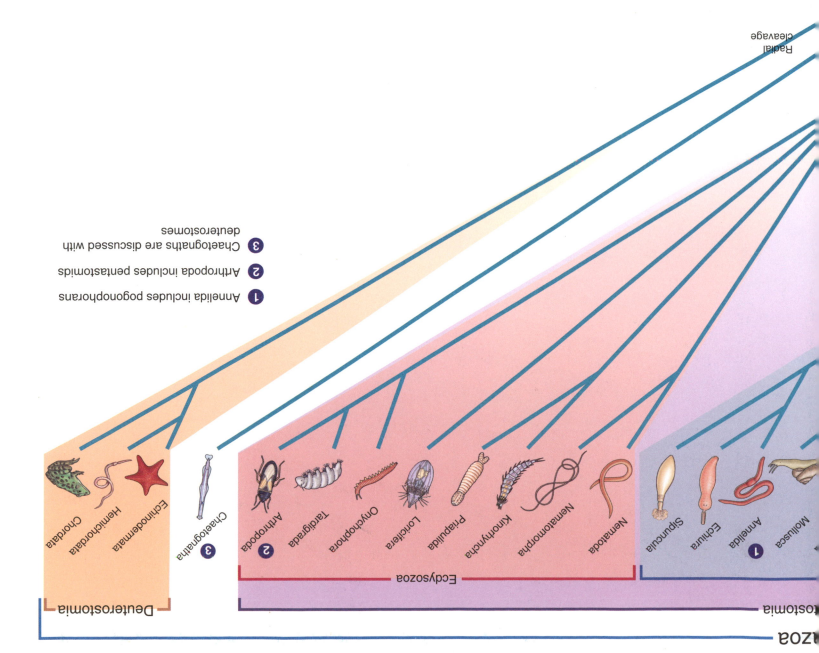

INTEGRATED PRINCIPLES OF
ZOOLOGY

FOURTEENTH EDITION

Cleveland P. Hickman, Jr.
WASHINGTON AND LEE UNIVERSITY

Larry S. Roberts
FLORIDA INTERNATIONAL UNIVERSITY

Susan L. Keen
UNIVERSITY OF CALIFORNIA AT DAVIS

Allan Larson
WASHINGTON UNIVERSITY

Helen I'Anson
WASHINGTON AND LEE UNIVERSITY

David J. Eisenhour
MOREHEAD STATE UNIVERSITY

Original Artwork by
WILLIAM C. OBER, M.D.
Washington and Lee University and Shoals Marine Laboratory
and
CLAIRE W. GARRISON, B.A.
Shoals Marine Laboratory, Cornell University

Boston   Burr Ridge, IL   Dubuque, IA   New York   San Francisco   St. Louis
Bangkok   Bogotá   Caracas   Kuala Lumpur   Lisbon   London   Madrid   Mexico City
Milan   Montreal   New Delhi   Santiago   Seoul   Singapore   Sydney   Taipei   Toronto

INTEGRATED PRINCIPLES OF ZOOLOGY, FOURTEENTH EDITION

Published by McGraw-Hill, a business unit of The McGraw-Hill Companies, Inc., 1221 Avenue of the Americas, New York, NY 10020. Copyright © 2008 by The McGraw-Hill Companies, Inc. All rights reserved. Previous editions 2006, 2004, 2001, and 1997. No part of this publication may be reproduced or distributed in any form or by any means, or stored in a database or retrieval system, without the prior written consent of The McGraw-Hill Companies, Inc., including, but not limited to, in any network or other electronic storage or transmission, or broadcast for distance learning.

Some ancillaries, including electronic and print components, may not be available to customers outside the United States.

 This book is printed on recycled, acid-free paper containing 10% postconsumer waste.

2 3 4 5 6 7 8 9 0 DOW/DOW 0 9 8

ISBN 978–0–07–297004–3
MHID 0–07–297004–9

Publisher: *Janice Roerig-Blong*
Executive Editor: *Patrick E. Reidy*
Developmental Editor: *Debra A. Henricks*
Senior Marketing Manager: *Tami Petsche*
Project Manager: *April R. Southwood*
Senior Production Supervisor: *Laura Fuller*
Lead Media Project Manager: *Jodi K. Banowetz*
Media Producer: *Daniel M. Wallace*
Associate Design Coordinator: *Brenda A. Rolwes*
Cover Designer: *Studio Montage, St. Louis, Missouri*
Senior Photo Research Coordinator: *John C. Leland*
Photo Research: *Mary Reeg*
Supplement Producer: *Melissa M. Leick*
Compositor: *Laserwords Private Limited*
Typeface: *10/12 Garamond*
Printer: *R. R. Donnelley Willard, OH*
Front cover image: *Polar Bear; © Digital Vision*
Back cover images: *Polar bear paws; © Creatas/PunchStock, Polar bear standing on the ice; © Geostock/ Getty Images*

The credits section for this book begins on page 880 and is considered an extension of the copyright page.

Library of Congress Cataloging-in-Publication Data

Integrated principles of zoology / Cleveland P. Hickman, Jr. ... [et al.]. – 14th ed.
  p. cm.
  Includes index.
  ISBN 978–0–07–297004–3 — ISBN 0–07–297004–9 (hard copy : alk. paper) 1. Zoology. I. Hickman, Cleveland P.
  QL47.2.H54 2008
  590–dc22

2007024506

CONTENTS IN BRIEF

iii

TABLE OF CONTENTS

PART THREE

Diversity of Animal Life

PART FOUR

Activity of Life

PART FIVE

Animals and Their Environments

CLEVELAND P. HICKMAN, JR.

Cleveland P. Hickman, Jr., Professor Emeritus of Biology at Washington and Lee University in Lexington, Virginia, has taught zoology and animal physiology for more than 30 years. He received his Ph.D. in comparative physiology from the University of British Columbia, Vancouver, B.C., in 1958 and taught animal physiology at the University of Alberta before moving to Washington and Lee University in 1967. He has published numerous articles and research papers in fish physiology, in addition to co-authoring these highly successful texts: *Integrated Principles of Zoology, Biology of Animals, Animal Diversity, Laboratory Studies in Animal Diversity,* and *Laboratory Studies in Integrated Principles of Zoology.*

Over the years Dr. Hickman has led many field trips to the Galápagos Islands. His current research is on intertidal zonation and marine invertebrate systematics in the Galápagos. He has published three field guides in the Galápagos Marine Life Series for the identification of echinoderms, marine molluscs, and marine crustaceans.

His interests include scuba diving, woodworking, and participating in chamber music ensembles.

Dr. Hickman can be contacted at: hickmanc@wlu.edu

LARRY S. ROBERTS

Larry S. Roberts, Professor Emeritus of Biology at Texas Tech University and an adjunct professor at Florida International University, has extensive experience teaching invertebrate zoology, marine biology, parasitology, and developmental biology. He received his Sc.D. in parasitology at the Johns Hopkins University and is the lead author of Schmidt and Roberts's *Foundations of Parasitology,* sixth edition. Dr. Roberts is also co-author of *Integrated Principles of Zoology, Biology of Animals,* and *Animal Diversity,* and is author of *The Underwater World of Sport Diving.*

Dr. Roberts has published many research articles and reviews. He has served as President of the American Society of Parasitologists, Southwestern Association of Parasitologists, and Southeastern Society of Parasitologists, and is a member of numerous other professional societies. Dr. Roberts also serves on the Editorial Board of the journal, *Parasitology Research.* His hobbies include scuba diving, underwater photography, and tropical horticulture.

Dr. Roberts can be contacted at: Lroberts1@compuserve.com

SUSAN KEEN

Susan Keen is a lecturer in the Section of Evolution and Ecology at the University of California at Davis. She received her Ph.D. in zoology from the University of California at Davis, following a M.Sc. from the University of Michigan at Ann Arbor. She is a native of Canada and obtained her undergraduate education at the University of British Columbia in Vancouver.

Dr. Keen is an invertebrate zoologist fascinated with jellyfish life histories. She has a particular interest in life cycles where both asexual and sexual phases of organisms are present, as they are in most jellyfishes. Her other research has included work on sessile marine invertebrate communities, spider populations, and Andean potato evolution.

Dr. Keen has been teaching evolution and animal diversity within the Introductory Biology series for 13 years. She enjoys all facets of the teaching process, from lectures and discussions to the design of effective laboratory exercises. In addition to her work with introductory biology, she offers seminars for the Davis Honors Challenge program, and for undergraduate and graduate students interested in teaching methods for biology. She was given an Excellence in Education Award from the Associated Students group at Davis in 2004. She attended the National Academies Summer Institute on Undergraduate Education in Biology in 2005, and was a National Academies Education Fellow in the Life Sciences for 2005–2006. Her interests include weight training, horseback riding, gardening, travel, and mystery novels.

Dr. Keen can be contacted at: slkeen@ucdavis.edu

ALLAN LARSON

Allan Larson is a professor at Washington University, St. Louis, MO. He received his Ph.D. in genetics at the University of California, Berkeley. His fields of specialization include evolutionary biology, molecular population genetics and systematics, and amphibian systematics. He teaches courses in introductory genetics, zoology, macroevolution, molecular evolution, and the history of evolutionary theory, and has organized and taught a special course in evolutionary biology for high-school teachers.

Dr. Larson has an active research laboratory that uses DNA sequences to examine evolutionary relationships among vertebrate species, especially in salamanders and lizards. The students in Dr. Larson's laboratory have participated in zoological field studies around the world, including projects in Africa, Asia, Australia, Madagascar, North America, South America, the Indo-Pacific Ocean, and the Caribbean Islands. Dr. Larson has authored numerous scientific publications, and has edited for the journals *The American Naturalist, Evolution, Journal of Experimental Zoology, Molecular Phylogenetics and Evolution,* and *Systematic Biology.* Dr. Larson serves as an academic advisor to undergraduate students and supervises the undergraduate biology curriculum at Washington University.

Dr. Larson can be contacted at: larson@wustl.edu

HELEN I'ANSON

Helen I'Anson, a native of England, is professor of biology at Washington and Lee University in Lexington, Virginia. She received her Ph.D. in physiology at the

University of Kentucky, Lexington, KY, and postdoctoral training at the University of Michigan, Ann Arbor, MI. She teaches courses in animal physiology, microanatomy, neuroendocrinology, general biology, and reproductive physiology. She has an active research program that focuses on the neural regulation of reproductive development. In particular, she is interested in how energy is partitioned in the developing animal, how signals from food and food storage depots are monitored by the brain, and how such signals are transduced to regulate reproductive activity at the onset of puberty in mammals.

Her interests include gardening, hiking, fishing, aromatherapy, music, and participating in choral ensembles.

Dr. I'Anson can be contacted at: iansonh@wlu.edu

DAVID J. EISENHOUR

David J. Eisenhour is an associate professor of biology at Morehead State University in Morehead, Kentucky. He received his Ph.D. in zoology from Southern Illinois University, Carbondale. He teaches courses in environmental science, human anatomy, general zoology, comparative anatomy, ichthyology, and vertebrate zoology. David has an active research program that focuses on systematics, conservation biology, and natural history of North American freshwater fishes. He has a particular interest in the diversity of Kentucky's fishes and is writing a book about that subject. He and his graduate students have authored several publications. David serves as an academic advisor to prepharmacy students.

His interests include fishing, landscaping, home remodeling, and entertaining his three young children, who, along with his wife, are enthusiastic participants in fieldwork.

Dr. Eisenhour can be contacted at: d.eisenhour@morehead-st.edu

Integrated Principles of Zoology is a college text designed for an introductory course in zoology. This fourteenth edition, as with previous editions, describes the diversity of animal life and the fascinating adaptations that enable animals to inhabit so many ecological niches.

We retain in this revision the basic organization of the thirteenth edition and its distinctive features, especially emphasis on the principles of evolution and zoological science. Also retained are several pedagogical features that have made previous editions easily accessible to students: opening chapter dialogues drawn from the chapter's theme; chapter summaries and review questions to aid student comprehension and study; concise and visually appealing illustrations; in-text derivations of generic names; chapter notes and essays that enhance the text by offering interesting sidelights to the narrative; literature citations; and an extensive glossary providing pronunciation, derivation, and definition of terms used in the text.

NEW TO THE FOURTEENTH EDITION

The authors welcome to the fourteenth edition Susan Keen, who supervised this revision. Many improvements are the direct result of Susan's new perspectives and those of many zoology instructors who submitted reviews of the thirteenth edition. We revised all chapters to streamline the writing and to incorporate new discoveries and literature citations. Our largest formal revision is to include a cladogram of animal phyla on the inside front cover of the book, and to reorder chapter contents in Part Three (Diversity of Animal Life) to match the arrangement of phyla on the cladogram. Each chapter in Part Three begins with a small image of the zoological cladogram highlighting the phylum or phyla covered in the chapter, followed by an expanded cladogram of the contents of each major phylum. We place stronger emphasis on phylogenetic perspectives throughout the book. Material formerly presented separately as "biological contributions" and "characteristics" of phyla is consolidated in a boxed list of phylum "characteristics" for each chapter in Part Three. New photographs are added to illustrate animal diversity in many phyla.

Material new to the fourteenth edition expands and updates our coverage of eight major principles: (1) scientific process and the role of theory, (2) cellular systems and metabolism, (3) endosymbiotic theory of eukaryotic origins, (4) physiological and ecological systems, (5) populational processes and conservation, (6) evolutionary developmental biology, (7) phylogenetic tests of morphological homologies, and (8) taxonomy. Exciting new fossil discoveries and molecular phylogenies contribute important changes to the last three principles. The primary changes to each major principle are summarized here with references to the relevant chapters.

Scientific Process and the Role of Theory

Many changes throughout the book increase the integration of hypothetico-deductive methodology in discussing new discoveries and controversies. We begin in Chapter 1 with a more detailed explanation of the hypothetico-deductive method of science and the important contrast between the comparative method versus experimental biology as complementary means of testing hypotheses. The role of theory in science is illustrated explicitly using Darwin's theory of common descent in Chapter 6. Uses of Darwin's theory of common descent to test evolutionary hypotheses and to construct taxonomies get expanded treatment in Chapter 10, including a new conceptual distinction between classification and systematization and coverage of DNA barcoding in species identification.

Cellular Systems and Metabolism

We expand in Chapter 3 our coverage of the components of eukaryotic cells, the biological roles of subcellular structures, and specializations of cellular surfaces. Expanded molecular topics include pH (Chapter 2), prions as diseases of protein conformation (Chapter 2), lipid metabolism (Chapter 4), and accumulation of "junk" or "parasitic" DNA in animal genomes (Chapter 5). In Chapter 7, a new boxed essay reports the discovery of actively dividing germ cells in adult female mammals, and a revised boxed essay updates applications of cell biology to contraceptive medicine.

Endosymbiotic Theory of Eukaryotic Origins

The history of the endosymbiotic theory is presented in more detail, including the empirical testing of its original claims and its more recent expansion to cover a broader evolutionary domain (Chapter 2). Important molecular phylogenetic evidence for separate evolutionary origins of nuclear, mitochondrial, and chloroplast genomes is presented in the form of a new global "tree of life" relating prokaryotic and eukaryotic genomes (Chapter 10). The role of endosymbiosis in diversification of unicellular eukaryotes gets new coverage in Chapter 11, and evolutionary loss of mitochondria from some infectious unicellular eukaryotes is added to Chapter 2.

Physiological and Ecological Systems

Numerous revisions address organismal physiology and its ecological consequences, beginning with the addition of "movement" as a general characteristic of life in Chapter 1. We add new results on tracheal respiration in insects (Chapter 21), respiratory gas transport in terrestrial arthropods and in vertebrates (Chapter 31), and lung ventilation in vertebrates (Chapter 31). Also revised are the plans of vertebrate circulatory systems, coronary circulation, and excitation and control of the heart (Chapter 31). New material appears on regulation of food intake and of digestion (Chapter 32), digestive processes in the vertebrate small intestine (Chapter 32), and foregut fermentation in ruminant mammals (Chapter 28). Evolution of centralized nervous systems, chemoreception, mechanoreception and photoreception in invertebrates gets new coverage in Chapter 33, with expanded explanation of synapses and conduction of action potentials. Endocrinology of invertebrates is expanded, and vertebrate endocrinology is updated to include discussion of white adipose tissue as an endocrine organ, the pancreatic polypeptide (PP) hormone, and controversies regarding medicinal uses of anabolic steroids (Chapter 34). Invertebrate excretory systems, especially arthropod kidneys, get expanded coverage in Chapter 30. We cover regional endothermy in fishes (Chapter 24), and add new explanatory material on the importance of water and osmotic regulation, especially in marine fishes (Chapter 30). Revision of Chapter 35 updates our knowledge of susceptibility and resistance to disease, including acquired immune deficiency. We add a new section on cetacean echolocation (Chapter 28), greater explanation of frog mating systems (Chapter 25), and avian reproductive strategies, including extra-pair copulations (Chapter 27). We provide greater coverage of scientific controversies regarding bee communication, eusociality, and genetics of animal behavior (Chapter 36). Concepts of food chains and food webs are now distinguished, and quantitative data are added to illustrate them using ecological pyramids (Chapter 38).

Populational Processes and Conservation

Modes of speciation receive expanded coverage and explanation (Chapter 6), as do concepts of fitness and inclusive fitness (Chapter 6), and costs and benefits of sexual versus asexual reproduction (Chapter 7). Conservation of natural populations is updated, especially in fishes (Chapter 24), mammals (Chapter 36), and tuataras (Chapter 26). Historical biogeographic processes are illustrated with expanded coverage of explanations for Wallace's Line, the geographic contact between evolutionarily disparate faunas (Chapter 37).

Evolutionary Developmental Biology

This rapidly growing discipline gets updated coverage both in concept and application. New concepts of developmental modularity and evolvability join our general coverage of evolutionary biology in Chapter 6. We discuss in Chapter 8 new evidence that some sponges have two germ layers. Cnidarian development and life cycles get expanded coverage, and the diploblastic status of cnidarians and ctenophores is reconsidered in light of new phylogenetic results (Chapter 13). We provide

molecular-genetic interpretations of the diploblast-triploblast distinction (Chapter 13) and updated details of triploblastic development (Chapter 14). Insights from genomic and developmental studies offer new interpretations of metazoan origins (Chapter 12) and suggest that changes in the expression of a single gene underlie alternative developmental pathways of arthropod limbs (uniramous versus biramous; Chapter 19). Developmental differences among chaetognaths, protostomes, and deuterostomes are reevaluated in light of new phylogenetic evidence (Chapter 22). We restructure our general coverage of body plans (Chapter 9) and provide greater explanation of the complex development of gastropod torsion (Chapter 16).

Phylogenetic Tests of Morphological Homologies

New molecular phylogenies and fossil discoveries revise our interpretations of many homologies and reveal independent evolution of similar characters in different groups. In light of these issues, we expand our coverage of the concept of homoplasy in Chapter 10. Chapter 16 incorporates new evidence challenging homology of metamerism in annelids and molluscs and illustrating the scientific process in action. Evidence from *Hox* gene expression is used in Chapter 19 to homologize the cephalothorax of spiders with heads of other arthropods and to support phylogenetic evidence for multiple origins of uniramous limbs from biramous ones in phylum Arthropoda. Homology of diffuse epidermal nervous systems and tripartite coeloms of echinoderms and hemichordates and nonhomology of dorsal hollow nerve chords of hemichordates and chordates change our favored hypotheses for relationships among these groups (Chapter 22). Developmental comparisons demonstrate nonhomology of coelomic compartmentalization of lophophorates with that of echinoderms and hemichordates (Chapter 22). New data and interpretations revise inferred characteristics of the most recent chordate ancestor (Chapter 23), origin and diversification of amniotes and their adaptations for terrestrial life (Chapter 26), evolution of the mammalian middle ear (Chapter 28), and details of hominid morphological evolution (Chapter 28).

Taxonomy

New molecular-phylogenetic and fossil data reject some familiar taxa and suggest new ones. We discuss evidence for a sister-group relationship of choanoflagellates and metazoans in Chapter 12. Chapter 14 discusses new phylogenetic results that underlie recognition of phylum Acoelomorpha and a revised phylogenetic hypothesis for nemertine worms. Acanthocephalans now appear to descend from a rotiferan ancestor (Chapter 15). Clade Clitellata (oligochaetes and leeches), pogonophorans, and vestimentiferans descend from polychaete annelids according to new phylogenetic data, making polychaetes paraphyletic (Chapter 17). Chapter 18 presents new evidence for clade Panarthropoda (Onychophora, Tardigrada, and Arthropoda). Chapters 19 and 21 present evidence supporting recognition of clade Pancrustacea (crustaceans and hexapods) and rejection of arthropod subphylum Uniramia. Chapter 20 includes new evidence that hexapods derive from a crustacean ancestor, and Pentastomida is subsumed in Crustacea. Entognatha and Insecta form separate clades within subphylum Hexapoda (Chapter 21). We update recognition of insect orders in Chapter 21. We introduce in Chapter 22 clade Ambulacraria (Echinodermata and Hemichordata), which is likely the sister group of chordates (Chapter 23). The fossil genus *Haikouella* gets increased coverage and illustration as the likely sister taxon to craniates (Chapter 23). Changes to fish taxonomy include using the clade name Petromyzontida for lampreys and removing bichirs from chondrosteans (Chapter 24). Early tetrapod evolution is extensively revised with reference to new fossil discoveries, including the genus *Tiktaalik* (Chapter 25). We replace the traditional use of "Reptilia" with one including the traditional reptiles, birds, and all descendants of their most recent common ancestor (Chapter 26). Phylogenetic results place turtles in the clade Diapsida (Chapter 26), contrary to earlier hypotheses. Amphisbaenians are now included within lizards according to their phylogenetic position, and the section on relationships of snakes to lizards is expanded (Chapter 26). Chapter 27 includes fairly extensive revisions of avian taxonomy based upon phylogenetic results from DNA sequence data.

TEACHING AND LEARNING AIDS

To help students in **vocabulary development,** key words are boldfaced and derivations of technical and zoological terms are provided, along with generic names of animals where they first appear in the text. In this way students gradually become familiar with the more common roots that form many technical terms. An extensive **glossary** provides pronunciation, derivation, and definition of each term. Many new terms were added to the glossary or rewritten for this edition.

A distinctive feature of this text is a **prologue** for each chapter that highlights a theme or fact relating to the chapter. Some prologues present biological, particularly evolutionary, principles; those in Part Three on animal diversity illuminate distinguishing characteristics of the group presented in the chapter.

Chapter notes, which appear throughout the book, augment the text material and offer interesting sidelights without interrupting the narrative. We prepared many new notes for this edition and revised several existing notes.

To assist students in chapter review, each chapter ends with a **concise summary,** a list of **review questions,** and **annotated selected references.** The review questions enable a student to self-test retention and understanding of the more important chapter material.

Again, William C. Ober and Claire W. Garrison have strengthened the art program for this text with many new full-color paintings that replace older art, or that illustrate new material. Bill's artistic skills, knowledge of biology, and experience gained from an earlier career as a practicing physician have enriched this text through ten of its editions. Claire practiced pediatric and obstetric nursing before turning to scientific illustration as a full-time career. Texts illustrated by Bill and Claire have received national recognition and won awards from the Association of Medical Illustrators, American Institute of Graphic Arts, Chicago Book Clinic, Printing Industries of America, and Bookbuilders West. They are also recipients of the Art Directors Award.

SUPPLEMENTS

For Students

Online Learning Center

The **Online Learning Center** for *Integrated Principles of Zoology* is a great place to review chapter material and to enhance your study routine. Visit **www.mhhe.com/hickmanipz14e** for access to the following online study tools:

- Chapter quizzing
- Key term flash cards
- Web links
- And more!

For Instructors

Online Learning Center

The *Integrated Principles of Zoology* **Online Learning Center (www.mhhe.com/hickmanipz14e)** offers a wealth of teaching and learning aids for instructors and students. Instructors will appreciate:

- A password-protected Instructor's Manual
- Access to the new online **Presentation Center** including illustrations, photographs, and tables from the text
- PowerPoint lecture outlines
- Links to professional resources
- Online activities for students such as chapter quizzing, key-term flash cards, web links, and more!

NEW! Presentation Center
www.mhhe.com/hickmanipz14e

Build Instructional Materials Wherever, Whenever, and However You Want! Accessed through the Online Learning Center, the **Presentation Center** is an online digital library containing assets such as art, photos, tables, PowerPoint image slides, and other media types that can be used to create customized lectures, visually enhanced tests and quizzes, compelling course websites, or attractive printed support materials.

Access to Your Book, Access to All Books! The **Presentation Center** library includes thousands of assets from many McGraw-Hill titles. This ever-growing resource gives instructors the power to utilize assets specific to an adopted textbook as well as content from all other books in the library.

Nothing Could Be Easier!

Presentation Center's dynamic search engine allows you to explore by discipline, course, textbook chapter, asset type, or keyword. Simply browse, select, and download the files you need to build engaging course materials. All assets are copyrighted by McGraw-Hill Higher Education but can be used by instructors for classroom purposes.

Instructor's Manual

This helpful ancillary provides chapter outlines, lecture enrichment suggestions, lesson plans, a list of changes from the previous edition, and source materials.

Test Bank

A computerized test bank utilizing testing software to create customized exams is available with this text. The user-friendly software allows instructors to search for questions by topic or format, edit existing questions or add new ones, and scramble questions to create multiple versions of the same test. Word files of the test bank questions are provided for those instructors who prefer to work outside the test-generator software.

Course Management Systems

Online content is available for a variety of course management systems including BlackBoard and WebCT.

Laboratory Studies in Integrated Principles of Zoology by Cleveland Hickman, Jr., and Lee B. Kats

Now in its fourteenth edition, this lab manual was written to accompany *Integrated Principles of Zoology,* and can be easily adapted to fit a variety of course plans.

Biology Digitized Video Clips

McGraw-Hill is pleased to offer digitized biology video clips on DVD! Licensed from some of the highest-quality science video producers in the world, these brief segments range from about five seconds to just under three minutes in length and cover all areas of general biology from cells to ecosystems. Engaging and informative, McGraw-Hill's digitized biology videos will help capture students' interest while illustrating key biological concepts and processes. Includes video clips on mitosis, Darwin's finches, cichlid mouth brooding, sponge reproduction, and much more!

ACKNOWLEDGMENTS

The authors extend their sincere thanks to the faculty reviewers whose numerous suggestions for improvement were of the greatest value in the revision process. Their experience with students of varying backgrounds, and their interest in and knowledge of the subject, helped to shape the text into its final form.

Kenneth Andrews, *East Central University*
Patricia Biesiot, *University of Southern Mississippi*
Mark Blackmore, *Valdosta State University*
Raymond Bogiatto, *California State University*
Roger Choate, *Oklahoma City Community College*
Jerry Cook, *Sam Houston State University*
Tamara Cook, *Sam Houston State University*
Michael Harvey, *Broward Community College*
Robert Hoyt, *Western Kentucky University*
Timothy Judd, *Southeast Missouri State University*
Robert L. Koenig, *Southwest Texas Jr. College*
William Kroll, *Loyola University of Chicago*
Sharyn Marks, *Humboldt State University*
R. Patrick Randolf, *University of California–Davis*
Anthony Stancampiano, *Oklahoma City Community College*
Rodney Thomas, *University of South Carolina*
Michael Toliver, *Eureka College*
Stim Wilcox, *Binghamton University*

The authors express their appreciation to the editors and support staff at McGraw-Hill Higher Education who made this project possible. Special thanks are due Patrick Reidy, Executive Editor, and Debra Henricks, Developmental Editor, who were the driving forces in piloting this text throughout its development. April Southwood, Project Manager, somehow kept authors, text, art, and production programs on schedule. John Leland oversaw the extensive photographic program and Brenda Rolwes managed the book's interior and cover design. We are indebted to them for their talents and dedication.

Although we make every effort to bring to you an error-free text, errors of many kinds inevitably find their way into a textbook of this scope and complexity. We will be grateful to readers who have comments or suggestions concerning content to send their remarks to Debra Henricks, Developmental Editor at debra_henricks@mcgraw-hill.com.

Cleveland P. Hickman, Jr.
Larry S. Roberts
Susan Keen
Allan Larson
Helen I'Anson
David J. Eisenhour

Introduction to Living Animals

1

A tube anemone (cerianthid, Botruanthus benedini) from the eastern Pacific.

1

Life: Biological Principles and the Science of Zoology

Zoologist studying the behavior of yellow baboons (Papio cynocephalus) *in the Amboseli Reserve, Kenya.*

The Uses of Principles

We gain knowledge of the animal world by actively applying important guiding principles to our investigations. Just as the exploration of outer space is both guided and limited by available technologies, exploration of the animal world depends critically on our questions, methods, and principles. The body of knowledge that we call zoology makes sense only when the principles that we use to construct it are clear.

The principles of modern zoology have a long history and many sources. Some principles derive from laws of physics and chemistry, which all living systems obey. Others derive from the scientific method, which tells us that our hypotheses regarding the animal world are useless unless they guide us to gather data that potentially can refute them. Many important principles derive from

previous studies of the living world, of which animals are one part. Principles of heredity, variation, and organic evolution guide the study of life from the simplest unicellular forms to the most complex animals, fungi, and plants. Because life shares a common evolutionary origin, principles learned from the study of one group often pertain to other groups as well. By tracing the origins of our operating principles, we see that zoologists are not an island unto themselves but part of a larger scientific community.

We begin our study of zoology not by focusing narrowly within the animal world, but by searching broadly for our most basic principles and their diverse sources. These principles simultaneously guide our studies of animals and integrate those studies into the broader context of human knowledge.

Zoology, the scientific study of animal life, builds on centuries of human inquiry into the animal world. Mythologies of nearly every human culture attempt to solve the mysteries of animal life and its origin. Zoologists now confront these same mysteries with the most advanced methods and technologies developed by all branches of science. We start by documenting the diversity of animal life and organizing it in a systematic way. This complex and exciting process builds on the contributions of thousands of zoologists working in all dimensions of the biosphere (Figure 1.1). We strive through this work to understand how animal diversity originated and how animals perform the basic processes of life that permit them to occupy diverse environments.

This chapter introduces the fundamental properties of animal life, the methodological principles on which their study is based, and two important theories that guide our research: (1) the theory of evolution, which is the central organizing

A

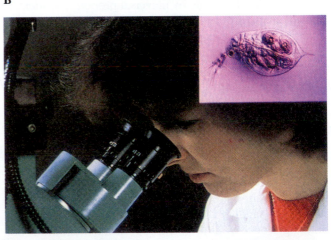

B

C

D

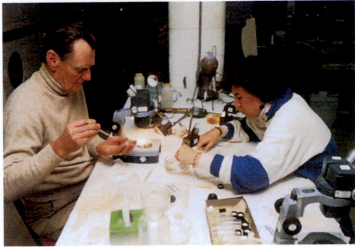

E

Figure 1.1

A few of the many dimensions of zoological research.
A, Observing moray eels in Maui, Hawaii. **B,** Working with tranquilized polar bears. **C,** Banding mallard ducks.
D, Observing *Daphnia pulex* (×150) microscopically.
E, Separating growth stages of crab larvae at a marine laboratory.

principle of biology, and (2) the chromosomal theory of inheritance, which guides our study of heredity and variation in animals. These theories unify our knowledge of the animal world.

FUNDAMENTAL PROPERTIES OF LIFE

Does Life Have Defining Properties?

We begin with the difficult question, What is life? Although many attempts have been made to define life, simple definitions are doomed to failure. When we try to give life a simple definition, we look for fixed properties maintained throughout life's history. However, the properties that life exhibits today (pp. 4–9) are very different from those present at its origin. The history of life shows extensive and ongoing change, which we call *evolution*. As the genealogy of life progressed and branched from the earliest living form to the millions of species alive today, new properties evolved and passed from parents to their offspring. Through this process, living systems have generated many rare and spectacular features that have no counterparts in the nonliving world. Unexpected properties emerge on many different lineages in life's evolutionary history, producing the great organismal diversity observed today.

We might try to define life by universal properties evident at its origin. Replication of molecules, for example, can be traced to life's origin and represents one of life's universal properties. Defining life in this manner faces the major problem that these are the properties most likely to be shared by some nonliving forms. To study the origin of life, we must ask how organic molecules acquired the ability for precise replication. But where do we draw the line between those replicative processes that characterize life and those that are merely general chemical features of the matter from which life arose? Replication of complex crystalline structures in nonliving chemical assemblages might be confused, for example, with the replicative molecular properties associated with life. If we define life using only the most advanced characteristics of the highly evolved living systems observed today, the nonliving world would not intrude on our definition, but we would eliminate the early forms of life from which all others descended and which give life its historical unity.

Ultimately our definition of life must be based on the common history of life on earth. Life's history of descent with modification gives it an identity and continuity that separates it from the nonliving world. We can trace this common history backward through time from the diverse forms observed today and in the fossil record to their common ancestor that arose in the atmosphere of the primitive earth (see Chapter 2). All organisms forming part of this long history of hereditary descent from life's common ancestor are included in our concept of life.

We do not force life into a simple definition, but we can readily identify the living world through its history of common evolutionary descent. Many remarkable properties have arisen during life's history and are observed in various combinations among living forms. These properties, discussed in the next section, clearly identify their possessors as part of the unified historical entity called life. All such features occur in the most highly evolved forms of life, such as those that compose the animal kingdom. Because they are so important for maintenance and functioning of living forms that possess them, these properties should persist through life's future evolutionary history.

General Properties of Living Systems

The most outstanding general features in life's history include chemical uniqueness; complexity and hierarchical organization; reproduction (heredity and variation); possession of a genetic program; metabolism; development; environmental interaction; and movement.

1. **Chemical uniqueness.** *Living systems demonstrate a unique and complex molecular organization.* Living systems assemble large molecules, known as macromolecules, that are far more complex than the small molecules of nonliving matter. These macromolecules are composed of the same kinds of atoms and chemical bonds that occur in nonliving matter and they obey all fundamental laws of chemistry; it is only the complex organizational structure of these macromolecules that makes them unique. We recognize four major categories of biological macromolecules: nucleic acids, proteins, carbohydrates, and lipids (see Chapter 2). These categories differ in the structures of their component parts, the kinds of chemical bonds that link their subunits together, and their functions in living systems.

   The general structures of these macromolecules evolved and stabilized early in the history of life. With some modifications, these same general structures are found in every form of life today. Proteins, for example, contain about 20 specific kinds of amino acid subunits linked together by peptide bonds in a linear sequence (Figure 1.2). Additional bonds occurring between amino acids that are not adjacent to each other in the protein chain give the protein a complex, three-dimensional structure (see Figures 1.2 and 2.15). A typical protein contains several hundred amino acid subunits. Despite the stability of this basic protein structure, the ordering of the different amino acids in the protein molecule is subject to enormous variation. This variation underlies much of the diversity that we observe among different kinds of living forms. The nucleic acids, carbohydrates, and lipids likewise contain characteristic bonds that link variable subunits (see Chapter 2). This organization gives living systems both a biochemical unity and great potential diversity.

2. **Complexity and hierarchical organization.** *Living systems demonstrate a unique and complex hierarchical organization.* Nonliving matter is organized at least into atoms and molecules and often has a higher degree of organization as well. However, atoms and molecules are combined into patterns in the living world that do not exist in the nonliving world. In living systems, we find a hierarchy of levels that includes, in ascending order of complexity, macromolecules, cells, organisms, populations, and species (Figure 1.3). Each level builds on the level below it

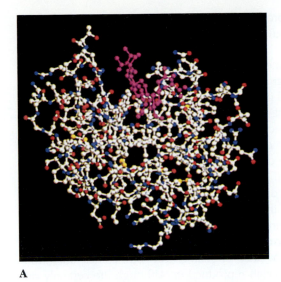

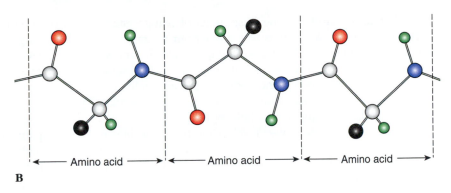

Figure 1.2

A computer simulation of the three-dimensional structure of the lysozyme protein **(A)**, which is used by animals to destroy bacteria. The protein is a linear string of molecular subunits called amino acids, connected as shown in **B**, which fold in a three-dimensional pattern to form the active protein. The white balls correspond to carbon atoms, the red balls to oxygen, the blue balls to nitrogen, the yellow balls to sulfur, the green balls to hydrogen, and the black balls **(B)** to molecular groups formed by various combinations of carbon, oxygen, nitrogen, hydrogen, and sulfur atoms that differ among amino acids. Hydrogen atoms are not shown in **A**. The purple molecule in **A** is a structure from the bacterial cell wall that is broken by lysozyme.

and has its own internal structure, which is also often hierarchical. Within the cell, for example, macromolecules are compounded into structures such as ribosomes, chromosomes, and membranes, and these are likewise combined in various ways to form even more complex subcellular structures called organelles, such as mitochondria (see Chapters 3 and 4). The organismal level also has a hierarchical substructure; cells combine to form tissues, which combine to form organs, which likewise combine to form organ systems (see Chapter 9).

Cells (Figure 1.4) are the smallest units of the biological hierarchy that are semiautonomous in their ability to conduct basic functions, including reproduction. Replication of molecules and subcellular components occurs only within a cellular context, not independently. Cells

Figure 1.3

Volvox globator (see pp. 229–231) is a multicellular chlorophytan that illustrates three different levels of the biological hierarchy: cellular, organismal, and populational. Each individual spheroid (organism) contains cells embedded in a gelatinous matrix. The larger cells function in reproduction, and the smaller ones perform the general metabolic functions of the organism. The individual spheroids together form a population.

Figure 1.4

Electron micrograph of ciliated epithelial cells and mucus-secreting cells (see pp. 192–195). Cells are the basic building blocks of living organisms.

are therefore considered the basic units of living systems (see Chapter 3). We can isolate cells from an organism and cause them to grow and to multiply under laboratory conditions in the presence of nutrients alone. This semi-autonomous replication is not possible for any individual molecules or subcellular components, which require additional cellular constituents for their reproduction.

Each successively higher level of the biological hierarchy is composed of units of the preceding lower level in the hierarchy. An important characteristic of this hierarchy is that the properties of any given level cannot be inferred even from the most complete knowledge of the properties of its component parts. A physiological feature, such as blood pressure, is a property of the organismal level; it is impossible to predict someone's blood pressure simply by knowing the physical characteristics of individual cells of the body. Likewise, systems of social interaction, as observed in bees, occur at the populational level; it is not possible to infer properties of this social system by studying individual bees in isolation.

The appearance of new characteristics at a given level of organization is called **emergence,** and these characteristics are called **emergent properties.** These properties arise from interactions among the component parts of a system. For this reason, we must study all levels directly, each one being the focus of a different subfield of biology (molecular biology; cell biology; organismal anatomy, physiology and genetics; population biology; Table 1.1). Emergent properties expressed at a particular level of the biological hierarchy are certainly influenced and restricted by properties of the lower-level components. For example, it would be impossible for a population of organisms that lack hearing to develop a spoken language. Nonetheless, properties of parts of a living system do not rigidly

determine properties of the whole. Many different spoken languages have emerged in human culture from the same basic anatomical structures that permit hearing and speech. The freedom of the parts to interact in different ways makes possible a great diversity of potential emergent properties at each level of the biological hierarchy.

Different levels of the biological hierarchy and their particular emergent properties are built by evolution. Before multicellular organisms evolved, there was no distinction between the organismal and cellular levels, and this distinction is still absent from single-celled organisms (see Chapter 11). The diversity of emergent properties that we see at all levels of the biological hierarchy contributes to the difficulty of giving life a simple definition or description.

3. **Reproduction.** *Living systems can reproduce themselves.* Life does not arise spontaneously but comes only from prior life, through reproduction. Although life certainly originated from nonliving matter at least once (see Chapter 2), this origin featured enormously long periods of time and conditions very different from the current biosphere. At each level of the biological hierarchy, living forms reproduce to generate others like themselves (Figure 1.5). Genes are replicated to produce new genes. Cells divide to produce new cells. Organisms reproduce, sexually or asexually, to produce new organisms (see Chapter 7). Populations may become fragmented to produce new populations, and species may split to produce new species through a process called speciation. Reproduction at any hierarchical level usually features an increase in numbers. Individual genes, cells, organisms, populations, or species may fail to reproduce themselves, but reproduction is nonetheless an expected property of these individuals.

Reproduction at each of these levels shows the complementary, and yet apparently contradictory, phenomena

TABLE 1.1

Different Hierarchical Levels of Biological Complexity That Display Reproduction, Variation, and Heredity

Level	Timescale of Reproduction	Fields of Study	Methods of Study	Some Emergent Properties
Cell	Hours (mammalian cell = ~16 hours)	Cell biology	Microscopy (light, electron), biochemistry	Chromosomal replication (meiosis, mitosis), synthesis of macromolecules (DNA, RNA, proteins, lipids, polysaccharides)
Organism	Hours to days (unicellular); days to years (multicellular)	Organismal anatomy, physiology, genetics	Dissection, genetic crosses, clinical studies, physiological experimentation	Structure, functions and coordination of tissues, organs and organ systems (blood pressure, body temperature, sensory perception, feeding)
Population	Up to thousands of years	Population biology, population genetics, ecology	Statistical analysis of variation, abundance, geographical distribution	Social structures, systems of mating, age distribution of organisms, levels of variation, action of natural selection
Species	Thousands to millions of years	Systematics and evolutionary biology, community ecology	Study of reproductive barriers, phylogeny, paleontology, ecological interactions	Method of reproduction, reproductive barriers

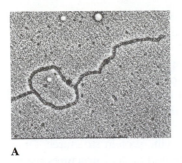

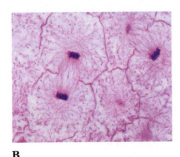

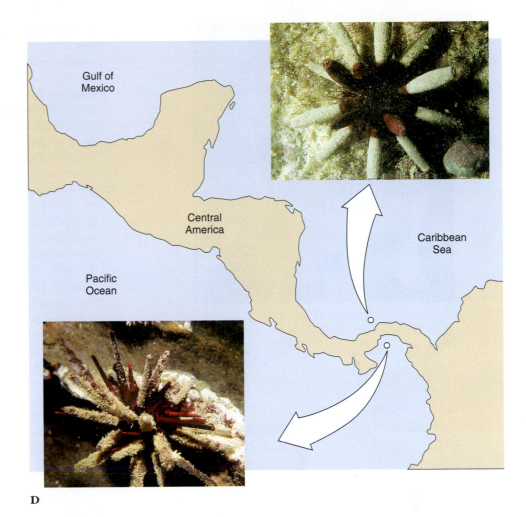

A

B

C

D

Figure 1.5

Reproductive processes observed at four different levels of biological complexity. **A,** Molecular level—electron micrograph of a replicating DNA molecule. **B,** Cellular level—micrograph of cell division at mitotic telophase. **C,** Organismal level—a king snake hatching. **D,** Species level—formation of new species in the sea urchin (*Eucidaris*) after geographic separation of Caribbean (*E. tribuloides*) and Pacific (*E. thouarsi*) populations by the formation of a land bridge.

of **heredity** and **variation.** Heredity is the faithful transmission of traits from parents to offspring, usually (but not necessarily) observed at the organismal level. Variation is the production of *differences* among the traits of different individuals. In a reproductive process, properties of descendants resemble those of their parents to varying degrees but usually are not identical to them. Replication of deoxyribonucleic acid (DNA) occurs with high fidelity, but errors occur at repeatable rates. Cell division is exceptionally precise, especially with regard to the nuclear material, but chromosomal changes occur nonetheless at measurable rates. Organismal reproduction likewise demonstrates both heredity and variation, the latter most obvious in sexually reproducing forms. Production of new populations and species also demonstrates conservation of some properties and changes of others. Two closely related frog species may have similar mating calls but differ in the rhythm of repeated sounds.

Interaction of heredity and variation in the reproductive process is the basis for organic evolution (see Chapter 6). If

heredity were perfect, living systems would never change; if variation were uncontrolled by heredity, biological systems would lack the stability that allows them to persist through time.

4. **Possession of a genetic program.** *A genetic program provides fidelity of inheritance* (Figure 1.6). Structures of the protein molecules needed for organismal development and functioning are encoded in **nucleic acids** (see Chapter 5). For animals and most other organisms, genetic information is contained in **DNA.** DNA is a very long, linear chain of subunits called nucleotides, each of which contains a sugar phosphate (deoxyribose phosphate) and one of four nitrogenous bases (adenine, cytosine, guanine, or thymine, abbreviated A, C, G, and T, respectively). The sequence of nucleotide bases contains a code for the order of amino acids in the protein specified by the DNA molecule. The correspondence between the sequence of bases in DNA and the sequence of amino acids in a protein is called the **genetic code.**

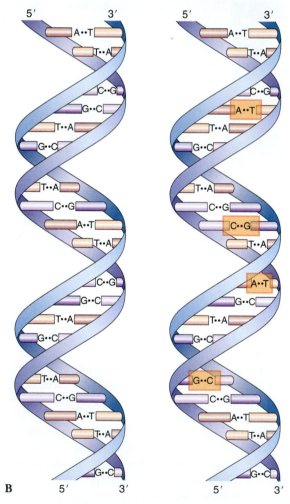

Figure 1.6

James Watson and Francis Crick with a model of the DNA double helix **(A)**. Genetic information is coded in the nucleotide base sequence inside the DNA molecule. Genetic variation is shown **(B)** in DNA molecules that are similar in base sequence but differ from each other at four positions. Such differences can encode alternative traits, such as different eye colors.

The genetic code arose early in the evolutionary history of life, and the same code occurs in bacteria and in the nuclear genomes of almost all animals and plants. The near constancy of this code among living forms provides strong evidence for a single origin of life. The genetic code has undergone very little evolutionary change since its origin because an alteration would disrupt the structure of nearly every protein, which would in turn severely disrupt cellular functions that require very specific protein structures. Only in the rare instance that the altered protein structures maintain their cellular functions would such a change be able to survive and be reproduced. Evolutionary change in the genetic code has occurred in the DNA contained in animal mitochondria, the organelles that regulate cellular energy. The genetic code in animal mitochondrial DNA therefore is slightly different from the standard code of nuclear and bacterial DNA. Because mitochondrial DNA specifies far fewer proteins than nuclear DNA, the likelihood of getting a change in the code that maintains cellular functions is greater there than in the nucleus.

5. **Metabolism.** *Living organisms maintain themselves by acquiring nutrients from their environments* (Figure 1.7). The nutrients are used to obtain chemical energy and molecular components for building and maintaining the living system (see Chapter 4). We call these essential chemical processes **metabolism.** They include digestion, acquisition of energy (respiration), and synthesis of molecules and structures. Metabolism is often viewed as an interaction of destructive (catabolic) and constructive (anabolic) reactions. The most fundamental anabolic and catabolic chemical processes used by living systems arose early in the evolutionary history of life, and all living forms share them. These reactions include synthesis of carbohydrates, lipids, nucleic acids, and proteins and their constituent parts and cleavage of chemical bonds to recover energy stored in them. In animals, many fundamental metabolic reactions occur at the cellular level, often in specific organelles found throughout the animal kingdom. Cellular respiration occurs, for example, in mitochondria. Cellular and nuclear membranes regulate metabolism by controlling the movement of molecules across the cellular and nuclear boundaries, respectively. The study of complex metabolic functions is called **physiology.** We devote a large portion of this book to describing and comparing the diverse tissues, organs, and organ systems that different groups of animals have evolved to perform the basic physiological functions of life (see Chapters 11 through 36).

6. **Development.** *All organisms pass through a characteristic life cycle.* Development describes the characteristic changes that an organism undergoes from its origin (usually the fertilization of an egg by sperm) to its final adult form (see Chapter 8). Development usually features changes in size and shape, and differentiation of structures within an organism. Even the simplest one-celled organisms grow in size and replicate their component parts until they divide into two or more cells. Multicellular organisms undergo more dramatic changes during their lives. Different developmental

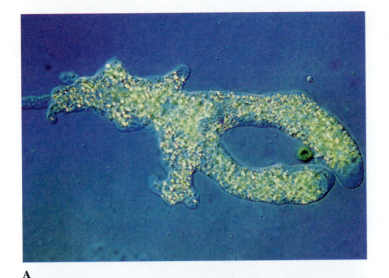

A

A

B

Figure 1.8

A, Adult monarch butterfly emerging from its pupal case. **B,** Fully formed adult monarch butterfly.

B

Figure 1.7

Feeding processes illustrated by **(A)** an ameba surrounding food and **(B)** a chameleon capturing insect prey with its projectile tongue.

stages of some multicellular forms are so dissimilar that they are hardly recognizable as belonging to the same species. Embryos are distinctly different from juvenile and adult forms into which they develop. Even postembryonic development of some organisms includes stages dramatically different from each other. The transformation that occurs from one stage to another is called **metamorphosis.** There is little resemblance, for example, among the egg, larval, pupal, and adult stages of metamorphic insects (Figure 1.8). Among animals, early stages of development are often more similar among organisms of related species than are later developmental stages. In our survey of animal diversity, we describe all stages of observed life histories but concentrate on adult stages in which diversity tends to be greatest.

7. **Environmental interaction.** *All animals interact with their environments.* The study of organismal interaction

with an environment is called **ecology.** Of special interest are the factors that influence geographic distribution and abundance of animals (see Chapters 37 and 38). The science of ecology reveals how an organism perceives environmental stimuli and responds in appropriate ways by adjusting its metabolism and physiology (Figure 1.9). All organisms respond to environmental stimuli, a property called **irritability.** The stimulus and response may be simple, such as a unicellular organism moving from or toward a light source or away from a noxious substance, or it may be quite complex, such as a bird responding to a complicated series of signals in a mating ritual (see Chapter 36). Life and environment are inseparable. We cannot isolate the evolutionary history of a lineage of organisms from the environments in which it occurred.

8. **Movement.** *Living systems and their parts show precise and controlled movements arising from within the system.* The energy that living systems extract from their environments permits them to initiate controlled movements. Such movements at the cellular level are essential for reproduction, growth, and many responses to stimuli in all living forms and for development in multicellular ones. Autonomous movement reaches great diversity in animals, and much of this book comprises descriptions of animal movement and the many adaptations that animals have evolved for locomotion. On a larger scale, entire populations or species may disperse from one geographic location to another one over time through their powers of movement. Movement characteristic of nonliving matter, such as that of particles in solution, radioactive decay of nuclei, and eruption of volcanoes is not precisely controlled by the moving objects themselves and often involves forces entirely external to them. The adaptive and often purposeful movements initiated by living systems are absent from the nonliving world.

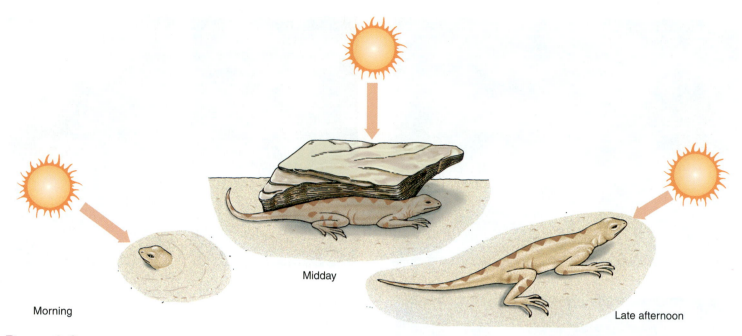

Morning

Midday

Late afternoon

Figure 1.9
A lizard regulates its body temperature by choosing different locations (microhabitats) at different times of day.

Life Obeys Physical Laws

To untrained observers, these eight properties of life may appear to violate basic laws of physics. Vitalism, the idea that life is endowed with a mystical vital force that violates physical and chemical laws, was once widely advocated. Biological research has consistently rejected vitalism, showing instead that all living systems obey basic laws of physics and chemistry. Laws governing energy and its transformations (thermodynamics) are particularly important for understanding life (see Chapter 4). The **first law of thermodynamics** is the law of conservation of energy. Energy is neither created nor destroyed but can be transformed from one form to another. All aspects of life require energy and its transformation. The energy to support life on earth flows from the fusion reactions in our sun and reaches the earth as light and heat. Sunlight captured by green plants and cyanobacteria is transformed by photosynthesis into chemical bonds. Energy in chemical bonds is a form of potential energy released when the bond is broken; the energy is used to perform numerous cellular tasks. Energy transformed and stored in plants is then used by animals that eat the plants, and these animals may in turn provide energy for other animals that eat them.

The **second law of thermodynamics** states that physical systems tend to proceed toward a state of greater disorder, or **entropy.** Energy obtained and stored by plants is subsequently released by various mechanisms and finally dissipated as heat. The complex molecular organization in living cells is attained and maintained only as long as energy fuels the organization. The ultimate fate of materials in the cells is degradation and dissipation of their chemical bond energy as heat. The process of evolution whereby organismal complexity can increase over time may appear at first to violate the second law of thermodynamics, but it does not. Organismal complexity is achieved and maintained only by the constant use and dissipation of energy flowing into the biosphere from the sun. Survival, growth, and reproduction of animals require energy that comes from breaking complex food molecules into simple organic waste. The processes by which animals acquire energy through nutrition and respiration reveal themselves to us through the many physiological sciences.

ZOOLOGY AS A PART OF BIOLOGY

Animals form a distinct branch on the evolutionary tree of life. It is a large and old branch that originated in the Precambrian seas over 600 million years ago. Animals form part of an even larger limb known as **eukaryotes,** organisms whose cells contain membrane-enclosed nuclei. This larger limb includes plants, fungi and numerous unicellular forms. Perhaps the most distinctive characteristic of the animals as a group is their means of nutrition, which consists of eating other organisms. Evolution has elaborated this basic way of life through diverse systems for capturing and processing a wide array of food items and for locomotion.

Animals are distinguished also by the absence of characteristics that have evolved in other eukaryotes. Plants, for example, use light energy to produce organic compounds (photosynthesis), and they have evolved rigid cell walls that surround their cell membranes; photosynthesis and cell walls are absent from animals. Fungi acquire nutrition by absorption of small organic molecules from their environments, and their body plan contains tubular filaments called *hyphae;* these structures are absent from the animal kingdom.

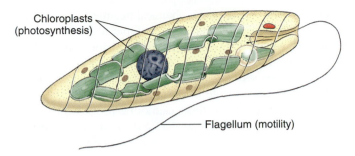

Figure 1.10
Some organisms, such as the single-celled *Euglena* (shown here) and *Volvox* (see Figure 1.3), combine properties that distinguish animals (locomotion) and plants (photosynthetic ability).

Some organisms combine properties of animals and plants. For example, *Euglena* (Figure 1.10) is a motile, single-celled organism that resembles plants in being photosynthetic, but it resembles animals in its ability to eat food particles. *Euglena* is part of a separate eukaryotic lineage that diverged from those of plants and animals early in the evolutionary history of eukaryotes. *Euglena* and other unicellular eukaryotes are sometimes grouped as the kingdom Protista, although this kingdom is an arbitrary grouping of unrelated lineages that violates taxonomic principles (see Chapter 10).

The fundamental structural and developmental features evolved by the animal kingdom are presented in Chapters 8 and 9.

PRINCIPLES OF SCIENCE

Nature of Science

We stated in the first sentence of this chapter that zoology is the scientific study of animals. A basic understanding of zoology therefore requires an understanding of what science is, what it is not, and how knowledge is gained using the scientific method.

Science is a way of asking questions about the natural world and sometimes obtaining precise answers to them. Although science, in the modern sense, has arisen recently in human history (within the last 200 years or so), the tradition of asking questions about the natural world is an ancient one. In this section, we examine the methodology that zoology shares with science as a whole. These features distinguish sciences from activities that we exclude from the realm of science, such as art and religion.

Despite an enormous impact of science on our lives, many people have only a minimal understanding of the nature of science. For example, on March 19, 1981, the governor of Arkansas signed into law the Balanced Treatment for Creation-Science and Evolution-Science Act (Act 590 of 1981). This act falsely presented "creation-science" as a valid scientific endeavor. "Creation-science" is actually a religious position advocated by a minority of the American religious community, and it does not qualify as science. The enactment of this law led to a historic lawsuit tried in December 1981 in the court of Judge William R. Overton, U.S.

District Court, Eastern District of Arkansas. The suit was brought by the American Civil Liberties Union on behalf of 23 plaintiffs, including religious leaders and groups representing several denominations, individual parents, and educational associations. The plaintiffs contended that the law was a violation of the First Amendment to the U.S. Constitution, which prohibits "establishment of religion" by government. This prohibition includes passing a law that would aid one religion or prefer one religion over another. On January 5, 1982, Judge Overton permanently enjoined the State of Arkansas from enforcing Act 590.

Considerable testimony during the trial dealt with the nature of science. Some witnesses defined science simply, if not very informatively, as "what is accepted by the scientific community" and "what scientists do." However, on the basis of other testimony by scientists, Judge Overton was able to state explicitly these essential characteristics of science:

1. It is guided by natural law.
2. It has to be explanatory by reference to natural law.
3. It is testable against the observable world.
4. Its conclusions are tentative and therefore not necessarily the final word.
5. It is falsifiable.

Pursuit of scientific knowledge must be guided by the physical and chemical laws that govern the state of existence. Scientific knowledge must explain what is observed by reference to natural law without requiring intervention of a supernatural being or force. We must be able to observe events in the real world, directly or indirectly, to test hypotheses about nature. If we draw a conclusion relative to some event, we must be ready always to discard or to modify our conclusion if further observations contradict it. As Judge Overton stated, "While anybody is free to approach a scientific inquiry in any fashion they choose, they cannot properly describe the methodology used as scientific if they start with a conclusion and refuse to change it regardless of the evidence developed during the course of the investigation." Science is separate from religion, and the results of science do not favor one religious position over another.

Unfortunately, the religious position formerly called "creation-science" has reappeared in American politics with the name "intelligent-design theory." We are forced once again to defend the teaching of science against this scientifically meaningless dogma.

Scientific Method

These essential criteria of science form the **hypothetico-deductive method.** The first step of this method is the generation of hypotheses or potential answers to the question being asked. These hypotheses are usually based on prior observations of nature or derived from theories based on such observations. Scientific hypotheses often constitute general statements about nature that may explain a large number of diverse observations. Darwin's hypothesis of natural selection, for example, explains the observations that many different species have properties that adapt them to their environments.

On the basis of the hypothesis, a scientist must make a prediction about future observations. The scientist must say, "If my hypothesis is a valid explanation of past observations, then future observations ought to have certain characteristics." The best hypotheses are those that make many predictions which, if found erroneous, will lead to rejection, or falsification, of the hypothesis.

The scientific method may be summarized as a series of steps:

1. Observation
2. Question
3. Hypothesis
4. Empirical test
5. Conclusions
6. Publication

Observations illustrated in Figure 1.1A-E form a critical first step in evaluating the life histories of natural populations. For example, observations of crab larvae shown in Figure 1.1E might cause the observer to question whether rate of larval growth is higher in undisturbed populations than in ones exposed to a chemical pollutant. A null hypothesis is then generated to permit an empirical test. A null hypothesis is one worded in a way that would permit data to reject it if it is false. In this case, the null hypothesis is that larval growth rates for crabs in undisturbed habitats are the same as those in polluted habitats. The investigator then performs an empirical test by gathering data on larval growth rates in a set of undisturbed crab populations and a set of populations subjected to the chemical pollutant. Ideally, the undisturbed populations and the chemically treated populations are equivalent for all conditions except presence of the chemical in question. If measurements show consistent differences in growth rate between the two sets of populations, the null hypothesis is rejected. One then concludes that the chemical pollutant does alter larval growth rates. A statistical test is usually needed to ensure that the differences between the two groups are greater than would be expected from chance fluctuations alone. If the null hypothesis cannot be rejected, one concludes that the data do not show any effect of the chemical treatment. The results of the study are then published to communicate findings to other researchers, who may repeat the results, perhaps using additional populations of the same or a different species. Conclusions of the initial study then serve as the observations for further questions and hypotheses to reiterate the scientific process.

Note that a null hypothesis cannot be proved correct using the scientific method. If the available data are compatible with it, the hypothesis serves as a guide for collecting additional data that potentially might reject it. Our most successful hypotheses are the ones that make specific predictions confirmed by large numbers of empirical tests.

The hypothesis of natural selection was invoked to explain variation observed in British moth populations (Figure 1.11). In industrial areas of England having heavy air pollution, many populations of moths contain primarily darkly pigmented (melanic) individuals, whereas moth populations inhabiting clean forests show a much higher frequency of lightly pigmented individuals. The hypothesis suggests that moths can survive most effectively by matching their surroundings, thereby remaining invisible to birds that seek to eat them. Experimental studies have shown that, consistent with this hypothesis, birds are able to locate and then to eat moths that do not match their surroundings. Birds in the same area frequently fail to find moths that match their surroundings, leaving them to reproduce and to increase their numbers relative to conspicuous moths. Another testable prediction of the hypothesis of natural selection is that when polluted areas are cleaned, the moth populations should demonstrate an increase in frequency of lightly pigmented individuals. Observations of such populations confirmed the result predicted by natural selection.

If a hypothesis is very powerful in explaining a wide variety of related phenomena, it attains the status of a **theory.** Natural selection is a good example. Our example of the use of natural selection to explain observed pigmentation patterns in moth populations is only one of many phenomena to which natural selection applies. Natural selection provides a potential explanation for the occurrence of many different traits distributed among virtually all animal species. Each of these instances constitutes a

A

B

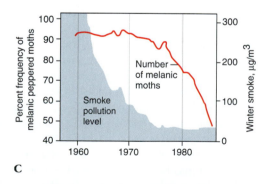
C

Figure 1.11

Light and melanic forms of the peppered moth, *Biston betularia* on **A,** a lichen-covered tree in unpolluted countryside and **B,** a soot-covered tree near industrial Birmingham, England. These color variants have a simple genetic basis. **C,** Recent decline in the frequency of the melanic form of the peppered moth with falling air pollution in industrial areas of England. The frequency of the melanic form still exceeded 90% in 1960, when smoke and sulfur dioxide emissions were still high. Later, as emissions fell and light-colored lichens began to grow again on the tree trunks, the melanic form became more conspicuous to predators. By 1986, only 50% of the moths were still of the melanic form, the rest having been replaced by the light form.

Ethics in Animal Research

The use of animals to serve human needs raises challenging ethical questions. Most controversial is the issue of animal use in biomedical and behavioral research and in testing commercial products.

Congress has passed a series of amendments to the Federal Animal Welfare Act, a body of laws covering care of vertebrate animals in laboratories and other facilities. These amendments are known as the three R's: *Reduction* in the number of animals needed for research; *Refinement* of techniques that might cause stress or suffering; *Replacement* of live animals with simulations or cell cultures whenever possible. As a result, the total number of animals used each year in research and in testing of commercial products has declined. Developments in cellular and molecular biology also have contributed to a decreased use of animals for research and testing. An animal rights movement has created an awareness of the needs of animals used in research and has stimulated researchers to discover more humane alternatives.

Computers and culturing of cells can substitute for experiments on animals only when the basic principles involved are well known. When the principles themselves are being scrutinized and tested, computer modeling is not sufficient. The National Research Council concedes that although the search for alternatives to animals in research and testing will continue, "the chance that alternatives will completely replace animals in the foreseeable future is nil." Realistic immediate goals, however, are reduction in number of animals used, replacement of mammals with other vertebrates, and refinement of experimental procedures to reduce discomfort of the animals being tested.

Medical and veterinary progress depends on research using animals. Every drug and vaccine developed to improve the human condition has been tested first on animals. Research using animals has enabled medical science to eliminate smallpox and polio from at least some parts of the world, and to immunize against diseases previously common and often deadly, including diphtheria, mumps, and rubella. It also has helped to create treatments for cancer, diabetes, heart disease, and depression, and to develop surgical procedures including heart surgery, blood transfusions, and cataract removal. AIDS research is wholly dependent on studies using animals. The similarity of simian AIDS, identified in rhesus monkeys, to human AIDS has permitted the disease in monkeys to serve as a model for the human disease. Recent work indicates that cats, too, may be useful models for development of an AIDS vaccine. Skin grafting experiments, first done with cattle and later with other animals, opened a new era in immunological research with vast contributions to treatment of disease in humans and other animals.

Research using animals also has benefited *other animals* through the development of veterinary cures. The vaccines for feline leukemia and canine parvovirus were first introduced to other cats and dogs. Many other vaccinations for serious diseases of animals were developed through research on animals—for example, rabies, distemper, anthrax, hepatitis, and tetanus. No endangered species is used in general research (except to protect that species from total extinction). Thus, research using animals has provided enormous benefits to humans and other animals. Still, much remains to be learned about treatment of diseases such as cancer, AIDS, diabetes, and heart disease, and research with animals will be required for this purpose.

Despite the remarkable benefits produced by research on animals, advocates of animal rights consider the harm done to animals in some research unethical. The most extreme animal-rights activists advocate total abolition of all forms of research using animals. The scientific community is deeply concerned about the impact of such attacks on the ability of scientists to conduct important experiments that will benefit people and animals. If we are justified to use animals for food and fiber and as pets, are we not justified in experimentation to benefit human welfare when these studies are conducted humanely and ethically?

The Association for Assessment and Accreditation of Laboratory Animal Care International supports the use of animals to advance medicine and science when nonanimal alternatives are not available and when animals are treated in an ethical and humane way. Accreditation by this organization allows research institutions to demonstrate excellence in their standards of animal care. Nearly all major institutions receiving funding from the National Institutes of Health have sought and received this accreditation. See the website at *www.aaalac.org* for more information on accreditation of laboratory animal care.

References on Animal-Research Ethics

Commission on Life Sciences, National Research Council. 1988. Use of laboratory animals in biomedical and behavioral research. Washington, D.C., National Academy Press. *Statement of national policy on guidelines for use of animals in biomedical research. Includes a chapter on benefits derived from use of animals.*

Groves, J. M. 1997. Hearts and minds: the controversy over laboratory animals. Philadelphia, Pennsylvania, Temple University Press. *Thoughtful review of the controversy by an activist who conducted extensive interviews with activists and animal-research supporters.*

Paul, E. F., and J. Paul, eds. 2001. Why animal experimentation matters: the use of animals in medical research. New Brunswick, New Jersey, Social Philosophy and Policy Foundation, and Transaction Publishers. *Essays by scientists, historians, and philosophers that express a defense of animal experimentation, demonstrating its moral acceptability and historical importance.*

specific hypothesis generated from the theory of natural selection. Note, however, that falsification of a specific hypothesis does not necessarily lead to rejection of the theory as a whole. Natural selection may fail to explain origins of human behavior, for example, but it provides an excellent explanation for many structural modifications of the pentadactyl (five-fingered) vertebrate limb for diverse functions. Scientists test many subsidiary hypotheses of their major theories to ask whether their theories are generally applicable. Most useful are theories that explain the largest array of different natural phenomena.

We emphasize that the meaning of the word "theory," when used by scientists, is not "speculation" as it is in ordinary English usage. Failure to make this distinction has been prominent in creationist challenges to evolution. The creationists have spoken of evolution as "only a theory," as if it were little better than a guess. In fact, the theory of evolution is supported by such massive evidence that biologists view repudiation of evolution as tantamount to repudiation of reason. Nonetheless, evolution, along with all other theories in science, is not proven in a mathematical sense, but it is testable, tentative, and falsifiable. Powerful theories that guide extensive research are called **paradigms.** The history of science shows that even major paradigms are subject to refutation and replacement when they fail to account for our observations of the natural world. They are then replaced by new paradigms in a process called a **scientific revolution.** For example, prior to the 1800s, animal species were studied as if they were specially created entities whose essential properties remained unchanged through time. Darwin's theories led to a scientific revolution that replaced these views with the evolutionary paradigm. The evolutionary paradigm has guided biological research for more than 140 years, and to date there is no scientific evidence that falsifies it; it has strong explanatory power and continues to guide active inquiry into the natural world. Evolutionary theory is generally accepted as the cornerstone of biology.

Chemists and physicists often use the term "law" to denote highly corroborated theories that appear to apply without exception to the physical world. Such laws are considered uniform throughout time and space. Because the biological world is temporally and spatially bounded, and because evolutionary change has produced an enormous diversity of forms with different emergent properties at multiple levels (Table 1.1), biologists now avoid using the term law for their theories. Nearly all of the biological laws proposed in the past have been found to apply only to some of life's diverse forms and not to all. Mendel's laws of inheritance, for example, do not apply to bacteria and often are violated even in animal and plant species that usually follow them. Darwin's theories of perpetual change and common descent of living forms (p. 15) are perhaps the only statements that one meaningfully might call laws of biology.

Experimental versus Evolutionary Sciences

The many questions asked about the animal world since Aristotle can be grouped into two major categories.* The first category seeks to understand the **proximate** or **immediate causes** that underlie the functioning of biological systems at a particular time and place. These include the problems of explaining how animals perform their metabolic, physiological, and behavioral functions at the molecular, cellular, organismal, and even populational levels. For example, how is genetic information expressed to guide the synthesis of proteins? What causes cells to divide to produce new cells? How does population density affect the physiology and behavior of organisms?

---

*Mayr, E. 1985. Chapter 25 in D. Kohn, ed. *The Darwinian Heritage.* Princeton, Princeton University Press.

The biological sciences that investigate proximate causes are called **experimental sciences,** and they proceed using the experimental method. Our goal is to test our understanding of a biological system. We predict the results of an experimental disturbance of the system based on our current understanding of it. If our understanding is correct, then the predicted outcome should occur. If, after the experimental disturbance, we see an unexpected outcome, we then discover that our understanding is incorrect or incomplete. Experimental conditions are repeated to eliminate chance occurrences that might produce erroneous conclusions. **Controls**—repetitions of the experimental procedure that lack the disturbance—are established to eliminate unknown factors that might bias the outcome of the experiment. The processes by which animals maintain a body temperature under different environmental conditions, digest their food, migrate to new habitats, or store energy are some additional examples of physiological phenomena studied by experiment (see Chapters 29 through 36). Subfields of biology that constitute experimental sciences include molecular biology, cell biology, endocrinology, developmental biology, and community ecology.

In contrast to questions concerning the proximate causes of biological systems are questions of the **ultimate causes** that have produced these systems and their distinctive characteristics through evolutionary time. For example, what are the evolutionary factors that caused some birds to acquire complex patterns of seasonal migration between temperate and tropical areas? Why do different species of animals have different numbers of chromosomes in their cells? Why do some animal species maintain complex social systems, whereas other species have solitary individuals?

The biological sciences that address questions of ultimate cause are called **evolutionary sciences,** and they proceed largely using the **comparative method** rather than experimentation. Characteristics of molecular biology, cell biology, organismal structure, development, and ecology are compared among related species to identify their patterns of variation. The patterns of similarity and dissimilarity are then used to test hypotheses of relatedness, and thereby to reconstruct the evolutionary tree that relates the species being studied. Recent advances in DNA sequencing technology permit detailed tests of relationships among all animal species. The evolutionary tree is then used to examine hypotheses of the evolutionary origins of the diverse molecular, cellular, organismal, and populational properties observed in the animal world. Clearly, evolutionary sciences rely on results of experimental sciences as a starting point. Evolutionary sciences include comparative biochemistry, molecular evolution, comparative cell biology, comparative anatomy, comparative physiology, and phylogenetic systematics.

A scientist's use of the phrase "ultimate cause," unlike Aristotle's usage, does not imply a preconceived goal for natural phenomena. An argument that nature has a predetermined goal, such as evolution of the human mind, is termed teleological. **Teleology** is the mistaken notion that evolution of living organisms is guided by purpose toward an optimal design. A major success of Darwinian evolutionary theory is its rejection of teleology in explaining biological diversification.

THEORIES OF EVOLUTION AND HEREDITY

We turn now to a specific consideration of the two major paradigms that guide zoological research today: Darwin's theory of evolution and the chromosomal theory of inheritance.

Darwin's Theory of Evolution

Darwin's theory of evolution is now over 140 years old (see Chapter 6). Darwin articulated the complete theory when he published his famous book *On the Origin of Species by Means of Natural Selection* in England in 1859 (Figure 1.12). Biologists today are frequently asked, "What is Darwinism?" and "Do biologists still accept Darwin's theory of evolution?" These questions cannot be given simple answers, because Darwinism encompasses several different, although mutually compatible, theories. Professor Ernst Mayr of Harvard University argued that Darwinism should be viewed as five major theories. These five theories have somewhat different origins and different fates and cannot be treated as only a single statement. The theories are (1) perpetual change, (2) common descent, (3) multiplication of species, (4) gradualism, and (5) natural selection. The first three theories are generally accepted as having universal application throughout the living world. The theories of gradualism and natural selection are controversial among evolutionists, although both are strongly advocated by a large portion of the evolutionary community and are important components of the Darwinian evolutionary paradigm. Gradualism and natural selection are clearly part of the evolutionary process, but their explanatory power might not be as widespread as Darwin intended. Legitimate scientific controversies regarding gradualism and natural selection often are misrepresented by creationists as challenges to the first three theories listed, although the validity of those first three theories is strongly supported by all relevant observations.

1. **Perpetual change.** This is the basic theory of evolution on which the others are based. It states that the living world is neither constant nor perpetually cycling, but is always changing. The properties of organisms undergo transformation across generations throughout time. This theory originated in antiquity but did not gain widespread acceptance until Darwin advocated it in the context of his other four theories. "Perpetual change" is documented by the fossil record, which clearly refutes creationists' claims for a recent origin of all living forms. Because it has withstood repeated testing and is supported by an overwhelming number of observations, we now regard "perpetual change" as a scientific fact.

2. **Common descent.** The second Darwinian theory, "common descent," states that all forms of life descended from a common ancestor through a branching of lineages (Figure 1.13). The opposing argument, that the different forms of life arose independently and descended to the present

Figure 1.12

Modern evolutionary theory is strongly identified with Charles Robert Darwin, who, with Alfred Russel Wallace, provided the first credible explanation of evolution. This photograph of Darwin was taken in 1854 when he was 45 years old. His most famous book, *On the Origin of Species,* appeared five years later.

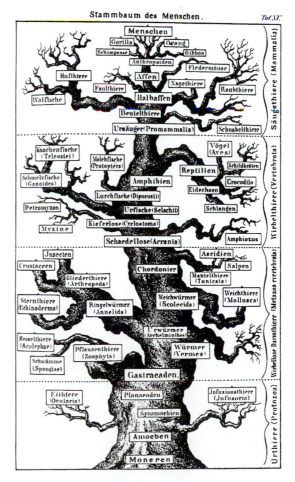

Figure 1.13

An early tree of life drawn in 1874 by the German biologist, Ernst Haeckel, who was strongly influenced by Darwin's theory of common descent. Many of the phylogenetic hypotheses shown in this tree, including the unilateral progression of evolution toward humans (= Menschen, *top*), have been refuted.

in linear, unbranched genealogies, has been refuted by comparative studies of organismal form, cell structure, and macromolecular structures (including those of the genetic material, DNA). All of these studies confirm the theory that life's history has the structure of a branching evolutionary tree, called a **phylogeny.** Species that share recent common ancestry have more similar features at all levels than do species whose most recent common ancestor is an ancient one. Much current research is guided by Darwin's theory of common descent toward reconstructing life's phylogeny using the patterns of similarity and dissimilarity observed among species. The resulting phylogeny serves as the basis for our taxonomic classification of animals (see Chapter 10).

3. **Multiplication of species.** Darwin's third theory states that the evolutionary process produces new species by splitting and transforming older ones. Species are now generally viewed as reproductively distinct populations of organisms that usually but not always differ from each other in organismal form. Once species are fully formed, interbreeding among members of different species does not occur or is too restricted to permit the species' lineages to merge. Evolutionists generally agree that the splitting and transformation of lineages produces new species, although there is still much controversy concerning details of this process (see Chapter 6) and the precise meaning of the term "species" (see Chapter 10). Much active scientific research examines historical processes that generate new species.

4. **Gradualism.** Gradualism states that the large differences in anatomical traits that characterize diverse species originate through the accumulation of many small incremental changes over very long periods of time. This theory is important because genetic changes having very large effects on organismal form are usually harmful to an organism. It is possible, however, that some genetic variants that have large effects are nonetheless sufficiently beneficial to be favored by natural selection. Therefore, although gradual evolution is known to occur, it may not explain the origins of all structural differences that we observe among species (Figure 1.14). Scientists are still actively studying this question.

5. **Natural selection.** Natural selection, Darwin's most famous theory, rests on three propositions. First, there is variation among organisms (within populations) for anatomical, behavioral, and physiological traits. Second, the variation is at least partly heritable so that offspring tend to resemble their parents. Third, organisms with different variant forms are expected to leave different numbers of offspring to future generations. Variants that permit their possessors most effectively to exploit their environments will preferentially survive to be transmitted to future generations. Over many generations, favorable new traits will spread throughout a population. Accumulation of such changes leads, over long periods of time, to production of new organismal characteristics and new species. Natural selection is therefore a creative process that generates novel forms from the small individual variations that occur among organisms within a population.

Figure 1.14

Gradualism provides a plausible explanation for the origins of different bill shapes in the Hawaiian honeycreepers shown here. This theory has been challenged, however, as an explanation of the evolution of such structures as vertebrate scales, feathers, and hair from a common ancestral structure. The geneticist Richard Goldschmidt viewed the latter forms as unbridgeable by any gradual transformation series.

Figure 1.15
According to Darwinian evolutionary theory, the different forms of these vertebrate forelimbs were molded by natural selection to adapt them for different functions. We show in later chapters that, despite these adaptive differences, these limbs share basic structural similarities.

Natural selection explains why organisms are constructed to meet the demands of their environments, a phenomenon called **adaptation** (Figure 1.15). Adaptation is the expected result of a process that accumulates the most favorable variants occurring in a population throughout long periods of evolutionary time. Adaptation was viewed previously as strong evidence against evolution, and Darwin's theory of natural selection was therefore important for convincing people that a natural process, capable of being studied scientifically, could produce new species. The demonstration that natural processes could produce adaptation was important to the eventual acceptance of all five Darwinian theories.

Darwin's theory of natural selection faced a major obstacle when it was first proposed: it lacked a successful theory of heredity. People assumed incorrectly that heredity was a blending process, and that any favorable new variant appearing in a population therefore would be lost. The new variant arises initially in a single organism, and that organism therefore must mate with one lacking the favorable new trait. Under blending inheritance, the organism's offspring would then have only a diluted form of the favorable trait. These offspring likewise would mate with others that lack the favorable trait. With its effects diluted by half each generation, the trait eventually would cease to exist. Natural selection would be completely ineffective in this situation.

Darwin was never able to counter this criticism successfully. It did not occur to Darwin that hereditary factors could be discrete and nonblending and that a new genetic variant therefore could persist unaltered from one generation to the next. This principle is called **particulate inheritance.** It was established after 1900 with the discovery of Gregor Mendel's genetic experiments, and it was eventually incorporated into what we now call the **chromosomal theory of inheritance.** We use the term **neo-Darwinism** to describe Darwin's theories as modified by incorporating this theory of inheritance.

Mendelian Heredity and the Chromosomal Theory of Inheritance
The chromosomal theory of inheritance is the foundation for current studies of genetics and evolution in animals (see Chapters 5 and 6). This theory comes from the consolidation of research done in the fields of genetics, which was founded by the experimental work of Gregor Mendel (Figure 1.16), and cell biology.

Genetic Approach
The genetic approach consists of mating or "crossing" populations of organisms that are true-breeding for alternative traits, and then following hereditary transmission of those traits through subsequent generations. "True-breeding" means that a population maintains across generations only one of the alternative traits when propagated in isolation from other populations. For example, most populations of fruit flies produce only red-eyed individuals, generation after generation, regardless of the environments in which they are raised; such strains are true-breeding for red eyes. Some laboratory strains of fruit flies produce only white-eyed individuals and are therefore true-breeding for white eyes (p. 88).

Gregor Mendel studied the transmission of seven variable features in garden peas, crossing populations that were true-breeding for alternative traits (for example, tall versus short plants). In the first generation (called the F_1 generation, for "filial"), only one of the alternative parental traits was observed; there was no indication of blending of the parental traits. In the example, the offspring (called F_1 *hybrids* because they represent a cross between two different forms) formed by crossing the tall and short plants were tall, regardless of whether the tall trait was inherited from the male or the female parent. These F_1 hybrids were allowed to self-pollinate, and both parental traits were found among their offspring (called the F_2 generation), although the trait observed in the F_1 hybrids (tall plants in this example) was

A

Figure 1.16

A, Gregor Johann Mendel. **B,** The monastery in Brno, Czech Republic, now a museum, where Mendel performed his experiments with garden peas.

B

three times more common than the other trait. Again, there was no indication of blending of the parental traits (Figure 1.17).

Mendel's experiments showed that the effects of a genetic factor can be masked in a hybrid individual, but that these factors are not physically altered during the transmission process. He postulated that variable traits are specified by paired hereditary factors, which we now call "genes." When **gametes** (eggs or sperm) are produced, the two genes controlling a particular

feature are segregated from each other and each gamete receives only one of them. Fertilization restores the paired condition. If an organism possesses different forms of the paired genes for a feature, only one of them is expressed in its appearance, but both genes nonetheless are transmitted unaltered in equal numbers to the gametes produced. Transmission of these genes is particulate, not blending. Mendel observed that inheritance of one pair of traits is independent of inheritance of other paired

Figure 1.17

Different predictions of particulate versus blending inheritance regarding the outcome of Mendel's crosses of tall and short plants. The prediction of particulate inheritance is upheld and the prediction of blending inheritance is falsified by the results of the experiments. The reciprocal experiments (crossing short female parents with tall male parents) produced similar results. (P_1 = parental generation; F_1 = first filial generation; F_2 = second filial generation.)

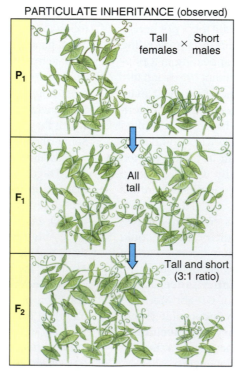

PARTICULATE INHERITANCE (observed)

P_1 — Tall females × Short males

F_1 — All tall

F_2 — Tall and short (3:1 ratio)

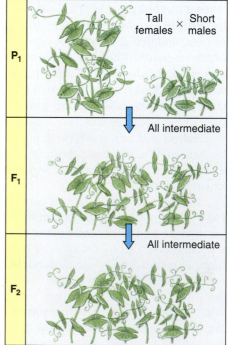

BLENDING INHERITANCE (not observed)

P_1 — Tall females × Short males

F_1 — All intermediate

F_2 — All intermediate

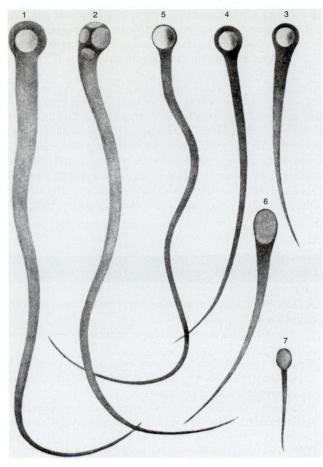

Figure 1.18
An early nineteenth-century micrographic drawing of sperm from (1) guinea pig, (2) white mouse, (3) hedgehog, (4) horse, (5) cat, (6) ram, and (7) dog. Some biologists initially interpreted these as parasitic worms in the semen, but in 1824, Jean Prévost and Jean Dumas correctly identified their role in egg fertilization.

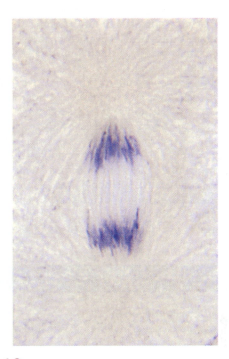

Figure 1.19
Paired chromosomes being separated before nuclear division in the process of forming gametes.

traits. We now know, however, that not all pairs of traits are inherited independently of each other; different traits that tend to be inherited together are said to be genetically linked (p. 88). Numerous studies, particularly of the fruit fly, *Drosophila melanogaster,* have shown that principles of inheritance discovered initially in plants apply also to animals.

Contributions of Cell Biology

Improvements in microscopes during the 1800s permitted cytologists to study the production of gametes by direct observation of reproductive tissues. Interpreting the observations was initially difficult, however. Some prominent biologists hypothesized, for example, that sperm were parasitic worms in semen (Figure 1.18). This hypothesis was soon falsified, and the true nature of gametes was clarified. As the precursors of gametes prepare to divide early in gamete production, the nuclear material condenses to reveal discrete, elongate structures called chromosomes. Chromosomes occur in pairs that are usually similar but not identical in appearance and informational content. The number of chromosomal pairs varies among species. One member of each pair is derived from the female parent and the other from the male parent. Paired chromosomes are physically associated and then segregated into different daughter cells during cell division prior to gamete formation (Figure 1.19). Each resulting gamete receives one chromosome from each pair. Different pairs of chromosomes are sorted into gametes independently of each other. Because the behavior of chromosomal material during gamete formation parallels that postulated for Mendel's genes, Sutton and Boveri in 1903 through 1904 hypothesized that chromosomes were the physical bearers of genetic material. This hypothesis met with extreme skepticism when first proposed. A long series of tests designed to falsify it nonetheless showed that its predictions were upheld. The chromosomal theory of inheritance is now well established.

SUMMARY

Zoology is the scientific study of animals, and it is part of biology, the scientific study of life. Animals and life in general can be identified by attributes that they have acquired over their long evolutionary histories. The most outstanding attributes of life include chemical uniqueness, complexity and hierarchical organization, reproduction, possession of a genetic program, metabolism, development, interaction with the environment, and movement. Biological systems comprise a hierarchy of integrative levels (molecular, cellular, organismal, populational, and species levels), each of which demonstrates a number of specific emergent properties.

Science is characterized by the acquisition of knowledge by constructing and then testing hypotheses through observations of the natural world. Science is guided by natural law, and its hypotheses are testable, tentative, and falsifiable. Zoological sciences can be subdivided into two categories, experimental sciences and evolutionary sciences. Experimental sciences use the experimental method to ask how animals perform their basic metabolic, developmental, behavioral, and reproductive functions, including investigations of their molecular, cellular, and populational systems. Evolutionary sciences use the comparative method to reconstruct the history of life, and then use that history to understand how diverse species and their molecular, cellular, organismal, and populational properties arose through evolutionary time. Hypotheses that withstand

repeated testing and therefore explain many diverse phenomena gain the status of a theory. Powerful theories that guide extensive research are called "paradigms." The major paradigms that guide the study of zoology are Darwin's theory of evolution and the chromosomal theory of inheritance.

The principles given in this chapter illustrate the unity of biological science. All components of biological systems are guided by natural laws and are constrained by those laws. Living organisms arise only from other living organisms, just as new cells can be produced only from preexisting cells. Reproductive processes occur at all levels of the biological hierarchy and demonstrate both heredity and variation. Interaction of heredity and variation at all levels of the biological hierarchy produces evolutionary change and has generated the great diversity of animal life documented throughout this book.

REVIEW QUESTIONS

1. Why is life difficult to define?
2. What are the basic chemical differences that distinguish living from nonliving systems?
3. Describe the hierarchical organization of life. How does this organization lead to the emergence of new properties at different levels of biological complexity?
4. What is the relationship between heredity and variation in reproducing biological systems?
5. Describe how evolution of complex organisms is compatible with the second law of thermodynamics.
6. What are the essential characteristics of science? Describe how evolutionary studies fit these characteristics whereas "scientific creationism" or "intelligent-design theory" does not.
7. Use studies of natural selection in British moth populations to illustrate the hypothetico-deductive method of science.

8. How do we distinguish the terms hypothesis, theory, paradigm, and scientific fact?
9. How do biologists distinguish experimental and evolutionary sciences?
10. What are Darwin's five theories of evolution (as identified by Ernst Mayr)? Which are accepted as fact and which continue to stir controversy among biologists?
11. What major obstacle confronted Darwin's theory of natural selection when it was first proposed? How was this obstacle overcome?
12. How does neo-Darwinism differ from Darwinism?
13. Describe the respective contributions of the genetic approach and cell biology to formulating the chromosomal theory of inheritance.

SELECTED REFERENCES

Futuyma, D. J. 1995. Science on trial: the case for evolution. Sunderland, Massachusetts, Sinauer Associates, Inc. *A defense of evolutionary biology as the exclusive scientific approach to the study of life's diversity.*

Kitcher, P. 1982. Abusing science: the case against creationism. Cambridge, Massachusetts, MIT Press. *A treatise on how knowledge is gained in science and why creationism does not qualify as science. Note that the position refuted as "scientific creationism" in this book is equivalent in content to the position more recently termed "intelligent-design theory."*

Kuhn, T. S. 1970. The structure of scientific revolutions, ed. 2, enlarged. Chicago, University of Chicago Press. *An influential and controversial commentary on the process of science.*

Mayr, E. 1982. The growth of biological thought: diversity, evolution and inheritance. Cambridge, Massachusetts, The Belknap Press of Harvard University Press. *An interpretive history of biology with special reference to genetics and evolution.*

Medawar, P. B. 1989. Induction and intuition in scientific thought. London, Methuen & Company. *A commentary on the basic philosophy and methodology of science.*

Moore, J. A. 1993. Science as a way of knowing: the foundations of modern biology. Cambridge, Massachusetts, Harvard University Press. *A lively, wide-ranging account of the history of biological thought and the workings of life.*

Perutz, M. F. 1989. Is science necessary? Essays on science and scientists. New York, E. P. Dutton. *A general discussion of the utility of science.*

Pigliucci, M. 2002. Denying evolution: creationism, scientism, and the nature of science. Sunderland, Massachusetts, Sinauer Associates, Inc. *A critique of science education and the public perception of science.*

Rennie, J. 2002. 15 answers to creationist nonsense. Sci. Am. 287:78–85 (July). *A guide to the most common arguments used by creationists against evolutionary biology, with concise explanations of the scientific flaws of creationists' claims.*

ONLINE LEARNING CENTER

Visit www.mhhe.com/hickmanipz14e for chapter quizzing, key term flash cards, web links, and more!

The Origin and Chemistry of Life

Earth's abundant supply of water was critical for the origin of life.

Spontaneous Generation of Life?

From ancient times, people commonly thought that life arose repeatedly by spontaneous generation from nonliving material in addition to parental reproduction. For example, frogs appeared to arise from damp earth, mice from putrefied matter, insects from dew, and maggots from decaying meat. Warmth, moisture, sunlight, and even starlight often were mentioned as factors that encouraged spontaneous generation of living organisms.

Among the efforts to synthesize organisms in the laboratory is a recipe for making mice, given by the Belgian plant nutritionist Jean Baptiste van Helmont (1648). "If you press a piece of underwear soiled with sweat together with some wheat in an open jar, after about 21 days the odor changes and the ferment . . . changes the wheat into mice. But what is more remarkable is that the mice which came out of the wheat and underwear were not small mice, not even miniature adults or aborted mice, but adult mice emerge!"

In 1861, the great French scientist Louis Pasteur convinced scientists that living organisms cannot arise spontaneously from nonliving matter. In his famous experiments, Pasteur introduced fermentable material into a flask with a long S-shaped neck that was open to air. The flask and its contents were then boiled for a long time to kill any microorganisms that might be present. Afterward the flask was cooled and left undisturbed. No fermentation occurred because all organisms that entered the open end were deposited in the neck and did not reach the fermentable material. When the neck of the flask was removed, microorganisms in the air promptly entered the fermentable material and proliferated. Pasteur concluded that life could not originate in the absence of previously existing organisms and their reproductive elements, such as eggs and spores. Announcing his results to the French Academy, Pasteur proclaimed, "Never will the doctrine of spontaneous generation arise from this mortal blow."

All living organisms share a common ancestor, most likely a population of colonial microorganisms that lived almost 4 billion years ago. This common ancestor was itself the product of a long period of prebiotic assembly of nonliving matter, including organic molecules and water, to form self-replicating units. All living organisms retain a fundamental chemical composition inherited from their ancient common ancestor.

According to the big-bang model, the universe originated from a primeval fireball and has been expanding and cooling since its inception 10 to 20 billion years ago. The sun and planets formed approximately 4.6 billion years ago from a spherical cloud of cosmic dust and gases. The cloud collapsed under the influence of its own gravity into a rotating disc. As material in the central part of the disc condensed to form the sun, gravitational energy was released as radiation. The pressure of this outwardly directed radiation prevented a collapse of the nebula into the sun. The material left behind cooled and eventually produced the planets, including earth (Figure 2.1).

In the 1920s, Russian biochemist Alexander I. Oparin and British biologist J. B. S. Haldane independently proposed that life originated on earth after an inconceivably long period of "abiogenic molecular evolution." Rather than arguing that the first living organisms miraculously originated all at once, a notion that formerly discouraged scientific inquiry, Oparin and Haldane argued that the simplest form of life arose gradually by the progressive assembly of small molecules into more complex organic molecules. Molecules capable of self-replication eventually would be produced, ultimately leading to assembly of living microorganisms.

WATER AND LIFE

The origin and maintenance of life on earth depend critically upon water. Water is the most abundant of all compounds in cells, forming 60% to 90% of most living organisms. Water has several extraordinary properties that explain its essential role in living systems and their origin. These properties result largely from hydrogen bonds that form between its molecules (Figure 2.2).

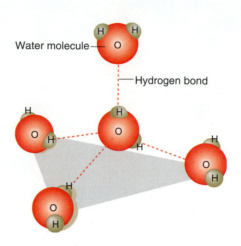

Figure 2.2
Geometry of water molecules. Each water molecule is linked by hydrogen bonds (*dashed lines*) to four other molecules. If imaginary lines connect the water molecules as shown, a tetrahedron is obtained.

Figure 2.1
Solar system showing narrow range of thermal conditions suitable for life.

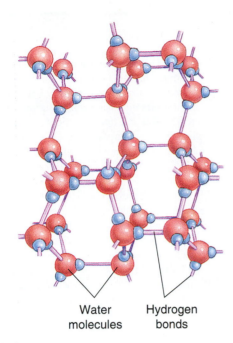

Figure 2.3

When water freezes at 0° C, the four partial charges of each atom in the molecule interact with the opposite charges of atoms in other water molecules. The hydrogen bonds between all the molecules form a crystal-like lattice structure, and the molecules are farther apart (and thus less dense) than when some of the molecules have not formed hydrogen bonds at 4° C.

Water has a **high specific heat capacity:** 1 calorie* is required to elevate the temperature of 1 g of water 1° C, a higher thermal capacity than any other liquid except ammonia. Much of this heat energy is used to rupture some hydrogen bonds in addition to increasing the kinetic energy (molecular movement), and thus the temperature, of the water. Water's high thermal capacity greatly moderates environmental temperature changes, thereby protecting living organisms from extreme thermal fluctuation. Water also has a **high heat of vaporization,** requiring more than 500 calories to convert 1 g of liquid water to water vapor. All hydrogen bonds between a water molecule and its neighbors must be ruptured before that water molecule can escape the surface and enter the air. For terrestrial animals (and plants), cooling produced by evaporation of water is important for expelling excess heat.

Another property of water important for life is its **unique density behavior** during changes of temperature. Most liquids become denser with decreasing temperature. Water, however, reaches its maximum density at 4° C *while still a liquid,* then becomes less dense with further cooling (Figure 2.3). Therefore, ice *floats* rather than sinking to the bottoms of lakes and ponds. If

---

*A calorie is defined as the amount of heat required to heat 1 g of water from 14.5° to 15.5° C. Although the calorie is the traditional unit of heat widely used in publications and tables, it is not part of the International System of Units (the SI system) which uses the joule (J) as the energy unit (1 cal = 4.184 J).

Figure 2.4

Because of hydrogen bonds between water molecules at the water-air interface, the water molecules cling together and create a high surface tension. Thus some insects, such as this water strider, can literally walk on water.

ice were denser than liquid water, bodies of water would freeze solid from the bottom upward in winter and might not melt completely in summer. Such conditions would severely limit aquatic life. In ice, water molecules form an extensive, open, crystal-like network supported by hydrogen bonds that connect all molecules. The molecules in this lattice are farther apart, and thus less dense, than in liquid water at 4° C.

Water has **high surface tension,** exceeding that of any other liquid but mercury. Hydrogen bonding among water molecules produces a cohesiveness important for maintaining protoplasmic form and movement. The resulting surface tension creates an ecological niche (see p. 826) for insects, such as water striders and whirligig beetles, that skate on the surfaces of ponds (Figure 2.4). Despite its high surface tension, water has **low viscosity,** permitting movement of blood through minute capillaries and of cytoplasm inside cellular boundaries.

Water is an excellent **solvent.** Salts dissolve more extensively in water than in any other solvent. This property results from the dipolar nature of water, which causes it to orient around charged particles dissolved in it. When, for example, crystalline NaCl dissolves in water, the Na^+ and Cl^- ions separate (Figure 2.5). The negative zones of the water dipoles attract the Na^+ ions while the positive zones attract the Cl^- ions. This orientation keeps the ions separated, promoting their dissociation. Solvents lacking this dipolar character are less effective at keeping the ions separated. Binding of water to dissolved protein molecules is essential to the proper functioning of many proteins.

Water also participates in many chemical reactions in living organisms. Many compounds are split into smaller pieces by the addition of a molecule of water, a process called **hydrolysis.** Likewise, larger compounds may be synthesized from smaller components by the reverse of hydrolysis, called **condensation reactions.**

$$R - R + H_2O \xrightarrow{\text{Hydrolysis}} R - OH + H - R$$

$$R - OH + H - R \xrightarrow{\text{Condensation}} R - R + H_2O$$

pH of Water Solutions

In pure liquid water (= distilled water), a small fraction of the water molecules split into ions of hydrogen (H^+) and hydroxide (OH^-); the concentration of both ions is 10^{-7} moles/liter. An acidic substance, when dissolved in water, contributes H^+ ions to solution, thereby increasing their concentration and causing an excess of H^+ ions over OH^- ions in solution. A basic substance does the reverse, contributing OH^- ions to the solution and making OH^- ions more common than H^+ ions. The degree to which a solution is acidic or basic is critical for most cellular processes and requires precise quantification and control; the structure and function of dissolved proteins, for example, depend critically on the concentration of H^+ in the solution.

The **pH** scale quantifies the degree to which a solution is acidic or basic. The scale ranges from 0 to 14 and represents the additive inverse of the logarithm (base 10) of the H^+ concentration (in moles/liter) of the solution. Pure liquid water therefore has a pH of 7 (H^+ concentration = 10^{-7} moles/liter). A solution with pH = 6.0 has an H^+ concentration ten times higher than that of pure water and is acidic, whereas a solution with pH = 8.0 has an H^+ concentration ten times lower than pure water and is basic. A concentrated strong acid, such as hydrochloric acid (HCl, known commercially as "muriatic acid" used to clean masonry) has an H^+ concentration of ~$1 = 10^0$ mole/liter, giving a pH of 0 (a concentration of H^+ 10,000,000 times that of pure water). A concentrated base, such as sodium hydroxide (NaOH, used commercially in liquid drain cleaners) has an H^+ concentration of approximately 10^{-14} mole/liter, giving a pH of 14.

A buffer is a dissolved substance (solute) that causes a solution to resist changes in pH because the buffer can remove added H^+ and OH^- ions from solution by binding them into compounds. Dissolved carbon dioxide in the form of bicarbonate (HCO_3^-) is a buffer that helps to protect human blood (pH = 7.3 to 7.5) from changes in pH. H^+ ions are removed from solution when they react with bicarbonate ions to form carbonic acid, which then dissociates into carbon dioxide and water. The excess carbon dioxide is removed during exhalation (p. 703). OH^- ions are removed from solution when this reaction is reversed, forming bicarbonate and hydrogen ions. The excess bicarbonate ions are secreted in the urine (p. 676), and the hydrogen ions serve to increase blood pH back to normal levels. Severe health problems occur if the pH of blood drops to 7 or rises to 7.8.

Because water is critical to the support of life, the continuing search for extraterrestrial life usually begins with a search for water. Plans for a human outpost on the moon likewise depend upon finding water there. As we write, NASA is planning to crash a space probe into the moon in 2009 in a search for ice; the moon's south pole is a prime candidate for a human outpost if ice is found there.

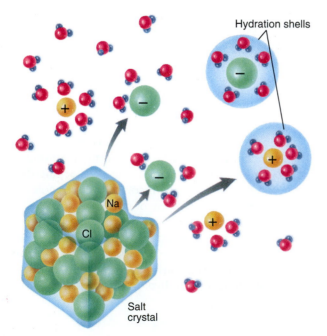

Figure 2.5

When a crystal of sodium chloride dissolves in water, the negative ends of the dipolar molecules of water surround the Na^+ ions, while the positive ends of water molecules face the Cl^- ions. The ions are thus separated and do not reenter the salt lattice.

ORGANIC MOLECULAR STRUCTURE OF LIVING SYSTEMS

Chemical evolution in the prebiotic environment produced simple organic compounds that ultimately formed the building blocks of living cells. The term "organic" refers broadly to compounds that contain carbon. Many also contain hydrogen, oxygen, nitrogen, sulfur, phosphorus, salts, and other elements. Carbon has a great ability to bond with other carbon atoms in chains of varying lengths and configurations. Carbon-to-carbon combinations introduce the possibility of enormous complexity and variety into molecular structure. More than a million organic compounds are known.

We review the kinds of organic molecules found in living systems, followed by further discussion of their origins in earth's primitive reducing atmosphere.

Carbohydrates: Nature's Most Abundant Organic Substance

Carbohydrates are compounds of carbon, hydrogen, and oxygen. These elements usually occur in the ratio of 1 C: 2 H: 1 O and are grouped as H—C—OH. Carbohydrates function in protoplasm mainly as structural elements and as a source of chemical energy. Glucose is the most important of these energy-storing carbohydrates. Familiar examples of carbohydrates include sugars, starches, and cellulose (the woody structure of plants). Cellulose occurs on earth in greater quantities than all other organic materials combined. Carbohydrates are synthesized by green plants from water and carbon dioxide, with the aid of solar energy. This process, called **photosynthesis,** is a reaction upon which all life depends, for it is the starting point in the formation of food.

Carbohydrates are usually grouped into the following three classes: (1) **monosaccharides,** or simple sugars; (2) **disaccharides,** or double sugars; and (3) **polysaccharides,** or complex sugars. Simple sugars have a single carbon chain containing 4 carbons (tetroses), 5 carbons (pentoses), or 6 carbons (hexoses). Other simple sugars have up to 10 carbons, but these sugars are not biologically important. Simple sugars, such as glucose, galactose, and fructose, all contain a free sugar group,

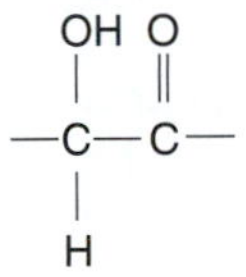

in which the double-bonded O may be attached to the terminal or nonterminal carbons of a chain. The hexose **glucose** (also called dextrose) is particularly important to the living world. Glucose is often shown as a straight chain (Figure 2.6A), but in water it forms a cyclic compound (Figure 2.6B). The "chair" diagram (Figure 2.7) of glucose best represents its true configuration, but all forms of glucose, however represented, are chemically equivalent. Other hexoses of biological significance include galactose and fructose, which are compared with glucose in Figure 2.8.

Disaccharides are double sugars formed by bonding two simple sugars. An example is maltose (malt sugar), composed of two glucose molecules. As shown in Figure 2.9, the two glucose molecules are joined by removing a molecule of water, causing the sharing of an oxygen atom by the two sugars. All disaccharides are formed in this manner. Two other common

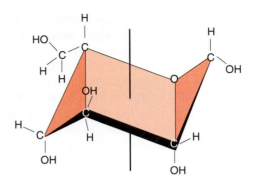

Figure 2.7
"Chair" representation of a glucose molecule.

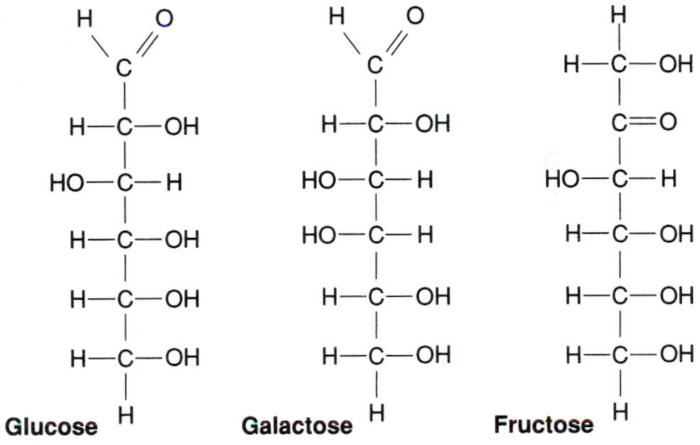

Glucose   **Galactose**   **Fructose**

Figure 2.8
These three hexoses are the most common monosaccharides.

A

B

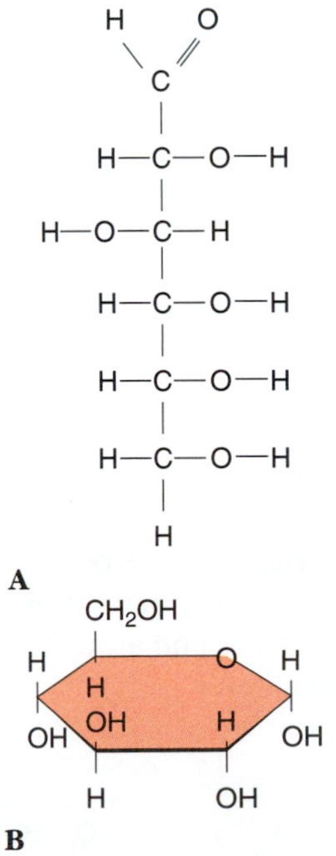

Figure 2.6
Two ways of depicting the simple sugar glucose. In **A,** the carbon atoms are shown in open-chain form. When dissolved in water, glucose tends to assume a ring form as in **B.** In this ring model the carbon atoms located at each turn in the ring are usually not shown.

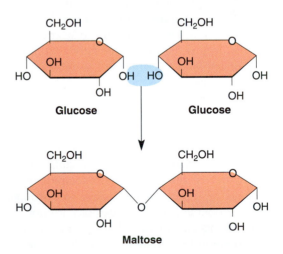

Figure 2.9
Formation of a double sugar (disaccharide maltose) from two glucose molecules with the removal of one molecule of water.

disaccharides are sucrose (ordinary cane, or table, sugar), formed by the linkage of glucose and fructose, and lactose (milk sugar), composed of glucose and galactose.

Polysaccharides are composed of many molecules of simple sugars (usually glucose) linked in long chains called polymers. Their empirical formula is usually written $(C_6H_{10}O_5)_n$, where n

designates the number of simple-sugar subunits in the polymer. Starch is the common polymer in which sugar is stored in most plants and is an important food for animals. **Chitin** is an important structural polysaccharide in the exoskeletons of insects and other arthropods (p. 404). **Glycogen** is an important polymer for storing sugar in animals. It is stored mainly in liver and muscle cells in vertebrates. When needed, glycogen is converted to glucose and delivered by blood to the tissues. Another polymer is **cellulose,** the principal structural carbohydrate of plants.

Lipids: Fuel Storage and Building Material

Lipids are fats and fatlike substances. They are molecules of low polarity; consequently, they are virtually insoluble in water but are soluble in organic solvents, such as acetone and ether. The three principal groups of lipids are neutral fats, phospholipids, and steroids.

Neutral Fats

The neutral or "true" fats are major fuels of animals. Stored fat is derived either directly from dietary fat or indirectly from dietary carbohydrates that the body has converted to fat for storage. Fats are oxidized and released into the bloodstream as needed to meet tissue demands, especially those of active muscle.

Neutral fats include triglycerides, which contain glycerol and three molecules of fatty acids. Neutral fats are therefore esters, a combination of an alcohol (glycerol) and an acid. Fatty acids in triglycerides are simply long-chain monocarboxylic acids; they vary in size but are commonly 14 to 24 carbons long. Production of a typical fat by the union of glycerol and stearic acid is shown in Figure 2.10A. In this reaction, three fatty acid molecules are seen to have united with OH groups of glycerol to form stearin (a neutral fat) plus three molecules of water.

Most triglycerides contain two or three different fatty acids attached to glycerol, and bear ponderous names such as myristoyl stearoyl glycerol (Figure 2.10B). The fatty acids in this triglyceride are **saturated;** every carbon within the chain holds two hydrogen atoms. Saturated fats, more common in animals than in plants, are usually solid at room temperature. **Unsaturated** fatty acids, typical of plant oils, have two or more carbon atoms joined by double bonds; the carbons are not "saturated" with hydrogen atoms and are available to form bonds with other atoms. Two common unsaturated fatty acids are oleic acid and linoleic acid (Figure 2.11). Plant fats, such as peanut oil and corn oil, tend to be liquid at room temperature.

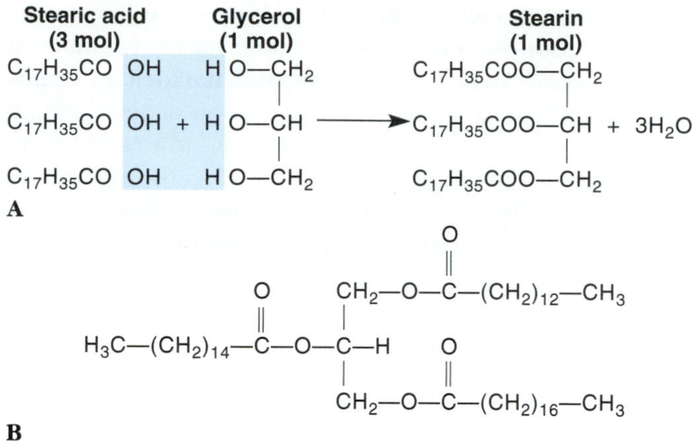

Figure 2.10

Neutral fats. **A,** Formation of a neutral fat from three molecules of stearic acid (a fatty acid) and glycerol. **B,** A neutral fat bearing three different fatty acids.

Phospholipids

Unlike fats that are fuels and serve no structural roles in the cell, phospholipids are important components of the molecular organization of tissues, especially membranes. They resemble triglycerides in structure, except that one of the three fatty acids is replaced by phosphoric acid and an organic base. An example is lecithin, an important phospholipid of nerve membranes (Figure 2.12). Because the phosphate group on phospholipids is charged and polar and therefore soluble in water, and the remainder of the molecule is nonpolar, phospholipids can bridge two environments and bind water-soluble molecules, such as proteins, to water-insoluble materials.

Steroids

Steroids are complex alcohols. Although they are structurally unlike fats, they have fatlike properties. The steroids are a large group of biologically important molecules, including cholesterol (Figure 2.13), vitamin D3, many adrenocortical hormones, and sex hormones.

Amino Acids and Proteins

Proteins are large, complex molecules composed of 20 kinds of amino acids (Figure 2.14). The amino acids are linked by **peptide bonds** to form long, chainlike polymers. In the formation of a peptide bond, the carboxyl group of one amino acid is linked by a covalent bond to the amino group of another, with elimination of water, as shown here:

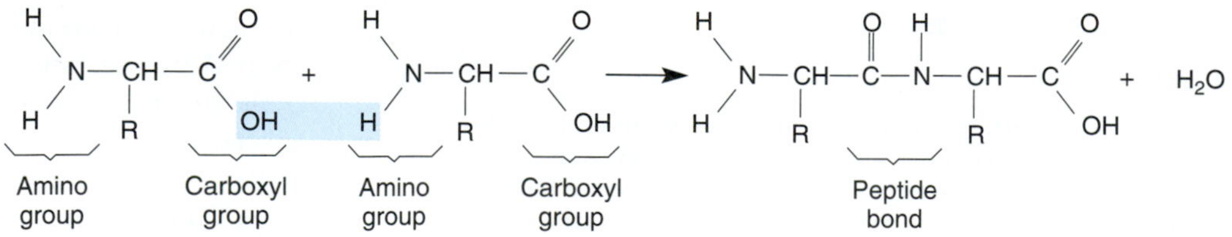

$$CH_3-(CH_2)_7-CH=CH-(CH_2)_7-COOH$$
Oleic acid

$$CH_3-(CH_2)_4-CH=CH-CH_2-CH=CH-(CH_2)_7-COOH$$
Linoleic acid

Figure 2.11

Unsaturated fatty acids. Oleic acid has one double bond and linoleic acid has two double bonds. The remainder of both acids is saturated.

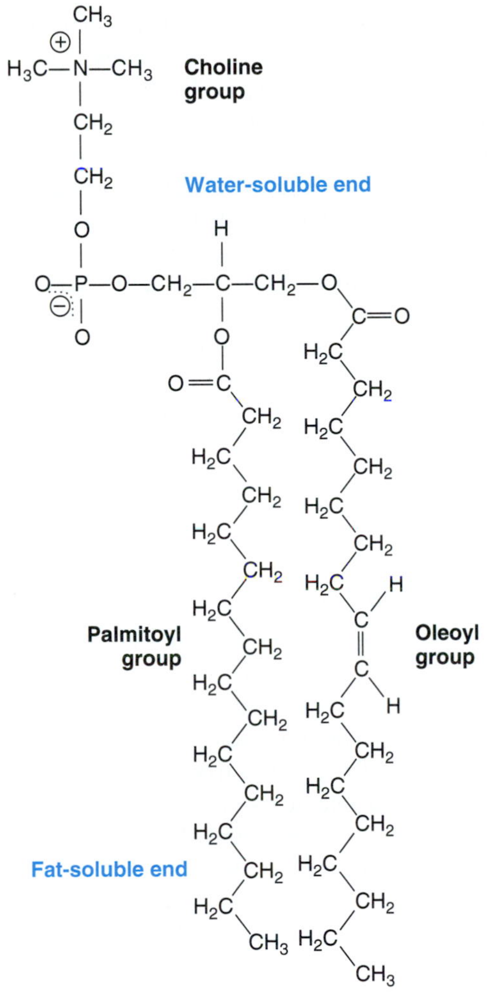

Figure 2.12

Lecithin (phosphatidyl choline), an important phospholipid of nerve membranes.

Figure 2.13

Cholesterol, a steroid. All steroids have a basic skeleton of four rings (three 6-carbon rings and one 5-carbon ring) with various side groups attached.

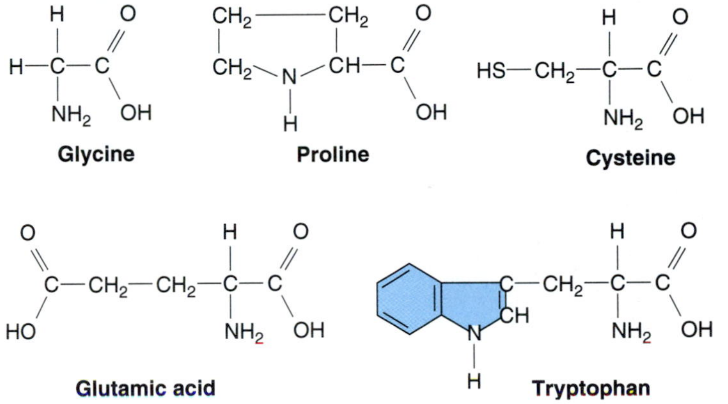

Figure 2.14

Five of the twenty kinds of amino acids.

recognize four levels of protein organization called primary, secondary, tertiary, and quaternary structures.

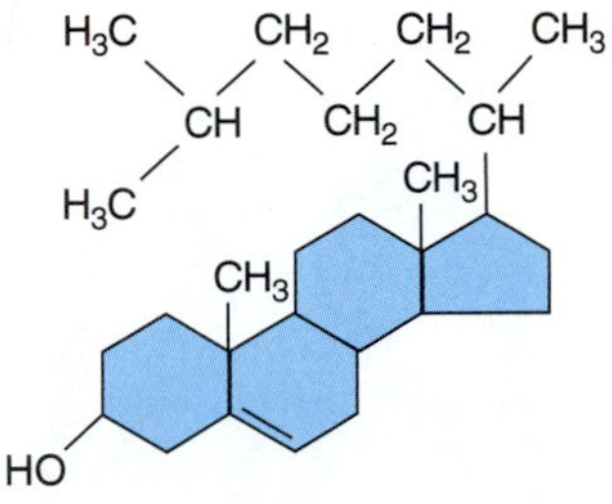

The **primary structure** of a protein is the sequence of amino acids composing the polypeptide chain. Because bonds between the amino acids in the chain can form only a limited number of stable angles, certain recurring structural patterns are assumed by the chain. These bond angles generate the **secondary structure,** such as the **alpha-helix,** which makes helical turns in a clockwise direction like a screw (Figure 2.15). The spirals of the chains are stabilized by hydrogen bonds, usually between a hydrogen atom of one amino acid and the peptide-bond oxygen of another amino acid from an adjacent turn of the helix. The helical and other configurations formed by the polypeptide chain bend and fold, giving the protein its complex, yet stable, three-dimensional **tertiary structure** (Figure 2.15). The folded chains are stabilized by chemical bonds between pairs of amino acids from different parts of the polypeptide chain. These bonds form between "side groups," parts of the amino acid not involved in a peptide bond. An example is the **disulfide bond,** a covalent bond between the sulfur atoms in two cysteine amino acids that are brought together by folds in the polypeptide chain. Also stabilizing the tertiary structure of proteins are hydrogen bonds, ionic bonds, and hydrophobic bonds.

The combination of two amino acids by a peptide bond forms a dipeptide having a free amino group on one end and a free carboxyl group on the other; therefore, additional amino acids can be joined until a long chain is produced. The 20 different kinds of amino acids can be arranged in an enormous variety of sequences of up to several hundred amino acid units, accounting for the large diversity of proteins found among living organisms.

A protein is not just a long string of amino acids; it is a highly organized molecule. For convenience, biochemists

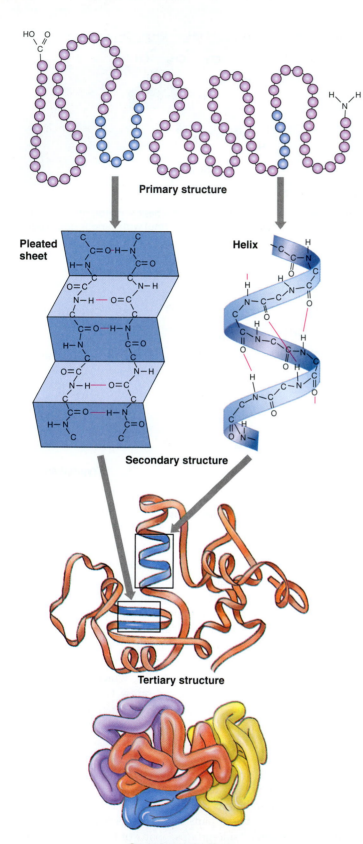

Primary structure

Pleated sheet

Helix

Secondary structure

Tertiary structure

Quaternary structure

Figure 2.15

Structure of proteins. The amino acid sequence of a protein (*primary structure*) encourages the formation of hydrogen bonds between nearby amino acids, producing coils and foldbacks (the *secondary structure*). Bends and helices cause the chain to fold back on itself in a complex manner (*tertiary structure*). Individual polypeptide chains of some proteins aggregate to form a functional molecule composed of several subunits (*quaternary structure*).

The term **quaternary structure** describes proteins that contain more than one polypeptide chain. For example, hemoglobin (the oxygen-carrying substance in blood) of higher vertebrates is composed of four polypeptide subunits held together in a single protein molecule (Figure 2.15).

Proteins perform many functions in living organisms. They form the structural framework of protoplasm and many cellular components. Many proteins function as **enzymes,** the biological catalysts required for almost every reaction in the body. Enzymes lower the activation energy required for specific reactions and enable life processes to proceed at moderate temperatures rather than requiring high temperatures. Enzymes control the reactions by which food is digested, absorbed, and metabolized. They promote the synthesis of structural materials for growth and to replace those lost by wear. They determine the release of energy used in respiration, growth, muscle contraction, physical and mental activities, and many other activities. Enzyme action is described in Chapter 4 (p. 60).

A **prion** is an infectious protein particle in which a protein of the host organism is contorted into an abnormal three-dimensional structure. Upon infection, the prion causes its host's normal copies of the protein to be refolded into the abnormal form, with pathological results. In "mad cow disease," a prion infection severely damages brain tissues and is fatal. Fatal neurological diseases associated with transmissible prions occur also in people (for example, kuru), and in sheep and goats (scrapie).

Nucleic Acids

Nucleic acids are complex polymeric molecules whose sequence of nitrogenous bases encodes the genetic information necessary for biological inheritance. They store directions for the synthesis of enzymes and other proteins, and are the only molecules that can (with the help of the right enzymes) replicate themselves. The two kinds of nucleic acids in cells are **deoxyribonucleic acid (DNA)** and **ribonucleic acid (RNA).** They are polymers of repeated units called **nucleotides,** each of which contains a sugar, a nitrogenous base, and a phosphate group. In addition to their role in nucleic acids, nucleotides have an important role as transporters of chemical energy in cellular metabolism (p. 62). Because the structure of nucleic acids is crucial to the mechanism of inheritance and protein synthesis, detailed information on nucleic acids is presented in Chapter 5 (p. 91).

CHEMICAL EVOLUTION

Both Haldane and Oparin proposed that earth's primitive atmosphere consisted of simple compounds such as water, molecular hydrogen, methane, and ammonia, but lacked oxygen gas

(O_2, also called "molecular oxygen"). The nature of the primeval atmosphere is critical for understanding life's origin. The organic compounds that compose living organisms are neither synthesized outside cells nor stable in the presence of molecular oxygen, which is abundant in the atmosphere today. The best evidence indicates, however, that the primitive atmosphere contained not more than a trace of molecular oxygen. The primeval atmosphere therefore was a reducing one, consisting primarily of molecules in which hydrogen exceeds oxygen; methane (CH_4) and ammonia (NH_3), for example, constitute fully reduced compounds. Such compounds are called "reducing" because they tend to donate electrons to other compounds, thereby "reducing" those compounds (p. 64). During this time, the earth was bombarded by large (100 km diameter) comets and meteorites, generating heat that repeatedly vaporized its oceans.

This reducing atmosphere was conducive to the prebiotic synthesis that led to life's beginnings, although totally unsuited for the organisms alive today. Haldane and Oparin proposed that ultraviolet radiation of such a gas mixture caused many organic substances, such as sugars and amino acids, to form. Haldane proposed that the early organic molecules accumulated in the primitive oceans to form a "hot dilute soup." In this primordial broth, carbohydrates, fats, proteins, and nucleic acids could have assembled to form the earliest structures capable of guiding their own replication.

If the simple gaseous compounds present in the early atmosphere are mixed with methane and ammonia in a closed glass system and kept at room temperature, they never react chemically with each other. To produce a chemical reaction, a continuous source of **free energy** sufficient to overcome reaction-activation barriers must be supplied. Ultraviolet light from the sun must have been intense on earth before the accumulation of atmospheric oxygen; ozone, a three-atom form of oxygen located high in the atmosphere, now blocks much of the ultraviolet radiation from reaching the earth's surface. Electrical discharges could have provided further energy for chemical evolution. Although the total amount of electrical energy released by lightning is small compared with solar energy, nearly all of the energy of lightning is effective in synthesizing organic compounds in a reducing atmosphere. A single flash of lightning through a reducing atmosphere generates a large amount of organic matter. Thunderstorms may have been one of the most important sources of energy for organic synthesis.

Widespread volcanic activity is another possible source of energy. One hypothesis maintains, for example, that life did not originate on the surface of the earth, but deep beneath the sea in or around **hydrothermal vents** (p. 816). Hydrothermal vents are submarine hot springs; seawater seeps through cracks in the seafloor until the water comes close to hot magma. The water is then superheated and expelled forcibly, carrying various dissolved molecules from the superheated rocks. These molecules include hydrogen sulfide, methane, iron ions, and sulfide ions. Hydrothermal vents have been discovered in several locations beneath the deep sea, and they would have been much more widely prevalent on the early earth. Interestingly, many heat- and sulfur-loving bacteria grow in hot springs today.

Prebiotic Synthesis of Small Organic Molecules

The Oparin-Haldane hypothesis stimulated experimental work to test the hypothesis that organic compounds characteristic of life could be formed from the simpler molecules present in the prebiotic environment. In 1953, Stanley Miller and Harold Urey in Chicago successfully simulated the conditions thought to prevail on the primitive earth. Miller built an apparatus designed to circulate a mixture of methane, hydrogen, ammonia, and water past an electric spark (Figure 2.16). Water in the flask was boiled to produce steam that helped to circulate the gases. The products formed in the electrical discharge (representing lightning) were condensed in the condenser and collected in the U-tube and small flask (representing an ocean).

After a week of continuous sparking, approximately 15% of the carbon from the reducing "atmosphere" had been converted into organic compounds that collected in the "ocean." The most

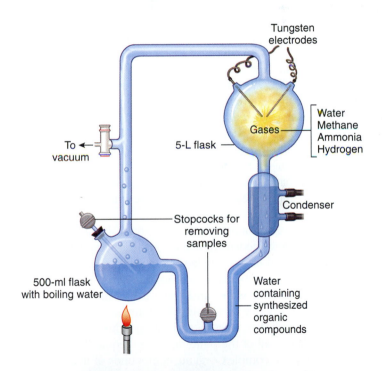

Figure 2.16
Dr. S. L. Miller with a replica of the apparatus used in his 1953 experiment on the synthesis of amino acids with an electric spark in a strongly reducing atmosphere.

striking finding was that many compounds related to life were synthesized. These compounds included four of the amino acids commonly found in proteins, urea, and several simple fatty acids. We can appreciate the astonishing nature of this synthesis when we consider that there are thousands of known organic compounds with structures no more complex than those of the amino acids formed. Yet in Miller's synthesis, most of the relatively few substances formed were compounds found in living organisms. This result was surely no coincidence, and it suggests that prebiotic synthesis on the primitive earth may have occurred under conditions not greatly different from those that Miller simulated.

Miller's experiments have been criticized in light of current opinion that the early atmosphere on earth was quite different from Miller's strongly reducing simulated atmosphere. Nevertheless, Miller's work stimulated many other investigators to repeat and to extend his experiment. Amino acids were found to be synthesized in many different kinds of gas mixtures that were heated (volcanic heat), irradiated with ultraviolet light (solar radiation), or subjected to electrical discharge (lightning). The only conditions required to produce amino acids were that the gas mixture be reducing and that it be subjected violently to a source of energy. In other experiments, electrical discharges were passed through mixtures of carbon monoxide, nitrogen, and water, yielding amino acids and nitrogenous bases. Although reaction rates were much slower than in atmospheres containing methane and ammonia, and yields were poor in comparison, these experiments support the hypothesis that the chemical beginnings of life can occur in atmospheres that are only mildly reducing. The need for methane and ammonia, however, led to proposals that these substances might have been introduced by comets or meteorites, or that they were synthesized near the hydrothermal vents.

Thus the experiments of many scientists have shown that highly reactive intermediate molecules such as hydrogen cyanide, formaldehyde, and cyanoacetylene are formed when a reducing mixture of gases is subjected to a violent energy source. These molecules react with water and ammonia or nitrogen to form more complex organic molecules, including amino acids, fatty acids, urea, aldehydes, sugars, and nitrogenous bases (purines and pyrimidines), all of the building blocks required for the synthesis of the most complex organic compounds of living matter. Further evidence for the natural abiotic synthesis of amino acids comes from finding amino acids in meteorites, such as the Murchison meteorite that landed in Australia in 1969.

Formation of Polymers

The next stage in chemical evolution involved the joining of amino acids, nitrogenous bases, and sugars to yield larger molecules, such as proteins and nucleic acids. Such synthesis does not occur easily in dilute solutions, because excess water drives reactions toward decomposition (hydrolysis). Although the primitive ocean might have been called a "primordial soup," it was probably a rather dilute one containing organic material that was approximately one-tenth to one-third as concentrated as chicken bouillon.

Need for Concentration

Prebiotic synthesis must have occurred in restricted areas where concentrations of the reactants were high. Violent weather on the primitive earth would have created enormous dust storms; impacts of meteorites would have lofted great amounts of dust into the atmosphere. The dust particles could have become foci of water droplets. Salt concentration in the particles could have been high and provided a concentrated medium for chemical reactions. Alternatively, perhaps the surface of the earth was too warm to have oceans but not too hot for a damp surface. This condition would have resulted from constant rain and rapid evaporation. Thus, the earth's surface could have become coated with organic molecules, an "incredible scum." Prebiotic molecules might have been concentrated by adsorption on the surface of clay and other minerals. Clay can concentrate and condense large amounts of organic molecules. The surface of iron pyrite (FeS_2) also has been suggested as a site for the evolution of biochemical pathways. The positively charged surface of pyrite would attract a variety of negative ions, which would bind to its surface. Furthermore, pyrite is abundant around hydrothermal vents, compatible with the hydrothermal-vent hypothesis.

Thermal Condensations

Most biological polymerizations are condensation (dehydration) reactions, in which monomers are linked together by the removal of water (p. 23). In living systems, condensation reactions always occur in an aqueous (cellular) environment containing appropriate enzymes. Without enzymes and energy supplied by ATP, macromolecules (proteins and nucleic acids) of living systems soon decompose into their constituent monomers.

Dehydration reactions could have occurred without enzymes in primitive earth conditions by thermal condensation. The simplest dehydration is accomplished by driving water from solids by direct heating. For example, heating a mixture of all 20 amino acids to 180° C produces a good yield of polypeptides.

The thermal synthesis of polypeptides to form "proteinoids" has been studied extensively by the American scientist Sidney Fox. He showed that heating dry mixtures of amino acids and then mixing the resulting polymers with water forms small spherical bodies. These proteinoid microspheres (Figure 2.17) possess certain characteristics of living systems. Each is not more than 2 μm in diameter, comparable in size and shape to spherical bacteria. The outer walls of the microspheres appear to have a double layer, and they show osmotic properties and selective diffusion. They may grow by accretion or proliferate by budding like bacteria. Proteinoids might have been used to assemble the first cells from macromolecular precursors. Formation of these polymers requires conditions likely to have occurred only in volcanoes. Organic polymers might have condensed on or in volcanoes and then, wetted by rain or dew, reacted further in solution to form polypeptides or polynucleotides.

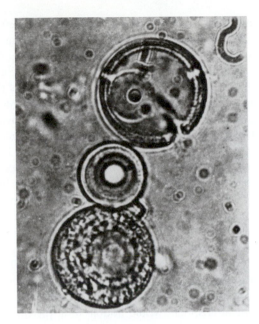

Figure 2.17

Electron micrograph of proteinoid microspheres. These proteinlike bodies can be produced in the laboratory from polyamino acids and may represent precellular forms. They have definite internal ultrastructure. (×1700)

ORIGIN OF LIVING SYSTEMS

The fossil record reveals that life existed 3.8 billion years ago; therefore, the origin of the earliest form of life can be estimated at approximately 4 billion years BP. The first living organisms were protocells, autonomous membrane-bound units with a complex functional organization that permitted the essential activity of self-reproduction. The primitive chemical systems that we have described lack this essential property. The principal problem in understanding the origin of life is explaining how primitive chemical systems could have become organized into living, autonomous, self-reproducing cells.

As we have seen, a lengthy chemical evolution on the primitive earth produced several molecular components of living forms. In a later stage of evolution, nucleic acids (DNA and RNA) began to behave as simple genetic systems that directed the synthesis of proteins, especially enzymes. However, this conclusion leads to a troublesome chicken-egg paradox: (1) How could nucleic acids have appeared without the enzymes that catalyze their synthesis? (2) How could enzymes have evolved without nucleic acids to encode their amino acid sequence? These questions come from a long-accepted notion that only proteins could act as enzymes. Startling evidence presented in the 1980s indicates that RNA in some instances has catalytic activity.

Catalytic RNA (ribozymes) can mediate processing of messenger RNA (removal of introns, p. 95), and can catalyze formation of peptide bonds. Strong evidence suggests that translation of mRNA by ribosomes (p. 95) is catalyzed by their RNA, not protein, content.

Therefore the earliest enzymes could have been RNA, and the earliest self-replicating molecules could have been RNA. Investigators are now calling this stage the "RNA world." Nonetheless,

proteins have several important advantages over RNA as catalysts, and DNA is a more stable carrier of genetic information than RNA. The first protocells containing protein enzymes and DNA should have been selectively favored over those with only RNA.

Once this protocellular stage of organization was reached, natural selection (pp. 124–126) would have acted on these primitive self-replicating systems. This stage was critical. Before this stage, biogenesis was shaped by the favorable environmental conditions on the primitive earth and by the nature of the reacting elements themselves. When self-replicating systems became responsive to natural selection, they began to evolve. The more rapidly replicating and more successful systems were favored, thereby gradually evolving efficient replicators. Evolution of the genetic code and fully directed protein synthesis followed. The system now meets the requirements for being the common ancestor of all living organisms.

Origin of Metabolism

Living cells today are organized systems with complex and highly ordered sequences of enzyme-mediated reactions. How did such vastly complex metabolic schemes develop? The exact history of this phase of life's evolution is unknown. We present here a model of the simplest sequence of events that could explain the origin of observed metabolic properties of living systems.

Organisms that can synthesize their food from inorganic sources using light or another source of energy are called **autotrophs** (Gr. *autos,* self, + *trophos,* feeder) (Figure 2.18). Organisms

Figure 2.18

Koala, a heterotroph, feeding on a eucalyptus tree, an autotroph. All heterotrophs depend for their nutrients directly or indirectly on autotrophs, which capture the sun's energy to synthesize their own nutrients.

lacking this ability must obtain their food supplies directly from the environment and are called **heterotrophs** (Gr. *heteros,* another, + *trophos,* feeder). The earliest postulated microorganisms are sometimes called **primary heterotrophs** because they relied on environmental sources for their food and existed prior to the evolution of any autotrophs. They were probably anaerobic organisms similar to bacteria of the genus *Clostridium.* Because chemical evolution had supplied generous stores of organic nutrients in the prebiotic soup, the earliest organisms would not have been required to synthesize their own food.

Autotrophs would have had a tremendous selective advantage over the primary heterotrophs in areas where organic nutrients became depleted. Evolution of autotrophic organisms most likely required acquisition of enzymatic activities to catalyze conversion of inorganic molecules to more complex ones, such as carbohydrates. The numerous enzymes of cellular metabolism appeared when cells started to utilize proteins for catalytic functions.

Carl Woese challenges the traditional view that the first organisms were primary heterotrophs. He finds it easier to visualize membrane-associated molecular aggregates that absorbed visible light and converted it with some efficiency into chemical energy. Thus the first organisms would have been autotrophs. Woese also suggests that the earliest "metabolism" comprised numerous chemical reactions catalyzed by nonprotein cofactors (substances necessary for the function of many of the protein enzymes in living cells). These cofactors would have been associated with membranes.

Appearance of Photosynthesis and Oxidative Metabolism

Autotrophy evolved in the form of photosynthesis. In photosynthesis, hydrogen atoms obtained from water react with carbon dioxide obtained from the atmosphere to generate sugars and molecular oxygen. Energy is stored in the form of covalent bonds between carbon atoms in the sugar molecule. Sugars provide nutrition to the organism and molecular oxygen is released into the atmosphere.

$$6CO_2 + 6H_2O \xrightarrow{\text{light}} C_6H_{12}O_6 + 6O_2$$

This equation summarizes the many reactions now known to occur in photosynthesis. Undoubtedly these reactions did not appear simultaneously, and other reduced compounds, such as hydrogen sulfide (H_2S), probably were the early sources of hydrogen.

Gradually, oxygen produced by photosynthesis accumulated in the atmosphere. When atmospheric oxygen reached approximately 1% of its current level, ozone began to accumulate and to absorb ultraviolet radiation, thereby greatly restricting the amount of ultraviolet light that reached the earth. Land and surface waters then were occupied by photosynthetic organisms, thereby increasing oxygen production.

Accumulation of atmospheric oxygen would interfere with anaerobic cellular metabolism that had evolved in the primitive

reducing atmosphere. As the atmosphere slowly accumulated oxygen gas (O_2), a new and highly efficient kind of metabolism appeared: **oxidative (aerobic) metabolism.** By using available oxygen as a terminal electron acceptor (p. 68) and completely oxidizing glucose to carbon dioxide and water, much of the bond energy stored by photosynthesis could be recovered. Most living forms became completely dependent upon oxidative metabolism.

Our atmosphere today is strongly oxidizing. It contains 78% molecular nitrogen, approximately 21% free oxygen, 1% argon, and 0.03% carbon dioxide. Although the time course for production of atmospheric oxygen is much debated, the most important source of oxygen is photosynthesis. Almost all oxygen currently produced comes from cyanobacteria (blue-green algae), eukaryotic algae, and plants. Each day these organisms combine approximately 400 million tons of carbon dioxide with 70 million tons of hydrogen to produce 1.1 billion tons of oxygen. Oceans are a major source of oxygen. Almost all oxygen produced today is consumed by organisms for respiration; otherwise, the amount of oxygen in the atmosphere would double in approximately 3000 years. Because Precambrian fossil cyanobacteria resemble modern cyanobacteria, it is reasonable to suppose that oxygen entering the early atmosphere came from their photosynthesis.

PRECAMBRIAN LIFE

The Precambrian period covers the geological time before the beginning of the Cambrian period some 570 to 600 million years BP. Most major animal phyla appear in the fossil record within a few million years at the beginning of the Cambrian period. This appearance has been called the "Cambrian explosion" because before this time, fossil deposits are mostly devoid of any organisms more complex than single-celled bacteria. Comparative molecular studies (p. 206) now suggest that the rarity of Precambrian fossils may represent poor fossilization rather than absence of animal diversity from the Precambrian period. Nonetheless, animals make a relatively late appearance in the history of life on earth. What were the early forms of life that generated both the oxidizing atmosphere critical for animal evolution and the evolutionary lineage from which animals would arise?

Prokaryotes and the Age of Cyanobacteria (Blue-Green Algae)

The earliest bacterium-like organisms proliferated, giving rise to a great variety of forms, some of which were capable of photosynthesis. From these arose the oxygen-producing **cyanobacteria** approximately 3 billion years ago.

Bacteria are called **prokaryotes,** meaning literally "before the nucleus." They contain a single, large molecule of DNA not located in a membrane-bound nucleus, but found in a nuclear region, or **nucleoid.** The DNA is not complexed with histone proteins, and prokaryotes lack membranous organelles such as mitochondria, plastids, Golgi apparatus, and endoplasmic reticulum (see Chapter 3). During cell division, the nucleoid divides and replicates of the cell's DNA are distributed to the daughter cells. Prokaryotes lack the chromosomal organization and chromosomal (mitotic) division seen in animals, fungi, and plants.

The name "algae" is misleading because it suggests a relationship to eukaryotic algae, and many scientists prefer the alternative name "cyanobacteria" rather than "blue-green algae." These organisms were responsible for producing an oxygen-rich atmosphere that replaced earth's primitive reducing atmosphere. Studies of biochemical reactions in extant cyanobacteria suggest that they evolved in a time of fluctuating oxygen concentration. For example, although they can tolerate atmospheric concentrations of oxygen (21%), the optimum concentration for many of their metabolic reactions is only 10%.

Bacteria and especially cyanobacteria ruled earth's oceans unchallenged for 1 to 2 billion years. The cyanobacteria reached the zenith of their success approximately 1 billion years BP, when filamentous forms produced great floating mats on the oceans' surfaces. This long period of cyanobacterial dominance, encompassing approximately two-thirds of the history of life, has been called with justification the "age of blue-green algae." Bacteria and cyanobacteria are so completely different from forms of life that evolved later that they were placed in a separate taxonomic kingdom, Monera.

Carl Woese and his colleagues at the University of Illinois discovered that the prokaryotes actually comprise at least two distinct lines of descent: the Eubacteria ("true" bacteria) and the Archaebacteria also called Archaea (p. 213). Although these two groups of bacteria look very much alike when viewed with the electron microscope, they are biochemically distinct. Archaebacteria differ fundamentally from bacteria in cellular metabolism, and their cell walls lack muramic acid, which is present in the cell walls of all Eubacteria. The most compelling evidence for differentiating these two groups comes from the use of one of the newest and most powerful tools at the disposal of the evolutionist, sequencing of nucleic acids (see note). Woese found that Archaebacteria differ fundamentally from other bacteria in the sequence of bases in ribosomal RNA (p. 95). Woese considers the Archaebacteria so distinct from the true bacteria that they should be considered a separate taxonomic kingdom, Archaea. The Monera then comprise only the true bacteria (see pp. 212–213 for further discussion and criticism of this taxonomy).

Appearance of Eukaryotes

Eukaryotes ("true nucleus"; Figure 2.19) have cells with membrane-bound nuclei containing **chromosomes** composed of **chromatin.** Constituents of eukaryotic chromatin include proteins called **histones** and RNA, in addition to DNA. Some nonhistone proteins are found associated with both prokaryotic DNA and eukaryotic chromosomes. Eukaryotes are generally larger than prokaryotes and contain much more DNA. Cellular division usually is by some form of mitosis. Within their cells are numerous membranous organelles, including mitochondria, in which the enzymes for oxidative metabolism are packaged. Eukaryotes include animals, fungi, plants, and numerous single-celled forms formerly called "protozoans" or "protists." Fossil evidence suggests that single-celled eukaryotes arose at least 1.5 billion years ago (Figure 2.20).

Molecular sequencing has emerged as a very successful approach to unraveling the ancient genealogies of living forms. The sequences of nucleotides in the DNA of an organism's genes are a record of evolutionary relationships, because every gene that exists today is an evolved copy of a gene that existed millions or even billions of years ago. Genes become altered by mutations through the course of time, but vestiges of the original gene usually persist. With modern techniques, one can determine the sequence of nucleotides in an entire molecule of DNA or in short segments of the molecule. When corresponding genes are compared between two different organisms, the extent to which the genes differ can be correlated with the time elapsed since the two organisms diverged from a common ancestor. Similar comparisons are made with RNA and proteins. These methods also permit scientists to synthesize long-extinct genes and proteins and to measure biochemical properties of the extinct proteins.

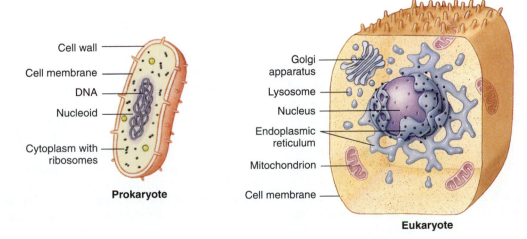

Cell wall

Cell membrane

DNA

Nucleoid

Cytoplasm with ribosomes

Prokaryote

Golgi apparatus

Lysosome

Nucleus

Endoplasmic reticulum

Mitochondrion

Cell membrane

Eukaryote

Figure 2.19

Comparison of prokaryotic and eukaryotic cells. Prokaryotic cells are about one-tenth the size of eukaryotic cells.

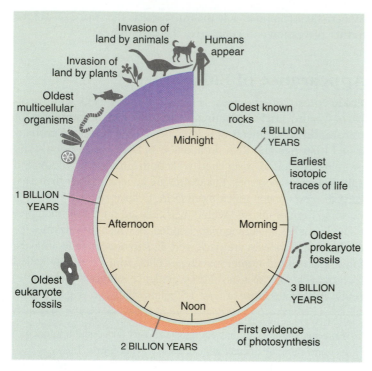

Figure 2.20

The clock of biological time. A billion seconds ago it was 1961, and most students using this text had not yet been born. A billion minutes ago the Roman empire was at its zenith. A billion hours ago Neanderthals were alive. A billion days ago the first bipedal hominids walked the earth. A billion months ago the dinosaurs were at the climax of their radiation. A billion years ago no animal had ever walked on the surface of the earth.

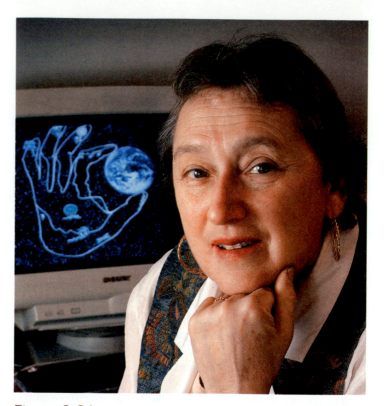

Figure 2.21

Dr. Lynn Margulis, whose endosymbiotic theory of the origins of mitochondria and chloroplasts is strongly supported by molecular evolutionary studies.

Because the organizational complexity of eukaryotes is much greater than that of prokaryotes, it is difficult to visualize how a eukaryote could have arisen from any known prokaryote. The American biologist Lynn Margulis (Figure 2.21) and others have proposed that eukaryotes did not arise from any single prokaryote but were derived from a **symbiosis** ("life together") of two or more types of bacteria. Mitochondria and plastids (photosynthetic organelles found only in plant cells), for example, each contain their own DNA (apart from the nucleus of the cell), which has some prokaryotic characteristics.

Nuclei, plastids, and mitochondria each contain genes encoding ribosomal RNA. Comparisons of the sequences of bases of these genes show that the nuclear, plastid, and mitochondrial DNAs represent distinct evolutionary lineages. Plastid and mitochondrial DNAs are closer in their evolutionary history to bacterial DNAs than to the eukaryotic nuclear DNA. Plastids are closest evolutionarily to cyanobacteria, and mitochondria are closest to another group of bacteria (purple bacteria), consistent with the symbiotic hypothesis of eukaryotic origins. Mitochondria contain the enzymes of oxidative metabolism, and plastids (a plastid with chlorophyll is a chloroplast) conduct photosynthesis. It is easy to see how a host cell able to accommodate such guests in its cytoplasm would have gained enormous evolutionary success.

The endosymbiotic theory proposes that a population ancestral to eukaryotic cells, derived from and resembling **anaerobic** (lacking oxidative metabolism) bacteria, evolved a nucleus and other intracellular membranes (p. 41) from infoldings of the cell membrane. Cells of this population acquired, by ingestion or parasitism, aerobic bacteria that avoided digestion and came to reside in the host cell's cytoplasm (p. 41). The endosymbiotic aerobic bacteria would have metabolized oxygen, which is toxic for their anaerobic host, and the anaerobic host cell would have given its aerobic residents food and physical protection. This mutually beneficial relationship would produce selection for the host cells and their residents to evolve a means of making their relationship a permanent one. Among the evolutionary outcomes of this selection would be compactness of the endosymbiont and its loss of genes redundant with those of its host (or the reverse).

Data collected to test this proposed mechanism show that its conditions are reasonable ones. Fossil data show that both aerobic and anaerobic bacteria were well established by 2.5 billion years ago, and that cells containing nuclei and internal membranes first appeared at this time. Some anaerobic, nucleated forms that lack mitochondria are alive today, including the human parasite *Giardia intestinalis,* although these forms probably represent descendants of lineages that formerly had mitochondria and lost them rather than lineages whose ancestry never featured mitochondria. Eukaryotic cells containing mitochondria are evident approximately 1.2 billion years ago. Bacteria have been introduced experimentally into single-celled eukaryotes and propagated as a symbiotic unit for many generations. Such experiments have shown further that the host cell can become dependent upon its resident bacteria for proteins whose functions formerly were performed by the host population prior to the experimental endosymbiosis.

In addition to claiming that mitochondria and plastids originated as bacterial symbionts, Lynn Margulis argues that eukaryote flagella, cilia (locomotory structures), and even the spindle of mitosis came from a kind of bacterium like a spirochete. Indeed, she suggests that this association (the spirochete with its new host cell) made evolution of mitosis possible. Margulis's evidence that organelles are former partners of an ancestral cell is now accepted by most biologists. Such merging of disparate organisms to produce evolutionarily novel forms is called symbiogenesis.

The first eukaryotes were undoubtedly unicellular, and many were photosynthetic autotrophs. Some of these forms lost their photosynthetic ability and became heterotrophs, feeding on eukaryotic autotrophs and prokaryotes. As cyanobacteria were cropped, their dense filamentous mats began to thin, providing space for other organisms. Carnivores appeared and fed on herbivores. Soon a balanced ecosystem of carnivores, herbivores, and primary producers appeared. By freeing space, cropping herbivores encouraged a greater diversity of producers, which in turn promoted evolution of new and more specialized croppers. An ecological pyramid developed with carnivores at the top of the food chain (p. 834).

The burst of evolutionary activity that followed at the end of the Precambrian period and beginning of the Cambrian period was unprecedented. Some investigators hypothesize that the explanation for the "Cambrian explosion" lies in the accumulation of oxygen in the atmosphere to a critical threshold level. Larger, multicellular animals required the increased efficiency of oxidative metabolism; these pathways could not be supported under conditions of limiting oxygen concentration.

SUMMARY

Living organisms show a remarkable uniformity in their chemical constituents and metabolism, reflecting their common descent from an ancient ancestor.

Life on earth could not have appeared without water, the primary component of living cells. The unique structure of water and its ability to form hydrogen bonds between adjacent water molecules are responsible for its special properties: solvency, high heat capacity, boiling point, surface tension, and lower density as a solid than as a liquid.

Life also depends critically on the chemistry of carbon. Carbon is especially versatile in bonding with itself and with other atoms, and it is the only element capable of forming the large molecules found in living organisms. Carbohydrates are composed primarily of carbon, hydrogen, and oxygen grouped as H—C—OH. The simplest carbohydrates are sugars, which serve as immediate sources of energy in living systems. Monosaccharides, or simple sugars, may bond together to form disaccharides or polysaccharides, which serve as storage forms of sugar or perform structural roles. Lipids constitute another class of large molecules featuring chains of carbon compounds; fats exist principally as neutral fats, phospholipids, and steroids. Proteins are large molecules composed of amino acids linked by peptide bonds. Many proteins function as enzymes that catalyze biological reactions. Each kind of protein has a characteristic primary, secondary, tertiary, and often, quaternary structure critical for its functioning. Nucleic acids are polymers of nucleotide units, each composed of a sugar, a nitrogenous base, and a phosphate group. They contain the material of inheritance and function in protein synthesis.

Experiments by Louis Pasteur in the 1860s convinced scientists that organisms do not arise repeatedly from inorganic matter. About 60 years later, A. I. Oparin and J. B. S. Haldane provided an explanation for how a common ancestor of all living forms could have arisen from nonliving matter almost 4 billion years ago. The origin of life followed a long period of "abiogenic molecular evolution" on earth in which organic molecules slowly accumulated in a "primordial soup." The atmosphere of the primitive earth was reducing, with little or no free oxygen present. Ultraviolet radiation, electrical discharges of lightning, or energy from hydrothermal vents could have provided energy for early formation of organic molecules. Stanley Miller and Harold Urey demonstrated the plausibility of the Oparin-Haldane hypothesis by simple but ingenious experiments. The concentration of reactants necessary for early synthesis of organic molecules might have been provided by damp surfaces, clay particles, iron pyrite, or other conditions. RNA might have been the primordial biomolecule, performing the functions of both genetic coding of information and catalysis. When self-replicating systems became established, evolution by natural selection could have increased their diversity and complexity.

The first organisms are hypothesized to have been primary heterotrophs, living on energy stored in molecules dissolved in a primordial soup. Later evolution produced autotrophic organisms, which can synthesize their own organic nutrients (carbohydrates) from inorganic materials. Autotrophs are better protected than heterotrophs from depletion of organic compounds from their environments. Molecular oxygen began to accumulate in the atmosphere as an end product of photosynthesis, an autotrophic process that produces sugars and oxygen by reacting water and carbon dioxide. Cyanobacteria appear to be primarily responsible for generation of atmospheric oxygen early in life's history.

All bacteria are prokaryotes, organisms that lack a membrane-bound nucleus and other organelles in their cytoplasm. The prokaryotes consist of two genetically distinct groups, Archaebacteria and Monera.

The eukaryotes apparently arose from symbiotic unions of two or more types of prokaryotes. The genetic material (DNA) of eukaryotes is borne in a membrane-bound nucleus, and also in mitochondria and sometimes plastids. Mitochondria and plastids have resemblances to bacteria, and their DNA is more closely allied to that of certain bacteria than to eukaryotic nuclear genomes.

REVIEW QUESTIONS

1. Explain each of these properties of water, and describe how each is conferred by the dipolar nature of a water molecule: high specific heat capacity; high heat of vaporization; unique density behavior; high surface tension; good solvent for ions of salts.
2. What was the composition of the earth's atmosphere at the time of the origin of life, and how did it differ from the atmosphere of today?
3. Regarding the experiments of Miller and Urey described in this chapter, explain what constituted the following in each case: observations, hypothesis, deduction, prediction, data, control. (The scientific method is described on pp. 11–14.)
4. Explain the significance of the Miller-Urey experiments.
5. Name three different sources of energy that could have powered reactions on early earth to form organic compounds.
6. By what mechanism might organic molecules have been concentrated in a prebiotic world so that further reactions could occur?
7. Name two simple carbohydrates, two storage carbohydrates, and a structural carbohydrate.
8. What characteristic differences in molecular structure distinguish lipids and carbohydrates?
9. Explain the difference between the primary, secondary, tertiary, and quaternary structures of a protein.
10. What are the important nucleic acids in a cell, and of what units are they constructed?
11. Distinguish among the following: primary heterotroph, autotroph, secondary heterotroph.
12. What is the source of oxygen in the present-day atmosphere, and what is its metabolic significance to most organisms living today?
13. Distinguish prokaryotes from eukaryotes.
14. Describe Margulis's view on the origin of eukaryotes from prokaryotes.
15. What was the "Cambrian explosion" and how might you explain it?

SELECTED REFERENCES

Berg, J. M., T. L. Tymoczko, and L. Stryer. 2007. Biochemistry, ed. 6. New York, W. H. Freeman. *A current and thorough textbook of biochemistry.*

Conway, Morris, S. 1993. The fossil record and the early evolution of the Metazoa. Nature **361:**219–225. *An important summary correlating fossil and molecular evidence.*

Fenchel, T. 2002. Origin and early evolution of life. Oxford, Oxford Univ. Press. *A review of current theories of the origin and early diversification of living forms.*

Gesteland, R. F., and J. F. Atkins, editors. 1999. The RNA world. Cold Spring Harbor, New York, Cold Spring Harbor Laboratory Press. *Evidence that there was a period when RNA served in both catalysis and transmission of genetic information.*

Kasting, J. F. 1993. Earth's early atmosphere. Science **259:**920–926. *Most investigators agree that there was little or no oxygen in the atmosphere of early earth and that there was a significant increase about 2 billion years ago.*

Knoll, A. H. 1991. End of the Proterozoic Eon. Sci. Am. **265:**64–73 (Oct.). *Multicellular animals probably originated only after oxygen in the atmosphere accumulated to a critical level.*

Lehninger, A. L., D. L. Nelson, and M. M. Cox. 2005. Lehninger principles of biochemistry, ed. 4. New York, Worth Publishers, Inc. *Clearly presented advanced textbook in biochemistry.*

Lodish, H., A. Berk, S. L. Zipursky, P. Matsudira, D. Baltimore, and J. Darnell. 2004. Molecular cell biology, ed. 5. New York, W. H. Freeman & Co. *Thorough treatment; begins with fundamentals such as energy, chemical reactions, bonds, pH, and biomolecules, then proceeds to advanced molecular biology.*

Margulis, L. 1998. Symbiotic planet: a new look at evolution. New York, Basic Books. *An important discussion of symbiogenesis in evolution.*

Orgel, L. E. 1994. The origin of life on the earth. Sci. Am. **271:**77–83 (Oct.). *There is growing evidence for an RNA world, but difficult questions are unanswered.*

Rand, R. P. 1992. Raising water to new heights. Science **256:**618. *Cites some ways that water can affect function of protein molecules.*

Wainright, P. O., G. Hinkle, M. L. Sogin, and S. L. Stickel. 1993. Monophyletic origins of the Metazoa: an evolutionary link with fungi. Science **260:**940–942. *Molecular evidence that multicellular animals share a more recent common ancestor with fungi than with plants or other eukaryotes.*

Waldrop, M. M. 1992. Finding RNA makes proteins gives "RNA world" a big boost. Science **256:**1396–1397. *Reports on the significance of Noller et al. (1992, Science **256:**1416), who found that ribosomal RNA can catalyze the formation of peptide bonds and hence proteins.*

ONLINE LEARNING CENTER

Visit www.mhhe.com/hickmanipz14e for chapter quizzing, key term flash cards, web links, and more!

3

Cells as Units of Life

A humpback whale, Megaptera novaeangliae, *leaps from the water.*

The Fabric of Life

It is a remarkable fact that living forms, from amebas and unicellular algae to whales and giant redwood trees, are formed from a single type of building unit: cells. All animals and plants are composed of cells and cell products. Thus the cell theory is another great unifying concept of biology.

New cells come from division of preexisting cells, and the activity of a multicellular organism as a whole is the sum of activities and interactions of its constituent cells. Energy is required to support life's activities, and virtually all of it flows from sunlight that is captured by green plants and algae and transformed by photosynthesis into chemical bond energy. Chemical bond energy is a form of potential energy that can be released when the bond is broken; the energy is used to perform electrical, mechanical, and osmotic tasks in the cell. Ultimately, all energy is dissipated into heat. This is in accord with the second law of thermodynamics, which states that there is a tendency in nature to proceed toward a state of greater molecular disorder, or entropy. Thus the high degree of molecular organization in living cells is attained and maintained only as long as energy fuels the organization.

CELL CONCEPT

More than 300 years ago the English scientist and inventor Robert Hooke, using a primitive compound microscope, observed boxlike cavities in slices of cork and leaves. He called these compartments "little boxes or cells." In the years that followed Hooke's first demonstration of the remarkable powers of the microscope to the Royal Society of London in 1663, biologists gradually began to realize that cells were far more than simple containers filled with "juices."

Cells are the fabric of life (Figure 3.1). Even the most primitive cells are enormously complex structures that form the basic units of all living organisms. All tissues and organs are composed of cells. In a human an estimated 60 trillion cells interact, each performing its specialized role in an organized partnership. In single-celled organisms all functions of life are performed within the confines of one microscopic package. There is no life without cells. The idea that a cell represents the basic structural and functional unit of life is an important unifying concept of biology.

With the exception of some eggs, which are the largest cells (in volume) known, cells are small and mostly invisible to the unaided eye. Consequently, our understanding of cells paralleled technical advances in the resolving power of microscopes. The Dutch microscopist Antoni van Leeuwenhoek sent letters to the Royal Society of London containing detailed descriptions of the numerous organisms he had observed using high-quality single lenses that he had made (1673 to 1723). In the early nineteenth century, improved design of microscopes permitted biologists to see separate objects only 1 μm apart. This advance was quickly followed by new discoveries that laid the groundwork for the **cell theory**—a theory stating that all living organisms are composed of cells.

In 1838 Matthias Schleiden, a German botanist, announced that all plant tissue was composed of cells. A year later one of his countrymen, Theodor Schwann, described animal cells as being similar to plant cells, an understanding that had been long delayed because animal cells are bounded only by a nearly invisible plasma membrane rather than a distinct cell wall characteristic

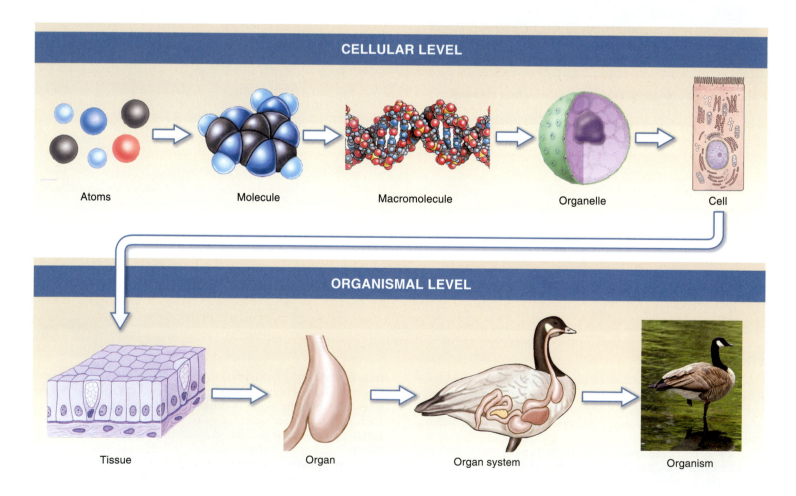

CELLULAR LEVEL

Atoms   Molecule   Macromolecule   Organelle   Cell

ORGANISMAL LEVEL

Tissue   Organ   Organ system   Organism

Figure 3.1

Biological organization from simple atoms to complex organisms. Atoms from molecules and macromolecules are assembled into organelles within each cell. Cells are grouped into tissues, organs, and organ systems to form a complex multicellular organism.

of plant cells. Schleiden and Schwann are thus credited with the unifying cell theory that ushered in a new era of productive exploration in cell biology. Another German, Rudolf Virchow, recognized that all cells came from preexisting cells (1858).

In 1840 J. Purkinje introduced the term **protoplasm** to describe cell contents. Protoplasm was at first thought to be a granular, gel-like mixture with special and elusive life properties of its own; cells were viewed as bags of thick soup containing a nucleus. Later the interior of cells became increasingly visible as microscopes were improved and better tissue-sectioning and staining techniques were introduced. Rather than being a uniform granular soup, a cell's interior is composed of numerous **cellular organelles,** each performing a specific function in the life of a cell. Today we realize that the components of a cell are so highly organized, structurally and functionally, that describing its contents as "protoplasm" is like describing the contents of an automobile engine as "autoplasm."

How Cells Are Studied

Light microscopes, with all their variations and modifications, have contributed more to biological investigation than any other instrument. They have been powerful exploratory tools for 300 years, and they continue to be so more than 50 years after invention of the electron microscope. However, electron microscopy has vastly enhanced our appreciation of the delicate internal organization of cells, and modern biochemical, immunological, physical, and molecular techniques have contributed enormously to our understanding of cell structure and function.

Electron microscopes employ high voltages to direct a beam of electrons through or at the surface of objects examined. The wavelength of the electron beam is approximately 0.00001 that of ordinary white light, thus permitting far greater magnification and resolution.

In preparation for viewing using the transmission electron microscope, specimens are cut into extremely thin sections (10 nm to 100 nm thick) and treated with "electron stains" (ions of elements such as osmium, lead, and uranium) to increase contrast between different structures. Electrons pass through a specimen, and images are seen on a fluorescent screen and photographed (Figure 3.2).

In contrast, specimens prepared for scanning electron microscopy are not sectioned, and electrons do not pass through them. The whole specimen is coated with an electron-dense material and bombarded with electrons, causing some electrons to be reflected back and secondary electrons to be emitted. An apparent three-dimensional image is recorded in the photograph. Although the magnification capability of scanning instruments is not as great as transmission microscopes, much has been learned about surface features of organisms and cells, as well as internal, membrane-bound structures. Examples of scanning electron micrographs are shown on pp. 145, 163, and 686.

A still greater level of resolution can be achieved with X-ray crystallography and nuclear magnetic resonance (NMR) spectroscopy. These techniques reveal the shapes of biomolecules and

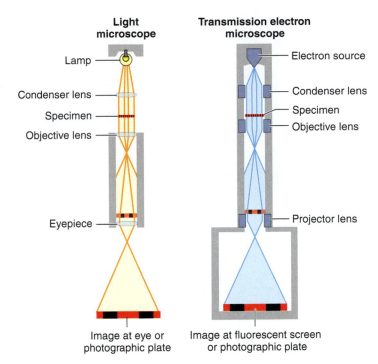

Figure 3.2

Comparison of optical paths of light and transmission electron microscopes. To facilitate comparison, the scheme of the light microscope has been inverted from its usual orientation with light source below and image above. In an electron microscope the lenses are magnets to focus the beam of electrons.

the relationships among atoms within them. Both techniques are laborious, but NMR spectroscopy does not require purification and crystallization of a substance, and molecules can be observed in solution.

Advances in techniques of cell study (cytology) are not limited to improvements in microscopes but include new methods of tissue preparation, staining for microscopic study, and the great contributions of modern biochemistry and molecular biology. For example, the various organelles of cells have differing, characteristic densities. Cells can be disrupted with most of the organelles remaining intact, then centrifuged in a density gradient (Figure 3.3), and relatively pure preparations of each organelle may be recovered. Thus the biochemical functions of various organelles may be studied separately. DNA and various types of RNA can be extracted and studied. Many enzymes can be purified and their characteristics determined. We use radioactive isotopes to study many metabolic reactions and pathways in cells. Modern chromatographic techniques can separate chemically similar intermediates and products. A particular cellular protein can be extracted and purified, and specific antibodies (see p. 775) can be prepared against the protein. When the antibody is complexed with a fluorescent substance and the complex is used to "stain" cells, the complex binds to the protein of interest, and its precise location in cells can be determined.

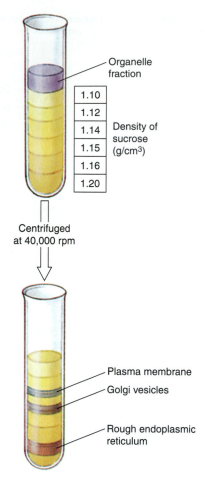

Figure 3.3

Separation of cell organelles in a density gradient by ultracentrifugation. The gradient is formed by layering sucrose solutions in a centrifuge tube, then carefully placing a preparation of mixed organelles on top. The tube is centrifuged at about 40,000 revolutions per minute for several hours, and organelles become separated down the tube according to their density.

ORGANIZATION OF CELLS

If we were to restrict our study of cells to fixed and sectioned tissues, we would be left with the erroneous impression that cells are static, quiescent, rigid structures. In fact, a cell's interior is in a constant state of upheaval. Most cells are continually changing shape; their organelles twist and regroup in a cytoplasm teeming with starch granules, fat droplets, and vesicles of various sorts. This description is derived from studies of living cell cultures with time-lapse photography and video. If we could see the swift shuttling of molecular traffic through gates in the cell membrane and the metabolic energy transformations within cell organelles, we would have an even stronger impression of internal turmoil. However, cells are anything but bundles of disorganized activity. There is order and harmony in cell functioning. Studying this dynamic phenomenon through a microscope, we realize that, as we gradually comprehend more and more about these units of life, we are gaining a greater understanding of the nature of life itself.

Prokaryotic and Eukaryotic Cells

We already described the radically different cell plan of prokaryotes and eukaryotes (p. 32). A fundamental distinction, expressed in their names, is that prokaryotes lack the membrane-bound nucleus present in all eukaryotic cells. Among other differences, eukaryotic cells have many membranous organelles (specialized structures that perform particular functions within cells) (Table 3.1).

Despite these differences, which are of paramount importance in cell studies, prokaryotes and eukaryotes have much in common. Both have DNA, use the same genetic code, and synthesize proteins. Many specific molecules such as ATP perform similar roles in both. These fundamental similarities imply common ancestry. The upcoming discussion is restricted to eukaryotic cells, of which all animals are composed.

TABLE 3.1

Comparison of Prokaryotic and Eukaryotic Cells

Characteristic	Prokaryotic Cell	Eukaryotic Cell
Cell size	Mostly small (1–10 μm)	Mostly large (10–100 μm)
Genetic system	DNA with some DNA-binding protein; simple, circular DNA molecule in nucleoid; nucleoid is not membrane bound	DNA complexed with DNA-binding proteins in complex linear chromosomes within nucleus with membranous envelope; circular mitochondrial and chloroplast DNA
Cell division	Direct by binary fission or budding; no mitosis	Some form of mitosis; centrioles in many; mitotic spindle present
Sexual system	Absent in most; highly modified if present	Present in most; male and female partners; gametes that fuse to form zygote
Nutrition	Absorption by most; photosynthesis by some	Absorption, ingestion, photosynthesis by some
Energy metabolism	No mitochondria; oxidative enzymes bound to cell membrane, not packaged separately; great variation in metabolic pattern	Mitochondria present; oxidative enzymes packaged therein; more unified pattern of oxidative metabolism
Intracellular movement	None	Cytoplasmic streaming, phagocytosis, pinocytosis
Flagella/cilia	If present, not with "9 + 2" microtubular pattern	With "9 + 2" microtubular pattern
Cell wall	Contains disaccharide chains cross-linked with peptides	If present, not with disaccharide polymers linked with peptides

Components of Eukaryotic Cells and Their Functions

Typically, eukaryotic cells are enclosed within a thin, selectively permeable **plasma membrane** (Figure 3.4). The most prominent organelle is a spherical or ovoid **nucleus,** enclosed within *two* membranes to form a double-layered **nuclear envelope** (Figure 3.4). Cellular material located between the cell membrane and nuclear envelope is collectively called **cytoplasm.** Within the cytoplasm are many organelles, such as mitochondria, Golgi complexes, centrioles, and endoplasmic reticulum. Plant cells typically contain **plastids,** some of which are photosynthetic organelles, and plant cells bear a cell wall containing cellulose outside the cell membrane.

The **fluid-mosaic model** is the currently accepted concept describing plasma membrane structure. By electron microscopy, a plasma membrane appears as two dark lines, each approximately 3 nm thick, at each side of a light zone (Figure 3.5). The entire membrane is 8 to 10 nm thick. This image is the result of a phospholipid bilayer, two layers of phospholipid molecules, all oriented with their water-soluble (hydrophilic) ends toward the outside and their fat-soluble portions (hydrophobic) toward the inside of the membrane (Figure 3.6). An important characteristic of the phospholipid bilayer is that it is fluidlike, giving the membrane flexibility and allowing the phospholipid molecules to move sideways freely within their own monolayer. Molecules of cholesterol are interspersed in the lipid portion of the bilayer (Figure 3.6). They make the membrane even less permeable to water-soluble ions and molecules and decrease membrane flexibility.

Glycoproteins (proteins with carbohydrates attached) are essential components of plasma membranes (Figure 3.6). Some of these proteins transport substances such as charged ions across the membrane (see p. 46, membrane function). Others act as specific receptors for various molecules or as highly specific cell markers. For example, the self/nonself recognition that enables the immune system to react to invaders (see Chapter 35) is based on proteins of this type. Some aggregations of protein molecules form pores or channels through which small polar molecules may enter (see gap junctions, p. 46, and channels, p. 49). Like the phospholipid molecules, most glycoproteins can move laterally in the membrane, although more slowly.

Nuclear envelopes contain less cholesterol than cell membranes, and pores in the envelope (Figure 3.7) allow molecules to move between nucleus and cytoplasm. Nuclei contain linear **chromosomes** suspended in **nucleoplasm.** The chromosomes are normally loosely condensed, flexible strands of **chromatin,** composed of a complex of DNA, and DNA-binding proteins. Chromosomal DNA carries the genetic information encoding cellular RNA and protein molecules (see Chapter 5). The linear chromosomes only become condensed and visible as discrete structures during cell division (see p. 52 for mitosis and p. 77 for meiosis). **Nucleoli** are specialized parts of certain chromosomes that stain in a characteristically dark manner. They carry multiple copies of the DNA information used to synthesize ribosomal

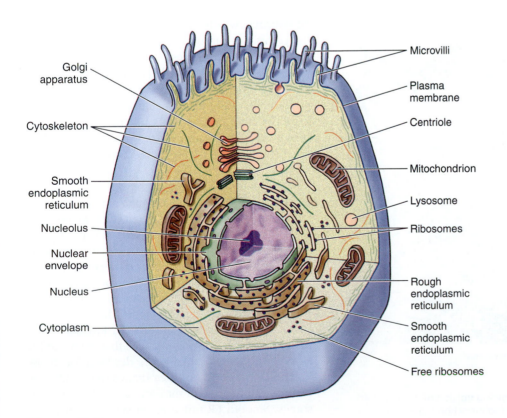

Figure 3.4

Generalized cell with principal organelles, as might be seen with the electron microscope.

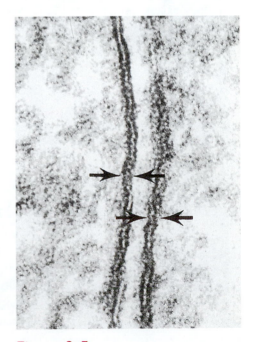

Figure 3.5

Plasma membranes of two adjacent cells. Each membrane (*between arrows*) shows a typical dark-light-dark staining pattern. (×325,000)

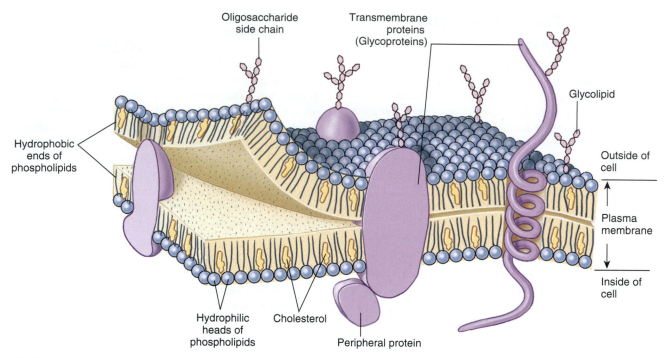

Figure 3.6

Diagram illustrating fluid-mosaic model of a plasma membrane.

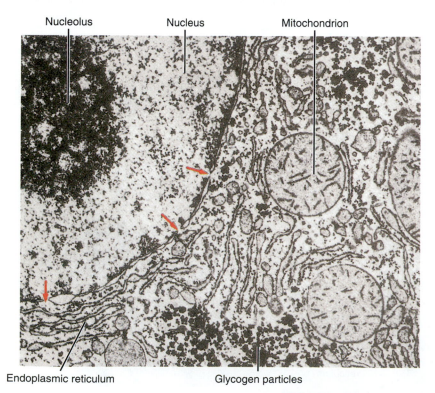

Figure 3.7

Electron micrograph of part of hepatic cell of rat showing portion of nucleus (*left*) and surrounding cytoplasm. Endoplasmic reticulum and mitochondria are visible in cytoplasm, and pores (*arrows*) can be seen in nuclear envelope. (×14,000)

RNA. After transcription from DNA, ribosomal RNA combines with protein to form the two subunits of **ribosomes,** which leave the nucleolus and pass to the cytoplasm through pores in the nuclear envelope. Ribosomes are sites of polypeptide or protein synthesis. They perform this function free, within the cytoplasm, when manufacturing polypeptides for use in the cytoplasm or nucleus. Alternatively, they become attached to **endoplasmic reticulum (ER)** when manufacturing polypeptides destined for the plasma membrane, lysosomes, or for export.

The outer membrane of the nuclear envelope is continuous with a cytoplasmic membranous system called endoplasmic reticulum (ER) (Figures 3.7 and 3.8). The space between the membranes of the nuclear envelope communicates with the space between the ER membranes (**cisterna,** pl., **cisternae).** The ER membranes may be covered on their outer surfaces with ribosomes and are thus designated **rough ER,** or they may lack ribosomal covering and be called **smooth ER.** Ribosomes on rough ER synthesize polypeptides that enter the ER cisternae or membrane and are destined for incorporation into the plasma membrane (Figure 3.9), for export from the cell, or they are bound for the lysosomes. Smooth ER functions in synthesis of lipids and phospholipids.

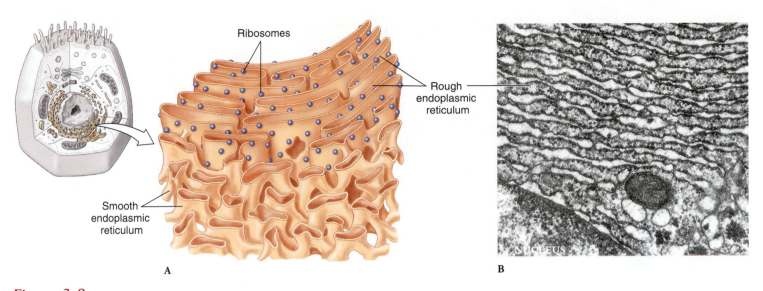

Figure 3.8

Endoplasmic reticulum. **A,** Endoplasmic reticulum is continuous with the nuclear envelope. It may have associated ribosomes (rough endoplasmic reticulum) or not (smooth endoplasmic reticulum). **B,** Electron micrograph showing rough endoplasmic reticulum. (×28,000)

The **Golgi complex** (Figures 3.9 and 3.10) is composed of a stack of membranous vesicles that function in storage, modification, and packaging of polypeptide and protein products produced by rough ER. The vesicles do not synthesize polypeptide or protein but may add complex carbohydrates to the molecules. Small vesicles of ER membrane containing polypeptide or protein detach and then fuse with sacs on the *cis* or "forming face" of a Golgi complex. After modification, the polypeptides or proteins become incorporated into vesicles that detach from the *trans* or "maturing face" of the complex (Figures 3.9 and 3.10). The contents of some of these vesicles may be expelled to the outside of the cell, as secretory products such as from a glandular cell. Some may carry integral polypeptides or proteins for incorporation into the plasma membrane, such as

receptors or carrier proteins. Others may contain enzymes that remain in the same cell that produces them. Such vesicles are called **lysosomes** (literally "loosening body," a body capable of causing lysis, or disintegration). Enzymes that they contain are involved in the breakdown of foreign material, including bacteria engulfed by a cell. Lysosomes also are capable of breaking down injured or diseased cells and worn-out cellular components. Their enzymes are so powerful that they kill the cell that formed them if enough of the lysosome membranes rupture. In normal cells the enzymes remain safely enclosed within the protective vesicle membranes. Lysosomal vesicles may pour their enzymes into a larger membrane-bound body containing an ingested food particle, a **food vacuole** or **phagosome** (see Figure 3.21).

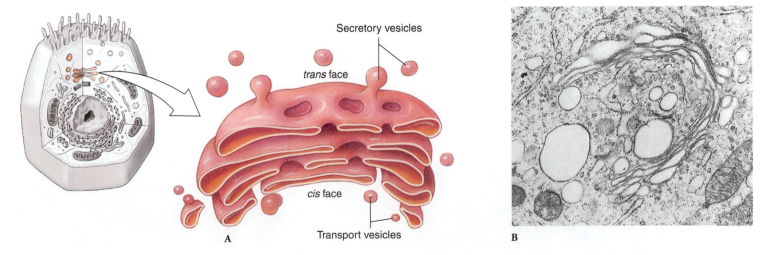

Figure 3.9

Golgi complex (=Golgi body, Golgi apparatus). **A,** The smooth cisternae of the Golgi complex have enzymes that modify polypeptides or proteins synthesized by the rough endoplasmic reticulum. **B,** Electron micrograph of a Golgi complex. (×46,000)

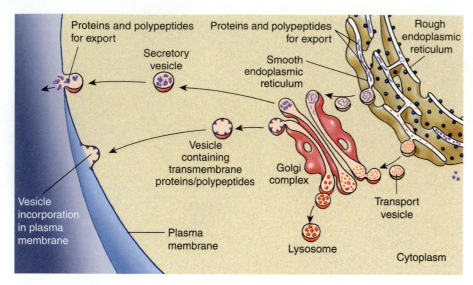

Figure 3.10

System for assembling, isolating, and secreting polypeptides and proteins for export in a eukaryotic cell, for lysosomes, or for incorporation into plasma membrane.

Mitochondria (sing., **mitochondrion**) (Figure 3.11) are conspicuous organelles present in nearly all eukaryotic cells. They are diverse in size, number, and shape; some are rodlike, and others are nearly spherical. They may be scattered uniformly throughout the cytoplasm or localized near cell surfaces and other regions of high metabolic activity. A mitochondrion is composed of a double membrane. The outer membrane is smooth, whereas the inner membrane is folded into numerous platelike or fingerlike projections called **cristae** (sing., **crista;** Figure 3.11), which increase the internal surface area where chemical reactions occur. These characteristic features make mitochondria easy to identify among organelles. Mitochondria are often called "powerhouses of cells," because enzymes located on the cristae catalyze the energy-yielding steps of aerobic metabolism (see Figure 4.14, p. 64). ATP (adenosine triphosphate), the most important energy-transfer molecule of all cells, is produced in this organelle. Mitochondria are self-replicating. They have a tiny, circular genome, much like genomes of prokaryotes except much smaller, which contains DNA specifying some, but not all, proteins of a mitochondrion.

Eukaryotic cells characteristically have a system of tubules and filaments that form a **cytoskeleton** (Figures 3.12 and 3.13). These provide support and maintain the form of cells, and in many cells, they provide a means of locomotion and translocation of macromolecules and organelles within the cell. The cytoskeleton is composed of microfilaments, microtubules, and intermediate filaments. **Microfilaments** are thin, linear structures, first observed distinctly in muscle cells, where they are responsible for the ability of a cell to contract. They are made of a protein called **actin.** Several dozen other proteins are known that bind with actin and determine its configuration and behavior in particular cells. One of these is **myosin,** whose interaction with actin causes contraction in muscle and other cells (p. 658). Actin microfilaments provide a means for movement of molecules and organelles through the cytoplasm, as well as movement of messenger RNA (p. 93) from the nucleus to particular positions within the cytoplasm. Actin and actin-binding proteins are also important in movement of vesicles between the ER, Golgi complex, and plasma membrane or lysosomes. **Microtubules,** somewhat larger than microfilaments, are tubular structures composed of a protein called **tubulin** (Figure 3.13). Each tubulin molecule is actually a doublet composed of two globular proteins. The molecules are attached head-to-tail to form a strand, and 13 strands aggregate to form a microtubule. Because the tubulin subunits in a microtubule are always attached head-to-tail, the ends of the microtubule differ chemically and functionally. One end (called the plus end) both adds and deletes tubulin subunits more rapidly than the other end (the minus end). Microtubules play a vital role in moving chromosomes toward daughter cells during cell division (p. 52), and they are important in intracellular architecture, organization, and transport. In addition, microtubules form essential parts of the structures of cilia and flagella (see next section). Microtubules radiate from a microtubule organizing center, the **centrosome,** near the nucleus. Centrosomes are not membrane bound. Within centrosomes are found a pair of **centrioles** (Figures 3.4 and 3.14), which are themselves

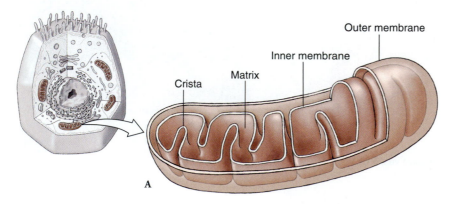

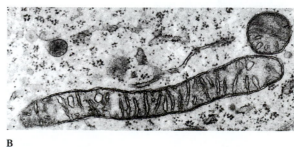

Figure 3.11

Mitochondria. **A,** Structure of a typical mitochondrion. **B,** Electron micrograph of mitochondria in cross and longitudinal section. (×30,000)

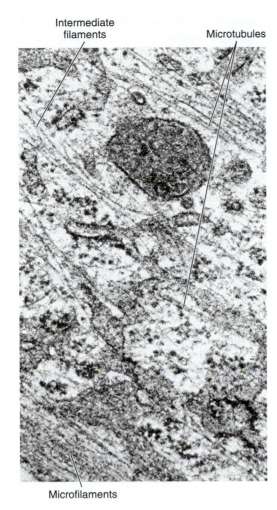

Figure 3.12

Cytoskeleton of a cell, showing its complex nature. Three visible cytoskeletal elements, in order of increasing diameter, are microfilaments, intermediate filaments, and microtubules. (×66,600)

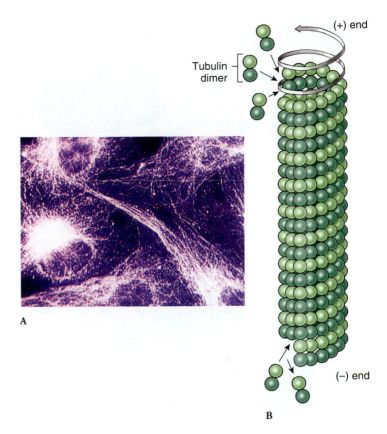

Figure 3.13

A, The microtubules in kidney cells of a baby hamster have been rendered visible by treatment with a preparation of fluorescent proteins that specifically bind to tubulin. **B,** A microtubule is composed of 13 strands of tubulin molecules, and each molecule is a dimer. Tubulin dimers are added to and removed from the (+) end of the microtubule more rapidly than at the (−) end.

composed of microtubules. Each centriole of a pair lies at right angles to the other and is a short cylinder of nine triplets of microtubules. They replicate before cell division. Although cells of higher plants do not have centrioles, a microtubule organizing center is present. **Intermediate filaments** are larger than microfilaments but smaller than microtubules. There are five biochemically distinct types of intermediate filaments, and their composition and arrangement depend on the cell type in which they occur. The type of intermediate filament is often determined in cancerous cells so that the original cell type can be identified. Knowing the particular cell type can often help in determining treatment options.

Surfaces of Cells and Their Specializations

The free surface of epithelial cells (cells that cover the surface of a structure or line a tube or cavity; see p. 195) sometimes bears either **cilia** or **flagella** (sing., **cilium, flagellum**). These are motile extensions of the cell surface that sweep materials past the cell. In many single-celled organisms and some small multicellular forms, they propel the entire organism through a liquid medium (see pp. 239, 282). Flagella provide the means of locomotion for male reproductive cells (sperm) of most animals (see p. 146) and many plants.

Cilia and flagella have different beating patterns (see p. 656), but their internal structure is the same. With few exceptions, the internal structures of locomotory cilia and flagella are composed of a long cylinder of nine pairs of microtubules enclosing a central pair (see Figure 29.11). At the base of each cilium or flagellum is a **basal body (kinetosome),** which is identical in structure to a single centriole.

Many cells move neither by cilia nor flagella but by **ameboid movement** using **pseudopodia.** Some groups of unicellular eukaryotes (p. 224), migrating cells in embryos of multicellular animals, and some cells of adult multicellular animals, such as white blood cells, show ameboid movement. Cytoplasmic streaming through the action of actin microfilaments extends a lobe (pseudopodium) outward from the surface of the cell. Continued streaming in the direction of a pseudopodium brings cytoplasmic organelles into the lobe and accomplishes movement of the entire cell. Some specialized pseudopodia have cores of microtubules (p. 227), and

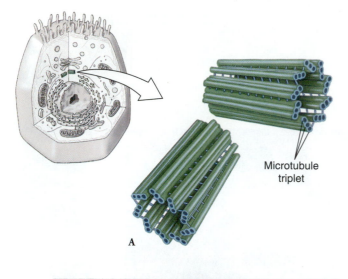

Microtubule triplet

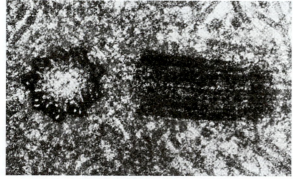

A

B

Figure 3.14

The centrosome. **A,** Each centrosome contains a pair of centrioles and each centriole is composed of nine triplets of microtubules arranged as a cylinder. **B,** Electron micrograph of a pair of centrioles, one in longitudinal (*right*) and one in cross section (*left*). The normal orientation of centrioles is at right angles to each other.

movement is effected by assembly and disassembly of the tubulin subunits.

Cells covering the surface of a structure (epithelial cells) or cells packed together in a tissue may have specialized junctional complexes between them. Nearest the free surface, the membranes of two cells next to each other appear to fuse, forming a **tight junction** (Figure 3.15). They are formed from rows of transmembrane proteins that bind tightly between adjacent cells. Tight junctions function as seals to prevent the passage of molecules between cells from one side of a layer of cells to another, because there is usually a space of about 20 nm between the plasma membranes of adjacent cells. The number of rows of transmembrane proteins in the tight junction determines how closely adjacent cells are sealed to each other. Tight junctions between intestinal cells, for example, force molecules from the intestinal contents to pass through epithelial cells during absorption, rather than between them. **Adhesion junctions** (Figure 3.15) occur just beneath tight junctions. These anchoring junctions are similar to tight junctions in that they encircle the cell. They are different from tight junctions in that they do

not seal adjacent cells to each other. Rather, the transmembrane proteins link together across a small intercellular space. Inside adjacent cells, the transmembrane proteins are attached to actin microfilaments and thus attach the cytoskeletons of adjacent cells to each other. Modified adhesion junctions are found between cardiac muscle cells, and these hold the cells together as the heart beats throughout the life of an organism (p. 694). At various points beneath tight junctions and adhesion junctions, in epithelial cells, small ellipsoid discs occur, within the plasma membrane in each cell. These appear to act as "spot-welds" and are called **desmosomes** (Figure 3.15). From each desmosome a tuft of intermediate filaments extends into the cytoplasm, and transmembrane linker proteins extend through the plasma membrane into the intercellular space to bind the discs of adjacent cells together. Desmosomes are not seals but seem to increase the strength of the tissue. Many are found between the cells of the skin in vertebrates (p. 645). **Hemidesmosomes** (Figure 3.15) are found at the base of cells and anchor them to underlying connective tissue layers. **Gap junctions** (Figure 3.15), rather than serving as points of attachment, provide a means of intercellular communication. They form tiny canals between cells, so that their cytoplasm becomes continuous, and small molecules and ions can pass from one cell to the other. Gap junctions may occur between cells of epithelial, nervous, and muscle tissues.

Another specialization of cell surfaces is the "lacing together" of adjacent cell surfaces where plasma membranes of the cells infold and interdigitate very much like a zipper. These infoldings are especially common in epithelial cells of kidney tubules and serve to increase the surface area of the cells for absorption or secretion. The distal or apical boundaries of some epithelial cells, as seen by electron microscopy, show regularly arranged **microvilli** (sing., **microvillus**). They are small, fingerlike projections consisting of tubelike evaginations of the plasma membrane with a core of cytoplasm containing bundles of actin microfilaments (Figures 3.15 and 3.16). They are seen clearly in the lining of the intestine where they greatly increase the absorptive and digestive surface. Such specializations appear as brush borders by light microscopy.

Membrane Function

The incredibly thin, yet sturdy, plasma membrane that encloses every cell is vitally important in maintaining cellular integrity. Once erroneously considered to be a rather static entity that defined cell boundaries and kept cell contents in place, plasma membranes (also called the plasmalemma) form dynamic structures having remarkable activity and selectivity. They are a permeability barrier that separates the interior from the external environment. They regulate the flow of molecules into and out of the cell, and provide many of the unique functional properties of specialized cells.

Membranes inside a cell surround a variety of organelles. Indeed, a cell is a system of membranes that divide it into numerous compartments. It has been estimated that if all membranes present in one gram of liver tissue were spread out flat, they would cover 30 square meters! Internal membranes share

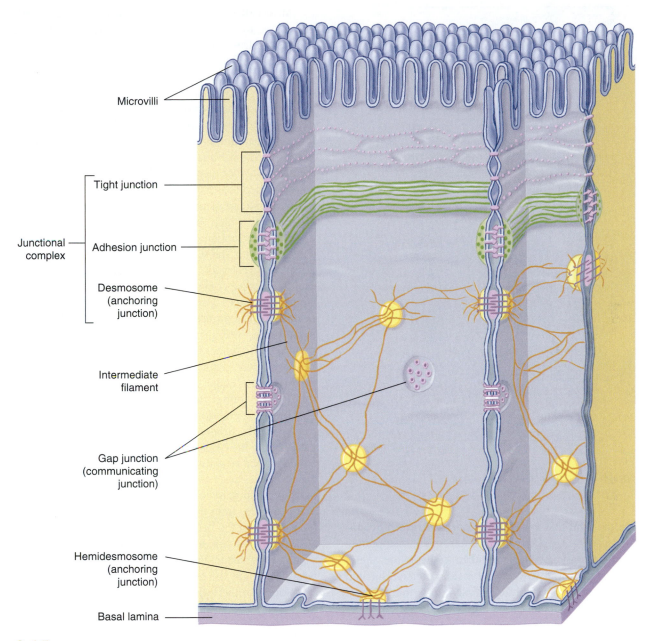

Microvilli

Tight junction

Junctional
complex

Adhesion junction

Desmosome
(anchoring
junction)

Intermediate
filament

Gap junction
(communicating
junction)

Hemidesmosome
(anchoring
junction)

Basal lamina

Figure 3.15

Junction types and locations are shown in columnar epithelial cells. Actin microfilaments (shown in green) and intermediate filaments (shown in orange) attach adhesion junctions and desmosomes, respectively, to the cytoskeleton.

many structural features of plasma membranes and are the site for many of a cell's enzymatic reactions.

A plasma membrane acts as a selective gatekeeper for entrance and exit of many substances involved in cell metabolism. Some substances can pass through with ease, others enter slowly and with difficulty, and still others cannot enter at all. Because conditions outside a cell are different from and more variable than conditions within a cell, it is necessary that passage of substances across the membrane be rigorously controlled.

We recognize three principal ways that a substance may enter across a cell membrane: (1) by **diffusion** along a concentration gradient; (2) by a **mediated transport system,** in which

the substance binds to a specific site on a transmembrane protein that assists it across the membrane; and (3) by **endocytosis,** in which the substance is enclosed within a vesicle that forms from the membrane surface and detaches inside the cell.

Diffusion and Osmosis

Diffusion is a movement of particles from an area of higher concentration to an area of lower concentration of the particles or molecules, thus tending to equalize the concentration throughout the area of diffusion. If a living cell surrounded by a membrane is immersed in a solution having a higher concentration of solute

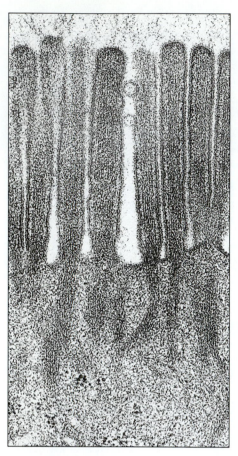

Figure 3.16
Electron micrograph of microvilli. (×59,000)

Most cell membranes are **selectively permeable,** that is, permeable to water but variably permeable or impermeable to solutes. In free diffusion it is this selectiveness that regulates molecular traffic. As a rule, gases (such as oxygen and carbon dioxide), urea, and lipid-soluble solutes (such as fats, fatlike substances, and alcohol; see p. 26) are the only solutes that can diffuse through biological membranes with any degree of freedom. Because many water-soluble molecules readily pass through membranes, such movements cannot be explained by simple diffusion. Sugars, many electrolytes, and macromolecules are moved across membranes by carrier-mediated processes that are described in the next section.

If we place a membrane between two unequal concentrations of solutes to which the membrane is impermeable, water flows through the membrane from the more dilute to the more concentrated solution. The water molecules move across the membrane down a concentration gradient from an area where the *water* molecules are more concentrated to an area on the other side of the membrane where they are less concentrated. This is **osmosis.**

We can demonstrate osmosis in a simple experiment by tying a selectively permeable membrane tightly over the end of a funnel. The funnel is filled with a salt solution and placed in a beaker of pure water so that the water levels inside and outside the funnel are equal. In a short time the water level in the glass tube of the funnel rises, indicating a net movement of water through the membrane into the salt solution (Figure 3.17).

Inside the funnel are salt molecules, as well as water molecules, while the beaker contains only water molecules. Thus the concentration of water is less inside the funnel because some of the available space is occupied by the nondiffusible salt ions. A concentration gradient exists for water molecules in the system. Water diffuses from the region of greater concentration of water (pure water in the beaker) to the region of lesser water concentration (salt solution inside the funnel).

As water enters the salt solution, the fluid level in the funnel rises. Eventually the pressure produced by the increasing weight of solution in the funnel pushes water molecules out as fast as they enter. The level in the funnel becomes stationary and the

molecules than the fluid inside the cell, a **concentration gradient** instantly exists between the two fluids across the membrane. Assuming that the membrane is **permeable** to the solute, there is a net movement of solute toward the inside, the side having the lower concentration. The solute diffuses "downhill" across the membrane until its concentrations on each side are equal.

Figure 3.17
Simple membrane osmometer. **A,** The end of a tube containing a salt solution is closed at one end by a selectively permeable membrane. The membrane is permeable to water but not to salt. **B,** When the tube is immersed in pure water, water molecules diffuse through the membrane into the tube. Water molecules are in higher concentration in the beaker because they are diluted inside the tube by salt ions. Because the salt cannot diffuse out through the membrane, the volume of fluid inside the tube increases, and the level rises.
C, When the weight of the column of water inside the tube exerts a downward force (hydrostatic pressure) causing water molecules to leave through the membrane in equal number to those that enter (osmotic pressure), the volume of fluid inside the tube stops rising. At this point the hydrostatic pressure is equivalent to the osmotic pressure.

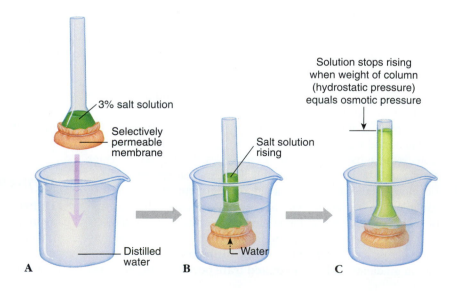

system is in equilibrium. The **osmotic pressure** of the solution is equivalent to the **hydrostatic pressure** necessary to prevent further net entry of water.

The concept of osmotic pressure is not without problems. A solution reveals an osmotic "pressure" only when it is separated from solvent by a selectively permeable membrane. It can be disconcerting to think of an isolated bottle of salt solution as having "pressure" much as compressed gas in a bottle (*hydrostatic* pressure) would have. Furthermore, the osmotic pressure is really the hydrostatic pressure that must be applied to a solution to keep it from gaining water *if* the solution were separated from pure water by a selectively permeable membrane. Consequently, biologists frequently use the term **osmotic potential** rather than osmotic pressure. However, since the term "osmotic pressure" is so firmly fixed in our vocabulary, it is necessary to understand the usage despite its potential confusion.

The concept of osmosis is very important in understanding how animals control their internal fluid and solute environment (see Chapter 30). For example, marine bony fishes maintain a solute concentration in their blood about one-third of that in seawater; they are **hypoosmotic** to seawater. If a fish swims into a river mouth and then up a freshwater stream, as salmon do, it would pass through a region where its blood solutes were equal in concentration to those in its environment **(isosmotic),** then enter freshwater, where its blood solutes were **hyperosmotic** to those in its environment. It must have physiological mechanisms to avoid net loss of water in the sea and gain of water in the river.

Diffusion Through Channels

Water and dissolved ions, since they are charged, cannot diffuse through the phospholipid component of the plasma membrane. Instead, they pass through specialized pores or channels created by transmembrane proteins. Ions and water move through these channels by diffusion. Ion channels allow specific ions of a certain size and charge to diffuse through them. They may allow ion diffusion at all times or they may be **gated channels,** requiring a signal to open or close them. Gated ion channels may open or close when a signaling molecule binds to a specific binding site on the transmembrane protein

(chemically-gated ion channels; Figure 3.18A) or when the ionic charge across a plasma membrane changes **(voltage-gated ion channels;** Figure 3.18B). Ion diffusion through channels is the basis of signaling mechanisms in the nervous system (see Chapter 33, p. 729) and in muscles (see Chapter 29, p. 660). Water channels are called **aquaporins,** and several different types have been discovered. They are especially important in the digestive system for absorption of water from food (see Chapter 32, p. 720), and in the kidney for water reabsorption during urine formation (see Chapter 30, p. 679).

Carrier-Mediated Transport

We have seen that a plasma membrane is an effective barrier to free diffusion of most molecules of biological significance, yet it is essential that such materials enter and leave a cell. Nutrients such as sugars and materials for growth such as amino acids must enter a cell, and wastes of metabolism must leave. Such molecules are moved across a membrane by special transmembrane proteins called **transporters,** or carriers. Transporters enable solute molecules to cross the phospholipid bilayer (Figure 3.19A). Transporters are usually quite specific, recognizing and transporting only a limited group of chemical substances or perhaps even a single substance.

At high concentrations of solute, mediated transport systems show a saturation effect. This means simply that the rate of influx reaches a plateau beyond which increasing the solute concentration has no further effect on influx rate (Figure 3.19B). This is evidence that the number of transporters available in a membrane is limited. When all transporters become occupied by solutes, the rate of transport is at a maximum and it cannot be increased. Simple diffusion shows no such limitation; the greater the difference in solute concentrations on the two sides of the membrane, the faster the influx.

Two distinctly different kinds of mediated transport mechanisms are recognized: (1) **facilitated diffusion,** in which a transporter assists a molecule to diffuse through the membrane that it cannot otherwise penetrate, and (2) **active transport,** in which energy is supplied to the transporter system to transport molecules in the direction opposite to a concentration gradient (Figure 3.20). Facilitated diffusion therefore differs from active

Figure 3.18

Gated channels require a signal to open (or close) them. **A,** Chemically-gated ion channels open (or close) when a signaling molecule binds to a specific binding site on the transmembrane protein. **B,** Voltage-gated ion channels open (or close) when the ionic charge across the membrane changes.

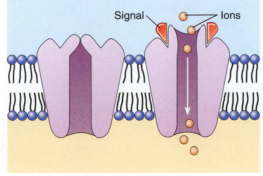

A Chemically-gated ion channel

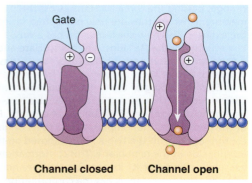

B Voltage-gated ion channel

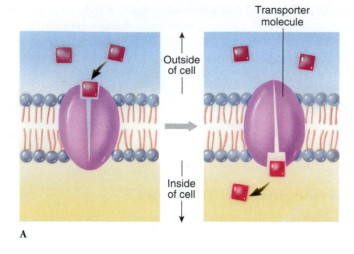

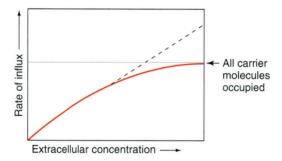

Figure 3.19

Facilitated transport. **A,** A transporter protein molecule binds with a molecule to be transported (substrate) on one side of a plasma membrane, changes shape, and releases the molecule on the other side. Facilitated transport takes place in the direction of a concentration gradient. **B,** Rate of transport increases with increasing substrate concentration until all transporter molecules are occupied.

transport in that it sponsors movement only in a downhill direction (in the direction of a concentration gradient) and requires no metabolic energy to drive the transport system.

In many animals facilitated diffusion aids in transport of glucose (blood sugar) into body cells that oxidize it as a principal energy source for the synthesis of ATP. The concentration of glucose is greater in blood than in the cells that consume it, favoring inward diffusion, but glucose is a water-soluble molecule that does not, by itself, penetrate cell membranes rapidly enough to support the metabolism of many cells; the carrier-mediated transport system increases the inward flow of glucose.

In active transport, molecules are moved uphill against the forces of passive diffusion. Active transport always involves an expenditure of energy (from ATP) because materials are transported against a concentration gradient. Among the most important active-transport systems in all animals are those that maintain sodium and potassium ion gradients between cells and the surrounding extracellular fluid or external environment. Most animal cells require a high internal concentration of potassium ions for protein synthesis at the ribosome and for certain enzymatic functions. The potassium ion concentration may be 20 to 50 times greater inside a cell than outside. Sodium ions, on the other hand, may be 10 times more concentrated outside a cell than inside. This sodium gradient forms the basis for electrical signal generation within the nervous system of animals (see Chapter 33, p. 728). Both of these ionic gradients are maintained by active transport of potassium ions into and sodium ions out of the cell. In many cells outward transport of sodium is linked to inward transport of potassium; this same transporter molecule does both. As much as 10% to 40% of all energy produced by cells is consumed by the **sodium-potassium exchange pump** (Figure 3.20).

Figure 3.20

Sodium-potassium pump, powered by bond energy of ATP, maintains the normal gradients of these ions across the cell membrane. The pump works by a series of conformational changes in the transporter: *Step 1.* Three ions of Na$^+$ bind to the interior end of the transporter, producing a conformational (shape) change in the protein complex. *Step 2.* The complex binds a molecule of ATP, cleaves it, and phosphate binds to the complex. *Step 3.* The binding of the phosphate group induces a second conformational change, passing the three Na$^+$ ions across the membrane, where they are now positioned facing the exterior. This new conformation has a very low affinity for the Na$^+$ ions, which dissociate and diffuse away, but it has a high affinity for K$^+$ ions and binds two of them as soon as it is free of the Na$^+$ ions. *Step 4.* Binding of the K$^+$ ions leads to another conformational change in the complex, this time leading to dissociation of the bound phosphate. Freed of the phosphate, the complex reverts to its original conformation, with the two K$^+$ ions exposed on the interior side of the membrane. This conformation has a low affinity for K$^+$ ions so that they are now released, and the complex has the conformation it started with, having a high affinity for Na$^+$ ions.

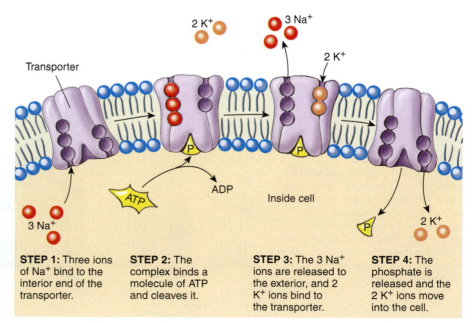

STEP 1: Three ions of Na$^+$ bind to the interior end of the transporter.

STEP 2: The complex binds a molecule of ATP and cleaves it.

STEP 3: The 3 Na$^+$ ions are released to the exterior, and 2 K$^+$ ions bind to the transporter.

STEP 4: The phosphate is released and the 2 K$^+$ ions move into the cell.

Endocytosis

Endocytosis, the ingestion of material by cells, is a collective term that describes three similar processes: phagocytosis, pinocytosis, and receptor-mediated endocytosis (Figure 3.21). They are pathways for specifically internalizing solid particles, small molecules and ions, and macromolecules, respectively. All require energy and thus may be considered forms of active transport.

Phagocytosis, which literally means "cell eating," is a common method of feeding among protozoa and lower metazoa. It is also the way in which white blood cells (leukocytes) engulf cellular debris and uninvited microbes or other pathogens in the blood. By phagocytosis, an area of the plasma membrane, coated externally with specific receptors and internally with actin and actin-binding proteins, forms a pocket that engulfs the solid material. The membrane-enclosed vesicle, a food vacuole or phagosome, then detaches from the cell surface and moves into the cytoplasm where it fuses with lysosomes, and its contents are digested by lysosomal enzymes.

Pinocytosis is similar to phagocytosis except that small areas of the surface membrane are invaginated into cells to form tiny vesicles. The invaginated pits and vesicles are called **caveolae** (ka-vee´o-lee). Specific binding receptors for the molecule or ion to be internalized are concentrated on the cell surface of caveolae. Pinocytosis apparently functions for intake of at least some vitamins, and similar mechanisms may be important in translocating substances from one side of a cell to the other (see "exocytosis," next heading) and internalizing signal molecules, such as some hormones or growth factors.

Receptor-mediated endocytosis is a specific mechanism for bringing large molecules within the cell. Proteins of the plasma membrane specifically bind particular molecules (termed **ligands** in this process), which may be present in the extracellular fluid in very low concentrations. The invaginations of the cell surface that bear the receptors are coated within the cell with a protein called **clathrin;** hence, they are described as **clathrin-coated pits.** As a clathrin-coated pit with its receptor-bound ligand invaginates and is brought within the cell, it is uncoated, the receptor and the ligand are dissociated, and the receptor and membrane material are recycled back to the surface membrane. Lysosomes fuse with the remaining vesicle, now called an **endosome,** and its contents are digested and absorbed into the cytoplasm. Some important proteins, peptide hormones, and cholesterol are brought into cells in this manner.

In phagocytosis, pinocytosis, and receptor-mediated endocytosis some amount of extracellular fluid is necessarily trapped in the vesicle and nonspecifically brought within the cell. We describe this as **bulk-phase endocytosis.**

Exocytosis

Just as materials can be brought into a cell by invagination and formation of a vesicle, the membrane of a vesicle can fuse with the plasma membrane and extrude its contents to the surrounding medium. This is the process of **exocytosis.** This process occurs in various cells to remove undigestible residues of substances brought in by endocytosis, to secrete substances such as hormones (Figure 3.10), and to transport a substance completely across a cellular barrier **(transcytosis),** as we just mentioned. For example, a substance may be picked up on one side of the wall of a blood vessel by pinocytosis, moved across the cell, and released by exocytosis.

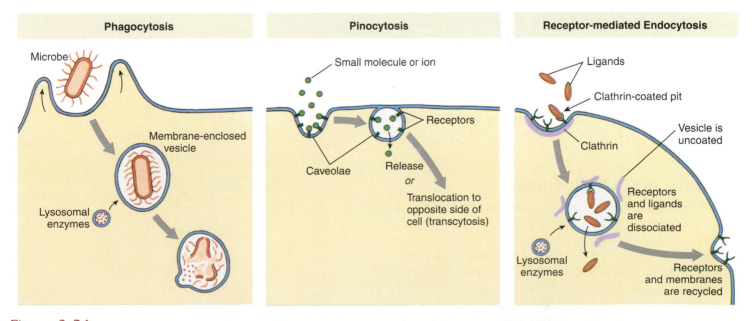

Figure 3.21

Three types of endocytosis. In phagocytosis the cell membrane binds to a large particle and extends to engulf it, forming a membrane-enclosed vesicle, a food vacuole or phagosome. In pinocytosis small areas of cell membrane, bearing specific receptors for a small molecule or ion, invaginate to form caveolae. Receptor-mediated endocytosis is a mechanism for selective uptake of large molecules in clathrin-coated pits. Binding of the ligand to the receptor on the surface membrane stimulates invagination of pits.

Actin and actin-binding proteins are now to be essential cytoskeletal components in the processes of endocytosis and exocytosis.

MITOSIS AND CELL DIVISION

All cells arise from the division of preexisting cells. All cells found in most multicellular organisms originated from the division of a single cell, a **zygote,** which is the product of union (fertilization) of an **egg** and a **sperm (gametes).** Cell division provides the basis for one form of growth, for both sexual and asexual reproduction, and for transmission of hereditary qualities from one cell generation to another cell generation.

In the formation of **body cells (somatic cells)** the process of nuclear division is **mitosis.** By mitosis each "daughter cell" is ensured a complete set of genetic instructions. Mitosis is a delivery system for distributing the chromosomes and the DNA they contain to continuing cell generations. Thus, a single zygote divides by mitosis to produce a multicellular organism, and damaged cells are replaced by mitosis during wound healing. As an animal grows, its somatic cells differentiate and assume different functions and appearances because of differential gene action. Although most of the genes in specialized cells remain silent and unexpressed throughout the lives of those cells, every cell possesses a complete genetic complement. Mitosis ensures equality of genetic potential; later, other processes direct the orderly expression of genes during embryonic development by selecting from the genetic instructions that each cell contains. (These fundamental properties of cells of multicellular organisms are discussed further in Chapter 8.)

In animals that reproduce **asexually** (see Chapter 7), mitosis is the only mechanism for the transfer of genetic information from parent to progeny, and thus the progeny are genetically identical to the parents in this case. In animals that reproduce **sexually** (see Chapter 7), the parents must produce **sex cells** (gametes or germ cells) that contain only half the usual number of chromosomes, so that progeny formed by the union of gametes will not contain double the parental number of chromosomes. This requires a special type of *reductional* division called **meiosis,** described in Chapter 5 (p. 77).

Structure of Chromosomes

As mentioned on page 41, DNA in eukaryotic cells occurs in chromatin, a complex of DNA with associated protein. Chromatin is organized into a number of discrete linear bodies called **chromosomes** (color bodies), so named because they stain deeply with certain biological dyes. In cells that are not dividing, chromatin is loosely organized and dispersed, so that individual chromosomes cannot be distinguished by light microscopy (see Figure 3.24, Interphase). Before division the chromatin becomes more compact, and chromosomes can be recognized and their individual morphological characteristics determined. They are of varied lengths and shapes, some bent and some rodlike. Their number is constant for a species, and every body cell (but not

the germ cells) has the same number of chromosomes regardless of a cell's function. A human, for example, has 46 chromosomes in each somatic cell.

During mitosis (nuclear division) chromosomes shorten further and become increasingly condensed and distinct, and each assumes a shape partly characterized by the position of a constriction, the **centromere** (Figure 3.22). The centromere is the location of the **kinetochore,** a disc of proteins that binds with microtubules of the spindle that forms during mitosis.

When chromosomes become condensed, the DNA is inaccessible, such that transcription (see Chapter 5, p. 93) cannot occur. Chromosomal condensation does, however, enable a cell to distribute chromosomal material efficiently and equally to daughter cells during cell division.

Phases in Mitosis

There are two distinct stages of cell division: division of nuclear chromosomes **(mitosis)** and division of cytoplasm **(cytokinesis).** Mitosis (that is, chromosomal segregation) is certainly the most obvious and complex part of cell division and that of greatest interest to cytologists. Cytokinesis normally immediately follows mitosis, although occasionally the nucleus may divide a number of times without a corresponding division of the cytoplasm. In such a case the resulting mass of protoplasm containing many nuclei is called a **multinucleate cell.** An example is the giant resorptive cell type of bone (osteoclast), which may contain 15 to 20 nuclei. Sometimes a multinucleate mass is formed by cell fusion rather than nuclear proliferation. This arrangement is called a **syncytium.** An example is vertebrate skeletal muscle, which is composed of multinucleate fibers formed by fusion of numerous embryonic cells.

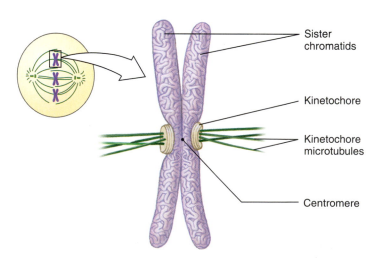

Sister chromatids

Kinetochore

Kinetochore microtubules

Centromere

Figure 3.22

Structure of a metaphase chromosome. Sister chromatids are still attached at their centromere. Each chromatid has a kinetochore, to which the kinetochore microtubules or fibers are attached. Kinetochore microtubules from each chromatid run to one of the centrosomes, which are located at opposite poles.

Mitosis is artificially divided into four successive stages or phases, although one stage merges into the next without sharp lines of transition. These phases are prophase, metaphase, anaphase, and telophase (Figures 3.23 and 3.24). When cells are not actively dividing, they are in interphase, a major part of the cell cycle described in some detail on page 54.

Prophase

At the beginning of prophase, centrosomes (along with their centrioles) replicate, the nuclear envelope disintegrates, and the two centrosomes migrate to opposite poles of the cell (Figure 3.23). At the same time, microtubules are manufactured between the two centrosomes to form a football-shaped **spindle,** so named because of its resemblance to nineteenth-century wooden spindles, used to twist thread together in spinning. Other microtubules radiate outward from each centrosome to form **asters.** The asters will develop into the microtubular portion of the cytoskeleton in each new daughter cell formed during cell division.

At this time the diffuse nuclear chromatin condenses to form visible chromosomes. These actually consist of two identical sister **chromatids** (Figure 3.22) formed by DNA replication (see Chapter 5, p. 93) during interphase and joined at their centromere. Dynamic spindle microtubules repeatedly extend and retract from each centrosome. When a microtubule encounters a kinetochore, it binds to the kinetochore, ceases

extending and retracting, and is now called a **kinetochore microtubule** or **fiber.** Thus, centrosomes send out "feelers" to find chromosomes.

Metaphase

Each centromere has two kinetochores, and each of the kinetochores is attached to one of the centrosomes by a kinetochore fiber. By a kind of tug-of-war during metaphase, the condensed sister chromatids are moved to the middle of the nuclear region, called the **metaphase plate** (Figures 3.23 and 3.24). The centromeres line up precisely in this region with the arms of the sister chromatids trailing off randomly in various directions.

Anaphase

The single centromere that has held two sister chromatids together now splits so that the two sister chromatids separate to become two independent chromosomes, each with its own centromere. The chromosomes move toward their respective poles, pulled by their kinetochore fibers. The arms of each chromosome trail behind as the microtubules shorten to drag a complete set of chromosomes toward each pole of the cell (Figures 3.23 and 3.24). Present evidence indicates that the force moving the chromosomes is disassembly of the tubulin subunits at the kinetochore end of the each microtubule.

As chromosomes approach their respective centrosomes, the centrosomes move farther apart as the microtubules are gradually disassembled.

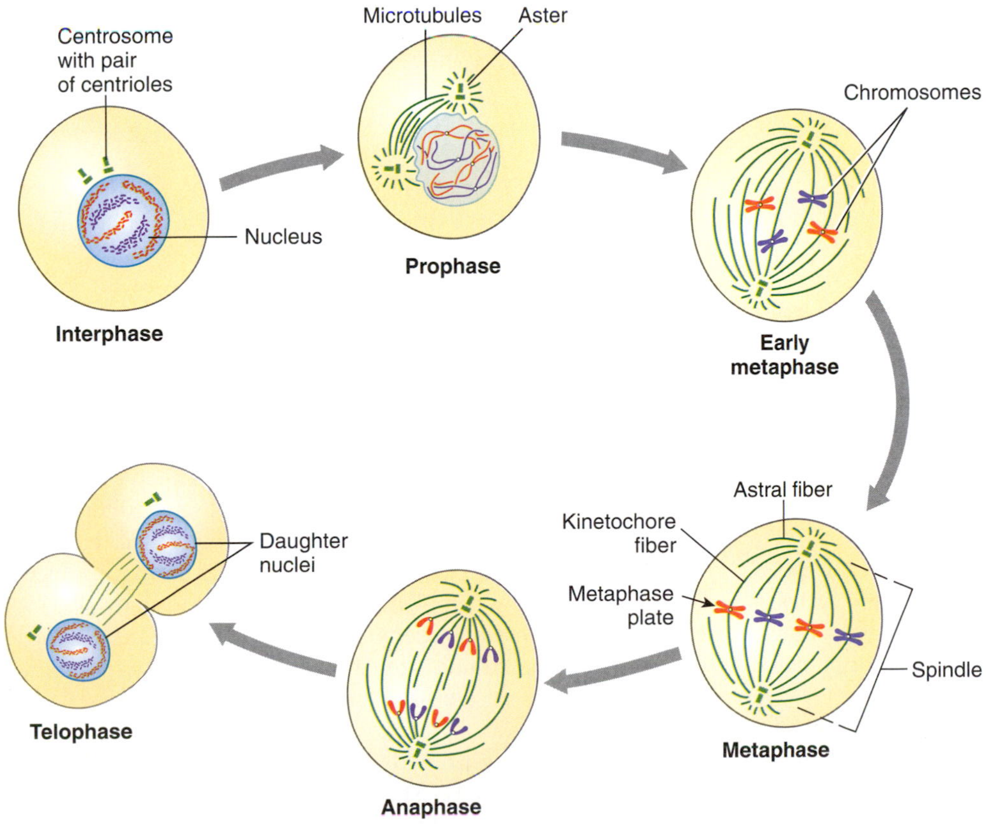

Telophase

When daughter chromosomes reach their respective poles, telophase has begun (Figures 3.23 and 3.24). The daughter chromosomes are crowded together and stain intensely with histological stains. Spindle fibers disappear and the chromosomes lose their identity, reverting to a diffuse chromatin network characteristic of an interphase nucleus. Finally, nuclear membranes reappear around the two daughter nuclei.

Cytokinesis: Cytoplasmic Division

During the final stages of nuclear division a **cleavage furrow** appears on the surface of a dividing cell and encircles it at the midline of the spindle (Figures 3.23 and 3.24). The cleavage furrow deepens and pinches the plasma membrane as though it were being tightened by an invisible

Figure 3.23
Stages of mitosis, showing division of a cell with two pairs of chromosomes. One chromosome of each pair is shown in red.

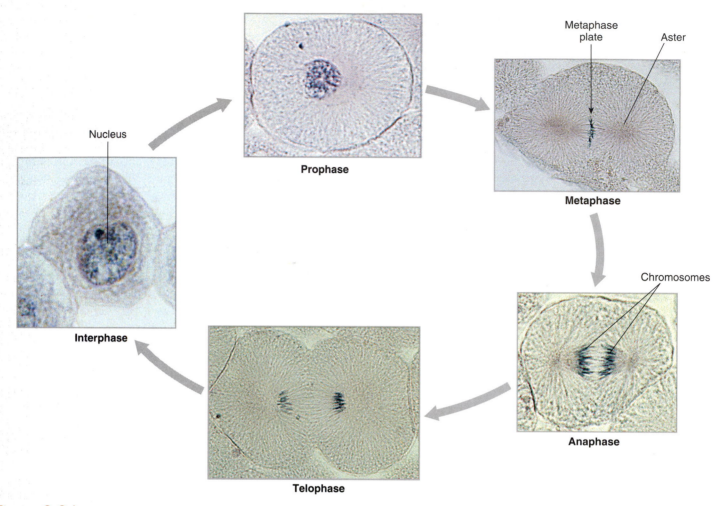

Figure 3.24
Stages of mitosis in whitefish.

rubber band. Actin microfilaments are present just beneath the surface in the furrow between the cells. Interaction with myosin and other actin-binding proteins, similar to muscle contraction mechanisms (p. 658), draws the furrow inward. Finally, the infolding edges of the plasma membrane meet and fuse, completing cell division.

Cell Cycle

Cycles are conspicuous attributes of life. The descent of a species through time is in a very real sense a sequence of life cycles. Similarly, cells undergo cycles of growth and replication as they repeatedly divide. A cell cycle is the interval between one cell division and the next (Figure 3.25).

Actual nuclear division, or mitosis, occupies only about 5% to 10% of the cell cycle; the rest of the cell's time is spent in **interphase**, the stage between nuclear divisions. For many years it was thought that interphase was a period of rest, because nuclei appeared inactive when observed by ordinary light microscopy. In the early 1950s new techniques for revealing DNA replication in nuclei were introduced at the same time that biologists

came to appreciate fully the significance of DNA as the genetic material. It was then discovered that DNA replication occurred during the interphase stage. Further studies revealed that many other protein and nucleic acid components essential to normal cell function, growth, and division were synthesized during the seemingly quiescent interphase period.

Replication of DNA occurs during a phase called the S phase (period of synthesis). In mammalian cells in tissue culture, the S phase lasts about six of the 18 to 24 hours required to complete one cell cycle. In this phase both strands of a DNA molecule must replicate; new complementary partners are synthesized for each strand so that two identical DNA molecules are produced from the original strand (see Chapter 5, p. 93). These complementary partners are the sister chromatids that are separated during the next mitosis.

The S phase is preceded and succeeded by G_1 and G_2 phases, respectively (G stands for "gap"), during which no DNA synthesis is occurring. For most cells, G_1 is an important preparatory stage for the replication of DNA that follows. During G_1, transfer RNA, ribosomes, messenger RNA, and many enzymes are synthesized. During G_2, spindle and aster proteins

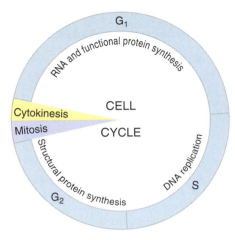

Figure 3.25

Cell cycle, showing relative duration of recognized phases. S, G_1, and G_2 are phases within interphase; S, synthesis of DNA; G_1, presynthetic phase; G_2, postsynthetic phase. After mitosis and cytokinesis the cell may go into an arrested, quiescent stage known as G_0. Actual duration of the cycle and the different phases varies considerably in different cell types.

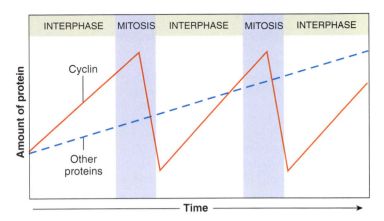

Figure 3.26

Variations in the level of cyclin in dividing cells of early sea urchin embryos. Cyclin binds with its cyclin-dependent kinase to activate the enzyme.

are synthesized in preparation for chromosome separation during mitosis. G_1 is typically of longer duration than G_2, although there is much variation in different cell types. Embryonic cells divide very rapidly because there is no cell growth between divisions, only subdivision of mass. DNA synthesis may proceed a hundred times more rapidly in embryonic cells than in adult cells, and the G_1 phase is very shortened. As an organism develops, the cycle of most of its cells lengthens, and many cells may be arrested for long periods of time in G_1 and enter a nonproliferative or quiescent phase called G_0. Most neurons or nerve cells, for example, divide no further and are essentially in a permanent G_0.

The events in cell cycles are exquisitely regulated. Transitions during cell cycles are mediated by enzymes called **cyclin-dependent kinases (cdk's)** and regulatory protein subunits that activate them, called **cyclins.** In general, kinases are enzymes that add phosphate groups to other proteins to activate or inactivate them, and kinases themselves may require activation. Cdk's become active only when they are bound to the appropriate cyclin, and cyclins are synthesized and degraded during each cell cycle (Figure 3.26). It seems likely that phosphorylation and dephosphorylation of specific cdk's and their interaction with phase-specific cyclins regulates the passage from one cell cycle phase to the next. Current research focuses on the checkpoints that regulate this phase to phase passage since disregulation of these mechanisms has been implicated in cancer.

Flux of Cells

Cell division is important for growth, for replacement of cells lost to natural attrition and for wound healing. Cell division is especially rapid during early development of an organism. At birth a human infant has about 2 trillion cells from repeated division of

a single fertilized egg or zygote. This immense number could be attained by just 42 cell divisions, with each generation dividing once every six to seven days. With only five more cell divisions, the cell number would increase to approximately 60 trillion, the number of cells in a human adult weighing 75 kg. Of course no organism develops in this machinelike manner. Cell division is rapid during early embryonic development, then slows with age. Furthermore, different cell populations divide at widely different rates. In some the average period between divisions is measured in hours, whereas in others it is measured in days, months, or even years. Some cells in the central nervous system stop dividing after the early months of fetal development and generally persist without further division for the life of the individual. Muscle cells also stop dividing during the third month of fetal development, and most future growth depends on enlargement of fibers already present.

In other tissues that are subject to abrasion, lost cells must be constantly replaced. It is estimated that in humans about 1% to 2% of all body cells—a total of 100 billion—are shed daily. Mechanical rubbing wears away the outer cells of the skin, and food in the alimentary canal removes lining epithelial cells. In addition, the restricted life cycle of blood cells involves enormous numbers of replacements. Such lost cells are replaced by mitosis.

Normal development, however, does entail cell death in which cells are not replaced. As cells age, they accumulate damage from destructive oxidizing agents and eventually die. Other cells undergo a programmed cell death, or **apoptosis** (a-puh-TOE-sis) (Gr. *apo-*, from, away from; + *ptosis*, a falling), which is in many cases necessary for continued health and development of an organism. For example, during embryonic development of vertebrates, fingers and toes develop as tissues between them die, excess immune cells that would attack the body's own tissues "commit suicide," and nerve cells die to create cerebral convolutions. Apoptosis consists of a well-coordinated and predictable series of events: the cells shrink, disintegrate, and their components are absorbed up by surrounding cells.

SUMMARY

Cells are the basic structural and functional units of all living organisms. Eukaryotic cells differ from the prokaryotic cells of bacteria and Archaebacteria in several respects, the most distinctive of which is presence of a membrane-bound nucleus containing hereditary material composed of DNA, bound to proteins to form chromatin. Chromatin consists of flexible, linear chromosomes, which become condensed and visible only during cell division.

Cells are surrounded by a plasma membrane that regulates the flow of molecules between the cell and its surroundings. The nucleus, enclosed by a double membrane, contains chromatin, associated proteins, and one or more nucleoli. Outside the nuclear envelope is cell cytoplasm, subdivided by a membranous network, the endoplasmic reticulum. Among the organelles within cells are the Golgi complex, mitochondria, lysosomes, and other membrane-bound vesicles. The cytoskeleton is composed of microfilaments (actin), microtubules (tubulin), and intermediate filaments (several types). Cilia and flagella are hairlike, motile appendages that contain microtubules. Ameboid movement by pseudopodia operates by means of the assembly and disassembly of actin microfilaments. Tight junctions, adhesion junctions, desmosomes, and gap junctions are structurally and functionally distinct connections between cells.

Membranes in a cell are composed of a phospholipid bilayer and other materials including cholesterol and transmembrane proteins. Hydrophilic ends of the phospholipid molecules are on the outer and inner surfaces of membranes, and the fatty acid portions are directed inward, toward each other, to form a hydrophobic core.

Substances can enter cells by diffusion, mediated transport, and endocytosis. Osmosis is diffusion of water through channels in a selectively permeable membrane as a result of osmotic pressure. Solutes to which the membrane is impermeable require channels or a transporter molecule to traverse the membrane. Water and ions move through open channels by diffusion (in the direction of a concentration gradient). Transport-mediated systems include facilitated diffusion and active transport (against a concentration gradient, which requires energy). Endocytosis includes bringing droplets (pinocytosis) or particles (phagocytosis) into a cell. In exocytosis the process of endocytosis is reversed.

The cell cycle in eukaryotes includes mitosis, or division of the nuclear chromosomes, and cytokinesis, the division of the cytoplasm, and interphase. During interphase, G_1, S, and G_2 phases are recognized, and the S phase is the time when DNA is synthesized (the chromosomes are replicated).

Cell division is necessary for the production of new cells from preexisting cells and is the basis for growth in multicellular organisms. During this process, replicated nuclear chromosomes divide by mitosis followed by cytoplasmic division or cytokinesis.

The four stages of mitosis are prophase, metaphase, anaphase, and telophase. In prophase, replicated chromosomes composed of sister chromatids condense into recognizable bodies. A spindle forms between the centrosomes as they separate to opposite poles of the cell. At the end of prophase the nuclear envelope disintegrates, and the kinetochores of each chromosome become attached to the centrosomes by microtubules (kinetochore fibers). At metaphase the sister chromatids are moved to the center of the cell, held there by the kinetochore fibers. At anaphase the centromeres divide, and the sister chromatids are pulled apart by the attached kinetochore fibers of the mitotic spindle. At telophase the sister chromatids, now called chromosomes, gather in the position of the nucleus in each cell and revert to a diffuse chromatin network. A nuclear membrane reappears, and cytokinesis occurs. At the end of mitosis and cytokinesis, two cells genetically identical to the parent cell have been produced.

Cells divide rapidly during embryonic development, then more slowly with age. Some cells continue to divide throughout the life of an animal to replace cells lost by attrition and wear, whereas others, such as nerve and muscle cells, complete their division during early development and many do not divide again. Some cells undergo a programmed cell death, or apoptosis.

REVIEW QUESTIONS

1. Explain the difference (in principle) between a light microscope and a transmission electron microscope.
2. Briefly describe the structure and function of each of the following: plasma membrane, chromatin, nucleus, nucleolus, rough endoplasmic reticulum (rough ER), Golgi complex, lysosomes, mitochondria, microfilaments, microtubules, intermediate filaments, centrioles, basal body (kinetosome), tight junction, gap junction, desmosome, glycoprotein, microvilli.
3. Name two functions each for actin and for tubulin.
4. Distinguish among cilia, flagella, and pseudopodia.
5. What are the functions of each of the main constituents of the plasma membrane?
6. Our current concept of the plasma membrane is known as the fluid-mosaic model. Why?
7. You place some red blood cells in a solution and observe that they swell and burst. You place some cells in another solution, and they shrink and become wrinkled. Explain what has happened in each case.
8. Explain why a beaker containing a salt solution, placed on a table in your classroom, can have a high osmotic pressure, yet be subjected to a hydrostatic pressure of only one atmosphere.
9. The plasma membrane is an effective barrier to molecular movement across it, yet many substances do enter and leave the cell. Explain the mechanisms through which this is accomplished and comment on the energy requirements of these mechanisms.
10. Distinguish among phagocytosis, pinocytosis, receptor-mediated endocytosis, and exocytosis.
11. Define the following: chromosome, centromere, centrosome, kinetochore, mitosis, cytokinesis, syncytium.
12. Explain phases of the cell cycle, and comment on important cellular processes that characterize each phase. What is G_0?
13. Name the stages of mitosis in order, and describe the behavior and structure of the chromosomes at each stage.
14. Briefly describe ways that cells may die during the normal life of a multicellular organism.

SELECTED REFERENCES

Alberts, B., D. Bray, K. Hopkin, A. Johnson, J. Lewis, M. Raff, K. Roberts, and P. Walter. 2003. Essential cell biology, ed. 2. New York, Garland Science Publishing. *A well-written text describing more detailed cellular mechanisms.*

Kaksonen, M., C. P. Toret, and D. G. Drubin. 2006. Harnessing action dynamics for clathrin-coated endocytosis. Nature Reviews, Molecular Cell Biology **7**:404–414. *While receptor-mediated endocytosis is the focus of this article, the role of actin in phagocytosis and pinocytosis are discussed.*

Roth, R. 2006. Clathrin-mediated endocytosis before fluorescent proteins. Nature Reviews, Molecular Cell Biology **7**:63–68. *An excellent discussion of the history of the scientific timeline underlying the dynamic process of receptor-mediated endocytosis.*

Wolfe, S. L. 1995. Introduction to cell and molecular biology, ed. 1. Belmont, CA, Thomson Brooks/Core Publishers. *Good coverage of cell structure and function for those interested in getting more detail.*

ONLINE LEARNING CENTER

Visit www.mhhe.com/hickmanipz14e for chapter quizzing, key term flash cards, web links, and more!

4

Cellular Metabolism

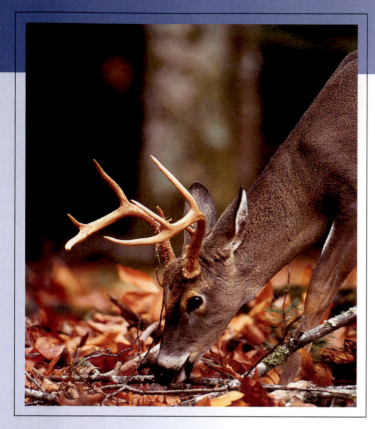

White-tailed deer (Odocoileus virginianus) *foraging for acorns.*

Deferring the Second Law

Living systems appear to contradict the second law of thermodynamics, which states that energy in the universe has direction and that it has been, and always will be, running down. In effect all forms of energy inevitably will be degraded to heat. This increase in disorder, or randomness, in any closed system is termed entropy. Living systems, however, *decrease* their entropy by *increasing* the molecular orderliness of their structure. An organism becomes vastly more complex during its development from fertilized egg to adult. The second law of thermodynamics, however, applies to closed systems, and living organisms are not closed systems. Animals grow and maintain themselves by borrowing free energy from the environment. When a deer feasts on the acorns and beechnuts of summer, it transfers potential energy, stored as chemical bond energy in the nuts' tissues,

to its own body. Then, in step-by-step sequences called biochemical pathways, this energy is gradually released to fuel the deer's many activities. In effect, the deer decreases its own internal entropy by increasing the entropy of its food. The orderly structure of the deer is not permanent, however, but will be dissipated when it dies.

The ultimate source of this energy for the deer—and for almost all life on earth—is the sun (Figure 4.1). Sunlight is captured by green plants, which fortunately accumulate enough chemical bond energy to sustain both themselves and animals that feed on them. Thus the second law is not violated; it is simply held at bay by life on earth, which uses the continuous flow of solar energy to maintain a biosphere of high internal order, at least for the period of time that life exists.

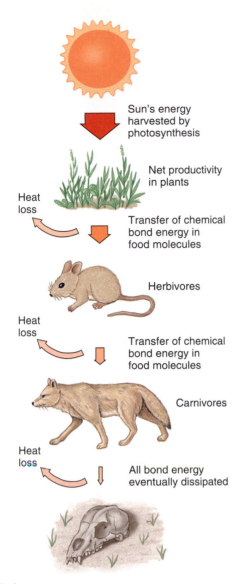

Figure 4.1

Solar energy sustains virtually all life on earth. With each energy transfer, however, about 90% of the energy is lost as heat.

All cells must obtain energy, synthesize their own internal structure, control much of their own activity, and guard their boundaries. **Cellular metabolism** refers to the collective chemical processes that occur within living cells to accomplish these activities. Although the enormous number of reactions in their aggregate are extremely complex, the central metabolic routes through which matter and energy are channeled appear to be conserved by the majority of living organisms.

ENERGY AND THE LAWS OF THERMODYNAMICS

The concept of energy is fundamental to all life processes. We usually express energy as the capacity to do work, to bring about change. Yet energy is a somewhat abstract quantity that is difficult to define and elusive to measure. Energy cannot be seen; it can be identified only by how it affects matter.

Energy can exist in either of two states: kinetic or potential. **Kinetic energy** is the energy of motion. **Potential energy** is stored energy, energy that is not doing work but has the capacity to do so. Energy can be transformed from one state to another. Especially important for living organisms is chemical energy, a form of potential energy stored in the chemical bonds of molecules. Chemical energy can be tapped when bonds are rearranged to release kinetic energy. Much of the work done by living organisms involves conversion of potential energy to kinetic energy.

Conversion of one form of energy to another is governed by the two laws of thermodynamics. The **first law of thermodynamics** states that energy cannot be created or destroyed. It can change from one form to another, but the total amount of energy remains the same. In short, energy is conserved. If we burn gasoline in an engine, we do not create new energy but merely convert the chemical energy in gasoline to another form, in this example, mechanical energy and heat. The **second law of thermodynamics,** introduced in the prologue to this chapter, concerns the transformation of energy. This fundamental law states that a closed system moves toward increasing disorder, or entropy, as energy is dissipated from the system (Figure 4.2). Living systems, however, are open systems that not only maintain their organization but also increase it, as during the development of an animal from egg to adult.

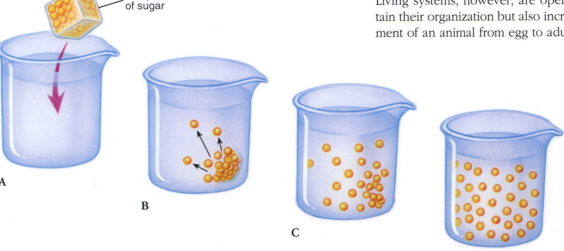

Figure 4.2

Diffusion of a solute through a solution, an example of entropy. When the solute (sugar molecules) is first introduced into a solution, the system is ordered and unstable (**B**). Without energy to maintain this order, the solute particles become distributed into solution, reaching a state of disorder (equilibrium) (**D**). Entropy has increased from left diagram to right diagram.

Free Energy

To describe the energy changes of chemical reactions, biochemists use the concept of **free energy.** Free energy is simply the energy in a system available for doing work. In a molecule, free energy equals the energy present in chemical bonds minus the energy that cannot be used. The majority of reactions in cells release free energy and are called **exergonic** (Gr. *ex,* out, + *ergon,* work). Such reactions are spontaneous and always proceed "downhill" since free energy is lost from the system. Thus:

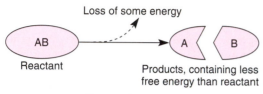

Exergonic reaction

However, many important reactions in cells require the addition of free energy and are said to be **endergonic** (Gr. *endon,* within, + *ergon,* work). Such reactions have to be "pushed uphill" because the products store more energy than the reactants.

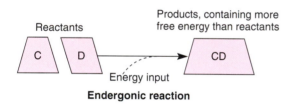

Endergonic reaction

Described on page 62, ATP is a ubiquitous, energy-rich intermediate used by organisms to power important uphill reactions such as those required for active transport of molecules across membranes (see Chapter 3, p. 49) and cellular synthesis.

THE ROLE OF ENZYMES

Enzymes and Activation Energy

For any reaction to occur, even exergonic ones that tend to proceed spontaneously, chemical bonds first must be destabilized. Some energy, termed the **activation energy,** must be supplied before the bond is stressed enough to break. Only then will an overall loss of free energy and formation of reaction products occur. This requirement can be likened to the energy needed to push a ball over the crest of a hill before it will roll spontaneously down the other side, the ball liberating its energy as it descends (Figure 4.3, top panel).

One way to activate chemical reactants is to raise the temperature. This increases the rate of molecular collisions and pushes chemical bonds apart. Thus, heat can impart the necessary activation energy to make a reaction proceed. However metabolic reactions must occur at biologically tolerable temperatures, which are usually too low to allow reactions to proceed

at a rate capable of sustaining life. Instead, living systems have evolved a different strategy: they employ **catalysts.**

Catalysts are chemical substances that accelerate reaction rates without affecting the products of the reaction and without being altered or destroyed by the reaction. A catalyst cannot make an energetically impossible reaction happen; it simply accelerates a reaction that would proceed at a very slow rate otherwise.

Enzymes are catalysts of the living world. The special catalytic talent of an enzyme is its power to reduce the amount of activation energy required for a reaction. In effect, an enzyme steers the reaction through one or more intermediate steps, each of which requires much less activation energy than that required for a single-step reaction (Figure 4.3). Note that enzymes do not supply the activation energy. Instead they lower the activation energy barrier, making a reaction more likely to proceed. Enzymes affect only the reaction rate. They do not in any way alter the free energy change of a reaction, nor do they change the proportions of reactants and products in a reaction.

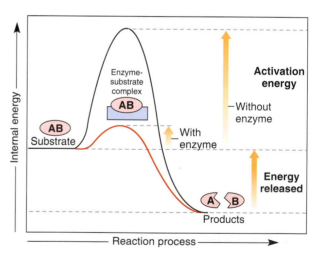

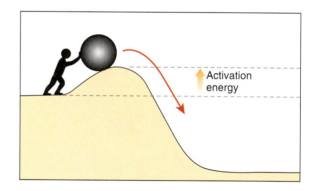

Figure 4.3

Energy changes during enzyme catalysis of a substrate. The overall reaction proceeds with a net release of energy (exergonic). In the absence of an enzyme, substrate is stable because of the large amount of activation energy needed to disrupt strong chemical bonds. The enzyme reduces the energy barrier by forming a chemical intermediate with a much lower internal energy state.

Nature of Enzymes

Enzymes are complex molecules that vary in size from small, simple proteins with a molecular weight of 10,000 to highly complex molecules with molecular weights up to 1 million. Many enzymes are pure proteins—highly folded and interlinked chains of amino acids. Other enzymes require participation of small nonprotein groups called **cofactors** to perform their enzymatic function. In some cases these cofactors are metallic ions (such as ions of iron, copper, zinc, magnesium, potassium, and calcium) that form a functional part of the enzyme. Examples are carbonic anhydrase (see Chapter 31, p. 705), which contains zinc; the cytochromes (some enzymes of the electron transport chain, p. 67), which contain iron; and troponin (a muscle contraction enzyme, see Chapter 29, p. 658), which contains calcium. Another class of cofactors, called **coenzymes,** is organic. Coenzymes contain groups derived from vitamins, most of which must be supplied in the diet. All B-complex vitamins are coenzymatic compounds. Since animals have lost the ability to synthesize the vitamin components of coenzymes, it is obvious that a vitamin deficiency can be serious. However, unlike dietary fuels and nutrients that must be replaced after they are burned or assembled into structural materials, vitamin components of coenzymes are recovered in their original form and are used repeatedly. Examples of coenzymes that contain vitamins are nicotinamide adenine dinucleotide (NAD), which contains the vitamin nicotinic acid (niacin); coenzyme A, which contains the vitamin pantothenic acid; and flavin adenine dinucleotide (FAD), which contains riboflavin (vitamin B_2).

Action of Enzymes

An enzyme functions by associating in a highly specific way with its **substrate,** the molecule whose reaction it catalyzes.

Enzymes bear an active site located within a cleft or pocket and that contains a unique molecular configuration. The active site has a flexible surface that enfolds and conforms to the substrate (Figure 4.4). The binding of enzyme to substrate forms an **enzyme-substrate complex (ES complex),** in which the substrate is secured by covalent bonds to one or more points in the active site of the enzyme. The ES complex is not strong and will quickly dissociate, but during this fleeting moment the enzyme provides a unique chemical environment that stresses certain chemical bonds in the substrate so that much less energy is required to complete the reaction.

If the formation of an enzyme-substrate complex is so rapidly followed by dissociation, how can biochemists be certain that an ES complex exists? The original evidence offered by Leonor Michaelis in 1913 is that, when the substrate concentration is increased while the enzyme concentration is held constant, the reaction rate reaches a maximum velocity. This *saturation effect* is interpreted to mean that all catalytic sites become filled at high substrate concentration. A saturation effect is not seen in uncatalyzed reactions. Other evidence includes the observation that the ES complex displays unique spectroscopic characteristics not displayed by either the enzyme or the substrate alone. Furthermore, some ES complexes can be isolated in pure form, and at least one kind (nucleic acids and their polymerase enzymes) has been directly visualized by electron microscopy.

Enzymes that engage in crucial energy-providing reactions of cells often proceed constantly and usually operate in sets rather than in isolation. For example, conversion of glucose to carbon dioxide and water proceeds through 19 reactions, each requiring a specific enzyme. Such crucial enzymes occur in

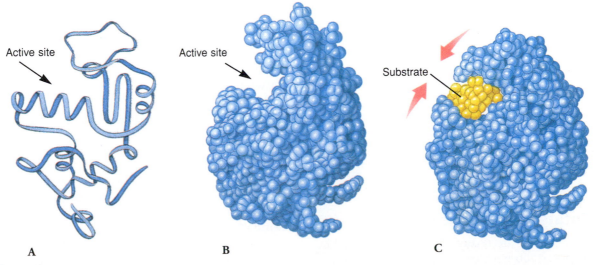

Figure 4.4

How an enzyme works. The ribbon model **(A)** and the space-filling model **(B)** show that the enzyme lysozyme bears a pocket containing the active site. When a chain of sugars (substrate) enters the pocket **(C)**, the protein enzyme changes shape slightly so that the pocket enfolds the substrate and conforms to its shape. This positions the active site (amino acids in the protein) next to a bond between adjacent sugars in the chain, causing the sugar chain to break.

relatively high concentrations in cells, and they may implement quite complex and highly integrated enzymatic sequences. One enzyme performs the first step, then another enzyme binds to the product and catalyzes the next step. This process continues until the end of the enzymatic pathway is reached. The reactions are said to be coupled. Coupled reactions are explained in a section on chemical energy transfer by ATP (see this page).

Specificity of Enzymes

One of the most distinctive attributes of enzymes is their high specificity. Specificity is a consequence of the exact molecular fit required between enzyme and substrate. Furthermore, an enzyme catalyzes only one reaction. Unlike reactions performed in an organic chemist's laboratory, no side reactions or by-products result. Specificity of both substrate and reaction is obviously essential to prevent a cell from being swamped with useless by-products.

However, there is some variation in degree of specificity. Some enzymes catalyze the oxidation (dehydrogenation) of only one substrate. For example, succinic dehydrogenase catalyzes the oxidation of succinic acid only (see the Krebs cycle, p. 66). Others, such as proteases (for example, pepsin and trypsin, released into the digestive tract during digestion, pp. 716 and 719), act on almost any protein, although each protease has its particular point of attack in the protein (Figure 4.5). Usually an enzyme binds one substrate molecule at a time, catalyzes its chemical change, releases the product, and then repeats the process with another substrate molecule. An enzyme may repeat this process billions of times until it is finally worn out (after a few hours to several years) and is degraded by scavenger enzymes in the cell. Some enzymes undergo successive catalytic cycles at speeds of up to a million cycles per minute, but most operate at slower rates. Many enzymes are repeatedly activated and inactivated; several mechanisms for regulating enzyme activity are well known (p. 72).

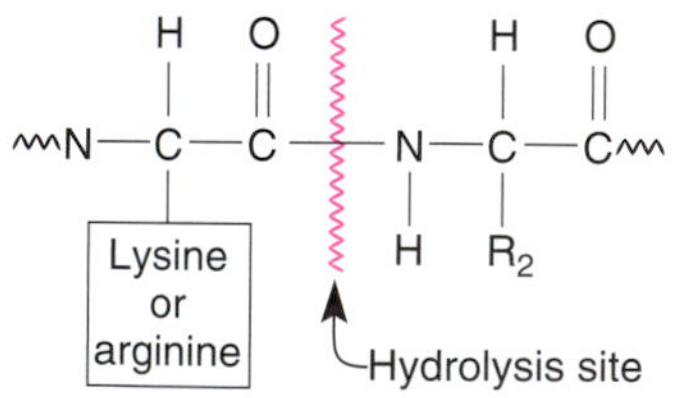

Figure 4.5

High specificity of trypsin. It splits only peptide bonds adjacent to lysine or arginine.

Enzyme-Catalyzed Reactions

Enzyme-catalyzed reactions are reversible, which is signified by double arrows between substrate and products. For example:

$$\text{Fumaric acid} + H_2O \rightleftharpoons \text{Malic acid}$$

However, for various reasons reactions catalyzed by most enzymes tend to go predominantly in one direction. For example,

the proteolytic enzyme pepsin degrades proteins into amino acids (a **catabolic** reaction), but it does not accelerate the rebuilding of amino acids into any significant amount of protein (an **anabolic** reaction). The same is true of most enzymes that catalyze the cleavage of large molecules such as nucleic acids, polysaccharides, lipids, and proteins. There is usually one set of reactions and enzymes that degrade them (catabolism; Gr. *kata,* down, + *bole,* throw), but they must be resynthesized by a different set of reactions catalyzed by different enzymes (anabolism; Gr. *ana,* up, + *bole,* throw).

The net **direction** of any chemical reaction depends on the relative energy contents of the substances involved. If there is little change in chemical bond energy of substrate and products, the reaction is more easily reversible. However, if large quantities of energy are released as the reaction proceeds in one direction, more energy must be provided in some way to drive the reaction in the reverse direction. For this reason many if not most enzyme-catalyzed reactions are in practice irreversible unless the reaction is coupled to another one that makes energy available. In cells both reversible and irreversible reactions are combined in complex ways to make possible both synthesis and degradation.

Hydrolysis literally means "breaking with water." In hydrolysis reactions, a molecule is cleaved by the addition of water at the cleavage site. A hydrogen is attached to one subunit and a hydroxyl (—OH) unit is attached to another. This breaks the covalent bond between subunits. Hydrolysis is the opposite of condensation (water-losing) reactions in which subunits of molecules are linked together by removal of water. Macromolecules are built by condensation reactions.

CHEMICAL ENERGY TRANSFER BY ATP

We have seen that endergonic reactions are those that do not proceed spontaneously because their products require an input of free energy. However, an endergonic reaction may be driven by coupling the energy-requiring reaction with an energy-yielding reaction. ATP is one of the most common intermediates in **coupled reactions,** and because it can drive such energetically unfavorable reactions, it is of central importance in metabolic processes.

An ATP molecule consists of adenosine (the purine adenine and the 5-carbon sugar ribose) and a triphosphate group (Figures 4.6 and 4.7). Most free energy in ATP resides in the triphosphate group, especially in two **phosphoanhydride bonds** between the three phosphate groups called **"high-energy bonds."** Usually, only the most exposed high-energy bond is hydrolyzed to release free energy when ATP is converted to adenosine diphosphate (ADP) and inorganic phosphate.

$$ATP + H_2O \rightarrow ADP + P_i$$

where P_i represents inorganic phosphate (i = inorganic). The high-energy groups in ATP are often designated by the

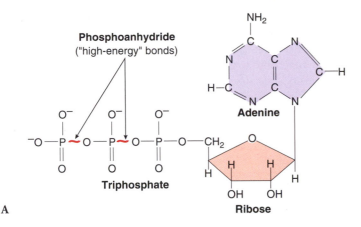

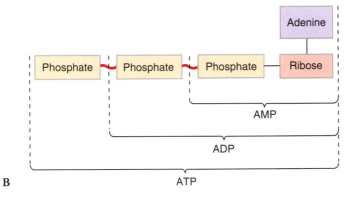

Figure 4.6

A, Structure of ATP. **B,** ATP formation from ADP and AMP. ATP: adenosine triphosphate; ADP: adenosine diphosphate; AMP: adenosine monophosphate.

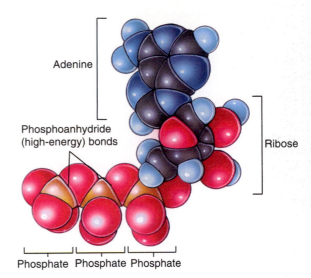

Figure 4.7

Space-filling model of ATP. In this model, carbon is shown in black; nitrogen in blue; oxygen in red; and phosphorus in orange.

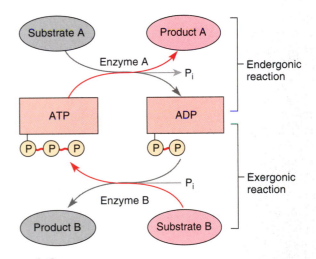

Figure 4.8

A coupled reaction. The endergonic conversion of substrate A to product A will not occur spontaneously but requires an input of energy from another reaction involving a large release of energy. ATP is the intermediate through which the energy is shuttled.

"tilde" symbol ~ (Figure 4.6). A high-energy phosphate bond is shown as ~P and a low-energy bond (such as the bond linking the triphosphate group to adenosine) as —P. Thus, ATP may be symbolized as A—P~P~P and ADP as A—P~P.

The way in which ATP can drive a coupled reaction is shown in Figure 4.8. A coupled reaction is really a system involving two reactions linked by an energy shuttle (ATP). The conversion of substrate A to product A is endergonic because the product contains more free energy than the substrate. Therefore energy must be supplied by coupling the reaction to an exergonic one, the conversion of substrate B to product B. Substrate B in this reaction is commonly called a **fuel** (for example, glucose or a lipid). Bond energy released in reaction B is transferred to ADP, which in turn is converted to ATP. ATP now contributes its phosphate-bond energy to reaction A, and ADP and P_i are produced again.

The high-energy bonds of ATP are actually rather weak, unstable bonds. Because they are unstable, the energy of ATP is readily released when ATP is hydrolyzed in cellular reactions. Note that ATP is an **energy-coupling agent** and *not* a fuel. It is not a storehouse of energy set aside for some future need. Rather it is produced by one set of reactions and is almost immediately consumed by another set. ATP is formed as it is needed, primarily by oxidative processes in mitochondria. Oxygen is not

consumed unless ADP and phosphate molecules are available, and these do not become available until ATP is hydrolyzed by some energy-consuming process. *Metabolism is therefore mostly self-regulating.*

CELLULAR RESPIRATION

How Electron Transport Is Used to Trap Chemical Bond Energy

Having seen that ATP is the one common energy denominator by which most cellular machines are powered, we must ask how this energy is captured from fuel substrates. This question directs

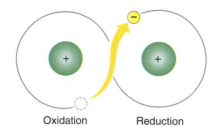

Oxidation   Reduction

Figure 4.9

A redox pair. The molecule at left is oxidized by the loss of an electron. The molecule at right is reduced by gaining an electron.

us to an important generalization: *all cells obtain their chemical energy requirements from oxidation-reduction reactions.* This means that in the degradation of fuel molecules, hydrogen atoms (electrons and protons) are passed from electron donors to electron acceptors with a release of energy. A portion of this energy can be trapped and used to form the high-energy bonds of molecules such as ATP.

An oxidation-reduction ("redox") reaction involves a transfer of electrons from an electron donor (the reducing agent) to an electron acceptor (the oxidizing agent). As soon as the electron donor loses its electrons, it becomes oxidized. As soon as the electron acceptor accepts electrons, it becomes reduced (Figure 4.9). In other words, a reducing agent becomes oxidized when it reduces another compound, and an oxidizing agent becomes reduced when it oxidizes another compound. Thus for every oxidation there must be a corresponding reduction.

In an oxidation-reduction reaction the electron donor and electron acceptor form a redox pair:

$$\text{Electron donor} \rightleftharpoons e^- + \text{Electron acceptor} + \text{Energy}$$

Electron donor $\rightleftharpoons$ e⁻ +   Electron acceptor + Energy
(reducing agent;   (oxidizing agent;
becomes oxidized)   becomes reduced)

When electrons are accepted by the oxidizing agent, energy is liberated because the electrons move to a more stable position.

ATP may be produced in a cell when electrons flow through a series of carriers. Each carrier is reduced as it accepts electrons and then is reoxidized when it passes electrons to the next carrier in the series. By transferring electrons stepwise in this manner, energy is gradually released, and ATP is produced. Ultimately, the electrons are transferred to a **final electron acceptor.** The nature of this final acceptor is the key that determines the overall efficiency of cellular metabolism.

Aerobic Versus Anaerobic Metabolism

Heterotrophs (organisms that cannot synthesize their own food but must obtain nutrients from the environment, including animals, fungi, and many single-celled organisms) are divided into two groups based on their overall efficiency of energy production during cellular metabolism: **aerobes,** those that use molecular oxygen as the final electron acceptor, and **anaerobes,** those that employ another molecule as the final electron acceptor.

As discussed in Chapter 2, life originated in the absence of oxygen, and an abundance of atmospheric oxygen was produced only after photosynthetic organisms (autotrophs) evolved. Some strictly anaerobic organisms still exist and indeed play important roles in specialized habitats. However, evolution has favored aerobic metabolism, not only because oxygen became available, but also because aerobic metabolism is vastly more efficient in energy production than anaerobic metabolism. In the absence of oxygen, only a very small fraction of the bond energy present in nutrients can be released. For example, when an anaerobic microorganism degrades glucose, the final electron acceptor (such as pyruvic acid) still contains most of the energy of the original glucose molecule. An aerobic organism on the other hand, using oxygen as the final electron acceptor, completely oxidizes glucose to carbon dioxide and water. Almost 20 times as much energy is released when glucose is completely oxidized as when it is degraded only to the stage of pyruvic acid. Thus, an obvious advantage of aerobic metabolism is that a much smaller quantity of food is required to maintain a given rate of metabolism.

Overview of Respiration

Aerobic metabolism is more familiarly called **cellular respiration,** defined as the oxidation of fuel molecules to produce energy with molecular oxygen as the final electron acceptor. We emphasize that oxidation of fuel molecules describes the *removal of electrons* and *not* the direct combination of molecular oxygen with fuel molecules. Let us look at this process in general before considering it in more detail.

Hans Krebs, the British biochemist who contributed so much to our understanding of respiration, described three stages in the complete oxidation of fuel molecules to carbon dioxide and water (Figure 4.10). In stage I, food passing through the intestinal tract are digested into small molecules that can be absorbed into the circulation. There is no useful energy yield during digestion, which is discussed in Chapter 32. In stage II, also called **glycolysis,** most of the digested food is converted into two 3-carbon units (pyruvic acid) in the cell cytoplasm. The pyruvic acid molecules then enter mitochondria, where in another reaction they join with a coenzyme (coenzyme A or CoA) to form acetyl coenzyme A or acetyl-CoA. Some ATP is generated in stage II, but the yield is small compared with that obtained in stage III of respiration. In stage III the final oxidation of fuel molecules occurs, with a large yield of ATP. This stage takes place entirely in mitochondria. Acetyl-CoA is channeled into the Krebs cycle where the acetyl group is completely oxidized to carbon dioxide. Electrons released from acetyl groups are transferred to special carriers that pass them to electron acceptor compounds in the electron transport chain. At the end of the chain the electrons (and the protons accompanying them) are accepted by molecular oxygen to form water.

Glycolysis

We begin our journey through the stages of respiration with glycolysis, a nearly universal pathway in living organisms that converts glucose into pyruvic acid. In a series of reactions occurring

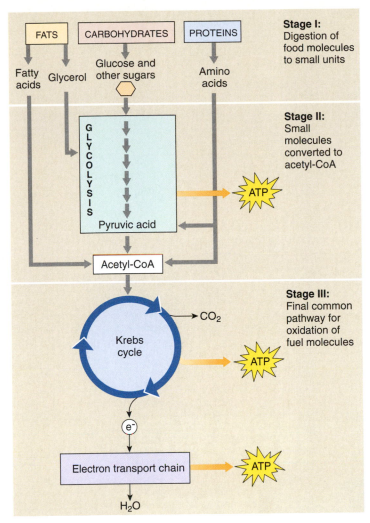

Figure 4.10

Overview of cellular respiration, showing the three stages in the complete oxidation of food molecules to carbon dioxide and water.

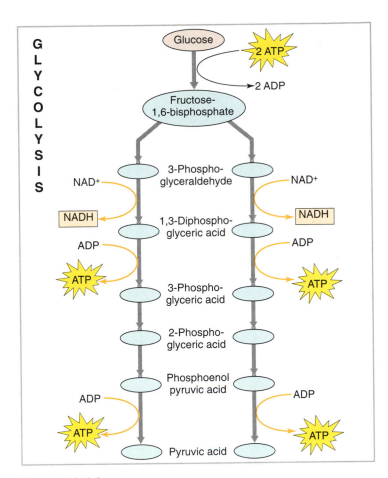

Figure 4.11

Glycolysis. Glucose is phosphorylated in two steps and raised to a higher energy level. High-energy fructose-1,6-bisphosphate is split into triose phosphates that are oxidized exergonically to pyruvic acid, yielding ATP and NADH.

in the cell cytosol, glucose and other 6-carbon monosaccharides are split into 3-carbon molecules of **pyruvic acid** (Figure 4.11). A single oxidation occurs during glycolysis, and each molecule of glucose yields two molecules of ATP. In this pathway the carbohydrate molecule is phosphorylated twice by ATP, first to glucose-6-phosphate (not shown in Figure 4.11) and then to fructose-1,6-bisphosphate. The fuel has now been "primed" with phosphate groups in this uphill portion of glycolysis and is sufficiently reactive to enable subsequent reactions to proceed. This is a kind of deficit financing required for an ultimate energy return many times greater than the original energy investment.

In the downhill portion of glycolysis, fructose-1,6-bisphosphate is cleaved into two 3-carbon sugars, which undergo an oxidation (electrons are removed), with the electrons and one of the hydrogen ions being accepted by **nicotinamide adenine dinucleotide (NAD+,** a derivative of the vitamin niacin) to produce a reduced form called **NADH.** NADH serves as a carrier molecule to convey high-energy electrons to the final electron transport chain, where ATP is produced.

The two 3-carbon sugars next undergo a series of reactions, ending with the formation of two molecules of pyruvic acid (Figure 4.11). In two of these steps, a molecule of ATP is produced. In other words, each 3-carbon sugar yields two ATP molecules, and since there are two 3-carbon sugars, four ATP molecules are generated. Recalling that two ATP molecules were used to prime the glucose initially, the net yield at this point is two ATP molecules. The 10 enzymatically catalyzed reactions in glycolysis can be summarized as:

$$\text{Glucose} + 2 \text{ ADP} + 2 \text{ P}_i + 2 \text{ NAD}^+ \rightarrow 2 \text{ pyruvic acid} + 2 \text{ NADH} + 2 \text{ ATP}$$

Acetyl-CoA: Strategic Intermediate in Respiration

In aerobic metabolism the two molecules of pyruvic acid formed during glycolysis enter a mitochondrion. There, each molecule is oxidized, and one of the carbons is released as carbon dioxide (Figure 4.12). The 2-carbon residue condenses with **coenzyme A (CoA)** to form **acetyl coenzyme A,** or **acetyl-CoA,** and an NADH molecule is also produced.

Pyruvic acid is the undissociated form of the acid:

$$CH_3 - \overset{\overset{\displaystyle O}{\|}}{C} - COOH$$

Under physiological conditions pyruvic acid typically dissociates

into pyruvate $(CH_3 - \overset{\overset{\displaystyle O}{\|}}{C} - COO^-)$ and H^+. It is correct to use either term in describing this and other organic acids (such as lactic acid, or lactate) in metabolism.

Acetyl-CoA is a critically important compound. Its oxidation in the Krebs cycle provides energized electrons to generate ATP, and it is also a crucial intermediate in lipid metabolism (p. 70).

Krebs Cycle: Oxidation of Acetyl-CoA

Degradation (oxidation) of the 2-carbon acetyl group of acetyl-CoA occurs within the mitochondrial matrix in a cyclic sequence

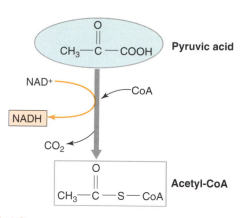

Figure 4.12

Formation of acetyl-CoA from pyruvic acid.

called the **Krebs cycle** (also called citric acid cycle and tricarboxylic acid cycle [TCA cycle]) (Figure 4.13). Acetyl-CoA condenses with a 4-carbon acid (oxaloacetic acid), releasing CoA to react again with more pyruvic acid. Through a cyclic series of reactions in the Krebs cycle, the two carbons from the acetyl group

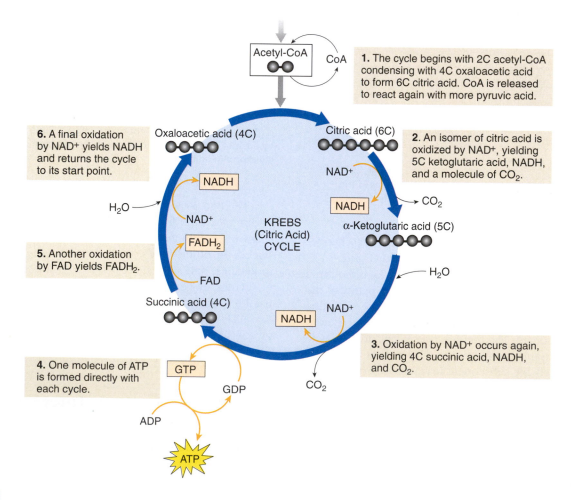

Figure 4.13

The Krebs cycle in outline form, showing production of three molecules of reduced NAD, one molecule of reduced FAD, one molecule of ATP, and two molecules of carbon dioxide. The molecules of NADH and $FADH_2$ will yield 11 molecules of ATP when oxidized in the electron transport system.

are released as carbon dioxide, and oxaloacetic acid is regenerated. Hydrogen ions and electrons in the oxidations transfer to NAD^+ and to FAD (flavine adenine dinucleotide, another electron acceptor), and a pyrophosphate bond is generated in the form of guanosine triphosphate (GTP). This high-energy phosphate readily transfers to ADP to form ATP. Thus, the overall products of the Krebs cycle are CO_2, ATP, NADH, and $FADH_2$:

$$\text{Acetyl unit} + 3\text{ NAD}^+ + \text{FAD} + \text{ADP} + P_i \rightarrow$$
$$2\text{ CO}_2 + 3\text{ NADH} + \text{FADH}_2 + \text{ATP}$$

The molecules of NADH and $FADH_2$ formed yield 11 molecules of ATP when oxidized in the electron transport chain. The other molecules in the cycle behave as intermediate reactants and products which are continuously regenerated as the cycle turns.

Aerobic cellular respiration uses oxygen as the final electron acceptor and releases carbon dioxide and water from the complete oxidation of fuels. The carbon dioxide that we, and other aerobic organisms, produce is removed from our bodies into the atmosphere during external respiration (see Chapter 31, p. 698). Fortunately for us and other aerobes, oxygen is continuously produced by cyanobacteria (blue-green algae), eukaryotic algae, and plants by the process of photosynthesis. In this process, hydrogen atoms obtained from water react with carbon dioxide from the atmosphere to generate sugars and molecular oxygen. Thus, a balance between oxygen used and produced, and carbon dioxide produced and used, is obtained across our planet. Unfortunately, excessive production of carbon dioxide due to human industrialization, and decreased production of oxygen due to our continuous removal of the world's forests, are threatening this delicate balance. Carbon dioxide levels continue to rise, leading to global atmospheric warming caused by the "greenhouse effect" (see Chapter 37, p. 807).

Electron Transport Chain

Transfer of hydrogen ions and electrons from NADH and $FADH_2$ to the final electron acceptor, molecular oxygen, is accomplished in an elaborate electron transport chain embedded in the inner membrane of mitochondria (Figure 4.14, see also p. 44). Each carrier molecule in the chain (labeled I to IV in Figure 4.14) is a large transmembrane protein-based complex that accepts and releases electrons at lower energy levels than the carrier preceding it in the chain. As electrons pass from one carrier molecule to the next, free energy is released. Some of this energy is used to transport H^+ ions across the inner mitochondrial membrane and in this way creates a H^+ gradient across the membrane. The H^+ gradient produced drives the synthesis of ATP. This process is called chemiosmotic coupling (Figure 4.14). According to this model, as electrons contributed by NADH and $FADH_2$ are carried down the electron transport chain, they activate proton transporting molecules which move protons (hydrogen ions) outward and into the space between the two mitochondrial membranes. This causes the proton concentration in this intermembrane space to

rise, producing a diffusion gradient that is used to drive the protons back into the mitochondrial matrix through special proton channels. These channels are ATP-forming transmembrane protein complexes (ATP synthase) that use the inward movement of protons to induce the formation of ATP. By this means, oxidation of one NADH yields three ATP molecules. $FADH_2$ from the Krebs cycle enters the electron transport chain at a lower level than NADH and so yields two ATP molecules. This method of energy capture is called **oxidative phosphorylation** because the formation of high-energy phosphate is coupled to oxygen consumption, and these reactions depend on demand for ATP by other metabolic activities within the cell.

Efficiency of Oxidative Phosphorylation

We can now calculate the ATP yield from the complete oxidation of glucose (Figure 4.15). The overall reaction is:

$$\text{Glucose} + 2\text{ ATP} + 36\text{ ADP} + 36\text{ P} + 6\text{ O}_2 \rightarrow$$
$$6\text{ CO}_2 + 2\text{ ADP} + 36\text{ ATP} + 6\text{ H}_2\text{O}$$

ATP has been generated at several points along the way (Table 4.1). The cytoplasmic NADH generated in glycolysis requires a molecule of ATP to fuel transport of each molecule of NADH into a mitochondrion; therefore, each NADH from glycolysis yields only two ATP (total of four), compared with the three ATP per NADH (total of six) formed within mitochondria. Accounting for the two ATP used in the priming reactions in glycolysis, the net yield may be as high as 36 molecules of ATP per molecule of glucose. The yield of 36 ATP is a theoretical maximum because some of the H^+ gradient produced by electron transport may be used for other functions, such as transporting substances in and out of the mitochondrion. Overall efficiency of aerobic oxidation of glucose is about 38%, comparing very favorably with human-designed energy conversion systems, which seldom exceed 5% to 10% efficiency.

The capacity for oxidative phosphorylation is also increased by the elaborate folding of the inner mitochondrial membrane (the cristae shown in Figure 4.14 and labeled in Figure 3.11, p. 44), providing a much greater surface area for more electron transport chain and ATP synthase proteins.

Anaerobic Glycolysis: Generating ATP Without Oxygen

Under anaerobic conditions, glucose and other 6-carbon sugars are first converted stepwise to a pair of 3-carbon pyruvic acid molecules during glycolysis, described on page 64 (see also Figure 4.11). This series of reactions yields two molecules of ATP and two molecules of NADH. In the absence of molecular oxygen, further oxidation of pyruvic acid cannot occur because without oxygen as the final electron acceptor in the electron transport chain, the Krebs cycle and electron transport chain cannot operate and cannot, therefore, reoxidize the NADH produced in glycolysis. The problem is neatly solved in most animal

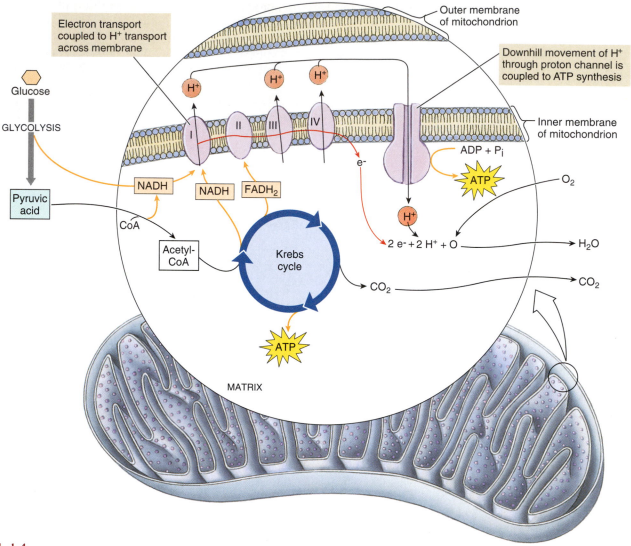

Figure 4.14

Oxidative phosphorylation. Most of the ATP in living organisms is produced in the electron transport chain. Electrons removed from fuel molecules in cellular oxidations (glycolysis and the Krebs cycle) flow through the electron transport chain, the major components of which are four transmembrane protein complexes (I, II, III, and IV). Electron energy is tapped by the major complexes and used to push H^+ outward across the inner mitochondrial membrane. The H^+ gradient created drives H^+ inward through proton channels (ATP synthase) that couple H^+ movement to ATP synthesis.

TABLE 4.1

Calculation of Total ATP Molecules Generated in Respiration

ATP Generated	Source
4	Directly in glycolysis
2	As GTP (→ATP) in Krebs cycle
4	From NADH in glycolysis
6	From NADH produced in pyruvic acid to acetyl-CoA reaction
4	From reduced FAD in Krebs cycle
18	From NADH produced in Krebs cycle
38 Total	
−2	Used in priming reactions in glycolysis
36 Net	

cells by reducing pyruvic acid to lactic acid (Figure 4.16). Pyruvic acid becomes the final electron acceptor and lactic acid the end product of anaerobic glycolysis. This step converts NADH to NAD^+, effectively freeing it to recycle and pick up more H^+ and electrons. In **alcoholic fermentation** (as in yeast, for example) the steps are identical to glycolysis down to pyruvic acid. One of its carbons is then released as carbon dioxide, and the resulting 2-carbon compound is reduced to ethanol, thus regenerating the NAD^+.

Anaerobic glycolysis is only one-eighteenth as efficient as complete oxidation of glucose to carbon dioxide and water, but its key virtue is that it provides *some* high-energy phosphate in situations in which oxygen is absent or in short supply. Many microorganisms live in places where oxygen is severely depleted, such as waterlogged soil, in mud of a lake or sea bottom, or within a decaying carcass. Vertebrate skeletal muscle may rely heavily on glycolysis during short

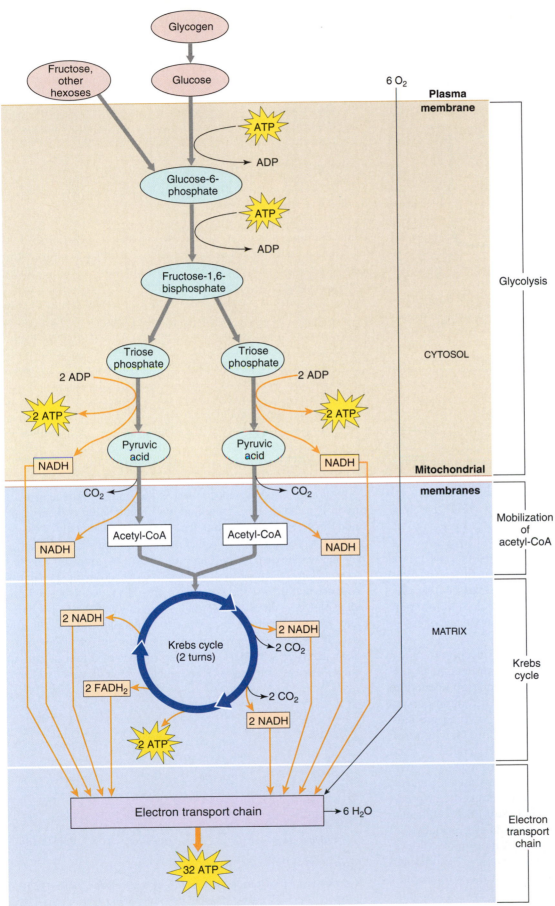

Figure 4.15

Pathway for oxidation of glucose and other carbohydrates. Glucose is degraded to pyruvic acid by cytoplasmic enzymes (glycolytic pathway). Acetyl-CoA is formed from pyruvic acid and is fed into the Krebs cycle. An acetyl-CoA molecule (two carbons) is oxidized to two molecules of carbon dioxide with each turn of the cycle. Pairs of electrons are removed from the carbon skeleton of the substrate at several points in the pathway and are carried by oxidizing agents NADH or $FADH_2$ to the electron transport chain where 32 molecules of ATP are generated. Four molecules of ATP are also generated by substrate phosphorylation in the glycolytic pathway, and two molecules of ATP (initially GTP) are formed in the Krebs cycle. This yields a total of 38 molecules of ATP (36 molecules net) per glucose molecule. Molecular oxygen is involved only at the very end of the pathway as the final electron acceptor at the end of the electron transport chain to yield water.

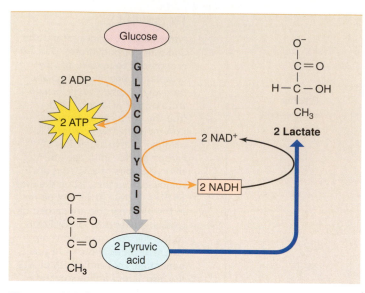

Figure 4.16

Anaerobic glycolysis, a process that proceeds in the absence of oxygen. Glucose is broken down to two molecules of pyruvic acid, with a net production of two molecules of ATP. Pyruvic acid, the final electron acceptor for the hydrogen ions and electrons released during pyruvic acid formation, is converted to lactic acid. Hydrogen and electrons are recycled through the carrier, NAD⁺.

bursts of activity when contraction is so rapid and powerful that oxygen delivery to tissues cannot supply energy demands by oxidative phosphorylation alone. At such times an animal has no choice but to supplement oxidative phosphorylation with anaerobic glycolysis. One kind of muscle fiber (white muscle) has few mitochondria and primarily uses anaerobic glycolysis for ATP production (see Chapter 29, p. 661). In all muscle types, intense or strenuous activity is followed by a period of increased oxygen consumption as lactic acid, the end product of anaerobic glycolysis, diffuses from muscle to the liver where it is metabolized. Because oxygen consumption increases following heavy activity, the animal is said to have acquired an **oxygen debt** during such activity, which is repaid when activity ceases, and accumulated lactic acid is metabolized.

Some animals rely heavily on anaerobic glycolysis during normal activities. For example, diving birds and mammals use glycolysis almost entirely to give them the energy needed to sustain long dives without breathing (that is, without requiring oxygen). Salmon would never reach their spawning grounds were it not for anaerobic glycolysis providing almost all of the ATP used in the powerful muscular bursts needed to carry them up rapids and falls. Many parasitic animals have dispensed with oxidative phosphorylation entirely at some stages of their life cycles. They secrete relatively reduced end products of their energy metabolism, such as succinic acid, acetic acid, and propionic acid. These compounds are produced in mitochondrial reactions that derive several more molecules of ATP than does the path from glycolysis to lactic acid, although such sequences are still far less efficient than the aerobic electron transport chain.

METABOLISM OF LIPIDS

The first step in the breakdown of a triglyceride is its hydrolysis to glycerol and three fatty acid molecules (Figure 4.17). Glycerol is phosphorylated and enters the glycolytic pathway (see Figure 4.10).

The remainder of the triglyceride molecule consists of fatty acids. For example, an abundant naturally occurring fatty acid is **stearic acid.**

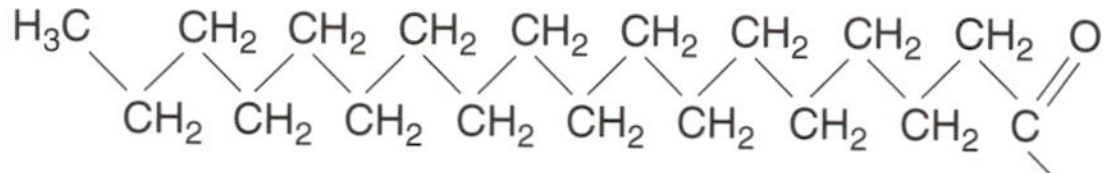

Stearic acid

The long hydrocarbon chain of a fatty acid is broken down by oxidation, two carbons at a time; these are released from the end of the molecule as acetyl-CoA. Although two high-energy phosphate bonds are required to prime each 2-carbon fragment, energy is derived both from the reduction of NAD⁺ and FAD to NADH and FADH₂, respectively, and from the acetyl group as it is degraded in the Krebs cycle. The complete oxidation of one molecule of 18-carbon stearic acid nets 146 ATP molecules. By comparison, three molecules of glucose (also totaling 18 carbons) yield 108 ATP molecules. Since there are three fatty acids in each triglyceride molecule, a total of 440 ATP molecules are formed. An additional 22 molecules of ATP are generated in the breakdown of glycerol, giving a grand total of 462 molecules of ATP—little wonder that fat is considered the king of animal fuels! Fats are more concentrated fuels than carbohydrates, because fats are almost pure hydrocarbons; they contain more hydrogen per carbon atom than sugars do, and it is the energized electrons of hydrogen that generate high-energy bonds when they are carried through the mitochondrial electron transport chain.

Fat stores are derived principally from surplus fats and carbohydrates in the diet. Acetyl-CoA is the source of carbon atoms used to build fatty acids. Because all major classes of organic molecules (carbohydrates, fats, and proteins) can be degraded to acetyl-CoA, all can be converted into stored fat. The biosynthetic pathway for fatty acids resembles a reversal of the catabolic pathway already described, but it requires an entirely different set of enzymes. From acetyl-CoA, the fatty acid chain is assembled two carbons at a time. Because fatty acids release energy when they are oxidized,

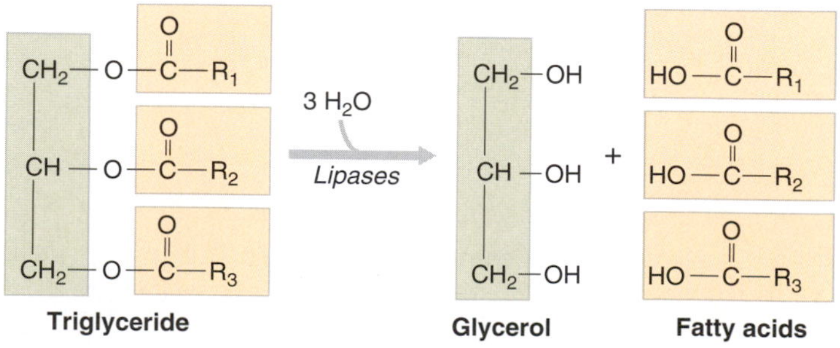

Triglyceride   Glycerol   Fatty acids

Figure 4.17

Hydrolysis of a triglyceride (neutral fat) by intracellular lipase. The R groups of each fatty acid represent a hydrocarbon chain.

they obviously require an input of energy for their synthesis. This energy is provided principally by electron energy from glucose degradation. Thus the total ATP derived from oxidation of a molecule of triglyceride is not as great as calculated, because varying amounts of energy are required for synthesis and storage.

Stored fats are the greatest reserve fuel in the body. Most usable fat resides in white adipose tissue composed of specialized cells packed with globules of triglycerides. White adipose tissue is widely distributed in the abdominal cavity, in muscles, around deep blood vessels and large organs (for example, heart and kidneys), and especially under the skin. Women average about 30% more fat than men, which is largely responsible for differences in shape between males and females. Humans can only too easily deposit large quantities of fat, generating hazards to health.

Physiological and psychological aspects of obesity are now being investigated by many researchers. There is increasing evidence that food intake, and therefore the amount of fat deposition, is regulated by feeding centers located in the brain (lateral and ventral hypothalamus and brain stem). The set point of these regions determines normal food intake and body weight for an individual, which may be maintained above or below what is considered normal for humans. Although evidence is accumulating for a genetic component to obesity, the epidemic proportions of obesity in the United States are more easily explained by lifestyle and feeding habits. Other developed countries show a similar, but less pronounced, trend toward development of an obesity problem.

Research also reveals that lipid metabolism in obese individuals appears to be abnormal compared to lean individuals. This research has resulted in the development of drugs that act at various stages of lipid metabolism, such as decreasing lipid digestion and absorption from the digestive tract, or increasing metabolism of lipids once they have been absorbed into the body.

METABOLISM OF PROTEINS

Since proteins are composed of amino acids, of which 20 kinds commonly occur (p. 26), the central topic of our consideration is amino acid metabolism. Amino acid metabolism is complex. Each of the 20 amino acids requires a separate pathway for biosynthesis and degradation. Amino acids are precursors to tissue proteins, enzymes, nucleic acids, and other nitrogenous constituents that form the fabric of cells. The central purpose of carbohydrate and fat oxidation is to provide energy, much of which is needed to construct and maintain these vital macromolecules.

Let us begin with the **amino acid pool** in blood and extracellular fluid from which the tissues draw their requirements. When animals eat proteins, most are digested in the digestive tract, releasing their constituent amino acids, which are then absorbed (Figure 4.18). Tissue proteins also are hydrolyzed during normal growth, repair, and tissue restructuring; their amino acids join those derived from protein found in food to enter the amino acid pool. A portion of the amino acid pool is used to rebuild tissue proteins, but most animals ingest a surplus of protein. Since amino acids are not excreted as such in any significant amounts, they must be disposed of in some other way. In fact, amino acids can be and are metabolized through oxidative pathways to yield high-energy phosphate. In short, excess proteins serve as fuel as do carbohydrates and fats. Their importance as fuel obviously depends on the nature of the diet. In carnivores that ingest a diet of almost pure protein and fat, nearly half of their high-energy phosphate comes from amino acid oxidation.

Before an amino acid molecule may enter the fuel depot, nitrogen must be removed by deamination (the amino group splits to form ammonia and a keto acid) or by transamination (the amino group is transferred to a keto acid to yield a new amino acid). Thus amino acid degradation yields two main products, carbon skeletons and ammonia, which are handled in different ways. Once nitrogen atoms are removed, the carbon skeletons

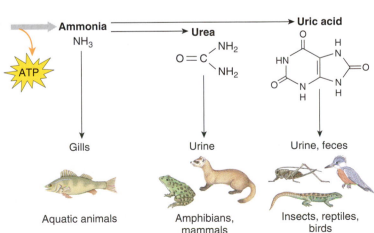

Figure 4.18
Fate of dietary protein.

of amino acids can be completely oxidized, usually by way of pyruvic acid or acetic acid. These residues then enter routes used by carbohydrate and fat metabolism (see Figure 4.10).

The other product of amino acid degradation is ammonia. Ammonia is highly toxic because it inhibits respiration by reacting with α-ketoglutaric acid to form glutamic acid (an amino acid), and effectively removes α-ketoglutarate from the Krebs cycle (see Figure 4.13). Disposal of ammonia offers little problem to aquatic animals because it is soluble and readily diffuses into the surrounding medium, often through respiratory surfaces. Terrestrial animals cannot get rid of ammonia so conveniently and must detoxify it by converting it to a relatively nontoxic compound. The two principal compounds formed are **urea** and **uric acid,** although a variety of other detoxified forms of ammonia are excreted by different animals. Among vertebrates, amphibians and especially mammals produce urea. Reptiles and birds, as well as many terrestrial invertebrates, produce uric acid (the excretion of uric acid by insects and birds is described on pp. 451 and 597, respectively).

The key feature that determines choice of nitrogenous waste is availability of water in the environment. When water is abundant, the chief nitrogenous waste is ammonia. When water is restricted, it is urea. Animals living in truly arid habitats use uric acid. Uric acid is highly insoluble and easily precipitates from solution, allowing its removal in solid form. Embryos of birds and reptiles benefit greatly from excretion of nitrogenous waste as uric acid, because waste cannot be eliminated through their eggshells. During embryonic development, harmless, solid uric acid is retained in one of the extraembryonic membranes. When a hatchling emerges into its new world, accumulated uric acid, along with the shell and membranes that supported development, is discarded.

MANAGEMENT OF METABOLISM

The complex pattern of enzymatic reactions that constitutes metabolism cannot be explained entirely in terms of physicochemical laws or chance happenings. Although some enzymes appear to function automatically, the activity of others is rigidly controlled. In the former case, suppose that the function of an enzyme is to convert A to B. If B is removed by conversion into another compound, the enzyme tends to restore the original ratio of B to A. Since many enzymes act reversibly, either synthesis or degradation may result. For example, an excess of an intermediate in the Krebs cycle would contribute to glycogen

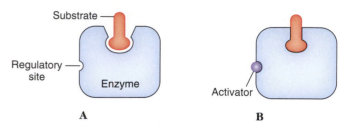

Figure 4.19

Enzyme regulation. **A,** The active site of an enzyme may only loosely fit its substrate in the absence of an activator. **B,** With the regulatory site of the enzyme occupied by an activator, the enzyme binds the substrate, and the site becomes catalytically active.

synthesis; a depletion of such a metabolite would lead to glycogen breakdown. This automatic compensation (equilibration) is not, however, sufficient to explain regulation of metabolism.

Mechanisms exist for critically regulating enzymes in both *quantity* and *activity.* In bacteria, genes leading to synthesis of an enzyme are switched on or off, depending on the presence or absence of a substrate molecule. In this way the *quantity* of an enzyme is controlled. It is a relatively imprecise process.

Mechanisms that alter activity of enzymes can quickly and finely adjust metabolic pathways to changing conditions in a cell. The presence or increase in concentration of some molecules can alter the shape (conformation) of particular enzymes, thus activating or inhibiting the enzyme (Figure 4.19). For example, phosphofructokinase, which catalyzes phosphorylation of glucose-6-phosphate to fructose-1,6-bisphosphate (see Figure 4.15), is inhibited by high concentrations of ATP or citric acid. Their presence means that a sufficient amount of precursors has reached the Krebs cycle and additional glucose is not needed. In some cases, the final end product of a particular metabolic pathway inhibits the first enzyme in the pathway. This method is termed **feedback inhibition.**

As well as being subject to alteration in physical shape, many enzymes exist in both an active and an inactive form. These forms may be chemically different. For example, one common way to activate or inactivate an enzyme is to add a phosphate group to the molecule, thus changing its conformational shape and either exposing or blocking the enzyme's active site. Enzymes that degrade glycogen (phosphorylase) and synthesize it (synthase) are both found in active and inactive forms. Conditions that activate phosphorylase tend to inactivate synthase and vice versa.

SUMMARY

Living systems are subject to the same laws of thermodynamics that govern nonliving systems. The first law states that energy cannot be destroyed, although it may change form. The second law states that the structure of systems proceeds toward total randomness, or increasing entropy, as energy is dissipated. Solar energy trapped by photosynthesis as chemical bond energy is passed through the food chain where it is used for biosynthesis, active transport, and motion, before finally being dissipated as heat. Living organisms are able to decrease their entropy and to maintain high internal order because the biosphere is an open system from which energy can be captured and used. Energy available for use in biochemical reactions is termed "free energy."

Enzymes are usually proteins, often associated with nonprotein cofactors, that vastly accelerate rates of chemical reactions in living systems. An enzyme acts by temporarily binding its reactant (substrate) onto an active site in a highly specific fit. In this configuration, internal activation energy barriers are lowered enough to modify the substrate, and the enzyme is restored to its original form.

Cells use the energy stored in chemical bonds of organic fuels by degrading fuels through a series of enzymatically controlled steps. This bond energy is transferred to ATP and packaged in the form of "high-energy" phosphate bonds. ATP is produced as it is required in cells to power various synthetic, secretory, and mechanical processes.

Glucose is an important source of energy for cells. In aerobic metabolism (respiration), the 6-carbon glucose is split into two 3-carbon molecules of pyruvic acid. Pyruvic acid is decarboxylated to form 2-carbon acetyl-CoA, a strategic intermediate that enters the Krebs cycle. Acetyl-CoA can also be derived from breakdown of fat. In the Krebs cycle, acetyl-CoA is oxidized in a series of reactions to carbon dioxide, yielding, in the course of the reactions, energized electrons that are passed to electron acceptor molecules (NAD^+ and FAD). In the final stage, the energized electrons are passed along an electron transport chain consisting of a series of electron carriers located in the inner membranes of mitochondria. A hydrogen gradient is produced as electrons are passed from carrier to carrier and finally to oxygen, and ATP is generated as the hydrogen ions flow down their electrochemical gradient through ATP synthase molecules located in the inner mitochondrial membrane. A net total of 36 molecules of ATP may be generated from one molecule of glucose.

In the absence of oxygen (anaerobic glycolysis), glucose is degraded to two 3-carbon molecules of lactic acid, yielding two molecules of ATP. Although anaerobic glycolysis is vastly less efficient than aerobic metabolism, it provides essential energy for muscle contraction when heavy energy expenditure outstrips the oxygen-delivery system of an animal; it also is the only source of energy generation for microorganisms living in oxygen-free environments.

Triglycerides (neutral fats) are especially rich depots of metabolic energy because the fatty acids of which they are composed are highly reduced and free of water. Fatty acids are degraded by sequential removal of 2-carbon units, which enter the Krebs cycle through acetyl-CoA.

Amino acids in excess of requirements for synthesis of proteins and other biomolecules are used as fuel. They are degraded by deamination or transamination to yield ammonia and carbon skeletons. The latter enter the Krebs cycle to be oxidized. Ammonia is a highly toxic waste product that aquatic animals quickly expel, often through respiratory surfaces. Terrestrial animals, however, convert ammonia into much less toxic compounds, urea or uric acid, for disposal.

Integration of metabolic pathways is finely regulated by mechanisms that control both amount and activity of enzymes. The quantity of some enzymes is regulated by certain molecules that switch on or off enzyme synthesis. Enzyme activity may be altered by the presence or absence of metabolites that cause conformational changes in enzymes and thus improve or diminish their effectiveness as catalysts.

REVIEW QUESTIONS

1. State the first and second laws of thermodynamics. Living systems may appear to violate the second law of thermodynamics because living things maintain a high degree of organization despite a universal trend toward increasing disorganization. What is the explanation for this apparent paradox?
2. Explain what is meant by "free energy" in a system. Will a reaction that proceeds spontaneously have a positive or negative change in free energy?
3. Many biochemical reactions proceed slowly unless the energy barrier to the reaction is lowered. How is this accomplished in living systems?
4. What happens in the formation of an enzyme-substrate complex that favors the disruption of substrate bonds?
5. What is meant by a "high-energy bond," and why might the production of molecules with such bonds be useful to living organisms?
6. Although ATP supplies energy to an endergonic reaction, why is it not considered a fuel?
7. What is an oxidation-reduction reaction and why are such reactions considered so important in cellular metabolism?
8. Give an example of a final electron acceptor found in aerobic and anaerobic organisms. Why is aerobic metabolism more efficient than anaerobic metabolism?
9. Why must glucose be "primed" with a high-energy phosphate bond before it can be degraded in the glycolytic pathway?
10. What happens to the electrons removed during the oxidation of triose phosphates during glycolysis?
11. Why is acetyl-CoA considered a "strategic intermediate" in respiration?
12. Why are oxygen atoms important in oxidative phosphorylation? What are the consequences if they are absent for a short period of time in tissues that routinely use oxidative phosphorylation to produce useful energy?
13. Explain how animals can generate ATP *without* oxygen. Given that anaerobic glycolysis is much less efficient than oxidative phosphorylation, why has anaerobic glycolysis not been discarded during animal evolution?

14. Why are animal fats sometimes called "the king of fuels"? What is the significance of acetyl-CoA to lipid metabolism?

15. The breakdown of amino acids yields two products: ammonia and carbon skeletons. What happens to these products?

16. Explain the relationship between the amount of water in an animal's environment and the kind of nitrogenous waste it produces.

17. Explain three ways that enzymes may be regulated in cells.

SELECTED REFERENCES

Alberts, B., D. Bray, K. Hopkin, A. Johnson, J. Lewis, M. Raff, K. Roberts, and P. Walter. 2003. Essential cell biology, ed. 2. New York, Garland Science Publishing. *Provides a more in-depth and well-written description of cellular metabolism.*

Berg, J., J. Tymoczko, and L. Stryer. 2002. Biochemistry, ed. 5. San Francisco, W. H. Freeman & Company. *One of the best undergraduate biochemistry texts.*

Lodish, H., A. Berk, S. L. Zipursky, P. Matsudaira, D. Baltimore, and J. Darnell. 2000. Molecular cell biology, ed. 4. San Francisco, W. H. Freeman & Company. *Chapter 16 is a comprehensive, well-illustrated treatment of energy metabolism.*

Wolfe, S. L. 1995. Introduction to cell and molecular biology. Belmont, CA. Thomson Brooks/Cole Publishers. *Covers the same topics as Wolfe's big book, but in less detail.*

ONLINE LEARNING CENTER

Visit www.mhhe.com/hickmanipz14e for chapter quizzing, key term flash cards, web links, and more!

PART TWO

Continuity and Evolution of Animal Life

2

A female Cardinalis cardinalis *(left) and a female* Cardinalis sinuatus *(right).*

Genetics: A Review

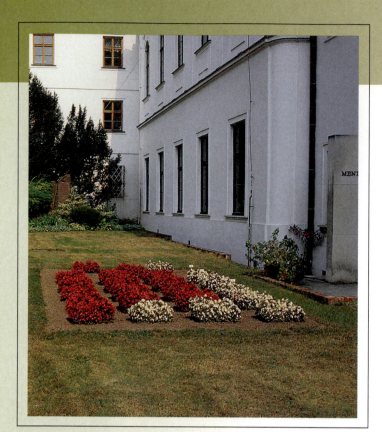

The site of Gregor Mendel's experimental garden, Brno, Czech Republic.

A Code for All Life

The principle of hereditary transmission is a central tenet of life on earth: all organisms inherit a structural and functional organization from their progenitors. What is inherited by an offspring is not an exact copy of the parent but a set of coded instructions that a developing organism uses to construct a body resembling its parents. These instructions are in the form of genes, the fundamental units of inheritance. One of the great triumphs of modern biology was the discovery in 1953 by James Watson and Francis Crick of the nature of the coded instructions in genes. The genetic material (deoxyribonucleic acid, DNA) is composed of nitrogenous bases arranged on a chemical chain of sugar-phosphate units. The genetic code lies in the linear order or sequence of bases in the DNA strand.

Because the DNA molecules replicate and pass from generation to generation, genetic variations can persist and spread in a population. Such molecular alterations, called mutations, are the ultimate source of biological variation and the raw material of evolution.

A basic principle of modern evolutionary theory is that organisms attain their diversity through hereditary modifications of populations. All known lineages of plants and animals are related by descent from common ancestral populations.

Heredity establishes the continuity of living forms. Although offspring and parents in a particular generation may look different, there is nonetheless a genetic continuity that runs from generation to generation for any species of plant or animal. An offspring inherits from its parents a set of coded information **(genes),** which a fertilized egg uses, together with environmental factors, to guide its development into an adult bearing unique physical characteristics. Each generation passes to the next the instructions required for maintaining continuity of life.

The gene is the unit entity of inheritance, the germinal basis for every characteristic that appears in an organism. The study of what genes are, how they are transmitted, and how they work is the science of genetics. It is a science that reveals the underlying causes of *resemblance,* as seen in the remarkable fidelity of reproduction, and of *variation,* the working material for organic evolution. All living forms use the same information storage, transfer, and translation system, which explains the stability of all life and reveals its descent from a common ancestral form. This is one of the most important unifying concepts of biology.

MENDEL'S INVESTIGATIONS

The first person to formulate the principles of heredity was Gregor Johann Mendel (1822 to 1884) (Figure 5.1 and p. 18), an Augustinian monk living in Brünn (Brno), Moravia. Brünn was then part of Austria but now lies in the eastern part of the Czech Republic. While conducting breeding experiments in a small monastery garden from 1856 to 1864, Mendel examined with great care the progeny of many thousands of plants. He presented in elegant simplicity the laws governing transmission of characters from parents to offspring. His discoveries, published in 1866, were of great significance, coming just after Darwin's publication of *On the Origin of Species by Means of Natural Selection.* Yet Mendel's discoveries remained unappreciated and forgotten until 1900—35 years after the completion of the work and 16 years after Mendel's death.

Mendel chose garden peas for his classic experiments because they had pure strains differing from each other by discrete characters. For example, some varieties were definitely dwarf and others tall; some strains produced smooth seeds and others wrinkled seeds (Figure 5.1). Mendel studied single characters that displayed sharply contrasting traits. He carefully avoided mere quantitative, continuously varying characteristics. A second reason for selecting peas was that they were self-fertilizing but subject to experimental cross-fertilization.

A giant advance in chromosomal genetics was made when the American geneticist Thomas Hunt Morgan and his colleagues selected a species of fruit fly, *Drosophila melanogaster,* for their studies (1910–1920). Flies were cheaply and easily reared in bottles in the laboratory, fed on a simple medium of bananas and yeast. Most importantly, they produced a new generation every 10 days, enabling Morgan to collect data at least 25 times more rapidly than with organisms that take longer to mature, such as garden peas. Morgan's work led to the mapping of genes on chromosomes and founded the discipline of cytogenetics.

Mendel crossed varieties having contrasting traits, making crosses for each of the seven characters shown in Figure 5.1. He removed the stamens (male part, containing the pollen) from a flower to prevent self-fertilization and then placed on the stigma (female part of flower) pollen from the flower of a plant true-breeding for the contrasting trait. Pollination from other sources such as wind and insects was rare and did not affect his results. Offspring from these crosses are called hybrids, meaning that they contain genetic information from two different parental strains. He collected seeds from the cross-fertilized flowers, planted these hybrid seeds, and examined the resulting plants for the contrasting traits being studied. These hybrid plants then produced offspring by self-pollination.

Mendel knew nothing of the cytological basis of heredity, since chromosomes and genes were not yet discovered. Although we can admire Mendel's power of intellect in his discovery of the principles of inheritance without knowledge of chromosomes, these principles are easier to understand if we first review chromosomal behavior, especially in meiosis.

CHROMOSOMAL BASIS OF INHERITANCE

In sexually reproducing organisms, special **sex cells,** or **gametes** (ova and sperm), transmit genetic information from parents to offspring. A scientific explanation of genetic principles required a study of germ cells and their behavior, and correlations between their transmission and certain visible results of inheritance. Nuclei of sex cells, especially the chromosomes, were early suspected of furnishing the real answer to the hereditary mechanism. Chromosomes are apparently the only entities transmitted in equal quantities from both parents to offspring.

When Mendel's laws were rediscovered in 1900, their parallelism with the cytological behavior of chromosomes was obvious. Later experiments showed that chromosomes carried hereditary material.

Meiosis: Reduction Division of Gametes

Although animal species differ greatly in the characteristic numbers, sizes, and shapes of chromosomes present in their body cells, a common feature is that chromosomes occur in pairs. The two members of a chromosomal pair contain similar genes encoding the same set of characteristics and usually, but not always, have the same size and shape. The members of such a pair are called **homologous** chromosomes; each individual member of a pair is called a **homolog.** One homolog comes from the mother and the other from the father. Meiosis is a special pair of cell divisions in which the genetic material replicates once followed by two rounds of cell division (Figure 5.2). The result is a set of four daughter cells, each of which has only *one* member of each homologous chromosome pair. The chromosomes present in a meiotic daughter cell or gamete are collectively called a single set of chromosomes. The number of chromosomes in a single set, which varies among species, is called the **haploid** (*n*) number of chromosomes. When

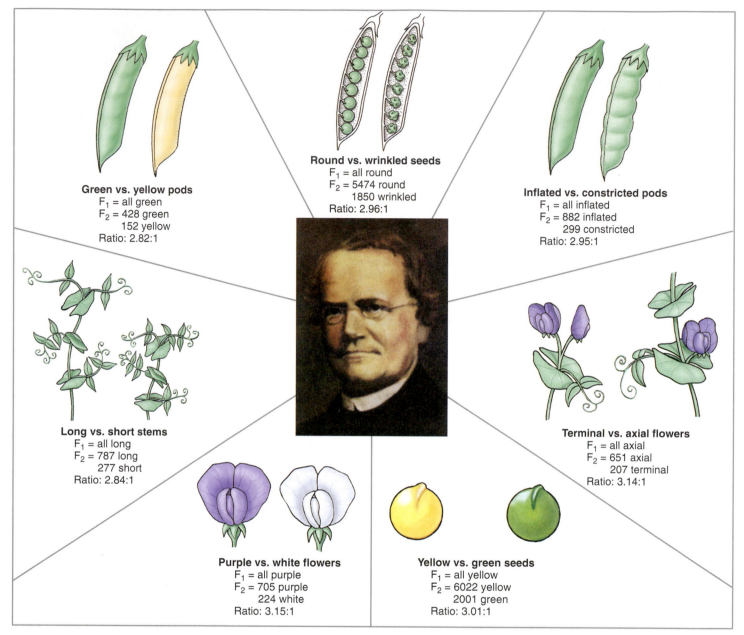

Green vs. yellow pods
F_1 = all green
F_2 = 428 green
      152 yellow
Ratio: 2.82:1

Round vs. wrinkled seeds
F_1 = all round
F_2 = 5474 round
      1850 wrinkled
Ratio: 2.96:1

Inflated vs. constricted pods
F_1 = all inflated
F_2 = 882 inflated
      299 constricted
Ratio: 2.95:1

Long vs. short stems
F_1 = all long
F_2 = 787 long
      277 short
Ratio: 2.84:1

Terminal vs. axial flowers
F_1 = all axial
F_2 = 651 axial
      207 terminal
Ratio: 3.14:1

Purple vs. white flowers
F_1 = all purple
F_2 = 705 purple
      224 white
Ratio: 3.15:1

Yellow vs. green seeds
F_1 = all yellow
F_2 = 6022 yellow
      2001 green
Ratio: 3.01:1

Figure 5.1

Seven experiments on which Gregor Mendel based his postulates. These are the results of monohybrid crosses for first and second generations.

a pair of gametes unites in fertilization, each gamete contributes its set of chromosomes to the newly formed cell, called a **zygote,** which has two complete sets of chromosomes. The number of chromosomes in two complete sets is called the **diploid** ($2n$) number. In humans the zygotes and all body cells normally have the diploid number ($2n$), or 46 chromosomes; the gametes have the haploid number (n), or 23, and meiosis reduces the number of chromosomes per cell from diploid to haploid.

Thus each cell normally has two copies of each gene coding for a given trait, one on each of the homologous chromosomes. Alternative forms of genes for the same trait are **allelic** forms, or **alleles.** Sometimes only one of the alleles has a visible effect on the organism, although both are present in each cell, and either

may be passed to progeny as a result of meiosis and subsequent fertilization.

Alleles are alternative forms of the same gene that have arisen by mutation of the DNA sequence. Like a baseball team with several pitchers, only one of whom can occupy the pitcher's mound at a time, only one allele can occupy a chromosomal locus (position). Alternative alleles for the locus may be on homologous chromosomes of a single individual, making that individual heterozygous for the gene in question. Numerous allelic forms of a gene may be found among different individuals in a population, a condition called "multiple alleles" (p. 85).

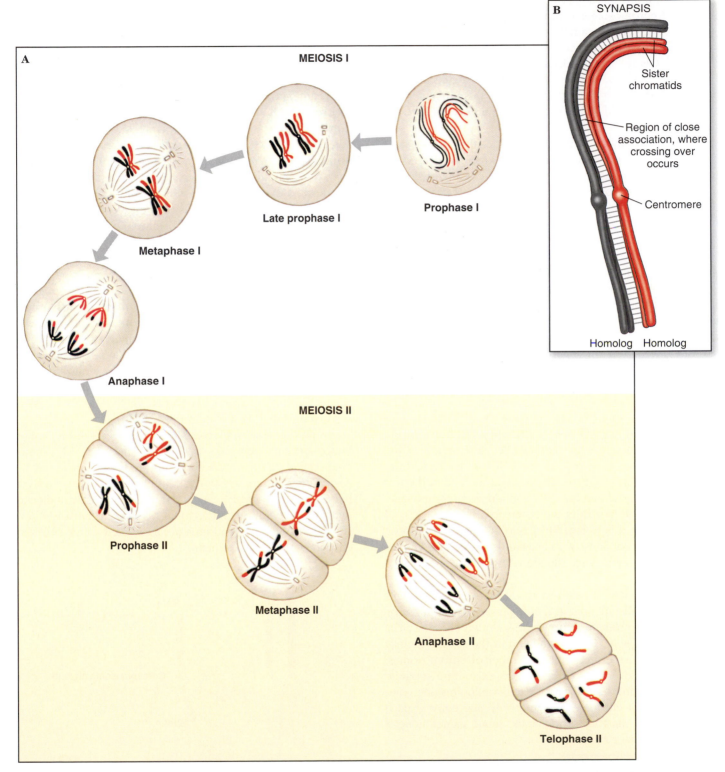

Figure 5.2

A, Meiosis in a sex cell with two pairs of chromosomes. Prophase I, homologous chromosomes come to lie with side-to-side contact, or synapsis, forming bivalents. A bivalent comprises a pair of homologous chromosomes, with each of the chromosomes containing a pair of identical chromatids joined by a centromere. Metaphase I, bivalents align at the spindle equator. Anaphase I, chromosomes of former bivalents are pulled toward opposite poles. Prophase II, daughter cells contain one of each homologous chromosome (haploid) but each chromosome is in replicated form (two chromatids attached at a centromere). Metaphase II, chromosomes align at the spindle equator. Anaphase II, chromatids of each chromosome separate. Telophase II, four haploid cells (gametes) formed, each with unreplicated chromosomes (one chromatid per chromosome). **B,** Synapsis occurs in prophase I, in which homologous chromosomes can break and exchange corresponding portions. The labelled sister chromatids and region of close association extend the full length of the bivalent.

During an individual's growth, all dividing cells contain the double set of chromosomes (mitosis is described on p. 52). In the reproductive organs, gametes (germ cells) are formed after meiosis, which *separates* the chromosomes of each homologous pair. Without this reductional division, the union of ovum (egg) and sperm would produce an individual with twice as many chromosomes as the parents. Continuation of this process in just a few generations could yield astronomical numbers of chromosomes per cell.

Most unique features of meiosis occur during prophase of the first meiotic division (Figure 5.2). Prior to meiosis, each chromosome has already replicated to form two chromatids joined at one point, the centromere. The two members of each pair of homologous chromosomes make side-by-side contact **(synapsis)** to form a **bivalent,** which permits genetic recombination between the paired homologous chromosomes (p. 89). Each bivalent is composed of two pairs of chromatids (each pair is a **dyad,** sister chromatids held together at their centromere), or *four* future chromosomes, and is thus called a **tetrad.** The position or location of any gene on a chromosome is the gene **locus** (pl., **loci**), and in synapsis all gene loci on a chromatid normally lie exactly opposite the corresponding loci on the sister chromatid and both chromatids of the homologous chromosome. Toward the end of prophase, the chromosomes shorten and thicken and then enter the first meiotic division.

In contrast to mitosis, the centromeres holding the chromatids together *do not divide* at anaphase. As a result, each of the dyads is pulled toward one of the opposite poles of the cell by microtubules of the division spindle. At telophase of the first meiotic division, each pole of the cell has one dyad from each tetrad formed at prophase. Therefore at the end of the first meiotic division, the daughter cells contain *one* chromosome of *each* homologous pair from the parent cell, so that the total chromosome number is reduced to haploid. However, because each chromosome contains two chromatids joined at a centromere, each cell contains twice the amount of DNA present in a gamete.

The second meiotic division more closely resembles events in mitosis. The dyads are split at the beginning of anaphase by division of their centromeres, and single-stranded chromosomes move toward each pole. Thus by the end of the second meiotic division, the cells have the haploid number of chromosomes, and each chromatid of the original tetrad exists in a separate nucleus. Four products are formed, each containing one complete haploid set of chromosomes and only one copy of each gene. Only one of the four products in female gametogenesis becomes a functional gamete (p. 146).

Sex Determination

Before the importance of chromosomes in heredity was realized in the early 1900s, genetic control of gender was totally unknown. The first scientific clue to chromosomal determination of sex came in 1902 when C. McClung observed that bugs (Hemiptera) produced two kinds of sperm in approximately equal numbers. One kind contained among its regular set of chromosomes a so-called accessory chromosome lacking in the other kind of sperm. Since all eggs of these species had the same number of haploid chromosomes, half the sperm would have the same number of chromosomes as the eggs, and half of them would have one chromosome less. When an egg was fertilized by a spermatozoon carrying the accessory (sex) chromosome, the resulting offspring was a female; when fertilized by a spermatozoon without an accessory chromosome, the offspring was a male. Therefore a distinction was made between sex chromosomes, which determine sex (and sex-linked traits); and **autosomes,** the remaining chromosomes, which do not influence sex. The particular type of sex determination just described is often called the XX-XO type, which indicates that females have two X chromosomes and males only one X chromosome (the O indicates absence of the chromosome). The XX-XO method of sex determination is depicted in Figure 5.3.

Later, other types of sex determination were discovered. In humans and many other animals each sex contains the same number of chromosomes; however, the sex chromosomes (XX) are alike in females but unlike (XY) in males. Hence a human egg contains 22 autosomes + 1 X chromosome. Sperm are of two kinds; half carry 22 autosomes + 1 X and half bear 22 autosomes + 1 Y. The Y chromosome is much smaller than the X and carries very little genetic information. At fertilization, when the 2 X chromosomes come together, offspring are female; when X and Y come together, offspring are male. The XX-XY kind of sex determination is shown in Figure 5.4.

A third type of sex determination is found in birds, moths, butterflies, and some fish, in which the male has 2 X (or sometimes called ZZ) chromosomes and the female an X and Y (or ZW). Finally, there are both invertebrates (p. 380) and vertebrates (p. 571) in which sex is determined by environmental or behavioral conditions rather than by sex chromosomes, or by genetic loci whose variation is not associated with visible difference in chromosomal structure.

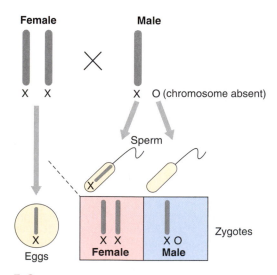

Figure 5.3

XX-XO sex determination. Only the sex chromosomes are shown.

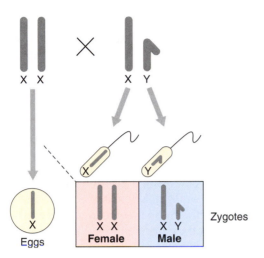

Figure 5.4
XX-XY sex determination. Only the sex chromosomes are shown.

In the case of X and Y chromosomes, homologous chromosomes are unlike in size and shape. Therefore, they do not both carry the same genes. Genes of the X chromosome often do not have allelic counterparts on the diminutive Y chromosome. This fact is very important in sex-linked inheritance (p. 87).

MENDELIAN LAWS OF INHERITANCE

Mendel's First Law

Mendel's **law of segregation** states that *in the formation of gametes, paired factors that may specify alternative phenotypes (visible traits) separate so that each gamete receives only one member of the pair.* In one of Mendel's original experiments, he pollinated pure-line tall plants with the pollen of pure-line dwarf plants. Thus the visible characteristics, or **phenotypes,** of the parents were tall and dwarf. Mendel found that all progeny in the first generation (F_1) were tall, just as tall as the tall parents of the cross. The reciprocal cross—dwarf plants pollinated with tall plants—gave the same result. The tall phenotype appeared in all progeny no matter which way the cross was made. Obviously, this kind of inheritance was not a blending of two traits, because none of the progeny was intermediate in size.

Next Mendel self-fertilized ("selfed") the tall F_1 plants and raised several hundred progeny, the second (F_2) generation. This time, *both* tall and dwarf plants appeared. Again, there was no blending (no plants of intermediate size), but the appearance of dwarf plants from all tall parental plants was surprising. The dwarf trait, seen in half of the grandparents but not in the parents, had reappeared. When he counted the actual number of tall and dwarf plants in the F_2 generation, he discovered that there were almost exactly three times more tall plants than dwarf ones.

Mendel then repeated this experiment for the six other contrasting traits that he had chosen, and in every case he obtained ratios very close to 3:1 (see Figure 5.1). At this point

it must have been clear to Mendel that he was dealing with hereditary determinants for the contrasting traits that did not blend when brought together. Even though the dwarf trait disappeared in the F_1 generation, it reappeared fully expressed in the F_2 generation. He realized that the F_1 generation plants carried determinants (which he called "factors") of both tall and dwarf parents, even though only the tall trait was visible in the F_1 generation.

Mendel called the tall factor **dominant** and the short **recessive.** Similarly, the other pairs of traits that he studied showed dominance and recessiveness. Whenever a dominant factor is present, the recessive one is not visible. The recessive trait appears only when both factors are recessive, or in other words, in a pure condition.

In representing his crosses, Mendel used letters as symbols; a capital letter denotes a dominant trait, and the corresponding lowercase letter denotes its recessive alternative. Modern geneticists still often follow this custom. Thus the factors for pure tall plants might be represented by *T/T,* the pure recessive by *t/t,* and the mix, or hybrid, of the two plants by *T/t.* The slash mark indicates that the alleles are on homologous chromosomes. The zygote bears the complete genetic constitution of the organism. All gametes produced by *T/T* must necessarily be *T,* whereas those produced by *t/t* must be *t.* Therefore a zygote produced by union of the two must be *T/t,* or a **heterozygote.** On the other hand, the pure tall plants (*T/T*) and pure dwarf plants (*t/t*) are **homozygotes,** meaning that the paired factors are alike on the homologous chromosomes and represent copies of the same allele. A cross involving variation at only a single locus is called a **monohybrid cross.**

In the cross between tall and dwarf plants there were two phenotypes: tall and dwarf. On the basis of genetic formulas there are three *hereditary* types: *T/T, T/t,* and *t/t.* These are called **genotypes.** A genotype is an allelic combination present in a diploid organism (*T/T, T/t,* or *t/t*), and the phenotype is the corresponding appearance of the organism (tall or dwarf).

One of Mendel's original crosses (tall plant and dwarf plant) could be represented as follows:

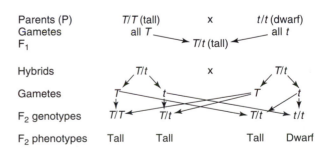

All possible combinations of F_1 gametes in the F_2 zygotes yield a 3:1 phenotypic ratio and a 1:2:1 genotypic ratio. It is convenient in such crosses to use the checkerboard method devised by Punnett (Punnett square) for representing the various combinations resulting from a cross. In the F_2 cross this scheme would apply:

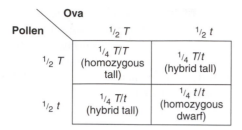

Ratio: 3 tall to 1 dwarf

The next step was an important one because it enabled Mendel to test his hypothesis that every plant contained non-blending factors from both parents. He self-fertilized the plants in the F_2 generation; the pollen of a flower fertilized the stigma of the same flower. The results showed that self-pollinated F_2 dwarf plants produced only dwarf plants, whereas one-third of the F_2 tall plants produced tall and the other two-thirds produced both tall and dwarf in the ratio of 3:1, just as the F_1 plants had done. Genotypes and phenotypes were as follows:

F_2 plants: Tall $\begin{cases} \frac{1}{4}\ T/T \xrightarrow{\text{Selfed}} \text{all } T/T \text{ (homozygous tall)} \\ \frac{1}{2}\ T/t \xrightarrow{\text{Selfed}} 1\ T/T:2\ T/t:1\ t/t \text{ (3 tall: 1 dwarf)} \end{cases}$

   Dwarf $\frac{1}{4}\ t/t \xrightarrow{\text{Selfed}} \text{all } t/t \text{ (homozygous dwarf)}$

This experiment showed that the dwarf plants were pure because they at all times gave rise to short plants when self-pollinated; the tall plants contained both pure tall and hybrid tall. It also demonstrated that, although the dwarf trait disappeared in the F_1 plants, which were all tall, dwarfness appeared in the F_2 plants.

Mendel reasoned that the factors for tallness and dwarfness were units that did not blend when they were together in a hybrid individual. The F_1 generation contained both of these units or factors, but when these plants formed their germ cells, the factors separated so that each germ cell had only one factor. In a pure-breeding plant both factors were alike; in a hybrid they were different. He concluded that individual germ cells were always pure with respect to a pair of contrasting factors, even when the germ cells were formed from hybrid individuals possessing both contrasting factors.

This idea formed the basis for Mendel's law of segregation, which states that whenever two factors are brought together in a hybrid, they segregate into separate gametes produced by the hybrid. The paired factors of the parent pass with equal frequency to the gametes. We now understand that the factors segregate because they occur on different chromosomes of a homologous pair, but the gametes receive only one chromosome of each pair in meiosis. Thus in current usage the law of segregation refers to the parting of homologous chromosomes during meiosis.

Mendel's great contribution was his quantitative approach to inheritance. His approach marks the birth of genetics, because before Mendel, people assumed that traits were blended like mixing together two colors of paint, a notion that unfortunately still lingers in the minds of many and was a problem for Darwin's theory of natural selection when he first proposed it (p. 17). If traits blended, variability would be lost in hybridization. With particulate inheritance, different alleles remain intact through the hereditary process and can be resorted like particles.

In not reporting conflicting findings, which must surely have arisen as they do in any original research, Mendel has been accused of "cooking" his results. The chances are, however, that he carefully avoided ambiguous material to strengthen his central message. Mendel's results have withstood repeated testing by other researchers, which confirms their scientific integrity.

Testcross

When an allele is dominant, heterozygous individuals containing that allele are identical in phenotype to individuals homozygous for it. Therefore one cannot determine the genotypes of these individuals just by observing their phenotypes. For instance, in Mendel's experiment of tall and dwarf traits, it is impossible to determine the genetic constitution of the tall plants of the F_2 generation by mere inspection of the tall plants. Three-fourths of this generation are tall, but which ones are heterozygotes?

As Mendel reasoned, the test is to cross the questionable individuals with pure recessives. If the tall plant is homozygous, all offspring in such a testcross are tall, thus:

Parents T/T (tall) x t/t (dwarf)

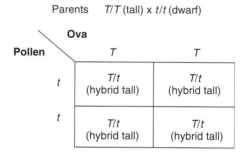

All of the offspring are T/t (hybrid tall). If the tall plant is heterozygous, half of the offspring are tall and half dwarf, thus:

Parents T/t (hybrid tall) x t/t (dwarf)

Pollen	Ova	
	T	t
t	T/t (hybrid tall)	t/t (homozygous dwarf)
t	T/t (hybrid tall)	t/t (homozygous dwarf)

The **testcross** is often used in modern genetics to assess the genetic constitution of offspring and to make desirable homozygous stocks of animals and plants.

Intermediate Inheritance

In some cases neither allele is completely dominant over the other, and the heterozygous phenotype is distinct from those of the parents, often intermediate between them. This is called

intermediate inheritance, or **incomplete dominance.** In the four-o'clock flower (*Mirabilis*), two allelic variants determine red versus pink or white flowers; homozygotes are red or white flowered, but heterozygotes have pink flowers. In a certain strain of chickens, a cross between those with black and splashed white feathers produces offspring that are not gray but a distinctive color called Andalusian blue (Figure 5.5). In each case, if the F_1s are crossed, the F_2s have a ratio of 1:2:1 in colors, or 1 red: 2 pink: 1 white in four-o'clock flowers and 1 black: 2 blue: 1 white for Andalusian chickens. This phenomenon can be illustrated for the chickens as follows:

Parents	B/B	(black feathers)	χ		B'/B'	(white feathers)
Gametes	all B				all B'	
F_1			B/B' (all blue)			
Crossing hybrids		B/B'	χ		B/B'	
Gametes		B, B'			B, B'	
F_2 genotypes	B/B		B/B'	B/B'		B'/B'
F_2 phenotypes	Black		Blue	Blue		White

When neither of the alleles is recessive, it is customary to represent both by capital letters and to distinguish them by the addition of a "prime" sign (B') or by superscript letters, for example, B^b (equals black feathers) and B^w (equals white feathers).

In this kind of cross, the heterozygous *phenotype* is indeed a blending of both parental types. It is easy to see how such observations would encourage the notion of blending inheritance. However, in the cross of black and white chickens or red and white flowers, only the hybrid phenotype is a blend; its hereditary factors do not blend and homozygous offspring breed true to the parental phenotypes.

Mendel's Second Law

Mendel's second law pertains to studies of two pairs of hereditary factors at the same time. For example, does the inheritance of factors for yellow versus green seeds influence the inheritance of factors for tall versus short plants when the strains being crossed differ for both seed color and plant height? Mendel performed crossing experiments between pea strains that differ by two or more phenotypic characters controlled by variation at different genes located on different chromosomes. According to Mendel's **law of independent assortment,** *genes located on different pairs of homologous chromosomes assort independently during meiosis.*

Mendel had already established that tall plants were dominant to dwarf. He also noted that crosses between plants bearing yellow seeds and plants bearing green seeds produced plants with yellow seeds in the F_1 generation; therefore yellow was dominant to green. The next step was to make a cross between

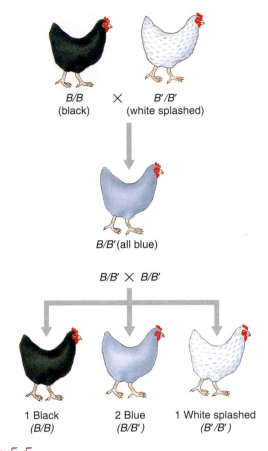

Figure 5.5
Cross between chickens with black and splashed white feathers. Black and white are homozygous; Andalusian blue is heterozygous.

plants differing in these two characteristics. When a tall plant with yellow seeds ($T/T\ Y/Y$) was crossed with a dwarf plant with green seeds ($t/t\ y/y$), the F_1 plants were tall and yellow as expected ($T/t\ Y/y$).

The F_1 hybrids were then self-fertilized giving the F_2 results shown in Figure 5.6.

Parents	$T/T\ Y/Y$	χ		$t/t\ y/y$
	(tall, yellow)			(dwarf, green)
Gametes	all TY			all ty
F_1		$T/t\ Y/y$		
		(tall, yellow)		

Mendel already knew that a cross between two plants bearing a single pair of alleles of the genotype T/t would yield a 3:1 ratio. Similarly, a cross between two plants with the genotypes Y/y would yield the same 3:1 ratio. If we examine *only* the tall and dwarf phenotypes expected in the outcome of the dihybrid experiment, they produce a ratio of 12 tall to 4 dwarf, which reduces to a ratio of 3:1. Likewise, a total of 12 plants have yellow seeds for every 4 plants that have green—again a 3:1 ratio. Thus the monohybrid ratio prevails for both traits when they are considered independently. The 9:3:3:1 ratio is nothing more than a combination of the two 3:1 ratios.

$$3:1 \times 3:1 = 9:3:3:1$$

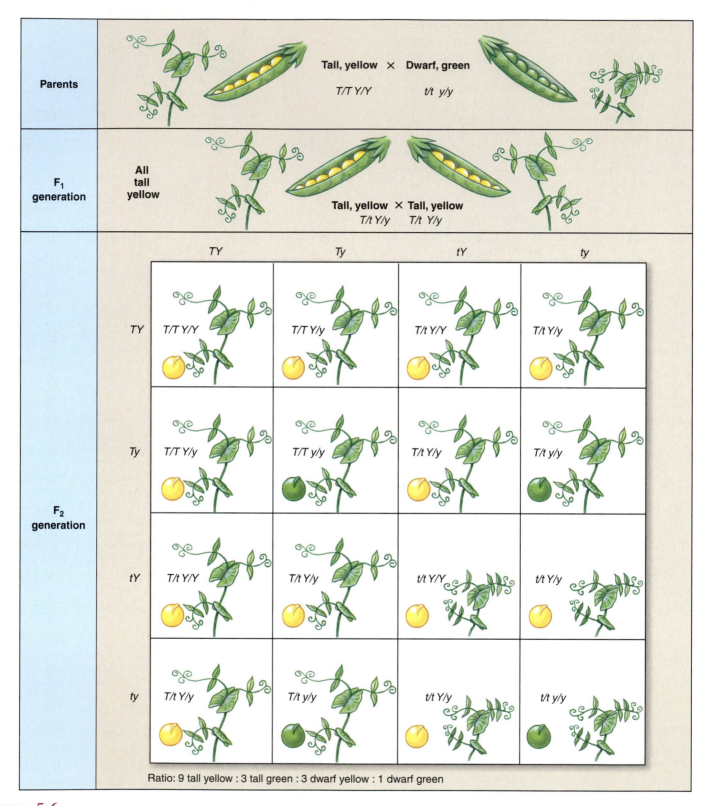

Ratio: 9 tall yellow : 3 tall green : 3 dwarf yellow : 1 dwarf green

Figure 5.6
Punnett square method for determining ratios of genotypes and phenotypes expected in a dihybrid cross for independently assorting genes.

When one of the alleles is unknown, it can be designated by a dash (*T/—*). This designation is used also when it is immaterial whether the genotype is heterozygous or homozygous, as when we count all of a genetically dominant phenotype. The dash could be either *T* or *t*.

The F_2 genotypes and phenotypes are as follows:

```
1  T/T  Y/Y ⎫
2  T/t  Y/Y ⎬  9 T/—Y/—    9 Tall yellow
2  T/T  Y/y ⎪
4  T/t  Y/y ⎭

1  T/T  y/y ⎫
2  T/t  y/y ⎬  3 T/—y/y    3 Tall green

1  t/t  Y/Y ⎫
2  t/t  Y/y ⎬  3 t/t—Y/—   3 Dwarf yellow

1  t/t  y/y    1 t/t y/y    1 Dwarf green
```

The results of this experiment show that segregation of alleles for plant height is entirely independent of segregation of alleles for seed color. Thus another way to state Mendel's law of independent assortment is that *paired copies of two different genes located on different (= nonhomologous) chromosomes segregate independently of one another*. The reason is that during meiosis the member of any pair of homologous chromosomes transmitted to a gamete is independent of which member of any other pair of chromosomes it receives. Of course, if the genes were close together on the same chromosome, they would assort together (be linked) unless crossing over occurred. Genes located very far apart on the same chromosome show independent assortment because crossing over occurs between them in nearly every meiosis. Linked genes and crossing over are discussed on p. 88.

One way to estimate proportions of progeny expected to have a given genotype or phenotype is to construct a Punnett square. With a monohybrid cross, this is easy; with a dihybrid cross, a Punnett square is laborious; and with a trihybrid cross, it is very tedious. We can make such estimates more easily by taking advantage of simple probability calculations. The basic assumption is that the genotypes of gametes of one sex have a chance of uniting with the genotypes of gametes of the other sex in proportion to the numbers of each present. This is generally true when the sample size is large enough, and the actual numbers observed come close to those predicted by the laws of probability.

We define probability, which is the expected frequency of an event, as follows:

$$\text{Probability (p)} = \frac{\text{Number of times an event happens}}{\text{Total number of trials or possibilities for the event to happen}}$$

For example, the probability (p) of a coin falling heads when tossed is 1/2, because the coin has two sides. The probability of rolling a three on a die is 1/6, because the die has six sides.

The probability of independent events occurring together (ordered events) involves the **product rule,** which is simply the product of their individual probabilities. When two coins are tossed together, the probability of getting two heads is $1/2 \times 1/2 = 1/4$, or 1 chance in 4. The probability of rolling two threes simultaneously with two dice is as follows:

$$\text{Probability of two threes} = 1/6 \times 1/6 = 1/36$$

We can use the product rule to predict the ratios of inheritance in monohybrid or dihybrid (or larger) crosses if the genes sort independently in the gametes (as they did in all of Mendel's experiments) (Table 5.1).

Note, however, that a small sample size may give a result quite different from that predicted. Thus if we tossed the coin three times and it fell heads each time, we would not be surprised. If we tossed the coin 1000 times and the number of heads diverged greatly from 500, we would strongly suspect something wrong with the coin. However, probability has no "memory." The probability of a coin toss yielding heads remains 1/2, no matter how many times the coin was tossed previously or results of the tosses.

Multiple Alleles

On page 78 we defined alleles as alternate forms of a gene. Whereas an individual can have no more than two alleles at a given locus (one each on each chromosome of the homologous pair, p. 78), many more dissimilar alleles can exist in a population. An example is the set of multiple alleles that affects coat color in rabbits. The different alleles are C (normal color), c^{ch} (chinchilla color), c^b (Himalayan color), and c (albino). The four alleles form a dominance series with C dominant over everything. The dominant allele is always written to the left and the recessive to the right:

$$C/c^b = \text{Normal color}$$
$$c^{ch}/c^b = \text{Chinchilla color}$$
$$c^b/c = \text{Himalayan color}$$
$$c/c = \text{albino}$$

Multiple alleles arise through mutations at the same gene locus at different times. Any gene can mutate (p. 100) if given time and thus can show many different alleles at the same locus.

Gene Interaction

The types of crosses previously described are simple in that the character variation results from the action of a single gene with one phenotypic effect. However, many genes have more than a single effect on organismal phenotypes, a phenomenon called **pleiotropy.** A gene whose variation influences eye color, for instance, could at the same time influence the development of other characters. An allele at one locus can mask or prevent the expression of an allele at another locus acting on the same trait, a phenomenon called **epistasis.** Another case of gene interaction

---

TABLE 5.1

Use of Product Rule for Determining Genotypic and Phenotypic Ratios in a Dihybrid Cross for Independently Assorting Genes

Parents' genotypes	T/t Y/y		×	T/t Y/y
Equivalent monohybrid crosses	T/t × T/t		and	Y/y × Y/y
Genotype ratios in F₁s of monohybrid crosses		1/4 T/T		1/4 Y/Y
		2/4 T/t		2/4 Y/y
		1/4 t/t		1/4 y/y

Combine two monohybrid ratios
to determine dihybrid genotype ratios

1/4 T/T	×	⎰ 1/4 Y/Y = 1/16 T/T Y/Y
		⎨ 2/4 Y/y = 2/16 T/T Y/y
		⎱ 1/4 y/y = 1/16 T/T y/y

2/4 T/t	×	⎰ 1/4 Y/Y = 2/16 T/t Y/Y
		⎨ 2/4 Y/y = 4/16 T/t Y/y
		⎱ 1/4 y/y = 2/16 T/t y/y

1/4 t/t	×	⎰ 1/4 Y/Y = 1/16 t/t Y/Y
		⎨ 2/4 Y/y = 2/16 t/t Y/y
		⎱ 1/4 y/y = 1/16 t/t y/y

Phenotype ratios in F₁s of monohybrid crosses

3/4 T/— (tall), 1/4 t/t (dwarf)
3/4 Y/— (yellow), 1/4 y/y (green)

Combine two monohybrid ratios
to determine phenotype ratios

3/4 T/—	×	⎰ 3/4 Y/— = 9/16 T/—Y/— (tall, yellow)
		⎱ 1/4 y/y = 3/16 T/—y/y (tall, green)

1/4 t/t	×	⎰ 3/4 Y/— = 3/16 t/t Y/— (dwarf, yellow)
		⎱ 1/4 y/y = 1/16 t/t y/y (dwarf, green)

Therefore phenotype ratios = 9 tall, yellow: 3 tall, green: 3 dwarf, yellow: 1 dwarf, green

---

is one in which several sets of alleles produce a cumulative effect on the same character.

Many cases are known in which the variation of a character results from two or more genes. Mendel probably did not appreciate the real significance of the genotype, as contrasted with the visible character—the phenotype. We now know that many different genes can affect a single phenotype (**polygenic inheritance**).

Several characters in humans are polygenic. In such cases the characters, instead of having discrete alternative phenotypes, show continuous variation between two extremes. Each of several genes has an allele that adds (+) and one that fails to add (−) an incremental dose to the value of the phenotype. This dosage-dependent inheritance is sometimes called **quantitative inheritance.** In this kind of inheritance the children are often phenotypically intermediate between the two parents. Variation at multiple genes influences phenotypic variation, but the different allelic forms of each gene remain unaltered as discrete hereditary factors as they are sorted into various genotypes. As the number of variable genes affecting a quantitative phenotype increases, intermediate conditions between the opposite extreme values of the phenotype become more continuous.

One illustration of such a type is the degree of pigmentation in matings between people having dark and light skin tones. The cumulative genes in such matings have a quantitative expression. Three or four genes are probably involved in skin pigmentation, but we simplify our explanation by referring to only two pairs of independently assorting genes. Thus a person with very dark pigment has two genes for pigmentation on separate chromosomes (A/A B/B). Each dominant allele contributes one unit of pigment. A person with very light pigment has alleles (a/a b/b) that contribute no color. (Freckles that commonly appear in the skin of very light people represent pigment contributed by entirely separate genes.) The offspring of very dark and very light parents would have an intermediate skin color (A/a B/b).

Children of parents having intermediate skin color show a range of skin color, depending on the number of genes for pigmentation that they inherit. Their skin color ranges from very dark (A/A B/B), to dark (A/A B/b or A/a B/B), intermediate (A/A b/b or A/a B/b or a/a B/B), light (A/a b/b or a/a B/b), to very light (a/a b/b). It is thus possible for parents heterozygous for skin color to produce children with darker or lighter colors than themselves.

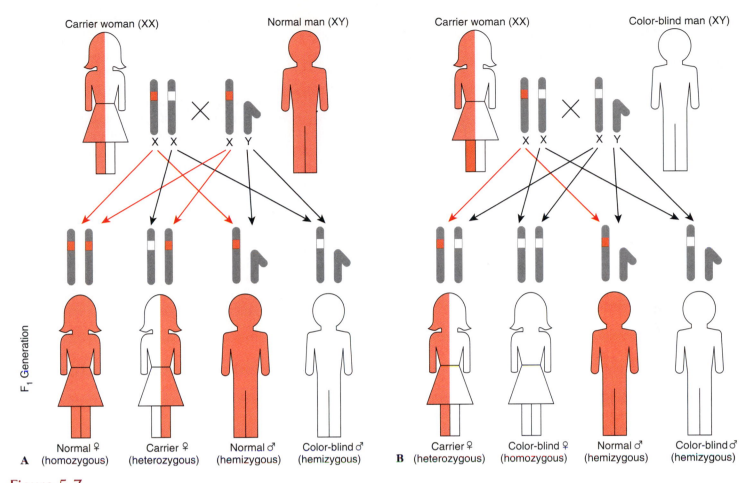

Figure 5.7

Sex-linked inheritance of red-green color blindness in humans. **A,** Carrier mother and normal father produce color blindness in one-half of their sons but in none of their daughters. **B,** Half of both sons and daughters of carrier mother and color-blind father are color blind.

Inheritance of eye color in humans is another example of gene interaction. One allele (*B*) determines whether pigment is present in the front layer of the iris. This allele is dominant over the allele for the absence of pigment (*b*). The genotypes *B/B* and *B/b* pigment generally produce brown eyes, and *b/b* produces blue eyes. However, these phenotypes are greatly affected by many modifier genes influencing, for example, the amount of pigment present, the tone of the pigment, and its distribution. Thus a person with *B/b* may even have blue eyes if modifier genes determine a lack of pigment, thus explaining the rare instances of a brown-eyed child of blue-eyed parents.

Sex-Linked Inheritance

It is known that inheritance of some characters depends on the sex of the parent carrying the gene and the sex of the offspring. One of the best-known sex-linked traits of humans is hemophilia (see Chapter 31, p. 690). Another example is red-green color blindness in which red and green colors are indistinguishable to varying degrees. Color-blind men greatly outnumber color-blind women.

When color blindness does appear in women, their fathers are color blind. Furthermore, if a woman with normal vision who is a carrier of color blindness (a **carrier** is heterozygous for the gene and is phenotypically normal) bears sons, half of them are likely to be color blind, regardless of whether the father had normal or affected vision. How are these observations explained?

Color blindness and hemophilia defects are recessive traits carried on the X chromosome. They are phenotypically expressed either when both genes are defective in the female or when only one defective gene is present in the male. The inheritance pattern of these defects is shown for color blindness in Figure 5.7. When the mother is a carrier and the father is normal, half of the sons but none of the daughters are color blind. However, if the father is color blind and the mother is a carrier, half of the sons *and* half of the daughters are color blind (on the average and in a large sample). It is easy to understand then why such defects are much more prevalent in males: a single sex-linked recessive gene in the male has a visible effect because he has only one X chromosome. What would be the outcome of a mating between a homozygous normal woman and a color-blind man?

Another example of a sex-linked character was discovered by Thomas Hunt Morgan (1910) in *Drosophila.* Normal eye

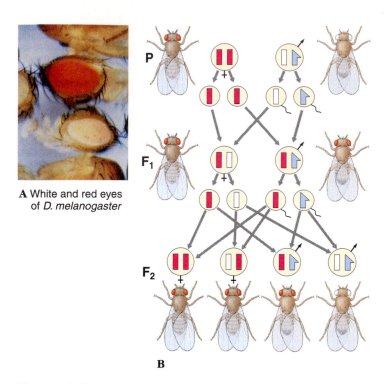

A White and red eyes of *D. melanogaster*

B

Figure 5.8

Sex-linked inheritance of eye color in fruit fly *Drosophila melanogaster.* **A,** White and red eyes of *D. melanogaster.* **B,** Genes for eye color are carried on X chromosome; Y carries no genes for eye color. Normal red is dominant to white. Homozygous red-eyed female mated with white-eyed male gives all red-eyed in F_1. F_2 ratios from F_1 cross are one homozygous red-eyed female and one heterozygous red-eyed female to one red-eyed male and one white-eyed male.

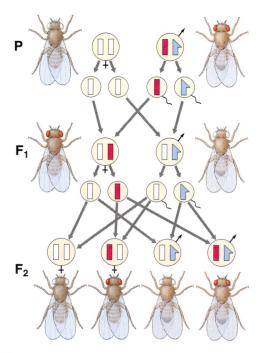

Figure 5.9

Reciprocal cross of Figure 5.8 (homozygous white-eyed female with red-eyed male) gives white-eyed males and red-eyed females in F_1. F_2 shows equal numbers of red-eyed and white-eyed females and red-eyed and white-eyed males.

color of this fly is red, but mutations for white eyes do occur (Figure 5.8). A gene for eye color is carried on the X chromosome. If true-breeding white-eyed males and red-eyed females are crossed, all F_1 offspring have red eyes because this trait is dominant (Figure 5.8). If these F_1 offspring are interbred, all F_2 females have red eyes; half of the males have red eyes and the other half have white eyes. No white-eyed females are found in this generation; only males have the recessive character (white eyes). The allele for white eyes is recessive and should affect eye color only in a homozygous condition. However, since the male has only one X chromosome (the Y does not carry a gene for eye color), white eyes appear whenever the X chromosome carries the allele for this trait. Males are said to be **hemizygous** (only one copy of a genetic locus is present) for traits carried on the X chromosome.

If the reciprocal cross is made in which females are white eyed and males red eyed, all F_1 females are red eyed and all males are white eyed (Figure 5.9). If these F_1 offspring are interbred, the F_2 generation shows equal numbers of red-eyed and white-eyed males and females.

Autosomal Linkage and Crossing Over

Linkage

Since Mendel's laws were rediscovered in 1900, it became clear that, contrary to Mendel's second law, not all factors segregate independently. Indeed, many traits are inherited together. Since the number of chromosomes in any organism is relatively small compared with the number of traits, each chromosome must contain many genes. All genes present on a chromosome are said to be **linked.** Linkage simply means that the genes are on the same chromosome, and all genes present on homologous chromosomes belong to the same linkage groups. Therefore there should be as many linkage groups as there are chromosome pairs.

Geneticists commonly use the word "linkage" in two somewhat different meanings. Sex linkage refers to inheritance of a trait on the sex chromosomes, and thus its phenotypic expression depends on the sex of the organism and the factors already discussed. Autosomal linkage, or simply, linkage, refers to inheritance of genes on a given autosomal chromosome. Letters used to represent such genes are normally written without a slash mark between them, indicating that they are on the same chromosome. For example, *AB/ab* shows that genes *A* and *B* are on the same chromosome. Interestingly, Mendel studied seven characteristics of garden peas that assorted independently because they are on seven different chromosomes. If he had studied eight characteristics, he would not have found independent assortment in two of the traits because garden peas have only seven pairs of homologous chromosomes.

In *Drosophila,* in which this principle has been studied most extensively, there are four linkage groups that correspond to the four pairs of chromosomes found in these fruit flies. Usually,

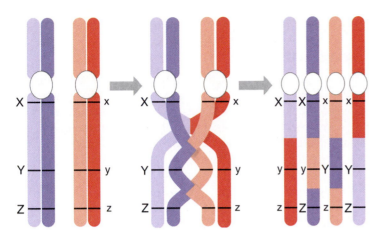

Figure 5.10

Crossing over during meiosis. Nonsister chromatids exchange portions, so that none of the resulting gametes is genetically the same as any other. Gene X is farther from gene Y than Y is from Z; therefore, gene X is more frequently separated from Y in crossing over than Y is from Z.

small chromosomes have small linkage groups, and large chromosomes have large groups.

Crossing Over

Linkage, however, is usually not complete. If we perform an experiment in which animals such as *Drosophila* are crossed, we find that linked traits separate in some percentage of the offspring. Separation of alleles located on the same chromosome occurs because of **crossing over.**

During the protracted prophase of the first meiotic division, paired homologous chromosomes break and exchange equivalent portions; genes "cross over" from one chromosome to its homolog, and vice versa (Figure 5.10). Each chromosome consists of two sister chromatids held together by means of a proteinaceous structure called a **synaptonemal complex.** Breaks and exchanges occur at corresponding points on nonsister chromatids. (Breaks and exchanges also occur between sister chromatids but usually have no genetic significance because sister chromatids are identical.) Crossing over is a means for exchanging genes between homologous chromosomes and thus greatly increases the amount of genetic recombination. The frequency of crossing over varies depending on the species, but usually at least one and often several crossovers occur each time chromosomes pair.

Because the frequency of recombination is proportional to the distance between loci, the relative linear position of each locus can be determined. Genes located far apart on very large chromosomes may assort independently because the probability of a crossover occurring between them in each meiosis is close to 100%. Such genes are found to be carried on the same chromosome only because each one is genetically linked to additional genes located physically between them on the chromosome. Laborious genetic experiments over many years have produced gene maps that indicate the positions of more than

500 genes distributed on the four chromosomes of *Drosophila melanogaster.*

Chromosomal Aberrations

Structural and numerical deviations from the norm that affect many genes at once are called chromosomal aberrations. They are sometimes called chromosomal mutations, but most cytogeneticists prefer to use the term "mutation" to refer to qualitative changes within a gene; gene mutations are discussed on page 100.

Despite the incredible precision of meiosis, chromosomal aberrations do occur, and they are more common than one might think. They are responsible for great economic benefit in agriculture. Unfortunately, they are responsible also for many human genetic malformations. It is estimated that five out of every 1000 humans are born with *serious* genetic defects attributable to chromosomal anomalies. An even greater number of embryos with chromosomal defects abort spontaneously, far more than ever reach birth.

Changes in chromosome numbers are called **euploidy** when there is the addition or deletion of whole sets of chromosomes and **aneuploidy** when a single chromosome is added to or subtracted from a set. A "set" of chromosomes contains one member of each homologous pair as would be present in the nucleus of a gamete. The most common kind of euploidy is **polyploidy,** the carrying of three or more sets of chromosomes by an organism. An organism with three or more complete sets of chromosomes is called a polyploid. Such aberrations are much more common in plants than in animals. Animals are much less tolerant of chromosomal aberrations, especially those in which sex determination requires a delicate balance between the numbers of sex chromosomes and autosomes. Many domestic plant species are polyploid (cotton, wheat, apples, oats, tobacco, and others), and perhaps 40% of flowering plants may have originated in this manner. Horticulturists favor polyploids because they often have more intensely colored flowers and more vigorous vegetative growth.

Aneuploidy is usually caused by failure of chromosomes to separate during meiosis **(nondisjunction).** If a pair of chromosomes fails to separate during the first or second meiotic divisions, both members go to one pole and none to the other. This condition results in at least one gamete or polar body having $n - 1$ chromosomes and another having $n + 1$ chromosomes. If an $n - 1$ gamete is fertilized by a normal n gamete, the result is a **monosomic** animal. Survival is rare because the lack of one chromosome gives an uneven balance of genetic instructions. **Trisomy,** the result of the fusion of a normal n gamete and an $n + 1$ gamete, is much more common, and several kinds of trisomic conditions are known in humans. Perhaps the most familiar is **trisomy 21,** or **Down syndrome.** As the name indicates, it involves an extra chromosome 21 combined with the chromosome pair 21, and it is caused by nondisjunction of that pair during meiosis. It occurs spontaneously, and there is seldom any family history of the abnormality. However, the risk of its appearance rises dramatically with increasing age of the mother;

it occurs 40 times more often in women over 40 years old than among women between the ages of 20 and 30. In cases where maternal age is not a factor, 20% to 25% of trisomy 21 results from nondisjunction during spermatogenesis; it is paternal in origin and is apparently independent of the father's age.

A *syndrome* is a group of symptoms associated with a particular disease or abnormality, although every symptom is not necessarily shown by every patient with the condition. An English physician, John Langdon Down, described in 1866 the syndrome that we now know is caused by trisomy 21. Because of Down's belief that the facial features of affected individuals were mongoloid in appearance, the condition has been called mongolism. The resemblances are superficial, however, and currently accepted names are trisomy 21 and Down syndrome. Among the numerous characteristics of the condition, the most disabling is severe mental retardation. This, as well as other conditions caused by chromosomal aberrations and several other birth defects, can be diagnosed *prenatally* by a procedure involving *amniocentesis*. The physician inserts a hypodermic needle through the abdominal wall of the mother and into fluids surrounding the fetus (*not into* the fetus) and withdraws some of the fluid, which contains some fetal cells. The cells are grown in culture, their chromosomes are examined, and other tests done. If a severe birth defect is found, the mother has the option of having an abortion performed. As an extra "bonus," the sex of the fetus is learned after amniocentesis. How? Alternatively, determination of concentrations of certain substances in the maternal serum, which is less invasive than amniocentesis, can detect about 60% of Down syndrome fetuses. Ultrasound scanning may be more than 80% accurate.

In all diploid species, normal development requires exactly two of each kind of autosome (not sex chromosomes). Nondisjunction can cause trisomies of other chromosomes, but because these lead to imbalance of many gene products, they almost always cause death before or soon after birth. However, each cell requires only one functional X chromosome (the other is inactivated in females). Nondisjunction of sex chromosomes is better tolerated but usually causes sterility and abnormalities of sex organs. For example, a human with XXY (Klinefelter syndrome) is a phenotypic male, usually infertile and with some female sexual characteristics. Presence of only one X (and no Y) is usually lethal in embryos, but the occasional live birth produces a phenotypic female with a variety of developmental abnormalities (Turner syndrome).

Structural aberrations involve whole sets of genes within a chromosome. A portion of a chromosome may be reversed, placing the linear arrangement of genes in reverse order **(inversion);** nonhomologous chromosomes may exchange sections **(translocation);** entire blocks of genes may be lost **(deletion),** usually causing serious developmental defects; or an extra section of chromosome may attach to a normal chromosome **(duplication).** These structural changes often produce phenotypic changes. Duplications, although rare, are important for evolution because they supply additional genetic information that may enable new functions.

GENE THEORY

Gene Concept

The term "gene" (Gr. *genos,* descent) was coined by W. Johannsen in 1909 for the hereditary factors of Mendel. Initially, genes were thought to be indivisible subunits of the chromosomes on which they occurred. Later studies with multiple mutant alleles demonstrated that alleles are in fact divisible by recombination; *portions* of a gene are separable. Furthermore, parts of many genes in eukaryotes are separated by sections of DNA that do not specify a part of the finished product **(introns).**

As the chief unit of genetic information, genes encode products essential for specifying the basic architecture of every cell, details of protein synthesis, cell division, and, directly or indirectly, the entire metabolic function of the cell. Because of their ability to mutate, to be assorted and shuffled in different combinations, genes are important units of heredity and variation in evolution. Genes maintain their identities for many generations despite mutational changes in some parts of their structure.

One Gene–One Enzyme Hypothesis

Since genes act to produce different phenotypes, we may infer that their action follows the scheme: gene → gene product → phenotypic expression. Furthermore, we may suspect that the gene product is usually a protein, because proteins act as enzymes, antibodies, hormones, and structural elements throughout the body.

The first clear, well-documented study to link genes and enzymes was performed on the common bread mold *Neurospora* by Beadle and Tatum in the early 1940s. This organism was ideally suited to a study of gene function for several reasons: these molds are much simpler to handle than fruit flies, they grow readily in well-defined chemical media, and they are haploid organisms unencumbered with dominance relationships among alleles. Furthermore, mutations were readily induced by irradiation with ultraviolet light. Ultraviolet-light-induced mutants, grown and tested in specific nutrient media, had single-gene mutations that were inherited. Each mutant strain was defective in one enzyme, which prevented that strain from synthesizing one or more complex molecules. The ability to synthesize a particular molecule was controlled by a single gene.

From these experiments Beadle and Tatum made an important and exciting formulation: *one gene produces one enzyme.* For this work they received the Nobel Prize for Physiology or Medicine in 1958. The new hypothesis was soon validated by research on many biosynthetic pathways. Hundreds of inherited disorders, including dozens of human hereditary diseases, are caused by single mutant genes causing loss of a specific enzyme. We now know that a particular protein may contain several chains of amino acids (polypeptides), each one specified by a different gene, and not all proteins specified by genes are enzymes (for example, structural proteins, antibodies, transport proteins, and hormones). Furthermore, genes directing the synthesis of various kinds of RNA were not included in Beadle and Tatum's formulation. Therefore a gene now may be defined more inclusively as *a nucleic acid sequence (usually DNA) that encodes a functional polypeptide or RNA sequence.*

STORAGE AND TRANSFER OF GENETIC INFORMATION

Nucleic Acids: Molecular Basis of Inheritance

Cells contain two kinds of nucleic acids: deoxyribonucleic acid (DNA), which is the genetic material, and ribonucleic acid (RNA), which functions in protein synthesis. Both DNA and RNA are polymers built of repeated units called **nucleotides.** Each nucleotide contains three parts: a **sugar,** a **nitrogenous base,** and a **phosphate group.** The sugar is a pentose (5-carbon) sugar; in DNA it is **deoxyribose** and in RNA it is **ribose** (Figure 5.11).

Nitrogenous bases of nucleotides are also of two types: pyrimidines, whose characteristic structure is a single, 6-membered ring, and purines, which contain two fused rings. Purines and pyrimidines contain nitrogen as well as carbon in their rings, which is why they are called "nitrogenous" bases. The purines in both RNA and DNA are adenine and guanine (Table 5.2). The pyrimidines in DNA are thymine and cytosine, and in RNA they are uracil and cytosine. Carbon atoms in the bases are numbered (for identification) according to standard biochemical notation (Figure 5.12). Carbons in ribose and deoxyribose are also numbered, but to distinguish them from the carbons in the bases, numbers for carbons in the sugars are given prime signs (see Figure 5.11).

The sugar, phosphate group, and nitrogenous base are linked as shown in the generalized scheme for a nucleotide:

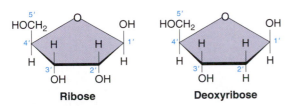

Figure 5.11 appears below; the generalized nucleotide scheme with labels Phosphate, Sugar, Nitrogenous base.

Figure 5.11

Ribose and deoxyribose, the pentose sugars of nucleic acids. A carbon atom lies in each of the four corners of the pentagon (labeled 1' to 4'). Ribose has a hydroxyl group (—OH) and a hydrogen on the number 2' carbon; deoxyribose has two hydrogens at this position.

	DNA	RNA
TABLE 5.2 Chemical Components of DNA and RNA		
Purines	Adenine	Adenine
	Guanine	Guanine
Pyrimidines	Cytosine	Cytosine
	Thymine	Uracil
Sugar	2-Deoxyribose	Ribose
Phosphate	Phosphoric acid	Phosphoric acid

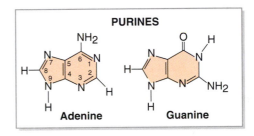

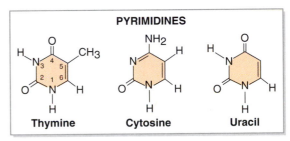

Figure 5.12

Purines and pyrimidines of DNA and RNA.

In DNA the "backbone" of the molecule is built of phosphoric acid and deoxyribose; to this backbone are attached the nitrogenous bases (Figure 5.13). The **5′ end** of the backbone has a free phosphate group on the **5′** carbon of the ribose, and the **3′ end** has a free hydroxyl group on the **3′** carbon. However, one of the most interesting and important discoveries about the nucleic acids is that DNA is not a single polynucleotide chain; it has *two* complementary chains that are precisely cross-linked by specific hydrogen bonding between purine and pyrimidine bases. The number of adenines equals the number of thymines, and the number of guanines equals the number of cytosines. This fact suggested a pairing of bases: adenine with thymine (AT) and guanine with cytosine (GC) (see Figures 1.6 and 5.14).

The result is a ladder structure (Figure 5.15). The upright portions are the sugar-phosphate backbones, and the connecting rungs are the paired nitrogenous bases, AT or GC. However, the ladder is twisted into a **double helix** with approximately 10 base pairs for each complete turn of the helix (Figure 5.16). The two DNA strands run in opposite directions **(antiparallel),** and the 5′ end of one strand is opposite the 3′ end of the other (Figure 5.16). The two strands are also **complementary**—the

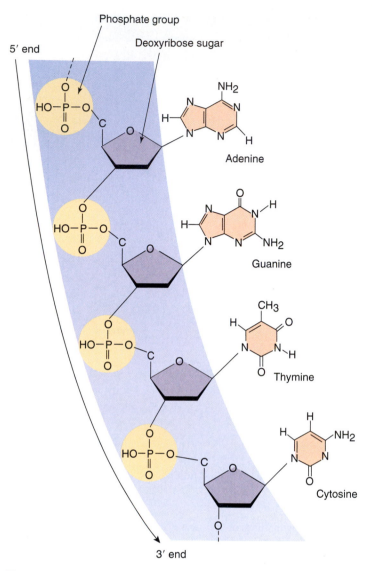

Figure 5.13

Section of a strand of DNA. Polynucleotide chain is built of a "backbone" of phosphoric acid and deoxyribose sugar molecules. Each sugar holds a nitrogenous base. Shown from top to bottom are adenine, guanine, thymine, and cytosine.

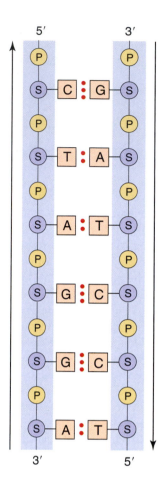

Figure 5.15

DNA, showing how the complementary pairing of bases between the sugar-phosphate "backbones" keeps the double helix at a constant diameter for the entire length of the molecule. Dots represent the three hydrogen bonds between each cytosine and guanine and the two hydrogen bonds between each adenine and thymine.

sequence of bases along one strand specifies the sequence of bases along the other strand.

The structure of DNA is widely considered the single most important biological discovery of the twentieth century. It was based on X-ray diffraction studies of Maurice H. F. Wilkins and Rosalind Franklin and on ingenious proposals of Francis H. C. Crick and James D. Watson published in 1953. Watson, Crick, and Wilkins were later awarded the Nobel Prize for Physiology or Medicine for their momentous work. Rosalind Franklin was not included because she died prior to the award.

RNA is similar to DNA in structure except that it consists of a *single* polynucleotide chain (except in some viruses), has ribose instead of deoxyribose, and has uracil instead of thymine.

Figure 5.14

Positions of hydrogen bonds between thymine and adenine and between cytosine and guanine in DNA.

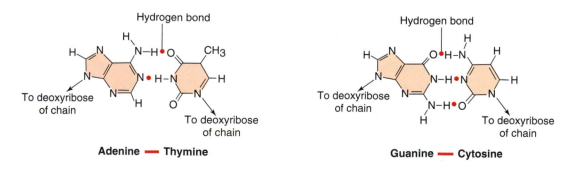

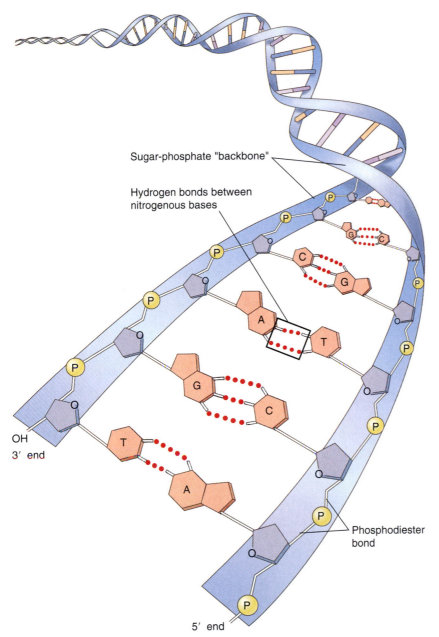

Sugar-phosphate "backbone"

Hydrogen bonds between
nitrogenous bases

OH
3′ end

Phosphodiester
bond

5′ end

Figure 5.16
DNA molecule.

Ribosomal, transfer, and messenger RNAs are the most abundant and well-known types (function described on pp. 95–96), but many structural and regulatory RNAs, such as micro RNAs, are known.

Every time a cell divides, the structure of DNA must be precisely copied in the daughter cells. This is called **replication** (Figure 5.17). During replication, the two strands of the double helix unwind, and each separated strand serves as a **template** against which a complementary strand is synthesized. An enzyme (DNA polymerase) catalyzes assembly of a new strand of polynucleotides with a thymine group going opposite the adenine group in the template strand, a guanine group opposite the

cytosine group, and the two reverse conditions. DNA polymerase synthesizes new strands only in the direction of 5′ to 3′. Because the parent DNA strands are antiparallel, one of which runs 5′ to 3′ and the other running 3′ to 5′, synthesis along one of the strands is continuous, and the other must be formed in a series of fragments, each of which begins with a 5′ end running toward a 3′ end (Figure 5.17).

DNA Coding by Base Sequence

Because DNA is the genetic material and contains a linear sequence of base pairs, an obvious extension of the Watson-Crick model is that the sequence of base pairs in DNA codes for, and is colinear with, the sequence of amino acids in a protein. The coding hypothesis must explain how a string of four different bases—a four-letter alphabet—could specify the sequence of 20 different amino acids.

In the coding procedure, obviously there cannot be a 1:1 correspondence between four bases and 20 amino acids. If a coding unit (often called a word, or **codon**) were two bases, only 16 words (4^2) could be formed, which could not specify 20 amino acids. Therefore the codon must contain at least three bases or three letters, because 64 possible words (4^3) could be formed by four bases when taken as triplets. A triplet code permits considerable redundancy of triplets (codons), because DNA encodes just 20 amino acids. Later work confirmed that nearly all amino acids are specified by more than one triplet code (Table 5.3).

DNA shows surprising stability, both in prokaryotes and in eukaryotes. Interestingly, it is susceptible to damage by harmful chemicals in the environment and by radiation. Such damage is usually not permanent, because cells have an efficient repair system. Various types of damage and repair are known, one of which is **excision repair.** Ultraviolet irradiation often damages DNA by linking adjacent pyrimidines by covalent bonds (dimerize), preventing transcription and replication. A series of several enzymes "recognizes" the damaged strand and excises the pair of dimerized pyrimidines and several bases following them. DNA polymerase then synthesizes the missing strand using the remaining one as a template, according to the base-pairing rules, and the enzyme **DNA ligase** joins the end of the new strand to the old one.

Transcription and the Role of Messenger RNA

Information is coded in DNA, but DNA does not participate directly in protein synthesis. The intermediary molecule between DNA and protein is another nucleic acid called **messenger RNA (mRNA).** The triplet codes in DNA are **transcribed** into mRNA, with uracil substituting for thymine (Table 5.3).

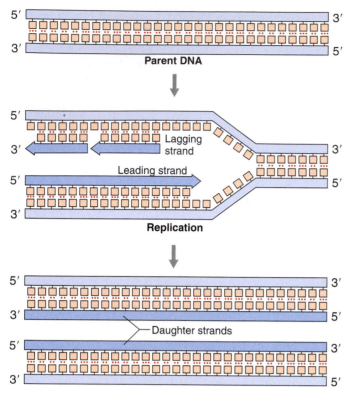

Figure 5.17

Replication of DNA. Parent strands of DNA part, and DNA polymerase synthesizes daughter strands using the base sequence of parent strands as a template. Because synthesis always proceeds in a 5′ to 3′ direction, synthesis of one strand is continuous, and the other strand must be synthesized as a series of fragments.

Ribosomal, transfer, and messenger RNAs are transcribed directly from DNA, each encoded by different sets of genes. RNA is formed as a complementary copy of one strand of the appropriate gene using an enzyme called **RNA polymerase.** (In eukaryotes each type of RNA [ribosomal, transfer, and messenger] is transcribed by a different type of RNA polymerase.) The RNA contains a sequence of bases that complements the bases in one of the two DNA strands, just as the DNA strands complement each other. Thus A in the template DNA strand is replaced by U in RNA; C is replaced by G; G is replaced by C; and T is replaced by A. Only one of the two chains is used as the template for RNA synthesis (Figure 5.18). A codon is referenced as the sequence of bases present in a mRNA molecule (Table 5.3), which is complementary and antiparallel to the template DNA strand (often called the "sense" strand) from which it is made. The DNA strand not used as a template during transcription of a gene is called the "antisense" strand.

A bacterial gene is encoded on a continuous stretch of DNA, transcribed into mRNA, and then translated (see the next section). The hypothesis that eukaryotic genes had a similar structure was rejected by the surprising discovery that some stretches of DNA are transcribed in the nucleus but are not found in the corresponding mRNA in the cytoplasm. In other words, pieces of the nuclear mRNA were removed in the nucleus before the finished mRNA was transported to the cytoplasm (Figure 5.19). Thus many genes are split, interrupted by sequences of bases that do not code for the final product, and mRNA transcribed from them must be edited or "matured" before translation in the cytoplasm. The intervening segments of DNA are called **introns,** and those that encode part of

TABLE 5.3

The Genetic Code: Amino Acids Specified by Codons of Messenger RNA

First Letter		Second Letter							Third Letter
		U		C		A		G	
U	UUU	Phenylalanine	UCU	Serine	UAU	Tyrosine	UGU	Cysteine	U
	UUC		UCC		UAC		UGC		C
	UUA	Leucine	UCA		UAA	End chain	UGA	End chain	A
	UUG		UCG		UAG		UGG	Tryptophane	G
C	CUU	Leucine	CCU	Proline	CAU	Histidine	CGU	Arginine	U
	CUC		CCC		CAC		CGC		C
	CUA		CCA		CAA	Glutamine	CGA		A
	CUG		CCG		CAG		CGG		G
A	AUU	Isoleucine	ACU	Threonine	AAU	Asparagine	AGU	Serine	U
	AUC		ACC		AAC		AGC		C
	AUA		ACA		AAA	Lysine	AGA	Arginine	A
	AUG	Methionine*	ACG		AAG		AGG		G
G	GUU	Valine	GCU	Alanine	GAU	Aspartic acid	GGU	Glycine	U
	GUC		GCC		GAC		GGC		C
	GUA		GCA		GAA	Glutamic acid	GGA		A
	GUG		GCG		GAG		GGG		G

*Also, begin polypeptide chain.

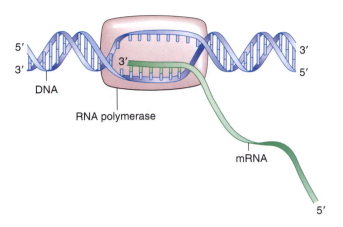

Figure 5.18

Transcription of mRNA from a DNA template. Transcription is similar for mRNA, rRNA, and tRNA except that each kind of RNA uses a different form of the enzyme, RNA polymerase. This diagram shows transcription midway to completion. Transcription began by unwinding the DNA helix, annealing of a RNA primer to the template strand of DNA, and extension of the primer at its 3' end by adding nucleotides (not shown) complementary to the sequence of bases in the template DNA strand. The primer is at the 5' end of the mRNA, which continues to grow in length by adding nucleotides at its annealed 3' end. When transcription is finished, the mRNA will detach completely from the DNA template.

the mature RNA and are translated into protein are called **exons.** Before mRNA leaves the nucleus, a methylated guanine "cap" is added at the 5' end, and a tail of adenine nucleotides (poly-*A*) is often added at the 3' end (Figure 5.19). The cap and the poly-*A* tail distinguish mRNA from other kinds of RNA molecules.

In mammals, genes coding for histones and for interferons are on continuous stretches of DNA. However, we now know that genes coding for many proteins are split. In lymphocyte differentiation the parts of the split genes coding for immunoglobulins are actually *rearranged* during development, so that different proteins result from subsequent transcription and translation. This rearrangement partly explains the enormous diversity

of antibodies manufactured by descendants of the lymphocytes (p. 775).

Base sequences in some introns are complementary to other base sequences in the intron, suggesting that the intron could fold so that complementary sequences would pair. This folding may be necessary to control proper alignment of intron boundaries before splicing. Most surprising of all is the discovery that, in some cases, RNA can "self-catalyze" the excision of introns. The ends of the intron join; the intron thus becomes a small circle of RNA, and the exons are spliced together. This process does not fit the classical definition of an enzyme or other catalyst because the molecule itself is changed by the reaction.

Translation: Final Stage in Information Transfer

The **translation** process occurs on **ribosomes,** granular structures composed of protein and **ribosomal RNA (rRNA).** Ribosomal RNA contains a large and a small subunit, and the small subunit comes to lie in a depression of a large subunit to form a functional ribosome (Figure 5.20). Messenger RNA molecules attach themselves to ribosomes to form a messenger RNA-ribosome complex. Because only a short section of mRNA makes contact with a single ribosome, the mRNA usually has several ribosomes attached along its length, each one at a different stage of synthesizing the encoded polypeptide. The entire complex, called a **polyribosome** or **polysome,** allows several molecules of the same kind of polypeptide to be synthesized concurrently, one on each ribosome of the polysome (Figure 5.20).

Assembly of polypeptides on the mRNA-ribosome complex requires another kind of RNA called **transfer RNA (tRNA).** Transfer RNAs have a complex secondary structure of folded stems and loops, often illustrated in the form of a cloverleaf (Figure 5.21), although the three-dimensional shape is somewhat different. Molecules of tRNA collect free amino acids from the cytoplasm and deliver them to the polysome, where they are assembled into a polypeptide. There are special tRNA molecules

Figure 5.19

Expression of ovalbumin gene of chicken. The entire gene of 7700 base pairs is transcribed to form the primary mRNA, then the 5' cap of methyl guanine and the 3' polyadenylate tail are added. After the introns are spliced out, the mature mRNA is transferred to the cytoplasm.

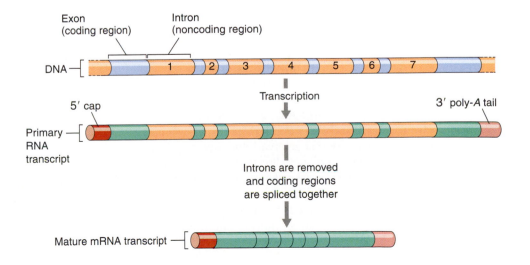

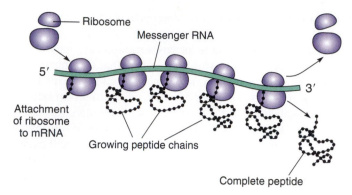

Figure 5.20

How the polypeptide chain is formed. As ribosomes move along messenger RNA in a 5′ to 3′ direction, amino acids are added stepwise to form the polypeptide chain.

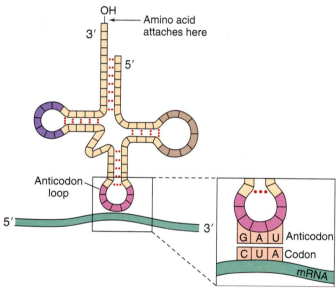

Figure 5.21

Diagram of a tRNA molecule. The anticodon loop bears bases complementary to those in the mRNA codon. The other two loops function in binding to the ribosome in polypeptide synthesis. The amino acid is added to the free single-stranded 3′ end by tRNA synthetase.

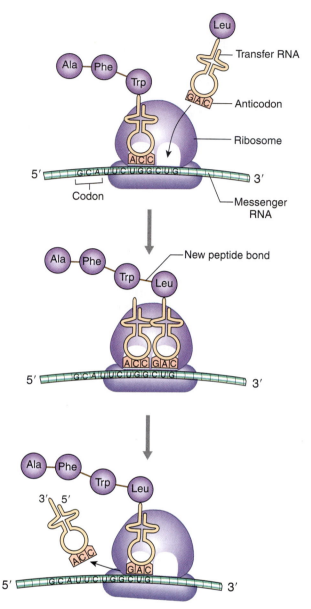

Figure 5.22

Formation of polypeptide chain on messenger RNA. As a ribosome moves down the messenger RNA molecule, transfer RNA molecules with attached amino acids enter the ribosome (top). Amino acids are joined together into a polypeptide chain, and transfer RNA molecules leave the ribosome (bottom).

for every amino acid. Furthermore, each tRNA is accompanied by a specific tRNA synthetase. Transfer RNA synthetases are enzymes that attach the correct amino acid to the terminal adenine on the 3′ end of each tRNA by a process called **charging.**

On the cloverleaf-shaped molecule of tRNA, a special sequence of three bases (the **anticodon**) is exposed in just the right way to form base pairs with complementary bases (the codon) in the mRNA. The codons are read and polypeptides assembled along the mRNA in a 5′ to 3′ direction. The anticodon of each tRNA is the key to the correct ordering of amino acids in the polypeptide being assembled.

For example, alanine is assembled into a polypeptide when it is signaled by the codon GCG in an mRNA. The translation

is accomplished by alanine tRNA in which the anticodon is CGC. An alanine tRNA is first charged with alanine by its tRNA synthetase. The alanine-tRNA complex enters the ribosome where it fits precisely into the right place on the mRNA strand. Then the next charged tRNA specified by the mRNA code (glycine tRNA, for example) enters the ribosome and attaches itself beside the alanine tRNA. The two amino acids are united by a peptide bond, and the alanine tRNA then detaches from the ribosome. The process continues stepwise as the polypeptide chain is built (Figure 5.22). A polypeptide of 500 amino acids can be assembled in less than 30 seconds.

Regulation of Gene Expression

In Chapter 8 we show how the orderly differentiation of an organism from fertilized ovum to adult requires expression of genetic material at every stage of development. Developmental biologists have provided convincing evidence that every cell in a developing embryo is genetically equivalent. Thus as tissues differentiate (change developmentally), each one uses only a part of the genetic instruction present in every cell. Genes express themselves only at certain times and not at others. Indeed, most of the genes are inactive at any given moment in a particular cell or tissue. The problem in development is to explain how, if every cell has a full gene complement, certain genes are "turned on" to produce proteins required for a particular developmental stage while other genes remain silent.

Although developmental changes bring the question of gene activation clearly into focus, gene regulation is necessary throughout an organism's existence. The cellular enzyme systems that control all functional processes obviously require genetic regulation because enzymes have powerful effects even in minute amounts. Enzyme synthesis must respond to the influences of supply and demand.

Gene Regulation in Eukaryotes

Several metabolic stages in eukaryotic cells may serve as control points for gene expression. Transcriptional and translational control are the primary stages for control of gene expression in animals, with gene rearrangement also used in some cases.

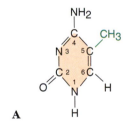

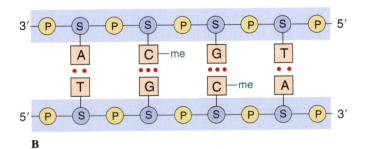

Figure 5.23

Some genes in eukaryotes are turned off by methylation of some cytosine residues in the chain. **A,** Structure of 5-methyl cytosine. **B,** Cytosine residues next to guanine are those that are methylated in a strand, thus allowing both strands to be symmetrically methylated.

Transcriptional Control   Control of transcription is probably the most important mechanism for regulating gene expression. **Transcription factors** are molecules that can have a positive or a negative effect on transcription of RNA from the DNA of target genes. The factors in some cases act within cells that produce them and in other cases are transported to different parts of the body prior to action. Examples of transcription factors are steroid receptors when bound to a steroid hormone. Steroid hormones produced by endocrine glands elsewhere in the body enter a target cell and bind with a receptor protein in the nucleus. The steroid-receptor complex then binds with DNA near the target gene (p. 755). Progesterone, for example, binds with a nuclear receptor in cells of the chicken oviduct; the hormone-receptor complex then activates transcription of genes encoding egg albumin and other substances.

An important mechanism for silencing genes is methylation of cytosine bases; a methyl group (CH_3—) binds the carbon in the 5 position in the cytosine ring (Figure 5.23A). This usually happens when the cytosine is next to a guanine base; thus, the bases in the complementary DNA strand would also be a cytosine and a guanine (Figure 5.23B). When the DNA is replicated, an enzyme recognizes the CG sequence and quickly methylates the daughter strand, keeping the gene inactive.

Translational Control   Genes can be transcribed and the mRNA sequestered so that translation is delayed. Development of eggs of many animals commonly uses this mechanism. Oocytes accumulate large quantities of messenger RNA during their development; then, fertilization activates metabolism and initiates translation of maternal mRNA.

Gene Rearrangement   Vertebrates contain cells called lymphocytes that bear genes encoding proteins called antibodies (p. 775). Each type of antibody binds only a particular foreign substance (antigen). Because the number of different antigens is enormous, diversity of antibody genes must be equally great. One source of this diversity is rearrangement of DNA sequences coding for antibodies during development of lymphocytes.

Molecular Genetics

Progress in our understanding of genetic mechanisms on the molecular level, as discussed in the last few pages, has been almost breathtaking in the last few years. We expect many more discoveries in the near future. This progress results from many biochemical techniques now used in molecular biology. We describe briefly the most important techniques.

Recombinant DNA

An important tool in this technology is a series of enzymes called **restriction endonucleases.** Each of these enzymes, derived from bacteria, cleaves double-stranded DNA at particular sites determined by their base sequences. Many of these endonucleases cut the DNA strands so that one has several bases projecting farther

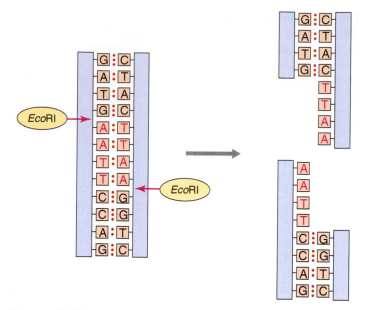

Figure 5.24

Action of restriction endonuclease, *Eco*RI. Such enzymes recognize specific base sequences that are palindromic (a palindrome is a word spelled the same backward and forward). *Eco*RI leaves "sticky ends," which anneal to other DNA fragments cleaved by the same enzyme. The strands are joined by DNA ligase.

than the other strand (Figure 5.24), leaving what are called "sticky ends." When these DNA fragments are mixed with others that have been cleaved by the same endonuclease, their sticky ends tend to anneal (join) by the rules of complementary base pairing. The ends are sealed into their new position by the enzyme **DNA ligase** in a process called **ligation.**

Besides their chromosomes, most prokaryotic and at least some eukaryotic cells have small circles of double-stranded DNA called *plasmids*. Although constituting only 1% to 3% of the bacterial genome, they may carry important genetic information, for example, resistance to an antibiotic. Plastids in plant cells (for example, chloroplasts) and mitochondria, found in most eukaryotic cells, are self-replicating and have their own complement of DNA in the form of small circles reminiscent of plasmids. The DNA of mitochondria codes for some mitochondrial proteins, whereas other mitochondrial proteins are specified by nuclear genes.

If DNA from a foreign source (such as a mammal) is ligated into a plasmid (see preceding note), the product is **recombinant DNA.** To produce the recombinant DNA in large quantities, the modified plasmid must be cloned in bacteria. The bacteria are treated with dilute calcium chloride to make them more susceptible to entry by the recombinant DNA, but plasmids do not enter most bacterial cells. Bacterial cells that have acquired the recombinant DNA can be identified if the plasmid has a marker, for example, resistance to an antibiotic. Then, only bacteria that can grow in the presence of the antibiotic are ones that have

absorbed the recombinant DNA. Some bacteriophages (bacterial viruses) also are used as carriers for recombinant DNA. Plasmids and bacteriophages that carry recombinant DNA are called **vectors.** The vectors retain the ability to replicate in the bacterial cells; therefore the recombinant insert is produced in large quantities, a process called amplification.

A clone is a collection of individuals or cells all derived by asexual reproduction from a single individual. When we speak of cloning a gene or plasmid in bacteria, we mean that we isolate a colony or group of bacteria derived from a single ancestor into which the gene or plasmid was inserted. Cloning is used to obtain large quantities of a gene that has been ligated into a bacterial plasmid.

Polymerase Chain Reaction

Recent advances permit a specific gene to be cloned enzymatically from any organism as long as part of the sequence of that gene is known. The technique is called the **polymerase chain reaction (PCR).** Two short chains of nucleotides called primers are synthesized; primers are complementary to different DNA strands in the known sequence at opposite ends of the gene to be cloned. A large excess of each primer is added to a sample of DNA from the organism, and the mixture is heated to separate the double helix into single strands. When the mixture is cooled, there is a much greater probability that each strand of the gene of interest will anneal to a primer than to the other strand of the gene—because the primer is present in a much higher concentration. A heat-stable DNA polymerase and the four deoxyribonucleotide triphosphates are added to the reaction mixture. DNA synthesis proceeds from the 3′ end of each primer, extending the primer in the 5′ to 3′ direction. Primers are designed so that the free 3′ end of each faces toward the gene whose sequence is to be cloned. Entire new complementary strands are synthesized, and the number of copies of the gene has doubled (Figure 5.25). The reaction mixture is then reheated and cooled again to allow more primers to bind original and new copies of each strand. With each cycle of DNA synthesis, the number of copies of the gene doubles. Since each cycle can take less than five minutes, the number of copies of a gene can increase from one to over one million in less than two hours! The PCR allows cloning a known gene from an individual patient, identification of a drop of dried blood at a crime scene, or cloning DNA of a 40,000-year-old woolly mammoth.

Recombinant DNA technology and PCR are currently being used to engineer crop plants, including soybeans, cotton, rice, corn, and tomato. Transgenic mice are commonly used in research, and gene therapy for human genetic diseases is being developed.

Genomics and Proteomics

The scientific field of mapping, sequencing, and analyzing genomes is now called **genomics.** Some researchers divide

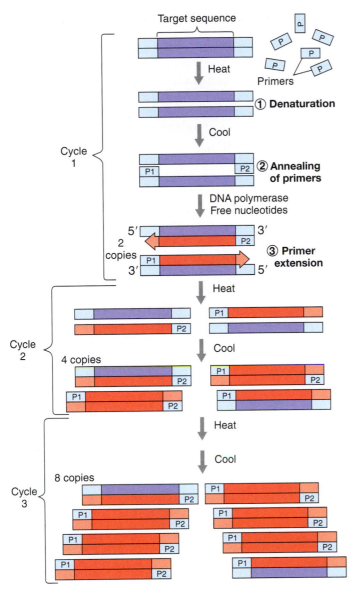

Figure 5.25

Steps in the polymerase chain reaction (PCR). Note that two different primers are required, one for each end of the target sequence.

technical improvements to make it possible by the twenty-second century. In fact, development and improvement of automated sequencers, as well as competition between the publicly supported Human Genome Sequencing Consortium and a large group of privately supported scientists (Celera Genomics and collaborators) led to publication of draft sequences in 2001!

Whether determination of the draft sequence was "the greatest scientific discovery of our time," as claimed by Davies's book (in Selected References), is debatable. Nevertheless, it was very exciting and yielded many surprises. For example, the human genome has fewer genes than thought previously, with 21,724 genes currently known. Only 5% of the 28% of the genome that is actually transcribed into RNA encodes protein. More than half the DNA present is repeated sequences of several types, including 45% in parasitic DNA elements. Parasitic DNA (also called "selfish" and "junk" DNA) is DNA that seems to serve no cellular or organismal function except its own propagation, but it may have utility in ways not yet clear.

Animal species vary by several orders of magnitude in the total amount of DNA present in their nuclear genomes (from slightly less than 10^8 to 10^{11} base pairs in a haploid gamete nucleus). At the low end are sponges (p. 248), some single-celled forms (see Chapter 11) and some arthropods (see Chapter 19), although the latter two groups include a wide range of genome sizes exceeding 10^9 in single-celled forms and reaching 10^{10} base pairs in arthropods. Most vertebrates have genomes of approximately 10^9 base pairs, but salamanders (p. 548), caecilians (p. 548), and lungfishes (p. 529) have genome sizes exceeding 10^{10} base pairs, with some salamanders reaching 10^{11} base pairs. Large genomes should not be considered advantageous, however, because most of the difference in genome size is contributed by accumulation of large amounts of the "selfish" or "junk" DNA in the larger genomes rather than DNA sequences useful to cellular metabolism and organismal function. The metabolic demands of replicating large amounts of DNA and the physical demands of housing it within the cell nucleus produce selection against accumulation of too much parasitic DNA in the genome. Animal groups with the largest genomes are likely those most able to tolerate accumulation of large amounts of parasitic DNA in their nuclear genomes without harming cellular and organismal functions. Because the emphasis of our book is on organismal biology, our review of genetics concentrates on genes that have clear roles in cellular and organismal functions, although these genes are a small minority of the DNA sequences present in animal nuclear genomes. Some DNA sequences considered useless to the organism exhibit variation that is useful in studies of population genetics (p. 126) and evolutionary relationships among species.

genomic analysis into "structural genomics" (mapping and sequencing) and "functional genomics" (development of genome-wide or system-wide experimental approaches to understand gene function).

In the 1970s Allan Maxam and Walter Gilbert in the United States and Frederick Sanger in England reported practical techniques for identifying the sequence of bases in DNA. By 1984 and 1985 scientists proposed to sequence and to map the entire human genome, an effort called the Human Genome Project. It was a most ambitious undertaking: the genome was estimated at 50,000 to 100,000 genes and regulatory subunits encoded in a linear sequence of about 3 to 6 billion pairs of bases. Using techniques available in 1988, it would have taken until 2700 to sequence the genome completely, but biologists then expected

A thousand human diseases, such as cystic fibrosis and Huntington's chorea, result from defects in single genes. Almost 300 disease-associated genes are known. Information developed from knowledge of gene sequences can permit new diagnostic tests, treatments, possible preventive strategies, and advances in molecular understanding of genetic diseases. However, to realize such benefits it is not sufficient simply to know the sequence of amino acids encoded by a nucleotide sequence in a gene. The human genome is responsible for hundreds of thousands of different proteins **(proteome).** The polypeptide encoded by a gene may be cleaved into separate functional parts or associated with polypeptides coded by other genes to produce diverse protein functions. Many scientists are now engaged in the difficult field of **proteomics:** to identify all the proteins in a cell, tissue, or organism; to determine how the proteins interact to accomplish their functions; and to outline the folding structures of the proteins.

GENETIC SOURCES OF PHENOTYPIC VARIATION

The creative force of evolution is natural selection acting on biological variation. Without variability among individuals, there could be no continued adaptation to a changing environment and no evolution (see Chapter 6). Although natural selection acts on varying organismal phenotypes, phenotypic variation within a population in a particular environment is often caused by variation in genotype. Preservation of favored phenotypes by natural selection therefore increases the abundance in a population of alleles associated with favored phenotypes, leading to adaptive evolution of the population. Through this process, a population evolves organismal phenotypes molded for effective use of environmental resources; such phenotypes are termed adaptations.

There are several genetic sources of phenotypic variation, all of which involve mutation at individual genes and the combining of the resulting alleles at variable genes into gametes and zygotes. Independent assortment of chromosomes during meiosis is a random process that creates new chromosomal combinations in gametes. In addition, chromosomal crossing over during meiosis allows recombination of linked genes between homologous chromosomes, further increasing variability. Random fusion of gametes from both parents also produces variation.

Thus sexual reproduction multiplies variation and provides the diversity and plasticity necessary for a species to survive environmental change. Sexual reproduction with its sequence of gene segregation and recombination across generations is what geneticist T. Dobzhansky called the "master adaptation" that makes all other evolutionary adaptations more accessible.

Although sexual reproduction reshuffles and amplifies whatever genetic diversity exists in a population, *new* genetic variation happens through gene mutations, chromosomal aberrations, and possibly by participation of parasitic DNA.

Gene Mutations

Gene mutations are chemicophysical changes that alter the sequence of bases in DNA. These mutations are studied directly by determining the DNA sequence and indirectly through their effects on organismal phenotype, if such effects occur. Some mutations produce a codon substitution as in the

human condition called **sickle cell anemia.** Homozygotes for the sickle cell allele often die before the age of 30 because the ability of their red blood cells to carry oxygen is greatly impaired by substitution of only a single amino acid in their hemoglobin. Other mutations involve deletion of one or more bases or insertion of additional bases into a DNA chain. Translation of mRNA is thus shifted, producing codons that specify incorrect amino acids and usually a nonfunctional or dysfunctional protein product.

Once a gene is mutated, it faithfully reproduces its new form. Many mutations are harmful; many are neither helpful nor harmful, and sometimes mutations are advantageous. Helpful mutations are of great significance to evolution because they furnish new possibilities with which natural selection can build adaptations. Natural selection determines which new alleles merit survival; the environment imposes a screening process that accumulates beneficial and eliminates harmful alleles.

When an allele of a gene is mutated to a new allele, the new form tends to be recessive, and its effects are normally masked by its partner allele. Only in the homozygous condition can such mutant alleles influence phenotype. Thus a population carries a reservoir of mutant recessive alleles, some of which are homozygous lethals but which are rarely present in the homozygous condition. Inbreeding encourages formation of homozygotes and increases the probability of recessive mutants being expressed in the phenotype.

Most mutations are destined for a brief existence. There are cases, however, in which mutations harmful or neutral under one set of environmental conditions become helpful under a different set. The earth's changing environment has provided numerous opportunities for favoring new gene mutations, as evidenced by the great diversity of animal life.

Frequency of Mutations

Although mutation occurs randomly with respect to an organism's needs, different mutation rates prevail at different loci. Some *kinds* of mutations are more likely to occur than others, and individual genes differ considerably in length. A long gene (more base pairs) is more likely to have a mutation than a short gene. Nevertheless, it is possible to estimate average spontaneous rates of mutation for different organisms and traits.

Genes are extremely stable. In the well-studied fruit fly, *Drosophila melanogaster,* there is approximately one detectable mutation per 10,000 loci (rate of 0.01% per locus per generation). The rate for humans is one per 10,000 to one per 100,000 loci per generation. If we accept the latter, more conservative figure, then a single normal allele is expected to undergo 100,000 generations before it is mutated. However, since human chromosomes contain approximately 21,724 loci, about every third person carries a new mutation. Similarly, each ovum or spermatozoon contains, on average, one mutant allele.

Since most mutations are deleterious, these statistics are anything but cheerful. Fortunately, most mutant genes are recessive and are not expressed in heterozygotes. Only a few by chance will increase enough in frequency for homozygotes to be produced.

MOLECULAR GENETICS OF CANCER

The crucial defect in cancer cells is that they proliferate in an unrestrained manner **(neoplastic growth).** The mechanism that controls the rate of division of normal cells has somehow been lost, and cancer cells multiply much more rapidly, invading other tissues in the body. Cancer cells originate from normal cells that lose their constraint on division and become dedifferentiated (less specialized) to some degree. Thus there are many kinds of cancer, depending on the original founder cells of the tumor. The change in many cancerous cells, perhaps all, has a genetic basis, and investigation of the genetic damage that causes cancer is now a major thrust of cancer research.

Oncogenes and Tumor-Suppressor Genes

We now recognize that specific genetic changes occurring in a particular clone of cells produce cancer. These genetic changes include alterations in numerous genes of two types, **oncogenes** and **tumor-suppressor genes.**

Oncogenes (Gr. *onkos,* bulk, mass, + *genos,* descent) occur normally in cells, and in their normal form they are called **proto-oncogenes.** One of these encodes a protein called **Ras.** Ras protein is a guanosine triphosphatase (GTPase) located just beneath the cell membrane. When a receptor on the cell surface binds a growth factor, Ras is activated and initiates a cascade of reactions causing cell division. The oncogene form encodes a protein that initiates the cell-division cascade even when the growth factor is absent from the surface receptor.

Of the many ways that cellular DNA can sustain damage, the three most important are ionizing radiation, ultraviolet radiation, and chemical mutagens. The high energy of ionizing radiation (X rays and gamma rays) causes electrons to be ejected from the atoms it encounters, producing ionized atoms with unpaired electrons (free radicals). The free radicals (principally from water) are highly reactive chemically, and they react with molecules in the cell, including DNA. Some damaged DNA is repaired, but if the repair is inaccurate, a mutation results. Ultraviolet radiation is of much lower energy than ionizing radiation and does not produce free radicals; it is absorbed by pyrimidines in DNA and causes formation of a double covalent bond between the adjacent pyrimidines. UV repair mechanisms can also be inaccurate. Chemical mutagens react with the DNA bases and cause mispairing during replication.

Gene products of tumor-suppressor genes act as a constraint on cell proliferation. One such product is called **p53** (for "53-kilodalton protein," a reference to its molecular weight). Mutations in the gene encoding p53 occur in about half of the 6.5 million cases of human cancer diagnosed each year. Normal p53 has several crucial functions, depending on the circumstances of the cell. It can trigger apoptosis (p. 55), act as a transcription activator or repressor (turning genes on or off), control progression from G_1 to S phase in the cell cycle, and promote repair of damaged DNA. Many of the mutations known in p53 interfere with its binding to DNA and thus its function.

SUMMARY

In sexual animals genetic material is distributed to offspring via gametes (ova and sperm), produced by meiosis. Each somatic cell in an organism has two chromosomes of each kind (homologous chromosomes) and is thus diploid.

Meiosis separates homologous chromosomes, so that each gamete has half the somatic chromosome number (haploid). In the first meiotic division, centromeres do not divide, and each daughter cell receives one of each pair of replicated homologous chromosomes with sister chromatids still attached to the centromere. At the beginning of the first meiotic division, replicated homologous chromosomes come to lie alongside each other (synapsis), forming a bivalent. The gene loci on one set of chromatids lie opposite the corresponding loci on the homologous chromatids. Portions of adjacent chromatids can exchange with the nonsister chromatids (crossing over) to produce new genetic combinations. At the second meiotic division, the centromeres divide, completing the reduction in chromosome number and amount of DNA. The diploid number is restored when male and female gametes fuse to form a zygote.

Gender is determined in many animals by the sex chromosomes; in humans, fruit flies, and many other animals, females have two X chromosomes, and males have an X and a Y.

Genes are the unit entities that influence all characteristics of an organism and are inherited by offspring from their parents. Allelic variants of genes might be dominant, recessive, or intermediate; a recessive allele in the heterozygous genotype will not be expressed in the phenotype but requires the homozygous condition for overt expression. In a monohybrid cross involving a dominant allele and a recessive alternative allele (both parents homozygous), the F_1 generation will be all heterozygous, whereas F_2 genotypes will occur in a 1:2:1 ratio, and phenotypes in a 3:1 ratio. This result demonstrates Mendel's law of segregation. Heterozygotes in intermediate inheritance show phenotypes distinct from homozygous phenotypes, sometimes intermediate forms, with corresponding alterations in phenotypic ratios.

Dihybrid crosses (in which genes for two different characteristics are carried on separate pairs of homologous chromosomes) demonstrate Mendel's law of independent assortment, and phenotypic ratios are 9:3:3:1 with dominant and recessive characters. Expected ratios in crosses of two or more characters are calculated from laws of probability.

Genes can have more than two alleles in a population, and different combinations of alleles can produce different phenotypic effects. Alleles of different genes can interact in producing a phenotype, as in polygenic inheritance and epistasis, in which one gene affects the expression of another gene.

A gene on the X chromosome shows sex-linked inheritance and produces an effect in males, even if a recessive allele is present, because the Y chromosome does not carry a corresponding allele.

All genes on a given autosomal chromosome are linked, and their variants do not assort independently unless they are very far apart on the chromosome, in which case crossing over occurs between them in nearly every meiosis. Crossing over increases the amount of genetic recombination in a population.

Occasionally, a pair of homologous chromosomes fails to separate in meiosis causing the gametes to get one chromosome too many or too few. Resulting zygotes usually do not survive; humans with $2n + 1$ chromosomes sometimes live, but they have serious abnormalities, such as Down syndrome.

Nucleic acids in the cell are DNA and RNA, which are large polymers of nucleotides composed of a nitrogenous base, pentose sugar, and phosphate group. The nitrogenous bases in DNA are adenine (A), guanine (G), thymine (T), and cytosine (C), and those in RNA are the same except that uracil (U) is substituted for thymine. DNA is a double-stranded, helical molecule in which the bases extend toward each other from the sugar-phosphate backbone: A always pairs with T and G with C. The strands are antiparallel and complementary, being held in place by hydrogen bonds between the paired bases. In DNA replication the strands part, and the enzyme DNA polymerase synthesizes a new strand along each parental strand, using the parental strand as a template.

A gene can encode either a ribosomal RNA (rRNA), a transfer RNA (tRNA), or a messenger RNA (mRNA); a gene of the latter kind specifies the sequence of amino acids in a polypeptide (one gene–one polypeptide hypothesis). In mRNA, each triplet of three bases specifies a particular amino acid.

Proteins are synthesized by transcription of DNA into the base sequence of a molecule of messenger RNA (mRNA), which functions in concert with ribosomes (containing ribosomal RNA [rRNA] and protein) and transfer RNA (tRNA). Ribosomes attach to the strand of mRNA and move along it, assembling the amino acid sequence of the protein. Each amino acid is brought into position for assembly by a molecule of tRNA, which itself bears a base sequence (anticodon) complementary to the respective codons of the mRNA. In eukaryotic nuclear DNA the sequences of bases in DNA coding for amino acids in a protein (exons) are interrupted by intervening sequences (introns). The introns are removed from the primary mRNA before it leaves the nucleus, and the protein is synthesized in the cytoplasm.

Genes, and the synthesis of the products for which they are responsible, must be regulated: turned on or off in response to varying environmental conditions or cell differentiation. Gene regulation in eukaryotes occurs at several levels, with control of transcription being a particularly important point of regulation.

Molecular genetic methods have made spectacular advances possible. Restriction endonucleases cleave DNA at specific base sequences, and DNA from different sources can be rejoined to form recombinant DNA. Combining mammalian with plasmid or viral DNA, a mammalian gene can be introduced into bacterial cells, which then multiply and produce many copies of the mammalian gene. The polymerase chain reaction (PCR) is used to clone specific genes if the base sequence of short pieces of DNA surrounding the gene are known. Draft sequences of the human genome were published in 2001. Among many exciting results was a revision of the number of genes to 21,724, down from previous estimates of 100,000. These genes are responsible for hundreds of thousands of proteins in a typical cell.

A mutation is a physicochemical alteration in the bases of DNA that may change the phenotypic effect of a gene. Although rare and usually detrimental to survival and reproduction of an organism, mutations are occasionally beneficial and may be accumulated in populations by natural selection.

Cancer (neoplastic growth) results from genetic changes in a clone of cells allowing unrestrained proliferation of those cells. Oncogenes (such as the gene coding for Ras protein) and inactivation of tumor-suppressor genes (such as that coding for p53 protein) are implicated in many cancers.

REVIEW QUESTIONS

1. What is the relationship between homologous chromosomes, copies of a gene, and alleles?
2. Describe or diagram the sequence of events in meiosis (both divisions).
3. What are the designations of the sex chromosomes in males of bugs, humans, and butterflies?
4. How do the chromosomal mechanisms determining sex differ in the three taxa in question 3?
5. Diagram by Punnett square a cross between individuals with the following genotypes: $A/a \times A/a$; $A/a\ B/b \times A/a\ B/b$.
6. Concisely state Mendel's law of segregation and his law of independent assortment.
7. Assuming brown eyes (B) are dominant over blue eyes (b), determine the genotypes of all the following individuals. The blue-eyed son of two brown-eyed parents marries a brown-eyed woman whose mother was brown eyed and whose father was blue eyed. Their child is blue eyed.
8. Recall that red color (R) in four-o'clock flowers is incompletely dominant over white (R'). In the following crosses, give the genotypes of the gametes produced by each parent and the flower color of the offspring: $R/R' \times R/R'$; $R'/R' \times R/R'$; $R/R \times R/R'$; $R/R \times R'/R'$.

9. A brown male mouse is mated with two female black mice. In several litters of young, the first female has had 48 black and the second female has had 14 black and 11 brown young. Can you deduce the pattern of inheritance of coat color and the genotypes of the parents?
10. Rough coat (R) is dominant over smooth coat (r) in guinea pigs, and black coat (B) is dominant over white (b). If a homozygous rough black is mated with a homozygous smooth white, give the appearance of each of the following: F_1; F_2; offspring of F_1 mated with smooth, white parent; offspring of F_1 mated with rough, black parent.
11. Assume right-handedness (R) is genetically dominant over left-handedness (r) in humans, and that brown eyes (B) are genetically dominant over blue (b). A right-handed, blue-eyed man marries a right-handed, brown-eyed woman. Their two children are (1) right handed, blue eyed and (2) left handed, brown eyed. The man marries again, and this time the woman is right handed and brown eyed. They have 10 children, all right handed and brown eyed. What are the probable genotypes of the man and his two wives?
12. In *Drosophila melanogaster,* red eyes are dominant to white and the variation for this characteristic is on the X

chromosome. Vestigial wings (v) are recessive to normal (V) for an autosomal gene. What will be the appearance of offspring of the following crosses: $X^w/X^w\ V/v \times X^w/Y\ v/v$, $X^w/X^w\ V/v \times X^w/Y\ V/v$.

13. Assume that color blindness is a recessive character on the X chromosome. A man and woman with normal vision have the following offspring: daughter with normal vision who has one color-blind son and one normal son; daughter with normal vision who has six normal sons; and a color-blind son who has a daughter with normal vision. What are the probable genotypes of all individuals?

14. Distinguish the following: euploidy, aneuploidy, and polyploidy; monosomy and trisomy.

15. Name the purines and pyrimidines in DNA and tell which pairs occur in the double helix. What are the purines and pyrimidines in RNA and to what are they complementary in DNA?

16. Explain how DNA is replicated.

17. Why could a codon not consist of only two bases?

18. Explain the transcription and processing of mRNA in the nucleus.

19. Explain the role of mRNA, tRNA, and rRNA in polypeptide synthesis.

20. What are four ways that genes can be regulated in eukaryotes?

21. In modern molecular genetics, what is recombinant DNA, and how is it prepared?

22. Name three sources of genetic recombination that contribute to phenotypic variation.

23. Distinguish between proto-oncogene and oncogene. Describe two mechanisms by which genetic change causes cancer?

24. What are Ras protein and p53? How can mutations in the genes for these proteins contribute to cancer?

25. Outline the essential steps in the polymerase chain reaction.

26. Draft sequences of the human genome have been published. What are some interesting observations based on the results? What are some potential benefits? What is the proteome?

SELECTED REFERENCES

Burt, A., and R. Trivers. 2006. Genes in conflict: the biology of selfish genetic elements. Cambridge, Massachusetts, Belknap Press of Harvard Univ. Press. *A thorough discussion of the means by which some genes acquire the ability to promote their own transmission or expression at the expense of others.*

Conery, J. S., and M. Lynch. 2000. The evolutionary fate and consequences of duplicate genes. Science **290:**1151–1155. *Gene duplications are an important source of genetic variation.*

Davies, K. 2001. Cracking the genome: inside the race to unlock human DNA. Craig Venter, Francis Collins, James Watson, and the story of the greatest scientific discovery of our time. New York, The Free Press. *Fascinating story of competition between the Human Genome Sequencing Project and Craig Venter's Celera Genomics. Of course, the genome is not "cracked" until its meaning is deciphered, and we are a long way from that.*

Dhand, R. 2000. Functional genomics. Nature **405:**819. *Introduction to a number of articles on functional genomics.*

Ezzell, C. 2002. Proteins rule. Sci. Am. **286:**40–47 (April). *An excellent explanation of the current status and problems of proteomics.*

Futreal, P. A., A. Kasprzyk, E. Birney, J. C. Mullikin, R. Wooster, and M. R. Stratton. 2001. Cancer and genomics. Nature **409:**850–852. *Good list of cancer genes.*

Griffiths, A. J., S. R. Wessler, R. C. Lewontin, W. M. Gelbart, D. T. Suzuki and J. H. Miller. 2005. Introduction to genetic analysis, ed. 8. New York, W. H. Freeman & Company. *A good and recent general genetics text.*

Hartwell, L. H., L. Hood, M. L. Goldberg, A. E. Reynolds, L. M. Silver, and R. C. Veres. 2004. Genetics: from genes to genomes, ed. 2. Boston, Massachusetts, McGraw-Hill Higher Education. *A good and recent text of genetics and genomics.*

Jimenez-Sanchez, G., B. Childs, and D. Valle. 2001. Human disease genes. Nature **409:**853–855. *They found "striking correlation between the function of the gene product and features of disease. . . ."*

Lewin, B. 2004. Genes VIII. Upper Saddle River, New Jersey, Pearson/Prentice Hall. *Thorough, up-to-date coverage on molecular biology of genes.*

Mange, E. J., and A. P. Mange. 1999. Basic human genetics, ed. 2. Sunderland, Massachusetts, Sinauer Associates. *A readable, introductory text concentrating on the genetics of the animal species of greatest concern to most of us.*

Mullis, K. B. 1990. The unusual origin of the polymerase chain reaction. Sci. Am. **262:**56–65 (April). *How the author had the idea for the simple production of unlimited copies of DNA while driving through the mountains of California.*

Pennisi, E. 2000. Genomics comes of age. Science **290:**2220–2221. *Determination of genomes of a number of organisms is "Breakthrough of the Year."*

Roberts, L. 2001. A history of the Human Genome Project. Science **291:**chart. *Each major discovery from double helix to full genome (1953–2001). Has glossary.*

The International Human Genome Mapping Consortium. 2001. A physical map of the human genome. Nature **409:**934–941. *The draft sequence published by the publicly supported consortium.*

Venter, J. C., M. D. Adams, E. W. Myers, P. W. Li, R. J. Mural, and 265 others. 2001. The sequence of the human genome. Science **291:**1304–1351. *The draft sequence published by Celera Genomics and its collaborators.*

Watson, J. D., and A. Berry. 2003. DNA: the secret of life. New York, Alfred A. Knopf. *An exciting account of the history and applications of genetics.*

ONLINE LEARNING CENTER

Visit www.mhhe.com/hickmanipz14e for chapter quizzing, key term flash cards, web links, and more!

Organic Evolution

A trilobite fossilized in Paleozoic rock.

A Legacy of Change

Life's history is a legacy of perpetual change. Despite the apparent permanence of the natural world, change characterizes all things on earth and in the universe. Earth's rock strata record the irreversible, historical change that we call organic evolution. Countless kinds of animals and plants have flourished and disappeared, leaving behind a sparse fossil record of their existence. Many, but not all, have left living descendants that bear some resemblance to them.

Life's changes are observed and measured in many ways. On a short evolutionary timescale, we see changes in the frequencies of different genetic traits within populations. Evolutionary changes in the relative frequencies of light- and dark-colored moths were observed within a single human lifetime in the polluted towns of industrial England. The formation of new species and dramatic changes in organismal form, as illustrated by the evolutionary diversification of Hawaiian birds, requires longer timescales covering 100,000 to 1 million years. Major evolutionary trends and episodic mass extinctions occur on even larger timescales, covering tens of millions of years. The fossil record of horses through the past 50 million years shows a series of different species replacing older ones through time and ending with the horses alive today. The fossil record of marine invertebrates shows us a series of mass extinctions separated by intervals of approximately 26 million years.

Because every feature of life as we know it today is a product of evolution, biologists consider organic evolution the keystone of all biological knowledge.

In Chapter 1, we introduced Darwinian evolutionary theory as the dominant paradigm of biology. Charles Robert Darwin and Alfred Russel Wallace (Figure 6.1) first established evolution as a powerful scientific theory. Today the reality of organic evolution can be denied only by abandoning reason. As the noted English biologist Sir Julian Huxley wrote, "Charles Darwin effected the greatest of all revolutions in human thought, greater than Einstein's or Freud's or even Newton's, by simultaneously establishing the fact and discovering the mechanism of organic evolution." Darwinian theory helps us to understand both the genetics of populations and long-term trends in the fossil record. Darwin and Wallace did not originate the basic idea of organic evolution, which has an ancient history. We review the history of evolutionary thinking as it led to Darwin's theory, evidence supporting it, and changes to the theory that have produced our modern synthetic theory of evolution.

ORIGINS OF DARWINIAN EVOLUTIONARY THEORY

Pre-Darwinian Evolutionary Ideas

Before the eighteenth century, speculation on origins of species rested on mythology and superstition, not on anything resembling a testable scientific theory. Creation myths often described the world remaining constant after a short period of creation. Nevertheless, some people approached the idea that nature has a long history of perpetual and irreversible change.

Early Greek philosophers, notably Xenophanes, Empedocles, and Aristotle, developed an early idea of evolutionary change. They recognized fossils as evidence for former life that they believed had been destroyed by natural catastrophe. Despite their intellectual inquiry, the Greeks failed to establish an evolutionary concept, and the issue declined well before the rise of Christianity. The opportunity for evolutionary thinking became even more restricted as a biblical account of the earth's creation became accepted as a tenet of faith. The year 4004 B.C. was fixed by Archbishop James Ussher (mid-seventeenth century) as the date of life's creation. Evolutionary views were considered rebellious and heretical, but they refused to die. The French naturalist Georges Louis Buffon (1707 to 1788) stressed the influence of environment on the modifications of animal form. He also extended the age of the earth to 70,000 years.

Lamarckism: The First Scientific Explanation of Evolution

French biologist Jean Baptiste de Lamarck (1744 to 1829; Figure 6.2) authored the first complete explanation of evolution in 1809, the year of Darwin's birth. He made a convincing case that fossils were remains of extinct animals. Lamarck's proposed evolutionary mechanism, **inheritance of acquired characteristics,** was engagingly simple: organisms, by striving to meet the demands of their environments, acquire adaptations and pass them by heredity to their offspring. According to Lamarck, the giraffe evolved its long neck because its ancestors lengthened their necks by stretching to obtain food and then passed the lengthened neck to their offspring. Over many generations, these changes accumulated to produce the long necks of modern giraffes.

We call Lamarck's concept of evolution **transformational,** because it claims that as individual organisms transform their characteristics through the use and disuse of parts, heredity makes corresponding adjustments to produce evolution. We now reject transformational theories because genetic studies show that traits acquired by an organism during its lifetime, such as strengthened muscles, are not inherited by offspring. Darwin's evolutionary theory differs from Lamarck's in being a **variational** theory, based on the distribution of genetic variation in populations. Evolutionary change is caused by differential survival and reproduction among organisms that differ in hereditary traits, not by inheritance of acquired characteristics.

Charles Lyell and Uniformitarianism

The geologist Sir Charles Lyell (1797 to 1875; Figure 6.3) established in his *Principles of Geology* (1830 to 1833) the principle of uniformitarianism. Uniformitarianism encompasses two important principles that guide scientific study of the history of nature: (1) that the laws of physics and chemistry have not changed throughout the history of the earth, and (2) that past geological events occurred by natural processes similar to those observed today. Lyell showed that natural forces, acting over long periods

A   B

Figure 6.1

Founders of the theory of evolution by natural selection. **A,** Charles Robert Darwin (1809 to 1882), as he appeared in 1881, the year before his death. **B,** Alfred Russel Wallace (1823 to 1913) in 1895. Darwin and Wallace independently developed the same theory. A letter and essay from Wallace written to Darwin in 1858 spurred Darwin into writing *On The Origin of Species,* published in 1859.

Figure 6.2

Jean Baptiste de Lamarck (1744 to 1829), French naturalist who offered the first scientific explanation of evolution. Lamarck's hypothesis that evolution proceeds by inheritance of acquired characteristics has been rejected and replaced by neo-Darwinian theories.

Figure 6.3

Sir Charles Lyell (1797 to 1875), English geologist and friend of Darwin. His book *Principles of Geology* greatly influenced Darwin during Darwin's formative period. This photograph was made about 1856.

of time, could explain the formation of fossil-bearing rocks. Lyell's geological studies led him to conclude that the earth's age must be measured in millions of years. These principles were important for discrediting miraculous and supernatural explanations of the history of nature and replacing them with scientific explanations. Lyell also stressed the gradual nature of geological changes that occur through time, and he argued further that such changes have no inherent tendency to occur in any particular direction. Both of these claims left important marks on Darwin's evolutionary theory.

Darwin's Great Voyage of Discovery

"After having been twice driven back by heavy southwestern gales, Her Majesty's ship *Beagle,* a ten-gun brig, under the command of Captain Robert FitzRoy, R.N., sailed from Devonport on the 27th of December, 1831." Thus began Charles Darwin's account of the historic five-year voyage of the *Beagle* around the world (Figure 6.4). Darwin, not quite 23 years old, had been asked to accompany Captain FitzRoy on the *Beagle,* a small vessel only 90 feet in length, which was about to depart on an

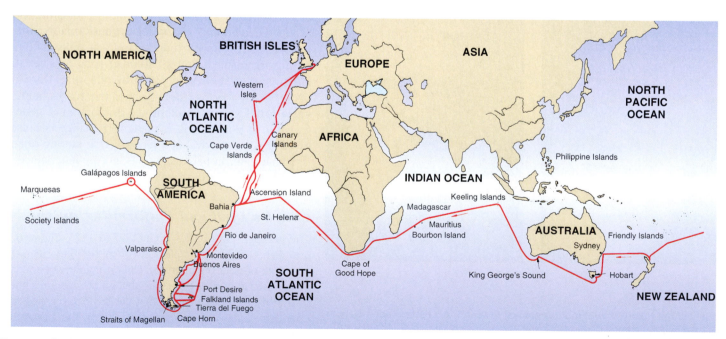

Figure 6.4

Five-year voyage of H.M.S. *Beagle.*

A

B

Figure 6.5

Charles Darwin and H.M.S. *Beagle*. **A,** Darwin in 1840, four years after the *Beagle* returned to England, and a year after his marriage to his cousin, Emma Wedgwood. **B,** The H.M.S. *Beagle* sails in Beagle Channel, Tierra del Fuego, on the southern tip of South America in 1833. The watercolor was painted by Conrad Martens, one of two official artists on the voyage of the *Beagle*.

extensive surveying voyage to South America and the Pacific (Figure 6.5). It was the beginning of the most important scientific voyage of the nineteenth century.

During the voyage (1831 to 1836), Darwin endured sea-sickness and the erratic companionship of Captain FitzRoy, but Darwin's youthful physical strength and early training as a naturalist equipped him for his work. The *Beagle* made many stops along the coasts of South America and adjacent islands. Darwin made extensive collections and observations on the fauna and flora of these regions. He unearthed numerous fossils of animals long extinct and noted the resemblance between fossils of the South American pampas and the known fossils of North America. In the Andes he encountered seashells embedded in rocks at 13,000 feet. He experienced a severe earthquake and watched mountain torrents that relentlessly wore away the earth. These observations, and his reading of Lyell's *Principles of Geology* during the voyage, strengthened his conviction that natural forces could explain the geological features of the earth.

In mid-September of 1835, the *Beagle* arrived at the Galápagos Islands, a volcanic archipelago straddling the equator 600 miles west of Ecuador (Figure 6.6). The fame of the islands stems from their oceanic isolation and rugged volcanic terrain. Circled by capricious currents, surrounded by shores of twisted lava bearing skeletal brushwood baked by the equatorial sun, inhabited by strange reptiles and by convicts stranded by the Ecuadorian government, the islands had few admirers among mariners. By the middle of the seventeenth century, the islands were known to Spaniards as "Las Islas Galápagos"—the tortoise islands. The giant tortoises, used for food first by buccaneers and later by American and British whalers, sealers, and ships of war, were the islands' principal attraction. At the time of Darwin's visit, the tortoises already were heavily exploited.

During the *Beagle's* five-week visit to the Galápagos, Darwin documented the unique character of the Galápagos plants and animals, including the giant tortoises, marine iguanas, mockingbirds, and ground finches. Darwin later described these studies as the "origin of all my views."

Darwin was struck by the fact that, although the Galápagos Islands and the Cape Verde Islands (visited earlier in this voyage of the *Beagle*) were similar in climate and topography, Galápagos plants and animals were related to those of the South American mainland and were entirely different from the African-derived forms of the Cape Verde Islands. Each Galápagos Island often contained a unique species related to forms on other Galápagos Islands. In short, Galápagos life must have originated in continental South America and then undergone modification in the various environmental conditions of the different islands. He

Figure 6.6

The Galápagos Islands viewed from the rim of a volcano.

Figure 6.7

Darwin's study at Down House in Kent, England, is preserved today much as it was when Darwin wrote *On The Origin of Species.*

concluded that living forms were neither divinely created nor immutable; they were, in fact, products of evolution.

On October 2, 1836, the *Beagle* returned to England, where Darwin conducted most of his scientific work (Figure 6.7). Most of Darwin's extensive collections had preceded him there, as had notebooks and diaries kept during the cruise. Darwin's journal, published three years after the *Beagle's* return to England, was an instant success and required two additional printings within the first year. Darwin later revised his journal as *The Voyage of the Beagle*, one of the most lasting and popular travel books.

The main product of Darwin's voyage, his theory of evolution, would continue to develop for more than 20 years after the *Beagle's* return. In 1838, he "happened to read for amusement" an essay on populations by T. R. Malthus (1766 to 1834), who stated that animal and plant populations, including human populations, tend to increase beyond the capacity of the environment to support them. Darwin already had been gathering information on artificial selection of animals under domestication. Darwin was especially fascinated by artificial breeds of pigeons. Many pigeon breeds differed so much in appearance and behavior that they would be considered different species if found in nature, yet all had clearly been derived from a single wild species, the rock pigeon (*Columba livia*). After reading Malthus's article, Darwin realized that a process of selection in nature, a "struggle for existence" because of overpopulation, could be a powerful force for evolution of wild species.

Darwin allowed the idea to develop in his own mind until it was presented in 1844 in a still-unpublished essay. Finally in 1856, he began to assemble his voluminous data into a work on the origin of species. He expected to write four volumes, a very big book, "as perfect as I can make it." However, his plans took an unexpected turn.

In 1858, he received a manuscript from Alfred Russel Wallace (1823 to 1913), an English naturalist in Malaya with whom he corresponded. Darwin was stunned to find that in a few pages, Wallace summarized the main points of the natural selection theory on which Darwin had been working for two decades. Rather than to withhold his own work in favor of Wallace as he was inclined to do, Darwin was persuaded by two close friends, the geologist Lyell and the botanist Hooker, to publish his views in a brief statement that would appear together with Wallace's paper in the *Journal of the Linnean Society*. Portions of both papers were read to an unimpressed audience on July 1, 1858.

For the next year, Darwin worked urgently to prepare an "abstract" of the planned four-volume work. This book was published in November 1859, with the title *On the Origin of Species by Means of Natural Selection, or the Preservation of Favoured Races in the Struggle for Life*. The 1250 copies of the first printing sold the first day! The book instantly generated a storm that has never abated. Darwin's views were to have extraordinary consequences on scientific and religious beliefs and remain among the greatest intellectual achievements of all time.

"Whenever I have found that I have blundered, or that my work has been imperfect, and when I have been contemptuously criticized, and even when I have been overpraised, so that I have felt mortified, it has been my greatest comfort to say hundreds of times to myself that 'I have worked as hard and as well as I could, and no man can do more than this.'" *Charles Darwin, in his autobiography, 1876.*

Once Darwin's caution had been swept away by the publication of *On the Origin of Species,* he entered an incredibly productive period of evolutionary thinking for the next 23 years, producing book after book. He died on April 19, 1882, and was buried in Westminster Abbey. The little *Beagle* had already disappeared, having been retired in 1870 and sold for scrap.

DARWINIAN EVOLUTIONARY THEORY: THE EVIDENCE

Perpetual Change

The main premise underlying Darwinian evolution is that the living world is neither constant nor perpetually cycling, but always changing. Perpetual change in the form and diversity of animal life throughout its 600- to 700-million-year history is seen most directly in the fossil record. A **fossil** is a remnant of past life uncovered from the crust of the earth (Figure 6.8). Some fossils constitute complete remains (insects in amber and mammoths), actual hard parts (teeth and bones), and petrified skeletal parts infiltrated with silica or other minerals (ostracoderms and molluscs). Other fossils include molds, casts, impressions, and fossil excrement (coprolites). In addition to documenting organismal evolution, fossils reveal profound changes in the earth's environment, including major changes in the distributions of lands and seas. Because many organisms left no fossils, a complete record of the past is always beyond our reach; nonetheless, discovery of new fossils and reinterpretation of familiar ones expand our knowledge of how the form and diversity of animals changed through geological time.

Fossil remains may on rare occasions include soft tissues preserved so well that recognizable cellular organelles are revealed by electron microscopy! Insects are frequently found entombed in amber, the fossilized resin of trees. One study of a fly entombed in 40-million-year-old amber revealed structures corresponding to muscle fibers, nuclei, ribosomes, lipid droplets, endoplasmic reticulum, and mitochondria (Figure 6.8D). This extreme case of mummification probably occurred because chemicals in the plant sap diffused into the embalmed insect's tissues.

Interpreting the Fossil Record

The fossil record is biased because preservation is selective. Vertebrate skeletal parts and invertebrates with shells and other hard structures left the best records (Figure 6.8). Soft-bodied animals, including jellyfishes and most worms, are fossilized only under very unusual circumstances such as those that formed the Burgess Shale of British Columbia (Figure 6.9). Exceptionally favorable conditions for fossilization produced the Precambrian fossil bed of South Australia, the tar pits of Rancho La Brea (Hancock Park, Los Angeles), the great dinosaur beds (Alberta, Canada, and Jensen, Utah; Figure 6.10), and the Yunnan and Lianoning provinces of China.

Fossils are deposited in stratified layers with new deposits forming on top of older ones. If left undisturbed, which is rare, a sequence is preserved with the ages of fossils being directly proportional to their depth in the stratified layers. Characteristic fossils often serve to identify particular layers. Certain widespread marine invertebrate fossils, including various foraminiferans (p. 242) and echinoderms (p. 472), are such good indicators of specific geological periods that they are called "index," or "guide," fossils. Unfortunately, the layers are usually tilted or show faults (cracks). Old deposits exposed by erosion may be covered with new deposits in a different plane. When exposed to tremendous pressures or heat, stratified sedimentary rock metamorphoses into crystalline quartzite, slate, or marble, which destroys fossils.

Stratigraphy for two major groups of African antelopes and its evolutionary interpretation are shown on Figure 6.11. Species in this group are identified by characteristic sizes and shapes of horns, which form much of the fossil record of this group. Solid vertical lines in Figure 6.11 denote the temporal distributions of species determined by presence of their characteristic horns in rock strata of various ages. Red lines denote the fossil records of living species, and gray lines denote the fossil records of extinct species. The dotted gray lines show the inferred relationships among living and fossil species based on their sharing of homologous structural features.

Geological Time

Long before the earth's age was known, geologists divided its history into a table of succeeding events based on the ordered layers of sedimentary rock. The "law of stratigraphy" produced a relative dating with the oldest layers at the bottom and the youngest at the top of the sequence. Time was divided into eons, eras, periods, and epochs as shown on the endpaper inside the back cover of this book. Time during the last eon (Phanerozoic) is expressed in eras (for example, Cenozoic), periods (for example, Tertiary), epochs (for example, Paleocene), and sometimes smaller divisions of an epoch.

In the late 1940s, radiometric dating methods were developed for determining the absolute age in years of rock formations. Several independent methods are now used, all based on

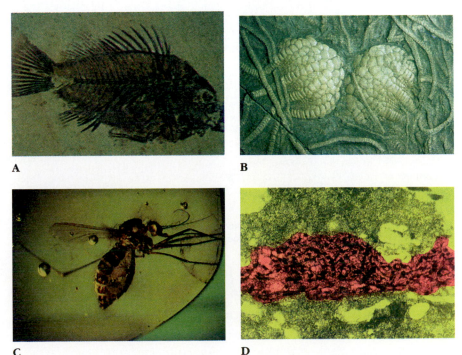

Figure 6.8

Four examples of fossil material. **A,** Fish fossil from rocks of the Green River Formation, Wyoming. Such fish swam here during the Eocene epoch of the Tertiary period, approximately 55 million years ago. **B,** Stalked crinoids (class Crinoidea, p. 487) from 85-million-year-old Cretaceous rocks. The fossil record of these echinoderms shows that they reached their peak millions of years earlier and began a slow decline to the present. **C,** An insect fossil that got stuck in the resin of a tree 40 million years ago and that has since hardened into amber. **D,** Electron micrograph of tissue from a fly fossilized as shown in **C;** the nucleus of a cell is marked in red.

Figure 6.9

A, Fossil trilobites visible at the Burgess Shale Quarry, British Columbia.
B, Animals of the Cambrian period, approximately 580 million years ago, as reconstructed from fossils preserved in the Burgess Shale of British Columbia, Canada. The main new body plans that appeared rather abruptly at this time established the body plans of animals familiar to us today.

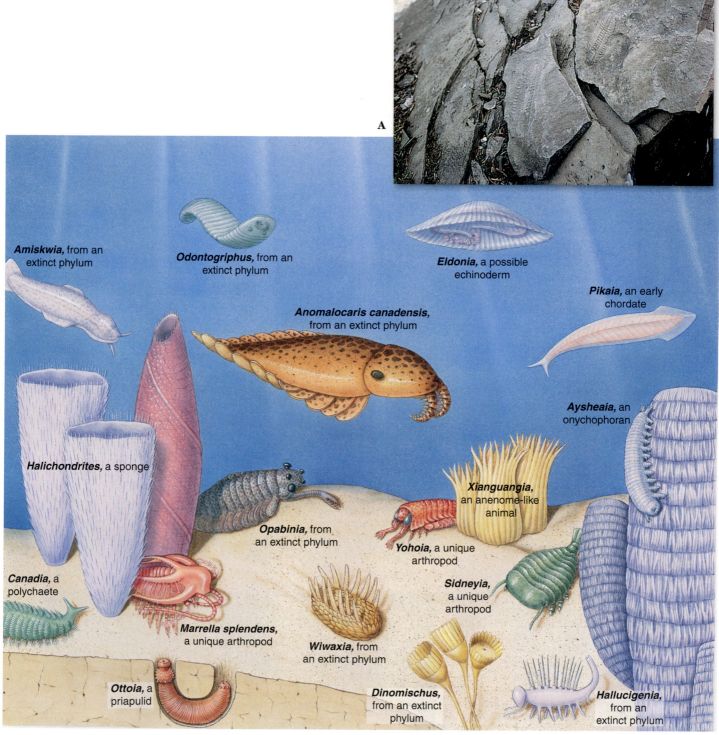

Figure 6.10
A dinosaur skeleton partially excavated from rock at Dinosaur Provincial Park, Alberta.

the radioactive decay of naturally occurring elements into other elements. These "radioactive clocks" are independent of pressure and temperature changes and therefore are not affected by often violent earth-building activities.

One method, potassium-argon dating, uses the decay of potassium-40 (^{40}K) to argon-40 (^{40}Ar) (12%) and calcium-40 (^{40}Ca) (88%). The half-life of potassium-40 is 1.3 billion years; half of the original atoms will decay in 1.3 billion years, and half of the remaining atoms will be gone at the end of the next 1.3 billion years. This decay continues until all radioactive potassium-40 atoms are gone. To measure the age of the rock, one calculates the ratio of remaining potassium-40 atoms to the amount of potassium-40 originally there (the remaining potassium-40 atoms plus the argon-40 and calcium-40 into which other potassium-40 atoms have decayed). Several such isotopes exist for dating purposes, some for dating the age of the earth itself. One of the most useful radioactive clocks depends on the decay of uranium into lead. With this method, rocks over 2 billion years old can be dated with a probable error of less than 1%.

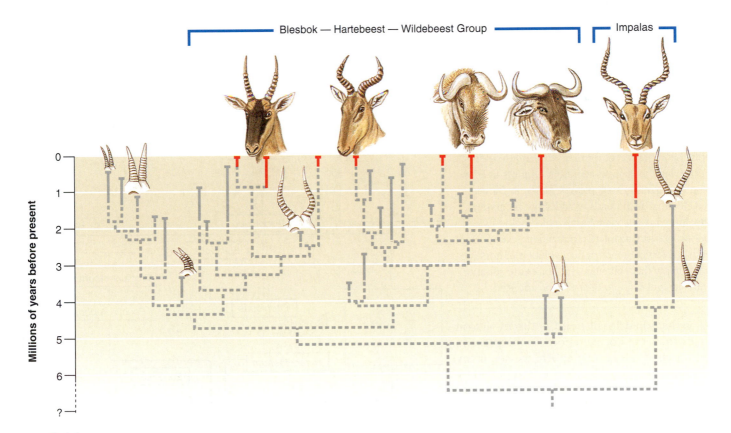

Figure 6.11
Stratigraphic record and inferred evolutionary relationships among alcelaphine (blesboks, hartebeests, wildebeests) and aepycerotine (impalas) antelopes in Africa. Species in this group are identified by characteristic sizes and shapes of horns found in rock strata of various ages. Solid vertical lines show the temporal distribution of species in rock strata whose ages are shown on the scale at the left side of the figure. Red lines show the temporal distributions of living species, and gray lines show the temporal distributions of extinct species in rock strata. Dotted gray lines show the inferred relationships among species based on their sharing of homologous structural features. The relative constancy of horn structure within species through geological time is consistent with the theory of punctuated equilibrium (p. 123). This fossil record shows that rates of speciation and extinction are higher for alcelaphine antelopes than for impalas.

The fossil record of macroscopic organisms begins near the start of the Cambrian period of the Paleozoic era, approximately 600 million years BP. Geological time before the Cambrian is called the Precambrian era or Proterozoic eon. Although the Precambrian era occupies 85% of all geological time, it has received much less attention than later eras, partly because oil, which provides the commercial incentive for much geological work, seldom exists in Precambrian formations. The Precambrian era contains well-preserved fossils of bacteria and algae, and casts of jellyfishes, sponge spicules, soft corals, segmented flatworms, and worm trails. Most, but not all, are microscopic fossils.

Evolutionary Trends

The fossil record allows us to view evolutionary change across the broadest scale of time. Species arise and go extinct repeatedly throughout the geological history recorded by the fossil record. Animal species typically survive approximately 1 to 10 million years, although their durations are highly variable. When we study patterns of species or taxon replacement through time, we observe trends. Trends are directional changes in the characteristic features or patterns of diversity in a group of organisms. Fossil trends clearly demonstrate Darwin's principle of perpetual change.

Our use of the phrase "evolutionary trend" does not imply that more recent forms are superior to older ones or that the changes represent progress in adaptation or organismal complexity. Although Darwin predicted that such trends would show progressive adaptation, many contemporary paleontologists consider progressive adaptation rare among evolutionary trends. Observed trends in the evolution of horses do not imply that contemporary horses are superior in any general sense to their Eocene ancestors.

A well-studied fossil trend is the evolution of horses from the Eocene epoch to the present. Looking back at the Eocene epoch, we see many different genera and species of horses replaced by others through time (Figure 6.12). George Gaylord Simpson (p. 208) showed that this trend is compatible with Darwinian evolutionary theory. The three characteristics that show the clearest trends in horse evolution are body size, foot structure, and tooth structure. Compared to modern horses, those of extinct genera were small; their teeth had a relatively small grinding surface, and their feet had a

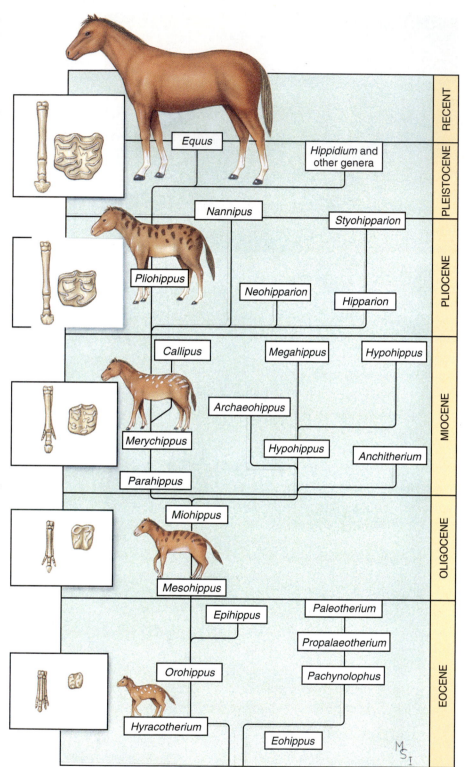

Figure 6.12

A reconstruction of genera of horses from Eocene to present. Evolutionary trends toward increased size, elaboration of molars, and loss of toes are shown together with a hypothetical phylogeny of extant and fossil genera.

relatively large number of toes (four). Throughout the subsequent Oligocene, Miocene, Pliocene, and Pleistocene epochs are continuing patterns of new genera arising and old ones going extinct. In each case, a net increase in body size, expansion of the grinding surface of the teeth, and reduction in the number of toes occurred. As the number of toes was reduced, the central digit became increasingly more prominent in the foot, and eventually only this central digit remained.

The fossil record shows a net change not only in the characteristics of horses but also variation in the numbers of different horse genera (and numbers of species) through time. The many horse genera of past epochs have been lost to extinction, leaving only a single survivor, *Equus*. Evolutionary trends in diversity are observed in fossils of many different groups of animals (Figure 6.13).

Trends in fossil diversity through time are produced by different rates of species formation versus extinction through time. Why do some lineages generate large numbers of new species whereas others generate relatively few? Why do different lineages undergo higher or lower rates of extinction (of species, genera, or taxonomic families) throughout evolutionary time? To answer these questions, we must turn to Darwin's other four theories of evolution. Regardless of how we answer these questions, however, the observed trends in animal diversity clearly illustrate Darwin's principle of perpetual change. Because the remaining four theories of Darwinism rely on the theory of perpetual change, evidence supporting these theories strengthens Darwin's theory of perpetual change.

Common Descent

Darwin proposed that all plants and animals have descended from an ancestral form into which life was first breathed. Life's history is depicted as a branching tree, called a **phylogeny.** Pre-Darwinian evolutionists, including Lamarck, advocated multiple independent origins of life, each of which gave rise to lineages that changed through time without extensive branching. Like all good scientific theories, common descent makes several important predictions that can be tested and potentially used to reject it. According to this theory, we should be able to trace the genealogies of all modern species backward until they converge on ancestral lineages shared with other species, both living and extinct.

We should be able to continue this process, moving farther backward through evolutionary time, until we reach the primordial ancestor of all life on earth. All forms of life, including many extinct forms that represent dead branches, will connect to this tree somewhere. Although reconstructing the history of life in this manner may seem almost impossible, phylogenetic research has been extraordinarily successful. How has this difficult task been accomplished?

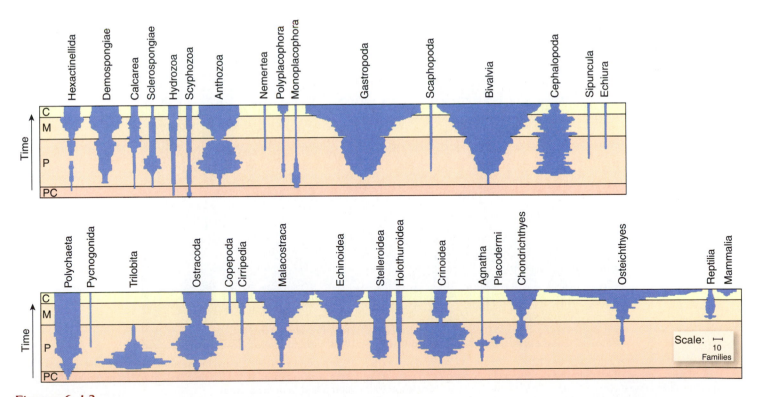

Figure 6.13
Diversity profiles of taxonomic families from different animal groups in the fossil record. The scale marks the Precambrian (PC), Paleozoic (P), Mesozoic (M), and Cenozoic (C) eras. The relative number of families is indicated from the width of the profile.

The Power of a Theory

Darwin's theory of common descent illustrates the scientific importance of general theories that give unified explanations to diverse kinds of data. Darwin proposed his theory of descent with modification of all living forms because it explained the patterns of similarity and dissimilarity among organisms in anatomical structures and cellular organization.

Anatomical similarities between humans and apes led Darwin to propose that humans and apes share more recent common ancestry with each other than they do with any other species. Darwin was unaware that his theory, a century later, would provide the primary explanation for similarities and dissimilarities among species in the structures of their chromosomes, sequences of amino acids in homologous proteins, and sequences of bases in homologous genomic DNA.

The accompanying figure shows photographs of a complete haploid set of chromosomes from each of four ape species: human (*Homo sapiens*), bonobo (the pygmy chimpanzee, *Pan paniscus*), gorilla (*Gorilla gorilla*), and orangutan (*Pongo pygmaeus*). Each chromosome in the human genome has a corresponding chromosome with similar structure and gene content in the genomes of other ape species. The most obvious difference between human and ape chromosomes is that the large second chromosome in the human nuclear genome was formed evolutionarily by a fusion of two smaller chromosomes characteristic of the ape genomes. Detailed study of the human and other ape chromosomes shows remarkable correspondence between them in genic content and organization.

Ape chromosomes are more similar to each other than they are to chromosomes of any other animals.

Comparison of DNA and protein sequences among apes likewise confirms their close genetic relationships, with humans and the two chimpanzee species being closer to each other than any of these species are to other apes. DNA sequences from the nuclear and mitochondrial genomes independently support the close relationships among ape species and especially the grouping of humans and chimpanzees as close relatives. Homologous DNA sequences of humans and chimpanzees are approximately 99% similar in base sequence.

Studies of variation in chromosomal structure, mitochondrial DNA sequences, and nuclear DNA sequences produced multiple independent data sets, each one potentially capable of rejecting Darwin's theory of common descent. Darwin's theory would be rejected, for example, if the chromosomal structures and DNA sequences of apes were no more similar to each other than to those of other animals. The data in this case support rather than reject predictions of Darwin's theory. The ability of Darwin's theory of common descent to make precise predictions of genetic similarities among these and other species, and to have those predictions confirmed by numerous empirical studies, illustrates its great strength. As new kinds of biological data have become available, the scope and strength of Darwin's theory of common descent have increased enormously. Indeed, nothing in biology makes sense in the absence of this powerful explanatory theory.

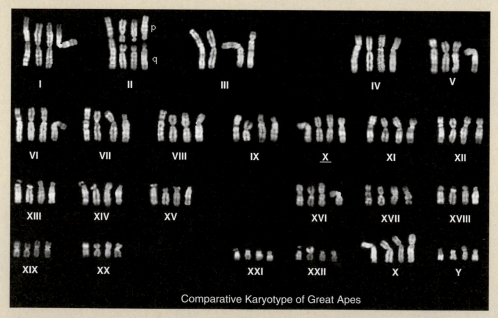

Comparative Karyotype of Great Apes

The human haploid genome contains 22 autosomes (I–XXII) and a sex chromosome (X or Y). The human chromosome is shown first in each group of four, followed by the corresponding chromosomes of bonobo, gorilla, and orangutan, in that order. Note that the chromatin of human chromosome II corresponds to that of two smaller chromosomes (marked p and q) in other apes.

Homology and Phylogenetic Reconstruction

Darwin recognized the major source of evidence for common descent in the concept of **homology.** Darwin's contemporary, Richard Owen (1804 to 1892), used this term to denote "the same organ in different organisms under every variety of form and function." A classic example of homology is the limb skeleton of vertebrates. Bones of vertebrate limbs maintain characteristic structures and patterns of connection despite diverse modifications for different functions (Figure 6.14). According to Darwin's theory of common descent, the structures that we call homologies represent characteristics inherited with some modification from a corresponding feature in a common ancestor.

Darwin devoted an entire book, *The Descent of Man and Selection in Relation to Sex,* largely to the idea that humans share common descent with apes and other animals. This idea was repugnant to many Victorians, who responded with predictable outrage (Figure 6.15). Darwin built his case mostly on anatomical comparisons revealing homology between humans and apes. To Darwin, the close resemblances between apes and humans could be explained only by common descent.

Throughout the history of all forms of life, evolutionary processes generate new characteristics that are then inherited by subsequent generations. Every time a new feature arises on an evolving lineage, we see the origin of a new homology. That homology gets transmitted to all descendant lineages unless it is subsequently lost. The pattern formed by the sharing of homologies among species provides evidence for common descent and allows us to reconstruct the branching evolutionary history of life. We can illustrate such evidence using a phylogenetic tree for a group of large, ground-dwelling birds (Figure 6.16). A new skeletal homology arises on each of the lineages shown (descriptions of specific homologies are not included because they are highly technical). The different groups of species located at the tips of the branches contain different combinations of these homologies, which reflect ancestry. For example, ostriches show homologies 1 through 5 and 8, whereas kiwis show homologies 1, 2, 13, and 15. Branches of the tree combine these species into a **nested hierarchy** of groups within groups (see Chapter 10). Smaller groups (species grouped near terminal branches) are contained within larger ones (species grouped by basal branches, including the trunk of the tree). If we erase the tree structure but retain patterns of homology observed in the living species, we are able to reconstruct the branching structure of the entire tree. Evolutionists test the theory of common descent by observing patterns of homology present within all groups of organisms. The pattern formed by all homologies taken together should specify a single branching tree that represents the evolutionary genealogy of all living organisms.

The nested hierarchical structure of homology is so pervasive in the living world that it forms the basis for our systematic classification of all forms of life (genera grouped into families, families grouped into orders, and other categories). Hierarchical classification even preceded Darwin's theory

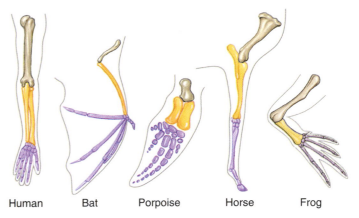

Human   Bat   Porpoise   Horse   Frog

Figure 6.14

Forelimbs of five vertebrates show skeletal homologies: *brown,* humerus; *orange,* radius and ulna; *purple,* "hand" (carpals, metacarpals, and phalanges). Clear homologies of bones and patterns of connection are evident despite evolutionary modification for different functions.

Figure 6.15

This 1873 advertisement for Merchant's Gargling Oil ridicules Darwin's theory of the common descent of humans and apes, which received only limited acceptance by the general public during Darwin's lifetime.

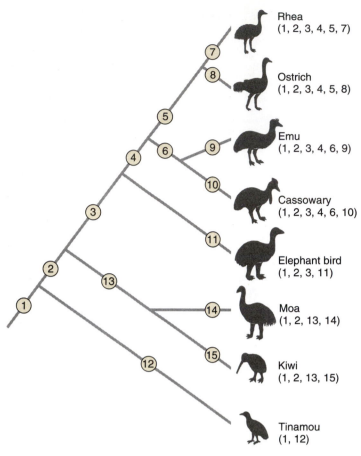

Rhea
(1, 2, 3, 4, 5, 7)

Ostrich
(1, 2, 3, 4, 5, 8)

Emu
(1, 2, 3, 4, 6, 9)

Cassowary
(1, 2, 3, 4, 6, 10)

Elephant bird
(1, 2, 3, 11)

Moa
(1, 2, 13, 14)

Kiwi
(1, 2, 13, 15)

Tinamou
(1, 12)

Figure 6.16

The phylogenetic pattern specified by 15 homologous structures in the skeletons of a group of flightless birds. Homologous features are numbered 1 through 15 and are marked both on the branches of the tree on which they arose and on the birds that have them. If you were to erase the tree structure, you would be able to reconstruct it without error from the distributions of homologous features shown for the birds at the terminal branches.

because this pattern is so evident, but it was not explained scientifically before Darwin. Once the idea of common descent was accepted, biologists began investigating the structural, molecular, and chromosomal homologies of animal groups. Taken together, the nested hierarchical patterns uncovered by these studies have permitted us to reconstruct evolutionary trees of many groups and to continue investigating others. Use of Darwin's theory of common descent to reconstruct the evolutionary history of life and to classify animals is the subject of Chapter 10.

Note that the earlier evolutionary hypothesis that life arose many times, forming unbranched lineages, predicts linear sequences of evolutionary change with no nested hierarchy of homologies among species. Because we do observe nested hierarchies of homologies, that hypothesis is rejected. Note also that because the creationist argument is not a scientific hypothesis, it can make no testable predictions about any pattern of homology and therefore fails to meet the criteria of a scientific theory of animal diversity.

Characters of different organisms that perform similar functions are not necessarily homologous. The wings of bats and birds, although homologous as vertebrate forelimbs, are not homologous as wings. The most recent common ancestor of bats and birds had forelimbs, but the forelimbs were not in the form of wings. Wings of bats and birds evolved independently and have only superficial similarity in their flight structures.

Bat wings are formed by skin stretched over elongated digits, whereas bird wings are formed by feathers attached along the forelimb. Such functionally similar but nonhomologous structures are often termed analogues.

Ontogeny, Phylogeny, and Recapitulation

Ontogeny is the history of the development of an organism through its entire life. Early developmental and embryological features contribute greatly to our knowledge of homology and common descent. Comparative studies of ontogeny show how the evolutionary alteration of developmental timing generates new characters, thereby producing evolutionary divergence among lineages.

The German zoologist Ernst Haeckel, a contemporary of Darwin, proposed that each successive stage in the development of an individual represented one of the adult forms that appeared in its evolutionary history. The human embryo with gill depressions in the neck corresponded, for example, to the adult appearance of a fishlike ancestor. On this basis Haeckel gave his generalization: *ontogeny (individual development) recapitulates (repeats) phylogeny (evolutionary descent)*. This notion later became known simply as **recapitulation** or the **biogenetic law.** Haeckel based his biogenetic law on the flawed premise that evolutionary change occurs by successively adding new features onto the end of an unaltered ancestral ontogeny while condensing the ancestral ontogeny into earlier developmental stages. This notion was based on Lamarck's concept of the inheritance of acquired characteristics (p. 105).

The nineteenth-century embryologist, K. E. von Baer, gave a more satisfactory explanation of the relationship between ontogeny and phylogeny. He argued that early developmental features were simply more widely shared among different animal groups than later ones. Figure 6.17 shows, for example, the early embryological similarities of organisms whose adult forms are very different (see Figure 8.21, p. 176). The adults of animals with relatively short and simple ontogenies often resemble pre-adult stages of other animals whose ontogeny is more elaborate, but embryos of descendants do not necessarily resemble the adults of their ancestors. Even early development undergoes evolutionary divergence among lineages, however, and it is not as stable as von Baer proposed.

We now know many parallels between ontogeny and phylogeny, but features of an ancestral ontogeny can be shifted either to earlier or later stages in descendant ontogenies. Evolutionary change in timing of development is called **heterochrony,**

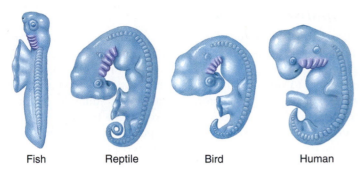

Figure 6.17

Fish   Reptile   Bird   Human

Comparison of gill arches of different embryos. All are shown separated from the yolk sac. Note the remarkable similarity of the four embryos at this early stage in development.

a term initially used by Haeckel to denote exceptions to recapitulation. If a descendant's ontogeny extends beyond its ancestral one, new characters can be added late in development, beyond the point at which development would have ended in an evolutionary ancestor. Features observed in the ancestor often are moved to earlier stages of development in this process, and ontogeny therefore does recapitulate phylogeny to some degree. Ontogeny also can be shortened during evolution, however. Terminal stages of the ancestor's ontogeny can be deleted, causing adults of descendants to resemble pre-adult stages of their ancestors (Figure 6.18). This outcome reverses the parallel between ontogeny and phylogeny (reverse recapitulation) producing **paedomorphosis** (the retention of ancestral juvenile characters by descendant adults). Because lengthening or shortening of ontogeny can change different parts of the body independently, we often see a mosaic of different kinds of developmental evolutionary change occurring concurrently. Therefore, cases in which an entire ontogeny recapitulates phylogeny are rare.

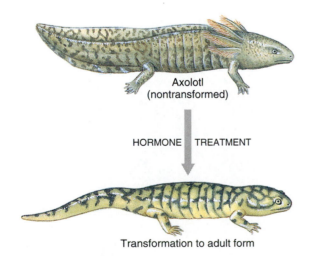

Axolotl
(nontransformed)

HORMONE | TREATMENT

Transformation to adult form

Figure 6.18

Aquatic and terrestrial forms of axolotls. Axolotls retain the juvenile, aquatic morphology (*top*) throughout their lives unless forced to metamorphose (*bottom*) by hormone treatment. Axolotls evolved from metamorphosing ancestors, an example of paedomorphosis.

Developmental Modularity and Evolvability

Evolutionary innovations occur not only by simple changes in rates of developmental processes but by changes in the physical location in the body where a process is activated. **Heterotopy** is the term traditionally used to describe a change in the physical location of a developmental process in an organism's body. For such a change to be successful, the developmental process must be compartmentalized into semiautonomous modules whose expression can be activated in new locations.

An interesting example of modularity and heterotopy occurs in some geckos. Geckos as a group typically have toepads, adhesive structures on the ventral side of the toes that permit climbing and clinging to smooth surfaces. Toepads consist of modified scales containing long protrusions, called setae, which can be molded to the surface of a substrate. A module responsible for toepad development is expressed in an unusual gecko species not only on the toes but also on the ventral side of the tip of the tail. This species has thereby acquired an additional adhesive appendage by ectopic expression of a standard developmental module.

Modularity is evident also in the homeotic mutations of the fruit fly, *Drosophila melanogaster*. Such mutations can substitute a developmental module for a leg in place of one normally specifying an antenna, thereby producing a fly with a pair of legs on its head. Another homeotic mutation in fruit flies transforms the balancer organs on the thorax into a second pair of wings; the balancer module is replaced by activation of the wing module, which in flies is activated normally only in a more anterior portion of the thorax.

Modularity is important in explaining some major evolutionary changes, such as evolution of tetrapod limbs (p. 546). The evolutionary transition from finlike limbs to the standard tetrapod limbs occurred by activating at the site of limb formation a set of homeobox genes (p. 173) whose expression pattern evolved initially as a module for forming part of the vertebral column. Shared patterns of gene expression between the vertebral column and the forelimbs and hindlimbs of tetrapods revealed the genetic and developmental mechanics of this module.

The term **evolvability** has been introduced recently to denote the great evolutionary opportunities created by having semiautonomous developmental modules whose expression can be moved from one part of the body to another. An evolving lineage that contains a large modular developmental toolkit can "experiment" with the construction of many new structures, some of which will persist and give rise to new homologies.

Multiplication of Species

Multiplication of species through time is a logical corollary to Darwin's theory of common descent. A branch point on the evolutionary tree means that an ancestral species has split into two different ones. Darwin's theory postulates that genetic variation present within a species, especially variation that occurs between geographically separated populations, provides the material from which new species are produced. Because evolution is a branching process, the total number of species produced by evolution increases through time, although most of these species eventually become extinct without leaving descendant species. A major challenge for evolutionists is to discover the process by which an ancestral species "branches" to form two or more descendant species.

Before we explore multiplication of species, we must decide what we mean by "species." As explained in Chapter 10, no consensus exists regarding definition of species. Most biologists agree, however, that important criteria for recognizing species include (1) descent of all members from a common ancestral population forming a **lineage** of ancestor-descendant populations, (2) reproductive compatibility (ability to interbreed) within and reproductive incompatibility between species for sexually reproducing animals, and (3) maintenance within species of genotypic and phenotypic cohesion (lack of abrupt differences among populations in allelic frequencies and organismal characteristics). The criterion of reproductive compatibility has received the greatest attention in studies of species formation, also called **speciation.**

Biological features that prevent different species from interbreeding are called **reproductive barriers.** The primary problem of speciation is to discover how two initially compatible populations evolve reproductive barriers that cause them to become distinct, separately evolving lineages. How do populations diverge from each other in their reproductive properties while maintaining complete reproductive compatibility within each population?

Reproductive barriers between populations usually evolve gradually. Evolution of reproductive barriers requires that diverging populations must be kept physically separate for long periods of time. If diverging populations reunite before reproductive barriers have evolved, interbreeding occurs between the populations and they merge. Speciation by gradual divergence in animals may require extraordinarily long periods of time, perhaps 10,000 to 100,000 years or more. Geographical isolation followed by gradual divergence is the most effective way for reproductive barriers to evolve, and many evolutionists consider geographical separation a prerequisite for branching speciation.

Geographical barriers between populations are not the same thing as reproductive barriers. Geographical barriers refer to spatial separation of two populations. They prevent gene exchange and are usually a precondition for speciation. Reproductive barriers result from evolution and refer to various behavioral, physical, physiological, and ecological factors that prevent interbreeding between different species. Behavioral barriers often evolve faster than other kinds of reproductive barriers. Geographical barriers do not guarantee that reproductive barriers will evolve. Reproductive barriers are most likely to evolve under conditions that include small population size, a favorable combination of selective factors, and long periods of geographical isolation. One or both of a pair of geographically isolated populations may become extinct prior to evolution of reproductive barriers between them. Over the vast span of geological time, however, conditions sufficient for speciation have occurred millions of times.

Allopatric Speciation

Allopatric ("in another land") populations of a species are those that occupy separate geographical areas. Because of their geographical separation, they cannot interbreed, but would be expected to do so if the geographic barriers between them were removed. If populations are allopatric immediately preceding and during evolution of reproductive barriers between them, the resulting speciation is called **allopatric speciation** or geographic speciation. The separated populations evolve independently and adapt to their respective environments, generating reproductive barriers between them as a result of their separate evolutionary paths. Because their genetic variation arises and changes independently, physically separated populations will diverge genetically even if their environments remain very similar. Environmental change between populations also can promote genetic divergence between them by favoring different phenotypes in the separated populations. Ernst Mayr (Figure 6.19) has contributed greatly to our knowledge of allopatric speciation through his studies of speciation in birds.

Allopatric speciation begins when a species splits into two or more geographically separated populations. This splitting can happen in either of two ways: by **vicariant speciation** or by a **founder event.** Vicariant speciation is initiated when climatic or geological changes fragment a species's habitat, producing impenetrable barriers that separate different populations geographically. For example, a mammalian species inhabiting a lowland forest could be divided by uplifting of a mountain barrier, sinking and flooding of a geological fault, or climatic changes that cause prairie or desert conditions to encroach on the forest. Formation of the isthmus of Panama separated populations of the sea urchin genus *Eucidaris*, leading to formation of the pair of species shown in Figure 1.5D.

Figure 6.19

Ernst Mayr (1904 to 2005), a major contributor to our knowledge of speciation and of evolution in general.

Vicariant speciation has two important consequences. Although the ancestral population is fragmented, individual fragments are usually left fairly intact. The vicariant process itself does not induce genetic change by reducing populations to a small size or by transporting them to unfamiliar environments. Another important consequence is that the same vicariant events may fragment several different species simultaneously. For example, fragmentation of a lowland forest most likely would disrupt numerous and diverse species, including salamanders, frogs, snails, and many other forest dwellers. Indeed, the same geographic patterns are observed among closely related species in different groups of organisms whose habitats are similar. Such patterns provide strong evidence for vicariant speciation.

An alternative means of initiating allopatric speciation is for a small number of individuals to disperse to a distant place where no other members of their species occur. The dispersing individuals may establish a new population in what is called a founder event. Allopatric speciation caused by founder events has been observed, for example, in the native fruit flies of Hawaii. Hawaii contains numerous patches of forest separated by volcanic lava flows. On rare occasions, strong winds can transport a few flies from one forest to another, geographically isolated forest where the flies are able to start a new population. Sometimes, a single fertilized female may found a new population. Unlike what happens in vicariant speciation, the new population initially has a very small size, which can cause its genetic structure to change quickly and dramatically from that of its ancestral population (see p. 129). When this event happens, phenotypic characteristics that were stable in the ancestral population often reveal unprecedented variation in the new population. As the newly expressed variation is sorted by natural selection, large changes in phenotype and reproductive properties occur, hastening evolution of reproductive barriers between the ancestral and newly founded populations.

Surprisingly, we often learn most about the genetics of allopatric speciation from cases in which formerly separated populations regain geographic contact following evolution of incipient reproductive barriers that are not absolute. The occurrence of mating between divergent populations is called **hybridization** and offspring of these matings are called **hybrids** (Figure 6.20). By studying the genetics of hybrid populations, we can identify the genetic bases of reproductive barriers.

Biologists often distinguish between reproductive barriers that impair fertilization (premating barriers) and those that impair growth and development, survival, or reproduction of hybrid individuals (postmating barriers). Premating barriers cause members of divergent populations either not to recognize each other as potential mates or not to complete the mating ritual successfully. Details of the horn structures of African antelopes (Figure 6.11) are important for recognizing members of the same species as potential mates. In some other cases, female and male genitalia of the different populations may be incompatible or the gametes may be incapable of fusing to form a zygote. In others, premating barriers may be strictly behavioral, with members of different species being otherwise nearly identical in phenotype. Different species that are indistinguishable in organismal appearance are called **sibling species.** Sibling species arise when allopatric populations diverge in the seasonal timing of reproduction or in auditory, behavioral, or chemical signals required for mating. Evolutionary divergence in these features can produce effective premating barriers without obvious changes in organismal appearance. Sibling species occur in groups as diverse as ciliates, flies, and salamanders.

Nonallopatric Speciation

Can speciation ever occur without prior geographic separation of populations? Allopatric speciation may seem an unlikely

---

The term *founder event* in its most general usage denotes a dispersal of organisms from an ancestral population across a geographic barrier to start a new, allopatric population. A founder event does not always directly cause important changes in the genetic constitution of the new population relative to the old one, although such changes are expected when the number of founding individuals is very small (less than 5 to 10 individuals, for example) and the ancestral population has a large amount of genetic variation. A change in the genetic constitution of a newly founded population because of a small number of founders is termed a *founder effect*, which includes a population bottleneck (p. 129). If a founder effect is so profound that selection acts in new ways on reproductively important characters, the founder event can induce speciation. *Founder-induced speciation* describes the subset of founder events in which a founder effect hastens species-level divergence of the newly founded population. Speciation in Hawaiian *Drosophila* as described in the text illustrates founder-induced speciation. Excluded from founder-induced speciation are founder events whose role in speciation is strictly to establish a new allopatric population capable of independent evolutionary change.

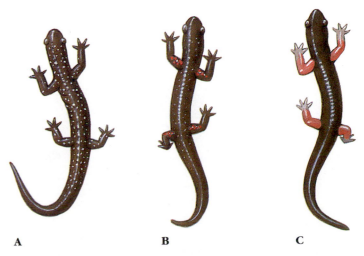

A   **B**   **C**

Figure 6.20
Pure and hybrid salamanders. Hybrids are intermediate in appearance between parental populations. **A,** Pure white-spotted *Plethodon teyahalee*; **B,** a hybrid between white-spotted *P. teyahalee* and red-legged *P. shermani*, intermediate in appearance for both spotting and leg color; **C,** pure red-legged *P. shermani*.

explanation for situations where many closely related species occur together in restricted areas that have no traces of physical barriers to animal dispersal. For example, several large lakes around the world contain very large numbers of closely related species of fish. The great lakes of Africa (Lake Malawi, Lake Tanganyika, and Lake Victoria) each contain many species of cichlid fishes that are found nowhere else. Likewise, Lake Baikal in Siberia contains many different species of sculpins that occur nowhere else in the world (Figure 6.21). It is difficult to conclude that these species arose anywhere other than in the lakes they inhabit, and yet those lakes are young on an evolutionary timescale and have no obvious environmental barriers that would fragment fish populations.

To explain speciation of fish in freshwater lakes and other examples like these, **sympatric** ("same land") **speciation** has been hypothesized. According to this hypothesis, different individuals within a species become specialized for occupying different components of the environment. By seeking and using very specific habitats in a single geographic area, different populations achieve sufficient physical and adaptive separation to evolve reproductive barriers. For example, cichlid species of African lakes are very different from each other in their feeding specializations. In many parasitic organisms, particularly parasitic insects, different populations may use different host species, thereby providing the physical separation necessary for reproductive barriers to evolve. Supposed cases of sympatric speciation have been criticized, however, because the reproductive distinctness of the different populations often

is not well demonstrated, so that we may not be observing formation of distinct evolutionary lineages that will become different species.

The occurrence of sudden sympatric speciation is perhaps most likely among higher plants. Between one-third and one-half of flowering plant species may have evolved by polyploidy (doubling of chromosome numbers), without prior geographic isolation of populations. In animals, however, speciation through polyploidy is an exceptional event.

Another possible mode of speciation, termed **parapatric speciation**, is geographically intermediate between allopatric and sympatric speciation. Two species are called parapatric with respect to each other if their geographic ranges are primarily allopatric but make contact along a borderline that neither species successfully crosses. In parapatric speciation, a geographically continuous ancestral species evolves within its range a borderline across which populations evolve species-level differences while maintaining geographic contact along the border.

The simplest model of parapatric speciation is one in which a change in environmental conditions splits the geographic range of a species into two environmentally distinct but geographically adjoining parts. An increase in temperature on a Caribbean island, for example, might cause part of a wet tropical forest to become a dry, sandy one. A lizard species occupying the formerly wet forest might become divided into geographically adjacent wet and dry forest populations. Unlike vicariant allopatric speciation, however, the populations in the different habitat types are not isolated by a physical barrier but maintain genetic

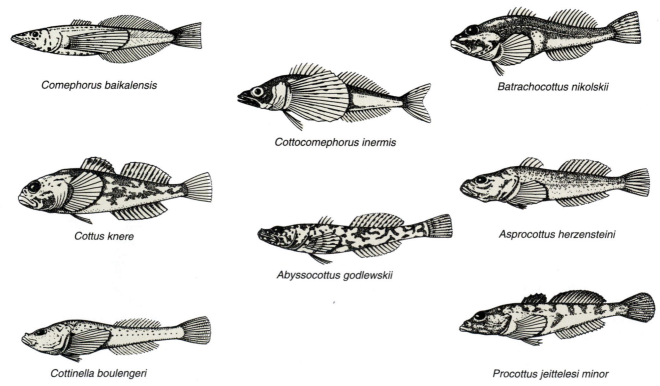

Comephorus baikalensis

Cottocomephorus inermis

Batrachocottus nikolskii

Cottus knere

Abyssocottus godlewskii

Asprocottus herzensteini

Cottinella boulengeri

Procottus jeittelesi minor

Figure 6.21

The sculpins of Lake Baikal, products of speciation that occurred within a single lake.

interactions along the geographic borderline between the different habitat types. The disparity in environmental conditions across the border, nonetheless, causes the populations to evolve as separate lineages adapted to different environments despite some gene exchange between them.

A parapatric distribution of species does not necessarily imply that speciation occurred parapatrically. Most cases of parapatrically distributed species show evidence of past allopatry, with subsequent removal of a geographic barrier permitting the two species to make geographic contact but with each species excluding the other one from its geographic territory.

The prevalence of parapatric speciation is controversial. This model of speciation predicts that parapatrically distributed populations differ mainly in adaptive features associated with the observed environmental differences but show relative homogeneity for other genetic variation. Comparisons of parapatrically distributed populations, including those of lizards occupying different forest types on Caribbean islands, often show extensive divergence for molecular variation unrelated to adaptive differentiation of the populations; such results are better explained by vicariant allopatric speciation than by parapatric speciation. In some cases, geological evidence shows that what is now a single island was physically fragmented into separate islands during warm periods when the sea level was higher than it is now; such evidence likewise favors the interpretation of allopatric speciation for parapatrically distributed species whose geographic contact occurs in formerly inundated areas.

Adaptive Radiation

The production of several ecologically diverse species from a common ancestral species is called **adaptive radiation,** especially when many disparate species arise within a short interval of geological time (a few million years). Some of our best examples of adaptive radiation are associated with lakes and young islands, which provide new evolutionary opportunities for aquatic and terrestrial organisms, respectively. Oceanic islands formed by volcanoes are initially devoid of life. They are gradually colonized by plants and animals from a continent or from other islands in separate founder events. The founders encounter ideal situations for evolutionary diversification, because environmental resources that were heavily exploited by other species on the mainland are free for colonization on the sparsely populated island. Archipelagoes, such as the Galápagos Islands, greatly increase opportunities for both founder events and ecological diversification. The entire archipelago is isolated from the continent and each island is geographically isolated from the others by sea; moreover, each island is different from every other one in its physical, climatic, and biotic characteristics.

Galápagos finches illustrate adaptive radiation on an oceanic archipelago (Figures 6.22 and 6.23). Galápagos finches (the name "Darwin's finches" was popularized in the 1940s by the British ornithologist David Lack) are closely related to each other, but each species differs from others in size and shape of the beak and in feeding habits. If the finches were specially created, it would require the strangest kind of coincidence for 13 similar kinds of finches to be created on the Galápagos Islands and nowhere else. Darwin's finches descended from a single ancestral population that arrived from the mainland and subsequently colonized the different islands of the Galápagos archipelago. The finches underwent adaptive radiation, occupying habitats that on the mainland were denied to them by the presence of other species better able to exploit those habitats. Galápagos finches thus assumed characteristics of mainland birds as diverse and unfinchlike as warblers and woodpeckers. A fourteenth Darwin's finch, found on isolated Cocos Island far north of the Galápagos archipelago, is similar in appearance to the Galápagos finches and almost certainly descended from the same ancestral founder.

Gradualism

Darwin's theory of gradualism opposes arguments for the sudden origin of species. Small differences, resembling those that we observe among organisms within populations today, are the raw material from which the different major forms of life evolved. This theory shares with Lyell's uniformitarianism the notion that we must not explain past changes by invoking unusual catastrophic events that are not observed today. If new species originated in single, catastrophic events, we should be able to see such events happening today and we do not. Instead, what we usually observe in natural populations are small, continuous changes in phenotypes. Such continuous changes can produce major differences among species only by accumulating over many thousands to millions of years. A simple statement of Darwin's theory of gradualism is that accumulation of quantitative changes leads to qualitative change.

Mayr (see Figure 6.19) makes an important distinction between populational gradualism and phenotypic gradualism. **Populational gradualism** states that new traits become established in a population by increasing their frequency initially from a small fraction of the population to a majority of the population. Populational gradualism is well established and is not controversial. **Phenotypic gradualism** states that new traits, even those that are strikingly different from ancestral ones, are produced in a series of small, incremental steps.

Phenotypic Gradualism

Phenotypic gradualism was controversial when Darwin first proposed it, and it is still controversial. Not all phenotypic changes are small, incremental ones. Some mutations that appear during artificial breeding change the phenotype substantially in a single mutational step. Such mutations traditionally are called "sports." Sports that produce dwarfing are observed in many species, including humans, dogs, and sheep, and have been used by animal breeders to achieve desired results; for example, a sport that deforms the limbs was used to produce Ancon sheep, which cannot jump hedges and are therefore easily contained (Figure 6.24). Many colleagues of Darwin who accepted his other theories considered phenotypic gradualism too extreme. If sporting mutations can be used in animal breeding, why must

Figure 6.22

Model for evolution of the 13 Darwin's finches on the Galápagos Islands. The model postulates three steps: (1) Immigrant finches from the South American mainland reach the Galápagos and colonize an island; (2) once the population becomes established, finches disperse to other islands where they adapt to new conditions and change genetically; and (3) after a period of isolation, secondary contact is established between different populations. The two populations are then recognized as separate species if they cannot interbreed successfully.

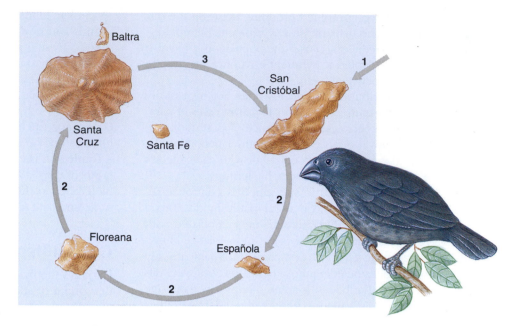

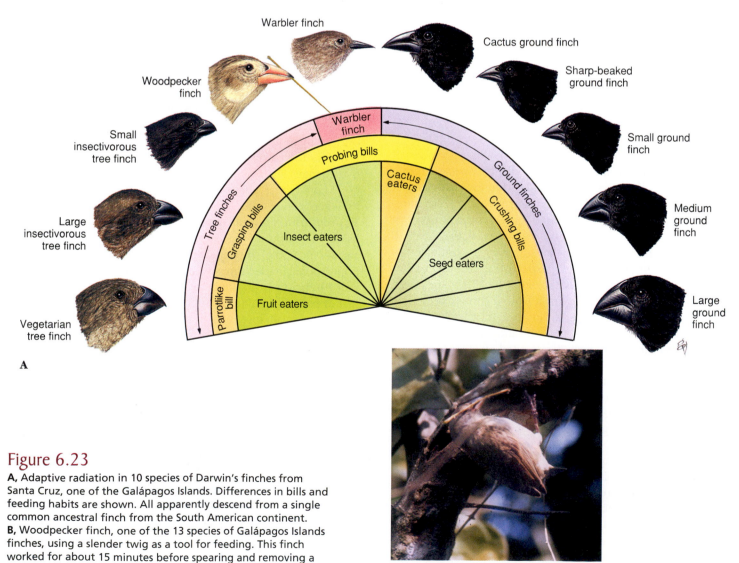

Figure 6.23

A, Adaptive radiation in 10 species of Darwin's finches from Santa Cruz, one of the Galápagos Islands. Differences in bills and feeding habits are shown. All apparently descend from a single common ancestral finch from the South American continent. B, Woodpecker finch, one of the 13 species of Galápagos Islands finches, using a slender twig as a tool for feeding. This finch worked for about 15 minutes before spearing and removing a wood roach from a break in the tree.

Figure 6.24
The Ancon breed of sheep arose from a "sporting mutation" that caused dwarfing of legs. Many of his contemporaries criticized Darwin for his claim that such mutations are not important for evolution by natural selection.

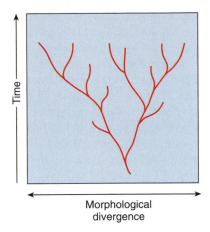

Figure 6.25
The phyletic gradualist model of evolutionary change in morphology, viewed as proceeding more or less steadily through geological time (millions of years). Bifurcations followed by gradual divergence led to speciation. Note that most morphological change accumulates incrementally within species lineages between branch points, which are not accompanied by unusually large amounts of morphological change.

we exclude them from our evolutionary theory? In favor of gradualism, Darwin and others have replied that sporting mutations always have negative side-effects that would cause selection to eliminate them from natural populations. Indeed, it is questionable whether Ancon sheep, despite their attractiveness to farmers, would propagate successfully in the presence of their long-legged relatives without human intervention. A mutation of large effect appears responsible, however, for an adaptive bill size polymorphism in an African finch species (*Pyrenestes ostrinus*) in which large-billed forms eat hard seeds and small-billed forms eat softer ones. Recent work in evolutionary developmental genetics (p. 174) illustrates the continuing controversy surrounding phenotypic gradualism.

Punctuated Equilibrium

When we view Darwinian gradualism on a geological time-scale, we may expect to find in the fossil record a long series of intermediate forms connecting the phenotypes of ancestral and descendant populations (Figure 6.25). This predicted pattern is called **phyletic gradualism.** Darwin recognized that phyletic gradualism is not often revealed by the fossil record. Studies conducted since Darwin's time generally have not revealed the continuous series of fossils predicted by phyletic gradualism. Is the theory of gradualism therefore refuted by the fossil record? Darwin and others claim that it is not, because the fossil record is too imperfect to preserve transitional series. Although evolution is a slow process by our standards, it is rapid relative to the rate at which good fossil deposits accumulate. Others have argued, however, that abrupt origins and extinctions of species in the fossil record force us to conclude that phyletic gradualism is rare.

Niles Eldredge and Stephen Jay Gould proposed **punctuated equilibrium** to explain the discontinuous evolutionary changes observed throughout geological time. Punctuated equilibrium

states that phenotypic evolution is concentrated in relatively brief events of branching speciation, followed by much longer intervals of morphological evolutionary stasis (Figure 6.26). Speciation is an episodic event, having a duration of approximately 10,000 to 100,000 years. Because species may survive for 5 to 10 million years, the speciation event is a "geological instant," representing 1% or less of a species's life span. Ten thousand years is plenty of time, however, for Darwinian gradual evolution to accomplish dramatic changes. A small fraction of the evolutionary history of a group therefore contributes most of the morphological evolutionary change that we observe. Punctuated equilibrium contrasts with the views of paleontologist George Simpson, who attributed only moderate rates of morphological

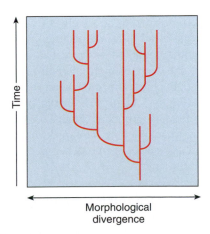

Figure 6.26
The punctuated equilibrium model sees morphological evolutionary change being concentrated in relatively rapid bursts of branching speciation (lateral lines) followed by prolonged periods of no change throughout geological time (millions of years).

evolution to branching speciation and expected most morphological change to accumulate gradually in the "phyletic" mode between events of branching speciation.

Founder-induced allopatric speciation provides a possible explanation for punctuated equilibria. Remember that founder-induced speciation requires the breaking of genetic equilibrium in a small, geographically isolated population. Such small populations have very little chance of being preserved in the fossil record. After a new genetic equilibrium forms and stabilizes, the new population may increase in size, thereby increasing the chance that some of its members will be preserved as fossils. Founder-induced speciation cannot be the exclusive cause of punctuated equilibrium, however, because punctuated equilibrium may be observed in groups where speciation by founder events is unlikely.

Evolutionists who lamented the imperfect state of the fossil record were treated in 1981 to the opening of an uncensored page of fossil history in Africa. Peter Williamson, a British paleontologist working in fossil beds 400 m deep near Lake Turkana, documented a remarkably clear record of speciation in freshwater snails. The geology of the Lake Turkana basin reveals a history of instability. Earthquakes, volcanic eruptions, and climatic changes caused the waters episodically to rise and to fall, sometimes by hundreds of feet. Thirteen lineages of snails show long periods of stability interrupted by relatively brief periods of rapid change in shell shape when snail populations were fragmented by receding waters. These populations diverged to produce new species that then remained unchanged through thick deposits before becoming extinct and being replaced by descendant species. The transitions occurred within 5000 to 50,000 years. In the few meters of sediment where speciation occurred, transitional forms were visible. Williamson's study conforms well to the punctuated equilibrium model of Eldredge and Gould.

Natural Selection

Natural selection is the major process by which evolution occurs in Darwin's theory of evolution. It gives us a natural explanation for the origins of **adaptation,** including all developmental, behavioral, anatomical, and physiological attributes that enhance an organism's ability to use environmental resources to survive and to reproduce. Evolution of color patterns that conceal moths from predators (Figure 1.11, p. 12), and of bills adapted to different modes of feeding in finches (Figure 6.23), illustrate natural selection leading to adaptation. Darwin developed his theory of natural selection as a series of five observations and three inferences drawn from them:

Observation 1—Organisms have great potential fertility.
All populations produce large numbers of gametes and potentially large numbers of offspring each generation. Population size would increase exponentially at an enormous rate if all individuals that were produced each generation survived and reproduced. Darwin calculated that, even in slow-breeding animals such as elephants, a single pair breeding from age 30 to 90 and having only six young could produce 19 million descendants in 750 years.

Observation 2—Natural populations normally remain constant in size, except for minor fluctuations.
Natural populations fluctuate in size across generations and sometimes go extinct, but no natural populations show the continued exponential growth that their reproductive biology theoretically could sustain.

Observation 3—Natural resources are limited.
Exponential growth of a natural population would require unlimited natural resources to provide food and habitat for the expanding population, but natural resources are finite.

Inference 1—A continuing *struggle* for existence exists among members of a population.
Survivors represent only a part, usually a very small part, of the individuals produced each generation. Darwin wrote in *On The Origin of Species* that "it is the doctrine of Malthus applied with manifold force to the whole animal and vegetable kingdoms." The struggle for food, shelter, and space becomes increasingly severe as overpopulation develops.

Observation 4—Populations show *variation* among organisms.
No two individuals are exactly alike. They differ in size, color, physiology, behavior, and many other ways.

Observation 5—Some variation is *heritable*.
Darwin noted that offspring tend to resemble their parents, although he did not understand how. The hereditary mechanism discovered by Gregor Mendel would be applied to Darwin's theory many years later.

Inference 2—Varying organisms show *differential survival and reproduction* favoring advantageous traits (= natural selection).
Survival in the struggle for existence is not random with respect to hereditary variation present in the population. Some traits give their possessors an advantage in using the environment for effective survival and reproduction. Survivors transmit their favored traits to offspring, thereby causing those traits to accumulate in the population.

Inference 3—Over many generations, natural selection generates new adaptations and new species.
The differential reproduction of varying organisms gradually transforms species and causes long-term "improvement" of populations. Darwin knew that people often use hereditary variation to produce useful new breeds of livestock and plants. *Natural* selection acting over millions of years should be even more effective in producing new types than the *artificial* selection imposed during a human lifetime. Natural selection acting independently on geographically separated populations would cause them to diverge from each other, thereby generating reproductive barriers that lead to speciation.

Darwin's Explanatory Model of Evolution by Natural Selection

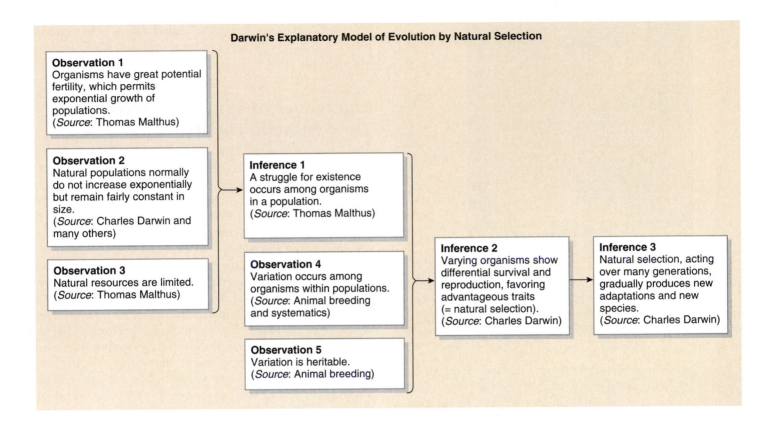

The popular phrase "survival of the fittest" was not originated by Darwin but was coined a few years earlier by the British philosopher Herbert Spencer, who anticipated some of Darwin's principles of evolution. Unfortunately the phrase later came to be coupled with unbridled aggression and violence in a bloody, competitive world. In fact, natural selection operates through many other characteristics of living organisms. The fittest animal may be one that enhances the living conditions of its population. Fighting prowess is only one of several means toward survival and reproductive advantage.

Natural selection may be considered a two-step process with a random component and a nonrandom component. Production of variation among organisms is the random component. The mutational process does not preferentially generate traits that are favorable to the organism; new variants are probably more likely to be unfavorable. The nonrandom component is the survival of different traits. This differential survival is determined by the effectiveness of different traits in permitting their possessors to use environmental resources to survive and to reproduce. The phenomenon of differential survival and reproduction among varying organisms is now called **sorting** and should not be equated with natural selection. We now know that even random processes (genetic drift, p. 127) can produce sorting among varying organisms. When selection operates, sorting occurs *because certain traits give their possessors advantages in survival and reproduction* relative to others that lack those traits. Selection is therefore a specific cause of sorting.

Darwin's theory of natural selection has been challenged repeatedly. One challenge claims that directed (nonrandom) variation governs evolutionary change. In the decades around 1900, diverse evolutionary hypotheses collectively called **orthogenesis** proposed that variation has momentum that forces a lineage to evolve in a particular direction that is not always adaptive. The extinct Irish elk was a popular example of orthogenesis. Newly produced variation was considered biased toward enlarging their antlers, thereby generating an evolutionary momentum for producing larger antlers. Natural selection was considered ineffective at stopping the antlers eventually from becoming so large and cumbersome that they forced the Irish elk into extinction (Figure 6.27). Orthogenesis explained apparently nonadaptive evolutionary trends that supposedly forced species into decline. Because extinction is the expected evolutionary fate of most species, disappearance of the Irish elk is not extraordinary and probably not related to large antlers. Subsequent genetic research on the nature of variation clearly has rejected the genetic predictions of orthogenesis.

Another recurring criticism of natural selection is that it cannot generate new structures or species but can only modify old ones. Most structures in their early evolutionary stages could not have performed the biological roles that the fully formed structures perform, and it is therefore unclear how natural selection could have favored them. What use is half a wing or the rudiment of a feather for a flying bird? To answer this criticism, we propose that many structures evolved initially for purposes different from the ones they have today. Rudimentary feathers would have been useful in thermoregulation, for example. The

Figure 6.27

Irish elk, a fossil species that once was used to support the orthogenetic idea that momentum in variation caused the antlers to become so large that the species was forced into extinction.

feathers later became useful for flying after they incidentally acquired aerodynamic properties. Natural selection then could act to improve the usefulness of feathers for flying. **Exaptation** denotes the utility of a structure for a biological role that was not part of the structure's evolutionary origin. Exaptation contrasts with adaptation, which implies that a structure arose by natural selection for a particular biological role. Bird feathers are therefore adaptations for thermoregulation but exaptations for flight. Because structural changes that separate members of different species are similar in kind to variation that we observe within species, it is reasonable to propose that selection can produce new species.

REVISIONS OF DARWIN'S THEORY

Neo-Darwinism

The most serious weakness in Darwin's theory was his failure to identify correctly the mechanism of inheritance. Darwin saw heredity as a blending phenomenon in which the hereditary factors of parents melded together in their offspring. Darwin also invoked the Lamarckian hypothesis that an organism could alter its heredity through use and disuse of body parts and through the direct influence of the environment. August Weismann rejected Lamarckian inheritance by showing experimentally that modifications of an organism during its lifetime do not change its heredity (see Chapter 5), and he revised Darwin's theory accordingly. We now use the term **neo-Darwinism** to denote Darwin's theory as revised by Weismann.

Mendelian genetics eventually clarified the particulate inheritance that Darwin's theory of natural selection required (p. 18). Ironically, when Mendel's work was rediscovered in 1900, it was considered antagonistic to Darwin's theory of natural selection. When mutations were discovered in the early 1900s, most geneticists thought that they produced new species in single

large steps. These geneticists relegated natural selection to the role of executioner, a negative force that merely eliminated the obviously unfit.

Emergence of Modern Darwinism: The Synthetic Theory

In the 1930s a new generation of geneticists began to reevaluate Darwin's theory from a mathematical perspective. These were population geneticists, scientists who studied variation in natural populations using statistics and mathematical models. Gradually, a new comprehensive theory emerged that brought together population genetics, paleontology, biogeography, embryology, systematics, and animal behavior in a Darwinian framework.

Population geneticists study evolution as a change in the genetic composition of populations. With the establishment of population genetics, evolutionary biology became divided into two different subfields. **Microevolution** pertains to evolutionary changes in frequencies of different allelic forms of genes (p. 78) within populations. **Macroevolution** refers to evolution on a grand scale, encompassing the origins of new organismal structures and designs, evolutionary trends, adaptive radiation, phylogenetic relationships of species, and mass extinction. Macroevolutionary research is based in systematics and the comparative method (p. 206). Following the evolutionary synthesis, both macroevolution and microevolution have operated firmly within the tradition of neo-Darwinism, and both have expanded Darwinian theory in important ways.

MICROEVOLUTION: GENETIC VARIATION AND CHANGE WITHIN SPECIES

Microevolution is the study of genetic change occurring within natural populations. Occurrence of different allelic forms of a gene in a population is called **polymorphism.** All alleles of all genes possessed by members of a population collectively form the **gene pool** of that population. The amount of polymorphism present in large populations is potentially enormous, because at observed mutation rates, many different alleles are expected for all genes.

Population geneticists study polymorphism by identifying the different allelic forms of a gene present in a population and then measuring the relative frequencies of the different alleles in the population. The relative frequency of a particular allelic form of a gene in a population is called its **allelic frequency.** For example, in the human population, there are three different allelic forms of the gene encoding the ABO blood types (p. 782). Using the symbol I to denote the gene encoding the ABO blood types, I^A and I^B denote genetically codominant alleles encoding blood types A and B, respectively. Allele i is a recessive allele encoding blood group O. Therefore genotypes $I^A I^A$ and $I^A i$ produce type A blood, genotypes $I^B I^B$ and $I^B i$ produce type B blood, genotype $I^A I^B$ produces type AB blood, and genotype ii produces type O blood. Because each individual contains two copies of this gene, the total number of copies present in the population is twice the number of individuals. What fraction of this total is represented by each of the three different allelic forms? In France, we find the following allelic frequencies: $I^A = .46$, $I^B = .14$, and $i = .40$. In Russia, the corresponding allelic frequencies differ ($I^A = .38$, $I^B = .28$, and $i = .34$), demonstrating microevolutionary divergence between these populations (Figure 6.28). Although alleles I^A and I^B are dominant to i, i is nearly as frequent as I^A and exceeds the frequency of I^B in both populations. Dominance describes the *phenotypic effect* of an allele in heterozygous individuals, not its relative abundance in a population of individuals. We will demonstrate that Mendelian inheritance and dominance do not alter allelic frequencies directly or produce evolutionary change in a population.

Genetic Equilibrium

In many human populations, genetically recessive traits, including the O blood type, blond hair, and blue eyes, are very common. Why have not the genetically dominant alternatives gradually supplanted these recessive traits? It is a common misconception that a characteristic associated with a dominant allele increases in frequency because of its genetic dominance. This misconception is refuted by a principle called **Hardy-Weinberg equilibrium** (see box next page), which forms the foundation for population genetics. According to this theorem, the hereditary process alone does not produce evolutionary change. In large biparental populations, allelic frequencies and genotypic ratios attain an equilibrium in one generation and *remain constant* thereafter *unless* disturbed by recurring mutations, natural selection, migration, nonrandom mating, or genetic drift (random sorting). Such disturbances are the sources of microevolutionary change.

A rare allele, according to this principle, does not disappear from a large population merely because it is rare. Certain rare traits, such as albinism and cystic fibrosis, persist for many generations. For example, albinism in humans is caused by a rare recessive allele a. Only one person in 20,000 is an albino, and this individual must be homozygous (a/a) for the recessive allele. Obviously the population contains many carriers, people with normal pigmentation who are heterozygous (A/a) for albinism. What is their frequency? A convenient way to calculate the frequencies of genotypes in a population is with the binomial expansion of $(p + q)^2$ (see box next page). We let p represent the allelic frequency of A and q the allelic frequency of a.

Assuming that mating is random, the distribution of genotypic frequencies is $p^2 = A/A$, $2pq = A/a$, and $q^2 = a/a$. Only the frequency of genotype a/a is known with certainty, 1/20,000; therefore:

$$q^2 = 1/20,000$$
$$q = (1/20,000)^{1/2} = 1/141$$
$$p = 1 - q = 140/141$$

The frequency of carriers is:

$$A/a = 2pq = 2 \times 140/141 \times 1/141 = 1/70$$

One person in every 70 is a carrier! Although the recessive trait is rare, the recessive allele is surprisingly common in the population. There is a message here for anyone proposing to eliminate a "bad" recessive allele from a population by controlling reproduction. It is practically impossible. Because only the homozygous recessive individuals reveal the phenotype against which artificial selection could act (by sterilization, for example), the allele would persist through heterozygous carriers. For a recessive allele present in 2 of every 100 persons (but homozygous in only 1 in 10,000 persons), it would require 50 generations of complete selection against the homozygotes just to reduce its frequency to one in 100 persons.

How Genetic Equilibrium Is Upset

Genetic equilibrium is disturbed in natural populations by (1) random genetic drift, (2) nonrandom mating, (3) recurring mutation, (4) migration, (5) natural selection, and interactions among these factors. Recurring mutation is the ultimate source of variability in all populations, but it usually requires interaction with one or more of the other factors to upset genetic equilibrium. We consider these other factors individually.

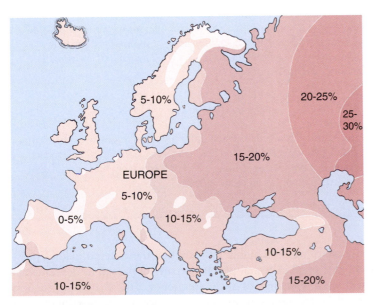

Figure 6.28

Frequencies of the blood-type B allele among humans in Europe. The allele is more common in the east and rarer in the west. The allele may have arisen in the east and gradually diffused westward through the genetic continuity of human populations. This allele has no known selective advantage; its changing frequency probably represents the effects of random genetic drift.

Genetic Drift

Some species, such as cheetahs (Figure 6.29), contain very little genetic variation, probably because their ancestral lineages

Hardy-Weinberg Equilibrium: Why the Hereditary Process Does Not Change Allelic Frequencies

The Hardy-Weinberg law is a logical consequence of Mendel's first law of segregation and expresses the tendency toward equilibrium inherent in Mendelian heredity.

Let us select for our example a population having a single locus bearing just two alleles T and t. The phenotypic expression of this gene might be, for example, the ability to taste a chemical compound called phenylthiocarbamide. Individuals in the population will be of three genotypes for this locus, T/T, T/t (both tasters), and t/t (non-tasters). In a sample of 100 individuals, let us suppose that we have 20 of T/T genotype, 40 of T/t genotype, and 40 of t/t genotype. We could then make a table showing the allelic frequencies (remember that every individual has two copies of the gene):

Genotype	Number of Individuals	Copies of the T Allele	Copies of the t Allele
T/T	20	40	
T/t	40	40	40
t/t	40		80
Total	100	80	120

Of the 200 copies, the proportion of the T allele is 80/200 = 0.4 (40%), and the proportion of the t allele is 120/200 = 0.6 (60%). It is customary to use "p" and "q" to represent the two allelic frequencies. The genetically dominant allele is represented by p, and the genetically recessive by q. Thus:

$$p = \text{frequency of } T = 0.4$$
$$q = \text{frequency of } t = 0.6$$
$$\text{Therefore } p + q = 1$$

Having calculated allelic frequencies in the sample, let us determine whether these frequencies will change spontaneously in a new generation of the population. Assuming that mating is random (gametes are sampled independently in pairs), each individual will contribute an equal number of gametes to the "common pool" from which the next generation is formed. Frequencies of gametes in the "pool" then will equal the allelic frequencies in the sample: 40% of the gametes will be T, and 60% will be t (ratio of 0.4:0.6). Both ova and sperm will, of course, show the same frequencies. The next generation is formed:

Sperm	Ova		
	$T = 0.4$		$t = 0.6$
$T = 0.4$	$T/T = 0.16$		$T/t = 0.24$
$t = 0.6$	$T/t = 0.24$		$t/t = 0.36$

Collecting genotypes, we have:

$$\text{frequency of } T/T = 0.16$$
$$\text{frequency of } T/t = 0.48$$
$$\text{frequency of } t/t = 0.36$$

Next, we determine the values of p and q from the randomly mated populations. From the table above, we see that the frequency of T will be the sum of genotypes T/T, which is 0.16, and one-half of the genotype T/t, which is 0.24:

$$T(p) = 0.16 + .5(0.48) = 0.4$$

Similarly, the frequency of t will be the sum of genotypes t/t, which is 0.36, and one-half the genotype T/t, which is 0.24:

$$t(p) = 0.36 + .5(0.48) = 0.6$$

The new generation bears exactly the same allelic frequencies as the parent population! Note that there has been no increase in the frequency of the genetically dominant allele T. Thus, *in a freely interbreeding, sexually reproducing population, the frequency of each allele would remain constant generation after generation in the absence of natural selection, migration, recurring mutation, and genetic drift* (see text). A mathematically minded reader will recognize that the genotype frequencies T/T, T/t, and t/t are actually a binomial expansion of $(p + q)^2$:

$$(p + q)^2 = p^2 + 2pq + q^2 = 1$$

A statistically minded reader will note that the equilibrium calculations give *expected* frequencies, which are unlikely to be realized exactly in a population of finite size. For this reason, finite population size is a cause of evolutionary change.

passed through periods when the total number of individuals in the population was very small. A small population clearly cannot contain large amounts of genetic variation. Each individual organism has at most two different allelic forms of each gene, and a single breeding pair contains at most four different allelic forms of each gene. Suppose that we have such a breeding pair. We know from Mendelian genetics (see Chapter 5) that chance decides which of the different allelic forms of a gene gets passed to offspring. It is therefore possible by chance alone that one or two of the parental alleles in this example will not be passed to

any offspring. It is highly unlikely that the different alleles present in a small ancestral population are all passed to descendants without any change of allelic frequency. This chance fluctuation in allelic frequency from one generation to the next, including loss of alleles from the population, is called **genetic drift.**

Genetic drift occurs to some degree in all populations of finite size. Perfect constancy of allelic frequencies, as predicted by Hardy-Weinberg equilibrium, occurs only in infinitely large populations, and such populations occur only in mathematical models. All populations of animals are finite and therefore experience

Figure 6.29
Cheetahs, a species whose genetic variability has been depleted to very low levels because of small population size in the past.

some effect of genetic drift, which becomes greater, on average, as population size declines. Genetic drift erodes genetic variability of a population. If population size remains small for many generations in a row, genetic variation can be greatly depleted. This loss is harmful to a species's evolutionary success because it restricts potential genetic responses to environmental change. Indeed, biologists are concerned that cheetah populations may have insufficient variation for continued survival.

A large reduction in the size of a population that increases evolutionary change by genetic drift is commonly called a bottleneck. A bottleneck associated with the founding of a new geographic population is called a founder effect and may be associated with formation of a new species (p. 119).

Nonrandom Mating

If mating is nonrandom, genotypic frequencies will deviate from Hardy-Weinberg expectations. For example, if two different alleles of a gene are equally frequent ($p = q = .5$), we expect half of the genotypes to be heterozygous ($2pq = 2[.5][.5] = .5$) and one-quarter to be homozygous for each of the respective alleles ($p^2 = q^2 = [.5]^2 = .25$). If we have **positive assortative mating,** individuals mate preferentially with others of the same genotype, such as albinos mating with other albinos. Matings among individuals homozygous for the same allele generate offspring that are homozygous like themselves. Matings among individuals heterozygous for the same pair of alleles produce on average 50% heterozygous offspring and 50% homozygous offspring (25% of each alternative type) each generation. Positive assortative mating increases the frequency of homozygous genotypes and decreases the frequency of heterozygous genotypes in a population but does not change allelic frequencies.

Preferential mating among close relatives also increases homozygosity and is called **inbreeding.** Whereas positive assortative mating usually affects one or a few traits, inbreeding simultaneously affects all variable traits. Strong inbreeding greatly increases chances that rare recessive alleles will become homozygous and be expressed.

Because inbreeding and genetic drift are both promoted by small population size, they are often confused with each other. Their effects are very different, however. Inbreeding alone cannot change allelic frequencies in the population, only the ways that alleles are combined into genotypes. Genetic drift changes allelic frequencies and consequently also changes genotypic frequencies. Even very large populations have the potential for being highly inbred if there is a behavioral preference for mating with close relatives, although this situation rarely occurs in animals. Genetic drift, however, will be relatively weak in very large populations.

Inbreeding has surfaced as a serious problem in zoos holding small populations of rare mammals. Matings of close relatives tend to bring together genes from a common ancestor and increase the probability that two copies of a deleterious gene will come together in the same organism. The result is "inbreeding depression." Our management solution is to enlarge genetic diversity by bringing together captive animals from different zoos or by introducing new stock from wild populations if possible. Paradoxically, where zoo populations are extremely small and no wild stock can be obtained, deliberate inbreeding is recommended. This procedure selects for genes that tolerate inbreeding; deleterious genes disappear if they kill animals homozygous for them.

Migration

Migration prevents different populations of a species from diverging. If a large species is divided into many small populations, genetic drift and selection acting separately in the different populations can produce evolutionary divergence among them. A small amount of migration in each generation keeps the different populations from becoming too distinct genetically. For example, the French and Russian populations whose ABO allele frequencies were discussed previously show some genetic divergence, but their genetic connection through intervening populations by continuing migration prevents them from becoming completely distinct.

Natural Selection

Natural selection can change both allelic frequencies and genotypic frequencies in a population. Although the effects of selection are often reported for particular polymorphic genes, we must stress that natural selection acts on the whole animal, not on isolated traits. An organism that possesses a superior combination of traits will be favored. An animal may have traits that confer no advantage or even a disadvantage, but it is successful overall if its combination of traits is favorable. When we claim that a genotype at a particular gene has a higher **relative fitness** than others, we state that on average that genotype confers an advantage in survival and reproduction in the population. If alternative genotypes have unequal probabilities of survival and reproduction, Hardy-Weinberg equilibrium is upset.

Using the genetic theory of natural selection, one can measure relative **fitness** values associated with different genotypes in a population. Geneticists often use W to denote the expected average fitness of a genotype in a population, with the genotype of highest fitness given a value of one and fitnesses of other genotypes indicated as fractions.

We illustrate measurement of fitness using genetic variation associated with the disease sickle-cell anemia in human populations. Considering only the alleles for normal hemoglobin (A) and sickle-cell hemoglobin (S) for the beta-hemoglobin gene in human populations (p. 100), the possible genotypes are AA, AS, and SS. Measurements of viability of individuals of these three genotypes in nonmalarial environments give a fitness value of 1 to genotypes AA and AS and a fitness of 0.2 to genotype SS. People having the SS genotype, who are susceptible to severe anemia, are expected to contribute only 20% as many offspring to the next generation on average as are individuals having the AA or AS genotypes. In malarial environments, genotype AS has the highest fitness (=1); genotype AA has a slightly decreased fitness (=0.9) because these individuals have a greater incidence of malaria than AS individuals, and SS has a low fitness (=0.2) because of anemia. From these measured fitness values and knowledge of the frequencies of alleles in a population and its system of mating, one can calculate the **average effect** that an allele has on the phenotype of relative fitness in that population. In the example of sickle-cell anemia, the average effect of allele S on fitness in a malarial environment is a balance between the strongly negative effect it has when homozygous and the positive effect that it has when heterozygous with allele A.

In Chapter 36, we discuss the related concept of **inclusive fitness**. The average effect of an allele on fitness is expressed not only by its direct contribution to the fitness of its possessors but by aid that its possessors give to close relatives, who are likely also to contain copies of the allele. The term "inclusive fitness" pertains to cases where the average effect of an allele would be calculated incorrectly if only its direct effects on fitness were measured.

Some traits and combinations of traits are advantageous for certain aspects of an organism's survival or reproduction and disadvantageous for others. Darwin used the term **sexual selection** to denote the selection of traits that are advantageous for obtaining mates but not for survival. Bright colors and elaborate feathers can enhance a male bird's competitive ability in obtaining mates while simultaneously increasing his visibility to predators (Figure 6.30). Environmental changes, such as extinction of a predator population, can alter the selective value of different traits. The action of selection on character variation is therefore very complex.

Figure 6.30

A pair of wood ducks. Brightly colored feathers of male birds probably confer no survival advantage and might even be harmful by alerting predators. Such colors nonetheless confer advantage in attracting mates, which overcomes, on average, the negative consequences of these colors for survival. Darwin used the term "sexual selection" to denote evolution of traits that give an individual an advantage in reproduction, even if the traits are neutral or harmful for survival.

evolution of a species. Interaction of genetic drift and selection in different populations permits many different genetic combinations of many polymorphic genes to be tested against natural selection. Migration among populations permits particularly favorable new genetic combinations to spread throughout the species as a whole. Interaction of selection, genetic drift, and migration in this example produces evolutionary change qualitatively different from what would result if any of these three factors acted alone. Geneticist Sewall Wright called this interaction *shifting balance* because it permits a population to explore different adaptive combinations of variable traits. Natural selection, genetic drift, mutation, nonrandom mating, and migration interact in natural populations to create an enormous opportunity for evolutionary change; perpetual stability, as predicted by Hardy-Weinberg equilibrium, almost never occurs across any significant amount of evolutionary time.

Measuring Genetic Variation Within Populations

How do we measure the genetic variation that occurs in natural populations? Genetic dominance, interactions between alleles of different genes, and environmental effects on a phenotype make it difficult to quantify genetic variation indirectly by observing organismal phenotypes. Variability can be quantified, however, at the molecular level.

Protein Polymorphism

Different allelic forms of genes encode proteins that often differ slightly in their amino acid sequence. This phenomenon is called

Interactions of Selection, Drift, and Migration

Subdivision of a species into small populations that exchange migrants is an optimal situation for promoting rapid adaptive

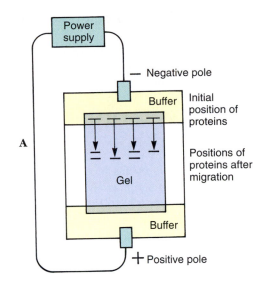

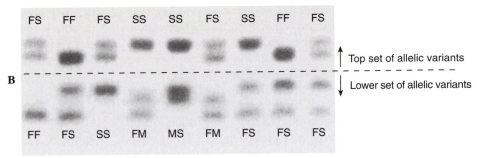

Figure 6.31

Study of genetic variation in proteins using gel electrophoresis. **A,** An electrophoretic apparatus separates allelic variants of proteins that differ in charge because of differences in their sequence of amino acids. **B,** Genetic variation in the protein leucine aminopeptidase for nine brown snails, *Helix aspersa.* Two different sets of allelic variants are revealed. The top set contains two alleles [denoted fast (F) and slow (S) according to their relative movement in the electric field]. Individuals homozygous for the fast allele show only a single fast band on the gel (FF), those homozygous for the slow allele show only a single slow band (SS), and heterozygous individuals have both bands (FS). The lower set contains three different alleles denoted fast (F), medium (M), and slow (S). Note that no individuals shown are homozygous for the medium (M) allele.

protein polymorphism. If these differences affect the protein's net electric charge, the different allelic forms can be separated using protein electrophoresis (Figure 6.31). We can identify the genotypes of particular individuals for protein-coding genes and measure allelic frequencies in a population.

Over the last 40 years, geneticists using this approach have discovered far more variation than was previously expected. Despite the high levels of polymorphism discovered using protein electrophoresis (Table 6.1), these studies underestimate both protein polymorphism and the total genetic variation present in a population. For example, protein polymorphism that does not involve charge differences is not detected. Furthermore, because the genetic code is degenerate (more than one codon for most amino acids, p. 94), protein polymorphism does not reveal all of the genetic variation present in protein-coding genes. Genetic changes that do not alter protein structure sometimes alter patterns of protein synthesis during development and can be very important to an organism. When all kinds of variation are considered, it is evident that most species have an enormous potential for further evolutionary change.

Quantitative Variation

Quantitative traits are those that show continuous variation with no obvious pattern of Mendelian segregation in their inheritance. The values of the trait in offspring often are intermediate between the values in the parents. Such traits are influenced by variation at many genes, each of which follows Mendelian inheritance and contributes a small, incremental amount to the total phenotype. Examples of traits that show quantitative variation include tail length in mice, length of a leg segment in grasshoppers, number of gill rakers in sunfishes, number of peas in pods, and height of adult males of the human species. When trait

TABLE 6.1

Values of Polymorphism (P) and Heterozygosity (H) for Various Animals and Plants as Measured Using Protein Electrophoresis

(a) Species	Number of Proteins	P*	H*
Humans	71	0.28	0.067
Northern elephant seal	24	0.0	0.0
Horseshoe crab	25	0.25	0.057
Elephant	32	0.29	0.089
Drosophila pseudoobscura	24	0.42	0.12
Barley	28	0.30	0.003
Tree frog	27	0.41	0.074

(b) Taxa	Number of Species	P*	H*
Plants	—	0.31	0.10
Insects (excluding *Drosophila*)	23	0.33	0.074
Drosophila	43	0.43	0.14
Amphibians	13	0.27	0.079
Reptiles	17	0.22	0.047
Birds	7	0.15	0.047
Mammals	46	0.15	0.036
Average		0.27	0.078

Source: Data from P.W. Hedrick, *Population biology.* Jones and Bartlett, Boston, 1984.
*P, the average number of alleles per gene per species; H, the proportion of heterozygous genes per individual.

values are graphed with respect to frequency distribution, they often approximate a normal, or bell-shaped, probability curve (Figure 6.32A). Most individuals fall near the average; fewer fall somewhat above or below the average, and extremes form the "tails" of the frequency curve with increasing rarity. Usually, the larger the population sample, the more closely the frequency distribution resembles a normal curve.

Selection can act on quantitative traits to produce three different kinds of evolutionary response (Figure 6.32B, C, and D). One outcome is to favor average values of the trait and to disfavor extreme ones; this outcome is called **stabilizing selection** (Figure 6.32B). **Directional selection** favors a phenotypic value either above or below the average and causes the population average to shift toward the favored value over time (Figure 6.32C). When we think about natural selection producing evolutionary change, it is usually directional selection that we have in mind, although we must remember that this is not

the only possibility. A third alternative is **disruptive selection** in which two different extreme phenotypes are simultaneously favored, but their average is disfavored (Figure 6.32D). The population then becomes bimodal, meaning that two very different phenotypic values predominate.

MACROEVOLUTION: MAJOR EVOLUTIONARY EVENTS

Macroevolution describes large-scale events in organic evolution. Speciation links macroevolution and microevolution. Major trends in the fossil record (see Figures 6.11, 6.12, and 6.13) are clearly within the realm of macroevolution. Patterns and processes of macroevolutionary change emerge from those of microevolution, but they acquire some degree of autonomy in doing so. The emergence of new adaptations and species, and the varying rates of speciation and extinction observed in the fossil record go beyond the fluctuations of allelic frequencies within populations.

Stephen Jay Gould recognized three different "tiers" of time at which we observe distinct evolutionary processes. The first tier constitutes the timescale of population genetic processes, from tens to thousands of years. The second tier covers millions of years, the scale on which rates of speciation and extinction are measured and compared among different groups of organisms. Punctuated equilibrium is a theory of the second tier, explaining the occurrence of speciation and morphological change and their association over millions of years. The third tier covers tens to hundreds of millions of years, and is marked by occurrence of episodic mass extinctions. In the fossil record of marine organisms, mass extinctions recur at intervals of approximately 26 million years. Five of these mass extinctions have been particularly disastrous (Figure 6.33). The study of long-term changes in animal diversity focuses on the third-tier timescale (see Figures 6.13 and 6.33).

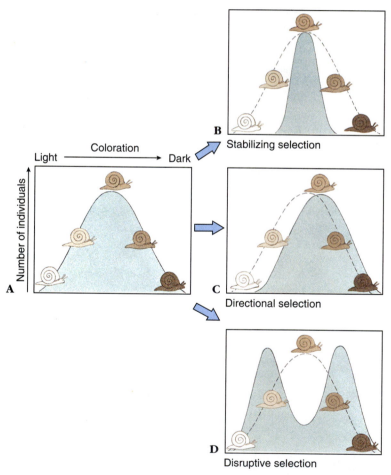

Figure 6.32

Responses to selection on a continuous (polygenic) character, coloration in a snail. **A,** The frequency distribution of coloration before selection. **B,** Stabilizing selection culls extreme variants from the population, in this case eliminating individuals that are unusually light or dark, thereby stabilizing the mean. **C,** Directional selection shifts the population mean, in this case by favoring darkly colored variants. **D,** Disruptive selection favors both extremes but not the mean; the mean is unchanged but the population no longer has a bell-shaped distribution of phenotypes.

Speciation and Extinction Through Geological Time

Evolutionary change at the second tier provides a new perspective on Darwin's theory of natural selection. Although a species may persist for many millions of years, it ultimately has two possible evolutionary fates: it may give rise to new species or become extinct without leaving descendants. Rates of speciation and extinction vary among lineages, and lineages that have the highest speciation rates and lowest extinction rates produce the greatest number of living species. The characteristics of a species may make it more or less likely than others to undergo speciation or extinction events. Because many characteristics are passed from ancestral to descendant species (analogous to heredity at the organismal level), lineages whose characteristics increase the probability of speciation and confer resistance to extinction should dominate the living world. This species-level process that produces differential rates of speciation and extinction among lineages is analogous in many ways

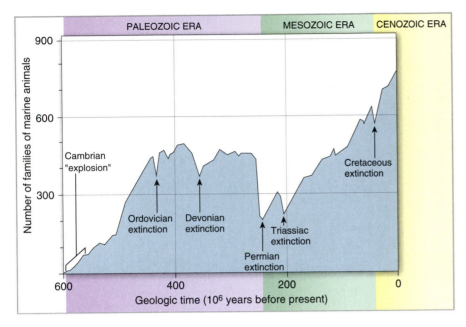

Figure 6.33

Changes in numbers of families of marine animals through time from the Cambrian period to the present. Sharp drops represent five major extinctions of skeletonized marine animals. Note that despite the extinctions, the overall number of marine families has increased to the present.

to natural selection. It represents an expansion of Darwin's theory of natural selection. This expansion is particularly important for macroevolution if one accepts the theory of punctuated equilibrium, which states that the evolutionarily important variation occurs primarily among rather than within species.

Species selection encompasses the differential survival and multiplication of species through geological time based on variation among lineages, especially in emergent, species-level properties. These species-level properties include mating rituals, social structuring, migration patterns, geographic distribution, and all other properties that emerge at the species level (see p. 6). Descendant species usually resemble their ancestors in these properties. For example, a "harem" system of mating in which a single male and several females compose a breeding unit characterizes some mammalian lineages but not others. We expect speciation rates to be enhanced by social systems that promote founding of new populations by small numbers of individuals. Certain social systems may increase the likelihood that a species will survive environmental challenges through cooperative action. Such properties would be favored by species selection over geological time.

Differential speciation and extinction among lineages also can be caused by variation in organismal-level properties (such as specialized versus generalized feeding) rather than species-level properties (see p. 6). Organisms that specialize in eating a restricted range of foods, for example, may be subjected more readily than generalized feeders to geographic isolation among populations, because areas where their preferred food is scarce or absent will function as geographic barriers to dispersal. Such geographic isolation could generate more frequent opportunities for speciation to occur throughout geological time. The fossil

records of two major groups of African antelopes suggest this result (Figure 6.11). A group of specialized grazers that contains blesboks, hartebeests, and wildebeests shows high speciation and extinction rates. Since the late Miocene, 33 extinct and 7 living species are known, representing at least 18 events of branching speciation and 12 terminal extinctions. In contrast, a group of generalist grazers and browsers that contains impalas shows neither branching speciation nor terminal extinction during this same interval of time. Interestingly, although these two lineages differ greatly in speciation rates, extinction rates, and species diversity, they do not differ significantly in total number of individual animals alive today.

Paleontologist Elisabeth Vrba, whose research produced the results in Figure 6.11, uses the term **effect macroevolution** to describe differential speciation and extinction rates among lineages caused by organismal-level properties. She reserves the term species selection for cases where species-level emergent properties are of primary importance. Some other evolutionary paleontologists consider effect macroevolution a subset of species selection because fitness differences occur among different species lineages rather than among varying organisms within species.

Mass Extinctions

When we study evolutionary change on an even larger timescale, we observe episodic events in which large numbers of taxa go extinct simultaneously. These events are called **mass extinctions** (see Figure 6.33). The most cataclysmic of these extinction episodes happened about 225 million years ago, when at least half of the families of shallow-water marine invertebrates, and fully 90% of marine invertebrate species disappeared within a few million years. This event was the **Permian extinction.** The **Cretaceous extinction,** which occurred about 65 million years ago, marked the end of dinosaurs, as well as numerous marine invertebrates and many small reptilian taxa.

Causes of mass extinctions and evolutionary timing of mass extinctions at intervals of approximately 26 million years are difficult to explain. Some people have proposed biological explanations for these episodic mass extinctions and others consider many mass extinctions artifacts of our statistical and taxonomic analyses. Walter Alvarez proposed that the earth was periodically bombarded by asteroids, causing these mass extinctions (Figure 6.34). The drastic effects of such bombardment of a planet were observed in July 1994 when fragments of Comet Shoemaker-Levy 9 bombarded Jupiter. The first fragment to hit Jupiter was estimated to have the force of 10 million hydrogen bombs. Twenty additional fragments hit Jupiter within the following week, one of which was 25 times more powerful than the first fragment. This bombardment was the most violent event in the recorded history of the solar system. A similar bombardment on earth would send

Figure 6.34

Twin craters of Clearwater Lakes in Canada show that multiple impacts on the earth are not as unlikely as they might seem. Evidence suggests that at least two impacts within a short time were responsible for the Cretaceous mass extinction.

debris into the atmosphere, blocking sunlight and causing drastic changes of climate. Temperature changes would challenge ecological tolerances of many species. Alvarez's hypothesis is being tested in several ways, including a search for impact craters left by asteroids and for altered mineral content of rock strata where mass extinctions occurred. Atypical concentrations of the rare-earth element iridium in strata at the Cretaceous-Tertiary boundary imply that this element entered the earth's atmosphere through asteroid bombardment.

Sometimes, lineages favored by species selection are unusually susceptible to mass extinction. Climatic changes produced by the hypothesized asteroid bombardments could produce selective challenges very different from those encountered at other times in the earth's history. Selective discrimination of particular biological traits by events of mass extinction is termed **catastrophic species selection.** For example, mammals survived the end-Cretaceous mass extinction that destroyed the dinosaurs and other prominent vertebrate and invertebrate groups. Following this event, mammals were able to use environmental resources that previously had been denied them, leading to their adaptive radiation.

Natural selection, species selection, and catastrophic species selection interact to produce the macroevolutionary trends seen in the fossil record. Studies of these interacting causal processes have made modern evolutionary paleontology an active and exciting field.

SUMMARY

Organic evolution explains the diversity of living organisms as the historical outcome of gradual change from previously existing forms. Evolutionary theory is strongly identified with Charles Robert Darwin, who presented the first credible explanation for evolutionary change. Darwin derived much of the material used to construct his theory from his experiences on a five-year voyage around the world aboard the H.M.S. *Beagle*.

Darwin's evolutionary theory has five major components. Its most basic proposition is *perpetual change*, the theory that the world is neither constant nor perpetually cycling but is steadily undergoing irreversible change. The fossil record amply demonstrates perpetual change in the continuing fluctuation of animal form and diversity following the Cambrian explosion 600 million years ago. Darwin's theory of *common descent* states that all organisms descend from a common ancestor through a branching of genealogical lineages. This theory explains morphological homologies among organisms as characteristics inherited with modification from a corresponding feature in their common evolutionary ancestor. Patterns of homology formed by common descent with modification permit us to classify organisms according to their evolutionary relationships.

Changes in the timing of developmental processes, termed heterochrony, and changes in their physical location within the body plan, termed heterotopy, explain the evolution of new morphological homologies. A developmental evolutionary module is a set of developmental processes and associated genes that can be expressed as a unit at different parts of the body to produce different structures with some shared developmental properties. Evolution of limbs in terrestrial vertebrates occurred by expressing at the limb bud a set of developmental processes that evolved initially to construct part of the vertebral column. Evolvability denotes the potential of a lineage to evolve new morphological features by using a set of developmental modules as an evolutionary toolkit.

A corollary of common descent is the *multiplication of species* through evolutionary time. Allopatric speciation denotes the evolution of reproductive barriers between geographically separated populations to generate new species. In some animals, especially parasitic insects that specialize on different host species, speciation may occur without geographical isolation, which is called sympatric speciation. Intermediate between allopatric speciation and sympatric speciation is a third mode, parapatric speciation, in which an environmental change splits a species into two environmentally distinct parts that maintain contact along a geographic borderline as they diverge to become separate species.

Adaptive radiation is the proliferation of many adaptively diverse species from a single ancestral lineage within a relatively short period of evolutionary time, such as a few million years. Oceanic archipelagoes, such as the Galápagos Islands, are particularly conducive to adaptive radiation of terrestrial organisms.

Darwin's theory of *gradualism* states that large phenotypic differences between species are produced by accumulation through evolutionary time of many individually small changes. Gradualism is still controversial. Mutations that have large effects on an organism have been useful in animal breeding, leading some to dispute Darwin's claim that such mutations are not important in evolution. On a macroevolutionary perspective, punctuated equilibrium states that most evolutionary change occurs in relatively brief events of branching speciation, separated by long intervals in which little phenotypic change accumulates.

Darwin's fifth major statement is that *natural selection* is the guiding force of evolution. This principle is founded on observations that all species overproduce their kind, causing a struggle for the limited resources that support existence. Because no two organisms are exactly alike, and because variable traits are at least partially heritable, those organisms whose hereditary endowment enhances their use of resources for survival and reproduction contribute disproportionately to the next generation. Over many generations, the sorting of variation by selection produces new species and new adaptations.

Mutations are the ultimate source of all new variation on which selection acts. Darwin's theory emphasizes that variation is produced at random with respect to an organism's needs and that differential survival and reproduction provide the direction for evolutionary change. Darwin's theory of natural selection was modified around 1900 and in subsequent decades by correction of his genetic errors. This modified theory is called neo-Darwinism.

Population geneticists discovered the principles by which genetic properties of populations change through time. A particularly important discovery, known as Hardy-Weinberg equilibrium, showed that the hereditary process itself does not change the genetic composition of populations. Important sources of evolutionary change include mutation, genetic drift, nonrandom mating, migration, natural selection, and their interactions.

Neo-Darwinism, as elaborated by population genetics, formed the basis for the Synthetic Theory of the 1930s and 1940s. Genetics, natural history, paleobiology, and systematics were unified by the common goal of expanding our knowledge of Darwinian evolution. Microevolution comprises studies of genetic change within contemporary populations. These studies show that most natural populations contain enormous amounts of variation. Macroevolution comprises studies of evolutionary change on a geological timescale. Macroevolutionary studies measure rates of speciation, extinction, and changes of diversity through time. These studies have expanded Darwinian evolutionary theory to include higher-level processes that regulate rates of speciation and extinction among lineages, including species selection and catastrophic species selection.

REVIEW QUESTIONS

1. Briefly summarize Lamarck's concept of the evolutionary process. What is wrong with this concept?
2. What is "uniformitarianism"? How did it influence Darwin's evolutionary theory?
3. Why was the *Beagle*'s journey so important to Darwin's thinking?
4. What was the key idea contained in Malthus's essay on populations that was to help Darwin formulate his theory of natural selection?
5. Explain how each of the following contributes to Darwin's evolutionary theory: fossils; geographic distributions of closely related animals; homology; animal classification.
6. How do modern evolutionists view the relationship between ontogeny and phylogeny? Explain how the observation of paedomorphosis conflicts with Haeckel's "biogenetic law."
7. What are the important differences between the vicariant and founder-event modes of allopatric speciation?
8. What are reproductive barriers? How do premating and postmating barriers differ?
9. Under what conditions is sympatric speciation proposed?
10. What is the main evolutionary lesson provided by Darwin's finches on the Galápagos Islands?
11. How is the observation of "sporting mutations" in animal breeding used to challenge Darwin's theory of gradualism? Why did Darwin reject such mutations as having little evolutionary importance?
12. What does the theory of punctuated equilibrium state about the occurrence of speciation throughout geological time? What observation led to this theory?
13. Describe the observations and inferences that compose Darwin's theory of natural selection.
14. Identify the random and nonrandom components of Darwin's theory of natural selection.
15. Describe some recurring criticisms of Darwin's theory of natural selection. How can these criticisms be refuted?
16. It is a common but mistaken belief that because some alleles are dominant and others are recessive, the dominants will eventually replace all the recessives in a population. How does the Hardy-Weinberg equilibrium refute this notion?
17. Assume that you are sampling a trait in animal populations; the trait is controlled by a single allelic pair A and a, and you can distinguish all three phenotypes AA, Aa, and aa (intermediate inheritance). Your sample includes:

Population	AA	Aa	aa	TOTAL
I	300	500	200	1000
II	400	400	200	1000

Calculate the distribution of phenotypes in each population as expected under Hardy-Weinberg equilibrium. Is population I in equilibrium? Is population II in equilibrium?
18. If after studying a population for a trait determined by a single pair of alleles you find that the population is not in equilibrium, what possible reasons might explain the lack of equilibrium?
19. Explain why genetic drift is more powerful in small populations.
20. Describe how the effects of genetic drift and natural selection can interact in a subdivided species.
21. Is it easier for selection to remove a deleterious recessive allele from a randomly mating population or a highly inbred population? Why?
22. Distinguish between microevolution and macroevolution, and describe some evolutionary processes evident only at the macroevolutionary level.

SELECTED REFERENCES

Avise, J. C. 2004. Molecular markers, natural history, and evolution, ed. 2. Sunderland, Massachusetts, Sinauer Associates. *An exciting and readable account of using molecular studies to help us understand evolution.*

Browne, E. J. 2002. Charles Darwin: a biography. New York, Knopf. *A comprehensive two-volume biography of Darwin completed in 2002.*

Conner, J. K., and D. L. Hartl. 2004. A primer of ecological genetics. Sunderland, Massachusetts, Sinauer Associates. *An introductory text on population genetics.*

Coyne, J. A., and H. A. Orr. 2004. Speciation. Sunderland, Massachusetts, Sinauer Associates. *A detailed coverage of speciation with emphasis on controversies in this field.*

Darwin, C. 1859. On the origin of species by means of natural selection, or the preservation of favoured races in the struggle for life. London, John Murray. *There were five subsequent editions by the author.*

Freeman, S., and J. C. Herron. 2007. Evolutionary analysis, ed. 4. Upper Saddle River, New Jersey, Pearson/Prentice Hall. *An introductory textbook on evolutionary biology designed for undergraduate biology majors.*

Futuyma, D. J. 2005. Evolution. Sunderland, Massachusetts, Sinauer Associates. *A very thorough introductory textbook on evolution.*

Gould, S. J. 2002. The structure of evolutionary theory. Cambridge, Massachusetts, Belknap Press of Harvard University Press. *A provocative discussion of what fossils tell us about the nature of life's evolutionary history.*

Graur, D., and W. H. Li. 2000. Fundamentals of molecular evolution. Sunderland, Massachusetts, Sinauer Associates. *A current textbook on molecular evolution.*

Hall, B. K. 1998. Evolutionary developmental biology. New York, Chapman and Hall. *An excellent textbook on the emerging field of developmental evolutionary biology.*

Hartl, D. L., and A. G. Clark. 2007. Principles of population genetics, ed. 4. Sunderland, Massachusetts, Sinauer Associates. *A current textbook on population genetics.*

Jablonski, D. 2005. Mass extinctions and macroevolution. Paleobiology **31:** S192–210. *A recent review of extinction theory.*

Levinton, J. S. 2001. Genetics, paleontology and macroevolution, ed. 2. Cambridge, U.K., Cambridge University Press. *A provocative discussion on the Darwinian basis of macroevolutionary theory.*

Mayr, E. 2001. What evolution is. New York, Basic Books. *A general survey of evolution by a leading evolutionary biologist.*

Mousseau, T. A., B. Sinervo, and J. Endler (eds.). 2000. Adaptive genetic variation in the wild. Oxford, U.K., Oxford University Press. *Detailed examples of adaptively important genetic variation in natural populations.*

Ruse, M. 1998. Philosophy of biology. Amherst, New York, Prometheus Books. *A collection of essays on evolutionary biology, including information on the Arkansas Balanced Treatment for Creation-Science and Evolution-Science Act.*

Stokstad, E. 2001. Exquisite Chinese fossils add new pages to book of life. Science **291:**232–236. *Exciting new fossil discoveries help to complete our understanding of life's evolutionary history. Some related articles immediately follow this one.*

Templeton, A. R. 2006. Population genetics and microevolutionary theory. Hoboken, New Jersey, Wiley-Liss. *An insightful treatment of evolutionary theory at the population level.*

West-Eberhard, M. J. 2003. Developmental plasticity and evolution. Oxford, U.K., Oxford University Press. *A provocative discussion of the role of development and modularity in evolution.*

ONLINE LEARNING CENTER

Visit www.mhhe.com/hickmanipz14e for chapter quizzing, key term flash cards, web links, and more!

The Reproductive Process

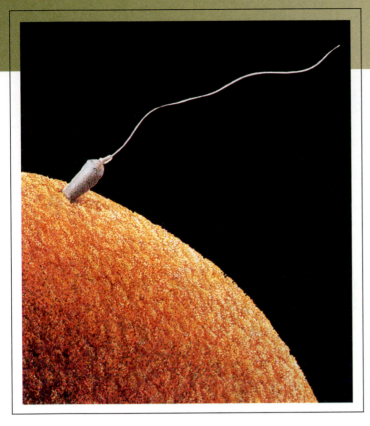

Human egg and sperm at the moment of fertilization.

"Omne vivum ex ovo"

In 1651, late in a long life, William Harvey, the English physiologist who earlier had founded experimental physiology by explaining the circuit of blood, published a treatise on reproduction. He asserted that all life developed from the egg—*omne vivum ex ovo*. This was insightful, since Harvey had no means for visualizing eggs of many animals, in particular the microscopic mammalian egg, which is no larger than a speck of dust to the unaided eye. Further, argued Harvey, eggs are launched into their developmental course by some influence from semen, a conclusion that was either remarkably perceptive or a lucky guess, since sperm also were invisible to Harvey. Such ideas differed sharply from existing notions of biogenesis, which saw life springing from many sources of which

eggs were but one. Harvey was describing characteristics of sexual reproduction in which two parents, male and female, must produce gametes that fuse to become a new individual.

Despite the importance of Harvey's aphorism that all life arises from eggs, it was not wholly correct. Life springs from reproduction of preexisting life, and reproduction may not be restricted to eggs and sperm. Asexual reproduction, the creation of new, genetically identical individuals by budding or fragmentation or fission from a single parent, is common, indeed characteristic, among some phyla. Nevertheless, most animals have found sex the winning strategy, probably because sexual reproduction promotes diversity, enhancing long-term survival of the lineage in a world of perpetual change.

Reproduction is one of the ubiquitous properties of life. Evolution is inextricably linked to reproduction, because the ceaseless replacement of aging predecessors with new life gives animal populations the means to adapt to a changing environment. In this chapter we distinguish asexual and sexual reproduction and explore the reasons why, for multicellular animals at least, sexual reproduction appears to offer important advantages over asexual. We then consider, in turn, the origin and maturation of germ cells; plan of reproductive systems; reproductive patterns in animals; and, finally, the endocrine events that orchestrate reproduction.

NATURE OF THE REPRODUCTIVE PROCESS

Two modes of reproduction are recognized: asexual and sexual. In **asexual** reproduction (Figure 7.1A and B) there is only one parent and with no special reproductive organs or cells. Each organism is capable of producing genetically identical copies of itself as soon as it becomes an adult. The production of copies is marvelously simple, direct, and typically rapid. **Sexual** reproduction (Figure 7.1C and D) as a rule involves two parents, each of which contributes special **germ cells** (**gametes** or **sex cells**) that in union (fertilization) develop into a new individual. The **zygote** formed from this union receives genetic material from both parents, and the combination of genes (p. 90) produces a genetically unique individual, bearing characteristics of the species but also bearing traits that make it different from its parents. Sexual reproduction, by recombining parental characters, multiplies variations and makes possible a richer and more diversified evolution.

Mechanisms for interchange of genes between individuals are more limited in organisms with only asexual reproduction. Of course, in asexual organisms that are haploid (bear only one set of genes, p. 77), mutations are immediately expressed and evolution proceeds quickly. In sexual animals, on the other hand, a gene mutation is often not expressed immediately, since it may be masked by its normal partner on the homologous chromosome. (Homologous chromosomes, discussed on p. 77, are those that pair during meiosis and carry genes encoding the same characteristics.) There is only a remote chance that both members of a gene pair will mutate in the same way at the same moment and thus be expressed immediately.

Asexual Reproduction: Reproduction without Gametes

Asexual reproduction (Figure 7.1A and B) is the production of individuals without gametes (eggs or sperm). It includes a number of distinct processes, all without involving sex or a second parent. Offspring produced by asexual reproduction all have the same genotype (unless mutations occur) and are called **clones.**

Asexual reproduction appears in bacteria and unicellular eukaryotes and in many invertebrate phyla, such as cnidarians, bryozoans, annelids, echinoderms, and hemichordates. In

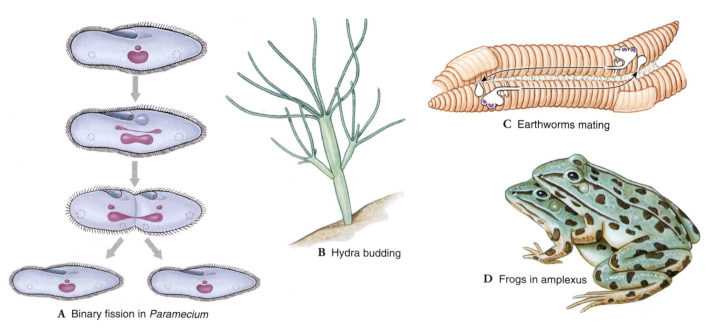

A Binary fission in *Paramecium*

B Hydra budding

C Earthworms mating

D Frogs in amplexus

Figure 7.1

Examples of asexual and sexual reproduction in animals. **A,** Binary fission in *Paramecium,* a single-celled eukaryote, results in two individuals. **B,** Budding, a simple form of asexual reproduction as shown in a hydra, a radiate animal. The buds, shown growing out of the central, parent hydra, eventually detach themselves and grow into fully formed individuals. **C,** Earthworms reproduce sexually, but are hermaphroditic, with each individual bearing both male and female organs. Each earthworm passes sperm from genital pores along grooves to seminal receptacles of its mate. **D,** Frogs, here in mating position (amplexus), represent bisexual reproduction, the most common form of sexual reproduction involving separate male and female individuals.

animal phyla in which asexual reproduction occurs, most members also employ sexual reproduction. In these groups, asexual reproduction ensures rapid increase in numbers when development and differentiation of the organism has not advanced to the point of forming gametes. Asexual reproduction is absent among vertebrates (although some forms of parthenogenesis have been interpreted as asexual by some authors; see p. 140).

It would be a mistake to conclude that asexual reproduction is in any way a "defective" form of reproduction relegated to the minute forms of life. Given the facts of their abundance, that they have persisted on earth for 3.5 billion years, and that they form the base of the food chain on which all higher forms depend, single-celled asexual organisms are both resoundingly abundant and supremely important. For these forms the advantages of asexual reproduction are its rapidity (many bacteria divide every half hour) and simplicity (no germ cells to produce and no time and energy expended in finding a mate).

The basic forms of asexual reproduction are fission (binary and multiple), budding, gemmulation, and fragmentation.

Binary fission is common among bacteria and protozoa (Figure 7.1A). In binary fission the body of the unicellular parent divides by mitosis (p. 52) into two approximately equal parts, each of which grows into an individual similar to the parent. Binary fission may be lengthwise, as in flagellate protozoa, or transverse, as in ciliate protozoa. In **multiple fission, or schizogony,** the nucleus divides repeatedly before division of the cytoplasm, producing many daughter cells simultaneously. Spore formation, called sporogony, is a form of multiple fission common among some parasitic protozoa, for example, malarial parasites.

Budding is an unequal division of an organism. A new individual arises as an outgrowth (bud) from its parent, develops organs like those of the parent, and then detaches itself. Budding occurs in several animal phyla and is especially prominent in cnidarians (Figure 7.1B).

Gemmulation is the formation of a new individual from an aggregation of cells surrounded by a resistant capsule, called a gemmule. In many freshwater sponges, gemmules develop in the fall and survive the winter in the dried or frozen body of the parent. In spring, the enclosed cells become active, emerge from the capsule, and grow into a new sponge.

In **fragmentation** a multicellular animal breaks into two or more parts, with each fragment capable of becoming a complete individual. Many invertebrates can reproduce asexually by simply breaking into two parts and then regenerating the missing parts of the fragments, for example, most anemones and many hydroids. Many echinoderms can regenerate lost parts, but this is not the same as reproduction by fragmentation.

Sexual Reproduction: Reproduction with Gametes

Sexual reproduction is the production of individuals from gametes. It includes **bisexual** (or **biparental**) reproduction as the most common form, involving two separate individuals. **Hermaphroditism** and **parthenogenesis** are less common forms of sexual reproduction.

Bisexual Reproduction

Bisexual reproduction is the *production of offspring formed by the union of gametes from two genetically different parents* (Figures 7.1C and D, and 7.2). Offspring will thus have a new genotype different from either parent. Individuals sharing parenthood are characteristically of different **sexes,** male and female (there are exceptions among sexually reproducing organisms, such as bacteria and some protozoa in which sexes are lacking). Each has its own reproductive system and produces only one kind of germ cell, spermatozoon or ovum, rarely both. Nearly all vertebrates and many invertebrates have separate sexes, and such a condition is called **dioecious** (Gr. *di,* two, + *oikos,* house). Individual animals that have both male and female reproductive organs are called **monoecious** (Gr. *monos,* single, + *oikos,* house). These animals are called **hermaphrodites** (from a combination of the names of the Greek god Hermes and goddess Aphrodite); this form of reproduction is described on page 140.

Distinctions between male and female are based, not on any differences in parental size or appearance, but on the size and mobility of the gametes they produce. The **ovum** (egg), produced by the female, is large (because of stored yolk to sustain early development), nonmotile, and produced in relatively small numbers. The **spermatozoon** (sperm), produced by the male, is small, motile, and produced in enormous numbers. Each sperm is a stripped-down package of highly condensed genetic material designed for the single purpose of reaching and fertilizing an egg.

There is another crucial event that distinguishes sexual from asexual reproduction: **meiosis,** a distinctive type of gamete-producing nuclear division (described in detail on p. 77). Meiosis differs from ordinary cell division (mitosis) in being a double division. Chromosomes split once, but the cell divides *twice,* producing four cells, each with half the original number of chromosomes (the **haploid** number). Meiosis is followed by **fertilization** in which two haploid gametes are combined to restore the normal (**diploid**) chromosomal number of the species.

The new cell (zygote), which now begins to divide by mitosis (described on p. 52), has equal numbers of chromosomes from each parent and is a unique individual bearing a recombination of parental characteristics. Genetic recombination is the great strength of sexual reproduction that keeps feeding new genetic combinations into the population.

Many unicellular organisms reproduce both sexually and asexually. When sexual reproduction does occur, it may or may not involve male and female gametes. Sometimes two mature sexual parent cells join together to exchange nuclear material or merge cytoplasm (**conjugation,** p. 232 in Chapter 11). Distinct sexes do not exist in these cases.

The male-female distinction is more clearly evident in most animals. Organs that produce germ cells are called **gonads.** The gonad that produces sperm is a **testis** (see Figure 7.12) and

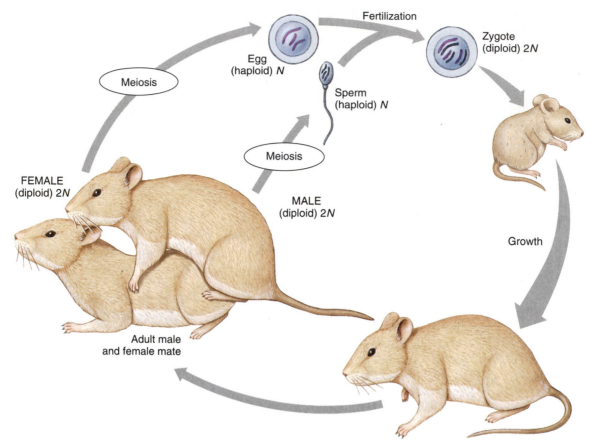

Figure 7.2

A sexual life cycle. The life cycle begins with haploid germ cells, formed by meiosis, combining to form a diploid zygote, which grows by mitosis to an adult. Most of the life cycle is spent as a diploid organism.

that which forms eggs is an **ovary** (see Figure 7.13). Gonads represent the **primary sex organs,** the only sex organs found in certain groups of animals. Most metazoa, however, have various **accessory sex organs** (such as penis, vagina, uterine tubes, and uterus) that transfer and receive germ cells. In the primary sex organs germ cells undergo many complicated changes during their development, the details of which are described on pages 143–146.

Hermaphroditism

Animals that have both male and female organs in the same individual are called **hermaphrodites,** and the condition is called **hermaphroditism.** In contrast to the dioecious state of separate sexes, hermaphrodites are **monoecious,** meaning that the same organism bears both male and female organs. Many sessile, burrowing, or endoparasitic invertebrate animals are hermaphoditic (for example, most flatworms, some hydroids and annelids, and all barnacles and pulmonate snails), as well as a few vertebrates (some fishes). Some hermaphrodites fertilize themselves, but most avoid self-fertilization by exchanging germ cells with another member of the same species (Figures 7.1C and 7.3). An advantage is that with every individual producing

eggs, a hermaphroditic species could potentially produce twice as many offspring as could a dioecious species in which half the individuals are nonproductive males. Some fishes are **sequential hermaphrodites,** in which a genetically programmed sex change occurs within an individual organism. In many species of reef fishes, for example, wrasses, an animal begins life as either a female or a male (depending on the species) but later becomes the opposite sex.

Parthenogenesis

Parthenogenesis ("virgin origin") is the development of an embryo from an unfertilized egg or one in which the male and female nuclei fail to unite following fertilization. There are many patterns of parthenogenesis. In one type, called **ameiotic parthenogenesis,** no meiosis occurs, and the egg is formed by mitotic cell division. This "asexual" form of parthenogenesis occurs in some species of flatworms, rotifers, crustaceans, insects, and probably others. In these cases, the offspring are clones of the parent because, without meiosis, the parent's chromosomal complement is passed intact to offspring.

In **meiotic parthenogenesis** a haploid ovum is formed by meiosis, and it may or may not be activated by the influence of

Figure 7.3
Hermaphroditic earthworms mating. Earthworms are "simultaneous" hermaphrodites; during mating each partner passes sperm from genital pores along grooves to seminal receptacles of its mate. They are held together by mucous secretions during this process.

a male's sperm. For example, in some species of fishes, a female may be inseminated by a male of the same or related species, but the sperm serves only to activate the egg; the male's genetic material is rejected before it can penetrate the egg (**gynogenesis**). In several species of flatworms, rotifers, annelids, mites, and insects, the haploid egg begins development spontaneously; no males are required to stimulate activation of an ovum. The diploid condition may be restored by chromosomal duplication or by autogamy (rejoining of haploid nuclei). A variant of this type of parthenogenesis occurs in many bees, wasps, and ants. In honey bees, for example, the queen bee can either fertilize eggs as she lays them or allow them to pass unfertilized. Fertilized eggs become diploid females (queens or workers), and unfertilized eggs develop parthenogenetically to become haploid males (drones); this type of sex determination is known as **haplodiploidy.** In some animals meiosis may be so severely modified that offspring are clones of the parent. Certain populations of whiptail lizards of the American southwest are clones consisting solely of females (Cole, 1984).

Parthenogenesis is surprisingly widespread in animals. It is an abbreviation of the usual steps of bisexual reproduction. It may have evolved to avoid the problem—which may be great in some animals—of bringing together males and females at the right moment for successful fertilization. The disadvantage of parthenogenesis is that if the environment should suddenly change parthenogenetic species have limited capacity to shift gene combinations to adapt to any new conditions. Bisexual species, by recombining parental characteristics, have a better chance of producing variant offspring that can utilize new environments.

Occasionally claims arise that spontaneous parthenogenetic development to term has occurred in humans. A British investigation of about 100 cases in which the mother denied having had intercourse revealed that in nearly every case the child possessed characteristics

not present in the mother, and consequently must have had a father. Nevertheless, mammalian eggs very rarely spontaneously start developing into embryos without fertilization. In certain strains of mice, such embryos will develop into fetuses and then die. The most remarkable instance of parthenogenetic development among vertebrates occurs in turkeys in which ova of certain strains, selected for their ability to develop without sperm, grow to reproducing adults.

Why Do So Many Animals Reproduce Sexually Rather Than Asexually?

Because sexual reproduction is so nearly universal among animals, it might be inferred to be highly advantageous. Yet it is easier to list disadvantages to sex than advantages. Sexual reproduction is complicated, requires more time, and uses much more energy than asexual reproduction. Males may waste valuable energy in competition for a mate and often possess sexual characteristics that can be detrimental to survival—for example, the elongated tail feathers of peacocks. Mating partners must come together, and this can be a disadvantage in sparsely populated areas for some species. Males and females must also coordinate their activities to produce young. Many biologists believe that an even more troublesome problem is the "cost of meiosis." A female that reproduces asexually passes all of her genes to her offspring, but when she reproduces sexually the genome is divided during meiosis and only half her genes flow to the next generation. Another cost is wastage in production of males, many of which fail to reproduce and thus consume resources that could be applied to production of females. Whiptail lizards of the American southwest offer a fascinating example of the potential advantage of parthenogenesis (discussed on p. 140). When unisexual and bisexual species of the same genus are reared under similar conditions in the laboratory, the population of the unisexual species grows more quickly because all unisexual lizards (all females) deposit eggs, whereas only 50% of the bisexual lizards do so (Figure 7.4).

Clearly, the costs of sexual reproduction are substantial. How are they offset? Biologists have disputed this question for years. One hypothesis suggests that sexual reproduction, with its separation and recombination of genetic material, keeps producing novel genotypes that *in times of environmental change* may be advantageous to survival and thus the organism may live to reproduce, whereas most others die. An often quoted example is the rapidly changing environment of organisms that is produced by parasites continuously evolving new mechanisms of attack and therefore favoring recombination of their hosts. Variability, advocates of this viewpoint argue, is sexual reproduction's trump card. Another hypothesis suggests that sexual recombination provides a means for the spread of beneficial mutations without a population being held back by deleterious ones. Experimental support for this hypothesis has recently been provided using the fruit fly, *Drosophila,* in which beneficial mutations increased to a greater degree in sexual populations compared with clonal (asexual) ones. These hypotheses are not mutually exclusive,

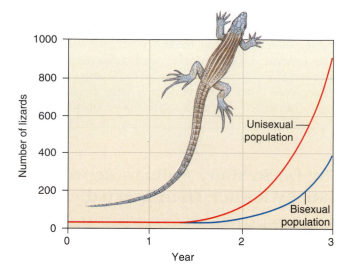

Figure 7.4

Comparison of the growth of a population of unisexual whiptail lizards with a population of bisexual lizards. Because all individuals of the unisexual population are females, all produce eggs, whereas only half the bisexual population are egg-producing females. By the end of the third year the unisexual lizards are more than twice as numerous as the bisexual ones.

however, and both provide possible explanations for the evolution of sexual reproduction.

There still remains the question of why sexual reproduction has been maintained in spite of its costs. Considerable evidence suggests that asexual reproduction is most successful in colonizing new environments. When habitats are empty what matters most is rapid reproduction; variability and increased fitness provided by beneficial genetic recombination matter little. As habitats become more crowded, competition between species for resources increases. Selection becomes more intense, and genetic variability—new beneficial genotypes produced by recombination in sexual reproduction—furnishes the diversity that permits a population to resist extinction. Therefore, on a geological timescale, asexual lineages, because they lack genetic flexibility, may be more prone to extinction than sexual lineages. Sexual reproduction is therefore favored by species selection (species selection is described on p. 133). There are many invertebrates that use both sexual and asexual reproduction, thus enjoying the advantages each has to offer.

THE ORIGIN AND MATURATION OF GERM CELLS

Many sexually reproducing organisms are composed of nonreproductive **somatic cells,** which are differentiated for specialized functions and die with the individual, and **germ cells,** which form the gametes: eggs and sperm. Germ cells provide continuity of life between generations, the **germ cell line.** Germ cells, or their precursors, the **primordial germ cells,** develop at the beginning of embryonic development (described in Chapter 8), usually in the endoderm, and migrate to the gonads. Here

they develop into eggs or sperm. The other cells of the gonads are somatic cells. They cannot form eggs or sperm, but they are necessary for support, protection, and nourishment of the germ cells during their development **(gametogenesis).**

A traceable germ cell line, as present in vertebrates, is also distinguishable in some invertebrates, such as nematodes and arthropods. In many invertebrates, however, germ cells develop directly from somatic cells at some period in the life of an individual.

Migration of Germ Cells

In vertebrates, the actual tissue from which gonads arise appears in early development as a pair of **genital ridges,** growing into the coelom from the dorsal coelomic lining on each side of the hindgut near the anterior end of the kidney (mesonephros).

Surprisingly perhaps, primordial germ cells do not arise in the developing gonad, but in the yolk-sac endoderm (p. 175). From studies with frogs and toads, it has been possible to trace the germ cell line back to the fertilized egg, in which a localized area of germinal cytoplasm (called **germ plasm**) can be identified in the vegetal pole of the uncleaved egg mass. This material can be followed through subsequent cell divisions of the embryo until it becomes situated in primordial germ cells in gut endoderm. From here the cells migrate by ameboid movement to the genital ridges, located on either side of the hindgut. A similar migration of primordial germ cells occurs in mammals (Figure 7.5). Primordial germ cells are the future stock of gametes for an animal. Once in the genital ridges and during subsequent gonadal development, germ cells begin to divide by mitosis, increasing their numbers from a few dozen to several thousand.

Sex Determination

At first gonads are sexually indifferent. In normal human males, a "male-determining gene" on the Y chromosome called **SRY (sex-determining region Y)** organizes the developing gonad into a testis instead of an ovary. Once formed, the testis secretes the steroid **testosterone.** This hormone, and its metabolite, **dihydrotestosterone (DHT),** masculinizes the fetus, causing the differentiation of penis, scrotum, and the male ducts and glands. It also destroys the incipient breast primordia, but leaves behind the nipples that are a reminder of the indifferent ground plan from which both sexes develop. Testosterone is also responsible for the masculinization of the brain, but it does so indirectly. Surprisingly, testosterone is enzymatically converted to estrogen in the brain, and it is **estrogen** that determines the organization of the brain for male-typical behavior.

Biologists have often stated that in mammals the indifferent gonad has an inherent tendency to become an ovary. Classic experiments performed in rabbits provide support for the idea that the female is the default sex during development. Removal of the fetal gonads before they have differentiated will invariably produce a female with uterine tubes, uterus, and vagina, even if the rabbit is a genetic male. Localization in 1994 of a region on the X chromosome named **DDS (dosage-sensitive sex reversal)** or

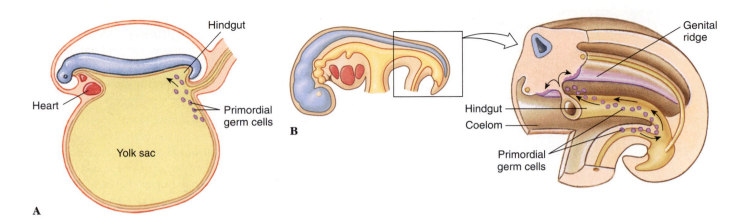

Figure 7.5

Migration of mammalian primordial germ cells. **A,** From the yolk sac the primordial germ cells migrate toward the region where the hindgut develops. **B,** Later-stage embryo in which the hindgut is more developed. Enlarged figure shows the germ cells migrating through the hindgut and into the genital ridges. In human embryos, migration is complete by the end of the fifth week of gestation.

SRVX (sex-reversing X), which promotes ovary formation, has challenged this view. In addition, the presence of such a region may help to explain feminization in some XY males. It is clear, however, that absence of testosterone in a genetic female embryo promotes development of female sexual organs: vagina, clitoris, and uterus. The developing female brain does require special protection from the effects of estrogen because, as mentioned earlier, estrogen causes masculinization of the brain. In rats, a blood protein (alpha-fetoprotein) binds to estrogen and keeps the hormone from reaching the developing female brain. This does not appear to be the case in humans, however, and even though circulating fetal estrogen levels can be quite high, the developing female brain does not become masculinized. One

possible explanation is that the level of brain estrogen receptors in the developing female brain is low, and therefore, high levels of circulating estrogen would have no effect.

The genetics of sex determination are treated in Chapter 5 (p. 80). Sex determination is strictly chromosomal in mammals, birds, amphibians, most reptiles, and probably most fishes. However, some fishes and reptiles lack sex chromosomes altogether; in these groups, gender is determined by nongenetic factors such as temperature or behavior. In crocodilians, many turtles, and some lizards the incubation temperature of the nest determines the sex ratio probably by indirectly activating and/or suppressing genes that direct development of the animals' sex organs. Alligator eggs, for example, incubated at low temperature all become females; those incubated at higher temperature all become males (Figure 7.6). Sex determination of many fishes is behavior dependent. Most of these species are hermaphroditic, possessing both male and female gonads. Sensory stimuli from the animal's social environment determine whether it will be male or female.

Gametogenesis

Mature gametes are produced by a process called gametogenesis. Although the same essential processes are involved in maturation of both sperm and eggs in vertebrates, there are some important differences. Gametogenesis in testes is called **spermatogenesis,** and in ovaries, **oogenesis.**

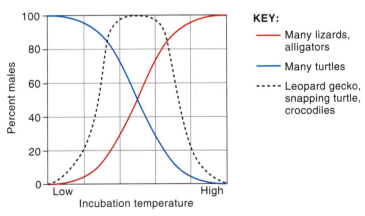

Figure 7.6

Temperature-dependent sex determination. In many reptiles that lack sex chromosomes incubation temperature of the nest determines gender. The graph shows that embryos of many turtles develop into males at low temperature, whereas embryos of many lizards and alligators become males at high temperatures. Embryos of crocodiles, leopard geckos, and snapping turtles become males at intermediate temperatures, and become females at higher or lower temperatures.

Source: Data from David Crews, "Animal Sexuality," Scientific American 270(1):108–114, January 1994.

Spermatogenesis

The walls of the seminiferous tubules contain differentiating germ cells arranged in a stratified layer five to eight cells deep (Figure 7.7). Germ cells develop in close contact with large **Sertoli** (sustentacular) **cells,** which extend from the periphery of the seminiferous tubules to the lumen and provide nourishment during germ-cell development and differentiation (Figure 7.8). The outermost layers contain **spermatogonia,** diploid cells that

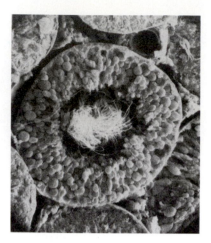

Figure 7.7

Section of a seminiferous tubule containing male germ cells. More than 200 meters long, highly coiled seminiferous tubules are packed in each human testis. This scanning electron micrograph reveals, in the tubule's central cavity, numerous tails of mature spermatozoa that have differentiated from germ cells in the periphery of the tubule. (×525)

From R. G. Kessel and R. H. Kardon, Tissues and Organs: A Text-Atlas of Scanning Electron Microscopy, *1979, W. H. Freeman and Co.*

have increased in number by mitosis. Each spermatogonium increases in size and becomes a **primary spermatocyte.** Each primary spermatocyte then undergoes the first meiotic division, as described in Chapter 5 (p. 79), to become two **secondary spermatocytes** (Figure 7.8).

For every structure in the reproductive system of males or females, there is a homologous structure in the other. This happens because during early development male and female characteristics begin to differentiate from the embryonic genital ridge, and two duct systems develop, which at first are identical in both sexes. Under the influence of sex hormones, the genital ridge develops into the testes of males and the ovaries of females. One duct system (mesonephric or Wolffian) becomes ducts of the testes in males and regresses in females. The other duct (paramesonephric or Müllerian) develops into the uterine tubes, uterus, and vagina of females and regresses in males. Similarly, the clitoris and labia of females are homologous to the penis and scrotum of males, since they develop from the same embryonic structures.

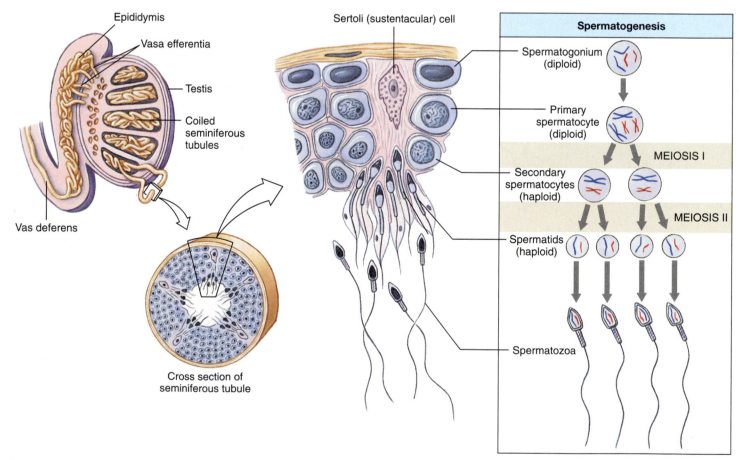

Figure 7.8

Spermatogenesis. Section of seminiferous tubule showing spermatogenesis. Germ cells develop within the recesses of large Sertoli (sustentacular) cells, that extend from the periphery of seminiferous tubules to their lumen, and that provide nourishment to the germ cells. Stem germ cells from which sperm differentiate are the spermatogonia, diploid cells located peripherally in the tubule. These divide by mitosis to produce either more spermatogonia or primary spermatocytes. Meiosis begins when primary spermatocytes divide to produce haploid secondary spermatocytes with double-stranded chromosomes. The second meiotic division forms four haploid spermatids with single-stranded chromosomes. As sperm develop, they are gradually pushed toward the lumen of the seminiferous tubule.

Each secondary spermatocyte enters the second meiotic division without intervention of a resting period. In the two steps of meiosis each primary spermatocyte gives rise to four **spermatids,** each containing the haploid number (23 in humans) of chromosomes. A spermatid usually contains a combination of his parents' chromosomes but may contain all chromosomes that the male inherited from his mother or from his father. Without further divisions the spermatids are transformed into mature **spermatozoa** or **(sperm)** (Figure 7.8). Modifications include great reduction of cytoplasm, condensation of the nucleus into a head, formation of a middle piece containing mitochondria, and a whiplike, flagellar tail for locomotion (Figures 7.8 and 7.9). The head consists of a nucleus containing the chromosomes for heredity and an **acrosome,** a distinctive feature of nearly all metazoa (exceptions are teleost fishes and certain invertebrates). In many species, both invertebrate and vertebrate, the acrosome contains enzymes that are released to clear an entrance through the layers that surround an egg. In mammals at least, one of the enzymes is hyaluronidase, which allows a sperm to penetrate the follicular cells surrounding an egg. A striking feature of many invertebrate spermatozoa is the acrosome filament, an extension of varying length in different species that projects suddenly from the sperm head when the latter first contacts the surface of an egg. Fusion of the egg and sperm plasma membranes is the initial event of fertilization (see Contact and Recognition between Egg and Sperm, p. 160).

The total length of a human sperm is 50 to 70 μm. Some toads have sperm that exceed 2 mm (2000 μm) in length (Figure 7.9) and are easily visible to the unaided eye. Most sperm, however, are microscopic in size (see p. 159 for an early seventeenth-century drawing of mammalian sperm, interpreted by biologists of the time as parasitic worms in the semen). In all sexually reproducing animals the number of sperm in males is far greater than the number of eggs in corresponding females. The number of eggs produced is correlated with the chances of young to hatch and to reach maturity.

Oogenesis

Early germ cells in the ovary, called **oogonia,** increase in number by mitosis. Each oogonium contains the diploid number of chromosomes. After the oogonia cease to increase in number, they grow in size and become **primary oocytes** (Figure 7.10). Before the first meiotic division, the chromosomes in each primary oocyte meet in pairs, paternal and maternal homologues, just as in spermatogenesis. When the first maturation (reduction) division occurs, the cytoplasm is divided unequally. One of the two daughter cells, the **secondary oocyte,** is large and receives most of the cytoplasm; the other is very small and is called the **first polar body** (Figure 7.10). Each of these daughter cells, however, has received half of the chromosomes.

In the second meiotic division, the secondary oocyte divides into a large **ootid** and a small polar body. If the first polar body also divides in this division, which sometimes happens, there are three polar bodies and one ootid (Figure 7.10). The ootid develops into a functional **ovum.** Polar bodies are nonfunctional, and they disintegrate. Formation of nonfunctional polar bodies is necessary to enable an egg to dispose of excess chromosomes,

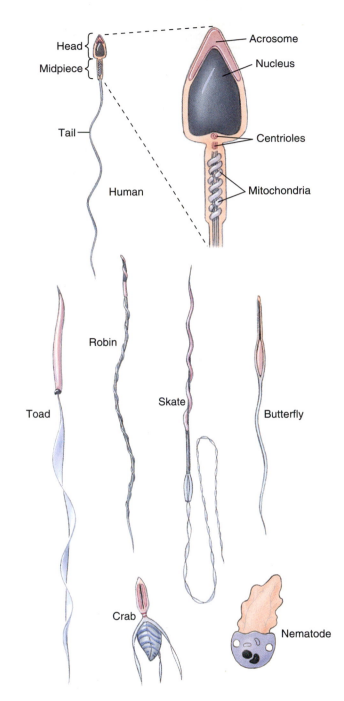

Figure 7.9
Examples of vertebrate and invertebrate sperm. The head and midpiece region of the human sperm is shown in more detail.

and the unequal cytoplasmic division makes possible a large cell with the cytoplasm containing a full set of cytoplasmic components needed for early development. Thus a mature ovum has N (haploid) number of chromosomes, the same as a sperm. However, each primary oocyte gives rise to only *one* functional gamete instead of four as in spermatogenesis.

In most vertebrates and many invertebrates the egg does not actually complete meiotic division before fertilization occurs. The general rule is that development is arrested during prophase I of the first meiotic division (in the primary oocyte phase). Meiosis resumes and is completed either at the time of ovulation (birds

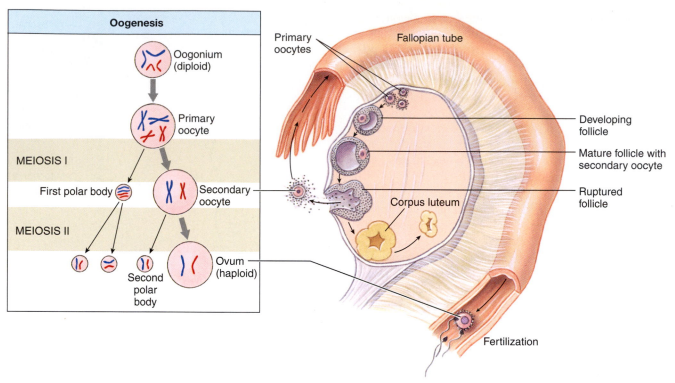

Figure 7.10

Oogenesis in humans. Early germ cells (oogonia) increase by mitosis during embryonic development to form diploid primary oocytes. After puberty, each menstrual month a diploid primary oocyte divides in the first meiotic division into a haploid secondary oocyte and a haploid polar body. If the secondary oocyte is fertilized, it enters the second meiotic division. The double-stranded chromosomes separate into a large ootid and small second polar body. The ootid develops into an ovum. Both ovum and second polar body now contain the N amount of DNA. Fusion of the haploid egg nucleus with a haploid sperm nucleus produces a diploid (2N) zygote.

and most mammals) or shortly after fertilization (many invertebrates, teleost fishes, amphibians, and reptiles). In humans, the ova begin the first meiotic division at about the thirteenth week of fetal development. Then their development arrests in prophase I as the primary oocyte until puberty, at which time one of these primary oocytes typically develops into a secondary oocyte each menstrual month. Thus, in humans meiosis II is completed only when the secondary oocyte is penetrated by a spermatozoon.

In many animals, the most obvious feature of egg maturation is deposition of yolk. Yolk, usually stored as granules or more organized platelets, is not a definite chemical substance but may be lipid or protein or both. Yolk may be synthesized within an egg from raw materials supplied by surrounding follicle cells, or preformed lipid or protein yolk may be transferred by pinocytosis from follicle cells to the oocyte.

Eggs also contain a large amount of mRNA that in not translated (p. 95) into polypeptides/proteins until fertilization triggers activation of these previously quiescent mRNA molecules. At this time the newly formed polypeptides/proteins begin to orchestrate the developmental process (see Chapter 8, p. 160).

Enormous accumulation of yolk granules, other nutrients (glycogen and lipid droplets), and quiescent mRNA cause an egg to grow well beyond the normal limits that force ordinary body (somatic) cells to divide. A young frog oocyte 50 µm in diameter, for example, grows to 1500 µm in diameter when mature after 3 years of growth in the ovary, and its volume is increased by a factor of 27,000. Bird eggs attain even greater absolute size; a hen egg increases 200 times in volume in only the last 6 to 14 days of rapid growth preceding ovulation.

Thus eggs are remarkable exceptions to the otherwise universal rule that organisms are composed of relatively minute cellular units. An egg's large size creates a problematic surface area-to-cell volume ratio, since everything that enters and leaves the ovum (nutrients, respiratory gases, wastes, and so on) must pass through the cell membrane. As the egg becomes larger, the available surface per unit of cytoplasmic volume (mass) becomes smaller. As we would anticipate, the metabolic rate of an egg gradually diminishes until a secondary oocyte or ovum (depending on the species) is in suspended animation until fertilization.

REPRODUCTIVE PATTERNS

The great majority of invertebrates, as well as many vertebrates, lay their eggs outside the body for development; these animals are called **oviparous** ("egg-birth"). Fertilization may be either internal (eggs are fertilized inside the body of a female before she lays them) or external (eggs are fertilized by a male after a female lays them). While many oviparous animals simply abandon their eggs rather indiscriminately, others display extreme care in finding places that will provide immediate and suitable sources of food for the young when they hatch.

Some animals retain eggs in their body (in the oviduct or uterus) while they develop, with embryos deriving all their nourishment from yolk stored within the egg. These animals are called **ovoviviparous** ("egg-live-birth"). Ovoviviparity occurs in several invertebrate groups (for example, various annelids, brachiopods, insects, and gastropod molluscs) and is common among certain fishes (p. 538) and reptiles (p. 580).

In the third pattern, **viviparous** ("live-birth"), eggs develop in the oviduct or uterus with embryos deriving their nourishment directly from the mother. Usually some kind of intimate anatomical relationship is established between developing embryos and their mother. In both ovoviviparity and viviparity, fertilization must be internal (within the body of the female) and the mother gives birth to young usually in a more advanced stage of development. Viviparity is confined mostly to lizards, snakes, mammals, and elasmobranch fishes, although viviparous invertebrates (scorpions, for example) and amphibians are known. Development of embryos within a mother's body, whether ovoviviparous or viviparous, obviously affords more protection to the offspring than egg-laying.

STRUCTURE OF REPRODUCTIVE SYSTEMS

The basic components of reproductive systems are similar in sexual animals, although differences in reproductive habits and methods of fertilization have produced many variations. Sexual systems consist of two components: (1) **primary organs,** which are the gonads that produce sperm and eggs and sex hormones; and (2) **accessory organs,** which assist the gonads in formation and delivery of gametes, and may also serve to support the embryo. They are of great variety, and include gonoducts (sperm ducts and oviducts), accessory organs for transferring spermatozoa into the female, storage organs for spermatozoa or yolk, packaging systems for eggs, and nutritional organs such as yolk glands and placenta.

Invertebrate Reproductive Systems

Invertebrates that transfer sperm from male to female for internal fertilization require organs and plumbing to facilitate this function that may be as complex as those of any vertebrate. In contrast, reproductive systems of invertebrates that simply release their gametes into the water for external fertilization may be little more than centers for gametogenesis. Polychaete annelids, for example, have no permanent reproductive organs. Gametes arise by proliferation of cells lining the body cavity. When mature the gametes are released through coelomic or nephridial ducts or, in some species, may exit through ruptures in the body wall.

Insects have separate sexes (dioecious), practice internal fertilization by copulation and insemination, and consequently have complex reproductive systems (Figure 7.11). Sperm from the testes pass through sperm ducts to seminal vesicles (where the sperm are stored) and then through a single ejaculatory duct to a penis. Seminal fluid from one or more accessory glands is added to the semen in the ejaculatory duct. Females have a pair of ovaries formed from a series of egg tubes (ovarioles). Mature ova pass through oviducts to a common genital chamber and then to a short copulatory bursa (vagina). In most insects, the male transfers sperm by inserting the penis directly into the female's genital bursa (vagina); from here they migrate, and are stored in a seminal receptacle. Often a single mating provides sufficient sperm to last the reproductive life of a female.

Vertebrate Reproductive Systems

In vertebrates the reproductive and excretory systems are together called the **urogenital system** because of their close anatomical connection, especially in males. This association is

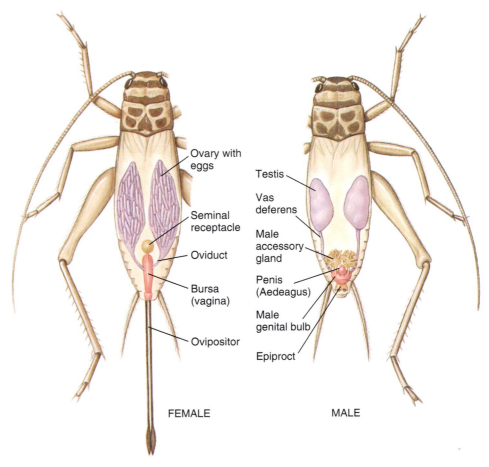

FEMALE

- Ovary with eggs
- Seminal receptacle
- Oviduct
- Bursa (vagina)
- Ovipositor

MALE

- Testis
- Vas deferens
- Male accessory gland
- Penis (Aedeagus)
- Male genital bulb
- Epiproct

Figure 7.11

Reproductive system of crickets. Sperm from the paired testes of males pass through sperm tubes (vas deferens) to an ejaculatory duct housed in the penis. In females, eggs from the ovaries pass through oviducts to the genital bursa. At mating sperm enclosed in a membranous sac (spermatophore) formed by the secretions of the accessory gland are deposited in the genital bursa of the female, then migrate to her seminal receptacle where they are stored. The female controls the release of a few sperm to fertilize her eggs at the moment they are laid, using the needlelike ovipositor to deposit the eggs in the soil.

very striking during embryonic development. In male fishes and amphibians the duct that drains the kidney (**opisthonephric duct** or **Wolffian duct**) also serves as a sperm duct (see p. 674). In male reptiles, birds, and mammals in which the kidney develops its own independent duct (**ureter**) to carry away waste, the old **mesonephric duct** becomes exclusively a sperm duct or **vas deferens.** In all these forms, with the exception of most mammals, the ducts open into a **cloaca** (derived, appropriately, from the Latin meaning "sewer"), a common chamber into which intestinal, reproductive, and excretory canals empty. Almost all placental mammals have no cloaca; instead the urogenital system has its own opening separate from the anal opening. In females, the **uterine duct** or **oviduct** is an independent duct that opens into the cloaca in animals having a cloaca.

Male Reproductive System

The male reproductive system of vertebrates, such as that of human males (Figure 7.12) includes testes, vasa efferentia, vas deferens, accessory glands, and (in some birds and reptiles, and all mammals) a penis.

Paired **testes** are the sites of sperm production. Each testis is composed of numerous **seminiferous tubules,** in which the sperm develop (Figure 7.8). The sperm are surrounded by **Sertoli cells** (or **sustentacular cells**), which nourish the developing sperm. Between the tubules are **interstitial cells** (or **Leydig cells**), which produce the male sex hormone (**testosterone**). In most mammals the two testes are housed permanently in a saclike scrotum suspended outside the abdominal cavity, or the testes descend into the scrotum during the breeding season.

This odd arrangement provides an environment of slightly lower temperature, since in most mammals (including humans) viable sperm do not form at temperatures maintained within the body. In marine mammals and all other vertebrates the testes are positioned permanently within the abdomen.

The sperm travel from the seminiferous tubules to the **vasa efferentia,** small tubes passing to a coiled **epididymis** (one for each testis), where final sperm maturation occurs and then to a **vas deferens,** the ejaculatory duct (Figures 7.8 and 7.12). In mammals the vas deferens joins the **urethra,** a duct that carries both sperm and urinary products through the **penis,** or external intromittent organ.

Most aquatic vertebrates have no need for a penis, since sperm and eggs are liberated into the water in close proximity to each other. However, in terrestrial (and some aquatic) vertebrates that bear their young alive or enclose the egg within a shell, sperm must be transferred to the female. Few birds have a true penis (examples of exceptions are the ostrich and the Argentine lake duck) and the mating process simply involves presenting cloaca to cloaca. Most reptiles and mammals have a true penis. In mammals the normally flaccid organ becomes erect when engorged with blood. Some mammals possess a bone in the penis (baculum), which presumably helps with rigidity.

In most mammals three sets of accessory glands open into the reproductive channels: a pair of **seminal vesicles,** a single **prostate gland,** and the pair of **bulbourethral glands** (Figure 7.12). Fluid secreted by these glands furnishes food to the sperm, lubricates the female reproductive tract for sperm, and counteracts the acidity of the vagina so that the sperm retain their viability longer after being deposited in the female.

Female Reproductive System

The ovaries of female vertebrates produce both ova and female sex hormones (estrogens and progesterone). In all jawed vertebrates, mature ova from each ovary enter the funnel-like opening of a **uterine tube** or **oviduct,** which typically has a fringed margin (fimbriae) that envelops the ovary at the time of ovulation. The terminal end of the uterine tube is unspecialized in most fishes and amphibians, but in cartilaginous fishes, reptiles, and birds that produce a large, shelled egg, special regions have developed for production of albumin and shell. In amniotes (reptiles, birds, and mammals; see Amniotes and the Amniotic Egg, p. 175) the terminal portion of the uterine tube is expanded into a muscular **uterus** in which shelled eggs are held before laying or in which embryos complete their development. In placental mammals, the

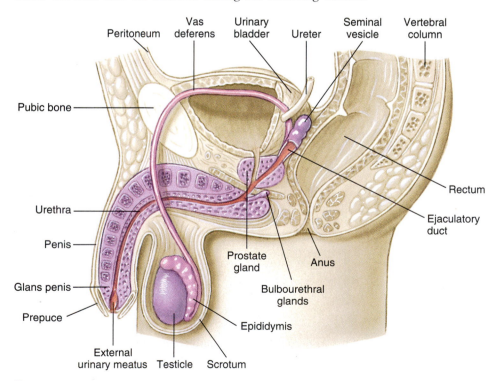

Figure 7.12
Human male reproductive system showing the reproductive structures in sagittal view.

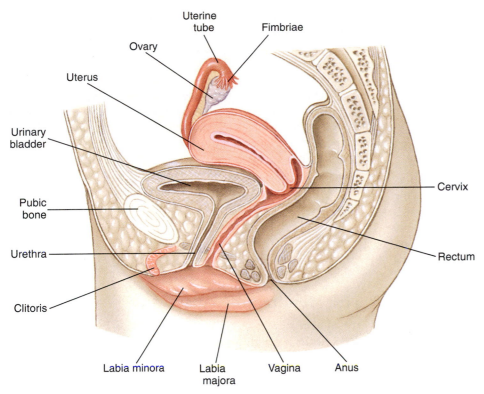

Figure 7.13
Human female reproductive system showing the pelvis in sagittal section.

walls of the uterus establish a close vascular association with the embryonic membranes through a **placenta** (see p. 177).

The paired ovaries of the human female (Figure 7.13), slightly smaller than the male testes, contain many thousands of oocytes. Each oocyte develops within a **follicle** that enlarges and finally ruptures to release a secondary oocyte (Figure 7.10). During a woman's fertile years, except following fertilization, approximately 13 oocytes mature each year, and usually the ovaries alternate in releasing oocytes. Because a woman is fertile for only about 30 years, of the approximately 400,000 primary oocytes in her ovaries at birth, only 300 to 400 have a chance to reach maturity; the others degenerate and are resorbed.

A long-held principle of mammalian reproductive biology has been that in the male, germ cell lines continue to remain functional and form sperm throughout adult life, while females possess a finite number of germ cells, and oocyte production ceases at birth. Indeed, we have just described human follicular development as such a case, where the primary oocytes present at birth provide her only source of follicles. Recently, an exciting discovery in mice has challenged this reproductive dogma. Juvenile and adult mouse ovaries have been shown to possess actively dividing germ cells that replenish the oocyte pool. If this finding can be extended to other mammalian species, it will have significant implications in the management of endangered species, where assisted reproductive techniques could be used to expand pools of oocytes that might mean the difference between extinction and survival.

The **uterine tubes,** or **oviducts,** are lined with cilia for propelling the egg away from the ovary from which it was released. The two ducts open into the upper corners of the **uterus,** or womb, which is specialized for housing the embryo during its intrauterine existence. It consists of thick muscular walls, many blood vessels, and a specialized lining: the **endometrium.** The uterus varies among different mammals, and in many it is designed to hold more than one developing embryo. Ancestrally it was paired but is fused to form one large chamber in many eutherian mammals.

The **vagina** is a muscular tube adapted to receive the male's penis and serves as the birth canal during expulsion of a fetus from the uterus. Where vagina and uterus meet, the uterus projects down into the vagina to form a **cervix.**

The external genitalia of human females, or **vulva,** include folds of skin, the **labia majora** and **labia minora,** and a small erectile organ, the **clitoris** (the female homolog of the glans penis of males). The opening into the vagina is often reduced in size in the virgin state by a membrane, the **hymen,** although in today's more physically active females, this membrane may be much reduced in extent.

ENDOCRINE EVENTS THAT ORCHESTRATE REPRODUCTION

Hormonal Control of Timing of Reproductive Cycles

From fish to mammals, reproduction in vertebrates is usually a seasonal or cyclic activity. Timing is crucial, because offspring should appear when food is available and other environmental conditions are optimal for survival. The sexual reproductive process is controlled by hormones, which are regulated by environmental cues, such as food intake, and seasonal changes in photoperiod, rainfall, or temperature, and by social cues. A region within the forebrain called the hypothalamus (p. 758) regulates the release of anterior pituitary gland hormones, which in turn stimulate tissues of the gonads (neurosecretion and the pituitary gland are described in Chapter 34). This hormonal system controls development of the gonads, accessory sex structures, and secondary sexual characteristics (see next section), as well as timing of reproduction.

The cyclic reproductive patterns of female mammals are of two types: **estrous cycle,** characteristic of most mammals, and **menstrual cycle,** characteristic only of the anthropoid primates (monkeys, apes, and humans). These two cycles differ in two important ways. First, in estrous cycles, females are receptive to males only during brief periods of **estrus,** or "heat," whereas in the menstrual cycle receptivity may occur throughout the cycle.

Second, a menstrual cycle, but not an estrous cycle, ends with breakdown and discharge of the inner portion of the uterus (endometrium). In an estrous cycle, each cycle ends with the endometrium simply reverting to its original state, without the discharge characteristic of the menstrual cycle.

Gonadal Steroids and Their Control

The ovaries of female vertebrates produce two kinds of steroid sex hormones—**estrogens** and **progesterone** (Figure 7.14). There are three kinds of estrogens: estradiol, estrone and estriol, of which estradiol is secreted in the highest amounts during reproductive cycles. Estrogens are responsible for development of female accessory sex structures (oviducts, uterus, and vagina) and for stimulating female reproductive activity. Secondary sex characters, those characteristics that are not primarily involved in formation and delivery of ova (or sperm in males), but that are essential for behavioral and functional success of reproduction, are also controlled or maintained by estrogens. Secondary sexual characteristics include distinctive skin or feather coloration, bone development, body size and, in mammals, initial development of the mammary glands. In female mammals, both estrogen and progesterone are responsible for preparing the uterus to receive a developing embryo. These hormones are controlled by **anterior pituitary gonadotropins: follicle-stimulating hormone (FSH),** and **luteinizing hormone (LH)** (Figure 7.15). The release of these two gonadotropins are in turn governed by **gonadotropin-releasing hormone (GnRH)** produced by neurosecretory cells in the **hypothalamus** (see p. 758 and Table 34.1). Through this control system environmental factors such as light, nutrition, and stress may influence reproductive cycles. Estrogens and progesterone feed back to the hypothalamus and anterior pituitary to keep secretion of GnRH, FSH, and LH in check (see Chapter 34, for a discussion of negative feedback of hormones).

The male sex steroid, **testosterone** (Figure 7.14), is manufactured by the **interstitial cells** of the testes. Testosterone, and its metabolite, **dihydrotestosterone (DHT),** are necessary for the growth and development of the male accessory sex structures (penis, sperm ducts, and glands), development of secondary male sex characters (such as bone and muscle growth, male plumage or pelage coloration, antlers in deer, and, in humans, voice quality), and male sexual behavior. Development of the testes and secretion of testosterone is controlled by FSH and LH,

the same anterior pituitary hormones that regulate the female reproductive cycle, and ultimately by GnRH from the hypothalamus. Like estrogens and progesterone in the female, testosterone and DHT feed back to the hypothalamus and anterior pituitary to regulate secretion of GnRH, FSH, and LH.

Recent identification of a peptide in the hypothalamus of birds and mammals that inhibits the secretion of GnRH and LH has led some scientists to believe that a gonadotropin-inhibiting hormone has finally been discovered. Further study is necessary, however, before we can be sure that this peptide antagonizes GnRH in all physiological conditions.

Both the ovary and testes secrete a peptide hormone, **inhibin,** which is secreted by the developing follicles in the female and by the **Sertoli cells** (or sustentacular cells) in the male. This hormone is an additional regulator of FSH secretion from the anterior pituitary in a negative feedback manner.

The Menstrual Cycle

The human menstrual cycle (L. *mensis,* month) consists of two distinct phases within the ovary: follicular phase and luteal phase, and three distinct phases within the uterus: menstrual phase, proliferative phase, and secretory phase (Figure 7.15). Menstruation (the "period") signals the **menstrual phase,** when part of the lining of the uterus (endometrium) degenerates and sloughs off, producing the menstrual discharge. Meanwhile, the **follicular phase** within the ovary is occurring, and by day 3 of the cycle blood levels of FSH and LH begin to rise slowly, prompting some of the ovarian follicles to begin growing and to secrete estrogen. As estrogen levels in the blood increase, the uterine endometrium heals and begins to thicken, and uterine glands within the endometrium enlarge **(proliferative phase).** By day 10 most of the ovarian follicles that began to develop at day 3 now degenerate (become **atretic**), leaving only one (sometimes two or three) to continue developing until it appears as a bulge on the surface of the ovary. This is a mature follicle or **graafian follicle.** During the latter part of the follicular phase, the graafian follicle secretes more estrogen, and also inhibin. As the levels of inhibin rise, the levels of FSH fall.

At day 13 or 14 in the cycle, the now high levels of estrogen from the graafian follicle stimulate a surge of GnRH from the hypothalamus, which induces a surge of LH (and to a lesser extent, FSH) from the anterior pituitary. The LH surge causes the graafian follicle to rupture **(ovulation),** releasing an oocyte

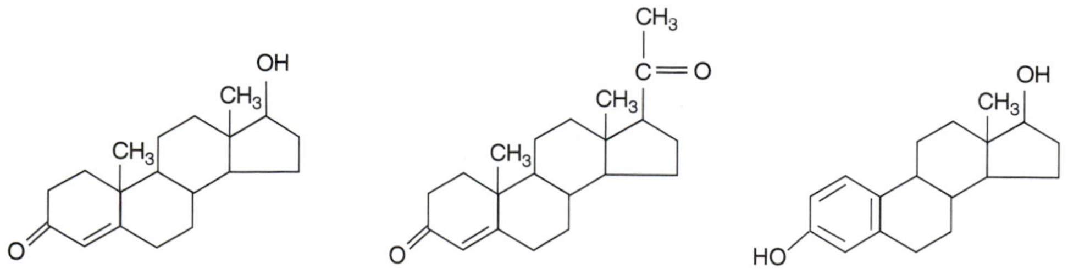

Testosterone   **Progesterone**   **Estradiol-17β**

Figure 7.14

Sex hormones. These three sex hormones show the basic four-ring steroid structure. The main female sex hormone, estradiol (an estrogen) is a C_{18} (18-carbon) steroid with an aromatic A ring (first ring to left). The main male sex hormone testosterone (an androgen) is a C_{19} steroid with a carbonyl group (C=O) on the A ring. The female sex hormone progesterone is a C_{21} steroid, also bearing a carbonyl group on the A ring.

Figure 7.15

Human menstrual cycle, showing changes in blood-hormone levels and uterine endometrium during the 28-day ovarian cycle. FSH promotes maturation of ovarian follicles, which secrete estrogen. Estrogen prepares the uterine endometrium and causes a surge in LH, which in turn causes ovulation and stimulates the corpus luteum to secrete progesterone and estrogen. Progesterone and estrogen production will persist only if an ovum is fertilized; without pregnancy progesterone and estrogen levels decline and menstruation follows.

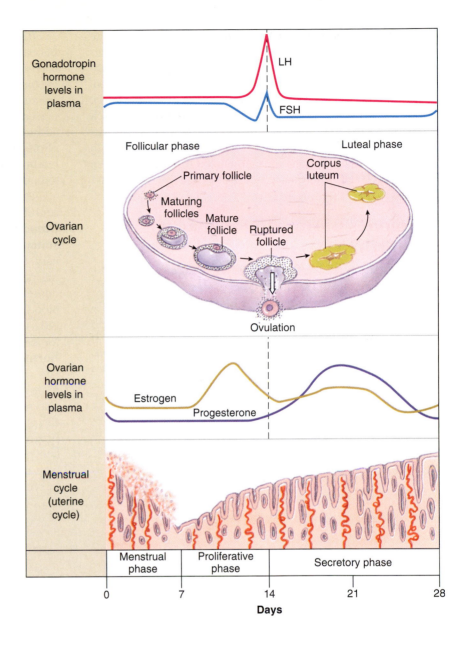

from the ovary. The oocyte remains viable for approximately 12 hours, during which time it may be fertilized by a sperm. During the ovarian **luteal phase,** a **corpus luteum** ("yellow body" for its appearance in cow ovaries) forms from the remains of the ruptured follicle that released the oocyte at ovulation (Figures 7.10 and 7.15). The corpus luteum, responding to continued stimulation by LH, becomes a transitory endocrine gland that secretes progesterone (and estrogen in primates). Progesterone ("before carrying [gestation]"), as its name implies, stimulates the uterus to undergo final maturational changes that prepare it for gestation **(secretory phase).** The uterus is now fully ready to house and to nourish an embryo. If fertilization has *not* occurred, the corpus luteum degenerates, and its hormones are no longer secreted. Since the uterine lining (endometrium) depends on progesterone and estrogen for its maintenance, their declining levels cause the uterine lining to deteriorate, leading to menstrual discharge of the next cycle.

Oral contraceptives (the "Pill") usually are combined preparations of estrogen and progesterone that act to decrease the output of pituitary gonadotropins FSH and LH. This prevents the ovarian follicles from ripening fully and usually prevents ovulation from occurring. Oral contraceptives are highly effective, with a failure rate of less than 1% if the treatment procedure is followed properly. More recently, estrogen and progesterone are administered as a once per month injection (Lunelle), as a skin patch (Ortho Evra), or as a vaginal ring (NuvaRing). Progesterone also acts on the reproductive tract as a whole, making it inhospitable for sperm and any fertilized oocyte. This mechanism has been exploited in progesterone-only contraceptives ("mini-pill," Depo-Provera), which may not block follicular development or ovulation, and in the "morning after pill," a postcoital emergency contraceptive that is now available over the counter in the United States for women aged 18 years or older.

GnRH from the hypothalamus, and LH and FSH from the anterior pituitary, are controlled by **negative feedback** of ovarian steroids (and inhibin). This negative feedback occurs throughout the menstrual cycle, except for a few days before ovulation. As just mentioned, ovulation is due to the *high levels of estrogen* causing a surge of GnRH, LH (and FSH). Such **positive feedback** mechanisms are rare in the body, since they move events away from stable set points. (Feedback mechanisms are described in Chapter 34, p. 756). This event is terminated by ovulation when estrogen levels fall as an oocyte is released from the follicle.

Hormones of Human Pregnancy and Birth

If fertilization occurs, it normally does so in the first third of the uterine tube **(ampulla).** The **zygote** travels from here to the uterus, dividing by mitosis to form a **blastocyst** (see Chapter 8, p. 177) by the time it reaches the uterus. The developing blastocyst adheres to the uterine surface after about 6 days and embeds itself in the endometrium. This process is called **implantation.** Growth of the embryo continues, producing a spherically shaped **trophoblast.** This embryonic stage contains three distinct tissue layers, the amnion, chorion, and embryo proper, the inner cell mass (see Figure 8.25, p. 178). The **chorion** becomes the source of **human chorionic gonadotropin (hCG),** which appears in the bloodstream soon after implantation. hCG stimulates the corpus luteum to continue to synthesize and to release both estrogen and progesterone (Figure 7.16).

The placenta forms the point of attachment between trophoblast and uterus (evolution and development of the placenta is described in Chapter 8, p. 177). Besides serving as a medium for the transfer of materials between maternal and fetal bloodstreams, the placenta also serves as an endocrine gland. The placenta continues to secrete hCG and also produces estrogen (mainly estriol) and progesterone. After about the third month of pregnancy, the corpus luteum degenerates in some mammals, but by then the placenta itself is the main source of both progesterone and estrogen (Figure 7.17).

Preparation of the mammary glands for secretion of milk requires two additional hormones, **prolactin (PRL)** and **human placental lactogen (hPL)** (or **human chorionic somatomammotropin**). PRL is produced by the anterior pituitary, but in nonpregnant women its secretion is inhibited. During pregnancy, elevated levels of progesterone and estrogen depress the inhibitory signal, and PRL begins to appear in the blood. PRL is also produced by the placenta during pregnancy. PRL, in combination with hPL, prepare the mammary glands for secretion. hPL, together with **human placental growth hormone (hPGH)** and maternal growth hormone, also stimulate an increase in available nutrients in the mother, so that more are provided to the developing embryo. The placenta also secretes β-endorphin and other endogenous opioids (see Chapter 33, p. 743) that regulate appetite and mood during pregnancy. Opioids may also contribute to a sense of wellbeing and help to alleviate some of the discomfort associated with the later months of pregnancy. Later the placenta begins to synthesize a peptide hormone called **relaxin;** this hormone

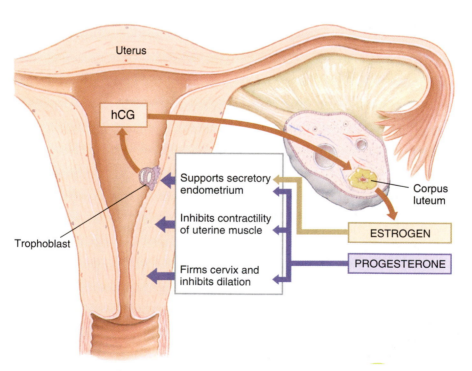

Figure 7.16

The multiple roles of progesterone and estrogen in normal human pregnancy. After implantation of an embryo in the uterus, the trophoblast (the future embryo and placenta) secretes human chorionic gonadotropin (hCG), which maintains the corpus luteum until the placenta, at about the seventh week of pregnancy, begins producing the sex hormones progesterone and estrogen.

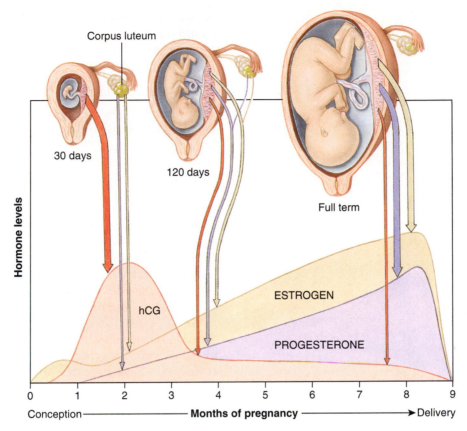

Figure 7.17

Hormone levels released from the corpus luteum and placenta during pregnancy. The width of the arrows suggests the relative amounts of hormone released; hCG (human chorionic gonadotropin) is produced solely by the placenta. Synthesis of progesterone and estrogen shifts during pregnancy from the corpus luteum to the placenta.

allows some expansion of the pelvis by increasing the flexibility of the pubic symphysis, and also dilates the cervix in preparation for delivery.

Birth, or **parturition,** occurs after approximately 9 months in humans and begins with a series of strong, rhythmic contractions of the uterine musculature, called **labor.** The exact signal that triggers birth is not fully understood in humans, but **placental corticotropin-releasing hormone (CRH)** appears to initiate the birth process. Just before birth, secretion of estrogen, which stimulates uterine contractions, rises sharply, while the level of progesterone, which inhibits uterine contractions, declines (Figure 7.17). This removes the "progesterone block" that keeps the uterus quiescent throughout pregnancy. **Prostaglandins,** a large group of hormones (long-chain fatty acid derivatives), also increase at this time, making the uterus more "irritable" (see Chapter 34, p. 762, for more on prostaglandins). Finally, stretching of the cervix sets in motion neural reflexes that stimulate secretion of **oxytocin** from the posterior pituitary. Oxytocin also stimulates uterine smooth muscle, leading to stronger and more frequent labor contractions. Secretion of oxytocin during childbirth is another example of **positive feedback.** This time the event is terminated by birth of the baby.

Childbirth occurs in three stages. In the first stage the cervix is enlarged by pressure from the baby in its bag of amniotic fluid,

which may be ruptured at this time (**dilation;** Figure 7.18B). In the second stage, the baby is forced out of the uterus and through the vagina to the outside (**expulsion;** Figure 7.18C). In the third stage, the placenta, or **afterbirth,** is expelled from the mother's body, usually within 10 minutes after the baby is born (**placental delivery;** Figure 7.18D).

Miscarriages during pregnancy, or spontaneous abortions, are quite common and serve as a mechanism to reject prenatal abnormalities such as chromosomal damage and other genetic errors, exposure to drugs or toxins, immune irregularities, or improper hormonal priming of the uterus. Modern hormonal tests show that about 30% of fertile zygotes are spontaneously aborted before or right after implantation; such miscarriages are unknown to the mother or are expressed as a slightly late menstrual period. Another 20% of established pregnancies end in miscarriage (those known to the mother), giving a spontaneous abortion rate of about 50%.

After birth, secretion of milk is triggered when the infant sucks on its mother's nipple. This leads to a reflex release of oxytocin from the posterior pituitary; when oxytocin reaches the mammary glands it causes contraction of smooth muscles lining ducts and sinuses of the mammary glands and ejection of milk.

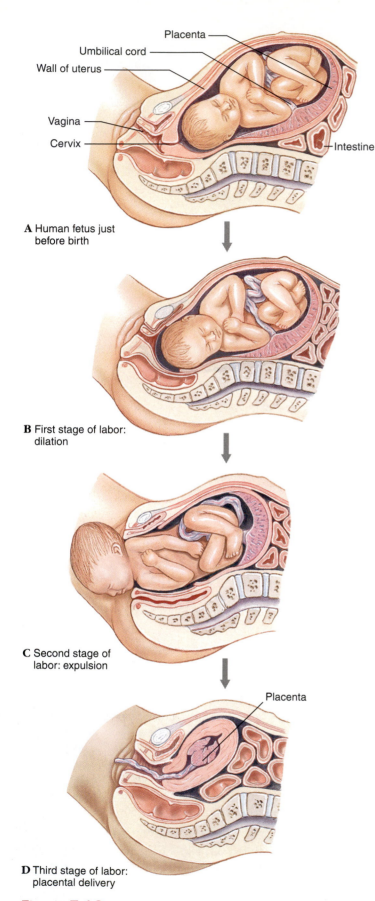

A Human fetus just before birth

B First stage of labor: dilation

C Second stage of labor: expulsion

D Third stage of labor: placental delivery

Figure 7.18
Birth, or parturition, in humans.

Suckling also stimulates release of prolactin from the anterior pituitary, which stimulates continued production of milk by the mammary glands.

Multiple Births

Many mammals give birth to more than one offspring at a time or to a litter (**multiparous**), each member of which has come from a separate egg. There are some mammals, however, that have only one offspring at a time (**uniparous**), although occasionally they may have more than one. The armadillo (*Dasypus*) is almost unique among mammals in giving birth to four offspring at one time—all of the same sex, either male or female, and all derived from the same zygote.

Human twins may come from one zygote (**identical**, or **monozygotic** twins; Figure 7.19A) or two zygotes (**nonidentical, dizygotic**, or **fraternal** twins; Figure 7.19B). Fraternal twins do not resemble each other any more than other children born separately in the same family, but identical twins are, of course, strikingly alike and always of the same sex. Triplets, quadruplets, and quintuplets may include a pair of identical twins. The other babies in such multiple births usually come from separate zygotes. About 33% of identical twins have separate placentas, indicating that the blastomeres separated at an early, possibly the two-cell, stage (Figure 7.19A, *top*). All other identical twins share a common placenta, indicating that splitting occurred after formation of the inner cell mass (see Figure 8.25 on p. 180). If splitting were to happen after placenta formation, but before the amnion forms, the twins would have individual amniotic sacs (Figure 7.19A, *middle*), as observed in the great majority of identical twins. Finally, a very small percentage of identical twins share one amniotic sac and a single placenta (Figure 7.19A, *bottom*), indicating that separation occurred after day 9 of pregnancy, by which time the amnion has formed. In these cases, the twins are at risk of becoming conjoined, a condition known as Siamese twinning. Embryologically, each member of fraternal twins has its own placenta and amnion (Figure 7.19B).

The frequency of twin births in comparison to single births is approximately 1 in 86, that of triplets 1 in 86^2, and that of quadruplets approximately 1 in 86^3. Frequency of identical twin births to all births is about the same the world over, whereas frequency of fraternal births varies with race and country. In the United States, three-fourths of all twin births are dizygotic (fraternal), whereas in Japan only about one-fourth are dizygotic. The tendency for fraternal twinning (but apparently not identical twinning) seems to run in family lines; fraternal twinning (but not identical twinning) also increases in frequency as mothers get older.

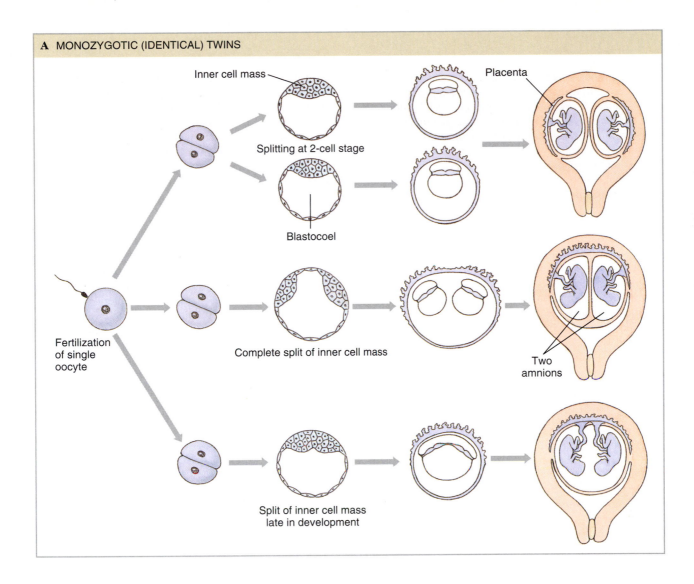

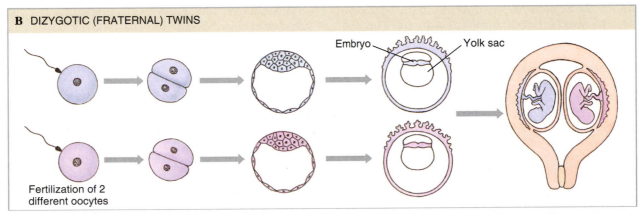

Figure 7.19

Formation of human twins. **A,** Monozygotic (identical) twin formation. **B,** Dizygotic (fraternal) twin formation. See text for explanation.

SUMMARY

Reproduction is the production of new life and provides an opportunity for evolution to occur. Asexual reproduction is a rapid and direct process by which a single organism produces genetically identical copies of itself. It may occur by fission, budding, gemmulation, or fragmentation. Sexual reproduction involves production of germ cells (sex cells or gametes), usually by two parents (bisexual reproduction), which combine by fertilization to form a zygote that develops into a new individual. Germ cells are formed by meiosis, reducing the number of chromosomes to haploid, and the diploid chromosome number is restored at fertilization. Sexual reproduction recombines parental characters and thus reshuffles and amplifies genetic diversity. Genetic recombination is important for evolution. Two alternatives to typical bisexual reproduction are hermaphroditism, the presence of both male and female organs in the same individual, and parthenogenesis, the development of an unfertilized egg.

Sexual reproduction exacts heavy costs in time and energy, requires cooperative investments in mating, and causes a 50% loss of genetic representation of each parent in the offspring. The classical view of why sex is needed is that it maintains variable offspring within the population, which may help the population to survive environmental change.

In vertebrates the primordial germ cells arise in the yolk-sac endoderm, then migrate to the gonad. In mammals, a gonad becomes a testis in response to masculinizing signals encoded on the Y chromosome of the male, and the reproductive tract masculinizes in response to circulating male sex steroids. Female reproductive structures (ovary, uterine tubes, uterus, and vagina) develop in the absence of signals encoded on the Y chromosome, although recent data suggests that a female-determining region on the X chromosome has a role in differentiation of female reproductive organs.

Germ cells mature in the gonads by a process called gametogenesis (spermatogenesis in males and oogenesis in females), involving both mitosis and meiosis. In spermatogenesis, each primary spermatocyte gives rise by meiosis and growth to four motile sperm, each bearing the haploid number of chromosomes. In oogenesis, each primary oocyte gives rise to only one mature, nonmotile, haploid ovum. The remaining nuclear material is discarded in polar bodies. During oogenesis an egg accumulates large food reserves within its cytoplasm.

Sexual reproductive systems vary enormously in complexity, ranging from some invertebrates, such as polychaete worms that lack any permanent reproductive structures to the complex systems of vertebrates and many invertebrates consisting of permanent gonads and various accessory structures for transferring, packaging, and nourishing gametes and embryos.

The male reproductive system of humans includes testes, composed of seminiferous tubules in which millions of sperm develop, and a duct system (vasa efferentia and vas deferens) that joins the urethra, glands (seminal vesicles, prostate, bulbourethral), and penis. The human female system includes ovaries, containing thousands of eggs within follicles; oviducts; uterus; and vagina.

The seasonal or cyclic nature of reproduction in vertebrates has required evolution of precise hormonal mechanisms that control production of germ cells, signal readiness for mating, and prepare ducts and glands for successful fertilization of eggs. Neurosecretory centers within the hypothalamus of the brain secrete gonadotropin-releasing hormone (GnRH), which stimulates endocrine cells of the anterior pituitary to release follicle-stimulating hormone (FSH) and luteinizing hormone (LH), which in turn stimulate the gonads. Estrogens and progesterone in females, and testosterone and dihydrotestosterone (DHT) in males, control the growth of accessory sex structures and secondary sex characteristics, in addition to feeding back to the hypothalamus and anterior pituitary to regulate GnRH, FSH, and LH secretion.

In the human menstrual cycle, estrogen induces the initial proliferation of uterine endometrium. A surge in GnRH and LH, induced by rising estrogen levels from the developing follicle(s), midway in the cycle causes ovulation and the corpus luteum to secrete progesterone (and estrogen in humans), which completes preparation of the uterus for implantation. If an egg is fertilized, pregnancy is maintained by hormones produced by the placenta and mother. Human chorionic gonadotropin (hCG) maintains secretion of progesterone and estrogen from the corpus luteum, while the placenta grows and eventually secretes estrogen, progesterone, hCG, human placental lactogen (hPL), human placental growth hormone (hPGH), prolactin (PRL), endogenous opioids, placental corticotropin-releasing hormone (CRH), and relaxin. Estrogen, progesterone, PRL, and hPL, as well as maternal prolactin, induce development of the mammary glands in preparation for lactation. hPL, hPGH, and maternal growth hormone also increase nutrient availability for the developing embryo.

Birth or parturition (at least in most mammals) appears to be initiated by release of placental CRH. In addition, a decrease in progesterone and an increase in estrogen levels occur so that the uterine muscle begins to contract. Oxytocin (from the posterior pituitary) and uterine prostaglandins continue this process until the fetus (followed by the placenta) is expelled. Placental relaxin makes the birth process easier by enabling expansion of the pelvis and dilation of the cervix.

Multiple births in mammals may result from division of one zygote, producing identical, monozygotic twins, or from separate zygotes, producing fraternal, dizygotic twins. Identical twins in humans may have separate placentas, or (most commonly) they may share a common placenta but have individual amniotic sacs.

REVIEW QUESTIONS

1. Define asexual reproduction, and describe four forms of asexual reproduction in invertebrates.
2. Define sexual reproduction and explain why meiosis contributes to one of its great strengths.

3. Explain why genetic mutations in asexual organisms lead to much more rapid evolutionary change than do genetic mutations in sexual forms. Why might harmful mutations be more deleterious to asexual organisms compared with sexual organisms?

4. Define two alternatives to bisexual reproduction— hermaphroditism and parthenogenesis—and offer a specific example of each from the animal kingdom. What is the difference between ameiotic and meiotic parthenogenesis?

5. Define the terms dioecious and monoecious. Can either of these terms be used to describe a hermaphrodite?

6. A paradox of sexual reproduction is that despite being widespread in nature, the question of why it exists at all is still unresolved. What are some disadvantages of sex? What are some consequences of sex that make it so important?

7. What is a germ cell line? How do germ cells pass from one generation to the next?

8. Explain how a spermatogonium, containing a diploid number of chromosomes, develops into four functional sperm, each containing a haploid number of chromosomes. In what significant way(s) does oogenesis differ from spermatogenesis?

9. Define, and distinguish among, the terms oviparous, ovoviviparous, and viviparous.

10. Name the general location and give the function of the following reproductive structures: seminiferous tubules, vas deferens, urethra, seminal vesicles, prostate gland, bulbourethral glands, mature follicle, oviducts, uterus, vagina, endometrium.

11. How do the two kinds of mammalian reproductive cycles— estrous and menstrual—differ from each other?

12. What are the male sex hormones and what are their functions?

13. Explain how the female hormones GnRH, FSH, LH, and estrogen interact during the menstrual cycle to induce ovulation and, subsequently, formation of the corpus luteum.

14. Explain the function of the corpus luteum in the menstrual cycle. If fertilization of the ovulated egg happens, what endocrine events occur to support pregnancy?

15. Describe the role of pregnancy hormones during human pregnancy. What hormones prepare the mammary glands for lactation and what hormones continue to be important during this process?

16. If identical human twins develop from separate placentas, when must the embryo have separated? When must separation have occurred if the twins share a common placenta but develop within separate amnions?

SELECTED REFERENCES

Cole, C. J. 1984. Unisexual lizards. Sci. Am. **250:**94–100 (Jan.). *Some populations of whiptail lizards from the American southwest consist only of females that reproduce by virgin birth.*

Crews, D. 1994. Animal sexuality. Sci. Am. **270:**108–114 (Jan.). *Sex is determined genetically in mammals and most other vertebrates, but not in many reptiles and fishes, which lack sex chromosomes. The author describes nongenetic sex determination and suggests a new framework for understanding the origin of sexuality.*

Crow, J. F. 1994. Advantages of sexual reproduction. Developmental Genetics **15:**205–213. *An excellent discussion of the advantages and disadvantages of sexual reproduction with a critique of the various hypotheses presented on this issue. Very readable.*

Forsyth, A. 1986. A natural history of sex: the ecology and evolution of sexual behavior. New York, Charles Scribner's Sons. *Engagingly written, factually accurate account of the sex lives of animals from unicellular organisms to humans, abounding in imagery and analogy. Highly recommended.*

Johnson, J., J. Cannling, T. Kaneko, J. P. Pru, and J. L. Tilly. 2004. Germline stem cells and follicular renewal in the postnatal mammalian ovary. Nature **428:**145–150. *Exciting new evidence that female mammals possess a renewable germ cell line, refuting an age-old hypothesis of reproductive biology.*

Johnson, M. H., and B. J. Everitt. 2000. Essential reproduction, ed. 5. Oxford, U.K., Blackwell Sciences Ltd. *Excellent coverage of reproductive physiology with emphasis on humans.*

Jones, R. E. 2006. Human reproductive biology, ed. 3. San Diego, Academic Press. *Thorough treatment of human reproductive physiology.*

Kinsley, C. H., and K. G. Lambert. 2006. The maternal brain. Sci. Am. **294:**72–79. *This excellent review discusses how the hormones secreted during pregnancy, and lactation in mammals appears to confer long-lasting benefits to the brain that alter skills and behavior associated with better parental care.*

Kriegsfeld, L. J., D. F. Mei, G. E. Bentley, Y. Ubuka, A. O. Mason, K. Inoue, K. Ukena, K. Tsutsui, and R. Silver. 2006. Identification and characterization of a gonadotropin-inhibitory system in the brains of mammals. Proceedings of the National Academy of Science **103:**2410–2415. *An original research paper that presents evidence of a gonadotropin-inhibiting hormone that suppresses the reproductive axis.*

Lee, D. M., R. R. Yeoman, D. E. Battaglia, R. L. Stouffer, M. B. Zelinski-Wooten, J. W. Fanton, and D. P. Wolf. 2004. Live birth after ovarian tissue transplant. Nature **428:**137–138. *New hope in the future for cancer patients that are made prematurely sterile is provided by the recent news of successful ovarian tissue transplants in monkeys.*

Lombardi, J. 1998. Comparative vertebrate reproduction. Boston, Kluwer Academic Publishers. *Comprehensive coverage of vertebrate reproductive physiology.*

Maxwell, K. 1994. The sex imperative: an evolutionary tale of sexual survival. New York, Plenum Press. *Witty survey of sex in the animal kingdom.*

Michod, R. E. 1995. Eros and evolution: a natural philosophy of sex. Reading, Massachusetts, Addison-Wesley Publishing Company. *In this engaging book, the author argues that sex evolved as a way of coping with genetic errors and avoiding homozygosity.*

Piñón, R. 2002. Biology of human reproduction. Sausalito, University Science Books. *An updated examination of human reproductive physiology.*

Ridley, M. 2001. The advantages of sex. www.pbs.org/wgbh/evolution/sex/advantage/ *An essay adapted from a New Scientist publication (4 Dec, 1993) summarizing hypotheses proposed for the evolution of sex.*

ONLINE LEARNING CENTER

Visit www.mhhe.com/hickmanipz14e for chapter quizzing, key term flash cards, web links and more!

Principles
of Development

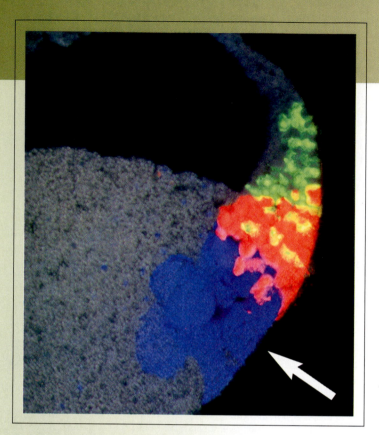

Spemann organizer cells (color) migrating from the dorsal lip (arrow) of a gastrula.

The Primary Organizer

During the first half of the twentieth century, experiments by the German embryologist Hans Spemann (1869 to 1941) and his student, Hilde Pröscholdt Mangold (1898 to 1924), ushered in the first of two golden ages of embryology. Working with salamanders, they found that tissue transplanted from one embryo into another could induce development of a complete organ, such as an eyeball, at the site of the transplant. This phenomenon is called embryonic induction. Mangold later discovered that one particular tissue, the dorsal lip from an embryonic stage called the gastrula, could induce the development of an entirely new salamander joined to the host salamander at the site of the transplant. (This work earned Spemann the Nobel Prize in Physiology or Medicine in 1935, but Hilde Mangold had died in a household accident only a few weeks after her research was published.) Spemann designated this dorsal lip tissue the **primary organizer,** now often called the **Spemann organizer.** Recent advances in molecular biology have inaugurated the second golden age of embryology,

still in progress. During this current golden age we are beginning to understand that induction is due to secretion of certain molecules that trigger or repress the activity of combinations of genes in nearby cells. For example, cells of the Spemann organizer migrate over the dorsal midline, secreting proteins with names like noggin, chordin, and follistatin. These proteins allow nearby cells to develop into the nervous system and other tissues along the middle of the back, and those tissues in turn release other proteins that induce development of other parts of the body. Such organizer proteins do not occur only in salamanders; remarkably similar proteins function in development of other vertebrates and even invertebrates. Because all animals appear to share similar molecular mechanisms for development, it may now be possible to understand how changes in such developmental controls led to the evolution of the great variety of animals. Research in this area has given rise to the exciting new field called evolutionary developmental biology.

How is it possible that a tiny, spherical fertilized human egg, scarcely visible to the naked eye, can develop into a fully formed, unique person, consisting of thousands of billions of cells, each cell performing a predestined functional or structural role? How is this marvelous unfolding controlled? Clearly all information needed must originate from the nucleus and in the surrounding cytoplasm. But knowing where the control system lies is very different from understanding how it guides the conversion of a fertilized egg into a fully differentiated animal. Despite intense scrutiny by thousands of scientists over many decades, it seemed until very recently that developmental biology, almost alone among the biological sciences, lacked a satisfactory explanatory theory. This now has changed. During the last two decades the combination of genetics and evolution with modern techniques of cellular and molecular biology has provided the long-sought explanation of animal development. Causal relationships between development and evolution have also become the focus of research. We do at last appear to have a conceptual framework to account for development.

EARLY CONCEPTS: PREFORMATION VERSUS EPIGENESIS

Early scientists and laypeople alike speculated at length about the mystery of development long before the process was submitted to modern techniques of biochemistry, molecular biology, tissue culture, and electron microscopy. An early and persistent idea was that young animals were preformed in eggs and that development was simply a matter of unfolding what was already there. Some claimed they could actually see a miniature adult in the egg or sperm (Figure 8.1). Even the more cautious argued that all parts of the embryo were in the egg, but too small and transparent to be seen. The concept of **preformation** was strongly advocated by most seventeenth- and eighteenth-century naturalist-philosophers.

In 1759 German embryologist Kaspar Friedrich Wolff clearly showed that in the earliest developmental stages of the chick, there was no preformed individual, only undifferentiated granular material that became arranged into layers. These layers continued to thicken in some areas, to become thinner in others, to fold, and to segment, until the body of the embryo appeared. Wolff called this process **epigenesis** ("origin upon or after"), an idea that a fertilized egg contains building material only, somehow assembled by an unknown directing force. Current ideas of development are essentially epigenetic in concept, although we know far more about what directs growth and differentiation.

Development describes the progressive changes in an individual from its beginning to maturity (Figure 8.2). In sexual multicellular organisms, development usually begins with a fertilized egg that divides mitotically to produce a many-celled embryo. These cells then undergo extensive rearrangements and interact with one another to generate an animal's body plan and all of the many kinds of specialized cells in its body. This generation of cellular diversity does not occur all at once, but emerges sequentially by a **hierarchy of developmental decisions.** The many familiar cell types that make up the body do not simply "unfold" at some point, but arise from conditions created in preceding stages. At each stage of development new structures arise from the interaction of less committed rudiments. Each interaction is increasingly restrictive, and the decision made at each stage in the hierarchy further limits developmental fate. Once cells embark on a course of differentiation, they become irrevocably committed to that course. They no longer depend on the stage that preceded them, nor do they have the option

Figure 8.1
Preformed human infant in sperm as imagined by seventeenth-century Dutch histologist Niklaas Hartsoeker, one of the first to observe sperm, using a microscope of his own construction. Other remarkable pictures published during this period depicted the figure sometimes wearing a nightcap!

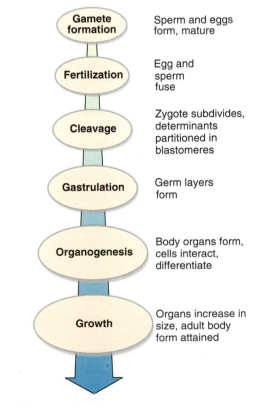

Gamete formation	Sperm and eggs form, mature
Fertilization	Egg and sperm fuse
Cleavage	Zygote subdivides, determinants partitioned in blastomeres
Gastrulation	Germ layers form
Organogenesis	Body organs form, cells interact, differentiate
Growth	Organs increase in size, adult body form attained

Figure 8.2
Key events in animal development.

of becoming something different. Once a structure becomes committed it is said to be **determined.** Thus the hierarchy of commitment is progressive and it is usually irreversible. The two basic processes responsible for this progressive subdivision are **cytoplasmic localization** and **induction.** We discuss both processes as we proceed through this chapter.

FERTILIZATION

The initial event in development in sexual reproduction is **fertilization,** the union of male and female gametes to form a **zygote.** Fertilization accomplishes two things: it provides for recombination of paternal and maternal genes, thus restoring the original diploid number of chromosomes characteristic of a species, and it activates the egg to begin development. However, sperm are not always required for development. Eggs of some species can be artificially induced to initiate development without sperm fertilization (artificial parthenogenesis), but in the great majority of cases an embryo will not be able to progress very far down the developmental path before lethal developmental abnormalities arise. However, some species have natural parthenogenesis (p. 140). Of these, some have eggs that develop normally in the absence of sperm. In other species (some fishes and salamanders), sperm is required for egg activation, but the sperm contributes no genetic material. Thus neither sperm contact nor the paternal genome is always essential for egg activation.

Oocyte Maturation

During oogenesis, described in the preceding chapter, an egg prepares itself for fertilization and for the beginning of development. Whereas a sperm eliminates all its cytoplasm and condenses its nucleus to the smallest possible dimensions, an egg grows in size by accumulating yolk reserves to support future growth. An egg's cytoplasm also contains vast amounts of messenger RNA, ribosomes, transfer RNA, and other elements required for protein synthesis. In addition, eggs of most species contain **morphogenetic determinants** that direct activation and repression of specific genes later in postfertilization development. The nucleus also grows rapidly in size during egg maturation, becoming bloated with RNA and so changed in appearance that it is given a special name, the **germinal vesicle.**

Most of this intense preparation occurs during an arrested stage of meiosis. In mammals, for example, it occurs during the prolonged prophase of the first meiotic division. The oocyte is now poised to resume meiotic divisions that are essential to produce a haploid female pronucleus that will join a male haploid pronucleus at fertilization. After resumption of meiosis, the egg rids itself of excess chromosomal material in the form of polar bodies (described in Chapter 7, p. 145). A vast amount of synthetic activity has preceded this stage. The oocyte is now a highly structured system, provided with a dowry which, after fertilization, will support nutritional requirements of the embryo and direct its development through cleavage.

Fertilization and Activation

Our current understanding of fertilization and activation derives in large part from more than a century of research on marine invertebrates, especially sea urchins. Sea urchins produce large numbers of eggs and sperm, which can be combined in the laboratory for study. Fertilization also has been studied in many vertebrates and, more recently, in mammals, using sperm and eggs of mice, hamsters, and rabbits.

Contact and Recognition Between Egg and Sperm

Most marine invertebrates and many marine fishes simply release their gametes into the ocean. Although an egg is a large target for a sperm, the enormous dispersing effect of the ocean and limited swimming range of a spermatozoon conspire against an egg and a sperm coming together by chance encounter. To improve likelihood of contact, eggs of numerous marine species release a chemotactic factor that attracts sperm to eggs. The chemotactic molecule is species-specific, attracting to eggs only sperm of the same species.

In sea urchin eggs, sperm first penetrate a jelly layer surrounding the egg, then contact an egg's vitelline envelope, a thin membrane lying just above the egg plasma membrane (Figure 8.3). At this point, egg-recognition proteins on the acrosomal process of the sperm (Figure 8.4) bind to species-specific sperm receptors on the vitelline envelope. This mechanism ensures that an egg recognizes only sperm of the same species. This is important in the marine environment where many closely related species may be spawning at the same time. Similar recognition proteins have been found on sperm of vertebrate species (including mammals) and presumably are a universal property of animals.

Prevention of Polyspermy

At the point of sperm contact with the egg vitelline envelope a **fertilization cone** appears into which the sperm head is later drawn (Figure 8.4). This event is followed immediately by

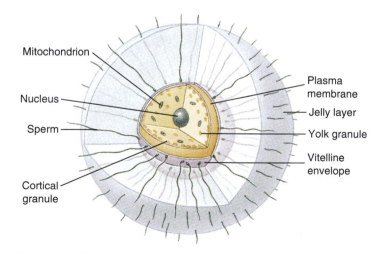

Mitochondrion

Nucleus

Sperm

Cortical granule

Plasma membrane

Jelly layer

Yolk granule

Vitelline envelope

Figure 8.3
Structure of sea urchin egg during fertilization.

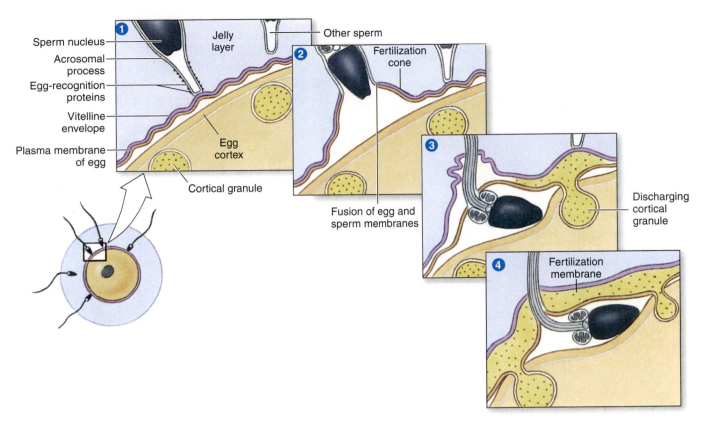

Figure 8.4
Sequence of events during sperm contact and penetration of a sea urchin egg.

important changes in the egg surface that block entrance of additional sperm, which, in marine eggs especially, may quickly surround the egg in swarming numbers (Figure 8.5). Entrance of more than one sperm, called **polyspermy,** must be prevented because union of more than two haploid nuclei would be ruinous for normal development. In a sea urchin egg, contact of the first sperm with the egg membrane is instantly followed by an electrical potential change in the egg membrane that prevents additional sperm from fusing with the membrane. This event, called the **fast block,** is followed immediately by the **cortical reaction,** in which thousands of enzyme-rich cortical granules, located just beneath the egg membrane, fuse with the membrane and release their contents into the space between the egg membrane and the overlying vitelline envelope (see Figure 8.4). The cortical reaction creates an osmotic gradient, causing water to rush into this space, elevating the envelope and lifting away all sperm bound to it, except the one sperm that has successfully fused with the egg membrane. One of the cortical granule enzymes causes the vitelline envelope to harden, and it is now called a **fertilization membrane.** The block to polyspermy is complete. Timing of these early events is summarized in Figure 8.6. Mammals have a similar security system that is erected within seconds after the first sperm fuses with the egg membrane.

Fusion of Pronuclei and Egg Activation

Once sperm and egg membranes have fused, the sperm loses its flagellum, which disintegrates. Its nuclear envelope then breaks apart, allowing the sperm chromatin to expand from its extremely condensed state. The enlarged sperm nucleus, now called a **pronucleus,** migrates inward to contact the female pronucleus. Their fusion forms the diploid **zygote nucleus.** Nuclear fusion takes only about 12 minutes in sea urchin eggs (Figure 8.6), but requires about 12 hours in mammals.

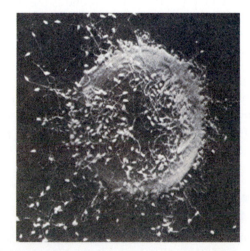

Figure 8.5
Binding of sperm to the surface of a sea urchin egg. Only one sperm penetrates the egg surface, the others being blocked from entrance by rapid changes in the egg membranes. Unsuccessful sperm are soon lifted away from the egg surface by a newly formed fertilization membrane.

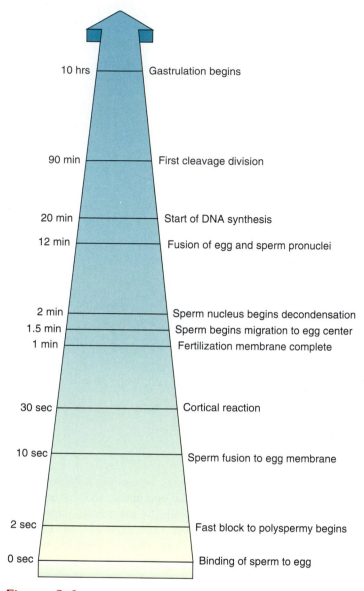

Time	Event
10 hrs	Gastrulation begins
90 min	First cleavage division
20 min	Start of DNA synthesis
12 min	Fusion of egg and sperm pronuclei
2 min	Sperm nucleus begins decondensation
1.5 min	Sperm begins migration to egg center
1 min	Fertilization membrane complete
30 sec	Cortical reaction
10 sec	Sperm fusion to egg membrane
2 sec	Fast block to polyspermy begins
0 sec	Binding of sperm to egg

Figure 8.6
Timing of events during fertilization and early development in a sea urchin.

Fertilization sets in motion several important changes in the cytoplasm of an egg—now called a zygote—that prepare it for cleavage. Inhibitors that had blocked metabolism and kept the egg quiescent, in suspended-animation, are removed. Fertilization is immediately followed by a burst of DNA and protein synthesis, the latter utilizing the abundant supply of messenger RNA previously stored in the egg cytoplasm. Fertilization also initiates an almost complete reorganization of the cytoplasm within which are morphogenetic determinants that activate or repress specific genes as development proceeds. Movement of cytoplasm repositions the determinants into new and correct spatial arrangements that are essential for proper development. The zygote now enters cleavage.

In animal eggs, fertilization induces an increase in the amount of free calcium ions inside the egg cytoplasm. This increase in intracellular free calcium regulates later developmental events and is essential for normal development to occur in all taxa studied, but the mechanisms controlling calcium levels vary. In some taxa, calcium ions are released from intracellular stores, whereas in others calcium enters the egg from outside via voltage-gated calcium channels (see Chapter 3, p. 49). Some organisms combine both mechanisms. The calcium signal can occur in a single pulse, as it does in jellyfish, starfish, and frog zygotes, or in a series of closely spaced pulses identified in ribbon worms, polychaetes, and mammals. Researchers once thought that the calcium signaling pattern might vary as part of the developmental dichotomy between protostomes and deuterostomes, but this is not the case. In even the short list of taxa just given, the two chordate deuterostomes exhibit different calcium release patterns, suggesting that the differing patterns are more likely related to the number and duration of developmental events that require calcium signaling.

CLEAVAGE AND EARLY DEVELOPMENT

During cleavage the embryo divides repeatedly to convert the large, unwieldy cytoplasmic mass into a large cluster of small, maneuverable cells called **blastomeres.** No growth occurs during this period, only subdivision of mass, which continues until normal **somatic** cell size is attained. At the end of cleavage the zygote has been divided into many hundreds or thousands of cells and the blastula stage is formed.

Before cleavage begins, an animal-vegetal axis is visible on the embryo. This axis exists because yolk, nutrition for the developing embryo, occurs only at one end, establishing **polarity** in the embryo. The yolk-rich end is the **vegetal pole** and the other end is the **animal pole** (Figure 8.7B); the animal pole contains mostly cytoplasm and very little yolk. The animal-vegetal axis provides a landmark or reference point on the embryo. Cleavage is generally an orderly sequence of cell divisions so that one cell divides to form two cells, these each divide to form four cells, the four make eight cells, and the process continues. During each division, a distinct cleavage furrow is visible in the cell. This **cleavage furrow** can be parallel or perpendicular to the animal-vegetal axis.

How Amount and Distribution of Yolk Affect Cleavage

The amount of yolk at the vegetal pole varies among taxa. Eggs with very little yolk, evenly distributed throughout the egg (Figure 8.7A, C, and E), are called **isolecithal** (Gr. *isos,* equal + *lekithos,* yolk). **Mesolecithal** (Gr. *mesos,* middle + *lekithos,* yolk) eggs have a moderate amount of yolk concentrated at the vegetal pole (Figure 8.7B), whereas **telolecithal**

RADIAL HOLOBLASTIC CLEAVAGE

A Sea star: Isolecithal egg

B Frog: Mesolecithal egg

Animal pole

Vegetal pole Gray crescent

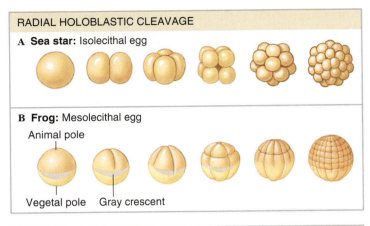

SPIRAL HOLOBLASTIC CLEAVAGE

C Nemertean worm: Isolecithal egg

DISCOIDAL MEROBLASTIC CLEAVAGE

D Chick: Telolecithal egg

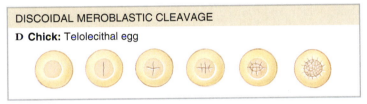

ROTATIONAL HOLOBLASTIC

E Mouse: Isolecithal egg

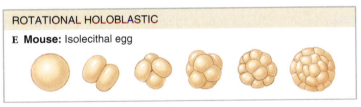

Figure 8.7
Cleavage stages in sea star, frog, nemertean worm, chick, and mouse. Yellow regions represent yolk in each diagram.

eggs (Gr. *telos,* end + *lekithos,* yolk) contain an abundance of yolk densely concentrated at the vegetal pole of the egg (Figure 8.7D). **Centrolecithal** eggs have a large, centrally located, mass of yolk.

The presence of yolk disrupts cleavage to varying degrees; when little yolk is present, cleavage furrows extend completely through the egg in **holoblastic** (Gr. *holo,* whole + *blastos,* germ) cleavage (Figure 8.7A, B, C, and E). When much yolk is present, cleavage is **meroblastic** (Gr. *meros,* part + *blastos,* germ), with cells sitting atop a mass of undivided yolk (Figure 8.7D). Meroblastic cleavage is incomplete because cleavage furrows cannot cut through the heavy concentration of yolk, but instead stop at the border between the cytoplasm and the yolk below.

Holoblastic cleavage occurs in isolecithal eggs and is present in echinoderms, tunicates, cephalochordates, nemerteans, and most molluscs, as well as in marsupial and placental mammals, including humans (Figure 8.7A, C, and E). Mesolecithal eggs also cleave holoblastically, but cleavage proceeds more slowly in the presence of yolk, leaving the vegetal region with a few large, yolk-filled cells, whereas the animal region has many small cells. Amphibian eggs (Figure 8.7B) illustrate this process.

Meroblastic cleavage occurs in telolecithal and centrolecithal eggs. In telolecithal eggs of birds, reptiles, most fishes, a few amphibians, cephalopod molluscs, and monotreme mammals, cleavage is restricted to cytoplasm in a narrow disc on top of the yolk (see chick development in Figure 8.7D). In centrolecithal eggs of insects and many other arthropods, cytoplasmic cleavage is limited to a surface layer of yolk-free cytoplasm, whereas the yolk-rich inner cytoplasm remains uncleaved (see Figure 8.15).

The function of yolk is to nourish the embryo. When much yolk is present, as in telolecithal eggs, young exhibit **direct development,** going from an embryo to a miniature adult. When little yolk is present, as in isolecithal or mesolecithal eggs, young develop into various larval stages capable of feeding themselves. In this **indirect development,** larvae differ from adults and must metamorphose into adult bodies (Figure 8.8). There is another way to compensate for absence of yolk: In most mammals, the mother nourishes the embryos by means of a placenta.

What Can We Learn From Development?

Biologists study development for different reasons. Some studies focus on understanding how the zygote, a single large cell, can produce the multitude of body parts in an organism.

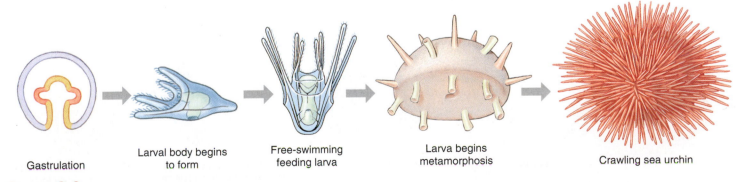

Gastrulation   Larval body begins to form   Free-swimming feeding larva   Larva begins metamorphosis   Crawling sea urchin

Figure 8.8
Indirect development in a sea urchin. After gastrulation, a free-swimming larva develops; it feeds and grows in ocean surface waters. The larva will metamorphose into a tiny bottom-dwelling sea urchin; the urchin feeds and grows, reaching sexual maturity in this body form.

Understanding the mechanisms of development requires knowledge of how cleavage partitions cytoplasm, how different cells interact, and how gene expression proceeds. These topics are covered on pages 170–174.

Another reason to study development is to search for commonalities among organisms. Commonalities in mechanisms of development are discussed on page 174, but there are also commonalities among organisms in the sequence of developmental events. All multicellular animals begin as zygotes and all go through cleavage and some subsequent developmental stages. Embryos of sponges, snails, and frogs diverge at some point to produce different adults. When does this divergence occur? All zygotes do not cleave in the same way; do certain types of cleavage characterize particular animal groups? Types of cleavage do characterize particular groups of animals, but cleavage type occurs along with other developmental features to form a suite of characters. Therefore, an overview of a developmental sequence is needed to explain other characters in the suite.

Based on these character suites, the 34 multicellular animal phyla fall into several distinct groups. Rather than attempting to grasp the details of 34 phyla, we can understand these phyla as variations on a much smaller number of developmental themes. The character suites are discussed on page 166, and in Chapter 9.

AN OVERVIEW OF DEVELOPMENT FOLLOWING CLEAVAGE

Blastulation

Cleavage subdivides the mass of the zygote until a cluster of cells called a **blastula** (Gr. *blastos,* germ, + *ule,* little) is formed (Figure 8.9). In mammals, the cluster of cells is called a blastocyst (see Figure 8.13E). In most animals, the cells are arranged around a central fluid-filled cavity (Figure 8.9) called a **blastocoel** (Gr. *blastos,* germ, + *koilos,* cavity). (A hollow blastula can be called a coeloblastula to distinguish it from a solid stereoblastula; the general account here assumes the blastula is hollow.) In the blastula stage, the embryo consists of a few hundred to several thousand cells poised for further development. There has been a great increase in total DNA content because each of the

many daughter cell nuclei formed by chromosomal replication at mitosis contains as much DNA as the original zygote nucleus. The whole embryo, however, is no larger than the zygote.

Formation of a blastula stage, with its one layer of germ cells, occurs in all multicellular animals. In most animals, development continues beyond the blastula to form one or two more germ layers in a gastrula stage. Sponges were previously thought to complete embryogenesis with only a single layer of blastula cells, but recent work shows that cell migrations produce external and internal layers in embryos of at least some sponges (see Figure 12.12). Whether such cell layers are homologous to the true germ layers of other organisms is debated. The germ layers ultimately produce all structures of the adult body; the germ layer derivatives for vertebrates are shown in Figure 8.26.

Gastrulation and Formation of Two Germ Layers

Gastrulation converts the spherical blastula into a more complex configuration and forms a second germ layer (Figure 8.9). There is variation in the process (see pp. 167–170). In some cases a second layer forms as cells migrate inward without making an internal cavity, but generally one side of the blastula bends inward in a process called invagination. This inward bending continues until the surface of the bending region extends about one-third of the way into the blastocoel, forming a new internal cavity (Figure 8.9). Picture a sphere being pushed inward on one side—the inward region forms a pouch. The internal pouch is the gut cavity, called an **archenteron** (Gr. *archae,* old, + *enteron,* gut) or a **gastrocoel** (Gr. *gaster,* stomach, + *koilos,* cavity). It sits inside the now-reduced blastocoel. The opening to the gut, where the inward bending began, is the **blastopore** (Gr. *blastos,* germ, + *poros,* hole).

The **gastrula** (Gr. *gaster,* stomach, + *ule,* little) stage has two layers: an outer layer of cells surrounding the blastocoel, called **ectoderm** (Gr. *ecto,* out, + *deros,* skin), and an inner layer of cells lining the gut, called **endoderm** (Gr. *endon,* inside, + *deros,* skin). In forming a mental picture of the developmental process, remember that cavities or spaces can only be defined by their boundaries. Thus, the gut cavity is a space defined by the layer of cells surrounding it (Figure 8.9).

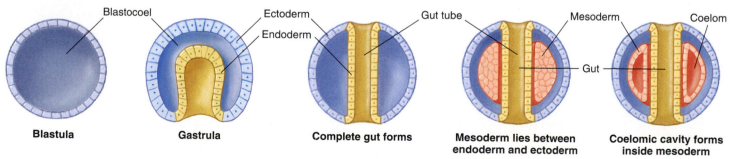

| Blastula | Gastrula | Complete gut forms | Mesoderm lies between endoderm and ectoderm | Coelomic cavity forms inside mesoderm |

Figure 8.9

A generalized developmental sequence showing formation of three germ layers and two body cavities that persist into adulthood.

This gut opens only at the blastopore; it is called a blind or **incomplete gut.** Anything consumed by an animal with a blind gut must either be completely digested, or the undigested parts egested through the mouth. Certain animals, sea anemones and flatworms, for example, have a blind gut, sometimes called a gastrovascular cavity. However, most animals have a **complete gut** with a second opening, the anus (Figure 8.9). The blastopore becomes the mouth in organisms with one suite of developmental characters, but it becomes the anus in organisms with another such suite of characters (see Figure 8.10).

Formation of a Complete Gut

When a complete gut forms, the inward movement of the archenteron continues until the end of the archenteron meets the ectodermal wall of the gastrula. The archenteron cavity extends through the animal, and the ectoderm and endoderm layers join together. This joining produces an endodermal tube, the gut, surrounded by the blastocoel, inside an ectodermal tube,

the body wall (Figure 8.9). The endodermal tube now has two openings, the blastopore and a second, unnamed, opening that formed when the archenteron tube merged with the ectoderm (Figure 8.9).

Formation of Mesoderm, a Third Germ Layer

The vast majority of multicellular animals proceed from a blastula to a gastrula, producing two germ layers. In one of many quirks of biological terminology, there is no term for organisms with only a single germ cell layer, but animals with two germ layers are called **diploblastic** (Gr. *diploos,* twofold, + *blastos,* germ). Diploblastic animals include sea anemones and comb jellies. Most animals have a third germ layer and are **triploblastic** (L. *tres,* three, + *blastos,* germ).

The third layer, **mesoderm** (Gr. *mesos,* middle, + *deros,* skin), eventually lies between the ectoderm and the endoderm (Figure 8.9). Mesoderm can form in two ways: cells arise from a ventral area near the lip of the blastopore and proliferate into the space between the archenteron and outer body wall (see Figure 8.13C), or the central region of the archenteron wall pushes outward into the space between the archenteron and outer body wall (see Figure 8.13A). Regardless of method, initial cells of mesoderm come from endoderm. (In a few groups, such as amphibians, part of a third layer of cells is made from ectoderm; this is called **ectomesoderm** (Gr. *ecto,* out, + *mesos,* middle, + *deros,* skin) to distinguish it from true, endodermally derived, mesoderm).

At the end of gastrulation, ectoderm covers the embryo, and mesoderm and endoderm have been brought inside (Figure 8.9). As a result, cells have new positions and new neighbors, so interactions among cells and germ layers then generate more of the body plan.

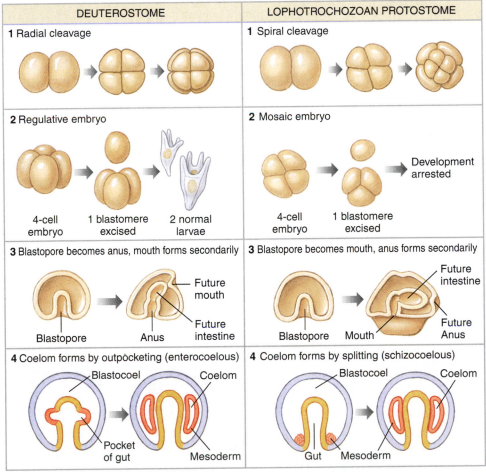

DEUTEROSTOME	LOPHOTROCHOZOAN PROTOSTOME
1 Radial cleavage	**1** Spiral cleavage
2 Regulative embryo 4-cell embryo 1 blastomere excised 2 normal larvae	**2** Mosaic embryo Development arrested 4-cell embryo 1 blastomere excised
3 Blastopore becomes anus, mouth forms secondarily Future mouth Future intestine Blastopore Anus	**3** Blastopore becomes mouth, anus forms secondarily Future intestine Future Anus Blastopore Mouth
4 Coelom forms by outpocketing (enterocoelous) Blastocoel Coelom Pocket of gut	**4** Coelom forms by splitting (schizocoelous) Blastocoel Coelom Gut Mesoderm

Figure 8.10

Developmental tendencies of lophotrochozoan protostomes (flatworms, annelids, molluscs, etc.) and deuterostomes. These tendencies are much modified in some groups, for example vertebrates. Cleavage in mammals is rotational rather than radial; in reptiles, birds, and many fishes cleavage is discoidal. Vertebrates have also evolved a derived form of coelom formation that is basically schizocoelous.

Formation of the Coelom

A **coelom** (Gr. *koilos,* cavity) is a body cavity completely surrounded by mesoderm; the band of mesoderm with its internal coelom lies inside the space previously occupied by the blastocoel (Figure 8.9). How did this happen? During gastrulation, the blastocoel is filled, partially or completely, with mesoderm. The coelomic cavity appears inside the mesoderm by one of two methods: **schizocoely** or **enterocoely.** These methods are discussed on pages 166 and 169. A coelom made by schizocoely is functionally equivalent to a coelom made by enterocoely. The method by which the coelom forms is an inherited character useful for grouping organisms into the suites of developmental characters mentioned earlier.

When coelom formation is complete, the body has three germ layers and two cavities (Figure 8.9). One cavity is the gut cavity and the other is the fluid-filled coelomic cavity. The coelom, surrounded by its mesodermal walls, has completely filled the blastocoel. Mesoderm around the coelom will eventually produce layers of muscles, among other structures.

SUITES OF DEVELOPMENTAL CHARACTERS

There are two major groups of triploblastic animals, **protostomes** and **deuterostomes.** The groups are identified by a suite of four developmental characters: (1) radial or spiral positioning of cells as they cleave, (2) regulative or mosaic cleavage of cytoplasm, (3) fate of the blastopore to become mouth or anus, and (4) schizocoelous or enterocoelous formation of a coelom. Snails and earthworms, among others, belong to the protostomes. Sea stars, fishes, and frogs, among others, belong to the deuterostomes.

Deuterostome Development

Cleavage Patterns

Radial cleavage (Figure 8.10) is so named because the embryonic cells are arranged in radial symmetry around the animal-vegetal axis. In radial cleavage of sea stars, the first cleavage plane passes right through the animal-vegetal axis, yielding two identical daughter cells (blastomeres). For the second cleavage division, furrows form simultaneously in both blastomeres, and these are oriented parallel to the animal-vegetal axis (but perpendicular to the first cleavage furrow). Cleavage furrows next form simultaneously in the four daughter blastomeres, this time oriented perpendicular to the animal-vegetal axis, yielding two tiers of four cells each. An upper tier of cells sits directly atop the tier of cells below it (Figure 8.10). Subsequent cleavages yield an embryo composed of several tiers of cells.

A second characteristic of cleavage concerns the fates of isolated blastomeres and the cytoplasm they contain. This issue did not come to light until biologists undertook developmental experiments with embryos early in cleavage. Imagine a four-cell embryo (Figure 8.10). All cells in an organism are ultimately derived from these four cells, but when are the products of each cell decided? If one cell is removed from the mass, can the other cells continue developing to produce a normal organism?

Most deuterostomes have **regulative development** where the fate of a cell depends on its interactions with neighboring cells, rather than on what piece of cytoplasm it acquired during cleavage. In these embryos, at least early in development, each cell is able to produce an entire embryo if separated from the other cells (Figure 8.11). In other words, an early blastomere

originally has the ability to follow more than one path of differentiation, but its interaction with other cells restricts its fate. If a blastomere is removed from an early embryo, the remaining blastomeres can alter their normal fates to compensate for the missing blastomere and produce a complete organism. This adaptability is termed regulative development.

Fate of Blastopore

A **deuterostome** (Gr. *deuteros,* second, + *stoma,* mouth) embryo develops through the blastula and gastrula stages, and forms a complete gut. The blastopore becomes the anus, and a second, unnamed, opening becomes the mouth, as is indicated by root words in the name of this group.

Coelom Formation

The final deuterostome feature concerns the origin of the coelom. In **enterocoely** (Gr. *enteron,* gut, + *koilos,* cavity), both mesoderm and coelom are made at the same time. In enterocoely, gastrulation begins with one side of the blastula bending inward to form the archenteron or gut cavity. As the archenteron continues to elongate inward, the sides of the archenteron push outward, expanding into a pouchlike coelomic compartment (see Figure 8.10). The coelomic compartment pinches off to form a mesodermally bound space surrounding the gut (see Figure 8.10). Fluid collects in this space. Notice that the cells

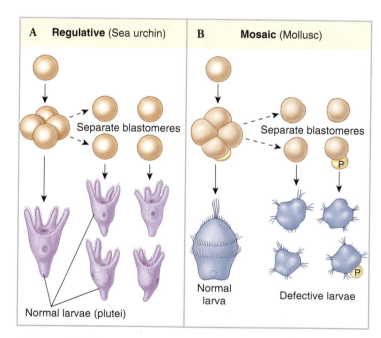

Figure 8.11

Regulative and mosaic cleavage. **A,** Regulative cleavage. Each of the early blastomeres (such as that of a sea urchin) when separated from the others develops into a small pluteus larva. **B,** Mosaic cleavage. In a mollusc, when blastomeres are separated, each gives rise to only a part of an embryo. The larger size of one defective larva is the result of the formation of a polar lobe (P) composed of clear cytoplasm of the vegetal pole, which this blastomere alone receives.

that form the coelom during enterocoely come from a different region of the endoderm than do those making the coelom during schizocoely (see Figure 8.10).

Examples of Deuterostome Development

The general outline of deuterostome development just given varies in some of its details depending upon the animal being studied. The presence of large amounts of yolk in some embryos further complicates the developmental sequence. A few examples of specific developmental sequences illustrate this variation.

Variations in Deuterostome Cleavage The typical deuterostome pattern is radial cleavage, but ascidian chordates (also called tunicates) exhibit **bilateral cleavage.** In ascidian eggs, the anteroposterior axis is established prior to fertilization by asymmetrical distribution of several cytoplasmic components (Figure 8.12). The first cleavage furrow passes through the animal-vegetal axis, dividing the asymmetrically distributed cytoplasm equally between the first two blastomeres. Thus, this first cleavage division separates the embryo into its future right and left sides, establishing its bilateral symmetry (hence the name bilateral holoblastic cleavage). Each successive division orients itself to this plane of symmetry, and the half-embryo formed on one side of the first cleavage is the mirror image of the half embryo on the other side.

Most mammals possess isolecithal eggs and a unique cleavage pattern called **rotational cleavage,** so called because of the orientation of blastomeres with respect to each other during the second cleavage division (see mouse development in Figure 8.7E). Cleavage in mammals is slower than in any other animal group. In humans, the first division is completed about 36 hours after fertilization (compared with about an hour and a half in sea urchins), and the next divisions follow at 12- to 24-hour intervals. As in most other animals, the first cleavage plane runs through the animal-vegetal axis to yield a two-cell embryo. However, during the second cleavage one of these blastomeres divides meridionally (through the animal-vegetal axis) while the other divides equatorially (perpendicular to the animal-vegetal axis). Thus, the cleavage plane in one blastomere is rotated 90 degrees with respect to the cleavage plane of the other blastomere (hence the name rotational cleavage). Furthermore, early divisions are asynchronous; not all blastomeres divide at the same time. Thus, mammalian embryos may

not increase regularly from two to four to eight blastomeres, but often contain odd numbers of cells. After the third division, the cells suddenly close into a tightly packed configuration, which is stabilized by tight junctions that form between outermost cells of the embryo. These outer cells form the **trophoblast.** The trophoblast is not part of the embryo proper but will form the embryonic portion of the placenta when the embryo implants in the uterine wall. Cells that actually give rise to the embryo proper form from the inner cells, called the **inner cell mass** (see blastula stage in Figure 8.13E).

Telolecithal eggs of reptiles, birds, and most fish divide by **discoidal cleavage.** Because of the great mass of yolk in these eggs, cleavage is confined to a small disc of cytoplasm lying atop a mound of yolk (see chick development in Figure 8.7D). Early cleavage furrows carve this cytoplasmic disc to yield a single layer of cells called the blastoderm. Further cleavages divide the blastoderm into five to six layers of cells (Figure 8.13D).

Variations in Deuterostome Gastrulation In sea stars, gastrulation begins when the entire vegetal area of the blastula flattens to form a **vegetal plate** (a sheet of epithelial tissue). This event is followed by a process called **invagination,** in which the vegetal plate bends inward and extends about one-third of the way into the blastocoel, forming the archenteron (Figure 8.13A). Coelomic formation is typical of enterocoely. As the archenteron continues to elongate toward the animal pole, and its anterior end expands into two pouchlike **coelomic vesicles,** which pinch off to form left and right coelomic compartments (Figure 8.13A).

The **ectoderm** gives rise to the epithelium of the body surface and to the nervous system. The **endoderm** gives rise to the epithelial lining of the digestive tube. The outpocketing of the archenteron is the origin of **mesoderm.** This third germ layer will form the muscular system, reproductive system, peritoneum (lining of the coelomic compartments), and the calcareous plates of the sea star's endoskeleton.

Frogs are deuterostomes with radial cleavage but morphogenetic movements of gastrulation are greatly influenced by the mass of inert yolk in the vegetal half of the embryo. Cleavage divisions are slowed in this half so that the resulting blastula consists of many small cells in the animal half and a few large cells in the vegetal half (see Figures 8.7B and 8.13B). Gastrulation in amphibians begins when cells located at the future dorsal side of the embryo invaginate to form a slitlike blastopore. Thus, as in sea stars, invagination initiates archenteron formation, but amphibian gastrulation begins in the marginal zone of the blastula, where animal and vegetal hemispheres come together, and where there is less yolk than in the vegetal region. Gastrulation progresses as sheets of cells in the marginal zone turn inward over the blastopore lip and move inside the gastrula to form mesoderm and endoderm (see opening figure of this chapter, p. 158). The three germ layers now formed are the primary structural layers that play crucial roles in further differentiation of the embryo.

Figure 8.12
Bilateral cleavage in tunicate embryos. The first cleavage division divides the asymmetrically distributed cytoplasm evenly between the first two blastomeres, establishing the future right and left sides of the adult animal. Bilateral symmetry of the embryo is maintained through subsequent cleavage divisions.

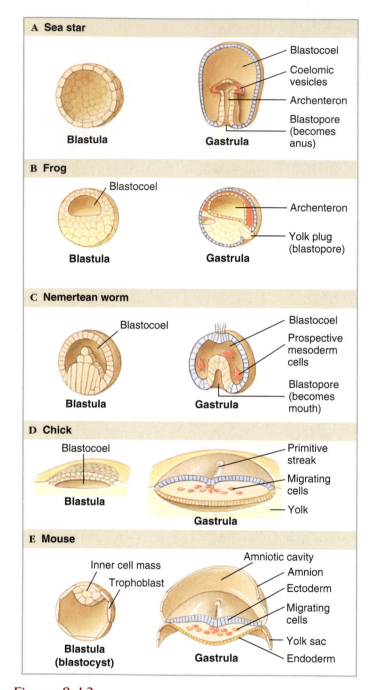

Figure 8.13

Blastula and gastrula stages in embryos of sea star, frog, nemertean worm, chick, and mouse.

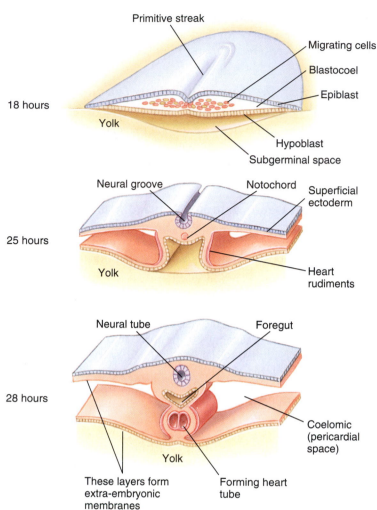

Figure 8.14

Gastrulation in a chick. Transverse sections through the heart-forming region of the chick show development at 18, 25, and 28 hours of incubation.

Cells of the epiblast move as a sheet toward the primitive streak, then roll over the edge and migrate as individual cells into the blastocoel. These migrating cells separate into two streams. One stream of cells moves deeper (displacing the hypoblast along the midline) and forms endoderm. The other stream moves between the epiblast and hypoblast to form mesoderm. Cells on the surface of the embryo compose the ectoderm. The embryo now has three germ layers, at this point arranged as sheetlike layers with ectoderm on top and endoderm at the bottom. This arrangement changes, however, when all three germ layers lift from the underlying yolk (Figure 8.14), then fold under to form a three-layered embryo that is pinched off from the yolk except for a stalk attachment to the yolk at midbody (see Figure 8.22).

Gastrulation in mammals is remarkably similar to gastrulation in reptiles and birds (see Figure 8.13E). Gastrulation movements in the inner cell mass produce a primitive streak. Epiblast cells move medially through the primitive streak into the blastocoel, and individual cells then migrate laterally through the blastocoel to form mesoderm and endoderm. Endoderm cells

In bird and reptile embryos (see Figure 8.13D), gastrulation begins with a thickening of the blastoderm at the caudal end of the embryo, which migrates forward to form a **primitive streak** (Figure 8.14). The primitive streak becomes the anteroposterior axis of the embryo and the center of early growth. The primitive streak is homologous to the blastopore of frog embryos, but in chicks it does not open into the gut cavity because of the obstructing mass of yolk. The blastoderm consists of two layers (epiblast and hypoblast) with a blastocoel between them.

(derived from the hypoblast) form a yolk sac devoid of yolk (since mammalian embryos derive nutrients directly from the mother via the placenta).

Amphibians, reptiles, and birds, which have moderate to large amounts of yolk concentrated in the vegetal region of the egg, have evolved derived gastrulation patterns in which the yolk does not participate in gastrulation. Yolk is an impediment to gastrulation and consequently the gastrulation process occurs around (amphibians) or on top (reptiles and birds) of the vegetal yolk. Mammalian eggs are isolecithal, and thus one might expect them to have a gastrulation pattern similar to that of sea stars. Instead they have a pattern more suited to telolecithal eggs. The best explanation for this feature of mammalian egg development is common ancestry with birds and reptiles. Reptiles, birds, and mammals share a common ancestor whose eggs were telolecithal. Thus, all three groups inherited their gastrulation patterns from this common ancestor, and mammals subsequently evolved isolecithal eggs but retained the telolecithal gastrulation pattern.

A further developmental complication in vertebrates is that coelom formation occurs by a modified form of schizocoely (see Figure 8.10), not enterocoely. The nonvertebrate chordates form the coelom by enterocoely, as is typical of deuterostomes.

Protostome Development

Cleavage Patterns

Spiral cleavage (see Figure 8.10) occurs in most protostomes. It differs from radial cleavage in two important ways. Rather than dividing parallel or perpendicular to the animal-vegetal axis, blastomeres cleave obliquely (approximately 45-degree angle) to this axis and typically produce quartets of cells that come to lie, not on top of but in the furrows between cells of the underlying layer. The upper layer of cells appears offset (shifted in a spiral fashion) from the lower (see Figure 8.10). In addition, spirally cleaving blastomeres pack themselves tightly together much like a group of soap bubbles, rather than just lightly contacting each other as do many radially cleaving blastomeres (see Figure 8.10).

Mosaic development characterizes most protostomes (see Figure 8.10). In mosaic development, cell fate is determined by the distribution of certain proteins and messenger RNAs, called **morphogenetic determinants,** in the egg cytoplasm. As cleavage occurs, these morphogenetic determinants are partitioned among the cells unequally. When a particular blastomere is isolated from the rest of the embryo, it still forms the characteristic structures decided by the morphogenetic determinants it contains (see Figure 8.11). In the absence of a particular blastomere, the animal lacks those structures normally formed by that blastomere, so it cannot develop normally. This pattern is called mosaic development because the embryo seems to be a mosaic of self-differentiating parts.

Fate of Blastopore

A **protostome** (Gr. *protos,* first, + *stoma,* mouth) is so named because the blastopore becomes the mouth, and the second, unnamed, opening becomes the anus.

Coelom Formation

In protostomes, a mesodermal band of tissue surrounding the gut forms before a coelom is made. If present, the inner coelomic cavity is made by **schizocoely.** To form mesoderm, endodermal cells arise ventrally at the lip of the blastopore (see Figure 8.10) and move, via **ingression,** into the space between the walls of the archenteron (endoderm) and outer body wall (ectoderm). These cells divide and deposit new cells, called mesodermal precursors, between the two existing cell layers (see Figure 8.13C). The proliferating cells become the mesoderm. Meticulous cell-lineage studies by embryologists established that in many organisms with spiral cleavage, for example, flatworms, snails, and related organisms, these mesodermal precursors arise from one large blastomere, called the 4d cell, that is present in a 29- to 64-cell embryo.

Some protostomes do not develop a coelom. Flatworms, like *Planaria,* develop to an early gastrula stage and then form a mesodermal layer as just described. Mesoderm completely fills the blastocoel and a coelom never forms (see Figure 9.3). Animals without a coelom are called **acoelomate.** In other protostomes, mesoderm lines only one side of the blastocoel, leaving a fluid-filled blastocoel next to the gut (see Figure 9.3). The fluid-filled cavity surrounding the gut is called a **pseudocoelom** (Gr. *pseudés,* false, + *koilos,* cavity); it is bordered on the inner edge by the endodermal gut lining, and on the outer edge by a layer of mesoderm next to the ectoderm. Thus, a pseudocoelom has mesoderm on only one side, whereas a true **coelom** is a fluid-filled cavity completely surrounded by mesoderm (see Figure 9.3). Acoelomate and pseudocoelomate body plans are discussed in more detail in Chapter 9.

For **coelomate** protostomes, like earthworms and snails, the mesodermal layer forms as just described, and a coelom is made by **schizocoely** (Gr. *schizein,* to split, + *koilos,* cavity). A coelom arises, as the name suggests, when the mesodermal band around the gut splits open centrally (see Figure 8.10). Fluid collects in the coelom.

Examples of Protostome Development

The protostomes are divided into two clades. One clade, **lophotrochozoan protostomes,** contains segmented worms, molluscs (snails, slugs, clams, octopus and their kin) and several less familiar taxa. The name of this clade refers to two features present in some members of the group: a horseshoe-shaped whorl of tentacles called a **lophophore** (see pp. 324–328), and a **trochophore larva** (see p. 337). Lophotrochozoans have the four protostome features described previously (see Figure 8.10). They typically form mesoderm from the embryonic 4d cell.

The other clade, **ecdysozoan protostomes,** includes arthropods (insects, spiders, crabs, and related organisms), roundworms, and other taxa which molt the exoskeleton. The name of this clade refers to shedding of the cuticle, **ecdysis** (Gr. *ekdyo,* take off, strip).

Variations in Protostome Cleavage   Spiral cleavage is typical of protostomes, but one highly specialized class of molluscs,

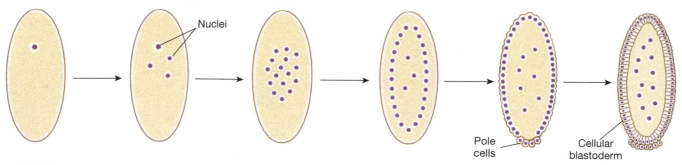

Figure 8.15

Superficial cleavage in a *Drosophila* embryo. The zygote nucleus at first divides repeatedly in the yolk-rich endoplasm by mitosis without cytokinesis. After several rounds of mitosis, most nuclei migrate to the surface where they are separated by cytokinesis into separate cells. Some nuclei migrate to the posterior pole to form the primordial germ cells, called pole cells. Several nuclei remain in the endoplasm where they will regulate breakdown of yolk products. The cellular blastoderm stage corresponds to the blastula stage of other embryos.

cephalopods, has bilateral cleavage like that of ascidian chordates (see p. 167 and Figure 8.12). Octopus, squid, and cuttlefish, among others, are cephalopods.

Many ecdysozoans do not exhibit spiral cleavage; in some, cleavage appears radial, and in others, such as insects, cleavage is neither spiral nor radial.

Centrolecithal eggs of insects undergo **superficial cleavage** (Figure 8.15) where the centrally located mass of yolk restricts cleavage to the cytoplasmic rim of the egg. This pattern is highly unusual because cytoplasmic cleavage (cytokinesis) does not occur until after many rounds of nuclear division. After roughly eight rounds of mitosis in the absence of cytoplasmic division (yielding 256 nuclei), the nuclei migrate to the yolk-free periphery of the egg. A few nuclei at the posterior end of the egg become surrounded by cytoplasm to form pole cells, which give rise to germ cells of the adult. Next, the entire egg cell membrane folds inward, partitioning each nucleus into a single cell, and yielding a layer of cells at the periphery surrounding the mass of yolk (Figure 8.15). Because yolk is an impediment to cleavage, this pattern avoids cleaving the yolk and instead confines cytoplasmic division to small regions of yolk-free cytoplasm.

Variations in Protostome Gastrulation   In most protostomes, mesodermal cells all derive from the 4d cell (see p. 169). However, in some nemertean worms (see Figure 8.13C), mesoderm derives from an earlier blastomere. Mesodermal origins are difficult to determine in many ecdysozoan protostomes due to the modified pattern of cleavage.

MECHANISMS OF DEVELOPMENT

Nuclear Equivalence

How does a developing embryo generate a multitude of cell types of a complete multicellular organism from the starting point of a single diploid nucleus of a zygote? To many nineteenth-century embryologists there seemed only one acceptable answer: as cell division ensued, hereditary material had to be parceled

unequally to daughter cells. In this view, the genome gradually became broken into smaller and smaller units until finally only the information required to impart the characteristics of a single cell type remained. This became known as the Roux-Weismann hypothesis, after the two German embryologists who developed the concept.

However, in 1892 Hans Driesch discovered that if he mechanically shook apart a two-celled sea urchin into separate cells, both half-embryos developed into normal larvae. Driesch concluded that both cells contained all genetic information of the original zygote. Still, this experiment did not settle the argument, because many embryologists thought that even if all cells contained complete genomes, the nuclei might become progressively modified in some way to dispense with the information not used in forming differentiated cells.

The efforts of Hans Driesch to disrupt egg development are poetically described by Peattie: "Behold Driesch grinding the eggs of Loeb's favorite sea urchin up between plates of glass, pounding and breaking and deforming them in every way. And when he ceased from thus abusing them, they proceeded with their orderly and normal development. Is any machine conceivable, Driesch asks, which could thus be torn down . . . have its parts all disarranged and transposed, and still have them act normally? One cannot imagine it. But of the living egg, fertilized or not, we can say that there lie latent within it all the potentialities presumed by Aristotle, and all of the sculptor's dream of form, yes, and the very power in the sculptor's arm." From Peattie, D. C. 1935. *An Almanac for Moderns.* New York, G. P. Putnam's Sons.

Around the turn of the century Hans Spemann introduced a new approach to testing the Roux-Weismann hypothesis. Spemann placed minute ligatures of human hair around salamander zygotes just as they were about to divide, constricting them until they were almost, but not quite, separated into two halves. The nucleus lay in one half of the partially divided zygote; the other side was anucleate, containing only cytoplasm. The zygote then completed its first cleavage division on the side containing the nucleus; the anucleate side remained undivided. Eventually, when the nucleated side had divided into about 16 cells, one of the cleavage nuclei would wander across the narrow cytoplasmic bridge to the anucleate side. Immediately this side began to divide and developed normally.

Sometimes, however, Spemann observed that the nucleated half of the embryo developed only into an abnormal ball of "belly" tissue. The explanation, Spemann discovered, depended on the presence of the gray crescent, a pigment-free area shown in Figure 8.7B. The gray crescent is required for normal development because it is the precursor of the Spemann organizer discussed in the opening essay on page 158.

Spemann's experiment demonstrated that every blastomere contains sufficient genetic information for the development of a complete animal. In 1938 he suggested another experiment that would demonstrate that even somatic cells of an adult contain a complete genome. The experiment, which Spemann characterized as being "somewhat fantastical" at that time, would be to remove the nucleus of an egg cell and replace it with the nucleus from a somatic cell from a different individual. If all cells contained the same genetic information as a zygote, then the embryo should develop into an individual that is genetically identical to the animal from which the nucleus was obtained. It took several decades to solve the technical difficulties, but the experiment was successfully performed on amphibians, and today it is done in a variety of mammals. The procedure is now familiarly called **cloning.** One of the most famous cloned mammals, Dolly the sheep, got the genetic material in her nuclei from the mammary glands of a six-year-old ewe.

If all nuclei are equivalent, what causes some cells to develop into neurons while others develop into skeletal muscle? In most animals (excluding insects), there are two major ways by which cells become committed to particular developmental fates: (1) cytoplasmic partitioning of determinative molecules during cleavage and (2) interaction with neighboring cells (inductive interactions). All animals use both of these mechanisms to some extent to specify different cell types. However, in some animals cytoplasmic specification is dominant, whereas others rely predominantly on inductive interactions.

Cytoplasmic Specification

A fertilized egg contains cytoplasmic components that are unequally distributed within the egg. These different cytoplasmic components are thought to contain morphogenetic determinants that control commitment of a cell to a particular cell type. These morphogenetic determinants are partitioned among different blastomeres as a result of cleavage, and the developmental fate of each cell becomes specified by the type of cytoplasm it acquires during development (see mosaic development, p. 169).

This process is especially striking (and easily visualized) in some tunicate species in which the fertilized egg contains as many as five differently colored types of cytoplasm (see Figure 8.12). These differently pigmented cytoplasms are segregated into different blastomeres, which then proceed to form distinct tissues or organs. For example, yellow cytoplasm gives

rise to muscle cells while gray equatorial cytoplasm produces notochord and neural tube. Clear cytoplasm produces larval epidermis and gray vegetal cytoplasm gives rise to the gut.

Embryonic Induction

Induction, the capacity of some cells to evoke a specific developmental response in others, is a widespread phenomenon in development. The classic experiments, cited in the opening essay on page 158, were reported by Hans Spemann and Hilde Mangold in 1924. When a piece of dorsal blastopore lip from a salamander gastrula was transplanted into a ventral or lateral position of another salamander gastrula, it invaginated and developed a notochord and somites. It also induced the *host* ectoderm to form a neural tube. Eventually a whole system of organs developed where the graft was placed, and then grew into a nearly complete secondary embryo (Figure 8.16). This creature was composed partly of grafted tissue and partly of induced host tissue.

It was soon found that *only* grafts from the dorsal lip of the blastopore were capable of inducing the formation of a complete or nearly complete secondary embryo. This area corresponds to the presumptive areas of notochord, somites, and prechordal plate (see p. 500). It was also found that only ectoderm of the host would develop a nervous system in the graft and that the reactive ability was greatest at the early gastrula stage and declined as the recipient embryo got older.

Spemann designated the dorsal lip area the **primary organizer** because it was the only tissue capable of inducing the

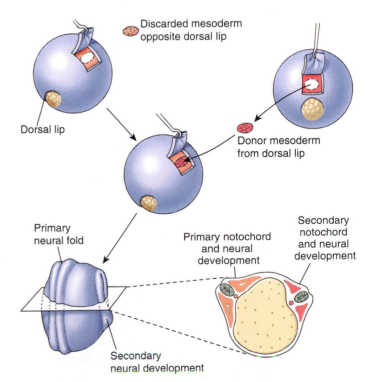

Figure 8.16
The Spemann-Mangold primary organizer experiment.

development of a secondary embryo in the host. It is now often called the Spemann organizer. Spemann also termed this inductive event **primary induction** because he considered it the first inductive event in development. Subsequent studies showed that many other cell types originate by later inductive interactions, a process called **secondary induction.**

Usually cells that have differentiated act as inductors for adjacent undifferentiated cells. Timing is important. Once a primary inductor sets in motion a specific developmental pattern in some cells, numerous secondary inductions follow. What emerges is a sequential pattern of development involving not only inductions but cell movement, changes in adhesive properties of cells, and cell proliferation. There is no "hard-wired" master control panel directing development, but rather a sequence of local patterns in which one step in development is a subunit of another. In showing that each step in the developmental hierarchy is a necessary preliminary for the next, Hans Spemann's induction experiments were among the most significant events in experimental embryology.

GENE EXPRESSION DURING DEVELOPMENT

Since every cell with few exceptions receives the same genetic material, cytoplasmic specification and induction must involve the activation of different combinations of genes in different cells. Understanding development is therefore ultimately a problem of understanding the genetics involved. It is not surprising that developmental genetics was first studied in the geneticists' favorite model organism, the fruit fly *Drosophila.* These studies have been repeated in several other model animals, such as the nematode worm *Caenorhabditis elegans,* zebra fish *Danio rerio,* frog *Xenopus laevis,* chick *Gallus gallus,* and mouse *Mus musculus.* This research suggests that epigenesis proceeds in three general stages: pattern formation, determination of position in the body, and induction of limbs and organs appropriate for that position. Each stage is guided by gradients of gene products that function as **morphogens.**

Pattern Formation

The first step in organizing development of an embryo is pattern formation: determination of the front-to-rear (anteroposterior), left-to-right, and back-to-front (dorsoventral) axes. As Spemann demonstrated in salamanders, the anteroposterior axis of the embryo is determined by the Spemann organizer, located in the gray crescent of a zygote. In *Drosophila* the anteroposterior axis is determined even before an egg is fertilized. Christiane Nüsslein-Volhard and her colleagues in Germany found that this determination is due to a gradient of mRNA that is secreted into the egg by nurse cells in the mother. The end of the egg that receives the highest level of this mRNA is fated to become the anterior of the embryo and eventually of the adult. The mRNA is transcribed from a gene called *bicoid* (pronounced BICK-oyd) in the nurse cells. After an egg is fertilized, *bicoid*

mRNA is translated into a protein morphogen called bicoid (not italicized) that binds to certain other genes. The products of these genes in turn activate others in a cascade that ultimately causes the production of an anteroposterior gradient. *Bicoid* is one of about 30 maternal genes that control pattern formation in an embryo. Some of these determine the dorsoventral axis. The gene *short gastrulation* leads to development of ventral structures, such as the nerve cord.

One of the most exciting discoveries in developmental genetics has been that the developmental genes of vertebrates and many other animals are similar to those of *Drosophila;* they are conserved over a wide range of animals. A gene similar to *bicoid* is also important in pattern formation in vertebrates. In vertebrates, however, the gene, called *Pitx2,* determines positioning of certain internal organs to either the left or right side of the body. Mutations in *Pitx2* in frogs, chicks, and mice can place the heart and stomach on the right instead of the left side. Such mutations may explain a reversal of organ position that sometimes occurs in humans. *Pitx2* is in turn activated by a protein produced by the gene *sonic hedgehog (Shh),* which is similar to a *Drosophila* gene called *hedgehog.* (The name *hedgehog* refers to the bristly appearance of fruit flies lacking the gene. The *"sonic"* comes from the video-game character Sonic the Hedgehog.) In vertebrates, *sonic hedgehog* is active in the left side only at the anterior end of the primitive streak (see Figure 8.13). *Short gastrulation* also has a counterpart in vertebrates—the gene *chordin,* which produces one of the proteins from the Spemann organizer.

In *Drosophila,* as well as other arthropods, annelid worms, chordates, and a few other groups, one important aspect of pattern formation along the anteroposterior axis is **segmentation,** also called **metamerism.** Segmentation is a division of the body into discrete segments or metameres (see Fig 9.6, p. 195). The segments are identical early in development, but later activation of different combinations of genes causes each segment to form different structures. For example, the anterior segment of insect embryos will form antennae, eyes, and mouthparts, while segments farther back will form legs. Segments are obvious in insects, but in fishes segmentation is apparent only in somites that produce such structures as vertebrae and repeated muscle bands (myomeres) (see Figure 24.24, p. 531). In *Drosophila* the number and orientation of segments is controlled by **segmentation genes.** There are three classes of segmentation genes: gap, pair-rule, and segment-polarity. **Gap genes** are activated first and divide an embryo into regions such as head, thorax, and abdomen. **Pair-rule genes** divide these regions into segments. Finally, **segment-polarity genes,** such as *hedgehog,* organize the anterior-to-posterior structures within each segment.

Homeotic and *Hox* Genes

Segmentation genes apparently regulate expression of other genes, ensuring that they are active only in appropriate segments. Such segment-specific genes are called homeotic genes. Mutations in homeotic genes, called **homeotic mutations,** place appendages or other structures in the wrong part of the body. For example, in *Drosophila* the homeotic gene *Antennapedia,*

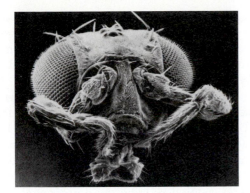

Figure 8.17

Head of a fruit fly with a pair of legs growing out of head sockets where antennae normally grow. The *Antennapedia* homeotic gene normally specifies the second thoracic segment (with legs), but the dominant mutation of this gene leads to this bizarre phenotype.

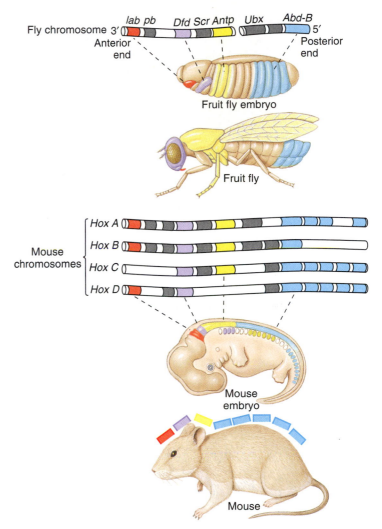

Figure 8.18

Homology of *Hox* genes in insects and mammals. These genes in both insects (fruit fly) and mammals (mouse) control the subdivision of the embryo into regions of different developmental fates along the anterior-posterior axis. The homeobox-containing genes lie on a single chromosome of the fruit fly and on four separate chromosomes in the mouse. Clearly defined homologies between the two, and the parts of the body in which they are expressed, are shown in color. The open boxes denote areas where it is difficult to identify specific homologies between the two. The *Hox* genes shown here are only a small subset of all the homeobox genes.

which helps trigger development of legs, is normally active only in the thorax. If the *Antennapedia* gene is activated by a homeotic mutation in the head of a maggot, the adult will have legs in place of antennae (Figure 8.17). *Antennapedia* and some other homeotic genes, as well as many other genes involved in development, include a sequence of 180 DNA base pairs, called the **homeobox.** The homeobox produces the part of a protein that attaches to the DNA of other genes, activating or blocking their expression.

Several other homeotic and nonhomeotic genes that are clustered close to *Antennapedia* on the same chromosome in *Drosophila* also include a homeobox. Genes in this cluster are called *Hom* genes. *Hom* genes do not encode specific limbs and organs. Instead, they function by specifying the location in the body along the anteroposterior axis. Intriguingly, the order of the *Hom* genes within the cluster on the chromosome is the same as the order in which they are expressed along the length of the body (Figure 8.18). One of the most exciting discoveries of the late twentieth century was that genes similar to *Hom* genes of *Drosophila* occur in other insects, as well as in chordates and unsegmented animals such as hydra and nematode worms. They also occur in plants and yeasts, and perhaps in all eukaryotes. These genes in organisms other than *Drosophila* were called *Hox* genes, but now all such genes are usually called *Hox* genes. Most *Hox* genes occur in a cluster on one chromosome. Mammals have four clusters, each on a different chromosome, with from 9 to 11 *Hox* genes each. As in *Drosophia,* the sequence of *Hox* genes within a cluster is the same as the front-to-rear order in which they are expressed in the body.

Morphogenesis of Limbs and Organs

Hox and other homeobox genes also play a role in shaping individual organs and limbs. As shown in Figures 8.18 and 8.19, for example, regions of the brain and identity of somites are specified by particular *Hox* and homeobox genes. Many other developmental genes that are also involved in pattern formation for the entire body also help shape individual limbs and organs by producing gradients of morphogens. One example, which has been studied by Cheryll Tickle and her coworkers at University College in London, is formation and development of limb buds in chicks. They have found that a new limb bud can be induced to grow from the side of a chick by implanting a bead soaked in fibroblast growth factor (FGF). This result implies that limbs are normally induced to develop by activation of the gene for FGF in appropriate parts of the body. Whether the limb bud develops into a wing or a leg depends on whether the FGF is applied toward the front or the rear of the chick.

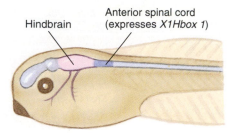

Control tadpole

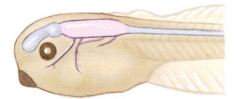

**Tadpole injected with antibodies
to *X1Hbox 1* protein**

Figure 8.19

How the inhibition of a homeodomain regulatory protein alters normal development of the central nervous system of a frog tadpole. When the protein (encoded by a homeobox DNA sequence known as *X1Hbox 1*) was inactivated by antibodies directed against it, the area that should have become anterior spinal cord transformed into hindbrain instead.

FGF also plays a role in shaping the limb. It is secreted by cells in an **apical ectodermal ridge** at the end of the limb bud. FGF acts as a morphogen that forms a gradient from the apical ectodermal ridge to the base of a limb bud. This gradient helps establish a proximodistal axis—one of three axes that guide development of a limb (Figure 8.20). Fingers or toes develop at the end of the proximodistal axis with the highest level of FGF. An anteroposterior axis is established by a gradient of sonic hedgehog and ensures that fingers or toes develop in the appropriate order. Finally, Wnt7a, a protein produced by a gene that is similar to the segment-polarity gene *wingless* in *Drosophila,* helps determine the dorsoventral axis. Wnt7a makes the dorsal side of the wing or foot different from the ventral side.

Evolutionary Developmental Biology

Zoologists have always looked to embryology for clues to the evolutionary history, or phylogeny, of animals. Developmental features such as the number of germ layers and the fate of the blastopore do suggest evolutionary relationships among different phyla. Advances in development genetics, such as those just described in the previous section, have made the relationship between development and evolution even closer and have given rise to an exciting new field called evolutionary

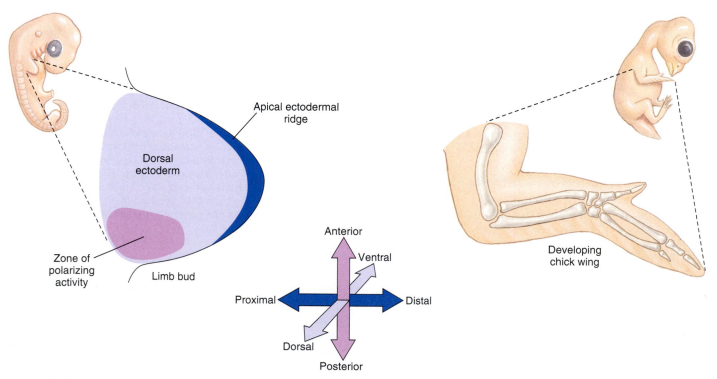

Figure 8.20

Morphogenesis in a vertebrate limb bud. The skeleton of a mature chicken limb is shown for orientation. Three axes are established in the limb bud: a proximal-distal axis by fibroblast growth factor (FGF) from the apical ectodermal ridge; an anterior-posterior axis by sonic hedgehog protein from the zone of polarizing activity; and a dorsal-ventral axis by Wnt7a protein from dorsal ectoderm.

developmental biology. Evolutionary developmental biology, often nicknamed evo-devo, is based on a realization that evolution is essentially a process in which organisms become different as a result of changes in the genetic control of development. The fact that the genes that control development are similar in animals as different as fruit flies and mice offers hope that we can reconstruct the evolutionary history of animals by understanding how functioning of those genes came to differ. Evolutionary developmental biology has already contributed several exciting concepts to our thinking about animal evolution, but the field is so new that it would be premature to accept these concepts as established. It is best to state them as questions for further study.

Are the body plans of all bilaterally symmetric animals fundamentally similar? As noted on page 172, *chordin,* one of the genes responsible for development of the nervous system in the dorsal part of a frog, is similar to *short gastrulation,* which is necessary for development of the ventral nerve cord in *Drosophila*. In addition, the gene *decapentaplegic* promotes dorsal development in *Drosophila,* and the similar gene *bone morphogenetic protein-4* promotes ventral development in frogs. In other words, insects and amphibians, whose body plans look so different, actually share a similar control of dorsoventral patterning, except that one is upside down compared with the other. This finding has prompted a reappraisal of an idea first proposed by the French naturalist Etienne Geoffroy St. Hilaire in 1822 after he noticed that in a dissected lobster on its back the nerve cord was above the gut, and the heart was below it, as in a vertebrate in its normal position. The idea that a vertebrate is like an inverted invertebrate was quickly rejected, but now biologists are once more considering whether the body plans of protostomes and deuterostomes are simply inverted relative to each other.

Can the anatomy of extinct ancestral species be inferred from the developmental genes shared by their descendants? The fact that dorsoventral patterning is similar in protostomes and deuterostomes suggests that the most recent common ancestor of these two branches had a similar dorsoventral patterning with a heart and nervous system separated by the gut. One can also infer from the similarity in *Hom/Hox* clusters in insects and chordates that the most recent common ancestor of protostomes and deuterostomes may have been segmented and that its segments differentiated by similar genes. It may also have had at least rudimentary eyes, judging from the fact that similar genes, *eyeless/Pax-6,* are involved in eye formation in a wide range of both protostomes and deuterostomes.

Instead of evolution proceeding by the gradual accumulation of numerous small mutations, could it proceed by relatively few mutations in a few developmental genes? The fact that formation of legs or eyes can be induced by a mutation in one gene suggests that these and other organs develop as modules (see p. 173). If so, then entire limbs and organs could have been lost or acquired during evolution as a result of one or a few mutations, which would challenge Darwin's theory of

gradualism (p. 121). If this is correct, then the apparently rapid evolution of numerous groups of animals during the few million years of the Cambrian explosion and at other times is more easily explained. Instead of requiring mutations in numerous genes, each with a small effect, evolution of different groups could be a result of changes in timing, number, or expression of relatively few developmental genes.

VERTEBRATE DEVELOPMENT

The Common Vertebrate Heritage

A prominent outcome of shared ancestry of vertebrates is their common pattern of development. This common pattern is best seen in the remarkable similarity of postgastrula vertebrate embryos (Figure 8.21). The likeness occurs at a brief moment in the development of vertebrates when shared chordate hallmarks of dorsal neural tube, notochord, pharyngeal gill pouches with aortic arches, ventral heart, and postanal tail are present at about the same stage of development. Their moment of similarity—when the embryos seem almost interchangeable—is all the more extraordinary considering the great variety of eggs and widely different types of early development that have converged toward a common design. Then, as development continues, the embryos diverge in pace and direction, becoming recognizable as members of their class, then their order, then family, and finally their species. The important contribution of early vertebrate development to our understanding of homology and evolutionary common descent is described in Chapter 6 in the section on Ontogeny, Phylogeny, and Recapitulation, page 116.

Amniotes and the Amniotic Egg

Reptiles, birds, and mammals form a monophyletic grouping of vertebrates called **amniotes,** so named because their embryos develop within a membranous sac, the **amnion.** The amnion is one of four **extraembryonic membranes** that compose a sophisticated support system within the **amniotic egg** (Figure 8.22), which evolved when the first amniotes appeared in the late Paleozoic era.

The **amnion** is a fluid-filled sac that encloses the embryo and provides an aqueous environment in which the embryo floats, protected from mechanical shock and adhesions.

Evolution of the second extraembryonic membrane, the **yolk sac,** actually predates appearance of amniotes many millions of years. The yolk sac with its enclosed yolk is a conspicuous feature of all fish embryos. After hatching, a growing fish larva depends on the remaining yolk provisions to sustain it until it can begin to feed itself (Figure 8.23). The yolk sac functions differently in animals that give live birth. In many viviparous vertebrates from diverse groups, the yolk sac becomes vascular and intimately associated with the mother's reproductive tract, allowing transfer of nutrients and respiratory gases

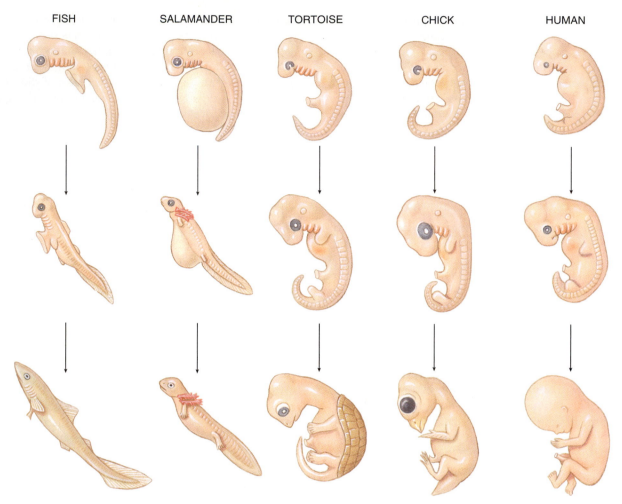

FISH   SALAMANDER   TORTOISE   CHICK   HUMAN

Figure 8.21

Early vertebrate embryos drawn from photographs. Embryos as diverse as fish, salamander, tortoise, bird, and human show remarkable similarity following gastrulation. At this stage (top row) they reveal features common to the entire subphylum Vertebrata. As development proceeds they diverge, each becoming increasingly recognizable as belonging to a specific class, order, family, and finally, species.

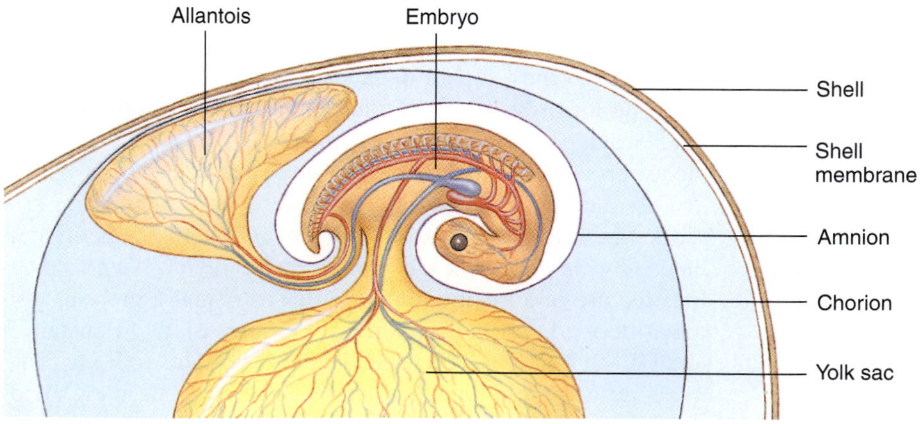

Allantois   Embryo

Shell

Shell membrane

Amnion

Chorion

Yolk sac

Figure 8.22

Amniotic egg at an early stage of development showing a chick embryo and its extraembryonic membranes.

between mother and fetus. Thus, a yolk sac placenta is formed. The mass of yolk is an extraembryonic structure because it is not a part of the embryo proper, and the yolk sac is an extraembryonic membrane because it is an accessory structure that develops outside the embryo and is discarded after the yolk is consumed.

The **allantois** is a sac that grows out of the hindgut and serves as a repository for metabolic wastes during development. It also functions as a respiratory surface for exchange of oxygen and carbon dioxide.

The **chorion** lies just beneath the eggshell and completely encloses the rest of the embryonic system. As the embryo grows and its need for oxygen increases, the allantois and chorion fuse to form the **chorioallantoic membrane.** This double membrane has a rich vascular network connected to the embryonic circulation.

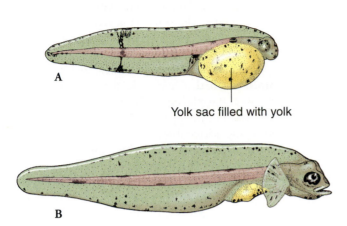

Figure 8.23
Fish larvae showing yolk sac. **A,** The one-day-old larva of a marine flounder has a large yolk sac. **B,** After 10 days of growth the larva has developed mouth, sensory organs, and a primitive digestive tract. With its yolk supply now exhausted, it must capture food to grow and survive.

Lying just beneath the porous shell, the vascular chorioallantois serves as a provisional "lung" across which oxygen and carbon dioxide can freely exchange. Thus an amniotic egg provides a complete life-support system for the embryo, enclosed by a tough outer shell. The amniotic egg is one of the most important adaptations to have evolved in vertebrates.

The evolution of a shelled amniotic egg made internal fertilization a reproductive requirement. A male must introduce sperm directly into the female reproductive tract, since sperm must reach and fertilize the egg before the eggshell is wrapped around it.

The Mammalian Placenta and Early Mammalian Development

Rather than developing within an eggshell like most other vertebrates most mammalian embryos evolved the strategy of developing within the mother's body. We have already seen that mammalian gastrulation closely parallels that of egg-laying amniotes. The earliest mammals were egg layers, and even today some mammals retain this primitive character; **monotremes** (duck-billed platypus and spiny anteater) lay large yolky eggs that closely resemble bird eggs. In **marsupials** (pouched mammals such as opossums and kangaroos), embryos develop for a time within the mother's uterus, but an embryo does not "take root" in the uterine wall, and consequently it receives little nourishment from the mother before birth. The young of marsupials are born at an early stage of development and continue developing sheltered in a pouch in the mother's abdominal wall, nourished with milk (reproduction in marsupials is described on pp. 628–629).

All other mammals, composing 94% of class Mammalia, are **placental mammals.** These mammals have evolved a **placenta,** a remarkable fetal structure through which an embryo is nourished. Evolution of this fetal organ required substantial restructuring, not only of extraembryonic membranes to form the placenta but also of the maternal oviduct, part of which had to expand into long-term housing for embryos, the **uterus.** Despite these modifications, development of extraembryonic membranes in placental mammals is remarkably similar to their development in egg-laying amniotes (compare Figures 8.22 and 8.24). In fact, for some nonmammalian vertebrates that give live birth, the extraembryonic membranes form a placenta. Some viviparous lizards and snakes have either a yolk sac placenta or a chorioallantoic placenta or both.

One of the most intriguing questions the placenta presents is, why is it not immunologically rejected by the mother? Both placenta and embryo are genetically alien to the mother because they contain proteins (called major histocompatibility proteins, p. 775) that differ from those of the mother. We would expect uterine tissues to reject the embryo just as the mother would reject an organ transplanted from her own child. The placenta is a uniquely successful foreign transplant, or **allograft,** because it has evolved measures for suppressing the immune response that normally would be mounted against it and the fetus by the mother. Experiments suggest that the chorion produces proteins and lymphocytes that block the normal immune response by suppressing formation of specific antibodies by the mother.

Early stages of mammalian cleavage, shown in Figure 8.13E, occur while a **blastocyst** is traveling down the oviduct toward the uterus, propelled by ciliary action and muscular peristalsis. When a human blastocyst is about six

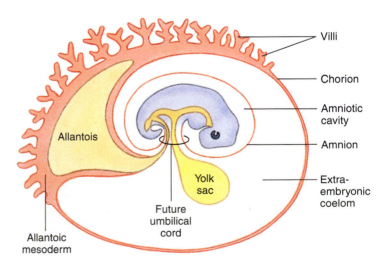

Figure 8.24
Generalized diagram of extraembryonic membranes of a mammal, showing how their development parallels that of a chick (compare with Figure 8.22). Most extraembryonic membranes of mammals have been redirected to new functions.

days old and composed of about 100 cells, it contacts the uterine endometrium (uterine lining) (Figure 8.25). On contact, the trophoblast cells proliferate rapidly and produce enzymes that break down the epithelium of the uterine endometrium. These changes allow the blastocyst to implant in the endometrium. By the eleventh or twelfth day the blastocyst is completely buried and surrounded by a pool of maternal blood. The trophoblast thickens, sending out thousands of tiny, fingerlike projections, the **chorionic villi.** These projections sink like roots into the uterine endometrium after the embryo implants. As development proceeds and embryonic demands for nutrients and gas exchange increase, the great proliferation of chorionic villi vastly increases the total surface area of the placenta. Although a human placenta at term measures only 18 cm (7 inches) across, its total absorbing surface is approximately 13 square meters—50 times the surface area of the skin of the newborn infant.

Since a mammalian embryo is protected and nourished through the placenta rather than with stored yolk, what happens to the four extraembryonic membranes it has inherited from early amniotes? The amnion remains unchanged, a protective water jacket in which the embryo floats. A fluid-filled yolk sac is also retained, although it contains no yolk. It has acquired a new function: during early development it is the source of stem cells that give rise to blood, lymphoid cells, and gametes. These stem cells later migrate into the developing embryo. In organisms such as racoons and mice, a heavily vascularized yolk sac implants in the uterus, along with the typical placenta. The two remaining extraembryonic membranes, allantois and the chorion, are recommitted to new functions. The allantois is no longer needed for storage of metabolic wastes. Instead it contributes to the **umbilical cord,** which links the embryo physically and functionally with the placenta. The chorion, the outermost membrane, forms most of the placenta itself. The rest of the placenta is formed by the adjacent uterine endometrium.

The embryo grows rapidly, and in humans all major organs of the body have begun their formation by the end of the fourth week of development. The embryo is now about 5 mm in length and weighs approximately 0.02 g. During the first two weeks of development **(germinal period)** the embryo is quite resistant to outside influences. However, during the next eight weeks, when all major organs are being established and body shape is forming **(embryonic period),** an embryo is more sensitive to disturbances that might cause malformations (such as exposure to alcohol or drugs taken by the mother) than at any other time in its development. The embryo becomes a **fetus** at approximately two months after fertilization. This ushers in the **fetal period,** which is primarily a growth phase, although organ systems (especially the nervous and endocrine systems) continue to differentiate. The fetus grows from approximately 28 mm and 2.7 g at 60 days to approximately 350 mm and 3000 g at term (nine months).

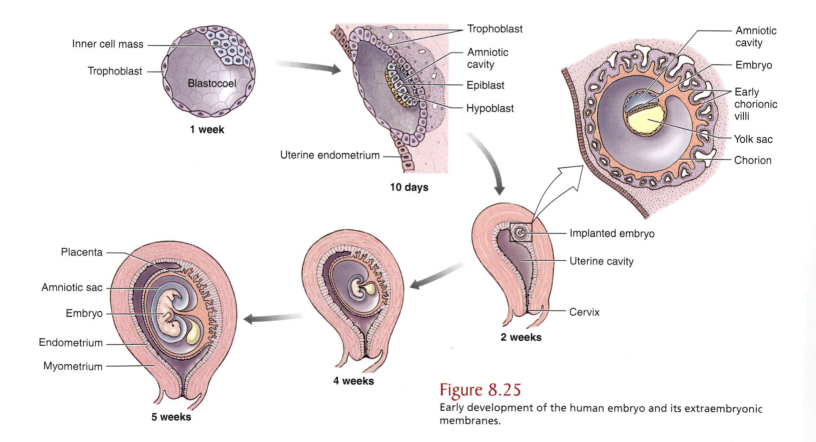

Figure 8.25

Early development of the human embryo and its extraembryonic membranes.

DEVELOPMENT OF SYSTEMS AND ORGANS

During vertebrate gastrulation the three germ layers are formed. These differentiate, as we have seen, first into primordial cell masses and then into specific organs and tissues. During this process, cells become increasingly committed to specific directions of differentiation. Derivatives of the three germ layers are diagrammed in Figure 8.26.

Assignment of early embryonic layers to specific "germ layers" (not to be confused with "germ cells," which are the eggs and sperm) is for the convenience of embryologists and is of no concern to the embryo. Whereas the three germ layers normally differentiate to form the tissue and organs described here, it is not the germ layer itself that determines differentiation, but rather the precise position of an embryonic cell with relation to other cells.

Derivatives of Ectoderm: Nervous System and Nerve Growth

The brain, spinal cord, and nearly all outer epithelial structures of the body develop from primitive ectoderm. They are among the earliest organs to appear. Just above the notochord, the ectoderm thickens to form a **neural plate.** The edges of this plate rise up, fold, and join together at the top to create an elongated, hollow **neural tube.** The neural tube gives rise to most of the nervous system: anteriorly it enlarges and differentiates into the brain and cranial nerves; posteriorly it forms the spinal cord and spinal motor nerves. Much of the rest of the peripheral nervous system is derived from **neural crest cells,** which pinch off from the neural tube before it closes (Figure 8.27). Among the multitude of different cell types and structures that originate with the neural crest are portions of the cranial nerves, pigment cells, cartilage and bone of most of the skull (including jaws), ganglia of the autonomic nervous

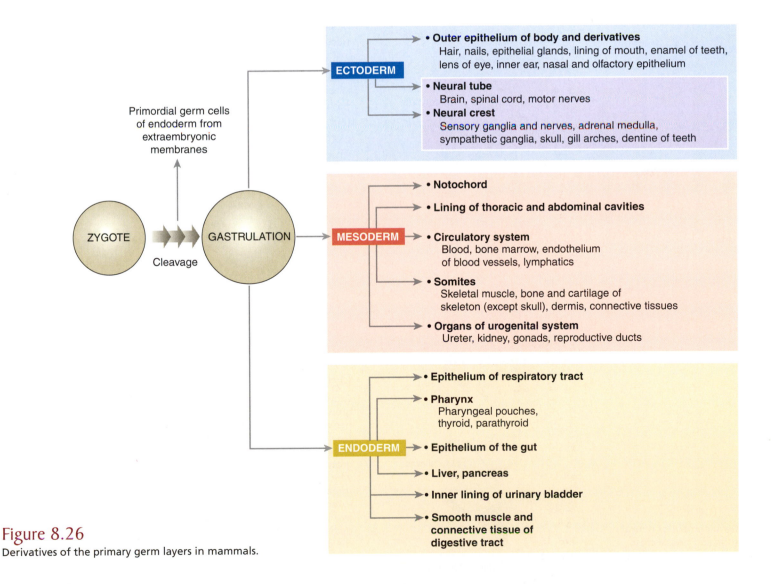

Figure 8.26
Derivatives of the primary germ layers in mammals.

system, medulla of the adrenal gland, and contributions to several other endocrine glands. Neural crest tissue is unique to vertebrates and was probably of prime importance in evolution of the vertebrate head and jaws.

How are the billions of nerve axons in the body formed? What directs their growth? Biologists were intrigued with these questions, which seemed to have no easy solutions. Because a single nerve axon may be more than a meter in length (for example, motor nerves running from the spinal cord to the toes), it seemed impossible that a single cell could reach out so far. The answer had to await the development of one of the most powerful tools available to biologists, the cell culture technique.

In 1907 embryologist Ross G. Harrison discovered that he could culture living neuroblasts (embryonic nerve cells) for weeks outside the body by placing them in a drop of frog lymph hung from the underside of a cover slip. Watching nerves grow for periods of days, he saw that each axon was an outgrowth of a single cell. As the axon extended outward, materials for growth flowed down the axon center to the growing tip (growth cone) where they were incorporated into new protoplasm (Figure 8.28).

The second question—what directs nerve growth—has taken longer to unravel. An idea held well into the 1940s was that nerve growth is a random, diffuse process. A major hypothesis proposed that the nervous system developed as an equipotential network, or blank slate, that later would be shaped by usage into a functional system. The nervous system just seemed too incredibly complex for us to imagine that nerve fibers could find their way selectively to so many predetermined destinations, yet it appears that this is exactly what they do! Research with invertebrate nervous systems indicated that each of the billions of nerve cell axons acquires a distinct identity that somehow directs it along a specific pathway to its destination. Many years ago Harrison observed that a growing nerve axon terminated in a growth cone, from which extend numerous tiny threadlike pseudopodial processes (filopodia) (Figure 8.28). Research has shown that the growth cone is steered by an array of guidance molecules secreted along the pathway and by the axon's target. This chemical guidance system, which must, of course, be genetically directed, is just one example of the amazing flexibility that characterizes the entire process of differentiation.

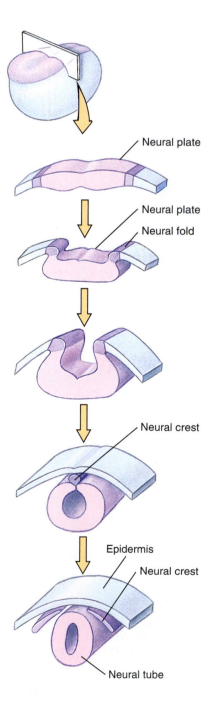

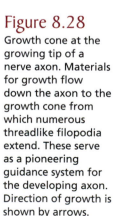

Figure 8.27

Development of neural tube and neural crest cells from neural plate ectoderm.

Neural plate

Neural plate

Neural fold

Neural crest

Epidermis

Neural crest

Neural tube

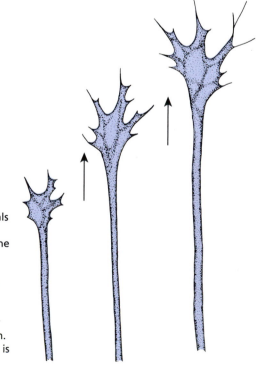

Figure 8.28

Growth cone at the growing tip of a nerve axon. Materials for growth flow down the axon to the growth cone from which numerous threadlike filopodia extend. These serve as a pioneering guidance system for the developing axon. Direction of growth is shown by arrows.

The tissue culture technique developed by Ross G. Harrison is now used extensively by scientists in all fields of active biomedical research, not just by developmental biologists. The great impact of the technique has been felt only in recent years. Harrison was twice considered for the Nobel Prize (1917 and 1933), but he failed ever to receive the award because, ironically, the tissue culture method was then considered "of rather limited value."

Derivatives of Endoderm: Digestive Tube and Survival of Gill Arches

In frog embryos the primitive gut makes its appearance during gastrulation with the formation of the **archenteron.** From this simple endodermal cavity develop the lining of the digestive tract, lining of the pharynx and lungs, most of the liver and pancreas, the thyroid and parathyroid glands, and the thymus (see Figure 8.26).

In other vertebrates the **alimentary canal** develops from the primitive gut and is folded off from the yolk sac by growth and folding of the body wall (Figure 8.29). The ends of the tube open to the exterior and are lined with ectoderm, whereas the rest of the tube is lined with endoderm. **Lungs, liver,** and **pancreas** arise from the foregut.

Among the most intriguing derivatives of the digestive tract are the pharyngeal pouches, which make their appearance in the early embryonic stages of all vertebrates (see Figure 8.21). During development the endodermally-lined pharyngeal pouches interact with overlying ectoderm to form gill arches. In fishes, gill arches develop into gills and supportive structures and serve as respiratory organs. When early vertebrates moved onto land, gills were unsuitable for aerial respiration and respiratory function was performed by independently evolved lungs.

Why then do gill arches persist in embryos of terrestrial vertebrates? Although gill arches serve no respiratory function in either embryos or adults of terrestrial vertebrates, they are necessary primordia for a variety of other structures. For example, the first

arch and its endoderm-lined pouch (the space between adjacent arches) form the upper and lower jaws and inner ear of vertebrates. The second, third, and fourth gill pouches contribute to the tonsils, parathyroid glands, and thymus. We can understand then why gill arches and other fishlike structures appear in early mammalian embryos. Their original function has been abandoned, but the structures are retained for new uses. The great conservatism of early embryonic development has conveniently provided us with a telescoped view of the origins of new adaptations.

Derivatives of Mesoderm: Support, Movement, and Beating Heart

The mesoderm forms most skeletal and muscular tissues, the circulatory system, and urinary and reproductive organs (see Figure 8.26). As vertebrates have increased in size and complexity, mesodermally derived supportive, movement, and transport structures have become an even greater proportion of the body.

Most **muscles** arise from the mesoderm along each side of the neural tube (Figure 8.30). This mesoderm divides into a linear series of blocklike somites (38 in humans), which by splitting, fusion, and migration become the axial skeleton, dermis of the dorsal skin, and muscles of the back, body wall, and limbs.

Mesoderm gives rise to the first functional organ, the embryonic heart. Guided by the underlying endoderm, two clusters of precardiac mesodermal cells move amebalike into position on either side of the developing gut. These clusters differentiate into a pair of double-walled tubes, which later fuse to form a single, thin tube (see Figure 8.14, p. 168).

As the cells group together, the first twitchings are evident. In a chick embryo, a favorite animal for experimental embryological studies, the primitive heart begins to beat on the second day of the 21-day incubation period; it begins beating before any true blood vessels have formed and before there is any blood to pump. As the ventricle primordium develops, the spontaneous cellular twitchings become coordinated into a feeble but rhythmical beat. New heart chambers, each with a beat faster than its predecessor, then develop.

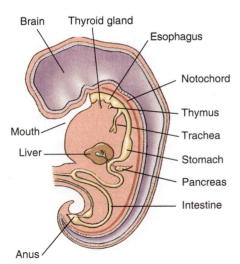

Brain
Thyroid gland
Esophagus
Notochord
Thymus
Mouth
Trachea
Liver
Stomach
Pancreas
Intestine
Anus

Figure 8.29
Derivatives of the alimentary canal of a human embryo.

Figure 8.30
Human embryo showing somites, which differentiate into skeletal muscles and axial skeleton.

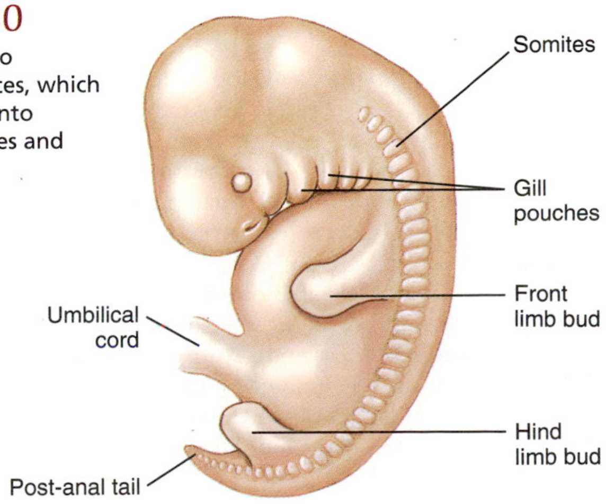

Somites
Gill pouches
Umbilical cord
Front limb bud
Hind limb bud
Post-anal tail

Finally a specialized area of heart muscle called the **sino-atrial (SA) node** develops and takes command of the entire heartbeat (the role of the SA node in the excitation of the heart is described on p. 694). The SA node becomes the heart's primary **pacemaker.** As the heart builds a strong and efficient beat, vascular channels open within the embryo and across the yolk. Within the vessels are the first primitive blood cells suspended in plasma.

Early development of the heart and circulation is crucial to continued embryonic development, because without circulation an embryo could not obtain materials for growth. Food is absorbed from the yolk and carried to the embryonic body, oxygen is delivered to all tissues, and carbon dioxide and other wastes are carried away. An embryo is totally dependent on these extraembryonic support systems, and the circulation is the vital link between them.

SUMMARY

Developmental biology encompasses the emergence of order and complexity during the development of a new individual from a fertilized egg, and the control of this process. The early preformation concept of development gave way in the eighteenth century to the theory of epigenesis, which holds that development is the progressive appearance of new structures that arise as the products of antecedent development. Fertilization of an egg by a sperm restores the diploid number of chromosomes and activates the egg for development. Both sperm and egg have evolved devices to promote efficient fertilization. The sperm is a highly condensed haploid nucleus provided with a locomotory flagellum. Many eggs release chemical sperm attractants, most have surface receptors that recognize and bind only with sperm of their own species, and all have developed devices to prevent polyspermy.

During cleavage an embryo divides rapidly and usually synchronously, producing a multicellular blastula. Cleavage is greatly influenced by quantity and distribution of yolk in the egg. Eggs with little yolk, such as those of many marine invertebrates, divide completely (holoblastic) and usually have indirect development with a larval stage interposed between the embryo and adult. Eggs having an abundance of yolk, such as those of birds, reptiles, and most arthropods divide only partially (meroblastic), and birds and reptiles have no larval stage.

Based on several developmental characteristics, bilateral metazoan animals are divided into two major groups. The Protostomia have mosaic cleavage and the mouth forms at or near the embryonic blastopore. The Deuterostomia have regulative cleavage and the mouth forms secondarily and not from the blastopore.

At gastrulation, cells on an embryo's surface move inward to form germ layers (endoderm, ectoderm, mesoderm) and the embryonic body plan. Like cleavage, gastrulation is much influenced by the quantity of yolk.

Despite the different developmental fates of embryonic cells, every cell contains a complete genome and thus the same nuclear information. Early development through cleavage is governed by cytoplasmic determinants derived from the maternal genome and placed in the egg cortex. As gastrulation approaches, control gradually shifts from maternal to embryonic as an embryo's own nuclear genes begin transcribing mRNA.

Harmonious differentiation of tissues proceeds in three general stages: pattern formation, determination of position in the body, and induction of limbs and organs appropriate for each position. Each stage is guided by morphogens. Pattern formation refers to determination of the anteroposterior, dorsoventral, and left-to-right body axes. In amphibians the anteroposterior axis is established by morphogens such as chordin from the Spemann organizer in the gray crescent of the zygote. In *Drosophila* that axis is determined by the morphogen bicoid, which is transcribed from maternal mRNA deposited at the anterior of the egg. In these and other segmented animals, such morphogens activate genes that divide the body into head, thorax and abdomen, and then into correctly oriented segments. The structures appropriate to each segment are then induced by homeotic genes, which are characterized by a particular sequence of DNA bases called the homeobox. Mutations in homeotic genes result in the development of inappropriate structures on a segment: legs on the head, for example.

The anteroposterior axis of an embryo is determined by homeotic and other homeobox-containing genes contained in one or more clusters on particular chromosomes. These genes, called *Hox* genes, occur not only in *Drosophila* and amphibians, but apparently in all animals. Each *Hox* gene is active in a particular region of the body, depending on its position within the cluster. Dorsoventral and left-right axes are similarly determined by morphogens that are produced only in the appropriate regions of the embryo. Similarly, morphogens guide the development of limbs along three body axes. Morphogens have been found to be remarkably similar in animals as different as *Drosophila* and amphibians. This realization has given rise to the field of evolutionary developmental biology, which is based on the idea that the evolution of the enormous variety of animals is the result of changes in the position and timing of relatively few genes that control development.

The postgastrula stage of vertebrate development represents a remarkable conservation of morphology when jawed vertebrates from fish to humans exhibit features common to all. As development proceeds, species-specific characteristics are formed.

Amniotes are terrestrial vertebrates that develop extraembryonic membranes during embryonic life. The four membranes are amnion, allantois, chorion, and yolk sac, each serving a specific life-support function for the embryo that develops within a self-contained egg (as in birds and most reptiles) or within the maternal uterus (mammals).

Mammalian embryos are nourished by a placenta, a complex fetal-maternal structure that develops in the uterine wall. During pregnancy the placenta becomes an independent nutritive, endocrine, and regulatory organ for the embryo.

Germ layers formed at gastrulation differentiate into tissues and organs. The ectoderm gives rise to skin and nervous system; endoderm gives rise to alimentary canal, pharynx, lungs, and certain glands; and mesoderm forms muscular, skeletal, circulatory, reproductive, and excretory organs.

REVIEW QUESTIONS

1. What is meant by epigenesis? How did Kaspar Friedrich Wolff's concept of epigenesis differ from the early notion of preformation?
2. How is an egg (oocyte) prepared during oogenesis for fertilization? Why is preparation essential to development?
3. Describe events that follow contact of a spermatozoon with an egg. What is polyspermy and how is it prevented?
4. What is meant by the term "activation" in embryology?
5. How does amount of yolk affect cleavage? Compare cleavage in a sea star with that in a bird.
6. What is the difference between radial and spiral cleavage?
7. What other developmental hallmarks are often associated with spiral or radial cleavage?
8. What is indirect development?
9. Using sea star embryos as an example, describe gastrulation. Explain how the mass of inert yolk affects gastrulation in frog and bird embryos.
10. What is the difference between schizocoelous and enterocoelous origins of a coelom?
11. Describe two different experimental approaches that serve as evidence for nuclear equivalence in animal embryos.
12. What is meant by "induction" in embryology? Describe the famous organizer experiment of Spemann and Mangold and explain its significance.
13. What are homeotic genes and what is the "homeobox" contained in such genes? What is the function of the homeobox? What are *Hox* genes? What is the significance of their apparently universal occurrence in animals?
14. What is the embryological evidence that vertebrates form a monophyletic group?
15. What are the four extraembryonic membranes of amniotic eggs of birds and reptiles and what is the function of each membrane?
16. What is the fate of the four extraembryonic membranes in embryos of placental mammals?
17. Explain what the "growth cone" that Ross Harrison observed at the ends of growing nerve fibers does to influence direction of nerve growth.
18. Name two organ system derivatives of each of the three germ layers.
19. What developmental characters are used to divide animals between protostome and deuterostome groups (clades)?

SELECTED REFERENCES

Carroll, S. B., J. K. Grenier, and S. D. Weatherbee. 2005. From DNA to diversity: molecular genetics and the evolution of animal design, ed. 2. Malden, Massachusetts, Blackwell Publishing. *Animal body plans develop through a hierarchy of gene interactions. As these interactions are understood, biologists seek commonalities in the "genetic toolkit" across a wide variety of taxa.*

Cibelli, J. B., R. P. Lanza, and M. D. West. 2002. The first human cloned embryo. Sci. Am. **286:**44–51 (Jan.). *Describes the first cloning of human embryos—but only to the 6-cell stage. Many scientists remain skeptical.*

Degnan, B. M., S. P. Leys, and C. Larroux. 2005. Sponge development and antiquity of animal pattern formation. Integr. Comp. Biol. **45:**335–341. *After blastula formation in a demosponge embryo, cell migration produces a two-layered gastrula that develops a third layer before becoming a free-swimming larva. If this pattern is typical, it suggests that both blastula and gastrula stages were present in ancestral metazoans.*

Gilbert, S. F. 2003. Developmental biology, ed. 7. Sunderland, Massachusetts, Sinauer Associates. *Combines descriptive and mechanistic aspects; good selection of examples from many animal groups.*

Gilbert, S. F., and A. M. Raunio (eds.). 1997. Embryology: constructing the organism. Sunderland, Massachusetts, Sinauer Associates. *The embryology of numerous animal groups.*

Goodman, C. S., and M. J. Bastiani. 1984. How embryonic nerve cells recognize one another. Sci. Am. **251:**58–66 (Dec.). *Research with insect larvae shows that developing neurons follow pathways having specific molecular labels.*

Leys, S. P., and D. Eerkes-Medrano. 2005. Gastrulation in calcareous sponges: in search of Haeckel's gastraea. Integr. Comp. Biol. **45:**342–351. *Ingression of cells during embryogenesis produces two germ layers (interpreted as gastrulation) in* Sycon, *a calcareous sponge. The ancestral pattern of gastrulation may be via ingression rather than by invagination, which occurs much later during larval metamorphosis.*

Nüsslein-Volhard, C. 1996. Gradients that organize embryo development. Sci. Am. **275:**54–61 (Aug.). *An account of the author's Nobel Prize–winning research.*

Rosenberg, K. R., and W. R. Trevathan. 2001. The evolution of human birth. Sci. Am. **285:**72–77 (Nov.). *Examines reasons why humans are the only primates to seek assistance during childbirth.*

Wolpert, L. 1991. The triumph of the embryo. Oxford, Oxford University Press. *Written for the nonspecialist, this engaging book is rich in detail and insight for all biologists interested in the development of life.*

ONLINE LEARNING CENTER

Visit www.mhhe.com/hickmanipz14e for chapter quizzing, key term flash cards, web links and more.

PART THREE

Diversity of Animal Life

3

A view of coral reef biodiversity.

Architectural Pattern of an Animal

*Cnidarian polyps have radial symmetry and cell-tissue grade of organization (*Dendronephthya *sp.).*

New Designs for Living

Zoologists today recognize 34 phyla of multicellular animals, each phylum characterized by a distinctive body plan and biological properties that set it apart from all other phyla. All are survivors of perhaps 100 phyla that appeared 600 million years ago during the Cambrian explosion, the most important evolutionary event in the geological history of life. Within the space of a few million years, virtually all major body plans that we see today, together with many other novel plans that we know only from the fossil record, were established. Entering a world sparse in species and mostly free of competition, these new life-forms diversified, producing new themes in animal architecture.

Later bursts of speciation that followed major extinction events produced mainly variations on established themes.

Established themes, in the form of distinctive body plans, are passed down a lineage from an ancestral population to its descendants; molluscs carry a hard shell, bird forelimbs make wings. These ancestral traits limit the morphological scope of descendants no matter what their lifestyle. Although penguin bodies are modified for an aquatic life, the wings and feathers of their bird ancestors might never adapt as well as fish fins and scales. Despite structural and functional evolution, new forms are constrained by the architecture of their ancestors.

The English satirist Samuel Butler proclaimed that the human body was merely "a pair of pincers set over a bellows and a stewpan and the whole thing fixed upon stilts." Most people less cynical than Butler would agree that the body is a triumph of intricate, living architecture. Less obvious, perhaps, is that the architecture of humans and most other animals conforms to the same well-defined plan. The basic uniformity of biological organization derives from the common ancestry of animals and from their basic cellular construction. Despite vast differences of structural complexity of organisms ranging from unicellular forms to humans, all share an intrinsic material design and fundamental functional plan. In this introduction to the diversity chapters (Chapters 11 through 28), we consider the limited number of body plans that underlie the apparent diversity of animal form and examine some of the common architectural themes that animals share.

HIERARCHICAL ORGANIZATION OF ANIMAL COMPLEXITY

Among the different unicellular and metazoan groups, we recognize five major grades of organization (Table 9.1). Each grade is more complex than the one preceding, and builds on it in a hierarchical manner.

The unicellular groups are the simplest eukaryotic organisms and represent the *protoplasmic* grade of organization. They are nonetheless complete organisms that perform all of the basic functions of life as seen in more complex animals. Within the confines of their cell, they show remarkable organization and division of labor, possessing distinct supportive structures, locomotor devices, fibrils, and simple sensory structures. The diversity observed among unicellular organisms is achieved

TABLE 9.1

Grades of Organization in Organismal Complexity

1. *Protoplasmic grade of organization.* Protoplasmic organization characterizes unicellular organisms. All life functions are confined within the boundaries of a single cell, the fundamental unit of life. Within a cell, protoplasm is differentiated into organelles capable of performing specialized functions.

2. *Cellular grade of organization.* Cellular organization is an aggregation of cells that are functionally differentiated. A division of labor is evident, so that some cells are concerned with, for example, reproduction, and others with nutrition. Some flagellates, such as *Volvox,* that have distinct somatic and reproductive cells are placed at the cellular level of organization. Many authorities also place sponges at this level.

3. *Cell-tissue grade of organization.* A step beyond the preceding is an aggregation of similar cells into definite patterns or layers and organized to perform a common function, to form a **tissue**. Sponges are considered by some authorities to belong to this grade, although jellyfishes and their relatives (Cnidaria) more clearly demonstrate the tissue plan. Both groups are still largely of the cellular grade of organization because most cells are scattered and not organized into tissues. An excellent example of a tissue in cnidarians is the **nerve net,** in which nerve cells and their processes form a definite tissue structure, with the function of coordination.

4. *Tissue-organ grade of organization.* An aggregation of tissues into organs is a further step in complexity. Organs are usually composed of more than one kind of tissue and have a more specialized function than tissues. This is the organizational level of flatworms (Platyhelminthes), in which well-defined organs such as eyespots, proboscis, and reproductive organs occur. In flatworms, the reproductive organs transcend the tissue-organ grade and are organized into a reproductive system.

5. *Organ-system grade of organization.* When organs work together to perform some function, we have the highest level of organization—an organ system. Systems are associated with basic body functions such as circulation, respiration, and digestion. The simplest animals having this type of organization are nemertean worms, which have a complete digestive system distinct from the circulatory system. Most animal phyla demonstrate this type of organization.

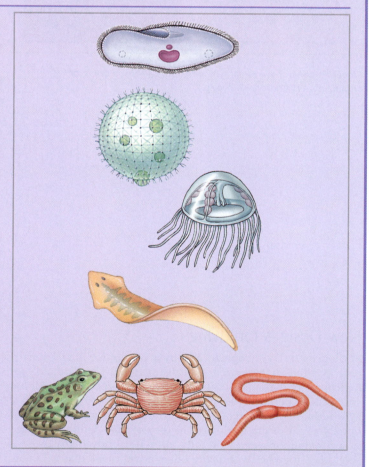

by varying the architectural patterns of subcellular structures, organelles, and the cell as a whole (see Chapter 11).

The **metazoa,** or multicellular animals, evolved greater structural complexity by combining cells into larger units. A metazoan cell is a specialized part of the whole organism and, unlike a unicellular organism, it is not capable of independent existence. Cells of a multicellular organism are specialized for performing the various tasks accomplished by subcellular elements in unicellular forms. The simplest metazoans show the *cellular* grade of organization in which cells demonstrate division of labor but are not strongly associated to perform a specific collective function (Table 9.1). In the more complex *cell-tissue* grade of organization, cells are grouped together and perform their common functions as a highly coordinated unit called a **tissue.** Animals at or beyond the cell-tissue grade of organization are termed **eumetazoans.** In animals of the *tissue-organ* grade of organization, tissues are assembled into still larger functional units called **organs.** Usually one type of tissue carries the burden of an organ's chief function, as muscle tissue does in the heart; other tissues—epithelial, connective, and nervous—perform supportive roles. The chief functional cells of an organ are called **parenchyma** (pa-ren´ka-ma; Gr. *para,* beside, + *enchyma,* infusion). The supportive tissues are its **stroma** (Gr. bedding). For instance, in the vertebrate pancreas the secreting cells are the parenchyma; capsule and connective tissue framework represent stroma.

Most metazoa have an additional level of complexity in which different organs operate together as **organ systems.** Eleven different kinds of organ systems are observed in metazoans: skeletal, muscular, integumentary, digestive, respiratory, circulatory, excretory, nervous, endocrine, immune, and reproductive. The great evolutionary diversity of these organ systems is covered in Chapters 14 through 28.

ANIMAL BODY PLANS

As described in the prologue to this chapter, the ancestral body plan constrains the form of its descendant lineage. Animal body plans differ in the grade of organization, in body symmetry, in the number of embryonic germ layers, and in the number of body cavities. Body symmetry can be generally determined from the external appearance of an animal, but other features of a body plan typically require a more detailed examination.

Animal Symmetry

Symmetry refers to balanced proportions, or correspondence in size and shape of parts on opposite sides of a median plane.

Spherical symmetry means that any plane passing through the center divides a body into equivalent, or mirrored, halves (Figure 9.1, *left*). This type of symmetry is found chiefly among some unicellular forms and is rare in animals. Spherical forms are best suited for floating and rolling.

Radial symmetry (Figure 9.1, *middle*) applies to forms that can be divided into similar halves by more than two planes passing through the longitudinal axis. These are tubular, vase, or bowl shapes found in some sponges and in hydras, jellyfish, sea urchins, and related groups, in which one end of the longitudinal axis is usually the mouth (the **oral** surface). In sessile forms, such as hydra and sea anemones, the basal attachment disc is the **aboral** surface. A variant form is **biradial symmetry** in which, because of some part that is single or paired rather than radial, only two planes passing through the longitudinal axis produce mirrored halves. Comb jellies (phylum Ctenophora, p. 282), which are globular but have a pair of tentacles, are an example. Radial and biradial animals are usually sessile, freely floating, or weakly swimming. Radial animals, with no anterior or posterior end, can interact with their environment in all directions—an advantage to sessile or free-floating forms with feeding structures arranged to snare prey approaching from any direction.

The two phyla that are primarily radial as adults, Cnidaria and Ctenophora, have been called the **Radiata,** although phylogenetic results suggest that this group is not monophyletic (p. 285). Echinoderms (sea stars and their kin) are primarily bilateral animals (their larvae are bilateral) that have become secondarily radial as adults.

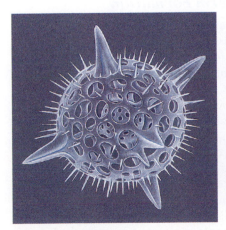

Spherical symmetry

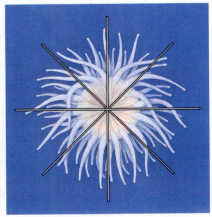

Radial symmetry

Bilateral symmetry

Figure 9.1
Animal symmetry. Illustrated are animals showing spherical, radial, and bilateral symmetry.

Bilateral symmetry applies to animals that can be divided along a sagittal plane into two mirrored portions—right and left halves (Figure 9.1, *right*). The appearance of bilateral symmetry in animal evolution was a major innovation, because bilateral animals are much better fitted for directional (forward) movement than are radially symmetrical animals. Bilateral animals form a monophyletic group of phyla called the **Bilateria.**

Bilateral symmetry is strongly associated with **cephalization,** differentiation of a head. The concentration of nervous tissue and sense organs in a head bestows obvious advantages to an animal moving through its environment head first. This is the most efficient positioning of organs for sensing the environment and responding to it. Usually the mouth of an animal is located on the head as well, since so much of an animal's activity is concerned with procuring food. Cephalization is always accompanied by differentiation along an anteroposterior axis, although the evolution of this axis preceded cephalization.

Some convenient terms used for locating regions of bilaterally symmetrical animals (Figure 9.2) are **anterior,** used to designate the head end; **posterior,** the opposite or tail end; **dorsal,** the back side; and **ventral,** the front or belly side. **Medial** refers to the midline of the body; **lateral,** to the sides. **Distal** parts are farther from the middle of the body; **proximal** parts are nearer. A **frontal plane** (sometimes called coronal plane) divides a bilateral body into dorsal and ventral halves by running through the anteroposterior axis and the right-left axis at right angles to the **sagittal plane,** the plane dividing an animal into right and left halves. A **transverse plane** (also called a cross section) would cut through a dorsoventral and a right-left axis at right angles to both the sagittal and frontal planes and would separate anterior and posterior portions (Figure 9.2). In vertebrates **pectoral** refers to the chest region or area associated with the anterior pair of appendages, and **pelvic** refers to the hip region or area associated with the posterior pair of appendages.

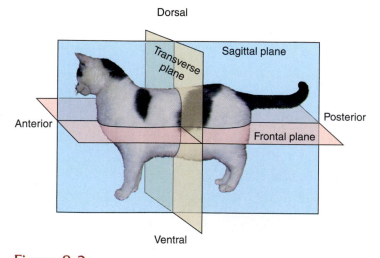

Dorsal

Transverse plane

Sagittal plane

Anterior

Posterior

Frontal plane

Ventral

Figure 9.2

The planes of symmetry as illustrated by a bilaterally symmetrical animal.

Body Cavities and Germ Layers

A body cavity is an internal space. The most obvious example is a gut cavity, but the vast majority of animals have a second, less obvious, cavity outside the gut. When this second cavity is fluid-filled, it may cushion and protect the gut from forces exerted on the body. In some animals, such as an earthworm, it also forms part of a hydrostatic skeleton used in locomotion.

Animals differ in the presence and number of body cavities. Sponges, at the cellular grade of organization, have no body cavities, not even a gut cavity. If sponges share the same developmental sequence as other metazoans, why do they lack a gut cavity? Where in the developmental sequence does a gut form? Sponges, like all metazoans, develop from a zygote to a blastula stage. A typical spherical blastula is composed of a layer of cells surrounding a fluid-filled cavity (see Figure 8.9). This cavity, a **blastocoel,** has no external opening, so it could not serve as a gut. In sponges, after the formation of a blastula, the cells reorganize to form an adult animal, although some researchers argue that a gastrula stage does occur before reorganization (see Figure 9.5, *upper pathway* and Chapter 12).

In animals other than sponges, development proceeds from a blastula to a **gastrula** stage, as one side of the blastula pushes inward, making a depression (see Figure 9.3). The depression becomes a gut cavity, also called a **gastrocoel** or **archenteron.** The external opening to the depression is the **blastopore;** it typically becomes the adult mouth or anus. The gut lining is **endoderm** and the outer layer of cells, surrounding the blastocoel, is **ectoderm** (Figure 9.3). The embryo now has two cavities, a gut and a blastocoel. Animals such as sea anemones and jellyfish develop from these two germ layers and are called **diploblastic** (Figure 9.5, *upper pathway*). They typically have radial symmetry as adult animals. The fluid-filled blastocoel persists in diploblasts, but in others it is filled with a third germ layer, **mesoderm.** Animals that possess ectoderm, mesoderm, and endoderm are termed **triploblastic** and the majority are bilaterally symmetrical.

Methods of Mesoderm Formation

Cells forming mesoderm are derived from endoderm, but there are two ways a middle tissue layer of mesoderm can form. In protostomes, mesoderm forms as endodermal cells from near the blastopore migrate into the blastocoel (Figure 9.3A). Following this event, three different body plans—acoelomate, pseudocoelomate, and coelomate—are possible (Figure 9.3A).

In the **acoelomate** plan, mesodermal cells completely fill the blastocoel, leaving a gut as the only body cavity (Figure 9.3A). The region between the ectodermal epidermis and endodermal digestive tract is filled with a spongy mass of space-filling cells, the **parenchyma** (Figure 9.4). Parenchyma is derived from embryonic connective tissue and is important in assimilation and transport of food and in disposal of metabolic wastes.

In the **pseudocoelomate** plan, mesodermal cells line the outer edge of the blastocoel, leaving two body cavities: a persistent blastocoel and a gut cavity (Figures 9.3A and 9.4). The

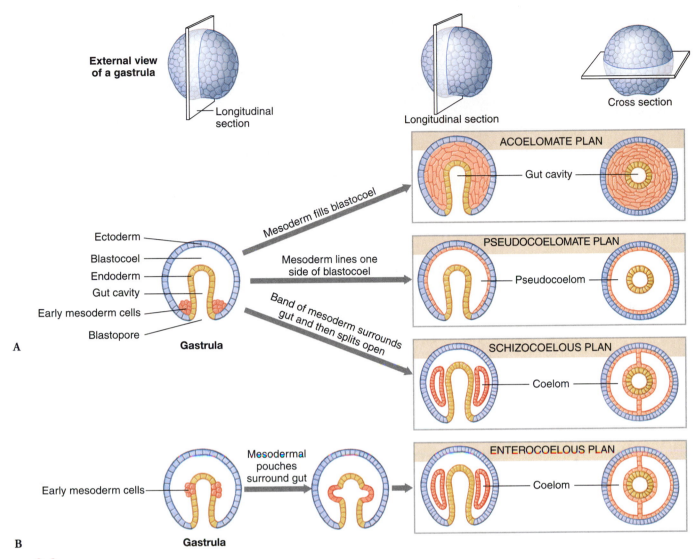

Figure 9.3

Mesoderm resides in different parts of the gastrula during formation of acoelomate, pseudocoelomate, and schizocoelous body plans **(A)**. Mesoderm and a coelom form together in the enterocoelus plan **(B)**.

blastocoel is now called a **pseudocoelom;** the name means false coelom in reference to mesoderm only partially surrounding the cavity, instead of completely surrounding it, as in a true **coelom.**

In the **schizocoelous** plan, mesodermal cells fill the blastocoel, forming a solid band of tissue around the gut. Then, through programmed cell death, space opens *inside* the mesodermal band (Figure 9.3A). This new space is a coelom. The embryo has two body cavities, a gut and a coelom.

In deuterostomes, mesoderm forms by an **enterocoelous** plan, where cells from the central portion of the gut lining begin to grow outward as pouches, expanding into the blastocoel (Figure 9.3B). The expanding pouch walls form a mesodermal ring. As the pouches move outward, they enclose a space. The space becomes a coelomic cavity or coelom. Eventually the pouches pinch off from the gut lining, completely enclosing a coelom bounded by mesoderm on all sides. The coelom completely fills the blastocoel. The embryo has two body cavities, a gut and a coelom.

A coelom made by **enterocoely** is functionally equivalent to a coelom made by **schizocoely,** and are represented as such in the **eucoelomate** body plan (Figure 9.4). Both kinds of coelomic cavities are bounded by mesoderm and lined with a **peritoneum,** a thin cellular membrane derived from mesoderm (Figure 9.4). Mesodermal **mesenteries** suspend organs in the coelom (Figure 9.4). A pseudocoelom lacks a peritoneum.

Developmental Origins of Body Plans in Triploblasts

Triploblastic animals follow one of several major developmental pathways to form a blastula from a zygote (Figure 9.5). The most common pathways are by spiral or radial cleavage (see Figure 8.10, p. 167).

Radial cleavage is typically accompanied by three other traits: The blastopore becomes an anus and a new opening makes the mouth, the coelom forms via enterocoely, and

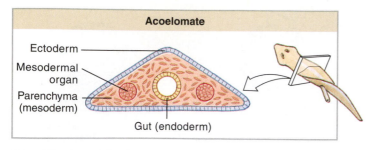

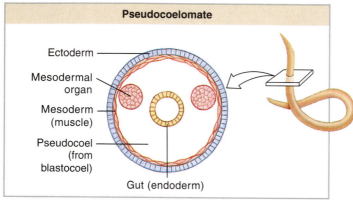

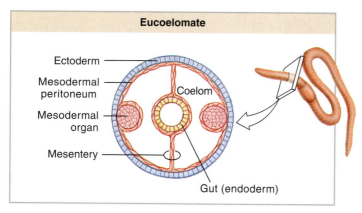

Figure 9.4

Acoelomate, pseudocoelomate, and eucoelomate body plans are shown as cross sections of representative animals. Note the relative positions of parenchyma, peritoneum, and body organs.

cleavage is regulative (see Figure 8.10, p. 165). Animals with these features are called deuterostomes (Figure 9.5, *lower pathway*); this group includes sea urchins and chordates.

Spiral cleavage produces an embryo whose developmental patterns contrast with those described for deuterostomes: the blastopore becomes the mouth, cleavage is mosaic (see Figure 8.10, p. 165), and mesoderm forms from a particular cell in the embryo, the 4d cell (see p. 169). The body may become acoelomate, pseudocoelomate, or coelomate, depending on the taxa (Figure 9.5, *central pathway*). If a coelom is present, it is made via schizocoely. Animals with these features are called lophotrochozoan protostomes; this group includes molluscs, segmented worms, and other taxa (Figure 9.5).

Lophotrochozoans are distinguished from ecdysozoan protostomes (not shown in Figure 9.5) where a range of cleavage patterns have been described. These include spiral cleavage, a

superficial cleavage pattern in which nuclei proliferate within common cytoplasm prior to separation by multiple cytoplasmic divisions (see Figure 8.15, p. 170) and a pattern initially resembling radial cleavage. Ecdysozoans may be coelomate or pseudocoelomate. Insects, crabs, and nematodes are among the ecdysozoans.

A Complete Gut Design and Segmentation

A few diploblasts and triploblasts have a blind or incomplete gut where food must enter and exit the same opening, but the majority of forms possess a complete gut (Figure 9.5). A complete gut makes possible a one-way flow of food from mouth to anus. A body constructed in this way is essentially a gut tube within another body tube. A tube-within-a-tube design has proved to be very versatile; members of the most common animal phyla, both invertebrate and vertebrate, have this plan.

Segmentation, also called metamerism, is another common feature of metazoans. Segmentation is a serial repetition of similar body segments along the longitudinal axis of the body. Each segment is called a **metamere,** or **somite.** In forms such as earthworms and other annelids (Figure 9.6), in which metamerism is most clearly represented, the segmental arrangement includes both external and internal structures of several systems. There is repetition of muscles, blood vessels, nerves, and setae of locomotion. Some other organs, such as those of sex, may be repeated in only a few segments. Evolutionary changes have obscured much of the segmentation in many animals, including humans.

The appearance of segmentation in body plans was a highly significant evolutionary event. Segmentation permits greater body mobility and complexity of structure and function. Its potential is amply displayed in phylum Arthropoda, the largest assemblage of animals on earth. Segmentation is found in phylum Chordata in addition to Annelida and Arthropoda (Figure 9.6), although superficial segmentation of ectoderm and body wall may appear among diverse groups of animals. The importance and potential of segmentation is discussed in Chapters 17 and 18.

COMPONENTS OF METAZOAN BODIES

Metazoan bodies consist of cellular components, derived from the three embryonic germ layers—ectoderm, mesoderm, and endoderm—as well as extracellular components.

Extracellular Components

Metazoan animals contain two important noncellular components: body fluids and extracellular structural elements. In all eumetazoans, body fluids are subdivided into two fluid "compartments": those that occupy **intracellular space,** within the body's cells, and those that occupy **extracellular space,** outside the cells. In animals with closed vascular systems

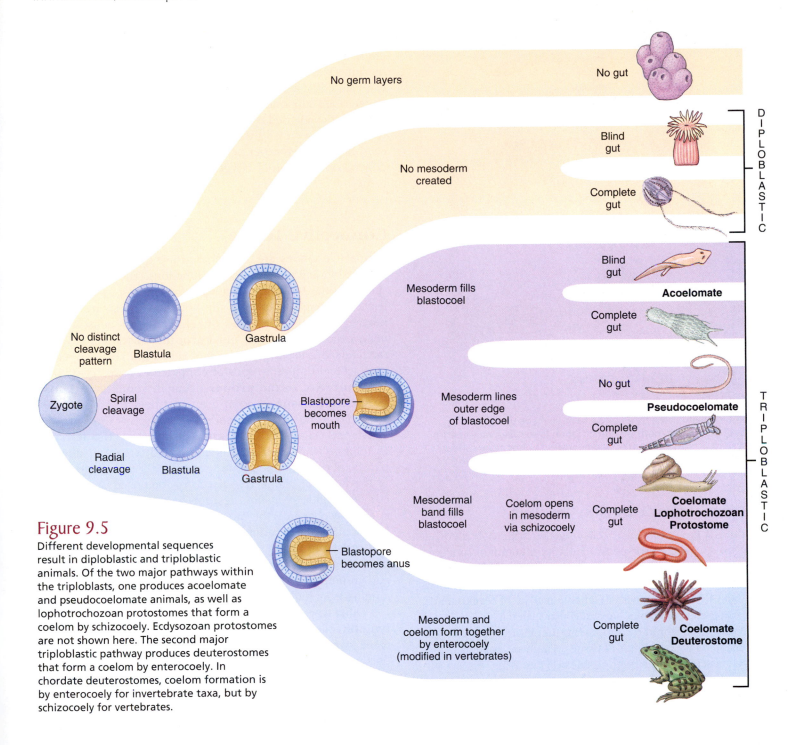

Figure 9.5
Different developmental sequences
result in diploblastic and triploblastic
animals. Of the two major pathways within
the triploblasts, one produces acoelomate
and pseudocoelomate animals, as well as
lophotrochozoan protostomes that form a
coelom by schizocoely. Ecdysozoan protostomes
are not shown here. The second major
triploblastic pathway produces deuterostomes
that form a coelom by enterocoely. In
chordate deuterostomes, coelom formation is
by enterocoely for invertebrate taxa, but by
schizocoely for vertebrates.

(such as segmented worms and vertebrates), the extracellular
fluids are subdivided further into **blood plasma** (the fluid
portion of blood) and **interstitial fluid.** Interstitial fluid,
also called tissue fluid, occupies the space surrounding cells.
Many invertebrates have open blood systems, however, with
no true separation of blood plasma from interstitial fluid. We
explore these relationships further in Chapter 31.

The term "intercellular," meaning "between cells," should not be
confused with the term "intracellular," meaning "within cells."

Extracellular structural elements are the supportive material
of the organism, including loose connective tissue (especially
well developed in vertebrates but present in all metazoa), cartilage (molluscs and chordates), bone (vertebrates), and cuticle
(arthropods, nematodes, annelids, and others). These elements
provide mechanical stability and protection (see Chapter 29).
In some instances, they act also as a depot of materials for
exchange between the cells and the interstitial fluid, and serve
as a medium for extracellular reactions. We describe diversity of
extracellular structural elements characteristic of different groups
of animals in Chapters 15 through 28.

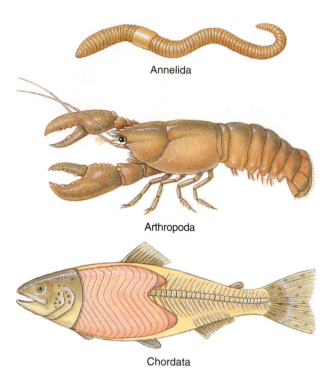

Annelida

Arthropoda

Chordata

Figure 9.6

Segmented phyla. These three phyla have all made use of an important principle in nature: segmentation (also called metamerism), or repetition of structural units. Segmentation brings more varied specialization because segments, especially in arthropods, have become modified for different functions.

Cellular Components: Tissues

A **tissue** is a group of similar cells (together with associated cell products) specialized for performance of a common function. The study of tissues is called **histology** (Gr. *histos,* tissue, + *logos,* discourse) or microanatomy. All cells in metazoan animals form tissues. Sometimes cells of a tissue may be of several kinds, and some tissues have much extracellular material.

During embryonic development, the germ layers become differentiated into four kinds of tissues. These are epithelial, connective, muscular, and nervous tissues (Figure 9.7). This is a surprisingly short list of only four basic tissue types that are able to meet the diverse requirements of animal life.

Epithelial Tissue

An **epithelium** (pl., epithelia) is a sheet of cells that covers an external or internal surface. Outside the body, epithelium forms a protective covering. Inside, epithelium lines all organs of the body cavity, as well as ducts and passageways through which various materials and secretions move. Thus, ions and molecules must pass through epithelial cells as they move to and from all other cells of the body. Consequently a large variety of transport molecules are located on epithelial cell membranes (see Chapter 3). Epithelial cells are also modified into glands that produce lubricating mucus or specialized products such as hormones or enzymes.

Epithelia are classified by cell form and number of cell layers. Simple epithelia (a single layer of cells; Figure 9.8) are found in all metazoan animals, while stratified epithelia (many cell layers; Figure 9.9) are mostly restricted to vertebrates. All types of epithelia are supported by an underlying basement membrane, which is a condensed region of ground substance of connective tissue, but is secreted by both epithelial and connective tissue cells. Blood vessels never penetrate into epithelial tissues, which depend on diffusion of oxygen and nutrients from underlying tissues.

Connective Tissue

Connective tissues are a diverse group of tissues that serve various binding and supportive functions. They are so widespread in the body that removal of other tissues would still leave the complete form of the body clearly apparent. Connective tissue is composed of relatively few cells, a great many extracellular fibers, and a **ground substance,** in which the fibers are suspended (together called **matrix**). We recognize several different types of connective tissue. Two kinds of **connective tissue proper** occur in vertebrates. **Loose connective tissue** is composed of fibers and both fixed and wandering cells suspended in a viscous fluid ground substance (Figure 9.10). **Dense connective tissue,** such as tendons and ligaments, is composed largely of densely packed fibers and little ground substance (Figure 9.10). Much of the fibers of connective tissue is composed of **collagen** (Gr. *kolla,* glue, + *genos,* descent), a protein of great tensile strength. Collagen is the most abundant protein in the animal kingdom, found in animal bodies wherever both flexibility and resistance to stretching are required. Connective tissue of invertebrates, as in vertebrates, consists of cells, fibers, and ground substance, and show a wide diversity of structure ranging from highly cellular to acellular histologies.

Other types of specialized connective tissue include **blood, lymph** (collectively considered vascular tissue), **adipose** (fat) tissue, **cartilage,** and **bone.** Vascular tissue is composed of distinctive cells in a fluid ground substance, the plasma. Vascular tissue lacks fibers under normal conditions. Blood composition is discussed in Chapter 31.

Cartilage is a semirigid form of connective tissue with closely packed fibers embedded in a gel-like ground substance (Figure 9.10). **Bone** is a calcified connective tissue containing calcium salts organized around collagen fibers (Figure 9.10). Structure of cartilage and bone is discussed in the section on skeletons in Chapter 29.

Muscular Tissue

Muscle is the most abundant tissue in the body of most animals. It originates (with few exceptions) from mesoderm, and its unit is the cell or **muscle fiber,** specialized for contraction. When viewed with a light microscope, **striated muscle** appears transversely striped (striated), with alternating dark and light bands (Figure 9.11). In vertebrates we recognize two types of striated

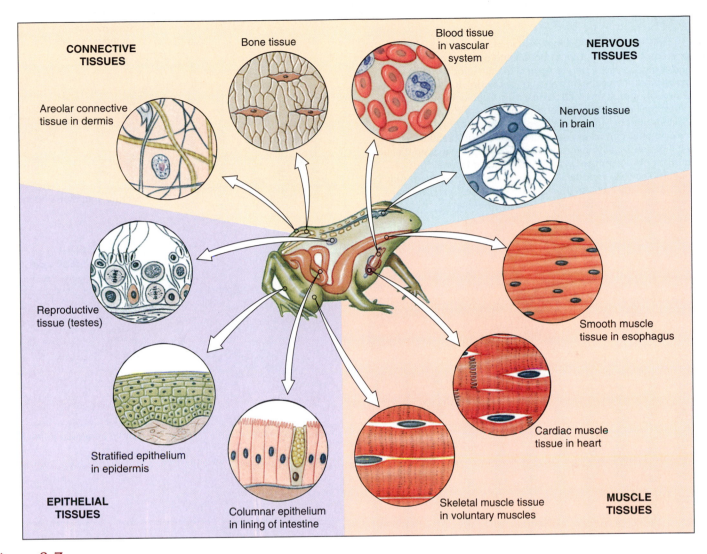

Figure 9.7
Types of tissues in a vertebrate, showing examples of where different tissues are located in a frog.

muscle: **skeletal** and **cardiac muscle.** In invertebrates, a third striated muscle type called **obliquely striated muscle** has been described. **Smooth** (or visceral) **muscle,** which lacks the characteristic alternating bands of the striated type is found in both invertebrates and vertebrates, although major ultrastructural differences have been described between them (Figure 9.11). Unspecialized cytoplasm of muscles is called **sarcoplasm,** and contractile elements within the fiber are **myofibrils.** Muscular movement is covered in Chapter 29.

Nervous Tissue

Nervous tissue is specialized for reception of stimuli and conduction of impulses from one region to another. Two basic types of cells in nervous tissue are **neurons** (Gr. nerve), the basic functional unit of nervous systems, and **neuroglia** (nu-rog´le-a; Gr. nerve, + *glia,* glue), a variety of nonnervous cells that insulate neuron membranes and serve various supportive functions. Figure 9.12 shows functional anatomy of a typical nerve cell. The functional roles of nervous tissue are treated in Chapter 33.

COMPLEXITY AND BODY SIZE

The most complex grades of metazoan organization permit and to some extent even promote evolution of large body size (Figure 9.13). Large size confers several important physical and ecological consequences for an organism. As animals become larger, the body surface increases much more slowly than body volume because surface area increases as the square of body length (length^2), whereas volume (and therefore mass) increases as the cube of body length (length^3). In other words, a large animal has less surface area relative to its volume than does a small animal of the same shape. The surface area of a large animal may be inadequate for respiration and nutrition by cells located deep within its body. There are two possible solutions to this problem. One solution is to fold or invaginate the body surface to increase the surface area or, as exploited by flatworms, flatten the body into a ribbon or disc so that no internal space is far from the surface. This solution allows a body to become large without internal complexity. However, most large animals adopted a second solution; they developed internal transport

Figure 9.8

Types of simple epithelium. **A, Simple squamous epithelium,** composed of flattened cells that form a continuous lining of blood capillaries, lungs, and other surfaces where it permits the diffusion of gases and transport of other molecules into and out of cavities. **B, Simple cuboidal epithelium** is composed of short, boxlike cells. Cuboidal epithelium usually lines small ducts and tubules, such as those of the kidney and salivary glands, and may have active secretory or absorptive functions. **C, Simple columnar epithelium** resembles cuboidal epithelium, but the cells are taller and usually have elongate nuclei. This type of epithelium is found on highly absorptive surfaces such as the intestinal tract of most animals. The cells often bear minute, fingerlike projections called microvilli that greatly increase the absorptive surface. In some organs, such as the female reproductive tract, cells may be ciliated.

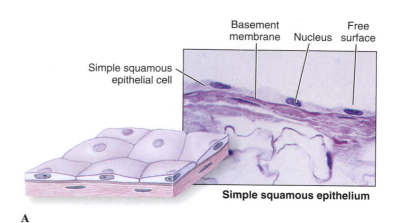

Simple squamous epithelium

A

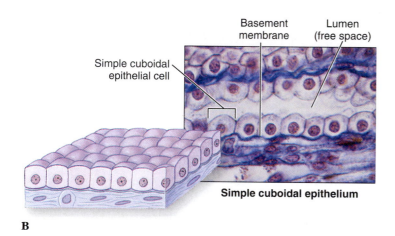

Simple cuboidal epithelium

B

Simple columnar epithelium

C

systems to shuttle nutrients, gases, and waste products between cells and the external environment.

Larger size buffers an animal against environmental fluctuations; it provides greater protection against predation and enhances offensive tactics; and it permits a more efficient use of metabolic energy. A large mammal uses more oxygen than a small mammal, but the cost of maintaining its body temperature is less per gram of weight for a large mammal than for a small one. Large animals also can move at less energy cost than can small animals. For example, a large mammal uses more oxygen in running than does a small mammal, but the energy cost of moving 1 g of its body over a given distance is much less for a large mammal than for a small one (Figure 9.14). For all of these reasons, ecological opportunities of larger animals are very different from those of small ones. In subsequent chapters we describe the extensive adaptive radiations observed in taxa of large animals.

The tendency for maximum body size to increase within lines of descent is known as "Cope's law of phyletic increase," named after nineteenth-century American paleontologist and naturalist Edward Drinker Cope. Cope noted that lineages begin with small organisms that give rise to larger and ultimately to giant forms. Large forms frequently become extinct, providing opportunities for new lineages, which in turn evolve larger forms. Cope's rule holds well for many nonflying vertebrates and invertebrate groups, even though Cope's Lamarckian explanation for the trend—that organisms evolved from an inner urge to attain a higher state of being (and larger size)—was preposterous. Many animal taxa contain lineages that show evolutionary miniaturization, in contrast to Cope's rule (for example, the insects).

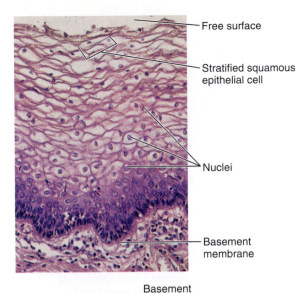

Free surface

Stratified squamous
epithelial cell

Nuclei

Basement
membrane

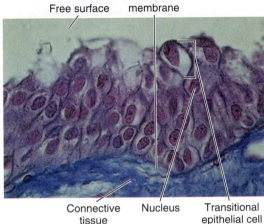

Free surface

Basement
membrane

Connective
tissue

Nucleus

Transitional
epithelial cell

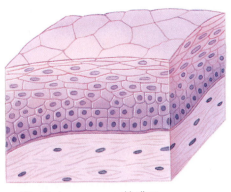

Stratified squamous epithelium

Stratified squamous epithelium
consists of two to many layers of
cells adapted to withstand mild
mechanical abrasion and distortion.
The basal layer of cells undergoes
continuous mitotic divisions,
producing cells that are pushed
toward the surface where they
are sloughed off and replaced by
new cells from beneath. This type
of epithelium lines the oral cavity,
esophagus, and anal canal of many
vertebrates, and the vagina of
mammals.

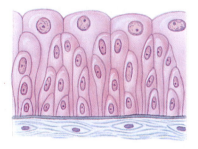

Transitional epithelium — unstretched

Transitional epithelium is a type
of stratified epithelium specialized
to accommodate great stretching.
This type of epithelium is found
in the urinary tract and bladder of
vertebrates. In the relaxed state it
appears to be four or five cell layers
thick, but when stretched it appears
to have only two or three layers of
extremely flattened cells.

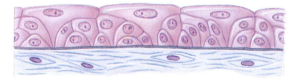

Transitional epithelium — stretched

Figure 9.9
Types of stratified epithelium.

SUMMARY

From the relatively simple organisms that mark the beginnings of life on earth, animal evolution has produced more intricately organized forms. Organelles are integrated into cells, cells into tissues, tissues into organs, and organs into systems. Whereas a unicellular organism performs all life functions within the confines of a single cell, a multicellular animal is an organization of subordinate units united at successive levels.

Every organism has an inherited body plan described in terms of body symmetry, number of embryonic germ layers, grade of organization, and number of body cavities. The majority of animals exhibit bilateral symmetry, but spherical and radial symmetry occur in some groups. Most animals are triploblastic and develop from three embryonic germ layers, but cnidarians and a few other forms are diploblastic. Sponges lack germ layers and possess a cellular grade of organization. Most animals have the tissue grade of organization.

All animals other than sponges have a gut cavity. Most animals have a second cavity that surrounds the gut. The second cavity may be a pseudocoelom or a coelom. There are two taxon-specific patterns of coelom formation, schizocoely and enterocoely.

Triploblastic animals are divided among deuterostomes and protostomes according to their particular developmental sequence. Protostomes are further divided into lophotrochozoan and ecdysozoan forms on the basis of more detailed features of development.

A metazoan body consists of cells, most of which are functionally specialized; body fluids, divided into intracellular and extracellular fluid compartments; and extracellular structural elements, which are fibers or formless materials that serve various structural functions in the extracellular space. The cells of metazoa develop into various tissues; basic types are epithelial, connective, muscular, and nervous. Tissues are organized into larger functional units called organs, and organs are associated to form systems.

One correlate of increased anatomical complexity is an increase in body size, which offers certain advantages such as more effective predation, reduced energy cost of locomotion, and improved homeostasis.

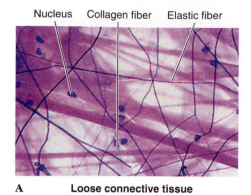

A   **Loose connective tissue**

Nucleus   Collagen fiber   Elastic fiber

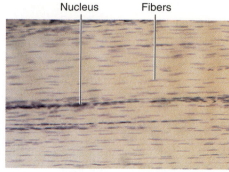

B   **Dense connective tissue**

Nucleus   Fibers

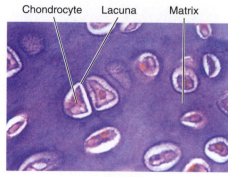

C   **Cartilage**

Chondrocyte   Lacuna   Matrix

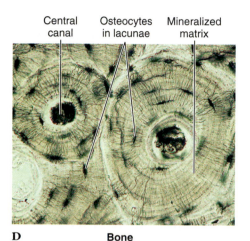

Central canal   Osteocytes in lacunae   Mineralized matrix

D   **Bone**

Figure 9.10

Types of connective tissue. **A, Loose connective tissue,** also called areolar connective tissue, is the "packing material" of the body that anchors blood vessels, nerves, and body organs. It contains fibroblasts that synthesize the fibers and ground substance of connective tissue and wandering macrophages that phagocytize pathogens or damaged cells. The different fiber types include collagen fibers (thick and red in micrograph) and thin elastic fibers (black and branching in micrograph) formed of the protein elastin. **B, Dense connective tissue** forms tendons, ligaments, and fasciae (fa´sha), the latter arranged as sheets or bands of tissue surrounding skeletal muscle. In a tendon (shown here) the collagenous fibers are extremely long and tightly packed together. **C, Cartilage** is a vertebrate connective tissue composed of a firm matrix containing cells (chondrocytes) located in small pockets called lacunae, and collagen and/or elastic fibers (depending on type of cartilage). In hyaline cartilage shown here, both collagen fibers and matrix are stained uniformly purple and cannot be distinguished one from the other. Because cartilage lacks a blood supply, all nutrients and waste materials must diffuse through the ground substance from surrounding tissues. **D, Bone,** strongest of vertebrate connective tissues, contains mineralized collagen fibers. Small pockets (lacunae) within the matrix contain bone cells, called osteocytes. The osteocytes communicate with each other by means of a tiny network of channels called canaliculi. Blood vessels, extensive in bone, are located in larger channels, including central canals. Bone undergoes continuous remodeling during an animal's life, and can repair itself following even extensive damage.

Nuclei of smooth muscle cells

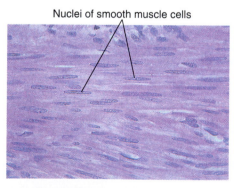

Smooth muscle is nonstriated muscle found in both invertebrates and vertebrates. Smooth muscle cells are long, and tapering, each containing a single nucleus. Smooth muscle is the most common type of muscle in invertebrates in which it serves as body wall musculature and surrounds ducts and sphincters. In vertebrates, smooth muscle surrounds blood vessels and internal organs such as intestine and uterus. It is called involuntary muscle in vertebrates because its contraction is usually not consciously controlled.

Skeletal muscle fiber   Nucleus   Striations

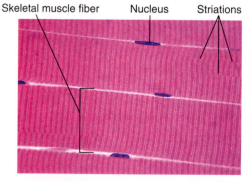

Skeletal muscle is a type of striated muscle found in both invertebrates and vertebrates. It is composed of extremely long, cylindrical fibers, which are multinucleate cells that may reach from one end of the muscle to the other. Viewed through the light microscope, the cells appear to have a series of stripes, called striations, running across them. Skeletal muscle is called voluntary muscle (in vertebrates) because it contracts when stimulated by nerves under conscious central nervous system control.

Striations   Intercalated discs   Nucleus

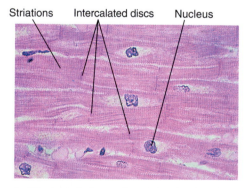

Cardiac muscle is another type of striated muscle found only in the vertebrate heart. The cells are much shorter than those of skeletal muscle and have only one nucleus per cell (uninucleate). Cardiac muscle tissue is a branching network of fibers with individual cells interconnected by junctional complexes called intercalated discs. Cardiac muscle is considered involuntary muscle because it does not require nerve activity to stimulate contraction. Instead, heart rate is controlled by specialized pacemaker cells located in the heart itself. However, autonomic nerves from the brain may alter pacemaker activity.

Figure 9.11

Types of muscle tissue.

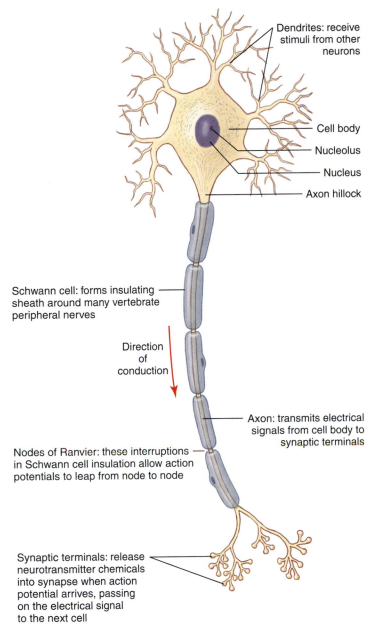

Dendrites: receive stimuli from other neurons

Cell body

Nucleolus

Nucleus

Axon hillock

Schwann cell: forms insulating sheath around many vertebrate peripheral nerves

Direction of conduction

Axon: transmits electrical signals from cell body to synaptic terminals

Nodes of Ranvier: these interruptions in Schwann cell insulation allow action potentials to leap from node to node

Synaptic terminals: release neurotransmitter chemicals into synapse when action potential arrives, passing on the electrical signal to the next cell

Figure 9.12

Functional anatomy of a neuron. From the nucleated cell body, or **soma,** extend one or more **dendrites** (Gr. *dendron,* tree), which receive electrical signals from receptors or other nerve cells, and a single **axon** that carries signals away from the cell body to other nerve cells or to an effector organ. The axon is often called a **nerve fiber.** Nerves are separated from other nerves or from effector organs by specialized junctions called synapses.

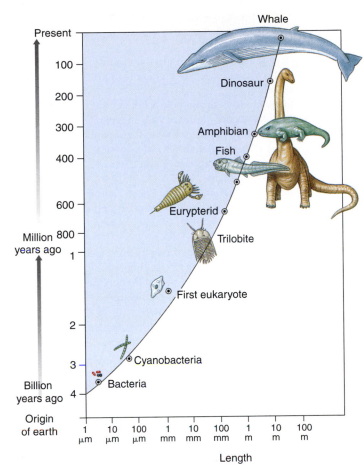

Figure 9.13

Graph showing the evolution of size (length) increase in organisms at different periods of life on earth. Note that both scales are logarithmic.

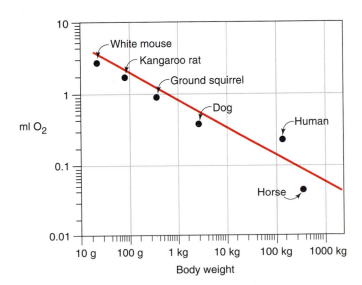

Figure 9.14

Net cost of running for mammals of various sizes. Each point represents the cost (measured in rate of oxygen consumption) of moving 1 g of body over 1 km. Cost decreases with increasing body size.

REVIEW QUESTIONS

1. Name the five grades of organization in organismal complexity and explain how each successive grade is more complex than the one preceding it.
2. Can you suggest why, during the evolutionary history of animals, there has been a tendency for maximum body size to increase? Do you think it inevitable that complexity should increase along with body size? Why or why not?
3. What is the meaning of the terms parenchyma and stroma as they relate to body organs?
4. Body fluids of eumetazoan animals are separated into fluid "compartments." Name these compartments and explain how compartmentalization may differ in animals with open and closed circulatory systems.
5. What are the four major types of tissues in metazoans?
6. How would you distinguish between simple and stratified epithelium? What characteristic of stratified epithelium might explain why it, rather than simple epithelium, is found lining the oral cavity, esophagus, and vagina?
7. What three elements are present in all connective tissue? Give some examples of different types of connective tissue.
8. What are three muscle tissue types found among animals? Explain how each is specialized for particular functions.
9. Describe the principal structural and functional features of a neuron.
10. Match the animal group with its body plan:

____ Unicellular	a. Nematode
____ Cell aggregate	b. Vertebrate
____ Blind sac, acoelomate	c. Protozoan
____ Tube-within-a-tube, pseudocoelomate	d. Flatworm
	e. Sponge
____ Tube-within-a-tube, eucoelomate	f. Arthropod
	g. Nemertean

11. Distinguish among spherical, radial, biradial, and bilateral symmetry.
12. Use the following terms to identify regions on your body and on the body of a frog: anterior, posterior, dorsal, ventral, lateral, distal, proximal.
13. How would frontal, sagittal, and transverse planes divide your body?
14. What is meant by segmentation? Name three phyla showing segmentation.

SELECTED REFERENCES

Arthur, W. 1997. The origin of animal body plans. Cambridge, U.K., Cambridge University Press. *Explores genetic, developmental, and population-level processes involved in the evolution of the 35 or so body plans that arose in the geological past.*

Baguna, J., and M. Ruitort. 2004. The dawn of bilaterian animals: the case of acoelomorph flatworms. Bioessays **26:**1046–1057. *Goes through the hypotheses for the transition between radial and bilateral body plans and the evidence for them.*

Cole, A. G., and B. K. Hall. 2004. The nature and significance of invertebrate cartilages revisited: distribution and histology of cartilage and cartilage-like tissues within the Metazoa. Zoology **107:**261–273. *Beautifully illustrated and well-written discussion of these tissues.*

Kessel, R. G. 1998. Basic medical histology: the biology of cells, tissues and organs. New York, Oxford University Press. *A current textbook of animal histology.*

Martindale, M. Q., J. R. Finnerty, and J. Q. Henry. 2002. The Radiata and the evolutionary origins of the bilaterian body plan. Molecular Phylogenetics and Evolution **24:**358–365. *Examines germ layers and symmetry in cnidarians and ctenophores to reconstruct the origin of the bilaterally symmetrical animals.*

McGowan, C. 1999. A practical guide to vertebrate mechanics. New York, Cambridge University Press. *Using many examples from his earlier book,* Diatoms to dinosaurs, *the author describes principles of biomechanics that underlie functional anatomy. Includes practical experiments and laboratory exercises.*

Royuela, M., B. Fraile, M. I. Arenas, and R. Paniagua. 2000. Characterization of several invertebrate muscle cell types: a comparison with vertebrate muscles. Microsc. Res. Tech. **41:**107–115. *Examines the ultrastructure of smooth, striated and oblique striated muscle in order to establish a clearer classification criteria.*

Welsch, U., and V. Storch. 1976. Comparative animal cytology and histology. London, Sidgwick & Jackson. *Comparative histology with good treatment of invertebrates.*

Willmer, P. 1990. Invertebrate relationships: patterns in animal evolution. Cambridge, U.K., Cambridge University Press. *Chapter 2 is an excellent discussion of animal symmetry, developmental patterns, origin of body cavities, and segmentation.*

ONLINE LEARNING CENTER

Visit www.mhhe.com/hickmanipz14e for chapter quizzing, key term flash cards, web links and more!

Taxonomy and Phylogeny of Animals

Molluscan shells from the collection of Jean Baptiste de Lamarck (1744 to 1829).

Order in Diversity

Evolution has produced a great diversity of species in the animal kingdom. Zoologists have named more than 1.5 million species of animals, and thousands more are described each year. Some zoologists estimate that species named so far constitute less than 20% of all living animals and less than 1% of all those that have existed.

Despite its magnitude, the diversity of animals is not without limits. Many conceivable forms do not exist in nature, as our myths of minotaurs and winged horses show. Animal diversity is not random but has definite order. Characteristic features of humans and cattle never occur together in a single organism as they do in the mythical minotaurs; nor do characteristic wings of birds and bodies of horses occur together naturally as they do in the mythical horse, Pegasus. Humans, cattle, birds, and horses are distinct groups of animals, yet they do share some important features, including vertebrae and homeothermy, that separate them from even more dissimilar forms such as insects and flatworms.

All human cultures classify familiar animals according to patterns in animal diversity. These classifications have many purposes. Some societies classify animals according to their usefulness or destructiveness to human endeavors; others may group animals according to their roles in mythology. Biologists organize animal diversity in a nested hierarchy of groups within groups according to evolutionary relationships as revealed by ordered patterns in their sharing of homologous features. This ordering is called a "natural system" because it reflects relationships that exist among animals in nature, outside the context of human activity. A systematic zoologists has three major goals: to discover all species of animals, to reconstruct their evolutionary relationships, and to communicate those relationships by constructing an informative taxonomic system.

Darwin's theory of common descent (Chapters 1 and 6) is the underlying principle that guides our search for order in the diversity of animal life. Our science of **taxonomy** ("arrangement law") produces a formal system for naming and grouping species to communicate this order. Animals that have very recent common ancestry share many features in common and are grouped most closely in our taxonomic classification. Taxonomy is part of the broader science of systematics, or comparative biology, in which studies of variation among animal populations are used to understand their evolutionary relationships. The study of taxonomy predates evolutionary biology, however, and many taxonomic practices are remnants of a pre-evolutionary world view. Adjusting our taxonomic system to accommodate evolution has produced many problems and controversies. Taxonomy has reached an unusually active and controversial point in its development in which several alternative taxonomic systems are competing for use. To understand this controversy, it is necessary first to review the history of animal taxonomy.

LINNAEUS AND TAXONOMY

The Greek philosopher and biologist Aristotle was the first to classify organisms according to their structural similarities. The flowering of systematics in the eighteenth century culminated in the work of Carolus Linnaeus (Figure 10.1), who designed our current scheme of classification.

Linnaeus was a Swedish botanist at the University of Uppsala. He had a great talent for collecting and classifying objects, especially flowers. Linnaeus produced an extensive system of classification for both plants and animals. This scheme, published in his great work, *Systema Naturae,* used morphology (the comparative study of organismal form) for arranging specimens in collections. He divided the animal kingdom into species and gave each one a distinctive name. He grouped species into genera, genera into

Figure 10.1
Carolus Linnaeus (1707 to 1778). This portrait was made of Linnaeus at age 68, three years before his death.

orders, and orders into "classes" (we use quotation marks or a capital letter to distinguish "class" as a formal taxonomic rank from its broader meaning as a group of organisms that share a common essential property). Because his knowledge of animals was limited, his lower categories, such as genera, often were very broad and included animals that are only distantly related. Much of his classification is now drastically altered, but the basic principle of his scheme is still followed.

Linnaeus's scheme of arranging organisms into an ascending series of groups of ever-increasing inclusiveness is a **hierarchical system** of classification. Major **taxa** (sing., **taxon**), into which organisms are grouped were given one of several standard **taxonomic ranks** to indicate the general degree of inclusiveness of the group. The hierarchy of taxonomic ranks has been expanded considerably since Linnaeus's time (Table 10.1). It now includes seven mandatory ranks for the animal kingdom, in descending

TABLE 10.1

Examples of Taxonomic Categories to Which Representative Animals Belong

Linnaean Rank	Human	Gorilla	Southern Leopard Frog	Katydid
Kingdom	Animalia	Animalia	Animalia	Animalia
Phylum	Chordata	Chordata	Chordata	Arthropoda
Subphylum	Vertebrata	Vertebrata	Vertebrata	Uniramia
Class	Mammalia	Mammalia	Amphibia	Insecta
Subclass	Eutheria	Eutheria	—	Pterygota
Order	Primates	Primates	Anura	Orthoptera
Suborder	Anthropoidea	Anthropoidea	—	Ensifera
Family	Hominidae	Hominidae	Ranidae	Tettigoniidae
Subfamily	—	—	Raninae	Phaneropterinae
Genus	*Homo*	*Gorilla*	*Rana*	*Scudderia*
Species	*Homo sapiens*	*Gorilla gorilla*	*Rana sphenocephala*	*Scudderia furcata*
Subspecies	—	—	—	*Scudderia furcata furcata*

The hierarchical taxonomy of four species (human, gorilla, Southern leopard frog, and katydid). Higher taxa generally are more inclusive than lower-level taxa, although taxa at two different levels may be equivalent in content. Closely related species are united at a lower point in the hierarchy than are distantly related species. For example, humans and gorillas are united at the level of the family (Hominidae) and above; they are united with the Southern leopard frog at the subphylum level (Vertebrata) and with the katydid at the kingdom level (Animalia). Mandatory Linnaean ranks are shown in bold type.

series: kingdom, phylum, "class," order, family, genus, and species. All organisms must be placed into at least seven taxa, one at each of the mandatory ranks. Taxonomists have the option of subdividing these seven ranks further to recognize more than seven taxa (superfamily, subfamily, superorder, suborder, etc.) for any particular group of organisms. In all, more than 30 taxonomic ranks are recognized. For very large and complex groups, such as fishes and insects, these additional ranks are needed to express different degrees of evolutionary divergence. Unfortunately, they also make the system more complex.

Introduction of evolutionary theory into animal taxonomy has changed the taxonomist's role from one of classification to **systematization.** Classification denotes the construction of classes, groupings of organisms that possess a common feature, called an essence, used to define the class. Organisms that possess the essential feature are members of the class by definition, and those that lack it are excluded. Because evolving species are subject always to change, the static nature of classes makes them a poor basis for a taxonomy of living systems. The activity of a taxonomist whose groupings of species represent units of common evolutionary descent is systematization, not classification. Species placed into a taxonomic group include the most recent common ancestor of the group and its descendants and thus form a branch of the phylogenetic tree of life. The species of a group thus formed represent a system of common descent, not a class defined by possession of an essential characteristic.

Because organismal characteristics are inherited from ancestral to descendant species, character variation is used to diagnose systems of common descent, but there is no requirement that an essential character be maintained throughout the system for its recognition as a taxon. The role of morphological or other features in systematization is therefore fundamentally different from the role of such characters in classification. In classification, a taxonomist asks whether a species being classified contains the defining feature(s) of a particular taxonomic class; in systematization, a taxonomist asks whether the characteristics of a species confirm or reject the hypothesis that it descends from the most recent common ancestor of a particular taxon. For example, tetrapod vertebrates descend from a common ancestor that had four limbs, a condition retained in most but not all of its descendants. Although they lack limbs, caecilians (p. 548) and snakes (p. 575) are tetrapods because they are parts of this system of common descent; other morphological and molecular characters group them respectively with living amphibians and lizards.

Although the hierarchical structure of Linnaean classification is retained in current taxonomy, the taxa are groupings of species related by evolutionary descent with modification, as diagnosed by sharing of homologous characters. As one moves up the taxonomic hierarchy from a species toward more inclusive groups, each taxon represents the descendants of an earlier ancestor, a larger branch of the tree of life.

Linnaeus's system for naming species is known as **binomial nomenclature.** Each species has a latinized name composed of two words (hence binomial) printed in italics (or underlined if handwritten or typed). The first word names the **genus,** which is capitalized; the second word is the **species epithet,** which is peculiar to the species within the genus and is written in lowercase (see Table 10.1). The great communicative value of Latin species names is that they are used consistently by scientists in all countries and languages; they are much more precise than "common names," which vary culturally and geographically.

The genus name is always a noun, and the species epithet is usually an adjective that must agree in gender with the genus. For instance, the scientific name of the common robin is *Turdus migratorius* (L. *turdus,* thrush; *migratorius,* of migratory habit). The species epithet never stands alone; the complete binomial must be used to name a species. Names of genera must refer only to single groups of organisms; the same name cannot be given to two different genera of animals. The same species epithet may be used in different genera, however, to denote different species. For example, the scientific name of the white-breasted nuthatch is *Sitta carolinensis.* The species epithet *"carolinensis"* is used in other genera for the species *Poecile carolinensis* (Carolina chickadee) and *Anolis carolinensis* (green anole, a lizard) to mean "of Carolina." All ranks above the species are designated using uninomial nouns, written with a capital initial letter.

Sometimes a species is divided into subspecies using a trinomial nomenclature (see katydid example, Table 10.1, and salamander example, Figure 10.2); such species are called **polytypic.** The generic, specific, and subspecific names are printed in italics (underlined if handwritten or typed). A polytypic species contains one subspecies whose subspecific name is a repetition of the species epithet and one or more additional subspecies whose names differ. Thus, to distinguish geographic variants of *Ensatina eschscholtzii,* one subspecies is named *Ensatina eschscholtzii eschscholtzii,* and different subspecies names are used for each of six other subspecies (Figure 10.2). Both the genus name and species epithet may be abbreviated as shown in Figure 10.2. Formal recognition of subspecies has lost popularity among taxonomists because subspecies are often based on minor differences in appearance that do not necessarily diagnose evolutionarily distinct units. When further study reveals that named subspecies are distinct evolutionary lineages, the subspecies are then often recognized as full species; indeed, many authors argue that the subspecies of *Ensatina eschscholtzii* are in fact separate species. Subspecies designations, therefore, should be viewed as tentative statements indicating that the species status of the populations needs further investigation.

SPECIES

While discussing Darwin's book, *On the Origin of Species,* in 1859, Thomas Henry Huxley asked, "In the first place, what is a species? The question is a simple one, but the right answer to it is hard to find, even if we appeal to those who should know most about it." We have used the term "species" so far as if it had a simple and unambiguous meaning. Actually, Huxley's commentary is as valid today as it was over 140 years ago. Our concepts of species have become more sophisticated, but the diversity of different concepts and disagreements surrounding their use are as evident now as in Darwin's time.

Despite widespread disagreement about the nature of species, biologists have repeatedly used certain criteria for identifying species. First, **common descent** is central to nearly all modern concepts of species. Members of a species must trace their ancestry to a common ancestral population although not necessarily to a single pair of parents. Species are thus historical entities. A second criterion is that species must be the **smallest distinct groupings** of organisms sharing patterns of ancestry and descent; otherwise, it would be difficult to separate species from higher taxa whose members also share common descent. Morphological characters traditionally have been important in identifying such groupings, but chromosomal and molecular characters now are used extensively for this purpose. A third important criterion is that of **reproductive community.** Members of a species must form a reproductive community that excludes members of other species. For sexually reproducing populations, interbreeding is critical for maintaining a reproductive community. For organisms whose reproduction is strictly asexual, reproductive community entails occupation of a particular ecological habitat in a particular place so that a reproducing population responds as a unit to evolutionary forces such as natural selection and genetic drift.

Any species has a distribution through space, its **geographic range,** and a distribution through time, its **evolutionary duration.** Species differ greatly from each other in both dimensions. Species having very large geographic ranges or worldwide distributions are called **cosmopolitan,** whereas those with very restricted geographic distributions are called **endemic.** If a species were restricted to a single point in space and time, we would have little difficulty recognizing it, and nearly every species concept would lead us to the same decision. We have little difficulty distinguishing from each other the different species of animals that we can find living in our local park or woods. However, when we compare a local population to similar but not identical populations located hundreds of miles away, it may be hard to determine whether these populations represent parts of a single species or different species (Figure 10.2).

Throughout the evolutionary duration of a species, its geographic range can change many times. A geographic range may be either continuous or disjunct, the latter having breaks within it where the species is absent. Suppose that we find two similar but not identical populations living

300 miles apart with no related populations between them. Are we observing a single species with a disjunct distribution or two different but closely related species? Suppose that these populations have been separated historically for 50,000 years. Is this enough time for them to have evolved separate reproductive communities, or can we still view them as parts of the same reproductive community? Clear answers to such questions are very hard to find. Differences among species concepts relate to solving these problems.

Typological Species Concept

Before Darwin, a species was considered a distinct and immutable entity. Species were defined by fixed, essential features (usually morphological) considered as a divinely created pattern or archetype. This practice constitutes the **typological (or morphological) species concept.** Scientists recognized species formally by designating a **type specimen** that was labeled and

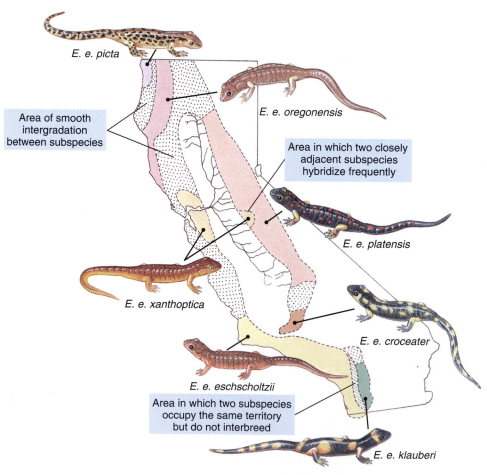

Area of smooth intergradation between subspecies

Area in which two closely adjacent subspecies hybridize frequently

Area in which two subspecies occupy the same territory but do not interbreed

E. e. picta

E. e. oregonensis

E. e. platensis

E. e. xanthoptica

E. e. croceater

E. e. eschscholtzii

E. e. klauberi

Figure 10.2

Geographic variation of color patterns in the salamander genus *Ensatina*. The species status of these populations has puzzled taxonomists for generations and continues to do so. Current taxonomy recognizes only a single species (*Ensatina eschscholtzii*) divided into subspecies as shown. Hybridization is evident between most adjacent populations, but studies of variation in proteins and DNA show large amounts of genetic divergence among populations. Furthermore, populations of the subspecies *E. e. eschscholtzii* and *E. e. klauberi* can overlap geographically without interbreeding.

deposited in a museum to represent the ideal form or morphology for the species (Figure 10.3). When scientists obtained additional specimens and wanted to assign them to a species, the type specimens of described species were consulted. The new specimens were assigned to a previously described species if they possessed the essential features of its type specimen. Small differences from the type specimen were considered accidental imperfections. Large differences from existing type specimens would lead a scientist to describe a new species with its own type specimen. In this manner, the living world was categorized into species.

Evolutionists discarded the typological species concept, but some of its traditions remain. Scientists still name species by describing type specimens deposited in museums, and the type specimen formally bears the name of the species. Organismal morphology is likewise still important in recognizing species; however, species are no longer viewed as classes of organisms defined by possession of certain morphological features. The basis of the evolutionary world view is that species are historical entities whose properties are subject always to change. Variation that we observe among organisms within a species is not an imperfect manifestation of an eternal "type"; the type itself is only an abstraction taken from the very real and important variation present within the species. A type is at best an average form that changes as organismal variation is sorted through time by natural selection. A type specimen serves only as a guide to the general morphological features that one may expect to find in a particular species as we observe it today.

The person who first describes a type specimen and publishes the name of a species is called the authority. This person's name and date of publication are often written after the species name. Thus, *Didelphis marsupialis* Linnaeus, 1758, tells us that Linnaeus was the first person to publish the species name of the opossum. Sometimes, the generic status of a species is revised following its initial description. In this case, the name of the authority is presented in parentheses. The Nile monitor lizard is denoted *Varanus niloticus* (Linnaeus, 1766) because the species originally was named by Linnaeus as *Lacerta nilotica,* and subsequently placed into a different genus.

Biological Species Concept

The most influential concept of species inspired by Darwinian evolutionary theory is the **biological species concept** formulated by Theodosius Dobzhansky and Ernst Mayr. This concept emerged during the evolutionary synthesis of the 1930s and 1940s from earlier ideas, and it has been refined and reworded several times since then. In 1982, Mayr stated the biological species concept as follows: *"A species is a reproductive community of populations (reproductively isolated from others) that occupies a specific niche in nature."* Note that a species is identified here according to reproductive properties of populations, not according to possession of any specific organismal characteristics.

Figure 10.3

Specimens of birds from the Smithsonian Institution (Washington D.C.), including birds originally collected by John J. Audubon, Theodore Roosevelt, John Gould, and Charles Darwin.

A species is an **interbreeding population** of individuals having common descent and sharing intergrading characteristics. Studies of populational variation in organismal morphology, chromosomal structure, and molecular genetic features are very useful for evaluating the geographical boundaries of interbreeding populations in nature. The criterion of the "niche" (see Chapter 38) recognizes that members of a reproductive community are expected also to have common ecological properties.

Because a reproductive community should maintain genetic cohesiveness, we expect organismal variation to be relatively smooth and continuous within species and discontinuous between them. Although the biological species is based on reproductive properties of populations rather than organismal morphology, morphology nonetheless can help us to diagnose biological species. Sometimes species status can be evaluated directly by conducting breeding experiments. Controlled breeding is practical only in a minority of cases, however, and our decisions regarding species membership usually are made by studying character variation. Variation in molecular characters is very useful for identifying geographical boundaries of reproductive communities. Molecular studies have revealed the occurrence of cryptic or **sibling species** (p. 119), which are too similar in morphology to be diagnosed as separate species by morphological characters alone.

The biological species concept has received strong criticism because of several perceived problems. First, the concept lacks an explicit temporal dimension. It provides a means for diagnosing species status of contemporary populations but gives little guidance regarding the species status of ancestral populations relative to their evolutionary descendants. Proponents of the biological species concept often disagree on the degree of reproductive isolation necessary for considering two populations separate species, thereby revealing some ambiguity in the concept. For example, should occurrence of limited hybridization between populations in a small geographic area cause them to be considered a single species despite evolutionary differences between them? Another problem is that because the biological species concept emphasizes interbreeding as the criterion of reproductive community, it denies the existence of species in groups of organisms that reproduce only asexually. It is common systematic practice, however, to describe species in all groups of organisms, regardless of whether reproduction is sexual or asexual.

Evolutionary Species Concept

The time dimension creates obvious problems for the biological species concept. How do we assign fossil specimens to biological species that are recognized today? If we trace a lineage backward through time, how far must we go before we have crossed a species boundary? If we could follow the unbroken genealogical chain of populations backward through time to the point where two sister species converge on their common ancestor, we would need to cross at least one species boundary somewhere. It would be very hard to decide, however, where to draw a sharp line between the two species.

To address this problem, the **evolutionary species concept** was proposed by Simpson in the 1940s to add an evolutionary time dimension to the biological species concept. This concept persists in a modified form today. A current definition of the evolutionary species is *a single lineage of ancestor-descendant populations that maintains its identity from other such lineages and that has its own evolutionary tendencies and historical fate.* Note that the criterion of common descent is retained here in the need for a lineage to have a distinct historical identity. Reproductive cohesion is the means by which a species maintains its identity from other such lineages and keeps its evolutionary fate separate from other species. The same kinds of diagnostic features discussed for the biological species concept are relevant for identifying evolutionary species, although in most cases only morphological features are available from fossils. Unlike the biological species concept, the evolutionary species concept applies both to sexually and asexually reproducing forms. As long as continuity of diagnostic features is maintained by the evolving lineage, it is recognized as a species. Abrupt changes in diagnostic features mark the boundaries of different species in evolutionary time.

Phylogenetic Species Concept

The last concept that we present is the **phylogenetic species concept.** The phylogenetic species concept is defined as an *irreducible (basal) grouping of organisms diagnosably distinct from other such groupings and within which there is a parental pattern of ancestry and descent.* This concept emphasizes most strongly the criterion of common descent. Both asexual and sexual groups are covered.

A phylogenetic species is a single population lineage with no detectable branching. The main difference in practice between the evolutionary and phylogenetic species concepts is that the latter emphasizes recognizing as separate species the smallest groupings of organisms that have undergone independent evolutionary change. The evolutionary species concept would group into a single species geographically disjunct populations that demonstrate some phylogenetic divergence but are judged similar in their "evolutionary tendencies," whereas the phylogenetic species concept would treat them as separate species. In general, a greater number of species would be described using the phylogenetic species concept than any other species concept, and many taxonomists consider it impractical for this reason. For strict adherence to cladistic systematics (p. 209), the phylogenetic species concept is ideal because only this concept guarantees strictly monophyletic units at the species level.

The phylogenetic species concept intentionally disregards details of evolutionary process and gives us a criterion that allows us to describe species without first needing to conduct detailed studies on evolutionary processes. Advocates of the phylogenetic species concept do not necessarily disregard the importance of studying evolutionary process. They argue, however, that the first step in studying evolutionary process is to have a clear picture of life's history. To accomplish this task, the pattern of common descent must be reconstructed in the greatest detail possible by starting with the smallest taxonomic units that have a history of common descent distinct from other such units.

Dynamism of Species Concepts

Current disagreements concerning concepts of species should not be considered discouraging. Whenever a field of scientific investigation enters a phase of dynamic growth, old concepts are reevaluated and either refined or replaced with newer, more progressive ones. The active debate occurring within systematics shows that this field has acquired unprecedented activity and importance in biology. Just as Thomas Henry Huxley's time was one of enormous advances in biology, so is the present time. Both times are marked by fundamental reconsiderations of the meaning of species. We cannot predict which concepts of species will remain useful 10 years from now. Researchers whose main interests are branching of evolutionary lineages, evolution of reproductive barriers among populations (p. 118), or ecological properties of species may favor different species concepts. The conflicts among the current concepts will lead us into the future. In many cases, different concepts agree on the locations of species boundaries, and disagreements identify particularly interesting cases of evolution in action. Understanding the conflicting perspectives, rather than learning a single species concept, is therefore of greatest importance for people now entering the study of zoology.

DNA Barcoding of Species

DNA barcoding is a technique for identifying organisms to species using sequence information from a standard gene present in all animals. The mitochondrial gene encoding cytochrome *c* oxidase subunit 1 (*COI*), which contains about 650 nucleotide base pairs, is a standard "barcode" region for animals. DNA sequences of *COI* usually vary among individuals of the same species but not extensively, so that variation within a species is much smaller than differences among species. DNA barcoding is applied to specimens in nature by taking a small DNA sample from blood or another expendable tissue. The method is useful also for specimens in natural-history museums, zoos, aquaria, and frozen-tissue collections. DNA sequences from such sources are checked against a public reference library of species identifiers to assign unknown specimens to known species. DNA barcoding does not solve the controversies regarding use of different species concepts, but it often permits the origin of a specimen to be identified to a particular local population, which is valuable information regardless of the species status that a taxonomist assigns to that population.

TAXONOMIC CHARACTERS AND PHYLOGENETIC RECONSTRUCTION

A major goal of systematics is to infer the evolutionary tree or **phylogeny** that relates all extant and extinct species. This task is accomplished by identifying organismal features, formally called **characters,** that vary among species. A character is any feature that the taxonomist uses to study variation within and among species. Taxonomists find characters by observing patterns of similarity among organisms in morphological, chromosomal, and molecular features (see p. 206), and less frequently in behavioral and ecological ones. Phylogenetic analysis depends upon finding among organisms shared features that are inherited from a common ancestor. Character similarity that results from common ancestry is called **homology** (see Chapter 6). Similarity does not always reflect common ancestry, however. Independent evolutionary origin of similar features on different lineages produces patterns of similarity among organisms that do not reflect common descent; this occurrence complicates the work of taxonomists. Character similarity that misrepresents common descent is called nonhomologous similarity or **homoplasy.** Endothermy of birds and mammals is an example of homoplasy; this condition arose separately in ancestral lineages of birds and mammals. Variation in other characters shows that birds and mammals are not each other's closest relatives (p. 499). For an example of molecular homoplasy, see the interpretation of character 41 (p. 211) in the boxed essay, Phylogenies from DNA Sequences.

Using Character Variation to Reconstruct Phylogeny

To infer the phylogeny of a group using characters that vary among its members, the first step is to determine which variant form of each character was present in the common ancestor of the entire group. This character state is called **ancestral** for the group as a whole. We presume that all other variant forms of the character arose later within the group, and these are called evolutionarily **derived character states.** Determining the **polarity** of a character refers to identifying which one of its contrasting

states is ancestral and which one(s) derived. For example, if we consider as a character the dentition of amniotic vertebrates (reptiles, birds, and mammals), presence versus absence of teeth in the jaws constitute alternative character states. Teeth are absent from modern birds but present in the other amniotes. To evaluate the polarity of this character, we must determine which character state, presence or absence of teeth, characterized the most recent common ancestor of amniotes and which state was derived subsequently within amniotes.

The method used to examine the polarity of a variable character is called **outgroup comparison.** We consult an additional group of organisms, called an **outgroup,** that is phylogenetically close but not within the group being studied. We infer that any character state found both within the group being studied and in the outgroup is ancestral for the study group. Amphibians and different groups of bony fishes constitute appropriate outgroups to the amniotes for polarizing variation in dentition of amniotes. Teeth are usually present in amphibians and bony fishes; therefore, we infer that presence of teeth is ancestral for amniotes and absence of teeth is derived. The polarity of this character indicates that teeth were lost in the ancestral lineage of all modern birds. Polarity of characters is evaluated most effectively when several different outgroups are used. All character states found in the study group that are absent from appropriate outgroups are considered derived.

Organisms or species that share derived character states form subsets within the study group called **clades** (Gr. *klados,* branch). A derived character shared by the members of a clade is formally called a **synapomorphy** (Gr. *synapsis,* joining together, + *morphē,* form) of that clade. Taxonomists use synapomorphies as evidence of homology to infer that a particular group of organisms forms a clade. Among extant amniotes, absence of teeth and presence of feathers are synapomorphies that identify the birds as a clade. A clade corresponds to a unit of evolutionary common descent; it includes all descendants of a particular ancestral lineage. The pattern formed by the derived states of all characters within our study group takes the form of a **nested hierarchy** of clades within clades. The goal is to identify all of the different clades nested within the study group, which would give a complete account of the patterns of common descent among species in the group.

Character states ancestral for a taxon are often called **plesiomorphic** for that taxon, and the sharing of ancestral states among organisms is termed **symplesiomorphy.** Unlike synapomorphies, however, symplesiomorphies do not provide useful information on nesting of clades within clades. In the example just given, we found that presence of teeth in jaws was plesiomorphic for amniotes. If we grouped together mammalian and reptilian groups, which possess teeth, to the exclusion of modern birds, we would not obtain a valid clade. Birds also descend from all common ancestors of reptiles and mammals and must be included within any clade that includes all reptiles and mammals. Errors in determining polarity of characters therefore clearly can produce errors in inference of phylogeny. It is important to note, however, that character states that are plesiomorphic at one taxonomic level can be synapomorphies at a more inclusive level. For example, the presence of jaws bearing teeth is a synapomorphy of gnathostome vertebrates (p. 511), a group that includes amniotes plus amphibians, bony fishes, and cartilaginous fishes, although teeth have been lost in birds and some other gnathostomes. The goal of phylogenetic analysis therefore can be restated as one of finding the appropriate taxonomic level at which any given character state is a synapomorphy. The character state is then used at that level to identify a clade.

The nested hierarchy of clades is presented as a branching diagram called a **cladogram** (Figure 10.4; see also Figure 6.16, and try to reconstruct this cladogram using only the sharing of numbered synapomorphies among the bird species). Taxonomists often make a technical distinction between a cladogram and a **phylogenetic tree.** The branches of a cladogram are only a formal device for indicating the nested hierarchy of clades within clades. The cladogram is not strictly equivalent to a phylogenetic tree, whose branches represent real lineages that occurred in the evolutionary past. To obtain a phylogenetic tree, we must add to the cladogram important additional interpretations concerning ancestors, the durations of evolutionary lineages, or the amounts of evolutionary change that occurred on the lineages. A cladogram is often used, however, as a first approximation of the branching structure of the corresponding phylogenetic tree.

Sources of Phylogenetic Information

We find characters used to construct cladograms in comparative morphology (including embryology), comparative cytology, and comparative biochemistry. **Comparative morphology** examines the varying shapes and sizes of organismal structures, including their developmental origins. Both macroscopic and microscopic characters are used, including details of cellular structure revealed by histology. As seen in Chapters 23 through 28, the variable structures of skull bones, limb bones, and integument (scales, hair, feathers) are particularly important for reconstructing the phylogeny of vertebrates. Comparative morphology uses specimens obtained from both living organisms and fossilized remains. **Comparative biochemistry** uses sequences of amino acids in proteins and the sequences of nucleotides in nucleic acids (see Chapter 5) to identify variable characters for constructing a cladogram (Figure 10.5). Direct sequencing of DNA is regularly

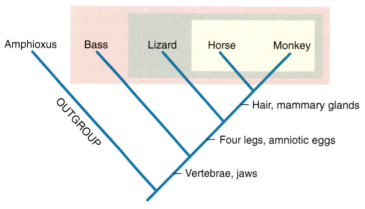

Figure 10.4

A cladogram as a nested hierarchy of taxa among five sampled chordate groups (Amphioxus, bass, lizard, horse, monkey). Amphioxus is the outgroup, and the study group comprises the four vertebrates. Four characters that vary among vertebrates are used to generate a simple cladogram: presence versus absence of four legs, amniotic eggs, hair, and mammary glands. For all four characters, absence is the ancestral state in vertebrates because this is the condition found in the outgroup, Amphioxus; for each character, presence is the derived state in vertebrates. Because they share presence of four legs and amniotic eggs as synapomorphies, the lizard, horse, and monkey form a clade relative to the bass. This clade is subdivided further by two synapomorphies (presence of hair and mammary glands) that unite the horse and monkey relative to the lizard. We know from comparisons involving even more distantly related animals that presence of vertebrae and jaws constitute synapomorphies of vertebrates and that Amphioxus, which lacks these features, falls outside the vertebrate clade.

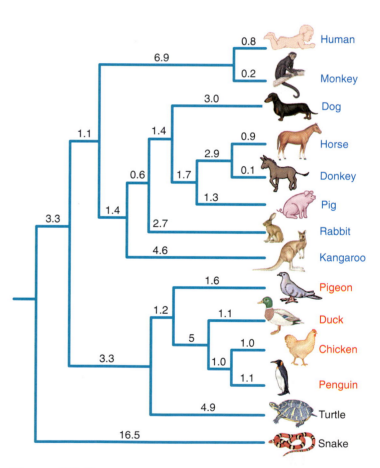

Figure 10.5

An early phylogenetic tree of representative amniotes based on inferred base substitutions in the gene that encodes the respiratory protein, cytochrome c. Numbers on the branches are the expected numbers of mutational changes that occurred in this gene along the different evolutionary lineages. Publication of this tree by Fitch and Margoliash in 1967 was influential in convincing systematists that molecular sequences contain phylogenetic information. Subsequent work confirms some hypotheses, including monophyly of mammals (blue) and birds (red) while rejecting others; kangaroo, for example, should be outside a branch containing all other mammals sampled.

applied to phylogenetic studies; however, comparisons of protein sequences are usually indirect, involving immunological or allozymic (see Figure 6.31) methods, or inferences from DNA sequences of protein-coding genes. Recent studies show that comparative biochemistry can be applied to some fossils in addition to living organisms. **Comparative cytology** uses variation in the numbers, shapes, and sizes of chromosomes and their parts (see Chapter 3 and p. 114) to obtain variable characters for constructing cladograms. Comparative cytology is used almost exclusively on living rather than fossilized organisms.

To add an evolutionary timescale necessary for producing a phylogenetic tree, we must consult the fossil record. We can look for the earliest appearance in fossils of derived morphological characters to estimate the ages of clades distinguished by those characters. The age of a fossil showing the derived characters of a particular clade is determined by radioactive dating (p. 109). An example of a phylogenetic tree constructed using these methods is Figure 25.3, page 547.

We can use comparative biochemical data to estimate the ages of different lineages on a phylogenetic tree. Some protein and DNA sequences undergo approximately linear rates of divergence through evolutionary time. The age of the most recent common ancestor of two species is therefore proportional to the differences measured between their proteins and DNA sequences. We calibrate evolution of proteins and DNA sequences by measuring their divergence between species whose most recent common ancestor has been dated using fossils. We then use the molecular evolutionary calibration to estimate ages of other branches on the phylogenetic tree.

THEORIES OF TAXONOMY

A theory of taxonomy establishes the principles that we use to recognize and to rank taxonomic groups. There are two currently popular theories of taxonomy: (1) traditional evolutionary taxonomy and (2) phylogenetic systematics (cladistics). Both are based on evolutionary principles. These two theories differ, however, on how evolutionary principles are used. These differences have important implications for how we use a taxonomy to study evolutionary processes.

The relationship between a taxonomic group and a phylogenetic tree or cladogram is important for both theories. This relationship can take one of three forms: **monophyly, paraphyly,** or **polyphyly** (Figure 10.6). A taxon is monophyletic if it includes the most recent common ancestor of the group and all descendants of that ancestor (Figure 10.6A). A taxon is paraphyletic if it includes the most recent common ancestor of all members of a group and some but not all descendants of that ancestor (Figure 10.6B). A taxon is polyphyletic if it does not include the most recent common ancestor of all members of a group; this condition requires that the group has had at least two separate evolutionary origins, usually requiring independent evolutionary acquisition of similar features (Figure 10.6C). Both evolutionary and cladistic taxonomy accept monophyletic groups and reject polyphyletic groups. They differ on acceptance of paraphyletic groups, however, and this difference has important evolutionary implications.

Traditional Evolutionary Taxonomy

Traditional **evolutionary taxonomy** incorporates two different evolutionary principles for recognizing and ranking higher taxa: (1) common descent and (2) amount of adaptive evolutionary change, as shown on a phylogenetic tree. Evolutionary taxa must have a single evolutionary origin, and must show unique adaptive features.

The mammalian paleontologist George Gaylord Simpson (Figure 10.7) and Ernst Mayr (see Figure 6.19) were highly influential in developing and formalizing the procedures of evolutionary taxonomy. According to Simpson and Mayr, a particular

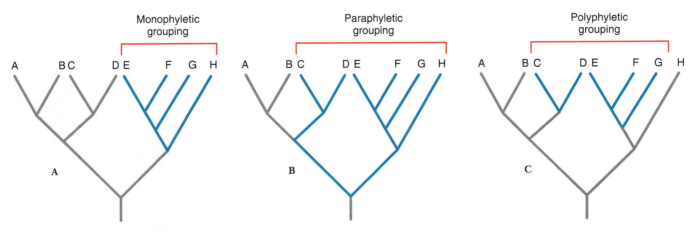

Figure 10.6

Relationships between phylogeny and taxonomic groups illustrated for a hypothetical phylogeny of eight species (A through H). **A,** *Monophyly*—a monophyletic group contains the most recent common ancestor of all members of the group and all of its descendants. **B,** *Paraphyly*—a paraphyletic group contains the most recent common ancestor of all members of the group and some but not all of its descendants. **C,** *Polyphyly*—a polyphyletic group does not contain the most recent common ancestor of all members of the group, thereby requiring that the group have at least two separate phylogenetic origins.

Figure 10.7

George Gaylord Simpson (1902 to 1984) formulated the principles of evolutionary taxonomy.

branch on an evolutionary tree is considered a higher taxon if it represents a distinct **adaptive zone.** Simpson describes an adaptive zone as "a characteristic reaction and mutual relationship between environment and organism, a way of life and not a place where life is led." By entering a new adaptive zone through a fundamental change in organismal structure and behavior, an evolving population can use environmental resources in a completely new way.

A taxon that constitutes a distinct adaptive zone is termed a **grade.** Simpson gives the example of penguins as a distinct adaptive zone within birds. The lineage immediately ancestral to all penguins underwent fundamental changes in the form of the body and wings to switch from aerial to aquatic locomotion (Figure 10.8). Aquatic birds that can fly both in air and underwater are somewhat intermediate in habitat, morphology, and behavior between aerial and aquatic adaptive zones. Nonetheless, the obvious modifications of the wings and body of penguins for swimming represent a new grade of organization. Penguins are therefore recognized as a distinct taxon within birds, the family Spheniscidae. The broader the adaptive zone when fully occupied by a group of organisms, the higher the rank given to the corresponding taxon.

Evolutionary taxa may be either monophyletic or paraphyletic. Recognition of paraphyletic taxa requires, however, that our taxonomies distort patterns of common descent. An evolutionary taxonomy of the anthropoid primates provides a good example (Figure 10.9). This

taxonomy places humans (genus *Homo*) and their immediate fossil ancestors in the family Hominidae, and it places the chimpanzees (genus *Pan*), gorillas (genus *Gorilla*), and orangutans (genus *Pongo*) in the family Pongidae. However, the pongid genera *Pan* and *Gorilla* share more recent common ancestry with the Hominidae than they do with the remaining pongid genus, *Pongo*. This arrangement makes the family Pongidae paraphyletic because it does not include humans, who also descend from the most recent common ancestor of all pongids (Figure 10.9). Evolutionary taxonomists nonetheless recognize the pongid genera as a single, family-level grade of arboreal, herbivorous primates having limited mental capacity; in other words, they show the same family-level adaptive zone. Humans are terrestrial, omnivorous primates who have greatly expanded mental and cultural attributes, thereby forming a distinct adaptive zone at the taxonomic level of the family. Unfortunately, if we want our taxa to constitute adaptive zones, we compromise our ability to present common descent effectively.

Traditional evolutionary taxonomy has been challenged from two opposite directions. One challenge states that because phylogenetic trees can be very difficult to obtain, it is impractical to base our taxonomic system on common descent and adaptive evolution. We are told that our taxonomy should represent a more easily measured feature, the overall similarity of organisms evaluated without regard to phylogeny. This principle is called **phenetic taxonomy.** Phenetic taxonomy contributed some useful analytical methods but did not have a strong impact on animal taxonomy, and scientific interest in this approach has declined. Despite the difficulties of reconstructing phylogeny, zoologists still consider this endeavor a central goal of their systematic work, and they are unwilling to compromise this goal for methodological purposes.

A   **B**

Figure 10.8

A, Penguin. **B,** Diving petrel. Penguins (avian family Spheniscidae) were recognized by George G. Simpson as a distinct adaptive zone within birds because of their adaptations for submarine flight. Simpson believed that the adaptive zone ancestral to penguins resembled that of diving petrels, which display adaptations for combined aerial and aquatic flight. Adaptive zones of penguins and diving petrels are distinct enough to be recognized taxonomically as different families within a common order (Ciconiiformes).

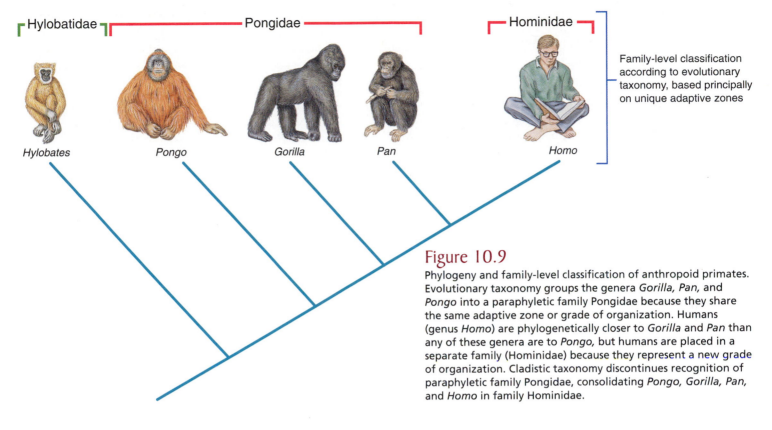

Family-level classification according to evolutionary taxonomy, based principally on unique adaptive zones

Figure 10.9

Phylogeny and family-level classification of anthropoid primates. Evolutionary taxonomy groups the genera *Gorilla, Pan,* and *Pongo* into a paraphyletic family Pongidae because they share the same adaptive zone or grade of organization. Humans (genus *Homo*) are phylogenetically closer to *Gorilla* and *Pan* than any of these genera are to *Pongo,* but humans are placed in a separate family (Hominidae) because they represent a new grade of organization. Cladistic taxonomy discontinues recognition of paraphyletic family Pongidae, consolidating *Pongo, Gorilla, Pan,* and *Homo* in family Hominidae.

Phylogenetic Systematics/Cladistics

A second and stronger challenge to evolutionary taxonomy is one known as **phylogenetic systematics** or **cladistics.** As the first name implies, this approach emphasizes the criterion of common descent and, as the second name implies, it is based on the cladogram of the group being classified. This approach to taxonomy was first proposed in 1950 by the German entomologist, Willi Hennig (Figure 10.10), and therefore is sometimes called "Hennigian systematics." All taxa recognized by Hennig's cladistic system must be monophyletic. We saw on Figure 10.9 how evolutionary taxonomists' recognition of the primate families Hominidae and Pongidae distorts genealogical relationships to emphasize adaptive uniqueness of the Hominidae. Because the most recent common ancestor of the paraphyletic family Pongidae is also an ancestor of the Hominidae, recognition of the Pongidae is incompatible with cladistic taxonomy. To avoid paraphyly, cladistic taxonomists have discontinued use of the traditional family Pongidae, placing chimpanzees, gorillas, and orangutans with humans in the family Hominidae. We adopt the cladistic classification in this book.

    Disagreement on the validity of paraphyletic groups may seem trivial at first, but its important consequences become clear when we discuss evolution. For example, claims that amphibians evolved from bony fish, that birds evolved from reptiles, or that humans evolved from apes might be made by an evolutionary taxonomist but are meaningless to a cladist. We imply by these statements that a descendant group (amphibians, birds, or humans) evolved from part of an ancestral group (bony fish, reptiles, and apes, respectively) to which the descendant does not belong. This usage automatically makes the ancestral group paraphyletic, and indeed bony fish, reptiles, and apes as traditionally recognized are paraphyletic groups. How are such paraphyletic groups recognized? Do they share distinguishing features not shared by the descendant group?

Figure 10.10

Willi Hennig (1913 to 1976), German entomologist who formulated the principles of phylogenetic systematics/cladistics.

Phylogenies from DNA Sequences

A simple example illustrates cladistic analysis of DNA sequence data to examine phylogenetic relationships among species. The study group in this example contains three species of chameleons, two from the island of Madagascar (*Brookesia theili* and *B. brygooi*) and one from Equatorial Guinea (*Chamaeleo feae*). The outgroup is a lizard of genus *Uromastyx*, which is a distant relative of chameleons. Do the molecular data in this example confirm or reject the prior taxonomic hypothesis that the two Madagascan chameleons are more closely related to each other than either one is to the Equatorial Guinean species?

The molecular information in this example comes from a piece of the mitochondrial DNA sequence (57 bases) for each species. Each sequence encodes amino acids 221–239 of a protein called "NADH dehydrogenase subunit 2" in the species from which it was obtained. These DNA base sequences are aligned and numbered as:

```
                    10        20        30        40        50
                    |         |         |         |         |
Uromastyx    AAACCTTAAAAGACACCACAACCATATGAACAACAACACCAACAATCAGCACACTAC
B. theili    AAACACTACAAAATATAACAACTGCATGAACAACATCAACCACAGCAAACATTTTAC
B. brygooi   AAACACTACAAGACATAACAACAGCATGAACTACTTCAACAACAGCAAATATTACAC
C. feae      AAACCCTACGAGACGCAACAACAATATGATCCACTTCCCCCACAACAAACACAATTT
```

Each column in the aligned sequences constitutes a character that takes one of four states: A, C, G, or T (a fifth possible state, absence of a base, is not observed in this example). Only characters that vary among the three chameleon species potentially contain information on which pair of species is most closely related. Twenty-three of the 57 aligned bases show variation *among chameleons*, as shown here in bold letters:

```
                    10        20        30        40        50
                    |         |         |         |         |
Uromastyx    AAACCTTAAAAGACACCACAACCATATGAACAACAACACCAACAATCAGCACACTAC
B. theili    AAACACTACAAAATATAACAACTGCATGAACAACATCAACCACAGCAAACATTTTAC
B. brygooi   AAACACTACAAGACATAACAACAGCATGAACTACTTCAACAACAGCAAATATTACAC
C. feae      AAACCCTACGAGACGCAACAACAATATGATCCACTTCCCCCACAACAAACACAATTT
```

To be useful for constructing a cladogram, a character must demonstrate sharing of derived characters (=synapomorphy). Which of these 23 characters demonstrate synapomorphies for chameleons? For each of the 23 variable characters, we must ask whether one of the states observed in chameleons is shared with the outgroup, *Uromastyx*. If so, this state is judged ancestral for chameleons and the alternative state(s) derived. Derived characters are identified for 21 of the 23 characters just identified; derived states are shown in blue:

```
                    10        20        30        40        50
                    |         |         |         |         |
Uromastyx    AAACCTTAAAAGACACCACAACCATATGAACAACAACACCAACAATCAGCACACTAC
B. theili    AAACACTACAAAATATAACAACTGCATGAACAACATCAACCACAGCAAACATTTTAC
B. brygooi   AAACACTACAAGACATAACAACAGCATGAACTACTTCAACAACAGCAAATATTACAC
C. feae      AAACCCTACGAGACGCAACAACAATATGATCCACTTCCCCCACAACAAACACAATTT
```

Note that polarity is ambiguous for two variable characters (at positions 23 and 54) whose alternative states in chameleons are not observed in the outgroup.

Of the characters showing derived states, 10 of them show synapomorphies among chameleons. These characters are marked here with numbers 1, 2, or 3 below the appropriate column.

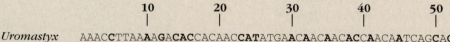

```
                    10        20        30        40        50
                    |         |         |         |         |
Uromastyx    AAACCTTAAAAGACACCACAACCATATGAACAACAACACCAACAATCAGCACACTAC
B. theili    AAACACTACAAAATATAACAACTGCATGAACAACATCAACCACAGCAAACATTTTAC
B. brygooi   AAACACTACAAGACATAACAACAGCATGAACTACTTCAACAACAGCAAATATTACAC
C. feae      AAACCCTACGAGACGCAACAACAATATGATCCACTTCCCCCACAACAAACACAATTT
             1         1         11        2   1 3   1     11
```

Paraphyletic groups are usually defined in a negative manner. They are distinguished only by lacking features found in a particular descendant group, because any traits that they share from their common ancestry are symplesiomorphies present also in the excluded descendants (unless secondarily lost). For example, apes are those "higher" primates that are not humans. Likewise, fish are those vertebrates that lack the distinguishing characteristics of tetrapods (amphibians and amniotes). What does it mean then to say that humans evolved from apes? To an evolutionary taxonomist, apes and humans are different adaptive zones or grades of organization; to say that humans evolved from apes states that bipedal, tailless organisms of large brain capacity evolved from arboreal, tailed organisms of smaller brain capacity. To a cladist, however, the statement

The eight characters marked 1 show synapomorphies grouping the two Madagascan species (*Brookesia theili* and *B. brygooi*) to the exclusion of the Equatorial Guinean species, *Chamaeleo feae*. We can represent these relationships as a cladogram:

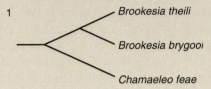

We can explain evolution of all characters favoring this cladogram by placing a single mutational change on the branch ancestral to the two *Brookesia* species. This is the simplest explanation for evolutionary change of these characters.

Characters marked 2 and 3 disagree with our cladogram and favor alternative relationships as shown here:

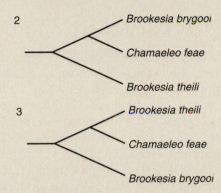

To explain evolutionary changes in characters favoring cladograms 2 or 3 using cladogram 1, we need at least two changes per character. Likewise, if we try to explain evolution of characters favoring cladogram 1 on cladograms 2 or 3, we need at least two changes for each of these characters. These two diagrams show the minimum numbers of changes required for character 5 (which favors cladogram 1) and character 41 (which favors cladogram 3) on cladogram 1; the ancestral state of each character is shown at the root of the tree and the states observed in each species at the tips of the branches:

character 5 (1 change)

character 41 (2 changes)

Systematists often use a principle called **parsimony** to resolve conflicts among taxonomic characters, as seen here. We choose as our best working hypothesis the cladogram that requires the smallest total amount of character change. In our example, cladogram 1 is favored by parsimony. For all 10 phylogenetically informative characters, cladogram 1 requires a total of 12 changes of character state (one for each of the 8 characters favoring it and two for each of the other 2 characters). Cladograms 2 and 3 each require at least 19 character-state changes, 7 steps longer than cladogram 1. By choosing cladogram 1, we claim that characters favoring cladograms 2 and 3 show homoplasy in their evolution.

The molecular sequences shown in this example therefore confirm predictions of the prior hypothesis, based on appearance and geography of these chameleons, that the *Brookesia* species shared a common ancestor with each other more recently than either one did with *Chamaeleo feae*.

As a further exercise, you should convince yourself that the 12 characters that vary among chameleons but which do not demonstrate unambiguous sharing of derived states are equally compatible with each of the three possible cladograms. For each character, find the minimum total number of changes that must occur to explain its evolution on each cladogram. You will see, if you do this exercise correctly, that the three cladograms do not differ in minimum numbers of changes required for each of these characters. For this reason, the characters are phylogenetically uninformative by the parsimony criterion.

Data from Townsend, T., and A. Larson. 2002. Molecular phylogenetics and mitochondrial genomic evolution in the Chamaeleonidae (Reptilia, Squamata). Molecular Phylogenetics and Evolution 23:22–36.

that humans evolved from apes says essentially that humans evolved from an arbitrary grouping of species that lack the distinctive characteristics of humans, a trivial statement that conveys no useful information. To a cladist, any statement that a particular monophyletic group descends from a paraphyletic one is nothing more than a claim that the descendant group evolved from something that it is not. Extinct ancestral groups are always paraphyletic because they exclude a descendant that shares their most recent common ancestor. Although many such groups have been recognized by evolutionary taxonomists, none are recognized by cladists.

Zoologists often construct paraphyletic groups because they are interested in a terminal, monophyletic group (such as humans), and they want to ask questions about its ancestry.

It is often convenient to lump together organisms whose features are considered approximately equally distant from the group of interest and to ignore their own unique features. It is significant in this regard that humans have never been placed in a paraphyletic group, whereas most other organisms have been. Apes, reptiles, fishes, and invertebrates are all terms that traditionally designate paraphyletic groups formed by combining various "side branches" found when human ancestry is traced backward through the tree of life. Such a taxonomy can give the erroneous impression that all of evolution is a progressive march toward humanity or, within other groups, a progressive march toward whatever species humans designate most "advanced." Such thinking is a relic of pre-Darwinian views that there is a linear scale of nature having "primitive" creatures at the bottom and humans near the top just below angels. Darwin's theory of common descent states, however, that evolution is a branching process with no linear scale of increasing perfection along a single branch. Nearly every branch contains its own combination of ancestral and derived features. In cladistics, this perspective is emphasized by recognizing taxa only by their own unique properties and not grouping organisms only because they lack the unique properties found in related groups.

Fortunately, there is a convenient way to express the common descent of groups without constructing paraphyletic taxa. It is done by finding what is called the **sister group** of the taxon of interest to us. Two different monophyletic taxa are each other's sister group if they share common ancestry more recently than either one does with any other taxa. The sister group of humans appears to be chimpanzees, with gorillas forming the sister group to humans and chimpanzees combined. Orangutans are the sister group of a clade that includes humans, chimpanzees, and gorillas; gibbons form the sister group of the clade that includes orangutans, chimpanzees, gorillas, and humans (see Figure 10.9).

Current State of Animal Taxonomy

The formal taxonomy of animals that we use today was established using the principles of evolutionary systematics and has been revised recently in part using the principles of cladistics. Introduction of cladistic principles initially replaces paraphyletic groups with monophyletic subgroups while leaving the remaining taxonomy mostly unchanged. A thorough revision of taxonomy along cladistic principles, however, will require profound changes, one of which almost certainly will be abandonment of Linnaean ranks. A new taxonomic system called PhyloCode is being developed as an alternative to Linnaean taxonomy; this system replaces Linnaean ranks with codes that denote the nested hierarchy of monophyletic groups conveyed by a cladogram. In our coverage of animal taxonomy, we try to use taxa that are monophyletic and therefore consistent with criteria of both evolutionary and cladistic taxonomy. We continue, however, to use Linnaean ranks. For familiar taxa that are clearly paraphyletic grades, we note this fact and suggest alternative taxonomic schemes that contain only monophyletic taxa.

In discussing patterns of descent, we avoid statements such as "mammals evolved from reptiles" that imply paraphyly and instead specify appropriate sister-group relationships. We avoid referring to groups of organisms as being primitive, advanced, specialized, or generalized because all groups of animals contain combinations of primitive, advanced, specialized, and generalized features; these terms are best restricted to describing specific characteristics and not an entire group.

Revision of taxonomy according to cladistic principles can cause confusion. In addition to new taxonomic names, we see old ones used in unfamiliar ways. For example, cladistic use of "bony fishes" includes amphibians and amniotes (including reptilian groups, birds, and mammals) in addition to finned, aquatic animals that we normally term "fish." Cladistic use of "reptiles" includes birds in addition to snakes, lizards, turtles, and crocodilians; however, it excludes some fossil forms, such as synapsids, that were traditionally placed in Reptilia (see Chapters 26 through 28). Taxonomists must be very careful to specify when using these seemingly familiar terms whether the traditional evolutionary taxa or newer cladistic taxa are being referenced.

MAJOR DIVISIONS OF LIFE

From Aristotle's time to the late 1800s, every living organism was assigned to one of two kingdoms: plant or animal. However, the two-kingdom system had serious problems. Although it was easy to place rooted, photosynthetic organisms such as trees and herbs among the plants and to place food-ingesting, motile forms such as insects, fishes, and mammals among the animals, unicellular organisms presented difficulties (see Chapter 11). Some forms were claimed both for the plant kingdom by botanists and for the animal kingdom by zoologists. An example is *Euglena* (p. 225), which is motile, like animals, but has chlorophyll and photosynthesis, like plants. Other groups, such as bacteria, were assigned rather arbitrarily to the plant kingdom.

Several alternative systems have been proposed to solve the problem of classifying unicellular forms. In 1866 Haeckel proposed the new kingdom Protista to include all single-celled organisms. At first bacteria and cyanobacteria (blue-green algae), forms that lack nuclei bounded by a membrane, were included with nucleated unicellular organisms. Finally, important differences were recognized between the anucleate bacteria and cyanobacteria (prokaryotes) and all other organisms that have membrane-bound nuclei (eukaryotes). In 1969 R. H. Whittaker proposed a five-kingdom system that incorporated the basic prokaryote-eukaryote distinction. The kingdom Monera contained the prokaryotes. The kingdom Protista contained the unicellular eukaryotic organisms (protozoa and unicellular eukaryotic algae). Multicellular organisms were split into three kingdoms by mode of nutrition and other fundamental differences in organization. The kingdom Plantae included multicellular photosynthesizing organisms, higher plants, and multicellular algae. Kingdom Fungi contained molds, yeasts, and fungi that obtain their food by absorption. Invertebrates (except the

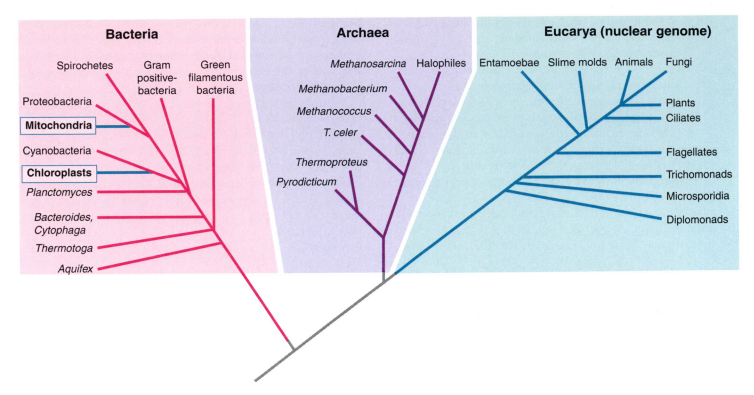

Figure 10.11

Phylogenetic overview of the three domains of life, Archaea, Bacteria and Eucarya, based on analysis of genes encoding ribosomal RNA. Because of their endosymbiotic origin (p. 34), organellar genomes of domain Eucarya (mitochondria, chloroplasts) are phylogenetically within the Bacteria rather than the clade that includes all eukaryotic nuclear genomes. Organisms of domain Eucarya therefore include cellular components of disparate evolutionary origins.

protozoa) and vertebrates compose the kingdom Animalia. Most of these forms ingest their food and digest it internally, although some parasitic forms are absorptive.

These different systems were proposed without regard to the phylogenetic relationships needed to construct evolutionary or cladistic taxonomies. The oldest phylogenetic events in the history of life have been obscure because the different forms of life share very few characters that can be compared among them to reconstruct phylogeny. Recently, however, a cladistic classification of all life-forms has been proposed based on phylogenetic information obtained from molecular data (the nucleotide base sequence of DNA encoding ribosomal RNA). According to this tree (Figure 10.11), Woese, Kandler, and Wheelis (1990) recognized three monophyletic **domains** above the kingdom level: Eucarya (all eukaryotes), Bacteria (the true bacteria), and Archaea (prokaryotes differing from bacteria in membrane structure and ribosomal RNA sequences). They did not divide Eucarya into kingdoms, although if we retain Whittaker's kingdoms Plantae, Animalia, and Fungi, Protista becomes a paraphyletic group (Figure 10.11). To maintain a cladistic classification, Protista must be discontinued by recognizing as separate kingdoms all of the labelled branches of Eucarya as shown in Figure 10.11.

Until a few years ago, animal-like protistans were traditionally studied in zoology courses as animal phylum Protozoa. Given current knowledge and the principles of phylogenetic systematics, this taxonomy commits two errors; "protozoa" are neither animals nor are they a valid monophyletic taxon at any level. Kingdom Protista is likewise invalid because it is not monophyletic. Animal-like protistans, now divided into seven or more phyla, are nonetheless of interest to students of zoology because they provide an important phylogenetic context for the study of animal diversity.

MAJOR SUBDIVISIONS OF THE ANIMAL KINGDOM

The phylum is the largest formal taxonomic category in the Linnaean classification of the animal kingdom. Metazoan phyla are often grouped together to produce additional, informal taxa intermediate between the phylum and the animal kingdom. Phylogenetic relationships among metazoan phyla have been particularly difficult to resolve both by morphological and molecular characters. Traditional groupings based on embryological and anatomical characters that may reveal phylogenetic affinities are:

Branch A (Mesozoa): phylum Mesozoa, the mesozoa
Branch B (Parazoa): phylum Porifera, the sponges, and phylum Placozoa
Branch C (Eumetazoa): all other phyla
    Grade I (Radiata): phyla Cnidaria, Ctenophora
    Grade II (Bilateria): all other phyla

Division A (Protostomia): characteristics in Figure 10.12
  Acoelomates: phyla Platyhelminthes,
    Gnathostomulida, Nemertea
  Pseudocoelomates: phyla Rotifera, Gastrotricha,
    Kinorhyncha, Nematoda, Nematomorpha,
    Acanthocephala, Entoprocta, Priapulida, Loricifera
  Eucoelomates: phyla Mollusca, Annelida, Arthropoda,
    Echiurida, Sipunculida, Tardigrada, Onychophora
Division B (Deuterostomia): characteristics in Figure 10.12
  phyla Phoronida, Ectoprocta, Chaetognatha,
  Brachiopoda, Echinodermata, Hemichordata, Chordata

Molecular phylogenetic results have challenged the placement of mesozoans (Branch A), suggesting that they might be derived from protostomia (Branch C, Grade II, Division A). As in the

outline, bilateral animals are customarily divided into protostomes and deuterostomes by their embryological development (Figure 10.12). However, some of the phyla are difficult to place into one of these two categories because they possess some characteristics of each group (see Chapter 15).

Molecular phylogenetic studies have challenged traditional classification of the Bilateria, but results are not yet strong enough to present a precise hypothesis of phylogenetic relationships among metazoan phyla. Molecular phylogenetic results place four phyla classified above as deuterostomes (Brachiopoda, Chaetognatha, Ectoprocta, and Phoronida) in the Protostomia. This is the scheme shown in Figure 10.12. Furthermore, the traditional major groupings of protostome phyla (acoelomates, pseudocoelomates, and eucoelomates) appear not to be monophyletic. Instead, protostomes are divided into two major

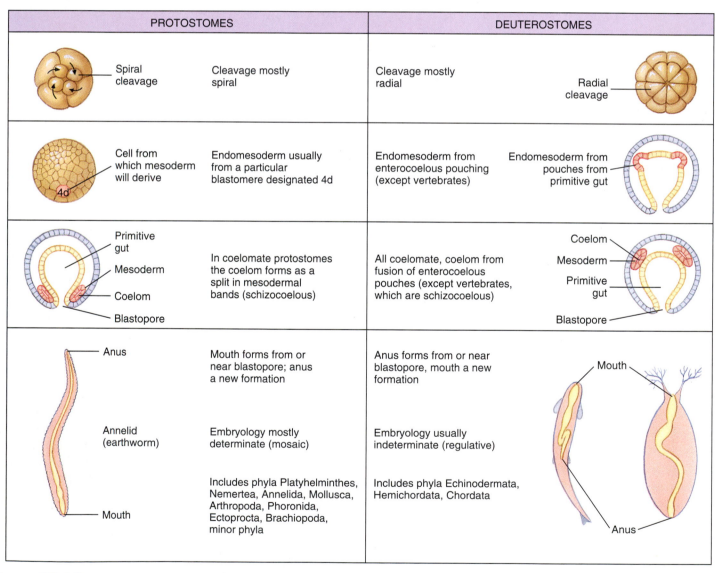

Figure 10.12

Basis for the distinction between divisions of bilateral animals. Traditional classifications often place phyla Brachiopoda, Ectoprocta, and Phoronida with deuterostomes, but recent molecular phylogenetic analyses place them with protostomes as shown here. Phylum Chaetognatha is of uncertain phylogenetic affinity and might lie outside these two groups.

monophyletic groups called the Lophotrochozoa and Ecdysozoa. Reclassification of the Bilateria is summarized:

Grade II: Bilateria
    Division A (Protostomia):
        Lophotrochozoa: phyla Platyhelminthes, Nemertea, Rotifera, Gastrotricha, Acanthocephala, Mollusca, Annelida, Echiurida, Sipunculida, Phoronida, Ectoprocta, Entoprocta, Gnathostomulida,

Chaetognatha, Brachiopoda
        Ecdysozoa: phyla Kinorhyncha, Nematoda, Nematomorpha, Priapulida, Arthropoda, Tardigrada, Onychophora, Loricifera
    Division B (Deuterostomia): phyla Chordata, Hemichordata, Echinodermata

Although further study is needed to confirm these new groupings, we use them to organize our survey of animal diversity.

SUMMARY

Animal systematics has three major goals: (1) to identify all species of animals, (2) to evaluate evolutionary relationships among animal species, and (3) to group animal species in a hierarchy of taxonomic groups (taxa) that conveys evolutionary relationships. Taxa are ranked to denote increasing inclusiveness as follows: species, genus, family, order, "class," phylum, and kingdom. All of these ranks can be subdivided to signify taxa that are intermediate between them. Names of species are binomial, with the first name designating the genus to which the species belongs (capitalized) followed by a species epithet (lowercase), both written in italics. Taxa at all other ranks are given single capitalized but nonitalicized names.

The biological species concept has guided the recognition of most animal species. A biological species is defined as a reproductive community of populations (reproductively isolated from others) that occupies a specific niche in nature. It is not immutable through time but changes during the course of evolution. Because the biological species concept may be difficult to apply in spatial and temporal dimensions, and because it excludes asexually reproducing forms, alternative concepts have been proposed. These alternatives include the evolutionary species concept and the phylogenetic species concept. No single concept of species is universally accepted by all zoologists, but zoologists agree that a species should constitute a population lineage with a history of evolutionary descent separate from other such lineages. Because species lineages are expected to differ from each other in the DNA sequence of the rapidly evolving mitochondrial gene *COI*, this gene sequence is used as a diagnostic "barcode" to assign specimens to species.

Two major schools of taxonomy are currently active. Traditional evolutionary taxonomy groups species into higher taxa according to the joint criteria of common descent and adaptive evolution; such taxa have a single evolutionary origin and occupy a distinctive adaptive zone. A second approach, called phylogenetic systematics or cladistics, emphasizes common descent exclusively in grouping species into higher taxa. Only monophyletic taxa (those having a single evolutionary origin and containing all descendants of the group's most recent common ancestor) are used in cladistics. In addition to monophyletic taxa, evolutionary taxonomy recognizes some taxa that are paraphyletic (having a single evolutionary origin but excluding some descendants of the most recent common ancestor of the group). Both schools of taxonomy exclude polyphyletic taxa (those having more than one evolutionary origin).

Both evolutionary taxonomy and cladistics require that patterns of common descent among species be assessed before higher taxa are recognized. Comparative morphology (including development), cytology, and biochemistry are used to reconstruct nested hierarchical relationships among taxa that reflect the branching of evolutionary lineages through time. The fossil record provides estimates of the ages of evolutionary lineages. Comparative studies and the fossil record jointly permit us to reconstruct a phylogenetic tree representing the evolutionary history of the animal kingdom.

Traditionally, all living forms were placed into two kingdoms (animal and plant) but more recently, a five-kingdom system (animals, plants, fungi, protistans, and monerans) has been followed. Neither of these systems conforms to the principles of evolutionary or cladistic taxonomy because they place single-celled organisms into either paraphyletic or polyphyletic groups. Based on our current knowledge of the phylogenetic tree of life, "protozoa" do not form a monophyletic group and they do not belong within the animal kingdom.

Phylogenetic relationships among animal phyla have been clarified by molecular phylogenetic studies, although many of these higher-level groupings remain tentative. Particularly controversial is the grouping of bilaterally symmetrical animals into clades Deuterostomia, Protostomia, Ecdysozoa, and Lophotrochozoa.

REVIEW QUESTIONS

1. List in order, from most inclusive to least inclusive, the principal categories (taxa) in Linnaean classification as currently applied to animals.
2. Explain why the system for naming species that originated with Linnaeus is "binomial."
3. How does the biological species concept differ from earlier typological concepts of a species? Why do evolutionary biologists prefer it to typological species concepts?
4. What problems have been identified with the biological species concept? How do other species concepts attempt to overcome these problems?
5. How are taxonomic characters recognized? How are such characters used to construct a cladogram?
6. How do monophyletic, paraphyletic, and polyphyletic taxa differ? How do these differences affect the validity of such taxa for both evolutionary and cladistic taxonomies?

7. How many different clades of two or more species are possible for species A–H shown in Figure 10.6A?
8. What is the difference between a cladogram and a phylogenetic tree? Given a cladogram for a group of species, what additional information is needed to obtain a phylogenetic tree?
9. How would cladists and evolutionary taxonomists differ in their interpretations of the statement that humans evolved from apes, which evolved from monkeys?

10. What taxonomic practices based on the typological species concept are retained in systematics today? How has their interpretation changed?
11. What are the five kingdoms distinguished by Whittaker? How does their recognition conflict with the principles of cladistic taxonomy?

SELECTED REFERENCES

Aguinaldo, A. M. A., J. M. Turbeville, L. S. Linford, M. C. Rivera, J. R. Garey, R. A. Raff, and J. A. Lake. 1997. Evidence for a clade of nematodes, arthropods and other moulting animals. Nature **387:**489–493. *This molecular phylogenetic study challenges traditional classification of the Bilateria.*

Avise, J. C. 2006. Evolutionary pathways in nature: a phylogenetic approach. Cambridge, U.K., Cambridge University Press. *A current treatment of phylogenetic knowledge.*

Ereshefsky, M. (ed.). 1992. The units of evolution. Cambridge, Massachusetts, MIT Press. *A thorough coverage of concepts of species, including reprints of important papers on the subject.*

Ereshefsky, M. 2001. The poverty of the Linnaean hierarchy. Cambridge, U.K., Cambridge University Press. *A philosophical critique of Linnaean taxonomy illustrating its problems with cladistic taxonomy.*

Felsenstein, J. 2002. Inferring phylogenies. Sunderland, Massachusetts, Sinauer Associates. *A thorough coverage of phylogenetic methods.*

Hall, B. K. 1994. Homology: the hierarchical basis of comparative biology. San Diego, Academic Press. *A collection of papers discussing the many dimensions of homology, the central concept of comparative biology and systematics.*

Hull, D. L. 1988. Science as a process. Chicago, University of Chicago Press. *A study of the working methods and interactions of systematists, containing a thorough review of the principles of evolutionary, phenetic, and cladistic taxonomy.*

Maddison, W. P., and D. R. Maddison. 2003. MacClade version 4.06. Sunderland, Massachusetts, Sinauer Associates, Inc. *A computer program for the MacIntosh that conducts phylogenetic analyses of systematic characters. The instruction manual stands alone as an excellent introduction to phylogenetic procedures. The computer program is user-friendly and excellent for instruction in addition to serving as a tool for analyzing real data.*

Mayr, E., and P. D. Ashlock. 1991. Principles of systematic zoology. New York, McGraw-Hill. *A detailed survey of systematic principles as applied to animals.*

Panchen, A. L. 1992. Classification, evolution, and the nature of biology. New York, Cambridge University Press. *Excellent explanations of the methods and philosophical foundations of biological classification.*

Swofford, D. 2002. Phylogenetic analysis using parsimony (and other methods) PAUP* version 4. Sunderland, Massachusetts, Sinauer Associates. *A powerful computer package for constructing phylogenetic trees from data.*

Valentine, J. W. 2004. On the origin of phyla. Chicago, Illinois, University of Chicago Press. *Phylogenetic analysis of animal phyla in a paleontological perspective.*

Wagner, G. P. (ed.). 2001. The character concept in evolutionary biology. San Diego, Academic Press. *A thorough coverage of evolutionary character concepts.*

Woese, C. R., O. Kandler, and M. L. Wheelis. 1990. Towards a natural system of organisms: proposal for the domains Archaea, Bacteria, and Eucarya. Proceedings of the National Academy of Sciences, USA, **87:**4576–4579. *Proposed cladistic classification for the major taxonomic divisions of life.*

ONLINE LEARNING CENTER

Visit www.mhhe.com/hickmanipz14e for chapter quizzing, key term flash cards, web links and more!

11

Protozoan Groups

• UNICELLULAR EUKARYOTES

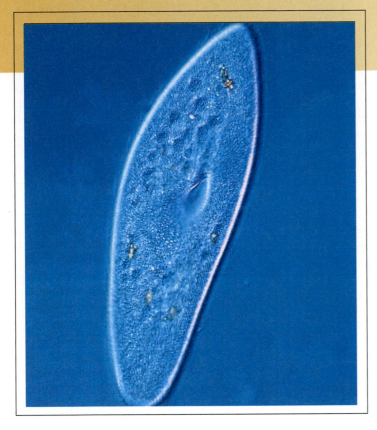

A paramecium.

Emergence of Eukaryotes and a New Life Pattern

The first reasonable evidence for life on earth dates from approximately 3.5 billion years ago. The first cells were prokaryotic bacteria-like organisms. The early prokaryotes diversified greatly over an enormous time span; their prokaryotic descendants now belong to two groups, the Eubacteria and the Archaea. One lineage within the ancient prokaryotes also gave rise to the first eukaryotic form. The key steps in the evolution of a eukaryotic cell from a prokaryotic ancestor involved symbiogenesis, a process whereby one prokaryote engulfed, but did not digest, another. The engulfed cell was eventually reduced to an organelle inside the host cell. The eukaryotic products of symbiogenesis include mitochondria and plastids.

A mitochondrion originated from an aerobic prokaryote capable of deriving energy in the presence of environmental oxygen. An anaerobic bacterium that engulfed such an aerobic form gained the capacity to grow in an oxygen-rich environment. The engulfed aerobic bacterium persisted inside the cell as a mitochondrion with its own genetic material. Over evolutionary time, most, but not all, genes from the mitochondrion came to reside in the host cell

nucleus. Almost all present-day eukaryotes have mitochondria and are aerobic.

The eukaryotic plastid originated when a cell engulfed a photosynthetic bacterium. When a prokaryote is engulfed and modified to become a eukaryotic organelle, we say the organelle developed via primary endosymbiosis. The chloroplasts in red algae, and in green algae and multicellular plants, arose this way. However, in some cases, a eukaryotic cell may obtain plastids from another eukaryote. This is secondary endosymbiosis. Two similar cells may have formed very differently, so it is not easy to untangle the evolutionary relationships among the diverse array of unicellular forms we now see.

The assemblage of eukaryotic unicellular organisms is collectively called protozoa. The inclusion of "zoa" in the name refers to two animal-like features: the absence of a cell wall, and the presence of at least one motile stage in the life cycle. However, the plant-animal distinction is not easily made in unicellular forms because many motile unicells carry photosynthetic plastids. The myriad of ways to live as a unicellular organism is fascinating, beguiling, and a little bewildering.

A protozoan, or unicellular eukaryote, is a complete organism in which all life activities occur within the limits of a single plasma membrane. Unicellular eukaryotes are found wherever life exists. They are highly adaptable and easily distributed from place to place. They require moisture, whether they live in marine or freshwater habitats, soil, decaying organic matter, or plants and animals. They may be sessile or free swimming, and they form a large part of the floating plankton. The same species are often found widely separated in time as well as in space. Some species may have spanned geological eras exceeding 100 million years.

Despite their wide distribution, many protozoa can live successfully only within narrow environmental ranges. Species adaptations vary greatly, and successions of species frequently occur as environmental conditions change.

Protozoa play an enormous role in the economy of nature. Their fantastic numbers are attested by the gigantic ocean soil deposits formed over millions of years by their skeletons. About 10,000 species of unicellular eukaryotes are symbiotic in or on animals or plants, sometimes even other protozoa. The relationship may be **mutualistic** (both partners benefit), **commensalistic** (one partner benefits, no effect on the other), or **parasitic** (one partner benefits at the expense of the other), depending on the species involved. Parasitic forms cause some of the most important diseases of humans and domestic animals.

HOW DO WE DEFINE PROTOZOAN GROUPS?

For many years, all protozoans were placed within a single phylum comprised of eukaryotic unicells, but phylogenetic studies showed this group was not monophyletic. Evidence suggests that the origin of the first eukaryote was followed by great diversification, leading some biologists to predict that more than 60 monophyletic eukaryotic clades will eventually emerge. The well-supported clade Opisthokonta (Figure 11.1) includes unicellular choanoflagellates, multicellular animals (metazoans) and fungi, among others (see p. 228). Like the Opisthokonta, the clade Viridiplantae has both unicellular and multicellular members; this group contains green algae, the bryophytes, and the vascular plants. The remaining eukaryotic clades contain less well-known organisms, many of which were considered protozoans.

Protozoans and their relatives have been given several names. Protozoans are usually unicellular, so the name Protoctista was initiated to include unicellular and closely related multicellular organisms in one group. However, protoctistan is far less commonly used than the names protist and protozoan. Protist is a general term that does not distinguish between plantlike and animal-like unicells, whereas protozoan was intended for a subset of animal-like unicellular organisms.

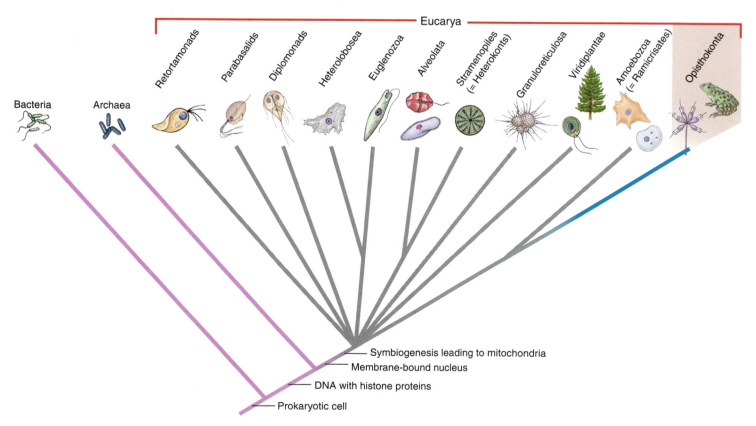

Figure 11.1

Cladogram showing two major prokaryotic branches and diversification of eukaryotes. Major eukaryotic clades containing protists are shown, but several clades of amebas and other forms are not shown. The order of branching remains to be determined for most clades. The very large opisthokont clade contains choanoflagellates, fungi, and all multicellular animals.

The two concepts, plantlike and animal-like, refer in part to the way that food is gathered. Plants are typically **autotrophic,** meaning that they synthesize their own organic constituents from inorganic substrates. Photosynthesis is one kind of autotrophy. Animals are typically **heterotrophic,** meaning that they obtain organic molecules synthesized by other organisms. Heterotrophic protozoa may ingest their food in a soluble form or in a particulate form. Particulate food is acquired by **phagocytosis** via an infolding or invagination of the cell membrane to surround a visible food particle (Figure 11.2). Heterotrophs that feed on visible particles are **phagotrophs** or **holozoic** feeders, whereas those that ingest soluble food are **osmotrophs** or **saprozoic** feeders.

A distinction between plants and animals on the basis of nutrition works well for multicellular forms, but the plant-animal distinction is not so clear among unicells. Autotrophic protozoa (phototrophs) use light energy to synthesize their organic molecules, but they often practice phagotrophy and osmotrophy as well. Even among heterotrophs, few are exclusively either phagotrophic or osmotrophic. A single class Euglenoidea (phylum Euglenozoa) contains some forms that are mainly phototrophs, some that are mainly osmotrophs, and some that are mainly phagotrophs. Species of *Euglena* show considerable variety in nutritional capability. Some species require certain preformed organic molecules, even though they are autotrophs, and some lose their chloroplasts if maintained in darkness, thus becoming permanent osmotrophs. The mode of nutrition employed by unicellular organisms is opportunistic and highly variable, even within a single species, so nutritional features have proved unreliable for defining protozoans, or protozoan subgroups.

Originally, the means of locomotion was used to distinguish three of the four classes in the traditional phylum Protozoa. Members of a parasitic class, once called Sporozoa, lack a distinct locomotory structure, but share an organelle capable of invading host cells. Members of the other three traditional protozoan classes differ in means of locomotion: flagellates (Figure 11.3) use **flagella,** ciliates (Figure 11.4) travel via a ciliated body surface, and amebas extend their **pseudopodia** (Figure 11.5) to move.

Typically, a flagellate has a few long flagella, and a ciliate has many short **cilia,** but no real morphological distinction exists between cilia and flagella. Some investigators have preferred to

Figure 11.3
One flagellum is clearly visible in the lower left of this photograph of *Euglena*.

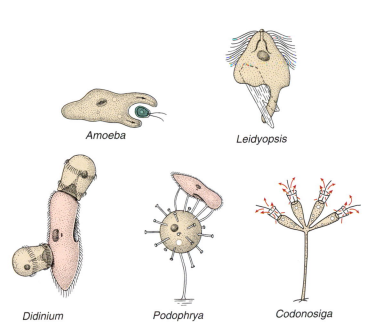

Amoeba   *Leidyopsis*

Didinium   *Podophrya*   *Codonosiga*

Figure 11.2
Some feeding methods among protozoa. *Amoeba* surrounds a small flagellate with pseudopodia. *Leidyopsis,* a flagellate living in the intestine of termites, forms pseudopodia and ingests wood chips. *Didinium,* a ciliate, feeds only on *Paramecium,* which it swallows through a temporary cytostome in its anterior end. Sometimes more than one *Didinium* feed on the same *Paramecium. Podophrya* is a suctorian ciliophoran. Its tentacles attach to its prey and suck prey cytoplasm into the body of the *Podophrya,* where it is pinched off to form food vacuoles. *Codonosiga,* a sessile flagellate with a collar of microvilli, feeds on particles suspended in the water drawn through its collar by the beat of its flagellum. Technically, all of these methods are types of phagocytosis.

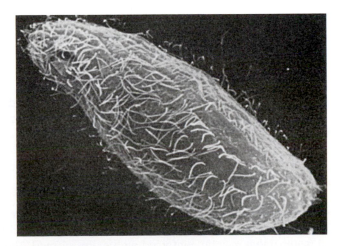

Figure 11.4
Scanning electron micrograph of a free-living ciliate *Tetrahymena thermophila* showing rows of cilia (×2000). Beating of flagella either pushes or pulls the organism through its medium, while cilia propel the organism by a "rowing" mechanism. Their structure is similar, whether viewed by scanning or transmission electron microscopy.

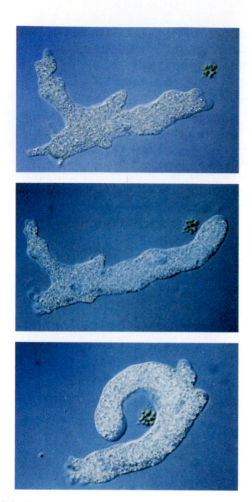

Figure 11.5

Ameboid movement. At top and center, the ameba extends a pseudopodium toward a *Pandorina* colony. At bottom, the ameba surrounds the *Pandorina* before engulfing it by phagocytosis.

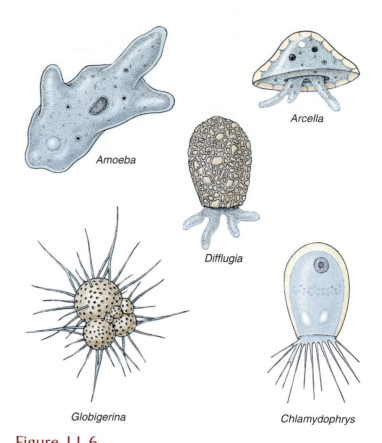

Figure 11.6

Examples of amebas. *Amoeba, Difflugia,* and *Arcella* have lobopodia; *Chlamydophrys* has filopodia; and the foraminiferan *Globigerina* bears reticulopodia.

call them both undulipodia (L. dim. of *unda,* a wave, 1 Gr. *podos,* a foot). However, a cilium propels water parallel to the surface to which the cilium is attached, whereas a flagellum propels water parallel to the main axis of the flagellum.

Amebas are able to assume a variety of body forms (Figure 11.5) due to flowing cell cytoplasm. The cytoplasm can be extended outward in pseudopodia of various shapes: **lobopodia** are blunt-tipped, **filipodia** are thin and sharply pointed, **rhizopodia** are branched filaments, and **reticulopodia** are branched filaments that merge to form a netlike structure (Figure 11.6). **Axopodia** are thin, pointed pseudopodia that contain a central longitudinal (axial) filament of microtubules (Figure 11.7).

Amebas that make shells are called **testate** (Figure 11.6). *Arcella* and *Difflugia* have their delicate plasma membrane covered with a protective **test** or shell of secreted siliceous or chitinoid material that may be reinforced with grains of sand. They move by means of pseudopodia that project from openings in the shell (Figure 11.6). Some very abundant shelled amebas are known as foraminiferans (*Globigerina,* Figure 11.6) or radiolarians (Figure 11.8). The name heliozoan (Figure 11.7) refers to freshwater amebas with axopodia; they may be testate or not. Amebas without shells are called naked.

To understand relationships among the wide range of unicellular forms, the Society of Protozoologists examined an enormous amount of research on protozoan structure, life history, and physiology, publishing, in 1980, a new classification of protozoa recognizing seven phyla. Three phyla contained the most well-known organisms: Apicomplexa held the sporozoans and related forms, Ciliophora held the ciliates, and Sarcomastigophora contained the amebas and the flagellates. Amebas and flagellates were grouped together because some flagellates could form pseudopodia, some species of ameba had flagellated stages, and at least one supposed ameba was really a flagellate without a flagellum. The phylum Sarcomastigophora was divided into two subphyla: Sarcodina contained the amebas and Mastigophora contained the flagellates. Mastigophorans were distinguished as plantlike (Phytomastigophorea) or animal-like (Zoomastigophorea). Our previous discussion of feeding mode as a taxonomic character should lead the reader to suspect that the Mastigophora was not a monophyletic group. However, the names remain quite descriptive and one easily deciphers a phytomastigophoran as a flagellate with plastids.

Molecular analyses, using sequences of bases in genes, particularly the gene encoding the small subunit of ribosomal RNA (p. 95), along with genes encoding some proteins, have revolutionized our concepts of phylogenetic affinities in protozoans, and indeed, all eukaryotes. The new clade names given to branches in a molecular phylogeny make it difficult for those already familiar with protozoan taxa to recognize group members, but retaining only the old

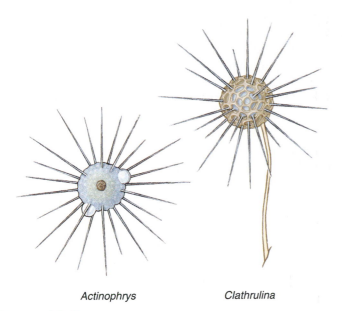

Figure 11.7

Actinophrys and *Clathrulina* are amebas with axopodia.

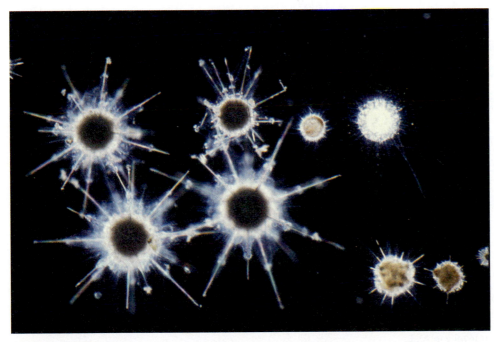

Figure 11.8

Some shelled amebas, like those shown here, are commonly called radiolarians.

names makes an informed reading of new research impossible. So, in the phylogenetic section at the end of this chapter, we retain the system of phyla outlined in comprehensive monographs such as Hausmann and Hülsmann (1996) and used in recent texts such as Roberts and Janovy (2005). However, we also use some recently erected clade names[1] as we discuss particular protozoan groups.

Some traditional names do not represent monophyletic groups. Molecular analyses show that the ameboid body form

[1]Patterson, D. J. 1999. *Amer. Nat.* **154** (supplement):S96–S124. Baldauf, S. L., A. J. Roger, I. Wenk-Siefert, and F. W. Doolittle. 2000. *Science* **290**:972–976.

has evolved independently several times, as has the shell. So animals called heliozoans are divided among five clades, and those called radiolarians are divided among three, according to some workers. Among the testate amebas, only foraminiferans appear to be a monophyletic group; they now belong in a clade called the Granuloreticulosa.

Despite the diversity of form, protozoans do demonstrate a basic body plan or grade—a single eukaryotic cell—and they amply demonstrate the enormous adaptive potential of that grade. Over 64,000 species have been named, and over half of these are fossils. Some workers estimate there may be 250,000 protozoan species. Although they are unicellular, protozoa are functionally complete organisms with many complicated, microanatomical structures. Their various organelles tend to be more specialized than those of the average cell in a multicellular organism. Particular organelles may perform as skeletons, sensory structures, conducting mechanisms, and other functions. These organelles bear closer scrutiny because of their functional importance and because differences in organelle structure can provide homologous characters on which to base taxonomic categories.

FORM AND FUNCTION

Locomotion

Cilia and Flagella

A cilium or flagellum has considerable internal structure. Each flagellum or cilium contains nine pairs of longitudinal microtubules arranged in a circle around a central pair (Figure 11.9), and this is true for all motile flagella and cilia in the animal kingdom, with a few notable exceptions. This "9 + 2" tube of microtubules in a flagellum or cilium is its **axoneme;** an axoneme is covered by a membrane continuous with the cell membrane covering the rest of the organism. At about the point where an axoneme enters the cell proper, the central pair of microtubules ends at a small plate within the circle of nine pairs (Figure 11.9A). Also at about that point, another microtubule joins each of the nine pairs, so that these form a short tube extending from the base of the flagellum into the cell. The tube consists of nine *triplets* of microtubules and is known as a **kinetosome (or basal body).** Kinetosomes are exactly the same in structure as **centrioles** that organize mitotic spindles during cell division (see Figure 3.14, p. 46). Centrioles of some flagellates may give rise to kinetosomes, or kinetosomes may function as centrioles. All typical flagella and cilia have a kinetosome at their base, regardless of whether they are borne by a protozoan or metazoan cell. Many small metazoans use cilia not only for locomotion but also to create water currents for their feeding and respiration. Ciliary movement is vital to many species in such functions as handling food, reproduction, excretion, and osmoregulation (as in flame cells, p. 296).

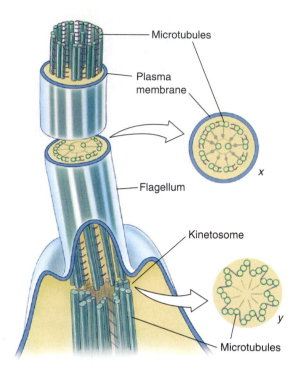

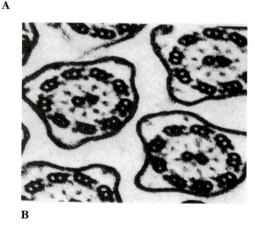

Shear resistance, causing the axoneme to bend when the filaments slide past each other, is provided by "spokes" from each doublet to the central pair of fibrils. These spokes are visible in electron micrographs. Direct evidence for the sliding microtubule hypothesis was obtained by attaching tiny gold beads to axonemal microtubules and observing their movement microscopically.

Pseudopodia

Pseudopodia are extensions of the cell cytoplasm used in locomotion (Figure 11.10). The cytoplasm is not homogeneous; sometimes peripheral and central areas of cytoplasm can be distinguished as **ectoplasm** and **endoplasm** (see Figure 11.10). Endoplasm appears more granular and contains the nucleus and cytoplasmic organelles. Ectoplasm appears more transparent (hyaline) by light microscopy, and it bears the bases of the cilia or flagella. Ectoplasm is often more rigid and is in the gel state of a colloid, whereas the more fluid endoplasm is in the sol state.

Colloidal systems are permanent suspensions of finely divided particles that do not precipitate, such as milk, blood, starch, soap, ink, and gelatin. Colloids in living systems are commonly proteins, lipids, and polysaccharides suspended in the watery fluid of cells (cytoplasm). Such systems may undergo sol-gel transformations, depending on whether the fluid or particulate components become continuous. In the sol state of cytoplasm, solids are suspended in a liquid, and in the semisolid gel state, liquid is suspended in a solid.

Figure 11.9

A, A flagellum illustrating the central axoneme, which is composed of nine pairs of microtubules plus a central pair. The axoneme is enclosed within the cell membrane. The central pair of microtubules ends near the level of the cell surface in a basal plate (axosome). The peripheral microtubules continue inward for a short distance to compose two of each of the triplets in the kinetosome (basal body) (at level *y* in **A**). **B,** Electron micrograph of a section through several cilia, corresponding to level *x* in **A**. (×133,000)

The current explanation for ciliary and flagellar movement is the **sliding microtubule hypothesis.** The movement is powered by a release of chemical bond energy in ATP (p. 62). Two little arms composed of the protein, dynein, are visible in electron micrographs on each of the pairs of peripheral tubules in the axoneme (level *x* in Figure 11.9), and these bear the enzyme adenosine triphosphatase (ATPase), which cleaves the ATP. When bond energy in ATP is released, the arms "walk along" one of the filaments in the adjacent pair, causing it to slide relative to the other filament in the pair.

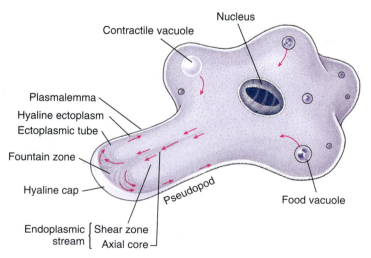

Figure 11.10

Ameba in active locomotion. Arrows indicate the direction of streaming endoplasm. The first sign of a new pseudopodium is thickening of the ectoplasm to form a clear hyaline cap, into which the fluid endoplasm flows. As the endoplasm reaches the forward tip, it fountains out and is converted into ectoplasm, forming a stiff outer tube that lengthens as the forward flow continues. Posteriorly the ectoplasm is converted into fluid endoplasm, replenishing the flow. Substratum is necessary for ameboid movement.

Characteristics of Unicellular Eukaryotes

1. **Unicellular;** some colonial, and some with multicellular stages in their life cycles
2. **Mostly microscopic,** although some are large enough to be seen with the unaided eye
3. All symmetries represented in the group; shape variable or constant (oval, spherical, or other)
4. **No germ layer present**
5. No organs or tissues, but **specialized organelles** are found; nucleus single or multiple
6. Free-living, mutualism, commensalism, parasitism all represented in the groups
7. Locomotion by **pseudopodia, flagella, cilia,** and direct cell movements; some sessile
8. Some provided with a **simple endoskeleton** or **exoskeleton,** but most are naked
9. **Nutrition of all types:** autotrophic (manufacturing own nutrients by photosynthesis), heterotrophic (depending on other plants or animals for food), saprozoic (using nutrients dissolved in the surrounding medium)
10. Aquatic or terrestrial habitat; free-living or symbiotic mode of life
11. Reproduction **asexually** by fission, budding, and cysts and **sexually** by conjugation or by syngamy (union of male and female gametes to form a zygote)
12. The simplest example of **division of labor between cells** is seen in certain colonial protozoa that have both somatic and reproductive zooids (individuals) in the colony.

Pseudopodia vary in composition and are of several types. The most familiar are **lobopodia** (Figures 11.5 and 11.10), which are rather large, blunt extensions of the cell body containing both endoplasm and ectoplasm. Some amebas characteristically do not extend individual pseudopodia, but move the whole body with pseudopodial motion; this movement is known as the **limax** form (for a genus of slugs, *Limax*). **Filopodia** are thin extensions, usually branching, and containing only ectoplasm. They occur in some amebas, such as *Euglypha* (Figure 11.17). **Reticulopodia** (see Figure 11.6) are distinguished from filopodia in that reticulopodia repeatedly rejoin to form a netlike mesh, although some protozoologists consider the distinction between filopodia and reticulopodia artificial. Members of superclass Actinopoda have **axopodia** (Figure 11.11), which are long, thin pseudopodia supported by axial rods of microtubules (Figure 11.11). The microtubules are arranged in a definite spiral or geometrical array, depending on the species, and constitute the axoneme of the axopod. Axopodia can be extended or retracted, apparently by addition or removal of microtubular material. Since the tips can adhere to the substrate, the organism can progress by a rolling motion, shortening the axonemes in front and extending those in the rear. Cytoplasm can flow along the axonemes, toward the body on one side and in the reverse direction on the other.

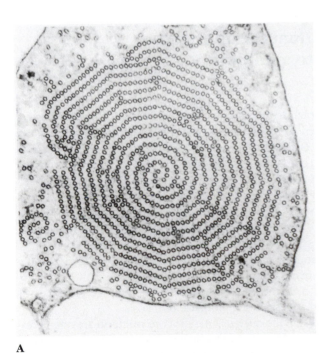

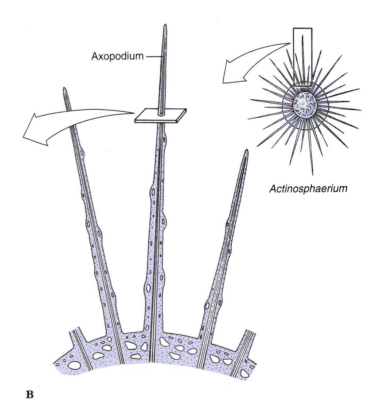

Axopodium

Actinosphaerium

A   **B**

Figure 11.11

A, Electron micrograph of axopodium (from *Actinosphaerium nucleofilum*) in cross section. **B,** Diagram of axopodium to show orientation of **A.** The axoneme of an axopodium is composed of an array of microtubules, which may vary from three to many in number depending on the species. Some species can extend or retract their axopodia quite rapidly. (×99,000)

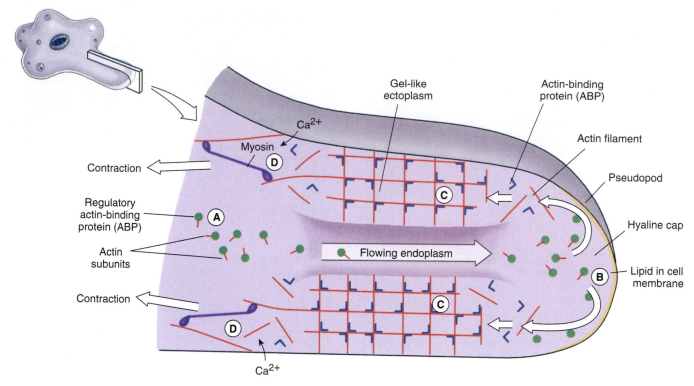

Figure 11.12

Mechanism of pseudopodial movement. In endoplasm, actin subunits are bound to regulatory actin-binding proteins that keep them from assembling **(A)**. Upon stimulation, hydrostatic force carries the subunits through a weakened gel to the hyaline cap. The actin subunits are freed from the regulatory proteins by lipids in the cell membrane **(B)**. Subunits quickly assemble into filaments and, upon interaction with actin-binding protein (ABP), form gel-like ectoplasm **(C)**. At the trailing edge, calcium ions activate an ABP that releases actin filaments from the gel, loosening the network enough that myosin molecules can pull on it **(D)**. Subunits pass up through the tube of ectoplasm to be reused.

Although pseudopodia are the chief means of locomotion in amebas, they can be formed by a variety of flagellate protozoa, as well as by ameboid cells of many animals. In fact, much defense against disease in the human body depends on ameboid white blood cells, and ameboid cells in many other animals, vertebrate and invertebrate, play similar roles.

When a typical lobopodium begins to form, an extension of ectoplasm called a **hyaline cap** appears, and endoplasm begins to flow toward and into the hyaline cap (Figures 11.10 and 11.12). The flowing endoplasm contains actin subunits attached to regulatory, actin-binding proteins (ABPs) that prevent actin from polymerizing. As endoplasm flows into the hyaline cap, it spreads to the periphery. Interaction with phospholipids in the cell membrane releases the actin subunits from their regulatory binding proteins and allows them to polymerize into actin filaments. The actin filaments become cross-linked to each other by another ABP to form a semisolid gel, transforming the ectoplasm into a tube through which the fluid endoplasm flows as the pseudopodium extends. Near the trailing edge of the gel, calcium ions activate an ABP that releases actin filaments from the gel and permits myosin to associate with and to pull these actin filaments. Thus contraction at the trailing edge creates a pressure that forces the fluid endoplasm, along with its now-dissociated actin subunits, back toward the hyaline cap.

Functional Components of Protozoan Cells

Nucleus

As in other eukaryotes, the nucleus is a membrane-bound structure whose interior communicates with the cytoplasm by small pores. Within the nucleus the genetic material (DNA) is borne on chromosomes. Except during cell division, chromosomes are not usually condensed in a form that can be distinguished, although during fixation of the cells for light microscopy, chromosomal material (chromatin) often clumps together irregularly, leaving some areas within the nucleus relatively clear. This appearance is described as **vesicular** and is characteristic of many protozoan nuclei (Figure 11.13). Condensations of chromatin may be distributed around the periphery of the nucleus or internally in distinct patterns. In most dinoflagellates (p. 236) chromosomes are visible through interphase as they would appear during prophase of mitosis.

Also within the nucleus, one or more **nucleoli** are often present (Figures 11.13 and 11.21). Characters such as the persistence of nucleoli during mitosis are useful in identifying protozoan clades.

Macronuclei of ciliates are described as **compact** or **condensed** because the chromatin material is more finely dispersed and clear areas cannot be observed with the light microscope (Figure 11.15).

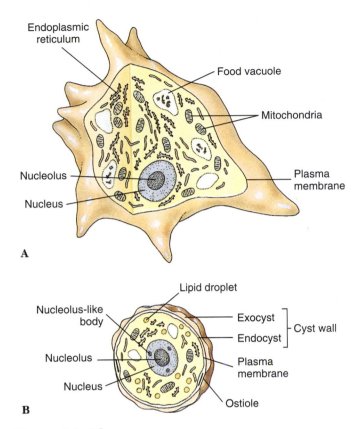

A

B

Figure 11.13
Structure of *Acanthamoeba palestinensis*. **A,** Active, feeding form. **B,** Cyst.

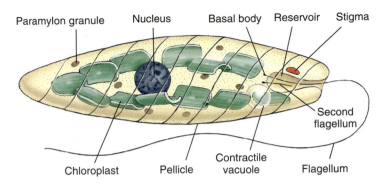

Figure 11.14
Euglena viridis. Features shown are a combination of those visible in living and stained preparations.

Mitochondria

A mitochondrion is an organelle used in energy production where oxygen serves as the terminal electron acceptor (see p. 67). It contains DNA. Cristae, the internal membranes of a mitochondrion (Figure 11.13), are of variable form, being flat, tubular, discoid, or branched (ramifying). The form of cristae is considered a homologous character and, in conjunction with other morphological features, is used to describe protozoan clades. In cells without mitochondria, **hydrogenosomes** may be present. Hydrogenosomes function in the absence of oxygen and are assumed to have evolved from mitochondria. **Kinetoplasts** are also assumed to be mitochondrial derivatives, but they work in association with a kinetosome, an organelle at the base of a flagellum.

Golgi Apparatus

The Golgi apparatus is part of the secretory system of the endoplasmic reticulum. Golgi bodies are also called **dictyosomes** in protozoan literature. **Parabasal bodies** are similar structures with potentially similar functions.

Plastids

Plastids are organelles containing a variety of photosynthetic pigments. The original addition of a plastid to eukaryotic cells occurred when a cyanobacterium was engulfed and not digested. **Chloroplasts** (Figure 11.14) contain different versions of chlorophylls (*a, b,* or *c*), but other kinds of plastids contain other pigments. For example, red algal plastids contain phycobilins. Particular pigments shared among unicellular eukaryotes may indicate shared ancestry, but plastids could also have been gained by secondary endosymbiosis.

Extrusomes

This general term refers to membrane-bound organelles in protozoans that are used to extrude something from the cell. The wide variety of structures extruded suggests that not all extrusomes are homologous. The ciliate **trichocyst** (p. 233) is an extrusome.

Nutrition

Holozoic nutrition implies phagocytosis (see Figure 11.2), in which an infolding or invagination of the cell membrane surrounds a food particle. As the invagination extends farther into the cell, it is pinched off at the surface (see Figure 3.21). The food particle thus is contained in an intracellular, membrane-bound vesicle, a **food vacuole** or **phagosome.** Lysosomes, small vesicles containing digestive enzymes, fuse with the phagosome and pour their contents into it, where digestion begins. As digested products are absorbed across the vacuolar membrane, the phagosome becomes smaller. Any undigestible material may be released to the outside by exocytosis, the vacuole again fusing with the cell-surface membrane. In most ciliates, many flagellates, and many apicomplexans, the site of phagocytosis is a definite mouth structure, the **cytostome** (Figure 11.15). In amebas, phagocytosis can occur at almost any point by envelopment of a particle with pseudopodia. Particles must be ingested through the opening of the test, or shell, in amebas that have tests. Flagellates may form a temporary cytostome, usually in a characteristic position, or they may have a permanent cytostome with specialized structure. Many ciliates have a characteristic structure for expulsion of waste matter, the **cytopyge** or **cytoproct,** found in a characteristic location. In some, the cytopyge also serves as the site for expulsion of the contents of the contractile vacuole.

Figure 11.15

Left, enlarged section of a contractile vacuole (water expulsion vesicle) of *Paramecium*. Water is apparently collected by endoplasmic reticulum, emptied into feeder canals and then into the vesicle. The vesicle contracts to empty its contents to the outside, thus serving as an osmoregulatory organelle. *Right, Paramecium,* showing cytopharynx, food vacuoles, and nuclei.

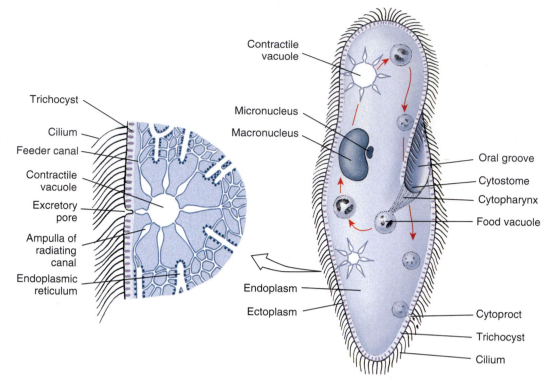

Saprozoic feeding may be by pinocytosis or by transport of solutes directly across the outer cell membrane. Pinocytosis and transport across a cell membrane are discussed on page 51. Direct transport across a membrane may be by diffusion, facilitated transport, or active transport. Diffusion is probably of little or no importance in nutrition of protozoa, except possibly in some endosymbiotic species. Some important food molecules, such as glucose and amino acids, may be brought into a cell by facilitated diffusion and active transport.

It has been shown that a stimulatory substance, or "inducer," must be present in the surrounding medium for many protozoa to initiate pinocytosis. Several proteins act as inducers, as can some salts and other substances; it appears that the inducer must be a positively charged molecule. Pinocytosis takes place at the inner end of the cytopharynx in protozoa possessing that structure.

Excretion and Osmoregulation

Vacuoles can be seen by light microscopy in the cytoplasm of many protozoa. Some of these vacuoles periodically fill with a fluid substance that is then expelled. Evidence is strong that these **contractile vacuoles** (Figures 11.10, 11.14, and 11.15) function principally in osmoregulation. They are more prevalent and fill and empty more frequently in freshwater protozoa than in marine and endosymbiotic species, where their surrounding medium would be more nearly isosmotic (having the same osmotic pressure) to their cytoplasm. Smaller species, which have a greater surface-to-volume ratio, generally have more rapid filling and

expulsion rates in their contractile vacuoles. Excretion of metabolic wastes, on the other hand, is almost entirely by diffusion. The main end product of nitrogen metabolism is ammonia, which readily diffuses from the small bodies of protozoa.

Although it seems clear that contractile vacuoles function to remove excess water that has entered cytoplasm by osmosis, a reasonable mechanism for such removal has been elusive. A recent hypothesis suggests that proton pumps (p. 67) on the vacuolar surface and on tubules radiating from it actively transport H^+ and cotransport bicarbonate (HCO_3^-) (Figure 11.16), which are osmotically active particles. As these particles accumulate within a vacuole, water would be drawn into the vacuole. Fluid within the vacuole would remain isosmotic to the cytoplasm. Then as the vacuole finally joins its membrane to the surface membrane and empties its contents to the outside, it would expel water, H^+, and HCO_3^-. These ions can be replaced readily by action of carbonic anhydrase on CO_2 and H_2O. Carbonic anhydrase is present in the cytoplasm of amebas.

Some ciliates, such as *Blepharisma,* have contractile vacuoles with structure and filling mechanisms apparently similar to those described for amebas. Others, such as *Paramecium,* have more complex contractile vacuoles. Such vacuoles are located in a specific position beneath the cell membrane, with an "excretory" pore leading to the outside, and surrounded by ampullae of about six feeder canals (Figure 11.15). Feeder canals, in turn, are surrounded by fine tubules about 20 nm in diameter, which connect with the canals during filling of ampullae and at their lower ends connect with the tubular system of endoplasmic reticulum. Ampullae and contractile vacuoles are surrounded by bundles of fibrils, which may function in contraction of these structures. Contraction of ampullae fills the vacuole. When the

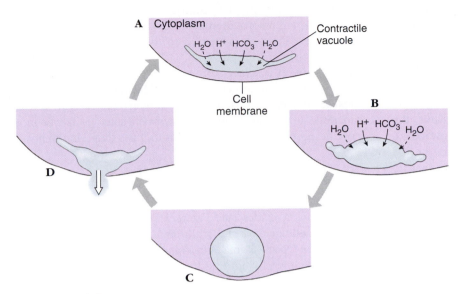

Figure 11.16

Proposed mechanism for operation of contractile vacuoles. **A, B,** Vacuoles are composed of a system of cisternae and tubules. Proton pumps in their membranes transport H^+ and cotransport HCO_3^- into the vacuoles. Water diffuses in passively to maintain an osmotic pressure equal to that in the cytoplasm. When the vacuole fills **C,** its membrane fuses with the cell's surface membrane, expelling water, H^+, and HCO_3^-. **D,** Protons and bicarbonate ions are replaced readily by action of carbonic anhydrase on carbon dioxide and water.

vacuole contracts to discharge its contents to the outside, the ampullae become disconnected from the vacuole, so that back-flow is prevented. Tubules, ampullae, or vacuoles may be sup-plied with proton pumps to draw water into their lumens by the mechanism already described.

Reproduction

Sexual phenomena occur widely among protozoa, and sexual processes may precede certain phases of asexual reproduction, but embryonic development does not occur; protozoa do not have embryos. The essential features of sexual processes include a reduction division of the chromosome number to half (diploid number to haploid number), the development of sex cells (gam-etes) or at least gamete nuclei, and usually a fusion of gamete nuclei (p. 236).

Fission

The cell multiplication process that produces more individuals in protozoa is called fission. The most common type of fission is **binary,** in which two essentially identical individuals result (Figure 11.17). When a progeny cell is considerably smaller than the parent and then grows to adult size, the process is called **budding.** Budding occurs in some ciliates. In **multiple fission,** division of the cytoplasm (cytokinesis) is preceded by several nuclear divisions, so that a number of individuals are produced almost simultaneously (see Figure 11.31). Multiple fis-sion, or **schizogony,** is common among the Apicomplexa and

some amebas. If the multiple fission is preceded by or associated with union of gametes, it is called **sporogony.**

The foregoing types of division are accompa-nied by some form of mitosis (p. 52). However, this mitosis is often somewhat unlike that found in metazoans. For example, the nuclear membrane often persists through mitosis, and the microtubular spindle may be formed within the nuclear mem-brane. Centrioles have not been observed in nuclear division of ciliates; the nuclear membrane persists in micronuclear mitosis, with the spindle within the nucleus. The macronucleus of ciliates seems simply to elongate, to constrict, and to divide without any recognizable mitotic phenomena **(amitosis).**

Sexual Processes

Although all protozoa reproduce asexually, and some are apparently exclusively asexual, the wide-spread occurrence of sex among protozoa testifies to its importance as a means of genetic recombi-nation. Gamete nuclei, or pronuclei, which fuse in

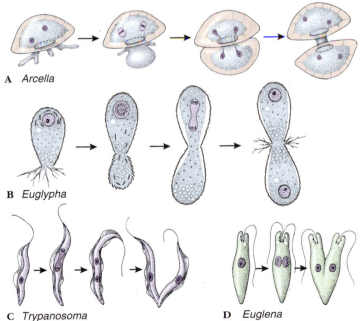

A *Arcella*

B *Euglypha*

C *Trypanosoma*   **D** *Euglena*

Figure 11.17

Binary fission in some amebas and flagellates. **A,** The two nuclei of *Arcella* divide as some of its cytoplasm is extruded and begins to secrete a new test for the daughter cell. **B,** The test of another ameba, *Euglypha,* is constructed of secreted platelets. Secretion of platelets for the daughter cell is begun before cytoplasm begins to move out of the aperture. As these are used to construct the test of the daughter cell, the nucleus divides. **C,** *Trypanosoma* has a kinetoplast (part of the mitochondrion) near the kinetosome of its flagellum close to its posterior end in the stage shown. All of these parts must be replicated before the cell divides. **D,** Division of *Euglena.* Compare **C** and **D** with Figure 11.27, fission in a ciliophoran.

fertilization to restore the diploid number of chromosomes, are usually borne in special gametic cells. When gametes all look alike, they are called **isogametes,** but most species have two dissimilar types, or **anisogametes.**

In animals meiosis usually occurs during or just before gamete formation (called gametic meiosis, p. 139). Such is indeed the case in Ciliophora and some flagellated and amebic groups. However, in other flagellated groups and in Apicomplexa, the first divisions *after* fertilization are meiotic **(zygotic meiosis),** and all individuals produced asexually (mitotically) in the life cycle up to the next zygote are haploid. Most protozoa that do not reproduce sexually probably are haploid, although demonstration of ploidy is difficult in the absence of meiosis. In some amebas (foraminiferans) haploid and diploid generations alternate **(intermediary meiosis),** a phenomenon widespread among plants.

Fertilization of an individual gamete by another is **syngamy,** but some sexual phenomena in protozoa do not involve syngamy. Examples are **autogamy,** in which gametic nuclei arise by meiosis and fuse to form a zygote within the same organism that produced them, and **conjugation,** in which an exchange of gametic nuclei occurs between paired organisms (conjugants). We describe conjugation further in the discussion of *Paramecium.*

Encystment and Excystment

Although separated from their external environment only by their delicate plasma membrane, unicellular forms are amazingly successful in habitats frequently subjected to extremely harsh conditions. Survival under harsh conditions surely is related to the ability to form **cysts,** dormant forms marked by possession of resistant external coverings and a complete shutdown of metabolic machinery. Cyst formation is also important to many parasitic forms that must survive a harsh environment between hosts (Figure 11.13). However, some parasites do not form cysts, apparently depending on direct transfer from one host to another. Reproductive phases such as fission, budding, and syngamy may occur in cysts of some species. Encystment has not been found in *Paramecium,* and it is rare or absent in marine forms.

Cysts of some soil-inhabiting and freshwater protozoa have amazing durability. Cysts of the soil ciliate *Colpoda* can survive 12 days in liquid nitrogen and 3 hours at 100°C. Survival of *Colpoda* cysts in dried soil has been shown for up to 38 years, and those of a certain small flagellate *(Podo)* can survive up to 49 years! Not all cysts are so sturdy, however. Those of *Entamoeba histolytica* tolerate gastric acidity but not desiccation, temperature above 50°C, or sunlight.

The conditions stimulating encystment are incompletely understood, although in some cases cyst formation is cyclic, occurring at a certain stage in the life cycle. In most free-living forms, adverse environmental change favors encystment. Such conditions may include food deficiency, desiccation, increased environmental osmotic pressure, decreased oxygen concentration, or change in pH or temperature.

During encystment a number of organelles, such as cilia or flagella, are resorbed, and the Golgi apparatus secretes cyst wall material, which is carried to the surface in vesicles and extruded.

Although the exact stimulus for excystation (escape from cysts) is usually unknown, a return of favorable conditions initiates excystment for those protozoa in which the cysts are a resistant stage. In parasitic forms the excystment stimulus may be more specific, requiring conditions similar to those found in the host.

MAJOR PROTOZOAN TAXA

The evolution of a eukaryotic cell was followed by diversification into many clades (see Figure 11.1), some of which contain both unicellular and multicellular forms. Clades of this type include the Opisthokonta, Viridiplantae, and the red algal clade, traditionally the phylum Rhodophyta. Rhodophyta is considered a plant clade because its members have plastids, are not heterotrophic, and lack flagellated stages (no motile sperm) in the life cycle. The clades we discuss further contain some members traditionally considered protozoans, so Viridiplantae and Opisthokonta are included, but Rhodophyta is not.

Opisthokonta

The Opisthokonta is a clade characterized by a combination of flattened mitochondrial cristae and one posterior flagellum on flagellated cells, if such cells exist. Recent protein sequence comparisons among taxa have also identified a short sequence of amino acids from one protein (elongation factor 1-alpha) that is shared by both unicellular and multicellular clade members. Relationships among clade members as suggested from sequence data from several proteins are shown in Figure 11.18.

The Opisthokonta contains metazoans and fungi as well as some unicellular taxa traditionally considered protozoans. The best-known unicells in this group are the microsporidians and choanoflagellates. Microsporidians are intracellular parasites now recognized as specialized fungi. Choanoflagellates (Figure 11.18) are solitary or colonial protozoans considered the most likely sister taxon to the metazoans. They are used to test hypotheses of how multicellular animals arose, specifically to identify features of the most recent common ancestor of animals and their closest unicellular relatives. We discuss them with sponges (phylum Porifera; p. 247) because of the strong resemblance between choanoflagellate cells and sponge choanocytes.

The Opisthokonta also contains less well-known unicells such as ichthyosporeans (animal parasites sometimes called DRIPs), nucleariid amebas, corallochytreans, and ministeriid amebas.

Stramenopiles

Members of the clade Stramenopiles have tubular mitochondrial cristae. Like opisthokonts, they may have flagellated cells, but stramenopiles are heterokont (Gr. *hetero,* different, + *kontos,*

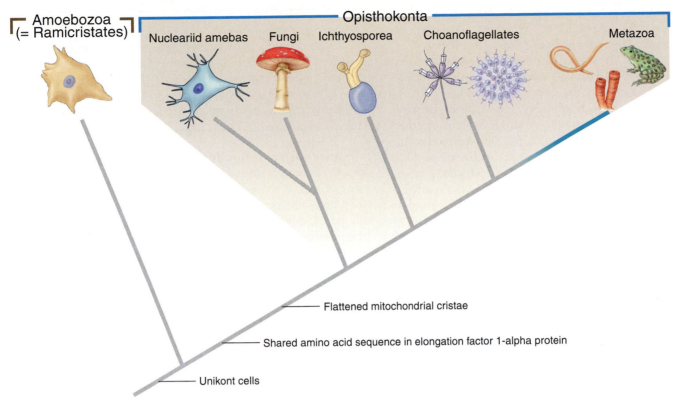

Figure 11.18

One hypothesis regarding relationships among some members of Opisthokonta: Choanoflagellates are shown as the sister taxon to Metazoa. Choanoflagellates shown are *Codonosiga* on the left and *Proterospongia* on the right.

pole) flagellates. They have two different flagella, both inserted at the cell anterior, instead of the posterior as in opisthokonts (Gr. *opisth,* posterior). In heterokonts, the forward directed flagellum is long and hairy, whereas the other is short, smooth, and trails behind the cell. This clade is sometimes called Heterokonta; the name stramenopile (L. *stramen,* straw, + *pile,* hair) refers to three-part tubular hairs covering the flagellum. This clade contains brown algae, yellow algae, and diatoms, all plantlike forms collecting energy with plastids, but animal-like forms are also present. The opalinids, a group of animal parasites once thought to be modified ciliates, and some heliozoans (see p. 243) are among the organisms placed in Stramenopiles.

Viridiplantae

The clade Viridiplantae contains unicellular and multicellular green algae, bryophytes, and vascular plants. Chloroplasts contain chlorophylls *a* and *b*. The flagellated plantlike branch of this lineage was once placed in class Phytomastigophorea by zoologists. However, other biologists placed the unicellular and multicellular green algae together in phylum Chlorophyta.

Phylum Chlorophyta

This group contains autotrophic single-celled algae such as *Chlamydomonas* (Figure 11.19) and colonial forms such as *Gonium* (Figure 11.19) and *Volvox* (Figure 11.20). *Volvox*

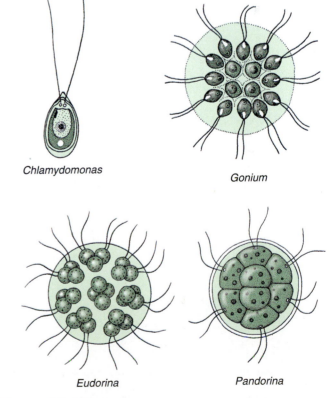

Chlamydomonas

Gonium

Eudorina

Pandorina

Figure 11.19

Examples of phylum Chlorophyta. They are all photoautotrophs.

is often studied in introductory courses because its mode of development is somewhat similar to embryonic development of some metazoans. The basic form of *Volvox,* a hollow ball of cells, is reminiscent of the metazoan blastula, leading some to suggest that the first metazoan was a nonphotosynthetic flagellate similar to *Volvox* in body design.

Volvox (Figure 11.20) is a green, hollow sphere that may reach a diameter of 0.5 to 1 mm. A single organism contains many thousands of cells (up to 50,000) embedded in the gelatinous surface of a jelly ball. Each cell is much like a euglenid (p. 231), with a nucleus, a pair of flagella, a large chloroplast, and a red **stigma.** A stigma is a shallow pigment cup that allows light from only one direction to strike a light-sensitive receptor. Adjacent cells are connected with each other by cytoplasmic strands. At one pole (usually in front as the colony moves), the stigmata are a little larger. Coordinated action of the flagella causes the colony to move by rolling over and over.

In *Volvox* we have a division of labor to the extent that most of the cells are somatic cells concerned with nutrition and locomotion, and a few germ cells located in the posterior half are responsible for reproduction. Reproduction is asexual or sexual. In either case only certain cells located around the equator or in the posterior half contribute to the next generation.

The original polarity of cells in *Volvox* is such that their flagella are protruding into the interior cavity of the developing organism. To move the flagella on the outside so that locomotion is possible, the entire spheroid must turn itself inside out. This process, called inversion, is *very unusual.* Of all other living organisms, only the sponges (phylum Porifera) have a comparable developmental process.

Asexual reproduction in *Volvox* occurs by repeated mitotic division of one of the germ cells to form a hollow sphere of cells, with the flagellated ends of the cells inside. The sphere then turns itself inside out to form a daughter colony similar to the parent colony. Several daughter colonies are formed inside the parent colony before they escape by rupture of the parent.

In **sexual reproduction** some of the cells differentiate into **macrogametes** or **microgametes** (Figure 11.20). Macrogametes are fewer and larger and are loaded with food for nourishment of the young organism. Microgametes, by repeated division, form bundles or balls of small flagellated sperm that leave the mother organism when they mature and swim to find a mature ovum. After fertilization, the zygote secretes a hard, spiny, protective shell around itself. When released by the rupture of a parent, a zygote remains quiescent during the winter. Within its shell the zygote undergoes repeated division, producing a small organism

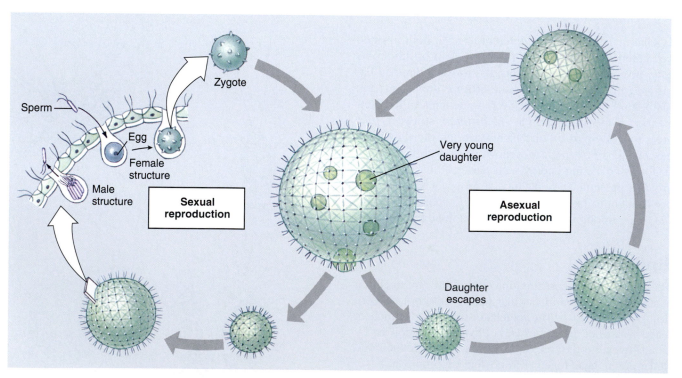

Figure 11.20

Life cycle of *Volvox.* Asexual reproduction occurs in spring and summer when specialized diploid reproductive cells divide to form young organisms that remain in the mother organism until large enough to escape. Sexual reproduction occurs largely in autumn when haploid sex cells develop. The fertilized ova may encyst and so survive the winter, developing into a mature asexual organism in the spring. In some species the organisms have separate sexes; in others both eggs and sperm are produced in the same organism.

that breaks out in the spring. A number of asexual generations may follow, during the summer, before sexual reproduction occurs again.

The order to which *Volvox* belongs (Volvocida) includes many freshwater flagellates, mostly green, with a cellulose cell wall through which two short flagella project. Many are colonial forms (Figure 11.19, *Pandorina, Eudorina, Gonium*), in which a single organism contains more than one cell but separate somatic and reproductive types do not exist.

Phylum Euglenozoa

The Euglenozoa (Figure 11.21) is generally considered a monophyletic group, based on the shared persistence of the nucleoli during mitosis, and the presence of discoid mitochondrial cristae. Members of this phylum have a series of longitudinal microtubules just beneath the cell membrane that help to stiffen the membrane into a **pellicle.** The phylum is divided into two subphyla, the Euglenida and the Kinetoplasta. Kinetoplastans are named for the presence of a unique organelle, the kinetoplast. This modified mitochondrion, associated with a kinetosome, carries a large disc of DNA. Kinetoplastans are all parasites, living in plants and animals.

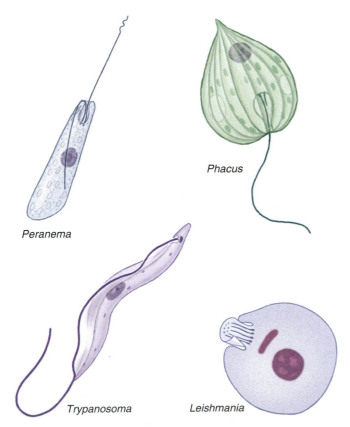

Figure 11.21

Examples of phylum Euglenozoa. *Peranema* is a colorless, free-living phagotroph, and *Phacus* is a green, free-living photoautotroph. *Trypanosoma* and *Leishmania* are parasitic, and some species cause serious diseases of humans and domestic animals. *Leishmania* is shown as its intracellular form, without an external flagellum.

Subphylum Euglenida

Euglenids, formerly in the Phytomastigophorea, have chloroplasts with chlorophyll *b*. These chloroplasts are surrounded by a double membrane and are likely to have arisen via secondary endosymbiosis.

Euglena viridis (Figure 11.14) is a representative flagellate commonly studied in introductory zoology courses. Its natural habitat is freshwater streams and ponds where there is considerable vegetation. The organisms are spindle shaped and about 60 μm long, but some species of *Euglena* are smaller and some larger (*E. oxyuris* is 500 μm long). Just beneath the outer membrane of *Euglena* are proteinaceous strips and microtubules that form a pellicle. In *Euglena* the pellicle is flexible enough to permit bending, but in other euglenids it may be more rigid. A flagellum extends from a flask-shaped **reservoir** at the anterior end, and another, short flagellum ends within the reservoir. A kinetosome occurs at the base of each flagellum, and a **contractile vacuole** empties into the reservoir. A red eyespot, or stigma, apparently functions in orientation to light. Within the cytoplasm are oval chloroplasts that bear chlorophyll and give the organism its greenish color. **Paramylon granules** of various shapes are masses of a starchlike food storage material.

Nutrition of *Euglena* is normally autotrophic (holophytic), but if kept in the dark the organism uses saprozoic nutrition, absorbing nutrients through its body surface. Mutants of *Euglena* can be produced that have permanently lost their photosynthetic ability. Although *Euglena* does not ingest solid food, some euglenids are phagotrophic. *Peranema* has a cytostome that opens alongside its flagellar reservoir.

Euglena reproduces by binary fission and can encyst to survive adverse environmental conditions.

Subphylum Kinetoplasta

Some of the most important protozoan parasites are kinetoplastans. Many of them belong to the genus *Trypanosoma* (Gr. *trypanon,* auger, + *soma,* body) (Figure 11.21) and live in the blood of fish, amphibians, reptiles, birds, and mammals. Some are nonpathogenic, but others produce severe diseases in humans and domestic animals. *Trypanosoma brucei gambiense* and *T. brucei rhodesiense* cause African sleeping sickness in humans, and *T. brucei brucei* causes a related disease in domestic animals. Trypanosomes are transmitted by tsetse flies (*Glossina* spp.). *Trypanosoma b. rhodesiense,* the more virulent of the sleeping sickness trypanosomes, and *T. b. brucei* have natural reservoirs (antelope and other wild mammals) that are apparently not harmed by the parasites. Some 10,000 new cases of human sleeping sickness are diagnosed each year, of which about half are fatal, and many of the remainder sustain permanent brain damage.

Trypanosoma cruzi causes Chagas' disease in humans in Central America and South America. It is transmitted by "kissing bugs" (Triatominae), a name arising from the bug's habit of biting its sleeping victim on the face. Acute Chagas' disease is most common and severe among children less than five years old, while the chronic disease is seen most often in adults. Symptoms are primarily a result of central and peripheral nervous dysfunction. Two to three million people in South and Central America show chronic Chagas' disease, and 45,000 of these die each year.

Several species of *Leishmania* (Figure 11.21) cause disease in humans. Infection with some species may cause a serious visceral disease affecting especially the liver and spleen; others can cause disfiguring lesions in the mucous membranes of the nose and throat, and the least serious result is a skin ulcer. *Leishmania* spp. are transmitted by sand flies. Visceral leishmaniasis and cutaneous leishmaniasis are common in parts of Africa and Asia, and the mucocutaneous form occurs in Central America and South America.

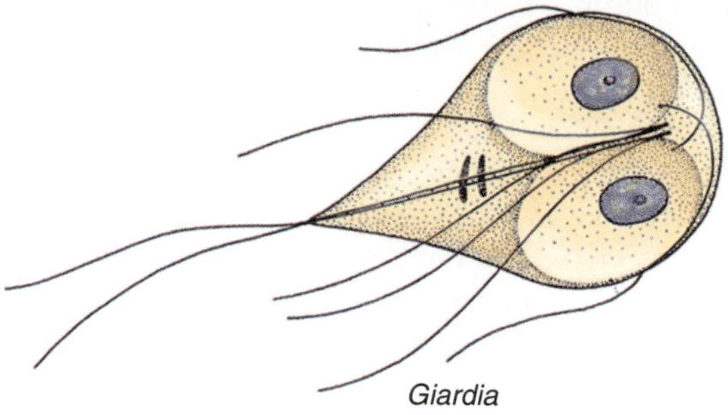

Phylum Retortamonada and the Diplomonads

This phylum is divided into two clades: Retortamonads and Diplomonads. Retortamonads include commensal and parasitic unicells, such as *Chilomastix* and *Retortamonas*. They lack mitochondria and Golgi bodies, so biologists wondered whether they branched from the main eukaryotic lineage before the mitochondrial symbiosis. Diplomonads, once a subgroup of retortamonds, also lack mitochondria, and were proposed as an early-diverging branch of the eukaryotic lineage. However, recent work showing that mitochondrial genes are present in the cell nucleus[2] makes it much more likely that the absence of mitochondria is a secondary loss, instead of primary absence.

Giardia, a diplomonad, is a well-studied parasite (Figure 11.22). Some species live in the human digestive tract, but others occur in birds or amphibians. It is often asymptomatic but may cause a rather discomfiting, but not fatal, diarrhea. Cysts are passed in the feces, and new hosts are infected by ingestion of cysts, often in contaminated water.

Giardia lamblia is commonly transmitted through water supplies contaminated with sewage. The same species, however, lives in a variety of mammals other than humans. Beavers seem to be an important source of infection in mountains of the western United States. When one has hiked for miles in the wild on a hot day, it can be very tempting to fill a canteen and drink from a crystal-clear beaver pond. Many cases of infection are acquired that way.

Alveolata

The alveolate clade, sometimes called a superphylum, contains three traditional phyla united by the shared presence of **alveoli,** membrane-bound sacs that lie beneath the cell membrane. In the Ciliophora (Figure 11.23), the alveoli produce pellicles; in the Dinoflagellata, a group of armored flagellates (Figure 11.29), the alveoli produce thecal plates, and in the Apicomplexa, containing

[2]Roger, A. J. 1999. Amer. Nat. **154** (supplement):S146–S163.

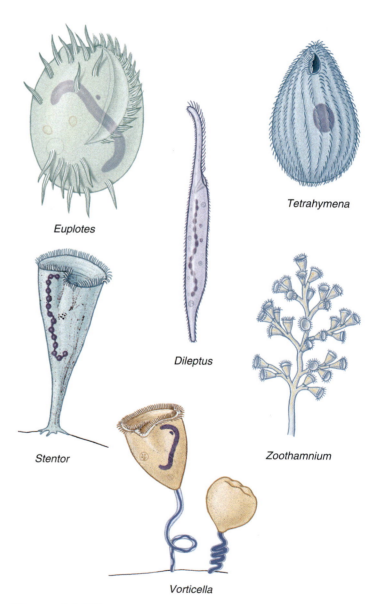

Figure 11.22

Giardia lamblia often causes diarrhea in humans.

Figure 11.23

Some representative ciliates. *Euplotes* have stiff cirri used for crawling about. Contractile fibrils in ectoplasm of *Stentor* and in stalks of *Vorticella* allow great expansion and contraction. Note the macronuclei, long and curved in *Euplotes* and *Vorticella*, shaped like a string of beads in *Stentor*.

intracellular parasitic species previously called sporozoans (see Figure 11.30), the alveoli have structural functions.

Phylum Ciliophora

Ciliates (Figure 11.23) are so named because the body surface is covered with cilia that beat in a coordinated rhythmical manner. The arrangement of cilia varies within the phylum and some ciliates lack cilia as adults, although they are present at other stages in the life cycle. In general, ciliates are larger than most other protozoa, but they range from 10 μm to 3 mm in length. Most ciliates are free-living in freshwater or marine habitats, but commensal and parasitic forms do occur. They are usually solitary and motile, but some are sessile and others are colonial. Ciliates are the most structurally complex of all protozoans, exhibiting a wide range of specializations.

The pellicle of ciliates may consist only of a cell membrane or in some species may form a thickened armor. Cilia are short and usually arranged in longitudinal or diagonal rows. Cilia may cover the surface of the organism or may be restricted to the oral region or to certain bands. In some forms cilia are fused into a sheet called an **undulating membrane** or into smaller **membranelles,** both used to propel food into the **cytopharynx** (gullet). In other forms there may be fused cilia forming stiffened tufts called **cirri,** often used in locomotion by the creeping ciliates (Figure 11.23).

An apparently structural system of fibers, in addition to the kinetosomes, forms the **infraciliature,** just beneath the pellicle (Figure 11.24). Each cilium terminates beneath the pellicle in its kinetosome, and from each kinetosome a fibril arises and passes beneath the row of cilia, joining with the other fibrils of that row. The cilia, kinetosomes, and other fibrils of that ciliary row form what is called a **kinety** (Figure 11.24). All ciliates seem to have kinety systems, even those that lack cilia at some stage. The infraciliature apparently does not coordinate ciliary beat, as

formerly thought. Coordination of ciliary movement seems to be by waves of depolarization of the cell membrane moving down the organism, similar to a nerve impulse.

Ciliates are always multinucleate, possessing at least one **macronucleus** and one **micronucleus,** but varying from one to many of either type. The macronuclei are apparently responsible for metabolic and developmental functions and for maintaining all the visible traits, such as the pellicular apparatus. Macronuclei vary in shape among the different species (Figures 11.15 and 11.23). Micronuclei participate in sexual reproduction and give rise to macronuclei after exchange of micronuclear material between individuals. Micronuclei divide mitotically, and macronuclei divide amitotically (see p. 227).

Some ciliates have small bodies in their ectoplasm between the bases of the cilia. Examples are **trichocysts** (Figures 11.15 and 11.24) and **toxicysts.** Upon mechanical or chemical stimulation, these bodies explosively expel a long, threadlike structure. The mechanism of expulsion is unknown. The function of trichocysts is thought to be defensive. When attacked by a *Didinium,* a paramecium expels its trichocysts but to no avail. Toxicysts, however, release a poison that paralyzes the prey of carnivorous ciliates. Toxicysts are structurally quite distinct from trichocysts. Many dinoflagellates also have trichocysts.

Most ciliates are holozoic. Most of them possess a cytostome (mouth) that in some forms is a simple opening and in others is connected to a gullet or ciliated groove. The mouth in some is strengthened with stiff, rodlike trichites for swallowing larger prey; in others, such as paramecia, ciliary water currents carry microscopic food particles toward the mouth. *Didinium* has a proboscis for engulfing paramecia on which it feeds (see Figure 11.2). Suctorians paralyze their prey and then ingest the contents through tubelike tentacles by a complex feeding mechanism that apparently combines phagocytosis with a sliding filament action of microtubules in the tentacles (see Figure 11.2).

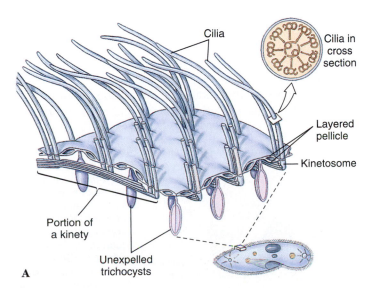

B Expelled trichocyst

Figure 11.24

Infraciliature and associated structures in ciliates. **A,** Structure of the pellicle and its relation to the infraciliature system. **B,** Expelled trichocyst.

Suctorians   Suctorians are ciliates in which the young possess cilia and are free swimming, and the adults grow a stalk for attachment, become sessile, and lose their cilia. They have no cytostome but feed by long, slender, tubelike tentacles. The suctorian captures living prey, usually a ciliate, by the tip of one or more tentacles and paralyzes it. The cytoplasm of the prey then flows through the attached tentacles, by a complex feeding mechanism that apparently combines phagocytosis with a sliding filament action of microtubules in the tentacles (Figure 11.2). Food vacuoles form in the feeding suctorian.

One of the best places to find freshwater suctorians is in algae that grow on the carapace of turtles. Common genera of suctorians found there are *Anarma* (without stalk or test) and *Squalorophrya* (with stalk and test). Other freshwater representatives are *Podophrya* (see Figure 11.2) and *Dendrosoma*. *Acinetopsis* and *Ephelota* are saltwater forms.

Suctorian parasites include *Trichophrya*, whose species occur on various invertebrates and freshwater fish; *Allantosoma*, which live in the intestine of certain mammals; and *Sphaerophrya*, which are found in *Stentor*.

Symbiotic Ciliates   Many symbiotic ciliates live as commensals, but some can be harmful to their hosts. *Balantidium coli* lives in the large intestine of humans, pigs, rats, and many other mammals (Figure 11.25). There seem to be host-specific strains, and the organism is not easily transmitted from one species to another. Transmission is by fecal contamination of food or water. Usually the organisms are not pathogenic, but in humans they sometimes invade the intestinal lining and cause a dysentery similar to that caused by *Entamoeba histolytica* (p. 242). The disease can be serious and even fatal. Infections are common in parts of Europe, Asia, and Africa but are rare in the United States.

Other species of ciliates live in other hosts. *Entodinium* (Figure 11.25) belongs to a group that has very complex structure and lives in the digestive tract of ruminants, where they may be very abundant. *Nyctotherus* live in the colon of frogs and toads. In aquarium and wild freshwater fishes, *Ichthyophthirius* causes a disease known to many fish culturists as "ick." Untreated, it can cause much loss of exotic fishes.

Free-Living Ciliates   Among the more striking and familiar ciliates are *Stentor* (Gr. herald with a loud voice), trumpet shaped and solitary, with a beadshaped macronucleus

(Figure 11.23); *Vorticella* (L. dim. of *vortex,* a whirlpool), bell-shaped and attached by a contractile stalk (Figure 11.23); and *Euplotes* (Gr. *eu,* true, good, + *ploter,* swimmer) with a flattened body and groups of fused cilia (cirri) that function as legs. Paramecia are usually abundant in ponds or sluggish streams containing aquatic plants and decaying organic matter. We discuss *Paramecium* in more detail, as a representative free-living ciliate.

***Form and Function in* Paramecium**   Paramecia are often described as slipper shaped. *Paramecium caudatum* is 150 to 300 µm in length and is blunt anteriorly and somewhat pointed posteriorly (see Figure 11.15). The organism has an asymmetrical appearance because of the **oral groove,** a depression that runs obliquely backward on the ventral side.

The **pellicle** is a clear, elastic membrane that may be ornamented by ridges or papilla-like projections (Figure 11.24), and its entire surface is covered with cilia arranged in lengthwise rows. Just below the pellicle is the thin clear **ectoplasm** that surrounds the larger mass of granular **endoplasm.** Embedded in ectoplasm just below the surface are spindle-shaped **trichocysts** (Figure 11.24), which alternate with the bases of cilia. The infraciliature can be seen only with special fixing and staining methods.

A **cytostome** at the end of the oral groove leads into a tubular **cytopharynx,** or **gullet.** Along the gullet an undulating membrane of modified cilia keeps food moving. Fecal material is discharged through a **cytoproct** posterior to the oral groove (see Figure 11.15). Within the endoplasm are food vacuoles containing food in various stages of digestion. There are two **contractile vacuoles,** each consisting of a central space surrounded by several **radiating canals** (see Figure 11.15) that collect fluid and empty it into the central vacuole. We describe excretion and osmoregulation on page 226.

Paramecium caudatum has two nuclei: a large kidney-shaped **macronucleus** and a smaller **micronucleus** fitted into the depression of the former. These can usually be seen only in stained specimens. The number of micronuclei varies in different species; for example, *P. multimicronucleatum* may have as many as seven.

Paramecia are holozoic, living on bacteria, algae, and other small organisms. Cilia in the oral groove sweep food particles in the water into the cytostome, from which they are carried into the cytopharynx by the undulating membrane. From the cytopharynx food is collected into a food vacuole that is constricted

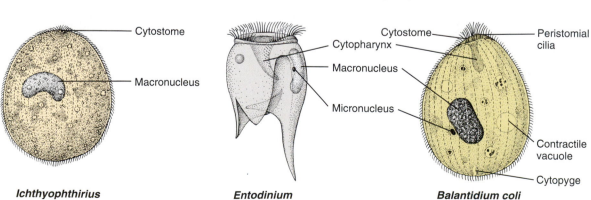

Figure 11.25

Some symbiotic ciliates. *Balantidium coli* is a parasite of humans and other mammals. *Ichthyophthirius* causes a common disease in aquarium and wild freshwater fishes. *Entodinium* is found in the rumen of cows and sheep.

Ichthyophthirius — Cytostome, Macronucleus

Entodinium — Cytostome, Cytopharynx, Macronucleus, Micronucleus

Balantidium coli — Peristomial cilia, Contractile vacuole, Cytopyge into the endoplasm. Food vacuoles circulate in a definite course through the cytoplasm while the food is being digested by enzymes from the endoplasm. Indigestible parts of the food are ejected through the cytoproct.

The body is elastic, allowing it to bend and to squeeze through narrow places. Its cilia can beat either forward or backward, so that the organism can swim in either direction. The cilia beat obliquely, causing the organism to rotate on its long axis. In the oral groove the cilia are longer and beat more vigorously than the others so that the anterior end swerves aborally. As a result of these factors, the organism moves forward in a spiral path (Figure 11.26A).

When a ciliate, such as a paramecium, contacts a barrier or a disturbing chemical stimulus, it reverses its cilia, backs up a short distance, and swerves the anterior end as it pivots on its posterior end. This behavior is called an **avoiding reaction** (Figure 11.26B). A paramecium may continue to change its direction to keep itself away from a noxious stimulus, and it may react in a similar fashion to keep itself within the zone of an attractant. A paramecium may also change its swimming speed. How does a paramecium "know" when to change directions or swimming speed? Interestingly, reactions of the organism depend on effects of the stimulus on the electrical potential difference across its cell membrane. Paramecia slightly hyperpolarize in attractants and depolarize in repellents that produce the avoiding reaction. Hyperpolarization increases the rate of the forward ciliary beat, and depolarization results in ciliary reversal and backward swimming.

Locomotor responses, by which an organism more or less continuously orients itself with respect to a stimulus, are called *taxes* (sing. *taxis*). Movement toward the stimulus is a positive taxis; movement away is a negative taxis. Some examples are thermotaxis, response to heat; phototaxis, response to light; thigmotaxis, response to contact; chemotaxis, response to chemical substances; rheotaxis, response to currents of air or water; galvanotaxis, response to constant electric current; and geotaxis, response to gravity. Some stimuli do not cause an orienting response but simply a change in movement: more rapid movement, more frequent random turning, or slowing or cessation of movement. Such responses are called kineses. Is the avoiding reaction of a paramecium a taxis or a kinesis?

***Reproduction in* Paramecium**   Paramecia reproduce only by binary fission across kineties (ciliary rows) but have certain forms of sexual phenomena called conjugation and autogamy.

In **binary fission** the micronucleus divides mitotically into two daughter micronuclei, which move to opposite ends of the cell (Figure 11.27). The macronucleus elongates and divides amitotically.

Conjugation occurs at intervals in ciliates. Conjugation is the temporary union of two individuals to exchange chromosomal material (Figure 11.28). During the union the macronucleus

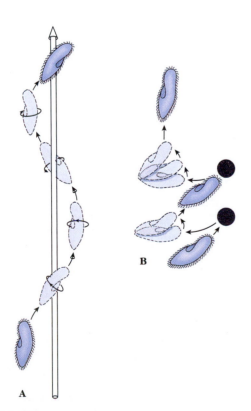

Figure 11.26

A, Spiral path of swimming *Paramecium.* **B,** Avoidance reaction of *Paramecium.*

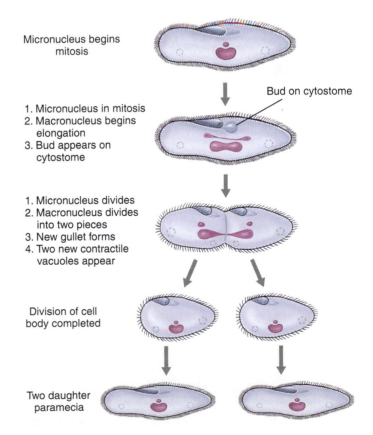

Micronucleus begins mitosis

1. Micronucleus in mitosis
2. Macronucleus begins elongation
3. Bud appears on cytostome

Bud on cytostome

1. Micronucleus divides
2. Macronucleus divides into two pieces
3. New gullet forms
4. Two new contractile vacuoles appear

Division of cell body completed

Two daughter paramecia

Figure 11.27

Binary fission in a ciliophoran *(Paramecium).* Division is across rows of cilia.

disintegrates and the micronucleus of each individual undergoes meiosis, giving rise to four haploid micronuclei, three of which degenerate (Figure 11.28A to C). The remaining micronucleus then divides into two haploid pronuclei, one of which is exchanged

A Two *Paramecium* individuals come into contact on their oral surface.

Micronucleus (2n)

Macronucleus

B The micronuclei divide by meiosis to produce four haploid micronuclei. Macronuclei degenerate.

C Three micronuclei degenerate; the remaining micronucleus divides to form "male" and "female" pronuclei.

D Male pronuclei are exchanged between conjugants.

E Male and female pronuclei fuse to make a diploid nucleus, and individuals separate.

F Three sets of mitotic divisions produce eight micronuclei; four of these become macronuclei while three degenerate.

G The remaining micronucleus divides twice as does the cell, producing four daughter cells.

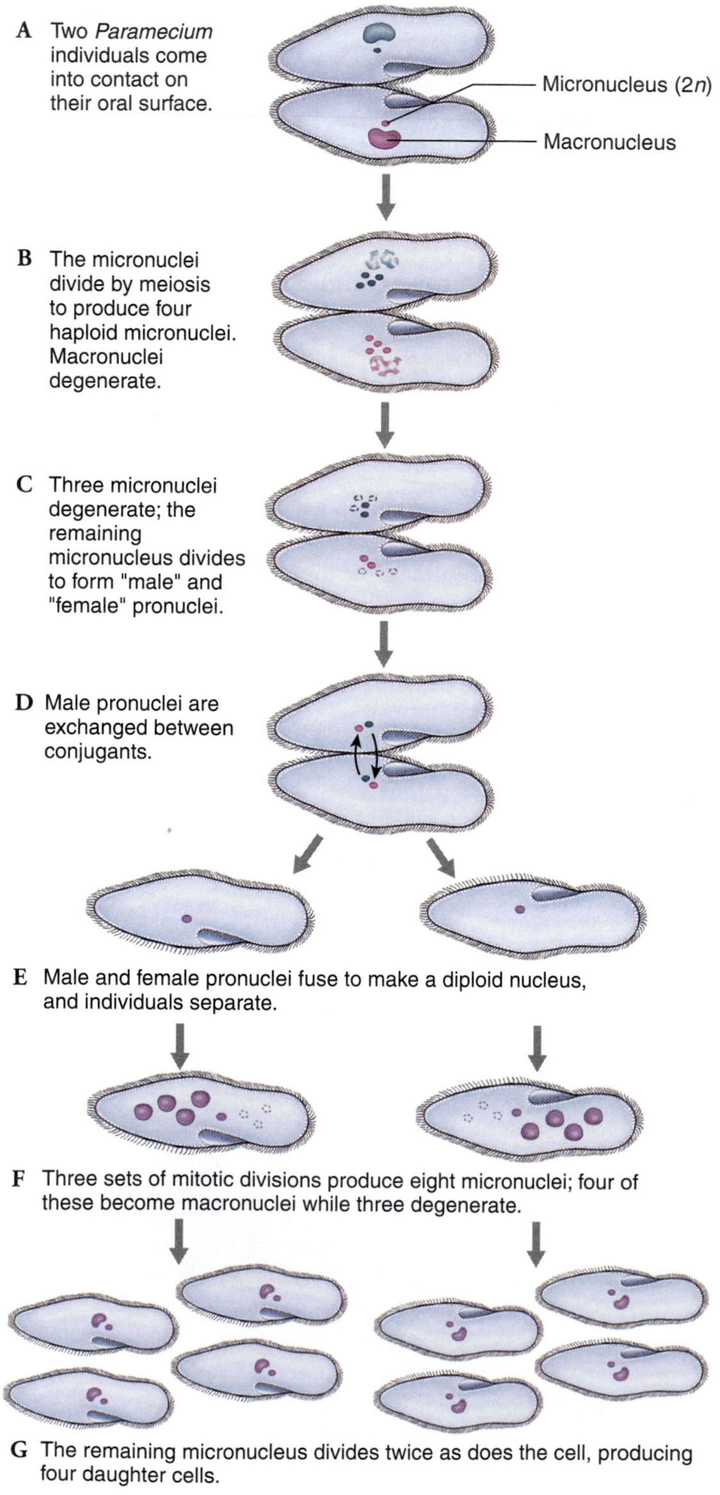

Figure 11.28

Scheme of conjugation in *Paramecium*.

with the other conjugant. The pronuclei fuse to restore the diploid number of chromosomes, followed by several more nuclear events detailed in Figure 11.28. Following this complicated process, the organisms may continue to reproduce by binary fission without conjugation.

The result of conjugation is similar to that of zygote formation, for each exconjugant contains hereditary material from two individuals. The advantage of sexual reproduction is that it permits gene recombinations, thus increasing genetic variation in the population. Although ciliates in clone cultures can apparently reproduce repeatedly and indefinitely without conjugation, the stock seems eventually to lose vigor. Conjugation restores vitality to a stock. Seasonal changes or a deteriorating environment usually stimulate sexual reproduction.

Autogamy is a process of self-fertilization similar to conjugation except that there is no exchange of nuclei. After the disintegration of the macronucleus and the meiotic divisions of the micronucleus, two haploid pronuclei fuse to form a synkaryon that is completely homozygous (see Chapter 5, p. 81).

Phylum Dinoflagellata

Dinoflagellates are another group formerly included by zoologists among Phytomastigophorea, and about half are photoautotrophic with chromoplasts bearing chlorophyll. The rest are colorless and heterotrophic. Ancestral dinoflagellates probably were heterotrophic, and some acquired chloroplasts by endosymbiosis from a variety of algal sources. Ecologically, some species are among the most important primary producers in marine environments. They commonly have two flagella, one equatorial and one longitudinal, each borne at least partially in grooves on the body (Figure 11.29). The body may be naked or covered by

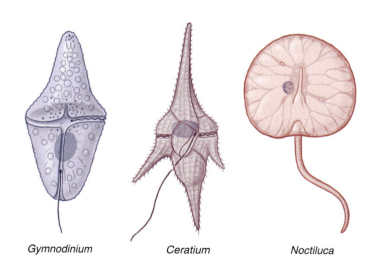

Gymnodinium   *Ceratium*   *Noctiluca*

Figure 11.29

Examples of phylum Dinoflagellata. *Gymnodinium* bears no cellulose plates. Some members of its family are autotrophic and some phagotrophic. *Ceratium* bears plates and is both autotrophic and phagotrophic. *Noctiluca* is entirely phagotrophic, can be very large (more than 1 mm wide), and has a large tentacle involved in feeding.

cellulose plates or valves. Many species can ingest prey through a mouth region between the plates near the posterior area of the body. *Ceratium* (Figure 11.29), for example, has a thick covering with long spines, into which the body extends, but it can catch food with posterior pseudopodia and ingest it between the flexible plates in the posterior groove. *Noctiluca* (Figure 11.29), a colorless dinoflagellate, is a voracious predator and has a long, motile tentacle, near the base of which its single, short flagellum emerges. *Noctiluca* is one of many marine organisms that can produce light (bioluminescence).

Several groups of autotrophic flagellates are planktonic primary producers (p. 834) in freshwater and marine environments; however, dinoflagellates are the most important, particularly in the sea. Zooxanthellae are dinoflagellates that live in mutualistic association in tissues of certain invertebrates, including other protozoa, sea anemones, horny and stony corals, and clams. The association with stony corals is of ecological and economic importance because only corals with symbiotic zooxanthellae can form coral reefs (see Chapter 13).

Dinoflagellates can damage other organisms, such as when they produce a "red tide." Although this name originally was applied to situations in which the organisms reproduced in such profusion (producing a "bloom") that the water turned red from their color, any instance of a bloom producing detectable levels of toxic substances is now called a red tide. The water may be red, brown, yellow, or not remarkably colored at all. The toxic substances are apparently not harmful to the organisms that produce them, but they may be highly poisonous to fish and other marine life. Several different types of dinoflagellates and one species of cyanobacterium have been responsible for red tides. Red tides have caused considerable economic losses to the shellfish industry. Another flagellate produces a toxin concentrated in the food chain, especially in large, coral reef fishes. The illness produced in humans after eating such fish is called ciguatera.

Pfiesteria piscicida is one of several related dinoflagellate species that may affect fish in brackish waters along the Atlantic coast, south of North Carolina. Much of the time *Pfiesteria* feeds on algae and bacteria, but something in the excreta of large fish schools causes it to release a powerful, short-lived, toxin. The toxin may stun or kill fish, often creating skin lesions. *Pfiesteria* has flagellated and ameboid forms among its more than 20 body types; some forms feed on fish tissues and blood. Although it does not have chloroplasts, it may sequester choroplasts from its algal prey and gain energy from them over the short term. This fascinating group of species was discovered in 1988.

Phylum Apicomplexa

All apicomplexans are endoparasites, and their hosts include many animal phyla. The presence of a certain combination of organelles, the **apical complex,** distinguishes this phylum (Figure 11.30A). The apical complex is usually present only in certain developmental stages of the organisms; for example,

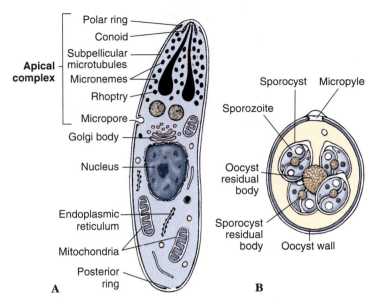

Figure 11.30

A, Diagram of an apicomplexan sporozoite or merozoite at the electron-microscope level, illustrating the apical complex. The polar ring, conoid, micronemes, rhoptries, subpellicular microtubules, and micropore (cytostome) are all considered components of the apical complex. **B,** Infective oocyst of *Eimeria*. The oocyst is the resistant stage and has undergone multiple fission after zygote formation (sporogony).

merozoites and **sporozoites** (Figure 11.31). Some structures, especially the **rhoptries** and **micronemes,** apparently aid in penetrating the host's cells or tissues.

Locomotor organelles are less obvious in this group than in other protozoa. Pseudopodia occur in some intracellular stages, and gametes of some species are flagellated. Tiny contractile fibrils can form waves of contraction across the body surfaces to propel the organism through a liquid medium.

The life cycle usually includes both asexual and sexual reproduction, and sometimes an invertebrate intermediate host. At some point in the life cycle, the organisms develop a **spore (oocyst),** which is infective for the next host and is often protected by a resistant coat. In the traditional phylum Protozoa, apicomplexans were in class Sporozoa, so the name sporozoan is sometimes applied here, but it can also be used for unrelated spore-forming taxa.

Class Coccidea   Coccidia are intracellular parasites in invertebrates and vertebrates, and the group includes species of very great medical and veterinary importance. We discuss three examples: *Eimeria,* which generally affects birds; *Toxoplasma,* which causes toxoplasmosis, a disease affecting cats and humans; and *Plasmodium,* the organism that causes malaria.

Eimeria Species.   The name "coccidiosis" is generally applied only to infections with *Eimeria* or *Isospora.* Humans can be infected with species of *Isospora,* but there is usually little disease. However, *Isospora* infections can be very serious in AIDS patients. Some species of *Eimeria* may cause serious disease in some domestic animals. Symptoms usually include severe diarrhea or dysentery.

Eimeria tenella is often fatal to young fowl, producing severe pathogenesis in the intestine. The organisms undergo schizogony (p. 227) in the intestinal cells, finally producing gametes. After fertilization the zygote forms an oocyst that exits its host via the feces (Figure 11.30B). Sporogony occurs within the oocyst outside the host, producing eight sporozoites in each oocyst. Infection occurs when a new host accidentally ingests a sporulated oocyst and the sporozoites are released by digestive enzymes.

Toxoplasma gondii.   A similar life cycle occurs in *Toxoplasma gondii,* a parasite of cats, but this species produces extraintestinal stages as well. When rodents, cattle, sheep, humans, many other mammals, or even birds, ingest sporozoites, the sporozoites cross from the intestine and begin rapid, asexual reproduction in a variety of tissues. As the host mounts an immune response, reproduction of the zoites slows, and they become enclosed in tough **tissue cysts.** The zoites, now called **bradyzoites,** accumulate in large numbers in each tissue cyst. Bradyzoites are infective for other hosts, including cats, where they can initiate the intestinal cycle in a cat that eats infected prey. Bradyzoites can remain viable and infective for months or years, and it is estimated that one-third of the world's human population carries tissue cysts containing bradyzoites in their body. The normal route of infection for humans is apparently consumption of infected meat that is insufficiently cooked.

In humans *Toxoplasma* causes little or no ill effects except in AIDS patients or in women infected during pregnancy, particularly in the first trimester. Such infection greatly increases the chances of a birth defect in the baby; it is now believed that 2% of all mental retardation in the United States is a result of congenital toxoplasmosis. Toxoplasmosis can also be a serious disease in persons who are immunosuppressed, either by drugs or AIDS. In such patients rupture of a tissue cyst, which would be contained easily in a person with a normal immune system, becomes a source of life-threatening infection.

Plasmodium: *The Malarial Organism.*   The best known coccidians are *Plasmodium* spp., causative organisms of the most important infectious disease of humans: **malaria.** Malaria is a very serious disease, difficult to control and widespread, particularly in tropical and subtropical countries. Four species of *Plasmodium* infect humans: *P. falciparum, P. vivax, P. malariae,* and *P. ovale.* Although each species produces its own peculiar clinical picture, all four have similar cycles of development in their hosts (Figure 11.31).

The parasite is carried by mosquitoes (*Anopheles*), and sporozoites are injected into a human with the insect's saliva during its bite. Sporozoites penetrate liver cells and initiate schizogony. In *P. falciparum* a single sporozoite produces up to 40,000 merozoites by schizogony. The products of this division then enter other liver cells to repeat the schizogonous cycle, or in *P. falciparum* they penetrate red blood cells after only one cycle in the liver. The period when the parasites are in the liver is the **incubation period,** and it lasts from 6 to 15 days, depending on the species of *Plasmodium.*

Merozoites released as a result of liver schizogony enter red blood cells, where they begin a series of schizogonous cycles. When they enter red blood cells, they become ameboid **trophozoites,** feeding on hemoglobin. The end product of the parasite's digestion of hemoglobin is a dark, insoluble pigment: **hemozoin.** Hemozoin accumulates in the host cell, is released when the next generation of merozoites is produced, and eventually accumulates in the liver, spleen, or other organs. A trophozoite within a red blood cell grows and undergoes schizogony, producing 6 to 36 merozoites, which, depending on the species, burst forth to infect new red cells. When a red blood cell containing merozoites bursts, it releases the parasite's metabolic products, which have accumulated there. Release of these foreign substances into the patient's circulation causes the chills and fever characteristic of malaria.

Some 16% or more of adults in the United States are infected with *Toxoplasma gondii;* we have no symptoms because the parasite is held in check by our immune systems. However, *T. gondii* is one of the most important opportunistic infections in AIDS patients. The latent infection is activated in between 5% and 15% of AIDS patients, often in the brain, with serious consequences. Another coccidian, *Cryptosporidium parvum,* first was reported in humans in 1976. We now recognize it as a major cause of diarrheal disease worldwide, especially in children in tropical countries. Waterborne outbreaks have occurred in the United States, and the diarrhea can be life-threatening in immunocompromised patients (such as those with AIDS). Infection rates for 2005 were about 3 cases per 100,000 persons. The latest coccidian pathogen to emerge has been *Cyclospora cayetanensis.* U.S. infection rates for 2005 were about 0.2 cases per 100,000 persons, with diarrhea being the most common symptom of infection. Infection usually occurs via ingestion of contaminated food or water.

Since the populations of schizonts maturing in red blood cells are synchronized to some degree, the episodes of chills and fever have a periodicity characteristic of the particular species of *Plasmodium.* In *P. vivax* (benign tertian) malaria and *P. ovale* malaria, episodes occur every 48 hours; in *P. malariae* (quartan) malaria, every 72 hours; and in *P. falciparum* (malignant tertian) malaria, about every 48 hours, although synchrony is less well defined in this species. People usually recover from infections with the first three species, but mortality is high in untreated cases of *P. falciparum* infection. Sometimes grave complications, such as **cerebral malaria,** occur. Unfortunately, *P. falciparum* is the most common species, accounting for 50% of all malaria in the world. Certain genes, for example the gene for sickle cell hemoglobin (p. 100 and p. 704), confer some resistance to malaria on people that carry them.

After some cycles of schizogony in red blood cells, infection of new cells by some of the merozoites causes production of **microgametocytes** and **macrogametocytes** rather than another generation of merozoites. When gametocytes are ingested by a mosquito feeding on a patient's blood, they mature

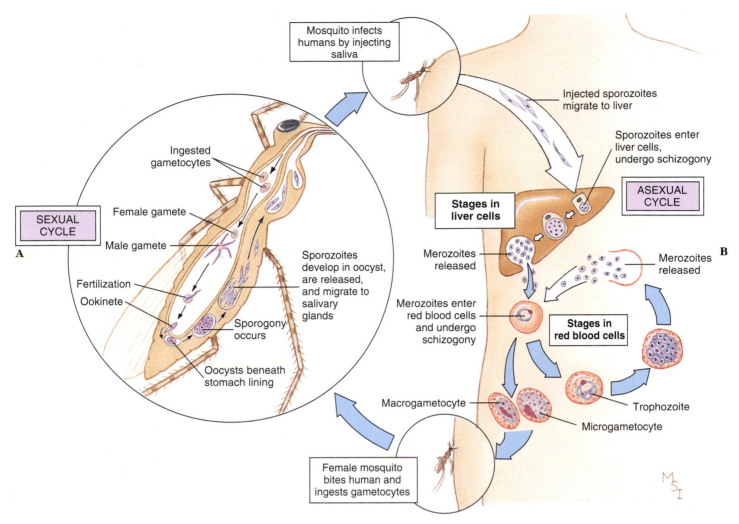

Figure 11.31

Life cycle of *Plasmodium vivax,* one of the protozoa (class Coccidia) that causes malaria in humans. **A,** Sexual cycle produces sporozoites in body of mosquito. Meiosis occurs just after zygote formation (zygotic meiosis). **B,** Sporozoites infect a human and reproduce asexually, first in liver cells and then in red blood cells. Malaria is spread by *Anopheles* mosquito, which ingests gametocytes along with human blood, then, when biting another victim, leaves sporozoites in new wound.

into **gametes,** and fertilization occurs. The zygote becomes a motile **ookinete,** which penetrates the stomach wall of the mosquito and becomes an **oocyst.** Within the oocyst, sporogony occurs, and thousands of **sporozoites** are produced. The oocyst ruptures, and the sporozoites migrate to the salivary glands, from which they are transferred to a human by a bite of the mosquito. Development in a mosquito requires 7 to 18 days but may be longer in cool weather.

Forty-one percent of the earth's people live in malarial regions. Elimination of mosquitoes and their breeding places by insecticides, drainage, and other methods has been effective in controlling malaria in some areas. However, difficulties in performing such activities in remote areas and areas suffering civil unrest, and acquisition of resistance to insecticides by mosquitoes and to antimalarial drugs by *Plasmodium* (especially *P. falciparum*), mean that malaria will be a serious disease of humans for a long time to come. Global estimates of deaths caused by malaria range from 700,000 to over 2 million, with 75% of such deaths being African children.

Other species of *Plasmodium* parasitize birds, reptiles, and mammals. Those of birds are transmitted chiefly by *Culex* mosquitoes.

A *disease* is any illness or disorder that can be recognized by a given set of signs and symptoms. *Epidemiology* is the study of all factors that influence transmission, geographic distribution, incidence, and prevalence of a disease. Epidemiology of parasitic diseases often involves poor sanitation and contamination of water or food with infectious stages. That is not the case with arthropod-borne diseases, such as malaria. Transmission and distribution of malaria depend on presence of a suitable *Anopheles* species, as well as its breeding, feeding, and resting habits. The climate (whether the mosquito can breed and feed throughout the year) is important, as are the prevalence of infected humans (especially asymptomatic individuals). It has nothing to do with improper waste disposal or poverty.

Classification of Protozoan Phyla (Unicellular Eukaryotes)

This classification primarily follows Hausmann and Hülsmann (1996) and is abridged from Roberts and Janovy (2005). With few exceptions, we are including only taxa of examples discussed in this chapter.

Much strong evidence indicates that phylum Sarcomastigophora and its subphyla are no longer tenable. Newer monographs consider amebas as belonging to several taxa with various affinities, not all yet determined. Organisms previously assigned to phylum Sarcomastigophora, subphylum Sarcodina, should be placed in at least two phyla, if not more. Nevertheless, amebas fall into a number of fairly recognizable morphological groups, which we will use for the convenience of readers and not assign such groups to specific taxonomic levels.

Phylum Chlorophyta (klor-of´i-ta) (Gr. *chlōros*, green, + *phyton*, plant). Unicellular and multicellular algae; photosynthetic pigments of chlorophyll *a* and *b*, reserve food is starch (characters in common with "higher" plants: bryophytes and vascular plants); all with biflagellated stages; flagella of equal length and smooth; mostly free-living photoautotrophs. Examples: *Chlamydomonas, Volvox*. Members of this phylum are placed in clade Viridiplantae.

Phylum Retortamonada (re-tor´ta-mo´nad-a) (L. *retorqueo*, to twist back, + *monas*, single, unit). Mitochondria and Golgi bodies lacking; three anterior and one recurrent (running toward posterior) flagellum lying in a groove; intestinal parasites or free living in anoxic environments. This phylum is divided into two clades with two genera in clade Retortamonads.

    Class Diplomonadea (di´plo-mon-a´de-a) (Gr. *diploos*, double, + L. *monas*, unit). One or two karyomastigonts (group of kinetosomes with a nucleus); individual mastigonts with one to four flagella; mitotic spindle within nucleus; cysts present; free living or parasitic. Nine genera of diplomonads comprise clade Diplomonads.

      Order Diplomonadida (di´plo-mon-a´di-da). Two karyomastigonts, each with four flagella, one recurrent; with variety of microtubular bands. Example: *Giardia*.

Phylum Axostylata (ak-so-sty-la´ta) (Gr. *axōn*, axle, + *stylos*, style, stake). With an axostyle made of microtubules.

    Class Parabasalea (par´a-bas-al´e-a) (Gr. *para*, beside, + *basis*, base). With very large Golgi bodies associated with karymastigont; up to thousands of flagella. *Trichomonas* and two other forms comprise clade Parabasalids.

      Order Trichomonadida (tri´ko-mon-a´di-da) (Gr. *trichos*, hair, + *monas*, unit). Typically at least some kinetosomes associated with rootlet filaments characteristic of trichomonads; parabasal body present; division spindle extranuclear; hydrogenosomes present; no sexual reproduction; true cysts rare; all parasitic. Examples: *Dientamoeba, Trichomonas*.

Phylum Euglenozoa (yu-glen-a-zo´a) (Gr. *eu-*, good, true, + *glēnē*, cavity, socket, + *zöon*, animal). With cortical microtubules; flagella often with paraxial rod (rodlike structure accompanying axoneme in flagellum); mitochondria with discoid cristae; nucleoli persist during mitosis. This phylum is synonomous with clade Euglenozoa.

    Subphylum Euglenida (yu-glen´i-da) With pellicular microtubules that stiffen pellicle.

      Class Euglenoidea (yu-glen-oyd´e-a) (Gr. *eu-*, good, true, + *glēnē*, cavity, socket, + *-ōideos*, form of, type of). Two heterokont flagella (flagella with different structures) arising from apical reservoir; some species with light-sensitive stigma and chloroplasts. Example: *Euglena*.

    Subphylum Kinetoplasta (ky-neet´o-plas´ta) (Gr. *kinētos*, to move, + *plastos*, molded, formed). With a unique mitochondrion containing a large disc of DNA; paraxial rod.

      Class Trypanosomatidea (try-pan´o-som-a-tid´e-a) (Gr. *trypanon*, a borer, + *sōma*, the body). One or two flagella arising from pocket; flagella typically with paraxial rod that parallels axoneme; single mitochondrion (nonfunctional in some forms) extending length of body as tube, hoop, or network of branching tubes, usually with single conspicuous

Parabasalids

The parabasalid clade contains some members of the phylum Axostylata. Members of this phylum have a stiffening rod composed of microtubules, the **axostyle,** that extends along the longitudinal axis of their body. Parabasalids, traditionally part of the class Parabasalea, possess a modified region of the Golgi apparatus called a parabasal body.

Much of the work on parabasalid structure has been done on species of *Trichomonas,* a disease-causing organism for humans and other animals. Some trichomonads (Figure 11.32) are of medical or veterinary importance. *Trichomonas vaginalis* infects the urogenital tract of humans and is sexually transmitted. It produces no symptoms in males but is the most common cause of vaginitis in females. **Pentatrichomonas hominis** lives in the cecum and colon of humans and *Trichomonas tenax* lives in the mouth; they apparently cause no disease. Other species of Trichomonadida are widely distributed through vertebrates of all classes and many invertebrates.

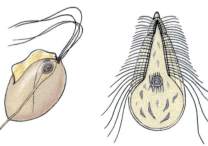
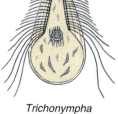
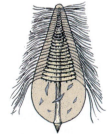

Trichomonas   *Trichonympha*   *Spirotrichonympha*

Figure 11.32

These three animals belong to the parabasalid clade. *Trichomonas vaginalis* is transmitted sexually and is a frequent cause of vaginitis in humans. *Trichonympha* and *Spirotrichonympha* are mutualistic symbionts in termites.

DNA-containing kinetoplast located near flagellar kinetosomes; Golgi body typically in region of flagellar pocket, not connected to kinetosomes and flagella; all parasitic. Examples: *Leishmania, Trypanosoma.*

Phylum Apicomplexa (ap´i-compleks´a) (L. *apex,* tip or summit, + *complex,* twisted around). Characteristic set of organelles (apical complex) associated with anterior end present in some developmental stages; cilia and flagella absent except for flagellated microgametes in some groups; cysts often present; all parasitic. This phylum is within clade Alveolata.

Class Gregarinea (gre-ga-ryn´e-a) (L. *gregarius,* belonging to a herd or flock). Mature gamonts (individuals that produce gametes) large, extracellular; gametes usually alike in shape and size; zygotes forming oocysts within gametocysts; parasites of digestive tract or body cavity of invertebrates; life cycle usually with one host. Examples: *Monocystis, Gregarina.*

Class Coccidea (kok-sid´e-a) (Gr. *kokkos,* kernel, grain). Mature gamonts small, typically intracellular; life cycle typically with merogony, gametogony, and sporogony; most species live inside vertebrates. Examples: *Cryptosporidium, Cyclospora, Eimeria, Toxoplasma, Plasmodium, Babesia.*

Phylum Ciliophora (sil-i-of´-or-a) (L. *cilium,* eyelash, + Gr. *phora,* bearing). Cilia or ciliary organelles in at least one stage of life cycle; two types of nuclei, with rare exception; binary fission across rows of cilia, budding and multiple fission also occur; sexuality involving conjugation, autogamy, and cytogamy; nutrition heterotrophic; contractile vacuole typically present; most species free living, but many commensal, some parasitic. (This is a very large group, now divided by the Society of Protozoologists classification into three classes and numerous orders and suborders. The classes are separated on the basis of technical characteristics of the ciliary patterns, especially around the cytostome, the development of the cytostome, and other characteristics.) Examples: *Paramecium, Colpoda, Tetrahymena, Balantidium, Stentor, Blepharisma, Epidinium, Euplotes, Vorticella, Carchesium,* *Trichodina, Podophrya, Ephelota.* This phylum is within clade Alveolata.

Phylum Dinoflagellata (dy´no-fla-jel-at´a) (Gr. *dinos,* whirling, + *flagellum,* little whip). Typically with two flagella, one transverse and one trailing; body usually grooved transversely and longitudinally, each groove containing a flagellum; chromoplasts usually yellow or dark brown, occasionally green or blue-green, bearing chlorophylls *a* and *c;* nucleus unique among eukaryotes in having chromosomes that lack or have low levels of histones; mitosis intranuclear; body form sometimes of spherical unicells, colonies, or simple filaments; sexual reproduction present; members free living, planktonic, parasitic, or mutualistic. Examples: *Zooxanthella, Ceratium, Noctiluca, Ptychodiscus.* This phylum is within clade Alveolata.

Amebas Although members of the former Sarcodina do not form a monophyletic group, we are considering them under this informal heading for the sake of descriptive simplicity. Amebas move by pseudopodia or locomotive protoplasmic flow without discrete pseudopodia; flagella, when present, usually restricted to developmental or other temporary stages; body naked or with external or internal test or skeleton; asexual reproduction by fission; sexuality, if present, associated with flagellated or, more rarely, ameboid gametes; most free living.

Rhizopodans Locomotion by lobopodia, filopodia (thin pseudopodia that often branch but do not rejoin), or by protoplasmic flow without production of discrete pseudopodia. Examples: *Amoeba, Entamoeba, Difflugia, Arcella, Chlamydophrys.* These amebas are divided among several clades.

Granuloreticulosans Locomotion by reticulopodia (thin pseudopodia that branch and often rejoin [anastomose]); includes foraminiferans. Examples: *Globigerina, Vertebralina.* Clade Granuloreticulosa contains these animals.

Actinopodans Locomotion by axopodia; includes radiolarians and heliozoans. Examples: *Actinophrys, Clathrulina.* These amebas are divided among several clades.

Amebas

Amebas are found in both fresh- and salt-water and in moist soils. Some are planktonic; some prefer a substratum. A few are parasitic. Most amebas reproduce by binary fission. Sporulation and budding occur in some.

Nutrition in amebas is holozoic; they ingest and digest liquid or solid foods. Most amebas are omnivorous, living on algae, bacteria, protozoa, rotifers, and other microscopic organisms. An ameba may ingest food at any part of its body surface merely by producing a pseudopodium to enclose the food (phagocytosis). The enclosed food particle, along with some environmental water, becomes a food vacuole, which is carried by the streaming movements of endoplasm. As digestion occurs within the vacuole by enzymatic action, water and digested materials pass into the cytoplasm. Undigested particles are eliminated through the cell membrane.

The shape of pseudopodia formed by each species of ameba has been used as a character for classification. In particular, the presence of axopodia (see p. 223), supported by axial rods of microtubules, was used to distinguish the actinopods from the other amebas, the nonactinopods. These descriptive terms are still in use.

Nonactinopod Amebas

Nonactinopod amebas may form lobopodia, filopodia, or rhizopodia (see p. 220). There are many species of rhizopod amebas; for example, the large *Chaos carolinense,* and the smaller amebas, *Amoeba verrucosa,* with short pseudopodia, and *A. radiosa,* with many slender pseudopodia. Rhizopodia are seen also in *Amoeba proteus,* the most commonly studied species of ameba.

Amoeba proteus lives in slow streams and ponds of clear water, often in shallow water on aquatic vegetation or on sides of ledges. They are rarely found free in water, for they require a substratum on which to crawl. They have an irregular shape because lobopodia may be formed at any point on their bodies.

They are colorless and about 250 to 600 µm in greatest diameter. They are bounded only by a plasma membrane. Ectoplasm and endoplasm are prominent. Organelles such as **nucleus, contractile vacuole, food vacuoles,** and small **vesicles** can be observed easily with a light microscope. Amebas live on algae, protozoa, rotifers, and even other amebas, upon which they feed by phagocytosis. An ameba can live for many days without food but decreases in volume during starvation. The time necessary for digestion by a food vacuole varies with the kind of food but is usually around 15 to 30 hours. When an ameba reaches full size, it divides by binary fission with typical mitosis.

How Do We Classify Amebas?   Classification of the amebas is in flux as researchers try to combine morphological characters, like the shape of pseudopodia or of mitochondrial cristae, with molecular data, such as protein sequences. A taxonomic group named on the basis of one feature may not be coincident with a group formed on the basis of another feature. Still, some patterns seem to emerge from combined data sets for two non-actinopod groups.

One group of amebas, alternately called Heterolobosea, amoeboflagellates, or schizopyrenids, has both ameboid and flagellated stages in the life cycle. The group is exemplified by *Naegleria fowleri,* a free-living organism from hot pools that causes amebic meningitis (primary amebic meningoencephalitis) if it enters humans. Organisms that cause human disease are frequently used as representatives in phylogenetic studies because they are more readily available than their wild counterparts.

Members of the Heterolobosea possess discoid mitochondrial cristae, as do members of the Euglenozoa, so the name Discicristata was proposed for the group composed of these two taxa. A close relationship between these taxa also emerged from protein sequence comparisons. However, a new group name may be required by the apparent presence of discoid mitochondrial cristae in organisms outside these two taxa.

Amebas forming lobopodia are grouped together as the Lobosa, exemplified by *Acanthamoebae* (Figure 11.13). This group also includes members of the Entamoebidae. They have branched mitochondrial cristae, a feature shared with slime molds (Mycetozoa). On the basis of branched or ramifying mitochondrial cristae, Lobosa and Mycetozoa were united as the Ramicristate clade. Protein sequence comparisons also united Mycetozoa and Lobosa, but workers called the combined group Amoebozoa. The protein sequence comparisons place the Amoebozoa/Ramicristates as the sister taxon to the Opisthokonta.

Entamoebidae   The entozoic amebas, those living inside humans or other animals, are members of clade Lobosa. They have branched pseudopodia, making them rhizopod amebas. Like several other protozoan taxa, they lack mitochondria.

There are many entozoic amebas, most of which live in intestines of humans or other animals. *Entamoeba histolytica* is the most important rhizopodan parasite of humans. It lives in the large intestine and on occasion can invade the intestinal wall by secreting enzymes that attack the intestinal lining. If this occurs, a serious and sometimes fatal amebic dysentery may result. The organisms may be carried by the blood to the liver and other organs and cause abscesses there. Many infected persons show few or no symptoms but are carriers, passing cysts in their feces. Diagnosis is complicated by the existence of a nonpathogenic species, *E. dispar,* which is morphologically identical to *E. histolytica.* Infection is spread by contaminated water or food containing cysts. *Entamoeba histolytica* is found throughout the world, but clinical amebiasis is most prevalent in tropical and subtropical areas.

Other species of *Entamoeba* found in humans are *E. coli* in the intestine and *E. gingivalis* in the mouth. Neither of these species is known to cause disease.

Another group of entozoic amebas is the Endamoebae. Examples include *Endamoeba blattae,* an endocommensal in the gut of cockroaches, as well as related species living in termites. Some evidence suggests that these animals are not closely related to the Entamoebae.

Granuloreticulosa   In this clade of amebas, slender pseudopodia extend through openings in the test, then branch and run together to form a protoplasmic net **(reticulopodia)** in which they ensnare their prey. Here captured prey is digested, and digested products are carried into the interior by flowing protoplasm.

Most reticulopods are foraminiferans, or forams. They are an ancient group of shelled amebas found in all oceans, with a few in fresh and brackish water. Most foraminiferans live on the ocean floor in incredible numbers, having perhaps the largest biomass of any animal group on earth. Their tests are of numerous types (Figures 11.6 and 11.33). Most tests are many chambered and are made of calcium carbonate, although they sometimes use silica, silt, and other foreign materials. Life cycles of foraminiferans are

A   **B**

Figure 11.33

A, Living foraminiferan, showing thin pseudopodia extending from test. **B,** Test of foraminiferan, *Vertebralina striata.* Foraminiferans are ameboid marine protozoa that secrete a calcareous, many-chambered test in which to live and then extrude protoplasm through pores to form a layer over the outside. The animal begins with one chamber, and as it grows, it secretes a succession of new and larger chambers, continuing this process throughout life. Many foraminiferans are planktonic, and when they die, their shells are added to the ooze on the ocean's bottom.

complex, for they have multiple fission and alternation of haploid and diploid generations (intermediary meiosis).

Foraminiferans have existed since Precambrian times and have left excellent fossil records. In many instances their hard shells are preserved unaltered. Many extinct species closely resemble those of the present day. They were especially abundant during the Cretaceous and Tertiary periods. Some were among the largest protozoa that have ever existed, measuring up to 100 mm (about 4 in) or more in diameter.

For untold millions of years tests of dead foraminiferans have been sinking to the bottom of the ocean, building up a characteristic ooze rich in lime and silica. About one-third of the seafloor is covered with shells of the genus *Globigerina*. This ooze is especially abundant in the Atlantic Ocean.

Of equal interest and of greater practical importance are the limestone and chalk deposits that were laid down by the accumulation of foraminiferans when sea covered what is now land. Later, through a rise in the ocean floor and other geological changes, this sedimentary rock emerged as dry land. The chalk deposits of many areas of England, including the White Cliffs of Dover, formed this way. The great pyramids of Egypt were made from stone quarried from limestone beds that were formed by a very large foraminiferan population that flourished during the early Tertiary period.

Since fossil foraminiferans can be found in well drillings, their identification is often important to oil geologists for identifying rock strata.

Actinopod Amebas

This polyphyletic group of amebas has axopod pseudopodia (Figures 11.7 and 11.11). The descriptive names heliozoan and radiolarian apply to some of these amebas, but taxa formerly placed in each group are now separated taxonomically, with heliozoans divided among five clades and radiolarians divided among three.

The name heliozoan refers to freshwater amebas with or without tests (see Figure 11.7). Examples are *Actinosphaerium*, which is about 1 mm in diameter and can be seen with the unassisted eye, and *Actinophrys* (see Figure 11.7), only 50 μm in diameter; neither has a test. *Clathrulina* (see Figure 11.7) secretes a latticed test.

The term radiolarian refers to marine testate amebas with intricate specialized skeletons of great beauty (Figure 11.34). The oldest known protozoa are marine actinopodans. Radiolarians are nearly all pelagic (live in open water). Most of them are planktonic in shallow water, although some live in deep water. The body is divided by a central capsule that separates inner and outer zones of cytoplasm. The central capsule, which may be spherical, ovoid, or branched, is perforated to allow cytoplasmic continuity. The skeleton is made of silica, strontium sulfate, or a combination of silica and organic matter and usually has a radial arrangement of spines that extend through the capsule from the center of the body. At the surface a shell may be fused with the spines. Around the capsule is a frothy mass of cytoplasm from which axopodia arise (p. 223). These are sticky to catch prey, which are carried by the streaming protoplasm to the central capsule to be digested. The ectoplasm on one side of the axial rod moves outward, or toward the tip, while on the other side it moves inward, or toward the test.

Radiolarians may have one or many nuclei. Their life history is not completely known, but binary fission, budding, and sporulation have been observed in them.

Radiolarians are among the oldest known protozoa because their relatively insoluble siliceous shells make durable fossils. They are usually found at great depths (4600 to 6100 meters), mainly in the Pacific and Indian oceans. Radiolarian ooze probably covers about 5 to 8 million square kilometers to a thickness of 700 to 4000 m. Under certain conditions, radiolarian ooze forms rocks (chert). Many fossil radiolarians occur in Tertiary rocks of California, and like foraminiferans, the identification of particular species is important to oil geologists for determining the age of rock strata.

PHYLOGENY AND ADAPTIVE DIVERSIFICATION

Phylogeny

Molecular evidence has almost completely revised our phylogeny of unicellular eukaryotes. It now seems that the ancestral eukaryote diversified into many morphologically distinct clades, although the branching order for diversification is still poorly understood. Many characters for phylogenetic analyses come from structural features of protozoan organelles. However, one must be able to distinguish an ancient organelle, formed through symbioses among prokaryotes, from a more recently acquired organelle formed through secondary symbioses among eukaryotes. The absence of an organelle such as a mitochondrion can be informative, but only if we have a way to distinguish whether mitochondria were present and later lost, or never present at all. Detailed studies of nuclear genomes and gene products—for example, mitochondrial enzymes produced by the nuclear genes—can distinguish between the primary absence of a structure and its

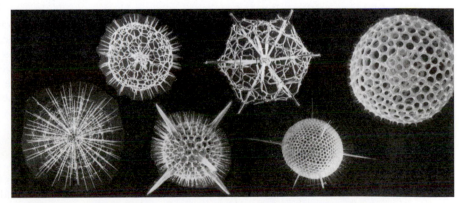

Figure 11.34

Types of radiolarian tests. In his study of these beautiful forms collected on the famous *Challenger* expedition of 1872 to 1876, Haeckel proposed our present concepts of symmetry.

secondary loss. It is now assumed that all amitochondriate protozoans had ancestors with mitochondria.

Plastids were another variable protozoan character that held promise for elucidating phylogenetic relationships. However, the presence of particular plastids in a wide variety of seemingly unrelated single-celled and multicellular eukaryotes created confusion until it eventually became clear that the primary endosymbiotic event with a cyanobacterium was followed by secondary and tertiary endosymbiotic events that transferred plastids among eukaryotic lineages. A disentangled pathway of endosymbiont transfers, in combination with results from new molecular data sets, suggests that many eukaryotic lineages now can be combined into a few eukaryotic supergroups. Members of some supergroups are shown in Figure 11.1: the stramenopiles and the alveolates are combined as the supergroup Chromalveolates; Opisthokonta and Amoebozoa are combined as Unikonts; the name refers to the single flagellum on flagellated cells. Two more supergroups, not shown in Figure 11.1, are created by combining taxa we have discussed with some groups not included in this text: Viridiplantae is combined with the red algal clade and glaucophytes to form the supergroup Plantae; Granuloreticulosans are joined with radiolarians and other organisms called cercozoans in the supergroup Rhizaria. There is weak support for a fifth supergroup, the Excavates, whose members share an unusual feeding groove. If this group is validated by further research, it will include the five leftmost clades in Figure 11.1 (Retortamonads, Parabasalids, Diplomonads, Heterolobosea, and Euglenozoa) along with other taxa not discussed here. Assuming that these supergroups survive further scrutiny, the next step is to determine the branching order among them. One hypothesis already under discussion is that the Unikonts are the sister taxon to all the other groups combined.

Adaptive Diversification

We have described some of the wide range of adaptations of protozoan groups in this chapter. Amebas range from bottom-dwelling, naked species to planktonic forms such as foraminiferans and radiolarians with beautiful, intricate tests. There are many symbiotic species of amebas. Flagellated forms likewise show adaptations for a similarly wide range of habitats, with the added variation of photosynthetic ability in many groups.

Within a single-cell body plan, the division of labor and specialization of organelles are carried furthest by ciliates. These have become the most complex of all protozoa. Specializations for intracellular parasitism have been adopted by Apicomplexa.

SUMMARY

"Animal-like," single-celled organisms were formerly assigned to the phylum Protozoa. It is now recognized that the "phylum" was composed of numerous taxa that do not form a monophyletic group. We use the terms *protozoa* and *protozoan* informally to refer to all these highly diverse organisms. They demonstrate the great adaptive potential of the basic body plan: a single eukaryotic cell. They occupy a vast array of niches and habitats. Many species have complex and specialized organelles.

All protozoa have one or more nuclei, and these often appear vesicular with light microscopy. Macronuclei of ciliates are compact. Nucleoli are often evident in the nuclei. Many protozoa have organelles similar to those found in metazoan cells.

Pseudopodial or ameboid movement is a locomotory and food-gathering mechanism in protozoa and plays a vital role as a defense mechanism in metazoa. It is accomplished by assembly of actin subunits into filaments and interaction of actin filaments with actin-binding proteins and myosin, and it requires expenditure of energy from ATP. Ciliary movement is likewise important in both protozoa and metazoa. Currently, the most widely accepted mechanism to account for ciliary movement is the sliding-microtubule hypothesis.

Various protozoa feed by holophytic, holozoic, or saprozoic means. The excess water that enters their bodies is expelled by contractile vacuoles (water-expulsion vesicles). Respiration and waste elimination are through the body surface. Protozoa can reproduce asexually by binary fission, multiple fission, and budding; sexual processes are common. Cyst formation to withstand adverse environmental conditions is an important adaptation in many protozoa.

The evolution of a eukaryotic cell was followed by diversification of lineages to form morphologically disparate clades, some of which contain both unicellular and multicellular forms. Major taxa discussed are identified partly on the basis of molecular characters and contain subsets of animals from traditional phyla. Members of several phyla have photoautotrophic species, including Chlorophyta, Euglenozoa, and Dinoflagellata. Some of these are very important planktonic organisms. Euglenozoa includes many nonphotosynthetic species, and some of these cause serious diseases of humans, such as African sleeping sickness and Chagas' disease. Apicomplexa are all parasitic, including *Plasmodium,* which causes malaria. Ciliophora move by means of cilia or ciliary organelles. They are a large and diverse group, and many are complex in structure. Amebas move by pseudopodia and are now assigned to a number of phyla.

REVIEW QUESTIONS

1. Explain why a protozoan may be very complex, even though it is composed of only one cell.
2. Distinguish among the following protozoan groups: Euglenozoa, Apicomplexa, Ciliophora, Dinoflagellata.
3. Distinguish vesicular and compact nuclei.
4. Explain the transitions of endoplasm and ectoplasm in ameboid movement. What is a current hypothesis regarding the role of actin in ameboid movement?
5. Distinguish lobopodia, filipodia, reticulopodia, and axopodia.

6. Contrast the structure of an axoneme of a cilium with that of a kinetosome.
7. What is the sliding-microtubule hypothesis?
8. Explain how protozoa eat, digest their food, osmoregulate, and respire.
9. Distinguish the following: binary fission, budding, multiple fission, and sexual and asexual reproduction.
10. What is the survival value of encystment?
11. Contrast and give an example of autotrophic and heterotrophic protozoa.
12. Name three kinds of amebas, and tell where they are found (their habitats).
13. Outline the general life cycle of malarial organisms. What explains the resurgence of malaria in recent years?
14. What is the public-health importance of *Toxoplasma,* and how do humans become infected with it? What is the public health importance of *Cryptosporidium* and *Cyclospora?*
15. Define the following with reference to ciliates: macronucleus, micronucleus, pellicle, undulating membrane, cirri, infraciliature, trichocysts, conjugation.
16. Outline the steps in conjugation of ciliates.
17. Explain why protozoans are neither plants nor animals.
18. Distinguish primary endosymbiogenesis from secondary endosymbiogenesis.

SELECTED REFERENCES

Allen, R. D. 1987. The microtubule as an intracellular engine. Sci. Am. **256:**42–49 (Feb.). *The action of microtubules accounts for the movement of chromosomes in mitosis and pseudopodial movement of filopodia and reticulopodia.*

Baldauf, S. L., A. J. Roger, I. Senk-Siefert, and W. F. Doolittle. 2000. A kingdom-level phylogeny of eukaryotes based on combined protein data. Science **290:**972–976. *They contend that combining sequence data for genes encoding several proteins indicates that there are 15 kingdoms of organisms.*

Burkholder, J. M. 2002. *Pfiesteria:* the toxic *Pfiesteria* complex. In G. Bitton (ed.), Encyclopedia of environmental microbiology pp. 2431–2447. New York, Wiley Publishers. *A nice summary of recent work on habitat and life cycles of* Pfiesteria, *including its effects on fish, shellfish, and humans.*

Cavalier-Smith, T. 1999. Principles of protein and lipid targeting in secondary symbiogenesis: euglenoid, dinoflagellate, and sporozoan plastid origins and the eukaryote family tree. J. Euk. Microbiol. **46:**347–366. *Many organisms are the products of secondary symbiogenesis (a eukaryote is consumed by another eukaryote, both products of primary symbiogenesis, and symbiont becomes an organelle), but tertiary symbiogenesis also has occurred (product of secondary symbiogenesis itself becomes a symbiont . . . and organelle).*

Harper, J. T., E. Waanders, and P. J. Keeling. 2005. On the monophyly of chromalveolates using a six-protein phylogeny of eukaryotes. Int. J. Syst. Evol. Microbiol. **55:**487–496. *Outlines support for a large clade uniting stramenopiles and alveolates.*

Harrison, G. 1978. Mosquitoes, malaria, and man: a history of the hostilities since 1880. New York, E. P. Dutton. *A fascinating story, well told.*

Hausmann, K., and N. Hülsmann. 1996. Protozoology. New York, Thieme Medical Publishers, Inc. *This was the most up-to-date, comprehensive treatment available before the release of Lee et al. (2000).*

Keeling, P. J. 2004. Diversity and evolutionary history of plastids and their hosts. Am. J. Bot. **91:**1481–1493. *A lucid description of plastid evolution and the evidence for primary, secondary, and tertiary endosymbioses.*

Keeling, P. J., G. Burger, D. J. Durnford, B. F. Lang, R. W. Lee, R. E. Pearlman, A. J. Roger, and M. W. Gray. 2005. The tree of eukaryotes. Trends Ecol. Evol. **20:**670–676. *Support for five eukaryotic supergroups is presented.*

Keeling, P. J., M. A. Luker, and J. D. Palmer. 2000. Evidence from beta-tubulin phylogeny that microsporidia evolved from within the fungi. Mol. Biol. Evol. **17:**23–31. *Microsporidians are shown to be a fungal subgroup, not a separate eukaryotic lineage.*

Lee, J. J., G. F. Leedale, and P. Bradbury (eds). 2000. An illustrated guide to the protozoa, ed. 2, 1432 pp., 2 vols. Lawrence, Kansas, Society of Protozoologists. *This long-awaited guide appeared in 2002. It is an essential reference for students of protozoa.*

Margulis, L., and K. V. Schwartz. 1998. Five kingdoms: an illustrated guide to life on earth, ed. 3. New York, W. H. Freeman and Company. *Although classification schemes in this book are not current, it has good descriptions of many taxa, clear descriptions of basic morphology, and useful photographs and drawings.*

Patterson, D. J. 1999. The diversity of eukaryotes. Amer. Nat. **154** (supplement):S96–S124. *Patterson provides morphological descriptions and synapomorphies for many clades containing protozoans.*

Roberts, L. S., and J. J. Janovy, Jr. 2005. Foundations of parasitology, ed. 7. Dubuque, Iowa, McGraw-Hill Higher Education. *Up-to-date and readable information on parasitic protozoa.*

Roger, A. J. 1999. Reconstructing early events in eukaryotic evolution. Amer. Nat. **154** (supplement):S146–S163. *Methods in determining whether absence of mitochondria is primary or due to a secondary loss are discussed here.*

Sleigh, M. A. 1989. Protozoa and other protists. London, Edward Arnold. *Extensively updated version of the author's* The biology of protozoa.

Steenkamp, E. T., J. Wright, and S. L. Baldauf. 2006. The protistan origins of animals and fungi. Mol. Biol. Evol. **23:**93–106. *Opisthokonta is a well-supported clade whose members share a short amino acid sequence in elongation factor 1-alpha.*

ONLINE LEARNING CENTER

Visit www.mhhe.com/hickmanipz14e for chapter quizzing, key term flash cards, web links and more!

12

A Caribbean demosponge, *Aplysina fistularis.*

Sponges and Placozoans

- PHYLUM PORIFERA: SPONGES
- PHYLUM PLACOZOA

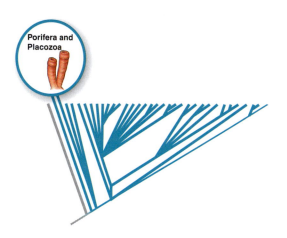

The Origins of Multicellularity

Sponges are the simplest multicellular animals. Because cells are the elementary units of life, organisms larger than unicellular organisms arose as aggregates of such building units. Nature experimented with producing larger organisms without cellular differentiation—certain large, single-celled marine algae, for example—but such examples are rarities. There are many advantages to multicellularity as opposed to simply increasing the mass of a single cell. Because cell surfaces exchange molecules with the environment, dividing a mass into smaller units greatly increases the surface area available for metabolic activities. It is impossible to maintain a workable surface-to-mass ratio by simply increasing the size of a single-celled

organism. Thus multicellularity is a highly adaptive path toward increasing body size.

Strangely, while sponges are multicellular, their organization is quite distinct from other metazoans. A sponge body is an assemblage of cells embedded in an extracellular matrix and supported by a skeleton of minute needlelike spicules and protein. Because sponges neither look nor behave like other animals, they were not completely accepted as animals by zoologists until well into the nineteenth century. Nonetheless, molecular evidence demonstrates that sponges are phylogenetically grouped with other metazoa.

ORIGIN OF METAZOA

Evolution of a eukaryotic cell was followed by diversification into many lineages (see Figure 11.1). Modern descendants of these lineages include unicellular protozoans (see Chapter 11), as well as colonial and multicellular plants and animals. We collectively call multicellular animals **metazoans.** Metazoans fall within the opisthokont clade (see Figure 11.18) along with fungi, choanoflagellates, and a few other groups. There is still debate over which taxon forms the sister group to the metazoans, but several phylogenies using molecular characters place choanoflagellates in this position.

Choanoflagellates are solitary or colonial aquatic eukaryotes, where each cell carries a flagellum surrounded by a collar of microvilli. Beating of the flagellum draws water into the collar, where microvilli collect tiny particles, typically bacteria. Many choanoflagellates are sessile and attached to hard surfaces, although one species attaches to floating diatom colonies allowing it to feed in midwater, even though it does not swim. Swimming does occur in *Proterospongia,* an unusual colonial form that propels itself through the water using flagella.

Choanoflagellate cells are noteworthy because they strongly resemble sponge feeding cells called choanocytes (see p. 248). It is very interesting to find a collared cell used in filter feeding in a colonial protozoan and in a sponge, whose ancestral lineage represents an early divergence from the lineage of all other multicellular animals (see cladogram on inside front cover). Was a sponge choanocyte inherited from a common ancestor with choanoflagellates? Arguments against this hypothesis include the observation that choanocytes occur only in adult sponges and are not part of the early developmental sequence. Instead, flagellated cells without collars develop into choanocytes after larval metamorphosis. Collar cells also occur in certain corals and some echinoderms, so if they were part of the earliest metazoan lineage, this morphology has been lost or suppressed in most taxa. Despite these objections, there is another clear link between choanoflagellates and metazoans: proteins used by colonial choanoflagellates for cell communication and adhesion are homologous to those that metazoans use in cell-to-cell signaling.[1] The morphology of the first metazoan is still a subject of debate.

In one approach to metazoan origins, researchers hypothesize transitional forms between presumed protozoan ancestors and simple metazoans. Clearly, our choice of particular protozoans as starting points, as well as particular metazoans we select as endpoints, will determine the hypothesized steps in evolution. Of two well-known evolutionary schemes, one starts with a multinucleate ciliate protozoan, and the other with a colony of flagellate protozoans similar to *Volvox,* but lacking photosynthetic abilities.

Proponents of the **syncytical ciliate hypothesis** hypothesize that metazoans arose from an ancestor shared with the single-celled ciliates. The common ancestor of metazoans acquired multiple nuclei within a single cell membrane and later became compartmentalized into the multicellular condition. It is assumed that the body form of the ancestor resembled that of modern ciliates and thus tended toward bilateral symmetry. Therefore the earliest metazoans would have been bilateral and similar to some extant flatworms. There are several objections to this hypothesis. It ignores embryology of the flatworms in which nothing similar to cellularization occurs; it does not explain the presence of flagellated sperm in metazoans; and, perhaps more important, it implies that the radial symmetry of cnidarians is derived from a primary bilateral symmetry.

The **colonial flagellate hypothesis**—first proposed by Haeckel in 1874—is the classical scheme, which, with various revisions, still has many followers. According to this hypothesis, metazoans descended from ancestors characterized by a hollow, spherical, colony of flagellated cells. Individual cells within the colony became differentiated for specific functional roles (reproductive cells, nerve cells, somatic cells, and so on), thus subordinating cellular independence to welfare of the colony as a whole. The colonial ancestral form was at first radially symmetrical and reminiscent of a blastula stage of development. This hypothetical ancestor was called a blastaea. Drawing on the developmental sequence of extant animals as a model, Haeckel hypothesized that ancestral forms similar to a gastrula may have existed. These ancestors were called gastraea. Cnidarians, with their radial symmetry, could have evolved from this form.

Most hypotheses for metazoan origins assume that all metazoans form a **monophyletic** group. Suggestions that sponges, cnidarians, and ctenophores evolved separately from triploblastic metazoans are not supported by molecular evidence. Data from comparisons of small-subunit ribosomal RNA sequences and from similarities in complex biochemical pathways across metazoans indicate that metazoans did not have a **polyphyletic origin.**

Molecular evidence does not support the syncytial ciliate hypothesis because ciliates are placed in a clade distinct from the opisthokonts. The placement of metazoans in Opisthokonta with choanoflagellates, such as *Codonosiga* and *Proterospongia* (see Figure 11.18), provides general support for the colonial flagellate hypothesis.

However, recent approaches to the problem of metazoan origins pay less attention to morphological transitions in favor of characterizing the regulatory components of the first metazoan genome. As already mentioned, the genetic instructions for cell-signaling proteins predate the transition from unicellular to multicellular forms. What other cell transmitters or morphogens did the first metazoan possess? One way to discover the answer is to compare the genomes or proteomes of simple metazoans, such as sponges, with those of more complex taxa.

Adult sponges have very simple bodies; they are aggregations of several different cell types, including choanocytes, held together by an extracellular matrix. A sponge body is not symmetrical; it has neither a mouth nor a digestive tract. Thus, we expect it to have a simple genetic architecture, perhaps reminiscent of the first animals. Surprisingly, the sponge genome contains many elements that code for parts of the

[1]King, N., C. T. Hittinger, and S. B. Carroll. 2003. Science **301**:361–363.

regulatory pathways of more complex metazoans, including proteins involved in spatial patterning, like those that specify an anterior and posterior pole in the larva. This discovery has led some biologists to hypothesize that modern sponges are less morphologically complex than were their ancestors.

Similar hypotheses have been applied to another phylum of simple animals discussed in this chapter. Members of Placozoa (see p. 257) have the smallest nuclear genome, and the largest mitochondrial genome, of any known metazoan. Their circular mitochondrial genome shares some features with metazoan outgroups, including chytrid fungi and choanoflagellates, but also has derived metazoan features.

It is worth remembering that organisms we see now are products of millions of years of evolution since the ancestors of their clades diverged from those of other metazoans. We expect to find genes unique to each phylum, along with those homologous to genes in metazoan outgroups, and to genes shared with more complex metazoans. Gene functions may have changed as new morphologies evolved, and there is much yet to be understood about the modern forms of the two phyla we discuss here. The body of a placozoan is at least as puzzling as that of a sponge; one cannot find heads or tails in either of them.

PHYLUM PORIFERA: SPONGES

Most animals move to search for food, but a sessile sponge (Figure 12.1) draws food and water into its body instead. The entrance of water through myriads of tiny pores is reflected in the phylum name, Porifera (po-rif'-er-a) (L. *porus,* pore, + *fera,* bearing). The sponge uses a flagellated "collar cell," the **choanocyte,** to move water (Figure 12.2). The beating of many tiny flagella, one per choanocyte, draws water past each cell, bringing in food and oxygen, as well as carrying away wastes. The sponge body is designed as an efficient aquatic filter for removing suspended particles from the surrounding water.

Most of the approximately 15,000 sponge species are marine; a few exist in **brackish** water, and some 150 species live in

Characteristics of Phylum Porifera

1. Multicellular; body an aggregation of several types of cells differentiated for various functions, some of which are organized into **incipient tissues** of a low level of integration
2. Body with pores (ostia), canals, and chambers that form a unique system of **water currents** on which sponges depend for food and oxygen
3. Mostly marine; all aquatic
4. Radial symmetry or none
5. Outer surface of flat pinacocytes; most interior surfaces lined with flagellated collar cells (choanocytes) that create water currents; a gelatinous protein matrix called mesohyl contains amebocytes of various types and skeletal elements
6. Skeletal structure of fibrillar collagen (a protein) and calcareous or siliceous crystalline spicules, often combined with variously modified collagen (spongin)
7. No organs or true tissues; digestion intracellular; excretion and respiration by diffusion
8. Reactions to stimuli apparently local and independent in cellular sponges, but electrical signals in syncytial glass sponges; nervous system probably absent
9. All adults sessile and attached to substratum
10. Asexual reproduction by buds or gemmules and sexual reproduction by eggs and sperm; free-swimming flagellated larvae in most

freshwater. Marine sponges are abundant in all seas and at all depths. Sponges vary in size from a few millimeters to the great loggerhead sponges, which may exceed 2 m in diameter. Many sponge species are brightly colored because of pigments in their dermal cells. Red, yellow, orange, green, and purple sponges are not uncommon.

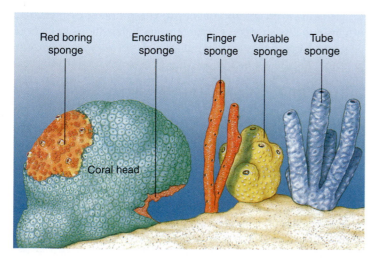

Figure 12.1

Some growth habits and forms of sponges.

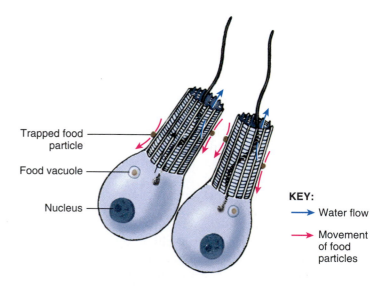

Figure 12.2

Sponge choanocytes have a collar of microvilli surrounding a flagellum. Beating of the flagellum draws water through the collar (blue arrows) where food is trapped on microvilli (red arrows).

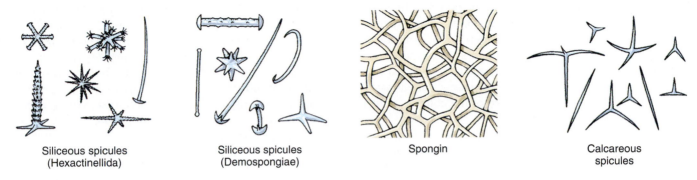

Figure 12.3
Diverse forms of spicules, many amazingly complex and beautiful, support a sponge body. Spongin fibers provide support in some sponges.

Although their embryos are free swimming, adult sponges are always attached, usually to rocks, shells, corals, or other submerged objects. Some bore holes into shells or rocks; others even grow on sand or mud. Some sponges, including the simplest, appear radially symmetrical, but many are quite irregular in shape. Some stand erect, some are branched or lobed, and others are low, even encrusting, in form (Figure 12.1). Their growth patterns often depend on shape of the substratum, direction and speed of water currents, and availability of space, so that the same species may differ markedly in appearance under different environmental conditions. Sponges in calm waters may grow taller and straighter than those in rapidly moving waters.

Many animals such as crabs, nudibranchs, mites, bryozoans, and fishes live as commensals or parasites in or on sponges. Larger sponges particularly tend to harbor a great variety of invertebrate **commensals.** Sponges also grow on many other living animals, such as molluscs, barnacles, brachiopods, corals, or hydroids. Some crabs attach pieces of sponge to their carapace for camouflage and for protection against predators. Although some reef fishes do graze on shallow-water sponges, most potential predators find sampling a sponge quite unpleasant. This antipredator effect is due to the sponge's often-noxious odor and elaborate skeletal framework.

The skeletal framework of a sponge can be fibrous and/or rigid. When present, the rigid skeleton consists of calcareous or siliceous support structures called **spicules** (Figure 12.3). The fibrous part of the skeleton comes from collagen fibrils in the intercellular matrix of all sponges. One form of collagen is traditionally known as **spongin** (Figure 12.3). Collagen comes in several types differing in chemical composition and form (for example, fibers, filaments, or masses surrounding spicules).

Sponges are an ancient group, with an abundant fossil record extending back to the early Cambrian period and even, according to some claims, the Precambrian. Classification is based on spicule form and chemical composition. Living poriferans traditionally have been assigned to three classes: Calcarea, Hexactinellida, and Demospongiae (Figure 12.4). Members of Calcarea typically have spicules of crystalline calcium carbonate with one, three, or four rays. Hexactinellids are glass sponges with six-rayed siliceous spicules, where the six

rays are arranged in three planes at right angles to each other. Members of Demospongiae have a skeleton of siliceous spicules, or spongin, or both. A fourth class (Sclerospongiae) was erected to contain sponges with a massive calcareous skeleton and siliceous spicules. Some zoologists maintain that known species of sclerosponges can be placed in the traditional classes of sponges (Calcarea and Demospongiae); thus we do not need a new class.

Form and Function
Sponges feed primarily by collecting suspended particles from water pumped through internal canal systems. Water enters

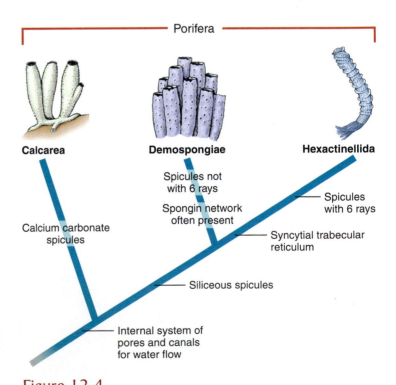

Figure 12.4
Cladogram depicting evolutionary relationships among the three classes of sponges with living representatives.

canals through a multitude of tiny incurrent pores in the outer layer of cells, a **pinacoderm.** Incurrent pores, called **dermal ostia** (Figure 12.5), have an average diameter of 50 μm. Inside the body, water is directed past the choanocytes, where food particles are collected on the choanocyte collar (Figure 12.2). The collar is made of many fingerlike projections, called microvilli, spaced about 0.1 μm apart. The use of the collar as a filter is one form of **suspension feeding.**

Sponges nonselectively consume food particles (bits of detritus, planktonic organisms and bacteria) sized between 0.1 μm and 50 μm. The smallest particles, accounting for about 80% of the particulate organic carbon, are taken into choanocytes by **phagocytosis.** Choanocytes may acquire protein molecules by **pinocytosis.** Two other cell types, **pinacocytes** and **archaeocytes,**

play a role in sponge feeding (see p. 253). Sponges may also absorb dissolved nutrients from the water.

The capture of food depends on the movement of water through the body. There are three main designs for the sponge body differing in the placement of the choanocytes. In the simplest **asconoid** system, the choanocytes lie in a large chamber called the **spongocoel;** in the **syconoid** system, the choanocytes lie in canals; and, in the **leuconoid** system, the choanocytes lie in distinct chambers (Figure 12.5). These three designs demonstrate an increase in complexity and efficiency of the water-pumping system, but they do not imply an evolutionary sequence. The leuconoid grade of construction is of clear adaptive value; it has the highest proportion of flagellated surface area for a given volume of cell tissue, so it efficiently meets

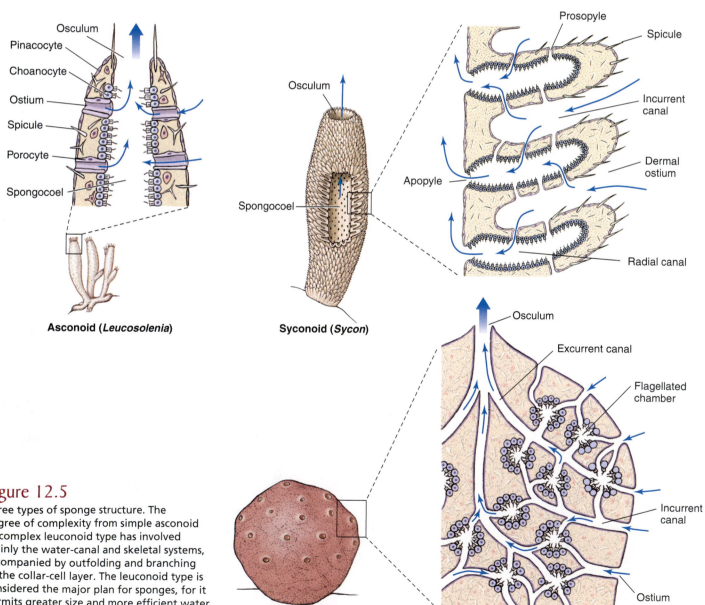

Asconoid (*Leucosolenia*)

Syconoid (*Sycon*)

Leuconoid (*Euspongia*)

Figure 12.5

Three types of sponge structure. The degree of complexity from simple asconoid to complex leuconoid type has involved mainly the water-canal and skeletal systems, accompanied by outfolding and branching of the collar-cell layer. The leuconoid type is considered the major plan for sponges, for it permits greater size and more efficient water circulation.

food demands. This leuconoid grade has evolved independently many times in sponges.

Types of Canal Systems

Asconoids Asconoid sponges have the simplest organization. Water is drawn into the sponge through microscopic dermal pores by the beating of large numbers of flagella on the choanocytes. These choanocytes line the internal cavity known as the spongocoel. As the choanocytes filter the water and extract food particles from it, used water is expelled through a single large osculum (Figure 12.5). This design has distinct limitations because choanocytes line the spongocoel and can collect food only from water directly adjacent to the spongocoel wall. Were the spongocoel to be large, most of the water and food in its central cavity would be inaccessible to choanocytes. Thus, asconoid sponges are small and tube-shaped. As an example, examine *Leucosolenia* (Gr. *leukos,* white, + *solen,* pipe) where slender, tubular individuals grow in groups attached by a common stolon, or stem, to objects in shallow seawater (Figure 12.5). *Clathrina* (L. *clathri,* latticework), another asconoid, has bright yellow, intertwined tubes (Figure 12.6). Asconoids are found only in the class Calcarea.

Syconoids Syconoid sponges look somewhat like larger editions of asconoids. They have a tubular body and single osculum, but the body wall, which is really the spongocoel lining, is thicker and more complex than that of asconoids. The lining has been folded outward to make choanocyte-lined canals (Figure 12.5). Folding the body wall into canals increases the surface area of the wall and thus increases the surface area covered by choanocytes. The canals are of small diameter compared with an asconoid spongocoel, so most of the water in a canal is accessible to choanocytes.

Water enters the syconoid body through dermal ostia that lead into incurrent canals. It then filters through tiny openings,

or **prosopyles,** into the **radial canals** (Figure 12.7). Here food is ingested by the choanocytes. The beating of the choanocytes's flagella forces the used water through internal pores, or **apopyles,** into the spongocoel. Notice that food capture does not occur in the syconoid spongocoel, so it is lined with epithelial-type cells rather than the flagellated cells present in asconoids. After the used water reaches the spongocoel, it exits the body through an **osculum.** As an example, examine *Sycon* (Gr. *sykon,* a fig), in Figure 12.5.

During development, syconoid sponges pass through an asconoid stage, following which flagellated canals form by evagination of the body wall. This developmental pattern provides evidence that syconoid sponges were derived from an ancestor with an asconoid body plan, but the syconoid condition is not homologous among all the sponges that possess it. Syconoids are found in class Calcarea and in some members of class Hexactinellida.

Leuconoids Leuconoid organization is the most complex of the sponge types and permits an increase in sponge size. In the leuconoid design, the surface area of the food-collecting regions with choanocytes is greatly increased; here the choanocytes line the walls of small chambers where they can effectively filter all the water present (Figure 12.5). The sponge body is composed

Figure 12.6
Clathrina canariensis (class Calcarea) is common on Caribbean reefs in caves and under ledges.

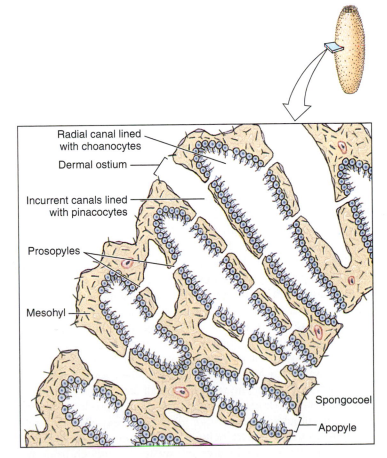

Radial canal lined with choanocytes

Dermal ostium

Incurrent canals lined with pinacocytes

Prosopyles

Mesohyl

Spongocoel

Apopyle

Figure 12.7
Cross section through wall of sponge *Sycon,* showing choanocytes in canals within the wall. Notice choanocytes do not line spongocoel.

Figure 12.8

This orange demosponge, *Mycale laevis,* often grows beneath platelike colonies of the stony coral, *Montastrea annularis.* The large oscula of the leuconoid canal system are seen at the edges of the plates. Unlike some other sponges, *Mycale* does not burrow into the coral skeleton and may actually protect coral from invasion by more destructive species. Pinkish radioles of a Christmas tree worm, *Spirobranchus giganteus* (phylum Annelida, class Polychaeta) also project from the coral colony. An unidentified reddish sponge can be seen to the right of the Christmas tree worm.

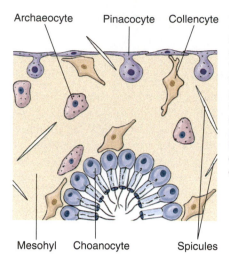

Figure 12.9

Small section through sponge wall, showing four types of sponge cells. Pinacocytes are protective and contractile; choanocytes create water currents and engulf food particles; archaeocytes have a variety of functions; collencytes secrete collagen.

of an enormous number of these tiny chambers. Clusters of flagellated chambers are filled from incurrent canals and discharge water into excurrent canals that eventually lead to an osculum (Figure 12.8).

A sponge pumps a remarkable amount of water. *Leuconia* (Gr. *leukos,* white), for example, is a small leuconoid sponge about 10 cm tall and 1 cm in diameter. It is estimated that water enters through some 81,000 incurrent canals at a velocity of 0.1 cm/second in each canal. However, because water passes into flagellated chambers with a greater cross-sectional area than the entry canals, water flow through the chambers slows to 0.001 cm/second. Such a flow rate allows ample opportunity for food capture by choanocytes. *Leuconia* has more than 2 million flagellated chambers where food collection occurs.

After food is removed, the used water is pooled to form an exit stream. The exit stream, containing the entire volume of water that entered the sponge over the myriad incurrent canals, leaves the sponge through an exit pore whose cross-sectional area is many times less than the total cross-sectional area of all the incurrent canals. The relatively small size of the exit pore, together with the large volume of used water, produces a very high exit velocity. In *Leuconia,* all water is expelled through a single osculum at a velocity of 8.5 cm/second—a jet force capable of carrying used water and wastes far enough from the sponge to avoid refiltering.

Some large sponges can filter 1500 liters of water a day, but unlike *Leuconia,* most leuconoids form large masses with numerous oscula (Figures 12.5 and 12.8), so that water exits from many local sites on the sponge. Most sponges are of the leuconoid type; leuconoid bodies account for most species within class Calcarea and are the most common types in other classes.

Types of Cells in the Sponge Body

Sponge cells are loosely arranged in a gelatinous matrix called **mesohyl,** or **mesenchyme** (Figure 12.9). The mesohyl is the connective tissue of sponges; in it are found various fibrils, skeletal elements, and ameboid cells. The absence of tissues or organs means that all fundamental processes must occur at the level of individual cells. Respiration and excretion occur by diffusion in each cell and, in freshwater sponges, excess water is expelled via contractile vacuoles in archaeocytes and choanocytes.

The visible activities and responses in sponges, other than propulsion of water, are alterations in shape, local contractions, propagating contractions, and closing and opening of incurrent and excurrent pores. Incurrent pores may close in response to heavy sediment in the water or other conditions that reduce the efficiency of feeding. The most common response is closure of the oscula. These movements are very slow, but the fact that there are responses of the whole body in animals that lack organization above the cellular level is puzzling. Apparently excitation spreads from cell to cell by an unknown mechanism; suggested mechanisms include mechanical stimuli and signaling molecules, possibly hormones. Electrical communication across the syncytial tissue of hexactinellid sponges (see p. 256) has been demonstrated, but nothing similar has been found in cellular sponges. Some zoologists point to the possibility of coordination by means of substances carried in the water currents, and others have tried, not very successfully, to demonstrate the presence of nerve cells. Although nerve cells have not been found, several other types of cells do occur.

Choanocytes   Choanocytes, which line flagellated canals and chambers, are ovoid cells with one end embedded in mesohyl and the other exposed. The exposed end bears a flagellum

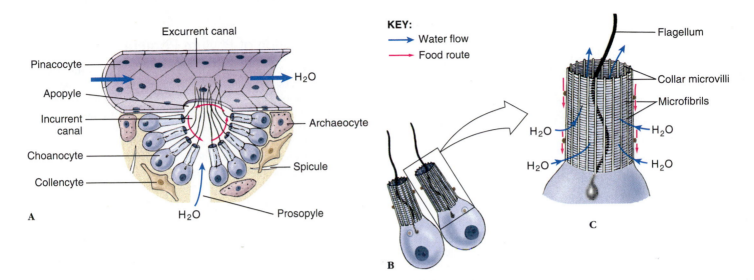

Figure 12.10

Food trapping by sponge cells. **A,** Cutaway section of canals showing cellular structure and direction of water flow. **B,** Two choanocytes and **C,** structure of the collar. Small red arrows indicate movement of food particles.

surrounded by a collar (Figures 12.9 and 12.10). The collar is composed of adjacent microvilli, connected to each other by delicate microfibrils, forming a fine filtering device for straining food particles from water (Figure 12.10B). The beating flagellum pulls water through the sievelike collar and forces it out through the open top of the collar. Particles too large to enter the collar become trapped in secreted mucus and slide down the collar to the base where they are phagocytized by the cell body. Larger particles have already been screened out by the small size of the dermal pores and prosopyles. Food engulfed by the cells is passed to a neighboring archaeocyte for digestion. Thus, digestion is entirely **intracellular,** so there is no extracellular gut cavity. Choanocytes also have a role in sexual reproduction.

Archaeocytes   Archaeocytes are ameboid cells that move in the mesohyl (Figures 12.9 and 12.10) and perform a number of functions. They can phagocytize particles at the pinacoderm and receive particles for digestion from choanocytes. Archaeocytes apparently can differentiate into any of the other types of more specialized cells in the sponge. Some, called **sclerocytes,** secrete spicules. Others, called **spongocytes,** secrete the spongin fibers of the skeleton, and **collencytes** secrete fibrillar collagen (p. 192). **Lophocytes** secrete large quantities of collagen but are distinguishable morphologically from collencytes.

Pinacocytes   The nearest approach to a true tissue in sponges is arrangement of the **pinacocyte** cells of the **pinacoderm** (Figures 12.9 and 12.10). A true tissue is a grouping of cells specialized for one function; a true tissue epithelium consists of a layer of specialized cells resting on a basal membrane. Pinacocytes are thin, flat, epithelial-like cells that cover the exterior surface and some interior surfaces of a sponge. Some are T-shaped with their cell bodies extending into the mesohyl. A layer of pinacocytes does not constitute an epithelium because a basal membrane is lacking in sponges. However, the pinacoderm is sufficiently specialized to be called an incipient tissue by some (see p. 192).

Pinacocytes may take up food particles by phagocytosis at the sponge surface. Pinacocytes are somewhat contractile and help to regulate surface area of a sponge. Some pinacocytes are modified as contractile **myocytes,** which are usually arranged in circular bands around oscula or pores, where they help regulate rate of water flow.

Cell Independence: Regeneration and Somatic Embryogenesis

Sponges have a tremendous ability to repair injuries and to restore lost parts, a process called regeneration. Regeneration does not imply reorganization of the entire animal, but only of the wounded portion. However, a complete reorganization of the structure and function of participating cells or bits of tissue does occur in somatic embryogenesis. If a sponge is cut into small fragments, or if the cells of a sponge are entirely dissociated and are allowed to fall into small groups, or aggregates, entire new sponges can develop from these fragments or aggregates of cells. This process has been termed **somatic embryogenesis.** Somatic embryogenesis involves a complete reorganization of the structure and functions of participating cells or bits of tissue. Isolated from influence of adjoining cells, they can realize their own potential to change in shape or function as they develop into a new organism.

Much experimental work has been done in this field. The process of reorganization appears to differ in sponges of differing complexity. There is still some controversy concerning just what mechanisms cause adhesion of the cells and the share that each type of cell plays in the formative process.

Regeneration following fragmentation is one means of asexual reproduction, a process whereby the genotype of the

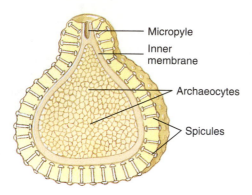

Micropyle

Inner
membrane

Archaeocytes

Spicules

Figure 12.11

Section through a gemmule of a freshwater sponge (Spongillidae). Gemmules are a mechanism for survival of the harsh conditions of winter. On return of favorable conditions, the archaeocytes exit through the micropyle to form a new sponge. The archaeocytes of the gemmule give rise to all cell types of the new sponge structure.

existing sponge is copied into other physiologically distinct sponge bodies. Asexual reproduction can also occur by bud formation. **External buds,** after reaching a certain size, may become detached from the parent and float away to form new sponges, or they may remain to form colonies. **Internal buds,** or **gemmules** (Figure 12.11), are formed in freshwater sponges and some marine sponges. Here, archaeocytes collect in the mesohyl and become surrounded by a tough spongin coat incorporating siliceous spicules. When the parent animal dies, the gemmules survive and remain dormant, preserving the species during periods of freezing or severe drought. Later, cells in the gemmules escape through a special opening, the **micropyle,** and develop into new sponges. Gemmulation in freshwater sponges (Spongillidae) is thus an adaptation to changing seasons. Gemmules are also a means of colonizing new habitats, since they can spread by streams or animal carriers. What prevents gemmules from hatching during the season of formation rather than remaining dormant? Some species secrete a substance that inhibits early germination of gemmules, and gemmules do not germinate as long as they are held in the body of the parent. Other species undergo maturation at low temperatures (as in winter) before they germinate. Gemmules in marine sponges also seem to be an adaptation to pass the cold of winter; they are the only form in which *Haliclona loosanoffi* exists during the colder parts of the year in the northern part of its range.

Sexual Reproduction

In **sexual reproduction** most sponges are **monoecious** (have both male and female sex cells in one individual). Sperm sometimes arise from transformation of choanocytes. In Calcarea and at least some Demospongiae, oocytes also develop from choanocytes; in other demosponges gametes apparently are derived from archaeocytes. Most sponges are viviparous; after fertilization the zygote is retained in and derives nourishment from its

parent, and a ciliated larva is released. In such sponges, sperm are released into the water by one individual and taken into the canal system of another. There choanocytes phagocytize the sperm; then the choanocytes transform into carrier cells, which carry the sperm through the mesohyl to oocytes. Other sponges are oviparous, and both oocytes and sperm are expelled into the water. The free-swimming larva of most sponges is a solid-bodied **parenchymula** (Figure 12.12A), although six other larval types exist, and some sponges exhibit direct development. The outwardly directed, flagellated cells of the parenchymula migrate to the interior after the larva settles and become choanocytes in the flagellated chambers.

Calcarea and a few Demospongiae have a very strange developmental pattern. A hollow blastula, called an stomoblastula (Figure 12.12B), develops, with flagellated cells toward the interior. The blastula then turns *inside out* **(inversion),** the flagellated ends of the cells becoming directed to the outside! Flagellated cells **(micromeres)** of the amphiblastula larva are at the anterior end, and larger, nonflagellated cells **(macromeres)** are at the posterior end. In contrast to other metazoan embryos, the micromeres invaginate into and are overgrown by the macromeres at metamorphosis during settlement. The flagellated micromeres become choanocytes, archeocytes, and collencytes of the new sponge, and the nonflagellated cells give rise to pinacoderm and sclerocytes.

Class Calcarea (Calcispongiae)

Calcarea (also called Calcispongiae) are calcareous sponges, so called because their spicules are composed of calcium carbonate. Spicules are straight (monaxons) or have three or four rays. These sponges tend to be small—10 cm or less in height—and

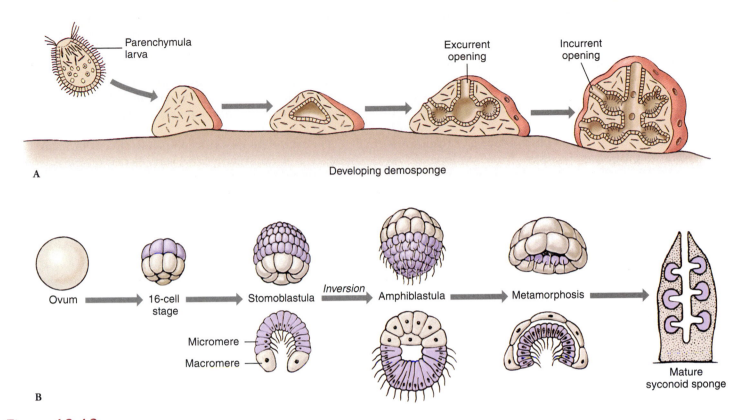

Figure 12.12

A, Development of demosponges. **B,** Development of the calcareous syconoid sponge *Sycon*.

tubular or vase-shaped. They may be asconoid, syconoid, or leuconoid in structure. Though many are drab in color, some are
bright yellow, red, green, or lavender. *Leucosolenia* and *Sycon*
(often called *Scypha* or *Grantia* by biological supply companies)
are marine shallow-water forms commonly studied in the laboratory (Figure 12.5). *Leucosolenia* is a small asconoid sponge
that grows in branching colonies, usually arising from a network of horizontal, stolonlike tubes (Figure 12.13). *Clathrina* is
small with intertwined tubes (Figure 12.6). *Sycon* is a solitary
sponge that may live singly or form clusters by budding. The
vase-shaped, typically syconoid animal is 1 to 3 cm long, with
a fringe of straight spicules around the osculum to discourage
small animals from entering.

Class Hexactinellida (Hyalospongiae): Glass Sponges

Glass sponges form class Hexactinellida (or Hyalospongiae).
Nearly all are deep-sea forms that are collected by dredging.
Most are radially symmetrical, with vase- or funnel-shaped bodies usually attached by stalks of root spicules to a substratum
(Figure 12.13, *Euplectella*) (N. L. from Gr. *euplektos*, well plaited).
They range from 7.5 cm to more than 1.3 m in length. Their distinguishing features include a skeleton of six-rayed siliceous spicules that are commonly bound together into a network forming a
glasslike structure.

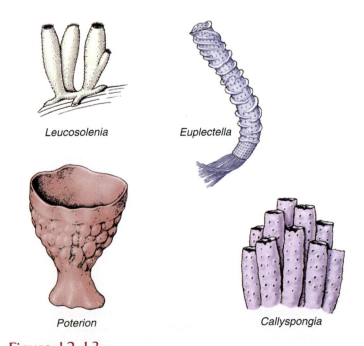

Figure 12.13

Some sponge body forms. *Euplectella* is in Hexactinellida, *Poterion*
and *Callyspongia* are members of Demospongiae, and *Leucosolenia*
is in Calcarea.

Their tissue structure differs so dramatically from other sponges that some scientists advocate placing hexactinellids in a subphylum separate from other sponges. The body of hexactinellids is composed of a single, continuous syncytial tissue (a tissue not divided into separate cells) called a **trabecular reticulum.** The trabecular reticulum is the largest, continuous syncytial tissue known in Metazoa. It is bilayered and encloses a thin, collagenous mesohyl between the two layers, as well as cellular elements such as archaeocytes, sclerocytes, and **choanoblasts.** Choanoblasts are associated with flagellated chambers, where the layers of the trabecular reticulum separate into a **primary reticulum** (incurrent side) and a **secondary reticulum** (excurrent, or atrial side) (Figure 12.14). The spherical choanoblasts are borne by the primary reticulum, and each choanoblast has one or more processes extending to **collar bodies,** the bases of which are also supported by the primary reticulum. Each collar body with its flagellum extends into the flagellated chamber through an opening in the secondary reticulum. Water is drawn into the space between primary and secondary reticula through prosopyles in the primary reticulum, then through the collars into the lumen of the flagellated chamber. Collar bodies do not participate in phagocytosis, which is accomplished by the primary and secondary reticula.

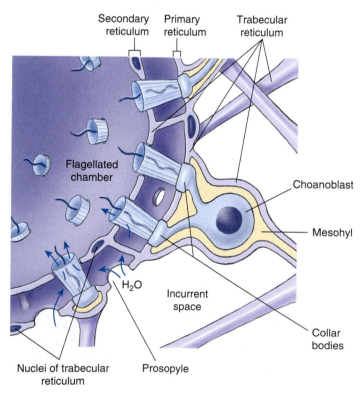

Figure 12.14

Diagram of part of a flagellated chamber of hexactinellids. The primary and secondary reticula are branches of the trabecular reticulum, which is syncytial. Cell bodies of the choanoblasts and their processes are borne by the primary reticulum and are embedded in a thin, collagenous mesohyl. Processes of the choanoblasts end in collar bodies, whose collars extend up through the secondary reticulum. Flagellar action propels water (*arrows*) to be filtered through the mesh of collar microvilli (see Figure 12.10).

The syncytial nature of these unusual sponges might suggest a syncytial origin for metazoans, but the details of development refute this idea. The tissue of the reticulum forms after typical embryonic cleavage and blastula formation. Following the 32-cell cleavage stage, new cells remain connected via cytoplasmic bridges, and the syncytium forms through a combination of cell fusion and envelopment. Thus, the animal is initially cellular.

The latticelike network of spicules found in many glass sponges is of exquisite beauty, such as that of *Euplectella,* or Venus's flower basket (Figure 12.13), a classic example of Hexactinellida.

Class Demospongiae

This group contains 95% of living sponge species, including most large sponges. Spicules are siliceous, but are not six rayed. Spicules may be bound together by spongin, or may be absent. So-called bath sponges, *Spongia* and *Hippospongia,* belong to the group called horny sponges, which have spongin skeletons and lack siliceous spicules entirely. All members of the class are leuconoid, and all are marine, except for members of the freshwater family Spongillidae.

Marine Demospongiae are quite varied and may be quite striking in color and shape (Figure 12.15). Some are encrusting; some are tall and fingerlike; some are low and spreading; some bore into shells; and some are shaped like fans, vases, cushions, or balls (Figure 12.15). Loggerhead sponges may grow several meters in diameter.

Freshwater sponges are widely distributed in well-oxygenated ponds and streams, where they encrust plant stems and old pieces of submerged wood. They may resemble a bit of wrinkled scum, be pitted with pores, and be brownish or greenish in color. Common genera are *Spongilla* (L. *spongia,* from Gr. *spongos,* sponge) and *Myenia.* Freshwater sponges are most common in midsummer, although some are more easily found in the fall. They do reproduce sexually, but existing genotypes may also reappear annually from gemmules. In late autumn, the sponge body dies and disintegrates, leaving the asexually formed gemmules to overwinter and begin the next year's population.

Phylogeny and Adaptive Diversification
Phylogeny

Sponges originated before the Cambrian period. Two groups of calcareous spongelike organisms occupied early Paleozoic reefs. The Devonian period saw rapid development of many glass sponges. Phylogenetic studies[2] using sequence data from large subunit rRNA, small subunit rRNA, and protein kinase C, indicate that sponges with calcareous spicules in the class Calcarea belong to a separate clade from those with spicules made of

[2]Borchiellini, C., M. Manuel, E. Alivon, N. Boury-Esnault, J. Vacelet, and Y. Le Parco. 2001. J. Evol. Biol. **14:**171–179. Medina, M., A. G. Collins, J. D. Silberman, and M. L. Sogin 2001. Proc. Nat. Acad. Sci., USA **98:**9707–9712.

A   B   C

Figure 12.15

Marine Demospongiae on Caribbean coral reefs. **A,** *Pseudoceratina* crassa is a colorful sponge growing at moderate depths. **B,** *Aplysina fistularis* is tall and tubular. **C,** *Monanchora unguifera* with commensal brittle star, *Ophiothrix suensoni* (phylum Echinodermata, class Ophiuroidea).

silica in classes Demospongaie and Hexactinellida. Two potential placements emerge for calcareous sponges: In one, calcareous sponges are the sister taxon to a clade of siliceous sponges, as we show in Figure 12.4, and in another, the phylum Porifera is paraphyletic because the calcareous "sponges" are more closely related to other metazoan taxa than they are to siliceous sponges. Clarification will require more data.

Adaptive Diversification

Porifera are a highly successful group that includes several thousand species in a variety of marine and freshwater habitats. Their diversification centers largely on their unique water-current system and its various degrees of complexity. However, within the silicious Demospongaie, a new feeding mode has evolved for a family of sponges inhabiting nutrient-poor deep-water caves. These deep-water sponges have a fine coating of tiny hooklike spicules over their highly branched bodies. The spicule layer entangles the appendages of tiny crustaceans swimming near the surface of the sponge. Later, filaments of the sponge body grow over prey, enveloping and digesting them. These sponges are carnivores, not suspension feeders, although some of them may augment their diets with nutrients obtained from symbiotic methanotrophic bacteria. The presence of the typical silicious spicules clearly identifies these animals as sponges, but they lack choanocytes and internal canals.

The loss of choanocytes in these species is doubtless disturbing for students learning to identify sponges, but students of evolution should be fascinated by it. The convoluted pathway taken by one branch of the sponge lineage clearly illustrates the nondirectional nature of evolution. To colonize such a nutrient-poor habitat initially, the ancestors of this group must have had at least one alternative feeding system, either carnivory or chemoautotrophy, already in place. Presumably, after the alternative method of food capture was in use, the choanocytes and internal canals were no longer formed. If there are further body modifications in this lineage, we might eventually not recognize the descendants as sponges. Imagine how the lineage would look if spicules were lost in favor of a greater reliance on the bacterial symbionts, and you will begin to understand why it is sometimes hard to trace morphological evolution or to identify the closest relatives of certain animals.

PHYLUM PLACOZOA

The phylum Placozoa (Gr. *plax, plakos,* tablet, plate, + *zōon,* animal) was proposed in 1971 by K. G. Grell to contain a single species, *Trichoplax adhaerens* (Figure 12.16A), a tiny (2 to 3 mm) marine form. The body is platelike and has no symmetry, no organs, and no muscular or nervous system. It also lacks both a basal lamina beneath the epidermis and an extracellular matrix, two features that were considered metazoan hallmarks. The body of a placozoan is composed of a dorsal epithelium of cover cells and shiny spheres and a thick ventral epithelium containing monociliated cells (cylinder cells) and nonciliated gland cells (Figure 12.16B). The space between the epithelia contains fibrous "cells" within a contractile syncytium. There are four cell types distinguished morphologically, but gene-expression studies suggest the presence of a fifth type.

Placozoans glide over their food, secrete digestive enzymes on it, and then absorb the products. In the laboratory, they feed on organic matter and small algae.

The life cycle of placozoans is not completely known. They divide asexually and produce "swarmer" stages by budding. Although sexual reproduction has not been observed, eggs occur in laboratory animals. Genetic studies of placozoans from around the world show that eight distinct lineages equivalent to species exist, although they cannot be distinguished morphologically.

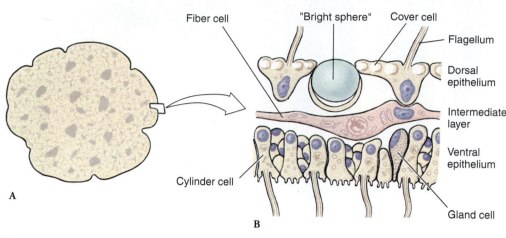

Figure 12.16

A, *Trichoplax adhaerens* is a marine, platelike animal only 2 to 3 mm in diameter. **B,** Section through *Trichoplax adhaerens,* showing histological structure.

Sexual reproduction has been inferred from molecular evidence of genetic diversity within a clade.

Grell considers *Trichoplax* diploblastic (see p. 192), with dorsal epithelium representing ectoderm and ventral epithelium representing endoderm because of its nutritive function. Recent gene-expression studies support these homologies. The origin of the middle fibrous layer is currently under study. As this group becomes better understood, the branching order for Placozoa and the two diploblastic phyla (see Chapter 13) may be clear relatively soon. At present we depict branching of placozoans, cnidarians, and ctenophorans as a polytomy (see cladogram on inside front cover).

SUMMARY

Sponges (phylum Porifera) are an abundant marine group with some freshwater representatives. They have various specialized cells, but these are not organized into tissues or organs. They depend on the flagellar beat of their choanocytes to circulate water through their bodies for food gathering and respiratory gas exchange. They are supported by secreted skeletons of fibrillar collagen, collagen in the form of large fibers or filaments (spongin), calcareous or siliceous spicules, or a combination of spicules and spongin in most species.

Sponges reproduce asexually by budding, fragmentation, and gemmules (internal buds). Most sponges are monoecious but produce sperm and oocytes at different times. Embryogenesis is unusual, with a migration of flagellated cells at the surface to the interior (parenchymella) or the production of an amphiblastula with inversion and growth of macromeres over micromeres. Sponges have great regenerative abilities.

Sponges are an ancient group, seemingly remote phylogenetically from other metazoa, but molecular evidence suggests that they are the sister group to Eumetazoa. Their adaptive diversification is centered on elaboration of the water circulation and filter-feeding system, except for one family of sponges where filter-feeding has been replaced by carnivory and reliance on bacterial symbionts for extra nutrition.

Phylum Placozoa is represented by a small platelike marine organism. It has only two cell layers with a fibrous syncytial layer between them. Some workers hypothesize that these layers are homologous to ectoderm and endoderm of more complex metazoans. Genetic studies indicate that there are eight species of placozoans.

REVIEW QUESTIONS

1. Briefly describe and contrast the syncytial ciliate hypothesis, the colonial flagellate hypothesis, and the polyphyletic origin of the metazoa. Which hypothesis seems most compatible with available data?
2. Give eight characteristics of sponges.
3. Briefly describe asconoid, syconoid, and leuconoid body types in sponges.
4. What sponge body type is most efficient and makes possible the largest body size?
5. Define the following: ostia, osculum, spongocoel, apopyles, prosopyles, spicules.
6. Define the following: pinacocytes, choanocytes, archaeocytes, sclerocytes, spongocytes, collencytes.

7. What material is found in the skeleton of all sponges?
8. Describe the skeletons of each class of sponges.
9. Describe how sponges feed, respire, and excrete.
10. What is a gemmule?
11. Why are glass sponges distinguished from sponges with cellular bodies?
12. Describe possible ancestors to sponges. Justify your answer.
13. Describe the body plan of Placozoa.
14. What features make placozoans interesting from a phylogenetic perspective?

SELECTED REFERENCES

Bergquist, P. R. 1978. Sponges. Berkeley, University of California Press. *Excellent monograph on sponge structure, classification, evolution, and general biology.*

Bond, C. 1997. Keeping up with the sponges. Nat. Hist. **106**:22–25. *Sponges are not fixed in permanent position; at least some can crawl on their substrate.* Haliclona loosanoffi *can move over 4 mm/day.*

Borchiellini, C., M. Manuel, E. Alivon, N. Boury-Esnault, J. Vacelet, and Y. Le Parco. 2001. Sponge paraphyly and the origin of Metazoa. J. Evol. Biol. **14**:171–179. *Results of this study suggest that members of class Calcarea are more closely related to other metazoans than to siliceous sponges.*

Grell, K. G. 1982. Placozoa. In S. P. Parker (ed.), Synopsis and classification of living organisms, vol. 1. New York, McGraw-Hill Book Company. *Synopsis of placozoan characteristics.*

Hooper, J. N. A., and R. W. M. van Soest. (eds.) 2002. Systema Porifera: a guide to the classification of sponges. New York, Kluwer Academic/Plenum. *A large and comprehensive work on sponge systematics and biology.*

King, N., C. T. Hittinger, and S. B. Carroll. 2003. Evolution of key cell signaling and adhesion protein families predates the origin of animals. Science **301**:361–363. *Cells in multicellular animals must aggregate and communicate. Proteins responsible for these functions in metazoans are homologous to those in choanoflagellates.*

Leys, S. P., and A. E. Ereskovsky. 2006. Embryogenesis and larval differentiation in sponges. Can. J. Zool. **84**:262–287. *A review of sponge development with clearly explained terms and excellent photomicrographs.*

Leys, S. P., and R. W. Meech. 2006. Physiology of coordination in sponges. Can. J. Zool. **84**:288–306. *Current research shows how sponge cells communicate.*

Medina, M., A. G. Collins, J. D. Silberman, and M. L. Sogin 2001. Evaluating hypotheses of basal animal phylogeny using complete sequences of large and small subunit rRNA. Proc. Nat. Acad. Sci., USA **98**:9707–9712. *Molecular data support placement of calcareous sponges in a clade distinct from siliceous sponges.*

Nielsen, C. 1995. Animal evolution: interrelationships of the living phyla. Oxford, Oxford University Press. *Several schemes for metazoan evolution using hypothetical ancestral forms are outlined here.*

Schierwater, B. 2005. My favorite animal, *Trichoplax adhaerens*. Bioessays **27**:1294–1302. *A personal description of the author's fascination with this animal.*

Vacelet, J., and N. Boury-Esnault. 1995. Carnivorous sponges. Nature **373**:333–335. *A fascinating article on feeding in these sponges. Later work demonstrates that symbiotic methanotrophic bacteria provide a second source of nutrition for the sponges.*

Vogel, S. 1981. Life in moving fluids: the physical biology of flow. Princeton, Princeton University Press. *A clear general discussion of how water flow influences animal design, with specific reference to water movement in the sponge body.*

Wood, R. 1990. Reef-building sponges. Am. Sci. **78**:224–235. *The author presents evidence that known sclerosponges belong to either the Calcarea or the Demospongiae and that a separate class Sclerospongiae is not needed.*

Wyeth, R. C. 1999. Video and electron microscopy of particle feeding in sandwich cultures of the hexactinellid sponge, *Rhabdocalyptus dawsoni*. Invert. Biol. **118**:236–242. *Phagocytosis is not by choanoblasts but by trabecular reticulum, especially primary reticulum. He places Hexactinellida in subphylum Sygmplasma and the rest of Porifera in subphylum Cellularia.*

ONLINE LEARNING CENTER

Visit www.mhhe.com/hickmanipz14e for chapter quizzing, key term flash cards, web links, and more!

13

Radiate Animals

- PHYLUM CNIDARIA
- PHYLUM CTENOPHORA

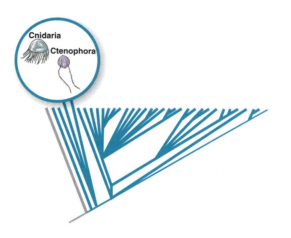

Tentacles of the coral Tubastraea coccinea *from the Caribbean.*

A Fearsome Tiny Weapon

Although members of phylum Cnidaria are more highly organized than sponges, they are still relatively simple animals. Most are sessile; those that are unattached, such as jellyfish, can swim only feebly. None can chase their prey. Indeed, we might easily get the false impression that cnidarians provide easy meals for other animals. The truth is, however, many cnidarians are very effective predators that are able to kill and eat prey that are much more highly organized, swift, and intelligent. They manage these feats because they possess tentacles that bristle with tiny, remarkably sophisticated weapons called nematocysts.

As it is secreted within the cell that contains it, the nematocyst is endowed with potential energy to power its discharge. It is as though a factory manufactured a gun, cocked and ready with a bullet in its chamber, as it rolls off the assembly line. Like a cocked gun, a completed nematocyst requires only a small stimulus to make it fire. Rather than a bullet, a tiny thread bursts from a nematocyst. Achieving a velocity of 2 meters/sec and an acceleration of $40,000 \times$ gravity, it instantly penetrates its prey and injects a paralyzing toxin. A small animal unlucky enough to brush against one of the tentacles is suddenly speared with hundreds or even thousands of nematocysts and quickly immobilized. Some nematocyst threads can penetrate human skin, causing sensations ranging from minor irritation to great pain, even death, depending on the species. A nematocyst is a fearsome, but wondrous, tiny weapon.

The two phyla discussed in this chapter are diploblastic (see cladogram on front inside cover), meaning that they have two embryonic cell layers, ectoderm and endoderm, from which adult structures develop. Two layers are produced as the embryo develops from a single-layered blastula to a gastrula (see Chapters 8 and 9). In adult diploblasts, the epidermis is derived from ectoderm, and the gut cavity lining or gastrodermis is derived from endoderm; this body plan is in marked contrast to that of adult sponges, where there are neither cell layers nor a gut cavity.

A new stage of development, gastrulation, characterizes diploblasts and produces the cell layers of adult animals. Therefore, one expects no evidence of cell layers at any stage of development in sponges or placozoans. However, as mentioned in Chapter 12, recent work on sponge development suggests that cell layers do develop in sponge larvae, but disappear as adults become a nonlayered aggregate of different cell types. The developmental sequence for placozoans is not known, but some biologists consider the two adult layers equivalent to derivatives of ectoderm and endoderm. Thus, it may be appropriate to add more phyla to the diploblast category if stages other than the adult are considered, or new homologies are established.

Currently, the diploblastic phyla are Cnidaria and Ctenophora. Adult organisms of both groups exhibit radial or biradial symmetry (see p. 187) and are not cephalized. Familiar cnidarians are sea anemones and jellyfishes, and some readers may know ctenophorans as comb jellies or sea walnuts.

PHYLUM CNIDARIA

Phylum Cnidaria (ny-dar′e-a) (Gr. *knide*, nettle, + L. *aria* [pl. suffix], like or connected with) is an interesting group of more than 9000 species. It includes some of nature's strangest and loveliest creatures: branching, plantlike hydroids; flowerlike sea anemones; jellyfishes; and those architects of the ocean floor, horny corals (sea whips, sea fans, and others) and stony corals whose thousands of years of calcareous house-building have produced great reefs and coral islands (p. 279).

The phylum takes its name from cells called **cnidocytes,** which contain organelles (cnidae) characteristic of the phylum. The most common type of cnida is the **nematocyst** described in the opening essay. Cnidocytes are formed only by cnidarians, but some ctenophores, molluscs, and flatworms eat hydroids bearing nematocysts, then store and use these stinging structures for their own defense.

Cnidarians are an ancient group with the longest fossil history of any metazoan, reaching back more than 700 million years. They are widespread in marine habitats, and there are a few in freshwater. Cnidarians are most abundant in shallow marine habitats, especially in warm temperatures and tropical regions. There are no terrestrial species. Colonial hydroids are usually found attached to mollusc shells, rocks, wharves, and other animals in shallow coastal water, but some species live at great depths. Floating and free-swimming medusae occur in open seas and lakes, often far from shore. Animals such as the Portuguese man-of-war and *Velella* (L. *velum*, veil, + *ellus*, dim. suffix) have floats or sails by which the wind carries them. Although they are mostly sessile, or at best, fairly slow moving or slow swimming, cnidarians are quite efficient predators of organisms that are much swifter and more complex.

Cnidarians sometimes live symbiotically with other animals, often as commensals on the shell or other surface of their host. Certain hydroids (Figure 13.1) and sea anemones commonly live on snail shells inhabited by hermit crabs, providing the crabs some protection from predators. Algal cells frequently live as mutuals in the tissues of cnidarians, notably in some freshwater hydras and in reef-building corals. The presence of the algae in reef-building corals limits the occurrence of coral reefs to relatively shallow, clear water where there is sufficient light for the photosynthetic requirements of the algae. These kinds of corals are an essential component of coral reefs, and reefs are extremely important habitats for many other species of invertebrates and vertebrates in tropical waters. Coral reefs are discussed further on page 279.

Although many cnidarians have little economic importance, reef-building corals are an important exception. Fish and other animals associated with reefs provide substantial amounts of food

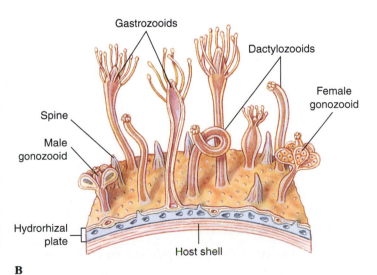

A   **B**

Figure 13.1

A, A hermit crab with its cnidarian mutuals. The host snail shell is blanketed with polyps of the hydrozoan *Hydractinia milleri*. The crab gets some protection from predation by the cnidarians, and the cnidarians get a free ride and bits of food from their host's meals. **B,** Portion of a colony of *Hydractinia*, showing the types of zooids and the stolon (hydrorhiza) from which they grow.

Labels in Figure 13.1B: Gastrozooids, Dactylozooids, Female gonozooid, Spine, Male gonozooid, Hydrorhizal plate, Host shell

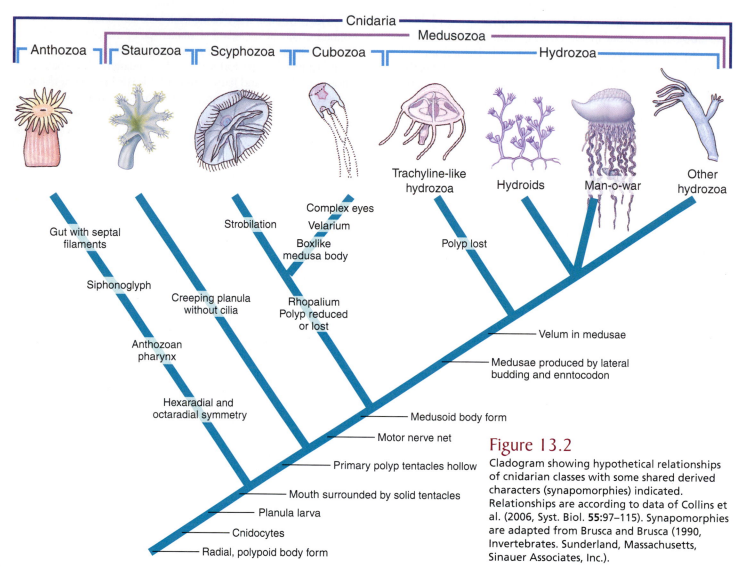

Figure 13.2
Cladogram showing hypothetical relationships of cnidarian classes with some shared derived characters (synapomorphies) indicated. Relationships are according to data of Collins et al. (2006, Syst. Biol. **55**:97–115). Synapomorphies are adapted from Brusca and Brusca (1990, Invertebrates. Sunderland, Massachusetts, Sinauer Associates, Inc.).

for humans, and reefs are of economic value as tourist attractions. Precious coral is used for making jewelry and ornaments, and coral rock serves for building purposes.

Four classes of Cnidaria were traditionally recognized (Figure 13.2): Hydrozoa (most variable class, including hydroids, fire corals, Portuguese man-of-war, and others), Scyphozoa ("true" jellyfishes), Cubozoa (cube jellyfishes), and Anthozoa (largest class, including sea anemones, stony corals, soft corals, and others). A fifth class, Staurozoa, has been proposed because recent phylogenies show that stauromedusans do not belong within the Scyphozoa. These odd animals do not make medusae, but the polyp body is topped by a medusa-like region (see p. 273).

Form and Function

Dimorphism and Polymorphism in Cnidarians

One of the most interesting—and sometimes puzzling—aspects of this phylum is the dimorphism and often polymorphism displayed by many of its members. All cnidarian forms fit into one of two morphological types (**dimorphism**): a **polyp,** or

hydroid form, which is adapted to a sedentary or sessile life, and a **medusa,** or jellyfish form, which is adapted for a floating or free-swimming existence (Figure 13.3).

Superficially the polyp and medusa seem very different, but actually each has retained the saclike body plan characteristic of the phylum (Figure 13.3). A medusa is essentially an unattached polyp with the tubular portion widened and flattened into a bell shape.

Polyps   Most polyps have tubular bodies. A mouth surrounded by tentacles defines the oral end of the body. The mouth leads into a blind gut or **gastrovascular cavity** (Figure 13.3). The aboral end of the polyp is usually attached to a substratum by a pedal disc or other device.

Polyps may reproduce asexually by budding, fission, or pedal laceration. In **budding,** a knob of tissue forms on the side of an existing polyp and develops a functional mouth and tentacles (see Figure 13.14). If a bud detaches from the polyp that made it, a clone is formed. If a bud stays attached to the polyp that made it, a colony will form and food may be shared through a common gastrovascular cavity (Figures 13.1 and 13.7). Polyps that do not bud are solitary; others form clones or colonies. The

Characteristics of Phylum Cnidaria

1. **Cnidocytes** present, typically housing stinging organelles called **nematocysts**
2. Entirely aquatic, some in freshwater, but most marine
3. **Radial symmetry** or biradial symmetry around a longitudinal axis with oral and aboral ends; no definite head
4. Two types of individuals: **polyps** and **medusae**
5. Adult body two-layered **(diploblastic)** with epidermis and gastrodermis derived from embryonic ectoderm and endoderm, respectively
6. Mesoglea, an extracellular matrix ("jelly") lies between body layers; amount of mesoglea variable; mesoglea with cells and connective tissue from ectoderm in some
7. Incomplete gut called **gastrovascular cavity;** often branched or divided with septa
8. **Extracellular digestion** in gastrovascular cavity and intracellular digestion in gastrodermal cells
9. Extensible tentacles usually encircle mouth or oral region
10. Muscular contractions via **epitheliomuscular cells,** which form an outer layer of longitudinal fibers at base of epidermis and an inner layer of circular fibers at base of gastrodermis; modifications of plan in hydrozoan medusa (independent ectodermal muscle fibers) and other complex cnidarians
11. Sense organs include well-developed statocysts (organs of balance) and ocelli (photosensitive organs); complex eyes in members of Cubozoa
12. **Nerve net** with symmetrical and asymmetrical synapses; diffuse conduction; two nerve rings in hydrozoan medusae
10. Asexual reproduction by budding in polyps forms clones and colonies; some colonies exhibit **polymorphism[1]**
11. Sexual reproduction by gametes in all medusae and some polyps; monoecious or dioecious; holoblastic indeterminate cleavage; planula larval form
12. No excretory or respiratory system
13. No coelomic cavity

[1]Note that polymorphism here refers to more than one structural form of individual within a species, as contrasted with the use of the word in genetics (p. 127), in which it refers to different allelic forms of a gene in a population.

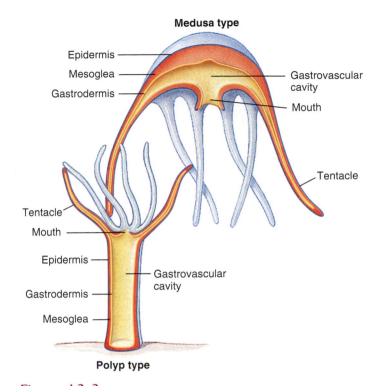

Medusa type

Epidermis
Mesoglea
Gastrodermis

Gastrovascular cavity
Mouth

Tentacle

Tentacle
Mouth
Epidermis
Gastrovascular cavity
Gastrodermis
Mesoglea

Polyp type

Figure 13.3
Comparison between polyp and medusa types of individuals.

Medusae   Medusae are usually free swimming and have bell- or umbrella-shaped bodies (Figures 13.3 and 13.10). They often exhibit tetramerous symmetry where body parts are arranged in fours. The mouth is usually centered on the concave (subumbrellar) side, and it may be pulled downward into frilly lobes that extend a long way beneath the umbrella or bell (Figure 13.17). Tentacles extend outward from the rim of the umbrella. Medusae have sensory structures for orientation (statocysts) and light reception (ocelli). Sensory information is integrated with motor response by a nerve ring at the base of the bell; two such rings are present in hydrozoan medusae (see Figure 13.11).

Medusae of class Scyphozoa are often called scyphomedusae, whereas those of class Hydrozoa are hydromedusae. Hydromedusae differ from scyphomedusae by the presence of a velum, a shelflike fold of tissue from the bottom of the bell that extends into the bell. By reducing the cross-sectional area at the bottom of the bell (see Figure 13.11), the velum increases the exit velocity of water from the bell, making each pulsation more efficient.

Life Cycles

In a cnidarian life cycle, polyps and medusae play different roles. The particular sequence of forms in the life cycle varies among cnidarian classes, but in general, a zygote develops into a motile planula larva. The planula settles on a hard surface and metamorphoses into a polyp. A polyp may make other polyps asexually, but eventually it produces free-swimming medusae by asexual reproduction (see Figures 13.7 and 13.19). Polyps produce medusae by budding, or other specialized methods like **strobilation** (see p. 272). Medusae reproduce sexually and are dioecious.

distinction between colonies and clones is sometimes blurred when a colony fragments.

A shared gastrovascular cavity permits polyp specialization. Many colonies include several morphologically distinct polyps, each specialized for a certain function, such as feeding, reproduction, or defense (Figure 13.1). Such colonies exhibit **polymorphism** (not to be confused with the population-genetic use of this term introduced in Chapter 6). In class Hydrozoa, feeding polyps or **hydranths,** are easily distinguished from reproductive polyps, or **gonangia,** by the absence of tentacles in gonangia. Gonangia typically make medusae.

Other methods of asexual reproduction in polyps are fission, where an individual divides in half as one side of the polyp pulls away from the other side, or pedal laceration, where tissue torn from the pedal disc develops into tiny new polyps. Pedal laceration and fission are common in sea anemones in class Anthozoa.

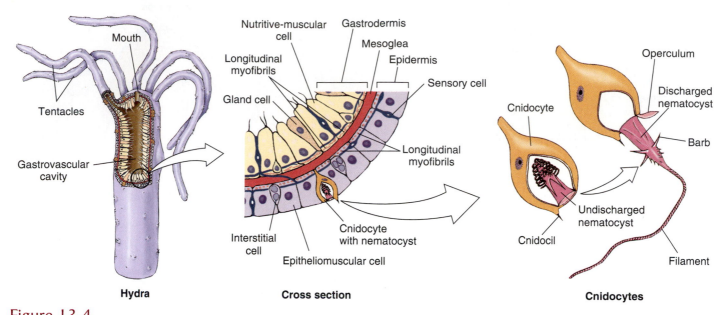

Hydra   **Cross section**   **Cnidocytes**

Figure 13.4

At right, structure of a stinging cell. Center, portion of the body wall of a hydra. Cnidocytes, which contain the nematocysts, arise in the epidermis from interstitial cells.

A life cycle that contains both an attached polyp and a swimming medusa permits organisms to take advantage of both pelagic (open water) and benthic (bottom) environments. Such life cycles occur in true jellyfishes of class Scyphozoa where the medusa is large and conspicuous and the polyps are typically very small. Most hydroids of class Hydrozoa also feature a sessile polyp stage, often in colonies, and a pelagic medusa stage.

However, there are many variations on the typical pattern. In some hydrozoans, the polyp colony is not sessile, but drifts across the ocean surface. The Portuguese man-of-war, *Physalia*, is one such drifter, using an inflated polyp as a gas-filled float (see Figure 13.15). Other colonies are collections of both polyps and medusae where pulsating bells propel the colony through the water.

Several life cycles do not include medusae. Anthozoans are presumed to have diverged from an ancestor of the other cnidarians before the medusa evolved in the latter branch (see Figure 13.2), but other cnidarians, including the hydrozoan *Hydra*, probably lost the medusa secondarily. The mechanism of loss is not clear in *Hydra*, but in other hydrozoans, a pattern of loss can be inferred from a comparison of modern forms. Most hydrozoans release medusae that later make gametes, but a few forms make medusae without releasing them from the colony. Gametes then form in the gonads of the medusae retained by the polyp colony. In some species only a short cuplike form surrounds the gonads (see Figure 13.9), and in others gonads develop right on the polyp colony with no trace of a medusa body. The latter organisms likely represent an extreme form of medusa retention and reduction.

Body Wall

The cnidarian body comprises an outer epidermis, derived from ectoderm, and an inner gastrodermis, derived from endoderm, with mesoglea between them (Figure 13.3). The gastrodermis lines the gut cavity and functions mainly in digestion. In polyps of the solitary hydrozoan, *Hydra*, the epidermal layer contains several cell types (Figure 13.4), including epitheliomuscular, interstitial, gland, sensory, and nerve cells (see pp. 269–270), as well as cnidocytes (see p. 265). Cnidarian bodies extend, contract, bend, and pulse, all in the absence of true mesodermally derived muscle cells. Instead, epitheliomuscular cells form most of the epidermis and serve both for covering and for muscular contraction (Figure 13.5). The bases of most such cells are extended parallel to the tentacle or body axis and contain myofibrils; they form the functional equivalent of a layer of longitudinal muscle next to the mesoglea. Contraction of these fibrils shortens the body or tentacles.

The mesoglea lies between the epidermis and the gastrodermis and is attached to both layers (Figure 13.3). It is gelatinous, or jellylike, and both epidermal and gastrodermal cells send processes into it. In polyps, it is a continuous layer extending over both body and tentacles, thinnest in the tentacles and thickest in the stalk portion. This arrangement allows the pedal region at

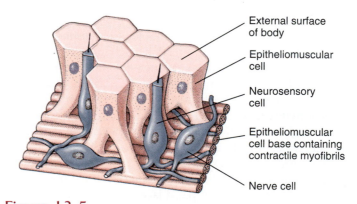

Figure 13.5

Epitheliomuscular and nerve cells in hydra.

the base of the animal to withstand great mechanical strain and gives the tentacles more flexibility.

The mesoglea helps to support the body and acts as a type of elastic skeleton. In class Anthozoa, the mesoglea is substantial and possesses ameboid cells. The mesogleal layer is also very thick in scyphozoan medusae and contains ameboid cells and fibers. The medusa bell has a fairly firm consistency, despite the fact that mesogleal jelly is 95% to 96% water. The buoyant mass of mesogleal "jelly" gives medusae the common name jellyfishes. The mesoglea is much thinner in the bells of hydromedusae and lacks ameboid cells or fibers.

Cnidocytes

As the opening essay attests, many cnidarians are very effective predators on prey larger and more intelligent than themselves. Such efficient predation is made possible by amply arming the tentacles with a unique cell type, the cnidocyte (Figure 13.4). Cnidocytes are borne in invaginations of ectodermal cells (Figure 13.4) and, in some forms, in endodermal cells. Each cnidocyte produces one of over 20 kinds of distinctive organelles called **cnidae** (Figure 13.6) that are discharged from the cell. During its development, a cnidoctye is properly called a **cnidoblast.** Once its cnida has been discharged, a cnidocyte is absorbed and replaced.

One type of cnida, the **nematocyst** (Figure 13.4), is used to inject a toxin for prey capture and defense. Nematocysts are tiny capsules composed of material similar to chitin and containing a

Figure 13.6

A, Several types of cnidae shown after discharge. At bottom are two views of a type that does not impale prey; it recoils like a spring, catching any small part of the prey in the path of the recoiling thread. **B,** Fired and unfired chidae from *Corynactis californica*.

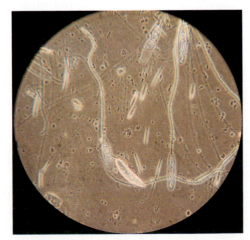

B

A

coiled tubular "thread" or filament, which is a continuation of the narrowed end of the capsule. This end of the capsule is covered by a little lid, or **operculum.** The inside of the undischarged thread may bear tiny barbs, or spines. Not all cnidae have barbs or inject poison. Some kinds, for example, do not penetrate prey but rapidly recoil like a spring after discharge, grasping and holding any part of the prey caught in the coil (Figure 13.6). Adhesive cnidae usually do not discharge in food capture, but are used in attachment and locomotion.

Except in Anthozoa, cnidocytes are equipped with a trigger-like **cnidocil,** which is a modified cilium. Anthozoan cnidocytes have a somewhat different ciliary mechanoreceptor. In some sea anemones, and perhaps other cnidarians, small organic molecules from the prey "tune" the mechanoreceptors, sensitizing them to the frequency of vibration caused by prey swimming. Tactile stimulation causes the nematocyst to discharge.

The mechanism of nematocyst discharge is remarkable. Evidence indicates that discharge is due to a combination of tensional forces generated during nematocyst formation and to an astonishingly high osmotic pressure within the nematocyst: 140 atmospheres. When stimulated to discharge, the high internal osmotic pressure causes water to rush into the capsule. The operculum opens, and the rapidly increasing *hydrostatic pressure* within the capsule forces the thread out with great force, turning inside out as it goes. At the everting end of the thread, the barbs flick to the outside like tiny switchblades. This minute but awesome weapon then injects poison when it penetrates prey.

Note again the distinction between osmotic and hydrostatic pressure (p. 49). The nematocyst is never required actually to contain 140 atmospheres of hydrostatic pressure within itself; such a hydrostatic pressure would doubtless cause it to explode. As the water rushes inside during discharge, the osmotic pressure falls rapidly, while the hydrostatic pressure rapidly increases.

Nematocysts of most cnidarians are not harmful to humans and are a nuisance at worst. However, the stings of a Portuguese man-of-war (see Figure 13.15) and certain jellyfishes are quite painful and sometimes dangerous (see note, p. 273).

Feeding and Digestion

Polyps are typically carnivorous, catching prey with their tentacles, and passing them through the mouth into the gastrovascular cavity for digestion. In *Hydra,* the tentacles are hollow and the tentacle cavity communicates with the gastrovascular cavity. Inside the gastrovascular cavity, gland cells discharge enzymes on the food to begin extracellular digestion, but intracellular digestion occurs in the gastrodermis (see p. 264).

The polyps of a hydrozoan colony capture and digest prey extracellularly, then pass a digestive broth into the common gastrovascular cavity where intracellular digestion occurs (see p. 266). In hydromedusae, both food type and digestive system are similar to that of the polyp. However, the body is oriented with the mouth facing downward in the center of the bell; the mouth is at the end of a tube called the **manubrium** (see Figure 13.11).

Scyphomedusae are typically larger than hydromedusae, but their basic form is similar. The mouth edge is extended as a manubrium, often with four frilly oral arms, sometimes called mouth lobes, used in capturing and ingesting prey (see Figure 13.19).

Anthozoan polyps, such as sea anemones, are carnivorous, feeding on fish or almost any animals of suitable size. They can expand and stretch their tentacles in search of small vertebrates and invertebrates, which they overpower with tentacles and nematocysts and carry to the mouth. A few species feed on minute forms caught by ciliary currents instead of eating large prey. Corals supplement their nutrition by collecting carbon from their algal symbionts (see p. 280).

Nerve Net

The nerve net of cnidarians is one of the best examples of a diffuse nervous system. This plexus of nerve cells is found both at the base of the epidermis and at the base of the gastrodermis, forming two interconnected nerve nets. Nerve processes (axons) end on other nerve cells at synapses or at junctions with sensory cells or effector organs (nematocysts or epitheliomuscular cells). Nerve action potentials are transmitted from one cell to another by release of a neurotransmitter from small vesicles on one side of the synapse or junction (p. 731). One-way transmission between nerve cells in higher animals is ensured because the vesicles are located on only one side of the synapse. However, cnidarian nerve nets are peculiar in that many of the synapses have vesicles of neurotransmitters on both sides, allowing transmission across the synapse in either direction. Another peculiarity of cnidarian nerves is the absence of any sheathing material (myelin) on the axons.

Nerve cells of the net have synapses with slender sensory cells that receive external stimuli, and the nerve cells have junctions with epitheliomuscular cells and nematocysts. Together with the contractile fibers of the epitheliomuscular cells, the sensory-nerve cell net combination is often termed a **neuromuscular system,** an important landmark in the evolution of nervous systems. The nerve net arose early in metazoan evolution, and it has never been completely lost phylogenetically. Annelids have it in their digestive systems. In the human digestive system it is represented by nerve plexuses in the musculature. Rhythmical peristaltic movements of the stomach and intestine are coordinated by this counterpart of the cnidarian nerve net.

Cnidarians do not have a local concentration of nerve cells that would approximate a central nervous system. However, some have argued that the nerve net and ring system in cnidarian medusae is as effective as a central nervous system when processing and responding to stimuli from three-dimensional surroundings. In scyphomedusae and the medusae of cubozoans, nerves are grouped in marginal sense organs, called **rhopalia,** that house chemoreceptors, statocysts, and often ocelli. The nerve nets form two or more systems, including a fast-conducting system to coordinate swimming movements and a slower one to coordinate movements of tentacles. In hydromedusae, two nerve rings that lie at the margin of the bell are formed by condensing the epidermal nerve net. Nerve rings process information from the sense organs and respond by changing swimming direction, pulsation rate, and position of tentacles.

Class Hydrozoa

The majority of Hydrozoa are marine and colonial in form, and a typical life cycle includes both an asexual polyp and a sexual medusa stage, as exemplified by a colonial marine hydroid such as *Obelia* (Gr. *obelias,* round cake).

Hydroid Colonies

A typical hydroid has a base, a stalk, and one or more terminal zooids. The base by which colonial hydroids attach to the substratum is a rootlike stolon, or **hydrorhiza** (see Figure 13.1), which gives rise to one or more stalks called **hydrocauli.** The living cellular part of the hydrocaulus is a tubular **coenosarc** (Figure 13.7), composed of the three typical cnidarian layers surrounding the coelenteron (gastrovascular cavity). The protective covering of the hydrocaulus is a nonliving chitinous sheath, or **perisarc.** Attached to the hydrocaulus are individual polyp animals, or zooids. Most zooids are feeding polyps called **hydranths,** or **gastrozooids.** They may be tubular, bottle-shaped, or vaselike, but all have a terminal mouth and a circle of tentacles. In **thecate** forms, such as *Obelia,* the perisarc continues as a protective cup around the polyp into which it can withdraw for protection (Figure 13.7). In others the polyp is **(athecate)** naked (Figure 13.8). In some forms the perisarc is an inconspicuous, thin film.

Hydranths capture and ingest prey, such as tiny crustaceans, worms, and larvae, thus providing nutrition for the entire colony. After partial extracellular digestion in a hydranth, the digestive broth passes along the common gastrovascular cavity where it is absorbed by gastrodermal cells, and intracellular digestion occurs.

Circulation within the gastrovascular cavity is a function of the ciliated gastrodermis but is also aided by rhythmical contractions and pulsations of the body.

Colonial hydroids bud off new individuals, thus increasing the size of the colony. New feeding polyps arise by budding, and medusa buds also arise on the colony. In *Obelia* these medusae bud from a reproductive polyp called a **gonangium.** Young medusae leave the colony as free-swimming individuals that mature and produce gametes (eggs and sperm) (Figure 13.7). In some species medusae remain attached to the colony and shed their gametes there. In other species medusae never develop and gametes are shed by male and female gonophores (Figure 13.9). Embryonation of the zygote produces a ciliated planula larva that swims freely for a time. Then it attaches to a substratum to develop into a minute polyp that gives rise, by asexual budding, to the hydroid colony, thus completing the life cycle.

Hydroid medusae are usually smaller than scyphozoan medusae, ranging from 2 to 3 mm to several centimeters in diameter (Figure 13.10). The margin of the bell projects inward as a shelflike **velum,** which partly closes the open side of the bell and is used in swimming (Figure 13.11). Muscular pulsations that alternately fill and empty the bell propel the animal forward, aboral side first, with a weak "jet propulsion." Tentacles attached to the bell margin are rich in nematocysts.

The mouth opening at the end of a suspended **manubrium** leads to a stomach and four radial canals that connect with a ring canal around the margin. This ring canal connects with the

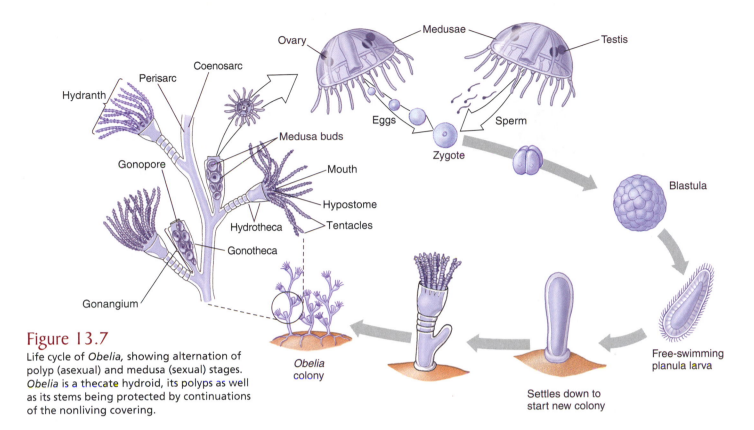

Figure 13.7
Life cycle of *Obelia*, showing alternation of polyp (asexual) and medusa (sexual) stages. *Obelia* is a thecate hydroid, its polyps as well as its stems being protected by continuations of the nonliving covering.

hollow tentacles. Thus the gastrovascular cavity is continuous from mouth to tentacles, and gastrodermis lines the entire system. Nutrition is similar to that of hydranths.

The nerve net is usually concentrated into two nerve rings at the base of the velum. The bell margin has a liberal supply of sensory cells. It usually also bears two kinds of specialized sense organs: **statocysts,** which are small organs of equilibrium (Figure 13.11B), and **ocelli,** which are light-sensitive organs.

The roles played by ectoderm and endoderm during the formation of hydromedusae have been investigated in one species (*Podocoryne carnea*). Here, as is typical for a hydrozoan, medusa buds are produced on the sides of gonangia by lateral budding. The buds have three cell layers: ectoderm, endoderm, and a unique derivative of ectoderm called the **entocodon.** Portions of the entocodon differentiate into smooth and striated muscles. Further smooth muscles in the velum and tentacles originate from ectoderm. The reader may recall that cnidarians lack true mesodermally derived muscle, using epitheliomusclular cells for contraction of polyps and of nonhydrozoan medusae. Thus the presence of smooth and striated muscles in hydrozoan medusae is surprising, as is the ectodermal origin of these muscles. The potential significance of this finding is discussed on page 285.

Freshwater Medusae

The freshwater medusa *Craspedacusta sowberii* (Figure 13.12) (order Hydroida) may have evolved from marine ancestors in the Yangtze River of China. Probably introduced with shipments

Figure 13.8
Athecate hydroids. **A,** *Ectopleura integra*, a solitary polyp with naked hydranths and gonophores. **B,** *Corymorpha* is a solitary hydroid that produces free-swimming medusae, each with a single trailing tentacle.

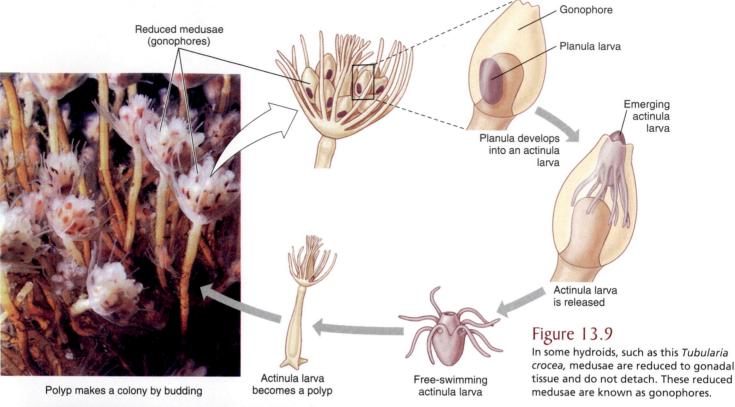

Reduced medusae (gonophores)

Gonophore

Planula larva

Planula develops into an actinula larva

Emerging actinula larva

Actinula larva is released

Polyp makes a colony by budding

Actinula larva becomes a polyp

Free-swimming actinula larva

Figure 13.9

In some hydroids, such as this *Tubularia crocea*, medusae are reduced to gonadal tissue and do not detach. These reduced medusae are known as gonophores.

Figure 13.10

Bell medusa, *Polyorchis penicillatus*, medusa stage of an unknown attached polyp.

of aquatic plants, this interesting form has now been found in many parts of Europe, throughout the United States, and in parts of Canada. Medusae may attain a diameter of 20 mm.

The polyp phase of this animal is tiny (2 mm) and has a very simple form with no perisarc and no tentacles. It occurs in colonies of a few polyps. For a long time its relation to the medusa was not recognized, and thus the polyp was given a name of its own, *Microhydra ryderi*. On the basis of its relationship to the jellyfish and the law of priority, both polyp and medusa should be called *Craspedacusta* (N.L. *craspedon,* velum, + Gr. *kystis,* bladder).

The polyp has three methods of asexual reproduction, as shown in Figure 13.12.

Hydra: A Solitary Freshwater Hydrozoan

Common freshwater hydras (Figure 13.13) live on the underside of aquatic leaves and lily pads in cool, clean freshwater of pools and streams. The hydra family is found throughout the world, with 16 species occurring in North America. Members of this family have been well studied, and much is known about their habits and body plan.

The body of a hydra can extend to a length of 25 to 30 mm or can contract to a tiny, gelatinous mass. It is a cylindrical tube with the aboral end drawn out into a slender stalk, ending in a **basal** (or pedal) **disc** for attachment. Unlike colonial polyps hydras can move about freely by gliding on a basal disc, aided by mucous secretions. They may even turn end over end or detach themselves and, by forming a gas bubble on the basal disc, float to the surface.

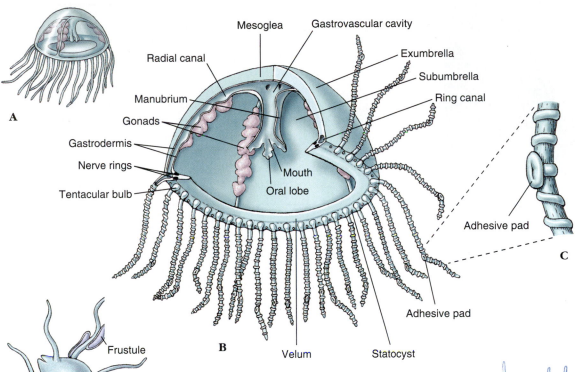

Figure 13.11

Structure of *Gonionemus.*
A, Medusa with typical
tetramerous arrangement.
B, Cutaway view showing
morphology. **C,** Portion of
a tentacle with its adhesive
pad and ridges
of nematocysts. **D,** Tiny
polyp, or hydroid stage, that
develops from the planula
larva. It can produce more
polyps by budding (frustules)
or produce medusa buds.

Feeding and Digestion   Hydras feed on a variety of small crustaceans, insect larvae, and annelid worms. The mouth, located on a conical elevation called the **hypostome,** is encircled by 6 to 10 hollow tentacles that, like the body, can greatly extend when the animal is hungry.

The mouth opens into the **gastrovascular cavity** which communicates with cavities in the tentacles. A hydra awaits its prey with tentacles extended (Figure 13.13). The food organism that brushes against its tentacles may find itself harpooned by scores of nematocysts that render it helpless, even though it may be larger than the hydra. The tentacles move toward the mouth, which slowly widens. Well moistened with mucous secretions, the mouth glides over and around the prey, totally engulfing it.

The activator that actually causes the mouth to open is the reduced form of glutathione, which occurs to some extent in all living cells. Glutathione escapes from the prey through wounds made by nematocysts, but only animals releasing enough of the chemical to activate a feeding response are eaten by a hydra. This mechanism explains how a hydra distinguishes between *Daphnia,* which it relishes, and some other forms that it refuses. If glutathione is placed in water containing hydras, each hydra will go through the motions of feeding even though no prey is present.

Inside the gastrovascular cavity, gland cells discharge enzymes on the food. Digestion is extracellular, but many food particles are drawn by pseudopodia into nutritive-muscular cells of the gastrodermis, where intracellular digestion occurs.

Nutritive-muscular cells are usually tall columnar cells and have laterally extended bases containing myofibrils. The myofibrils run at right angles to the body or tentacle axis and so form a

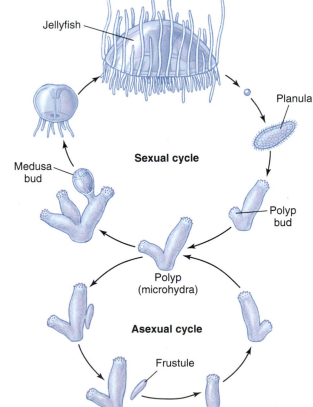

Figure 13.12

Life cycle of *Craspedacusta,* a freshwater hydrozoan. The polyp
has three methods of asexual reproduction: by budding off new
individuals, which may remain attached to the parent (colony
formation); by constricting off nonciliated planula-like larvae
(frustules), which can move around and give rise to new polyps; and
by producing medusa buds, which develop into sexual jellyfish.

Figure 13.13
Hydra catches an unwary water flea with the nematocysts of its tentacles. This hydra already contains one water flea eaten previously.

circular muscle layer. However, this muscle layer in hydras is very weak, and longitudinal extension of the body and tentacles is achieved mostly by increasing the volume of water in the gastrovascular cavity. Water is brought in through the mouth by beating of cilia on the nutritive-muscular cells. Thus, water in the gastrovascular cavity serves as a **hydrostatic skeleton.** The two cilia on the free end of each cell also serve to circulate food and fluids in the digestive cavity. The cells often contain large numbers of food vacuoles. Gastrodermal cells in green hydras *(Chlorohydra)* (Gr. *chloros,* green, + *hydra,* a mythical nine-headed monster slain by Hercules) bear green algae (zoo-chlorellae, phylum Chlorophyta), which give the hydras their color. This existence is probably a case of symbiotic mutualism, since the algae use the respiratory carbon dioxide from the hydra to form organic compounds useful to the host. Algae receive shelter and probably other physiological requirements in return.

Interstitial cells are scattered among the bases of the nutritive cells. They transform into other types of cells when the need arises.

Cnidocytes are not present in the gastrodermis.

Epidermis
The epidermal layer contains epitheliomuscular, interstitial, gland, cnidocyte, and sensory and nerve cells. **Epitheliomuscular cells** compose most of the epidermis and serve both for covering and for muscular contraction (see Figure 13.5). The bases of most of these cells are extended parallel to the tentacle or body axis and contain myofibrils, thus forming a layer of longitudinal muscle next to the mesoglea. Contraction of these fibrils shortens the body or tentacles.

Interstitial cells are undifferentiated stem cells found among the bases of the epitheliomuscular cells. Differentiation of interstitial cells produces cnidoblasts, sex cells, buds, nerve cells, and others, but generally not epitheliomuscular cells (which reproduce themselves).

Gland cells are tall cells, located around the basal disc and mouth, that secrete an adhesive substance for attachment and sometimes a gas bubble for floating (see Figure 13.4).

Over 230 years ago, Abraham Trembley was astonished to discover that isolated sections of the stalk of hydra could regenerate and each become a complete animal. Since then, over 2000 investigations of hydra have been published, and the organism has become a classic model for morphological differentiation. The mechanisms governing morphogenesis have great practical importance, and the simplicity of hydra lends itself to these investigations. Substances controlling development (morphogens), such as those determining which end of a cut stalk will develop a mouth and tentacles, have been discovered, and they may be present in the cells in extremely low concentrations (10^{-10} M).

Cnidocytes occur throughout the epidermis. Hydras have three functional types of cnidae: those that penetrate prey and inject poison (penetrants, see Figure 13.4), those that recoil and entangle prey (volvents), and those that secrete an adhesive substance used in locomotion and attachment (glutinants).

Sensory cells are scattered among the other epidermal cells, especially near the mouth and tentacles and on the basal disc. The free end of each sensory cell bears a flagellum, which is the sensory receptor for chemical and tactile stimuli. The other end branches into fine processes that synapse with nerve cells.

Nerve cells of the epidermis are generally multipolar (have many processes), although in more highly organized cnidarians the cells may be bipolar (with two processes). Their processes (axons) form synapses with sensory cells and other nerve cells and junctions with epitheliomuscular cells and cnidocytes. There are both one-way (morphologically asymmetrical) and two-way synapses with other nerve cells.

Reproduction
Hydras reproduce sexually and asexually. In asexual reproduction, buds appear as outpocketings of the body wall and develop into young hydras that eventually detach from the parent. Most species are dioecious. Temporary gonads (Figure 13.14) usually appear in autumn, stimulated by lower temperatures and perhaps also by reduced aeration of stagnant waters. Testes or ovaries, when present, appear as rounded projections on the surface of the body (Figure 13.14). Eggs in the ovary usually mature one at a time and are fertilized by sperm shed into the water.

Zygotes undergo holoblastic cleavage to form a hollow blastula. The inner part of the blastula delaminates to form the endoderm, and the mesoglea is laid down between ectoderm and endoderm. A cyst forms around the embryo before it breaks loose from the parent, enabling it to survive winter. Young hydras hatch in spring when weather is favorable.

Other Hydrozoans
Members of orders Siphonophora and Chondrophora are among the most specialized Hydrozoa. They usually form polymorphic swimming or floating colonies containing modified medusae and polyps.

There are several types of polyp individuals. Gastrozooids are feeding polyps with a single long tentacle arising from the

Figure 13.14
Hydra with developing bud and ovary.

base of each. Some of these long, stinging tentacles become separated from the feeding polyp and are called **dactylozooids,** or fishing tentacles. These tentacles sting prey and lift it to the lips of feeding polyps. Among the modified medusoid individuals are the **gonophores,** which are little more than sacs containing either ovaries or testes.

Physalia (Gr. *physallis,* bladder), the Portuguese man-of-war (Figure 13.15), is a colony with a rainbow-hued float of blues and pinks that carries it along the surface waters of tropical seas. Many are blown to shore on the eastern coast of the United States. The long, graceful tentacles, actually zooids, are laden with nematocysts and are capable of inflicting painful stings. The float, called a **pneumatophore,** is believed to have expanded from the original larval polyp. It contains a sac arising from the body wall and is filled with a gas similar to air. The float acts as a type of nurse-carrier for future generations of individuals that bud from it and hang suspended in the water. Some siphonophores, such as *Stephalia* and *Nectalia,* possess swimming bells as well as a float.

An interesting mutualistic relationship exists between *Physalia* and a small fish called *Nomeus* (Gr. herdsman) that swims among the tentacles with perfect safety. Why the fish is not stung to death by its host's nematocysts is unclear, but like the anemone fish to be discussed later, *Nomeus* is probably protected by a skin mucus that does not stimulate nematocyst discharge.

Other hydrozoans secrete massive calcareous skeletons that resemble true corals (Figure 13.16). They are sometimes called **hydrocorals.**

Figure 13.15
A Portuguese man-of-war colony, *Physalia physalis* (order Siphonophora, class Hydrozoa). Colonies often drift onto southern ocean beaches, where a hazard to bathers. Each colony of medusa and polyp types is integrated to act as one individual. As many as 1000 zooids may be found in one colony. The nematocysts secrete a powerful neurotoxin.

Class Scyphozoa

Class Scyphozoa (si-fo-zo'a) (Gr. *skyphos,* cup) includes most of the larger jellyfishes, or "cup animals." A few scyphozoans, such as *Cyanea* (Gr. *kyanos,* dark-blue substance), may attain a bell diameter exceeding 2 m and tentacles 60 to 70 m long (Figure 13.17), but most range from 2 to 40 cm in diameter. Most drift or swim in the open sea, some even at depths of 3000 m. Movement is by rhythmical pulsations of the bell.

Bells of different species vary in depth from a shallow saucer shape to a deep helmet or goblet shape, but a velum is never present. Tentacles around the bell, or umbrella, may be numerous or few, and they may be short as in *Aurelia* (L. *aurum,* gold; Figure 13.18) or long as in *Cyanea.* The margin of the umbrella is scalloped, usually with each indentation, or notch bearing a pair of **lappets,** and between them is a sense organ called a **rhopalium** (tentaculocyst). *Aurelia* has eight such notches. Some scyphozoans have 4, others 16. Each rhopalium is club-shaped and contains a hollow statocyst for equilibrium and one or two pits lined with sensory epithelium. In some species the rhopalia also bear ocelli.

The **nervous system** in scyphozoans is a nerve net, with a subumbrella net that controls bell pulsations and another, more diffuse net that controls local reactions such as feeding.

Tentacles, manubrium, and often the entire body surface are well supplied with nematocysts that can deliver painful stings. However, the primary function of scyphozoan nematocysts is not to attack humans but to paralyze prey animals, which are conveyed to the mouth lobes with the help of other tentacles or by bending of the umbrella margin.

The mouth is centered on the subumbrella side. The manubrium usually forms four frilly **oral arms** that are used in capturing and ingesting prey. The mouth leads into a stomach.

Figure 13.16

These hydrozoans form calcareous skeletons that resemble true coral. **A,** *Stylaster roseus* (order Stylasterina) occurs commonly in caves and crevices in coral reefs. These fragile colonies branch in only a single plane and may be white, pink, purple, red, or red with white tips. **B,** Species of *Millepora* (order Milleporina) form branching or platelike colonies and often grow over the horny skeleton of gorgonians (see p. 280), as is shown here. They have a generous supply of powerful nematocysts that produce a burning sensation on human skin, justly earning the common name fire coral. The inset photo shows the extended tentacles.

Internally, extending out from the stomach of scyphozoans are four **gastric pouches** in which gastrodermis extends down in little tentacle-like projections called **gastric filaments.** These filaments are covered with nematocysts to quiet any prey that may still be struggling. Gastric filaments are lacking in

Figure 13.17

Giant jellyfish, *Cyanea capillata* (order Semaeostomeae, class Scyphozoa). A North Atlantic species of *Cyanea* reaches a bell diameter exceeding 2 m. It is known as the "sea blubber" by fishermen.

hydromedusae. A complex system of **radial canals** branches outward from the pouches to a **ring canal** in the margin and forms a part of the gastrovascular cavity.

Aurelia, the familiar moon jelly (Figure 13.18), feeds on small planktonic animals. Its medusae, 7 to 10 cm in diameter, are commonly found in waters off both the east and west coasts of the United States. The bell has relatively short tentacles, not used in prey capture. Food items are caught in mucus on the umbrella surface, and are carried to "food pockets" on the umbrella margin by cilia. From there, ciliated oral lobes move food to the gastrovascular cavity. Cilia in the gastrodermis layer keep a current of water moving to bring food and oxygen into the stomach and to expel wastes.

Sexes are separate, with gonads located in the gastric pouches. Fertilization is internal, with sperm being carried by ciliary currents into the gastric pouch of females. Zygotes may develop in seawater or may be brooded in folds of the oral arms. The ciliated planula larva becomes attached and develops into a **scyphistoma,** a hydralike form (Figure 13.19) that may bud to produce a polyp clone. By a process of **strobilation** the scyphistoma of *Aurelia* forms a series of saucerlike buds, **ephyrae,** and is now called a **strobila** (Figure 13.19). When ephyrae break loose, they grow into mature jellyfish.

Figure 13.18

Moon jellyfish *Aurelia aurita* (class Scyphozoa) is cosmopolitan in distribution. It feeds on planktonic organisms caught in mucus on its umbrella.

The life cycle just described is typical of scyphozoans, but there is variation within this class. In a few species, a larva develops directly into a medusa, and the polyp stage is absent.

The scyphozoans *Cassiopeia* and *Rhizostoma* also exhibit odd body forms. Visitors to Florida often notice the "upside-down jellyfish," *Cassiopeia* (L. mythical queen of Ethiopia) because it is typically found lying on its "back" in strongly sunlit shallow lagoons, in contrast to the usual swimming habit of medusae. It also has an unusual, highly branched, mouth. A similar mouth form can be seen in *Rhizostoma* (Gr. *rhiza,* root, + *stoma,* mouth) from colder waters. Both animals belong to a group of scyphozoans without tentacles on the umbrella margin and with a characteristic oral arm structure. During development, edges of the oral lobes fold over and fuse, forming canals (**arms** or **brachial canals**) that become highly branched. These canals open to the surface at frequent intervals by pores called "mouths"; the original mouth is obliterated in the fusion of the oral lobes. Planktonic organisms caught in the mucus of the frilly oral arms are transported by cilia to the mouths and then up the brachial canals to the gastric cavity. *Cassiopeia's* umbrella margin contracts about 20 times a minute, creating water currents to bring plankton into contact with the mucus and nematocysts of its oral lobes. Its tissues are abundantly supplied with symbiotic dinoflagellates (p. 237) (**zooxanthellae**). As they lie sunning themselves in shallow water, *Cassiopeia* are thus reminiscent of large flowers in more ways than one.

Class Staurozoa

Animals in this class are commonly called stauromedusans and were previously considered unusual scyphozoans, even though their life cycle does not include a medusa phase. The solitary polyp body is stalked (Figure 13.20) and uses an adhesive disc to attach to seaweeds and other objects on the sea bottom. The top of the polyp resembles a medusa, although previous interpretations have noted that the bottom of the "medusa" resembles a polyp. The top of the polyp has eight extensions ("arms"), ending in tentacle clusters, surrounding the mouth. Polyps reproduce sexually. The nonswimming planula develops directly into a new polyp.

Class Cubozoa

The Cubozoa once were considered an order (Cubomedusae) of Scyphozoa. The medusoid is the predominant form (Figure 13.21); the polypoid is inconspicuous and in most cases unknown. Some cubozoan medusae may range up to 25 cm tall, but most are about 2 to 3 cm. In transverse section the bells are almost square. A tentacle or group of tentacles is found at each corner of the square at the umbrella margin. The base of each tentacle is differentiated into a flattened, tough blade called a **pedalium** (Figure 13.21). Rhopalia are present, each housing six eyes in addition to other sense organs. There are two copies of each of three kinds of eyes: two forms of ocelli, and a sophisticated camera-type eye with a cornea and cellular lens. The umbrella margin is not scalloped, and the subumbrella edge turns inward to form a **velarium.** The velarium functions as a velum does in hydrozoan medusae, increasing swimming efficiency, but it differs structurally. Cubomedusae are strong swimmers and voracious predators, feeding mostly on fish in near-shore areas, such as mangrove swamps. Stings of some species can be fatal to humans.

The complete life cycle is known for only one species, *Tripedalia cystophora* (L. *tri,* three, + Gr. *pedalion,* rudder). The polyp is tiny (1 mm tall), solitary, and sessile. New polyps bud laterally, detach, and creep away. Polyps do not produce ephyrae but metamorphose directly into medusae.

Chironex fleckeri (Gr. *cheir,* hand, + *nexis,* swimming) is a large cubomedusa known as the sea wasp. Its stings are quite dangerous and sometimes fatal. Most fatal stings have been reported from tropical Australian waters, usually following quite massive stings. Witnesses have described victims as being covered with "yards and yards of sticky wet string." Stings are very painful, and death, if it is to occur, ensues within a matter of minutes. If death does not occur within 20 minutes after stinging, complete recovery is likely.

Class Anthozoa

Anthozoans, or "flower animals," are polyps with a flowerlike appearance (Figure 13.22). There is no medusa stage. Anthozoa are all marine and occur in both deep and shallow water and in polar seas as well as tropical seas. They vary greatly in size and may be solitary or colonial. Many forms are supported by skeletons.

The class has three subclasses: **Hexacorallia** (or **Zoantharia**), containing sea anemones, hard corals, and others; **Ceriantipatharia,** containing only tube anemones and thorny corals; and **Octocorallia** (or **Alcyonaria**), containing soft and horny corals, such as sea fans, sea pens, sea pansies, and others. Zoantharians and

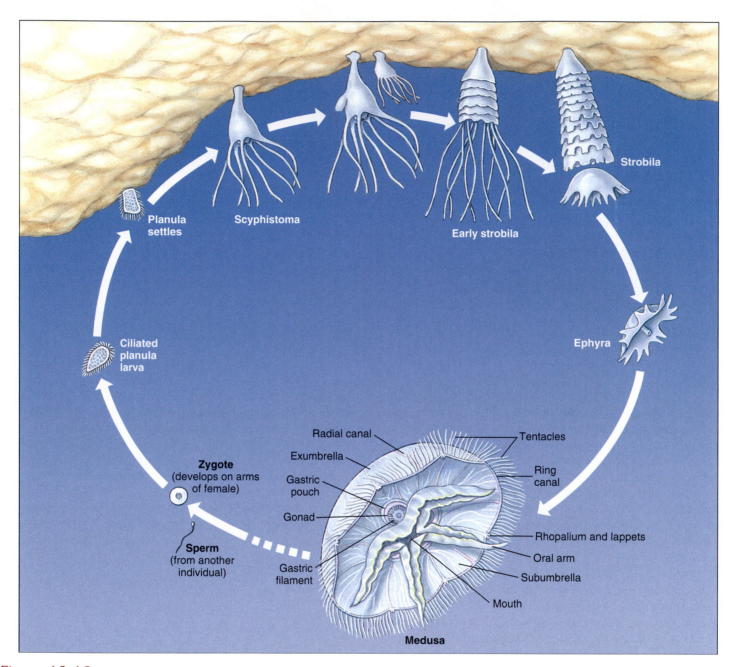

Planula settles

Scyphistoma

Early strobila

Strobila

Ephyra

Ciliated planula larva

Zygote
(develops on arms of female)

Sperm
(from another individual)

Radial canal

Exumbrella

Gastric pouch

Gonad

Gastric filament

Tentacles

Ring canal

Rhopalium and lappets

Oral arm

Subumbrella

Mouth

Medusa

Figure 13.19

Life cycle of *Aurelia*, a marine scyphozoan medusa.

ceriantipatharians have a **hexamerous** plan (of six or multiples of six) or polymerous symmetry and have simple tubular tentacles arranged in one or more circlets on the oral disc. Octocorallians are **octomerous** (built on a plan of eight) and always have eight pinnate (featherlike) tentacles arranged around the margin of the oral disc (Figure 13.23).

The gastrovascular cavity is large and partitioned by septa, or mesenteries, which are inward extensions of the body wall. Where one septum extends into the gastrovascular cavity from the body wall, another extends from the diametrically opposite side; thus, they are said to be **coupled.** In Hexacorallia, the septa are not only coupled, they are also **paired** (Figure 13.24). The muscular arrangement varies among different

groups, but usually features circular muscles in the body wall and longitudinal and transverse muscles in the septa.

The mesoglea is a mesenchyme containing ameboid cells. A general tendency toward biradial symmetry in the septal arrangement occurs in the shape of the mouth and pharynx. There are no special organs for respiration or excretion.

Sea Anemones

Sea anemone (order Actiniaria) polyps are larger and heavier than hydrozoan polyps (see Figure 13.22). Most range from 5 mm or less to 100 mm in diameter, and from 5 mm to 200 mm long, but some grow much larger. Some sea anemones are quite

Figure 13.20
Thaumatoscyphus hexaradiatus are an example of class Staurozoa.

Figure 13.22
Sea anemones are the familiar and colorful "flower animals" of tide pools, rocks, and pilings of the intertidal zone. Most, however, are subtidal, their beauty seldom revealed to human eyes. These are rose anemones. *Tealia piscivora* (subclass Hexacorallia, class Anthozoa).

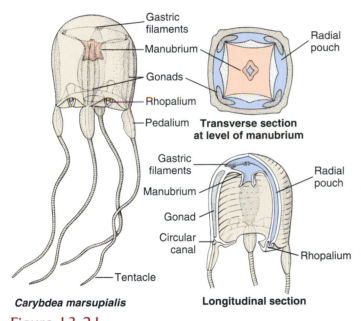

Gastric filaments
Manubrium
Gonads
Rhopalium
Pedalium
Radial pouch

Transverse section at level of manubrium

Gastric filaments
Manubrium
Gonad
Circular canal
Radial pouch
Rhopalium
Tentacle

Longitudinal section

Carybdea marsupialis

Figure 13.21
Carybdea, a cubozoan medusa.

A   B

Figure 13.23
A, Orange sea pen *Ptilosarcus gurneyi* (order Pennatulacea, class Anthozoa). Sea pens are colonial forms that inhabit soft bottoms. The base of the fleshy body of the primary polyp is buried in the bottom. It gives rise to numerous secondary, branching polyps. **B,** Close-up of a gorgonian. The pinnate tentacles characteristic of subclass Octocorallia are apparent.

colorful. Anemones occur in coastal areas all over the world, especially in warmer waters. They attach by means of their pedal discs to shells, rocks, timber, or whatever submerged substrata they can find. Some burrow in mud or sand.

Sea anemones are cylindrical in form with a crown of tentacles arranged in one or more circles around the mouth of the flat **oral disc** (Figure 13.24). The slit-shaped mouth leads into a **pharynx.** At one or both ends of the mouth is a ciliated groove called a **siphonoglyph,** which extends into the pharynx. The siphonoglyph creates a water current directed into the pharynx. Cilia elsewhere on the pharynx direct water outward. Currents thus created supply oxygen and remove wastes. They also help to maintain an internal fluid pressure, providing a hydrostatic skeleton that serves in lieu of a true skeleton as a support for opposing muscles.

The pharynx leads into a large **gastrovascular cavity** that is divided into six radial chambers by means of six pairs of **primary (complete) septa,** or **mesenteries,** extending vertically from the body wall to the pharynx (Figure 13.24). Openings between chambers (septal perforations) in the upper part of the pharyngeal region aid water circulation. Smaller **(incomplete) septa** partially subdivide the large chambers and provide a means of increasing the surface area of the gastrovascular cavity. The free edge of each incomplete septum forms a type of sinuous cord called a **septal filament,** which contains nematocysts and

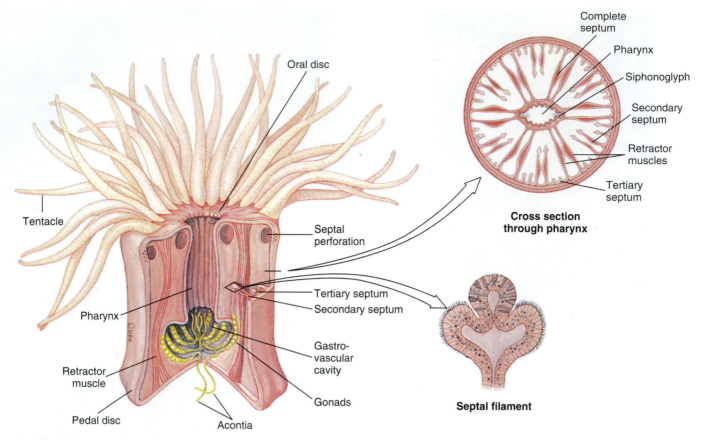

Figure 13.24

Structure of a sea anemone. Free edges of septa and acontia threads are equipped with nematocysts to complete paralyzation of prey begun by the tentacles.

gland cells for digestion. In some anemones (such as *Metridium*) the lower ends of the septal filaments are prolonged into **acontia threads,** also provided with nematocysts and gland cells, which can be protruded through the mouth or through pores in the body wall to aid prey capture or defense. The pores also aid in rapid discharge of water from the body when the animal is endangered and contracts to a small size.

Sea anemones are carnivorous, feeding on fish or almost any live (and sometimes dead) animals of suitable size. Some species live on minute forms caught by ciliary currents.

Feeding behavior in many zoantharians is under chemical control. Some respond to reduced glutathione. In certain others two compounds are involved: asparagine, the feeding activator, causes a bending of tentacles toward the mouth; then reduced glutathione induces swallowing of food.

Muscles are well developed in sea anemones, but the arrangement is quite different from that in Hydrozoa. Longitudinal fibers of the epidermis occur only in the tentacles and oral disc of most species. The strong longitudinal muscles of the column are gastrodermal and located in the septa (Figure 13.24). Gastrodermal circular muscles in the column are well developed.

Most anemones can glide slowly along a substrate on their pedal discs. They can expand and stretch their tentacles in search of small vertebrates and invertebrates, which they overpower with tentacles and nematocysts and carry to their mouth. When

disturbed, sea anemones contract and withdraw their tentacles and oral discs. Some anemones swim to a limited extent by rhythmical bending movements, which may permit escape from enemies such as sea stars and nudibranchs. *Stomphia,* for example, at the touch of a predatory sea star, detaches its pedal disc and creeps or swims to escape (Figure 13.25). This escape reaction is elicited not only by the touch of the star but also by exposure to liquids exuded by the star or to crude extracts made from its tissues. The sea star exudes steroid saponins that are toxic and irritating to most invertebrates. Extracts from nudibranchs also can provoke this reaction in some sea anemones.

Anemones form some interesting mutualistic relationships with other organisms. Many species harbor symbiotic dinoflagellates (zooxanthellae) within their tissues, similar to the hard coral–zooxanthellae association (p. 280), and the anemones profit from the product of algal photosynthesis. Some anemones habitually attach to the shells occupied by certain hermit crabs. The hermit encourages the relationship and, finding its favorite species, which it recognizes by touch, it massages the anemone until it detaches. The hermit crab holds the anemone against its own shell until the anemone is firmly attached. The crab derives some protection against predators by the anemone. The anemone gets free transportation and particles of food dropped by the hermit crab.

Certain damselfishes (anemone fishes) (family Pomacentridae) form associations with large anemones, especially in tropical

A   B

Figure 13.25

A. A sea anemone that swims. **B.** When attacked by a predatory sea star *Dermasterias,* the anemone *Stomphia didemon* (subclass Hexacorallia, class Anthozoa) detaches from the bottom and rolls or swims spasmodically to a safer location.

Indo-Pacific waters (Figure 13.26). An unknown property of the skin mucus of the fish causes an anemone's nematocysts not to discharge, but if some other fish is so unfortunate as to brush the anemone's tentacles, it is likely to become a meal. The anemone obviously provides shelter for the anemone fish, and the fish may help ventilate the anemone by its movements, keep the anemone free of sediment, and even lure an unwary victim to seek the same shelter.

Sexes are separate in some sea anemones, whereas others are hermaphroditic. Monoecious species are **protandrous** (produce sperm first, then eggs). Gonads are arranged on the margins of the septa, and fertilization occurs externally or in the gastrovascular cavity. The zygote develops into a ciliated larva. Asexual reproduction commonly occurs by **pedal laceration** or by longitudinal fission, occasionally by transverse fission or by budding. In pedal laceration, small pieces of the pedal disc break off as the animal moves, and each of these regenerates a small anemone.

Figure 13.26

Orangefin anemone fish (*Amphiprion chrysopterus*) nestles in the tentacles of its sea anemone host. Anemone fishes do not elicit stings from their hosts but may lure unsuspecting other fish to become meals for the anemone.

Hexacorallian Corals

Hexacorallian corals belong to the order Scleractinia, sometimes called true or stony corals. Stony corals might be described as miniature sea anemones that live in calcareous cups that they have secreted (Figures 13.27 and 13.28). Like that of anemones, a coral polyp's gastrovascular cavity is subdivided by septa arranged in multiples of six (hexamerous) and its hollow tentacles surround the mouth, but there is no siphonoglyph.

Instead of a pedal disc, the epidermis at the base of the column secretes a limy skeletal cup, including sclerosepta, which project into the polyp between its true septa (Figure 13.28). Living polyps can retract into the safety of their cup when not feeding. Since the skeleton is secreted below the living tissue rather than within it, the calcareous material is an exoskeleton. In many colonial corals, the skeleton may become massive, building up over many years, with the living coral forming a sheet of tissue over the surface (Figure 13.29). The gastrovascular cavities of the polyps are all connected through this sheet of tissue.

Three other small orders of Zoantharia are recognized.

Tube Anemones and Thorny Corals

Members of subclass Ceriantipatharia have unpaired septa. Tube anemones (order Ceriantharia) (Figure 13.30) are solitary and live buried to the level of the oral disc in soft sediments. They occupy tubes constructed of secreted mucus and threads of nematocyst-like organelles, into which they can withdraw. Thorny or black corals (order Antipatharia) (Figure 13.31) are colonial and attached to a firm substratum. Their skeleton is of a horny material and has thorns. Both of these orders are small in numbers of species and are limited to warmer waters of the sea.

Octocorallian Corals

Octocorallia (Alcyonaria) have strict octomerous symmetry, with eight pinnate tentacles and eight unpaired, complete septa (see Figure 13.23). They are all colonial, and gastrovascular

A

B

C

Figure 13.27

A, Cup coral *Tubastrea* sp. Its polyps form clumps resembling groups of sea anemones. Although often found on coral reefs, *Tubastrea* is not a reef-building coral (ahermatypic) and has no symbiotic zooxanthellae in its tissues. **B,** Polyps of *Montastrea cavernosa* are tightly withdrawn during daytime but open to feed at night, as in **C** (subclass Hexacorallia).

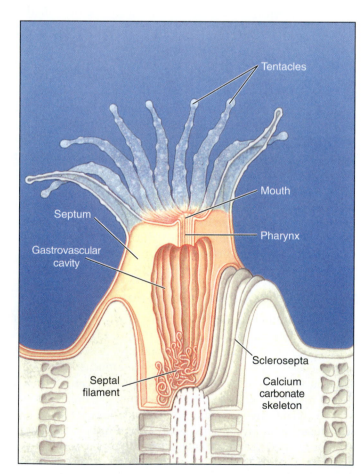

Figure 13.28

Polyp of a hexacorallian coral (order Scleractinia) showing calcareous cup (exoskeleton), gastrovascular cavity, sclerosepta, septa, and septal filaments.

Figure 13.29

Boulder star coral, *Montastrea annularis,* (subclass Hexacorallia, class Anthozoa). Colonies can grow up to 10 feet (3 m) high.

Figure 13.30
A tube anemone (subclass Ceriantipatharia, order Ceriantharia) extends from its tube at night. Its oral disc bears long tentacles around the margin and short tentacles around the mouth.

cavities of the polyps communicate through a system of gastrodermal tubes called **solenia** (Figure 13.32). The solenia run through an extensive mesoglea (**coenenchyme**) in most octocorallians, and the surface of the colony is covered by epidermis. The skeleton is secreted in the coenenchyme and contains limy spicules, fused spicules, or a horny protein, often in combination. Thus the skeletal support of most alcyonarians is an endoskeleton. The variation in pattern among the species of octocorallians lends great variety to the form of the colonies: from soft corals such as *Dendronephthya* (Figure 13.33), with their spicules scattered through the coenenchyme, to the tough, axial supports of sea fans and other gorgonian corals (Figure 13.34), to the fused spicules of organ-pipe coral.

Tentacle   Spiny skeleton

Enlargement of single polyp
B

A

Figure 13.31
A, Colony of *Antipathes,* a black or thorny coral (order Antipatharia, subclass Ceriantipatharia, class Anthozoa). Most abundant in deep waters in the tropics, black corals secrete a tough, proteinaceous skeleton that can be worked into jewelry. **B,** The polyps of Antipatharia have six simple, nonretractile tentacles. The spiny processes in the skeleton are the origin of the common name thorny corals.

Renilla (L. *ren,* kidney, + *illa,* suffix), the sea pansy, is a colony reminiscent of a pansy flower. Its polyps are embedded in the fleshy upper side, and a short stalk that supports the colony is embedded in the seafloor. *Ptilosarcus* (Gr. *ptilon,* feather, + *sarkos,* flesh), a sea pen, belongs to the same order and may reach a length of 50 cm (see Figure 13.23).

The graceful beauty of octocorallians—in hues of yellow, red, orange, and purple—helps to create the "submarine gardens" of the coral reefs.

Coral Reefs

Most students have seen photographs or movies giving a glimpse of the vibrant color and life found on coral reefs, and some may have been fortunate enough to visit a reef. Coral reefs are among the most productive of all ecosystems, and they have a diversity of life-forms rivaled only by tropical rain forests. They are large formations of calcium carbonate (limestone) in shallow tropical seas deposited by living organisms over thousands of years; living plants and animals are confined to the top layer of reefs where they add calcium carbonate to that deposited by their predecessors. The

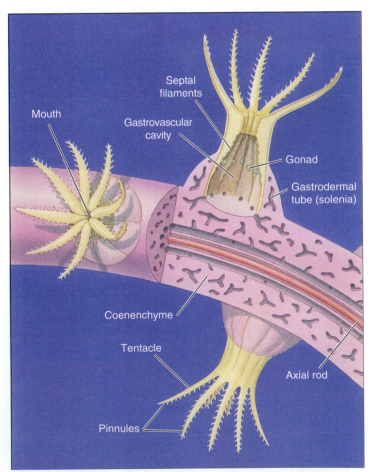

Septal filaments

Mouth

Gastrovascular cavity

Gonad

Gastrodermal tube (solenia)

Coenenchyme

Tentacle

Axial rod

Pinnules

Figure 13.32
Polyps of an octocorallian coral. Note the eight pinnate tentacles, coenenchyme, and solenia. They have an endoskeleton of limy spicules often with a horny protein, which may be in the form of an axial rod.

Figure 13.33

A soft coral, *Dendronephthya* sp. (order Alcyonacea, subclass Octocorallia, class Anthozoa), on a Pacific coral reef. The showy hues of this soft coral vary from pink and yellow to bright red and contribute much color to Indo-Pacific reefs.

However, hermatypic (Gr. *herma,* support, mound, + *typos,* type) corals seem essential to formation of large reefs, since such reefs do not occur where these corals cannot live.

Hermatypic corals require warmth, light, and the salinity of undiluted seawater. These requirements limit coral reefs to shallow waters between 30 degrees north and 30 degrees south latitude and excludes them from areas with upwelling of cold water or areas near major river outflows with attendant low salinity and high turbidity. These corals require light because they have mutualistic dinoflagellates (zooxanthellae) living in their tissues. The microscopic zooxanthellae are very important to the corals; their photosynthesis and fixation of carbon dioxide furnish food molecules for their hosts, they recycle phosphorus and nitrogenous waste compounds that otherwise would be lost, and they enhance the ability of the coral to deposit calcium carbonate.

most important organisms that precipitate calcium carbonate from seawater to form reefs are the scleractinian, **hermatypic** (reef-building) **corals** (Figure 13.29) and **coralline algae.** Not only do coralline algae contribute to the total mass of calcium carbonate, but their precipitation of the substance helps to hold the reef together. Some octocorallians and hydrozoa (especially *Millepora* [L. *mille,* thousand, + *porus,* pore] spp., "fire coral," see Figure 13.16B) contribute in some measure to the calcareous material, and an enormous variety of other organisms contributes small amounts.

Because zooxanthellae are vital to hermatypic corals, and water absorbs light, hermatypic corals rarely live at depths greater than 100 feet (30 m). Interestingly, some deposits of coral reef limestone, particularly around Pacific islands and atolls, reach great thickness—even thousands of feet. Clearly, corals and other organisms could not have grown from the bottom in the abyssal blackness of deep sea and reached shallow water where light could penetrate. Charles Darwin was the first to realize that such reefs began their growth in *shallow* water around volcanic islands; then as the islands slowly sank beneath the sea, the growth of the reefs kept pace with the rate of sinking, thus producing deep deposits.

A   B   C

Figure 13.34

Colonial gorgonian, or horny, corals (order Gorgonacea, subclass Octocorallia, class Anthozoa) are conspicuous components of reef faunas. These examples are from the western Pacific. **A,** Red gorgonian *Melithaea* sp. **B,** A sea fan, *Subergorgia mollis.* **C,** Red whip coral, *Ellisella* sp.

Classification of Phylum Cnidaria

Strong molecular and morphological evidence now indicates that members of the former phylum Myxozoa, commonly occurring fish parasites, are highly derived cnidarians.[1] At this time, we cannot place them with confidence in the following classification; it is possible that they are hydrozoans or a separate class.

Class Hydrozoa (hi-dro-zo′á) (Gr. *hydra,* water serpent, + *zōon,* animal). Solitary or colonial; asexual polyps and sexual medusae, although one type may be suppressed; hydranths with no mesenteries; medusae (when present) with a velum; both freshwater and marine. Examples: *Hydra, Obelia, Physalia, Tubularia.*

Class Scyphozoa (si-fo-zo′a) (Gr. *skyphos,* cup, + *zoōn,* animal). Solitary; polyp stage reduced or absent; bell-shaped medusae without velum; gelatinous mesoglea much enlarged; margin of bell or umbrella typically with eight notches that are provided with sense organs; all marine. Examples: *Aurelia, Cassiopeia, Rhizostoma.*

Class Staurozoa (stò-ro-zōá) (Gr. *stauros,* a cross, + *zōon,* animal). Solitary; polyps only; medusa absent; polyp surface extended into eight clusters of tentacles surrounding mouth; attachment via adhesive disc; all marine. Examples: *Haliclystis, Lucernaria.*

[1]Siddall, M. E., et al. 1995. J. Parasitol. **81:**961–967.

Class Cubozoa (ku′bo-zo′a) (Gr. *kybos,* a cube, + *zōon,* animal). Solitary; polyp stage reduced; bell-shaped medusae square in cross section, with tentacle or group of tentacles hanging from a bladelike pedalium at each corner of the umbrella; margin of umbrella entire, without velum but with velarium; all marine. Examples: *Tripedalia, Carybdea, Chironex, Chiropsalmus.*

Class Anthozoa (an-tho-zo′a) (Gr. *anthos,* flower, + *zōon,* animal). All polyps; no medusae; solitary or colonial; gastrovascular cavity subdivided by at least eight mesenteries or septa bearing nematocysts; gonads endodermal; all marine.

   Subclass Hexacorallia (heks′a-ko-ral′e-a) (Gr. *hex,* six, + *korallion,* coral) **(Zoantharia).** With simple unbranched tentacles; mesenteries in pairs; sea anemones, hard corals, and others. Examples: *Metridium, Anthopleura, Tealia, Astrangia, Acropora.*

   Subclass Ceriantipatharia (se-re-an-tip′a-tha′re-a) (N. L. combination of Ceriantharia and Antipatharia). With simple unbranched tentacles; mesenteries unpaired; tube anemones and black or thorny corals. Examples: *Cerianthus, Antipathes, Stichopathes.*

   Subclass Octocorallia (ok′to-ko-ral′e-a) (L. *octo,* eight + Gr. *korallion,* coral) **(Alcyonaria).** With eight pinnate tentacles; eight complete, unpaired mesenteries; soft and horny corals. Examples: *Tubipora, Alcyonium, Gorgonia, Plexaura, Renilla.*

Several types of reefs are commonly recognized. **Fringing reefs** are close to a landmass with either no lagoon or a narrow lagoon between reef and shore. A **barrier reef** (Figure 13.35) runs roughly parallel to shore and has a wider and deeper lagoon than does a fringing reef. **Atolls** are reefs that encircle a lagoon but not an island. These types of reefs typically slope rather steeply into deep water at their seaward edge. **Patch** or **bank reefs** occur some distance back from the steep, seaward slope in lagoons of barrier reefs or atolls. The so-called Great Barrier Reef, extending 2027 km long and up to 145 km from shore off the northeast coast of Australia, is actually a complex of reef types.

Fringing, barrier, and atoll reefs all have distinguishable zones characterized by different groups of corals and other animals. The side of the reef facing the sea is the **reef front** or **fore reef slope** (Figure 13.35). The reef front is parallel to the shore and perpendicular to the predominant direction of wave travel. It slopes downward into deeper water, sometimes gently at first, then precipitously. Characteristic assemblages of scleractinian corals grow deep on the slope, high near the crest, and in intermediate zones. In shallow water or slightly emergent at the top of the reef front is a **reef crest.** The upper front and the crest bear the greatest force of waves and must absorb great energy during storms. Pieces of coral and other organisms are broken off at such times and thrown shoreward onto the **reef flat,** which slopes down into the lagoon. The reef flat thus accumulates calcareous material eventually ground into coral sand. The sand is stabilized by growth of plants such as turtle grass and coralline algae and ultimately becomes cemented into the mass of the reef by precipitation of carbonates. A reef is not an unbroken wall facing the sea but is

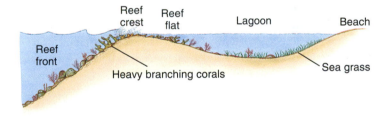

A

B

Figure 13.35

A, Profile of a barrier reef. **B,** Portion of an atoll from the air. Reef slope plunges into deep water at left (dark blue), lagoon at right.

highly irregular, with grooves, caves, crevices, channels through from the flat to the front, and deep, cup-shaped holes ("blue holes"). Octocorallians grow in areas more protected from the full force of waves, as well as on the flat and the deeper areas of the fore reef slope. Many other kinds of organisms inhabit cryptic locations such as caves and crevices.

Enormous numbers of species and individuals of invertebrate groups and fishes populate the reef ecosystem. For example, there are 300 *common* species of fishes on Caribbean reefs and more than 1200 on the Great Barrier Reef complex of Australia. It is marvelous that such diversity and productivity can be maintained, since reefs are washed by nutrient-poor waves of the open ocean. Although relatively little nutrient enters the ecosystem, little is lost because the interacting organisms are so efficient in recycling. The corals even feed on feces of fish swimming over them!

Despite their great intrinsic and economic value, coral reefs in many areas are threatened by a variety of factors, mostly of human origin. These include enrichment with nutrients (from sewage and runoff of agricultural fertilizer from nearby land) and overfishing of herbivorous fishes, both of which contribute to overgrowth of multicellular algae. Agricultural pesticides, sediment from tilled fields and dredging, and oil spills degrade reefs. Such environmental stresses kill corals directly, or make them more susceptible to the numerous coral diseases that have been observed in recent years. Coral reefs are apparently suffering from effects of global warming. When their surrounding water becomes too warm, corals expel their zooxanthellae (coral "bleaching") for unknown reasons. Instances of coral bleaching are increasingly common around the world. Furthermore, higher atmospheric concentrations of carbon dioxide (from burning hydrocarbon fuels) tends to acidify ocean water, which makes precipitation of $CaCO_3$ by corals more difficult metabolically.

PHYLUM CTENOPHORA

Ctenophora (te-nof'o-ra) (Gr. *kteis, ktenos,* comb, + *phora,* pl. of bearing) is composed of about 150 species. All are marine forms occurring in all seas but especially in warm waters. They take their name from eight rows of comblike plates used for locomotion. Common names for ctenophores are "sea walnuts" and "comb jellies."

Except for a few creeping and sessile forms, ctenophores are free-swimming. Although these feeble swimmers are more common in surface waters, ctenophores sometimes occur at considerable depths. They are often at the mercy of tides and strong currents, but they avoid storms by swimming into deeper water. In calm water they may rest vertically with little movement, but when moving they use their ciliated comb plates to propel themselves mouth-end forward.

From an examination of *Pleurobrachia* (Figure 13.36), it is clear that biradial symmetry results from the presence of two tentacles on the body. There is no head, but an oral-aboral axis is present. The mouth leads from a pharynx into a branched digestive

A *Pleurobrachia*   B *Mnemiopsis*

Figure 13.36

A, Comb jelly *Pleurobrachia* sp. (order Cydippida). Its fragile beauty is especially evident at night when it luminesces from its comb rows. **B,** *Mnemiopsis* sp. (order Lobata).

tract ending with an anal pore. The body is transparent and has a gelatinous layer, derived from embryonic ectoderm and endoderm, between the two adult tissue layers. The gelatinous layer contains an extensive set of muscle fibers; fiber patterns are radial, as well as in meridional and latitudinal bands around the body. Muscle fibers are also present in the extensible tentacles.

Ctenophore tentacles capture small planktonic organisms, typically crustaceans such as copepods, from the surrounding waters. Extended tentacles trail in the water, and passing prey are caught by epidermal glue cells called **colloblasts** (Figure 13.37C). Colloblasts contain an adhesive material discharged on contact with prey; the adhesive binds to the prey and the rest of the colloblast cell remains attached to the tentacle. Food-laden tentacles are wiped across the mouth.

Ctenophores with short tentacles may collect food on the ciliated body surface. Ctenophores without tentacles may feed on other gelatinous animals such as medusae, salps (see p. 503), or other ctenophores. Entire prey may be consumed or small parts, such as tentacles, removed. Some ctenophores that feed on cnidarians collect undischarged cnidocytes from their prey and incorporate them into epidermal tissue as a defense mechanism. The ctenophore *Haekelia rubra* (named after Ernst Haeckel, nineteenth-century German zoologist) consumes hydromedusae tentacles in this way.

Ctenophores were previously divided between two classes: Tentaculata and Nuda. Based on evidence that the classes are not monophyletic groups, most biologists discuss ctenophore diversity using seven orders below the class level. Morphological and molecular evidence suggest that one common order (Cydippida) is polyphyletic. One family within Cydippida appears related to members of the order Beroida (see Figure 13.39), whereas others may form new clades. Until new classes have been formulated, we will not discuss ctenophore subgroups.

A fundamental understanding of the ctenophore body plan can be gained from consideration of *Pleurobrachia* and a few other examples.

Representative Type: *Pleurobrachia*

Pleurobrachia (Gr. *pleuron,* side, + L. *brachia,* arms) is about 1.5 to 2 cm in diameter (Figure 13.36). The oral pole bears the mouth opening, and the aboral pole has a sensory organ, the **statocyst.**

Comb Plates

On the surface are eight equally spaced bands called **comb rows,** which extend as meridians from the aboral pole and end before reaching the oral pole (Figure 13.37). Each band consists of transverse plates of long fused cilia called **comb plates** (Figure 13.37D). Ctenophores are propelled by beating of cilia on the comb plates. The beat in each row starts at the aboral end and proceeds successively along the combs to the oral end. All eight rows normally beat in unison. The animal is thus driven forward with the mouth in advance. The animal can swim backward by reversing the direction of the wave.

Tentacles

The two **tentacles** are long, solid and very extensible, and they can be retracted into a pair of **tentacle sheaths.** When completely extended, they may measure 15 cm in length. The surface of the tentacles bears **colloblasts,** or glue cells (Figure 13.37C), which secrete a sticky substance for catching and holding small animals.

Body Wall

The cellular layers of ctenophores are generally similar to those of cnidarians. Between the epidermis and gastrodermis is a gelatinous **collenchyme** that fills most of the interior of the body and contains muscle cells and ameboid cells. Muscle cells are distinct and are not contractile portions of epitheliomuscular cells (in contrast to Cnidaria).

Digestive System and Feeding

The **gastrovascular system** comprises a mouth, a pharynx, a stomach, and a system of gastrovascular canals that branch through the jelly to extend to the comb plates, tentacular sheaths,

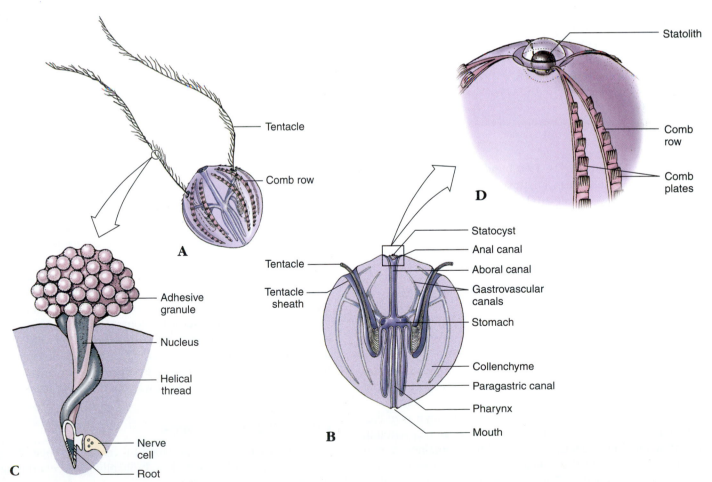

Figure 13.37

Comb jelly *Pleurobrachia,* a ctenophore. **A,** External view. **B,** Hemisection. **C,** Colloblast, an adhesive cell characteristic of ctenophores. **D,** Portion of comb rows showing comb plates, each composed of transverse rows of long fused cilia.

Characteristics of Phylum Ctenophora

1. Eight rows of combs (ctenes) arranged radially around body
2. **Colloblasts,** adhesive cells used in prey capture, present in most
3. Entirely marine
4. Symmetry **biradial;** arrangement of internal canals and position of the paired tentacles change the radial symmetry into a combination of radial and bilateral
5. Body ellipsoidal or spherical in shape with oral and aboral ends; no definite head
6. Adult body with gelatinous middle layer containing muscle cells; derivation of middle cellular layer controversial (ectodermal vs. endodermal) affecting status as diploblastic or triploblastic
7. Complete gut; mouth opens into pharynx; gut with a series of branching **gastrovascular canals;** gut terminates at **anal pore;** wastes exit via anal pore and mouth
8. **Extracellular digestion** in pharynx
9. Two extensible tentacles occur in most
10. Muscular contractions via **muscle fibers (cells),** not epitheliomuscular cells
11. Nervous system consisting of a subepidermal plexus concentrated around the mouth and beneath the comb plate rows; an **aboral sense organ** (statocyst)
12. Reproduction monoecious in most; gonads (endodermal origin) on the walls of the digestive canals, which are under the rows of comb plates; mosaic or regulative cleavage within embryos; cydippid larva
13. No respiratory system
14. No coelomic cavity

Figure 13.38
A cydippid larva.

and elsewhere (Figure 13.37). Two blind canals terminate near the mouth, and an aboral canal that passes near the statocyst and then divides into two small **anal canals** through which undigested material is expelled. Digestion is both extracellular and intracellular.

Respiration and Excretion

Respiration and excretion occur through the body surface.

Nervous and Sensory Systems

Ctenophores have a nervous system similar to that of cnidarians. It features a subepidermal plexus, which is concentrated under each comb plate, but no central control as is found in more complex animals.

The sense organ at the aboral pole is a statocyst (see Figure 13.37B and D). Tufts of cilia support a calcareous statolith, with the whole being enclosed in a bell-like container. Alterations in the position of the animal change the pressure of the statolith on the tufts of cilia. The sense organ coordinates beating of the comb rows but does not trigger their beat.

The epidermis of ctenophores bears abundant sensory cells to detect chemical and other stimuli. When a ctenophore contacts an unfavorable stimulus, it often reverses the beat of its comb plates and moves backward. Comb plates are very sensitive to touch, which often causes them to withdraw into the animal.

Reproduction and Development

Pleurobrachia, like most other ctenophores, is monoecious. Gonads are located on the lining of the gastrovascular canals under the comb plates. Fertilized eggs are discharged through the epidermis into the water.

Cleavage in ctenophores varies among cell lineages. Some lineages are determinate (mosaic), since the parts of the animal that will be formed by each blastomere are determined early in embryogenesis. If one of these blastomeres is removed in the early stages, the resulting embryo will be deficient. Other cell lineages are like those of cnidarians where development is regulative (indeterminate). The free-swimming **cydippid larva** (Figure 13.38) develops gradually into an adult without metamorphosis.

Other Ctenophores

Ctenophores are fragile and beautiful creatures. Their transparent bodies glisten like fine glass, brilliantly iridescent during the day and luminescent at night.

One of the most striking ctenophores is *Beroe* (L. a nymph), which may be more than 100 mm in length and 50 mm in breadth (Figure 13.39A). It is conical or thimble-shaped and flattened in

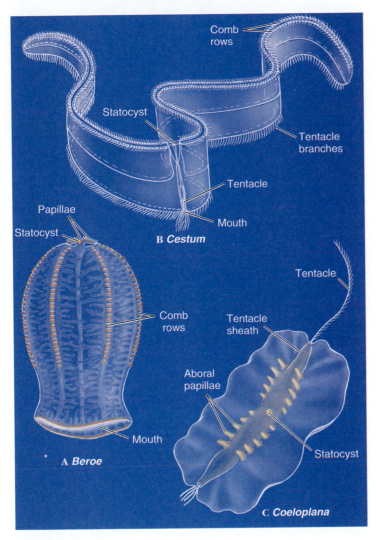

Figure 13.39

Diversity among phylum Ctenophora. **A,** *Beroe* sp. (order Beroida).
B, *Cestum* sp. (order Cestida). **C,** *Coeloplana* sp. (order Platyctenea).

the tentacular plane. The tentacular plane in *Beroe* is defined as where the tentacles would have been, because it has a large mouth but no tentacles. The animal is pink or rusty brown. Its body wall is covered with an extensive network of canals formed by union of the paragastric and meridional canals.

Highly modified ctenophores such as *Cestum* (L. *cestus,* girdle) use sinuous body movements as well as their comb plates in locomotion. Venus's girdle (*Cestum,* Figure 13.39B) is highly compressed in the tentacular plane. Bandlike, it may be more than 1 m long and presents a graceful appearance as it swims in the oral direction. The highly modified *Ctenoplana* (Gr. *ktenos,* comb, + L. *planus,* flat) and *Coeloplana* (Gr. *koilos,* hollow, + L. *planus,* flat) (Figure 13.39C) are rare but interesting because they have disc-shaped bodies flattened in the oral-aboral axis and are adapted for creeping rather than swimming. A common ctenophore along the Atlantic and Gulf coasts is *Mnemiopsis* (Gr. *mneme,* memory, + *opsis,* appearance) (see Figure 13.36B), which has a laterally compressed body with two large oral lobes and unsheathed tentacles.

Nearly all ctenophores emit flashes of luminescence at night, especially such forms as *Mnemiopsis* (see Figure 13.36B). The vivid flashes of light seen at night in southern seas are often caused by members of this phylum.

Since the 1980s population explosions of *Mnemiopsis leidyi* in the Black and Azov Seas have led to catastrophic declines in fisheries there. Inadvertently introduced from the coast of the Americas with ballast water of ships, the ctenophores feed on zooplankton, including small crustaceans and eggs and larvae of fish. The normally inoffensive *M. leidyi* is kept in check in the Atlantic by certain specialized predators, but introduction of such predators into the Black Sea carries its own dangers. However, accidental introduction of the predatory ctenophore *Beroe ovata* into the Black Sea seems to have resulted first in a decline of *M. leidyi* and then in the loss of the new predator.

PHYLOGENY AND ADAPTIVE DIVERSIFICATION

Phylogeny of the Diploblasts

The two phyla discussed in this chapter were traditionally considered diploblastic, radially symmetrical, animals whose body plans were distinct from both the sponges and the triploblastic, bilaterally symmetrical, animals that comprise the rest of the Metazoa. The hard and fast distinctions between the diploblastic and triploblastic conditions are increasingly blurred by detailed morphological studies and by studies of gene expression. Both cnidarians and ctenophores have a gelatinous middle layer surrounded by an outer (epidermal) layer derived from ectoderm and by an inner gut lining derived from endoderm. This body plan is clearly diploblastic, but the presence of cells within the middle gelatinous layer is problematic. If cells of the middle layer derive from endoderm, then they represent a true mesodermal layer of the type seen in triploblasts. If cells in the middle layer derive from ectoderm, then derivation of the middle layer is not the same as in the majority of triploblasts; some workers refer to this layer as ectomesoderm.

In most cnidarians, there are relatively few cells within the mesoglea, so the diploblastic nature of this group has not been much debated. However, the formation of the extensive entocodon layer during development of hydrozoan medusae has led to the suggestion that some cnidarians are triploblastic. Further increasing controversy is the fact that one of the products of the entocodon is striated muscle. Smooth and striated muscles are considered true muscle cells, unlike the contractile epitheliomuscular cells of other cnidarians. In triploblasts, true muscles are produced by mesodermal cells, yet the hydrozoan entocodon is ectodermal in origin, as are other smooth muscle cells present in hydrozoan medusae.

In triploblastic mesoderm, certain genes are expressed during muscle formation; probes of gene expression in hydrozoan medusae showed that genes homologous to those of triploblast mesoderm were expressed in diploblast endoderm. This is not surprising because mesoderm is an endodermal derivative. However, it is surprising also to find one gene associated with mesodermal muscle expressed during the formation of ectodermal muscle in hydrozoans. What does it all mean? Given that true muscle formation from ectoderm occurs in one part of the life cycle of one cnidarian class, it may well represent an independent origin of muscle in one branch of the diploblast lineage, but interpretation of these results is far from settled.

Recent reexamination of the development of ctenophores has led to the observation that muscle cells in the middle layer originate from endodermal cells, rather than ectodermal cells, as initially reported. If further studies confirm the endodermal origin of ctenophore muscle cells, then ctenophores would be triploblastic in the same sense as the bilaterally symmetrical animals.

It might seem that the designation of body symmetry as radial or bilateral would be more straightforward than the issue of number of embryonic layers. However, this too is much debated. The adult cnidarian is clearly radially symmetrical, as is the adult ctenophore. However, studies of the cnidarian planula larva show that it swims with one end consistently moving forward. If the forward end is designated as "anterior," then the larva has a distinct anterior-posterior axis. The planula larva settles onto a hard substratum with the forward-swimming end serving as the attached end. The trailing end of the larva becomes the oral end of the developing polyp.

Recall that *Hox* genes are highly conserved throughout almost all Metazoa and control expression of other genes determining body axis and morphogenesis along the body axis (see p. 172). Cnidarians do not have as many anterior, central, and posterior *Hox* genes as do most triploblasts, but they do have some genes homologous to anterior and posterior *Hox* genes (central *Hox* genes are lacking). Where are the genes homologous to the anterior *Hox* genes of triploblasts expressed in a sea anemone polyp? Are they expressed in the oral or aboral end? They are expressed in the oral end of the polyp. These results are puzzling; did the radially symmetrical cnidarians have a bilaterally symmetrical ancestor, or does the genetic potential for bilateral symmetry predate the bilateral body plan? At present, the answer is not clear.

The reader may have noticed another curiosity in the above account: the forward-swimming end of the larva attaches to the substratum at metamorphosis and becomes the aboral end of the polyp. The aboral end of the polyp is where the posterior *Hox* gene expression is seen. Does this mean that larval orientation is inversely related to polyp orientation? No one knows, but in sponges, where the adult animal has no distinct body axis at all, the larva also has one end that swims forward. With which end does it attach to the substratum? In the sponge *Sycon raphanus,* larvae usually attach with the forward-swimming end, but sometimes with the trailing end, and occasionally with the side of the larva. In most of the triploblastic bilaterally symmetrical animals, the anteroposterior axis of the adult is already obvious in the larval stage, so there is little basis for comparison with sponges and cnidarians. Given the preceding discussion, it is perhaps not surprising that the branching order for the diploblastic phyla is not yet determined. We depict a polytomy for cnidarian, ctenophoran, and placozoan branches.

Cnidarian Phylogeny

Potential antecedents of those hallmark organelles of cnidarians, nematocysts, can be found among some single-celled groups, for example, trichocysts and toxicysts in ciliates and trichocysts in dinoflagellates (see p. 233). In fact, some dinoflagellates have organelles that are strikingly similar in structure to nematocysts.

Relationships among cnidarian classes are still controversial. A fascinating area for speculation is the structure of an ancestral cnidarian life cycle: Which came first, the polyp or the medusa? Of two important hypotheses, one postulates that the basal cnidarian was a trachyline-like hydrozoan with a medusa stage, the other that the basal cnidarian was an anthozoan polyp without a medusa in the life cycle.

If the ancestral cnidarians had life cycles similar to trachyline-like hydrozoans, a larval form would metamorphose directly into a medusa without an intervening polyp. Under this hypothesis, a polyp phase was added later in evolutionary history, explaining why some biologists refer to a polyp as a second larval stage. However, molecular evidence suggests that Anthozoa is the sister taxon to the rest of phylum Cnidaria (see Figure 13.2). Development of medusae would then become a synapomorphy of the other classes, with a subsequent loss of a polyp stage in ancestors of Trachylina. One feature that fits well with this hypothesis is the shared possession of a linear mitochondrial genome in groups with medusae: Anthozoans and all the other metazoans have a circular mitochondrial genome, considered the ancestral condition. We illustrate the taxon Medusozoa as combining all classes with medusae in the life cycle.

Adaptive Diversification

Neither phylum has deviated far from its basic structural plan. In Cnidaria, both polyp and medusa are constructed on the same scheme but medusae have expanded sensory and locomotor capacities.

Nonetheless, cnidarians have achieved large numbers of individuals and species, demonstrating a surprising degree of diversity considering the simplicity of their basic body plan. They are efficient predators, many feeding on prey quite large in relation to themselves. Some are adapted for feeding on small particles. The colonial form of life is well explored, with some colonies growing to great size among corals, and others, such as siphonophores, showing astonishing polymorphism and specialization of individuals within a colony.

Ctenophores have adhered to the arrangement of the comb plates and their biradial symmetry, but they vary in body shape, and presence or absence of tentacles. A few have adopted a creeping or sessile habit.

SUMMARY

Phyla Cnidaria and Ctenophora have a primary radial symmetry; radial symmetry is an advantage for sessile or free-floating organisms because environmental stimuli come from all directions equally. Cnidaria are surprisingly efficient predators because they possess stinging organelles called nematocysts. Both phyla are essentially diploblastic (some triploblastic, depending on the definition of mesoderm), with a body wall composed of epidermis and gastrodermis separated by a mesoglea. The digestive-respiratory (gastrovascular) cavity has a mouth and no anus in cnidarians, but an anal pore is present in ctenophores. Cnidarians are at the tissue level of organization. They have two basic body types (polypoid and medusoid), and in many hydrozoans and scyphozoans the life cycle involves both an asexually reproducing polyp and a sexually reproducing medusa.

That unique organelle, the cnida, is produced by a cnidoblast (which becomes the cnidocyte) and is coiled within a capsule. When discharged, some types of cnidae called nematocysts penetrate prey and inject poison. Discharge is effected by a change in permeability of the capsule and an increase in internal hydrostatic pressure because of the high osmotic pressure within the capsule.

Most hydrozoans are colonial and marine, but the freshwater hydras are commonly demonstrated in class laboratories. They have a typical polypoid form but are not colonial and have no medusoid stage. Most marine hydrozoans are in the form of a branching colony of many polyps (hydranths). Hydrozoan medusae may be free-swimming or remain attached to their colony.

Scyphozoans are typical jellyfishes, in which the medusa is the dominant body form, and many have an inconspicuous polypoid stage. A new class, Staurozoa, has been erected to contain staurodusans, formerly part of Scyphozoa. Cubozoans are predominantly medusoid. They include the dangerous sea wasps.

Anthozoans are all marine and are polypoid; there is no medusoid stage. The most important subclasses are Hexacorallia (with hexamerous or polymerous symmetry) and Octocorallia (with octomerous symmetry). The largest hexacorallian orders contain sea anemones, which are solitary and do not have a skeleton, and stony corals, which are mostly colonial and secrete a calcareous exoskeleton. Stony corals are a critical component in coral reefs, which are habitats of great beauty, productivity, and ecological and economic value. Octocorallia contain the soft and horny corals, many of which are important and beautiful components of coral reefs.

Ctenophora are biradial and swim by means of eight comb rows. Colloblasts, with which they capture small prey, characterize the phylum.

REVIEW QUESTIONS

1. Explain the selective advantage of radial symmetry for sessile and free-floating animals.
2. What characteristics of phylum Cnidaria are most important in distinguishing it from other phyla?
3. Name and distinguish the taxonomic classes in phylum Cnidaria.
4. Distinguish between polyp and medusa forms.
5. Explain the mechanism of nematocyst discharge. How can a hydrostatic pressure of one atmosphere be maintained within the nematocyst until it receives an expulsion stimulus?
6. What is an unusual feature of the nervous system of cnidarians?
7. In what way is a hydra atypical as a hydrozoan?
8. Name and give functions of the main cell types in the epidermis and in the gastrodermis of hydra.
9. What stimulates feeding behavior in hydras?
10. Define the following with regard to hydroids: hydrorhiza, hydrocaulus, coensosarc, perisarc, hydranth, gonangium, manubrium.
11. Give an example of a highly polymorphic, floating, colonial hydrozoan.
12. Distinguish the following from each other: statocyst and rhopalium; scyphomedusae and hydromedusae; scyphistoma, strobila, and ephyrae; velum, velarium, and pedalium; Hexacorallia and Octocorallia.
13. Define the following with regard to sea anemones: siphonoglyph; primary septa or mesenteries; incomplete septa; septal filaments; acontia threads; pedal laceration.
14. Describe three specific interactions of anemones with nonprey organisms.
15. Contrast the skeletons of hexacorallian and alcyonarian corals.
16. Coral reefs generally are limited in geographic distribution to shallow marine waters. How do you explain this observation?
17. Specifically, what kinds of organisms are most important in deposition of calcium carbonate on coral reefs?
18. How do zooxanthellae contribute to the welfare of hermatypic corals?
19. Distinguish each of the following from each other: fringing reefs; barrier reefs; atolls; patch or bank reefs.
20. What characteristics of Ctenophora are most important in distinguishing it from other phyla?
21. How do ctenophores swim, and how do they obtain food?
22. Compare cnidarians and ctenophores, giving five ways in which they resemble each other and five ways in which they differ.
23. Cnidarians and ctenophores are considered diploblastic, but why might some biologists label them triploblastic?

SELECTED REFERENCES

Buddemeier, R. W., and S. V. Smith. 1999. Coral adaptation and acclimatization: a most ingenious paradox. Am. Zool. **39:**1–9. *First of a series of papers in this issue dealing with effects of climatic and temperature changes on coral reefs.*

Coates, M. M. 2003. Visual ecology and functional morphology of Cubozoa (Cnidaria). Integr. Comp. Biol. **43:**542–548. *Collected information on the three different kinds of eyes in cubozoans.*

Collins, A. G., P. Schuchert, A. C. Marques, T. Jankowski, M. Medina, and B. Schierwater. 2006. Medusozoan phylogeny and character evolution clarified by new large and small subunit rDNA and an assessment of the utility of phylogenetic mixture models. Syst. Biol. **55:**97–115. *Authors produce a working cladogram of major cnidarian taxa, including class Staurozoa.*

Crossland, C. J., B. G. Hatcher, and S. V. Smith. 1991. Role of coral reefs in global ocean production. Coral Reefs **10:**55–64. *Because of extensive recycling of nutrients within reefs, their net energy production for export is relatively minor. However, they play a major role in inorganic carbon precipitation by biologically-mediated processes.*

Finnerty, J. R., K. Pang, P. Burton, D. Paulson, and M. Q. Martindale. 2004. Origins of bilateral symmetry: *Hox* and Dpp expression in a sea anemone. Science **304:**1335–1337. *Suggested homology between sea anemone oral end and anterior region of triploblasts based on* Hox *gene expression. Letters and comments that followed this article offer a full discussion of issues.*

Kenchington, R., and G. Kelleher. 1992. Crown-of-thorns starfish management conundrums. Coral Reefs **11:**53–56. *The first article of an entire issue on the starfish:* Acanthaster planci, *a predator of corals. Another entire issue was devoted to this predator in 1990 (p. 447).*

Lesser, M. P. 1997. Oxidative stress causes coral bleaching during exposure to elevated temperatures. Coral Reefs **16:**187–192. *Evidence that reactive types of oxygen molecules, perhaps produced by the zooxanthellae, cause cell damage and expulsion of zooxanthellae. Stress on the zooxanthellae may be induced by increased temperature or UV irradiation.*

Martindale, M. Q., K. Pang, and J. R. Finnerty, J. R. 2004. Investigating the origins of triploblasty: "mesodermal" gene expression in a diploblastic animal, the sea anemone *Nematostella vectensis* (phylum Cnidaria; class Anthozoa). Development **131:**2463–2474. *A discussion of the entocodon problem and putative mesoderm in cnidarians with illustrations and photos showing gene expression during development.*

Pennisi, E. 1998. New threat seen from carbon dioxide. Science **279:**989. *Increase in atmospheric CO_2 is acidifying ocean water, making it more difficult for corals to deposit $CaCO_3$. If CO_2 doubles in the next 70 years, as expected, reef formation will decline by 40%, and by 75% if CO_2 doubles again.*

Podar, M., S. H. D. Haddock, M. I. Sogin, and G. R. Harbison. 2001. A molecular phylogenetic framework for the phylum Ctenophora using 18S rRNA genes. Mol. Phylogen. Evol. **21:**218–230. *Similar evolutionary relationships among ctenophore orders appear from morphological and molecular studies.*

Rosenberg, E., and Y. Loya, 1999. *Vibrio shiloi* is the etiological (causative) agent of *Oculina patagonica* bleaching: general implications. Reef Encounter **25:**8–10. *These investigators believe all coral bleaching is due to bacteria, not just that of* O. patagonica. *The bacterium that they report* (Vibrio shiloi) *needs high temperatures.*

ONLINE LEARNING CENTER

Visit www.mhhe.com/hickmanipz14e for chapter quizzing, key term flash cards, web links, and more!

Thysanozoon nigropapillosum, *a marine turbellarian (order Polycladida).*

Flatworms, Mesozoans, and Ribbon Worms

- PHYLUM ACOELOMORPHA
- PHYLUM PLATYHELMINTHES
- PHYLUM MESOZOA
- PHYLUM NEMERTEA

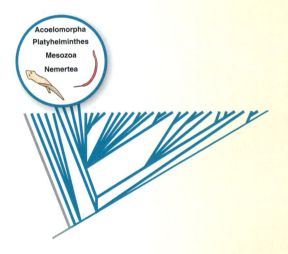

Getting Ahead

For most cnidarians and ctenophores one side of the animal is just as important as any other for snaring prey coming from any direction. But if an animal is active in seeking food, shelter, home sites, and reproductive mates, it requires a different set of strategies and a new body organization. Active, directed movement is most efficient with an elongated body form with head (anterior) and tail (posterior) ends. In addition, one side of the body is kept up (dorsal) and the other side, specialized for locomotion, is kept down (ventral).

The result is a bilaterally symmetrical animal in which the body can be divided along only one plane of symmetry to yield two halves that are mirror images of each other. Furthermore, because it is better to determine where one is going than where one has been, sense organs and centers for nervous control have come to be concentrated on the head. This process is called cephalization. Cephalization and primary bilateral symmetry occur together in almost all animals more complex than sponges, cnidarians, and ctenophores.

Four phyla are discussed in this chapter. In contrast to diploblasts, members of three of the four phyla have **bilaterally symmetrical, triploblastic** bodies. This body plan is present in modern or ancestral forms of all metazoans yet to be discussed. The body is triploblastic because a middle germ layer, mesoderm, derived from endoderm, is present. Three germ layers—ectoderm, endoderm, and mesoderm—produce all adult body structures (see p. 179 for typical derivatives of these layers).

Members of the phylum Mesozoa do not have clearly defined body layers, and their development does not include gastrulation. However, these animals are highly specialized parasites, so some researchers argue that the simple bodies in modern forms were derived from more complex free-living ancestors. Molecular studies indicate that mesozoans possess some genetic and biochemical markers present in triploblastic animals. Their position on the cladogram on the inside front cover reflects this evidence.

Members of two of the bilaterally symmetrical triploblastic phyla discussed here have **acoelomate** (Gr. *a*, not, + *koilōma*, cavity) bodies. A coelom is a cavity that develops *entirely within* mesoderm (see p. 165). Acoelomate bodies do not have a coelom. Readers may recall that diploblasts also lack a coelom but were not called acoelomate; the term is used only for animals possessing mesoderm. Acoelomate taxa do not form a monophyletic group in most analyses, so we use the term only to describe a particular body plan.

Typical acoelomates have only one internal space, the digestive cavity (Figure 14.1). The region between the epidermis and the digestive cavity lining is filled with a cellular, mesodermally derived **parenchyma.** Parenchyma is a form of packing tissue containing more cells and fibers, and less extracellular matrix (ECM), than the mesoglea of cnidarians. Organs are another derivative of mesoderm that increases internal complexity in triploblasts. We see this complexity in members of the acoelomate phyla Acoelomorpha and Platyhelminthes.

Some members of Acoelomorpha are atypical acoelomates because they lack a digestive cavity. In these small worms food particles enter through the mouth and move into a cellular or syncytial mass derived from endoderm. A temporary digestive cavity may form within the endoderm.

A typical acoelomate animal has a gut cavity, lined with endodermally-derived cells, and surrounded by a mass of tissue derived from mesoderm. A typical **coelomate** animal has a gut cavity surrounded by a mesodermal mass that houses a fluid-filled **coelom.** As described in Chapter 9, there are two ways that a coelom can form. In schizocoely, a coelom forms when a solid band of mesoderm surrounding the gut splits open, forming a space where fluid collects. In enterocoely, a coelom forms as endoderm lining the gut pushes outward, enclosing a coelomic cavity.

Each of the two methods of coelom formation co-occurs with other developmental characters to form character suites (see Figure 8.10) defining two metazoan clades: **Protostomia** and **Deuterostomia** (see cladogram on inside front cover). Members of Protostomia have spiral or centrolecithal cleavage, but not radial cleavage. Cleavage is mosaic, not regulative. The embryonic blastopore becomes the mouth, not the anus as in deuterostomes, and when a coelom is present, it forms by schizocoely, not by enterocoely. Most triploblastic metazoan phyla belong to one of these clades. Platyhelminthes are acoelomate protostomes, whereas members of phylum Nemertea, discussed at the end of this chapter, are coelomate protostomes with organ systems. However, this chapter begins with phylum Acoelomorpha, whose members are presumed to have diverged from the main line of bilateral metazoan evolution prior to evolution of protostomes or deuterostomes (see cladogram on inside front cover).

PHYLUM ACOELOMORPHA

Acoelomorphs (Figure 14.2) are small flat worms less than 5 mm in length. The word "worm" is loosely applied to elongate bilateral invertebrate animals without appendages. At one time zoologists considered worms (Vermes) a group in their own right. This group included a highly diverse assortment of forms that were eventually distributed among various phyla. However, zoologists still refer to many animal taxa as flatworms, ribbon worms, roundworms, and segmented worms.

Acoelomorph flatworms typically live in marine sediments, although a few forms are pelagic. There are some species in brackish water. Most acoelomorphs are free-living, but some are symbiotic and others parasitic. The group contains approximately 350 species.

Members of phylum Acoelomorpha were formerly placed in class Turbellaria within phylum Platyhelminthes (see p. 292). Two turbellarian orders, Acoela and Nemertodermatida, now represent two subgroups in phylum Acoelomorpha.

Acoelomorphs have a cellular ciliated epidermis. The parenchyma layer contains a small amount of ECM and circular, longitudinal, and diagonal muscles.

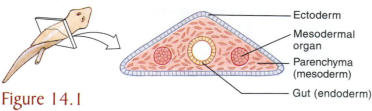

Figure 14.1
Diagram of acoelomate body plan (cross section).

Ectoderm
Mesodermal organ
Parenchyma (mesoderm)
Gut (endoderm)

Figure 14.2
Acoelomorph worms, *Waminoa* sp., on a bubble coral, *Plerogyra sinuosa*.

The digestive system of some acoelomorphs opens from a mouth to a tubelike pharynx followed by a sacklike gut. There is no anus. In many acoels, the gut and pharynx are absent, so the mouth leads into a mass of endodermally derived cells or an endodermally derived syncytial mass (Figure 14.3). When food is passed into temporary spaces, gastrodermal phagocytotic cells digest food intracellularly.

Acoelomorphs are monoecious. The female reproductive organ produces gametes and nutrition for the young at the same time; the resultant yolk-filled eggs are called **endolecithal** eggs. After fertilization, some or all cleavage events produce a duet-spiral pattern of new cells. The duet-spiral pattern may be a defining morphological feature for acoelomorphs, but further study is required for confirmation.

Other defining features proposed for acoelomorphs are biochemical (patterns of neurotransmitters) or rely on details of cellular ultrastructure such as formation of a network of interconnecting rootlets from epidermal cilia.

Acoelomorphs have a distinct anteroposterior axis, but the diffuse collection of nerve cells at the anterior end of the body lacks ganglia typical of a "true" brain. Acoelomorphs have a radial arrangement of nerves in the body, instead of a ladderlike pattern seen in flatworms within phylum Platyhelminthes. Acoelomorph statocysts differ in structure from those of platyhelminths.

Phylogeny of Acoelomorpha

Several phylogenetic studies using molecular characters (for example, mitochondrial genome and myosin II genes) describe acoelomorphs as early diverging, bilaterally symmetrical

Characteristics of Phylum Acoelomorpha

1. Rootlets of epidermal cilia form an interconnecting network
2. Entirely aquatic, some in brackish water, but most in marine sediments
3. Most free-living, some commensal, some parasitic
4. **Bilateral symmetry;** anterior concentration of nerve cells; **body flattened dorsoventrally**
5. Adult body three-layered **(triploblastic)**
6. Body acoelomate, ECM reduced
7. Epidermis cellular
8. Gut absent, or if present, gut incomplete and sacklike
9. **Mesodermal** muscle cells produce longitudinal, circular, and diagonal muscles
10. Diffuse system of anterior neurons connected to radially arranged nerve cords
11. Sense organs include statocysts (organs of balance) and ocelli
12. Asexual reproduction by fragmentation
13. Monoecious sexual reproduction via well-developed gonads, ducts, and accessory organs, internal fertilization; duet-spiral cleavage
14. No excretory or respiratory system

triploblasts. Acoelomorphs have only four or five *Hox* genes, unlike free-living members of Platyhelminthes, which have seven or eight such genes.

CLADES WITHIN PROTOSTOMIA

Most triploblastic metazoans are divided among two large clades or superphyla: Protostomia and Deuterostomia (see cladogram on inside front cover). Division into these two groups is based largely on features of development (see p. 166), although the two groups are also recovered in most phylogenies using molecular characters.

The Protostomia is divided into two large clades: **Lophotrochozoa** and **Ecdysozoa.** Platyhelminthes is the first protostome phylum discussed; it and the remaining phyla included in this chapter belong to Lophotrochozoa. The set of phyla now considered to be lophotrochozoans first appeared as a clade in molecular phylogenies. Prior to the construction of these phylogenies, biologists distinguished groups within the protostomes on the basis of body plan. Acoelomate taxa were assumed to be closely related to each other, as were coelomate protostomes.

Molecular phylogenies grouped acoelomate and coelomate taxa together within the protostomes, instead dividing protostomes into two subsets with distinctive molecular signatures. Some morphological characters were shared by members of each subset. Members of **Ecdysozoa** possess a cuticle that is molted as their bodies grow. Members of **Lophotrochozoa** share either an odd horse-shoe shaped feeding structure, the **lophophore** (see p. 324), or a particular larval form called the **trochophore** (see p. 337).

Trochophore larvae are minute, translucent, and roughly top-shaped (see Figure 16.7). They have a prominent circlet of cilia and sometimes one or two accessory circlets. Trochophores occur in

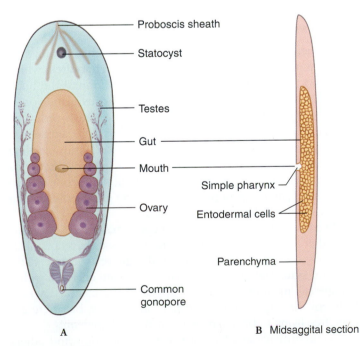

Figure 14.3

A, Generalized acoelomorph flatworm. **B,** Midsagittal section showing gut cavity filled with endodermal cells.

the early development of many marine members of Annelida and Mollusca and are assumed ancestral for these groups. Trochophore-like larvae occur in some marine members of Platyhelminthes, Nemertea, Echiura, and Sipunculida, among others.

PHYLUM PLATYHELMINTHES

Members of phylum Platyhelminthes (Gr. *platys,* flat, + *helmins,* worm), are commonly called flatworms. They range in size from a millimeter or less to some tapeworms that are many meters in length, although most are 1 to 3 cm. Their bodies may be slender, broadly leaflike, or long and ribbonlike. The phylum contains free-living forms, such as the common planarian (Figure 14.4), as well as parasitic flukes and tapeworms.

Because there is no single unique characteristic (synapomorphy) for the phylum as a whole, some authorities argue that the phylum Platyhelminthes is not a valid monophyletic group. However, there is a defining feature for a large parasitic clade within Platyhelminthes. The parasites share an external body covering, called a **syncytial tegument,** or **neodermis,** which contrasts with the cellular ciliated epidermis of most free-living forms. Some morphological features of the free-living flatworms suggest shared ancestry with the parasitic forms. Pending resolution of the intense debate on the nature of this group of worms, we continue to present it as a phylum.

A

B

Figure 14.4
A, Stained planarian. **B,** *Bipalium,* a terrestrial flatworm.

Platyhelminthes is divided into four classes (Figure 14.5): Turbellaria, Trematoda, Monogenea, and Cestoda. Class Turbellaria contains the free-living flatworms, along with some symbiotic and parasitic forms. Most turbellarians are bottom dwellers in marine or freshwater, living under stones or other hard objects. Freshwater planarians can be found in streams, pools, and even hot springs. Terrestrial flatworms are limited to moist places under stones or logs. Class Turbellaria is depicted as a paraphyletic taxon (see p. 294) and awaits a complete revision.

All members of classes Monogenea, Trematoda (flukes), and Cestoda (tapeworms) are parasitic. Most Monogenea are ectoparasites, but all trematodes and cestodes are endoparasitic. Many species have indirect life cycles with more than one host; the first host is often an invertebrate, and the final host is usually a vertebrate. Humans serve as hosts for a number of species.

Many animals covered in this chapter and Chapters 11, 15, 17, 18, 19, 20, and 21 are parasites. People have suffered greatly through the centuries from their parasites and those of their domestic animals. Fleas and bacteria conspired to destroy a third of the European population in the seventeenth century, and malaria, schistosomiasis, and African sleeping sickness have sent millions to their graves. Even today, after successful campaigns against yellow fever, malaria, and hookworm infections in many parts of the world, parasitic diseases in association with nutritional deficiencies are the primary killers of people. Civil wars and environmental changes have led to resurgences in malaria, trypanosomiasis, and leishmaniasis, and global prevalences of intestinal roundworms are unchanged in the last 50 years.

Form and Function

Epidermis, Muscles

Most turbellarians have a cellular, ciliated epidermis resting on a basement membrane. It contains rod-shaped **rhabdites,** which swell and form a protective mucous sheath around the body when discharged with water. Single-cell mucous glands open on the surface of the epidermis (Figure 14.6). Most orders of turbellarians have **dual-gland** adhesive organs in the epidermis. These organs consist of three cell types: viscid and releasing gland cells and anchor cells (Figure 14.7). Secretions of the viscid gland cells apparently fasten microvilli of the anchor cells to the substrate, and secretions of the releasing gland cells provide a quick, chemical detaching mechanism.

In contrast to the ciliated cellular epidermis of most turbellarians, adult members of the three parasitic classes have a nonciliated body covering called a **syncytial tegument** (Figure 14.8). The term **syncytial** means that many nuclei are enclosed within a single cell membrane. It might appear that a completely new body covering appeared in the parasitic classes, but there are some free-living turbellarians with an atypical epidermis.

Some turbellarians have a syncytial epidermis and others have a syncytial "insunk" epidermis where cell bodies (containing nuclei) are located beneath the basement membrane of the

Characteristics of Phylum Platyhelminthes

1. No clear defining feature
2. In marine, freshwater, and moist terrestrial habitats
3. Turbellarian flatworms are mostly free living; classes Monogenea, Trematoda, and Cestoda entirely parasitic
4. **Bilateral symmetry;** definite polarity of anterior and posterior ends; **body flattened dorsoventrally**
5. Adult body three-layered **(triploblastic)**
6. Body acoelomate
7. Epidermis may be cellular or syncytial (ciliated in some); **rhabdites** in epidermis of most Turbellaria; epidermis a syncytial **tegument** in Monogenea, Trematoda, Cestoda, and some Turbellaria
8. Gut incomplete, may be branched, absent in cestodes
9. Muscular system primarily of a sheath form and of mesodermal origin; layers of circular, longitudinal, and sometimes oblique fibers beneath the epidermis
10. Nervous system consisting of a **pair of anterior ganglia** with **longitudinal nerve cords** connected by transverse nerves and located in the mesenchyme in most forms
11. Sense organs include statocysts (organs of balance) and ocelli
12. Asexual reproduction by fragmentation and other methods as part of complex parasite life cycles
13. Most forms monoecious; reproductive system complex, usually with well-developed gonads, ducts, and accessory organs; internal fertilization; development direct in free-swimming forms and those with single hosts; complicated life cycle often involving several hosts in many internal parasites
14. Excretory system of two lateral canals with branches bearing **flame cells (protonephridia);** lacking in some forms
15. Respiratory, circulatory, and skeletal systems lacking; lymph channels with free cells in some trematodes

epidermis. Cell bodies communicate with the surface cytoplasm (distal cytoplasm) by sending extensions upward. These extensions fuse to form the syncytial covering, much as they do in syncytial tegument formation. The term "insunk" is a misnomer because the surface cytoplasm arises from upward extensions of cell bodies, not from cell bodies sinking beneath the basement membrane.

Adults of all members of Trematoda, Monogenea, and Cestoda possess a syncytial covering that entirely lacks cilia and is designated a tegument (Figure 14.8). The larval forms of many of these groups are ciliated, but the ciliated covering is shed once a host is contacted. Epidermal shedding has been suggested as a means of avoiding host immune response. Development of tegument occurs as several surface layers of the epidermis are shed; eventually fused cytoplasmic extensions from cell bodies below the basement membrane become the surface covering of the body. The tegument is sometimes called the **neodermis,** and its shared presence among the parasites is the basis for uniting trematodes, monogeneans, and cestodes in clade **Neodermata** (see Figure 14.5).

The tegument is resistant to the immune system of the host in endoparasites, and it resists digestive juices in tapeworms and others that dwell in a host gut. The syncytial nature of the tegument might render it more resistant because there are no penetrable junctions between cells. The tegument can be both absorptive and secretory. The tegument of one tapeworm has been shown to release enzymes that reduce the effectiveness of the host's digestive system. Tapeworm tegument absorbs nutrients from the host's digestive cavity—tapeworms have neither mouth nor gut.

In the body wall below the basement membrane of flatworms are layers of **muscle fibers** that run circularly, longitudinally, and diagonally. A meshwork of **parenchyma** cells, developed from mesoderm, fills the spaces between muscles and visceral organs. Parenchyma cells in some, perhaps all, flatworms are not a separate cell type but are the noncontractile portions of muscle cells.

Nutrition and Digestion

In general, platyhelminth digestive systems include a mouth, a pharynx, and an intestine (Figure 14.9). In turbellarians, such as the planarian, *Dugesia,* the pharynx is enclosed in a **pharyngeal sheath** (Figure 14.9) and opens posteriorly just inside the mouth, through which it can extend. The intestine has three many-branched trunks, one anterior and two posterior. The whole forms a **gastrovascular cavity** lined with columnar epithelium (Figure 14.9).

Planarians are mainly carnivorous, feeding largely on small crustaceans, nematodes, rotifers, and insects. They can detect food from some distance by means of chemoreceptors. They entangle prey in mucous secretions from the mucous glands and rhabdites. A planarian grips prey with its anterior end, wraps its body around the prey item, extends its pharynx and sucks up food in small amounts.

Intestinal secretions contain proteolytic enzymes for some **extracellular digestion.** Bits of food are sucked into the intestine, where phagocytic cells of the gastrodermis complete digestion **(intracellular).** Undigested food is egested through the pharynx. Monogeneans and trematodes graze on host cells, feeding on cellular debris and body fluids. The mouth of trematodes and monogeneans usually opens at or near the anterior end of their body into a muscular, nonextensible pharynx (Figures 14.10 and 14.18). Posteriorly, their esophagus opens into a blindly ending intestine, which is commonly Y-shaped but may be highly branched or unbranched, depending on the species.

Because cestodes have no digestive tract, they must depend on host digestion, and absorption is confined to small molecules from the host's digestive tract.

Excretion and Osmoregulation

Excretory systems remove wastes from the body, whereas osmoregulatory systems control water balance. Osmoregulatory systems are very common in freshwater animals where concentration gradients between internal fluids and the external environment cause bloating as water enters across the body's

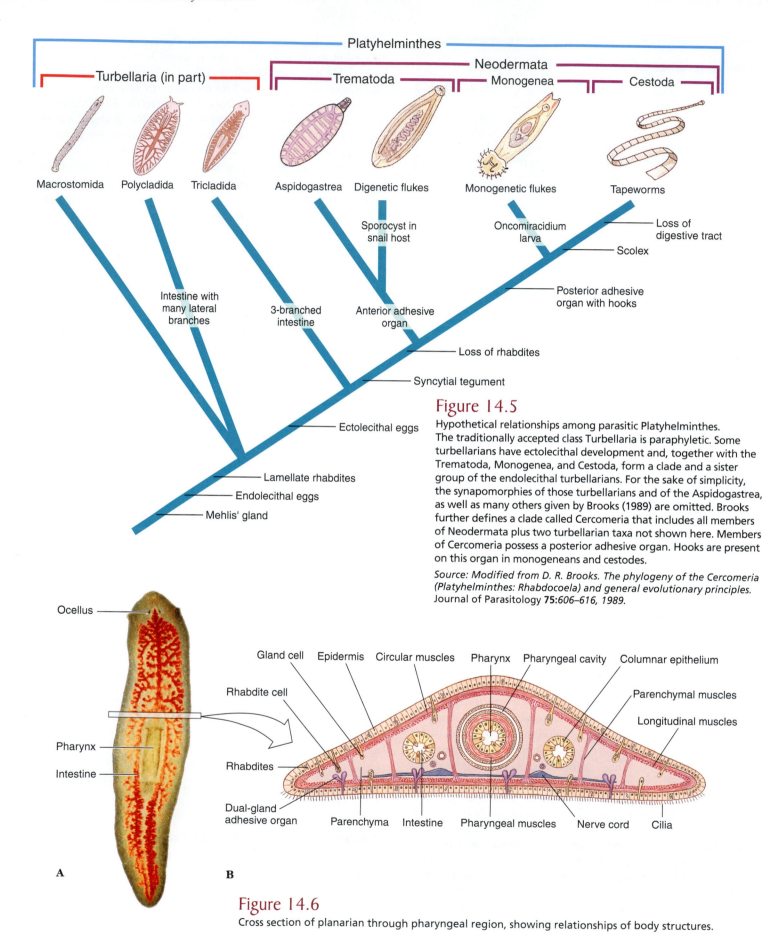

Figure 14.5

Hypothetical relationships among parasitic Platyhelminthes. The traditionally accepted class Turbellaria is paraphyletic. Some turbellarians have ectolecithal development and, together with the Trematoda, Monogenea, and Cestoda, form a clade and a sister group of the endolecithal turbellarians. For the sake of simplicity, the synapomorphies of those turbellarians and of the Aspidogastrea, as well as many others given by Brooks (1989) are omitted. Brooks further defines a clade called Cercomeria that includes all members of Neodermata plus two turbellarian taxa not shown here. Members of Cercomeria possess a posterior adhesive organ. Hooks are present on this organ in monogeneans and cestodes.

Source: Modified from D. R. Brooks. The phylogeny of the Cercomeria (Platyhelminthes: Rhabdocoela) and general evolutionary principles. Journal of Parasitology 75:606–616, 1989.

Figure 14.6

Cross section of planarian through pharyngeal region, showing relationships of body structures.

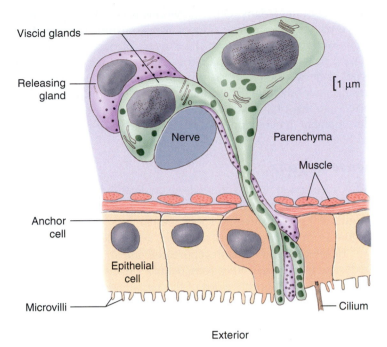

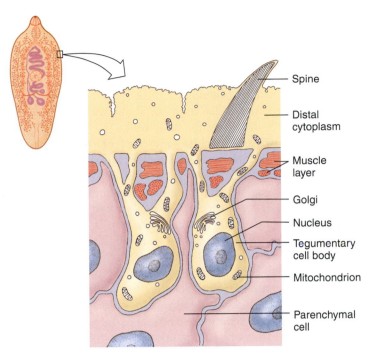

Figure 14.7

Reconstruction of dual-gland adhesive organ of the turbellarian *Haplopharynx* sp. There are two viscid glands and one releasing gland, which lie beneath the body wall. The anchor cell lies within the epidermis, and one of the viscid glands and the releasing gland are in contact with a nerve.

Figure 14.8

Diagrammatic drawing of the structure of the tegument of a trematode *Fasciola hepatica*.

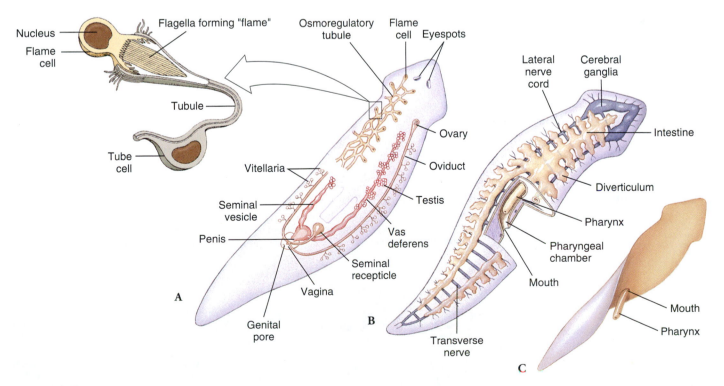

Figure 14.9

Structure of a planarian. **A,** Reproductive and osmoregulatory systems, shown in part. Inset at left is enlargement of flame cell. **B,** Digestive tract and ladder-type nervous system. Pharynx is shown in resting position. **C,** Pharynx extended through ventral mouth.

permeable membranes (see p. 668). Excess water is typically removed via osmoregulatory systems. Sometimes osmoregulation and excretion are combined when waste products are dissolved in water shed from the body. Flatworms have a system of **protonephridia** (Figure 14.9A) that could be used for excretion or osmoregulation. Although a small amount of ammonia is excreted via protonephridia, most metabolic wastes are largely removed by diffusion across the body wall.

Flatworm **protonephridia** (excretory or osmoregulatory organs closed at the inner end) have **flame cells** (Figure 14.9A). A flame cell is cup-shaped with a tuft of flagella extending from the inner face of the cup. In some turbellarians and in all Neodermata, the protonephridia form a **weir** (Old English *wer*, a fence placed in a stream to catch fish); the rim of the cup is elongated into fingerlike projections that extend between similar projections of a tubule cell. The space (lumen) enclosed by the tubule cell continues into collecting ducts that finally open to the outside by pores. Beating flagella (resembling a flickering flame) drive fluid down the collecting ducts and provide a negative pressure to draw fluid through the delicate interlacing projections of the weir. The wall of the duct beyond the flame cell commonly bears folds or microvilli that probably function in resorption of certain ions or molecules.

In planarians, collecting ducts join and rejoin into a network along each side of the animal (Figure 14.9) and may empty through many nephridiopores. This system is mainly osmoregulatory because it is reduced or absent in marine turbellarians, which do not have to expel excess water.

Flame cell protonephridia are present also in the parasitic taxa. Monogeneans usually have two excretory pores opening laterally, near the anterior. Collecting ducts of trematodes empty into an excretory bladder that opens to the exterior by a terminal pore (Figure 14.10). In cestodes there are two main excretory canals on each side that are continuous through the entire length of the worm (see Figure 14.22). They join in the last segment (proglottid, see p. 303) to form an excretory bladder that opens by a terminal pore. When the terminal proglottid is shed, the two canals open separately.

Nervous System

The most primitive flatworm nervous system, found in some turbellarians, is a **subepidermal nerve plexus** resembling the nerve net of cnidarians. Other flatworms have, in addition to a nerve plexus, one to five pairs of **longitudinal nerve cords** lying under the muscle layer. Freshwater planarians have one ventral pair (Figure 14.9B). Connecting nerves form a "ladder-type" pattern. Their brain is a bilobed mass of ganglion cells arising anteriorly from the ventral nerve cords. Neurons are organized into sensory, motor, and association types—an important development in evolution of nervous systems.

Sense Organs

Active locomotion in flatworms has favored not only cephalization in the nervous system but also further evolution of sense organs. **Ocelli,** or light-sensitive eyespots, are common in turbellarians (Figure 14.9A), monogeneans, and larval trematodes.

Tactile cells and chemoreceptive cells are abundant over the body, and in planarians they form definite organs on the auricles (the earlike lobes on the sides of the head). Some species also have statocysts for equilibrium and rheoreceptors for sensing direction of the water current. Sensory endings are abundant around the oral sucker of trematodes and holdfast organ (scolex, p. 303) of cestodes and around genital pores in both groups.

Reproduction and Regeneration

Many turbellarians reproduce both asexually (by fission) and sexually. Asexually, freshwater planarians merely constrict behind the pharynx and separate into two animals, each of which regenerates the missing parts—a quick means of population increase. Reduced population density may cause an increase in the rate of fissioning. In some fissioning forms individuals may remain temporarily attached, forming chains of zooids (Figure 14.11).

The considerable powers of regeneration in planarians have provided an interesting system for experimental studies of development. For example, a piece excised from the middle of a planarian can regenerate both a new head and a new tail. However, the piece retains its original polarity: a head grows at the anterior end and a tail at the posterior end. An extract of heads added to a culture medium containing headless worms will prevent regeneration of new heads, suggesting that substances in one region will suppress regeneration of the same region at another level of the body.

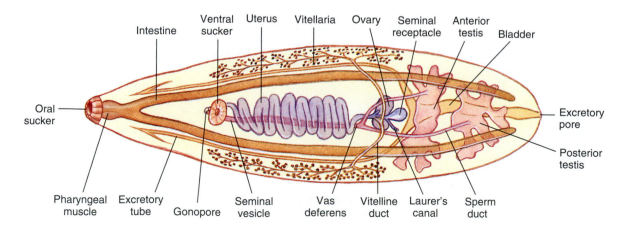

Figure 14.10
Structure of human liver fluke *Clonorchis sinenesis.*

Intestine · Ventral sucker · Uterus · Vitellaria · Ovary · Seminal receptacle · Anterior testis · Bladder · Oral sucker · Excretory pore · Posterior testis · Pharyngeal muscle · Excretory tube · Gonopore · Seminal vesicle · Vas deferens · Vitelline duct · Laurer's canal · Sperm duct

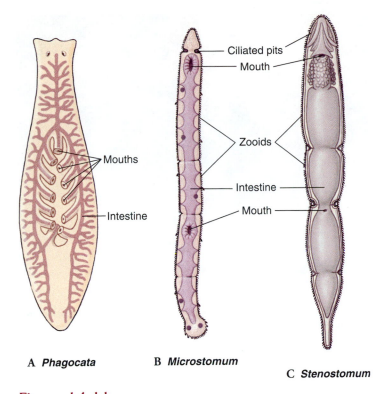

Figure 14.11

Some small freshwater turbellarians. **A,** *Phagocata* has numerous pharynges. **B** and **C,** Incomplete fission results for a time in a series of attached zooids.

Trematodes undergo asexual reproduction in their intermediate hosts, snails. Details of their astonishing life cycles are described on page 298. Some juvenile cestodes show asexual reproduction, budding off hundreds, or in some cases, even millions, of offspring (p. 306).

Almost all flatworms are monoecious (hermaphroditic) but practice cross-fertilization. In some turbellarians the yolk for nutrition of a developing embryo is contained within the egg cell itself **(endolecithal),** and embryogenesis shows spiral determinate cleavage typical of protostomes (p. 169). Possession of endolecithal eggs is considered ancestral for flatworms. Trematodes, monogeneans, cestodes, and many of the turbellarian groups share a derived condition in which female gametes contain little or no yolk, and yolk is contributed by cells released from separate organs called **vitellaria.** Yolk cells are conducted toward a juncture with the **oviduct** by **vitelline ducts** (Figures 14.9 and 14.10). Usually a number of yolk cells surround the zygote within the eggshell; thus development is **ectolecithal.** Cleavage is affected in such a way that a spiral pattern cannot be distinguished. The entire package consisting of yolk cells and zygote, surrounded by the eggshell, moves into the **uterus** and finally is released through a common genital pore or a separate uterine pore (Figure 14.10).

Access to the yolk in ectolecithal eggs is problematic for the developing embryo, but the outermost epidermal layers of some embryos grow outward to encompass the yolk. As the outermost epidermal layer is shed during development, successive inner layers enclose and utilize yolk. It has been hypothesized that the shedding of epidermal layers to permit yolk intake in ectolecithal turbellarians formed the evolutionary basis for the shedding of larval epidermal layers as the syncytial tegument forms.

Male reproductive organs include one, two, or more **testes** connected to **vasa efferentia** that join to become a single **vas deferens.** The vas deferens commonly leads into a **seminal vesicle** and hence to a papilla-like **penis** or an extensible copulatory organ called a **cirrus.**

During breeding season turbellarians develop both male and female organs, which usually open through a common genital pore (Figure 14.9A). After copulation one or more fertilized eggs and some yolk cells become enclosed in a small cocoon. The cocoons are attached by little stalks to the underside of stones or plants. Embryos emerge as juveniles that resemble mature adults. In some marine forms embryos develop into ciliated free-swimming larvae very similar to the trochophore larvae of other members of Lophotrochozoa.

Monogeneans hatch as free-swimming larvae that attach to the next host and develop into juveniles. Larval trematodes emerge from the eggshell as ciliated larvae that penetrate a snail intermediate host, or they may hatch only after being eaten by a snail. Most cestodes hatch only after being consumed by an intermediate host. Many different animals serve as intermediate hosts, and depending on the species of tapeworm may require one or more specific intermediate hosts to complete the life cycle.

Class Turbellaria

Turbellarians are mostly free-living worms that range in length from 5 mm or less to 50 cm. They can be found under objects in marine, freshwater, and terrestrial habitats. There are about six species of terrestrial turbellarians in the United States. Their mouth is on the ventral side and leads into a gut cavity, often via a pharynx. Turbellarians are often distinguished on the basis of the form of the gut (present or absent; simple or branched; pattern of branching) and pharynx (simple; folded; bulbous). Except for order Polycladida (Gr. *poly,* many, + *klados,* branch), turbellarians with endolecithal eggs have a simple gut and a simple pharynx. In a few turbellarians there is no recognizable pharynx. Polyclads have a folded pharynx and a gut with many branches (Figure 14.12). Polyclads include many marine forms of moderate to large size (3 to more than 40 mm) (Figure 14.13), and a highly branched intestine is correlated with larger size in turbellarians. Members of order Tricladida (Gr. *treis,* three, + *klados,* branch), which are ectolecithal and include freshwater planaria, have a three-branched intestine (Figure 14.12).

Turbellarians typically are creeping forms that combine muscular with ciliary movements to achieve locomotion. Very small planaria swim by means of their cilia. Others move by gliding, head slightly raised, over a slime track secreted by the marginal adhesive glands. Beating of epidermal cilia in the slime track moves the animal forward, while rhythmical muscular waves can be seen passing backward from the head. Large polyclads and terrestrial turbellarians crawl by muscular undulations, much in the manner of a snail.

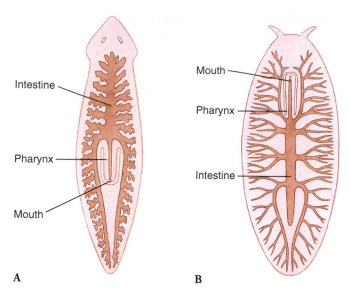

Figure 14.12

Intestinal pattern of two orders of turbellarians. **A,** Tricladida. **B,** Polycladida.

Figure 14.13

Pseudobiceros hancockanus, a marine polyclad turbellarian. Marine polyclads are often large and beautifully colored. The orange polyps of *Tubastrea aurea,* an ahermatypic coral, and *Aplidium cratiferum,* a colonial tunicate (see Chapter 23) that looks something like cartilage, are also in the photograph.

Class Trematoda

Trematodes are all parasitic flukes, and as adults they are almost all found as endoparasites of vertebrates. They are chiefly leaf-like in form with one or more suckers but lack the opisthaptor present in monogenean flukes (p. 303).

Other structural adaptations for parasitism are apparent: various penetration glands or glands to produce cyst material, organs for adhesion such as suckers and hooks, and increased reproductive capacity. Otherwise, trematodes share several characteristics with ectolecithal turbellarians, such as a well-developed alimentary canal (but with the mouth at the anterior, or cephalic, end) and similar reproductive, excretory, and nervous systems, as well as a musculature and parenchyma that are only slightly modified from those of turbellarians. Sense organs are poorly developed.

Of the subclasses of Trematoda, subclass Aspidogastrea is the least well known. Most parasites in this group have only a single host, usually a mollusc. If there is a second host, it is usually a fish or turtle. Subclass Digenea (Gr. *dis,* double, + *genos,* race) is the largest and best known, with many species of medical and economic importance.

Subclass Digenea

With rare exceptions, digeneans have a complex life cycle, the first **(intermediate)** host being a mollusc and the **definitive** host (the host in which sexual reproduction occurs, sometimes called the **final** host) being a vertebrate. In some species a second, and sometimes even a third, intermediate host intervenes. The group has radiated greatly, and its members parasitize almost all kinds of vertebrate hosts. Digeneans inhabit, according to species, a wide variety of sites in their hosts: all parts of the digestive tract, respiratory tract, circulatory system, urinary tract, and reproductive tract.

Among the world's most amazing biological phenomena are digenean life cycles. Although cycles of different species vary widely in detail, a typical example would include an adult, egg

(shelled embryo), miracidium, sporocyst, redia, cercaria, and metacercaria stages (Figure 14.14). The shelled embryo or larva usually passes from the definitive host in excreta and must reach water to develop further. There, it hatches to a free-swimming, ciliated larva, the **miracidium.** The miracidium penetrates the tissues of a snail, where it transforms into a **sporocyst.** Sporocysts reproduce asexually to yield either more sporocysts or a number of **rediae.** Rediae, in turn, reproduce asexually to produce more rediae or to produce **cercariae.** In this way a single egg can give rise to an enormous number of progeny. Cercariae emerge from the snail and can either penetrate the final host directly (for example, the blood fluke *Schistosoma mansoni*), penetrate a second intermediate host (for example, the lung fluke *Paragonimus westermani*), or encyst on aquatic vegetation (for example, the intestinal fluke *Fasciolopsis buski*). At this point, cercariae develop into **metacercariae,** which are essentially juvenile flukes. When the metacercariae are eaten by the final host, the juveniles migrate to the site of final infection and grow into adults.

Some of the most serious parasites of humans and domestic animals belong to Digenea (Table 14.1). The first digenean life cycle revealed was that of *Fasciola hepatica* (L. *fasciola,* a small bundle, band), which causes "liver rot" in sheep and other ruminants. Adult flukes live in the bile passage of the liver, and eggs are passed in feces. After hatching, a miracidium penetrates a snail to become a sporocyst. There are two generations of rediae, and the cercaria encysts on vegetation. When infested vegetation is eaten by a sheep or other ruminant (or sometimes humans), the metacercariae excyst and grow into young flukes.

Clonorchis sinensis: **Liver Fluke in Humans.** *Clonorchis* (Gr. *clon,* branch, + *orchis,* testis) is the most important liver fluke of humans and is common in many regions of eastern Asia, especially in China, Southeast Asia, and Japan. Cats, dogs, and pigs are also often infected.

Classification of Phylum Platyhelminthes

Class Turbellaria (tur′bel-lar′e-a) (L. *turbellae* [pl.], stir, bustle, + *aria,* like or connected with): **turbellarians.** Usually free-living forms with soft, flattened bodies; covered with ciliated epidermis containing secreting cells and rodlike bodies (rhabdites); mouth usually on ventral surface sometimes near center of body; no body cavity except intercellular lacunae in parenchyma; mostly hermaphroditic, but some have asexual fission. A paraphyletic taxon awaiting revision. Examples: *Dugesia* (planaria), *Microstomum, Planocera.*

Class Trematoda (trem′a-to′da) (Gr. *trema,* with holes, + *eidos,* form): **digenetic flukes.** Body of adults covered with a syncytial tegument without cilia; leaflike or cylindrical in shape; usually with oral and ventral suckers, no hooks; alimentary canal usually with two main branches; mostly monoecious; development indirect, with first host a mollusc, final host usually a vertebrate; parasitic in all classes of vertebrates. Examples: *Fasciola, Clonorchis, Schistosoma.*

Class Monogenea (mon′o-gen′e-a) (Gr. *mono,* single, + *gene,* origin, birth): **monogenetic flukes.** Body of adults covered with a syncytial tegument without cilia; body usually leaflike to cylindrical in shape; posterior attachment organ with hooks, suckers, or clamps, usually in combination; monoecious; development direct, with single host and usually with free-swimming, ciliated larva; all parasitic, mostly on skin or gills of fish. Examples: *Dactylogyrus, Polystoma, Gyrodactylus.*

Class Cestoda (ses-to′da) (Gr. *kestos,* girdle, + *eidos,* form): **tapeworms.** Body of adults covered with nonciliated, syncytial tegument; general form of body tapelike; scolex with suckers or hooks, sometimes both, for attachment; body usually divided into series of proglottids; no digestive organs; usually monoecious; larva with hooks; parasitic in digestive tract of all classes of vertebrates; development indirect with two or more hosts; first host may be vertebrate or invertebrate. Examples: *Diphyllobothrium, Hymenolepis, Taenia.*

Structure. The worms vary from 10 to 20 mm in length (Figure 14.10). Their structure is typical of many trematodes in most respects. They have an **oral sucker** and a **ventral sucker.** The **digestive system** consists of a pharynx, a muscular esophagus, and two long, unbranched intestinal ceca. The **excretory system** consists of two protonephridial tubules, with branches lined with flame cells. The two tubules unite to form a single median bladder that opens to the outside. The nervous system, like that of other flatworms, consists of two cerebral ganglia connected to longitudinal cords that have transverse connectives.

As is common in flukes, about 80% of the body is devoted to reproduction. The **reproductive system** is hermaphroditic and complex. They have two branched **testes** that unite to form a single **vas deferens,** which widens into a **seminal vesicle.** The seminal vesicle leads into an **ejaculatory duct,** which terminates at the genital opening. The female system contains a branched **ovary** with a short **oviduct,** which is joined by ducts from the **seminal receptacle** and **vitellaria** at an **ootype.** The ootype is surrounded by a glandular mass, **Mehlis's gland,** of uncertain function. From Mehlis's gland the much-convoluted **uterus** runs to the genital

pore. Cross-fertilization between individuals is usual, and sperm are stored in the seminal receptacle. When an oocyte is released from the ovary, it is joined by a sperm and a group of vitelline cells and is fertilized. The vitelline cells release a proteinaceous shell material, which is stabilized by a chemical reaction; the Mehlis's gland secretions are added, and the egg passes into the uterus.

Life Cycle. The normal habitat of the adults is in the bile passageways of humans and other fish-eating mammals (Figure 14.14). Eggs, each containing a complete miracidium, are shed into water with the feces but do not hatch until they are ingested by the snail *Parafossarulus* or related genera. Eggs, however, may live for some weeks in water. In a snail a miracidium enters the tissues and transforms into a sporocyst, which produces one generation of rediae. A redia is elongated, with an alimentary canal, a nervous system, an excretory system, and many germ cells in the process of development. Rediae pass into the liver of the snail where the germ cells continue embryonation and give rise to tadpolelike cercariae. These two asexual stages in the intermediate host allow a single miracidium to produce as many as 250,000 infective cercariae.

Cercariae escape into the water and swim until they encounter a fish of family Cyprinidae. They then bore under scales into the fish's muscles. Here cercariae lose their tails and encyst as metacercariae. If a mammal eats raw or undercooked infected fish, the metacercarial cyst dissolves in the intestine, and young flukes apparently migrate up the bile duct, where they become adults. There the flukes may live for 15 to 30 years.

The effect of these flukes on a person depends mainly on the extent of the infection but includes abdominal pain and other abdominal symptoms. A heavy infection can cause a pronounced cirrhosis of the liver and death. Cases are diagnosed through fecal examinations. Destruction of snails that carry larval stages has been used as a method of control. However, the simplest method to avoid infection is to make sure that all fish used as food is thoroughly cooked.

***Schistosoma:* Blood Flukes.** **Schistosomiasis,** infection with blood flukes of genus *Schistosoma* (Gr. *schistos,* divided, + *soma,* body), ranks among the major infectious diseases in the world, with 200 million people infected. The disease is widely prevalent over much of Africa and parts of South America, West Indies, Middle East, and Far East. The old generic name for the worms was *Bilharzia* (from Theodor Bilharz, German parasitologist who discovered *Schistosoma haematobium*), and the infection was called bilharziasis, a name still used in many areas.

Blood flukes differ from most other flukes in being dioecious and having the two branches of the digestive tube united into a single tube in the posterior part of the body. Males are broader and heavier and have a large, ventral groove, the **gynecophoric canal,** posterior to the ventral sucker. The gynecophoric canal embraces the long, slender female (Figure 14.15).

Three species account for most schistosomiasis in humans: *S. mansoni,* which lives primarily in veins draining the large intestine; *S. japonicum,* which occurs mostly in veins of the small intestine; and *S. haematobium,* which lives in veins of the urinary bladder. *Schistosoma mansoni* is common in parts of Africa, Brazil, northern South America, and the West Indies; species of

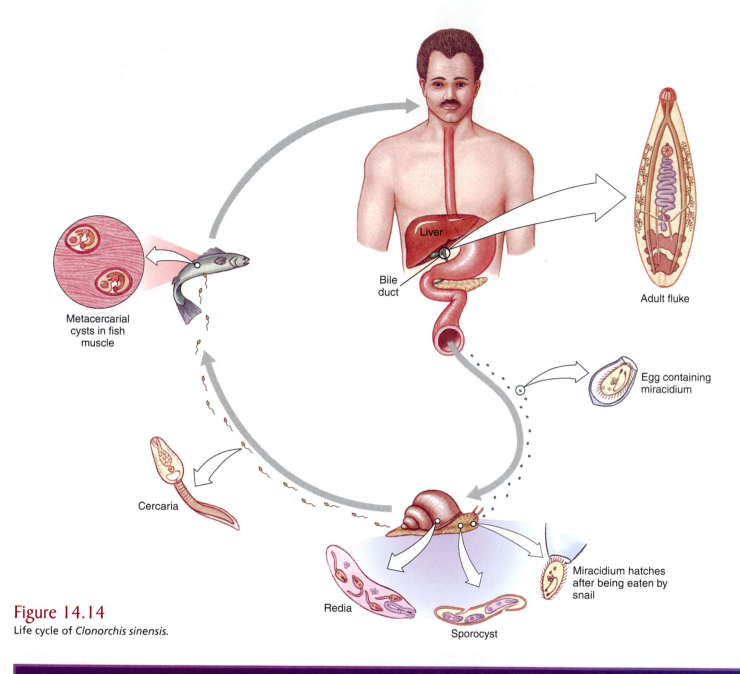

Figure 14.14

Life cycle of *Clonorchis sinensis*.

TABLE 14.1

Examples of Flukes Infecting Humans

Common and Scientific Names	Means of Infection; Distribution and Prevalence in Humans
Blood flukes (*Schistosoma* spp.); three widely prevalent species, others reported	Cercariae in water penetrate skin; 200 million people infected with one or more species
S. mansoni	Africa, South and Central America
S. haematobium	Africa
S. japonicum	Eastern Asia
Chinese liver flukes (*Clonorchis sinensis*)	Eating metacercariae in raw fish; about 30 million cases in eastern Asia
Lung flukes (*Paragonimus* spp.), seven species, most prevalent is *P. westermani*	Eating metacercariae in raw freshwater crabs, crayfish; Asia and Oceania, sub-Saharan Africa, South and Central America; several million cases in Asia
Intestinal fluke (*Fasciolopsis buski*)	Eating metacercariae on aquatic vegetation; 10 million cases in eastern Asia
Sheep liver fluke (*Fasciola hepatica*)	Eating metacercariae on aquatic vegetation; widely prevalent in sheep and cattle, occasional in humans

Figure 14.15

A, Adult male and female *Schistosoma japonicum* in copulation. The male has a long gynecophoric canal that holds the female. Humans are usually hosts of adult parasites, found mainly in Africa but also in South America and elsewhere. Humans become infected by wading or bathing in cercaria-infested waters. **B,** Life cycle of *Schistosoma mansoni.*

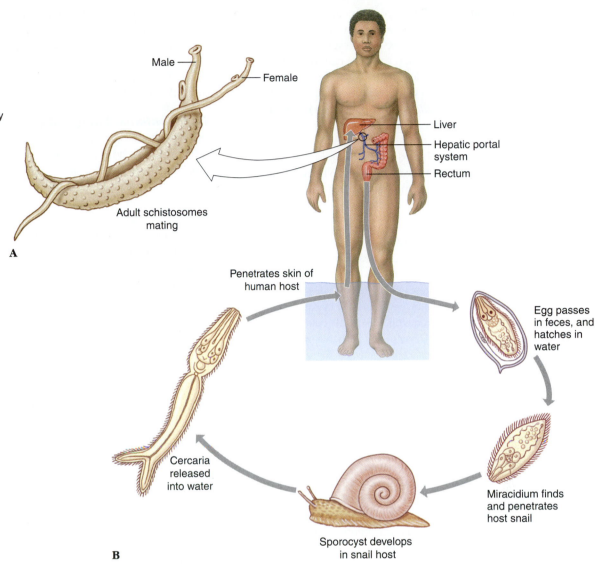

Male
Female
Adult schistosomes mating
A

Liver
Hepatic portal system
Rectum

Penetrates skin of human host

Egg passes in feces, and hatches in water

Cercaria released into water

Miracidium finds and penetrates host snail

Sporocyst develops in snail host

B

Biomphalaria are the principal snail intermediate hosts. *Schistosoma haematobium* is widely prevalent in Africa, using snails of the genera *Bulinus* and *Physopsis* as the main intermediate hosts. *Schistosoma japonicum* is confined to the Far East, and its hosts are several species of *Oncomelania.*

Unfortunately, some projects intended to raise the standard of living in some tropical countries, such as the Aswan High Dam in Egypt, have increased the prevalence of schistosomiasis by creating more habitats for the snail intermediate hosts. Before the dam was constructed, the 500 miles of the Nile River between Aswan and Cairo was subjected to annual floods; alternate flooding and drying killed many snails. Four years after dam completion, prevalence of schistosomiasis had increased sevenfold along that segment of the river. Prevalence in fishermen around the lake above the dam increased from a very low level to 76%.

The life cycle of blood flukes is similar in all species. Eggs are discharged in human feces or urine; if they get into water, they hatch as ciliated miracidia, which must contact the required

kind of snail within a few hours to survive. In the snail, they transform into sporocysts, which produce another generation of sporocysts. Daughter sporocysts give rise to cercariae directly, without formation of rediae. Cercariae escape from the snail and swim until they contact bare human skin. They penetrate the skin, shedding their tails in the process, and reach a blood vessel where they enter the circulatory system. There is no metacercarial stage. The young schistosomes make their way to the hepatic portal system of blood vessels and undergo a period of development in the liver before migrating to their characteristic sites. As eggs are released by adult females, they are somehow extruded through the wall of veins and through the gut or bladder lining, to be voided with feces or urine, according to species. Many eggs do not make this difficult transit and are swept by blood flow back to the liver or other areas, where they become centers of inflammation and tissue reaction (see Figure 35.7).

The parasite's eggs produce the main ill effects of schistosomiasis. With *S. mansoni* and *S. japonicum*, eggs in the intestinal wall cause ulceration, abscesses, and bloody diarrhea with abdominal pain. Similarly, *S. haematobium* causes ulceration of the bladder wall with bloody urine and pain on urination. Eggs swept to the liver or other sites cause symptoms associated with the organs where they lodge. When they are caught in the capillary bed of the liver, they impede circulation and cause cirrhosis, a fibrotic reaction that interferes with liver function (Figure 14.16). Of the three species, *S. haematobium* is considered least serious and *S. japonicum* most severe. The prognosis is poor in heavy infections of *S. japonicum* without early treatment.

Control is best achieved by educating people to dispose of body wastes hygienically and to avoid exposure to contaminated water. These are difficult problems for poverty-stricken people living under unsanitary, crowded conditions.

Although proper disposal of body wastes is the best control for schistosomiasis, other strategies are being pursued with varying success: chemotherapy, vector control, and vaccination. Development of a vaccine is the subject of much research, but an effective vaccine is not yet available. Vector control by environmental management and by biological means appears promising. Biological controls include introduction of species of snails, crayfish, and fish that prey on the snail vectors. However, biological control attempts for other species have often been fraught with unexpected ecological impacts. In some cases, the biological control has been more of a problem in the long run than the pest species it was supposed to control. Many biologists consider such introductions an extreme risk that should be avoided.

Schistosome Dermatitis (Swimmer's Itch).

Various species of schistosomes in several genera cause a rash or dermatitis when their cercariae penetrate hosts that are unsuitable for further development. Cercariae of several genera whose normal hosts are North American birds cause dermatitis in bathers in northern lakes. Severity of the rash increases with an increasing number of contacts with the organisms, or sensitization. After penetration, cercariae are attacked and killed by the host's immune mechanisms, and they release allergenic substances, causing itching. The condition is more an annoyance than a serious threat to health, but there may be economic losses to vacation trade around infested lakes.

Paragonimus: Lung Flukes.

Several species of *Paragonimus* (Gr. *para*, beside, + *gonimos*, generative), a fluke that lives in the lungs of its host, are known from a variety of mammals. *Paragonimus westermani* (Figure 14.17) from East Asia and the southwest Pacific parasitizes a number of wild carnivores, humans, pigs, and rodents. Its eggs are coughed up in the sputum, swallowed, then eliminated with feces. Zygotes develop in the water, and the miracidium penetrates a snail host. Within the snail, miracidia give rise to sporocysts, which in turn develop into rediae. Cercariae form in rediae and are then shed into the water or ingested directly by freshwater crabs that prey on infected snails. Metacercariae develop in the crabs, and the infection is acquired by eating raw or undercooked crab meat. The infection causes respiratory symptoms, with breathing difficulties and chronic cough. Fatal cases are common. A closely related species, *P. kellicotti,* occurs in mink and similar animals in North America, but only one human case has been recorded. Its metacercariae live in crayfish.

Some Other Trematodes.

Fasciolopsis buski (L. *fasciola*, small bundle, + Gr. *opsis*, appearance) parasitizes the intestine of humans and pigs in India and China. Larval stages occur in several species of planorbid snails, and cercariae encyst on water chestnuts, an aquatic vegetation eaten raw by humans and pigs.

Leucochloridium is noted for its remarkable sporocysts. Snails (*Succinea*) eat vegetation infected with eggs from bird droppings. Sporocysts become much enlarged and branched, and cercariae

Figure 14.16

Cut surface of a liver showing severe fibrosis. The patient was a 27-year-old man who died from hematemesis (vomiting blood) associated with spleen and liver enlargement. Over 180 pairs of adult *Schistosoma mansoni* were counted at autopsy.

Courtesy A. W. Cheever/From H. Zaiman, A Pictorial Presentation of Parasites.

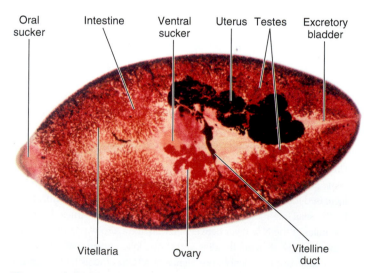

Figure 14.17

Lung fluke *Paragonimus westermani.* Adults are up to 2 cm long. Eggs discharged in sputum or feces hatch into free-swimming miracidia that enter snails. Cercariae from snails enter freshwater crabs and encyst in soft tissues. Humans are infected by eating poorly cooked crabs or by drinking water containing larvae freed from dead crabs.

encyst within the sporocyst. Sporocysts enter the snail's head and tentacles, become brightly striped with orange and green bands, and pulsate at frequent intervals. Birds are attracted by the enlarged and pulsating tentacles, eat the snails, and so complete the life cycle.

Class Monogenea

Monogenetic flukes were traditionally placed as an order of Trematoda, but morphological and molecular data confirm that they are sufficiently different to be classified separately. Cladistic analysis places them closer to the Cestoda, and some authorities now argue that cestodes and monogenean flukes are sister taxa, both having a posterior attachment organ with hooks. Monogeneans are all parasites, primarily of gills and external surfaces of fish. A few are found in the urinary bladder of frogs and turtles, and one parasitizes the eye of a hippopotamus. Although widespread and common, monogeneans seem to cause little damage to their hosts under natural conditions. However, like numerous other fish pathogens, they become a serious threat when their hosts are crowded together, as, for example, in fish farming.

Life cycles of monogeneans are direct, with a single host. The egg hatches to produce a ciliated larva, called an **oncomiracidium,** that attaches to its host. The oncomiracidium bears hooks on its posterior, which in many species become the hooks on the large posterior attachment organ **(opisthaptor)** of the adult (Figure 14.18).

Because monogeneans must cling to the host and withstand the force of water flow over the gills or skin, adaptive diversification has produced a wide array of opisthaptors in different species. Opisthaptors may bear large and small hooks, suckers, and clamps, often in combination with each other.

Class Cestoda

Cestoda, or tapeworms, differ in many respects from the preceding classes. They usually have long flat bodies composed of a **scolex,** for attachment to the host, followed by a linear series of reproductive units or **proglottids** (Figure 14.19). The scolex, or holdfast, is usually provided with suckers or suckerlike organs and often with hooks or spiny tentacles as well (Figure 14.19).

Tapeworms entirely lack a digestive system but do have well-developed muscles, and their excretory system and nervous system are somewhat similar to those of other flatworms. They have no special sense organs but do have sensory endings in the tegument that are modified cilia (Figure 14.20).

As in Monogenea and Trematoda, no external, motile cilia occur in adults, and the tegument is of a distal cytoplasm with

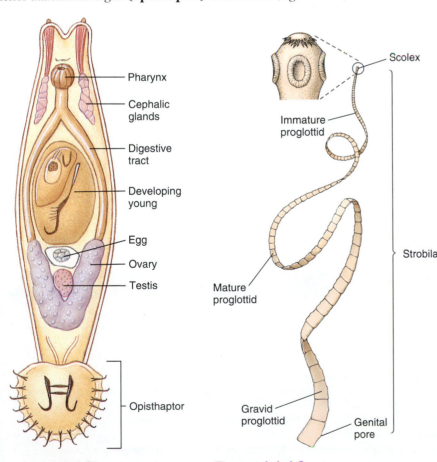

Figure 14.18
A monogenetic fluke *Gyrodactylus cylindriformis,* ventral view.

Figure 14.19
A tapeworm, showing strobila and scolex. The scolex is the organ of attachment.

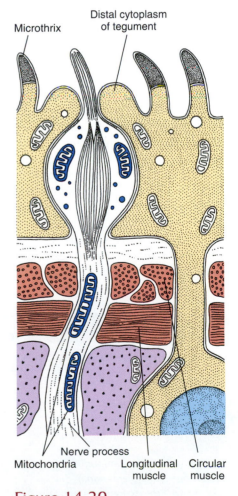

Figure 14.20
Schematic drawing of a longitudinal section through a sensory ending in the tegument of *Echinococcus granulosus.*

sunken cell bodies beneath the superficial muscle layer (Figure 14.20). In contrast to monogeneans and trematodes, however, the entire surface of cestodes is covered with minute projections similar to microvilli of the vertebrate small intestine (p. 48). These **microtriches** (sing. **microthrix**) greatly enlarge the surface area of the tegument, which is a vital adaptation for a tapeworm since it must absorb all its nutrients across its tegument.

Subclass Eucestoda contains the great majority of species of Cestoda. The main body of tapeworms, the chain of proglottids, is called a **strobila** (see Figure 14.19). Typically, there is a **germinative zone** just behind the scolex where new proglottids are formed. As younger proglottids are differentiated in front of it, each individual proglottid moves outward in the strobila, and its gonads mature.

Unlike most other flatworms, many eucestodes are known to self-fertilize, although mutual cross-fertilization remains the norm when mates are available. Each proglottid contains a complete male and female reproductive system, and during mutual cross-fertilization, sperm from each strobila is transferred to the other. However, many tapeworms are known to double back upon themselves so that two proglottids from the same individual may fertilize one another. The shelled embryos form in the uterus of the proglottid, and they are expelled through a uterine pore or the entire proglottid is shed from the worm as it breaks free at zones of muscle weakness between each proglottid.

The tapeworm body is unusual because of the absence of many typical landmarks. There is no head. The scolex, used for attachment, is a remnant of the *posterior* part of the ancestral body. Cestodes and monogeneans thus share a posterior attachment organ with hooks. Absence of the gut and absorption of nutrients by the tegument have already been discussed.

Some zoologists have maintained that the proglottid formation of cestodes represents "true" segmentation (metamerism), but we do not support this view. Segmentation of tapeworms is best considered a replication of sex organs to increase reproductive capacity and is not homologous to the metamerism found in Annelida, Arthropoda, and Chordata (see pp. 190 and 381).

More than 1000 species of tapeworms are known to parasitologists. With rare exceptions, cestodes require at least two hosts, and adults are parasites in the digestive tract of vertebrates. Often their intermediate host is an invertebrate. Collectively these animals are capable of infecting almost all vertebrate species. Normally, adult tapeworms do little harm to their hosts. The most common tapeworms found in humans are listed in Table 14.2. More detailed descriptions of tapeworm life cycles can be found in accounts of several common species to follow.

GUTLESS WONDER

Though lacking skeletal strengths
Which we associate with most
Large forms, tapeworms go to great
Lengths to take the measure of a host.

Monotonous body sections
In a limp mass-production line
Have nervous and excretory connections

And the means to sexually combine
And to coddle countless progeny
But no longer have the guts
To digest for themselves or live free
Or know a meal from soup to nuts.

Copyright 1975 by John M. Burns. Reprinted by permission of the author from *BioGraffiti: A Natural Selection* by John M. Burns. Republished as a paperback by W. W. Norton & Company, Inc., 1981.

Taenia saginata: Beef Tapeworm

Structure   *Taenia saginata* (Gr. *tainia,* band, ribbon) is called the beef tapeworm, but it lives as an adult in the human intestine. Juvenile forms occur primarily in intermuscular tissue of cattle. A mature adult may reach a length of 10 m or more. Its scolex has four suckers for attachment to the intestinal wall, but no hooks. A short neck connects the scolex to the strobila, which may have as many as 2000 proglottids. Gravid proglottids bear shelled, infective larvae (Figure 14.21) which become detached and pass in feces.

Although tapeworms lack true metamerism, there is repetition of the reproductive and excretory systems in each proglottid. Excretory canals in the scolex are continued the length of the body by a pair of dorsolateral, and a pair of ventrolateral, excretory

---

TABLE 14.2

Common Cestodes of Humans

Common and Scientific Name	Means of Infection; Prevalence in Humans
Beef tapeworm (*Taenia saginata*)	Eating rare beef; most common of all tapeworms in humans
Pork tapeworm (*Taenia solium*)	Eating rare pork; less common than *T. saginata*
Fish tapeworm (*Diphyllobothrium latum*)	Eating rare or poorly cooked fish; fairly common in Great Lakes region of United States, and other areas of world where raw fish is eaten
Dog tapeworm (*Dipylidium caninum*)	Unhygienic habits of children (juveniles in flea and louse); moderate frequency
Dwarf tapeworm (*Hymenolepis nana*)	Juveniles in flour beetles; common
Unilocular hydatid (*Echinococcus granulosus*)	Cysts of juveniles in humans; infection by contact with dogs; common wherever humans are in close relationship with dogs and ruminants
Multilocular hydatid (*Echinococcus multilocularis*)	Cysts of juveniles in humans; infection by contact with foxes; less common than unilocular hydatid

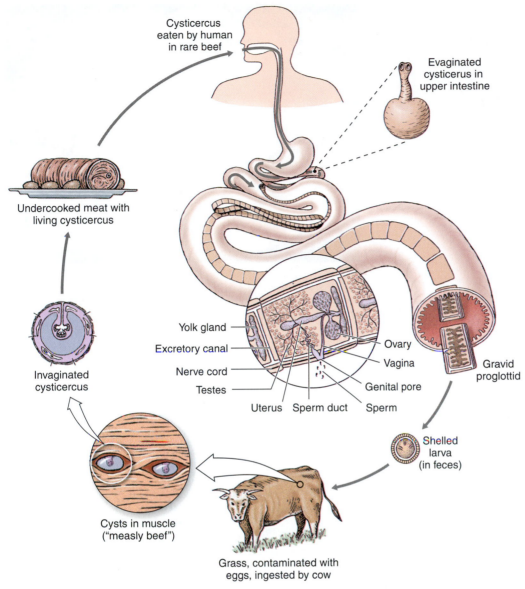

Figure 14.21

Life cycle of beef tapeworm, *Taenia saginata*. Ripe proglottids detach in the human intestine, leave the body in feces, crawl onto grass, and are ingested by cattle. Eggs hatch in the cow's intestine, freeing oncospheres, which penetrate into muscles and encyst, developing into "bladder worms." A human eats infected rare beef, and cysticercus is freed in intestine where it attaches to the intestinal wall, forms a strobila, and matures.

canals. These paired canals are connected by a transverse canal near the posterior end of each proglottid. Two longitudinal **nerve cords** from a **nerve ring** in the scolex also run back through each proglottid (Figure 14.22). Attached to the excretory ducts are flame cells. Each mature proglottid also contains muscles and parenchyma as well as a complete set of male and female organs similar to those of a trematode.

In this group of tapeworms, vitellaria are typically a single, compact **vitelline gland** located just posterior to the ovaries. When gravid proglottids break off and pass out with the feces, they usually crawl out of the fecal mass and onto vegetation nearby. There they may be eaten by grazing cattle. A proglottid ruptures as it dries, further scattering the embryos on soil and

grass. Embryos may remain viable on grass for as long as 5 months.

Life Cycle   Shelled larvae (**oncospheres**) swallowed by cattle hatch and use their hooks to burrow through the intestinal wall into blood or lymph vessels and finally reach voluntary muscle, where they encyst to become **bladder worms** (juveniles called **cysticerci**). There the juveniles develop an invaginated scolex but remain quiescent. When raw or undercooked "measly" meat is eaten by a suitable host, the cyst wall dissolves, the scolex evaginates and attaches to the intestinal mucosa, and new proglottids begin to develop. It takes 2 to 3 weeks for a mature worm to form. When a person is infected with one of these tapeworms, numerous gravid proglottids are expelled daily, sometimes even crawling out the anus. Humans become infected by eating rare roast beef, steaks, and barbecues. Considering that about 1% of American cattle are infected, that 20% of all cattle slaughtered are not federally inspected, and that even in inspected meat one-fourth of infections are missed, it is not surprising that tapeworm infection is fairly common. Infection is easily avoided when meat is thoroughly cooked.

Some Other Tapeworms

Taenia solium: Pork Tapeworm.

Adult *Taenia solium* (Gr. *tainia*, band, ribbon) live in the human small intestine whereas juveniles occur in the muscles of pigs. The scolex has both suckers and hooks arranged on its tip (see Figure 14.19), the **rostellum.** The life history of this worm is similar to that of the beef tapeworm, except that people become infected by eating insufficiently cooked pork.

Taenia solium is much more dangerous than *T. saginata* because cysticerci, as well as adults, can develop in humans. If someone accidentally ingests eggs or proglottids the liberated embryos migrate to any of several organs and form cysticerci (Figure 14.23). The condition is called **cysticercosis.** Common sites are eyes and brain, and infection in such locations can cause blindness, serious neurological symptoms, or death.

Diphyllobothrium latum: Fish Tapeworm.   Adult *Diphyllobothrium* (Gr. *dis,* double, + *phyllon,* leaf, + *bothrion,* hole,

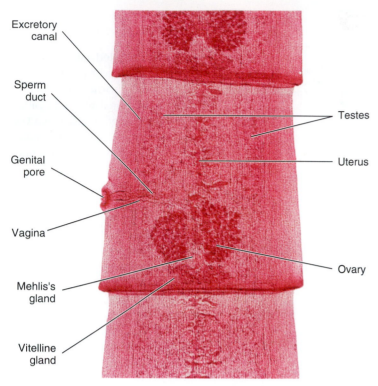

Figure 14.22

Mature proglottid of *Taenia pisiformis,* a dog tapeworm. Portions of two other proglottids also shown.

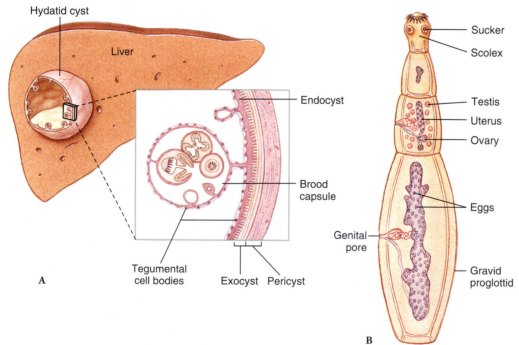

Figure 14.24

Echinococcus granulosus, a dog tapeworm, which may be dangerous to humans. **A,** Early hydatid cyst or bladder-worm stage found in cattle, sheep, hogs, and sometimes humans produces hydatid disease. Humans acquire disease by unsanitary habits in association with dogs. When eggs are ingested, liberated larvae encyst in the liver, lungs, or other organs. Brood capsules containing scolices are formed from the inner layer of each cyst. The cyst enlarges, developing other cysts with brood pouches. It may grow for years to the size of a basketball, necessitating surgery. **B,** The adult tapeworm lives in intestine of a dog or other carnivore.

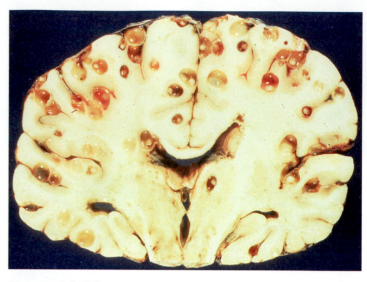

Figure 14.23

Section through the brain of a person who died of cerebral cysticercosis, an infection with cysticerci of *Taenia solium.*

trench) live in the intestines of humans, dogs, cats, and other mammals; immature stages develop in crustaceans and fish. With a length up to 20 m, it is the largest cestode that infects humans. Fish tapeworm infections can occur anywhere in the world where people commonly eat raw fish; in the United States infections are most common in the Great Lakes region. In Finland the worm can cause a serious anemia not apparent in other areas.

Echinococcus granulosus: **Unilocular Hydatid.** *Echinococcus granulosus* (Gr. *echinos,* hedgehog, + *kokkos,* kernel) (Figure 14.24B), a dog tapeworm, causes hydatidosis, a very serious human disease in many parts of the world. Adult worms develop in canines, and juveniles grow in more than 40 species of mammals, including humans, monkeys, sheep, reindeer, and cattle. Thus humans may serve as a dead-end host for this tapeworm. The juvenile stage is a special kind of cysticercus called a **hydatid cyst** (Gr. *hydatis,* watery vesicle). It grows slowly but for a long time—up to 20 years—reaching the size of a basketball in an unrestricted site such as the liver. If a hydatid grows in a critical location, such as heart or central nervous system, serious symptoms may appear sooner. The main cyst maintains a single (or unilocular) chamber, but daughter cysts bud off, and each contains thousands of scolices. Each scolex produces a worm when eaten by a canine. The only treatment is surgical removal of the hydatid.

Phylogeny and Adaptive Diversification of Platyhelminthes

Phylogeny

Although class Turbellaria is clearly paraphyletic, we retain the taxon because presentation based on thorough cladistic analysis would require introduction of many more taxa and characteristics beyond the scope of this book and not yet common in zoological literature. For example, ectolecithal turbellarians should be allied with trematodes, monogeneans, and cestodes as the sister group to endolecithal turbellarians (see Figure 14.5).

Several synapomorphies, including the unique architecture of the tegument, indicate that neodermatans (trematodes, monogeneans, and cestodes) form a monophyletic group, and monophyly of Neodermata is supported by sequence data from several different molecular markers.[1]

Features of the most recent common ancestor of all bilaterian animals have been much debated. Some investigators propose that a **planuloid ancestor** (perhaps one very similar to the planula larva of cnidarians) may have given rise to one branch of descendants that were sessile or free-floating and radial, which became the Cnidaria, and another branch that acquired a creeping habit and bilateral symmetry. Bilateral symmetry is a selective advantage for creeping or swimming animals because sensory structures are concentrated on the anterior end (cephalization), which is the end that first encounters environmental stimuli. This ancestral form would have had a simple body with a blind gut, perhaps much like the body of an acoelomorph flatworm.

Adaptive Diversification

Whether Platyhelminthes is a valid monophyletic group remains a subject of debate, although there is little argument that Turbellaria is a paraphyletic group awaiting revision. There can be little doubt that the flatworm body plan has been successful, and descendants of ancient flatworms have diversified extensively. The descendants of those ancient flatworms have been particularly successful as parasites, and many groups of flatworms have become highly specialized for a parasitic existence.

PHYLUM MESOZOA

The name Mesozoa (mes-o-zo'a) (Gr. *mesos,* in the middle, + *zōon,* animal) was coined by an early investigator (van Beneden, 1876) who considered the group a "missing link" between protozoa and metazoa. These minute, ciliated, wormlike animals represent an extremely simple level of organization. All mesozoans live as parasites or symbionts in marine invertebrates, and the majority of them are only 0.5 to 7 mm in length. Most are

composed of only 20 to 30 cells arranged basically in two layers. The layers are not homologous to the germ layers of other metazoans.

The two classes of mesozoans, Rhombozoa and Orthonectida, differ so much from each other that some authorities place them in separate phyla.

Rhombozoans (Gr. *rhombos,* a spinning top, + *zōon,* animal) live in kidneys of benthic cephalopods (bottom-dwelling octopuses, cuttlefishes, and squids) where they apparently cause no harm to the host. Adults, called **vermiforms** (or nematogens), are long and slender (Figure 14.25). Their inner, reproductive cells give rise to vermiform larvae that grow and then reproduce. When a population becomes crowded, reproductive cells of some adults develop into gonadlike structures producing male and female gametes. Zygotes grow into minute (0.04 mm) ciliated infusoriform larvae (Figure 14.25B), quite unlike the parent. These larvae are shed with host urine into the seawater. The next part of the life cycle is unknown because infusoriform larvae are not immediately infective to a new host.

Orthonectids (Gr. *orthos,* straight, + *nektos,* swimming) (Figure 14.26) parasitize a variety of invertebrates, such as brittle stars, bivalve molluscs, polychaetes, and nemerteans. Their life cycles involve sexual and asexual phases, and the asexual stage is quite different from that of rhombozoans. It consists of a multinucleated mass called a **plasmodium,** which by division ultimately gives rise to males and females.

Phylogeny of Mesozoans

There is still much to learn about these mysterious little parasites, but probably one of the most intriguing questions is the place of mesozoans in the evolutionary picture. Some investigators consider them primitive or degenerate flatworms and even place them in phylum Platyhelminthes. Present molecular evidence groups mesozoans with flatworms in superphylum Lophotrochozoa. However, a molecular phylogeny that included an orthonectid and two species from a rhombozoan subgroup, the dicyemids, did not show members of the two classes to be sister taxa, so the phylum may not be monophyletic.

PHYLUM NEMERTEA (RHYNCHOCOELA)

Nemerteans (nem-er'te-ans) are often called ribbon worms. Their name (Gr. *Nemertes,* one of the Nereids, unerring one) refers to the unerring aim of the proboscis, a long muscular tube (Figures 14.27 and 14.28) that can be thrust out swiftly to grasp their prey. The phylum is also called Rhynchocoela (ring'ko-se'la) (Gr. *rhynchos,* beak, + *koilos,* hollow), which also refers to the proboscis. They are thread-shaped or ribbon-shaped bilaterally symmetrical, triploblastic worms; There are about 1000 species in the group; nearly all are marine.

Nemertean worms are usually less than 20 cm long, although a few are several meters in length (Figure 14.29). *Lineus*

[1]Telford, et al. 2003. Proc. R. Soc. London B **270:**1077–1083.

Figure 14.25

Two methods of reproduction by mesozoans. **A,** Asexual development of vermiform larvae from reproductive cells in the axial cell of an adult. **B,** Under crowded conditions in the host kidney, reproductive cells develop into gonads with gametes that produce infusoriform dispersal larvae that emerge in the host urine.

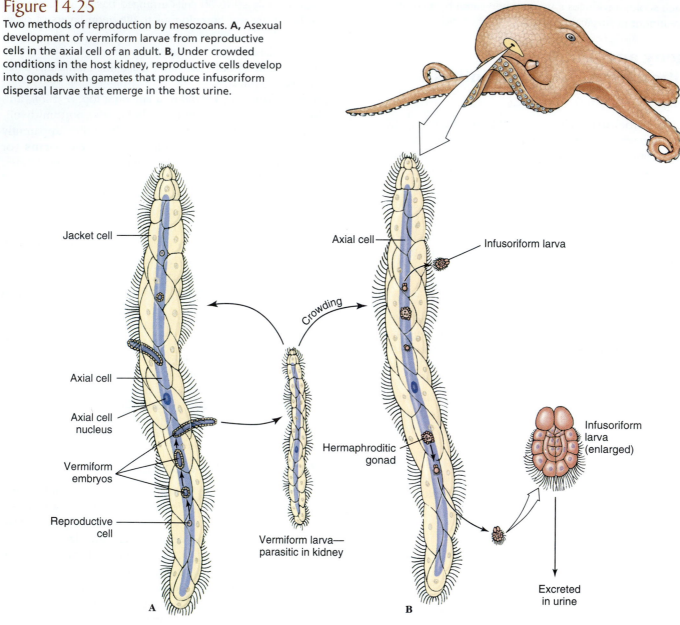

Jacket cell

Axial cell

Axial cell

Axial cell nucleus

Vermiform embryos

Reproductive cell

A

Crowding

Vermiform larva—parasitic in kidney

Axial cell

Infusoriform larva

Hermaphroditic gonad

Infusoriform larva (enlarged)

Excreted in urine

B

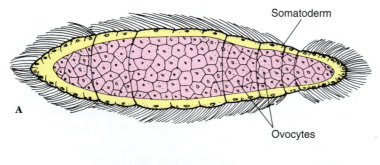

Somatoderm

Ovocytes

A

B

Figure 14.26

A, Female and, **B,** male orthonectid (*Rhopalura*). This mesozoan parasitizes such forms as flatworms, molluscs, annelids, and brittle stars. The structure is a single layer of ciliated epithelial cells surrounding an inner mass of sex cells.

Figure 14.27

Ribbon worm *Amphiporus bimaculatus* (phylum Nemertea) is 6 to 10 cm long, but other species range up to several meters. The proboscis of this specimen is partially extended at the top; the head is marked by two brown spots. The animal is photographed on an algal frond.

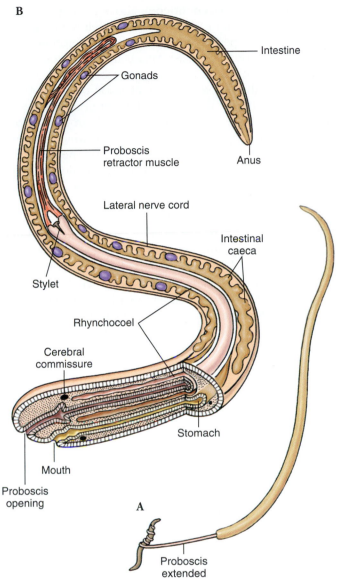

Figure 14.28

A, *Amphiporus,* with proboscis extended to catch prey. **B,** Structure of female nemertean worm *Amphiporus* (diagrammatic). Dorsal view to show proboscis.

longissimus (L. *linea,* line) was reportedly able to stretch up to 60 m in length! Their colors can be bright, although most are dull or pallid. Some live in secreted gelatinous tubes.

With few exceptions, the general body plan of nemerteans is similar to that of turbellarians. The epidermis of nemerteans is ciliated and has many gland cells. There are flame cells in the excretory system. Rhabdites have been found in several nemerteans, including *Lineus,* but some work indicates that they are not homologous to rhabdites in flatworms. Nemerteans also differ from flatworms in their reproductive system. Ribbon worms are mostly dioecious. In marine forms there is a ciliated larva that has some resemblance to trochophore larvae found in annelids and molluscs.

Nemerteans show some derived features absent from flatworms. The most obvious of these is the eversible **proboscis** and its sheath, for which there are no counterparts within any

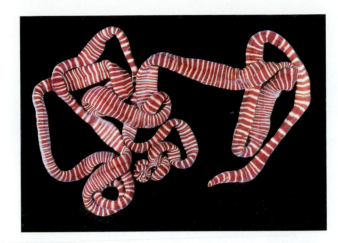

Figure 14.29

Baseodiscus is a genus of nemerteans whose members typically measure several meters in length. This *B. mexicanus* is from the Galápagos Islands.

other phylum. In the odd genus *Gorgonorhynchus* (Gr. *Gorgo,* name of a female monster of terrible aspect, + *rhynchos,* beak, snout) the proboscis is divided into many proboscides, which appear as a mass of wormlike structures when everted. Another difference is the presence of an **anus** in adults, producing a **complete digestive system.** A digestive system with an anus is more efficient because ejection of waste materials back through the mouth is not necessary. Ingestion and defecation can occur simultaneously. Nemerteans are also the simplest animals to have a closed loop **blood-vascular system.**

A few nemerteans occur in moist soil and freshwater. *Prostoma rubrum* (Gr. *pro,* before, in front of, + *stoma,* month) which is 20 mm or less in length, is a well-known freshwater species. The larger number of nemertean species are marine; at low tide they are often coiled under stones. It seems probable that they are active at high tide and quiescent at low tide. Some nemerteans such as *Cerebratulus* (L. *cerebrum,* brain, + *ulus,* dim. suffix) often live in empty mollusc shells. Small species often live among seaweed, or they may be found swimming near the surface of the water. Nemerteans are often secured by dredging at depths of 5 to 8 m or deeper.

Although a few species are commensals or scavengers, most ribbon worms are active predators on small invertebrates. A few species are specialized egg predators (considered ectoparasites) on brachyuran crabs, and in high numbers can consume all the embryos in their host's clutch.

Form and Function

Many nemerteans are difficult to examine because they are so long and fragile. *Amphiporus* (Gr. *amphi,* on both sides, + *poros,* pore), a genus of smaller forms that ranges from 2 to 10 cm in length, is fairly typical of nemertean structure (Figure 14.28). Its body wall consists of ciliated epidermis and layers of circular and longitudinal muscles (Figure 14.30). Locomotion consists largely of gliding over a slime track, although larger species move by muscular contractions. Some large species are even capable of undulatory swimming when threatened.

Characteristics of Phylum Nemertea

1. An **eversible proboscis,** which lies free in a cavity (rhynchocoel) above the alimentary canal unique to nemerteans
2. In marine, freshwater, and moist terrestrial habitats
3. Nemerteans are mostly free-living, with a few parasitic species
4. Bilateral symmetry; highly contractile body that is cylindrical anteriorly and flattened posteriorly
5. Body **triploblastic;** adult parenchyma partly gelatinous
6. Rhynchocoel is a true coelomic cavity, but its unusual position and function as part of the proboscis mechanism leads many to question whether it is homologous to the coelom of other protostomes
7. Epidermis with cilia and gland cells; rhabdites in some
8. **Complete digestive system** (mouth to anus)
9. Body-wall musculature of outer circular and inner longitudinal layers with diagonal fibers between the two; sometimes another circular layer inside the longitudinal layer
10. Nervous system usually a four-lobed brain connected to paired longitudinal nerve trunks or, in some, middorsal and midventral trunks
11. Sensory **ciliated pits** or **head slits** on each side of head, which communicate between the outside and the brain; tactile organs and ocelli (in some)
12. Asexual reproduction by fragmentation
13. Sexes separate with simple gonads; few hermaphrodites; **pilidium larvae** in some
14. Excretory system of two coiled canals, which are branched with **flame cells**
15. **Blood-vascular system with two or three longitudinal trunks**
16. No respiratory system

cavity of its own, the **rhynchocoel,** above the digestive tract (but not connected with it). The proboscis itself is a long, blind muscular tube that opens at the anterior end at a proboscis pore above the mouth (Figure 14.28). Muscular pressure on fluid in the rhynchocoel causes the long tubular proboscis to be everted rapidly through the proboscis pore. Eversion of the proboscis exposes a sharp barb, called a stylet (absent in some nemerteans). The sticky, slime-covered proboscis coils around the prey and stabs it (often repeatedly) with the stylet, while pouring a toxic secretion on the prey (Figure 14.28). The neurotoxin in some species has recently been identified as tetrodotoxin, commonly known as the poison in puffer fishes. Then, retracting its proboscis, a nemertean draws the prey near its mouth and the subdued prey is swallowed whole.

Nemerteans have a true circulatory system, and blood flow is maintained by a combination of the contractile walls of the vessels and general body movements. As a result the flow is irregular and often reverses direction in the vessels. Two to many flame-bulb protonephridia are closely associated with the circulatory system, so that their function appears to be truly excretory (for disposal of metabolic wastes), in contrast to their apparently osmoregulatory role in Platyhelminthes.

Nemerteans have a pair of nerve ganglia, and one or more pairs of longitudinal nerve cords are connected by transverse nerves.

Some species reproduce asexually by fragmentation and regeneration. Nemerteans show a surprising range of sexual reproductive strategies. Most species are dioecious and fertilization is often external, although many exceptions are known: Some species are hermaphroditic, some have internal fertilization, and some even have ovoviviparous development.

The mouth is anterior and ventral, and the digestive tract is complete, extending the full length of the body and ending at an anus. There are usually no muscles in the gut wall itself; instead cilia move food through the intestine. Digestion is largely extracellular in the gut lumen.

The favorite prey of most nemerteans is annelids and other small invertebrates. Their diets may be highly specialized or extremely varied, depending on the species. Some species appear able to detect prey only when they physically bump into it, whereas others are capable of tracking prey over great distances. When prey is encountered, nemerteans seize it with a proboscis that lies in an interior

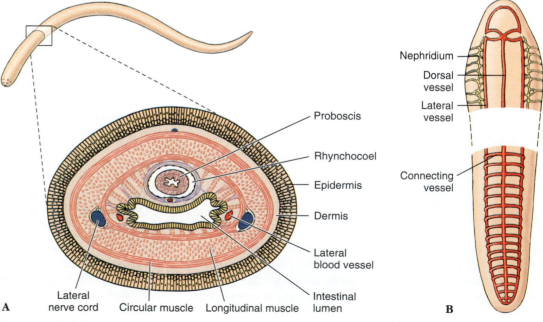

Figure 14.30

A, Diagrammatic cross section of female nemertean worm. **B,** Excretory and circulatory systems of nemertean worm. Flame bulbs along nephridial canal are closely associated with lateral blood vessels.

Classification of Phylum Nemertea

Class Enopla (en'o-pla) (Gr. *enoplos,* armed). Proboscis usually armed with stylets; mouth opens in front of brain. Examples: *Amphiporus, Prostoma.*

Class Anopla (an'o-pla) (Gr. *anoplos,* unarmed). Proboscis lacks stylets; mouth opens below or posterior to brain. Examples: *Cerebratulus, Tubulanus, Lineus.* Class Anopla is contentious because some authorities consider it is a paraphyletic group.

Phylogeny of Nemertea

There has been some debate about the phylogenetic position of nemerteans. Like many other lophotrochozoan protostomes, nemerteans exhibit spiral cleavage. Later development varies across the phylum, with some work showing typical formation of mesoderm from endoderm, as well as instances of mesodermal formation from ectoderm.

Nemerteans make a variety of larval forms, and in some species all stages of development occur within an egg case. The evolutionary relationship between the various larval forms and the typical trochophore has been much discussed. Some similarities exist, but a distinctive central ring of cilia (a prototroch) is not present. However, a new developmental study of an early diverging species found that a band of cilia does form around the larva, although it later degenerates. The brief existence of a ciliary band provides evidence that a trochophore larva was present in ancestral nemerteans, and supports their placement in Lophotrochozoa.

A second area of controversy is the nature of the nemertean body plan. In the introduction to this chapter, we distinguished coelomate and acoelomate body plans. Are nemerteans acoelomate or coelomate animals? The rhynchocoel is an internal cavity lined by mesoderm that forms by schizocoely, making it a true coelom. However, a typical coelom (p. 165) forms a fluid-filled cavity around the digestive tract. The rhynchocoel instead lies above the digestive tract, extending about three-quarters the length of the body from the anterior end. The rhynchocoel differs from a typical coelom in both position and function. A typical coelom surrounds, cushions, and protects the gut, but it also forms part of the hydrostatic skeleton, stiffening when surrounding muscles contract. A rhynchocoel is fluid-filled and surrounded by muscles; muscular contraction increases hydrostatic pressure and eventually everts the proboscis. We leave the reader, in the company of future generations of biologists, to ponder whether the protostome coelom and the rhynchocoel are homologous structures.

SUMMARY

Acoelomorpha and Platyhelminthes are among the simplest bilaterally symmetrical forms, a condition of adaptive value for actively crawling or swimming animals. They are acoelomate. They are triploblastic and at the organ-system level of organization. Members of Acoelomorpha have very simple nervous and digestive systems; some lack a gut entirely.

The body surface of turbellarians is usually a cellular epithelium, at least in part ciliated, containing mucous cells and rod-shaped rhabdites that function together in locomotion. Members of all other classes of flatworms are covered by a nonciliated, syncytial tegument with cell bodies beneath superficial muscle layers. Digestion is extracellular and intracellular in most; cestodes must absorb predigested nutrients across their tegument because they have no digestive tract. Osmoregulation is accomplished by flame-cell protonephridia, and removal of metabolic wastes and respiration occur by diffusion across the body wall. Flatworms have a ladder-type nervous system with motor, sensory, and association neurons. Most flatworms are hermaphroditic, and asexual reproduction occurs in some groups.

Class Turbellaria is a paraphyletic group that includes mostly free-living and carnivorous members. Digenetic trematodes have a mollusc intermediate host and almost always a vertebrate definitive host. The extensive asexual reproduction that occurs in their intermediate host helps to increase the chances that some of their offspring will reach a definitive host. Aside from the tegument, digeneans share many basic structural characteristics with turbellarians. Digenea includes a number of important parasites of humans and domestic animals. Digeneans contrast with Monogenea, which are important ectoparasites of fishes and have a direct life cycle (without intermediate hosts).

Cestodes (tapeworms) generally have a scolex at their posterior end, followed by a long chain of proglottids, each of which contains a complete set of reproductive organs of both sexes. The anterior end of the body has been lost evolutionarily. Cestodes live as adults in the digestive tract of vertebrates. They have microvillus-like microtriches on their tegument, which increase its surface area for absorption. Shelled larvae are passed in the feces, and juveniles develop in a vertebrate or invertebrate intermediate host.

Flatworms and the cnidarians both likely evolved from a common planuloid ancestor, some of whose descendants became sessile or free-floating and radial (cnidarians), while others became creeping and bilateral (flatworms).

Sequence analysis of rDNA, as well as some developmental and morphological criteria, suggest that Acoelomorpha, heretofore considered an order of turbellarians, diverged from a common ancestor shared with other Bilateria and is the sister group of all other bilateral phyla.

Members of phylum Mesozoa are very simply organized animals that are parasitic in kidneys of cephalopod molluscs (class Rhombozoa) and in several other invertebrate groups (class Orthonectida). They have only two cell layers, but these are not homologous to the germ layers of higher metazoans. They have a complicated life history that is still incompletely known. Their simple organization may have been derived from a more complex platyhelminth-like ancestor.

Members of Nemertea have a complete digestive system with an anus and a true circulatory system. They are free-living, mostly marine, and they capture prey by ensnaring it with their long, eversible proboscis. The proboscis cavity, the rhynchocoel, appears to be a coelomic cavity.

REVIEW QUESTIONS

1. Why is bilateral symmetry of adaptive value for actively motile animals?
2. Match the terms in the right column with the classes in the left column:

    _____ Turbellaria   a. Endoparasitic
    _____ Monogenea   b. Free-living and commensal
    _____ Trematoda   c. Ectoparasitic
    _____ Cestoda

3. Describe the body plan of a typical turbellarian.
4. Distinguish two mechanisms by which flatworms supply yolk for their embryos. Which system is evolutionarily ancestral for flatworms and which one is derived?
5. What do planarians (triclad flatworms) eat, and how do they digest it?
6. Briefly describe the osmoregulatory system, the nervous system, and the sense organs of turbellarians, trematodes, and cestodes.
7. Contrast asexual reproduction in triclad turbellarians, Trematoda, and Cestoda.
8. Contrast a typical life cycle of a monogenean with that of a digenetic trematode.
9. Describe and contrast the tegument of most turbellarians and the other classes of platyhelminths. Does the tegument provide evidence that trematodes, monogeneans, and cestodes form a clade within Platyhelminthes? Why?
10. Answer the following questions with respect to both *Clonorchis* and *Schistosoma:* (a) how do humans become infected? (b) what is the general geographical distribution? (c) what are the main disease conditions produced?
11. Why is *Taenia solium* a more dangerous infection than *Taenia saginata?*
12. What are two cestodes for which humans can serve as intermediate hosts?
13. Define each of the following with reference to cestodes: scolex, microtriches, proglottids, strobila.
14. Give three differences between nemerteans and platyhelminths.
15. Recent evidence suggests that members of Acoelomorpha constitute a sister group for all other Bilateria. How do members of this group differ from typical protostomes?
16. Nemerteans possess a body cavity formed by schizocoely. What is this cavity, and how is it different from the cavity in most coelomate animals?
17. How do mesozoans differ morphologically from the other phyla discussed in this chapter?

SELECTED REFERENCES

Brooks, D. R. 1989. The phylogeny of the Cercomeria (Platyhelminthes: Rhabdocoela) and general evolutionary principles. J. Parasitol. **75:**606–616. *Cladistic analysis of parasitic flatworms.*

Baguñà, J., and M. Ruitort. 2004. The dawn of bilaterian animals: the case of acoelomorph flatworms. Bioessays **26:**1046–1057. *Genetic and morphological evidence for acoelomorphs as early diverging bilaterally symmetrical animals.*

Desowitz, R. S. 1981. New Guinea tapeworms and Jewish grandmothers. New York, W. W. Norton & Company. *Accounts of parasites and parasitic diseases of humans. Entertaining and instructive. Recommended for all students.*

Hanelt, B., D. Van Schyndel, C. M. Adema, L. L. Lewis, and E. S. Loker. 1996. The phylogenetic position of *Rhopalura ophiocomae* (Orthonectida) based on 18S ribosomal DNA sequence analysis. Mol. Biol. Evol. **13:**1187–1191. *Orthonectid mesozoans align with triploblastic animals and do not form the sister taxon to rhombozoans.*

Kobayashi, M., H. Furuya, and P. W. H. Holland. 1999. Dicyemids are higher animals. Nature. **401:**762. *Sequence analysis of the gene for Hox protein is evidence that mesozoans are members of superphylum Lophotrochozoa and that they are derived from a more complex ancestor that has undergone simplification during its parasitic evolution. They ". . . are not basal and primitive animals and should not be excluded from Metazoa."*

Livaitis, M. K., and K. Rohde. 1999. A molecular test of platyhelminth phylogeny: inferences from partial 28S rDNA sequences. Invert. Biol. **118:**42–56. *This report does not support a basal position for Acoela and presents evidence that Monogenea is paraphyletic.*

Roberts, L. S., and J. Janovy, Jr. 2005. G. D. Schmidt and L. S. Roberts's foundations of parasitology, ed. 7. Dubuque, Iowa, McGraw-Hill Higher Education. *Up-to-date and readable accounts of parasitic flatworms.*

Ruiz-Trillo, I., M. Ruitort, H. M. Fourcade, J. Baguñà, and J. L. Boore. 2004. Mitochondrial genome data support the basal position of Acoelomorpha and the polyphyly of the Platyhelminthes. Mol. Phylogen. Evol. **33:**321–332. *Evidence that Acoelomorpha is the sister taxon to the remaining Bilateria.*

Strickland, G. T. 2000. Hunter's tropical medicine and emerging infectious diseases, ed. 8. Philadelphia, W. B. Saunders Company. *A valuable source of information on parasites of medical importance.*

Telford, M. J., A. E. Lockyear, C. Cartwright-Finch, and D. T. J. Littlewood. 2003. Combined large and small subunit ribosomal RNA phylogenies support a basal position of the acoelomate flatworms. Proc. R. Soc. London B **270:**1077–1083. *An up-to-date review of the molecular evidence that led to removal of the Acoela from Platyhelminthes.*

Turbeville, J. M. 2002. Progress in nemertean biology: development and phylogeny. Integ. and Comp. Biol. **42:**692–703. *A discussion of nemertean features that suggest shared ancestry with flatworms and other lophotrochozoan phyla.*

Tyler, S., and M. S. Tyler. 1997. Origin of the epidermis in parasitic Platyhelminthes. Int. J. Parasit. **27:**715–738. *Descriptions of epidermal replacement as tegument forms in Neodermata.*

ONLINE LEARNING CENTER

Visit www.mhhe.com/hickmanipz14e for chapter quizzing, key term flash cards, web links, and more!

Ectoprocts and other animals fouling a boat bottom.

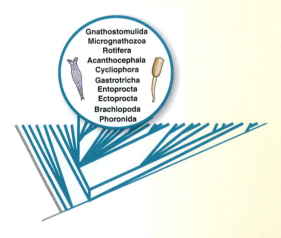

Gnathiferans and Smaller Lophotrochozoans

- PHYLUM GNATHOSTOMULIDA
- PHYLUM MICROGNATHOZOA
- PHYLUM ROTIFERA
- PHYLUM ACANTHOCEPHALA
- PHYLUM CYCLIOPHORA
- PHYLUM GASTROTRICHA
- PHYLUM ENTOPROCTA
- PHYLUM ECTOPROCTA
- PHYLUM BRACHIOPODA
- PHYLUM PHORONIDA

Gnathostomulida
Micrognathozoa
Rotifera
Acanthocephala
Cycliophora
Gastrotricha
Entoprocta
Ectoprocta
Brachiopoda
Phoronida

Some Evolutionary Experiments

During the Cambrian period, about 535 to 530 million years ago, a most fertile time occurred in evolutionary history. For over 3 billion years before this time, evolution had forged little more than prokaryotes and unicellular eukaryotes. Then, within the space of a few million years, all of the major phyla of macroscopic invertebrates, and probably all of the smaller phyla, became established. This was the Cambrian explosion, the greatest evolutionary "bang" the world has known. In fact, the fossil record suggests that more phyla existed in the Paleozoic Era than exist now, but some disappeared during major extinction events that punctuated the evolution of life on earth since that time. Greatest of these disruptions was the Permian extinction about 230 million years ago.

Evolution has produced many "experimental models." Some of these models failed because they were unable to survive in changing conditions. Others flourished, and their descendants are the dominant species and individuals inhabiting the world today. Still others produced a small number of species, some of which persist. Members of these phyla are interesting because they occupy specialized habitats—for example, living between sand grains—and often have unusual morphologies. Relationships among these groups have been and continue to be the subject of considerable controversy.

Protostomia is a large clade, sometimes called a superphylum, whose members share some developmental characters including formation of the mouth from the embryonic blastopore. Protostome phyla are divided between two large clades: Lophotrochozoa and Ecdysozoa. Lophotrochozoan embryos typically exhibit spiral mosaic cleavage.

Most of the lophotrochozoan phyla described in Chapter 14 exhibit an acoelomate body plan. Some lophotrochozoan phyla described in this chapter also have this plan, but others possess **pseudocoelomate** bodies (Gr. *pseudo,* false, + *koilōma,* cavity).

A pseudocoelomate body contains an internal cavity surrounding the gut, but this cavity is *not* completely lined with mesoderm, as it would be in a coelomate animal (Figure 15.1). A pseudocoelom is an embryonic blastocoel that persists throughout development, leading some to describe animals with this body plan as blastocoelomates. There is a mesodermal layer on the outer edge of the cavity, but the endodermal gut lining forms the inner boundary of the pseudocoelom (see p. 189).

The pseudocoel may be filled with fluid, or it may contain a gelatinous matrix with some mesenchymal cells. It shares some functions of a coelom: space for development and differentiation of digestive, excretory, and reproductive systems, a simple means of circulation or distribution of materials throughout the body, a storage place for waste products to be discharged to the outside by excretory ducts, and a hydrostatic support.

Many pseudocoelomate animals are quite small, so the most likely function of the pseudocoel in these animals is to permit internal circulation in the absence of a true circulatory system.

Four lophotrochozoan phyla discussed in this chapter belong to a small clade whose ancestors possessed complex cuticular jaws. The clade is called Gnathifera, and its members are phyla Gnathostomulida, Micrognathozoa, Rotifera, and Acanthocephala (Figure 15.2).

Six other lophotrochozoan phyla are also included in this chapter. Members of Gastrotricha are tiny aquatic animals that may be closely related to gnathiferans. Placement of Cycliophora and Entoprocta is subject to debate, but phylogenies based on molecular characters place them within Lophotrochozoa, along with three taxa that bear the eponymous **lophophore:** Ectoprocta, Brachiopoda, and Phoronida. This horseshoe-shaped feeding structure covered in tentacles has sometimes been used to unite these three phyla into a clade of lophophorates, but not all researchers agree that the structure is homologous. As the body plans and developmental patterns of these small, but fascinating, animals are reexamined, their phylogenetic placement may change.

CLADE GNATHIFERA

Gnathiferans, other than acanthocephalans, possess small cuticular jaws with a homologous microstructure (Figure 15.2). The number of pairs of such jaws varies within the clade. Members of Gnathostomulida, Micrognathozoa, and Rotifera are tiny, free-living, aquatic animals. Acanthocephalans are wormlike endoparasites living as adults in fishes or other vertebrates.

Rotifera and Acanthocephala are presumed sister taxa, together forming a clade called Syndermata. Their close relationship first appeared in molecular phylogenies and led morphologists to examine acanthocephalans anew, searching for evidence that these parasites were highly derived rotifers. There is little external similarity between free-swimming rotifers and endoparasitic worms, but members of both groups have a **eutelic** syncytial epidermis. Eutely refers to constancy in the numbers of nuclei present, as illustrated by the constant numbers of nuclei in various organs of one species of rotifer: E. Martini (1912) reported that he always found 183 nuclei in the brain, 39 in the stomach, and 172 in the coronal epithelium. Despite the shared structure of the epidermis, the union of two morphologically disparate taxa into clade Syndermata is still controversial.

PHYLUM GNATHOSTOMULIDA

Gnathostomulids are delicate wormlike animals less than 2 mm long (Figure 15.3). The first known species of Gnathostomulida (nath'o-sto-myu'lid-a) (Gr. *gnathos,* jaw, + *stoma,* mouth, +

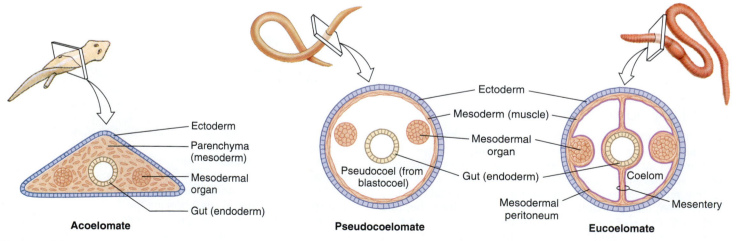

Figure 15.1
Acoelomate, pseudocoelomate, and eucoelomate body plans.

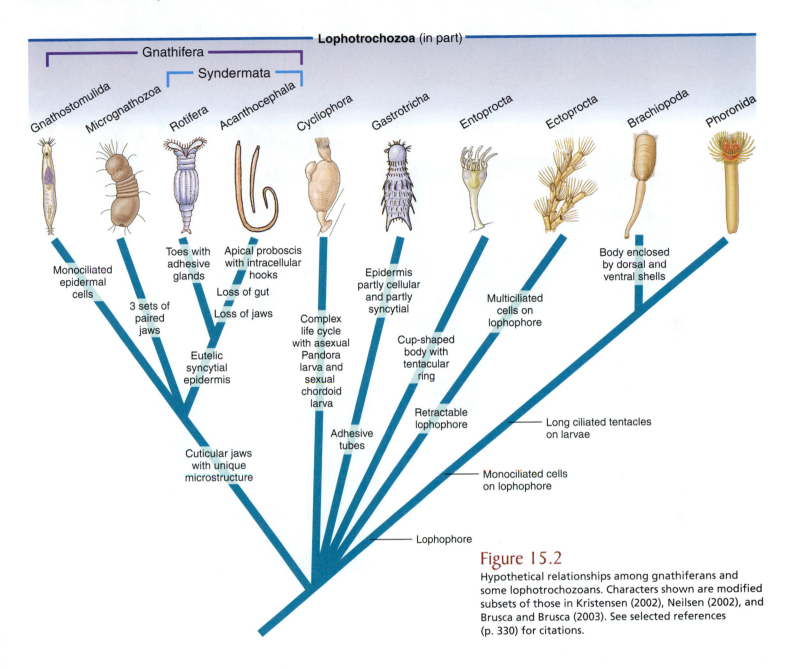

Figure 15.2

Hypothetical relationships among gnathiferans and some lophotrochozoans. Characters shown are modified subsets of those in Kristensen (2002), Neilsen (2002), and Brusca and Brusca (2003). See selected references (p. 330) for citations.

L. *ulus,* dim. suffix) was observed in 1928 in the Baltic, but its description was not published until 1956. Since then jaw worms have been found in many parts of the world, including the Atlantic coast of the United States, and over 80 species in 18 genera have been described.

Gnathostomulids live in interstitial spaces of very fine sandy coastal sediments and silt from the intertidal to depths of several hundred meters. They can endure conditions of very low oxygen. They often occur in large numbers and frequently in association with gastrotrichs, nematodes, ciliates, tardigrades, and other small forms.

Gnathostomulids can glide, swim in loops and spirals, and bend the head from side to side. The epidermis is ciliated, but each epidermal cell has only one cilium, a condition rarely found in lophotrochozoans other than some gastrotrichs (p. 321). The nervous system is only partially described, but appears to be primarily associated with a host of sensory cilia and ciliary pits on the head.

Gnathostomulids feed by scraping bacteria and fungi from the substratum with a pair of jaws on the pharynx. The pharynx leads into a simple blind gut. Some morphologists have suggested that a tissue strand connecting the posterior gut to the epidermis is a remnant of an ancestral complete gut, but this conjecture requires more support.

The body is acoelomate with a poorly developed parenchyma layer. There is no circulatory system, so gnathostomulids probably rely on diffusion for circulation, excretion, and gas exchange.

Description of the reproductive systems and mating behavior of these worms is far from complete. Gnathostomulids are primarily protandric or simultaneous hermaphrodites engaging in mutual cross-fertilization that occurs internally. Fertilized animals each appear to produce a single zygote developing via spiral cleavage.

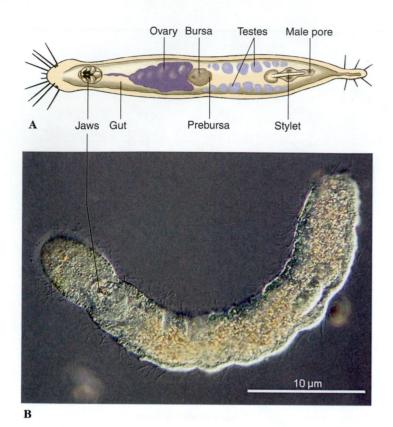

Figure 15.3

A, *Gnathostomula jenneri* (phylum Gnathostomulida) is a tiny member of the interstitial fauna between grains of sand or mud. Species in this family are among the most commonly encountered jaw worms, found in shallow water and down to depths of several hundred meters. **B,** *Gnathostomula paradoxa* is abundant in sediments near burrows of marine polychaetes in the North Sea. Its ecology is very similar to that of *G. jenneri* from the North American Atlantic coast.

PHYLUM MICROGNATHOZOA

The first and only micrognathozoan species, *Limnognathia maerski,* was collected from Greenland in 1994 but not formally described until 2000. Micrognathozoans are tiny animals that are interstitial (living between sand grains) and about 142 μm long. The body consists of a two-part head, a thorax, and an abdomen with a short tail (Figure 15.4). The cellular epidermis has dorsal plates but lacks plates ventrally. These animals move using cilia and also possess a unique ventral ciliary adhesive pad that produces glue.

There are three pairs of complex jaws. The mouth leads into a relatively simple gut. An anus opens to the outside only periodically. There are two pairs of protonephridia.

The reproductive system is not well understood. Only female reproductive organs have been found, so perhaps the animals reproduce parthenogenetically. Cleavage and subsequent development have not been studied.

PHYLUM ROTIFERA

Rotifera (ro-tif′e-ra) (L. *rota,* wheel, + *fera,* those that bear) derive their name from the characteristic ciliated crown, or **corona,** that, when beating, often gives the impression of rotating wheels

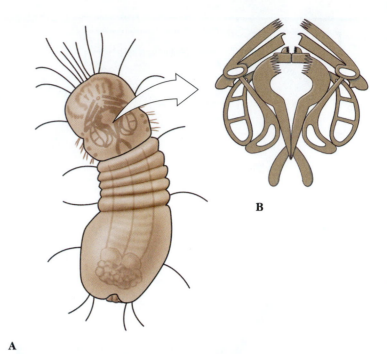

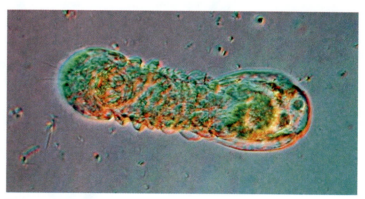

Figure 15.4

A, *Limnognathia maerski,* a micrognathozoan. **B,** Detail of complex jaws. **C,** A living specimen. This animal was found on moss in a freshwater spring on Disko Island, Greenland. It swims, or crawls, consuming bacteria, blue-green algae, and diatoms.

(Figure 15.5). Rotifers range from 40 μm to 3 mm in length, but most are between 100 and 500 μm long. There are about 2000 species of rotifers.

Rotifers are adapted to many ecological conditions. Most species are benthic, living on the bottom or in vegetation of ponds or along the shores of freshwater lakes where they swim or crawl on the vegetation. Some species live in the water film between sand grains of beaches (meiofauna). Pelagic forms (Figure 15.6B) are common in surface waters of freshwater lakes and ponds. Some rotifers are epizoic (live on the body of another animal) or parasitic.

Some rotifers have bizarre shapes (Figure 15.6). Their shapes are often correlated with their mode of life. Floaters are usually globular and saclike; creepers and swimmers are somewhat elongated and wormlike; and sessile types are

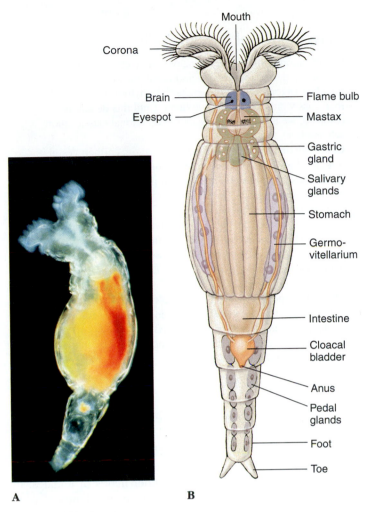

A   **B**

Figure 15.5

A. A live *Philodina*, a common rotifer; **B.** Structure of *Philodina*.

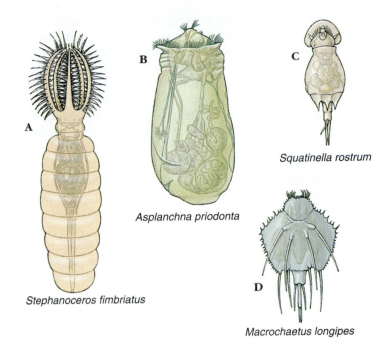

Stephanoceros fimbriatus

Asplanchna priodonta

Squatinella rostrum

Macrochaetus longipes

Figure 15.6

Variety of form in rotifers, **A**, *Stephanoceros* has five long, fingerlike coronal lobes with whorls of short bristles. It catches its prey by closing its funnel when food organisms swim into it, and the bristly lobes prevent prey from escaping. **B**, *Asplanchna* is a pelagic, predatory genus with no foot. **C**, *Squatinella* has a semicircular nonretractable, transparent hoodlike extension covering the head. **D**, *Machrochaetus* is dorsoventrally flattened.

commonly vaselike, with a thickened outer epidermis (lorica). Some are colonial.

Many species of rotifers can endure long periods of desiccation, during which they resemble grains of sand. Desiccated rotifers are very tolerant of environmental extremes. For example, some moss-dwelling species have been dried for up to 4 years before reviving upon addition of water. Other rotifers have survived temperatures as cold as −272° C before being successfully revived.

Form and Function

External Features

A rotifer's body comprises a head bearing a ciliated corona, a trunk, and a posterior tail, or foot. Except for the corona, the body is nonciliated and covered with a cuticle. One of the best-known genera is *Philodina* (Gr. *philos,* fond of, + *dinos,* whirling) (Figure 15.5).

Their ciliated corona, or crown, surrounds a nonciliated central area of their head, which may bear sensory bristles or papillae. A head's appearance depends on which of several types of corona it has—usually a circlet of some sort, or a pair of trochal

(coronal) discs (the term *trochal* comes from a Greek word meaning wheel). Cilia on the corona beat in succession, giving the appearance of a revolving wheel or pair of wheels. Their **mouth** is located in the corona on the midventral side. Coronal cilia function in both locomotion and feeding.

The trunk may be elongated, as in *Philodina* (Figure 15.5), or saccular in shape (see Figure 15.6). The trunk contains visceral organs and often bears sensory antennae. The body wall of many species is superficially ringed to simulate segmentation. Although some rotifers have a true, secreted cuticle, all have a fibrous layer within their epidermis. The fibrous layer in some is quite thick and forms a caselike **lorica,** which is often arranged in plates or rings.

Their **foot** is narrower and usually bears one to four **toes.** The cuticle of the foot may be ringed so that it is telescopically retractile. The foot is an attachment organ and contains **pedal glands** that secrete an adhesive material used by both sessile and creeping forms. It is tapered gradually in some forms (Figure 15.5) and sharply set off in others (Figure 15.6). In swimming pelagic forms, the foot is usually reduced. Rotifers can move by creeping with leechlike movements aided by the foot, or by swimming with the coronal cilia, or both.

Internal Features

Underneath the cuticle is a **syncytial epidermis,** which secretes the cuticle, and bands of **subepidermal muscles,** some circular, some longitudinal, and some running through the pseudocoel to visceral organs. The **pseudocoel** is large, occupying the space between body wall and viscera. It is filled with fluid, some of the muscle bands, and a network of mesenchymal ameboid cells.

The digestive system is complete. Some rotifers feed by sweeping minute organic particles or algae toward the mouth by beating the coronal cilia. Cilia dispose of larger unsuitable particles. Their pharynx **(mastax)** is fitted with a muscular portion equipped with hard jaws **(trophi)** for sucking in and grinding food particles. The mastax can be a crushing and grinding form among suspension feeders or a grasping and piercing form in predatory species. The constantly chewing mastax is often a distinguishing feature of these tiny animals. Carnivorous species feed on protozoa and small metazoans, which they capture by trapping or grasping. Trappers have a funnel-shaped area around the mouth. When small prey swim into the funnel, the lobes fold inward to capture and to hold them until they are drawn into the mouth and pharynx. Hunters have trophi that are projected and used like forceps to seize prey, bring it back into the pharynx, and then pierce it or break it so that edible parts may be recovered and the rest discarded. **Salivary** and **gastric glands** are believed to secrete enzymes for extracellular digestion. Absorption occurs in the stomach.

The excretory system typically consists of a pair of **protonephridial tubules,** each with several **flame cells,** that empty into a common bladder. The bladder, by pulsating, empties into a **cloaca**—into which the intestine and oviducts also empty. The fairly rapid pulsation of the protonephridia—one to four times per minute—would indicate that protonephridia are important osmoregulatory organs. Water apparently enters through the mouth rather than across the epidermis; even marine species empty their bladder at frequent intervals.

The nervous system contains a bilobed **brain,** dorsal to the mastax in the "neck" region of the body, which sends paired nerves to the sense organs, mastax, muscles, and viscera. Sensory organs include paired **eyespots** (in some species such as *Philodina*), sensory bristles and papillae, and ciliated pits and dorsal antennae.

Reproduction

Rotifers are dioecious, and males are usually smaller than females. However, despite having separate sexes, males are entirely unknown in the class Bdelloidea, and in the Monogononta they seem to occur only for a few weeks of the year.

The female reproductive system in the Bdelloidea and Monogononta consists of combined ovaries and yolk glands **(germovitellaria)** and oviducts that open into the cloaca. Yolk is supplied to developing ova by flow-through cytoplasmic bridges, rather than as separate yolk cells as in ectolecithal Platythelminthes.

Mictic (Gr. *miktos*, mixed, blended) refers to the capacity of haploid eggs to be fertilized (that is, "mixed") with the male's sperm nucleus to form a diploid embryo. Amictic ("without mixing") eggs are diploid and develop parthenogenetically.

Classification of Phylum Rotifera

Classification of the rotifers remains a subject of debate. Some authorities demote Seisonidea and Bdelloidea to orders within the class Digonata. Others demote the phylum Acanthocephala to a class within the phylum Rotifera. Until this debate is resolved, we continue to present the traditional classification scheme here.

Class Seisonidea (sy'son-id'e-a) (Gr. *seison,* earthen vessel, + *eidos,* form). Marine; elongated form; corona vestigial; sexes similar in size and form; females with pair of ovaries and no vitellaria; single genus *(Seison)* with two species; epizoic on gills of a crustacean *(Nebalia).*

Class Bdelloidea (del-oyd'e-a) (Gr. *bdella,* leech, + *eidos,* form). Swimming or creeping forms; anterior end retractile; corona usually with pair of trochal discs; males unknown; parthenogenetic; two germovitellaria. Examples: *Philodina* (Figure 15.5), *Rotaria.*

Class Monogononta (mon'o-go-non'ta) (Gr. *monos,* one, + *gonos,* primary sex gland). Swimming or sessile forms; single germovitellarium; males reduced in size; eggs of three types (amictic, mictic, dormant). Examples: *Asplanchna* (Figure 15.6B), *Epiphanes.*

In Bdelloidea (*Philodina,* for example), all females are parthenogenetic and produce diploid eggs that hatch into diploid females. These females reach maturity in a few days. In class Seisonidea females produce haploid eggs that must be fertilized and that develop into either males or females. In Monogononta, however, females produce two kinds of eggs (Figure 15.7). During most of the year diploid females produce thin-shelled, **diploid amictic eggs.** Amictic eggs develop parthenogenetically into diploid (amictic) females. However, such rotifers often live in temporary ponds or streams and are cyclic in their reproductive patterns. Any one of several environmental factors—for example, crowding, diet, or photo-period (according to species)—may induce amictic eggs to develop into diploid mictic females that produce thin-shelled **haploid eggs.** If these eggs are not fertilized, they develop into haploid males. But if fertilized, the eggs, called **mictic** eggs, develop a thick, resistant shell and become dormant. They survive over winter ("winter eggs") or until environmental conditions are again suitable, at which time they hatch into amictic females. Dormant eggs are often dispersed by winds or birds, which may explain the peculiar distribution patterns of rotifers.

The male reproductive system includes a single testis and a ciliated sperm duct that runs to a genital pore (males usually lack a cloaca). The end of the sperm duct is specialized as a copulatory organ. Copulation is usually by hypodermic impregnation; the penis can penetrate any part of a female's body wall and inject sperm directly into her pseudocoel. The zygote undergoes modified spiral cleavage.

Females hatch with adult features, needing only a few days' growth to reach maturity. Males often do not grow and are sexually mature at hatching.

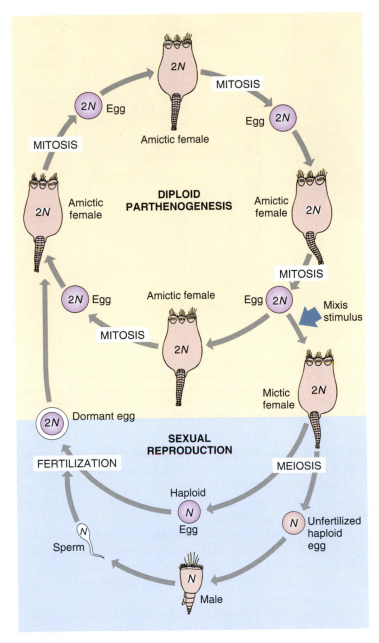

Figure 15.7

Reproduction of some rotifers (class Monogononta) is parthenogenetic during the part of the year when environmental conditions are suitable. In response to certain stimuli, females begin to produce haploid (*N*) eggs. If haploid eggs are not fertilized, they hatch into haploid males. Males provide sperm to fertilize other haploid eggs, which then develop into diploid (2*N*), dormant eggs that can resist the rigors of winter. When suitable conditions return, dormant eggs resume development, and a female hatches.

Phylogeny of Rotifera

Rotifers are a cosmopolitan group of about 2000 species, some of which occur throughout the world. However, recent molecular work has begun to question the taxonomic affinity of some of these groups, and worldwide distributions of some are apparently an artifact of morphological similarity rather than taxonomic relatedness. Rotifers are most common in freshwater, but many species also live in brackish water or even damp soils or mosses. In contrast, strictly marine species are rather few in number.

According to the traditional classification scheme on page 318, Rotifera has three classes, but some authorities demote Seisonidea and Bdelloidea to orders within a class called Digonata. Others divide the phylum into two classes: one containing the seisonids, and the other containing bdelloids and monogononts under the name Eurotatoria.

In some molecular phylogenies, spiny-headed worms called acanthocephalans (see below) arise within Rotifera. The idea that these specialized endoparasites are highly derived rotifers[1] is controversial, but if this placement is substantiated by other data sets, phylum Acanthocephala will be demoted to a class within Rotifera. At present, we depict Acanthocephala as the sister taxon to Rotifera.

PHYLUM ACANTHOCEPHALA

Members of phylum Acanthocephala (a-kan'tho-sef'a-la) (Gr. *akantha*, spine or thorn, + *kephalē*, head) are commonly called "spiny-headed worms." The phylum derives its name from one of its most distinctive features, a cylindrical, invaginable proboscis bearing rows of recurved spines, by which it attaches itself to the intestine of its host (Figure 15.8). The phylum is cosmopolitan, and more than 1100 species are known, most of which parasitize fish, birds, and mammals. All acanthocephalans are endoparasitic, living as adults in the intestine of vertebrates. Larvae of spiny-headed worms develop in arthropods, either crustaceans or insects, depending on the species.

Various species range in size from less than 2 mm to more than 1 m in length. Females of a species are usually larger than males. Their body is usually bilaterally flattened, with numerous transverse wrinkles. Worms are typically cream color but may absorb yellow or brown pigments from the intestinal contents.

Form and Function

In life the body is somewhat flattened, although students may see turgid and cylindrical specimens that were treated with tap water before fixation (Figure 15.8C).

The body wall is syncytial, and its surface is covered by minute crypts 4 to 6 µm deep, which greatly increase the surface area of the tegument. About 80% of the thickness of the tegument is the radial fiber zone, which contains a **lacunar system** of ramifying fluid-filled canals (Figure 15.8A and B). Exchange of gases, nutrients, and wastes occurs primarily across the body wall by diffusion. Within the body, diffusion is facilitated by the lacunar system. Curiously, body-wall muscles are tubelike and filled with fluid. Tubes in the muscles are continuous with the lacunar system; therefore, circulation of lacunar fluid may well bring nutrients to and remove wastes from the muscles. There is no heart or other circulatory system, and contraction of the muscles moves lacunar fluid through the canals and muscles. Thus, the lacunar fluid, which also permeates most tissues of

[1]Welch, M. D. B. 2000. Invert. Biol. **119**:17–26.

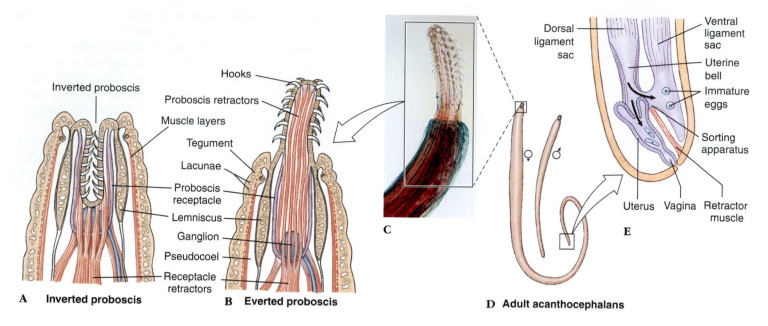

Figure 15.8

Structure of a spiny-headed worm (phylum Acanthocephala). **A** and **B,** Eversible spiny proboscis by which the parasite attaches to the intestine of its host, often doing great damage. Because they lack a digestive tract, food is absorbed through the tegument. **C,** Live acanthocephalan, *Leptorhynchoides thecatus,* with proboscis everted. **D,** Male is typically smaller than female. **E,** Scheme of the genital selective apparatus of a female acanthocephalan. It is a unique device for separating immature from mature fertilized eggs. Eggs containing larvae enter the uterine bell and pass on to the uterus and exterior. Immature eggs are shunted into the ventral ligament sac or into the pseudocoel to undergo further development.

the body, appears to serve as an unusual circulatory system in these animals. Both longitudinal and circular body-wall muscles are present.

Their proboscis, which bears rows of recurved hooks, is attached to the neck region (Figure 15.8) and can be inverted into a **proboscis receptacle** by retractor muscles. Attached to the neck region (but not within the proboscis) are two elongated hydraulic sacs (lemnisci) that may serve as reservoirs of lacunar fluid from the proboscis when that organ is invaginated or aid in gas exchange between the body and the proboscis; their exact function remains unknown, however.

There is no respiratory system. When present, the excretory system consists of a pair of **protonephridia** with flame cells. These unite to form a common tube opening into the sperm duct or uterus.

Their nervous system has a central ganglion within the proboscis receptacle and nerves to the proboscis and body. There are sensory endings on the proboscis and genital bursa. However, like many obligate endoparasites, both the nervous system and sense organs of these animals are greatly reduced.

Acanthocephalans have no digestive tract, and they must absorb all nutrients through their tegument. They can absorb various molecules by specific membrane-transport mechanisms, and other substances can cross the tegumental membrane by pinocytosis. Their tegument bears some enzymes, such as peptidases, which can cleave several dipeptides, and the amino acids are then absorbed by the worm. Like cestodes (p. 303), acanthocephalans require host dietary carbohydrate, but their mechanism for absorption of glucose is different. As glucose is

absorbed, it is rapidly phosphorylated and compartmentalized, so that a metabolic "sink" is created into which glucose from the surrounding medium can flow. Glucose diffuses down the concentration gradient into the worm because it is constantly removed as soon as it enters.

Acanthocephalans are dioecious. Males have a pair of testes, each with a vas deferens, and a common ejaculatory duct that ends in a small penis. During copulation sperm are ejected into the vagina, travel up the genital duct, and escape into the pseudocoel of the female.

In females the ovarian tissue in the ligament sac breaks into **ovarian balls** that are released from the genital ligaments, or ligament sacs, and float free in the pseudocoel. One of the ligament sacs leads to a funnel-shaped **uterine bell** that receives the developing shelled embryos and passes them to the uterus (Figure 15.8). An interesting and unique selective apparatus operates here. Fully developed embryos are slightly longer than immature ones, and they are actively selected and passed into the uterus, while immature eggs are rejected and retained for further maturation. The shelled embryos, which are discharged in feces of their vertebrate host, do not hatch until eaten by an intermediate host.

No species is normally a parasite of humans, although species that usually occur in other hosts infect humans occasionally. *Macracanthorhynchus hirudinaceus* (Gr. *makros,* long, large, + *akantha,* spine, thorn, + *rhynchos,* beak) occurs throughout the world in the small intestine of pigs and sometimes in other mammals. For *M. hirudinaceus* the intermediate host is any of several species of soil-inhabiting beetle larvae, especially scarabeids. Grubs of the June beetle (*Phyllophaga*) are frequent hosts. Here the larva (**acanthor**) burrows through the intestine and develops into a juvenile (**cystacanth**) in the insect's hemocoel. Pigs become infected by eating the grubs. Acanthocephalans penetrate the intestinal wall with their spiny proboscis to attach to the host. In many cases there is remarkably little inflammation, but in some species the inflammatory response of the host is intense. Infection with these worms can cause great pain, particularly if the gut wall is completely perforated. Multiple infections may do considerable damage to a pig's intestine, and perforations can occur.

Phylogeny of Acanthocephala

Based largely on the shape and organization of spines on the proboscis, acanthocephalans are traditionally divided into three classes: Archiacanthocephala, Eoacanthocephala, and Palaeacanthocephala. Recent molecular work suggests that the phylum status of this group is unwarranted, and in fact the acanthocephalans are a class of highly derived rotifers. This finding has stirred considerable debate among invertebrate zoologists. If acanthocephalans evolved from within Rotifera, it should be possible to outline some steps in the evolution of parasitism from a rotifer ancestor to an acanthocephalan. However, there is still debate as to which rotifer class contains animals most closely related to acanthocephalans. Shared morphological features have been identified in members of both Seisonidea and Eurotatoria. New molecular characters favor a particular evolutionary pathway, so we may eventually explain how these parasites arose and which animals were ancestral hosts.

PHYLUM CYCLIOPHORA

In December 1995, P. Funch and R. M. Kristensen reported their discovery of some very strange little creatures clinging to the mouthparts of the Norway lobster *(Nephrops norvegicus)*. The animals were tiny, only 0.35 mm long and 0.10 mm wide, and did not fit into any known phylum. They were named *Symbion pandora*, the first members of phylum Cycliophora (Figure 15.9). Two other species have since been found on other species of lobsters, but they have not been formally described.

Figure 15.9
Symbion pandora, a cycliophoran living on setae on the mouthparts of lobsters.

Cycliophorans have a very specialized habitat: they live on mouthparts of marine decapod crustaceans in the Northern Hemisphere. They attach to bristles on the mouthparts with an adhesive disc on the end of an acellular stalk. They feed by collecting bacteria, or bits of food dropped from their lobster host, on a ring of compound cilia that surrounds the mouth.

The body plan is relatively simple. The mouth leads into a U-shaped gut ending with an anus that opens outside the ciliated ring. The body is acoelomate. The epidermis is cellular and surrounded by a cuticle.

The life cycle has sexual and asexual phases. Feeding animals make internal buds, called Pandora larvae, which become new feeding individuals upon release. Clone-members quickly occupy vacant areas on the lobster mouthparts. Internal budding is also used to make a new feeding and digestive system for a feeding animal—the existing system degenerates and is replaced by one from the internal bud.

As a prelude to sexual reproduction, male or female larvae are made. A male larva is released from a feeding individual and settles atop another animal housing a female larva. A male larva produces secondary males with reproductive organs; internal fertilization occurs as one secondary male mates with a female larva leaving the body of a feeding animal. Once the egg in the female is fertilized, a chordoid larva develops inside the body of its mother, consuming it. The chordoid larva swims to a new lobster host where it makes a feeding animal by internal budding. The feeding animal then makes a clone of feeders by internal budding.

Relationships to other phyla are quite controversial. Funch and Kristensen consider the organisms to be protostomes, and most analyses place them in Lophotrochozoa, sometimes within or allied to Gnathifera.

PHYLUM GASTROTRICHA

Gastrotricha (gas-tro-tri′ka) (N. L. fr. Gr. *gaster, gastros,* stomach or belly, + *thrix, trichos,* hair) includes small, ventrally flattened animals usually less than 1 mm in length. The largest species of gastrotrichs can reach lengths of about 3 mm. Superficially, gastrotrichs may appear somewhat like rotifers but lacking a corona and mastax and having a characteristically bristly or scaly body. They are usually found gliding on the substrate, or the surface of an aquatic plant or animal, by means of their ventral cilia, or they compose part of the meiofauna in interstitial spaces between substrate particles.

Gastrotrichs occur in fresh, brackish, and salt water. The 450 or so species are about equally divided between these environments. Many species are cosmopolitan, but only a few occur in both freshwater and the sea. Much is yet to be learned about their distribution and biology.

Form and Function

A gastrotrich (Figures 15.10 and 15.11) is usually elongated, with a convex dorsal surface bearing a pattern of bristles, spines, or scales, and a flattened ciliated ventral surface. Cells on the ventral surface may be monociliated or multiciliated. The head is

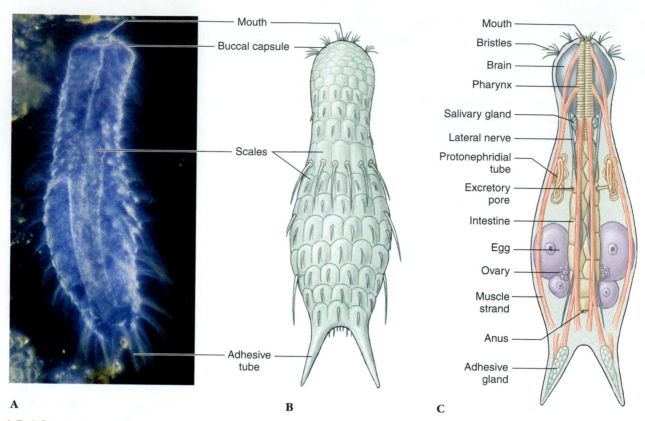

Mouth
Buccal capsule
Scales
Adhesive tube

Mouth
Bristles
Brain
Pharynx
Salivary gland
Lateral nerve
Protonephridial tube
Excretory pore
Intestine
Egg
Ovary
Muscle strand
Anus
Adhesive gland

A   B   C

Figure 15.10

A, Live *Chaetonotus simrothic,* a common gastrotrich. **B,** Dorsal surface. **C,** Internal structure, ventral view.

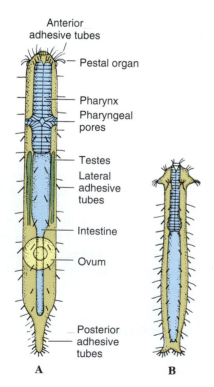

Anterior adhesive tubes
Pestal organ
Pharynx
Pharyngeal pores
Testes
Lateral adhesive tubes
Intestine
Ovum
Posterior adhesive tubes

A   B

Figure 15.11

Gastrotrichs in order Macrodasyida. **A,** *Macrodasys.* **B,** *Turbanella.*

often lobed and ciliated, and the tail end may be greatly elongated or forked in some species.

A partially syncytial epidermis is found beneath the cuticle; it has some cellular regions. Longitudinal muscles are better developed than are circular ones, and in most cases they are unstriated. Adhesive tubes secrete a substance for attachment. A dual-gland system for attachment and release resembles that described for Turbellaria (p. 292).

No specialized respiratory or circulatory structures occur in gastrotrichs; gas exchange is by simple diffusion in these tiny animals. At least some species appear capable of anaerobic respiration. Their digestive system is complete and comprises a mouth, a muscular pharynx, a stomach-intestine, and an anus (Figure 15.10C). Food is largely algae, protozoa, bacteria, and detritus, which are directed to their mouth by their head cilia. Digestion appears to be extracellular, although little is known about the exact mechanisms of digestion and nutrient absorption. Protonephridia are equipped with **solenocytes** rather than flame cells. Solenocytes have a single flagellum enclosed in a cylinder of cytoplasmic rods as opposed to the many flagella found in flame bulbs. There is no body cavity in gastrotrichs, and the internal organs are all packed tightly into the compact body.

Their nervous system includes a brain near the pharynx and a pair of lateral nerve trunks. Sensory structures are similar to those in rotifers, except that eyespots are generally lacking although some species have pigmented eyespots (ocelli) in the brain. Sensory bristles, often concentrated on the head, are modified from cilia and are primarily tactile.

Gastrotrichs are typically hermaphroditic, although the male system of some is so rudimentary that they are functionally parthenogenetic females. Like rotifers, some gastrotrichs produce thin-walled, rapidly developing eggs and thick-shelled, dormant eggs. The thick-shelled eggs can withstand harsh environments and may survive dormancy for some years. Cleavage is not well studied but appears to be radial. Development is direct, and juveniles have the same form as adults. Growth and maturation are often rapid, and newly hatched juveniles usually reach sexual maturity within just a few days.

PHYLUM ENTOPROCTA

Entoprocta (en'to-prok'ta) (Gr. *entos,* within, + *proktos,* anus) is a small phylum of about 150 species of tiny, sessile animals that superficially resemble hydroid cnidarians but have ciliated tentacles that tend to roll inward (Figure 15.12B and C). Most entoprocts are microscopic, and none is more than 5 mm long. They may be solitary or colonial, but all are stalked and sessile. All are ciliary feeders.

With the exception of *Urnatella* (L. *urna,* urn, + *ellus,* dim. suffix), which occurs in freshwater, all entoprocts are marine forms that have a wide distribution from polar regions to tropics. Most marine species are restricted to coastal and brackish waters and often grow on shells and algae. Some are commensals on marine annelid worms. Entoprocts occur from the intertidal to depths of around 500 m. Freshwater entoprocts occur on the underside of rocks in running water. *U. gracilis* is the only common freshwater species in North America (Figure 15.12A).

Form and Function

The body, or **calyx,** of an entoproct is cup-shaped, bears a crown, or circle, of ciliated tentacles, and may be attached to a substratum by a single stalk and an attachment disc with adhesive glands, as in the solitary *Loxosoma* and *Loxosomella* (Gr. *loxos,* crooked, + *soma,* body) (Figure 15.12B), or by two or more stalks in colonial forms. Movement is usually restricted in entoprocts, but *Loxosoma,* which lives in the tubes of marine annelids, is quite active, moving over the annelid and its tube freely.

The body wall comprises a cuticle, cellular epidermis, and longitudinal muscles. The tentacles and stalk are continuations of the body wall. The 8 to 30 tentacles forming the crown are ciliated on their lateral and inner surfaces, and each can move individually. Tentacles can roll inward to cover and to protect the mouth and anus but cannot be retracted into the calyx. The gut is U-shaped and ciliated, and both the mouth and the anus open within the circle of tentacles. Entoprocts are **ciliary suspension feeders.** Long cilia on the sides of the tentacles keep a current of water containing protozoa, diatoms, and particles of detritus moving inward between the tentacles. Short cilia on the inner surfaces of the tentacles capture food and direct it downward toward the mouth. Digestion and absorption occur within the stomach and intestine before wastes are discharged from the anus.

The pseudocoel is largely filled with a gelatinous parenchyma in which is embedded a pair of flame bulb protonephridia and their ducts, which unite and empty near the mouth. A well-developed **nerve ganglion** occurs on the ventral side of the stomach, and the body surface bears sensory bristles and pits. Circulatory and respiratory organs are absent. Exchange

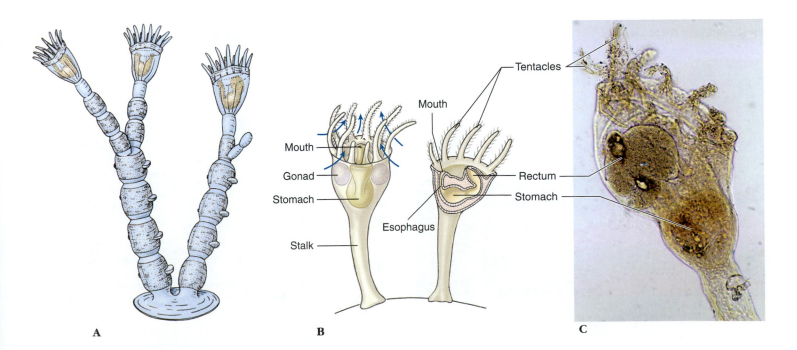

Figure 15.12
A, *Urnatella,* a freshwater entoproct, forms small colonies of two or three stalks from a basal plate. **B,** *Loxosomella,* a solitary entoproct. Both solitary and colonial entoprocts can reproduce asexually by budding, as well as sexually. **C,** A live *Loxosomella.*

of gases occurs across the body surface, probably much of it through the tentacles.

Some species are dioecious, but many are monoecious, most often protandrous hermaphrodites, where the gonad at first produces sperm and later eggs. Colonial forms may have monoecious or dioecious zooids, and colonies can even contain zooids of both sexes. The gonoducts open within the circle of tentacles.

Fertilized eggs develop in a depression, or brood pouch, between the gonopore and the anus. Entoprocts have a modified spiral cleavage pattern with mosaic blastomeres. The embryo gastrulates by invagination. Mesoderm develops from the 4d cell. The trochophore-like larva (see p. 337) is ciliated and free-swimming. It has an apical tuft of cilia at the anterior end and a ciliated girdle around the ventral margin of the body. Eventually the larva settles to a substratum and metamorphoses into an adult zooid.

LOPHOPHORATES

The final three phyla in this chapter are the most controversial taxa placed within Protostomia. Evidence that Ectoprocta, Brachiopoda, and Phoronida belong within the lophotrochozoan subgroup of protostomes (see Figure 15.2) comes from sequence analysis of the genes encoding small-subunit ribosomal RNA. Some developmental data are consistent with the molecular data: in phoronids, the blastopore becomes the mouth, as is typical for protostomes, but in brachiopods, the blastopore disappears and both mouth and anus develop from new openings. Ectoproct cleavage appears to be mosaic, another protostome feature.

Other aspects of development support placement of these taxa within Deuterostomia: cleavage is radial in all three phyla, and each has a coelom formed by enterocoely. As in some deuterostomes, the coelom is divided into three parts: **protocoel, mesocoel,** and **metacoel.**

Members of all three phyla possess a feeding device called a **lophophore** (Gr. *lophos,* crest or tuft, + *phorein,* to bear). A lophophore is a unique arrangement of ciliated tentacles borne on a ridge or fold of the body wall (Figure 15.13). The tentacles are hollow and contain an extension of the mesocoel. Thus, the thin ciliated tentacles are not only an efficient feeding device, but also a respiratory device permitting gas exchange between the surrounding water and the internal coelomic fluid. The lophophore and its crown of tentacles can usually be extended for feeding or withdrawn for protection.

The gut is U-shaped, with the mouth opening inside the lophophore ring, and the anus opening outside the ring (Figures 15.13 and 15.18). A flap of tissue, the epistome, covers the mouth and contains an extension of the protocoel. The regions of the body housing the mesocoel and metacoel are called the mesosome and metasome, respectively. In Ectoprocta, the fluid-filled mesocoel and metacoel cavities are part of the hydraulic system for lophophore extension. In other groups the metacoel houses the viscera.

At various points in history, Ectoprocta, Brachiopoda, and Phoronida have been considered protostomes with some deuterostome characteristics or deuterostomes with some protostome characteristics. Their current placement as lophotrochozoan protostomes reflects acceptance of molecular data in the face of conflicting and inconsistent morphological data.

In some classification schemes, the shared presence of a lophophore is used to unite all three phyla into a lophophorate clade. We do not present this scheme. Both molecular and morphological evidence support the placement of Brachiopoda and Phoronida as each other's closest relatives, but the position of Ectoprocta is less clear. It may belong with the other two phyla, but a detailed analysis of lophophore function[2] suggests otherwise. A lophophore is well suited to capturing suspended

[2]Nielsen, C. 2002. Integ. and Comp. Biol. **42:**685–691.

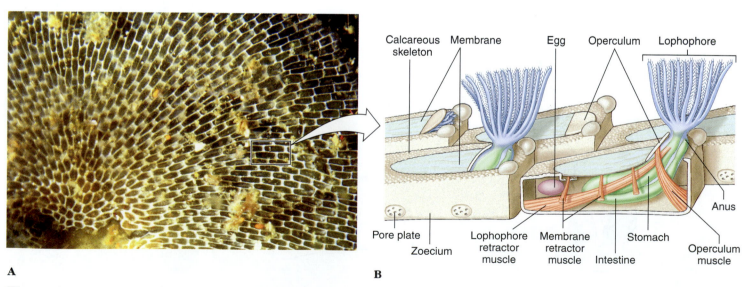

A B

Figure 15.13

A, Skeletal remains of a colony of *Membranipora,* a marine encrusting bryozoan (Ectoprocta). Each little oblong zoecium is the calcareous former home of a zooid. **B,** Portion of a colony of an encrusting bryozoan. Two zooids are shown extended from their chambers, the zoecia. The tiny zooids pop up to feed with their tentacular crown, then quickly withdraw at the slightest disturbance. The mouth is inside the lophophore ring, but the anus lies outside.

food particles from the surrounding water—is the ectoproct lophophore convergent to that of brachiopods and phoronids? Much remains to be discovered about these odd animals.

PHYLUM ECTOPROCTA (BRYOZOA)

Ectoprocta (ek-to-prok′ta) (Gr. *ektos,* outside, + *proktos,* anus) contains aquatic animals that often encrust hard surfaces. Most species are sessile, but some slide slowly, and others crawl actively, across the surfaces they inhabit. With very few exceptions, they are colony builders. Each member of a colony is small, typically less than 0.5 mm. Colony members, called **zooids,** feed by extending their lophophores into surrounding water to collect tiny particles. Zooids secrete small containers in which they live and thus form an exoskeleton (Figure 15.13). The exoskeleton or **zoecium,** may, according to species, be gelatinous, chitinous, or stiffened with calcium and possibly also impregnated with sand. Its shape may be boxlike, vaselike, oval, or tubular.

Ectoprocts have left a rich fossil record since the Ordovician period and are diverse and abundant today. There are about 4500 living species of ectoprocts. They live in both freshwater and marine habitats but largely in shallow waters.

Some colonies form limy encrustations on seaweed, shells, and rocks; others form fuzzy or shrubby growths or erect, branching colonies that look like seaweed (Figure 15.14). Some ectoprocts might easily be mistaken for hydroids but can be distinguished under a microscope by presence of an anus (Figure 15.13). Although zooids are minute, the colonies are often several centimeters in diameter; some encrusting colonies may be a meter or more in width, and erect forms may reach 30 cm or more in height. Marine forms exploit all kinds of firm surfaces, such as shells, rocks, large brown algae, mangrove roots, ship

bottoms, and even the bottoms of icebergs! Freshwater ectoprocts may form mosslike colonies on stems of plants or on rocks, usually in shallow ponds or pools. In some freshwater forms individuals are borne on finely branching stolons that form delicate tracings on the underside of rocks or plants. Other freshwater ectoprocts are embedded in large masses of gelatinous material.

Ectoprocts have long been called bryozoans, or moss animals (Gr. *bryon,* moss, + *zōon,* animal), a term that originally included Entoprocta also. However, because entoprocts have their anus located within the tentacular crown, they are commonly separated from ectoprocts, which, like other lophophorates, have their anus outside the circle of tentacles. Many authors continue to use the name "Bryozoa" but exclude entoprocts from the group.

Form and Function

Each member of a colony lives in a tiny chamber, called a zoecium, which is secreted by its epidermis (Figure 15.13). Each **zooid** consists of a feeding polypide and a case-forming cystid. A **polypide** includes the lophophore, digestive tract, muscles, and nerve centers. A **cystid** includes the body wall of an animal, together with its secreted exoskeleton.

Polypides live a type of jack-in-the-box existence, popping up to feed but, at the slightest disturbance, quickly withdrawing into their little chamber, which often has a tiny trapdoor (operculum) that shuts to conceal its inhabitant (Figure 15.13). To extend its tentacular crown, certain muscles contract, which increases hydrostatic pressure within the body cavity and pushes the lophophore out by a hydraulic mechanism. Other muscles can contract to withdraw the crown to safety with great speed.

When feeding, an animal extends its lophophore and spreads its tentacles to form a funnel. Cilia on the tentacles draw water into the funnel and out between the tentacles. Food particles

Figure 15.14

Colonies of marine ectoprocts. **A,** The zooids are extended in this lacy colony of *Membranipora tuberculata* **B,** *Bugula neritina* has upright, branching colonies.

A

B

A

B

Figure 15.15

A, Ciliated lophophore of *Electra pilosa,* a marine ectoproct. The thin central tube is the base of a vibraculum, a modified zooid that sweeps the colony surface. **B,** *Plumatella repens,* a freshwater bryozoan (phylum Ectoprocta). It grows on the underside of rocks and on vegetation in lakes, ponds, and streams.

caught by cilia in the funnel are drawn into the mouth, both by a pumping action of the muscular pharynx and by action of cilia along the length of the tentacles and in the pharynx itself. Undesirable particles can be rejected by reversing the ciliary action, by drawing the tentacles close together, or by retracting the whole lophophore into the zoecium. The lophophore ridge tends to be circular in marine ectoprocts (Figure 15.15A) and U-shaped in freshwater species (Figure 15.15B). A septum divides the mesocoel in the lophophore from the larger posterior metacoel. A protocoel and epistome occur only in freshwater ectoprocts.

Digestion in the ciliated, U-shaped digestive tract begins extracellularly in the stomach and is completed intracellularly in the intestine. Respiratory, vascular, and excretory organs are absent. Gaseous exchange is through the body surface, and since ectoprocts are small, coelomic fluid is adequate for internal transport. Pores in the walls between adjoining zooids permit exchange of materials throughout the colony by way of the coelomic fluid. Coelomocytes engulf and store waste materials. A ganglionic mass and a nerve ring surround the pharynx, but no specialized sense organs are present.

Most colonies contain only feeding individuals, but specialized zooids incapable of feeding (collectively called

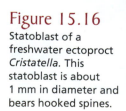

Figure 15.16

Statoblast of a freshwater ectoproct *Cristatella.* This statoblast is about 1 mm in diameter and bears hooked spines.

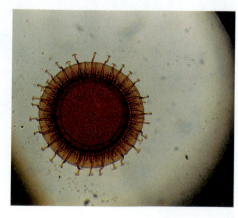

heterozooids) occur in some species. One type of modified zooid (called an *avicularium*) resembles a bird beak that snaps at small invading organisms that might foul a colony. Another type (called a *vibraculum*) has a long bristle that apparently helps to sweep away foreign particles (Figure 15.15A).

Most ectoprocts are hermaphroditic. Some species shed eggs into seawater, but most brood their eggs, some within the coelom and some externally in a special brood chamber called an ovicell, which is a modified zoecium in which an embryo develops. In some cases many embryos proliferate asexually from that initial embryo in a process called **polyembryony.** Cleavage is radial but apparently mosaic. Little is known of mesoderm derivation. Larvae of nonbrooding species have a functional gut and swim for a few months before settling; larvae of brooding species do not feed and settle after a brief free-swimming existence. They attach to the substratum by secretions from an **adhesive sac,** then metamorphose to their adult form.

Each colony begins from this single metamorphosed primary zooid, which is called an **ancestrula.** The ancestrula then undergoes asexual budding to produce the many zooids of a colony. Freshwater ectoprocts have another type of budding that produces **statoblasts** (Figure 15.16), which are hard, resistant capsules containing a mass of germinative cells. Statoblasts are formed during summer and fall. When a colony dies in late autumn, statoblasts remain, and in spring they give rise to new polypides and eventually to new colonies.

PHYLUM BRACHIOPODA

Brachiopoda (brak-i-op′o-da) (Gr. *brachiōn,* arm, + *pous, podos,* foot), or lamp shells, are an ancient group. Although about 325 species are now living, some 12,000 fossil species, which once flourished in Paleozoic and Mesozoic seas, have been described. Modern forms have changed little from early ones. Genus *Lingula* (L. tongue) (Figure 15.17A) is considered a "living fossil," having existed virtually unchanged since Ordovician times. Most modern brachiopod shells range between 5 and 80 mm in length, but some fossil forms reached 30 cm.

Brachiopods are attached, bottom-dwelling, marine forms that mostly prefer shallow water, although they are known from nearly all ocean depths. Externally brachiopods resemble bivalved molluscs in having two calcareous shell valves secreted by a mantle. They were, in fact, classified with molluscs until the middle of the nineteenth century, and their name refers to the arms of the **lophophore,** which were thought homologous to the mollusc foot. Brachiopods, however, have dorsal and ventral valves instead of right and left lateral valves as do bivalve molluscs and, unlike bivalves, most of them are attached to a substrate either directly or by means of a fleshy stalk called a **pedicel.** Some, such as *Lingula,* live in vertical burrows in sand or mud. Muscles open and close the valves and provide movement for the stalk and tentacles.

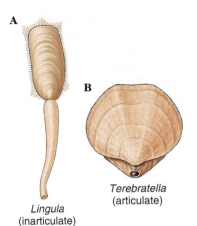

Figure 15.17

Brachiopods. **A,** *Lingula,* an inarticulate brachiopod that normally occupies a burrow. The contractile pedicel can withdraw the body into the burrow. **B,** An articulate brachiopod, *Terebratella.* The valves have a tooth-and-socket articulation, and a short pedicel projects through the pedicel valve to attach to the substratum.

Terebratella
(articulate)

Lingula
(inarticulate)

In most brachiopods the ventral (pedicel) valve is slightly larger than the dorsal (brachial) valve, and one end projects in the form of a short, pointed beak perforated where the fleshy stalk passes through the shell to attach to the substratum (Figure 15.17B). In many the pedicel valve is shaped like a classic oil lamp of ancient Greece and Rome, so that brachiopods came to be called "lamp shells."

Shell structure distinguishes the two classes of branchiopods. Shell valves of Articulata have a connecting hinge with an interlocking tooth-and-socket arrangement, as in *Terebratella* (L. *terebratus,* a boring, + *ella,* dim. suffix); those of Inarticulata lack the hinge and are held together by muscles only, as in *Lingula* and *Glottidia* (Gr. *glōttidos,* mouth of windpipe).

Their body occupies only the posterior part of the space between the valves (Figure 15.18), and extensions of the body wall form mantle lobes that line and secrete the shell. Their large horseshoe-shaped lophophore in the anterior mantle cavity bears long, ciliated tentacles used in respiration and feeding. Ciliary water currents carry food particles between the gaping valves and over the lophophore. Tentacles catch food particles, and

ciliated grooves carry the particles along the arm of the lophophore to their mouth. Rejection tracts carry unwanted particles to the mantle lobe where they are swept out in ciliary currents. Organic detritus and some algae are apparently primary food sources. A brachiopod's lophophore not only creates food currents, as do other lophophorates, but also seems to absorb dissolved nutrients directly from environmental seawater.

As in other lophophorates, the posterior metacoel bears the viscera. One or two pairs of nephridia open into the coelom and empty into the mantle cavity. Coelomocytes, which ingest particulate wastes, are expelled by nephridia. There is an open circulatory system with a contractile heart. Lophophore and mantle are probably the chief sites of gaseous exchange. There is a nerve ring with a small dorsal and a larger ventral ganglion.

Most species have separate sexes, and temporary gonads discharge gametes through the nephridia. Most fertilization is external, but a few species brood their eggs and young.

Cleavage is radial, and coelom and mesoderm formation in at least some brachiopods is enterocoelic. The blastopore closes, but its relationship to the mouth is uncertain. In articulates, metamorphosis of larvae occurs after they have attached by a pedicel. In inarticulates, juveniles resemble a minute brachiopod with a coiled pedicel in the mantle cavity. There is no metamorphosis. As a larva settles, its pedicel attaches to the substratum, and adult existence begins.

PHYLUM PHORONIDA

Phylum Phoronida (fo-ron′i-da) (L. *Phoronis,* in mythology, surname of Io, who was turned into a white heifer) contains approximately 20 species of small, wormlike animals. Most live on the substrate below shallow coastal waters, especially in temperate seas. They range from a few millimeters to 30 cm in length. Each worm secretes a leathery or chitinous tube in which it lies free, but which it never leaves. The tubes may be anchored singly or

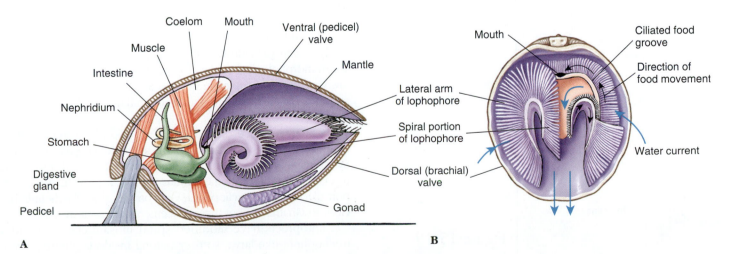

Figure 15.18

Phylum Brachiopoda. **A,** An articulate brachiopod (longitudinal section). Note that its pedicel emerges from the ventral valve, so that when it is attached to a substrate, an articulate brachiopod is "upside down," with its ventral valve on top and its dorsal valve below. **B,** Feeding and respiratory currents. Large (blue) arrows show water flow over lophophore; small (black) arrows indicate food movement toward mouth in ciliated food groove.

in a tangled mass on rocks, shells, or pilings or buried in sand. They thrust out the tentacles on the lophophore for feeding, but a disturbed animal can withdraw completely into its tube.

A lophophore has two parallel ridges curved in a horseshoe-shape, the bend located ventrally and the mouth lying between the two ridges (Figure 15.19). Horns of the ridges often coil into twin spirals. Each ridge carries hollow ciliated tentacles, which, like the ridges themselves, are extensions of the body wall.

Cilia on the tentacles direct a water current toward a groove between the two ridges, which leads toward the mouth. Plankton and detritus caught in this current become entangled in mucus and are carried by cilia to the mouth. The anus lies dorsal to the mouth, outside the lophophore, flanked on each side by a nephridiopore. Water leaving the lophophore passes over the anus and nephridiopores, carrying away wastes. Cilia in the stomach area of the U-shaped gut aid food movement.

The body wall consists of cuticle, epidermis, and both longitudinal and circular muscles. The protocoel is present as a small cavity in the epistome; it connects to the mesocoel along the lateral aspects of the epistome (Figure 15.19). A septum separates the metacoel from the mesocoel. Phoronids have an extensive system of contractile blood vessels in a functionally but not technically closed circulatory system. They have no heart. Their blood contains hemoglobin within nucleated cells. There is a pair of metanephridia. A nerve ring sends nerves to tentacles and body wall, but the system is diffuse and lacks a distinct ganglion that could be called a brain. A single giant motor fiber lies in the epidermis, and an epidermal nerve plexus supplies the body wall and epidermis.

There are both monoecious (the majority) and dioecious species of Phoronida, and at least two species reproduce asexually. Fertilization may be internal or external, but contrary to early reports cleavage is radial. Coelom formation is by a highly modified enterocoelous route, but the blastopore becomes the mouth. A free-swimming, ciliated larva, called an actinotroch, which sinks to the bottom, metamorphoses into an adult, secretes a tube, and becomes sessile.

PHYLOGENY

Evidence from a sequence analysis of the small-subunit 18S ribosomal genes suggests that some time after ancestral deuterostomes diverged from ancestral protostomes in the Precambrian, protostomes split again into two large groups (or superphyla): Ecdysozoa, containing phyla that go through a series of molts during development, and Lophotrochozoa, including lophophorate phyla and phyla many of whose larvae are trochophore-like (see p. 291).[3]

Most lophotrochozoans share some developmental features such as spiral mosaic cleavage and formation of the mouth from the embryonic blastopore, but there is no common body plan. Lophotrochozoa contains acoelomate, pseudocoelomate, and coelomate members. Of those animals already described, platyhelminths, gnathostomulids, gastrotrichs, and cycliophorans are acoelomate, whereas pseudocoelomate members include rotifers, acanthocephalans, and entoprocts. Nemertean worms and the three lophophorate phyla are all coelomate animals, but in nemerteans the coelom surrounds the proboscis and differs greatly from the tripartite coelom in ectoprocts, brachiopods, and phoronids.

The lophotrochozoan protostomes are a heterogeneous group for which evolutionary branching order remains to be determined. Many group members described in this chapter are small and relatively poorly known. Clade Gnathifera represents one hypothesis for relationships among four phyla: Gnathostomulida, Micrognathozoa, Rotifera, and Acanthocephala. Members of the first three taxa share complex cuticular jaws, whereas acanthocephalans are assumed to be descended from ancestors that possessed such jaws.

Acanthocephalans are highly specialized parasites with a unique, and likely ancient morphology. However, DNA sequence analysis can provide hypothetical phylogenetic relationships when morphological or developmental similarities between taxa are virtually or completely absent. Such analysis has led to the startling conclusion that acanthocephalans are highly derived rotifers.[4]

Gene sequence data place Acanthocephala and Rotifera together as clade Syndermata, sharing a eutelic syncytial epidermis. Syndermata is placed with Micrognathozoa and Gnathostomulida in clade Gnathifera. Some similar morphological features support placement of Cycliophora, Gastrotricha, and Platyhelminthes close to Gnathifera, but these placements are controversial.

Entoprocts have modified spiral mosaic cleavage and a trochophore-like larva, so they belong inside Lophotrochozoa, but outside clade Gnathifera.

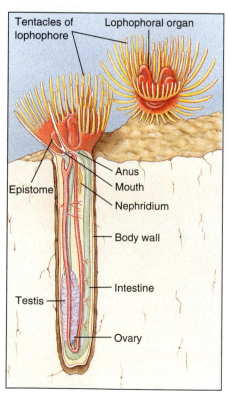

Figure 15.19

Internal structure of *Phoronis* (phylum Phoronida), in diagrammatic vertical section.

Labels in figure:
Tentacles of lophophore
Lophophoral organ
Epistome
Anus
Mouth
Nephridium
Body wall
Intestine
Testis
Ovary

[3]Balavoine, G., and A. Adoutte. 1998. Science **280**:397–398.
[4]Welch, W. D. B. 2000 Invert. Biol. **199**:17–26.

Entoprocts were once included with phylum Ectoprocta in a phylum called Bryozoa, but ectoprocts are true coelomate animals, and most zoologists prefer to place them in a separate group. Ectoprocts are still often called bryozoans. Sequence analysis places both entoprocts and ectoprocts among lophotrochozoan phyla.

Ectoprocts, brachiopods, and phoronids share a lophophore and a tripartite coelom, but other features are mixtures of developmental traits from both protostomes and deuterostomes. As already discussed (p. 324), there is debate over whether the lophophorates form a clade, and whether the group members, individually or collectively, belong within Protostomia or Deuterostomia.

Division of the coelom into three parts (**trimerous, or tripartite**) is a feature shared with deuterostomes, but the character must be convergent if lophophorates are protostomes. Furthermore, some authors question the trimerous nature and homologies of the coelom in some lophophorates (for example, whether the space in the epistome of inarticulate brachiopods is a protocoel, whether the mesocoel and metacoel in brachiopods are homologous to these spaces in other lophophorates, and whether the body coelum of ectoprocts is homologous to that of brachiopods and phoronids).

With such large issues unresolved, little can be said about diversification within Lophotrochozoa. The ten phyla just described have become the focus of more intense study in recent years; perhaps evolutionary relationships among these phyla will be clearer soon. The phyla to be discussed in Chapters 16 and 17 are clearly members of Lophotrochozoa and possess developmental features typical of the group.

SUMMARY

Lophotrochozoans do not share a body plan; instead, group members are acoelomate, pseudocoelomate, and coelomate.

Clade Gnathifera contains four phyla whose common ancestor is hypothesized to possess cuticular jaws with a unique microstructure. Included phyla are Gnathostomulida, Micrognathozoa, Rotifera, and Acanthocephala.

Gnathostomulida is a curious phylum containing tiny worm-like animals living among sand grains and silt. The animals do not have anus.

Micrognathozoa consists of a single species of tiny animals living between sand grains. These animals have three pairs of complex jaws similar to those of rotifers and gnathostomulids.

Phylum Rotifera is composed of small, mostly freshwater species with a ciliated corona, which creates currents of water to draw planktonic food toward the mouth. Their mouth opens into a muscular pharynx, or mastax, which is equipped with jaws. The Bdelloidea are obligate parthenogens, and males appear not to exist in this group.

Acanthocephalans are all parasitic in the intestine of vertebrates as adults, and their juvenile stages develop in arthropods. They have an anterior, invaginable proboscis armed with spines, which they embed in the intestinal wall of their host. They do not have a digestive tract and so must absorb all nutrients across their tegument. Molecular evidence and a shared eutelic syncytial epidermis suggest a phylogenetic affinity of acanthocephalans and rotifers, and therefore a gnathiferan origin of acanthocephalans, which requires an evolutionary loss of jaws in an ancestral acanthocephalan lineage.

Cycliophorans are very tiny animals living on the setae of mouthparts of lobsters. They have complex life cycles with sexual and asexual phases.

Gastrotrichs are also tiny aquatic animals. They have ventrally flattened bodies with bristles or scales. They move by cilia or adhesive glands.

Entoprocts are small sessile aquatic animals with a cup-shaped body on a small stalk. They have a crown of ciliated feeding tentacles encircling both the mouth and anus.

Ectoprocta, Brachiopoda, and Phoronida all bear a lophophore, which is a crown of ciliated tentacles surrounding the mouth but not the anus and containing an extension of the mesocoel. They are sessile as adults, have a U-shaped digestive tract, and have a free-swimming larva. The lophophore functions as both a respiratory and a feeding structure, its cilia creating water currents from which food particles are filtered.

Ectoprocts are abundant in marine habitats, living on a variety of submerged substrata, and a number of species are common in freshwater. Ectoprocts are colonial, and although each individual is quite small, colonies are commonly several centimeters or more in width or height. Each individual lives in a chamber (zoecium), which is a secreted exoskeleton of chitinous, calcium carbonate, or gelatinous material.

Brachiopods were very abundant in the Paleozoic era but have been declining in numbers and species since the early Mesozoic era. Their bodies and lophophores are covered by a mantle, which secretes a dorsal and a ventral valve (shell). They are usually attached to the substrate directly or by means of a pedicel.

Phoronida are the least common lophophorates, living in tubes mostly in shallow coastal waters. They thrust the lophophore out of the tube for feeding.

REVIEW QUESTIONS

1. What are some adaptive advantages of a pseudocoel compared to the acoelomate condition?
2. Explain the difference between a true coelom and a pseudocoel.
3. What character unites members of clade Gnathifera?
4. What characters are used to unite rotifers and acanthocephalans as members of clade Syndermata?
5. How are the valves of a brachiopod oriented in terms of the dorsal-ventral axis?
6. What habitat is shared by micrognathozoans and gnathostomulids?
7. Where would you look if you had to find a cycliophoran?
8. How does an entoproct differ from an ectoproct?

9. What is the normal size of a rotifer; where is it found; and what are its major features?

10. Explain the difference between mictic and amictic eggs of rotifers. What is the adaptive value of each?

11. What is eutely?

12. Describe the major features of the acanthocephalan body.

13. How do acanthocephalans get food?

14. The evolutionary ancestry of acanthocephalans is particularly obscure. Describe some characters of acanthocephalans that make it surprising that they could be highly derived rotifers.

15. About how big are gastrotrichs, gnathostomulids, and micrognathozoans?

16. What are distinguishing characteristics of entoprocts?

17. What characters do lophophorate phyla have in common? What characters distinguish them from each other?

18. Define each of the following: lophophore, zoecium, zooid, polypide, cystid, statoblasts.

19. What are some protostome characters found among lophophorates? What are their deuterostome characters?

20. How is the lophophore of ectoprocts extended?

SELECTED REFERENCES

Balavoine, G., and A. Adoutte. 1998. One or three Cambrian radiations? Science **280:**397–398. *Discusses radiation into superphyla Ecdysozoa, Lophotrochozoa, and Deuterostomia.*

Brusca, R. C., and G. J. Brusca. 2003. Invertebrates. ed. 2. Sunderland, Massachusetts, Sinauer Associates, Inc. *A comprehensive invertebrate text.*

Cohen, B. L., and A. Weydmann. 2005. Molecular evidence that phoronids are a subtaxon of brachiopods (Brachiopoda : Phoronata) and that genetic divergence of metazoan phyla began long before the early Cambrian. Org. Divers. Evol. **5:**253–273. *Indicates that phoronids arose within Brachiopoda and should no longer be considered a phylum.*

Conway Morris, S., B. L. Cohen, A. B. Gawthrop, and T. Cavalier-Smith. 1996. Lophophorate phylogeny. Science **272:**282–283. *These authors urged caution in acceptance of the taxon Lophotrochozoa proposed by Halanych et al. (1995).*

Funch, P., and R. M. Kristensen. 1995. Cycliophora is a new phylum with affinities to Entoprocta and Ectoprocta. Nature **378:**711–714. *The first description of Symbion pandora.*

Giribet, G., M. V. Sorenson, P. Funch, R. M. Kristensen, and W. Sterrer. 2004. Investigations into the phylogenetic position of Micrognathozoa using four molecular loci. Cladistics **20:**1–13. *Research supports the placement of micrognathozoans outside any known phylum.*

Halanych, K. M., J. D. Bacheller, A. M. A. Aguinaldo, S. M. Liva, D. M. Hillis, and J. A. Lake. 1995. Evidence from 18S ribosomal DNA that lophophorates are protostome animals. Science **267:**1641–1643. *Despite much morphological and developmental evidence that lophophorates are deuterostomes, they clustered with annelids and molluscs in this analysis. The authors proposed Lophotrochozoa, defined as the last common ancestor of lophophorate taxa, annelids, and molluscs, and all descendants of that ancestor.*

Halanych, K. M., and Y. Passamaneck. 2001. A brief review of metazoan phylogeny and future prospects in Hox-research. Am. Zool. **41:**629–639. *A good review of the arguments for and against the lophotrochozoa and ecdysozoa hypotheses.*

Helfenbein, K. G., and J. L. Boore. 2004. The mitochondrial genome of *Phoronis architecta*—Comparisons demonstrate that phoronids are Lophotrochozoan protostomes. Mol. Biol. Evol. **21:**153–157. *Analysis of the mitochondrial DNA sequence shows a gene arrangement very similar to that of a chiton.*

Kristensen, R. M. 2002. An introduction to Loricifera, Cycliophora, and Micrognathozoa. Integ. and Comp. Biol. **42:**641–651. *A clear and informative description of these little-known animal groups.*

Neilsen, C. 2002. The phylogenetic position of Entoprocta, Ectoprocta, Phoronida, and Brachiopoda. Integ. and Comp. Biol. **42:**685–691. *Presents evidence that lophophorates do not form a monophyletic group and that phoronids and brachiopods are deuterostomes.*

Rieger, R. M., and S. Tyler. 1995. Sister-group relationship of Gnathostomulida and Rotifera-Acanthocephala. Invert. Biol. **114:**186–188. *Evidence that gnathostomulids are the sister group of a clade containing rotifers and acanthocephalans.*

Wallace, R. L. 2002. Rotifers: exquisite metazoans. Integ. and Comp. Biol. **42:**660–667. *This paper summarizes recent work on rotifers, but assumes basic knowledge of the group.*

Welch, M. D. B. 2000. Evidence from a protein-coding gene that acanthocephalans are rotifers. Invert. Biol. **119:**17–26. *Sequence analysis of a gene coding for a heat-shock protein supports a position of acanthocephalans within Rotifera. Other molecular and morphological evidence is cited that supports this position.*

ONLINE LEARNING CENTER

Visit www.mhhe.com/hickmanipz14e for chapter quizzing, key term flash cards, web links, and more!

Molluscs
• PHYLUM MOLLUSCA

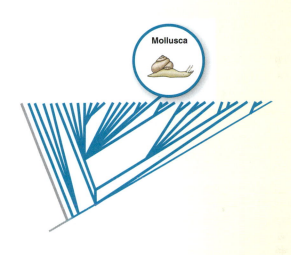

Fluted giant clam, Tridacna maxima.

A Significant Space

Long ago in the Precambrian era, the most complex animals populating the seas were acoelomate. The simplest, and probably the first, mode of achieving a fluid-filled space within the body was retention of the embryonic blastocoel, as in pseudocoelomates. This was not the best evolutionary solution because, for example, the organs lay loose in the body cavity.

Some descendants of Precambrian acoelomate organisms evolved a more elegant arrangement: a fluid-filled space *within* the mesoderm, the *coelom*. This meant that the space was lined with mesoderm and the organs were suspended by mesodermal membranes, the *mesenteries*. Mesenteries provided an ideal location for networks of blood vessels, and the suspended alimentary canal could become more muscular, more highly specialized, and more diversified without interfering with other organs.

Development of a coelom was a major step in the evolution of larger and more complex forms. All major groups in chapters to follow are coelomates. In some, a large fluid-filled coelom surrounded by muscles becomes a hydrostatic skeleton that permits rapid shape change and efficient burrowing. However, the heavy molluscan shell makes shape change impossible, and the molluscan coelom is relatively small.

MOLLUSCS

Mollusca (mol-lus′ka) (L. *molluscus,* soft) is one of the largest animal phyla after Arthropoda. There are over 90,000 living species and some 70,000 fossil species. Molluscs are coelomate lophotrochozoan protostomes, and as such they develop via spiral mosaic cleavage and make a coelom by schizocoely. The ancestral larval stage is a trochophore, but development is variously modified within the classes.

The name Mollusca indicates one of their distinctive characteristics, a soft body. This very diverse group (Figure 16.1) includes chitons, tusk shells, snails, slugs, nudibranchs, sea butterflies, clams, mussels, oysters, squids, octopuses, and nautiluses. The group ranges from fairly simple organisms to some of the most complex of invertebrates; sizes range from almost microscopic to the giant squid *Architeuthis.* These huge molluscs may grow to nearly 20 m long, including their tentacles. They may weigh up to 900 kg (1980 pounds). The shells of some giant clams, *Tridacna gigas,* which inhabit Indo-Pacific coral reefs, reach 1.5 m in length and weigh more than 250 kg. These are extremes, however, for probably 80% of all molluscs are less than 10 cm in maximum shell size. The phylum includes some of the most sluggish and some of the swiftest and most active invertebrates. It includes herbivorous grazers, predaceous carnivores, filter feeders, detritus feeders, and parasites.

Molluscs are found in a great range of habitats, from the tropics to polar seas, at altitudes exceeding 7000 m, in ponds, lakes, and streams, on mud flats, in pounding surf, and in open ocean from the surface to abyssal depths. They represent a variety of lifestyles, including bottom feeders, burrowers, borers, and pelagic forms.

According to fossil evidence, molluscs originated in the sea, and most of them have remained there. Much of their evolution occurred along the shores, where food was abundant and habitats were varied. Only bivalves and gastropods moved into brackish and freshwater habitats. As filter feeders, bivalves were unable to leave aquatic surroundings. Only slugs and snails (gastropods) actually invaded the land. Terrestrial snails are limited in their range by their need for humidity, shelter, and presence of calcium in the soil.

Humans exploit molluscs in a variety of ways. Many kinds of molluscs are used as food. Pearl buttons are obtained from shells of bivalves. The Mississippi and Missouri river basins and artificial propagation furnish material for this industry in the United States. Pearls, both natural and cultured, are produced in the shells of clams and oysters, most of them in a marine oyster, *Meleagrina,* found around eastern Asia.

Some molluscs are considered pests because of the damage they cause. Burrowing shipworms, which are bivalves of several species (see Figure 16.32), do great damage to wooden ships and wharves. To prevent the ravages of shipworms, wharves must be either creosoted or built of concrete (unfortunately, some shipworms ignore creosote, and some bivalves bore into concrete). Snails and slugs frequently damage garden and other vegetation. In addition, snails often serve as intermediate hosts for serious parasites of humans and domestic animals. Boring snails of genus *Urosalpinx* rival sea stars in destroying oysters.

In this chapter we explore the various major groups of molluscs (Figure 16.2) including those groups having limited diversity (classes Caudofoveata, Solenogastres, Monoplacophora, and Scaphopoda). Members of class Polyplacophora (chitons) are common to abundant marine animals, especially in the intertidal zone. Bivalves (class Bivalvia) have evolved many species, both marine and freshwater. Class Cephalopoda (squids, cuttlefish,

A   **B**   **C**

Figure 16.1

Molluscs: a diversity of life-forms. The basic body plan of this ancient group has become variously adapted for different habitats.
A, A chiton *(Tonicella lineata),* class Polyplacophora. **B,** A marine snail *(Calliostoma annulata),* class Gastropoda. **C,** A nudibranch *(Chromodoris* sp.) class Gastropoda. **D,** Pacific giant clam *(Panope abrupta),* with siphons to the left, class Bivalvia. **E,** An octopus *(Octopus briareus),* class Cephalopoda, forages at night on a Caribbean coral reef.

D   **E**

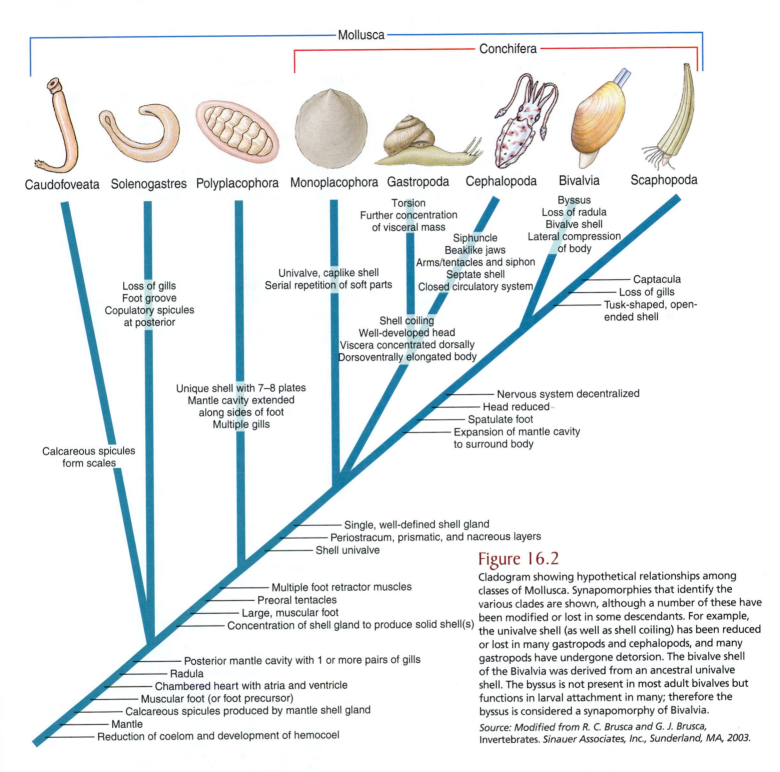

Figure 16.2

Cladogram showing hypothetical relationships among classes of Mollusca. Synapomorphies that identify the various clades are shown, although a number of these have been modified or lost in some descendants. For example, the univalve shell (as well as shell coiling) has been reduced or lost in many gastropods and cephalopods, and many gastropods have undergone detorsion. The bivalve shell of the Bivalvia was derived from an ancestral univalve shell. The byssus is not present in most adult bivalves but functions in larval attachment in many; therefore the byssus is considered a synapomorphy of Bivalvia.

Source: Modified from R. C. Brusca and G. J. Brusca, Invertebrates. *Sinauer Associates, Inc., Sunderland, MA, 2003.*

octopuses, and their kin) contains the largest and most intelligent of all invertebrates. Most abundant and widespread of molluscs, however, are snails and their relatives (class Gastropoda). Although enormously diverse, molluscs have in common a basic body plan (pp. 334–336). The coelom in molluscs is limited to a space around the heart, and perhaps around the gonads and part of the kidneys. Although it develops embryonically in a manner similar to the coelom of annelids (see ch. 17), the functional consequences of this space are quite different because it is not used in locomotion.

FORM AND FUNCTION

The enormous variety, great beauty, and easy availability of shells of molluscs have made shell collecting a popular pastime. However, many amateur shell collectors, even though able to name hundreds of the shells that grace our beaches, know very little about the living animals that created those shells and once lived in them. Reduced to its simplest dimensions, the mollusc body plan may be said to consist of a **head-foot** portion and a **visceral mass** portion (Figure 16.3). The head-foot is the more active area, containing the feeding, cephalic sensory, and locomotor organs. It depends primarily on muscular action for its function. The visceral mass is the portion containing digestive, circulatory, respiratory, and reproductive organs, and it depends primarily on ciliary tracts for its functioning. Two folds of skin, outgrowths of the dorsal body wall, form a protective **mantle,** which encloses a space between the mantle and body wall called the **mantle cavity.** The mantle cavity houses **gills (ctenidia)** or a lung, and in some molluscs the mantle secretes a protective **shell** over the visceral mass. Modifications of the structures that make up the head-foot and the visceral mass produce the great diversity of patterns observed in Mollusca. Greater emphasis on either the head-foot portion or the visceral mass portion can be observed in various classes of molluscs.

Head-Foot

Most molluscs have well-developed heads, which bear their mouth and some specialized sensory organs. Photosensory receptors range from fairly simple ones to the complex eyes of cephalopods. Tentacles are often present. Within the mouth is a structure unique to molluscs, the radula, and usually posterior to the mouth is the chief locomotor organ, or foot.

Radula

The radula is a rasping, protrusible, tonguelike organ found in all molluscs except bivalves and most solenogasters. It is used for feeding and consists of a ribbonlike membrane on which are mounted rows of tiny teeth that point backward (Figure 16.4). Complex muscles move the radula and its supporting cartilages **(odontophore)** in and out of the mouth while the membrane is partly rotated over the tips of the cartilages. There may be a few or as many as 250,000 teeth, which, when protruded, can scrape, pierce, tear, or cut. The usual function of the radula is twofold: to rasp fine particles of food material from hard surfaces and to serve as a conveyor belt for carrying particles in a continuous stream toward the digestive tract. As the radula wears away anteriorly, new rows of teeth are continuously replaced by secretion at its posterior end. The pattern and number of teeth in a row are specific for each species and are used in the classification of molluscs. Very interesting radular specializations, such as for boring through hard materials or for harpooning prey, are found in some forms.

Foot

The molluscan foot (see Figure 16.3) may be variously adapted for locomotion, for attachment to a substratum, or for a combination of

Characteristics of Phylum Mollusca

1. Dorsal body wall forms pair of folds called the **mantle,** which encloses the **mantle cavity,** is modified into **gills** or **lungs,** and secretes the **shell** (shell absent in some); ventral body wall specialized as a muscular **foot,** variously modified but used chiefly for locomotion; radula in mouth
2. Live in marine, freshwater, and terrestrial habitats
3. Free-living or occasionally parasitic
4. Body bilaterally symmetrical (bilateral asymmetry in some); unsegmented; often with definite head
5. Triploblastic body
6. **Coelom** limited mainly to area around heart, and perhaps lumen of gonads, part of kidneys, and occasionally part of the intestine
7. Surface epithelium usually ciliated and bearing mucous glands and sensory nerve endings
8. Complex digestive system; rasping organ **(radula)** usually present; anus usually emptying into mantle cavity; internal and external **ciliary tracts** often of great functional importance
9. Circular, diagonal, and longitudinal muscles in the body wall; mantle and foot highly muscular in some classes (for example cephalopods and gastropods)
10. Nervous system of paired cerebral, pleural, pedal, and visceral ganglia, with nerve cords and subepidermal plexus; ganglia centralized in nerve ring in gastropods and cephalopods
11. Sensory organs of touch, smell, taste, equilibrium, and vision (in some); the highly developed direct **eye** (photosensitive cells in retina face light source) of cephalopods is similar to the indirect eye (photosensitive cells face away from light source) of vertebrates but arises as a skin derivative in contrast to the brain eye of vertebrates
12. No asexual reproduction
13. Both **monoecious** and **dioecious** forms; **spiral cleavage;** ancestral larva a **trochophore,** many with a **veliger** larva, some with direct development
14. One or two kidneys **(metanephridia)** opening into the pericardial cavity and usually emptying into the mantle cavity
15. Gaseous exchange by **gills, lungs, mantle,** or **body surface**
16. **Open circulatory system** (secondarily closed in cephalopods) of heart (usually three chambered), blood vessels, and sinuses; respiratory pigments in blood

Figure 16.3

Generalized mollusc. Although this construct is often presented as a "hypothetical ancestral mollusc (HAM)," most experts now reject this interpretation. For example, the molluscan ancestor probably was covered with calcareous spicules, rather than a univalve shell. Such a diagram is useful, however, to facilitate description of the general body plan of molluscs.

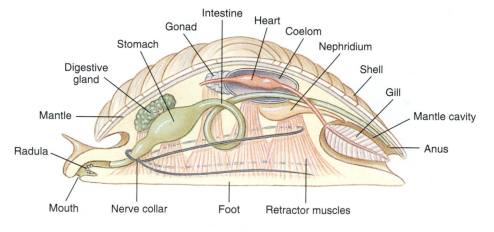

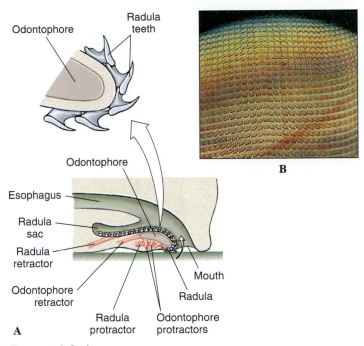

Figure 16.4

A, Diagrammatic longitudinal section of a gastropod head showing a radula and radula sac. The radula moves back and forth over the odontophore cartilage. As the animal grazes, the mouth opens, the odontophore is thrust forward, the radula gives a strong scrape backward bringing food into the pharynx, and the mouth closes. The sequence is repeated rhythmically. As the radula ribbon wears out anteriorly, it is continually replaced posteriorly. **B,** Radula of a snail prepared for microscopic examination.

functions. It is usually a ventral, solelike structure in which waves of muscular contraction effect a creeping locomotion. However, there are many modifications, such as the attachment disc of limpets, the laterally compressed "hatchet foot" of bivalves, or the siphon for jet propulsion in squids and octopuses. Secreted mucus is often used as an aid to adhesion or as a slime tract by small molluscs that glide on cilia.

In snails and bivalves the foot is extended from the body hydraulically, by engorgement with blood. Burrowing forms can extend the foot into the mud or sand, enlarge it with blood pressure, then use the engorged foot as an anchor to draw the body forward. In pelagic (free-swimming) forms the foot may be modified into winglike parapodia, or thin, mobile fins for swimming.

Visceral Mass

Mantle and Mantle Cavity

The mantle is a sheath of skin, extending from the visceral mass, that hangs down on each side of the body, protecting the soft parts and creating between itself and the visceral mass a space called the mantle cavity. The outer surface of the mantle secretes the shell.

The mantle cavity (see Figure 16.3) plays an enormous role in the life of a mollusc. It usually houses respiratory organs (gills or lung), which develop from the mantle, and the mantle's own exposed surface serves also for gaseous exchange. Products from the digestive, excretory, and reproductive systems are emptied into the mantle cavity. In aquatic molluscs a continuous current of water, kept moving by surface cilia or by muscular pumping, brings in oxygen and, in some forms, food. This same water current also flushes out wastes and carries reproductive products out to the environment. In aquatic forms the mantle is usually equipped with sensory receptors for sampling environmental water. In cephalopods (squids and octopuses) the muscular mantle and its cavity create jet propulsion used in locomotion. Many molluscs can withdraw their head or foot into the mantle cavity, which is surrounded by the shell, for protection.

In the simplest form, a mollusc ctenidium (gill) consists of a long, flattened axis extending from the wall of the mantle cavity (Figure 16.5). Many leaflike gill filaments project from the central axis. Water is propelled by cilia between gill filaments, and blood diffuses from an afferent vessel in the central axis through the filament to an efferent vessel. Direction of blood movement is opposite to the direction of water movement, thus establishing a countercurrent exchange mechanism (see p. 532). The two ctenidia are located on opposite sides of the mantle cavity and are arranged so that the cavity is functionally divided into an incurrent chamber and an excurrent chamber. The basic arrangement of gills is variously modified in many molluscs.

Shell

The shell of a mollusc, when present, is secreted by the mantle and is lined by it. Typically there are three layers (Figure 16.6A). The **periostracum** is the outer organic layer, composed of an organic substance called conchiolin, which consists of quinone-tanned protein. It helps to protect underlying calcareous layers from erosion by boring organisms. It is secreted by a fold of the mantle edge, and growth occurs only at the margin of the shell. On the older parts of the shell, periostracum often becomes worn away. The middle **prismatic layer** is composed of densely packed prisms of calcium carbonate (either aragonite or calcite) laid down in a protein matrix. It is secreted by the glandular margin of the mantle, and increase in shell size occurs at the shell margin as the animal grows. The inner **nacreous layer** of the shell lies next to the mantle and is secreted continuously by the mantle surface, so that it increases in thickness during the life of the animal. The calcareous nacre is laid down in thin layers. Very thin and wavy layers produce the iridescent mother-of-pearl found in abalones (*Haliotis*), chambered nautiluses (*Nautilus*), and many bivalves. Such shells may have 450 to 5000 fine parallel layers of crystalline calcium carbonate for each centimeter of thickness.

There is great variation in shell structure among molluscs. Freshwater molluscs usually have a thick periostracum that gives some protection against acids produced in the water by decay of leaf litter. In many marine molluscs the periostracum is relatively thin, and in some it is absent. Calcium for the shell comes from environmental water or soil or from food. The first shell appears during the larval period and grows continuously throughout life.

Internal Structure and Function

Gas exchange occurs in specialized respiratory organs such as ctenidia, secondary gills and lungs, as well as the body surface,

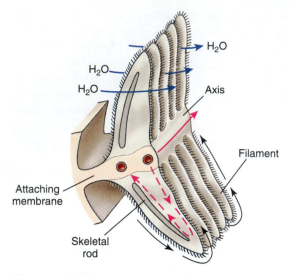

Figure 16.5

Primitive condition of mollusc ctenidium. Circulation of water between gill filaments is by cilia, and blood diffuses through the filament from the afferent vessel to the efferent vessel. Black arrows are ciliary cleansing currents.

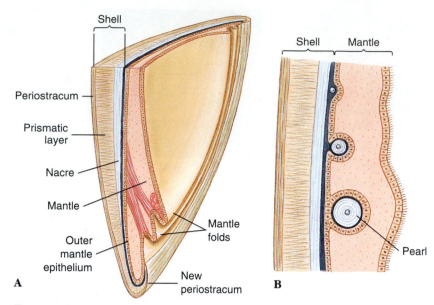

Figure 16.6

A, Diagrammatic vertical section of shell and mantle of a bivalve. The outer mantle epithelium secretes the shell; the inner epithelium is usually ciliated. **B,** Formation of pearl between mantle and shell as a parasite or bit of sand under the mantle becomes covered with nacre.

particularly the mantle. There is an **open circulatory system** with a pumping heart, blood vessels, and blood sinuses. In an open circulatory system blood is not entirely contained within blood vessels; rather it flows through vessels in some parts of the body and enters open sinuses in other parts. An open circulatory system is less efficient at supplying oxygen to all tissues in the body, so it is common in slow-moving animals. Insects are a notable exception, but in these animals oxygen is distributed by the tracheal system, not by the circulatory system. In a closed circulatory system, blood moves to and from tissues within blood vessels. Most cephalopods have a closed circulatory system with heart, vessels, and capillaries.

The digestive tract is complex and highly specialized, according to feeding habits of the various molluscs, and is usually equipped with extensive ciliary tracts. Most molluscs have a pair of kidneys (**metanephridia,** a type of nephridium in which the inner end opens into the coelom by a **nephrostome**). Ducts of the kidneys in many forms also serve for discharge of eggs and sperm.

The **nervous system** consists of several pairs of ganglia with connecting nerve cords, and it is generally simpler than that of annelids and arthropods. The nervous system contains neurosecretory cells that, at least in certain air-breathing snails, produce a growth hormone and function in osmoregulation. There are various types of highly specialized sense organs.

Reproduction and Life History

Most molluscs are dioecious, although some are hermaphroditic. The free-swimming **trochophore** larva that emerges from the egg in many molluscs (Figure 16.7) is remarkably similar to that seen in annelids. Direct metamorphosis of a trochophore into a small juvenile, as in chitons, is viewed as ancestral for molluscs. However,

in many molluscan groups (especially gastropods and bivalves) the trochophore stage is followed by a uniquely molluscan larval stage called a **veliger.** The free-swimming veliger (Figure 16.8) has the beginnings of a foot, shell, and mantle. In many molluscs the trochophore stage occurs in the egg, and a veliger hatches to become the only free-swimming stage. Cephalopods, some freshwater bivalves, and freshwater and some marine snails have no free-swimming larvae; instead, juveniles hatch directly from eggs.

Trochophore larvae (Figure 16.7) are minute, translucent, more or less top-shaped, and have a prominent circlet of cilia (prototroch) and sometimes one or two accessory circlets. They are found in molluscs and annelids exhibiting the ancestral embryonic development pattern and are usually considered homologous between the two phyla. Some form of trochophore-like larva also occurs in marine turbellarians, nemertines, brachiopods, phoronids, sipunculids, and echiurids, and together with recent molecular evidence, it suggests a phylogenetic grouping of these phyla. Based on developmental and molecular evidence, some zoologists unite them in a taxon called Trochozoa within superphylum Lophotrochozoa.

CLASSES OF MOLLUSCS

For more than 50 years five classes of living molluscs were recognized: Amphineura, Gastropoda, Scaphopoda, Bivalvia (also called Pelecypoda), and Cephalopoda. Discovery of *Neopilina* in the 1950s added another class (Monoplacophora), and Hyman[1]

---

[1]Hyman, L. H. 1967. The Invertebrates, vol. VI. New York, McGraw-Hill Book Company.

Figure 16.7

A, Generalized trochophore larva. Molluscs and annelids with an ancestral pattern of embryonic development have trochophore larvae, as do several other phyla. **B,** Trochophore of a Christmas tree worm, *Spirobranches spinosus* (Annelida).

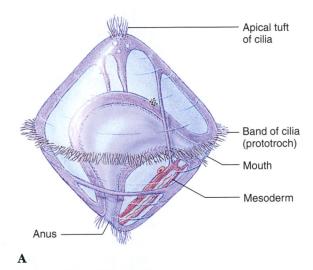

Apical tuft of cilia

Band of cilia (prototroch)

Mouth

Mesoderm

Anus

A

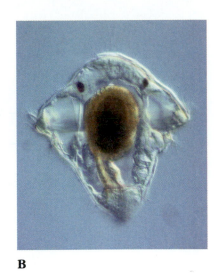

B

Figure 16.8

Veliger of a snail, *Pedicularia,* swimming. The adults are parasitic on corals. The ciliated process (velum) develops from the prototroch of the trochophore (Figure 16.7).

contended that solenogasters and chitons were separate classes (Aplacophora and Polyplacophora), lapsing the name Amphineura. Subsequently, Aplacophora was divided into the sister groups Caudofoveata and Solenogastres.[2] Members of both groups are wormlike and shell-less with calcareous scales or spicules in their integument. They have reduced heads and lack nephridia. In spite of these similarities, there are important differences between groups.

Class Caudofoveata

Members of class Caudofoveata comprise about 120 species of wormlike, marine organisms ranging from 2 to 140 mm in length (see Figure 16.2). They are mostly burrowers and orient themselves vertically, with the terminal mantle cavity and gills at the entrance of the burrow. They feed primarily on microorganisms

and detritus. They possess an oral shield, an organ apparently associated with food selection and intake, as well as a radula. They have one pair of gills. They are dioecious. The body plan of caudofoveates may have more features in common with the ancestor of molluscs than any other living group. This class is sometimes called Chaetodermomorpha.

Class Solenogastres

Solenogasters (see Figure 16.2) are a small group of about 250 species of marine animals similar to caudofoveates. Solenogasters, however, usually have no radula and no gills (although secondary respiratory structures may be present). Their foot is represented by a midventral, narrow furrow, the pedal groove. They are hermaphroditic. Solenogasters are bottom-dwellers, and often live and feed on cnidarians. This class is sometimes called Neomeniomorpha.

Class Monoplacophora

Monoplacophora were long thought to be extinct; they were known only from Paleozoic shells. However, in 1952 living specimens of *Neopilina* (Gr. *neo,* new, + *pilos,* felt cap) were dredged up from the ocean bottom near the west coast of Costa Rica. About 25 species of monoplacophorans are now known. These molluscs are small and have a low, rounded shell and a creeping foot (Figure 16.9). The mouth bears a characteristic radula. They superficially resemble limpets, but unlike most other molluscs, have some serially repeated organs. These animals have three to six pairs of gills, two pairs of auricles, three to seven pairs of metanephridia, one or two pairs of gonads, and a ladderlike nervous system with 10 pairs of pedal nerves. Such serial repetition occurs to a more limited extent in chitons. Why should there be repeated sets of body structures in these animals? Body structures repeat in each segment of an annelid worm (see p. 362). Are the repeated structures indications that molluscs had a segmented (metameric) ancestor? In the past, some biologists thought so, but current research indicates that *Neopilina* shows pseudometamerism, and that molluscs did not have a metameric ancestor.

[2]Boss, K. J. 1982. Mollusca. In S. P. Parker, ed., *Synopsis and Classification of Living Organisms,* vol. 1. New York, McGraw-Hill Book Company.

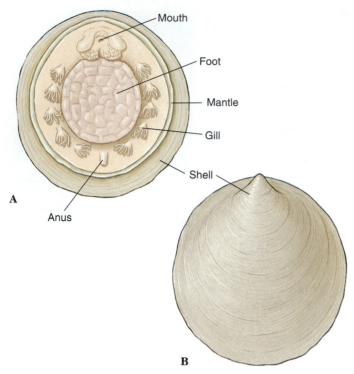

A

Mouth

Foot

Mantle

Gill

Shell

Anus

B

Figure 16.9

Neopilina, class Monoplacophora. Living specimens range from 3 mm to about 3 cm in length. **A,** Ventral view. **B,** Dorsal view.

Class Polyplacophora: Chitons

Chitons (Gr. coat of mail, tunic) (Figures 16.10 and 16.11) represent a somewhat more diverse molluscan group with about 1000 currently described species. They are rather flattened dorsoventrally and have a convex dorsal surface that bears seven or eight

articulating limy plates, or valves, hence their name Polyplacophora ("many plate bearers"). The plates overlap posteriorly and are usually dull colored to match the rocks to which chitons cling. Their head and cephalic sensory organs are reduced, but photosensitive structures **(esthetes),** which have the form of eyes in some chitons, pierce the plates.

Most chitons are small (2 to 5 cm); the largest, *Cryptochiton* (Gr. *crypto,* hidden, + *chiton,* coat of mail), rarely exceeds 30 cm. They prefer rocky surfaces in intertidal regions, although some live at great depths. Most chitons are stay-at-home organisms, straying only very short distances for feeding. Most feed by projecting the radula outward from the mouth to scrape algae from rocks. Scraping is aided by radular teeth reinforced with the iron-containing mineral, magnetite. However, the chiton *Placiphorella velata* is an unusual predatory species that uses a specialized head flap to capture small invertebrate prey. A chiton clings tenaciously to its rock with its broad, flat foot. If detached, it can roll up like an armadillo for protection.

The mantle forms a girdle around the margin of the plates, and in some species mantle folds cover part or all of the plates. Compared with other molluscan classes, the mantle cavity of polyplacophorans is extended along the side of the foot, and the gills are more numerous. The gills are suspended from the roof of the mantle cavity along each side of the broad ventral foot. With the foot and the mantle margin adhering tightly to the substrate, these grooves become closed chambers, open only at the ends. Water enters the grooves anteriorly, flows across the gills bringing a continuous supply of oxygen, and leaves posteriorly. At low tide the margins of the mantle can be tightly pressed to the substratum to diminish water loss, but in some circumstances, the mantle margins can be held open for limited air breathing. A pair of **osphradia** (chemoreceptive sense organs for sampling water) are found in the mantle grooves near the anus of many chitons.

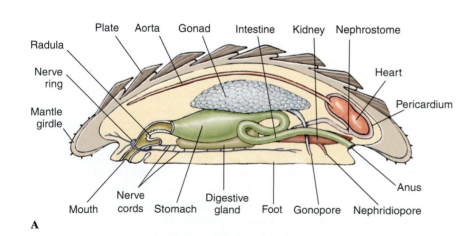

A

Radula

Nerve ring

Mantle girdle

Plate   Aorta   Gonad   Intestine   Kidney   Nephrostome

Heart

Pericardium

Mouth   Nerve cords   Stomach   Digestive gland   Foot   Gonopore   Nephridiopore   Anus

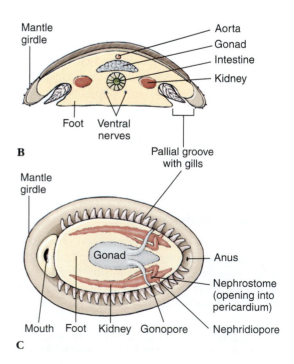

B

Mantle girdle

Foot   Ventral nerves

Aorta

Gonad

Intestine

Kidney

Pallial groove with gills

C

Mantle girdle

Mouth   Foot   Kidney   Gonopore

Gonad

Anus

Nephrostome (opening into pericardium)

Nephridiopore

Figure 16.10

Anatomy of a chiton (class Polyplacophora). **A,** Longitudinal section. **B,** Transverse section. **C,** External ventral view.

Figure 16.11

Mossy chiton, *Mopalia muscosa*. The upper surface of the mantle, or "girdle," is covered with hairs and bristles, an adaptation for defense.

Blood pumped by the three-chambered heart reaches the gills by way of an aorta and sinuses. A pair of kidneys (metanephridia) carries waste from the pericardial cavity to the exterior. Two pairs of longitudinal nerve cords are connected in the buccal region.

Sexes are separate in most chitons, and trochophore larvae metamorphose directly into juveniles, without an intervening veliger stage.

Class Scaphopoda

Scaphopoda, commonly called tusk shells or tooth shells, are benthic marine molluscs found from the subtidal zone to over 6000 m depth. They have a slender body covered with a mantle and a tubular shell open at both ends. In scaphopods the molluscan body plan has taken a new direction, with the mantle wrapped around the viscera and fused to form a tube. There are about 900 living species of scaphopods, and most are 2.5 to 5 cm long, although they range from 4 mm to 25 cm long.

The foot, which protrudes through the larger end of the shell, is used to burrow into mud or sand, always leaving the small end of the shell exposed to the water above (Figure 16.12). Respiratory water circulates through the mantle cavity both by movements of the foot and ciliary action. Gills are absent, and gaseous exchange therefore occurs in the mantle. Most food is detritus and protozoa from the substratum. It is caught on cilia of the foot or on the mucus-covered, ciliated adhesive knobs of the long tentacles (**captacula**) extending from the head and is conveyed to the nearby mouth. A radula carries food to a crushing gizzard. The captacula may serve some sensory function, but eyes, tentacles, and osphradia typical of many other molluscs are lacking.

Sexes are separate, and the larva is a trochophore.

Class Gastropoda

Among molluscs, class Gastropoda is by far the largest and most diverse, containing over 70,000 living and more than 15,000 fossil species. It contains so much diversity that there is no single general term in our language that can apply to it. It contains snails, limpets, slugs, whelks, conchs, periwinkles, sea slugs, sea hares, and sea butterflies. These forms range from marine molluscs to terrestrial, air-breathing snails and slugs. Gastropods are usually sluggish, sedentary animals because most of them have heavy shells and slow locomotion. Some are specialized for climbing, swimming, or burrowing. Shells are their chief defense.

The shell, when present, is always of one piece (**univalve**) and may be coiled or uncoiled. Starting at the **apex,** which contains the oldest and smallest **whorl,** the whorls become successively larger and spiral around the central axis, or **columella** (Figure 16.13). The shell may be right handed (**dextral**) or left handed (**sinistral**), depending on the direction of coiling. Direction of coiling is genetically controlled and dextral shells are far more common. Many snails have an **operculum,** a plate made of tanned protein that covers the shell **aperture** when the body is withdrawn into the shell.

Gastropods range from microscopic forms to giant marine forms such as *Pleuroploca gigantea,* a snail with a shell up to 60 cm long, and sea hares, *Aplysia* (see Figure 16.22), some

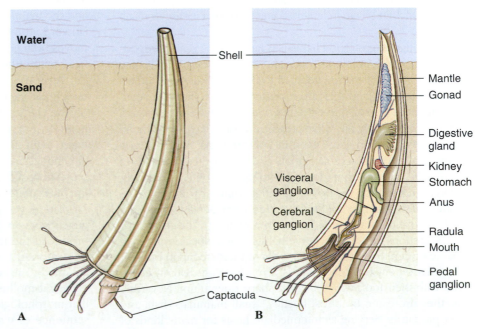

Figure 16.12

The tusk shell, *Dentalium* (class Scaphopoda). **A,** It burrows into soft mud or sand and feeds by means of its prehensile tentacles (captacula). Respiratory currents of water are drawn in by ciliary action through the small open end of the shell, then expelled through the same opening by muscular action. **B,** Internal anatomy of *Dentalium.*

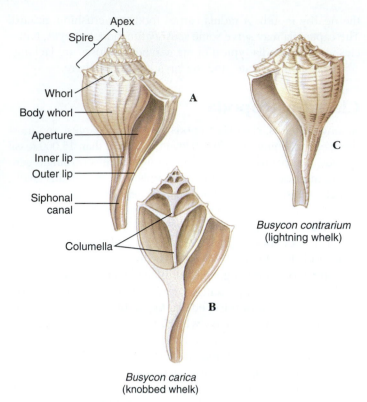

Figure 16.13

Shell of the whelk *Busycon*. **A** and **B**, *Busycon carica*, a dextral, or right-handed, shell. A dextral shell has the aperture on the right side when the shell is held with the apex up and the aperture facing the observer. **C**, *B. contrarium*, a sinistral, or left-handed, shell.

species of which reach 1 m in length. Most gastropods, however, are between 1 and 8 cm in length. Some fossil species are as much as 2 m long.

The range of gastropod habitats is large. In the sea gastropods are common both in littoral zones and at great depths, and some are even pelagic. Some are adapted to brackish water and others to freshwater. On land they are restricted by such factors as mineral content of the soil and extremes of temperature, dryness, and acidity. Even so, they are widespread, and some have been found at great altitudes and some even in polar regions. Snails occupy all kinds of habitats: in small pools or large bodies of water, in woodlands, in pastures, under rocks, in mosses, on cliffs, in trees, underground, and on the bodies of other animals. They have successfully undertaken every mode of life except aerial locomotion.

Gastropods may be protected by shells, by distasteful or toxic secretions, and by secretive habits. Some species are even capable of deploying the stinging cells of their cnidarian prey for their own defense. A few such as *Strombus* can deal an active blow with their foot, which bears a sharp operculum. Nevertheless, they are eaten by birds, beetles, small mammals, fish, and other predators. Serving as intermediate hosts for many kinds of parasites, especially trematodes (p. 298), snails are often harmed by larval stages of parasites.

There are three gastropod subclasses: Prosobranchia, Opisthobranchia, and Pulmonata. They are described on page 345, but subclass names are commonly used to discuss particular animals. Familiar prosobranchs include periwinkles, limpets, whelks,

conchs, abalones, and cowries. Marine sea slugs, sea hares, and nudibranchs are often called opisthobranchs. Pulmonates include most land and freshwater snails and slugs.

Form and Function

Torsion   Gastropod development varies with the particular group under discussion, but in general there is a trochophore larval stage followed by a veliger larval stage where the shell first forms. The veliger has two ciliated velar lobes, used in swimming, and the developing foot is visible (Figure 16.14). The mouth is anterior and the anus is posterior initially, but the relative positions of the shell, digestive tract and anus, nerves that lie along both sides of the digestive tract, and the mantle cavity containing the gills, all change in a process called **torsion.**

Torsion is usually described as a two-step process. In the first step, an asymmetrical foot retractor muscle contracts and pulls the shell and enclosed viscera (containing organs of the body) 90 degrees counterclockwise, relative to the head. This movement brings the anus from the posterior to the right side of the body (Figure 16.14). Typical descriptions state that movement of the shell accompanies visceral movement, but recent detailed studies have shown that movement of the shell is independent of visceral movements. The first movements of the shell rotate it between 90 and 180 degrees into a position that will persist into adulthood. It was previously assumed that the mantle cavity, which houses both the gills and the anus in adult animals, moved with the anus in the first 90 degrees of torsion. However, new studies have shown that the mantle cavity develops on the right side of the body near the anus, but is initially separate from it. The anus and mantle cavity usually move farther to the right and the mantle cavity is remodeled to encompass the anus. In a slower and more variable series of changes, the digestive tract moves both laterally and dorsally so that the anus lies above the head within the mantle cavity (Figure 16.14).

After torsion, the anus and mantle cavity open above the mouth and head. The left gill, kidney, and heart atrium are now on the right side, whereas the original right gill, kidney, and heart atrium are now on the left, and the nerve cords form a figure eight. Because of the space available in the mantle cavity, the animal's sensitive head end can now be withdrawn into the protection of the shell, with the tougher foot, and when present the operculum, forming a barrier to the outside.

The developmental sequence just described is called ontogenetic torsion. Evolutionary torsion is the series of changes that produced the modern torted gastropod body from the ancestral untorted form. The hypothetical ancestral gastropod was assumed to have a posterior mantle cavity like the hypothetical ancestral mollusc (see Figure 16.3). It has long been assumed that morphological changes in ontogenetic torsion represent the sequence of evolutionary changes. However, new studies of development in several kinds of gastropods suggest a different scenario; researchers hypothesize that the ancestral gastropod had two lateral mantle cavities, much like those in *Neopilina* (see Figure 16.9) and in chitons (see Figure 16.10). A single mantle cavity over the head may have arisen when the left lateral mantle cavity was lost and the right cavity expanded toward the

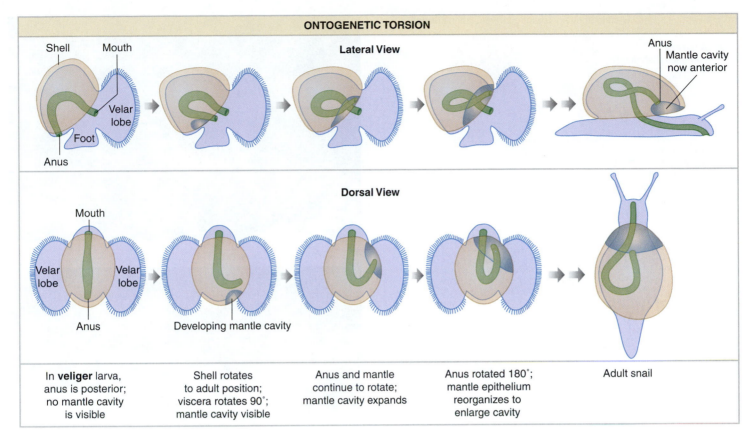

Figure 16.14
Ontogenetic torsion in a gastropod veliger larva.

middle of the body after the first 90 degrees of torsion. Careful study of ontogenetic torsion shows that asynchronous displacements of the shell, visceral mass and anus, and the mantle cavity are possible, although some features move together in some taxa. Torsion has been reinterpreted as a conserved anatomical stage, where the shell has moved to the adult position and the anus and mantle cavity are on the right side of the body, rather than a conserved process of change.[3]

Varying degrees of **detorsion** are seen in opisthobranchs and pulmonates, and the anus opens to the right side or even to the posterior. However, both of these groups were derived from torted ancestors.

The curious arrangement that results from torsion poses a serious sanitation problem by creating the possibility of wastes being washed back over the gills **(fouling)** and causes us to wonder what strong evolutionary pressures selected for such a strange realignment of body structures. Several explanations have been proposed, none entirely satisfying. For example, sense organs of the mantle cavity (osphradia) would better sample water when turned in the direction of travel. Certainly the consequences of torsion and the resulting need to avoid fouling have been very important in the subsequent evolution of gastropods. These consequences cannot be explored, however, until we describe another unusual feature of gastropods—coiling.

Coiling   Coiling, or spiral winding, of the shell and visceral mass is not the same as torsion. Coiling may occur in the larval stage at the same time as torsion, but the fossil record shows that coiling was a separate evolutionary event and originated in gastropods earlier than did torsion. Nevertheless, all living gastropods have descended from coiled, torted ancestors, whether or not they now show these characteristics.

Early gastropods had a bilaterally symmetrical **planospiral** shell, in which all whorls lay in a single plane (Figure 16.15A). Such a shell was not very compact, since each whorl had to lie completely outside the preceding one. The compactness problem of a planospiral shell was solved by the **conispiral** shape, in which each succeeding whorl is at the side of the preceding one (Figure 16.15B). However, this shape was clearly unbalanced, hanging as it was with much weight over to one side. Better weight distribution was achieved by shifting the shell upward and posteriorly, with the shell axis oblique to the longitudinal axis of the foot (Figure 16.15C). The weight and bulk of the main body whorl, the largest whorl of the shell, pressed on the right side of the mantle cavity, however, and apparently interfered with the organs on that side. Accordingly, the gill, atrium, and kidney of the right side have been lost in most living gastropods, leading to a condition of *bilateral asymmetry*. Curiously, a few modern species have secondarily returned to a planospiral shell form.

Although loss of the right gill was probably an adaptation to the mechanics of carrying a coiled shell, that condition displayed in most modern prosobranchs made possible a way to avoid the

[3]Page, L. R. 2003. J. Exp. Zool. Part B: **297B**:11–26.

Cleft in shell and mantle for ventilating current

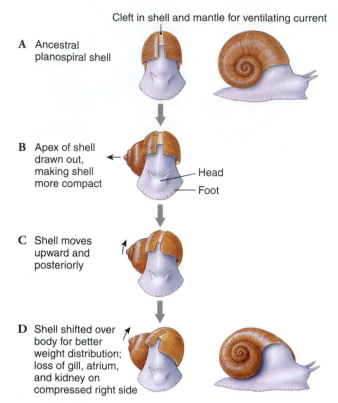

A Ancestral planospiral shell

B Apex of shell drawn out, making shell more compact
— Head
— Foot

C Shell moves upward and posteriorly

D Shell shifted over body for better weight distribution; loss of gill, atrium, and kidney on compressed right side

Figure 16.15

Evolution of shell in gastropods. **A,** Earliest coiled shells were planospiral, each whorl lying completely outside the preceding whorl. **B,** Better compactness was achieved by snails in which each whorl lay partially to the side of the preceding whorl. **C** and **D,** Better weight distribution resulted when shell was moved upward and posteriorly. However, some modern forms have reevolved the planospiral form.

fouling problem caused by torsion. Water is brought into the left side of the mantle cavity and out the right side, carrying with it wastes from the anus and nephridiopore, which lie near the right side. Ways in which fouling is avoided in other gastropods are mentioned on pages 345–346.

Feeding Habits   Feeding habits of gastropods are as varied as their shapes and habitats, but all include use of some adaptation of the radula. Most gastropods are herbivorous, rasping particles of algae from hard surfaces. Some herbivores are grazers, some are browsers, and some are planktonic feeders. *Haliotis,* the abalone (Figure 16.16A), holds seaweed with its foot and breaks off pieces with its radula. Land snails forage at night for green vegetation.

Some snails, such as *Bullia* and *Buccinum,* are scavengers living on dead and decaying flesh; others are carnivores that tear their prey with radular teeth. *Melongena* feeds on clams, especially *Tagelus,* the razor clam, thrusting its proboscis between the gaping shell valves. *Fasciolaria* and *Polinices* (Figure 16.16B) feed on a variety of molluscs, preferably bivalves. *Urosalpinx cinerea,* oyster borers, drill holes through the shells of oysters. Their radula, bearing three longitudinal rows of teeth, is used first to begin the drilling action; then the snails glide forward, evert an accessory boring organ through a pore in the anterior

A

B

Figure 16.16

A, Red abalone, *Haliotis rufescens.* This huge, limpetlike snail is prized as food and extensively marketed. Abalones are strict vegetarians, feeding especially on sea lettuce and kelp. **B,** Moon snail, *Polinices lewisii.* A common inhabitant of West Coast sand flats, the moon snail is a predator of clams and mussels. It uses its radula to drill neat holes through its victim's shell, through which the proboscis is then extended to eat the bivalve's fleshy body.

sole of their foot, and hold it against the oyster's shell, using a chemical agent to soften the shell. Short periods of rasping alternate with long periods of chemical activity until a neat round hole is completed. With its proboscis inserted through the hole, a snail may feed continuously for hours or days, using its radula to tear away the soft flesh. *Urosalpinx* is attracted to its prey at some distance by sensing some chemical, probably one released in metabolic wastes of the prey.

Cyphoma gibbosum (see Figure 16.21B) and related species live and feed on gorgonians (phylum Cnidaria, Chapter 13) in shallow, tropical coral reefs. These snails are commonly known as flamingo tongues. During normal activity their brightly colored mantle entirely envelops the shell, but it can be quickly withdrawn into the shell aperture when the animal is disturbed.

Members of genus *Conus* (Figure 16.17) feed on fish, worms, and molluscs. Their radula is highly modified for prey capture.

Figure 16.17

Conus extends its long, wormlike proboscis **(A)**. When a fish attempts to consume this tasty morsel, the *Conus* stings it in the mouth and kills it. The snail engulfs the fish with its distensible stomach **(B)**, then regurgitates the scales and bones some hours later.

A gland charges the radular teeth with a highly toxic venom. When *Conus* senses the presence of its prey, a single radular tooth slides into position at the tip of the proboscis. Upon striking the prey, the proboscis expels a tooth like a harpoon, and the venom immediately paralyzes the prey. This is an effective adaptation for a slowly moving predator to prevent escape of a swiftly moving prey. Some species of *Conus* can deliver very painful stings, and in several species the sting is lethal to humans. The venom consists of a series of toxic peptides, and each *Conus* species carries peptides **(conotoxins)** that are specific for the neuroreceptors of its preferred prey. Conotoxins have become valuable tools in research on the various receptors and ion channels of nerve cells.

Some gastropods, such as the queen conch (*Strombus gigas*), feed on organic deposits on the sand or mud. Others collect the same sort of organic debris but can digest only microorganisms contained in it. Some sessile gastropods, such as some limpets, are ciliary feeders that use gill cilia to draw in particulate matter, roll it into a mucous ball, and carry it to their mouth. Some sea butterflies secrete a mucous net to catch small planktonic forms; then they draw the web into their mouth.

After maceration by the radula or by some grinding device, such as a gizzard in the sea hare *Aplysia,* digestion is usually extracellular in the lumen of the stomach or digestive glands. In ciliary feeders the stomachs are sorting regions, and most digestion is intracellular in digestive glands.

Internal Form and Function   Respiration in most gastropods is performed by a **ctenidium** (two ctenidia is the primitive condition, found in some prosobranchs) located in the mantle cavity, though some aquatic forms lack gills and instead depend on the mantle and skin. After some prosobranchs lost one gill, most of them lost half of the remaining one, and the central axis became attached to the wall of the mantle cavity (Figure 16.18). Thus they attained the most efficient gill arrangement for the way the water circulated through the mantle cavity (in one side and out the other).

Pulmonates lack gills, but have a highly vascular area in their mantle that serves as a **lung** (Figure 16.19). Most of the mantle margin seals to the back of the animal, and the lung opens to the outside by a small opening called a **pneumostome.** The mantle cavity fills with air by contraction of the mantle floor. Many aquatic pulmonates must surface to expel a bubble of gas from their lung. To inhale, they curl the edge of the mantle around the pneumostome to form a siphon.

Most gastropods have a single nephridium (kidney). The circulatory and nervous systems are well developed (Figure 16.19). The latter incorporates three pairs of ganglia connected by nerves. Sense organs include eyes or simple photoreceptors, statocysts, tactile organs, and chemoreceptors. The simplest type of gastropod eye is simply a cuplike indentation in the skin lined with pigmented photoreceptor cells. In many gastropods

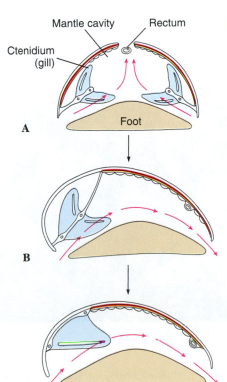

Figure 16.18

Evolution of ctenidia in gastropods, **A,** Primitive condition with two ctenidia and excurrent water leaving the mantle cavity by a dorsal slit or hole. **B,** Condition after one ctenidium had been lost. **C,** Derived condition found in most marine gastropods, in which filaments on one side of remaining gill are lost, and axis is attached to the mantle wall.

Figure 16.19

Anatomy of a pulmonate snail.

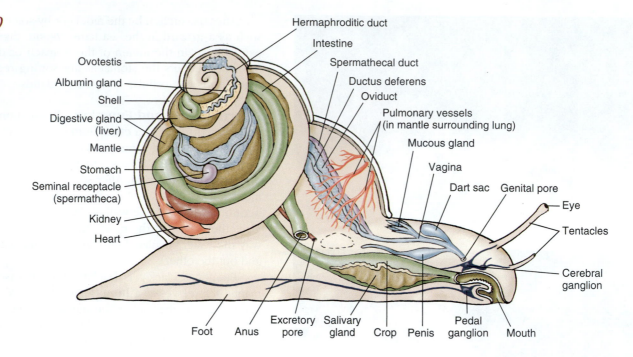

Labels: Ovotestis, Albumin gland, Shell, Digestive gland (liver), Mantle, Stomach, Seminal receptacle (spermatheca), Kidney, Heart, Hermaphroditic duct, Intestine, Spermathecal duct, Ductus deferens, Oviduct, Pulmonary vessels (in mantle surrounding lung), Mucous gland, Vagina, Dart sac, Genital pore, Eye, Tentacles, Cerebral ganglion, Foot, Anus, Excretory pore, Salivary gland, Crop, Penis, Pedal ganglion, Mouth

the eyecup contains a lens covered with a cornea. A sensory area called an **osphradium,** located at the base of the incurrent siphon of most gastropods, is chemosensory in some forms, although its function may be mechanoreceptive in some and remains unknown in others.

There are both dioecious and monoecious gastropods. Many gastropods perform courtship ceremonies. During copulation in monoecious species there is an exchange of spermatozoa or spermatophores (bundles of sperm). Many terrestrial pulmonates eject a dart from a dart sac (Figure 16.19) into the partner's body to heighten excitement before copulation. After copulation each partner deposits its eggs in shallow burrows in the ground. Gastropods with the most primitive reproductive characteristics discharge ova and sperm into seawater where fertilization occurs, and embryos soon hatch as free-swimming trochophore larvae. In most gastropods fertilization is internal.

Fertilized eggs encased in transparent shells may be emitted singly to float among the plankton or may be laid in gelatinous layers attached to a substratum. Some marine forms enclose their eggs, either in small groups or in large numbers, in tough egg capsules, or in a wide variety of egg cases (Figure 16.20). Offspring generally emerge as veliger larvae (see Figure 16.8), or they may spend the veliger stage in the case or capsule and emerge as young snails. Some species, including many freshwater snails, are ovoviviparous, brooding their eggs and young in their oviduct.

A

B

Figure 16.20

Eggs of marine gastropods. **A,** The wrinkled whelk, *Nucella lamellosa,* lays egg cases resembling grains of wheat; each contains hundreds of eggs.
B, Egg ribbon of a dorid nudibranch.

Major Groups of Gastropods

Traditional classification of class Gastropoda recognizes three subclasses: Prosobranchia, the largest subclass, almost all of which are marine; Opisthobranchia, an assemblage including sea slugs, sea hares, nudibranchs, and canoe shells, all marine; and Pulmonata, containing most freshwater and terrestrial species. Currently, gastropod taxonomy is in flux. Evidence suggests that Prosobranchia is paraphyletic. Opisthobranchia may or may not be paraphyletic, but Opisthobranchia and Pulmonata together apparently form a monophyletic grouping. The number of subclasses within the Gastropoda and the relationships among them remain subjects of considerable controversy. For convenience and organization, we continue to use the words "prosobranchs" and "opisthobranchs," recognizing that they may not represent valid taxa.

Prosobranchs   This group contains most marine snails and some freshwater and terrestrial gastropods. The mantle cavity is anterior as a result of torsion, with the gill or gills lying in front of the heart. Water enters the left side and exits from the right side, and the edge of the mantle often extends into a long siphon to separate incurrent from excurrent flow. In prosobranchs with two gills (for example, the abalone *Haliotis* and keyhole limpets *Diodora*, Figures 16.16A and 16.21A), fouling is avoided by having the excurrent water go up and out through one or more holes in the shell above the mantle cavity.

Prosobranchs have one pair of tentacles. Sexes are usually separate. An operculum is often present.

They range in size from periwinkles and small limpets (*Patella* and *Diodora*) (Figure 16.21A) to horse conchs (*Pleuroploca*) which grow shells up to 60 cm in length, making them the largest gastropods in the Atlantic Ocean. Familiar examples of prosobranchs include abalones (*Haliotis*), which have an ear-shaped shell; whelks (*Busycon*), which lay their eggs in double-edged, disc-shaped capsules attached to a cord a meter long; common periwinkles (*Littorina*); moon snails (*Polinices*); oyster borers (*Urosalpinx*), which bore into oysters and suck their juices; rock shells (*Murex*), a European species that was used to make the royal purple of the ancient Romans; and some freshwater forms (*Goniobasis* and *Viviparus*).

Opisthobranchs   Opisthobranchs are an odd assemblage of molluscs that include sea slugs, sea hares, sea butterflies, and bubble shells. They are nearly all marine; most are shallow-water forms, hiding under stones and seaweed; a few are pelagic. Currently nine or more orders of opisthobranchs are recognized. Opisthobranchs show partial or complete detorsion; thus the anus and gill (if present) are displaced to the right side or rear of the body. Clearly, the fouling problem is obviated if the anus is moved away from the head toward the posterior. Two pairs of tentacles are usually found, and the second pair is often further modified (**rhinophores,** Figure 16.22), with platelike folds that apparently increase the area for chemoreception. Their shell is typically reduced or absent. All are monoecious.

Sea hares (*Aplysia,* Figure 16.22), have large, earlike anterior tentacles and vestigial shells. In pteropods or sea butterflies (*Corolla* and *Clione*) the foot is modified into fins for swimming; thus, they are pelagic.

Nudibranchs are carnivorous and often brightly colored (Figure 16.23). Plumed sea slugs (Aeolidae) which feed primarily on sea anemones and hydroids, have elongate papillae (**cerata**) covering their back. They ingest their prey's nematocysts and transport the nematocysts undischarged to the tips of their cerata. There the nematocysts are placed in cnidosacs that open to the outside, and the aeolid can use these highjacked nematocysts for its own defense. *Hermissenda* is one of the more common West Coast nudibranchs.

Sacoglossan sea slugs are characterized by a radula with a single tooth per row that is used to pierce algal cells, allowing the slug to suck the contents. Similar to their aeolid cousins, some sacoglossans can steal functional organelles from their prey for their own benefit. In fact, many species have evolved special branches of the gut that run throughout the body; photosynthetic plastids from their algal food are directed into these branches rather than being digested, and they continue to function for quite some time. Some carnivorous nudibranchs likewise take advantage of intact zooxanthellae from their cnidarian prey (p. 280). This ability to usurp the photosynthetic machinery of their prey has led to the nickname "solar-powered sea slugs" for some species (for example, *Elysia crispata*).

Figure 16.21

A, *Diodora aspera,* a gastropod with a hole in its apex through which water leaves the mantle cavity. **B,** Flamingo tongues, *Cyphoma gibbosum,* are showy inhabitants of Caribbean coral reefs, where they are associated with gorgonians. These snails have a smooth, creamy, orange to pink shell that is normally covered by the brightly marked mantle.

A

B

Rhinophore   Oral tentacle

A

B

Figure 16.22

A, The sea hare, *Aplysia dactylomela,* crawls and swims across a tropical seagrass bed, assisted by large, winglike parapodia, here curled above the body. **B,** When attacked, sea hares squirt a copious protective secretion from their "purple gland" in the mantle cavity.

Figure 16.23

Phyllidia ocellata, a nudibranch. Like other *Phyllidia* spp., it has a firm body with dense calcareous spicules and bears its gills along the sides, between its mantle and foot.

A

Pneumostome

B

Figure 16.24

A, Pulmonate land snail. Note two pairs of tentacles; the second, larger pair bears the eyes. **B,** Banana slug, *Ariolimax columbianus.* Note pneumostome.

Pulmonates   Pulmonates include land and most freshwater snails and slugs (and a few brackish and saltwater forms). They have lost their ancestral ctenidia, but their vascularized mantle wall has become a lung, which fills with air by contraction of the mantle floor (some aquatic species have developed secondary gills in the mantle cavity). The anus and nephridiopore open near the pneumostome, and waste is expelled forcibly with air or water from the lung. Pulmonates show some detorsion. They are monoecious. Aquatic species have one pair of nonretractile tentacles, at the base of which are eyes; land forms have two pairs of tentacles, with the posterior pair bearing eyes (Figure 16.24).

Class Bivalvia (Pelecypoda)

Bivalvia are also known as Pelecypoda (pel-e-sip'o-da), or "hatchet-footed" animals, as their name implies (Gr. *pelekys,* hatchet, + *pous, podos,* foot). They are bivalved molluscs that include mussels, clams, scallops, oysters, and shipworms (Figures 16.25 to 16.29) and they range in size from tiny seed shells 1 to 2 mm in length to giant South Pacific clams, *Tridacna,* which may reach more than 1 m in length and as much as 225 kg (500 pounds) in weight (see Figure 16.33). Most bivalves are sedentary **filter feeders** that depend on currents produced by cilia on their gills to gather food materials. Unlike gastropods, they have no head, no radula, and very little cephalization.

Most bivalves are marine, but many live in brackish water and in streams, ponds, and lakes.

Figure 16.25

Bivalve molluscs. **A,** Mussels, *Mytilus edulis,* occur in northern oceans around the world; they form dense beds in the intertidal zone. A host of marine creatures live protected beneath attached mussels. **B,** Scallops *(Chlamys opercularis)* swim to escape attack by starfish *(Asterias rubens).* When alarmed, these most agile of bivalves swim by clapping the two shell valves together.

A

B

Freshwater clams were once abundant and diverse in streams throughout the eastern United States, but they are now easily the most jeopardized group of animals in the country. Of more than 300 species once present, nearly two dozen are extinct, more than 60 are considered endangered, and as many as 100 more are threatened. A combination of causes is responsible, of which the damming and impoundment of rivers is likely the most important. Pollution and sedimentation from mining, industry, and agriculture are other important culprits. Poaching to supply the cultured pearl industry is also a significant contributor. And in addition, introductions of exotic species make the problem worse. For example, the prolific zebra mussels (see note, p. 349) attach in great numbers to the native clams, exhausting food supplies (phytoplankton) in the surrounding water.

Form and Function

Shell   Bivalves are laterally compressed, and their two shells **(valves)** are held together dorsally by a hinge ligament that causes the valves to gape ventrally. The valves are drawn together by adductor muscles that work in opposition to the hinge ligament (Figure 16.26C and D). The **umbo** is the oldest part of the shell, and growth occurs in concentric lines around it (Figure 16.26A).

Pearl production is a by-product of a protective device used by the animals when a foreign object (grain of sand, parasite, or other) becomes lodged between the shell and mantle (see Figure 16.6). The mantle secretes many layers of nacre around the irritating object. Pearls are cultured by inserting particles of nacre, usually taken from the shells of freshwater clams, between the shell and mantle of a certain species of oyster and by keeping the oysters in enclosures for several years. *Meleagrina* is an oyster used extensively by the Japanese for pearl culture.

Body and Mantle   The **visceral mass** is suspended from the dorsal midline, and the muscular foot is attached to the visceral mass anteroventrally (Figure 16.27). The ctenidia hang down on each side, each covered by a fold of the mantle. The posterior edges of the mantle folds are modified to form dorsal excurrent and ventral incurrent openings (Figure 16.28A). In some marine bivalves the mantle is drawn out into long muscular siphons that allow the clam to burrow into the mud or sand and to extend the siphons to the water above (Figure 16.28B to D).

Locomotion   Bivalves initiate movement by extending a slender muscular foot between the valves (Figure 16.28D). They pump blood into their foot, causing it to swell and to act as an anchor in the mud or sand; then longitudinal muscles contract to shorten the foot and pull the animal forward.

Scallops and file shells swim with a jerky motion by clapping their valves together to create a sort of jet propulsion. The mantle edges can direct the stream of expelled water, so that the animals can swim in virtually any direction (Figures 16.25B and 16.29).

Gills   Gaseous exchange occurs through both mantle and gills. Gills of most bivalves are highly modified for filter-feeding; they are derived from primitive ctenidia by a great lengthening of filaments on each side of the central axis (see Figure 16.27). As ends of long filaments became folded back toward the central axis, ctenidial filaments took the shape of a long, slender W. Filaments lying beside each other became joined by ciliary junctions or tissue fusions, forming platelike **lamellae** with many vertical water tubes inside. Thus water enters the incurrent siphon, propelled by ciliary action, then enters the water tubes through pores between the filaments in the lamellae, proceeds dorsally into a common **suprabranchial chamber** (Figure 16.30), and then out the excurrent aperture.

Figure 16.26

Tagelus plebius, stubby razor clam (class Bivalvia). **A,** External view of left valve. **B,** Inside of right shell showing scars where muscles were attached. The mantle was attached at its insertion area. **C** and **D,** Sections showing function of adductor muscles and hinge ligament. In **C,** the adductor muscle is relaxed, allowing the hinge ligament to pull the valves apart. In **D,** the adductor muscle is contracted, pulling the valves together.

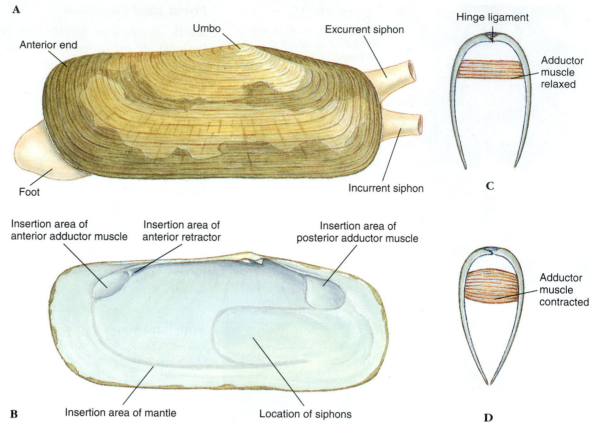

A

Anterior end — Umbo — Excurrent siphon — Hinge ligament — Adductor muscle relaxed

Foot — Incurrent siphon — **C**

B — Insertion area of anterior adductor muscle — Insertion area of anterior retractor — Insertion area of posterior adductor muscle — Adductor muscle contracted — **D**

Insertion area of mantle — Location of siphons

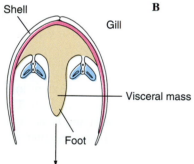

Shell — Gill — Visceral mass — Foot

B

Figure 16.27

Sections through a bivalve shell and body, showing relative positions of visceral mass and foot. Evolution of bivalve ctenidia: By a great lengthening of individual filaments, ctenidia became adapted for filter-feeding and separated the incurrent chamber from the excurrent, suprabranchial chamber.

Feeding

Most bivalves are filter feeders. Respiratory currents bring both oxygen and organic materials to the gills where ciliary tracts direct currents to the tiny pores of the gills. Gland cells on the gills and labial palps secrete copious amounts of mucus, which entangles particles suspended in water going through gill pores. These mucous masses slide down the outside of the gills toward food grooves at the lower edge of the gills (Figure 16.31). Heavier particles of sediment drop off the gills as a result of gravitational pull, but smaller particles travel along the food grooves toward the labial palps. The palps, being also grooved and ciliated, sort the particles and direct tasty ones encased in the mucous mass into the mouth.

Some bivalves, such as *Nucula* and *Yoldia,* are deposit feeders and have long proboscides attached to the labial palps (see Figure 16.28C). These can be protruded onto sand or mud to collect food particles, in addition to particles attracted by gill currents.

Shipworms (Figure 16.32) burrow in wood and feed on particles they excavate. Symbiotic bacteria live in a special organ in the bivalve and produce cellulase to digest wood. Other bivalves such as giant clams gain much of their nutrition as adults from the photosynthetic products of symbiotic dinoflagellates living in their mantle tissue (Figure 16.33).

Septibranchs, another group of bivalves, draw small crustaceans or bits of organic debris into the mantle cavity by sudden inflow of water created by the pumping action of a muscular septum in the mantle cavity.

Figure 16.29

Representing a group that has evolved from burrowing ancestors, the surface-dwelling scallop *Aequipecten irradians* has developed sensory organs along its mantle edges (tentacles and a series of blue eyes).

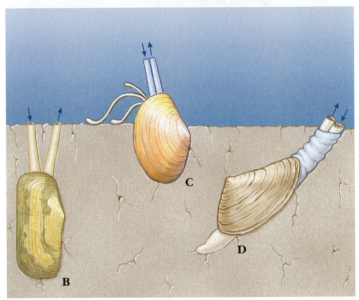

Figure 16.28

Adaptations of siphons in bivalves. **A,** In northwest ugly clams *Entodesma saxicola,* incurrent and excurrent siphons are clearly visible. **B** to **D,** In many marine forms the mantle is drawn out into long siphons. In **A, B,** and **D,** the incurrent siphon brings in both food and oxygen. In **C,** *Yoldia,* the siphons are respiratory; long ciliated palps feel about over the mud surface and convey food to the mouth.

Internal Structure and Function   The floor of the stomach of filter-feeding bivalves is folded into ciliary tracts for sorting a continuous stream of particles. In most bivalves a cylindrical **style sac** opening into the stomach secretes a gelatinous rod called a **crystalline style,** which projects into the stomach and is kept whirling by means of cilia in the style sac (Figure 16.34). Rotation of the style helps to dissolve its surface layers, freeing digestive enzymes (especially amylase) that it contains, and to roll the mucous food mass. Dislodged particles are sorted, and suitable ones are directed to the digestive gland or engulfed by amebocytes. Further digestion is intracellular.

The three-chambered heart, which lies in the pericardial cavity (see Figure 16.31), has two atria and a ventricle and beats slowly, ranging from 0.2 to 30 times per minute. Part of the blood is oxygenated in the mantle and returns to the ventricle through the atria; the rest circulates through sinuses and passes in a vein to the kidneys, from there to the gills for oxygenation, and back to the atria.

A pair of U-shaped kidneys (nephridial tubules) lies just ventral and posterior to the heart (Figure 16.31B). The glandular portion of each tubule opens into the pericardium; the bladder portion empties into the suprabranchial chamber.

Zebra mussels, *Dreissena polymorpha,* are a recent and disastrous biological introduction into North America. They were apparently picked up as veligers with ballast water by one or more ships in freshwater ports in northern Europe and then expelled between Lake Huron and Lake Erie in 1986. This 4 cm bivalve spread throughout the Great Lakes by 1990, and by 1994 it was as far south on the Mississippi as New Orleans, as far north as Duluth, Minnesota, and as far east as the Hudson River in New York. It attaches to any firm surface and filter-feeds on phytoplankton. Populations rapidly increase in size. They foul water-intake pipes of municipal and industrial plants, impede intake of water for municipal supplies, and have far-reaching effects on the ecosystem (see note, p. 347). Zebra mussels will cost billions of dollars to control if they can be controlled at all.

Another freshwater clam, *Corbicula fluminea,* was introduced into the United States from Asia more than 50 years ago by unknown means. Despite efforts to control *Corbicula* that cost over a billion dollars per year, it is now a pest throughout most of the continental United States, infesting water systems and clogging pipes.

The nervous system consists of three pairs of widely separated ganglia connected by commissures and a system of nerves. Sense organs are poorly developed. They include a pair of statocysts in the foot, a pair of osphradia of uncertain function in the

Figure 16.30

Section through heart region of a freshwater clam to show relation of circulatory and respiratory systems. Respiratory water currents: water is drawn in by cilia, enters gill pores, and then passes up water tubes to suprabranchial chambers and out excurrent aperture. Blood in gills exchanges carbon dioxide for oxygen. Blood circulation: ventricle pumps blood forward to sinuses of foot and viscera, and posteriorly to mantle sinuses. Blood returns from mantle to atria; it returns from viscera to the kidney, and then goes to the gills, and finally to the atria.

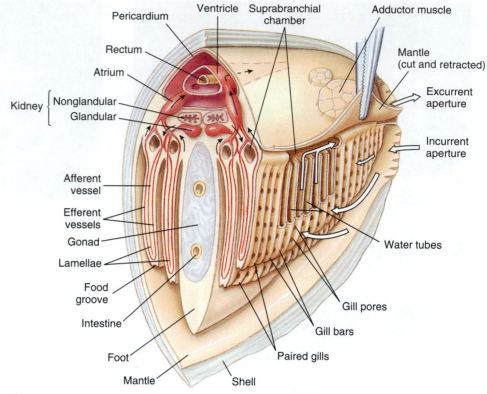

Figure 16.31

A, Feeding mechanism of freshwater clam. Left valve and mantle are removed. Water enters the mantle cavity posteriorly and is drawn forward by ciliary action to the gills and palps. As water enters the tiny openings of the gills, food particles are sieved out and caught in strings of mucus that are carried by cilia to the palps and directed to the mouth. Sand and debris drop into the mantle cavity and are removed by cilia. **B,** Clam anatomy.

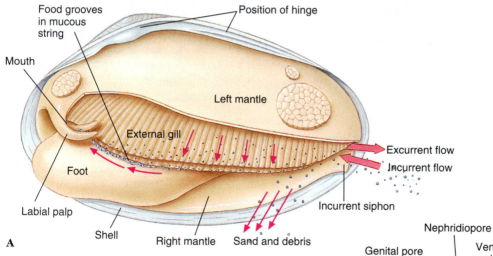

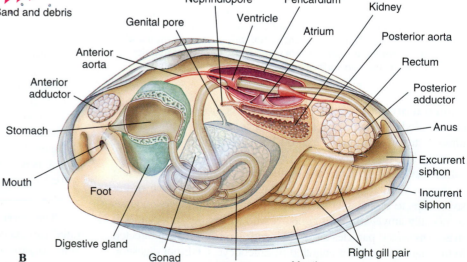

A

B

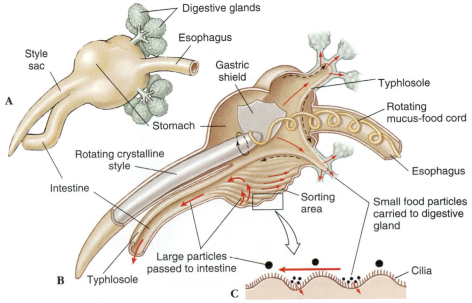

Figure 16.33

Clam *(Tridacna gigas)* lies buried in coral rock with greatly enlarged siphonal area visible. These tissues are richly colored and bear enormous numbers of symbiotic single-celled algae (zooxanthellae) that provide much of the clam's nutriment.

Figure 16.32

A, Shipworms are bivalves that burrow in wood, causing great damage to unprotected wooden hulls and piers. They are nicknamed "termites of the sea." **B,** The two small, anterior valves, seen at left, are used as rasping organs to extend the burrow.

Figure 16.34

Stomach and crystalline style of ciliary-feeding clam. **A,** External view of stomach and style sac. **B,** Transverse section showing direction of food movements. Food particles in incoming water are caught in a cord of mucus that is kept rotating by the crystalline style. Ridged sorting areas direct large particles to the intestine and small food particles to digestive glands. **C,** Sorting action of cilia.

mantle cavity, tactile cells, and sometimes simple pigment cells on the mantle. Scallops (*Aequipecten, Chlamys*) have a row of small blue eyes along each mantle edge (see Figure 16.29). Each eye has a cornea, lens, retina, and pigmented layer. Tentacles on the margin of the mantle of *Aequipecten* and *Lima* have tactile and chemoreceptor cells.

Reproduction and Development   Sexes are usually separate. Gametes are discharged into the suprabranchial chamber to be carried out with the excurrent flow. An oyster may produce 50 million eggs in a single season. In most bivalves fertilization is external. The embryo develops into trochophore, veliger, and spat stages (Figure 16.35).

In most freshwater clams fertilization is internal. Eggs drop into the water tubes of the gills where they are fertilized by sperm entering with the incurrent flow (see Figure 16.31). They develop there into a bivalved **glochidium larva** stage, which is a specialized veliger (Figure 16.36A). Glochidia need to attach themselves to specific fish hosts and live parasitically for several weeks to complete their development. Various mussel species have unique tactics for getting their larvae in contact with a suitable fish host species. Some simply discharge glochidia into the water column; if they come into contact with a suitable passing fish or amphibian, they can attach to the gills or skin and complete their development. In other species, the mantle flap of brooding females, in which the glochidia are held in a gelatinous packet called a **conglutinate,** has a size and shape unique to each mussel species. This mantle flap is often used as a lure to bring potential host species into contact with the glochidia. For example, the conglutinate of a gravid female pocketbook mussel (*Lampsilis ovata*) grows to resemble a small fish (Figure 16.36B). This mantle flap is then wriggled like a fishing lure to attract nearby bass, which serve as host for the glochidia. When a hungry bass strikes the mantle, instead of a meal it gets a mouthful of glochidia, which promptly attach to the fishes' gills.

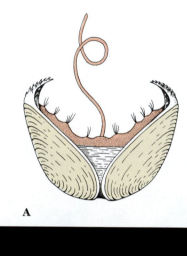

A

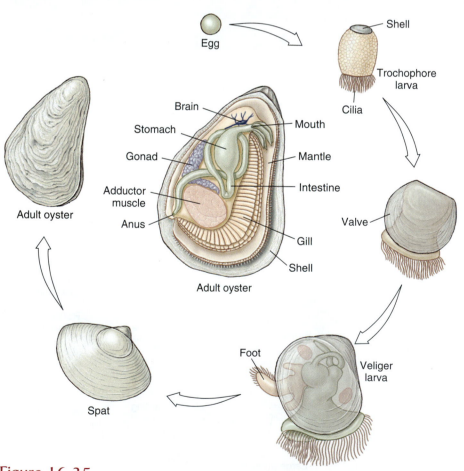

Figure 16.35

Life cycle of oysters. Oyster larvae swim about for approximately 2 weeks before settling down for attachment to become spats. Oysters take about 4 years to grow to commercial size.

B

Figure 16.36

A, Glochidium, or larval form, for some freshwater clams. When larvae are released from the mother's brood pouch, they may become attached to a fish's gill by clamping their valves closed. They remain as parasites on the fish for several weeks. Their size is approximately 0.3 mm. **B,** Some clams have adaptations that help their glochidia find a host. The mantle edge of this female pocketbook mussel (*Lampsilis ovata*) mimics a small minnow, complete with eye. When a smallmouth bass comes to dine, it gets doused with glochidia.

After encysting on a suitable host to complete development, the juveniles detach and sink to the bottom to begin independent lives. Larval "hitchhiking" helps distribute a form whose locomotion is very limited while also preventing larvae from being swept downstream out of the lake.

Boring Many bivalves can burrow into mud or sand, but some have evolved a mechanism for burrowing into much harder substances, such as wood or stone.

Teredo, Bankia, and some other genera are called shipworms. They can be very destructive to wooden ships and wharves. These strange little clams have a long, wormlike appearance, with a pair of slender posterior siphons that keep water flowing over the gills, and a pair of small globular valves on the anterior end with which they burrow (Figure 16.32). The valves have microscopic teeth that function as very effective wood rasps. The animals extend their burrows with an unceasing rasping motion of the valves. This motion sends a continuous flow of fine wood particles into the digestive tract where they are attacked by cellulase produced by symbiotic bacteria. Interestingly, these bacteria also fix nitrogen, an important property for their hosts, which live on a diet (wood) high in carbon but deficient in nitrogen.

Some clams bore into rock. The piddock *(Pholas)* bores into limestone, shale, sandstone, and sometimes wood or peat. It has strong valves that bear spines, which it uses to cut away the rock gradually while anchoring itself with its foot. *Pholas* may grow to 15 cm long and make rock burrows up to 30 cm long.

Class Cephalopoda

Cephalopoda (Gr. *kephalē,* head, + *pous, podos,* foot) include squids, octopuses, nautiluses, devilfish, and cuttlefish. All are marine, and all are active predators.

Their modified foot is concentrated in the head region. It takes the form of a funnel for expelling water from the mantle cavity, and the anterior margin is drawn out into a circle or crown of arms or tentacles.

Cephalopods range upward in size from 2 or 3 cm. The common squid of markets, *Loligo,* is about 30 cm long. Giant squids, *Architeuthis,* at up to almost 60 ft in length and weighing nearly a ton, are the largest invertebrates known.

Fossil records of cephalopods go back to Cambrian times. The earliest shells were straight cones; others were curved or coiled, culminating in the coiled shell similar to that of the modern *Nautilus,* the only remaining member of the once flourishing nautiloids (Figure 16.37). Cephalopods without shells or with internal shells (such as octopuses and squids) apparently evolved from some early straight-shelled ancestor. Many ammonoids, which are extinct, had quite elaborate shells (Figure 16.37C).

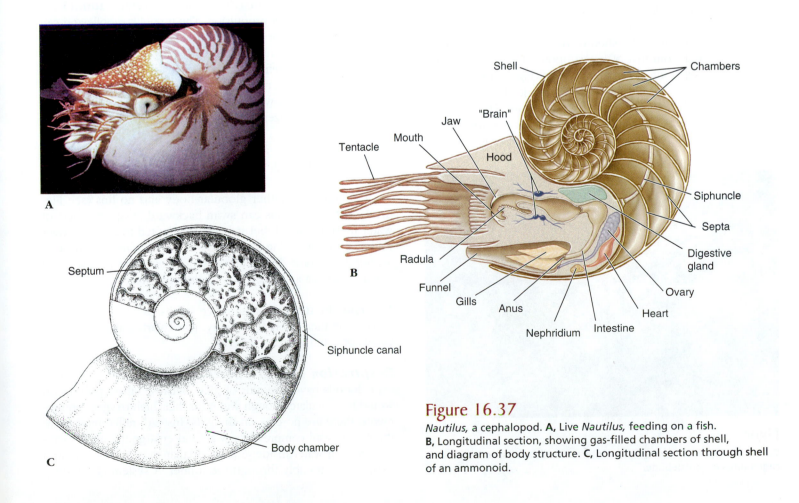

Figure 16.37

Nautilus, a cephalopod. **A,** Live *Nautilus,* feeding on a fish. **B,** Longitudinal section, showing gas-filled chambers of shell, and diagram of body structure. **C,** Longitudinal section through shell of an ammonoid.

The natural history of some cephalopods is fairly well known. They are marine animals and appear sensitive to the degree of salinity. Few are found in the Baltic Sea, where the water has a low salt content. Cephalopods are found at various depths. Octopuses are often seen in the intertidal zone, lurking among rocks and crevices, but occasionally they are found at great depths. The more active squids are rarely found in very shallow water, and some have been taken at depths of 5000 m. *Nautilus* is usually found near the bottom in water 50 to 560 m deep, near islands in the southwestern Pacific.

The enormous giant squid, *Architeuthis,* is very poorly known because no one has ever studied a living specimen. The anatomy has been described from stranded animals, from those captured in nets of fishermen, and from specimens found in the stomach of sperm whales. The mantle length is 5 to 6 m, and the head is up to 1 m. They have the largest eyes in the animal kingdom: up to 25 cm (10 in) in diameter. They apparently eat fish and other squids, and they are an important food item for sperm whales. They are thought to live on or near the sea bottom at a depth of 1000 m, but some have been observed swimming at the surface.

Form and Function

Shell   Although early nautiloid and ammonoid shells were heavy, they were made buoyant by a series of **gas chambers,** as is that of *Nautilus* (Figure 16.37B), enabling the animal to maintain neutral buoyancy. The shell of *Nautilus,* although coiled, is quite different from that of a gastropod. The shell is divided by transverse septa into internal chambers (Figure 16.37B), only the last inhabited by the living animal. As it grows, it moves forward, secreting behind its body a new septum. The chambers are connected by a cord of living tissue called a **siphuncle,** which extends from the visceral mass. Cuttlefishes (Figure 16.38)

Figure 16.38
Cuttlefish, *Sepia latimanus,* has an internal shell familiar to keepers of caged birds as "cuttlebone."

also have a small, curved shell, but it is entirely enclosed by the mantle. In squids most of the shell has disappeared, leaving only a thin, proteinaceous strip called a pen, which is enclosed by the mantle. In *Octopus* (Gr. *oktos,* eight, + *pous, podos,* foot) the shell has disappeared entirely.

After a member of genus *Nautilus* secretes a new septum, the new chamber is filled with fluid similar in ionic composition to that of *Nautilus's* blood (and of seawater). Fluid removal involves active secretion of ions into tiny intercellular spaces in the siphuncular epithelium, so that a very high local osmotic pressure is produced, and water is drawn out of the chamber by osmosis. The gas in the chamber is just the respiratory gas from the siphuncle tissue that diffuses into the chamber as fluid is removed. Thus gas pressure in the chamber is 1 atmosphere or less because it is in equilibrium with gases dissolved in the seawater surrounding the *Nautilus,* which are in turn in equilibrium with air at the surface of the sea, despite the fact that the *Nautilus* may be swimming at 400 m beneath the surface. That the shell can withstand implosion by the surrounding 41 atmospheres (about 600 pounds per square inch), and that the siphuncle can remove water against this pressure are marvelous feats of natural engineering!

Locomotion   Cephalopods swim by forcefully expelling water from the mantle cavity through a ventral **funnel** (or **siphon**)—a sort of jet propulsion. The funnel is mobile and can be pointed forward or backward to control direction; the force of water expulsion controls speed.

Squids and cuttlefishes are excellent swimmers. The squid body is streamlined and built for speed (Figure 16.39). Cuttlefishes swim more slowly. The lateral fins of squids and cuttlefishes serve as stabilizers, but they are held close to the body for rapid swimming.

Nautilus is active at night; its gas-filled chambers keep the shell upright. Although not as fast as squids, it moves surprisingly well.

Octopus has a rather globular body and no fins (see Figure 16.1E). An octopus can swim backward by spurting jets of water from its funnel, but it is better adapted to crawling over rocks and coral, using suction discs on its arms to pull or to anchor itself. Some deep-water octopods have the arms webbed like an umbrella and swim in a medusa-like fashion (p. 271).

Internal Features   The active habits of cephalopods are reflected in their internal anatomy, particularly their respiratory, circulatory, and nervous systems.

Respiration and Circulation.   Except for nautiloids, cephalopods have one pair of gills. Because ciliary propulsion would not circulate enough water for their high oxygen requirements, there are no cilia on the gills. Instead, radial muscles in the mantle wall compress the wall and enlarge the mantle cavity, drawing water inside. Strong circular muscles contract and expel water forcibly through the funnel. A system of one-way

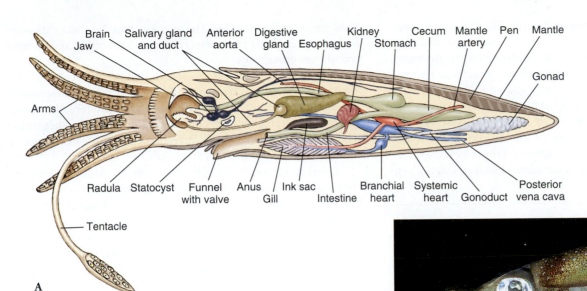

Figure 16.39
A, Lateral view of squid anatomy, with the left half of the mantle removed. **B,** Reef squid *Sepioteuthis lessoniana*.

valves prevents water from being taken in through the funnel or expelled around the mantle margin.

Likewise, the open circulatory system of ancestral molluscs would be inadequate for cephalopods. Their circulatory system has evolved into a closed network of vessels, and capillaries conduct blood through the gill filaments. Furthermore, the molluscan plan of circulation places the entire systemic circulation before the blood reaches the gills (in contrast to vertebrates, in which the blood leaves the heart and goes directly to the gills or lungs). This functional problem was solved by the development of **accessory** or **branchial hearts** (Figure 16.39A) at the base of each gill to increase the pressure of blood going through the capillaries there.

Nervous and Sensory Systems.
Nervous and sensory systems are more elaborate in cephalopods than in other molluscs. The brain, the largest in any invertebrate, consists of several lobes with millions of nerve cells. Squids have giant nerve fibers (among the largest known in the animal kingdom), which are activated when the animal is alarmed and initiate maximal contractions of the mantle muscles for a speedy escape.

Squid nerves played an important role in early biophysical studies. Our current understanding of transmission of action potentials along and between nerve fibers (see Chapter 33) is based primarily on work performed using the giant nerve fibers of squids, *Loligo* spp. A. Hodgkin and A. Huxley received the Nobel Prize in Physiology or Medicine, 1963, for their achievements in this field.

Sense organs are well developed. Except for *Nautilus*, which has relatively simple eyes, cephalopods have highly complex eyes with cornea, lens, chambers, and retina (Figure 16.40).

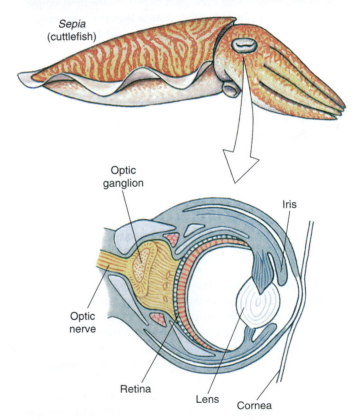

Figure 16.40
Eye of a cuttlefish *(Sepia)*. The structure of cephalopod eyes is very similar to that of eyes of vertebrates (see item 11, p. 334).

Orientation of the eyes is controlled by the statocysts, which are larger and more complex than in other molluscs. The eyes are held in a constant relation to gravity, so that the slit-shaped pupils are always in a horizontal position.

Most cephalopods are apparently colorblind, but their eyesight is excellent and their visual acuity underwater far surpasses our own. They can also be taught to discriminate between shapes—for example, a square and a rectangle—and to remember such a discrimination for a considerable time. Experimenters find it easy to modify octopod behavior patterns by devices of reward and punishment. Octopods are capable of observational learning; when one octopus observes another being rewarded by making a correct choice, the observer learns which choice is rewarded and consistently makes the same selection when given the opportunity.

When similar structures that are not inherited from a common ancestor evolve in different ways in unrelated animals, we call it **convergence,** or **convergent evolution.** For many years cephalopod eyes and vertebrate eyes have been cited as a marvelous example of convergent evolution. Cephalopod and vertebrate eyes are similar in many details of structure but differ in development. Compound eyes of arthropods (pp. 427, 747), differing in both structure and development, were viewed as examples of other, independently derived eyes in animals. Now we recognize that all triploblastic animals with eyes, even those with the most simple eyespots, such as platyhelminths, share at least two conserved genes: that for rhodopsin, a visual pigment, and *Pax 6,* now sometimes called the "master control gene for eye morphogenesis." Once these two genes originated, natural selection eventually produced the specialized organs of vertebrates, molluscs, and arthropods.

Octopods use their arms for tactile exploration and can discriminate between textures by feel but apparently not between shapes. Their arms are well supplied with both tactile and chemoreceptor cells. Cephalopods seem to lack a sense of hearing.

Communication   Little is known of social behavior of nautiloids or deep-water cephalopods, but inshore and littoral forms such as *Sepia, Sepioteuthis, Loligo,* and *Octopus* have been studied extensively. Although their tactile sense is well developed and they have some chemical sensitivity, visual signals are the predominant means of communication. These signals consist of a host of movements of the arms, fins, and body, as well as many color changes. Movements may range from minor body motions to exaggerated spreading, curling, raising, or lowering of some or all of the arms. Color changes are effected by **chromatophores,** cells in the skin that contain pigment granules (see Chapter 29, p. 647). Tiny muscle cells surround each elastic chromatophore, whose contractions pull the cell boundary of the chromatophore outward, causing it to expand greatly. As the cell expands, the pigment becomes dispersed, changing the color pattern of the animal. When the muscles relax, chromatophores return to their original size, and pigment becomes concentrated again. By means of the chromatophores, which are under nervous and probably hormonal control, an elaborate system of changes in color and pattern is possible, including general darkening or lightening; flushes of pink, yellow, or lavender; and formation of bars, stripes,

spots, or irregular blotches. These colors may be used variously as danger signals, as protective coloring, in courtship rituals, and probably in other ways.

By assuming different color patterns of different parts of the body, a squid can transmit three or four different messages *simultaneously* to different individuals and in different directions, and it can instantaneously change any or all of the messages. Probably no other system of communication in invertebrates can convey so much information so rapidly.

Deep-water cephalopods may have to depend more on chemical or tactile senses than their littoral or surface cousins, but they also produce their own type of visual signals, for they have evolved many elaborate luminescent organs.

Most cephalopods other than nautiloids have another protective device. An ink sac that empties into the rectum contains an **ink gland** that secretes **sepia,** a dark fluid containing the pigment melanin, into the sac. When the animal is alarmed, it releases a cloud of ink, which may hang in the water as a blob or be contorted by water currents. The animal quickly departs from the scene, leaving the ink as a decoy to the predator.

Reproduction   Sexes are separate in cephalopods. Spermatozoa are encased in spermatophores and stored in a sac that opens into the mantle cavity. One arm of adult males is modified as an intromittent organ, called a **hectocotylus,** used to pluck a spermatophore from his own mantle cavity and insert it into the mantle cavity of a female near the oviduct opening (Figure 16.41). Before copulation males often undergo color displays, apparently directed against rival males. Eggs are fertilized as they leave the oviduct and are then usually attached to stones or other objects. Some octopods tend their eggs. Females of *Argonauta,* the paper nautilus, secrete a fluted "shell," or capsule, in which eggs develop.

The large yolky eggs undergo meroblastic cleavage. During embryonic development, the head and foot become

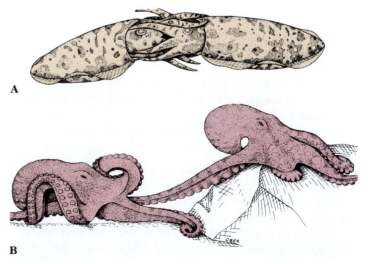

A

B

Figure 16.41

Copulation in cephalopods. **A,** Mating cuttlefishes. **B,** Male octopus uses modified arm to deposit spermatophores in female mantle cavity to fertilize her eggs. Octopuses often tend their eggs during development.

indistinguishable. The ring around the mouth, which bears the arms, or tentacles, may be derived from the anterior part of the foot. Juveniles hatch from eggs; no free-swimming larva exists in cephalopods.

Major Groups of Cephalopods

There are three subclasses of cephalopods: Nautiloidea, which have two pairs of gills; the entirely extinct Ammonoidea; and Coleoidea, which have one pair of gills. Nautiloidea populated Paleozoic and Mesozoic seas, but there survives only one genus, *Nautilus* (see Figure 16.37), of which there are five species. *Nautilus's* head, with its 60 to 90 or more tentacles, can be extended from the opening of the body compartment of its shell. Its tentacles have no suckers but are made adhesive by secretions. They are used in searching for, sensing, and grasping food. Beneath its head is the funnel. Mantle, mantle cavity, and visceral mass are sheltered by the shell.

Ammonoids were widely prevalent in the Mesozoic era but became extinct by the end of the Cretaceous period. They had chambered shells analogous to nautiloids, but the septa were more complex, and the septal sutures (where septa contact the inside of the shell) were frilled (compare shells in Figure 16.37B and C). The reasons for their extinction remain a mystery. Present evidence suggests that they were gone before the asteroid bombardment at the end of the Cretaceous period (inside back cover), whereas some nautiloids, which some ammonoids closely resembled, survive to the present.

Subclass Coleoidea includes all living cephalopods except *Nautilus*. The classification of living cephalopods is a subject of debate, but most authorities place the octopuses and vampire squids together in the superorder Octopodiformes, whereas squids, cuttlefish, and their relatives are grouped into the superorder Decapodiformes. Members of order Sepioidea (cuttlefishes and their relatives) have a rounded or compressed, bulky body bearing fins (Figure 16.38). They have eight arms and two tentacles. Both arms and tentacles have suckers, but tentacles bear suckers only at their ends. Members of the orders Myopsida and Degopsida (squids, Figure 16.39) have a more cylindrical body but also have eight arms and two tentacles. Order Vampyromorpha (vampire squid) contains only a single, deep-water species. Members of order Octopoda have eight arms and no tentacles (Figure 16.1E). Their bodies are short and saclike, with no fins. The suckers in squids are stalked (pendunculated), with hardened rims bearing teeth; in octopuses the suckers are sessile and rimless.

PHYLOGENY AND ADAPTIVE DIVERSIFICATION

The first molluscs probably arose during Precambrian times because fossils attributed to Mollusca have been found in geological strata as old as the early Cambrian period. On the basis of such shared features as spiral cleavage, mesoderm from the 4d blastomere, and trochophore larva, many zoologists argue that Mollusca are protostomes, allied with the annelids in Lophotrochozoa. Opinions differ, however, as to the exact nature of the relationship among lophotrochozoans. Some characters suggest that molluscs and annelids are sister taxa, but we do not depict a branching order for these taxa.

Annelid worms have a developmental pattern very similar to that of molluscs, but the annelid body is metameric, composed of serially repeated segments, whereas there are no true segments in molluscs. Both annelids and molluscs are coelomate protostomes, but the coelom is greatly reduced in molluscs as compared with annelids. Opinions differ as to whether molluscs were derived from a wormlike ancestor independent of annelids, share an ancestor with annelids after the advent of the coelom, or share a segmented common ancestor with annelids. Until the lophotrochozoan phylogeny is better resolved, it will not be possible to determine whether molluscs and annelids shared a coelomate ancestor.

The hypothesis that annelids and molluscs shared a segmented ancestor is strengthened if the repeated body parts present in *Neopilina* (class Monoplacophora), and in some chitons, can be considered evidence of metamerism. However, recent morphological and developmental studies indicate that these parts are not remnants of an ancestral metameric body. A new perspective on the evolution of repeated parts (gills and muscles) comes from analysis of a new molluscan cladogram. This cladogram was based on molecular characters from a wide range of molluscs, including a monoplacophoran.[4] The cladogram places monoplacophorans as the sister taxon to chitons; it unites the two taxa with repeated body parts in a clade called Serialia. Further, clade Serialia does not branch from the base of the molluscan tree, as it would if the ancestral mollusc were segmented. Instead, clade Serialia branches from the molluscan lineage after the wormlike caudofoveates and solenogastres, indicating that the repeated structures are derived molluscan features, not ancestral features.

Some researchers have noted that annelids are not the only segmented animals. Arthropods also have segmented bodies, but molecular sequence data place arthropods in clade Ecdysozoa, not in clade Lophotrochozoa with annelids and molluscs. This means that arthropods are more distantly related to annelids and to molluscs than either group is to the other. The third segmented group, the chordates, is placed within the deuterostome clade. The segmented phyla are not closely related to each other according to our current understanding of metazoan phylogeny.

Did segmentation originate independently within the three metameric taxa? At present, there is no clear answer, but several hypotheses are currently under consideration. One hypothesis suggests that genes for segmentation were present in basal bilaterians and have been suppressed many times. Another suggests that the two segmented protostomes, annelids and arthropods, are sister taxa, but this hypothesis conflicts with the recent placement of annelids and arthropods in different clades within Protostomia. Several scientists are currently studying the detailed mechanisms that produce segments in annelids and arthropods, as well as the formation of repeated body parts in certain molluscs. Differences in biochemical pathways and developmental steps that produce

[4]Giribet, G., et al. 2006. Proc. Natl. Acad. Sci. USA **103**:7723–7728.

Classification of Phylum Mollusca

Class Caudofoveata (kaw'do-fo-ve-at' a) (L. *cauda*, tail, + *fovea*, small pit). Wormlike; shell, head, and excretory organs absent; radula usually present; mantle with chitinous cuticle and calcareous scales; oral pedal shield near anterior mouth; mantle cavity at posterior end with pair of gills; sexes separate; formerly united with solenogasters in class Aplacophora. Examples: *Chaetoderma, Limifossor*.

Class Solenogastres (so-len'o-gas' trez) (Gr. *solen*, pipe, + *gaster*, stomach): **solenogasters.** Wormlike; shell, head, and excretory organs absent; radula present or absent; mantle usually covered with calcareous scales or spicules; rudimentary mantle cavity posterior, without true gills, but sometimes with secondary respiratory structures; foot represented by long, narrow, ventral pedal groove; hermaphroditic. Example: *Neomenia*.

Class Monoplacophora (mon'o-pla-kof'o-ra) (Gr. *monos*, one, + *plax*, plate, + *phora*, bearing). Body bilaterally symmetrical with a broad flat foot; a single limpetlike shell; mantle cavity with three to six pairs of gills; large coelomic cavities; radula present; three to seven pairs of nephridia, two of which are gonoducts; separate sexes. Example: *Neopilina* (Figure 16.9).

Class Polyplacophora (pol'y-pla-kof'o-ra) (Gr. *polys*, many, several, + *plax*, plate, + *phora*, bearing): **chitons.** Elongated, dorsoventrally flattened body with reduced head; bilaterally symmetrical; radula present; shell of seven or eight dorsal plates; foot broad and flat; gills multiple along sides of body between foot and mantle edge; sexes usually separate, with a trochophore but no veliger larva. Examples: *Mopalia* (Figure 16.11), *Tonicella* (Figure 16.1A).

Class Scaphopoda (ska-fop'o-da) (Gr. *skaphē*, trough, boat, + *pous, podos*, foot): **tusk shells.** Body enclosed in a one-piece tubular shell open at both ends; conical foot; mouth with radula and contractile tentacles (captacula); head absent; mantle for respiration; sexes separate; trochophore larva. Example: *Dentalium* (Figure 16.12).

Class Gastropoda (gas-trop'o-da) (Gr. *gaster*, stomach, + *pous, podos*, foot): **snails and slugs.** Body asymmetrical and shows effects of torsion; body usually in a coiled shell (shell uncoiled or absent in some); head well developed, with radula; foot large and flat; one or two gills, or with mantle modified into secondary gills or a lung; most with single atrium and single nephridium; nervous system with cerebral, pleural, pedal, and visceral ganglia; dioecious or monoecious, some with trochophore, typically with veliger, some without pelagic larva. Examples: *Busycon, Polinices* (Figure 16.16B), *Physa, Helix, Aplysia* (Figure 16.22).

Class Bivalvia (bi-val've-a) (L. *bi*, two, + *valva*, folding door, valve) **(Pelecypoda): bivalves.** Body enclosed in a two-lobed mantle; shell of two lateral valves of variable size and form, with dorsal hinge; head greatly reduced, but mouth with labial palps; no radula; no cephalic eyes, a few with eyes on mantle margin; foot usually wedge shaped; gills platelike; sexes usually separate, typically with trochophore and veliger larvae. Examples: *Anodonta, Venus, Tagelus* (Figure 16.26), *Teredo* (Figure 16.32).

Class Cephalopoda (sef'a-lop'o-da) (Gr. *kephalē*, head, + *pous, podos*, foot): **squids, cuttlefish, nautilus, and octopuses.** Shell often reduced or absent; head well developed with eyes and a radula; head with arms or tentacles; foot modified into siphon; nervous system of well-developed ganglia, centralized to form a brain; sexes separate, with direct development. Examples: *Sepioteuthis* (Figure 16.39), *Octopus* (Figure 16.1E), *Sepia* (Figure 16.38).

---

segmented bodies across taxa would support the hypothesis that segmentation arose several times independently.

Fossils are remains of past life uncovered from the crust of the earth (see Chapter 6). They can be actual parts or products of animals (teeth, bones, shells, and so on), petrified skeletal parts, molds, casts, impressions, footprints, and others. Soft and fleshy parts rarely leave recognizable fossils. Therefore we have no good record of molluscs before they had shells, and there can be some doubt that certain early fossil shells are really remains of molluscs, particularly if the group they represent is now extinct (for example, the Hyolitha). The issue of how to define a mollusc from hard parts alone was emphasized by Yochelson (1978, Malacologia **17:**165), who said, "If scaphopods were extinct and soft parts were unknown, would they be called mollusks? I think not."

A "hypothetical ancestral mollusc" (Figure 16.3) was long viewed as representing the original mollusc ancestor, but neither a solid shell nor a broad, crawling foot are now considered universal characters for Mollusca (Figure 16.42). The primitive ancestral mollusc was probably a small (about 1 mm) more or less wormlike organism with a ventral gliding surface and a dorsal mantle. It may have had a chitinous cuticle and calcareous scales. It probably had a posterior mantle cavity with two gills, a radula, a ladderlike nervous system, and an open circulatory system with a heart. There remains debate about whether among living molluscs the primitive condition is most nearly approached by caudofoveates or solenogasters. However the fossil record for caudofaveates goes back only to the Silurian (about 440 MYA) and we have no real fossil record for the solenogasters. In contrast, some monoplacophoran groups (Heliconelloida) have fossil records stretching back to the earliest Cambrian (about 510 MYA). Despite the poor fossil record of these shell-less groups, both aplacophoran classes probably diverged from primitive ancestors before the development of a solid shell, a distinct head with sensory organs, and a ventral muscularized foot. Polyplacophorans probably also diverged early from the main lines of molluscan evolution before the veliger was established as the larva (see Figure 16.2). Some workers believe that shells of polyplacophorans are not homologous to shells of other molluscs because they differ structurally and developmentally.

There remains debate over the exact relationships of the molluscan classes to one another, but most zoologists favor Gastropoda and Cephalopoda forming the sister group to Monoplacophora (see Figure 16.2). Both gastropods and cephalopods have a greatly expanded visceral mass. The mantle cavity was moved to the right side by torsion in gastropods, but in cephalopods the mantle cavity was extended

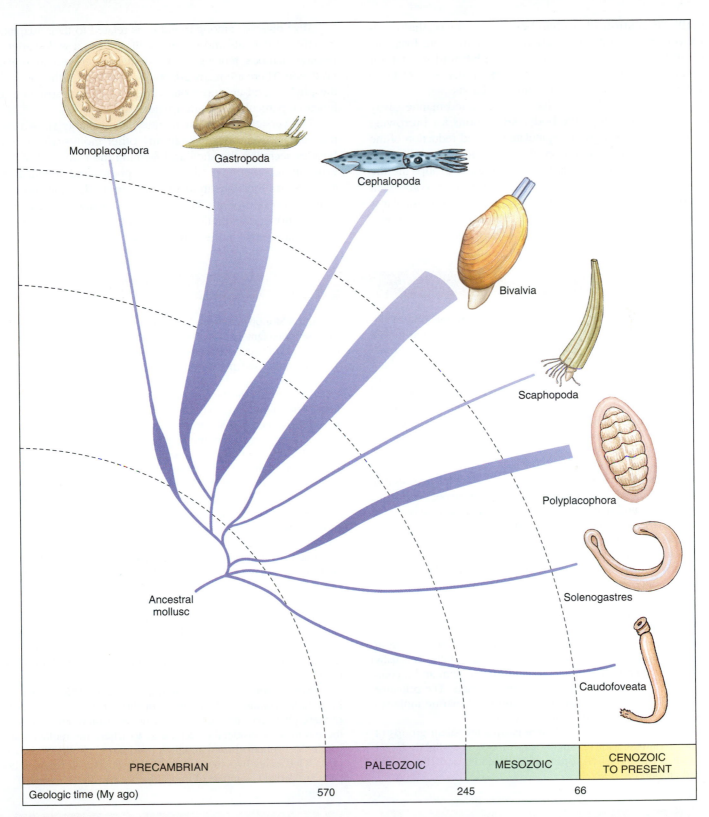

Monoplacophora

Gastropoda

Cephalopoda

Bivalvia

Scaphopoda

Polyplacophora

Solenogastres

Caudofoveata

Ancestral mollusc

| PRECAMBRIAN | PALEOZOIC | MESOZOIC | CENOZOIC TO PRESENT |

Geologic time (My ago)   570   245   66

Figure 16.42

Classes of Mollusca, showing their derivations and relative abundance.

ventrally. Evolution of a chambered shell in cephalopods was a very important contribution to their freedom from the substratum and their ability to swim. Elaboration of their respiratory, circulatory, and nervous systems is correlated with their active predatory and swimming habits.

Scaphopods and bivalves have an expanded mantle cavity that essentially envelops the body. Adaptations for burrowing characterize these groups: the spatulate foot and reduction of the head and sense organs. However, there is some debate still about whether morphological similarities among these groups are the result of shared ancestry or shared life habits (that is, convergent evolution). The classification of bivalve molluscs in particular is a subject of intense debate, and few authorities can agree on a scheme of nomenclature or taxonomic relation within this group.

Most diversity among molluscs is related to their adaptation to different habitats and modes of life and to a wide variety of feeding methods, ranging from sedentary filter-feeding to active predation. There are many adaptations for food gathering within the phylum and an enormous variety in radular structure and function, particularly among gastropods.

The versatile glandular mantle has probably shown more plastic adaptive capacity than any other molluscan structure. Besides secreting the shell and forming the mantle cavity, it is variously modified into gills, lungs, siphons, and apertures, and it sometimes functions in locomotion, in the feeding processes, or in a sensory capacity. The shell, too, has undergone a variety of evolutionary adaptations making molluscs one of the most successful groups on the planet today.

SUMMARY

Mollusca is the largest lophotrochozoan phylum and one of the largest and most diverse of all phyla, its members ranging in size from very small organisms to the largest of invertebrates. The basic body divisions of molluscs are head-foot and visceral mass, which is usually covered by a shell. The majority are marine, but some are freshwater, and a few are terrestrial. They occupy a wide variety of niches. A number are economically important, and a few are medically important as hosts of parasites.

Molluscs are coelomate although their coelom is limited to the area around the heart, gonads, and occasionally part of the intestine. Evolutionary development of a coelom was important because it enabled better organization of visceral organs and, in many of the animals that have it, but not molluscs, an efficient hydrostatic skeleton.

The mantle and mantle cavity are important characteristics of molluscs. The mantle secretes the shell and overlies a part of the visceral mass to form a cavity housing the gills. The mantle cavity has been modified into a lung in some molluscs. The foot is usually a ventral, solelike, locomotory organ, but it may be variously modified, as in cephalopods, where it has become arms and a funnel. The radula is found in all molluscs except bivalves and many solenogasters and is a protrusible, tonguelike organ with teeth used in feeding. Except in cephalopods, which have secondarily developed a closed circulatory system, the circulatory system of molluscs is open, with a heart and blood sinuses. Molluscs usually have a pair of nephridia connecting with the coelom and a complex nervous system with a variety of sense organs. The primitive larva of molluscs is the trochophore, and most marine molluscs also have a more advanced larva, the veliger.

Classes Caudofoveata and Solenogastres are small groups of wormlike molluscs with no shell. Scaphopoda is a slightly larger class with a tubular shell, open at both ends, and the mantle wrapped around the body.

Class Monoplacophora is a tiny, univalve marine group showing pseudometamerism. Polyplacophora are more common, marine organisms with shells in the form of a series of seven or eight plates. They are rather sedentary animals with a row of gills along each side of their foot.

Gastropoda are the most successful and largest class of molluscs. Their interesting evolutionary history includes a torted stage where anus and head are at the same end, as well as coiling, which represents an elongation and spiraling of the visceral mass. Torsion has led to the problem of fouling, which is the release of excreta over the head and in front of the gills, and this has been solved in various ways among different gastropods. Among the solutions to fouling are bringing water into one side of the mantle cavity and out the other (many gastropods), some degree of detorsion (opisthobranchs), and conversion of the mantle cavity into a lung (pulmonates).

Class Bivalvia are marine and freshwater, and they have their shell divided into two valves joined by a dorsal ligament and held together by an adductor muscle. Most of them are filter feeders, drawing water through their gills by ciliary action.

Members of class Cephalopoda are all predators and many can swim rapidly. Their arms or tentacles capture prey by adhesive secretions or by suckers. They swim by forcefully expelling water from their mantle cavity through a funnel, which was derived from the foot.

There is both embryological and molecular evidence that molluscs share a common ancestor with annelids more recently than either of these phyla do with arthropods or deuterostome phyla. However, there remains considerable debate as to where the molluscs arose within the Lophotrochozoa and their relationship to other protostome phyla. One recent hypothesis states that molluscs with repeated body parts are derived members of the clade, not ancestral forms.

REVIEW QUESTIONS

1. Members of such a large and diverse phylum as Mollusca impact humans in many ways. Discuss this statement.
2. How does a molluscan coelom develop embryologically? Why was the evolutionary development of a coelom important?
3. What are characteristics of Mollusca that distinguish it from other phyla?
4. Briefly describe characteristics of the hypothetical ancestral mollusc, and tell how each class of molluscs (Caudofoveata,

Solenogastres, Monoplacophora, Polyplacophora, Scaphopoda, Gastropoda, Bivalvia, Cephalopoda) differs from the ancestral condition with respect to each of the following: shell, radula, foot, mantle cavity and gills, circulatory system, and head.

5. Define the following: ctenidia, odontophore, periostracum, prismatic layer, nacreous layer, metanephridia, nephrostome, trochophore, veliger, glochidium, osphradium.

6. Briefly describe the habitat and habits of a typical chiton.

7. Define the following with respect to gastropods: operculum, columella, torsion, fouling, bilateral asymmetry, rhinophore, pneumostome.

8. What functional problem results from torsion? How have gastropods evolved to avoid this problem?

9. Gastropods have diversified enormously. Illustrate this statement by describing variations in feeding habits found in gastropods.

10. Distinguish between opisthobranchs and pulmonates.

11. Briefly describe how a typical bivalve feeds and how it burrows.

12. How is the ctenidium modified from the ancestral form in a typical bivalve?

13. What is the function of the siphuncle of cephalopods?

14. Describe how cephalopods swim and how they eat.

15. Describe adaptations in the circulatory and neurosensory systems of cephalopods that are particularly valuable for actively swimming, predaceous animals.

16. Distinguish between ammonoids and nautiloids.

17. Which other invertebrate groups are likely to be the closest relatives of molluscs? What evidence supports and contradicts these relationships?

SELECTED REFERENCES

Abbott, R. T., and P. A. Morris. 2001. R. T. Peterson (ed.). A field guide to shells: Atlantic coasts and the West Indies, ed. 5. Boston, Houghton Mifflin Company. *An excellent revision of a popular handbook.*

Barinaga, M. 1990. Science digests the secrets of voracious killer snails. Science **249**:250–251. *Describes current research on the toxins produced by cone snails.*

Bergström, J. 1989. The origin of animal phyla and the new phylum Procoelomata. Lethaia **22**:259–269. *Argues that Caudofoveata are the only surviving members of Procoelomata, putative ancestral sclerite-bearing early Cambrian metazoan.*

Fleischman, J. 1997. Mass extinctions come to Ohio. Discover **18**(5):84–90. *Of the 300 species of freshwater bivalves in the Mississippi River basin, 161 are extinct or endangered.*

Gehring, W. J., and I. Kazuho. 1999. *Pax 6:* mastering eye morphogenesis and eye evolution. Trends Genet. **15**:371–377. *The authors discuss morphogenetic pathways by which various animal eyes could have evolved from a common ancestral type of photoreceptive cell.*

Giribet, G., A. Okusu, A. R. Lindgren, S. W. Huff, M. Schrödl, and M. Nishiguchi. 2006. Evidence for a clade composed of molluscs with serially repeated structures: monoplacophorans are related to chitons. Proc. Natl. Acad. Sci. USA **103**:7723–7728. *Molluscs with repeated structures are derived taxa and not likely to have inherited repeated structures from a segmented ancestor.*

Gosline, J. M., and M. D. DeMont. 1985. Jet-propelled swimming in squids. Sci. Am. **252**:96–103 (Jan.). *Mechanics of swimming in squid are analyzed; elasticity of collagen in mantle increases efficiency.*

Gould, S. J. 1994. Common pathways of illumination. Nat. Hist. **103**:10–20. Pax-6 *gene controls eye morphogenesis in* Drosophila *and vertebrates.*

Hanlon, R. T., and J. B. Messenger. 1996. Cephalopod behaviour. Cambridge, U.K., Cambridge University Press. *Intended for nonspecialists and specialists.*

Haszprunar, G. 2000. Is the Aplacophora monophyletic? A cladistic point of view. Amer. Malac. Bull. **15**:115–130. *Asserts that solenogasters are the sister group to all extant molluscan groups, including the caudofaveates.*

Holloway, M. 2000. Cuttlefish say it with skin. Nat. Hist. **109**(3):70–76. *Cuttlefish and other cephalopods can change texture and color of their skin with astonishing speed. Fifty-four components of cuttlefish "vocabulary" have been described, including color display, skin texture, and a variety of arm and fin signals.*

Page, L. R. 2003. Gastropod ontogenetic torsion: developmental remnants of an ancient evolutionary change in body plan. J. Exp. Zool. Part B: **297B**:11–26. *A torted body is the result of asynchronous movements of the shell and viscera, accompanied by remodeling of mantle epithelium.*

Ward, P. D. 1998. Coils of time. Discover **19**(3):100–106. *Present Nautilus has apparently existed essentially unchanged for 100 million years, and all other species known were derived from it, including king nautilus (Allonautilus), a recent derivation.*

Woodruff, D. S., and M. Mulvey. 1997. Neotropical schistosomiasis: African affinities of the host snail *Biomphalaria glabrata* (Gastropoda: Planorbidae). Biol. J. Linn. Soc. **60**:505–516. *The pulmonate snail* Biomphalaria glabrata *is the intermediate host in the New World for* Schistosoma mansoni, *an important trematode of humans (p. 299). Allozyme analysis shows that* B. glabrata *clusters with African species rather than the neotropical ones. Thus, when* S. mansoni *was brought to New World in African slaves, it found a compatible host.*

Zorpette, G. 1996. Mussel mayhem, continued. Sci. Am. **275**:22–23 (Aug.). *Some benefits, though dubious, of the zebra mussel invasion have been described, but these are outweighted by the problems created.*

ONLINE LEARNING CENTER

Visit www.mhhe.com/hickmanipz14e for chapter quizzing, key term flash cards, web links, and more!

Annelids and Allied Taxa

- PHYLUM ANNELIDA, INCLUDING POGONOPHORANS (SIBOGLINIDS)
- PHYLUM ECHIURA
- PHYLUM SIPUNCULA

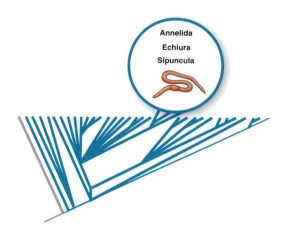

Annelida
Echiura
Sipuncula

Chloeia *sp., a polychaete.*

Dividing the Body

Although a fluid-filled coelom provided an efficient hydrostatic skeleton for burrowing, precise control of body movements was probably difficult for the earliest coelomates. The force of muscle contraction in one area was carried throughout the body by the fluid in the undivided coelom. In contrast, there were distinct coelomic compartments within the bodies of ancestral annelids. Compartments, known as segments or metameres, were separated from neighbors by partitions called septa. Septa permitted each fluid-filled segment to respond individually to local muscle contraction—one segment could be long and thin and another short and round. Annelids illustrate segmentation, or *metamerism;* their bodies are composed of serially repeated units. Each unit contains components of most organ systems, such as circulatory, nervous, and excretory systems.

The evolutionary advent of metamerism was significant because it made possible much greater complexity in structure and function. Metamerism increased burrowing efficiency by permitting the independent movement of separate segments. Fine control of movements allowed, in turn, the evolution of a more sophisticated nervous system. Moreover, repetition of body parts gave the organisms a built-in redundancy that provided a safety factor: if one segment should fail, others could still function. Thus an injury to one part would not necessarily be fatal.

The evolutionary potential of the metameric body plan is amply demonstrated by the large and diverse phyla Annelida, Arthropoda and Chordata, which likely represent three separate evolutionary origins of metamerism.

The wormlike animal phyla described in this chapter are coelomate protostomes belonging to subgroup Lophotrochozoa. They develop by spiral mosaic cleavage, form mesoderm from derivatives of the 4d cell, make a coelom by schizocoely, and share a trochophore as the ancestral larval form. Three phyla are discussed: Annelida, Echiura, and Sipuncula.

Members of phylum Annelida are segmented worms living in marine, freshwater, and moist terrestrial habitats. Marine bristle worms, leeches, and the familiar earthworms belong to this group. Annelida also now includes pogonophoran and vestimentiferan worms, formerly either placed together in phylum Pogonophora, or placed in distinct phyla: Pogonophora and Vestimentifera. These deep-ocean worms belong in clade Siboglinidae within class Polychaeta.

Worms in phylum Echiura and phylum Sipuncula are benthic marine animals with unsegmented bodies. Several phylogenetic studies using molecular sequence data place echiurans within phylum Annelida as a derived group of polychaetes where segmentation has been lost, but this placement is not universally accepted. We depict echiurans as the sister taxon to Annelida, and sipunculans as the sister taxon to a clade composed of Annelida and Echiura (Figure 17.1).

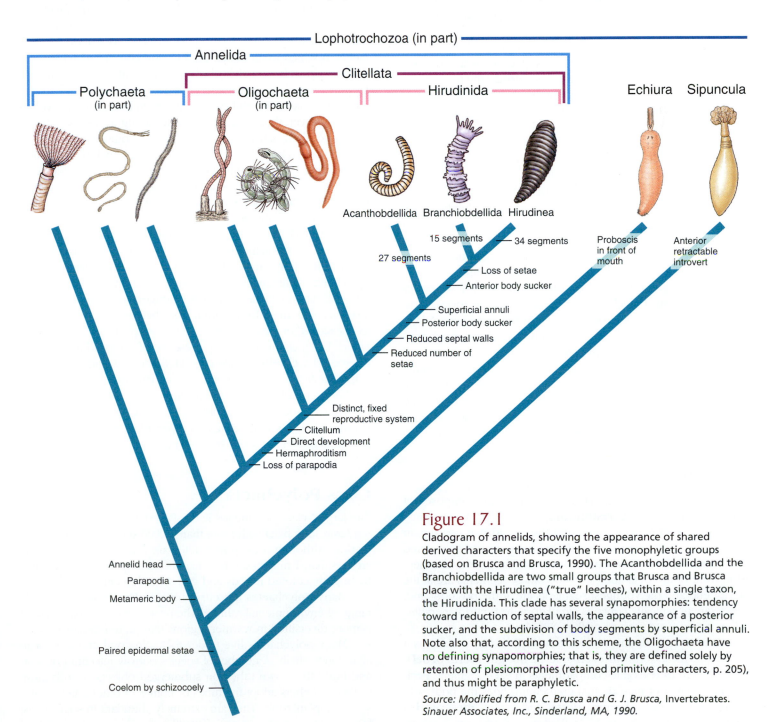

Figure 17.1

Cladogram of annelids, showing the appearance of shared derived characters that specify the five monophyletic groups (based on Brusca and Brusca, 1990). The Acanthobdellida and the Branchiobdellida are two small groups that Brusca and Brusca place with the Hirudinea ("true" leeches), within a single taxon, the Hirudinida. This clade has several synapomorphies: tendency toward reduction of septal walls, the appearance of a posterior sucker, and the subdivision of body segments by superficial annuli. Note also that, according to this scheme, the Oligochaeta have no defining synapomorphies; that is, they are defined solely by retention of plesiomorphies (retained primitive characters, p. 205), and thus might be paraphyletic.

Source: Modified from R. C. Brusca and G. J. Brusca, Invertebrates. *Sinauer Associates, Inc., Sinderland, MA, 1990.*

PHYLUM ANNELIDA, INCLUDING POGONOPHORANS (SIBOGLINIDS)

Phylum Annelida (an-nel′i-da) (L. *annelus,* little ring, + *ida,* pl. suffix) consists of the segmented worms. It is a diverse phylum, numbering approximately 15,000 species, the most familiar of which are earthworms and freshwater worms (class Oligochaeta) and leeches (class Hirudinida). However, approximately two-thirds of the phylum comprises marine worms (class Polychaeta), which are less familiar to most people. Some polychaetes are grotesque in appearance whereas others are graceful and beautiful. They include clamworms, plumed worms, parchment worms, scaleworms, lugworms, and many others.

Annelida are worms whose bodies are divided into similar **segments,** (also called metameres) arranged in linear series and externally marked by circular rings called **annuli** (the name of the phylum refers to this characteristic). Body segmentation **(metamerism)** is a division of the body into a series of segments, each of which contains similar components of all major organ systems. In annelids the segments are delimited internally by septa.

Annelids are sometimes called "bristle worms" because, with the exception of leeches, most annelids bear tiny chitinous bristles called **setae** (L. *seta,* hair or bristle). Short needlelike setae help anchor segments during locomotion and long, hairlike setae aid aquatic forms in swimming. Since many annelids burrow or live in secreted tubes, stiff setae also aid in preventing the worm from being pulled out or washed out of its home. Robins know from experience how effective earthworms' setae are.

Annelids have a worldwide distribution, and a few species are cosmopolitan. Polychaetes are chiefly marine forms. Most are benthic, but some live pelagic lives in the open seas. Oligochaetes and leeches occur predominantly in freshwater or terrestrial soils. Some freshwater species burrow in mud and sand and others among submerged vegetation. Many leeches are predators, specialized for piercing their prey and feeding on blood or soft tissues. A few leeches are marine, but most live in freshwater or in damp regions. Suckers are typically found at both ends of the body for attachment to the substratum or to their prey.

Body Plan

The annelid body typically has a two-part head, composed of a **prostomium** and a **peristomium** followed by a segmented body and a terminal portion called the **pygidium** bearing an anus (Figure 17.2). The head and pygidium are not considered to be segments. New segments differentiate during development just in front of the pygidium; thus the oldest segments are at the anterior end and the youngest segments are at the posterior end. Each segment typically contains circulatory, respiratory, nervous, and excretory structures, as well as a coelom.

In most annelids the coelom develops embryonically as a split in the mesoderm on each side of the gut **(schizocoel),** forming a pair of coelomic compartments in each segment. **Peritoneum** (a layer of mesodermal epithelium) lines the body wall of each compartment, forming dorsal and ventral **mesenteries** that cover all organs (Figure 17.3). Peritonea of adjacent segments

meet to form **septa,** which are perforated by the gut and longitudinal blood vessels. The body wall surrounding the peritoneum and coelom contains strong circular and longitudinal muscles adapted for swimming, crawling, and burrowing (Figure 17.3).

Except in leeches, the coelom of most annelids is filled with fluid and serves as a **hydrostatic skeleton.** Because the volume of fluid in a coelomic compartment is essentially constant, contraction of the longitudinal body-wall muscles causes a segment to shorten and to become larger in diameter, whereas contraction of the circular muscles causes it to lengthen and become thinner. The presence of septa means that widening and elongation occur in restricted areas; crawling motions are produced by alternating waves of contraction by longitudinal and circular muscles passing down the body (peristaltic contractions). Segments in which longitudinal muscles are contracted widen and anchor themselves against the substrate while other segments, in which circular muscles are contracted, elongate and stretch forward. Forces powerful enough for rapid burrowing as well as locomotion can thus be generated. Swimming forms use undulatory rather then peristaltic movements in locomotion.

An annelid body has a thin outer layer of nonchitinous cuticle surrounding the epidermis (Figure 17.3). Paired epidermal setae (Figures 17.2 and 17.17) are ancestral for annelids, although they have been reduced or lost in some. The annelid digestive system is not segmented: the gut runs the length of the body perforating each septum (Figure 17.3). Longitudinal dorsal and ventral blood vessels follow the same path, as does the ventral nerve cord.

Traditionally, annelids are divided among three classes: Polychaeta, Oligochaeta, and Hirudinida. Polychaeta is a paraphyletic class because ancestors of oligochaetes and hirudineans (leeches) arose from within polychaetes. Oligochaetes and leeches together form a monophyletic group called Clitellata (see Figure 17.1), characterized by presence of a reproductive structure called a **clitellum** (see p. 371). Some authorities now consider Clitellata to be an annelid class containing oligochaetes and leeches as orders, but we retain the three classes and consider Clitellata a clade whose members are class Oligochaeta and class Hirudinida. Class Oligochaeta is a paraphyletic group because ancestors of leeches arose from within it.

Class Polychaeta

The largest class of annelids is the Polychaeta (Gr. *polys,* many, + *chaitē,* long hair) with more than 10,000 species, most of them marine. Although most polychaetes are 5 to 10 cm long, some are less than 1 mm, and others may be as long as 3 m. They may be brightly colored in reds and greens, iridescent, or dull.

Many polychaetes are euryhaline and can tolerate a wide range of environmental salinity. The freshwater polychaete fauna is more diversified in warmer regions than in temperate zones.

Many polychaetes live under rocks, in coral crevices, or in abandoned shells. A number of species burrow into mud or sand and build their own tubes on submerged objects or in bottom sediment. Others adopt the tubes or homes of other animals, and some are planktonic. They are extremely abundant in some areas; for example, a square meter of mudflat may contain thousands of

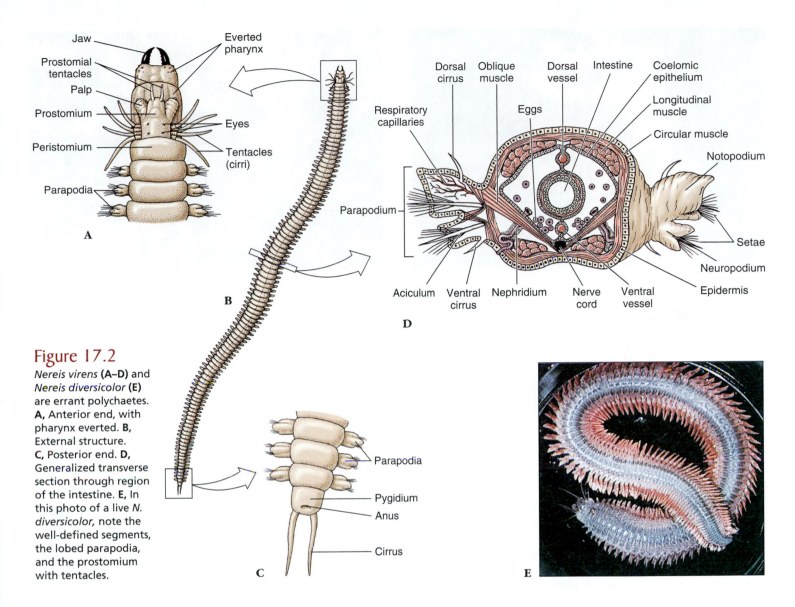

Figure 17.2

Nereis virens (**A–D**) and *Nereis diversicolor* (**E**) are errant polychaetes. **A,** Anterior end, with pharynx everted. **B,** External structure. **C,** Posterior end. **D,** Generalized transverse section through region of the intestine. **E,** In this photo of a live *N. diversicolor,* note the well-defined segments, the lobed parapodia, and the prostomium with tentacles.

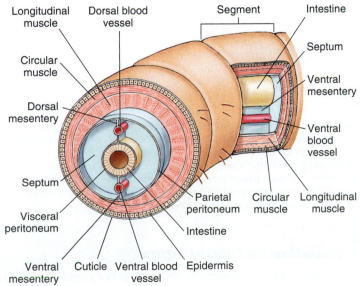

Figure 17.3

Annelid body plan.

polychaetes. They play a significant part in marine food chains because they are eaten by fish, crustaceans, hydroids, and many other predators.

Polychaetes differ from other annelids in having a well-differentiated head with specialized sense organs; paired appendages, called **parapodia,** on most segments; and no clitellum (Figure 17.2). As their name implies, they have many setae, usually arranged in bundles on the parapodia. They exhibit the most pronounced differentiation of body segments and specialization of sensory organs found in annelids (see p. 367).

Polychaetes are often divided into two morphological groups based on their activity: sedentary polychaetes and errant (free-moving) polychaetes. Sedentary polychaetes spend much or all of their time in tubes or permanent burrows. Many of them, especially those that live in tubes, have elaborate devices for feeding and respiration (Figure 17.4). Errant polychaetes (L. *errare,* to wander), include free-swimming pelagic forms, active burrowers, crawlers, and tubeworms that only leave their tubes for feeding or breeding. Most of these, like clam worms in the genus *Nereis* (Gr. name of a sea nymph) (see Figure 17.2), are predatory and equipped

A

B

Figure 17.4

Tube-dwelling sedentary polychaetes. **A,** Christmas-tree worm, *Spirobranchus giganteus,* live in a calcareous tube. On its head are two whorls of modified tentacles (radioles) used to collect suspended food particles from the surrounding water. Notice the finely branched filters visible on the edge of one radiole. **B,** Sabellid polychaetes, *Bispira brunnea,* live in leathery tubes.

with jaws or teeth. They have an eversible, muscular pharynx armed with teeth that can be thrust out with surprising speed to capture prey.

Form and Function

A polychaete typically has a **prostomium,** which may or may not be retractile and which often bears eyes, tentacles, and sensory palps (see Figure 17.2). The **peristomium** surrounds the mouth and may bear setae, palps, or, in predatory forms, chitinous jaws. Ciliary feeders may bear a crown of tentacles that can be opened like a fan or withdrawn into the tube (Figure 17.4).

The polychaete trunk is segmented, and most segments bear parapodia, which may have lobes, cirri, setae, and other parts on them (see Figure 17.2). Parapodia are used in crawling, swimming, or for anchoring the animal in its tube. They usually serve as the chief respiratory organs, although some polychaetes also have gills. *Amphitrite* (Gr. a mythical sea nymph), for example, has three pairs of branched gills and long extensible tentacles (Figure 17.5). *Arenicola* (L. *arena,* sand, + *colo,* inhabit), the burrowing lugworm (Figure 17.6), has paired gills on certain segments.

Nutrition

A polychaete's digestive system consists of a foregut, a midgut, and a hindgut. The foregut includes a stomodeum, a pharynx, and an anterior esophagus. It is lined with cuticle, and the jaws, where present, are constructed of cuticular protein. The more anterior portions of the midgut secrete digestive enzymes but absorption takes place toward the posterior end. A short hindgut connects the midgut to the exterior via the anus, which is on the pygidium.

Errant polychaetes are typically predators and scavengers. Sedentary polychaetes feed on suspended particles, or they may be deposit feeders, consuming particles on or in the sediment.

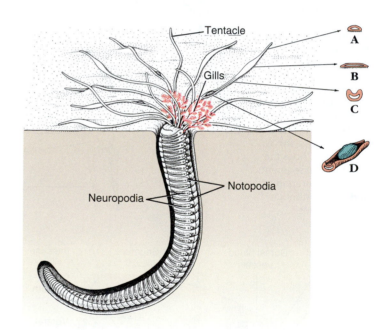

Figure 17.5

Amphitrite, which builds its tubes in mud or sand, extends long grooved tentacles out over the mud to pick up bits of organic matter. The smallest particles are moved along food grooves by cilia, larger particles by peristaltic movement. Its plumelike gills are blood red. **A,** Section through exploratory end of tentacle. **B,** Section through tentacle in area adhering to substratum. **C,** Section showing ciliary groove. **D,** Particle being carried toward mouth.

Circulation and Respiration

Polychaetes show considerable diversity in both circulatory and respiratory structures. As previously mentioned, parapodia and gills serve for gaseous exchange in various species. However,

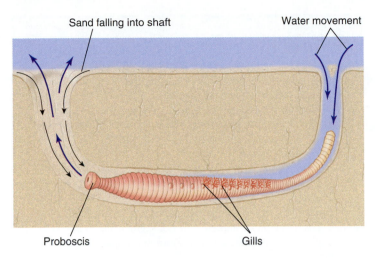

Figure 17.6
Arenicola, the lugworm, lives in a U-shaped burrow in intertidal mudflats. It burrows by successive eversions and retractions of its proboscis. By peristaltic movements it keeps water filtering through the sand. The worm then ingests the food-laden sand.

in some polychaetes there are no special organs for respiration, and gaseous exchange takes place across the body surface.

The circulatory pattern varies greatly. In *Nereis* a dorsal longitudinal vessel carries blood anteriorly, and a ventral longitudinal vessel conducts it posteriorly (see Figure 17.2D). Blood flows between these two vessels via segmental networks in the parapodia, septa, and around the intestine. In the burrowing predatory worm *Glycera* (Gr. *Glykera*, a feminine proper name) the circulatory system is reduced and joins directly with the coelom. Septa are incomplete, and thus the coelomic fluid assumes the function of circulation.

Many polychaetes have respiratory pigments such as hemoglobin, chlorocruorin, or hemerythrin (p. 704).

Excretion

Excretory organs consist of protonephridia and mixed proto- and metanephridia in some, but most polychaetes have metanephridia (see Figure 17.2). There is one pair per segment, with the inner end of each **(nephrostome)** opening into a coelomic compartment. Coelomic fluid passes into the nephrostome, and selective resorption occurs along the nephridial duct (see Figure 17.18).

Nervous System and Sense Organs

Organization of the central nervous system in polychaetes follows the basic annelid plan (see Figure 17.19). Dorsal cerebral ganglia connect with a subpharyngeal ganglion via a circumpharyngeal connective. A double ventral nerve cord courses the length of the worm, with metamerically arranged ganglia.

Sense organs are highly developed in polychaetes and include eyes, nuchal organs, and statocysts. Eyes, when present, may range from simple eyespots to well-developed organs. Eyes are most conspicuous in errant worms. Usually the eyes are retinal cups, with rodlike photoreceptor cells (lining the cup wall) directed toward the

lumen of the cup. The highest degree of eye development occurs in the family Alciopidae, which has large, image-resolving eyes similar in structure to those of some cephalopod molluscs (see Figure 16.40, p. 355), with cornea, lens, retina, and retinal pigments. Alciopid eyes also have accessory retinas, a characteristic independently evolved by deep-sea fishes and some deep-sea cephalopods. The accessory retinas of alciopids are sensitive to different wavelengths. The eyes of these pelagic animals may be well adapted to function because penetration by the different wavelengths of light varies with depth. Studies with electroencephalograms show that they are sensitive to dim light of the deep sea. Nuchal organs are ciliated sensory pits or slits that appear to be chemoreceptive, an important factor in food gathering. Some burrowing and tube-building polychaetes have statocysts that function in body orientation.

Reproduction and Development

Polychaetes have no permanent sex organs, and they usually have separate sexes. Reproductive systems are simple: Gonads appear as temporary swellings of the peritoneum and shed their gametes into the coelom. The gametes are then carried to the outside through gonoducts, through the metanephridia, or by rupture of the body wall. Fertilization is external, and the early larva is a trochophore (see Figure 16.7).

Some polychaetes live most of the year as sexually immature animals called atokes, but during the breeding season a portion of the body becomes sexually mature and swollen with gametes (Figure 17.7). An example is the palolo worm, which lives in burrows among coral reefs. During the swarming period, the sexually mature portions, now called epitokes, break off and swim to the surface. Just before sunrise, the sea is literally covered with them, and at sunrise they burst, freeing eggs and sperm for fertilization. Anterior portions of the worms regenerate new posterior sections. Swarming is of great adaptive value because the synchronous maturation of all the epitokes ensures the maximum number of fertilized eggs. However, this reproductive strategy is very hazardous; many types of predators have a feast on the swarming worms. In the meantime, the atoke remains safely in its burrow to produce another epitoke at the next cycle. In some polychaetes, epitokes arise from atokes by asexual budding (Figure 17.8) and become complete worms.

Representative Polychaetes

Clam Worms: *Nereis*   Clam worms (see Figure 17.2), or sand worms as they are sometimes called, are errant polychaetes that live in mucous-lined burrows in or near low tide. Sometimes they are found in temporary hiding places, such as under stones, where they stay with their bodies covered and their heads protruding. They are most active at night, when they wiggle out of their hiding places and swim or crawl over the sand in search of food.

The body, containing about 200 segments, may grow to 30 or 40 cm in length. The head is composed of a prostomium and a peristomium. The prostomium bears a pair of stubby palps,

Figure 17.7

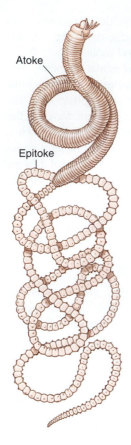

Atoke

Epitoke

Eunice viridis, the Samoan palolo worm. The posterior segments make up the epitokal region, consisting of segments packed with gametes. Each segment has an eyespot on the ventral side. Once a year the worms swarm, and the epitokes detach, rise to the surface, and discharge their ripe gametes, leaving the water milky. By the next breeding season, the epitokes are regenerated.

sensitive to touch and taste; a pair of short sensory tentacles; and two pairs of small dorsal eyes that are light sensitive. The peristomium bears the ventral mouth, a pair of chitinous jaws, and four pairs of sensory tentacles (see Figure 17.2A).

Each parapodium has two lobes: a dorsal **notopodium** and a ventral **neuropodium** (see Figure 17.2D) that bear setae with many blood vessels. Parapodia are used for both creeping and swimming and are controlled by oblique muscles that run from the midventral line to the parapodia in each segment. The worm swims by lateral undulatory movement of the body. It can dart through the water with considerable speed. These undulatory movements can also be used to suck water into or pump it out of the burrow.

Clam worms feed on small animals, other worms, and a variety of larval forms. They seize food with their chitinous jaws, which they protrude through the mouth when they evert their pharynx. Food is swallowed as the worm withdraws its pharynx. Movement of food through the alimentary canal is by peristalsis.

Characteristics of Phylum Annelida

1. Unique annelid head and paired epidermal setae present (lost in leeches); parapodia present in the ancestral condition
2. Marine, freshwater, and terrestrial
3. Most free-living, some symbiotic, some ectoparasitic
4. Body bilaterally symmetrical, **metameric,** often with distinct head
5. Triploblastic body
6. Coelom (schizocoel) well developed and divided by septa, except in leeches; coelomic fluid functions as hydrostatic skeleton
7. Epithelium secretes outer transparent moist cuticle
8. Digestive system complete and not segmentally arranged
9. Body wall with outer circular and inner longitudinal muscle layers
10. Nervous system with a double ventral nerve cord and a pair of ganglia with lateral nerves in each segment; brain a pair of dorsal cerebral ganglia with connectives to ventral nerve cord
11. Sensory system of tactile organs, taste buds, statocysts (in some), photoreceptor cells, and eyes with lenses (in some); specialization of head region into differentiated organs, such as tentacles, palps, and eyespots of polychaetes
12. Asexual reproduction by fission and fragmentation; capable of complete regeneration
13. Hermaphroditic or separate sexes; larvae, if present, are trochophore type; asexual reproduction by budding in some; spiral cleavage and mosaic development
14. Excretory system typically a **pair of nephridia for each segment; nephridia** remove waste from blood as well as from coelom
15. Respiratory gas exchange through skin, **gills,** or **parapodia**
16. **Circulatory system closed** with muscular blood vessels and aortic arches ("hearts") for pumping blood, segmentally arranged; respiratory pigments (hemoglobin, hemerythrin, or chlorocruorin) often present; amebocytes in blood plasma

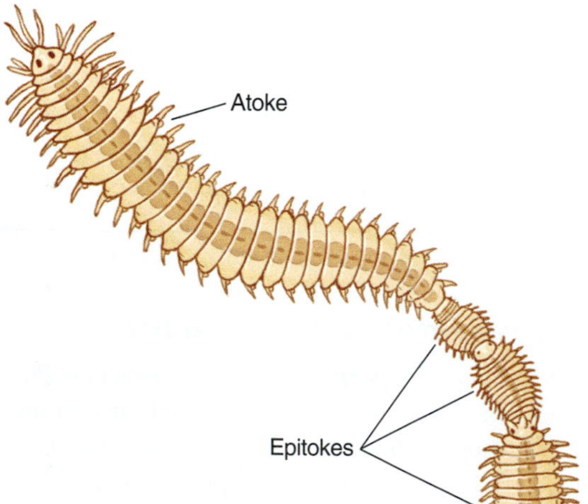

Figure 17.8

Atoke

Epitokes

Rather than transforming a portion of its body into an epitoke, *Autolytus prolifer* asexually buds off complete worms from its posterior end that become sexual epitokes.

Scale Worms   Scale worms (Figure 17.9) are members of the family Polynoidae (Gr. *Polynoē,* daughter of Nereus and Doris, a sea god and goddess), one of the most diverse, abundant, and widespread of polychaete families. Their flattened bodies are covered with broad scales, modified from dorsal parts of the parapodia. Most species are of modest size, but some are enormous (up to 190 mm long and 100 mm wide). They are carnivorous and eat a wide variety of animals. Many are commensal, living in burrows of other polychaetes or in association with cnidarians, molluscs, or echinoderms.

Fireworms   *Hermodice carunculata* (Gr. *herma,* reef, + *dex,* a worm found in wood) (Figure 17.10) and related species are called fireworms because their hollow, brittle setae contain a poisonous secretion. The setae puncture a hand that touches them, and then break off in the wound to cause skin irritation. Fireworms feed on corals, gorgonians, and other cnidarians.

Tubeworms   Polychaete tube-dwellers secrete many types of tubes. Some are parchmentlike or leathery (see Figure 17.4B);

Figure 17.9

A scale worm, *Hesperonoe adventor,* normally lives as a commensal in the tubes of *Urechis* (phylum Echiura, p. 379).

Figure 17.10

A fireworm, *Hermodice carunculata,* feeds on gorgonians and stony corals. Its setae are like tiny glass fibers and serve to ward off predators.

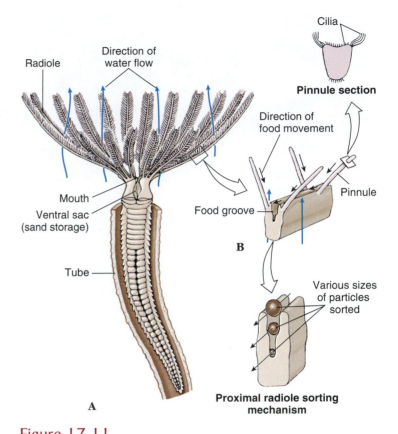

Figure 17.11

Sabella, a polychaete ciliary feeder, extends its crown of feeding radioles from its leathery secreted tube, reinforced with sand and debris. **A,** Anterior view of the crown. Cilia direct small food particles along grooved radioles to mouth and discard larger particles. Sand grains are directed to storage sacs and later are used in tube building. **B,** Distal portion of radiole showing ciliary tracts of pinnules and food grooves.

some are firm, calcareous tubes attached to rocks or other surfaces (see Figure 17.4A); and some are simply grains of sand or bits of shell or seaweed cemented together with mucous secretions. Many species burrow in sand or mud, lining their burrows with mucus (see Figure 17.6).

Most sedentary tube and burrow dwellers are particle feeders, using cilia or mucus to obtain food, typically plankton and detritus. Some deposit feeders, like *Amphitrite* (see Figure 17.5), protrude their heads above the mud and extend long tentacles over the surface to find food. Cilia and mucus on the tentacles entrap particles found on the sea bottom and move them toward the mouth. Lugworms, *Arenicola,* use an interesting combination of suspension and deposit feeding. They live in a U-shaped burrow through which, by peristaltic movements, they cause water to flow. Food particles are trapped by the sand at the front of the burrow, and *Arenicola* then ingests the food-laden sand (see Figure 17.6).

Fanworms, or "featherduster" worms, are beautiful tubeworms, fascinating to watch as they emerge from their secreted tubes and unfurl their lovely tentacular crowns to feed (see Figure 17.4). A slight disturbance, sometimes even a passing shadow, causes them to duck back quickly into the safety of the homes. Food drawn to the feathery arms, or **radioles,** by ciliary action is trapped in mucus and carried down ciliated food grooves to the mouth (Figure 17.11). Particles too large for the food grooves pass along the margins of the food grooves and fall away before they reach the mouth. Only small particles of food enter the mouth; sand grains are stored in a sac to be used later in enlarging the tube.

The parchment worm, *Chaetopterus* (Gr. *chaite⁻,* long hair, + *pteron,* wing), feeds on suspended particles by an entirely different mechanism (Figure 17.12). It lives in a U-shaped, parchmentlike tube buried, except for the tapered ends, in sand or mud along the shore. The worm attaches to the side of the tube by ventral suckers. Fans (modified parapodia on segments 14 to 16) pump water through the tube by rhythmical movements. A pair of enlarged parapodia on segment 12 secretes a

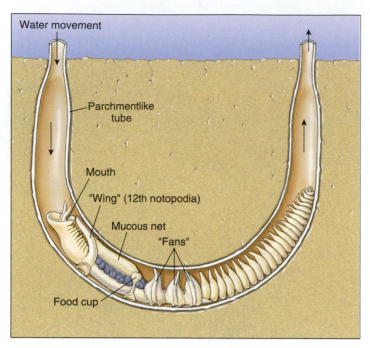

Figure 17.12

Chaetopterus, a sedentary polychaete, lives in a U-shaped tube in the sea bottom. It pumps water through the parchmentlike tube (of which one-half has been cut away here) with its three pistonlike fans. The fans beat 60 times per minute to keep water currents moving. The winglike notopodia of the twelfth segment continuously secrete a mucous net that strains out food particles. As the net fills with food, the food cup rolls it into a ball, and when the ball is large enough (about 3 mm), the food cup bends forward and deposits the ball in a ciliated groove to be carried to the mouth and swallowed.

long mucous net that reaches back to a small food cup just in front of the fans. All water passing through the tube is filtered through this mucous net, the end of which is rolled into a ball by cilia in the cup. When the ball is about the size of a BB shot (about 3 mm diameter), the fans stop beating and the ball of food and mucus is rolled forward by ciliary action to the mouth and swallowed.

Clade Siboglinidae (Pogonophorans)   Members of former phylum Pogonophora (po′go-nof′e-ra) (Gr. *pōgōn*, beard, +*phora*, bearing), or beardworms, were entirely unknown before the twentieth century. The first specimens to be described were collected from deep-sea dredgings off the coast of Indonesia in 1900. They have since been discovered in several seas, including the western Atlantic off the U.S. eastern coast. Some 150 species have been described so far. Most species are less than 1 mm in diameter but can be 10 to 75 cm in length.

Most siboglinids live in mud on the ocean floor, at depths of 100 to 10,000 m. This location accounts for their delayed discovery, for they are obtained only by dredging. They are sessile animals that secrete very long chitinous tubes in which they live, and probably only extend the anterior end of their body for absorbing nutrients. The tubes are generally oriented upright in bottom sediments. A tube can be three to four times the length of the animal, which can move up or down inside its tube but cannot turn around.

Beardworms have a long, cylindrical body covered with cuticle. Cuticle, epidermis, and circular and longitudinal muscles compose the body wall. The body is divided into a short anterior **forepart;** a long, very slender **trunk;** and a small, segmented **opisthosoma** (Figure 17.13). Paired epidermal setae are present on the trunk and opisthosoma. At the anterior end of the body, a cephalic lobe bears from 1 to 260 long tentacles (the "beard" that gives this phylum its name), depending on species. Tentacles are hollow extensions of the coelom and bear minute pinnules. For a part or all of their length, tentacles lie parallel with each other, enclosing a cylindrical intertentacular space into which the pinnules project (Figure 17.14).

Siboglinids are remarkable in having no mouth or digestive tract, making their mode of nutrition a puzzling matter. They absorb some nutrients dissolved in seawater, such as glucose, amino acids, and fatty acids, through the pinnules and microvilli of their tentacles. Most of their energy, however, apparently is derived from a mutualistic association with chemoautotrophic bacteria. These bacteria oxidize hydrogen sulfide to provide energy to produce organic compounds from carbon dioxide. Siboglinids bear the bacteria in an organ called a **trophosome,**

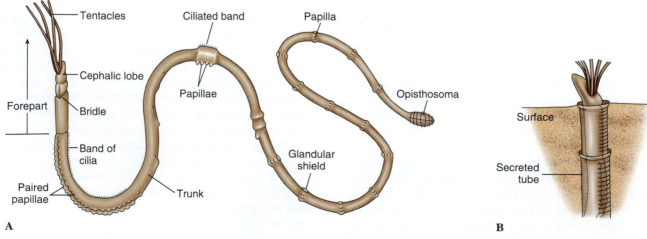

Figure 17.13

Diagram of a typical siboglinid. **A,** External features. In life, the body is much more elongated than shown in this diagram. **B,** Position in tube.

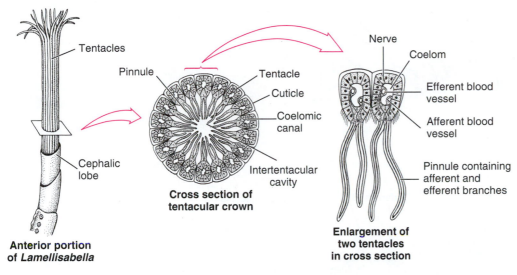

Cross section of tentacular crown

Enlargement of two tentacles in cross section

Anterior portion of *Lamellisabella*

Figure 17.14

Cross section of tentacular crown of sibioglinid *Lamellisabella*. Tentacles arise from ventral side of forepart at base of cephalic lobe. Tentacles (which vary in number in different species) enclose a cylindrical space, with the pinnules forming a kind of nutrient uptake network. Food molecules may be absorbed into the blood supply of tentacles and pinnules.

which is derived embryonically from the midgut (all traces of foregut and hindgut are absent in adults).

There is a well-developed closed circulatory system. Photoreceptors are similar to those of other annelids.

Among the most amazing animals found in deep-water, Pacific rift communities (see Chapter 38, p. 839) are vestimentiferans, *Riftia pachyptila*. These giant beardworms live around deep-water hydrothermal vents and grow up to 3 m long and 5 cm in diameter (Figure 17.15). The trophosome of other sibioglinids is confined to the posterior part of the trunk, which is buried in sulfide-rich sediments, but the trophosome of *Riftia* occupies most of its large trunk. It has a much larger supply of hydrogen sulfide, enough to nourish its large body, in the effluent of the hydrothermal vents.

Figure 17.15

A colony of giant beardworms (vestimentiferans, clade Siboglinidae) at great depth near a hydrothermal vent along the Galápagos Trench, eastern Pacific Ocean.

Sexes are separate, with a pair of gonads and a pair of gonoducts in the trunk section. Little developmental work has been done on these deep-sea worms, but research suggests that cleavage is unequal and atypical. It seems to be closer to spiral than to radial. Development of the apparent coelom is schizocoelic, not enterocoelic as was originally described. The worm-shaped embryo is ciliated but a poor swimmer. It is probably carried by water currents until it settles.

Clade Clitellata

Clade Clitellata contains earthworms, and their relatives, in class Oligochaeta and leeches in class Hirudinida. Members of this clade share a unique reproductive structure called a **clitellum.** The clitellum is a ring of secretory cells in the epidermis that appears on the worm's exterior as a fat band around the body about one-third of the body length from the anterior end. The clitellum is always visible in oligochaetes, but it appears only during the reproductive season in leeches. Members of Clitellata are derived annelids that lack parapodia, presumably an evolutionary loss from a polychaete ancestor. Clitellates are all hermaphroditic (monoecious) animals that exhibit direct development: Young develop inside a cocoon secreted by the clitellum, so no trochophore larva is visible. Small worms emerge from cocoons.

Class Oligochaeta

More than 3000 species of oligochaetes are found in a great variety of sizes and habitats. They include the familiar earthworms and many species that live in freshwater. Most are terrestrial or freshwater forms, but some are parasitic, and a few live in marine or brackish water.

With few exceptions, oligochaetes bear setae, which may be long or short, straight or curved, blunt or needlelike, or arranged singly or in bundles. Whatever the type, setae are less numerous in oligochaetes than in polychaetes, as is implied by the class name, which means "few long hairs." Aquatic forms usually have longer setae than do earthworms.

Form and Function The main features of an oligochaete body are described with reference to the familiar earthworm. The circulatory system and excretory structures described in earthworms are typical of annelids in general, but the digestive and nervous systems have aspects specific to oligochaetes.

Earthworms, sometimes called "night crawlers," burrow in moist rich soil, and usually live in branched, interconnected tunnels. The species commonly studied in laboratories is *Lumbricus terrestris* (L. *lubricum,* earthworm). It ranges in size from 12 to 30 cm long (Figure 17.16), but is small in comparison to giant tropical forms whose 4-m-long bodies may comprise 150 to upward of 250 segments.

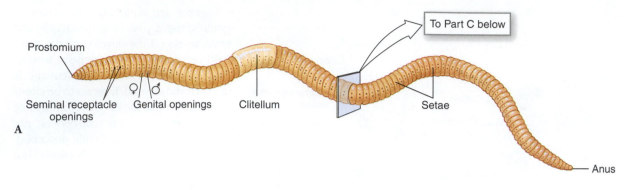

A

Prostomium

♀|♂

Seminal receptacle openings   Genital openings   Clitellum   Setae

To Part C below

Anus

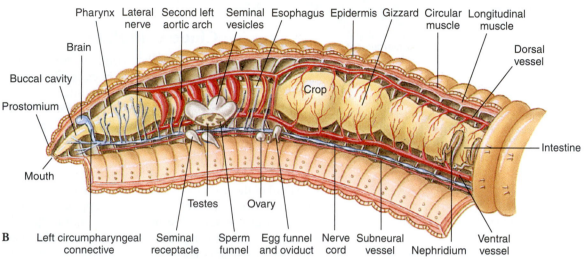

Pharynx   Lateral nerve   Second left aortic arch   Seminal vesicles   Esophagus   Epidermis   Gizzard   Circular muscle   Longitudinal muscle

Brain

Dorsal vessel

Buccal cavity

Prostomium

Crop

Mouth

Intestine

B   Left circumpharyngeal connective   Seminal receptacle   Sperm funnel   Egg funnel and oviduct   Nerve cord   Subneural vessel   Nephridium   Ventral vessel

Testes   Ovary

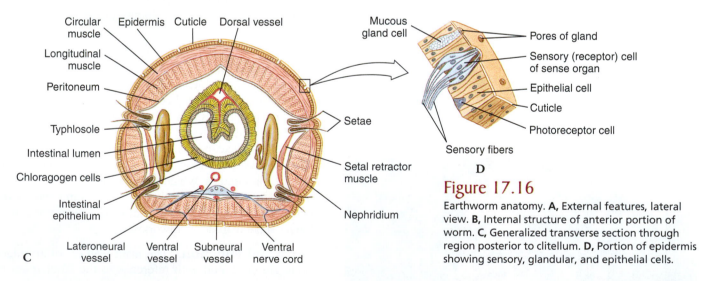

Circular muscle   Epidermis   Cuticle   Dorsal vessel

Longitudinal muscle

Peritoneum

Typhlosole

Intestinal lumen

Chloragogen cells

Intestinal epithelium

Lateroneural vessel   Ventral vessel   Subneural vessel   Ventral nerve cord

C

Setae

Setal retractor muscle

Nephridium

Mucous gland cell

Pores of gland

Sensory (receptor) cell of sense organ

Epithelial cell

Cuticle

Photoreceptor cell

Sensory fibers

D

Figure 17.16

Earthworm anatomy. **A,** External features, lateral view. **B,** Internal structure of anterior portion of worm. **C,** Generalized transverse section through region posterior to clitellum. **D,** Portion of epidermis showing sensory, glandular, and epithelial cells.

Earthworms normally emerge at night, but in damp rainy weather they stay near the surface, often with mouth or anus protruding from the burrow. In very dry weather they may burrow several feet underground, coil in a slime chamber and become dormant.

Earthworms use peristaltic movement: Contractions of circular muscles in the anterior end lengthen the body, pushing the anterior end forward where it anchors. Anchoring is accomplished by contraction of the longitudinal muscles in forward segments—these segments become short and wide, pushing against the sides of the burrow. As they do so, bristlelike rods called setae project outward through small pores in the cuticle. Setae dig into the walls of the burrow to anchor the forward segments; contractions of longitudinal muscles then shorten the rest of the body, pulling the posterior end up behind the anchored anterior region. As waves of extension and contraction pass along the entire body, it gradually moves forward.

The paired epidermal setae of oligochaetes are set in a sac within the body wall and moved by muscles (Figure 17.17), as

Figure 17.17

Seta with its muscle attachments showing relation to adjacent structures. Setae lost by wear are replaced by new ones, which develop from formative cells.

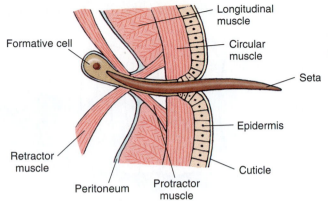

they are in polychaetes. However, oligochaetes do not have parapodia; instead the setae extend directly out of the body wall on each segment. In most earthworms each segment bears four pairs of chitinous setae (Figure 17.16C), although there may be more than 100 such setae per segment in some oligochaetes.

Aristotle called earthworms the "intestines of the soil." Some 22 centuries later Charles Darwin published his observations in his classic *The Formation of Vegetable Mould Through the Action of Worms.* He showed how worms enrich soil by bringing subsoil to the surface and mixing it with topsoil. An earthworm can ingest its own weight in soil every 24 hours, and Darwin estimated that from 10 to 18 tons of dry earth per acre pass through their intestine annually, thus bringing potassium and phosphorus from the subsoil and also adding nitrogenous products to the soil from their own metabolism. They also drag leaves, twigs, and organic substances into their burrows closer to the roots of plants. Their activities are vitally important in aerating soil. Darwin's views were at odds with his contemporaries, who thought earthworms were harmful to plants. But recent research has amply confirmed Darwin's findings, and earthworm management is now practiced in many countries.

Nutrition Most oligochaetes are scavengers. Earthworms feed mainly on decaying organic matter, bits of leaves and vegetation, refuse, and animal matter. After being moistened by secretions from the mouth, food is drawn inward by the sucking action of their muscular pharynx. The liplike prostomium aids in manipulating food into position. Calcium from soil swallowed with food tends to produce a high blood calcium level. **Calciferous glands** along the esophagus secrete calcium ions into the gut and so reduce the calcium ion concentration of their blood. Calciferous glands also function in regulating acid-base balance of body fluids.

Leaving the esophagus, food is stored temporarily in the thin-walled **crop** before being passed on to the **gizzard,** which grinds food into small pieces. Digestion and absorption occur in the **intestine.** The wall of the intestine is infolded dorsally to form a **typhlosole,** which greatly increases the absorptive and digestive surface (Figure 17.16C).

Surrounding the intestine and dorsal vessel and filling much of the typhlosole is a layer of yellowish **chloragogen tissue** (Gr. *chlōros,* green, + *agōgē,* a carrying away). This tissue serves

as a center for synthesis of glycogen and fat, a function roughly equivalent to that of liver cells. When full of fat, chloragogen cells are released into the coelom where they float freely as cells called **eleocytes** (Gr. *elaio,* oil, + *kytos,* hollow vessel [cell]), which transport materials to the body tissues. Eleocytes can pass from segment to segment and may accumulate around wounds and regenerating areas, where they break down and release their contents into the coelom. Chloragogen cells also function in excretion.

Circulation and Respiration Annelids have a double transport system: coelomic fluid and a closed circulatory system. Food, wastes, and respiratory gases are carried by both coelomic fluid and blood in varying degrees. Blood circulates in a closed system of vessels, which includes capillary systems in the tissues. Five main blood trunks run lengthwise through the body.

A single **dorsal vessel** runs above the alimentary canal from the pharynx to the anus. It is a pumping organ, provided with valves, and it functions as a true heart. This vessel receives blood from vessels of the body wall and digestive tract and pumps it anteriorly into five pairs of **aortic arches.** The function of aortic arches is to maintain a steady pressure of blood in the ventral vessel.

A single **ventral vessel** serves as an aorta. It receives blood from the aortic arches and delivers it to the brain and rest of the body, providing segmental vessels to the walls, nephridia, and digestive tract.

Their blood contains colorless ameboid cells and a dissolved respiratory pigment, **hemoglobin** (p. 704). The blood of some annelids may have respiratory pigments other than hemoglobin, as noted on page 367.

Earthworms have no special respiratory organs, but gaseous exchange occurs across their moist skin.

Excretion Each segment (except the first three and the last one) bears a pair of **metanephridia.** Each metanephridium occupies parts of two successive segments (Figure 17.18). A ciliated funnel, the **nephrostome,** lies just anterior to an intersegmental septum and leads by a small ciliated tubule through the septum into the segment behind, where it connects with the main part of the nephridium. Several complex loops of increasing size compose the nephridial duct, which terminates in a bladderlike structure leading to an opening, the **nephridiopore.** The nephridiopore opens to the outside near the ventral row of setae. By means of cilia, wastes from the coelom are drawn into the nephrostome and tubule, where they are joined by salts and organic wastes transported from blood capillaries in the glandular part of the nephridium. Waste is discharged to the outside through a nephridiopore.

Aquatic oligochaetes excrete ammonia; terrestrial oligochaetes usually excrete the much less toxic urea. *Lumbricus* produces both, the level of urea depending somewhat on environmental conditions. Both urea and ammonia are produced by chloragogen cells, which may break off and enter the metanephridia directly, or their products may be

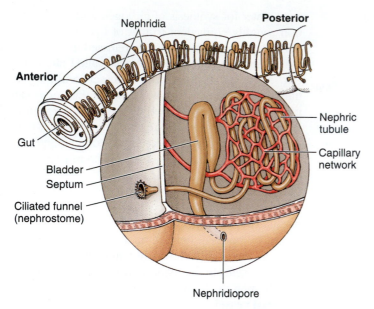

Figure 17.18

Nephridium of earthworm. Wastes are drawn into the ciliated nephrostome in one segment, then passed through the loops of the nephridium, and expelled through the nephridiopore of the next segment.

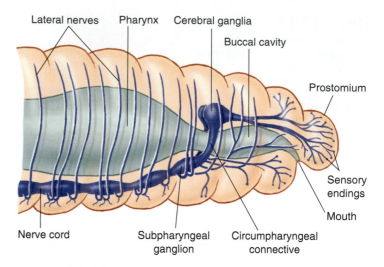

Figure 17.19

Anterior portion of earthworm and its nervous system. Note concentration of sensory endings in this region.

carried by the blood. Some nitrogenous waste is eliminated through the body surface.

Oligochaetes are largely freshwater animals, and even such terrestrial forms as earthworms must exist in a moist environment. Osmoregulation is a function of the body surface and the nephridia, as well as the gut and dorsal pores. *Lumbricus* will gain weight when placed in tap water and lose it when returned to soil. Salts as well as water can pass across the integument, salts apparently being actively transported.

Nervous System and Sense Organs

The nervous system in earthworms (Figure 17.19) consists of a central system and peripheral nerves. The central system reflects the typical annelid pattern: a pair of **cerebral ganglia** (the "brain") above the pharynx, a pair of **connectives** passing around the pharynx connecting the brain with the first pair of ganglia in the nerve cord; a solid **ventral nerve cord,** really double, running along the floor of the coelom to the last segment; and a pair of fused ganglia on the nerve cord in each segment. Each pair of fused ganglia provides nerves to the body structures, which contain both sensory and motor fibers.

Neurosecretory cells have been found in the brain and ganglia of both oligochaetes and polychaetes. They are endocrine in function and secrete neurohormones concerned with the regulation of reproduction, secondary sex characteristics, and regeneration.

For rapid escape movements most annelids have from one to several very large axons commonly called **giant axons** (Figure 17.20), or giant fibers, located in the ventral nerve cord. Their large diameter increases rate of conduction (see p. 730) and makes possible simultaneous contractions of muscles in many segments.

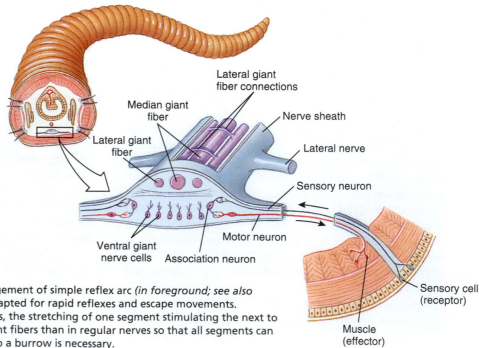

Figure 17.20

Portion of nerve cord of earthworm showing arrangement of simple reflex arc *(in foreground; see also p. 734)* and the three dorsal giant fibers that are adapted for rapid reflexes and escape movements. Ordinary crawling involves a succession of reflex acts, the stretching of one segment stimulating the next to stretch. Impulses are transmitted much faster in giant fibers than in regular nerves so that all segments can contract simultaneously when quick withdrawal into a burrow is necessary.

In the dorsal median giant fiber of *Lumbricus,* which is 90 to 160 μm in diameter, speed of conduction has been estimated at 20 to 45 m/second, several times faster than in ordinary neurons of this species. This is also much faster than in polychaete giant fibers, probably because in earthworms the giant fibers are enclosed in myelinated sheaths, which insulate them.

Simple sense organs are distributed all over the body. Earthworms have no eyes but do have many lens-shaped photoreceptors in their epidermis. Most oligochaetes are negatively phototactic to strong light but positively phototactic to weak light. Many single-celled sense organs are widely distributed in the epidermis. What are presumably chemoreceptors are most numerous on the prostomium. In the integument are many free nerve endings which are probably tactile in nature.

General Behavior   Earthworms are among the most defenseless of creatures, yet their abundance and wide distribution indicate their ability to thrive. Although they have no specialized sense organs, they are sensitive to many stimuli. They react positively to mechanical stimuli when such stimuli are moderate and negatively to a strong stimulus (such as footfall near them), which causes them to retire quickly into their burrows. They react to light, which they avoid unless it is very weak. Chemical responses aid them in the choice of food.

Chemical as well as tactile responses are very important to earthworms. They not only must sample the organic content of soil to find food, but also must sense its texture, acidity, and calcium content.

Experiments show that earthworms have some learning ability. They can be taught to avoid an electric shock, and thus can develop an association reflex. Darwin credited earthworms with a great deal of intelligence because they pulled leaves into their burrows by the narrow end, the easiest way for drawing a leaf-shaped object into a small hole. Darwin assumed that seizure of leaves by worms did not result from random handling or from chance but was deliberate. However, investigations since Darwin's time have shown that the process is mainly one of trial and error, for earthworms often seize a leaf several times before getting it right.

Reproduction and Development   Earthworms are monoecious (hermaphroditic); both male and female organs are found in the same animal (see Figure 17.16B). In *Lumbricus* reproductive systems are found in segments 9 to 15. Two pairs of small testes and two pairs of sperm funnels are surrounded by three pairs of large seminal vesicles. Immature sperm from the testes mature in seminal vesicles, then pass into sperm funnels and down sperm ducts to the male genital pores in segment 15, where they are expelled during copulation. Eggs are discharged by a pair of small ovaries into the coelomic cavity, where ciliated funnels of the oviducts carry them outside through female genital pores on segment 14. Two pairs of seminal receptacles in segments 9 and 10 receive and store sperm from the mate during copulation.

Reproduction in earthworms may occur throughout the year as long as warm, moist weather prevails at night (Figure 17.21).

When mating, worms extend their anterior ends from their burrows and bring their ventral surfaces together (Figure 17.21). Their surfaces are held together by mucus secreted by the **clitellum** (L. *clitellae,* packsaddle) and by special ventral setae, which penetrate each other's bodies in the regions of contact. After discharge, sperm travel to seminal receptacles of the other worm via its seminal grooves. After copulation each worm secretes first a mucous tube and then a tough, chitinlike band that forms a **cocoon** around its clitellum. As the cocoon passes forward, eggs from the oviducts, albumin from skin glands, and sperm from the mate (stored in the seminal receptacles) pour into it. Fertilization of eggs then occurs within the cocoon. When the cocoon slips past the anterior end of the worm, its ends close, producing a sealed, lemon-shaped body. Embryogenesis occurs within the cocoon, and the form that hatches from the egg is a young worm similar to the adult. Thus development is direct with no metamorphosis. Juveniles do not develop a clitellum until they are sexually mature.

Representative Oligochaetes   Freshwater oligochaetes usually are smaller and have more conspicuous setae than earthworms. They are more mobile than earthworms and tend to have better-developed sense organs. Most are benthic forms that crawl on the substrate or burrow in soft mud. Aquatic oligochaetes are an important food source for fishes. A few are ectoparasitic.

Some of the more common freshwater oligochaetes are the 1 mm long *Aeolosoma* (Gr. *aiolos,* quick-moving, + *soma,* body) (Figure 17.22B); the 10 to 25 mm long *Stylaria* (Gr. *stylos,* pillar) (Figure 17.22A); the 5 to 10 mm long *Dero* (Gr. *dere,* neck or throat) (Figure 17.22D). The common 30 to 40 mm long *Tubifex* (L. *tubus,* tube, + *faciens,* to make or do) (Figure 17.22C) is reddish and lives with its head in mud at the bottom of ponds and its tail waving in the water. *Tubifex* is an alternate host necessary in the life cycle of *Myxobolus cerebralis,* a parasite that causes a very serious condition called whirling disease in rainbow trout in North America. Some oligochaetes, such as *Aeolosoma,* may asexually form chains of zooids by transverse fission (Figure 17.22B).

Class Hirudinida: Leeches

Class Hirudinida is divided into three orders, Hirudinea, the "true" leeches, and two others that merit mention here because their members are morphological intermediates between oligochaetes and true leeches (see Figure 17.1). Oligochaetes have variable numbers of segments, segments bear setae, and there are no suckers on the body. True leeches have 34 segments, entirely lack setae, and possess anterior and posterior suckers. Members of order Acanthobdellida have 27 segments, bear setae on the first five segments, and have a posterior sucker. Members of order Branchiobdellida have 14 or 15 segments, no setae, and an anterior sucker. Branchiobdellids are commensal or parasitic on crayfish. Hereafter, leech refers to members of order Hirudinea.

Leeches occur predominantly in freshwater habitats, but a few are marine, and some have even adapted to terrestrial life in warm, moist places. They are more abundant in tropical countries than in temperate zones.

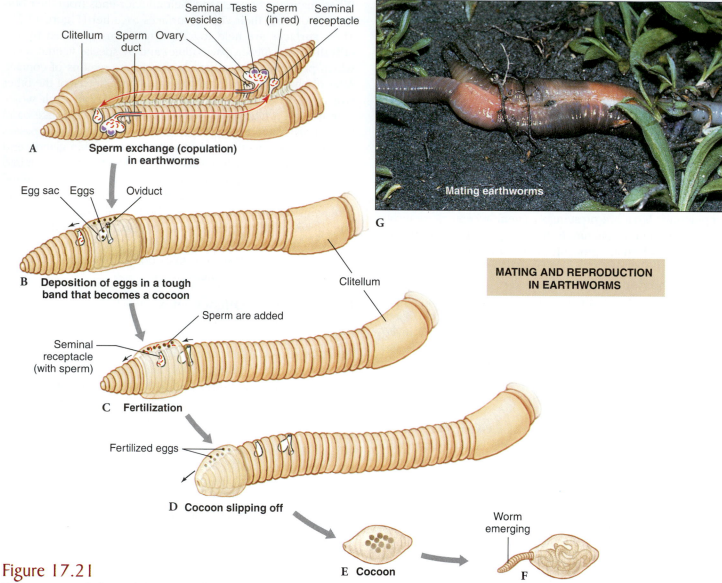

Figure 17.21

Earthworm copulation and formation of egg cocoons. **A,** Mutual insemination; sperm from genital pore (segment 15) pass along seminal grooves to seminal receptacles (segments 9 and 10) of each mate. **B** and **C,** After worms separate, the clitellum secretes first a mucous tube and then a tough band that forms a cocoon. The developing cocoon passes forward to receive eggs from oviducts and sperm from seminal receptacles. **D,** As cocoon slips off over anterior end, its ends close and seal. **E,** Cocoon is deposited near burrow entrance. **F,** Young worms emerge in 2 to 3 weeks. **G,** Two earthworms in copulation. Their anterior ends point in opposite directions as their ventral surfaces are held together by mucous bands secreted by the clitella.

Most leeches are between 2 and 6 cm in length, but some, including "medicinal" leeches, reach 20 cm. The giant of all is the Amazonian *Haementeria* (Gr. *haimateros,* bloody) (Figure 17.23), which reaches 30 cm.

Leeches are usually flattened dorsoventrally and exhibit a variety of patterns and colors: black, brown, red, or olive green. Many leeches live as carnivores on small invertebrates; some are temporary parasites; and some are permanent parasites, never leaving their host. Some leeches attack human beings and are a nuisance to outdoor enthusiasts.

Like oligochaetes, leeches are hermaphroditic and have a clitellum, which appears only during breeding season. The clitellum secretes a cocoon for reception of eggs.

Form and Function Unlike other annelids, leeches have a fixed number of segments but they appear to have many more because each segment is marked by transverse grooves to form superficial rings (Figure 17.24).

Unlike other annelids, leeches lack distinct coelomic compartments. In all but one species the septa have disappeared, and the coelomic cavity is filled with connective tissue and a system of spaces called **lacunae.** The coelomic lacunae form a regular system of channels filled with coelomic fluid, which in some leeches serves as an auxiliary circulatory system.

Leeches are more highly specialized than oligochaetes. They have lost the setae used by oligochaetes in locomotion and have developed suckers for attachment while sucking

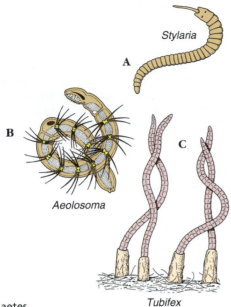

Figure 17.22

Some freshwater oligochaetes.
A, *Stylaria* has the prostomium
drawn out into a long snout. **B,**
Aeolosoma uses cilia around the
mouth to sweep in food particles,
and it buds off new individuals
asexually. **C,** *Tubifex* lives head
down in long tubes. **D,** *Dero* has
ciliated anal gills.

Figure 17.23

The world's largest leech, *Haementeria ghilianii,* on the arm of Dr. Roy
K. Sawyer, who found it in French Guiana, South America.

blood (their gut is specialized for storage of large quantities
of blood). Most leeches crawl with looping movements of the
body, by attaching first one sucker and then the other and pulling the body along the surface. Aquatic leeches swim with a
graceful undulatory movement.

Nutrition   Leeches are popularly considered parasitic, but
many are predaceous. Most freshwater leeches are active predators or scavengers equipped with a proboscis that can be
extended to ingest small invertebrates or to take blood from
cold-blooded vertebrates. Some can force their pharynx or
proboscis into soft tissues such as the gills of fish. Some terrestrial leeches feed on insect larvae, earthworms, and slugs,
which they hold by an oral sucker while using a strong sucking
pharynx to ingest food. Other terrestrial forms climb bushes
or trees to reach warm-blooded vertebrates such as birds or
mammals.

Most leeches are fluid feeders. Many prefer to feed on tissue fluids and blood pumped from open wounds. Some freshwater leeches are true bloodsuckers, preying on cattle, horses,
humans, and other mammals. True bloodsuckers, which include
the so-called medicinal leech, *Hirudo medicinalis* (L. *hirudo,* a
leech) (Figure 17.25), have cutting plates, or chitinous "jaws," for
cutting through tough skin. Some parasitic leeches leave their
hosts only during the breeding season, and certain fish parasites
are permanently parasitic, depositing their cocoons on their host
fish. However, even the true bloodsuckers rarely remain on the
host for a long period of time.

For centuries "medicinal leeches" *(Hirudo medicinalis)* were used
for bloodletting because of the mistaken idea that a host of bodily
disorders and fevers were caused by an excess of blood. A 10- to
12-cm-long leech can extend to a much greater length when distended with blood, and the amount of blood it can suck is considerable. Leech collecting and leech culture in ponds were practiced
in Europe on a commercial scale during the nineteenth century.
Wordsworth's poem "The Leech-Gatherer" was based on this use
of leeches.

Leeches are once again being used medically. When fingers,
toes, or ears are severed, microsurgeons can reconnect arteries but
not all the more delicate veins. Leeches are used to relieve conges-
tion until the veins can grow back into the healing appendage.

Respiration and Excretion   Gas exchange occurs only
through the skin except in some fish leeches, which have gills.
There are 10 to 17 pairs of nephridia, in addition to coelomocytes
and certain other specialized cells that also may be involved in
excretory functions.

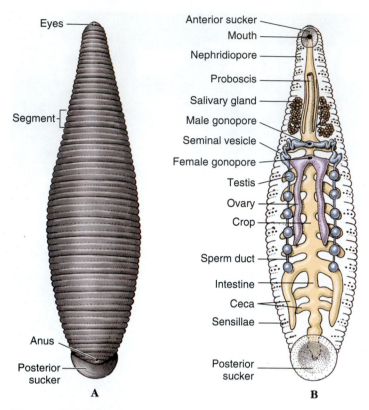

Figure 17.24
Structure of a leech, *Placobdella*. **A**, External appearance, dorsal view. **B**, Internal structure, ventral view.

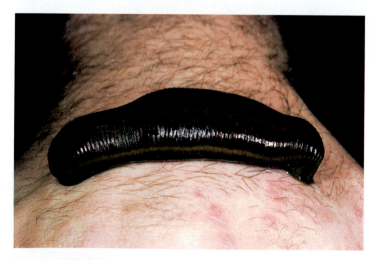

Figure 17.25
Hirudo medicinalis feeding on blood from human arm.

Nervous and Sensory Systems   Leeches have two "brains": one is anterior and composed of six pairs of fused ganglia (forming a ring around the pharynx), the other is posterior and composed of seven pairs of fused ganglia. An additional 21 pairs of segmental ganglia occur along the double nerve cord. In addition to free sensory nerve endings and photoreceptor cells in the epidermis, there is a row of sense organs, called **sensillae,** in the central annulus of each segment. Pigment-cup ocelli also are present in many species.

Leeches are highly sensitive to stimuli associated with the presence of a prey or host. They are attracted by and will attempt to attach to an object smeared with appropriate host substances, such as fish scales, oil secretions, or sweat. Those that feed on the blood of mammals are attracted by warmth; terrestrial haemadipsids of the tropics will converge on a person standing in one place.

Reproduction   Leeches are hermaphroditic but cross-fertilize during copulation. Sperm are transferred by a penis or by hypodermic impregnation (a spermatophore is expelled from one worm and penetrates the integument of the other). After copulation their clitellum secretes a cocoon that receives eggs and sperm. Leeches may bury their cocoons in mud, attach them to submerged objects, or, in terrestrial species, place them in damp soil. Development is similar to that of oligochaetes.

Circulation   The coelom of leeches has been reduced by the invasion of connective tissue and, in some, by a proliferation of chloragogen tissue, to a system of coelomic sinuses and channels. Some orders of leeches retain a typical oligochaete circulatory system, and in these the coelomic sinuses act as an auxiliary blood-vascular system. In other orders the traditional blood vessels are lacking and the system of coelomic sinuses forms the only blood-vascular system. In those orders contractions of certain longitudinal channels provide propulsion for the blood (the equivalent of coelomic fluid).

PHYLUM ECHIURA

Phylum Echiura (ek-ee-yur'a) (Gr. *echis*, viper, serpent, +*oura* tail, + *ida*, pl. suffix) consists of about 140 species of marine worms that burrow into mud or sand, live in empty snail shells or sand-dollar tests, or rocky crevices. They are found in all oceans—most commonly in littoral zones of warm waters—but some are found in polar waters or dredged from depths of up to 10,000 m. They vary in length from a few millimeters to 40 or 50 cm.

Echiurans are cylindrical and somewhat sausage-shaped (Figure 17.26). Anterior to the mouth is a flattened, extensible proboscis, which cannot be retracted into the trunk. Echiurans are often called "spoon worms" because of the shape of the contracted proboscis in some species. The nervous system of echiurans is fairly simple with a ventral nerve cord that runs the length of the trunk and continues dorsally into the proboscis. The proboscis has a ciliated groove leading to the mouth. While they lie buried, the proboscis can extend out over the mud for exploration and deposit-feeding (Figure 17.27). Most species gather very small particles of detritus and move them along the proboscis by cilia; larger particles are moved by a combination of cilia and muscular action or by muscular action alone. Unwanted particles can be rejected along the route to the mouth. The proboscis is

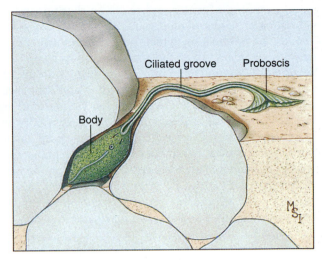

Figure 17.27

Bonellia (phylum Echiura) is a detritus feeder. Lying in its burrow, it explores the surface with its long proboscis, which picks up organic particles and carries them along a ciliated groove to the mouth.

short in some forms and long in others. *Bonellia,* which is only 8 cm long, can extend its proboscis up to 2 m.

One common form, *Urechis* (Gr. *oura,* tail, + *echis,* viper, serpent), has a very short proboscis and lives in a U-shaped burrow in which it secretes a funnel-shaped mucous net. It pumps water through the net, capturing bacteria and fine particulate material in it. *Urechis* periodically swallows the food-laden net. *Lissomyema* (Gr. *lissos,* smooth, + *mys,* muscle) lives in empty gastropod shells in which it constructs galleries irrigated by

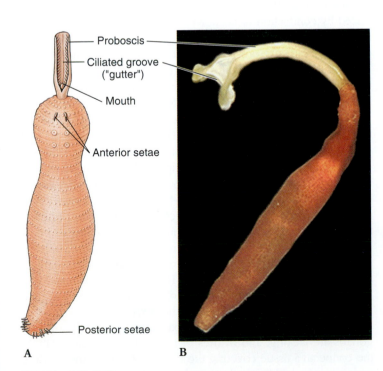

Figure 17.26

A, *Echiurus,* an echiuran common on both Atlantic and Pacific coasts of North America. **B,** *Anelassorhynchus,* an echiuran of the tropical Pacific. The shape of their proboscis lends them the common name of "spoon worms."

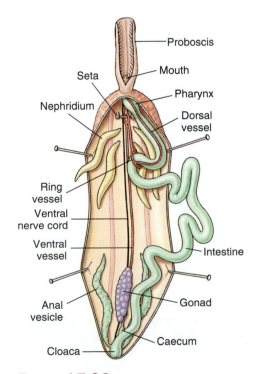

Figure 17.28

Internal anatomy of an echiuran.

rhythmical pumping of water and feeds on detritus and the organic coating of sand and mud gathered by this process.

Cuticle and epithelium, which may be smooth or ornamented with papillae, cover the muscular body wall. There may be a pair of anterior setae or a row of bristles around the posterior end (see Figure 17.26). The coelom is large. The digestive tract is long and coiled and terminates at the posterior end (Figure 17.28). A pair of anal sacs may have an excretory and osmoregulatory function. Most echiurans have a closed circulatory system with colorless blood but contain hemoglobin in coelomic corpuscles and certain body cells. There are one to many pairs of nephridia, which serve mainly as gonoducts in some species. Gas exchange probably occurs primarily in the hindgut, which is continually filled and emptied by cloacal irrigation.

In some species sexual dimorphism is pronounced, with the female being much the larger of the two. *Bonellia* has an extreme sexual dimorphism, and tiny males live on the body of the female or in her nephridia. Determination of sex in *Bonellia* is very interesting. Free-swimming larvae are sexually undifferentiated. Those that settle on the proboscis of a female become males (1 to 3 mm long). About 20 males are usually found in a single female. Larvae that do not contact a female proboscis metamorphose into females. The stimulus for development into males is apparently a hormone produced by a female's proboscis.

Sexes are separate, with gonads being produced by special regions of the peritoneum in each sex. Mature sex cells break loose from these gonadal regions and leave the body cavity by way of the nephridia. Fertilization is usually external.

Early cleavage and trochophore stages are very similar to those of annelids and sipunculans. The trochophore stage, which may last from a few days to 3 months, according to species, is followed by gradual metamorphosis to a wormlike adult.

PHYLUM SIPUNCULA

Phylum Sipuncula (sigh-pun′kyu-la) (L. *sipunculus,* little siphon) consists of about 250 species of benthic marine worms, at depths ranging from the intertidal to over 5000 m. They live sedentary lives in burrows in mud or sand, occupy borrowed snail shells, or live in coral crevices or among vegetation. Some species construct their own rock burrows by chemical and perhaps mechanical means. More than half of the species are restricted to tropical zones. Some are tiny, slender worms, but the majority range from 3 to 10 cm in length. Some are commonly known as "peanut worms" because, when disturbed, they can contract to a peanut shape (Figure 17.29).

Sipunculans have no segmentation or setae. They are most easily recognized by a slender retractile **introvert,** or **proboscis,** which is continually and rapidly being run in and out of the anterior end. Walls of the **trunk** are muscular. When the introvert is everted, the mouth can be seen at its tip surrounded by a crown of ciliated tentacles. Little is known about the details of sipunculan feeding. Some species appear to be deposit feeders

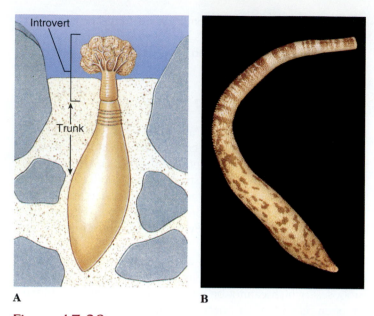

A   **B**

Figure 17.29

Sipunculans. *Themiste* **(A)** and *Phascolosoma* **(B)** are both burrowing genera of cosmopolitan distribution.

or detritivores, whereas others appear to be suspension feeders. Some nutrition may also come from dissolved organic compounds directly from the water column. Undisturbed sipunculans usually extend the anterior end from their burrow or hiding place and stretch out their tentacles to explore and to feed. Organic matter collected in mucus on the tentacles is moved to the mouth by ciliary action. The introvert is extended by hydrostatic pressure produced by contraction of body-wall muscles against the coelomic fluid. The lumen of the hollow tentacles is not connected to the coelom but rather to one or two blind, tubular compensation sacs that lie along their esophagus (Figure 17.30). These sacs receive fluid from the tentacles when the introvert is retracted. Retraction is effected by special retractor muscles. The surface of the introvert is often rough because of surface spines, hooks, or papillae.

There is a large, fluid-filled coelom traversed by muscle and connective-tissue fibers. Their digestive tract is a long tube that doubles back on itself to form a U-shape and ends in an anus near the base of the introvert (Figure 17.30). A pair of large nephridia opens to the outside to expel waste-filled coelomic amebocytes; the nephridia also serve as gonoducts. Circulatory and respiratory systems are lacking, but coelomic fluid contains red corpuscles that have a respiratory pigment, hemerythrin, used in transportation of oxygen. Gas exchange appears to occur largely across the tentacles and introvert. Their nervous system has a bilobed cerebral ganglion just behind the tentacles and a ventral nerve cord extending the length of the body.

With only a few exceptions, sexes are separate. Permanent gonads are lacking, and ovaries or testes develop seasonally in the connective tissue covering the origins of one or more of the retractor muscles. Sex cells are released through the nephridia. The larval form is usually a trochophore. Asexual reproduction also occurs by transverse fission, the posterior one-fifth of the parent constricting off to become a new individual in some species.

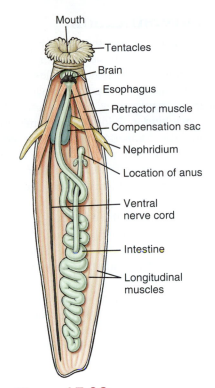

Mouth
Tentacles
Brain
Esophagus
Retractor muscle
Compensation sac
Nephridium
Location of anus
Ventral nerve cord
Intestine
Longitudinal muscles

Figure 17.30
Internal structure of *Sipunculus*.

EVOLUTIONARY SIGNIFICANCE OF METAMERISM

No truly satisfactory explanation has yet been given for the origins of segmentation and the coelom, although the subject has stimulated much speculation and debate. All classical explanations have had important arguments leveled against them, and more than one may be correct, or none, as suggested by R. B. Clark.[1] Clark stressed the functional and evolutionary significance of these features to the earliest animals that possessed them. He argued forcefully that the adaptive value of a coelom was as a **hydrostatic skeleton** in a burrowing animal. Thus contraction of muscles in one part of the animal could act antagonistically on muscles in another part by transmission of the force of contraction through the enclosed constant volume of fluid in the coelom.

Although the original function of the coelom may have served to facilitate burrowing in the substrate, certain other advantages accrued to its possessors. For example, coelomic fluid would have acted as a circulatory fluid for nutrients and wastes, making large numbers of flame cells distributed throughout the tissues unnecessary. Gametes could be stored in the spacious coelom for simultaneous release by all individuals in the population (thus enhancing chances of fertilization). Such a synchronous release of gametes would have selected for greater

nervous and endocrine control. The coelom may have evolved in response to different selective pressures in protostomes and deuterostomes.

The origins of a metameric (segmented) body are at least as puzzling as the origins of the coelom. True metamerism occurs in annelids, arthropods, and chordates. The placement of the annelids and arthropods in Protostomia and of the chordates in Deuterostomia makes it unlikely that segmentation is homologous among these three taxa. Within the protostomes, annelids are placed in clade Lophotrochozoa, whereas arthropods are in clade Ecdysozoa. In both clades most phyla are not segmented, again making it unlikely that members of these two phyla inherited a segmented body plan from a common ancestor. Annelids and molluscs have very similar developmental programs leading to a trochophore larva, but the annelid trochophore develops a series of segments as it grows, whereas the mollusc trochophore does not grow in this way (see discussion p. 336).

It is possible that all bilaterally symmetrical metazoans shared a segmented ancestor and that segmentation genes were suppressed in most lineages, but preliminary studies of the details of how segments form (genetic control and chemical signaling) in different phyla do not support this hypothesis.[2] Instead, current evidence supports the hypothesis that segmentation arose independently multiple times.

The selective advantage of a segmented body for annelids appears to lie in the efficiency of burrowing made possible by shape change in individual coelomic compartments of the hydrostatic skeleton. However, this explanation cannot be extended to the arthropods because, as Chapters 19, 20, and 21 describe, the rigid exoskeleton of the arthropods prohibits shape change among segments, and the coelom is small in comparison to that of annelids. Clearly, there is much to learn about metamerism.

PHYLOGENY AND ADAPTIVE DIVERSIFICATION

Phylogeny

Annelids and molluscs share many developmental features, so they were presumed by many biologists to be very closely related, perhaps sister taxa. However, the shared developmental features are likely to be retained ancestral features for lophotrochozoan protostomes.

Pogonophoran and vestimentiferan worms were once placed outside phylum Annelida, but they have been reinterpreted as derived members of class Polychaeta and are now placed in clade Siboglinidae within this class. Only a small portion of the siboglinid body is segmented.

Two other groups of worms, sipunculids and echiurans, are closely related to annelids according to phylogenies using molecular characters. Some of these phylogenies place

---

[1]Clark, R. B. 1964. Dynamics in metazoan evolution. The origin of the coelom and segments. Oxford, U.K., Clarendon Press.

[2]Seaver, E. C. 2003. Int. J. Dev. Biol. 47:583–595.

echiurans within Annelida, as derived polychaetes that have lost segmentation. There are serially repeated structures, such as nerve-cord ganglia and mucous glands in echiuran larvae, and serially repeated nephridia in echiuran adults. Some biologists interpret these repeated structures as remnants from a segmented ancestor and place echiurans within Annelida. The presence of paired epidermal setae in some echiuran species provides strong support for placing echiurans within Annelida. One recently developed phylogenetic tree placed echiurans near capitellid polychaetes; both taxa dwell in sediments. If this result is supported by more studies, Echiura, like Pogonophora, may no longer be a valid phylum.

The placement of Sipuncula is more contentious than that of Echiura. Sipunculans are not metameric and they do not have setae. Larval development is similar to that of annelids, molluscs, and echiurans. Further study, especially where molecular characters are used, may clarify the position of these worms within Lophotrochozoa. For the present, we depict them as the sister taxon to a clade of annelids and echiurans.

Within phylum Annelida, class Polychaeta is a paraphyletic group because we have evidence that ancestral clitellates arose from within Polychaeta.

Adaptive Diversification

Annelids are an ancient group that has undergone extensive adaptive diversification. The basic body structure, particularly of polychaetes, lends itself to almost endless modification. As marine worms, polychaetes have a wide range of habitats.

A basic adaptive feature in evolution of annelids is their septal arrangement, resulting in fluid-filled coelomic compartments. Fluid pressure in these compartments is used to create a hydrostatic skeleton, which in turn permits precise movements such as burrowing and swimming. Powerful circular and longitudinal muscles can flex, shorten, and lengthen the body.

Feeding adaptations show great variation, from the sucking pharynx of oligochaetes and the chitinous jaws of carnivorous polychaetes to the specialized tentacles and radioles of particle feeders. The evolution of a trophosome to house the chemoautotrophic bacteria that provide nutrients to siboglinids is an adaptation to deep-sea life.

In polychaetes the parapodia have been adapted in many ways and for a variety of functions, chiefly locomotion and respiration.

In leeches many adaptations (such as suckers, cutting jaws, pumping pharynx, and distensible gut) relate to their predatory and bloodsucking habits.

SUMMARY

Phylum Annelida is a large, cosmopolitan group containing marine polychaetes, earthworms and freshwater oligochaetes, and leeches. Certainly the most important structural innovation underlying diversification of this group is metamerism (segmentation), a division of the body into a series of similar segments, each of which contains a repeated arrangement of many organs and systems. The coelom also is highly developed in annelids, and this, together with the septal arrangement of fluid-filled compartments and a well-developed body-wall musculature, is an effective hydrostatic skeleton for precise burrowing and swimming movements. Further segmented specialization occurs in arthropods, the subjects of Chapters 19, 20, and 21.

Polychaetes, the largest class of annelids, are mostly marine. On each segment they have many setae, which are borne on paired parapodia. Parapodia show a wide variety of adaptations among polychaetes, including specialization for swimming, respiration, crawling, maintaining position in a burrow, pumping water through a burrow, and accessory feeding. Some polychaetes are mostly predaceous and have an eversible pharynx with jaws. Other polychaetes rarely leave the burrows or tubes in which they live. Several types of deposit- and filter-feeding are known among members of this group. Polychaetes are dioecious, have a reproductive system lacking a clitellum, external fertilization, and a trochophore larva.

Siboglinids live in tubes on the deep-ocean floor, and they are metameric. They have no mouth or digestive tract but apparently absorb some nutrient by the crown of tentacles at their anterior end. Much of their energy is due to chemoautotrophy of bacteria in their trophosome.

Clade Clitellata encompasses class Oligochaeta and class Hirudinida. Class Oligochaeta contains earthworms and many freshwater forms; they have a small number of setae per segment (compared with Polychaeta) and no parapodia. They have a closed circulatory system, and the dorsal blood vessel is the main pumping organ. Paired nephridia occur in most segments. Earthworms contain the typical annelid nervous system: dorsal cerebral ganglia connected to a double, ventral nerve cord with segmental ganglia running the length of the worm. Oligochaetes are hermaphroditic and practice cross-fertilization. The clitellum plays an important role in reproduction, including secretion of mucus to surround the worms during copulation and secretion of a cocoon to receive eggs and sperm and in which embryonation occurs. A small, juvenile worm hatches from the cocoon.

Leeches (class Hirudinida) are mostly freshwater, although a few are marine and a few are terrestrial. They feed mostly on fluids; many are predators, some are temporary parasites, and a few are permanent parasites. The hermaphroditic leeches reproduce in a fashion similar to that of oligochaetes, with cross-fertilization and cocoon formation by the clitellum.

Echiurans are burrowing marine worms, and most are deposit feeders, with a proboscis anterior to their mouth. Some species bear epidermal setae. They lack segmentation. The validity of this group as a phylum is a subject of debate.

Sipunculans are small, burrowing marine worms with an eversible introvert at their anterior end. The introvert bears tentacles used for deposit feeding. Sipunculans are not segmented.

Embryological evidence places annelids with molluscs and arthropods in the Protostomia. Recent molecular evidence suggests that annelids and molluscs are more closely related to each other (in Lophotrochozoa) than either phylum is to arthropods (in Ecdysozoa). Echiurans are closely related to annelids and may have arisin within this phylum. Sipunculans are also allied to annelids, but also share certain features with molluscs.

REVIEW QUESTIONS

1. What characteristics of phylum Annelida distinguish it from other phyla?
2. How are members of clade Clitellata distinguished from polychetes?
3. Describe the annelid body plan, including body wall, segments, coelom and its compartments, and coelomic lining.
4. Explain how the hydrostatic skeleton of annelids helps them to burrow. How is the efficiency for burrowing increased by segmentation?
5. Describe three ways that various polychaetes obtain food.
6. Define each of the following: prostomium, peristomium, pygidium, radioles, parapodium.
7. Explain functions of each of the following in earthworms: pharynx, calciferous glands, crop, gizzard, typhlosole, chloragogen tissue.
8. Compare the main features of each of the following in each class of annelids: circulatory system, nervous system, excretory system.
9. Describe the function of the clitellum and cocoon.
10. How are freshwater oligochaetes generally different from earthworms?
11. Describe the ways in which leeches obtain food.
12. What is the largest siboglinid known, and how is it nourished?
13. What features are shared between annelids and echiurans?
14. Where does a sipunculan live, and how does it collect food?
15. What was the evolutionary significance of segmentation and the coelom to its earliest possessors?

SELECTED REFERENCES

Childress, J. J., H. Felbeck, and G. N. Somero. 1987. Symbiosis in the deep sea. Sci. Am. **256:**114–120 (May). *The amazing story of how the animals around deep-sea vents, including* Riftia pachyptila, *absorb hydrogen sulfide and transport it to their mutualistic bacteria. For most animals, hydrogen sulfide is highly toxic.*
Cutler, E. B. 1995. The Sipuncula. Their systematics, biology, and evolution. Ithaca, New York, Cornell University Press. *The author tried to "bring together everything known about" sipunculans.*
Davis, G. K., and N. H. Patel. 2000. The origin and evolution of segmentation. Trends Genet. **15:**M68–M72. *Discussion of segmentation with a focus on arthropods.*
Fischer, A., and U. Fischer. 1995. On the life-style and life-cycle of the luminescent polychaete *Odontosyllis enopla* (Annelida: Polychaeta). Invert. Biol. **114:**236–247. *If epitokes of this species survive their spawning swarm, they can return to a benthic existence.*
Halanych, K. M., T. D. Dahlgren, and D. McHugh. 2002. Unsegmented annelids? Possible origins of four lophotrochozoan worm taxa. Integ. and Comp. Biol. **42:**678–684. *A nice summary of current morphological and molecular studies on classification of pogonophorans, echiurids, myzostomids, and sipunculans.*
Lent, C. M., and M. H. Dickinson. 1988. The neurobiology of feeding in leeches. Sci. Am. **258:**98–103 (June). *Feeding behavior in leeches is controlled by a single neurotransmitter (serotonin).*
McClintock, J. 2001. Blood suckers. Discover **22:**56–61 (Dec.). *Describes modern medical uses for leeches.*
McHugh, D. 2000. Molecular phylogeny of Annelida. Can. J. Zool. **78:**1873–1884. *Descriptions of monophyletic groups within Annelida supported by molecular data.*
Menon, J., and A. J. Arp. 1998. Ultrastructural evidence of detoxification in the alimentary canal of *Urechis caupo.* Invert. Biol. **117:**307–317. *This curious echiuran has detoxification bodies in its gut cells and epithelial cells that allow it to live in a highly toxic sulfide environment.*
Mirsky, S. 2000. When good hippos go bad. Sci. Am. **282:**28 (Jan.). Placobdelloides jaegerskioeldi *is a parasitic leech that breeds only in the rectum of hippopotomuses.*
Patel, N. H. 2003. The ancestry of segmentation. Dev. Cell **5:**2–4. *Explores the idea that segmentation is an ancestral feature of all bilaterally symmetrical animals.*
Pernet, B. 2000. A scaleworm's setal snorkel. Invert. Biol. **119:**147–151. Sthenelais berkeleyi *is an apparently rare but large (20 cm) polychaete that buries its body in sediment and communicates with water above just by its anterior end. Ciliary movement on parapodia pumps water into the burrow for ventilation. The worm remains immobile for long periods, except when prey comes near; it then rapidly everts its pharynx to capture prey.*
Rouse, G. W. 2001. A cladistic analysis of Siboglinidae Caullery, 1914 (Polychaeta: Annelida): Formerly the phyla Pogonophora and Vestimentifera. Zool. J. Linn. Soc. **132:**55–80. *Diagnostic features of Siboglinidae and its subgroups are provided.*
Seaver, E. C. 2003. Segmentation: mono- or polyphyletic. Int. J. Dev. Biol. **47:**583–595. *Preliminary comparisons of the segmentation process in annelids, arthropods, and chordates suggest that annelids and arthropods do not share mechanisms of segmentation, but vertebrates and arthropods may share some mechanisms.*
Winnepenninckx, B. M. H., Y. Van de Peer, and T. Backeljau. 1998. Metazoan relationships on the basis of 18S rRNA sequences: A few years later . . . Am. Zool. **38:**888–906. *Their calculations and analysis support monophyly of Clitellata but cast doubt on monophyly of Polychaeta.*

ONLINE LEARNING CENTER

Visit www.mhhe.com/hickmanipz14e for chapter quizzing, key term flash cards, web links, and more!

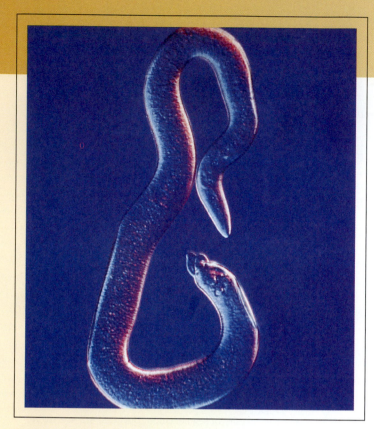

Male Trichinella spiralis, *a nematode.*

Smaller Ecdysozoans

- PHYLUM NEMATODA
- PHYLUM NEMATOMORPHA
- PHYLUM KINORHYNCHA
- PHYLUM PRIAPULIDA
- PHYLUM LORICIFERA
- PHYLUM ONYCHOPHORA
- PHYLUM TARDIGRADA

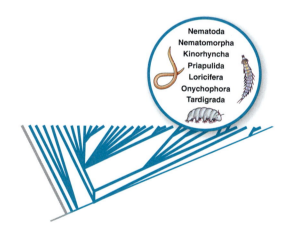

Nematoda
Nematomorpha
Kinorhyncha
Priapulida
Loricifera
Onychophora
Tardigrada

A World of Nematodes

Without any doubt, nematodes are the most important pseudocoelomate animals, in terms of both numbers and their impact on humans. Nematodes are abundant over most of the world, yet most people are only occasionally aware of them as parasites of humans or of their pets. We are not aware of the millions of these worms in the soil, in ocean and freshwater habitats, in plants, and in all kinds of animals. Their dramatic abundance moved N. A. Cobb[1] to write in 1914:

> If all the matter in the universe except the nematodes were swept away, our world would still be dimly recognizable, and if, as disembodied spirits, we could then investigate it,

we should find its mountains, hills, vales, rivers, lakes, and oceans represented by a thin film of nematodes. The location of towns would be decipherable, since for every massing of human beings there would be a corresponding massing of certain nematodes. Trees would still stand in ghostly rows representing our streets and highways. The location of the various plants and animals would still be decipherable, and, had we sufficient knowledge, in many cases even their species could be determined by an examination of their erstwhile nematode parasites.

[1]From N. A. Cobb. 1914. Yearbook of the United States Department of Agriculture, p. 472.

P rotostome animals include flatworms, roundworms, molluscs, annelids, and arthropods, among many other taxa (see cladogram on inside front cover). Many protostomes, such as annelids, roundworms, and arthropods, possess a **cuticle,** a nonliving external layer secreted by the epidermis. A firm cuticle surrounding the body wall, like that present in roundworms and arthropods, restricts growth. In such animals, the cuticle is molted, and the outer layer shed via **ecdysis,** as the body increases in size.

Protostome phyla are divided between two large clades: Lophotrochozoa and Ecdysozoa. Ecdysozoa (Figure 18.1) comprises those taxa that molt the cuticle as they grow. Where it has been studied, molting is regulated by the hormone **ecdysone;** biologists assume that a homologous set of biochemical steps regulates molting among all ecdysozoans. Ecdysozoan taxa, other than loriciferans, were first united as a clade in phylogenies based on molecular characters.

As was the case with lophotrochozoan phyla, ecdysozoans do not share a common body plan. Members of Nematoda, Nematomorpha, and Kinorhyncha have pseudocoelomate bodies. Members of Priapulida have not been carefully studied, but are assumed to be pseudocoelomate. The pseudocoelom is used as a hydrostatic skeleton in nematodes, kinorhynchs, and priapulids. Within Loricifera, species apparently vary in body plan: some are described as pseudocoelomate and others appear acoelomate. Members of clade Panarthropoda have coelomate bodies, but their coeloms are quite reduced in size as compared with those of annelids. Panarthropoda is an enormous group of animals, containing three phyla: Onychophora, Tardigrada, and Arthropoda.

Arthropoda is the largest phylum in terms of numbers of described species and forms the subject of Chapters 19, 20, and 21. This chapter describes all other ecdysozoan phyla.

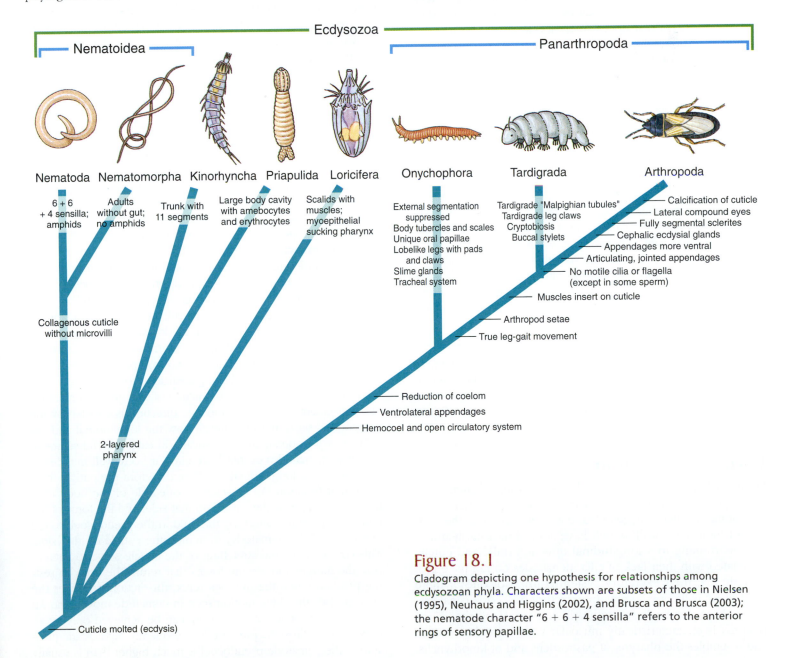

Figure 18.1

Cladogram depicting one hypothesis for relationships among ecdysozoan phyla. Characters shown are subsets of those in Nielsen (1995), Neuhaus and Higgins (2002), and Brusca and Brusca (2003); the nematode character "6 + 6 + 4 sensilla" refers to the anterior rings of sensory papillae.

PHYLUM NEMATODA: ROUNDWORMS

Approximately 25,000 species of Nematoda (nem-a-to´da) (Gr., *nematos*, thread) have been named, but many authorities now prefer Nemata for the name of this phylum. It has been estimated that if all species were known, the number might be nearer 500,000. They live in the sea, in freshwater, and in soil, from polar regions to the tropics, and from mountaintops to the depths of the sea. Good topsoil may contain billions of nematodes per acre. Nematodes also parasitize virtually every type of animal and many plants. Effects of nematode infestation on crops, domestic animals, and humans make this phylum one of the most important of all parasitic animal groups.

Free-living nematodes feed on bacteria, yeasts, fungal hyphae, and algae. They may be saprozoic or coprozoic (live in fecal material). Predatory species may eat rotifers, tardigrades, small annelids, and other nematodes. Many species feed on plant juices from higher plants, which they penetrate, sometimes causing agricultural damage of great proportions. Nematodes themselves may be prey for mites, insect larvae, and even nematode-capturing fungi. *Caenorhabditis elegans,* a free-living nematode, is easy to culture in the laboratory and has become an invaluable model for studies of developmental biology.

In 1963 Sydney Brenner started studying a free-living nematode, *Caenorhabditis elegans,* the beginning of some extremely fruitful research. Now this small worm has become one of the most important experimental models in biology. The origin and lineage of all the cells in its body (959) have been traced from zygote to adult, and the complete "wiring diagram" of its nervous system is known—all neurons and all connections between them. Its genome has been completely mapped, and scientists have sequenced its entire genome of 3 million bases comprising 19,820 genes. Many basic discoveries of gene function, such as genes encoding proteins essential for programmed cell death, have been made and will be made using *C. elegans.*

Virtually every species of vertebrate and many invertebrates serve as hosts for one or more types of parasitic nematodes. Nematode parasites in humans cause much discomfort, disease, and death, and in domestic animals they are a source of great economic loss.

Form and Function

Distinguishing characteristics of this large group of animals are their cylindrical shape; their flexible, nonliving cuticle; their lack of motile cilia or flagella (except in one species); the muscles of their body wall, which have several unusual features, such as running in a longitudinal direction only, and eutely. Correlated with their lack of cilia, nematodes do not have protonephridia; their excretory system consists of one or more large gland cells opening by an excretory pore, or a canal system without gland cells, or both cells and canals together. Their pharynx is characteristically muscular with a triradiate lumen and resembles the pharynx of gastrotrichs and of kinorhynchs.

Most nematode worms are less than 5 cm long, and many are microscopic, but some parasitic nematodes are more than 1 m in length.

Their outer body covering is a relatively thick, noncellular **cuticle** secreted by the underlying epidermis **(hypodermis).** This cuticle is shed during juvenile growth stages, which is one of the characters that places nematodes in the Ecdysozoa. The hypodermis is syncytial, and its nuclei are located in four **hypodermal cords** that project inward (Figure 18.2). Dorsal and ventral hypodermal cords bear longitudinal dorsal and ventral nerves, and the lateral cords bear excretory canals. The cuticle is of great functional importance to the worm, serving to contain the high **hydrostatic pressure** (turgor) exerted by fluid in the pseudocoel and protecting the worm from hostile environments such as dry soils or the digestive tracts of their hosts. The several layers of the cuticle are primarily of **collagen,** a structural protein also abundant in vertebrate connective tissue. Three of the layers are composed of crisscrossing fibers, which confer some longitudinal elasticity on the worm but severely limit its capacity for lateral expansion.

Body-wall muscles of nematodes are very unusual. They lie beneath the hypodermis (epidermal syncytium) and contract longitudinally only. There are no circular muscles in the body wall. The muscles are arranged in four bands, or quadrants, separated by the four hypodermal cords (Figure 18.2). Each muscle cell has a contractile **fibrillar** portion (or **spindle**) and a noncontractile **sarcoplasmic** portion (cell body). The spindle is distal and abuts the hypodermis, and the cell body projects into the pseudocoel. The spindle is striated with bands of actin and myosin, reminiscent of vertebrate skeletal muscle (see Figure 9.11, p. 196, and p. 656). The cell bodies contain the nuclei and are a major depot for glycogen storage in the worm. From each cell body a process or **muscle arm** extends either to the ventral or the dorsal nerve. Although not unique to nematodes, this arrangement is very unusual; in most animals nerve processes (axons, p. 727) extend to the muscle, rather than the other way around.

The fluid-filled pseudocoel, in which the internal organs lie, constitutes a hydrostatic skeleton. Hydrostatic skeletons, found in many invertebrates, lend support by transmitting the force of muscle contraction to the enclosed, noncompressible fluid. Normally, muscles are arranged antagonistically, so that movement is effected in one direction by contraction of one group of muscles, and movement back in the opposite direction is effected by the antagonistic set of muscles. Recall how the longitudinal and circular muscles operate antagonistically in each annelid segment. However, nematodes do not have circular body-wall muscles to antagonize the longitudinal muscles; therefore the cuticle must serve that function. As muscles on one side of the body contract, they compress the cuticle on that side, and the force of the contraction is transmitted (by the fluid in the pseudocoel) to the other side of the nematode, stretching the cuticle on that side. This compression and stretching of the cuticle serve to antagonize the muscle and are the forces that return the body to resting position when the muscles relax; this action produces the characteristic thrashing motion seen in nematode movement. An increase in efficiency of this system can be achieved only by an increase in hydrostatic pressure. Consequently, hydrostatic pressure in the nematode pseudocoel is much higher than is usually

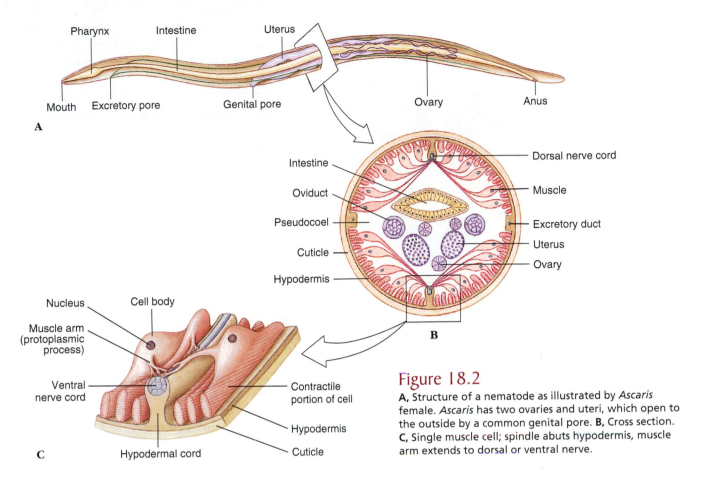

Figure 18.2

A, Structure of a nematode as illustrated by *Ascaris* female. *Ascaris* has two ovaries and uteri, which open to the outside by a common genital pore. **B,** Cross section. **C,** Single muscle cell; spindle abuts hypodermis, muscle arm extends to dorsal or ventral nerve.

found in other kinds of animals that have hydrostatic skeletons but that also have antagonistic muscle groups.

The alimentary canal of nematodes consists of a mouth (Figure 18.2), a muscular pharynx, a long nonmuscular intestine, a short rectum, and a terminal anus. Food is sucked into the pharynx when the muscles in its anterior portion contract rapidly and open the lumen. Relaxation of the muscles anterior to the food mass closes the lumen of the pharynx, forcing the food posteriorly toward the intestine. The intestine is one cell-layer thick. Food matter moves posteriorly by body movements and by additional food being passed into the intestine from the pharynx. Defecation is accomplished by muscles that simply pull the anus open, and expulsive force is provided by the high pseudocoelomic pressure that surrounds the gut.

Adults of many parasitic nematodes have an anaerobic energy metabolism; thus, a Krebs cycle and cytochrome system characteristic of aerobic metabolism are absent. They derive energy through glycolysis and probably through some incompletely known electron-transport sequences. Interestingly, some free-living nematodes and free-living stages of parasitic nematodes are obligate aerobes and have a Krebs cycle and cytochrome system.

A **ring of nerve tissue and ganglia** around the pharynx gives rise to small nerves to the anterior end and to two **nerve cords,** one dorsal and one ventral. **Sensory papillae** are concentrated around the head and tail. The **amphids** (Figure 18.3) are a pair of somewhat more complex sensory organs that open on each side of the head at about the same level as the cephalic circle of papillae. The amphidial opening leads into a deep

cuticular pit with sensory endings of modified cilia. Amphids are usually reduced in nematode parasites of animals, but most parasitic nematodes bear a bilateral pair of **phasmids** near the posterior end. They are rather similar in structure to amphids.

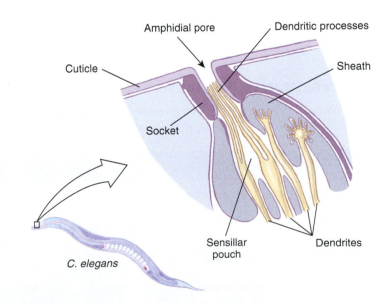

Figure 18.3

Diagram of an amphid in *Caenorhabditis elegans.*

Redrawn from Wright, K. A. 1980. Nematode sense organs. In B. M. Zuckerman (ed.), Nematodes as biological models, Vol. 2, Aging and other model systems. Copyright © Academic Press, New York.

Figure 18.4

A, Cross section of a male nematode. **B,** Posterior end of a male nematode.

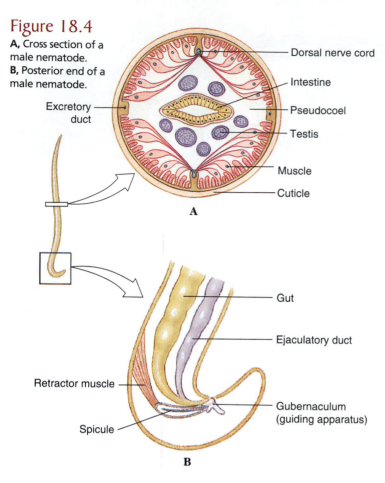

Most nematodes are dioecious. Males are smaller than females, and their posterior end usually bears a pair of **copulatory spicules** (Figure 18.4). Fertilization is internal, and eggs are usually stored in the uterus until deposition. Development among free-living forms is typically direct. The four juvenile stages are each separated by a molt, or shedding, of the cuticle. Many parasitic nematodes have free-living juvenile stages. Others require an intermediate host to complete their life cycles.

Representative Nematode Parasites

As mentioned on page 386, nearly all vertebrates and many invertebrates are parasitized by nematodes. A number of these are very important pathogens of humans and domestic animals. A few nematodes are common in humans in North America (Table 18.1), but they and many others usually abound in tropical countries. Space permits mention of only a few in this discussion.

Copulatory spicules of male nematodes are not true intromittent organs, since they do not conduct sperm, but are another adaptation to cope with high internal hydrostatic pressure. Spicules must hold the vulva of a female open while the ejaculatory muscles overcome the hydrostatic pressure in the female and rapidly inject sperm into her reproductive tract. Furthermore, nematode spermatozoa are unique among those studied in the animal kingdom in that they lack a flagellum and acrosome. Within a female's reproductive tract, sperm become ameboid and move by pseudopods. Could this be another adaptation to the high hydrostatic pressure in the pseudocoel?

Ascaris lumbricoides: The Large Roundworm of Humans

Because of its size and availability, *Ascaris* (Gr. *askaris,* intestinal worm) is usually selected as a type for study in zoology, as well as in experimental work. Thus it is probable that parasitologists know more about structure, physiology, and biochemistry of *Ascaris* than of any other nematode. This genus includes several species. One of the most common, *A. megalocephala,* is found in the intestine of horses. *Ascaris lumbricoides* (Figure 18.5) is one of the most common nematode parasites found in humans; recent surveys have shown a prevalence of up to 25% in some areas of the southeastern United States, and more than 1.27 *billion* people are infected worldwide. The large roundworm of pigs, *A. suum,* is morphologically close to *A. lumbricoides,* and they were long considered the same species.

A female *Ascaris* may lay 200,000 eggs a day, carried by the host's feces. Given suitable soil conditions, embryos develop into infective juveniles within 2 weeks. Direct sunlight and high temperatures are rapidly lethal, but the eggs have an amazing tolerance to other adverse conditions, such as desiccation or lack of oxygen. Shelled juveniles can remain viable for many months or even years in soil. Infection usually occurs when eggs are ingested with uncooked vegetables or when children put soiled fingers

TABLE 18.1

Common Parasitic Nematodes of Humans in North America

Common and Scientific Names	Mode of Infection; Prevalence
Hookworm (*Ancylostoma duodenale* and *Necator americanus*)	Contact in soil with juveniles that burrow into skin; common in southern states
Pinworm (*Enterobius vermicularis*)	Inhalation of dust with ova and by contamination with fingers; most common worm parasite in United States
Intestinal roundworm (*Ascaris lumbricoides*)	Ingestion of embryonated ova in contaminated food; common in rural areas of Appalachia and southeastern states
Trichina worm (*Trichinella* spp.)	Ingestion of infected muscle; occasional in humans throughout North America
Whipworm (*Trichuris trichiura*)	Ingestion of contaminated food or by unhygienic habits; usually common wherever *Ascaris* is found

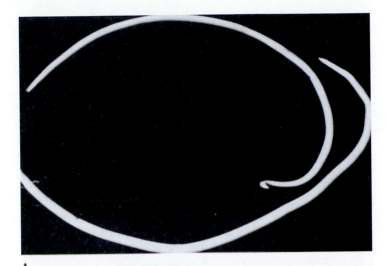

Figure 18.5

A, Intestinal roundworm *Ascaris lumbricoides*, male and female. Male, *top,* is smaller and has characteristic sharp kink in the end of the tail. Females of this large nematode may be over 30 cm long. **B,** Intestine of a pig, nearly completely blocked by *Ascaris suum*. Such heavy infections are also fairly common with *A. lumbricoides* in humans.

cause a serious pneumonia at this stage. On reaching the pharynx, juveniles are swallowed, passed through the stomach, and finally mature about 2 months after the eggs were ingested. In the intestine, where they feed on intestinal contents, worms cause abdominal symptoms and allergic reactions, and in large numbers they may cause intestinal blockage. Parasitism by *Ascaris* is rarely fatal, but death can occur if the intestine is blocked by a heavy infestation. Perforation of the intestine with resultant peritonitis is not uncommon, and wandering worms may occasionally emerge from the anus or throat or may enter the trachea or eustachian tubes and middle ears. Infection rates tend to be highest in children, and males tend to be more heavily infected than females, presumably because boys are more likely to ingest dirt.

Other ascarids are common in wild and domestic animals. Species of *Toxocara,* for example, are found in dogs and cats. Their life cycle is generally similar to that of *Ascaris,* but juveniles often do not complete their tissue migration in adult dogs, remaining in the host's body in a stage of arrested development. Pregnancy in a female dog, however, stimulates juvenile worms to wander, and they infect the embryos in the uterus. Puppies are then born with worms. These ascarids also survive in humans but do not complete their development, leading to an occasionally serious condition in children known as *visceral larva migrans*. This is a good argument for pet owners to practice immediate hygienic disposal of canine wastes!

Hookworms

Hookworms are so named because the anterior end curves dorsally, suggesting a hook. The most common species is *Necator americanus* (L. *necator,* killer), whose females are up to 11 mm long. Males can reach 9 mm in length. Large plates in their mouths (Figure 18.6) cut into the intestinal mucosa of the host where they suck blood and pump it through their intestine, partially digesting it and absorbing the nutrients. They suck much more blood than they need for food, and heavy infections cause anemia in patients. Hookworm disease in children may result in retarded mental and physical growth and a general loss of energy.

or toys in their mouths. Unsanitary defecation habits "seed" the soil or drinking water, and viable eggs remain long after all signs of the fecal matter have disappeared. Thus infection rates tend to be highest in areas where waste treatment practices do not control these factors.

When a host swallows embryonated eggs, the tiny juveniles hatch. They burrow through the intestinal wall into veins or lymph vessels and are carried through the heart to the lungs. There they break out into alveoli and are carried up to the trachea. If the infection is large, they may

Plates

Figure 18.6

A, Mouth of hookworm displaying cutting plates. **B,** Section through anterior end of hookworm attached to dog intestine. Note cutting plates pinching off mucosa from which the thick muscular pharynx sucks blood. Esophageal glands secrete anticoagulant to prevent blood from clotting.

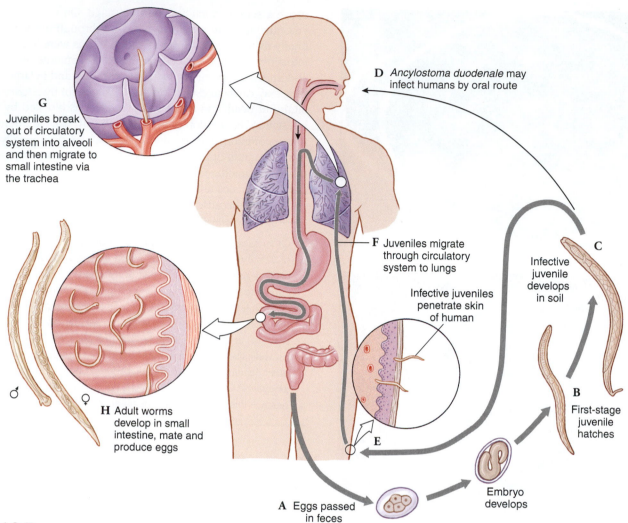

D *Ancylostoma duodenale* may infect humans by oral route

G Juveniles break out of circulatory system into alveoli and then migrate to small intestine via the trachea

F Juveniles migrate through circulatory system to lungs

Infective juvenile develops in soil

C

Infective juveniles penetrate skin of human

♂   ♀

H Adult worms develop in small intestine, mate and produce eggs

E

B First-stage juvenile hatches

Embryo develops

A Eggs passed in feces

Figure 18.7

The life cycle of hookworms: a shelled embryo develops into a first-stage juvenile which is followed by two molts. The resulting third-stage juvenile enters developmental arrest until it reaches a new host (**A** to **C**). Human infection may be via the mouth (**D**) or skin (**E**). Juveniles migrate through the circulatory system to lungs (**F**), enter alveoli (**G**), and then reach the intestine where they mate (**H**).

Drawing by William Ober and Claire Garrison.

Eggs pass in the feces, and juveniles hatch in the soil, where they live on bacteria (Figure 18.7). When human skin comes in contact with infected soil, infective juveniles burrow through the skin to the blood, reach the lungs and finally the intestine in a manner similar to that described for *Ascaris.*

Trichina Worm

Trichinella spiralis (Gr. *trichinos,* of hair, + *-ella,* diminutive) is one of several species of tiny nematodes responsible for the potentially lethal disease trichinosis. Adult worms burrow in the mucosa of the small intestine where females produce living young. Juveniles penetrate blood vessels and are carried throughout the body, where they may be found in almost any tissue or body space. Eventually, they penetrate skeletal muscle cells, becoming one of the largest known intracellular parasites. Juveniles cause astonishing redirection of gene expression in their host cell, which loses its striations and becomes a **nurse cell** that nourishes the worm (Figure 18.8). When raw or poorly

Figure 18.8

Muscle infected with trichina worm *Trichinella spiralis.* The juveniles lie within muscle cells that the worms have induced to transform into nurse cells (commonly called cysts). An inflammatory reaction occurs around the nurse cells. Juveniles may live 10 to 20 years, and nurse cells eventually may calcify.

cooked meat containing encysted juveniles is swallowed, the worms are liberated into the intestine where they mature.

Trichinella spp. can infect a wide variety of mammals in addition to humans, including hogs, rats, cats, and dogs. Hogs become infected by eating garbage containing pork scraps with juveniles or by eating infected rats. In addition to *T. spiralis*, we now know there are four other sibling species in the genus. They differ in geographic distribution, infectivity to different host species, and freezing resistance.

Heavy infections may cause death, but lighter infections are more common—about 12 cases are discovered annually in the United States, but infection is still common in other parts of the world.

Pinworms

Pinworms, *Enterobius vermicularis* (Gr. *enteron,* intestine, + *bios,* life), cause relatively little disease, but they are the most common nematode parasites in the United States, estimated at 30% of all children and 16% of adults. Adult parasites (Figure 18.9) live in the large intestine and cecum. Females, up to about 12 mm in length, migrate to the anal region at night to lay their eggs (Figure 18.9). Scratching the resultant itch effectively contaminates hands and bedclothes. Eggs develop rapidly and become infective within 6 hours at body temperature. When they are swallowed, they hatch in the duodenum, and the worms mature in the large intestine.

Diagnosis of most intestinal roundworms is usually made by examination of a small bit of feces under the microscope and finding characteristic eggs. However, pinworm eggs are not often found in the feces because the female deposits them on the skin around the anus. The "Scotch tape method" is more effective. The sticky side of cellulose tape is applied around the anus to collect the eggs, then the tape is placed on a glass slide and examined under a microscope. Several drugs are effective against this parasite, but all members of a family should be treated at the same time because the worms easily spread through a household.

Classification of Phylum Nematoda

The traditional classification is based on the work of Kampfer, et al.

Class Secernentea (= Phasmida) Amphids ventrally coiled or derived therefrom; three esophageal glands; some with phasmids; both free-living and parasitic forms. Examples: *Caenorhabditis, Ascaris, Enterobius, Necator, Wuchereria.*

Class Adenophorea (= Aphasmida) Amphids generally well-developed, pocketlike; five or more esophageal glands; phasmids absent; excretory system lacking lateral canals, formed of single, ventral, glandular cells, or entirely absent; mostly free-living, but includes some parasites. Examples: *Dioctophyme, Trichinella, Trichuris.*

Classification of nematodes is somewhat more satisfactory at the order and superfamily level; division into classes relies on characteristics that are not striking and that are difficult for novices to distinguish. Argument exists about monophyly of the nematodes (Adamson[1]), but some molecular work supports the traditional classes (Kampfer[2]). A recent molecular phylogeny divides nematodes among 12 clades.[3]

[1]Adamson, M. 1987. Can. J. Zool. **65:**1478–1482.
[2]Kampfer, S., et al. 1998. Invert. Biol. **117:**29–36.
[3]Holterman, T., et al. 2006. Mol. Biol. Evol. **23:**1792–1800.

Filarial Worms

At least eight species of filarial nematodes infect humans, and some of these are major causes of disease. Some 120 million people in tropical countries are infected with *Wuchereria bancrofti* (named for Otto Wucherer) or *Brugia malayi* (named for S. L. Brug), which places these species among the scourges of humanity. The worms live in the lymphatic system, and females are as long as 10 cm. Disease symptoms are associated with inflammation and obstruction of the lymphatic system. Females release live young, tiny **microfilariae,** into the blood and lymphatic system (Figure 18.10). As they feed, mosquitos ingest microfilariae, which develop inside the mosquitos to the infective stage. They escape from the mosquito when it is feeding again on a human and penetrate the wound made by the mosquito bite.

The dramatic manifestations of elephantiasis are produced occasionally after long and repeated exposure to reinfection. The condition is marked by an excessive growth of connective tissue and enormous swelling of affected parts, such as the scrotum, legs, arms, and more rarely, the vulva and breasts (Figure 18.11).

Another filarial worm causes river blindness (onchocerciasis) and is carried by black flies. It infects more than 37 million people in parts of Africa, Arabia, Central America, and South America.

The most common filarial worm in the United States is probably the dog heartworm, *Dirofilaria immitis* (Figure 18.12). Carried by mosquitos, it also can infect other canids, cats,

Figure 18.9

Pinworms, *Enterobius vermicularis*. **A,** Female worm from human large intestine (slightly flattened in preparation), magnified about 20 times. **B,** Group of pinworm eggs, which are usually discharged at night around the anus of the host, who, by scratching during sleep, gets fingernails and clothing contaminated.

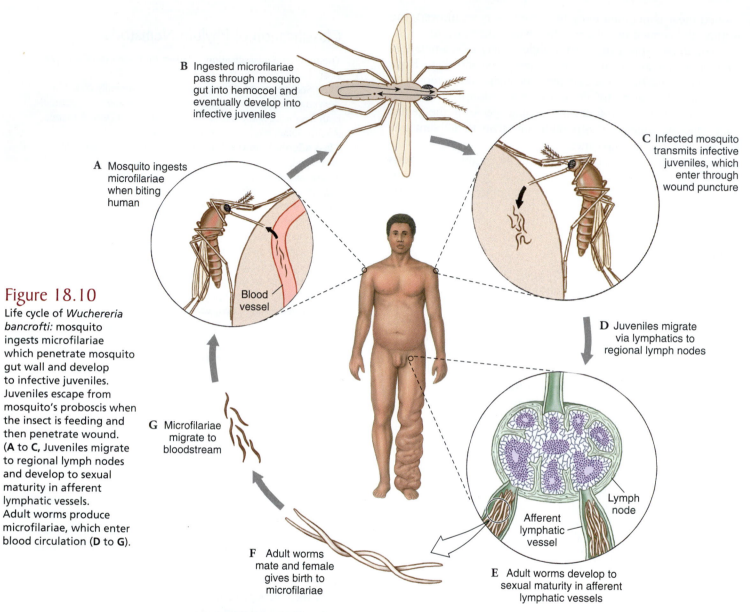

B Ingested microfilariae pass through mosquito gut into hemocoel and eventually develop into infective juveniles

A Mosquito ingests microfilariae when biting human

C Infected mosquito transmits infective juveniles, which enter through wound puncture

Blood vessel

D Juveniles migrate via lymphatics to regional lymph nodes

G Microfilariae migrate to bloodstream

Lymph node

Afferent lymphatic vessel

F Adult worms mate and female gives birth to microfilariae

E Adult worms develop to sexual maturity in afferent lymphatic vessels

Figure 18.10

Life cycle of *Wuchereria bancrofti:* mosquito ingests microfilariae which penetrate mosquito gut wall and develop to infective juveniles. Juveniles escape from mosquito's proboscis when the insect is feeding and then penetrate wound. (**A** to **C**, Juveniles migrate to regional lymph nodes and develop to sexual maturity in afferent lymphatic vessels. Adult worms produce microfilariae, which enter blood circulation (**D** to **G**).

Figure 18.11

Elephantiasis of leg caused by adult filarial worms of *Wuchereria bancrofti,* which live in lymph passages and block the flow of lymph. Tiny juveniles, called microfilariae, are ingested with blood meal of mosquitos, where they develop to infective stage and are transmitted to a new host.

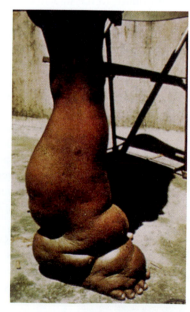

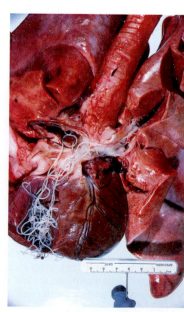

Figure 18.12

Dirofilaria immitis in right ventricle, extending up into right and left pulmonary arteries of an eight-year-old Irish setter.

ferrets, sea lions, and occasionally humans. Along the Atlantic and Gulf Coast states and northward along the Mississippi River throughout the midwestern states, prevalence in dogs is up to 45%. It occurs in other states at a lower prevalence. This worm causes a very serious disease among dogs, and no responsible owner should fail to provide "heartworm pills" for a dog during mosquito season.

PHYLUM NEMATOMORPHA

The popular name for Nematomorpha (nem′a-to-mor′fa) (Gr. *nema, nematos,* thread, + *morphē,* form) is "horsehair worms," based on an old superstition that the worms arise from horsehairs that happen to fall into water. The worms do look something like hairs from a horse's tail. They were long included within Nematoda because both groups share the structure of the cuticle, presence of epidermal cords, longitudinal muscles only, and pattern of nervous system. They are currently placed as the sister taxon to nematodes.

About 320 species of horsehair worms have been named. Worldwide in distribution, they are free-living as adults and parasitic in arthropods as juveniles. Adults live almost anywhere in wet to moist surroundings if oxygen is adequate.

Form and Function

Horsehair worms are extremely long and slender, with a cylindrical body. They are generally about 0.5 to 3 mm in diameter but can be up to 1 m in overall length. Their anterior end is usually rounded, and their posterior end is rounded or has two or three caudal lobes (Figure 18.13).

Their body wall is much like that of nematodes: a secreted cuticle, a hypodermis, and musculature of **longitudinal muscles** only.

Their digestive system is vestigial. The pharynx is a solid cord of cells, and the intestine does not open to the cloaca. Larval forms absorb food from their arthropod hosts through their body wall. Until recently, adults were thought to live entirely on stored nutrients. Recent research has shown that adults absorb organic molecules through their vestigial gut and body wall in much the same way as juveniles.

Circulatory, respiratory, and excretory systems are lacking and probably occur on a primarily cellular level. However, very little is known about the physiology of these worms. There is a nerve ring around the pharynx and a midventral nerve cord.

Life cycles of nematomorphs are poorly known. In the cosmopolitan genus, *Gordius* (named for an ancient king who tied an intricate knot), juveniles may encyst on vegetation likely to be eaten by a grasshopper or other arthropod. Gordiid larval stages also have hooks or stylets that may be used to bore into a host, perhaps via the integument or the gut lining. In other cases, the gordiid may infect the host via its drinking water. Larvae encyst in the host; in some cases, it seems that development continues after the first host is eaten by a second host. In the marine nematomorph, *Nectonema* (Gr. *nektos,* swimming, + *nema,* a thread), juveniles occur in hermit crabs and other crabs.

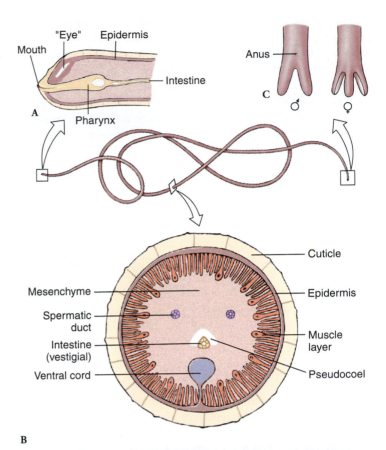

Figure 18.13

Structure of *Paragordius,* a nematomorph. **A,** Longitudinal section through the anterior end. **B,** Transverse section. **C,** Posterior end of male and female worms. Nematomorphs, or "horsehair worms," are very long and very thin. Their pharynx is usually a solid cord of cells and appears nonfunctional. *Paragordius,* whose pharynx opens through to the intestine, is unusual in this respect and also in the possession of a photosensory organ ("eye"). **D,** *Paragordius tricuspidatus* emerges from the body of a European cricket, *Nemobius sylvestris.*

After several months in the hemocoel (p. 396) of an arthropod host, juveniles complete a single molt and emerge into water as mature adults. If the host is a terrestrial insect, the parasite somehow stimulates the host to seek water. Worms do not emerge from the host unless water is nearby.

Nematomorphs are dioecious. Adults wriggle slowly once in water, with males typically being more active than females. In both sexes, gonads empty into a cloaca through gonoducts. Females discharge their eggs into water in long strings.

PHYLUM KINORHYNCHA

Kinorhyncha (kin'o-ring'ka) (Gr. *kinein,* to move, + *rhynchos,* beak) are marine worms a little larger than rotifers and gastrotrichs but usually not more than 1 mm long. This phylum has also been called Echinodera, meaning spiny necked. About 179 species have been described to date.

Kinorhynchs are cosmopolitan, living from pole to pole, from intertidal areas to 8000 m in depth. Most live in mud or sandy mud, but some have been found in algal holdfasts, sponges, or other invertebrates.

Form and Function

The body of kinorhynchs is divided into head, neck, and trunk regions. The trunk has 11 segments, marked externally by spines and cuticular plates. (Figure 18.14). The retractile head sometimes called an introvert, has five to seven circlets of spines with a small retractile proboscis. The spines, called scalids, function in locomotion, chemoreception, and mechanoreception. Each contains 10 or fewer monociliary sensory cells. The body is flat ventrally and arched dorsally. The body wall is composed of a chitinous cuticle, a cellular epidermis, and longitudinal epidermal cords, much like those of nematodes. The arrangement of muscles is correlated with the segments, and unlike the nematodes, kinorhynchs have circular, longitudinal, and diagonal muscle bands.

A kinorhynch cannot swim. In silt and mud where it commonly lives, it burrows by extending the head into the mud and anchoring it with spines. Extension of the head takes place as trunk muscles increase hydrostatic pressure on the small amount of fluid in the pseudocoel. After extension it draws its body forward until its head is retracted into its body. When disturbed, a kinorhynch draws in its head and protects it with a closing apparatus of cuticular plates on the neck, or on the neck and trunk.

Their digestive system is complete, with a mouth at the tip of a proboscis, followed by a pharynx, an esophagus, a nonciliated midgut, and a cuticle-lined hindgut, as well as an anus. Kinorhynchs feed on diatoms or by digesting organic material from the surface of mud particles through which they burrow.

Their pseudocoel is filled with amebocytes and organs leaving little fluid space. A multinucleated solenocyte protonephridium on each side of the gut between the eighth and ninth segments serves as their excretory system.

The nervous system is in contact with the epidermis, with a multilobed brain encircling their pharynx, and with a ventral ganglionated nerve cord extending throughout the body. Sense organs are represented by sensory bristles and by eyespots in some.

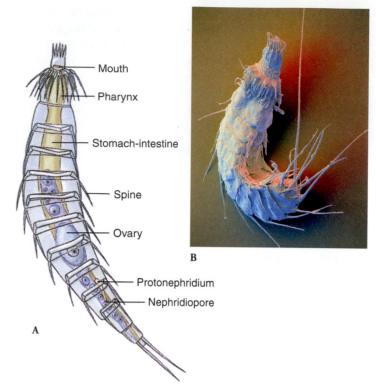

Figure 18.14

A, *Echinoderes,* a kinorhynch, is a minute marine worm. Segmentation is superficial. The head, with its circle of spines, is retractile. **B,** Colored scanning electron micrograph (SEM) of kinorhynch *Antigomonas sp.*

Sexes are separate, with paired gonads and gonoducts. There is a series of about six juvenile stages and a definitive, nonmolting adult. No asexual reproduction has been found.

PHYLUM PRIAPULIDA

Priapulida (pri'a-pyu'li-da) (Gr. *priapos,* phallus, + *ida,* pl. suffix) are a small group (only 16 species) of marine worms found chiefly in colder waters of both hemispheres. They have been reported along the Atlantic coast from Massachusetts to Greenland and along the Pacific coast from California to Alaska. They live in mud and sand of the seafloor and range from intertidal zones to depths of several thousand meters. *Tubiluchus* (L. *tubulus,* dim. of *tubus,* waterpipe) is a minute detritus feeder adapted to interstitial life in warm coralline sediments. *Maccabeus* (named for a Judean patriot who died in 160 B.C.) is a tiny tube-dweller discovered in muddy Mediterranean bottoms.

Form and Function

Priapulids have cylindrical bodies, most less than 12 to 15 cm long, but *Halicryptus higginsi* is up to 39 cm in length. Most are burrowing predaceous animals that feed on soft-bodied invertebrates such as polychaete worms (p. 364). They usually orient themselves upright in mud with their mouth at the surface. However, *Tubiluchus* feeds on organic detritus in the sediments around coral reefs. They are adapted for burrowing by body contractions.

Figure 18.15

A, Major internal structures of *Priapulus*. **B**, *Priapulus caudatus* from Lurefjord, Norway.

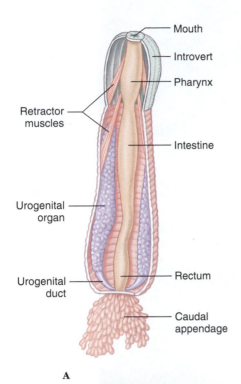

Mouth
Introvert
Pharynx
Retractor muscles
Intestine
Urogenital organ
Rectum
Urogenital duct
Caudal appendage

A

B

The body includes an introvert, trunk, and usually one or two caudal appendages (Figure 18.15). Their retractable introvert is ornamented with papillae and ends with rows of curved spines (scalids) that surround their mouth. Scalids have sensory and locomotory functions. Extension of the introvert occurs as circular muscles increase hydrostatic pressure on the internal fluid-filled cavity. Derivation of the cavity is not clear. The eversible pharynx is used for capturing small, soft-bodied prey. *Maccabeus* has a crown of brachial tentacles around its mouth.

Their trunk is not metameric but is superficially divided into 30 to 100 rings and is covered with tubercles and spines. The tubercles are probably sensory in function. The anus and urogenital pores are located at the posterior end of the trunk. Caudal appendages are hollow stems believed to be respiratory and probably chemoreceptive in function. A chitinous cuticle, molted periodically throughout life, covers their body.

Their digestive system contains a muscular pharynx and a straight intestine and rectum (Figure 18.15). There is a nerve ring around the pharynx and a midventral nerve cord. Amebocytes occur in the fluids of the body cavity and, at least in some species have corpuscles containing a respiratory pigment called hemerythrin.

Sexes are separate, although males have never been discovered for *Maccabeus*. Paired urogenital organs each contain a gonad and clusters of solenocytes, both connected to a protonephridial tubule that carries both gametes and excretory products outside the body. Fertilization is external. Embryology is poorly known. In *Meiopriapulus* development is direct, and females brood their developing embryos. In most species, the zygote appears to undergo radial cleavage and to develop into a loricate larva. Larvae of *Priapulus* dig into mud and become detritus feeders.

PHYLUM LORICIFERA

Loricifera (L., *lorica*, corselet, + Gr., *phora*, bearing) are a very recently described phylum of animals (1983) with only 11 currently described species and approximately 80 undescribed species. The tiny animals (ranging from 0.10 to 0.50 mm long) have a protective external case (lorica) and live in spaces between grains of marine gravel, to which they cling tightly. Although they were first described from specimens collected off the coast of France, they are apparently distributed worldwide. Most species have been found in coarse marine sediments at depths of 300–450 m, although one species was recently collected from 8000 m.

Form and Function

The loriciferan body has five regions: mouth cone, head or introvert, neck, thorax, and abdomen. There are nine circlets of scalids on the introvert. Scalids are similar to those of kinorhynchs and serve locomotory and sensory functions. The covering on the abdomen, a lorica, may have thick cuticular plates, or it may be thin and folded. The entire forepart of their body can be retracted into the circular lorica (Figure 18.16). Their diet is unknown but there is speculation that they feed on bacteria. Their brain fills most of the head, and oral spines are innervated by nerves from the brain and other ganglia. The body cavity has been described as a pseudocoel in some species, but other species are considered acoelomate.

Loriciferans are dioecious with dimorphic males and females. Copulation occurs, but life cycles are not well known. There is a distinct larval phase called a Higgins larva. Three species within genus *Rugiloricus* have life cycles that differ in the number of larval stages. In one species, a Higgins larva molts into an adult; in another a Higgins larva molts into a second stage, which molts into an adult, and in a third species the life cycle is more complex as a Higgins larva is followed by parthenogenetic stages. Higgins larvae themselves also differ in form with benthic larvae having toes and pelagic larvae lacking toes.

Figure 18.16

A, Dorsal view of adult loriciferan, *Nanaloricus mysticus,* showing internal features. **B,** Live animal, 0.3 mm.

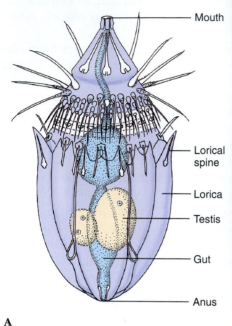

- Mouth
- Lorical spine
- Lorica
- Testis
- Gut
- Anus

A

B

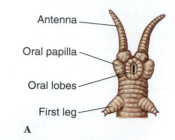

- Antenna
- Oral papilla
- Oral lobes
- First leg

A

B

Figure 18.17

Peripatus, a caterpillar-like onychophoran that has characteristics in common with both annelids and arthropods. **A,** Ventral view of head. **B,** In natural habitat.

CLADE PANARTHROPODA

Panarthropoda contains Arthropoda and two allied phyla, Onychophora and Tardigrada. In these taxa, the coelom is reduced and a hemocoel develops. In onychophorans and arthropods, a coelom develops by schizocoely, but coelom formation has been described as enterocoelic in tardigrades. In all three phyla, the main coelomic cavity later fuses with the blastocoel to form a new cavity called a **hemocoel,** or a mixocoel. The hemocoel is lined by an extracellular matrix, not the mesodermal peritoneum that originally lined the coelom. Blood from the open circulatory system enters the hemocoel and surrounds the internal organs. A muscular heart is present, but tubular blood vessels occur in only part of the body; blood enters and leaves the hemocoel through blood vessels. There may be small coelomic cavities surrounding a few organs in other parts of the body.

Phylum Onychophora

Members of phylum Onychophora (on-y-kof′o-ra) (Gr. *onyx,* claw, + *pherein,* to bear) are commonly called "velvet worms," or "walking worms." There are approximately 70 species of caterpillar-like animals, ranging from about 0.5 to 15 cm in length. They live in rain forests and other moist, leafy habitats in tropical and subtropical regions and in some temperate regions of the Southern Hemisphere. Most velvet worms are predaceous, feeding on caterpillars, insects, snails, and worms. Some onychophorans live in termite nests and feed on termites.

Their fossil record shows that they have changed little in their 500-million-year history. A fossil form, *Aysheaia,* discovered in the Burgess Shale deposit of British Columbia and dating back to mid-Cambrian times, is very much like modern onychophorans (see Figure 6.9, p. 110). Onychophorans were probably far more common at one time than they are now. Today they are terrestrial and extremely retiring, becoming active only at night or when the air is nearly saturated with moisture.

Form and Function

External Features   Onychophorans are more or less cylindrical and show no external segmentation except for the paired appendages (Figure 18.17). The skin is soft, velvety, and covered with a thin, flexible cuticle that contains protein and chitin. In structure and chemical composition it resembles arthropod cuticle; however, it never hardens like arthropod cuticle, and it is molted in patches rather than all at one time. The body is studded with tiny **tubercles,** some of which bear sensory bristles. The color may be green, blue, orange, dark gray, or black, and minute scales on the tubercles give the body an iridescent and velvety appearance. The head bears a pair of large **antennae,** each with an annelid-like eye at the base. The ventral mouth has a pair of clawlike **mandibles** and is flanked by a pair of **oral papillae,** which can expel a slimy defensive secretion (Figure 18.17).

Their 14 to 43 pairs of **unjointed legs** are short, stubby, and clawed. Onychophorans crawl by passing waves of contraction from anterior to posterior. When a body region extends, the legs lift up and move forward. The legs are more ventrally located than are parapodia of annelids.

Internal Features   The body wall is muscular like that of annelids. The body cavity is a **hemocoel,** imperfectly divided into compartments, or sinuses, much like those of arthropods (see p. 424). **Slime glands** on each side of the body cavity open on the oral papillae. When disturbed by a predator, the animal can launch two streams of sticky fluid from these glands up to 30 cm. Hardening rapidly, this adhesive can entangle the would-be predator and hold it firmly for the onycophoran to consume at its leisure.

The mouth, surrounded by lobes of skin, contains a dorsal tooth and a pair of lateral mandibles for grasping and cutting prey. There is a muscular pharynx and a straight digestive tract (Figure 18.18).

Each leg-bearing body segment contains a pair of **nephridia,** each nephridium with a vesicle, ciliated funnel and duct, and nephridiopore opening at the base of a leg. Absorptive cells in the midgut excrete crystalline uric acid, and certain pericardial cells function as nephrocytes, storing excretory products taken from the blood.

Figure 18.18
Internal anatomy of an onychophoran.

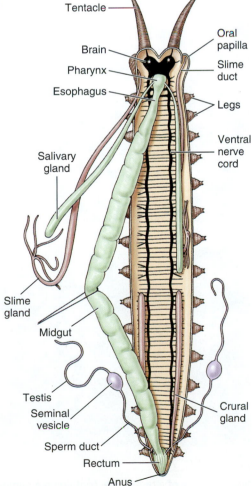

For respiration there is a **tracheal system** that connects to all parts of the body and communicates with the outside by many openings, or **spiracles,** scattered all over the body. Onychophorans cannot close their spiracles to prevent water loss, so although the tracheae are efficient, these animals are restricted to moist habitats. Their tracheal system is somewhat different from that of arthropods and probably originated independently.

The open circulatory system has, in the pericardial sinus, a dorsal, tubular heart with a pair of ostia in each segment.

The nervous system of onycophorans is organized much like a ladder with paired ventral nerve cords running close to the top of each row of legs connected by commisures running across the width of the body. Nerves to antennae and head region extend from the brain, and ganglionic swellings at the base of each leg supply nerves to legs and body wall. Sense organs include a relatively well-developed eye, taste spines around the mouth, tactile papillae on the integument, and hygroscopic receptors that orient the animal toward water vapor. Although these animals were assumed to have a limited behavioral repertoire, recent work has shown social behavior and group hunting in an Australian species.

With the exception of one known parthenogenic species, onychophorans are dioecious, with paired reproductive organs. Little is known about the mating habits of these animals, but in some species a portion of the uterus is expanded as a seminal receptacle, presumably for copulation. In at least one species the male deposits spermatophores, seemingly at random, on the back of the female. White blood cells then dissolve the skin beneath the spermatophores. Sperm can then enter the body cavity and migrate in the blood to the ovaries to fertilize eggs. Onychophorans may be oviparous, ovoviviparous, or viviparous. Only two Australian genera are oviparous, laying shell-covered eggs in moist places. In all other onychophorans eggs develop in the uterus, and live young leave the mother's body. In some species there is a placental attachment between mother and young (viviparous); in others young develop in the uterus without attachment (ovoviviparous). Nonplacental species typically have eggs with a large amount of yolk, the eggs cleave superficially, in a manner similar to arthropods. When little yolk is present, cleavage is complete.

Phylum Tardigrada

Tardigrada (tar-di-gray'da) (L. *tardus,* slow, + *gradus,* step), or "water bears," are minute organisms usually less than a millimeter in length. Most of the 900 described species are terrestrial forms that live in the water film surrounding mosses and lichens or damp soils. Some live in freshwater algae or mosses or in bottom debris, and some are marine, usually inhabiting interstitial spaces between sand grains, in both deep and shallow seawater. They share many characteristics with arthropods.

They have an elongated, cylindrical, or a long oval body that is unsegmented. The head is merely the anterior part of the trunk. The trunk bears four pairs of short, stubby, unjointed legs, each armed with four to eight claws (Figure 18.19). They are covered by a nonchitinous cuticle that is molted along with

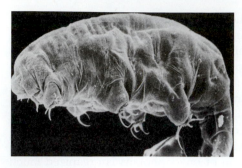

Figure 18.19

Scanning electron micrograph of an aquatic tardigrade, *Pseudobiotus*.

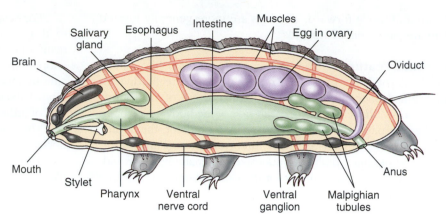

Figure 18.20

Internal anatomy of a tardigrade.

claws and buccal apparatus four or more times in the life history. Some, such as *Echiniscus,* expel feces when molting, leaving the feces in the discarded cuticle.

The mouth of tardigrades opens into a buccal tube that empties into a muscular pharynx adapted for sucking (Figure 18.20). Two needlelike stylets flanking the buccal tube can be protruded through the mouth. These stylets pierce plant or animal cells, and the pharynx sucks in the liquid contents. Some tardigrades suck body juices of nematodes, rotifers, and other small animals, while others are parasitic on larger animals such as sea cucumbers or barnacles. At the junction of intestine and rectum, three glands, thought to be excretory and often called Malpighian tubules, empty into the digestive system. Cilia are absent.

Most of the body cavity is a hemocoel, with their true coelom restricted to the gonadal cavity. There are no circulatory or respiratory systems, gaseous exchange occurring by diffusion through the body surface.

Their muscular system consists of a number of long muscle bands, usually comprised of one or a few large muscle cells each. Circular muscles are absent, but hydrostatic pressure of the body fluid may act as a skeleton. Being unable to swim (with one exception), water bears crawl awkwardly, clinging to the substrate with their claws.

Their brain is relatively large and covers most of the dorsal surface of the pharynx. Circumpharyngeal connectives link it to the subpharyngeal ganglion, from which the double ventral nerve cord extends posteriorly as a chain of four ganglia that appear to control the four pairs of legs.

Sexes are separate in tardigrades. In some freshwater and moss-dwelling species, males are unknown and parthenogenesis seems to be the rule. Some species also have dwarf males, but in most tardigrades that have been studied males and females occur with approximately equal frequency. In some species sperm is deposited directly into the female's seminal receptacle or cloaca during copulation; in others sperm is injected into the body cavity by piercing the cuticle. Eggs of some species are highly ornate (Figure 18.21). Egg laying, like defecation, apparently occurs primarily at molting, when the volume of coelomic fluid is reduced. Females of some species cement their eggs to

a submerged object, whereas others deposit eggs in the molted cuticle (Figure 18.22). In some such cases, fertilization is indirect and males gather around the old cuticle containing unfertilized eggs and shed sperm into it.

Detailed research on the development of tardigrades is lacking, but cleavage is apparently complete. A stereogastrula is

Figure 18.21

Scanning electron micrograph of a highly ornate egg of a tardigrade, *Macrobiotus hufelandii*.

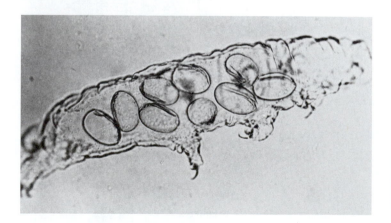

Figure 18.22

Molted cuticle of a tardigrade, containing a number of fertilized eggs.

formed. Five pairs of coelomic pouches appear, reminiscent of the enterocoelous development of many deuterostomes. However, all but the last pair, which fuse to form the gonad, disappear during development, and the gonocoel is the only true coelomic space left in adults. Development is direct and rapid. After about 14 days, juveniles use their claws to break free of the egg. At this time the number of cells in the body is relatively fixed, and growth occurs primarily by increase in cell size rather than cell number.

One of the most intriguing features of terrestrial tardigrades is their capacity to enter a state of suspended animation, called cryptobiosis, during which metabolism is virtually imperceptible; such an organism can withstand prolonged harsh environmental conditions. Under gradual drying conditions, water content of the body decreases from 85% to only 3%, movement ceases, and the body becomes barrel-shaped. In a cryptobiotic state tardigrades can resist temperature extremes from +149° C to −272° C, ionizing radiation, oxygen deficiency, preservatives such as ether and absolute alcohol, and other adverse conditions and may survive for years. Activity resumes when moisture is again available. Some nematodes and rotifers also can undergo cryptobiosis.

PHYLOGENY

Evolutionary relationships among ecdysozoans are not well understood. Members of this clade do not share a common cleavage pattern. Cleavage in nematodes and nematomorphs is described as unique, or not obviously spiral or radial. In priapulids, it is somewhat similar to radial cleavage. Cleavage has not been studied in kinorhynchs, loriciferans, and tardigrades. In onychophoran eggs containing large amounts of yolk, the cytoplasm does not cleave, but nuclei do divide. Development is similar to that in arthropods with centrolecithal eggs (see p. 172). In onychophoran eggs with little yolk, cleavage is complete (holoblastic), but the cleavage pattern varies, appearing spiral in some taxa and radial in others.

In the absence of developmental characters, branch order is not defined for all ecdysozoans, but roundworms, phylum Nematoda, are united with horsehair worms, phylum Nematamorpha, in clade Nematoidea (see Figure 18.1). Recent phylogenies place the two phyla as sister taxa sharing a collagenous cuticle.

Phylum Kinorhyncha is shown as the sister taxon to phylum Priapulida based on the shared two-layered pharynx. Kinorhynchs have mouthparts (oral styles on a noninversible mouth cone) similar to those of loriciferans, but loriciferans also share some morphological features with larval nematomorphs and with priapulids. Some workers erect clade Scalidophora to contain kinorhynchs, priapulids, and loriciferans, but more work is needed on these animals before a branch order is specified.

Clade Panarthropoda unites three phyla whose evolutionary association has been clear for a long time. Velvet worms, phylum Onychophora, are the sister taxon to a clade comprising arthropods and tardigrades.

Onychophoran characteristics shared with arthropods include a tubular heart and hemocoel with open circulatory system, presence of tracheae (probably not homologous), absence of ectodermal cilia,

and large size of their brain. Onychophorans also share a number of characteristics with annelids: metamerically arranged nephridia, muscular body wall, pigment cup ocelli, and ciliated reproductive ducts. Unique characteristics include oral papillae, slime glands, body tubercles, and suppression of external segmentation.

Some authors have advocated that onychophorans should be included with myriapods and insects in phylum Arthropoda. However, most authors believe that their differences warrant keeping them in a separate phylum (see Figure 18.1). Sequence analysis supports placement of Onychophora in clade Panarthropoda, with a second branch containing both arthropods and tardigrades.

Tardigrades have some similarities to rotifers, particularly in their reproduction and their cryptobiotic tendencies, and some authors have called them pseudocoelomates. Their embryogenesis, however, would seem to put them among coelomates. The enterocoelic origin of the mesoderm is a deuterostome characteristic, but five pouches form, some of which fuse while others disappear, unlike the pattern in typical deuterostomes. Other authors identify several important synapomorphies that suggest grouping them with arthropods (see Figure 18.1). DNA sequence analysis supports alignment with arthropods in Ecdysozoa. Tardigrades and arthropods also share two morphological features: arthropod-type setae and muscles that insert on the cuticle (see Figure 18.1).

Reconstructing the evolutionary history of life is a fascinating pursuit, but biologists lack developmental and morphological information for many taxa, as is clear from comments presented here. Many of the less well-known taxa are very small animals living in obscure habitats, for example, the spaces between sand grains, but until all character states can be described for these animals, our knowledge of both ecdysozoan and metazoan phylogeny will remain incomplete.

Adaptive Diversification

Arthropods aside, certainly the most impressive adaptive diversification in this group of phyla is shown by nematodes. They are by far the most numerous in terms of both individuals and species, and they have been able to adapt to almost every habitat available to animal life. Their basic pseudocoelomate body plan, with the cuticle, hydrostatic skeleton, and longitudinal muscles, has proved generalized and plastic enough to adapt to an enormous variety of physical conditions. Free-living lineages gave rise to parasitic forms on at least several occasions, and virtually all potential hosts have been exploited. A recent phylogeny indicates that plant-parasitic forms have arisen from fungal-feeding ancestors in three independent evolutionary events. All types of life cycle occur: from simple and direct to complex, with intermediate hosts; from normal dioecious reproduction to parthenogenesis, hermaphroditism, and alternation of free-living and parasitic generations. A major factor contributing to evolutionary opportunism of nematodes has been their extraordinary capacity to survive suboptimal conditions, for example, developmental arrests in many free-living and animal parasitic species and ability to undergo cryptobiosis (survival in harsh conditions by assuming a very low metabolic rate) in many free-living and plant parasitic species.

SUMMARY

Phyla covered in this chapter possess a range of body plans. Analysis of nucleotide similarities in the gene for 18S small-subunit rDNA provides evidence that they belong to superphylum Ecdysozoa. All members of this clade molt their cuticles.

Arthropods aside, Nematoda is the largest and most important of these phyla, and although only 25,000 species are described currently, estimates suggest there may be as many as 500,000 species alive today. They are more or less cylindrical, tapering at the ends, and covered with a tough, secreted cuticle. Their body-wall muscles are longitudinal only, and to function well in locomotion, such an arrangement must enclose a volume of fluid in the pseudocoel at high hydrostatic pressure. This fact of nematode life has a profound effect on most of their other physiological functions, for example, ingestion of food, egestion of feces, excretion, copulation, and others. Most nematodes are dioecious, and there are four juvenile stages, each separated by a molt of the cuticle. Almost all invertebrate and vertebrate animals and many plants have nematode parasites, and many other nematodes are free-living in soil and aquatic habitats. Some parasitic nematodes have part of their life cycle free-living, some undergo a tissue migration in their host, and some have an intermediate host in their life cycle. Some parasitic nematodes cause severe diseases in humans and other animals.

Nematomorpha or horsehair worms superficially resemble nematodes and have parasitic juvenile stages in arthropods, followed by a free-living, aquatic adult stage.

Kinorhyncha and Loricifera are small phyla of tiny, aquatic pseudocoelomates. Kinorhynchs anchor and then pull themselves by spines on their head. Loriciferans can withdraw their bodies into their lorica. Priapulids are marine burrowing worms of moderate size.

Clade Panarthropoda contains onychophorans, tardigrades, and arthropods. They have open circulatory systems with a hemocoel.

Onychophora are caterpillar-like animals found in humid, mostly tropical habitats. They are segmented and crawl by means of a series of unjointed, clawed appendages.

Tardigrades are minute animals, mostly terrestrial, living in the water film that surrounds mosses and lichens. They have eight unjointed legs and a nonchitinous cuticle. They can undergo cryptobiosis, withstanding adverse conditions for long periods.

REVIEW QUESTIONS

1. What is a cuticle?
2. Define ecdysis.
3. What is a hydrostatic skeleton?
4. Distinguish a solenocyte from a flame-cell protonephridium.
5. Explain two peculiar features of the body-wall muscles in nematodes.
6. What feature of body-wall muscles in nematodes requires a high hydrostatic pressure in the pseudocoelomic fluid for efficient function?
7. Explain the interaction of cuticle, body-wall muscles, and pseudocoelomic fluid in locomotion of nematodes.
8. Explain how the high pseudocoelomic pressure affects feeding and defecation in nematodes.
9. Outline the life cycle of each of the following: *Ascaris lumbricoides,* hookworm, *Enterobius vermicularis, Trichinella spiralis, Wuchereria bancrofti.*
10. Where in the human body are adults of each species in question 9 found?
11. Outline the life cycle of a gordiid nematomorph.
12. How are nematodes and nematomorphs alike, and how are they different?
13. Where do kinorhynchs live?
14. Describe the introvert of a loriciferan and a priapulid.
15. How is a hemocoel different from a true coelom?
16. In what sense is a hemocoel part of the circulatory system?
17. In what habitats would you encounter tardigrades?
18. How does cryptobiosis in tardigrades increase the likelihood of survival?
19. Describe the two major protostome clades and give a defining feature for each.
20. List the predominant body plan (acoelomate, pseudocoelomate, or coelomate) for members of each protostome phylum and discuss how our picture of protostome evolution would change if each body plan were a homologous character.

SELECTED REFERENCES

Aguinaldo, A. M. A., J. M. Turbeville, L. S. Linford, M. C. Rivera, J. J. F. R. Garey, R. A. Raff, and J. A. Lake. 1997. Evidence for a clade of nematodes, arthropods and other moulting animals. Nature **387:**489–493. *Sequence analysis to support a superphylum Ecdysozoa.*

Balavoine, G., and A. Adoutte. 1998. One or three Cambrian radiations? Science **280:**397–398. *Discusses radiation into superphyla Ecdysozoa, Lophotrochozoa, and Deuterostomia.*

Bird, A. F., and J. Bird. 1991. The structure of nematodes, ed. 2. New York, Academic Press. *The most authoritative reference available on nematode morphology. Highly recommended.*

Brusca, R. C., and G. J. Brusca. 2003. Invertebrates, ed. 2. Sunderland, Massachusetts, Sinauer Associates, Inc. *A Comprehensive invetebrate text.*

Chan, M.-S. 1997. The global burden of intestinal nematode infections—fifty years on. Parasitol. Today **13:**438–443. *According to this author, most recent estimates are 1.273 billion infections (24% prevalence) with* Ascaris, *0.902 billion (17% prevalence) with* Trichuris, *and 1.277 billion (24% prevalence) with hookworms. Worldwide prevalence of these nematodes has remained essentially unchanged in 50 years!*

Despommier, D. D. 1990. *Trichinella spiralis:* the worm that would be virus. Parasitol. Today **6:**193–196. *Juveniles of* Trichinella *are among the largest of all intracellular parasites.*

Dopazo, H., and J. Dopazo. 2005. Genome-scale evidence of the nematode-arthropod clade. Genome Biol. **6:**R41. *Phylogenetic trees constructed* *under the hypothesis that a coelomate body is homologous are compared with those assuming ecdysis is a homologous trait.*

Duke, B. O. L. 1990. Onchocerciasis (river blindness)—can it be eradicated? Parasitol. Today **6:**82–84. *Despite the introduction of a very effective drug, the author predicts that this parasite will not be eradicated in the foreseeable future.*

Garey, J. R., M. Krotec, D. R. Nelson, and J. Brooks. 1996. Molecular analysis supports a tardigrade-arthropod association. Invert. Biol. **115:**79–88. *Relationship of tardigrades and arthropods based on morphological characters is supported by sequence analysis of the gene encoding small-subunit rRNA.*

Gould, S. J. 1995. Of tongue worms, velvet worms, and water bears. Natural History **104**(1):6–15. *Intriguing essay on affinities of Pentastomida, Onychophora, and Tardigrada and how they, along with larger phyla, were products of the Cambrian explosion.*

Halanych, K. M., and Y. Passamaneck. 2001. A brief review of metazoan phylogeny and future prospects in Hox-research1. Am. Zool. **41:**629–639. *A good review of the arguments for and against the lophotrochozoa and ecdysozoa hypotheses.*

Holterman, M., A. van der Wurff, S. van den Elsen, H. van Megen, T. Bongers, O. Holovachov, J. Bakker, and J. Helder. 2006. Phylum-wide analysis of SSU rDNA reveals deep phylogenetic relationships among nematodes and accelerated evolution toward crown clades. Mol. Biol. Evol. **23:**1792–1800. *A new nematode phylogeny shows that plant-parasitic species arose from fungivorous ancestors in three lineages.*

Neuhaus, B., and R. P. Higgins. 2002. Ultrastructure, biology, and phylogenetic relationships of Kinorhyncha. Integ. and Comp. Biol. **42:**619–632. *A detailed summary of the biology and morphology of these animals.*

Nielsen, C. 1995. Animal evolution: Interrelaionships of the living phyla. Oxford University Press, New York. *The author proposes homologous features for many lesser-known taxa.*

Ogilvie, B. M., M. E. Selkirk, and R. M. Maizels. 1990. The molecular revolution and nematode parasitology: yesterday, today, and tomorrow. J. Parasitol. **76:**607–618. *Modern molecular biology has wrought enormous changes in investigations on nematodes.*

Poinar, G. O., Jr. 1983. The natural history of nematodes. Englewood Cliffs, New Jersey, Prentice-Hall, Inc. *Contains a great deal of information about these fascinating worms.*

Reinhard, J., and D. M. Rowell. 2005. Social behaviour in an Australian velvet worm, *Euperipatoides rowelli* (Onychophora: Peripatopsidae). J. Zool. **267:**1–7. *This velvet worm hunts collectively and has an organized social structure where one female is dominant.*

Taylor, M. J., and A. Hoerauf. 1999. *Wolbachia* bacteria of filarial nematodes. Parasitol. Today **15:**437–442. *All filarial parasites of humans have endosymbiotic* Wolbachia, *and most filarial nematodes of all kinds are infected. Nematodes can be "cured" by treatment with the antibiotic tetracycline. If cured, they cannot reproduce. Bacteria are apparently passed vertically from females to offspring.*

ONLINE LEARNING CENTER

Visit www.mhhe.com/hickmanipz14e for chapter quizzing, key term flash cards, web links, and more!

19

Trilobites, Chelicerates, and Myriapods

- PHYLUM ARTHROPODA
- SUBPHYLUM TRILOBITA
- SUBPHYLUM CHELICERATA
- SUBPHYLUM MYRIAPODA

A scorpion.

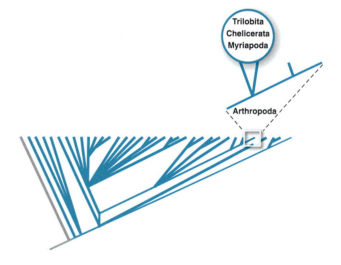

A Suit of Armor

Sometime, somewhere in the Precambrian era, a major milestone occurred in the evolution of life on earth. The soft cuticle in the segmented ancestor of animals we now call arthropods was hardened by deposition of additional protein and an inert polysaccharide called chitin. The cuticular exoskeleton offered some protection against predators and other environmental hazards, and it conferred on its possessors a formidable array of other selective advantages. For example, a hardened cuticle provided a more secure site for muscle attachment, allowed adjacent segments and joints to function as levers, and vastly improved the potential for rapid locomotion, including flight. Of course, a suit of armor could not be uniformly hard; the animal would be as immobile as the rusted Tin Woodsman in the *Wizard of Oz*. Stiff sections of cuticle were separated from each other by thin, flexible sections, which formed sutures and joints. The cuticular exoskeleton had enormous evolutionary potential. Jointed extensions on each segment became appendages.

As the hardened cuticle evolved, or perhaps concurrently with it, many other changes took place in the bodies and life cycles of protoarthropods. Growth required a sequence of cuticular molts controlled by hormones. Coelomic compartments reduced their hydrostatic skeletal function, perhaps causing a regression of the coelom and its replacement with an open system of sinuses (hemocoel). Motile cilia were lost. These changes and others are called "arthropodization." Some zoologists argue that all changes in arthropodization followed from the development of a cuticular exoskeleton. If several different ancestors had independently evolved a cuticular exoskeleton, then they independently would have evolved the suite of characters that we call arthropodization. If this were the case, the huge phylum Arthropoda would be polyphyletic. We agree with other zoologists who find the weight of evidence supports single-phylum status.

PHYLUM ARTHROPODA

Phylum Arthropoda (ar-throp'o-da) (Gr. *arthron,* joint, + *pous, podos,* foot) is currently the most species-diverse phylum in the animal kingdom, composed of more than three-fourths of all known species. Approximately 1,100,000 species of arthropods are recorded, and it is likely that as many more remain to be classified. (In fact, based on surveys of insect faunas in the canopies of rain forests, many estimates of yet undescribed species are much higher.) Arthropods include spiders, scorpions, ticks, mites, crustaceans, millipedes, centipedes, insects, and other less well-known groups. In addition, there is a rich fossil record extending to the very late Precambrian period.

Few arthropods exceed 60 cm in length, and most are much smaller. However, Paleozoic eurypterids reached up to 3 m and some ancient dragonfly-like insects (Protodonata) had wingspans approaching 1 m. Currently, the largest arthropod, a Japanese crab *Macrocheira* (Gr. *makros,* large, + *cheir,* hand), spans approximately 4 m; the smallest is a parasitic mite *Demodex* (Gr. *dēmos,* people, + *dex,* a wood worm), which is less than 0.1 mm long.

Arthropods are usually active, energetic animals. They utilize all modes of feeding—carnivory, herbivory, and omnivory—although most are herbivorous. Many aquatic arthropods are omnivorous or depend on algae for nourishment, and most land forms live chiefly on plants. In diversity of ecological distribution, arthropods have no rivals.

Although many terrestrial arthropods compete with humans for food and spread serious vertebrate diseases, they are essential for the pollination of many food plants, and they also serve as food in the ecosystem, yield drugs, and generate products such as silk, honey, beeswax, and dyes.

Arthropods are more widely distributed throughout all regions of the earth's biosphere than are members of any other eukaryotic phylum. They occur in all types of environment from the deepest ocean depths to very high elevations, and from the tropics far into both northern and southern polar regions. Different species are adapted for life in the air; on land; in fresh, brackish, and marine waters; and in or on the bodies of plants and other animals. Some species live in places where no other animal could survive.

Relationships Among Arthropod Subgroups

Arthropods are ecdysozoan protostomes belonging to clade Panarthropoda (see Figure 18.1). They have segmented bodies, a chitinous cuticle often containing calcium, and jointed appendages. The critical stiffening of the cuticle to form a jointed exoskeleton is sometimes called "arthropodization."

Arthropods diversified greatly, but it is relatively easy to identify particular body plans characterizing arthropod subgroups. For example, centipedes and millipedes have trunks composed of repeated similar segments, whereas spiders have two distinct body regions and lack repeated segments. Arthropoda is divided into several **subphyla** based on our current understanding of the relationships among subgroups.

Traditionally, centipedes, millipedes, and related forms called pauropods and symphylans, were grouped with the insects in subphylum Uniramia. Members of Uniramia all possessed **uniramous** appendages—those with a single branch—as opposed to **biramous** appendages, which have two branches (Figure 19.1). Phylogenies constructed using molecular data did not support Uniramia as a monophyletic group. Further, as the genetic basis for the uniramian versus biramian appendage was better understood (see p. 439), it became increasingly unlikely that all uniramous appendages were inherited from a single common ancestor with such appendages.

Currently, five arthropod subphyla are defined. Centipedes, millipedes, pauropods, and symphylans are placed in subphylum **Myriapoda.** Insects are placed in subphylum **Hexapoda.** Spiders, ticks, horseshoe crabs, and their relatives form subphylum **Chelicerata.** Lobsters, crabs, barnacles, and many others form subphylum **Crustacea.** We include tongue worms, members of former phylum Pentastomida, in Crustacea. The extinct trilobites are placed in subphylum **Trilobita.**

Relationships among the subphyla are controversial. One hypothesis assumes that all arthropods possessing a particular mouthpart, called a **mandible** (Figure 19.1), belong to a single clade, Mandibulata. This clade includes members of Myriapoda, Hexapoda, and Crustacea. Arthropods that do not have mandibles possess **chelicerae** (Figure 19.1), as exemplified by spiders. Thus, according to the "mandibulate hypothesis," myriapods, hexapods, and crustaceans are more closely related to each other than are any of them to chelicerates. Critics of the mandibulate hypothesis argue that the mandibles in each group are so different from each other that they could not be homologous. Mandibles of crustaceans are multijointed with chewing or biting surfaces on the mandible bases (gnathobasic mandible), whereas those of myriapods and hexapods have a single joint with the biting surface on the distal edge (entire-limb mandible). There are also some differences in the muscles controlling the two types. Proponents of the mandibulate hypothesis respond that the 550-million-year history of the mandibulates makes possible the evolution of diverse mandibles from an ancestral type.

We assume that subphylum Trilobita was the earliest diverging arthropod subgroup. We depict subphylum Crustacea as the sister taxon of subphylum Hexapoda, but do not specify a branching order for subphylum Myriapoda, subphylum Chelicerata, or the combined branch with hexapods and crustaceans (Figure 19.2). Evidence of a close relationship between hexapods and crustaceans emerged from several phylogenetic studies using molecular characters; these studies prompted a reevaluation of the morphological characters in members of both taxa. We unite subphylum Crustacea with subphylum Hexapoda in clade Pancrustacea. The exact nature of the close relationship between these two subphyla is at issue and is discussed in Chapters 20 and 21.

Following a general introduction to the arthropods in this chapter, we cover three subphyla: Trilobita, Chelicerata, and Myriapoda. Chapter 20 is devoted to subphylum Crustacea and Chapter 21 to subphylum Hexapoda.

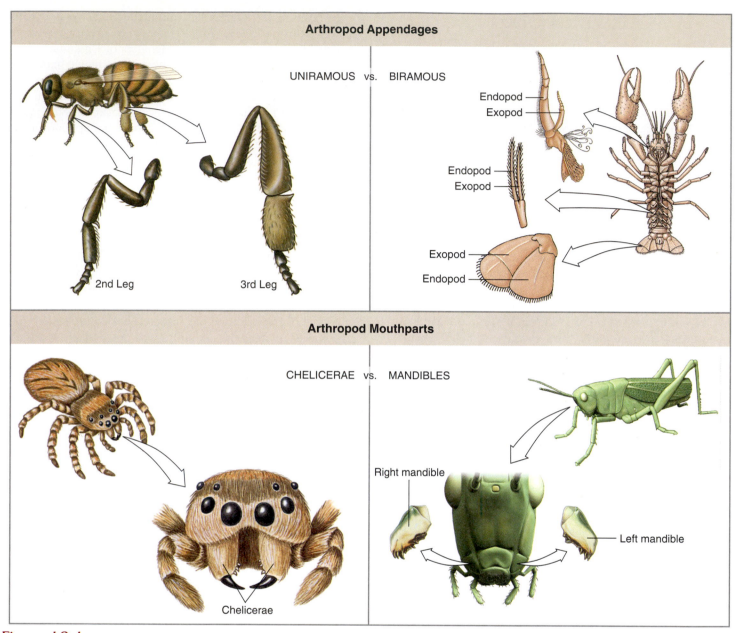

Figure 19.1

Two important arthropod characters: Appendages may be uniramous (honey bee leg) or biramous (lobster limbs); mouthparts may include chelicerae (spider) or mandibles (grasshopper). Note that presence or absence of gills is unrelated to appendage form.

Why Have Arthropods Achieved Such Great Diversity and Abundance?

Arthropods exhibit great diversity (number of species), wide distribution, variety of habitats and feeding habits, and have an uncanny genetic predisposition for adaptation to changing conditions. In our discussion we briefly summarize some structural and physiological patterns that have aided their rise to dominance.

1. **A versatile exoskeleton.** Arthropods possess an exoskeleton that is highly protective without sacrificing flexibility or mobility. This skeleton is the **cuticle,** an outer covering secreted by the underlying epidermis. The cuticle consists of an inner, relatively thick, **procuticle** and an outer, relatively

thin, **epicuticle** (see Figure 19.3). Both the procuticle and the epicuticle each consist of several layers (lamina). The outer epicuticle is composed of protein, often with lipids. The protein is stabilized and hardened by chemical cross-linking, called **sclerotization,** which increases its protective ability. In many insects the outermost layer of epicuticle is composed of waxes that reduce water loss.

The procuticle is divided into an **exocuticle,** which is secreted before a molt, and an **endocuticle,** which is secreted after molting. Both of these layers are composed of **chitin** bound with protein. Chitin is a tough, nitrogenous polysaccharide that is insoluble in water, alkalis, and weak acids. The procuticle is not only flexible and lightweight,

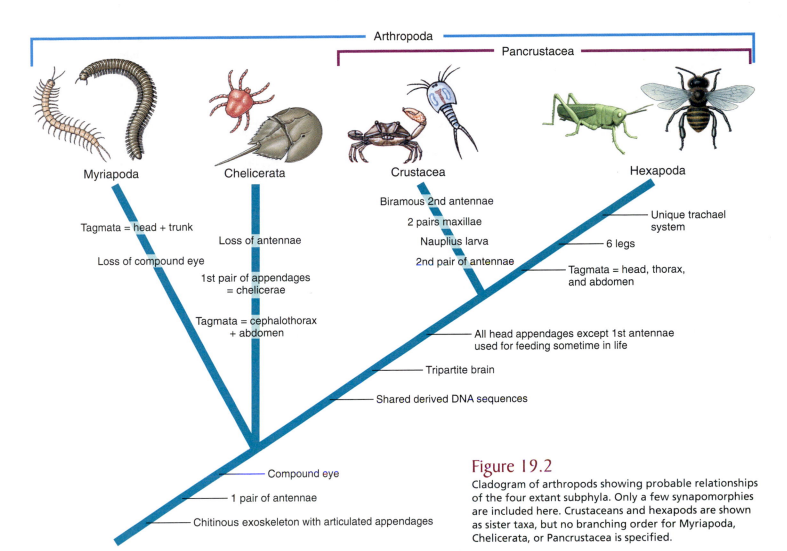

Figure 19.2

Cladogram of arthropods showing probable relationships of the four extant subphyla. Only a few synapomorphies are included here. Crustaceans and hexapods are shown as sister taxa, but no branching order for Myriapoda, Chelicerata, or Pancrustacea is specified.

but also affords protection against dehydration and other biological and physical stresses. In insects chitin makes up about 50% of the procuticle, with the remainder being protein. In some crustaceans, chitin is 60% to 80% of the procuticle; additionally, most crustaceans possess some areas of the procuticle impregnated with **calcium salts.** The addition of calcium salts reduces flexibility, but increases strength. In the hard shells of lobsters and crabs, for example, this calcification is extreme.

The cuticle may be soft and permeable or may form a veritable coat of armor. Between body segments and between the segments of appendages the cuticle is thin and flexible, creating movable joints and permitting free movements. In crustaceans and insects the cuticle forms ingrowths **(apodemes)** to which muscles attach. Cuticle may also line foregut and hindgut, line and support the tracheae and be adapted for biting mouthparts, sensory organs, copulatory organs, and ornamental purposes. It is indeed a versatile material.

The nonexpansible cuticular exoskeleton does, however, impose important restrictions on growth. To grow, an arthropod must shed its outer covering at intervals and grow a larger one—a process called **molting.** The process of molting terminates in the actual shedding of the skin, or **ecdysis.** Arthropods

Figure 19.3

Structure of crustacean cuticle.

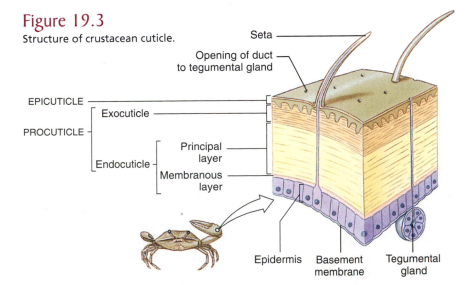

Characteristics of Phylum Arthropoda

1. **Jointed appendages;** ancestrally, one pair to each segment, but number often reduced; appendages often modified for specialized functions
2. Living in marine, freshwater, and terrestrial habitats; many capable of flight
3. Free-living and parasitic taxa
4. Bilateral symmetry; **segmented body** divided into functional groups called **tagmata:** head and trunk; head, thorax, and abdomen; or cephalothorax and abdomen; definite head
5. Triploblastic body
6. **Reduced coelom** in adult; most of body cavity consisting of hemocoel (sinuses, or spaces, in the tissues) filled with blood
7. **Cuticular exoskeleton;** containing protein, lipid, chitin, and often calcium carbonate secreted by underlying epidermis and shed (molted) at intervals; although **chitin** occurs in a few groups other than arthropods, its use better developed in arthropods
8. **Complete digestive system;** mouthparts modified from ancestral appendages and adapted for different methods of feeding; alimentary canal shows great specialization by having, in various arthropods, chitinous teeth, compartments, and gastric ossicles
9. **Complex muscular system,** with exoskeleton for attachment, **striated muscles** for rapid actions, smooth muscles for visceral organs; no cilia
10. **Nervous system** similar to that of annelids, with dorsal brain connected by a ring around the gullet to a double nerve chain of ventral ganglia; fusion of ganglia in some species
11. Well-developed sensory organs; behavioral patterns much more complex than those of most invertebrates, with wider occurrence of **social organization**
12. Parthenogenesis in some taxa
13. **Sexes usually separate,** with paired reproductive organs and ducts; usually internal fertilization; oviparous, ovoviviparous, or viviparous; often with **metamorphosis**
14. Paired excretory glands called **coxal, antennal,** or **maxillary glands** present in some; others with excretory organs called **Malpighian tubules**
15. Respiration by **body surface, gills, tracheae** (air tubes), or **book lungs**
16. **Open circulatory system,** with dorsal **contractile heart,** arteries, and hemocoel (blood sinuses)

may molt many times before reaching adulthood, and some continue to molt after that. More details of the molting process are given for crustaceans (p. 428) and for insects (p. 455). An exoskeleton is also relatively heavy and becomes proportionately heavier as body size increases. Terrestrial arthropods may be limited in body size because of this relationship.

2. **Segmentation and appendages provide for more efficient locomotion.** The ancestral arthropod body plan was likely a linear series of similar segments, each with a pair of jointed appendages. However, extant groups exhibit a wide variety of segments and appendages. There has been a tendency for segments to combine or to fuse into functional groups, called **tagmata** (sing., **tagma**), which have

specialized purposes. Spider bodies, for example, have two tagmata. Appendages are frequently differentiated and specialized for pronounced division of labor. Limb segments are essentially hollow levers moved by internal muscles, most of which are striated, providing rapid action. The appendages have sensory hairs (as well as bristles and spines) and may be modified and adapted for sensory functions, food handling, swift and efficient walking, and swimming.

3. **Air piped directly to cells.** Most terrestrial arthropods have a highly efficient tracheal system of air tubes, which delivers oxygen directly to the tissues and cells and makes a high metabolic rate possible during periods of intense activity. This system also tends to limit body size. Aquatic arthropods breathe mainly by some form of internal or external gill system.

4. **Highly developed sensory organs.** Sensory organs are found in great variety, from the compound (mosaic) eye to those accomplishing touch, smell, hearing, balancing, and chemical reception. Arthropods are keenly alert to what happens in their environment.

5. **Complex behavior patterns.** Arthropods exceed most other invertebrates in complexity and organization of their activities. Innate (unlearned) behavior unquestionably controls much of what they do, but learning also plays an important part in the lives of many species.

6. **Limiting intraspecific competition through metamorphosis.** Many arthropods pass through metamorphic changes, including a larval form quite different from the adult in structure. Larval forms often are adapted for eating food different from that of adults and occupy a different space, resulting in less competition within a species.

SUBPHYLUM TRILOBITA

Trilobites probably had their beginnings before the Cambrian period, in which they flourished. They have been extinct for 200 million years, but were abundant during the Cambrian and Ordovician periods. Their name refers to the trilobed shape of the body in cross section, caused by a pair of longitudinal grooves. They were dorsoventrally flattened bottom dwellers and probably scavengers (Figure 19.4A). Most of them could roll up like pill bugs, and they ranged from 2 to 67 cm in length. Despite their antiquity, they were highly specialized arthropods.

Their exoskeleton contained chitin, strengthened in some areas by calcium carbonate. There were three tagmata in the body: head (also called cephalon), trunk, and pygidium. Their cephalon was one piece but showed signs of ancestral segmentation; their trunk had a variable number of segments; and segments of the pygidium, at the posterior end, were fused into a plate. Their cephalon bore a pair of antennae, compound eyes, a mouth, and four pairs of leglike appendages. There were no true mouthparts (ancestrally derived from jointed appendages), but a hypostome (page 412) likely served in feeding. Each body segment except the last also bore a pair of biramous appendages. One of the branches had a fringe of filaments that may have served as gills.

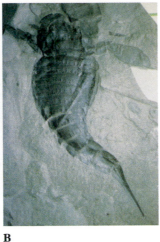

A   **B**

Figure 19.4

Fossils of early arthropods. **A,** Trilobite fossils, dorsal view. These animals were abundant in mid-Cambrian period. **B,** Eurypterid fossil. Eurypterids flourished in Europe and North America from Ordovician to Permian periods.

SUBPHYLUM CHELICERATA

Chelicerate arthropods are an ancient group that includes eurypterids (extinct), horseshoe crabs, spiders, ticks and mites, scorpions, sea spiders and other less well-known groups such as sun scorpions and whip scorpions. Their bodies are composed of two tagmata: a cephalothorax or prosoma, and an abdomen, or opisthosoma. They are characterized by six pairs of cephalothoracic appendages that include a pair of chelicerae (mouthparts), a pair of pedipalps, and four pairs of walking legs. They have no antennae. Most chelicerates suck liquid food from their prey. There are three chelicerate classes (Figure 19.5).

Class Merostomata

Class Merostomata contains eurypterids, all now extinct, and xiphosurids, or horseshoe crabs, an ancient group sometimes called "living fossils."

Subclass Eurypterida

The eurypterids, or giant water scorpions (see Figure 19.4B) were the largest of all fossil arthropods, some reaching a length of 3 m. Their fossils occur in rocks from the Cambrian to the

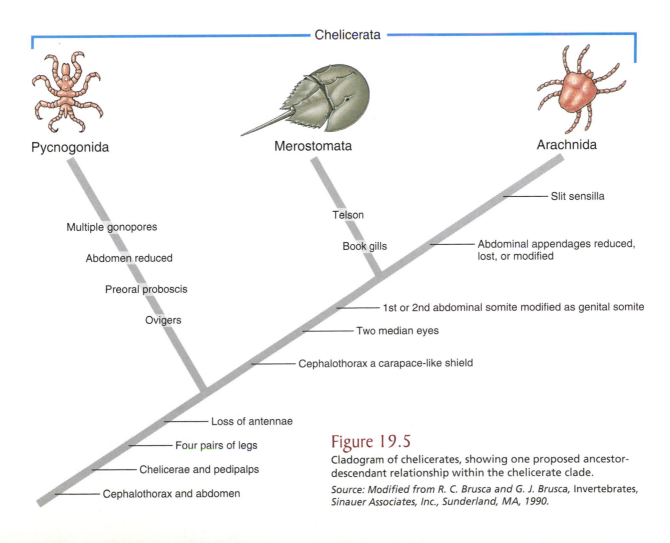

Figure 19.5

Cladogram of chelicerates, showing one proposed ancestor-descendant relationship within the chelicerate clade.

Source: Modified from R. C. Brusca and G. J. Brusca, Invertebrates, *Sinauer Associates, Inc., Sunderland, MA, 1990.*

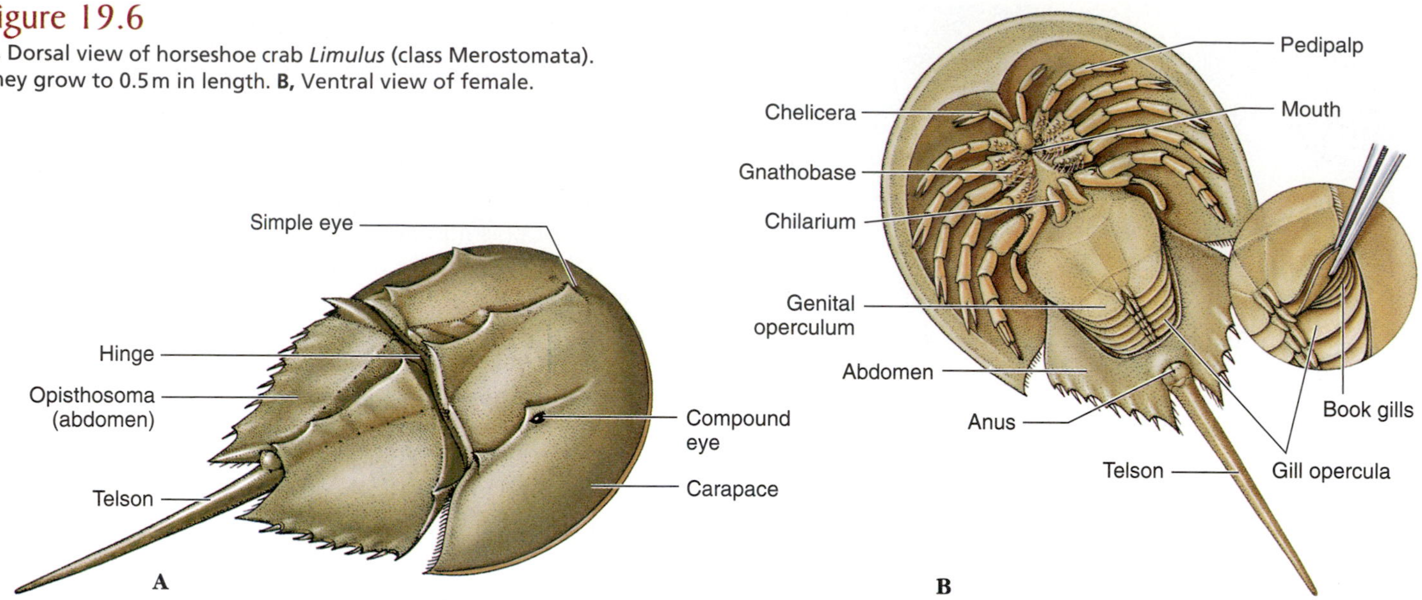

Figure 19.6

A, Dorsal view of horseshoe crab *Limulus* (class Merostomata). They grow to 0.5 m in length. **B,** Ventral view of female.

Simple eye

Hinge

Opisthosoma (abdomen)

Telson

Compound eye

Carapace

A

Pedipalp

Chelicera

Gnathobase

Chilarium

Mouth

Genital operculum

Abdomen

Anus

Telson

Book gills

Gill opercula

B

Permian periods. They had many features resembling those of marine horseshoe crabs (Figure 19.6) as well as those of scorpions. Their heads had six fused segments and bore both simple and compound eyes as well as chelicerae and pedipalps. They also had four pairs of walking legs, and their abdomen had 12 segments and a spikelike telson.

Eurypterids were the dominant predators of their time and some had anterior appendages modified into large, crushing claws. It is possible that development of dermal armor in early fishes (pp. 510–511) resulted from selection pressure of eurypterid predation.

Subclass Xiphosurida: Horseshoe Crabs

Xiphosurids are an ancient marine group that dates from the Cambrian period. Our common horseshoe crab *Limulus* (L. *limus,* sidelong, askew) (Figure 19.6) goes back practically unchanged to the Triassic period. Only three genera (four species) survive today: *Limulus,* which lives in shallow water along the North American Atlantic coast (including the Gulf coast down through Texas and Mexico); *Carcinoscorpius* (Gr. *karkinos,* crab, + *skorpiōn,* scorpion), along the southern shore of Japan; and *Tachypleus* (Gr. *tachys,* swift, + *pleutēs,* sailor), in the East Indies and along the coast of southern Asia. They usually live in shallow water.

Xiphosurids have an unsegmented, horseshoe-shaped **carapace** (hard dorsal shield) and a broad abdomen, which has a long **telson,** or tailpiece. Their cephalothorax bears a pair of chelicerae, one pair of pedipalps, and four pairs of walking legs, whereas their abdomen has six pairs of broad, thin appendages that are fused in the median line (Figure 19.6). On five abdominal appendages, **book gills** (flat, leaflike gills) occur under the gill opercula. There are two lateral, rudimentary eyes and two simple eyes on the carapace. The horseshoe crab swims by means of its abdominal plates and can walk with its walking legs. It feeds at night on worms and small molluscs, which it seizes with its chelicerae and walking legs.

During the mating season horseshoe crabs come to shore by the thousands at high tide to mate. A female burrows into sand where she lays eggs, with one or more smaller males following closely to add sperm to the nest before the female covers it with sand. American *Limulus* mate and lay eggs during high tides of full and new moons in spring and summer. Eggs are warmed by the sun and protected from waves until young larvae hatch and return to the sea, carried by another high tide. Larvae are segmented and are often called "trilobite larvae" because they resemble trilobites.

Class Pycnogonida: Sea Spiders

About 1000 species of sea spiders occupy marine habitats ranging from shallow, coastal waters to deep-ocean basins. Some sea spiders are only a few millimeters long, but others are much larger with legspans up to nearly 0.75 m. They have small, thin bodies and usually four pairs of long, thin walking legs. In addition, they have a feature unique among arthropods: segments are duplicated in some groups, so that they possess five or six pairs of legs instead of the four pairs normally characteristic of chelicerates. Males of many species bear a subsidiary pair of legs **(ovigers)** (Figure 19.7) on which they carry developing eggs, and ovigers are often absent in females. Many species also are equipped with chelicerae and palps. Chelicerae are sometimes called chelifores in this group.

The small head (cephalon) has a raised projection with two pairs of simple eyes. The mouth is at the tip of a long **proboscis,** which sucks juices from cnidarians and soft-bodied animals. Their circulatory system is limited to a simple dorsal heart, and excretory and respiratory systems are absent. The long, thin body and legs provide a large surface area, in proportion to body volume, that is evidently sufficient for diffusion of gases and wastes. Because of the small size of the body, the digestive system and gonads have branches that extend into the legs.

Sea spiders occur in all oceans, but they are most abundant in polar waters. *Pycnogonum* (Figure 19.7B) is a common

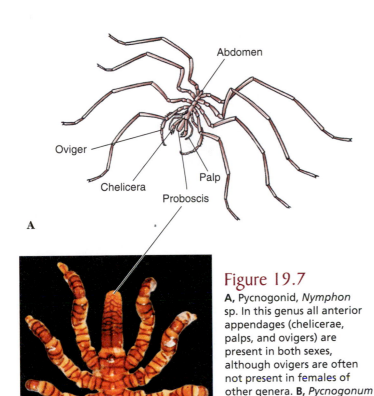

Figure 19.7

A, Pycnogonid, *Nymphon* sp. In this genus all anterior appendages (chelicerae, palps, and ovigers) are present in both sexes, although ovigers are often not present in females of other genera. **B,** *Pycnogonum hancockii,* a pycnogonid with relatively short legs. Females of this genus have neither chelicerae nor ovigers and males have ovigers.

intertidal genus on both Atlantic and Pacific coasts of the United States; it has relatively short, heavy legs. *Nymphon* (Figure 19.7A) is the largest genus of pycnogonids, with over 200 species. It occurs from subtidal depths to 6800 m in all oceans except the Black and Baltic seas.

Some research suggests that pycnogonids belonged to an early diverging arthropod lineage outside any of the subphyla but morphological and molecular evidence strongly supports the placement of pycnogonids in the Chelicerata (see Phylogeny section).

Class Arachnida

Arachnids (Gr. *arachnē,* spider) exhibit enormous anatomical variation. In addition to spiders, the group includes scorpions, pseudoscorpions, whip scorpions, ticks, mites, daddy longlegs (harvestmen), and others. There are many differences among these taxa with respect to form and appendages. They are mostly free-living and are most common in warm, dry regions.

Arachnids have become extremely diverse: More than 80,000 species have been described to date. They were among the first arthropods to move into terrestrial habitats. For example, scorpions are among Silurian fossils, and by the end of the Paleozoic period mites and spiders had appeared.

All arachnids have two tagmata: a cephalothorax (head and thorax) and an abdomen, which may or may not be segmented. The abdomen houses the reproductive organs and respiratory organs such as tracheae and book lungs. The cephalothorax

usually bears a pair of chelicerae, a pair of pedipalps, and four pairs of walking legs (Figure 19.8). Most arachnids are predaceous and have fangs, claws, venom glands, or stingers; fangs are modified chelicerae, whereas claws (chelae) are modified pedipalps. They usually have a strong sucking pharynx with which they ingest the fluids and soft tissues from the bodies of their prey. Among their interesting adaptations are spinning glands of spiders.

Most arachnids are harmless to humans and actually do much good by destroying injurious insects. Arachnids typically feed by releasing digestive enzymes over or into their prey and then sucking the predigested liquid. A few, such as black widow and brown recluse spiders, can give dangerous bites. Stings of scorpions may be quite painful, and those of a few species can be fatal. Some ticks and mites are carriers of diseases as well as causes of annoyance and painful irritations. Certain mites damage a number of important food and ornamental plants by sucking their juices.

Several smaller orders are not included in our discussion.

Order Araneae: Spiders

Spiders are a large group of arachnids comprising about 40,000 species distributed throughout the world. The spider body is compact: a **cephalothorax (prosoma)** and **abdomen (opisthosoma),** both unsegmented and joined by a slender pedicel. A few spiders have a segmented abdomen, which is considered an ancestral character.

Anterior appendages include a pair of **chelicerae** (Figure 19.8), which have terminal **fangs** through which run ducts from venom glands, and a pair of leglike **pedipalps,** which have sensory function and are also used by males to transfer sperm. The basal parts of pedipalps may be used to manipulate food (Figure 19.8). Four pairs of **walking legs** terminate in claws.

All spiders are predaceous, feeding largely on insects, which they effectively dispatch with poison from their fangs. Some spiders chase prey; others ambush them; and many trap them in a net of silk. After a spider seizes prey with its chelicerae and injects venom, it liquefies the prey's tissues with digestive fluid and sucks the resulting broth into its stomach. Spiders with teeth

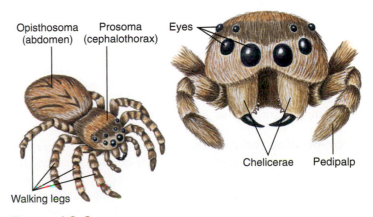

Figure 19.8

External anatomy of a jumping spider, with anterior view of head (at *right*).

Figure 19.9
Spider, internal anatomy.

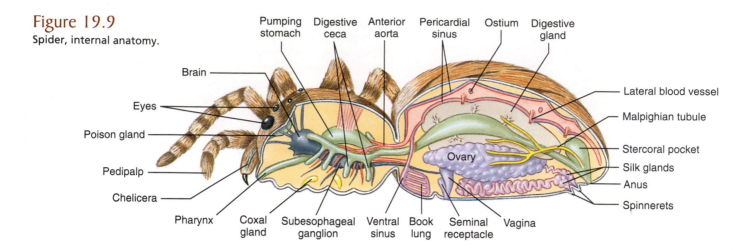

at the bases of chelicerae crush or chew prey, aiding digestion by enzymes from their mouth.

Spiders breathe by means of book lungs or tracheae or both. Book lungs consist of many parallel air pockets extending into a blood-filled chamber (Figure 19.9). Air enters the chamber by a slit in the body wall. Tracheae form a system of air tubes that carry air directly to the blood from an opening called a spiracle. The tracheae are similar to those in insects (p. 449) but are much less extensive and have evolved independently in both arthropod lineages. The tracheal systems of arthropods thus represent a case of massive evolutionary convergence.

Spiders and insects have also independently evolved a unique **excretory system of Malpighian tubules** (Figure 19.9), which work in conjunction with specialized resorptive cells in the intestinal epithelium. Potassium and other solutes and waste materials are secreted into the tubules, which drain the fluid, or "urine," into the intestine. Resorptive cells recapture most potassium and water, leaving behind such wastes as uric acid. This recycling of water and potassium allows species living in dry environments to conserve body fluids by producing a nearly dry mixture of urine and feces. Many spiders also have **coxal glands,** which are modified nephridia that open at the coxa, or base, of the first and third walking legs.

Spiders usually have eight **simple eyes,** each with a lens, optic rods, and a retina (Figure 19.8). They are used chiefly for perception of moving objects, but some, such as those of hunting and jumping spiders, may form images. Since a spider's vision is often poor, its awareness of its environment depends largely on cuticular mechanoreceptors, such as **sensory setae** (sensilla). Fine setae covering the legs can detect vibrations in the web, struggling prey, or even air movements.

Web-Spinning Habits   The ability to spin silk is central to a spider's life, as it is in some other arachnids such as tetranychid spider mites. Two or three pairs of spinnerets containing hundreds of microscopic tubes run to special abdominal **silk glands** (Figure 19.9). A scleroprotein secretion emitted as a liquid from the spinnerets hardens to form a silk thread. Threads of spider silk are stronger than steel threads of the same diameter and second in strength only to fused quartz fibers.

Many species of spiders spin silk webs. The kind of net varies among species. Some webs are simple and consist merely of a few strands of silk radiating out from a spider's burrow or place of retreat. Others spin beautiful, geometrical orb webs. However, spiders use silk threads for many additional purposes: nest lining, sperm webs or egg sacs, bridge lines, draglines, warning threads, molting threads, attachment discs, nursery webs, and for wrapping prey items (Figure 19.10). Not all spiders spin webs for traps. Some spiders throw a sticky bolus of silk to capture their prey. Others, such as wolf spiders, jumping spiders (see Figure 19.8), and fisher spiders (Figure 19.11), simply chase and catch their prey. These spiders likely lost the ability to produce silk for prey capture.

Figure 19.10
Grasshopper, snared and helpless in the web of a golden garden spider (*Argiope aurantia*), is wrapped in silk while still alive. If the spider is not hungry, the prize will be saved for a later meal.

Figure 19.11
Fisher spider, *Dolomedes triton,* feeds on a minnow. This handsome spider feeds mostly on aquatic and terrestrial insects but occasionally captures small fishes and tadpoles. It pulls its paralyzed victim from the water, pumps in digestive enzymes, then sucks out the predigested contents.

Reproduction   A courtship ritual visually precedes mating. Before mating, a male spins a small web, deposits a drop of sperm on it, and then picks up the sperm to be stored in special cavities of his pedipalps. When he mates, he inserts his pedipalps into the female genital opening to store the sperm in his mate's seminal receptacles. A female lays her eggs in a silken net, which she may carry or attach to a web or plant. A cocoon may contain hundreds of eggs, which hatch in approximately two weeks. Young usually remain in the egg sac for a few weeks and molt once before leaving it. The number of molts may vary, but typically ranges between four and twelve before adulthood is reached.

Are Spiders Really Dangerous?   It is amazing that such small and innocuous creatures have generated so much unreasonable fear in human minds. Spiders are timid creatures that, rather than being dangerous enemies to humans, are actually allies in the continuing battle with insects and other arthropod pests. Venom produced to kill prey is usually harmless to humans. The most poisonous spiders bite only when threatened or when defending their eggs or young. Even American tarantulas (Figure 19.12), despite their fearsome size, are *not* dangerous. They rarely bite, and their bite is about as serious as a bee sting.

There are, however, two genera in the United States that can give severe or even fatal bites: *Latrodectus* (L. *latro,* robber, + *dectes,* biter; **black widow,** five species) and *Loxosceles* (Gr. *loxos,* crooked, + *skelos,* leg; **brown recluse,** 13 species). Black widows are moderate to small in size and shiny black, usually with a bright orange or red spot, commonly in the shape of an hourglass, on the underside of their abdomen (Figure 19.13A). Their venom is neurotoxic, acting on the nervous system. About four or five of every 1000 reported bites are fatal.

Brown recluse spiders are brown and bear a violin-shaped dorsal stripe on their cephalothorax (Figure 19.13B). Their venom is hemolytic rather than neurotoxic, producing death of tissues and skin surrounding the bite. Their bite can be mild to serious and occasionally fatal to small children and older individuals.

A   B

Figure 19.13

A, Black widow spider, *Latrodectus mactans,* suspended on her web. Note the red "hourglass" on the ventral side of her abdomen.
B, Brown recluse spider, *Loxosceles reclusa,* is a small venomous spider. Note the small violin-shaped marking on its cephalothorax. The venom is hemolytic and dangerous.

Some spiders in other parts of the world are also dangerous, for example, funnelweb spiders *Atrax* spp. in Australia. Most dangerous of all are spiders in the South and Central American genus *Phoneutria.* They are large (10 to 12 cm leg span) and quite aggressive. Their venom is among the most pharmacologically toxic of spider venoms, and their bites cause intense pain, neurotoxic effects, sweating, acute allergic reaction, and nonsexual enlargement of the penis.

W. S. Bristowe (*The World of Spiders.* 1971 Rev. ed. London, Collins) estimated that at certain seasons a field in Sussex, England (that had been undisturbed for several years) had a population of 2 million spiders to the acre. He concluded that so many spiders could not successfully compete except for the many specialized adaptations they had evolved. These include adaptations to cold and heat, wet and dry conditions, and light and darkness.

Some spiders capture large insects, some only small ones; web-builders snare mostly flying insects, whereas hunters seek those that live on the ground. Some lay eggs in the spring, others in the late summer. Some feed by day, others by night, and some have developed flavors that are distasteful to birds or to certain predatory insects. As it is with spiders, so has it been with other arthropods; their adaptations are many and diverse and contribute in no small way to their long success.

Figure 19.12

A tarantula, *Brachypelma vagans.*

Order Scorpiones: Scorpions

Scorpions are perhaps the most ancient of terrestrial arthropods and comprise about 1400 species worldwide. Although scorpions are more common in tropical and subtropical regions, some occur in temperate zones. Scorpions are generally secretive, hiding in burrows or under objects by day and feeding at night. They feed largely on insects and spiders, which they seize with their pedipalps and shred with their chelicerae.

A   **B**

Figure 19.14

A, An emperor scorpion (order Scorpiones), *Paninus imperator,* with young, which stay with the mother until their first molt. **B,** Harvestmen, *Mitopus* sp. (order Opiliones). Harvestmen run rapidly on their stiltlike legs. They are especially noticeable during the harvesting season, hence the common name.

Sand-dwelling scorpions locate prey by sensing surface waves generated by the movements of insects on or in the sand. These waves are detected by compound slit sensilla located on the last segment of the legs. A scorpion can locate a burrowing cockroach 50 cm away and reach it in three or four quick movements.

Scorpion tagmata are a rather short **cephalothorax,** which bears chelicerae, pedipalps, legs, a pair of large median eyes, and usually two to five pairs of small lateral eyes; a **preabdomen** (or **mesosoma**) of seven segments; and a long slender **postabdomen** (or **metasoma**) of five segments, which ends in a stinging apparatus (Figure 19.14A). Their chelicerae are small; their pedipalps are large and chelate (pincerlike); and the four pairs of walking legs are long and eight-jointed.

On the ventral side of the abdomen are curious comblike **pectines,** which serve as tactile organs for exploring the ground and for sex recognition. The stinger on the last segment consists of a bulbous base and a curved barb that injects venom. Venom of most species is not harmful to humans but may produce a painful swelling. However, the sting of certain species of *Androctonus* in Africa and *Centruroides* (Gr. *kenteō,* to prick, + *oura,* tail, + *oides,* form) in Mexico can be fatal unless antivenom is administered. In general, larger species tend to be less venomous than smaller species and rely on their greater strength to overpower prey.

Scorpions perform a complex mating dance, the male holding the female's chelae as he steps back and forth. He kneads her chelicerae with his own and, in some species, stings her on her pedipalp or on the edge of her cephalothorax. The stinging action is slow and deliberate, and the stinger remains in the female's body for several minutes. Both individuals remain motionless during that time. Finally, the male deposits a spermatophore and pulls the female over it until the sperm mass is taken into the female orifice. Scorpions are truly viviparous; females brood their young within their reproductive tract. After several months to a year of development anywhere from 1 to over 100 young are produced, depending on the species. The young, only a few millimeters long, crawl onto their mother's back until after their

first molt (Figure 19.14A). They mature in 1 to 8 years and may live for as long as 15 years.

Order Opiliones: Harvestmen

Harvestmen (Figure 19.14B), often known as "daddy longlegs," are common throughout the world and comprise about 5000 species. They are easily distinguished from spiders: their abdomen and cephalothorax are rounded and broadly joined, without the constriction of a pedicel; their abdomen shows external segmentation; and they have only two eyes, mounted on a tubercle on their cephalothorax. They have four pairs of long, spindly legs that end in tiny claws. They can cast off one or more of these legs without apparent ill effect if they are grasped by a predator (or human hand). The ends of their chelicerae are pincerlike, and, while carnivorous, they are often scavengers as well.

Harvestmen are not venomous and are harmless to humans. Odoriferous glands that open on the cephalothorax deter some predators with their noxious secretions. Other than some mites, opilionids are unique among arachnids in having a penis for direct sperm transfer; all are oviparous.

Traditionally allied with Acari, more recent studies indicate that Opiliones forms a clade with scorpions and two smaller orders. They are the sister group of scorpions.

Order Acari: Ticks and Mites

Members of order Acari are without doubt the most medically and economically important group of arachnids. They far exceed other orders in numbers of individuals and species. Although about 40,000 species have been described, some authorities estimate that from 500,000 to 1 million species exist. Hundreds of individuals of several species of mites may be found in a small portion of leaf mold in forests. They occur throughout the world in both terrestrial and aquatic habitats, even extending into such inhospitable regions as deserts, polar areas, and hot springs. Many acarines are parasitic during one or more stages of their life cycle.

Most mites are 1 mm or less in length. Ticks, which are only one suborder of Acari, range from a few millimeters to occasionally 3 cm. A tick may become enormously distended with blood after feeding on its host.

Acarines differ from all other arachnids in having complete fusion of the cephalothorax and abdomen, with no sign of external division or segmentation (Figure 19.15). They carry their mouthparts on a little anterior projection, the **capitulum,** which consists mainly of the feeding appendages surrounding the mouth. On each side of their mouth is a chelicera, which functions in piercing, tearing, or gripping food. The form of the chelicerae varies greatly in different families. Lateral to the chelicerae is a pair of segmented pedipalps, which also vary greatly in form and function related to feeding. Ventrally the bases of the pedipalps fuse to form a **hypostome,** whereas a **rostrum,**

A   **B**

Figure 19.15

A, Wood tick, *Dermacentor variabilis* (order Acari). Larvae, nymphs, and adults are all ectoparasites or micropredators that drop off their hosts to molt to the next stage. **B,** Red velvet (harvest) mite, *Trombidium* sp. As with chiggers *(Trombicula),* only larvae of *Trombidium* are ectoparasites. Nymphs and adults are free-living and feed on insect eggs and small invertebrates.

Figure 19.17

Damage to *Chamaedorea* sp. palm caused by mites of the family Tetranychidae (order Acari). Over 130 species of this family occur in North America, and some are serious agricultural pests. Mites pierce plant cells and suck out contents, giving leaves the mottled appearance shown here.

or **tectum,** extends dorsally over their mouth. Adult mites and ticks usually have four pairs of legs, although there may be only one to three in some specialized forms.

Most acarines transfer sperm directly, but many species use a spermatophore. A larva with six legs hatches from the egg, and one or more eight-legged nymphal stages follow before the adult stage is reached.

Many species of mites are entirely free-living. *Dermatophagoides farinae* (Gr. *dermatos,* skin, + *phagō,* to eat, + *eidos,* likeness of form) (Figure 19.16) and related species are denizens of house dust all over the world, sometimes causing allergies and dermatoses. There are some marine mites, but most aquatic species live in freshwater. They have long, hairlike setae on their legs for swimming, and their larvae may be parasitic on aquatic invertebrates. Such abundant organisms must be important ecologically, but many acarines have more direct effects on our food supply and health. Spider mites (family Tetranychidae) are serious agricultural pests on fruit trees, cotton, clover, and many other plants. They suck the contents of plant cells, producing a mottled appearance on the leaves (Figure 19.17), and construct a protective web from silk glands opening near the base of the chelicerae. Larvae of genus *Trombicula* are called **chiggers** or

redbugs. They feed on the dermal tissues of terrestrial vertebrates, including humans, and may cause an irritating dermatitis, but they do not burrow or remain attached to the host. Some species of chiggers transmit a disease called Asiatic scrub typhus. Hair follicle mites, *Demodex* (Figure 19.18), are apparently nonpathogenic in humans; they infect most of us although we are unaware of them. In some cases they may produce a mild dermatitis. Other species of *Demodex* and other genera of mites cause mange in domestic animals. Human itch mites, *Sarcoptes scabiei* (Figure 19.19), cause intense itching as they burrow beneath the skin. Infestations of these mites were very common during World War II because of the crowded conditions under which people were forced to live.

Figure 19.16

Scanning electron micrograph of house dust mite, *Dermatophagoides farinae.*

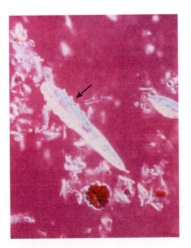

Figure 19.18

Demodex folliculorum, human follicle mite.

Figure 19.19

Sarcoptes scabiei, human itch mite.

Figure 19.20

Boophilus annulatus, a tick that carries Texas cattle fever.

The inflamed welt and intense itching that follows a chigger bite is not the result of the chigger burrowing into the skin, as is popularly believed. Rather a chigger bites through the skin with its chelicerae and injects a salivary secretion containing powerful enzymes that liquefy skin cells. Human skin responds defensively by forming a hardened tube that the larva uses as a drinking straw and through which it gorges itself with host cells and fluid. Scratching usually removes the chigger but leaves the tube, which is a source of irritation for several days.

In addition to disease conditions that they themselves cause, ticks are among the world's premier disease vectors, ranking second only to mosquitos. They surpass other arthropods in carrying a great variety of infectious agents including apicomplexans, rickettsial, viral, bacterial, and fungal organisms. Species of *Ixodes* carry the most common arthropod-borne infection in the United States, Lyme disease (see note). Species of *Dermacentor* (see Figure 19.15A) and other ticks transmit Rocky Mountain spotted fever, a poorly named disease because most cases occur in the eastern United States. *Dermacentor* also transmits tularemia and agents of several other diseases. Texas cattle fever, also called red-water fever, is caused by a protozoan parasite transmitted by cattle ticks, *Boophilus annulatus* (Figure 19.20). Many more examples could be cited.

An epidemic of arthritis occurred in the 1970s in the town of Lyme, Connecticut. Subsequently known as Lyme disease, it is caused by a bacterium and carried by ticks of the genus *Ixodes*. There are now thousands of cases a year in Europe and North America, and other cases have been reported from Japan, Australia, and South Africa. Many people bitten by infected ticks recover spontaneously or do not get the disease. Others, if not treated at an early stage with appropriate antibiotics, develop a chronic, disabling disease.

SUBPHYLUM MYRIAPODA

The term "myriapod," meaning "many footed," describes members of four classes in subphylum Myriapoda that have evolved a pattern of two tagmata—head and trunk—with paired appendages on most or all trunk segments. Myriapods include Chilopoda (centipedes), Diplopoda (millipedes), Pauropoda (pauropods), and Symphyla (symphylans) (Figure 19.21).

Myriapods use tracheae to carry respiratory gases directly to and from all body cells in a manner similar to onychophorans (p. 397) and some arachnids, but tracheal systems have likely evolved independently in each group.

Excretion is usually by Malpighian tubules, but these have evolved independently of Malpighian tubules found in Chelicerata.

Class Chilopoda

Chilopoda (ki-lop′o-da) (Gr. *cheilos,* margin, lip, + *pous, podos,* foot), or centipedes, are land forms with somewhat flattened bodies. Centipedes prefer moist places such as under logs, bark, and stones. They are very agile carnivores, living on cockroaches and other insects, and earthworms. They kill their prey with their venom claws and then chew it with their mandibles. The largest centipede in the world, *Scolopendra gigantea,* is nearly 30 cm in length. Common house centipedes *Scutigera* (L. *scutum,* shield, + *gera,* bearing), which have 15 pairs of legs, are much smaller and often seen scurrying around bathrooms and damp cellars, where they catch insects. Most species of centipedes are harmless to humans, although many tropical centipedes are dangerous. There are about 3,000 species worldwide.

Centipede bodies may contain from a few to 177 segments (Figure 19.22). Each segment, except the one behind the head and the last two in the body, bears a pair of jointed legs. Appendages of the first body segment are modified to form venom claws. The last pair of legs is longer than the others and serves a sensory function.

The head appendages are similar to those of an insect (Figure 19.22B). There are a pair of antennae, a pair of mandibles, and one or two pairs of maxillae. A pair of eyes on the dorsal side of the head consists of groups of ocelli.

The digestive system is a straight tube into which salivary glands empty at the anterior end. Two pairs of Malpighian tubules empty into the hind part of the intestine. There is an elongated heart with a pair of arteries to each segment. The heart has a series of ostia to provide for return of blood to the heart from the hemocoel. Respiration is by means of a tracheal system of branched air tubes that come from a pair of spiracles in each segment. The nervous system is typically arthropodan, and there is also a visceral nervous system.

Sexes are separate, with unpaired gonads and paired ducts. Some centipedes lay eggs and others are viviparous. The young are similar in form to adults and do not undergo metamorphosis.

Class Diplopoda

Diplopoda (Gr. *diploo,* double, two, + *pous, podos,* foot) are commonly called millipedes, which literally means "thousand feet" (Figure 19.23A). Millipedes are not as active as centipedes: They walk with a slow, graceful motion, not wriggling as centipedes do. They prefer dark, moist places under logs or stones. Most millipedes are herbivorous, feeding on decayed plant matter, although sometimes they eat living plants. Millipedes are slow-moving animals and may roll into a coil when disturbed. Many millipedes also protect themselves from predation by

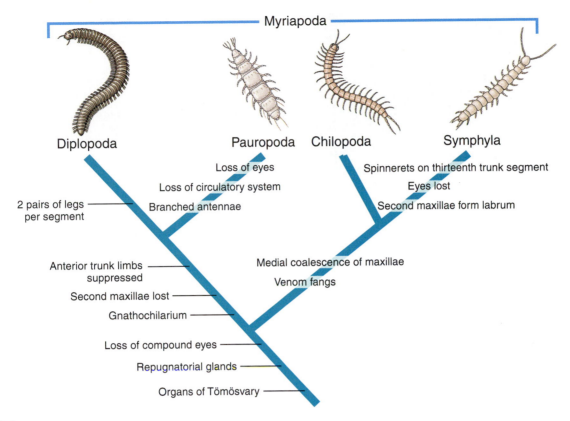

Figure 19.21

Cladogram showing hypothetical relationships of myriapods. Organs of Tömösvary are unique sensory organs opening at the bases of the antennae, and repugnatorial glands, located on certain segments or legs, secrete an obnoxious substance for defense. The gnathochilarium is formed in diplopods and pauropods by fusion of the first maxillae, and the collum is the collarlike tergite of the first trunk segment.

Source: Modified from R. C. Brusca and G. J. Brusca, Invertebrates, *Sinauer Associates, Inc., Sunderland, MA, 1990.*

secreting toxic or repellent fluids from special glands (**repugnatorial glands**) positioned along the sides of the body. Common examples of this class are *Spirobolus* and *Julus*, both of which have wide distribution. There are more than 10,000 species of millipedes worldwide.

The cylindrical body of a millipede is formed by 25 to more than 100 segments. Their short thorax consists of four segments, each bearing one pair of legs. Each abdominal segment has two pairs of legs, leading to the impression of a thousand feet. The millipede exoskeleton is reinforced with calcium carbonate.

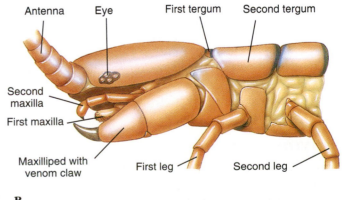

A

B

Figure 19.22

A, A centipede, *Scolopendra* (class Chilopoda) from the Amazon Basin, Peru. Most segments have one pair of appendages each. First segment bears a pair of venom claws, which in some species can inflict serious wounds. Centipedes are carnivorous. **B,** Head of centipede.

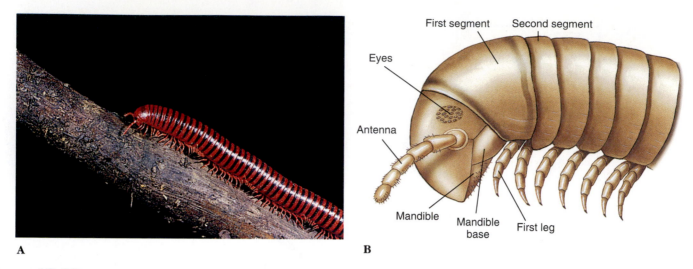

First segment   Second segment

Eyes

Antenna

Mandible   Mandible   First leg
base

A   B

Figure 19.23

A, A tropical millipede with warning coloration. Note the typical doubling of appendages on most segments, hence diplosegments. **B,** Head of millipede.

Their head bears two clumps of simple eyes and a pair each of antennae, mandibles, and maxillae (Figure 19.23B). The general body structures are similar to those of centipedes. Two pairs of spiracles on each abdominal segment open into air chambers that connect to tracheal air tubes. There are two genital apertures toward the anterior end.

In most millipedes the appendages of the seventh segment are specialized as copulatory organs. After millipedes copulate, females lay eggs in a nest and guard them carefully. Interestingly, larval forms have only one pair of legs to each segment.

Class Pauropoda

Pauropoda (Gr. *pauros,* small, + *pous, podos,* foot) are a group of minute (2 mm or less), soft-bodied myriapods, numbering almost 500 species. Although widely distributed, pauropods are the least well known myriapods. They live in moist soil, leaf litter, or decaying vegetation and under bark and debris. Representative genera are *Pauropus* and *Allopauropus.*

Pauropods have a small head with branched antennae and no true eyes, but they have a pair of sense organs that resemble eyes (Figure 19.24A). Their 12 trunk segments usually bear nine pairs of legs (none on the first or the last two segments). They have only one tergal (dorsal) plate covering each two segments.

Tracheae, spiracles, and circulatory system are lacking. Pauropods are probably most closely related to diplopods.

Class Symphyla

Symphyla (Gr. *sym,* together, + *phylon,* tribe) are small (2 to 10 mm) and have centipede-like bodies (Figure 19.24B). They live in humus, leaf mold, and debris. *Scutigerella* (L. dim. of *Scutigera*) are often pests on vegetables and flowers, particularly in greenhouses. They are soft bodied, with 14 segments, 12 of which bear legs and one a pair of spinnerets. The antennae are long and unbranched.

The mating behavior of *Scutigerella* is unusual. Males place a spermatophore at the end of a stalk. When a female finds it, she takes it into her mouth, storing sperm in special buccal pouches. Then she removes eggs from her gonopore with her mouth and attaches them to moss or lichen, or to walls of crevices, smearing them during handling with some of the semen, thereby fertilizing them. Young at first have only six or seven pairs of legs and development is direct.

Symphylans are eyeless but have sensory pits at the bases of the antennae. Their tracheal system is limited to a pair of spiracles on their head and tracheal tubes to anterior segments only. Only 160 species are described.

A   B

Figure 19.24

A, Pauropod. Pauropods are minute, whitish myriapods with three-branched antennae and nine pairs of legs. They live in leaf litter and under stones. They are eyeless but have sense organs that resemble eyes. **B,** *Scutigerella,* a symphylan, is a minute whitish myriapod that is sometimes a greenhouse pest.

PHYLOGENY AND ADAPTIVE DIVERSIFICATION

Phylogeny

Extant arthropods are divided among four subphyla. Relationships among subphyla are subject to debate, but the taxon Pancrustacea, which contains hexapods and crustaceans, is well supported. Which subphylum is the sister taxon to Pancrustacea? According to the mandibulate hypothesis, Myriapoda is grouped with Pancrustacea, but phylogenies where molecular characters are used rarely support this branching pattern. There is more support for placement of Myriapoda as the sister taxon for Chelicerata, and for these two groups together to form the sister taxon for Pancrustacea. As more genetic data are combined with morphological characters in phylogenetic studies, relationships among subphyla will become clearer.

Biologists assume that the ancestral arthropod had a segmented body with one pair of appendages per segment. During evolution, adjacent segments fused to make body regions (tagmata). How many segments contributed to a head in each group of arthropods? *Hox* gene studies indicate that the first five segments, at least, fused to form the head tagma in all four extant subphyla. It is surprising to find the same pattern of fusion in chelicerates as in other subphyla because a head is not immediately obvious in a chelicerate. Spider bodies have two tagmata: prosoma, or cephalothorax, and opisthosoma, or abdomen. Is the head part of the prosoma? *Hox* gene comparisons indicate that the entire prosoma corresponds to the head of other arthropods.

Studies of the heads of pycnogonids were used to detect the phylogenetic position of these odd animals. Sea spiders have spindly bodies and unusual chelicerae. There was speculation that pycnogonids were not chelicerates at all, but instead were the sister taxon of all other arthropods. In the earliest fossil arthropods, appendages emerge from the first head segment, but in spiders and horseshoe crabs, chelicerae and the nerves that control them originate from the second segment during early development. Initial studies of nerve patterns in larval sea spiders indicated that their chelicerae arose, and were controlled, from the first segment. If this result had been confirmed, pycnogonids would be considered the sister taxon to all other arthropods. However, subsequent studies using *Hox* gene expression to define segment boundaries did not support this result. Sea spiders remain within subphylum Chelicerata. They and all living arthropods have head appendages that arise from the region of the head that corresponds to the second segment.

Another controversial area of arthropod biology where genetic studies have proved helpful lies in the evolution and antiquity of uniramous and biramous appendages. Hexapods and myriapods have uniramous appendages, but trilobites and some crustaceans have biramous appendages. If the ancestral appendage were biramous, then the switch to uniramous appendages might have occurred in one lineage whose descendants now carry this trait. Such reasoning led biologists to group hexapods with myriapods, but phylogenies using molecular characters

repeatedly placed hexapods with crustaceans. Is it likely that the uniramous limb evolved more than once? This question would be more easily answered if the genetic basis of limb structure were understood. Studies of the genetic determination of limb

Classification of Phylum Arthropoda

Subphylum Trilobita (tri′lo-bi′ta) (Gr. *tri,* three, + *lobos,* lobe): **trilobites.** All extinct forms; Cambrian to Carboniferous; body divided by two longitudinal furrows into three lobes; distinct head, trunk, and abdomen, biramous (two-branched) appendages.

Subphylum Chelicerata (ke-lis′e-ra′ta) (Gr. *chēlē,* claw, + *keras,* horn, + *ata,* group suffix): **eurypterids, horseshoe crabs, spiders, ticks.** First pair of appendages modified to form chelicerae; pair of pedipalps and four pairs of legs; no antennae; no mandibles; cephalothorax and abdomen usually unsegmented.

Subphylum Myriapoda (mir-ee-ap′o-da) (Gr. *myrias,* a myriad, + *pous, podus,* foot); **myriapods.** All appendages uniramous; head appendages consisting of one pair of antennae, one pair of mandibles, and one or two pairs of maxillae.

Subphylum Crustacea (crus-ta′she-a) (L. *crusta,* shell, + *acea,* group suffix): **crustaceans.** Mostly aquatic, with gills; cephalothorax usually with dorsal carapace; biramous appendages, modified for various functions; head appendages consisting of two pairs of antennae, one pair of mandibles, and two pairs of maxillae; development primitively with nauplius stage (see classification of crustaceans, p. 438).

Subphylum Hexapoda (hek-′sap′oda) (Gr. *hex,* six, + *pous, podus,* foot): **hexapods.** Body with distinct head, thorax, and abdomen; pair of antennae; mouthparts modified for different food habits; head of six fused segments; thorax of three segments; abdomen with variable number, usually 11 somites; thorax with two pairs of wings (sometimes one pair or none) and three pairs of jointed legs; separate sexes; usually oviparous; gradual or abrupt metamorphosis.

Classification of Subphylum Chelicerata

Class Merostomata (mer′o-sto′ma-ta) (Gr. *mēros,* thigh, + *stoma,* mouth, + *ata,* group suffix): **aquatic chelicerates.** Cephalothorax and abdomen; compound lateral eyes; appendages with gills; sharp telson; subclasses Eurypterida (all extinct) and Xiphosurida, horseshoe crabs. Example: *Limulus.*

Class Pycnogonida (pik′no-gon′i-da) (Gr. *pyknos,* compact, + *gony,* knee, angle): **sea spiders.** Small (3 to 4 mm), but some reach 500 mm; body chiefly cephalothorax; tiny abdomen; usually four pairs of long walking legs (some with five or six pairs); mouth on long proboscis; four simple eyes; no respiratory or excretory system. Example: *Pycnogonum.*

Class Arachnida (ar-ack′ni-da) (Gr. *arachnē,* spider): **scorpions, spiders, mites, ticks, harvestmen.** Four pairs of legs; segmented or unsegmented abdomen with or without appendages and generally distinct from cephalothorax; respiration by gills, tracheae, or book lungs; excretion by Malpighian tubules and/or coxal glands; dorsal bilobed brain connected to ventral ganglionic mass with nerves, simple eyes; chiefly oviparous; no true metamorphosis. Examples: *Argiope, Centruroides.*

Classification of Subphylum Myriapoda

Class Diplopoda (di-plop′o-da) (Gr. *diploos,* double, + *pous, podos,* foot): **millipedes.** Body almost cylindrical; head with short antennae and simple eyes; body with variable number of segments; short legs, usually two pairs of legs to a segment; oviparous. Examples: *Julus, Spirobolus.*

Class Chilopoda (ki-lop′o-da) (Gr. *cheilos,* lip, + *pous, podos,* foot): **centipedes.** Dorsoventrally flattened body; variable number of segments, each with one pair of legs; one pair of long antennae; oviparous. Examples: *Cermatia, Lithobius, Geophilus.*

Class Pauropoda (pau-ro′po-da) (Gr. *pauros,* small, + *pous, podos,* foot): **pauropods.** Minute (1 to 1.5 mm); cylindrical body consisting of double segments and bearing 9 or 10 pairs of legs; no eyes. Example: *Pauropus.*

Class Symphyla (sym′fy-la) (Gr. *syn,* together, + *phylē,* tribe): **garden centipedes.** Slender (1 to 8 mm) with long, filiform antennae; body consisting of 15 to 22 segments with 10 to 12 pairs of legs; no eyes. Example: *Scutigerella.*

branching show that modulation of expression of one gene (*Distal-less,* or *Dll*) determines the number of limb branches (see p. 439). Gene expression can be modified within a lineage, so the number of limb branches present is unlikely to be homologous.

The number of appendages per segment is another variable character within Arthropoda. The ancestral arthropod is assumed to have had one pair per segment. Millipedes, in class Diplopoda, have two pairs of appendages on most trunk segments. Did the millipede pattern originate by the repeated fusion of two ancestral segments? Perhaps it did, but expression of the *Distal-less* gene might also have a role here. Larval millipedes have only one pair of appendages per segment.

Adaptive Diversification

Arthropods demonstrate multiple evolutionary trends toward pronounced tagmatization by differentiation or fusion of segments, giving rise to such combinations of tagmata as head and trunk; head, thorax, and abdomen; or cephalothorax (fused head and thorax) and abdomen. The ancestral arthropod condition is to have similar appendages on each segment. More derived forms have appendages specialized for specific functions, or some segments that lack appendages entirely.

Much of the amazing diversity in arthropods seems to have evolved because of modification and specialization of their cuticular exoskeleton and their jointed appendages, resulting in a wide variety of locomotor and feeding adaptations.

While adaptations and specializations made possible by the cuticular exoskeleton of arthropods and other morphological and behavioral features may have fostered high diversity, another important factor ensuring the incredible evolutionary success of the arthropods was undoubtedly their small body size, which allowed them many more types of specialized niches than would be available for larger organisms.

SUMMARY

Arthropoda is the largest, most abundant and diverse phylum of animals. Arthropods are segmented, coelomate, ecdysozoan protostomes with well-developed organ systems. Most show marked tagmatization. They occur in virtually all habitats capable of supporting life. Perhaps more than any other single factor, prevalence of arthropods is explained by adaptations made possible by their cuticular exoskeleton and small size. Other important elements are jointed appendages, tracheal respiration, efficient sensory organs, complex behavior, and metamorphosis.

Trilobites were a dominant Paleozoic subphylum, now extinct. Members of subphylum Chelicerata have no antennae, and their main feeding appendages are chelicerae. In addition, they have a pair of pedipalps (which may be similar to the walking legs) and four pairs of walking legs. Class Merostomata includes the extinct eurypterids and the ancient, although still extant, horseshoe crabs. Class Pycnogonida contains the sea spiders, which are odd little animals with a large suctorial proboscis and vestigial abdomen. The great majority of living chelicerates are in class Arachnida: spiders (order Araneae), scorpions (order Scorpiones), harvestmen (order Opiliones), ticks and mites (order Acari), and others.

Tagmata of most spiders (cephalothorax and abdomen) show no external segmentation and are joined by a waistlike pedicel. Spiders are predaceous, and their chelicerae are provided with venom glands for paralyzing or killing prey. They breathe by book lungs, tracheae, or both. Spiders can spin silk, which they use for a variety of purposes, including in some cases webs for trapping prey.

Distinctive characters of scorpions are their large, clawlike pedipalps and their clearly segmented abdomen, which bears a terminal stinging apparatus. Harvestmen have small, ovoid bodies with very long, slender legs. Their abdomen is segmented and broadly joined to their cephalothorax.

The cephalothorax and abdomen of ticks and mites are completely fused, and mouthparts are borne on an anterior capitulum. Like spiders, some mites can spin silk. They are the most numerous of any arachnids; some are important carriers of disease, and others are serious plant pests.

Members of subphylum Myriapoda have a head followed by a series of trunk segments. The most familiar myriapods are predatory centipedes and herbivorous millipedes.

REVIEW QUESTIONS

1. What are important distinguishing features of arthropods?
2. Name the subphyla of arthropods, and give a few examples of each.
3. Briefly discuss the contribution of the cuticle to the success of arthropods, and name some other factors that have contributed to their success.
4. What is a trilobite?
5. What appendages are characteristic of chelicerates?
6. Briefly describe the distinguishing morphological features of each of the following: eurypterids, horseshoe crabs, pycnogonids.
7. Why are horseshoe crabs in the same subphylum as spiders?
8. What are the tagmata of arachnids, and which tagmata bear appendages?
9. Describe the mechanism of each of the following with respect to spiders: feeding, excretion, sensory reception, web-spinning, reproduction.
10. What are the most important spiders in the United States that are dangerous to humans? How does their venom work?
11. Distinguish each of the following orders from each other: Araneae, Scorpiones, Opiliones, Acari.
12. Discuss the economic and medical importance of members of order Acari to human well-being.
13. How do centipedes capture and subdue prey?

SELECTED REFERENCES

Averof, M. 1998. Evolutionary biology: origin of the spider's head. Nature **395:**436–437. *Summary of research on homology of the head across arthropod subphyla.*

Bowman, A. S., J. W. Dillwith, J. R. Sauer. 1996. Tick salivary prostaglandins: presence, origin and significance. Parasitol. Today **12:**388–396. *Tick prostaglandins act as immunosuppressants, anticoagulants, and analgesics. They allow the tick to feed over an extended time without the blood clotting, an inflammatory reaction occurring, or the host dislodging them.*

Foelix, R. F. 1996. Biology of spiders. New York, Oxford University Press. *Attractive, comprehensive book with extensive references; of interest to both amateurs and professionals.*

Hubbell, S. 1997. Trouble with honeybees. Nat. Hist. **106:**32–43. *Parasitic mites (Varroa jacobsoni on bee larvae and Acarapis woodi in the trachea of adults) cause serious losses among honey bees.*

Hwang, U. W., M. Friedrich, D. Tautz, C. J. Park, and W. Kim. 2001. Mitochondrial protein phylogeny joins myriapods with chelicerates. Nature **413:**154–157. *A sister taxon relationship between myriapods and chelicerates emerges from this study.*

Jager, M., J. Murienne, C. Clabaut, J. Deutsch, H. Le Guyader, and M. Manuel. 2006. Homology of arthropod anterior appendages revealed by *Hox* gene expression in a sea spider. Nature **441:**506–508. *Segment boundaries in sea spider heads show that chelifores (chelicerae) originate from the second head segment.*

Lane, R. P., and R. W. Crosskey (eds). 1993. Medical insects and arachnids. London, Chapman and Hall. *This is the best book currently available on medical entomology.*

Luoma, J. R. 2001. The removable feast. Audubon **103**(3):48–54. *During May and June large numbers of horseshoe crabs ascend the shores of U.S. Atlantic states to breed and lay eggs. Since the 1980s they have been heavily harvested to be chopped up and used for bait. This practice has led to serious declines in* Limulus *populations, with accompanying declines in populations of migrating shore birds that feed on* Limulus *eggs.*

Mallatt, J., J. R. Garey, and J. W. Shultz. 2004. Ecdysozoan phylogeny and Bayesian inference: first use of nearly complete 28s and 18s rRNA gene sequences to classify arthropods and their kin. Mol. Phylogen. Evol. **31:**178–191. *Results indicate that the Crustacea is paraphyletic without hexapods, but that Pancrustacea is a monophyletic group, that chelicerates and myriapods are sister taxa, and that Panarthropoda is a monophyletic group. There was no support for a mandibulate clade.*

McDaniel, B. 1979. How to know the ticks and mites. Dubuque, Iowa, William C. Brown Publishers. *Useful, well-illustrated keys to genera and higher categories of ticks and mites in the United States.*

Ostfeld, R. S. 1997. The ecology of Lyme-disease risk. Am. Sci. **85:**338–346. *Lyme disease, caused by a bacterium transmitted by ticks, has been reported in 48 of the 50 United States and seems to be increasing in frequency and geographic range.*

Polis, G. A. (ed). 1990. The biology of scorpions. Stanford, California, Stanford University Press. *The editor brings together a readable summary of what is known about scorpions.*

Schultz, J. W. 1990. Evolutionary morphology and phylogeny of Arachnida. Cladistics **6:**1–38. *A cladistic analysis of arachnid orders based on morphological data; this study disrupted the traditional views that scorpions are the sister group of other arachnids or were the sister group of eurypterids.*

Shear, W. A. 1994. Untangling the evolution of the web. Am. Sci. **82:**256–266. *Fossil spider webs are nonexistent. Evolution of the web must be studied by comparing modern spider webs to each other and correlating studies of spider anatomy.*

Suter, R. B. 1999. Walking on water. Am. Sci. **87:**154–159. *Fishing spiders (Dolomedes) depend on surface tension to walk on water.*

Weaver, D. C. 1999. Mysterious fevers. Discover **20:**37–40. *Ehrlichiosis is caused by a bacterial parasite of white blood cells transmitted by ticks.*

ONLINE LEARNING CENTER

Visit www.mhhe.com/hickmanipz14e for chapter quizzing, key term flash cards, web links, and more!

Crustaceans

- PHYLUM ARTHROPODA
- SUBPHYLUM CRUSTACEA

A Sally Lightfoot crab, Grapsus grapsus, *from the Galápagos Islands.*

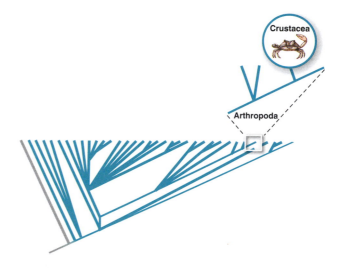

"Insects of the Sea"

The Crustacea (L. *crusta,* shell) are named after the hard shell that most crustaceans bear. Over 67,000 species have been described, and several times that number probably exist. Most familiar to people are the edible ones, for example, lobsters, crayfishes, shrimps, and crabs. In addition to these "crusty" crustaceans, there is an astonishing array of less familiar forms such as copepods, ostracods, water fleas, whale lice, tadpole shrimp, and krill. They fill a wide variety of ecological roles and show enormous variation in morphological characteristics, making a satisfactory description of the group as a whole very difficult.

We live in the Age of Arthropods, notwithstanding our anthropocentric attachment to our tradition of calling the current era the Age of Mammals. Together, insects and crustaceans compose more than 80% of all named animal species. Just as insects pervade terrestrial habitats (more than a million named species and countless trillions of individuals), crustaceans abound in oceans, lakes, and rivers. Some walk, some burrow, and some (such as barnacles) are sessile. Some swim upright, others swim upside down, and many are delicate microscopic forms that drift as plankton in oceans or in lakes. Indeed, it is probable that some of the most abundant animals in the world are members of the copepod genus *Calanus*. In recognition of their dominance of marine habitats, it is understandable that crustaceans have been called "insects of the sea."

Extant arthropods are divided among four subphyla (see Figure 19.2). Crustacea and Hexapoda share five derived features and are united in clade Pancrustacea (Figure 20.1). We depict crustaceans and hexapods as sister taxa, but some phylogenies using molecular characters support the hypothesis that hexapods arose from *within* the crustacean lineage. If the same pattern emerges from studies using other genes, it will be phylogenetically correct to refer to insects as "terrestrial crustaceans." Our description of crustaceans as "insects of the sea" in the prologue to this chapter describes only the ecological role of these animals.

Crustaceans are divided among five classes (Figure 20.1), although preliminary phylogenies using molecular characters do not support the monophyly of all classes. We have placed members of the former phylum Pentastomida in class Maxillopoda,

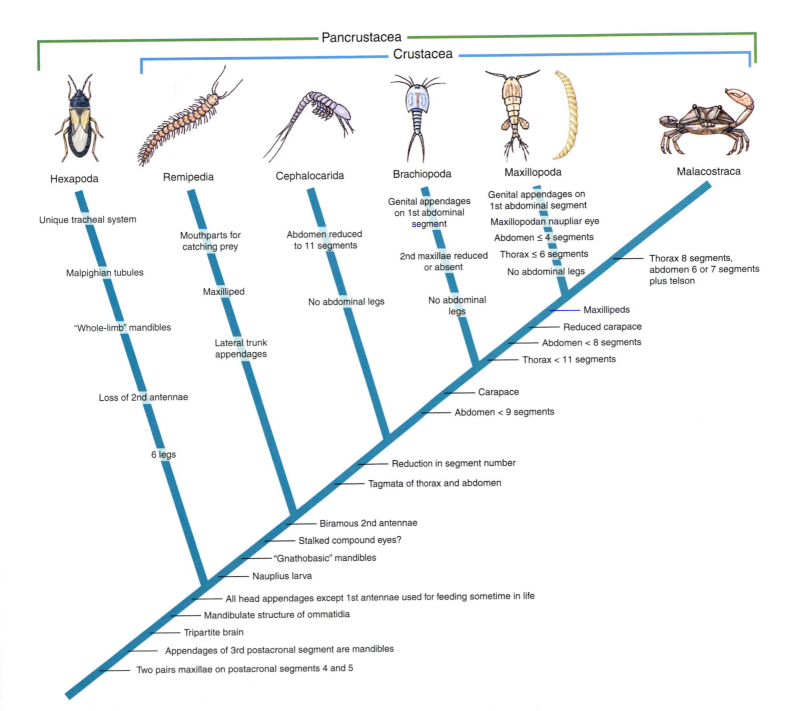

Figure 20.1

Cladogram showing hypothetical ancestor-descendant relationships of hexapods and classes of crustaceans. Hexapods and crustaceans are hypothetical sister groups evolving from a common ancestor defined by numerous shared derived characteristics. Characters followed by a question mark may be ancestral rather than shared derived features. The acron is the anterior region of the head and is not counted as a segment.

subclass Pentastomida. Pentastomids, often called tongue worms, are parasites of vertebrates, living in lungs or nasal cavities. They are closely related to fish lice in subclass Branchiura.

Features of crustacean classes and subclasses are discussed following a general introduction to crustacean biology.

SUBPHYLUM CRUSTACEA

General Nature of a Crustacean

Crustacea are mainly marine; however, there are many freshwater and a few terrestrial species. Crustaceans differ from other arthropods in a variety of ways, but the distinguishing characteristic is that crustaceans are the only arthropods with **two pairs of antennae.** In addition to two pairs of antennae and a pair of **mandibles,** crustaceans have two pairs of **maxillae** on the head, followed by a pair of appendages on each body segment. In some crustaceans not all segments bear appendages. All appendages, except perhaps the first antennae, are ancestrally **biramous** (two main branches), and at least some appendages of present-day adults show that condition. Organs specialized for respiration, if present, function as **gills.**

Most crustaceans have between 16 and 20 segments, but some forms have 60 segments or more. A larger number of segments is an ancestral feature. The more derived condition is to have fewer segments and increased tagmatization (see p. 406). Major tagmata are head, thorax, and abdomen. In most Crustacea one or more thoracic segments are fused with the head to form a **cephalothorax.** Tagmata are not homologous throughout the subphylum (or even within some classes) because in different groups different segments may have fused to form what we now call, for example, a head or a cephalothorax.

By far the largest class of crustaceans is the class Malacostraca, which includes lobsters, crabs, shrimps, beach hoppers, sow bugs, and many others. These species show a surprisingly constant arrangement of body segments and tagmata, which is considered the ancestral plan of this class (Figure 20.2). This typical body plan has a head of five (six embryonically) fused segments, a thorax of eight segments, and an abdomen of six segments (seven in a few species). At the anterior end is a nonsegmented **rostrum** and at the posterior end is a nonsegmented **telson,** which with the last abdominal segment and its **uropods** forms a tail fan in many forms.

In many crustaceans the dorsal cuticle of the head may extend posteriorly and around the sides of the animal to cover or be fused with some or all of the thoracic and abdominal segments. This covering is called a **carapace** (Figure 20.2). In some groups the carapace forms clamshell-like valves that cover most or all of the body. In decapods (including lobsters, shrimp, crabs, and others), the carapace covers the entire cephalothorax but not the abdomen.

Form and Function

Because of their size and easy availability, large crustaceans such as crayfishes have been well studied and are commonly included in introductory laboratory courses. Hence many of the comments here apply specifically to crayfishes and their relatives.

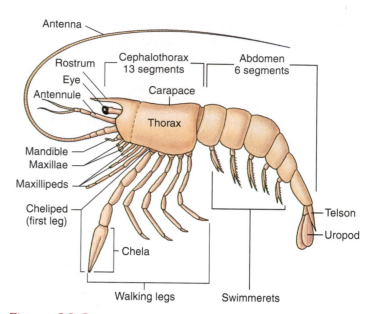

Figure 20.2
Archetypical plan of Malacostraca. The two maxillae and three maxillipeds have been separated diagrammatically to illustrate the general plan.

External Features

Bodies of crustaceans are covered with a secreted cuticle composed of chitin, protein, and calcareous material. The harder, heavy plates of larger crustaceans are particularly high in calcareous deposits. The hard protective covering is soft and thin at the joints between segments, allowing flexibility of movement. The carapace, if present, covers much or all of the cephalothorax; in decapods such as crayfishes, all head and thoracic segments are enclosed dorsally by the carapace. Each segment not enclosed by the carapace is covered by a dorsal cuticular plate, or **tergum** (Figure 20.3A), and a ventral transverse bar, or **sternum,** lies between the segmental appendages (Figure 20.3B). The abdomen terminates in a telson that bears the anus.

Position of the **gonopores** varies according to sex and group of crustaceans. They may be on or at the base of a pair of appendages, at the terminal end of the body, or on segments without legs. For example, in crayfishes openings of the vasa deferentia are on the median side at the base of the fifth pair of walking legs, and those of the oviducts are at the base of the third pair. In females an opening to the seminal receptacle is usually located in the midventral line between the fourth and fifth pairs of walking legs.

Appendages   Members of classes Malacostraca (for example, crayfishes) and Remipedia typically have a pair of jointed appendages on each segment (Figure 20.3B), although abdominal segments in the other classes typically do not bear appendages. Considerable specialization is evident in appendages of derived crustaceans such as crayfishes. The basic, biramous plan is illustrated by a crayfish appendage such as a maxilliped, a thoracic limb modified to become a feeding appendage (Figure 20.4). The basal portion, or **protopod,** bears a lateral **exopod** and a medial

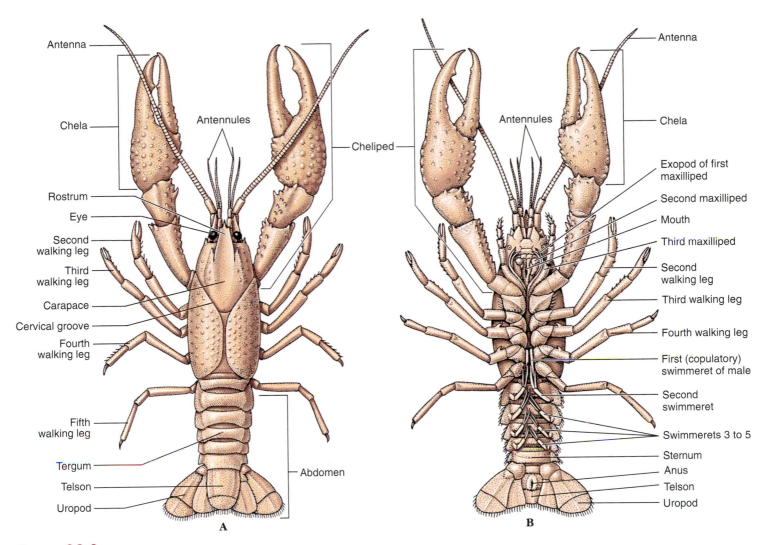

Figure 20.3
External structure of crayfishes. **A,** Dorsal view. **B,** Ventral view.

endopod. The protopod is composed of two parts (**basis** and **coxa**), whereas the exopod and endopod have from one to several parts each. Variations on the basic form exist (Figure 20.5). Some appendages, such as the walking legs of crayfishes, have become secondarily uniramous. Medial or lateral processes, called **endites** and **exites,** respectively, sometimes occur on crustacean limbs. An exite on the protopod is called an **epipod,** which is often modified as a gill. Table 20.1 shows how various appendages have become modified from the presumed ancestral biramous plan and now perform disparate functions.

Terminology applied by various workers to crustacean appendages is not uniform. At least two systems are in wide use. Alternative terms to those we have used, for example, are protopodite, exopodite, endopodite, basipodite, coxopodite, and epipodite. The first and second pairs of antennae may be termed antennules and antennae, and first and second maxillae are often called maxillules and maxillae. A rose by any other name . . .

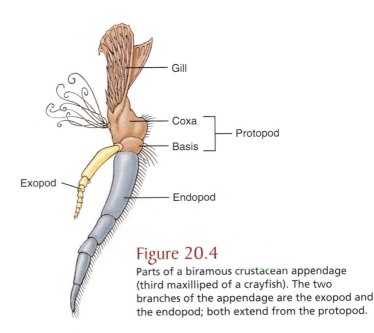

Figure 20.4
Parts of a biramous crustacean appendage (third maxilliped of a crayfish). The two branches of the appendage are the exopod and the endopod; both extend from the protopod.

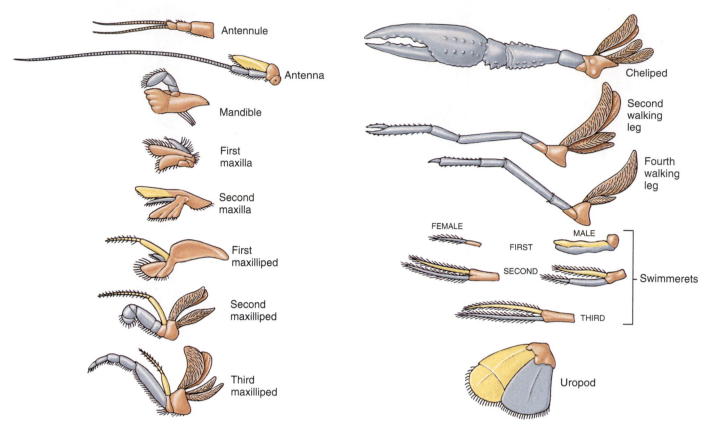

Figure 20.5

Appendages of a crayfish showing variations on the basic biramous plan, as found in a swimmeret. Protopod, brown; endopod, blue; exopod, yellow.

Structures that share a similar basic plan and have descended from a common ancestral form are said to be homologous, whether they have the same function or not. Since specialized walking legs, mouthparts, chelipeds, and swimmerets have all developed from a common biramous appendage (that has become modified to perform different functions), they are all homologous to each other—a condition called **serial homology.** Limbs ancestrally were all very similar, but during structural evolution some branches have been reduced, some lost, some greatly altered, and some new parts added. Crayfishes and their allies possess the most elaborate serial homology in the animal kingdom, with 17 distinct but serially homologous types of appendages (Table 20.1). For example, compare the size of the chela of the cheliped to the tiny claw (chela) of the second walking leg in Figure 20.5.

Internal Features

The muscular and nervous systems in the thorax and abdomen clearly show segmentation, but there are marked modifications in other systems. Most changes involve concentration of parts in a particular region or else reduction or complete loss of parts.

Hemocoel   The major body space in arthropods is not a coelom, but a persistent blastocoel that becomes a blood-filled **hemocoel** (see p. 396). In crustaceans the only coelomic compartments remaining are the end sacs of excretory organs and the space around the gonads.

Muscular System   Striated muscles form a considerable part of the body of most Crustacea. Muscles are usually arranged in antagonistic groups: **flexors,** which draw a part toward the body, and **extensors,** which extend the part outward. The abdomen of a crayfish has powerful flexors (Figure 20.6), which are used when the animal swims backward with a burst of speed to escape from predators.

Respiratory System   Gas exchange in smaller crustaceans occurs across thin areas of cuticle (for example, in the appendages) or the entire body, and specialized structures for gas exchange may be absent. Larger crustaceans have gills, which are delicate, featherlike projections with very thin cuticle. In decapods the sides of the carapace enclose the gill cavity, which is open anteriorly and ventrally (Figure 20.7). Gills may project from the pleural wall into the gill cavity, from the articulation of thoracic legs with the body, or from thoracic coxae. The latter two types are typical of crayfishes. The "bailer," a part of the second maxilla, draws water over the gill filaments, into the gill cavity at the bases of the legs, and out of the gill cavity at the anterior.

Circulatory System   Crustaceans and other arthropods have an "open" or lacunar type of circulatory system. This means that there are no veins and no separation of blood from interstitial

TABLE 20.1

Crayfish Appendages

Appendage	Protopod	Endopod	Exopod	Function
First antenna (antennule)	3 segments, statocyst in base	Many-jointed feeler	Many-jointed feeler	Touch, taste, equilibrium
Second antenna (antenna)	2 segments, excretory pore in base	Long, many-jointed feeler	Thin, pointed blade	Touch, taste
Mandible	2 segments, heavy jaw and base of palp	2 distal segments of palp	Absent	Crushing food
First maxilla (maxillule)	2 segments, with 2 thin endites	Small unjointed lamella	Absent	Food handling
Second maxilla (maxilla)	2 segments, with 2 endites and 1 scaphognathite (epipod)	1 small pointed segment	Part of scaphognathite (bailer)	Drawing currents of water into gills
First maxilliped	2 medial plates and epipod	2 small segments	1 basal segment, plus many-jointed filament	Touch, taste, food handling
Second maxilliped	2 segments plus gill (epipod)	5 short segments	2 slender segments	Touch, taste, food handling
Third maxilliped	2 segments plus gill (epipod)	5 larger segments	2 slender segments	Touch, taste, food handling
First walking leg (cheliped)	2 segments plus gill (epipod)	5 segments with heavy pincer (chela)	Absent	Offense and defense
Second walking leg	2 segments plus gill (epipod)	5 segments plus small pincer	Absent	Walking and prehension
Third walking leg	2 segments plus gill (epipod); genital pore in female	5 segments plus small pincer	Absent	Walking and prehension
Fourth walking leg	2 segments plus gill (epipod)	5 segments, no pincer	Absent	Walking
Fifth walking leg	2 segments; genital pore in male; no gill	5 segments, no pincer	Absent	Walking
First swimmeret	In female reduced or absent; in male fused with endopod to form tube			In male, transferring sperm to female
Second swimmeret Male	Structure modified for transfer of sperm to female	Structure modified for transfer of sperm to female		
Female	2 segments	Jointed filament	Jointed filament	Creating water currents; carrying eggs and young
Third, fourth, and fifth swimmerets	2 short segments	Jointed filament	Jointed filament	Creating water currents; in female carrying eggs and young
Uropod	1 short, broad segment	Flat, oval plate	Flat, oval plate; divided into 2 parts with hinge	Swimming; egg protection in female

fluid, as there is in animals with closed systems. Hemolymph (blood) leaves the heart through arteries, circulates through the hemocoel, and returns to venous **sinuses,** or spaces, instead of veins before it enters the heart.

A dorsal heart is the chief propulsive organ. It is a single-chambered sac of striated muscle. Hemolymph enters the heart from the surrounding **pericardial sinus** through paired ostia, with valves that prevent backflow into the sinus (Figure 20.7). From the heart hemolymph enters one or more arteries. Valves in the arteries prevent a backflow of hemolymph. Small arteries empty into tissue sinuses, which in turn often discharge into a large **sternal sinus** (Figure 20.7).

From there, afferent sinus channels carry hemolymph to the gills, if present, for oxygen and carbon dioxide exchange. Hemolymph then returns to the pericardial sinus by efferent channels (Figure 20.7).

Hemolymph in arthropods may be colorless, reddish, or bluish, as in many Crustacea. Hemocyanin, a copper-containing

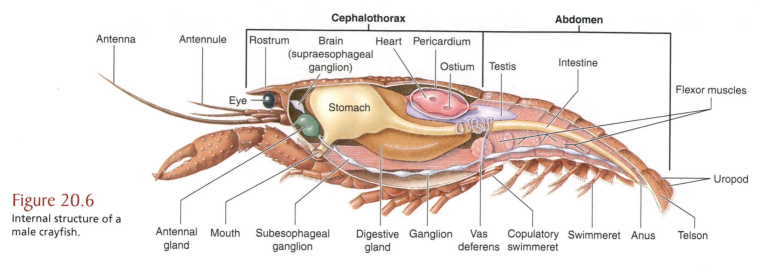

Figure 20.6
Internal structure of a male crayfish.

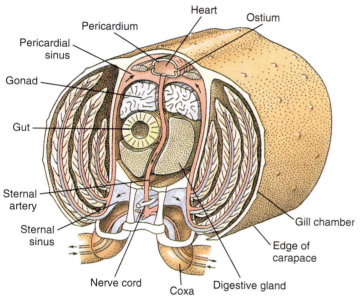

Figure 20.7
Diagrammatical cross section through heart region of a crayfish showing direction of blood flow in this "open" blood system. Heart pumps blood to body tissues through arteries, which empty into tissue sinuses. Returning blood enters sternal sinus, then goes through gills for gas exchange, and finally back to pericardial sinus by efferent channels. Note absence of veins.

respiratory pigment, or hemoglobin, an iron-containing pigment, may be carried in solution. Hemolymph has the property of clotting, which prevents its loss in minor injuries. Some ameboid cells release a thrombinlike coagulant that precipitates clotting.

Excretory System   Excretory organs of adult crustaceans are a pair of tubular structures located in the ventral part of their head anterior to the esophagus (Figure 20.6). They are called **antennal glands** or **maxillary glands,** depending on whether they open at the base of the antennae or at the base of the second maxillae. A few adult crustaceans have both. Excretory organs of decapods are antennal glands, also called green glands

in this group. Crustaceans do not have Malpighian tubules, the excretory organs of spiders and insects.

The **end sac** of the antennal gland consists of a small vesicle **(saccule)** and a spongy mass called a **labyrinth.** The labyrinth connects by an **excretory tubule** to a dorsal **bladder,** which opens to the exterior by a pore on the ventral surface of the basal antennal segment (Figure 20.8). Hydrostatic pressure within the hemocoel provides force for filtration of fluid into the end sac. Filtrate is excreted as urine after the resorption of salts, amino acids, glucose, and some water.

Excretion of nitrogenous wastes (mostly ammonia) takes place by diffusion across thin areas of cuticle, especially gills. The so-called "excretory organs" function principally to regulate ionic and osmotic composition of body fluids. Freshwater crustaceans, such as crayfishes, are constantly threatened with dilution of their blood

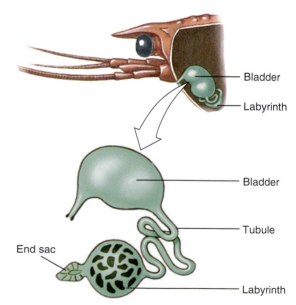

Figure 20.8
Scheme of antennal gland (green gland) of crayfishes. (In natural position organ is much folded.) Some crustaceans lack a labyrinth, and the excretory tubule (nephridial canal) is a much-coiled tube.

by water, which diffuses across the gills and other water-permeable surfaces. Antennal glands, by forming a dilute, low-salt urine, act as an effective "flood-control" device. Some Na⁺ and Cl⁻ are lost in the urine, but this loss is compensated by active absorption of dissolved salt by the gills. In marine crustaceans, such as lobsters and crabs, the antennal glands functions to adjust salt composition of hemolymph by selective modification of salt content from urine. In these forms urine remains isosmotic to the blood.

Nervous and Sensory Systems   Nervous systems of crustaceans and annelids have much in common, although those of crustaceans have more fusion of ganglia (Figure 20.6). The brain consists of a pair of **supraesophageal ganglia** that supplies nerves to the eyes and two pairs of antennae. It is joined by connectives to the **subesophageal ganglion,** a fusion of at least five pairs of ganglia that supply nerves to mouth, appendages, esophagus, and antennal glands. The double ventral nerve cord has a pair of ganglia for each segment and nerves serving the appendages, muscles, and other parts. In addition to this central system, there may be a sympathetic nervous system associated with the digestive tract.

Crustaceans have well-developed sense organs. The largest sense organs of crayfishes are eyes and statocysts. Widely distributed over the body are **tactile hairs,** delicate projections of cuticle that are especially abundant on chelae, mouthparts, and telson. Chemical senses of taste and smell reside in receptors on antennae, mouthparts, and other places.

A saclike **statocyst,** opening to the surface by a dorsal pore, is found on the basal segment of each first antenna of crayfishes. Statocysts contain a ridge that bears sensory setae formed from the chitinous lining and grains of sand that serve as **statoliths.** Whenever the animal changes its position, corresponding changes in the position of the grains on the sensory setae are relayed as stimuli to the brain, and the animal can adjust itself accordingly. Each molt (ecdysis) of cuticle results in loss of the cuticular lining of statocysts and the sand grains that they contain. New grains are acquired through the dorsal pore after ecdysis.

Eyes in many crustaceans are compound, composed of many photoreceptor units called **ommatidia** (Figure 20.9). Covering the rounded surface of each eye is a transparent area of cuticle, the **cornea,** which is divided into many small squares or hexagons known as facets. These facets form the outer faces of the ommatidia. Each ommatidium behaves like a tiny eye and contains several kinds of cells arranged in a columnar fashion (Figure 20.9). Black pigment cells are found between adjacent ommatidia and the movement of pigment in an arthropod compound eye permits it to adjust to different amounts of light. There are three sets of pigment cells in each ommatidium: distal retinal, proximal retinal, and reflecting; these are so arranged that they can form a collar or sleeve around each ommatidium. For strong light (day adaptation) the distal retinal pigment moves inward and meets the outward-moving proximal retinal pigment so that a complete pigment sleeve forms around the ommatidium (Figure 20.9). In this condition only rays that strike the cornea directly will reach the photoreceptor cells (retinuli), for each ommatidium is shielded from others. Thus each ommatidium will see only a limited area

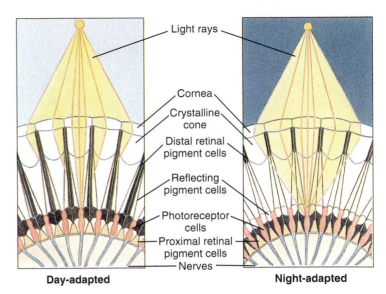

Day-adapted   Night-adapted

Figure 20.9

Portion of compound eye of an arthropod showing migration of pigment in ommatidia for day and night vision. Five ommatidia are represented in each diagram. In daytime each ommatidium is surrounded by a dark pigment collar so that each ommatidium is stimulated only by light rays that enter its own cornea (mosaic vision); in nighttime, pigment forms incomplete collars and light rays can spread to adjacent ommatidia (continuous, or superposition, image).

Light rays
Cornea
Crystalline cone
Distal retinal pigment cells
Reflecting pigment cells
Photoreceptor cells
Proximal retinal pigment cells
Nerves

of the field of vision (a mosaic, or **apposition,** image). In dim light distal and proximal pigments separate so that light rays, with the aid of reflecting pigment cells, have a chance to spread to adjacent ommatidia and to form a continuous, or **superposition,** image. This second type of vision is less precise but takes maximum advantage of the limited amount of light received.

Reproduction, Life Cycles, and Endocrine Function

Most crustaceans have separate sexes, and there are various specializations for copulation among different groups. Barnacles are monoecious but generally practice cross-fertilization. In some ostracods and harpacticoid copepods males are scarce, and reproduction is usually parthenogenetic. Most crustaceans brood their eggs in some manner: branchiopods and barnacles have special brood chambers, copepods have brood sacs attached to the sides of their abdomen (see Figure 20.16), and many malacostracans carry eggs and young attached to their abdominal appendages.

Crayfishes have direct development: there is no larval form. A tiny juvenile with the same form as the adult and a complete set of appendages and segments hatches from the egg. However, development is indirect in most crustaceans, and larvae quite unlike adults in structure and appearance hatch from eggs. Change from larva ultimately to an adult is called **metamorphosis.** The ancestral and most widely occurring larva in Crustacea is the **nauplius** (Figures 20.10 and 20.22). Nauplii bear only three pairs of appendages: uniramous first antennules, biramous antennae, and biramous mandibles. All function as swimming appendages at this stage. Subsequent development may involve a gradual

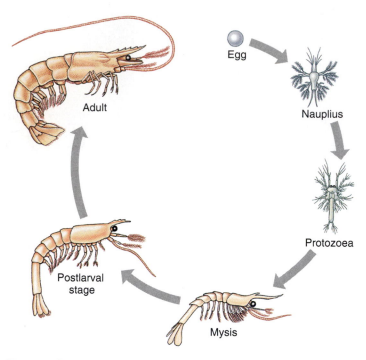

Figure 20.10

Life cycle of a Gulf shrimp, *Farfantepenaeus*. Penaeids spawn at depths of 40 to 90 m. Young larval forms are planktonic and move inshore to water of lower salinity to develop as benthic juveniles and adults. Older shrimp return to deeper water offshore.

change to adult body form, and appendages and segments are added through a series of molts. However, transformation to the adult form may involve more abrupt changes. For example, metamorphosis of a barnacle proceeds from a free-swimming nauplius to a larva with a bivalve carapace called a cyprid and finally to a sessile adult with calcareous plates.

Molting and Ecdysis Molting, the physiological process of making a larger cuticle, and ecdysis (ek′dī-sis) (Gr. *ekdyein,* to strip off), the shedding of the cuticle, are necessary for the body to increase in size because the exoskeleton is nonliving and does not grow as the animal grows. Much of a crustacean's functioning, including its reproduction, behavior, and many metabolic processes, is directly affected by the physiology of the molting cycle.

Cuticle, which is secreted by underlying epidermis, has several layers (Figures 19.3 and 20.11). The outermost is **epicuticle,** a very thin layer of lipid-impregnated protein. The bulk of cuticle is the several layers of **procuticle:** (1) **exocuticle,** which lies just beneath the epicuticle and contains protein, calcium salts, and chitin; (2) **endocuticle,** which itself is composed of (3) a **principal layer,** which contains more chitin and less protein and is heavily calcified, and (4) an uncalcified **membranous layer,** a relatively thin layer of chitin and protein.

Molting animals grow in the **intermolt** phases, or instars, with soft tissues increasing in size until there is no free space within the cuticle. When the body fills the cuticle, the animal enters the **premolt** phase. Growth occurs over a much longer time period than is apparent from examining the external size of the animal.

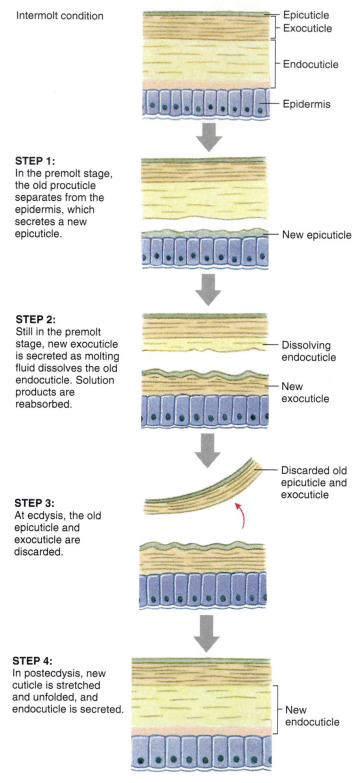

Intermolt condition
— Epicuticle
— Exocuticle
— Endocuticle
— Epidermis

STEP 1:
In the premolt stage, the old procuticle separates from the epidermis, which secretes a new epicuticle.

— New epicuticle

STEP 2:
Still in the premolt stage, new exocuticle is secreted as molting fluid dissolves the old endocuticle. Solution products are reabsorbed.

— Dissolving endocuticle

— New exocuticle

STEP 3:
At ecdysis, the old epicuticle and exocuticle are discarded.

— Discarded old epicuticle and exocuticle

STEP 4:
In postecdysis, new cuticle is stretched and unfolded, and endocuticle is secreted.

— New endocuticle

Figure 20.11

Cuticle secretion and resorption in ecdysis.

During the molting process and some time before actual ecdysis, epidermal cells enlarge considerably. They separate from the membranous layer, secrete a new epicuticle, and begin secreting a new exocuticle (Figure 20.11). Enzymes are released into the area above the new epicuticle. These enzymes begin to dissolve old endocuticle, and soluble products are resorbed and stored within the body of the crustacean. Some calcium salts are stored as **gastroliths** (mineral accretions) in the walls of the stomach. Finally, only exocuticle and epicuticle of the old cuticle remain, underlain by new epicuticle and new exocuticle. The animal swallows water, which it absorbs through its gut, and its blood volume increases greatly. Internal pressure causes the cuticle to split along preformed lines of weakness in the cuticle, and the animal pulls itself out of its old exoskeleton (Figure 20.12). Following this is a stretching of the still soft new cuticle, deposition of new endocuticle, redeposition of salvaged inorganic salts and other constituents, and hardening of the new cuticle. During the period of molting, the animal is defenseless and remains hidden and quiescent.

When a crustacean is young, ecdysis must occur frequently to allow growth, and the molting cycle is relatively short. As the animal approaches maturity, intermolt periods become progressively longer, and in some species molting ceases entirely. During intermolt periods, increase in tissue mass occurs as living tissue replaces water.

Hormonal Control of the Ecdysis Cycle
Although ecdysis is hormonally controlled, the cycle is often initiated by environmental stimuli perceived by the central nervous system. Such stimuli may include temperature, day length, and humidity (in the case of land crabs) or a combination of environmental signals. The signal from the central nervous system decreases production of a **molt-inhibiting hormone** by the **X-organ.** (The X-organ is a group of neurosecretory cells in the medulla terminalis of the brain.) In crayfishes and other decapods, the medulla terminalis is found in the eyestalk. The hormone is carried in axons of the X-organ to the **sinus gland** (which itself is probably not glandular in function), also in the eyestalk, where it is released into the hemolymph.

A drop in level of molt-inhibiting hormone promotes release of a **molting hormone** from the **Y-organs.** Y-organs lie beneath the epidermis near the adductor muscles of the mandibles, and they are homologous to prothoracic glands of insects, which produce the hormone ecdysone. Action of molting hormone is to initiate processes leading to ecdysis. Once initiated, the cycle proceeds automatically without further action of hormones from either X- or Y-organs.

Other Endocrine Functions   Removal of eyestalks accelerates molting; in addition, crustaceans whose eyestalks have been removed can no longer adjust their body coloration to match the background conditions. It was discovered long ago that this defect was caused not by loss of vision but by loss of hormones in the eyestalks. The body color of crustaceans is largely a result of pigments in special branched cells (chromatophores) in the epidermis. Concentration of pigment granules in the center of the cells causes lightening, and dispersal of pigment throughout the cells causes darkening. Pigment behavior is controlled by hormones from neurosecretory cells in the eyestalk, as is migration of retinal pigment for light and dark adaptation in the eyes (see Figure 20.9).

Neurosecretory cells are nerve cells that are modified for secretion of hormones. They are widespread in invertebrates and also occur in vertebrates. Cells in the vertebrate hypothalamus and posterior pituitary are good examples (see p. 759).

Release of neurosecretory material from the pericardial organs in the wall of the pericardium causes an increase in the rate and amplitude of the heartbeat.

Androgenic glands, first discovered in an amphipod (*Orchestia,* a common beach hopper), occur in male malacostracans. Unlike most other endocrine organs in crustaceans, these are not neurosecretory organs. Their secretion stimulates expression of male sexual characteristics. Young malacostracans have rudimentary androgenic glands, but in females these glands fail to develop. If they are artificially implanted in a female, her ovaries transform to testes and begin to produce sperm, and her appendages begin to acquire male characteristics at the next

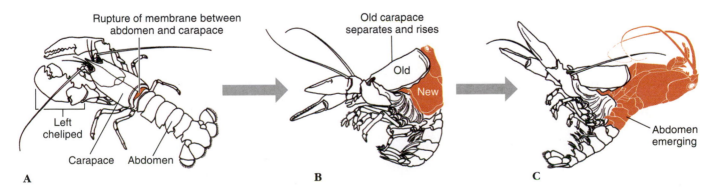

Figure 20.12
Molting sequence in a lobster, *Homarus americanus.* **A,** Membrane between carapace and abdomen ruptures, and carapace begins slow elevation. This step may take up to 2 hours. **B** and **C,** Head, thorax, and finally abdomen withdraw. This process usually takes no more than 15 minutes. Immediately after ecdysis, chelipeds are desiccated and body is very soft. Lobster continues rapid absorption of water so that within 12 hours the body increases about 20% in length and 50% in weight. Tissue water will be replaced by protein in succeeding weeks.

molt. In isopods the androgenic glands are found in testes; in all other malacostracans they are between muscles of the coxopods of the last thoracic legs and partly attached near ends of the vasa deferentia. Although females do not possess organs similar to androgenic glands, their ovaries produce one or two hormones that influence secondary sexual characteristics.

Hormones that influence other body processes in Crustacea also may be present, and evidence suggests that a neurosecretory substance produced in the eyestalk regulates the level of blood sugar.

Feeding Habits

Feeding habits and adaptations for feeding vary greatly among crustaceans. Many forms can shift from one type of feeding to another depending on environment and food availability, but all use the same fundamental set of mouthparts. Mandibles and maxillae function to ingest food; maxillipeds hold and crush food. Walking legs, particularly chelipeds, serve in food capture in predaceous forms.

Many crustaceans, both large and small, are predatory, and some have interesting adaptations for killing prey. For example, one kind of mantis shrimp carries, on one of its walking legs, a specialized digit that can be drawn into a groove and released suddenly to pierce passing prey. Pistol shrimps (*Alpheus* spp.) have an enormously enlarged chela that can be cocked like the hammer of a gun and snapped shut at great speed, forming a cavitation bubble that implodes with a force sufficient to stun their prey.

Food sources of **suspension feeders** range from plankton and detritus to bacteria. **Predators** consume larvae, worms, crustaceans, snails, and fishes. **Scavengers** eat dead animal and plant matter. Suspension feeders, such as fairy shrimps, water fleas, and barnacles, use their legs, which bear a thick fringe of setae, to create water currents that sweep food particles through the setae. Mud shrimps (*Upogebia* spp.) use long setae on their first two pairs of thoracic appendages to strain food from water circulated through their burrow by movements of their swimmerets.

Crayfishes have a two-part stomach (Figure 20.13). The first part contains a **gastric mill** in which food, already shred by their mandibles, can be further ground by three calcareous teeth into particles fine enough to pass through a filter of setae

in the second part; food particles then pass into the intestine for chemical digestion.

A BRIEF SURVEY OF CRUSTACEANS

Crustaceans are an extensive group of over 67,000 species worldwide with many subdivisions. They have many structures, habitats, and modes of living. Some are much larger than crayfishes; others are smaller, even microscopic. Some are highly developed and specialized; others have simpler organization.

Readers should realize that the following summary of crustaceans and the classification on page 438 are misleadingly brief. Although we mention all classes, a complete presentation of taxa in the hierarchy below class level would require coverage well beyond the scope of this textbook.

Class Remipedia

Remipedia (Figure 20.14A) is a very small class of Crustacea. The 10 species described so far have come from caves with connections to the sea. Remipedes have some ancestral crustacean features. There are 25 to 38 trunk segments (thorax and abdomen), all bearing paired, biramous, swimming appendages that are essentially alike. Antennules are biramous. Both pairs of maxillae and a pair of maxillipeds, however, are prehensile and apparently adapted for feeding. The shape of the swimming appendages is similar to that found in Copepoda, but unlike copepods and cephalocarids, swimming legs are directed laterally rather than ventrally.

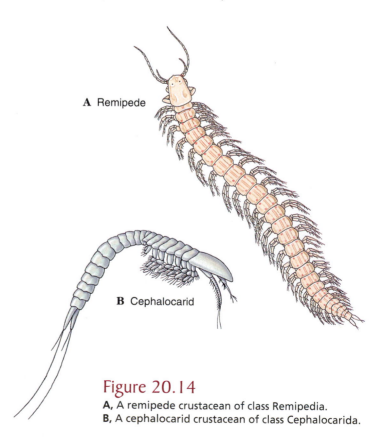

A Remipede

B Cephalocarid

Figure 20.14
A, A remipede crustacean of class Remipedia.
B, A cephalocarid crustacean of class Cephalocarida.

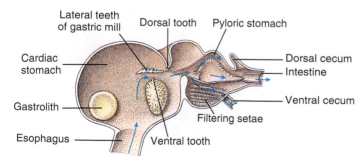

Figure 20.13
Malacostracan stomach showing gastric "mill" and directions of food movements. Mill has chitinous ridges, or teeth, for mastication, and setae for straining food before it passes into the pyloric stomach.

Class Cephalocarida

Cephalocarida (Figure 20.14B) is also a small group, with only nine species known. Cephalocarids occur along both coasts of the United States, in the West Indies, and in Japan. They are 2 to 3 mm long and live in bottom sediments from the intertidal zone to a depth of 300 m. Some features are ancestral: thoracic limbs are very similar to each other, and second maxillae are similar to thoracic limbs. Cephalocarids have no eyes, carapace, or abdominal appendages. True hermaphrodites, they are unique among Arthropoda in discharging both eggs and sperm through a common duct.

Class Branchiopoda

There are over 10,000 species of Branchiopoda, which represent a crustacean type with some ancestral characters. Three orders are recognized: **Anostraca** (fairy shrimp and brine shrimp, Figure 20.15B), with no carapace; **Notostraca** (tadpole shrimp, Figure 20.15A), whose carapace forms a large dorsal shield; and **Diplostraca** (water fleas, Figure 20.15C), which typically have a carapace that encloses the body but not the head, or a carapace that encloses the entire body (clam shrimps). Branchiopods have flattened and leaflike **phyllopodia**—legs that serve as the chief respiratory organs (hence the name branchiopods). Most branchiopods also use their legs for suspension feeding, and in groups other than cladocerans, they use their legs for locomotion as well.

Most branchiopods are freshwater forms. Most important and abundant are water fleas (cladocerans), which often form a large proportion of freshwater zooplankton. Their interesting means of reproduction is reminiscent of that of some rotifers (see Chapter 15). During summer cladocerans often produce only females, by parthenogenesis, rapidly increasing the population. With onset of unfavorable conditions, some males are produced, and eggs that must be fertilized are produced by normal meiosis (production of overwintering, fertilized eggs is termed ephipia). Fertilized eggs are highly resistant to cold and desiccation, and they are very important to the survival of the overwintering population and for passive transfer to new habitats. Most cladocerans have direct development, whereas other branchiopods have gradual metamorphosis.

Class Ostracoda

Members of Ostracoda are, like diplostracans, enclosed in a bivalve carapace and resemble tiny clams, 0.25 to 10 mm long (Figure 20.16A). They are commonly called mussel shrimp or seed shrimp; they have a worldwide distribution and are important in aquatic food webs. Ostracods show considerable fusion of trunk segments, obscuring division between the thorax and abdomen. The trunk has one to three pairs of limbs, with the number of thoracic appendages reduced to two or none. Feeding and locomotion are principally by use of the head appendages. Most ostracods are benthic or climb on plants, but some are planktonic or burrowing, and a few are parasitic. Feeding habits are diverse; there are particle, plant, and carrion feeders and predators. They are widespread in both marine and freshwater habitats. Most of the 6,000 known species are dioecious, but some are parthenogenetic. Some bizarre male mussel shrimps emit light and may synchronize their flashing to attract females. Development is by gradual metamorphosis. There are thousands of extant species and over 10,000 ostracod fossil

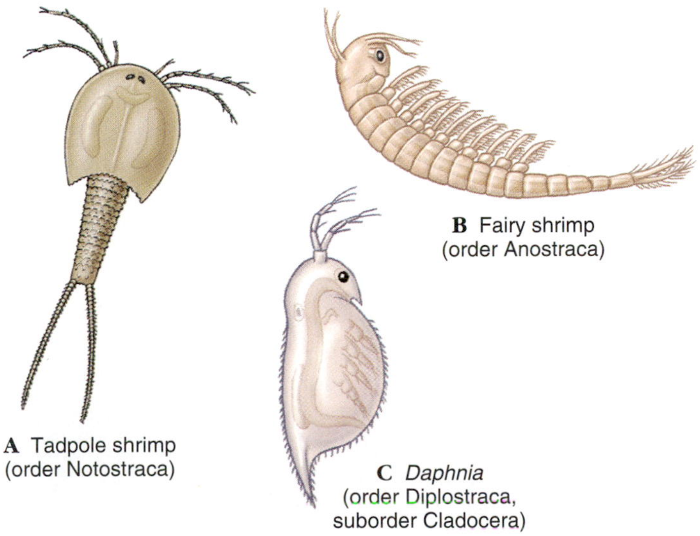

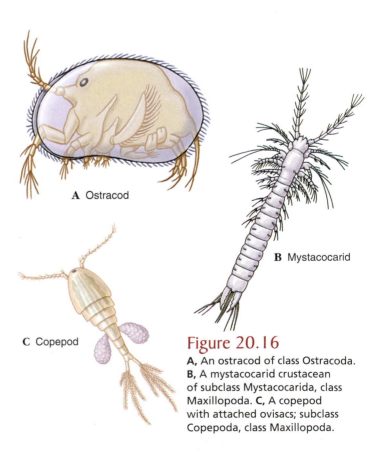

A Tadpole shrimp (order Notostraca)
B Fairy shrimp (order Anostraca)
C *Daphnia* (order Diplostraca, suborder Cladocera)

Figure 20.15
Animals in **A, B,** and **C** are members of class Branchiopoda.

A Ostracod
B Mystacocarid
C Copepod

Figure 20.16
A, An ostracod of class Ostracoda. **B,** A mystacocarid crustacean of subclass Mystacocarida, class Maxillopoda. **C,** A copepod with attached ovisacs; subclass Copepoda, class Maxillopoda.

species whose presence in certain rock strata often serve as important indicators of oil deposits.

Class Maxillopoda

Class Maxillopoda (10,000 species worldwide) includes a number of crustacean groups traditionally considered classes that form a monophyletic group within Crustacea. They basically have five cephalic, six thoracic, and usually four abdominal segments plus a telson, but reductions of the above are common. There are no typical appendages on the abdomen. The eye of the nauplius (when present) has a unique structure and is called a **maxillopodan eye.**

Subclass Mystacocarida

Mystacocarida is a class of tiny crustaceans (less than 0.5 mm long) that live in interstitial water between sand grains of marine beaches (Figure 20.16B). Only 10 species have been described, but mystacocarids are widely distributed through many parts of the world.

Subclass Copepoda

This group is third only to Malacostraca in number of species, and their collective biomass exceeds billions of metric tons throughout the marine and fresh waters of the world. Copepods are small (usually a few millimeters or less in length) and rather elongate, tapering toward the posterior. They lack a carapace and retain a simple, median, nauplius (maxillopodan) eye in adults (Figure 20.16C). They have a single pair of uniramous maxillipeds and four pairs of rather flattened, biramous, thoracic swimming appendages. The fifth pair of legs is reduced. The posterior part of the body is usually separated from the anterior, appendage-bearing portion by a major articulation. Antennules are often longer than other appendages and used in swimming. Copepoda have become very diverse and evolutionarily enterprising, with large numbers of symbiotic as well as free-living species. Many parasites are highly modified, and adults may be so highly modified (and may depart so far from the description just given) that they can hardly be recognized as arthropods, let alone crustaceans.

Ecologically, free-living copepods are of extreme importance, often dominating the primary consumer level (p. 834) in aquatic communities. In many marine localities the copepod *Calanus* is the most abundant organism in zooplankton and has the greatest proportion of total biomass (p. 835). In other localities it may be surpassed in biomass only by euphausids (p. 436). *Calanus* is an important dietary component of such economically and ecologically important fish as herring, menhaden, and sardines. This genus is also important to the larvae of larger fish and (along with euphausids) is an important food item for some whales and sharks that are filter feeders. Other genera commonly occur in marine zooplankton, and some forms such as *Cyclops* and *Diaptomus* may form an important component of freshwater plankton. Many species of copepods are parasites of a wide variety of other

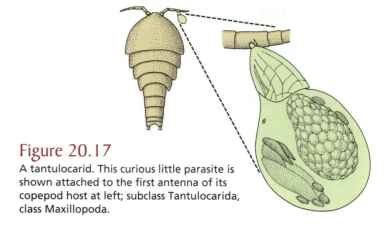

Figure 20.17

A tantulocarid. This curious little parasite is shown attached to the first antenna of its copepod host at left; subclass Tantulocarida, class Maxillopoda.

marine invertebrates and marine and freshwater fish, and can be of economic importance. Some species of free-living copepods serve as intermediate hosts for parasites of humans, such as *Diphyllobothrium* (a tapeworm) and *Dracunculus* (a nematode), as well as of other animals.

Development in copepods is indirect, and some highly modified parasites show striking metamorphoses.

Subclass Tantulocarida

Tantulocarida (Figure 20.17) is the most recently described class (here considered a subclass) of crustaceans (1983). Only about 12 species are known so far. They are tiny (0.15 to 0.2 mm) copepod-like ectoparasites of other deep-sea benthic crustaceans. They have no recognizable head appendages except one pair of antennae on sexual females. Their life cycle is not known with certainty, but present evidence suggests that there is a parthenogenetic cycle and a bisexual cycle with fertilization. **Tantulus** larvae penetrate the cuticle of their hosts by a mouth tube. Then their abdomen and all thoracic limbs are lost during metamorphosis to an adult. Alone among maxillopodans, juveniles bear six to seven abdominal segments, but other evidence supports inclusion in this class.

Subclass Branchiura

Branchiurans are a small group of primarily fish ectoparasites whose mouthparts are modified for sucking (Figure 20.18). Members of this group are usually between 5 and 10 mm long and may be found on marine or freshwater fish. They typically have a broad, shieldlike carapace, compound eyes, four biramous thoracic appendages for swimming, and a short, unsegmented abdomen. Second maxillae have become modified as suction cups, enabling the parasites to move on their fish host or even from fish to fish. Heavily infested fish may get fungal infections and die. There is no nauplius, and young resemble adults except in size and degree of development of appendages.

Figure 20.18

Fish louse; subclass Branchiura, class Maxillopoda.

Subclass *Pentastomida*

Members of former phylum Pentastomida (pen-ta-stom′i-da) (Gr. *pente*, five, + *stoma*, mouth), or tongue worms, consist of about 130 species of wormlike parasites of the respiratory system of vertebrates. Adult pentastomids live mostly in lungs of reptiles, such as snakes, lizards, and crocodiles, but one species, *Reighardia sternae*, lives in air sacs of terns and gulls, and another, *Linguatula serrata* (Gr. *lingua*, tongue), lives in the nasopharynx of canines and felines (and occasionally humans). Although more common in tropical areas, they also occur in North America, Europe, and Australia.

Adults range from 1 to 13 cm in length. Transverse rings give their bodies a segmented appearance (Figure 20.19). Their body is covered with a nonchitinous and highly porous cuticle that is molted periodically during larval stages. The anterior end may bear five short protuberances (hence the name Pentastomida). Four of these bear chitinous claws, and the fifth bears the mouth (Figure 20.20). There is a simple straight digestive system, adapted for sucking blood from the host. The nervous system, similar to that other arthropods, has paired ganglia along the ventral nerve cord. The only sense organs appear to be papillae. There are no circulatory, excretory, or respiratory organs.

Sexes are separate, and females are usually larger than males. A female may produce several million eggs, which pass up the trachea of the host, are swallowed, and exit with feces. Larvae hatch as oval, tailed creatures with four stumpy legs. Most pentastomid life cycles require an intermediate vertebrate host such as a fish, a reptile, or, rarely, a mammal, that is eaten by the definitive vertebrate host. After ingestion by an intermediate host, larvae penetrate the intestine, migrate randomly in the body, and finally metamorphose into

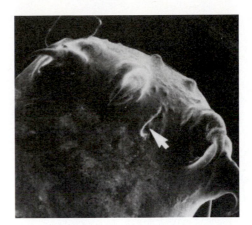

Figure 20.20
Anterior end of a pentastome. Note both the mouth (*arrow*), between the middle hooks, and the apical sensory papillae.

nymphs. After growth and several molts, a nymph finally becomes encapsulated and dormant. When eaten by a final host, a juvenile finds its way to a lung, feeds on blood and tissue, and matures.

Several species have been found encysted in humans, the most common being *Armillifer armillatus* (L. *armilla,* ring, bracelet, + *fero,* to bear), but usually they cause few symptoms. *Linguatula serrata* is a cause of nasopharyngeal pentastomiasis, or "halzoun," a disease of humans in the Middle East and India.

Subclass *Cirripedia*

Cirripedia includes barnacles (order Thoracica), which are usually enclosed in a shell of calcareous plates, as well as three smaller orders of burrowing or parasitic forms. Barnacles are sessile as adults and may be attached to the substrate by a stalk (gooseneck barnacles) (Figure 20.21B) or directly (acorn barnacles) (Figure 20.21A). Typically their carapace (mantle) surrounds their body and secretes a shell of calcareous plates. The head is reduced, they have no abdomen, and the thoracic legs are long, many-jointed cirri with hairlike setae. The cirri are extended through an opening between the calcareous plates to filter out small particles on which the animal feeds (Figure 20.21). Although all barnacles are marine, they are often found in the intertidal zone and are therefore exposed to drying and sometimes freshwater for some periods of time. For example, *Semibalanus balanoides* can tolerate below-freezing temperatures in the Arctic tidal zone and can survive exposed on its rocky substrate for up to nine hours in the summer. During these periods the aperture between the plates closes to a very narrow slit.

Barnacles frequently foul ship bottoms by settling and growing there. So great may be their number that the speed of the ship may be reduced by 30% to 40%, necessitating drydocking the ship to remove them. They may also live atop whales (see Figure 20.26).

Most nonparasitic barnacles are hermaphroditic and undergo a striking metamorphosis during development. Most hatch as nauplii, which soon become cyprid larvae, so called because of their resemblance to an ostracod genus *Cypris*. They have a bivalve carapace and compound eyes. Cyprids attach to the substrate by means of their first antennae, which have adhesive glands, and begin their metamorphosis. This involves several

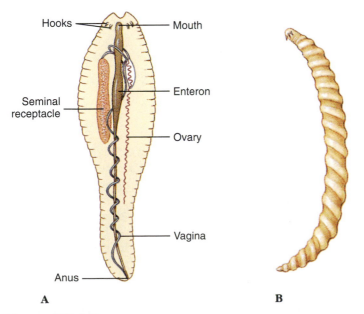

A   B

Figure 20.19
Two pentastomids. **A,** *Linguatula,* found in nasal passages of carnivorous mammals. Female is shown with some internal structures. **B,** Female *Armillifer,* a pentastomid with pronounced body rings. In parts of Africa and Asia, humans are parasitized by immature stages; adults (10 cm long or more) live in lungs of snakes. Human infection may occur from eating snakes or from contaminated food or water.

Labels (Figure 20.19A): Hooks, Mouth, Seminal receptacle, Enteron, Ovary, Vagina, Anus

A

B

Figure 20.21

Barnacles; order Thoracica, subclass Cirripedia, class Maxillopoda. **A,** Acorn barnacles, *Balanus balanoides,* on an intertidal rock await the return of the tide. **B,** Common gooseneck barnacles, *Lepas anatifera.* Note the feeding legs, or cirri, on *Lepas.* Barnacles attach themselves to a variety of firm substrates, including rocks, pilings, and boat bottoms.

dramatic changes, including secretion of calcareous plates, loss of eyes, and transformation of swimming appendages to cirri.

Members of order Rhizocephala, such as *Sacculina,* are highly modified parasites of crabs. These barnacles are dioecious. Like other cirripedes, they start life as nauplii and then become cyprid larvae, but when they find a host, females of most species metamorphose into a **kentrogon** (Gr. *kentron,* a point, spine, + *gonos,* progeny) which injects cells of the parasite into the hemocoel of its crab host (Figure 20.22). Eventually, rootlike absorptive processes grow throughout the crab's body, and the parasite reproductive structures become externalized between cephalothorax and reflexed abdomen of the crab. Males at the cyprid stage attach to the external female brood chamber.

The exact position at which reproductive structures become externalized from the crab's body is of great adaptive value for rhizocephalan parasites. Because a crab's egg mass (if it had one) would be borne in this position, a crab treats the parasite as if it were a mass of the crab's own eggs. It protects, ventilates, and grooms its parasite and actually assists in the parasite's reproduction by performing spawning behavior at the appropriate time. The crab's grooming is necessary for continued good health of the parasite. But what if the rhizocephalan's larva is so unlucky as to infect a male crab? No problem. During the parasite's internal growth in the male crab, it castrates its host, and the crab becomes structurally and behaviorally like a female!

Class Malacostraca

Malacostraca, with over 20,000 species worldwide, is the largest class of Crustacea and shows great diversity. The diversity is reflected by the higher classification of the group, which includes three subclasses, 14 orders, and many suborders, infraorders, and superfamilies. We confine our coverage to a few of the most important orders. We described the characteristic body plan of malacostracans on page 422.

Order Isopoda

Isopods are one of the few crustacean groups to have successfully invaded terrestrial habitats in addition to freshwater and seawater habitats and the only crustaceans to have become truly terrestrial.

They are typically dorsoventrally flattened, lack a carapace, and have sessile compound eyes. Maxillipeds are the first pair of thoracic limbs; other thoracic limbs lack exopods and are similar. Abdominal appendages bear gills or lunglike organs called pseudotracheae and, except for uropods, also are similar to each other (hence the name isopods). Many species have the ability to roll into a tight ball for protection.

Common land forms are the sow bugs, or pill bugs (*Porcellio* and *Armadillidium,* Figure 20.23A), which live under stones and in damp places. Although they are terrestrial, they lack an efficient cuticular covering and other adaptations possessed by insects to conserve water; therefore they must live in moist environments (for example, under wet logs or rocks). *Caecidotea*

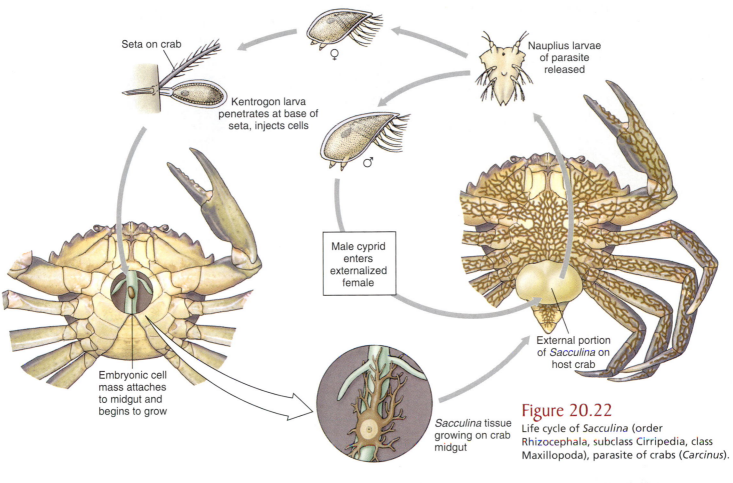

Seta on crab

Kentrogon larva
penetrates at base of
seta, injects cells

Embryonic cell
mass attaches
to midgut and
begins to grow

Male cyprid
enters
externalized
female

Nauplius larvae
of parasite
released

Sacculina tissue
growing on crab
midgut

External portion
of *Sacculina* on
host crab

Figure 20.22
Life cycle of *Sacculina* (order
Rhizocephala, subclass Cirripedia, class
Maxillopoda), parasite of crabs (*Carcinus*).

Figure 20.23
A, Four pill bugs, *Armadillidium vulgare*
(order Isopoda, class Malacostraca),
common terrestrial forms. **B,** Freshwater
sow bug, *Caecidotea* sp., an aquatic isopod.

Figure 20.24
An isopod parasite (*Anilocra* sp.) on a coney (*Cephalopholis fulvus*)
inhabiting a Caribbean coral reef (order Isopoda, class Malacostraca).

(Figure 20.23B) is a common freshwater form found under rocks
and among aquatic plants. *Ligia* is a common marine form that
scurries across the beach or rocky shore. Some isopods are parasites of fishes (Figure 20.24) or crustaceans.

Development is essentially direct but may be highly metamorphic in specialized parasites.

Order Amphipoda

Amphipods resemble isopods in that they lack a carapace, and
have sessile compound eyes and only one pair of maxillipeds
(Figure 20.25). However, they are usually compressed laterally,

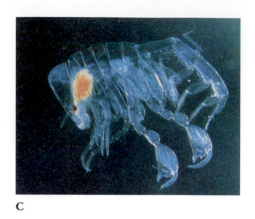

A

B

C

Figure 20.25

Marine amphipods. **A,** Free-swimming amphipod, *Anisogammarus* sp. **B,** Skeleton shrimp, *Caprella* sp., shown on a bryozoan colony, resemble praying mantids. **C,** *Phronima,* a marine pelagic amphipod, takes over the tunic of a salp (subphylum Urochordata, see Chapter 23). Swimming by means of its abdominal swimmerets, which protrude from the opening of the barrel-shaped tunic, the amphipod maneuvers to catch its prey. The tunic is not seen (order Amphipoda, class Malacostraca).

and their gills are in the typical thoracic position. Furthermore, their thoracic and abdominal limbs are each arranged in two or more groups that differ in form and function. For example, one group of abdominal legs may be for swimming and another group for jumping. There are many marine amphipods, including some beach-dwelling forms (for example, *Orchestia,* a beach hopper), numerous freshwater genera (*Hyalella* and *Gammarus*), and a few parasites (Figure 20.26). Development is direct and without a true metamorphosis.

Order Euphausiacea

Euphausiacea is a group of only about 90 species, but they are important as oceanic plankton known as "krill" (Figure 20.27).

Figure 20.27

Meganyctiphanes (order Euphausiacea, class Malacostraca) "northern krill."

A

B

Figure 20.26

A, Head and mouth of a healthy California grey whale, *Eschrichtius robustus,* bearing its characteristic heavy load of barnacles (order Thoracica, subclass Cirripedia, class Maxillopoda) and cyamid parasites (order Amphipoda, class Malacostraca) (*arrows*). Note yellowish plates of baleen in mouth (p. 640). **B,** Cyamid parasites of grey whale. Unlike most amphipods, these are dorsoventrally flattened. They have sharp, grasping claws on their legs.

Figure 20.28

Decapod crustaceans. **A,** A bright orange tropical rock crab, *Grapsus grapsus,* is a conspicuous exception to the rule that most crabs bear cryptic coloration. **B,** A hermit crab, *Elassochirus gilli,* which has a soft abdominal exoskeleton, lives in a snail shell that it carries about and into which it can withdraw for protection. **C,** A male fiddler crab, *Uca* sp., uses its enlarged cheliped to wave territorial displays and in threat and combat. **D,** A red night shrimp, *Rhynchocinetes rigens,* prowls caves and overhangs of coral reefs, but only at night. **E,** A spiny lobster, *Panulirus argus* (shown here), and the northern lobster, *Homarus americanus,* are consumed with gusto by many people (order Decapoda, class Malacostraca).

They are about 3 to 6 cm long, have a carapace that is fused with all thoracic segments but does not entirely enclose their gills. They have no maxillipeds, but have thoracic limbs with exopods. Most are bioluminescent, with a light-producing substance in an organ called a **photophore.** Some species may occur in enormous swarms, covering up to 45 m² and extending up to 500 m in one direction. They form a major portion of the diet of baleen whales and many fishes. Eggs hatch as nauplii, and development is indirect and metamorphic.

Order Decapoda

Decapods have three pairs of maxillipeds and five pairs of walking legs. In crabs the first pair of walking legs is modified to form pincers **(chelae),** but the second and third pairs may also be chelate, as in crayfishes, lobsters, and most shrimps. Decapods range in size from a few millimeters to the largest of all arthropods, Japanese spider crabs, whose chelae span 4 m from tip to tip. Crayfishes, lobsters, crabs, and "true" shrimp belong in this group (Figures 20.28 and 20.29). There are about 18,000 species of decapods, and the order is extremely diverse. They are very important ecologically and economically, and numerous species are relished as food.

Figure 20.29

Sponge crab, *Dromidia antillensis.* This crab is one of several species that deliberately mask themselves with material from their immediate environment (order Decapoda, class Malacostraca).

Classification of Subphylum Crustacea

Higher classification of Crustacea is complex and subject to change as new data become available. This listing relies on several sources, omitting many smaller taxa.

Class Remipedia (remə-'pədēə) (L. *remipedes,* oar-footed). No carapace; one-segmented protopods; biramous antennules and antennae; all trunk appendages similar; cephalic appendages large and raptorial; maxilliped segment fused to head; trunk unregionalized. Example: *Speleonectes.*

Class Cephalocarida (se'fə-'lō ka'rədə) (Gr. *kephalē,* head, + *karis,* shrimp, + *ida,* pl. suffix). No carapace; phyllopodia, one-segmented protopods; uniramous antennules and biramous antennae; compound eyes lacking; no abdominal appendages; maxilliped similar to thoracic leg. Example: *Hutchinsoniella.*

Class Branchiopoda (bran-kēä'pōdə) (Gr. *branchia,* gills, + *pous, podos,* foot). Phyllopodia; carapace present or absent; no maxillipeds; antennules reduced; compound eyes present; no abdominal appendages; maxillae reduced.

    Order Anostraca (ənästrəkə) (Gr. *an-,* prefix meaning without, + *ostrakon,* shell): **fairy shrimp and brine shrimp.** No carapace; no abdominal appendages; uniramous antennae. Examples: *Artemia, Branchinecta.*

    Order Notostraca (nō'tästrəkə) (Gr. *nōtos,* the back, + *ostrakon,* shell): **tadpole shrimp.** Carapace forming large dorsal shield; abdominal appendages present, reduced posteriorly; antennae vestigial. Examples: *Triops, Lepidurus.*

    Order Diplostraca (diplōstrəkə) (Gr. *diploos,* double, + *ostrakon,* shell): **water fleas** (cladocerans) and **clam shrimps** (conchostracans). Carapace folded, usually enclosing trunk but not head (cladocerans) or enclosing entire body (conchostracans); biramous antennae. Examples: *Daphnia, Leptodora, Lynceus.*

Class Ostracoda (ästrə'kōdə) (Gr. *ostrakodes,* having a shell): **ostracods.** Bivalve carapace entirely encloses body; body unsegmented or indistinctly segmented; no more than two pairs of trunk appendages. Examples: *Cypris, Cypridina, Gigantocypris.*

Class Maxillopoda (mak'sila'pōdə) (L. *maxilla,* the jawbone, + *pous, podos,* a foot). Usually five cephalic, six thoracic, and four abdominal segments plus a telson, but reductions common; no typical appendages on abdomen; nauplial eye of unique structure (maxillopodan eye); carapace present or absent.

    Subclass Mystacocarida (mistàkō-'karədə) (Gr. *mystax,* mustache, + *karis,* shrimp, + *ida,* pl. suffix): **mustache shrimps.** No carapace; body of cephalon and ten-segmented trunk; telson with clawlike caudal rami; cephalic appendages nearly identical, but antennae and mandibles biramous, other head appendages uniramous; second through fifth trunk segments with short, single-segment appendages. Example: *Derocheilocaris.*

    Subclass Copepoda (kōpe-'pədə) (Gr. *kōpē,* oar, + *pous, podos,* foot): **copepods.** No carapace; thorax typically of seven segments, of which first and sometimes second fuse with head to form cephalothorax; antennules uniramous; antennae bi- or uniramous; four to five pairs swimming legs; parasitic forms often highly modified. Examples: *Cyclops, Diaptomus, Calanus, Ergasilus, Lernaea, Salmincola, Caligus.*

    Subclass Tantulocarida (tan'tü'lōkarədə) (L. *tantulus,* so small, + *caris,* shrimp). No recognizable cephalic appendages except antennae on sexual female; solid median cephalic stylet; six free thoracic segments, each with pair of appendages, anterior five biramous; six abdominal segments; minute copepod-like ectoparasites. Examples: *Basipodella, Deoterthron.*

    Subclass Branchiura (bran-'kēyurə) (Gr. *branchia,* gills, + *ura,* tail): **fish lice.** Body oval, head and most of trunk covered by flattened carapace, incompletely fused to first thoracic segment; thorax with four pairs of appendages, biramous; abdomen unsegmented, bilobed; eyes compound; antennae and antennules reduced; maxillules often forming suctoral discs. Examples: *Argulus, Chonopeltis.*

    Subclass Pentastomida (pen-ta-stom'i-da) (Gr. *pente,* five, + *stoma,* mouth): **pentastomids.** Wormlike unsegmented body with five short anterior protuberances, four bear claws and the fifth bears the sucking mouth. Examples: *Armillifer, Linguatula.*

    Subclass Cirripedia (sirə-'pēdēə) (L. *cirrus,* curl of hair, + *pes, pedis,* foot): **barnacles.** Sessile or parasitic as adults; head reduced and abdomen rudimentary; paired compound eyes absent; body segmentation indistinct; usually hermaphroditic; in free-living forms carapace becomes mantle, which secretes calcareous plates; antennules become organs of attachment, then disappear. Examples: *Balanus, Policipes, Sacculina.*

Class Malacostraca (malə-'kä-strəkə) (Gr. *malakos,* soft, + *ostrakon,* shell). Usually with eight segments in thorax and six plus telson in abdomen; all segments with appendages; antennules often biramous; first one to three thoracic appendages often maxillipeds; carapace covering head and part or all of thorax, sometimes absent; gills usually thoracic epipods.

    Order Isopoda (īso-'pōdə) (Gr. *isos,* equal, + *pous, podos,* foot): **isopods.** No carapace; antennules usually uniramous, sometimes vestigial; eyes sessile (not stalked); gills on abdominal appendages; body commonly dorsoventrally flattened; second thoracic appendages usually not prehensile. Examples: *Armadillidium, Caecidotea, Ligia, Porcellio.*

    Order Amphipoda (am-'fi-'pōdə) (Gr. *amphis,* on both sides, + *pous, podos,* foot): **amphipods.** No carapace; antennules often biramous; eyes usually sessile; gills on thoracic coxae; second and third thoracic limbs usually prehensile; typically bilaterally compressed body form. Examples: *Orchestia, Hyalella, Gammarus.*

    Order Euphausiacea (yü-foz-ē-āshēə) (Gr. *eu,* well, + *phausi,* shining bright, + L. *acea,* suffix, pertaining to): **krill.** Carapace fused to all thoracic segments but not entirely enclosing gills, no maxillipeds; all thoracic limbs with exopods. Example: *Meganyctiphanes.*

    Order Decapoda (də'ka-'pōdə) (Gr. *deka,* ten, + *pous, podos,* foot): **shrimps, crabs, lobsters.** All thoracic segments fused with and covered by carapace; eyes on stalks; first three pairs of thoracic appendages modified to maxillipeds. Examples: *Farfantepenaeus (= Penaeus), Cancer, Pagurus, Grapsus, Homarus, Panulirus.*

Crabs, especially, exist in a great variety of forms. Although resembling crayfishes, they differ from the latter in having a broader cephalothorax and reduced abdomen. Familiar examples along the seashore are hermit crabs (Figure 20.28B), which live in snail shells (because their abdomens are not protected by the same heavy exoskeleton as are the anterior parts); fiddler crabs, *Uca* (Figure 20.28C), which burrow in sand just below the high-tide level and emerge to run over the sand while the tide is out; spider crabs such as *Libinia;* interesting decorator crabs *Dromidia,* and others, which cover their carapaces with sponges and sea anemones for protective camouflage (Figure 20.29).

PHYLOGENY AND ADAPTIVE DIVERSIFICATION

Phylogeny

Among Crustacea, Remipedia seem to have the most ancestral characters (see Figure 20.1): They have a long body, with no tagmatization behind the head, a double ventral nerve cord, and serially arranged digestive ceca. Fossils of a puzzling arthropod from the Mississippian period seem to be the sister group of remipedians, and their morphology suggested one mechanism for the origin of biramous appendages. They have *two pairs* of uniramous limbs on each segment. Thus, it was suggested that each crustacean segment represents two ancestral segments that fused ("diplopodous condition," as seen in Diplopoda, p. 414), and that biramous appendages derived from fusion of both limbs on an ancestral diplopodous segment. However, it is now known that modulation in expression of the *Distal-less (Dll)* gene determines location of distal ends of anthropod limbs. In each primordial (embryonic) biramous appendage, the gene product of

Dll can be observed in two groups of cells, each of which will become a branch of the limb. In a uniramous limb primordium, there is only one such group of cells, and in primodia of phyllopodous limbs (as in class Branchiopoda), there are as many groups expressing *Dll* as there are limb branches.

The wormlike pentastomids were placed in Ecdysozoa near arthropods because their larval form resembles tardigrade larvae, their cuticle is molted, and there are other similarities in sperm morphology and larval appendages. Phylogenies based on sequences of ribosomal RNA genes indicate that pentastomids are crustaceans. A recent study of gene arrangements and base sequences of mitochondrial DNA confirmed this result. Pentastomids are now considered highly derived crustaceans, placed in class Maxillopoda near fish lice (subclass Branchiura).

Adaptive Diversification

The level of adaptive diversification demonstrated by the crustaceans is great, with the exploitation of virtually all aquatic resources. They are unquestionably the dominant arthropod group in marine environments, and they share dominance of freshwater habitats with insects. Invasions of terrestrial environments have been much more limited, with isopods being the only notable success. There are a few other terrestrial examples, such as land crabs. The most diverse class is Malacostraca, and the most abundant groups are Copepoda and Ostracoda. Members of both taxa include planktonic suspension feeders and numerous scavengers. Copepods have been particularly successful as parasites of both vertebrates and invertebrates, and it is clear that present parasitic copepods are products of numerous invasions of such niches.

SUMMARY

Crustacea is a large, primarily aquatic subphylum. In addition to a pair of mandibles, crustaceans have common two pairs of antennae and two pairs of maxillae. Their tagmata are a head and trunk or a head, thorax, and abdomen. Many have a carapace. Crustaceans' appendages are ancestrally biramous.

All arthropods must periodically cast off their old cuticle (ecdysis) and grow in size before the newly secreted cuticle hardens. Premolt and postmolt periods are hormonally controlled, as are several other processes, such as change in body color and expression of sexual characteristics.

Feeding habits vary greatly in Crustacea, and there are many forms of predators, scavengers, suspension feeders, and parasites. Respiration is through the body surface or by gills, and excretory organs take the form of maxillary or antennal glands. Circulation, as in other arthropods, is through an open system of sinuses (hemocoel), and a dorsal, tubular heart is the chief pumping organ. Most crustaceans have compound eyes composed of units called ommatidia. Sexes are usually separate.

Class Branchiopoda is characterized by phyllopodia and contains, among others, order Diplostraca, which is ecologically

important as zooplankton. Within class Maxillopoda, members of subclass Copepoda lack a carapace and abdominal appendages. They are abundant and are among the most important of the primary consumers in many freshwater and marine ecosystems. Many are parasitic. Most members of subclass Cirripedia (barnacles) are sessile as adults, secrete a shell of calcareous plates, and filter-feed by means of their thoracic appendages. Subclass Branchiura contains fish lice. Closely related to fish lice are tongue worms; they are parasitic in the lungs and nasal cavities of vertebrates. These members of former phylum Pentastomida now comprise subclass Pentastomida in class Maxillopoda.

Malacostraca is the largest and most diverse crustacean class, and the most important orders are Isopoda, Amphipoda, Euphausiacea, and Decapoda. All have both abdominal and thoracic appendages. Isopods lack a carapace and are usually dorsoventrally flattened. Amphipods also lack a carapace but are usually laterally flattened. Euphausiaceans are important oceanic plankton called krill. Decapods include crabs, shrimps, lobsters, crayfishes, and others; they have five pairs of walking legs (including chelipeds) on their thorax.

REVIEW QUESTIONS

1. What are the tagmata and appendages on the head of crustaceans? What other important characteristics of Crustacea distinguish them from other arthropods?
2. Define each of the following: tergum, sternum, telson, protopod, exopod, endopod, epipod, endite, and exite.
3. What is meant by homologous structures? What is meant by serial homology, and how do crustaceans show serial homology?
4. What is a carapace?
5. Briefly describe respiration and circulation in crayfishes.
6. Briefly describe the function of antennal and maxillary glands in Crustacea.
7. How does a crayfish detect changes in position?
8. What is the photoreceptor unit of a compound eye? How does this unit adjust to varying amounts of light?
9. What is a nauplius? What is the difference between direct and indirect development in Crustacea?
10. Describe the molting process in Crustacea, including the action of hormones and the process of ecdysis.
11. Which classes and subclasses of Crustacea (Branchiopoda, Ostracoda, Copepoda, Cirripedia, and Malacostraca) are most diverse? Most numerous? Distinguish them from each other.
12. Compare and contrast Isopoda, Amphipoda, Euphausiacea, and Decapoda.
13. What is the significance of Remipedia to the hypotheses concerning the origin of crustaceans?
14. Briefly explain the genetic determination of biramous and uniramous appendages.
15. What is a tongue worm, and where would it be found?

SELECTED REFERENCES

Bliss, D. E. (editor-in-chief). 1982–1985. The biology of Crustacea, vols. 1–10. New York, Academic Press, Inc. *This series is a standard reference for all aspects of crustacean biology.*

Boore, J. L., D. V. Lavrov, and W. M. Brown. 1998. Gene translocation links insects and crustaceans. Nature **392:**667–668. *A single mitochondrial gene translocation, indicative of a recent common ancestor, is shared by insects and crustaceans, but not present in chelicerates or myriapods.*

Boyd, C. E., and J. W. Clay. 1998. Shrimp aquaculture and the environment. Sci. Am. **278:**58–65 (June). *Shrimp aquaculture can have adverse consequences on the environment (pollution).*

Chang, E. S., S. A. Chang, and E. P. Mulder. 2001. Hormones in the lives of crustaceans: An overview. Am. Zool. **41:**1090–1097. *Summary of research into hormone function in the American lobster.*

Galant, R., and S. B. Carroll. 2002. Evolution of a transcriptional repression domain in an insect Hox protein. Nature **415:**910–913. *There are levels of a protein (Ultrabithorax, Ubx), encoded by a* Hox *gene, in abdomens of insects, where they repress expression of another gene,* Distal-less (Dll), *which is required for limb information. Crustacean abdomens and onychophorans have high Ubx but can form limbs on their abdomen, showing that Ubx is a conditional repressor in those groups.*

Giribet, G., G. D. Edgecombe, and W. C. Wheeler. 2001. Arthropod phylogeny based on eight molecular loci and morphology. Nature **413:**157–161. *Support for Crustacea and Insecta as sister groups in a mandibulate clade.*

Gould, S. J. 1996. Triumph of the root-heads. Nat. Hist. **105:**10–17. *An informative essay on parasite-host coevolution using* Sacculina *as an example.*

Holden, C. 1997. Green crabs advance north. Science. **276:**203. *A report on the advance of European green crab (*Carcinus maenas*) up the west coast of the United States.*

Huys, R., G. A. Boxhall, and R. J. Lincoln. 1993. The tantulocaridan life cycle: the circle closed? J. Crust. Biol. **13:**432–442. *The current*

hypothesis of a parthenogenetic cycle alternating with a cycle that includes fertilization in these bizarre little creatures.

Lavrov, D. L., W. M. Brown, and J. L. Boore. 2004. Phylogenetic position of Pentastomida and (pan)crustacean relationships. Proc. R. Soc. Lond. Ser. B. **271:**537–544. *Pentastomids are maxillopod crustaceans, probably closely related to fish lice.*

Laufer, H., and W. J. Biggers. 2001. Unifying concepts learned from methyl farnesoate for invertebrate reproduction and postembryonic development. Am. Zool. **41:**442–457. *Methyl farnesoate performs similar functions in crustaceans as juvenile hormone does in insects.*

Martin, J. W., and G. E. Davis. 2001. An updated classification of the recent Crustacea. Los Angeles, Natural History Museum of Los Angeles County Science Series 39. 124 pp.

Panganiban, G., A. Sebring, L. Nagy, and S. Carroll. 1995. The development of crustacean limbs and the evolution of arthropods. Science **270:**1363–1366. *Probing for particular homeotic gene products suggests that all arthropods derive from a common ancestor and that biramous and uniramous limbs derive from modulation of* Distal-less (Dll) *gene expression.*

Storch, V., and B. G. M. Jamieson. 1992. Further spermatological evidence for including the Pentastomida (tongue worms) in the Crustacea. Int. J. Parasitol. **22:**95–108. *Morphological and developmental data to support the placement of pentastomids as derived crustaceans rather than a distinct phylum.*

Versluis, M., B. Schmitz, A. von der Heydt, and D. Lohse. 2000. How snapping shrimp snap: through captivating bubbles. Science **289:**2114–2117. *Snapping of their chela is strong enough to cause cavitation bubbles. Implosion of the bubbles stuns prey.*

Zill, S. N., and E.-A., Seyfarth. 1996. Exoskeletal sensors for walking. Sci. Am. **275:** 86–90 (July). *Cockroaches, crabs, and spiders have sensors in the exoskeleton of their legs that act as biological strain gauges.*

ONLINE LEARNING CENTER

Visit www.mhhe.com/hickmanipz14e for chapter quizzing, key term flash cards, web links, and more!

Hexapods

- PHYLUM ARTHROPODA
- SUBPHYLUM HEXAPODA

The majority of animal species is composed of insects.

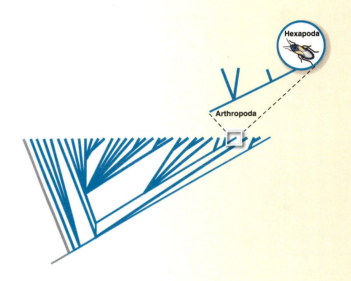

A Winning Combination

Humans suffer staggering economic losses to insects. Locust outbreaks in Africa seem a thing of the past to many today, but this is far from true. Locust populations fluctuate between quiet phases, where they cover only 16 million square kilometers in 30 African countries, and plague phases, where they cover 29 million square kilometers of land in 60 countries. A swarm of locusts. *Schistocerca gregaria,* contains 40 to 80 million insects per square kilometer. In peak phases, they cover 20% of the earth's land surface and affect the livelihood of one-tenth of the earth's population. The last plague phase was 1986–1989, but the Food and Agriculture Organization (FAO) of the United Nations monitors and maps population sizes continuously to respond quickly to outbreaks (http://www.fao.org/ag/locusts/en/info/info/faq/index.html).

In the western United States and Canada, an outbreak of mountain pine beetles in the 1980s and 1990s killed pines on huge acreages, and the 1973 to 1985 outbreak of spruce budworm in fir/spruce forests killed millions of conifer trees. Since its introduction in the 1920s, a fungus that causes Dutch elm disease, mainly transmitted by European bark beetles, has virtually obliterated American elm trees in North America. Since 2004, another alien invader, the emerald ash borer, a beetle, threatens the ash trees of North America.

These examples remind us of our ceaseless struggle with the dominant group of animals on earth today: insects. Insects far outnumber all other species of animals in the world combined, and numbers of individuals are equally enormous. Some scientists have estimated that there are 200 million insects for every single human alive today! Insects have an unmatched ability to adapt to all land environments and to virtually all climates. Many have exploited freshwater and shoreline habitats, and have evolved extraordinary abilities to survive adverse environmental conditions.

ubphylum Hexapoda is named for the presence of **six legs** in members of the group. All legs are **uniramous.** Hexapods have **three tagmata**—head, thorax, and abdomen—with appendages on the head and thorax. Abdominal appendages are greatly reduced or absent. There are two classes within Hexapoda: Entognatha and Insecta (Figure 21.1).

Entognatha is a small group whose members have the bases of mouthparts enclosed within the head capsule. There are three orders of entognathans. Members of Protura and Diplura are tiny, eyeless, and inhabit soils or dark, damp places where they are rarely noticed. Members of Collembola are commonly called

springtails because of their ability to leap; an animal 4 mm long may leap 20 times its body length. Collembolans live in soil, in decaying plant matter, on freshwater pond surfaces, and along the seashore. They can be very abundant, reaching millions per hectare in some soils, but like other entognathans, their small size makes them less visible to the casual observer.

Insecta is an enormous class whose members have ecto-gnathous mouthparts, where the bases of mouthparts lie outside the head capsule. Winged insects are called pterygotes, and wingless insects are called apterygotes. Class Insecta contains one group whose members diverged from ancestors of

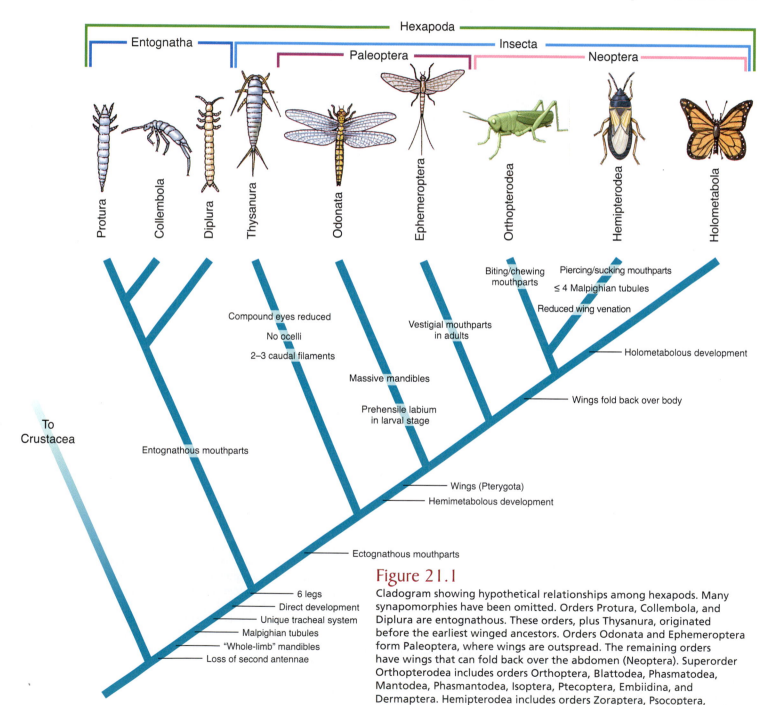

Figure 21.1

Cladogram showing hypothetical relationships among hexapods. Many synapomorphies have been omitted. Orders Protura, Collembola, and Diplura are entognathous. These orders, plus Thysanura, originated before the earliest winged ancestors. Orders Odonata and Ephemeroptera form Paleoptera, where wings are outspread. The remaining orders have wings that can fold back over the abdomen (Neoptera). Superorder Orthopterodea includes orders Orthoptera, Blattodea, Phasmatodea, Mantodea, Phasmantodea, Isoptera, Ptecoptera, Embiidina, and Dermaptera. Hemipterodea includes orders Zoraptera, Psocoptera, Hemiptera, Thysanoptera, and Phthiraptera; and superorder Holometabola encompasses all holometabolous orders.

the wingless order Thysanura, which forms the sister taxon to all other insects. Insect wings evolved in a common ancestor of the latter clade (Figure 21.1). Thysanurans are called primitively wingless to distinguish them from orders whose members do not have wings now, but whose ancestors were winged.

CLASS INSECTA

Insecta (L. *insectus,* cut into) is the most diverse and abundant of all groups of arthropods. There are more species of insects than species of all other animals combined. The number of insect species classified is currently at 1.1 million, but experts estimate that as many as 30 million species may exist. There is also striking evidence of continuing and sometimes rapid evolution among living insects.

It is difficult to appreciate fully the ecological, medical, and economic significance of this extensive group. The study of insects (**entomology**) occupies the time and resources of skilled men and women all over the world. The struggle between humans and their insect competitors seems to be endless, yet paradoxically insects have so interwoven themselves into the economy of nature in so many useful roles that most terrestrial ecosystems would collapse without them.

Insects differ from other arthropods in having ectognathous mouthparts and usually **two pairs of wings** on the thoracic region of the body, although some have one pair of wings or none. Insects range in size from less than 1 mm to 20 cm in length, the majority being less than 2.5 cm long. Some of the largest insects live in tropical areas.

Distribution

Insects are among the most abundant and widespread of all land animals. They have spread into practically all habitats that support life except the sea. Relatively few are truly marine, but some are common in intertidal zones. Marine water striders (*Halobates*), which live on the *surface* of the ocean, are the only marine invertebrates that live on the sea-air interface. Insects are common in brackish water, in salt marshes, and on sandy beaches. They are abundant in freshwater, in soil, in forests (especially the tropical forest canopy), and they are found even in deserts and wastelands, on mountaintops, and as parasites in and on plants and animals.

Their wide distribution is made possible by their powers of flight and their highly adaptable nature. Insects evolved wings and invaded the air 250 million years before flying reptiles, birds, or mammals. In most cases they can easily surmount barriers that are virtually impassable to many other animals. Their small size allows them to be carried by currents of both wind and water to far regions. Their well-protected eggs can withstand rigorous conditions and can be carried long distances by birds and other animals. Their agility and ecological aggressiveness enable them to occupy every possible niche in a habitat. No single pattern of biological adaptation can be applied to them.

Adaptability

Insects have shown an amazing evolutionary adaptability, as evidenced by their wide distribution and enormous diversity of species. Most of their structural modifications are in their wings, legs, antennae, mouthparts, and alimentary canals. Such wide diversity enables this vigorous group to use all available food and shelter resources. Some are parasitic, some suck the sap of plants, some chew the foliage of plants, some are predaceous, and some live on the blood of various animals. Within these different groups, specialization occurs, so that a particular kind of insect will eat, for instance, leaves of only one kind of plant. This specificity of eating habits lessens competition with other species and to a great extent accounts for their biological diversity.

Insects are well adapted to dry and desert regions. Their hard and protective exoskeleton limits evaporation. Some insects also extract most of the water from food, fecal material, and by-products of cell metabolism.

External Form and Function

Insects show a remarkable variety of morphological characteristics, but, as in other arthropods, the exoskeleton is formed of a complex system of plates called **sclerites,** connected by concealed, flexible hinge joints. Muscles between sclerites enable insects to make precise movements. Rigidity of their exoskeleton is attributable to unique scleroproteins and not to its chitin component. It is waterproof and its lightness makes flying possible. By contrast, the cuticle of crustaceans is stiffened mostly by minerals.

Insects are much more homogeneous in tagmatization than are Crustacea. Insect tagmata comprise a head, thorax, and abdomen. The cuticle of each body segment typically is composed of four plates (sclerites): a dorsal notum (tergum), a ventral sternum, and a pair of lateral pleura. Pleura of abdominal segments are often partially membranous rather than sclerotized. Some insects are fairly generalized in body structure; some are highly specialized. Grasshoppers, or locusts, are a generalized type often used in laboratories to demonstrate general features of insects (Figure 21.2).

The head usually bears a pair of relatively large compound eyes, a pair of antennae, and usually three ocelli (Figure 21.2). Antennae, which vary greatly in size and form (Figure 21.3), act as tactile organs, olfactory organs, and in some cases as auditory organs. Mouthparts, formed from specially hardened cuticle, typically consist of a labrum, a pair each of mandibles and maxillae, a labium, and a tonguelike hypopharynx. The type of mouthparts that an insect possesses determines how it feeds. We discuss some of these modifications later in this chapter.

The thorax comprises three segments: prothorax, mesothorax, and metathorax, each bearing a pair of legs (Figure 21.2). In most insects the mesothorax and metathorax also bear a pair of wings. Wings are cuticular extensions formed by the epidermis. They consist of a double membrane containing veins of thicker cuticle that serve to expand the wings after eclosion from the pupa and to strengthen the wings aerodynamically. Although these veins vary in their patterns among different taxa, they are relatively constant within a family, genus, or species and serve as one means of classification and identification.

Legs of insects often are modified for special purposes. Many terrestrial forms have walking legs with terminal pads and claws. These pads may be sticky for walking upside down, as

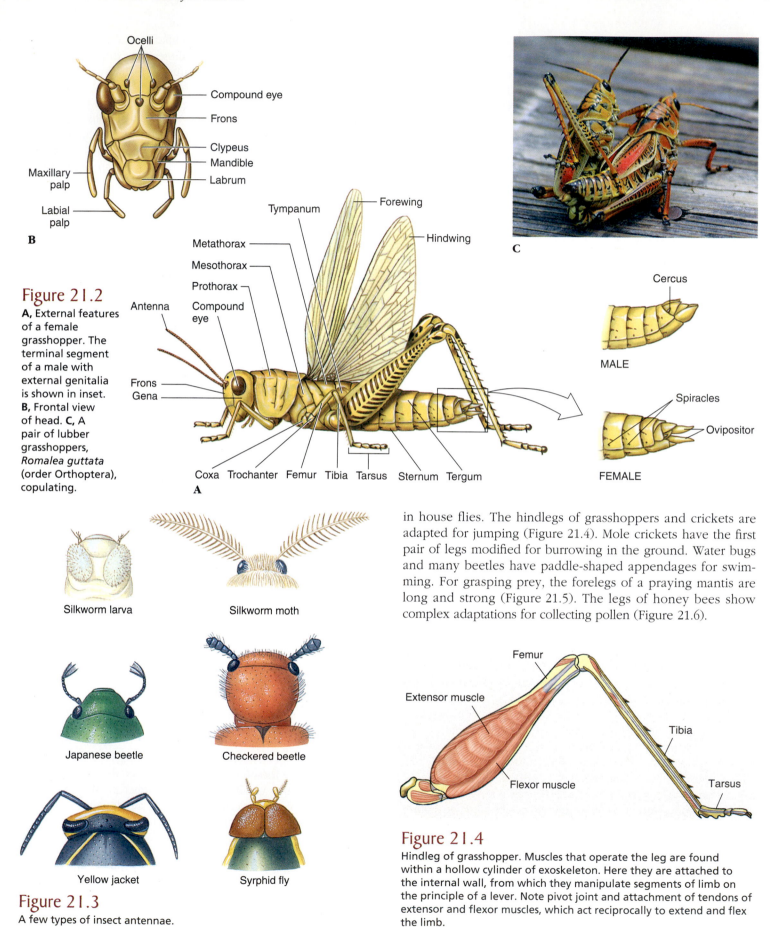

Figure 21.2
A, External features of a female grasshopper. The terminal segment of a male with external genitalia is shown in inset. **B,** Frontal view of head. **C,** A pair of lubber grasshoppers, *Romalea guttata* (order Orthoptera), copulating.

Labels in figure A (head, frontal view B): Ocelli, Compound eye, Frons, Clypeus, Mandible, Labrum, Maxillary palp, Labial palp

Labels in figure A (lateral view): Antenna, Compound eye, Prothorax, Mesothorax, Metathorax, Tympanum, Forewing, Hindwing, Frons, Gena, Coxa, Trochanter, Femur, Tibia, Tarsus, Sternum, Tergum

Labels (male/female abdomen): Cercus, MALE, Spiracles, Ovipositor, FEMALE

Figure 21.3
A few types of insect antennae.

Labels: Silkworm larva, Silkworm moth, Japanese beetle, Checkered beetle, Yellow jacket, Syrphid fly

in house flies. The hindlegs of grasshoppers and crickets are adapted for jumping (Figure 21.4). Mole crickets have the first pair of legs modified for burrowing in the ground. Water bugs and many beetles have paddle-shaped appendages for swimming. For grasping prey, the forelegs of a praying mantis are long and strong (Figure 21.5). The legs of honey bees show complex adaptations for collecting pollen (Figure 21.6).

Figure 21.4
Hindleg of grasshopper. Muscles that operate the leg are found within a hollow cylinder of exoskeleton. Here they are attached to the internal wall, from which they manipulate segments of limb on the principle of a lever. Note pivot joint and attachment of tendons of extensor and flexor muscles, which act reciprocally to extend and flex the limb.

Labels: Femur, Extensor muscle, Flexor muscle, Tibia, Tarsus

A

B

Figure 21.5

A, Praying mantis (order Mantodea) feeding on an insect. **B,** Praying mantis laying eggs.

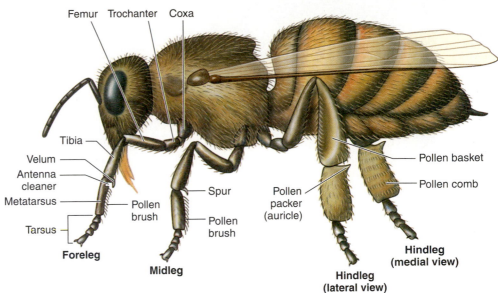

Figure 21.6

Adaptive legs of worker honey bee. In the foreleg, the toothed indentation covered with the velum combs out the antennae. The spur on the middle leg removes wax from wax glands on the abdomen. Pollen brushes on the front and middle legs comb off pollen picked up on body hairs and deposit it on the pollen brushes of the hindlegs. Long hairs of the pollen comb on the hindleg remove pollen from the comb of the opposite leg; then the auricle (pollen packer) presses it into a pollen basket when the leg joint is flexed back. A bee carries her load in both baskets to the hive and pushes pollen into a cell, to be cared for by other workers.

The insect abdomen comprises 9 to 11 segments; the eleventh, when present, bears a pair of **cerci** (appendages at the posterior end). Larval or nymphal forms may have a variety of abdominal appendages, but these appendages are lacking in adults. The genitalia emerge from segments 8 and 9 of the abdomen (Figure 21.2A), they are often useful in identification and classification.

There are innumerable variations in body form among insects. Beetles are usually thick and plump (see Figure 21.7A); damselflies, ant lions, and walking sticks are long and slender (Figure 21.7B); many aquatic beetles are streamlined; butterflies have the broadest wings of all; and cockroaches are flat, adapted to living in crevices. The ovipositor of female ichneumon wasps is extremely long (Figure 21.8), whereas the anal cerci form forceps in earwigs but are long and many-jointed in stoneflies and mayflies. The hymenopteran stinger is a modified ovipositor. Antennae are long in cockroaches and katydids, short in dragonflies and most beetles, knobbed in butterflies, and plumed in some moths. Many other dramatic variations exist (see Figure 21.3). Perhaps most amazing is the fact that mouthparts, antennae, legs, cerci, and ovipositors are all modified appendages.

Locomotion

Walking   When walking, most insects use a triangle of legs involving the first and last leg of one side together with the middle leg of the opposite side. In this way, terrestrial insects

A

B

Figure 21.7

A, A giant horned beetle, *Diloboderus abderus* (order Coleoptera), from Uruguay. Though the ferocious-looking processes from the head and thorax might appear to be for pinching or stabbing an opponent, they actually are used to lift or pry up a rival of the same species away from resources. **B,** Walking sticks, *Diapheromera femorata* (order Phasmatodea), mating. The species is common in much of North America. It is wingless, and despite its camouflage as a twig, it is eaten by numerous predators.

Figure 21.8

An ichneumon wasp with the end of the abdomen raised to thrust her long ovipositor into wood to find a tunnel made by the larva of a wood wasp or wood-boring beetle. She can bore 13 mm or more into the wood to lay her eggs in the larva of a wood-boring beetle, which will become host for the ichneumon larvae. Other ichneumon species attack spiders, moths, flies, crickets, caterpillars, and other insects.

keep at least three of their six legs on the ground at all times—a tripod arrangement that bestows stability.

Some insects, such as water striders, *Gerris* (L. *gero*, to carry), can walk on the surface of water. A water strider has on its footpads nonwetting hairs that do not break the surface film of water but merely indent it. As it skates along on its two pairs of posterior legs, *Gerris* uses its reduced and toothed pair of prothoracic legs to capture and hold prey. Water striders exhibit unusual cleaning behavior and may do complete flips on the water surface in an attempt to dislodge debris from their thoracic terga (Figure 21.9). Bodies of marine water striders, *Halobates* (Gr. *halos*, the sea, + *bātes*, one that treads), excellent surfers on rough ocean waves, are further protected by a water-repellent coat of close-set hairs shaped like thick hooks.

Power of Flight   Insects are the only invertebrates that can fly, and they share the power of flight with birds and flying mammals. However, their wings evolved in a manner different from limb buds of birds and mammals and are not homologous to them. Insect wings are formed by outgrowths from the body wall

of the mesothoracic and metathoracic segments and are composed of cuticle. Recent fossil evidence suggests that insects may have evolved fully functional wings over 400 million years ago.

Most insects have two pairs of wings, but Diptera (true flies) have only one pair (Figure 21.10), the hindwings being represented by a pair of tiny **halteres** (balancers) that vibrate and are responsible for equilibrium during flight. Males of order Strepsiptera have only a hind pair of wings and an anterior pair of "halteres." Males of scale insects also have one pair of wings but no halteres. Some insects are ancestrally (for example, silverfish) or secondarily (for example, fleas) wingless. Reproductive female ants shed their wings after their nuptial flight (males die), and reproductive male and female termites have wings, but workers in both cases are wingless. Lice and fleas are always wingless.

Wings may be thin and membranous, as in flies and many other groups (see Figure 21.8); thick and horny, such as the front wings of beetles (see Figure 21.7A); parchmentlike, such as the front wings of grasshoppers; covered with fine scales, as in butterflies and moths; or covered with hairs, as in caddis flies.

Wing movements are controlled by a complex of muscles in the thorax. **Direct flight muscles** are attached to a part of the wing itself. **Indirect flight muscles** are not attached to the wing and cause wing movement by altering the shape of the thorax. The wing is hinged at the thoracic tergum and also slightly laterally on a pleural process, which acts as a fulcrum (Figure 21.11). In most insects, the upstroke of the wing is effected by contracting indirect muscles that pull the tergum down toward the sternum (Figure 21.11A). Dragonflies and cockroaches accomplish the downstroke by contracting direct muscles attached to the wings lateral to the pleural fulcrum. In Hymenoptera and Diptera (see p. 464) all major flight muscles are indirect. The downstroke occurs when the sternotergal muscles (muscles inserted on sternum and tergum) relax and longitudinal muscles of the thorax contract

Figure 21.9

Water strider, *Gerris* sp. (order Hemiptera). The animal is supported on its long, slender legs by the water's surface tension.

Figure 21.10

House fly, *Musca domestica* (order Diptera). House flies can become contaminated with over 100 human pathogens, which may be transferred to human and animal food by direct contact, regurgitated food, and feces.

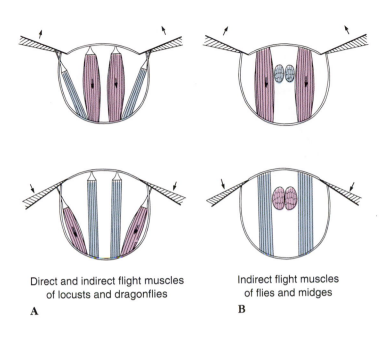

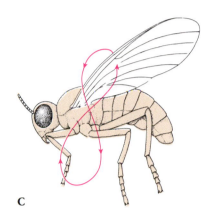

Figure 21.11

A, Flight muscles of insects such as cockroaches, in which upstroke is by indirect muscles and downstroke is by direct muscles. **B,** In insects such as flies and bees, both upstroke and downstroke are by indirect muscles. **C,** The figure-eight path followed by the wing of a flying insect during the upstroke and downstroke.

Direct and indirect flight muscles of locusts and dragonflies

A

Indirect flight muscles of flies and midges

B

C

and arch the tergum (Figure 21.11B), pulling up the tergal articulations relative to the pleura. The downstroke in beetles and grasshoppers involves both direct and indirect muscles.

Insectan flight-muscle contraction has two basic types of neural control: **synchronous** and **asynchronous.** Larger insects such as dragonflies and butterflies have wings with synchronous muscles, in which a single nerve impulse stimulates a muscle contraction and thus one wing stroke. Wings with asynchronous muscles occur in Hymenoptera, Diptera, Coleoptera, and some Hemiptera, see pp. 463–464. Their mechanism of action is complex and depends on the storage of potential energy in resilient parts of the thoracic cuticle. As one set of muscles contracts (moving the wing in one direction), they also stretch the antagonistic set of muscles, causing them to contract (and move the wing in the other direction). Because muscle contractions are not in phase with nervous stimulation, only occasional nerve impulses are necessary to keep the muscles contracting and relaxing. Thus extremely rapid wing beats are possible. For example, butterflies (with synchronous muscles) may beat as few as four times per second. Insects with asynchronous muscles, however, such as flies and bees, may produce 100 beats per second or more. The fruit fly, *Drosophila* (Gr. *drosos,* dew, + *philos,* loving), can fly at 300 beats per second, and midges have been clocked at more than 1000 beats per second.

Obviously flying entails more than a simple flapping of wings; a forward thrust is necessary. As indirect flight muscles alternate rhythmically to raise and to lower the wings, direct flight muscles alter the angle of the wings so that they act as lifting airfoils during both upstroke and downstroke, twisting the leading edge of the wings downward during the downstroke and upward during the upstroke. A figure-eight movement (Figure 21.11C) results, spilling air from the trailing edges of the wings. The quality of forward thrust depends, of course, on several factors, such as variations in wing venation, wing load

(grams of body weight divided by total wing area), how much the wings are tilted, and wing length and shape.

Flight speeds vary tremendously. The fastest flyers usually have narrow, fast-moving wings with a strong tilt and a strong figure-eight component. Sphinx moths and horse flies achieve approximately 48 km (30 miles) per hour and dragonflies approximately 40 km (25 miles) per hour. Some insects are capable of long continuous flights. Migrating monarch butterflies, *Danaus plexippus* (Gr. after Danaus, mythical king of Arabia) (see Figure 21.27A), travel hundreds to thousands of miles south in the fall, flying at a speed of approximately 10 km (6 miles) per hour, to reach their overwintering roosts in Mexico and California.

Internal Form and Function

Nutrition

The digestive system (Figure 21.12; see also Figure 32.9, p. 714) consists of a foregut (mouth with salivary glands, esophagus, crop for storage, and gizzard for grinding in some); a midgut (stomach and gastric ceca); and a hindgut (intestine, rectum, and anus). Some digestion may occur in the crop as food mixes with enzymes from the saliva, but no absorption occurs there. The main site for digestion and absorption is the midgut, and the ceca may increase the digestive and absorptive area. Little absorption of nutrients occurs in the hindgut (with certain exceptions, such as wood-eating termites), but this is a major area for resorption of water and some ions (see p. 451).

Most insects feed on plant juices and plant tissues **(phytophagous** or **herbivorous).** Some insects feed on specific plants; others, such as grasshoppers, eat almost any plant. Caterpillars of many moths and butterflies eat foliage of only certain plants. Certain species of ants and termites cultivate fungus gardens as a source of food.

Figure 21.12
Internal structure of female grasshopper.

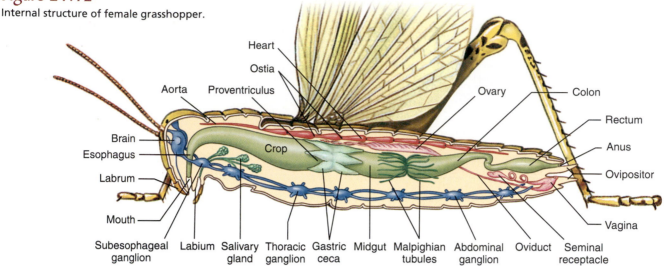

Heart
Ostia
Aorta   Proventriculus
Ovary   Colon
Rectum
Brain
Anus
Esophagus
Crop
Labrum
Ovipositor
Vagina
Mouth
Subesophageal   Labium   Salivary   Thoracic   Gastric   Midgut   Malpighian   Abdominal   Oviduct   Seminal
ganglion   gland   ganglion   ceca   tubules   ganglion   receptacle

Many beetles and larvae of many insects live on dead animals **(saprophagous).** Some insects are **predaceous,** catching and eating other insects as well as other types of animals (see Figure 21.5). However, the so-called predaceous diving beetle, *Cybister fimbriolatus* (Gr. *kybistēr,* diver), is not as predaceous as once supposed, but is largely a scavenger.

Many insects are **parasitic** as adults, as larvae, or, in some cases, both juveniles and adults are parasites. For example, fleas (Figure 21.13) live on blood of mammals as adults, but their larvae are free-living scavengers. Lice (Figures 21.14 and 21.15) are parasitic throughout their life cycle. Many parasitic insects are themselves parasitized by other insects, a condition called **hyperparasitism.** Larvae of many varieties of wasps live and complete much of their metamophosis inside the bodies of spiders or other insects (Figure 21.16), consuming their hosts and eventually killing them. Because they always kill their hosts, they are known as **parasitoids** (a lethal type of parasite). Parasitoid insects are enormously important in controlling populations of other insects.

For each type of feeding, mouthparts are adapted in a specialized way. **Sucking mouthparts** usually form a tube and can easily pierce the tissues of plants or animals. Mosquitos (order Diptera) demonstrate this arrangement well. Their mandibles,

maxillae, hypopharynx, and labrum-epipharynx are elongated into needlelike stylets, together forming a **fascicle** (Figure 21.17C), which pierces the skin of their prey to enter a blood vessel. The hypopharynx bears a salivary duct, and the labrum-epipharynx forms a food channel. The labrum forms a sheath for the fascicle that bends back during feeding (Figure 21.17C). In honey bees

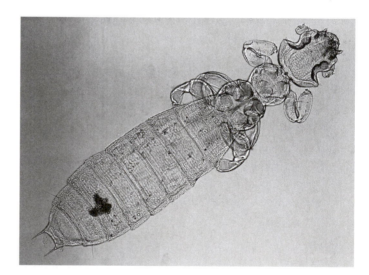

Figure 21.14
Gliricola porcelli (order Phthiraptera), a chewing louse of guinea pigs. Antennae are normally held in the deep grooves on the sides of the head.

Figure 21.13
Female human flea, *Pulex irritans* (order Siphonaptera).

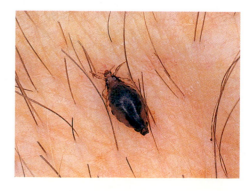

Figure 21.15
The head and body louse of humans, *Pediculus humanus* (order Phthiraptera), feeding.

A   B

Figure 21.16

A, Larval stage of the tomato hornworm, *Manduca sexta* (order Lepidoptera). The more than 100 species of North American sphinx moths are strong fliers and mostly nocturnal feeders. Their larvae are called hornworms because of the large, fleshy posterior spine.
B, Hornworm parasitized by a tiny wasp, *Apanteles* (a parasitoid), which laid its eggs inside the caterpillar. The wasp larvae have emerged, and their pupae are formed on the caterpillar's skin. Young wasps emerge in 5 to 10 days, but the caterpillar dies.

the labium forms a flexible and contractile "tongue" covered with many hairs. When a bee plunges its proboscis into nectar, the tip of the tongue bends upward and moves back and forth rapidly. Liquid enters the tube by capillary action and is drawn inside continuously by a pumping pharynx. In adult butterflies and moths, mandibles are usually absent (they are always present in larvae), and the maxillae form a long sucking proboscis (Figure 21.17D) for drawing nectar from flowers. At rest the proboscis coils into a flat spiral. While feeding the proboscis extends and fluid is pumped inside by pharyngeal muscles.

House flies, blow flies, and fruit flies have **sponging** and **lapping mouthparts** (Figure 21.17E). At the apex of the labium is a pair of large, soft lobes with grooves on the lower surface that serve as food channels. These flies lap liquid food or liquefy food first with salivary secretions. Horse flies not only sponge surface liquids but pierce the skin with slender, tapering mandibles, and then absorb blood.

Chewing mouthparts such as those of grasshoppers and many other herbivorous insects are adapted for seizing and crushing food (Figure 21.17A); those of most carnivorous insects are sharp and pointed for piercing their prey. Mandibles of chewing insects are strong, toothed plates whose edges can bite or tear while the maxillae hold the food and pass it toward the mouth. Enzymes secreted by the salivary glands provide chemical action to aid the chewing process.

Circulation

A tubular heart creates a peristaltic wave (see Figure 21.12) that moves hemolymph (blood) forward through the only blood vessel, a dorsal aorta. Accessory pulsatory organs help move hemolymph into the wings and legs, and flow is also facilitated by body movements. Hemolymph consists of plasma and amebocytes and apparently has little role in oxygen transport in most insects, but hemoglobin occurs in the hemolymph of some species (especially aquatic immatures occupying environments of low oxygen tension) and functions in oxygen transport.

Gas Exchange

Terrestrial animals require efficient respiratory systems that permit rapid oxygen and carbon dioxide exchange but at the same time restrict water loss. In insects this is the function of the **tracheal system,** an extensive network of thin-walled tubes that branch into every part of the body (Figure 21.18). The tracheal system of insects evolved independently of that of other arthropodan groups such as spiders. The tracheal trunks open to the outside by **spiracles,** usually two pairs on the thorax and seven or eight pairs on the abdomen. A spiracle may be merely a hole in the integument, as in primitively wingless insects, but there is usually a valve or some other closing mechanism that reduces water loss. The evolution of a tracheal system with valves must have been very important in enabling insects to move into drier habitats. Spiracles may also possess a filtering device such as a sieve plate or a set of interlocking bristles that prevents the entrance of water, parasites, or dust into the tracheae.

Tracheae are composed of a single layer of cells and are lined with cuticle that is shed during molts along with the outer cuticle. Spiral thickenings of cuticle (called **taenidia**) support the tracheae and prevent their collapse. Tracheae branch out into smaller tubes, ending in very fine, fluid-filled tubules called **tracheoles** (lined with cuticle, but not shed at ecdysis), which branch into a fine network over the cells. Large insects may have tracheae several millimeters in diameter that taper down to 1 to 2 μm. Tracheoles then taper to 0.5 to 0.1 μm in diameter. In one stage of silkworm larvae it is estimated that there are 1.5 million tracheoles! Some lepidopterous (moths and butterflies) larvae have an abdominal mass of tracheoles that forms the structural and physiological equivalent of a vertebrate lung. Scarcely any living cell is more than a few micrometers away from a tracheole. In fact, the ends of some tracheoles actually indent the membranes of cells that they supply, so that they terminate close to mitochondria. The tracheal system affords efficient transport usually without use of oxygen-carrying pigments in hemolymph, although hemoglobin is present in some.

The tracheal system may also include **air sacs,** which are apparently dilated tracheae without taenidia (Figure 21.18A). They are thin walled and flexible and are most common in the body cavity although they sometimes occur in appendages. Air sacs may allow internal organs to change in volume during growth without changing the shape of the insect, and they reduce the weight of large insects. However, in many insects the air sacs increase the volume of air inspired and expired. Muscular movements in the abdomen draw air into the tracheae and expand the sacs, which collapse on expiration. In some insects—locusts, for example—additional pumping is provided by telescoping the abdomen, pumping with the prothorax, or thrusting the head forward and backward.

Recent studies of insect respiration using X-rays have shown that tracheal expansion and compression also occur in response to movements of jaw or limb muscles. Contraction of these muscles increases pressure inside the exoskeleton, and this elevated

Figure 21.17

Four types of insect mouthparts. **A,** Chewing mouthparts of a grasshopper. **B** and **C,** Sucking mouthparts of a mosquito. Parts of piercing fascicle are shown in cross section (**C**). **D,** Sucking mouthparts of a butterfly. Mandibles are absent, and maxillae form a long proboscis. **E,** Sponging mouthparts of a house fly. A pair of large lobes with grooves on their lower surface are on the end of the labium.

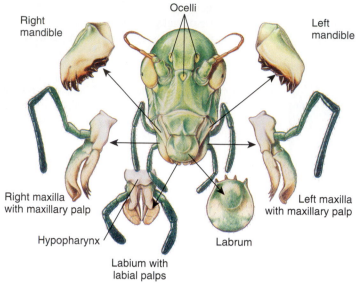

A   GRASSHOPPER

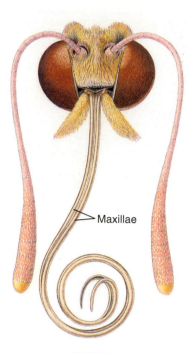

D   BUTTERFLY

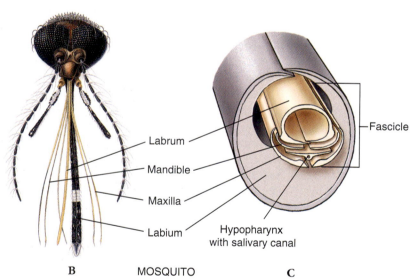

B   MOSQUITO

C

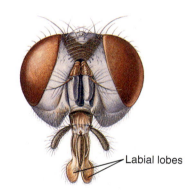

E   HOUSE FLY

pressure causes contraction of tracheae, effectively permitting insects to exhale. When muscles involved in tracheal compression relax, tracheae expand due to the recoil by taenidial rings. If tracheae contract when spiracles are closed, increased internal pressure enhances oxygen diffusion to cells.

In some very small insects, gas transport occurs entirely by diffusion along a concentration gradient. Consumption of oxygen causes a reduced pressure in their tracheae that pulls air inward through the spiracles.

The tracheal system is an adaptation for air breathing, but many insects (nymphs, larvae, and adults) live in water. In small, soft-bodied aquatic nymphs, gaseous exchange may occur by diffusion through the body wall, usually into and out of a tracheal network just under the integument. Aquatic nymphs of stoneflies, mayflies, and damselflies have a variety of **tracheal gills,** which are thin extensions of the body wall containing a rich tracheal supply. Gills of dragonfly nymphs are ridges in the rectum (rectal gills) where gas exchange occurs as water enters and leaves.

Although diving beetles, *Dytiscus* (Gr. *dytikos,* able to swim), can fly, they spend most of their life in water as excellent swimmers. How do they, and other aquatic insects, respire? They use an "artificial gill" in the form of a bubble of air (a plastron) held under the first pair of wings. The bubble is kept stable by a layer of hairs on top of the abdomen and is in contact with spiracles on the abdomen. Oxygen from the bubble diffuses into their tracheae and is replaced by diffusion of oxygen from the surrounding water. However, nitrogen from the bubble diffuses into the water, slowly decreasing the size of the bubble; therefore, diving beetles must surface every few hours to replace the air. Mosquito larvae are not good swimmers but live just below the surface, protruding short breathing tubes like snorkels to the surface for air (see Figure 21.23B). Spreading oil on the water, a favorite method of mosquito control, clogs their tracheae with oil and suffocates the larvae. "Rat-tailed maggots" of certain syrphid flies have an extensible tail that can stretch as much as 15 cm to the water surface.

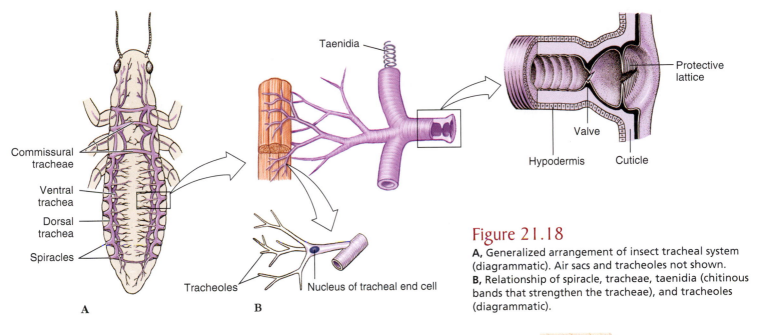

Figure 21.18

A, Generalized arrangement of insect tracheal system (diagrammatic). Air sacs and tracheoles not shown. **B,** Relationship of spiracle, tracheae, taenidia (chitinous bands that strengthen the tracheae), and tracheoles (diagrammatic).

Excretion and Water Balance

Insects and spiders have each evolved independently a unique excretory system consisting of **Malpighian tubules** that operate in conjunction with specialized glands in the wall of the rectum. Malpighian tubules, variable in number, are thin, elastic, blind tubules attached to the juncture between the midgut and hindgut (Figures 21.12 and 21.19A). Free ends of the tubules lie in the hemocoel and are bathed in hemolymph.

The mechanism of urine formation in Malpighian tubules of herbivorous insects appears to depend on a proton pump that adds hydrogen ions to the lumen of the tubule. Hydrogen ions are then exchanged for potassium ions (Figure 21.19B). This primary secretion of ions pulls water with it by osmosis to produce a potassium-rich fluid. Other solutes and waste materials also are secreted or diffuse into the tubule. The predominant waste product of nitrogen metabolism in most insects is uric acid, which is virtually insoluble in water (see p. 673). Uric acid enters the upper end of tubules, where the pH is slightly alkaline, as relatively soluble potassium and urate (abbreviated KHUr in Figure 21.19). As forming urine passes into the lower end of tubules, potassium combines with carbon dioxide and is reabsorbed as potassium bicarbonate ($KHCO_3$). As a result the pH of the fluid becomes acidic (pH 6.6), and insoluble uric acid (HUr) precipitates. As urine drains into the intestine and passes through the hindgut, specialized rectal glands reabsorb chloride, sodium (and in some cases potassium), and water.

Since water requirements vary among different types of insects, this ability to cycle water and salts is very important. Insects living in dry environments may resorb nearly all water from the rectum, producing a nearly dry mixture of urine and feces. However, freshwater larvae need to excrete water as well as to conserve salts. Insects that feed on dry grains need to conserve water and to excrete salt. In contrast, leaf-feeding insects ingest and excrete quantities of fluid. For example, aphids (p. 463) pass the excess fluid in the form of a sweetish material

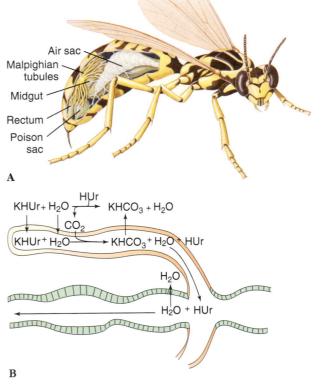

Figure 21.19

Malpighian tubules of insect. **A,** Malpighian tubules are located at the juncture of the midgut and hindgut (rectum) as shown in the cutaway view of a wasp. **B,** Function of Malpighian tubules. Hydrogen ions are actively exchanged for potassium ions in the upper tubules. Water and potassium acid urate (KHUr) follow. Potassium is resorbed with water and other solutes in the rectum.

called **honeydew,** which is relished by other insects, especially ants (see Figure 21.33A). Honeydew promotes growth of sooty mold (fungus) on leaves of infested plants and "rains" on cars parked beneath infested trees.

Nervous System

The nervous system in general resembles that of larger crustaceans, with a similar tendency toward the fusion of ganglia (see Figure 21.12). A number of insects have a giant fiber system. There is also a stomodeal nervous system that corresponds in function to the autonomic nervous system of vertebrates. Neurosecretory cells located in various parts of the brain have an endocrine function, but, except for their role in molting and metamorphosis, little is known of their activity.

Sense Organs

Along with neuromuscular coordination, insects have unusually keen sensory perception. Their sense organs are mostly microscopic and are located chiefly in the body wall. Each type usually responds to a specific stimulus including mechanical, auditory, chemical, visual, and other stimuli.

Mechanoreception Mechanical stimuli (those involving touch, pressure, and vibration) are detected by **sensilla.** A sensillum may be simply a **seta,** or hairlike process, connected with a nerve cell; a nerve ending just under the cuticle and lacking a seta, or a more complex organ (scolophorous organ) consisting of sensory cells with their endings attached to the body wall. Such organs are widely distributed over the antennae, legs, and body.

Auditory Reception Very sensitive setae (hair sensilla) or tympanal organs may detect specific frequencies of airborne sounds. In tympanal organs a number of sensory cells (ranging from a few to hundreds) extend to a very thin tympanic membrane that encloses an air space in which vibrations are detected. Tympanal organs occur in certain Orthoptera (Figure 21.2), Hemiptera, and Lepidoptera. Most insects are fairly insensitive to airborne sounds but can detect vibrations reaching them through the substrate. Organs on the legs usually detect vibrations of the substrate. Some nocturnal moths (for example, family Noctuidae) can detect ultrasonic pulses emitted by bats for echolocation (p. 627) and dive to the ground when they perceive bats.

Chemoreception Chemoreceptors (for taste or smell) are usually bundles of sensory cell processes often located in sensory pits. These are often on mouthparts, but in many insects they are also on the antennae, and butterflies, moths, and flies also have them on the tarsi of the legs. Chemical sense is generally keen, and some insects can detect certain odors for several kilometers. Many patterns of insect behavior such as feeding, mating, habitat selection, and host-parasite relations are mediated through chemical senses. These senses also play a crucial role in

responses of insects to artificial repellents and attractants. For example, an increase in carbon dioxide concentration, such as would be caused by a potential host nearby, causes a resting mosquito to begin flying, then it follows gradients of warmth and moisture and other cues to find its host. Diethyl toluamide **(DEET),** a repellent, apparently blocks the mosquito's ability to sense lactic acid, thus preventing host location.

Visual Reception Insect eyes are of two types, simple and compound. **Simple eyes** are found in some nymphs and larvae and in many adults. Most insects have three ocelli on their head. Honey bees probably use ocelli to monitor light intensity and photoperiod (length of day) but not to form images.

Most adult insects have **compound eyes,** which may cover much of the head. They consist of thousands of ommatidia—6300 in the eye of a honey bee, for example. The structure of the compound eye is similar to that of crustaceans (Figure 21.20). An insect such as a honey bee can see simultaneously in almost all directions around its body, but it is more myopic than humans, and images, even of nearby objects, are probably fuzzy. However, most flying insects rate much higher than humans in flicker-fusion tests. Flickers of light become fused in human eyes at a frequency of 45 to 55 per second, but bees and blow flies can distinguish as many as 200 to 300 separate flashes of light per second. This is undoubtedly advantageous in analyzing a fast-changing landscape during flight.

A bee can distinguish colors, but its sensitivity begins in the ultraviolet range, which human eyes cannot see. Although uniform in color to our perception, bee-pollinated flowers often have petals with lines and angular shapes that differ in ultraviolet (UV) light absorption and reflection. The lines and shapes of UV absorption acts as a "nectar guide," leading bees to nectar in the flower. Many insects, such as butterflies, also have vision sensitive to red wavelengths, but honey bees are red-blind.

Other Senses Insects also have well-developed senses for temperature, especially on the antennae and legs, and for

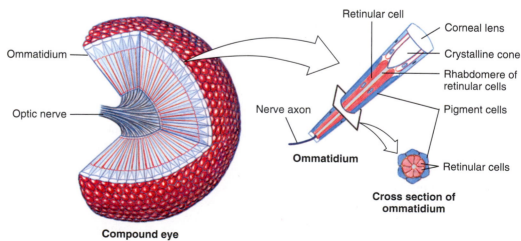

Figure 21.20
Compound eye of an insect. A single ommatidium is shown enlarged to the right.

humidity, as well as for proprioception (sensation of muscle stretch and body position), gravity, and other physical properties.

Neuromuscular Coordination

Insects are active creatures with excellent neuromuscular coordination. Arthropod muscles are typically cross-striated, just as vertebrate skeletal muscles are. A flea can leap a distance of 100 times its own length, and an ant can carry in its jaws a load greater than its own weight. This sounds as though insect muscle were stronger than that of other animals. Actually, however, the force that a particular muscle can exert is related directly to its cross-sectional area, not its length. Based on maximum load moved per square centimeter of cross section, the strength of insect muscle is relatively the same as that of vertebrate muscle. The illusion of great strength of insects (and other small animals) is simply a consequence of small body size.

In terms of proportionate body length, a flea's jump would be the equivalent of a 6-foot human executing a standing high jump of 600 feet! Actually, a flea's muscles are not entirely responsible for its jump; they cannot contract rapidly enough to reach the required acceleration. Fleas depend on pads of *resilin,* a protein with unusual elastic properties, which is also found in wing-hinge ligaments of many other insects. Resilin releases 97% of its stored energy on returning from a stretched position, compared with only 85% in most commercial rubber. When a flea prepares to jump, it rotates its hind femurs and compresses the resilin pads, then engages a "catch" mechanism. In effect, it has cocked itself. To take off, the flea needs to exert a relatively small muscular action to unhook the catches, allowing the resilin to expand.

Reproduction

Parthenogenesis occurs prominently in life cycles of some Hemiptera and Hymenoptera (see insect orders, pp. 463–464), but sexual reproduction is the norm for insects. Sexes are separate and various means are used to attract mates. A female moth releases a powerful pheromone that males can detect for a great distance. Fireflies use flashes of light; some insects find each other by sounds or color signals and by various kinds of courtship behavior.

Once a mate has been attracted, fertilization is usually internal. Sperm may be released directly or packaged into spermatophores. During the evolutionary transition of ancestral insects from marine to terrestrial life, spermatophores were widely used. Spermatophores may be transferred without copulation, as in silverfish, where a male deposits a spermatophore on the ground, then spins signal threads to guide a female to it. Alternatively, spermatophores may be deposited in the female vagina (see Figure 21.12) during copulation; in many cases, especially in butterflies, nutrients are also passed to the female via the spermatophore. Copulation (Figure 21.21) evolved much later than indirect sperm transfer using spermatophores.

A

B

Figure 21.21

Copulation in insects (see also Figures 21.2B and 21.7B). **A,** *Omura congrua* (order Orthoptera) are a kind of grasshopper found in Brazil. **B,** Bluet damselflies, *Enallagma* sp. (order Odonata), are common throughout North America. Here, a male still grasps a female after copulation. The female (white abdomen) lays eggs in the water.

Usually sperm are stored in the seminal receptacle of a female in numbers sufficient to fertilize more than one batch of eggs. Many insects mate only once during their lifetime, but others, such as male damselflies, copulate several times per day.

Insects usually lay a great many eggs. A queen honey bee, for example, may lay more than 1 million eggs during her lifetime. On the other hand, some flies are viviparous and bring forth only a single offspring at a time. Insects that make no provision for care of their young may lay many more eggs than do insects that provide for their young or those that have a very short life cycle.

Most species lay their eggs in a particular habitat to which visual, chemical, or other cues guide them. Butterflies and moths lay their eggs on the specific kind of plant on which the caterpillar must feed. For example, a tiger moth may look for a pigweed, a sphinx moth for a tomato or tobacco plant, and a monarch butterfly for a milkweed plant (Figure 21.22). Insects whose immature stages are aquatic characteristically lay their eggs in water (Figure 21.23). A tiny braconid wasp lays her eggs on the caterpillar of the sphinx moth where the larvae will feed internally. After feeding and growing within the caterpillar, the braconid larvae emerge from the host and pupate externally within tiny cocoons (Figure 21.23). An ichneumon wasp, with unerring accuracy, seeks a certain kind of larva in which her young will live as

parasitoids. Her long ovipositor may have to penetrate 1 to 2 cm of wood to find a larva of a wood wasp or a wood-boring beetle in which she will deposit her eggs (see Figure 21.8).

Metamorphosis and Growth

Early development occurs within eggs, and hatching young escape from eggs in various ways. During postembryonic development most insects change in form, undergoing **metamorphosis** (see Figure 21.22). During this period they must undergo a series of molts to grow, and each stage between molts is called an **instar.**

Although metamorphosis occurs in many animals, insects illustrate it more dramatically than any other group. The transformation, for instance, of a hickory horned devil caterpillar into a beautiful royal walnut moth represents an astonishing morphological change. In insects metamorphosis is associated with the evolution of wings, which are restricted to the reproductive stage. Adults, in effect, have become mating and dispersal stages for those species possessing flight.

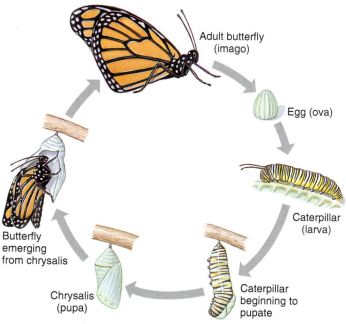

Figure 21.22

Complete (holometabolous) metamorphosis in a butterfly, *Danaus plexippus*. Eggs hatch to produce first of several larval instars. Last larval instar molts to become a pupa. Adult emerges at pupal molt.

Ametabolous (Direct) Development

A few insects, such as silverfish and springtails, undergo direct development. The young, or juveniles, are similar to adults except in size and sexual maturation. Stages are egg, juvenile, and adult. Primitively wingless insects are among those with this type of development.

Hemimetabolous (Incomplete) Metamorphosis

Some insects undergo a **hemimetabolous** (Gr. *bemi,* half, + *metabolē,* change), or gradual (incomplete), metamorphosis. These include grasshoppers, cicadas, mantids, and terrestrial bugs, which have terrestrial young, and mayflies, stoneflies, dragonflies, and aquatic bugs that lay their eggs in water and whose young are aquatic. The young are called **nymphs,** and their wings develop externally as budlike outgrowths in early instars and increase in size as the animal grows by successive molts and becomes a winged **adult** (Figures 21.24 and 21.25). Aquatic nymphs in some orders have tracheal gills or other modifications for aquatic life (Figure 21.26). The stages of hemimetabolous insects are egg, nymph (several instars), and adult (see Figure 21.25).

The *biological* meaning of the word "bug" is a great deal more restrictive than in common English usage. People often refer to all insects as "bugs," even extending the word to include such non-animals as bacteria, viruses, and glitches in computer programs. Strictly speaking, however, a bug is a member of order Hemiptera and nothing else.

Holometabolous (Complete) Metamorphosis

Approximately 88% of insects undergo a complete or **holometabolous** (Gr. *holo,* complete, + *metabolē,* change) metamorphosis, which separates physiological processes of growth

A   B

Figure 21.23

A, Mosquito *Culex* (order Diptera) lays her eggs in small packets or rafts on the surface of standing or slowly moving water. **B,** Mosquito larvae are the familiar wrigglers of ponds and ditches. To breathe, they hang head down, with respiratory tubes projecting through the surface film of water. Motion of vibratile tufts of fine hairs on the head brings a constant supply of food to these filter feeders.

A   B

Figure 21.24

A, Ecdysis in a cicada, *Tibicen davisi* (order Hemiptera). The old cuticle splits along a dorsal midline as a result of increased blood pressure and of air forced into the thorax by muscle contraction. The emerging insect is pale, and its new cuticle is soft. The wings will be expanded by blood pumped into veins, and the insect enlarges by taking in air. **B,** An adult *Tibicen davisi*.

Physiology of Metamorphosis

Hormones regulate metamorphosis in insects. Major endocrine organs that are involved in development are the **brain, prothoracic (ecdysial) glands, corpora cardiaca,** and **corpora allata** (Figure 34.4, p. 757).

The intercerebral part of the brain and the ganglia of the nerve cord contain several groups of neurosecretory cells that produce a brain hormone called **prothoracicotropic hormone (PTTH).** These neurosecretory cells send their axons to paired organs behind the brain, called corpora cardiaca, which serve as storage and release organs for PTTH (and also produce other hormones). PTTH is carried via the hemolymph to the prothoracic gland, a glandular organ in the head or prothorax produces **molting hormone,** or **ecdysone** (ek-dĭ′sōn) in response to PTTH. Ecdysone sets in motion certain processes that lead to molting and discarding the old cuticle (ecdysis).

Larval molting persists as long as **juvenile hormone** from the corpora allata is present in sufficient amounts, along with molting hormone in the hemolymph. Under these conditions, each molt produces a larger larva (see Figure 34.4).

In later instars, the corpora allata release progressively less juvenile hormone. When juvenile hormone is at a very low level, a larva molts to become a pupa (instead of a *larger* larva), and likewise, the cessation of juvenile hormone production in a pupa leads to an adult at the next molt (metamorphosis). Control of development is the same in hemimetabolous insects, except that there is no pupa, and cessation of juvenile hormone production occurs in the final nymphal instar. The corpora allata again become active in adult insects, in which juvenile hormone is important in normal sexual reproduction and gamete formation. The prothoracic glands degenerate in adults of most insects, and adults do not molt.

(larva) from those of differentiation (pupa) and reproduction (adult) (see Figure 21.22). In effect, each stage functions without competition with the other stages, for larvae often live in entirely different surroundings and eat different foods from adults. The wormlike **larvae,** which usually have chewing mouthparts, are known as caterpillars, maggots, bagworms, fuzzy worms, or grubs. After a series of instars, a larva molts to enter a transitional stage called a **pupa.** Pupae are usually inactive and enveloped by a case, which may take several different forms. Pupae are nonfeeding and are the stage in which many insects also pass the winter. The **adult** insect emerges (ecloses) from the overwintering pupa in the spring, revealing an insect with crumpled, miniature wings. In a short time its wings expand and harden, and the insect is on its way. The stages of a holometabolous insect, then, are egg, larva (several instars), pupa, and adult (see Figure 21.22). Adults undergo no further molting.

Insect hormones have been the focus of fascinating experiments. For example, if the corpora allata (and thus juvenile hormone) are removed surgically from a larva, the following molt is a metamorphosis. Conversely, if the corpora allata from a young larva are transplanted into a final larval instar, the latter can be converted into a giant larva, because metamorphosis to a pupa cannot occur.

Figure 21.25

Life history of a hemimetabolous insect.

Egg   ◄—————— Nymphs ——————►   Adult chinch bug

an insect, such signals forecast adverse conditions to come, for example, lengthening or shortening of days. Thus photoperiod, or day length, is often a signal that initiates diapause. After diapause is initiated, another environmental signal is usually required to end it. Such signal may be return of favorable temperature after a prolonged period of cold or an occasion of rain after a dry period, as in a desert.

Diapause always occurs at the end of an active growth stage of the molting cycle so that, when the diapause period ends an insect is ready for another molt. One species of the ant *Myrmica* reaches the third instar stage in late summer. Many larvae do not develop beyond this point until the following spring, even if temperatures are mild or if the larvae are kept in a warm laboratory. Depending on the species, insects may enter diapause at any stage of their life cycle.

Figure 21.26

A, Stonefly, *Perla* sp. (order Plecoptera). **B,** Ten-spot dragonfly, *Libellula pulchella* (order Odonata). **C,** Nymph of dragonfly. Both stoneflies and dragonflies have aquatic nymphs that undergo gradual metamorphosis.

Defense

Insects as a group display many colors. This is especially true of butterflies, moths, and beetles. Even within a species the color pattern may vary in a seasonal way, and colors may differ between males and females. Some color patterns and body forms in insects are highly adaptive in evasion of predation, such as those for **mimicry** (imitation of a noxious species by a palatable one, Figure 21.27), **aposematic coloration** (warning coloration to advertise noxious qualities), and **crypsis** (camouflage in shape or coloration to escape notice, Figure 21.28).

Diapause

Many animals, including many types of insects, undergo a period of dormancy in their annual life cycle. In temperate zones there may be a period of winter dormancy, called hibernation, or a period of summer dormancy, called estivation, or both. There are periods in the life cycle of many insects when eggs, larvae, pupae, or even adults remain dormant for a long time because external conditions are too harsh or unfavorable for survival in states of normal activity. Thus the life cycle is synchronized with periods of suitable environmental conditions and abundance of food. Most insects enter a dormant state when some factor of the environment, such as temperature, becomes unfavorable, and dormancy continues until conditions again become favorable.

However, some species have a prolonged arrest of growth that occurs regardless of environment, whether or not favorable conditions prevail. This type of dormancy is called **diapause** (di′a-poz) (Gr. *dia,* through, dividing into two parts, + *pausis,* a stopping), and it is an important adaptation to survive adverse environmental conditions. Diapause is internally controlled in each species and sometimes varies between subspecies within a species, but it is usually initiated by a particular signal. In the environment of

Besides color, insects have other methods of protecting themselves. The cuticular exoskeleton affords good protection for many. Some, such as stink bugs, have repulsive odors and tastes; others protect themselves by a good offense, for many are very aggressive and fight (for example, bees and ants); and still others are swift in running for cover when danger threatens.

Many insects practice chemical warfare in a variety of ingenious ways. Some repel an assault by virtue of their bad taste, odor, or poisonous properties; others use chemical exudates that mechanically prevent a predator from attacking. Caterpillars of some monarch butterflies (Figure 21.27) assimilate cardiac glycosides from certain species of milkweed (family Asclepiadaceae); this substance confers unpalatability on larvae and adults and induces vomiting in some (but not all) of their avian predators.

Figure 21.27

Mimicry in butterflies.
A, Monarch butterfly is distasteful to, and avoided by some birds because as a caterpillar it fed on the acrid milkweed. **B,** The monarch is mimicked by the smaller viceroy butterfly, *Limenitis archippus,* which feeds on willows and may (or may not) be tasteful to birds, but is not eaten because it so closely resembles the monarch in color and markings. This kind of mimicry is called Batesian mimicry, although recent evidence argues for Müllerian mimicry.

A

B

C

Figure 21.28

Crypsis (camouflage) in insects. **A,** *Estigena pardalis* (order Lepidoptera) in Java resembles a dead leaf. **B,** Bizarre processes from the thorax of a treehopper from Mexico, *Sphongophorus* sp. (order Hemiptera), masquerade as parts of the twig on which it feeds. **C,** Broken outlines and color of a katydid (*Dysonia* sp., order Orthoptera) in Costa Rica give it the appearance of the leaves on which it has been feeding.

Bombardier beetles, on the other hand, produce an irritating spray, which they aim accurately at attacking ants or other enemies.

Behavior and Communication

The keen sensory perceptions of insects make them extremely responsive to many stimuli. Stimuli may be internal (physiological) or external (environmental), and responses are governed by both the physiological state of the animal and the pattern of nerve pathways traveled by the impulses. Many responses are simple, such as orientation toward or away from a stimulus, for example, avoidance of light by a cockroach, or attraction of carrion flies to the odor of dead flesh.

Much behavior of insects, however, is not a simple matter of orientation but involves a complex series of responses. A pair of dung beetles chew a bit of dung, roll it into a ball, and roll the ball laboriously to where they intend to bury it after laying their eggs in it (Figure 21.29). Cicadas slit the bark of a twig and then lay an egg in each of the slits. Female potter wasps *Eumenes* scoop clay into pellets, carry them one by one to a building site, and fashion them into dainty little narrow-necked clay pots, into each of which the wasp lays an egg. Then the mother wasp hunts and paralyzes a number of caterpillars, pokes them into the opening of a pot, and seals the opening with clay. Each egg, in its own protective pot, hatches to find a well-stocked larder of food.

Figure 21.29

Dung beetles *Canthon pilularis* (order Coleoptera), chew off a bit of dung, form it into a ball, and then roll it to where they will bury it in soil. One beetle pushes while the other pulls. Eggs are laid in the ball, and larvae feed on the dung. Dung beetles are black, an inch or less in length, and common in pasture fields.

Much of such behavior is innate; however, learning is more important than was once believed. The potter wasp, for example, must learn where she has left her pots if she is to return to fill them with caterpillars one at a time. Social insects, which have been studied extensively, are capable of most basic forms of learning used by mammals. An exception is insight learning. Apparently insects, when faced with a new problem, cannot reorganize their memories to construct a new response.

Some insects can memorize and perform in sequence tasks involving multiple signals in various sensory areas. Worker honey bees have been trained to walk through mazes that involved five turns in sequence, using such clues as color of a marker, distance between two spots, or angle of a turn. The same is true of ants. Workers of one species of *Formica* learned a six-point maze at a rate only two or three times slower than that of laboratory rats. Foraging trips of ants and bees often wind and loop in a circuitous route, but once a forager has found food, the return trip is relatively direct. One investigator suggests that the continuous series of calculations necessary to figure the angles, directions, distance, and speed of the trip and to convert it into a direct return could involve a stopwatch, compass, and integral vector calculus. How an insect does this is unknown.

Insects communicate with each other by means of chemical, visual, auditory, and tactile signals. **Chemical** signals take the form of **pheromones,** which are substances secreted by one individual that affect behavior or physiological processes of another individual. Many pheromones have been described. Like hormones, pheromones are effective in minute quantities. Known functions of various pheromones include attraction of the opposite sex, release of certain behavior patterns (for example, aggregation pheromones to enable mass attack of bark beetles on a tree or for overwintering of ladybug beetles), to fend off aggression, to mark trails and territories, and to signal alarms. Social insects, such as bees, ants, wasps, and termites, can recognize a nestmate—or an alien in the nest—by means of identification pheromones. Social parasites escape detection—and certain destruction—by imitating or duplicating pheromones produced by members of their host colony. Pheromones determine caste in termites and to some extent in ants and bees. They are a primary integrating force in populations of social insects.

Many insect pheromones have been isolated and identified. Traps baited with pheromones have been used for many years to monitor insects of economic importance. They can be used to detect presence of an insect, such as a new arrival from a neighboring area (tracking spread of European gypsy moth in the United States or presence of European corn ear worms in a field), or to monitor changes in population levels. Use of pheromone traps has become an important tool to detect potential outbreaks, allowing sufficient time to plan remedial action.

Sound production and **reception** in insects have been studied extensively, and although a sense of hearing is not present in all insects, this means of communication is meaningful to those insects that use it. Sounds serve as warning devices, advertisement of territorial claims, or courtship songs. Sounds of crickets and grasshoppers seem to be concerned with courtship and aggression. Male crickets scrape the modified edges of their forewings together to produce their characteristic chirping. The long, drawn-out sound of male cicadas, a call to attract females, is produced by vibrating membranes in a pair of organs on the ventral side of the basal abdominal segment.

There are many forms of **tactile communication,** such as tapping, stroking, grasping, and antennae touching, which evoke responses varying from recognition to recruitment and alarm. Certain kinds of flies, springtails, and beetles manufacture their own **visual signals** in the form of **bioluminescence.** Best known of luminescent beetles are fireflies, or lightningbugs (which are neither flies nor bugs, but beetles), in which a flash of light helps to locate a prospective mate. Each species has its own characteristic flashing rhythm produced on the ventral side of the last abdominal segments. Females flash an answer to the species-specific pattern to attract males. This interesting "love call" has been adopted by species of *Photuris,* which prey on male fireflies of other species they attract (Figure 21.30).

Social Behavior

Insects rank very high in the animal kingdom in their organization of social groups, and cooperation within more complex groups depends heavily on chemical and tactile communication. Social communities are not all complex, however. Some community groups are temporary and uncoordinated, as are hibernating associations of carpenter bees or feeding gatherings of aphids. Some are coordinated for only brief periods, and some cooperate more fully, such as tent caterpillars, *Malacosoma,* which join in building a home web and feeding net. However, all of these are open communities with limited social behavior.

In the **eusocial** Hymenoptera (honey bees and ants) and Isoptera (termites) a complex social life is necessary for perpetuation of the species. They involve all stages of the life cycle, communities are usually ongoing, all activities are collective, and there is reciprocal communication and division of labor. The society usually demonstrates polymorphism, or **caste** differentiation.

Honey bees have one of the most complex organizations in the insect world. Instead of lasting one season, their organization continues for an indefinite period. As many as 60,000 to 70,000 honey bees may live in a single hive. Of these, there are three castes: a single sexually mature female, or **queen;** a few hundred **drones,** which are sexually mature males; and the rest are **workers,** which are sexually inactive genetic females (Figure 21.31).

Workers take care of young, secrete wax with which they build the six-sided cells of honeycomb, gather nectar from flowers, manufacture honey, collect pollen, and ventilate and guard the hive. Several drones fertilize a queen during her nuptial (mating) flight, at which time enough sperm is stored in her seminal receptacle to last her a lifetime. Drones die after mating, and those remaining in the hive at summer's end are driven out by workers and starve.

Castes are determined partly by fertilization and partly by what is fed to larvae. Drones develop parthenogenetically from unfertilized eggs (and consequently are haploid); queens and workers develop from fertilized eggs (and thus are diploid; see haplodiploidy, p. 142). Female larvae that will become queens

Figure 21.30

Firefly femme fatale, *Photuris versicolor,* eating a male *Photinus tanytoxus,* which she has attracted with false mating signals.

Figure 21.31

Queen bee surrounded by her court. The queen is the only egg layer in the colony. The attendants, attracted by her pheromones, constantly lick her body. As food and the queen's pheromones are transferred from these bees to others, the queen's presence is communicated throughout the colony.

are fed royal jelly, a secretion from the salivary glands of nurse workers. Royal jelly differs from the "worker jelly" fed to ordinary larvae, but components in it that are essential for queen determination have not yet been identified. Honey and pollen are added to worker diet about the third day of larval life. Pheromones in "queen substance," which is produced by a queen's mandibular glands, prevent female workers from maturing sexually. Workers produce royal jelly only when the level of "queen substance" pheromone in the colony drops, most typically due to overcrowding. This change also occurs when the queen becomes too old, dies, or is removed. Then workers' ovaries develop, and they start enlarging a larval cell and feeding a larva royal jelly to produce a new queen. Production of a new queen may be followed by swarming where the old queen leaves with part of the colony.

Honey bees have evolved an efficient system of communication by which, through certain body movements, their scouts inform workers of the location and quantity of food sources (Figure 36.23, p. 800).

Termite colonies contain several castes, consisting of fertile individuals, both males and females, and immature individuals (Figure 21.32). Some fertile individuals may have wings and may leave the colony, mate, lose their wings, and as **king** and **queen** start a new colony. Wingless fertile individuals may under certain conditions substitute for the king or queen. Immature members are wingless and become **workers** and **soldiers.** Soldiers have large heads and mandibles and serve to defend the colony. As in bees and ants, extrinsic factors cause caste differentiation. Reproductive individuals and soldiers secrete inhibiting pheromones that pass throughout the colony to nymphs through a mutual feeding process, called **trophallaxis,** so that they become sterile workers. Workers also produce pheromones, and if the level of "worker substance" or "soldier substance" falls, as might happen after an attack by marauding predators, for example, the next generation produces compensating proportions of the appropriate caste.

Ants also have highly organized societies. Superficially, they resemble termites, but they are related to bees and wasps in the order Hymenoptera and can be distinguished easily. In contrast to termites, ants are usually dark in color, are hard bodied, and have a constriction posterior to their first abdominal somite. Antennae of ants are elbowed, while those of termites are threadlike or resemble a string of beads (moniliform).

A   B

Figure 21.32

A, Termite workers, *Reticulitermes flavipes* (order Isoptera), eating yellow pine. Workers are wingless immature animals that tend the nest. **B,** Termite queen becomes a distended egg-laying machine. The queen and several workers and soldiers are shown here.

A

B

Figure 21.33

A, An ant (order Hymenoptera) tending a group of aphids (order Homoptera). The aphids feed copiously on plant juices and excrete the excess as a clear liquid rich in carbohydrates ("honeydew"), which is cherished as a food by ants. **B,** A weaver ant nest in Australia.

In ant colonies males die soon after mating and the queen either starts her own new colony or joins some established colony and does the egg laying. Sterile females are wingless workers and soldiers that do work of the colony: gather food, care for young, and protect the colony. In many larger colonies there may be two or three types of individuals within each caste.

Ants have evolved some striking patterns of "economic" behavior, such as making slaves, farming fungi, herding "ant cows" (aphids or other members of Sternorrhyncha) (Figure 21.33A), sewing their nests together with silk (Figure 21.33B), and using tools.

INSECTS AND HUMAN WELFARE

Beneficial Insects

Although most of us think of insects primarily as pests, all terrestrial life, including human life, would have great difficulty surviving if all insects were suddenly to disappear. Some

A   B   C

Figure 21.34

Some beneficial insects. **A,** A predaceous stink bug (order Hemiptera) feeds on a caterpillar. Note the sucking proboscis of the bug. **B,** A ladybug beetle (order Coleoptera). Adults (and larvae of most species) feed voraciously on plant pests such as mites, aphids, scale insects, and thrips. **C,** A parasitic wasp (*Larra bicolor*) attacking a mole cricket. The wasp drives the cricket from its burrow, then stings and paralyzes it. After the wasp deposits her eggs, the mole cricket recovers and resumes an active life—until it is killed by developing wasp larvae.

produce useful materials: honey and beeswax from bees, silk from silkworms, and shellac from a wax secreted by lac insects. More important, however, insects are necessary for the cross-fertilization of many crops. Bees pollinate almost $14 billion worth of food crops per year in the United States alone, and this figure does not include pollination of forage crops for livestock or pollination by other insects.

Very early in their evolution, insects and flowering plants formed a relationship of mutual adaptations (**coevolution**) that have been to each other's advantage. Insects exploit flowers for food, and flowers exploit insects for pollination. Each floral development of petal and sepal arrangement is correlated with the sensory array of certain pollinating insects. Among these mutual adaptations are amazing devices of allurements, traps, specialized structures, nectar guides (p. 452), and precise timing.

Many predaceous insects, such as tiger beetles, aphid lions, ant lions, praying mantids, and ladybug beetles, destroy harmful insects (Figure 21.34). Parasitoid insects are very important in controlling populations of many harmful insects. Dead animals are quickly consumed by maggots hatched from eggs laid in carcasses. In turn, insects serve as important sources of food for many other animals.

Harmful Insects

Harmful insects include those that eat and destroy plants and fruits, such as grasshoppers, chinch bugs, corn borers, boll weevils, grain weevils, San Jose scale, and hundreds of others (Figure 21.35). Nearly every cultivated crop has several insect pests. People expend enormous resources in all agricultural

A   B   C

Figure 21.35

Some insect pests. **A,** Japanese beetles, *Popillia japonica* (order Coleoptera), are serious pests of fruit trees and ornamental shrubs. They were introduced into the United States from Japan in 1917. **B,** Longtailed mealybug, *Pseudococcus longispinus* (order Hemiptera). Many mealybugs are pests of commercially valuable plants. **C,** Corn ear worms, *Heliothis zea* (order Lepidoptera). An even more serious pest of corn is the infamous corn borer, an import from Europe in 1908 or 1909.

activities, in forestry, and in the food industry to counter insects and the damage they cause. Outbreaks of bark beetles or defoliators such as spruce budworms, hemlock woolly adelgids, and gypsy moths have caused tremendous economic losses and have become a major element in determining the composition of forests in the United States. Gypsy moths, introduced into the United States in 1869 in an ill-advised attempt to breed a better silkworm, have spread throughout the northeast as far south as Virginia and as far west as Minnesota. They defoliate forests in years when there are outbreaks. In 1981, they defoliated 13 million acres in 17 northeastern states.

Ten percent of all arthropod species are parasitic insects, or insects that in essence are "micropredators" because they attack but do not remain on their hosts. Lice, bloodsucking flies, warble flies, bot flies, and many others attack humans or domestic animals or both. Malaria, carried by *Anopheles* mosquitos (Figure 21.36), is still one of the world's major diseases, infecting hundreds of millions of people each year and causing millions of deaths. Mosquitos also transmit yellow fever and lymphatic filariasis. Fleas carry plague, which at times in history has killed significant portions of human populations. House flies are vectors of typhoid, as are lice for typhus fever; tsetse flies carry African sleeping sickness; and bloodsucking bugs, *Rhodnius* and related genera, transmit Chagas's disease. The newest viral plague to hit North America, the West Nile virus, is carried by over 40 species of mosquito, especially *Culex,* and infects humans, some other mammals, and over 75 species of birds, some of which serve as reservoirs for the virus.

There is tremendous destruction of food, clothing, and property by weevils, cockroaches, ants, clothes moths, termites, and carpet beetles. Not the least of insect pests are bedbugs, *Cimex,* bloodsucking hemipterous insects whose association with humans originated from bats that shared their caves early in human evolution. Bedbug infestation is rising throughout the developed world for unknown reasons. Proposed explanations include increased importation of bugs by travelers and reluctance to use insecticides.

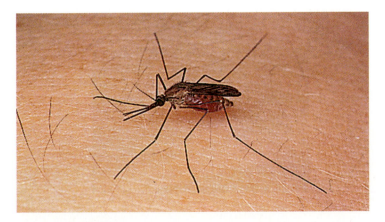

Figure 21.36
A mosquito, *Anopheles quadrimaculatus* (order Diptera). *Anopheles* spp. are vectors of malaria.

> ### Classification of Subphylum Hexapoda
>
> **Class Entognatha** (en'tog-natha) (Gr. *entos*, within, inside, + *gnathos*, jaw): **entognaths.** Base of mouthparts lies within head capsule; mandibles have one articulation. Example: *Entomobrya.*
> **Class Insecta** (in-sek'ta) (L. *insectus*, cut into): **insects.** Bases of mouthparts exposed and exiting head capsule; mandibles usually have two regions of articulation. Examples: *Drosophila, Bombus, Anopheles* (insect orders listed on pp. 462–464).

Control of Insects

Because all insects are an integral part of the ecological communities to which they belong, their total destruction would obviously do more harm than good. All terrestrial food webs would be seriously disturbed or destroyed. The beneficial role of insects in our environment is often overlooked, and in our zeal to control the pests we spray the landscape indiscriminately with extremely effective "broad-spectrum" insecticides that eradicate good, as well as harmful, insects. We have also found, to our dismay, that many chemical insecticides persist in the environment and accumulate as residues in the bodies of animals higher in the food chain, including ourselves. Furthermore, many insects have evolved a resistance to insecticides in common use. Honey bees are especially susceptible to insecticides, and resistance mainly is developed by harmful insects.

In recent years, methods of control other than chemical insecticides have been under intense investigation, experimentation, and development. Economics, concern for the environment, and consumer demand are causing thousands of farmers across the United States to use alternatives to strict dependence on chemicals.

Many organisms useful in the biological control of insects are currently being used or are under scientific investigation. These organisms include bacterial, viral, and fungal pathogens. For example, a bacterium, *Bacillus thuringiensis,* produces a toxin that is quite effective in control of lepidopteran pests (cabbage looper, imported cabbage worm, tomato worm, gypsy moth). Other strains of *B. thuringiensis* have toxins that kill insects in other orders, and the species diversity of target insects is being widened by techniques of genetic engineering. Genes coding for the toxin produced by *B. thuringiensis* also have been introduced into other bacteria and even into the plants themselves, which makes the plants resistant to insect attack. Many of our grain crops, especially corn, now possess genes that express protein toxic to specific insect pests, thereby obviating the need for pesticides. However, some insects have now evolved resistance to the *B. thuringiensis* toxin.

Classification of Class Entognatha and Class Insecta

Entomologists do not all agree on names of orders or on the limits of each order. Some choose to combine and others to divide the groups. However, this synopsis of orders is one that is rather widely accepted.

I. Class Entognatha

Order Protura (pro-tu′ra) (Gr. *protos,* first, + *oura,* tail). Minute (1 to 1.5 mm); no eyes or antennae; appendages on abdomen as well as thorax; live in soil and dark, humid places; direct development.

Order Diplura (dip-lu′ra) (Gr. *diploos,* double, + *oura,* tail): **japygids** and **campodeids.** Usually less than 10 mm; pale, eyeless; a pair of long terminal filaments or pair of caudal forceps; live in damp humus or rotting logs; development direct.

Order Collembola (col-lem′bo-la) (Gr. *kolla,* glue, + *embolon,* peg, wedge): **springtails, jumping bristletails,** and **snow fleas.** Small (5 mm or less); compound eyes lacking, eye patches of 1 to several lateral ocelli; respiration by trachea or body surface; a springing organ folded under the abdomen for leaping; abundant in soil; sometimes swarm on pond surface film or on snowbanks in spring; development direct.

II. Class Insecta

Subclass Apterygota

Order Thysanura (thy-sa-nu′ra) (Gr. *thysanos,* tassel, + *oura,* tail): **silverfish** (Figure 21.37) and **bristletails.** Small to medium size; large eyes; long antennae; three long terminal cerci; live under stones and leaves and around human habitations; development direct.

Subclass Pterygota

1. Infraclass Paleoptera

Order Ephemeroptera (e-fem-er-op′ter-a) (Gr. *ephēmeros,* lasting but a day, + *pteron,* wing): **mayflies** (Figure 21.38). Wings membranous; forewings larger than hindwings; adult mouthparts vestigial; nymphs aquatic, with lateral tracheal gills.

Order Odonata (o-do-na′ta) (Gr. *odontos,* tooth, + *ata,* characterized by): **dragonflies, damselflies** (Figure 21.21B and 21.26B). Large; membranous wings are long, narrow, net veined, and similar in size; long and slender body; aquatic nymphs with gills and prehensile labium for capture of prey.

2. Infraclass Neoptera

Order Orthoptera (or-thop′ter-a) (Gr. *orthos,* straight, + *pteron,* wing): **grasshoppers** (Figure 21.2), **locusts, crickets, katydids.** Wings, when present, with forewings thickened and hindwings folded like a fan under forewings; chewing mouthparts.

Order Blattodea (blə-′tō-dēə) (L. *blatta,* cockroach, + Gr. *eidos,* form, + *ea,* characterized by): **cockroaches.** Common insects in tropical areas; often in houses in northern areas; oval, flattened bodies may exceed 5 cm in length; tarsi with 5 segments; wings typically present, often reduced.

Order Phasmatodea (faz-mə-′tō-dēə) (Gr. *phasma,* apparition, + *eidos,* form, + *ea,* characterized by): **walkingsticks** (Figure 21.7B)

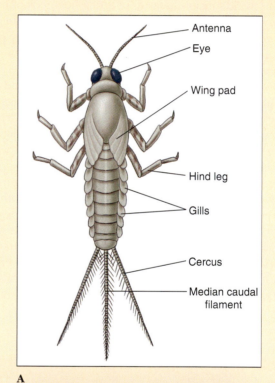

A

Figure 21.37
Silverfish *Lepisma* (order Thysanura) is often found in homes.

B

Figure 21.38
Mayfly (order Ephemeroptera). **A,** Nymph. **B,** Adult.

Classification of Class Entognatha and Class Insecta

and **leaf insects**. Bodies elongated and sticklike or flattened and laterally expanded; herbivorous, tropical forms may be very large (up to 30 cm).

Order Mantodea (man-ˈtō-dēə) (Gr. *mantis*, soothsayer, + *eidos*, form, + *ea*, characterized by): **mantids** (Figure 21.5). Bodies elongated with raptorial front legs; predatory; may reach 10 cm in length.

Order Mantophasmatodea (man-ˈtō-faz-mə-ˈtō-dēə) (an amalgamation of the order names for praying mantids [Mantodea] and walkingsticks [Phasmatodea]): **gladiators.** Secondarily wingless; chewing mouthparts; resemble a combination of a praying mantis and a walkingstick; nocturnal predators, on insects and spiders; described in 2002; rare, found in Africa; 6 to 8 species.

Order Dermaptera (der-mapˈter-a) (Gr. *derma*, skin, + *pteron*, wing): **earwigs.** Very short forewings; large and membranous hindwings folded under forewings when at rest; chewing mouthparts; forcepslike cerci.

Order Plecoptera (ple-kopˈter-a) (Gr. *plekein*, to twist, + *pteron*, wing): **stoneflies** (Figure 21.26A). Membranous wings; larger and fanlike hindwings; aquatic nymph with tufts of tracheal gills.

Order Isoptera (i-sopˈter-a) (Gr. *isos*, equal, + *pteron*, wing): **termites** (Figure 21.32). Small; membranous, narrow wings similar in size with few veins; wings shed at maturity; erroneously called "white ants"; distinguishable from true ants by broad union of thorax and abdomen; complex social organization.

Order Embiidina (em-bē-əˈdi-nə) (Gr. *embios*, lively, + *eidos*, form, + *ina*, resembling): **webspinners.** Small; male wings membranous, narrow, and similar in size; wingless females; chewing mouthparts; colonial; make silk-lined channels in tropical soil.

Order Psocoptera (so-copˈter-a) (Gr. *psoco*, rub away, + *pteron*, wing) **(Corrodentia): psocids, book lice, bark lice.** Body usually small, may be as large as 10 mm; membranous, narrow wings with few veins, usually held rooflike over abdomen when at rest; some wingless species; found in books, bark, bird nests, on foliage.

Order Zoraptera (zo-rapˈter-a) (Gr. *zōros*, pure, + *apteryos*, wingless): **zorapterans.** As large as 2.5 mm; membranous, narrow wings usually shed at maturity; colonial and termitelike.

Order Phthiraptera (thī–rapˈ-ter-a) (Gr. *phteir*, louse, + *apteros*, wingless): **lice.** Wingless ectoparasites adapted for clinging to warm-blooded hosts. **Sucking lice** (Figure 21.15) in former order Anoplura now constitute suborder Anoplura, mouthparts adapted for piercing and sucking, includes head lice, crab lice, and body lice. **Chewing lice** (Figure 21.14) in former order Mallophaga now divided among three suborders.

Order Thysanoptera (thy-sa-nopˈter-a) (Gr. *thysanos*, tassel, + *pteron*, wing): **thrips.** Length 0.5 to 5 mm (a few longer); wings, if present, long, very narrow, with few veins, and fringed with long hairs; sucking mouthparts; destructive plant-eaters, but some feed on insects.

Order Hemiptera (he-mipˈter-a) (Gr. *hemi*, half, + *pteron*, wing). Members have unique mouthparts specialized for piercing and sucking. Hemiptera is divided into three suborders: Heteroptera, Auchenorrhyncha, and Sternorrhyncha. Heteroptera contains **true bugs**; size 2 to 100 mm; wings present or absent; forewings with basal portion thickened and partly sclerotized; apical portion membranous; hindwings membranous; at rest, wings held flat over abdomen; many with odorous scent glands; includes water scorpions, water striders (Figure 21.9), bedbugs, squash bugs, assassin bugs, chinch bugs, stink bugs, plant bugs, lace bugs, and many others. Auchenorrhyncha contains **hoppers** (Figures 21.30 and 21.39) and **cicadas** (Figure 21.26); four wings typical if wings are present. Sternorrhyncha contains **whiteflies, psyllids, aphids, mealybugs** (Figure 21.35B), and **scale insects;** four wings typical if wings are present; often with complex life histories; many species are plant pests.

Order Neuroptera (neu-ropˈter-a) (Gr. *neuron*, nerve, + *pteron*, wing): **dobsonflies, ant lions** (Figure 21.40), **lacewings.** Medium to large size; similar, membranous wings with many cross veins; chewing mouthparts; dobsonflies with greatly enlarged mandibles in males, and with aquatic larvae; ant lion larvae (doodlebugs) make craters in sand to trap ants.

Order Coleoptera (ko-le-opˈter-a) (Gr. *koleos*, sheath, + *pteron*, wing): **beetles** (Figures 21.7A, 21.29, 21.35A), including **fireflies** (Figure 21.30) and **weevils** (Figure 21.41D). The largest order of animals in the world with 250,000 described species; front wings (elytra) thick, hard, opaque; membranous hindwings folded under front wings at rest; mouthparts for biting and chewing; includes

Figure 21.39
Oak treehoppers, *Platycotis vittata* (order Hemiptera).

Figure 21.40
Adult ant lion (order Neuroptera).

Classification of Class Entognatha and Class Insecta

ground beetles, carrion beetles, whirligig beetles, darkling beetles, stag beetles, dung beetles, diving beetles, boll weevils, and many others.

Order Strepsiptera (strep-sip′ter-a) (Gr. *strepsis,* a turning, + *pteron,* wing): **stylops** or **twisted wing parasites.** Females wingless, without eyes or antennae; males with vestigial forewings and fan-shaped hindwings; females and larvae parasites of bees, wasps, and other insects.

Order Mecoptera (me-kop′ter-a) (Gr. *mekos,* length, + *pteron,* wing): **scorpionflies.** Small to medium size; wings long, slender, with many veins; at rest, wings held rooflike over back; males have scorpion-like clasping organ at end of abdomen; carnivorous; live in most woodlands.

Order Lepidoptera (lep-i-dop′ter-a) (Gr. *lepidos,* scale, + *pteron,* wing): **butterflies** (Figure 21.41A) and **moths.** Membranous wings covered with overlapping scales; wings coupled or overlapping; mouthparts a sucking tube, coiled when not in use; larvae (caterpillars) with chewing mandibles for plant eating, stubby prolegs on abdomen, and silk glands for spinning cocoons; antennae knobbed in butterflies, and usually filamentous (sometimes plumed) in moths (Figure 21.41B).

Order Diptera (dip′ter-a) (Gr. *di,* two, + *pteron,* wing): **true flies** (Figure 21.10). Single pair of wings, membranous and narrow; hindwings reduced to inconspicuous balancers (halteres); sucking mouthparts or adapted for sponging, lapping, or piercing; legless larvae (maggots); includes crane flies, mosquitos, moth flies, midges, fruit flies, flesh flies, house flies, horse flies, bot flies, blow flies, and many others.

Order Trichoptera (tri-kop′ter-a) (Gr. *trichos,* hair, + *pteron,* wing): **caddisflies.** Small, soft bodies; wings well veined and partially scaled, hairy, folded rooflike over hairy body; chewing mouthparts, mandibles much reduced; aquatic larvae of many species construct cases of leaves, sand, gravel, bits of shell, or plant matter, bound together with secreted silk or cement; some make silk feeding nets attached to rocks in stream.

Order Siphonaptera (si-fon-ap′ter-a) (Gr. *siphon,* a siphon, + *apteros,* wingless): **fleas** (Figure 21.13). Small; wingless; bodies laterally compressed; legs adapted for leaping; ectoparasites or micropredators of birds and mammals; larvae legless, maggotlike scavengers.

Order Hymenoptera (hi-men-op′ter-a) (Gr. *hymen,* membrane, + *pteron,* wing): **ants, bees, wasps** (Figure 21.41C). Very small to large; membranous, narrow wings coupled distally; subordinate hindwings; mouthparts for chewing and lapping liquids; ovipositor sometimes modified into stinger, piercer, or saw (Figure 21.8); both social and solitary species, most larvae legless, blind, and maggotlike.

Figure 21.41

A, *Papilio krishna* (order Lepidoptera) is a beautiful swallowtail butterfly from India. Members of the Papilionidae grace many areas of the world, both tropical and temperate, including North America. Compare knobbed antennae with plumed antennae in **B,** *Rothschildia jacobaea,* a saturniid moth from Brazil. *Hyalophora cecropia* is a common saturniid in North America. **C,** Paper wasps (order Hymenoptera) attending their pupae and larvae. **D,** *Curculio proboscideus,* the chestnut weevil, is a member of the largest family (Curculionidae) of the largest insect order (Coleoptera). This family includes many serious agricultural pests.

A number of viruses and fungi that have potential as insecticides have been isolated. Difficulties and expense in rearing these agents have been overcome in certain cases, and some are now commercially produced.

Introduction of natural predators or parasitoids of insect pests has met with some success. In the United States vedalia beetles from Australia help to control cottony-cushion scale on citrus plants, and numerous instances of control by use of insect parasitoids are recorded.

Another approach to biological control is to interfere with reproduction or behavior of insect pests with sterile males or with naturally occurring organic compounds that act as hormones or pheromones. Such research, although very promising, is slow because of our limited understanding of insect behavior and the problems of isolating and identifying complex compounds that are produced in such minute amounts. Nevertheless, pheromones will probably play an important role in biological pest control in the future.

A systems approach called **integrated pest management** is practiced with many crops. This approach involves mixing all possible, practical techniques to contain pest infestations at a tolerable level; for example, cultivating techniques (resistant plant varieties, crop rotation, tillage techniques, timing of sowing, planting or harvesting, and others), use of biological controls, and sparse use of insecticides.

The sterile-male approach has been used effectively in eradicating screwworm flies, a livestock pest. Large numbers of male insects, sterilized by irradiation in the pupal stage, are introduced into the natural population; females that mate with the sterile flies lay infertile eggs.

PHYLOGENY AND ADAPTIVE DIVERSIFICATION

Our understanding of evolutionary relationships among the arthropod subphyla has changed greatly over the last decade. Subphylum Uniramia, which united myriapods and hexapods, was predicated on the assumption that uniramous appendages were a shared derived character (synapomorphy) that united these groups to the exclusion of all other arthropods. However, Uniramia did not emerge as a monophyletic taxon in phylogenies based on molecular characters, which grouped hexapods and crustaceans as a clade. Crustaceans have biramous appendages, so such a large difference in appendage form was initially considered strong evidence against the hypothesis of a sister-taxon relationship of hexapods and crustaceans. However, once the genetic basis of branched appendages was better understood, it was clear that changes in the number of branches on an appendage could evolve by a relatively simple change in gene expression. Members of former subphylum Uniramia were then divided between subphylum Myriapoda and subphylum Hexapoda.

Hexapods are now united with members of subphylum Crustacea in clade Pancrustacea. Both taxa have mandibles, but it is unclear whether pancrustaceans are more closely related to myriapods, a mandibulate taxon, or to chelicerates, which lack mandibles.

While clade Pancrustacea is well supported, the nature of the relationship between crustacean and hexapod subgroups is subject to debate. Some phylogenies indicate a sister-taxon relationship between crustaceans and hexapods, as we show in Figure 19.2, but others indicate that hexapods arose *within* Crustacea. If this result is supported by future studies, subphylum Crustacea will be paraphyletic unless it is redefined to include hexapods. In phylogenies where hexapods originate within Crustacea, they appear closest to branchiopod, cephaplocarid, and remipedian crustaceans. The next decade of research should clarify the evolutionary position of Hexapoda within Pancrustacea.

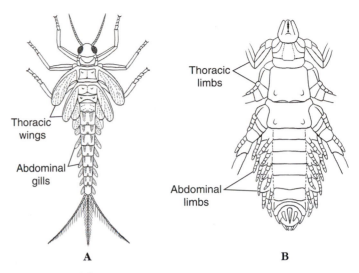

Figure 21.42

A, A Paleozoic "mayfly" nymph with thoracic winglets and abdominal gills. The thoracic winglets could have been precursors of wings. **B,** A Paleozic insect with multiramous thoracic legs and vestigal, multiramous abdominal limbs.

Within Hexapoda, class Entognatha is shown as the sister taxon to class Insecta in Figure 21.1, but some research indicates that entognathous mouthparts may have evolved several times, and that some entognathans may be closer to insects than they are to other entognathans.

Insect fossils, although not abundant, have been found in numbers sufficient to give a general idea of the evolutionary history of the class. Although several groups of marine arthropods, such as trilobites, crustaceans, and xiphosurans, were present in the Cambrian period, the first terrestrial arthropods—scorpions and millipedes—did not appear until the Silurian period. The first insects, which were wingless, date from the Devonian period, although a Silurian fossil has been tentatively identified as an insect.

It is likely that ancestral insects had a head and trunk of many similar segments, most or all of which bore limbs. Early fossil insects had small abdominal appendages (and apparently some multiramous appendages, Figure 21.42), and some modern, apterygote (primitively wingless) orders have abdominal styli that are considered vestigial legs. We now understand that absence of abdominal legs in most insects results from a pattern of expression of certain *Hox* genes that prevents expression of the *Distal-less* gene in the abdomen of insects but not crustaceans or onychophorans (p. 439).[1,2]

The evolutionary origin of insect wings has long been a puzzle. Recent evidence based on fossil insect mandibles suggests that winged insects were in existence about 400 million years ago. By the Carboniferous period, several orders of winged insects (Paleoptera) most of which are now extinct, had appeared. The

[1]Galant, R., and S. B. Carroll. 2002. Evolution of a transcriptional repression domain in an insect Hox protein. Nature **415**:910–913.
[2]Ronshaugen, M., N. McGinnis, and W. McGinnis. 2002. Hox protein mutation and macroevolution of the insect body plan. Nature **415**:914–917.

adaptive value of wings for flight is clear, but such structures did not spring into existence fully developed. One hypothesis was that wings developed from lateral thoracic expansions that were useful in gliding. However, this hypothesis did not explain an origin or function of articulations and neuromusculature in the protowings that would provide raw material for selection and eventual evolution of flapping wings to support flight. An alternative hypothesis is that ancestral flying insects were derived from aquatic insects or insects with aquatic juveniles that bore external gills on their thorax from which wings could have been derived. Thoracic and abdominal gills on Paleozoic insects were apparently articulated and movable, capable of ventilation and swimming movements. They may have provided the morphological structures for "prowings." The evolution of broadly attached thoracic prowings (incapable of providing flight) in semi-aquatic insects would have increased the body temperature of *basking* insects possessing them over "wingless" forms. The subsequent expansion of these thoracic prowings for behavioral temperature regulation (basking) could easily have provided the morphological stage required for the evolution of truly functional wings (Figure 21.43) large enough to support flight.

The ancestral winged insect gave rise to three lines, which differed in their ability to flex their wings. Two of these (Odonata and Ephemeroptera) have outspread wings or hold their wings vertically over the abdomen. The other line has wings that can fold back horizontally over the abdomen; it branched into three groups, all of which were present by the Permian period. One group with hemimetabolous metamorphosis, chewing mouthparts, and cerci includes Orthoptera, Dermaptera, Isoptera, and Embiidina; another group with hemimetabolous metamorphosis and a tendency toward sucking mouthparts includes Thysanoptera, Hemiptera, and perhaps also Psocoptera, Zoraptera, and Phthiraptera, although there is some disagreement among authorities about the last group. Insects with holometabolous metamorphosis have the most specialized life history, and these apparently form a clade that includes the remaining neopterous orders (for example, Lepidoptera, Diptera, Hymenoptera).

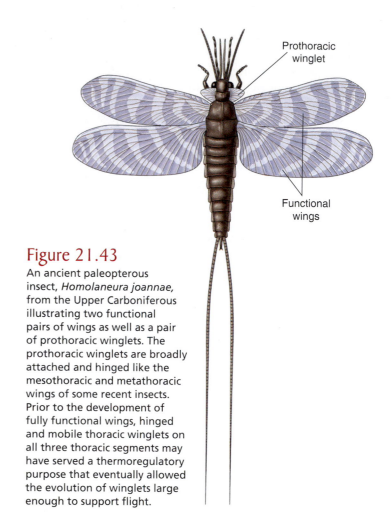

Figure 21.43

An ancient paleopterous insect, *Homolaneura joannae*, from the Upper Carboniferous illustrating two functional pairs of wings as well as a pair of prothoracic winglets. The prothoracic winglets are broadly attached and hinged like the mesothoracic and metathoracic wings of some recent insects. Prior to the development of fully functional wings, hinged and mobile thoracic winglets on all three thoracic segments may have served a thermoregulatory purpose that eventually allowed the evolution of winglets large enough to support flight.

Adaptive properties of insects have been stressed throughout this chapter. The directions and ranges of their adaptive radiation, both structurally and physiologically, have been amazingly varied. Whether it be in the area of habitat, feeding adaptations, means of locomotion, reproduction, or general mode of living, the adaptive achievements of insects are truly remarkable.

SUMMARY

Members of subphylum Hexapoda have uniramous appendages and bear one pair of antennae, a pair of mandibles, and two pairs of maxillae on the head. Tagmata are head, thorax, and abdomen. Hexapods have three pairs of jointed thoracic legs. There are two classes: Entognatha and Insecta. Entognatha contains three orders of tiny animals commonly inhabiting soils.

Insecta is the largest class of the world's largest phylum. Insects have mouthparts whose bases extend out of the head capsule.

The evolutionary success of insects in terrestrial habitats is largely explained by features such as wings (present in most), waterproofing their cuticle and other mechanisms to minimize water loss, and the ability to become dormant during adverse conditions.

Most insects bear two pairs of wings on their thorax, although some have one pair and some are primitively or secondarily wingless. Wing movements in some insects are controlled by direct

flight muscles, which insert directly on the base of the wings in the thorax, whereas others have indirect flight muscles, which move the wings by changing the shape of the thorax. Each contraction of synchronous flight muscles requires a nerve impulse, while asynchronous flight muscles contract many times for each nerve impulse.

Feeding habits vary greatly among insects, and there is an enormous variety of specialization of mouthparts reflecting the particular feeding habits of a given insect. They breathe by means of a tracheal system, which is a system of tubes that opens by spiracles on the thorax and abdomen. Excretory organs are Malpighian tubules floating freely in the hemocoel.

Sexes are separate in insects, and fertilization is usually internal. Almost all insects undergo metamorphosis during development. In hemimetabolous (incomplete) metamorphosis, larval instars are called immatures or nymphs, and adults emerge at the last nymphal molt. In holometabolous (complete) metamorphosis, the last larval molt gives rise to a nonfeeding stage (pupa). A winged adult emerges at the final, pupal, molt. Both types of metamorphosis are hormonally controlled.

Insects are important to human welfare, particularly because they pollinate food and forage crop plants, control populations of other, harmful insects by predation and parasitism, and serve as food for other animals. Many insects are harmful to human interests because they inflict great damage on crops, food, forests, clothing, and property, and many are carriers of important diseases affecting humans and domestic animals.

Hexapods are united with crustaceans in clade Pancrustacea. Entognathans either form or include the sister taxon of insects. They—like the basal insects—are wingless. Legs have been lost from the abdomen of insects, being now confined to each of the three thoracic segments. Wings of ancestral winged insects may have been derived from external gills of aquatic nymphs or adults, and there may have been a stage in which tergal expansions (or prowings), whether articulated or not, served to enhance the efficiency of behavioral thermoregulation until a size was reached to form efficient aerofoils for truly functional wings.

Adaptive diversity and the numbers of both species and individuals in Insecta are enormous.

REVIEW QUESTIONS

1. What characteristics of hexapods distinguish them from *all* other arthropods?
2. How are insects distinguished from other hexapods?
3. Explain why indirect flight muscles can beat much more rapidly than direct flight muscles.
4. How do insects walk?
5. What major parts form an insect's gut, and how do these parts function?
6. Describe three different types of mouthparts found in insects, and tell how they are adapted for feeding on different foods.
7. Describe the tracheal system of a typical insect and explain why it is able to function efficiently without oxygen-carrying pigments in the hemolymph. Why would a tracheal system not be suitable for humans?
8. Describe the unique excretory system of insects. How is uric acid formed?
9. Describe sensory receptors that insects have to various stimuli.
10. Explain the difference between holometabolous and hemimetabolous metamorphosis in insects, including stages of each.
11. Describe hormonal control of metamorphosis in insects, including the action of each hormone and where each is produced.
12. What is diapause, and what is its adaptive value?
13. Briefly describe three features that insects have evolved to avoid predation.
14. Describe and give an example of each of four ways that insects use to communicate with each other.
15. What castes are found in honey bees and in termites, and what is the function of each?
16. What are mechanisms of caste determination in honey bees and termites?
17. What is trophallaxis? What function(s) does it serve in termites?
18. Name several ways in which insects are beneficial to humans and several ways they are detrimental.
19. In what ways are detrimental insects controlled? What is integrated pest management?
20. What are the most probable characteristics of the most recent common ancestor of insects? What major lineages descended from it?
21. What is a plausible scenario for evolution of wings and flying insects?

SELECTED REFERENCES

Arnett, R. H., Jr., and M. C. Thomas, eds. 2000. American beetles, vol. 1. Boca Raton, Florida, CRC Press.

Arnett, R. H., Jr., M. C. Thomas, P. E. Skelley, and J. H. Franks, eds. 2002. American beetles, vol. 2. Polyphaga: Scarabaeoidea through Curculionoidea. Boca Raton, Florida, CRC Press. *These two volumes present recent detailed keys to families of American beetles; the second edition of volume 1 (2001) includes additions and corrections.*

Beckage, N. E. 1997. The parasitic wasp's secret weapon. Sci. Am. **277:**82–87 (Nov.). *This parasitoid wasp carries a virus that invades the host insect when the wasp lays its eggs, then paralyzes the host.*

Bennet-Clark, H. C. 1998. How cicadas make their noise. Sci. Am. **278:**58–61 (May). *Male cicadas are the loudest known insects.*

Berenbaum, M. R. 1995. Bugs in the system. Reading, Massachusetts, Addison-Wesley Publishing Company. *How insects impact human affairs. Well written for a wide audience, highly recommended.*

Downs, A. M. R., K. A. Stafford, and G. C. Coles. 1999. Head lice: Prevalence in schoolchildren and insecticide resistance. Parasitol. Today **15:**1–4. *This report is mainly concerned with England, but head lice are one of the most common parasites of American schoolchildren.*

Douglas, M. M. 1981. Thermoregulatory significance of thoracic lobes in the evolution of insect wings. Science **211:**84–86. *Evolution of broadly attached thoracic winglets could have increased the body temperature excess of ancient wingless insects by 55% over that of forms without winglets. Subsequent expansion of winglets for behavioral thermoregulation could have provided the morphological stage required for evolution of functional wings.*

Engel, M. S., and D. A. Grimaldi. 2004. New light shed on the oldest insect. Nature **427:**627–630. *Silurian fossil mandibles indicate that insects and entognathous hexapods had evolved by this period.*

Gullan, P. J., and P. S. Cranston. 2005. The insects: an outline of entomology, ed. 3. Malden, Massachusetts, Blackwell Publishing. *An easy-to-use text for general entomology with good figures, modern phylogenies, and data on biogeography.*

Hayashi, A. M. 1999. Attack of the fire ants. Sci. Am. **280:**26, 28 (Feb.). *Fire ants have invaded Galápagos, Melanesia, and West Africa, where they may be blinding elephants and otherwise disrupting the ecosystem.*

Heinrich, B., and H. Esch. 1994. Thermoregulation in bees. Am. Sci. **82:**164–170. *Fascinating behavioral and physiological adaptations for* *increasing and decreasing body temperature allow bees to function in a surprisingly wide range of environmental temperatures.*

Hölldobler, B. H., and E. O. Wilson. 1990. The ants. Cambridge, Massachusetts, Harvard University Press. *The fascinating story of social organization in ants.*

Hubbell, S. 1997. Trouble with honey bees. Nat. Hist. **106**(4):32–43. *Infection of honey bees with* Varroa *mites is a big problem for beekeepers.*

Johnson, N. F., and C. Triplehorn. 2005. Borror and DeLong's introduction to the study of insects, ed. 7. Belmont, California, Brooks/Cole Publishing Company. *An up-to-date reference text for the study of insects; keys are included, but some require specialized knowledge of morphological characters. See Arnett and Thomas for keys to the American beetle families.*

Kingsolver, J. G., and M. A. R. Koehl. 1985. Aerodynamics, thermoregulation, and the evolution of insect wings: differential scaling and evolutionary change. Evolution **39**(3):488–504. *This work supports the thermoregulatory hypothesis proposed by Douglas by showing that as winglets approach the maximum benefit for temperature regulation they also reach the minimum size required for either gliding and/or flapping flight.*

Kukalova-Peck, J. 1978. Origin and evolution of insect wings and their relation to metamorphosis, as documented by the fossil record. Journal of Morphology **156**:53–126. *A thorough examination of the paleontological evidence of the paleopterous insects and a re-examination of the stages that may have led to the evolution of insect wings and metamorphosis.*

Levine, M. 2002. How insects lose their limbs. Nature **415**:848–849. *A Hox gene product in the abdomen of insects inhibits action of another gene product that is necessary for limb formation.*

Raff, R. A. 1996. The shape of life: genes, development, and the evolution of animal form. Chicago, University of Chicago Press. *Includes a good account of how insect wings may have evolved.*

Regier, J. C., J. W. Shultz, and R. E. Kambic. 2005. Pancrustacean phylogeny: hexapods are terrestrial crustaceans and maxillopods are not monophyletic. Proc. R. Soc. Lond. Ser. B. **272**:395–401. *Hexapods are closely related to Branchiopoda, Cephalocarida, and Remipedia.*

Tallamy, D. W. 1999. Child care among the insects. Sci. Am. **280**:72–77 (Jan.). *Most insects render no assistance to eggs or young, but females, and in a few cases, males, of a variety of species, care for and protect eggs and young.*

Topoff, H. 1990. Slave-making ants. Am. Sci. **78**:520–528. *An amazing type of social parasitism in which certain species of ants raid the colonies of related species, abduct their pupae, then exploit them to do all the work in the host colony.*

Westneat, M. W., O. Betz, R. W. Blob, K. Fezzaa, W. J. Cooper, and W. Lee. 2003. Tracheal respiration in insects visualized with synchrotron X-ray imaging. Science **299**:558–560. *Tracheae compress actively, exhaling air, but expand passively.*

Whiting, R. 2004. Phylogenetic relationships and evolution of insects, pp. 330–344. In J. Cracraft and M. J. Donoghue, eds., Assembling the tree of life. New York, Oxford University Press. *A detailed discussion of current hypotheses for insect evolution, including the suggestion that entognathous hexapods do not form a monophyletic group, unlike ectognathous forms.*

ONLINE LEARNING CENTER

Visit www.mhhe.com/hickmanipz14e for chapter quizzing, key term flash cards, web links, and more!

22

A mass of sea stars *(Pisaster ochraceus) above the waterline at low tide.*

Chaetognaths, Echinoderms, and Hemichordates

- PHYLUM CHAETOGNATHA
- PHYLUM ECHINODERMATA
- PHYLUM HEMICHORDATA

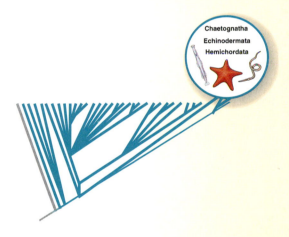

A Design to Puzzle the Zoologist

Libbie Hyman, a distinguished American zoologist, once described echinoderms as a "noble group especially designed to puzzle the zoologist." With a combination of characteristics that should delight the most avid reader of science fiction, echinoderms would seem to confirm Lord Byron's observation that

*Tis strange—but true;
for truth is always strange;
Stranger than fiction.*

Despite the adaptive value of bilaterality for free-moving animals, and the merits of radial symmetry for sessile animals, echinoderms confounded the rules by becoming free moving but radial. That they evolved from a bilateral ancestor there can be no doubt, for their larvae are bilateral. They undergo a bizarre metamorphosis to a radial adult in which there is a 90° reorientation in body axis.

A compartment of the coelom has been transformed in echinoderms into a unique water-vascular system that uses hydraulic power to operate a multitude of tiny tube feet used in food gathering and locomotion. Dermal ossicles may fuse together to invest an echinoderm in armor or may be reduced to microscopic bodies. Many echinoderms have miniature jawlike pincers (pedicellariae) scattered on their body surface, often stalked and sometimes equipped with poison glands.

This constellation of characteristics is unique in the animal kingdom. Despite the vast amount of research that has been devoted to them, we are still far from understanding many aspects of echinoderm biology. Only recently have we discovered the position of the anterior-posterior body axis in sea stars—a feature so obvious in almost all other animals.

The triploblastic metazoans are divided into two large clades: Protostomia and Deuterostomia. Clades are characterized by a combination of molecular and morphological features, with many morphological features visible in early development. Classical developmental characters associated with protostomes are spiral mosaic cleavage; formation of the mouth from the embryonic blastopore (protostomy); and formation of a coelom by schizocoely, when a coelom is present. Classical deuterostome developmental features are radial regulative cleavage; formation of the mouth from a second opening (deuterostomy); and coelom formation by enterocoely. All deuterostomes are coelomate.

Members of some phyla possess all the developmental characters in each suite: marine annelids and molluscs are classical protostomes, and echinoderms are classical deuterostomes. However, readers of the preceding chapters will be aware that there are taxa within Protostomia that do not exhibit all protostome features. As well, the two protostome subdivisions, Lophotrochozoa and Ecdysozoa, differ in terms of the classical traits.

Most lophotrochozoans, other than the problematic lophophorates, exhibit several of these traits. Most lophotrochozoans also share a trochophore-like larval form which is not present in members of Ecdysozoa.

Ecdysozoan cleavage is spiral in a few groups, but many have eggs with substantial amounts of yolk, so only nuclei divide (superficial cleavage). The shared feature for all well-studied ecdysozoans is the ability to molt their cuticles.

Despite the progress made in charting the path of metazoan evolution, there are some phyla whose body plans seemed to represent a "mix-and-match" strategy; they have both protostome and deuterostome features. The lophophorates have already been mentioned, and here we include another phylum, Chaetognatha, whose evolutionary position is still much debated. The chaetognaths are pelagic marine predators commonly called arrow worms. They are an ancient group with representatives among the Burgess Shale fossils, so they are more than 500 million years old. Fossil forms are strikingly similar to modern forms in external appearance. This fascinating phylum has been placed in Deuterostomia, in Protostomia, and outside the two groups.

We depict Chaetognatha as outside the protostome and deuterostome clades (see foldout cladogram and Figure 22.1),

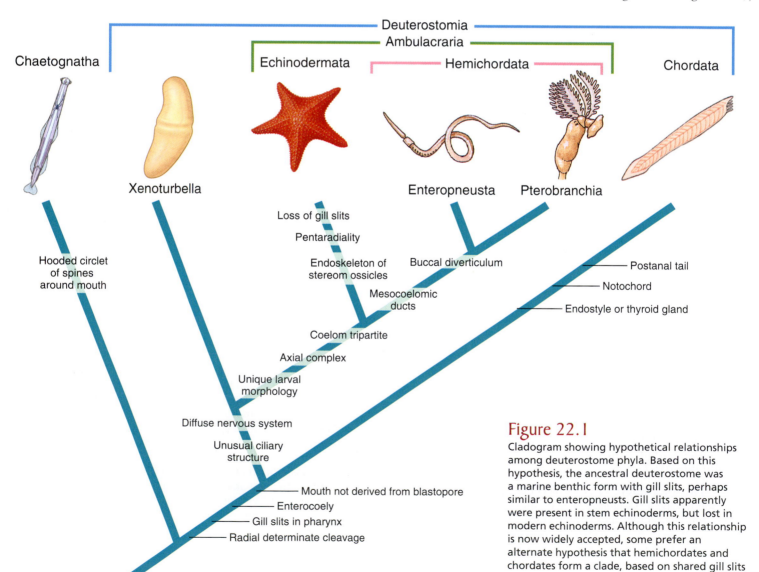

Figure 22.1

Cladogram showing hypothetical relationships among deuterostome phyla. Based on this hypothesis, the ancestral deuterostome was a marine benthic form with gill slits, perhaps similar to enteropneusts. Gill slits apparently were present in stem echinoderms, but lost in modern echinoderms. Although this relationship is now widely accepted, some prefer an alternate hypothesis that hemichordates and chordates form a clade, based on shared gill slits and dorsal, hollow nerve cord.

pending new studies. Chaetognaths are discussed in this chapter, followed by two deuterostome phyla, Echinodermata and Hemichordata, now considered sister taxa in clade Ambulacraria (Figure 22.1).

PHYLUM CHAETOGNATHA

A common name for chaetognaths is arrow worms. They are all marine animals and most are highly specialized for their planktonic existence. The name Chaetognatha (ke-tog′na-tha) (Gr. *chaitē,* long flowing hair, + *gnathos,* jaw) refers to the sickle-shaped bristles on each side of their mouth. This is not a large group, and only about 100 species are known. Their small, straight bodies resemble miniature torpedoes, or darts, ranging from less than 1 to about 12 cm in length (Figure 22.2).

With the exception of *Spadella* (Gr. *spadix,* palm frond, +*ella,* dim. suffix), a benthic genus, and a few species that live close to the deep-ocean floor, arrow worms are all adapted for a planktonic existence. They usually swim to the surface at night and descend during daytime. Much of the time they drift passively, but they can dart forward in swift spurts, using their caudal fin and longitudinal muscles—a fact that no doubt contributes to their success as planktonic predators. Horizontal fins bordering the trunk serve largely as stabilizers and are used in flotation rather than in active swimming.

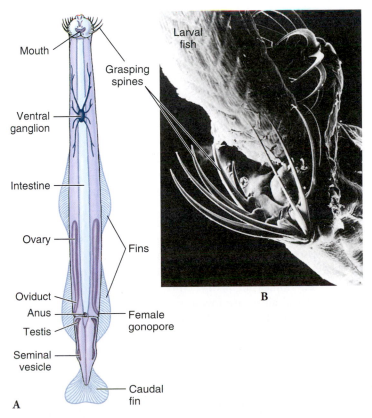

Figure 22.2

Arrow worms. **A,** Internal structure of *Sagitta.* **B,** Scanning electron micrograph of a juvenile arrow worm, *Flaccisagitta hexaptera* (35 mm length) eating a larval fish.

Form and Function

Arrow worms are unsegmented, and their body includes a head, trunk, and postanal tail (Figure 22.2A). Beneath the head is a large depression, called a vestibule, leading to the mouth. The vestibule contains teeth and is flanked on both sides by curved chitinous spines used in seizing prey. A pair of eyes is located dorsally. A peculiar hood formed from a fold of the neck can be drawn forward over the head and spines. When a chaetognath captures prey, it retracts its hood, and its teeth and raptorial spines spread apart and then snap shut with startling speed. Arrow worms are voracious predators, living on planktonic forms, especially copepods, but including a variety of other planktonic crustaceans, small fish, and even other chaetognaths (Figure 22.2B). When they are abundant, as they often are, they may have a substantial ecological impact. Most species are highly mobile and nearly transparent, characteristics likely of adaptive value in their role as planktonic predators.

A thin cuticle covers the body, and their epidermis is single layered except along the sides, where it is stratified in a thick layer. These are the only invertebrates with a many-layered epidermis.

Arrow worms have a complete digestive system and a well-developed coelom. The coelom is used as a hydrostatic skeleton. There is a nervous system with a nerve ring connecting a cerebral ganglion above the esophagus to a number of lateral ganglia and a large ventral ganglion. Sense organs include eyes, sensory bristles, and possibly a unique U-shaped ciliary loop that extends over their neck from the back of the head. The exact function of this loop remains unknown, but it may detect vibrations, water currents, or may be chemosensory. Chaetognaths use vibrations of the sensory bristles to detect prey. Respiratory and excretory systems are entirely lacking and these processes appear to be accomplished solely by diffusion. Recently, a previously unknown and loosely organized hemal system has been described for chaetognaths.

Arrow worms are hermaphroditic with either reciprocal or self-fertilization. Eggs of *Sagitta* (L. arrow) bear a coat of jelly and are planktonic. Eggs of other arrow worms may be released to sink to the substratum during development, attached to stationary objects, or attached to the parent and carried while they develop. Juveniles develop directly without metamorphosis.

Chaetognath embryogenesis suggests deuterostome affinities. The mouth does not arise from the blastopore, and the coelom develops by enterocoely. However, the exact nature of coelom formation remains contentious. Some authorities state that embryonic coelom formation is clearly enterocoelous, while others argue that it differs from that of typical deuterostomes because their coelom is formed by a backward extension from the archenteron rather than by pinched-off coelomic sacs. There is no true peritoneum lining the coelom.

Chaetognaths were described as having a tripartite coelom like that present in members of Ambulacraria, but apparently the third section of the coelom is merely a partition between male and female gonads in these hermaphrodites.

Early descriptions of chaetognath cleavage describe it as radial, complete and equal, but more recent studies dispute this description, finding instead that cleavage planes in four-cell embryos are similar to those of crustaceans and nematodes.

Chaetognaths and nematodes both lack circular muscles and have a similar arrangement of longitudinal muscles. Some phylogenies based on nucleotide sequences placed chaetognaths within Ecdysozoa, but there have been no reports that the thin cuticle is molted.

One phylogenetic study using *Hox* genes as characters suggested that chaetognaths branched from the metazoan lineage before the protostome/deuterostome split. The most recent phylogenies place them within protostomes as the sister taxon to all other members. Protostome features present in adult animals include the ventral nerve cord. Adult characters and nucleotide sequence data are at odds with developmental data, so the evolutionary position of chaetognaths remains in doubt.

CLADE AMBULACRARIA

Ambulacraria is a superphylum that contains two deuterostome phyla: Echinodermata and Hemichordata (Figure 22.1). Echinoderms, including sea stars, brittle stars, and sea cucumbers, are familiar to many people, but hemichordates, known as acorn worms and pterobranchs, are much less familiar. Aside from classical deuterostome features, members of Ambulacraria share a three-part (tripartite) coelom, similar larval forms, and an axial complex (a highly specialized metanephridium).

The sister taxon to Ambulacraria shown in Figure 22.1 is *Xenoturbella*. This genus contains two species, *Xenoturbella bocki*, and *X. westbladi*; it is sometimes described as phylum Xenoturbellida. The animals, first described in 1949, are wormlike, ciliated, reach lengths of 4 cm, and live in North Sea mud. *Xenoturbella* feeds on bivalve eggs.

The body has few morphological features other than a blind gut. A coelom and excretory structures are absent. The body is not cephalized, but there is a diffuse, netlike nervous system reminiscent of that in flatworms. Muscles are present. There are no structured gonads, but sexual reproduction occurs.

Why is this animal placed in Deuterostomia? *Xenoturbella* has been allied with both flatworms and molluscs, but when included in phylogenies using molecular characters, the genus lies within Deuterostomia. Its placement as the sister taxon to Ambulacraria is based on the shared presence of unusual ciliary structures and the diffuse nervous system. If this position is supported by later studies, we must assume that many ancestral features have been lost in *Xenoturbella*.

PHYLUM ECHINODERMATA

Echinoderms are marine forms and include sea stars (also called starfishes), brittle stars, sea urchins, sea cucumbers, and sea lilies. They represent a bizarre group sharply distinguished from all other animals. The name Echinodermata (e-ki'no-der'ma-ta) (L. *echinatus,* prickly, + Gr.*derma*, skin, +*ata*, characterized by) is derived from their external spines or protuberances. All members of the phylum have a calcareous endoskeleton either in the form of plates or represented by scattered tiny ossicles.

The most noticeable characteristics of echinoderms are (1) spiny endoskeleton of plates, (2) water-vascular system, (3) pedicellariae, (4) dermal branchiae, and (5) basic pentaradial symmetry in the adults. No other group with such complex organ systems has radial symmetry.

Echinoderms are an ancient group of animals extending back to at least the Cambrian period. *Arkarua,* an enigmatic fossil from the Vendian (560 million years ago) has been identified as the earliest known echinoderm (Gehling, 1987), but many authorities consider this identification inconclusive. Despite an excellent fossil record, the origin and early evolution of echinoderms are still obscure. It seems clear that they descend from bilateral ancestors because their larvae are bilateral but become radially symmetrical later in development. Many zoologists believe that early echinoderms were sessile and evolved radiality as an adaptation to sessile existence. Bilaterality is of adaptive value to animals that travel directionally through their environment, while radiality is of value to animals whose environment meets them on all sides equally. Hence, the body plan of present-day echinoderms seems to have been derived from one that was attached to the seafloor, had radial symmetry and radiating grooves (ambulacra) for food gathering, and had an upward-facing oral side. Attached forms likely were once plentiful, but only about 80 species, all in class Crinoidea, still survive (Figure 22.3). Oddly, conditions have favored survival of their free-moving descendants, although they are still quite radial, and among them are some of the most abundant marine animals. Nevertheless, in the exception that proves the rule (that bilaterality is adaptive for free-moving animals), at least three groups of echinoderms (sea cucumbers and two groups of sea urchins) have secondarily evolved a superficial bilateral organization (although there remains a pentaradial organization of skeletal and most organ systems).

Most echinoderms have no ability to osmoregulate and thus rarely venture into brackish waters. They occur in all oceans of the world and at all depths, from intertidal to abyssal regions. Often the most common animals in deep ocean are echinoderms. The most abundant species found in the Philippine Trench (10,540 m) was a sea cucumber. Echinoderms are virtually all bottom dwellers, although there are a few pelagic species.

No parasitic echinoderms are known, but a few are commensals. On the other hand, a wide variety of other animals make their homes in or on echinoderms, including parasitic or commensal algae, protozoa, ctenophores, turbellarians, cirripedes, copepods, decapods, snails, clams, polychaetes, fishes, and other echinoderms.

Asteroids, or sea stars (Figure 22.4), are commonly found on hard, rocky surfaces, but numerous species are at home on sandy or soft substrates. Some species are particle feeders, but many are predators, feeding particularly on sedentary or sessile prey, since sea stars themselves are relatively slow moving.

Ophiuroids—brittle stars, or serpent stars (see Figures 22.14 and 22.17)—are by far the most active echinoderms, they move

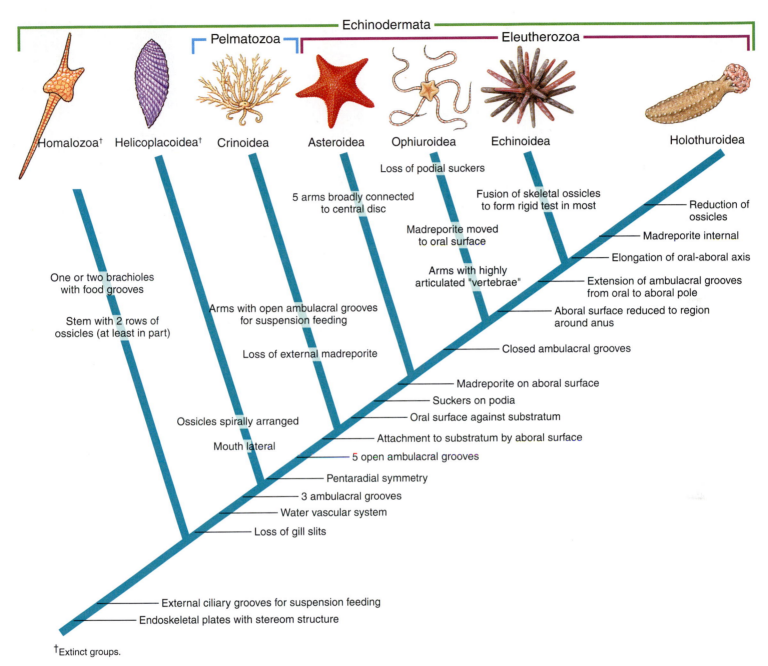

Figure 22.3

Cladogram showing hypothetical relationships among echinoderm groups. The extinct Homalozoa (carpoids), which were not radial in symmetry but had stereom endoskeletal plates, represent an early split from echinoderms. The extinct helicoplacoids had three ambulacral grooves that wound around their bodies in spiral fashion, and are hypothesized to be the sister group of modern echinoderms. Evolution of pentaradial symmetry was an adaptation to sessile existence and is a synapomorphy for modern echinoderms. The cladogram presented here depicts ophiuroids as haven risen separately from asteroids, after evolution of closed ambulacral grooves; possession of five arms would have arisen independently in these groups. An alternate scenario, also well supported, unites Asteroidea and Ophiuroidea in a clade, with five arms being a synapomorphy and independent evolution of closed ambulacral grooves in ophiuroids and the common ancestor of echinoids and holothuroids.

by bending their jointed muscular arms, rather than by walking on tube feet. A few species are reported to have swimming ability, and some burrow. They may be scavengers, browsers, deposit or filter feeders, or predators. Some are commensal in large sponges, in whose water canals they may live in great numbers.

Holothurians, or sea cucumbers (see Figure 22.24), are widely prevalent in all seas. Many are found on sandy or mucky bottoms, where they lie concealed. Compared with other echinoderms, holothurians are greatly extended in the oral-aboral axis. They are oriented with that axis more or less parallel to the substrate and lying on one side. Most are suspension or deposit feeders.

Echinoids, or sea urchins (see Figure 22.18), are adapted for living on the ocean floor and always keep their oral surface in

Characteristics of Phylum Echinodermata

1. Unique **water-vascular system** of coelomic origin extends from body surface as series of tentacle-like projections **(podia,** or **tube feet)** protracted by increase of fluid pressure within them; opening to exterior **(madreporite** or **hydropore)** usually present
2. Living in marine habitats
3. Free-living taxa
4. Body unsegmented (nonmetameric) with **pentaradial symmetry;** body rounded, cylindrical, or star-shaped, with five or more radiating areas, or **ambulacra,** alternating with interambulacral areas; no head
5. Triploblastic body
6. Coelom extensive, forming perivisceral cavity and cavity of water-vascular system; coelom of enterocoelous type; coelomic fluid with amebocytes
7. **Endoskeleton** of **dermal calcareous ossicles** with **spines** or of calcareous **spicules** in dermis; covered by epidermis (ciliated in most); **pedicellariae** (in some)
8. Digestive system usually complete; axial or coiled; anus absent in ophiuroids
9. Skeletal elements connected by ligaments of mutable collagenous tissue under neural control, ligaments can be "locked" into rigid posture or relaxed to allow free movement at will; locomotion by **tube feet,** which project from **ambulacral areas,** by movement of spines, or by movement of arms, which project from central disc of body
10. Nervous system with circumoral ring and radial nerves; usually two or three systems of networks located at different levels in the body, varying in degree of development according to group
11. **No brain;** few specialized sensory organs; sensory system of tactile and chemoreceptors, podia, terminal tentacles, photoreceptors, and statocysts
12. Autotomy and regeneration of lost parts conspicuous; asexual reproduction by fragmentation in some
13. Sexes separate (except a few hermaphroditic) with large gonads, single in holothuroids but multiple in most; simple ducts, with no elaborate copulatory apparatus or secondary sexual structures; fertilization usually external; eggs brooded in some; development through **free-swimming, bilateral, larval stages** (some with direct development); metamorphosis to radial adult or subadult form; radial cleavage and regulative development
14. **Excretory organs absent**
15. Respiration by **papulae, tube feet, respiratory tree** (holothuroids), and **bursae** (ophiuroids)
16. Blood-vascular system **(hemal system)** much reduced, playing little if any role in circulation, and surrounded by extensions of coelom **(perihemal sinuses);** main circulation of body fluids (coelomic fluids) by peritoneal cilia

Figure 22.4

Some sea stars (class Asteroidea) from the Pacific. **A,** Pincushion star, *Culcita navaeguineae,* preys on coral polyps and also eats other small organisms and detritus. **B,** *Choriaster granulatus* scavenges dead animals on shallow Pacific reefs. **C,** *Tosia queenslandensis* from the Great Barrier Reef System browses encrusting organisms. **D,** Crown-of-thorns star, *Acanthaster planci,* is a major coral predator (see note, p. 478).

Crinoids (see Figure 22.29) stretch their arms outward and up like a flower's petals and feed on plankton and suspended particles. Most living species become detached from their stems as adults, but they nevertheless spend most of their time on the substrate, fastened by aboral appendages called cirri.

A zoologist who admires the fascinating structure and function of echinoderms can share with a layperson an admiration of the beauty of their symmetry, often enhanced by bright colors. Many species are rather drab, but others may be orange, red, purple, blue, and often multicolored.

Because of the spiny nature of their structure, echinoderms are not often prey of other animals—except other echinoderms (sea stars). Some fishes have strong teeth and other adaptations that enable them to feed on echinoderms. A few mammals, such as sea otters, feed on sea urchins. In scattered parts of the world, humans relish sea urchin gonads, either raw or roasted on the half shell. Trepang, the cooked, protein-rich body wall of certain large sea cucumbers, is a delicacy in many east Asian countries. Unfortunately, the intense, often illegal, fishery for sea cucumbers has severely depleted their populations in many areas of the tropical world.

contact with the substratum. Regular urchins, which are radially symmetrical, feed chiefly on algae or detritus, while irregulars, which are secondarily bilateral, feed on small particles. "Regular" sea urchins prefer hard substrates, but sand dollars and heart urchins ("irregular" urchins) are usually found on sand.

Sea stars feed on a variety of molluscs, crustaceans, and other invertebrates. In some areas they may perform an important ecological role as a top carnivore in a community. Their chief economic impact is on clams and oysters. A single starfish can eat as many as a dozen oysters or clams in a day. To rid shellfish beds of these pests, quicklime is sometimes spread over areas where they abound. Quicklime damages the delicate epidermal membrane, destroying the dermal branchiae and ultimately the animal itself. Unfortunately, other soft-bodied invertebrates are also damaged. However, oysters remain with their shells tightly closed until the quicklime is degraded.

Echinoderms have been widely used in developmental studies because their gametes are usually abundant and easy to collect and raise in a laboratory. Investigators can follow embryonic developmental stages with great accuracy. We know more about molecular biology of sea urchin development than that of almost any other embryonic system. Artificial parthenogenesis was first discovered in sea urchin eggs, when it was found that, by treating eggs with hypertonic seawater or subjecting them to a variety of other stimuli, development would proceed without sperm.

Class Asteroidea

Sea stars, often called starfishes, demonstrate basic features of echinoderm structure and function very well. There are about 1500 living species and they are easily obtainable. Thus we will consider them first, then comment on major differences shown by other groups.

Sea stars are familiar along the shoreline where large numbers may aggregate on rocks. Sometimes they cling so tenaciously that they are difficult to dislodge without tearing off some tube feet. They also live on muddy or sandy substrates and among coral reefs. They are often brightly colored and range in size from a centimeter in greatest diameter to about a meter across. *Asterias* (Gr. *asteros,* a star) is a common genus of the east coast of the United States and is commonly studied in zoology laboratories. *Pisaster* (Gr. *pisos,* a pea, + *asteros,* a star, p. 469) is common on the west coast of the United States, as is *Dermasterias* (Gr. *dermatos,* skin, leather, + *asteros,* a star), the leather star.

Form and Function

External Features Sea stars are composed of a central disc that merges gradually with the tapering arms (rays). The body is somewhat flattened, flexible, and covered with a ciliated, pigmented epidermis. Their mouth is centered on the under, or oral, side, surrounded by a soft peristomial membrane. An **ambulacrum** (pl., **ambulacra,** L. *ambulacrum,* a covered way, an alley, a walk planted with trees) or **ambulacral area,** runs from the mouth on the oral side of each arm to the tip of the arm. Sea stars typically have five arms, but they may have more (Figure 22.4D), and there are as many ambulacral areas as there are arms. An **ambulacral groove** is found along the middle of each ambulacral area, and the groove is bordered by rows of **tube feet (podia)** (Figure 22.5). These in turn are usually protected by movable **spines.** A large **radial nerve** can be seen in the center of each ambulacral groove (Figure 22.6C), between the rows of tube feet. The nerve is very superficially located, covered only by thin epidermis. Under the nerve is an extension of the coelom and the radial canal of the water-vascular system, all of which are external to the underlying ossicles (Figure 22.6C). In all other classes of living echinoderms except crinoids, these structures are covered by ossicles or other dermal tissue; thus ambulacral grooves in asteroids and crinoids are said to be *open,* and those of the other groups are *closed.*

The aboral surface is usually rough and spiny, although spines of many species are flattened, so that the surface appears smooth (see Figure 22.4C). Around the bases of spines are groups of minute, pincerlike **pedicellariae,** bearing tiny jaws manipulated by muscles (Figure 22.7). These jaws keep the body surface free of debris, protect papulae, and sometimes aid

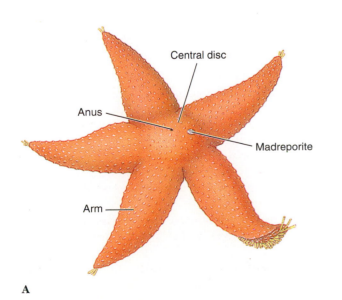

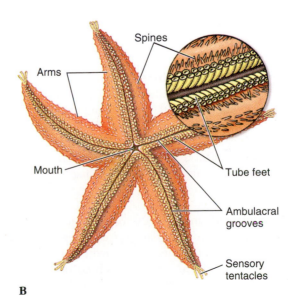

A **B**

Figure 22.5
External anatomy of asteroid. **A,** Aboral view. **B,** Oral view.

Figure 22.6

A, Internal anatomy of a sea star. **B,** Water-vascular system. Podia penetrate between ossicles. (Polian vesicles are not present in *Asterias.*) **C,** Cross section of arm at level of gonads, illustrating open ambulacral grooves.

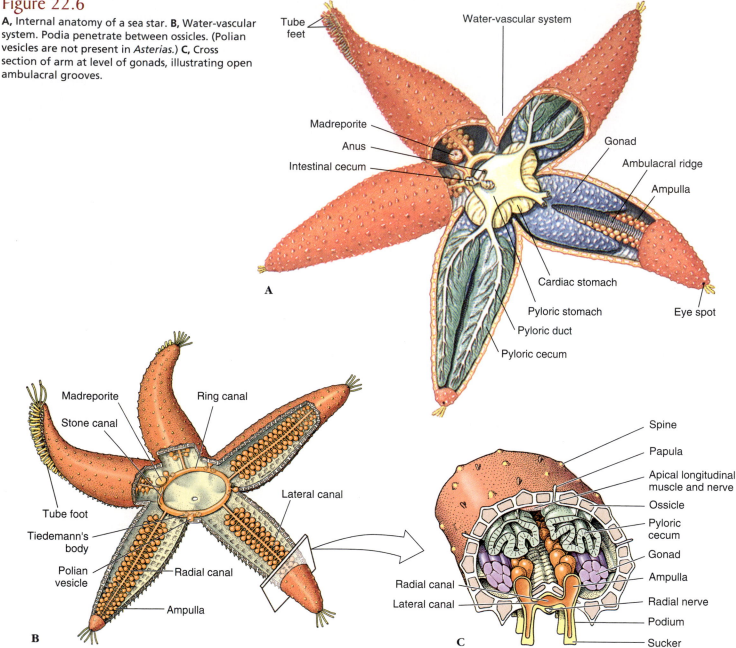

in food capture. **Papulae (dermal branchiae** or **skin gills)** are soft delicate projections of the coelomic cavity, covered only with epidermis and lined internally with peritoneum; they extend out through spaces between ossicles and are involved with respiration (Figures 22.6C and 22.7F). Also on the aboral side are an inconspicuous anus and a conspicuous circular **madreporite** (see Figure 22.5A), a calcareous sieve leading to the water-vascular system.

Endoskeleton   Beneath the epidermis of sea stars is a mesodermal endoskeleton of small calcareous plates, or **ossicles,** bound together with connective tissue. This connective tissue is an unusual form of mutable collagen, termed **catch collagen,** that is under neurological control. Catch collagen can be

changed from a "liquid" to a "solid" form very quickly when stimulated by the nervous system. This characteristic provides echinoderms with some unique mechanical properties, perhaps most importantly the ability to hold various postures without muscular effort. From the ossicles project spines and tubercles that form the spiny surface. Ossicles are penetrated by a meshwork of spaces, usually filled with fibers and dermal cells. This internal meshwork structure is described as **stereom** and is unique to echinoderms.

Muscles in the body wall move the rays and can partially close the ambulacral grooves by drawing their margins together.

Coelom, Excretion, and Respiration   Coelomic compartments of larval echinoderms give rise to several structures in

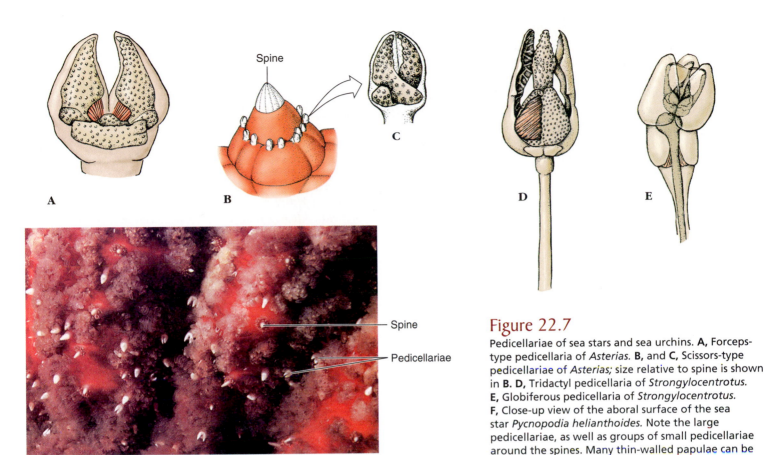

Figure 22.7
Pedicellariae of sea stars and sea urchins. **A,** Forceps-type pedicellaria of *Asterias*. **B,** and **C,** Scissors-type pedicellariae of *Asterias;* size relative to spine is shown in **B. D,** Tridactyl pedicellaria of *Strongylocentrotus*. **E,** Globiferous pedicellaria of *Strongylocentrotus*. **F,** Close-up view of the aboral surface of the sea star *Pycnopodia helianthoides*. Note the large pedicellariae, as well as groups of small pedicellariae around the spines. Many thin-walled papulae can be seen.

adults, one of which is a spacious body coelom filled with fluid. This fluid contains amebocytes (coelomocytes), bathes internal organs, and projects into papulae. The ciliated peritoneal lining of the coelom circulates the fluid around the body cavity and into the papulae. Exchange of respiratory gases and excretion of nitrogenous waste, principally ammonia, occur by diffusion through the thin walls of papulae and tube feet. Some wastes may be engulfed by coelomocytes, which migrate through the epithelium of papulae or tube feet to the exterior.

Water-Vascular System   The water-vascular system is another coelomic compartment unique to echinoderms. It is a set of canals and specialized tube feet that, together with the dermal ossicles, forms a hydraulic system. In sea stars the primary functions of the water-vascular system are locomotion and food gathering, in addition to respiration and excretion.

Structurally, the water-vascular system opens to the outside through small pores in the madreporite. The madreporite of asteroids is on the aboral surface (see Figure 22.5A) and leads into a **stone canal,** which descends toward a **ring canal** around the mouth (see Figure 22.6B). **Radial canals** diverge from the ring canal, one into the ambulacral groove of each ray. Also attached to the ring canal are four or five pairs of folded, pouch-like **Tiedemann's bodies** and from one to five **polian vesicles** (polian vesicles are absent in some sea stars, such as *Asterias*). Tiedemann's bodies may produce coelomocytes, and polian

vesicles are apparently for fluid storage and regulation of the internal pressure within the water-vascular system.

A series of small **lateral canals,** each with a one-way valve, connects the radial canal to the cylindrical podia, or tube feet, along the sides of the ambulacral groove in each ray. Each podium is a hollow, muscular tube, the inner end of which is a muscular sac, or **ampulla,** which lies within the body coelom (see Figure 22.6A and C), and the outer end of which usually bears a **sucker.** Some species lack suckers. Podia pass to the outside between ossicles in the ambulacral groove.

The water-vascular system operates hydraulically and is an effective locomotor mechanism. The basic design applies muscular pressure to coelomic fluid in tube feet to stiffen them for walking. The ampulla at the top of the tube foot serves as a fluid reservoir. Each tube foot has in its walls connective tissue that maintains the cylinder at a relatively constant diameter. Contraction of muscles in the ampulla forces fluid into the podium, extending it. Valves in the lateral canals prevent backflow of fluid into the radial canals. Conversely, contraction of the longitudinal muscles in the tube foot retracts the podium, forcing fluid back into the ampulla. Contraction of muscles in one side of the podium bends the organ toward that side. Small muscles at the end of the tube foot can raise the middle of the disclike end, creating suction when the end is applied to a firm substrate. It has been estimated that by combining mucous adhesion with suction, a single podium can exert a pull equal to 0.25 to 0.3 newtons. Coordinated action of all or many of the tube feet is sufficient to draw the animal up a vertical surface or over rocks. The ability to move while firmly adhering to the substrate is a clear advantage to an animal living in a sometimes wave-churned environment.

On a soft surface, such as muck or sand, suckers are ineffective (numerous sand-dwelling species have no suckers), so the tube feet are employed as legs. Locomotion becomes mainly a stepping process. Most sea stars can move only a few centimeters per minute, but some very active ones—*Pycnopodia* (Gr. *pyknos,* compact, dense, + *pous, podos,* foot) (Figure 22.8B), for example—can move 75 to 100 cm per minute. When inverted, a sea star bends its rays until some tubes reach the substratum and attach as an anchor; then it slowly rolls over.

Tube feet are innervated by the central nervous system (ectoneural and hyponeural systems). Nervous coordination enables tube feet to move in a single direction, although not in unison, so that the sea star may progress. If the radial nerve in an arm is cut, podia in that arm lose coordination, although they can still function. If the circumoral nerve ring is cut, podia in all arms become uncoordinated, and movement ceases.

Feeding and Digestive System   The mouth on the oral side leads through a short esophagus to a large stomach in the central disc. The lower (cardiac) part of the stomach can be everted through the mouth during feeding (Figure 22.9), and excessive eversion is prevented by gastric ligaments. The upper (pyloric) part is smaller and connects by ducts to a pair of large **pyloric ceca (digestive glands)** in each arm (see Figure 22.6A). Digestion is mostly extracellular, although some intracellular digestion may occur in the ceca. A short intestine leads aborally from the pyloric stomach, and there are usually a few small, saclike **intestinal ceca** (see Figure 22.6A). The anus is inconspicuous, and some sea stars lack an intestine and anus.

Many sea stars are carnivorous and feed on molluscs, crustaceans, polychaetes, echinoderms, other invertebrates, and sometimes small fishes. Sea stars consume a wide range of food items,

but many show particular preferences. Some select brittle stars, sea urchins, or sand dollars, swallow them whole and later regurgitate undigestible ossicles and spines (see Figure 22.8B). Some attack other sea stars, and if they are small compared with their prey, they may attack and begin eating at the end of one arm.

Since 1963 there have been numerous reports of increasing numbers of the crown-of-thorns starfish (*Acanthaster planci* [Gr. *akantha,* thorn, + *asteros,* star]) (Figure 22.4D) that were damaging large areas of coral reef in the Pacific Ocean. Crown-of-thorns stars feed on coral polyps, and they sometimes occur in large aggregations, or "herds." There is some evidence that outbreaks have occurred in the past, but an increase in frequency during the past 40 years suggests that some human activity may be affecting the sea stars. Of reefs surveyed in 2002, 12% had outbreaks, compared with 1988 when 10% had outbreaks, resulting in extensive damage. Efforts to control these organisms are very expensive and of questionable effectiveness. The controversy continues, especially in Australia, where it is exacerbated by extensive media coverage.

Some asteroids feed heavily on molluscs (Figure 22.8A), and *Asterias* is a significant predator on commercially important clams and oysters. When feeding on a bivalve, a sea star will wrap itself around its prey, attaching its podia to the valves, and then exert a steady pull, using its feet in relays. A force of some 12.75 newtons can thus be exerted. This would be roughly equivalent to a human attempting to lift a 1000-lb weight with one hand. In half an hour or so the bivalve's adductor muscles fatigue and relax. With a very small gap available, the star inserts its soft everted stomach into the space between the valves and wraps it around the soft parts of the bivalve to begin digestion. After feeding, the sea star draws its stomach inward by contraction of its stomach muscles and relaxation of body-wall muscles.

Some sea stars feed on small particles, either entirely or in addition to carnivorous feeding. Plankton and other organic particles coming in contact with an animal's surface are carried by epidermal cilia to the ambulacral grooves and then to the mouth.

Hemal System   The so-called hemal system is not very well developed in asteroids, and its function in all echinoderms is unclear. The hemal system has little to do with circulation of body fluids. It is a system of tissue strands enclosing unlined sinuses and is itself enclosed in another coelomic compartment, or **perihemal channels** (Figure 22.9). Research with at least one sea star shows that absorbed nutrients appear in the hemal system within a few hours of feeding, and eventually concentrate in the gonads and podia. Thus, the hemal system appears to play a role in distributing digested nutrients. It also includes the **axial complex,** which pressure-filters blood vascular fluids (Figure 22.9).

Nervous System   The nervous system consists of three units at different levels in the disc and arms. Chief of these systems is an **oral (ectoneural)** system composed of a **nerve ring** around

A   **B**

Figure 22.8

A, *Orthasterias koehleri* eating a clam. **B,** This *Pycnopodia helianthoides* has been overturned while eating a large sea urchin, *Strongylocentrotus franciscanus.* This sea star has 20 to 24 arms and can range up to 1 m in diameter (arm tip to arm tip).

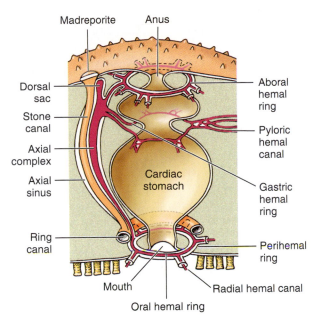

Figure 22.9

Hemal system of asteroids. The main perihemal channel is the thin-walled axial sinus, which encloses both the axial organ and stone canal. Other features of the hemal system are shown.

Figure 22.10

Pacific sea star, *Echinaster luzonicus,* can reproduce itself by splitting across the disc, then regenerating missing arms. The one shown here has evidently regenerated six arms from the longer one at top left.

the mouth and a main **radial nerve** into each arm. It appears to coordinate the tube feet. A **deep (hyponeural)** system lies aboral to the oral system, and an **aboral** system consists of a ring around the anus and radial nerves along the roof of each ray. An **epidermal nerve plexus** or nerve net freely connects these systems with the body wall and related structures. The epidermal plexus coordinates responses of the dermal branchiae to tactile stimulation—the only instance known in echinoderms in which coordination occurs through a nerve net.

Sense organs are not well developed. Tactile organs and other sensory cells are scattered over the surface, and an ocellus is at the tip of each arm. Their reactions are mainly to touch, temperature, chemicals, and differences in light intensity. Sea stars are usually more active at night.

Reproductive System, Regeneration, and Autotomy

Most sea stars have separate sexes. A pair of gonads lies in each interradial space (see Figure 22.6A). Fertilization is external and occurs in early summer when eggs and sperm are shed into the water. A secretion from neurosecretory cells located on the radial nerves stimulates maturation and shedding of asteroid eggs.

Echinoderms can regenerate lost parts. Sea-star arms can regenerate readily, even if all are lost. Sea stars also have the power of autotomy and can cast off an injured arm near its base. Regeneration of a new arm may take several months.

Some species can regenerate a complete new sea star (Figure 22.10) from a detached arm. For most asteroids to regenerate, the detached arm must contain a part (about one-fifth) of the central disc. In some species, however, asexual reproduction from single detached arms with no trace of the central disc is a common means of asexual reproduction (for example, *Linckia*).

In former times fishermen used to dispatch sea stars they collected from their oyster beds by chopping them in half with a hatchet—a worse than futile activity. Some sea stars reproduce asexually under normal conditions by cleaving their central disc, each part regenerating the rest of the disc and missing arms.

Development   Development is quite varied among different lineages of sea stars. Some species produce benthic egg masses in which juveniles develop. Other species produce eggs that are brooded, either under the oral side of the animal or in specialized aboral structures, and development is direct. Some species even have viviparous brooding of young in the gonad of the adults. However, most sea stars produce free-swimming planktonic larvae. Even here there is variation, and some species provision their young with sufficient yolk to develop without feeding in the water column, whereas others require a prolonged period of feeding to gain sufficient energy to metamorphose into the adult body form.

Early embryogenesis shows a typical ancestral deuterostome pattern (see Figures 8.7A and 8.11A). Gastrulation is by invagination, and the anterior end of the archenteron pinches off to become a coelomic cavity, which expands in a U-shape to fill the blastocoel. Each leg of the U, at the posterior, constricts to become a separate vesicle, and these eventually give rise to the main coelomic compartments of the body (metacoels, called **somatocoels** in echinoderms). The anterior portion of the U undergoes subdivision to form protocoels and mesocoels (called **axocoels** and **hydrocoels** in echinoderms) (Figure 22.11). The left hydrocoel will become the water-vascular system, and the left axocoel will give rise to the stone canal and perihemal channels. The right axocoel and hydrocoel will disappear. The free-swimming larva has cilia arranged in bands and is called a **bipinnaria** (Figure 22.2A). These ciliated tracts become extended into larval arms. Soon the larva grows three adhesive arms and a sucker at its anterior end and is then called a **brachiolaria** (Figure 22.12B). At that time it attaches to the substratum, forms a temporary attachment stalk, and undergoes metamorphosis.

Metamorphosis involves a dramatic reorganization of a bilateral larva into a radial juvenile. The anteroposterior axis of the larva is lost, and *what was the left side becomes the oral surface,*

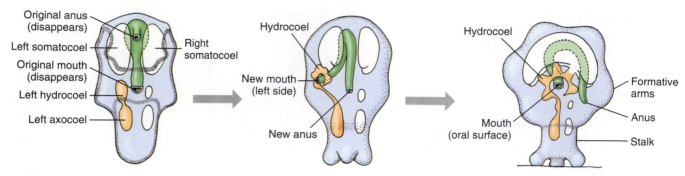

Figure 22.11

Asteroid metamorphosis. The left somatocoel becomes the oral coelom, and the right somatocoel becomes the aboral coelom. The left hydrocoel becomes the water-vascular system and the left axocoel the stone canal and perihemal channels. The right axocoel and hydrocoel are lost.

and the larval right side becomes the aboral surface (see Figure 22.11). Correspondingly, larval mouth and anus disappear, and a new mouth and anus form on what were originally the left and right sides, respectively. The portion of the anterior coelomic compartment from the left side expands to form the ring canal of the water-vascular system around the mouth, and then it grows branches to form radial canals. As the short, stubby arms and the first podia appear, the animal detaches from its stalk and begins life as a young sea star. A number of regulatory genes found in bilateral animals are conserved in echinoderms and have surprisingly similar functions. For example, *Distal-less* and its homolog in vertebrates regulate outgrowth of limbs in these animals; its homolog in echinoderms is active in development of tube feet.

Sea Daisies   Strange little (less than 1 cm diameter), disc-shaped animals (Figure 22.13) were discovered in water over 1000 m deep off New Zealand. Sometimes called sea daisies, they were originally described (1986) as a new class of echinoderms Concentricycloidea. Only two species are known so far.

Most zoologists now agree that they are highly derived spinulosid asteroids. Phylogenetic analysis of rDNA places them within Asteroidea.

Sea daisies are pentaradial in symmetry but have no arms. Their tube feet are located around the periphery of the disc, rather than along ambulacral areas, as in other echinoderms. Their water-vascular system includes two concentric ring canals; the outer ring may represent radial canals since podia arise from it. A hydropore, homologous to the madreporite, connects the inner ring canal to the aboral surface. One species has no digestive tract; its oral surface is covered by a membranous **velum,** by which it apparently absorbs nutrients. The other species has a shallow, saclike stomach but no intestine or anus.

Class Ophiuroidea

Brittle stars are largest of the major groups of echinoderms with over 2000 extant species, and they are probably the most abundant also. They abound in all types of benthic marine habitats, even carpeting the abyssal sea bottom in many areas.

Form and Function

Apart from a typical possession of five arms, brittle stars are surprisingly different from asteroids. Arms of brittle stars are slender and sharply set off from the central disc (Figure 22.14). They have no pedicellariae or papulae, and their ambulacral grooves are closed and covered with arm ossicles. Their tube feet are without suckers; they aid in feeding but are of limited use in locomotion. In contrast to asteroids, the madreporite of ophiuroids is located on the oral surface, on one of the oral shield ossicles (Figure 22.15). Ampullae are absent from podia, and force for protrusion of a podium is generated by a proximal muscular portion of the podium.

Each jointed arm consists of a column of articulated ossicles (so-called **vertebrae**), connected by muscles and covered by plates. Locomotion is by arm movement. Arms are moved forward in pairs and are placed against the substratum, while one (any one) is extended forward or trailed behind, and the animal is pulled or pushed along in a jerky fashion.

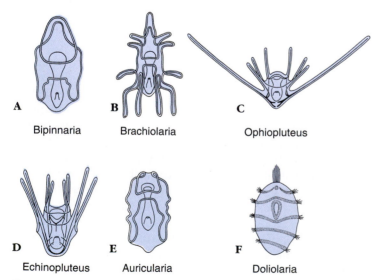

A   Bipinnaria
B   Brachiolaria
C   Ophiopluteus
D   Echinopluteus
E   Auricularia
F   Doliolaria

Figure 22.12

Larvae of echinoderms. **A,** Bipinnaria of asteroids. **B,** Brachiolaria of asteroids. **C,** Ophiopluteus of ophiuroids. **D,** Echinopluteus of echinoids. **E,** Auricularia of holothuroids. **F,** Doliolaria of crinoids.

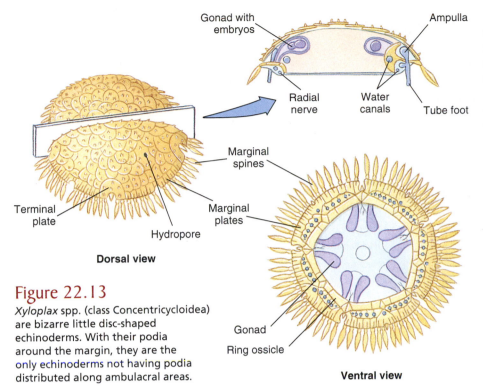

Dorsal view

Ventral view

Figure 22.13

Xyloplax spp. (class Concentricycloidea) are bizarre little disc-shaped echinoderms. With their podia around the margin, they are the only echinoderms not having podia distributed along ambulacral areas.

Figure 22.14

A, Brittle star, *Ophiura lutkeni* (class Ophiuroidea). Brittle stars do not use their tube feet for locomotion but can move rapidly (for an echinoderm) by means of their arms. **B,** Basket star, *Astrophyton muricatum* (class Ophiuroidea). Basket stars extend their many-branched arms to filter-feed, usually at night. They show a strongly negative phototropic response.

A

B

Five movable plates that serve as jaws surround the mouth (Figure 22.15). There is no anus. Their skin is leathery, with dermal plates and spines arranged in characteristic patterns. Surface cilia are mostly lacking.

Visceral organs are confined to the central disc, since their rays are too slender to contain them (Figure 22.16). Their stomach is saclike, and there is no intestine. Indigestible material is cast out of the mouth.

Five pairs of invaginations called **bursae** open toward the oral surface by genital slits at the bases of the arms. Water circulates in and out of these sacs for exchange of gases. On the coelomic wall of each bursa are small gonads that discharge their ripe sex cells into the bursa. Gametes pass through the genital slits into the water for fertilization (Figure 22.17A).

Sexes are usually separate; a few ophiuroids are hermaphroditic. Some brood their young in the bursae; the young escape through the genital slits or by rupturing the aboral disc. Most species produce a free-swimming larva called an ophiopluteus, and its ciliated bands extend onto delicate, beautiful larval arms (see Figure 22.12C). During metamorphosis to a juvenile, there is no temporarily attached phase, as there is in asteroids.

Water-vascular, nervous, and hemal systems are similar to those of sea stars. Each arm contains a small coelom, a radial nerve, and a radial canal of the water-vascular system.

Biology

Brittle stars tend to be secretive, living on hard substrates in locations where little or no light penetrates. They are often negatively phototropic and insinuate themselves into small crevices between rocks, becoming more active at night. They are commonly fully exposed in the permanent darkness of deep seas. Ossicles in arms of at least some photosensitive ophiuroids show a remarkable adaptation for photoreception. Tiny, rounded structures on their aboral surface serve as microlenses, focusing light on nerve bundles just beneath them. Related species that are indifferent to light have no such structures.

Ophiuroids feed on a variety of small particles, either browsing food from the bottom or suspension feeding. Podia are important in transferring food to the mouth. Some brittle stars extend arms into the water and catch suspended particles in mucous strands between arm spines. Basket stars perch on corals, extending their branched arms to capture plankton (Figure 22.17B)

Some ophiuroids are carnivorous, and at least one species is a fish specialist that assumes an ambush posture with the central disc held off the substrate. When an unsuspecting fish enters the "shelter" beneath the central disc, the star twists abruptly to trap the fish in a spiral cylinder formed by the spiny arms.

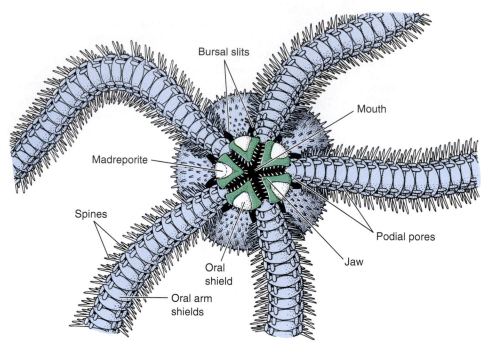

Figure 22.15

Oral view of spiny brittle star, *Ophiothrix*.

Regeneration and autotomy are even more pronounced in brittle stars than in sea stars. Many seem very fragile, releasing an arm or even part of the disc at the slightest provocation. Some can reproduce asexually by cleaving the disc; each new individual then regenerates missing parts.

Some common ophiuroids along the coast of the United States are *Amphipholis* (Gr. *amphi,* both sides of, + *pholis,*

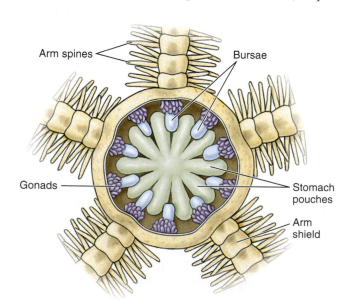

Figure 22.16

Ophiuroid with aboral disc wall cut away to show principal internal structures. Bursae are fluid-filled sacs in which water constantly circulates for respiration. They also serve as brood chambers. Only bases of arms are shown.

horny scale) (viviparous and hermaphroditic), *Ophioderma* (Gr. *ophis,* snake, + *dermatos,* skin), *Ophiothrix* (Gr. *ophis,* snake, + *thrix,* hair), and *Ophiura* (Gr. *ophis,* snake, + *oura,* tail) (see Figure 22.14). Basket stars *Gorgonocephalus* (Gr. *Gorgo,* name of a female monster of terrible aspect, + *kephalē,* a head) (Figure 22.17B) and *Astrophyton* (Gr. *asteros,* star, + *phyton,* creature, animal) (see Figure 22.14B) have arms that branch repeatedly. Most ophiuroids are drab, but some are attractive, with bright color patterns (Figure 22.17A).

Class Echinoidea

There are roughly 950 living species of echinoids, which generally have a compact body enclosed in an endoskeletal test, or shell. Dermal ossicles, which have become closely fitting plates, form a test. Echinoids lack arms, but their tests reflect a typical pentamerous plan of echinoderms in their five ambulacral areas. These ambulacral regions are visible as five "spiky" bands in Figure 22.18E. A close look at the arrangement of tube feet in Figure 22.19 also reveals the ambulacral regions. The most notable modification of the ancestral body plan is that the oral surface which bears the tube feet and faces the substratum in sea stars, has expanded around to the aboral side, so that ambulacral areas extend up to an area close to the anus (**periproct**).

A majority of living species of sea urchins are "regular"; they have a hemispherical shape, radial symmetry, and medium to long spines (Figures 22.18 and 22.19). Sand dollars (Figure 22.20) and heart urchins (Figure 22.21) are "irregular" because members of their orders have become secondarily bilateral; their spines are usually very short. Regular urchins move by means of their tube feet, with some assistance from their spines, and irregular

A   B

Figure 22.17

A, This brittle star, *Ophiopholis aculeata,* has its bursae swollen with eggs, which it is ready to expel. The arms have been broken and are regenerating. **B,** Oral view of a basket star, *Gorgonocephalus eucnemis,* showing pentaradial symmetry.

Figure 22.18
Diversity among regular sea urchins (class Echinoidea). **A,** Ten-lined pencil urchin, *Eucidaris metularia* in the Red Sea. Members of this order have many ancestral characters and have survived since the Paleozoic era. They may be closest in resemblance to the common ancestor of all other extant echinoids. **B,** Slate-pencil urchin, *Heterocentrotus mammilatus.* The large, triangular spines of this urchin were formerly used for writing on slates. **C,** Aboral spines of the intertidal urchin, *Colobocentrotus atratus,* are flattened and mushroom shaped, while the marginal spines are wedge shaped, giving the animal a streamlined form to withstand pounding surf. **D,** *Diadema antillarum* is a common species in the West Indies and Florida. **E,** *Astropyga magnifica* is one of the most spectacularly colored sea urchins, with bright-blue spots along its interambulacral areas.

urchins move chiefly by their spines (Figure 22.20). Some echinoids are quite colorful, and some have greatly reduced tests. These "soft-tested" urchins often have bright warning coloration and their pedicellariae deliver painful toxins.

Echinoids have a wide distribution in all seas, from intertidal regions to deep oceans. Regular urchins often prefer rocky or hard substrates, whereas sand dollars and heart urchins like to burrow into a sandy substrate. Distributed along one or both coasts of North America are common genera of regular urchins (*Arbacia* [Gr. *Arbakēs,* first king of Media], *Strongylocentrotus*

Figure 22.19
Purple sea urchin, *Strongylocentrotus purpuratus,* is common along the Pacific coast of North America where there is heavy wave action.

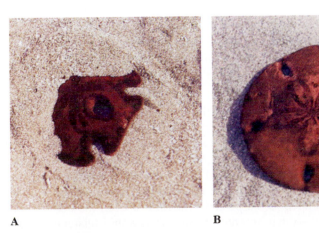

Figure 22.20
Two sand dollar species. **A,** *Encope grandis* as normally found burrowing near the surface on a sandy bottom. **B,** Removed from the sand. The short spines and petaloids on the aboral surface of this *Encope micropora* are easily seen.

[Gr. *strongylos,* round, compact, + *kentron,* point, spine] [Figure 22.19], and *Lytechinus* [Gr. *lytos,* dissolvable, broken, + *echinos,* sea urchin]) and sand dollars (*Dendraster* [Gr. *dendron,* tree, stick, + *asteros,* star] and *Echinarachnius* [Gr. *echinos,* sea urchin, + *arachnē,* spider]). The West Indies–Florida region is rich in echinoderms, including echinoids, of which *Diadema* (Gr. *diadeō,* to bind around), with its long, needle-sharp spines, is a notable example (see Figure 22.18D; see note on this page).

Form and Function

In general, an echinoid test is a compact skeleton of 10 double rows of plates that bear movable, stiff spines (Figure 22.22). The plates are sutured firmly. During periods of rapid growth, the plates may not keep pace with the soft tissue, producing somewhat loose sutures. The five pairs of ambulacral rows are homologous to the five arms of sea stars and have pores (Figure 22.22B) through which long tube feet extend. The plates bear small tubercles on which the round ends of spines articulate as ball-and-socket joints. Spines are moved by small muscles around the bases.

There are several kinds of pedicellariae, most common of which are three jawed and are mounted on long stalks (see Figure 22.7D and E). Pedicellariae help keep the body clean, especially by preventing marine larvae from settling on the body surface. Pedicellariae of many species bear venom glands, and their toxin paralyzes small prey.

Diadema antillarum is not nearly as prominent as it once was. In January 1983, an epidemic swept through the Caribbean and along the Florida Keys. Its cause has never been determined, but it decimated the *Diadema* population, leaving less than 5% of the original numbers. Other species of sea urchins were unaffected. However, various types of algae, formerly grazed heavily by *Diadema* have increased greatly on the reefs, and *Diadema* populations have not recovered. This abundance of algae has had a disastrous effect on coral reefs around Jamaica. Herbivorous fish around that island had been chronically overharvested, and then, after the *Diadema* epidemic, there was nothing left to control algal overgrowth. Coral reefs around Jamaica have been largely destroyed as a result. As of this writing, a modest recovery in *Diadema* populations is occurring in parts of the Caribbean.

Five converging teeth surround the mouth of regular urchins. In some sea urchins branched gills (modified podia) encircle the peristome (region around the mouth). Genital pores and madreporite are located aborally in the periproct region which surrounds the anus (Figure 22.22). Sand dollars also have teeth, and their mouth is located at about the center of the oral side, but their anus has shifted to the posterior margin or even the oral side of the disc, so that an anteroposterior axis and bilateral symmetry can be recognized. Bilateral symmetry is even more accentuated in heart urchins, with their anus near the posterior on the oral side and their mouth moved away from the oral pole toward the anterior (see Figure 22.21).

Aboral pole   Periproct

A   B   Mouth

Figure 22.21

An irregular echinoid, *Meoma,* one of the largest heart urchins (test up to 18 cm). *Meoma* occurs in the West Indies and from the Gulf of California to the Galápagos Islands. **A,** Aboral view. Anterior ambulacral area is not modified as a petaloid in heart urchins, although it is in sand dollars. **B,** Oral view. Note curved mouth at anterior end and periproct at posterior end.

Inside the test (Figure 22.22) are the coiled digestive system and a complex chewing mechanism (in regular urchins and in sand dollars), called **Aristotle's lantern** (Figure 22.23), to which the teeth are attached. A ciliated **siphon** connects the esophagus to the intestine and enables water to bypass the stomach to concentrate food for digestion in the intestine. Sea urchins are largely omnivorous, but their primary diet often consists mostly of algae and other organic material, which they graze with their teeth. Sand dollars have short club-shaped spines that move sand and its organic contents over the aboral surface and down the sides. Fine food particles drop between the spines, and ciliated tracts on the oral side carry the particles to their mouth.

Hemal and nervous systems are basically similar to those of asteroids. Ambulacral grooves are closed, and radial canals of the water-vascular system run just beneath the test, one in each ambulacral radius (see Figure 22.22). Ampullae for the podia are within the test, and each ampulla usually communicates with its podium by *two* canals through pores in the ambulacral plate; consequently, such pores in the plates are in pairs.

Peristomial gills, where present, are of little importance in respiratory gas exchange, this function being performed principally by other podia. Although the gills appear to provide some oxygen to the muscles associated with Aristotle's lantern, they seem to function primarily to accommodate pressure changes in the pharyngeal coelom during feeding movements of the lantern complex. In irregular urchins respiratory podia are thin walled, flattened, or lobulate and are arranged in ambulacral fields called **petaloids** on the aboral surface. Petaloids form what after appears to be a flower pattern atop sand dollars (see Figure 22.20). Irregular urchins also have short, suckered, single-pored podia in ambulacral and sometimes interambulacral areas; these podia function in food handling.

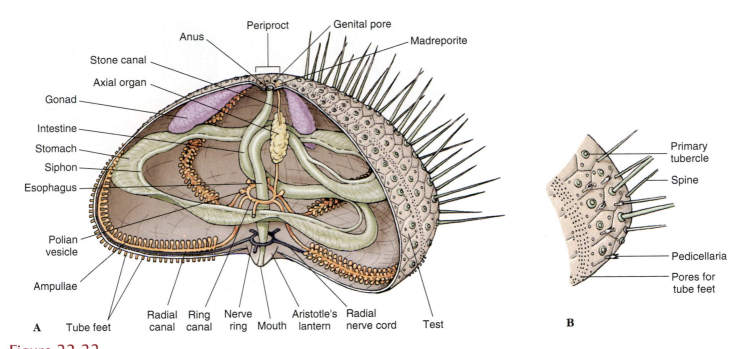

Figure 22.22

A, Internal structure of a sea urchin; water-vascular system in tan. B, Detail of portion of endoskeleton.

Sexes are separate, and both eggs and sperm are shed into the sea for external fertilization. Some, such as certain pencil urchins, brood their young in depressions between the spines. **Echinopluteus larvae** (Figure 22.12D) of nonbrooding echinoids may live a planktonic existence for several months and then metamorphose quickly into young urchins.

Class Holothuroidea

In a phylum characterized by odd animals, class Holothuroidea (sea cucumbers) contains members that both structurally and physiologically are among the strangest. These animals bear a

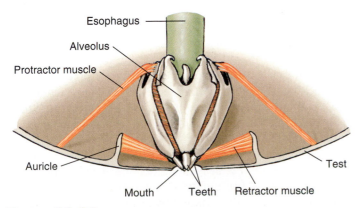

Figure 22.23

Aristotle's lantern, a complex mechanism used by sea urchins for masticating their food. Five pairs of retractor muscles draw the lantern and teeth up into the test; five pairs of protractors push the lantern down and expose the teeth. Other muscles produce a variety of movements. Only major skeletal parts and muscles are shown in this diagram.

remarkable resemblance to the vegetable after which they are named (see Figure 22.24). Compared with other echinoderms, holothurians are greatly elongated in the oral-aboral axis, and ossicles are much reduced in most; consequently, these animals are soft bodied. Some species crawl on the sea floor, others are found beneath rocks, and some are burrowers.

There are approximately 1150 living species of holothuroids. Common species along the east coast of North America are *Cucumaria frondosa* (L. *cucumis,* cucumber), *Sclerodactyla briareus* (Gr. *skleros,* hard, + *daktylos,* finger) (see Figure 22.26), and the translucent, burrowing *Leptosynapta* (Gr. *leptos,* slender, + *synapsis,* joining together). Along the Pacific coast there are several species of *Cucumaria* (see Figure 22.24C) and the striking reddish brown *Parastichopus* (Gr. *para,* beside, + *stichos,* line or row, + *pous, podos,* foot) (see Figure 22.24A), with very large papillae.

Form and Function

Their body wall is usually leathery, with tiny ossicles embedded in it (Figure 22.25), although a few species have large ossicles forming a dermal armor (see Figure 22.24B). Because of the elongate body form of sea cucumbers, they characteristically lie on one side. The body wall contains circular and longitudinal muscles along the ambulacra.

In some species locomotor tube feet are equally distributed to the five ambulacral areas (see Figure 22.24C) or all over the body, but most have well-developed tube feet only in the ambulacra normally applied to the substratum (see Figure 22.24A and B). Thus a secondary bilaterality is present, albeit of quite different origin from that of irregular urchins. The side applied to the substratum has three ambulacra and is called a **sole;** tube feet in the dorsal ambulacral areas, if present, are usually without suckers

A   **B**   **C**

Figure 22.24

Sea cucumbers (class Holothuroidea). **A,** Common along the Pacific coast of North America, *Parastichopus californicus* grows to 50 cm in length. Its tube feet on the dorsal side are reduced to papillae and warts. **B,** In sharp contrast to most sea cucumbers, the surface ossicles of *Psolus chitonoides* are developed into a platelike armor. The ventral surface is a flat, soft, creeping sole, and the mouth (surrounded by tentacles) and anus are turned dorsally. **C,** Tube feet are found in all ambulacral areas of *Cucumaria miniata* but are better developed on its ventral side, shown here.

and may be modified as sensory papillae. All tube feet, except oral tentacles, may be absent in burrowing forms. Oral tentacles are 10 to 30 retractile, modified tube feet around the mouth.

A sea cucumber's coelomic cavity is spacious and fluid filled and has many coelomocytes. The fluid-filled coelom now serves as a hydrostatic skeleton. Dermal ossicles are small and not connected to each other, so they no longer form an endoskeleton.

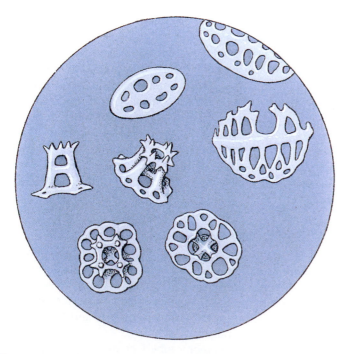

Figure 22.25

Ossicles of sea cucumbers are usually microscopic bodies buried in leathery dermis. They can be extracted from this tissue with commercial bleach and are important taxonomic characteristics. The ossicles shown here, called tables, buttons, and plates, are from *Holothuria difficilis*. They illustrate the meshwork (stereom) structure observed in ossicles of all echinoderms at some stage in their development (×250).

The digestive system empties posteriorly into a muscular **cloaca** (Figure 22.26). A **respiratory tree** composed of two long, many-branched tubes also empties into the cloaca, which pumps seawater into it. The respiratory tree serves for both respiration and excretion and is not present in any other group of living echinoderms. Gas exchange also occurs through skin and tube feet.

The hemal system is more well developed in holothurians than in other echinoderms. Their water-vascular system is peculiar in that the madreporite lies free in the coelom.

Sexes are usually separate, but some holothurians are hermaphroditic. Among echinoderms, only sea cucumbers have a single gonad. The gonad is usually in the form of one or two clusters of tubules that join at the gonoduct. Fertilization is external, and the free-swimming larva is called an **auricularia** (see Figure 22.12E). Some species brood their young either inside their body or somewhere on the body surface.

Biology

Sea cucumbers are sluggish, moving partly by means of their ventral tube feet and partly by waves of contraction in the muscular body wall. Most sedentary species trap suspended food particles in mucus of their outstretched oral tentacles or pick up particles from the surrounding surface. They then stuff their tentacles into the pharynx, one by one, ingesting captured food (Figure 22.27A). Others crawl along the substrate grazing the bottom with their tentacles (Figure 22.27B).

Sea cucumbers have a peculiar power of what appears to be self-mutilation but is in reality a defense mechanism. When irritated or subjected to unfavorable conditions, many species can cast out a part of their viscera by a strong muscular contraction that may either rupture the body wall or evert its contents through their anus. Lost parts are soon regenerated. Certain species have Cuvierian tubules which are attached to the posterior part of the respiratory tree and when expelled may entangle an enemy (Figure 22.27C). These tubules become long and sticky after expulsion, and some contain toxins.

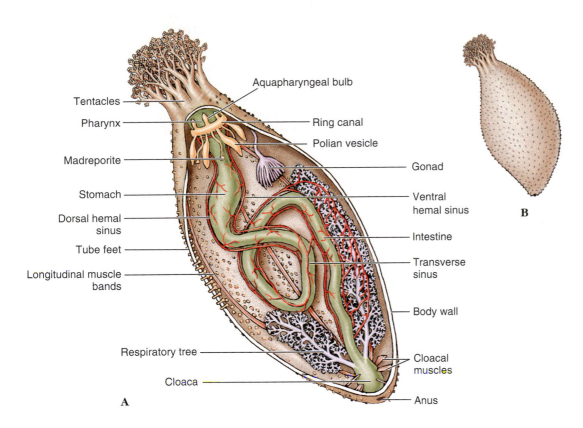

Figure 22.26
Anatomy of a sea cucumber *Sclerodactyla*. **A,** Internal. *Red,* hemal system. **B,** External.

An interesting commensal relationship exists between certain sea cucumbers and the pealfish, *Carapus,* which uses the cloaca and respiratory tree of the sea cucumber as shelter.

Class Crinoidea

Crinoids include about 625 species of sea lilies and feather stars. As fossil records reveal, crinoids were once far more numerous than they are now. They differ from other echinoderms by being attached during a substantial part of their lives. Sea lilies have a flower-shaped body at the tip of an attached stalk (Figure 22.28). Feather stars have long, many-branched arms, and adults are free-moving, though they may remain in the same spot for long periods (Figure 22.29). During metamorphosis feather stars become sessile and stalked, but after several months they detach and become free-moving. Many crinoids are deep-water forms, but feather stars may inhabit shallow waters, especially in Indo-Pacific and West-Indian–Caribbean regions, where the largest numbers of species occur.

Figure 22.27

A, *Eupentacta quinquesemita* extends its tentacles to collect particulate matter in the water, then puts them one by one into its mouth and cleans the food from them. **B,** Moplike tentacles of *Parastichopus californicus* are used for deposit feeding on the bottom. **C,** *Bohadschia argus* expels its Cuvierian tubules, modified parts of its respiratory tree, when it is disturbed. These sticky strands, containing a toxin, discourage potential predators.

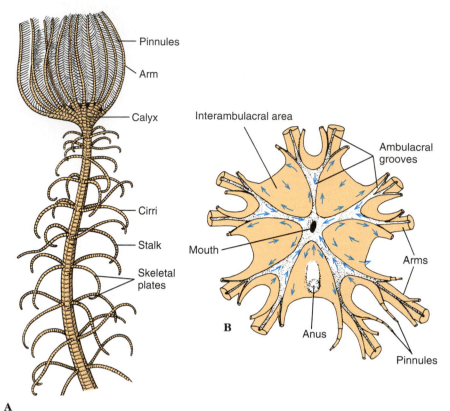

Figure 22.28

Crinoid structure. **A,** Sea lily (stalked crinoid) with portion of stalk. Modern crinoid stalks rarely exceed 60 cm, but fossil forms were as much as 20 m long. **B,** Oral view of calyx of a crinoid, *Antedon,* showing direction of ciliary food currents. Ambulacral grooves with podia extend from mouth along arms and branching pinnules. Food particles touching podia are tossed into ambulacral grooves and carried, tangled in mucus, by strong ciliary currents toward mouth. Particles falling on interambulacral areas are carried by cilia first toward mouth and then outward and finally dropped off the edge, thus keeping the oral disc clean.

Figure 22.29

Comantheria briareus are crinoids found on Pacific coral reefs. They extend their arms into the water to catch food particles both during the day and at night.

Form and Function

Their body disc, or **calyx,** is covered with a leathery skin **(tegmen)** containing calcareous plates. Epidermis is poorly developed. Five flexible arms branch to form many more arms, each with many lateral **pinnules** arranged like barbs on a feather (Figure 22.28). Calyx and arms together are called the **crown.** Sessile forms have a long, jointed **stalk** attached to the aboral side of the body. This stalk is composed of plates, appears jointed, and may bear **cirri.** Madreporite, spines, and pedicellariae are absent.

Their upper (oral) surface bears a mouth, which opens into a short esophagus, from which the long intestine with diverticula proceeds aborally for a distance and then makes a complete turn to an **anus,** which may be on a raised cone (Figure 22.28B). **Ambulacral grooves** are open and ciliated and serve to carry food to their mouth (Figure 22.28B). Simple tube feet without suckers line the ambulacral grooves, which extend into the pinnules. With the aid of tube feet and mucous strands, crinoids collect small organisms from the surrounding waters.

Their water-vascular system follows the basic echinoderm plan. However, the system functions entirely on existing coelomic fluid. There is no madreporite to allow exchange of fluid with the surrounding environment. The nervous system has an oral ring and a radial nerve that runs to each arm. The aboral or entoneural system is more highly developed in crinoids than in most other echinoderms. This system innervates the podia, which proliferate along the lateral pinnules, providing both feeding and sensory functions. Additional sensory organs are scanty and simple.

Sexes are separate. Gonads are simply masses of cells in the genital cavity of arms and pinnules. Gametes escape without ducts through a rupture in pinnule walls. Brooding occurs in some forms. **Doliolaria** larvae (Figure 22.12F) are free swimming for a time before they become attached and metamorphose.

PHYLOGENY AND ADAPTIVE DIVERSIFICATION

Phylogeny

Echinoderms left an extensive fossil record and evolved about 25 anatomically distinct body forms which account for 20 currently recognized classes. Most of these became extinct by the end of the Paleozoic, and only five classes survive today. Despite their fossil record, numerous contesting hypotheses on their phylogeny have been proposed. Based on their bilateral larvae, many invertebrate zoologists argue that their ancestors were bilateral and that their coelom had three pairs of spaces (trimeric or tripartite). Some investigators have proposed that radial symmetry arose in a free-moving echinoderm ancestor and that sessile

Classification of Phylum Echinodermata

There are about 7000 living and 20,000 extinct or fossil species of Echinodermata. The traditional classification placed all free-moving forms that were oriented with oral side down in subphylum Eleutherozoa, containing most living species. The other subphylum, Pelmatozoa, contained mostly forms with stems and oral side up; most extinct classes and living Crinoidea belong to this group. Although alternative schemes have strong supporters, cladistic analysis provides evidence that the two traditional subphyla are monophyletic. This list includes only groups with living members.

Subphylum Pelmatozoa (pel-ma'to-zo'a) (Gr. *pelmatos;* a stalk, + *zōon,* animal). Body in form of cup or calyx, borne on aboral stalk during part or all of life; oral surface directed upward; open ambulacral grooves; madreporite absent; both mouth and anus on oral surface; several fossil classes plus living Crinoidea.

Class Crinoidea (krin-oi'de-a) (Gr. *krinon,* lily; + *eidos,* form; + *ea,* characterized by): **sea lilies** and **feather stars.** Five arms branching at base and bearing pinnules; ciliated ambulacral grooves on oral surface with tentacle-like tube feet for food gathering; spines, madreporite, and pedicellariae absent. Examples: *Antedon, Comantheria* (Figure 22.29).

Subphylum Eleutherozoa (e-lu'ther-o-zo'a) (Gr. *eleutheros,* free, not bound, + *zōon,* animal). Body form star-shaped, globular, discoidal, or cucumber-shaped; oral surface directed toward substratum or oral-aboral axis parallel to substratum; body with or without arms; ambulacral grooves open or closed.

Class Asteroidea (as'ter-oy'de-a) (Gr. *aster,* star, + *eidos,* form, + *ea,* characterized by): **sea stars (starfish).** Star-shaped, with arms not sharply marked off from central disc; ambulacral grooves open, with tube feet on oral side; tube feet often with suckers; anus and madreporite aboral; pedicellariae present.

Examples: *Asterias, Pisaster* (p. 469). This group includes former members of **class Concentricycloidea** (kon-sen'tri-sy-kloy'de-a) (L. *cum,* together, + *centrum,* center [having a common center], + Gr. *kyklos,* circle, + *eidos,* form, + *ea,* characterized by): **sea daisies.** Disc-shaped body, with marginal spines but no arms; concentrically arranged skeletal plates; ring of suckerless podia near body margin; hydropore present; gut present or absent, no anus. Example: *Xyloplax* (Figure 22.13).

Class Ophiuroidea (o'fe-u-roy'de-a) (Gr. *ophis,* snake, + *oura,* tail, + *eidos,* form, + *ea,* characterized by): **brittle stars** and **basket stars.** Star-shaped, with arms sharply marked off from central disc; ambulacral grooves closed, covered by ossicles; tube feet without suckers and not used for locomotion; pedicellariae absent; anus absent. Examples: *Ophiura* (Figure 22.14A), *Gorgonocephalus* (Figure 22.17B).

Class Echinoidea (ek'i-noy'de-a) (Gr. *echinos,* sea urchin, hedgehog, + *eidos,* form, + *ea,* characterized by): **sea urchins, sea biscuits,** and **sand dollars.** More or less globular or disc-shaped, with no arms; compact skeleton or test with closely fitting plates; movable spines; ambulacral grooves closed; tube feet with suckers; pedicellariae present. Examples: *Arbacia, Strongylocentrotus* (Figure 22.19), *Lytechinus, Mellita.*

Class Holothuroidea (hol'o-thu-roy'de-a) (Gr. *holothourion,* sea cucumber, + *eidos,* form, + *ea,* characterized by): **sea cucumbers.** Cucumber-shaped, with no arms; spines absent; microscopic ossicles embedded in thick muscular wall; anus present; ambulacral grooves closed; tube feet with suckers; circumoral tentacles (modified tube feet); pedicellariae absent; madreporite internal. Examples: *Sclerodactyla, Parastichopus, Cucumaria* (Figure 22.24C).

---

groups were derived several times independently from free-moving ancestors. However, this view does not consider the adaptive significance of radial symmetry for a sessile existence.

The more traditional view is that the first echinoderms were sessile, became radial as an adaptation to that existence, and then gave rise to the free-moving groups. Figure 22.3 is consistent with this hypothesis. It depicts endoskeletal plates with stereom structure and the presence of external ciliary grooves for feeding as early echinoderm (or pre-echinoderm) developments. Extinct carpoids (Figures 22.30A) had stereom ossicles but were not radially symmetrical, and the status of their water-vascular system, if any, is uncertain. Some investigators regard carpoids as a separate subphylum of echinoderms (Homalozoa) and consider them closer to chordates. Fossil helicoplacoids (Figures 22.30B) show evidence of three, true ambulacral grooves, and their mouth was on the side of their body. We show both groups as early diverging echinoderms (Figure 22.3).

Attachment to a substratum by their aboral surface would have selected for radial symmetry, explaining the origin of subphylum Pelmatozoa whose living members are crinoids. Both Cystoidea (extinct) and Crinoidea were primitively attached to a substratum by an aboral stalk. An ancestor that became

free-moving and applied its oral surface to the substratum would have given rise to subphylum Eleutherozoa.

Phylogeny within Eleutherozoa is controversial. Most investigators agree that echinoids and holothuroids form a clade, but opinions diverge on the relationship of ophiuroids and asteroids. Figure 22.3 illustrates the view that ophiuroids arose after closure of their ambulacral grooves, but this scheme treats evolution of five ambulacral rays (arms) in ophiuroids and asteroids as independently evolved. Alternatively, if ophiuroids and asteroids are a single clade, then closed ambulacral grooves must have been lost in the asteroids, or evolved separately in ophiuroids and in the common ancestor of echinoids and holothuroids.

Adaptive Diversification

Diversification of echinoderms has been limited by their most important characters: radial symmetry, water-vascular system, and dermal endoskeleton. If their ancestors had a brain and specialized sense organs, these were lost in adoption of radial symmetry. Only now are gene expression studies beginning to help researchers identify structures such as the anterior-posterior axis in adult echinoderms. The best evidence currently available suggests that

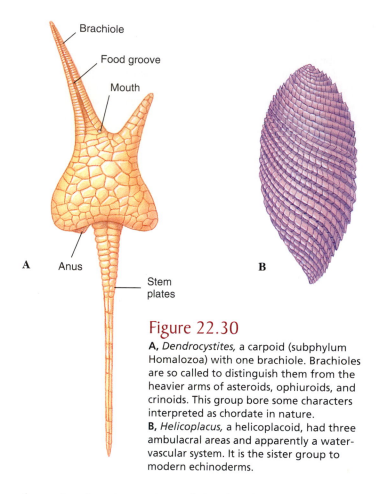

Figure 22.30

A, *Dendrocystites,* a carpoid (subphylum Homalozoa) with one brachiole. Brachioles are so called to distinguish them from the heavier arms of asteroids, ophiuroids, and crinoids. This group bore some characters interpreted as chordate in nature.
B, *Helicoplacus,* a helicoplacoid, had three ambulacral areas and apparently a water-vascular system. It is the sister group to modern echinoderms.

Characteristics of Phylum Hemichordata

1. Body divided into **proboscis, collar,** and **trunk; buccal diverticulum** in posterior part of proboscis
2. Enteropneusta free-moving and of burrowing habits; pterobranchs sessile, mostly colonial, living in secreted tubes
3. Free-living
4. Bilaterally symmetrical, soft bodied; wormlike or short and compact with stalk for attachment
5. Triploblastic
6. Single coelomic pouch in proboscis, but paired pouches in collar and trunk
7. Ciliated epidermis
8. Digestive system complete
9. Longitudinal and circular muscles in body wall in some
10. A subepidermal nerve plexus thickened to form dorsal and ventral nerve cords, with a ring connective in the collar; some species with hollow **dorsal nerve cord**
11. Sensory neurons in proboscis likely function in chemoreception
12. Colonies form by asexual budding in pterobranchs; asexual reproduction by fragmentation in enteropneusts
13. Sexes separate in Enteropneusta, with gonads projecting into body cavity; tornaria larva in some Enteropneusta
14. A single **glomerulus** connected to blood vessels may have excretory function and is considered a metanephridium
15. Respiratory system of **gill slits** (few or none in pterobranchs) connecting the pharynx with outside
16. Circulatory system of dorsal and ventral vessels and dorsal heart

---

the oral surface is anterior and the aboral surface is posterior. Under this hypothesis, the arms represent lateral growth zones.

Echinoderms have diversified within the benthic habitat. There are large numbers of creeping forms with filter-feeding, deposit-feeding, scavenging, and herbivorous habits, and very rare pelagic forms. In this light the relative success of asteroids as predators is impressive and probably attributable to the extent to which they have exploited the hydraulic mechanism of their tube feet.

The basic body plan of echinoderms appears to have severely limited their evolutionary opportunities to become parasites. Indeed, the most mobile of echinoderms, ophiuroids, which are also the ones most able to insert their bodies into small spaces, are the only group with significant numbers of commensal species.

PHYLUM HEMICHORDATA

Hemichordata (hem'i-kor-da'ta) (Gr. *hemi,* half, + *chorda,* string, cord) are marine animals formerly considered a subphylum of chordates, based on their possession of gill slits and a rudimentary notochord. However, the so-called hemichordate notochord is really a buccal diverticulum (called a **stomochord,** meaning "mouth-cord") and not homologous to a chordate notochord, so hemichordates are ranked as a separate phylum.

Hemichordates are vermiform bottom dwellers, living usually in shallow waters. Some colonial species live in secreted

tubes. Most are sedentary or sessile. Their distribution is almost cosmopolitan, but their secretive habits and fragile bodies make collecting them difficult.

There are two classes. Members of class Enteropneusta (Gr. *enteron,* intestine, + *pneustikos,* of, or for, breathing) (acorn worms) range from 20 mm to 2.5 m in length. Members of class Pterobranchia (Gr. *pteron,* wing, + *branchia,* gills) are smaller, usually 1 to 12 mm, not including their stalk. About 75 species of enteropneusts and three small genera of pterobranchs are recognized.

Hemichordates have a tripartite coelom.

Class Enteropneusta

Enteropneusts, or acorn worms, are sluggish, wormlike animals that live in burrows or under stones, usually in mud or sand flats of intertidal zones. *Balanoglossus* (Gr. *balanos,* acorn, + *glōssa,* tongue) and *Saccoglossus* (Gr. *sakkos,* sac, strainer, + *glōssa,* tongue) (Figure 22.31) are common genera.

Form and Function

Their mucus-covered body is divided into three distinct regions: a tonguelike proboscis, a short collar, and a long trunk (protosome, mesosome, and metasome).

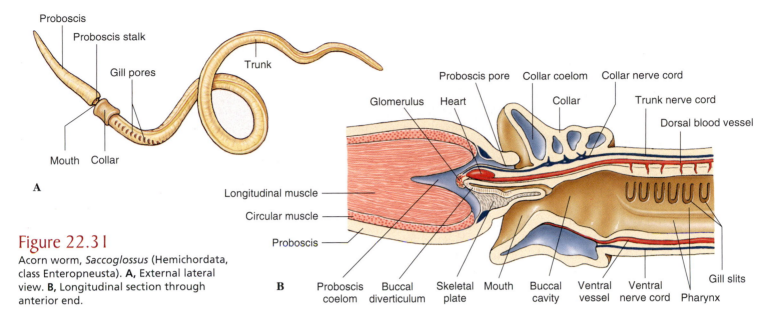

Figure 22.31

Acorn worm, *Saccoglossus* (Hemichordata, class Enteropneusta). **A,** External lateral view. **B,** Longitudinal section through anterior end.

Proboscis   Their proboscis is the active part of the animal. It probes its surroundings, collecting food in mucus strands on its surface. Cilia carry particles to the groove at the edge of the collar, direct them to the mouth on the underside, and then the particles are swallowed. Large particles can be rejected by covering the mouth with the edge of the collar (Figure 22.32).

Burrow dwellers use their proboscis to excavate, thrusting it into mud or sand and allowing cilia and mucus to move sand backward. They also may ingest sand or mud as they go, extracting its organic contents. They build U-shaped, mucus-lined burrows, usually with two openings 10 to 30 cm apart and with the base of the U 50 to 75 cm below the surface. They can thrust

their proboscis out the front opening for feeding. Defecation at the back opening builds characteristic spiral mounds of feces that leave a telltale clue to locations of burrows.

In the posterior end of their proboscis is a small coelomic sac (protocoel) into which extends a **buccal diverticulum,** a slender, blindly ending pouch of the gut that reaches forward into the buccal region and was formerly considered a notochord. A slender canal connects the protocoel with a **proboscis pore** to the outside (Figure 22.31B). Paired coelomic cavities in the collar also open by pores. By pulling water through these pores into the coelomic sacs, their proboscis and collar can be stiffened to aid in burrowing. Contraction of body musculature then forces excess water out through the gill slits, reducing the hydrostatic pressure and allowing the animal to move forward.

Branchial System   A row of **gill pores** is located dorsolaterally on each side of the trunk just behind the collar (Figure 22.32A). Pores open from a series of gill chambers that in turn connect with a series of U-shaped **gill slits** in the sides of the pharynx (Figure 22.31B). There are no gills on the gill slits, but some respiratory gaseous exchange occurs in the vascular branchial epithelium, as well as in the body surface. Ciliary currents keep a fresh supply of water moving from their mouth through the pharynx and out the gill slits and branchial chambers to the outside.

Feeding and the Digestive System   Hemichordates are largely ciliary-mucus feeders. Food is caught in mucus on the proboscis and collar. Cilia move the food to the ventral part of the pharynx and esophagus, then to the intestine where digestion and absorption occur.

Circulatory and Excretory Systems   A middorsal vessel carries the colorless blood forward above the gut. In the collar the vessel expands into a sinus and a heart vesicle above the buccal diverticulum. Blood then enters a network of blood sinuses called the **glomerulus,** which partially surrounds these

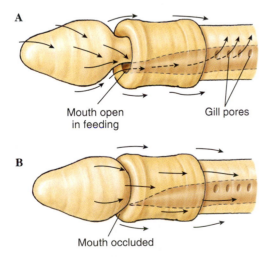

Figure 22.32

Food currents of enteropneust hemichordate. **A,** Side view of acorn worm with mouth open, showing direction of currents created by cilia on proboscis and collar. Food particles are directed toward mouth and digestive tract. Rejected particles move toward outside of collar. Water leaves through gill pores. **B,** When mouth is occluded, all particles are rejected and passed onto the collar. Nonburrowing and some burrowing hemichordates use this feeding method.

structures (Figure 22.31B). The glomerulus has an excretory function and is homologous to the echinoderm axial complex. Blood travels posteriorly through a ventral vessel below the gut, passing through extensive sinuses to the gut and body wall.

Nervous and Sensory Systems   Their nervous system consists mostly of a subepithelial network, or plexus, of nerve cells and fibers to which processes of epithelial cells are attached. Thickenings of this net form dorsal and ventral nerve cords that are united posterior to the collar by a ring connective. The dorsal cord continues into the collar and furnishes many fibers to the plexus of the proboscis. The dorsal nerve cord **(neurochord)** is formed by an invagination of the ectoderm and is hollow in some species. This striking similarity to the pattern in chordates is taken as evidence for homology with the dorsal nerve cord of the Chordata, but not all researchers accept this interpretation. The neurochord contains giant nerve cells with processes running to nerve trunks. This nerve plexus system is quite reminiscent of that of cnidarians and echinoderms.

Sensory receptors include neurosensory cells throughout the epidermis (especially in the proboscis, a preoral ciliary organ that may be chemoreceptive) and photoreceptor cells.

Reproductive System and Development   Sexes are separate in enteropneusts. Although most species reproduce only sexually, at least one species undergoes asexual reproduction. A dorsolateral row of gonads runs along each side of the anterior trunk. Fertilization is external, and in some species a ciliated **tornaria** larva develops and at certain stages is so similar to an echinoderm bipinnaria that it was once considered an echinoderm larva (Figure 22.33). The familiar *Saccoglossus* of American waters has direct development without a tornaria stage.

Class Pterobranchia

The basic plan of class Pterobranchia is similar to that of Enteropneusta, but certain structural differences are correlated with the sedentary lifestyle of pterobranchs. The first pterobranch ever reported was obtained by the famed *Challenger* expedition

of 1872 to 1876. Although first placed among Polyzoa (Entoprocta and Ectoprocta), its affinities to hemichordates were later recognized. Only three genera (*Atubaria, Cephalodiscus,* and *Rhabdopleura*) are known.

Pterobranchs are small animals, usually within a range of 1 to 7 mm in length, although the stalk may be longer. Many individuals of *Cephalodiscus* (Gr. *kephalē*, head, + *diskos*, disc) (Figure 22.34) live together in collagenous tubes, which often form an anastomosing system. Zooids are not connected, however, and live independently in the tubes. Through apertures in these tubes, they extend their crown of tentacles. They are attached to the tube walls by extensible stalks that can retract the owners back into the tubes when necessary.

The body of *Cephalodiscus* is divided into the three regions—proboscis, collar, and trunk—characteristic of hemichordates. There is only one pair of gill slits, and the alimentary canal is U-shaped, with their anus near their mouth. The proboscis is shield-shaped. At the proboscis base are five to nine pairs of branching arms with tentacles containing an extension of the coelomic compartment of the mesosome, as in a lophophore. Ciliated grooves on the tentacles and arms collect food. Some species are dioecious, and others are monoecious. Asexual reproduction by budding also may occur.

In *Rhabdopleura* (Gr. *rhabdos*, rod, + *pleura*, a rib, the side), which is smaller than *Cephalodiscus,* individuals remain together to form a colony of zooids connected by a stolon and enclosed in secreted tubes (Figure 22.35). The collar in these forms bears two branching arms. No gill slits or glomeruli are present. New individuals are produced by budding from a creeping basal stolon, which branches on a substratum. No pterobranch has a tubular nerve cord in the collar, but otherwise their nervous system is similar to that of Enteropneusta.

Fossil graptolites of the middle Paleozoic era often are placed as an extinct class under Hemichordata. They are important index fossils of Ordovician and Silurian geological strata. Alignment of graptolites with hemichordates has been very controversial, but discovery of an organism that seems to be a living graptolite lends strong support to the hypothesis. It has been described as a new species of pterobranch, called *Cephalodiscus graptolitoides.*

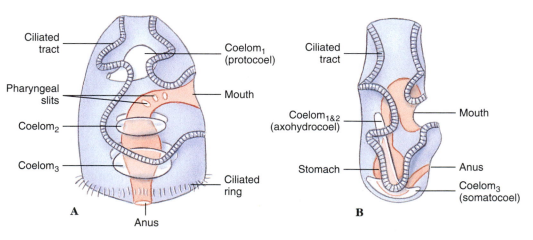

Figure 22.33
Comparison of a hemichordate tornaria **(A)** to an echinoderm bipinnaria **(B)**.

Figure 22.34

Cephalodiscus, a pterobranch hemichordate. These tiny (5 to 7 mm) forms live in tubes in which they can move freely. Ciliated tentacles and arms direct currents of food and water toward mouth.

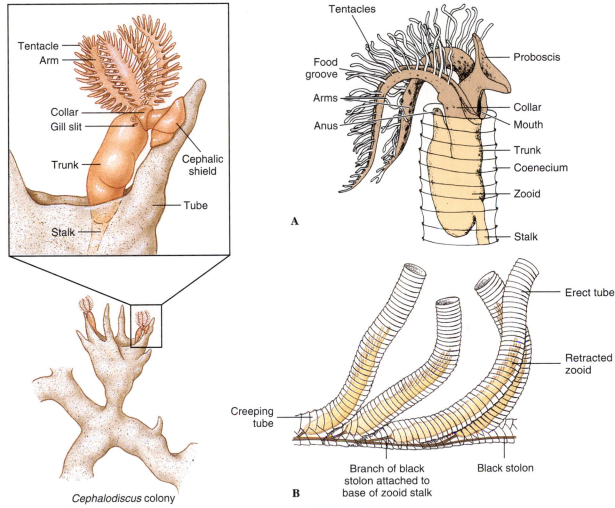

Cephalodiscus colony

Figure 22.35

A, *Rhabdopleura,* a pterobranch hemichordate in its tube. Individuals live in branching tubes connected by stolons, and protrude ciliated tentacles for feeding. **B,** Portion of a colony.

PHYLOGENY AND ADAPTIVE DIVERSIFICATION

Phylogeny

Hemichordate phylogeny has long been puzzling. Hemichordates share characters with both echinoderms and chordates. With chordates they share pharyngeal slits. If hemichordates are the sister taxon to echinoderms, as described by the Ambulacraria hypothesis (see Figure 22.1), then gill slits are an ancestral feature for deuterostomes. Gill slits are presumed lost in a lineage ancestral to all echinoderms, although some researchers find evidence for gill slits in the extinct carpoid echinoderms. Thus, the loss of gill slits occurred before the lineages with extant members branched from those of other echinoderms (Figure 22.3).

The Ambulacraria hypothesis unites echinoderms and hemichordates on the basis of a shared diffuse epidermal nervous system. It is therefore unlikely that the short dorsal, somewhat hollow, nerve cord in the collar zone of hemichordates is homologous to the dorsal hollow nerve cord of chordates.

Another phylogenetically important character is the shared tripartite coelom in hemichordates and echinoderms. This feature is now assumed to characterize members of Ambulacraria. Reexamination of development of the coelomic compartments in lophophorates and chaetognaths indicates that compartments in these animals are not homologous to those of Ambulacraria. The buccal diverticulum in the hemichordate mouth cavity, long thought homologous to the notochord of chordates, is now considered a synapomorphy of hemichordates themselves.

Early embryogenesis of hemichordates is remarkably like that of echinoderms, and early tornaria larvae are almost identical to bipinnaria larvae of asteroids, suggesting that echinoderms form the sister group of hemichordates (see Figure 22.1). Sequence analysis of 18S rDNA suggests that Enteropneusta is not a monophyletic group, and that hemichordates are the sister group to echinoderms.* This, and other more recent, analyses place chordates as sister taxa to the echinoderm and hemichordate clade within the Deuterostomia.

Adaptive Diversification

It was long thought that because of their sessile lives and their habitat in secreted tubes in ocean bottoms, where conditions are fairly stable, pterobranchs had undergone little adaptive

*(Cameron et al., 2000).

divergence from the most recent common ancestor of hemi-chordates. They have a tentacular type of ciliary feeding which was presumed ancestral for hemichordates. Enteropneusts, on the other hand, although sluggish, are more active than ptero-branchs. Lacking tentaculated arms, they use a proboscis to trap small organisms in mucus, or they eat sand as they burrow and digest organic sediments from the sand in what was presumed to be a derived mode of feeding. However, recent molecular evidence suggests that pterobranchs are derived from within the enteropneust lineage, making this scenario unlikely.

Some authorities also recognize the monotypic class Planc-tosphaeroidea. However, most invertebrate zoologists agree that this animal is likely a hemichordate larva, despite the fact that it cannot be currently linked to any specific adult.

SUMMARY

Arrow worms (phylum Chaetognatha) are a small group but an impor-tant component of marine plankton. They have a well-developed coe-lom and are effective predators, catching other planktonic organisms with the teeth and chitinous spines around their mouth.

Phylum Echinodermata shows characteristics of the Deuterosto-mia division of the animal kingdom. They are an important marine group sharply distinguished from other phyla of animals. They have pentaradial symmetry but were derived from bilateral ancestors.

Sea stars (class Asteroidea) can be used to illustrate echino-derms. Sea stars usually have five arms, which merge gradually with a central disc. Like other echinoderms, they have no head and few specialized sensory organs. Their mouth is directed toward the sub-stratum. They have stereom dermal ossicles, respiratory papulae, and open ambulacral grooves. Many sea stars have pedicellariae. Their water-vascular system is an elaborate hydraulic system derived embryonically from one of their coelomic compartments. Along the ambulacral areas, branches of the water-vascular system (tube feet) are important in locomotion, food gathering, respiration, and excre-tion. Many sea stars are predators, whereas others feed on small particles. Sexes are separate, and reproductive systems are simple. The bilateral, free-swimming larva becomes attached, transforms to a radial juvenile, then detaches and becomes a motile sea star.

Arms of brittle stars (class Ophiuroidea) are slender and sharply set off from the central disc. Ophiuroids have no pedicellariae or ampullae and their ambulacral grooves are closed. Their tube feet have no suckers, and their madreporite is on the oral side. They crawl by means of arm movements, and their tube feet function in food gathering.

Dermal ossicles of most sea urchins (class Echinoidea) are closely fitting plates, the body is compact, and there are no arms. Ambulacral areas are closed and extend around their body toward the aboral pole. Sea urchins move by means of tube feet or by their spines. Some urchins (sand dollars and heart urchins) have returned to adult bilateral symmetry.

Dermal ossicles in sea cucumbers (class Holothuroidea) are very small; therefore the body wall is soft. Their ambulacral areas also are closed and extend toward the aboral pole. Holothuroids are greatly elongated in the oral-aboral axis and lie on their side. Because cer-tain ambulacral areas are characteristically against the substratum, sea cucumbers have also undergone some return to bilateral sym-metry. Tube feet around the mouth are modified into tentacles, with which they feed. They have an internal respiratory tree, and their madreporite hangs free in the coelom.

Sea lilies and feather stars (class Crinoidea) are the only group of living echinoderms, other than asteroids, with open ambulacral grooves. They are mucociliary particle feeders and lie with their oral side up.

Sea daisies (former class Concentricycloidea) are an enigmatic group now placed within class Asteroidea. They are circular in shape, have marginal tube feet, and two concentric ring canals in their water-vascular system.

Ancestors of echinoderms were probably bilaterally symmetri-cal, and they probably evolved through a sessile stage that became radially symmetrical and then gave rise to free-moving forms.

Members of phylum Hemichordata are marine wormlike forms formerly considered chordates because their buccal diverticulum was thought to be homologous to a notochord. Like chordates, most have paired gill slits. Divisions of their body (proboscis, collar, trunk) contain typical deuterostome coelomic compartments (proto-coel, mesocoel, metacoel). Hemichordate class Enteropneusta con-tains burrowing worms that capture food using mucus and cilia on their proboscis. Members of class Pterobranchia are tube dwellers that also use mucus and cilia to feed, but they capture food on ten-tacles. Hemichordates are important phylogenetically because they show affinities with chordates and echinoderms. Together with the echinoderms, they form clade Ambulacraria. Recent work suggests that Ambulacraria, together with the odd wormlike *Xenoturbella*, forms the sister group of chordates.

REVIEW QUESTIONS

1. What constellation of characteristics possessed by echinoderms is found in no other phylum?
2. How do we know that echinoderms were derived from an ancestor with bilateral symmetry?
3. Distinguish the following groups of echinoderms from each other: Crinoidea, Asteroidea, Ophiuroidea, Echinoidea, Holothuroidea.
4. What is an ambulacrum, and what is the difference between open and closed ambulacral grooves?
5. Trace or make a rough copy of Figure 22.6B without labels; then from memory label the parts of the water-vascular system of sea stars.
6. Briefly explain the mechanism of action of a sea star's tube foot.
7. What structures are involved in the following functions in sea stars? Briefly describe the action of each: respiration, feeding and digestion, excretion, reproduction.

8. Compare the structures and functions in question 7 as they are found in brittle stars, sea urchins, sea cucumbers, and crinoids.
9. Briefly describe development in sea stars, including metamorphosis.
10. Match groups in the left column with *all* correct answers in the right column.

    _____ Crinoidea   a. Closed ambulacral grooves
    _____ Asteroidea   b. Oral surface generally upward
    _____ Ophiuroidea   c. With arms
    _____ Echinoidea   d. Without arms
    _____ Holothuroidea   e. Approximately globular or disc-shaped
                                      f. Elongated in oral-aboral axis
                                      g. With pedicellariae
                                      h. Madreporite internal
                                      i. Madreporite on oral plate

11. Define the following: pedicellariae, madreporite, respiratory tree, Aristotle's lantern, papulae, Cuvierian tubules.
12. What evidence suggests that ancestral echinoderms were sessile?
13. Give four examples of how echinoderms are important to humans.
14. What is a major difference in function of the coelom in holothurians compared with other echinoderms?
15. Describe a reason for the hypothesis that the ancestor of eleutherozoan groups was a radial, sessile organism.
16. What characteristics do Hemichordata share with Echinodermata, and how do the two phyla differ?
17. Distinguish Enteropneusta from Pterobranchia.
18. How do chaetognaths feed?

SELECTED REFERENCES

Aizenberg, J., A. Tkachenkoo, S. Weiner, L. Addadi, and G. Hendler. 2001. Calcitic microlenses as part of the photoreceptor system in brittle stars. Nature **412**:819–822. *Tiny bumps on stereom ossicles in arms serve as microlenses to focus light on nerve photoreceptors.*

Ball, E. E., and D. J. Miller. 2006. Phylogeny: the continuing classificatory conundrum of chaetognaths. Curr. Biol. **16**:R593–R596. *Nice summary of chaetognath biology and two recent phylogenetic studies in the same issue.*

Baker, A. N., F. W. E. Rowe, and H. E. S. Clark. 1986. A new class of Echinodermata from New Zealand. Nature **321**:862–864. *Describes the strange sea daisies.*

Bourlat, S. J., T. Juliusdottir, C. J. Lowe, R. Freeman, J. Aronowicz, M. Kirschner, E. S. Lander, M. Thorndyke, H. Nakano, A. B. Kohn, A. Heyland, L. L. Moroz, R. R. Copley, and M. J. Telford. 2006. Deuterostome phylogeny reveals monophyletic chordates and the new phylum Xenoturbellida. Nature **444**:85–88. *The sister taxon to the chordates is Ambulacraria plus Xenoturbella; characters uniting clades within Deuterostomia are given.*

Bourlat, S. J., C. Nielsen, A. E. Lockyer, D. T. J. Littlewood, and M. J. Telford. 2003. *Xenoturbella* is a deuterostome that eats molluscs. Nature **424**:925–928. *Explanation for molluscan genetic material associated with* Xenoturbella *in some phylogenetic studies.*

Cameron, C. B., J. R. Garey, and B. J. Swalla. 2000. Evolution of the chordate body plan: new insights from phylogenetic analyses of deuterostome phyla. Proc. Nat. Acad. Sci. **97**(9):4469–4474. *Molecular sequence data suggests that enteropneusts are paraphyletic and pterobranchs have evolved from an enteropneust-like ancestor.*

Gilbert, S. F. 2006. Developmental biology, ed. 8. Sunderland, Massachusetts, Sinauer Associates. *Any modern text in developmental biology, such as this one, provides a multitude of examples in which studies on echinoderms have contributed (and continue to contribute) to our knowledge of development.*

Halanych, K. M. 1996. Testing hypothesis of chaetognath origins: long branches revealed by 18S ribosomal DNA. Syst. Biol. **45**:223–246. *Analysis suggests chaetognaths are most closely related to nematodes.*

Hendler, G., J. E. Miller, D. L. Pawson, and P. M. Kier. 1995. Sea stars, sea urchins, and allies: Echinoderms of Florida and the Caribbean. Washington, Smithsonian Institution Press. *An excellent field guide for echinoderm identification.*

Hickman, C. P., Jr. 1998. A field guide to sea stars and other echinoderms of Galápagos. Lexington, VA, Sugar Spring Press. *Provides descriptions and nice photographs of members of classes Asteroidea, Ophiuroidea, Echinoidea, and Holothuroidea in the Galápagos Islands.*

Hughes, T. P. 1994. Catastrophes, phase shifts, and large-scale degradation of a Caribbean coral reef. Science **265**:1547–1551. *Describes the sequence of events, including the die-off of sea urchins, leading to the destruction of the coral reefs around Jamaica.*

Israelsson, O., and G. E. Budd. 2005. Eggs and embryos in *Xenoturbella* (phylum uncertain) are not ingested prey. Dev. Genes Evol. **215**:358–363. *Eggs inside* Xenoturbella *were produced in situ and are not prey items.*

Lane, D. J. W. 1996. A crown-of-thorns outbreak in the eastern Indonesian Archipelago, February 1996. Coral Reefs **15**:209–210. *This is the first report of an outbreak of* Acanthaster planci *in Indonesia. It includes a good photograph of an aggregation of these sea stars.*

Mooi, R., and B. David. 1998. Evolution within a bizarre phylum: homologies of the first echinoderms. Am. Zool. **38**:965–974. *These authors contend that "The familiarity of a sea star or a sea urchin belies their overall weirdness." They describe the Etraxial/Axial Theory (EAT) of echinoderm skeletal homologies.*

Smith, A. B., K. J. Peterson, G. Wray, and D. T. J. Littlewood. 2003. From bilateral symmetry to pentaradiality: the phylogeny of hemichordates and echinoderms, pp. 365–383. In J. Cracraft and M. J. Donohue, eds. Assembling the tree of life. New York, Oxford University Press. *An excellent discussion of recent changes in deuterostome phylogeny.*

Svitii, K. W. 1993. It's alive, and it's a graptolite. Discover **14**(7):18–19. *Short account of the discovery of a "living fossil."* Cephalodiscus graptolitoides.

Woodley, J. D., P. M. H. Gayle, and N. Judd. 1999. Sea-urchins exert top-down control of macroalgae on Jamaican coral reefs (2). Coral Reefs **18**:193. *In areas where* Tripneustes *(another sea urchin) have invaded fore reefs there is much less macroalgae, and such areas present a better chance that corals can recolonize. Diadema recovery has been slow.*

Wray, G. A., and R. A. Raff. 1998. Body builders of the sea. Nat. Hist. **107**:38–47. *Regulatory genes in bilateral animals have assumed new but analogous roles in radial echinoderms.*

ONLINE LEARNING CENTER

Visit www.mhhe.com/hickmanipz14e for chapter quizzing, key term flash cards, web links, and more!

23

Two amphioxus in feeding posture.

Chordates

- GENERAL CHARACTERISTICS, PROTOCHORDATES, AND ORIGIN OF THE EARLY VERTEBRATES

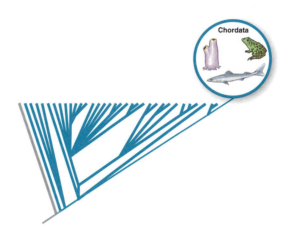

It's a Long Way From Amphioxus

Along the more southern coasts of North America, half buried in sand on the seafloor, lives a small fishlike translucent animal quietly filtering organic particles from seawater. Inconspicuous, of no commercial value and largely unknown, this creature is nonetheless one of the famous animals of classical zoology. It is amphioxus, an animal that wonderfully exhibits the five distinctive hallmarks of the phylum Chordata—(1) dorsal, tubular nerve cord overlying (2) a supportive notochord, (3) pharyngeal slits and (4) endostyle for filter-feeding, and (5) a postanal tail for propulsion—all wrapped up in one creature with textbook simplicity. Amphioxus is an animal that might have been designed by a zoologist for the classroom. During the nineteenth century, with interest in vertebrate ancestry running high, amphioxus was considered by many to resemble closely the direct ancestor of vertebrates. Its exalted position was later acknowledged by Philip Pope in a poem sung to the tune of "Tipperary." It ends with the refrain:

> *It's a long way from amphioxus*
> *It's a long way to us.*
> *It's a long way from amphioxus*
> *To the meanest human cuss.*
> *Well, it's good-bye to fins and gill slits*
> *And its welcome lungs and hair,*
> *It's a long, long way from amphioxus*
> *But we all came from there.*

But amphioxus' place in the sun was not to endure. For one thing, amphioxus lacks one of the most important of vertebrate characteristics, a distinct head with special sense organs and the equipment for shifting to an active predatory mode of life. Absence of a head, together with several specialized features, suggests to zoologists today that amphioxus represents an early departure from vertebrate ancestry. It seems that we are a very long way indeed from amphioxus. Nevertheless, while amphioxus is denied the vertebrate ancestral award, it resembles the chordate condition immediately preceding the origin of vertebrates more closely than any other living animal we know.

THE CHORDATES

Animals most familiar to most people belong to phylum Chordata (kor-da'ta) (L. *chorda,* cord). Humans are members and share the characteristic from which the phylum derives its name—the **notochord** (Gr. *nōton,* back, + L. *chorda,* cord) (Figure 23.1). All members of the phylum possess this structure, either restricted to early development or present throughout life. The notochord is a rodlike, semirigid body of cells enclosed by a fibrous sheath, which extends, in most cases, the length of the body just ventral to the central nervous system. Its primary purpose is to stiffen the body, providing skeletal scaffolding for the attachment of swimming muscles.

The structural plan of chordates shares features of many nonchordate invertebrates, such as bilateral symmetry, anteroposterior axis, coelom, tube-within-a-tube arrangement, metamerism, and cephalization. However, the exact phylogenetic position of chordates within the animal kingdom is unclear.

Two possible lines of descent have been proposed. Earlier speculations that focused on the arthropod-annelid-mollusc group (Protostomia branch) of the invertebrates have fallen from favor. It is now believed that only members of the echinoderm-hemichordate assemblage (Deuterostomia branch) deserve serious consideration as the chordate sister group. Chordates share with other deuterostomes several important characteristics: radial cleavage (p. 166), anus derived from the first embryonic opening (blastopore) and mouth derived from an opening of secondary

Characteristics of Phylum Chordata

1. Bilateral symmetry; segmented body; three germ layers; well-developed coelom
2. **Notochord** (a skeletal rod) present at some stage in the life cycle
3. **Single, dorsal, tubular nerve cord;** anterior end of cord usually enlarged to form brain
4. **Pharyngeal pouches** present at some stage in the life cycle; in aquatic chordates these develop into pharyngeal slits
5. **Endostyle** in floor of pharynx or a **thyroid gland** derived from the endostyle
6. **Postanal tail** projecting beyond the anus at some stage but may or may not persist
7. Complete digestive system
8. **Segmentation,** if present, restricted to outer body wall, head, and tail and not extending into coelom

origin, and a coelom primitively formed by fusion of enterocoelous pouches (although in most vertebrates coelom formation is schizocoelus, but independently derived, as an accommodation for their large yolks). These uniquely shared characteristics indicate a natural unity among the Deuterostomia.

Phylum Chordata shows a more fundamental unity of organs and organ systems than do other phyla. Ecologically, chordates are among the most adaptable of organic forms and are able to occupy most kinds of habitat. They illustrate perhaps better than any other animal group the basic evolutionary processes of origin of new structures, adaptive strategies, and adaptive radiation.

Traditional and Cladistic Classification of the Chordates

Traditional Linnaean classification of chordates (p. 509) provides a convenient way to indicate the taxa included in each major group. However, in cladistic usage, some of the traditional taxa, such as Agnatha and Reptilia, are no longer recognized. Such taxa do not satisfy the requirement of cladistics that only **monophyletic** groups are valid taxonomic entities, that is, groups that contain all known descendants of a single common ancestor. The reptiles, for example, are considered a **paraphyletic** grouping because this group does not contain all of the descendants of their most recent common ancestor (p. 568). The common ancestor of reptiles as traditionally recognized is also the ancestor of birds. As shown in the cladogram (see Figure 23.3), reptiles, birds, and mammals compose a monophyletic group called Amniota, so named because all develop from an egg having special extraembryonic membranes, one of which is the amnion. Therefore according to cladistics, reptiles can only be used as a term of convenience to refer to amniotes that are not birds or mammals; there are no derived characters that unite reptiles to the exclusion of birds and mammals. The reasons why nonmonophyletic groups are not used in cladistic taxonomy are explained in Chapter 10 (p. 209).

The phylogenetic tree of chordates (Figure 23.2) and the cladogram of chordates (Figure 23.3) provide different kinds of

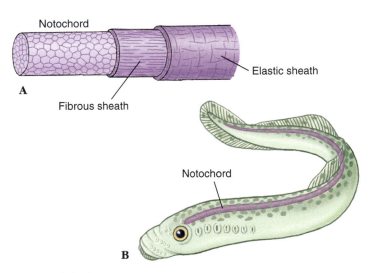

Figure 23.1

A, Structure of the notochord and its surrounding sheaths. Cells of the notochord proper are thick walled, pressed together closely, and filled with semifluid. Stiffness is caused mainly by turgidity of fluid-filled cells and surrounding connective tissue sheaths. This primitive type of endoskeleton is characteristic of all chordates at some stage of the life cycle. The notochord provides longitudinal stiffening of the main body axis, a base for trunk muscles, and an axis around which the vertebral column develops. **B,** In hagfishes and lampreys it persists throughout life, but in other vertebrates it is largely replaced by vertebrae. In mammals, remnants are found in nuclei pulposi of intervertebral discs. The method of notochord formation is different in the various groups of animals. In amphioxus it originates from endoderm; in birds and mammals it arises as an anterior outgrowth of the embryonic primitive streak.

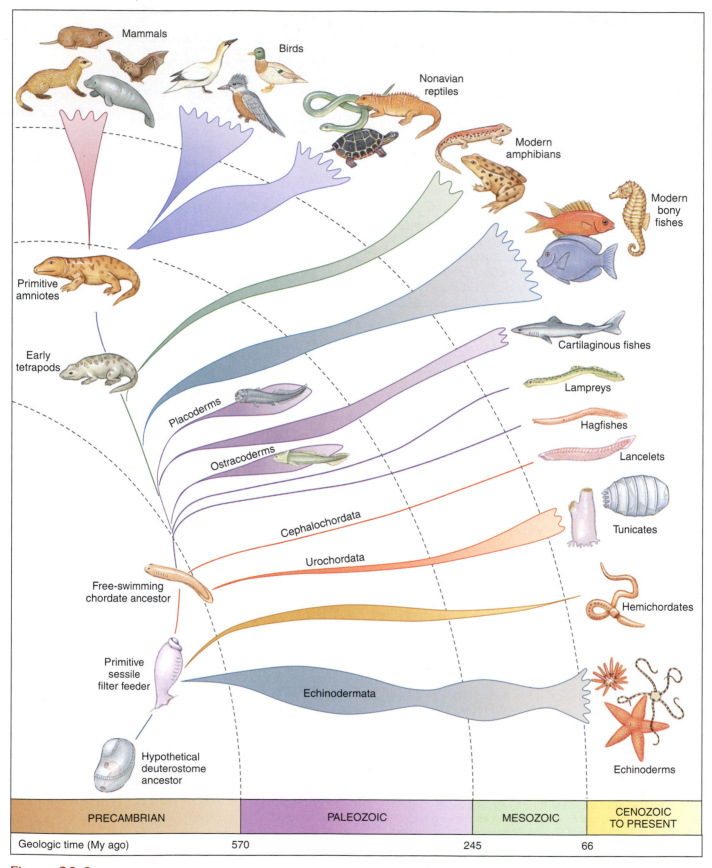

Figure 23.2

Phylogenetic tree of the chordates, suggesting probable origin and relationships. Other schemes have been suggested and are possible. The relative abundance in numbers of species of each group through geological time, as indicated by the fossil record, is suggested by the bulging and thinning of that group's line of descent.

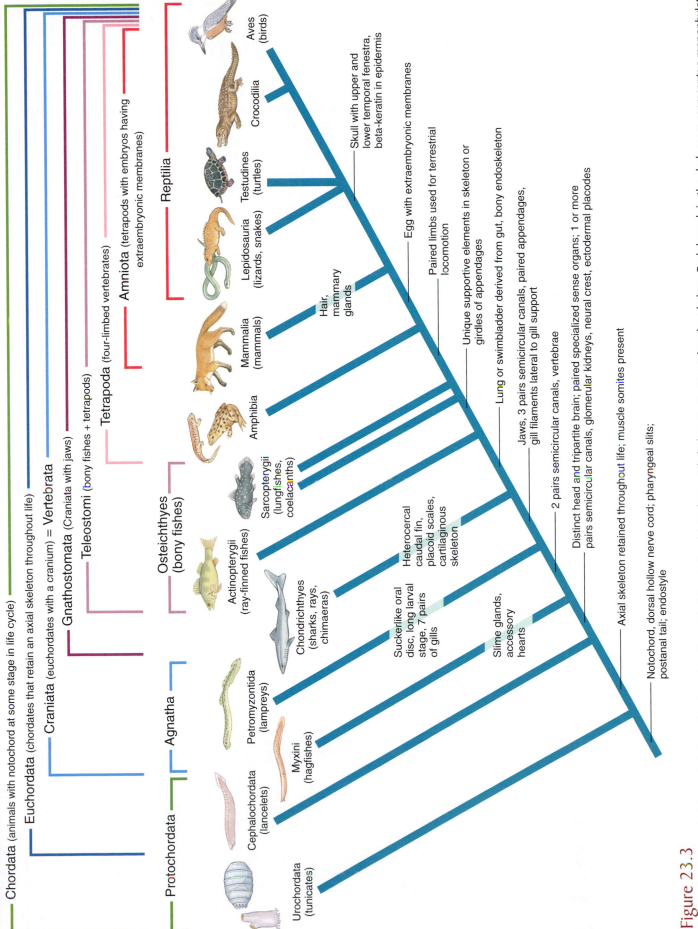

Figure 23.3

Cladogram of living members of phylum Chordata showing probable relationships of monophyletic groups composing the phylum. Each branch in the cladogram represents a monophyletic group. Some derived character states that identify the branchings are shown at right of the branch points. Nesting brackets across the top of the cladogram identify monophyletic groupings within the phylum. The term Craniata, although commonly equated with Vertebrata, is preferred by many authorities because it recognizes that some jawless vertebrates have a cranium but no vertebrae. The lower set of brackets identify the traditional groupings Protochordata, Agnatha, Osteichthyes, and Reptilia. These paraphyletic groups are not recognized in cladistic treatments, but are shown because of widespread use.

Chordata (animals with notochord at some stage in life cycle)

Euchordata (chordates that retain an axial skeleton throughout life)

Craniata (euchordates with a cranium) = Vertebrata

Gnathostomata (Craniata with jaws)

Teleostomi (bony fishes + tetrapods)

Tetrapoda (four-limbed vertebrates)

Amniota (tetrapods with embryos having extraembryonic membranes)

Reptilia

Vertebrata (Craniata with jaws)

Protochordata

Agnatha

Osteichthyes (bony fishes)

Urochordata (tunicates)

Cephalochordata (lancelets)

Myxini (hagfishes)

Petromyzontida (lampreys)

Chondrichthyes (sharks, rays, chimaeras)

Actinopterygii (ray-finned fishes)

Sarcopterygii (lungfishes, coelacanths)

Amphibia

Mammalia (mammals)

Lepidosauria (lizards, snakes)

Testudines (turtles)

Crocodilia

Aves (birds)

Skull with upper and lower temporal fenestra, beta-keratin in epidermis

Egg with extraembryonic membranes

Paired limbs used for terrestrial locomotion

Unique supportive elements in skeleton or girdles of appendages

Hair, mammary glands

Lung or swimbladder derived from gut; bony endoskeleton

Jaws, 3 pairs semicircular canals, paired appendages, gill filaments lateral to gill support

2 pairs semicircular canals, vertebrae

Heterocercal caudal fin, placoid scales, cartilaginous skeleton

Distinct head and tripartite brain; paired specialized sense organs; 1 or more pairs semicircular canals, glomerular kidneys, neural crest, ectodermal placodes

Suckerlike oral disc, long larval stage, 7 pairs of gills

Slime glands, accessory hearts

Axial skeleton retained throughout life; muscle somites present

Notochord, dorsal hollow nerve cord; pharyngeal slits; postanal tail; endostyle

information. The cladogram shows a nested hierarchy of taxa grouped by their sharing of derived characters. These characters may be morphological, physiological, embryological, behavioral, chromosomal, or molecular in nature. By contrast, the branches of a phylogenetic tree are intended to represent real lineages that occurred in the evolutionary past. Geological information regarding ages of lineages is added to information from the cladogram to generate a phylogenetic tree for the same taxa.

In our treatment of chordates, we have retained the traditional Linnaean classification (p. 509) because subfields of zoology are organized according to this scheme and because the alternative—thorough revision following cladistic principles—would require extensive change and abandonment of familiar rankings. However, we have tried to use monophyletic taxa as much as possible, because such usage is necessary to reconstruct the evolution of morphological characters in chordates.

Several traditional divisions of phylum Chordata used in Linnaean classifications are shown in Table 23.1. A fundamental separation is Protochordata from Vertebrata. Since the former lack a well-developed head, they also are called Acraniata. All vertebrates have a well-developed skull enclosing the brain and are called Craniata. We should note that some cladistic classifications exclude Myxini (hagfishes) from the group Vertebrata because they lack vertebrae, although retaining them in Craniata since they do have a cranium. The vertebrates (craniates) may be variously subdivided into groups based on shared characteristics. Two such subdivisions shown in Table 23.1 are: (1) Agnatha, vertebrates lacking jaws (hagfishes and lampreys), and Gnathostomata, vertebrates having jaws (all other vertebrates), and (2) Amniota, vertebrates whose embryos develop within a fluid-filled sac, the amnion (reptiles, birds, and mammals), and Anamniota, vertebrates lacking this adaptation (fishes and amphibians). The Gnathostomata in turn can be subdivided into Pisces, jawed vertebrates with appendages, if any, in the form of fins; and Tetrapoda (Gr. *tetras,* four, + *podos,* foot), jawed vertebrates with appendages, if any, in the form of limbs. Note that several of these groupings are paraphyletic (Protochordata, Acraniata, Agnatha, Anamniota, Pisces) and consequently are not accepted in cladistic classifications. Accepted monophyletic taxa are shown at the top of the cladogram in Figure 23.3 as a nested hierarchy of increasingly more inclusive groupings.

FIVE CHORDATE HALLMARKS

Five distinctive characteristics that, taken together, set chordates apart from all other phyla are **notochord, dorsal tubular nerve cord, pharyngeal pouches** or **slits, endostyle,** and **postanal tail.** These characteristics are always found at some embryonic stage, although they may be altered or may disappear in later stages of the life cycle. All but pharyngeal pouches or slits are unique to chordates; hemichordates also have pharyngeal slits. A dorsal nerve cord is present in some hemichordates, but its homology to that of chordates is uncertain.

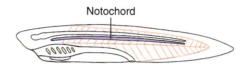

Notochord

The notochord is a flexible, rodlike structure, extending the length of the body. It is the first part of the endoskeleton to appear in the embryo. The notochord is an axis for muscle attachment, and because it can bend without shortening, it permits undulatory movements of the body. In most protochordates and in jawless vertebrates, the notochord persists throughout life (see Figure 23.1). In all vertebrates except hagfishes a series of cartilaginous or bony vertebrae are formed from mesenchymal cells derived from blocks of mesodermal cells (somites) lateral to the notochord. In most vertebrates, the notochord is displaced by vertebrae, although remains of the notochord may persist between or within the vertebrae.

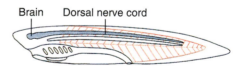

Dorsal Tubular Nerve Cord

In most invertebrate phyla that have a nerve cord, it is ventral to the alimentary canal and is solid, but in chordates the single cord is dorsal to the alimentary canal and is a tube (although the

TABLE 23.1

Traditional Divisions of the Phylum Chordata

Urochordata (tunicates)	Cephalo- chordata (lancelets)	Myxini (hagfishes)	Petromy- zontida (lampreys)	Chondrich- thyes (sharks)	Osteichthyes (bony fishes)	Amphibia (amphibians)	Reptilia (reptiles)	Aves (birds)	Mammalia (mammals)

←———————————————————————————————— Chordata ——————————————————————————————————→

←— Protochordata —→ ←——————————————————————— Vertebrata ———————————————————————→

←— Acraniata —→ ←———————————————————————— Craniata ————————————————————————→

←———— Agnatha ————→ ←——————————————— Gnathostomata ———————————————→

←——— Pisces ———→ | ←——————— Tetrapoda ———————→

←———————————— Anamniota ————————————→ | ←———— Amniota ————→

hollow center may be nearly obliterated during growth). The anterior end becomes enlarged to form the brain in vertebrates. The hollow cord is produced in the embryo by infolding of ectodermal cells on the dorsal side of the body above the notochord. Among vertebrates, the nerve cord passes through the protective neural arches of the vertebrae, and the brain is surrounded by a bony or cartilaginous cranium.

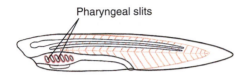

Pharyngeal slits

Pharyngeal Pouches and Slits

Pharyngeal slits are openings that lead from the pharyngeal cavity to the outside. They are formed by inpocketing of the outside ectoderm (pharyngeal grooves) and evagination, or outpocketing, of the endodermal lining of the pharynx (pharyngeal pouches). In aquatic chordates, the two pockets break through the pharyngeal cavity where they meet to form the pharyngeal slit. In amniotes some pockets may not break through the pharyngeal cavity and only grooves are formed instead of slits. In tetrapod (four-footed) vertebrates the pharyngeal pouches give rise to several different structures, including the Eustachian tube, middle ear cavity, tonsils, and parathyroid glands (see p. 181).

The perforated pharynx evolved as a filter-feeding apparatus and is used as such in protochordates. Water with suspended food particles is drawn by ciliary action through the mouth and flows out through pharyngeal slits where food is trapped in mucus. In vertebrates, ciliary action is replaced by a muscular pump that drives water through the pharynx by expanding and contracting the pharyngeal cavity. Also modified were the aortic arches that carry blood through the pharyngeal bars. In protochordates these are simple vessels surrounded by connective tissue. Early fishes added a capillary network having only thin, gas-permeable walls, thus improving efficiency of gas transfer between blood and the water outside. These adaptations led to the evolution of **internal gills,** completing conversion of the pharynx from a filter-feeding apparatus in protochordates to a respiratory organ in aquatic vertebrates.

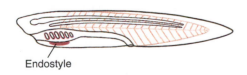

Endostyle

Endostyle or Thyroid Gland

Until recently, the endostyle was not recognized as a chordate character. However, it or its derivative, the thyroid gland, is found in all chordates, but in no other animals. The endostyle, in the pharyngeal floor, secretes mucus that traps small food particles brought into the pharyngeal cavity. An endostyle is found in protochordates and lamprey larvae. Some cells in the endostyle

secrete iodinated proteins. These cells are homologous with the iodinated-hormone-secreting thyroid gland of adult lampreys and the remainder of vertebrates. In primitive chordates, endostyle and perforated pharynx work together to create an efficient filter-feeding apparatus.

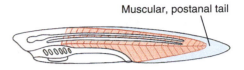

Muscular, postanal tail

Postanal Tail

The postanal tail, together with somatic musculature and the stiffening notochord, provides the motility that larval tunicates and amphioxus need for their free-swimming existence. As a structure added to the body behind the end of the digestive tract, it clearly has evolved specifically for propulsion in water. Its efficiency is later increased in fishes with the addition of fins. The tail is evident in humans only as a vestige (the coccyx, a series of small vertebrae at the end of the spinal column) but most other mammals have a waggable tail as adults.

ANCESTRY AND EVOLUTION

Since the mid-nineteenth century when Darwin's theory of common descent became the focal point for recognizing relationships among groups of living organisms, zoologists have debated the question of chordate origins. It has been difficult to reconstruct the evolutionary history of the earliest chordates because they were probably soft-bodied creatures that had little chance of being preserved as fossils. Although more Cambrian chordates recently have been discovered (see pp. 507–510), the fossil record is sparse. Consequently, such reconstructions largely come from the study of living organisms, especially from an analysis of early developmental stages, which tend to be more evolutionarily conserved than the differentiated adult forms they become.

Most early efforts to identify kinship of chordates to other phyla are now recognized as based on similarities related to analogy rather than homology. Analogous structures are those that perform similar functions but have different origins (such as wings of birds and butterflies). Homologous structures, on the other hand, share a common origin but may look different (at least superficially) and perform quite different functions. For example, all vertebrate forelimbs are homologous because they are derived from a pentadactyl limb of the same ancestor, even though they may be modified as differently as a human's arm and a bird's wing. Homologous structures share a genetic heritage; analogous structures do not. Obviously, only homologous similarities reveal common ancestry.

Zoologists at first speculated that chordates evolved within the protostome lineage (annelids and arthropods) but discarded such ideas when they realized that supposed morphological similarities were not homologous. Early in the twentieth century when further theorizing became rooted in developmental patterns of animals, it became apparent that the chordates must have originated within the deuterostome branch of the animal kingdom. As explained in Chapter 8 (p. 166 and Figure 8.10), Deuterostomia, a grouping that includes the echinoderms, hemichordates, and chordates, has several important embryological features, as well as shared gene sequences, that clearly separate it from Protostomia and establish its monophyly. Thus deuterostomes are almost certainly a natural grouping of interrelated animals that have their common origin in ancient Precambrian seas. Several lines of anatomical, developmental, and molecular evidence suggest that somewhat later, at the base of the Cambrian period some 570 million years ago, the first distinctive chordates arose from a lineage related to echinoderms and hemichordates (Figure 23.2). Evidence from phylogenetic analysis of gene sequences, development, and morphology strongly suggest that a clade containing both echinoderms and hemichordates is the sister group of chordates (see Figure 22.1, p. 470). Information about the biology of the earliest chordates can be gleaned from examination of the two living chordate groups that are not vertebrates, Urochordata and Cephalochordata.

SUBPHYLUM UROCHORDATA (TUNICATA)

Urochordates ("tail-chordates"), more commonly called tunicates, include about 1600 species. They are found in all seas from near shoreline to great depths. Most are sessile as adults, although some are free-living. The name "tunicate" is suggested by the usually tough, nonliving **tunic** that surrounds the animal and contains cellulose (Figure 23.4). As adults, tunicates are highly specialized chordates, for in most species only the larval form, which resembles a microscopic tadpole, bears all the chordate hallmarks. During adult metamorphosis, the notochord (which, in the larva, is restricted to the tail, hence the group name Urochordata) and tail disappear, while the dorsal nerve cord becomes reduced to a single ganglion.

Urochordata is divided into three classes: **Ascidiacea** (Gr. *askiolion,* little bag, + *acea,* suffix), **Appendicularia** (L. *appendic,* hang to), and **Thaliacea** (Gr. *thalia,* luxuriance, + *acea,* suffix). Members of Ascidiacea are by far the most common, diverse, and best known. They are often called "sea squirts" because some species forcefully discharge a jet of water from the excurrent siphon when irritated. All but a few ascidian species are sessile animals, attached to rocks or other hard substrates such as pilings or bottoms of ships. In many areas, they are among the most abundant of intertidal animals.

Ascidians may be solitary, colonial, or compound. Each of the solitary and colonial forms has its own test, but among compound forms many individuals may share the same test (Figure 23.5). In some compound ascidians each member has its own incurrent siphon, but the excurrent opening is common to the group.

Solitary ascidians (see Figure 23.4) are usually spherical or cylindrical forms. Lining the tunic is an inner membrane, the **mantle.** On the outside are two projections: an **incurrent siphon,** or oral siphon, which corresponds to the anterior end

of the body, and an **excurrent siphon,** or atrial siphon, that marks the dorsal side. When a sea squirt is expanded, water enters the incurrent siphon and passes into a capacious ciliated **pharynx** that is minutely perforated by slits to form an elaborate basketwork. Water passes through the slits into an **atrial cavity** and out through the excurrent siphon.

Feeding depends on formation of a mucous net that is secreted by a glandular groove, the **endostyle,** located along the midventral side of the pharynx. Cilia on bars of the pharynx pull the mucus into a sheet that spreads dorsally across the inner face of the pharynx. Food particles brought in the incurrent opening are trapped on the mucous net, which is then worked into a rope and carried posteriorly by cilia into the esophagus and stomach. Nutrients are absorbed in the midgut and indigestible wastes are discharged from the anus, located near the excurrent siphon.

The circulatory system consists of a ventral heart and two large vessels, one on either side of the heart; these vessels connect to a diffuse system of smaller vessels and spaces serving the pharyngeal basket (where respiratory exchange occurs), the digestive organs, gonads, and other structures. An odd feature found in no other chordate is that the heart drives the blood first in one direction for a few beats, then pauses, reverses its action, and drives the blood in the opposite direction for a few beats. Another remarkable feature is the presence of strikingly high amounts of rare elements in the blood, such as vanadium and niobium. The vanadium concentration in the sea squirt *Ciona* may reach 2 million times its concentration in seawater. The function of these rare metals in the blood is a mystery.

The nervous system is restricted to a **nerve ganglion** and plexus of nerves that lie on the dorsal side of the pharynx.

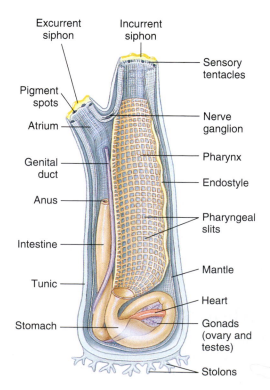

Figure 23.4

Structure of a common tunicate, *Ciona* sp.

B

Figure 23.5

A, Two yellow-edged sea squirts, *Rhopalaea* sp., on a Philippine reef. Note the large incurrent siphon and the smaller excurrent siphon for each animal. **B,** Seven colonies of compound tunicates, *Atriolum robustum,* on a Pacific reef. Individuals in a colony share a common tunic (yellow), but each has a separate incurrent (oral) siphon. Each colony has a single, large excurrent (atrial) siphon on top.

A

Beneath the nerve ganglion is located the **subneural gland,** connected by a duct to the pharynx.

Sea squirts are hermaphroditic, with usually a single ovary and a single testis in the same animal. Germ cells are carried by ducts into the atrial cavity, and then into the surrounding water where fertilization occurs.

Of the five chief characteristics of chordates, adult sea squirts have only two: pharyngeal slits and endostyle. However, the larval form reveals the secret of their true relationship. The tadpole larva (Figure 23.6) is an elongate, transparent form with all five chordate characteristics: notochord, hollow dorsal nerve cord, propulsive postanal tail, and a large pharynx with endostyle and pharyngeal slits. The larva does not feed but swims for some hours or days before fastening itself vertically by adhesive papillae to a solid object. It then undergoes a dramatic metamorphosis

(Figure 23.6) to become a sessile adult, so modified as to become almost unrecognizable as a chordate.

Tunicates of class Thaliacea, known as thaliaceans or salps, are barrel- or lemon-shaped pelagic forms with transparent, gelatinous bodies that, despite the considerable size that some species reach, are nearly invisible in sunlit surface waters. They occur singly or in colonial chains that may reach several meters in length (Figure 23.7). The cylindrical thaliacean body is typically surrounded by bands of circular muscle, with incurrent and excurrent siphons at opposite ends. Water pumped through the body by muscular contraction (rather than by cilia as in ascidians) is used for locomotion by a sort of jet propulsion, for respiration, and as a source of particulate food that is filtered on mucous surfaces. Many are provided with luminous organs, which give a brilliant light at night. Most of the body is hollow, with the viscera forming a compact mass on the ventral side.

The life histories of thaliaceans are often complex and are adapted to respond to sudden increases in their food supply. The appearance of a phytoplankton bloom, for example, is met by an explosive population increase leading to extremely high density of thaliaceans. Common forms include *Doliolum* and *Salpa,* both of which reproduce by an alternation of sexual and asexual generations.

The third tunicate class, the Appendicularia (Larvacea in some classifications) are curious larvalike pelagic creatures shaped like a bent tadpole. The name Larvacea refers to their resemblance to the larval stages of other tunicates. They feed by a method unique in the animal world. Each builds a delicate house, a transparent hollow sphere of mucus interlaced with filters and passages

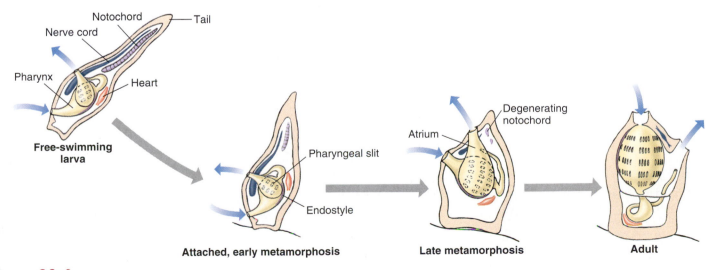

Figure 23.6

Metamorphosis of a solitary ascidian from a free-swimming larval stage.

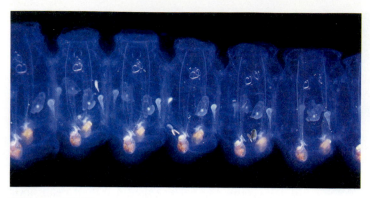

Figure 23.7

Salps. The transparent individuals of this delicate, planktonic species are grouped in a chain. Visible within each individual is an opaque gonad, an opaque gut, and a long serrated gill bar.

SUBPHYLUM CEPHALOCHORDATA

Cephalochordates are lancelets: slender, laterally compressed, translucent animals about 3 to 7 cm in length (Figure 23.9) that inhabit the sandy bottoms of coastal waters around the world. Lancelets originally bore the generic name *Amphioxus* (Gr. *amphi,* both ends, + *oxys,* sharp), later surrendered by priority to *Branchiostoma* (Gr. *branchia,* gills, + *stoma,* mouth). Amphioxus is still used, however, as a convenient common name for all of approximately 29 species in this diminutive subphylum. Five species of amphioxus occur in North American coastal waters.

Amphioxus is especially interesting because it has the five distinctive characteristics of chordates in simple form. Water enters the mouth, driven by cilia in the buccal cavity and pharynx, then passes through numerous pharyngeal slits where food is trapped in mucus secreted by the endostyle, and is then moved by cilia into the intestine. Here the smallest food particles are separated from the mucus and passed into the **hepatic cecum** where they are phagocytized and digested intracellularly. Food is moved through the gut by means of cilia, which are concentrated in a darkly staining area called the ileocolic ring (Figure 23.9), rather than by muscular contractions as in vertebrates. As in tunicates, filtered water passes first into an **atrium,** then leaves the body by an **atriopore** (equivalent to the excurrent siphon of tunicates).

The closed circulatory system is complex for so simple a chordate. The flow pattern is remarkably similar to that of fishes, although there is no heart. Blood is pumped forward in the **ventral aorta** by peristaltic-like contractions of the vessel wall, then passes upward through branchial arteries (aortic arches) in the pharyngeal bars to paired **dorsal aortas** which join to become a single dorsal aorta. From here blood is distributed to body tissues by microcirculation and then is collected in veins, which return it to the ventral aorta. Lacking both erythrocytes and hemoglobin, their blood is thought to transport nutrients but to play little role in gas exchange. There are no gills specialized for respiration in the pharynx; gas exchange occurs over the surface of the body.

The nervous system is centered around a hollow nerve cord lying above the notochord. Pairs of **spinal nerve roots** emerge at each trunk myomeric (muscle) segment. Sense organs are simple, including an anterior, unpaired **ocellus** that functions as a photoreceptor. Although the anterior end of the nerve cord is not enlarged into the characteristic vertebrate brain, it is apparently homologous to parts of the vertebrate brain.

Sexes are separate. Gametes are released in the atrium, then pass

through which water enters (Figure 23.8). Tiny phytoplankton and bacteria trapped on a feeding filter inside the house are drawn into the animal's mouth through a strawlike tube. When the filters become clogged with waste, which happens about every 4 hours, the appendicularian abandons its house and builds a new house, a process that takes only a few minutes. Like thaliaceans, appendicularians can quickly build up dense populations when food is abundant. At such times scuba diving among the houses, which are about the size of walnuts, is likened to swimming through a snowstorm! Appendicularians are paedomorphic; they are sexually mature animals that have retained the larval body form of their evolutionary ancestors (see the note explaining paedomorphosis on p. 508).

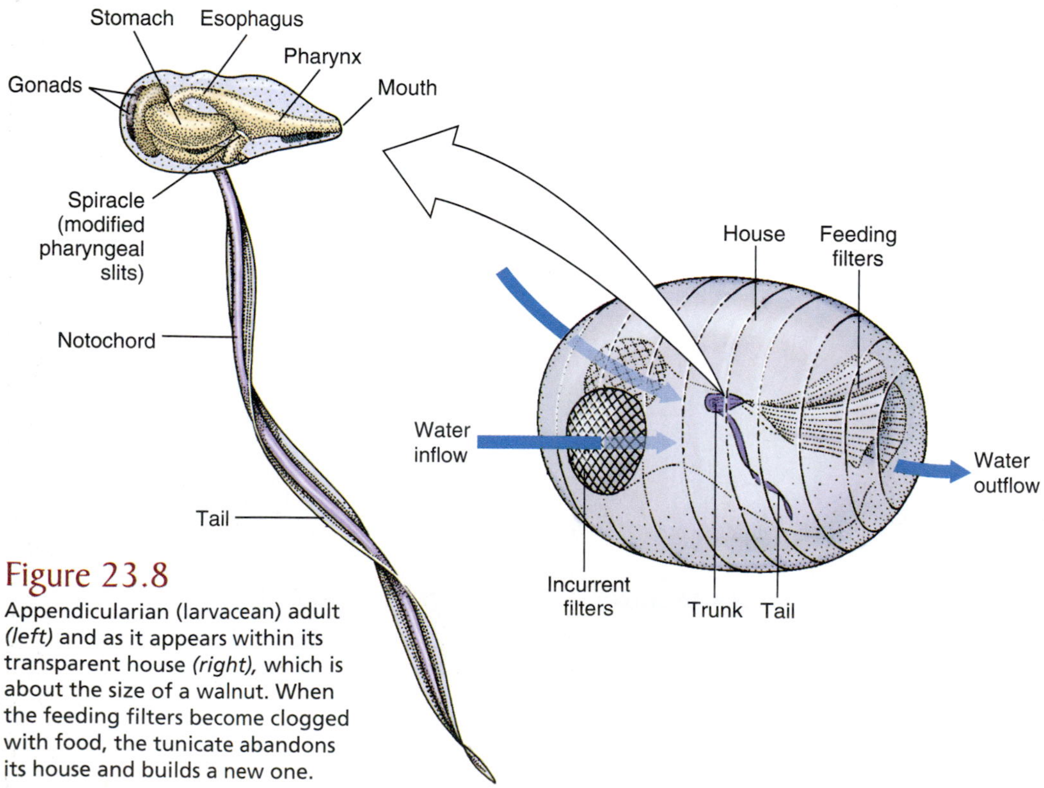

Stomach   Esophagus
Pharynx
Gonads
Mouth
Spiracle
(modified
pharyngeal
slits)
Notochord
Water
inflow
House   Feeding
filters
Tail
Water
outflow
Incurrent
filters
Trunk   Tail

Figure 23.8

Appendicularian (larvacean) adult *(left)* and as it appears within its transparent house *(right),* which is about the size of a walnut. When the feeding filters become clogged with food, the tunicate abandons its house and builds a new one.

Figure 23.9

Amphioxus. This interesting bottom-dwelling cephalochordate illustrates the five distinctive chordate characteristics (notochord, dorsal nerve cord, pharyngeal slits, endostyle, and postanal tail). The vertebrate ancestor is thought to have had a similar body plan. **A,** Living amphioxus in typical position for filter-feeding. Note the oral hood with tentacles surrounding the mouth. **B,** Internal structure.

A

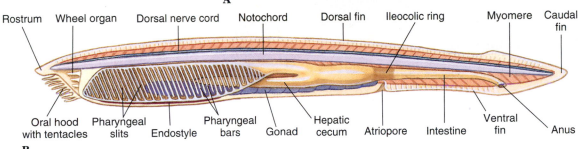

B

through the atriopore to the outside where fertilization occurs. Cleavage is total (holoblastic) and a gastrula is formed by invagination. Larvae hatch soon after egg deposition and gradually assume the shape of adults.

No other chordate shows the basic diagnostic chordate characteristics as clearly as amphioxus. In addition to the five chordate anatomical hallmarks, amphioxus possesses several structural features that resemble the vertebrate plan. Among these are a hepatic cecum, a diverticulum that resembles the vertebrate pancreas in secreting digestive enzymes and the liver in storing glycogen, **segmented trunk musculature,** and the basic circulatory plan of more advanced chordates. As discussed on page 508, most zoologists consider amphioxus a living descendant of an ancestor that gave rise to both cephalochordates and vertebrates. Therefore cephalochordates are, in cladistic terms, the living sister group of vertebrates (see Figure 23.3).

SUBPHYLUM VERTEBRATA (CRANIATA)

The third subphylum of chordates is the large and diverse Vertebrata, the subject of Chapters 24 through 28. This monophyletic group shares the basic chordate characteristics with the other two subphyla, but in addition it exhibits a number of novel characters that the others do not share. The alternative name of the subphylum, Craniata, more accurately describes the group since all have a cranium (bony or cartilaginous braincase) but some jawless fishes lack vertebrae.

Adaptations That Have Guided Early Vertebrate Evolution

The earliest vertebrates were substantially larger and considerably more active than the protochordates. Increased speed and mobility resulted from modifications of skeletal structures and muscles. The higher activity level and size of vertebrates also requires structures specialized in the location, capture, and digestion of food and adaptations designed to support a high metabolic rate.

Musculoskeletal Modifications

Most vertebrates possess both an exoskeleton and endoskeleton of cartilage or bone. The endoskeleton permits almost unlimited body size with much greater economy of building materials than the exoskeleton of arthropods. Some vertebrates have become the most massive organisms on earth. The endoskeleton forms an excellent jointed scaffolding for attachment of segmented muscles. The segmented body muscles (myomeres) changed from the V-shaped muscles of cephalochordates to the W-shaped muscles of vertebrates. This increased complexity of folding in the myomeres provides powerful control over an extended length of the body. Also unique to vertebrates are the presence of fin rays of dermal origin, aiding in swimming.

The endoskeleton probably was composed initially of cartilage that later gave way to bone. Cartilage, with its fast growth and flexibility, is ideal for constructing the first skeletal framework of all vertebrate embryos. The endoskeleton of living hagfishes, lampreys, sharks and their kin, and even in some "bony"

Characteristics of Subphylum Vertebrata

1. Chief diagnostic features of chordates—**notochord, dorsal tubular nerve cord, pharyngeal pouches, endostyle** or **thyroid gland,** and **postanal tail**—all present at some stage of the life cycle
2. **Integument** of two divisions, an outer epidermis of stratified epithelium from ectoderm and an inner dermis of connective tissue from mesoderm; many modifications of skin, such as glands, scales, feathers, claws, horns, and hair
3. Distinctive cartilage or bone **endoskeleton** consisting of vertebral column (except in hagfishes, which lack vertebrae) and a head skeleton (cranium and pharyngeal skeleton) derived largely from **neural crest cells**
4. **Muscular pharynx;** in fishes pharyngeal pouches open to the outside as slits and bear gills; in tetrapods pharyngeal pouches are sources of several glands
5. Complex, W-shaped muscle segments or **myomeres** to provide movement
6. Complete, **muscularized digestive tract** with distinct liver and pancreas
7. Circulatory system consisting of a **ventral heart** of multiple chambers; closed blood vessel system of arteries, veins, and capillaries; blood containing **erythrocytes** with **hemoglobin;** paired aortic arches connecting ventral and dorsal aortas and giving off branches to the gills among aquatic vertebrates; in terrestrial forms; aortic arches modified into pulmonary and systemic circuits
8. Well-developed **coelom** divided into a pericardial cavity and a pleuroperitoneal cavity
9. Excretory system consisting of **paired, glomerular kidneys** provided with ducts to drain waste to the cloaca
10. Highly differentiated **tripartite brain;** 10 or 12 pairs of **cranial nerves;** a pair of spinal nerves for each primitive myotome; **paired special sense organs** derived from **epidermal placodes**
11. **Endocrine system** of ductless glands scattered throughout the body
12. Nearly always separate sexes; each sex containing gonads with ducts that discharge their products either into the cloaca or into special openings near the anus
13. Most vertebrates with two pairs of appendages supported by limb girdles and appendicular skeleton

fishes, such as sturgeons, is mostly cartilage. Bone may have been adaptive in early vertebrates in several ways. The presence of bone in the skin of ostracoderms and other ancient fishes certainly provided protection from predators, although there are some more important benefits of bone. The structural strength of bone is superior to cartilage, making it ideal for muscle attachment in areas of high mechanical stress. One of the most interesting ideas is that the function associated with the origin of bone was for mineral regulation. Phosphorus and calcium are used for many physiological processes and are in particularly high demand in organisms with high metabolic rates. Storage and regulation of calcium and phosphorus ions were likely important functions of bone in the earliest vertebrates.

We should note that most vertebrates possess an extensive exoskeleton (one that develops from the skin), although it is highly modified in many forms. Some of the most primitive fishes, including ostracoderms and placoderms, were partly covered in a bony, dermal armor. This armor is modified as scales in later fishes. Many of the bones encasing the brain of advanced vertebrates actually develop from cells that originate from the dermis! Most vertebrates are further protected with keratinized structures derived from the epidermis, such as reptilian scales, hair, feathers, claws, and horns.

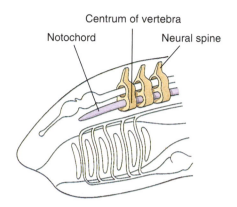

Physiology Upgrade

Vertebrates have modifications to the digestive, respiratory, circulatory, and excretory systems that meet an increased metabolic demand. The perforated pharynx evolved in early chordates as a filter-feeding device, which uses cilia and mucus to move water and to trap small suspended food particles. In vertebrates, the addition of muscles to the pharynx created a powerful pump for moving water. With the origin of highly vascularized gills, the function of the pharynx shifted to primarily gas exchange. Changes in the gut, including a shift from movement of food by ciliary action to muscular action and addition of accessory digestive glands, the liver and pancreas, managed the increased amount of food ingested. A ventral three-chambered heart consisting of a sinus venosus, atrium, and ventricle, and erythrocytes with hemoglobin enhanced transportation of nutrients, gases, and other substances. Protochordates have no distinct kidneys, but vertebrates possess paired, glomerular kidneys that remove metabolic waste products and regulate body fluids and ions.

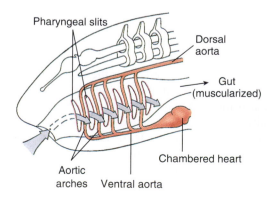

New Head, Brain, and Sensory Systems

When vertebrate ancestors shifted from filter-feeding to active predation, new sensory, motor, and integrative controls became essential for location and capture of larger prey. The anterior end of the nerve cord became enlarged as a **tripartite brain** (forebrain, midbrain, and hindbrain) and protected by a cartilaginous or bony cranium. Paired special sense organs designed for distance reception evolved. These included eyes with lenses and inverted retinas; pressure receptors, such as paired inner ears designed for equilibrium and sound reception; chemical receptors including taste and exquisitely sensitive olfactory organs; lateral-line receptors for detecting water vibrations; and electroreceptors for detecting electrical currents that signal prey.

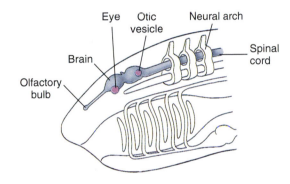

Neural Crest, Ectodermal Placodes, and Hox Genes

Development of the vertebrate head and special sense organs was largely the result of two embryonic innovations present only in vertebrates: **neural crest** and **ectodermal placodes.** The neural crest, a population of ectodermal cells lying along the length of the embryonic neural tube (see Figure 8.27 and p. 182), contributes to the formation of many different structures, among them most of the cranium, pharyngeal skeleton, tooth dentine, some cranial nerves, ganglia, some endocrine glands, and Schwann cells. In addition they may regulate the development of adjacent tissue, such as tooth enamel and pharyngeal muscles (branchiomeres). The ectodermal placodes (Gr. *placo,* plate) are platelike ectodermal thickenings that appear on either side of the neural tube. These give rise to the olfactory epithelium, lens of the eye, inner ear epithelium, some ganglia and cranial nerves, lateral-line mechanoreceptors, and electroreceptors. The placodes also induce the formation of taste buds. Thus the vertebrate head with its sensory structures located adjacent to the mouth (later equipped with prey-capturing jaws), stemmed from the creation of new cell types.

Recent studies of the distribution of homeobox genes that control the body plan of chordate embryos (homeobox genes are described on pp. 171–172) suggest that *Hox* genes were duplicated at about the time of the origin of vertebrates. One copy of *Hox* genes is found in amphioxus and other invertebrates whereas living gnathostomes have four copies. Perhaps these additional copies of genes that control body plan provided genetic material free to evolve a more complex kind of animal.

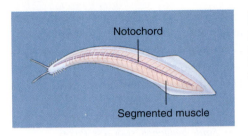

Figure 23.10
Pikaia, an early chordate from the Burgess Shale of British Columbia, Canada.

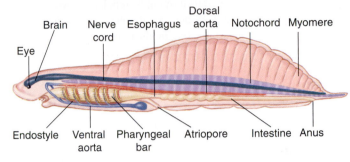

Figure 23.11
Haikouella, a chordate with several vertebrate features from early Cambrian shales of Haikou, China. It has been hypothesized to be the sister group of vertebrates (craniates).

The Search for the Ancestral Vertebrate

Most of the early Paleozoic vertebrate fossils, the jawless ostracoderms (p. 510), share many novel features of organ system development with living vertebrates. These organ systems must have originated in an early vertebrate or invertebrate chordate lineage. Fossil invertebrate chordates are rare and known primarily from two fossil beds—the well-known middle Cambrian Burgess Shale of Canada (p. 110) and the recently discovered early Cambrian fossil beds of Chengjiang and Haikou, China. An ascidian tunicate and *Yunnanozoon,* a probable cephalochordate, are known from Chengjiang. Slightly better known is *Pikaia,* a ribbon-shaped, somewhat fishlike creature about 5 cm in length discovered in the Burgess Shale (Figure 23.10). The presence of V-shaped myomeres and a notochord clearly identifies *Pikaia* as a chordate. The superficial resemblance of *Pikaia* to living amphioxus suggests that it may be an early cephalochordate.

A wealth of information about the origin of vertebrates is provided by *Haikouella lanceolata,* a small fishlike creature known from over 300 fossil specimens recently discovered in 530-million-year-old sediments near Haikou. It possessed several characters that clearly identify it as a chordate, including notochord, pharynx, and dorsal nerve cord, but also had characters, including pharyngeal muscles, paired eyes, and an enlarged brain, that are characteristic of vertebrates (Figure 23.11). However, it is not a vertebrate, because the fossils lack evidence of several diagnostic vertebrate traits, including a cranium, an ear, and a distinct telencephalon (anterior region of the forebrain). Thus, it is transitional in morphology between cephalochordates and vertebrates. Jon Mallatt, Jun-yuan Chen, and colleagues, who have studied these fossils extensively, hypothesized *Haikouella* to be the sister taxon of vertebrates. Despite recent fossil discoveries of early chordates, many speculations regarding vertebrate ancestry have focused on the living protochordates, in part because they are much better known than the fossil forms.

Butler and Hodos provided an explanation of how the paired eyes of vertebrates evolved from the unpaired, median ocellus of an amphioxus-like ancestor. The homeotic gene *Pax-6* is responsible for the formation of an eye-producing region near the midbrain. Products of another gene, *Sonic hedgehog,* suppress *Pax-6* expression at the midline, thus forming paired, lateral eyes. In mice, genetic manipulations that produce an absence of *Sonic hedgehog* result in formation of an unpaired, median eye.

Garstang's Hypothesis of Chordate Larval Evolution

The chordates have pursued two paths in their early evolution, one path leading to the sedentary urochordates, the other to active, mobile cephalochordates and vertebrates. One hypothesis, proposed in 1928 by Walter Garstang of England, suggested that the chordate ancestral lineage retained into adulthood the larval form of sessile tunicate-like animals. The tadpole larva of tunicates does indeed bear all the right attributes to qualify as a possible vertebrate ancestor: notochord, hollow dorsal nerve cord, pharyngeal slits, endostyle, and postanal tail. At some point, Garstang suggested, the tadpole larva failed to metamorphose into an adult tunicate, instead developing gonads and reproducing in the larval stage. With continued evolution, a new group of free-swimming animals appeared, the ancestors of cephalochordates and vertebrates (Figure 23.12).

Garstang called this process **paedomorphosis** (Gr. *pais,* child + *morphē,* form), a term describing the evolutionary retention of juvenile or larval traits in the adult body. Garstang suggested that evolution may occur in larval stages of animals, in

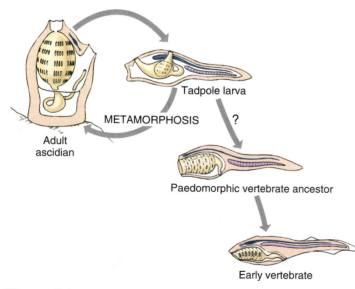

Metamorphosis
Tadpole larva
Adult ascidian
?
Paedomorphic vertebrate ancestor
Early vertebrate

Figure 23.12

Garstang's hypothesis of larval evolution. Adult tunicates live on the sea floor but produce a free-swimming tadpole larva. According to this hypothesis, more than 550 million years ago, some larvae began to reproduce in the swimming stage. These evolved into the first vertebrates.

this case leading to the vertebrate lineage. Paedomorphosis is a well-known phenomenon in several different animal groups (paedomorphosis in amphibians is described on pp. 550–551). Although long popular, Garstang's hypothesis has been challenged recently. First, phylogenies generated from molecular data, along with information from fossils, suggest that the ancestor of deuterostomes was free-swimming. Second, additional phylogenies from molecular data show that the sessile ascidians represent a derived body form, and that free-swimming appendicularians are most similar in body form to ancestral chordates.

Paedomorphosis, the displacement of ancestral larval or juvenile features into a descendant adult, can be produced by three different evolutionary-developmental processes: neoteny, progenesis, and postdisplacement. In neoteny, the growth rate of body form is slowed so that the animal does not attain the ancestral adult form when it reaches maturity. Progenesis is the precocious maturation of gonads in a larval (or juvenile) body that then stops growing and never attains the adult body form. In postdisplacement, the onset of a developmental process is delayed relative to reproductive maturation, so that the ancestral adult form is not attained by the time of reproductive maturation. Neoteny, progenesis and postdisplacement thus describe different ways in which paedomorphosis can happen. Biologists use the inclusive term paedomorphosis to describe results of these evolutionary-developmental processes.

Position of Amphioxus

Zoologists consider the cephalochordate amphioxus the closest living relative of vertebrates. Cephalochordates share several characters with vertebrates that are absent from tunicates, including segmented myomeres, dorsal and ventral aortas, branchial or aortic arches, and podocytes, specialized excretory cells. However, as noted in the prologue to the chapter (p. 496), amphioxus is unlike the most recent common ancestor of vertebrates because it lacks the tripartite brain, chambered heart, special sensory organs, muscular gut and pharynx, and neural crest tissue inferred to have been present in that ancestor. In addition, the larger fins of some extinct cephalochordates suggest that they were more free-swimming than modern amphioxus.

Despite these specializations most zoologists believe that amphioxus has largely retained the body structure of the immediate prevertebrate condition. Thus cephalochordates are probably the closest living relative of vertebrates (see Figure 23.3) (the fossil *Haikouella* is probably the sister taxon of vertebrates).

The Ammocoete Larva of Lampreys as a Model of the Primitive Vertebrate Body Plan

Lampreys (jawless fishes of the class Petromyzontida, discussed in Chapter 24) have a freshwater larval stage called the **ammocoete** (Figure 23.13). In body form, appearance, life habit, and many anatomical details, the ammocoete larva resembles amphioxus. In fact, lamprey larvae were given the genus name *Ammocoetes*

Traditional Linnean Classification of Living Members of Phylum Chordata

Phylum Chordata

Subphylum Urochordata (u'ro-kor-da'ta) (Gr. *oura*, tail, + L. *chorda*, cord, + *ata*, characterized by) **(Tunicata): tunicates.** Notochord and nerve cord in free-swimming larva only; ascidian adults sessile, encased in tunic. About 1600 species.

Subphylum Cephalochordata (sef'a-lo-kor-da'ta) (Gr. *kephalē*, head, + L. *chorda,* cord): **lancelets (amphioxus).** Notochord, nerve cord, and postanal tail persist throughout life; fishlike in form. 29 species.

Subphylum Vertebrata (ver'te-bra'ta) (L. *vertebratus*, backboned) **(Craniata): vertebrates.** Bony or cartilaginous cranium surrounding tripartite brain; well-developed head with paired sense organs; usually with vertebrae; heart present, with multiple chambers, muscularized digestive tract, paired kidneys.

    Superclass Agnatha (ag'na-tha) (Gr. *a*, without, + *gnathos,* jaw): **hagfishes, lampreys.** Without true jaws or paired appendages. A paraphyletic group.

        Class Myxini (mik-sin'y) (Gr. *myxa*, slime): **hagfishes.** Mouth with four pairs of tentacles; buccal funnel absent; 1 to 16 pairs of external gill openings; vertebrae absent; slime glands present. About 70 species.

        Class Petromyzontida (pet'trō-mī-zon'ti-də) (Gr. *petros*, stone, + *myzon,* sucking): **lampreys.** Mouth surrounded by keratinized teeth but no barbels, buccal funnel present; seven pairs of external gill openings; vertebrae present only as neural arches. 38 species.

    Superclass Gnathostomata (na'tho-sto'ma-ta) (Gr. *gnathos*, jaw, + *stoma,* mouth): **jawed fishes, tetrapods.** With jaws and (usually) paired appendages.

        Class Chondrichthyes (kon-drik'thee-eez) (Gr. *chondros*, cartilage, + *ichthys*, fish): **sharks, skates, rays, chimaeras.** Cartilaginous skeleton; intestine with spiral valve; claspers present in males; no swim bladder. About 970 species.

        Class Actinopterygii (ak'ti-nop-te-rij'ee-i) Gr. *aktis*, ray, + *pteryx*, fin, wing): **ray-finned fishes.** Skeleton ossified; single gill opening covered by operculum; paired fins supported primarily by dermal rays; appendage musculature within body; swim bladder mainly a hydrostatic organ, if present; atrium and ventricle not divided. About 27,000 species.

        Class Sarcopterygii (sar-cop-te-rij'ee-i) (Gr. *sarkos*, flesh, + *pteryx*, fin, wing): **lobe-finned fishes.** Skeleton ossified, single gill opening covered by operculum; paired fins with sturdy internal skeleton and musculature within appendage; diphycercal tail; intestine with spiral valve; usually with lunglike swim bladder; atrium and ventricle at least partly divided. 8 species. Paraphyletic unless tetrapods are included.

        Class Amphibia (am-fib'e-a) (Gr. *amphi,* both or double, + *bios,* life): **amphibians.** Ectothermic tetrapods; respiration by lungs, gills, or skin; development through larval stage; skin moist, containing mucous glands, and lacking scales. About 5500 species.

        Class Reptilia (rep-til'e-a) (L. *repere,* to creep): **reptiles.** Ectothermic tetrapods possessing lungs; embryo develops within shelled egg; no larval stage; skin dry, lacking mucous glands, and covered by epidermal scales. A paraphyletic group unless birds are included. About 8100 species.

        Class Aves (ay'veez) (L. pl. of *avis,* bird): **birds.** Endothermic vertebrates with front limbs modified for flight; body covered with feathers; scales on feet. About 9700 species.

        Class Mammalia (ma-may'lee-a) (L. *mamma,* breast): **mammals.** Endothermic vertebrates possessing mammary glands; body more or less covered with hair; brain large, with neocortex; three middle ear bones. About 4800 species.

(Gr. *ammos,* sand, + *koitē,* bed, referring to the preferred larval habitat) in the nineteenth century when it was erroneously thought to be an adult cephalochordate, closely allied with amphioxus. Ammocoete larvae are so different from adult lampreys that the mistake is understandable; the exact relationship was not explained until metamorphosis into an adult lamprey was observed.

Ammocoete larvae have a long, slender body with an oral hood surrounding the mouth much like amphioxus (Figure 23.13).

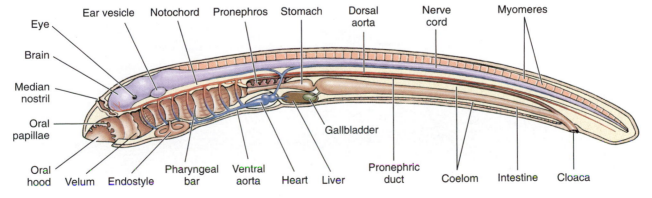

Figure 23.13

Ammocoete, freshwater larval stage of a lamprey. Although they resemble amphioxus in many ways, ammocoetes have a well-developed brain, paired eyes, pronephric kidneys, heart, and other features lacking in amphioxus but representative of the vertebrate body plan.

Ammocoetes are filter feeders, but instead of drawing water by ciliary action into the pharynx as amphioxus does, ammocoetes produce a feeding current by muscular pumping action much like modern fishes. The arrangement of body muscle into myomeres, the presence of a notochord serving as chief skeletal axis, and the plan of the circulatory system all resemble these features in amphioxus.

Ammocoetes do have several characteristics lacking in amphioxus that are homologous to those of vertebrates. These include a chambered heart, tripartite brain, paired special sense organs derived from ectodermal placodes, and pituitary gland. The kidney is pronephric (pp. 673–674) and conforms to the basic vertebrate plan. Instead of the numerous pharyngeal slits of amphioxus, there are only seven pairs of pharyngeal pouches and slits in ammocoetes. From pharyngeal bars separating the pharyngeal slits project gill filaments bearing secondary lamellae much like the more extensive gills of modern fishes (see Figure 24.29, p. 533). Ammocoetes also have a true liver replacing the hepatic cecum of amphioxus, a gall-bladder, and pancreatic tissue (but no distinct pancreatic gland).

Ammocoetes display the most primitive condition for these characteristics of any living vertebrate. They clearly illustrate many shared derived characters of vertebrates that are obscured in the development of other vertebrates. They may approach most closely the supposed body plan of the ancestral vertebrate.

The Earliest Vertebrates

The earliest known vertebrate fossils, until recently, were armored jawless fishes called **ostracoderms** (os-trak′o-derm) (Gr. *ostrakon*, shell, + *derma*, skin) from late Cambrian and Ordovician deposits. During the last 10 years researchers have discovered a number of 530-million-year-old fossils in the amazing Chengjiang deposits belonging to two (possibly only one) fishlike vertebrates:

Myllokunmingia (Gr. *myllo*, sea fish, + Kunming, a city in China) and *Haikouichthys* (Haikou, a city in China, + Gr. *ichthy*, fish). These fossils push back the origin of vertebrates to at least the early Cambrian. These fossils showed many typical vertebrate characters including a heart, paired eyes, otic (ear) capsules, and what has been interpreted as rudimentary vertebrae.

The earliest ostracoderms were armored with bone in their dermis and lacked paired fins that later fishes found so important for stability (Figure 23.14). The swimming movements of one of the early groups, the **heterostracans** (Gr. *heteros*, different, + *ostrakon*, shell) must have been imprecise, although sufficient to propel them along the ocean bottom where they searched for food. With fixed circular or slitlike mouth openings they may have filtered small food particles from the water or ocean bottom. However, unlike the ciliary filter-feeding protochordates, ostracoderms sucked water into the pharynx by muscular pumping, an important innovation that suggests to some authorities that ostracoderms may have been active predators that fed on soft-bodied animals.

The term "ostracoderm" does not denote a natural evolutionary assemblage but rather is a term of convenience for describing several groups of heavily armored extinct jawless fishes.

During the Devonian period, the heterostracans underwent a major radiation, producing numerous peculiar-looking forms. Without ever evolving paired fins or jaws, these earliest vertebrates flourished for 150 million years until becoming extinct near the end of the Devonian period.

Coexisting with heterostracans throughout much of the Devonian period were **osteostracans** (Gr. *osteon*, bone, + *ostrakon*, shell). Osteostracans had paired pectoral fins, an innovation that functioned to improve swimming efficiency by controlling

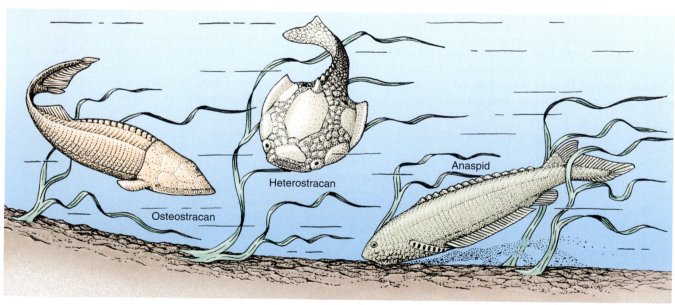

Figure 23.14

Three ostracoderms, jawless fishes of Silurian and Devonian times. They are shown as they might have appeared while searching for food on the floor of a Devonian sea. All were probably filter feeders, but employed a strong pharyngeal pump to circulate water rather than the much more limiting mode of ciliary feeding used by their protochordate ancestors (presumably resembling amphioxus for this feature).

yaw, pitch, and roll. A typical osteostracan, such as *Cephalaspis* (Gr. *kephalē,* head, + *aspis,* shield) (Figure 23.14), was a small animal, seldom exceeding 30 cm in length. It was covered with a heavy, dermal armor of cellular bone, including a single-piece head shield. Examination of internal features of the braincase reveal a sophisticated nervous system and sense organs, similar to those of modern lampreys.

Another group of ostracoderms, the **anaspids** (Figure 23.14), were more streamlined than other ostracoderms. These and other ostracoderms enjoyed an impressive radiation in the Silurian and Devonian periods. However, all ostracoderms became extinct by the end of the Devonian period.

For decades geologists have used strange, microscopic tooth-like fossils called **conodonts** (Gr. *kōnos,* cone, + *odontos,* tooth) to date Paleozoic marine sediments without having any idea what kind of creature originally possessed these elements. The discovery in the early 1980s of fossils of complete conodont animals has changed this situation. With their phosphatized toothlike elements, myomeres, notochord, and paired eye and otic capsules, conodonts clearly belong to the vertebrate clade (Figure 23.15). Although their exact position in this clade is unclear, they are important in understanding the evolution of early vertebrates.

The Swedish paleozoologist Erik Stensiö was the first to approach fossil anatomy with the same painstaking attention to minute detail that morphologists have long applied to the anatomical study of living fishes. He developed novel and exacting methods for gradually grinding away a fossil, a few micrometers at a time, to reveal internal features. He was able to reconstruct not only bone anatomy, but nerves, blood vessels, and muscles in numerous groups of Paleozoic and early Mesozoic fishes. His innovative methods are widely used today by paleozoologists.

Early Jawed Vertebrates

All jawed vertebrates, whether extinct or living, are collectively called **gnathostomes** ("jaw mouth") in contrast to the jawless vertebrates, the **agnathans** ("without jaw"). Gnathostomes are a monophyletic group since presence of jaws is a derived character state shared by all jawed fishes and tetrapods. Agnathans, however, are defined principally by the absence of jaws, a character that is not unique to jawless fishes since jaws are lacking in vertebrate ancestors. Thus, Agnatha is paraphyletic.

The origin of jaws was one of the most important events in vertebrate evolution. The utility of jaws is obvious: they allow predation on large and active forms of food not available to jawless vertebrates and permit manipulation of objects. Ample evidence suggests that jaws arose through modifications of the first or second of the serially repeated cartilaginous gill arches. But how did this mandibular arch change from a function of gill support and ventilation to one of feeding as jaws? Expansion of this arch and evolution of new, associated muscles may have first assisted gill ventilation, perhaps to meet the increasing metabolic demands of early vertebrates. Once enlarged and equipped with extra muscles, the first pharyngeal arch could have easily been modified to serve as jaws. Evidence for this remarkable transformation includes, first, both gill arches and jaws form from upper and lower bars that bend forward and are hinged in the middle (Figure 23.16). Second, both gill arches and jaws are derived from neural crest cells. Third, the jaw musculature is homologous to the original gill support musculature as evidenced by cranial nerve distribution. Nearly as remarkable as this drastic morphological remodeling is the subsequent evolutionary fate of jawbone elements—their transformation into ear ossicles of the mammalian middle ear (see the note on p. 745).

An additional feature characteristic of all gnathostomes is the presence of paired pectoral and pelvic appendages in the form of fins or limbs. These likely originated as stabilizers to check yaw, pitch, and roll generated during active swimming. The fin-fold hypothesis has been proposed to explain the origin of paired fins.

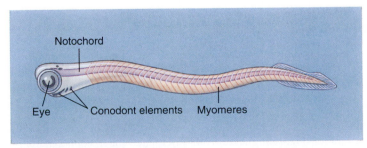

Figure 23.15
Restoration of a living conodont. Conodonts superficially resembled amphioxus, but they possessed a much greater degree of encephalization (paired eyes and otic [ear] capsules) and bonelike mineralized elements—all indicating that conodonts were chordates and probably vertebrates. Conodont elements are considered to be part of a food-handling apparatus.

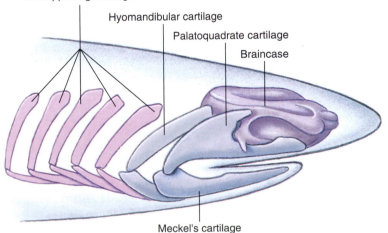

Figure 23.16
How vertebrates got their jaws. Resemblance between jaws and gill supports of primitive fishes, such as this Carboniferous shark, suggests that the upper jaw (palatoquadrate) and lower jaw (Meckel's cartilage) evolved from structures that originally functioned as gill supports. The gill supports immediately behind the jaws are hinged like jaws and served to link the jaws to the braincase. Relics of this transformation are seen during the development of modern sharks.

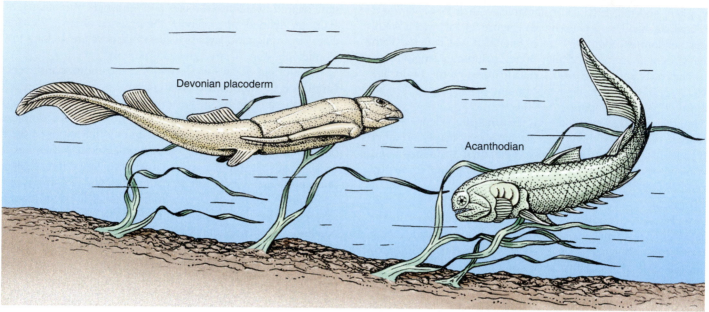

Figure 23.17

Early jawed fishes of the Devonian period, 400 million years ago. Shown are a placoderm *(left)* and an acanthodian *(right)*. Jaws and gill supports from which the jaws evolved develop from neural crest cells, a diagnostic character of vertebrates. Most placoderms were bottom dwellers that fed on benthic animals although some were active predators. Acanthodians carried less armor than placoderms and had large, anteriorly placed eyes and prominent spines on paired fins. Most were marine but several species lived in freshwater.

According to this hypothesis, paired fins arose from paired continuous, ventrolateral folds or fin-forming zones. The addition of skeletal supports in the fins served to enhance their properties of providing stability during swimming. Evidence for this hypothesis is found in the paired flaps of *Myllokunmingia* and anaspids and in the multiple paired fins of acanthodians, also described in this section. However, pectoral fins appear in the fossil record before pelvic fins, suggesting a more complex evolutionary scenario. In one fish lineage the muscle and skeletal supports in the paired fins became strengthened, allowing them to become adapted for locomotion on land as limbs. The origin of jaws and paired appendages may be linked to a second *Hox* duplication, near the origin of the gnathostomes. Both jaws and paired fins were major innovations in vertebrate evolution, among the most important reasons for the subsequent major radiations of vertebrates that produced the modern fishes and all tetrapods, including you, the reader of this book.

Among the first jawed vertebrates were the heavily armored **placoderms** (plak′o-derm) (Gr. *plax,* plate, + *derma,* skin). These first appear in the fossil record in the early Silurian period (Figure 23.17). Placoderms evolved a great variety of forms, some very large (one was 10 m in length!) and grotesque in appearance. They were armored fish covered with diamond-shaped scales or with large plates of bone. All became extinct by the end of the Devonian period and appear to have left no descendants. However, **acanthodians,** a group of early jawed fishes known from the Silurian to Permian periods, and characterized by large, anteriorly set eyes and fins with large spines (Figure 23.17), are included in a clade that underwent a great radiation into the bony fishes that dominate the waters of the world today.

SUMMARY

Phylum Chordata is named for the rodlike notochord that forms a stiffening body axis at some stage in the life cycle of every chordate. All chordates share five distinctive hallmarks that set them apart from all other phyla: notochord, dorsal tubular nerve cord, pharyngeal pouches, endostyle, and postanal tail. Two of the three chordate subphyla are invertebrates and lack a well-developed head. They are Urochordata (tunicates), most of which are sessile as adults but all of which have a free-swimming larval stage, and Cephalochordata (lancelets), fishlike forms that include the famous amphioxus.

Chordates have evolutionary affinities to echinoderms and hemichordates, although their precise origins have been controversial. Most zoologists now consider the chordate ancestor to have been a small, free-swimming, filter-feeding creature.

Subphylum Vertebrata includes the backboned members of the animal kingdom (hagfishes actually lack vertebrae but are included with the Vertebrata by tradition because they share numerous homologies with vertebrates). As a group vertebrates are characterized by having a well-developed head, and by their comparatively large size, high degree of motility, and a distinctive body plan that embodies several distinguishing features that permitted the exceptional adaptive radiation of the group. Most important of these are the living endoskeleton that allows continuous growth and provides a sturdy framework for efficient muscle action and attachment, a muscular pharynx with slits and gills (lost or greatly modified in terrestrial vertebrates) with vastly increased respiratory efficiency, a muscularized gut, chambered heart, and glomerular kidneys for meeting higher metabolic demands, and an advanced nervous system with a distinct brain and paired sense organs. Insight into the early evolution of vertebrates is provided by examination of several fossil forms, including *Haikouella,* conodonts, and ostracoderm fishes, and ammocoetes, the larval form of living lampreys. Evolution of jaws and paired appendages likely contributed to the incredible success of one group of vertebrates, the gnathostomes.

REVIEW QUESTIONS

1. What characteristics are shared by the three deuterostome phyla that indicate a monophyletic group of interrelated animals?
2. Explain how the use of a cladistic classification for the vertebrates results in important regroupings of the traditional vertebrate taxa (refer to Figure 23.3). Why are certain traditional groupings such as Osteichthyes and Agnatha not recognized in cladistic usage?
3. Name five hallmarks shared by all chordates, and explain the function of each.
4. In debating the question of chordate origins, zoologists eventually agreed that chordates must have evolved within the deuterostome assemblage rather than from a protostome group as earlier argued. What embryological evidences support this view?
5. Offer a description of an adult tunicate that would identify it as a chordate, yet distinguish it from any other chordate group.
6. Amphioxus long has been of interest to zoologists searching for a vertebrate ancestor. Explain why amphioxus captured such interest and why it no longer is considered to resemble closely the most recent common ancestor of all vertebrates.
7. Both sea squirts (urochordates) and lancelets (cephalochordates) are filter-feeding organisms. Describe the filter-feeding apparatus of a sea squirt and explain in what ways its mode of feeding is similar to, and different from, that of amphioxus.
8. Explain why it is necessary to know the life history of a tunicate to understand why tunicates are chordates.
9. List three groups of adaptations that guided vertebrate evolution, and explain how each has contributed to the success of vertebrates.
10. In 1928 Walter Garstang hypothesized that tunicates resemble the ancestral stock of the vertebrates. Explain this hypothesis.
11. What is the phylogenetic position of *Haikouella,* and what evidence supports its placement there?
12. Distinguish between ostracoderms and placoderms. What important evolutionary advances did each contribute to vertebrate evolution? What are conodonts?
13. Explain how zoologists think the vertebrate jaw evolved.

SELECTED REFERENCES

Ahlberg, P. E. 2001. Major events in early vertebrate evolution. London, Taylor & Francis. *Evolution of vertebrates up to the split of major jawed fish groups, incorporating molecular, fossil, and embryological data. Many important contributions, but some conclusions are controversial.*

Alldredge, A. 1976. Appendicularians. Sci. Am. **235:**94–102 (July). *Describes the biology of larvaceans, which build delicate houses for trapping food.*

Bowler, P. J. 1996. Life's splendid drama: evolutionary biology and the reconstruction of life's ancestry 1860–1940. Chicago, University of Chicago Press. *Thorough and eloquent exploration of scientific debates over reconstruction of history of life on earth; chapter 4 treats theories of chordate and vertebrate origins.*

Carroll, R. L. 1997. Patterns and processes of vertebrate evolution. New York, Cambridge University Press. *A comprehensive analysis of the evolutionary processes that have influenced large-scale changes in vertebrate evolution.*

Donoghue, P. C. J., P. L. Forey, and R. J. Aldridge. 2000. Conodont affinity and chordate phylogeny. Biol. Rev. **75:**191–251. *In this summary of early chordate evolution, the authors provide evidence that conodonts are vertebrates and that lampreys and hagfishes do not form a monophyletic group.*

Forey, P., and P. Janvier. 1994. Evolution of the early vertebrates. Am. Sci. **82:**554–565. *Summarizes the biology and evolution of many groups of ostracoderms and other primitive craniates.*

Gee, H. 1996. Before the backbone: views on the origin of the vertebrates. New York, Chapman & Hall. *Outstanding review of the many vertebrate origin hypotheses. Gee links much of the recent genetic, developmental, and molecular evidence in his discussion.*

Gould, S. J. 1989. Wonderful life: the Burgess Shale and the nature of history. New York, W.W. Norton & Company. *In this book describing the marvelous Cambrian fossils of the Burgess Shale, Gould "saves the best for last" by inserting an epilogue on* Pikaia, *the first known chordate.*

Jeffries, R. P. S. 1986. The ancestry of the vertebrates. Cambridge, Cambridge University Press. *This book is an excellent summary of the deuterostome groups and of the various competing hypotheses of vertebrate ancestry.*

Long, J. A. 1995. The rise of fishes: 500 million years of evolution. Baltimore, The Johns Hopkins University Press. *An authoritative, liberally illustrated evolutionary history of fishes.*

Mallatt, J., and J.-Y. Chen. 2003. Fossil sister group of craniates: predicted and found. J. Morph. **258:**1–31. *Reevaluation of* Haikouella *fossils revealed several features that suggest it is the sister taxon to craniates.*

Maisey, J. G. 1996. Discovering fossil fishes. New York, Henry Holt & Company. *Handsomely illustrated chronology of fish evolution with cladistic analysis of evolutionary relationships.*

Shimeld, S. M., and P. W. Holland. 2000. Vertebrate innovations. Proc. Natl. Acad. Sci. **97:**4449–4452. *Focuses on developmental characters, including neural crest, ectodermal placodes, and Hox genes.*

Stokes, M. D., and N. D. Holland. 1998. The lancelet. Am. Sci. **86**(6):552–560. *Describes the historical role of amphioxus in early hypotheses of vertebrate ancestry and summarizes recent molecular data that has rekindled interest in amphioxus.*

ONLINE LEARNING CENTER

Visit www.mhhe.com/hickmanipz14e for chapter quizzing, key term flash cards, web links, and more!

24

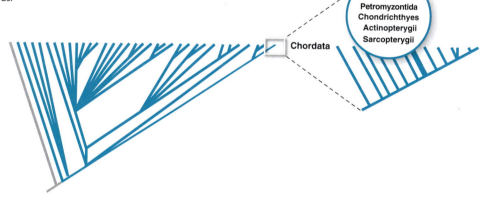

Hammerhead shark, *Sphyrna*, near the Galápagos Islands.

Fishes

- PHYLUM CHORDATA
- CLASS MYXINI
- CLASS PETROMYZONTIDA
- CLASS CHONDRICHTHYES
- CLASS ACTINOPTERYGII
- CLASS SARCOPTERYGII

Myxini
Petromyzontida
Chondrichthyes
Actinopterygii
Sarcopterygii

Chordata

What Is a Fish?

In common (and especially older) usage, the term "fish" denotes a mixed assortment of water-dwelling animals. We speak of jellyfish, cuttlefish, starfish, crayfish, and shellfish, knowing full well that when we use the word "fish" in such combinations, we are not referring to a true fish. In earlier times, even biologists did not make such a distinction. Sixteenth-century natural historians classified seals, whales, amphibians, crocodiles, even hippopotamuses, as well as a host of aquatic invertebrates, as fishes. Later biologists were more discriminating, eliminating first the invertebrates and then the amphibians, reptiles, and mammals from the narrowing concept of a fish. Today we recognize a fish as an aquatic vertebrate with gills, appendages, if present, in the form of fins, and usually a skin with scales of dermal origin. Even this modern concept of the term "fish" is used for convenience, not as a taxonomic unit. Fishes are not a monophyletic group, because the ancestor of land vertebrates (tetrapods) is found within one group of fishes (the sarcopterygians). Thus, fishes can be defined in an evolutionary sense as all vertebrates that are not tetrapods. Because fishes live in habitats that are less accessible to humans than terrestrial habitats, people have rarely appreciated the remarkable diversity of these vertebrates. Nevertheless, whether appreciated by humans or not, the world's fishes have enjoyed an effusive proliferation that has produced an estimated 28,000 living species—more than all other species of vertebrates combined—with adaptations that have fitted them to almost every conceivable aquatic environment. No other animal group threatens their domination of the world's seas, lakes, and streams.

The life of a fish is bound to its body form. Their mastery of stream, lake, and ocean is revealed in the many ways that fishes have harmonized their life design to the physical properties of their aquatic surroundings. Suspended in a medium that is 800 times more dense than air, a trout or pike can remain motionless, varying its neutral buoyancy by adding or removing air from its swim bladder. It may dart forward or at angles, using its fins as brakes and tilting rudders. With excellent organs for salt and water exchange, fishes can steady and finely tune their body fluid composition in their freshwater or seawater environment. Their gills are the most effective respiratory devices in the animal kingdom for extracting oxygen from a medium that contains less than 1/20 as much oxygen as air. Fishes have excellent olfactory and visual senses and a unique lateral-line system, which has an exquisite sensitivity to water currents and vibrations. Thus in mastering the physical problems of their element, early fishes evolved a basic body plan and set of physiological strategies that both shaped and constrained the evolution of their descendants.

ANCESTRY AND RELATIONSHIPS OF MAJOR GROUPS OF FISHES

Fishes are a vast array of distantly related gill-breathing aquatic vertebrates with fins. They are the most ancient and most diverse of the clade Vertebrata, constituting five of the nine living vertebrate classes and about half of the approximately 55,000 vertebrate species.

Fishes are of ancient ancestry, having descended from an unknown free-swimming protochordate ancestor about 550 million years ago (hypotheses of chordate and vertebrate origins are discussed in Chapter 23). The earliest vertebrates were an assemblage of jawless **agnathan** fishes, including the ostracoderms (see Figure 23.14, p. 510). One group of ostracoderms gave rise to the jawed **gnathostomes.**

The use of *fishes* as the plural form of *fish* may sound odd to most people accustomed to using *fish* in both the singular and the plural. *Fish* refers to one or more individuals of the same species; *fishes* refers to more than one species.

The jawless agnathans include along with the extinct ostracoderms the living **hagfishes** and **lampreys,** fishes adapted as scavengers or parasites. Although hagfishes have no vertebrae and lampreys have only rudimentary vertebrae, they nevertheless are included with the subphylum Vertebrata because they have a cranium and many other vertebrate homologies. Although hagfishes and lampreys superficially look much alike, they are in fact so different from each other that they have been assigned to separate taxonomic classes by zoologists.

All remaining fishes have paired appendages and jaws and are included, along with tetrapods (land vertebrates) in the monophyletic group of gnathostomes. They appear in the fossil record in the late Silurian period with fully formed jaws, and no forms intermediate between agnathans and gnathostomes

are known. By the Devonian period, the Age of Fishes, several distinct groups of jawed fishes were common. One of these, the placoderms (p. 512), became extinct in the following Carboniferous period, leaving no descendants. A second group, **cartilaginous fishes** of class Chondrichthyes (sharks, rays, and chimaeras), lost the heavy dermal armor of early jawed fishes and adopted cartilage rather than bone for the skeleton. Most chondrichthyans are active predators with sharklike or raylike body forms that changed only slightly over the ages. As a group, sharks and their kin flourished during the Devonian and Carboniferous periods of the Paleozoic era but declined dangerously close to extinction at the end of the Paleozoic. They recovered in the early Mesozoic and radiated to form the modest but thoroughly successful assemblage of modern sharks and rays (Figure 24.1).

The other two groups of gnathostome fishes, **acanthodians** (p. 512) and **bony fishes,** were abundant and diverse in the Devonian period. Acanthodians resembled bony fishes but were distinguished by having heavy spines on all fins except the caudal fin. They became extinct in the lower Permian period. Although phylogenetic affinities of the acanthodians are much debated, many authors consider them the sister group of bony fishes. **Bony fishes** (Osteichthyes, Figure 24.2) are the most abundant fishes today. We can recognize two distinct groups of bony fishes. Of these two, by far the most diverse are **ray-finned fishes** (class Actinopterygii), which radiated to form most modern bony fishes. The other group, the **lobe-finned fishes** (class Sarcopterygii), contains few fish species today but includes the sister group of the tetrapods. Lobe-finned fishes are represented today by **lungfishes** and **coelacanths**—meager remnants of important lineages that flourished in the Devonian period (see Figure 24.1). A classification of the major fish taxa is on page 540.

LIVING JAWLESS FISHES

Living jawless fishes include approximately 108 species divided between two classes: Myxini (hagfishes) with about 70 species and Petromyzontida (lampreys) with 38 species (Figures 24.3 and 24.4). Members of both groups lack jaws, internal ossification, scales, and paired fins, and both groups share porelike gill openings and an eel-like body form.

Based on this morphological similarity, these two groups formerly were united under the name "Cyclostomata," a grouping shown to be paraphyletic when morphological characters were analyzed with cladistic methods (p. 209). Lampreys have many characters, including vertebrae, extrinsic eye muscles, at least two semicircular canals, and cerebellum, which they uniquely share with gnathostomes. Interestingly, recent analysis of molecular characters shows hagfishes and lampreys forming a monophyletic unit. This grouping, inconsistent with the morphological data, is not supported by most zoologists, and this "cyclostome" hypothesis needs further testing. Thus we take the view that hagfishes form the sister group of a clade that includes lampreys and gnathostomes (see Figure 24.2).

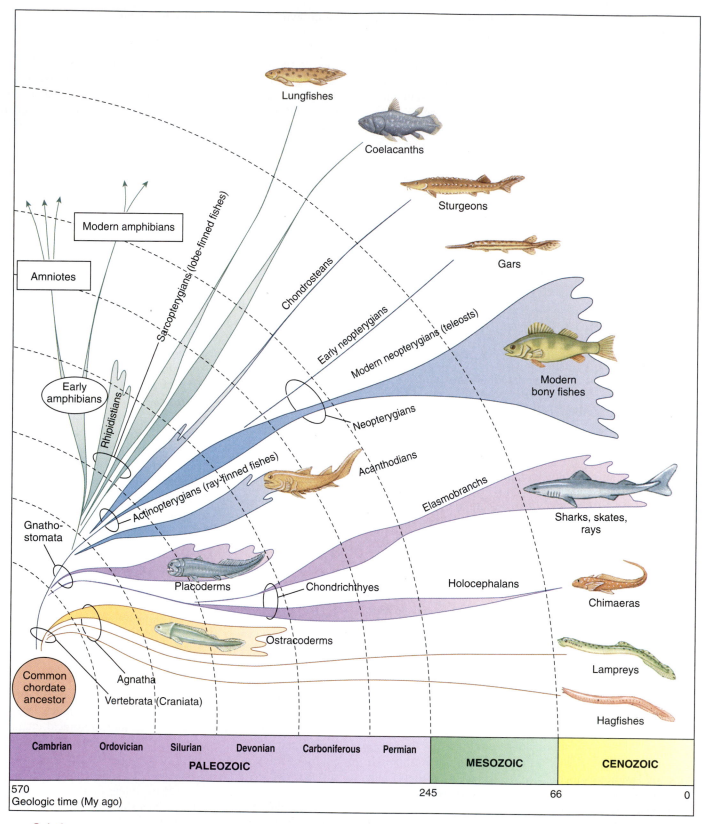

Figure 24.1

Graphic representation of the family tree of fishes, showing evolution of major groups through geological time. Numerous lineages of extinct fishes are not shown. Widened areas in the lines of descent indicate periods of adaptive radiation and relative number of species in each group. The lobe-finned fishes (sarcopterygians), for example, flourished in the Devonian period, but declined and are today represented by only four surviving genera (lungfishes and coelacanths). Homologies shared by sarcopterygians and tetrapods suggest that they form a clade. Sharks and rays radiated during the Carboniferous period, declined in the Permian, then radiated again in the Mesozoic era. Johnny-come-latelies in fish evolution are the spectacularly diverse modern fishes, or teleosts, which make up most living fishes.

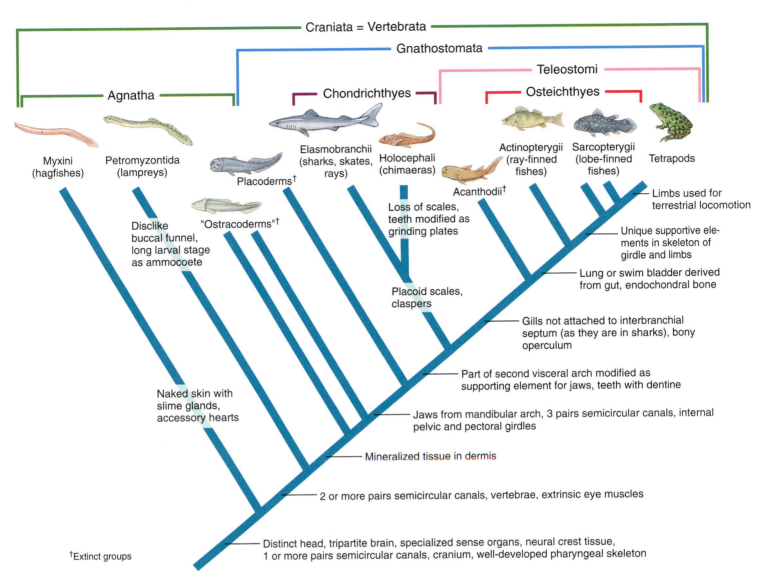

Figure 24.2

Cladogram of the fishes, showing the probable relationships of major monophyletic fish taxa. Several alternative relationships have been proposed. Extinct groups are designated by a dagger (†). Some shared derived characters are shown to the right of branch points. The groups Agnatha and Osteichthyes, although paraphyletic structural grades considered undesirable in cladistic classification, are conveniently recognized in systematics because they share broad structural and functional patterns of organization.

Class Myxini: Hagfishes

Hagfishes are an entirely marine group that feeds on annelids, molluscs, crustaceans, and dead or dying fishes. Thus they are not parasitic like lampreys but are scavengers and predators. There are about 70 species of hagfishes, of which the best known in North America are the Atlantic hagfish, *Myxine glutinosa* (Gr. *myxa*, slime) (see Figure 24.3), and the Pacific hagfish, *Eptatretus stoutii* (N. L. *ept*, Gr. *hepta*, seven, + *tretos*, perforated). Although almost completely blind, hagfishes are quickly attracted to food, especially dead or dying fishes, by their keenly developed senses of smell and touch. A hagfish enters a dead or dying animal through an orifice or by digging into the body. Using two toothed, keratinized plates on its tongue that fold together in a pincerlike action, the hagfish rasps bits of flesh from its prey. For extra leverage, the hagfish often ties a knot in its tail, then passes the knot forward along its body until it is pressed securely against the side of its prey (see Figure 24.3D).

While the strange features of hagfishes fascinate, hagfishes have not endeared themselves to commercial fishermen. In earlier days of commercial fishing mainly by gill nets and set lines, hagfish often bit into the bodies of captured fish and devoured the contents, leaving behind a useless sack of skin and bones. But as large and efficient trawls came into use, hagfishes ceased to be an important pest. Recently the commercial fishing industry "turned the tables" and began targeting hagfishes as a source of leather for golf bags and boots. Fishing pressure has been so intense that some species have greatly declined.

Hagfishes are renowned for their ability to generate enormous quantities of slime. If disturbed or roughly handled, a

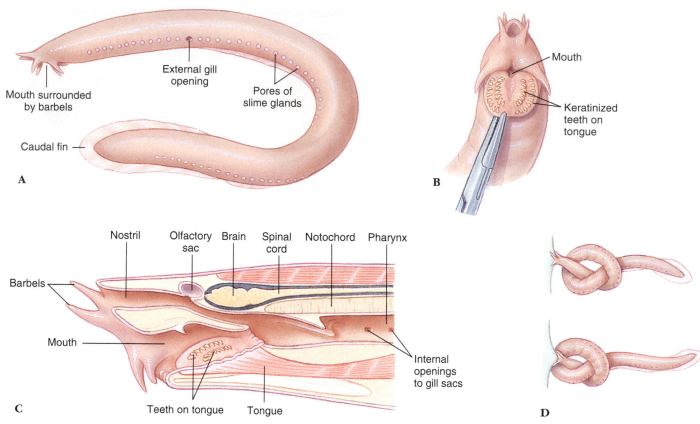

Figure 24.3

Atlantic hagfish, *Myxine glutinosa* (class Myxini). **A,** External anatomy; **B,** Ventral view of head, showing keratinized teeth used to grasp food during feeding; **C,** Sagittal section of head region (note retracted position of rasping tongue and internal openings into a row of gill sacs); **D,** Hagfish knotting, showing how it obtains leverage to tear flesh from prey.

hagfish exudes a milky fluid from special glands positioned along its body. On contact with seawater, the fluid forms a slime so slippery that the animal is almost impossible to grasp.

Figure 24.4

Sea lamprey, *Petromyzon marinus,* feeding on body fluids of a dying fish.

Unlike any other vertebrate, the body fluids of hagfishes are in osmotic equilibrium with seawater, as are most marine invertebrates. Hagfishes have several other anatomical and physiological peculiarities, including a low-pressure circulatory system served by three accessory hearts in addition to the main heart positioned behind the gills.

The reproductive biology of hagfishes remains largely a mystery, despite a still unclaimed prize offered more than 100 years ago by the Copenhagen Academy of Science for information on the animal's breeding habits. It is known that females, which in some species outnumber males 100 to one, produce small numbers of surprisingly large, yolky eggs 2 to 7 cm in diameter depending on the species. There is no larval stage.

Class Petromyzontida: Lampreys

All lampreys of the Northern Hemisphere belong to family Petromyzontidae (Gr. *petros,* stone, + *myzon,* sucking). The group name refers to a lamprey's habit of grasping a stone with its mouth to hold its position in current. The destructive marine lamprey, *Petromyzon marinus,* occurs on both sides of the Atlantic Ocean (in America and Europe) and may attain a length of 1 m (Figure 24.4). *Lampetra* (L. *lambo,* to lick or lap up) also has a wide distribution in North America and Eurasia and ranges from 15 to 60 cm long.

Characteristics of Class Myxini

1. Body slender, eel-like, rounded, with **naked skin containing slime glands**
2. **No paired appendages,** no dorsal fin (the caudal fin extends anteriorly along the dorsal surface)
3. **Fibrous and cartilaginous skeleton;** notochord persistent
4. Biting mouth with two rows of eversible teeth, but no jaws
5. Heart with sinus venosus, atrium, and ventricle; **accessory hearts,** aortic arches in gill region
6. Five to 16 pairs of gills with a variable number of gill openings
7. **Pronephric** and segmented **mesonephric kidneys;** marine, **body fluids isosmotic with seawater**
8. Digestive system **without stomach;** no spiral valve or cilia in intestinal tract
9. Dorsal nerve cord with differentiated brain; **no cerebellum;** 10 pairs of cranial nerves; dorsal and ventral nerve roots united
10. Sense organs of taste, smell, and hearing; **eyes degenerate; one pair semicircular canals**
11. Sexes separate (ovaries and testes in same individual but only one is functional); external fertilization; large yolky eggs, **no larval stage**

There are 20 species of lampreys in North America. About half of these belong to the nonparasitic brook type; the others are parasitic. The genus *Ichthyomyzon* (Gr. *ichthyos,* fish, + *myzon,* sucking), which includes three parasitic and three nonparasitic species, is restricted to eastern North America. On the west coast of North America the chief marine form is *Lampetra tridentata,* commonly sold as *P. marinus* by biological supply companies.

All lampreys ascend freshwater streams to breed. Marine forms are anadromous (Gr. *anadromos,* running upward); that is, they leave the sea where they spend their adult lives to swim up streams to spawn. In North America all lampreys spawn in winter or spring. Males begin building a nest and are joined later by females. Using their oral discs to lift stones and pebbles and vigorous body vibrations to sweep away light debris, they form an oval depression (Figure 24.5). At spawning, with the female attached to a rock to maintain her position over the nest, the male attaches to the dorsal side of her head. As eggs are shed into the nest, they are fertilized by the male. The sticky eggs adhere to pebbles in the nest and quickly become covered with sand. Adults die soon after spawning.

Eggs hatch in about 2 weeks, releasing small larvae called **ammocoetes,** which are so unlike their parents that early biologists thought they were a separate species. The larva bears a remarkable resemblance to amphioxus and possesses the basic

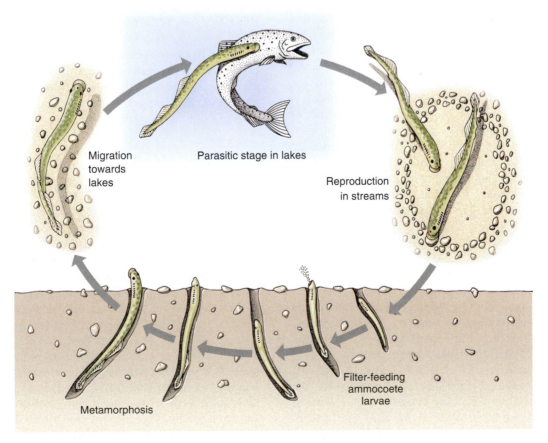

Figure 24.5
Life cycle of the "landlocked" form of the sea lamprey *Petromyzon marinus.*

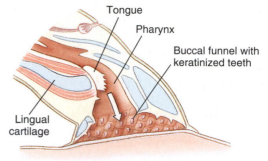

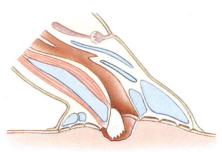

Attachment to fish with keratinized teeth and suction

Tongue protruded for rasping flesh

Figure 24.6

How a lamprey uses its keratinized tongue to feed. After firmly attaching to a fish by its buccal funnel, the protrusible tongue rapidly rasps an opening through the fish's integument. Body fluid, abraded skin, and muscle are eaten.

chordate characteristics in such simplified and easily visualized form that it has been considered a chordate archetype (p. 509). After absorbing the remainder of their yolk supply, young ammocoetes, now about 7 mm long, leave the nest gravel and drift downstream to burrow in a suitable sandy, low-current area. The larvae live as suspension feeders while growing slowly for 3 to 7 or more years, then rapidly metamorphose into adults. This change involves the eruption of eyes, replacement of the hood by the oral disc with keratinized teeth, enlargement of fins, maturation of gonads, and modification of the gill openings.

Parasitic lampreys either migrate to the sea, if marine, or remain in freshwater, where they attach themselves by their suckerlike mouth to a fish and, with their sharp keratinized teeth, rasp through the flesh and suck out body fluids (Figure 24.6). To promote the flow of blood, the lamprey injects an anticoagulant into the wound. When gorged, the lamprey releases its hold but leaves the fish with a large, gaping wound that can be fatal. Parasitic freshwater adults live 1 to 2 years before spawning and then die; anadromous forms live 2 to 3 years.

Nonparasitic lampreys do not feed after emerging as adults and their digestive tract degenerates to a nonfunctional strand of tissue. Within a few months they spawn and die.

Invasion of the Great Lakes by the sea lamprey, *Petromyzon marinus,* in this century has had a devastating effect on fisheries. No lampreys were present in the Great Lakes west of Niagara Falls until the Welland Ship Canal was deepened between 1913 and 1918, allowing lampreys to bypass the Falls. Moving first through Lake Erie to Lakes Huron, Michigan, and Superior, sea lampreys, accompanied by overfishing, caused a total collapse of a multimillion-dollar lake trout fishery in the early 1950s. Rainbow trout, lake whitefish, lake herring, chubs, and other species were destroyed in turn. After reaching peak abundance in 1951 in Lakes Huron and Michigan and in 1961 in Lake Superior, sea

lampreys began to decline, due in part to depletion of their food and in part to expensive control measures (mainly chemical larvacides placed in selective spawning streams). Lake trout, aided by a restocking program, are now recovering, but wounding rates are still high in some lakes.

CLASS CHONDRICHTHYES: CARTILAGINOUS FISHES

There are about 970 living species in class Chondrichthyes, an ancient group that appeared in the Devonian period. Although a much smaller and less diverse assemblage than bony fishes, their impressive combination of well-developed sense organs, powerful jaws and swimming musculature, and predaceous habits ensures them a secure and lasting place in the aquatic community. One of their distinctive features is a cartilaginous skeleton. Although calcification may be extensive in their skeletons, bone is entirely absent throughout the class—a curious evolutionary feature, since Chondrichthyes are derived from ancestors having well-developed bone. Although bone was lost in Chondrichthyes, possibly through a process of neoteny (p. 508), phosphatized mineral tissues were retained in teeth, scales, and spines. Almost all chondrichthyans are marine; only 28 species live primarily in freshwater.

Except for whales, sharks include the largest living vertebrates. The larger sharks may reach 12 m in length. Dogfish sharks commonly studied in zoological laboratories rarely exceed 1 m.

Subclass Elasmobranchii: Sharks, Skates, and Rays

The 13 living orders of elasmobranchs number about 937 species. Coastal waters are dominated by ground sharks, order Carcharhiniformes, which contains typical-looking sharks such as tiger and bull sharks and more bizarre forms, including hammerheads (Figure 24.7). Order Lamniformes contains several large, pelagic sharks dangerous to humans, including great white and mako sharks. Dogfish sharks, familiar to generations of comparative anatomy students, are in order Squaliformes. Skates belong to the order Rajiformes, and several groups of rays (stingrays, eagle rays, manta rays, and devil rays) belong to the order Myliobatiformes.

Although most sharks are by nature timid and cautious, some of them are dangerous to humans. There are numerous authenticated cases of shark attacks by great white sharks, *Carcharodon* (Gr. *karcharos,* sharp, + *odous,* tooth) (reaching 6 m); mako sharks, *Isurus* (Gr. *is,* equal, + *ouros,* tail); tiger sharks, *Galeocerdo* (Gr. *galeos,* shark, + *kerdō,* fox); bull sharks, *Carcharhinus leucas* (Gr. *Karcharos,* sharp, + *rhinos,* nose); and hammerhead sharks, *Sphyrna* (Gr. *sphyra,* hammer). More shark casualties have been reported from tropical and temperate waters of the Australian region than from any other. During World War II there were several reports of mass shark attacks on victims of ship sinkings in tropical waters.

The worldwide shark fishery is experiencing unprecedented pressure, driven by the high price of shark fins for shark-fin soup, an Asian delicacy (which commonly sells for up to $100.00 per bowl). Coastal shark populations in general have declined so rapidly that "finning" has been outlawed in the United States; other countries, too, are setting quotas to protect threatened shark populations. Even in the Marine Resources Reserve of the Galápagos Islands, one of the world's exceptional wild places, tens of thousands of sharks have been killed illegally for the Asian shark-fin market. Contributing to the threatened collapse of shark fisheries worldwide is the low fecundity of sharks and the long time required by most sharks to reach sexual maturity; some species take as long as 35 years.

Form and Function

Although to most people sharks have a sinister appearance and fearsome reputation, they are at the same time among the most gracefully streamlined of all fishes. The body of a dogfish shark (Figure 24.8) is fusiform (spindle-shaped). The asymmetrical heterocercal tail, in which the vertebral column turns upward and extends into the dorsal lobe of the caudal fin, provides thrust and some lift as it sweeps back and forth. There are paired **pectoral** and **pelvic** fins supported by appendicular skeletons, one or two

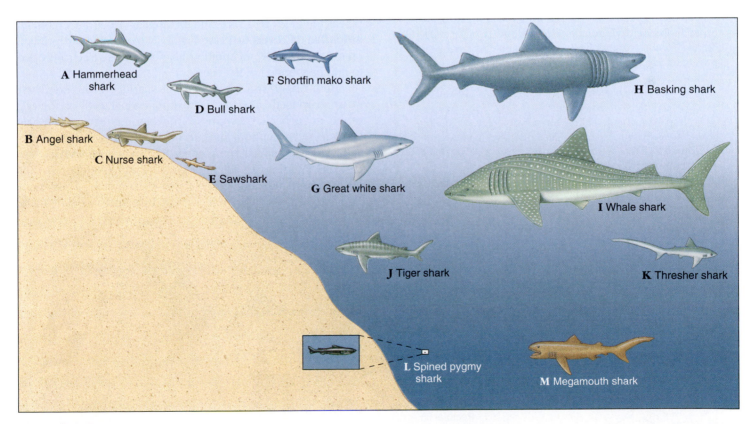

Figure 24.7

Diversity in sharks of the subclass Elasmobranchii: **A,** hammerhead shark, *Sphyrna;* **B,** angel shark, *Squatina;* **C,** nurse shark, *Ginglymostoma cirratum;* **D,** bull shark, *Carcharhinus leucas;* **E,** sawshark, *Pristiophorus;* **F,** shortfin mako shark, *Isurus oxyrinchus;* **G,** great white shark, *Carcharodon carcharias;* **H,** basking shark, *Cetorhinus maximus;* **I,** whale shark, *Rhincodon typus;* **J,** tiger shark, *Galeocerdo cuvier;* **K,** thresher shark, *Alopias vulpinus;* **L,** spined pygmy shark, *Squaliolus laticaudus;* and **M,** megamouth shark, *Megachasma pelagios.*

Characteristics of Class Chondrichthyes

1. Large (average about 2 m), **body fusiform,** or dorsoventrally depressed, with a **heterocercal** caudal fin (diphycercal in chimaeras) (see Figure 24.16); paired pectoral and pelvic fins; pelvic fins in male modified as "claspers"

2. **Mouth ventral;** two olfactory sacs that do not open into the mouth cavity in elasmobranchs; nostrils open into mouth cavity in chimaeras; jaws present

3. Skin with **placoid scales** (see Figure 24.17) or naked; teeth of modified placoid scales and **polyphyodont** in elasmobranchs; teeth modified as grinding plates in chimaeras

4. **Endoskeleton entirely cartilaginous;** notochord persistent but reduced; vertebrae complete and separate (vertebrae present but centra absent in chimaeras)

5. Digestive system with J-shaped stomach (stomach absent in chimaeras); **intestine with spiral valve;** often with large oil-filled liver for buoyancy

6. Circulatory system of several pairs of aortic arches; single circulation; hepatic portal and renal portal systems; heart with sinus venosus, atrium, ventricle, and conus arteriosus

7. Respiration by means of five to seven pairs of gills leading to **exposed gill slits** in elasmobranchs; four pairs of gills covered by an operculum in chimaeras

8. No swim bladder or lung

9. Opisthonephric kidney and rectal gland; blood isosmotic or slightly hyperosmotic to seawater; **high concentrations of urea and trimethylamine oxide in blood**

10. Brain of two olfactory lobes, two cerebral hemispheres, two optic lobes, cerebellum, medulla oblongata; 10 pairs of cranial nerves; **three pairs of semicircular canals;** senses of smell, vibration reception (lateral-line system), vision, and electroreception well-developed

11. Sexes separate; gonads paired; reproductive ducts open into cloaca (separate urogenital and anal openings in chimaeras); oviparous, ovoviviparous, or viviparous; direct development; **fertilization internal**

median **dorsal** fins (each with a spine in *Squalus* [L. a kind of sea fish]), and a median **caudal** fin. A median **anal** fin is present in most sharks, including the smooth dogfish, *Mustelus* (L. *mustela,* weasel). In males, the medial part of the pelvic fin is modified to form a **clasper,** which is used in copulation. Paired **nostrils** (blind pouches) are ventral and anterior to the mouth (Figure 24.9). The lateral eyes are lidless, and behind each eye a spiracle (remnant of the first gill slit) is usually present. Five gill slits are found anterior to each pectoral fin. The tough, leathery skin is covered with tooth-like, dermal **placoid scales** arranged to reduce the turbulence of water flowing along the body surface during swimming.

Sharks are well equipped for their predatory life. They track their prey using highly sensitive senses in an orderly sequence. Sharks may initially detect prey from a kilometer or more away with their large olfactory organs, capable of detecting chemicals as low as 1 part per 10 billion. The laterally placed nostrils of hammerhead sharks (see Figure 24.7) may enhance odor localization by improving stereo-olfaction. Prey also may be located from long distances by sensing low-frequency vibrations with mechanoreceptors in the **lateral-line system.** This system is composed of special receptor organs **(neuromasts)** in interconnected tubes and pores extending along the sides of the body and over the head (Figure 24.10). At closer range a shark switches to vision as the primary method of tracking prey. Contrary to popular belief, most sharks have excellent vision, even in dimly lit waters. During the final stage of attack, sharks are guided to their prey by the bioelectric fields that surround all animals. Electroreceptors, the **ampullae of Lorenzini** (see Figure 24.9), are located primarily on the shark's head. In addition, sharks may use electroreception to find prey buried in the sand.

Both upper and lower jaws of sharks are provided with many sharp teeth. The front row of functional teeth on the edge of the jaw is backed by rows of developing teeth that replace worn teeth throughout the life of the shark (see Figures 24.8 and 24.9). The mouth cavity opens into a large **pharynx,** which contains openings to separate gill slits and spiracles. A short, wide esophagus runs to the J-shaped stomach. A **liver** and **pancreas** open into a short, straight **intestine,** which contains

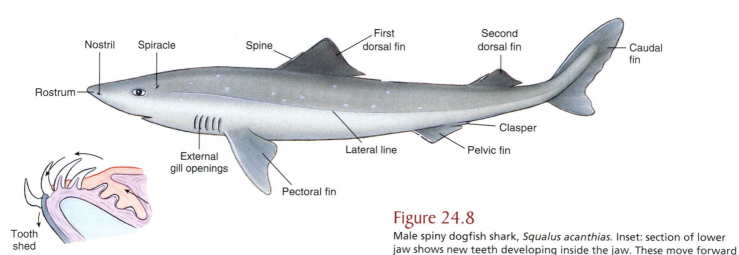

Figure 24.8

Male spiny dogfish shark, *Squalus acanthias.* Inset: section of lower jaw shows new teeth developing inside the jaw. These move forward to replace lost teeth. Rate of replacement varies among species.

Figure 24.9
Head of sand tiger shark, *Carcharias* sp. Note the series of successional teeth. Also visible below the eye are ampullae of Lorenzini.

the **spiral valve** that slows passage of food and increases the absorptive surface (Figure 24.11). Attached to a short rectum is a **rectal gland,** unique to chondrichthyans, which secretes a colorless fluid containing a high concentration of sodium chloride. The rectal gland assists the **opisthonephric kidney** in regulating salt concentration of the blood. Chambers of the **heart** are arranged in tandem formation, and blood circulates in the same pattern seen in other gill-breathing vertebrates (Figure 24.11). Blood leaving the heart via the ventral aorta enters capillary beds in the gills where oxygen is absorbed, then circulated to the rest of the body via the dorsal aorta, without first reentering the heart (see Figure 31.10A, p. 692).

All chondrichthyans have internal fertilization, but maternal support of embryos is highly variable. Some sharks and all skates lay large, yolky eggs immediately after fertilization; these species are termed **oviparous.** Some deposit their eggs in a horny capsule called a "mermaid's purse," which often is provided with

tendrils that wrap around the first firm object it contacts, much like tendrils of grape vines. Embryos are nourished from the yolk for a long period—6 to 9 months in some, as much as 2 years in one species—before hatching as miniature replicas of adults. Many sharks, however, retain embryos in their reproductive tract for prolonged periods. Many are **ovoviviparous** (lecithotrophic viviparous) species, which retain developing young in the uterus while they are nourished by contents of their yolk sac until born. Still other species have true **viviparous** reproduction. In these, embryos receive nourishment from the maternal bloodstream through a **placenta,** or from nutritive secretions, "uterine milk," produced by the mother. Some sharks, including sand tigers, exhibit a grisly type of reproduction in which embryos receive additional nutrition by eating eggs and siblings. The evolution of retention of embryos by many elasmobranchs was an important innovation that contributed to their success. Regardless of the initial amount of maternal support, all parental care ends once eggs are laid or young are born.

Marine elasmobranchs have developed an interesting solution to the physiological problem of living in a salty medium. To prevent water from being drawn out of the body osmotically, elasmobranchs retain nitrogenous compounds, especially urea and trimethylamine oxide, in their extracellular fluid. These solutes, combined with the blood salts, raise the blood solute concentration to exceed slightly that of seawater, eliminating an osmotic inequality between their bodies and surrounding seawater.

More than half of all elasmobranchs are rays, a group that includes skates, electric rays, sawfishes, stingrays, and manta rays. Most are specialized for bottom dwelling, with a dorsoventrally flattened body and greatly enlarged pectoral fins, which they move in a wavelike fashion to propel themselves (Figure 24.12). Gill openings are on the underside of the head, but the large spiracles are on top. Respiratory water enters through these spiracles to prevent clogging the gills, because the mouth is often buried in sand. Teeth are adapted for crushing prey: molluscs, crustaceans, and an occasional small fish.

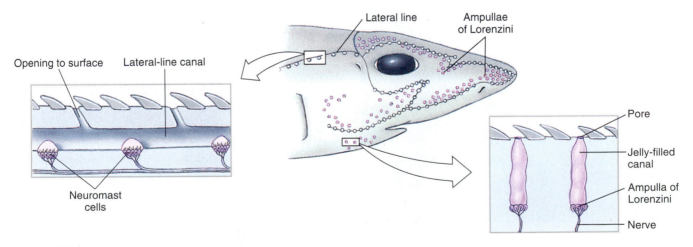

Figure 24.10
Sensory canals and receptors in a shark. The ampullae of Lorenzini respond to weak electric fields, and possibly to temperature, water pressure, and salinity. The lateral-line sensors, called neuromasts, are sensitive to disturbances in the water, enabling a shark to detect nearby objects by reflected waves in the water.

Figure 24.11

Internal anatomy of a dogfish shark, *Squalus acanthias.*

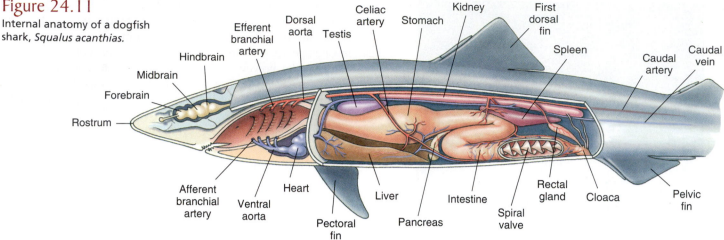

Stingrays have a slender and whiplike tail armed with one or more saw-edged spines with venom glands at the base. Wounds from the spines are excruciatingly painful, and may heal slowly and with complications. Electric rays are sluggish fish with large electric organs on each side of their head (Figure 24.13). Each organ is composed of numerous vertical stacks of disclike cells connected in parallel so that when all cells discharge simultaneously, a high-amperage current is produced that flows out into the surrounding water. The voltage produced is relatively low (50 volts) but power output may be almost one kilowatt—quite sufficient to stun prey or discourage predators. Electric rays were

A

B

Figure 24.12

Skates and rays are specialized for life on the seafloor. Both clearnose skates, *Raja eglanteria* (A), and southern stingrays, *Dasyatis americana* (B), are flattened dorsoventrally and move by undulations of winglike pectoral fins. This stingray (B) is followed by a pilot fish.

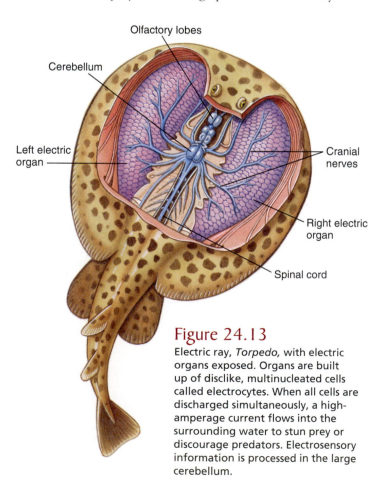

Figure 24.13

Electric ray, *Torpedo*, with electric organs exposed. Organs are built up of disclike, multinucleated cells called electrocytes. When all cells are discharged simultaneously, a high-amperage current flows into the surrounding water to stun prey or discourage predators. Electrosensory information is processed in the large cerebellum.

Figure 24.14
Spotted ratfish, *Hydrolagus collei,* of North American west coast. This species is one of the most handsome of chimaeras, which tend toward bizarre appearances.

used by ancient Egyptians for a form of electrotherapy in the treatment of afflictions such as arthritis and gout.

Subclass Holocephali: Chimaeras

Members of the small subclass Holocephali, distinguished by such suggestive names as ratfish (Figure 24.14), rabbitfish, spookfish, and ghostfish, are remnants of a line that diverged from the shark lineage at least 360 million years ago. Fossil chimaeras (ky-meer′-uz) first occurred in the Devonian period, reached their zenith in the Cretaceous and early Tertiary periods (120 million to 50 million years ago), and have declined ever since. Today there are only about 33 species extant.

Anatomically chimaeras have several features linking them to elasmobranchs, but they possess a suite of unique characters, too. Instead of a toothed mouth, their jaws bear large flat plates. The upper jaw is completely fused to the cranium, a most unusual feature in fishes. Their food includes seaweed, molluscs, echinoderms, crustaceans, and fishes—a surprisingly mixed diet for such a specialized grinding dentition. Chimaeras are not commercial species and are seldom caught. Despite their bizarre shape, they are beautifully colored with a pearly iridescence.

OSTEICHTHYES: BONY FISHES

Origin, Evolution, and Diversity

In the early to middle Silurian, a lineage of fishes with bony endoskeletons gave rise to a clade of vertebrates that contains 96% of living fishes and all living tetrapods. Fishes of this clade have traditionally been termed "bony fishes" **(Osteichthyes)**; bony fishes and tetrapods are united by the presence of **endochondral bone** (bone that replaces cartilage developmentally, p. 650), presence of lungs or a swim bladder derived from the gut, and several cranial and dental characters. Because traditional usage of Osteichthyes does not describe a monophyletic (natural) group (see Figure 24.2), most recent classifications, including the one presented on page 540, do not recognize this

term as a valid taxon. Rather, it is used as a term of convenience to describe vertebrates with endochondral bone that are conventionally termed "bony fishes."

Fossils of the earliest bony fishes show similarities in several craniopharyngeal structures, including a bony operculum and branchiostegal rays, with acanthodians (p. 512 and Figure 23.17), indicating they likely descended from a common ancestor. By the middle Devonian bony fishes already had radiated extensively into two major groups, with adaptations that fitted them for every aquatic habitat except the most inhospitable. One of these groups, ray-finned fishes (class Actinopterygii), includes modern bony fishes (Figure 24.15), the most species-rich group of living vertebrates. A second group, lobe-finned fishes (class Sarcopterygii), is represented today by only eight fishlike vertebrates, the lungfishes and coelacanths (see Figures 24.22 and 24.23); however, it includes the sister group of land vertebrates (tetrapods).

Several key adaptations contributed to the radiation of bony fishes. They have an **operculum** over the gill composed of bony plates and attached to a series of muscles. This feature increases respiratory efficiency because outward rotation of the operculum creates a negative pressure so that water would be drawn across the gills, as well as pushed across by the mouth pump (see Figure 31.20). A gas-filled derivative of the esophagus provides an additional means of gas exchange in hypoxic waters and an efficient means for achieving neutral buoyancy. In fishes that use these pouches primarily for gas exchange, the pouches are called lungs, while in fishes that use these pouches primarily for buoyancy, the pouches are called swim bladders (pp. 531–532). Progressive specialization of jaw musculature and skeletal elements involved in feeding is another key feature of bony fish evolution.

Class Actinopterygii: Ray-Finned Fishes

Ray-finned fishes are an enormous assemblage containing all our familiar bony fishes—almost 27,000 species. Earliest actinopterygians, known as **palaeoniscids** (pay′lee-o-nis′ids), were small fishes, with large eyes, heterocercal caudal fin (Figure 24.16), and thick, interlocking scales with an outer layer of **ganoin** (Figure 24.17). These fishes had a single dorsal fin and numerous bony rays derived from scales stacked end to end, distinctively different in appearance from the lobe-finned fishes with which they shared the Devonian waters. Palaeoniscids are represented by fossil fragments as early as the late Silurian, and flourished throughout the late Paleozoic era, during the same period that ostracoderms, placoderms, and acanthodians disappeared and sarcopterygians declined in abundance (see Figure 24.1).

From those earliest ray-finned fishes arose several clades. Bichirs, in the clade Cladistia, have lungs, heavy ganoid scales, and other characteristics similar to those of palaeoniscids (Figure 24.18). The 16 species of bichirs live in freshwaters of Africa. A second group are the **chondrosteans** (Gr. *chondros,* cartilage, + *osteon,* bone), represented by 27 species of freshwater and anadromous sturgeons and paddlefishes (Figure 24.18). Nearly all chondrosteans have experienced severe population declines from dam construction, overfishing, and pollution.

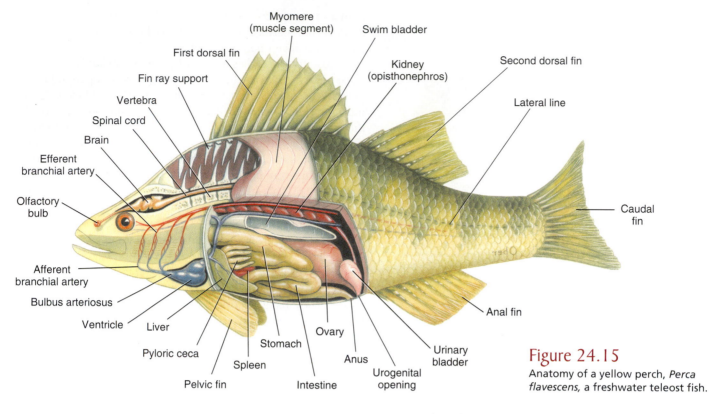

Figure 24.15
Anatomy of a yellow perch, *Perca flavescens*, a freshwater teleost fish.

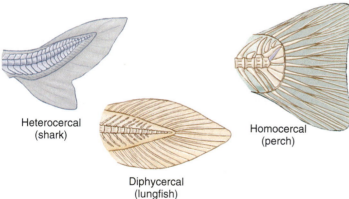

Figure 24.16
Types of caudal fins among fishes.

The third major group of ray-finned fishes to emerge from palaeoniscid stock were the **neopterygians** (Gr. *neos*, new, + *pteryx*, fin). Neopterygians appeared in the late Permian and radiated extensively during the Mesozoic era (see Figure 24.1). During the Mesozoic one lineage gave rise to a secondary radiation that led to modern bony fishes, the teleosts. There are two surviving genera of early neopterygians, the bowfin, *Amia* (Gr. tunalike fish), of shallow, weedy waters of the Great Lakes and Mississippi River basin, and gars, *Lepisosteus* (Gr. *lepidos*, scale, + *osteon*, bone), of eastern and southern North America (Figure 24.19). The seven species of gars are large, ambush predators with elongate bodies and jaws filled with needlelike teeth. Gars and bowfin may surface to gulp air, filling their vascularized swim bladder with air to supplement oxygen obtained in the gills.

The major clade of neopterygians are **teleosts** (Gr. *teleos*, perfect, + *osteon*, bone), the modern bony fishes (see Figure 24.15). Teleost diversity is astounding, with almost 27,000 described species, representing about 96% of all living fishes or about half of all vertebrates (Figure 24.20). In addition, it has been estimated that there are an additional 5,000 to 10,000 undescribed species. Although most of the 200 or so new species of teleosts described each year are from poorly sampled areas such as South America or deep oceanic waters, several new species are described each year from areas as well known as the freshwaters of North America! Teleosts range in size from 7 mm adult minnows to 17 m oarfish and 900 kg, 4.5 m blue marlin (Figure 24.20). These fishes occupy almost every conceivable habitat, from elevations up to 5200 m in Tibet to 8000 m below the surface of the ocean. Some species live in hot springs at 44°C, while others live under the Antarctic ice at −2°C. They may live in lakes with salt concentrations three times that of seawater, caves of total darkness, swamps devoid of oxygen, or even make extended excursions onto land, as do mudskippers (Figure 24.20).

Several morphological trends in the teleost lineage allowed it to diversify into this truly incredible variety of habitats and forms. Heavy dermal armor of primitive ray-finned fishes was replaced by light, thin, flexible **cycloid** and **ctenoid** scales (see Figure 24.17). Some teleosts, such as most eels and catfishes, completely lack scales. Increased mobility and speed that resulted from loss of heavy armor improved predator avoidance and feeding efficiency. Changes in the fins of teleosts increased maneuverability and speed and allowed fins to serve a variety of other functions. The symmetrical shape of the **homocercal** tail (see Figure 24.16) of most teleosts focused musculature

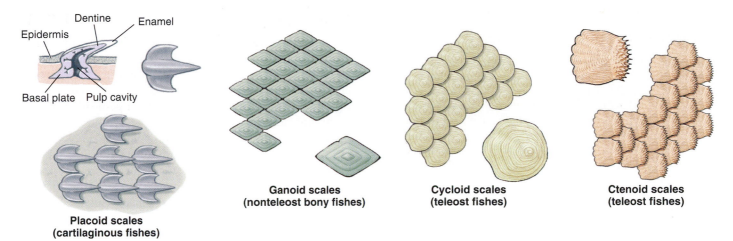

Dentine
Epidermis — Enamel

Basal plate — Pulp cavity

**Placoid scales
(cartilaginous fishes)**

**Ganoid scales
(nonteleost bony fishes)**

**Cycloid scales
(teleost fishes)**

**Ctenoid scales
(teleost fishes)**

Figure 24.17

Types of fish scales. Placoid scales are small, conical toothlike structures characteristic of Chondrichthyes. Diamond-shaped ganoid scales, present in early bony fishes such as the gar, are composed of layers of silvery enamel (ganoin) on the upper surface and bone on the lower. Teleosts have either cycloid or ctenoid scales. These are thin and flexible and are arranged in overlapping rows.

Figure 24.18

Primitive ray-finned fishes of class Actinopterygii. **A,** Bichir, *Polypterus bichir,* of equatorial West Africa. It is a nocturnal predator. **B,** Atlantic sturgeon, *Acipenser oxyrhynchus* (now uncommon), of Atlantic coastal rivers. **C,** Paddlefish, *Polyodon spathula,* of the Mississippi River basin reaches 2 m and 80 kg.

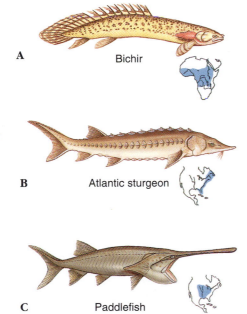

A — Bichir

B — Atlantic sturgeon

C — Paddlefish

A

B

Figure 24.19

Nonteleost neopterygian fishes. **A,** Bowfin, *Amia calva.* **B,** Longnose gar, *Lepisosteus osseus.* They frequent slow-moving streams and swamps of eastern North America where they may hang motionless in the water, ready to snatch passing fishes.

contractions on the tail, permitting greater speed. The dorsal fin shifted from a fixed keel that primarily prevented rolling, to a flexible and highly specialized structure in derived teleosts (see Figure 24.15). These changes in the morphology of fins were useful for camouflage, braking and other complex movements, streamlining, and social communication. Bizarre modifications of the dorsal fin include the lure of anglerfishes, venom-delivering spines of scorpionfishes, and suctorial disc of sharksuckers (Figure 24.20). Teleost lineages demonstrated an increasingly fine control of gas resorption and secretion in the swim bladder. Control of buoyancy likely coevolved with fin modifications to refine locomotion. Several anatomical modifications improved feeding efficiency. Changes in jaw suspension enabled the orobranchial cavity to expand rapidly, creating a highly sophisticated suction device. Rapid jaw protrusion by sliding the upper jaw forward

increased final attack velocity. Gill arches of many teleosts diversified into powerful **pharyngeal jaws** for chewing, grinding, and crushing. With so many innovations teleosts have become the most diverse of fishes.

Characteristics of Class Actinopterygii

1. **Skeleton with bone of endochondral origin;** caudal fin heterocercal in ancestral forms, usually **homocercal** in descendant forms (Figure 24.16); skin with mucous glands and embedded dermal scales (Figure 24.21); scales **ganoid** in ancestral forms, scales **cycloid, ctenoid** or absent in derived forms (Figure 24.17)
2. Paired and median fins present, **supported by long dermal rays (lepidotrichia);** muscles controlling fin movement within body
3. Jaws present; teeth usually present with enamaloid covering; olfactory sacs do not open into mouth; spiral valve present in ancestral forms, absent in derived forms
4. Respiration primarily by gills supported by arches and covered with an **operculum**
5. **Swim bladder** often present with or without a duct connecting to esophagus, usually functioning in buoyancy
6. Circulation consisting of a heart with a sinus venosus, an undivided atrium, and an undivided ventricle; single circulation; typically four aortic arches; nucleated erythrocytes
7. Excretory system of paired opisthonephric kidneys; sexes usually separate; fertilization usually external; larval forms may differ greatly from adults
8. Nervous system of a brain with small cerebrum, optic lobes, and cerebellum; 10 pairs of cranial nerves; three pairs of semicircular canals

Class Sarcopterygii: Lobe-Finned Fishes

The ancestor of tetrapods is found within a group of extinct sarcopterygian fishes called **rhipidistians,** which included several lineages that flourished in freshwaters and shallow coastal areas in the late Paleozoic. Rhipidistians, such as *Eusthenopteron* (see Figure 25.2, p. 546), were cylindrical, large-headed fishes with fleshy fins, and presumably lungs. The evolution of tetrapods from rhipidistians is discussed in Chapter 25.

All early sarcopterygians had lungs as well as gills, and a tail of the **heterocercal** type. However, during the Paleozoic the orientation of the vertebral column changed so that the tail became a symmetrical **diphycercal** tail (see Figure 24.16). These sarcopterygians had powerful jaws; heavy, enameled scales with a dentinelike material called **cosmine;** and strong, fleshy, paired lobed fins that may have been used to clamber over benthic substrates filled with woody debris. The sarcopterygian clade today is represented by only eight fish species: six species of lungfishes and two species of coelacanths (Figures 24.22 and 24.23).

Of the three surviving genera of lungfishes, most similar to early forms is *Neoceratodus* (Gr. *neos,* new, + *keratos,* horn, + *odes,* form), the living Australian lungfish, which may attain a length of 1.5 m (Figure 24.22). This lungfish, unlike its relatives, normally relies on gill respiration, and cannot survive long out of water. The South American lungfish, *Lepidosiren* (L. *lepidus,* pretty, + *siren,* mythical mermaid), and the African lungfish, *Protopterus* (Gr. *prōtos,* first, + *pteron,* wing), can live out of water for long periods of time. *Protopterus* lives in African streams and ponds that may dry during the dry season, with

A

B

C

D

Figure 24.20

Diversity among teleosts. **A,** Blue marlin, *Makaira nigricans,* one of the largest teleosts. **B,** Mudskippers, *Periophthalmus* sp., make extensive excursions on land to graze on algae and capture insects; they build nests in which the young hatch and are guarded by the mother. **C,** Protective coloration of the flamboyant lionfish, *Pterois* sp., advises caution; the dorsal spines are venomous. **D,** The sucking disc on the sharksucker, *Echeneis naucrates,* is a modification of the dorsal fin.

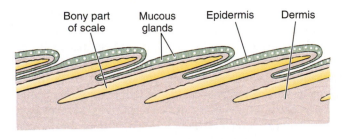

Figure 24.21
Section through the skin of a bony fish, showing the overlapping scales (yellow). The scales lie in the dermis and are covered by epidermis.

oxygen include a lung for respiratory gas exchange and partially separated pulmonary and systemic cardiovascular circuits.

Coelacanths also arose in the Devonian period, radiated somewhat, and reached their peak of diversity in the Mesozoic era. At the end of the Mesozoic era they nearly disappeared but left one remarkable surviving genus, *Latimeria* (Figure 24.23). Because the last coelacanths were believed to have become extinct 70 million years ago, the scientific world was astonished when the remains of a coelacanth were found on a dredge off the coast of South Africa in 1938. An intensive search to locate more

Australian lungfish   African lungfishes   South American lungfish

Figure 24.22
Lungfishes are lobe-finned fishes of class Sarcopterygii. The Australian lungfish, *Neoceratodus forsteri,* is the least specialized of three lungfish genera. The African lungfishes, *Protopterus* sp., are best adapted of the three for remaining dormant in mucus-lined cocoons breathing air during prolonged periods of drought.

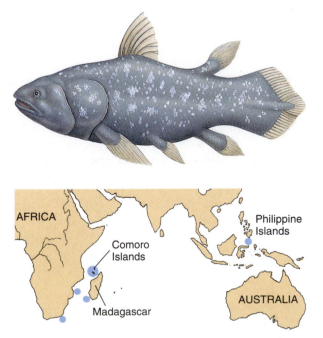

Figure 24.23
The coelacanth genus *Latimeria* is a surviving marine relict of a group of lobe-finned fishes that flourished some 350 million years ago.

specimens was successful off the coast of the Comoro Islands. There fishermen occasionally catch them at great depths with hand lines, providing specimens for research. This was the only known population of *Latimeria* until 1998, when the scientific world again was surprised by the capture of a new species of coelacanth in Sulawesi, Indonesia, 10,000 km from the Comoros!

The "modern" marine coelacanths are descendants of the Devonian freshwater stock. The tail is diphycercal (see Figure 24.16) but possesses a small lobe between the upper and lower caudal lobes, producing a three-pronged structure (Figure 24.23).

Coelacanths are a deep metallic blue with irregular white or brassy flecks, providing camouflage against the dark lava-cave reefs they inhabit. Young are born fully formed after hatching internally from eggs 9 cm in diameter—the largest among bony fishes.

STRUCTURAL AND FUNCTIONAL ADAPTATIONS OF FISHES

Locomotion in Water

To the human eye, some fishes appear capable of swimming at extremely high speeds. But our judgment is unconsciously tempered by our own experience that water is a highly resistant medium through which to move. Most fishes, such as a trout or a minnow, can swim maximally about 10 body lengths per second, obviously an impressive performance by human standards. When

their mud beds baked hard by the hot tropical sun. The fish burrows down at the approach of the dry season and secretes a copious slime mixed with mud to form a hard cocoon in which it estivates until rains return. Adaptations for using atmospheric

Characteristics of Class Sarcopterygii

1. **Skeleton with bone of endochondral origin;** caudal fin **diphycercal** in living representatives, heterocercal in ancestral forms; skin with embedded dermal scales (Figure 24.21) with a layer of dentinelike material, **cosmine,** in ancestral forms
2. Paired and median fins present; paired fins with a single basal skeletal element and short dermal rays; muscles that move paired fins located on appendage
3. Jaws present; teeth are covered with true enamel and typically are crushing plates restricted to palate; olfactory sacs paired, may or may not open into mouth; intestine with spiral valve
4. Gills supported by bony arches and covered with an **operculum**
5. Swim bladder vascularized and used for respiration and buoyancy (fat-filled in coelacanths)
6. Circulation consisting of heart with a sinus venosus, two atria, a partly divided ventricle, and a conus arteriosus; **double circulation** with pulmonary and systemic circuits; characteristically five aortic arches
7. Nervous system with a cerebrum, a cerebellum, and optic lobes; 10 pairs of cranial nerves; three pairs of semicircular canals
8. Sexes separate; fertilization external or internal

these speeds are translated into kilometers per hour it means that a 30 cm (1 foot) trout can swim only about 10.4 km (6.5 miles) per hour. As a general rule, the larger the fish the faster it can swim.

Measuring fish cruising speeds accurately is best done in a "fish wheel," a large ring-shaped channel filled with water that is turned at a speed equal and opposite to that of the fish. Much more difficult to measure are the sudden bursts of speed that most fish can make to capture prey or to avoid being captured. A hooked bluefin tuna was once "clocked" at 66 km per hour (41 mph); swordfish and marlin are thought to be capable of incredible bursts of speed approaching, or even exceeding, 110 km per hour (68 mph). Such high speeds can be sustained for no more than 1 to 5 seconds.

The propulsive mechanism of a fish is its trunk and tail musculature. The axial, locomotory musculature is composed of zigzag bands, called **myomeres.** Muscle fibers in each myomere are relatively short and connect the tough connective tissue partitions that separate each myomere from the next. On the surface myomeres take the shape of a W lying on its side (Figure 24.24) but internally the bands are complexly folded and nested so that the pull of each myomere extends over several vertebrae. This arrangement produces more power and finer control of movement since many myomeres are involved in bending a given segment of the body.

Understanding how fishes swim can be approached by studying the motion of a very flexible fish such as an eel (Figure 24.25). The movement is serpentine, not unlike that of a snake, with waves of contraction moving backward along its body by alternate contraction of myomeres on either side. The anterior end of the body bends less than the posterior end, so that each undulation increases in amplitude as it travels along the body. While undulations move backward, bending of the body pushes laterally against the water, producing a **reactive force** that is directed forward, but at an angle. It can be analyzed as having two components: **thrust,** which is used to overcome drag and propels the fish forward, and **lateral force,** which tends to make the fish's head "yaw," or deviate from the course in the same direction as its tail. This side-to-side head movement is very obvious in a swimming eel or shark, but many fishes have a large, rigid head with enough surface resistance to minimize yaw.

The movement of an eel is reasonably efficient at low speed, but its body shape generates too much frictional drag for rapid swimming. Fishes that swim rapidly, such as trout, are less flexible and limit body undulations mostly to the caudal region (Figure 24.25). Muscle force generated in the large anterior muscle mass is transferred through tendons to the relatively nonmuscular caudal peduncle and tail where thrust is generated. This form of swimming reaches its highest development in tunas, whose bodies do not flex at all. Virtually all thrust is derived from powerful beats of the caudal fin (Figure 24.26). Many fast oceanic fishes such as marlin, swordfish, amberjacks, and wahoo have swept-back caudal fins shaped like a sickle. Such fins are the aquatic counterpart of the high aspect ratio wings of the swiftest birds (p. 601).

The body temperature of most fishes is the same as their environment, because any heat generated internally quickly is lost into the surrounding water. However, some fishes, such as tunas (Figure 24.26) and mako sharks, maintain a high temperature in their swimming muscles and viscera—as much as 10°C warmer than surrounding water. Marlins (see Figure 24.20A) and other billfishes elevate the temperature of their brain and retina. Research by F. G. Carey and others explains how these fishes accomplish this kind of thermoregulation, called **regional endothermy.** Heat is generated as a by-product of various activities, including digestion and swimming, or for billfishes, by a specialized heat-generating organ beneath the brain. This heat is conserved with a rete mirabile, a parallel bundle of blood vessels arranged to provide a countercurrent flow of blood (p. 682). High temperatures apparently promote powerful swimming and enhance digestive and nervous system activity. Fishes with regional endothermy are the fastest in the world.

Swimming is the most economical form of animal locomotion, largely because aquatic animals are almost perfectly supported by their medium and need to expend little energy to overcome the force of gravity. If we compare the energy cost per kilogram of body weight of traveling 1 km by different forms of locomotion, we find swimming costs only 0.39 kcal (salmon) as compared with 1.45 kcal for flying (gull) and 5.43 kcal for walking (ground squirrel). However, part of the unfinished business of biology is understanding how fish and aquatic mammals are

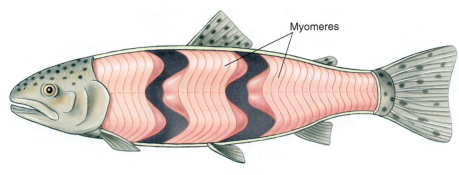

Figure 24.24
Trunk musculature of a teleost fish, partly dissected to show internal arrangement of the muscle bands (myomeres). The myomeres are folded into a complex, nested grouping, an arrangement that favors stronger and more controlled swimming.

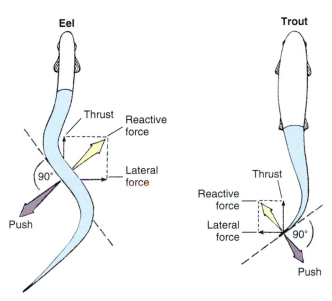

Figure 24.25
Movements of swimming fishes, showing the forces developed by an eel-shaped and spindle-shaped fish.

Source: From Vertebrate Life, by Pough et al., 1996. Reprinted by permission of Prentice-Hall, Inc., Upper Saddle River, NJ.

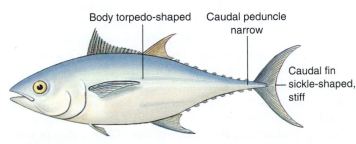

Figure 24.26
Bluefin tuna, *Thunnus thynnus*, showing adaptations for fast swimming. Powerful trunk muscles pull on the slender caudal peduncle. Since the body does not bend, all of the thrust comes from beats of the stiff, sickle-shaped caudal fin.

able to move through water while creating almost no turbulence. The secret lies in the way aquatic animals bend their bodies and fins (or flukes) to swim and in the friction-reducing properties of the body surface.

Neutral Buoyancy and the Swim Bladder

All fishes are slightly heavier than water because their skeletons and other tissues contain heavy elements present only in trace amounts in natural waters. To keep from sinking, sharks, which lack a swim bladder, must always keep moving forward in the water. The asymmetrical (heterocercal) tail of a shark provides lift as it sweeps through the water, and its broad head and flat pectoral fins (see Figure 24.8) act as angled planes to provide additional lift. Sharks also are aided in buoyancy by having very large livers containing a special fatty hydrocarbon called **squalene** with a density of only 0.86 grams per milliliter. The liver thus acts like a large sack of buoyant oil that helps to compensate the shark's heavy body.

By far the most efficient flotation device is a gas-filled space. The **swim bladder** serves this purpose in bony fishes (Figure 24.27). It arose from the paired lungs of primitive Devonian bony fishes. Lungs were probably a ubiquitous feature of Devonian freshwater bony fishes when, as we have seen, warm, swampy habitats would have made such an accessory respiratory structure advantageous. Swim bladders are present in most pelagic bony fishes but are absent in tunas, most abyssal fishes, and most bottom dwellers, such as flounders and sculpins.

Without a swim bladder, bony fishes sink because their tissues are denser than water. To achieve neutral buoyancy, they displace additional water by a volume of gas in a swim bladder, thus adjusting their total density to match that of the surrounding water. This adjustment allows fishes with a swim bladder to remain suspended indefinitely at any depth with no muscular effort. Unlike bone, blood, and other tissues, gas is compressible and changes volume as a fish changes its depth. If a fish swims to a greater depth, the greater pressure exerted by the surrounding water compresses the gas in the swim bladder, so that the fish becomes less buoyant and begins to sink. The volume of gas in the swim bladder must be increased to establish a new equilibrium buoyancy. When a fish swims upward, gas in the bladder expands because of the reduced surrounding water pressure, making the fish more buoyant. Unless gas is removed, the fish will continue to ascend with increasing speed as the swim bladder continues to expand.

Gas may be removed from the swim bladder in two ways. The more primitive **physostomous** (Gr., *phys*, bladder, + *stoma*, mouth) fishes (trout, for example) have a pneumatic duct that connects the swim bladder to the esophagus, through which they may expel air. More derived teleosts exhibit the **physoclistous** (Gr., *phys*, bladder, + *clist*, closed) condition in which the pneumatic duct is lost in adults. In physoclistous fishes, gas is

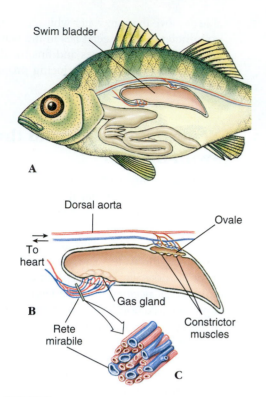

Figure 24.27

A, Swim bladder of a teleost fish. The swim bladder lies in the coelom just beneath the vertebral column. **B,** Gas is secreted into the swim bladder by the gas gland. Gas from the blood is moved into the gas gland by the rete mirabile, a complex array of tightly-packed capillaries that act as a countercurrent multiplier to increase oxygen concentration. The arrangement of venous and arterial capillaries in the rete is shown in **C.** To release gas during ascent, a muscular valve opens, allowing gas to enter the ovale from which the gas is removed by diffusion into the blood.

absorbed by blood from the **ovale,** a vascularized area of the swim bladder (Figure 24.27). Both types of fishes require gas to be secreted into the swim bladder from the blood, although a few shallow-water-inhabiting physostomes may gulp air to fill their swim bladder.

Physiologists who were at first baffled by the secretion mechanism now understand how it operates. In brief, the gas gland secretes lactic acid, which enters the blood, causing a localized high acidity in the **rete mirabile** that forces hemoglobin to release its load of oxygen. The capillaries in the rete are arranged in parallel, creating a countercurrent multiplier (p. 677), allowing oxygen to reach high concentrations in the gas gland and to diffuse into the swim bladder. The final gas pressure attained in the swim bladder depends on the length of the rete capillaries; they are relatively short in fishes living near the surface, but are extremely long in deep-sea fishes.

The amazing effectiveness of this device is exemplified by a fish living at a depth of 2400 m (8000 feet). To keep the bladder inflated at that depth, the gas inside (mostly oxygen, but also variable amounts of nitrogen, carbon dioxide, argon, and even some carbon monoxide) must have a pressure exceeding 240 atmospheres, which is much greater than the pressure in a fully

charged steel gas cylinder. Yet the oxygen pressure in the fish's blood cannot exceed 0.2 atmosphere—in equilibrium with the oxygen pressure in the atmosphere at the sea surface.

Hearing and Weberian Ossicles

Fishes, like other vertebrates, detect sounds as vibrations in the inner ear. Detecting these vibrations is difficult for aquatic vertebrates because their bodies are nearly the same density as the surrounding water, letting sound waves pass through the fish's body nearly undetected.

A particularly elegant solution to this problem is found in the ostariophysans, a group of teleosts that contains the minnows, characins, suckers, and catfishes. The ostariophysans include about 7900 species and are usually the dominant fishes in diversity and abundance in freshwater habitats. Their success may be partly due to their **Weberian ossicles,** a set of small bones, which allow them to hear faint sounds over a much broader range of frequency than other teleosts. Reception of sound begins at the swim bladder, which can be easily vibrated because it is air-filled. Sound vibrations are transmitted from the swim bladder to the inner ear by the Weberian ossicles (Figure 24.28). This system has some similarity to the tympanum and middle ear bones of mammals (pp. 744–745), but evolved independently. Adaptations to improve hearing are not confined to ostariophysans. For example, herrings and anchovies have anterior expansions of the swim bladder that directly contact the skull. The importance of the swim bladder in these fishes is demonstrated by experiments in which the swim bladder is artificially deflated, reducing sensitivity to sounds.

Respiration

Fish gills are composed of thin filaments, each covered with a thin epidermal membrane folded repeatedly into platelike **lamellae** (Figure 24.29). These are richly supplied with blood vessels. The gills are located inside the pharyngeal cavity and are covered with a movable flap, the **operculum.** This arrangement provides excellent protection to delicate gill filaments, streamlines the body, and makes possible a pumping system for moving water through the mouth, across the gills, and out the operculum. Instead of opercular flaps as in bony fishes, elasmobranchs have a series of **gill slits** (see Figure 24.8) out of which the water flows. In both elasmobranchs and bony fishes the branchial mechanism is arranged to pump water continuously and smoothly over the gills, although to an observer it appears that fish breathing is pulsatile. The flow of water is opposite the direction of blood flow (countercurrent flow), the best arrangement for extracting the greatest possible amount of oxygen from water. Some bony fishes can remove as much as 85% of the dissolved oxygen from water passing over their gills. Very active fishes, such as herring and mackerel, can obtain sufficient water for their high oxygen demands only by swimming forward continuously to force water into their open mouth and across their gills. This process is called ram ventilation. Such fishes would be asphyxiated if placed in an aquarium that restricts free-swimming movements, even if the water is saturated with oxygen.

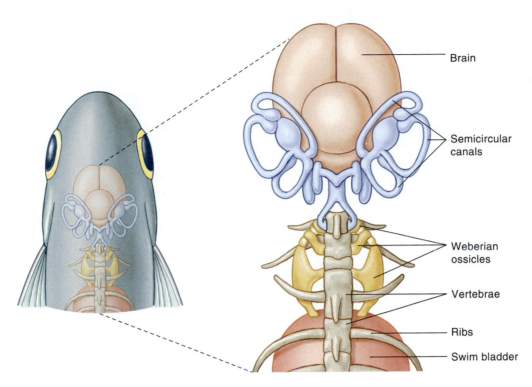

Figure 24.28
Weberian ossicles are small bones that transmit sound vibrations received in the swim bladder to the inner ear. Teleosts with this apparatus can detect faint sounds over a much broader range of frequency than other fishes.

A surprising number of fishes can live out of water for varying lengths of time by breathing air. Several devices are employed by different fishes. We already have described the lungs of lungfishes, gars, and the extinct rhipidistians. Freshwater eels often make overland excursions during rainy weather, using the skin as a major respiratory surface. Electric eels, *Electrophorus* (Gr. *ēlektron,* something bright, + *phoros,* to bear), have degenerate gills and must supplement gill respiration by gulping air through a vascular mouth cavity. One of the best air breathers of all is the Indian climbing perch, *Anabas* (Gr. *anabainō,* to go up), which spends most of its time on land near the water's edge, breathing air through special air chambers above much-reduced gills.

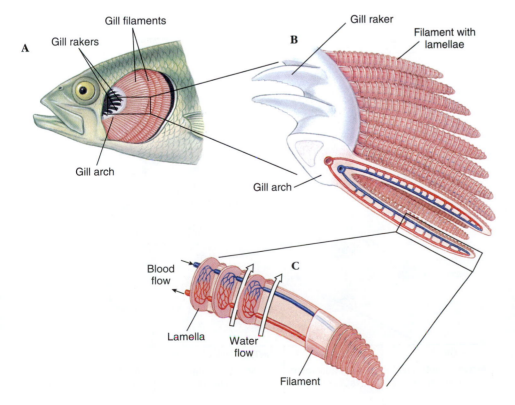

Figure 24.29
Gills of fish. The bony, protective flap covering the gills (operculum) has been removed, **A,** to reveal branchial chamber containing the gills. There are four gill arches on each side, each bearing numerous filaments. A portion of a gill arch **(B)** shows gill rakers that project forward to strain out food and debris, and gill filaments that project to the rear. A single gill filament **(C)** is dissected to show the blood capillaries within the platelike lamellae. Direction of water flow *(large arrows)* is opposite the direction of blood flow.

Osmotic Regulation

Freshwater is an extremely dilute medium with a salt concentration (0.001 to 0.005 gram moles per liter [M]) much below that of the blood of freshwater fishes (0.2 to 0.3 M). Water therefore tends to enter their bodies osmotically, and salt is lost by diffusion outward. Although the scaled and mucus-covered body surface is almost totally impermeable to water, water gain and salt loss do occur across thin membranes of the gills. Freshwater fishes are **hyperosmotic regulators** with several defenses against these problems (Figure 24.30). First, excess water is pumped out by the **opisthonephric** kidneys (p. 673), which are capable of forming very dilute urine. Second, special **salt-absorbing cells** located in the gill epithelium actively move salt ions, principally sodium and chloride, from water to the blood. This absorption, together with salt present in the fish's food, replaces diffusive salt loss. These mechanisms are so efficient that a freshwater fish devotes only a small part of its total energy maintaining itself in osmotic balance.

Perhaps 90% of all bony fishes are restricted to either a freshwater or a seawater habitat because they are incapable of osmotic regulation in the "wrong" habitat. Most freshwater fishes quickly die if placed in seawater, as will marine fishes placed in freshwater. However, some 10% of all teleosts can pass back and forth with ease between both habitats. These **euryhaline fishes** (Gr. *eurys*, broad, + *hals*, salt) are of two types: those such as many flounders, sculpins, and killifish that live in estuaries or certain intertidal areas where the salinity fluctuates throughout the day; and those such as salmon, shad, and eels, that spend part of their life cycle in freshwater and part in seawater.

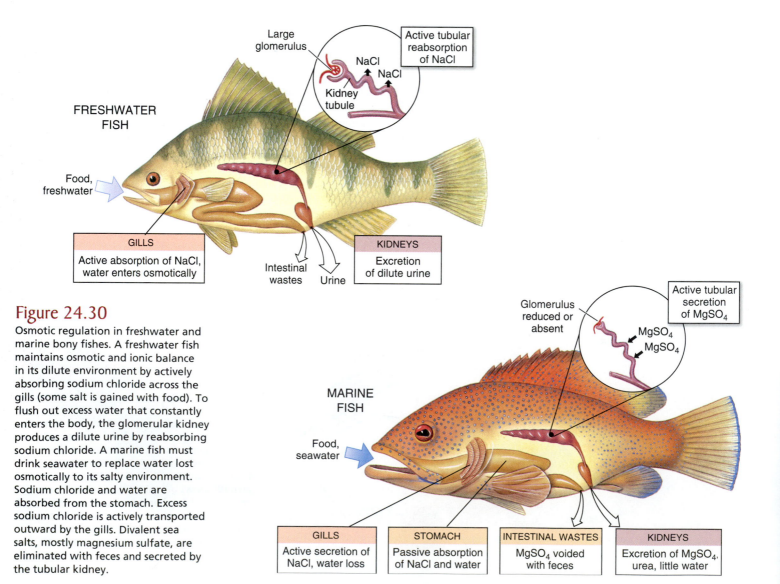

Figure 24.30

Osmotic regulation in freshwater and marine bony fishes. A freshwater fish maintains osmotic and ionic balance in its dilute environment by actively absorbing sodium chloride across the gills (some salt is gained with food). To flush out excess water that constantly enters the body, the glomerular kidney produces a dilute urine by reabsorbing sodium chloride. A marine fish must drink seawater to replace water lost osmotically to its salty environment. Sodium chloride and water are absorbed from the stomach. Excess sodium chloride is actively transported outward by the gills. Divalent sea salts, mostly magnesium sulfate, are eliminated with feces and secreted by the tubular kidney.

Marine bony fishes are **hypoosmotic regulators** that encounter a completely different set of problems. Having a much lower blood salt concentration (0.3 to 0.4 M) than the seawater around them (about 1 M), they tend to lose water and gain salt. A marine teleost fish quite literally risks drying out, much like a desert mammal deprived of water. To compensate for water loss, a marine teleost drinks seawater (Figure 24.30). Excess salt accompanying the seawater is disposed in multiple ways. Major sea salt ions (sodium, chloride, and potassium) are carried by the blood to the gills where they are secreted outward by special **salt-secretory cells.** The remaining sea salt ions, mostly magnesium, sulfate, and calcium, are voided with feces or excreted by the kidneys. Unlike a freshwater fish's kidneys, which form urine by the usual filtration-resorption sequence typical of most vertebrate kidneys (pp. 673–677), a marine fish's kidneys excrete divalent ions by tubular secretion. Since very little if any filtrate is formed, the glomeruli have lost their importance and disappeared in some marine teleosts. Pipefishes and the frogfish shown in Figure 24.32 are examples of "aglomerular" marine fishes.

Feeding Behavior

For any fish, feeding is one of its main concerns in day-to-day living. Although many a luckless angler would swear otherwise, a fish devotes more time and energy to eating, or searching for food to eat, than to anything else. Throughout the long evolution of fishes, there has been unrelenting selective pressure for those adaptations that enable a fish to win the eat-or-be-eaten contest. Certainly the most far-reaching single event was the evolution of jaws (p. 511). Jaws freed fishes from a largely passive filter-feeding existence, enabling them to adopt a predatory mode of life. Improved means of capturing larger prey demanded stronger muscles, more agile movement, better balance, and improved special senses. More than any other aspect of its life habit, feeding behavior shapes the fish.

Most fishes are **carnivores** and prey on a myriad of animal foods from zooplankton and insect larvae to large vertebrates. Some deep-sea fishes are capable of eating victims nearly twice their own size—an adaptation for life in a world where meals are infrequent. Most advanced ray-finned fishes cannot masticate their food as we can because doing so would block the current of water across the gills. Some, however, such as wolf eels (Figure 24.31), have molarlike teeth in their jaws for crushing prey, which may include hard-bodied crustaceans. Others that do grind their food use powerful pharyngeal teeth in their throat. Most carnivorous fish swallow their prey whole, using sharp-pointed teeth in their jaws and on the roof of the mouth to seize prey. The incompressibility of water assists many large-mouthed predators in capturing prey. When the mouth is suddenly opened, water rushes in, sweeping the victim inside (Figure 24.32).

A second group of fishes are **herbivores** that eat plants and macroalgae. Plant eaters are relatively uncommon among fishes, but are crucial intermediates in the food chain in some habitats. Plant eaters are most common in coral reefs (parrotfishes, damselfishes, and surgeonfishes) and in tropical freshwater habitats (some minnows, characins, and catfishes).

Figure 24.31
Wolf eel, *Anarrhichthys ocellatus,* feeding on a sea cucumber it has captured and pulled to the opening of its den.

Figure 24.32
Longlure frogfish, *Antennarius multiocellatus,* awaits its meal. Above its head swings a modified dorsal fin spine ending in a fleshy tentacle that contracts and expands in a convincing wormlike manner. When a fish approaches the alluring bait, the huge mouth opens suddenly, creating a strong inward current that sweeps the prey inside. The entire process takes only 4 milliseconds!

Suspension feeders that crop the abundant microorganisms of the sea form a third and diverse group of fishes ranging from fish larvae to basking sharks. However, the most characteristic group of plankton feeders are herringlike fishes (menhaden, herring, anchovies, capelin, pilchards, and others), mostly **pelagic** (open-sea dwellers) fishes that travel in large schools.

Both phytoplankton and smaller zooplankton are strained from the water with sievelike gill rakers (see Figure 32.1, p. 710). Because plankton feeders are the most abundant of all marine fishes, they are important food for numerous larger but less abundant carnivores. Many freshwater fishes also depend on plankton for food.

Other groups of fishes include **scavengers,** such as hagfishes, that consume dead and dying animals, and **detritivores,** such as some suckers and minnows, that consume fine, particulate organic matter. Some fishes use a **parasitic** mode of feeding in which they consume parts of other live fishes. Examples of these include lampreys (p. 518) and the candiru, *Vandellia,* a tiny elongate catfish that feeds on the gill epithelia of host fishes. Finally it should be noted that most fishes, though specialized for a narrow diet, may use other foods when available.

Digestion in most fishes follows the vertebrate plan. Except in several fishes that lack distinct stomachs, food proceeds from stomach to tubular intestine, which tends to be short in carnivores (see Figure 24.15) but may be extremely long and coiled in herbivorous and detritivorous forms. In herbivorous grass carps, for example, the intestine may be nine times the body length, an adaptation for the lengthy digestion required for plant carbohydrates. In carnivores, some protein digestion may be initiated in the acid medium of the stomach, but the principal function of the stomach is to store often large and infrequent meals while awaiting their reception by the intestine.

Digestion and absorption proceed simultaneously in the intestine. A curious feature of ray-finned fishes, especially teleosts, is the presence of numerous **pyloric ceca** (see Figure 24.15) found in no other vertebrate group. Their primary function appears to be fat absorption, although all classes of digestive enzymes (protein-, carbohydrate-, and fat-splitting) are secreted there.

Migration

Freshwater Eels

For centuries naturalists had been puzzled about the life history of freshwater eels, *Anguilla* (an-gwil'la) (L. eel), a common and commercially important species of coastal streams of the North Atlantic. Eels are **catadromous** (Gr. *kata,* down, + *dromos,* running), meaning that they spend most of their lives in freshwater but migrate to the sea to spawn. Each fall, large numbers of eels were seen swimming down rivers toward the sea, but no adults ever returned. Each spring countless numbers of young eels, called "elvers" (Figure 24.33), each about the size of a wooden matchstick, appeared in coastal rivers and began swimming upstream. Beyond the assumption that eels must spawn somewhere at sea, location of their breeding grounds was completely unknown.

The first clue was provided by two Italian scientists, Grassi and Calandruccio, who in 1896 reported that elvers were not larval eels but rather were relatively advanced juveniles. True larval eels, they discovered, were tiny, leaf-shaped, completely transparent creatures that bore little resemblance to an eel. They had been called **leptocephali** (Gr. *leptos,* slender, + *kephal¯e,* head) by early naturalists, who never suspected their true identity. In 1905 Johann Schmidt, supported by the Danish government, began a systematic study of eel biology, which he continued until his death in 1933. With cooperation of captains of commercial vessels plying the Atlantic, thousands of leptocephali were caught in different areas of the Atlantic with plankton nets Schmidt supplied. By noting where larvae in different stages of development were captured, Schmidt and his colleagues eventually reconstructed the spawning migrations.

When adult eels leave the coastal rivers of Europe and North America, they swim steadily and apparently at great depth for 1 to 2 months until they reach the Sargasso Sea, a vast area of warm oceanic water southeast of Bermuda (Figure 24.33). Here, at depths of 300 m or more, the eels spawn and die. Minute larvae then begin an incredible journey back to the streams of Europe and North America. Since the Sargasso Sea is much closer to the American coastline, American eel larvae make their journey in only about eight months, compared to three years for European eel larvae. Males typically remain in brackish water of coastal rivers, while females migrate as far as several hundred kilometers upstream. After 8 to 15 years of growth, females, now 1 m long, return to the ocean to join the smaller males in the journey back to the spawning grounds in the Sargasso Sea.

Recent enzyme electrophoretic analysis of eel larvae confirmed not only the existence of separate European and American species but also Schmidt's belief that the European and American eels spawn in partially overlapping areas of the Sargasso Sea.

Homing Salmon

The life history of salmon is nearly as remarkable as that of freshwater eels and certainly has received far more popular attention. Salmon are **anadromous** (Gr. *anadromos,* running upward); they spend their adult lives at sea but return to freshwater to spawn. Atlantic salmon, *Salmo salar* (L. *salmo,* salmon, + *sal,* salt), and Pacific salmon (six species in the genus *Oncorhynchus* [on-ko-rink'us] [Gr. *onkos,* hook, + *rhynchos,* snout]) have this practice, but there are important differences among the seven species. Atlantic salmon may make repeated upstream spawning runs. The six Pacific salmon species (sockeye, coho, pink, Chinook, chum, and Japanese masu) each make a single spawning run (Figure 24.34), after which they die.

The virtually infallible homing instinct of the Pacific species is legendary: after migrating downstream as a smolt, (a juvenile stage) a sockeye salmon ranges many hundreds of miles over the Pacific for nearly 4 years, grows to 2 to 5 kg in weight, and then returns almost unerringly to spawn in the headwaters of its parent stream. Some straying does occur and is an important means of increasing gene flow and populating new streams.

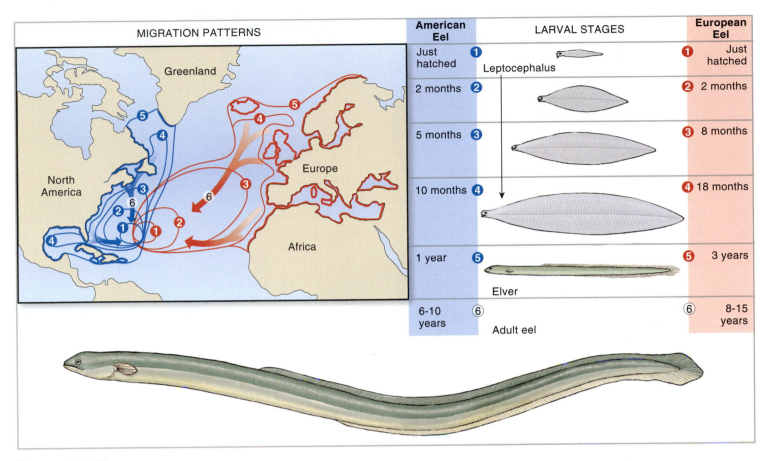

American Eel		LARVAL STAGES	European Eel	
Just hatched	❶	*Leptocephalus*	❶	Just hatched
2 months	❷		❷	2 months
5 months	❸		❸	8 months
10 months	❹		❹	18 months
1 year	❺	*Elver*	❺	3 years
6-10 years	⑥	*Adult eel*	⑥	8-15 years

Figure 24.33

Life histories of the European eel, *Anguilla anguilla,* and American eel, *Anguilla rostrata.* Migration patterns of European species are shown in red. Migration patterns of American species are shown in blue. Boxed numbers refer to stages of development. Note that the American eel completes its larval metamorphosis and sea journey in one year. It requires nearly three years for the European eel to complete its much longer journey.

Salmon runs in the Pacific Northwest have been devastated by a lethal combination of spawning stream degradation by logging, pollution and, especially, by more than 50 hydroelectric dams, which obstruct upstream migration of adult salmon and kill downstream migrants as they pass through the dams' power-generating turbines. In addition, the chain of reservoirs behind the dams, which has converted the Columbia and Snake Rivers into a series of lakes, increases mortality of young salmon migrating downstream by slowing their passage to the sea. The result is that the annual run of wild salmon is today only about 3% of the 10 to 16 million fish that ascended the rivers 150 years ago. While recovery plans have been delayed by the power industry, environmental groups argue that in the long run losing the salmon will be more expensive to the regional economy than making changes now that will allow salmon stocks to recover.

Experiments by A. D. Hasler and others have shown that homing salmon are guided upstream by the characteristic odor of their parent stream. When salmon finally reach the spawning beds of their parents (where they themselves were hatched), they spawn and die. The following spring, newly hatched fry transform into smolts before and during the downstream

Figure 24.34

Migrating Pacific sockeye salmon (*Oncorhynchus nerka*).

migration. At this time they are imprinted (p. 791) with the distinctive odor of the stream, which is apparently a mosaic of compounds released by the characteristic vegetation and soil in the watershed of the parent stream. They also seem to imprint on odors of other streams they pass while migrating downriver and use these odors in reverse sequence as a map during the upriver migration as returning adults.

How do salmon find their way to the mouth of a coastal river from the trackless miles of open ocean? Salmon move hundreds of miles away from the coast, much too far to be able to detect the odor of their parent stream. Experiments suggest that some migrating fish, like birds, can navigate by orienting to the position of the sun. However, migrant salmon can navigate on cloudy days and at night, indicating that sun navigation, if used at all, cannot be a salmon's only navigational cue. Fish also (again, like birds) appear able to detect and to navigate to the earth's magnetic field. Finally, fishery biologists concede that salmon may not require precise navigational abilities at all, but instead may use ocean currents, temperature gradients, and food availability to reach the general coastal area where "their" river is located. From this point, they would navigate by their imprinted odor map, making correct turns at each stream junction until they reach their natal stream.

Reproduction and Growth

In a group as diverse as fishes, it is no surprise to find extraordinary variations on the basic theme of sexual reproduction. Most fishes favor a simple theme: they are **dioecious,** with **external fertilization** and **external development** of their eggs and embryos (oviparity). However, as tropical fish enthusiasts are well aware, the ever-popular ovoviviparous guppies and mollies of home aquaria bear their young alive after development in the ovarian cavity of the mother (Figure 24.35). As described earlier in this chapter (p. 523), some viviparous sharks develop a kind of placental attachment through which the young are nourished during gestation.

Let us return to the much more common oviparous mode of reproduction. Many marine fishes are extraordinarily profligate egg producers. Males and females come together in great schools and release vast numbers of gametes into the water to drift with currents. Large female cod may release 4 to 6 million eggs at a single spawning. Less than one in a million eggs will survive the numerous perils of the ocean to reach reproductive maturity.

Unlike the minute, buoyant, transparent eggs of pelagic marine teleosts, those of many near-shore bottom-dwelling (benthic) species are larger, typically yolky, nonbuoyant, and adhesive. Some bury their eggs, many attach them to vegetation, some deposit them in nests, and some even incubate them in their mouths (Figure 24.36). Many benthic spawners guard their eggs. Intruders expecting an easy meal of eggs may be met with a vivid and often belligerent display by the guard, which is almost always male.

Freshwater fishes almost invariably produce nonbuoyant eggs. Those, such as perch, that provide no parental care simply scatter their myriads of eggs among weeds or along the sediment. Freshwater fishes that do provide egg care, such as bullhead catfishes and some darters, produce fewer, larger eggs that enjoy a better chance for survival.

Elaborate preliminaries to mating are the rule for freshwater fishes. A female Pacific salmon, for example, performs a ritualized mating "dance" with her breeding partner after arriving at the spawning bed in a fast-flowing, gravel-bottomed stream (Figure 24.37). She then turns on her side and scoops a nest hole with her tail. As eggs are laid by the female, they are fertilized by a male (Figure 24.37). After the female covers the eggs with gravel, the exhausted fish dies.

Soon after an egg of an oviparous species is laid and fertilized, it absorbs water and the outer layer hardens. Cleavage follows, and a blastoderm forms, astride a relatively enormous yolk mass. Soon the yolk mass is enclosed by the developing blastoderm, which then begins to assume a fishlike shape. Many fish hatch as larvae, carrying a semitransparent sac of yolk, which provides their food supply until the mouth and digestive tract have developed. The larvae then begin searching for their own food. After a period of growth a larva undergoes a metamorphosis, especially dramatic in many marine species, including eels (see Figure 24.33). Body shape is refashioned, fin and color patterns change, and the animal becomes a juvenile bearing the unmistakable definitive body form of its species.

Figure 24.36

Male banded jawfish, *Opistognathus macrognathus,* orally brooding its eggs. The male retrieves the female's eggs and incubates the eggs until they hatch. During brief periods when the jawfish is feeding, the eggs are left in the burrow.

Figure 24.35

Rainbow surfperch, *Hypsurus caryi,* giving birth. All of the West Coast surfperches (family Embiotocidae) are ovoviviparous.

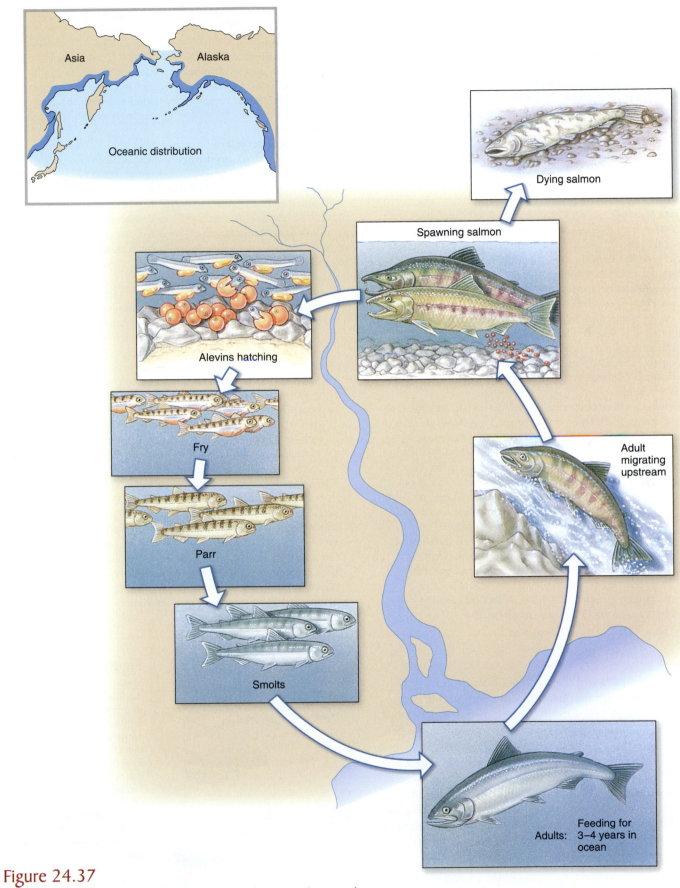

Figure 24.37
Spawning Pacific salmon, *Oncorhynchus*, and development of eggs and young.

Classification of Living Fishes

The following Linnaean classification of major fish taxa follows that of Nelson (2006). The probable relationships of these traditional groupings together with the major extinct groups of fishes are shown in a cladogram in Figure 24.2. Other schemes of classification have been proposed. Because of the difficulty of determining relationships among the numerous living and fossil species, we can appreciate why fish classification has undergone, and will continue to undergo, continuous revision.

Phylum Chordata

Subphylum Vertebrata (Craniata)

Superclass Myxinomorphi

Class Myxini (mik-sin′y) (Gr. *myxa*, slime): **hagfishes.** No jaws or paired fins; mouth with four pairs of tentacles; buccal funnel absent; 1 to 16 pairs of external gill openings; vertebrae absent; slime glands present. Examples: *Myxine, Epaptretus;* about 70 species, marine.

Superclass Petromyzontomorphi

Class Petromyzontida (pet′tr⁻o-m⁻i-zon′ti-də) (Gr. *petros*, stone, + *myzon*, sucking): **lampreys.** No jaws or paired fins; mouth surrounded by keratinized teeth but no barbels, buccal funnel present; seven pairs of external gill openings; vertebrae present only as neural arches. Examples: *Petromyzon, Ichthyomyzon, Lampetra;* 38 species, freshwater and anadromous.

Superclass Gnathostomata (na′tho-sto′-ma-ta) (Gr. *gnathos*, jaw, + *stoma*, mouth). Jaws present; paired appendages present (secondarily lost in a few forms); three pairs of semicircular canals; notochord partly or completely replaced by centra.

Class Chondrichthyes (kon-drik′thee-eez) (Gr. *chondros*, cartilage + *ichthys*, fish): **cartilaginous fishes.** Cartilaginous skeleton; teeth not fused to jaws and usually replaced; no swim bladder; intestine with spiral valve; claspers present in males.

Subclass Elasmobranchii (e-laz′mo-bran′kee′i) (Gr. *elasmos*, plated, + *branchia*, gills): **sharks, skates, and rays.** Placoid scales or derivatives (scutes and spines) usually present; five to seven gill arches and gill slits in separate clefts along pharynx; upper jaw not fused to cranium. Examples: *Squalus, Raja, Charcarodon, Sphyrna;* about 937 species, mostly marine.

Subclass Holocephali (hol′o-sef′a-li) (Gr. *holos*, entire, + *kephalē*, head): **chimaeras, ratfishes.** Scales absent; four gill slits covered by operculum; jaws with tooth plates; accessory clasping organ (tentaculum) in males; upper jaw fused to cranium. Examples: *Chimaera, Hydrolagus;* about 33 species, marine.

Class Actinopterygii (ak′ti-nop-te-rij′ee-i) (Gr. *aktis*, ray, + *pteryx*, fin, wing): **ray-finned fishes.** Skeleton ossified; single gill opening covered by operculum; paired fins supported primarily by dermal rays; limb musculature within body; swim bladder mainly a hydrostatic organ, if present; atrium and ventricle not divided; teeth with enameloid covering.

Subclass Cladistia (clə-dis′tē-a) (Gr. *cladi*, branch): **bichirs.** Rhombic ganoid scales; lungs; spiracle present; dorsal fin consisting of 5 to 18 finlets. Examples: *Polypterus;* about 16 species, freshwater.

Subclass Chondrostei (kon-dros′tē-i) (Gr. *chondros*, cartilage, + *osteon*, bone): **paddlefishes, sturgeons.** Skeleton primarily cartilage; caudal fin heterocercal; large scutes or tiny ganoid scales present; spiracle usually present; more fin rays that ray supports. Examples: *Polyodon, Acipenser;* 29 species, freshwater and anadromous.

Subclass Neopterygii (nee′op-te-rij′ee-i) (Gr. *neo*, new, + *pteryx*, fin, wing): **gars, bowfin, teleosts.** Skeleton primarily bone; caudal fin usually homocercal; scales cycloid, ctenoid, absent, or rarely, ganoid. Fin ray number equal to their supports in dorsal and anal fins. Examples: *Amia, Lepisosteus, Anguilla, Oncorhynchus, Perca;* about 27,000 species, nearly all aquatic habitats.

Class Sarcopterygii (sar-cop-te-rij′ee-i) (Gr. *sarkos*, flesh, + *pteryx*, fin, wing): **lobe-finned fishes.** Skeleton ossified; single gill opening covered by operculum; paired fins with sturdy internal skeleton and musculature within appendage; diphycercal tail; intestine with spiral valve; usually with lungs; atrium and ventricle at least partly divided; teeth with enamel covering. Examples: *Latimeria* (coelacanths); *Neoceratodus, Lepidosiren, Protopterus* (lungfishes); 8 species, marine and freshwater. Not monophyletic unless tetrapods are included.

Growth is temperature dependent. Consequently, fish living in temperate regions grow rapidly in summer when temperatures are high and food is abundant but nearly stop growing in winter. Annual rings in the scales, otoliths, and other bony parts reflect this seasonal growth (Figure 24.38), a distinctive record of convenience to fishery biologists who wish to determine a fish's age. Unlike birds and mammals, which stop growing after reaching adult size, most fishes after attaining reproductive maturity continue to grow for as long as they live. This may be a selective advantage, since the larger the fish, the more gametes it produces and the greater its contribution to future generations.

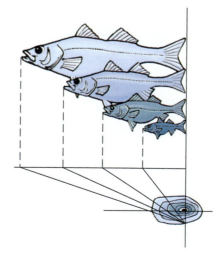

Figure 24.38

Scale growth. Fish scales disclose seasonal changes in growth rate. Growth is interrupted during winter, producing year marks (annuli). Each year's increment in scale growth is a ratio to the annual increase in body length. Otoliths (ear stones) and certain bones can also be used in some species to determine age and growth rate.

SUMMARY

Fishes are poikilothermic, gill-breathing aquatic vertebrates with fins for appendages. They include the oldest vertebrate groups, having originated from an unknown chordate ancestor in the Cambrian period or possibly earlier. Five classes of living fishes are recognized. The jawless hagfishes (class Myxini) and lampreys (class Petromyzontida), have an eel-like body form without paired fins, a cartilaginous skeleton, a notochord that persists throughout life, and a disclike mouth adapted for sucking or biting. All other vertebrates have jaws, a major development in vertebrate evolution. Members of class Chondrichthyes (sharks, rays, skates, and chimaeras) have a cartilaginous skeleton (a degenerative feature), paired fins, excellent sensory organs, and an active, characteristically predaceous habit. Bony fishes may be divided into two classes. Lobe-finned fishes of class Sarcopterygii, represented today by lungfishes and coelacanths, form a paraphyletic group if tetrapods are excluded, as done in traditional classification. Terrestrial vertebrates arose from one lineage within this group. The second is ray-finned fishes (class Actinopterygii), a huge and diverse modern assemblage containing nearly all familiar freshwater and marine fishes. Modifications of the skeletal and muscular systems in this group increased locomotion and feeding efficiency.

Modern bony fishes (teleosts) have radiated to form approximately 27,000 species that reveal an enormous diversity of adaptations, body form, behavior, and habitat preference. Most fishes swim by undulatory contractions of the body muscles, which generate thrust (propulsive force) and lateral force. Eel-like fishes oscillate the whole body, but in more rapid swimmers undulations are limited to the caudal region or caudal fin alone.

Most pelagic bony fishes achieve neutral buoyancy in water using a gas-filled swim bladder, the most effective gas-secreting device known in the animal kingdom. Sensitivity to sounds may be enhanced by Weberian ossicles that transmit sounds from the swim bladder to the inner ear. Gills of fishes, having efficient countercurrent flow between water and blood, facilitate high rates of oxygen exchange. All fishes show well-developed osmotic and ionic regulation, achieved principally by the kidneys and gills.

With the exception of agnathans, all fishes have jaws that are variously modified for carnivorous, herbivorous, planktivorous, and detritivorous feeding modes.

Many fishes are migratory, and some, such as catadromous freshwater eels and anadromous salmon, make remarkable migrations of great length and precision. Fishes reveal an extraordinary range of sexual reproductive strategies. Most fishes are oviparous, but ovoviviparous and viviparous fishes are not uncommon. Reproductive investment may be in large numbers of eggs with low survival (many marine fishes) or in fewer eggs with greater parental care for better survival (freshwater fishes).

REVIEW QUESTIONS

1. Provide a brief description of the fishes citing characteristics that would distinguish them from all other animals.
2. What characteristics distinguish hagfishes and lampreys from all other fishes? How do they differ in morphology from each other?
3. Describe feeding behavior in hagfishes and lampreys. How do they differ?
4. Describe the life cycle of sea lampreys, *Petromyzon marinus,* and the history of their invasion of the Great Lakes.
5. In what ways are sharks well equipped for a predatory life habit?
6. What function does the lateral-line system serve? Where are receptors located?
7. Explain how bony fishes differ from sharks and rays in the following systems or features: skeleton, scales, buoyancy, respiration, and reproduction.
8. Match ray-finned fishes in the right column with the group to which each belongs in the left column:

   _____ Chondrosteans   a. Perch
   _____ Nonteleost   b. Sturgeon
           neopterygians   c. Gar
   _____ Teleosts   d. Salmon
                       e. Paddlefish
                       f. Bowfin
9. Make a cladogram that includes the following groups of fishes: chondrosteans, elasmobranchs, hagfishes, holocephalans, lampreys, lungfishes, teleosts. Add the following synapomorphies to the diagram: claspers, cranium, endochondral bone, fleshy fins, jaws, vertebrae.
10. List four characteristics of teleosts that contributed to their incredible diversity and abundance.
11. Only eight species of lobe-finned fishes are alive today, remnants of a group that flourished in the Devonian period of the Paleozoic. What morphological characteristics distinguish lobe-finned fishes? What is the literal meaning of Sarcopterygii, the class to which lobe-finned fishes belong?
12. Give the geographical locations of the three surviving genera of lungfishes and explain how they differ in their ability to survive out of water. Which of the three is the most similar to the earliest, fossil lungfishes?
13. Describe discovery of living coelacanths. What is the evolutionary significance of the group to which they belong?
14. Compare the swimming movements of eels with those of trout, and explain why the latter are more efficient for rapid locomotion.
15. Sharks and bony fishes approach or achieve neutral buoyancy in different ways. Describe the methods evolved in each group. Why must a teleost fish adjust the gas volume in its swim bladder when it swims upward or downward? How is gas volume adjusted?
16. What is meant by "countercurrent flow" as it applies to fish gills?
17. How do Weberian ossicles increase a fish's sensitivity to sounds?

18. Compare the osmotic problem and the mechanism of osmotic regulation in freshwater and marine bony fishes.
19. Two principal groups of fishes, with respect to feeding behavior, are the carnivores and the suspension feeders. How are these two groups adapted for their feeding behavior?
20. Describe the life cycle of freshwater eels. How does the life cycle of American eels differ from that of European eels?

21. How do adult Pacific salmon find their way back to their parent stream to spawn?
22. What mode of reproduction in fishes is described by each of the following terms: oviparous, ovoviviparous, viviparous?
23. Reproduction in marine pelagic fishes and in freshwater fishes is distinctively different. How and why do they differ?

SELECTED REFERENCES

Barton, M. 2007. Bond's biology of fishes, ed. 3. Belmont, California, Thomson Brooks/Cole. *A superior revision of Bond's text that emphasizes anatomy, physiology, and ecology.*

Carey, F. G. 1973. Fishes with warm bodies. Sci. Am. **228:**36–44 (Feb.). *Classic paper about how fishes with regional endothermy keep warm.*

Helfman, G. J., B. B. Collette, and D. E. Facey. 1997. The diversity of fishes. Malden, Massachusetts, Blackwell Science. *This delightful and information-packed textbook focuses on adaptation and diversity and is particularly strong in evolution, systematics, and history of fishes.*

Horn, M. H., and R. N. Gibson. 1988. Intertidal fishes. Sci. Am. **258:**64–70 (Jan.). *Describes the special adaptations of fishes living in a demanding environment.*

Long, J. A. 1995. The rise of fishes: 500 million years of evolution. Baltimore, The Johns Hopkins University Press. *A lavishly illustrated evolutionary history of fishes.*

Martini, F. H. 1998. Secrets of the slime hag. Sci. Am. **279:**70–75 (Oct.). *Biology of the most primitive living craniate.*

Moyle, P. B., and J. J. Cech, Jr. 2004. Fishes: an introduction to ichthyology, ed. 5. Englewood Cliffs, New Jersey, Prentice-Hall, Inc. *Textbook written in a lively style and stressing ecology rather than morphology.*

Nelson, J. S. 2006. Fishes of the world, ed. 4. New York, John Wiley & Sons, Inc. *Authoritative classification of all major groups of fishes.*

Page, L. M., and B. M. Burr. 1991. A field guide to freshwater fishes: North America north of Mexico. Boston, Houghton Mifflin. *Range maps and color illustrations for most species.*

Paxton, J. R., and W. N. Eschmeyer. 1998. Encyclopedia of fishes, ed. 2. San Diego, Academic Press. *Excellent authoritative reference that focuses on diversity and is spectacularly illustrated.*

Springer, V. G., and J. P. Gold. 1989. Sharks in question. Washington, Smithsonian Institution Press. *Morphology, biology, and diversity of sharks, richly illustrated.*

Webb, P. W. 1984. Form and function in fish swimming. Sci. Am. **251:**72–82 (July). *Specializations of fish for swimming and analysis of thrust generation.*

Weinberg, S. 2000. A fish caught in time: the search for the coelacanth. London, Fourth Estate. *The exciting history of the coelacanth discoveries.*

ONLINE LEARNING CENTER

Visit www.mhhe.com/hickmanipz14e for chapter quizzing, key term flash cards, web links, and more!

Early Tetrapods and Modern Amphibians

- PHYLUM CHORDATA
- CLASS AMPHIBIA

A pickeral frog, Rana palustris, *during metamorphosis.*

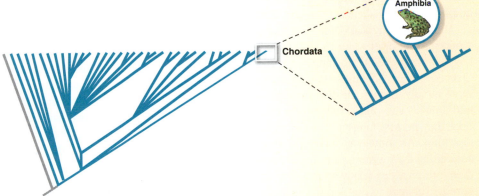

From Water to Land in Ontogeny and Phylogeny

A chorus of frogs beside a pond on a spring evening heralds one of nature's dramatic events. Mating frogs produce masses of eggs, which will hatch into limbless, gill-breathing, fishlike tadpole larvae that feed and grow. Then a remarkable transformation occurs. Hindlegs appear and gradually lengthen. The tail shortens. Larval teeth and gills are lost. Eyelids develop. Forelimbs emerge. In a few weeks the aquatic tadpole has completed its metamorphosis to an adult frog.

The evolutionary transition from water to land occurred not in weeks but over millions of years. A lengthy series of alterations cumulatively fitted the vertebrate body plan for life on land. The origin of land vertebrates is no less a remarkable feat for this fact—a feat that would have a poor chance of succeeding today because well-established terrestrial competitors would exclude poorly adapted transitional forms.

Amphibians include the only living vertebrates that have a transition from water to land in both their ontogeny and phylogeny. Even after 350 million years of evolution, amphibians remain quasiterrestrial, hovering between aquatic and land environments. This double life is expressed in their name. Even the amphibians best adapted for a terrestrial existence cannot stray far from moist conditions. Many, however, have developed ways to keep their eggs out of open water where their larvae would be exposed to enemies.

Adaptation for life on land is a major theme of the remaining vertebrate groups, which form a clade called the superclass Tetrapoda. Amphibians and amniotes (including nonavian reptiles, birds, and mammals) are the two major extant branches of **tetrapod** phylogeny, which originates in the Devonian period. Many tetrapod lineages have lost one or both of the two pairs of limbs that give the group its name. Amphibians are ectothermic, (p. 000) primitively quadrupedal tetrapods with glandular skin. Many depend on freshwater streams or pools for their reproduction. In this chapter, we review the origins of terrestrial vertebrates and discuss the amphibian branch in detail. We discuss the major amniote groups in Chapters 26 through 28.

MOVEMENT ONTO LAND

Movement from water to land is perhaps the most dramatic event in animal evolution because it involves invasion of a physically hazardous habitat. Life originated in water. Animals are mostly water in composition, and all cellular activities occur in water. Nevertheless, organisms invaded land, carrying their watery composition with them. Vascular plants, pulmonate snails, and tracheate arthropods made this transition much earlier than did vertebrates, and thus generated a food supply that terrestrial vertebrates eventually would use. Although invasion of land required modification of almost every organ system, aquatic and terrestrial vertebrates retain many structural and functional similarities. We see a transition between aquatic and terrestrial vertebrates most clearly today in the many living amphibians that make this transition during their own life histories.

Important physical differences that animals must accommodate when moving from water to land include those of (1) oxygen content, (2) density, (3) temperature regulation, and (4) habitat diversity. Oxygen is at least 20 times more abundant in air, and it diffuses much more rapidly through air than through water. Consequently, terrestrial animals can obtain oxygen far more easily than aquatic ones once they possess lungs and other appropriately adapted respiratory structures. Air, however, has approximately 1000 times less buoyant density than water and is approximately 50 times less viscous. It therefore provides relatively little support against gravity, requiring terrestrial animals to develop strong limbs and to remodel their skeleton to achieve adequate structural support. Air fluctuates in temperature more readily than does water, and terrestrial environments therefore experience harsh and unpredictable cycles of freezing, thawing, drying, and flooding. Terrestrial animals require behavioral and physiological strategies to protect themselves from thermal extremes.

Despite its hazards, the terrestrial environment offers a great variety of habitats including boreal, temperate, and tropical forests, grasslands, deserts, mountains, oceanic islands, and polar regions (pp. 808–812). Safe shelter for protection of vulnerable eggs and young is found much more readily in many of these terrestrial habitats than in aquatic ones.

EARLY EVOLUTION OF TERRESTRIAL VERTEBRATES

By the Devonian period, beginning about 400 million years ago, bony fishes had diversified to include many freshwater forms. An important combination of characteristics that evolved originally in aquatic habitats now gave its possessors some ability to explore terrestrial habitats. These characteristics included two structures that connected to the pharynx, an air-filled cavity, which functioned as a swim bladder, and paired internal nares (evolutionary origin shown on Figure 25.1), which functioned in chemoreception. On land, this combination of structures would be used to draw oxygen-rich air through the nares and into the air-filled cavity, whose surface would permit some respiratory gas exchange with body fluids. The bony elements of paired fins, modified for support and movement on underwater surfaces (evolutionary origin shown on Figure 25.1), gained sufficient strength to provide support and movement of the body on land.

The internal nares, air-filled cavity, and paired limbs of an aquatic tetrapod ancestor therefore were available for modification by later evolution to fit them for terrestrial breathing and support. The air-filled cavity illustrates an important evolutionary principle in which a structure that has evolved by natural selection for an initial utility or role is later recruited or "coopted" for a new role. Note that the air-filled cavities called "lungs" and "swim bladders" are homologous, with the specific terms distinguishing the structure's role for air breathing in terrestrial forms versus buoyancy during swimming in aquatic ones.

Freshwater habitats are inherently unstable, being prone to evaporation or depletion of the dissolved oxygen needed to support vertebrate life. It is therefore not surprising that multiple fish groups, given a combination of structures that could be coopted for terrestrial breathing and locomotion, evolved some degree of terrestriality. Mudskippers and lungfish are two familiar examples of evolution of terrestriality by fishes; however, only one such transition, occurring in the early Devonian period, provided the ancestral lineage of all tetrapod vertebrates. This lineage ultimately evolved the characteristic tetrapod adaptations to air breathing, including increased vascularization of the air-filled cavity with a rich capillary network to form a lung, and a **double circulation** to direct the deoxygenated blood into the lungs for oxygenation, and the oxygenated blood from the lungs to other body tissues.

Tetrapods evolved their limbs in an ancestral aquatic habitat during the Devonian period prior to their evolutionary movement onto land. Although fish fins may seem very different from the jointed limbs of tetrapods, an examination of bony elements of the paired fins of lobe-finned fishes shows that they broadly resemble homologous structures of amphibian limbs. In *Eusthenopteron,* a Devonian lobe-fin, we can recognize an upper arm bone (humerus) and two forearm bones (radius and ulna) as well as other elements that we homologize with wrist bones of tetrapods (Figure 25.2). *Eusthenopteron* could paddle itself through the bottom mud of pools with its fins, but it could not walk upright because backward and forward movement of the fins was limited to about 20 to 25 degrees. The recently discovered fossil genus *Tiktaalik* is morphologically intermediate between lobe-finned fishes and tetrapods. *Tiktaalik* probably inhabited shallow, oxygen-depleted streams or swamps, using its limbs to support the body while placing its snout above water to breathe air. This form also might have traversed land. *Acanthostega,* one of the earliest known Devonian tetrapods, had well-formed tetrapod limbs with clearly formed digits on both fore- and hindlimbs, but these limbs were too weakly constructed to hoist the animals' bodies off the surface for proper walking on land. *Ichthyostega,* however, with a fully

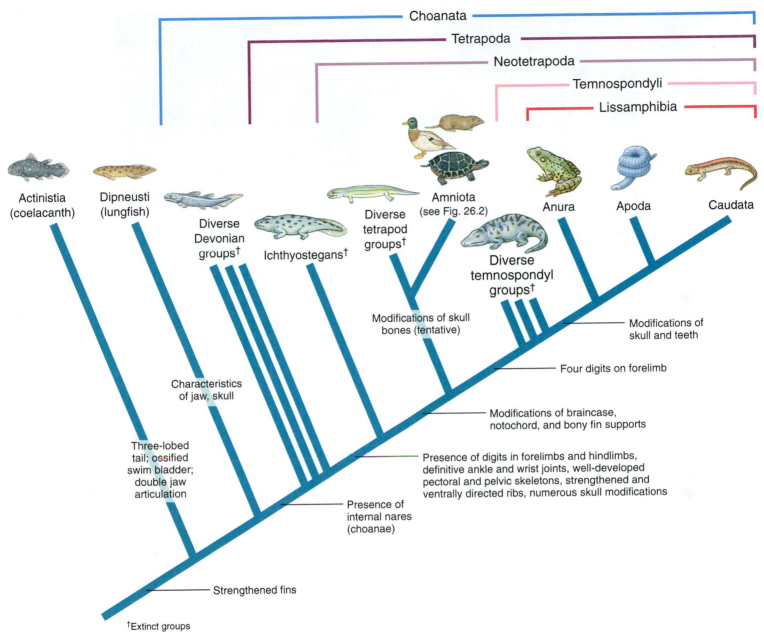

Figure 25.1

Tentative cladogram of the Tetrapoda with emphasis on descent of the amphibians. Especially controversial are the relationships of major tetrapod groups (Amniota, Temnospondyll and diverse early tetrapod groups) and outgroups (Actinistia, Dipneusti, extinct Devonian groups). All aspects of this cladogram are controversial, however, including relationships of the Lissamphibia. The relationships shown for the three groups of Lissamphibia are based on molecular evidence. Extinct groups are marked with a dagger symbol (†).

developed shoulder girdle, bulky limb bones, well-developed muscles, and other adaptations for terrestrial life, must have been able to pull itself onto land, although it probably did not walk very well.

Until the late 1980s zoologists thought that the early tetrapods had five fingers and five toes on their hands and feet, the basic pentadactyl plan of most living tetrapods. However, the more recently discovered fossils of Devonian tetrapods have more than five digits, indicating that the five-digit pattern became stabilized later in tetrapod evolution.

Evidence points to lobe-finned fishes as the closest relatives of tetrapods; in cladistic terms they contain the sister group of tetrapods

(Figures 25.1 and 25.3). Both lobe-finned fishes and early tetrapods such as *Acanthostega* and *Ichthyostega* shared several characteristics of their skull, teeth, and pectoral girdle. *Ichthyostega* (Gr. *ichthys*, fish, + *stegē*, roof, or covering, in reference to the roof of the skull, which was shaped like that of a fish) represents an early offshoot of tetrapod phylogeny that possessed several adaptations, in addition to jointed limbs, that equipped it for life on land. These include stronger vertebrae and associated muscles to support their body in air, new muscles to elevate their head, strengthened shoulder and hip girdles, protective rib cage, modified ear structure for detecting airborne sounds, foreshortening of the skull, and lengthening of the snout. *Ichthyostega*

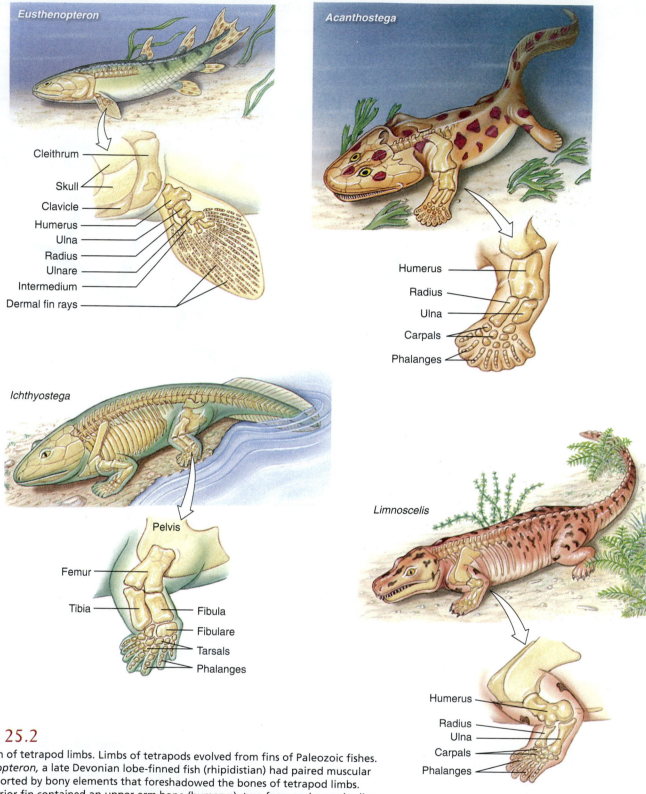

Figure 25.2

Evolution of tetrapod limbs. Limbs of tetrapods evolved from fins of Paleozoic fishes. *Eusthenopteron,* a late Devonian lobe-finned fish (rhipidistian) had paired muscular fins supported by bony elements that foreshadowed the bones of tetrapod limbs. The anterior fin contained an upper arm bone (humerus), two forearm bones (radius and ulna), and smaller elements homologous to wrist bones of tetrapods. As typical of fishes, the pectoral girdle, consisting of the cleithrum, clavicle, and other bones, was firmly attached to the skull. In *Acanthostega,* one of the earliest known Devonian tetrapods (appearing about 360 million years BP), dermal fin rays of the anterior appendage were replaced by eight fully evolved fingers. *Acanthostega* was probably exclusively aquatic because its limbs were too weak for travel on land. *Ichthyostega,* a contemporary of *Acanthostega,* had fully formed tetrapod limbs and must have been able to walk on land. The hindlimb bore seven toes (the number of front limb digits is unknown). *Limnoscelis,* an anthracosaur of the Carboniferous (about 300 million years BP) had five digits on both front and hindlimbs, the basic pentadactyl model that became the tetrapod standard.

nonetheless resembled aquatic forms in retaining a tail complete with fin rays and in having opercular (gill-covering) bones.

Bones of *Ichthyostega,* the most thoroughly studied of all early tetrapods, were first discovered on an East Greenland mountainside in 1897 by Swedish scientists looking for three explorers lost two years earlier during an ill-fated attempt to reach the North Pole by a hot-air balloon. Later expeditions by Gunnar Säve-Söderberg uncovered skulls of *Ichthyostega,* but Säve-Söderberg died at age 38 before he could study the skulls. After Swedish paleontologists returned to the Greenland site where they found the remainder of *Ichthyostega's* skeleton, Erik Jarvik, one of Säve-Söderberg's assistants, examined the skeleton in detail. This research became his life's work, producing a description of *Ichthyostega* that remains the most detailed of any Paleozoic tetrapod. Jarvik suffered a crippling stroke at age 88 in 1994, but had by then virtually completed an extensive monograph on *Ichthyostega,* which was published in 1996.

Evolutionary relationships of early tetrapod groups remain controversial. We present a tentative cladogram (Figure 25.1), which almost certainly will undergo revision as new data are reported. Several extinct lineages plus **Lissamphibia,** which contains modern amphibians, are placed in a group called **temnospondyls.** This group generally has only four digits on the forelimb rather than the five characteristic of most tetrapods. Lissamphibians arose during the Carboniferous and later diversified, probably by early Triassic, to produce ancestors of the three major groups of amphibians alive today, **frogs** (Anura or Salientia), **salamanders** (Caudata or Urodela), and **caecilians** (Apoda or Gymnophiona).

Two additional generally recognized but nonetheless controversial groupings of Carboniferous and Permian tetrapods, **lepospondyls** and **anthracosaurs,** are judged from skull structure to be closer to amniotes than to temnospondyls (Figure 25.3). We cover the amniote branch of tetrapod phylogeny in Chapters 26 through 28.

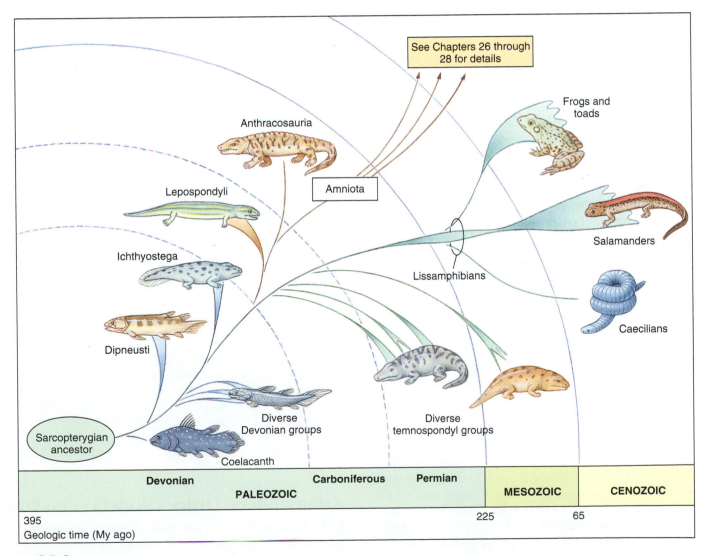

Figure 25.3

Early tetrapod evolution and the descent of amphibians. Tetrapods share most recent common ancestry with diverse Devonian groups. Amphibians share most recent common ancestry with diverse temnospondyls of the Carboniferous and Permian periods of the Paleozoic, and Triassic period of the Mesozoic.

MODERN AMPHIBIANS

The three living amphibian orders comprise more than 6000 species. Most share general adaptations for life on land, including skeletal strengthening. Amphibian larvae and some aquatic adult salamanders use the ancestral lateral-line system for sensory input, but metamorphosed adults use a redesigned olfactory epithelium to sense airborne odors and ears to detect sounds.

Nonetheless, most amphibians meet problems of independent life on land only halfway. In the ancestral amphibian life history, eggs are aquatic and hatch to produce an aquatic larval form that uses gills for respiration. A metamorphosis follows in which gills are lost. Metamorphosed amphibians use cutaneous respiration on land, and many have lungs, which are present throughout larval life and activated for breathing air at metamorphosis. Many amphibians retain this general pattern, but important exceptions include some salamanders that lack a complete metamorphosis and retain a permanently aquatic, larval morphology throughout life. Some caecilians, some frogs, and some other salamanders live entirely on land and have no aquatic larval phase. Both alternatives are evolutionarily derived conditions. Some frogs, salamanders, and caecilians that undergo a complete metamorphic life cycle nonetheless remain in water as adults rather than moving onto land during their metamorphosis.

Even the most terrestrial amphibians remain dependent on very moist environments. Their skin is thin, and it requires moisture for protection against desiccation in air. An intact frog loses water nearly as rapidly as a skinless frog. Amphibians also require moderately cool environments. As ectotherms, their body temperature is determined by and varies with the environment, greatly restricting where they can live. Cool and wet environments are especially important for reproduction. Eggs are not well protected from desiccation, and they must be shed directly into water or onto moist terrestrial surfaces.

Caecilians: Order Gymnophiona (Apoda)

The order Gymnophiona (jim'no-fy'o-na) (Gr. *gymnos,* naked, + *opineos,* of a snake) contains approximately 173 species of elongate, limbless, burrowing creatures commonly called **caecilians** (Figure 25.4). They occur in tropical forests of South America (their principal home), Africa, India, and Southeast Asia. Caecilians possess a long, slender body, small dermal scales in the skin of some, many vertebrae, long ribs, no limbs, and a terminal anus. Eyes are small, and most species are blind as adults. Special sensory tentacles occur on the snout. Because they are almost entirely burrowing or aquatic, caecilians are seldom observed. Their food consists mostly of worms and small invertebrates, which they find underground. Fertilization is internal, and males have a protrusible copulatory organ. Eggs usually are deposited in moist ground near water. Some species have aquatic larvae; larval development in other species occurs within the egg. In some species eggs are carefully guarded during their development in folds of the body. Viviparity also is common in some caecilians, with embryos obtaining nourishment by eating the wall of the oviduct.

Characteristics of Modern Amphibians

1. **Skeleton mostly bony,** with varying numbers of vertebrae; ribs present in some, absent or fused to vertebrae in others
2. Body forms vary greatly among species: salamanders usually have distinct head, neck, trunk, and tail; adult frogs have a compressed body with fused head and trunk and no intervening neck; caecilians have an elongated trunk not strongly demarcated from the head and a terminal anus
3. **Limbs usually four (quadrupedal)** in two pairs with associated shoulder/hip girdle, although some forms have a single pair of limbs and others no limbs; webbed feet often present; no true nails; **forelimb usually with four digits** but sometimes five and sometimes fewer
4. **Heart with a sinus venosus,** two atria, one ventricle, a conus arteriosus; **double circulation** through the heart in which pulmonary arteries and veins supply lungs (when present) and return oxygenated blood to heart; skin abundantly supplied with blood vessels
5. **Skin smooth, moist and glandular;** integument modified for **cutaneous respiration;** pigment cells (chromatophores) common and of considerable variety; **granular glands** associated with secretion of defensive compounds
6. Respiration by skin and in some forms by gills and/or lungs; presence of gills and lungs varies among species and by developmental stage of some species; forms with aquatic larvae lose gills at metamorphosis in frogs; many salamanders retain gills and an aquatic existence throughout life
7. Ectothermic, body temperature dependent upon environmental temperature and not modulated by metabolically generated heat
8. Excretory system of paired mesonephric or opisthonephric kidneys; urea main nitrogenous waste
9. Ear with **tympanic membrane** (eardrum) and **stapes** (columella) for transmitting vibrations to inner ear
10. For vision in air, cornea rather than lens is principal refractive surface for bending light; **eyelids** and **lachrymal glands** protect and wash eyes
11. Mouth usually large with small teeth in upper or both jaws; paired **internal nostrils** open into a nasal cavity lined with **olfactory epithelium** at anterior part of mouth cavity and enable breathing in lung-breathing forms
12. Ten pairs of cranial nerves
13. Separate sexes; fertilization mostly external in frogs and toads but internal via a spermatophore in most salamanders and caecilians; predominantly oviparous, some ovoviviparous or viviparous; metamorphosis usually present; **moderately yolky eggs** (mesolecithal) **with jellylike membrane coverings**

Salamanders: Order Urodela (Caudata)

Order Urodela (Gr. *oura,* tail, + *delos,* evident) comprises tailed amphibians, approximately 553 species of salamanders. Salamanders occur in almost all northern temperate regions of the world, and they are abundant and diverse in North America. Salamanders occur also in tropical areas of Central and northern

Figure 25.4

A, Female caecilian coiled around eggs in burrow. **B,** Pink-head caecilian (*Herpele multiplicata*), native to western Africa.

South America. Salamanders are typically small; most of the common North American salamanders are less than 15 cm long. Some aquatic forms are considerably longer, and Japanese giant salamanders sometimes exceed 1.5 m in length.

Most salamanders have limbs set at right angles to the trunk, with forelimbs and hindlimbs of approximately equal size. In some aquatic and burrowing forms, limbs are rudimentary or missing.

Salamanders are carnivorous both as larvae and adults, preying on worms, small arthropods, and small molluscs. Since their food is rich in proteins, they do not store great quantities of fat or glycogen. Like all amphibians, they are ectotherms with a low metabolic rate.

Respiration

Salamanders demonstrate an unusually diverse array of respiratory mechanisms. They share the general amphibian condition of having in their skin extensive vascular nets that serve respiratory exchange of oxygen and carbon dioxide. At various stages of their life history, salamanders also may have external gills, lungs, both, or neither gills nor lungs. Salamanders with an aquatic larval stage hatch with gills but lose them later if a metamorphosis occurs. Several lineages of salamanders have evolved permanently aquatic forms that fail to complete metamorphosis and retain their gills and finlike tail throughout life. Lungs, the most widespread respiratory organ of terrestrial vertebrates, are present from birth in salamanders that have them, and they become the primary means of respiration following metamorphosis.

Although we normally associate lungs with terrestrial organisms and gills with aquatic ones, salamander evolution has produced aquatic forms that breathe primarily with lungs and terrestrial forms that lack them completely. Amphiumas of the salamander family Amphiumidae have evolved a completely aquatic life history with a greatly reduced metamorphosis. Amphiumas nonetheless lose their gills before adulthood and then breathe primarily by lungs. They point their nostrils above the surface of the water to get air.

In contrast to the amphiumas, all species of the family Plethodontidae are lungless, and many of these species are

Life Cycles

Some salamanders are aquatic or terrestrial throughout their entire life cycle, but the ancestral condition is metamorphic, having aquatic larvae and terrestrial adults that live in moist places under stones and rotten logs. Eggs of most salamanders are fertilized internally; a female recovers in her cloaca a packet of sperm (**spermatophore**) deposited by a male on a leaf or stick (Figure 25.5). Aquatic species lay their eggs in clusters or stringy masses in water. Their eggs hatch to produce an aquatic larva having external gills and a finlike tail. Completely terrestrial species deposit eggs in small, grapelike clusters under logs or in excavations in soft moist earth, and in many species adults guard their eggs (Figure 25.6). Terrestrial species have **direct development:** they bypass the larval stage and hatch as miniature versions of their parents. The most complex of salamander life cycles occurs in some American newts, whose aquatic larvae metamorphose to form terrestrial juveniles that later metamorphose again to produce secondarily aquatic, breeding adults (Figure 25.7). Many newt populations skip the terrestrial "red eft" stage, however, remaining entirely aquatic.

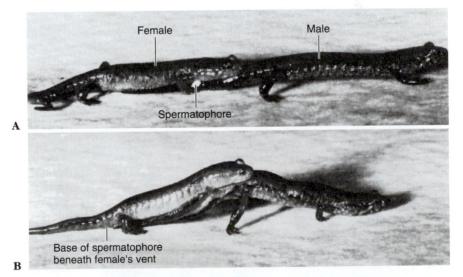

Figure 25.5

Courtship and sperm transfer in pygmy salamanders, *Desmognathus wrighti.* After judging the female's receptivity by the presence of her chin on his tail base, the male deposits a spermatophore on the ground, then moves forward a few paces.
A, The white mass of the sperm atop a gelatinous base is visible at the level of the female's forelimb. The male moves ahead, the female following until the spermatophore is at the level of her vent. **B,** The female has recovered the sperm mass in her vent, while the male arches his tail, tilting the female upward and presumably facilitating recovery of the sperm mass. The female later uses the sperm stored in her body to fertilize eggs internally before laying them.

Figure 25.6
Female dusky salamander (*Desmognathus* sp.) attending eggs. Some salamanders exercise parental care of eggs, which includes rotating eggs and protecting them from fungal infections and predation by various arthropods and other salamanders.

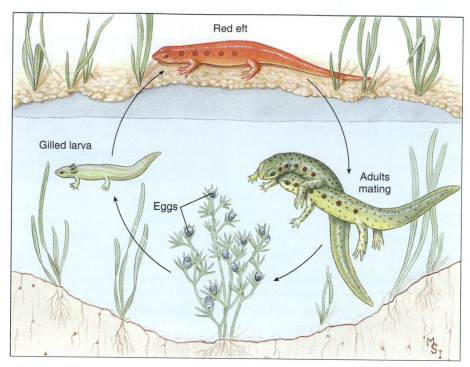

Figure 25.7
Life history of a red-spotted newt, *Notophthalmus viridescens* of the family Salamandridae. In many habitats the aquatic larva metamorphoses into a brightly colored "red eft" stage, which remains on land from 1 to 3 years before transforming into a secondarily aquatic adult.

strictly terrestrial. This large family contains more than 350 species, including many familiar North American salamanders (see Figures 25.5, 25.6, and 25.8). The efficiency of cutaneous respiration is increased by penetration of a capillary network into the epidermis or by thinning of the epidermis over superficial dermal capillaries. Cutaneous respiration is supplemented by pumping air through the mouth where respiratory gases are exchanged across the vascularized membranes of the buccal (mouth) cavity (buccopharyngeal breathing). Evolutionary loss of lungs probably occurred in an ancestral plethodontid lineage that occupied swift streams where lungs would have conferred too much buoyancy, and where water is so cool and well oxygenated that cutaneous respiration alone could support life. Some plethodontids have aquatic larvae whose gills

are lost at metamorphosis. Others retain a permanently larval form with gills throughout life. Many others are completely terrestrial and bear the distinction of being the only vertebrates to have neither lungs nor gills at any stage of their life history. It is odd that the most completely terrestrial salamanders evolved in a group that lacks lungs.

Paedomorphosis

A persistent phylogenetic trend observed in salamander evolution is for descendants to retain into adulthood features that were present only in pre-adult stages of their ancestors. Some characteristics of ancestral adult morphology are consequently eliminated. This condition is called **paedomorphosis** (Gr. "child form"; see Chapter 6, p. 117). The most dramatic form of paedomorphosis occurs in species that become sexually mature while retaining their gills, aquatic life habit, and other larval characteristics. These nonmetamorphic species are said to be **perennibranchiate** ("permanently gilled"). Mud puppies of genus *Necturus* (Figure 25.9), which live on bottoms of ponds and lakes, are an extreme example. These and many other salamanders are obligately perennibranchiate; they have never been observed to metamorphose under any conditions.

Some other species of salamanders reach sexual maturity with larval morphology but, unlike *Necturus,* may metamorphose to terrestrial forms under certain environmental conditions. Good examples are found in *Ambystoma* species from Mexico and the United States, which include naturally transforming types,

Figure 25.8
Longtail salamander, *Eurycea longicauda*, a common plethodontid salamander.

Figure 25.9

Paedomorphosis in salamanders. **A,** The mud puppy *Necturus* sp., is a permanently gilled (perennibranchiate) aquatic form. **B,** An axolotl (*Ambystoma mexicanum*) may remain permanently gilled, or, should its pond habitat evaporate, metamorphose to a terrestrial form that loses its gills and breathes by lungs. The axolotl shown is an albino form commonly used in laboratory experiments but uncommon in natural populations.

nontransforming types, and nontransforming types that can be made to transform when treated with the thyroid hormone, thyroxine (T_4). Gilled individuals are called **axolotls** (Figure 25.9). Their typical habitat consists of small ponds that can disappear through evaporation in dry weather. When its pond evaporates, an axolotl metamorphoses to a terrestrial form, losing its gills and breathing with lungs. It can then travel across land in search of new ponds, to which it must return to reproduce. Axolotls are forced to metamorphose artificially when they are treated with thyroxine (T_4). Thyroid hormones (T_3 and T_4) are essential for amphibian metamorphosis. The pituitary gland appears not to become fully active in nonmetamorphosing forms, thereby failing to release thyroid-stimulating hormone (TSH, p. 759), which is required to stimulate the thyroid gland to produce thyroid hormones.

Paedomorphosis takes many different forms in different salamanders. It may affect the body as a whole or may be restricted to one or a few structures. The amphiumas lose their gills and activate their lungs before maturity, but they retain many general features of larval body form. Paedomorphosis is important even in terrestrial plethodontids, which never have

an aquatic larval stage. One sees the effects of paedomorphosis, for example, in the shapes of the hands and feet of the tropical plethodontid genus *Bolitoglossa* (Figure 25.10). Ancestral morphology of *Bolitoglossa* features well-formed digits that grow outward from the pad of the hand or foot during development. Some species have enhanced their ability to climb smooth vegetation, such as banana trees, by halting growth of the digits and retaining throughout life a padlike foot. This padlike foot can produce adhesion and suction to attach the salamander to smooth vertical surfaces, and thereby serves an important adaptive function.

Frogs and Toads: Order Anura (Salientia)

The approximately 5283 species of frogs and toads that compose order Anura (Gr. *an*, without, + *oura*, tail) are for most people the most familiar amphibians. Anura is an old group, known from the Triassic period, 250 million years ago. Frogs and toads occupy a great variety of habitats. Their aquatic mode of reproduction and water-permeable skin prevent them from wandering too far from sources of water, however, and their ectothermy bars them from polar and subarctic habitats. The name of the order, Anura, refers to an obvious group characteristic, absence of tails in adults. Although all have a tailed larval stage during embryonic or larval development, only genus *Ascaphus* contains

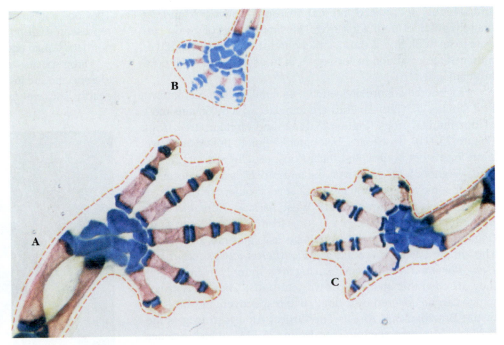

Figure 25.10

Foot structure of representatives of three different species of the tropical plethodontid salamander genus *Bolitoglossa*. These specimens have been treated chemically to clear the skin and muscles and to stain the bone red/pink and cartilage blue. The species having the most fully ossified and distinct digits **(A, C)** live primarily on the forest floor. The species having the padlike foot caused by restricted digital growth **(B)** climbs smooth leaves and stems using the foot surface to produce suction or adhesion for attachment. The padlike foot evolved by paedomorphosis; it was derived evolutionarily by truncating development of the body, which prevents full digital development.

Figure 25.11

Two common North American frogs. **A,** Bullfrog, *Rana catesbeiana,* largest American frog and mainstay of the frog-leg epicurean market (family Ranidae). **B,** Green tree frog, *Hyla cinerea,* a common inhabitant of swamps of the southeastern United States (family Hylidae). Note adhesive pads on the feet.

A

B

a tail-like structure in adults. Frogs and toads are specialized for jumping, as suggested by the alternative order name, Salientia, which means leaping.

We see in the appearance and life habit of their larvae further distinctions between Anura and Caudata. Eggs of most frogs hatch into a tadpole having a long, finned tail, both internal and external gills, no legs, specialized mouthparts for herbivorous feeding (salamander larvae and some tadpoles are carnivorous), and a highly specialized internal anatomy. They look and act entirely different from adult frogs. Metamorphosis of a frog tadpole to an adult frog is thus a striking transformation. The perennibranchiate condition never occurs in frogs and toads as it does in salamanders.

Frogs and toads are divided into 44 families. The best-known frog families in North America are Ranidae, which contains most of our familiar frogs (Figure 25.11A), and Hylidae, tree frogs (Figure 25.11B). True toads, belonging to family Bufonidae, have short legs, stout bodies, and thick skins, usually with prominent warts (Figure 25.12). However, the term "toad" is used informally also for some terrestrial members of several other families.

The largest anuran is the West African *Conraua goliath,* which is more than 30 cm long from tip of nose to anus (Figure 25.13). This giant eats animals as big as rats and ducks. The smallest frogs recorded are *Eleutherodactylus iberia* and *Psyllophryne didactyla,* measuring less than 1 cm in length; they are also the smallest known tetrapods. These tiny frogs, which can be covered by a dime, live respectively in Cuba and in the Brazilian rain forest. The largest American frog is the bullfrog, *Rana catesbeiana* (see Figure 25.11A), which reaches a head and body length of 20 cm.

Habitats and Distribution

Probably the most familiar frogs are the species of genus *Rana* (Gr. frog), found throughout temperate and tropical regions of the world except in New Zealand, the oceanic islands, and southern South America. They usually live near water, although some,

such as wood frogs, *R. sylvatica,* spend most of their time on damp forest floors. Wood frogs probably return to pools only for breeding in early spring. The larger bullfrogs, *R. catesbeiana,* and green frogs, *R. clamitans,* are nearly always found in or near permanent water or swamps. Leopard frogs, *R. pipiens* and related species, have a wider variety of habitats and are the most widespread of North American frogs; they are commonly used in biology laboratories and for classical electrophysiological research. They occur in some form in nearly every state, although sparingly represented along the Pacific coast. They also extend far into northern Canada and as far south as Panama.

Frog species are often patchy in distribution, being restricted to a specific stream or pool and absent or scarce in similar habitats nearby. Pickerel frogs (*R. palustris*), for example, are abundant

Figure 25.12

American toad, *Bufo americanus* (family Bufonidae). This principally nocturnal yet familiar amphibian feeds on large numbers of insect pests and on snails and earthworms. The rough skin contains numerous glands that produce a surprisingly poisonous milky fluid, providing excellent protection from a variety of potential predators.

Figure 25.13

Conraua (Gigantorana) goliath (family Ranidae) of West Africa, the world's largest frog. This specimen weighed 3.3 kg (approximately 7½ pounds).

only in certain localized regions. Recent studies have shown that many populations of frogs worldwide may be suffering declines in numbers and becoming even more geographically fragmented than usual.

Most larger frogs are solitary except during breeding season. During breeding periods most of them, especially males, are very noisy. Each male usually takes possession of a particular perch near water, where he may remain for hours or even days, trying to attract a female to that spot. At times frogs are mainly silent, and their presence is not detected until they are disturbed. When they enter the water, they dart swiftly to the bottom of the pool and kick the substrate to conceal themselves in a cloud of muddy water. In swimming, they hold their forelimbs near their body and kick backward with their webbed hindlimbs, propelling themselves forward. When they surface to breathe, only the head and foreparts are exposed, since they conceal themselves behind any available vegetation.

Amphibian populations are falling in various parts of the world, although many species continue to thrive. No single explanation fits all declines, although loss of habitat predominates. In some populations, changes are simply random fluctuations caused by periodic droughts and other naturally occurring phenomena. Frog and toad eggs exposed on the surfaces of ponds are especially sensitive to damage from ultraviolet radiation. Climatic changes that reduce water depth at oviposition sites increase ultraviolet exposure of embryos and make them more susceptible to fungal infection. Declines in population survival may include an increased incidence of malformed individuals, such as frogs with extra limbs. Malformed limbs are often associated with infection by trematodes (p. 298).

During winter months most frogs in temperate climates hibernate in the soft mud underlying pools and streams. Their life processes are at a very low ebb during hibernation, and the

energy that they need is derived from glycogen and fat stored in their bodies during the spring and summer. More terrestrial frogs, such as tree frogs, hibernate in humus of the forest floor. They tolerate low temperatures, and many actually survive freezing all extracellular fluid, representing 35% of the body water. Such frost-tolerant frogs prepare for winter by accumulating glucose and glycerol in body fluids, thereby protecting tissues from the normally damaging effects of ice-crystal formation.

Declines of some amphibian populations may be caused by other amphibians. While native American amphibians continue to disappear as wetlands are drained, an exotic frog introduced into southern California has found many favorable habitats there. African clawed frogs, *Xenopus laevis* (Figure 25.14), are voracious, aggressive, primarily aquatic frogs that rapidly displace native frogs and fishes from waterways. This species was introduced into North America in the 1940s when it was used extensively in human pregnancy tests. When more efficient tests appeared in the 1960s, some hospitals released surplus frogs into nearby streams, where these prolific breeders have flourished and are now considered pests. Similar results occurred when giant toads, *Bufo marinus* (to 23 cm in length), were introduced to Queensland, Australia, and southern Florida to control agricultural pests. They are rapidly spreading in Australia, producing numerous ecological changes, including displacement of native anurans.

Adult frogs have numerous enemies, such as snakes, aquatic birds, turtles, raccoons, and humans; fish prey on tadpoles, and only a few tadpoles survive to maturity. Although usually defenseless, many frogs and toads in the tropics and subtropics are aggressive, jumping and biting at predators. Some defend themselves by feigning death. Most anurans can inflate their lungs so that they are difficult to swallow. When disturbed along the margin of a pond or brook, a frog often remains quite still; when it thinks it is detected, it jumps, not always into the water where

Figure 25.14

African clawed frog, *Xenopus laevis*. The claws, an unusual feature in frogs, are on the hind feet. This frog has been introduced into California, where it is considered a serious pest.

enemies may lurk, but into grassy cover on the bank. When held, a frog may cease its struggles for an instant to put its captor off guard and then suddenly leap, voiding its urine. A frog's best protection is its ability to leap and, in some species, to use poison glands. Dendrobatid frogs use potent toxins for defense. Bull-frogs in captivity do not hesitate to snap at tormentors and can inflict painful bites.

Integument and Coloration

A frog's skin is thin and moist, and it is attached loosely to the body only at certain points. Histologically the skin contains two layers: an outer stratified **epidermis** and an inner spongy **dermis** (Figure 25.15). The epidermal layer, which is shed periodically when a frog or toad "molts," contains deposits of **keratin,** a tough, fibrous protein that limits abrasion and loss of water from the skin. More terrestrial amphibians such as toads have especially heavy deposits of keratin, although amphibian keratin is soft, unlike the hard keratin that forms scales, claws, feathers, horns, and hair of amniotes.

The epidermis produces two types of integumentary glands that grow into the loose dermal tissues below. Small **mucous** glands secrete a protective mucous waterproofing onto the skin surface, and large granular **serous** glands produce a whitish, watery poison highly irritating to predators. All amphibians produce a skin poison, but its effectiveness varies among species and with different predators. The extremely toxic poison of three species of *Phyllobates,* a genus of small South American dendrobatid frogs, is used by a western Colombian tribe to poison points of blowgun darts. Most species of the family Dendrobatidae produce toxic skin secretions, some of which are among the most lethal animal secretions known, more dangerous even than venoms of sea snakes or any of the most venomous arachnids.

A frog's skin color is produced, as in other amphibians, by special pigment cells, **chromatophores,** located mainly in the dermis. Amphibian chromatophores, like those of many other vertebrates, are branched cells containing pigment that may be concentrated in a small area or dispersed throughout the branching processes to control skin coloration (Figure 25.16; see also p. 648). Most amphibians have three types of chromatophores: uppermost in the dermis are **xanthophores,** containing

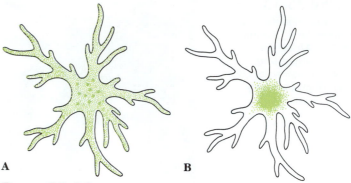

Figure 25.16

Pigment cells (chromatophores). **A,** Pigment dispersed. **B,** Pigment concentrated. The pigment cell does not contract or expand; color effects are produced by streaming of cytoplasm, carrying pigment granules into cell branches for maximum color effect or to the center of the cell for minimum effect. Control over dispersal or concentration of pigment is mostly by light stimuli acting through a pituitary hormone.

yellow, orange, or red pigments; beneath these lie **iridophores,** containing a silvery, light-reflecting pigment; and lowermost are **melanophores,** containing black or brown melanin. Iridophores act like tiny mirrors, reflecting light back through the xanthophores to produce the bright colors of many tropical frogs. Surprisingly perhaps, green hues so common in North American frogs are produced not by green pigment but by an interaction of xanthophores containing a yellow pigment and underlying iridophores that, by reflecting and scattering light (Tyndall scattering), produce a blue color. Blue light is filtered by the overlying yellow pigment and thus appears green. Many frogs can adjust their color to match their background and thus to camouflage themselves (Figure 25.17).

Skeletal and Muscular Systems

Amphibians, like other vertebrates, have a well-developed **endoskeleton** of bone and cartilage to support the body and its muscular movements. Movement onto land required limbs capable of supporting the body's weight and introduced a new set of stress and leverage forces. The entire musculoskeletal system of an adult frog is specialized for jumping and swimming by simultaneous extensor thrusts of the hindlimbs.

The amphibian vertebral column assumes a new role as a support from which the abdomen is suspended and to which limbs attach. Because anurans move with limbs rather than swimming with serial contractions of the trunk musculature, the vertebral column has lost much of the original flexibility characteristic of fishes. It is a rigid frame for transmitting force from the hindlimbs to the body. Anurans are further specialized by an extreme shortening of the body. Typical frogs have only nine trunk vertebrae and a rodlike **urostyle,** which represents several fused caudal vertebrae (coccyx) (Figure 25.18). The limbless caecilians, which do not share these specializations for tetrapod locomotion, may have as many as 285 vertebrae.

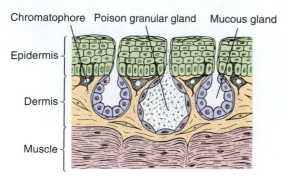

Chromatophore   Poison granular gland   Mucous gland

Epidermis

Dermis

Muscle

Figure 25.15

Section through frog skin.

Figure 25.17
Cryptic coloration of the gray tree frog, *Hyla versicolor.* Camouflage is so good that presence of this frog usually is disclosed only at night by its resonant, flutelike call.

easily imagine how selective pressures generated by terrestrial locomotion remodeled ancestral lobe-fins into limbs.

Limb muscles are presumably homologous to radial muscles that move the fins of fishes, but the muscular arrangement has become so complex in tetrapod limbs that its exact correspondence with fin musculature is unclear. Despite this complexity, we can identify two major groups of muscles on any limb: an anterior and ventral group that pulls the limb forward and toward the midline (protraction and adduction), and a second set of posterior and dorsal muscles that draws the limb back and away from the body (retraction and abduction).

Trunk musculature, which in fishes is segmentally organized into powerful muscular bands (myomeres, p. 530) for locomotion by lateral flexion, has been much modified during amphibian evolution. Dorsal (epaxial) muscles are arranged to support the head and to brace the vertebral column. Ventral (hypaxial) muscles are more developed in amphibians than in fishes, since they must support the viscera in air without the buoying assistance of water.

Respiration and Vocalization

Amphibians use three respiratory surfaces for gas exchange in air: skin (cutaneous breathing), mouth (buccal breathing), and lungs. Frogs and toads are more dependent on lung breathing than are salamanders; nevertheless, skin provides an important supplementary avenue for gas exchange in anurans, especially during hibernation in winter. Even when lung breathing predominates, carbon dioxide is lost primarily across the skin while oxygen is absorbed primarily across the lungs.

Lungs are supplied by pulmonary arteries (derived from the sixth aortic arches) and blood returns directly to the left atrium by pulmonary veins. Frog lungs are ovoid, elastic sacs with their inner surfaces divided into a network of septa that are subdivided into small terminal air chambers called faveoli. Faveoli of

A frog's skull differs greatly from those of other vertebrates; it is much lighter in weight, less ossified, flattened in profile, and has fewer bones. The front of the skull, containing the nose, eyes, and brain, is better developed, whereas the back of the skull, which contains the gill apparatus in fishes, is much reduced (see Figure 25.18).

Limbs are of typical tetrapod pattern in having three main joints (hip, knee, and ankle; or shoulder, elbow, and wrist). The foot is typically five-rayed (pentadactyl) and the hand is four-rayed, with both foot and hand having several joints in each digit (Figure 25.18). It is a repetitive system resembling the bone structure of lobe-fins, which are distinctly suggestive of amphibian limbs (see Figure 25.1). One can

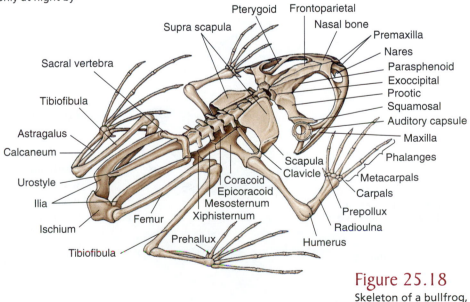

Figure 25.18
Skeleton of a bullfrog, *Rana catesbeiana.*

frog lungs are much larger than alveoli of amniote vertebrates, and consequently frog lungs have a smaller relative surface available for gas exchange: the respiratory surface of *Rana pipiens* is about 20 cm² per cubic centimeter of air contained, compared with 300 cm² for humans. The major challenge in lung evolution was not development of a good internal vascular surface, but rather the mechanism of moving air. A frog is a positive-pressure breather that fills its lungs by forcing air into them; this system contrasts with the negative-pressure system of amniotes. A frog's breathing is explained in Figure 25.19. One can easily follow this sequence in a living frog at rest: rhythmical throat movements of mouth breathing occur continuously before flank movements indicate that the lungs are being emptied and refilled.

Vocal cords, located in the **larynx,** or voice box, are much more developed in male frogs than in females. A frog produces sound by passing air back and forth over the vocal cords between the lungs and a large pair of sacs (vocal pouches) in the floor of the mouth. The vocal sacs also serve as effective resonators in males, which use their voices to attract mates. Most species have unique sound patterns. Spring peepers produce high-pitched sounds surprisingly strident for such tiny frogs. Green frogs produce banjolike calls, whereas those of leopard frogs are long and guttural, and bullfrogs produce resonant "jug-o-rum" calls.

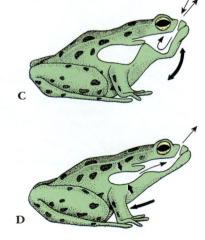

Figure 25.19

Breathing in a frog. Frogs are positive-pressure breathers that fill their lungs by forcing air into them. **A,** Floor of mouth is lowered, drawing air in through nostrils. **B,** With nostrils closed and glottis open, the frog forces air into its lungs by elevating floor of mouth. **C,** Mouth cavity rhythmically ventilates for a period. **D,** Lungs are emptied by contraction of body-wall musculature and by elastic recoil of lungs.

Circulation

Amphibian circulation is a closed system of arteries and veins serving a vast peripheral network of capillaries through which blood is forced by a single pressure pump, the heart. The principal differences in circuitry relative to fishes involve the shift from gill to lung breathing. Elimination of gills removes a major obstacle to blood flow from the arterial circuit, but lung breathing introduces two new evolutionary challenges. The first challenge is to provide a blood circuit to the lungs. As we have seen, this challenge was met by converting the sixth aortic arch into pulmonary arteries to serve the lungs and by developing new pulmonary veins for returning oxygenated blood to the heart (p. 693). The second evolutionary challenge was to separate pulmonary circulation from the rest of the body's circulation, so that oxygenated blood from the lungs is sent to the body and deoxygenated venous blood returning from the body is sent to the lungs. Solving this problem required a double circulation consisting of separate pulmonary and systemic circuits. Tetrapods solved the problem by evolving a partition down the center of the heart, creating a double pump, one for each circuit. However, partitioning is only partial in amphibians and most nonavian reptiles compared to birds and mammals, which have the most extensively divided hearts containing two atria and two ventricles.

Frog hearts (Figure 25.20) have two separate atria and a single undivided ventricle. Blood from the body (systemic circuit) first enters a large receiving chamber, the sinus venosus, which forces blood into the right atrium. The left atrium receives freshly oxygenated blood from the lungs and skin. The right and left atria contract asynchronously so that although the **ventricle** is undivided, blood remains mostly separated when it enters this chamber. When the ventricle contracts, oxygenated pulmonary blood enters the systemic circuit and deoxygenated systemic blood enters the pulmonary circuit. This separation is aided by a **spiral valve,** which divides the systemic and pulmonary flows in the **conus arteriosus** (Figure 25.20), and by different blood pressure in the pulmonary and systemic blood vessels leaving the conus arteriosus. The exact mechanism and precision of separation of oxygenated and deoxygenated blood in the conus arteriosus remain unclear.

Feeding and Digestion

Frogs are carnivorous as adults, as are most other amphibians, and they feed on insects, spiders, worms, slugs, snails, millipedes, and nearly anything else that moves and is small enough to swallow whole. They snap at moving prey with their protrusible tongue, which is attached to the front of the mouth and is free at the posterior end. The highly glandular free end of the tongue produces a sticky secretion that adheres to prey. When teeth are present on the premaxillae, maxillae, and vomers (roof of the mouth), they are used to prevent escape of prey, not for biting or chewing. The digestive tract is relatively short in adult amphibians, a characteristic of most carnivores, and it produces a variety of enzymes for digesting proteins, carbohydrates, and fats.

Classification of Class Amphibia

Order Gymnophiona (jim'no-fy'o-na) (Gr. *gymnos*, naked, + *ophioneos*, of a snake) **(Apoda): caecilians.** Body elongate; limbs and limb girdle absent; dermal scales present in skin of some; tail short or absent; 95 to 285 vertebrae; pantropical; 3 families, 33 genera, approximately 173 species.
Order Urodela (yur'uh-dēl'uh) (Gr. *oura*, tail, + *delos*, evident) **(Caudata): salamanders.** Body with head, trunk, and tail; no scales; usually two pairs of equal limbs; 10 to 60 vertebrae; predominantly Holarctic; 9 living families, 64 genera, approximately 553 species.
Order Anura (uh-nur'uh) (Gr. *an*, without, + *oura*, tail) **(Salientia): frogs, toads.** Head and trunk fused; no tail; no scales; two pairs of limbs; large mouth; lungs; 6 to 10 vertebrae including urostyle (coccyx); cosmopolitan, predominantly tropical; 44 families, 362 genera, approximately 5283 species.

Larval stages of anurans (tadpoles) are usually herbivorous, feeding on pond algae and other vegetable matter; they have a relatively long digestive tract because their bulky food must be submitted to time-consuming fermentation before useful products can be absorbed.

Nervous System and Special Senses

Three fundamental parts of the brain—forebrain (telencephalon), concerned with the sense of smell; midbrain (mesencephalon), concerned with vision; and hindbrain (rhombencephalon), concerned with hearing and balance—show dramatic evolutionary trends in tetrapods (p. 736). Cephalization increases with emphasis on information processing by the brain and a corresponding loss of independence of the spinal ganglia, which

are capable only of stereotyped reflexive behavior. Nonetheless, a headless frog preserves an amazing degree of purposive and highly coordinated behavior despite lacking central perception mechanisms and therefore being technically dead. With only the spinal cord intact, it maintains normal body posture and can accurately raise its leg to wipe an irritant from its skin. It will even use the opposite leg if the closer leg is held.

The forebrain (Figure 25.21) contains the olfactory center, which assumes greatly increased importance on land for detection of dilute airborne odors. The sense of smell is one of the dominant special senses in frogs. The remainder of the forebrain, or cerebrum, has little importance in amphibians. Instead, complex integrative activities of frogs occur in the midbrain optic lobes. The hindbrain is divided into an anterior cerebellum and a posterior medulla. The cerebellum (Figure 25.21), which coordinates equilibrium and movement, is not well developed in amphibians, which stay close to the ground or other surfaces and are not noted for dexterity of movement. All sensory neurons except those of vision and smell pass through the medulla, which is at the anterior end of the spinal cord. Here are located centers for auditory reflexes, respiration, swallowing, and vasomotor control.

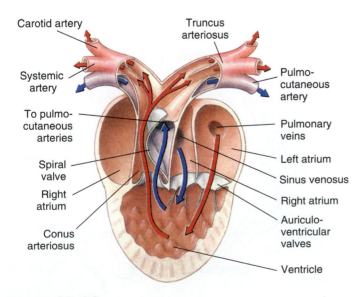

Figure 25.20
Structure of a frog heart. *Red arrows,* oxygenated blood. *Blue arrows,* deoxygenated blood.

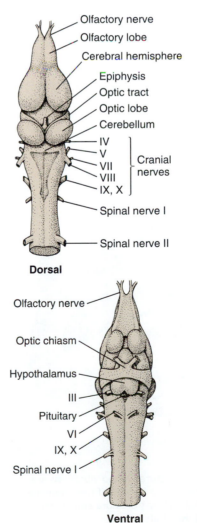

Figure 25.21
Brain of a frog, dorsal and ventral views.

Evolution of a semiterrestrial life by amphibians features a reordering of sensory receptor priorities on land. The pressure-sensitive lateral-line (acousticolateral) system of fishes remains only in aquatic larvae of amphibians and in a few strictly aquatic adult amphibian species. This system serves no purpose on land, because it relies on reflected pressure waves to detect and to localize objects in water. The task of detecting airborne sounds falls on the ear.

A frog's ear is by amniote standards a simple structure: a middle ear closed externally by a large **tympanic membrane** (eardrum) and containing a **columella** (stapes) that transmits vibrations to the inner ear (Figure 25.22). The latter contains a **utricle,** from which arise three semicircular canals, and a **saccule** bearing a diverticulum, or **lagena.** The lagena is partly covered with a **tectorial membrane,** which in its fine structure resembles the much more complex mammalian cochlea. In most frogs this structure senses low-frequency sound energy not greater than 4000 Hz (cycles per second); in bullfrogs the main frequency response is in the 100 to 200 Hz range, which matches the energy of a male frog's low-pitched call.

Vision is the dominant special sense in many amphibians (the mostly blind caecilians are obvious exceptions). Several modifications of ancestral aquatic eyes serve to adapt them for use in air. Lachrymal glands and eyelids keep eyes moist, wiped free of dust, and shielded from injury. Since the cornea is exposed to air, it is an important refractive surface, removing much of the burden from the lens of bending light rays and focusing an image on the retina. As in fishes, accommodation (adjusting focus for near and distant objects) occurs by moving the lens. Unlike eyes of most fishes, amphibian eyes at rest are adjusted for distant objects, and the lens is moved forward to focus on nearby objects.

Keeping a sharp image on the retina for approaching or receding objects requires accommodation, which occurs in several different ways among vertebrates. Eyes of bony fishes and lampreys are set for near vision in the resting stage; to focus on distant objects, the lens is moved backward. In amphibians, sharks, and snakes, the relaxed eye is focused on distant objects and the lens is moved *forward* to focus on nearby objects. In birds, mammals, and all nonavian reptiles except snakes, the lens accommodates by changing its *curvature* rather than by moving forward or backward. The resting eye in these forms is set for distant vision, and to focus on nearby objects the lens curvature is increased by being squeezed (or, in some, allowed to relax) into a rounded shape.

A **retina** contains both **rods and cones,** the latter providing frogs with color vision. The iris contains well-developed circular and radial muscles and can rapidly expand or contract the aperture (pupil) to accommodate changing illumination. The upper lid of the eye is fixed, but the lower one is folded into a transparent **nictitating membrane** capable of moving across the eye surface (Figure 25.23). Frogs and toads generally possess good vision, which is crucial for animals that rely on quick escape to avoid their numerous predators and on accurate movements to capture rapidly moving prey.

Other sensory receptors include tactile and chemical receptors in skin, taste buds on the tongue and palate, and a well-developed olfactory epithelium lining the nasal cavity.

Reproduction

Because frogs and toads are ectothermic, they breed, feed, and grow only during warm seasons. One of the first drives after the dormant period is breeding. In spring males call vociferously to attract females. When their eggs are mature, females enter

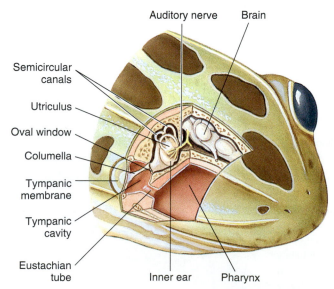

Figure 25.22

Cutaway of frog head showing ear structure. Sound vibrations are transmitted from the tympanic membrane by way of the columella to the inner ear. The eustachian tube allows pressure equilibration between the tympanic cavity and the pharynx.

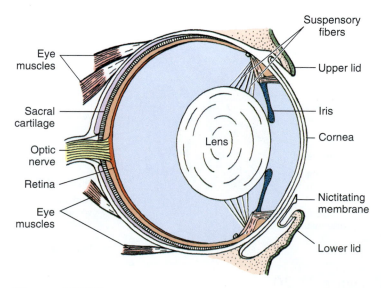

Figure 25.23

Amphibian eye.

water and are clasped by males in a process called **amplexus** (Figure 25.24) in which eggs are fertilized externally (after leaving the female's body). As a female lays eggs, a male discharges sperm over the eggs to fertilize them. After fertilization, the jelly layers absorb water and swell. Eggs are laid in large masses, usually anchored to vegetation.

A fertilized egg (zygote) begins development almost immediately (Figure 25.25). By repeated division (cleavage) an egg is converted into a hollow ball of cells (blastula). The blastula undergoes gastrulation (p. 164) and then continues to differentiate to form an embryo with a tail bud. At 2 to 21 days, depending on temperature, a tadpole hatches from the protective jelly coats that had surrounded the original fertilized egg.

At hatching, a tadpole has a distinct head, trunk, and a compressed tail. Its mouth is on the ventral side of the head and has keratinized jaws for scraping vegetation from hard objects. Behind the mouth is a ventral adhesive disc for clinging to objects. In

Figure 25.24
A male green tree frog, *Hyla cinerea*, clasps a larger female during breeding season in a South Carolina swamp. Clasping (amplexus) is maintained until the female deposits her eggs. Like most tree frogs, these are capable of rapid and marked color changes; the male here, normally green, has darkened during amplexus.

Figure 25.25
Life cycle of a leopard frog.

front of the mouth are two deep pits, which later become nostrils. Swellings on each side of the head later become external gills. There are three pairs of external gills, which later become internal gills covered by a flap of skin (operculum) on each side. On the right side the operculum completely fuses with the body wall, but on the left side a small opening, the spiracle (L. *spiraculum,* air hole) remains. Water flows through the spiracle after entering the mouth and passing the internal gills. Hindlimbs appear first during metamorphosis, while forelimbs remain temporarily hidden by folds of the operculum. The tail is resorbed. The intestine becomes much shorter. The mouth undergoes a transformation into the adult condition. Lungs develop, and gills are resorbed (Figure 25.25). Leopard frogs usually complete metamorphosis within 3 months, whereas bullfrogs take 2 or 3 years to complete the process.

The life history just described is typical of most temperate-zone anurans but only one of many alternative patterns in tropical anurans. Some remarkable reproductive strategies of tropical anurans are illustrated in Figure 25.26. Some species lay their eggs in foam masses that float on the surface of the water; some deposit their eggs on leaves overhanging ponds and streams into which the emerging tadpoles will drop; some lay their eggs in damp burrows; and others place their eggs in water trapped in tree cavities or in water-filled chambers of some bromeliads (epiphytic plants in the tropical forest canopy). While most frogs abandon their eggs, some, such as the tropical dendrobatids (a family that includes the poison-dart frogs), tend their eggs. When the tadpoles hatch, they squirm onto the parent's back to be carried for varying lengths of time (Figure 25.26C). Marsupial frogs carry their developing eggs in a pouch on the back (Figure 25.26A). In Surinam frogs (Figure 25.26B), the male and female do backward somersaults during mating, and eggs and sperm slide into the space between the mating pair. The male presses the fertilized eggs into the female's back, which develops a spongy incubating layer that eventually sloughs once young are hatched. In the species-rich tropical genus *Eleutherodactylus,* mating occurs on land, and eggs hatch directly into froglets; the aquatic larval stage is eliminated, freeing these frogs from an obligatory association with pools or streams.

Migration of frogs and toads depends upon their breeding habits. Males usually return to a pond or stream before females, which they then attract by calling. Some salamanders also have a strong homing instinct, returning each year to reproduce in the same pool, to which they are guided by olfactory cues. The initial stimulus for migration in many cases is attributable to a seasonal cycle in the gonads plus hormonal changes that increase frogs' sensitivity to changes in temperature and humidity.

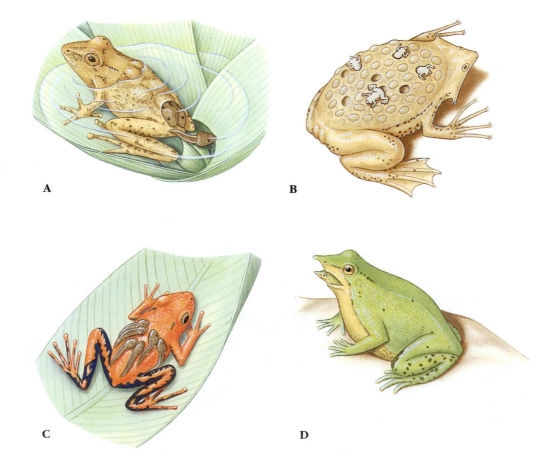

Figure 25.26
Unusual reproductive strategies of anurans.
A, Female South American pygmy marsupial frog, *Flectonotus pygmaeus,* carries developing larvae in a dorsal pouch. **B,** Female Surinam frog carries eggs embedded in specialized brooding pouches on the dorsum; froglets emerge and swim away when development is complete. **C,** Male poison arrow frog, *Phyllobates bicolor,* carries tadpoles adhering to its back.
D, Tadpoles of a male Darwin's frog, *Rhinoderma darwinii,* develop into froglets in its vocal pouch. When ready to emerge, a froglet crawls into the parent's mouth, which the parent opens to allow the froglet's escape.

A   B

C   D

SUMMARY

Amphibians are ectothermic, primitively quadrupedal vertebrates that have glandular skin and breathe by lungs, gills, and/or skin. They form one of two major branches of tetrapod phylogeny, the other one being amniotes. Modern amphibians comprise three major evolutionary groups. Caecilians (order Gymnophiona) are a small tropical group of limbless, elongate forms. Salamanders (order Urodela) are tailed amphibians that have retained the generalized four-limbed body plan of their Paleozoic ancestors. Frogs and toads (order Anura) are the largest group of modern amphibians, adults of which are specialized for a jumping mode of locomotion on land or in water.

Most amphibians have a biphasic life cycle, beginning with an aquatic larva that later metamorphoses to produce a terrestrial adult, which returns to water to lay eggs. Some frogs, salamanders, and caecilians have evolved direct development that omits the aquatic larval stage, and some caecilians have evolved viviparity. Salamanders are unique among amphibians in having evolved several perennibranchiate species that retain a permanently larval morphology throughout life, eliminating a terrestrial phase completely. The perennibranchiate condition is obligate in some species, but others metamorphose to a terrestrial form if their pond habitat evaporates.

Although amphibians have evolved adaptations to the aquatic phase of their life history, adaptations to their terrestrial existence are particularly noteworthy. Respiratory exchange of gases occurs across porous skin in all amphibians and is supplemented in most amphibians by lungs. Oddly, the most highly terrestrial salamanders lack lungs whereas some aquatic forms use lungs as their major respiratory structure. Terrestrial forms have strong skeletons with terrestrial adaptations evident in the structuring of the ribs, pectoral and pelvic girdles, and limbs. Derived features of amphibian auditory and visual systems and associated regions of the brain facilitate sensory perception on land.

Despite their adaptations for terrestrial life, adults and eggs of all amphibians require cool, moist environments if not actual pools or streams. Eggs and adult skin have no effective protection against very cold, hot, or dry conditions, greatly restricting adaptive diversification of amphibians to environments that have moderate temperatures and abundant water.

REVIEW QUESTIONS

1. How did the characteristic differences between aquatic and terrestrial environments influence the early evolution of tetrapods?
2. Describe the different modes of respiration used by amphibians. What paradox do amphiumas and terrestrial plethodontids present regarding the association of lungs with life on land?
3. Evolution of the tetrapod limb was one of the most important advances in vertebrate history. Describe the supposed sequence in its evolution.
4. Compare the general life-history patterns of salamanders with those of frogs. Which group shows a greater variety of evolutionary changes of the ancestral biphasic amphibian life cycle?
5. Give the literal meaning of the name Gymnophiona. What animals are included in this amphibian order, what do they look like, and where do they live?
6. What is the literal meaning of the order names Urodela and Anura? What major features distinguish members of these two orders from each other?
7. Describe the breeding behavior of a typical woodland salamander.
8. How has paedomorphosis been important to evolutionary diversification of salamanders?
9. Describe the integument of a frog. How are the various skin colors produced in frogs?
10. Describe amphibian circulation.
11. Explain how the forebrain, midbrain, hindbrain, and the sensory structures with which each brain division is concerned have developed to meet sensory requirements for amphibian life on land.
12. Briefly describe the reproductive behavior of frogs. In what important ways do frogs and salamanders differ in their reproduction?

SELECTED REFERENCES

Clack, J. A. 2002. Gaining ground: the origin and evolution of tetrapods. Bloomington, Indiana, Indiana University Press. *An authoritative account of paleontological evidence regarding tetrapod origins.*

Conant, R., and J. T. Collins. 1998. A field guide to reptiles and amphibians: eastern and central North America. The Peterson field guide series. Boston, Houghton Mifflin Company. *Updated version of a popular field guide; color illustrations and distribution maps for all species.*

Daeschler, E. B., N. H. Shubin, and F. A. Jenkins, Jr. 2006. A Devonian tetrapod-like fish and the evolution of the tetrapod body plan. Nature **440:**757–763. *Describes Tiktaalik, a fossil intermediate between lobe-fin fishes and tetrapods.*

Duellman, W. E., and L. R. Trueb. 1994. Biology of amphibians. Baltimore, Johns Hopkins University Press. *Important comprehensive sourcebook of information on amphibians, extensively referenced and illustrated.*

Frost, D. R., T. Grant, J. Faivovich, R. H. Bain, A. Haas, C. F. B. Haddad, R. O. De Sá, A. Channing, M. Wilkinson, S. C. Donnellan, C. J. Raxworthy, J. A. Campbell, B. L. Blotto, P. Moler, R. C. Drewes, R. A. Nussbaum, J. D. Lynch, D. M. Green, and W. C. Wheeler. 2006. The amphibian tree of life. Bull. Am. Mus. Nat. Hist. **297:**1–370. *A phylogeny for living amphibians derived from a large compilation of morphological and molecular characters.*

Halliday, T. R., and K. Adler (eds.). 2002. Firefly encyclopedia of reptiles and amphibians. Toronto, Canada, Firefly Books. *Excellent authoritative reference work with high-quality illustrations.*

Heatwole, H. (ed.). 1994–1995. Amphibian biology. Chipping Norton, NSW, Australia, Surrey Beatty and Sons. *A two-volume work containing extensive coverage of amphibian integuments and social systems.*

Jamieson, B. G. M. (ed.). 2003. Reproductive biology and phylogeny of Anura. Enfield, New Hampshire, Science Publishers, Inc. *Provides detailed coverage of reproductive biology and early evolutionary diversification of frogs and toads.*

Lannoo, M. (ed.). 2005. Amphibian declines: the conservation status of United States species. Berkeley, California, University of California Press. *A survey of conservation status of American amphibians.*

Lewis, S. 1989. Cane toads: an unnatural history. New York, Dolphin/ Doubleday. *Based on an amusing and informative film of the same title, this book describes the introduction of cane toads to Queensland, Australia and the unexpected consequences of their population explosion there. "If Monty Python teamed up with National Geographic, the result would be* Cane Toads.*"*

Petranka, J. W. 1998. Salamanders of the United States and Canada. Washington, DC, Smithsonian Institution Press. *A comprehensive coverage of life history and ecology of American and Canadian salamanders.*

Pough, F. H., R. M. Andrews, J. E. Cadle, M. L. Crump, A. H. Savitsky, and K. D. Wells. 2004. Herpetology, ed. 3. Upper Saddle River, New Jersey, Prentice-Hall. *A current general textbook of herpetology.*

Savage, J. M. 2002. The amphibians and reptiles of Costa Rica. Chicago, University of Chicago Press. *Costa Rica hosts a diversity of anuran, caecilian, and salamander species. Courses offered by the Organization for Tropical Studies give college students an opportunity to study this amphibian fauna.*

Sever, D. M. (ed.). 2003. Reproductive biology and phylogeny of Urodela (Amphibia). Enfield, New Hampshire, Science Publishers, Inc. *A thorough review of reproductive biology and evolutionary relationships among salamanders.*

Stebbins, R. C., and N. W. Cohen. 1995. A natural history of amphibians. Princeton, New Jersey, Princeton University Press. *Worldwide treatment of amphibian biology, emphasizing physiological adaptations, ecology, reproduction, behavior, and a concluding chapter on amphibian declines.*

Zug, G. R., L. J. Vitt, and J. P. Caldwell. 2001. Herpetology: an introduction to the biology of amphibians and reptiles. San Diego, Academic Press. *A current general textbook of herpetology.*

ONLINE LEARNING CENTER

Visit www.mhhe.com/hickmanipz14e for chapter quizzing, key term flash cards, web links, and more!

Amniote Origins and Nonavian Reptiles

- PHYLUM CHORDATA
- CLASS REPTILIA

Hatching Komodo lizard (Varanus komodoensis).

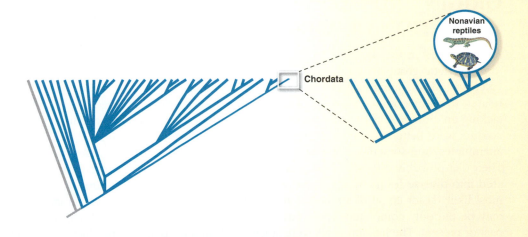

Enclosing the Pond

The amphibians, with well-developed limbs, redesigned sensory and respiratory systems, and modifications of the postcranial skeleton for supporting the body in air, have made a notable conquest of land. But, with shell-less eggs, thin, moist skin, and often gilled larvae, their development remains hazardously tied to water. The ancestor of a clade containing nonavian reptiles, birds, and mammals evolved an egg better adapted for dry terrestrial conditions. This shelled egg, perhaps more than any other adaptation, unshackled early reptiles from the aquatic environment by freeing the developmental process from dependence on aquatic or very moist terrestrial environments. In fact, the "pond-dwelling" stages were not eliminated but enclosed within a series of extraembryonic membranes that provided complete support for embryonic development. One membrane, the amnion, encloses a fluid-filled cavity, the "pond,"

within which the developing embryo floats. Another membranous sac, the allantois, serves both as a respiratory surface and as a chamber for the storage of nitrogenous wastes. Enclosing these membranes is a third membrane, the chorion, through which oxygen and carbon dioxide freely pass. Finally, surrounding and protecting everything is a porous, parchmentlike or leathery shell.

With the last ties to aquatic reproduction severed, conquest of land by vertebrates was ensured. Paleozoic tetrapods that developed this reproductive pattern were ancestors of a single, monophyletic assemblage called Amniota, named after the innermost of the three extraembryonic membranes, the amnion. Before the end of the Paleozoic era amniotes had diverged into multiple lineages that gave rise to all nonavian reptiles, birds, and mammals.

The paraphyletic class Reptilia (rep-til′e-a) (L. *repto*, to creep) includes nearly 8000 species (approximately 340 species in the United States and Canada) occupying a great variety of aquatic and terrestrial habitats, in many of which they are diverse and abundant. Nevertheless, reptiles are perhaps remembered best for what they once were, rather than for what they are now. The Age of Reptiles, which lasted for more than 165 million years, saw the appearance of a great radiation of reptilian lineages into a bewildering array of terrestrial and aquatic forms. Among these were herbivorous and carnivorous dinosaurs, many of huge stature and awesome appearance, that dominated animal life on land. Then, during a mass extinction at the end of the Mesozoic era, many reptilian lineages became extinct. Among the lineages to emerge from the Mesozoic extinction are those of today's reptiles. One of these lineages includes the two species of tuataras (*Sphenodon*) of New Zealand, sole survivors of a group that otherwise disappeared 100 million years ago. But other groups, especially lizards and snakes, have radiated since the Mesozoic extinction into diverse and abundant forms. The morphology, physiology, and behavior of some reptiles, especially lizards, probably are more similar to the first amniotes than any other living vertebrate group. In the next section we discuss the origin of amniotes, their radiation into diverse groups, and adaptations for life on dry land.

ORIGIN AND EARLY EVOLUTION OF AMNIOTES

As mentioned in the prologue to this chapter, amniotes are a monophyletic group that appeared and diversified in the late Paleozoic. Most zoologists agree that amniotes are most closely related to anthracosaurs, a group of **anamniotes** (vertebrates lacking an amnion) of the early Carboniferous period. Anthracosaurs were better adapted for terrestrial life than most other anamniotes, and sometimes initially mistaken for early reptiles. Their diet probably consisted mostly of insects, which had radiated into diverse forms by the Carboniferous. *Diadectes* is the most likely sister taxon of amniotes, and interestingly enough, may be the only completely herbivorous anamniotic tetrapod, past or present. The first amniotes were small and lizardlike, but quickly underwent an adaptive radiation by the early Permian period into forms diverse in morphology, feeding biology, and use of habitats (Figure 26.1).

Early diversification of amniotes produced three patterns of holes (fenestra) in the temporal region of the skull. **Anapsid** (Gr. *an*, without, + *apsis*, arch) skulls have no openings in the temporal area of the skull behind the **orbit** (opening in the skull for the eye); thus, the temporal region of the skull is completely roofed by dermal bones (Figure 26.2). This skull morphology was present in the earliest amniotes, and in one living group, the turtles, although the anapsid condition in turtles likely evolved secondarily, from ancestors having temporal fenestra. Two other amniote clades, Diapsida and Synapsida, represent separate evolutionary derivations from the ancestral anapsid condition.

The **diapsid** (Gr. *di*, double, + *apsis*, arch) skull has two temporal openings: one pair located low over the cheeks, and a second pair positioned above the lower pair, in the roof of the skull, and separated from the first by a bony arch (Figure 26.2). Diapsid skulls are present in birds and all amniotes traditionally considered "reptiles," except turtles (see Figure 26.1). In many living diapsids (lizards, snakes, and birds) one or both of the bony arches and openings have been lost, perhaps to facilitate skull kinesis (see Figure 26.12). The earliest diapsids gave rise to five morphologically distinct clades. The **lepidosaurs** include most of the living reptiles, including lizards, snakes, and tuataras. The **archosaurs** include dinosaurs, pterosaurs, and living crocodilians and birds. A third, smaller clade, the **sauropterygians,** includes several extinct aquatic groups, the most conspicuous of which are the large, long-necked plesiosaurs (see Figure 26.1). **Ichthyosaurs,** represented by extinct, aquatic dolphinlike forms (see Figure 26.1), form a fourth clade of diapsids. Placement of the last clade, the **turtles,** in the diapsid clade is controversial, although we treat turtles as highly modified members of the clade Diapsida here. Turtle morphology is a mix of ancestral and derived characters that has scarcely changed since they first appeared in the fossil record in the Triassic some 200 million years ago. Turtle skulls lack temporal fenestra and are often considered to be the only living descendants of parareptiles, an early anapsid group. However, other morphological and genetic evidence published over the past 15 years places turtles within the diapsid clade, suggesting that the two pairs of temporal fenestra characteristic of diapsids were lost early in turtle evolution. The relationship of turtles to other diapsids is unclear; morphology of the appendicular skeleton suggests affinities with lepidosaurs, but genetic evidence suggests affinities with archosaurs.

The third skull fenestration condition is **synapsid** (Gr. *syn*, together, + *apsis*, arch), characterized by a single pair of temporal openings located low on the cheeks and bordered by a bony arch (Figure 26.2). The synapsid condition occurs in a clade that includes mammals and their extinct relatives, the therapsids and pelycosaurs (see Figure 26.1). Synapsids were the first amniote group to undergo extensive adaptive radiation and were the dominant large amniotes of the late Paleozoic.

What was the functional significance of the temporal openings in early amniotes? In living forms these openings are occupied by large muscles that elevate (adduct) the lower jaw. Changes in jaw musculature might reflect a shift from suction feeding in aquatic vertebrates to a terrestrial feeding type that required larger muscles to produce more static pressure, for doing activities such as nipping plant material with the anterior teeth, or for grinding food with the posterior teeth. Amniotes exhibit considerably more variation in feeding biology than anamniotes, and herbivory is common in many amniote lineages. Although the functional significance of the evolution of temporal openings in amniotes is not fully understood, it is clear that expansion of the jaw musculature of amniotes must have been important. Even in fenestra-lacking turtle skulls, emarginations (notches) in the temporal area of the skull provide room for large jaw muscles.

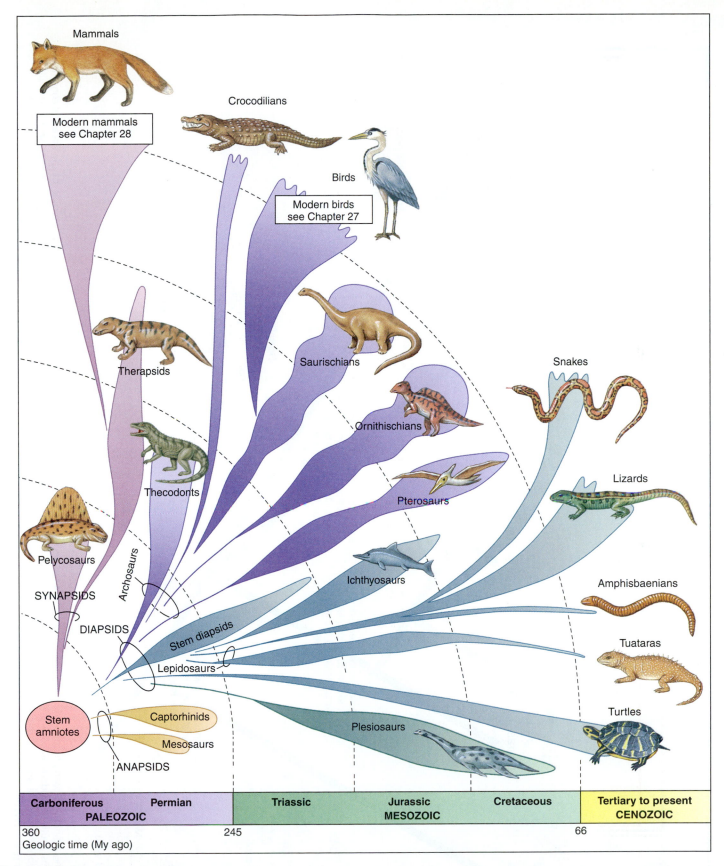

Figure 26.1

Evolution of amniotes. The evolutionary origin of amniotes occurred by evolution of an amniotic egg that made reproduction on land possible, although this egg may well have developed before the earliest amniotes had ventured far on land. The amniote assemblage, which includes nonavian reptiles, birds, and mammals, evolved from a lineage of small, lizardlike forms that retained the anapsid skull pattern of early tetrapods. First to diverge from the primitive stock was a lineage that evolved a skull pattern termed the synapsid condition. All other amniotes, including birds and all living nonavian reptiles except turtles, have a skull pattern known as diapsid. Turtles retain the primitive, anapsid skull pattern. The great Mesozoic radiation of reptiles may have resulted partly from the increased variety of ecological habitats that amniotes could exploit.

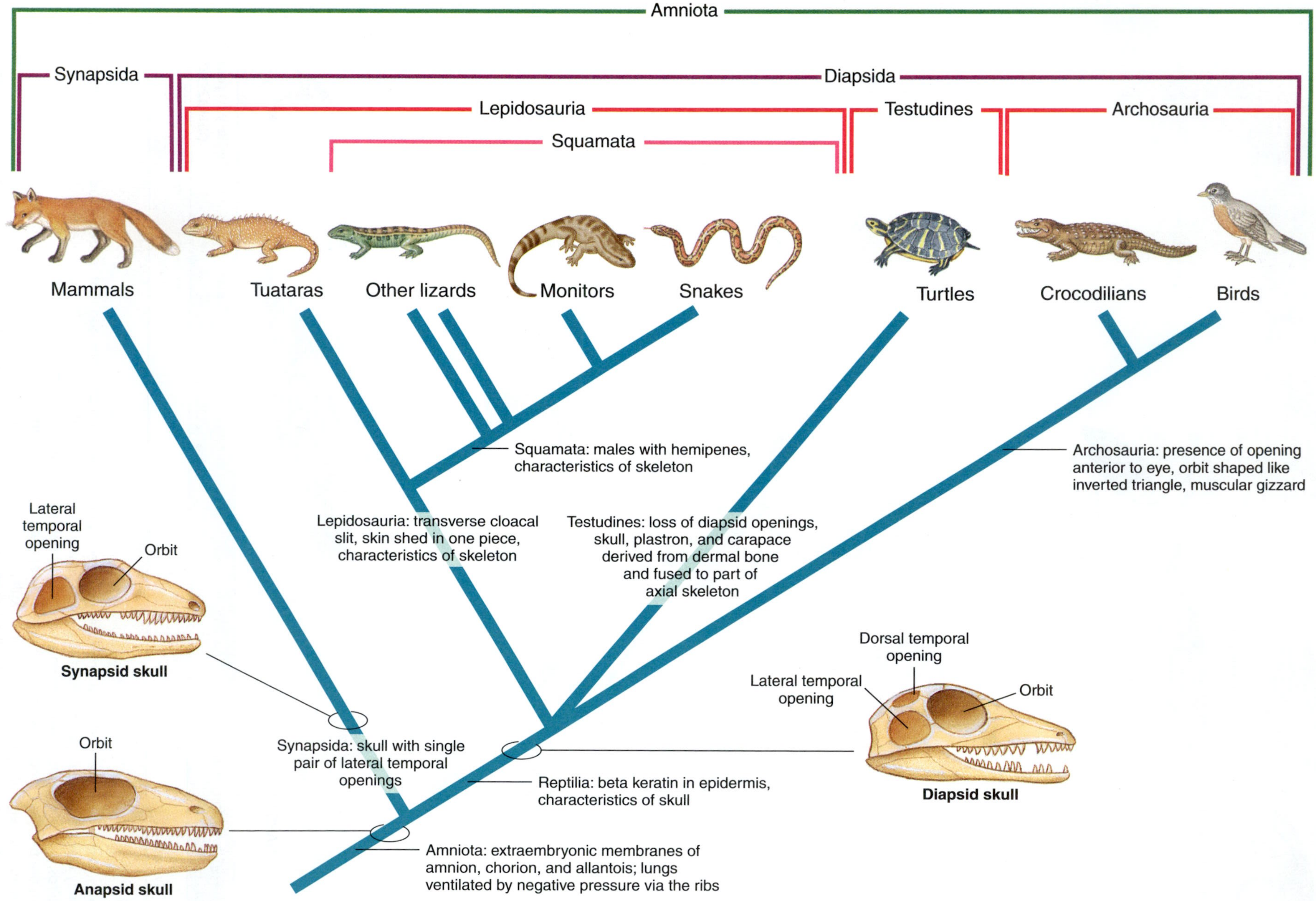
Amniota
Synapsida
Diapsida
Lepidosauria
Testudines
Archosauria
Squamata
Mammals
Tuataras
Other lizards
Monitors
Snakes
Turtles
Crocodilians
Birds
Squamata: males with hemipenes, characteristics of skeleton
Archosauria: presence of opening anterior to eye, orbit shaped like inverted triangle, muscular gizzard
Lepidosauria: transverse cloacal slit, skin shed in one piece, characteristics of skeleton
Testudines: loss of diapsid openings, skull, plastron, and carapace derived from dermal bone and fused to part of axial skeleton
Lateral temporal opening
Orbit
Synapsid skull
Dorsal temporal opening
Lateral temporal opening
Orbit
Diapsid skull
Orbit
Synapsida: skull with single pair of lateral temporal openings
Reptilia: beta keratin in epidermis, characteristics of skull
Anapsid skull
Amniota: extraembryonic membranes of amnion, chorion, and allantois; lungs ventilated by negative pressure via the ribs

Figure 26.2

Cladogram of living Amniota showing monophyletic groups. Some shared derived characters (synapomorphies) of the groups are given. The skulls represent the ancestral condition of the three groups. Skulls of modern diapsids and synapsids are often highly modified by loss or fusion of skull bones that obscures the ancestral condition. Representative skulls for anapsids are *Nyctiphruetus* of the upper Permian; for diapsids, *Youngina* of the upper Permian; for synapsids, *Aerosaurus,* a pelycosaur of the lower Permian. The relationship of turtles to other reptiles is controversial: some researchers consider them archosaurs; others consider them lepidosaurs or direct living descendents of anapsids.

Derived Characters of Amniotes

1. **Amniotic egg.** All amniotes have eggs with four extraembryonic membranes, the **amnion, allantois, chorion,** and **yolk sac** (Figure 26.3). The amnion encloses the embryo in fluid, cushioning the embryo and providing an aqueous medium for growth. Metabolic wastes are stored in a sac formed by the allantois. The chorion surrounds the entire contents of the egg, and like the allantois, is highly vascularized. Thus, the chorion and allantois form an efficient respiratory organ for removing carbon dioxide and acquiring oxygen. Most amniotic eggs are surrounded by a mineralized but often flexible shell, although many lizards and snakes and most mammals lack shelled eggs. The shell forms an important mechanical support, and especially for birds, a semipermeable barrier, which allows passage of gases but limits water loss. Like eggs of anamniotes, amniotic eggs have a yolk sac for nutrient storage (see pp. 175–177), although the yolk sac tends to be larger in amniotes. In marsupial and placental mammals, the yolk sac does not store yolk, but it may form a temporary or persistent yolk-sac placenta to transfer nutrients, gases, and wastes between the embryo and mother.

   How did the amniotic egg evolve? It is tempting to think of the amniotic egg as *the* land egg. However, many amphibians lay eggs on land and many amniotic eggs, such as those of turtles, must be buried in wet soil or deposited in areas of high humidity. Still, amniotic eggs can be laid in places too dry for any amphibian; clearly the evolution of amniotic eggs was a major factor in the success of tetrapods on land. Perhaps a more important selective advantage of the amniotic egg was that it permitted development of a larger, faster-growing embryo. Anamniote eggs are supported mainly by a thick jellylike layer. This jelly layer is inadequate to support large eggs and limits movement of oxygen into the eggs. One hypothesis suggests that a first step in the evolution of the amniotic egg was replacement of the jelly layer with a shell, which provided better support and movement of oxygen.

Furthermore, calcium deposited in the shell can be dissolved and absorbed by the growing embryo, to provide one of the raw materials needed for skeleton construction. This hypothesis is supported by physiological studies, which show that embryos of species with the smallest amniotic eggs have a metabolic rate about three times that of embryos of anamniotes with eggs of the same size.

All amniotes lack gilled larvae and have internal fertilization. These characteristics, along with the amniotic egg, eliminated the need for aquatic environments during reproduction. Shelled amniotic eggs require internal fertilization because sperm cannot penetrate the shell. Internal fertilization in amniotes is accomplished with a copulatory organ, except for tuataras and most birds, which transfer sperm by pressing their cloacas together. The most common copulatory organ in amniotes, a penis derived from the cloacal wall, appears to be another amniote innovation.

2. **Rib ventilation of the lungs.** Amphibians, like air-breathing fishes, fill their lungs by *pushing* air from the oral and pharyngeal cavities into the lungs (see Figure 31.22). In contrast, amniotes *draw* air into their lungs (**aspiration**) by expanding the thoracic cavity using costal (rib) muscles or pulling the liver (with other muscles) posterior. This shift from positive to negative ventilation probably influenced the evolution of amniote limbs. Early tetrapods used their rib muscles to make lateral undulations, causing a wriggling motion not unlike that of elongate fishes. Enlargement and mechanical rearrangement of the limbs enabled more efficient locomotion of amniotes, particularly for those with large bodies, while allowing rib muscles to serve a primary function in lung ventilation. However, we note that limbs are not required for terrestrial locomotion as some amniotes, such as snakes and amphisbaenians, are very successful with no limbs at all!

3. **Thicker and more waterproof skin.** Although the skin varies widely in structure among living amniotes and anamniote tetrapods, amniote skin tends to be more keratinized

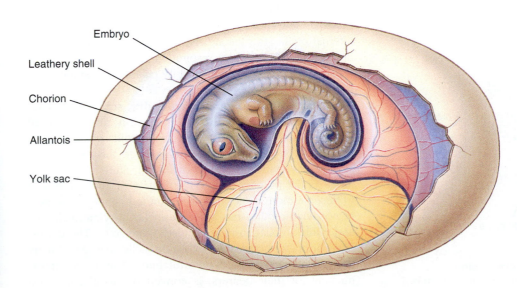

Figure 26.3

Amniotic egg. The embryo develops within the amnion and is cushioned by amniotic fluid. Food is provided by yolk from the yolk sac and metabolic wastes are deposited within the allantois. As development proceeds, the allantois fuses with the chorion, a membrane lying against the inner surface of the shell; both membranes are supplied with blood vessels that assist in the exchange of oxygen and carbon dioxide across the porous shell. Because this kind of egg is an enclosed, self-contained system, it is often called a "cleidoic" egg (Gr. *kleidoun,* to lock in).

and less permeable to water. A wide variety of structures composed of keratin, such as scales, hair, feathers, and claws, project from amniote skin, and the epidermis itself is more heavily keratinized. Keratin gives the skin protection from physical trauma, and lipids in the skin limit water loss through the skin. Reptiles uniquely have a form of keratin called beta keratin. This especially hard form of keratin is in reptile scales and scutes, bird feathers, and other reptile epidermal structures (Figure 26.4). Keratin and lipids limit the skin's ability to exchange respiratory gases—so, unlike most amphibians, few amniotes use their skin as a primary respiratory organ. Amniote gas exchange takes place primarily in the lungs, which have considerably more surface area than anamniote lungs.

Changes in Traditional Classification of Reptiles

With increasing use of cladistic methodology in zoology, and its insistence on hierarchical arrangement of monophyletic groups (see p. 209), important changes have been made in the classification of amniotes. As traditionally defined, class Reptilia includes snakes, lizards, tuataras, crocodilians, and turtles, in addition to several extinct groups, such as dinosaurs, plesiosaurs, pterosaurs, and the "mammal-like reptiles," but excludes birds. However, birds and reptiles share several derived characters, including several skull and ankle characteristics along with presence of a special type of keratin in the skin called beta keratin, which unites them as a monophyletic group (see Figure 26.2). Birds evolved from a group of dinosaurs called dromeosaurs (see p. 587). Thus, reptiles as traditionally defined form a **paraphyletic** group because they do not include all descendants of their most recent ancestor.

Crocodilians and birds are sister groups; they are more recently descended from a common ancestor than either is from any other living reptilian lineage. In other words, birds and crocodilians belong to a monophyletic group apart from other reptiles

and, according to the rules of cladistics, should be assigned to a clade that separates them from the remaining reptiles. This clade is in fact recognized; it is Archosauria (see Figures 26.1 and 26.2), a grouping that also includes the extinct dinosaurs and pterosaurs. Archosaurs, along with their sister group, the lepidosaurs (tuataras, lizards, and snakes), and turtles, form a monophyletic group that cladists call Reptilia. Note that the early pelycosaur and therapsid ancestors of mammals were formerly termed "mammal-like reptiles." However, they are not in the clade Reptilia and are no longer called "reptiles" (see p. 614 and Figure 28.2). Here we use Reptilia and reptiles in a cladistic fashion to include living amniote groups traditionally termed "reptiles" together with birds and all extinct groups more closely related to these than to mammals. The term "nonavian reptiles" is used here to refer to a paraphyletic group that includes the living turtles, lizards, snakes, tuataras, and crocodilians, and a number of extinct groups, including plesiosaurs, ichthyosaurs, pterosaurs, and dinosaurs. Nonavian reptiles are the subject of most of the remainder of this chapter; birds, which form the remainder of clade Reptilia, are discussed in Chapter 27.

CHARACTERISTICS OF NONAVIAN REPTILES THAT DISTINGUISH THEM FROM AMPHIBIANS

Amphibians and nonavian reptiles are both ectothermic tetrapods, relying on the environment for adjustment of their body temperature. However, they differ in several characteristics, many of which are directly related to the success of nonavian reptiles in dry, terrestrial environments.

1. **Nonavian reptiles have better developed lungs than those of amphibians.** Nonavian reptiles, like other amniotes, rely primarily on lungs for gas exchange. These lungs have more surface area than the lungs of amphibians and are ventilated by drawing air into the lungs, rather than the amphibian method of pushing air into lungs. Nonavian reptiles expand the thoracic cavity, drawing air into the lungs by expanding the rib cage (lizards and snakes), or moving internal organs (turtles and crocodiles). Pulmonary respiration is supplemented by respiration in the cloaca or pharynx in many aquatic turtles and by cutaneous respiration in sea snakes and turtles.

2. **Nonavian reptiles have tough, dry scaly skin that offers protection against desiccation and physical injury.** Amphibians must maintain a thin, moist skin to permit effective gas exchange. However, this skin makes amphibians vulnerable to dehydration. Shift from skin to lungs as the primary site of respiration permitted the skin to become more desiccation resistant in amniotes. The skin of nonavian reptiles has an epidermis of varying thickness, and a thick, collagen-rich dermis (Figure 26.4). The dermis has chromatophores, color-bearing cells that give many lizards and snakes their colorful hues. This layer, unfortunately for their bearers, is converted into alligator and

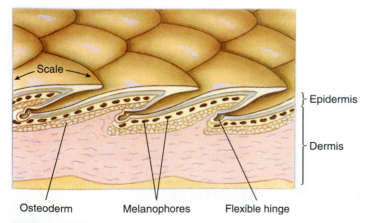

Figure 26.4

Section of the skin of a nonavian reptile showing overlapping, keratinized scales in the epidermis and bony osteoderms in the dermis.

Scale
Epidermis
Dermis
Osteoderm   Melanophores   Flexible hinge

snakeskin leather, esteemed for expensive pocketbooks and shoes. Resistance to desiccation primarily is provided by hydrophobic lipids in the epidermis. The epidermis also contains a hard form of keratin called beta keratin that is unique to the clade Reptilia. The characteristic scales of nonavian reptiles, formed mostly of beta keratin, provide protection against wear in terrestrial environments. The primarily epidermal scales are not homologous to the scales of fishes, which are mostly bony, dermal structures (see Figure 29.2, p. 647). In crocodilians, scales remain throughout life, growing gradually to replace wear. In other nonavian reptiles, such as lizards and snakes, new keratinized epidermis grows beneath the old, which is then shed at intervals. Turtles add new layers of keratin under the old layers of the platelike scutes, which are modified scales. Crocodiles and many lizards (skinks, for example), possess bony plates called osteoderms (Figure 26.4) located in the dermis, beneath the keratinized scales.

3. **The amniotic egg of nonavian reptiles permits rapid development of large young in relatively dry environments.** As described on page 567, the amniotic egg allows nonavian reptiles to reproduce outside water, although their eggs often require relatively high humidity to avoid desiccation. All eggs of nonavian reptiles are impregnated with calcium, but only turtles have eggs with rigid shells. The shells of other nonavian reptiles are leathery. For many species of nonavian reptiles, egg or embryo development occurs in a female's reproductive tract, providing even greater protection from predators and dehydration, and potential for the mother to manage the embryo's nutritional and other physiological needs.

4. **The jaws of nonavian reptiles are efficiently designed for applying crushing or gripping force to prey.** Jaws of fishes are designed for suction feeding or for quick closure, but once the prey is seized, little static force can be applied. The larger jaw muscles of nonavian reptiles, like other amniotes, have better mechanical advantage. In addition, the tongue is muscular and mobile, functioning to move food in the mouth for mastication and swallowing.

5. **Nonavian reptiles have an efficient and versatile circulatory system and higher blood pressure than amphibians.** In all nonavian reptiles, like other amniotes, the right atrium, which receives deoxygenated blood from the body, is completely partitioned from the left atrium, which receives oxygenated blood from the lungs. Crocodilians have two completely separated ventricles as well (Figure 26.5); in other nonavian reptiles the ventricle is incompletely partitioned into multiple chambers. Even in those forms with incomplete ventricular septa, flow patterns within the heart limit admixture of pulmonary (oxygen-rich) and systemic (oxygen-poor) blood; all nonavian reptiles therefore have two functionally separate circulations. The incomplete separation between the right and left sides of the heart provides an added benefit of permitting blood to bypass the lungs when pulmonary respiration is not occurring (for example, diving or estivation).

6. **Nonavian reptiles have efficient strategies for water conservation.** Most amphibians excrete their metabolic waste primarily as ammonia. Ammonia is toxic at relatively low concentrations and must be removed in a dilute solution. Because of the water required for excretion of ammonia, it is not adaptive for vertebrates occupying dry,

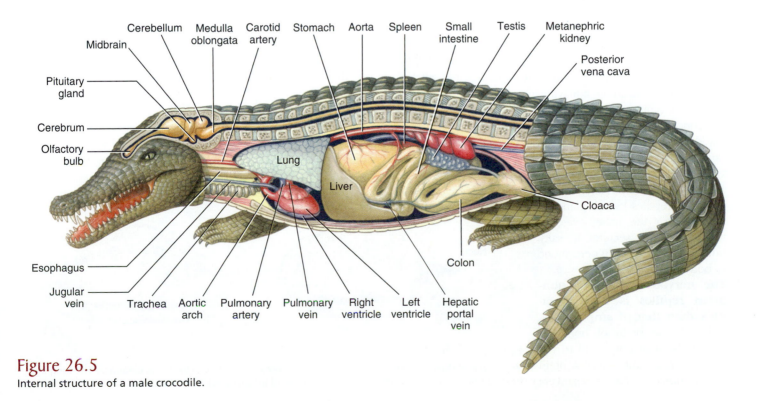

Figure 26.5
Internal structure of a male crocodile.

Characteristics of Nonavian Reptiles (Class Reptilia)

1. **Body covered with keratinized epidermal scales** and sometimes bony dermal plates; **integument with few glands**
2. **Two paired limbs, usually with five toes,** and adapted for climbing, running, or paddling; limbs vestigial or absent in snakes and some lizards
3. Skeleton well ossified; ribs with sternum (sternum absent in snakes) forming a complete thoracic basket; **skull with one occipital condyle**
4. Respiration primarily by lungs, which have high surface area and filled by **aspiration; no gills;** cloaca, pharynx, or skin used for respiration by some
5. Circulatory system functionally divided into **pulmonary and systemic circuits;** heart typically consisting of a sinus venosus, an atrium completely divided into two chambers, and a ventricle incompletely divided into three chambers; crocodilians with a sinus venosus, two atria, and two ventricles
6. Ectothermic; many thermoregulate behaviorally
7. **Metanephric kidney (paired); uric acid main nitrogenous waste**
8. Nervous system with optic lobes on dorsal side of brain; **12 pairs of cranial nerves** in addition to nervus terminalis; enlarged cerebrum
9. Sexes separate; **fertilization internal;** copulatory organ a **penis,** hemipenes, or rarely absent
10. **Eggs covered with calcareous or leathery shells; extraembryonic membranes (amnion, chorion, yolk sac, and allantois)** present during embryonic life; **no aquatic larval stages**

terrestrial environments. Nonavian reptiles excrete their nitrogenous wastes as uric acid. Uric acid is relatively nontoxic, so it can be concentrated without harmful effects. All amniotes, including nonavian reptiles, have metanephric kidneys, but nonavian reptiles (unlike mammals) cannot concentrate their urine in the kidneys. Thus the urinary bladder of nonavian reptiles receives undiluted urine. Water (and most salts) is resorbed in the bladder, and "urine" is voided as a semisolid mass of uric acid. Salts are removed in many nonavian reptiles by salt glands—located near the nose, eyes, or tongue—which secrete a salty fluid that is strongly hyperosmotic to body fluids.

7. **The nervous system of non-avian reptiles is more complex than that of amphibians.** Although the brain of nonavian reptiles is small compared to the brain of other amniotes, the brain of all amniotes has a relatively

enlarged cerebrum. Enlargement of the cerebrum is correlated with integration of sensory information and control of muscles during locomotion. Nonavian reptiles have particularly good vision, and many species display brilliant coloration. Smell is not as developed in most reptiles, but snakes and many lizards use a highly sensitive sense of smell to find prey and mates. In lizards and snakes, olfaction is assisted by a well-developed Jacobson's organ, a specialized olfactory chamber in the roof of the mouth.

CHARACTERISTICS AND NATURAL HISTORY OF REPTILIAN ORDERS

Order Testudines (Chelonia): Turtles

Turtles appear in the fossil record in the Upper Triassic, some 200 million years ago. From the Triassic, turtles survived to the present with very little change to their early morphology. They are enclosed in shells consisting of a dorsal **carapace** (Fr. from Sp., *carapacho,* covering) and a ventral **plastron** (Fr. breastplate). The shell is composed of two layers: an outer layer of keratin and an inner layer of bone. New layers of keratin are laid down beneath the old as the turtle grows and ages. The bony layer is a fusion of ribs, vertebrae, and many dermally ossifying elements (Figure 26.6). Unique among vertebrates, turtle limbs and limb girdles are located *inside* the ribs! Lacking teeth, a turtle's jaw is provided with tough, keratinized plates for gripping food (Figure 26.7).

Clumsy and unlikely as they appear to be within their protective shells, turtles are nonetheless a varied and ecologically diverse group that seems able to adjust to human presence.

The terms "turtle," "tortoise," and "terrapin" are applied variously to different members of the turtle order. In North American usage, they are all correctly called turtles. The term "tortoise" is frequently applied to land turtles, especially large forms. British usage of the terms is different: "tortoise" is the inclusive term, whereas "turtle" is applied only to the aquatic members.

One consequence of living in a rigid shell with fused ribs is that a turtle cannot expand its chest to breathe. Turtles solved

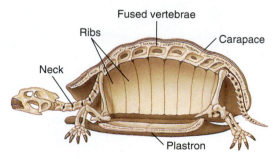

Fused vertebrae

Ribs

Carapace

Neck

Plastron

Figure 26.6

Skeleton and shell of a turtle, showing fusion of vertebrae and ribs with the carapace. The long and flexible neck allows the turtle to withdraw its head into its shell for protection.

Figure 26.7
Snapping turtle, *Chelydra serpentina,* showing the absence of teeth. Instead, the jaw edges are covered with a keratinized plate.

Figure 26.8
Mating Galápagos tortoises, *Geochelone elephantopus.* The male has a concave plastron that fits over the highly convex carapace of the female, helping to provide stability during mating. Males utter a roaring sound during mating, the only time they are known to emit vocalizations.

this problem by employing certain abdominal and pectoral muscles as a "diaphragm." Air is drawn in by increasing abdominal cavity volume by contracting limb flank muscles. Exhalation is also active and is accomplished by drawing the shoulder girdle back into the shell, thus compressing the viscera and forcing air out of the lungs. Breathing is visible as the bellowslike movements of the turtle's "limb pockets": folds of skin between the limbs and the rigid shell. Movements of the limbs during walking also help to ventilate the lungs. Many aquatic turtles gain enough oxygen by pumping water in and out of a vascularized mouth cavity or cloaca; this activity enables them to remain submerged for long periods when inactive. When active they must lung-breathe more frequently.

A turtle's brain, like that of other reptiles, is small, never exceeding 1% of body weight. The cerebrum, however, is larger than that of an amphibian, and turtles are able to learn a maze about as quickly as a rat. Turtles have a middle and an inner ear, but perception of sound is poor. Not unexpectedly, therefore, turtles are virtually mute (the biblical "voice of the turtle" refers to the turtledove, a bird), although many tortoises utter grunting or roaring sounds during mating (Figure 26.8). Compensating for poor hearing are a good sense of smell, acute vision, and color perception evidently as good as that of humans.

Turtles are oviparous. Fertilization is internal and all turtles, even marine forms, bury their shelled, amniotic eggs in the ground. Usually they exercise considerable care in constructing their nest, but once eggs are deposited and covered, the female deserts them. An interesting feature of turtle reproduction is that in some turtle families, as in all crocodilians and some lizards, nest temperature determines sex of the hatchlings. In turtles, low temperatures during incubation produce males and high temperatures produce females (Figure 26.9). All nonavian reptiles with temperature-dependent sex determination lack sex chromosomes.

Marine turtles, buoyed by their aquatic environment, may reach great size. Leatherbacks are largest, attaining a length of 2 m and weight of 725 kg. Green turtles (Figure 26.10), so named because of their greenish body fat, may exceed 360 kg, although most individuals of this economically valuable and heavily exploited species seldom live long enough to reach anything approaching this size. Some land tortoises may weigh several hundred kilograms, such as the giant tortoises (Figure 26.8) of the Galápagos Islands that so intrigued Darwin during his visit there in 1835. Most tortoises

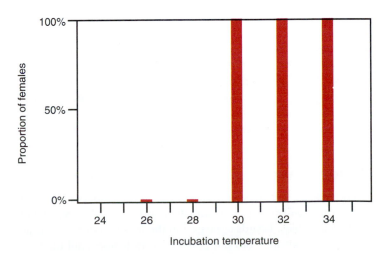

Figure 26.9
Temperature-dependent sex determination in the European pond turtle, *Emys orbicularis.* Eggs incubated at high temperatures produce females, while eggs incubated at low temperature produce males.

Figure 26.10

Green sea turtle, *Chelonia mydas*. Green turtles are herbivores that subsist on marine grasses and algae. Sea turtles range widely in the oceans, returning to land only to deposit their eggs. Sea turtles are found in all tropical oceans.

Figure 26.11

Alligator snapping turtle, *Macroclemys temmincki*, of the southeastern United States lies on the bottom, mouth agape, luring fishes and other unwary prey by undulating a pink, wormlike protrusion from its tongue. Any prey attempting to eat the bait is instantly captured in powerful jaws.

are rather slow moving; an hour of determined trudging carries a large Galápagos tortoise approximately 300 m (although they may move much more rapidly for short distances). Their low metabolism probably explains their longevity, for some are believed to live more than 150 years.

The shell, like a medieval coat of armor, offers obvious advantages. The head and appendages can be drawn in for protection. The familiar box tortoise, *Terrapene carolina*, has a plastron that is hinged, forming two movable parts that can be pulled up against the carapace so tightly that one can hardly force a knife blade between the shells. Some turtles, such as the large eastern snapping turtle (see Figure 26.7), have reduced shells, making complete withdrawal for protection quite impossible. Snappers, however, have another formidable defense, as their name implies. They are entirely carnivorous, living on fishes, frogs, waterfowl, or almost anything that comes within reach of their powerful jaws. An alligator snapper lures unwary fish into its mouth with a pink, wormlike extension of its tongue that serves as a "bait" (Figure 26.11). Alligator snappers are wholly aquatic and come ashore only to lay their eggs.

Order Squamata: Lizards and Snakes

Squamates are the most recent and diverse products of diapsid evolution, comprising approximately 95% of all known living nonavian reptiles. Lizards appeared in the fossil record as early as the Jurassic, but they did not begin their radiation until the Cretaceous period of the Mesozoic era when dinosaurs were at the climax of their radiation. Snakes appeared during the late Jurassic period, probably from a group of lizards whose descendants include the Gila monster and monitor lizards. Two specializations

in particular characterize snakes: extreme elongation of their body and accompanying displacement and rearrangement of internal organs; and specializations for eating large prey.

Viviparity in living nonavian reptiles is limited to squamates and has evolved in that clade about 100 separate times. Viviparity typically occurs by increasing the length of time eggs are kept within the oviduct. Developing young obtain nutrients from yolk sacs (**lecithotrophy** or ovoviviparity) or via the mother (**placentotrophy**), or some combination of both. Viviparous snakes and lizards are usually associated with cool climates. Viviparity may be an adaptation that permits squamates to regulate the temperature of their developing embryos, using **ectothermy,** for optimal development. Gravid mothers can move to favorable environments to keep their bodies, and their young, warm enough to ensure rapid development. Also, young retained in the mother's body are better protected from predation than deposited eggs.

Skulls of squamates are modified from the ancestral diapsid condition by loss of dermal bone ventral and posterior to the lower temporal opening. This modification has allowed evolution in most lizards and snakes of a mobile skull having movable joints. Such a skull is called a **kinetic skull.** The quadrate, which in other reptiles is fused to the skull, has a joint at its dorsal end, as well as its usual articulation with the lower jaw. In addition, joints in the palate and across the roof of the skull allow the snout to be tilted upward (Figure 26.12). Specialized mobility of the skull enables squamates to seize and to manipulate their prey; it also increases the effective closing force of the

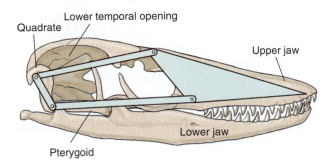

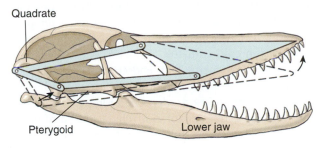

Figure 26.12

Kinetic diapsid skull of a modern lizard (monitor lizard, *Varanus* sp.) showing the joints that allow the snout and upper jaw to move on the rest of the skull. The quadrate can move at its dorsal end and ventrally at both the lower jaw and the pterygoid. The front part of the braincase is also flexible, allowing the snout to be raised. Note that the lower temporal opening is very large with no lower border; this modification of the diapsid condition, common in modern lizards, provides space for expansion of large jaw muscles. The upper temporal opening lies dorsal and medial to the postorbital-squamosal arch and is not visible in this drawing.

jaw musculature. The skull of snakes is even more kinetic than that of lizards. Such exceptional skull mobility is considered a major factor in diversification of lizards and snakes.

Traditionally, the order Squamates was divided into three suborders: Sauria (or Lacertilia) (lizards), Serpentes (snakes), and Amphisbaenia (amphisbaenians). Amphisbaenians now are considered modified lizards, and we discuss them with lizards. Snakes form a monophyletic group but uniquely share a common ancestor with varanid lizards, thus making Sauria paraphyletic. Here we use "lizard" and "Sauria" to describe all squamates that are not snakes.

Suborder Sauria: Lizards

Lizards are an extremely diverse group, including terrestrial, burrowing, aquatic, arboreal, and aerial members. Among the more familiar groups in this varied suborder are **geckos** (Gekkonidae) (Figure 26.13) small, agile, mostly nocturnal forms with adhesive toe pads that enable them to walk upside down and on vertical surfaces; **iguanids** (Iguanidae), including most familiar New World lizards, often brightly colored with ornamental crests, frills, and throat fans, and the remarkable marine iguana of the Galápagos Islands (Figure 26.14); **skinks** (Scincidae), with elongate bodies, an armor of tight-fitting osteoderms

Figure 26.13

Tokay, *Gekko gecko*, of Southeast Asia has a true voice and is named after the strident repeated *to-kay, to-kay* call.

(see Figure 26.4), and reduced limbs in many species; **monitors** (Varanidae), large, active predators that include the largest lizard, the Komodo dragon, *Varanus komodensis* (see photo on p. 563); and **chameleons** (Chamaeleonidae), a group of arboreal lizards, mostly of Africa and Madagascar. Chameleons are entertaining creatures that catch insects with a sticky-tipped tongue that can be projected accurately and rapidly to a distance greater than the length of their body (Figure 26.15). The great majority of lizards have four limbs and relatively short bodies, but in many the limbs are reduced, and a few such as glass lizards (Figure 26.16) are completely limbless.

Most lizards have movable eyelids, whereas a snake's eyes are permanently covered with a transparent cap. Lizards have keen vision for daylight (retinas rich in both cones and rods; see p. 750 for discussion of color vision), although one group, the nocturnal geckos, has retinas composed entirely of rods. Most lizards have an external ear that snakes lack. The inner ear of lizards is variable in structure, but as with other reptiles, hearing does

Figure 26.14

A large male marine iguana, *Amblyrhynchus cristatus,* of the Galápagos Islands, feeding underwater on algae. This is the only marine lizard in the world. It has special salt-removing glands in the eye orbits and long claws that enable it to cling to the bottom while feeding on small red and green algae, its principal diet. It may dive to depths exceeding 10 m (33 feet) and remain submerged more than 30 minutes.

Figure 26.15
A chameleon snares a dragonfly. After cautiously edging close to its target, the chameleon suddenly lunges forward, anchoring its tail and feet to the branch. A split second later, it launches its sticky-tipped, foot-long tongue to trap the prey. The eyes of this common European chameleon, *Chamaeleo chamaeleon,* are swiveled forward to provide binocular vision and excellent depth perception.

Figure 26.16
A glass lizard, *Ophisaurus* sp., of the southeastern United States. This legless lizard feels stiff and brittle to the touch and has an extremely long, fragile tail that readily fractures when the animal is struck or seized. Most specimens, such as this one, have only a partly regenerated tip to replace a much longer tail previously lost. Glass lizards can be readily distinguished from snakes by the deep, flexible groove running along each side of the body. They feed on worms, insects, spiders, birds' eggs, and small nonavian reptiles.

not play an important role in the lives of most lizards. Geckos are exceptions because males are strongly vocal (to announce territory and to discourage approach of other males) and they must, of course, hear their own vocalizations. Other species of lizards vocalize in defensive behavior.

Many lizards live in the world's hot and arid regions. Since their skin lacks glands, water loss by this avenue is much

reduced. They produce a semisolid urine with a high content of crystalline uric acid. This excellent mechanism for conserving water is found in other groups living successfully in arid habitats (birds, insects, and pulmonate snails). Some, such as Gila monsters of southwestern United States deserts, store fat in their tails, which they use during drought to provide energy and metabolic water (Figure 26.17). Gila monsters, and their close relatives, beaded lizards, are the only lizards capable of delivering a venomous bite.

Lizards, like nearly all nonavian reptiles, are ectotherms, adjusting their body temperature by moving among different microclimates (see Chapter 30, p. 680). Cold climates provide limited opportunities for ectotherms to raise their body temperature to preferred levels. As a result, there are relatively few nonavian reptile species in cold climates. However, because ectotherms use considerably less energy than endotherms, nonavian reptiles are successful in ecosystems with low productivity and warm climates, such as tropical deserts, dry, open forests, and grasslands. Thus, ectothermy is not an "inferior" characteristic of nonavian reptiles, but rather a successful strategy for coping with specific environmental challenges.

The amphisbaenians, or "worm lizards," are lizards highly specialized for a fossorial (burrowing) life. Amphisbaenia means "double walk," in reference to their peculiar ability to move backward nearly as effectively as forward. Amphisbaenians appear so different from other lizards that, until recently, they were placed in a separate suborder, Amphisbaenia. However, phylogenetic analyses of morphological and molecular data show that they are highly modified lizards, although their relationship to other lizard families is uncertain. Amphisbaenians have elongate, cylindrical bodies of nearly uniform diameter, and most lack any trace of external limbs (Figure 26.18). Eyes usually are hidden underneath the skin, and there are no external ear openings. Their skull is solidly built and either conical, or spade-shaped, which aids in tunneling through the soil. The skin is formed into numerous,

Figure 26.17
Gila monster, *Heloderma suspectum,* of southwestern United States desert regions and the related Mexican beaded lizard are the only venomous lizards known. These brightly colored, clumsy-looking lizards feed principally on birds' eggs, nesting birds, mammals, and insects. Unlike venomous snakes, the Gila monster secretes venom from glands in its lower jaw. The chewing bite is painful to humans but seldom fatal.

Figure 26.18

A worm lizard of the suborder Amphisbaenia. Worm lizards are burrowing forms with a solidly constructed skull used as a digging tool. The species pictured, *Amphisbaena alba,* is widely distributed in South America.

independently moving rings which can grip the soil, creating a movement not unlike that of earthworms. Amphisbaenians have an extensive distribution in South America and tropical Africa. In the United States, one species, *Rhineura floridana,* occurs in Florida where it is called the "graveyard snake."

Suborder Serpentes: Snakes

Snakes are limbless and usually lack both pectoral and pelvic girdles (the latter persists as a vestige in pythons, boas, and some other snakes). The numerous vertebrae of snakes, shorter and wider than those of tetrapods, permit quick lateral undulations through grass and over rough terrain. Ribs increase rigidity of the vertebral column, providing more resistance to lateral stresses. Elevation of the neural spine gives the numerous muscles more leverage.

Many lineages of lizards exhibit reduction or loss of limbs, but none of these lineages experienced the remarkable radiation of snakes. Because of the limited kinesis of lizard skulls (see Figure 26.12), they are able to consume only relatively small food items. In contrast, the highly kinetic skull and feeding apparatus of snakes, which enable them to eat prey several times their own diameter, are remarkable specializations, which may be responsible for their incredible success. Unlike lizard jaws, the two halves of the lower jaw (mandibles) are joined only by muscles and skin, allowing them to spread widely apart. Many skull bones are so loosely articulated that the entire skull can flex asymmetrically to accommodate oversized prey (Figure 26.19). Since a snake must keep breathing during the slow process of swallowing, its tracheal opening (glottis) is thrust forward between the two mandibles.

The cornea of a snake's eye, lacking a movable eyelid, is permanently protected with a transparent membrane called a spectacle, which, together with reduced eyeball mobility, gives snakes the cold, unblinking stare that many people find

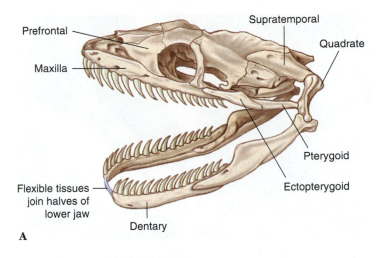

Figure 26.19

A, Lateral view of python skull. Each side of the extremely kinetic skull has several movable bones (labeled) which permit extraordinary movements of the jaw in feeding. The halves of the lower jaw are united by flexible soft tissues, permitting wide separation and independent movement of each side. **B,** The great mobility of the snake jaw and skull elements is evident in this snake swallowing an egg.

unnerving. Most snakes have relatively poor vision, arboreal snakes of the tropical forest being a conspicuous exception (Figure 26.20). Some arboreal snakes possess excellent binocular vision that helps them track prey through branches where scent trails would be impossible to follow.

Snakes have no external ears or tympanic membrane. This condition, together with absence of any obvious response to aerial sounds, led to a widespread opinion that snakes are totally deaf. But snakes do have internal ears, and recent work has shown quite clearly that within a limited range of low frequencies (100 to 700 Hz), hearing in snakes compares favorably with that of most lizards. Snakes are also quite sensitive to vibrations carried in the ground.

Nevertheless, most snakes employ chemical senses rather than vision and hearing to hunt their prey. In addition to the usual olfactory areas in the nose, which are not well developed, there are **Jacobson's organs** (vomeronasal organs), a pair of

Figure 26.20

Parrot snake, *Leptophis ahaetulla.* The slender body of this Central American tree snake is an adaptation for sliding along branches without weighing them down.

pitlike organs in the roof of the mouth. These are lined with richly innervated chemosensory epithelium. The forked tongue, flicking through the air, picks up scent molecules and conveys them to the mouth; the tongue is then drawn past Jacobson's organs. Information is then transmitted to the brain where scents are identified (Figure 26.21).

Snakes have evolved several solutions to the obvious problem of movement without limbs. The most typical pattern of movement is **lateral undulation** (Figure 26.22B). Movement follows an S-shaped path, with a snake propelling itself by exerting lateral force against surface irregularities. A snake seems to "flow," since the moving loops appear stationary with respect to the ground. Lateral undulatory movement is fast and efficient under most but not all circumstances. **Concertina movement** (Figure 26.22A) enables a snake to move in a narrow passage, as

when climbing a tree by using irregular channels in the bark. A snake extends forward while bracing S-shaped loops against the sides of the channel. To advance in a straight line, as when stalking prey, many heavy-bodied snakes employ **rectilinear movement.** Two or three sections of the body rest on the ground to support the snake's weight. Intervening sections are lifted free of the ground and pulled forward by muscles (shown in red in Figure 26.22C) that originate on ribs and insert on the ventral skin. Rectilinear movement is a slow but effective way of moving inconspicuously toward prey, even when there are no surface irregularities. **Side-winding** is a fourth form of movement that enables desert vipers to move with surprising speed across loose, sandy surfaces with minimum surface contact (Figure 26.22D). The sidewinder rattlesnake moves by throwing its body forward in loops with its body lying at an angle of about 60 degrees to its direction of travel.

Most snakes capture their prey by grasping it with their mouth and swallowing it while still alive. Swallowing a struggling, kicking, biting animal is dangerous, so most snakes that swallow prey alive specialize on smaller prey, such as worms, insects, fish, frogs, and, less frequently, small mammals. Many of these snakes, which may be quite fast, locate prey by actively foraging. Snakes that first kill their prey by constriction

Figure 26.21

A blacktail rattlesnake, *Crotalus molossus,* flicks its tongue to smell its surroundings. Scent particles trapped on the tongue's surface are transferred to Jacobson's organs, olfactory organs in the roof of the mouth. Note the heat-sensitive pit organ between the nostril and eye.

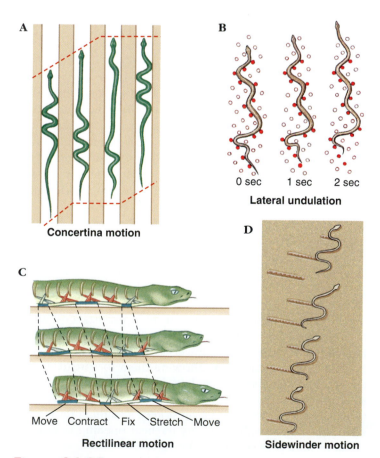

Figure 26.22

Snake locomotion. **A,** Concertina motion. **B,** Lateral undulation. **C,** Rectilinear motion. **D,** Sidewinder motion. Refer to text for explanation.

Figure 26.23
Nonvenomous African house snake, *Boaedon fuluginosus,* constricting a mouse before swallowing it.

Figure 26.24
Spectacled, or Indian, cobra, *Naja naja.* Cobras erect the front of the body and flatten the neck as a threat display and before attacking. Although a cobra's strike range is limited, all cobras are dangerous because of the extreme toxicity of their venom.

(Figure 26.23) often specialize on large mammalian prey. The largest constrictors are able to kill and to swallow prey as large as deer, leopards, and crocodilians. However, because muscle rearrangements that permit constricting prey also reduce speed of travel, most constrictors tend to ambush prey.

Other snakes kill their prey before swallowing by injecting it with venom. Less than 20% of all snakes are venomous, although venomous species outnumber nonvenomous species 4 to 1 in Australia. Venomous snakes are usually divided into five families, based in part on type of fangs. Vipers (family Viperidae) have highly developed, movable, tubular fangs at the front of their mouth. This family includes American pit vipers and Old World true vipers, which lack facial heat-sensing pits. Among the latter are the common European adder and African puff adder. A second family of venomous snakes (family Elapidae) has short, permanently erect fangs in the front of their mouth. In this group are cobras (Figure 26.24), mambas, coral snakes, and kraits. The highly venomous sea snakes are usually placed in a third family (Hydrophiidae). The fossorial mole vipers (family Atractaspididae) usually have fangs similar to vipers, although their relationship to other venomous snakes is not clear. The very large family Colubridae, which contains most familiar (and nonvenomous) snakes, does include a few snakes that have been responsible for human fatalities. Two examples are the African boomslang and the African twig snake, both rear-fanged snakes that normally use their venom to quiet struggling prey.

Snakes of subfamily Crotalinae within family Viperidae are called **pit vipers** because they possess special heat-sensitive pit organs (loreal pits) on their heads, located between their nostrils and eyes (Figures 26.21 and 26.25). All best-known North American venomous snakes are pit vipers, such as cottonmouths, copperheads, and rattlesnakes. The pits are supplied with a dense packing of free nerve endings from the fifth cranial nerve. These respond to radiant energy in the long-wave infrared

(5000 to 15,000 nm) and are especially sensitive to heat emitted by warm-bodied birds and mammals that are their food (infrared wavelengths of about 10,000 nm). Some measurements suggest that pit organs can distinguish temperature differences of only 0.003° C from a radiating surface. Pit vipers use pit organs to track warm-blooded prey and to aim strikes, which they can make as effectively in total darkness as in daylight. Boa constrictors and pythons also have heat receptors (in their lips), but the anatomy is quite different from that of pit vipers, suggesting that their heat receptors evolved independently.

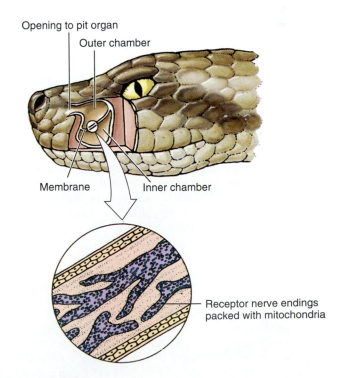

Opening to pit organ
Outer chamber
Membrane
Inner chamber
Receptor nerve endings packed with mitochondria

Figure 26.25
Pit organ of rattlesnake, a pit viper. Cutaway shows location of a deep membrane that divides the pit into inner and outer chambers. Heat-sensitive nerve endings are concentrated in the membrane.

The Mesozoic World of Dinosaurs

When in 1842 the English anatomist Richard Owen coined the term *dinosaur* ("fearfully-great lizard") to describe fossil Mesozoic reptiles of gigantic size, only three poorly known dinosaur genera were distinguished. But with new and marvelous fossil discoveries quickly following, by 1887 zoologists were able to distinguish two groups of dinosaurs based on differences in structure of the pelvic girdles. The Saurischia ("lizard-hipped") had a simple, three-pronged pelvis with hip bones arranged much as they are in other nonavian reptiles. The large bladelike ilium is attached to one or two sacral vertebrae. The pubis and ischium extend anteriorly and ventrally respectively, and all three bones meet at the hip socket, a deep opening on the side of the pelvis. The Ornithischia ("bird-hipped") had a somewhat more complex pelvis. The ilium and ischium were arranged similarly in ornithischians and saurischians, but the ornithischian pubis was a narrow, rod-shaped bone with anteriorly and posteriorly directed processes lying alongside the ischium. Oddly, while the ornithischian pelvis, as the name suggests, was similar to that of birds, birds evolved from the saurischian lineage.

Dinosaurs and their living descendants, the birds, are archosaurs ("ruling lizards"), a group that includes crocodiles and pterosaurs (refer to the classification of amniotes on p. 581). As traditionally recognized, dinosaurs are a paraphyletic group because they do not include birds, which are descended from the theropod dinosaur lineage.

From among the various archosaurian radiations of the Triassic there emerged a lineage with limbs drawn under the body to provide an upright posture. This lineage gave rise to the earliest dinosaurs of the late Triassic. In *Herrerasaurus,* a bipedal dinosaur from Argentina, we see one of the most distinctive characteristics of dinosaurs: walking upright on pillarlike legs, rather than on legs splayed outward as with modern amphibians and nonavian reptiles. This arrangement allowed the legs to support the great weight of the body while providing an efficient and rapid stride.

Two groups of saurischian dinosaurs have been proposed based on differences in feeding habits and locomotion: the carnivorous and bipedal theropods, and the herbivorous and quadrupedal sauropods. *Coelophysis* was an early theropod with a body form typical of all theropods: powerful hindlegs with three-toed feet; long, heavy counterbalancing tail; slender, grasping forelimbs; flexible neck; and a large head with jaws armed with daggerlike teeth. Large predators such as *Allosaurus,* common during the Jurassic, were replaced by even more massively built carnivores of the Cretaceous, such as *Tyrannosaurus,* which reached a length of 14.5 m (47 ft), stood nearly 6 m high, and weighed more than 7200 kg (8 tons). Not all predatory saurischians were massive; several were swift and nimble, such as *Velociraptor* ("speedy predator") of the Upper Cretaceous.

Herbivorous saurischians, the quadrupedal sauropods, appeared in the late Triassic. Although early sauropods were small- and medium-sized dinosaurs, those of the Jurassic and Cretaceous attained gigantic proportions, the largest terrestrial vertebrates ever to have lived. *Brachiosaurus* reached 25 m (82 ft) in length and may have weighed in excess of 30,000 kg (33 tons). Even larger sauropods have been discovered; *Argentinosaurus* was 40 m (132 ft) long and weighed at least 80,000 kg. With long necks and long front legs, sauropods were the first vertebrates adapted to feed on trees. They reached their greatest diversity in the Jurassic and began to decline in overall abundance and diversity during the Cretaceous.

The second group of dinosaurs, the Ornithischia, were all herbivorous. Although more varied, even grotesque, in appearance than saurischians, the ornithischians are united by several derived skeletal features that indicate common ancestry. The huge back-plated *Stegosaurus* of the Jurassic is a well-known example of armored ornithischians, which comprised two of the five major groups of ornithischians. Even more shielded with bony plates than stegosaurs were the heavily built ankylosaurs, "armored tanks" of the dinosaur world. As the Jurassic gave way to the Cretaceous, several groups of unarmored ornithischians appeared, although many bore impressive horns. The steady increase in ornithiscian diversity in the Cretaceous paralleled a concurrent gradual decline in giant sauropods, which had flourished in the Jurassic. *Triceratops* is representative of horned dinosaurs that were common in the Upper Cretaceous. Even more prominent in the Upper Cretaceous were the hadrosaurs, such as *Parasaurolophus,* which are believed to have lived in large herds. Many hadrosaurs had skulls elaborated with crests that probably functioned as vocal resonators to produce species-specific calls. The bipedal pachycephalosaurs, appearing during the late Cretaceous, had thick skulls used in pushing competitions or for ramming enemies.

Dinosaurs likely had considerably more complex parental care than most other reptilian groups. Support for dinosaurs as caregivers can be found by examining the phylogenetic relationships of archosaurs. Two living groups, birds and crocodilians, are members of the clade Archosauria in which dinosaurs are contained (Figure 26.1). Because both crocodilians and birds share well-developed parental care, it is likely that dinosaurs exhibited similar behavior. In addition, fossil nests of dinosaurs are known for several groups. In one case, a fossil adult of the small theropod *Oviraptor* was found with a nest of eggs. Originally, it was hypothesized that the adult was a predator on the eggs (*Oviraptor* means "egg seizer"). Later, an embryo in similar eggs was found and identified as *Oviraptor,* indicating that the adult was probably with its own eggs! Examination of baby *Maiasaura* (a hadrosaur) found in a nest revealed considerable wear on their teeth. This suggests that the babies had remained in a nest and were possibly fed by adults during part of their early life.

Sixty-five million years ago, the last Mesozoic dinosaurs became extinct, leaving birds and crocodilians as the only surviving lineages of archosaurs. The demise of dinosaurs coincided with a large asteroid impact on the Yucatan peninsula that would have produced worldwide environmental upheaval. An alternative explanation, supported by many paleontologists, suggests that extinctions were caused by changing climates and landforms at the close of the Cretaceous. However, these hypotheses do not explain why dinosaurs became extinct while most other vertebrate lineages persisted. We continue to be fascinated by the awe-inspiring, often staggeringly large creatures that dominated the Mesozoic era for 165 million years—an incomprehensibly long period of time. Today, inspired by clues from fossils and footprints from a lost world, scientists continue to piece together the puzzle of how various dinosaur groups arose, behaved, and diversified.

SAURISCHIANS

ORNITHISCHIANS

66 My ago

CRETACEOUS

Titanosaurus
12 m (40 ft)

Velociraptor
1.8 m (6 ft)

Parasaurolophus (duck-billed dinosaur)
10 m (33 ft)

Triceratops
9 m (30 ft)

144 My ago

JURASSIC

Brachiosaurus
25 m (82 ft)

Allosaurus
11 m (35 ft)

Stegosaurus
9 m (30 ft)

Ilium

Ischium

Pubis

208 My ago

TRIASSIC

Coelophysis
3 m (10 ft)

Pubis

Ilium

Ischium

Herrerasaurus 4 m (13 ft)
One of the oldest known
dinosaurs. Has characteristics
of both saurischians and
ornithischians.

245 My ago

All vipers have a pair of teeth, modified as fangs, on the maxillary bones. These lie in a membranous sheath when the mouth is closed. When a viper strikes, a special muscle and bone lever system erects the fangs as the mouth opens (Figure 26.26). Fangs are driven into the prey by the thrust of the strike, and venom is injected into the wound through a channel in the fangs. A viper immediately releases its prey after the bite and follows it until it is paralyzed or dead. Then the snake swallows its prey whole. The bite of a pit viper can be dangerous to humans, although in many instances the snake injects very little venom when it bites. Each year in the United States approximately 7000 bites from pit vipers are reported, resulting in only about 5 deaths.

Even the saliva of harmless snakes possesses limited toxic qualities, and it is logical that there was natural selection for this toxic tendency as snakes evolved. Snake venoms have traditionally been divided into two types. The **neurotoxic** type acts mainly on the nervous system, affecting the optic nerves (causing blindness) or the phrenic nerve of the diaphragm (causing paralysis of respiration). The **hemorrhagin** type breaks down red blood cells and blood vessels and produces extensive hemorrhaging of blood into tissue spaces. In fact, most snake venoms are complex mixtures of various fractions that attack different organs in specific ways; they seldom can be assigned categorically to one or the other of the traditional types. In addition, all venoms possess enzymes that accelerate digestion.

The toxicity of a venom is measured by the median lethal dose on laboratory animals (LD_{50}). By this standard the venoms of the Australian tiger snake and some of the sea snakes appear to be the most deadly. However, several larger snakes are more dangerous. The aggressive king cobra, which may exceed 5.5 m in length, is the largest and perhaps the most dangerous of all venomous snakes. It is estimated worldwide that 50,000 to 60,000 people die from snakebite each year. Most deaths occur in India, Pakistan, Myanmar, and nearby countries where poorly shod people frequently come into contact with venomous snakes and frequently do not get immediate medical attention following snakebite. The snakes primarily responsible for deaths in these areas are Russell's viper, saw-scaled viper, and several species of cobras.

The LD_{50} (median lethal dose) has been the standardized procedure for assaying toxicity of drugs; it was originally developed in the 1920s by pharmacologists. In practice, small samples of laboratory animals, usually mice, are exposed to a graded series of doses of the drug or toxin. The dose that kills 50% of the animals in the test period is recorded as the LD_{50}. Expensive and time consuming, this classical procedure is being replaced by alternative methods that greatly reduce the number of animals needed. Among these alternatives are cytotoxicity tests that evaluate the ability of test substances to kill cells, and toxikinetic procedures that measure interaction of a drug or toxin with a living system.

Most snakes are **oviparous** (L. *ovum*, egg, + *parere*, to bring forth) species that lay their shelled, elliptical eggs beneath rotten logs, under rocks, or in holes in the ground. Most of the remainder, including all American pit vipers, except the tropical bushmaster, are ovoviviparous (L. *ovum*, egg, + *vivus*, living, + *parere*, to bring forth), giving birth to well-formed young that are nourished by yolk while in the uterus. Very few snakes exhibit placental viviparity (L. *vivus*, living, + *parere*, to bring forth); in these snakes the placenta permits exchange of materials between the embryonic and maternal bloodstreams. Snakes are able to store sperm and can lay several clutches of fertile eggs at long intervals after a single mating.

Order Sphenodonta: Tuataras

The order Sphenodonta is represented by two living species of genus *Sphenodon* (Gr. *sphenos*, wedge, + *odontos*, tooth) of New Zealand (Figure 26.27). Tuataras are sole survivors of the sphenodontid lineage that radiated modestly during the early Mesozoic era but declined toward the end of the Mesozoic. Several species of tuataras were once widespread throughout the two main islands of New Zealand but the two living species are now restricted to small islets of Cook Strait and off the northeast coast of North Island. On some of these islands, under protection by the New Zealand government, they are prospering. Loss of the tuatara populations on the main islands of New Zealand was caused by humans intentionally or accidentally introducing nonnative animals, including rats, cats, dogs, and goats, which preyed upon tuataras and their eggs or destroyed their habitat. Tuataras are particularly vulnerable because they have slow growth and low reproductive rates.

Tuataras are lizardlike forms 66 cm long or less that live in burrows often shared with petrels. They are slow-growing animals with long lives; one is recorded to have lived 77 years.

Tuataras have captured the interest of zoologists because of numerous features that are almost identical to those of early Mesozoic diapsids that lived 200 million years ago. These features include a diapsid skull with two temporal openings bounded by

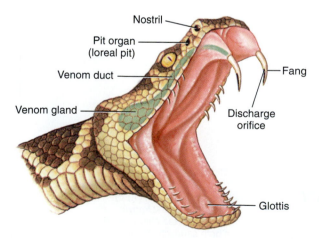

Figure 26.26
Head of rattlesnake showing the venom apparatus. The venom gland, a modified salivary gland, is connected by a duct to the hollow fang.

Nostril

Pit organ (loreal pit)

Venom duct

Venom gland

Fang

Discharge orifice

Glottis

Classification of Early Amniotes and Living Nonavian Reptiles

The following Linnaean classification is modified from Carroll (1988)[1] and agrees with the genealogical relationships of living non-avian reptiles shown in Figure 26.2. Relationships of turtles to other diapsids are controversial. Extinct groups are indicated by a dagger.

Subclass Anapsida[†] (a-nap′se-duh) (Gr. *an*, without, + *apsis*, arch): **anapsids.** Amniotes having some primitive features, such as a skull with no temporal opening.

    Order Captorhinida[†] (kap-to-rine′uh-duh) (Gr. *kapto*, to seize, + *rhinos*, nose). Amniotes of Carboniferous and early Permian.

Subclass Diapsida (di-ap′se-duh) (Gr. *di*, double, + *apsis*, arch): **diapsids.** Amniotes having a skull with two temporal openings.

    Order Testudines (tes-tu′din-eez) (L. *testudo*, tortoise) **(Chelonia): turtles.** Body in a bony case of dorsal carapace and ventral plastron; jaws with keratinized beaks instead of teeth; vertebrae and ribs fused to overlying carapace; temporal openings lost; approximately 300 species.

    Superorder Lepidosauria (lep-i-do-sor′ee-uh) (Gr. *lepidos*, scale, + *sauros*, lizard). Diapsid lineage appearing in the Triassic; characterized by sprawling posture; no bipedal specializations; diapsid skull often modified by loss of one or both temporal arches; transverse cloacal slit, skin shed in one piece.

        Order Squamata (skwa-ma′ta) (L. *squamatus*, scaly, + *ata*, characterized by): **snakes** and **lizards.** Skin of keratinized epidermal scales or plates, which is shed; quadrate movable; skull kinetic (except amphisbaenians); vertebrae usually concave in front; paired copulatory organs.

            Suborder Lacertilia (lay-sur-till′ee-uh) (L. *lacerta*, lizard) **(Sauria): lizards.** Body slender, usually with four limbs but many forms lack one or both pairs of limbs; rami of lower jaw fused; eyelids movable; amphisbaenians with eyes hidden beneath skin; external ear present; this paraphyletic suborder contains approximately 4800 species.

            Suborder Serpentes (sur-pen′tes) (L. *serpere*, to creep): **snakes.** Body elongate; limbs, ear openings, and middle ear absent; mandibles joined anteriorly by ligaments; eyelids fused into transparent spectacle; tongue forked and protrusible; left lung reduced or absent; approximately 2900 species.

        Order Sphenodonta (sfen′o-don′tuh) (Gr. *sphen*, wedge, + *odontos*, tooth): **tuataras.** Primitive diapsid skull; vertebrae biconcave; quadrate immovable; parietal eye present; two extant species in *Sphenodon*.

    Superorder Ichthyosauria[†] (ik′thee-o-sor′ee-uh) (Gr. *ichthys*, fish, + *sauros*, lizard). Mesozoic marine dolphin-shaped diapsids with large eyes and vertical tails.

    Superorder Sauropterygia[†] (sor-op-te-ig′ee-uh) (Gr. *sauros*, lizard, + *pteryginos*, winged). Mesozoic marine reptiles.

        Order Plesiosauria[†] (plees′ee-o-sor′ee-uh) (Gr. *plesios*, near, + *sauros*, lizard). Long-necked Mesozoic marine reptiles with paddlelike limbs.

    Superorder Archosauria (ark′uh-sor′ee-uh) (Gr. *arch¯on*, ruling, + *sauros*, lizard). Diapsid lineage appearing in the Triassic, tending toward bipedalism; orbit-shaped like an upside-down triangle; anteorbital fenestra (opening in the skull anterior to the orbit) and gizzard present; ventricle fully divided; parental care of young.

        Order Crocodilia (crok′uh-dil′ee-uh) (L. *crocodilus*, crocodile): **crocodilians.** Skull elongate and massive; nares terminal; secondary palate present; four-chambered heart; vertebrae usually concave in front; forelimbs usually with five digits; hindlimbs with four digits; quadrate immovable; advanced social behavior; 23 species.

        Order Pterosauria[†] (ter′uh-sor′ee-uh) (Gr. *pteron*, winged, + *sauros*, lizard). Flying, Mesozoic archosaurs with membranous wings.

        Order Saurischia (sor-ish′ee-uh) (Gr. *sauros*, lizard, + *ischion*, hip). Mesozoic dinosaurs; bipedal carnivores and mostly quadrupedal herbivores; ancestral (lizardlike) hip structure.

            Suborder Sauropodomorpha[†] (sor′uh-pod-uh-morf′uh) (Gr. *sauros*, lizard, + *podos*, foot, + *morphē*, form). Herbivorous saurischians including Mesozoic giants such as *Brachiosaurus, Apatosaurus,* and *Diplodocus.*

            Suborder Theropoda (the-ro′po-duh) (Gr. *thēr*, wild beast, + *podos*, foot). Carnivorous saurischians including huge predators such as *Tyrannosaurus* and small, agile predators such as *Deinonychus* and *Velociraptor.* Birds are descended from this lineage.

        Order Ornithischia[†] (orn′uh-thish′ee-uh) (Gr. *ornis*, bird, + *ischion*, hip). Mesozoic dinosaurs; bipedal and quadrupedal herbivores such as *Stegosaurus, Triceratops,* and *Parasaurolophus;* derived (birdlike) hip structure.

Subclass Synapsida (sin-ap′si-duh) (Gr. *syn*, together, + *apsis*, arch). Amniotes having skull with one pair of lateral temporal openings.

    Order Pelycosauria (pel′uh-ko-sor′ee-uh) (Gr. *pelyx*, wooden bowl, + *sauros*, lizard). Carboniferous and Permian synapsids with many ancestral amniote characteristics; therapsids are descended from this paraphyletic group.

    Order Therapsida (ther-ap′-si-duh) (Gr. *ther*, wild beast, + *apsis*, arch). Permian and Triassic synapsids with many mammal-like characteristics; living mammals are descended from this paraphyletic group.

[1]Carroll, R. L. 1988. *Vertebrate paleontology and evolution.* New York, W. H. Freeman and Company.

---

complete arches. Tuataras also bear a well-developed median parietal eye complete with elements of cornea, lens, and retina (although since it is buried beneath opaque skin this "third eye" can register only changes in light intensity). *Sphenodon* represents one of the slowest rates of morphological evolution known among vertebrates.

Order Crocodilia: Crocodiles and Alligators

Modern crocodilians are the only surviving reptiles of the archosaurian lineage that gave rise to the great Mesozoic radiation of dinosaurs and their kin and to birds. Although modern

Figure 26.27

Tuatara, *Sphenodon* sp., a living representative of order Sphenodonta. This "living fossil" reptile has, on top of the head, a well-developed parietal "eye" with retina, lens, and nervous connections to the brain. Although covered with scales, this third eye is sensitive to light. The parietal eye may have been an important sense organ in early reptiles. Tuataras are found today only on certain islands off the coastline of New Zealand.

crocodiles belong to a clade that began its radiation in the late Cretaceous period, they differ little in structural details from primitive crocodilians of the early Mesozoic. Having remained mostly unchanged for nearly 200 million years, crocodilians face an uncertain future in a world dominated by humans. Modern crocodilians are divided into three families: alligators and caimans, mostly a New World group; crocodiles, which are widely distributed and include the saltwater crocodile, one of the largest living nonavian reptiles; and gharials, represented by a single species in India and Burma.

All crocodilians have an elongate, robust, well-reinforced skull and massive jaw musculature arranged to provide a wide gape and rapid, powerful closure. Teeth are set in sockets, a type of dentition called **thecodont** that was typical of Mesozoic archosaurs as well as the earliest birds. Another adaptation, found in no other vertebrate except mammals, is a complete secondary palate. This innovation pushed the internal nares to the posterior, allowing a crocodilian to breathe when its mouth is filled with water or food (or both). Crocodilians also share with birds and mammals a four-chambered heart with completely divided atria and ventricles.

The estuarine crocodile, *Crocodylus porosus,* found in southern Asia, and the Nile crocodile, *C. niloticus,* (Figure 26.28A) grow to great size (adults weighing 1000 kg have been reported) and are swift and aggressive. Crocodiles are known to attack animals as large as cattle, deer, and people. Alligators (Figure 26.28B) are usually less aggressive than crocodiles and far less dangerous to humans. In the United States, *Alligator mississippiensis* (Figure 26.28B) is the only species of alligator; *Crocodylus acutus,* restricted to extreme southern Florida, is the only species of crocodile. Large alligators are powerful animals nevertheless,

and adults have almost no enemies but humans. The chink in their formidable armor is their developmental stages. Nests left unguarded by the mother are almost certain to be discovered and raided by any of several mammals that relish eggs, and young hatchlings may be devoured by large fishes or wading birds.

Male alligators make loud bellows during the mating season. Alligators and crocodiles are oviparous. Usually 20 to 50 eggs are laid in a mass of dead vegetation or buried in the sand and guarded by the mother. The mother hears vocalizations from hatching young and responds by opening the nest to allow the hatchlings to escape. As with many turtles and some lizards, incubation temperature of the eggs determines sex ratio of the offspring. However, unlike turtles (p. 571) low nest temperatures

A

B

Figure 26.28

Crocodilians. **A,** Nile crocodile, *Crocodylus niloticus,* basking. The fourth tooth of the lower jaw fits outside the slender upper jaw; alligators lack this feature. **B,** American alligator, *Alligator mississippiensis,* an increasingly noticeable resident of rivers, bayous, and swamps of the southeastern United States.

produce only females, whereas high nest temperatures produce only males. This results in highly unbalanced sex ratios in some areas. For example, in one study area in Louisiana, female hatchlings outnumbered males five to one.

Crocodiles and alligators can be distinguished on the basis of head morphology. Crocodiles have a relatively narrow snout, and when their mouths are closed, the fourth lower jaw tooth is visible where it fits into a notch on the upper jaw. Alligators generally have a broader snout, and their fourth lower jaw tooth is hidden when the mouth is closed (Figure 26.28). Gharials have very narrow snouts, and are largely fish eaters.

SUMMARY

Amniotes diverged from a group of early tetrapods during the late Paleozoic era, some 300 million years ago, and quickly radiated into diverse forms occupying a range of aquatic and terrestrial habitats. Their success as terrestrial vertebrates can be attributed to several adaptations, especially the amniotic egg. The amniotic egg, with its shell and four extraembryonic membranes—the amnion, allantois, chorion, and yolk sac—permits rapid development of embryos in terrestrial environments. Additional amniote adaptations that permit occupation of dry environments include a thick, water-resistant skin, excretion of urea or uric acid, and complex lungs ventilated by trunk muscles.

Before the end of the Paleozoic, amniotes diversified to form three groups distinguished by skull structure: anapsids, which lack temporal fenestra; synapsids, which have one pair of temporal fenestra, and diapsids, which have two pairs of temporal fenestra. Mammals evolved from early synapsids. Early diapsids gave rise to all living nonavian reptiles (tentatively including turtles) and to birds. There are no living descendants of Paleozoic anapsids. One clade of diapsids, the archosaurs, underwent a great, worldwide radiation during the Mesozoic into large and morphologically diverse forms, including the ichthyosaurs, plesiosaurs, pterosaurs, and dinosaurs. Although these are now extinct, descendants of some dinosaurs survived the great extinction at the end of the Mesozoic, and underwent their own diversification as birds.

Turtles (order Testudines), with their distinctive shells, have changed little in anatomy since the Triassic period. Turtles are a small group of long-lived terrestrial, semiaquatic, aquatic, and marine species. They lack teeth, and instead bear keratinized plates on their jaws. All are oviparous, and all, including marine forms, bury their eggs.

Lizards and snakes (order Squamata) represent 95% of all living nonavian reptiles. Lizards are a diversified and successful group, particularly in hot climates. They are distinguished from snakes by having united lower jaw halves, movable eyelids, and external ear openings. Lizards and snakes are ectothermic, regulating their temperature by moving among different microenvironments. Most lizards and snakes are oviparous, although viviparity is not uncommon, especially in cooler climates. Amphisbaenians are a small group of tropical lizards specialized for burrowing. They have ringed, usually limbless, bodies and a solid skull.

Snakes, which evolved from one group of lizards, are characterized by elongate, limbless bodies and a highly kinetic skull that allows them to swallow prey several times larger than the snake's diameter. Most snakes rely on chemical senses, including Jacobson's organ, to hunt prey, rather than on visual and auditory senses. Two groups of snakes (pit vipers and boids) have unique infrared-sensing organs for tracking prey. Some snakes swallow their prey alive; others kill their prey by constriction or with venom. Different groups of venomous snakes are distinguished by the anatomy and placement of their fangs.

The tuataras of New Zealand (order Sphenodonta) are relics and sole survivors of a group that otherwise disappeared 100 million years ago. They bear several features that are almost identical to those of Mesozoic fossil diapsids.

Crocodiles, alligators, and caimans (order Crocodilia) are the only living nonavian representatives of the archosaurian lineage that gave rise to the extinct dinosaurs and the living birds. Crocodilians have several adaptations for a carnivorous, semiaquatic life, including a massive skull with powerful jaws, and a secondary palate. They have the most complex social behavior of any living nonavian reptile.

REVIEW QUESTIONS

1. What are the four membranes associated with amniotic eggs? What is the function of each of these membranes?
2. How do the skin and respiratory systems of amniotes differ from their early tetrapod ancestors?
3. Amniotes are divided into three groups based on their skull morphologies. What are these three groups, and how do their skulls differ? Which living amniotes, if any, originated from each of these three groups?
4. Why are "reptiles," as traditionally defined, a paraphyletic group? How has cladistic taxonomy revised Reptilia to make it monophyletic?
5. Describe ways in which nonavian reptiles are more functionally or structurally suited for terrestriality than amphibians.
6. Describe the principal structural features of turtles that would distinguish them from any other nonavian reptile order.

7. How might nest temperature affect egg development in turtles? In crocodilians?
8. What is meant by a "kinetic" skull and what benefit does it confer? How are snakes able to eat large prey?
9. In what ways are the special senses of snakes similar to those of lizards, and in what ways have they evolved for specialized feeding strategies?
10. What are amphisbaenians? What morphological adaptations do they have to aid in burrowing?
11. Distinguish ornithischian and saurischian dinosaurs based on their hip anatomy. Which lineage gave rise to birds?
12. How do snakes and crocodilians breathe when their mouths are full of food?
13. What is the function of Jacobson's organ of snakes?
14. What is the function of the "pit" of pit vipers?

15. What is the difference in structure or location of the fangs of a rattlesnake, a cobra, and an African boomslang?
16. Most lizards and snakes are oviparous, but some are ovoviviparous or have placental viviparity. Compare these methods of reproduction in squamates. In what regions are viviparity most common?
17. Describe how a snake moves by lateral undulation. Why might this form of locomotion be inefficient on an unstable surface (such as sand) or a surface lacking irregularities? What forms of locomotion would work for a snake under these conditions?
18. Why are tuataras (*Sphenodon*) of special interest to biologists? Where would you have to go to see one in its natural habitat?
19. From which diapsid lineage have crocodilians descended? What other major fossil and living vertebrate groups belong to this same lineage? In what structural and behavioral ways do crocodilians differ from other living nonavian reptiles?

SELECTED REFERENCES

Alvarez, W., and F. Asaro. 1990. An extraterrestrial impact. Sci. Am. **263:**78–84 (Oct.). *This article and an accompanying article by V. E. Courtillot, "A volcanic eruption," present opposing interpretations of the cause of the Cretaceous mass extinction that led to the demise of the dinosaurs.*

Cogger, H. G., and R. G. Zweifel (eds.). 1998. Encyclopedia of reptiles and amphibians. San Diego, Academic Press. *This comprehensive, up-to-date, and lavishly illustrated volume was written by some of the best-known herpetologists in the field.*

Crews, D. 1994. Animal sexuality. Sci. Am. **270:**108–114 (Jan.). *The reproductive strategies of reptiles, including nongenetic sex determination, provide insights into the origins and functions of sexuality.*

Erickson, G. M. 1999. Breathing life into *Tyrannosaurus rex.* Sci. Am. **281:**42–49 (Sept.). *Present evidence suggests that* T. rex *was gregarious and acquired food by scavenging and predation.*

Greene, H. W. 1997. Snakes: The evolution of mystery in nature. Berkeley, University of California Press. *Beautiful photographs accompany a well-written volume for the scientist or novice.*

Halliday, T. R., and K. Adler (eds.). 1986. The encyclopedia of reptiles and amphibians. New York, Facts on File, Inc. *Comprehensive and beautifully illustrated treatment of reptilian groups with helpful introductory sections on origins and characteristics.*

King, G. 1996. Reptiles and herbivory. London, Chapman & Hall. *Explains the adaptations reptiles use in obtaining nutrients from a herbivorous diet.*

Lillywhite, H. B. 1988. Snakes, blood circulation and gravity. Sci. Am. **259:**92–98 (Dec.). *Even long snakes are able to maintain blood circulation when the body is extended vertically (head-up posture) through special circulatory reflexes that control blood pressure.*

Lohmann, K. J. 1992. How sea turtles navigate. Sci. Am. **266:**100–106 (Jan.). *Recent evidence suggests that sea turtles use the earth's magnetic field and the direction of ocean waves to navigate back to their natal beaches to nest.*

Mattison, C. 1995. The encyclopedia of snakes. New York, Facts on File, Inc. *Generously illustrated book treating evolution, physiology, behavior, and classification of snakes.*

Norman, D. 1991. Dinosaur! New York, Prentice-Hall. *Highly readable account of the life and evolution of dinosaurs, with fine illustrations.*

Paul, G. S. 2000. The Scientific American book of dinosaurs. New York, St. Martin's Press. *Essays emphasizing functional morphology, behavior, evolution, and extinction of dinosaurs.*

Pianka, E. R., and L. J. Vitt. 2003. Lizards: windows to the evolution of diversity. Berkeley, University of California Press. *Hundreds of color photographs highlight a treatment of the behavior and evolution of lizards.*

Pough, F. H., R. M. Andrews, J. E. Cadle, M. L. Crump, A. H. Savitzky, and K. D. Wells. 2003. Herpetology, ed. 3. Upper Saddle River, New Jersey, Prentice-Hall. *A comprehensive textbook treating diversity, physiology, behavior, ecology, and conservation of reptiles and amphibians.*

Sumida, S. S, and K. L. M. Martin (eds.). 1997. Amniote origins: completing the transition to land. San Diego, Academic. *Discusses the diversity, evolution, ecology, and adaptive morphology of early amniotes, with special emphasis on the origin of the amniotic egg.*

Zug, G. R., L. J. Vitt, and J. P. Caldwell. 2001. Herpetology: an introduction to the biology of amphibians and reptiles, ed. 2. San Diego, Academic Press. *A current general textbook of herpetology.*

ONLINE LEARNING CENTER

Visit www.mhhe.com/hickmanipz14e for chapter quizzing, key term flash cards, web links, and more!

Storks during night migration.

Birds

- PHYLUM CHORDATA
- CLASS AVES

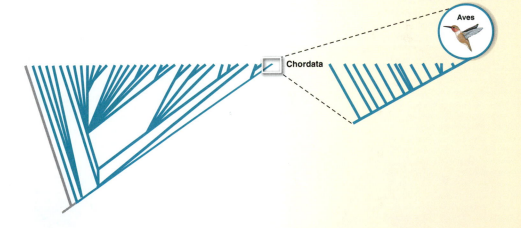

Long Trip to a Summer Home

Perhaps it was ordained that some birds, having mastered flight, would use this power to make the long seasonal migrations. Moving between southern wintering regions and northern summer breeding regions with long summer days and an abundance of insects provides parents with ample food to rear their young. Predators of birds are not so abundant in the far North, and a brief once-a-year appearance of vulnerable young birds does not encourage buildup of predator populations. Migration also vastly increases the amount of space available for breeding and reduces aggressive territorial behavior. Finally, migration favors homeostasis—the balancing of physiological processes that maintains internal stability—by allowing birds to avoid climatic extremes.

Still, the wonder of the migratory pageant remains, and there is much yet to learn about its mechanisms. What times migration, and what determines that each bird shall store sufficient fuel for the journey? How did the sometimes difficult migratory routes originate, and what cues do birds use in navigation? What was the origin of this instinctive force to follow the retreat of winter northward? It is instinct that drives the migratory waves in spring and fall, instinctive blind obedience that carries most birds successfully to their northern nests, while countless others fail and die, winnowed by the ever-challenging environment.

Of the vertebrates, birds of class Aves (ay'veez) (L. pl. of *avis,* bird) are the most noticeable, the most melodious, and many think the most beautiful. With more than 9700 species distributed over nearly the entire earth, birds outnumber any other vertebrate group except fishes. Birds live in forests and deserts, in mountains and prairies, and on all oceans. Four species are known to have visited the North Pole, and one, a skua, was seen at the South Pole. Some birds live in total darkness in caves, finding their way by echolocation, and others dive to depths greater than 45 m to prey on aquatic life. The "bee" hummingbird of Cuba, weighing in at only 1.8 g, is one of the smallest vertebrate endotherms.

The single unique feature that distinguishes birds from other living animals is their feathers. If an animal has feathers, it is a bird; if it lacks feathers, it is not a bird. Feathers also were present in some theropod dinosaurs, but these feathers were not capable of supporting flight.

There is great uniformity of structure among birds. Despite approximately 150 million years of evolution, during which they proliferated and adapted to specialized ways of life, we have no difficulty recognizing a living bird as a bird. In addition to feathers, all birds have forelimbs modified into wings (although not always used for flight); all have hindlimbs adapted for walking, swimming, or perching; all have keratinized beaks; and all lay eggs. The reason for this great structural and functional uniformity is that birds evolved into flying machines, which greatly restricts morphological diversity, so much more evident in other vertebrate classes.

A bird's entire anatomy is designed around flight. An airborne life for a large vertebrate is a highly demanding evolutionary challenge. A bird must, of course, have wings for support and propulsion. Bones must be light yet serve as a rigid airframe. The respiratory system must be highly efficient to meet the intense metabolic demands of flight and also serve as a thermoregulatory device to maintain a constant body temperature. A bird must have a rapid and efficient digestive system to process an energy-rich diet; it must have a high metabolic rate; and it must have a high-pressure circulatory system. Above all, birds must have a finely tuned nervous system and acute senses, especially superb vision, to handle the complex problems of headfirst, high-velocity flight.

ORIGIN AND RELATIONSHIPS

Approximately 147 million years ago, a flying animal drowned and settled to the bottom of a shallow marine lagoon in what is now Bavaria, Germany. It was rapidly covered with fine silt and eventually fossilized. There it remained until discovered in 1861 by a workman splitting slate in a limestone quarry. The fossil was approximately the size of a crow, with a skull not unlike that of modern birds except that the beaklike jaws bore small bony teeth set in sockets like those of dinosaurs (Figure 27.1). The skeleton was decidedly reptilian with a long bony tail, clawed fingers, and abdominal ribs. It might have been classified as a theropod dinosaur except that it carried an unmistakable imprint of **feathers,** those marvels of biological engineering that

A

B

Figure 27.1

Archaeopteryx, a 147-million-year-old ancestor of modern birds.
A, Cast of the second and most nearly perfect fossil of *Archaeopteryx,* which was discovered in a Bavarian stone quarry. Ten specimens of *Archaeopteryx* have been discovered, the most recent one described in 2007.
B, Reconstruction of *Archaeopteryx.*

only birds possess. *Archaeopteryx lithographica* (ar-kee-op′ter-ix lith-o-graf′e-ca, Gr., meaning "ancient wing inscribed in stone"), as the fossil was named, was an especially fortunate discovery because the fossil record of birds was disappointingly meager. The finding was also dramatic because it demonstrated beyond reasonable doubt the phylogenetic relatedness of birds and theropod dinosaurs.

Zoologists had long recognized the similarity of birds and reptiles. The skulls of birds and reptiles abut against the first neck vertebra by a single occipital condyle (a small bony knob; mammals have two such knobs). Birds and reptiles have a single middle ear bone, the stapes (mammals have three middle ear bones). Birds and reptiles have a lower jaw composed of five or six bones, whereas the lower jaw of mammals has one bone, the dentary. Birds and reptiles excrete their nitrogenous wastes as uric acid whereas mammals excrete theirs as urea. Birds and most reptiles lay similar yolked eggs with the early embryo developing on the surface by shallow cleavage divisions.

The distinguished English zoologist Thomas Henry Huxley was so impressed with these and many other anatomical and physiological affinites that he called birds "glorified reptiles" and classified them with a group of dinosaurs called theropods that displayed several birdlike characteristics (Figures 27.2 and 27.3). Theropod dinosaurs share many derived characters with birds, the most obvious of which is the elongate, mobile, S-shaped neck.

Dromeosaurs, a group of theropods that includes *Velociraptor,* share many additional derived characters with birds, including a furcula (fused clavicles) and lunate wrist bones that permit swiveling motions used in flight (Figure 27.3). Additional evidence linking birds to dromeosaurs comes from recently described fossils from late Jurassic and early Cretaceous deposits in Liaoning Province, China. These spectacular fossils, including *Protarchaeopteryx* and *Caudipteryx,* are dromeosaur-like theropods, but with feathers! It is unlikely these feathered dinosaurs could fly, however, as they had short forelimbs and symmetrical vaned feathers (the flight feathers of modern flying birds are asymmetrical). While these feathers could not have been used for powered flight, they may have been useful for controlling glides or jumps from trees. These primitive feathers, like modern feathers, may have been colorful and used in social displays. Additional theropod dinosaurs recently unearthed in China, such as *Sinosauropteryx,* are covered with filaments that appear to be homologous with feathers. The filamentous covering of these dinosaurs likely served as insulation and was a precursor to vaned feathers. Other fossils from Spain and Argentina of birds more derived than *Archaeopteryx* document the development of the keeled sternum and alula (see Figure 27.7), loss of teeth, and fusion of bones characteristic of modern birds. Clearly, a phylogenetic approach to classification would include birds with theropod dinosaurs. With this view, dinosaurs are not extinct—they are with us today as birds!

Living birds (Neornithes) are divided into two groups: (1) **Paleognathae** (Gr. *palaios,* ancient, + *gnathos,* jaw), the large flightless ostrichlike birds and the kiwis, often called ratite birds, which have a flat sternum with poorly developed pectoral muscles, and tinamous, and (2) **Neognathae** (Gr. *neos,* new, + *gnathos,* jaw), all other birds, nearly all of which are flying birds that have a keeled sternum to which powerful flight muscles attach. There are a number of flightless neognathus birds, some of which lack a keeled sternum (Figure 27.4). Flightlessness has appeared independently among many groups of birds; the fossil record reveals flightless wrens, pigeons, parrots, cranes, ducks, auks, and even a flightless owl. Penguins are flightless although they use their wings to "fly" through water (p. 208). Flightlessness almost always has evolved on islands where few terrestrial predators are found. Flightless birds living on continents today are the large paleognathids (ostrich, rhea, cassowary, emu), which can run fast enough to escape predators. An ostrich can run 70 km (42 miles) per hour, and claims of speeds of 96 km (60 miles) per hour have been made. The evolution and dispersal of flightless birds are discussed on pages 115 and 819, respectively.

The bodies of flightless birds are dramatically redesigned to remove all of the restrictions of flight. The keel of the sternum is lost, and heavy flight muscles (as much as 17% of the body weight of flying birds), as well as other specialized flight apparatus, disappear. Because body weight is no longer a constraint, flightless birds tend to become large. Several extinct flightless birds were enormous: giant moas of New Zealand weighed more than 225 kg (500 pounds) and elephantbirds of Madagascar, the largest bird that ever lived, probably weighed nearly 450 kg (about 1000 pounds) and stood nearly 2 m tall.

STRUCTURAL AND FUNCTIONAL ADAPTATIONS FOR FLIGHT

Just as an airplane must be designed and built according to rigid aerodynamic specifications if it is to fly, so too must birds meet stringent structural requirements if they are to stay airborne. All the special adaptations found in flying birds contribute to two things: more power and less weight. Flight by humans became possible when we developed an internal combustion engine and learned how to reduce the weight-to-power ratio to a critical point. Birds accomplished flight millions of years ago. Unlike airplanes, birds also must feed themselves, convert food into metabolic energy, escape predators, repair their own injuries, maintain a constant body temperature, and reproduce.

Feathers

Feathers are very lightweight, yet possess remarkable toughness and tensile strength. Most typical of bird feathers are **contour feathers,** vaned feathers that cover and streamline the bird's body. A contour feather consists of a hollow **quill,** or **calamus,** emerging from a skin follicle, and a **shaft,** or **rachis,** which is a

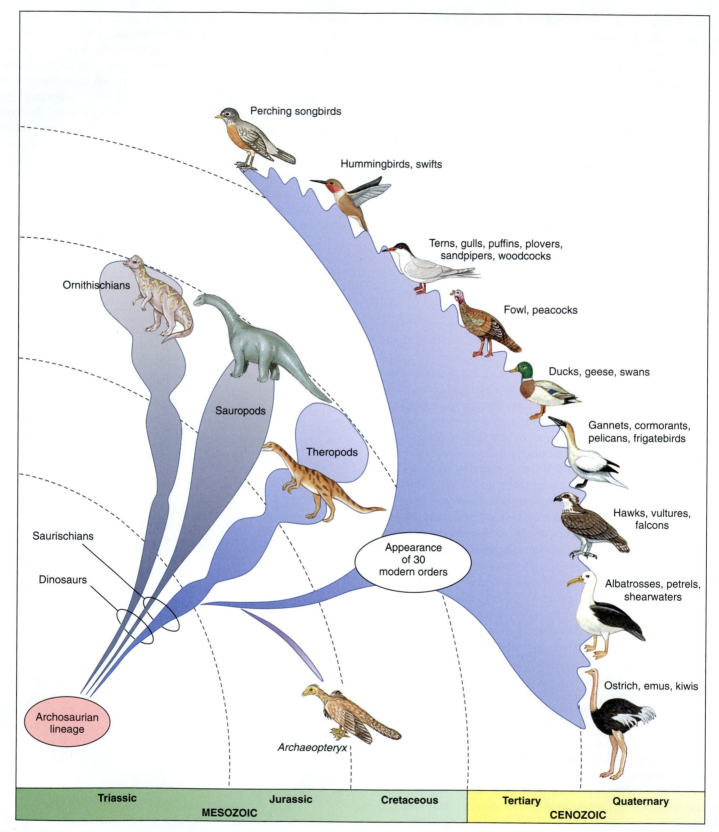

Perching songbirds

Hummingbirds, swifts

Terns, gulls, puffins, plovers, sandpipers, woodcocks

Ornithischians

Fowl, peacocks

Ducks, geese, swans

Sauropods

Gannets, cormorants, pelicans, frigatebirds

Theropods

Hawks, vultures, falcons

Saurischians

Appearance of 30 modern orders

Dinosaurs

Albatrosses, petrels, shearwaters

Archosaurian lineage

Ostrich, emus, kiwis

Archaeopteryx

Triassic	Jurassic	Cretaceous	Tertiary	Quaternary
MESOZOIC			CENOZOIC	

Figure 27.2

Evolution of modern birds. Of 30 living bird orders, 9 of the more important are shown. The earliest known bird, *Archaeopteryx,* lived in the Upper Jurassic, about 147 million years ago. *Archaeopteryx* uniquely shares many specialized aspects of its skeleton with the smaller theropod dinosaurs and is considered to have evolved within the theropod lineage. Evolution of modern bird orders occurred rapidly during the Cretaceous and early Tertiary periods.

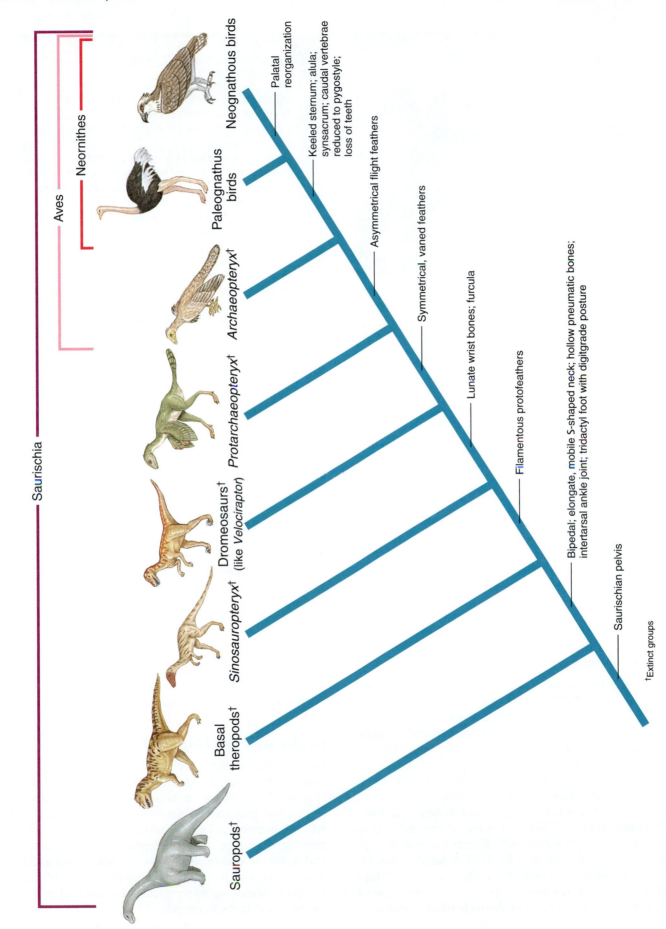

Figure 27.3
Cladogram of the Saurischia, showing the relationship of several taxa to modern birds. Shown are a few of the shared derived characters, mostly related to flight, that were used to construct the genealogy. The ornithischians are the sister group to the saurischians and all are members of the clade Archosauria (see Figures 26.1, p. 565, and 26.2, p. 566).

Figure 27.4

One of the strangest birds in a strange land, a flightless cormorant, *Nannopterum harrisi*, of the Galápagos Islands dries its wings after a fishing forage. It is a superb swimmer, propelling itself through the water with its feet to catch fish and octopuses. The flightless cormorant is an example of a carinate bird (having a keeled sternum) that has lost the keel and the ability to fly.

Characteristics of Class Aves

1. Body usually spindle-shaped, with four divisions: head, neck, trunk, and tail; **neck elongate and S-shaped**
2. **Forelimbs modified as wings;** hindlimbs adapted for perching, walking, or swimming; foot with four toes (2 or 3 toes in some)
3. Epidermal **covering of feathers** and **leg scales;** thin integument of epidermis and dermis; no sweat glands; oil or preen gland at base of tail; **pinna of ear rudimentary**
4. **Fully ossified skeleton with air cavities;** skull bones fused with **one occipital condyle;** skull diapsid with antorbital fenestra; each jaw covered with a keratinized sheath, forming a **beak; no teeth;** ribs with strengthening processes, the uncinate process attaching ribs with one another; **tail not elongate,** reduced to **pygostyle;** sternum usually well developed with keel; **single bone in middle ear**
5. Nervous system well developed, with 12 pairs of cranial nerves and brain with **large cerebellum and optic lobes**
6. Circulatory system of four-chambered heart with two atria and two ventricles; **separate pulmonary and systemic circuits; right aortic arch persisting;** nucleated red blood cells
7. **Endothermic**
8. Respiration by slightly expansible lungs **(parabronchi),** with thin **air sacs** among the visceral organs and skeleton; **syrinx (voice box)** near junction of trachea and bronchi
9. Excretory system of metanephric kidney; ureters open into cloaca; **no bladder;** semisolid urine; uric acid main nitrogenous waste
10. Sexes separate; testes paired, with the vas deferens opening into the cloaca; **females with functional left ovary and oviduct only;** copulatory organ (penis) in ducks, geese, paleognathids, and a few others
11. Fertilization internal; **amniotic eggs with much yolk and hard calcareous shells; incubation external;** young active at hatching **(precocial)** or helpless and naked **(altricial);** sex determination by chromosomes (females heterogametic)

continuation of the quill and bears numerous **barbs** (Figure 27.5). The barbs are arranged in closely parallel fashion and spread diagonally outward from both sides of the central shaft to form a flat, expansive, webbed surface, the **vane.** There may be several hundred barbs in a vane.

If a feather is examined with a microscope, each barb appears to be a miniature replica of the feather with numerous parallel filaments called **barbules** set in each side of the barb and spreading laterally from it. There may be 600 barbules on each side of a barb, adding up to more than 1 million barbules for the feather. Barbules of one barb overlap barbules of a neighboring barb in a herringbone pattern and are held together with great tenacity by tiny hooks. Should two adjoining barbs become separated—and considerable force is needed to pull the vane apart—they are instantly zipped together again by drawing the feather through the fingertips. A bird, of course, does this preening with its bill, and much of a bird's time is occupied with preening to keep its feathers in perfect condition.

Types of Feathers

Different types of bird feathers serve different functions. **Contour feathers** (Figure 27.5E) give the bird its outward form and are the type we have already described. Contour feathers that extend beyond the body and are used in flight are called **flight feathers. Down feathers** (Figure 27.5H) are soft tufts without a prominent rachis, hidden beneath contour feathers. They are soft because their barbules lack hooks. They are especially abundant on the breast and abdomen of water birds and on young quail and grouse and function principally to conserve heat. **Filoplume feathers** (Figure 27.5G) are hairlike, degenerate feathers; each is a weak shaft with a tuft of short barbs at the tip. They are the "hairs" of a plucked fowl. They have no known function. Bristles around the mouths of flycatchers and whippoorwills are probably modified filoplumes. A fourth type of highly modified feather, the **powder-down feather,** is found on

herons, bitterns, hawks, and parrots. Tips of these disintegrate as they grow, releasing a talclike powder that helps to waterproof the feathers and give them metallic luster.

Origin and Development

Like a reptile's scale to which it is homologous, a feather develops from an epidermal elevation overlying a nourishing dermal core (Figure 27.5A). However, rather than flattening like a scale, a feather bud rolls into a cylinder and sinks into the follicle from which it is growing. During growth, pigments (lipochromes and melanin) are added to the epidermal cells. As a feather enlarges and nears the end of its growth, the soft rachis and barbs are transformed into hard structures by deposition of keratin. The protective sheath splits apart, allowing the end of a feather to protrude and barbs to unfold.

Figure 27.5

Types of bird feathers and their development. **A** to **E,** Successive stages in development of a vaned, or contour, feather. Growth occurs within a protective sheath, **D,** that splits open when growth is complete, allowing the mature feather to spread flat. **F** to **H,** Other feather varieties, including a pheasant feather with aftershaft, **F,** filoplumes, **G,** and down feathers, **H.**

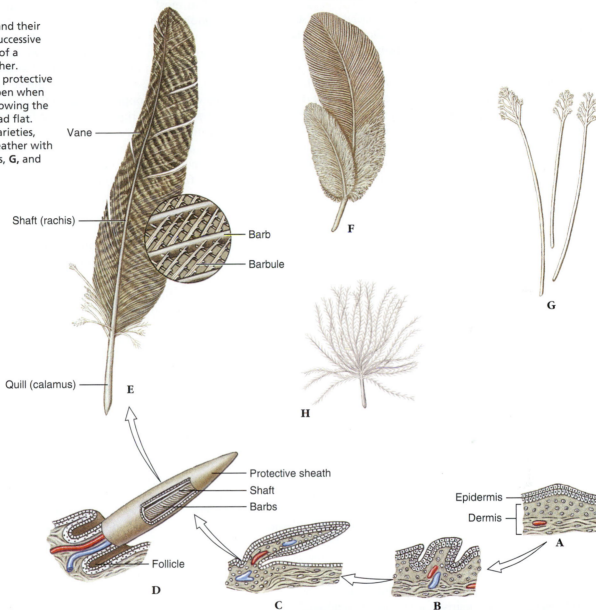

Molting

When fully grown, a feather, like mammalian hair, is a dead structure. Shedding, or molting, of feathers is a highly orderly process. Except in penguins, which molt all at once, feathers are discarded gradually to avoid appearance of bare spots. Flight and tail feathers are lost in exact pairs, one from each side, so that balance is maintained (Figure 27.6). Replacements emerge before the next pair is lost, and most birds can continue to fly unimpaired during the molting period; however, many water birds (ducks, geese, loons, and others) lose all their primary feathers at once and are grounded during the molt. Many prepare for molting by moving to isolated bodies of water where they can find food and more easily escape enemies. Nearly all birds molt at least once a year, usually in late summer after nesting season.

Figure 27.6

Osprey, *Pandion haliaetus* (order Falconiformes), lands while holding a freshly captured fish. Feathers are molted in sequence in exact pairs so that balance is maintained during flight.

The vivid color of feathers is of two kinds: pigmentary and structural. Red, orange, and yellow feathers are colored by pigments, called lipochromes, deposited in feather barbules as they are formed. Black, brown, red-brown, and gray colors are from a different pigment, melanin. Blue feathers of blue jays, indigo buntings, and bluebirds depend not on pigment but on scattering of shorter wavelengths of light by particles within the feather; these are structural colors. Blue feathers are usually underlain by melanin, which absorbs certain wavelengths, thus intensifying the blue. Such feathers look the same from any angle of view. Green colors are almost always a combination of yellow pigment and blue feather structure. Another kind of structural color is the beautiful iridescent color of many birds, which ranges from red, orange, copper, and gold to green, blue, and violet. Iridescent color is based on interference that causes light waves to reinforce, to weaken, or to cancel each other. Iridescent colors may change with the angle of view; quetzals, for example, look blue from one angle and green from another. Among vertebrates, only tropical reef fishes can vie with birds for intensity and vividness of color.

Skeleton

A major structural requirement for flight is a light, yet sturdy, skeleton (Figure 27.7A). As compared with the earliest known bird, *Archaeopteryx,* (Figure 27.7B) bones of modern birds are phenomenally light, delicate, and laced with air cavities. Such **pneumatized** bones (Figure 27.8) are nevertheless strong. The skeleton of a frigate bird with a 2.1 m (7-foot) wingspan weighs only 114 grams (4 ounces), less than the weight of all its feathers.

As archosaurs, birds evolved from ancestors with diapsid skulls (p. 566). However, skulls of modern birds are so specialized that it is difficult to see any trace of the original diapsid condition. A bird's skull is built lightly and mostly fused into one piece. The braincase and orbits are large to accommodate a bulging brain and large eyes needed for quick motor coordination and superior vision. Yet, a pigeon skull weighs only 0.21% of its body weight; by comparison a rat's skull weighs 1.25% of its body weight. As a whole, however, the skeleton of a bird is not lighter than that of a mammal of similar size. The difference is in distribution of mass: whereas the skull and pneumatized wing bones are especially light, the leg bones are heavier than those of mammals. This helps lower the bird's center of gravity as required for aerodynamic stability.

In *Archaeopteryx,* both jaws contained teeth set in sockets, an archosaurian characteristic. Modern birds are completely toothless, having instead a horny (keratinous) beak molded around the bony jaws. The mandible is a complex of several bones hinged to provide a double-jointed action that permits the mouth to gape widely. Most birds have kinetic skulls (kinetic skulls of lizards are described on p. 572). The attachment of the upper jaw to the skull is flexible; this allows the upper jaw to move slightly, thus increasing the gape. In some birds, such as parrots, the upper jaw is especially flexible because it is hinged to the skull.

The most distinctive feature of the vertebral column is its rigidity. Most vertebrae except the **cervicals** (neck vertebrae) are fused together. Most caudal vertebrae are fused into a **pygostyle** (see Figure 27.7A), while many of the remaining vertebrae in the trunk are fused as the **synsacrum.** These fused vertebrae and the pelvic girdle form a stiff but light framework to support the legs and to provide rigidity for flight. To assist in this rigidity, ribs are braced against each other with uncinate processes (see Figure 27.7A). Except in flightless birds, the sternum bears a large, thin keel that provides an attachment for powerful flight muscles. Fused clavicles form an elastic **furcula** that apparently stores energy as it flexes during wing beats. Examination of the anatomy of *Archaeopteryx* permits some insight to its flight abilities. The asymmetrical feathers and large furcula of *Archaeopteryx* lend strong support that it was a flying bird. However, as compared to modern birds, it probably was not a strong flier, because its small sternum offered little area for attachment of flight muscles (see Figure 27.7B).

Bones of the forelimbs are highly modified for flight. They are reduced in number, and several are fused together. Despite these alterations, a bird's wing is clearly a rearrangement of the basic vertebrate tetrapod limb from which it arose (p. 546), and all the elements—arm, forearm, wrist, and fingers—are represented in modified form (see Figure 27.7).

Muscular System

The locomotor muscles of wings are relatively massive to meet demands of flight. Largest of these is the **pectoralis,** which depresses the wings in flight. Its antagonist is the **supracoracoideus** muscle, which raises the wing (Figure 27.9). Surprisingly, perhaps, this latter muscle is not located on the backbone (anyone who has been served the back of a chicken knows that it offers little meat) but is positioned under the pectoralis on the breast. It is attached by a tendon to the upper side of the humerus of the wing so that it pulls from below by an ingenious "rope-and-pulley" arrangement. Both pectoralis and supracoracoideus are anchored to the keel of the sternum. Positioning the main muscle mass low in the body improves aerodynamic stability.

From the main leg muscle mass in the thigh, thin but strong tendons extend downward through sleevelike sheaths to the toes. Consequently the feet are nearly devoid of muscles, explaining the thin, delicate appearance of a bird's leg. This arrangement places the main muscle mass near a bird's center of gravity and at the same time allows great agility to the slender, lightweight feet. Because the feet are composed mostly of bone, tendon, and tough, scaly skin, they are highly resistant to damage from freezing. When a bird perches on a branch, an ingenious toe-locking mechanism (Figure 27.10) is activated, which prevents the bird from falling off its perch when asleep. The same mechanism causes the talons of a hawk or owl automatically to sink deeply into its prey as the legs bend under

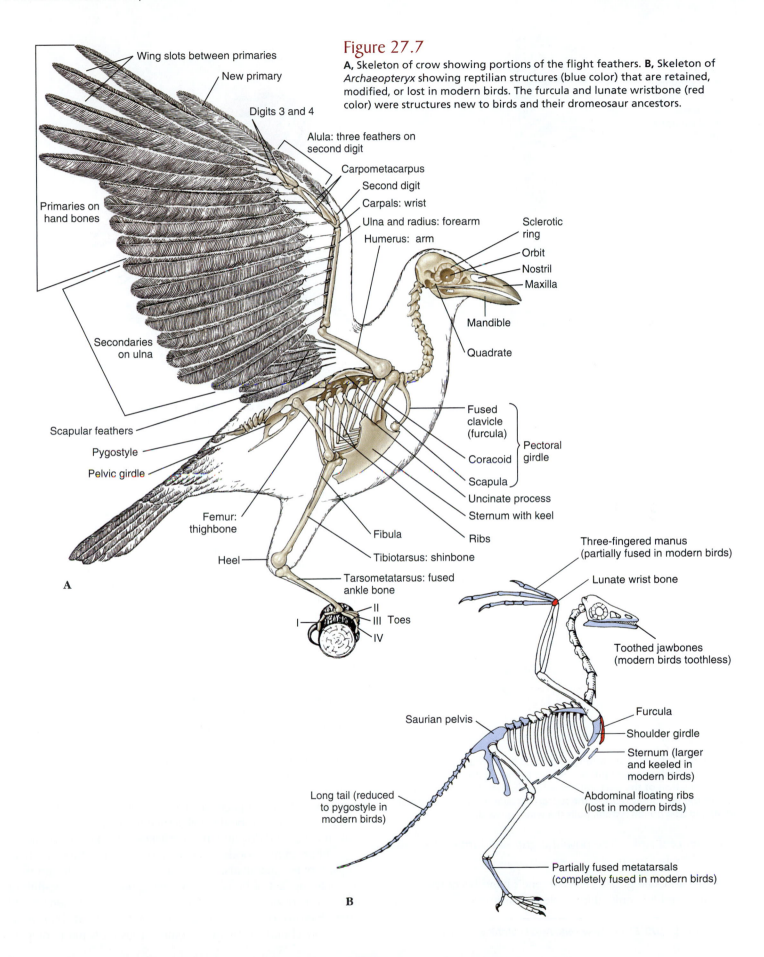

Figure 27.7

A, Skeleton of crow showing portions of the flight feathers. **B,** Skeleton of *Archaeopteryx* showing reptilian structures (blue color) that are retained, modified, or lost in modern birds. The furcula and lunate wristbone (red color) were structures new to birds and their dromeosaur ancestors.

Wing slots between primaries

New primary

Digits 3 and 4

Alula: three feathers on second digit

Carpometacarpus

Second digit

Carpals: wrist

Ulna and radius: forearm

Humerus: arm

Sclerotic ring

Orbit

Nostril

Maxilla

Mandible

Quadrate

Primaries on hand bones

Secondaries on ulna

Scapular feathers

Pygostyle

Pelvic girdle

Fused clavicle (furcula)

Coracoid

Scapula

Pectoral girdle

Uncinate process

Sternum with keel

Ribs

Femur: thighbone

Fibula

Tibiotarsus: shinbone

Heel

Tarsometatarsus: fused ankle bone

II
III Toes
I
IV

A

Three-fingered manus (partially fused in modern birds)

Lunate wrist bone

Toothed jawbones (modern birds toothless)

Saurian pelvis

Furcula

Shoulder girdle

Sternum (larger and keeled in modern birds)

Abdominal floating ribs (lost in modern birds)

Long tail (reduced to pygostyle in modern birds)

Partially fused metatarsals (completely fused in modern birds)

B

Figure 27.8

Hollow wing bone of a songbird showing stiffening struts and air spaces that replace bone marrow. Such "pneumatized" bones are remarkably light and strong.

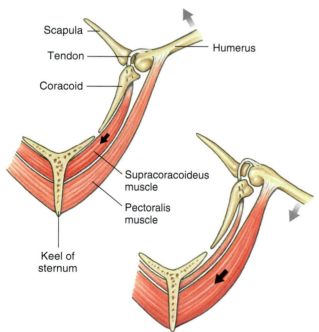

Scapula
Tendon
Coracoid
Humerus
Supracoracoideus muscle
Pectoralis muscle
Keel of sternum

Figure 27.9

Flight muscles of a bird are arranged to keep the center of gravity low in the body. Both major flight muscles are anchored on the sternum keel. Contraction of the pectoralis muscle pulls the wing downward. Then, as the pectoralis relaxes, the supracoracoideus muscle contracts and, acting as a pulley system, pulls the wing upward.

the impact of a strike. The powerful grip of a bird of prey was described by L. Brown[1]

> When an eagle grips in earnest, one's hand becomes numb, and it is quite impossible to tear it free, or to loosen

[1]From Brown, L. 1970. *Eagles.* New York, Arco Publishing.

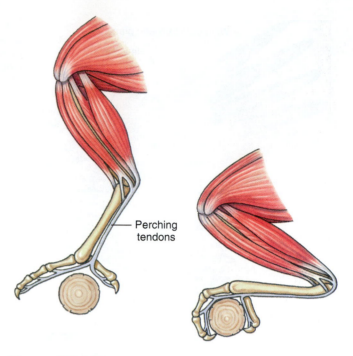

Perching tendons

Figure 27.10

Perching mechanism of a bird. When a bird settles on a branch, tendons automatically tighten, closing the toes around the perch.

the grip of the eagle's toes with the other hand. One just has to wait until the bird relents, and while waiting one has ample time to realize that an animal such as a rabbit would be quickly paralyzed, unable to draw breath, and perhaps pierced through and through by the talons in such a clutch.

Birds have lost the long reptilian tail, still fully evident in *Archae-opteryx,* and have substituted a pincushion-like muscle mound into which the tail feathers are rooted. It contains a perplexing array of tiny muscles, as many as 1000 in some species, which control the crucial tail feathers. The most complex muscular system of all is found in the neck of birds; the thin and stringy muscles, elaborately interwoven and subdivided, provide the bird's neck with the ultimate in vertebrate flexibility.

Food, Feeding, and Digestion

In their early evolution, most birds were carnivorous, feeding principally on insects, already well established on the earth's surface in both variety and numbers long before birds made their appearance. With the advantage of flight, birds could hunt insects on the wing and carry their assault to insect refuges mostly inaccessible to their earthbound tetrapod peers. Today, there is a bird to hunt nearly every insect; they probe the soil, search the bark, scrutinize every leaf and twig, and drill into insect galleries hidden in tree trunks.

Other animal foods (worms, molluscs, crustaceans, fish, frogs, reptiles, mammals, as well as other birds) all found their way into the diet of birds. A very large group, nearly one-fifth of all birds, feeds on nectar. Some birds are omnivores (often termed **euryphagous,** or "wide-eating" species) that will eat whatever is seasonally abundant. However, omnivorous birds must compete

Figure 27.11

Some bills of birds showing variety of adaptations.

Raven
Generalized bill

Cardinal
Seed cracker

American avocet
Worm burrow probe

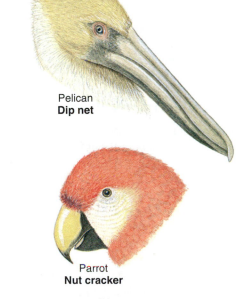

Pelican
Dip net

Parrot
Nut cracker

Flamingo
Zooplankton strainer

Anhinga
Fish spear

Eagle
Meat tearer

with numerous other omnivores for the same broad spectrum of food. Others are specialists (called **stenophagous,** or "narrow-eating" species) that have the pantry to themselves—but at a price. Should the food source be reduced or destroyed for some reason (disease or adverse weather, for example), their survival may be jeopardized.

The beaks of birds are strongly adapted to specialized food habits—from generalized types such as strong, pointed beaks of ravens, to grotesque, highly specialized ones in flamingoes, pelicans, and avocets (Figure 27.11). The beak of a woodpecker is a straight, hard, chisellike device. Anchored to a tree trunk with its tail serving as a brace, the woodpecker delivers powerful, rapid blows to excavate nest cavities or expose burrows of wood-boring insects. It then uses its long, flexible, barbed tongue to catch insects in their galleries. A woodpecker's skull is especially thick to absorb shock.

How much do birds eat? By a peculiar twist of reality, the commonplace "to eat like a bird" is supposed to signify a diminutive appetite. Yet birds, because of their intense metabolism, are voracious feeders. Small birds with their high metabolic rate eat more food relative to their body mass than large birds. This happens because oxygen consumption increases only about three-fourths as rapidly as body weight. For example, the resting metabolic rate (oxygen consumed per gram of body weight) of a hummingbird is 12 times that of a pigeon and 25 times that of a chicken. A 3 g hummingbird may eat 100% of its body weight in food each day, an 11 g blue tit about 30%, and a 1880 g domestic chicken, 3.4%. Obviously the weight of food consumed also depends on water content of the food, since water has no nutritive value. A 57 g Bohemian waxwing was estimated to eat 170 g of watery *Cotoneaster* berries in one day—three times its body weight! Seed-eaters of equivalent size might eat only 8 g of dry seeds per day.

Birds process their food rapidly and thoroughly with efficient digestive equipment. A shrike can digest a mouse in 3 hours, and berries will pass completely through the digestive tract of a thrush in just 30 minutes. Because birds lack teeth, foods that require grinding are reduced in the gizzard. The poorly developed salivary glands mainly secrete mucus for lubricating food. Many birds have an enlargement **(crop)** of the esophagus at its lower end that serves as a storage chamber.

In pigeons, doves, and some parrots, the crop not only stores food but also produces a lipid- and protein-rich "milk," composed of sloughed epithelial cells of the crop lining. For a few days after hatching, the helpless young are fed regurgitated crop milk by both parents.

The stomach proper consists of two compartments, a **proventriculus,** which secretes gastric juice, and the muscular **gizzard,** which is lined with keratinized plates that serve as millstones for grinding food. To assist in the grinding process, birds swallow coarse, gritty objects or pebbles, which lodge in the gizzard. Certain birds of prey such as owls form pellets of indigestible materials, for example, bones and fur, in the proventriculus by sloughing the gut lining to enclose this material and ejecting it through the mouth. At the junction of the small intestine with the colon are paired **ceca;** these are well developed in herbivorous birds in which they serve as fermentation chambers.

In young birds the dorsal wall of the cloaca bears the **bursa of Fabricius,** which processes the B lymphocytes that are important in the immune response (p. 774).

Circulatory System

The general plan of bird circulation is not greatly different from that of mammals, although their shared characteristics were evolved in parallel. Their four-chambered heart is large, with strong ventricular walls; thus, birds share with mammals a complete separation of respiratory and systemic circulations. However, the right aortic arch, instead of the left as in the mammals, leads to the dorsal aorta. The two jugular veins in the neck are connected by a cross vein, an adaptation for shunting blood from one jugular to the other as the head rotates. The brachial and pectoral arteries to the wings and breast are unusually large.

The heartbeat is extremely fast, and, as in mammals, there is an inverse relationship between heart rate and body weight. For example, a turkey has a heart rate at rest of about 93 beats per minute, a chicken has a resting rate of 250 beats per minute, and a blackcapped chickadee has 500 beats per minute when asleep, which may increase to a phenomenal 1000 beats per minute

during exercise. Blood pressure in birds is roughly equivalent to that in mammals of similar size.

A bird's blood contains **nucleated, biconvex erythrocytes.** (Mammals, the only other endothermic vertebrates, have enucleated, biconcave erythrocytes that are somewhat smaller than those of birds.) The **phagocytes,** or mobile ameboid cells of the blood, are very active and efficient in birds in repairing wounds and destroying microbes.

Respiratory System

The respiratory system of birds differs radically from the lungs of reptiles and mammals and is marvelously adapted for meeting the high metabolic demands of flight. In birds the finest branches of the bronchi, rather than ending in saclike alveoli as in mammals, are developed as tubelike **parabronchi** through which air flows continuously. The parabronchi form the lungs of birds. Also unique is an extensive system of nine interconnecting **air sacs** located in pairs in the thorax and abdomen and even extended by tiny tubes into the centers of the long bones (Figure 27.12). Air sacs connect to the lungs in such a way that most of the inspired air bypasses the lungs and flows directly into the posterior air sacs, which serve as reservoirs for fresh air. On expiration, this oxygenated air is passed through the lung and collected in the anterior air sacs. From there it flows directly to the outside. Thus, it takes two respiratory cycles for a single breath of air to pass through the respiratory system, allowing for continuous one-way flow through the respiratory exchange chambers, the parabronchi (Figure 27.12). The advantage of such a system is that an almost continuous stream of oxygenated air is passed through the richly vascularized parabronchi. Although many details of a bird's respiratory system are not yet understood, it is clearly the most efficient respiratory system of any terrestrial vertebrate.

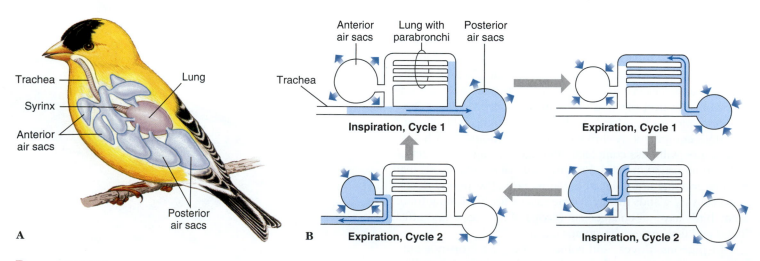

Figure 27.12

Respiratory system of a bird. **A,** Lungs and air sacs. One side of the bilateral air sac system is shown. **B,** Movement of a single volume of air through a bird's respiratory system. Two full respiratory cycles are required to move air through the system.

The remarkable efficiency of a bird's respiratory system is emphasized by bar-headed geese that routinely migrate over the Himalayan mountains and have been sighted flying over Mt. Everest (8848 meters or 29,141 feet) under conditions that are severely hypoxic to humans. They reach altitudes of 9000 meters in less than a day, without the acclimatization that is absolutely essential for humans even to approach the upper reaches of Mt. Everest.

In addition to performing its principal respiratory function, the air sac system helps to cool the bird during vigorous exercise. A pigeon, for example, produces about 27 times more heat when flying than when at rest. Air sacs have numerous diverticula that extend inside the larger pneumatic bones of the pectoral and pelvic girdles, wings, and legs. Because they contain warmed air, they provide considerable buoyancy to the bird.

Excretory System

Urine is formed in the relatively large paired metanephric kidneys by glomerular filtration followed by selective modification of the filtrate in the tubule (the details of this sequence are given on pp. 674–679). Urine passes by way of **ureters** to the **cloaca.** There is no urinary bladder.

Birds, like reptiles, excrete their nitrogenous wastes as uric acid, rather than urea. In shelled eggs, all excretory products must remain within the eggshell with the growing embryo. Uric acid crystallizes from solution and can be stored harmlessly within the eggshell. Because of uric acid's low solubility, a bird can excrete 1 g of uric acid in only 1.5 to 3 ml of water, whereas a mammal may require 60 ml of water to excrete 1 g of urea. The concentration of uric acid occurs almost entirely in the cloaca, where it is combined with fecal material, and the water reabsorbed.

Bird kidneys are much less efficient than mammalian kidneys in removal of solutes, primarily ions of sodium, potassium, and chloride. Most mammals can concentrate solutes to 4 to 8 times that of blood concentration, and some such as desert rodents can concentrate urine to nearly 25 times that of blood. By comparison, most birds concentrate solutes only slightly greater than that of blood (the best that any bird can concentrate is about 6 times that of blood).

To compensate for weak solute-concentrating ability of the kidney, some birds, especially marine birds that must excrete large salt loads from the food they eat and seawater they drink, use extrarenal mechanisms to remove salt from the body. **Salt glands,** one located above each eye of sea birds (Figure 27.13), can excrete highly concentrated solutions of sodium chloride, up to twice the concentration of seawater. The salt solution runs out the internal or external nostrils, giving gulls, petrels, and other sea birds a perpetual runny nose. The size of the salt gland in some birds depends on how much salt the bird takes in its diet. For example, a race of mallard ducks living a semimarine life in Greenland has salt glands 10 times larger than those of ordinary freshwater mallards.

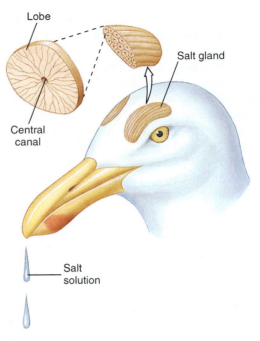

Figure 27.13

Salt glands of a marine bird (gull). One salt gland is located above each eye. Each gland consists of several lobes arranged in parallel. One lobe is shown in cross section, much enlarged. Salt is secreted into many radially arranged tubules, then flows into a central canal that leads into the nose.

Nervous and Sensory Systems

The design of a bird's nervous and sensory system reflects the complex problems of flight and a highly visible existence, in which it must gather food, mate, defend territory, incubate and rear young, and correctly distinguish friend from foe. The brain of a bird has well-developed **cerebral hemispheres, cerebellum,** and **optic lobes** (Figure 27.14). The **cerebral cortex**—chief coordinating center of a mammalian brain—is thin, unfissured, and poorly developed in birds. But the core of the cerebrum, the **dorsal ventricular ridge,** has enlarged into the principal integrative center, controlling such activities as eating, singing, flying, and all complex instinctive reproductive activities. Relatively intelligent birds, such as crows and parrots, have

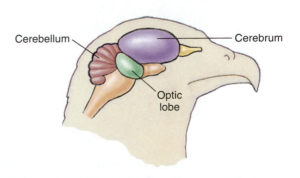

Figure 27.14

Bird brain showing principal divisions.

larger cerebral hemispheres than do less intelligent birds such as chickens and pigeons. The **cerebellum** is much larger in birds than in reptiles and serves as a crucial coordinating center where muscle-position sense, equilibrium sense, and visual cues are assembled and used to coordinate movement and balance. The **optic lobes,** laterally bulging structures of the midbrain, form a visual association apparatus comparable to the visual cortex of mammals.

The senses of smell and taste of some birds are poor, but are relatively well developed in many other birds, such as carnivorous birds, flightless birds, oceanic birds, and waterfowl. Birds have good hearing and superb vision, the keenest in the animal kingdom. As in mammals, a bird's ear consists of three regions: (1) **external ear,** a sound-conducting canal extending to the **eardrum,** (2) **middle ear,** containing a rodlike **columella** that transmits vibrations, and (3) **inner ear,** where the organ of hearing, the **cochlea,** is located. A bird's cochlea is much shorter than the coiled mammalian cochlea, yet birds can hear roughly the same range of sound frequencies as humans. However, birds do not hear high-frequency sounds as well as similar-sized mammals. Actually, a bird's ear far surpasses that of humans in capacity to distinguish differences in intensities and to respond to rapid fluctuations in pitch.

A bird's eye resembles that of other vertebrates in gross structure but is relatively larger, less spherical, and almost immobile; instead of turning their eyes, birds turn their heads with their long and flexible necks to scan the visual field. The light-sensitive **retina** (Figure 27.15) is generously equipped with rods (for dim light vision) and cones (for good acuity and color vision). Cones predominate in diurnal birds, and rods are more numerous in nocturnal birds. A distinctive feature of a bird's eye is the **pecten,** a highly vascularized organ attached to the retina near the optic nerve and jutting into the vitreous humor (Figure 27.15). The pecten is thought to provide nutrients and oxygen to the eye. On the anterior side of the eye is a **sclerotic**

ring of platelike bones that serve to strengthen and focus the large eye (see Figure 27.7).

The position of a bird's eyes in its head is correlated with its life habits. Herbivores that must avoid predators have eyes placed laterally to give a wide view of the world; predaceous birds such as hawks and owls have eyes directed to the front, allowing more binocular vision for better depth perception. In birds of prey and some others, the **fovea,** or region of keenest vision on the retina, is placed in a deep pit, which makes it necessary for the bird to focus exactly on the source. Many birds, moreover, have two foveae on the retina (Figure 27.15): the central one for sharp monocular views and the posterior one for binocular vision. Woodcocks can probably see binocularly both forward and backward. The visual acuity of a hawk is about 8 times that of a human (enabling it to see clearly a crouching rabbit more than a mile away), and an owl's ability to see in dim light is more than 10 times that of a human. Birds have good color vision, especially toward the red end of the spectrum.

Many birds can see into the ultraviolet, enabling them to view environmental features inaccessible to us but accessible to insects (such as flowers with ultraviolet-reflecting "nectar guides" that attract pollinating insects). Several species of ducks, hummingbirds, kingfishers, and passerines (songbirds) can see in the near ultraviolet (UV) down to 370 nm (the human eye filters out ultraviolet light below 400 nm). For what purpose do birds use their UV-sensitivity? Some, such as hummingbirds, may be attracted to nectar-guiding flowers, like insects. But, for the others, the benefit derived from UV-sensitivity is unknown.

FLIGHT

What prompted the evolution of flight in birds, the ability to rise free of earthbound concerns, as almost every human has dreamed of doing? The air was a relatively unexploited habitat stocked with flying insects for food. Flight also offered escape from terrestrial predators and opportunity to travel rapidly and widely to establish new breeding areas and to benefit from year-round favorable climate by migrating north and south with the seasons.

Two competing hypotheses of the origin of bird flight have been offered: birds began to fly either by climbing to a high place and gliding down, or by flapping their wings into the air from the ground. The first hypothesis, termed arboreal, or "trees down," has been long favored. Proponents of this view envision an arboreal ancestor of *Archaeopteryx* gliding from tree to tree, or perhaps "pouncing" on prey below using wings to control its attack. Modifications permitting lift and powered flight would be highly advantageous for this kind of life. Indeed, there are many arboreal squirrels and lizards that use gliding to move among trees. The kind of locomotion envisioned by proponents of the arboreal hypothesis is perhaps best exhibited in the kakapo, a living species of "flightless" New Zealand parrot, which climbs trees using its hindlimbs, and glides among trees using its wings and occasionally flaps to refine its glides. A weakness of this hypothesis is that the feathered dromeosaurs were primarily

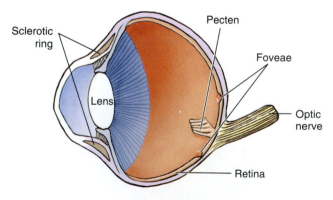

Figure 27.15

A hawk eye has all the structural components of a mammalian eye, plus a peculiar pleated structure, the pecten, believed to provide nourishment to the retina. The extraordinarily keen vision of hawks is attributed to the extreme density of cone cells in the foveae: 1.5 million per fovea compared to 0.2 million for humans.

ground-dwelling, although some of the smaller ones appear to have morphological adaptations that support climbing.

Proponents of the cursorial, or "ground-up," hypothesis suggest that the feathered wings of bipedal, ground-dwelling ancestors of the first flying birds may have been used as snares to capture insects or to refine aerodynamic control during leaps to capture flying insects. Thus, as the wings became larger, they would have been capable of powered flight. However, a ground launch requires working against gravity rather than enlisting its help! No living gliders launch from the ground and examples of ground-dwelling vertebrates that pursue flying insects are lacking. A slightly more convincing scenario is suggested by studies of chukar partridge chicks, which use wingbeats to assist running up steep inclines. Perhaps the ancestors of birds used wings to assist in climbing. Although the evidence is weighted toward the arboreal hypothesis, the debate about origin of flight has not been settled. Interestingly enough, feathers were certainly a requirement for bird flight, but were not required for powered flight in two other lineages, bats and extinct pterosaurs, which lack feathers and, notably, convincing flight-origin hypotheses of their own.

Bird Wing as a Lift Device

To fly, birds must generate lift forces greater than their own mass to become airborne and they must provide propulsion to move. They use their wings to provide both. In general, the distal part of the wing, the modified hand bones with the attached primaries, acts as a propeller to provide propulsion. Lift is provided by feathers in the more medial part of the wing, the secondaries, associated with the forearm. The wing is streamlined in cross section, with a slightly concave lower surface (**cambered**) and with small, tight-fitting feathers where the leading edge meets the air (Figure 27.16). Air slips smoothly over the wing, creating lift with minimum drag. Some lift is produced by positive pressure against the undersurface of the wing. But on the upper side, where the airstream must travel farther and faster over the convex surface, a negative pressure is created that provides more than two-thirds of the total lift.

The lift-to-drag ratio of an airfoil is determined by the angle of tilt (angle of attack) and the airspeed (Figure 27.16). At high speeds sufficient lift is generated so that the wing is held at a low angle of attack, creating less drag. As speed decreases, lift is increased by increasing the angle of attack, but drag forces also increase. When the angle of attack becomes too steep, usually around 15 degrees, turbulence appears on the upper surface, lift is destroyed, and stalling occurs. Stalling can be delayed or prevented by placing a **wing slot** along the leading edge; this structure directs a layer of rapidly moving air across the upper wing surface. Wing slots were and still are used in aircraft traveling at a low speed. In birds, two kinds of wing slots occur: (1) the **alula,** or group of small feathers on the thumb (see Figures 27.6 and 27.7), which provides a midwing slot, and (2) **slotting between the primary feathers,** which provides a wing-tip slot. In many songbirds, these together

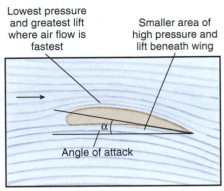

Air flow around wing

Lowest pressure and greatest lift where air flow is fastest

Smaller area of high pressure and lift beneath wing

α

Angle of attack

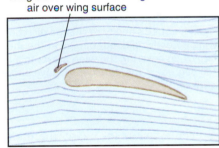

Lift-destroying turbulence

α

Stalling at low speed

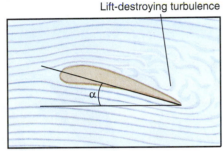

Wing slot directs fast-moving air over wing surface

Preventing stall with wing slots

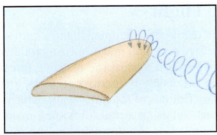

Formation of wing tip vortex

Figure 27.16

Air patterns formed by an airfoil, or wing, moving from right to left. At low speed the angle of attack (α) must increase to maintain lift but this increases the threat of stalling. The upper figures show how low-speed stalling can be prevented with wing slots. Wing tip vortex (*bottom*), a turbulence that tends to develop at high speeds, reduces flight efficiency. The effect is reduced in wings that sweep back and taper to a tip.

Figure 27.17

In normal flapping flight of strong fliers like ducks, the wings sweep downward and forward fully extended. Thrust is provided by the primary feathers at the wing tips. To begin the upbeat, the wing is bent, bringing it upward and backward. The wing then extends, ready for the next downbeat.

Figure 27.18

The secret of a hummingbird's ability to change direction instantly, or hang motionless in the air while sipping nectar from a flower, lies in its wing structure. The wing is nearly rigid, but hinged at the shoulder by a swivel joint and powered by a supracoracoideus muscle that is unusually large for the bird's size. When hovering the wing moves in a sculling motion. The leading edge of the wing moves forward on the forward stroke, then swivels nearly 180 degrees at the shoulder to move backward on the backstroke. The effect is to provide lift without propulsion on *both* forward and backward strokes.

provide stall-preventing slots for nearly the entire distal (and aerodynamically more important) half of the wing.

Flapping Flight

Two forces are required for flapping flight: a vertical *lifting* force to support the bird's weight, and a horizontal thrusting force to move the bird forward against the resistive forces of friction. Thrust is provided mainly by primary feathers at the wing tips, while secondary feathers of the inner wing, which do not move so far or so fast, act as an airfoil, providing mainly lift. Greatest power is applied on the downstroke. The primary feathers are bent upward and twist to a steep angle of attack, biting into the air like a propeller (Figure 27.17). The entire wing (and the bird's body) is pulled forward. On the upstroke, the primary feathers bend in the opposite direction so that their upper surfaces twist into a positive angle of attack to produce thrust, just as the lower surfaces did on the downstroke. A powered upstroke is essential for hovering flight, as in hummingbirds (Figure 27.18), and is important for fast, steep takeoffs by small birds with elliptical wings.

Basic Forms of Bird Wings

Bird wings vary in size and form because successful exploitation of different habitats has imposed special aerodynamic requirements. Four types of bird wings are easily recognized.[2]

Elliptical Wings

Birds that must maneuver in forested or brushy habitats, as do sparrows, warblers, doves, woodpeckers, and magpies (Figure 27.19A), have elliptical wings. This type has a **low aspect ratio** (ratio of length to average width). The wings of the highly maneuverable British Spitfire fighter plane of World War II fame conformed closely to the outline of a sparrows wing. Elliptical wings are slotted between the primary feathers; this arrangement helps to prevent stalling during sharp turns, low-speed flight, and frequent landing and takeoff. Each separated primary feather behaves as a narrow wing with a high angle of attack, providing high lift at low speed.

---

[2]Saville, D. B. O. 1957. Adaptive evolution in the avian wing. *Evolution* **11**:212–224.

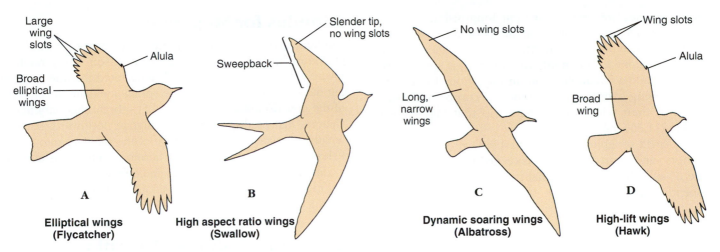

Figure 27.19
Four basic forms of bird wings.

High maneuverability of elliptical wings is exemplified by the tiny chickadee, which can change course within 0.03 second.

High Aspect Ratio Wings

Birds that feed during flight, such as swallows, hummingbirds, and swifts, or that make long migrations, such as plovers, sandpipers, terns, and gulls (Figure 27.19B), have wings that sweep back and taper to a slender tip. They are rather flat in section, have a **high aspect ratio,** and lack wing-tip slotting characteristic of elliptical wings. Sweepback and wide separation of wing tips reduce "tip vortex," a drag-creating turbulence that tends to develop at wing tips at faster speeds (see Figure 27.16). This type of wing is aerodynamically efficient for high-speed flight but cannot easily keep a bird airborne at low speeds. The fastest birds, such as sandpipers, clocked at 175 km (109 miles) per hour, belong to this group.

Dynamic Soaring Wings

Oceanic soaring birds, including albatrosses, shearwaters, and gannets (Figure 27.19C) also have high aspect ratio wings, shaped like those of sailplanes. Such long, narrow wings lack slots and are adapted for **dynamic soaring.** Dynamic soaring can only be done over seas with strong, reliable winds, and exploits different wind speeds near the ocean surface (slow) and well above the surface (fast). A bird that uses dynamic soaring begins a downwind glide from an elevated position, gaining speed as it descends. Near the surface of the ocean, it turns into the wind and rises into stronger winds. Although its velocity relative to the ocean slows, the strong winds over its wings provide the lift to keep it aloft.

High-Lift Wings

Vultures, hawks, eagles, owls, and ospreys (Figure 27.19D)—predators that carry heavy loads—have wings with slotting, alulas, and pronounced camber, all of which promote high lift at low speed. Wings of these birds have an aspect ratio intermediate between that of elliptical wings and high aspect ratio wings. Many of these birds are land soarers, with broad, slotted wings that provide the sensitive response and maneuverability required for static soaring in capricious air currents over land.

MIGRATION AND NAVIGATION

We described advantages of migration in the prologue to this chapter. Not all birds migrate, of course, but most North American and European species do, and the biannual journeys of some are truly extraordinary undertakings.

Migration Routes

Most migratory birds have well-established routes trending north and south. Since most birds (and other animals) breed in the Northern Hemisphere, where most of the earth's landmass is concentrated, most birds migrate south for the northern winter and north to nest during the northern summer. Of the 4000 or more species of migrant birds (a little less than half the total bird species), most breed in the more northern latitudes of the hemisphere. Some use different routes in the fall and spring (Figure 27.20). Some, especially certain aquatic species, complete their migratory routes in a very short time. Others, however, make a leisurely trip, often stopping along the way to feed. Some warblers are known to take 50 to 60 days to migrate from their winter quarters in Central America to their summer breeding areas in Canada. Many smaller species migrate at night and feed by day; others migrate chiefly in daytime; and many swimming and wading birds migrate by either day or night.

Many birds are known to follow landmarks, such as rivers and coastlines, but others do not hesitate to fly directly over large bodies of water in their routes. Some birds have very wide migration lanes, whereas others, such as certain sandpipers, are restricted to very narrow ones, keeping well to the coastlines because of their food requirements.

Some species are known for their long-distance migrations. Arctic terns, greatest globe spanners of all, breed north of the Arctic Circle during the northern summer, then migrate to Antarctic regions for the northern winter. This species also is known to take a circuitous route in migrations from North America, passing over to the coastlines of Europe and Africa and then to their winter quarters, a trip that may exceed 18,000 km (11,200 miles).

Many small songbirds, such as warblers, vireos, thrushes, flycatchers, and sparrows, also make great migratory treks (Figure 27.20). Migratory birds that nest in Europe or Central Asia spend the northern winter in Africa.

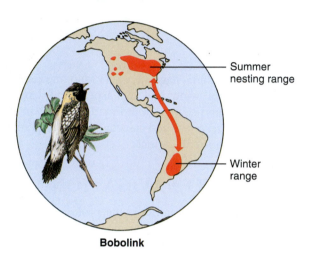

Bobolink

American golden plover

Figure 27.20

Migrations of bobolinks, *Dolichonyx oryzivorus,* and American golden plovers, *Pluvialis dominica.* Bobolinks commute 22,500 km (14,000 miles) each year between nesting sites in North America and their range in Argentina, where they spend the northern winters, a phenomenal feat for such a small bird. Although the breeding range has extended to colonies in western areas, these birds take no shortcuts but adhere to the ancestral seaboard route. American golden plovers fly a loop migration, striking out across the Atlantic in their southward autumnal migration but returning in the spring by way of Central America and the Mississippi Valley because ecological conditions are more favorable at that time.

Stimulus for Migration

Humans have known for centuries that onset of reproductive cycles of birds is closely related to season. Only relatively recently, however, has it been demonstrated that the lengthening days of late winter and early spring stimulate development of the gonads and accumulation of fat—both important internal changes that predispose birds to migrate northward. Long day length stimulates the anterior lobe of the pituitary into activity. Release of pituitary gonadotropic hormone in turn sets in motion a complex series of physiological and behavioral changes, resulting in gonadal growth, fat deposition, migration, courtship and mating behavior, and care of the young.

Direction Finding in Migration

Numerous experiments suggest that most birds navigate chiefly by sight. Birds recognize topographical landmarks and follow familiar migratory routes—a behavior assisted by flock migration, during which navigational resources and experience of older birds can be pooled. In addition to visual navigation, birds have a highly accurate sense of time. Numerous studies support an old, much debated hypothesis that birds can detect and navigate by the earth's magnetic field. The navigational abilities of birds are primarily instinctive, although they may require calibration with existing navigational landmarks. In addition, learning can play a role because a bird's navigational abilities may improve with experience.

In the early 1970s W. T. Keeton showed that flight bearings of homing pigeons were significantly disturbed by magnets attached to the birds' heads, or by minor fluctuations in the geomagnetic field. But until recently the nature and position of a magnetic receptor in pigeons remained a mystery. Deposits of a magnetic substance called magnetite (Fe_3O_4) have been discovered in the neck musculature of pigeons and migratory white-crowned sparrows. If this material is coupled to sensitive muscle receptors, as has been proposed, the structure could serve as a magnetic compass that would enable birds to detect and to orient their migrations to the earth's magnetic field.

Experiments by German ornithologists G. Kramer and E. Sauer and American ornithologist S. Emlen demonstrated convincingly that birds can navigate by celestial cues: the sun by day and the stars by night. Using special circular cages, Kramer concluded that birds maintain compass direction by referring to the sun, (Figure 27.21). This is called **sun-azimuth orientation** (*azimuth,* compass bearing of the sun). To use the sun as a compass birds must know the time of day because the sun's position changes throughout the day. By exposing birds to altered light cycles to shift their perception of daybreak, researchers showed that birds do use an internal clock in this fashion. Sauer's and Emlen's ingenious planetarium experiments strongly suggest that some birds, probably many, are able to detect and to navigate by the North Star axis around which the constellations appear to rotate.

When the birds grew to an age for migration, they were placed in cages under a normal night sky that allowed recording of the direction in which they tried to migrate. Birds that had seen only points of light during their development, with no rotation of the sky, showed no ability to detect direction and moved randomly. Birds that had developed seeing the normal sky rotated around the North Star oriented correctly for migration; and the group that developed seeing the sky rotated about Orion showed consistent orientation as if Betelgeuse were the North Star, even though now exposed to a normal night sky with stars rotating around the North Star. Thus, Emlen elegantly showed that these birds do not hatch with an innate sense of direction but must learn direction by seeing the sky rotate around a "pole" star!

Some remarkable feats of bird navigation still defy understanding. Most birds undoubtedly use a combination of environmental and innate cues to migrate. Migration is a rigorous undertaking. The target is often small, and natural selection relentlessly eliminates individuals making errors in migration, leaving only the best navigators to propagate the species.

SOCIAL BEHAVIOR AND REPRODUCTION

The adage says "birds of a feather flock together," and many birds are indeed highly social creatures. Especially during the breeding season, sea birds gather, often in enormous colonies, to nest and to rear young (Figure 27.22). Land birds, with some conspicuous exceptions, such as starlings and rooks, tend to be less gregarious than sea birds during breeding and to seek isolation for rearing their brood. Species that covet separation from their kind during breeding may aggregate for migration or feeding. Togetherness offers advantages: mutual protection from enemies, greater ease in finding mates, less opportunity for individual straying during migration, and mass huddling for protection against low night temperatures during migration. Certain species, such as pelicans (Figure 27.23), may use highly organized cooperative behavior to feed. At no time are the highly organized social interactions of birds more evident than during the breeding season, as they establish territorial claims, select mates, build nests, incubate and hatch their eggs, and rear their young.

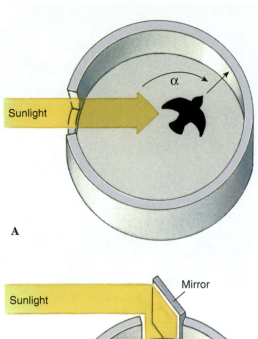
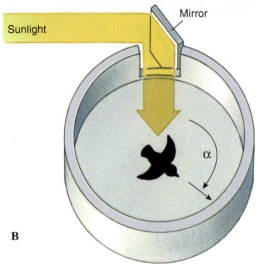

Figure 27.21
Gustav Kramer's experiments with sun-compass navigation in starlings. **A,** In a windowed, circular cage, the bird fluttered to align itself in the direction it would normally follow if it were free. **B,** When the true angle of the sun is deflected with a mirror, the bird maintains the same relative position to the sun. This shows that these birds use the sun as a compass. The bird navigates correctly throughout the day, changing its orientation to the sun as the sun moves across the sky.

In an elegant set of experiments designed to determine whether nocturnal migrants have an innate sense of direction or learn direction as nestlings, Stephen Emlen raised indigo buntings under three sets of conditions in a planetarium in which star patterns could be modified. One group of nestlings was allowed to see stars in a normal night sky rotating around the North Star. A second group of nestlings saw a night sky star pattern that was rotating around Betelgeuse, a bright star in the constellation Orion, as if Betelgeuse were the North Star. A third group of nestlings was raised seeing only points of light at night that did not rotate.

Figure 27.22
Part of a colony of northern gannets, *Morus bassanus,* showing extremely close spacing between pairs in this highly social bird. Order Pelecaniformes.

Figure 27.23

Cooperative feeding behavior by white pelicans, *Pelecanus onocrotalus.* **A,** Pelicans form a horseshoe to drive fish together. **B,** Then they plunge simultaneously to scoop fish in their huge bills. These photographs were taken 2 seconds apart.

Reproductive System

During most of the year the **testes** of males are tiny bean-shaped bodies. But during the breeding season they enlarge greatly, as much as 300 times their nonbreeding size. Since males of most species lack a penis, copulation is a matter of bringing cloacal surfaces into contact, usually while the male stands on the back of the female (Figure 27.24). Some swifts and hawks copulate in flight.

In females of most birds, only the left ovary and oviduct develop; those on the right dwindle to vestigial structures (Figure 27.25). Eggs discharged from the ovary enter the expanded end of the oviduct, the **infundibulum.** The oviduct runs posteriorly to the cloaca. While eggs are passing down the oviduct, **albumin,** or egg white, from special glands is added to them; farther down the oviduct, shell membrane, shell, and shell pigments are also secreted around the egg. Fertilization occurs in the upper oviduct several hours before layers of albumin, shell membranes, and shell are added. Sperm remain alive in a female's oviduct for many days after a single mating. Hen eggs show good fertility for 5 or 6 days after mating, but then fertility drops rapidly. However, an egg occasionally may be fertile as long as 30 days after separation of hens from a rooster.

Mating Systems

Two types of mating systems in animals are **monogamy,** in which an individual has only one mate and **polygamy,** in which an individual has more than one mate during a breeding period.

Figure 27.24

Copulation in waved albatrosses, *Diomeda irrorata.* In most bird species males lack a penis. A male copulates by standing on the back of a female, pressing his cloaca against that of the female, and passing sperm to the female.

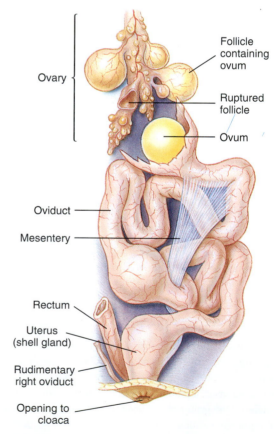

Figure 27.25

Reproductive system of a female bird. For most birds, only the left ovary and reproductive tract are functional. Structures on the right dwindle to vestiges.

Monogamy is rare in most animal groups but common in birds; more than 90% of birds are monogamous. In a few bird species, such as swans and geese, partners are chosen for life and often remain together throughout the year. Seasonal monogamy is more common as the great majority of migrant birds pair only during the breeding season, living independent lives the rest of the year and perhaps choosing a different mate the next breeding season.

One reason that monogamy is much more common among birds than among mammals is that male and female birds are equally adept at most aspects of parental care. Male mammals do not gestate the young and do not lactate and thus can provide little help in caring for the young. Female and male birds can alternate care of the nest and young, which permits one parent to be at the nest at all times. For some species, a female remains on the nest for months at a time, and is fed by the male. This constant attendance to the nest may be particularly important in species that would experience high loss of eggs or young to predators or rival birds if a nest were left unguarded. For many bird species, the high demands on a male to care for the young or his mate preclude the establishment of nests with additional females.

Although most birds have a monogamous mating system, either member of a pair may mate with an individual that is not the partner. Recent **DNA** analyses have shown most passerine species frequently are "unfaithful," engaging in extra-pair copulations. As a result, nests of many of these monogamous species contain a sizeable portion (30% or more) of young with fathers other than the attendant male. Why do individuals engage in extra-pair copulations? By mating with an individual of better genetic quality, fitness of the offspring can be improved. Also, mating with multiple partners increases genetic variation of the offspring. A male can father more offspring with additional partners, and the mate of the extra-pair female he has mated provides parental care to his offspring!

The most common form of polygamy in birds, when it occurs, is **polygyny** ("many females"), in which a male has more than one female mate. In many species of grouse, males gather in a collective display ground, the **lek**, which is divided into individual territories, each vigorously defended by a displaying male (Figure 27.26). There is nothing of value in a lek to the female except the male, and all he can offer are his genes, for only females care for the young. Usually a dominant male and several subordinate males occur in a lek. Competition among males for females is intense, but females appear to choose the dominant male for mating because, presumably, social rank correlates with genetic quality.

Polyandry ("many males"), in which a female mates with several males and the male incubates the eggs, is relatively rare in birds. It is practiced by several shorebird species, including spotted sandpipers, *Actitis macularia*. Female spotted sandpipers defend territories and mate with multiple males. Males incubate eggs within the female's territory and provide most parental care. This unusual reproductive strategy and clustering of individuals may be in response to high predation on spotted sandpiper nests.

Figure 27.26
Dominant male greater sage grouse, *Centrocercus urophasianus*, surrounded by several hens that have been attracted by his "booming" display.

Nesting and Care of Young

To produce offspring, all birds lay eggs that must be incubated by one or both parents. Most duties of incubation fall on females, although in many instances both parents share the task, and occasionally only males incubate the eggs.

Most birds build some form of nest in which to rear their young. Some birds simply lay their eggs on bare ground or rocks, making no pretense of nest building. Others build elaborate nests such as pendant nests constructed by orioles, delicate lichen-covered mud nests of hummingbirds (Figure 27.27) and flycatchers, chimney-shaped mud nests of cliff swallows, and floating nests of rednecked grebes. Most birds take considerable pains to conceal their nests from enemies. Woodpeckers, chickadees, bluebirds, and many others place their nests in tree hollows or other cavities; kingfishers excavate tunnels in the banks of streams for their nests; and birds of prey build high in lofty trees or on inaccessible cliffs. Nest parasites such as the brown-headed cowbird and the European cuckoo build no nests at all but simply lay their eggs in the nests of birds smaller than themselves. When the eggs hatch, the foster parents care for the cowbird young, which outcompete the host's own hatchlings.

The developmental state of newly hatched birds varies among species. **Precocial** young, such as quail, fowl, ducks, and most water birds, are covered with down when hatched and can run or swim as soon as their plumage is dry (Figure 27.28). The most precocial birds are brush turkeys or megapodes of Australia, which incubate eggs in sand and vegetation mounds like crocodilians. Their young can fly at hatching. However, most precocial young, even those able to leave the nest soon after hatching, are still fed or protected from predators by their parents for some time. **Altricial** hatchlings, in contrast, are naked and unable to see or to walk at birth and remain in the nest for a week or more. Parents of altricial species must carry food to

Figure 27.27

Anna's hummingbird, *Calypte anna,* feeding its young in its nest of fibers and down from plants, bound together with spiderwebs and camouflaged with lichens. The female builds the nest, incubates two pea-sized eggs, and rears the young with no assistance from a male. Anna's hummingbird is a common resident of California. It is the only hummingbird to overwinter in the United States.

their young almost constantly, for young birds may eat more than their weight each day. Many birds are not easily categorized as precocial or altricial because their young are intermediate in development at birth. For example, gulls and terns are born covered with down and with eyes open, but are unable to leave the nest for some time.

Although it may seem that precocial young have all the advantages, with their greater ability to find food and to escape predation, altricial birds have some advantages of their own. Because altricial birds lay relatively small eggs with minimal yolk supplies, a mother has a relatively small investment in her eggs and can easily replace eggs lost to predation or extreme weather

conditions. Altricial young also grow faster, perhaps due to the higher growth potential of immature tissue.

BIRD POPULATIONS

Bird populations, like those of other animals, vary in size from year to year. Snowy owls, for example, are subject to population cycles that closely follow cycles in their food supply, mainly rodents. Voles, mice, and lemmings in the north have a fairly regular 4-year cycle of abundance; at population peaks, predator populations of foxes, weasels, and buzzards, as well as snowy owls, increase because there is abundant food for rearing their young. After a crash in the rodent population, snowy owls move south, seeking alternative food supplies. They occasionally appear in large numbers in southern Canada and the northern United States, where their total absence of fear of humans makes them easy targets for thoughtless hunters.

Occasionally activities of people may cause spectacular changes in bird distribution. Both starlings (Figure 27.29) and house sparrows have been accidentally or deliberately introduced into numerous countries, to become the two most abundant bird species on earth, with the exception of domestic fowl.

Humans also are responsible for the extinction of many bird species. More than 80 species of birds have, since 1681, followed the last dodo to extinction. Most were victims of changes in their habitat or competition with introduced species. But several have been hunted to extinction, among them passenger pigeons,

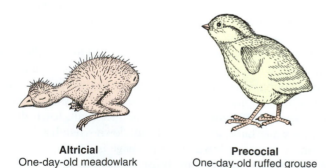

Altricial
One-day-old meadowlark

Precocial
One-day-old ruffed grouse

Figure 27.28

Comparison of one-day-old altricial and precocial young. The altricial meadowlark (*left*) hatches nearly naked, blind, and helpless. The precocial ruffed grouse (*right*) is covered with down, alert, strong legged, and able to feed itself.

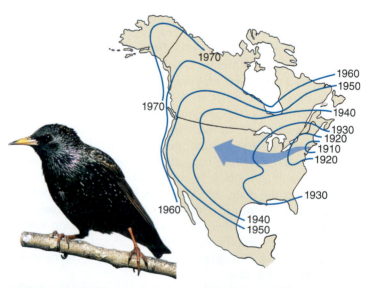

Figure 27.29

Colonization of North America by starlings, *Sturnus vulgaris,* after the introduction of 120 birds into Central Park in New York City in 1890. There are now perhaps 200 million starlings in the United States alone, testimony to the great reproductive potential of birds. Starlings are omnivorous, eating mostly insects in spring and summer and shifting to wild fruits in the fall.

Figure 27.30

Sport-shooting passenger pigeons in Louisiana during the nineteenth century. Relentless sport and market hunting, before establishment of state and federal hunting regulations, in addition to clearing of the hardwood forests that served as nesting habitats, eventually dropped the population too low to sustain colonial breeding. The last passenger pigeon died in captivity in 1914.

which only a century ago darkened the skies over North America in incredible numbers estimated in the billions (Figure 27.30).

Today, game bird hunting is a well-managed renewable resource in the United States and Canada, and while hunters kill millions of game birds each year, none of the 74 bird species legally hunted are endangered. Hunting interests, by acquiring large areas of wetlands for migratory bird refuges and sanctuaries, have contributed to the recovery of both game and nongame birds.

Lead poisoning of waterfowl is a side effect of hunting. Before long-delayed federal regulations went into effect in 1991, requiring the use of nonlead shot for all inland and coastal waterfowl hunting, shotguns scattered more than 3000 tons of lead each year in the United States alone. When waterfowl eat the pellets (which they mistake for seeds or grist), the pellets are ground and eroded in their gizzards, facilitating absorption of lead into their blood. Lead poisoning paralyzes or weakens birds, leading to death by starvation. Today, birds are still dying from ingesting lead shot that has accumulated over the years.

Of particular concern is the recent sharp decline of songbirds in the United States and southern Canada. Amateur birdwatchers and ornithologists have recorded that many songbird species that were abundant as recently as 40 years ago are now suddenly scarce. There are several reasons for the decline. Excessive fragmentation of forests throughout much of the United States has increased exposure of nests of forest-dwelling species to nest predators such as blue jays, raccoons, and opossums, and to nest parasites such as brown-headed cowbirds. House cats also kill millions of small birds every year. From a study of radio-collared farm cats in Wisconsin, researchers estimated that in that state alone, cats may kill 19 million songbirds in a single year.

The rapid loss of tropical forests—approximately 170,000 square kilometers each year, an area about the size of the state of Washington—is depriving some 250 species of songbird migrants of their wintering homes. Recent studies indicate that stressors on the wintering grounds are seriously decreasing the physiological condition of birds, particularly songbirds, prior to northward migration. Of all long-term threats facing songbird populations, tropical deforestation is the most serious and most intractable to change. If the rate of deforestation accelerates in the next few decades as expected, the world's tropical forests will have disappeared by 2040 (Terborgh, 1992).

Some birds, such as robins, house sparrows, and starlings, can accommodate these changes and may even thrive on them, but for most birds the changes are adverse. Terborgh (1992) warns that unless we take leadership in managing our natural resources wisely we soon could be facing the "silent spring" that Rachel Carson envisioned in 1962.

Classification of Living Birds of Class Aves

Class Aves contains about 9700 species distributed among 30 orders of living birds. Understanding the relationships of living birds, and consequently placing them in a classification, has been difficult because of the apparent rapid diversification of birds in the Cretaceous and early Tertiary. Prior to the study of Sibley and Alquest (1990), which used DNA hybridization, classification had been primarily based on morphological similarity. The classification proposed by Sibley and Alquest has been widely adopted, although their results suggested a number of surprising relationships, including a sister-group relationship of Anseriformes (ducks and geese) and ratites. New attempts at discovering the higher relationships of birds have utilized many kinds of data, but especially mitochondrial DNA (mtDNA) and nuclear DNA sequences. The avian classification and numbers of living bird species in orders that we present mostly follows that given by Gill (2006), which itself is based on Sibley and Alquest's study and many other, more recent, phylogenetic reconstructions. We note that determining higher relationships of birds is an active area of study, and our understanding of these relationships is likely to change in the near future.

Class Aves (L. *avis,* bird)

    Superorder Paleognathae (Gr. *palaios,* ancient, + *gnathos,* jaw). Modern birds with primitive archosaurian palate. Ratites, which include ostrich rheas, cassowaries, emus, and kiwis, (with unkeeled sternum) and tinamous (with keeled sternum).

        Order Struthioniformes (stroo′thi-on-i-for′meez) (L. *struthio,* ostrich, + *forma,* form): **ostrich.** The ostrich, *Struthio camelus* (Figure 27.31), of Africa, is the largest of living birds, with some specimens being 2.4 m tall and weighing 135 kg. The feet are provided with only two toes of unequal size covered with pads, which enable the birds to travel rapidly over sandy ground.

        Order Rheiformes (rē′i-for′meez) (Gr. *rhea,* rhea, + form): **rheas.** Two species of large, flightless birds found in grasslands of South America.

        Order Casuariiformes (cas′ū-ər′i-for-meez) (Mal. *casuar,* cassowary, + form): **cassowaries, emu.** The three species of cassowaries occupy forests of northern Australia and New Guinea. The emu is the second largest living bird species, and is confined to Australia. All are flightless.

        Order Dinornithiformes (din′or-nith′i-for-meez) (Gr. *din,* terrible, + Gr. *ornith,* bird, + form): **kiwis.** Kiwis, about the size of a domestic fowl, are unusual in having only the mere vestige of a wing. This order also includes the extinct, flightless moas, some of which reached 2 m at the shoulder. Three living species, all in New Zealand.

        Order Tinamiformes (tin-am′i-for′meez) (N.L. *Tinamus,* type genus, + form): **tinamous.** Ground-dwelling, grouselike birds of Central and South America. About 47 species.

    Superorder Neognathae (Gr. *neos,* new, + *gnathos,* jaw). Modern birds with flexible palate.

        Order Anseriformes (an′ser-i-for′meez) (L. *anser,* goose, + form): **swans, geese, ducks.** The members of this order have broad bills with filtering ridges at their margins, a foot web restricted to the front toes, and a long breastbone with a low keel. About 162 species, worldwide distribution.

        Order Galliformes (gal′li-for′meez) (L. *gallus,* cock, + form): **quail, grouse, pheasants, ptarmigan, turkeys,** **domestic fowl.** Chickenlike ground-nesting herbivores with strong beaks and heavy feet. The bobwhite quail, *Colinus virginianus,* is found all over the eastern half of the United States. The ruffed grouse, *Bonasa umbellus,* is found in about the same region, but in woods instead of the open pastures and grain fields, which the bobwhite frequents. About 290 species, worldwide distribution.

        Order Sphenisciformes (sfe-nis′i-for′meez) (Gr. *Sphēniskos,* dim. of *sphen,* wedge, from the shortness of the wings, + form): **penguins.** Web-footed marine swimmers of southern seas from Antarctica north to the Galápagos Islands. Although penguins are carinate birds, they use their wings as paddles for swimming rather than for flight. About 17 species.

        Order Gaviiformes (gay′vee-i-for′meez) (L. *gavia,* bird, probably sea mew, + form): **loons.** The five species of loons are remarkable swimmers and divers with short legs and heavy bodies. They live exclusively on fish and small aquatic forms. The familiar great northern diver, *Gavia immer,* is

Figure 27.31
Ostrich, *Struthio camelus,* of Africa, the largest of all living birds. Order Struthioniformes.

Classification of Living Birds of Class Aves

found mainly in northern waters of North America and Eurasia.

Order Podicipediformes (pod′i-si-ped′i-for′meez) (L. *podex*, rump, + *pes, pedis,* foot): **grebes.** These are short-legged divers with lobate-webbed toes. The pied-billed grebe, *Podilymbus podiceps,* is a familiar example of this order. Grebes are most common in old ponds where they build their raftlike floating nests. Twenty-two species, worldwide distribution.

Order Phoenicopteriformes (fēn′i-cop-ter′i-for′meez) (Gr. *phoenico,* reddish-purple, + *pter,* wing, + form): **flamingos** (Figure 27.32). Large, colorful, wading birds that use lamellae in their beaks to strain zooplankton from the water. Five species.

Order Procellariiformes (pro-sel-lar′ee-i-for′meez) (L. *procella,* tempest, + form): **albatrosses, petrels, fulmars, shearwaters.** All are marine birds with hooked beak and tubular nostrils. In wingspan (more than 3.6 m in some), albatrosses are the largest of flying birds. About 112 species, worldwide distribution.

Order Pelecaniformes (pel-e-can-i-for′meez) (Gr. *pelekan,* pelican, + form): **pelicans, cormorants, gannets, boobies, and others.** These are colonial fish-eaters with throat pouch and all four toes of each foot included within the web. About 65 species, worldwide distribution, especially in the tropics.

Order Ciconiiformes (si-ko′nee-i-for′meez) (L. *ciconia,* stork, + form): **herons, bitterns, storks, ibises, spoonbills, vultures.** These are long-necked, long-legged, mostly colonial waders and vultures. A familiar eastern North American representative is the great blue heron, *Ardea herodias,* which frequents marshes and ponds. About 116 species, worldwide distribution.

Order Falconiformes (fal′ko-ni-for′meez) (L. *falco,* falcon, + form): **eagles, hawks, falcons, condors, buzzards.** Diurnal birds of prey. All are strong fliers with keen vision and sharp, curved talons. About 304 species, worldwide distribution.

Order Gruiformes (groo′i-for′meez) (L. *grus,* crane, + form): **cranes, rails, coots, gallinules.** Mostly prairie and marsh breeders. About 212 species, worldwide distribution.

Order Charadriiformes (ka-rad′ree-i-for′meez) (N.L. *Charadrius,* genus of plovers, + form): **gulls** (Figure 27.33), **oyster catchers, plovers, sandpipers, terns, woodcocks, turnstones, lapwings, snipe, avocets, phalaropes, skuas, skimmers, auks, puffins.** All are shorebirds. They are strong fliers and are usually colonial. About 367 species, worldwide distribution.

Order Columbiformes (co-lum′bi-for′meez) (L. *columba,* dove, + form): **pigeons, doves.** All have short necks, short legs, and a short, slender bill. The flightless dodo, *Raphus cucullatus,* of the Mauritius Islands became extinct in 1681. About 308 species, worldwide distribution.

Order Psittaciformes (sit′ta-si-for′meez) (L. *psittacus,* parrot, + form): **parrots, parakeets.** Birds with hinged and movable upper beak, fleshy tongue. About 364 species, pantropical distribution.

Order Opisthocomiformes (o-pis′thō-co-mi-for′meez) (Gr. *opistho,* back, + L. *comos,* with long hair, + form): **hoatzin.**

Figure 27.32
Greater flamingos, *Phoenicopterus ruber,* on an alkaline lake in East Africa. Order Phoenicopteriformes.

The relationship of the single species in this order to other birds is uncertain. The young of this South American, herbivorous bird has large claws on its wings that it uses to climb trees.

Order Musophagiformes (myu′-so-fa-ji-for′meez) (L. *musa,* banana, + Gr. *phagō,* to eat, + form): **turacos.** Medium to large birds of dense forest or forest edge with a conspicuous patch of crimson on the spread wing. Bill brightly colored, wings short and rounded. Twenty-three species restricted to Africa.

Order Cuculiformes (ku-koo′li-for′meez) (L. *cuculus,* ccuckoo, + form): **cuckoos, roadrunners.** European cuckoos, *Cuculus canorus,* lay their eggs in nests of smaller birds, which rear the young cuckoos. American cuckoos, black-billed and yellow-billed, usually rear their own young. About 138 species, worldwide distribution.

Order Strigiformes (strij′i-for′meez) (L. *strix,* screech owl, + form): **owls.** Nocturnal predators with large eyes, powerful beaks and feet, and silent flight. About 180 species, worldwide distribution.

Figure 27.33
Laughing gulls, *Larus atricilla,* in flight. Order Charadriiformes.

Classification of Living Birds of Class Aves

Order Caprimulgiformes (kap′ri-mul′ji-for′meez) (L. *caprimulgus,* goatsucker, + form): **goatsuckers, nighthawks, whippoorwills.** Night and twilight feeders with small, weak legs and wide mouths fringed with bristles. Whippoorwills, *Antrostomus vociferus,* are common in the woods of the eastern states, and nighthawks, *Chordeiles minor,* are often seen and heard in the evening flying around city buildings. About 118 species, worldwide distribution.

Order Apodiformes (up-pod′i-for′meez) (Gr. *apous,* footless, + form): **swifts, hummingbirds.** These are small birds with short legs and rapid wingbeat. The familiar chimney swift, *Chaetura pelagia,* fastens its nest in chimneys by means of saliva. A swift found in China builds a nest of saliva that is used by Chinese people for soup making. Most species of hummingbirds are found in the tropics, but there are 14 species in the United States, of which only one, the ruby-throated hummingbird, is found in the eastern part of the country. About 429 species, worldwide distribution.

Order Coliiformes (ka-ly′i-for′meez) (Gr. *kolios,* green woodpecker, + form): **mousebirds.** Small crested birds of uncertain relationship. Six species restricted to southern Africa.

Order Trogoniformes (tro-gon′i-for′meez) (Gr. *trōgon,* gnawing, + form): **trogons.** Richly colored, long-tailed birds. About 39 species, pantropical distribution.

Order Coraciiformes (ka-ray′see-i-for′meez or kor′uh-sigh′uh-for′meez) (N.L. *coracii* from Gr. *korakias,* a kind of raven, + form): **kingfishers, hornbills, and others.** Birds with strong, prominent bills and which nest in cavities. In the eastern half of the United States, belted kingfishers, *Megaceryle alcyon,* are common along most waterways of any size. About 209 species, worldwide distribution.

Order Piciformes (pis′i-for′meez) (L. *picus,* woodpecker, + form): **woodpeckers, toucans, puffbirds, honeyguides.** Birds with highly specialized bills and having two toes extending forward and two backward. All nest in cavities. There are many species of woodpeckers in North America,

most common of which are flickers and downy, hairy, red-bellied, redheaded, and yellow-bellied woodpeckers. Largest is the pileated woodpecker, which is usually found in mature forests. About 398 species, worldwide distribution.

Order Passeriformes (pas′er-i-for′meez) (L. *passer,* sparrow, + form): **perching songbirds** (Figure 27.34). This is the largest order of birds, containing 56 families and 60% of all birds. Most have a highly developed syrinx. Their feet are adapted for perching on thin stems and twigs. The young are altricial. To this order belong many birds with beautiful songs such as thrushes, warblers, mockingbird, meadowlark, and hosts of others. Others of this order, such as swallows, magpie, starling, crows, raven, jays, nuthatches, and creepers, have no songs worthy of the name. About 5750 species, worldwide distribution.

Figure 27.34

Ground finch, *Geospiza fuliginosa,* one of the famous Darwin's finches of the Galápagos Islands. Order Passeriformes.

SUMMARY

The more than 9700 species of living birds are egg-laying, endothermic vertebrates with feathers and forelimbs modified as wings. Birds are closest phylogenetically to theropods, a group of Mesozoic dinosaurs with several birdlike characteristics. The oldest known fossil bird, *Archaeopteryx* from the Jurassic period of the Mesozoic era, had numerous reptilian characteristics and was almost identical to certain theropod dinosaurs except that it had feathers. It is probably the sister taxon of modern birds.

Adaptations of birds for flight are of two basic kinds: those reducing body weight and those promoting more power for flight. Feathers, the hallmark of birds, are complex derivatives of reptilian scales and combine lightness with strength, water repellency, and high insulative value. Body weight is further reduced by elimination of some bones, fusion of others (also providing rigidity for flight), and presence of hollow, air-filled spaces in many bones. The light, keratinized bill, replacing the heavy jaws and teeth of reptiles, serves as both hand and mouth for birds and is variously adapted for different feeding habits.

Adaptations that provide power for flight include high metabolic rate and body temperature coupled with an energy-rich diet; a highly efficient respiratory system consisting of a system of air sacs arranged to provide a constant, one-way flow of air through the lungs; powerful flight and leg muscles arranged to place muscle weight near the bird's center of gravity; and an efficient, high-pressure circulation.

Birds have keen eyesight, good hearing, and superb coordination for flight. The metanephric kidneys produce uric acid as the principal nitrogenous waste.

Birds fly by applying the same aerodynamic principles as an airplane and using similar equipment: wings for lift, support, and propulsion, a tail for steering and landing control, and wing slots for control at low flight speed. Flightlessness in birds is unusual but has evolved independently in several bird orders, usually on islands where terrestrial predators are absent; all are derived from flying ancestors. Arboreal and cursorial hypotheses for the origin of flight have been proposed. The arboreal hypothesis, currently favored by zoologists, proposes that wings and feathers were first used to glide down from trees and later modified for powered flight.

Bird migration refers to regular movements between summer nesting places and wintering regions. Spring migration to the north, where more food is available for nestlings, enhances reproductive success. Many cues are used for finding direction during migration, including innate sense of direction and ability to navigate by the sun, stars, or the earth's magnetic field.

The highly developed social behavior of birds is manifested in vivid courtship displays, mate selection, territorial behavior, and incubation of eggs and care of the young. Most birds have a monogamous mating system, although extra-pair copulations are common. Young hatch at various levels of development; altricial young are naked and helpless, while precocial young are feathered and able to walk and to feed themselves.

REVIEW QUESTIONS

1. Explain the significance of the discovery of *Archaeopteryx*. Why did this fossil demonstrate beyond reasonable doubt that birds are grouped phylogenetically with dinosaurs?
2. The special adaptations of birds contribute to two essentials for flight: more power and less weight. Explain how each of the following contributes to one or both of these two essentials: feathers, skeleton, muscle distribution, digestive system, circulatory system, respiratory system, excretory system, reproductive system.
3. How do marine birds rid themselves of excess salt?
4. In what ways are a bird's ears and eyes specialized for the demands of flight?
5. Explain how a bird's wing is designed to provide lift. What design features help to prevent stalling at low flight speeds?
6. Describe four basic forms of bird wings. How does wing shape correlate with flight speed and maneuverability?
7. Contrast the arboreal and cursorial hypotheses for the origin of flight in birds.
8. What are advantages of seasonal migration for birds?
9. Describe different navigational resources birds may use in long-distance migration.
10. What are some advantages of social aggregation among birds?
11. More than 90% of all bird species are monogamous. Explain why monogamy is much more common among birds than among mammals.
12. Briefly describe an example of polygyny and an example of polyandry among birds.
13. Define the terms precocial and altricial as they relate to birds.
14. Offer some examples of how human activities have affected bird populations.

SELECTED REFERENCES

Ackerman, J. 1998. Dinosaurs take wing. Nat. Geog. **194**(1):74–99. *Beautifully illustrated synopsis of dinosaur to bird evolution.*

Bennet, P. M., and I. E. F. Owens. 2002. Evolutionary ecology of birds: life histories, mating systems, and extinction. Oxford, UK, Oxford University Press. *A phylogenetic approach to understanding how natural and sexual selection have led to the incredible diversity of bird mating systems.*

Brooke, M., and T. Birkhead (eds.). 1991. The Cambridge encyclopedia of ornithology. New York, Cambridge University Press. *Comprehensive, richly illustrated treatment that includes a survey of all modern bird orders.*

Elphick, J. (ed.). 1995. The atlas of bird migration: tracing the great journeys of the world's birds. New York, Random House. *Lavishly illustrated collection of maps of birds' breeding and wintering areas, migration routes, and many facts about each bird's migration journey.*

Emlen, S. T. 1975. The stellar-orientation system of a migratory bird. Sci. Am. **233**:102–111 (Aug.). *Describes fascinating research with indigo buntings, revealing their ability to navigate by the center of celestial rotation at night.*

Feduccia, A. 1996. The origin and evolution of birds. New Haven, Yale University Press. *An updated successor to the author's* The Age of Birds *(1980) but more comprehensive; rich source of information on evolutionary relationships of birds.*

Gill, F. B. 2006. Ornithology, ed. 3. New York, W. B. Freeman and Company: *Popular, comprehensive, and accurate ornithology text.*

Norbert, U. M. 1990. Vertebrate flight. New York, Springer-Verlag. *Detailed review of the mechanics, physiology, morphology, ecology, and evolution of flight. Covers bats as well as birds.*

Padian, K., and L. M. Chiappe. 1998. The origin of birds and their flight. Sci. Am. **279**:38–47 (Feb.). *The authors argue that birds evolved from small, predatory dinosaurs that lived on the ground.*

Peterson, R. T. 2002. A field guide to the birds of eastern and western North America, ed. 5. Boston, Houghton Mifflin. *One of the best field guides for the region.*

Proctor, N. S., and P. J. Lynch. 1998. Manual of ornithology: avian structure and function. New Haven, Connecticut, Yale University Press. *A heavily illustrated ornithology text.*

Sibley, C. G., and J. E. Ahlquist. 1990. Phylogeny and classification of birds: a study in molecular evolution. New Haven, Yale University Press. *A comprehensive application of DNA annealing experiments to the problem of resolving avian phylogeny.*

Terborgh, J. 1992. Why American songbirds are vanishing. Sci. Am. **266**:98–104 (May). *The number of songbirds in North America has been dropping sharply. The author suggests reasons why.*

Terres, J. K. 1980. The Audubon Society encyclopedia of North American birds. New York, Alfred A. Knopf, Inc. *Comprehensive, authoritative, and richly illustrated.*

Waldvogel, J. A. 1990. The bird's eye view. Am. Sci. **78**:342–353 (July–Aug.). *Birds possess visual abilities unmatched by humans. So how can we know what they really see?*

Wellnhofer, P. 1990. *Archaeopteryx.* Sci. Am. **262**:70–77 (May). *Description of perhaps the most important fossil ever discovered.*

ONLINE LEARNING CENTER

Visit www.mhhe.com/hickmanipz14e for chapter quizzing, key term flash cards, web links, and more!

Juvenile grizzly bear, Ursus arctos horribilis.

Mammals

- PHYLUM CHORDATA
- CLASS MAMMALIA

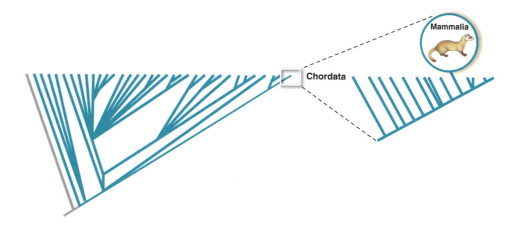

The Tell-Tale Hair

Hair evolved in a common ancestor of all mammals and has been retained to varying degrees in all species descended from it. Hair is therefore diagnostic for mammals; except in some pathological conditions, all mammals have hair at some point in their lives, and hair occurs in no other living organisms. Even those living mammals apparently without hair, such as whales, usually have a few hairs on their bodies. Mammalian hair has undergone numerous adaptive modifications for diverse uses. Mammals use hair for concealment, behavioral signaling, waterproofing and buoyancy; their hair may serve as sensitive vibrissae on their snouts or as prickly quills. Perhaps the most important use of their hair is thermal insulation, helping to maintain a high, constant body temperature in all climates, and thus support a high level of activity.

Mammals are among the most active animals, exhibiting speed and endurance in aquatic, aerial, and terrestrial habitats. They maintain this activity in nearly all environmental conditions, including the cool of night, baking deserts, frigid polar seas, and icy winters. Numerous evolutionary innovations underlie this exuberant adaptive diversification. In addition to hair, mammals uniquely possess a set of middle ear bones for transmitting sounds to the inner ear, mammary glands for nourishing newborns, a large brain with a unique covering of the cerebrum (the neocortex), a diaphragm for efficient ventilation of lungs, and adaptations for a highly developed sense of smell. Most mammals have an intrauterine, vascular placenta for feeding the embryo, specialized teeth and jaw musculature for processing food, and an upright gait for rapid and efficient locomotion.

Mammals, with their highly developed nervous system and numerous adaptations, occupy almost every environment on earth that supports life. Although not a large group (about 4800 species as compared with more than 9700 species of birds, 28,000 species of fishes, and 1,100,000 species of insects), class Mammalia (mam-may'lee-a) (L. *mamma,* breast) is among the most biologically differentiated groups in the animal kingdom. Mammals are exceedingly diverse in size, shape, form, and function. They range in size from the Kitti's hognosed bat in Thailand, weighing only 1.5 g, to blue whales, exceeding 130 metric tons.

Despite their adaptability and in some instances because of it, mammals have been influenced by human activity more than any other group of animals. We have domesticated numerous mammals for food and clothing, as beasts of burden, and as pets. We use millions of mammals each year in biomedical research. We have introduced alien mammals into new habitats, occasionally with benign results but more frequently with unexpected disaster. Although history provides numerous warnings, we continue to overcrop valuable wild stocks of mammals. The whaling industry has threatened itself with total collapse by exterminating its own resource—a classic example of self-destruction in the modern world, in which competing segments of an industry are intent only on reaping all they can today as though tomorrow's supply were of no concern whatever. In some cases destruction of a valuable mammalian resource has been deliberate, such as the officially sanctioned (and tragically successful) policy during the Indian wars of exterminating bison to drive the Plains Indians into starvation. Although commercial hunting has declined, the ever-increasing human population with accompanying destruction of wild habitats has harassed and disfigured mammalian faunas. In 2006, 510 species of mammals were listed as "critically endangered" or "endangered" by the International Union for the Conservation of Nature and Natural Resources (IUCN), including most cetaceans, cats (except domestic cats), otters, and primates (except humans).

We are becoming increasingly aware that our presence on this planet makes us responsible for the character of our natural environment. Since our welfare has been and continues to be closely related to that of the other mammals, it is clearly in our interest to preserve the natural environment of which all mammals, ourselves included, are a part. We need to remember that nature can do without humans but humans cannot exist without nature.

ORIGIN AND EVOLUTION OF MAMMALS

The evolutionary descent of mammals from their earliest amniote ancestors is perhaps the most fully documented transition in vertebrate history. From the fossil record, we can trace the derivation over 150 million years of endothermic, furry mammals from their small, ectothermic, hairless ancestors. Skull structures and especially teeth are the most abundant fossils, and it is largely from these structures that we identify the evolutionary descent of mammals.

The structure of the skull roof permits us to identify three major groups of amniotes that diverged in the Carboniferous period of the Paleozoic era, the **synapsids, anapsids,** and **diapsids.** The synapsid group, which includes the mammals and their ancestors, has a pair of temporal openings in the skull associated with attachment of jaw muscles (Figure 28.1B). Synapsids were the first amniote group to radiate widely into terrestrial habitats. As discussed in Chapter 26, the anapsid group is characterized by solid skulls and includes some of the earliest amniotes (Figure 28.1A). The diapsids have two pairs of temporal openings in the skull (Figure 28.1C; see also Figure 26.2, p. 566) and contain dinosaurs, lizards, snakes, crocodilians, birds, and their ancestors. Turtles have a skull with an anapsid morphology, but phylogenetic analysis places them in the diapsid clade, suggesting their skull morphology evolved independently.

The earliest synapsids radiated extensively into diverse herbivorous and carnivorous forms often collectively called **pelycosaurs** (Figures 28.2 and 28.3). These early synapsids were the most common and largest amniotes in the early Permian. Pelycosaurs share a general outward resemblance to lizards, but this resemblance is misleading. Pelycosaurs are not closely related to lizards, which are diapsids, nor are they a monophyletic group. From one group of early carnivorous synapsids arose the **therapsids** (Figure 28.3), the only synapsid group to survive beyond the Paleozoic. With therapsids we see for the first time an efficient erect gait with upright limbs positioned beneath the body, rather than spawled out to the sides of the body, as in lizards and primitive pelycosaurs. Since stability was reduced by raising the animal from the ground, the muscular coordination center of the brain, the cerebellum, assumed an expanded role. Modifications in the morphology of the therapsid skull and mandibular

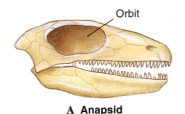

Figure 28.1

Skulls of early amniotes, showing the pattern of temporal openings that distinguish the three groups.

A Anapsid

B Synapsid

C Diapsid

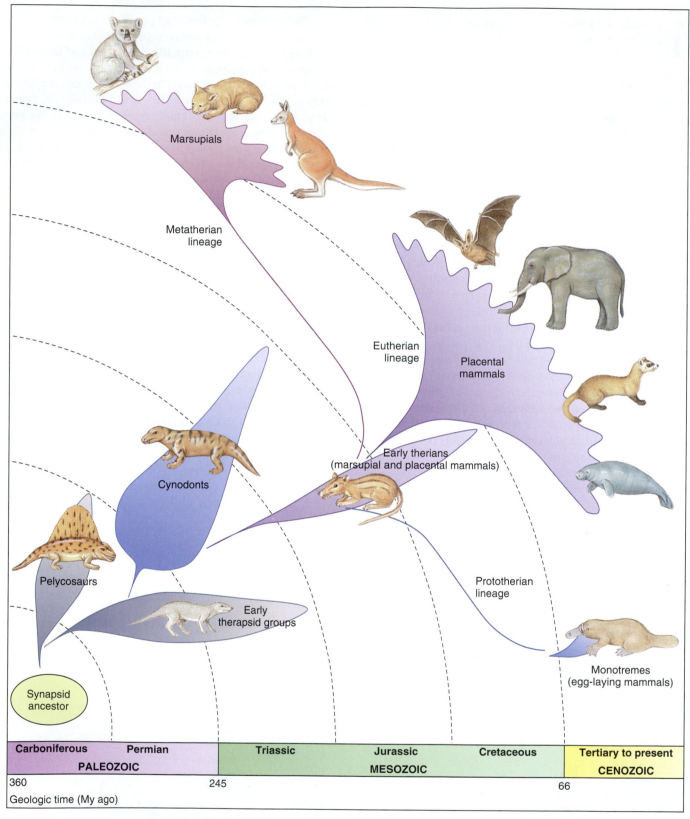

Figure 28.2

Evolution of major groups of synapsids. The synapsid lineage, characterized by lateral temporal openings in the skull, began with pelycosaurs, early mammal-like amniotes of the Permian. Pelycosaurs radiated extensively and evolved changes in jaws, teeth, and body form that presaged several mammalian characteristics. These trends continued in their successors, the therapsids, especially in cynodonts. One lineage of cynodonts gave rise in the Triassic to therians (marsupial and placental mammals). Fossil evidence, as currently interpreted, indicates that all three groups of living mammals—monotremes, marsupials, and placentals—are derived from the same cynodont lineage. The great radiation of modern placental orders occurred during the Cretaceous and Tertiary periods.

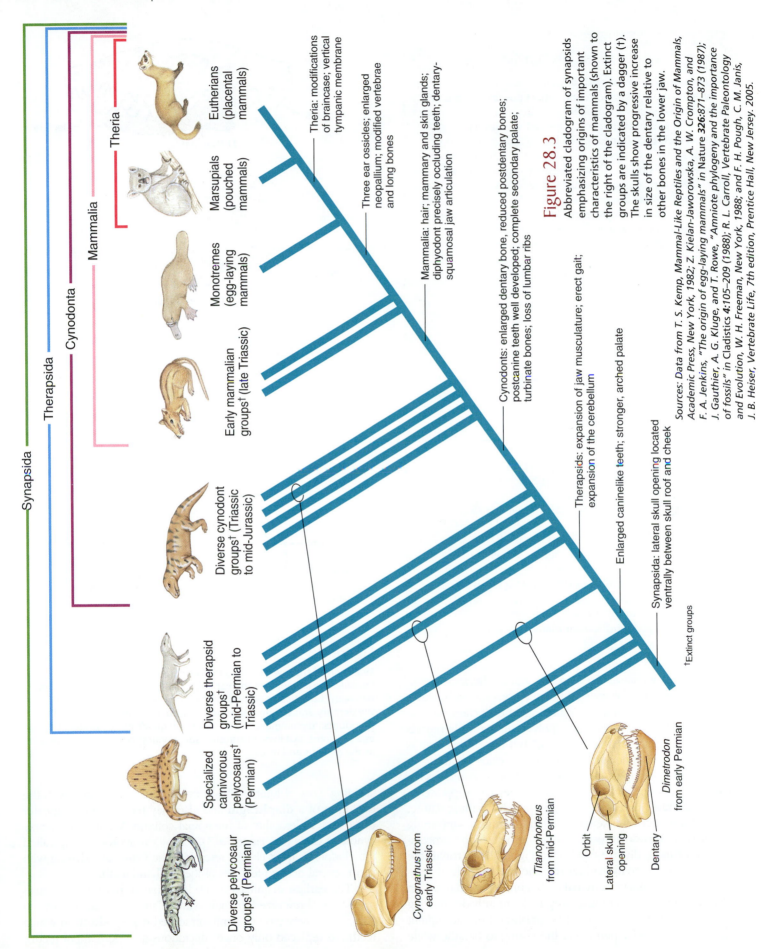

Eutherians (placental mammals)

Marsupials (pouched mammals)

Monotremes (egg-laying mammals)

Early mammalian groups† (late Triassic)

Diverse cynodont groups† (Triassic to mid-Jurassic)

Diverse therapsid groups† (mid-Permian to Triassic)

Specialized carnivorous pelycosaurs† (Permian)

Diverse pelycosaur groups† (Permian)

Theria

Mammalia

Cynodonta

Therapsida

Synapsida

Theria: modifications of braincase; vertical tympanic membrane

Three ear ossicles; enlarged neopallium; modified vertebrae and long bones

Mammalia: hair; mammary and skin glands; diphyodont precisely occluding teeth; dentary-squamosal jaw articulation

Cynodonts: enlarged dentary bone, reduced postdentary bones; postcanine teeth well developed; complete secondary palate; turbinate bones; loss of lumbar ribs

Therapsids: expansion of jaw musculature; erect gait; expansion of the cerebellum

Enlarged caninelike teeth; stronger, arched palate

Synapsida: lateral skull opening located ventrally between skull roof and cheek

†Extinct groups

Cynognathus from early Triassic

Titanophoneus from mid-Permian

Dimetrodon from early Permian

Orbit

Lateral skull opening

Dentary

Figure 28.3

Abbreviated cladogram of synapsids emphasizing origins of important characteristics of mammals (shown to the right of the cladogram). Extinct groups are indicated by a dagger (†). The skulls show progressive increase in size of the dentary relative to other bones in the lower jaw.

Sources: Data from T. S. Kemp, Mammal-Like Reptiles and the Origin of Mammals, Academic Press, New York, 1982; Z. Kielan-Jaworowska, A. W. Crompton, and F. A. Jenkins, "The origin of egg-laying mammals" in Nature 326:871–873 (1987); J. Gauthier, A. G. Kluge, and T. Rowe, "Amniote phylogeny and the importance of fossils" in Cladistics 4:105–209 (1988); R. L. Carroll, Vertebrate Paleontology and Evolution, W. H. Freeman, New York, 1988; and F. H. Pough, C. M. Janis, J. B. Heiser, Vertebrate Life, 7th edition, Prentice Hall, New Jersey, 2005.

Characteristics of Class Mammalia

1. **Body mostly covered with hair,** but reduced in some
2. **Integument** with **sweat, scent, sebaceous,** and **mammary glands,** underlain by a thick layer of fat
3. Skull with **two occipital condyles** and **secondary palate; turbinate bones** in nasal cavity; jaw joint between squamosal and dentary bones (Figure 28.4), middle ear with **three ossicles** (malleus, incus, stapes); **seven cervical vertebrae** (except some xenarthrans [edentates] and manatees); **pelvic bones fused**
4. Mouth with **diphyodont teeth** (milk, or deciduous, teeth replaced by a permanent set); teeth **heterodont** in most (varying in structure and function); lower jaw a **single enlarged bone (dentary)**
5. **Movable eyelids** and **fleshy external ears (pinnae)**
6. Circulatory system of a four-chambered heart (two atria and two ventricles); **persistent left aortic arch,** and **nonnucleated, biconcave red blood corpuscles**
7. Respiratory system of lungs with alveoli, and larynx; **secondary palate** (anterior bony palate and posterior continuation of soft tissue, the soft palate) separates air and food passages (Figure 28.5); **muscular diaphragm** for air exchange separates thoracic and abdominal cavities; convoluted **turbinate bones** in the nasal cavity for warming and moistening inspired air
8. Excretory system of metanephric kidneys with ureters that usually open into a bladder
9. Brain highly developed, especially **cerebral cortex;** 12 pairs of cranial nerves; olfactory sense highly developed
10. Endothermic and homeothermic
11. Cloaca present only in monotremes (present but shallow in marsupials)
12. Separate sexes; reproductive organs of a penis, testes (usually in a scrotum), ovaries, oviducts, and uterus; sex determination by chromosomes (males is heterogametic)
13. Internal fertilization; **embryos develop in a uterus** with **placental attachment** (except in monotremes); **fetal membranes (amnion, chorion, allantois)**
14. Young nourished by **milk from mammary glands**

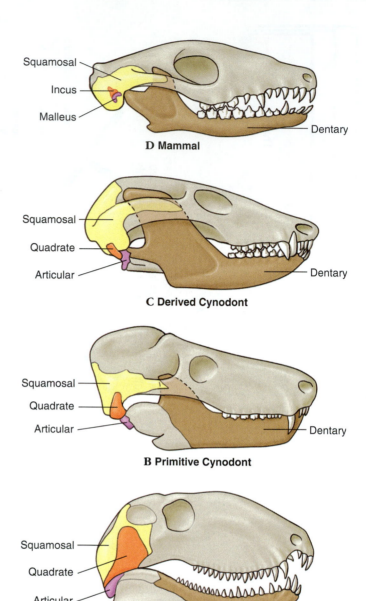

D Mammal

C Derived Cynodont

B Primitive Cynodont

A Pelycosaur

Figure 28.4

Evolution of the jaw joint and middle ear bones in the synapsid lineage. The jaw joint in the earliest synapsids, the pelycosaurs, was between the articular and quadrate bones. A new joint between the dentary and squamosal bones evolved in the most advanced cynodonts. In mammals, the articular and quadrate no longer function as a jaw joint, but instead transmit sound vibrations in the middle ear as the malleus and incus.

adductor muscles increased feeding efficiency. Therapsids radiated into numerous herbivorous and carnivorous forms; however most early forms disappeared during the great extinction event at the end of the Permian. Previously, pelycosaurs and therapsids have been called "mammal-like reptiles," but use of this term is inappropriate, because they are not part of the clade Reptilia.

One therapsid group to survive into the Mesozoic era was the **cynodonts.** Cynodonts evolved several features that supported a high metabolic rate: increased and specialized jaw musculature, permitting a stronger bite; several skeletal changes, supporting greater agility; **heterodont** teeth, permitting better food processing and use of more diverse foods (see Figure 28.4); **turbinate bones,** in the nasal cavity, aiding retention of body heat (Figure 28.5); and a secondary bony palate (Figure 28.5), enabling an animal to breathe while holding prey in its mouth or chewing food. The secondary palate would be important to subsequent mammalian evolution by permitting the young to breathe while

suckling. Loss of lumbar ribs in cynodonts is correlated with the evolution of a **diaphragm** and also may have provided greater dorsoventral flexibility of the spinal column. Within the diverse cynodont clade (see Figure 28.3), a small carnivorous group called trithelodontids most closely resembles the mammals, sharing with them several derived features of the skull and teeth.

The earliest mammals of the late Triassic period were small mouse- or shrew-sized animals with enlarged crania, redesigned jaws, and a new type of dentition, called **diphyodont,** in which teeth are replaced only once (deciduous and permanent teeth).

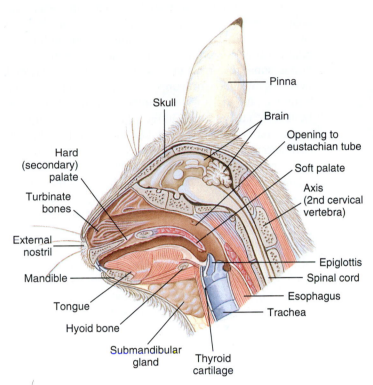

Figure 28.5
Sagittal section of the head of a rabbit.

This contrasts with the primitive amniote pattern of continual tooth replacement throughout life (polyphyodont teeth). One of the more amazing transformations involved the three middle ear bones, the malleus, incus, and stapes, which function to transmit sound vibrations in mammals (p. 745). The stapes, homologous to the columella or hyomandibula of other vertebrates, already functioned in hearing in early synapsids. The malleus and incus originated from the articular and quadrate, respectively, two bones that previously served as the jaw joint but became reduced in size (better to transmit sound vibrations) and relocated to the middle ear (see Figure 28.4). A new jaw joint was formed between the dentary and squamosal (temporal) bones. This dentary-squamosal joint is the defining characteristic for fossil mammals.

The earliest mammals were almost certainly endothermic, although their body temperature would have been rather lower than modern placental mammals. Hair was essential for insulation, and the presence of hair implies that sebaceous and sweat glands must also have evolved at this time to condition hair and to facilitate thermoregulation. The fossil record is silent on the appearance of mammary glands, but they must have evolved before the end of the Triassic. The young of early mammals probably hatched from eggs in a very immature condition, totally dependent on maternal milk, warmth, and protection. This mode of reproduction occurs today only in monotremes (a mammalian subgroup containing echidnas and platypus).

Oddly, early mammals of the mid-Triassic, having developed nearly all novel attributes of modern mammals, had to wait for another 150 million years before they could achieve their great diversity. While dinosaurs became diverse and abundant, all

nonmammalian synapsid groups became extinct. But mammals survived, first as shrewlike, probably nocturnal, creatures. Then, in the Cretaceous period, but especially during the Eocene epoch that began about 58 million years ago, modern mammals began to diversify rapidly. The great Cenozoic radiation of mammals is partly attributed to numerous habitats vacated by extinction of many amniote groups at the end of the Cretaceous. Mammalian radiation was almost certainly promoted by the facts that mammals were agile, endothermic, intelligent, adaptable, and gave birth to living young, which they protected and nourished from their own milk supply, thus dispensing with vulnerable eggs laid in nests.

Recent fossil discoveries and cladistic analyses have shed light on the origin of whales (order Cetacea), and illustrate the importance of using fossil evidence when answering phylogenetic questions. Although the traditional view linked whales with an extinct group of wolflike creatures called mesonychids, molecular analysis of living species placed whales as the sister group to the hippopotamuses, within the order of even-toed hoofed mammals (Artiodactylia). Recent fossil discoveries in Pakistan and elsewhere provide an almost unbroken record of early whale evolution. Particularly important are remains of the ankle bones, which are diagnostic for artiodactyls. The earliest whales have a pulleylike astragulus (an ankle bone), which clearly links whales to artiodactyls. Although cladistic analysis of these recent fossils tentatively suggests that whales are the sister group to all artiodactyls, and not just hippos, the lack of agreement among researchers emphasizes the need for additional fossil material and analyses to refine evolutionary hypotheses.

Class Mammalia includes 26 orders: one order containing monotremes, seven orders of marsupials, and 18 orders of placentals. A complete classification is on pages 637–640.

STRUCTURAL AND FUNCTIONAL ADAPTATIONS OF MAMMALS

Integument and Its Derivatives

Mammalian skin and especially its modifications distinguish mammals as a group. As the interface between an animal and its environment, the skin is strongly molded by an animal's way of life. In general the skin is thicker in mammals than in other classes of vertebrates, although as in all vertebrates it is composed of **epidermis** and **dermis** (Figure 28.6). The epidermis is thinner where it is well protected by hair, but in places that are subject to much contact and use, such as palms or soles, its outer layers become thick and cornified with keratin.

Hair

Hair is especially characteristic of mammals, although humans are not very hairy creatures and, in whales, hair is reduced to only a few sensory bristles on the snout. A hair grows from a hair follicle that, although epidermal in origin, is sunk into the dermis of the skin (Figure 28.6). A hair grows continuously by

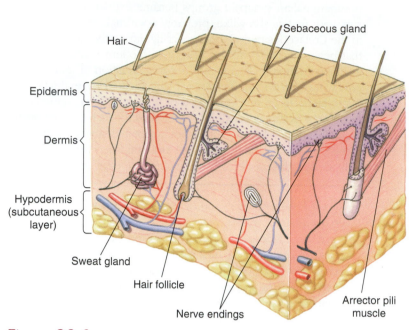

Figure 28.6

Structure of human skin (epidermis and dermis) and hypodermis, showing hair and glands.

rapid proliferation of cells in the follicle. As a hair shaft is pushed upward, new cells are carried away from their source of nourishment and die, filled with the same dense type of fibrous protein, called **keratin,** that constitutes nails, claws, hooves, and feathers. Thus, true hair, found only in mammals, is composed of dead, keratin-packed epidermal cells.

Mammals characteristically have two kinds of hair forming their **pelage** (fur coat): (1) dense and soft **underhair** for insulation and (2) coarse and longer **guard hair** for protection against wear and to provide coloration. Underhair traps a layer of insulating air. In aquatic mammals, such as fur seals, otters, and beavers, it is so dense that it is almost impossible to wet. In water, guard hairs become wet and adhere to each other, forming a protective blanket over the underhair (Figure 28.7).

When a hair reaches a certain length, it stops growing. Normally it remains in its follicle until a new growth starts, whereupon it falls out. In most mammals there are periodic molts of the entire coat. In humans, hair is shed and replaced throughout life (although balding males confirm that replacement is not assured!).

A hair is more than a strand of keratin. It consists of three layers: the medulla or pith in the center of the hair, the cortex with pigment granules next to the medulla, and the outer cuticle composed of imbricated scales. Hair of different mammals shows a considerable range of structure. It may be deficient in cortex, such as the brittle hair of deer, or it may be deficient in medulla, such as the hollow, air-filled hairs of wolverines. Hairs of rabbits and some others are scaled to interlock when pressed together. Curly hair, such as that of sheep, grows from curved follicles.

In the simplest cases, such as foxes and seals, the coat is shed once every summer. Most mammals have two annual molts, one in the spring and one in the fall. Summer coats are always much thinner than winter coats and in some mammals may be a different color. Several northern mustelid carnivores such as weasels have white winter coats and brown summer coats. It was once believed that the white inner pelage of arctic animals conserved body heat by reducing radiation loss; in fact, dark and white pelages radiate heat equally well. Winter white pelage of arctic animals is simply camouflage in a land of snow. The varying hare of North America has three annual molts: the white winter coat is replaced by a brownish gray summer coat, and this is replaced in autumn by a grayer coat, which is soon shed to reveal the winter white coat beneath (Figure 28.8). White fur of arctic mammals in winter (leukemism) is not to be confused with albinism, caused by a recessive gene that blocks pigment (melanin) formation. Albinos have red eyes and pinkish skin, whereas arctic animals in their winter coats have dark eyes and often dark-colored ear tips, noses, and tail tips.

Most mammals have somber colors that help disguise their presence. Often a species is marked with "salt-and-pepper" coloration or a disruptive pattern that helps make it inconspicuous in its natural surroundings. Examples are spots of leopards and fawns and stripes of tigers. Skunks advertise their presence with conspicuous warning coloration.

The hair of mammals has become modified to serve many purposes. Bristles of hogs, spines of porcupines and their kin, and vibrissae on the snouts of most mammals are examples. **Vibrissae,** commonly called "whiskers," are really sensory hairs that provide a tactile sense to many mammals. The slightest movement of a vibrissa generates impulses in sensory nerve

Figure 28.7

American beaver, *Castor canadensis,* gnawing on an aspen tree. This second largest rodent (the South American capybara is larger) has a heavy waterproof pelage consisting of long, tough guard hairs overlying the thick, silky underhair so valued in the fur trade. Order Rodentia, family Castoridae.

Figure 28.8

Snowshoe, or varying, hare, *Lepus americanus* in **A,** brown summer coat and, in **B,** white winter coat. In winter, extra hair growth on the hind feet broadens the animal's support in snow. Snowshoe hares are common residents of the taiga and are an important prey for lynxes, foxes, and other carnivores. Order Lagomorpha, family Leporidae.

A

B

Figure 28.9

Dogs are frequent victims of a porcupine's impressive quills. Unless removed (usually by a veterinarian) quills will continue to work their way deeper in the flesh causing great distress and may lead to the victim's death.

next set of antlers. For several years each new pair of antlers is larger and more elaborate than the previous set. Annual growth of antlers strains mineral metabolism, since during the growing season an older moose or elk must accumulate 50 or more pounds of calcium salts from its vegetable diet.

Horns of the pronghorn antelope (family Antilocapridae) are similar to true horns of bovids except that the keratinized portion is forked and shed annually. Giraffe horns are similar to antlers but retain their integumentary covering and are not shed.

endings that travel to special sensory areas in the brain. Vibrissae are especially long in nocturnal and burrowing animals.

Porcupines, hedgehogs, echidnas, and a few other mammals have developed an effective and dangerous spiny armor. When cornered, the common North American porcupine turns its back toward its attacker and lashes out with its barbed tail. The lightly attached quills break off at their bases when they enter the skin and, aided by backward-pointed hooks on their tips, work deeply into tissues. Dogs are frequent victims (Figure 28.9) but fishers, wolverines, and bobcats are able to flip a porcupine onto its back to expose vulnerable underparts.

Horns and Antlers

Several kinds of horns or hornlike structures are found in mammals. **True horns,** found in members of family Bovidae (for example, sheep and cattle), are hollow sheaths of keratinized epidermis that embrace a core of bone arising from the skull (see Figure 29.3). True horns are not shed, are not branched (although they may be greatly curved), grow continuously, and occur in both sexes.

Antlers of the deer family Cervidae are branched and composed of solid bone when mature. During their annual spring growth, antlers develop beneath a covering of highly vascular soft skin called **velvet** (Figure 28.10). Except for caribou (see Figure 28.17A), only males of the species produce antlers. When growth of antlers is complete just before the fall breeding season, blood vessels constrict and a stag removes the velvet by rubbing its antlers against trees. Antlers are shed after the breeding season. New buds appear a few months later to herald the

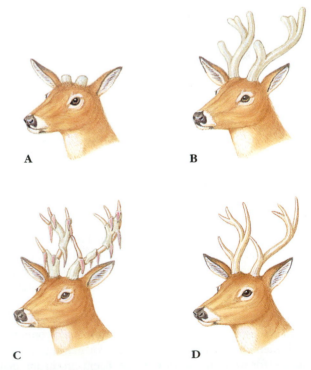

A   **B**

C   **D**

Figure 28.10

Annual growth of buck deer antlers. **A,** Antlers begin growth in late spring, stimulated by pituitary gonadotropins. **B,** Bone grows very rapidly until halted by a rapid rise in testosterone production by the testes. **C,** The skin (velvet) dies and sloughs off. **D,** Testosterone levels peak during the fall breeding season. Antlers are shed in January as testosterone levels subside.

Rhinoceros horn consists of hairlike keratinized filaments that arise from dermal papillae cemented together, but they are not attached to the skull.

Trade in rhino products, especially the horn, has pushed Asian and African rhinos to the brink of extinction. Rhino horn is valued in China as an agent for reducing fever, and for treating heart, liver, and skin disease, and in North India as an aphrodisiac. Such supposed medicinal values are totally without pharmacological basis. The principal use of rhino horns, however, is to fashion handles for ceremonial daggers called jambiyas in the Middle East. Between 1970 and 1997, horns from 22,350 rhinos were imported into northern Yemen alone. An international ban prohibiting the trade of rhino horn has reduced, but not eliminated, the now illegal trade, and populations continue to suffer.

Glands

Of all vertebrates, mammals have the greatest variety of integumentary glands. Most fall into one of four classes: sweat, scent, sebaceous, and mammary. All are derivatives of epidermis (see Figure 28.6).

Sweat glands are tubular, highly coiled glands that occur over much of the body surface in most mammals (see Figure 28.6). They are not present in other vertebrates. There are two kinds of sweat glands: eccrine and apocrine. **Eccrine glands** secrete a watery fluid that, if evaporated on the skin's surface, draws heat away from the skin and cools it. Eccrine glands occur in hairless regions, especially foot pads, in most mammals, although in horses and most primates they are scattered over the body. They are either reduced or absent in rodents, rabbits, and whales. **Apocrine glands** are larger than eccrine glands and have longer and more convoluted ducts. Their secretory coil is in the dermis and extends deep into the hypodermis. They always open into a hair follicle or where a hair once was. Apocrine glands develop near puberty and are restricted (in humans) to the axillae (armpits), mons pubis, breasts, prepuce, scrotum, and external auditory canals. In contrast to watery secretions of eccrine glands, apocrine secretions are milky fluids, whitish or yellow in color, that dry on the skin to form a film. Apocrine glands are not involved in heat regulation. Their activity is correlated with reproductive function and signal such things as cycle stage and receptivity in the female, and degree of maturity and relatedness of individuals within a species.

Scent glands occur in nearly all mammals. Their location and functions vary greatly. They are used for communication with members of the same species, for marking territorial boundaries, for warning, or for defense. Scent-producing glands are located in orbital, metatarsal, and interdigital regions (deer); behind the eyes and on the cheek (pica and woodchuck); penis (muskrats, beavers, and many canines); base of the tail (wolves and foxes); back of the head (dromedary); and anal region (skunks, minks, and weasels). The latter, the most odoriferous of all glands, open by ducts into the anus; their secretions can be discharged forcefully for 2 to 3 meters. During mating season many mammals give off strong scents for attracting the opposite sex. Humans also are endowed with scent glands. However we tend to dislike our own scent, a concern that has stimulated a lucrative deodorant industry to produce an endless output of soaps and odor-masking concoctions.

Sebaceous glands (see Figure 28.6) are intimately associated with hair follicles, although some are free and open directly onto the surface. The cellular lining of the gland is discharged in the secretory process and must be renewed for further secretion. These gland cells become distended with a fatty accumulation, then die, and are expelled as a greasy mixture called **sebum** into the hair follicle. Called a "polite fat" because it does not turn rancid, it serves as a dressing to keep skin and hair pliable and glossy. Most mammals have sebaceous glands over their entire body; in humans they are most numerous in the scalp and on the face.

Mammary glands, for which mammals are named, occur on all female mammals and in a rudimentary form on all male mammals. They develop by thickening the epidermis to form a milk line along each side of the abdomen in the embryo. On certain parts of these lines the mammae appear while the intervening parts of the ridge disappear. Mammary glands increase in size at maturity, becoming considerably larger during pregnancy and subsequent nursing of young. In human females, adipose tissue begins to accumulate around the mammary glands at puberty to form the breast. In most mammals milk is secreted from mammary glands via nipples or teats, but monotremes lack nipples and simply secrete milk into a depression on the mother's belly where it is lapped by the young.

Food and Feeding

Mammals exploit an enormous variety of food sources; some mammals require highly specialized diets, whereas others are opportunistic feeders that thrive on diversified diets. Food habits and physical structure are thus inextricably linked. A mammal's adaptations for attack and defense and its specializations for finding, capturing, chewing, swallowing, and digesting food all determine a mammal's anatomy and habits.

Teeth, perhaps more than any other single physical characteristic, reveal the life habit of a mammal (Figure 28.11). With certain exceptions (monotremes, anteaters, certain whales), all mammals have teeth, and their modifications are correlated with what the mammal eats.

As mammals evolved during the Mesozoic, major changes occurred in teeth and jaws. Unlike the uniform **homodont** dentition of the first synapsids, mammalian teeth became differentiated to perform specialized functions such as cutting, seizing, gnawing, tearing, grinding, and chewing. Teeth differentiated in this manner are called **heterodont.** Mammalian dentition is differentiated into four types: **incisors,** with simple crowns and sharp edges, used mainly for snipping or biting; **canines,** with long conical crowns, specialized for piercing; **premolars** and **molars** with compressed crowns and one or more cusps, suited for shearing, slicing, crushing, or grinding. The ancestral tooth formula for most mammals, which expresses the number of each tooth type in one-half of the upper and lower jaw, was I 3/3, C 1/1, PM 4/4, M 3/3 = 44. Members of order Insectivora (shrews), some omnivores, and carnivores come closest to this ancestral pattern (Figure 28.11).

Unlike most other vertebrates, mammals do not continuously replace their teeth throughout their lives. Most mammals grow just two sets of teeth: a temporary set, called **deciduous,** or **milk,** teeth, which is replaced by a permanent set when the skull has grown large enough to accommodate a full set. Only incisors, canines, and premolars are deciduous; molars are never replaced and the single permanent set must last a lifetime.

Feeding Specializations

The feeding, or trophic, apparatus of a mammal—teeth and jaws, tongue, and alimentary canal—are adapted to its particular feeding habits. Mammals are customarily divided among four basic trophic categories—insectivores, carnivores, omnivores, and herbivores—but many other feeding specializations have evolved in mammals, as in other living organisms, and the feeding habits of many mammals defy exact classification. The principal feeding specializations of mammals are shown in Figure 28.11.

Insectivorous mammals, such as shrews, moles, anteaters, and most bats, are small. They feed on insects, as well as a variety of other small invertebrates. Since insectivores eat little fibrous vegetable matter that requires prolonged fermentation, their intestinal tract tends to be short (Figure 28.12). Insectivorous mammals have teeth with pointed cusps, permitting them to puncture the exoskeleton or skin of prey. Because many omnivores and carnivores also consume insects, the insectivorous diet is distinctive mainly by its lack of plant material and vertebrates.

Herbivorous mammals that feed on grasses and other vegetation form two main groups: (1) **browsers** and **grazers,** such as ungulates (hooved mammals including horses,

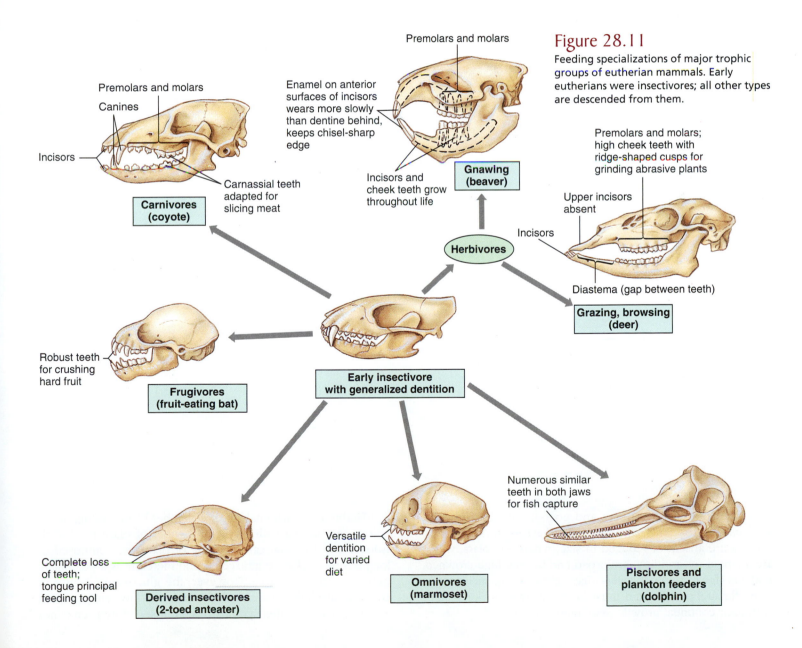

Figure 28.11
Feeding specializations of major trophic groups of eutherian mammals. Early eutherians were insectivores; all other types are descended from them.

Premolars and molars

Enamel on anterior surfaces of incisors wears more slowly than dentine behind, keeps chisel-sharp edge

Premolars and molars

Canines

Incisors

Carnassial teeth adapted for slicing meat

Carnivores (coyote)

Incisors and cheek teeth grow throughout life

Gnawing (beaver)

Premolars and molars; high cheek teeth with ridge-shaped cusps for grinding abrasive plants

Upper incisors absent

Incisors

Herbivores

Diastema (gap between teeth)

Grazing, browsing (deer)

Robust teeth for crushing hard fruit

Frugivores (fruit-eating bat)

Early insectivore with generalized dentition

Complete loss of teeth; tongue principal feeding tool

Derived insectivores (2-toed anteater)

Versatile dentition for varied diet

Omnivores (marmoset)

Numerous similar teeth in both jaws for fish capture

Piscivores and plankton feeders (dolphin)

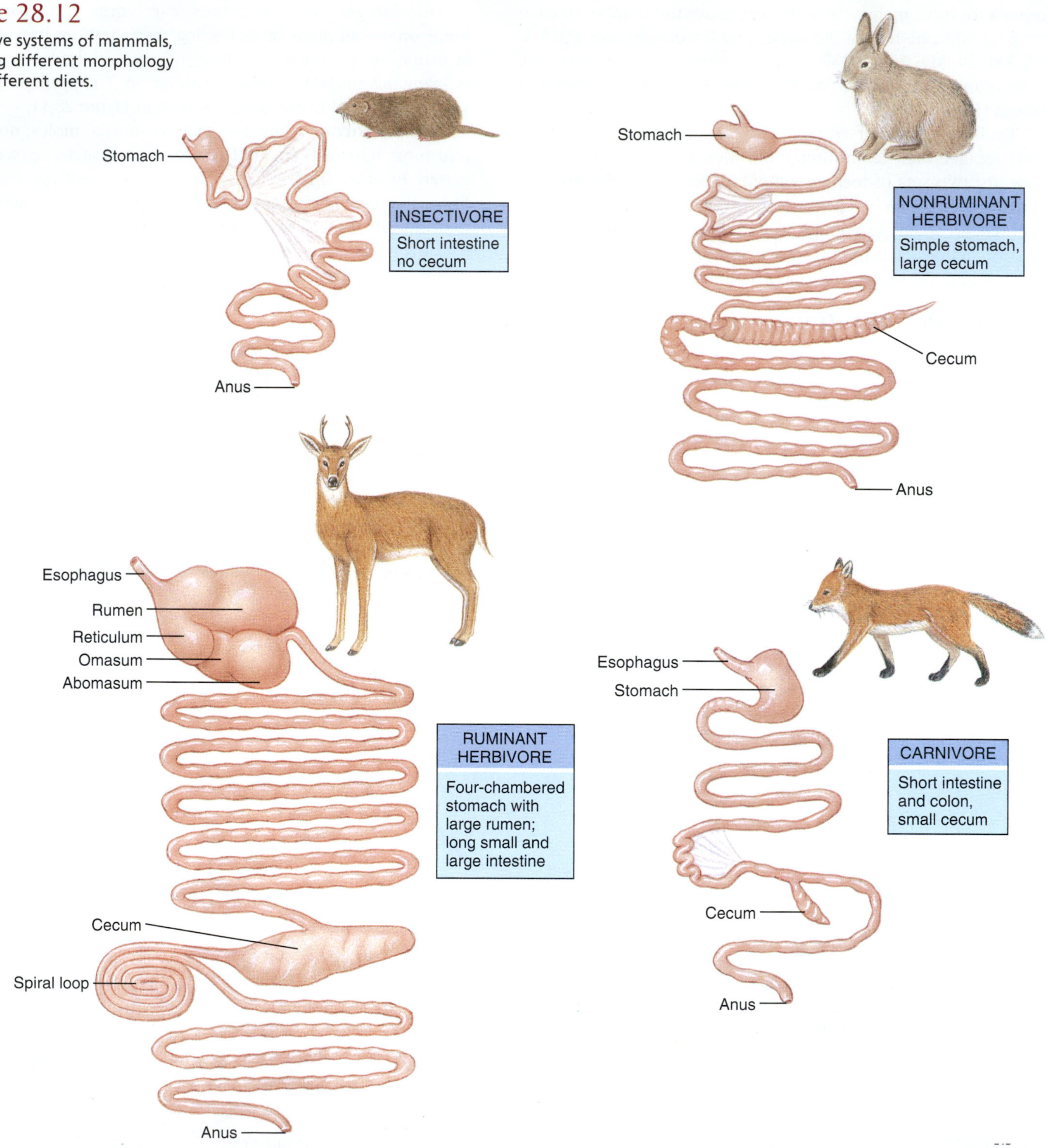

Figure 28.12

Digestive systems of mammals, showing different morphology with different diets.

INSECTIVORE
Short intestine no cecum

- Stomach
- Anus

NONRUMINANT HERBIVORE
Simple stomach, large cecum

- Stomach
- Cecum
- Anus

RUMINANT HERBIVORE
Four-chambered stomach with large rumen; long small and large intestine

- Esophagus
- Rumen
- Reticulum
- Omasum
- Abomasum
- Cecum
- Spiral loop
- Anus

CARNIVORE
Short intestine and colon, small cecum

- Esophagus
- Stomach
- Cecum
- Anus

deer, antelope, cattle, sheep, and goats); and (2) **gnawers,** such as many rodents, and rabbits and hares. In herbivores, canines are absent or reduced in size, whereas molars, which are adapted for grinding, are broad and usually high-crowned. Rodents (for example, beavers) have chisel-sharp incisors that grow throughout life and must be worn away to keep pace with their continual growth (see Figure 28.11).

Herbivorous mammals have a number of interesting adaptations for dealing with their fibrous diet of plant food. **Cellulose,** the structural carbohydrate of plants, is composed of long chains of glucose molecules, and therefore is a potentially nutritious food resource. However, the glucose molecules in cellulose are linked by chemical bonds that few enzymes can attack. No vertebrates synthesize cellulose-splitting enzymes (cellulases). Instead, herbivorous vertebrates harbor anaerobic bacteria and protozoa in fermentation chambers in their gut. Simple carbohydrates, proteins, and lipids produced by the microorganisms can be absorbed by the host, and the host can digest the microorganisms themselves.

Fermentation in some herbivores, such as horses, zebras, rabbits, elephants, some primates, and many rodents, occurs primarily in the colon and in a spacious side pocket, or diverticulum, called a **cecum** (Figure 28.12). Although some absorption happens in the colon and cecum, most fermentation occurs after the primary absorptive area (the small intestine), and many nutrients are lost in the feces. Rabbits and many rodents often eat their fecal pellets **(coprophagy),** giving the food a second pass through the gut to extract additional nutrients.

Ruminants (cattle, bison, buffalo, goats, antelopes, sheep, deer, giraffes, and okapis) have a huge **four-chambered stomach** (Figure 28.12). As a ruminant feeds, grass passes down the esophagus to the **rumen,** where it is digested by microorganisms and then formed into small balls of cud. At its leisure, the ruminant returns a cud to its mouth where the cud is deliberately chewed at length to crush the fiber. Swallowed again, food returns to the rumen where cellulolytic bacteria and protozoa continue fermentation. The pulp passes to the **reticulum,** then to the **omasum,** where water, soluble food, and microbial products are absorbed. The remainder proceeds to the **abomasum** ("true" acid stomach) and small intestine, where proteolytic enzymes are secreted and normal digestion occurs. Perhaps because ruminants are particularly good at extracting nutrients from forage, they are the primary large herbivores in ecosystems with a shortage of food, such as tundras and deserts.

Herbivores generally have large, long digestive tracts and must eat a considerable amount of plant food to survive. An African elephant weighing 6 tons must consume 135 to 150 kg (300 to 400 pounds) of rough fodder each day to obtain sufficient nourishment for life.

Carnivorous mammals feed mainly on herbivores. This group includes foxes, dogs, weasels, wolverines, fishers, and cats. Carnivores are well-equipped with biting and piercing teeth and powerful clawed limbs for killing their prey. Since their protein diet is more easily digested than the fibrous food of herbivores, their digestive tract is shorter and the cecum small or absent (Figure 28.12). Carnivores organize their feeding into discrete meals rather than feeding continuously (as do most herbivores) and therefore have much more leisure time.

Note that the terms "insectivores" and "carnivores" have two different uses in mammals: to describe diet and to denote specific taxonomic orders of mammals. For example, not all carnivores belong to order Carnivora (many marsupials and cetaceans are carnivorous) and not all members of order Carnivora are carnivorous. Order Carnivora contains many opportunistic feeders and some, such as pandas, are strict vegetarians.

In general, carnivores lead more active—and by human standards more interesting—lives than do herbivores. Since a carnivore

must find and catch its prey, there is a premium on intelligence; many carnivores, such as cats, are noted for their stealth and cunning in hunting prey (Figure 28.13). This has led to a selection of herbivores capable either of defending themselves or of detecting and escaping carnivores. Thus for herbivores, there has been a premium on keen senses, speed, and agility. Some herbivores, however, survive by virtue of their sheer size (rhinos, elephants) or by defensive group behavior (muskoxen).

Humans have changed the rules in the carnivore-herbivore contest. Carnivores, despite their intelligence, have suffered much from human presence and have been virtually exterminated in some areas. Small herbivores, on the other hand, with their potent reproductive ability, have consistently defeated our most ingenious efforts to banish them from our environment. The problem of rodent pests in agriculture has intensified (see Figure 28.30, and p. 632); we have removed carnivores, which served as the herbivores' natural population control, but have not been able to devise a suitable substitute.

Omnivorous mammals—pigs, raccoons, many rodents, bears, and most primates, including humans—use both plants and animals for food. Many carnivorous forms also eat fruits, berries, and grasses when hard pressed. Foxes, which usually feed on small rodents and birds, eat frozen apples, beechnuts, and corn when their normal food sources are scarce. Other mammals often considered herbivores, such as some rodents, have a mixed diet of insects, seeds, and fruit.

Many mammals cache food stores during periods of plenty. This habit is most pronounced in rodents, such as squirrels, chipmunks, gophers, and certain mice. Tree squirrels collect nuts, conifer seeds, and fungi and store these in caches for winter use. Often each item is hidden in a different place (scatter hoarding) and marked by a scent to assist relocation in the future. Some caches of chipmunks and red squirrels can be quite large (Figure 28.14).

Body Weight and Food Consumption

The relationship between body size and metabolic rate was discussed in relation to food consumption of birds (p. 595). The

Figure 28.13

Lionesses, *Panthera leo*, eating a wildebeest. Lacking stamina for a long chase, lions stalk prey and then charge suddenly, surprising their prey. Lions gorge themselves with their kill, then sleep and rest for periods as long as one week before eating again. Order Carnivora, family Felidae.

Figure 28.14

Eastern chipmunk, *Tamias striatus,* with cheek pouches stuffed with seeds to be carried to a hidden cache. It will try to store several liters of food for the winter. It hibernates but awakens periodically to eat some of its cached food. Order Rodentia, family Sciuridae.

smaller the mammal, the greater is its metabolic rate and the more it must consume relative to its body size (Figure 28.15). This happens because the metabolic rate of a mammal—and therefore the amount of food it must eat to sustain this metabolic rate—varies in rough proportion to the relative surface area rather than to body weight. Surface area is proportional to approximately 0.7 power of body weight, and the amount of food a mammal (or bird) eats also is roughly proportional to a 0.7 power of its body weight. For example, a 3 g mouse will consume *per gram body weight* five times more food than does a 10 kg dog and about 30 times more food than does a 5000 kg

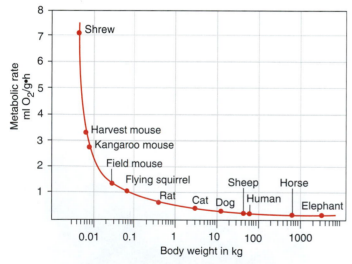

Figure 28.15

Relationship between body weight and metabolic rate for mammals. This relationship, often called the "mouse-to-elephant" curve, shows that metabolic rate is intense for small mammals like shrews and mice, and declines with increasing body weight of the species.

Source: From Eckert Animal Physiology: Mechanisms and Adaptations, *4/e by D. Randall, W. Burggren, K. French. © 1978, 1983, 1988, and 1997 by W. H. Freeman and Company. Used with permission.*

elephant. Thus small mammals (shrews, bats, and mice) must spend much more time hunting and eating food than do large mammals. The smallest shrews weighing only 2 g may eat more than their body weight each day and will starve to death in a few hours if deprived of food (Figure 28.16). In contrast, large carnivores can remain fat and healthy with only one meal every few days. Mountain lions are known to kill an average of one deer a week, although they will kill more frequently when game is abundant.

Migration

Migration is a more difficult undertaking for mammals than for birds or fishes, because terrestrial locomotion is more energetically expensive than swimming or flying. Not surprisingly, few terrestrial mammals make regular seasonal migrations, preferring instead to center their activities in a defined and limited home range. Nevertheless, there are some striking terrestrial mammalian migrations. More migrators are found in North America than on any other continent.

An example is barren-ground caribou of Canada and Alaska, which undertake direct and purposeful mass migrations spanning 160 to 1100 km (100 to 700 miles) twice annually (Figure 28.17). From winter ranges in boreal forests (taiga), they migrate rapidly in late winter and spring to calving ranges on the barren grounds (tundra). Calves are born in mid-June. As summer progresses, caribou are increasingly harassed by warble and nostril flies that bore into their flesh, by mosquitos that drink their blood (estimated at a liter per caribou each week during the height of the mosquito season), and by wolves that prey on their calves. They move southward in July and August, feeding little along the way. In September they reach the taiga and feed there almost continuously on low ground vegetation. Mating (rut) occurs in October.

Plains bison, before their deliberate near extinction by humans, made huge circular migrations to separate summer and winter ranges.

Figure 28.16

Shorttail shrew, *Blarina brevicauda,* eating a grasshopper. This tiny but fierce mammal, with a prodigious appetite for insects, mice, snails, and worms, spends most of its time underground and so is seldom seen. Shrews are believed to resemble insectivorous ancestors of placental mammals. Order Insectivora, family Soricidae.

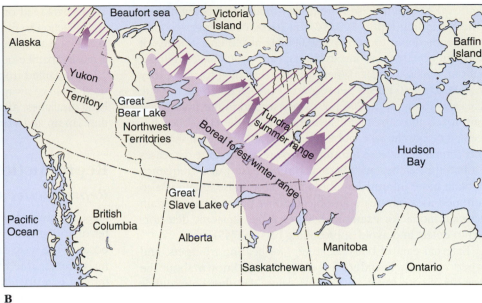

A   B

Figure 28.17

Barren-ground caribou, *Rangifer tarandus,* of Canada and Alaska. **A,** Adult male caribou in autumn pelage and antlers in velvet. **B,** Summer and winter ranges of some major caribou herds in Canada and Alaska (other herds not shown occur on Baffin Island and in western and central Alaska). Principal spring migration routes are indicated by arrows; routes vary considerably from year to year. The same species is known as reindeer in Europe. Order Artiodactyla, family Cervidae.

Caribou have suffered a drastic decline in numbers since early times when their population reached several million. By 1958 less than 200,000 remained in Canada. The decline has been attributed to several factors, including habitat alteration from exploration and development in the North, but especially to excessive hunting. For example the western Arctic herd in Alaska exceeded 250,000 caribou in 1970. Following five years of heavy unregulated hunting, a 1976 census revealed only about 75,000 animals remaining. After restricting hunting, the herd had increased to 340,000 by 1988 and 490,000 in 2006. However, this recovery is threatened by proposed expansion of oil extraction in several Alaskan wildlife refuges, including the Arctic National Wildlife Refuge. The Porcupine caribou herd, which calves on the Arctic National Wildlife Refuge, numbers about 123,000, down by about one-third from 1989.

The longest mammalian migrations are made by the oceanic seals and whales. Gray whales, for example, migrate between Alaska in summer and Baja California, Mexico, in winter, an annual migration of over 18,000 km (11,250 miles). One of the most remarkable migrations is that of northern fur seals, which breed on the Pribilof Islands approximately 300 km (185 miles) off the coast of Alaska and north of the Aleutian Islands. From wintering grounds off southern California females journey as much as 2800 km (1740 miles) across open ocean, arriving in the spring at the Pribilofs where they congregate in enormous numbers (Figure 28.18). Young are born within a few hours or days after arrival of the cows. Then the bulls, having already arrived and established territories, collect groups of cows, which they guard with vigilance. After calves have been nursed for approximately three months, cows and juveniles leave for their long migration southward. Bulls do not follow but remain in the Gulf of Alaska during the winter.

Although we might expect bats, the only winged mammals, to use their gift of flight to migrate, few of them do. Most spend winters in hibernation. Four species of American bats that migrate spend their summers in northern or western states and their winters in the southern United States or Mexico.

Flight and Echolocation

Many mammals scamper about in trees with amazing agility; some can glide from tree to tree (Figure 28.19) and one group, the bats, has full flight. Gliding or flying evolved independently in several groups of mammals, including marsupials, rodents, flying lemurs, and bats. Flying squirrels (Figure 28.19) actually glide rather than fly, using the gliding skin (patagium) that extends from the sides of the body.

Bats are mostly nocturnal or crepuscular (active at twilight) and thus hold a niche unoccupied by most birds. Their achievement is attributed to two features: flight and capacity to navigate by echolocation. Together these adaptations enable bats to fly and to avoid obstacles in absolute darkness, to locate and to catch insects with precision, and to find their way deep into caves (a habitat largely unexploited by other mammals and birds) where they sleep during the daytime hours.

Research has been concentrated on members of the family Vespertilionidae, to which most common North American bats belong. When in flight, bats emit short pulses 5 to 10 msec in duration in a narrow directed beam from the mouth or nose

(Figure 28.20). Each pulse is frequency modulated, being highest at the beginning, up to 100,000 Hz (hertz, cycles per second), and sweeps down to perhaps 30,000 Hz at the end. Sounds of this frequency are ultrasonic to human ears, which have an upper limit of about 20,000 Hz. When bats are searching for prey, they produce about 10 pulses per second. If prey is detected, the rate increases rapidly up to 200 pulses per second in the final phase of approach and capture. Pulses are spaced so the echo of each is received before the next pulse is emitted, an adaptation that prevents jamming. Because transmission-to-reception time decreases as a bat approaches an object, it can increase pulse frequency to obtain more information about an object. Pulse length is also shortened as a bat nears an object. Some prey of bats, certain nocturnal moths for example, have evolved ultrasonic detectors used to detect and avoid approaching bats (p. 745).

External ears of bats are large, like hearing trumpets, and shaped variously in different species. Less is known about the inner ear of bats, but it obviously receives the ultrasonic sounds emitted. A bat builds up a mental image of its surroundings from echo scanning that approaches the resolution of a visual image from eyes of diurnal animals.

Some bats, including the approximately 170 species of bats in suborder Megachiroptera, lack echolocation abilities. They are primarily nocturnal, although several species are diurnal.

These bats feed on fruit, flowers, and nectar, using large eyes and olfaction to find their food. Flowers of plants that are pollinated by bats open at night, are white or light in color, and emit a musky, batlike odor that nectar-feeding bats find attractive.

The famed tropical vampire bats have razor-sharp incisors used to shave away the epidermis of their prey, exposing underlying capillaries. After infusing an anticoagulant to aid blood flow, they lap up and store their meal in a specially modified stomach.

Reproduction

Reproductive Cycles

Most mammals have definite mating seasons, usually in winter or spring and timed so that birth and the rearing of young occur at the most favorable time of the year. Many male mammals are capable of fertile copulation at any time, but female fertility is restricted to a specific time during a periodic cycle, or **estrous cycle.** Females copulate with males only during a relatively brief period in this cycle known as heat or **estrus.** (Figure 28.21).

How often females are in estrus varies greatly among different mammals. Animals that have only a single estrus during their breeding season are called **monestrous;** those having a

---

Many insectivores (for example, shrews and tenrecs) use echolocation, but it is crudely developed as compared with bats. Toothed whales, however, have a highly developed capacity to locate objects by echolocation. Totally blind sperm whales in perfect health have been captured with food in their stomachs. Although the mechanism of sound production and reception remains imperfectly understood, it is thought that clicks are produced in the nasal passages as air is moved through nasal valves and associated air sacs while the blowhole is closed. The clicks are focused into a narrow beam by the melon, a lens-shaped, fatty body in the

forehead. Because of certain physical properties of water, it is necessary for toothed whales to emit very high frequency pulses—as high as 220,000 Hz. Returning echos are primarily received through the lower jaw, and channeled through oil-filled sinuses in the dentary (lower jawbone) to the inner ear. The inner ear is surrounded by a capsule of bone enclosing a fatty mixture that blocks sounds except for those transmitted through the lower jaw. Toothed whales can apparently determine size, shape, speed, direction, and density of objects in the water and know the position of every whale in their pod.

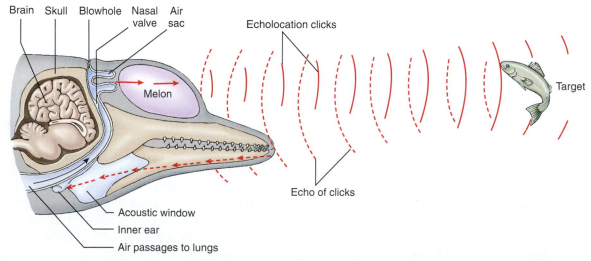

In dolphins, clicks are produced by moving air through the nasal canal and are focused into beams through the melon. Returning sounds primarily are received through the acoustic window, a posterior part of the mandibular with very thin bone, and channeled through oil in the mandible to the middle and inner ears.

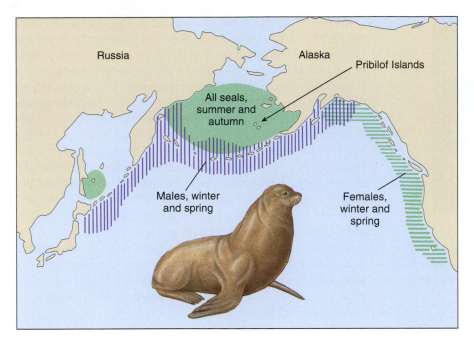

Figure 28.18
Annual migrations of northern fur seals, *Callorhinus ursinus,* showing the separate wintering grounds of males and females. Both males and females of the larger Pribilof population migrate in early summer to the Pribilof Islands, where females give birth to their pups and then mate with males. Order Carnivora, family Otariidae.

Figure 28.19
Northern flying squirrel, *Glaucomys sabrinus,* gliding in for a landing. Area of undersurface is nearly trebled when gliding skin is spread. Glides of 40 to 50 m are possible. Good maneuverability during flight is achieved by adjusting position of the gliding skin with special muscles. Flying squirrels are nocturnal and have superb night vision. Order Rodentia, family Sciuridae.

recurrence of estrus during their breeding season are called **polyestrous.** Dogs, foxes, and bats belong to the first group; field mice and squirrels are all polyestrous as are many mammals living in the more tropical regions of the earth. Old World monkeys and humans have a somewhat different cycle in which the postovulation period is terminated by **menstruation,** during which the, endometrium (lining of the uterus) collapses and is discharged with some blood through the vagina. This **menstrual cycle** is described in Chapter 7 (pp. 150–152).

A curious phenomenon that lengthens the gestation period of many mammals is delayed implantation. The blastocyst remains dormant while its implantation in the uterine wall is postponed for periods of a few weeks to several months. For many mammals (for example, bears, seals, weasels, badgers, bats, and many deer) delayed implantation is a device for extending gestation so the young are born at the time of year best for their survival.

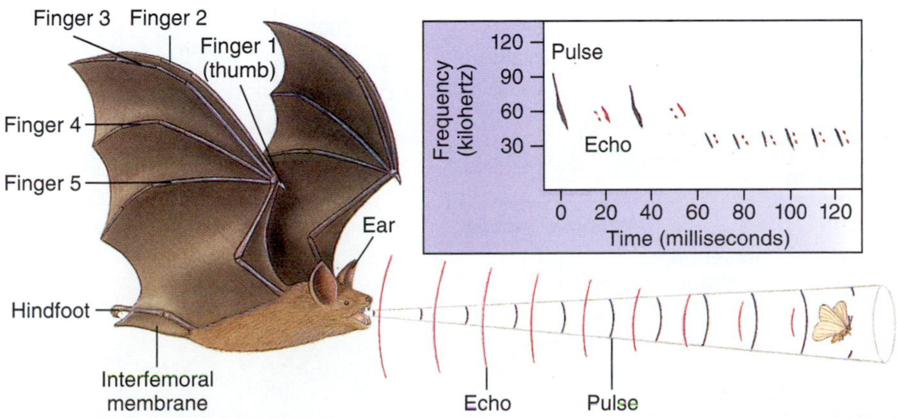

Reproductive Patterns

There are three different patterns of reproduction in mammals. One pattern is represented by egg-laying (oviparous) mammals, the **monotremes.** The duck-billed platypus has one breeding season each year. Ovulated eggs, usually two, are fertilized in the oviduct. Embryos continue to develop in the uterus for 10 to 12 days, where they are nourished by yolk supplies deposited prior to ovulation, and secretions from the mother. A thin, leathery shell is secreted around the embryos prior to the eggs being laid. The platypus lays its eggs in a burrow where they hatch at a relatively undeveloped state after about 12 days. Echidnas incubate their eggs in an abdominal pouch. After hatching, young feed on milk produced by the mother's mammary glands. Because monotremes have no nipples, young lap milk secreted onto the belly fur of the mother.

Figure 28.20
Echolocation of an insect by a little brown bat, *Myotis lucifugus.* Frequency modulated pulses are directed in a narrow beam from the bat's mouth. As the bat nears its prey, it emits shorter, lower signals at a faster rate. Order Chiroptera, family Vespertilionidae.

Figure 28.21

African lions, *Panthera leo,* mating. Lions breed at any season, although predominantly in spring and summer. During the short period a female is receptive, she may mate repeatedly. Three or four cubs are born after gestation of 100 days. Once the mother introduces the cubs into the pride, they are treated with affection by both adult males and females. Cubs go through an 18- to 24-month apprenticeship learning how to hunt and then are frequently driven from the pride to manage themselves. Order Carnivora, family Felidae.

Marsupials are pouched, viviparous mammals that exhibit a second pattern of reproduction. Although only eutherians are called "placental mammals," marsupials do have a primitive type of placenta, called a **choriovitelline** (or yolk sac) **placenta.** An embryo (blastocyst) of a marsupial is at first encapsulated by shell membranes and floats free for several days in the uterine fluid. After "hatching" from the shell membranes, embryos of most marsupials do not implant, or "take root" in the uterus as they would in eutherians, but erode shallow depressions in the uterine wall in which they lie and absorb nutrient secretions from the mucosa by way of a vascularized yolk sac. Gestation (the intrauterine period of development) is brief in marsupials, and therefore all marsupials give birth to tiny young that are effectively still embryos, both anatomically and physiologically. However, early birth is followed by a prolonged interval of lactation and parental care (Figure 28.22).

Although it is tempting to view the ephemeral choriovitelline placenta of marsupials as transitional between absence of a placenta in monotremes and the persistent **chorioallantoic placenta** of placental mammals, cladistic analysis challenges this hypothesis. All marsupials and placental mammals have a choriovitelline placenta, and a chorioallantoic placenta occurs in some marsupials. Perhaps the chorioallantoic placenta was present in the common ancestor of marsupials and placental mammals, but later was lost in most marsupials.

In red kangaroos (Figure 28.23) the first pregnancy of the season begins with a 33-day gestation, after which the young (joey) is born, crawls to the pouch without assistance from its mother, and attaches to a nipple. The mother immediately becomes pregnant again, but the presence of a suckling young in the pouch arrests development of the new embryo in the uterus at about the 100-cell stage. This period of arrest, called **embryonic diapause,** lasts approximately 235 days during which time the first joey is growing in the pouch. When the joey leaves the pouch, the uterine embryo resumes development and is born about a month later. The mother again becomes pregnant, but because the second joey is suckling, once again development of the new embryo is arrested. Meanwhile, the first joey returns to the pouch occasionally to suckle. At this point the mother has three young of different ages dependent upon her for nourishment: a joey on foot, a joey in the pouch, and a diapause embryo in the uterus. There are variations on this remarkable sequence—not all marsupials have developmental delays like kangaroos, and some do not even have pouches—but in all, the young are born at an extremely early stage of development and undergo prolonged development while dependent on a teat (Figure 28.24).

The third pattern of reproduction is that of viviparous **placental mammals,** the eutherians. In placentals, reproductive investment is in prolonged gestation, unlike marsupials in which reproductive investment is in prolonged lactation (Figure 28.22). Embryos remain in the uterus, nourished by food supplied initially by a choriovitelline placenta and later by a chorioallantoic type of placenta (described on pp. 177–178), an intimate connection between mother and young. Length of gestation is longer in placentals than marsupials, and in large mammals it is much longer (Figure 28.22). For example, mice have a gestation period of 21 days; rabbits and hares, 30 to 36 days; cats and dogs, 60 days; cattle, 280 days; and elephants, 22 months (the longest). Important

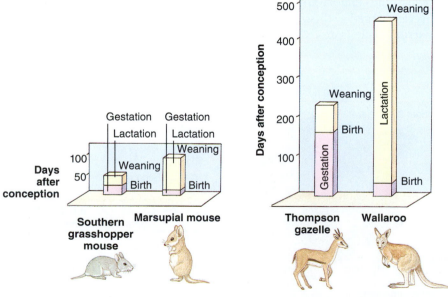

Figure 28.22

Comparison of gestation and lactation periods between matched pairs of ecologically similar species of marsupial and placental mammals. The graph shows marsupials have shorter intervals of gestation and much longer intervals of lactation than in similar species of placentals.

Figure 28.23

Kangaroos have a complicated reproductive pattern in which the mother may have three young in different stages of development dependent on her at once. Order Diprotodontia, family Macropodidae.

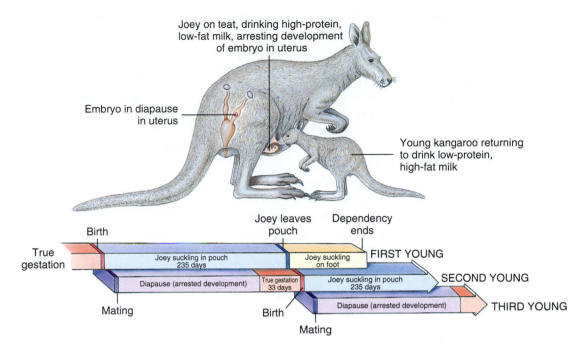

exceptions include baleen whales, the largest mammals, whose gestation period is only 12 months, and small bats no larger than mice, whose gestation period extends 4 to 5 months. The condition of the young at birth also varies. An antelope bears its young well furred, eyes open, and able to run. Newborn mice, however, are blind, naked, and helpless. We all know how long it takes a human baby to gain its footing. Human growth is in fact slower than that of any other mammal, and this is one of the distinctive attributes that sets us apart from other mammals.

Is the placental mode of reproduction superior to that of marsupials? The traditional view holds this to be true, citing the low species diversity and small geographic range of marsupials, and the success of introduced placentals in Australia at the expense of some marsupials. Clearly, placentals have the advantage of higher reproductive rate, and retaining young in a pouch is not possible for fully aquatic forms. However, marsupials may have some advantages of their own. Because marsupials invest less energy in newborns, more energy would be available to a marsupial for replacement of

lost young. This mode of reproduction may be advantageous in highly unpredictable climates, such as those of Australia. Regardless of these arguments, marsupials have succeeded alongside placentals in South and Central America, where they have experienced a modest radiation into about 80 species, and who could doubt the tenacity of the North American opossum?

The renowned fecundity of small rodents, and the effect of removing their natural predators, is felicitously expressed in this excerpt from Thornton Burgess's **Portrait of a Meadow Mouse.**

He's fecund to the nth degree
In fact this really seems to be
His one and only honest claim
To anything approaching fame.
In just twelve months, should all survive,
A million mice would be alive—
His progeny. And this, 'tis clear,
Is quite a record for a year.
Quite unsuspected, night and day
They eat the grass that would be hay.
On any meadow, in a year,
The loss is several tons, I fear,
Yet man, with prejudice for guide,
The checks that nature doth provide
Destroys. The meadow mouse survives
And on stupidity he thrives.

Figure 28.24

Southern opossums, *Didelphis marsupialis,* 15 days old, fastened to teats in mother's pouch. When born after a gestation period of only 12 days, they are the size of honey bees. They remain attached to the nipples for 50 to 60 days. Order Didelphimorpha, family Didelphidae.

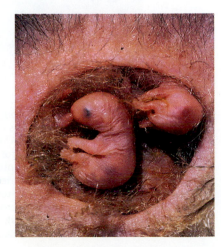

The number of young produced by mammals in a season depends on mortality rate, which, for some mammals such as mice, may be high at all age levels. Usually, the larger the animal, the smaller the number of young in a litter. Small rodents, which serve as prey for many carnivores, usually produce more than one litter of several young each season. Meadow voles, *Microtus pennsylvanicus,* are known to produce as many as 17 litters of four to nine young in a year. Most carnivores have but one litter of three to five young per year. Large mammals, such as elephants and horses, give birth to a single young with each pregnancy. An elephant produces, on average, four calves during her reproductive life of perhaps 50 years.

Territory and Home Range

Many mammals have territories—areas from which individuals of the same species are excluded. In fact, many wild mammals, like some humans are basically unfriendly to their own kind, especially so to their own sex during the breeding season. If a mammal dwells in a burrow or den, this area forms the center of its territory. If it has no fixed address, the territory is marked out, usually with the highly developed scent glands (p. 620). Territories vary greatly in size depending on the size of the animal and its feeding habits. Grizzly bears have territories of several square kilometers, which they guard zealously against all other grizzlies.

Mammals usually use natural features of their surroundings in staking their claims. These are marked with secretions from the scent glands or by urinating or defecating. When an intruder knowingly enters another's marked territory, it is immediately placed at a psychological disadvantage. Should a challenge follow, the intruder almost invariably breaks off the encounter in a submissive display characteristic for the species. Territoriality and aggressive and submissive displays are described in more detail in Chapter 36; pages 793–796.

A beaver colony is a family unit, and beavers are among several mammalian species in which males and females form strong monogamous bonds that last a lifetime. Because beavers invest considerable time and energy in constructing a lodge and dam and storing food for winter (Figure 28.25), the family, especially the adult male, vigorously defends its real estate against intruding beavers. Most of the work of building dams and lodges is undertaken by male beavers, but females assist when not occupied with their young.

An interesting exception to the strong territorial nature of most mammals is the prairie dog, which lives in large, friendly communities called prairie dog "towns" (Figure 28.26). When a

new litter has been reared, adults relinquish the old home to the young and move to the edge of the community to establish a new home. Such a practice is totally antithetical to the behavior of most mammals, which drive off the young when they are self-sufficient.

A **home range** of a mammal is a much larger foraging area surrounding a defended territory. Home ranges are not defended in the same way as are territories; home ranges may, in fact, overlap, producing a neutral zone used by owners of several territories for seeking food.

Mammalian Populations

A population of animals includes all members of a species that share a particular space and potentially interbreed (see Chapter 38). All mammals (like other organisms) live in ecological communities, each composed of numerous populations of different animal and plant species. Each species is affected by the activities of other species and by other changes, especially climatic, that occur. Thus populations are always changing in size. Populations of small mammals are lowest before the breeding season and greatest just after addition of new members. Beyond these expected changes in population size, mammalian populations may fluctuate from other causes.

Figure 28.26
Immature black-tailed prairie dogs, *Cynomys ludovicianus,* greeting adult. These highly social prairie dwellers are herbivores that serve as an important prey for many animals. They live in elaborate tunnel systems so closely interwoven that they form "towns" of as many as 1000 individuals. Towns are subdivided into family units, each with one or two males, several females, and their litters. Although prairie dogs display ownership of burrows with territorial calls, they are friendly with inhabitants of adjacent burrows. The name "prairie dog" derives from the sharp, doglike bark they make when danger threatens. Order Rodentia, family Sciuridae.

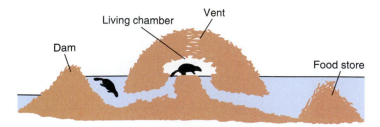

Figure 28.25
Each beaver colony constructs its own lodge in a pond created by damming a stream. Each year the mother bears four or five young; when the third litter arrives, 2-year-olds are driven out of the colony to establish new colonies elsewhere. Order Rodentia, family Castoridae.

Irregular fluctuations are commonly produced by variations in climate, such as unusually cold, hot, or dry weather, or by natural catastrophes, such as fires, hailstorms, and hurricanes. These are **density-independent** factors because they affect a population whether it is crowded or dispersed. However, the most spectacular fluctuations are **density dependent;** that is, they correlate with population crowding. These extrinsic limits to growth are discussed in Chapter 38; page 830.

In his book *The Arctic* (1974. Montreal, Infacor, Ltd.), Canadian naturalist Fred Bruemmer describes the growth of lemming populations in arctic Canada:

"After a population crash one sees few signs of lemmings; there may be only one to every 10 acres. The next year, they are evidently numerous; their runways snake beneath the tundra vegetation, and frequent piles of rice-sized droppings indicate the lemmings fare well. The third year one sees them everywhere. The fourth year, usually the peak year of their cycle, the populations explode. Now more than 150 lemmings may inhabit each acre of land and they honeycomb it with as many as 4000 burrows. Males meet frequently and fight instantly. Males pursue females and mate after a brief but ardent courtship. Everywhere one hears the squeak and chitter of the excited, irritable, crowded animals. At such times they may spill over the land in manic migrations."

Cycles of abundance are common among many rodent species. One of the best-known examples is the mass migrations of Scandinavian and arctic North American lemmings following population peaks. Lemmings (Figure 28.27) breed all year, although more in summer than in winter. The gestation period is only 21 days; young born at the beginning of summer are weaned in 14 days and are capable of reproducing by end of summer. At the peak of their population density, having devastated the vegetation by tunneling and grazing, lemmings begin long, mass migrations to find new undamaged habitats for food and space. They swim across streams and small lakes as they go but cannot distinguish these from large lakes, rivers, and the sea, in which they drown. Since lemmings are the main diet of many carnivorous mammals and birds, any change in lemming population density affects all their predators as well.

Varying hares (snowshoe rabbits, Figure 28.8) of North America show 10-year cycles in abundance. The well-known

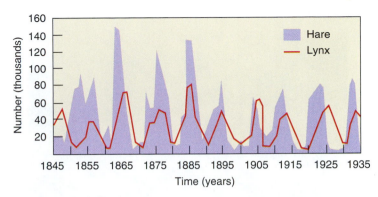

Figure 28.28

Changes in population size of varying hare and lynx in Canada as indicated by pelts received by the Hudson's Bay Company over a 90-year period. Abundance of lynx (predator) follows that of the hare (prey).

fecundity of rabbits enables them to produce litters of three or four young as many as five times per year. Their density may increase to 4000 hares competing for food in each square mile of northern forest. Predators (owls, minks, foxes, and especially lynxes) also increase (Figure 28.28). Then the population crashes precipitously for reasons that have long been a puzzle to scientists. Rabbits die in great numbers, not from lack of food or from an epidemic disease (as was once believed) but evidently from some density-dependent psychogenic cause. As crowding increases, hares become more aggressive, show signs of fear and defense, and stop breeding. The entire population reveals symptoms of pituitary-adrenal gland exhaustion, an endocrine imbalance called "shock disease," which results in death. These dramatic crashes are not well understood. Whatever the causes, population crashes that follow superabundance, although harsh, permit vegetation to recover, providing survivors with a much better chance for successful breeding.

HUMANS AND MAMMALS

Some 10,000 years ago, at the time people developed agricultural methods, they also began domestication of mammals. Dogs were certainly among the first to be domesticated, probably entering voluntarily into their human dependence. Dogs are an extremely adaptable and genetically plastic species derived from wolves. Much less genetically variable and certainly less social than dogs are domestic cats, probably derived from an African race of wildcat. Wildcats look like oversized domestic cats and are still widespread in Africa and Eurasia. Domestication of cattle, buffaloes, sheep, and pigs probably came much later. Beasts of burden—horses, camels, oxen, and llamas—probably were subdued by early nomadic peoples. Certain domestic species no longer exist as wild animals, for example, one-humped dromedary camels of North Africa and llamas and alpacas of South America. All truly domestic animals breed in captivity; many have been molded by selective breeding to yield characteristics desirable for human purposes.

Some mammals hold special positions as "domestic" animals. Elephants have never been truly domesticated because they will

Figure 28.27

Collared lemming, *Dicrostonyx* sp., a small rodent of the far north. Populations of lemmings fluctuate widely. Order Rodentia, family Cricetidae.

Figure 28.29

Herd of reindeer, *Rangifer tarandus,* during annual roundup by Laplanders in northern Sweden. The same species is known as caribou in North America. Order Artiodactyla, family Cervidae.

seldom breed in captivity. In Asia, adults are captured and submit to a life of toil with astonishing docility. Reindeer of northern Scandinavia are domesticated only in the sense that they are "owned" by nomadic peoples who follow them in their seasonal migrations (Figure 28.29). Elands of Africa are undergoing experimental domestication in several places. They are placid, gentle, and immune to native diseases and produce excellent meat.

Activities of mammals can in some instances conflict with human activities. Rodents and rabbits are capable of inflicting staggering damage to growing crops and stored food (Figure 28.30). We have provided an inviting forage for rodents with our agriculture and convenienced them further by removing their natural predators. Rodents also carry various diseases. Bubonic plague and typhus are carried by various rodents, including brown rats and prairie dogs. Tularemia (rabbit fever), is transmitted to humans by wood ticks carried by rabbits, woodchucks, muskrats, and other rodents. Rocky Mountain spotted fever is carried to humans by

Figure 28.30

Brown rat, *Rattus norvegicus,* living all too successfully beside human habitations. Brown rats not only cause great damage to food stores but also spread disease, including bubonic plague (a disease, carried by infected fleas, that greatly influenced human history in medieval Europe), typhus, infectious jaundice, *Salmonella* food poisoning, and rabies. Order Rodentia, family Muridae.

ticks from ground squirrels and dogs; Lyme disease is transmitted by ticks from white-tailed deer. Trichina worms and tapeworms are acquired by humans who eat the meat of infected hogs, cattle, and other mammals.

In the introduction to this chapter, we alluded to the discouraging exploitation of whales as one example of our inability to reconcile human needs with preservation of wildlife. Extermination of a species for commercial gain is so totally indefensible that no debate is required. Once a species is extinct, no amount of scientific or technical ingenuity will bring it back. What has taken millions of years to evolve can be destroyed in a decade of thoughtless exploitation. Many people are concerned with the awesome impact we have on wildlife, and there is more determination today to reverse a regrettable trend than ever before. If given a chance, mammals will usually make spectacular recoveries from human depredations, as have sea otters and saiga antelopes, both once in danger of extinction and now numerous.

HUMAN EVOLUTION

Darwin devoted an entire book, *The Descent of Man and Selection in Relation to Sex* (1871), largely to human evolution. The idea that humans shared common descent with apes and other animals was repugnant to the Victorian world, which responded with predictable outrage (see Figure 6.15, p. 115). Because at that time virtually no fossil evidence linking humans with apes existed, Darwin built his case mostly on anatomical comparisons between humans and apes. To Darwin, the close resemblances between apes and humans could be explained only by common descent.

The search for fossils, especially for a "missing link" that would provide a connection between apes and humans, began when two skeletons of Neandertals[1] were collected in the 1880s. Then in 1891, Eugene Dubois discovered the famous Java man (*Homo erectus*). The most spectacular discoveries, however, have been made in Africa, especially between 1967 and 1977, which American paleoanthropologist Donald C. Johanson calls the "golden decade." During this same period, comparative biochemical studies demonstrated that humans and chimpanzees are as similar genetically as are many sibling species. Comparative cytology provided evidence that chromosomes of humans and apes are homologous (p. 114). We are no longer searching for a "missing link" to establish common descent of humans and apes, our closest living relatives.

Evolutionary Radiation of Primates

Humans are primates, a fact that even pre-evolutionist Linnaeus recognized. All primates share certain significant characteristics: grasping fingers on all four limbs, flat fingernails instead of claws, and forward-pointing eyes with binocular vision and excellent depth perception. The accompanying

---

[1]The traditional spelling "Neanderthal" has been replaced by "Neandertal" in many recent publications. In old German, *thal* (valley) is pronounced with a hard *t,* rather than with a soft *th.* Modern spellings of many German words were changed to reflect their pronunciation, resulting in *thal* becoming *tal.*

synopsis will highlight probable relationships of major primate groups.

The earliest primate was probably a small, nocturnal animal similar in appearance to tree shrews. This ancestral primate lineage split into two lineages, one of which gave rise to lemurs and lorises, with wet noses (Strepsirhini); and the other to tarsiers (Figure 28.31), monkeys (Figure 28.32), and apes (Figure 28.33). Traditionally the lemurs, lorises, and tarsiers have been called the **prosimians,** a group that is now known to be paraphyletic, and the apes and monkeys have been called **simians** or **anthropoids.** Prosimians and many simians are arboreal (tree-dwellers), which is probably the ancestral lifestyle for both groups. Flexible limbs are essential for active animals moving through trees. Grasping hands and feet, in contrast to clawed feet of squirrels and other rodents, enable primates to grip limbs, to hang from branches, to seize and to manipulate food, and, most significantly, to use tools. Primates have highly developed sense organs, especially acute, binocular vision, and proper coordination of limb and finger muscles to assist their active arboreal life. Of course, sense organs are no better than the brain processing sensory information. Precise timing, judgment of distance, and alertness require a large cerebral cortex.

The earliest simian fossils appeared in Africa in late Eocene deposits, some 40 million years ago. Many of these primates became diurnal rather than nocturnal, making vision the dominant special sense, now enhanced by color perception. We recognize three major simian clades. These are (1) New World monkeys of Central and South America (ceboids; Figure 28.32A), including howler monkeys, spider monkeys, and tamarins, (2) Old World monkeys (cercopithecoids), including baboons (Figure 28.32B), mandrills, and colobus monkeys, and

Figure 28.32

Monkeys. **A,** Red howler monkeys, *Alovatta seniculus,* order Primates, family Cebidae, an example of New World monkeys. **B,** Olive baboons, *Papio homadryas,* order Primates, family Cercopithecidae, an example of Old World monkeys.

(3) apes (Figure 28.33). Old World monkeys and apes (including humans) are sister taxa, and together form the sister group of New World monkeys. In addition to their geographic separation, Old World monkeys differ from New World monkeys in lacking a grasping tail, while having close-set nostrils, better opposable, grasping thumbs, and more derived teeth. Apes differ from Old World monkeys in having a larger cerebrum, a more dorsally placed scapula, and loss of the tail. Humans, orangutans, gorillas, and chimpanzees are now recognized to belong to a single family, Hominidae, and are referred to here

Figure 28.31

A prosimian, the Mindanao tarsier, *Tarsius syrichta carbonarius,* of Mindanao Island in the Philippines. Order Primates, family Tarsiidae.

Figure 28.33

Gorillas, *Gorilla gorilla,* order Primates, family Hominidae, examples of apes.

as **hominids.** All fossil hominid species that are phylogenetically placed closer to living humans than to chimpanzees are referred to here as **humans.**

Apes first appear in 20-million-year-old fossils. At this time woodland savannas were arising in Africa, Europe, and North America. Perhaps motivated by greater abundance of food on the ground, these apes left the trees and became largely terrestrial.

Early Humans

The gradual replacement of forests with grasslands in eastern Africa during the late Pliocene provided an impetus for apes to adapt to an open environment, the savannas. Because of the benefits of standing upright (better view of predators, freeing of hands for using tools, defense, caring for young, and gathering food) emerging hominids gradually evolved upright posture. This important transition was an enormous leap because it required extensive redesigning of the skeleton and muscle attachments.

Evidence of the earliest humans of this period is extremely sparse. Yet in 2001 the desert sands of Chad yielded one of the most astonishing and important discoveries of modern paleontology, a remarkably complete skull of a human dated at nearly 7 million years ago. Named *Sahelanthropus tchadensis* ("Sahel hominid of Chad") by its discoverer, French paleontologist Michel Brunet, this creature is by far the most ancient human yet discovered (Figure 28.34). Although its brain is no larger than that of a chimp (between 320 and 380 cm^3), its relatively small canine teeth, massive brow ridges, and features of the face and basicranium suggest that the skull is that of a human. Until this skull was discovered, the earliest fossil placed closer to humans than chimpanzees was *Ardipithecus ramidus* from the sands of Ethiopia, originally dated at 4.4 million years (Figure 28.34). *Ardipithecus ramidus* is a mosaic of apelike and derived humanlike traits, with indirect (and controversial) evidence that it may have been bipedal. Between 1997 and 2001 additional fossils of *Ardipithecus* were discovered that extend its existence back to about 5.8 million years ago. *Sahelanthropus* and *Ardipithecus* were probably forest-dwellers, based on forest associated fossils found in the same deposits.

The most celebrated early human fossil is a 40% complete skeleton of a female *Australopithecus afarensis* (Figure 28.35). Unearthed in 1974 and named "Lucy" by its discoverer Donald Johanson, *A. afarensis* was a short, bipedal human with face and brain size slightly larger than that of a chimpanzee. This species was sexually dimorphic in size; females stood about 1 m (Figure 28.36) and males stood about 1.5 m. Their teeth suggest that they were primarily fruit-eaters, but likely incorporated some meat into their diet. The numerous fossils of *A. afarensis* date from 3.7 to 3.0 million years ago.

In 1995 *Australopithecus anamensis* was discovered in the Rift Valley of Kenya. Many researchers believe that this species, which lived between 4.2 and 3.9 million years ago, is an intermediate between *Ardipithecus* and *A. afarensis* (Lucy). The extremely humanlike lower leg bones of *Australopithecus anamensis* provide strong evidence that this species was bipedal.

In the last two decades there has been an explosion of australopithecine fossil finds, with eight putative species requiring interpretation. Some of these finds are known as gracile

australopithecines because of their relatively light build, especially in skull and teeth (although all were more robust than modern humans). The gracile australopithecines are generally considered more closely related to early *Homo* species and, by extension, of the lineage leading to modern humans. Following *A. anamensis* and Lucy (*Australopithecus afarensis*; 3.7 to 3 million years ago) there appeared the bipedal *Australopithecus africanus*, which lived between 3 and 2.3 million years ago. This species was similar to *A. afarensis*, but had a more humanlike face, slightly larger body, and a brain size about one-third as large as that of modern humans. In 1998 a partial skull, dated at 2.5 million years ago, was discovered in Ethiopia. Named *Australopithecus garhi*, this species has been suggested to be the missing link between *Australopithecus* and *Homo*, although its phylogenetic position is controversial.

Coexisting with the earliest species of *Homo* was a different lineage of large and robust australopithecines that existed between 2.5 and 1.2 million years ago. One of these was *Paranthropus robustus* (see Figure 28.34), which probably approached the size of a gorilla. The "robust" australopithecines were heavy jawed with skull crests and large back molars, used for chewing coarse roots and tubers. They are an extinct branch in hominid evolution and not part of our own lineage.

Emergence of *Homo*

Although researchers are divided over who the first members of *Homo* were, and indeed how to define the genus *Homo*, the earliest *known* species of the genus was *Homo habilis*, a fully erect hominid. *Homo habilis* ("handy man") is thought to have been about 127 cm (5 feet) tall and to have weighed about 45 kg (100 pounds), although females were probably smaller. This poorly known species was larger brained than the australopithecines and its brain was more humanlike in shape; one brain cast shows a bulge representing Broca's motor speech area, suggesting that *H. habilis* was capable at least of rudimentary speech. *Homo habilis* was the first species unquestionably to use simple stone tools. Many researchers consider the oldest *H. habilis* remains, from 2.4 to 1.8 million years ago, to represent a different species, *Homo rudolfensis*.

About 1.8 million years ago *Homo erectus* appeared, a large hominid standing 150 to 170 cm (5 to 5.5 feet) tall, with a low but distinct forehead and strong brow ridges. Its brain capacity was around 1000 cm^3, intermediate between the brain capacity of *H. habilis* and modern humans (see Figure 28.34). *Homo erectus* had a successful and complex culture and became the first hominid to migrate out of Africa, spreading throughout Asia. Their brain shape, revealed by casts of skulls, suggests that limited speech was possible, and charcoal deposits indicated that they controlled and made use of fire. The earliest *H. erectus* fossils often are called *H. ergaster*. Yet another amazing hominid find was announced in 2004: *Homo floresiensis*, a species only 1 m tall, from the island of Flores, Indonesia. This species arose from *H. erectus* and became extinct only about 13,000 years ago.

Homo sapiens: **Modern Humans**

Recent molecular genetic studies indicate that human populations have formed a single evolutionary lineage for the past

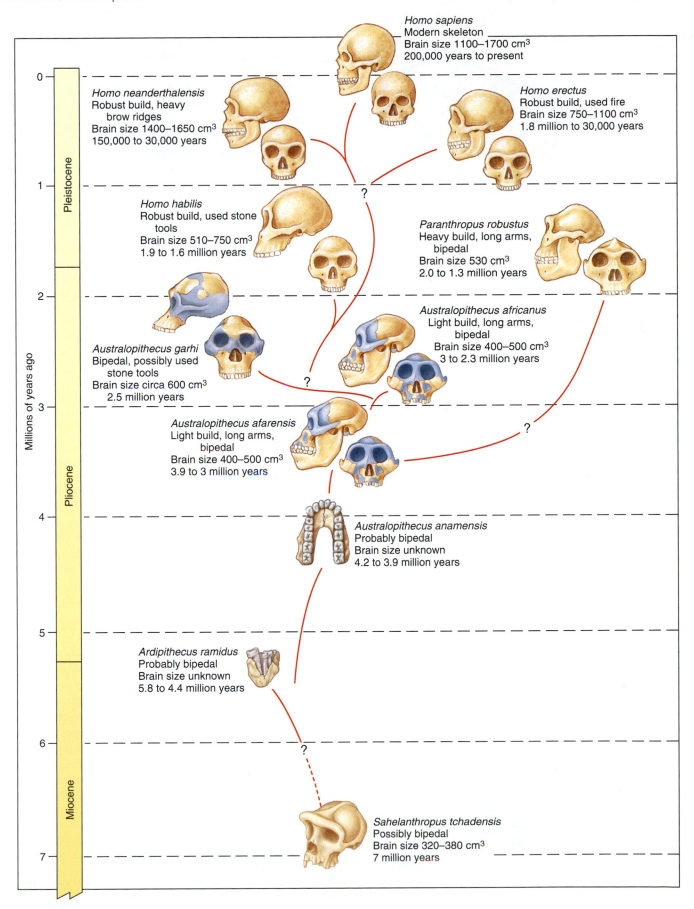

Figure 28.34

Hominid skulls, showing several of the best-known lines preceding modern humans (*Homo sapiens*).

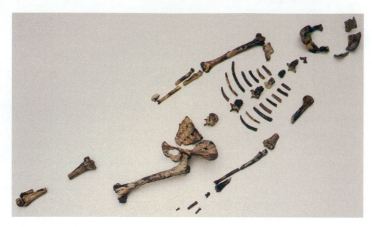

Figure 28.35

Lucy (*Australopithecus afarensis*), the most nearly complete skeleton of an early human ever found. Lucy is dated at 3.2 million years old. A nearly complete skull of *A. afarensis* was discovered in 1994.

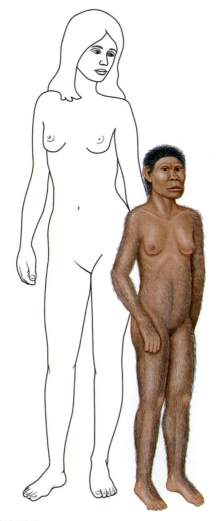

Figure 28.36

A reconstruction of the appearance of Lucy (*right*) compared with a modern human (*left*).

1.7 million years. During this time, populations on different continents have shown some geographic differentiation, but they maintained at least low levels of gene exchange and all have made genetic contributions to modern humans. Several major expansions of populations out of Africa occurred during this time.

Anthropologists have named fossil species to denote spatial and temporal variation in phenotypic characters within this lineage; however, this lineage might constitute a single species according to some species concepts. The earliest human remains originally classified as *Homo sapiens* ("wise man"), from 500,000 to 300,000 years ago, and from 800,000 years ago, now are identified by anthropologists as *H. heidelbergensis* and *H. antecessor*. One group of well-known humans, the **Neandertals,** arose about 150,000 years ago and occupied most of Europe and the Middle East. These morphologically robust humans were originally classified as a subspecies of *H. sapiens,* but recent evidence pointing to their distinctiveness has led anthropologists to recognize them as *H. neanderthalensis.* Neandertals had a brain size similar to that of modern humans and developed more sophisticated stone tools than *H. erectus.* Their robust, heavily muscled bodies allowed them to survive the cold climates of the Ice Age and to hunt large Pleistocene mammals, including woolly rhinoceros, bison, and mammoth. Burial of the dead is first known in Neandertals, and they possibly had complex rituals or religion. However, they lacked the complex art, technology, and culture of their successors.

Fossil and mitochondrial DNA (mtDNA) evidence indicates that characteristics of *H. sapiens,* as presently defined, arose in Africa about 200,000 years ago (Figure 28.34). All mtDNA can be traced to a single female, dubbed "Eve," who lived in Africa approximately 170,000 years ago. About 30,000 years ago Neandertals and remaining *H. erectus* disappeared, approximately 10,000 years after the first appearance of *H. sapiens* in Europe and Eastern Asia. Modern humans were tall people with a culture very different from that of Neandertals. Implement crafting developed rapidly, and human culture became enriched with aesthetics, artistry, and sophisticated language.

In closing our discussion of human evolution, it is important to note that recognition of species in *Homo* is based almost entirely on morphology. Recognition of three or more distinct species of *Homo* does not necessarily imply the occurrence of branching speciation in this lineage; it is possible that we are observing phyletic change within a single species through time, and using the species names only to denote different grades of evolution. At present there is only one human species alive today, an unusual situation when one considers that three to five species of humans have been present during nearly all of the last 4 million years.

The Unique Human Position

Biologically, *Homo sapiens* is a product of the same processes that have directed the evolution of every organism from the time of life's origin. Mutation, isolation, genetic drift, and natural

Classification of Living Mammalian Orders[1]

Class Mammalia

Subclass Prototheria (pro'to-thir'ee-a) (Gr. *prōtos,* first, + *thēr,* wild animal).

Infraclass Ornithodelphia (or'ni-tho-del'fee-a) (Gr. *ornis,* bird, + *delphys,* womb). Monotreme mammals.

Order Monotremata (mon'o-tre'ma-tah) (Gr. *monos,* single, + *trēma,* hole): **egg-laying (oviparous) mammals: duck-billed platypus, echidnas.** Three species in this order are from Australia, Tasmania, and New Guinea. The most noted member of the order is the duck-billed platypus, *Ornithorhynchus anatinus.* Spiny anteaters, or echidnas, *Tachyglossus,* have a long, narrow snout adapted for feeding on ants, their chief food.

Subclass Theria (thir'ee-a) (Gr. *thēr,* wild animal).

Infraclass Metatheria (met'a-thir'ee-a) (Gr. *meta,* after, + *thēr,* wild animal). Marsupial mammals.

Order Didelphimorphia (dy'del-fi-mor'fee-a) (Gr. *di,* two, + *delphi,* uterus, + *morph,* form): **American opossums.** These mammals, like other marsupials, are characterized by an abdominal pouch, or marsupium, in which they rear their young. Most species are found in Central and South America, but one species, the Virginia opossum, *Didelphis virginiana,* is widespread in North America; 66 species.

Order Paucituberculata (pos'see-tu-ber-cu-la'ta) (L. *pauci,* few, + *tuberculum,* knob): **shrew opossums.** Tiny, shrew-sized marsupials found in western South America; seven species.

Order Microbiotheria (my'cro-by-o-ther'ee-a) (Gr. *micro,* small, + *bio,* life, + *thēr,* wild animal): **Monito del Monte.** A South American mouse-sized marsupial that may be more closely related to Australian marsupials; one species.

Order Dasyuromorphia (das-ee-yur'o-mor'fee-a) (Gr. *dasy,* hairy, + *uro,* tail, + *morph,* form): **Australian carnivorous mammals.** In addition to a number of larger carnivores, this order includes a number of marsupial "mice," all of which are carnivorous. Confined to Australia, Tasmania, and New Guinea; 64 species.

Order Peramelemorphia (per'a-mel-e-mor'fee-a) (Gr. *per,* pouch, + *mel,* badger, + *morph,* form): **bandicoots.** Like placentals, members of this group have a chorioallantoic placenta and a high rate of reproduction for marsupials. Confined to Australia, Tasmania, and New Guinea; 22 species.

Order Notoryctemorphia (no'to-rict'te-mor'fee-a) (Gr. *not,* back, + *oryct,* digger, + *morph,* form): **marsupial moles.** Bizarre, semifossorial marsupials present in Australia; two species.

Order Diprotodontia (dy'pro-to-don'tee-a) (Gr. *di,* two, + *pro,* front, + *odont,* tooth): **koalas, wombats, possums, wallabies, kangaroos.** Diverse marsupial group containing some of the largest and most familiar marsupials. Present in Australia, Tasmania, New Guinea, and many islands of the East Indies; 131 species.

Infraclass Eutheria (yu-thir'ee-a) (Gr. *eu,* true, + *thēr,* wild animal). Placental mammals.

Order Insectivora (in-sec-tiv'o-ra) (L. *insectum,* an insect, + *vorare,* to devour): **insect-eating mammals: shrews, hedgehogs, tenrecs, moles.** Small, sharp-snouted animals that spend a great part of their lives underground. Distributed throughout the world except for Australia and New Zealand. Shrews are among the smallest mammals known; 440 species.

Order Macroscelidea (mak-ro-sa-lid'ee-a) (Gr. *makros,* large, + *skelos,* leg): **elephant shrews.** Secretive mammals with long legs, a snoutlike nose adapted for foraging for insects, large eyes. Widespread in Africa; 15 species.

Order Dermoptera (der-mop'ter-a) (Gr. *derma,* skin, + *pteron,* wing): **flying lemurs.** These are related to true bats and consist of a single genus *Galeopithecus.* They are not lemurs (which are primates) and cannot fly but glide like flying squirrels. They occur in the Malay peninsula in the East Indies; two species.

Order Chiroptera (ky-rop'ter-a) (Gr. *cheir,* hand, + *pteron,* wing): **bats.** Wings of bats, the only true flying mammals, are modified forelimbs in which the second to fifth digits are elongated to support a thin integumental membrane for flying. The first digit (thumb) is short with a claw. Common North American forms are little brown bats, *Myotis,* free-tailed bats, *Tadarida,* and big brown bats, *Eptesicus.* In Old World tropics fruit bats, or "flying foxes," *Pteropus,* are largest of all bats, with a wingspread of 1.2 to 1.5 m; they live chiefly on fruits; 977 species.

Order Scandentia (skan-dent'e-a) (L. *scandentis,* climbing): **tree shrews.** Small, squirrel-like mammals of the tropical rain forests of southern and southeastern Asia. Despite their name, many are not especially well-adapted for life in trees, and some are almost completely terrestrial; 16 species.

Order Primates (pry-may'teez) (L. *prima,* first): **prosimians, monkeys, apes, humans.** First in the animal kingdom in brain development, with especially large cerebral cortex. Most species are arboreal, apparently derived from tree-dwelling insectivores. It is believed that their tree-dwelling habits of agility in capturing food or avoiding enemies were largely responsible for their advances in brain structure. As a group they are generalized with five digits (usually provided with flat nails) on both forelimbs and hindlimbs. All except humans have their bodies covered with hair. Forelimbs are often adapted for grasping, as are the hindlimbs sometimes. The group is singularly lacking in claws, scales, horns, and hoofs. There are two suborders; 279 species.

Suborder Strepsirhini (strep'suh-ry-nee) (Gr. *strepsō,* to turn, twist, + *rhinos,* nose): **lemurs, aye-aye, lorises, pottos, bush babies.** Seven families of arboreal primates, formerly called prosimians,

---

[1]Based on Nowak, R. M. 1999. *Walker's Mammals of the World,* ed. 6, Baltimore, The Johns Hopkins University Press.

Classification of Living Mammalian Orders

concentrated on Madagascar, but with species in Africa, southeast Asia, and Malay peninsula. All have a wet, naked region (rhinarium) surrounding comma-shaped nostrils, a long nonprehensile tail, and a second toe provided with a claw. Their food is both plants and animals; 49 species.

Suborder Haplorhini (hap'lo-ry-nee) (Gr. *haploos*, single, simple, + *rhinos*, nose): **tarsiers, marmosets, New and Old World monkeys, gibbons, gorilla, chimpanzees, orangutan, humans.** Six families, all except tarsiers are in the clade Anthropoidea. Haplorhine primates have dry, hairy noses, ringed nostrils and differences in skull morphology that distinguish them from strepsirhine primates. Family **Tarsiidae** contains crepuscular and nocturnal tarsiers (Figure 28.31), with large, forward-facing eyes and reduced snout (five species). New World Monkeys, sometimes called platyrrhine monkeys because the nostrils are widely separated, are contained in two families: **Callitrichidae** (marmosets and tamarins; 35 species) and **Cebidae** (capuchin-like monkeys; 65 species). Callitrichids, which include the colorful lion tamarins, have prehensile hands and quadrupedal locomotion. Cebid monkeys are much larger than any callitrichid. They include capuchin monkeys, *Cebus*, spider monkeys, *Ateles*, and howler monkeys, *Alouatta*. Some cebids (including spider and howler monkeys) have prehensile tails, used like an additional hand for grasping and swinging.

Old World monkeys, termed catarrhine monkeys because their nostrils are set close together and open to the front, are placed in family **Cercopithecidae**, with 96 species. They include mandrills, *Mandrillus*, baboons, *Papio*, macaques, *Macaca*, and langurs, *Presbytis*. The thumb and large toe are opposable. Some have internal cheek pouches; none have prehensile tails. Family **Hylobatidae** contains gibbons and siamang (11 species of genus *Hylobates*), with arms much longer than legs, prehensile hands with fully opposable thumb, and locomotion by true brachiation. Family **Hominidae** contains four living genera and five species: *Gorilla* (one species), *Pan* (two species, chimpanzee and bonobo), *Pongo* (one species, orangutan), and *Homo* (one species, human). The first three of these four genera were formerly placed in the paraphyletic family Pongidae; family Hominidae contained only humans. This separation is not recognized by cladistic taxonomy because the most recent common ancestor of the family Pongidae is also the ancestor of humans.

Order Xenarthra (ze-nar'thra) (Gr. *xenos*, intrusive, + *arthron*, joint) (formerly Edentata [L. *edentatus*, toothless]): **anteaters, armadillos** (Figure 28.37), **sloths.** Either toothless (anteaters) or with simple, peglike teeth (sloths and armadillos). Most live in South and Central America, although nine-banded

Figure 28.37

Nine-banded armadillo, *Dasypus novemcinctus*. This nocturnal species occupies long tunnels during the day, which are dug by powerful, clawed forelimbs. Order Xenarthra, family Dasypodidae.

armadillos, *Dasypus novemcinctus*, are common in the southern United States and spreading north; 29 species.

Order Pholidota (fol'i-do'ta) (Gr. *pholis*, horny scale): **pangolins.** An odd group of mammals whose bodies are covered with overlapping horny scales formed from fused bundles of hair. Their home is in tropical Asia and Africa; seven species.

Order Lagomorpha (lag'o-mor'fa) (Gr. *lagos*, hare, + *morphē*, form): **rabbits, hares, pikas** (Figure 28.38). With long, constantly growing incisors, like rodents, but unlike rodents, they have an additional pair of peglike incisors growing behind the first pair. All lagomorphs are herbivores with cosmopolitan distribution; 81 species.

Order Rodentia (ro-den'che-a) (L. *rodere*, to gnaw): **gnawing mammals: squirrels, rats, woodchucks.** Most numerous of all mammals both in numbers

Figure 28.38

A pika, *Ochotona princeps*, atop a rockslide in Alaska. This little rat-sized mammal does not hibernate but prepares for winter by storing dried grasses beneath boulders. Order Lagomorpha, family Ochotonidae.

Figure 28.39
Grizzly bear, *Ursus arctos horribilis,* of Alaska. Grizzlies, once common in the lower 48 states, are now confined largely to wilderness areas. Order Carnivora, family Ursidae.

and species. Characterized by two pairs of chisel-like incisors that grow throughout life and are adapted for gnawing. With their impressive reproductive rate, adaptability, and capacity to invade nearly all terrestrial habitats, they are of great ecological importance. Important families of this order are **Sciuridae** (squirrels and woodchucks), **Muridae** (rats and house mice), **Castoridae** (beavers), **Erethizontidae** (porcupines), **Geomyidae** (pocket gophers), and **Cricetidae** (hamsters, deer mice, gerbils, voles, lemmings); 2052 species.

Order Carnivora (car-niv'o-ra) (L. *caro,* flesh, + *vorare,* to devour): **flesh-eating mammals: dogs, wolves, cats, bears** (Figure 28.39)**, weasels, seals, sea lions** (Figure 28.40)**, walruses.** All except the giant panda, have predatory habits, and their teeth are especially adapted for tearing flesh; most have canines for killing prey. They are distributed all over the world except in Australian and Antarctic regions where there are no native forms (besides seals). Among more familiar families are **Canidae** (dog family), containing dogs, wolves, foxes, and coyotes; **Felidae** (cat family), whose members include domestic cats, tigers, lions, cougars, and lynxes; **Ursidae** (bears); **Procyonidae** (raccoons); and **Mustelidae** (fur-bearing family), containing martens, skunks, weasels, otters, badgers, minks, and wolverines; **Otariidae** (eared seals), containing fur seals and sea lions; 280 species.

Order Tubulidentata (tu'byu-li-den-ta'ta) (L. *tubulus,* tube, + *dens,* tooth): **aardvark.** "Aardvark" is Dutch for earth pig, a peculiar animal with a piglike body found in Africa; one species.

Order Proboscidea (pro'ba-sid'e-a) (Gr. *proboskis,* elephant's trunk, from *pro,* before, + *boskein,* to feed): **proboscis mammals: elephants.** Largest of living land animals, with two upper incisors elongated as tusks, and well-developed molar teeth. Asiatic or Indian elephants, *Elephas maximus,* have long been partly domesticated and trained to do heavy tasks. Taming of African elephants, *Loxodonta africana,* is more difficult but was done extensively by ancient Carthaginians and Romans, who employed them in their armies; two species.

Order Hyracoidea (hy'ra-coi'de-a) (Gr. *hyrax,* shrew): **hyraxes** (coneys). Coneys are herbivores restricted to Africa and Syria. They have some resemblance to short-eared rabbits but have teeth like rhinoceroses, with hooves on their toes and pads on their feet. They have four toes on the front feet and three toes on the back; seven species.

Order Sirenia (sy-re'ne-a) (Gr. *seiren,* sea nymph): **sea cows and manatees.** Large, clumsy, aquatic mammals with large head, no hindlimbs, and forelimbs modified into flippers. The sea cow (dugong) of tropical coastlines of East Africa, Asia, and Australia and three species of manatees of the Caribbean area and Florida, Amazon River, and West Africa are the only living species. A fifth species, the large Steller's sea cow, was hunted to extinction by humans in the mid-eighteenth century; four species.

Order Perissodactyla (pe-ris'so-dak'ti-la) (Gr. *perissos,* odd, + *dactylos,* toe): **odd-toed hoofed mammals: horses, asses, zebras, tapirs, rhinoceroses.** Odd-toed hoofed mammals have an odd number of toes (one or three), each with a cornified hoof. Both Perissodactyla and Artiodactyla are often referred to as **ungulates** (L. *ungula,* hoof), or hoofed mammals, with teeth adapted for grinding plants. The horse family (Equidae), which also includes asses and zebras, has only one functional toe. Tapirs have a short proboscis formed from the upper lip and nose. The rhinoceros, *Rhinoceros,* includes several species found in Africa and Southeast Asia. All are herbivorous; 17 species.

Figure 28.40
A Galápagos sea lion bull, *Zalophus californianus,* barks to indicate his territorial ownership. Order Carnivora, family Otariidae.

Classification of Living Mammalian Orders

Order Artiodactyla (ar'te-o-dak'ti-la) (Gr. *artios,* even, + *daktylos,* toe): **even-toed hoofed mammals: swine, camels, deer and their allies, giraffes, hippopotamuses, antelopes, cattle, sheep, goats.** Most of these ungulates have two toes, although the hippopotamus and some others have four (Figure 28.41). Each toe is sheathed in a cornified hoof. Many, such as cattle, deer, and sheep have horns or antlers. Many are ruminants. Most are strictly herbivores, but some species, such as pigs, are omnivorous. The group is divided into nine living families and many extinct ones and includes some of the most valuable domestic animals. Artiodactyla is commonly divided into three suborders: the **Suina** (pigs, peccaries, and hippopotamuses), the **Tylopoda** (camels), and the **Ruminantia** (deer, giraffes, sheep, cattle); 221 species.

Order Cetacea (see-tay'she-a) (L. *cetus,* whale): **whales** (Figure 28.42), **dolphins, porpoises.**

Anterior limbs of cetaceans are modified into broad flippers; posterior limbs are absent. Some have a fleshy dorsal fin and the tail is divided into transverse fleshy flukes. Nostrils are represented by a single or double blowhole on top of the head. They have no hair except for a few on the muzzle, no skin glands except the mammary and those of the eye, no external ear, and small eyes. The order is divided into **toothed whales** (suborder Odontoceti), represented by dolphins, porpoises, and sperm whales; and **baleen whales** (suborder Mysticeti) represented by rorquals, right whales, and gray whales. Baleen whales are generally larger than toothed whales. The blue whale, a rorqual, is the heaviest animal that has ever lived. Rather than teeth, baleen whales have a straining device of whalebone (baleen) attached to the palate, used to filter plankton; 78 species.

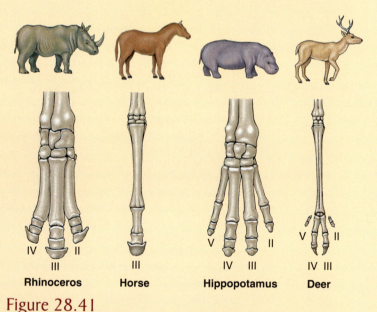

Figure 28.41

Odd-toed and even-toed ungulates. Rhinoceros and horse (order Perissodactyla) are odd-toed; hippopotamus and deer (order Artiodactyla) are even-toed. The lighter, faster mammals run on only one or two toes.

Figure 28.42

Humpback whale, *Megaptera novaeangliae,* breaching. Among the most acrobatic of whales, humpbacks appear to breach to stun fish schools or to communicate information to other pod members. Order Cetacea, family Balaenopteridae.

selection have operated for us as they have for other animals. Yet we have what no other animal species has, a nongenetic cultural evolution that provides a constant feedback between past and future experience. Our symbolic languages, capacities for conceptual thought, knowledge of our history, and abilities to manipulate our environment emerge from this nongenetic cultural endowment. Finally, we owe much of our cultural and intellectual achievements to our arboreal ancestry which bequeathed us with binocular vision, superb visuotactile discrimination, and manipulative skills in use of our hands. If the horse (with one toe instead of five fingers) had human mental capacity, could it have accomplished what humans have?

SUMMARY

Mammals are endothermic and homeothermic vertebrates whose bodies are insulated by hair and who nurse their young with milk. The approximately 4800 species of mammals are descended from the synapsid lineage of amniotes that arose during Carboniferous period of the Paleozoic era. Their evolution can be traced from pelycosaurs of the Permian period to therapsids of the late Permian and Triassic periods of the Mesozoic era. One group of therapsids, the cynodonts, gave rise during the Triassic to mammals. Mammalian evolution was accompanied by the appearance of many important derived characteristics, among these an enlarged brain with greater sensory integration, high metabolic rate, endothermy, heterodont teeth, and many changes in the skeleton that supported a more active life. Mammals diversified rapidly during the Tertiary period of the Cenozoic era.

Mammals are named for the glandular milk-secreting organs of females (rudimentary in males), a unique adaptation which, combined with prolonged parental care, buffers infants from demands of foraging for themselves and eases the transition to adulthood. Hair, the integumentary outgrowth that covers most mammals, serves variously for mechanical protection, thermal insulation, protective coloration, and waterproofing. Mammalian skin is rich in glands: sweat glands that function in evaporative cooling, scent glands used in social interactions, and sebaceous glands that secrete lubricating skin oil. All placental mammals have deciduous teeth that are replaced by permanent teeth (diphyodont dentition). Four kinds of teeth—incisors, canines, premolars, and molars—may be highly modified in different mammals for specialized feeding tasks, or they may be absent.

Food habits of mammals strongly influence their body form and physiology. Insectivores have pointed teeth for piercing the exoskeleton of insects and other small invertebrates. Herbivorous mammals have specialized teeth for grinding cellulose and silica-rich plants and have specialized regions of the gut for harboring bacteria that digest cellulose. Carnivorous mammals have adaptations, including specialized jaw muscles and teeth, for killing and processing their prey, mainly herbivorous mammals. Omnivores feed on both plant and animal foods and have a variety of tooth types.

Some marine, terrestrial, and aerial mammals migrate; some migrations, such as those of fur seals and caribou, are extensive. Migrations are usually made toward favorable climatic and optimal food and calving conditions, or to bring the sexes together for mating.

Mammals with true flight, the bats, are mainly nocturnal and thus avoid direct competition with birds. Most employ ultrasonic echolocation to navigate and to feed in darkness.

Living mammals with the most primitive reproductive characters are egg-laying monotremes of the Australian region. After hatching, the young are nourished with their mother's milk. All other mammals are viviparous. Embryos of marsupials have brief gestation periods, are born underdeveloped, and complete their early growth in the mother's pouch, nourished by milk. The remaining mammals are eutherians, mammals that develop a sophisticated placental attachment between mother and embryos through which embryos are nourished for a prolonged period.

Mammal populations fluctuate from both density-dependent and density-independent causes and some mammals, particularly rodents, may experience extreme cycles of abundance in population density. The unqualified success of mammals as a group cannot be attributed to greater organ system perfection, but rather to their impressive overall adaptability—the capacity to fit more perfectly in total organization to environmental conditions and thus exploit virtually every habitat on earth.

Darwinian evolutionary principles give us great insight into our own origins. Humans are primates, a mammalian group that descended from a shrewlike ancestor. The common ancestor of all modern primates was arboreal and had grasping fingers and forward-facing eyes capable of binocular vision. Primates radiated to form two lineages: (1) lemurs and lorises and (2) tarsiers, monkeys, and apes (including humans). Earliest humans appeared in Africa about 7 million years ago, and gave rise to several species of australopithecines, which persisted for over 3 million years. Australopithecines gave rise to, and coexisted with, *Homo habilis,* the first user of stone tools. *Homo erectus* appeared about 1.8 million years ago and was eventually replaced by modern humans, *Homo sapiens.*

REVIEW QUESTIONS

1. Describe the evolution of mammals, tracing their synapsid lineage from early amniote ancestors to true mammals. How would you distinguish pelycosaurs, early therapsids, cynodonts, and mammals?

2. Describe structural and functional adaptations that appeared in early amniotes that foreshadowed the mammalian body plan. Which mammalian attributes do you think were especially important to successful radiation of mammals?

3. Hair is believed to have evolved in therapsids as an adaptation for insulation, but modern mammals have adapted hair for several other purposes. Describe these.

4. What is distinctive about each of the following: horns of bovids, antlers of deer, and horns of rhinos? Describe the growth cycle of antlers.

5. Describe location and principal function(s) of each of the following skin glands: sweat glands (eccrine and apocrine), scent glands, sebaceous glands, and mammary glands.

6. Define "diphyodont" and "heterodont" and explain why both terms apply to mammalian dentition.

7. Describe food habits of insectivorous, herbivorous, carnivorous, and omnivorous mammals. Give the common names of some mammals belonging to each group.

8. Most herbivorous mammals depend on cellulose as their main energy source, yet no mammal synthesizes cellulose-splitting enzymes. How are the digestive tracts of mammals specialized for symbiotic digestion of cellulose?

9. How does fermentation differ between horses and cattle?

10. What is the relationship of body mass to metabolic rate of mammals?

11. Describe the annual migrations of barren-ground caribou and fur seals.
12. Explain what is distinctive about the life habit and mode of navigation in bats.
13. Describe and distinguish patterns of reproduction in monotremes, marsupials, and placental mammals. What aspects of mammalian reproduction are present in *all* mammals but in no other vertebrates?
14. Distinguish between territory and home range for mammals.
15. What is the difference between density-dependent and density-independent causes of fluctuations of the size of mammalian populations?
16. Describe the hare-lynx population cycle, considered a classic example of a prey-predator relationship (Figure 28.28). From your examination of the cycle, formulate a hypothesis to explain the oscillations.

17. What do the terms Theria, Metatheria, Eutheria, Monotremata, and Marsupialia mean? List mammals that are grouped under each taxon.
18. What anatomical characteristics set primates apart from other mammals?
19. What role does a fossil named "Lucy" play in the reconstruction of human evolutionary history?
20. In what ways do the genera *Australopithecus* and *Homo*, which coexisted for at least 1 million years, differ?
21. When did the different species of *Homo* appear and how did they differ socially?
22. What major attributes make the human position in animal evolution unique?

SELECTED REFERENCES

Feldhamer, G. A., L. C. Drickamer, S. H. Vessey, and J. F. Merritt. 2003. Mammalogy: adaptation, diversity, and ecology, ed. 2. Dubuque, Iowa, WCB/McGraw-Hill. *Modern well-illustrated textbook.*

Grzimek's encyclopedia of mammals. 1990. vol. 1–5. New York, McGraw-Hill Publishing Company. *Valuable source of information on all mammal orders.*

Johanson, D. C., and M. A. Edey. 1981. Lucy, the beginnings of humankind. New York, Simon & Schuster. *Entertaining account of Johanson's discovery of the famous, nearly complete* Australopithecus afarensis *skeleton.*

Macdonald, D. (ed.). 1984. The encyclopedia of mammals. New York, Facts on File Publications. *Coverage of all mammal orders and families, enhanced with fine photographs and color artwork.*

Nowak, R. M. 1999. Walker's mammals of the world, ed. 6. Baltimore, The Johns Hopkins University Press. *The definitive illustrated reference work on mammals, with descriptions of all extant and recently extinct species.*

Pilbeam, D., R. D. Martin, and D. Jones. 1994. Cambridge encyclopedia of human evolution. New York, Cambridge University Press. *Comprehensive and informative encyclopedia written for nonspecialists. Highly readable and highly recommended.*

Preston-Mafham, R., and K. Preston-Mafham. 1992. Primates of the world. New York, Facts on File Publications. *A small "primer" with high quality photographs and serviceable descriptions.*

Rice, J. A. (ed.). 1994. The marvelous mammalian parade. Nat. Hist. **103**(4):39–91. *A special multi-authored section on mammalian evolution.*

Rismiller, P. D., and R. S. Seymour. 1991. The echidna. Sci. Am. **294**:96–103 (Feb.). *Recent studies of this fascinating monotreme have revealed many secrets of its natural history and reproduction.*

Stringer, C. B. 1990. The emergence of modern humans. Sci. Am. **263**:98–104 (Dec.). *A review of the geographical origins of modern humans.*

Suga, N. 1990. Biosonar and neural computation in bats. Sci. Am. **262**:60–68 (June). *How the bat nervous system processes echolocation signals.*

Tattersall, I. 2001. How we came to be human. Sci. Amer. **285**:56–63 (Dec.). *How acquiring language and capacity for symbolic art sets* Homo sapiens *apart from Neandertals.*

Templeton, A. R. 2002. Out of Africa again and again. Nature **416**:45–51. *A comprehensive analysis of molecular genetic data indicating that humans have evolved as a cohesive lineage for the past 1.7 million years, with multiple migrations of populations out of Africa.*

Wong, K. 2002. The mammals that conquered the seas. Sci. Am. **286**:70–79 (May). *New fossils and DNA evidence help to unravel the evolutionary history of whales.*

ONLINE LEARNING CENTER

Visit www.mhhe.com/hickmanipz14e for chapter quizzing, key term flash cards, web links, and more!

PART FOUR

Activity of Life

Female broad-billed hummingbird sipping nectar.

29

Support, Protection, and Movement

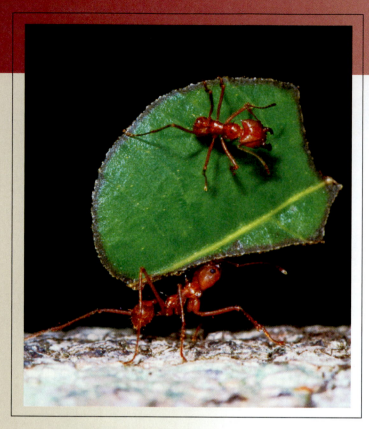

An ant carries with ease a flower petal that is heavier than the ant's body weight.

Of Grasshoppers and Superman

"A dog," remarked Galileo in the seventeenth century, "could probably carry two or three such dogs upon his back; but I believe that a horse could not carry even one of its own size." Galileo was referring to the principle of scaling, a procedure that allows us to understand the physical consequences of changing body size. A grasshopper can jump to a height of 50 times the length of its body, yet a man in a standing jump cannot clear an obstacle of his own height. Without an understanding of scaling, this comparison could easily lead us to expect something extraordinary about the musculatures of insects. To the authors of a nineteenth-century entomology text it seemed that "This wonderful strength of insects is doubtless the result of something peculiar in the structure and arrangement of their muscles, and principally their extraordinary power of contraction." But grasshopper muscles are in fact no more powerful than human muscles because *muscles of small and large animals exert the same force per cross-sectional area*. Grasshoppers leap high in proportion to their size because they are small, not because they possess extraordinary muscles.

The authors of this nineteenth-century text further suggested that if vertebrates had the powers of insects, they would surely have "caused the early desolation of the world." More probably, such powers would have led to their own desolation. Earthly mortals would need more than superhuman muscles were they to leap in the proportions of a grasshopper. They would require superhuman tendons, ligaments, and bones to withstand the stresses of mighty contractions, not to mention the crushing strains of landing again on earth at terminal velocity. The feats of Superman would be quite impossible were he built of the structural materials available to earthbound animals, rather than of the wondrous materials available to inhabitants of the mythical planet Krypton.

INTEGUMENT

The integument is the outer covering of the body, a protective wrapping that includes the skin and all structures derived from or associated with skin, such as hair, setae, scales, feathers, and horns. In most animals it is tough and pliable, providing mechanical protection against abrasion and puncture and forming an effective barrier against invasion of bacteria. It may also provide moisture proofing against fluid loss or gain. Skin helps protect underlying cells against the damaging action of the sun's ultraviolet rays. In addition to being a protective cover, the skin serves a variety of important regulatory functions. For example, in endothermic animals, most of the body's heat is lost through the skin; the skin contains mechanisms that cool the body when it is too hot and slow heat loss when the body is too cold. Skin contains sensory receptors that provide essential information about the immediate environment. It has excretory functions and in some animals respiratory functions as well. Through skin pigmentation some organisms can make themselves more or less conspicuous. Skin secretions can make an animal sexually attractive or repugnant or provide olfactory and/or phenomenal cues that influence behavioral interactions between individuals.

Invertebrate Integument

Many protozoa have only the delicate cell or plasma membranes for external coverings; others, such as *Paramecium,* have developed a protective pellicle. Most multicellular invertebrates, however, have more complex tissue coverings. The principal covering is a single-layered **epidermis.** Some invertebrates have added a secreted noncellular **cuticle** over the epidermis for additional protection.

The molluscan epidermis is delicate and soft and contains mucous glands, some of which secrete the calcium carbonate of the shell. Cephalopod molluscs (squids and octopuses) have developed a more complex integument, consisting from the surface inward of cuticle, simple epidermis, layer of connective tissue, layer of reflecting cells (iridocytes), and thicker layer of connective tissue.

Arthropods have the most complex of invertebrate integuments, providing not only protection but also skeletal support. Development of a firm exoskeleton and jointed appendages suitable for attachment of muscles has been a key feature in the extraordinary diversity of this phylum, largest of animal groups. The arthropod integument consists of a single-layered **epidermis** (more precisely called **hypodermis**), which secretes a complex cuticle of two zones (Figure 29.1A). The thicker inner zone, the **procuticle,** is composed of protein and chitin (a polysaccharide) laid down in layers (lamellae) much like veneers of plywood. The outer zone of cuticle, lying on the external surface above the procuticle, is the thin **epicuticle.** The epicuticle is a nonchitinous complex of proteins and lipids that provides a protective moisture-proofing barrier to the integument.

An arthropod cuticle may be a tough but soft and flexible layer, as it is in many microcrustaceans and insect larvae, or it may be hardened by one of two ways. In decapod crustaceans, for example, crabs and lobsters, the cuticle is stiffened by **calcification,** the deposition of calcium carbonate in the outer layers of the procuticle. In insects hardening occurs when protein molecules bond together with stabilizing cross-linkages within and between adjacent lamellae of the procuticle. The result of this process, called **sclerotization,** is formation of a highly resistant and insoluble protein, **sclerotin.** The arthropod cuticle is one of the toughest materials synthesized by animals; it is strongly resistant to pressure and tearing and can withstand boiling in concentrated alkali, yet it is light, having a specific mass of only 1.3 (1.3 times the weight of water).

When arthropods molt, epidermal cells first divide by mitosis. Enzymes secreted by the epidermis digest most of the procuticle. Digested materials are then absorbed and consequently not lost by the body. Then in the space beneath the old cuticle a new epicuticle and procuticle are formed. After the old cuticle is shed, the new cuticle is thickened and calcified or sclerotized.

Vertebrate Integument and Derivatives

The basic plan of the vertebrate integument, as exemplified by frog and human skin (Figure 29.1B and C), includes a thin, outer stratified epithelial layer, the **epidermis,** derived from ectoderm and an inner, thicker layer, the **dermis,** or true skin, which is of mesodermal origin. (Ectoderm and mesoderm are embryonic germ layers, described in Figure 8.26, p. 179.)

Although the epidermis is thin and appears simple in structure, it gives rise to most derivatives of the integument, such as hair, feathers, claws, and hooves. The epidermis is a stratified squamous epithelium (p. 195) consisting of several layers of cells. Cells of the basal part undergo frequent mitosis to renew layers that lie above. As outer layers of cells are displaced upward by new generations of cells beneath, an exceedingly tough, fibrous protein called **keratin** accumulates in the interior of the cells. Gradually, keratin replaces all metabolically active cytoplasm. The cell dies and is eventually shed, lifeless and scalelike. Such is the origin of dandruff as well as a significant fraction of household dust. This process is called **keratinization,** and the cell, thus transformed, has become **cornified.** Cornified cells, highly resistant to abrasion and water diffusion, form the outermost **stratum corneum.** This epidermal layer becomes especially thick in areas exposed to persistent pressure or wear such as calluses, foot pads of mammals, and the scales of reptiles and birds.

The dermis is a dense connective tissue layer (p. 196) containing blood vessels, collagenous fibers, nerves, pigment cells, fat cells, and connective tissue cells called fibroblasts. These elements support, cushion, and nourish the epidermis, which is devoid of blood vessels. In addition, other cells present in this connective tissue layer (macrophages, mast cells, and lymphocytes, see Chapter 35), provide a first line of defense if the outer epidermal layer is broken.

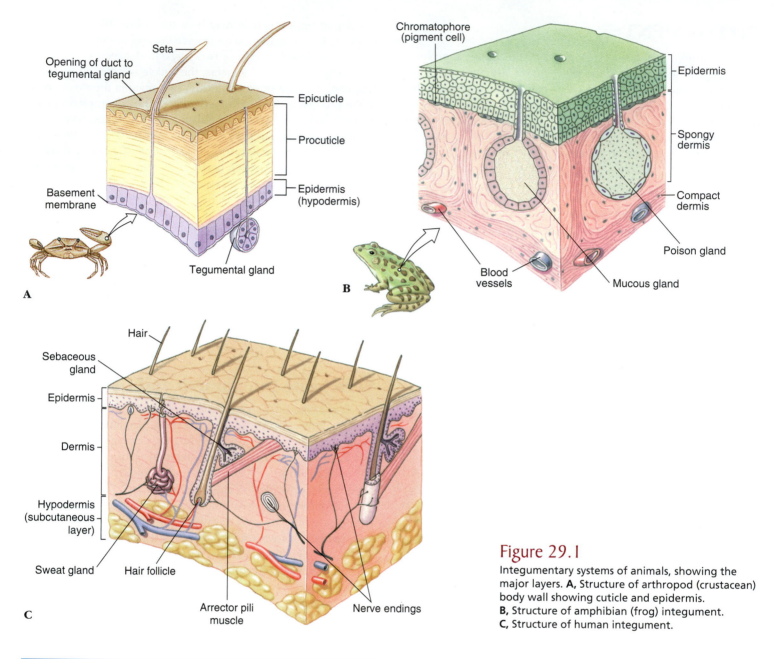

Figure 29.1

Integumentary systems of animals, showing the major layers. **A,** Structure of arthropod (crustacean) body wall showing cuticle and epidermis. **B,** Structure of amphibian (frog) integument. **C,** Structure of human integument.

Lizards, snakes, turtles, and crocodilians were among the first to exploit the adaptive possibilities of the remarkably tough protein keratin. Reptilian epidermal scale that develops from keratin is a much lighter and more flexible structure than the bony, dermal scale of fishes, yet it provides excellent protection from abrasion and desiccation (Figure 29.2). Scales may be overlapping structures, as in snakes and some lizards, or develop into plates, as in turtles and crocodilians. In birds, keratin found new uses. Feathers, beaks, and claws, as well as scales, are all epidermal structures composed of dense keratin. Mammals continued to capitalize on keratin's virtues by turning it into hair, hooves, claws, and nails. As a result of its keratin content, hair is by far the strongest material in the body. It has a tensile strength comparable to that of rolled aluminum and is nearly twice as strong, weight for weight, as the strongest bone.

The **dermis** may also contain true bony structures, of dermal origin. Heavy bony plates were common in ostracoderms and placoderms of the Paleozoic era (Figure 23.17, p. 512) and persist in some living fishes, such as sturgeons (Figure 24.19, p. 527). Scales of contemporary fishes are bony dermal structures that have evolved from the bony armor of the Paleozoic fishes but are much smaller and more flexible. Fish scales are thin bony slivers covered with a mucus-secreting epidermis (Figure 29.2). Most amphibians lack dermal bones in their skin, except for vestiges of dermal scales found in a few tropical caecilian species. In reptiles dermal bones provide the armor of crocodilians, the beaded skin of many lizards, and contribute to the shell of turtles. Dermal bone also gives rise to antlers, as well as the bony core of horns.

Structures such as claws, beaks, nails, and horns contain combinations of epidermal (keratinized) and dermal components. Their basic structure is the same, with a central bony core

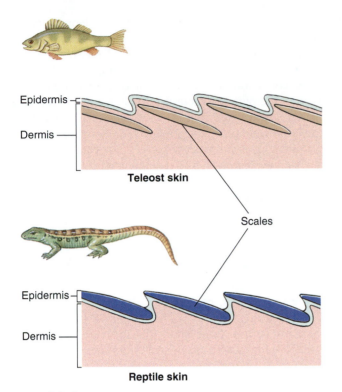

Figure 29.2

Integument of bony fishes and lizards. Bony (teleost) fishes have bony scales from dermis, and lizards have horny scales from epidermis. Thus they are not homologous structures. Dermal scales of fishes are retained throughout life. Since a new growth ring is added to each scale each year, fishery biologists use scales to tell the age of fishes. Epidermal scales of reptiles are shed periodically.

covered by a vascularized nutritive layer of the dermis, and an outer epithelial layer. This epithelial layer has a germinative component responsible for the continual growth of horns, hooves, claws, and beaks. The outer epithelial layer is keratinized. Overgrowth of these structures is prevented by constant wear and abrasion (Figure 29.3).

Animal Coloration

The colors of animals may be vivid and dramatic when serving as important recognition marks or as warning coloration, or they may be subdued or cryptic when used for camouflage. Integumentary color is usually produced by pigments, but in many insects and in some vertebrates, especially birds, certain colors are produced by the physical structure of the surface tissue, which reflects certain light wavelengths and eliminates others. Colors produced this way are called **structural color,** and they are responsible for the most beautifully iridescent and metallic hues in the animal kingdom. Many butterflies and beetles and a few fishes share with birds the distinction of being the earth's most resplendent animals. Certain structural colors of feathers are caused by minute, air-filled spaces or pores that reflect white light (white feathers) or some portion of the spectrum (for example, Tyndall blue coloration produced by scattering of light [see note, p. 592]). Iridescent colors that change hue as an animal's angle shifts with respect to an observer are produced when light is reflected from several layers of thin, transparent film. By phase interference, light waves reinforce, weaken, or eliminate each other to produce some of the purest and most brilliant colors we know.

More common than structural colors in animals are **pigments** (biochromes), an extremely varied group of large molecules that reflect light rays. In crustaceans and ectothermic vertebrates these pigments are contained in large cells with branching processes, called **chromatophores** (Figures 29.1 and 29.4A). The pigment may concentrate in the center of the cell in an aggregate too small to be visible, or it may disperse throughout the cell and its processes, providing maximum display. Chromatophores of cephalopod molluscs are entirely different (Figure 29.4B). Each is a small saclike cell filled with pigment granules and surrounded by muscle cells that, when contracted, stretch the whole cell into a pigmented sheet. When muscles relax, the elastic chromatophore quickly shrinks to a small sphere. With such pigment cells squids and octopuses can alter their color more rapidly than any other animal.

Most widespread of animal pigments are the **melanins,** a group of black or brown polymers responsible for various

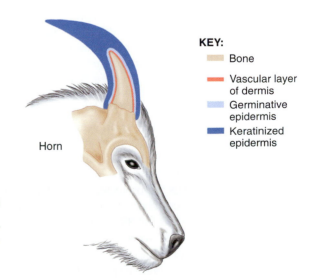

KEY:

- Bone
- Vascular layer of dermis
- Germinative epidermis
- Keratinized epidermis

Figure 29.3

Similarity of structure of integumentary derivatives. Claws, beaks, and horns are all built of similar combinations of epidermal (keratinized) and dermal components. A central bony core is covered by a vascularized nutritive layer of dermis. An outer epithelial layer has a basal germinative component which proliferates to allow these structures to grow continually. The thickened surface epithelium is keratinized or cornified. Note that the relative thickness of each component is not drawn to scale.

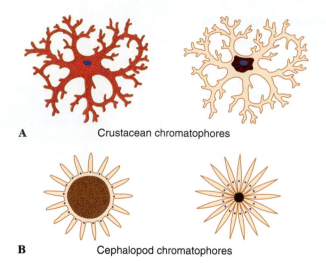

A   Crustacean chromatophores

B   Cephalopod chromatophores

Figure 29.4

Chromatophores. **A,** The crustacean chromatophore showing the pigment dispersed (*left*) and concentrated (*right*). Vertebrate chromatophores are similar. **B,** The cephalopod chromatophore is an elastic capsule surrounded by muscle fibers that, when contracted (*left*), stretch the capsule to expose the pigment.

earth-colored shades that most animals wear. Yellow and red colors are often caused by **carotenoid** pigments, which are frequently contained within special pigment cells called **xanthophores.** Most vertebrates are incapable of synthesizing their own carotenoid pigments but must obtain them directly or indirectly from plants. Two entirely different classes of pigments called ommochromes and pteridines are usually responsible for the yellow pigments of molluscs and arthropods. Green colors are rare; when they occur, they are usually produced by yellow pigment overlying blue structural color. **Iridophores,** a third type of chromatophore, contain crystals of guanine or some other purine, rather than pigment. They produce a silvery or metallic effect by reflecting light.

By vertebrate standards, mammals are a somber-colored group (pp. 618–619). Most mammals are largely color blind, a deficiency associated with a lack of bright colors in the group. Exceptions are the brilliantly colored skin patches of some baboons and mandrills. Significantly, primates have color vision and thus can appreciate such eye-catching ornaments. The muted colors of mammals are caused by melanin, which is deposited in growing hair by dermal melanophores.

Injurious Effects of Sunlight

The familiar vulnerability of human skin to sunburn reminds us of the potentially damaging effects of ultraviolet radiation on protoplasm. Many animals, such as flatworms, if exposed to the sun in shallow water are damaged or killed by ultraviolet radiation. Most land animals are protected from such damage by the screening action of special body coverings, for example, the cuticle of arthropods, scales of reptiles, and feathers and fur of birds and mammals. Humans, however, are "naked apes" that lack the furry protection of most other mammals. We must depend on thickening of the epidermis (particularly the outermost stratum corneum) and on epidermal pigmentation for protection.

Most ultraviolet radiation is absorbed in the epidermis, but about 10% penetrates the dermis. Damaged cells in both epidermis and dermis release histamine and other vasodilator substances that cause blood-vessel enlargement in the dermis and the characteristic red coloration of sunburn. Light skins suntan through the formation of the pigment **melanin** in the deeper epidermis and by "pigment darkening," that is, the photooxidative blackening of bleached pigment already present in the epidermis. Unfortunately, tanning does not bestow perfect protection. Sunlight still ages the skin prematurely, and tanning itself causes the skin to become dry and leathery. Sunlight also is responsible for approximately 1 million new cases of skin cancer annually in the United States alone, making skin cancer the most common of malignancies. There is now strong evidence that genetic mutations caused by high doses of sunlight received during the pre-adult years are responsible for skin cancers that appear after middle age.

SKELETAL SYSTEMS

Skeletons are supportive systems that provide rigidity to the body, surfaces for muscle attachment, and protection for vulnerable body organs. The familiar bone of the vertebrate skeleton is only one of several kinds of supportive and connective tissues serving various binding and weight-bearing functions, which are described in this section.

Hydrostatic Skeletons

Not all skeletons are rigid; many invertebrate groups use their body fluids as an internal hydrostatic skeleton. Muscles in the body wall of an earthworm, for example, have no firm base for attachment but develop muscular force by contracting against incompressible coelomic fluids, enclosed within a limited space, much like the hydraulic brake system of an automobile.

Alternate contractions of the circular and longitudinal muscles of the body wall enable a worm to become thin and then thicken, producing backward-moving waves of motion that propel the animal forward (Figure 29.5). Movement in earthworms and other annelids is helped by septa that separate their body into more or less independent compartments (Figure 17.3, p. 365). An obvious advantage is that if a worm is punctured or even cut into pieces, each part can still develop pressure and move. Worms that lack internal compartments, for example, the lugworm *Arenicola* (Figure 17.6, p. 367), are rendered helpless if body fluid is lost through a wound.

Many muscles not only produce movement but also provide a unique form of skeletal support. The elephant's trunk is an excellent example of a structure that lacks any obvious form of skeletal support, yet is capable of bending, twisting, elongating, and lifting heavy weights (Figure 29.6). An elephant's trunk, tongues of mammals and reptiles, and tentacles of cephalopod molluscs are examples of **muscular hydrostats.** Like hydrostatic skeletons of worms, muscular hydrostats work because they are composed of incompressible tissues that remain at constant volume. The remarkably diverse movements of muscular hydrostats depend on muscles arranged in complex patterns.

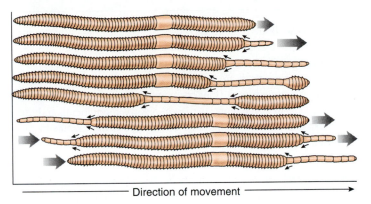

Figure 29.5

How an earthworm moves forward. When circular muscles contract, longitudinal muscles are stretched by internal fluid pressure and the worm elongates. Then, by alternate contraction of longitudinal and circular muscles, a wave of contraction passes from anterior to posterior. Bristlelike setae are extended to anchor the animal and prevent slippage as it moves forward.

Figure 29.6

Muscular trunk of an elephant, an example of a muscular hydrostat.

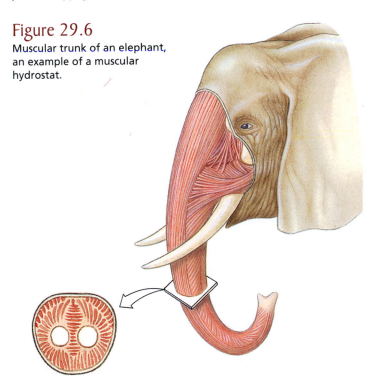

Rigid Skeletons

Rigid skeletons differ from hydrostatic skeletons in one fundamental way: rigid skeletons consist of rigid elements, usually jointed, to which muscles can attach. Since muscles can only contract and not actively lengthen, rigid skeletons provide the anchor points required by opposing sets of muscles, such as flexors and extensors, that allow movement in more than one direction.

There are two principal types of rigid skeletons: **exoskeleton,** typical of molluscs, arthropods and many other invertebrates; and **endoskeleton,** characteristic of echinoderms, vertebrates, and some cnidarians. An invertebrate's exoskeleton may be mainly protective, but it may also perform a vital role in locomotion. An exoskeleton may take the form of a shell, a spicule, or a calcareous, proteinaceous, or chitinous plate. It may be rigid, as in molluscs, or jointed and movable, as in arthropods. Unlike an endoskeleton, which grows with the animal, an exoskeleton is often a limiting coat of armor that must be periodically molted to make way for an enlarged replacement (molting and ecdysis in crustaceans is described on p. 428). Some invertebrate exoskeletons, such as the shells of snails and bivalves, grow with the animal.

The arthropod-type exoskeleton is perhaps a better arrangement for small animals than a vertebrate-type endoskeleton because a hollow cylindrical tube can support much more weight without collapsing than can a solid cylindrical rod of the same material and weight. Arthropods can thus enjoy both protection and structural support from their exoskeleton. For larger animals the hollow cylinder would be completely impractical. If made thick enough to support the body weight, it would be too heavy to lift; but if kept thin and light, it would be extremely sensitive to buckling or shattering on impact. Finally, can you imagine the sad plight of an animal the size of an elephant when it shed its exoskeleton to molt?

The vertebrate endoskeleton is formed inside the body and is composed of bone and cartilage, which are forms of specialized connective tissue (see p. 196). Bone not only supports and protects but is also the major body reservoir for calcium and phosphorus. In amniote vertebrates red blood cells, platelets, and white blood cells are formed in the bone marrow.

Notochord and Cartilage

The **notochord** (see Figure 23.1, p. 497) is a semirigid supportive axial rod of protochordates and all vertebrate larvae and embryos. It is composed of large, vacuolated cells and is surrounded by layers of elastic and fibrous sheaths. It is a stiffening device, preserving body shape during locomotion. Except in jawless vertebrates (lampreys and hagfishes), the notochord is surrounded or replaced by vertebrae during embryonic development.

Cartilage is a major skeletal element of some vertebrates. Jawless fishes (for example, lampreys) and elasmobranchs (sharks, skates, and rays) have purely cartilaginous skeletons, which is an evolutionarily derived feature, since their Paleozoic ancestors had bony skeletons. Other vertebrates as adults have principally bony skeletons with some cartilage interspersed. Cartilage is a soft, pliable tissue that resists compression. The basic form, **hyaline cartilage,** has a clear, glassy appearance (see Figure 9.10, p. 196). It is composed of cartilage cells (**chondrocytes**) surrounded by a firm complex protein-sugar gel interlaced with a meshwork of collagenous fibers. Blood vessels are virtually absent—the reason that sports injuries involving cartilage heal slowly. In addition to forming the cartilaginous skeleton of some vertebrates and that of all vertebrate embryos, hyaline cartilage forms articulating surfaces of many bone joints of most adult vertebrates and supporting tracheal, laryngeal, and bronchial rings of the respiratory system (see p. 701). Two other types of cartilage, elastic and fibrous, are similar to hyaline cartilage, except that in the case of elastic cartilage the fiber type is predominantly elastic, whereas in the case of fibrous cartilage numerous collagenous fiber bundles are present, often arranged in herringbone patterns.

Cartilage similar to hyaline cartilage occurs in some invertebrates, for example the radula of gastropod molluscs (p. 342) and lophophore of brachiopods (p. 326). The cartilage of cephalopod molluscs is of a special type with long, branching processes that resemble cells of vertebrate bone.

Bone

Bone is a living tissue that differs from other connective and supportive tissues by having significant deposits of inorganic calcium salts laid down in an extracellular matrix composed of collagenous fibers in a protein-sugar gel. Unlike cartilage, it is highly vascular. Its structural organization is such that bone has nearly the tensile strength of cast iron, yet is only one-third as heavy.

Bone is never formed in vacant space but is always laid down by replacement in areas occupied by some form of connective tissue. Most bone develops from cartilage and is called **endochondral** ("within-cartilage") or **replacement bone.** Embryonic cartilage is gradually eroded leaving it extensively honeycombed; bone-forming cells then invade these areas and begin depositing extracellular bone matrix, which becomes calcified, around strandlike remnants of the cartilage. A second type of bone is **intramembranous bone,** which develops directly from sheets of embryonic cells. Dermal bone, mentioned on page 646, is a type of intramembranous bone. In tetrapod vertebrates intramembranous bone is restricted mainly to bones of the face, cranium, and clavicle; the remainder of the skeleton is endochondral bone. Whatever the embryonic origin, once fully formed, endochondral and intramembranous bone look similar.

Fully formed bone, however, may vary in density. **Spongy** (or cancellous) **bone** consists of an open, interlacing framework of bony tissue, oriented to give maximum strength under normal stresses and strains that bone receives. All bone develops first as spongy bone, but some bones, through further deposition of bone matrix, become **compact.** Compact bone is dense, appearing solid to the unaided eye. Both spongy and compact bone are found in typical long bones of tetrapods (Figure 29.7).

Microscopic Structure of Bone   Compact bone is composed of a calcified bone matrix arranged in concentric rings. Between these rings are cavities **(lacunae)** containing bone cells **(osteocytes),** which are interconnected by many minute passages **(canaliculi).** These passages allow communication between bone cells, via gap junctions (p. 46), and serve to distribute nutrients throughout the bone. This entire organization of lacunae and canaliculi is arranged into an elongated cylinder called an **osteon** (also called **haversian system**) (Figure 29.7). Bone consists of bundles of osteons cemented together and interconnected with blood vessels and nerves although noncellular matrix predominates. As a result of the presence of blood vessels and nerves, bone breaks can heal rapidly, and bone diseases can be as painful as any other tissue disease.

Like muscle, bone is subject to "use and disuse." When we exercise our muscles, our bones respond by producing new bone tissue to give added strength. In fact, the bumps and processes to which muscles attach are produced by bone in response to the action of muscle forces. Conversely, when bones are not subject to stress, as in space flight, the body resorbs the mineral, and the bones become weak. Astronauts who spend many months in space must exercise to a greater degree than on earth to prevent such resorption and bone weakness.

Bone is a dynamic tissue, and bone remodelling and growth are complex restructuring processes, involving both destruction by bone-resorbing cells **(osteoclasts)** and deposition by bone-building cells **(osteoblasts).** Both processes occur simultaneously so that new osteons are formed as old ones are resorbed. The marrow cavity inside grows larger by bone resorption of the inner surface of the surrounding bone while new bone is laid down on the outer bone surface by bone deposition. Bone growth responds to several hormones, in particular **parathyroid hormone** from the parathyroid gland, which stimulates bone resorption, and **calcitonin** from the thyroid gland, which inhibits bone resorption. These two hormones, together with vitamin D_3, **1,25-dihydroxyvitamin D_3,** are responsible for maintaining a constant level of calcium in the blood. The effect of hormones on bone growth and resorption is described in more detail on page 764.

Following menopause, a woman loses 5% to 6% of her bone mass annually, often leading to the disease osteoporosis and increasing the risk of bone fractures. Dietary supplementation with calcium and vitamin D_3 has been advocated to prevent such losses, and together with exercise they can slow demineralization after menopause. Therapy with the female sex hormone estrogen (see p. 150) is often used in postmenopausal women, because ovarian production of estrogen drops significantly after menopause. More frequently, however, low doses of estrogen are usually accompanied by low doses of the female hormone progesterone, since this combination decreases the risk of breast and uterine cancer, which are significant side effects of estrogen therapy alone. **Bisphosphonates** are an alternative therapy to hormone replacement therapy (HRT) for women with history of breast or uterine cancer in the family. This class of drugs is nonhormonal and works by decreasing bone breakdown activity of osteoclasts. Finally, **selective estrogen-receptor modulators (SERMs)** are another type of treatment for osteoporosis. These are synthetic hormone replacement drugs that work by mimicking the effects of estrogen on bone but do not appear to increase the risk of breast or uterine cancer. Among animals, only humans are troubled with osteoporosis, perhaps a consequence of the long postreproductive life of the human species. Osteoporosis is traditionally considered a female problem, but it is currently estimated that one in five men are at risk for osteoporosis (http://www.iofbonehealth.org/).

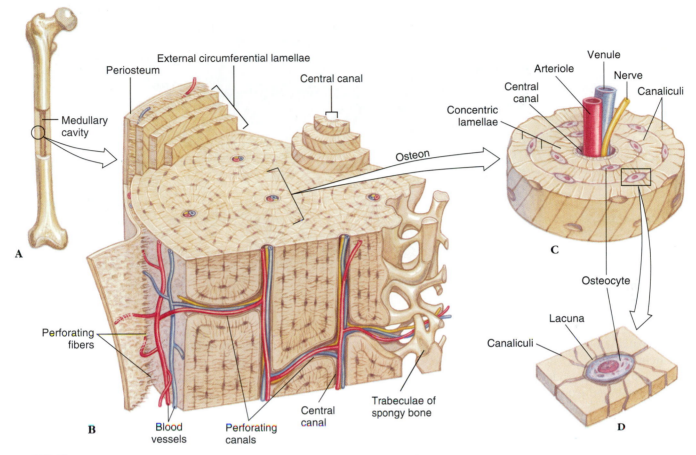

Figure 29.7

Structure of compact bone. **A,** Adult long bone with a cut into the medullary cavity. **B,** Enlarged section showing osteons, the basic histological unit of bone. **C,** Enlarged view of an osteon showing the concentric lamellae and the osteocytes (bone cells) arranged within lacunae. **D,** An osteocyte within a lacuna. Bone cells receive nutrients from the circulatory system via tiny canaliculi that interlace the calcified matrix. Bone cells are known as osteoblasts when they are building bone, but, in mature bone shown here, they become resting osteocytes. Bone is covered with dense connective tissue called periosteum.

Plan of the Vertebrate Skeleton

The vertebrate skeleton is composed of two main divisions: **axial skeleton,** which includes skull, vertebral column, sternum, and ribs, and **appendicular skeleton,** which includes the limbs (or fins or wings) and pectoral and pelvic girdles (Figures 29.8 and 29.9). Not surprisingly, the skeleton has undergone extensive remodeling in the course of vertebrate evolution. The move from water to land forced dramatic changes in body form. With increased cephalization, further concentration of brain, sense organs, and food-gathering apparatus in the head occurred and the skull became the most intricate portion of the skeleton. Some early fishes had as many as 180 skull bones (a source of frustration to paleontologists) but through loss of some bones and fusion of others, skull bones became greatly reduced in number during evolution of the tetrapods. Amphibians and lizards have 50 to 95, and mammals, 35 or fewer. Humans have 29.

The vertebral column is the main stiffening axis of the post-cranial skeleton. In fishes it serves much the same function as the notochord, providing points for muscle attachment and provides stiffness, preserving body shape during muscle contraction. With evolution of amphibious and terrestrial tetrapods, the vertebrate body was no longer buoyed by the aquatic environment. The vertebral column became structurally adapted to withstand new regional stresses transmitted to the column by the two pairs of appendages. In amniote tetrapods (reptiles, birds, and mammals), the vertebrae are differentiated into **cervical** (neck), **thoracic** (chest), **lumbar** (back), **sacral** (pelvic), and **caudal** (tail) vertebrae. In frogs, birds, and also in humans caudal vertebrae are reduced in number and size, and sacral vertebrae are fused. The number of vertebrae varies among the different vertebrates. Pythons seem to lead the list with more than 400. In humans (Figure 29.9) there are 33 in a young child, but in adults 5 are fused to form the **sacrum** and 4 to form the **coccyx.** Besides the sacrum and coccyx, humans have 7 cervical, 12 thoracic, and 5 lumbar vertebrae. The number of cervical vertebrae (7) is constant in nearly all mammals, whether the neck is short as in dolphins, or long as in giraffes.

In mammals, the first two cervical vertebrae, **atlas** and **axis,** are modified to support the skull and to permit pivotal movements. The atlas bears the globe of the head much as the

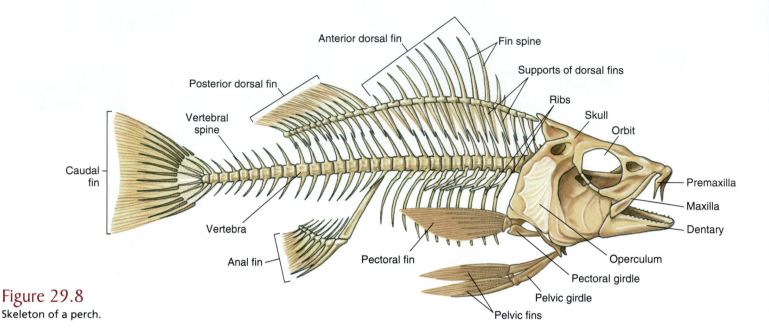

Figure 29.8
Skeleton of a perch.

mythological Atlas bore the earth on his shoulders. The axis, the second vertebra, permits the head to turn from side to side.

Ribs are long or short skeletal structures that articulate medially with vertebrae and extend into the body wall. Fishes have a pair of ribs for every vertebra (see Figure 29.8); they serve as stiffening elements in the connective-tissue septa that separate the muscle segments and thus improve the effectiveness of muscle contractions. Many fishes have both dorsal and ventral ribs, and some have numerous riblike intermuscular bones as well—all of which increase the difficulty and reduce the pleasure of eating certain kinds of fish. Other vertebrates have a reduced number of ribs, and some, such as the familiar leopard frog, have no ribs at all. In mammals the ribs together form the thoracic basket, which supports the chest wall and prevents collapse of the lungs. Mammals such as sloths have 24 pairs of ribs, whereas horses possess 18 pairs. Primates other than humans have 13 pairs of ribs; both male and female humans have 12 pairs, although approximately 1 person in 20 has a thirteenth pair.

Most vertebrates, fishes included, have paired appendages. All fishes except agnathans have thin pectoral and pelvic fins that are supported by the pectoral and pelvic girdles, respectively (see Figure 29.8). Tetrapods (except caecilians, some salamanders, snakes, and limbless lizards) have two pairs of **pentadactyl** (five-toed) limbs, also supported by girdles. The pentadactyl limb is similar in all tetrapods, alive and extinct; even when highly modified for various modes of life, the elements are rather easily homologized (the evolution of the pentadactyl limb is illustrated in Figure 25.2, p. 546).

Modifications of the basic pentadactyl limb for life in different environments often involve bone loss or fusion rather than addition of new bone. The ends of the appendages are more likely to be modified, such as the bony structures of the feet and hands. Horses and their relatives evolved a foot structure for fleetness by elongation of the third toe. In effect, a horse stands on its third fingernail (hoof), much like a ballet dancer standing on the tips of the toes. A bird wing is a good example of distal modification. A bird embryo bears 13 distinct wrist and hand bones (carpals and metacarpals), but most of these and the finger bones (phalanges) regress during development, leaving four bones in three digits in an adult (see p. 593). The proximal bones (humerus, radius, and ulna), however, are only slightly modified in the bird wing.

In nearly all tetrapods the pelvic girdle is firmly attached to the axial skeleton, since the greatest locomotory forces transmitted to the body come from the hindlimbs. The pectoral girdle, however, is much more loosely attached to the axial skeleton, providing the forelimbs with greater freedom for manipulative movements.

Effect of Body Size on Bone Stress

As Galileo realized in 1638, the ability of animals' limbs to support a load decreases as animals increase in size (chapter opening essay, p. 644). Imagine two animals, one twice as long as the other, that are proportionally identical. The larger animal is twice as long, twice as wide, and twice as tall as the smaller. The volume (and the weight) of the larger animal is eight times the volume of the smaller ($2 \times 2 \times 2 = 8$). However, the strength of the larger animal's legs is only four times the strength of the smaller, because bone, tendon, and muscle strength are proportional to cross-sectional area. So, as Galileo noted, eight times the weight would have to be carried by only four times the strength. Because the maximum strength of mammalian bone is rather uniform per unit of cross-sectional area, how can animals become larger without placing unbearable stresses on long limb bones? One obvious solution is to make bones stouter and therefore stronger. However, throughout much of their size range, bone shape in different sized mammals does not change much. Instead, mammals have adapted limb posture so that stresses are shifted to

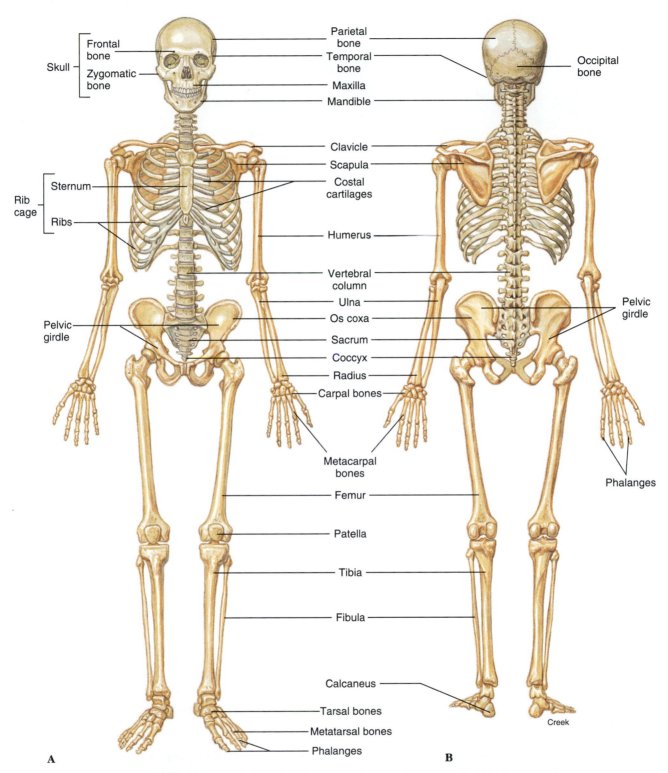

Skull
Frontal bone
Zygomatic bone
Parietal bone
Temporal bone
Maxilla
Mandible
Occipital bone

Clavicle
Scapula
Costal cartilages

Rib cage
Sternum
Ribs

Humerus

Vertebral column
Ulna
Os coxa
Sacrum
Coccyx
Radius
Carpal bones

Pelvic girdle

Pelvic girdle

Metacarpal bones

Phalanges

Femur

Patella

Tibia

Fibula

Calcaneus
Tarsal bones
Metatarsal bones
Phalanges

A

B

Creek

Figure 29.9

Human skeleton. **A,** Ventral view. **B,** Dorsal view. In comparison with other mammals, the human skeleton is a patchwork of primitive and specialized parts. Erect posture, produced by specialized changes in legs and pelvis, enabled the primitive arrangement of arms and hands (arboreal adaptation of human ancestors) to be used for manipulation of tools. Development of the skull and brain followed as a consequence of the premium natural selection put on dexterity and ability to appraise the environment.

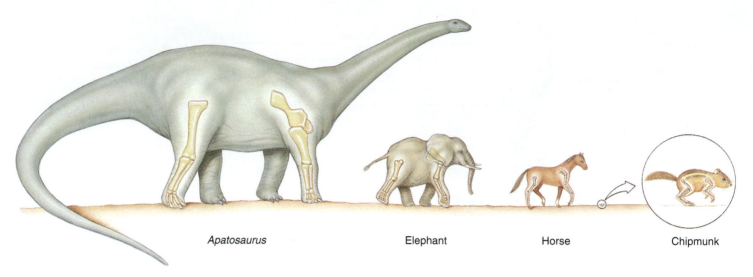

Apatosaurus   Elephant   Horse   Chipmunk

Figure 29.10

Comparison of postures in small and large mammals, showing the effect of scale. Because of its more upright posture, bone stresses in the horse are similar to those in the chipmunk. In mammals larger than horses (above about 300 kg), greatly increased stresses require that bones become exceedingly robust and that the animal lose agility.

align with the long axis of the bones, rather than transversely. Small animals the size of a chipmunk run in a crouched limb posture, whereas a large mammal such as a horse, has adopted an upright posture (Figure 29.10). Bones and muscles are capable of carrying far more weight when aligned more closely with the ground reaction force, as they are in a horse's leg. In this way, peak bone stresses during strenuous activity are no greater for a galloping horse than for a running chipmunk or dog.

For animals larger than horses, further mechanical advantage by changing limb posture is not possible because the limbs are fully upright. Instead, the long bones of an elephant weighing 2.5 metric tons, and those of the enormous dinosaur *Apatosaurus,* weighing an estimated 34 metric tons, are (were) extremely thick and robust (Figure 29.10), providing the safety factor these massive animals require(d). However, top running speeds of the largest terrestrial mammals decline with increasing size, and analysis of one of the most formidable dinosaurs, *Tyrannosaurus,* concluded that it could not run (Hutchinson and Garcia, 2002).

ANIMAL MOVEMENT

Movement is an important characteristic of animals. Animal movement occurs in many forms in animal tissues, ranging from barely discernible streaming of cytoplasm to extensive movements of powerful skeletal muscles. Most animal movement depends on a single fundamental mechanism: **contractile proteins,** which can change their form to allow relaxation and contraction. Contractile machinery is always composed of ultrafine fibrils arranged to contract when powered by **ATP.** By far the most important protein contractile system is the **actomyosin system,** composed of two proteins, **actin** and **myosin.** This is an almost universal biomechanical system found from protozoa to vertebrates; it performs a long list of diverse functional roles. Cilia and flagella, however, are composed of proteins other

than actin and myosin, and thus are exceptions to the rule. In this discussion we examine the three principal kinds of animal movement: ameboid, ciliary and flagellar, and muscular.

Ameboid Movement

Ameboid movement is a form of movement especially characteristic of amebas and other unicellular forms; it is also found in many wandering cells of metazoans, such as white blood cells, embryonic mesenchyme, and numerous other mobile cells that move through tissue spaces.

Research with a variety of ameboid cells, including pathogen-fighting phagocytes in blood, has produced a consensus model to explain extension and withdrawal of **pseudopodia** (false feet) and ameboid crawling. Optical studies of an ameba in movement suggest that the outer layer of nongranular, gel-like **ectoplasm** surrounds a more fluid core of **endoplasm** (see Figure 11.10, p. 222). Movement depends on actin and other regulatory proteins. According to this hypothesis (Stossel, 1994), as the pseudopod extends, hydrostatic pressure forces actin subunits in the flowing endoplasm into the pseudopod where they dissociate from regulatory proteins and are then able to assemble into a network to form the gel-like ectoplasm. At the trailing edge of the gel, where the network disassembles, freed actin interacts, in the presence of calcium ions, with myosin to create a contractile force that pulls the cell forward behind the extending pseudopod. Locomotion is assisted by membrane-adhesion proteins that attach temporarily to the substrate to provide traction, enabling the cell to crawl steadily forward (see Figure 11.12, p. 224).

Ciliary and Flagellar Movement

Cilia are minute, hairlike, motile processes that extend from surfaces of cells of many animals. They are a particularly distinctive feature of ciliate protistans, but except for nematodes in which motile cilia are absent and arthropods in which they are rare, cilia are found in all major groups of animals. Cilia perform many roles either in moving small organisms such as unicellular ciliates and ctenophores (see Figure 29.12B) through their aquatic environment or in propelling fluids and materials over epithelial surfaces of larger animals.

Cilia are of remarkably uniform diameter (0.2 to 0.5 μm) wherever they are found. Electron microscopy reveals that each cilium has at its base a **basal body** (kinetosome) that is structurally similar to a centriole (see Figure 3.14, p. 46). Each basal body gives rise to a peripheral circle of nine double microtubules arranged around two single microtubules in the center (Figure 29.11) forming the structural support and machinery for movement within each cilium. Each microtubule is composed of a spiral array of protein subunits called **tubulin** (see Figure 3.13B, p. 45). The microtubule doublets around the periphery are connected to each other and to the central pair of microtubules by a complex system of **microtubule-associated proteins** (MAPs). Also extending from each doublet is a pair of arms composed of the protein **dynein.** The dynein arms, which act as cross bridges between doublets, operate to produce a sliding force between microtubules. During ciliary movement, microtubules behave as "sliding filaments" that move past one another much like the sliding filaments of vertebrate skeletal muscle described in the discussion on the sliding filament hypothesis, page 658. During ciliary flexion, dynein arms link to adjacent microtubules, then swivel and release in repeated cycles, causing microtubules on the concave side to slide outward past microtubules on the convex side. This process increases curvature of the cilium. During the recovery stroke microtubules on the opposite side slide outward to bring the cilium back to its starting position.

A **flagellum** is a whiplike structure longer than a cilium and usually present singly or in small numbers at one end of a cell. They are found in many single-celled eukaryotes, in animal spermatozoa, and in sponges. Flagella have the same basic internal structure as cilia, although several exceptions to the 9 + 2 arrangement have been noted; for example, sperm tails of flatworms have but one central microtubule, and sperm tails of a mayfly have no central microtubule. The main difference between a cilium and a flagellum is in their beating pattern rather than in their structure. A flagellum beats symmetrically with snakelike undulations so that water is propelled parallel to the long axis of the flagellum. A cilium, in contrast, beats asymmetrically with a fast power stroke in one direction followed by a slow recovery during which the cilium bends as it returns to its original position (Figure 29.12A). Water is propelled parallel to the ciliated surface (Figure 29.12A and B).

Muscular Movement

Contractile tissue is most highly developed in muscle cells called **fibers.** Although muscle fibers themselves can do work only by contraction and cannot actively lengthen, they can be arranged in so many different configurations and combinations that almost any movement is possible.

Types of Vertebrate Muscle

Vertebrate muscle is broadly classified on the appearance of muscle cells (fibers) when viewed with a light microscope. Both **skeletal muscle** and **cardiac muscle** appear transversely striped **(striated),** with alternating dark and light bands (Figure 29.13), although unlike skeletal muscle, cardiac muscle

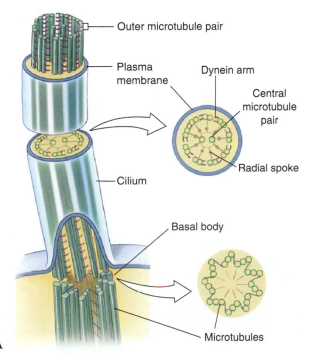

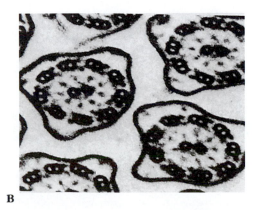

B

Figure 29.11

A, Longitudinal and cross section of a cilium showing the microtubules and microtubule-associated proteins (MAPs) of the 9 + 2 arrangement typical of both cilia and flagella. The central pair of microtubules end near the level of the cell surface. The peripheral microtubules continue inward for a short distance to compose two of each of the triplets in the basal body (kinetosome). **B,** Electron micrograph of section through several cilia. (×133,000)

A

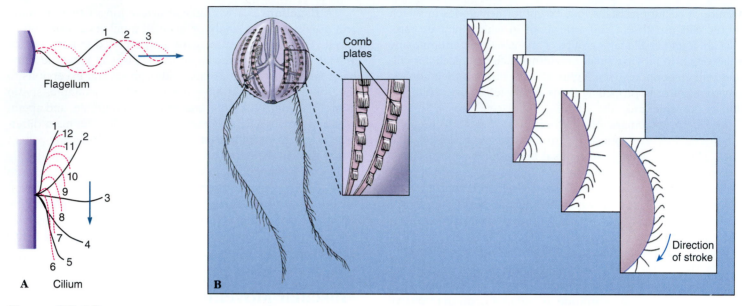

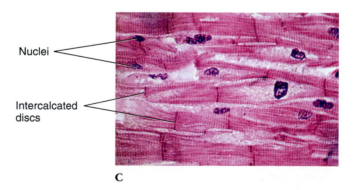

Figure 29.12

A, Flagellum beats in wavelike undulations, propelling water parallel to the main axis of itself. Cilium propels water in direction parallel to the cell surface. **B,** Movement of cilia in comb plates of a ctenophore. Note how the waves of beating comb plates pass down a comb row, opposite the direction of the power stroke of individual cilia. The movement of one comb plate lifts the plate below it and so triggers the next lower plate and so on.

is uninucleate and with branching cells. A third type of vertebrate muscle is **smooth** (or visceral) **muscle,** which lacks the characteristic alternating bands of the striated type.

Skeletal muscle is typically organized into sturdy, compact bundles or bands (Figure 29.13A). It is called skeletal muscle because it is attached to skeletal elements and is responsible for

movements of the trunk, appendages, respiratory organs, eyes, mouthparts, and other structures. Skeletal muscle **fibers** are extremely long, cylindrical, multinucleate cells that usually reach from one end of the muscle to the other. They are packed into bundles called **fascicles** (L. *fasciculus,* small bundle), which are enclosed by tough connective tissue (see Figure 29.14). The

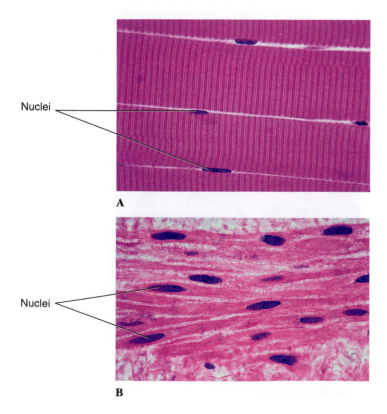

Figure 29.13

Photomicrographs of types of vertebrate muscle. **A,** Skeletal muscle (human) showing several striated fibers (cells) lying side by side. Note that nuclei are peripheral. **B,** Smooth muscle (human) showing absence of striations. Note elongate nuclei in the long fibers. **C,** Cardiac muscle (monkey) is striated, similar to skeletal muscle. Note the vertical bars, called intercalated discs, joining separate fibers end to end.

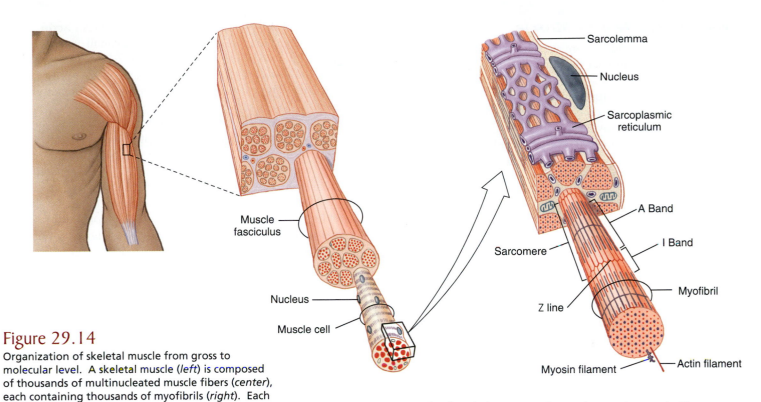

Figure 29.14

Organization of skeletal muscle from gross to molecular level. A skeletal muscle (*left*) is composed of thousands of multinucleated muscle fibers (*center*), each containing thousands of myofibrils (*right*). Each myofibril contains numerous myosin and actin filaments that interact to slide past each other during contraction to shorten the muscle. The sarcoplasmic reticulum is a network of modified endoplasmic reticulum tubules surrounding the myofibrils and serves as a reservoir for stored calcium that is released during each membrane depolarization, and initiates filament sliding during muscle contraction.

fascicles are in turn grouped into a discrete **muscle** surrounded by a thick connective tissue layer. Most skeletal muscles taper at their ends, where they connect to bones by tendons. Other muscles, such as the ventral abdominal muscles, are flattened sheets.

In most fishes, amphibians, and to some extent lizards and snakes, muscles are organized into segments that alternate with the vertebrae. Skeletal muscles of other vertebrates, by splitting, fusion, and shifting, have developed into specialized muscles best suited for manipulating jointed appendages that have evolved for locomotion on land. Skeletal muscle contracts powerfully and quickly but fatigues more rapidly than does smooth muscle. Skeletal muscle is sometimes called **voluntary muscle** because it is stimulated by motor neurons (see p. 727) under conscious control.

Muscles can only contract or shorten and relax to their original length; they provide movement only in one direction, and are therefore often grouped as an antagonistic set of muscles. Antagonistic muscles are functional opposites that oppose each other's action. For example, the biceps brachii on one side of the upper arm is opposed in its action by the triceps brachii on the opposite side of the arm. By contracting against each other, they balance and smooth rapid movements in two different directions.

Smooth muscle lacks the striations typical of skeletal muscle (Figure 29.13B). The cells are much smaller, tapering strands, each containing a single, central nucleus. The cells interdigitate with each other such that the tapered end of one cell lies close to the central nuclear region of the next. Smooth muscle cells are organized into sheets of muscle encircling cavities and tubular structures of the body, such as the walls of the alimentary canal, blood vessels, respiratory passages, and urinary and genital ducts. Smooth muscle is typically slow acting and can maintain prolonged contractions with very little energy expenditure. It is under the control of the autonomic nervous system (see p. 738), hormones and local mechanisms; thus, unlike skeletal muscle, its contractions are involuntary and unconscious. Smooth muscle functions by sustained contraction or relaxation. For example, smooth muscles push material through a tube, such as the intestine, by active contractions or they change the diameter of a tube to regulate fluid or air flow, such as in a blood vessel or airway.

Cardiac muscle, the seemingly tireless muscle of the vertebrate heart, combines certain characteristics of both skeletal and smooth muscle (Figure 29.13C). It is fast acting and striated like skeletal muscle, but contraction is under involuntary autonomic and hormonal control like smooth muscle. External control mechanisms serve only to modulate the intrinsic rate of contraction; the heartbeat originates within specialized cardiac muscle, and the heart continues to beat even if removed from the body (heart excitation is described on p. 694). Cardiac muscle is composed of closely opposed, but separate, uninucleate cell fibers joined to each other by junctional complexes (see Chapter 3, p. 46) within vertical bars called **intercalated discs.**

Types of Invertebrate Muscle

Smooth and striated muscles are also characteristic of invertebrate animals, as well as another type called oblique striated muscle. There are many variations of all three types and even instances in which structural and functional features of vertebrate smooth and striated muscle are combined. Striated muscle appears in invertebrate groups as diverse as cnidarians and arthropods. The thickest muscle fibers known, approximately 3 mm in diameter and 6 cm long, are those of giant barnacles and of Alaska king crabs living along the Pacific coast of North America. Such large muscle cells lend themselves well to physiological studies and are understandably popular with muscle physiologists.

In the limited space available to treat the great diversity of muscle structure and function in the invertebrate assemblage, we have selected for discussion two functional extremes: the specialized adductor muscles of molluscs and the fast flight muscles of insects.

Bivalve molluscan muscles contain fibers of two types. One kind is striated muscle that can contract rapidly, enabling the bivalve to snap shut its valves when disturbed. Scallops use these "fast" muscle fibers to swim in their awkward manner (see Figure 16.25B, p. 347). The second muscle type is smooth muscle, capable of slow, long-lasting contractions. Using these fibers, a bivalve can keep its valves tightly shut for hours or even days. Such adductor muscles use little metabolic energy and receive remarkably few neural signals to maintain the activated state. The contracted state has been likened to a "catch mechanism" involving a low rate of cross-bridge cycling (see p. 661) between contractile proteins within the muscle fiber with a low energy expenditure. However, research continues to clarify the mechanism, and similar mechanisms have been discovered in some types of vertebrate smooth muscle.

Insect flight muscles are virtually the functional antithesis of the slow, holding adductor muscles of bivalves. The wings of some small flies operate at frequencies greater than 1000 beats per second. The **fibrillar muscle,** which contracts at these frequencies—far greater than even the most active of vertebrate muscles—shows unique characteristics. It has very limited extensibility; the wing leverage system is arranged so that the muscles shorten only slightly during each downbeat of the wings. Furthermore, muscles and wings operate as a rapidly oscillating system in an elastic thorax (see Figure 21.11, p. 447). Since the muscles recoil elastically and are activated by stretch during flight, they receive excitatory neural signals only periodically rather than one signal per contraction; one reinforcement signal for every 20 or 30 contractions is enough to keep the system active. Insect flight muscles are described in more detail in Chapter 21 (pp. 446–447).

Structure of Striated Muscle

As mentioned on page 655, striated muscle is so named because of periodic bands, plainly visible under the light microscope, that pass across the widths of muscle cells. Each cell, or **fiber,** is a multinucleated tube containing numerous **myofibrils,** packed together and invested by a plasma membrane, the **sarcolemma** (Figure 29.14). The myofibril contains two types of **filaments,** composed of the protein **myosin,** and the protein **actin.** These are the contractile proteins of the muscle. Actin filaments are held together by a dense structure called the Z line. The functional unit of the myofibril, the **sarcomere,** extends between successive Z lines. These anatomical relationships are diagrammed in Figure 29.14.

Each myosin filament is composed of many myosin molecules packed together in an elongate bundle (Figure 29.15). Each myosin molecule contains two polypeptide chains, each having a club-shaped head (Figure 29.15A). They are lined up in two bundles to form a myosin filament. The two myosin bundles are held end to end at the center of each sarcomere so that the double heads of each myosin molecule face outward from the center of the filament and point toward the Z lines to which the actin filaments are attached (Figure 29.15B). The myosin heads act as binding sites for high-energy ATP and during muscle contraction they form molecular cross bridges that interact with the actin filaments.

Human muscle tissue develops before birth, and a newborn child's complement of skeletal muscle fibers is similar to that of an adult. Although an adult male weight lifter and a young boy have a similar number of muscle fibers, the weight lifter may be several times the boy's strength because repeated high-intensity, short-duration exercise has induced the synthesis of additional actin and myosin filaments. Each fiber has hypertrophied, becoming larger and stronger. This type of exercise favors hypertrophy of fast glycolytic fibers (see p. 661) that fatigue quickly. Endurance exercise such as long-distance running produces a very different response. Fast oxidative and intermediate fiber types are stimulated (see p. 661), and develop more mitochondria and myoglobin and become adapted for a high rate of oxidative phosphorylation. These changes, together with the development of more capillaries serving the fibers, lead to increased capacity for long-duration activity rather than increasing the strength of contraction.

Actin filaments are composed of a backbone of a double strand of the protein actin, twisted into a double helix. In addition, two actin-binding proteins, tropomyosin and troponin, form part of the actin filament complex. They are important in regulating the interaction of actin and myosin during muscle contraction. Two thin strands of **tropomyosin** lie near the grooves between the actin strands. Each tropomyosin strand is itself a double helix as shown in Figure 29.15C. **Troponin,** a complex of three globular proteins, is located at intervals along the actin filament. Troponin acts as a calcium-dependent switch that controls the contraction process.

The actin filament complexes extend outward from both sides of the Z line and overlap with myosin bundles toward the center of each sarcomere (Figures 29.15B and 29.16).

Sliding Filament Hypothesis of Muscle Contraction

In the 1950s the English physiologists A. F. Huxley and H. E. Huxley independently proposed the **sliding filament hypothesis**

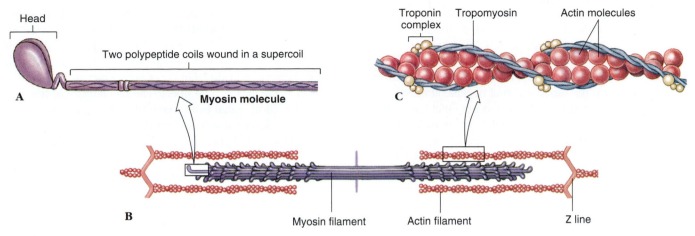

Figure 29.15

Molecular structure of actin and myosin filaments of skeletal muscle. **A,** The myosin molecule is composed of two polypeptides coiled together and expanded at their ends into a globular head. **B,** The myosin filament is composed of a bundle of myosin molecules with the globular heads extended outward toward the actin filaments on both sides. **C,** The actin filament consists of a double strand of actin surrounded by two tropomyosin strands. A globular protein complex, troponin, occurs in pairs at every seventh actin unit. Troponin is a calcium-dependent switch that controls the interaction between actin and myosin.

to explain striated muscle contraction. According to this hypothesis, the actin and myosin filaments become linked together by molecular cross bridges, which act as levers to pull the filaments past each other. During contraction, the club-shaped heads on the myosin filaments form cross bridges that move rapidly back and forth, alternately attaching to and releasing from receptor sites on the actin filaments, and drawing actin filaments past the myosin filaments in a ratchetlike action. As contraction continues, the Z lines are pulled closer together (Figure 29.16). Thus the sarcomere shortens. Because all sarcomere units shorten together, the whole muscle contracts. Relaxation is a passive process. When cross bridges between the actin and myosin filaments release, the sarcomeres are free to lengthen. This requires some force, which is supplied by recoil of elastic fibers within the connective tissue layers of the muscle (see p. 196) and by antagonistic muscles or the force of gravity.

Control of Contraction

Muscle contracts in response to nerve stimulation. If the nerve supply to a muscle is severed, the muscle **atrophies,** or wastes away. Skeletal muscle fibers are innervated by motor neurons whose cell bodies are located in the central nervous system (brain and spinal cord) (see p. 734). Each cell body gives rise to a motor axon that leaves the central nervous system to travel by way of a peripheral nerve trunk to a muscle where it branches repeatedly into many terminal branches. Each terminal branch innervates a single muscle fiber. Depending on the type of muscle, a single

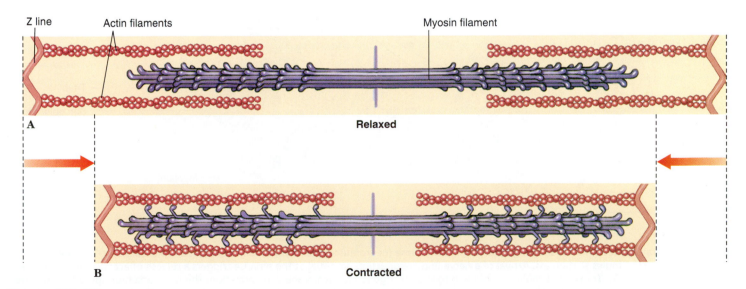

Figure 29.16

Sliding filament hypothesis, showing how actin and myosin filaments interact during contraction. **A,** Muscle relaxed. **B,** Muscle contracted.

motor axon may innervate as few as three or four muscle fibers (where very precise control is needed, such as the muscles that control eye movement) or as many as 2000 muscle fibers (where precise control is not required, such as large leg muscles). A motor neuron and all muscle fibers it innervates is called a **motor unit.** The motor unit is the functional unit of skeletal muscle control. When a motor neuron fires, the action potential passes to all fibers of the motor unit and each is stimulated to contract simultaneously. Total force exerted by a muscle depends on the number of motor units activated. Precise control of movement is achieved by varying the number of motor units activated at any one time. A smooth and steady increase in muscle tension is produced by increasing the number of motor units brought into play; this is called motor unit **recruitment.**

The Neuromuscular Junction

The place where a motor axon terminates on a muscle fiber is called the **neuromuscular** (or **myoneural**) **junction** (Figure 29.17). At the junction is a tiny gap, or **synaptic cleft,** that thinly separates a nerve terminal and muscle fiber. Close to the junction, the neuron stores a chemical, **acetylcholine,** in minute vesicles known as **synaptic vesicles.** Vesicles of acetylcholine are released into the synaptic cleft when a nerve signal or action potential (see Chapter 33, p. 728) reaches a synapse. Acetylcholine is a chemical mediator or **neurotransmitter** that diffuses

across the synaptic cleft and acts on the muscle fiber membrane, or sarcolemma, by binding to receptor sites and generating an electrical depolarization (see p. 728). The depolarization spreads rapidly over the muscle-fiber sarcolemma. Thus the synapse is a special chemical bridge that couples together the electrical activities of nerve and muscle fibers. The mechanism of transmission of an electrical signal from nerve fiber to muscle is very similar to signal transmission between two nerve fibers described on page 731 and in Figures 33.7 and 33.8 (p. 731 and p. 732).

Built into vertebrate skeletal muscle is an elaborate conduction system that serves to carry the depolarization from the neuromuscular junction to the densely packed filaments within the fiber. Along the surface of the sarcolemma are numerous invaginations of the membrane that project into the muscle fiber as a system of tubules, called **T-tubules** (Figure 29.17). The membrane depolarization passes down these T-tubules and into the muscle fiber. The T-tubules are closely associated with the **sarcoplasmic reticulum,** a system of modified endoplasmic reticulum (p. 42) that runs parallel to the actin and myosin filaments. The sarcoplasmic reticulum stores calcium and its release around the actin and myosin filaments enables the muscle fiber to contract.

Excitation-Contraction Coupling

How does electrical depolarization of the sarcolemma and T-tubules activate the contractile machinery? In resting,

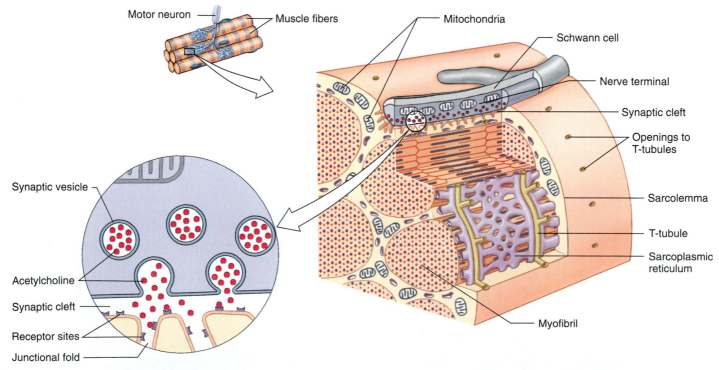

Figure 29.17

Section of vertebrate skeletal muscle showing a nerve-muscle synapse (neuromuscular or myoneural junction), sarcoplasmic reticulum, and connecting transverse tubules (T-tubules). Arrival of a nerve impulse or action potential at the synapse triggers a release of acetylcholine into the synaptic cleft (*inset at left*). Transmitter molecules binding to receptors generate membrane depolarization. This depolarization spreads across the sarcolemma, into the T-tubules, and to the sarcoplasmic reticulum where the sudden release of calcium sets in motion the contractile machinery of the myofibril.

unstimulated muscle, shortening does not occur because thin tropomyosin strands surrounding the actin filaments lie in a position that prevents the myosin heads from attaching to actin. When muscle is stimulated and the action potential is Transmitted down the T-tubules, the electrical depolarization stimulates the sarcoplasmic reticulum surrounding the fibrils to release calcium ions (Figure 29.18). The calcium binds to the actin-binding protein, troponin. Troponin immediately undergoes changes in shape that causes tropomyosin to move out of its blocking position, exposing active sites on the actin filaments. Myosin heads then bind to these sites, forming cross bridges between adjacent myosin and actin filaments. This sets in motion an attach-pull-release cycle, or **cross-bridge cycling,** that occurs in a series of steps as shown in Figure 29.18. Release of bond energy from ATP hydrolysis activates the myosin head, which swings 45 degrees, at the same time releasing a molecule of ADP. This is the power stroke that pulls the actin filament a distance of about 10 nm, and it comes to an end when phosphate is released and another ATP molecule binds to the myosin head, freeing it from the active site. Thus each cycle requires expenditure of energy in the form of ATP (Figure 29.18).

Shortening continues as long as nerve action potentials arrive at the neuromuscular junction and free calcium remains available around the actin and myosin filaments. The cross-bridge cycling can repeat again and again, 50 to 100 times per second, pulling actin and myosin filaments past each other. While the distance each sarcomere can shorten is very small, this distance is multiplied by the thousands of sarcomeres lying end to end in a muscle fiber. Consequently, a strongly contracting muscle may shorten by as much as one-third its resting length.

When stimulation stops, calcium is quickly pumped back into the sarcoplasmic reticulum. Troponin resumes its original configuration; tropomyosin moves back into its blocking position on actin, and the muscle relaxes.

Energy for Contraction

Muscle contraction requires large amounts of energy. ATP is the immediate source of energy, and its levels within muscle are held almost constant due to immediate replenishment from three main sources. Glucose is transported to muscle in the blood where it is catabolized during **aerobic metabolism** (see Chapter 4, p. 64) to produce ATP. **Glycogen** store within muscle can also supply glucose molecules for ATP production. Finally, muscles have an energy reserve in the form of **creatine phosphate.**

Glycogen is a polysaccharide chain of glucose molecules (p. 25) stored in both liver and muscle. Muscle has by far the larger store—three-fourths of all glycogen in the body is stored in muscle. As a supply of energy for contraction, glycogen has three important advantages: it is relatively abundant, it can be mobilized quickly, and it can provide energy under anoxic conditions. Enzymes convert glycogen to glucose-6-phosphate molecules, the first stage of glycolysis that leads to generation of ATP (p. 64).

Creatine phosphate is a high-energy phosphate compound that stores bond energy during periods of rest. As ADP is produced from ATP during contraction, creatine phosphate releases its stored bond energy to convert ADP to ATP. This reaction can be summarized as:

$$\text{Creatine Phosphate} + \text{ADP} \rightarrow \text{ATP} + \text{Creatine}$$

Some muscle types (**slow** and **fast oxidative fibers,** next section) rely heavily on glucose and oxygen supplies transported to muscle via the circulatory system. If muscular contraction is not too vigorous or too prolonged, glucose from the blood or released from glycogen can be completely oxidized to carbon dioxide and water by aerobic metabolism. During prolonged or heavy exercise, however, blood flow to the muscles, although greatly increased above resting levels, cannot supply oxygen to the mitochondria rapidly enough to complete oxidation of glucose. The contractile machinery then receives its energy largely by **anaerobic glycolysis,** a process that does not require oxygen (p. 67). The ability to take advantage of this anaerobic pathway, although not nearly as efficient as the aerobic one, is of great importance; without it, all forms of heavy muscular exertion would be impossible. Indeed, **fast glycolytic fibers** (next section) rely almost exclusively on anaerobic glycolysis to produce energy for contraction.

During anaerobic glycolysis, glucose is degraded to lactic acid with release of energy. Lactic acid accumulates in the muscle and diffuses rapidly into the general circulation. If muscular exertion continues, muscle fatigue occurs. This was originally attributed to the buildup of lactic acid causing enzyme inhibition. However, more recent data suggest that accumulation of inorganic phosphate causes muscle fatigue, at least in muscle types that rely heavily on creatine phosphate stores. Thus the anaerobic pathway is a self-limiting one, since continued heavy exertion leads to exhaustion. The muscles incur an **oxygen debt** because accumulated lactic acid must be oxidized by extra oxygen. After a period of exertion, oxygen consumption remains elevated until all lactic acid has been oxidized or consumed in resynthesis of glycogen.

Muscle Performance

Fast and Slow Fibers

Skeletal muscles of vertebrates consist of more than one type of fiber. **Slow oxidative fibers,** which are specialized for slow, sustained contractions without fatigue, are important in maintaining posture in terrestrial vertebrates. Such muscles are often called **red muscles** because they contain an extensive blood supply, a high density of mitochondria for supplying ATP via aerobic metabolism, and abundant stored myoglobin which supplies oxygen reserves, all of which give the muscle a red color.

Two kinds of **fast fibers,** capable of fast, powerful contractions are known. One kind of fast fiber (**fast glycolytic fiber**) lacks an efficient blood supply and a high density of mitochondria and myoglobin. Muscles composed of these fibers (often called **white muscles**) are usually pale in color, function anaerobically, and fatigue rapidly. The "white meat" of chicken is a familiar example. Members of the cat family have

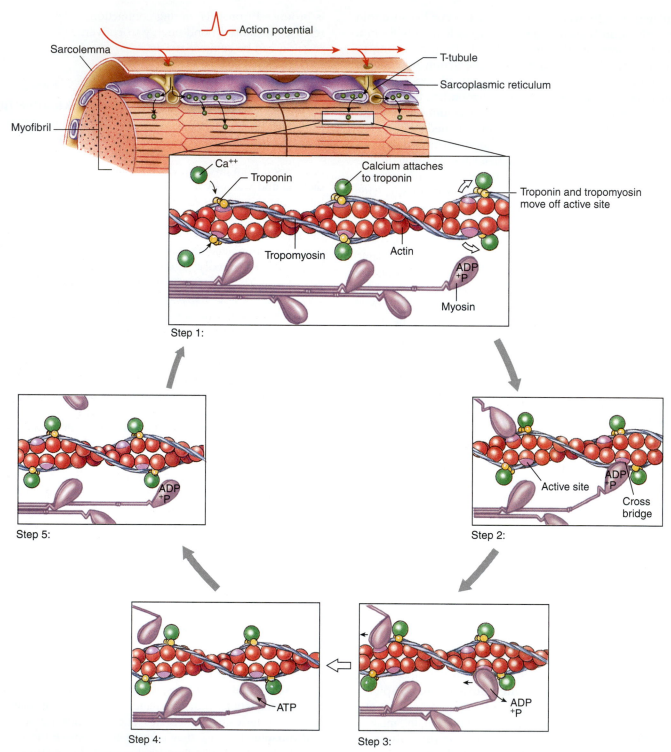

Figure 29.18

Excitation-contraction coupling in vertebrate skeletal muscle. **Step 1:** An action potential spreads along the sarcolemma and is conducted inward to the sarcoplasmic reticulum by way of T-tubules. Calcium ions are released from the sarcoplasmic reticulum and diffuse rapidly into the myofibrils, binding to troponin molecules on the actin filament. Troponin and tropomyosin molecules are moved away from the active sites. **Step 2:** Myosin cross bridges bind to the exposed active sites. **Step 3:** Using the energy stored in ATP, the myosin head swings toward the center of the sarcomere. ADP and a phosphate group are released. **Step 4:** The myosin head binds another ATP molecule; this frees the myosin head from the active site on actin. **Step 5:** The myosin head splits ATP, retaining the energy released as well as the ADP and the phosphate group. The cycle can now be repeated as long as calcium is present to open active sites on the actin molecules.

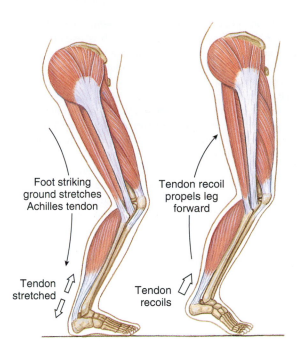

Figure 29.19

Energy storage in the Achilles tendon of human and kangaroo legs. During running, stretching of the Achilles tendon when the foot strikes the ground stores kinetic energy that is released to propel the leg forward.

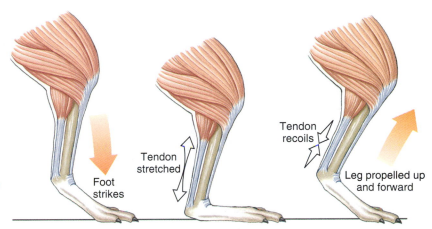

running muscles formed almost entirely of fast glycolytic fibers that operate anaerobically. During a chase, such muscles accumulate a substantial oxygen debt, which is replenished after the chase. For example, a cheetah after a high-speed chase lasting less than a minute will pant heavily for 30 to 40 minutes before its oxygen debt is relieved. Weight lifters favor activation and development of these muscle fibers and thus cannot sustain lifting heavy weights for long periods. The other kind of fast fiber (**fast oxidative fiber**) has an extensive blood supply and a high density of mitochondria and myoglobin, and functions largely aerobically. Some animals use these muscles for rapid, sustained activities. Most migratory birds, such as geese and swans, as well as dogs and ungulates (hoofed mammals), for example, have limb (or wing) muscles with a high percentage of fast oxidative fibers, and are capable of active locomotion for long periods of time. Most muscles possess a mixture of these different fiber types which permit a range of activity.

Importance of Tendons in Energy Storage

When mammals walk or run, much kinetic energy is stored from step to step as elastic strain energy in the tendons. For example, during running the Achilles tendon is stretched by a combination of downward force of the body on the foot and contraction of the calf muscles. The tendon then recoils, extending the foot while the muscle is still contracted, propelling the leg forward (Figure 29.19). An extreme example of this bouncing-ball principle is the bounding of a kangaroo (Figure 29.19). This type of movement uses far less energy than would be required if every step relied solely on alternate muscle contraction and relaxation.

There are many examples of elastic storage in the animal kingdom. It is used in the ballistic jumps of grasshoppers and fleas, in the wing hinges of flying insects, in the hinge ligaments of bivalve molluscs, and in the highly elastic large dorsal ligament (ligamentum nuchae) that helps to support the head of hoofed mammals.

SUMMARY

An animal is wrapped in a protective covering, the integument, which may be as simple as the delicate plasma membrane of an ameba or as complex as the skin of a mammal. The arthropod exoskeleton is the most complex of invertebrate integuments, consisting of a two-layered cuticle secreted by a single-layered epidermis. It may be hardened by calcification or sclerotization and must be molted at intervals to permit body growth. Vertebrate integument consists of two layers: the epidermis, which gives rise to various derivatives such as hair, feathers, and claws; and the dermis, which supports and nourishes the epidermis. It also is the origin of bony derivatives such as fish scales and deer antlers.

Integument color is of two kinds: structural color, produced by refraction or scattering of light by particles in the integument, and pigmentary color, produced by pigments that are usually confined to special pigment cells (chromatophores).

Skeletons are supportive systems that may be hydrostatic or rigid. The hydrostatic skeletons of several soft-walled invertebrate groups depend on body-wall muscles that contract against a noncompressible internal fluid of constant volume. In a similar manner, muscular hydrostats, such as the tongue of mammals and reptiles, and the trunk of elephants, rely on muscle bundles arranged in complex patterns to produce movement without either skeletal support or a liquid-filled cavity. Rigid skeletons have evolved with attached muscles that act with the supportive skeleton to produce movement. Arthropods have an external skeleton, which must be shed periodically to make possible an enlarged replacement. The vertebrates developed an internal skeleton, a framework formed of cartilage and/or bone, that can grow with the animal, while, in the case of bone, additionally serving as a reservoir of calcium and phosphate.

Animal movement, whether in the form of cytoplasmic streaming, ameboid movement, or the contraction of an organized muscle mass, depends on specialized contractile proteins. The most important of these is the actomyosin system, which is usually organized into elongate actin and myosin filaments that slide past one another during contraction. When a muscle is stimulated, an electrical depolarization is conducted into the muscle fibers through T-tubules to the sarcoplasmic reticulum, causing the release of calcium. Calcium binds to a protein-troponin complex associated with the actin filament. This binding causes tropomyosin to shift from its blocking position and allows the myosin heads to form cross bridges with the actin filament. Powered by ATP, the myosin heads swivel back and forth to pull the actin and myosin filaments past each other. Phosphate bond energy for contraction is supplied by carbohydrate fuels.

Vertebrate skeletal muscle consists of variable percentages of both slow fibers, used principally for sustained postural contractions, and fast fibers, used in locomotion. Tendons are important in locomotion because the kinetic energy stored in stretched tendons at one stage of a locomotory cycle is released at a subsequent stage.

REVIEW QUESTIONS

1. The arthropod exoskeleton is the most complex of invertebrate integuments. Describe its structure, and explain the difference in the way cuticle is hardened in crustaceans and in insects.

2. Distinguish between epidermis and dermis in vertebrate integument, and describe structural derivatives of these two layers.

3. What is the difference between structural colors and colors based on pigments? How do the chromatophores of vertebrates and cephalopod molluscs differ in structure and function?

4. As "naked apes" humans lack the protective investment of fur that shields other mammals from the damaging effects of sunlight. How does human skin respond to ultraviolet radiation in the short term and with continued exposure?

5. Hydrostatic skeletons have been defined as a mass of fluid enclosed within a muscular wall. How would you modify this definition to make it apply to a muscular hydrostat? Offer examples of both hydrostatic skeleton and muscular hydrostat.

6. One of the special qualities of vertebrate bone is that it is a living tissue that permits continuous remodeling. Explain how the structure of bone allows this remodeling to happen.

7. What is the difference between endochondral and membranous bone? Between spongy and compact bone?

8. Discuss the role of osteoclasts, osteoblasts, parathyroid hormone, and calcitonin in bone growth.

9. The laws of scaling tell us that doubling the length of an animal will increase its weight eightfold while the force its bones can bear increases only fourfold. What solutions to this problem have evolved that allow animals to become large, while maintaining bone stresses within margins of safety?

10. Name the major skeletal components included in the axial and in the appendicular skeleton.

11. An unexpected discovery from studies of ameboid movement is that the same proteins found in the contractile system of metazoan muscle—actin and myosin—are present in ameboid cells. Explain how these and other proteins are believed to interact during ameboid movement.

12. A "9 + 2" arrangement of microtubules is typical of both cilia and flagella. Explain how this system is thought to function to produce a bending motion. What is the difference between a cilium and a flagellum?

13. What functional features of molluscan smooth muscle and insect fibrillar muscle set them apart from typical vertebrate muscle?

14. The sliding filament model of skeletal muscle contraction assumes a sliding or slipping of interdigitating filaments of actin and myosin. Electron micrographs show that during contraction the actin and myosin filaments remain of constant length while the distance between Z lines shortens. Explain how this happens in terms of the molecular structure of the muscle filaments. What is the role of regulatory proteins (troponin and tropomyosin) in contraction?

15. While the sarcoplasmic reticulum of muscle was first described by nineteenth-century microscopists, its true significance was not appreciated until its intricate structure was revealed much later by the electron microscope. What could you tell a nineteenth-century microscopist to enlighten him or her about the structure of the sarcoplasmic reticulum and its role in the coupling of excitation and contraction?

16. The filaments of skeletal muscle are moved by free energy derived from the hydrolysis of ATP. During sustained muscle contraction ATP levels remain fairly constant, while levels of creatine phosphate fall. Explain why this is so. Under what circumstances is an oxygen debt incurred during muscle contraction?

17. During evolution, skeletal muscle became adapted to functional demands ranging from sudden, withdrawal movements of a startled worm, to the sustained contractions required to maintain mammalian posture, to supporting a long, fast chase across the African savanna. What are some of the fiber types in vertebrate muscle that evolved to support these kinds of activities?

SELECTED REFERENCES

Alexander, R. M. 1992. The human machine. New York, Columbia University Press. *Describes all kinds of human movement with the human body viewed as an engineered machine. Well chosen illustrations.*

Anderson, J. I., Schjerling, P., and Saltin, B. 2000. Muscle, genes and athletic performance. Sci. Am. **283:**48–55 (Sept.). *Good discussion of muscle structure and function related to human athletic performance.*

Hadley, N. F. 1986. The arthropod cuticle. Sci. Am. **255:**104–112 (July). *Describes properties of this complex covering that account for much of the adaptive success of arthropods.*

Hutchinson, J. R., and M. Garcia. 2002. *Tyrannosaurus* was not a fast runner. Nature **415:**1018–1021. *Analysis concludes that* Tyrannosaurus *would have had less than half enough leg muscle mass to run and could only walk.*

Leffell, D. J., and D. E. Brash. 1996. Sunlight and skin cancer. Sci. Am. **275:**52–59 (July). *Skin cancer that appears in older people begins with damage received decades earlier. Many cases are caused by a mutation in a single gene.*

Marx, J. 2004. Coming to grips with bone loss. Science **305:**1420–1422. *An interesting review of new therapies for prevention and treatment of osteoporosis.*

Randall, D., W. Burggren, and K. French. 2001. Eckert animal physiology: mechanisms and adaptations. New York, W. H. Freeman & Company. *A comprehensive and comparative treatment of animal physiology.*

Stossel, T. P. 1994. The machinery of cell crawling. Sci. Am. **271:**54–63 (Sept.). *Cell crawling depends on the orderly assembly and disassembly of an actin protein scaffold.*

Westerblad, H., D. G. Allen, and J. Lännergren. 2002. Muscle fatigue: lactic acid or inorganic phosphate the major cause? News Physiol. Sci. **17:**17–21. *Well-written review of recent data that provides an alternate explanation for muscle fatigue.*

Willmer, P., G. Stone, and I. Johnston. 2005. Environmental physiology of animals, ed. 2. Oxford, U.K., Blackwell Science Ltd. *Well-written information regarding environmental adaptations in both vertebrates and invertebrates.*

ONLINE LEARNING CENTER

Visit www.mhhe.com/hickmanipz14e for chapter quizzing, key term flash cards, web links, and more!

30

Homeostasis: Osmotic Regulation, Excretion, and Temperature Regulation

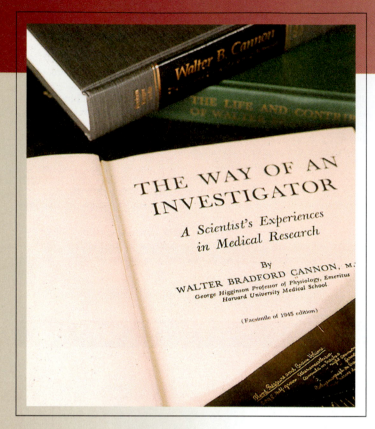

Title page of Walter B. Cannon's autobiography.

Homeostasis: Birth of a Concept

The tendency toward internal stabilization of the animal body was first recognized by Claude Bernard, great French physiologist of the nineteenth century who, through his studies of blood glucose and liver glycogen, first discovered internal secretions. After a lifetime of study and experimentation, Bernard gradually developed the principle for which he is best remembered, that of constancy of the internal environment, a principle that in time would pervade physiology and medicine. Years later, at Harvard University, American physiologist Walter B. Cannon (Figure 30.1) reshaped and restated Bernard's idea. From his studies of the nervous system and reactions to stress, he described the ceaseless balancing and rebalancing of physiological processes that maintain stability and restore the normal state when it has been disturbed. He also gave it a name: **homeostasis.** The term soon flooded the medical literature of the

1930s. Physicians spoke of getting their patients back into homeostasis. Even politicians and sociologists saw what they considered deep nonphysiological implications. Cannon enjoyed this broadened application of the concept and later suggested that democracy was the form of government that took a homeostatic middle course. Despite the enduring importance of the homeostasis concept, Cannon never received the Nobel Prize—one of several acknowledged oversights of the Nobel Committee. Late in life, Cannon expressed his ideas about scientific research in his autobiography, *The Way of an Investigator.* This engaging book describes the resourceful career of a homespun man whose life embodied the traits that favor successful research.

Figure 30.1
Walter Bradford Cannon (1871 to 1945), Harvard professor of physiology who coined the term "homeostasis" and developed the concept originated by French physiologist Claude Bernard (Figure 31.2, p. 688).

The concept of homeostasis, described in the chapter opening essay, permeates all physiological thinking and is the theme of this chapter and Chapter 31. Although this concept was first developed from studies with mammals, it applies to single-celled organisms as well as to multicellular organisms. Potential changes in the internal environment arise from two sources. First, metabolic activities require a constant supply of materials, such as oxygen, nutrients, and salts, that cells withdraw from their surroundings and that must be continuously replenished. Cellular activity also produces waste products that must be expelled. Second, the internal environment responds to changes in the organism's external environment. Changes from either source must be stabilized by the physiological mechanisms of homeostasis.

In more complex metazoans, cellular homeostasis is maintained by the coordinated activities of all systems of the body, except for the reproductive system. The various homeostatic activities are orchestrated by the circulatory, nervous, and endocrine systems, and also by organs that serve as sites of exchange with the external environment. These last include kidneys, lungs or gills, digestive tract, and integument. Through these organs oxygen, nutrients, minerals, and other constituents of body fluids enter, water is exchanged, heat is lost, and metabolic wastes are eliminated.

Thus, systems within an organism function in an integrated way to maintain a constant internal environment around a setpoint. Small deviations from this setpoint in variables such as pH, temperature, osmotic pressure, metabolic fuels (for example, glucose or fatty acids), carbon dioxide, and oxygen levels activate physiological mechanisms that return the variable to its setpoint by a process called **negative feedback** regulation (see Chapter 34, p. 756).

We look first at the problems of controlling the internal fluid environment of animals living in aquatic habitats. Next we briefly examine how these problems are solved by terrestrial animals and consider the function of the organs that regulate their internal state. Finally we look at strategies that have evolved for living in a world of changing temperatures.

WATER AND OSMOTIC REGULATION

Water and osmotic regulation provide a means for maintaining internal solute concentrations within a range that allows cellular function to proceed optimally. As discussed in Chapter 3 (p. 46), the selective permeability of cell membranes means that changes in ion concentrations on either side of the membrane will dramatically alter ion and water flow through the membrane. Cell volume will rise and fall if cells are exposed to a hypoosmotic or hyperosmotic environment, respectively, and both changes will produce negative effects on cellular metabolism. The concept of water and osmotic regulation applies to single-celled and multicellular animals alike; however, multicellular animals can control ion and water balance inside cells by regulating the ion and water content of the fluids that bathe them.

How Marine Invertebrates Meet Problems of Salt and Water Balance

Most marine invertebrates are in osmotic equilibrium with their seawater environment. With body surfaces permeable to salts and water, their body-fluid concentration rises or falls in conformity with changes in concentrations of seawater. Because such animals are incapable of regulating osmotic pressure of their body fluid, they are called **osmotic conformers.** Invertebrates living in the open sea are seldom exposed to osmotic fluctuations because the ocean is a highly stable environment. Oceanic invertebrates have, in fact, very limited abilities to withstand osmotic change. Should they be exposed to dilute seawater, they absorb water by osmosis and die quickly because their body's cells cannot tolerate dilution and are helpless to prevent it. These animals are restricted to living in a narrow salinity range and are said to be **stenohaline** (Gr. *stenos,* narrow, + *hals,* salt). An example is the marine spider crab, *Maia* (Figure 30.2).

Conditions along coasts and in estuaries and river mouths are much less constant than those of the open ocean. Here animals must cope with large and often abrupt changes in salinity as the tides ebb and flow and mix with freshwater draining from rivers. These animals are termed **euryhaline** (Gr. *eurys,* broad, + *hals,* salt), meaning that they can survive a wide range of salinity changes, mainly by demonstrating v arying powers of **osmotic regulation.** For example, a brackish-water shore crab, *Eriocheir,* can resist dilution of body fluids by dilute (brackish) seawater (Figure 30.2). Although the concentration of salts in the body fluids falls, it does so less rapidly than the fall in seawater concentration. This crab is a **hyperosmotic regulator,** meaning that it maintains its body fluids more concentrated (hence *hyper-*) than the surrounding water.

By regulating against excessive dilution, and thus protecting the cells from extreme changes, brackish-water shore crabs can live successfully in the physically unstable but biologically rich coastal environment. Nevertheless, with only a limited capacity for osmotic

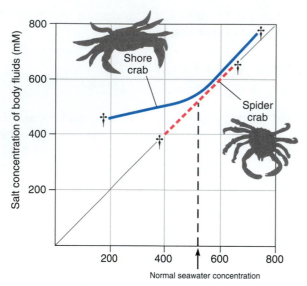

Figure 30.2

Salt concentration of body fluids of two crabs as affected by variations in seawater concentration. The 45-degree line represents equal concentration between body fluids and seawater. Since the spider crab, *Maia* sp., cannot regulate the salt concentration of its fluids, it conforms to whatever changes happen in the seawater (*dashed red line*). The shore crab, *Eriocheir* sp., however, can regulate osmotic concentration of its fluids to some degree in dilute seawater. For example, when seawater is 200 mM (millimolar), the shore crab's body fluid is approximately 430 mM (*solid blue line*). Crosses at ends of lines indicate limits of tolerance for each species beyond which they will die.

regulation, they will die if exposed to greatly diluted seawater. To understand how a brackish-water shore crab and other coastal invertebrates achieve hyperosmotic regulation, let us examine the problems they face. First, because a crab's body fluids are osmotically more concentrated than the dilute seawater outside, water flows into its body, especially across the thin, permeable membranes of the gills. As with the membrane osmometer containing a salt solution (p. 48), water diffuses inward because of higher solute concentration inside. For the crab, were this inflow of water allowed to continue unchecked, its body fluids would soon become diluted and unbalanced. The problem is solved by the kidneys (antennal glands located in the crab's head; see Figure 30.7 showing similar glands of a crayfish), which excrete excess water as dilute urine.

The second problem is salt loss. Again, because the animal is saltier than its environment, it cannot avoid loss of ions by outward diffusion across its gills. Salt is also lost in urine. To compensate for solute loss, special salt-secreting cells in the gills actively remove ions from dilute seawater and move them into the blood, thus maintaining the internal osmotic concentration. This is an **active transport** (p. 49) process that requires energy because ions must be transported against a concentration gradient from a lower salt concentration (in dilute seawater) to an higher one (in blood).

Invasion of Freshwater

Some 400 million years ago, during the Silurian and Lower Devonian periods, the major groups of jawed fishes began to penetrate into brackish-water estuaries and then gradually into freshwater rivers. Before them lay a new, unexploited habitat already stocked with food in the form of insects and other invertebrates, which had preceded them into freshwater. However, advantages that this new habitat offered were balanced by a tough physiological challenge: the necessity of developing effective osmotic regulation.

Freshwater animals face challenges similar to but more extreme than those of the shore crab just described. They must keep the salt concentration of their body fluids higher than that of the water in which they live. Water enters their bodies osmotically, and salt is lost by outward diffusion. Unlike the habitat of the brackish-water shore crab, freshwater is much more dilute than are coastal estuaries, and there is no retreat, no salty sanctuary into which a freshwater animal can retire for osmotic relief. It must and has become a permanent and highly efficient hyperosmotic regulator.

The scaled and mucus-covered body surface of a fish is nearly waterproof, yet remains flexible. In addition, freshwater fishes have several defenses against the problems of water gain and salt loss. First, water that inevitably enters the body by osmosis is pumped out by the kidney, which is capable of forming very dilute urine by reabsorbing sodium chloride (Figure 30.3A). Second, special salt-absorbing cells located in the gills transport salt ions, principally sodium and chloride (present in small quantities even in freshwater), from the water to the blood. This process, together with salt present in the fish's food, replaces diffusive salt loss. These mechanisms are so efficient that a freshwater fish devotes only a small part of its total energy expenditure to maintain osmotic balance.

Crayfishes, aquatic insect larvae, clams, and other freshwater animals are also hyperosmotic regulators and face the same hazards as freshwater fishes; they tend to gain too much water and lose too much salt. Like freshwater fishes, they excrete excess water as urine and replace lost salt by active ion transport across the gills.

Amphibians living in water also must compensate for salt loss by actively absorbing salt from the water (Figure 30.4). They use their skin for this purpose. Physiologists learned some years ago that pieces of frog skin continue to transport sodium and chloride actively for hours when removed and placed in a specially balanced salt solution. Fortunately for biologists, but unfortunately for frogs, these animals are so easily collected and maintained in the laboratory that frog skin became a favorite membrane system for studies of ion-transport phenomena.

Return of Fishes to the Sea

Bony fishes living in the oceans today are descendants of earlier freshwater bony fishes that moved back into the sea during the Triassic period approximately 200 million years ago. Over millions of years freshwater fishes established an ionic concentration in the body fluid equivalent to approximately one-third that of seawater. Body fluid of terrestrial vertebrates is remarkably similar to that of dilute seawater too, a fact that is undoubtedly related to their ancient marine heritage.

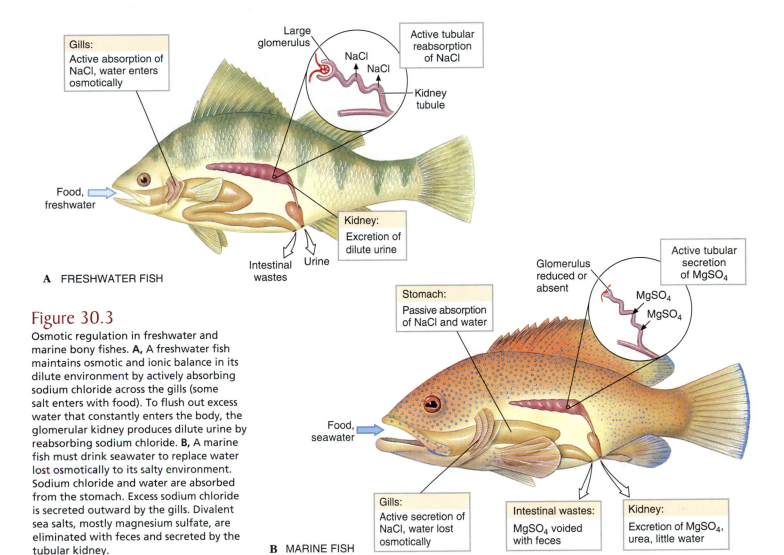

Figure 30.3

Osmotic regulation in freshwater and marine bony fishes. **A,** A freshwater fish maintains osmotic and ionic balance in its dilute environment by actively absorbing sodium chloride across the gills (some salt enters with food). To flush out excess water that constantly enters the body, the glomerular kidney produces dilute urine by reabsorbing sodium chloride. **B,** A marine fish must drink seawater to replace water lost osmotically to its salty environment. Sodium chloride and water are absorbed from the stomach. Excess sodium chloride is secreted outward by the gills. Divalent sea salts, mostly magnesium sulfate, are eliminated with feces and secreted by the tubular kidney.

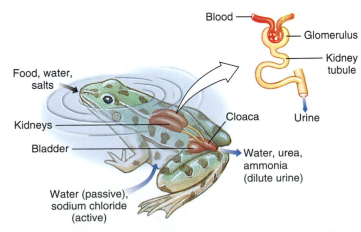

Figure 30.4

Exchange of water and solute in a frog. Water enters the highly permeable skin and is excreted by kidneys. The skin also actively transports ions (sodium chloride) from the environment. Kidneys form dilute urine by reabsorbing sodium chloride. Urine flows into the urinary bladder, where, during temporary storage, most of the remaining sodium chloride is removed and returned to the blood.

By expressing concentration of salt in seawater or body fluids in molarity, we are saying that the osmotic strength is equivalent to the molar concentration of an ideal solute having the same osmotic strength. In fact, seawater and animal body fluids are not ideal solutions because they contain electrolytes that dissociate in solution. A 1 M solution of sodium chloride (which dissociates in solution) has a much greater osmotic strength than a 1 M solution of glucose, an ideal solute (nonelectrolyte) that does not dissociate in solution. Consequently, biologists usually express osmotic strength of a biological solution in osmolarity rather than molarity. A 1 osmolar solution exerts the same osmotic pressure as a 1 M solution of a nonelectrolyte.

When some freshwater bony fishes of the Triassic period ventured back to the sea, they encountered a new set of problems. Having a much lower internal osmotic concentration than the seawater around them, they lost water and gained salt. Indeed a marine bony fish literally risks desiccation, much like a desert mammal deprived of water.

Marine bony fishes living today maintain the salt concentration of their body fluids at approximately one-third that of seawater (body fluids = 0.3 to 0.4 gram mole per liter [M]; seawater = 1M). They are **hypoosmotic regulators** because they maintain their body fluids at a lower concentration (hence *hypo-*) than their seawater environment. To compensate for water loss a marine bony fish drinks seawater (Figure 30.3B). This seawater is absorbed from the intestine, and the major sea salt, sodium chloride, is carried by the blood to the gills, where specialized salt-secreting cells transport it back into the surrounding sea. Ions remaining in the intestinal residue, especially magnesium, sulfate, and calcium, are voided with feces or excreted by kidneys. In this indirect way, marine fishes rid themselves of excess sea salts and replace water lost by osmosis. Samuel Taylor Coleridge's ancient mariner, surrounded by "water, water, everywhere, nor any drop to drink" undoubtedly would have been tormented even more had he known of the marine fishes' ingenious solution to thirst. A marine fish regulates the amount of seawater it drinks, consuming only enough to replace water loss and no more.

Sharks and rays (elasmobranchs) are almost totally marine, and achieve osmotic balance differently. The salt composition of shark's blood is similar to that of bony fishes, but the blood also carries a large content of the organic compounds urea and trimethylamine oxide (TMAO). Urea is a metabolic waste that most animals quickly excrete. The shark kidney, however, conserves urea, allowing it to accumulate in the blood and body tissues and raising the blood osmolarity to equal or slightly exceed that of seawater. With osmotic difference between blood and seawater eliminated, water balance is not a problem for sharks and other elasmobranchs; they are in osmotic equilibrium with their environment.

The high concentration of urea in the blood and body tissues of sharks and rays—more than 100 times as high as in mammals—could not be tolerated by most other vertebrates. In the latter, such high concentrations of urea disrupt peptide bonds of proteins, altering protein configuration. This would also occur in sharks and rays except that TMAO has the opposite effect and serves to stabilize proteins in the presence of high urea levels. So accommodated are elasmobranchs to urea that their tissues cannot function without it, and their heart will stop beating in its absence.

How Terrestrial Animals Maintain Salt and Water Balance

The problems of living in an aquatic environment seem small indeed compared with problems of life on land. Since animal bodies are mostly water, all metabolic activities proceed in water, and life itself was conceived in water, it might seem that animals were meant to stay in water. Yet many animals, like the plants preceding them, moved onto land, carrying their watery composition with them. Once on land, terrestrial animals continued their adaptive diversification, solving the threat of desiccation, until they became abundant even in some of the most arid parts of the earth.

Terrestrial animals lose water by evaporation from respiratory and body surfaces, excretion in urine, and elimination in feces. They replace such losses by water in food, drinking water when available, and retaining **metabolic water** formed in cells by oxidation of metabolic fuel molecules, such as carbohydrates and fats. Stored fat becomes an important source of metabolic water in diving mammals. Certain arthropods—for example, desert roaches, certain ticks and mites, and mealworms—are able to absorb water vapor directly from atmospheric air. In some desert rodents, metabolic water gain may constitute most of the animals' water intake.

Particularly revealing is a comparison of water balance in human beings, nondesert mammals that drink water, with that of kangaroo rats, desert rodents that may drink no water at all (Table 30.1). Kangaroo rats acquire all their water from their food: 90% is metabolic water derived from oxidation of fuel molecules (see Figure 4.10, p. 65) and 10% as free moisture in food. Although we eat foods with a much higher water content than the dry seeds of a kangaroo rat's diet, we still must drink half our total water requirement.

TABLE 30.1

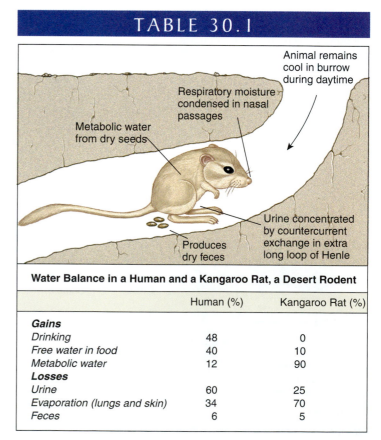

Water Balance in a Human and a Kangaroo Rat, a Desert Rodent

	Human (%)	Kangaroo Rat (%)
Gains		
Drinking	48	0
Free water in food	40	10
Metabolic water	12	90
Losses		
Urine	60	25
Evaporation (lungs and skin)	34	70
Feces	6	5

Source: Some data from K. Schmidt-Nielsen, *How animals work.* Cambridge University Press, 1972.

Given ample water to drink, humans can tolerate extremely high environmental temperatures while preventing a rise in body temperature. Our ability to keep cool by evaporation was impressively demonstrated more than 200 years ago by a British scientist who remained for 45 minutes in a room heated to 260° F (126° C). A steak he carried in with him was thoroughly cooked, but he remained uninjured and his body temperature did not rise. Sweating rates may exceed 3 liters of water per hour under such conditions and cannot be tolerated unless the lost water is replaced by drinking. Without water, a human continues to sweat unabatedly until the water deficit exceeds 10% of the body weight, when collapse occurs. With a water deficit of 12% a human is unable to swallow even if offered water, and death occurs when the water deficit reaches about 15% to 20%. Few people can survive more than a day or two in a desert without water. Thus humans are not physiologically well adapted for desert climates but may prosper there nonetheless by virtue of their technological culture.

Excretion of wastes presents a special problem in water conservation. The primary end product of protein breakdown is ammonia, a highly toxic material. Fishes easily excrete ammonia by diffusion across their gills, since there is an abundance of water to wash it away. Terrestrial insects, nonavian reptiles, and birds have no convenient way to rid themselves of toxic ammonia; instead, they convert it into uric acid, a nontoxic, almost insoluble compound. This conversion enables them to excrete a semisolid urine with little water loss. Uric acid as an end product has another important benefit. Nonavian reptiles and birds lay amniotic eggs enclosing their embryos (see Figure 26.3, p. 567), together with their stores of food and water, and wastes that accumulate during development. By converting ammonia to uric acid, a developing embryo's waste can be precipitated into solid crystals, which are stored harmlessly within the amniotic sac, inside the egg until hatching.

Marine birds and turtles have evolved an effective solution for excreting large loads of salt eaten with their food. Located above each eye is a special **salt gland** capable of excreting a highly concentrated solution of sodium chloride—up to twice the concentration of seawater. In birds the salt solution runs out the nares (see p. 597 and Figure 27.13). Marine lizards and turtles, like *Alice in Wonderland*'s Mock Turtle, shed their salt-gland secretion as salty tears. Salt glands are important accessory organs of salt excretion in these animals because their kidneys cannot produce a concentrated urine, as can mammalian kidneys.

INVERTEBRATE EXCRETORY STRUCTURES

Many protozoan groups and some freshwater sponges have special excretory organelles called contractile vacuoles. More complex invertebrates have excretory organs that are basically tubular structures forming urine by first producing an ultrafiltrate or fluid secretion from the blood. This fluid secretion enters the proximal end of the tubule and is modified continuously as it flows down the tubule. The final product is urine.

Contractile Vacuole

The tiny, spherical, intracellular vacuole of protozoa and freshwater sponges is not a true excretory organ, since ammonia and other nitrogenous wastes of metabolism readily diffuse into the surrounding water. The contractile vacuole of freshwater protozoa is an organ of water balance that expels excess water gained by osmosis. As water enters the protozoan, a vacuole grows and finally collapses as it empties its contents through a pore on the surface. The cycle is repeated rhythmically. Although the mechanism for filling the vacuole is not fully understood, recent research suggests that contractile vacuoles are surrounded by a network of membranous channels populated with numerous proton pumps (proton pumps are described in connection with the electron transport chain in Chapter 4, p. 67 and following). Proton pumps create H^+ and HCO^- gradients that draw water into the vacuole, forming an isosmotic solution. These ions are excreted when the vacuole empties (see Figure 11.16, p. 227).

Contractile vacuoles are common in freshwater protozoa, sponges, and radiate animals (such as hydra), but rare or absent in marine forms of these groups, which are isosmotic with seawater and consequently neither lose nor gain too much water.

Nephridium

The most common type of invertebrate excretory organ is the nephridium, a tubular structure designed to maintain appropriate osmotic balance. One of the simplest arrangements is the flame-cell system (or **protonephridium**) of acoelomates (flatworms) and some pseudocoelomates.

In planaria and other flatworms the protonephridial system takes the form of two highly branched duct systems distributed throughout the body (Figure 30.5). Fluid enters the system through specialized "flame cells." The rhythmical beat of the flagellar tuft within each flame cell, suggestive of a tiny flickering flame, creates a negative pressure that draws fluid into the tubular portion of the system. As the fluid moves down the tubule, water and metabolites valuable to the body are recovered by reabsorption, leaving wastes behind to be expelled through excretory pores that open at intervals on the body surface. Nitrogenous wastes (mainly ammonia) diffuse across the entire body surface.

The flame-cell system is extensively branched throughout a flatworm's body. Thus, these acoelomate animals have no necessity for a circulatory system to deliver wastes to a centralized excretory system (such as the kidneys of vertebrates and many invertebrates).

The protonephridium just described is a **closed** system. The tubules are closed on the inner end and urine is formed from a fluid that must first enter the tubules by being transported across flame cells. A more advanced type of nephridium is the **open,** or "true," nephridium **(metanephridium)** found in several eucoelomate phyla such as annelids (Figure 30.6), molluscs,

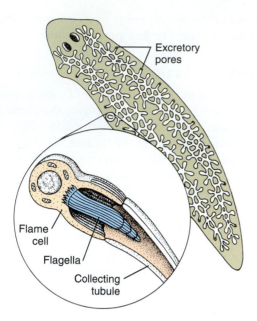

Figure 30.5

Flame-cell system of a flatworm. Body fluids collected by flame cells (protonephridia) are passed down a system of ducts to excretory pores on the body surface.

and several smaller phyla. A metanephridium is more advanced than a protonephridium in two important ways. First, the tubule is open at *both* ends, allowing fluid to be swept into the tubule through a ciliated funnel-like opening, the **nephrostome.** Second, a metanephridium is surrounded by a network of blood vessels that assists reclamation from the tubular fluid of water and valuable materials such as salts, sugars, and amino acids.

Despite these differences, the basic process of urine formation is the same in protonephridia and metanephridia: fluid enters and flows continuously through a tubule where the fluid is selectively modified by (1) withdrawing valuable solutes from it and returning these to the body (reabsorption) and (2) adding waste solutes to it (secretion). The sequence ensures removal of wastes from the body without loss of useful materials. Kidneys of vertebrates operate in basically the same way (p. 673).

Arthropod Kidneys

The paired **antennal glands** of crustaceans, located in the ventral part of the head (Figure 30.7), are an advanced design of the basic nephridial organ. However, they lack open nephrostomes. Instead, blood hydrostatic pressure forms a protein-free filtrate of the blood (ultrafiltrate) in the end sac. In the tubular portion of the gland, selective reabsorption of certain salts and active secretion of others modifies the filtrate as it moves toward the bladder. Thus crustaceans have excretory organs that are basically vertebrate-like in the functional sequence of urine formation.

Insects and spiders have a unique excretory system consisting of **Malpighian tubules** that operate in conjunction with specialized glands in the wall of the rectum (Figure 30.8). These thin, elastic, blind Malpighian tubules are closed and lack an arterial supply. Urine formation does not occur by filtration of body fluids as is the case in the nephridium, but is produced by

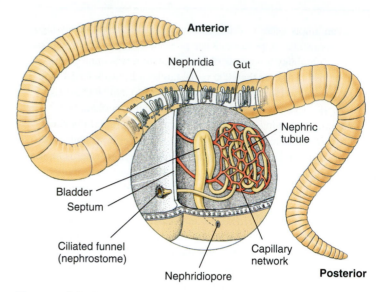

Figure 30.6

Excretory system of an earthworm. Each segment has a pair of large nephridia suspended in a fluid-filled coelom. Each nephridium occupies two segments because the ciliated funnel (nephrostome) drains the segment anterior to the segment containing the rest of the nephridium. The nephric tubule reabsorbs valuable materials from the tubular fluid into the capillary network.

tubular secretion mechanisms by the cells lining the Malpighian tubules that are bathed in hemolymph (blood). This process is initiated by active transport (p. 49) of hydrogen ions into the Malpighian tubular lumen. These hydrogen ions are then moved

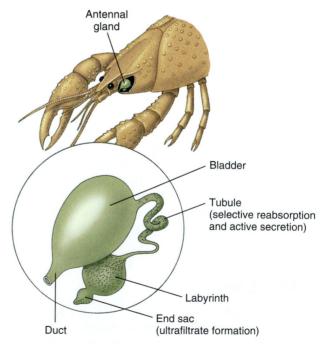

Figure 30.7

Crustacean antennal glands shown in a crayfish. These are filtration kidneys in which a filtrate of the blood is formed in the end sac. The ultrafiltrate is converted into urine as it passes down the tubule toward the bladder.

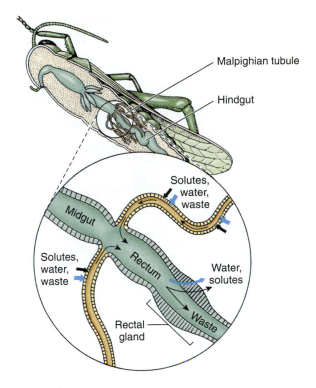

Figure 30.8
Malpighian tubules of insects. Malpighian tubules are located at the juncture of the midgut and hindgut (rectum). Solutes, especially potassium, are actively secreted into the tubules from the surrounding hemolymph (blood). Water, uric acid, and other wastes follow. This fluid drains into the rectum, where solutes (including potassium) and water are actively reabsorbed, leaving wastes to be excreted.

back into the cells lining the tubule using protein transporters that move hydrogen ions in exchange for sodium or potassium ions, and chloride ions follow passively. Herbivorous and omnivorous insects secrete primarily potassium into the tubular lumen. Carnivorous insects, such as blood-sucking mosquitos, initially secrete a fluid high in sodium that reflects the high salt content of blood plasma (see Chapter 31, p. 688). As the red blood cells are digested, the fluid content of sodium falls and becomes rich in potassium ions. The secretion of ions creates an osmotic pressure that draws water, solutes, and nitrogenous wastes, especially uric acid, into the tubule. Uric acid enters the blind-ending distal segment of the tubule as soluble potassium urate, which precipitates as insoluble uric acid in the proximal end of the tubule. Once the formative urine drains into the rectum, water and salts may be reabsorbed by specialized rectal glands, leaving behind uric acid, excess water, salts and other wastes, which are expelled in feces. Rectal glands of aquatic insect larval stages absorb solute, but little water, whereas blood-sucking insects can alter the amount of salt and water reabsorption during and between meals. The feces of blood-sucking insects will be high in water and excess salts during and immediately after a blood meal, but low in salts and water between meals. The Malpighian tubule excretory system is ideally suited for life in dry environments and has contributed to the adaptive diversification of insects on land.

VERTEBRATE KIDNEY

Ancestry and Embryology

From comparative studies of development, biologists believe that the kidney of the earliest vertebrates extended the length of the coelomic cavity and was composed of segmentally arranged tubules, each resembling an invertebrate nephridium. Each tubule opened at one end into the coelom by a nephrostome and at the other end into a common **archinephric duct.** This ancestral kidney is called an **archinephros** ("ancient kidney"), and we find a segmented kidney very similar to an archinephros in embryos of hagfishes and caecilians (Figure 30.9). Almost from the beginning, the reproductive system, which develops beside the excretory system from the same segmental blocks of trunk mesoderm, used the nephric ducts as a convenient conducting system for reproductive products. Thus even though the two systems have nothing functionally in common, they are closely associated in their use of common ducts (Figure 30.9).

Kidneys of living vertebrates developed from this primitive plan. During embryonic development of amniote vertebrates, kidneys develop in three successive stages: **pronephros, mesonephros,** and **metanephros** (Figure 30.9). Some, but not all, of these stages are observed in other vertebrate groups. In all vertebrate embryos, the pronephros is the first kidney to appear. It is located anteriorly in the body and becomes part of the persistent kidney only in adult hagfishes and a few bony fish species. In all other vertebrates the pronephros degenerates during development and is replaced by a more centrally located mesonephros. The mesonephros is the functional kidney of embryonic amniotes (nonavian reptiles, birds, and mammals). The mesonephros and metanephros, together called the **opisthonephros** (Figure 30.9), form the adult kidney of most fishes and amphibians.

The metanephros, characteristic of adult amniotes, is distinguished in several ways from the pronephros and mesonephros. It is more caudally located and it is a much larger, more compact structure containing a very large number of nephric tubules. It is drained by a new duct, the **ureter,** which developed when the old archinephric duct was relinquished to the reproductive system of the male for sperm transport. Thus three successive kidney types—pronephros, mesonephros, metanephros—succeed each other embryologically, and to some extent phylogenetically, in amniotes.

Vertebrate Kidney Function

The vertebrate kidney is part of many interconnected mechanisms that maintain cellular homeostasis, and it is the principal organ that regulates volume and composition of the internal fluid environment. While we commonly describe the vertebrate kidney as an organ of excretion, the removal of metabolic wastes is incidental to its osmoregulatory function.

The organization of kidneys differs somewhat in different groups of vertebrates, but in all the basic functional unit is the **nephron,** and urine is formed by three well-defined physiological processes: **filtration, reabsorption,** and **secretion.** This discussion focuses mainly on the mammalian kidney, which is the most completely understood osmoregulatory organ.

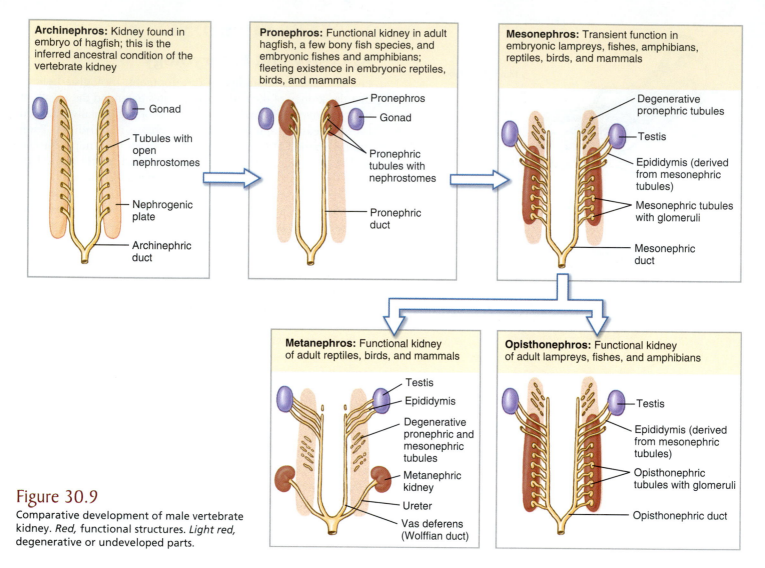

Figure 30.9

Comparative development of male vertebrate kidney. *Red*, functional structures. *Light red*, degenerative or undeveloped parts.

The two human kidneys are small organs comprising less than 1% of the body weight, yet they receive a remarkable 20% to 25% of the total cardiac output—1440 to 1800 liters of blood each day. This vast blood flow is channeled to approximately 2 million nephrons, which form the bulk of the two kidneys. Each nephron begins with an expanded chamber, the **Bowman's capsule,** containing a tuft of capillaries, the **glomerulus** (glo-mer′yoo-lus), which together are called the **renal corpuscle.** Blood pressure in the glomerular capillaries forces an almost protein-free **filtrate** into a Bowman's capsule and along a **renal tubule,** consisting of several segments that perform the functions of reabsorption and secretion in the process of urine formation. The filtrate passes first into a **proximal convoluted tubule,** then into a long, thin-walled **loop of Henle,** which extends deep into the inner portion of the kidney (the **medulla**) before returning to the outer portion (the **cortex**) where it becomes a **distal convoluted tubule.** From the distal tubule the fluid empties into a **collecting duct,** which drains into the **renal pelvis.** Here the urine is collected before being carried by the **ureter** to the **urinary bladder.** These anatomical relationships are shown in Figure 30.10.

Urine that leaves the collecting duct is very different from the filtrate produced in the renal corpuscle. During its travels through the renal tubule and collecting duct, both the composition and concentration of the original filtrate change. Some solutes such as all glucose and amino acids, and most sodium, have been reabsorbed while others, such as hydrogen ions and urea, have been concentrated in urine.

The nephron, with its pressure filter and tubule, is intimately associated with blood circulation (Figures 30.10 and 30.11). Blood from the aorta enters each kidney through a large **renal artery,** which divides into a branching system of smaller arteries. The arterial blood reaches the glomerulus through an **afferent arteriole** and leaves by an **efferent arteriole.** From the efferent arteriole the blood travels to an extensive capillary network that surrounds and supplies the proximal and distal convoluted tubules and the loop of Henle. This capillary network provides a means for the pickup and delivery of materials that are reabsorbed or secreted by the kidney tubules. From these capillaries blood is collected by veins that unite to form the **renal vein.** This vein returns blood to the posterior vena cava.

Glomerular Filtration

Let us now return to the glomerulus, where the process of urine formation begins. The glomerulus acts as a specialized mechanical filter in which an almost protein-free filtrate of the plasma

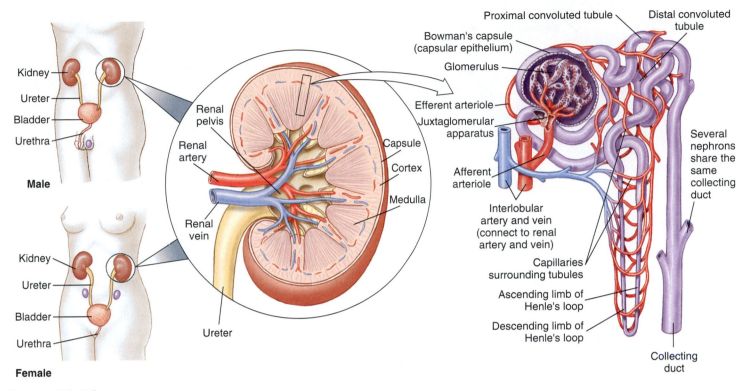

Figure 30.10

Urinary system of humans, with enlargements showing detail of the kidney and a single nephron.

is driven by the blood pressure across capillary walls and into the fluid-filled space of the Bowman's capsule. The diameter of

Figure 30.11

Scanning electron micrograph of a cast of the microcirculation of a mammalian kidney, showing several glomeruli and associated blood vessels. The Bowman's capsule, which normally surrounds each glomerulus, has been digested away in preparing the cast.

From R. G. Kessel and R. H. Kardon, Tissues and Organs: A Text-Atlas of Scanning Electron Microscopy, 1979, W. H. Freeman and Co.

the afferent arteriole entering the glomerulus is greater than that of the exiting efferent arteriole providing the high hydrostatic pressure that allows formation of the glomerular filtrate. Solute molecules small enough to pass through the filtration slits of the capillary wall are carried through by water in which they are dissolved. Red blood cells and almost all plasma proteins, however, are withheld because they are too large to pass through these pores (Figure 30.12).

The filtrate continues through the renal tubular system where it undergoes extensive modification before becoming urine. Human kidneys form approximately 180 liters (nearly 50 gallons) of filtrate each day, a volume many times exceeding the total blood volume. If this volume of water and the valuable nutrients and salts it contains were lost, the body would soon be depleted of these compounds. Depletion does not happen because nearly all of the filtrate is reabsorbed. The final urine volume in humans averages 1.2 liters per day.

Conversion of filtrate into urine involves two processes: (1) modification of the composition of the filtrate through tubular reabsorption and secretion, and (2) changes in the total osmotic concentration of urine through regulation of water excretion.

Tubular Reabsorption

Approximately 60% of the filtrate volume and virtually all of the glucose, amino acids, vitamins and other valuable nutrients are reabsorbed in the proximal convoluted tubule. Much of this reabsorption is by **active transport,** in which cellular energy is used to transport materials from tubular fluid to the surrounding

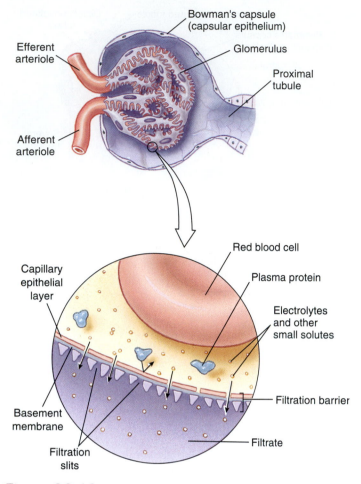

Figure 30.12

Renal corpuscle, showing (*enlargement*) the filtration of fluid through the glomerular capillary membrane. Water, electrolytes, and other small molecules pass the porous filtration barrier, but almost all plasma proteins are too large to pass the barrier. The filtrate is thus almost protein-free.

capillary network and thus into the blood circulation. Electrolytes such as sodium, potassium, calcium, bicarbonate, and phosphate are reabsorbed by ion pumps, which are carrier proteins driven by the hydrolysis of ATP (ion pumps are described on p. 50). Because an essential function of the kidney is to regulate plasma concentrations of electrolytes, all are individually reabsorbed by ion pumps specific for each electrolyte. Some are strongly reabsorbed and others weakly reabsorbed, depending on the body's ability to conserve each mineral. Some materials are passively reabsorbed. Negatively charged chloride ions, for example, passively accompany active reabsorption of positively charged sodium ions in the proximal convoluted tubule. Water, too, is withdrawn passively from the tubule, as it osmotically follows the active reabsorption of solutes.

For most substances there is an upper limit to the amount of substance that can be reabsorbed. This upper limit is termed the **transport maximum** (renal threshold) for that substance. For example, glucose normally is reabsorbed completely by the kidney because the transport maximum for glucose is well above the amount of glucose usually present in the glomerular filtrate. Should the plasma glucose concentration exceed this threshold level, as in the disease diabetes mellitus, glucose appears in the urine (Figure 30.13).

In the disease diabetes mellitus ("sweet running through"), glucose rises to abnormally high concentrations in the blood plasma (hyperglycemia) because the hormone insulin, which enables body cells to take up glucose, is deficient (see Chapter 34, p. 767). As blood glucose rises above a normal level of about 100 mg/100 ml of plasma, the concentration of glucose in the filtrate also rises, and more glucose must be reabsorbed by the proximal tubule. Eventually a point is reached (about 300 mg/100 ml of plasma) at which reabsorptive capacity of the tubular cells is saturated. Above this transport maximum for glucose, glucose spills into the urine. In untreated diabetes mellitus the victim's urine tastes sweet, thirst is unrelenting, and the body wastes away despite a large food intake. In England the disease for centuries was appropriately called the "pissing evil."

Unlike glucose or amino acids, most electrolytes are excreted in the urine in variable amounts. Reabsorption of sodium, the dominant cation in the plasma, illustrates the flexibility of the reabsorption process. The human kidney filters approximately 600 g of sodium every 24 hours. Nearly all of this sodium is reabsorbed, but the exact amount is matched precisely to sodium intake. With a normal sodium intake of 4 g per day, the kidney excretes 4 g and reabsorbs 596 g each day. A person on a low-salt

Figure 30.13

The mechanism for the tubular reabsorption of glucose can be likened to a conveyor belt running at constant speed. **A,** When glucose concentrations in the glomerular filtrate are low, all is reabsorbed. **B,** When glucose concentrations in the filtrate have reached the transport maximum, all carrier sites for glucose are occupied. If glucose levels rise further, **C,** as in the disease diabetes mellitus, some glucose escapes the carriers and appears in urine.

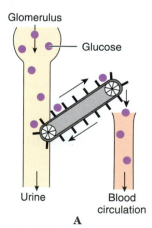
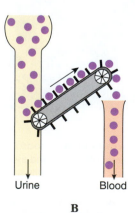
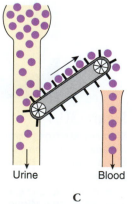

A   B   C

diet of 0.3 g of sodium per day still maintains salt balance because only 0.3 g escapes reabsorption. But with a very high salt intake, above 20 g per day, the kidney cannot excrete sodium as fast as it enters. The unexcreted sodium chloride causes retention of additional water in the body fluids, and the person begins to gain weight. (The salt intake of the average North American is about 6 to 18 g per day, approximately 20 times more than the body needs, and three times more than is considered acceptable for those predisposed to high blood pressure.)

A human kidney can adapt to excrete large quantities of salt (sodium chloride) under conditions of high salt intake. In societies accustomed to widespread use of foods heavily salted for preservation (for example, salted pork and salt herring) daily intakes may approach or even exceed 100 g. Body weight remains normal under such conditions. However, the acute ingestion of 20 to 40 g/day by volunteers unadapted to such large intakes of salt caused swelling of tissues, increase in body weight, and an increase in blood pressure.

Distal convoluted tubules make additional adjustments of filtrate composition. Sodium reabsorbed by the proximal convoluted tubule—some 85% of the total filtered—is obligatory reabsorption; this amount is reabsorbed independent of sodium intake. In the distal convoluted tubule, however, sodium reabsorption is controlled by **aldosterone,** a steroid hormone secreted by the adrenal gland (p. 766). Aldosterone increases both active reabsorption of sodium and secretion of potassium by the distal tubules; the hormone thus decreases loss of sodium and increases loss of potassium in the urine. The secretion of aldosterone is regulated (1) by the enzyme **renin,** produced by the **juxtaglomerular apparatus,** a complex of cells located at the junction of the afferent arteriole with the glomerulus (see Figure 30.10), and (2) by elevated blood potassium levels. Renin is released in response to a low blood sodium level, to low blood pressure (which can occur if the blood volume drops too low), or to low sodium in the glomerular filtrate. Renin then initiates a series of enzymatic events culminating in the production of **angiotensin,** a blood-borne hormone that has several related effects. First, it stimulates the release of aldosterone, which acts in turn to increase sodium reabsorption and potassium secretion by the distal tubule. Second, it increases the secretion of **antidiuretic hormone** (vasopressin, discussed on p. 679), which promotes water conservation by the kidney. Third, it increases blood pressure. Finally, it stimulates thirst, which also is stimulated by decreased blood volume or increased blood osmolarity. These actions of angiotensin tend to reverse the circumstances (low blood sodium and low blood pressure and/or blood volume) that triggered the secretion of renin. Sodium and water are conserved, and blood volume and blood pressure are restored to normal.

The flexibility of distal reabsorption of sodium varies considerably in different animals: it is restricted in humans but very broad in many rodents. These differences have appeared because selective pressures during evolution have resulted in rodents adapted for dry environments. They must conserve water and at the same time excrete considerable sodium. Humans, however, were not designed to accommodate the large salt appetites many have. Our closest relatives, the great apes, are vegetarians with an average salt intake of less than 0.5 g per day.

Tubular Secretion

In addition to reabsorbing materials from glomerular filtrate, the nephron can secrete materials across the tubular epithelium and *into* the filtrate. In this process, the reverse of tubular reabsorption, carrier proteins in the tubular epithelial cells selectively transport substances from blood in capillaries outside the tubule to the filtrate inside the tubule. Tubular secretion enables a kidney to increase the urine concentrations of materials to be excreted, such as hydrogen and potassium ions, drugs, and various foreign organic materials. The tubular epithelium is able to recognize foreign organic materials, such as ingested pharmaceutical drugs, because they are metabolized by the liver to form cationic or anionic molecules. These molecules are transported by the tubular epithelium, which has both cationic and anionic transporters in its membrane. The distal convoluted tubule is the site of most tubular secretion.

In the kidneys of bony marine fishes, nonavian reptiles, and birds, tubular secretion is a much more complex process than it is in mammalian kidneys. Marine bony fishes actively secrete large amounts of magnesium and sulfate, seawater salts that are by-products of their mode of osmotic regulation (see p. 669). Nonavian reptiles and birds excrete uric acid instead of urea as their major nitrogenous waste. The material is actively secreted by the tubular epithelium. Since uric acid is nearly insoluble, it forms crystals in the urine and requires little water for excretion. Thus excretion of uric acid is an important adaptation for water conservation.

Water Excretion

Kidneys closely regulate the osmotic pressure of the blood. When fluid intake is high, the kidney excretes dilute urine, saving salts and excreting water. When fluid intake is low, kidneys conserve water by forming concentrated urine. A dehydrated person can concentrate urine to approximately four times blood osmotic concentration. This important ability to concentrate urine enables us to excrete wastes with minimal water loss.

The capacity of mammalian and some avian kidneys to produce a concentrated urine involves an interaction between the loop of Henle and the collecting ducts. This interplay results in formation of an osmotic gradient in the kidney, as shown in Figure 30.14. In the cortex, interstitial fluid is isosmotic with blood, but deep in the medulla the osmotic concentration is four times greater than that of blood (in rodents and desert mammals that can produce highly concentrated urine the osmotic gradient is much greater than in humans). The high osmotic concentrations in the medulla are produced by an exchange of ions in the loop of Henle by **countercurrent multiplication.** "Countercurrent" refers to the opposite directions of fluid movement in the two limbs of the loop of Henle: down in the descending limb and up the ascending limb. "Multiplication" describes the increasing osmotic concentration in the medulla surrounding the loops of Henle and collecting ducts resulting from ion exchange between the two limbs of the loop.

The functional characteristics of this system are as follows. The descending limb of the loop of Henle is permeable to water but

Figure 30.14

Mechanism of urine concentration in mammals. Sodium and chloride diffuse or are pumped from the ascending limb of the loop of Henle, and water is withdrawn passively from the descending limb, which is impermeable to sodium chloride. Sodium chloride reabsorbed from the ascending limb of the loop of Henle and urea reabsorbed from the collecting duct raise the osmotic concentration in the kidney medulla, creating an osmotic gradient for controlled reabsorption of water from the collecting duct.

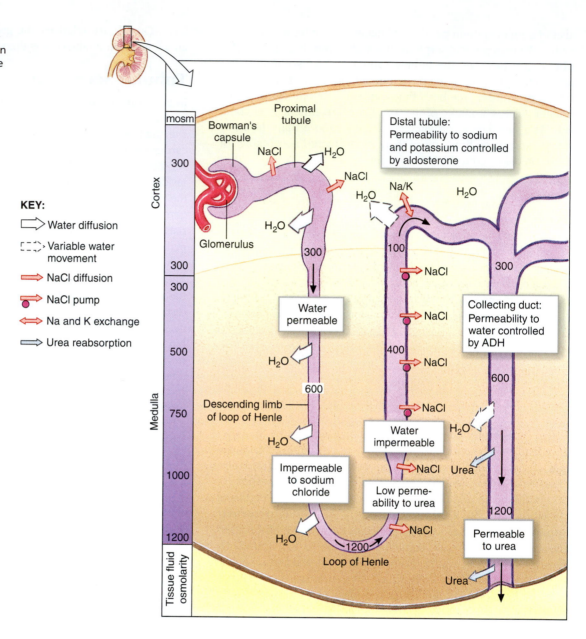

KEY:

⇒ Water diffusion

⇢ Variable water movement

⇒ NaCl diffusion

⊶⇒ NaCl pump

⇔ Na and K exchange

⇒ Urea reabsorption

impermeable to solutes. The ascending limb is nearly impermeable to water. Sodium chloride passively moves from the lower ascending limb and is actively transported from the thick portion of the ascending limb into the surrounding tissue fluid (Figure 30.14). As the interstitium surrounding the loop becomes more concentrated with solute, water is withdrawn from the descending limb by osmosis. The tubular fluid in the base of the loop, now more concentrated, moves up the ascending limb, where still more sodium chloride diffuses or is pumped out. In this way the effect of ion movement in the ascending limb is multiplied as more water is withdrawn from the descending limb and more concentrated fluid is presented to the ascending limb (Figures 30.14 and 30.15). The blood capillaries surrounding the loops of Henle, the **vasa recta,** are also arranged in a countercurrent fashion. This arrangement of blood vessels is important in maintaining the osmotic concentration gradient of the medulla and cortex.

Final adjustment of urine concentration occurs not in the loops of Henle but in the collecting ducts. Formative urine that

enters the distal tubule from the loop of Henle is dilute (because of salt withdrawal) and may be diluted still more by active reabsorption of more sodium chloride in the distal tubule. The formative urine, low in solutes but carrying urea, now flows down into the collecting duct. Because of the high concentration of solutes surrounding the collecting duct, water is withdrawn from the urine. As the urine becomes more concentrated, urea also diffuses out of it. The lower regions of the collecting duct are permeable to urea, and thus far, four different urea transporters have been discovered. Some of this urea flows back into the lower portion of the ascending limb of the loop of Henle, but since the loop of Henle is less permeable to urea, its concentrations build within the medulla tissue fluid. This buildup of urea contributes significantly to the high osmotic concentration of the medulla and to the countercurrent multiplication mechanism (Figure 30.15).

The amount of water reabsorbed and the final concentration of the urine depend on permeability of the walls of the distal

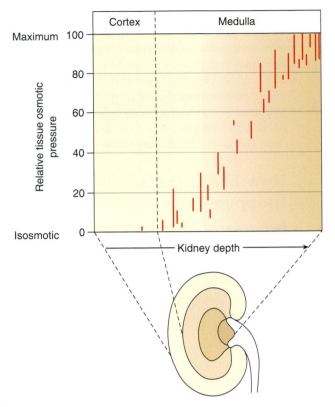

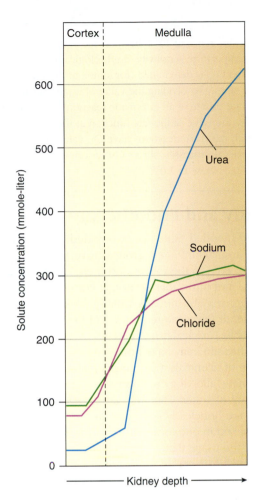

Figure 30.15

Osmotic concentration of tissue fluid in the mammalian kidney. Tissue fluid is isosmotic in the kidney cortex (*to left in diagram*) but osmotic concentration increases continuously through the medulla, reaching a maximum at the papilla where the urine drains into the ureter.

convoluted tubule and the collecting duct. This is controlled by the **antidiuretic hormone** (ADH, or vasopressin), which is released by the posterior pituitary gland (neurohypophysis, see p. 758). In turn, special receptors in the brain that constantly sense the osmotic pressure of body fluids govern the release of this hormone. When the blood osmotic pressure increases or blood volume decreases, as during dehydration, the pituitary gland releases more ADH. ADH increases permeability of the collecting duct by increasing the number of water channels in the collecting duct epithelial cells. Then, as the fluid in the collecting duct passes through the hyperosmotic region of the kidney medulla, water diffuses through the channels into the surrounding interstitial fluid and is carried away by the blood circulation. The urine loses water and becomes more concentrated. Given this sequence of events for dehydration, it is not difficult to anticipate how the system responds to overhydration: the pituitary stops releasing ADH, the channels in the collecting duct epithelial cells decrease in number, and a large volume of dilute urine is excreted.

The varying ability of different mammals to form a concentrated urine correlates closely with length of the loops of Henle. Beavers, which have no need to conserve water in their aquatic environment, have short loops and can concentrate their urine only to about twice the osmolarity of the blood. Humans, with relatively longer loops, can concentrate urine 4.2 times that of the blood. As we would anticipate, desert mammals have much

greater urine-concentrating powers. A camel can produce a urine 8 times the plasma concentration, a gerbil 14 times, and an Australian hopping mouse 22 times. In this creature, the greatest urine concentrator of all, the long loops of Henle extend to the tip of a long renal papilla that pushes out into the mouth of the ureter.

TEMPERATURE REGULATION

We have seen that a fundamental problem facing an animal is keeping its internal environment in a state that permits normal cell function. Biochemical activities are sensitive to the chemical environment and our discussion thus far has examined how the chemical environment is stabilized. Biochemical reactions are also extremely sensitive to temperature. All enzymes have an optimum temperature; at temperatures above or below this optimum, enzyme function is impaired. Temperature therefore is a severe constraint for animals, all of which must maintain biochemical stability. When body temperature drops too low, metabolic processes slow, reducing the amount of energy the animal can muster for activity and reproduction. If body temperature rises too high, metabolic reactions become unbalanced and enzymatic activity is hampered or even destroyed. Thus animals can succeed only in a restricted range of temperature, usually between 0° to 40° C. Animals must either find a habitat where they do not have to contend with temperature extremes, or they must develop means of stabilizing their metabolism independent of temperature extremes.

A temperature difference of 10° C has become a standard used to measure the temperature sensitivity of a biological function. This value, called the Q_{10}, is determined (for temperature intervals of exactly 10° C) simply by dividing the value of a rate function (such as metabolic rate or rate of an enzymatic reaction) at the higher temperature by the value of the rate function at the lower temperature. In general, metabolic reactions have Q_{10} values of about 2.0 to 3.0. Purely physical processes, such as diffusion, have much lower Q_{10} values, usually close to 1.0.

Ectothermy and Endothermy

The terms "cold-blooded" and "warm-blooded" have long been used to divide animals into two groups: invertebrates and vertebrates that feel cold to the touch, and those, such as humans, other mammals, and birds, that do not. It is true that body temperature of mammals and birds is usually (though not always) warmer than the air temperature, but a "cold-blooded" animal is not necessarily cold. Tropical fishes, and insects and nonavian reptiles basking in the sun, may have body temperatures equaling or surpassing those of mammals. Conversely, many "warm-blooded" mammals hibernate, allowing their body temperature to approach the freezing point of water. Thus the terms "warm-blooded" and "cold-blooded" are hopelessly subjective and nonspecific but are so firmly entrenched in our vocabulary that most biologists find it easier to accept the usage than to try to change it.

The term **poikilothermic** (body temperature that fluctuates with environmental temperature) and **homeothermic** (constant body temperature, regulated independent of environmental temperature) are frequently used by zoologists as alternatives to "cold-blooded" and "warm-blooded," respectively. These terms, which refer to variability of body temperature, are more precise and more informative, but still offer difficulties. For example, deep-sea fishes live in an environment having no perceptible temperature change. Even though their body temperature is absolutely stable, to call such fishes homeotherms would distort the intended application of the term. Furthermore, among homeothermic birds and mammals many allow their body temperature to change between day and night, or, as with hibernators, between seasons.

Physiologists prefer yet another way to describe body temperatures, one that reflects the fact that an animal's body temperature is a balance between heat gain and heat loss. All animals produce heat from cellular metabolism, but in most the heat is conducted away as fast as it is produced. In these animals, the **ectotherms**—and the overwhelming majority of animals belong to this group—body temperature is determined solely by the environment. Many ectotherms exploit their environment behaviorally to select areas of more favorable temperature (such as basking in the sun) but the source of energy used to increase body temperature comes from the environment, not from within the body. Alternatively some animals are able to generate and to retain enough metabolic heat to elevate their own body temperature to a high but stable level. Because the source of their body heat is internal, they are called **endotherms.** These favored few in the animal kingdom are the birds and mammals, as well as a few nonavian reptiles and fast-swimming fishes, and certain insects that are at least partially endothermic. Endothermy allows birds and mammals to stabilize their internal temperature, allowing biochemical processes and nervous system functions to proceed at steady high levels of activity. Endotherms may thus remain active in winter and exploit habitats denied to ectotherms. In reality, endotherms that have a large surface-to-volume ratio (are small) with subsequent high heat loss and/or a limited food availability tend toward decreased activity and hibernation in colder climates.

How Ectotherms Achieve Temperature Independence

Behavioral Adjustments

Although ectotherms cannot control their body temperature physiologically, many are able to regulate their body temperature behaviorally with considerable precision. Ectotherms often have the option of seeking areas in their environment where the temperature is favorable to their activities. Some ectotherms, such as desert lizards, exploit hour-to-hour changes in solar radiation to keep their body temperatures relatively constant (Figure 30.16). In the early morning they expose their head to absorb the sun's heat. By midmorning they emerge from their burrows and bask in the sun with their bodies flattened to absorb heat. As the day warms, they turn to face the sun to reduce exposure, and raise their bodies from the hot substrate. In the hottest part of the day they may retreat to their burrows or move into the shade. Later they emerge to bask as the sun sinks lower and the air temperature drops.

These behavioral patterns help to maintain a relatively steady body temperature of 36° to 39° C while the air temperature varies between 29° and 44° C. Some lizards can tolerate intense midday heat without shelter. The desert iguana of the southwestern United States prefers a body temperature of 42° C when active and can tolerate a rise to 47° C, a temperature that is lethal to all birds and mammals and most other lizards. The term "cold-blooded" clearly does not apply to these animals!

Metabolic Adjustments

Even without the help of the behavioral adjustments just described, most ectotherms can adjust their metabolic rates to the prevailing temperature such that the intensity of metabolism remains mostly unchanged. This is called **temperature compensation** and involves complex biochemical and cellular adjustments. These adjustments enable a fish or a salamander, for example, to benefit from almost the same level of activity in both warm and cold environments. Thus, whereas endotherms achieve metabolic homeostasis by maintaining their body temperature independent of environmental temperature, ectotherms accomplish much the same by directly maintaining their metabolism independent of body temperature. This metabolic regulation also is a form of homeostasis.

Temperature Regulation in Endotherms

Most mammals have body temperatures between 36° and 38° C, somewhat lower than those of birds, which range between 40°

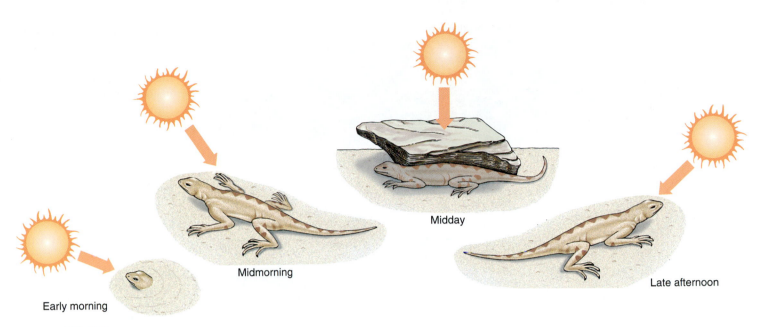

Figure 30.16

How a lizard regulates its body temperature behaviorally. In the morning the lizard absorbs the sun's heat through its head while keeping the rest of its body protected from cool morning air. Later it will emerge to bask. At noon, with its body temperature high, it seeks shade from the hot sun. When the air temperature drops in the late afternoon, it emerges and lies parallel to the sun's rays.

and 42° C. Constant temperature is maintained by a delicate balance between heat production and heat loss—not a simple matter when these animals are alternating between periods of rest and bursts of heat-producing activity.

Heat is produced by the animal's metabolism. This includes oxidation of foods, basal cellular metabolism, and muscular contraction. Because much of an endotherm's daily caloric intake is required to generate heat, especially in cold weather, the endotherm must eat more food than an ectotherm of the same size. Heat is lost by radiation, conduction, and convection (air movement) to a cooler environment and by evaporation of water (Figure 30.17). A bird or mammal can control both processes of heat production and heat loss within rather wide limits. If the animal becomes too cool, it can generate heat by increasing muscular activity (exercise or shivering) and by decreasing heat loss by increasing its insulation. If it becomes too warm, it decreases heat production and increases heat loss. We examine these processes with examples.

Adaptations for Hot Environments

Despite the harsh conditions of deserts—intense heat during the day, cold at night, and scarcity of water, vegetation, and cover—many kinds of animals live there successfully. The smaller desert mammals are mostly **fossorial** (living mainly in the ground) or **nocturnal** (active at night). The lower temperature and higher humidity of burrows help to reduce water loss by evaporation. As explained on page 670, desert animals such as the kangaroo rat and the American desert ground squirrels can, if necessary, derive the water they need from metabolism of their dry food, drinking no water at all. Such animals produce a highly concentrated urine and form almost completely dry feces.

Large desert ungulates (hooved mammals that regurgitate and chew their partially digested food) obviously cannot escape

desert heat by living in burrows. Animals such as camels and desert antelopes (gazelle, oryx, and eland) possess a number of adaptations for coping with heat and dehydration. Figure 30.18 shows those of the eland. Mechanisms for controlling water loss and preventing overheating are closely linked. The glossy, pallid

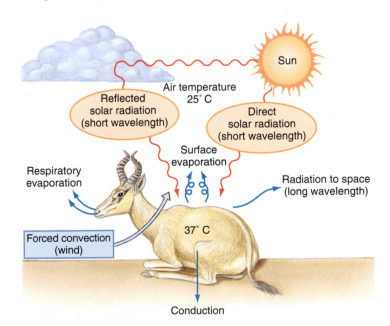

Figure 30.17

Exchange of heat between a large mammal and its environment on a warm day. Red arrows indicate sources of net heat gain by the animal (all radiation); blue arrows are avenues of net heat loss (evaporative cooling, conduction to the ground, longwave radiation into space, and forced convection by the wind). If air and ground temperatures were warmer than the animal, the arrows for forced convection, conduction, and radiation would be reversed. Then the animal could lose heat only by evaporative cooling.

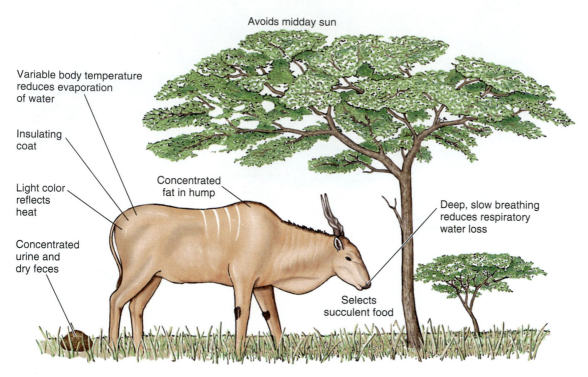

Avoids midday sun

Variable body temperature
reduces evaporation
of water

Insulating
coat

Concentrated
fat in hump

Light color
reflects
heat

Deep, slow breathing
reduces respiratory
water loss

Concentrated
urine and
dry feces

Selects
succulent food

Figure 30.18

Physiological and behavioral adaptations of the common eland for regulating temperature in the hot, arid savanna of Central Africa.

color of fur reflects direct sunlight, and fur itself is an excellent insulation that resists heat. Heat is lost by convection and conduction from the underside of elands where the fur is very thin. Fat tissue, an essential food reserve, is concentrated in a single hump on the back, instead of being uniformly distributed under the skin where it would impair loss of heat by radiation. Elands avoid evaporative water loss—the only means an animal has for cooling itself when the environmental temperature is higher than that of the body—by permitting their body temperature to drop during the cool night and then to rise slowly during the day as the body stores heat. Only when the body temperature reaches 41° C must elands prevent further rise through **evaporative cooling** by sweating and panting. Respiratory moisture is condensed and reabsorbed in nasal passages as air is breathed out. They conserve water by producing a concentrated urine and dry feces. Camels have all of these adaptations developed to a similar or even greater degree; they are perhaps the most perfectly adapted of all large desert mammals.

Adaptations for Cold Environments

In cold environments mammals and birds use two major mechanisms to maintain homeothermy: (1) **decreased conductance,** reduction of heat loss by increasing the effectiveness of the insulation, and (2) **increased heat production.**

In all mammals living in the cold regions of the earth, fur thickness increases in winter, sometimes by as much as 50%. Thick underhair is the principal insulating layer, whereas the longer and more visible guard hair serves as protection against wear and for protective coloration. However, unlike the well-insulated trunk of the body, the body extremities (legs, tail, ears,

nose) of arctic mammals are thinly insulated and exposed to rapid cooling. To prevent these parts from becoming major avenues of heat loss, they are allowed to cool to low temperatures, often approaching freezing point. The heat in the warm arterial blood is not lost from the body, however. Instead, a **countercurrent heat exchange** between the outgoing warm blood and the returning cold blood prevents heat loss. Arterial blood in the leg of an arctic mammal or bird passes in close contact with a network of small veins. Because arterial blood flow is opposite to that of returning venous blood, heat is exchanged very efficiently from artery to veins. When arterial blood reaches the foot it has transferred nearly all of its heat to the veins returning blood to the body core (Figure 30.19). Thus little heat is lost to the surrounding cold air from poorly insulated distal regions of the leg. Countercurrent heat exchangers in appendages also are common in aquatic mammals such as seals and whales, which have thinly insulated flippers that would be avenues of excessive heat loss in the absence of this heat-salvaging arrangement.

A consequence of peripheral heat exchange is that legs and feet of mammals and birds living in cold environments must function at low temperatures. Temperatures of the feet of arctic foxes and barren-ground caribou are just above freezing point; in fact, the temperature may be below 0°C in footpads and hooves. To keep feet supple and flexible at such low temperatures, fats in the extremities have very low melting points, perhaps 30°C lower than ordinary body fats.

In severely cold conditions all mammals can produce more heat by **augmented muscular activity** through exercise or shivering. We are all familiar with the effectiveness of both activities. A person can increase heat production as much as 18-fold by violent shivering when maximally stressed by cold. Another source of heat is increased oxidation of foods, especially from stores of brown fat (brown fat is described on p. 720). This mechanism is called **nonshivering thermogenesis.**

Small mammals the size of lemmings, voles, and mice meet the challenge of cold environments in a different way. Small mammals are not as well insulated as large mammals because thickness of fur is limited by the need to maintain mobility. Consequently these forms exploit the excellent insulating qualities of snow by living under it in runways on the forest floor, where incidentally, their food also is located. In this **subnivean environment** the temperature seldom drops below −5°C even

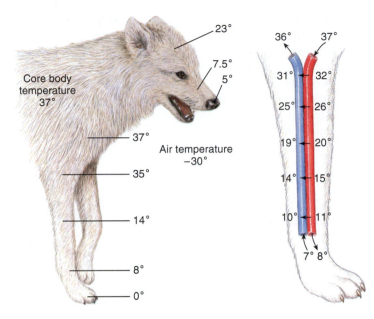

Figure 30.19

Countercurrent heat exchange in the leg of an arctic wolf. The diagram to the left shows how the extremities cool when the animal is exposed to low air temperatures. The diagram to the right depicts a portion of the front leg artery and vein, showing how heat is exchanged between arterial and venous blood. Thus, heat is shunted back into the body and conserved.

though the air temperature above may fall to −50° C. Snow insulation decreases thermal conductance from small mammals just as thick pelage does for large mammals. Living beneath the snow is really a type of avoidance response to cold.

Adaptive Hypothermia in Birds and Mammals

Endothermy is energetically expensive. Whereas an ectotherm can survive for weeks in a cold environment without eating, an endotherm must always have energy resources to supply its high metabolic rate. The problem is especially acute for small birds and mammals which, because of their intense metabolism, may require a daily intake of food approaching their own body weight to maintain homeothermy (food consumption by birds is related on p. 594, and by mammals on p. 620). It is not surprising then that a few small birds and mammals have evolved ways to abandon homeothermy for periods ranging from a few hours to several months, allowing their body temperature to fall until it approaches or equals the temperature of surrounding air.

Some very small mammals, such as bats, maintain high body temperatures when active but allow their body temperature to drop profoundly when inactive and asleep. This is called **daily torpor,** an adaptive hypothermia that provides enormous saving of energy to small endotherms that are never more than a few hours away from starvation at normal body temperatures. Hummingbirds also may drop their body temperature at night when food supplies are low (Figure 30.20).

Many small and medium-sized mammals in northern temperate regions solve the problem of winter scarcity of food and low temperature by entering a prolonged and controlled

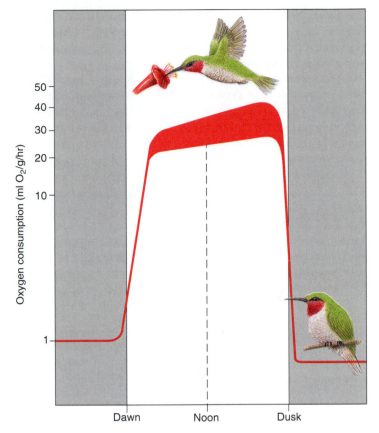

Figure 30.20

Torpor in hummingbirds. Body temperature and oxygen consumption (red line) are high when hummingbirds are active during the day but may drop to one-twentieth these levels during periods of food shortage. Torpor vastly lowers demands on the bird's limited energy reserves.

state of dormancy: **hibernation.** True hibernators, such as ground squirrels, jumping mice, marmots, and woodchucks (Figure 30.21), prepare for hibernation by storing body fat. Entry into hibernation is gradual. After a series of "test drops" during which body temperature decreases a few degrees and then returns to normal, the animal cools to within a degree or less of the ambient temperature. Metabolism decreases to a fraction of normal. In ground squirrels, for example, the respiratory rate

Figure 30.21

Hibernating woodchuck, *Marmota monax* (order Rodentia), in den exposed by road-building work sleeps on, unaware of the intrusion. Woodchucks begin hibernating in late September while the weather is still warm and may sleep six months. The animal is rigid and decidedly cold to the touch. Breathing is imperceptible, as slow as one breath every five minutes. Although it appears to be dead, it will awaken if the den temperature drops dangerously low.

decreases from a normal rate of 200 per minute to 4 or 5 per minute, and the heart rate from 150 to 5 beats per minute. During arousal a hibernator both shivers violently and employs non-shivering thermogenesis to produce heat.

Some mammals, such as bears, badgers, raccoons, and opossums, enter a state of prolonged sleep in winter with little or no decrease in body temperature. Prolonged sleep is not true hibernation. Bears of the northern forest sleep for several months. A bear's heart rate may decrease from 40 to 10 beats per minute, but body temperature remains normal and the bear is awakened if

sufficiently disturbed. One intrepid but reckless biologist narrowly escaped injury when he crawled into a den and attempted to measure the bear's rectal temperature with a thermometer!

Some invertebrates and vertebrates will enter a state of dormancy during the summer called **estivation** or "summer sleep." In this state breathing rates and metabolism decrease when temperatures are high, food is scarce, and when dehydration threatens. Some examples of animals that estivate are land snails, blue land crabs, African lungfishes, the desert tortoise, the pigmy mouse, and Colombian ground squirrels.

SUMMARY

Throughout life, matter and energy pass through the body, potentially disturbing the internal physiological state. Homeostasis, the ability of an organism to maintain internal stability despite such challenges, is a characteristic of all living systems. Homeostasis involves the coordinated activity of several physiological and biochemical mechanisms, and it is possible to relate some major events in animal evolution to increasing internal independence from the consequences of environmental change. In this chapter we have examined two aspects of homeostasis: (1) the varying ability of animals to stabilize the osmotic and chemical composition of the blood, and (2) the capacity of animals to regulate their temperatures in thermally challenging environments.

Most marine invertebrates must either depend on the osmotic stability of the ocean to which they conform, or be able to tolerate wide fluctuations in environmental salinity. Some of the latter show limited powers of osmotic regulation, the capacity to resist internal osmotic change through the evolution of specialized regulatory organs. All animals living in freshwater are hyperosmotic to their environment and have developed mechanisms for recovering salt from the environment and eliminating excess water that enters the body osmotically.

All vertebrate animals show excellent osmotic homeostasis. Marine bony fishes maintain their body fluids distinctly hypoosmotic to their environment by drinking seawater and physiologically distilling it. Elasmobranchs (sharks and rays) have adopted a strategy of osmotic homeostasis by retaining urea and trimethylamine oxide (TMAO) in the blood.

The kidney is the most important organ for regulating the chemical and osmotic composition of the blood. In all metazoa kidneys are some variation on a basic theme: a tubular structure that forms urine by introducing a fluid secretion or filtrate of the blood or interstitial fluid into a tubule in which it is selectively modified to form urine. Terrestrial vertebrates have especially sophisticated kidneys, since they must be able to regulate closely the water content of the blood by balancing gains and expenditures. The basic excretory unit is the nephron, composed of a glomerulus in which an ultrafiltrate of the blood is formed, and a long nephric tubule

in which the formative urine is selectively modified by the tubular epithelium. Water, salts, and other valuable materials pass by reabsorption to the peritubular circulation, and certain wastes pass by secretion from the circulation to the tubular urine. All mammals and some birds can produce urine more concentrated than blood by means of a countercurrent multiplier system localized in the loops of Henle, a specialization not found in other vertebrates.

Temperature has a profound effect on the rate of biochemical reactions and, consequently, on the metabolism and activity of all animals. Animals may be classified according to whether body temperature is variable (poikilothermic) or stable (homeothermic), or by the source of body heat, whether external (ectothermic) or internal (endothermic).

Ectotherms partially free themselves from thermal constraints by seeking habitats with favorable temperatures, by behavioral thermoregulation, or by adjusting their metabolism to the prevailing temperature through biochemical alterations.

Endothermic birds and mammals differ from ectotherms in having a much higher production of metabolic heat and a much lower conductance of heat from the body. They maintain constant body temperature by balancing heat production with loss.

Small mammals in hot environments for the most part escape intense heat and reduce evaporative water loss by burrowing. Large mammals employ several strategies for dealing with direct exposure to heat, including reflective insulation, heat storage by the body, and evaporative cooling.

Endotherms in cold environments maintain body temperature by decreasing heat loss with thickened pelage or plumage, by peripheral cooling, and by increasing heat production through shivering or nonshivering thermogenesis. Small endotherms may avoid exposure to low temperatures by living under the snow.

Adaptive hypothermia is a strategy used by small mammals and birds to blunt energy demands during periods of inactivity (daily torpor) or periods of prolonged cold and minimal food availability (hibernation). Some vertebrates and invertebrates enter a similar state during the summer when temperatures are high, food is scarce, and when dehydration threatens (estivation).

REVIEW QUESTIONS

1. Define homeostasis. What evolutionary advantages for a species might result from successful maintenance of internal homeostasis?
2. The problems of water balance may have arisen when early metazoan animals began invading estuaries and rivers. Describe the physiological challenges confronting marine

invertebrates entering freshwater and, using crustaceans as an example, suggest solutions to these challenges.
3. Distinguish the terms in the following pairs: osmotic conformity and osmotic regulation; stenohaline and euryhaline; hyperosmotic and hypoosmotic.

4. Young downstream salmon migrants moving from their freshwater natal streams into the sea leave an environment nearly free of salt to enter one containing three times as much salt as their body fluids. Describe osmotic challenges of each environment and suggest physiological adjustments salmon must make in moving from freshwater to the sea.

5. Most marine invertebrates are osmotic conformers. How does their body fluid differ from that of sharks and rays, which are also in near osmotic equilibrium with their environment?

6. What strategy does a kangaroo rat use that allows it to exist in the desert without drinking any water?

7. In what animals would you expect to find a salt gland? What is its function?

8. Relate the function of contractile vacuoles to the following experimental observations: to expel an amount of fluid equal in volume to the volume of the animal required 4 to 53 minutes for some freshwater protozoa, and between 2 and 5 hours for some marine species.

9. How does a protonephridium differ structurally and functionally from a true nephridium (metanephridium)? In what ways are they similar?

10. Describe the developmental stages of kidneys in amniotes. How does the developmental sequence for amniotes differ from that of amphibians and fishes?

11. In what ways does the nephridium of an earthworm parallel the human nephron in structure and function?

12. Describe what happens during the following stages in urine formation in the mammalian nephron: filtration, tubular reabsorption, tubular secretion.

13. Explain how the cycling of sodium chloride between the descending and ascending limbs of the loop of Henle in the mammalian kidney, and special permeability of these tubules, produces high osmotic concentrations in interstitial fluids in the kidney medulla. Explain the role of urea in producing high osmotic concentrations in the interstitial fluid of the medulla.

14. Explain how antidiuretic hormone (vasopressin) controls excretion of water in mammalian kidneys.

15. Define the following terms and comment on the limitations (if any) of each in describing the thermal relationships of animals to their environments: poikilothermy, homeothermy, ectothermy, endothermy.

16. Defend the statement: "Both ectotherms and endotherms achieve metabolic homeostasis in unstable thermal environments, but they do so by employing different physiological strategies."

17. Large mammals live successfully in deserts and in the arctic. Describe different adaptations mammals use to maintain homeothermy in each environment.

18. Explain why it is advantageous for certain small birds and mammals to abandon homeothermy during brief or extended periods of their lives.

SELECTED REFERENCES

Beyenbach, K. W. 2003. Transport mechanisms of diuresis in Malpighian tubules of insects. J. Exp. Biol. **206**:3845–3856. *An excellent review of Malpighian tubular secretion in blood-sucking insects compared with aquatic and terrestrial insects.*

Cossins, A. R., and K. Bowler. 1987. Temperature biology of animals. London, Chapman and Hall. *Comprehensive treatment of both ectotherms and endotherms.*

Heinrich, B. 1996. The thermal warriors: strategies of insect survival. Cambridge, Massachusetts, Harvard University Press. *Describes the many fascinating ways that insects respond to their temperature environment.*

Ianowski, J. P., and M. J. O'Donnell. 2004. Basolateral ion transport mechanisms during fluid secretion by *Drosophila* Malpighian tubules: Na⁺ recycling, Na⁺: K⁺: 2Cl⁻ cotransport and Cl⁻ conductance. J. Exp. Biol. **207**:2599–2609. *Research paper that presents data in support of a revised model for Malpighian tubular secretion in a terrestrial insect.*

Louw, G. N. 1993. Physiological animal ecology. New York, Longman Scientific & Technical. *Clearly presented survey with emphasis on thermoregulation and water relations in animals.*

Randall, D., W. Burggren, and K. French. 2001. Eckert animal physiology: mechanisms and adaptations. New York, W. H. Freeman & Company. *A comprehensive and comparative treatment of animal physiology.*

Sands, J. M. 1999. Urea transport: it's not just "freely diffusible" anymore. News Physiol. Sci. **14**:46–47. *Summarizes studies that have reported urea transport proteins in the collecting duct of the nephron.*

Schmidt-Nielsen, K. 1981. Countercurrent systems in animals. Sci. Am. **244**:118–128 (May). *Explains how countercurrent systems transfer heat, gases, or ions between fluids moving in opposite directions.*

Schultz, S. G. 1996. Homeostasis, Humpty Dumpty and integrative biology. News Physiol. Sci. **11**:238–246. *Describes the central role that homeostasis plays in the study of physiological systems.*

Smith, H. W. 1953. From fish to philosopher. Boston, Little, Brown & Company. *Classic account of vertebrate kidney evolution.*

Storey, K. B., and J. M. Storey. 1990. Frozen and alive. Sci. Am. **263**:92–97 (Dec.). *Explains how many animals have evolved strategies for surviving complete or almost complete freezing during the winter months.*

Willmer, P., G. Stone, and I. Johnston. 2005. Environmental physiology of animals, ed. 2. Oxford, Blackwell Science Ltd. *Well-written information regarding excretory systems in both vertebrates and invertebrates.*

ONLINE LEARNING CENTER

Visit www.mhhe.com/hickmanipz14e for chapter quizzing, key term flash cards, web links, and more!

Internal Fluids and Respiration

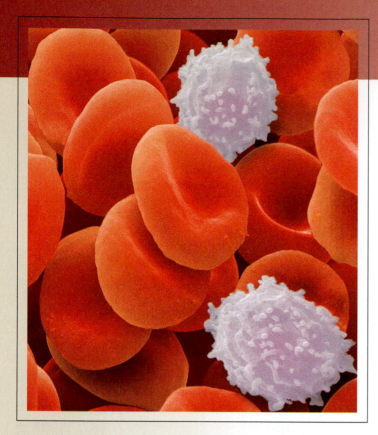

Scanning electron micrograph of blood cells.

William Harvey's Discovery

Ceaselessly, during a human life, the heart pumps blood through arteries, capillaries, and veins: about 5 liters per minute, until by the end of a normal life the heart has contracted some 2.5 billion times and pumped 300,000 tons of blood. When the heart stops its contractions, life also ends.

The crucial importance of the heart and its contractions for human life has been known since antiquity, probably almost as long as humans have existed. However, the circuit flow of blood, the notion that the heart pumps blood into arteries through the circulation and receives it back in veins became known only a few hundred years ago. The first correct description of blood flow by the English physician William Harvey initially received vigorous opposition when published in 1628.

Centuries earlier, the Greek anatomist Galen had taught that air enters the heart from the windpipe and that blood was able to pass from one ventricle to the other through "pores" in the interventricular septum of the heart. Galen also believed that blood first flowed out of the heart into all vessels, then returned—a kind of ebb and flow of blood. Even though there was almost nothing correct about this concept, it was still doggedly trusted at the time of Harvey's publication.

Harvey's conclusions were based on sound experimental evidence. He used a variety of animals for his experiments and chided human anatomists, saying that if only they had acquainted themselves with anatomy of lower vertebrates, they would have understood the blood's circuit. By tying ligatures on arteries, he noticed that the region between the heart and ligature swelled. When veins were tied off, the swelling occurred beyond the ligature. When blood vessels were cut, blood flowed in arteries from the cut end nearest the heart; the reverse happened in veins. By means of such experiments, Harvey discovered the correct scheme of blood circulation, even though he could not see the capillaries that connected the arterial and venous flows.

Single-celled organisms live in direct contact with their environments. They obtain nutrients and oxygen and release wastes directly across their cell surface. These organisms are so small that no special internal system of transport, beyond normal streaming movements of cytoplasm, is required. Even some simple multicellular forms, such as sponges, cnidarians, and flatworms, lack the internal complexity and metabolic demands that would require a circulatory system. Most other multicellular organisms, because of their increased size, activity, and complexity, need a specialized circulatory system to transport nutrients, waste products, and respiratory gases to and from all tissues of their body. In addition to serving these primary transport needs, circulatory systems have acquired additional functions; hormones (see Chapter 34) are moved from glands or cells that produce them, to target tissues where they act in concert with the nervous system (see Chapter 33) to integrate organismal function. Water, electrolytes, and the many other constituents of body fluids are distributed and exchanged between different organs and tissues. An effective response to disease and injury is vastly accelerated by an efficient circulatory system. Homeothermic birds and mammals depend heavily on blood circulation to conserve or to dissipate heat as required for maintenance of constant body temperature.

Gas exchange by diffusion across surface membranes alone is possible only for very small organisms less than 1 mm in diameter. For example, in single-celled organisms, oxygen is acquired and carbon dioxide is liberated in this way, since diffusion paths are short and the surface area of the organism is large relative to volume. As animals became larger and evolved a waterproof covering, specialized devices, such as lungs and gills, evolved to increase the effective surface for gas exchange. In addition, because gases diffuse so slowly through living tissue, a circulatory system was necessary to distribute gases to and from deep tissues of the body. Even these adaptations were inadequate for complex animals with high rates of cellular respiration. The solubility of oxygen in the blood plasma is so low that plasma alone cannot carry enough oxygen to support metabolic demands. With evolution of special oxygen-transporting blood proteins, such as hemoglobin, which seems to have evolved in conjunction with the circulatory system, the oxygen-carrying capacity of blood increased greatly. Thus what began as a simple and easily satisfied requirement resulted in evolution of several complex and essential respiratory and circulatory adaptations.

INTERNAL FLUID ENVIRONMENT

The body fluid of a single-celled organism is cellular cytoplasm, a liquid-gel substance in which the various membrane systems and organelles are suspended. In multicellular animals body fluids are divided into two main compartments, **intracellular** and **extracellular.** The intracellular compartment (also called intracellular fluid) is the collective fluid inside all the body's cells. The extracellular compartment (or extracellular fluid) is the fluid outside and surrounding the cells (Figure 31.1A). Thus the cells are bathed by their own aqueous environment, the extracellular fluid that buffers them from often harsh physical and chemical changes occurring outside the body. The importance of extracellular fluid was first emphasized by the great French physiologist Claude Bernard (Figure 31.2). In animals having closed circulatory systems (vertebrates, annelids, and a few other invertebrate groups; see p. 691) extracellular fluid is further subdivided into **blood plasma** and **interstitial (intercellular) fluid** (see Figure 31.1A). Blood vessels of a closed circulatory system contain plasma, whereas interstitial fluid, or tissue fluid as

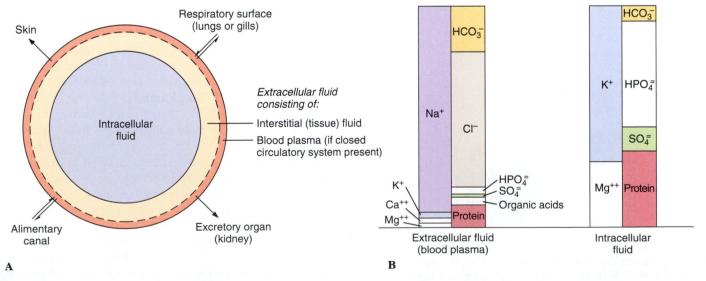

Figure 31.1

Fluid compartments of an animal's body. **A,** All body cells can be represented as belonging to a single large fluid compartment that is completely surrounded and protected by extracellular fluid (*milieu intérieur*). In animals with a closed circulatory system, this fluid is further subdivided into plasma and interstitial fluid. All exchanges with the environment occur across the plasma compartment. **B,** Typical electrolyte composition of extracellular and intracellular fluids. Total equivalent concentration of each major constituent is shown. Equal amounts of anions (negatively charged ions) and cations (positively charged ions) are in each fluid compartment. Note that sodium and chloride, major plasma electrolytes, are virtually absent from intracellular fluid (actually they are present in low concentration). Note the much higher concentration of protein inside cells.

it is sometimes called, occupies spaces surrounding the cells in the body. Nutrients and gases passing between vascular plasma and cells must traverse this narrow fluid separation. Interstitial fluid is constantly formed from plasma by movement of fluid to and from those microscopic vessels that are in close proximity to every cell (capillaries; see p. 696).

Composition of the Body Fluids

All these fluid spaces—plasma, interstitial, and intracellular—differ from each other in solute composition, but all have one feature in common: they are mostly water. Despite their firm appearance, animals are 70% to 90% water. Humans, for example, are approximately 70% water by weight. Of this, 50% is cell water, 15% is interstitial-fluid water, and the remaining 5% is in blood plasma. Plasma serves as the pathway of exchange between the cells of the body and the outside world. This exchange of respiratory gases, nutrients, and wastes is accomplished by specialized organs (for example, kidney, lung, gill, alimentary canal), as well as by skin (see Figure 31.1A).

Body fluids contain many inorganic and organic substances in solution. Principal among these are inorganic electrolytes and proteins. **Sodium, chloride,** and **bicarbonate ions** are the chief extracellular electrolytes, whereas **potassium, magnesium,** and **phosphate ions** and **proteins** are the major intracellular electrolytes (see Figure 31.1B). These differences are dramatic; they are always maintained despite continuous flow of materials into and out of cells of the body. The two subdivisions of extracellular fluid—plasma and interstitial fluid—have similar compositions except that plasma has more proteins, which are mostly too large to move from capillaries into interstitial fluid.

Figure 31.2

French physiologist Claude Bernard (1813 to 1878), one of the most influential of nineteenth-century physiologists. Bernard believed in the constancy of the *milieu intérieur* ("internal environment"), which is the extracellular fluid bathing the cells. He emphasized that it is through the *milieu intérieur* that foods and wastes and gases are exchanged and through which chemical messengers are distributed. He wrote, "The living organism does not really exist in the external environment (the outside air or water) but in the liquid *milieu intérieur* . . . that bathes the tissue elements."

COMPOSITION OF BLOOD

Among invertebrates that lack a circulatory system (such as flatworms and cnidarians) it is not possible to distinguish a true "blood." These organisms possess a clear, watery tissue fluid containing some phagocytic cells, a little protein, and a mixture of salts similar to seawater. The "blood" of invertebrates with open circulatory systems (p. 691) is more complex and is often called hemolymph (Gr. *haimo,* blood, + L. *lympha,* water). Invertebrates with closed circulatory systems (p. 691), on the other hand, maintain a clear separation between blood contained within blood vessels and tissue (interstitial) fluid surrounding blood vessels and in direct contact with cells.

In vertebrates, blood is a complex liquid tissue composed of plasma and formed elements or cellular components suspended in plasma.

The composition of mammalian blood follows:

Plasma—55% of blood

1. Water 90%
2. Dissolved solids, consisting of plasma proteins (albumin, globulins, fibrinogen), glucose, amino acids, electrolytes, various enzymes, antibodies, hormones, metabolic wastes, and traces of many other organic and inorganic materials
3. Dissolved gases, especially oxygen, carbon dioxide, and nitrogen

Formed elements (Figure 31.3)—45% of blood

1. Red blood cells (erythrocytes), containing hemoglobin for transport of oxygen and carbon dioxide
2. White blood cells (leukocytes), serving as scavengers and as defensive cells
3. Cell fragments (platelets in mammals) or cells (thrombocytes in other vertebrates) that function in blood coagulation

Plasma proteins are a diverse group of large and small proteins that perform numerous functions. The major protein groups are (1) **albumins,** the most abundant group, constituting 60% of the total, which help to keep plasma in osmotic equilibrium with the cells of the body; (2) **globulins,** a diverse group of high-molecular weight protein (35% of total) immunoglobulins (see p. 776) and various metal-binding proteins; and (3) **fibrinogen,** a very large protein that functions in blood coagulation. Blood **serum** is plasma minus the proteins involved in clot formation.

Red blood cells, or **erythrocytes,** are present in enormous numbers in blood, approximately 5.4 million per milliliter of blood in adult men and 4.8 million in adult women. In mammals and birds, red cells form continuously from large nucleated **erythroblasts** in red bone marrow (in other vertebrates kidneys and spleen are the principal sites of red blood cell production). During erythrocyte formation hemoglobin is synthesized and the precursor cells divide several times. In mammals the nucleus shrinks during development to become nonfunctional and eventually is lost from the cell by exocytosis (p. 51). The majority of cellular organelles are also lost, such as ribosomes, mitochondria, and most enzyme systems. What is left is a biconcave disc consisting of a baglike membrane packed with about 280 million molecules

FORMED ELEMENTS OF HUMAN BLOOD

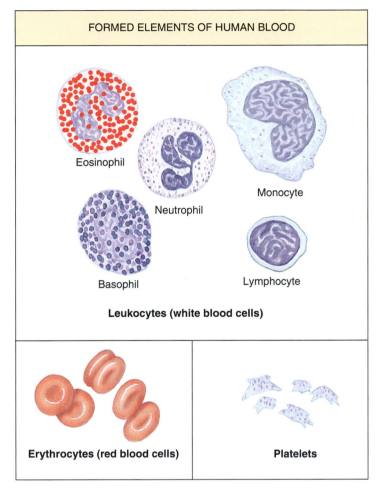

Eosinophil

Neutrophil

Monocyte

Basophil

Lymphocyte

Leukocytes (white blood cells)

Erythrocytes (red blood cells)

Platelets

Figure 31.3

Formed elements of human blood. Hemoglobin-containing red blood cells of humans and other mammals lack nuclei, but those of all other vertebrates have nuclei (see Figure 31.4B). Various leukocytes provide a system of protection for the body. Platelets are cell fragments that participate in the blood's clotting mechanism.

of the blood-transporting pigment hemoglobin. Approximately 33% of an erythrocyte by weight is hemoglobin. The biconcave shape (Figures 31.3 and 31.4A) is a mammalian innovation that provides a larger surface for gas diffusion than would a flat or spherical shape. All other vertebrates have nucleated erythrocytes that are usually ellipsoidal in shape (Figure 31.4B).

An erythrocyte enters the circulation for an average life span of approximately 4 months. During this time it may journey 11,000 km, squeezing repeatedly through the smallest blood vessels, capillaries, which are sometimes so narrow that the erythrocyte must bend to pass through. At last it fragments and is quickly engulfed by large scavenger cells called **macrophages** located in the liver, bone marrow, and spleen. Iron from the heme component of hemoglobin is salvaged to be used again; the rest of the heme is converted to **bilirubin,** a bile pigment. It is estimated that a human body produces 10 million erythrocytes and destroys another 10 million every second.

White blood cells, or **leukocytes,** form part of the immune system of the body (see Chapter 35). In adults they number only approximately 50,000 to 100,000 per milliliter of blood, a ratio of

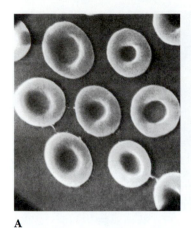
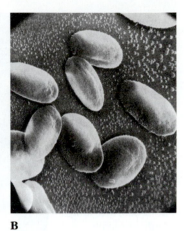

A   B

Figure 31.4

Mammalian and amphibian red blood cells. **A,** Erythrocytes of a gerbil are biconcave discs containing hemoglobin and no nucleus. **B,** Frog erythrocytes are convex discs, each containing a nucleus, which is plainly visible in the scanning electron micrograph as a bulge in the center of each cell. (Magnifications: mammalian erythrocytes, × 6300; frog erythrocytes, × 2400.)

1 white cell to 500 to 1000 red cells. There are several kinds of white blood cells: **granulocytes** (subdivided into **neutrophils, basophils,** and **eosinophils**), and **agranulocytes,** the lymphocytes and monocytes (see Figure 31.3).

Hemostasis: Prevention of Blood Loss

It is essential that animals have ways of preventing rapid loss of body fluids after an injury. In animals with closed circulatory systems (p. 691) blood is flowing and is under considerable hydrostatic pressure. Thus, it is especially vulnerable to hemorrhagic loss.

When a vessel is damaged, smooth muscle (p. 196) in the wall of the vessel contracts, which causes the vessel lumen to narrow, sometimes so strongly that blood flow is completely stopped. This simple but highly effective means of preventing hemorrhage is used by invertebrates and vertebrates alike. Beyond this first defense against blood loss, all vertebrates, as well as some larger, active invertebrates with high blood pressures, have in the blood special cellular elements and proteins that are capable of forming plugs, or clots, at an injury site.

In vertebrates **blood coagulation** is the dominant hemostatic defense. Blood clots form as a tangled network of fibers from one of the plasma proteins, **fibrinogen.** The transformation of fibrinogen into a **fibrin** meshwork (Figure 31.5) that entangles blood cells to form a gel-like clot is catalyzed by the enzyme thrombin. Thrombin is normally present in blood in an inactive form called **prothrombin,** which must be activated for coagulation to occur.

In this process, blood platelets (see Figure 31.3) and the damaged cells of blood vessels play a vital role. Platelets form in red bone marrow from large, multinucleated cells that regularly pinch off bits of their cytoplasm; thus they are fragments of cells. There are 150,000 to 300,000 platelets per cubic millimeter of blood. When the normally smooth inner surface of a blood vessel is disrupted, either by a break or by deposits of a

Figure 31.5

Human red blood cells trapped in fibrin clot. Clotting is initiated after tissue damage by disintegration of platelets in blood, resulting in a complex series of intravascular reactions that end with conversion of a plasma protein, fibrinogen, into long, tough, insoluble polymers of fibrin. Fibrin and entangled red blood cells form the blood clot, which arrests bleeding.

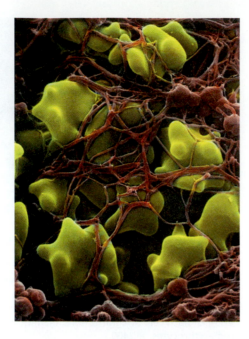

cholesterol-lipid material, platelets rapidly adhere to the surface and release **thromboplastin** and other clotting factors. These platelet derived factors, calcium ions, as well as tissue thromboplastin and other factors released from damaged tissue initiate conversion of prothrombin to active thrombin (Figure 31.6).

The catalytic sequence in this scheme is complex, involving a series of plasma protein factors, each normally inactive until activated by a previous factor in the sequence. The sequence behaves like a "cascade" with each reactant in the sequence leading to a large increase in the amount of the next reactant. At

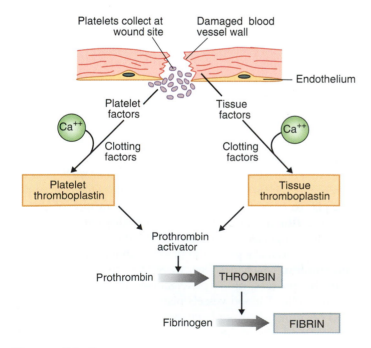

Figure 31.6

Stages in formation of fibrin.

least 13 different plasma coagulation factors have been identified. A deficiency of only a single factor can delay or prevent the clotting process. Why has such a complex clotting mechanism evolved? Probably it is necessary to provide a fail-safe system capable of responding to any kind of internal or external hemorrhage that might occur and yet is unlikely to be activated into forming dangerous intravascular clots in the absence of injury.

Several kinds of clotting abnormalities occur in humans. One of these, hemophilia, is a condition characterized by failure of blood to clot, so that even insignificant wounds can cause continuous severe bleeding and death. It is caused by a rare mutation (the condition occurs in about 1 in 10,000 males) on the X sex chromosome, resulting in an inherited lack of one of the platelet factors in males and in homozygous females. Called the "disease of kings," it once ran through several interrelated royal families of Europe, apparently having originated from a mutation in one of Queen Victoria's parents.

Hemophilia is one of the best known cases of sex-linked inheritance in humans (p. 87). Actually two different loci on the X chromosome are involved. Classic hemophilia (hemophilia A) accounts for about 80% of persons with the condition, and the remainder are caused by Christmas disease (hemophilia B). The recessive allele at each locus results in a deficiency of a different platelet factor.

CIRCULATION

Most animals have evolved mechanisms, in addition to simple diffusion, for transporting materials among various regions of their body. For sponges (see Figure 12.5, p. 250) and diploblasts (see Chapter 13) the water in which they live provides the medium for transport. Water, propelled by ciliary, flagellar, or body movements, passes through channels or compartments to facilitate the movement of food, respiratory gases, and wastes. True circulatory systems—containing vessels through which blood moves—are essential to animals so large or so active that diffusional processes alone cannot supply their oxygen needs. An animal's shape obviously is important. Flattened and leaflike acoelomate flatworms (see Chapter 14) survive without a circulatory system because the distance of any body part from the surface is short; respiratory gases and metabolic wastes transfer by simple diffusion even though many are relatively large animals.

A circulatory system having a full complement of components—pump, arterial distribution system, capillaries that interface with cells, venous reservoir and return system—is fully recognizable in annelid worms (see Chapter 17). In earthworms (Figure 31.7) there are two main vessels, a dorsal vessel carrying blood anteriorly, and a ventral vessel delivering blood posteriorly throughout the body by way of segmental vessels and a dense tissue capillary network. The dorsal vessel drives the blood forward by peristalsis (see p. 714) and thus serves as a heart. Five aortic arches that connect the dorsal and ventral vessels laterally are also contractile and serve as accessory hearts

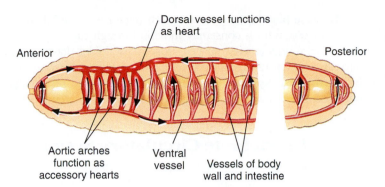

Figure 31.7

Blood flow through the closed vascular system of an earthworm.

to maintain a steady flow of blood into the ventral vessel and to the head region. Many smaller segmental vessels that deliver blood to tissue capillaries are actively contractile as well. We see then that there is no localized pump pushing the blood through a system of passive tubes; instead the power of contraction is widely distributed throughout the vascular system.

Open and Closed Circulations

The system just described is a **closed circulation** because the circulating fluid, **blood,** is confined to vessels throughout its journey through the vascular system. Many invertebrates have an **open circulation** in which there are no small blood vessels or capillaries interfacing with cells or connecting arteries with veins. In insects and other arthropods, in most molluscs, and in many smaller invertebrate groups blood sinuses, collectively called a **hemocoel,** replace capillary beds found in animals with closed systems. During development of the body cavity in these groups, the blastocoel is not completely obliterated by the mesoderm (Figure 31.8) and fuses with embryonic coelomic cavities.

Thus, the hemocoel is composed of the primary body cavity (persistent blastocoel) and secondary coelomic cavities through which blood (also called **hemolymph**) freely circulates (bottom diagrams in Figure 31.8). Since there is no separation of the extracellular fluid into blood plasma and lymph (as there is in a closed circulation), blood volume is large and may constitute 20% to 40% of body volume. By contrast, blood volume in animals with closed circulations (vertebrates, for example) is only about 5% to 10% of body volume.

In arthropods, the heart and all viscera lie in the hemocoel, bathed by blood (Figure 31.8). Blood enters the heart through valved openings, the ostia, and the heart's contractions, which resemble a forward-moving peristaltic wave, propel blood into a limited arterial system. Blood is distributed to the head and other organs, then escapes into the hemocoel. It is routed through the body and appendages by a system of baffles and longitudinal membranes (septa) before returning to the heart (Figure 31.9). Because blood pressure is very low in open systems, seldom exceeding 4 to 10 mm Hg, many arthropods have auxiliary hearts or contractile vessels to boost blood flow.

It is interesting to note that insects and many other terrestrial arthropods do not use their circulatory system for respiratory gas transport. Instead, a separate respiratory system has evolved for this purpose: a tracheal system in insects and some other terrestrial arthropods (centipedes, millipedes and some spiders, p. 449), pseudotracheae in terrestrial pill bugs, and book lungs in some spiders.

In animals with closed circulatory systems (most annelids, cephalopod molluscs, and all vertebrates) the coelom increases in size during embryonic development to obliterate the blastocoel and forms a secondary body cavity (top diagrams in Figure 31.8). A system of continuously connected blood vessels develops within the mesoderm. Closed systems have certain features in common. A **heart** pumps blood into **arteries** that branch and narrow into **arterioles** and then into a vast system

Figure 31.8

Diagrams showing how open and closed circulatory systems develop. The principal body cavity of arthropods is the hemocoel, which is formed by fusion of the primary blastocoel with the coelom.

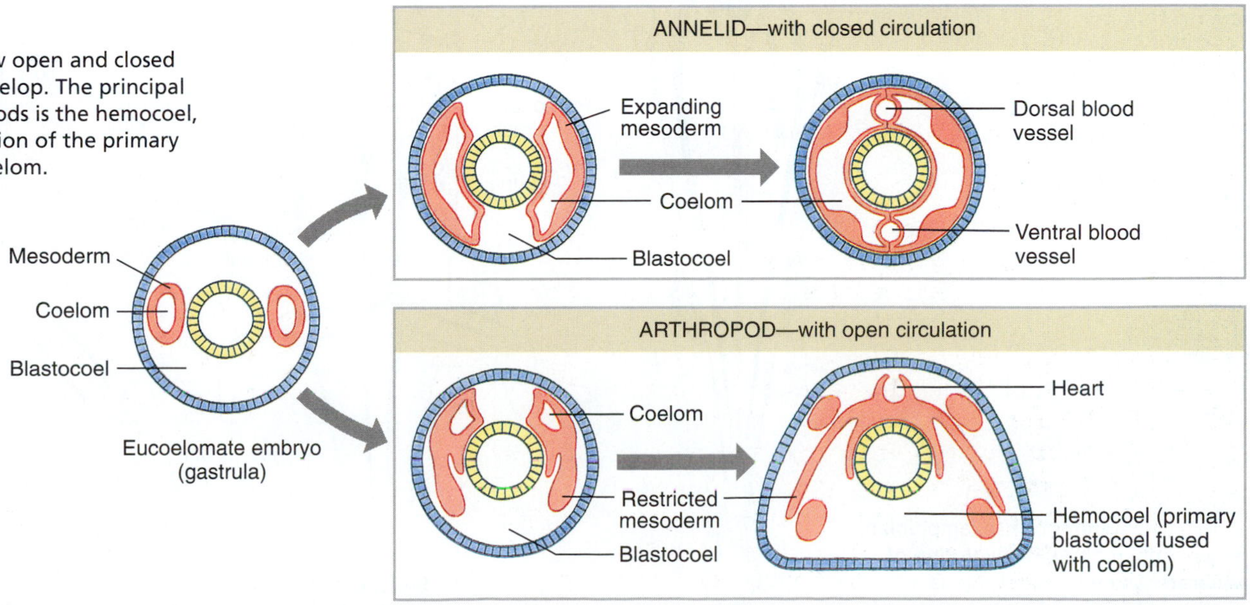

ANNELID—with closed circulation

Expanding mesoderm

Coelom

Blastocoel

Dorsal blood vessel

Ventral blood vessel

Mesoderm

Coelom

Blastocoel

Eucoelomate embryo (gastrula)

ARTHROPOD—with open circulation

Coelom

Restricted mesoderm

Blastocoel

Heart

Hemocoel (primary blastocoel fused with coelom)

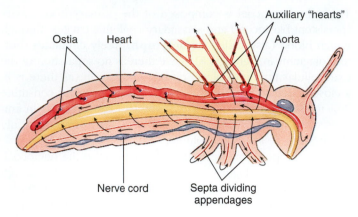

Figure 31.9
Circulatory system of an insect. Although the circulatory system is open, blood is directed through the appendages in channels formed by longitudinal septa. Arrows indicate the course of circulation.

of **capillaries** that interfaces with cells in body tissues. Blood leaving capillaries enters **venules** and then **veins** that return the blood to the heart. Capillary walls are thin (only one cell thick), permitting rapid rates of transfer of materials between blood and tissues. Closed systems are more suitable for large and active animals because blood can be moved rapidly to the most active tissues. In addition, flow to various organs can be readjusted to meet changing needs by varying the diameters of blood vessels.

Because blood pressures are much higher in closed than in open systems, fluid is constantly moving through capillary walls into the surrounding tissue spaces. Most of this fluid is drawn back into capillaries by osmosis (see p. 697). The remainder is recovered by a **lymphatic system** (p. 697), which has evolved separately but in conjunction with the high-pressure system of vertebrates.

Plan of Vertebrate Circulatory Systems

In vertebrates the principal differences in the blood-vascular system involve the gradual separation of the heart into two separate pumps as vertebrates evolved from aquatic life with gill breathing to fully terrestrial life with lung breathing. These changes are shown in Figure 31.10, which compares the circulation of fish, amphibians, and mammals.

A fish heart contains two main chambers in series, an **atrium** and a **ventricle.** The atrium is preceded by an enlarged chamber, the **sinus venosus,** which collects blood from the venous system to assure a smooth delivery of blood to the heart. Elasmobranchs have a fourth heart chamber, the **conus arteriosus,** which dampens blood pressure oscillations before blood flows into delicate blood capillaries. Teleost fish have a **bulbous arteriosus** that serves the same function. Blood makes a single circuit through a fish's vascular system; it is pumped from the heart to the gills, where it is oxygenated, then flows into the dorsal aorta to be distributed to body organs, and finally returns by

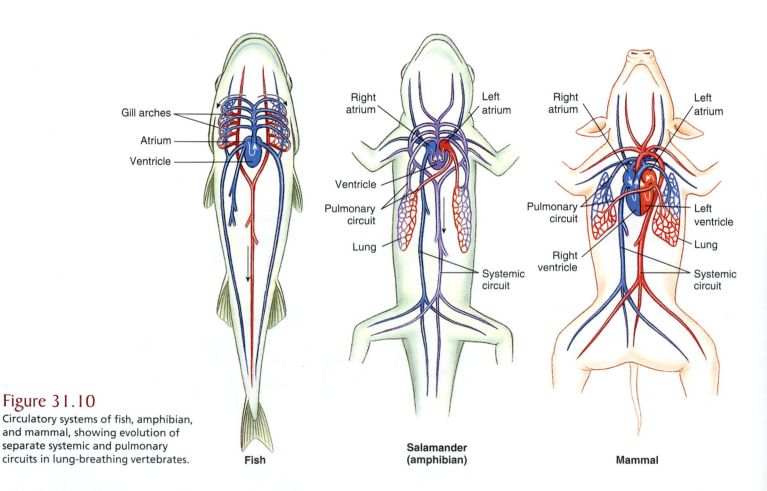

Figure 31.10
Circulatory systems of fish, amphibian, and mammal, showing evolution of separate systemic and pulmonary circuits in lung-breathing vertebrates.

Fish

Salamander (amphibian)

Mammal

veins to the heart. In this circuit the heart must provide sufficient pressure to push the blood through two sequential capillary systems, first that of the gills, and then that of the remainder of the body. The principal disadvantage of the single-circuit system is that gill capillaries offer so much resistance to blood flow that blood pressures to body tissues are greatly reduced.

With evolution of lungs instead of gills between the heart and aorta, vertebrates developed a high-pressure **double circulation**: a **systemic circuit** that provides oxygenated blood to the capillary beds of the body organs; and a **pulmonary circuit** that serves the lungs. The beginning of this major evolutionary change probably resembled the condition seen in lungfishes and amphibians. In modern amphibians (frogs, toads, salamanders) the atrium is completely separated by a partition into two atria (Figure 31.11). The right atrium receives venous blood from the body while the left atrium receives oxygenated blood from the lungs and skin. The ventricle is undivided, but venous and arterial blood remain mostly separate due to the spiral fold of the conus arteriosus, the arrangement of septa or folds in vessels leaving the heart (Figure 31.11), and differential blood pressures in the exiting vessels. A septum partially divides the ventricle in most nonavian reptiles, and it is complete in crocodilians, birds, and mammals (Figure 31.12). Systemic and pulmonary circuits are now separate circulations, each served by one half of a dual heart (Figure 31.12).

Mammalian Heart

The four-chambered mammalian heart (Figure 31.12) is a muscular organ located in the thorax and covered by a tough, fibrous sac, the **pericardium.** Blood returning from the lungs flows through the **pulmonary veins** and collects in the **left atrium,** passes into the **left ventricle,** and is pumped into the body (systemic) circulation via the **aorta.** Blood returning from the body flows through the **inferior** (posterior) and **superior** (anterior) **vena cava** into the **right atrium,** and passes into the **right ventricle,** which pumps it into the lungs via the **pulmonary arteries.** Backflow of blood in the heart is prevented by two sets of valves, formed as extensions of the inner wall of the heart, that open and close passively in response to pressure differences between the heart chambers. The **left atrioventricular** (bicuspid) and the **right atrioventricular** (tricuspid) valves separate the cavities of the atrium and ventricle in

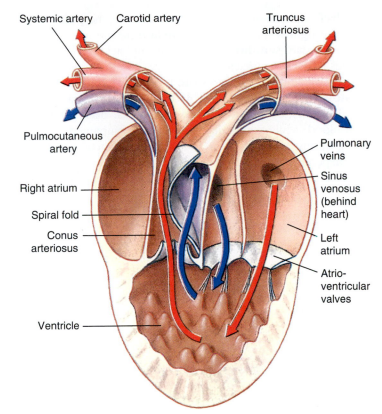

Figure 31.11

Route of blood through a frog heart. Atria are completely separated, and the spiral fold in the conus arteriosus helps to route blood to lungs and systemic circulation.

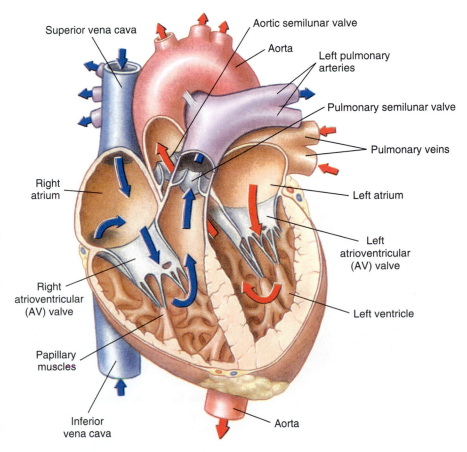

Figure 31.12

Human heart. Deoxygenated blood enters the right side of the heart and is pumped to the lungs. Oxygenated blood returning from the lungs enters the left side of the heart and is pumped to the body. The right ventricular wall is thinner than that of the left ventricle, a result of less muscular force generation to pump blood to such close structures as the lungs.

each half of the heart. Where the great arteries, the **pulmonary** from the right ventricle and the **aorta** from the left ventricle, leave the heart, **semilunar valves** prevent backflow from these arteries into the ventricles.

Contraction is called **systole** (sis′to-lee), and relaxation, **diastole** (dy-as′to-lee) (Figure 31.13). When the atria contract (atrial systole), the ventricles relax (ventricular diastole) and fill with blood, while ventricular systole is accompanied by atrial filling during atrial diastole. The rate of contraction or heartbeat depends on age, sex, and especially exercise. Exercise may increase **cardiac output** (volume of blood forced from either

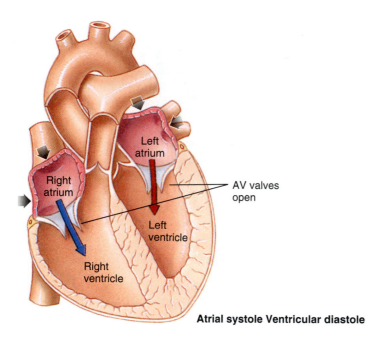

Atrial systole Ventricular diastole

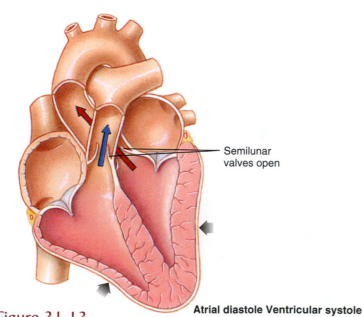

Atrial diastole Ventricular systole

Figure 31.13
Human heart in systole and diastole.

ventricle each minute) more than fivefold due to an increase in both heart **rate** and **stroke volume** (volume of blood forced from either ventricle per beat). Heart rates among vertebrates vary with level of metabolism and body size. Ectothermic codfish have a heart rate of approximately 30 beats per minute; endothermic rabbits of about the same weight have a rate of approximately 200 beats per minute. Small animals have higher heart rates than do large animals, reflecting the increase in metabolic rate that occurs with a decrease in body size (p. 624). For example, the heart rate in an elephant is 25 beats per minute, in a human 70 per minute, in a cat 125 per minute, in a mouse 400 per minute, and in the tiny 4 g shrew, the smallest mammal, the heart rate approaches a prodigious 800 beats per minute.

Excitation and Control of the Heart

The vertebrate heart is a muscular pump composed of **cardiac muscle.** Cardiac muscle resembles skeletal muscle—both are types of striated muscle—but cardiac cells are branched and joined end-to-end by junctional complexes (intercalated discs) to form a branching network (see Figure 29.13, p. 656). Unlike skeletal muscle, vertebrate cardiac muscle does not depend on nerve activity to initiate a contraction. Instead, regular contractions are established by specialized cardiac muscle cells, called **pacemaker cells.** In a nonavian reptile, bird, or mammal heart the pacemaker is in the **sinoatrial (SA) node,** a remnant of the ancestral sinus venosus still found in fish and amphibians. Electrical activity initiated in the pacemaker spreads over the muscle of the two atria and then, after a slight delay, to the secondary pacemaker, the atrioventricular (AV) node, at the top of the ventricles. At this point electrical activity is conducted rapidly through the **bundle of His** and left and right branching bundles to the apex of the ventricle and then continues through specialized fibers **(Purkinje fibers)** to the apex or "tip" of the ventricles (Figure 31.14). This arrangement allows the contraction to begin within cardiac cells in this region of the ventricles and spread upward to squeeze out the blood in the most efficient way; it also ensures that both ventricles contract simultaneously and with enough of a delay to allow the atria to fill before electrical activity is again initiated at the SA node. Structural specializations in Purkinje fibers, such as well-developed intercalated discs (see Figure 29.13, p. 656) and numerous gap junctions, facilitate rapid conduction through these fibers.

The **cardiac center** in the brain provides external regulation to the heart and is located in the medulla (p. 735). It connects to the heart by two sets of nerves, the parasympathetic **vagus** nerves, and applies a braking action to the heart rate when activated by the brain, while activation of the sympathetic nerves speeds it up. Both sets of nerves terminate in the SA node, thus guiding the activity of the pacemaker.

The cardiac center in turn receives sensory information about a variety of stimuli. Pressure receptors (sensitive to blood pressure) and chemical receptors (sensitive primarily to carbon dioxide and pH) are located at strategic points in the vascular system. The cardiac center uses this information to adjust heart rate and cardiac output in response to activity or changes in

between muscle fibers that the heart's own pumping action squeezes through sufficient oxygenated blood to the muscle. In larger fish and frogs, as well as in nonavian reptiles, birds, and mammals, however, the larger heart size and the thickness of the heart muscle require that the heart have its own vascular supply, the **coronary circulation.** Coronary arteries divide to form an extensive capillary network surrounding the muscle fibers and provide them with oxygen and nutrients. Heart muscle has an extremely high oxygen demand. Even at rest the heart removes 70% of oxygen from the blood, in contrast to most other body tissues, which remove only about 25%. Therefore, an increase in the work of the heart must be met by a massive increase in coronary blood flow—up to nine times the resting level during strenuous exercise. Any reduction in coronary circulation due to partial or complete blockage (coronary artery disease) may lead to a heart attack (myocardial infarction) in which heart cells may die from lack of oxygen.

Coronary artery disease (CAD) is currently the number one killer in the United States. Risk factors can be divided into those that cannot be modified and those that can be modified. Risk factors that cannot be modified include a family history of heart disease, being a male, being a postmenopausal female, or being over 45 years old. Risk factors that can be modified include smoking, high blood cholesterol levels, high blood pressure, uncontrolled diabetes, being overweight or obese, stress, high saturated fat and cholesterol diets, and a sedentary lifestyle. Reducing modifiable risk factors can significantly decrease the risk of CAD.

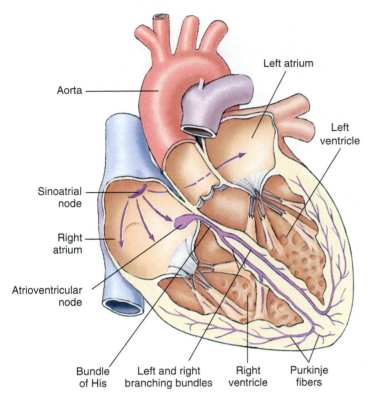

Figure 31.14

Mechanisms controlling heartbeat. Arrows indicate spread of excitation from the sinoatrial node, across the atria, to the atrioventricular node. The wave of excitation is then conducted very rapidly to ventricular muscle over the specialized conducting bundles and Purkinje fiber system.

body position. Feedback mechanisms thus control the heart and keep its activity constantly attuned to needs of the body.

Because the heartbeat is initiated in specialized muscle cells, vertebrate hearts, together with the hearts of molluscs and several other invertebrates, are called **myogenic** ("muscle origin") hearts. Although the nervous system does alter pacemaker activity to slow down or speed up heart rate, a myogenic heart beats spontaneously and involuntarily even if completely removed from the body. An isolated turtle or frog heart beats for hours if placed in a balanced salt solution. The myogenic activity of a human heart allows heart transplants to occur successfully even if a heart has been out of a body for several hours, especially if the heart's activity is slowed by cooling. Some invertebrates, for example decapod crustaceans, have **neurogenic** ("nerve origin") hearts. In these hearts a cardiac ganglion (group of nerve cell bodies) located on the heart serves as pacemaker. If this ganglion is separated from the heart, the heart stops beating, even though the ganglion itself remains rhythmically active.

Coronary Circulation

It is no surprise that an organ as active as the heart needs a generous blood supply of its own. The heart of small fish and frogs is small and the muscle is so thoroughly channeled with spaces

Arteries

All vessels leaving the heart are called arteries whether they carry oxygenated blood (aorta) or deoxygenated blood (pulmonary artery). To withstand high, pulsing pressures generated during ventricular systole, the largest arteries closest to the heart **(elastic arteries)** are invested with thick layers of elastic fibers, very little smooth muscle (p. 196), and tough inelastic connective tissue (Figure 31.15). The elasticity of these arteries allows them to stretch as the surge of blood leaves the heart during ventricular systole and then to recoil, compressing the fluid column, during ventricular diastole. This elasticity maintains the high blood pressure created by each heartbeat, and blood moves in a forward direction because semilunar valves at the entrances to these arteries impede backward flow (Figure 31.12). Thus the normal arterial pressure in humans varies only between 120 mm Hg (systole) and 80 mm Hg (diastole) (usually expressed as 120/80), rather than dropping to zero during diastole as we might expect in a fluid system with an intermittent pump. Arteries farther away from the heart possess more smooth muscle and less elastic fibers. These arteries, called **muscular arteries,** can increase or decrease their diameter, which results in dampening of the heartbeat-associated high pressure and flow oscillations before the blood reaches body organs.

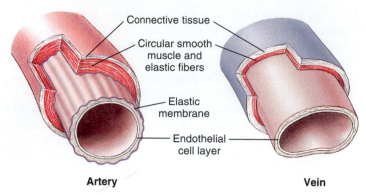

Figure 31.15

Artery and vein, showing tissue layers. Note greater thickness of the muscle layer in the artery. This layer has more elastic fibers in elastic arteries and more smooth muscle in muscular arteries.

Thickening and loss of elasticity in arteries is known as *arteriosclerosis*. When arteriosclerosis is characterized by fatty deposits of cholesterol in artery walls, the condition is *atherosclerosis*. Scientists now think that inflammation precedes the accumulation of fat in the arteries. Irregularities in the walls of blood vessels often cause blood to clot around them, forming a *thrombus*. When a bit of the thrombus breaks off and is carried by the blood to lodge elsewhere, it is an *embolus*. If the embolus blocks one of the coronary arteries, the person has a heart attack (a "coronary"). The portion of the heart muscle served by the branch of the coronary artery that is blocked is starved for oxygen. It may be replaced by scar tissue if the person survives. A thrombus may also occur within a coronary artery and this is a more common cause of heart attack.

As arteries branch and narrow into **arterioles,** the walls are composed primarily of only one or two layers of smooth muscle surrounding an epithelial layer (Figure 31.16). Contraction of this muscle narrows the arterioles and reduces the flow of blood to body organs, diverting it to where it is most needed. Blood must be pumped with a hydrostatic pressure sufficient to overcome resistance of the narrow passages through which it must flow. Consequently, large animals tend to have higher blood pressure than do small animals.

Blood pressure was first measured in 1733 by Stephan Hales, an English clergyman with unusual inventiveness and curiosity. He tied his mare, which was "to have been killed as unfit for service," on her back and exposed the femoral artery. This he cannulated with a brass tube, connecting it to a tall glass tube with the windpipe of a goose. The use of the windpipe was both imaginative and practical; it gave the apparatus flexibility "to avoid inconveniences that might arise if the mare struggled." The blood rose 8 feet in the glass tube and bobbed up and down with systolic and diastolic beats of the heart. The weight of the 8-foot column of blood was equal to the blood pressure. We now express blood pressure as the height of a column of mercury (Hg), which is 13.6 times heavier than water. Hales's figures, expressed in millimeters of mercury, indicate that he measured a blood pressure of 180 to 200 mm Hg, about normal for a horse.

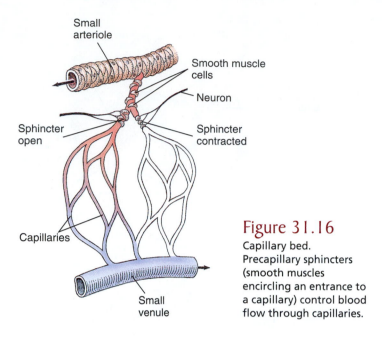

Figure 31.16

Capillary bed. Precapillary sphincters (smooth muscles encircling an entrance to a capillary) control blood flow through capillaries.

Today, blood pressure is measured in humans with an instrument called a **sphygmomanometer.** A cuff is inflated on the upper arm with air to a pressure sufficient to close the arteries in the arm. If a stethoscope is held over the brachial artery (in the crook of the elbow) and air is slowly released from the cuff, the first turbulent spurts of blood through the artery as it opens slightly can be heard. This is equivalent to systolic pressure. As pressure in the cuff decreases, the turbulent sound finally disappears as blood runs smoothly through the artery. The pressure at which the sound disappears is diastolic pressure. Digital sphygmomanometers are now more commonly used.

Capillaries

The Italian Marcello Malpighi was the first to describe capillaries in 1661, thus confirming the existence of the minute links between the arterial and venous systems that Harvey knew must exist but could not see. Malpighi studied capillaries of a living frog's lung, which is still one of the simplest and most vivid preparations for demonstrating capillary blood flow.

Capillaries are present in enormous numbers, forming extensive networks in nearly all tissues (Figure 31.16). In muscle there are more than 2000 per square millimeter (1,250,000 per square inch), but not all are open at the same time. Indeed, perhaps less than 1% are open in resting skeletal muscle. But when muscle is active, all capillaries may open to bring oxygen and nutrients to the working muscle fibers and to carry away metabolic wastes.

Capillaries are extremely narrow, averaging in mammals about 8 μm in diameter, which is only slightly wider than the red blood cells that must pass through them. Their walls are formed by a single layer of thin **endothelial** cells, held together by a delicate basement membrane and few connective-tissue fibers.

Capillary Exchange

Capillaries are quite permeable to small ions, nutrients, and water. Blood pressure within a capillary tends to force fluids out through or between the capillary endothelial cells and into the surrounding interstitial space (p. 687). Fluid may pass between the endothelial cells via water-filled clefts (approximately 4 nm wide) or through endothelial cells in pinocytotic vesicles (p. 51) that appear to shuttle material from one side of epithelial cells to the other. Lipid-soluble substances can diffuse easily through the plasma membranes of endothelial cells and into the interstitial fluid. Because larger molecules such as plasma proteins cannot pass through the capillary endothelial cell clefts, an almost protein-free filtrate is forced outward. This fluid movement is important in irrigating the interstitial space, in providing tissue cells with oxygen, glucose, amino acids, and other nutrients. For capillary exchange to be effective, fluids that leave the capillaries must at some point reenter the circulation and bring with them cellular metabolic wastes. If they did not, fluid would quickly accumulate in tissue spaces, causing swelling or edema. The delicate balance of fluid exchange across the capillary wall is achieved by the two opposing forces of hydrostatic (blood) pressure and osmotic pressure (Figure 31.17).

In a capillary, the blood pressure that pushes water molecules and solutes through the capillary endothelial cell clefts is greatest at the arteriolar end of the capillary and declines along its length as blood pressure falls (Figure 31.17). Opposing blood hydrostatic pressure is an osmotic pressure (p. 49) created by the proteins that cannot pass through the capillary endothelial cell clefts. This **colloid osmotic pressure,** which is about 25 mm Hg in mammalian plasma, tends to draw water back into the capillary from the tissue fluid. The result of these two opposing forces is that water and solutes tend to be filtered out of the arteriolar end of the capillary where hydrostatic pressure exceeds colloid osmotic pressure, and to be drawn in again at the venous end where colloid osmotic pressure exceeds hydrostatic pressure.

The amount of fluid filtered across the capillary endothelial cells fluctuates greatly among different capillaries. Usually outflow exceeds inflow, and some excess fluid remains in the interstitial spaces between tissue cells. This excess is collected and removed by **lymph capillaries** of the **lymphatic system** and eventually this fluid, called **lymph,** is returned to the circulatory system via larger lymph vessels.

Veins

Venules and veins into which the capillary blood drains for its return journey to the heart are thinner walled, less elastic, and of considerably larger diameter than their corresponding arteries and arterioles (see Figure 31.15). Blood pressure in the venous system is low, from approximately 10 mm Hg, where capillaries drain into venules, to almost zero in the right atrium. Because pressure is so low, venous return must get assistance from valves in the veins, body skeletal muscles surrounding the veins, suction created during diastole of the heart, and the rhythmical action of the lungs during breathing. Without these mechanisms, blood will pool in the lower extremities of a standing animal—a very real problem if humans must stand for long periods. Veins that lift blood from the extremities to the heart contain valves that divide the long column of blood into segments. Valves are formed as infoldings of the endothelial cell layer and underlying connective tissue. When skeletal muscles contract, as in even slight activity, veins are squeezed, and blood within them moves toward the heart because valves within the veins keep blood from flowing backward. The well-known risk of fainting while standing at stiff attention in hot weather usually can be prevented by deliberately pumping leg muscles. Negative pressure in the thorax created by inspiratory movements of the lungs also speeds venous return by sucking blood up the large vena cava into the heart.

Lymphatic System

The lymphatic system of vertebrates is an extensive network of thin-walled vessels that arise as blind-ended lymph capillaries in most tissues of the body. These unite to form a treelike structure of increasingly larger lymph vessels, which finally drain into veins in the lower neck (Figure 31.18). A principal function of the lymphatic system is to return to the blood the excess fluid (lymph) filtered across capillary endothelial cells into interstitial spaces. This fluid-filtrate, called lymph, is similar to plasma but has a much lower concentration of protein. Large molecules, especially fats absorbed from the gut, also reach the circulatory system by way of the lymphatic system. The rate of lymph flow is very low,

Figure 31.17

Fluid movement across the capillary endothelial cell clefts. At the arterial end of the capillary, hydrostatic (blood) pressure exceeds colloid osmotic pressure contributed by plasma proteins, and a plasma filtrate is forced out of the capillary. At the venous end, colloid osmotic pressure exceeds the hydrostatic pressure, and fluid is drawn into the capillary. In this way plasma nutrients are carried into the interstitial space where they can enter cells, and metabolic wastes from the cells are drawn into the plasma and carried away.

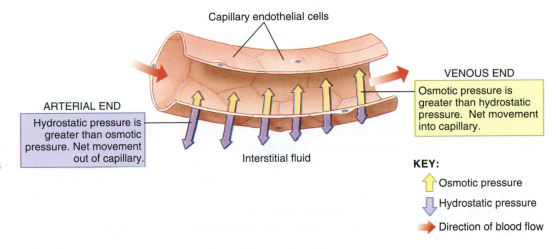

Capillary endothelial cells

VENOUS END
Osmotic pressure is greater than hydrostatic pressure. Net movement into capillary.

ARTERIAL END
Hydrostatic pressure is greater than osmotic pressure. Net movement out of capillary.

Interstitial fluid

KEY:
Osmotic pressure
Hydrostatic pressure
Direction of blood flow

Figure 31.18

Human lymphatic system, showing major vessels, **A**, and a detail of the intimate association between the blood and lymphatic capillaries, **B**.

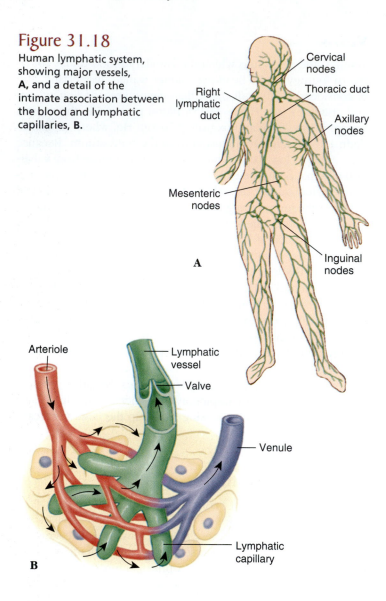

a minute fraction of blood flow, and larger lymph vessels possess valves similar in structure and function to those found in veins.

The lymphatic system also plays a central role in the body's defenses. Located at intervals along the lymph vessels are **lymph nodes** (Figure 31.18) that have several defense-related functions (see Chapter 35). Cells in the lymph glands such as macrophages remove foreign particles, especially bacteria, which might otherwise enter the general circulation. They are also centers (together with bone marrow and thymus gland) for production, maintenance, and distribution of lymphocytes—essential components of a body's defense mechanisms.

RESPIRATION

Energy from food is released by oxidative processes, usually with molecular oxygen as the terminal electron acceptor. Oxygen for this purpose is taken into the body across a respiratory surface. Physiologists distinguish two separate but interrelated respiratory processes: **cellular respiration,** the oxidative processes that occur within cells (p. 63), and **external respiration,** the exchange of oxygen and carbon dioxide between the organism and its

environment. In this section we describe external respiration and transport of gases from respiratory surfaces to body tissues.

Problems of Aquatic and Aerial Breathing

How an animal respires is determined largely by the nature of its environment. The two great arenas of animal evolution—water and land—are vastly different in their physical characteristics. The most obvious difference is that air contains far more oxygen—at least 20 times more—than does water. For example, water at 5° C (41° F) fully saturated with air contains approximately 9 ml of oxygen per liter (0.9%); by comparison air contains 209 ml of oxygen per 1000 ml (21%). The density and viscosity of water are approximately 800 and 50 times greater, respectively, than those of air. Furthermore, gas molecules diffuse 10,000 times more rapidly in air than in water. These differences mean that aquatic animals must have evolved very efficient ways of removing oxygen from water, yet even the most advanced fishes with highly efficient gills and pumping mechanisms may use as much as 20% of their energy just extracting oxygen from water. By comparison, the cost for mammals to breathe is only 1% to 2% of energy produced during resting metabolism.

Respiratory surfaces must be thin and always kept wet with a fine film of fluid to allow diffusion of gases across an aqueous phase between the environment and the underlying circulation. This is hardly a problem for aquatic animals, immersed as they are in water, but it is a challenge for air breathers. To keep respiratory membranes moist and protected from injury, air breathers have in general developed them as invaginations of the body surface and then added pumping mechanisms to move air in and out of the respiratory region. The lung is the best example of a successful solution to breathing on land. In general **evaginations** of the body surface, such as gills, are most suitable for aquatic respiration; **invaginations,** such as lungs and tracheae, are best for air breathing. We now consider some important examples of respiratory organs employed by animals.

Respiratory Organs

Gas Exchange by Direct Diffusion

Protozoa, sponges, cnidarians, and many worms respire by direct diffusion of gases between organism and environment. As noted at the beginning of this chapter this kind of **cutaneous respiration** is not adequate when the cellular mass exceeds approximately 1 mm in diameter. However, by greatly increasing the surface of the body relative to its mass, many multicellular animals can supply part or all of their oxygen requirements by direct diffusion. Flatworms are an example of this strategy.

Cutaneous respiration frequently supplements gill or lung breathing in larger animals such as amphibians and fishes. For example, an eel can exchange 60% of its oxygen and carbon dioxide through its highly vascular skin. During winter hibernation, frogs and even turtles exchange all their respiratory gases through their skin while submerged in ponds or springs. Most species of salamanders are lungless. Some lungless salamanders have larvae with gills, and gills persist in the adults of some, but adults of most species have neither lungs nor gills.

Gas Exchange Through Tubes: Tracheal Systems

Insects and certain other terrestrial arthropods (centipedes, millipedes, and some spiders) have a highly specialized type of respiratory system, in many respects the simplest, most direct, and most efficient respiratory system found in active animals. It is a branching system of tubes (**tracheae**) that extends to all parts of the body (Figure 31.19). The smallest end channels are fluid-filled **tracheoles**, less than 1 μm in diameter, that terminate in close association with the plasma membranes of body cells. Air enters and leaves the tracheal system through valvelike openings (**spiracles**) that may be closed to reduce water loss. A filter may also be present to decrease entrance of water, debris, or parasites (see Figure 21.18, p. 451). Some insects can ventilate the tracheal system with body movements; the familiar telescoping movement of the abdomen of bees on hot summer days is an example. Respiratory pigments are present in insect blood, but because the cells have a direct pipeline to the outside, bringing oxygen in and carrying carbon dioxide out, an insect's respiration is independent of its circulatory system. Consequently, insect blood plays no direct role in oxygen transport.

Efficient Exchange in Water: Gills

Gills or **branchia** of various types are effective respiratory devices for life in water. Gills may be simple **external** extensions of the body surface, such as **dermal papulae** of sea stars (p. 476) or **branchial tufts** (gills) of marine worms (p. 366) and aquatic amphibians

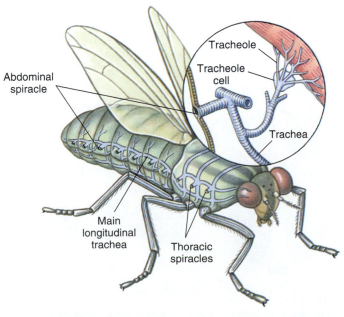

Figure 31.19

Tracheal system of insects. Air enters through spiracles, then travels through tracheae to reach tissues at tracheoles.

(p. 551). The dorsal lobe of paddlelike appendages called **parapodia** also serves as an external respiratory surface for some marine polychaete worms in which blood vessels branch through the parapodia close to the surface for enhanced gas exchange. (p. 365). Most efficient are **internal gills** of fishes (p. 533) and arthropods. Fish gills are thin filamentous structures, richly supplied with blood vessels arranged so that blood flow is opposite to the flow of water across the gills. This arrangement, called **countercurrent flow** (pp. 532 and 677), provides the greatest possible extraction of oxygen from water. Water flows over the gills in a steady stream, pulled and pushed by an efficient, two-valved, branchial pump composed of the mouth and opercular cavities (Figure 31.20). Gill ventilation is often assisted by the fish's forward movement through the water with its mouth open (ram ventilation).

Lungs

Gills are unsuitable for life in air because, when removed from the buoying water medium, gill filaments collapse, dry, and stick together; a fish out of water rapidly asphyxiates despite the abundance of oxygen around it. Consequently most air-breathing vertebrates possess lungs, highly vascularized internal cavities. Structures called lungs occur in certain invertebrates (pulmonate snails, scorpions, some spiders, some small crustaceans), but these structures are not homologous to vertebrate lungs and are not efficiently ventilated.

Lungs that can be ventilated by muscle movements to produce a rhythmic exchange of air are characteristic of terrestrial vertebrates. Most rudimentary of vertebrate lungs are those of lungfishes, which use them to supplement, or even to replace, gill respiration during periods of drought. Although of simple construction, a lungfish lung is supplied with a capillary network in its largely unfurrowed walls, a tubelike connection to the pharynx, and a primitive ventilating system for moving air in and out of the lung.

Amphibian lungs vary from simple, smooth-walled, baglike lungs of some salamanders to the subdivided lungs of frogs and toads (Figure 31.21). Total surface available for gas exchange is much increased in lungs of nonavian reptiles, which are subdivided into numerous interconnecting air sacs. Most elaborate of all are mammalian lungs containing millions of small sacs, called **alveoli** (Figures 31.21 and 31.23), each intimately associated with a rich vascular network. Human lungs have a total surface area of from 50 to 90 m²—50 times the area of the skin surface—and contain 1000 km of capillaries. A large surface area is essential for the high oxygen uptake required to support the elevated metabolic rate of endothermic mammals.

A disadvantage of lungs is that gas is exchanged between blood and air only in the alveoli and alveolar ducts, located at the ends of a branching tree of air tubes (trachea, bronchi, and bronchioles [Figures 31.21 and 31.23]). The volume of air in a lung's passageways where gas exchange does not occur is called "dead space." Unlike the efficient one-way flow of water across fish gills, air must enter and exit a lung through the same channel. After exhalation, the air tubes are filled with "used" air from the alveoli which, during the following inhalation, is pulled back into the lungs with fresh air. This air shuttles back and forth with each breath, adding to the difficulty of properly ventilating

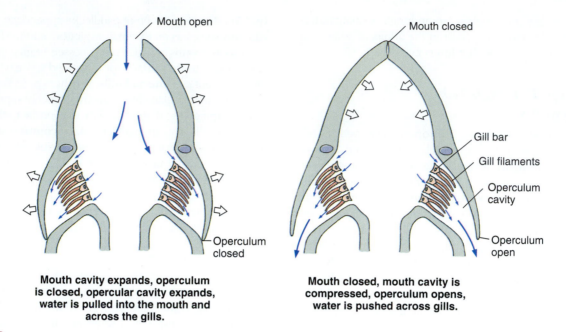

Mouth cavity expands, operculum is closed, opercular cavity expands, water is pulled into the mouth and across the gills.

Mouth closed, mouth cavity is compressed, operculum opens, water is pushed across gills.

Figure 31.20

How a fish ventilates its gills. Through the action of two skeletal muscle pumps, one in the mouth cavity, the other in the opercular cavity, water is drawn into the mouth, passes over the gills, and exits through the gill covers (opercular clefts).

lungs. In fact, lung ventilation in humans is so inefficient that in normal breathing only approximately one-sixth of the air in the lungs is replenished with each inspiration. Even after forced expiration, 20% to 35% of the air remains in the lungs.

In birds, lung efficiency is improved vastly by adding an extensive system of air sacs (Figure 31.21 and p. 596) that serve as air reservoirs during ventilation. On inspiration, about 25% of incoming air passes over the lung **parabronchi** (one-cell-thick air capillaries) where gas exchange occurs. The remaining 75% of incoming air bypasses the lungs to enter the air sacs (gas exchange does not occur here). At expiration some of this fresh air passes directly through the lung passages and into the lung parabronchi. Thus the parabronchi receive nearly fresh air during both inspiration and expiration. The beautifully designed bird lung is a result of selective pressures during evolution of flight with its high metabolic demands.

Amphibians employ a **positive pressure** action to force air into their lungs, unlike most nonavian reptiles, birds, and mammals, which ventilate their lungs by **negative pressure,** in which air is pulled into the lungs by expansion of the thoracic cavity. Frogs ventilate the lungs by first drawing air into the mouth through the **nostrils** (external nares). Then, closing the nostrils and raising the floor of the mouth, or buccal cavity, they drive air into the lungs (Figure 31.22). Much of the time, however, frogs rhythmically ventilate only the buccal cavity, a well-vascularized respiratory surface that supplements cutaneous and pulmonary respiration.

Structure and Function of the Mammalian Respiratory System

Air enters a mammalian respiratory system through nostrils (external nares), passes through a **nasal chamber,** lined with

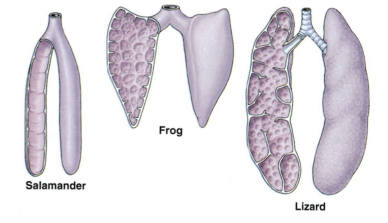

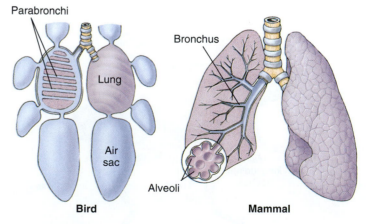

Figure 31.21

Variations among internal structures of lungs among vertebrate groups ranging from simple sacs with little exchange surface between blood and air spaces in amphibians to complex, lobulated structures, each with complex divisions and extensive exchange surfaces in birds and mammals.

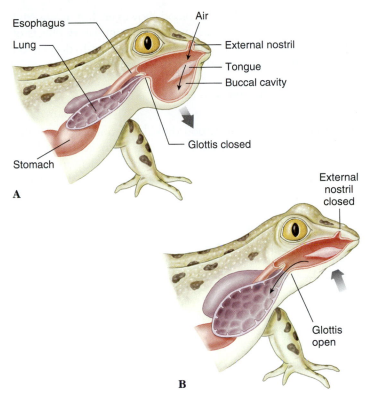

Figure 31.22

Breathing in frogs. Frogs, positive-pressure breathers, fill their lungs by forcing air into them. **A,** The buccal cavity floor is lowered, drawing air in through the nostrils. **B,** With nostrils closed and glottis open, frogs force air into lungs by elevating the buccal cavity floor. The buccal cavity may be ventilated rhythmically for a period before the lungs are emptied by contraction of body-wall musculature and by elastic recoil of lungs.

mucus-secreting epithelium, and then through **internal nares,** nasal openings connected to the **pharynx.** Here, where pathways of digestion and respiration cross, inhaled air leaves the pharynx by passing into a narrow opening, the **glottis,** while food enters the esophagus to pass to the stomach (see Figure 32.10, p. 715). The glottis opens into the **larynx,** or voice box, and then into the **trachea,** or windpipe. The trachea branches into two **bronchi,** one to each lung (Figure 31.23). Within the lungs each bronchus divides and subdivides into small tubes **(bronchioles)** that lead via **alveolar ducts** to the air sacs **(alveoli)** (Figure 31.23). The single-layered endothelial walls of the alveoli and alveolar ducts are thin and moist to facilitate exchange of gases between air and adjacent blood capillaries. Air passageways are lined with both mucus-secreting and ciliated epithelial cells, which play an important role in conditioning the air before it reaches the alveoli. Partial cartilage rings in the walls of the tracheae, bronchi, and even some of the larger bronchioles prevent those structures from collapsing.

In its passage to the air sacs, air undergoes three important changes: (1) it is filtered free from most dust and other foreign substances, (2) it is warmed to body temperature, and (3) it is saturated with moisture.

The lungs contain much elastic connective tissue. They are covered by a thin layer of tough epithelium known as the **visceral pleura.** Continuous with this layer is a similar layer, the **parietal pleura,** that lines the inner surface of the walls of the chest (Figure 31.23). The two layers of the pleura are in close contact, are lubricated by tissue fluid, and slide over one another as the lungs expand and contract. The "space" between the pleura, called the **pleural cavity,** maintains a partial vacuum, or negative **intrapleural pressure,** which helps keep the lungs expanded to fill the pleural cavity, and therefore no real pleural

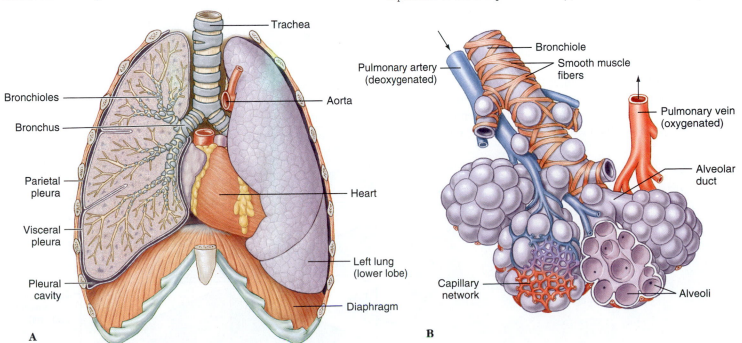

Figure 31.23

A, Lungs of human with right lung shown in section. **B,** Terminal portion of bronchiole showing air sacs with their blood supply. Arrows show direction of blood flow.

space exists. The chest cavity is bounded by the spine, ribs, and breastbone, and floored by the **diaphragm,** a dome-shaped, muscular partition between the chest cavity and abdomen. A muscular diaphragm occurs only in mammals.

Ventilating the lungs

The chest cavity is an air-tight chamber. During inspiration the ribs are pulled upward by contraction of the external intercostal muscles, and the diaphragm contracts and flattens. The resultant increase in volume of the chest cavity (Figure 31.24) causes intrapleural pressure to fall to a more negative value and air pressure in the lungs, **intrapulmonary pressure,** to fall below atmospheric pressure: air rushes in through passageways to equalize the pressure. **Tidal volume** is the amount of air (ml) that is moved during this process. Normal **expiration** is a less active process than inspiration. When the muscles relax, the ribs and diaphragm return to their original position, the chest cavity decreases in size, intrapulmonary and intrapleural pressures increase, the elastic lungs deflate, and air exits (Figure 31.24). During forced expiration which occurs during exercise, the ribs are pulled down and inward to a greater extent than during normal expiration by an additional set of muscles located between the ribs (internal intercostal muscles). Simultaneous contraction of abdominal muscles forces the diaphragm upward to a greater degree as it relaxes due to upward pressure from abdominal organs, especially the liver, below it. Together these mechanisms expel more air and also enhance inspiratory volume during the next breath.

How Breathing Is Coordinated

Breathing is normally involuntary and automatic but can come under voluntary control. Neurons in the medulla of the brain (p. 735) regulate normal, quiet breathing. They spontaneously produce rhythmical bursts that stimulate contraction of the diaphragm and external intercostal muscles during inspiration. However, respiration must adjust itself to changing requirements of the body for oxygen. Carbon dioxide rather than oxygen has the greatest effect on respiratory rate because under normal conditions arterial oxygen does not decline enough to stimulate chemical receptors **(chemoreceptors)** in the medulla of the brain. Even a small rise

in carbon dioxide level in the blood, however, has a powerful effect on respiratory activity. Actually, the stimulatory effects of carbon dioxide are due in part to an increase in hydrogen-ion concentration in cerebrospinal fluid bathing the brain.

$$CO_2 + H_2O \leftrightarrow H_2CO_3 \leftrightarrow H^+ + HCO_3^-$$

This reaction shows that carbon dioxide combines with water to form carbonic acid. Carbonic acid then dissociates to release hydrogen ions, making the cerebrospinal fluid more acidic, and stimulating respiratory chemoreceptors in the medulla of the brain. Both rate and depth of respiration increase. Peripheral chemoreceptors, located close to the heart and in the neck region, monitor peripheral changes in blood levels of carbon dioxide and hydrogen ions and send stimulatory signals to the medulla respiratory centers if these levels rise.

It is well known that swimmers can remain submerged much longer without breathing if they vigorously hyperventilate first to blow off carbon dioxide from the lungs, lowering circulating carbon dioxide levels in the blood and thereby delaying the overpowering urge to breathe. The practice is dangerous because blood oxygen is depleted while swimming just as rapidly as without prior hyperventilation, but this will not stimulate respiratory centers, and the swimmer may lose consciousness when the oxygen supply to the brain drops below a critical point. Several documented drownings among swimmers attempting long underwater swimming records have been caused by this practice.

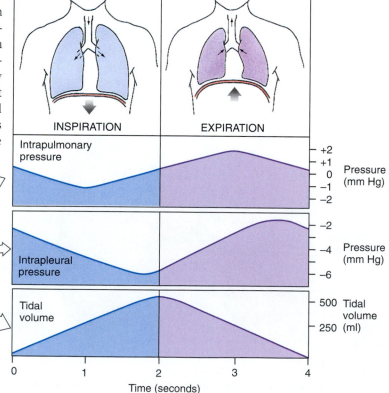

Figure 31.24
Mechanism of breathing in humans.

Gaseous Exchange in Lungs and Body Tissues: Diffusion and Partial Pressure

Air (the atmosphere) is a mixture of gases: about 71% nitrogen, 20.9% oxygen, in addition to fractional percentages of other gases, such as carbon dioxide (0.03%). Gravity attracts the atmospheric gases to the earth. At sea level the atmosphere exerts a pressure due to gravity equal to 760 mm Hg (atmospheric pressure; 1 atm). Because air is a mixture of gases, *part* of the 760 mm Hg pressure **(partial pressure)** is due to each component gas. For example, the partial pressure of oxygen is $0.209 \times 760 = 159$ mm, and that for carbon dioxide is $0.0003 \times 760 = 0.23$ mm in dry air. (In fact, atmospheric air is never completely dry, and the varying amount of water vapor present exerts a pressure in proportion to its concentration, like other gases.)

As soon as air enters the respiratory tract, its composition changes (Table 31.1, Figure 31.25). Inspired air becomes saturated with water vapor as it travels through the air-filled passageways toward the alveoli. When inspired air reaches the alveoli, it mixes with residual air remaining from the previous respiratory cycle. Partial pressure of oxygen drops and that of carbon dioxide rises. Upon expiration, air from the alveoli mixes with air in the dead space to produce still a different mixture (Table 31.1). Although no significant gas exchange occurs in the dead space, the air it contains is the first air to leave the body when expiration begins.

Because the partial pressure of oxygen in lung alveoli is greater (100 mm Hg) than it is in blood entering lung capillaries (40 mm Hg), oxygen diffuses into the capillaries. In a similar manner carbon dioxide in blood of lung capillaries has a higher concentration (46 mm Hg) than is present in lung alveoli (40 mm Hg), so carbon dioxide diffuses from the blood capillaries into the alveoli.

In tissues respiratory gases also move along their concentration gradients (Figure 31.25). Body cells continuously use and produce oxygen and carbon dioxide, respectively, so the partial pressure of oxygen in the arterial blood (100 mm Hg) entering the tissue bed is greater than in the tissues (0 to 30 mm Hg), and partial pressure of carbon dioxide in tissues (45 to 68 mm Hg) is greater than that in blood (40 mm Hg). In each case gases diffuse from a location of higher concentration to one of lower concentration.

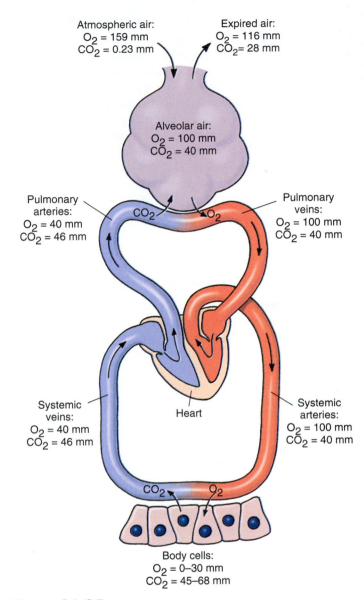

Figure 31.25

Exchange of respiratory gases in lungs and tissue cells. Numbers are partial pressures in millimeters of mercury (mm Hg).

TABLE 31.1

Partial Pressures and Gas Concentrations in Air and Body Fluids

	Nitrogen (N_2)	Oxygen (O_2)	Carbon Dioxide (CO_2)	Water Vapor (H_2O)
Inspired air (dry)	600 (79%)	159 (20.9%)	0.2 (0.03%)	—
Alveolar air (saturated)	573 (75.4%)	100 (13.2%)	40 (5.2%)	47 (6.2%)
Expired air (saturated)	569 (74.8%)	116 (15.3%)	28 (3.7%)	47 (6.2%)
Arterial blood	573	100	40	
Peripheral tissues	573	0–30	45–68	
Venous blood	573	40	46	

Note: Values expressed in millimeters of mercury (mm Hg). Percentages indicate proportion of total atmospheric pressure at sea level (760 mm Hg). Inspired air is shown as dry, although atmospheric air always contains variable amounts of water. If, for example, atmospheric air at 20° C were half saturated (relative humidity 50%), the partial pressures and percentages would be N₂ 593.5 (78.1%); O₂ 157 (20.6%); CO₂ 0.2 (0.03%); and H₂O 8.75 (1.1%).

Because of the weight of water, pressure increases the equivalent of 1 atmosphere for every 10 m of depth in seawater, and the pressure of the air supplied to a diver must be increased correspondingly so that it can be drawn into the lungs. Under the increased pressure, additional air dissolves in the blood, the amount depending on depth and time at depth of a dive. If a diver ascends slowly, the gas comes out of solution imperceptibly and is breathed out from the lungs. However, if the ascent is too rapid, the air comes out of solution and forms bubbles in the blood and other tissues, a condition known as *decompression sickness* or *the bends*. The result is painful and, if severe, can cause paralysis or death.

How Respiratory Gases Are Transported

In some invertebrates respiratory gases are simply carried dissolved in body fluids. However, solubility of oxygen is so low in water that this mode of transport is adequate only for animals with low rates of metabolism. For example, only approximately 1% of a human's oxygen requirement can be transported in this way. Consequently in many invertebrates and in virtually all vertebrates, nearly all oxygen and a significant amount of carbon dioxide are transported by special colored proteins, or **respiratory pigments,** in the blood. In all vertebrates these respiratory pigments are packaged into red blood cells.

The most widespread respiratory pigment in the animal kingdom is **hemoglobin,** a red, iron-containing protein present in all vertebrates and many invertebrates. Each molecule of hemoglobin is 5% **heme,** an iron-containing compound giving the red color to blood, and 95% **globin,** a colorless protein. The heme portion of hemoglobin has a great affinity for oxygen; each gram of hemoglobin can carry a maximum of approximately 1.3 ml of oxygen. Because there are approximately 15 g of hemoglobin in each 100 ml of blood, fully oxygenated blood contains approximately 20 ml of oxygen per 100 ml. Of course, for hemoglobin to be of value to the body it must hold oxygen in a reversible chemical combination so that it can be released to tissues. The actual amount of oxygen that combines with hemoglobin depends on the shape or conformation of the hemoglobin molecule, which is affected by several factors, including the concentration of oxygen itself. When the oxygen concentration is high, as it is in the capillaries of the lung alveoli, hemoglobin binds oxygen; in tissues where the prevailing oxygen partial pressure is low, hemoglobin releases its stored oxygen reserves (Figure 31.26).

We can express the oxygen-carrying capacity of hemoglobin relative to the surrounding oxygen concentration as **hemoglobin saturation curves** (also called oxygen dissociation curves [Figure 31.26]). As these curves show, the lower the surrounding oxygen partial pressure, the greater the quantity of oxygen released from hemoglobin. This important characteristic of hemoglobin allows more oxygen to be released to those tissues that have a high level of aerobic cellular respiration

(p. 64) and therefore need it most (those having the lowest partial pressure of oxygen).

Sickle cell anemia is an up-to-now incurable, inherited condition (p. 100) in which a single amino acid (glutamic acid) in normal hemoglobin (HbA) is replaced by a valine in sickle cell hemoglobin (HbS). The ability of HbS to carry oxygen is severely impaired, and erythrocytes tend to crumple during periods of oxygen stress (for example, during exercise). Capillaries become clogged with misshapen red cells; the affected area is very painful, and the tissue may die. About 1 in 10 African-Americans carry the trait (heterozygous). Heterozygotes do not have sickle cell anemia and live normal lives, but if both parents are heterozygous, each of their offspring has a 25% chance of inheriting the disease.

Another factor that affects hemoglobin saturation curves and therefore release of oxygen to tissues is the sensitivity of **oxyhemoglobin** (hemoglobin with bound oxygen) to carbon dioxide. Carbon dioxide shifts the hemoglobin saturation curve to the right (Figure 31.26B), a phenomenon called the **Bohr effect** after the Danish scientist who first described it. As carbon dioxide enters the blood from respiring tissues, it causes hemoglobin to unload more oxygen. The opposite event occurs in the lungs; as carbon dioxide diffuses from blood into alveolar space, the hemoglobin saturation curve shifts back to the left, allowing more oxygen to be loaded onto hemoglobin. Increased carbon dioxide in the blood lowers blood pH, as does addition of acid to blood (for example, lactic acid from exercising muscles, p. 661). A lower pH also shifts the hemoglobin saturation curve to the right and causes release of oxygen to active tissues (Figure 31.26B).

Although hemoglobin is the only vertebrate respiratory pigment, several other respiratory pigments are known among invertebrates. *Hemocyanin,* a blue, copper-containing protein, occurs in crustaceans and most molluscs. Among other pigments is *chlorocruorin* (klor-a-cru'-o-rin), a green-colored, iron-containing pigment found in four families of polychaete tubeworms. Its structure and oxygen-carrying capacity are very similar to those of hemoglobin, but it is carried free in the plasma rather than being enclosed in blood cells. *Hemerythrin* is a red pigment found in some polychaete worms. Although it contains iron, this metal is not present in a heme group (despite the name of the pigment!), and its oxygen-carrying capacity is poor compared to hemoglobin.

The same blood that transports oxygen to the tissues from the lungs must carry carbon dioxide back to the lungs on its return trip. However, unlike oxygen that is transported almost exclusively in combination with hemoglobin, carbon dioxide is transported in three different forms. A small fraction of the blood-borne carbon dioxide, only about 5%, is carried as gas physically dissolved in the plasma. The remainder diffuses

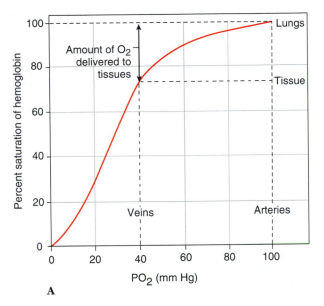

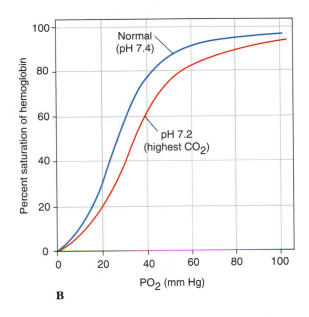

Figure 31.26

Hemoglobin saturation curves. Curves show how the amount of oxygen that can bind to hemoglobin is related to oxygen partial pressure (PO_2). **A,** At the higher partial pressure in the lungs, hemoglobin can load with more oxygen. In the tissues the oxygen concentration is less, so hemoglobin can carry less and unloads more oxygen. **B,** Hemoglobin is also sensitive to carbon dioxide partial pressure (Bohr effect and pH). As carbon dioxide enters blood from the tissues, it shifts the curve to the right, decreasing affinity of hemoglobin for oxygen. Thus the hemoglobin unloads more oxygen in the tissues where carbon dioxide concentration is higher or when pH is lower.

into red blood cells. In red blood cells, most carbon dioxide, approximately 70%, becomes carbonic acid through action of the enzyme carbonic anhydrase. Carbonic acid immediately dissociates into hydrogen ion and bicarbonate ion. We can summarize the entire reaction as follows:

$$CO_2 + H_2O \overset{\text{carbonic anhydrase}}{\rightleftarrows} H_2CO_3 \rightleftarrows H^+ + HCO_3^-$$

The hydrogen ions combine with hemoglobin to form **deoxyhemoglobin,** thus preventing a severe decrease in blood pH, and releasing oxygen simultaneously. Biocarbonate ions are transported out of the red blood cells in exchange for chloride ions **(the chloride shift).** Bicarbonate ions remain in solution in plasma because unlike carbon dioxide, bicarbonate is extremely soluble (Figure 31.27).

Figure 31.27

Transport of carbon dioxide in the blood. **A,** Carbon dioxide produced by cellular respiration diffuses from the tissues into plasma and red blood cells. Carbonic anhydrase in red blood cells catalyzes conversion of carbon dioxide into carbonic acid, then bicarbonate and hydrogen ions. The bicarbonate diffuses out of the cells, and diffusion inward of chloride ions maintains electrical balance. Hydrogen ions associate with hemoglobin. **B,** The lower partial pressure of carbon dioxide in the alveoli of the lungs favors reversal of these reactions.

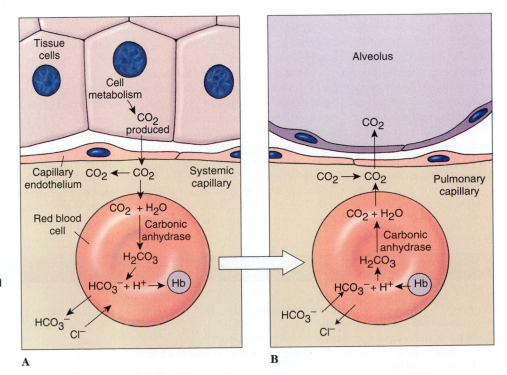

Another fraction of the carbon dioxide, approximately 25%, combines reversibly with hemoglobin. Carbon dioxide does not combine with the heme group but with amino groups of several amino acids to form a compound called **carbaminohemoglobin.**

All of these reactions are reversible. When blood reaches the lungs, bicarbonate is transported back into red blood cells (the chloride shift transporter reverses direction), it reacts with the hydrogen ions that are released from hemoglobin as oxygen is bound preferentially, and it is converted back into carbon dioxide. Carbon dioxide diffuses from red blood cells into plasma, along with carbon dioxide released from hemoglobin as it combines with oxygen. This carbon dioxide ultimately diffuses from plasma into alveolar air.

Unfortunately for humans and many other animals, hemoglobin has an affinity for carbon monoxide that is about 200 times greater than its affinity for oxygen. Consequently, even when carbon monoxide is present in the atmosphere at lower concentrations than oxygen, it tends to displace oxygen from hemoglobin to form a stable compound called carboxyhemoglobin. Air containing only 0.2% carbon monoxide may be fatal. Because of their higher respiratory rate, children and small animals are poisoned more rapidly than adults. Carbon monoxide is becoming an atmospheric contaminant of ever-increasing proportions as the world's population and industrialization continue to increase rapidly.

SUMMARY

Fluid in the body, whether intracellular, plasma, or interstitial, is mostly water, but contains many dissolved substances, including electrolytes and proteins. Vertebrate blood consists of fluid plasma and formed elements, including red and white blood cells and platelets. Plasma has many dissolved solids, as well as dissolved gases. Mammalian red blood cells lose their nucleus during development and contain the oxygen-carrying pigment, hemoglobin. White blood cells are important defensive elements. Platelets are vital in the process of clotting, necessary to prevent excess blood loss when a blood vessel is damaged. Platelets release a series of factors that activate prothrombin to thrombin, an enzyme that causes fibrinogen to change to the gel form, fibrin, forming the fibers of a blood clot.

In open circulatory systems, such as those of arthropods and most molluscs, blood moves from arteries into a hemocoel, which is a body cavity derived from the blastocoel and embryonic coelomic cavities. In closed circulatory systems, such as those of annelids, vertebrates, and cephalopod molluscs, the heart pumps blood into arteries, then into arterioles of smaller diameter, through a bed of fine capillaries, through venules, and finally through veins, which lead back to the heart. In fishes, which have a single atrium and a single ventricle, blood is pumped to gills and then directly to systemic capillaries throughout the body without first returning to the heart. With evolution of lungs, vertebrates developed a double circulation consisting of a systemic circuit serving the body, and a pulmonary circuit serving the lungs. To be fully efficient, this change required partitioning of both atrium and ventricle to form a double pump; partial partitioning occurs in lungfishes and amphibians which have two atria but an undivided ventricle, and is complete in crocodilians, birds, and mammals, which have two ventricles.

One-way flow of blood during the heart's contraction (ventricular systole) and relaxation (ventricular diastole) is assured by valves between the atria and ventricles and between the ventricles and pulmonary arteries and aorta. Although the heart can beat spontaneously, due to the presence of pacemaker cells, its rate is controlled by hormones and by nerves from the central nervous system. Heart muscle uses much oxygen and has a well-developed coronary blood circulation. The walls of arteries are thicker than those of veins, and the elastic connective tissue in the walls of larger arteries allows them to expand during ventricular systole and to recoil during ventricular diastole. Normal arterial blood pressure (hydrostatic) of humans in systole is 120 mm Hg and in diastole, 80 mm Hg. Because capillary endothelial cells possess narrow water-filled clefts between them, a protein-free filtrate crosses capillary walls, its movement determined by a balance between opposing forces of hydrostatic and protein osmotic pressure. Substances also leave and enter the blood via pinocytotic vesicles and diffusion (lipid-soluble molecules) through the endothelial cells. Tissue fluid that does not reenter the capillary system is collected by the lymphatic system (now called lymph) and returned to blood by lymph ducts.

Very small animals can depend on diffusion between the external environment and their tissues or cytoplasm for transport of respiratory gases, but larger animals require specialized organs, such as gills, tracheae, or lungs, for this function. Gills and lungs provide an increased surface area for exchange of respiratory gases between blood and environment. Many animals have special respiratory pigments and other mechanisms to help transport oxygen and carbon dioxide in blood. The most widespread respiratory pigment in the animal kingdom, hemoglobin, has a high affinity for oxygen at high oxygen concentrations but releases it at lower concentrations. Vertebrate hemoglobin, which is packaged in red blood cells, combines readily with oxygen in gills or lungs, then releases it in respiring body tissues where the oxygen partial pressure is low. Blood carries carbon dioxide from tissues to lungs as bicarbonate ion and dissolved gas in the plasma, and, in combination with hemoglobin, in red blood cells.

REVIEW QUESTIONS

1. Name the chief intracellular electrolytes and the chief extracellular electrolytes.
2. What is the fate of spent erythrocytes in the body?
3. Outline or briefly describe the sequence of events that leads to blood coagulation.

4. Two distinctly different styles of circulatory systems have evolved among animals: open and closed. What is "open" about an open circulatory system? Closed systems sometimes are cited as adaptive for actively moving animals with (at least at times) high metabolic demand. Can you suggest possible reasons for this assertion?

5. Place the following in correct order to describe the circuit of blood through the vascular system of a fish: ventricle, gill capillaries, sinus venosus, body tissue capillaries, atrium, dorsal aorta.

6. Trace the flow of blood through the heart of a mammal, naming the four chambers, their valves, and explaining where the blood entering each atrium comes from and where blood leaving each ventricle goes. When the ventricles contract, what prevents blood from reentering the atria? What factors cause blood to move forward at high pressure in the aorta?

7. Explain the origin and conduction of the excitation that leads to a heart contraction. Why is the vertebrate heart said to be myogenic? If the heart is myogenic, how do you account for alterations in rate of the heartbeat?

8. Define the terms systole and diastole. Distinguish atrial and ventricular systole and diastole.

9. Explain the movement of fluid across capillary endothelial cells. How does balance of hydrostatic pressure and colloid osmotic pressure determine direction of net fluid flow?

10. Blood pressure at the arterial end of capillaries is about 40 mm Hg in humans. If blood pressure at the venous end is about 15 mm Hg, and colloid osmotic pressure is 25 mm Hg throughout, what is the net effect on fluid movement between capillaries and tissue spaces?

11. Provide a brief description of the lymphatic system. What are its principal functions? Why is movement of lymph through the lymphatic system very slow?

12. What is an advantage of a fish's gills for breathing in water and a disadvantage for breathing on land?

13. Describe the tracheal system of insects. What is the advantage of such a system for a small animal?

14. Trace the route of inspired air in humans from the nostrils to the smallest chamber of the lungs. What is the "dead air space" of a mammalian lung and how does it affect the partial pressure of oxygen reaching the alveoli? How is this problem partially solved by bird respiratory systems?

15. The amount of time that scuba divers can spend underwater is limited by several factors, including time required to deplete the air supply in their tanks. To make their air last longer novice divers may be instructed to breathe slowly and exhale as much as possible on each breath. Can you suggest a reason why this behavior would lengthen a diver's air supply?

16. How does a frog ventilate its lungs? Contrast an amphibian's positive-pressure breathing with a mammal's negative-pressure breathing.

17. What is the role of carbon dioxide in the control of rate and depth of mammalian breathing?

18. The air pressure supplied to a scuba diver must equal that exerted by the surrounding seawater, and for each 10 m increase in depth, pressure of the surrounding seawater increases one full atmosphere. Assuming the partial pressure of oxygen in air at sea level (one atmosphere) is 0.209×760 mm Hg (= 159 mm Hg), what partial pressure of oxygen would a diver be breathing at a depth of 30 m?

19. Explain how oxygen is carried in blood, including specifically the role of hemoglobin. Answer the same question with regard to carbon dioxide transport.

20. The ability of hemoglobin to bind oxygen decreases with decreasing oxygen concentration and also decreases with increasing carbon dioxide concentration. What effect do these phenomena have on the delivery of oxygen to tissues?

SELECTED REFERENCES

Bartecchi, C. E. 1998. If you don't have a defibrillator. Sci. Am. **278**:91. *Describes "cough" and "thump" techniques that can be used instead of cardiopulmonary resuscitation (CPR).*

Brusca, R., and G. Brusca, 2003. Invertebrates, ed. 2. Sunderland, MA, Sinauer Associates. *Good discussion of origins of body cavities in invertebrates.*

Burggren, W. W. 1997. Identifying and evaluating patterns in cardiorespiratory physiology. Am. Zool. **37**:109–115. *One of several papers in this issue which is composed of a symposium on cardiorespiratory physiology.*

Eisenberg, M. S. 1998. Defibrillation: The spark of life. Sci. Am. **278**:86–90. *When the pacemaker loses its rhythm, heart muscle commences uncoordinated contractions. Application of a brief electrical shock from a defibrillator often can "reset" the pacemaker. Defibrillators have saved many lives.*

Hardison, R. 1999. The evolution of hemoglobin. Am. Sci. **87**:126–137. *Comparison of amino-acid sequences in hemoglobins from animals, plants, unicellular eukaryotes, and eubacteria suggests that all share a common ancestor early in organismal evolution.*

Jain, R. K., and P. F. Carmeliet. 2001. Vessels of death or life. Sci. Am. **285**:38–43. *Overview of blood vessel development (argiogenisis) and its role in many serious diseases.*

Kiberstis, P., and J. Marx. 1996. Cardiovascular medicine. Science **272**:663. *Introduction to a series of news and articles on current research on heart development, genetics of blood pressure, genetics of cardiovascular disease, mouse models of atherosclerosis, molecular therapies for vascular diseases, new drugs for stroke.*

Libby, P. 2002. Atherosclerosis: the new view. Sci. Am. **286**:47–55 (May). *Describes the most recent data on development of atherosclerosis.*

Lillywhite, H. B. 1988. Snakes, blood circulation and gravity. Sci. Am. **259**:92–98 (Dec.). *How a snake's vascular system is designed to counter the effects of gravity.*

Nucci, M. L., and A. Abuchowski. 1998. The search for blood substitutes. Sci. Am. **278**:73–77. *Shortage of blood supplies and risk of contamination have made the search for substitutes more urgent.*

Randall, D., W. Burggren, and K. French. 2001 Eckert animal physiology: mechanisms and adaptations. New York, W. H. Freeman & Company. *A readable, comparative approach to animal physiology.*

ONLINE LEARNING CENTER

Visit www.mhhe.com/hickmanipz14e for chapter quizzing, key term flash cards, web links, and more!

Digestion and Nutrition

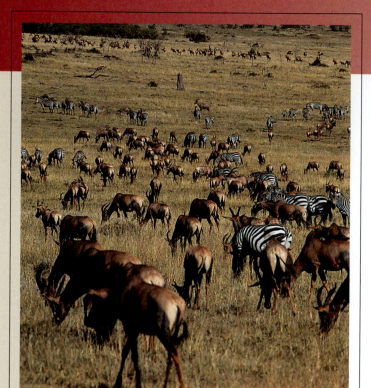

Topi and zebras on the African savannah.

A Consuming Cornucopia

Sir Walter Raleigh observed that the difference between a rich man and a poor man is that the former eats when he pleases while the latter eats when he can get food. Unlike the affluent for whom food acquisition requires only the selection of prepackaged foods at a well-stocked supermarket, the world's poor can appreciate that for them, as for the rest of the animal kingdom, procuring food is fundamental to survival. For most animals, eating is the main business of living.

Potential food is everywhere and little remains unexploited. Animals bite, chew, nibble, crush, graze, browse, shred, rasp, filter, engulf, enmesh, suck, and absorb foods of incredible variety. What an animal eats and how it eats profoundly affect an animal's feeding specialization, its behavior, its physiology, and its internal and external anatomy—in short, both its body form and its role in the web of life. The endless evolutionary jostling between predator and prey has provided adaptations for eating and adaptations for avoiding being eaten. By whatever means food may be secured, there is far less variation among animals in the subsequent digestive simplification of foods. Vertebrates and invertebrates alike use similar digestive enzymes. Even more uniform are the final biochemical pathways for nutrient use and energy transformation. The nourishment of animals is like a cornucopia in which the food flows in rather than out. A great diversity of foods procured by countless feeding adaptations streams into the mouth of the horn, is simplified, and finally used for organismal survival and reproduction.

All organisms require energy to maintain their highly ordered and complex structures. This energy is chemical bond energy released by transforming complex compounds acquired from an organism's environment into simpler ones.

The ultimate source of energy for life on earth is the sun. Sunlight is captured by chlorophyll molecules in green plants, which transform a portion of this energy into chemical bond energy (food energy). Green plants are **autotrophic** organisms, which require only inorganic compounds absorbed from their surroundings to provide the raw material for synthesis and growth. Most autotrophic organisms are the chlorophyll-bearing **phototrophs,** although some, the chemosynthetic bacteria, are **chemotrophs** that gain energy from inorganic chemical reactions.

Almost all animals are **heterotrophic organisms** that depend on already synthesized organic compounds of plants and other animals to obtain the materials they will use for growth, maintenance, and reproduction of their kind. The food of animals is normally the complex tissues of other organisms and is usually too bulky to be absorbed directly by cells. Thus, it must be digested into soluble molecules small enough to be used.

Animals may be divided into categories based on dietary habits. **Herbivorous** animals feed mainly on plant life. **Carnivorous** animals feed mainly on herbivores and other carnivores. **Omnivorous** animals eat both plants and animals. **Saprophagous** animals feed on decaying organic matter.

Ingestion of foods and their simplification by **digestion** are only initial steps in nutrition. Foods reduced by digestion to more simple, soluble, molecular forms are **absorbed** and **transported** to the body's tissues. There they are **assimilated** into the structures of cells. Oxygen is also transported to tissues, where food products are **oxidized,** or burned to yield energy and heat. Food not immediately used is **stored** for future use. Wastes produced by oxidation must be **excreted.** Food products unsuitable for digestion are **egested** in feces.

We first examine feeding adaptations of animals and then discuss digestion and absorption of food. We close with a consideration of nutritional requirements of animals.

FEEDING MECHANISMS

Few animals can absorb nutrients directly from their external environments. Exceptions are some blood parasites (p. 239), certain intestinal protozoan parasites (p. 242), and tapeworms (p. 303) and acanthocephalans, (p. 320) that nourish themselves on primary organic molecules absorbed directly across their body surfaces. These nutrients have already been digested by their host organism. Most animals, however, must work for their meals. They are active feeders that have evolved numerous specializations for obtaining food. With food procurement as one of the most potent driving forces in animal evolution, natural selection has placed a high priority on adaptations for exploiting new sources of food and the means of food capture and intake. In this brief discussion we consider some of the major food-gathering devices.

Feeding on Particulate Matter

Drifting microscopic particles abound in the upper hundred meters of ocean water. Most of this multitude is **plankton,** organisms too small to do anything but drift with the ocean's currents. The rest is organic debris, disintegrating remains of dead plants and animals. Although this oceanic swarm of plankton forms a rich life domain, it is unevenly distributed. The heaviest plankton growth occurs in estuaries and areas of upwelling, where there is an abundant nutrient supply. It is consumed by numerous larger animals, invertebrates and vertebrates, using a variety of feeding mechanisms.

One of the most important and widely employed methods for feeding is **suspension-feeding** (Figure 32.1). Most suspension feeders use ciliated surfaces to produce currents that draw drifting food particles into their mouths. Most suspension-feeding invertebrates, such as tube-dwelling polychaete worms, bivalve molluscs, hemichordates, and most protochordates, entrap particulate food on mucous sheets that convey the food into the digestive tract. Others, such as fairy shrimps, water fleas, and barnacles, use sweeping movements of their setae-fringed legs to create water currents and to entrap food, which is transferred to their mouth. Freshwater developmental stages of certain insect orders use fanlike arrangements of setae or spin silk nets to entrap food.

One form of suspension-feeding, often called **filter-feeding,** has evolved frequently as a secondary modification among representatives of groups that are primarily selective feeders. These animals possess filtering devices that strain food from water as it passes through. Examples include many microcrustaceans, fishes such as herring, menhaden, and basking sharks, certain birds such as flamingos, and the largest of all animals, baleen (whalebone) whales (Figure 32.1). The vital importance of one component of plankton, diatoms, in supporting a great pyramid of filter-feeding animals is stressed by N. J. Berrill:[1]

> A humpback whale . . . needs a ton of herring in its stomach to feel comfortably full—as many as five thousand individual fish. Each herring, in turn, may well have 6000 or 7000 small crustaceans in its own stomach, each of which contains as many as 130,000 diatoms. In other words, some 400 billion yellow-green diatoms sustain a single medium-sized whale for a few hours at most.

Another type of particulate feeding exploits deposits of disintegrated organic material (detritus) that accumulates on and in the substratum; this type is called **deposit-feeding.** Some deposit feeders, such as many annelids and some hemichordates, simply pass the substrate through their bodies, removing nutrients from it. Others, such as scaphopod molluscs, certain bivalve molluscs, and some sedentary and tube-dwelling polychaete worms, use appendages to gather organic deposits some distance from the body and move them toward the mouth (Figure 32.2).

Feeding on Food Masses

Among the most interesting animal adaptations are those that have evolved for procuring and manipulating solid food. Such

[1]Berrill, N. J. 1958. *You and the universe.* New York, Dodd, Mead & Co.

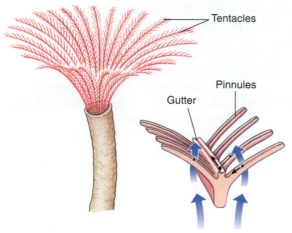

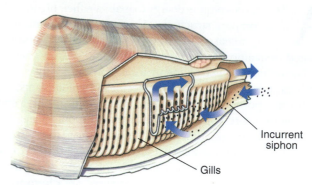

B Bivalve molluscs (class Bivalvia, phylum Mollusca) use their gills as feeding devices, as well as for respiratory gas exchange. Water currents created by cilia on the gills carry food particles into the incurrent siphon and between slits in the gills where they are entangled in a mucous sheet covering the gill surface. Ciliated food grooves then transport the particles to the mouth (not shown). Arrows indicate direction of water movement.

A Marine fan worms (class Polychaeta, phylum Annelida) have a crown of tentacles. Numerous cilia on the edges of the tentacles draw water (*blue arrows*) between pinnules where food particles are entrapped in mucus; particles are then carried down a "gutter" in the center of the tentacle to the mouth (*black arrows*).

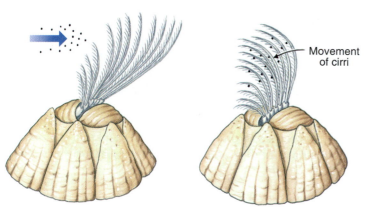

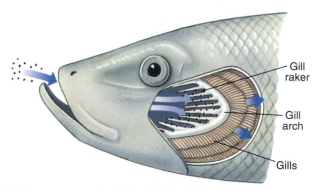

D Herring and other filter-feeding fishes (class Actinopterygii, phylum Chordata) use gill rakers that project forward from the gill arches into the pharyngeal cavity to strain plankton. Herring swim almost constantly, forcing water and suspended food into their mouth; food is strained by their gill rakers, and water passes through the gill openings.

C Barnacles (class Maxillopoda, subphylum Crustacea, phylum Arthropoda) sweep their thoracic appendages (cirri) through the water to trap plankton and other organic particles on fine bristles that fringe the cirri. Food is transferred to the barnacle's mouth by the first, short cirri.

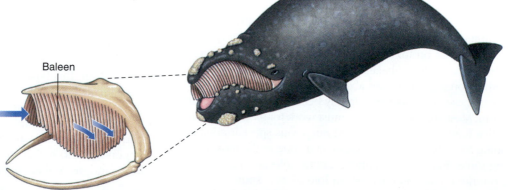

Figure 32.1

Some suspension and filter feeders and their feeding mechanisms.

E Whalebone whales (class Mammalia, phylum Chordata) filter out plankton, principally large crustaceans called krill, with whalebone, or baleen. Water enters the swimming whale's open mouth by force of the animal's forward motion and is strained out through more than 300 keratinized baleen plates that hang like a curtain from the roof of the mouth. Krill and other plankton caught in the baleen are periodically collected by the huge tongue and swallowed.

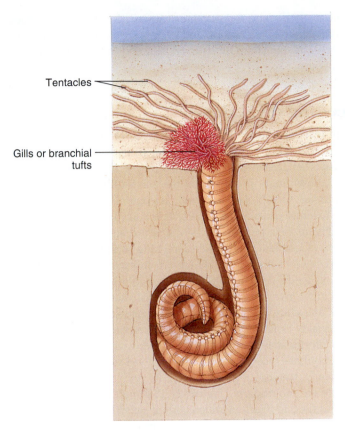

Figure 32.2

The annelid *Amphitrite* is a deposit feeder that lives in a mucus-lined burrow and extends long feeding tentacles in all directions across the surface. Food trapped on mucus is conveyed along the tentacles to the mouth.

Figure 32.3

This African egg-eating snake, *Dasypeltis,* subsists entirely on hard-shelled birds' eggs, which it swallows whole. Its special adaptations are reduced size and number of teeth, enormously expansible jaw provided with elastic ligaments, and teethlike vertebral spurs that puncture the shell. Shortly after the second photograph was taken, the snake punctured and collapsed the egg, swallowed its contents, and regurgitated the crushed shell.

adaptations and animals bearing them are largely shaped by what the animal eats.

Predators must locate, capture, hold, and swallow prey. Most carnivorous animals simply seize food and swallow it intact, although some employ toxins that paralyze or kill prey at time of capture. Although no true teeth appear among invertebrates, many have beaks or toothlike structures for biting and holding. A familiar example is the carnivorous polychaete *Nereis*, which possesses a muscular pharynx armed with chitinous jaws that can be everted with great speed to seize prey (see Figure 17.2, p. 365). Once a capture is made, the pharynx is retracted and prey swallowed. Fishes, amphibians, and nonavian reptiles use their teeth principally to grip prey and to prevent its escape until they can swallow it whole. Snakes and some fishes can swallow enormous meals. Gripping of prey, together with absence of limbs, is associated with some striking feeding adaptations in these groups: recurved teeth for seizing and holding prey and distensible jaws and stomachs to accommodate their large and infrequent meals (Figure 32.3). Birds lack teeth, but their bills are often provided with serrated edges or the upper bill is hooked for seizing and tearing prey (see Figure 27.11, p. 595).

Many invertebrates reduce food size by shredding devices (such as shredding mouthparts of many crustaceans) or by tearing devices (such as beaklike jaws of cephalopod molluscs; see

Figure 16.37, p. 353, and Figure 16.39, p. 355). Insects have three pairs of appendages on their heads that serve variously as jaws, chitinous teeth, chisels, tongues, or sucking tubes (see Figure 21.17, p. 450). Usually the first pair serves as crushing teeth; the second as grasping jaws; and the third, as a probing and tasting tongue.

True mastication, chewing or crushing of food as opposed to tearing, is found only among mammals. Mammals usually have four different types of teeth, each adapted for specific functions. **Incisors** are designed for biting, cutting, and stripping; **canines** are for seizing, piercing and tearing; **premolars** and **molars,** at the back of the jaw, are for grinding and crushing (Figure 32.4). This basic pattern is often greatly modified in animals having specialized food habits (Figure 32.5; see also Figure 28.11, p. 621). Herbivores usually lack canines but have well-developed molars with enamel ridges for grinding. Well-developed, self-sharpening incisors of rodents grow throughout life and must be worn away by gnawing to keep pace with growth. Some teeth have become so highly modified that they

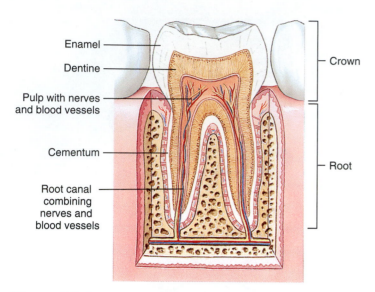

Figure 32.4

Structure of human molar tooth. A tooth is built of three layers of calcified tissue covering: enamel, which is 98% mineral and the hardest material in the body; dentine, which composes the mass of the tooth and is approximately 75% mineral; and cementum, which forms a thin covering over the dentine in the root of the tooth and is very similar to dense bone in composition. The pulp cavity contains loose connective tissue, blood vessels, nerves, and tooth-building cells.

are no longer useful for biting or chewing food. An elephant's tusk (Figure 32.6) is a modified upper incisor used for defense, attack, and rooting, and a male wild boar has modified canines used as weapons. Many feeding specializations of mammals are described on pages 621–623.

Herbivorous, or plant-eating, animals have evolved special devices for crushing and cutting plant material. Some invertebrates have scraping mouthparts, such as the radula of snails (see Figure 16.4, p. 335). Insects such as locusts have grinding and cutting mandibles; herbivorous mammals such as horses and cattle use wide, corrugated molars for grinding. All such mechanisms disrupt the tough cellulose cell wall to accelerate its digestion by intestinal microorganisms, as well as to release cell

contents for direct enzymatic breakdown. Thus herbivores digest food that carnivores cannot, and in doing so, convert plant material into organic compounds for consumption by carnivores and omnivores.

Feeding on Fluids

Fluid-feeding is especially characteristic of parasites, but it is practiced among many free-living forms as well. Some internal parasites (endoparasites) simply absorb the nutrient surrounding them, unwittingly provided by the host. Others bite and rasp host tissue, suck blood, and feed on contents of the host's intestine. External parasites (ectoparasites) such as leeches, lampreys (see Figure 24.6, p. 520), parasitic crustaceans, and insects use a variety of efficient piercing and sucking mouthparts to feed on blood or other body fluid. There are numerous arthropods that feed on fluids, for example, fleas, mosquitos, sucking lice, bedbugs, ticks and mites, to name some of the more troublesome ones. Many are vectors of serious diseases of humankind and thus qualify as far more than pesky annoyances.

Unfortunately for humans and other warm-blooded animals, the ubiquitous mosquito excels in its blood-sucking habit. Alighting gently, the mosquito punctures its prey with an array of six needlelike mouthparts (Figure 21.17B and C, p. 450). One of these is used to inject an anticoagulant saliva (responsible for the irritating itch that follows the "bite" and serving as a vector for microorganisms causing malaria, yellow fever, encephalitis, and other diseases); another mouthpart is a channel through which blood is sucked. It is of little comfort that only females dine on blood to obtain nutrients necessary for forming their eggs.

DIGESTION

In the process of digestion, which means literally "carrying asunder," organic foods are mechanically and chemically broken into small units for absorption. Although food solids consist principally of carbohydrates, proteins, and fats, the very components that form the body of the consumer, these components must first be reduced to their simplest molecular units before they can

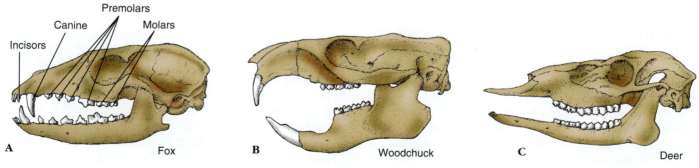

Figure 32.5

Mammalian dentition. **A,** Teeth of gray fox, a carnivore, showing the four types of teeth; **B,** Woodchuck, a rodent, has chisel-like incisors that continue to grow throughout life as they wear; **C,** White-tailed deer, a browsing ungulate, with flat premolars and molars bearing complex ridges suited for grinding.

Figure 32.6

An African elephant loosening soil from a salt lick with its tusk. Elephants use their powerful modified incisors in many ways to search for food and water: plowing the ground for roots, prying apart branches to reach the edible cambium, and drilling into dry riverbeds for water.

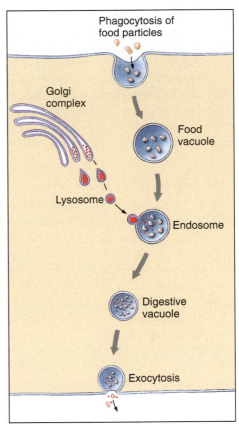

Figure 32.7

Intracellular digestion. Lysosomes containing digestive enzymes (lysozymes) are produced within a cell by the Golgi complex. Lysosomes fuse with food vacuoles and release enzymes that digest the enclosed food. Usable products of digestion are absorbed into the cytoplasm, and indigestible wastes are expelled.

be assimilated. Each animal reassembles some of these digested and absorbed units into organic compounds of the animal's own unique pattern.

In protozoa and sponges digestion is entirely **intracellular** (Figure 32.7). A food particle is enclosed within a food vacuole by phagocytosis (see p. 51). Digestive enzymes are added and the products of digestion, the simple sugars, amino acids, and other molecules, are absorbed into the cell cytoplasm where they may be used directly or, in the case of multicellular animals, may be transferred to other cells. Food wastes are simply extruded from the cell by exocytosis (p. 51).

There are important limitations to intracellular digestion. Only particles small enough to be phagocytized can be engulfed, and every cell must be capable of secreting all necessary enzymes, and of absorbing the products into their cytoplasm. These limitations were resolved with evolution of an **alimentary system** in which **extracellular** digestion of large food masses could occur. In extracellular digestion certain cells lining the **lumen** (cavity) of an alimentary canal specialize in forming various digestive secretions, such as enzymes, whereas other cells function largely, or entirely, in absorption. Many simpler metazoans, such as radiates, turbellarian flatworms, and ribbon worms (nemerteans), practice both intracellular and extracellular digestion. With evolution of greater complexity and appearance of complete mouth-to-anus alimentary systems, extracellular digestion became emphasized, together with increasing regional specialization of the digestive tract. For arthropods and vertebrates, digestion is almost entirely extracellular. Ingested food is exposed to various mechanical, chemical, and bacterial treatments, to different acidic and alkaline regions, and to digestive juices that are added at appropriate stages as food passes through the alimentary canal.

Action of Digestive Enzymes

Mechanical processes of cutting and grinding by teeth and muscular mixing by an intestinal tract are important in digestion. However, reduction of foods to small, absorbable units relies principally on chemical breakdown by **enzymes,** discussed in Chapter 4 (pp. 60–62). Digestive enzymes are **hydrolytic** enzymes, or **hydrolases,** so called because food molecules are split by a process of **hydrolysis,** breaking of a chemical bond by adding the components of water across it:

$$R\text{---}R + H_2O \xrightarrow[\text{enzyme}]{\text{digestive}} R\text{---}OH + H\text{---}R$$

In this general enzymatic reaction, R—R represents a larger food molecule split into two products, R—OH and R—H. Usually these reaction products must in turn be split repeatedly before the original molecule is reduced to its numerous subunits. Proteins, for example, are composed of hundreds, or even thousands, of interlinked amino acids, which must be completely separated before individual amino acids can be absorbed. Similarly, complex carbohydrates must be reduced to simple sugars. Fats (lipids) are reduced to molecules of glycerol, fatty acids, and monoglycerides (see Digestion in the Vertebrate Small Intestine for specific enzyme examples, p. 717). Some fats, unlike proteins and carbohydrates, may be absorbed without first being completely hydrolyzed, since they are able to diffuse through the plasma membrane of the cells lining the alimentary canal. There are specific enzymes for digestion of each class of organic compounds. These enzymes are located in specific regions of an alimentary canal in an "enzyme chain," in which one enzyme may complete what another has started. Products then move posteriorly for still further hydrolysis.

Motility in Alimentary Canal

Food is moved through digestive tracts by **cilia** or by specialized **musculature,** and often by both. Movement is usually by cilia in acoelomate and pseudocoelomate metazoa that lack the mesodermally derived gut musculature of true coelomates. Cilia move intestinal fluids and materials also in some eucoelomates, such as most molluscs, in which the coelom is weakly developed. In animals with well-developed coeloms, the gut is usually lined with two opposing layers of smooth muscle: a longitudinal layer, in which smooth muscle fibers run parallel with the length of the gut, and a circular layer, in which muscle fibers embrace the circumference of the gut (see Figure 32.13). A characteristic gut movement is **segmentation,** the alternate constriction of rings of smooth muscle of the intestine that constantly divide and squeeze contents back and forth (Figure 32.8A). Walter B. Cannon of homeostasis fame (p. 667), while still a medical student at Harvard in 1900, was the first to use X rays to

watch segmentation in experimental animals that had been fed suspensions of barium sulfate. Segmentation serves to mix food but does not move it through the gut. Another kind of muscular action, called **peristalsis,** moves food down the gut with waves of contraction of circular muscle behind and relaxation in front of the food mass (bolus) (Figure 32.8B).

ORGANIZATION AND REGIONAL FUNCTION OF ALIMENTARY CANALS

Metazoan alimentary canals can be divided into five major regions: (1) reception, (2) conduction and storage, (3) grinding and early digestion, (4) terminal digestion and absorption, and (5) water absorption and concentration of solids. Food progresses from one region to the next, allowing digestion to proceed in sequential stages (Figures 32.9 and 32.10; see also Figure 28.12, p. 622).

Receiving Region

The first region of an alimentary canal consists of devices for feeding and swallowing. These include **mouthparts** (for example,

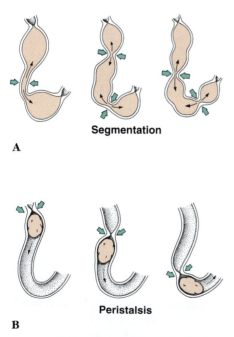

Segmentation

A

Peristalsis

B

Figure 32.8

Movement of intestinal contents by segmentation and peristalsis. **A,** Segmentational movements of food showing how constrictions squeeze food back and forth, mixing it with enzymes. The sequential mixing movements occur at about 1-second intervals. **B,** Peristaltic movement, showing how food is propelled forward by a traveling wave of contraction behind the food mass (bolus).

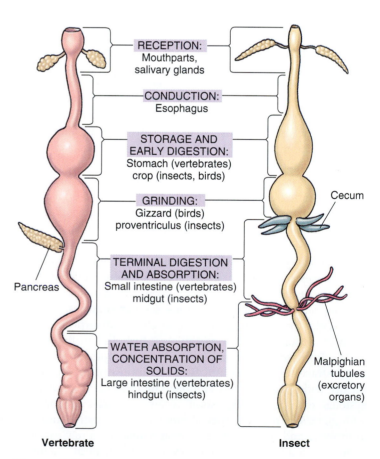

Vertebrate   **Insect**

Figure 32.9

Generalized digestive tracts of a vertebrate and an insect, showing the major functional regions of metazoan digestive systems.

mandibles, jaws, teeth, radula, bills), **buccal cavity,** or mouth, and muscular **pharynx,** or throat. Most metazoans other than suspension feeders have **salivary glands** (buccal glands) that produce lubricating secretions containing mucus to assist swallowing (Figure 32.9). Salivary glands often have other specialized functions such as secretion of toxic enzymes for quieting struggling prey and secretion of salivary enzymes to begin digestion. Salivary secretions of leeches, for example, are complex mixtures containing an anesthetic substance (making its bite nearly painless) and several enzymes that prevent blood coagulation and increase blood flow by dilating veins and dissolving cell adhesion molecules that bind cells together.

Salivary **amylase** is a carbohydrate-splitting enzyme that begins hydrolysis of plant and animal starches. It occurs only in certain herbivorous molluscs, some insects, and in primate mammals, including humans. Starches are long polymers of glucose, and salivary amylase does not completely hydrolyze it but breaks it mostly into two-glucose fragments called **maltose.** Some free glucose and longer fragments of starch are also produced. When the food bolus is swallowed, salivary amylase continues to act for some time, digesting perhaps half of the starch before the enzyme is inactivated by the acidic environment of the stomach. Further starch digestion resumes beyond the stomach in the intestine as the pH of the intestinal lumen increases.

A tongue is a vertebrate innovation, usually attached to the floor of the mouth, that assists in food manipulation and swallowing. Tongues are also used as chemosensors and possess taste buds that are used to determine palatability of foods (see Chapter 33, p. 741). It may be used for other purposes, however, such as food capture (for example, frogs, chameleons, woodpeckers, anteaters) or as an olfactory sensor (many lizards and snakes).

In humans, swallowing begins with the tongue pushing moistened food toward the pharynx. The nasal cavity closes reflexively by raising the soft palate. As food slides into the pharynx, the epiglottis tips down over the trachea, nearly closing it (Figure 32.10). Some particles of food may enter the opening of the trachea but contraction of laryngeal muscles prevents them from going farther. Once food is in the esophagus, peristaltic contraction of esophageal muscles forces it smoothly toward the stomach. The top one-third of the esophagus is surrounded by skeletal muscle as well as smooth muscle, so the act of swallowing is voluntary until the food has traveled past this upper region.

Conduction and Storage Region

The **esophagus** of vertebrates and many invertebrates serves to transfer food to the digestive region. In many invertebrates (annelids, insects, octopods) the esophagus is expanded into a **crop** (see Figure 32.9), used for food storage before digestion. Among vertebrates, only birds have a crop, which serves to store and to soften food (grain, for example) before it passes to the stomach, or to allow mild fermentation of food before it is regurgitated to feed nestlings.

Region of Grinding and Early Digestion

In most vertebrates, and in some invertebrates, the **stomach** provides initial digestion as well as storage and mixing of food with digestive juices. Mechanical breakdown of food, especially plant food with its tough cellulose cell walls, often continues in herbivorous animals by grinding and crushing

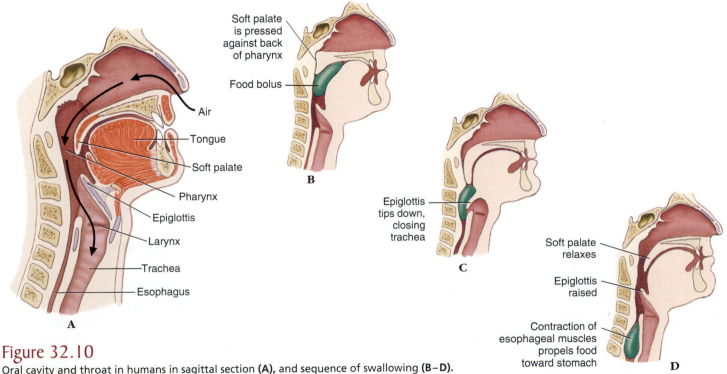

Figure 32.10

Oral cavity and throat in humans in sagittal section **(A)**, and sequence of swallowing **(B–D)**.

devices in the stomach. The muscular **gizzard** of terrestrial oligochaete worms and birds is assisted by stones and grit swallowed along with food or, in arthropods, by hardened linings (for example, chitinous teeth of an insect's proventriculus [see Figure 32.9], and calcareous teeth of gastric mills of crustaceans [see Figure 20.13, p. 430]).

Digestive diverticula—blind tubules or pouches arising from the main passage—often supplement stomachs of many invertebrates. They are usually lined with a multipurpose epithelium having cells specialized for secreting mucus or digestive enzymes, or absorption or storage. Examples include ceca of polychaete annelids, digestive glands of bivalve molluscs (see Figure 16.31, p. 350), hepatopancreas of crustaceans, and pyloric ceca of sea stars.

Herbivorous vertebrates have evolved several strategies for exploiting cellulose-splitting microorganisms to derive maximal nutrition from plant food. Despite its abundance on earth, cellulose that encloses plant cells can be digested only by an enzyme, **cellulase,** that has limited distribution in the living world. No metazoan animals can produce intestinal cellulase for direct digestion of cellulose. However many herbivorous metazoans harbor microorganisms (bacteria and protozoa) in their gut that do produce cellulase. These microorganisms ferment cellulose under anaerobic conditions in the gut, producing fatty acids and sugars that herbivores can use. While the ultimate fermentation machine is the multichambered stomach of the ruminant ungulates described on page 623, many other animals harbor microorganisms in other parts of the gut, such as their intestine proper or their cecum (see Figure 28.12, p. 622).

Stomachs of carnivorous and omnivorous vertebrates are typically a U-shaped muscular tube whose inner epithelial layer contains glands that produce proteolytic enzymes and strong acids, the latter an adaptation that probably arose for killing prey and halting bacterial activity. When food arrives at a stomach, a **cardiac sphincter** opens reflexively to allow food to enter, then closes to prevent regurgitation back into the esophagus. In humans, peristaltic waves pass over the filled stomach at a rate of approximately three each minute. Churning is most vigorous at the intestinal end where food is steadily released into the **duodenum,** the first region of the small intestine. A **pyloric sphincter** regulates flow of food into the intestine and prevents regurgitation into the stomach. Deep tubular glands in the stomach wall secrete **gastric juice,** in humans approximately 2 liters each day. Three types of cells line these glands: **goblet cells,** which secrete mucus, **chief cells,** which secrete **pepsinogen,** and **parietal** or **oxyntic cells,** which secrete **hydrochloric acid.** Pepsinogen is the precursor of **pepsin,** a **protease** (protein-splitting enzyme) produced from pepsinogen only in an acid medium (pH 1.6 to 2.4). This highly specific enzyme splits large proteins by preferentially splitting certain peptide bonds scattered along the peptide chain of a protein molecule. Although pepsin, because of its specificity, cannot completely degrade proteins, it effectively hydrolyzes them into smaller polypeptides. Other proteases that together can split all peptide bonds complete digestion of protein in the intestine. Pepsin is present in stomachs of nearly all vertebrates.

Rennin (not to be confused with renin, an enzyme produced by vertebrate kidneys, p. 677) is a milk-curdling enzyme found in the stomachs of ruminant mammals. It probably occurs in many other mammals. By clotting and precipitating milk proteins, it slows movement of milk through the stomach. Rennin extracted from stomachs of calves is used in making cheese. Human infants, lacking rennin, digest milk proteins with acidic pepsin, just as adults do.

Secretion of gastric juice is intermittent. Although a small volume of gastric juice is secreted continuously, even during prolonged periods of starvation, secretion normally increases when stimulated by sight and smell of food, by presence of food in the stomach, and by emotional states such as anxiety and anger.

That stomach mucosa is not digested by its own powerful acid secretions results from another gastric secretion, mucus, made of water, salts, and mucin, a highly viscous organic compound. Mucus coats and protects the mucosa from both chemical and mechanical injury. We should note that despite a popular misconception that an "acid stomach" is unhealthy, a notion promoted by advertising, stomach acidity is normal and essential. Sometimes, however, the protective mucous coating fails. This failure is often associated with an infection from a bacterium (*Helicobacter pylori*) that secretes toxins causing inflammation of the stomach's lining. This inflammation may lead to a stomach ulcer.

A most unusual and classic investigation of digestion was made by U.S. Army surgeon William Beaumont during the years 1825 to 1833. His subject was a young, hardliving French-Canadian voyageur named Alexis St. Martin, who in 1822 accidentally shot himself in the abdomen with a musket, the blast "blowing off integuments and muscles of the size of a man's hand, fracturing and carrying away the anterior half of the sixth rib, fracturing the fifth, lacerating the lower portion of the left lobe of the lungs, the diaphragm, and perforating the stomach." Miraculously the wound healed, but a permanent opening, or fistula, formed that permitted Beaumont to see directly into St. Martin's stomach (Figure 32.11). St. Martin became a permanent, although temperamental, patient in Beaumont's care, which included food and housing. Over a period of eight years, Beaumont was able to observe and record how the lining of the stomach changed under different psychological and physiological conditions, how foods changed during digestion, effects of emotional states on stomach motility, and many other facts about the digestive process of his famous patient.

Region of Terminal Digestion and Absorption: Intestine

The importance of an intestine varies widely among animal groups. In invertebrates that have extensive digestive diverticula in which food is digested and phagocytized, an intestine may serve only as a pathway for conducting wastes from the body. In other invertebrates with simple stomachs, and in all vertebrates, intestines are equipped for both digestion and absorption.

Figure 32.11

Dr. William Beaumont at Fort Mackinac, Michigan Territory, collecting gastric juice from Alexis St. Martin.

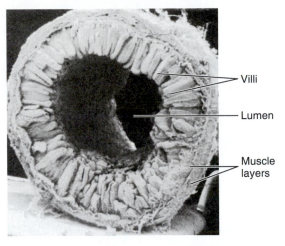

Villi

Lumen

Muscle layers

Figure 32.12

Scanning electron micrograph of a rat intestine showing numerous fingerlike villi that project into the lumen and vastly increase effective absorptive and secretory surface of the intestine. (×21)

From R. G. Kessel and R. H. Kardon, Tissues and Organs; A Text-Atlas of Scanning Electron Microscopy, 1979, W. H. Freeman and Co.

Devices for increasing the internal surface area of an intestine are highly developed in vertebrates, but are generally absent among invertebrates. Perhaps the most direct way to increase the absorptive surface of a gut is to increase its length. Coiling of their intestine is common among all vertebrate groups and reaches its highest development in mammals, in which the length of the intestine may exceed eight times the length of their body. Although a coiled intestine is rare among invertebrates, other strategies for increasing surface area sometimes occur. For example, the **typhlosole** of terrestrial oligochaete worms (see Figure 17.16, p. 372), an inward folding of the dorsal intestinal wall that runs the full length of the intestine, effectively increases internal surface area of a gut in a narrow body lacking space for a coiled intestine.

Lampreys and sharks have longitudinal or spiral folds in their intestine. Other vertebrates have developed elaborate folds (amphibians, nonavian reptiles, birds, and mammals) and minute fingerlike projections called **villi** (birds and mammals), which give the inner surface of fresh intestinal tissue the appearance of velvet (Figure 32.12). Electron microscopy reveals that each cell lining the intestinal cavity additionally is bordered by hundreds of short, delicate processes called **microvilli** (Figure 32.13C and D; see also Figure 3.16, p. 48). These processes, together with larger villi and intestinal folds, may increase the internal surface area of an intestine more than a thousand times as compared to a smooth cylinder of the same diameter. This elaborate surface greatly facilitates absorption of food molecules.

Digestion in the Vertebrate Small Intestine

Food is released into the small intestine through a **pyloric sphincter,** which relaxes at intervals to allow entry of acidic stomach contents into the initial segment of the small intestine, or **duodenum.** Two secretions pour into this region: **pancreatic juice** and **bile** (Figure 32.14). Both secretions have a high bicarbonate content, especially pancreatic juice, which effectively neutralizes gastric acid, raising the pH of the liquefied food mass, now called **chyme,** from 1.5 to 7 as it enters the duodenum. This change in pH is essential because all intestinal enzymes are effective only in a neutral or slightly alkaline medium.

Cells of intestinal mucosa (Figure 32.13A), like those of the stomach mucosa, are subjected to considerable wear and are constantly undergoing replacement. Cells deep in the crypt between adjacent villi divide rapidly and migrate up the villus. In mammals the cells reach the tip of the villus in about two days. There they are shed, along with their membrane enzymes, into the lumen at a rate of some 17 billion a day along the length of the human intestine. Before they are shed, however, these cells differentiate into absorptive cells that function to transport nutrients from the intestinal lumen into the network of blood and lymph vessels, once digestion is complete.

Bile The liver secretes bile into the **bile duct,** which drains into the upper intestine (duodenum). Between meals bile collects in the **gallbladder,** an expansible storage sac that releases bile when stimulated by the presence of fatty food in the duodenum. Bile contains water, bile salts, and pigments, but no enzymes (Figure 32.14). **Bile salts** (mainly sodium taurocholate and sodium glycocholate) are essential for digestion of fats. Fats, because of their tendency to remain in large, water-insoluble globules, are especially resistant to enzymatic digestion. Bile salts reduce surface tension of fat globules, allowing the churning action of the intestine to break fats into tiny droplets (emulsification). With total surface exposure of fat particles greatly increased, fat-splitting lipases are able to reach and to hydrolyze the triglyceride

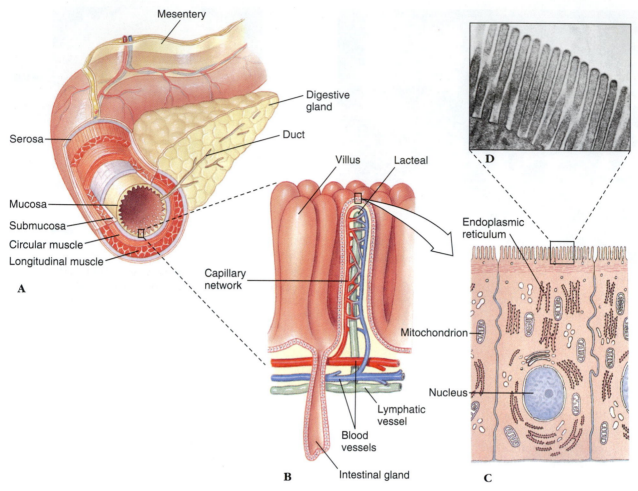

Figure 32.13

Organization of vertebrate digestive tract, showing **A,** the successive layers of mucosa, submucosa, muscle, and enveloping serosa; an enzyme-secreting digestive gland (e.g., pancreas); and the thin mesentery that positions the intestine within the body cavity. **B,** Portion of mucosal lining of intestine, showing fingerlike villi. **C,** Section of a single cell from the mucosal lining. **D,** Microvilli on surface of mucosal cell, rat intestine. (×16,400)

REGION	SECRETION	pH	COMPOSITION
Salivary glands	Saliva	6.5	Amylase Bicarbonate
Stomach	Gastric juice	1.5	Pepsinogen HCl Rennin in ruminant mammals
Liver and gallbladder	Bile	7–8	Bile salts and pigments Cholesterol
Pancreas	Pancreatic juice	7–8	Trypsin, Chymotrypsin Carboxypeptidase Lipase, Amylase Nucleases Bicarbonate
Small intestine	Membrane enzymes	7–8	Aminopeptidase Maltase Lactase Sucrase Alkaline Phosphatase Nucleotidase Nucleosidase

Figure 32.14

Secretions of a mammalian alimentary canal with the principal components and pH of each secretion.

molecules. The yellow-green color of bile is produced by **bile pigments,** breakdown products of hemoglobin from aged red blood cells. Bile pigments also give feces its characteristic color.

Bile production is only one of the liver's many functions. This highly versatile organ is a storehouse for glycogen, production center for plasma proteins, site of protein synthesis and detoxification of protein wastes, site for destruction of aged red blood cells, and center for metabolism of fat, amino acids, and carbohydrates.

Pancreatic Juice Pancreatic secretion of vertebrates contains several enzymes of major importance in digestion (Figure 32.14). Two powerful proteases, **trypsin** and **chymotrypsin,** continue enzymatic digestion of proteins begun by pepsin, which is now inactivated by the alkalinity of the intestine. Trypsin and chymotrypsin, like pepsin, are highly specific proteases that split peptide bonds deep inside a protein molecule. Hydrolysis of a peptide linkage may be shown as:

Pancreatic juice also contains **carboxypeptidase,** which removes amino acids from carboxyl ends of polypeptides; **pancreatic lipase,** which hydrolyzes fats into fatty acids and glycerol; **pancreatic amylase,** a starch-splitting enzyme identical to salivary amylase in its action; and **nucleases,** which degrade RNA and DNA to nucleotides.

Membrane Enzymes Cells lining the intestine have digestive enzymes embedded in their surface membrane that continue digestion of carbohydrates, proteins, and phosphate compounds (Figure 32.14). These enzymes of microvillus membranes (see Figure 32.13D) include **aminopeptidase,** which splits terminal amino acids from the amino end of short peptides, and several **disaccharidases,** enzymes that split 12-carbon sugar molecules into 6-carbon units. The disaccharidases include **maltase,** which splits maltose into two molecules of glucose; **sucrase,** which splits sucrose to fructose and glucose; and **lactase,** which breaks lactose (milk sugar) into glucose and galactose. Also present are **alkaline phosphatase,** an enzyme that attacks a variety of phosphate compounds, and **nucleotidases** and **nucleosidases,** which continue the breakdown of nucleotides into nucleosides and finally ribose and deoxyribose sugars, and purines and pyrimidies.

Although milk is the universal food of newborn mammals and one of the most complete human foods, many adult humans cannot digest milk because they are deficient in lactase, the enzyme that hydrolyzes lactose (milk sugar). Lactose intolerance is genetically determined. It is characterized by abdominal bloating, cramps, flatulence, and watery diarrhea, all appearing within 30 to 90 minutes after ingesting milk or its unfermented by-products. (Fermented dairy products, such as yogurt and cheese, create no intolerance problems.)

Northern Europeans and their descendants, which include a majority of European-Americans, are most tolerant of milk. Many other ethnic groups are generally intolerant to lactose, including the Japanese, Chinese, Inuit (Eskimos), South American Indians, and most Africans. Only about 30% of African-Americans are tolerant; those who are tolerant trace their ancestry to east and central Africa where dairying is traditional and tolerance to lactose is high.

Absorption

Little food is absorbed in the stomach because digestion is still incomplete and because of limited absorptive surface area. However, some materials, such as lipid-soluble drugs and alcohol, are absorbed mostly there, which contributes to their rapid action. Most digested food is absorbed from the small intestine where numerous finger-shaped villi and microvilli provide an enormous surface area through which materials can pass from the intestinal lumen into the circulation.

Carbohydrates are absorbed almost exclusively as simple sugars (monosaccharides, for example, glucose, fructose, and galactose) because the intestine is virtually impermeable to polysaccharides. Proteins are absorbed principally as their amino acid subunits, although a limited amount of small proteins or peptide fragments may be absorbed. Both active and passive transport processes transfer simple sugars and amino acids across the intestinal epithelium (see p. 49).

Immediately after a meal these materials are in such high concentration in the gut that they readily diffuse by facilitated transport into the blood, where their concentration is initially lower. However, if absorption were passive only, we would expect transfer to cease as soon as concentrations of a substance became equal on both sides of the intestinal epithelium. Passive transfer alone would permit valuable nutrients to be lost in feces. In fact, very little is lost because passive transfer is supplemented by an **active transport** (p. 49) mechanism located in epithelial cells that transfers digested food molecules into the blood. Materials thus are moved *against* their concentration gradient, a process requiring energy. Although not all food products are actively transported, those that are, such as glucose, galactose, and most amino acids, are carried by specific protein transporters for each kind of molecule.

As mentioned previously, fat droplets are emulsified by bile salts and then digested by pancreatic lipase. Triglycerides are broken into fatty acids and monoglycerides, which complex with bile salts to form minute droplets called **micelles.** When micelles contact the microvilli of the intestinal epithelium, fatty acids and monoglycerides are absorbed by simple diffusion. They then enter the endoplasmic reticulum of absorptive cells, where they are resynthesized into triglycerides before passing into **lacteals** (see Figure 32.13B). From the lacteals, fat droplets enter the lymphatic system (see Figure 31.18, p. 698) and eventually pass into the blood circulation through the thoracic duct. After a fatty meal, even a peanut butter sandwich, the presence of numerous fat droplets in the blood imparts a milky appearance to blood plasma. Triglyceride digestion, absorption, and metabolism are currently a major research focus and have led to the development of a number of antiobesity drugs used to control our body weight.

Region of Water Absorption and Concentration of Solids

The large intestine consolidates the indigestible remnants of digestion by reabsorption of water to form solid or semisolid feces for removal from the body by defecation. Reabsorption of water is of special significance in insects, especially those living in dry environments, which must (and do) conserve nearly all water entering the rectum. Specialized **rectal glands** absorb water and ions as needed, leaving behind fecal pellets that are almost completely dry. In nonavian reptiles and birds, which also produce nearly dry feces, most water is reabsorbed in the cloaca. A white pastelike feces is formed containing both indigestible food wastes and uric acid.

The colon of most vertebrates contains enormous numbers of bacteria, which first enter the sterile colon of a newborn infant with its food. In adult humans approximately one-third of the dry weight of feces is bacteria; these include harmless bacteria as well as bacteria that can cause serious illness should they escape into the abdomen or bloodstream. Normally the body's defenses prevent invasion of such bacteria. Bacteria degrade organic wastes in feces and provide some nutritional benefit by synthesizing certain vitamins (vitamin K and small quantities of some of B vitamins), which are absorbed by the body.

REGULATION OF FOOD INTAKE

Most animals unconsciously adjust intake of food to balance energy expenditure. If energy expenditure is increased by greater physical activity, more food is consumed. Most vertebrates, from fish to mammals, eat for calories rather than bulk because, if their diet is diluted with fiber, they respond by eating more. Similarly, intake is adjusted downward following a period of several days when caloric intake is too high.

Hunger centers located in the hypothalamus and brain stem of the brain (pages 735 and 736) regulate the intake of food. A drop in blood glucose level stimulates a craving for food. While most animals seem able to stabilize their weight at normal levels with ease, many humans cannot. Obesity is rising throughout the industrial world and is a major health problem in many countries today. According to recent surveys some 65%

of adults and 15% of children in the United States are either overweight or obese. In Canada, while the numbers are lower than in the United States, a similar increase to 23% (2004) from 14% (1978 to 1979) has been noted. Assessment of overweight relies on body mass index (BMI, weight in kilograms divided by the square of height in meters), waist circumference, and risk factor for diseases associated with obesity, such as type 2 diabetes, cardiovascular disease, and some cancers. A BMI of 25 or greater is considered overweight, whereas a BMI of 30 or greater is determined obese.

Some obese people do not eat significantly more food than thinner people, but rather they have an inherited genetic predisposition to gain weight on a high-fat or high-carbohydrate diet. The increase in fast-food meals, larger portion sizes, and more sedentary lifestyle is, however, associated with the prevalence of obesity in developed countries. Some obese people may also have a reduced capacity to burn excess calories by "diet-induced thermogenesis." Placental mammals are unique in having a dark adipose tissue called **brown fat,** specialized for generation of heat. Newborn mammals, including human infants, have much more brown fat than adults. In human infants brown fat is located in the chest, upper back, and near the kidneys. Abundant mitochondria in brown fat contain a protein called **uncoupling protein,** which acts to uncouple the production of ATP during oxidative phosphorylation (p. 67). Thermogenesis in brown fat is stimulated by excess food and by cold temperatures (**nonshivering thermogenesis,** see p. 682) and is activated by the sympathetic nervous system (p. 739), which responds to signals from the hypothalamus and brain stem. In people of average weight, increased caloric intake induces brown fat to dissipate excess energy as heat through the action of uncoupling protein. The Pima Indians of Arizona have low sympathetic nervous system activity, and this may contribute to the prevalence of obesity in this population. Current research is investigating this link between diet-induced thermogenesis and obesity in an attempt to provide new therapies to obese people.

Many mammals have two kinds of adipose tissue that perform completely different functions. **White adipose tissue,** the bulk of body fat, is adapted for storage of fat derived mainly from surplus fats and carbohydrates in the diet. It is distributed throughout the body, particularly in the deep layers of skin. **Brown adipose tissue** is highly specialized for mediating nonshivering and diet-induced thermogenesis rather than for the storage of fat. Brown fat, unique to placental mammals, is especially well developed in hibernating species of bats and rodents, but occurs also in many nonhibernating species such as rabbits, artiodactyls, carnivores, and primates (including humans). It is brown because it is packed with mitochondria containing large quantities of iron-bearing cytochrome molecules. In ordinary body cells, ATP is generated by a flow of electrons down the electron transport chain (p. 67). This ATP then powers various cellular processes. In brown fat cells heat is generated instead of ATP.

There are other reasons for obesity in addition to the fact that many people simply eat too much and get too little exercise. Fat stores are supervised by the hypothalamus and brain stem, which may be set at a point higher or lower than the norm. A high setting can be lowered somewhat by exercise, but as dieters are painfully aware, the body defends its fat stores with remarkable tenacity. In 1995, a hormone produced by fat cells was found to cure obesity in mutant mice lacking the gene that produces the hormone. The hormone, called **leptin,** appears to operate through a feedback system that tells the hypothalamus and brain stem how much fat the body carries. If levels are high, release of leptin by fat cells leads to diminished appetite and increased thermogenesis. The discovery of leptin has initiated a flurry of research on obesity and a resurgence of commercial interest in producing a weight-loss drug based on leptin. Unfortunately, most obese people do not respond to infusions of leptin, and actually produce higher than normal amounts themselves. Their brain appears to have become resistant to these high leptin levels and does not respond by decreasing appetite. Fat cells have now been shown to secrete a variety of hormones, and current research focuses on understanding how these multiple signals interact with short-term satiety signals produced during the digestive process (discussed in the next section) to regulate food intake and subsequently body weight.

Regulation of Digestion

The digestive process is coordinated by a family of hormones (see Chapter 34) produced by the body's most diffuse endocrine tissue, the gastrointestinal tract. These hormones are examples of the many substances produced by a vertebrate's body that have hormonal function, yet are not necessarily produced by discrete endocrine glands. Because of their diffuse origins gastrointestinal (GI) hormones have been difficult to isolate for study and only recently have they been researched in depth.

Among the principal GI hormones are gastrin, cholecystokinin (CCK), and secretin (Figure 32.15). **Gastrin** is a small polypeptide hormone produced by endocrine cells in the pyloric portion of the stomach. Gastrin is secreted in response to stimulation by parasympathetic nerve endings (vagus nerve), and when protein food enters the stomach. Its main actions are to stimulate hydrochloric acid secretion from parietal or oxyntic cells and to increase gastric motility. Gastrin is an unusual hormone because it exerts its action on the same organ from which it is secreted. **CCK** is also a polypeptide hormone, and it has a striking structural resemblance to gastrin, suggesting that the two arose by duplication of ancestral genes. CCK is secreted by endocrine cells in the walls of the upper small intestine in response to the presence of fatty acids and amino acids in the duodenum. It has at least three distinct functions. It stimulates gallbladder contraction and thus increases the flow of bile salts into the intestine; it stimulates an enzyme-rich secretion from the pancreas; and it acts on the brain stem to contribute a feeling of satiety after a meal, particularly one rich in fats. The first hormone to be discovered, **secretin** (see the opening essay for Chapter 34 on p. 753), is produced by endocrine cells in the duodenal wall. It is secreted in response to food and strong acid in the stomach and small intestine, and its principal action is to stimulate release of an alkaline pancreatic fluid that neutralizes

Figure 32.15

Three hormones of digestion. Shown are principal actions of the hormones gastrin, CCK (cholecystokinin), and secretin.

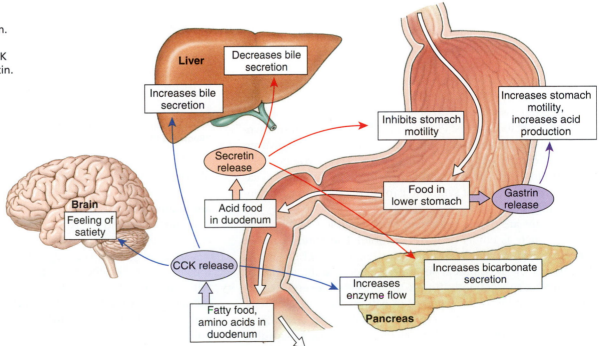

stomach acid as it enters the intestine. It also aids fat digestion by inhibiting gastric motility and increasing production of an alkaline bile secretion from the liver.

GI hormones continue to be isolated and their structure determined. So far, all are peptides, and many are present in both the GI tract and in the central nervous system. One of these is CCK, which has been found in high concentrations in the cerebral cortex and hypothalamus of mammals. By providing a feeling of satiety after eating (just mentioned) it may play some role in regulating appetite. Several other GI peptides, for example, vasoactive intestinal peptide (VIP), glucagon-like peptide 1 (GLP-1), pancreatic polypeptide (PP), gastric inhibitory peptide (GIP), ghrelin, and peptide YY (PYY) appear to play neurotransmitter roles in the brain. For example, ghrelin, PP, and PYY also appear to be short-term regulators of food intake. Ghrelin levels rise before a meal and appear to stimulate appetite, while PP and PYY levels rise during a meal and induce satiety. Much research now focuses on newly discovered peptides in the hope of finding a "magic bullet" to solve the current obesity crisis.

NUTRITIONAL REQUIREMENTS

Food of animals must include **carbohydrates, proteins, fats, water, mineral salts,** and **vitamins.** Carbohydrates and fats are required as fuels for energy and for the synthesis of various substances and structures. Proteins (actually the amino acids of which they are composed) are needed for synthesis of species-specific proteins and other nitrogen-containing compounds. Water is required as the solvent for body chemistry and as a major component of all fluids of the body. Inorganic salts are required as anions and cations of body fluids and tissues and form important structural and physiological components throughout the body. Vitamins are accessory factors from food that are often built into the structure of many enzymes.

A **vitamin** is a relatively simple organic compound that is not a carbohydrate, fat, protein, or mineral and is required in very small amounts in the diet for some specific cellular function. Vitamins are not sources of energy but function as coenzymes that are often associated with activity of important enzymes that serve vital metabolic roles. Plants and many microorganisms synthesize all the organic compounds they need; animals, however, have lost certain synthetic abilities during their evolution and depend ultimately on plants to supply these compounds. Vitamins therefore represent synthetic gaps in the metabolic machinery of animals.

Vitamins are usually classified as fat soluble (soluble in fat solvents such as ether) or water soluble. The **water-soluble vitamins** include the B complex and vitamin C (Table 32.1). Vitamins of the B complex, so grouped because the original B vitamin was subsequently found to consist of several distinct molecules, tend to occur together in nature. Almost all animals, vertebrate and invertebrate, require B vitamins; they are "universal" vitamins. A dietary need for vitamin C and the **fat-soluble vitamins** A, D₃, E, and K is mostly restricted to vertebrates,

TABLE 32.1	
Human Nutrient Requirements	

Water-Soluble Vitamins

Thiamine (B_1)	Folacin (folic acid)
Riboflavin (B_2)	Vitamin B_{12} (cobalamin)
Niacin (nicotinic acid)	Biotin
Pyridoxine (B_6)	Ascorbic acid (C)
Pantothenic acid	

Fat-Soluble Vitamins

A, D_3, E, and K

Minerals

Major	*Trace*
Calcium	Iron
Phosphorus	Fluorine
Sulfur	Zinc
Potassium	Copper
Chlorine	Silicon
Sodium	Vanadium
Magnesium	Tin
	Nickel
	Selenium
	Manganese
	Iodine
	Molybdenum
	Chromium
	Cobalt

Amino Acids

Phenylalanine	Methionine
Lysine	Tryptophan
Isoleucine	Threonine
Leucine	Arginine*
Valine	Histidine*

Polyunsaturated Fatty Acids

Arachidonic
Linoleic
Linolenic

*Required for normal growth of children.

although some are required by certain invertebrates. Even within groups of close relationship, requirements for vitamins are relative, not absolute. A rabbit does not require vitamin C, but guinea pigs and humans do. Some songbirds require vitamin A, but others do not.

The recognition years ago that many human diseases and those of domesticated animals were caused by or associated with dietary deficiencies led biologists to search for specific nutrients that would prevent such diseases. These studies eventually yielded a list of **essential nutrients** for people and other animal species studied. Essential nutrients are those needed for normal growth and maintenance and that *must* be supplied in the diet. In other words, it is "essential" that these nutrients be in the diet because an animal cannot synthesize them from other dietary constituents. Nearly 30 organic compounds (amino acids and vitamins) and 21 elements are essential for humans (Table 32.1). Considering that the body contains thousands of different organic compounds, the list in Table 32.1 is remarkably short. Animal cells are able to synthesize compounds of enormous variety and complexity from a small, select group of raw materials.

In the average diet of North Americans approximately 50% of the total calories (energy content) comes from carbohydrates and 40% comes from lipids. Proteins, essential as they are for structural needs, supply only a little more than 10% of total calories of an average diet of North Americans. Carbohydrates are widely consumed because they are more abundant and cheaper than proteins or lipids. Actually humans and many other animals can subsist on diets devoid of carbohydrates, provided sufficient total calories and essential nutrients are present. Inuits, before the decline of their native culture, lived on a diet that was high in fat and protein and very low in carbohydrate.

Lipids are needed principally to provide energy. However, at least three fatty acids are essential for humans because we cannot synthesize them. Much interest and research have been devoted to lipids in our diets because of the association between fatty diets, obesity, and the disease **atherosclerosis.** The matter is complex, but evidence suggests that atherosclerosis may occur when a diet is high in saturated lipids (lipids with no double bonds in the carbon chains of fatty acids) but low in polyunsaturated lipids (two or more double bonds in the carbon chains).

Atherosclerosis (Gr. *atheroma,* tumor containing gruel-like matter, + *sclerosis,* to harden) is a degenerative disease in which fatty substances are deposited in the lining of arteries, resulting in narrowing of the passage and eventual hardening and loss of elasticity. Current evidence suggests that inflammation of the artery wall precedes deposition of fat. Elevated cholesterol levels can fuel such inflammation. Cholesterol-lowering drugs are used by many humans in an effort to decrease the risk of cardiovascular diseases such as atherosclerosis and related conditions.

Proteins are expensive foods and limited in the diet. Proteins, of course, are not themselves the essential nutrients but rather contain essential amino acids. Of the 20 amino acids commonly found in proteins, eight and possibly 10 are essential to humans (Table 32.1). We can synthesize the rest from other amino acids. Generally, animal proteins have more essential amino acids than do proteins of plant origin. All eight essential amino acids must be present simultaneously in the diet for protein synthesis. If one or more is missing, use of the other amino acids will be reduced proportionately; they cannot be stored and are metabolized for energy. Thus heavy reliance on a single plant source as a diet will inevitably lead to protein deficiency. This problem can be corrected if two kinds of plant proteins having complementary

strengths in essential amino acids are ingested together. For example, a balanced protein diet can be prepared by mixing wheat flour, which is deficient only in lysine, with a legume (peas or beans), which is a good source of lysine but deficient in methionine and cysteine. Each plant complements the other by having adequate amounts of those amino acids that are deficient in the other.

Because animal proteins are rich in essential amino acids, they are in great demand in all countries. North Americans eat far more animal proteins than do Asians and Africans. In 2001 the annual per capita consumption of meat was 122 kg in the United States, 72 kg in Europe, 27 kg in Asia, 15 kg in North Africa, and 11 kg in sub-Saharan Africa.[2] In 2001–2003, 30% to 35% of calories in the diet of developed countries came from animal products. By comparison, in developing countries only 10% to 15% of calories came from animal sources.[2] North Americans consume approximately one-quarter of all beef produced in the world.

Two different types of severe food deficiency are recognized: marasmus, general undernourishment from a diet low in both calories and protein, and kwashiorkor, protein malnourishment from a diet adequate in calories but deficient in protein. Marasmus (Gr. *marasmos,* to waste away) is common in infants weaned too early and placed on low-calorie–low-protein diets; these children are listless, and their bodies waste away. Kwashiorkor is a West African word describing a disease a child gets when displaced from the breast by a newborn sibling. This disease is characterized by retarded growth, anemia, weak muscles, a bloated body with typical pot belly, acute diarrhea, susceptibility to infection, and high mortality.

Undernourishment and malnourishment rank as two of the world's oldest problems and remain major health problems today, afflicting an eighth of the human population. Growing children and pregnant and lactating women are especially vulnerable to the devastating effects of malnutrition. Cell proliferation and growth in the human brain are most rapid in the terminal months of gestation and the first year after birth. Adequate protein for neuron development is a requirement during this critical time to prevent neurological dysfunction. Brains of children who die of protein malnutrition during the first year of life have 15% to 20% fewer brain cells than those of normal children (Figure 32.16). Malnourished children who survive this period suffer permanent brain damage and cannot be helped by later corrective treatment (Figure 32.17). Studies suggest that poverty, with attendant lack of educational and medical resources, and lowered expectations, exacerbates effects of malnutrition by delaying intellectual development.[3]

---

[2]Food and Agriculture organization of the United Nations, http://www.fao.org/Faostat/
[3]Brown, J. L, and E. Pollitt. 1996. Malnutrition, poverty and intellectual development. Sci. Am. 274:38–43 (Feb.).

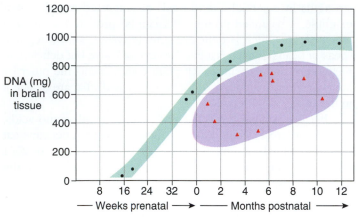

Figure 32.16

Effect of early malnutrition on cell number (measured as total DNA content) in a human brain. This graph shows that malnourished infants (*purple oval*) have far fewer brain cells than do normal infants (*green growth curve*).

Figure 32.17

Refugee child suffering severe malnutrition.

SUMMARY

Autotrophic organisms (mostly green plants), using inorganic compounds as raw materials, capture the energy of sunlight through photosynthesis and produce complex organic molecules. Heterotrophic organisms (bacteria, fungi, and animals) use organic compounds synthesized by plants, and chemical bond energy stored therein, for their own nutritional and energy needs.

A large group of animals with very different levels of complexity feed by filtering minute organisms and other particulate matter suspended in water. Others feed on organic detritus deposited in the substrate. Selective feeders, on the other hand, have evolved mechanisms for manipulating larger food masses, including various devices for seizing, scraping, boring, tearing, biting, and chewing. Fluid-feeding is characteristic of endoparasites, which may absorb food across the general body surface, and of ectoparasites, herbivores, and predators that have developed specialized mouthparts for piercing and sucking.

Digestion is a process of breaking food mechanically and chemically into molecular subunits for absorption. Digestion is intracellular in protozoan groups and sponges. In more complex metazoans it is supplemented, and finally replaced entirely, by extracellular digestion, which occurs in sequential stages in a tubular cavity, or alimentary canal. The mouth receives food, mixes it with lubricating saliva, then passes it down the esophagus to regions where food may be stored (crop), or ground (gizzard), or acidified and subjected to early digestion (vertebrate stomach). Among vertebrates, most digestion occurs in the small intestine. Enzymes from the pancreas and intestinal mucosa hydrolyze proteins, carbohydrates, fats, nucleic acids, and various phosphate compounds. The liver secretes bile, containing salts that emulsify fats. Once foods are digested,

their products are absorbed as molecular subunits (monosaccharides, amino acids, fatty acids and glycerol) into blood or lymph vessels of villi of the small intestine. The large intestine (colon) serves mainly to absorb water and minerals from the food wastes as they pass through it. It also contains symbiotic bacteria that produce certain vitamins.

Most animals balance food intake with energy expenditure. Food intake is regulated primarily by hunger centers located in the hypothalamus and brain stem. In mammals, should caloric intake exceed requirements for energy, excess calories normally are dissipated as heat in specialized brown fat tissue.

Several gastrointestinal hormones coordinate digestive functions. They include gastrin, which stimulates acid secretion by the stomach; CCK, which stimulates gallbladder and pancreatic secretion and leads to satiety; and secretin, which stimulates bicarbonate secretion from the pancreas and inhibits gastric motility. New GI hormones are being added to this list as they are discovered. For example, ghrelin stimulates appetite, while PP and PYY induce satiety.

All animals require a balanced diet containing both fuels (mainly carbohydrates and lipids) and structural and functional components (proteins, minerals, and vitamins). For every multicellular animal, certain amino acids, lipids, vitamins, and minerals are "essential" dietary factors that cannot be produced by an animal's own synthetic machinery. Animal proteins are better-balanced sources of amino acids than are plant proteins, which tend to lack one or more essential amino acids. Undernourishment and protein malnourishment are among the world's major health problems, afflicting millions of people. Ironically, obesity and diseases associated with this condition are major health problems in developed countries of the world.

REVIEW QUESTIONS

1. Distinguish between the following pairs of terms: autotrophic and heterotrophic; phototrophic and chemotrophic; herbivores and carnivores; omnivores and insectivores.

2. Suspension-feeding is one of the most important methods of feeding among animals. Explain the characteristics, advantages, and limitations of suspension-feeding, and name three different groups of animals that are suspension feeders.

3. An animal's feeding adaptations are an integral part of an animal's behavior and usually shape the appearance of the animal itself. Discuss contrasting feeding adaptations of carnivores and herbivores.

4. Explain how food is propelled through the digestive tract.

5. Compare intracellular with extracellular digestion and suggest why there has been a phylogenetic trend in some animals from intracellular to extracellular digestion.

6. Which structural modifications vastly increase the internal surface area of the intestine (both invertebrate and vertebrate), and why is this large surface area important?

7. Trace digestion and final absorption of a carbohydrate (starch) in the vertebrate gut, naming the carbohydrate-splitting enzymes, where they are found, the breakdown products of starch digestion, and in what form they are finally absorbed.

8. As in question 7, trace digestion and final absorption of a protein.

9. Explain how fats are emulsified and digested in the vertebrate gut. Explain how bile aids the digestive process even though it contains no enzymes. Provide an explanation for the following observation: fats are broken into fatty acids and monoglycerides in the intestinal lumen, but appear later in the blood as fat droplets.

10. Explain the phrase "diet-induced thermogenesis" and relate it to the problem of obesity in some people. What other factors may contribute to human obesity?

11. Name three hormones of the gastrointestinal tract and explain how they assist in the coordination of gastrointestinal function.

12. Name the basic classes of foods that serve mainly as (a) fuels and as (b) structural and functional components.

13. If vitamins are neither biochemically similar compounds nor sources of energy, what characteristics distinguish vitamins as a distinct group of nutrients? What are the water-soluble and the fat-soluble vitamins?

14. Why are some nutrients considered "essential" and others "nonessential" even though both types of nutrients are used in growth and tissue repair?

15. Explain the difference between saturated and unsaturated lipids, and comment on the current interest in these compounds as they relate to human health.

16. What is meant by "protein complementarity" among plant foods?

SELECTED REFERENCES

Bachman, E. S., H. Dhillon, C-Y. Zhang, S. Cinti, A. C. Bianco, B. K. Kobilka, and B. B. Lowell. 2002. βAR signalling required for diet-induced thermogenesis and obesity resistance. Science **297**:843–845. *This paper describes an important study performed in mice showing that diet-induced thermogenesis is regulated by the sympathetic nervous system.*

Blaser, M. J. 1996. The bacteria behind ulcers. Sci. Am. **274**:104–107 (Jan.). *We now know that most cases of stomach ulcers are caused by acid-loving microbes. At least one-third of the human population are infected although most do not become ill.*

Hill, J. O., H. R. Wyatt, G. W. Reed, and J. C. Peters. 2003. Obesity and the environment: where do we go from here? Science **299**:853–855. *Read this and related articles on the obesity crisis in a special issue of Science.*

Martins, I. J., and T. G. Redgrave. 2004. Obesity and post-prandial lipid metabolism. Feast or famine? J. Nut. Biochem. **15**:130–141. *An excellent review of the disease of obesity and current research in lipid metabolism.*

Milton, K. 1993. Diet and primate evolution. Sci. Am. **269**:86–93 (Aug.). *Studies with primates suggest that modern human diets often diverge greatly from those to which the human body may be adapted.*

Morrison, S. F. 2004. Central pathways controlling brown adipose tissue thermogenesis. News Physiol. Sci. **19**:67–74. *A readable review relating thermogenesis to energy expenditure.*

Randall, D., W. Burggren, and K. French. 2002. Eckert animal physiology: mechanisms and adaptations, ed. 5. New York, W. H. Freeman & Company. *A comprehensive and comparative treatment of animal physiology including an excellent section on digestion in animals.*

Ronti, T., G. Lupattelli, and E. Mannarino. 2006. The endocrine function of adipose tissue: an update. Clin. Endocrinol. **64**:355–365. *Summary of the signalling molecules secreted by fat stores and their role in energy balance.*

Sanderson, S. L., and R. Wassersug. 1990. Suspension-feeding vertebrates. Sci. Am. **262**:96–101 (Mar.). *A variety of vertebrates, some enormous in size, eat by filtering out small organisms from massive amounts of water passed through a feeding apparatus.*

Weindrach, R. 1996. Caloric restriction and aging. Sci. Am. **274**:46–52 (Jan.). *Organisms from single-celled protists to mammals live longer on well-balanced but low-calorie diets. The potential benefits for humans are examined.*

Willme, P., G. Stone, and I. Johnston. 2005. Environmental physiology of animals, ed. 2. Oxford, U.K., Blackwell Science Ltd. *Well-written information regarding environmental adaptations of both vertebrates and invertebrates.*

ONLINE LEARNING CENTER

Visit www.mhhe.com/hickmanipz14e for chapter quizzing, key term flash cards, web links, and more!

Nervous Coordination: Nervous System and Sense Organs

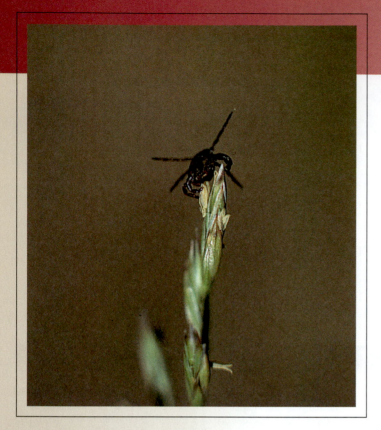

Wood tick on a grass stem awaits its host.

The Private World of the Senses

Information from the senses of vision, hearing, taste, olfaction, and touch continually assaults us. These classic five senses are supplemented by sensory inputs of cold, warmth, vibration, and pain, as well as by information from numerous internal sensory receptors that operate silently and automatically to keep our interior domain working smoothly.

The world revealed by our senses is uniquely human even if our individual senses are not. We share this exclusive world with no other animal, nor can we venture into the sensory world of any other animal except as an abstraction through our imagination.

The idea that each animal enjoys an unshared sensory world was first conceived by Jakob von Uexküll, a seldom cited German biologist of the early part of the last century. Von Uexküll asks us to try to enter the world of a tick through our imagination, supplemented by what we know of tick biology. It is a world of

temperature, of light and dark, and of the odor of butyric acid, a chemical common to all mammals. Insensitive to all other stimuli, the tick climbs up a blade of grass to wait, for years if necessary, for cues that will betray the presence of a potential host. Later, swollen with blood, she drops to the earth, lays her eggs, and dies. The tick's impoverished sensory world, devoid of sensory luxuries and fine-tuned by natural selection for the world she will encounter, has ensured her single goal, reproduction.

A bird and a bat may share for a moment precisely the same environment. The worlds of their perceptions, however, are vastly different, structured by the limitations of the sensory windows each employs and by the brain that garners and processes what it needs for survival. For one it is a world dominated by vision; for the other, echolocation. The world of each is alien to the other, just as their worlds are to us.

The nervous system originated in a fundamental property of life: **irritability,** the ability to respond to environmental stimuli (see Chapter 1, p. 9). The response may be simple, such as a protozoan moving to avoid a noxious substance, or quite complex, such as a vertebrate animal responding to elaborate signals of courtship. A protistan receives and responds to a stimulus, all within the confines of a single cell. Evolution of multicellularity and more complex levels of animal organization required increasingly complex mechanisms for communication between cells and organs. Relatively rapid communication is by **neural mechanisms** and involves propagated electrochemical signaling along and between cell membranes. The basic plan of a nervous system is to receive information from the external and internal environments, to encode this information, and to transmit and to process it for appropriate action. We examine these functions in this chapter. Relatively less rapid or long-term adjustments in animals are governed by **hormonal mechanisms,** the subject of Chapter 34.

NEURONS: FUNCTIONAL UNITS OF NERVOUS SYSTEMS

A **neuron,** or nerve cell, may assume many shapes, depending on its function and location; a typical kind is shown diagrammatically in Figure 33.1. From the nucleated cell body extend cytoplasmic processes of two types: one or more **dendrites** in all but the simplest neuron, and a single **axon.** As the name dendrite suggests (Gr. *dendron,* tree), these processes are often profusely branched. They, and the entire cell body surface, are the nerve cell's receptive apparatus, designed to receive information from several different sources at once. Some of these inputs are excitatory, causing a signal to be generated and propagated, others are inhibitory, making signal generation and propagation less likely.

The single axon (Gr. *axon,* axle), often a long fiber that may be meters in length in the largest mammals, is relatively uniform in diameter, and typically carries signals away from the cell body. In vertebrates and some complex invertebrates, the axon is often covered with an insulating sheath of **myelin,** which accelerates signal propagation.

Neurons are commonly classified as **afferent,** or **sensory; efferent,** or **motor;** and **interneurons,** which are neither sensory nor motor but interconnect neurons. Afferent and efferent neurons lie mostly outside the **central nervous system** (brain and nerve cord) in the **peripheral nervous system,** while interneurons, which in humans are 99% of all neurons in the body, lie entirely within the central nervous system. Afferent neurons are connected to **receptors.** Receptors function to convert external and internal environmental stimuli into nerve signals, which are carried by afferent neurons into the central nervous system. Here signals may be perceived as conscious sensation. Nerve signals also move to efferent neurons, which carry them via the peripheral nervous system to **effectors,** such as muscles or glands.

In vertebrates, nerve processes (usually axons) are often bundled together in a wrapping of connective tissue to form a **nerve** (Figure 33.2). Cell bodies of these nerve processes are located either in the central nervous system or in **ganglia,** which

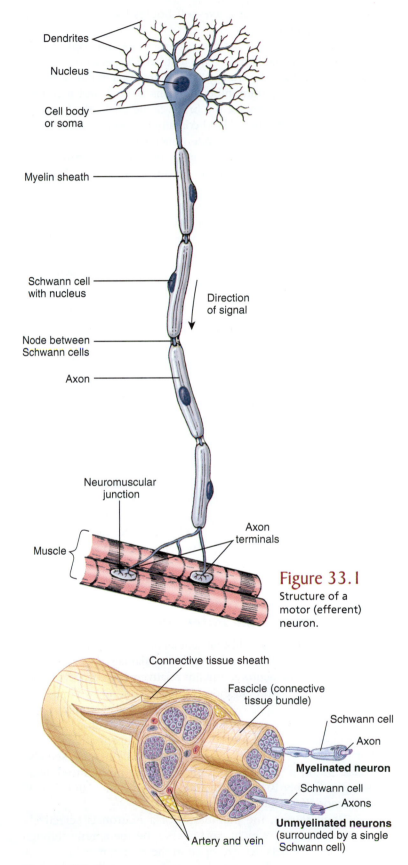

Figure 33.1

Structure of a motor (efferent) neuron.

Figure 33.2

Structure of a peripheral nerve showing nerve fibers surrounded by various layers of connective tissue. A nerve may contain thousands of both efferent and afferent fibers.

are discrete bundles of nerve-cell bodies located outside the central nervous system.

Surrounding neurons are non-nervous **neuroglial cells** (often simply called **"glial"** cells) that have a special relationship to neurons. Glial cells are extremely numerous in the vertebrate brain, where they outnumber neurons 10 to 1 and may form almost half the volume of the brain. Some glial cells form intimate insulating sheaths of lipid-containing **myelin** around nerve fibers. Vertebrate nerves are often enclosed by concentric rings of myelin, produced by special glial cells called **Schwann cells** (Figure 33.3) in the peripheral nervous system, and **oligodendrocytes** in the central nervous system. Certain glial cells, called **astrocytes** because of their radiating, starlike shape, serve as nutrient and ion reservoirs for neurons, as well as providing a scaffold during brain development, enabling migrating neurons to find their destinations from points of origin. Astrocytes, and smaller **microglial** cells, are essential for the regenerative process that follows brain injury. Unfortunately, astrocytes also participate in several diseases of the nervous system, including Parkinsonism, multiple sclerosis, and brain tumor development.

Nature of a Nerve Action Potential

A **nerve** signal or **action potential** is an electrochemical message of neurons, the common functional denominator of all nervous system activity. Despite the incredible complexity of nervous systems of many animals, nerve action potentials are basically alike in all neurons and in all animals. An action potential is an "all-or-none" phenomenon; either the fiber is conducting an action potential, or it is not. Because all action potentials are alike, the only way that a nerve fiber can vary its signal is by changing the frequency of signal conduction. Frequency change is the language of a nerve fiber. A fiber may conduct no action potentials at all or very few per second up to a maximum approaching 1000 per second. The higher the frequency (or rate) of conduction, the greater is the level of excitation.

Resting Membrane Potential

Membranes of neurons, like all cellular membranes, have a selective permeability that creates ionic imbalances. The interstitial fluid surrounding neurons contains relatively high concentrations of sodium (Na^+) and chloride (Cl^-) ions, but a low concentration of potassium ions (K^+) and large impermeable anions with negative charge, such as proteins. Inside the neuron, the ratio is reversed: K^+ and impermeable anion concentration is high, but Na^+ and Cl^- concentrations are low (Figure 33.4; see also Figure 31.1B, p. 687). These differences are pronounced; there is approximately 10 times more Na^+ outside than in and 25 to 30 times more K^+ inside than out.

When at rest, the membrane of a neuron is selectively permeable to K^+, which can traverse the membrane through potassium channels that are open in the resting membrane (see Chapter 3, p. 49). The permeability to Na^+ is nearly zero because Na^+ channels are closed in a resting membrane. Potassium ions tend to diffuse outward through the membrane, following the gradient of potassium concentration. Very quickly the positive charge

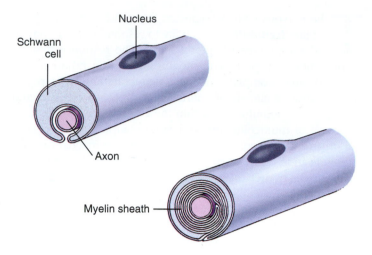

Figure 33.3

Development of the myelin sheath in a myelinated neuron of the peripheral nervous system. The whole Schwann cell grows around an axon, then rotates around it, enclosing the axon in a tight, multilayered sheath. The myelin sheath insulates a nerve axon and facilitates transmission of nerve signals or action potentials.

outside reaches a level that prevents any more K^+ from diffusing out of the axon (because like charges repel each other), and because large anions cannot pass through the membrane, positively charged potassium ions are drawn back into the cell. Now the resting membrane is at equilibrium, with an electrical gradient that exactly balances the concentration gradient. This **resting membrane potential** is usually −70 mV (millivolts), with the inside of the membrane negative with respect to the outside.

Sodium Pump

A resting cell membrane has a very low permeability to Na^+. Nevertheless, due to the high concentration gradient and electrical attraction, some Na^+ leaks through it into the cell, even in the resting condition. When the axon is active, during an action potential, Na^+ flows inward through open Na^+ channels with each passing signal. If not removed, the accumulation of Na^+ inside the axon would cause the resting membrane potential of the fiber to decay. This decay is prevented by **sodium pumps,** each a complex of protein subunits embedded in the plasma membrane of the axon (see Figure 3.20, p. 50). Each sodium pump uses energy from the hydrolysis of ATP to transport sodium from the inside to the outside of the membrane. The sodium pump in nerve axons, as in many other cell membranes, also moves K^+ into the axon while it is moving Na^+ out. Thus, it is a **sodium-potassium exchange pump** that helps to restore the ion gradients of both Na^+ and K^+. In addition, in the central nervous system, astrocytes (mentioned earlier on this page) help to maintain the correct balance of ions surrounding neurons by storing excess potassium produced during neuronal activity.

Action Potential

A nerve **action potential** is a rapidly moving change in electrical membrane potential (Figure 33.5). It is a very rapid and brief **depolarization** of the membrane of the nerve fiber. This means

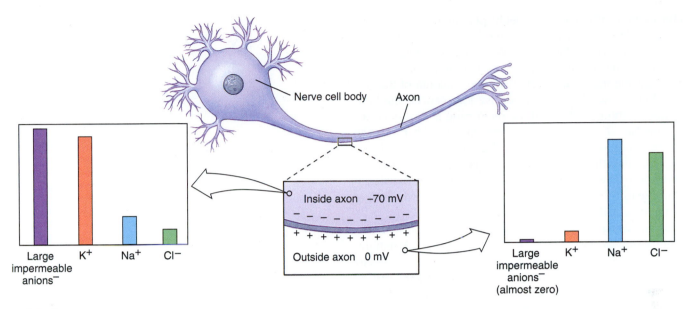

Figure 33.4

Ionic composition inside and outside a resting nerve cell. An active sodium-potassium exchange pump located in the cell membrane drives sodium to the outside, keeping its concentration low inside. Potassium concentration is high inside. Although the membrane is "leaky" to potassium, this ion is held inside by the repelling positive charge outside the membrane, and its attraction to large negatively charged anions inside the membrane, which cannot leave the cell.

that the membrane potential changes from rest (approximately −70 mV) in a positive direction and overshoots 0 mV to about +35 mV. In other words, the membrane potential reverses for an instant so that the outside becomes negative compared with the inside. Then, as the action potential moves ahead, the membrane returns to its normal resting membrane potential, ready to conduct another signal. The entire event occupies approximately a millisecond. Perhaps the most significant property of the nerve action potential is that it is **self-propagating;** once started the action potential moves along the nerve fiber automatically and unchanged in intensity, much like the burning of a fuse.

What causes the reversal of polarity in the cell membrane during passage of an action potential? We have seen that the resting membrane potential depends on the high membrane permeability to K+, some 50 to 70 times greater than the permeability to Na+. When the action potential arrives at a given point in a neuron membrane, the change in membrane potential causes **voltage-gated**

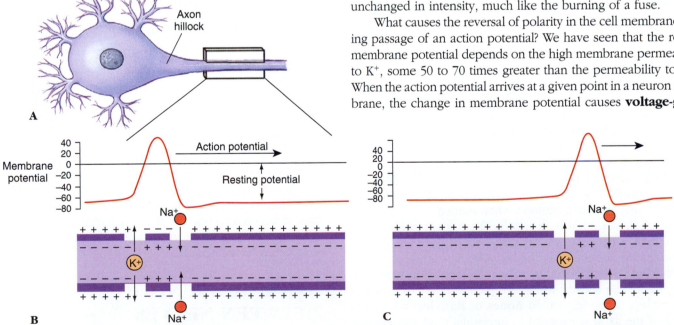

Figure 33.5

Conduction of an action potential or nerve signal. The action potential originates in the axon hillock of the neuron (A) and moves toward the right. B and C show the electrical event and associated changes in localized membrane permeability to sodium and potassium. The position of the action potential in C is shown about 4 milliseconds after B. When the action potential arrives at a point, voltage-gated sodium channels open, allowing sodium ions to enter. Sodium inflow reverses the membrane polarity, making the inner surface of the axon positive and the outside negative. Sodium channels then close and voltage-gated potassium channels open. Potassium ions move out and restore the normal resting potential. These events occur repeatedly along a membrane until the axon terminals are reached.

Na$^+$ channels (p. 49) to open suddenly, permitting Na$^+$ to diffuse into the axon from the outside, moving down the concentration gradient for Na$^+$. The voltage-gated Na$^+$ channels remain open for less than a millisecond. Only a very minute amount of Na$^+$ moves across the membrane—less than one-millionth of the Na$^+$ outside—but this sudden rush of positive ions cancels the local resting membrane potential and the membrane is **depolarized.** Then, as the Na$^+$ channels close, the membrane quickly regains its resting properties as K$^+$ ions quickly diffuse out through voltage-gated K$^+$ channels that open briefly in response to the membrane depolarization. The membrane becomes once again practically impermeable to Na$^+$, and permeable primarily to K$^+$ as the resting membrane potential is reestablished.

Thus, the rising phase of an action potential is associated with rapid influx (inward movement) of Na$^+$ (Figure 33.5). When the action potential reaches its peak, Na$^+$ permeability is restored to normal, and K$^+$ permeability briefly increases above the resting level and K$^+$ ions exit. Increased potassium permeability causes the action potential to drop rapidly toward the resting membrane level, during the **repolarization** phase. The membrane is now ready to transmit another action potential. These events occur at each point along the nerve fiber membrane, as the action potential is conducted from the axon hillock, where the axon potential originated, to the axon terminals (Figure 33.5).

High-Speed Conduction

Although ionic and electrical events associated with action potentials are much the same throughout the animal kingdom, conduction velocities vary enormously from nerve to nerve and from animal to animal—from as slow as 0.1 m/sec in sea anemones to as fast as 120 m/sec in some mammalian motor axons. The speed of conduction is highly correlated with diameter of the axon. Small axons conduct slowly because internal resistance to current flow is high. In most invertebrates, where fast conduction velocities are important for quick response, such as in locomotion to capture prey or to avoid capture, axon diameters are larger. Giant axons of squids are nearly 1 mm in diameter and carry impulses 10 times faster than ordinary axons in the same animal. A squid's giant axon innervates the animal's mantle musculature, which is used for powerful mantle contractions when it swims by jet propulsion. Similar giant axons enable earthworms, which are normally slow-moving animals, to withdraw almost instantaneously into their burrows when startled.

Although vertebrates do not possess giant axons, they achieve high conduction velocities by a cooperative relationship between axons and the investing layers of myelin laid down by the Schwann cells or oligodendrocytes described on page 728. Insulating myelin sheaths are interrupted at intervals of about 1 mm or less by nodes (called **nodes of Ranvier**) where the surface of the axon is exposed to interstitial fluid surrounding the nerve. In these **myelinated fibers** action potentials depolarize the axon membrane only at nodes because the myelin sheath prevents depolarization elsewhere (Figure 33.6). Ion pumps and channels that move ions across a membrane are concentrated in each node. Once an action potential starts down an axon, depolarization of the first node initiates an electrical current that flows to the neighboring node, causing it to depolarize and to trigger

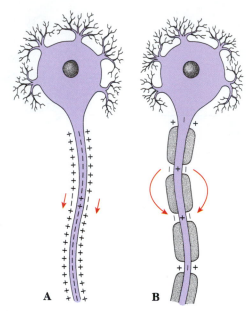

Figure 33.6
Action potential conduction in unmyelinated and myelinated fibers. In unmyelinated fibers (A), the action potential spread is continuous, and must depolarize, the entire length of the axon membrane. In myelinated fibers (B), the action potential leaps from node to node, bypassing the insulated portions of the fiber. This is saltatory conduction, which is much faster than continuous conduction.

an action potential. Thus the action potential leaps from node to node, a kind of conduction called **saltatory** (L. *salto,* to dance, leap). The gain in efficiency as compared with nonmyelinated axons is impressive. For example, a frog myelinated axon only 12 μm in diameter conducts nerve impulses at the same speed as a squid unmyelinated axon 350 μm in diameter.

Some invertebrates, including prawns and insects, also have fast axons invested with multiple layers of a myelin-like substance that is interrupted at intervals much like myelinated axons of vertebrates. Conduction rates, though not as fast as vertebrate saltatory conduction, are much faster than unmyelinated axons of the same diameter in other invertebrates.

Temperature also regulates conduction velocity in animals. Endotherms usually have a high conduction velocity since they maintain a constant body temperature (37° C in humans), whereas the conduction velocity in ectotherms fluctuates with environmental temperatures.

SYNAPSES: JUNCTIONS BETWEEN NERVES

When an action potential passes down an axon to its terminal, it must cross a small gap, or **synapse** (Gr. *synapsis,* contact, union), separating it from another neuron or an effector organ. Two distinct kinds of synapses are known: electrical and chemical.

Electrical synapses, although much less common than chemical synapses, have been demonstrated in both invertebrate and vertebrate groups. Electrical synapses are points at which ionic currents flow directly across a narrow gap junction (see Figure 3.15, p. 47) from one neuron to another. Electrical synapses show no time lag and consequently are important for escape reactions. Signals are bidirectional at many electrical synapses, but unidirectional ones have been demonstrated in Crustacea. Electrical synapses also have been observed in other excitable cell types, and form an important method of communication in the heart between cardiac muscle cells (p. 694) and between cells in smooth muscle tissue (for example, the uterus, p. 149).

Much more complex than electrical synapses are **chemical synapses,** which contain packets or vesicles of specialized chemicals called **neurotransmitters.** Neurons bringing action potentials toward chemical synapses are called **presynaptic neurons;** those carrying action potentials away are **postsynaptic neurons.** At a synapse, membranes are separated by a narrow gap, or **synaptic cleft,** having a width of approximately 20 nm.

The axon of most neurons divides at its end into many branches, each of which bears a synaptic knob or terminal that sits on dendrites or the cell body of the next neuron (Figure 33.7A). Because a single action potential coming down a nerve axon is transmitted along these many branches and synaptic endings, many impulses may converge on the cell body at one instant or may diverge onto more than one postsynaptic neuron. In addition,

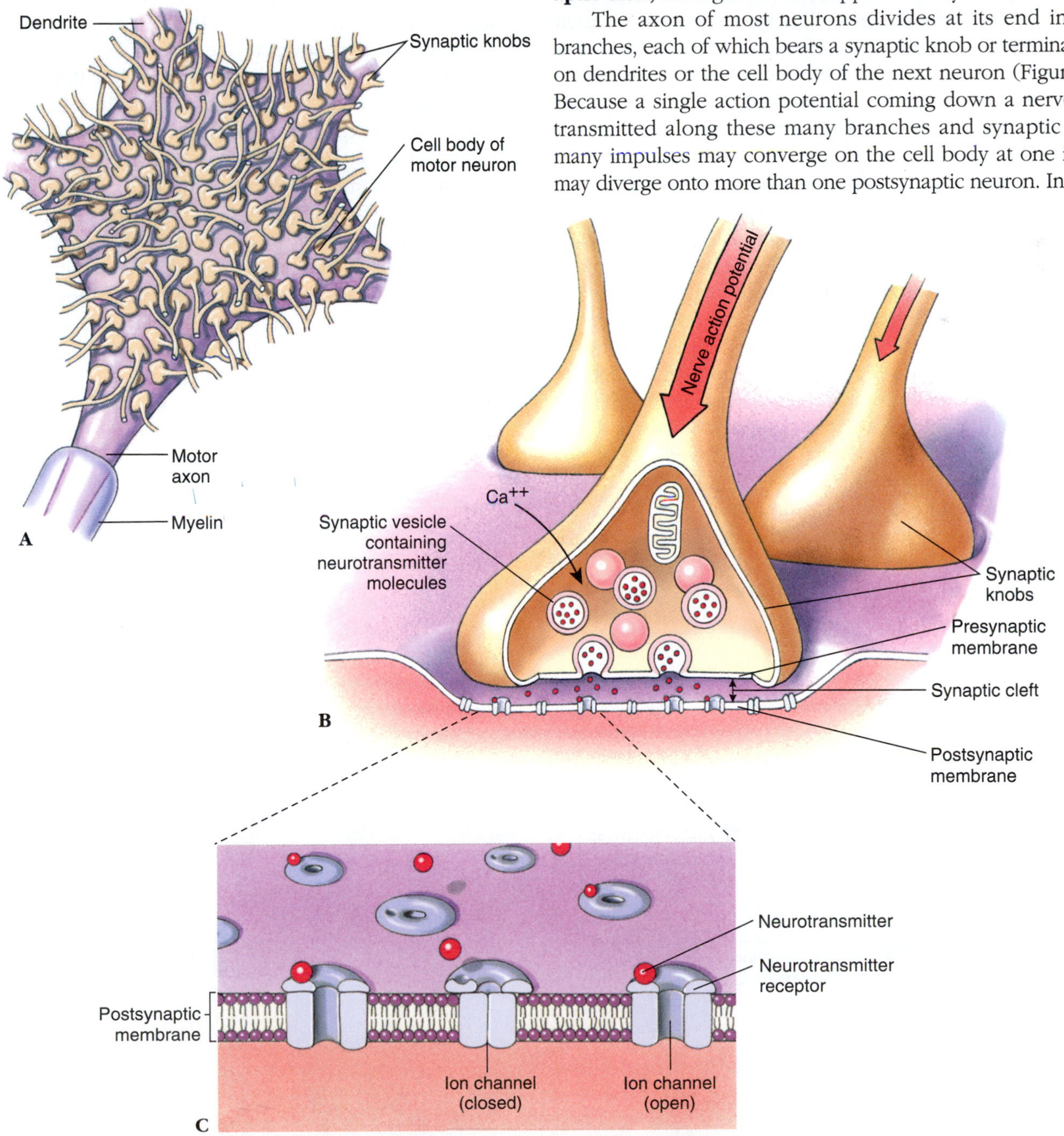

Figure 33.7

Transmission of action potentials across nerve synapses. **A,** A cell body of a motor nerve is shown with the axon terminals of interneurons. Each terminal ends in a synaptic knob; thousands of synaptic knobs may rest on a single nerve cell body and its dendrites. **B,** A synaptic knob enlarged 60 times more than in **A.** An action potential traveling down the axon causes movement of synaptic vesicles to the presynaptic membrane where exocytosis occurs, releasing neurotransmitter molecules into the cleft. **C,** Diagram of a synaptic cleft at the ultrastructural level. Upon vesicular exocytosis, neurotransmitter molecules move rapidly across the gap to bind briefly with chemically-gated ion channels in the postsynaptic membrane. Binding of neurotransmitter to its receptor produces a change in the potential of the postsynaptic membrane, in this case, caused by opening of the ion channels.

axon terminals of many neurons may almost cover a nerve cell body and its dendrites with thousands of synapses.

The 20-nm interstitial fluid-filled gap between presynaptic and postsynaptic membranes prevents action potentials from spreading directly to a postsynaptic neuron. Instead synaptic knobs secrete one or more specific neurotransmitters that communicate chemically with the postsynaptic cell. One of the most common neurotransmitters of the peripheral nervous system is **acetylcholine,** which illustrates typical synaptic transmission. Inside synaptic knobs are numerous tiny **synaptic vesicles,** each containing several thousand molecules of acetylcholine. When an action potential arrives at a terminal knob a sequence of events occurs as portrayed in Figures 33.7 and 33.8. An action potential causes an inward movement of calcium (Ca^+) ions through voltage-gated channels in the synaptic knob membrane and this induces exocytosis of some neurotransmitter-filled synaptic vesicles. Acetylcholine molecules diffuse into the gap in a fraction of a millisecond and bind briefly to receptor molecules on ion channels in the postsynaptic membrane. These **chemically-gated** channels (p. 49) open and ions flow through the channels while they remain open. This flow of ions creates a voltage change in the postsynaptic membrane. Whether this postsynaptic excitatory potential is large enough to trigger an action potential depends on how many acetylcholine molecules are released and how many channels are opened. Acetylcholine is rapidly destroyed by the enzyme **acetylcholinesterase,** which converts acetylcholine into acetate and choline. If not inactivated in this way, the neurotransmitter would continue to stimulate the postsynaptic membrane indefinitely as long as the ion channels remain open. Organophosphate insecticides (such as malathion) and certain military nerve gases are poisonous for precisely this reason; they

block acetylcholinesterase. The final step in the sequence is reabsorption of choline into the presynaptic terminal, resynthesis of acetylcholine and its storage in synaptic vesicles, ready to respond to another action potential.

Vertebrate and invertebrate nervous systems both have many different chemical neurotransmitters. Those that depolarize postsynaptic membranes are released at **excitatory synapses,** while those that move the resting membrane potential in a more negative direction **(hyperpolarization),** thereby stabilizing them against depolarization, are released at **inhibitory synapses.** Whether a neurotransmitter causes a postsynaptic excitatory or inhibitory potential depends on what specific ions flow through the chemically-gated channels to which they bind. Thus, neurotransmitters can be both excitatory and inhibitory. Such examples include acetylcholine, norepinephrine, dopamine, and serotonin. However, some neurotransmitters always appear to be inhibitory (for example, glycine and gamma aminobutyric acid [GABA]), while others seem always to be excitatory (for example, glutamate). Neurons in the central nervous system have both excitatory and inhibitory synapses among the hundreds or thousands of synaptic knobs on the dendrites and cell body of each neuron.

The net balance of all excitatory and inhibitory inputs received by a postsynaptic cell determines whether it generates an action potential (Figure 33.8). If many excitatory signals are received at one time, they can reduce the resting membrane potential enough in the postsynaptic membrane to elicit an action potential. Inhibitory signals, however, stabilize the postsynaptic membrane, making it less likely that an action potential will be generated. The synapse is a crucial part of the decision-making equipment of the central nervous system, modulating flow of information from one neuron to the next.

Figure 33.8

Sequence of events in synaptic transmission at an excitatory synapse.

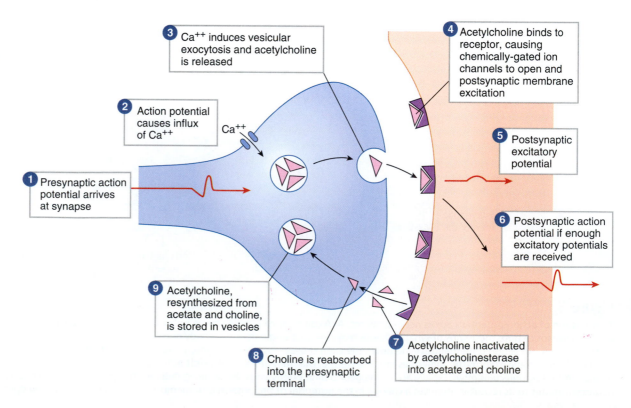

3 Ca^{++} induces vesicular exocytosis and acetylcholine is released

4 Acetylcholine binds to receptor, causing chemically-gated ion channels to open and postsynaptic membrane excitation

2 Action potential causes influx of Ca^{++}

Ca^{++}

5 Postsynaptic excitatory potential

1 Presynaptic action potential arrives at synapse

6 Postsynaptic action potential if enough excitatory potentials are received

9 Acetylcholine, resynthesized from acetate and choline, is stored in vesicles

8 Choline is reabsorbed into the presynaptic terminal

7 Acetylcholine inactivated by acetylcholinesterase into acetate and choline

EVOLUTION OF NERVOUS SYSTEMS

Invertebrates: Development of Centralized Nervous Systems

Various metazoan phyla reveal a progressive increase in complexity of nervous systems that probably reflects in a general way stages in evolution of nervous systems. The simplest pattern of invertebrate nervous systems is the nerve net of radially symmetrical animals, such as sea anemones, jellyfishes, hydras, and comb jellies (Figure 33.9A). A nerve net forms an extensive network in and under the epidermis over all the body. A signal starting in one part of this net spreads in all directions, since synapses in most radially symmetrical animals do not restrict transmission to one-way movement, as often occurs in animals with more complex nervous systems. There is evidence of organization into **reflex arcs** (p. 734) with branches of a nerve net connecting to sensory receptors in the epidermis and to epithelial cells that have contractile properties. Although most responses tend to be generalized, many are astonishingly complex for so simple a nervous system. Part of the nerve net is concentrated into two nerve rings in medusan forms of Cnidaria (see Figure 13.11, p. 269) and receives sensory input from **statocysts,** organs of balance (p. 746), **ocelli,** light-sensitive organs, as well as sensory cells detecting chemical and tactile stimuli. Nerve nets are found among vertebrates in nerve plexuses located, for example, in the intestinal wall; such nerve plexuses govern generalized intestinal movements such as peristalsis and segmentation (p. 714).

Bilateral nervous systems, the simplest of which occur in flatworms, represent a distinct increase in complexity over the nerve net of radiate animals. Recent evidence from the study of genetic mechanisms controlling brain development in insect and mouse embryos shows homology of regulatory gene families. These data suggest that a common ancestral brain utilizing these genes may have evolved prior to the protostome-deuterostome divergence. Flatworms have two anterior ganglia, composed of groups of nerve cell bodies from which two main nerve trunks run posteriorly, with lateral branches extending throughout the body (Figure 33.9B). This is the simplest nervous system showing differentiation into a **peripheral nervous system** (a communication network extending to all parts of the body) and a **central nervous system** (a concentration of nerve cell bodies), which coordinates everything. More complex invertebrates exhibit a more centralized nervous system (brain), with two longitudinal fused nerve cords and many ganglia. Elaborate nervous systems of annelids consist of a bilobed brain, a double nerve cord with segmental ganglia, and distinctive **afferent** (sensory) and **efferent** (motor) neurons (Figure 33.9C). Segmental ganglia are relay stations for coordinating regional activity.

The basic plan of molluscan nervous systems is a series of three pairs of well-defined ganglia, but cephalopods (such as octopuses and squids) have ganglia burgeoned into textured nervous centers of great complexity; those of octopuses contain more than 160 million cells. Sense organs, too, are highly developed. Consequently, cephalopod behavior far outstrips that of any other invertebrate, and they are capable of learning (p. 356).

The basic plan of arthropod nervous systems (Figure 33.9D) resembles that of annelids, but ganglia are larger and sense organs are much better developed. Social behavior is often elaborate, particularly in hymenopteran insects (bees, wasps, and ants), and most arthropods are capable of considerable manipulation of their environment. Despite the small size of an insect's brain, examples of learning have been documented in bees, wasps, ants, flies, locusts, and grasshoppers. The brain region associated with learning appears to be areas called **mushroom**

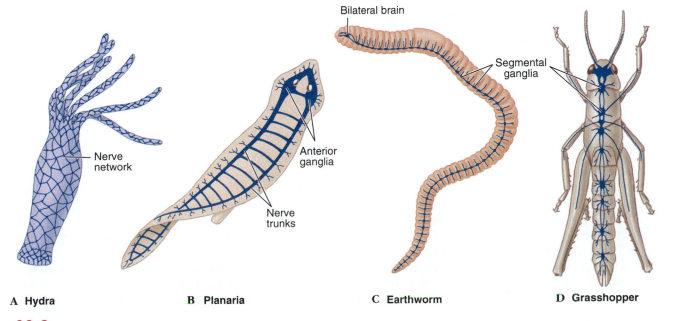

A Hydra   **B Planaria**   **C Earthworm**   **D Grasshopper**

Figure 33.9

Invertebrate nervous systems. **A,** Nerve net of radiates, the simplest neural organization. **B,** Flatworm system, the simplest linear-type nervous system of two nerves connected to a complex neuronal network. **C,** Annelid nervous system, organized into a bilobed brain and ventral cord with segmental ganglia. **D,** Arthropod nervous system, also segmental, with large ganglia and more elaborate sense organs.

bodies, which are largest in social insects, and experiments have shown changes in these structures with age and experience.

Vertebrates: Fruition of Encephalization

The basic plan of the vertebrate nervous system is a hollow, *dorsal* nerve cord terminating anteriorly in a large mass, or brain. This pattern contrasts with the nerve cord of bilateral invertebrates, which is solid and ventral to their alimentary canal. By far the most important trend in evolution of vertebrate nervous systems is the great elaboration of size, configuration, and functional capacity of the brain, a process called **encephalization.** Vertebrate encephalization has brought to full fruition several functional capabilities including fast responses, great capacity for storage of information, and enhanced complexity and flexibility of behavior. Another consequence of encephalization is the ability to form associations between past, present, and (at least in humans) future events.

Spinal Cord

The **brain** and **spinal cord** compose the central nervous system. During early embryonic development, the spinal cord and brain begin as an ectodermal neural groove, which by folding and enlarging becomes a long, hollow neural tube (see Figure 8.14, p. 168). The cephalic end enlarges to form brain vesicles, and the rest becomes the spinal cord. Unlike any invertebrate nerve cord, segmental nerves of spinal cords of vertebrates (31 pairs in humans) are separated into dorsal sensory roots and ventral motor roots. Sensory nerve cell bodies are gathered together into dorsal root (spinal) ganglia. Both dorsal (sensory) and ventral (motor) roots meet beyond the spinal cord to form a mixed spinal nerve (Figure 33.10).

The spinal cord encloses a central spinal canal and is additionally wrapped in three layers of membranes called **meninges** (men-in'jeez; Gr. *meningos,* membrane). In cross section the cord shows two zones (Figure 33.10). An inner zone of gray matter, resembling in shape the wings of a butterfly, contains the cell bodies of motor neurons and interconnecting interneurons. An outer zone of white matter contains bundles of axons and dendrites linking different levels of the cord with each other and with the brain.

Reflex Arc

Many neurons work in groups called **reflex arcs,** a fundamental unit of neural operation that has remained conserved during evolution of the nervous system. A reflex arc contains at least two neurons, but usually there are more. Parts of a typical reflex arc are (1) a **receptor,** a sense organ in skin, muscle, or another organ; (2) an **afferent,** or sensory, neuron, which carries impulses toward the central nervous system; (3) the **central nervous system,** where synaptic connections are made between sensory neurons and interneurons; (4) an **efferent,** or motor, neuron, which makes a synaptic connection with the interneuron and carries impulses from the central nervous system; and (5) an **effector,** by which an animal responds to environmental changes. Examples of effectors are muscles, glands, ciliated cells,

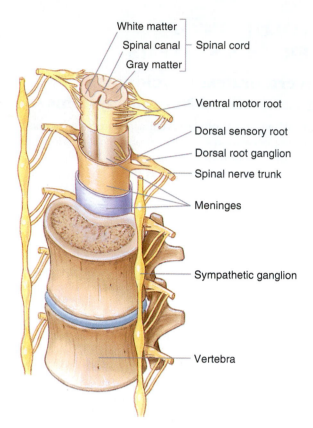

White matter
Spinal canal — Spinal cord
Gray matter

Ventral motor root
Dorsal sensory root
Dorsal root ganglion
Spinal nerve trunk
Meninges

Sympathetic ganglion

Vertebra

Figure 33.10

Human spinal cord and its protection. Two vertebrae show the position of the spinal cord, emerging spinal nerves, and the sympathetic trunk. The cord is wrapped by three layers of membrane (meninges) between two of which lies a protective bath of cerebrospinal fluid.

cnidocytes of cnidarians, electric organs of fishes, and pigmented cells called chromatophores (p. 647).

A reflex arc in vertebrates in its simplest form contains only two neurons—a sensory (afferent) neuron and a motor (efferent) neuron (for example, the "knee-jerk" or stretch reflex, Figure 33.11A). Usually, however, interneurons are interposed between sensory and motor neurons (Figure 33.11B). An interneuron may connect afferent and efferent neurons on the same side of the spinal cord or on opposite sides, or it may connect them on different levels of the spinal cord, either on the same or opposite sides.

A **reflex act** is a response to a stimulus acting over a reflex arc. It is involuntary, meaning that it is often not under the control of the will. For example, many vital processes of the body, such as control of breathing, heartbeat, diameter of blood vessels, and sweat secretion are reflex acts. Some reflex acts are innate; others are acquired through learning.

In almost any reflex act, a number of reflex arcs are involved. For instance, a single afferent sensory neuron may make synaptic connections with many efferent motor neurons. In a similar way an efferent neuron may receive signals from many afferent neurons. Afferent neurons also make connections with ascending sensory neurons, which travel in the white matter of the spinal cord, bringing information about peripheral reflexes to the brain. Reflex activity may then be modified by signals from the brain traveling along descending motor

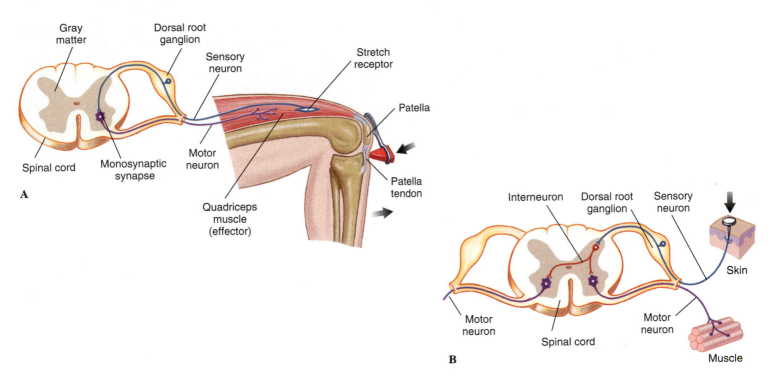

Figure 33.11

The reflex arc. **A,** The "knee-jerk" or stretch reflex, a simple reflex arc. Sudden pressure on the patellar ligament stretches muscles in the upper leg. Action potentials generated in stretch receptors are conducted along afferent (sensory) neurons to the spinal cord and relayed directly to an efferent (motor) nerve cell body. Action potentials pass along efferent neurons to leg muscles (effectors), stimulating them to contract. **B,** Multisynaptic relfex arc. A more common reflex arc includes interneurons between the sensory and motor neuron. Tack puncture is sensed by pain receptors in the skin and the signal is conducted along afferent fibers to the spinal cord where synaptic connections are made with interneurons. Here, an interneuron is shown making connections with motor neurons on both sides of the spinal cord, such that stimulation of muscle fibers in more than one part of the body (both legs, for example) allows coordination of muscle responses to the tack puncture.

neurons, which impinge on the final efferent motor neurons before they leave the spinal cord for the periphery.

Brain

Unlike the spinal cord, which has changed little in structure during vertebrate evolution, the brain has changed dramatically. The ancestral vertebrate brain of fishes and early tetrapods expanded to form a deeply fissured and enormously intricate brain in the lineage leading to mammals (Figure 33.12). It reaches its greatest complexity in the human brain, which contains some 35 billion neurons, each of which may receive information from tens of thousands of synapses at one time. The ratio between weight of the brain and that of the spinal cord affords a fair criterion of an animal's intelligence. In fishes and amphibians this ratio is approximately 1:1; in humans the ratio is 55:1—in other words, the brain is 55 times heavier than the spinal cord. Although the human brain is not the largest (the sperm whale's brain is seven times heavier) nor the most convoluted (that of the porpoise is even more folded), it is by all odds the best in overall performance. This "great ravelled knot," as the British physiologist Sir Charles Sherrington called the human brain, in fact may be so complex that it will never be able to understand its own function!

Brains of early vertebrates had three principal divisions: a forebrain, or **prosencephalon;** a midbrain, or **mesencephalon;** and a hindbrain, or **rhombencephalon** (Figure 33.13). Each part was concerned with one or more special sense: the forebrain

with smell, the midbrain with vision, and the hindbrain with hearing and balance. These primitive but very fundamental concerns of the brain have been in some instances amplified and in others reduced or overshadowed during continued evolution as sensory priorities were shaped by an animal's habitat and way of life.

Hindbrain　The **medulla oblongata,** the most posterior division of the brain, is really a conical continuation of the spinal cord (Figure 33.14A and B). The medulla, together with the more anterior midbrain, constitutes the "brain stem," an area that controls numerous vital and largely subconscious activities such as heartbeat, respiration, vascular tone, gastric secretions, and swallowing. The brain stem also contains centers that appear to integrate incoming peripheral information regarding satiety and feeding stimuli. The **pons,** also a part of the hindbrain, contains a thick bundle of fibers that carry impulses from one side of the cerebellum to the other, and also connects both medulla and cerebellum to other brain regions (Figure 33.14A and B).

The **cerebellum,** lying dorsal to the medulla, controls equilibrium, posture, and movement (Figure 33.14A and B). Its development is directly correlated with an animal's mode of locomotion, agility of limb movement, and balance. It is usually weakly developed in amphibians and nonavian reptiles, forms that live close to the ground, and well developed in the more agile bony fishes. It reaches its apogee in birds and mammals in which it is greatly expanded and folded. The cerebellum does

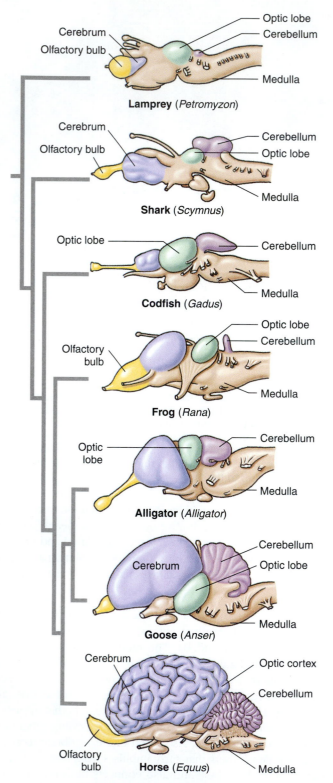

Figure 33.12

Evolution of the vertebrate brain. Note the progressive increase in size of the cerebrum. The cerebellum, concerned with equilibrium and motor coordination, is largest in animals whose balance and precise motor movements are well developed (fishes, birds, and mammals).

not initiate movement but operates as a precision error-control center, or servomechanism, that programs a movement initiated somewhere else, such as the motor cortex of the cerebrum (Figure 33.14A). Primates and especially humans, who possess a manual dexterity far surpassing that of other animals, have the most complex cerebellum. Movements of hands and fingers may involve cerebellar coordination of simultaneous contraction and relaxation of hundreds of individual muscles.

Midbrain   The midbrain (see Figure 33.13) consists mainly of the **tectum** (including the **optic lobes**), which contains nuclei that serve as centers for visual and auditory reflexes. (In neurophysiological usage a nucleus is a small aggregation of nerve cell bodies within the central nervous system.) The midbrain has undergone little evolutionary change in structure among vertebrates but has changed markedly in function. It mediates the most complex behavior of fishes and amphibians, integrating visual, tactile, and auditory information. Such functions have been gradually assumed by the forebrain in amniotes. In mammals, the midbrain is mainly a relay center for information on its way to higher brain centers.

Forebrain   Just anterior to the midbrain lie the **thalamus** and **hypothalamus,** the most posterior elements of the forebrain (Figure 33.14B). The egg-shaped thalamus is a major relay station that analyzes and passes sensory information to higher brain centers. In the hypothalamus are several "housekeeping" centers that regulate body temperature, water balance, appetite, and thirst—all functions concerned with maintenance of internal constancy (homeostasis). Neurosecretory cells located in the hypothalamus produce several neurohormones (described in Chapter 34). The hypothalamus also contains centers for regulating reproductive function and sexual behavior, and it participates in emotional behaviors.

The anterior portion of the forebrain, or **cerebrum** (Figure 33.14A and B), can be divided into two anatomically distinct areas, the **paleocortex** and **neocortex.** Originally concerned with smell, it became well developed in advanced fishes and early terrestrial vertebrates, which depend on this special sense. In mammals and especially in primates the paleocortex is a deep-lying area called a rhinencephalon ("nose brain"), because many of its functions depend on olfaction. Better known as the **limbic system,** it mediates several species-specific behaviors that relate to fulfilling needs such as feeding and sex. One region of the limbic system, the **hippocampus,** has been extensively studied as a site of spatial learning and memory. The hippocampus has gained notoriety because its neurons have mitotic capabilities in adults, a previously unknown occurrence in mammalian neurons.

Although a late arrival in vertebrate evolution, the neocortex completely overshadows the paleocortex and has become so expanded that it envelops much of the forebrain and all of the midbrain (Figure 33.14). Almost all integrative activities primitively assigned to the midbrain are transferred to the neocortex, or **cerebral cortex** as it is usually called.

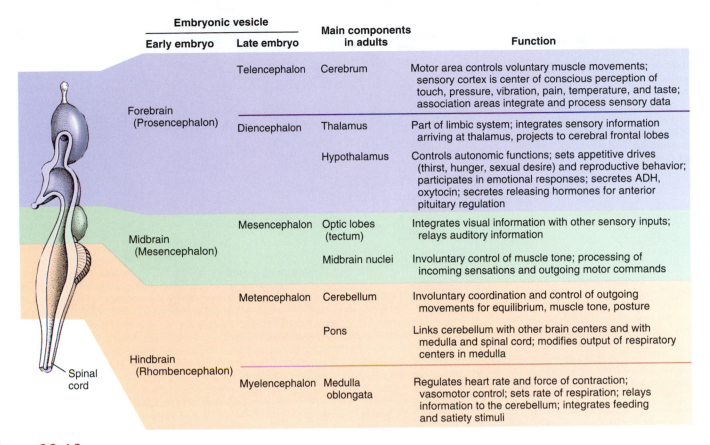

Embryonic vesicle		Main components in adults	Function
Early embryo	**Late embryo**		
Forebrain (Prosencephalon)	Telencephalon	Cerebrum	Motor area controls voluntary muscle movements; sensory cortex is center of conscious perception of touch, pressure, vibration, pain, temperature, and taste; association areas integrate and process sensory data
	Diencephalon	Thalamus	Part of limbic system; integrates sensory information arriving at thalamus, projects to cerebral frontal lobes
		Hypothalamus	Controls autonomic functions; sets appetitive drives (thirst, hunger, sexual desire) and reproductive behavior; participates in emotional responses; secretes ADH, oxytocin; secretes releasing hormones for anterior pituitary regulation
Midbrain (Mesencephalon)	Mesencephalon	Optic lobes (tectum)	Integrates visual information with other sensory inputs; relays auditory information
		Midbrain nuclei	Involuntary control of muscle tone; processing of incoming sensations and outgoing motor commands
Hindbrain (Rhombencephalon)	Metencephalon	Cerebellum	Involuntary coordination and control of outgoing movements for equilibrium, muscle tone, posture
		Pons	Links cerebellum with other brain centers and with medulla and spinal cord; modifies output of respiratory centers in medulla
	Myelencephalon	Medulla oblongata	Regulates heart rate and force of contraction; vasomotor control; sets rate of respiration; relays information to the cerebellum; integrates feeding and satiety stimuli

Spinal cord

Figure 33.13
Divisions of the vertebrate brain.

Functions in the cerebrum have been localized by direct stimulation of exposed brains of people and experimental animals, postmortem examination of persons suffering from various lesions, and surgical removal of specific brain areas in experimental animals. The cortex contains discrete motor and sensory areas (Figures 33.14 and 33.15). The motor areas control voluntary muscle movements, while the sensory cortex is the center of conscious perception of touch, pressure, pain, temperature, and taste. Vision, smell, hearing, and speech are purely sensory or motor regions located in specific areas on the cerebral lobes. In addition, there are large "silent" regions, called **association areas,** concerned with memory, judgment, reasoning, and other integrative functions. These regions are not directly connected to sense organs or muscles.

Thus in mammals, and especially in humans, separate parts of the brain mediate conscious and unconscious functions. The unconscious mind, all of the brain except the cerebral cortex, governs numerous vital functions that are removed from conscious control: respiration, blood pressure, heart rate, hunger, thirst, temperature balance, salt balance, sexual drive, and basic (sometimes irrational) emotions. The brain is also a complex endocrine gland that regulates and receives feedback from the body's endocrine system (see Chapter 34). The conscious mind, or cerebral cortex, is the site of higher mental activities (for example, planning and reasoning), memory, and integration of sensory information. Memory appears to transcend all parts of the brain rather than being a property of any particular part of the brain as once believed.

The right and left hemispheres of the cerebral cortex are bridged through the corpus callosum (see Figure 33.14B), a neural connection through which the two hemispheres are able to transfer information and coordinate mental activities. In humans, the two hemispheres of the brain are specialized for different functions: the left hemisphere for language development, mathematical and learning capabilities, and sequential thought processes; and the right hemisphere for spatial, musical, artistic, intuitive, and perceptual activities. In addition, each hemisphere controls the opposite side of the body. It has been known for a long time that even extensive damage to the right hemisphere may cause varying degrees of left-sided paralysis but has little effect on intellect and speech. Conversely, damage to the left hemisphere usually causes loss of speech and may have disastrous effects on intellect. Because these differences in brain symmetry and function exist at birth, they appear to be inborn rather than the result of developmental or environmental effects as once believed.

Hemispheric specialization has long been considered a unique human trait, but it was recently discovered in the brains of songbirds in which one side of the brain is specialized for song production.

Figure 33.14

A, External view of the human brain, showing lobes of the cerebrum and localization of major function of the cerebrum and cerebellum. **B,** Section through midline of the human brain showing one cerebral hemisphere of the cerebrum, the thalamus and hypothalamus of the forebrain, and the pons, medulla, and cerebellum of the hindbrain.

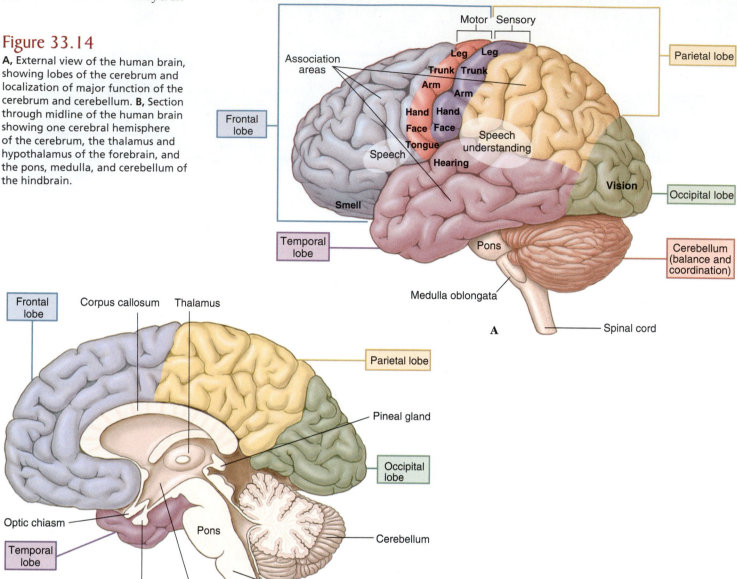

Although the large size of their brain undoubtedly makes humans the most intelligent of animals, it is apparent that they can do without much of it and still remain intelligent. Brain scans of persons with hydrocephalus (enlargement of the head as a result of pressure disturbances that cause the brain ventricles [fluid-filled cavities within the brain] to enlarge many times their normal size) show that although many such persons are functionally disabled, others are nearly normal. The cranium of one person with hydrocephalus was nearly filled with cerebrospinal fluid and the only remaining cerebral cortex was a thin layer of tissue, 1 mm thick, pressed against his cranium. Yet this young man, with only 5% of his brain, had achieved first-class honors in mathematics at a British university and was socially normal. This and other similarly dramatic observations suggest that there is enormous redundancy and spare capacity in corticocerebral function. It also suggests that the deep structures of the brain, which are relatively spared in hydrocephalus, may perform functions once attributed to the cortex.

Peripheral Nervous System

The peripheral nervous system includes all nervous tissue outside the central nervous system. It has two functional divisions: **sensory** or **afferent division,** which brings sensory information to the central nervous system, and **motor** or **efferent division,** which conveys motor commands to muscles and glands. The efferent division has two components: (1) **somatic nervous system,** which innervates skeletal muscle, and (2) **autonomic nervous system,** which innervates smooth muscle, cardiac muscle, and glands.

Autonomic Nervous System   The autonomic system governs involuntary, internal functions of the body that do not ordinarily affect consciousness, such as movements of alimentary canal and heart, contraction of smooth muscle of blood vessels, urinary bladder, iris of the eye, and others, plus secretions of various glands.

Autonomic nerves originate in the brain or spinal cord as do nerves of the somatic nervous system, but unlike the latter,

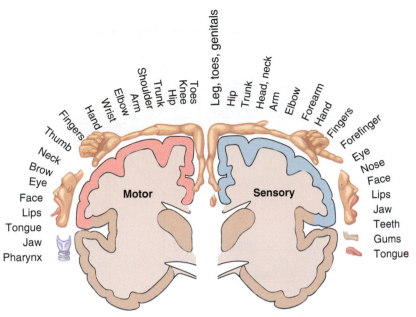

Figure 33.15

Arrangement of sensory and motor cortices shown in transverse section (see Figure 33.14 for surface view). Localizations of sensory terminations from different parts of the body are shown at right; origins of descending motor pathways are shown at left. The motor cortex lies in front of the sensory cortex, so the two are not superimposed. These maps were produced from the 1930s work of Canadian neurosurgeon Wilder Penfield. Recent research shows that the motor cortex is not as orderly as the map suggests; rather correspondence between cortical areas and areas of the body they control is more diffuse.

and are often embedded in tissue layers close to effector organs in the parasympathetic system (see Figure 33.16).

All *preganglionic* neurons, whether sympathetic or parasympathetic, release acetylcholine at their synapse with postganglionic cells. However, parasympathetic *postganglionic* neurons release acetylcholine at their endings, whereas sympathetic *postganglionic* neurons with a few exceptions release norepinephrine (also called noradrenaline). This difference is another important characteristic distinguishing the two parts of the autonomic nervous system.

As a general rule the parasympathetic division is associated with nonstressful activities, such as resting, eating, digestion, and urination. The sympathetic division is active under conditions of physical or emotional stress. Under such conditions heart rate increases, blood vessels to the skeletal muscles dilate, blood vessels in the viscera constrict, activity of the intestinal tract decreases, and metabolic rate increases. The importance of these responses in emergency reactions (sometimes called the fright, fight or flight response) are described in chapter 34 (p. 767). It should be noted, however, that the sympathetic division is active to some degree also during resting conditions in maintaining normal blood pressure and body temperature.

autonomic fibers consist of not one but two motor neurons (Figure 33.16). They synapse once after leaving the cord and before arriving at the effector organ. These synapses are located outside the spinal cord in ganglia. Axons passing from the cord to the ganglia are called preganglionic autonomic neurons; those passing from the ganglia to the effector organs are called postganglionic neurons.

Subdivisions of the autonomic system are the **parasympathetic** and **sympathetic** systems. Most organs in the body are innervated by both sympathetic and parasympathetic neurons, whose actions are antagonistic (Figure 33.17). If one neuron stimulates an activity, the other inhibits it. However, neither kind of nerve is exclusively excitatory or inhibitory. For example, parasympathetic neurons inhibit heartbeat but excite peristaltic movements of the intestine; sympathetic neurons increase heartbeat but inhibit intestinal peristaltic movement.

Parasympathetic neurons emerge from the central nervous system either in brain stem cranial nerves or in spinal nerves emerging from the sacral (pelvic) region of the spinal cord (Figures 33.16 and 33.17). In the sympathetic division nerve cell bodies of all preganglionic neurons are located in thoracic and upper lumbar areas of the spinal cord. Their neurons exit through the ventral roots of spinal nerves, separate from these, and go to sympathetic ganglia (Figures 33.10 and 33.17), which are paired and form a chain on each side of the spinal column.

Ganglia are usually remote from the effector organ in the sympathetic system (for example, sympathetic ganglion chain)

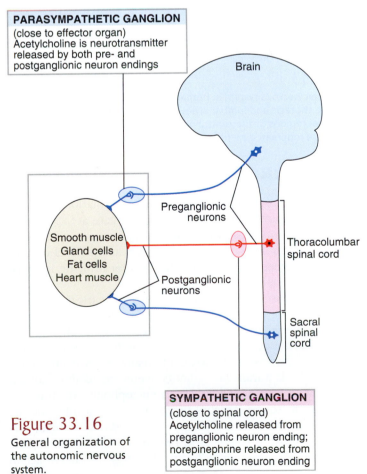

Figure 33.16

General organization of the autonomic nervous system.

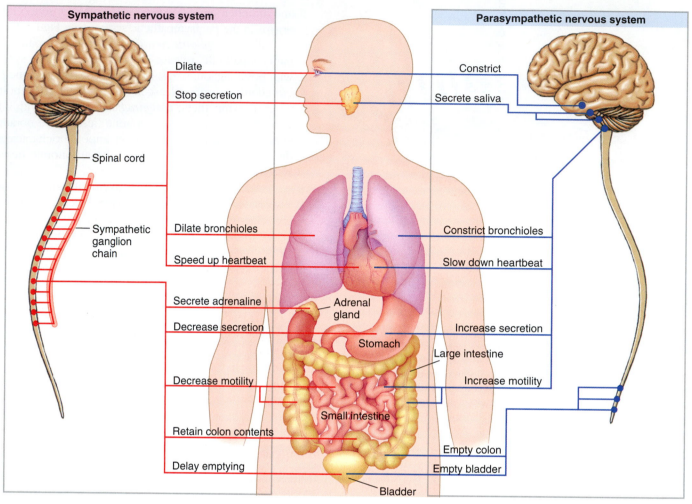

Sympathetic nervous system

Dilate

Stop secretion

Spinal cord

Sympathetic ganglion chain

Dilate bronchioles

Speed up heartbeat

Secrete adrenaline

Decrease secretion

Decrease motility

Retain colon contents

Delay emptying

Parasympathetic nervous system

Constrict

Secrete saliva

Constrict bronchioles

Slow down heartbeat

Adrenal gland

Increase secretion

Stomach

Large intestine

Increase motility

Small intestine

Empty colon

Empty bladder

Bladder

Figure 33.17

Autonomic nervous system in humans. Outflow of autonomic nerves from the central nervous system is shown at left. Sympathetic (*red*) outflow is from the thoracic and lumbar areas of the spinal cord by way of a chain of sympathetic ganglia. Parasympathetic (*blue*) outflow is from the cranial and sacral regions of the central nervous system; parasympathetic ganglia (not shown) are located in or adjacent to the organs innervated. Most organs are innervated by fibers from both sympathetic and parasympathetic divisions.

SENSE ORGANS

Animals require a constant inflow of information from the environment to regulate their lives. Sense organs are specialized sensory receptors designed for detecting environmental status and change. An animal's sense organs are its first level of environmental perception; they are channels for bringing information to the central nervous system.

A **stimulus** is some form of energy—electrical, mechanical, chemical, or radiant. Sensory receptors within a sense organ transform energy from a stimulus into nerve action potentials, the common language of the nervous system. In a very real sense, then, sense organs are biological transducers. A microphone, for example, is a transducer that converts mechanical (sound) energy into electrical energy. Like a microphone, which is sensitive only to sound, sensory receptors are, as a rule, specific for one kind of stimulus. Thus eyes respond only to light, ears to sound, pressure receptors to pressure, and chemoreceptors to chemicals, converting all forms of energy into nerve action

potentials that can be transmitted to the central nervous system, eliciting a response via the previously described reflex arc that is fundamental to all nervous systems.

Since all nerve action potentials are qualitatively alike, how do animals perceive and distinguish different sensations of varying stimuli? The answer is that real perception of sensation is performed by localized regions of the brain, where sensory receptors from each sense organ have their own hookup. This concept of "labeled lines" of communication to specific brain regions was first described in the 1830s by Johannes Müller who called this the **law of specific nerve energies.** Action potentials arriving at a particular sensory area of the brain can be interpreted in only one way. For example, pressure on the eye causes us to see "stars" or other visual patterns; mechanical distortion of the eye initiates action potentials in the optic nerve fibers that are perceived as light sensations. Although such a procedure probably could never be done, a deliberate surgical switching of optic and auditory nerves would cause the recipient literally to see thunder and hear lightning!

Classification of Receptors

Receptors are traditionally classified by their location. Those near the external surface, called **exteroceptors,** keep an animal informed about its external environment. Internal parts of the body are provided with **interoceptors,** which receive stimuli from internal organs. Muscles, tendons, and joints have **proprioceptors,** which are sensitive to changes in tension of muscles and provide an organism with a sense of body position. Sometimes receptors are classified by the form of energy to which the receptors respond, such as **chemical, mechanical, light,** or **thermal.**

Chemoreception

Chemoreception is the oldest and most universal sense in the animal kingdom. It probably guides behavior of animals more than any other sense. Unicellular forms use **contact chemical receptors** to locate food and adequately oxygenated water and to avoid harmful substances. These receptors elicit an orientation behavior, called **chemotaxis,** toward or away from a chemical source. Most metazoans have specialized **distance chemical receptors,** which are often developed to a remarkable degree of sensitivity. Distance chemoreception, usually called smell or olfaction, guides feeding behavior, location and selection of sexual mates, territorial and trail marking, and alarm reactions of numerous animals.

In all vertebrates and in insects, the senses of **taste** and **smell** are clearly distinguishable. Although there are similarities between taste and smell receptors, in general taste is more restricted in response and is less sensitive than smell. Central nervous system centers for taste and smell are located in different parts of the brain.

Insect chemoreceptors are located in sensory hairs called sensilla. Taste sensilla are present on the mouthparts, legs, wing margins, and ovipositor in females. They have a single pore at the tip and recognize four classes of compounds: sugar (attractive), bitter (repelling), salts, and water. Olfactory sensilla are located on the head on two pairs of olfactory organs: the antennae and the maxillary palps (p. 450). Pores on the cuticullar walls of these sensilla allow odorant and pheromone molecules from the environment to contact the olfactory receptor neurons.

Social insects and many other animals, including mammals, produce species-specific compounds, called **pheromones,** that constitute a highly developed chemical language. Pheromones are a diverse group of organic compounds that an animal releases to affect the physiology or behavior of another individual of the same species. Information regarding territory, societal hierarchy, sex and reproductive state are transmitted via this system. Ants, for example, are walking batteries of glands (Figure 33.18) that produce numerous chemical signals. These include releaser pheromones, such as alarm and trail pheromones, and primer pheromones, which alter endocrine and reproductive systems of different castes in a colony.

In vertebrates, taste receptors occur in the mouth cavity and especially on the tongue (Figure 33.19), where they provide a means for judging foods before they are swallowed. A **taste bud** consists of a cluster of receptor cells surrounded by supporting cells; it is provided with a small external pore through which slender tips of sensory cells project. Chemicals being tasted interact with specific receptor sites on microvilli of receptor cells. Taste sensations are categorized as sweet, salty, acid, bitter, and possibly umami (Japanese, roughly translated to mean "meaty" or "savory"). Although the mechanisms are different for each basic taste sensation, receptor cells are depolarized by the specific chemical and action potentials are generated. Contrary to what was originally thought, taste receptors can respond to different types of taste categories, although they may respond more strongly to one particular type. These action potentials are transmitted across chemical synapses (p. 731) and travel along sensory neurons to specific brain regions. Taste discrimination depends on assessment by the brain of the relative activity of the five different taste receptor subtypes. This assessment is similar to color vision in vertebrates, where a whole rainbow of colors can be distinguished by relative excitation of only three types of color photoreceptors (see p. 750). Because receptor cells are subject to abrasion by foods, taste buds have a short life (5 to 10 days in mammals) and are continually being replaced.

Although olfactory sense is a primal sense for many animals, used for identification of food, sexual mates, and predators, olfaction is most highly developed in mammals. Even humans, although a species not celebrated for detecting smells, can discriminate perhaps 20,000 different odors. A human nose can detect 1/25 of one-millionth of 1 mg of mercaptan, the odoriferous substance of skunks. Even so, our olfactory abilities compare poorly with those of other mammals that rely on olfaction for survival. A dog explores new surroundings with its nose much as we do with our eyes. A dog's nose is justifiably renowned; with some odorous sources a dog's nose is at least a million times more sensitive than ours. Dogs are assisted in their proficiency by having a nose located close to the ground where odors from passing creatures tend to linger.

Olfactory endings are located in a special epithelium covered by a thin film of mucus, positioned deep in the nasal cavity (Figure 33.20). Within this epithelium lie millions of olfactory neurons, each with several hairlike cilia protruding from the free end. Odor molecules entering the nose bind to receptor proteins located in the cilia; this binding generates action potentials that travel along axons to the olfactory bulb of the brain. From here odor information is sent to the olfactory cortex where odors are analyzed. Odor information is then projected to higher brain centers that influence emotions, thoughts, and behavior.

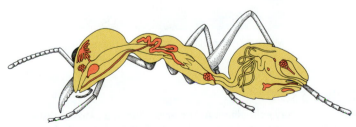

Figure 33.18
Pheromone-producing glands of an ant *(shown in orange)*.

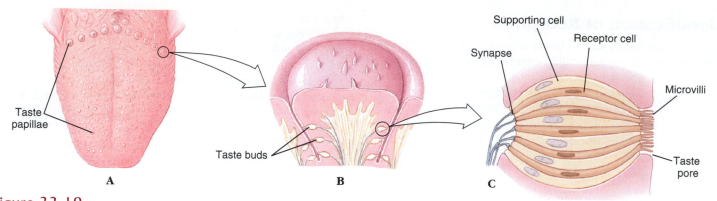

Figure 33.19

Taste receptors. **A,** Surface of human tongue showing location of taste papillae. **B,** Position of taste buds on a taste papilla. **C,** Structure of a taste bud.

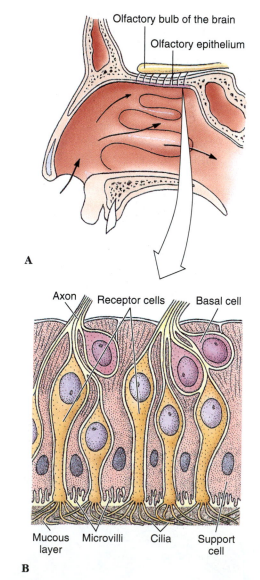

Using techniques of molecular genetics (p. 97), researchers discovered a large family of genes that appears to encode for odor reception in mammals (including humans). About 70 genes from the same family have been identified in the fruit fly, *Drosophila,* and some also in the nematode, *C. elegans.* This suggests that the gene family for olfaction is ancient and highly conserved through evolution. Each of the 1000 or so genes discovered in mammals encodes a separate type of odor receptor. Since mammals can detect at least 20,000 different odors, each receptor must respond to several odor molecules, and each odor molecule must bind with several types of receptors, each of which responds to a part of the molecule's structure. Brain-mapping techniques have shown that each olfactory neuron projects to a characteristic location on the olfactory bulb, providing a two-dimensional map that identifies which receptors have been activated in the nose. In addition, olfactory neurons expressing the same odor-receptor gene converge to a fixed olfactory bulb region, which might provide an explanation for the extremely high sensitivity of smell. Projected to the brain, odor information is recognized as a unique scent.

Because flavor of food depends on odors reaching the olfactory epithelium through the throat passage, taste and smell are easily confused. All "tastes" other than the five basic ones (sweet, sour, acid, salty, and possibly umami) result from flavor molecules reaching the olfactory epithelium in this manner. Food loses its appeal during a common cold because a stuffy nose blocks odors rising from the mouth.

Many terrestrial vertebrates possess an additional olfactory organ, the **vomeronasal organ (VNO** or **Jacobson's organ),** which responds to pheromones. The VNO is also lined with olfactory epithelium and is located in paired, blind-ending passages that open into the nasal or oral cavity. The VNO olfactory receptor cells respond to chemical signals carried by pheromones, which are chemically different from the odor molecules that stimulate nasal olfactory receptor cells. Equivocal information currently exists as to the presence of a functional VNO system in humans, and recent research has focused on the possibility of pheromonal communication in humans. The best established example is that of menstrual cycle synchronization that occurs in women living together in close association, such as in dormitories.

Figure 33.20

Human olfactory epithelium. **A,** The epithelium is a patch of tissue positioned in the roof of the nasal cavity. **B,** It is composed of supporting cells, basal cells, and olfactory receptor cells with cilia protruding from their free ends.

Mechanoreception

Mechanoreceptors are sensitive to quantitative forces such as touch, pressure, stretching, sound, vibration, and gravity—in short, they respond to motion. Animals interact with their environments, feed themselves, maintain normal postures, and walk, swim, or fly, using a steady flow of information from mechanoreceptors.

Touch

Invertebrates, especially insects, have many kinds of receptors sensitive to touch. Such receptors are well endowed with tactile hairs sensitive to both touch and vibrations. Superficial touch receptors of vertebrates are distributed over the body but tend to be concentrated in areas especially important for exploring and interpreting the environment. In most vertebrates these areas are the face and extremities of limbs. Of the more than half-million separate touch-sensitive spots on the surface of a human body, most occur on the tongue and fingertips, as might be expected based on the large portion of sensory cortex that receives information from these regions (see Figure 33.15). The simplest touch receptors are bare nerve endings in skin, but touch receptors assume varying shapes and sizes. Each hair follicle is crowded with receptors sensitive to touch.

Pacinian corpuscles, relatively large mechanoreceptors that register deep touch and pressure in mammalian skin, illustrate the general properties of mechanoreceptors. These corpuscles are common in deep layers of skin, connective tissue surrounding muscles and tendons, and abdominal mesenteries. Each corpuscle consists of a nerve terminus surrounded by a capsule of numerous, concentric, onionlike layers of connective tissue (Figure 33.21). Pressure at any point on a capsule distorts its nerve ending, producing a graded **receptor potential,** a local

flow of electric current similar to a postsynaptic excitatory potential (p. 732). Progressively stronger stimuli lead to correspondingly stronger receptor potentials until a **threshold current** is produced; this current initiates an action potential in a sensory nerve fiber. A second action potential is initiated at removal of the pressure, but not during the pressure. This response is called adaptation (not to be confused with the evolutionary meaning of this term [see Chapter 6]) and characterizes many touch receptors, which are admirably suited to detecting a sudden mechanical change but readily adapt to new conditions. We are aware of new pressures when we put on our shoes and clothing in the morning, but we are not reminded of these pressures all day.

Pain

Pain receptors are relatively unspecialized nerve fiber endings that respond to a variety of stimuli signaling possible or real damage to tissues. These free nerve endings also respond to other stimuli, such as mechanical movement of a tissue and temperature changes. Pain fibers respond to small peptides, such as substance P and bradykinins, which are released by injured cells. This type of response is termed *slow pain. Fast pain* responses (for example, a pin prick, cold or hot stimuli) are a more direct response of the nerve endings to mechanical or thermal stimuli.

Pain is a distress call from the body signaling some noxious stimulus or internal disorder. Although there is no cortical pain center, discrete areas have been located in the brain stem where pain messages from the periphery terminate. These areas contain two kinds of small peptides, endorphins and enkephalins, that are endogenous opiates with morphinelike or opiumlike activity. When released, they bind with specific opiate receptors in the midbrain. They are the body's own analgesics.

Just as pain is a sign of danger, sensory pleasure is a sign of a stimulus useful to the subject. Pleasure depends on the internal state of an animal and is judged with reference to homeostasis and some physiological set point. Pleasure states may be produced by release of endogenous opioids within the central nervous system.

Lateral-Line System of Fishes and Amphibians

A lateral line is a distant touch reception system for detecting wave vibrations and currents in water. Receptor cells, called **neuromasts,** are located on the body surface in aquatic amphibians and some fishes, but in many fishes they are located within canals running beneath the epidermis; these canals open at intervals to the surface (Figure 33.22). Each neuromast is a collection of **hair cells** with sensory endings, or cilia, embedded in a gelatinous, wedge-shaped mass, the **cupula.** The cupula projects into the center of the lateral-line canal so that it bends in response to any disturbance of water on the body surface. The lateral-line system is one of the principal sensory systems that guide fishes in their movements and in location of predators, prey, and social partners (p. 522).

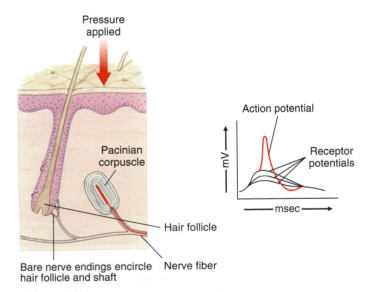

Figure 33.21

Response of pacinian corpuscle to applied pressure. Progressively stronger pressure produces stronger receptor potentials. When the threshold stimulus is reached, an all-or-none action potential is generated in the afferent nerve fiber.

The lateral line serves another function in some fish, the reception of small bioelectric signals (produced during heart and muscle activity) from other members of their species or from intruder or prey species. The **electroreceptor cells** are found in pores closely associated with the lateral-line system and in some species, such as the shark, are concentrated primarily on the head (see Figure 24.10, p. 523). In addition to receiving electric signals, some fish are able to generate weak or strong electric fields produced by **electric organs,** which are modified muscles located near the tail (for example, in some freshwater fish, such as the catfish and electric eel). Intruders or prey can be located as they produce a disturbance in the electric field. Potential mates are recognized in some species since the discharge frequency is different between the sexes. Fish with strong electric fields can use electrolocation both to locate and to stun their prey (electric eels). Others do not possess electroreceptors but have electric organs on either side of the head (for example, in marine electric rays; see p. 524) and use an electric current to stun prey.

Hair cells form an important sensory component of several mechanoreceptors found in both invertebrate (statocysts) and vertebrate (vestibular organs) organs of equilibrium, discussed on page 746.

Hearing

An ear is a specialized receptor for detecting sound waves in the surrounding environment. Because sound communication and reception are integral to lives of terrestrial vertebrates, we may be surprised to discover that most invertebrates inhabit a silent world. Only certain arthropod groups—crustaceans, spiders, and insects—have developed true sound-receptor organs. Even among insects, only locusts, cicadas, crickets, grasshoppers, and most moths possess ears, and these are of simple design: a pair of air pockets, each enclosed by a tympanic membrane that passes sound vibrations to sensory cells. Despite their spartan construction, insect ears are beautifully designed to detect the sound of a potential mate, a rival male, or a predator.

Especially interesting are ultrasonic detectors of certain nocturnal moths. These have evolved specifically to detect approaching bats and thus lessen the moth's chance of becoming a bat's evening meal (echolocation in bats is described on p. 625). Each moth ear possesses just two sensory receptors, known as A_1 and A_2 (Figure 33.23). The A_1 receptor responds to ultrasonic cries of a bat that is still too far away to detect the moth. As the bat approaches and its cries increase in intensity, the receptor fires more rapidly, informing the moth that the bat is coming nearer. Since the moth has two ears, its nervous system can determine the bat's position by comparing firing rates from the two ears. The moth's strategy is to fly away before the bat detects it. But if the bat continues its approach, the second (A_2) receptor in each ear, which responds only to high-intensity sounds, will fire. The moth responds immediately with an evasive maneuver, usually making a power dive to a bush or the ground where it is safe because the bat cannot distinguish a moth's echo from those of its surroundings.

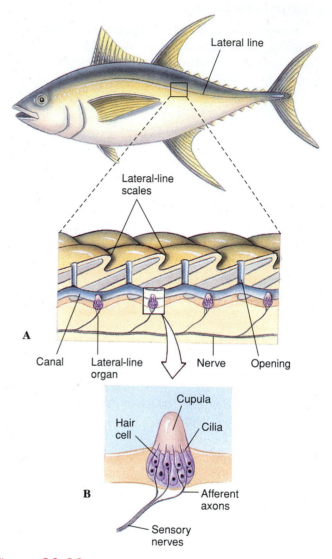

Figure 33.22

Lateral-line system. **A,** Lateral line of a bony fish with both exposed and hidden neuromasts. **B,** Structure of a neuromast (lateral-line organ).

In their evolution, vertebrate ears originated as a balance organ, or **labyrinth,** whose function is described in the next section. In all jawed vertebrates, from fishes to mammals, the labyrinth has a similar structure, consisting of two small chambers called the **saccule** and the **utricle,** and three **semicircular canals** (Figure 33.24). In fishes the base of the saccule is extended into a tiny pocket (**lagena**) that, during evolution of vertebrates, developed into the hearing receptor of tetrapods. With continued elaboration and elongation in birds and mammals, the fingerlike lagena was modified to form a **cochlea.**

A human ear (Figure 33.25) is representative of mammalian ears. The outer, or external, ear collects sound waves and funnels them through an **auditory canal** to an eardrum or **tympanic membrane** lying next to the middle ear. The middle ear is an air-filled chamber containing a remarkable chain of three tiny bones, or ossicles, known as the **malleus** (hammer), **incus** (anvil), and **stapes** (stirrup), named because of their fancied

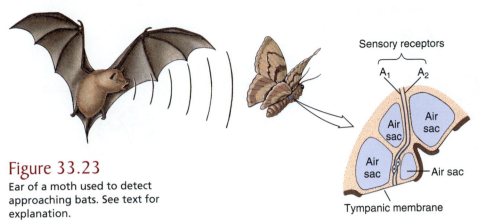

Within the inner ear is the organ of hearing, or **cochlea** (Gr. *cochlea,* snail's shell), which is coiled in mammals, making two and one half turns in humans (Figure 33.25B). The cochlea is divided longitudinally into three tubular canals running parallel with one another. This relationship is indicated in Figure 33.26. These canals become progressively smaller from the base of the cochlea to the apex. One of these canals is called the **vestibular canal;** its base is closed by the oval window. The **tympanic canal,** which is in communication with the vestibular canal at the tip of the cochlea, has its base closed by the **round window.** Between these two canals is a **cochlear duct,** which contains the **organ of Corti,** the actual sensory apparatus (Figure 33.25C and D). Within the organ of Corti are fine rows of hair cells that run lengthwise from the base to the tip of the cochlea. At least 24,000 hair cells are present in a human ear. The 80 to 100 "hairs" on each cell are actually microvilli and a single large cilium (see Chapter 3, p. 46, and Chapter 29, p. 654), which project into the endolymph of the cochlear canal. Each cell is connected with neurons of the auditory nerve. Hair cells rest on the **basilar membrane,** which separates the tympanic canal and cochlear duct, and they are covered by the **tectorial membrane,** lying directly above them (Figure 33.25D).

Figure 33.23

Ear of a moth used to detect approaching bats. See text for explanation.

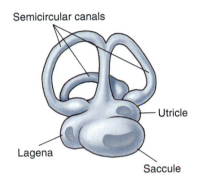

Figure 33.24

Vestibular apparatus of a teleost fish, containing three semicircular canals, which respond to angular acceleration; two balance organs (utricle and saccule), which are static receptors that signal the fish's position in relation to gravity; and a small chamber, the lagena, which is specialized for sound reception.

resemblance to these objects. These bones conduct sound waves across the middle ear (Figure 33.25B). The bridge of bones is so arranged that the force of sound waves pushing against the tympanic membrane is amplified as much as 90 times where the stapes contacts the **oval window** of the inner ear. Muscles attached to the middle ear bones contract when the ear receives very loud noises, providing the inner ear some protection from damage. The middle ear connects with the pharynx by means of an **eustachian tube,** which permits pressure equalization on both sides of the tympanic membrane.

When a sound wave strikes the ear, its energy is transmitted through the ossicles of the middle ear to the oval window, which oscillates back and forth, moving the fluid of the vestibular and tympanic canals (Figure 33.26). Because these fluids are noncompressible, an inward movement of the oval window produces a corresponding outward movement of the round window. The fluid oscillations also cause the basilar membrane with its hair cells to vibrate simultaneously.

According to the **place hypothesis of pitch discrimination** formulated by Georg von Békésy, different areas of the basilar membrane respond to different frequencies; for every sound frequency, there is a specific "place" on the basilar membrane where hair cells respond to that frequency (Figure 33.26). Initial displacement of the basilar membrane starts a wave traveling down the membrane, much as flipping a rope at one end starts a wave moving down the rope (Figure 33.27). The displacement wave increases in amplitude as it moves from the oval window toward the apex of the cochlea, reaching a maximum at the region of the basilar membrane where the natural frequency of the membrane corresponds to the sound frequency. Here, the membrane vibrates with such ease that the energy of the traveling wave is completely dissipated. Hair cells within the organ of Corti in that region are stimulated and action potentials conveyed to the axons of the auditory nerve. Isolated hair cells have been shown to respond to a particular band of frequencies depending on their location within the cochlea. Thus, action potentials that are carried by certain axons of the auditory nerve are interpreted by the hearing center as particular tones. **Loudness**

The origin of the three tiny bones of the mammalian middle ear—the malleus, incus, and stapes—is one of the most extraordinary and well-documented transitions in vertebrate evolution. Amphibians, nonavian reptiles, and birds have a single rodlike ear ossicle, the stapes (also called the columella), which originated as a jaw support (hyomandibular) as seen in fishes (see Figure 23.16, p. 511). With evolution of the earliest tetrapods, the braincase became firmly sutured to the skull, and the hyomandibular, no longer needed to brace the jaw, became converted into the stapes. In a similar way, the two additional ear ossicles of the mammalian middle ear—the malleus and incus—originated from parts of the jaw of early vertebrates. The quadrate bone of nonavian reptilian upper jaws became the incus, and the articular bone of the lower jaw became the malleus. Homology of nonavian reptilian jaw bones to mammalian ear bones is clearly documented in the fossil record and in embryological development of mammals.

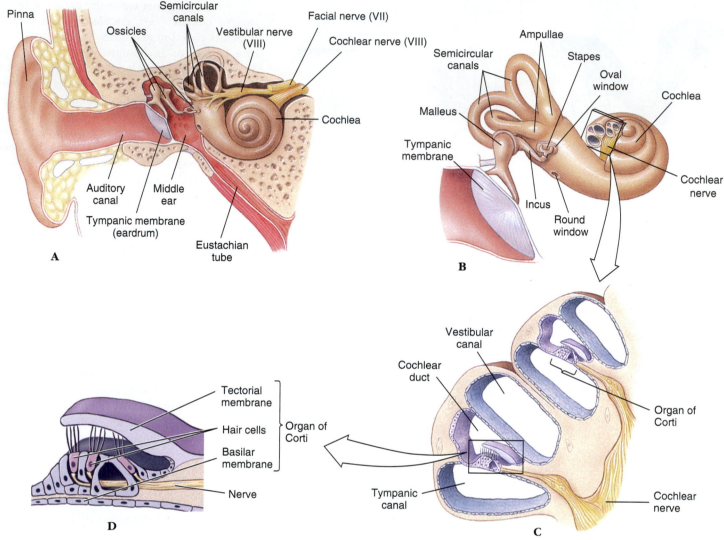

Figure 33.25

Human ear. **A,** Longitudinal section showing external, middle, and inner ear. **B,** Enlargement of middle ear and inner ear. The cochlea of the inner ear has been opened to show the arrangement of canals within. **C,** Enlarged cross section of cochlea showing the organ of Corti. **D,** Detail of ultrastructure of the organ of Corti.

of a tone depends on the number of hair cells stimulated, whereas **timbre,** or quality, of a tone is produced by the pattern of hair cells stimulated by sympathetic vibration. This latter characteristic of tone enables us to distinguish between different human voices and different musical instruments, although the notes in each case may be of the same pitch and loudness.

Most recent auditory research has focused on a more active role for hair cells within the organ of Corti. Experiments have demonstrated that outer hair cells may respond to sound waves by changing their length and thus mechanically altering the position of the basilar and tectorial membranes. Although a function of such movements is not yet established in vivo, it has been suggested that this active response of these receptor cells in the organ of Corti might increase both sensitivity and selectivity of hearing.

Equilibrium

In invertebrates, specialized sense organs for monitoring gravity and low-frequency vibrations often appear as **statocysts.** Each is a simple sac lined with hair cells and containing a heavy calcareous structure, the **statolith** (Figure 33.28). The delicate, hairlike filaments of sensory cells are activated by the shifting position of the statolith when the animal changes position. Statocysts occur in many invertebrate phyla from cnidarians to arthropods. All are built on similar principles.

The vertebrate organ of equilibrium is the **labyrinth,** or **vestibular organ.** It consists of two small chambers **(saccule** and **utricle)** and three **semicircular canals** (see Figure 33.25B). The utricle and saccule are static balance organs that, like invertebrate statocysts, give information about position of the head or body with respect to the force of gravity. As the head is tilted in one direction or another, stony accretions press on different

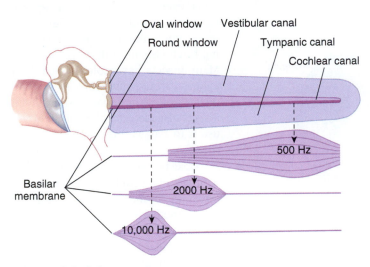

Figure 33.26

Frequency localization in the cochlea of the mammalian ear as it would appear with the cochlea stretched out. Sound waves transmitted to the oval window produce vibration waves that travel down the basilar membrane. High-frequency vibrations cause the membrane to resonate near the oval window. Low-frequency tones travel farther down the basilar membrane.

groups of hair cells; these cells send nerve action potentials to the brain, which interprets this information with reference to head position.

The semicircular canals of vertebrates are designed to respond to **rotational acceleration** and are relatively insensitive to linear acceleration. The three semicircular canals are at right angles to each other, one for each axis of rotation. They are filled with fluid (endolymph), and within each canal is a bulb-like enlargement, the **ampulla,** which contains hair cells. The hair cells are embedded in a gelatinous membrane, the **cupula,** which projects into the fluid. The cupula is similar in structure to the cupula of the lateral-line system of fishes (p. 743). When the

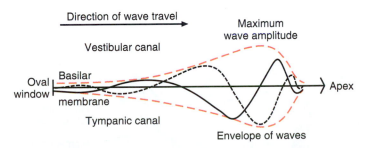

Figure 33.27

Traveling waves along the basilar membrane. The oval window is at left, and the cochlear apex at right. The two wave formations (*solid* and *dashed lines*) occur at separate instants of time. The curves in color represent the extreme displacements of the membrane by traveling waves as they reach their maximum amplitude where the natural frequency of the basilar membrane corresponds to the sound frequency. At this point along the basilar membrane, hair cells in the organ of Corti are stimulated.

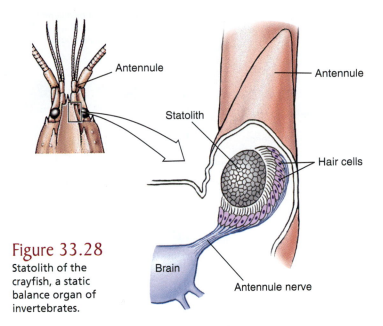

Figure 33.28

Statolith of the crayfish, a static balance organ of invertebrates.

head rotates, fluid in the canal at first tends not to move because of inertia. Since the cupula is attached, its free end is pulled in the direction opposite the direction of rotation (Figure 33.29). Bending of the cupula distorts and excites the hair cells embedded in it, increases the discharge rate of afferent nerve fibers leading from the ampulla to the brain, and produces a sensation of rotation. Since the three canals of each ear are in different planes, acceleration in any direction stimulates at least one ampulla.

Photoreception: Vision

Light-sensitive receptors are called **photoreceptors.** These receptors range from simple light-sensitive cells scattered randomly on the body surface of many invertebrates (dermal light sense) to the exquisitely developed camera-type eye of vertebrates and cephalopods. Eyespots of astonishingly advanced organization appear even in some unicellular forms. That of a dinoflagellate, *Nematodinium,* bears a lens, a light-gathering chamber, and a photoreceptive pigment cup—all developed within a single-celled organism (Figure 33.30). The dermal light receptors of many invertebrates are of much simpler design. They are far less sensitive than optic receptors, but they are important in locomotory orientation, pigment distribution in chromatophores, photoperiodic adjustment of reproductive cycles, and other behavioral changes.

More highly organized eyes, many capable of excellent image formation, are based on one of two different principles: either a single-lens, camera-type eye such as those of cephalopod molluscs and vertebrates; or a multifaceted (compound) eye as in arthropods. Arthropod **compound eyes** are composed of many independent visual units called **ommatidia** (Figure 33.31). Light enters each corneal lens and is absorbed by visual pigments in the rhabdomere of the retinular cells. These receptor cells depolarize and generate action potentials in the axon leaving each ommatidium. Eyes of bees contain about 15,000 of these

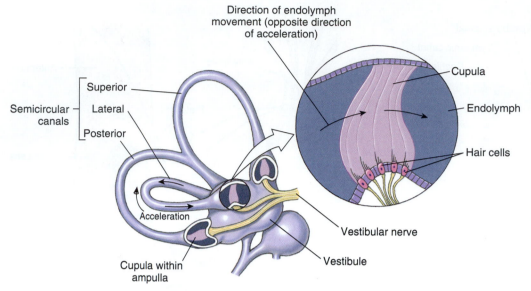

Figure 33.29

How the semicircular canals respond to angular acceleration. Because of inertia, endolymph in the semicircular canal corresponding to the plane of motion moves past the cupula in a direction opposite to that of angular acceleration. Movement of the cupula stimulates hair cells.

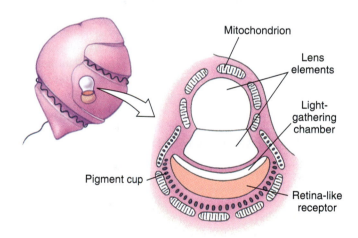

Figure 33.30

Eyespot of the dinoflagellate, *Nematodinium*.

Figure 33.31

Compound eye of an insect. A single ommatidium is shown enlarged at right.

units, each of which views a separate narrow sector of the visual field. Such eyes form a mosaic of images of varying brightness from the separate units. Many insects have color vision; honeybees can use ultraviolet light to see nectar guides in flowers. Many flying insects also detect polarized light and use it to navigate through their environments. Resolution (the ability to see objects sharply) is poor compared with that of a vertebrate eye. A fruit fly, for example, must be closer than 3 cm to see another fruit fly as anything but a single spot. However, a compound eye is especially well suited to detecting motion, as anyone who has tried to swat a fly knows.

Eyes of certain annelids, molluscs, and all vertebrates are built like a camera—or rather we should say that a camera is modeled somewhat after these eyes. A camera-type eye contains in the front a light-tight chamber and lens system, which focuses an image of the visual field on a light-sensitive surface (the retina) in the back (Figures 16.39 and 33.32).

The spherical eyeball is built of three layers: (1) a tough outer white **sclera** for support and protection, (2) middle **choroid coat,** containing blood vessels for nourishment, and (3) light-sensitive **retina** (Figure 33.32). The **cornea** is a transparent anterior modification of the sclera. A pigmented curtain, or **iris,** regulates the size of the light opening, or **pupil.** The pupil is usually a round or vertical slit-shaped opening in vertebrates, but a horizontal slit-shaped opening in cephalopods. Just behind the iris is a **lens,** a transparent, elastic oval disc that bends light rays to focus an image on the retina. **Ciliary muscles** are attached to and encircle the lens. In vertebrates they can alter curvature of the lens so

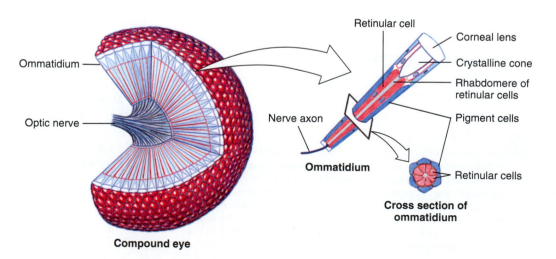

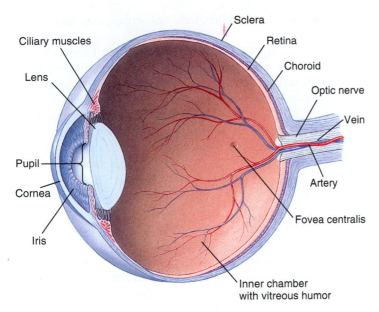

Figure 33.32
Structure of the human eye.

The **fovea centralis,** or **fovea,** the region of keenest vision, is located in the center of the retina (see Figure 33.32), in direct line with the center of the lens and cornea. It contains only cones, a vertebrate specialization for diurnal (daytime) vision. Acuity of an animal's eyes depends on the density of cones in the fovea. A human fovea and that of a lion contain approximately 150,000 cones per square millimeter, but many water and field birds have up to 1 million cones per square millimeter. Their eyes are as good as our eyes would be if aided by eight-power binoculars.

At the peripheral parts of the retina only rods are found. Rods are high-sensitivity receptors for dim light. At night, the cone-filled fovea is unresponsive to low levels of light and we become functionally color blind ("at night all cats are gray"). Under nocturnal conditions, the position of greatest visual acuity is not at the center of the fovea but at its edge. Thus it is easier to see a dim star at night by looking slightly to one side of it.

that images at varying distances from the eye can be focused on the retina. In cephalopods the ciliary muscles move the rigid lens closer or farther away from the retina to focus on images. In terrestrial vertebrates the cornea actually does most of the bending of light rays, whereas the lens adjusts focus for near and far objects. Between cornea and lens is an **outer chamber** filled with watery **aqueous humor;** between lens and retina is a much larger **inner chamber** filled with viscous **vitreous humor.**

In cephalopods the photoreceptor cells of the retina point forward and directly absorb incoming light; in vertebrates photoreceptor cells point backward and absorb light bouncing off the back of the eye. In vertebrates the retina is composed of several cell layers (Figure 33.33). The outermost layer, closest to the sclera, consists of pigment cells or **chromatophores.** Adjacent to this layer are the photoreceptors, **rods** and **cones.** Approximately 125 million rods and 1 million cones are present in each human eye. Cones are primarily concerned with color vision in ample light and rods with colorless vision in dim light. Next is a network of **intermediate neurons** (bipolar, horizontal, and amacrine cells) that process and relay visual information from the photoreceptors to ganglion cells whose axons form the optic nerve. The network permits much convergence, especially for rods. Information from several hundred rods may converge on a single ganglion cell, an adaptation that greatly increases the effectiveness of rods in dim light. Cones show very little convergence. By coordinating activities between different ganglion cells, and adjusting sensitivities of bipolar cells, horizontal and amacrine cells improve overall contrast and quality of a visual image.

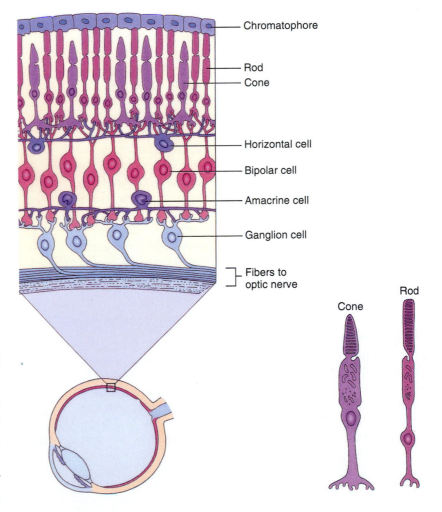

Figure 33.33
Structure of a primate retina, showing organization of intermediate neurons that connect photoreceptor cells to ganglion cells of the optic nerve.

One of several marvels of the vertebrate eye is its capacity to compress the enormous range of light intensities presented to it into a narrow range that can be handled by optic nerve fibers. Light intensity between a sunny noon and starlit night differs by more than 10 billion to 1. Rods quickly saturate with high light intensity, but cones do not; they shift their operating range with changing ambient light intensity so that a high-contrast image is perceived over a broad range of light conditions. This shift is made possible by complex interactions among the network of nerve cells that lie between the cones and the ganglion cells that generate retinal output to the brain.

Chemistry of Vision

Both rods and cones contain light-sensitive pigments known as **rhodopsins.** Each rhodopsin molecule consists of a large protein, **opsin,** an enzyme, and a small carotenoid molecule, **retinal,** a derivative of vitamin A. When a quantum of light strikes the photopigment and is absorbed by the rhodopsin molecule, retinal is isomerized, changing the shape of the molecule. This molecular change triggers the enzymatic activity of opsin, which sets in motion a biochemical sequence of several steps. This complex sequence behaves as an excitatory cascade that vastly amplifies the energy of a single photon to cause hyperpolarization (p. 732) of a rod or cone. This hyperpolarization signal is transmitted through the intermediate neurons and leads to depolarization and generation of action potentials in a ganglion cell. It is interesting to note that light reception in invertebrate eyes leads to *depolarization* of receptor cells, while similar light signals induce *hyperpolarization* in vertebrate receptor cells.

The amount of intact rhodopsin in a retina depends on the intensity of light reaching the eye. A dark-adapted eye contains much rhodopsin and is very sensitive to weak light. Conversely in a light-adapted eye, most of the rhodopsin is split into retinal and opsin. It takes approximately half an hour for a light-adapted eye to accommodate to darkness, while the rhodopsin level gradually increases.

Color vision

Cones function to perceive color and require 50 to 100 times more light for stimulation than do rods. Consequently, night vision is almost totally rod vision. Unlike humans, who have both day and night vision, some vertebrates specialize for one or the other. Strictly nocturnal animals, such as bats and owls, have pure rod retinas. Purely diurnal forms, such as common gray squirrels and some birds, have only cones; they are virtually blind at night.

In 1802 an English physician and physicist Thomas Young speculated that we see color by relative excitation of three kinds of photoreceptors: one each for red, green, and blue. In the 1960s Young's prescient hypothesis was eventually supported through the combined work of several groups of researchers. Humans have three types of cones, each containing a visual pigment that responds to a particular wavelength of light (Figure 33.34). Blue cones absorb the most light at 430 nm, green cones at 540 nm, and red cones at 575 nm. Variation in the structure of opsin produces the different visual pigments found in rods and the three types of cones. Colors are perceived by comparing levels of excitation of the three different kinds of cones. For example, a light having a wavelength of 530 nm would excite green cones 95%, red cones about 70%, and blue cones not at all. This comparison is made both in nerve circuits in the retina and in the visual cortex of the brain, and our brain interprets this combination as green.

Color vision occurs in some members of all vertebrate groups with the possible exception of amphibians. Bony fishes and birds have particularly good color vision. Surprisingly, most mammals are color blind; exceptions are primates and a few other species such as squirrels.

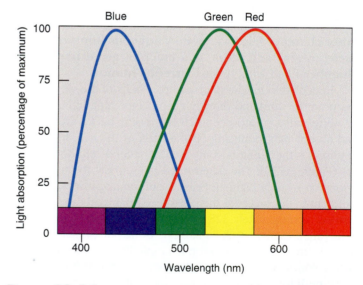

Figure 33.34

The absorption spectrum of human vision. Three types of visual pigments in cones absorb maximally at 430 nm (blue cones), 540 nm (green cones), and 575 nm (red cones).

SUMMARY

The nervous system is a rapid-communication system that interacts continuously with the endocrine system to control coordination of body function. The basic unit of nervous integration in all animals is the neuron, a highly specialized cell designed to conduct self-propagating electrical events, called action potentials, to other cells. Action potentials are transmitted from one neuron to another across synapses, which may be either electrical or chemical. The thin gap between neurons at chemical synapses is bridged by a chemical neurotransmitter molecule, which is released from the synaptic knob and can be either stimulatory or inhibitory.

The simplest organization of neurons into a system is the nerve net of cnidarians, basically a plexus of nerve cells that, with additions, is the basis of nervous systems of several invertebrate phyla. With the appearance of ganglia (nerve centers) in bilateral flatworms, nervous systems differentiated into central and peripheral divisions. Molluscs and arthropods possess a higher level of brain complexity than other invertebrates and are capable of learning. In vertebrates, the central nervous system consists of a brain and spinal cord. Fishes and amphibians have a three-part brain, whereas in mammals, the cerebral cortex has become a vastly enlarged multicomponent structure that has assumed the most important integrative activities of the nervous system. It completely overshadows the ancient brain, which is consigned to the role of relay center and to monitoring numerous unconscious but nonetheless vital functions such as breathing, blood pressure, and heart rate.

In humans the left cerebral hemisphere is usually specialized for language and mathematical skills while the right hemisphere is specialized for visual-spatial and musical skills.

The peripheral nervous system connects the central nervous system to receptors and effector organs. It is divided broadly into an afferent system, which conducts sensory signals to the central nervous system, and an efferent system, which conveys motor impulses to effector organs. This efferent system is subdivided into a somatic nervous system, which innervates skeletal muscle, and an autonomic nervous system, which innervates smooth and cardiac muscles and glands. The autonomic nervous system is subdivided into anatomically distinct sympathetic and parasympathetic systems, each of which sends fibers to most body organs. Generally the sympathetic system governs excitatory activities and the parasympathetic system governs maintenance and restoration of body resources.

Sensory organs are receptors designed especially to respond to internal or environmental change. The most primitive and ubiquitous sense is chemoreception. Chemoreceptors may be contact receptors such as the insect and vertebrate sense of taste, or distance receptors such as olfaction, which detects airborne molecules. In both cases, a specific chemical interacts with a particular receptor and results in impulses transmitted to, and interpreted by, the brain. In spite of the similarity between these two senses, the sense of olfaction is far more sensitive and complex.

Receptors for touch, pain, equilibrium, and hearing are all mechanical force receptors. Touch and pain receptors are characteristically simple structures, but hearing and equilibrium are highly specialized senses based on special hair cells that respond to mechanical deformation. Sound waves received by the ear are mechanically amplified and transmitted to the inner ear where different areas of the cochlea respond to different sound frequencies. Equilibrium receptors, also located in the inner ear in vertebrates, consist of two saclike static balance organs and three semicircular canals that detect rotational acceleration. Invertebrates monitor gravity and position using statocysts.

Vision receptors (photoreceptors) are associated with special pigment molecules that photochemically decompose in the presence of light and, in doing so, trigger nerve action potentials in optic fibers. The advanced compound eye of arthropods is especially well suited to detecting motion in the visual field. Cephalopods and vertebrates have a camera eye with focusing optics. Photoreceptor cells of the retina are of two kinds: rods, designed for high sensitivity with dim light, and cones, designed for color vision in daylight. Cones predominate in the fovea centralis of human eyes, the area of keenest vision. Rods are more abundant in peripheral areas of the retina.

REVIEW QUESTIONS

1. Define the following terms: neuron, axon, dendrite, myelin sheath, afferent neuron, efferent neuron, association neuron.
2. Glial cells far outnumber neurons and contribute roughly half the weight of the mammalian nervous system. Offer examples of functions glial cells perform in the peripheral nervous system and in the central nervous system.
3. The concentration of potassium ions inside a nerve cell membrane is higher than the concentration of sodium ions outside the membrane, yet the inside of the membrane (where the cation concentration is higher) is negative to the outside. Explain this observation in terms of permeability properties of the membrane.
4. What ionic and electrical changes occur during passage of an action potential along an axon?
5. Explain different ways in which invertebrates and vertebrates have achieved high velocities for conduction of action potentials. Can you suggest why the invertebrate solution would not be suitable for the homeothermic birds and mammals?
6. Why is the sodium pump *indirectly* important to the action potential and to maintaining the resting membrane potential?
7. Describe the microstructure of a chemical synapse. Summarize what happens when an action potential arrives at a synapse.
8. Describe the cnidarian (radiate) nervous system. How is a tendency toward centralization of the nervous system manifested in flatworms, annelids, molluscs, and arthropods?
9. How does a vertebrate spinal cord differ morphologically from nerve cords of invertebrates?
10. The knee-jerk reflex is often called a stretch reflex because a sharp tap on the patellar ligament stretches the quadriceps femoris, the extensor muscle of the leg. Describe the components and sequence of events that lead to a "knee jerk." Why is this reflex simpler than most reflex arcs? What is the difference between a reflex arc and a reflex act?
11. Name major functions associated with the following brain structures: medulla oblongata, cerebellum, tectum, thalamus, hypothalamus, cerebrum, limbic system.

12. What functional activities are associated with the left and the right hemispheres of the cerebral cortex?

13. What is the autonomic nervous system and what activities does it perform that distinguish it from the somatic nervous system? Why can the autonomic nervous system be described as a "two-neuron" system?

14. Give the meaning of the statement, "The idea that all sense organs behave as biological transducers is a unifying concept in sensory physiology."

15. Chemoreception in vertebrates and insects is mediated through clearly distinguishable senses of taste and smell. Contrast these two senses in humans in terms of anatomical location and nature of the receptors and sensitivity to chemical molecules. Describe chemoreception in insects.

16. What is the vomeronasal organ and what activity does it perform? Why is its functioning considered distinct from the sense of smell, but a component of the olfactory system of vertebrates?

17. Explain how ultrasonic detectors of certain nocturnal moths are adapted to help them escape an approaching bat.

18. Outline the place theory of pitch discrimination as an explanation of the human ear's ability to distinguish sounds of different frequencies.

19. Explain how the semicircular canals of the ear are designed to detect rotation of a human head in any directional plane.

20. Contrast the structure and functioning of the compound eye of arthropods with the camera-type eye of cephalopod molluscs and vertebrates.

21. Explain what happens when light strikes a dark-adapted rod that leads to the generation of a nerve impulse. What is the difference between rods and cones in their sensitivity to light?

22. In 1802 Thomas Young hypothesized that we see color because the retina contains three kinds of receptors. What evidence substantiates Young's hypothesis? How can we perceive any color in the visible spectrum when our retinas contain only three classes of color cones?

SELECTED REFERENCES

Axel, R. 1995. The molecular logic of smell. Sci. Am. **273:**154–159 (Oct.). *Describes a surprisingly large family of genes that encodes odor-detecting molecules. This and other findings help to illuminate how the nose and brain may perceive scents.*

Behmer, S. T. 2005. Learning in insects. Encyclopedia of Entomology. Dordrecht, The Netherlands, Kluwer Academic Publishers, pp. 1278–1283 *Provides an overview of insect learning, their abilities, and limitations.*

Changeux, J-P. 1993. Chemical signaling in the brain. Sci. Am. **269:**58–62 (Nov.). *Studies of the electric organs of fishes provide insights into how neurons in the human brain transmit information from one to the next.*

Delcomyn, F. 1998. Foundation of neurobiology. New York, W. H. Freeman & Company. *A good introduction to neuroscience that introduces diversity among animals and some interesting modern neurobiology research.*

Haddad, D., F. Schaupp, R. Brandt, G. Manz, R. Menzel, and A. Haase. 2004. NMR imaging of the honeybee brain. J. Insect Sci. **4:**7–13. *NMR microscopy reveals the 3-D structure of the honeybee brain in a noninvasive way. Excellent images illustrate the structures of this tiny brain that is capable of rapid neural processing, memory retention, and communication of those memories.*

Kammermeier, L., and H. Reichert. 2001. Evolution of the nervous system: common developmental genetic mechanisms for patterning invertebrate and vertebrate brains. Br. Res. Bull. **55:**675–682. *Discussion of the discovery of homology of genes regulating embryonic development of an insect* (Drosophila melanogaster) *and a mammalian (mouse) brain.*

McClintock, M. K. 2000. Human pheromones: primers, releasers, signalers, or modulates? In K. Wallen and J. E. Schneider, (eds.), Reproduction in context. Social and environmental influences on reproduction. Cambridge, Massachusetts, The MIT Press, pp. 355–420. *A discussion of pheromones with particular reference to equivocal information in humans.*

Nathan, P. 1997. The nervous system, ed. 4. London, Whurr Publications Ltd. *One of the best of several semipopular accounts of the nervous system.*

Nef, P. 1998. How we smell: the molecular and cellular bases of olfaction. News Physiol. Sci. **13:**1–5 (Feb.). *Describes three models for odor perception, each based on experimental data.*

Randall, D., W. Burggren, and K. French. 2001. Eckert animal physiology: mechanisms and adaptations. New York, W. H. Freeman & Company. *A comprehensive and comparative treatment of animal physiology, with particularly good information on the nervous systems and sense organs of animals.*

Smith, D. V., and R. F. Margolskee. 2001. Making sense of taste. Sci. Am. **284:**32–39 (March). *Describes the mechanism of taste reception in a very readable way.*

Ulfendahl, M., and A. Flock. 1998. Outer hair cells provide active tuning in the organ of Corti. News Physiol. Sci. **13:**107–111 (July). *Describes recent experiments that suggest a more active role for the sensory hair cells within the auditory system of mammals.*

ONLINE LEARNING CENTER

Visit www.mhhe.com/hickmanipz14e for chapter quizzing, key term flash cards, web links, and more!

34

Chemical Coordination: Endocrine System

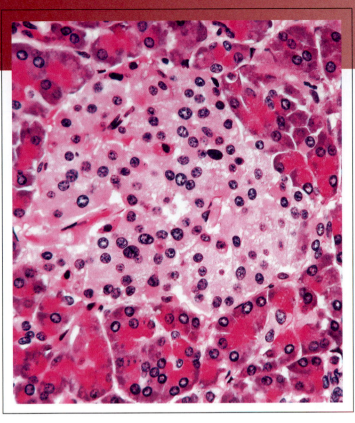

An endocrine islet of Langerhans, site of insulin and glucagon synthesis, within the human pancreas.

The Birth of Endocrinology

The birth date of endocrinology as a science is usually given as 1902, when two English physiologists, W. H. Bayliss and E. H. Starling (Figure 34.1), demonstrated the action of a hormone in a classic experiment that is still considered a model of the scientific method. Bayliss and Starling were interested in determining how the pancreas secreted its digestive juice into the small intestine at the proper time of the digestive process. They tested the hypothesis that acidic food entering the intestine triggered a nervous reflex that released pancreatic juice. To test this hypothesis, Bayliss and Starling cut away all nerves serving a tied-off loop of the small intestine of an anesthetized dog, leaving the isolated loop connected to the body only by its circulation. Injecting acid into the nerveless loop, they saw a pronounced flow of pancreatic juice. Thus, rather than a nervous reflex,

some chemical messenger had circulated through the blood from the intestine to the pancreas, causing the pancreas to secrete. Yet acid itself could not be the factor because it had no effect when injected directly into the circulation.

Bayliss and Starling then designed the crucial experiment that started the new science of endocrinology. Suspecting that the chemical messenger originated in the mucosal lining of the intestine, they prepared an extract of scrapings from the mucosa, injected it into the dog's circulation, and were rewarded with an abundant flow of pancreatic juice. They named the messenger present in the intestinal mucosa *secretin*. Later Starling coined the term **hormone** to describe all such chemical messengers, since he correctly surmised that secretin was only the first of many hormones awaiting discovery.

A   **B**

Figure 34.1

Founders of endocrinology. **A,** Sir William H. Bayliss (1860 to 1924). **B,** Ernest H. Starling (1866 to 1927).

The endocrine system, the second great integrative system controlling an animal's activities, communicates by chemical messengers called **hormones** (Gr. *hormōn,* to excite). The classical definition of hormones states that they are chemical compounds released into the blood in small amounts and transported by the circulatory system throughout the body to distant **target cells** where they initiate physiological responses.

Many hormones are secreted by **endocrine glands,** small, well-vascularized ductless glands composed of groups of cells arranged in cords or plates. Since endocrine glands have no ducts, their only connection with the rest of the body is by blood or other body fluid; they receive their raw materials from their extensive blood supply and secrete their finished hormonal products into it. **Exocrine glands,** in contrast, have ducts for discharging their secretions onto a free surface. Examples of exocrine glands are sweat glands and sebaceous glands of skin, salivary glands, and the various enzyme-secreting glands lining the walls of the stomach and intestine (see Chapter 32).

The classical definitions of hormones and endocrine glands just given, like so many other generalizations in biology, are changing as new information appears. Some hormones, such as certain neurosecretions, may never enter the general circulation. Furthermore, evidence suggests that many hormones, such as insulin and many hormones of digestion (see Chapter 32, p. 721), are synthesized in minute amounts in a variety of non-endocrine tissues (nerve cells, for example), and some, such as cytokines, are secreted by cells of the immune system (p. 777). Such hormones may function as **neurotransmitters** in the brain or as local tissue factors **(parahormones),** which stimulate cell growth or some biochemical process. Most hormones, however, are blood borne and therefore diffuse into every tissue space in the body.

Compared with nervous systems, endocrine systems are slow acting because of the time required for a hormone to reach the appropriate tissue, cross the capillary endothelium, and diffuse through tissue fluid to, and sometimes into, cells. The minimum response time is seconds and may be much longer. Hormonal responses in general are long lasting (minutes to days) whereas those under nervous control are short term (milliseconds to minutes). We expect to find endocrine control where a sustained effect is required, as in many metabolic, growth, and reproductive processes. Despite such differences, nervous and endocrine systems function without sharp separation as a single, interdependent system. Endocrine glands often receive directions from the brain. Conversely, many hormones act on the nervous system and significantly affect a wide array of animal behaviors.

Even though Bayliss and Starlig (see chapter opening essay) are considered the founders of endocrinology, the first formal experiment in endocrinology was actually performed in 1849 by Professor Arnold Adolph Berthold at the University of Gottingen. He conclusively demonstrated that a blood-borne signal was produced by the testes, and that this chemical was responsible for producing both physical and behavioral characteristics that distinguished an adult male rooster from immature chickens and adult, castrated male chickens (capons). Berthold castrated male chicks and divided them into three groups. He left one group of controls to grow normally without their testes, and he reimplanted the testes into the second group. The third group was implanted with testes from different chicks. As the chicks grew, he observed that the castrated group developed into capons, with no interest in hens, lacking rooster plumage and male aggressive behavior. The second and third groups of birds were indistinguishable from each other, with full male plumage, normal aggressive behavior and interest in hens. Berthold then killed the birds and dissected them. He found that the transplanted testes had developed their own blood supply and were functioning normally. From this classic experiment, Berthold concluded that because there was no nerve supply to the testes, the testes must secrete a blood-borne signal that produced all characteristics of maleness.

All hormones are low-level signals. Even when an endocrine gland is secreting maximally, its hormone is so greatly diluted by the large volume of blood it enters that its plasma concentration seldom exceeds 10^{-9} M (or one billionth of a 1M concentration). Some target cells respond to plasma concentrations of hormone as low as 10^{-12} M. Since hormones have far-reaching and often powerful influences on cells, it is evident that their effects are vastly amplified at the cellular level.

MECHANISMS OF HORMONE ACTION

Widespread distribution of hormones in an animal permits certain hormones, such as growth hormone from the vertebrate pituitary gland, to affect most, if not all, cells during specific stages of cellular differentiation. Whether hormones produce widespread or highly specific responses only in certain cells and at certain times depends on the presence of **receptor molecules** on or in target cells. A hormone will engage only those cells that display the receptor that, by virtue of its specific molecular shape, binds with the hormone molecule. Other cells are insensitive to the hormone's presence because they lack specific receptors. Hormones act through two kinds of receptors: **membrane-bound receptors** and **nuclear receptors.**

Membrane-Bound Receptors and the Second-Messenger Concept

Many hormones, such as most amino acid derivatives, and peptide hormones that are too large, or too polar, to pass through plasma membranes, bind to transmembrane proteins (see Figure 3.6, p. 42) that act as receptor sites present on the surfaces of target-cell membranes. The hormone and receptor form a complex that triggers a cascade of molecular events within a cell. The hormone thus behaves as a **first messenger** that causes activation of a **second-messenger** system in the cytoplasm. At least six different molecules have been identified as second messengers. Each works via a specific **kinase,** which causes activation or inactivation of rate-limiting enzymes that modify the direction and rate of cytoplasmic processes (Figure 34.2). Since many molecules are activated at each level in the second-messenger system cascade after a single hormone molecule has been bound, the message is amplified, perhaps many thousands of times.

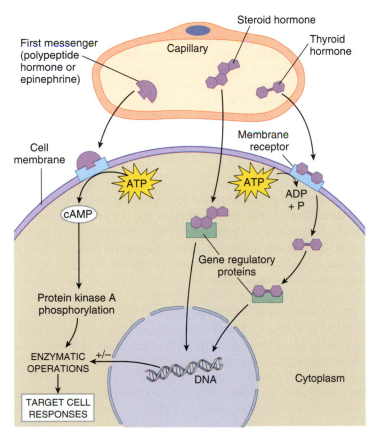

Figure 34.2

Mechanisms of hormone action. Peptide hormones and epinephrine act through second-messenger systems, as for example, cyclic AMP, shown here. The combination of hormone with a membrane receptor stimulates the enzyme adenylate cyclase to catalyze formation of cyclic AMP (second messenger). Thyroid hormones bind to a membrane receptor and are transported into the cell by active transport. There they bind to cytoplasmic receptors that are transported into the nucleus to alter gene transcription. Steroid hormones penetrate the cell membrane to combine with cytoplasmic or nuclear receptors that alter gene transcription.

Second-messenger systems known to participate in hormone actions are **cyclic AMP** (cAMP), **cyclic GMP** (cGMP), **Ca^{++}/calmodulin, inositoltrisphosphate** (IP_3), and **diacylglycerol** (DAG). Cyclic AMP was the first to be investigated, and has been shown to mediate actions of many peptide hormones, including parathyroid hormone, glucagon, adrenocorticotropic hormone (ACTH), thyrotropic hormone (TSH), melanophore-stimulating hormone (MSH), and vasopressin. It also mediates action of epinephrine (also called adrenaline), an amino acid derivative. Interestingly, the same hormone may activate different second-messenger systems in each type of target cell, such that a single hormone produces multiple actions within an animal.

Other membrane-bound receptors possess kinase activity themselves and are activated when the hormone binds to the receptor, for example, the insulin and insulin-like growth factor membrane receptors.

Nuclear Receptors

Unlike peptide hormones and epinephrine, which are much too large to pass through plasma membranes, **steroid hormones** (for example, estrogen, testosterone, and aldosterone), are lipid-soluble molecules that readily diffuse through plasma membranes. Once inside the cytoplasm of target cells, steroid hormones bind selectively to receptor molecules. While these receptor molecules may be located in either cytoplasm or nucleus, their ultimate site of activity is nuclear. The hormone-receptor complex, now known as a **gene regulatory protein,** then activates or inhibits specific genes. As a result, gene transcription is altered, since messenger RNA molecules are synthesized on specific sequences of DNA. Stimulation or inhibition of mRNA formation modifies production of key enzymes, thus setting in motion a hormone's observed effect (Figure 34.2). Thyroid hormones and the insect-molting hormone, ecdysone (a steroid, p. 26), also act through nuclear receptors. Thyroid hormones first bind to a transmembrane protein transport molecule (p. 49) that uses ATP to move the hormones into the cell.

Compared with peptide hormones that act *indirectly* through second-messenger systems, steroids and thyroid hormones have a *direct* effect on protein synthesis because they bind a nuclear receptor that modifies specific gene activity.

Recent evidence suggests that lipid-soluble hormones, such as estrogen, may also possess membrane-bound receptors that activate second-messenger systems in the same way as peptide hormones, providing multiple and complex control of target cells.

Control of Secretion Rates of Hormones

Hormones influence cellular functions by altering rates of many different biochemical processes. Many affect enzymatic activity and thus alter cellular metabolism, some change membrane permeability, some regulate synthesis of cellular proteins, and some stimulate release of hormones from other endocrine glands. Because these are all dynamic processes that must adapt to changing metabolic demands, they must be regulated, not merely activated, by appropriate hormones. This regulation is achieved by precisely controlled release of a hormone into the blood. Concentration of a hormone in body fluid depends on two factors: its rate of secretion and the rate at which it is inactivated and removed from the circulation. Consequently, if secretion is to be correctly controlled, an endocrine gland requires information about the level of its own hormone(s) in the plasma.

Most hormones are controlled by negative feedback systems that operate between glands secreting hormones and outputs or products of target cells (Figure 34.3). A feedback pattern is one in which output is constantly compared with a set point, like a thermostat. For example, CRH (corticotropin-releasing hormone), secreted by the hypothalamus, stimulates the pituitary (containing target cells) to release ACTH. ACTH stimulates the adrenal gland (containing target cells) to secrete cortisol. As the level of ACTH rises in the plasma, it acts on, or "feeds back" on, the hypothalamus to inhibit release of CRH. Similarly, as cortisol levels rise in the plasma, it "feeds back" on the hypothalamus and pituitary to inhibit release of both CRH and ACTH, respectively. Thus any deviation from the set point (a specific plasma level of each hormone) leads to corrective action in the opposite direction (Figure 34.3). Such a **negative feedback** system is highly effective in preventing extreme oscillations in hormonal output. However, hormonal feedback systems are more complex than a rigid "closed-loop" system such as a thermostat that controls a house's central heating system, because hormonal feedback may be altered by input from the nervous system or by metabolites or other hormones.

Extreme oscillations in hormone output do sometimes occur under natural conditions. However, because they have the potential to disrupt finely tuned homeostatic mechanisms, such extreme oscillations, as a result of **positive feedback,** are highly

regulated and possess an obvious shutoff mechanism. During positive feedback, the signal (or output of the system) feeds back to the control system and causes an increase in the initial signal. In this way the initial signal becomes progressively amplified to produce an explosive event. For example, hormones controlling parturition (childbirth) become elevated from their normal setpoint and are shut off by birth of offspring from the uterus (see Chapter 7).

INVERTEBRATE HORMONES

All invertebrate taxa produce hormones, and while cnidarians, nematodes, and annelids possess endocrine cells, endocrine glands appear in molluses and arthropods, with the latter being more complex. Invertebrate hormones are peptides (often neuropeptides), steroids, or terpenoids (lipid-soluble organic molecules), but peptides and neuropeptides are the most common among different invertebrate groups. Some hormones are vertebrate-like in structure and function (for example, steroids), but there is much greater diversity in invertebrate endocrine function than in vertebrates. Invertebrate hormones regulate color changes, growth, reproduction, and internal homeostatic mechanisms, such as metabolism, metabolic fuel levels, and osmoregulation.

In many metazoan phyla, the principal source of hormones is **neurosecretory cells,** specialized nerve cells capable of synthesizing and secreting hormones. Their products, called neurosecretory hormones, are discharged directly into the circulation, and serve as a crucial link between the nervous and endocrine systems.

Peptides and neuropeptides have been shown to regulate many physiological processes in invertebrates. In Crustacea, **cardioactive peptide** increases heart rate. Hormones that regulate metabolism of carbohydrates, fats, and amino acids belong to the **crustacean hyperglycaemic hormone** family (CHH) in Crustacea and to the **adipokinetic hormone** family in insects. **Diuretic hormones** stimulate the secretion of fluid in insect Malpighian tubules (p. 672). A family of small neuropeptides called **FMRFamide-related peptides** (FaRPs) appears to have evolved as early as bilateral symmetry, and their functions seem to have been conserved across phyla. Peptides of this family are known to regulate muscular tissues of the body, and digestive and reproductive processes in many invertebrates, as well as osmoregulatory processes in nematodes, annelids, molluscs, and insects, and arterial hemolymph flow in crustaceans. They are now being isolated and characterized in vertebrates. An extensively studied neurosecretory process in invertebrates is control of development and metamorphosis of insects. In insects, as in other arthropods, growth is a series of steps in which the rigid, nonexpansible exoskeleton is periodically discarded and replaced with a new, larger one. Most insects undergo a process of metamorphosis (p. 454), in which a series of juvenile stages, each requiring formation of a new exoskeleton, end with a molt.

Insect physiologists discovered that molting and metamorphosis are primarily controlled by interaction of two hormones,

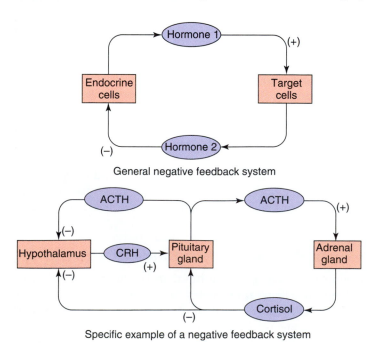

General negative feedback system

Specific example of a negative feedback system

Figure 34.3

Negative feedback systems.

one favoring growth and differentiation of adult structures and another favoring retention of juvenile structures. These two hormones are **molting hormone** or **ecdysone,** a steroid produced by the prothoracic gland, and **juvenile hormone,** a terpenoid produced by the corpora allata (Figure 34.4).

Ecdysone is controlled by **prothoracicotropic hormone** or **PTTH.** This hormone is a polypeptide (molecular weight about 5000) produced by neurosecretory cells of the brain, and transported by axons to the corpora cardiacum where it is stored. At intervals during juvenile growth, release of PTTH into the blood stimulates the prothoracic gland to secrete ecdysone. Ecdysone bound to its nuclear receptor acts directly on the chromosomes as a gene regulatory protein (Figure 34.2) to cause a molt, and subsequent development of adult structures. It is held in check, however, by juvenile hormone, which favors retention of juvenile characteristics. During juvenile life, juvenile hormone predominates and each molt yields another larger juvenile (Figure 34.4). Finally output of juvenile hormone decreases, allowing final metamorphosis to the adult stage.

Juvenile hormone appears to be important, at least in some insects, during **diapause** (or arrested development), which can occur at any stage of metamorphosis. Diapause usually occurs due to seasonal changes in environmental conditions, such as cold temperatures or changes in day length. In some insects high levels of juvenile hormone inhibit release of PTTH, and thus ecdysone levels remain low and development to the next stage is arrested. In other insects diapause is due to a decrease in brain neurosecretory activity and a direct reduction of PTTH, or to a direct effect of temperature on the prothoracic glands, causing decreased ecdysone secretion. Juvenile hormone is also present in adult insects where it is involved in regulation of egg development in females. In addition, low levels cause decreased reproductive function during adult diapause (or dormancy), which occurs during winter months in some insects.

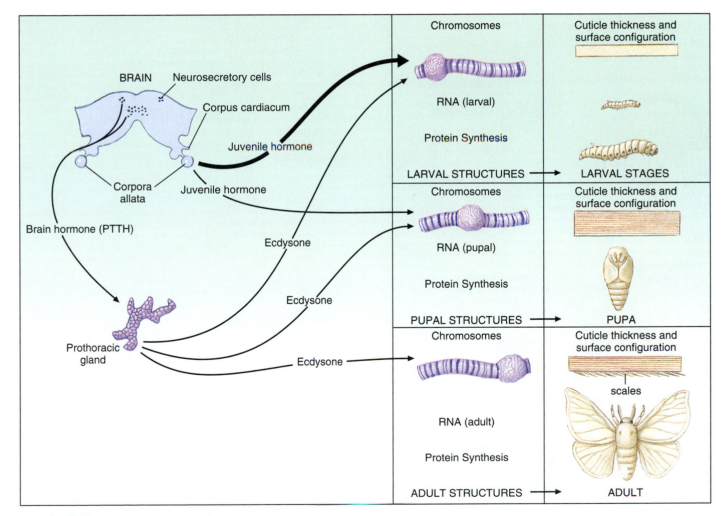

Figure 34.4

Endocrine control of molting in a moth, typical of insects having complete metamorphosis. Many moths mate in spring or summer, and eggs soon hatch into the first of several larval stages, called instars. After the final larval molt, the last and largest larva (caterpillar) spins a cocoon in which it pupates. The pupa overwinters, and an adult emerges in the spring to start a new generation. Juvenile hormone and ecdysone interact to control molting and pupation. Many genes are activated during metamorphosis, as seen by puffing of chromosomes (center column). Puffs form in sequence during successive molts. Changes in cuticle thickness and surface characteristics are shown at right.

Chemists have synthesized several potent analogs of juvenile hormone, which hold great promise as insecticides. Minute quantities of these synthetic analogs induce abnormal final molts or prolong or block development. Based on the extensive study of all aspects of endocrine regulation in insects, scientists have synthesized many different endocrine disruptors designed to regulate insect populations. Unlike chemical insecticides, they are highly specific, but given the conservation of hormone functions across invertebrate groups, they are less ecologically benign than originally thought.

VERTEBRATE ENDOCRINE GLANDS AND HORMONES

In the remainder of this chapter we describe some of the best understood and most important of vertebrate hormones. While this discussion is limited principally to a brief overview of mammalian hormonal mechanisms (since laboratory mammals and humans have always been the objects of the most intensive research), we note some important differences in functional roles of hormones among different vertebrate groups.

Hormones of the Hypothalamus and Pituitary Gland

The pituitary gland, or **hypophysis,** is a small gland (0.5 g in humans) lying in a well-protected position between the roof of the mouth and floor of the brain (Figure 34.5). It is a two-part gland having a double embryological origin. The **anterior pituitary** (adenohypophysis) is derived embryologically from the roof of the mouth. The **posterior pituitary** (neurohypophysis) arises from a ventral portion of the brain, the **hypothalamus,** and is connected to it by a stalk, the **infundibulum.** Although the anterior pituitary lacks any anatomical connection to the brain, it is functionally connected to it by a special portal circulatory system. A portal circulation is one that delivers blood from one capillary bed to another (Figures 34.5 and 34.6). In this case,

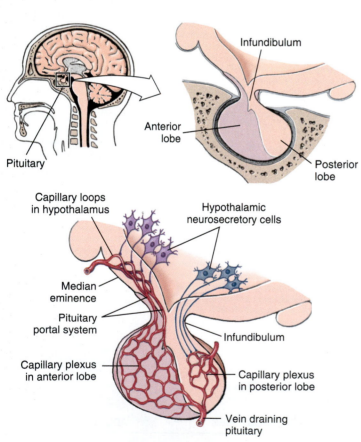

Figure 34.5

Human hypothalamus and pituitary gland. The posterior lobe is connected directly to the hypothalamus by axons of neurosecretory cells. The anterior lobe is indirectly connected to the hypothalamus by a portal circulation (shown in red) beginning in the base of the hypothalamus and ending in the anterior pituitary.

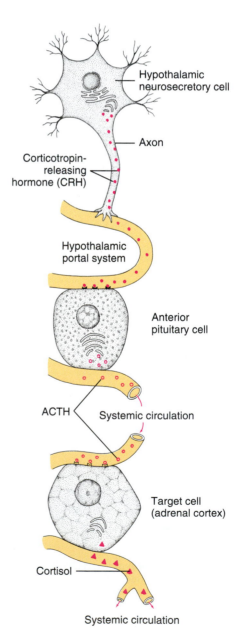

Figure 34.6

Relationship of hypothalamic, pituitary, and target-gland hormones. The hormone sequence controlling the release of cortisol from the adrenal cortex is used as an example.

the portal circulation provides a link between neurosecretory cells of the hypothalamus and endocrine cells of the anterior pituitary gland.

Hypothalamus and Neurosecretion

Because of the strategic importance of the pituitary in influencing most hormonal activities in the body, the pituitary was once called the "master gland." This description is not appropriate, however, because anterior pituitary hormones are regulated by a higher council, the neurosecretory centers of the hypothalamus. The hypothalamus is itself under ultimate control by inputs from other regions of the brain. The hypothalamus contains groups of neurosecretory cells, which are specialized nerve cells (Figures 34.5 and 34.6) that manufacture neurohormones called **releasing hormones** or **release-inhibiting hormones** (or "factors"). These neurohormones travel down nerve axons to their endings in the median eminence. Here they enter a capillary network to complete their journey to the anterior pituitary by way of the pituitary portal system. The hypothalamic hormones then stimulate or inhibit release of various anterior pituitary hormones. Several hypothalamic releasing and release-inhibiting hormones have been discovered, characterized chemically, and isolated in pure state (Table 34.1), although the identification and action of some of the hypothalamic hormones listed in Table 34.1 is still tentative.

Anterior Pituitary

The anterior pituitary consists of an **anterior lobe** (pars distalis) as shown in Figure 34.5, and an **intermediate lobe** (pars intermedia), which is absent in some animals (including humans). The anterior pituitary produces seven hormones, and in those animals with an intermediate lobe, all but one are released by the anterior lobe.

Four hormones of the anterior pituitary are **tropic hormones** (from the Greek *tropē,* to turn toward) that regulate other endocrine glands (Table 34.1). **Thyroid-stimulating hormone (TSH)** or **thyrotropin** stimulates production of thyroid hormones by the thyroid gland. Two tropic hormones, commonly called **gonadotropins,** act on the gonads (ovaries of females, testes of males). These are **follicle-stimulating hormone (FSH)** and **luteinizing hormone (LH).** FSH promotes egg production and secretion of estrogen in females, and supports sperm production in males. LH induces ovulation, corpus luteum production, and secretion of female sex steroids, progesterone and estrogen. In males, LH promotes production of male sex steroids (primarily testosterone). It once was called interstitial cell stimulating hormone (ICSH) in males, before it was discovered to be identical to LH in females. The hormonal control of reproduction is extensively discussed in Chapter 7. The fourth tropic hormone, **adrenocorticotropic hormone (ACTH),** increases production and secretion of steroid hormones from the adrenal cortex.

Prolactin and the structurally related **growth hormone (GH)** are proteins. Prolactin is essential for preparing mammary glands for lactation; after birth it is required for production of milk. Prolactin also is implicated in parental behavior in a wide variety of vertebrates. Beyond its more traditional role in reproductive processes, prolactin regulates water and electrolyte balance in many species. More recently, prolactin has been shown to be a chemical mediator of the immune system and is important in formation of new blood vessels (angiogenesis). Unlike tropic hormones, prolactin acts directly on its target tissues rather than through other hormones.

GH (also called somatotropin) performs a vital role in governing body growth through its stimulatory effect on cellular mitosis, on synthesis of messenger RNA and protein, and on metabolism, especially in new tissue of young vertebrates. Growth hormone acts directly on growth and metabolism, as well as indirectly through a polypeptide hormone, **insulin-like growth factor (IGF)** or somatomedin, produced by the liver.

The only anterior pituitary hormone produced by the intermediate lobe is **melanocyte-stimulating hormone (MSH).** In cartilaginous and bony fishes, amphibians, and nonavian reptiles, MSH is a direct-acting hormone that promotes dispersion of the pigment melanin within melanocytes, causing darkening of the skin. In birds and mammals, MSH is produced by cells in the anterior pituitary rather than the intermediate lobe, but its physiological function remains unclear. MSH appears unrelated to pigmentation in endotherms, although it will cause darkening of the skin in humans if injected into the circulation. Until recently, many endocrinologists thought MSH was a vestigial hormone in mammals, but interest has been rekindled by studies showing that it enhances memory and growth of the fetus. In addition, MSH has been isolated from specific regions of the hypothalamus, where it has been linked to regulation of ingestive behaviors and metabolism in adult mammals. Future studies will determine if a similar role exists for MSH during development. MSH and ACTH are derived from a precursor molecule (pro-opiomelanocortin or POMC) that is transcribed and translated from a single gene.

Posterior Pituitary

The hypothalamus is the source of two hormones of the posterior lobe of the pituitary (Table 34.1). They are formed in neurosecretory cells in the hypothalamus, whose axons extend down the infundibular stalk and into the posterior lobe. The hormones are secreted from axon terminals ending in close proximity to blood capillaries, which the hormones enter when released (see Figure 34.5). In a sense the posterior lobe is not a true endocrine gland, but a storage and release center for hormones manufactured entirely in the hypothalamus. The two posterior-lobe hormones of mammals, oxytocin and vasopressin, are chemically very much alike. Both are polypeptides consisting of eight amino acids (octapeptides, Figure 34.7). These hormones are among the fastest-acting hormones, since they are capable of producing a response within seconds of their release from the posterior lobe.

Oxytocin has two important specialized reproductive functions in adult female mammals. It stimulates contraction of

TABLE 34.1
Hormones of the Vertebrate Pituitary

	Hormone	Chemical Nature	Principal Action	Hypothalamic Controls
Adenohypophysis				
(Anterior lobe)	Thyroid-stimulating hormone (TSH)	Glycoprotein	Stimulates thyroid hormone synthesis and secretion	TSH-releasing hormone (TRH)
	Follicle-stimulating hormone (FSH)	Glycoprotein	Female: follicle maturation and estrogen synthesis Male: stimulates sperm production	Gonadotropin-releasing hormone (GnRH)[1]
	Luteinizing hormone (LH)	Glycoprotein	Female: stimulates ovulation, corpus luteum formation, estrogen and progesterone synthesis Male: testosterone secretion	Gonadotropin-releasing hormone (GnRH)[1] Gonadotropin release-inhibiting hormone (GnIH)[2]
	Prolactin (PRL)	Protein	Mammary gland growth, milk synthesis, immune response and angiogenesis in mammals, parental behavior, electrolyte and water balance in lower vertebrates	Dopamine (prolactin release-inhibiting hormone or PIH) Prolactin-releasing factor (PRF)?
	Growth hormone (GH) (somatotropin)	Protein	Stimulates soft tissue and bone growth, protein synthesis, mobilization of glycogen and fat stores	Growth hormone-releasing hormone (GHRH) Growth hormone release-inhibiting hormone (GHIH) or somatostatin
	Adrenocorticotropic hormone (ACTH)	Polypeptide	Stimulates glucocorticoid synthesis by adrenal cortex	Corticotropin-releasing hormone (CRH)
Intermediate lobe[3]	Melanocyte-stimulating hormone (MSH)	Polypeptide	Increased melanin synthesis by melanocytes in epidermis of ectotherms; function not clear in endotherms	Melanocyte-stimulating hormone-inhibiting hormone (MSHIH)
Neurohypophysis				
(Posterior lobe)	Oxytocin	Octapeptide	Milk ejection and uterine contractions, sexual behavior and pair bonding in monogamous species	
	Vasopressin[4] (antidiuretic hormone or ADH)	Octapeptide	Water reabsorption in mammalian kidneys	
	Vasotocin[5]	Octapeptide	Increases water reabsorption	

[1] One GnRH hormone regulates both FSH and LH, but some recent studies suggest a separate FSH-releasing hormone (FSH-RH).
[2] Recently discovered GnIH in birds and mammals.
[3] Birds and some mammals lack an intermediate lobe. In these forms, MSH is produced by the anterior lobe.
[4] In mammals.
[5] In all vertebrate classes except mammals, although additional, related hormones have been identified.

uterine smooth muscles during parturition (birth of young). In clinical practice, oxytocin is used to induce delivery during prolonged labor and to prevent uterine hemorrhage after birth. A second action of oxytocin is that of milk ejection by the mammary glands in response to suckling. Recent work also has established a role for oxytocin in pair-bonding behavior in both sexes in some monogamous vole species.

Vasopressin, the second posterior-lobe hormone, acts on collecting ducts of the kidney to increase water reabsorption and thus restrict urine flow, as already described on page 679. It is therefore often called **antidiuretic hormone** or **ADH.** Vasopressin also increases blood pressure through its generalized constrictor effect on smooth muscles of arterioles. Finally, vasopressin acts centrally to increase thirst, and therefore, drinking behavior.

All jawed vertebrates secrete two posterior-lobe hormones that are quite similar to those of mammals. All are octapeptides, but their structures vary because of amino-acid substitutions in three of eight amino-acid positions in the molecule.

Figure 34.7

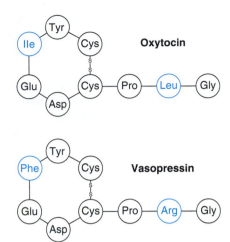

Posterior pituitary hormones of mammals. Both oxytocin and vasopressin consist of eight amino acids (the two sulfur-linked cysteine molecules are considered a single amino acid, cystine). Oxytocin and vasopressin are identical except for amino-acid substitutions in the blue positions. Abbreviations represent amino acids.

Of all posterior-lobe hormones, **vasotocin** (Table 34.1) has the widest phylogenetic distribution and is considered the parent hormone from which other octapeptides evolved. It occurs in all vertebrate classes except mammals. It is a water-balance hormone in amphibians, especially toads, in which it acts to conserve water by (1) increasing permeability of skin (to promote water absorption from the environment), (2) stimulating water reabsorption from the urinary bladder, and (3) decreasing urine flow. Action of vasotocin is best understood in amphibians, but it appears to play some water-conserving role in birds and nonavian reptiles as well.

Pineal Gland

In all vertebrates the dorsal part of the brain, the diencephalon (Figure 33.13, p. 737) gives rise to a saclike evagination called the pineal complex, which lies just beneath the skull in a midline position. In ectothermic vertebrates the pineal complex contains glandular tissue and a photoreceptive sensory organ involved in pigmentation responses and in light-dark biological rhythms. In lampreys, many amphibians, lizards, and tuataras (*Sphenodon*, p. 580), the median photoreceptive organ is so well developed, containing structures analogous to the lens and cornea of lateral eyes, that it is often called a third eye. In birds and mammals, the pineal complex is an entirely glandular structure called the **pineal gland.** The pineal gland produces the hormone **melatonin.** Melatonin secretion is strongly affected by exposure to light. Its production is lowest during daylight hours and highest at night. In nonmammalian vertebrates, the pineal gland is responsible for maintaining **circadian rhythms**—self-generated (endogenous) rhythms that are about 24 hours in length. A circadian rhythm serves as a biological clock for many physiological processes that follow a regular pattern.

In mammals, an area of the hypothalamus called the **suprachiasmatic nucleus** has become the primary circadian pacemaker, although the pineal gland still produces melatonin nightly and serves to reinforce the circadian rhythm of the suprachiasmatic nucleus. In birds and mammals in which seasonal rhythms in reproduction are regulated by **photoperiod,** melatonin plays a critical role in timing gonadal activity. In long-day breeders, such as horses, ferrets, hamsters, and deer mice, reduced light stimulation with shortening day length in the autumn increases melatonin secretion, and in these species reproductive activity is suppressed during winter months. Lengthening days in the spring have the opposite effect, and reproductive activities are resumed. Short-day breeders, such as white-tailed deer, silver fox, spotted skunk, and sheep, are stimulated by reduced day length in the autumn; increasing melatonin levels at this time are associated with increased reproductive activity. The role of melatonin is an indirect one in both cases since melatonin itself does not stimulate or inhibit the reproductive axis.

The pineal gland produces subtle and incompletely understood effects on circadian and annual rhythms in nonphotoperiodic mammals (such as humans). For example, melatonin secretion has been linked to a sleeping and eating disorder in humans known as seasonal affective disorder (SAD). Some people living in northern latitudes, where day lengths are short in winter and when melatonin production is elevated, become depressed in winter, sleep long periods, and may have eating binges. Often this wintertime disorder can be treated by exposure to sunlamps with full-spectrum light; such exposure depresses melatonin secretion by the pineal gland. Disturbed physiological rhythms associated with jet-lag, shift-work, and aging also are linked to inappropriate melatonin rhythms.

Brain Neuropeptides

The blurred distinction between endocrine and nervous systems is nowhere more evident than in the growing list of hormone-like neuropeptides that have been discovered in the central and peripheral nervous systems of vertebrates and invertebrates. In mammals, approximately 40 neuropeptides (short chains of amino acids) have been located using immunological labeling with antibodies and visualized in histological sections under a microscope, and the list is still growing. Many are known to lead double lives—to be capable of behaving both as hormones, carrying signals from gland cells to their targets, and as neurotransmitters, relaying signals between nerve cells. For example, both oxytocin and vasopressin have been discovered at widespread sites in the brain by immunological methods. Related to this discovery is the fascinating observation that people and experimental animals injected with minute quantities of vasopressin experience enhanced learning and improved memory. This effect of vasopressin in brain tissue is unrelated to its well-known antidiuretic function in the kidney (p. 679). Several hormones, such as gastrin and cholecystokinin (p. 721) (formerly considered to function only in the gastrointestinal tract), have been discovered in the cerebral cortex, hippocampus, and hypothalamus. In addition to its gastrointestinal actions, cholecystokinin is known to function in control of feeding and satiety and may serve other functions as a brain neuroregulator.

Among the dramatic developments in this field was the discovery in 1975 of endorphins and enkephalins, neuropeptides that bind with opiate receptors and influence perception of pleasure and pain (see note on p. 743). Endorphins and enkephalins are found also in brain circuits that modulate several other functions unrelated to pleasure and pain, such as control of blood pressure, body temperature, body movement, feeding, and reproduction. Even more intriguing, endorphins are derived from the same prohormone (POMC) that gives rise to the anterior pituitary hormones ACTH and MSH.

The radioimmunoassay technique developed by Solomon Berson and Rosalyn Yalow about 1960 revolutionized endocrinology and neurochemistry. First, antibodies to the hormone of interest (insulin, for example) are prepared by injecting a mammal, such as guinea pigs or rabbits, with the hormone. Then, a fixed amount of radioactively labeled insulin and unlabeled insulin antibodies is mixed with the sample of blood plasma to be measured. The native insulin in the blood plasma and the radioactive insulin compete for insulin antibodies. The more insulin present in the sample, the less radioactive insulin will bind to the antibodies. Bound and unbound insulin are then separated, and their radioactivities are measured together with those of appropriate standard solutions containing known amounts of insulin to determine the amount of insulin present in the blood sample. The method is so incredibly sensitive that it can measure the equivalent of a cube of sugar dissolved in one of the Great Lakes.

Prostaglandins and Cytokines

Prostaglandins

Prostaglandins are derivatives of long-chain unsaturated fatty acids that were discovered in seminal fluid in the 1930s. At first they were thought to be produced only by the prostate gland (hence the name) but now have been found in virtually all mammalian tissues. Prostaglandins often act as local hormones and have diverse actions on many different tissues, making generalizations about their effects difficult. Many of their effects, however, involve smooth muscle. In some tissues prostaglandins regulate vasodilation or vasoconstriction by their action on smooth muscle in walls of blood vessels. They are known to stimulate contraction of uterine smooth muscle during childbirth. There also is evidence that overproduction of uterine prostaglandins is responsible for painful symptoms of menstruation (dysmenorrhea) experienced by many women. Several inhibitors of prostaglandins that provide relief from these symptoms have now been approved as medicines. Among other actions of prostaglandins is their intensification of pain in damaged tissues, mediation of the inflammatory response, and involvement in fever.

Cytokines

For some years we have known that cells of the immune system communicate with each other and that this communication was crucial to the immune response. Now we understand that a large group of polypeptide hormones called **cytokines** (p. 777)

mediate communication between cells participating in an immune response. Cytokines can affect the cells that secrete them, affect nearby cells, and like other hormones, they can affect cells in distant locations. Their target cells bear specific receptors for the cytokine bound to the surface membrane. Cytokines coordinate a complex network, with some target cells being activated, stimulated to divide and often to secrete their own cytokines. The same cytokine that activates some cells may suppress division of other target cells. Cytokines also are involved in formation of blood, and more recently, their role in regulation of energy balance by the central nervous system is being explored.

Hormones of Metabolism

An important group of hormones adjusts the delicate balance of metabolic activities. Rates of chemical reactions within cells are often regulated by long sequences of enzymes (see Chapter 4, p. 60). Although such sequences are complex, each step in a pathway is mostly self-regulating as long as the equilibrium between substrate, enzyme, and product remains stable. However, hormones may alter activity of crucial enzymes in a metabolic process, thus accelerating or inhibiting the entire process. The most important hormones of metabolism are those of the thyroid, parathyroid, adrenal glands, and pancreas, as well as the previously mentioned growth hormone of the anterior pituitary. Finally, white adipose tissue must now be included as an endocrine organ since it secretes many peptides, called adipokines, some of which regulate metabolism.

Thyroid Hormones

The thyroid gland is a large endocrine gland located in the neck of all vertebrates. It is composed of thousands of tiny spherelike units, called follicles, where two thyroid hormones, **triiodothyronine** and **thyroxine** (T_3 and T_4, respectively) are synthesized, stored, and released into the bloodstream as needed. The size of the follicles, and amount of stored T_3 and T_4 they contain, depends on the activity of the gland (Figure 34.8). A third hormone, **calcitonin,** is also secreted by the thyroid gland in mammals; this hormone is discussed in the section on Calcium Metabolism (p. 764).

A unique characteristic of the thyroid is its ability to concentrate high levels of **iodine;** in most animals this single gland contains well over half the body store of iodine. Epithelial cells of thyroid follicles actively trap iodine from the blood and combine it with the amino acid tyrosine, creating the two thyroid hormones. T_3 contains three iodine atoms, whereas T_4 contains four iodine atoms. T_4 is formed in much greater amounts than T_3, but in many animals T_3 is the more physiologically active hormone. T_4 is now considered a precursor to T_3. The most important actions of T_3 and T_4 are to (1) promote normal growth and development of the nervous system of growing animals, and (2) stimulate metabolic rate.

Undersecretion of thyroid hormones in fishes, birds, and mammals dramatically impairs growth, especially of the nervous

Figure 34.8

Appearance of thyroid gland follicles viewed through the microscope (approximately ×350). When inactive, follicles are distended with colloid, the storage form of thyroid hormones, and epithelial cells are flattened. When active, the colloid disappears as thyroid hormones are secreted into the circulation, and epithelial cells become greatly enlarged.

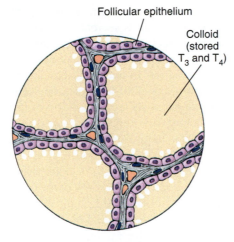

Follicular epithelium

Colloid (stored T_3 and T_4)

Inactive follicles

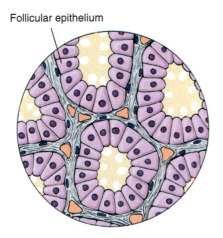

Follicular epithelium

Active follicles

system. A human **cretin,** a mentally retarded dwarf, is a result of thyroid malfunction from a very early age. Conversely, oversecretion of thyroid hormones causes precocious development in all vertebrates, although its effect is particularly prominent in fish and amphibians. In frogs and toads, transformation from aquatic herbivorous tadpole without lungs or legs to semiterrestrial or terrestrial carnivorous adult with lungs and four legs, occurs when the thyroid gland becomes active at the end of larval development. Stimulated by rising thyroid hormone levels of the blood, metamorphosis and climax occur (Figure 34.9). Growth of frogs after metamorphosis is directed by growth hormone.

In birds and mammals, the best known action of thyroid hormones is control of oxygen consumption and heat production. The thyroid maintains metabolic activity of homeotherms (birds and mammals) at a normal level. Oversecretion of thyroid hormones accelerates body processes by as much as 50%, causing irritability, nervousness, fast heart rate, intolerance of warm environments, and loss of body weight despite increased appetite. Undersecretion of thyroid hormones slows metabolic activities, leading to loss of mental alertness, slowing of heart rate, muscular weakness, increased sensitivity to cold, and weight gain. An important function of the thyroid gland is to promote adaptation to cold environments by increasing heat production. Thyroid hormones stimulate cells to produce more heat and store less chemical energy (ATP); in other words, thyroid hormones *reduce* efficiency of cellular oxidative phosphorylation (p. 67). Consequently many cold-adapted mammals have heartier appetites and eat more food in winter than in summer although their activity is about the same in both seasons. In winter, a larger portion of the food is being converted directly into body-warming heat.

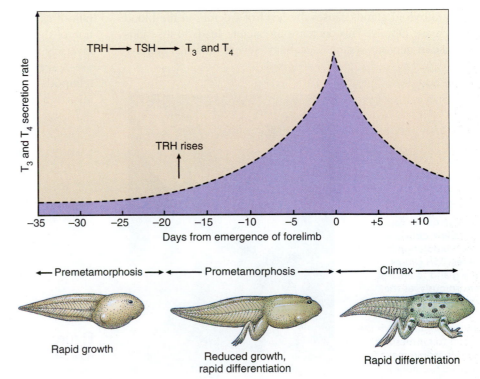

Figure 34.9

Effect of thyroid hormones (T_3 and T_4) on growth and metamorphosis of a frog. The release of TRH from the hypothalamus at the end of premetamorphosis sets in motion hormonal changes (increased TSH, T_3 and T_4) leading to metamorphosis. Thyroid hormone levels are maximal at the time forelimbs emerge.

Synthesis and release of thyroid hormones are governed by thyrotropic hormone (TSH) from the anterior pituitary gland (Table 34.1). TSH is regulated in turn by thyrotropin-releasing hormone (TRH) of the hypothalamus. As noted on page 760, TRH is part of a higher regulatory council that controls the tropic hormones of the anterior pituitary. TRH and TSH control thyroid activity in an excellent example of negative feedback (p. 756). It can be overridden, however, by neural stimuli, such as exposure to cold, which directly stimulates increased release of TRH and thus TSH.

Some years ago, a condition called goiter was common among people living in the Great Lakes region of the United States and Canada, as well as some other parts of the world, such as the Swiss Alps. This type of goiter is an enlargement of the thyroid gland caused by deficiency of iodine in the diet. Thus, TSH levels rise due to a decrease in thyroid hormone negative feedback. Overstimulation of the thyroid gland by TSH to produce thyroid hormone with insufficient iodine causes the gland to hypertrophy, sometimes so much that the entire neck region becomes swollen (Figure 34.10). Goiter caused by iodine deficiency is seldom seen in North America because of the widespread use of iodized salt. However, it is estimated that even today 200 million people worldwide experience varying degrees of goiter, mostly in high mountains of South America, Europe, and Asia.

Hormonal Regulation of Calcium Metabolism

Closely associated with the thyroid gland and in some animals buried within it are **parathyroid glands.** These tiny glands occur as two pairs in humans but vary in number and position in other vertebrates. They were discovered at the end of the nineteenth century when the fatal effects of "thyroidectomy" were traced to unknowing removal of parathyroid glands together with the thyroid gland. In birds and mammals, including humans, removal of the parathyroid glands causes the level of calcium in the blood to decrease rapidly. This decrease in calcium leads to a serious increase in nervous-system excitability, severe muscular spasms,

tetany, and finally death. Subsequently, it was discovered that parathyroid glands secrete a hormone, **parathyroid hormone (PTH),** which is essential to maintenance of calcium homeostasis. Calcium ions are extremely important for formation of healthy bones. In addition, they are needed for such functions as neurotransmitter and hormone release, muscle contraction, intracellular signaling, and blood clotting.

Before considering how hormones maintain calcium homeostasis, it is helpful to summarize mineral metabolism in bone, a densely packed storehouse of both calcium and phosphorus (see Chapter 29, p. 650 for bone structure and function). Bone contains approximately 98% of the calcium and 80% of the phosphorus in humans. Although bone is second only to teeth as the most durable material in the body (as evidenced by survival of fossil bones for millions of years) it is in a state of constant turnover in living vertebrates. Bone-building cells **(osteoblasts)** synthesize organic fibers and glycoproteins of bone matrix which later become mineralized with a form of calcium phosphate called hydroxyapatite. Bone-resorbing cells **(osteoclasts)** are giant cells that dissolve the bony matrix, releasing calcium and phosphate into the blood. These opposing activities allow bone constantly to remodel itself, especially in a growing animal, producing structural improvements to counter new mechanical stresses on the body. They additionally provide a vast and accessible reservoir of minerals that can be withdrawn as needed for general cellular requirements.

The level of calcium in the blood is maintained by three hormones that coordinate absorption, storage, and excretion of calcium ions. If blood calcium should decrease slightly, the parathyroid gland increases its secretion of **PTH.** This increase stimulates osteoclasts to dissolve bone adjacent to these cells, thus releasing calcium and phosphate into the bloodstream and returning blood calcium level to normal. PTH also decreases the rate of calcium excretion by the kidney and increases production of the hormone 1,25-dihydroxyvitamin D_3. PTH levels vary inversely with blood calcium level, as shown in Figure 34.11.

A second hormone involved in calcium metabolism in all tetrapods is derived from vitamin D_3. Vitamin D_3, like all vitamins, is a dietary requirement. But unlike other vitamins, vitamin

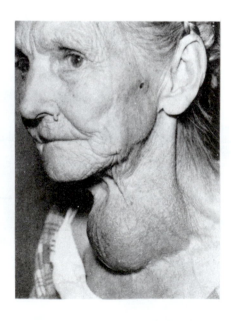

Figure 34.10

A large goiter caused by iodine deficiency. Overstimulation by excess TSH causes the gland to enlarge enormously, as the thyroid gland is stimulated to extract enough iodine from the blood to synthesize the body's requirement for thyroid hormones.

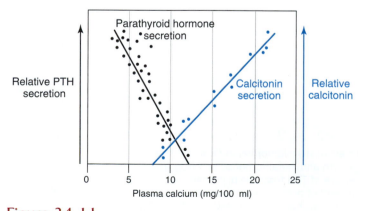

Figure 34.11

How rate of secretion of parathyroid hormone (PTH) and calcitonin respond to changes in blood-calcium level in a mammal.

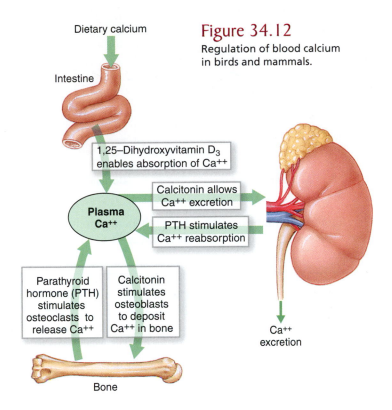

Figure 34.12
Regulation of blood calcium in birds and mammals.

Dietary calcium

Intestine

1,25–Dihydroxyvitamin D₃ enables absorption of Ca⁺⁺

Plasma Ca⁺⁺

Calcitonin allows Ca⁺⁺ excretion

PTH stimulates Ca⁺⁺ reabsorption

Parathyroid hormone (PTH) stimulates osteoclasts to release Ca⁺⁺

Calcitonin stimulates osteoblasts to deposit Ca⁺⁺ in bone

Ca⁺⁺ excretion

Bone

D_3 may also be synthesized in the skin from a precursor by irradiation with ultraviolet light from the sun. Vitamin D_3 is then converted in a two-step oxidation to a hormonal form, **1,25-dihydroxyvitamin D_3.** This steroid hormone is essential for active calcium absorption by the gut (Figure 34.12). Production of 1,25-dihydroxyvitamin D_3 is stimulated by low plasma phosphate as well as by an increase in PTH secretion.

In humans, a deficiency of vitamin D_3 causes rickets, a disease characterized by low blood calcium and weak, poorly calcified bones that tend to bend under postural and gravitational stresses. Rickets has been called a disease of northern winters, when sunlight is minimal. It once was common in the smoke-darkened cities of England and continental Europe.

A third calcium-regulating hormone, **calcitonin,** is secreted by specialized cells (C cells) in the thyroid gland of mammals and in the ultimobranchial glands of other vertebrates. Calcitonin

is released in response to elevated levels of calcium in the blood. It rapidly suppresses calcium withdrawal from bone, decreases intestinal absorption of calcium, and increases excretion of calcium by the kidneys. Calcitonin therefore protects the body against an increase in level of calcium in the blood, just as parathyroid hormone protects it from a decrease in blood calcium (Figure 34.12). Calcitonin has been identified in all vertebrate groups, but its importance is uncertain because replacement of calcitonin is not required for maintenance of calcium homeostasis, at least in humans, if the thyroid gland is surgically removed (also removing the C cells).

Hormones of the Adrenal Cortex

The mammalian adrenal gland is a double gland composed of two unrelated types of glandular tissue: an outer region of adrenocortical cells, or **cortex,** and an inner region of specialized cells, the **medulla** (Figure 34.13). In nonmammalian vertebrates homologs of adrenocortical and medullary cells are organized quite differently; they may be intermixed or distinct, but never arranged in a cortex-medulla relationship as in mammals.

At least 30 different compounds have been isolated from adrenocortical tissue, all of them closely related lipoidal compounds known as steroids. Only a few of these compounds are true steroid hormones; most are various intermediates in synthesis of steroid hormones from **cholesterol** (Figure 34.14). Corticosteroid hormones are commonly classified into two groups, according to their function: glucocorticoids and mineralocorticoids.

Glucocorticoids, such as **cortisol** (Figure 34.14) and **corticosterone,** are concerned with food metabolism, inflammation, and stress. They promote synthesis of glucose from compounds other than carbohydrates, particularly amino acids and fats. The overall effect of this process, called **gluconeogenesis,** is to increase the level of glucose in the blood, thus providing a quick energy source for muscle and nervous tissue. Glucocorticoids are also important in diminishing the immune response to various inflammatory conditions. Because several diseases of humans are inflammatory diseases (for example, allergies, hypersensitivity, and rheumatoid arthritis), these corticosteroids have important medical applications.

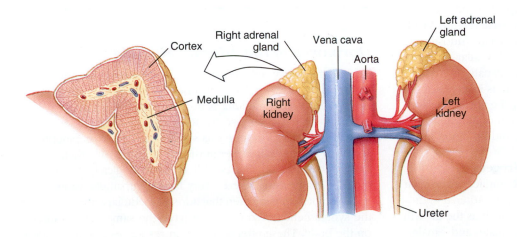

Cortex

Right adrenal gland

Vena cava

Aorta

Left adrenal gland

Medulla

Right kidney

Left kidney

Ureter

Figure 34.13
Paired adrenal glands of humans, showing gross structure and position on the upper poles of the kidneys. Steroid hormones are produced by the cortex. The sympathetic hormones epinephrine and norepinephrine are produced by the medulla.

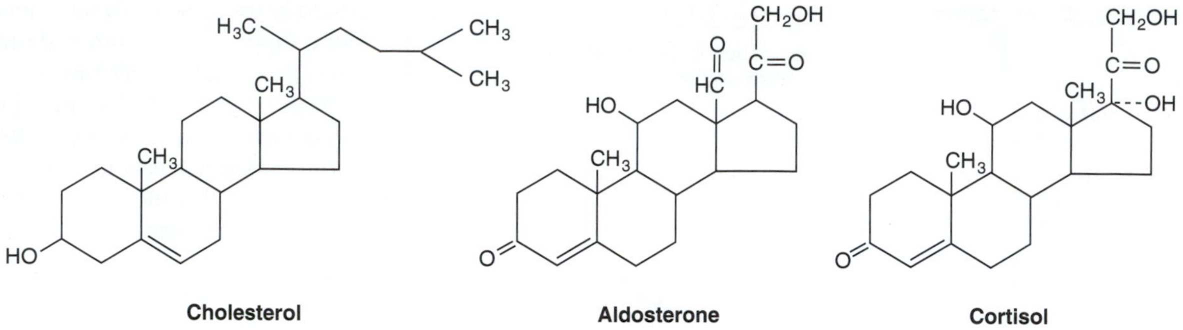

Figure 34.14

Hormones of the adrenal cortex. Cortisol (a glucocorticoid) and aldosterone (a mineralocorticoid) are two of several steroid hormones synthesized from cholesterol in the adrenal cortex.

Cholesterol   Aldosterone   Cortisol

Synthesis and secretion of glucocorticoids are controlled principally by ACTH of the anterior pituitary (see Figure 34.6), while ACTH is controlled in turn by corticotropin-releasing hormone (CRH) of the hypothalamus (Table 34.1). As with pituitary control of the thyroid, a negative feedback relationship exists between CRH, ACTH, and the adrenal cortex (see Figure 34.3). An increase in release of glucocorticoids suppresses output of CRH and ACTH; the resulting decline in blood level of CRH and ACTH then inhibits further release of glucocorticoids from the adrenal cortex. An opposite sequence of events happens should the blood level of glucocorticoids drop: CRH and ACTH output increases, which in turn stimulates secretion of glucocorticoids. CRH is known to mediate stressful stimuli through the adrenal axis.

Adrenal steroid hormones, especially the glucocorticoids, are remarkably effective in relieving symptoms of rheumatoid arthritis, allergies, and various disorders of connective tissue, skin, and blood. Following the report in 1948 by P. S. Hench and his colleagues at the Mayo Clinic that cortisone dramatically relieved pain and crippling effects of advanced arthritis, steroid hormones were hailed by the media as "wonder drugs." Optimism soon diminished, however, when it became apparent that severe side effects always attended long-term administration of antiinflammatory steroids. Steroid therapy lulls the adrenal cortex into inactivity and may permanently impair the body's capacity to produce its own steroids. Today steroid therapy is applied with caution, because the inflammatory response is a necessary part of the body's defenses.

Mineralocorticoids, the second group of corticosteroids, are those that regulate salt balance. **Aldosterone** (Figure 34.14) is by far the most important steroid of this group. Aldosterone promotes tubular reabsorption of sodium and tubular secretion of potassium by the kidneys. Since sodium usually is in short supply in diets of many animals and potassium is in excess, mineralocorticoids play vital roles in preserving the correct balance of blood electrolytes. The salt-regulating action of aldosterone is controlled by the renin-angiotensin system, and by blood levels of potassium ions described on page 677.

Adrenocortical tissue also produces **androgens** (Gr. *andros,* man, + *genesis,* origin), which, as the name implies, are similar in effect to the male sex hormone, testosterone. Adrenal androgens promote some developmental changes, such as the growth spurt, that occur just before puberty in human males and females.

Development of **anabolic steroids,** synthetic hormones related to testosterone, has led to widespread abuse of steroids among athletes.

Use of anabolic steroids by athletes became major news following Ben Johnson's drug-fueled win of the 100-meter race at the 1988 Olympics. Despite almost universal condemnation by Olympic, medical, and college sports authorities, an unscientific and clandestine program of experimentation with anabolic steroids has become popular with many amateur and professional athletes in many countries. These synthetics (and testosterone and its precursors) cause hypertrophy of skeletal muscle and improve performance that depends on strength. Unfortunately, they also have serious side effects, including testicular atrophy (and infertility), periods of irritability, abnormal liver function, and cardiovascular disease. Recent data suggest that steroid abuse among adolescents is increasing. Most users are male and in 2004, self-reporting estimates of use were 2.4% in tenth-graders and 3.4% in twelfth-graders. This represents a significant increase since 1991, when use was reported as 1.8% in tenth-graders and 2.1% in twelfth-graders (http://www.drugabuse.gov/Infofacts/Steroids.html). Use among professional athletes is not well documented, although anecdotal evidence suggests that such drugs are popular among athletes in a variety of sports.

The U.S. National Baseball Hall of Fame has shunned some star players because of their suspected use of steroids; a controversial case is that of Mark McGwire, whose record-breaking 70 home runs with the St. Louis Cardinals in 1998 would have ensured his election except for suspicion that his performance was enhanced by steroids.

Hormones of the Adrenal Medulla

Adrenal medullary cells secrete two structurally similar hormones: **epinephrine (adrenaline)** and **norepinephrine (noradrenaline).** The adrenal medulla is derived embryologically from the same tissue that gives rise to postganglionic sympathetic neurons of the autonomic nervous system (p. 739). Norepinephrine serves as a neurotransmitter at the endings of sympathetic nerve axons. Thus functionally, as well as embryologically, the adrenal medulla can be considered a very large sympathetic ganglion.

It is not surprising then that adrenal medullary hormones and the sympathetic nervous system have the same general effects on the body. These effects center on responses to emergencies, such as fear and strong emotional states, flight from danger, fighting, lack of oxygen, blood loss, and exposure to pain. Walter B. Cannon, of homeostasis fame (p. 667), termed these "fight or flight" responses that are appropriate for survival. We are familiar with the increased heart rate, tightening of the stomach, dry mouth, trembling muscles, general feeling of anxiety, and increased awareness that attends sudden fright or other strong emotional states. These effects are attributable to increased activity of the sympathetic nervous system and to rapid release into the blood of epinephrine and norepinephrine from the adrenal medulla. Activation of the adrenal medulla by the sympathetic nervous system prolongs the effects of sympathetic system activation.

Epinephrine and norepinephrine have many other effects of which we are not as aware, including constriction of arterioles (which, together with increased heart rate, increases blood pressure), mobilization of liver glycogen and fat stores to release glucose and fatty acids for energy, increased oxygen consumption and heat production, hastening of blood coagulation, and inhibition of the gastrointestinal tract. These changes prepare the body for emergencies and are activated in stressful conditions.

Hormones from Islet Cells of the Pancreas

The pancreas is both an exocrine and an endocrine organ (Figure 34.15). The exocrine portion produces pancreatic juice, a mixture of digestive enzymes and bicarbonate ions conveyed by a duct (or ducts) to the digestive tract (see Chapter 32). Scattered within the extensive exocrine portion of the pancreas are numerous small islets of tissue, called **islets of Langerhans** (Figure 34.15 and photograph on p. 753). This endocrine portion of the pancreas is only 1% to 2% of the total weight of the organ. The islets are without ducts and secrete their hormones directly into blood vessels that extend throughout the pancreas.

Three polypeptide hormones are secreted by different cell types within the islets: **insulin,** produced by **beta cells, glucagon,** produced by **alpha cells,** and a recently discovered hormone called **pancreatic polypeptide** (PP), also produced within the islets. Insulin and glucagon have antagonistic actions of great importance in metabolism of carbohydrates and fats. Meals rich in carbohydrates stimulate insulin release as blood glucose levels rise following digestion and absorption of the meal (p. 719). Insulin is essential for uptake of blood-borne glucose by cells, especially skeletal muscle cells. Insulin promotes entry of glucose into body cells through its action on a glucose transporter molecule found in cell membranes. Although insulin-dependent glucose transporter molecules have been demonstrated on neurons of the central nervous system, neurons do not require insulin for glucose uptake. This independence from insulin is very important because unlike other cells of the body, neurons almost exclusively use glucose as an energy source. The exact role of insulin-dependent glucose transporters in the brain is not clear, but insulin is important in central regulation of food intake and body weight. Cells of the rest of the body, however, require insulin to use glucose; without insulin the level of glucose in the blood rises to abnormally high levels, a condition called

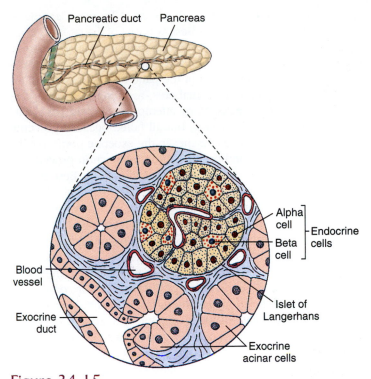

Figure 34.15

The pancreas is composed of two kinds of glandular tissue: exocrine acinar cells that secrete digestive juices that enter the intestine through the pancreatic duct, and endocrine islets of Langerhans. The islets of Langerhans secrete the hormones insulin, glucagon, and pancreatic polypeptide directly into the blood circulation.

hyperglycemia. When this level exceeds the transport maximum of the kidney (see note on p. 676) sugar (glucose) appears in urine. Insulin deficiency also inhibits uptake of amino acids by skeletal muscle, and fats and proteins in muscle are metabolized to provide energy. Body cells starve while the urine abounds in the very substance the body craves. The disease of insulin deficiency, called **diabetes mellitus type 1,** afflicts nearly 5% of humans in varying degrees of severity. If left untreated, it can lead to severe damage to kidneys, eyes, and blood vessels, and it can greatly shorten life expectancy. Humans can also develop insulin insensitivity or **diabetes mellitus type 2,** with similar symptoms to type 1 diabetes mellitus. This disease is occurring at an increasing rate as more individuals become overweight and obese (see Chapter 32, p. 720). In 2000, a global estimate of humans with type 2 diabetes was 171 million, with a projected increase to 366 million by 2030. An increase in exercise levels and a change in diet can help to decrease insulin levels and to alleviate symptoms in such individuals.

In 1982, insulin became the first hormone produced by genetic engineering (recombinant DNA technology, p. 98) to be marketed for human use. Recombinant insulin has the exact structure of human insulin and therefore will not stimulate an immune response (see Chapter 35, p. 775), which has often been a problem for diabetics receiving insulin purified from pig or cow pancreas.

The first extraction of insulin in 1921 by two Canadians, Frederick Banting and Charles Best, was one of the most dramatic and important events in the history of medicine. Many years earlier two German scientists, J. Von Mering and O. Minkowski, discovered that surgical removal of the pancreas of dogs invariably caused severe symptoms of diabetes, resulting in the animal's death within a few weeks. Many attempts were made to isolate the diabetes preventive factor, but all failed because powerful protein-splitting digestive enzymes in the exocrine portion of the pancreas destroyed the hormone during extraction procedures. Following a hunch, Banting, in collaboration with Best and his physiology professor J. J. R. Macleod, tied off the pancreatic ducts of several dogs. This caused the exocrine portion of the gland with its hormone-destroying enzyme to degenerate, but left the islets' tissues healthy long enough for Banting and Best to extract insulin successfully from these glands. Injected into another dog, the insulin immediately lowered the level of sugar in the blood (Figure 34.16). Their experiment paved the way for commercial extraction of insulin from slaughterhouse animals. It meant that millions of people with diabetes, previously fated to disability or death, could look forward to more normal lives.

Glucagon, the second hormone of the pancreas, has several effects on carbohydrate and fat metabolism that are opposite to the effects of insulin. Low blood glucose levels and absorption of amino acids into the blood following digestion (p. 719) stimulate glucagon secretion. For example, glucagon raises the blood-glucose level (by converting liver glycogen to glucose), whereas insulin lowers blood glucose. Glucagon and insulin do not have the same effects in all vertebrates, and in some, glucagon is lacking altogether.

The newest pancreatic hormone, PP, is released after a meal and reduces appetite. As yet its physiological role in energy metabolism is unknown, but since its administration to mice and humans reduces food intake, it is likely to become a focus of future research in the fight against obesity.

Figure 34.16

Charles H. Best and Sir Frederick Banting in 1921 with the first dog to be kept alive by insulin.

Growth Hormone and Metabolism

Growth hormone (GH) is a particularly important metabolic hormone in young animals during growth and development. It acts directly on long bones to promote cartilaginous growth and bone formation by cell division and protein synthesis, thus producing an increase in length and density of bone. GH also acts indirectly on growth via stimulation of **insulin-like growth factor (IGF)** or somatomedin release from the liver. This polypeptide hormone promotes mobilization of glycogen from liver stores and release of fat from adipose tissue stores necessary for growth processes. Thus, GH is considered a **diabetogenic hormone,** since oversecretion leads to an increase in blood glucose and can cause insulin insensitivity or diabetes mellitus type 2. If produced in excess, GH causes giantism. A deficiency of this hormone in a human child leads to dwarfism.

White Adipose Tissue as an Endocrine Organ

The discovery of the *ob* gene in 1994, which codes for the hormone, **leptin** (Gr. *leptos,* thin), produced by white fat cells (adipose tissue), began a period of intense research on adipose tissue. So far 26 peptides, **adipokines,** with autocrine, paracrine, and endocrine functions have been described.

Leptin is an important hormone that regulates eating behavior and long-term energy balance as part of a feedback system that informs the brain, particularly the hypothalamus and brain stem, of the energy status of the periphery. Evidence suggests that leptin is more important during times of lower food, and consequently, energy availability, since decreased fat stores secrete less leptin. At such times, the brain responds by directing available energy away from nonessential processes, such as reproduction, and stimulates increased foraging and feeding behaviors. Leptin has become immensely significant in the study of satiety signals and energy expenditure, since these studies relate to the overall problem of human obesity (p. 720). Blood-plasma levels of leptin mirror those of insulin, which also provides an important feedback signal to the brain regarding adipose tissue stores.

Of the long list of recently discovered adipokines, there are several, in addition to leptin, that appear to be involved in the regulation of energy balance. **Adiponectin** tends to lower blood glucose levels by increasing the effects of insulin on liver and skeletal muscle. Several studies also suggest that low blood adiponectin levels are linked with obesity and type 2 diabetes incidence. In addition, high blood adiponectin levels are associated with a decreased risk of coronary artery disease, which is due in part to adiponectin decreasing the likelihood of fatty deposits of cholesterol in the artery walls (p. 696). **Tumor necrosis factor-α (TNF-α),** as well as other cytokines, are secreted by adipose tissue, and high levels are associated with obesity-related insulin insensitivity. Currently these and several other adipokines, where the data are more controversial, are the subject of intense obesity-related research.

SUMMARY

Hormones are chemical messengers synthesized by special endocrine and other cells and transported by blood or other body fluid to target cells where they affect cell function by altering specific biochemical processes. Specificity of response is ensured by the presence on or in target cells of protein receptors that bind only selected hormones. Hormone effects are vastly amplified in target cells by acting through one of two basic mechanisms. Many hormones, including epinephrine, glucagon, vasopressin, and some hormones of the anterior pituitary, cause production of a "second messenger," such as cyclic AMP, that relays the hormone's message from a surface receptor to a cell's biochemical machinery. Steroid hormones and thyroid hormones operate through mainly cytoplasmic or nuclear receptors. A hormone-receptor complex is formed that alters protein synthesis by stimulating or inhibiting gene transcription.

Most invertebrate hormones are products of neurosecretory cells, although endocrine cells are found in cnidarians, nematodes, and annelids, and endocrine glands occur in molluscs and arthropods. Peptide, neuropeptide, steroid, and terpenoid hormones regulate many physiological processes. The best understood invertebrate endocrine system is that controlling molting and metamorphosis in insects. A juvenile insect grows by passing through a series of molts under control of two hormones, one (ecdysone) favoring molting to an adult and the other (juvenile hormone) favoring retention of juvenile characteristics. Ecdysone is controlled by a neurosecretory hormone (PTTH) from the brain. Juvenile hormone, ecdysone, and PTTH all play an important role in the regulation of diapause (arrested development) that can occur at any stage of metamorphosis, as well as in the adult (dormancy).

The vertebrate endocrine system is orchestrated by the hypothalamus. Release of all anterior pituitary hormones is primarily regulated by hypothalamic neurosecretory products called releasing hormones (or release-inhibiting hormones). The hypothalamus also produces two neurosecretory hormones, which are stored in and released from the posterior lobe of the pituitary. In mammals these two hormones are oxytocin, which stimulates milk production and uterine contractions during parturition; and vasopressin (antidiuretic hormone), which acts on the kidney to restrict urine production, causes vasoconstriction of blood vessels, and increases thirst. In amphibians, nonavian reptiles, and birds, vasotocin replaces vasopressin as the water-balance hormone.

The anterior lobe of the pituitary produces seven well-characterized hormones. Four of these are tropic hormones that regulate subservient endocrine glands: thyrotropic hormone (TSH), which controls secretion of thyroid hormones; adrenocorticotropic hormone (ACTH), which stimulates release of steroid hormones by the adrenal cortex, primarily the glucocorticoids, cortisol or corticosterone; and follicle-stimulating hormone (FSH) and luteinizing hormone (LH), which act on ovaries and testes. Three direct-acting hormones are (1) prolactin, which plays several diverse roles, including stimulation of milk production during lactation; (2) growth hormone, which governs body growth and metabolism; and (3) melanocyte-stimulating hormone (MSH), which controls melanocyte dispersion in ectothermic vertebrates.

The pineal gland, a derivative of the pineal complex of the diencephalon of the brain, produces the hormone melatonin. In many vertebrates, melatonin, which is released in response to darkness, maintains circadian rhythms. In birds and mammals that are seasonal breeders, melatonin level provides information regarding daylength, and thereby indirectly regulates seasonal reproductive activity.

Recent application of ultrasensitive radioimmunochemical techniques has revealed many neuropeptides in the brain, several of which behave as neurotransmitters in the brain but as hormones elsewhere in the body. The classical definition of a hormone has been modified to include other chemical messengers, such as prostaglandins and cytokines, which originate in sources other than clearly defined endocrine glands.

Several hormones play important roles in regulating cellular metabolic activities. Two thyroid hormones, triiodothyronine (T_3) and thyroxine (T_4), control growth, development of the nervous system, and cellular metabolism. Calcium metabolism is regulated principally by three hormones: parathyroid hormone from the parathyroid glands, a hormonal derivative of vitamin D_3, 1,25-dihydroxyvitamin D_3, and calcitonin from the C cells of thyroid gland or ultimobranchial glands. Parathyroid hormone and 1,25-dihydroxyvitamin D_3 increase plasma calcium levels; calcitonin decreases plasma calcium levels.

The principal steroid hormones of the adrenal cortex are glucocorticoids, which stimulate formation of glucose from nonglucose sources (gluconeogenesis), and mineralocorticoids, which regulate blood electrolyte balance. The adrenal medulla is the source of epinephrine and norepinephrine, which have many effects, including assisting the sympathetic nervous system in emergency responses. They also increase energy substrates in the blood for use in emergency situations.

Glucose metabolism is regulated by the antagonistic action of two pancreatic hormones. Insulin is needed for cellular use of blood glucose, it also increases storage of fats in adipose tissue and uptake of amino acids in muscle. Glucagon opposes the action of insulin. A third pancreatic hormone, pancreatic polypeptide, is released after meals and reduces apetite.

White adipose tissue is now considered an endocrine organ and secretes many peptides, called adipokines. Leptin feeds back to the hypothalamus to modulate food intake and long-term energy balance. Adiponectin tends to lower blood glucose levels and decreases the risk of cardiovascular disease, while tumor necrosis factor-α appears to increase the risk of obesity-related insulin insensitivity.

REVIEW QUESTIONS

1. Outline the famous experiment of Berthold that marks the first endocrine experiment. What might the hypothesis have been?
2. Provide definitions for the following: hormone, endocrine gland, exocrine gland, hormone receptor molecule.
3. Hormone receptor molecules are the key to understanding specificity of hormone action on target cells. Describe and distinguish between receptors located on the cell surface and those located in the nucleus of target cells. Name two hormones whose action is mediated through each type of receptor.

4. What is the importance of feedback systems in the control of hormonal output? Offer an example of a hormonal feedback pattern.

5. In which invertebrate phyla is endocrine function observed? Which invertebrate phyla possess endocrine glands? Give two examples of invertebrate hormones that regulate metabolism.

6. Explain how the three hormones involved in insect growth—ecdysone, juvenile hormone, and PTTH—interact in molting and metamorphosis.

7. Name seven hormones produced by the anterior pituitary gland. Why are four of these seven hormones called "tropic hormones"? Explain how secretion of the anterior pituitary hormones is controlled by neurosecretory cells in the hypothalamus.

8. Describe the chemical nature and function of two posterior-lobe hormones, oxytocin and vasopressin. What is distinctive about the way these neurosecretory hormones are secreted compared with the neurosecretory release and release-inhibiting hormones that control the anterior pituitary hormones?

9. What is the evolutionary origin of the pineal gland of birds and mammals? Explain the role of the pineal hormone, melatonin, in regulating seasonal reproductive rhythms in birds and mammals. Does melatonin have any function in humans?

10. What are endorphins and enkephalins? What are prostaglandins?

11. What are some functions of hormones called cytokines?

12. What are the two most important functions of the thyroid hormones?

13. Explain how you would interpret the graph in Figure 34.11 to show that PTH and calcitonin act in a complementary way to control blood-calcium level.

14. Describe the principal functions of the two major groups of adrenocortical steroids, the glucocorticoids, and the mineralocorticoids. To what extent do these names provide clues to their function?

15. Where are the hormones epinephrine and norepinephrine produced and what is their relationship to the sympathetic nervous system and its response to emergencies?

16. Explain the actions of two hormones of the islets of Langerhans on the level of glucose in the blood. What is the consequence of insulin insufficiency or insensitivity as in the disease diabetes mellitus?

17. What is the function of the hormone, leptin? Why has its discovery proved important in the area of regulation of feeding?

18. Why is adipose tissue now considered to be an endocrine organ?

SELECTED REFERENCES

Bentley, P. J. 1998. Comparative vertebrate endocrinology, ed. 3. Cambridge, Cambridge University Press. *Undergraduate text with good evolutionary perspective.*

Dockray, G. J. 2004. The expanding family of–RFamide related peptides and their effects on feeding behavior. Exp. Physiol. **89:**229–235. *This review describes an ever expanding family of neuropeptides (FaRPs) that appear to regulate a diverse number of endocrine functions in invertebrate and vertebrates.*

Gard, P. 1998 Human endocrinology. Bristol, Pennsylvania, Taylor & Francis Ltd. *Good treatment of human endocrinology.*

Griffin, J. E., and S. R. Ojeda. 2004. Textbook of endocrine physiology, ed. 5. Oxford, Oxford University Press. *Excellent upper-undergraduate and graduate text of human endocrinology.*

Hadley, M. E., and J. E. Levine. 2007. Endocrinology, ed. 6. Englewood Cliffs, New Jersey, Prentice-Hall, Inc. *Undergraduate-level textbook in vertebrate endocrinology.*

Kahn, S. E., R. L. Hull, and K. M. Utzschneider. 2006. Mechanisms linking obesity to insulin resistance and type 2 diabetes. Nature **444:**640–846. *This review discusses the endocrine mechanisms that can lead to type 2 diabetes in obese humans.*

Krajniak, K. G. 2005. Annelid endocrine disruptors and a survey of invertebrate FMRFamide-related peptides Integr. Comp. Biol. **45:**88–96. *A brief overview of functions of FMRFamide-related peptides (FaRPs) in invertebrates can be found in the second part of this review.*

LaFont, R. 2000. The endocrinology of invertebrates. Ecotoxicology **9:**41–57. *A well-written review of invertebrate endocrinology compared with vertebrate endocrine function.*

Lienhard, G. E., J. W. Slot, D. E. James, and M. M. Mueckler. 1992. How cells absorb glucose. Sci. Am. **266:**86–91 (Jan.). *How insulin regulates the function of a special transporter molecule that moves glucose across cell membranes.*

Oehlmann, J., and U. Schulte-Oehlmann. 2003. Endocrine disruption in invertebrates. Pure Appl. Chem. **75:**2207–2218. *Provides an overview of invertebrate hormones and the effects of endocrine disruptors in invertebrate groups.*

Ronti, T., G. Lupattelli, and E. Mannarino. 2006. The endocrine function of adipose tissue: an update. Clin. Endocrinol. **64:**355–365. *This review discusses recent evidence for white adipose tissue as an endocrine organ.*

Rosen, E. D., and B. M. Spiegelman. 2006. Nature **444:**847–853. *Provides a discussion of the biology of white adipose tissue and its role in endocrine regulation of energy balance.*

Tarrant, A. M. 2005. Endocrine-like signaling in cnidarians: current understanding and implications for ecophysiology. Integr. Comp. Biol. **45:**201–214. *Review endocrine function in cnidarians in comparison to other invertebrate and vertebrate endocrinology.*

Van Gaal, L. F., I. L. Mertens, and C. E. De Block. 2006. Mechanisms linking obesity with cardiovascular disease. Nature **444:**875–880. *Provides an excellent summary of the mechanisms by which obesity can lead to an elevated risk of cardiovascular disease.*

Woods, S. C., P. A. Rushing, and R. J. Seeley. 2001. Neuropeptides and the control of energy homeostasis. Nutrition and Brain **5:**93–115. *A readable discussion of food intake regulation and body weight control, incorporating our most recent hypotheses regarding leptin.*

ONLINE LEARNING CENTER

Visit www.mhhe.com/hickmanipz14e for chapter quizzing, key term flash cards, web links, and more!

Immunity

A leukocyte (orange) is attaching to and engulfing Bacillus cells (blue, rod-shaped). It will use enzymes to digest the bacteria. The whole process is known as phagocytosis.

The Language of Cells in Immunity

Cells of the immune system communicate with each other by mechanisms similar to that used by a peptide hormone (see Chapter 34). Target cells have receptors protruding through their plasma membrane that specifically bind the signal molecules and only those molecules. Binding of a signal causes changes in the receptor molecule (or in an associated membrane protein), and this initiates a cascade of activations involving protein kinases and phosphorylases (enzymes that transfer phosphate groups). Transcription factors are mobilized, and in the nucleus these factors initiate transcription of formerly inactive genes, leading to synthesis of products they encode (see Chapter 5).

One large group of signaling molecules of the immune system are called *cytokines*. Cytokines and their receptors perform an intricate and elaborate ballet of activation and regulation, causing some cells to proliferate, suppressing proliferation of others, and stimulating secretion of additional cytokines or other defense molecules. Precise signaling among cells and exact performance of their duties are essential to maintainence of health and defense against invading viruses, bacteria, and parasites and for prevention of unrestrained cell division, as in cancer. Successful establishment of invaders in our bodies depends on evasion or subversion of our immune system, and inappropriate response of immune cells may itself produce disease. We have learned to manipulate the immune response so that we can transplant organs between individuals, but progressive failure in immune cell communication results in profound disease, such as AIDS.

The immune system is located throughout an animal's body, and it is as crucial to survival as the respiratory, circulatory, nervous, skeletal, or any other system. Every animal's environment is filled with parasites and potential parasites: flatworms, nematodes, arthropods, single-celled eukaryotes, bacteria, and viruses. Whether any parasite can infect an animal, the **host,** and the resulting severity of disease depends largely on the host's defense system.

SUSCEPTIBILITY AND RESISTANCE

A host is **susceptible** to a parasite if the host cannot eliminate a parasite before the parasite becomes established. The host is **resistant** if its physiological status prevents establishment and survival of the parasite. Corresponding terms from the viewpoint of a parasite would be **infective** and **noninfective.**

These terms denote only the success or failure of infection, not the underlying mechanisms. Mechanisms that increase resistance (and correspondingly reduce susceptibility and infectivity) may involve either attributes of a host not related to active defense mechanisms or specific defense mechanisms mounted by a host in response to a foreign invader. It is important to remember that these terms are relative, not absolute; for example, an organism may be more or less resistant than another, and its resistance can vary, depending on age, health, and environmental exposure.

The term **immunity** is often used as a synonym for resistance. A more precise statement is that *an animal demonstrates immunity if it possesses cells or tissues capable of recognizing and protecting the animal against nonself invaders.* Most animals show some degree of **innate immunity,** defense that does not depend on prior exposure to the invader. In addition to having innate immunity, vertebrates develop **acquired immunity,** which is specific to a particular nonself material, requires time for its development, and occurs more quickly and vigorously on secondary response.

Frequently, resistance conferred by immune mechanisms is not complete. In some instances a host may recover clinically and be resistant to a specific challenge, but some parasites may remain and reproduce slowly, as in toxoplasmosis (p. 238), Chagas' disease (p. 231), and malaria (p. 238). This condition is called **premunition.**

INNATE DEFENSE MECHANISMS
Physical and Chemical Barriers

The unbroken surface of most animals provides a barrier to invading organisms. It may be tough and cornified, as in many terrestrial vertebrates, or sclerotized, as in arthropods. Soft outer surfaces are usually protected by a layer of mucus, which lubricates the surface and helps dislodge particles from it.

A variety of antimicrobial substances occur in secretions of animals. These include antimicrobial peptides discussed in the next section. In insects pathogens such as parasitic wasp eggs induce release of chemicals that form melanin to encapsulate the eggs. Chemical defenses present in many vertebrates include a low pH in the stomach and vagina and hydrolytic enzymes in secretions of the alimentary tract. Mucus is produced by mucous membranes lining the digestive and respiratory tract of vertebrates and contains parasiticidal substances such as **IgA** and **lysozyme.** IgA is a class of antibody (p. 776) that can cross cellular barriers easily and is an important protective agent in mucus of the intestinal epithelium. It is secreted onto the surface of cells lining the alimentary canal in response to invasion by specific bacteria. Thus, it is part of an animal's acquired immune response (p. 775). IgA is present also in saliva and sweat. Lysozyme is an enzyme that attacks the cell wall of many bacteria.

Various cells, including those involved in the acquired immune response, liberate protective compounds. A family of low-molecular-weight glycoproteins, the **interferons,** are released by a variety of eukaryotic cells in response to invasion by intracellular parasites (including viruses) and other stimuli. **Tumor necrosis factor (TNF)** is a member of the protein-signaling molecule family called **cytokines** (see Table 35.2) and is produced mainly by cells called **macrophages** (see p. 773), some **T lymphocytes** (see p. 774), and white adipose tissue (p. 768). TNF is a major mediator of inflammation (p. 780) and in sufficient concentration causes **fever.** Fever in mammals is one of the most common symptoms of infection. The protective role of fever, if any, remains unclear, but high body temperature may destabilize certain viruses and bacteria.

The intestine of most animals harbors a population of bacteria that seem not to be harmed by host defenses, nor do they elicit any protective defense response. In fact, the normal intestinal microflora in vertebrates tends to inhibit establishment of pathogenic microbes.

Substances in normal human milk can kill intestinal protozoa such as *Giardia lamblia* and *Entamoeba histolytica* (see Chapter 11), and these substances may be important in protecting infants against these and other infections. Antimicrobial elements in human breast milk include lysozyme, defensins (see next section), IgA, IgG (another class of antibody), interferons, and leukocytes (white blood cells, see also Chapter 31).

Some species of mammals are susceptible to infection with parasites such as *Schistosoma mansoni* (see Chapter 14), and others are partially or completely resistant. Without mediation by antibody, macrophages of more resistant species (rats, guinea pigs, rabbits) kill schistosome juveniles, but macrophages of susceptible species do not.

Complement is a series of proteins that are activated in sequence as a host response to invading organisms. Activation of complement by the **classical pathway** (so called only because it was discovered first) depends on antibody (p. 776) bound to the surface of the invading organism and so is an effector mechanism in the acquired immune response (p. 775). Complement activated by the **alternative pathway** is an important innate defense against invasion by bacteria and some fungi. Activation of this pathway is by interaction of complement proteins produced early in the cascade sequence with polysaccharides in the outer coating of the microorganism. Classical and alternative pathways share some, but not all, components. Both pathways rely on activation of the third component in the complement cascade (C3), and from this point on both pathways are the same. Active C3 initiates a cascade of activations, ultimately causing lysis of the invading cell. The host's own cells are not lysed because regulatory proteins rapidly inactivate the first active component of complement when it binds to host cells but not to foreign cells. The active C3 component also binds to invading target cells effectively tagging them for **phagocytosis.** The process of tagging pathogens for subsequent phagocytosis is called **opsonization.** Finally, active C3 attracts lymphocytes (p. 774) to the site of infection and enhances inflammation (p. 780). Complement-like proteins called Teps (thioester-containing proteins) have been discovered in insects and appear to function similar to the alternative pathway of the complement system.

Antimicrobial Peptides

Insects tend to be resistant to infection with many microbial pathogens. In the 1980s experiments showed that inoculation of moth larvae with bacteria caused a release of a barrage of antimicrobial agents that killed the bacteria, even without prior exposure to the invader. Since that time, hundreds of antimicrobial peptides have been described from a broad spectrum of animals, invertebrates and vertebrates. They are especially important at surfaces where organism meets environment, such as skin or mucous membranes. For example, epithelial glands in the frog skin secrete high concentrations of antimicrobial peptides at sites of irritation or injury. Antimicrobial peptides do not have such high specificity as does the acquired immune response of vertebrates, but rather each peptide is effective against a different category of microbe, for example gram-positive bacteria (bacteria that stain with "Gram stain"), gram-negative bacteria, and fungi. Release of the peptides is immediate in presence of the foreign organism and is not subject to prior immunizing experience with the microbe. Conventional antibiotics usually work by blocking a critical protein in an invading microbe, but these peptides interfere with internal signaling of a microbe or perforate its surface with holes.

Antimicrobial peptides in mammals have been called **defensins.** They do not harm cells of the organism from which they originate. Macrophages, neutrophils, eosinophils and cells around linings of intestinal, respiratory, and urogenital tracts secrete defensins in response to stimulation by molecules on the surface of microbes or, in some cases, to their metabolic products. Such molecules are conserved over a range of microbes but do not occur on host cells. Defensins may be chemotoxic to neutrophils, or they may enhance the inflammatory response (p. 780), or the acquired immune response (p. 775). Several neuropeptides (p. 761) and cytokines (p. 762) demonstrate antimicrobial activity.

Release of peptides begins when receptors on a cell's surface recognize a microbial molecule. Many of these receptors are known as **Toll** proteins or **Toll-like receptors (TLRs),** so called because they are located in a cell membrane where they receive signals from the outside. At least nine TLRs have been described in humans, each of which recognizes a specific pattern of molecules from a class of microbes. Activation of a particular TLR signals the nucleus to synthesize a peptide against that particular microbe. An ever-increasing list of non-TLRs is now being discovered that also function as innate immunity receptor systems.

Cellular Defenses: Phagocytosis

For defense against an invader an animal's cells must recognize when a substance does not belong; they must recognize *nonself.* **Phagocytosis** illustrates the process of nonself recognition, and it also serves as a process for removing senescent cells and cellular debris from the host. Phagocytosis occurs in almost all metazoa and is a feeding mechanism in many single-celled organisms (p. 225). A cell that has this ability is a **phagocyte.** Phagocytes engulf a particle within an invagination of the phagocyte's cell membrane (see Chapter 3, p. 51). The invagination becomes pinched off, and the particle becomes enclosed within an intracellular vacuole. Other cytoplasmic vesicles called **lysosomes** (see Chapter 3, p. 43) join with the particle-containing vacuole and provide digestive enzymes to destroy the particle. Lysosomes of many phagocytes also contain enzymes that catalyze production of cytotoxic **reactive oxygen intermediates (ROIs)** and **reactive nitrogen intermediates (RNIs).** Examples of ROIs are superoxide radical (O_2^-), hydrogen peroxide (H_2O_2), singlet oxygen ($^1O_2$), and hydroxyl radical (OH•). RNIs include nitric oxide (NO) and its oxidized forms, nitrite (NO_2^-) and nitrate (NO_3^-). All such intermediates are potentially toxic to invasive microorganisms or parasites.

Phagocytes and Other Defense Cells

Many invertebrates have specialized cells that function as itinerant troubleshooters within the body, acting to engulf or to encapsulate foreign material (see Table 35.1) and to repair wounds. Such cells are variously called amebocytes, hemocytes, or coelomocytes in different animals. If a foreign particle is small, it is engulfed by phagocytosis; but if it is larger than about 10 μm, it is usually encapsulated. Arthropods can encapsulate a foreign object by deposition of melanin around it, either from cells of the capsule or by precipitation from the hemolymph (blood).

In vertebrates several categories of cells are capable of phagocytosis. **Monocytes** arise from stem cells in the bone marrow (Figure 35.1) and give rise to **macrophages.** These cells are members of the **mononuclear phagocyte system** (formerly classified as the **reticuloendothelial system**), which are phagocytic cells stationed around the body. The mononuclear phagocyte system includes macrophages in connective tissue, lymph nodes, spleen, and lung; **Kupffer cells** in sinusoids of the liver; **osteoclasts** of bone; and **microglial cells** in the central nervous system. Macrophages also have important roles in the acquired immune response of vertebrates (see p. 775).

Some **polymorphonuclear leukocytes (PMNs),** a name that refers to the highly variable shape of their nucleus (see Figures 31.3 and 35.4), are circulating phagocytes in blood.

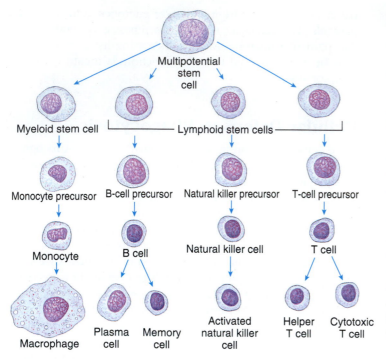

Figure 35.1

Lineage of some cells active in immune response. These cells, as well as red blood cells and other white blood cells are derived from multipotential stem cells in the bone marrow. B cells mature in bone marrow and are released into blood or lymph. Precursors of T cells complete their development in the thymus gland. Precursors of macrophages circulate in blood as monocytes.

Another name for these leukocytes is **granulocytes,** which alludes to the many small granules visible in their cytoplasm after treatment with appropriate stains. According to the staining properties of their granules, granulocytes are further subdivided into **neutrophils, eosinophils,** and **basophils.** Neutrophils are the most abundant (60% to 70% of the total leukocytes), and they provide the first line of phagocytic defense in an infection. Their lysosomes catalyze production of ROIs that are toxic to invading organisms. Eosinophils in normal blood account for about 2% to 5% of the total leukocytes, and basophils are the least numerous at about 0.5%. A high **eosinophilia** (eosinophil count in the blood) is often associated with allergic diseases and parasitic infections.

Several other kinds of cells, including basophils, are not important as phagocytes but are important cellular components of the defense system. **Mast cells** are basophil-like cells found in the dermis and other connective tissues (see Chapter 9, p. 196). When they are stimulated to do so (during inflammation, p. 780), basophils and mast cells release a number of pharmacologically active substances that affect surrounding cells. **Lymphocytes** (Figure 35.1), including **T lymphocytes (T cells)** and **B lymphocytes (B cells),** are crucial in the acquired immune response of vertebrates. **Natural killer (NK) cells** are lymphocyte-like cells that can kill virus-infected and tumor cells in absence of antibody. They release interferons, cytokines that activate other defense cells, as well as substances that lyse the target cell.

IMMUNITY IN INVERTEBRATES

One principal test of the ability of invertebrate tissues to recognize nonself, and by inference potential pathogens, is by grafting a piece of tissue from another individual of the same species (**allograft**) or a different species (**xenograft**) onto the host. If the graft grows in place with no host response, the host tissue is treating it as self, but if cell response and rejection of the graft occur, the host exhibits immune recognition. Most invertebrates, even simple sponges (p. 248), tested in this way reject xenografts; and almost all can reject allografts to some degree (Table 35.1). Interestingly, nemertines and molluscs apparently do not reject allografts. Even some animals with quite loose body organization, such as Porifera and Cnidaria, can reject allografts; this response may be an adaptation to avoid loss of integrity of an individual sponge or colony under conditions of crowding, with attendant danger of overgrowth or fusion with other individuals. Interestingly, sponges, cnidarians, annelids and insects (for example, American cockroaches, *Periplaneta americana*) reject allografts from the same source more quickly upon second exposure; thus they show at least short-term immunological memory.

Hemocytes of molluscs release degradative enzymes during phagocytosis and encapsulation, and antimicrobial substances occur in body fluids of a variety of invertebrates. Substances functioning as opsonins occur in annelids, insects, crustaceans, echinoderms, and molluscs.

Bacterial, viral, and fungal infections in some insects stimulate production of antimicrobial peptides (p. 773), but these peptides show broad-spectrum activity and are not specific for a single infective agent. Specific, induced responses that demonstrate memory upon challenge, previously considered a hallmark of acquired immunity of vertebrates, have been found in copepods and American cockroaches. Injection of prawns with a specific virus coat protein produces protection against the virus from which the coat protein was isolated. In addition, studies in the water flea and bumble bee show that immunity can be passed from one generation to the next.

Contact with infectious organisms can bring the defense systems of snails into enhanced levels of readiness that last for up to two months or more. Susceptibility of snail hosts of the trematode *Schistosoma mansoni* depends heavily on genotype of the snail. Excretory/secretory products of the trematode stimulate motility of hemocytes from resistant snails but inhibit motility of hemocytes from susceptible hosts. Hemocytes from resistant snails encapsulate the trematode larva and apparently kill it with superoxide and H_2O_2 and then destroy it by phagocytosis (Figure 35.2). It appears that the cytokine interleukin-1 occurs in resistant snails and is responsible for activating hemocytes.

Evidence from studies of innate immunity in invertebrates has begun to blur the lines between acquired and innate immune systems. Although many of the mechanisms are quite different, analogous phenomena of memory and specificity of response have now been found in invertebrates. These criteria have always been used to distinguish acquired and innate immunity. One key difference that is still recognized, however, is amplification of the immune response to secondary exposures through

TABLE 35.1
Some Invertebrate Leukocytes and Their Functions

Group	Cell Types and Functions	Phagocytosis	Encapsulation	Allograft Rejection	Xenograft Rejection
Sponges	Archaeocytes (wandering cells that differentiate into other cell types and can act as phagocytes)	+	+	+*	+*
Cnidarians	Amebocytes: "lymphocytes"	+		+	+
Nemertines	Agranular leukocytes; granular macrophagelike cells	+		−	±
Annelids	Basophilic amebocytes (accumulate as "brown bodies"), acidophilic granulocytes	+	+	+	+
Sipunculids	Several types	+	+	±	+
Insects	Several types, depending on family; e.g., plasmatocytes, granulocytes, spherule cells, coagulocytes (blood clotting)	+	+	−	±
Crustaceans	Granular phagocytes; refractile cells that lyse and release contents	+	+	−	+
Molluscs	Amebocytes	+	+	−	+
Echinoderms	Amebocytes, spherule cells, pigment cells, vibratile cells (blood clotting)	+	+	+	+
Tunicates	Many types, including phagocytes; "lymphocytes"	+	+	+	+

Data from Lackie, A. M. 1980. Parasitology **80**:393–412. (See Lackie's article for references.)
*Transplantation reactions occur, but the extent to which the leukocytes are involved is unknown.

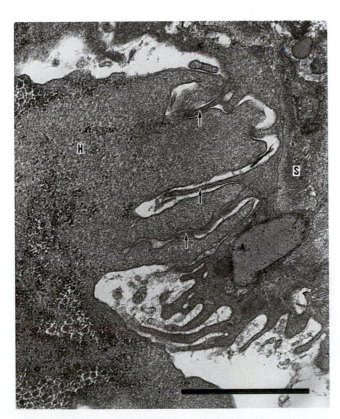

Figure 35.2

Electron micrograph of a hemocyte (H) from a schistosome-resistant strain of a snail attacking a *Schistosoma mansoni* larva (S) under in vitro conditions. Note the hemocyte processes apparently engaged in phagocytosis of portions of the larval tegument (*arrows*). (Scale bar = 1 μm.)

immune cell proliferation specific to the pathogen challenge. This phenomenon is central to acquired immune responses of vertebrates, which is discussed in the next section.

ACQUIRED IMMUNE RESPONSE IN VERTEBRATES

The specialized system of nonself recognition possessed by vertebrates produces increased resistance to *specific* foreign substances or invaders on repeated exposures. Investigations on mechanisms involved are currently intense, and our knowledge of them is increasing rapidly.

The immune response is stimulated by a specific foreign substance called an **antigen,** and an antigen is any substance that will stimulate an immune response. Antigens may be any of a variety of substances with a molecular weight over 3000. They are most commonly proteins and are usually (but not always) foreign to the host. The acquired immune response has two arms, called **humoral** and **cellular.** Humoral immunity is based on **antibodies,** which are both on cell surfaces and dissolved in blood and lymph, whereas cellular immunity is entirely associated with cell surfaces. There is extensive communication and interaction among cells of the two arms.

Basis of Self and Nonself Recognition
Major Histocompatibility Complex

We have known for many years that nonself recognition is very specific. If tissue from one individual is transplanted into another individual of the same species, the graft will grow for a time and then die as immunity against it rises. In the absence of drugs that modify the immune response, tissue grafts will grow successfully only if they are between identical twins or between individuals of highly inbred strains of animals. The molecular basis for this nonself recognition involves a specific group of proteins embedded in the cell surface. In vertebrates these proteins are coded by certain genes, now known as the **major histocompatibility complex (MHC).** MHC proteins are among the most variable known, and unrelated individuals almost always have different genes. There are two types of MHC proteins: class I and class II. Class I proteins occur on the surfaces of virtually all cells, whereas class II MHC proteins occur only on certain cells participating in immune responses, such as lymphocytes and macrophages.

The capability of an acquired immune response develops over a period of time in the early development of an organism. All substances present at the time the capacity develops are recognized as self in later life. Unfortunately, the system of self and nonself recognition sometimes fails, and an animal may begin to produce antibodies against some part of its own body. This condition leads to one of several known *autoimmune diseases,* such as rheumatoid arthritis, multiple sclerosis, lupus, and diabetes mellitus type 1.

Recognition Molecules

We discuss the role of MHC proteins in nonself recognition on page 778, but MHC proteins are not themselves the molecules that recognize foreign substances. This task falls to two basic types of molecules, the genes for which probably evolved from a common ancestor: **antibodies** and **T-cell receptors.** Each vertebrate individual has an enormous variety of antibodies and T-cell receptors, *each of which binds specifically* to one particular antigen (or part of an antigen), even though that antigen may have never been present in the body previously.

Antibodies

Antibodies are proteins called **immunoglobulins.** They are borne in the surface of B lymphocytes or secreted by cells **(plasma cells)** derived from B cells. The basic antibody molecule consists of four polypeptide strands: two identical light chains and two identical, longer heavy chains held together in a Y-shape by disulfide bonds and hydrogen bonds (Figure 35.3). The amino-acid sequence toward the two top ends of the Y varies in both the heavy and light chains, according to the specific antibody molecule (the **variable region**), and this variation determines with which antigen the antibody can bind. Each of

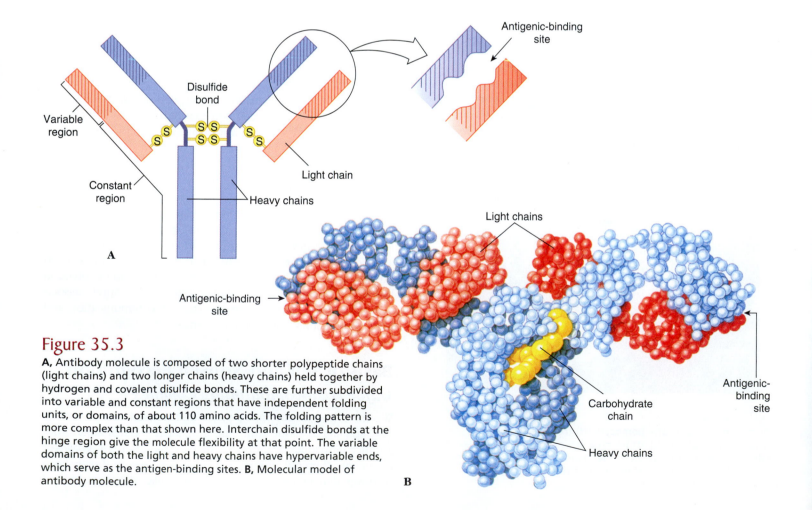

Figure 35.3

A, Antibody molecule is composed of two shorter polypeptide chains (light chains) and two longer chains (heavy chains) held together by hydrogen and covalent disulfide bonds. These are further subdivided into variable and constant regions that have independent folding units, or domains, of about 110 amino acids. The folding pattern is more complex than that shown here. Interchain disulfide bonds at the hinge region give the molecule flexibility at that point. The variable domains of both the light and heavy chains have hypervariable ends, which serve as the antigen-binding sites. **B,** Molecular model of antibody molecule.

the ends of the Y forms a cleft that acts as the antigen-binding site (Figure 35.3), and specificity of the molecule depends on the shape of the cleft and properties of the chemical groups that line its walls. The remainder of an antibody is called the **constant region.** The variable end of the antibody molecule is often called **Fab,** for **a**ntigen-**b**inding **f**ragment, and the constant end is known as **Fc,** for **c**rystallizable **f**ragment (Figure 35.3). The so-called constant region is not really constant: the light chains can be either of two types, and the heavy chains may be any of five types. The type of heavy chain determines the **class** of the antibodies, known as **IgM, IgG** (commonly called *gamma globulin*), **IgA, IgD,** and **IgE.** The class of the antibody determines the specific role of the antibody in the immune response (for example, whether the antibody is secreted or held on a cell surface) but not the antigen it recognizes.

A major problem of immunology is understanding how the mammalian genome could contain information needed to produce at least a million different antibodies. The answer seems to be that antibody genes occur in pieces, rather than as continuous stretches of DNA, and that the antigen-recognizing sites (variable regions) of heavy and light chains of antibody molecules are pieced together from information supplied by separate DNA sequences, which can be shuffled by recombination activating genes (RAGs) to increase diversity of the gene products. The immense repertoire of antibodies is achieved in part by complex gene rearrangements and in part by frequent somatic mutations that produce additional variation in protein structure of the variable regions of the heavy and light antibody chains. Analogous processes occur in the production of genes for T-cell receptors.

Functions of Antibody in Host Defense   Antibodies can mediate destruction of an invader (antigen) in a number of ways. A foreign antigen, for example, becomes coated with antibody molecules as their Fab regions become bound to it effectively immobilizing the antigen. Macrophages recognize the projecting Fc regions and are stimulated to phagocytose the antigen-antibody complex (opsonization). Antibodies also may be able to neutralize toxins that are secreted by an invader.

Another important process, particularly in destruction of bacterial cells, is interaction of antibodies with complement activated by the classical pathway. As noted on page 772, the first component in the classical pathway is activated by antibody bound to the surface of the invading organism. The end result in both classical and alternative complement pathways is the same, that is, lysis of a foreign cell. Both pathways also lead to opsonization or enhancement of inflammation (p. 780). Binding of complement to antigen-antibody complexes can facilitate clearance of these potentially harmful masses by phagocytic cells.

Antibody bound to the surface of an invader may trigger contact killing of the invader by host natural killer cells in what is called **antibody-dependent, cell-mediated cytotoxicity (ADCC).** Receptors for Fc of bound antibody on a microorganism or tumor cell cause natural killer cells to adhere to them and to secrete the cytotoxic contents of their vacuoles.

T-Cell Receptors

T-cell receptors are transmembrane proteins on the surfaces of T cells. Like antibodies, T-cell receptors have a constant region and a variable region. The constant region extends slightly into the cytoplasm and the variable region, which binds with specific antigens, extends outward. Most T cells also bear other transmembrane proteins closely linked to the T-cell receptors, which serve as **accessory** or **coreceptor** molecules. There are about 200 known **CD** (for **c**lusters of **d**ifferentiation proteins) **molecules** or **markers,** one of which, CD3, associates with the constant region of T-cell receptors. The other CD molecules bind to specific ligands on target cells.

Subsets of T Cells

Lymphocytes are **activated** when they are stimulated to move from their recognition phase, in which they simply bind with particular antigens, to a phase in which they proliferate and differentiate into cells that function to eliminate the antigens. We also speak of activation of effector cells, such as macrophages, when they are stimulated to perform their protective function.

Communication between cells in the immune response, regulation of the response, and certain effector functions are performed by different kinds of T cells (Figure 35.4). Although morphologically similar, subsets of T cells can be distinguished by characteristic proteins in their surface membranes. For example, cells with the coreceptor protein CD4 and CD28 are **T-helper cells** (or T_H). These cells secrete cytokines that modulate the activity of other types of lymphocytes and macrophages during an immune response. Some T_H cells (designated T_H1) activate cell-mediated immunity against bacterial and viral attack, and others (called T_H2) activate humoral immunity and antibody release (Figure 35.4).

Cytotoxic T lymphocytes (CTLs) are cells with the coreceptor protein $CD8^+$ that kill target cells expressing certain antigens. A CTL binds tightly to the target cell and secretes a protein that causes pores to form in the cell membrane, resulting in lysis. **T-suppressor cells** eventually suppress an immune response by inhibiting other T- and B-cell activity and **T-memory cells** provide antigen memory for activation during future immune responses.

Cytokines

The 1980s saw rapid advances in our knowledge of how cells of immunity communicate with each other. They do this by means of protein hormones called **cytokines** (Table 35.2; see also Chapter 34, p. 762). Cytokines can produce their effects on the same cells that produce them, on cells nearby, or on cells distant in the body from their production site. Recently, several cytokines including interferon-γ have been shown to possess antimicrobial activity.

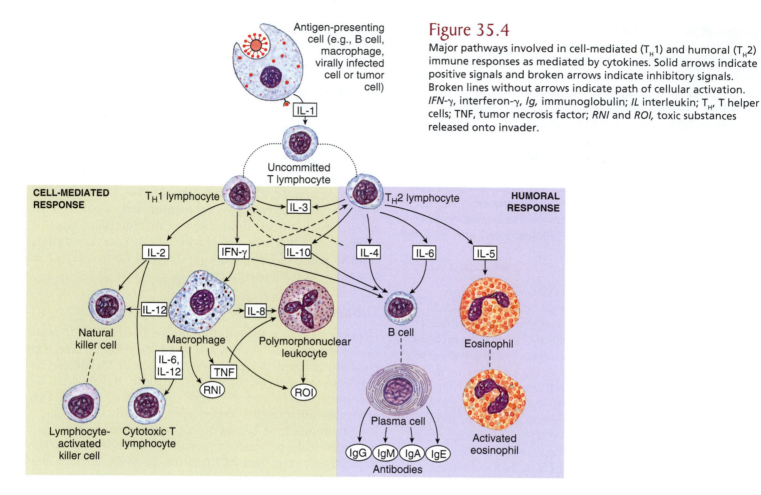

Figure 35.4

Major pathways involved in cell-mediated (T_H1) and humoral (T_H2) immune responses as mediated by cytokines. Solid arrows indicate positive signals and broken arrows indicate inhibitory signals. Broken lines without arrows indicate path of cellular activation. *IFN-γ*, interferon-γ; *Ig,* immunoglobulin; *IL* interleukin; T_H, T helper cells; TNF, tumor necrosis factor; *RNI* and *ROI,* toxic substances released onto invader.

Interleukins (ILs) were originally so called because they are synthesized by leukocytes and have their effect on leukocytes. We now know that some other kinds of cells can produce interleukins, and interleukins produced by leukocytes can affect other kinds of cells.

Generation of a Humoral Response: T_H2 Arm

When an antigen is introduced into a body, it binds to a specific antibody on the surface of an appropriate B cell, but this binding is usually not sufficient to activate the B cell to multiply. The B cell internalizes the antigen-antibody complex and then incorporates portions of the antigen into its own cell surface, bound in the cleft of MHC II protein (Figures 35.5 and 35.6). That portion of the antigen presented on the surface of the B cell (**antigen-presenting cell** or **APC**) is called the **epitope** (or **determinant**). T_H2 cells with a specific T-cell receptor for that particular epitope recognize the epitope bound to the MHC II protein. Binding of the T-cell receptor to the epitope–MHC II complex is enhanced by the coreceptor CD4, which itself binds to the constant portion of the MHC II protein (Figure 35.6). Bound CD4 molecule also transmits a stimulation signal to the interior of the T cell. Activation of the T cell further requires interaction of additional costimulatory signals (for

example, the CD40 molecule and receptor) from other proteins on the surface of the B and T cells. CD8 coreceptors function in a similar way on CTLs; they enhance binding of the T-cell receptor and transmit a stimulatory signal into the T cell.

Activated T_H2 cells secrete IL-4, IL-5, IL-6, and IL-10 (Table 35.2) which activate the B cell that has the same epitope and class II MHC protein on its surface (Figures 35.4 and 35.5). The B cell multiplies rapidly and produces many plasma cells, which secrete large quantities of antibody. The antibody binds to antigen and macrophages recognize this complex and are stimulated to engulf it (opsonization) (Figure 35.5). The antibody is secreted for a period of time, then the plasma cells die. Thus if we measure the concentration of antibody (**titer**) soon after the antigen is injected, we can detect little or none. Titer then rises rapidly as plasma cells secrete the antibody, and it may decrease somewhat as they die and antibody is degraded (Figure 35.7). However, if we give another dose of antigen (**challenge**), there is little or no lag, and antibody titer rises quicker to a higher level than after the first dose. This is the **secondary** or **anamnestic response,** and it occurs because some of the activated B cells gave rise to long-lived **memory cells.** There are many more memory cells present in the body than the original B lymphocyte with the appropriate antibody on its surface, and they rapidly multiply to produce additional plasma cells. Existence of an anamnestic response has great practical value because it is the basis for protective vaccines.

TABLE 35.2
Some Important Cytokines

Cytokine	Principal Source	Function
Interleukin-1 (IL-1)	Activated macrophages	Mediates inflammation, activates T cells, B cells, and macrophages
Interleukin-2 (IL-2)	T_H1 cells	Major growth factor for T and B cells, enhances cytolytic activity of natural killer cells, causing them to proliferate and become **lymphocyte-activated killer (LAK) cells**
Interleukin-3 (IL-3)	Activated T and B cells	Multilineage colony-stimulating factor; promotes growth and differentiation of all cell types in bone marrow
Interleukin-4 (IL-4)	Mostly by T_H2 cells	Growth factor for B cells, some T cells, and mast cells; promotes antibody secretion
Interleukin-5 (IL-5)	T_H2 cells	Activates eosinophils; acts with IL-2 and IL-4 to stimulate growth and differentiation of B cells
Interleukin-6 (IL-6)	Macrophages, endothelial cells, fibroblasts, and T_H2 cells	Important growth factor for B cells late in their differentiation, activates CTLs
Interleukin-8 (IL-8)	Antigen-activated T cells, macrophages, endothelial cells, fibroblasts, and platelets	Activating and chemotactic factor for neutrophils, and to a lesser extent, other PMNs
Interleukin-10 (IL-10)	T_H2 cells	Inhibits T_H1, NK, and macrophage cytokine synthesis, promotes B-cell proliferation
Interleukin-12 (IL-12)	Macrophages and B cells	Activates NK cells and T cells; potently induces production of IFN-γ, shifts immune response to T_H1
Transforming growth factor-β (TGF-β)	Macrophages, lymphocytes, and other cells	Inhibits lymphocyte proliferation, CTL and LAK cell generation, and macrophage cytokine production
Interferon-α (IFN-α)	Body cells under viral attack	Activates NK cells and macrophages
Interferon-β (IFN-β)	Body cells under viral attack	Activates NK cells and macrophages
Interferon-γ (IFN-γ)	T_H1 cells and LAK cells	Strong macrophage-activating factor; causes a variety of cells to express class II MHC molecules; promotes T- and B-cell differentiation; activates neutrophils; activates endothelial cells to allow lymphocytes to pass through walls of vessels; antimicrobial activity
Tumor necrosis factor (TNF)	Activated macrophages and T_H1 cells	Major mediator of inflammation; low concentrations activate endothelial cells, activate PMNs, stimulate macrophages and cytokine production (including IL-1, IL-6, IL-12, and TNF itself); higher concentrations cause increased synthesis of prostaglandins, resulting in fever

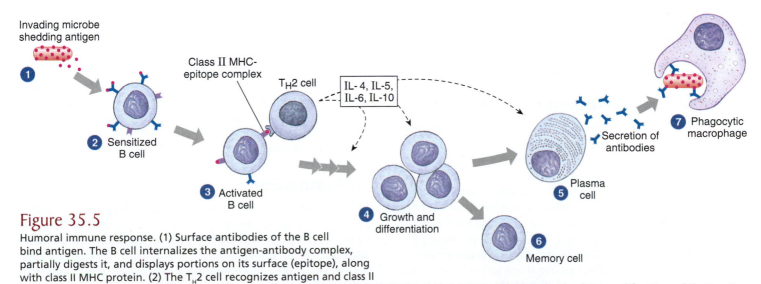

Figure 35.5

Humoral immune response. (1) Surface antibodies of the B cell bind antigen. The B cell internalizes the antigen-antibody complex, partially digests it, and displays portions on its surface (epitope), along with class II MHC protein. (2) The T_H2 cell recognizes antigen and class II protein on the B cell, is activated, and secretes interleukins 4, 5, 6, and 10 (IL-4, IL-5, IL-6, IL-10). (3) T_H2 then stimulates proliferation of the B cell, which carries antigen and class II protein on its surface. (4) IL-4, IL-5, IL-6, and IL-10 promote activation and differentiation of B cells into (5) many plasma cells that secrete antibodies. (6) Some B-cell progeny become memory cells. (7) Antibody produced by plasma cells binds to antigen and stimulates macrophages to consume antigen (opsonization).

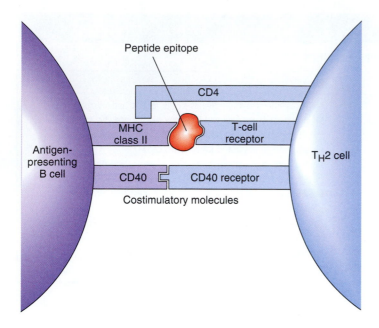

Figure 35.6

Interacting molecules during activation of a T$_H$2-helper cell by an antigen-presenting B cell.

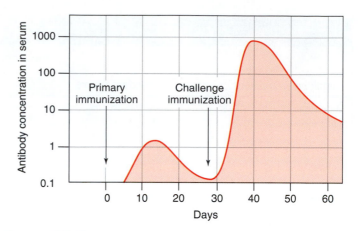

Figure 35.7

Typical antibody response after primary and challenge immunizations. The secondary response is a result of the large numbers of memory cells produced after the primary B-cell activation.

Many aspects of immunology have been greatly assisted by the discovery of a method for producing stable clones of cells that produce only one kind of antibody. Such monoclonal antibodies bind only to *one kind* of antigenic determinant (most proteins bear many different antigenic determinants and thus stimulate the body to produce complex mixtures of antibodies). *Monoclonal* antibodies are made by fusing normal antibody-producing plasma cells with a continuously growing plasma cell line, producing a hybrid of the normal cell with one that can divide indefinitely in culture. This cell line is called a *hybridoma*. Clones are selected from among the hybrids and are grown to become "factories" that produce almost unlimited quantities of one specific antibody. Hybridoma techniques discovered in 1975 have become one of the most important research tools for producing antibodies used by immunologists.

Cell-Mediated Response: T$_H$1 Arm

Many immune responses involve little, if any, antibody and depend on the action of cells only. In cell-mediated immunity the epitope of an antigen is also presented by APCs, but the T$_H$1 arm of the immune response is activated (see Figure 35.4). In this case APCs may be virally infected cells, tumor cells, or infected macrophages that have phagocytosed bacteria. T$_H$1 cells recognize the epitope-MHC II complex and become activated to release IL-2, TNF, and interferon-γ (INF-γ). IL-2 promotes the activity of activated B and T cells and enhances the cytotoxic activity of natural killer cells, causing them to proliferate and to become **lymphocyte-activated killer (LAK) cells.** LAK cells also release INF-γ. INF-γ is a potent macrophage-activating factor. It promotes B- and T-cell differentiation and B-cell proliferation and activates capillary endothelial cells so that lymphocytes

may pass through blood vessels into infected tissue areas. INF-γ is also a major inducer of nonspecific response called **inflammation.** TNF activates PMNs and stimulates macrophages and cytokine production. Cytotoxic T cells also interact with APC surface receptors and this, together with stimulation by IL-2 and INF-γ (secreted from activated T$_H$1 cells), causes them to proliferate and to secrete proteins that cause pores to form in the infected cells, resulting in lysis.

Transplantation of organs from one person to another requires immunosuppressing the recipient enough that the new organ is not rejected without leaving the patient defenseless against infection. Since discovery of a fungus-derived drug called cyclosporine, many organs, for example, kidneys, hearts, lungs, and livers can be transplanted. Cyclosporine inhibits IL-2 and affects CTLs more than T$_H$2 lymphocytes. It has no effect on other white blood cells or on healing mechanisms, so that a patient can still mount an immune response but not reject the transplant. However, the patient must continue to take cyclosporine because if the drug is stopped, the body will recognize the transplanted organ as foreign and reject it.

Like humoral immunity, cell-mediated immunity shows a secondary response due to large numbers of memory T cells produced from the original activation. For example, a second tissue graft (challenge) between the same donor and host is rejected much more quickly than the first.

Inflammation

Inflammation is a vital process in the innate immune response involved in mobilization of body defenses against an invading organism or other tissue damage and in repair of damage thereafter. The course of events in the inflammatory process is greatly influenced by the prior immunizing experience of the body with the invader and by the duration of the invader's presence or its persistence in the body. The innate processes by which an invader is actually destroyed, however, are themselves nonspecific but usually lead to activation of the acquired immune response. Manifestations of inflammation are **delayed type hypersensitivity (DTH)** and **immediate hypersensitivity,** depending on whether the response is cell mediated or antibody mediated.

The DTH reaction is a type of cell-mediated immunity in which the ultimate effectors are activated macrophages. The term *delayed type hypersensitivity* is derived from the fact that a period of 24 hours or more elapses from the time of antigen introduction until a response is observed in an immunized subject. This is because T_H1 cells with specific receptors in their surface for that particular antigen require some time to arrive at the antigen site, recognize epitopes displayed by APCs, and become activated to secrete IL-2, tumor necrosis factor (TNF), and IFN-γ. TNF causes endothelial cells of the blood vessels to express on their surface molecules to which leukocytes adhere: first neutrophils and then lymphocytes and monocytes. TNF also causes endothelium to secrete inflammatory cytokines such as IL-8, which increase the mobility of leukocytes and facilitate their passage through the endothelium. TNF and IFN-γ also cause endothelial cells to change shape, favoring leakage of macromolecules and passage of cells into the infected or damaged tissue. Escape of fibrinogen from blood vessels leads to conversion of fibrinogen to fibrin, and the area becomes swollen and firm.

As monocytes leave blood vessels, they become activated macrophages, which are the main effector cells of a DTH. They phagocytize a particulate antigen, secrete mediators that promote local inflammation, and cytokines and growth factors that promote healing. If the antigen is not destroyed and removed, its chronic presence leads to deposition of fibrous connective tissue, or **fibrosis.** Nodules of inflammatory tissue called **granulomas** may accumulate around persistent antigen and are found in numerous parasitic infections (Figure 35.8).

Immediate hypersensitivity is quite important in some parasitic infections. This reaction involves degranulation of mast cells in the area. Their surfaces bear receptors for Fc portions of antibody, especially IgE. Occupation of these sites by antigen-specific antibodies enhances degranulation of mast cells when the Fab portions bind the particular antigen. There is a rapid release of several mediators, such as histamine, that cause dilation of local blood vessels and increased vascular permeability. Escape of blood plasma into the surrounding tissue causes swelling **(wheal),** and engorgement of vessels with blood produces redness, the characteristic **flare** (Figure 35.9). If immediate hypersensitivity becomes more widespread in the body (systemic), it is called **anaphylaxis,** which may be fatal if not treated rapidly. During a more controlled inflammatory response, the purpose of swelling and change in permeability of capillaries is to allow antibodies and leukocytes to move from capillaries and easily to reach the invader. The first phagocytic line of defense is neutrophils, which may last a few days, then macrophages (either fixed or differentiated from monocytes) become predominant.

Some degree of cell death **(necrosis)** always occurs in inflammation, but necrosis may not be prominent if the inflammation is minor. When necrotic debris is confined within a localized area, pus (spent leukocytes and tissue fluid) may cause an increase in hydrostatic pressure, forming an **abscess.** An area of inflammation that opens out to a skin or mucous surface is an **ulcer.**

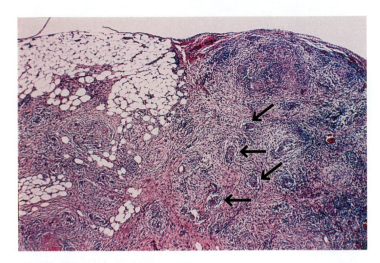

Figure 35.8
Granulomatous reaction around eggs (*arrows*) of *Schistosoma mansoni* in mesenteries.

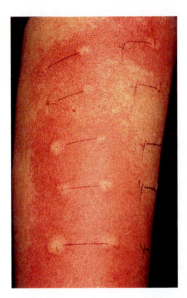

Figure 35.9
Wheal and flare reactions around sites of antigen injection for allergy testing.

Immediate hypersensitivity in humans is the basis for **allergies** and **asthma,** which are quite undesirable conditions, leading one to wonder why they evolved. Some scientists propose that the allergic response originally evolved to help the body ward off parasites because only allergens and parasite antigens stimulate production of large quantities of IgE. Avoidance of or reduction in effects of parasites would have conferred a selective advantage in human evolution. The hypothesis is that in absence of heavy parasitic challenge, the immune system is free to react against other substances, such as ragweed pollen. People now living where parasites remain abundant are less troubled with allergies than are those living in relatively parasite-free areas.

Acquired Immune Deficiency Syndrome (AIDS)

AIDS is an extremely serious disease in which the ability to mount an immune response is disabled severely. It is caused by **human immunodeficiency virus (HIV).** HIV preferentially invades and destroys T_H lymphocytes due to the virus expressing CD4 protein as a major surface receptor. Normally, T_H cells form 60% to 80% of the T-cell population; in AIDS their levels plummet to undetectable, so the humoral immune response is destroyed and the cell-mediated response is compromised.

The first case of AIDS was recognized in 1981, and the number of people infected with HIV/AIDS continues to rise. In 2001 approximately 35 million people were infected, and this number rose to 39 million people in 2005[1]. Only 10% of the world's population live in sub-Saharan Africa, but about two-thirds of all people infected with HIV/AIDS live in this region. HIV/AIDS is now rapidly spreading through the Indian subcontinent and Southeast Asia. In developing countries about 12,000 people become infected every day. HIV infection virtually always progresses to AIDS after a latent period of some years. Since T_H cell populations are virtually destroyed by HIV, AIDS patients are continuously plagued by infections with microbes and parasites that cause insignificant problems in persons with normal immune responses. If left untreated, AIDS is a terminal disease. There are now highly effective, though costly, drugs that can slow progression of the disease. Some drugs attack the enzyme needed by the virus to produce its DNA (for example, AZT), whereas others inhibit the enzymes required to assemble new virus (for example, protease inhibitors). Because the virus mutates (p. 100) to produce many different strains during progress of an infection, efforts to produce a vaccine have been fruitless.

BLOOD GROUP ANTIGENS

ABO Blood Types

Blood cells differ chemically from person to person, and when two incompatible blood types are mixed, **agglutination** (clumping together) of erythrocytes results. The basis of these chemical differences is naturally occurring antigens on the membranes of red blood cells. The best known of these inherited immune systems is the ABO blood group (p. 126). Antigens A and B are inherited as codominant alleles of a single gene. Homozygotes for a recessive allele at the same gene have type O blood, which lacks A and B antigens. Thus, as shown in Table 35.3, an individual with, for example, genes I^A/I^A or I^A/i develops A antigen (blood type A). The presence of an I^B gene produces B antigens (blood type B), and for the genotype I^A/I^B both A and B antigens develop on the erythrocytes (blood type AB). Epitopes of A and B also are present on the surfaces of many epithelial and most endothelial cells.

There is an odd feature about the ABO system. Normally we would expect that a type A individual would develop antibodies against type B cells only if cells bearing B epitopes were first introduced into the body. In fact, type A persons acquire anti-B antibodies soon after birth, even without exposure to type B cells. Similarly, type B individuals come to carry anti-A antibodies at a very early age. Type AB blood has neither anti-A nor anti-B antibodies (since if it did, it would destroy its own blood cells), and type O blood has both anti-A and anti-B antibodies. There is evidence that the antibodies develop as a response to A and B

epitopes on intestinal microorganisms when the intestine becomes colonized with bacteria after birth. Presumably, small and unnoticed infections with the bacteria occur. The antibodies thus produced cross-react with the A and B epitopes on erythrocytes.

We see then that the blood-group names identify their *antigen* content. Persons with type O blood are called universal donors because, lacking antigens, their blood can be infused into a person with any blood type. Even though it contains anti-A and anti-B antibodies, these are so diluted during transfusion that they do not react with A or B antigens in a recipient's blood. Persons with AB blood are universal recipients because they lack antibodies to A and B antigens. In practice, however, clinicians insist on matching blood types to prevent any possibility of incompatibility.

Rh Factor

Karl Landsteiner, an Austrian—later American—physician discovered ABO blood groups in 1900. In 1940, 10 years after receiving a Nobel Prize, he made still another famous discovery. This was a blood group called the Rh factor, named after rhesus monkeys, in which it was first found. Approximately 85% of white individuals in the United States have the factor (positive) and the other 15% do not (negative). The Rh factor is encoded by a dominant allele at a single gene. Rh-positive and Rh-negative bloods are incompatible; shock and even death may follow their mixing when Rh-positive blood is introduced into an Rh-negative person who has been sensitized by an earlier transfusion of Rh-positive blood. Rh incompatibility accounts for a peculiar and often fatal **hemolytic disease of the newborn (erythroblastosis fetalis).** If an Rh-negative mother has an Rh-positive baby (father is Rh-positive) she can become immunized by fetal blood during the birth process. Anti-Rh antibodies are predominately IgG and can cross the placenta during a subsequent pregnancy and agglutinate fetal blood. Erythroblastosis fetalis normally is not a problem in cases of ABO incompatibility between mother and fetus because antibodies to ABO antigens are primarily IgM and cannot cross the placenta.

The genetics of the Rh factor are very much more complicated than proposed when the factor was first discovered. Some authorities think that three genes located close together on the same chromosome are involved, whereas others adhere to a system of one gene with many alleles. In 1968 a revision of the single-gene concept listed 37 alleles necessary to explain the phenotypes then known. Furthermore, the frequency of the various alleles varies greatly between whites, Asians, and blacks.

Erythroblastosis fetalis can now be prevented by giving an Rh-negative mother anti-Rh antibodies just after the birth of her first child. These antibodies remain long enough to neutralize any Rh-positive fetal blood cells that may have entered her circulation, thus preventing her own antibody machinery from being stimulated to produce Rh-positive antibodies. Active, permanent immunity is blocked. The mother must be treated after every subsequent pregnancy (assuming the father is Rh-positive). If the mother has already developed an immunity, however, the baby may be saved by an immediate, massive transfusion of blood free of antibodies.

---

[1]UNAIDS 2006 Report on Global AIDS epidemic.

TABLE 35.3

Major Blood Groups

Blood Type	Genotype	Antigens on Red Blood Cells	Antibodies in Serum	Can Give Blood To	Can Receive Blood From	Frequency in United States (%)		
						Whites	Blacks	Asians
O	i/i	None	Anti-A and anti-B	All	O	45	48	31
A	I^A/I^A, I^A/i	A	Anti-B	A, AB	O, A	41	27	25
B	I^B/I^B, I^B/i	B	Anti-A	B, AB	O, B	10	21	34
AB	I^A/I^B	AB	None	AB	All	4	4	10

SUMMARY

A plethora of viral, prokaryotic, and eukaryotic parasites exists in every animal's environment, and a defense (immune) system is crucial to survival. Immunity can be defined concisely as possession of tissues capable of recognizing and protecting the animal against nonself invaders. Most animals have some amount of innate immunity, and vertebrates develop acquired immunity. The surface of most animals provides a physical barrier to invasion, and has a variety of antimicrobial substances in its body secretions.

Exposure of many animals, vertebrate and invertebrate, to many microorganisms stimulates their innate immune response. This response is based on release of antimicrobial peptides, is immediate, not requiring prior immunizing exposure, and is nonspecific but related to the category of invading microbe.

Phagocytes engulf particles and usually digest or kill them with enzymes and cytotoxic secretions. Many invertebrates have specialized cells that can perform defensive phagocytosis. Several kinds of vertebrate cells, especially macrophages and neutrophils, are important phagocytes, and cells of the mononuclear phagocyte system (formerly classified as the reticuloendothelial system) reside in various sites in the body. Eosinophils are important in allergies and many parasitic infections. Basophils, mast cells, T and B lymphocytes, and natural killer cells are not phagocytic but play vital roles in defense.

Many invertebrates show nonself recognition by rejection of xenografts or allografts or both. In some cases they may show enhanced response on repeated exposure. Bacterial, viral, and fungal infections stimulate the release of degradative enzymes, and antimicrobial peptides and activate phagocytes. Acquired responses that demonstrate memory and specificity have been demonstrated in some invertebrate phyla.

An acquired immune response is elicited by an antigen. Vertebrates demonstrate increased resistance to *specific* foreign substances (antigens) on repeated exposure, and the resistance is based on a vast number of specific recognition molecules:

antibodies and T-cell receptors. Nonself recognition depends on markers in cell surfaces known as major histocompatibility (MHC) proteins. Antibodies are borne on the surfaces of B lymphocytes (B cells) and in solution in the blood after secretion by the progeny of B cells, plasma cells. T-cell receptors occur only on the surfaces of T lymphocytes (T cells).

The cells of immunity communicate with each other and with other cells in the body by means of protein hormones called cytokines such as interleukins, tumor necrosis factor, and interferon-γ. The two arms of the vertebrate immune response are the humoral response (T_H2), involving antibodies, and the cell-mediated response (T_H1), involving cell surfaces only. When one arm is activated or stimulated, its cells produce cytokines that tend to suppress activity in the other arm. Activation of either arm requires that the antigen be phagocytosed by an APC (antigen-presenting cell, for example, a macrophage or a B cell), which partially digests the antigen and presents its determinant (epitope) on the surface of the APC along with an MHC class II protein. Virally infected cells and tumor cells also present antigen to T_H1 cells. Extensive communication by cytokines and activation (and suppression) of various cells in the response leads to production of specific antibody or proliferation of T cells with specific receptors that recognize an antigenic epitope. After the initial response, memory cells remain in the body and are responsible for enhanced response on next exposure to the antigen.

Damage to the immune response caused by HIV (human immunodeficiency virus) in production of AIDS (acquired immune deficiency syndrome) is due primarily to destruction of a crucial set of helper T cells, bearing the CD4 coreceptor on their surface.

Inflammation is an important part of the body's defense; it is greatly influenced by prior immunizing experience with an antigen.

People have genetically determined antigens in the surfaces of their red blood cells (ABO blood groups and others); blood types must be compatible in transfusions, or transfused blood will be agglutinated by antibodies in the recipient.

REVIEW QUESTIONS

1. Distinguish susceptibility from resistance, and innate from acquired immunity. Why are these traditionally recognized types of immunity now more difficult to distinguish?

2. What are some examples of innate defense mechanisms that are chemical in nature? What is complement?

3. After a phagocyte has engulfed a particle, what usually happens to the particle?

4. What are some important phagocytes in vertebrates?

5. What is the molecular basis of self and nonself recognition in vertebrates?

6. What is the difference between T cells and B cells?

7. What is a cytokine? What are some functions of cytokines?

8. Outline the sequence of events in a humoral immune response from the introduction of antigen to the production of antibody.

9. Define the following: plasma cell, secondary response, memory cell, complement, opsonization, titer, challenge, cytokine, natural killer cell, interleukin-2.

10. What are the functions of CD4 and CD8 proteins on the surface of T cells?

11. In general, what are consequences of activation of the T_H1 arm of the immune response? Activation of the T_H2 arm?

12. Distinguish between class I and class II MHC proteins.

13. Describe a typical inflammatory response.

14. What is a major mechanism by which HIV damages the immune system in AIDS?

15. Give the genotypes of each of the following blood types: A, B, O, AB. What happens when a person with type A gives blood to a person with type B? With type AB? With type O?

16. What causes hemolytic disease of the newborn (erythroblastosis fetalis)? Why does the condition not arise in cases of ABO incompatibility?

17. Give some evidence that cells of many invertebrates bear molecules on their surface that are specific to the species and even to a particular individual animal.

18. Give an example of immunological memory in invertebrates.

SELECTED REFERENCES

Aderem, A., and R. J. Ulevitch. 2000. Toll-like receptors in the induction of the innate immune response. Nature **406:**782–787. *A good review of the role of Toll-like receptors in the innate immune response.*

Alberts, B., A. Johnson, J. Lewis, M. Raff, K. Roberts, and P. Walter. 2002. Molecular biology of the cell, ed. 4. New York, Garland Publishing. *Concise discussion of complement system with clear figures.*

Cherry, S., and N. Silverman. 2006. Host-pathogen interactions in *Drosophila*: new tricks from an old friend. Nature Immnol. **7:**911–917. *Using an array of genetic tools and* Drosophila *as a model system, this review describes exciting new advances in the study of insect innate immune function.*

Devereux, G. 2006. The increase in the prevalence of asthma and allergy: food for thought. Nature Rev. **6:**869–874. *An interesting article that reviews current evidence for dietary changes causing an increase in asthma and allergies in developed countries.*

Flajnik, M. F., and L. Du Pasquier. 2004. Evolution of innate and adaptive immunity: can we draw a line? Trends Immunol. **25:**640–644. *Discusses how sequencing animal genomes has led to evidence of adaptive immunity genes and molecules in invertebrates and how the line between innate and adaptive immune systems is becoming blurred.*

Gartner, L. P., and J. L. Hiatt. 2001. Color textbook of histology. Philadelphia, W. B. Sanders Company. *Excellent chapters describing cells of the mammalian immune system and their functions.*

Gura, T. 2001. Innate immunity: ancient system gets new respect. Science **291:**2068–2071. *Innate immunity evolved early in animal evolution.*

Klotman, M. E., and T. L. Chang. 2006. Defensins in innate antiviral immunity. Nature Rev. **6:**447–456. *This review describes new data that suggest an antiviral role for mammalian defensins against both the virus and the host cell.*

Kurtz, J. 2005. Specific memory within innate immune systems. Trends Immunol. **26:**186–192. *An excellent review of evidence for memory within the innate immune system of invertebrates.*

Kurtz, J., and K. Franz. 2003. Evidence for memory in invertebrate immunity. Nature **425:**37–38. *Copepods show memory when infected with their natural parasite, a tapeworm.*

Letvin, N. L. 2006. Progress and obstacles in the development of an AIDS vaccine. Nature **6:**930–939. *Provides an excellent review of HIV biology and immune regulatory mechanisms, as well as a good discussion of progress toward an effective vaccine against AIDS.*

Litman, G. W., J. P. Cannon, and L. J. Dishaw. 2005. Reconstructing immune phylogeny: new perspectives. Nature Rev. Immunol. **5:**866–879. *Describes the interrelatedness of adaptive and innate immunity from an evolutionary perspective and provides a clear visualization of the similar selection pressures that have influenced pathogenicity and host protection mechanisms.*

Little, T. J., D. Hultmark, and A. F. Read. 2005. Invertebrate immunity and the limits of mechanistic immunology. Nature Immunol. **6:**651–654. *A review of some exciting data showing anticipatory behavior in the invertebrate innate immune system.*

Medzhitov, R., and C. A. Janeway Jr. 2002. Decoding the patterns of self and nonself by the innate immune system. Science **296:**298–300. *The several strategies employed by innate immune systems for discrimination of self and nonself.*

Roberts, J. P. 2004. Are HIV vaccines fighting fire with gasoline? The Scientist **18:**26–27. *Clear and concise summary of why a vaccine for HIV/AIDS is proving difficult to produce. Associated material includes recent information on vaccines that are currently being tested.*

Vizioli, J., and M. Salzet. 2002. Antimicrobial peptides from animals: focus on invertebrates. Trends Pharmacol. Sci. **23:**494–496. *A short review of invertebrate antimicrobial peptides and their possible use as antibiotics.*

Zasloff, M. 2002. Antimicrobial peptides of multicellular organisms. Nature **415:**389–395. *This short review discusses the molecules and mechanisms behind antimicrobial peptide activity in innate immune function and their possible use as antiinfective drugs.*

ONLINE LEARNING CENTER

Visit www.mhhe.com/hickmanipz14e for chapter quizzing, key term flash cards, web links, and more!

Animal Behavior

Sage grouse (Centrocercus urophasianus) *male displaying to females, North America.*

The Lengthening Shadow of One Person

Ralph Waldo Emerson said that an institution is the lengthening shadow of one person. For Charles Darwin the shadow is long indeed, for he brought into being entire fields of knowledge, such as evolution, ecology, and finally, after a long gestation, animal behavior. Above all, he altered how we think about ourselves, the earth we inhabit, and the animals that share it with us.

Charles Darwin, with the insight of genius, showed how natural selection would favor specialized behavioral patterns for survival. Darwin's pioneering book, *The Expression of the Emotions in Man and Animals*, published in 1872, mapped a strategy for behavioral research still used today. However, science in 1872 was unprepared for Darwin's central insight that behavioral patterns, no less than

bodily structures, are selected and have evolutionary histories. Another 60 years would pass before such concepts would flourish within behavioral science.

In 1973, the Nobel Prize in Physiology or Medicine was awarded to three pioneering zoologists, Karl von Frisch, Konrad Lorenz, and Niko Tinbergen (Figure 36.1). The citation stated that these three were the principal architects of the new science of ethology, the scientific study of animal behavior, particularly under natural conditions. It was the first time that any contributor to the behavioral sciences was so honored. The discipline of animal behavior had arrived.

A

B

C

Figure 36.1

Pioneers of the science of ethology. **A,** Konrad Lorenz (1903 to 1989). **B,** Karl von Frisch (1886 to 1982). **C,** Niko Tinbergen (1907 to 1988).

Behavioral biologists ask *how* animals behave and *why* they behave as they do. "How" questions are concerned with immediate or **proximate causation,** and are studied experimentally (p. 14). For example, a biologist might explain the singing of a male white-throated sparrow in the spring by hormonal or neural mechanisms. Such physiological or mechanistic causes of behavior are proximate factors. Proximate factors underlying animal behavior are covered in Chapters 33 and 34, which discuss nervous systems and hormonal coordination. Alternatively, a biologist might ask what function singing serves the sparrow, and then identify events in the ancestry of birds that led to springtime singing. These are "why" questions that focus on **ultimate causation,** the evolutionary origin and purpose of a behavior. Questions of ultimate causation are answered using comparative methodology (p. 14) applying phylogenetic analysis (p. 205) to understand evolutionary changes in behavior and their associated morphological and environmental contexts. The focus of this chapter is evolutionary explanation of animal behavior and the challenges of behavioral evolution for Darwin's theory of natural selection (p. 124).

The study of animal behavior has several different historical roots, and there is no universally accepted term for the whole subject. **Comparative psychology** emerged from efforts to find general laws of behavior that apply to many species, including humans. Replicable experimental approaches concentrated particularly on white rats, pigeons, dogs, and occasionally primates. Following criticisms that the discipline lacked an evolutionary perspective and focused too narrowly on white rats as a model for other organisms, many comparative psychologists developed phylogenetically based investigations, some of these conducted under natural conditions.

The aim of a second approach, **ethology,** is to describe the behavior of an animal in its *natural habitat.* Most ethologists have been zoologists who gather data by field observations and experiments. Nature provides the variables, which are manipulated using such approaches as presenting animal models, playing recordings of animal vocalizations, and altering habitats. Modern ethologists also test their hypotheses under controlled laboratory conditions. Laboratory results often direct further testing of hypotheses using observations of free-ranging animals in undisturbed natural environments.

Ethology emphasizes the importance of ultimate factors influencing behavior. A great contribution of von Frisch, Lorenz, and Tinbergen was to demonstrate that behavioral traits are measurable entities like anatomical or physiological traits. The central theme of ethology is that behavioral traits can be identified and measured, homologies determined, and their evolutionary histories investigated to provide causal explanations.

Much work by comparative psychologists and ethologists resides in the discipline of **behavioral ecology.** Behavioral ecologists determine how individuals behave to maximize reproductive and evolutionary success. Behavioral ecologists often concentrate on a particular aspect of behavior, such as mate choice, foraging, or parental investment. **Sociobiology,** the ethological study of social behavior, was formalized with the 1975 publication of E. O. Wilson's *Sociobiology: The New Synthesis.* Wilson describes social behavior as reciprocal communication of a cooperative nature (transcending mere sexual activity) that permits a group of organisms of the same species to become organized in a cooperative manner. In a complex system of social interactions, individuals are highly dependent on others for daily living. While social behavior appears in many groups of animals, Wilson identified four "pinnacles" of complex social behavior. These are (1) colonial invertebrates, such as the Portuguese man-of-war (p. 271), which is a composite of interdependent individual organisms; (2) social insects, such as ants, bees, and termites, which have developed sophisticated systems of communication; (3) nonhuman mammals, such as dolphins, elephants, and primates that have highly developed social systems; and (4) humans.

Wilson's inclusion of human behavior in sociobiology and his assumptions of genetic foundations for many human social behaviors have been strongly criticized. Complex systems of human social interactions, including religion, economic systems, and such objectionable characteristics as racism, sexism, and war, are emergent properties (p. 6) of human culture and its history. Is it meaningful to search for a specific genetic basis or justification for such phenomena? Many answer "no," and look instead to the field of sociology, rather than sociobiology, to help us understand complex, emergent properties of human societies.

DESCRIBING BEHAVIOR: PRINCIPLES OF CLASSICAL ETHOLOGY

Early ethologists sought to identify and to explain relatively invariant components of behavior shared by diverse animal species. From such studies emerged several concepts that were first popularized in Tinbergen's influential book, *The Study of Instinct* (1951).

Some basic concepts of animal behavior can be illustrated by the egg-retrieval response of greylag geese (Figure 36.2), described by Lorenz and Tinbergen in a famous paper in 1938. If Lorenz and Tinbergen presented a female greylag goose with an egg a short distance from her nest, she would rise, extend her neck until the bill was just over the egg and then bend her neck, pulling the egg carefully into the nest.

Although this behavior appeared to be rational, Tinbergen and Lorenz noticed that if they removed the egg after the goose had begun her retrieval, or if the egg being retrieved slipped away, the goose would continue the retrieval movement without

the egg until she was again settled comfortably on her nest. Then, seeing that the egg had not been retrieved, she would repeat the egg-rolling pattern.

Thus, the bird performed egg-rolling behavior as a program that, once initiated, had to be completed in a standard way. Lorenz and Tinbergen viewed egg-retrieval as a "fixed" pattern of behavior: a motor pattern mostly invariable in its performance. A behavior of this type, performed in an orderly, predictable sequence is called **stereotypical behavior.** Of course, stereotyped behavior may not occur identically on all occasions, but it should be recognizable, even when performed inappropriately. Further experiments by Tinbergen disclosed that the greylag goose was not particularly discriminating about what she retrieved. Almost any smooth and rounded object placed outside the nest would trigger egg-rolling behavior; even a small toy dog and a large yellow balloon were dutifully retrieved. When the goose settled down on such objects, they obviously did not feel right and she discarded them.

Lorenz and Tinbergen realized that an egg outside the nest must act as a stimulus, or trigger, to release egg-retrieval behavior. Lorenz termed the triggering stimulus a **releaser,** a simple stimulus in the environment that would trigger a certain innate behavior. Because the animal usually responded to some specific aspect of the releaser (sound, shape, or color, for example) the effective stimulus was called a **sign stimulus.** Ethologists have described hundreds of sign stimuli. In every case the response is highly predictable. For example, the alarm call of adult herring gulls always releases a crouching freeze response in their chicks. Certain nocturnal moths take evasive maneuvers or drop to the ground when they hear ultrasonic cries of bats that feed on them (p. 745); most other sounds do not release this response.

These examples illustrate the predictable and programmed nature of much animal behavior. This is even more evident when stereotyped behavior is released inappropriately. In spring a male three-spined stickleback, a small fish, selects a territory that it defends vigorously against other males. The underside of the male becomes bright red, and an approach of another red-bellied male releases a threat posture or even an aggressive attack. Tinbergen's suspicion that a male's red belly served as a releaser for aggression was reinforced when a passing red postal truck evoked attack behavior from males in his aquarium. Tinbergen then performed experiments using a series of models presented to the males. He found that they vigorously attacked any model bearing a red stripe, even a plump lump of wax with a red underside, yet

Figure 36.2

Egg-rolling behavior of the greylag goose, *Anser anser,* as studied by Lorenz and Tinbergen. In this stereotypical behavior, the egg outside the nest (*1*) is a sign stimulus for the goose to approach it (*2*) and to pull the egg toward the nest (*3–4*). The position shown in frame 4 is used to roll the egg toward the nest. The goose completes its return to the nest in this fashion even if the egg being retrieved rolls away.

Figure 36.3

Stickleback models used to study territorial behavior. The carefully made model of a stickleback (*left*), without a red belly, is attacked much less frequently by a territorial male stickleback than the four simple red-bellied models.

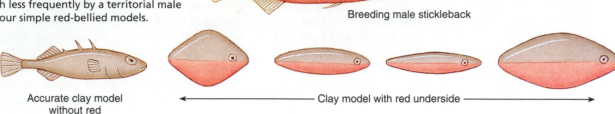

Breeding male stickleback

Accurate clay model
without red

← Clay model with red underside →

a model that closely resembled a male stickleback but lacked the red belly was attacked less frequently (Figure 36.3). Tinbergen discovered other examples of stereotyped behavior released by simple sign stimuli. Male English robins furiously attacked a bundle of red feathers placed in their territory but ignored a stuffed juvenile robin without red feathers (Figure 36.4).

We see in these examples costs to programmed behavior that leads to improper responses. Fortunately for red-bellied sticklebacks and red-breasted English robins, their aggressive response toward red works appropriately most of the time because red objects are uncommon in their environments. Why do these and other animals not use reasoning to choose an appropriate response? Under conditions that are relatively consistent and predictable, automatic preprogrammed responses may be most efficient. Even if they can or could, judging an appropriate response might take too much time. Releasers have the advantage of focusing an animal's attention on the relevant signal, and release of a preprogrammed stereotyped behavior enables an animal to respond rapidly when speed is essential for survival.

Figure 36.4

Two models of an English robin. The bundle of red feathers is attacked by male robins, whereas the stuffed juvenile bird (*right*) without a red breast is ignored.

From N. Tinbergen, The study of instinct, Oxford University Press, Oxford, England, 1951; modified from D. Lack, The life of the robin, H. F. & G. Witherby Ltd., London, England, 1943. Reprinted by permission of Oxford University Press.

CONTROL OF BEHAVIOR

Stereotyped behavior suggested to ethologists that they were observing inherited, or **innate,** behavior. Many kinds of preprogrammed behavior appear suddenly in an animal's ontogeny and are indistinguishable from similar behavior performed by older, experienced individuals. Orb-weaving spiders build their webs without practice, and male crickets court females without lessons from more experienced crickets or by learning from trial and error. To such behaviors the term innate, or instinctive, is applied.

Instinctive behaviors are, like organismal morphology, dependent on interactions between an organism and its environments during ontogeny. Although an instinct might appear rigid and fixed, instincts are products of evolutionary change and remain subject to further evolutionary change by selection. Dog breeds, for example, often have characteristic behavioral tendencies that differ among breeds according to the selection imposed by past breeders. Breeders of sheepdogs, for example, have constructed genetic combinations that reinforce behaviors useful in herding sheep and suppress behaviors destructive to this endeavor. Behavioral performance is modified further by training individual sheepdogs. Dogs of another breed might only rarely express behaviors conducive to sheep herding, yet all dogs share very recent common ancestry on evolutionary standards. Evolution of instincts is perhaps best conceived as a narrowing of behavioral repertoires so that the nervous system reinforces a particular subset of behaviors and limits use of alternatives. As environments and selective forces change, different behavioral modes will be genetically stabilized.

Many complex sequences of behavior in invertebrate animals are largely invariate in their execution and appear to follow precise rules without learning. Programmed behavior is important for survival, especially for animals that never know their parents. They must respond to stimuli immediately and correctly as they emerge. Animals with long lives and with parental care or other opportunities for social interactions nonetheless may improve or change their behavior by learning.

The Genetics of Behavior

Hereditary transmission of most innate behavior is complex, with many interacting genes and environmental factors influencing each behavioral trait. However, a few examples of behavioral differences within species show simple Mendelian transmission from parents to offspring. Perhaps the most convincing example

Figure 36.5

The genetics of hygienic behavior in female worker honey bees, as demonstrated by W. C. Rothenbuhler. The results are explained by assuming that there are two independently assorting genes, one associated with uncapping cells containing diseased larvae, and the other associated with removing diseased larvae from cells. Hemizygous male progeny are not shown. See text for further explanation.

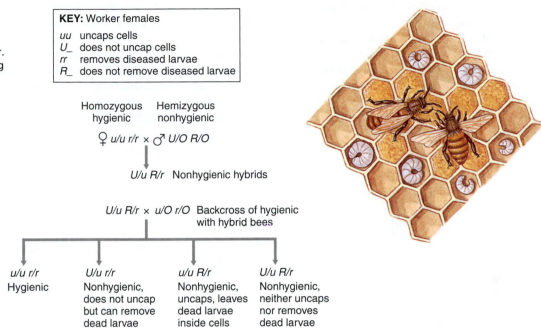

KEY: Worker females

- *uu*   uncaps cells
- *U_*   does not uncap cells
- *rr*   removes diseased larvae
- *R_*   does not remove diseased larvae

Homozygous   Hemizygous
hygienic   nonhygienic

♀ *u/u r/r* × ♂ *U/O R/O*

U/u R/r   Nonhygienic hybrids

U/u R/r × *u/O r/O*   Backcross of hygienic
with hybrid bees

u/u r/r	*U/u r/r*	*u/u R/r*	*U/u R/r*
Hygienic	Nonhygienic, does not uncap but can remove dead larvae	Nonhygienic, uncaps, leaves dead larvae inside cells	Nonhygienic, neither uncaps nor removes dead larvae

is the inheritance of hygienic behavior in bees. Honey bees are susceptible to a bacterial disease, American foulbrood (*Bacillus larvae*). A bee larva that catches foulbrood dies. If bees remove dead larvae from their hive they reduce the chance of an infection spreading.

Some strains of bees, called "hygienic," uncap hive cells containing rotting larvae and remove them from the hive. W. C. Rothenbuhler described two components to this behavior: first removal of cell caps, and second removal of larvae. Hygienic bees have homozygous recessive genotypes for two different genes. Uncapping behavior is performed by individuals homozygous for a recessive allele, *u,* at one gene, and removal behavior is performed by individuals homozygous for a recessive allele, *r,* at a second gene. When Rothenbuhler crossed hygienic bees (*u/u r/r*) with a nonhygienic strain (*U/U R/R*), he found that all hybrids (*U/u R/r*) were nonhygienic. Thus only workers having both genes in the homozygous recessive condition show the complete behavior. Next, Rothenbuhler performed a "backcross" between the hybrids and the hygienic parental strain. As we expect if hygienic behavior is transmitted by allelic variation of two genes, four different kinds of bees resulted (Figure 36.5). Approximately one-quarter of the bees were homozygous recessive for both *u* and *r* and showed the complete behavior: they both uncapped the cells and removed infected larvae. Another quarter of the offspring (*u/u R/r*) uncapped but did not remove dead bees. Another quarter (*U/u r/r*) did not uncap, but would remove the larvae if another worker uncapped the cells. Workers that were heterozygous for the dominant allele at both genes (*U/u R/r*) would not perform either part of the cleaning behavior (Figure 36.5). The results show clearly that each component of the cleaning behavior is associated with one gene segregating independently of the gene influencing the other behavioral component.

Most inherited behaviors do not show simple segregation and independence; instead, hybrids of subspecies or species commonly show intermediate or confused behavior. A classic study by W. C. Dilger on nest-building behavior in different species of lovebirds revealed such an outcome. Lovebirds are small parrots of the genus *Agapornis* (Figure 36.6). Each species has its own method of courtship and technique for carrying nesting material. Fischer's lovebirds, *A. personata fischeri,* cut long strips of nesting material from vegetation, then carry to the nest one strip at a time. Peach-faced lovebirds, *A. roseicollis,* carry several strips of torn nesting material simultaneously by tucking them into feathers of the lower back and rump. Dilger, who crossed

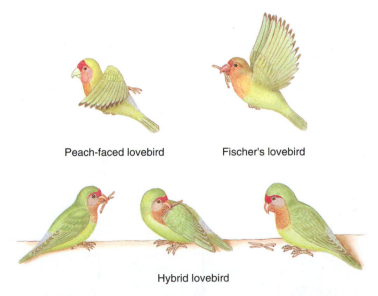

Peach-faced lovebird   Fischer's lovebird

Hybrid lovebird

Figure 36.6

Confused behavior in hybrid lovebirds, *Agapornis* sp. The peach-faced lovebird carries nest-building material tucked into its feathers; Fischer's lovebird carries nest-building material in its beak. The hybrids attempted both carrying methods; neither method was initially accomplished successfully.

the two species successfully, found that hybrids displayed a confused conflict between a tendency to carry material in the feathers (inherited from the peach-faced lovebirds) and a tendency to carry material in the bill (inherited from Fischer's lovebird) (Figure 36.6). Hybrids attempted both feather-tucking and bill-carrying but performed neither behavior correctly. Hybrids inherited a behavior intermediate between those of the parents. With experience hybrids improved their carrying ability by tending to carry the material in their bills, like Fischer's lovebird.

Learning and Diversity of Behavior

Another aspect of behavior is learning, which we define as modification of behavior through experience. An excellent model system for studying learning processes is the marine opistho-branch snail, *Aplysia* (Figure 36.7), a subject of intense experimentation by E. R. Kandel and his associates. Gills of *Aplysia* are partly covered by the mantle cavity and open to the outside by a siphon (Figure 36.8). If one prods the siphon, *Aplysia* withdraws its siphon and gills and folds them in the mantle cavity.

This simple protective response, called gill-withdrawal reflex, is repeated when *Aplysia* extends its siphon again. If the siphon is touched repeatedly, however, *Aplysia* decreases the gill-withdrawal response and ignores the stimulus. This behavioral modification illustrates a widespread form of learning called **habituation.** If *Aplysia* is given a noxious stimulus (for example, an electric shock) to the head at the same time the siphon is touched, it becomes **sensitized** to the stimulus and withdraws its gills as completely as it did before habituation occurred. Sensitization, then, can reverse previous habituation.

The nervous pathways of habituation and sensitization in *Aplysia* are known. Receptors in the siphon connect through sensory neurons (black pathways in Figure 36.8) to motor neurons (blue pathway in Figure 36.8) that control the gill-withdrawal muscles and muscles of the mantle cavity. Kandel found that repeated stimulation of the siphon diminished the release of synaptic transmitter from sensory neurons. Sensory neurons continue to fire when the siphon is probed but, with less neurotransmitter being released into the synapse, the system becomes less responsive.

Sensitization requires action of a different kind of neuron called a facilitating interneuron. These interneurons make connections between sensory neurons in the snail's head and motor neurons that control muscles of the gill and mantle (Figure 36.8). When sensory neurons in the head are stimulated by an electric shock, they fire facilitating interneurons, which end on the synaptic terminals of sensory neurons (red pathways in Figure 36.8). These endings in turn *increase* the amount of transmitter released by the siphon sensory neurons. This release increases excitation of the excitatory interneurons and motor neurons leading to the gill and mantle muscles. Motor neurons now fire more readily than before. The system is now sensitized because any stimulus to the siphon produces a strong gill-withdrawal response.

Kandel's studies indicate that strengthening or weakening of the gill-withdrawal reflex involves changes in levels of transmitter in existing synapses. However, more complex kinds of learning may involve formation of new neural pathways and connections, as well as changes in existing circuits.

Figure 36.7

The sea hare, *Aplysia* sp., an opisthobranch gastropod used in many neurophysiological and behavioral studies.

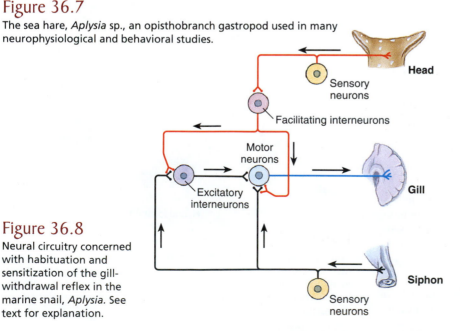

Figure 36.8

Neural circuitry concerned with habituation and sensitization of the gill-withdrawal reflex in the marine snail, *Aplysia*. See text for explanation.

Head

Sensory neurons

Facilitating interneurons

Motor neurons

Excitatory interneurons

Gill

Siphon

Sensory neurons

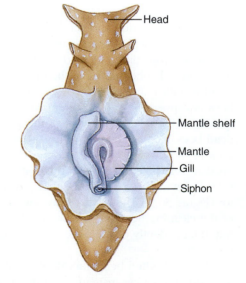

Head

Mantle shelf

Mantle

Gill

Siphon

Imprinting

Another kind of learned behavior is **imprinting,** the imposition of a stable behavior in a young animal by exposure to particular stimuli during a critical period in the animal's development. As soon as a newly hatched gosling or duckling is strong enough to walk, it follows its mother when away from the nest. After it has followed the mother for some time it follows no other animal. However, if the eggs are hatched in an incubator or if the mother is separated from the eggs as they hatch, the goslings follow the first large object they see. As they grow, the young geese prefer the artificial "mother" to anything else, including their true mother. The goslings are said to be imprinted on the artificial mother.

Imprinting was observed at least as early as the first century A.D. when the Roman naturalist Pliny the Elder wrote of "a goose which followed Lacydes as faithfully as a dog." Konrad Lorenz was the first to study imprinting systematically. When Lorenz hand-reared goslings, they formed an immediate and permanent attachment to him and waddled or swam after him wherever he went (Figure 36.9). They could no longer be induced to follow their own mother or another human being. Lorenz found that imprinting occurs in a brief sensitive period early in an individual's life and that once established the imprinted bond usually persists throughout life.

What imprinting shows is that a goose's brain (or the brain of numerous other birds and mammals that show imprinting-like behavior) accommodates the imprinting experience. Natural selection favors evolution of a brain that imprints in this way, in which following the mother and obeying her commands are important for survival. The fact that a gosling can be made to imprint to a mechanical toy duck or a person under artificial conditions is a cost that is tolerated because goslings seldom encounter these stimuli in their natural environments. The disadvantages of the system's simplicity are outweighed by the advantages of its reliability.

One final example completes our consideration of learning. Songbirds demonstrate robust sex differences in many behaviors. Males of many species of birds have characteristic territorial songs that identify the singers to other birds and announce territorial rights to other males of that species. Like many other songbirds, a male white-crowned sparrow must learn the song of its species by hearing the song of its father. If a sparrow is hand-reared in acoustic isolation in a laboratory, it develops an abnormal song (Figure 36.10); if an isolated bird is allowed to hear recordings of normal white-crowned sparrow songs during a critical period of 10 to 50 days after hatching, it learns to sing normally. It even imitates the local dialect that it hears.

Characteristics of the song are not determined by learning alone. If during the critical learning period, an isolated male white-crowned sparrow hears a recording of another species of sparrow, even a closely related one, it does not learn the song. It learns only the song appropriate to its own species. Thus although the song must be learned, the brain is constrained to recognize and to learn vocalizations produced by males of its species alone. Learning a wrong song would produce behavioral chaos, and natural selection favors a system that eliminates such errors. Another example of complex interactions between learned and innate factors is illustrated by navigation of seasonally migratory birds (p. 601).

Figure 36.9

Ducklings imprinted on Konrad Lorenz follow him as faithfully as they would a natural mother.

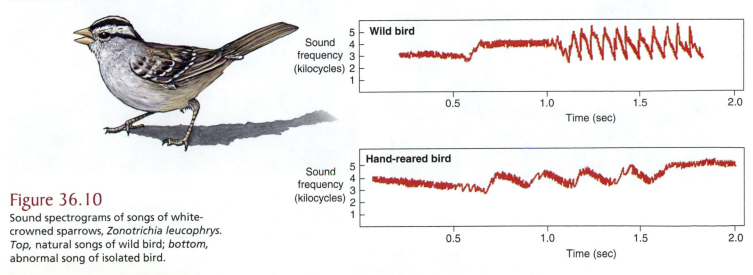

Figure 36.10

Sound spectrograms of songs of white-crowned sparrows, *Zonotrichia leucophrys.* *Top,* natural songs of wild bird; *bottom,* abnormal song of isolated bird.

SOCIAL BEHAVIOR

When we think of "social" animals we tend to think of highly structured honey bee colonies, herds of antelope grazing on the African plains (Figure 36.11), schools of herring, or flocks of starlings. Social behavior of animals *of the same species* living together is by no means limited to such obvious examples. Any interaction resulting from a response of one animal to another of the same species is a social behavior. Even a pair of rival males fighting for possession of a female displays a social interaction.

Social aggregations are only one kind of social behavior, and indeed not all aggregations of animals are social. Clouds of moths attracted to a light at night, barnacles attracted to a common float, or trout gathering in the coolest pool of a stream are nonsocial groupings of animals responding to environmental signals. Social aggregations depend on signals from the animals themselves, causing the animals to remain together and to influence one another.

Not all animals showing sociality are social to the same degree. While all sexually reproducing species must at least cooperate enough to achieve fertilization, some animals limit their adult sociality to breeding. Alternatively, swans, geese, albatrosses, and beavers, to name just a few, form strong monogamous bonds that last a lifetime. The most persistent social bonds usually form between mothers and their young and, for birds and mammals, these bonds usually terminate at fledging or weaning.

Figure 36.11
Mixed herd of topi and common zebra grazing on a savanna of tropical Africa.

Selective Consequences of Sociality

Social aggregations are beneficial for defense, both passive and active, from predators. Musk-oxen form a passive defensive circle when threatened by a wolf pack and are much less vulnerable than an individual facing wolves alone.

As an example of active defense, a breeding colony of gulls, alerted by alarm calls of a few, attack predators *en masse;* this collective attack discourages a predator more effectively than individual attacks. Members of a town of prairie dogs, although divided into social units called coteries, cooperate by warning each other with a special bark when danger threatens. Thus every individual in a social organization benefits from the eyes, ears, and noses of all other members of the group. Experimental tests using a wide variety of predators and prey support the notion that the larger the group, the less likely is an individual within the group to be eaten.

Sociality offers several potential benefits to reproduction. It facilitates encounters between males and females, which, for solitary animals, can consume much time and energy. Sociality also synchronizes reproductive behavior through the mutual stimulation that individuals give one another. Among colonial birds the sounds and displays of courting individuals set in motion reproductive endocrine changes in other individuals. Because there is more social stimulation, large colonies of gulls produce more young per nest than do small colonies. Furthermore, parental care that social animals provide their offspring increases survival of the brood (Figure 36.12). Social living provides opportunities for individuals to give aid and to share food with young other than their own. Such interactions within a social network have produced some intricate cooperative behavior among parents, their young, and their kin.

Of the many other advantages of social organization noted by ethologists, we mention only a few in this brief treatment: cooperation in hunting for food; huddling for mutual protection

Figure 36.12
An infant yellow baboon, *Papio cyanocephalus,* "jockey rides" its mother. Later, as the infant is weaned, the mother-infant bond weakens and the infant will be refused rides.

from severe weather; opportunities for division of labor, which is especially well developed in social insects; and the potential for learning and transmitting useful information through a society.

Observers of a seminatural colony of macaque monkeys in Japan recount an interesting example of acquiring and passing tradition in a society. The macaques were given sweet potatoes and wheat at a feeding station on the beach of an island colony. One day a young female named Imo was observed washing the sand off a sweet potato in seawater. The behavior was quickly imitated by Imo's playmates and later by Imo's mother. Still later when the young members of the troop became mothers they waded into the sea to wash their potatoes; their offspring imitated them without hesitation. The tradition became firmly established in the troop (Figure 36.13).

Some years later, Imo, as an adult, discovered that she could separate wheat from sand by tossing a handful of sandy wheat in the water; allowing the sand to sink, she could retrieve the floating wheat to eat. Again, within a few years, wheat-sifting became a tradition in the troop.

Imo's peers and social inferiors copied her innovations most readily. The adult males, her superiors in the social hierarchy, would not adopt the practice but continued laboriously to pick wet sand grains off their sweet potatoes and to scout the beach for single grains of wheat.

Acquisition of food-cleaning skills by Imo and her peers shows that a social setting provides opportunities for acquisition and sharing of complex learned behaviors, ones that transcend simple imprinting and habituation. Imo's food-cleaning behaviors demonstrate a conditioned response, learning that certain methods lead repeatedly to a desired result, plus reasoning and insight for evaluating which methods are useful for cleaning various food items.

Figure 36.13
Japanese macaque washing sweet potatoes. The tradition began when a young female named Imo began washing sand from the potatoes before eating them. Younger members of the troop quickly learned the behavior.

Social living also can have disadvantages as compared with a solitary existence for some animals. Species that survive by camouflage from potential predators profit by being dispersed. Large predators benefit from a solitary existence for a different reason, their requirement for a large supply of prey. Thus there is no overriding adaptive advantage to sociality that inevitably disfavors a solitary way of life. Advantages and disadvantages of sociality depend upon the ecological situation.

Ethologist Tim Clutton-Brock distinguishes **socially coordinated behavior,** in which an individual adjusts its actions to the presence of others to increase its own reproductive success directly, from **cooperative behavior,** in which an individual performs activities that benefit others because such behavior ultimately benefits that individual's genetic contributions to future generations. The former category includes agonistic and competitive behaviors, territoriality, and formation of various mating systems. The latter category includes cooperative foraging and breeding behaviors, and especially behaviors that may benefit an individual's close relatives or cause others to reciprocate a beneficial behavior. If an individual's presence in a group benefits its survival and reproductive success, selection should favor evolution of cooperative strategies.

Agonistic or Competitive Behavior

Animals may compete for food, water, sexual mates, or shelter when such requirements are limited and therefore worth a fight. Much of what animals do to resolve competition is called **aggression,** which we define as an offensive physical action, or threat, to force others to abandon something they own or might attain. Many ethologists consider aggression part of a somewhat more inclusive category called **agonistic** (Gr. contest) behavior, any activity related to fighting, whether it be aggression, defense, submission, or retreat.

Most aggressive encounters lack the violence that we usually associate with fighting. Many species possess specialized weapons such as sharpened teeth, beaks, claws, or horns used for protection from, or predation on, other species. Although potentially dangerous, such weapons are seldom used in any severely damaging way against members of *their own species.*

Animal aggression within a species seldom produces injury or death because animals have evolved many symbolic **ritualized threat displays** that carry mutually understood meanings to establish a dominance hierarchy within the population. A ritualized display is a behavior that has been modified through evolution to make it increasingly effective in serving a communicative function. Through **ritualization,** simple movements or traits become more intensive, conspicuous, or precise, and acquire the function of a signal to reduce misunderstanding. Fights over mates, food, or territory become ritualized jousts rather than bloody, no-holds-barred battles. When fiddler crabs (see Figure 20.28) spar for reproductive territory on intertidal sands, their large claws usually are only slightly opened. Even in intense fighting when claws are used, the crabs grasp each other in a way that prevents reciprocal injury. Male venomous snakes competing for a female engage in stylized bouts by winding themselves together; each attempts to butt the other's head with its own until one becomes pinned to the ground and retreats. The rivals do not bite each other. Males of many fish species contest territorial boundaries with lateral threat displays, puffing themselves to look as threatening as possible. The encounter is usually settled when either animal perceives itself obviously inferior in the social hierarchy, withdraws, and swims away. Rival giraffes engage in largely symbolic "necking" matches in which two males standing side by side wrap and unwrap their necks (Figure 36.14). Neither uses its potentially lethal hooves on the other, and neither is injured.

Thus animals fight as though programmed by rules that prevent serious injury. Fights between rival bighorn rams are spectacular to watch, and the sound of clashing horns may be heard for hundreds of meters (Figure 36.15), but the skull is so well protected by massive horns that injury occurs only by accident. Nevertheless, despite these constraints, aggressive encounters on occasion can be true fights to the death of a rival. If African male elephants are unable to resolve dominance conflicts painlessly with ritual postures, they resort to incredibly violent battles, with each trying to plunge its tusks into the most vulnerable parts of the opponent's body.

More commonly, however, the loser of a ritualized encounter may simply run away, or signal defeat by a specialized subordination ritual. A likely loser profits by communicating submission as quickly as possible, thereby avoiding a thrashing. Such submissive displays that signal the end of a fight can be almost the opposite of threat displays (Figure 36.16). In his book, *The Expression of the Emotions in Man and Animals* (1872), Charles Darwin described the seemingly opposite nature of threat

Figure 36.15
Male bighorn sheep, *Ovis canadensis,* fight for social dominance during the breeding season.

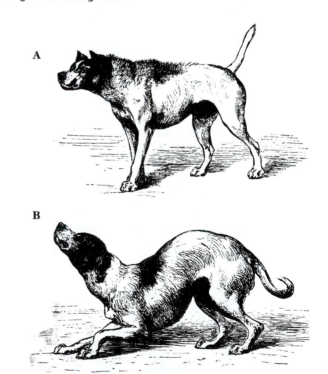

Figure 36.16
Darwin's illustration of his principle of antithesis as shown by the postures of dogs. **A,** A dog approaches another dog with hostile, aggressive intentions. **B,** The same dog is in a humble and conciliatory state of mind. The signals of aggressive display have been reversed.

Figure 36.14
Male Masai giraffes, *Giraffa camelopardalis,* fight for social dominance. Such fights are largely symbolic, seldom causing injury.

and appeasement displays as the "principle of antithesis." The principle remains accepted by ethologists today.

The winner of an aggressive competition is dominant to the loser, the subordinate. For the victor, dominance means enhanced access to all contested resources that contribute to reproductive success: food, mates, and territory. In a social species, dominance interactions often form a **dominance hierarchy.** One animal at the top wins encounters with all other members in the social group; the second in rank wins all but those with the top-ranking individual.

Such a simple, ordered hierarchy was first observed in chickens by Schjelderup-Ebbe, who called the hierarchy a "peck-order." Once social ranking is established, actual pecking diminishes and is replaced by threats, bluffs, and bows. Top hens and roosters get unquestioned access to food and water, dusting areas, and the roost. The system works because it reduces social tensions that would constantly surface if animals had to fight constantly over resources.

Not all dominance hierarchies have clear-cut dominant and subordinate individuals. In some hierarchies, dominant animals are frequently challenged by subordinates.

Subordinates in any social order may be expendable. In many systems they never reproduce and are often the first to die. During times of food scarcity, death of weaker members protects resources for stronger members. Rather than sharing food, the excess population is sacrificed. This sacrifice results from the individual advantage that stronger, dominant individuals possess during such circumstances.

Territoriality

Territorial ownership is another facet of sociality in animal populations. A territory is a fixed area whose occupant excludes intruders *of the same species* and sometimes other species. This exclusion involves defending the area from intruders and being conspicuous on the site. Territorial defense occurs in numerous animals: insects, crustaceans, fishes, amphibians, lizards, birds, and mammals, including humans.

Sometimes the space defended moves with the individual. Individual distance, as one example, can be observed in the spacing between swallows or pigeons on a wire, gulls on a beach, or people waiting for a bus.

Territoriality is generally an alternative to dominance behavior, although both systems can operate in the same species. A territorial system may work well when the population is low, but it may fail with increasing population density and be replaced with dominance hierarchies when all animals occupy a common space.

Like every other competitive endeavor, territoriality carries both costs and advantages. It is beneficial when it ensures access to limited resources, *unless* territorial boundaries cannot be maintained with little effort. Presumed benefits of a territory are, in fact, numerous: uncontested access to a foraging area; enhanced attractiveness to females thus reducing problems of pair-bonding, mating, and rearing young; reduced disease transmission; reduced vulnerability to predators. Advantages of holding a territory wane if an individual must spend most of its time in boundary disputes with neighbors and other intruders.

Most of the time and energy required for territoriality are expended when the territory is first established. Once boundaries

are located they tend to be respected, and aggressive behavior diminishes as territorial neighbors come to recognize each other. Indeed, neighbors may look so peaceful that an observer not present when the territories were established may conclude incorrectly that the animals are not territorial. A "beachmaster" sea lion (a dominant male with many females) seldom quarrels with neighbors, who have their own territories to defend. However, he must be constantly vigilant against bachelor bulls who challenge his mating privileges.

Birds are conspicuously territorial. Most male songbirds establish territories in early spring and defend these vigorously against males of the same species during spring and summer when mating and nesting occur. A male song sparrow, for example, has a territory of approximately three-fourths of an acre. In any given area, the number of song sparrows remains approximately the same each year. The population remains stable because the young occupy territories of adults that die. Any surplus in the song sparrow population is excluded from territories and thus not able to mate.

Sea birds such as gulls, gannets, boobies, and albatrosses occupy colonies that are divided into very small territories just large enough for nesting (Figure 36.17). Territories of these birds cannot include their fishing grounds, since they all forage at sea where their food is always shifting in location and shared by all.

Territorial behavior is not as prominent with mammals as it is with birds. Mammals are less mobile than birds, making it more difficult for them to patrol a territory for trespassers. Instead,

Figure 36.17

Gannet nesting colony. Note precise spacing of nests with each occupant just beyond pecking distance of its neighbors.

many mammals have **home ranges** (p. 630). A home range is the total area that an individual traverses in its activities. It is not an exclusive, defended preserve but overlaps home ranges of other individuals of the same species.

For example, home ranges of baboon troops overlap extensively, although a small part of each range becomes the recognized territory of each troop for its exclusive use. Home ranges may shift considerably among the seasons. A baboon troop may have to shift to a new range during the dry season to obtain water and better grass. Elephants, before their movements were restricted by humans, made long seasonal migrations across the African savanna to new feeding ranges. However, home ranges established for each season are remarkably consistent in size.

Mating Systems

Animals display diverse mating systems. Behavioral ecologists generally classify mating systems by the ways that males and females associate for mating. **Monogamy** is an association between one male and one female at a time. **Polygamy** is a general term that incorporates all multiple mating systems where females or males have more than one mate. **Polygyny** refers to a male that mates with more than one female. **Polyandry** is a system in which a female mates with more than one male. There are specific types of polygyny. **Resource-defense polygyny** occurs when males gain access to females indirectly by holding critical resources. For example, female bullfrogs prefer to mate with males who are larger and older. These males defend territories of higher quality than smaller males because their territories have favored temperatures or are free of predatory leeches. **Female-defense polygyny** occurs when females aggregate and, consequently, are defendable. Thus, when female elephant seals occupy a small island, dominant males can defend and mate with them relatively easily (Figure 36.18). This situation was previously called a "harem." **Male-dominance polygyny** occurs when females select mates from aggregations of males competing for an opportunity to mate. For example, some animals form **leks.** A lek is a communal display ground where males congregate to attract and to court females. Females choose and mate with the male having the most attractive qualities (Figure 36.19). Leks characterize some birds, including prairie chickens and sage grouse. In these systems, sexual selection (p. 130) is often intense, producing evolution of bizarre courtship rituals and exaggerated morphological traits.

Cooperative Behavior, Altruism, and Kin Selection

If, as Darwin suggested, animals should behave selfishly and strive to produce as many offspring as possible, why do some animals help others at some risk to themselves? As noted earlier (p. 792), general advantages of a group context may explain selection for cooperative behavior; however, some forms of cooperative behavior are so extreme that they would seem to require an additional explanation. Why do some individuals forego breeding to serve the reproductive success of others? Why do some individuals appear to sacrifice themselves so that other members of their group can survive? Until the mid 1960s, scientists had trouble explaining in Darwinian terms how such **altruistic behavior** might persist in a population.

Most altruistic behaviors formerly were explained using a **group-selection** argument. Group selectionists suggested that animals that helped others or that failed to mate did so for the

Figure 36.18

Two elephant seals, *Mirounga angustirostris,* fight to establish dominance. Males are much larger than females in this highly polygynous society.

Figure 36.19

Male sage grouse, *Centrocercus urophasianus,* displaying at its lek.

benefit of the group as a whole. Therefore, such behaviors produced increased survivorship of groups whose members behaved altruistically. According to proponents of this argument, selection occurs at the level of the group, not at the level of the individual as Darwin suggested. However, the group-selection argument as originally proposed by V. C. Wynne-Edwards in 1962 has been rejected by most behavioral ecologists for several reasons.

For example, if alleles associated with a risky altruistic behavior, such as giving calls to warn others of predators, were distributed randomly in a social group, those lacking such genes would flourish. They would be warned with no risk to themselves; their chances of reproduction would be greater and, in time, the "selfish" alleles would eliminate the altruistic ones from the group's gene pool.

In 1964, W. D. Hamilton, based largely on his studies of insects, proposed a new way to explain altruistic behavior by modifying Darwin's original concept of fitness. He reasoned that fitness is measured not just by the number of offspring produced but by an increase or decrease of particular alleles in the gene pool. Thus, an individual may act altruistically, even at great risk, if it helps to increase representation of its alleles in the gene pool. Alleles are shared by all relatives, including parents and offspring, brothers and sisters, cousins and other relations. Alleles that influence altruistic behavior among relatives would persist in future generations. Because the most closely related animals share the most genes by common descent, we expect that altruistic behavior would be most common among closely related individuals. Parental behaviors that benefit survival of offspring, as studied in birds (p. 605), are an obvious example. Thus, if everything else were equal, brothers who on average share half their alleles would be more likely to aid one another than they would a cousin who shares on average only 25% of their alleles. Hamilton's hypothesis based on this genetic explanation for altruism and cooperation is called **kin selection.** Essentially, kin selection operates by individuals assisting the survival and reproduction of other individuals who possess the same genes by common descent.

Hamilton's hypothesis revolutionized evolutionary and behavioral biology. The main criterion of Darwinian fitness in genetic models of natural selection is the relative number of an individual's alleles that are passed to future generations. Hamilton, however, developed the concept of **inclusive fitness,** which is the relative number of an individual's alleles that are passed to future generations either as a result of an individual's own reproductive success *or that of related individuals.* Thus, kin selection and inclusive fitness may explain many altruistic behaviors that have perplexed biologists for many years.

A good example of altruism and kin selection in nature is the remarkable cooperation and coordination among euscocial insects such as ants, bees, and wasps. Through haplodiploidy (p. 142), where males are haploid and females are diploid, full sisters are genetically related on average by 75% rather than

50% (Figure 36.20). Sisters are more closely related to each other than to their own daughters! Therefore, they cooperate with other members of their social group, forego breeding themselves and aid their queen to produce more sisters who are more closely related (75% related) than potential offspring (50% related). This explanation is challenged by molecular genetic discoveries that in many haplodiploid insects, reproductive females have multiple mates. The nonreproductive females who tend their mother's younger offspring indiscriminately are therefore not likely to be tending only their full siblings. An average relatedness between nonreproductive workers and their tended siblings would need to exceed 50% for kin selection to provide a convincing explanation for reproductive division of labor (often called **eusociality**) in Hymenoptera.

Female Belding's ground squirrels, found in the High Sierra of California, give alarm calls when a predator approaches (Figure 36.21). Alarm calling warns other members of the social group and is risky to the alarm caller. However, benefits to alarm calling outweigh risks because alarm callers are warning *related* individuals. Thus, alarm-calling behavior,

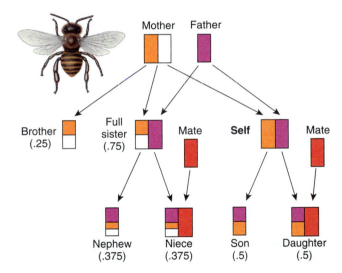

Figure 36.20

Haplodiploidy in honey bees, showing degrees of relatedness of a female worker bee (labeled Self) to individuals she might raise. In honey bees, as in other haplodiploid animals, diploid females develop from fertilized eggs, and males develop from unfertilized eggs. Each daughter of a male gets all his genes (purple bar) and *full* sisters receive an identical one half of their genome from the same father. Open bars represent other, unrelated alleles. Because full sisters also share half the genes they receive from their common mother (orange bar), the relatedness of Self to a full sister is 0.75, the average of 0.5 and 1.0. (In a diploid-diploid system as in humans, the relatedness of siblings is 0.5 because both paternally and maternally inherited genes have a 50% chance of being present in a sibling.) Note that relatedness of female workers to a brother is only 0.25, because brothers are fatherless. If the mother bee has mated with more than one male, average relatedness of a worker bee to her mother's immature daughters will vary between 0.5 and 0.75.

Figure 36.21

Belding's ground squirrel, *Spermophilus beldingi,* gives an alarm call to warn of an approaching predator. This risky behavior endangers the callers to benefit noncallers. Females whose young are nearby give more such alarm calls than do other individuals.

even if it is risky, may be favored by selection if it increases inclusive fitness of a caller.

Kinship theory suggests that animals may evolve an ability to recognize categories of relatives so that cooperation or aid-giving behavior is directed more efficiently toward relatives. Although kin-recognition behavior was discussed by Hamilton, little was known about it until almost 20 years after he wrote his influential papers. Through experimental studies we now know that animals of many kinds discriminate kin and non-kin: isopods, insects, fishes, frog and toad tadpoles, birds, squirrels, and monkeys. Individuals of some species can even discriminate between full siblings and half-siblings and between cousins and unrelated individuals. Thus, some species have a finely tuned ability to identify relatives of various degrees of relatedness. Cues used in kin recognition vary among species. Birds often use vocalizations, whereas chemical cues are common in other groups.

Because altruistic behavior occurs among unrelated individuals in many natural populations, kin-selection theory does not explain all altruistic behavior. The theory of **reciprocal altruism,** formulated initially by Robert Trivers, provides an additional Darwinian foundation for explaining altruistic behaviors among individuals, including ones who are not close relatives. According to this theory, an individual is selected to perform altruistic acts if doing so increases that individual's chances of receiving equally or more valuable favors from others. Reciprocal altruism is most likely to evolve in species that have stable social groupings whose members are mutually interdependent for defense, nutrition or reproduction, with frequent opportunities for altruistic interaction.

Darwinian selective explanations for social behavior have been investigated extensively using the mathematical theory of games to determine which behaviors may qualify as an **evolutionary stable strategy** or **ESS.** An ESS is expected to persist over long periods of evolutionary time because it prevails in competition with alternative strategies that might arise. An altruistic behavior would not be an ESS if it were subject to cheaters who could preferentially obtain altruistic behaviors from unrelated others without giving in return. Because altruistic behavior is hard to study in nature, results of ESS theory help to focus investigations on populations and behaviors that are most likely to show evolutionary stability. Ritualized displays of animal aggression (p. 793) are considered strong examples of evolutionary stable strategies because they prevent individuals from escalating conflicts to the point of serious injury and thereby enhance survivorship. Selection should favor ritualized displays over alternative behaviors that would subject an individual to violence. An important research topic is to determine how truthful are the displays used by animals to avoid violent conflict or to procure mates. Would it be an evolutionary stable strategy for organisms to evolve displays that deceptively overstate their actual strength or desirability for mating?

Gerald Wilkinson's studies of food sharing by vampire bats show that individuals do reciprocate altruistic behavior. Vampire bats aggregate in resting areas and leave those areas at night to obtain blood meals from large mammals, which are often difficult to find. A vampire bat that obtains a blood meal may perform the altruistic behavior of regurgitating blood to other hungry members of its group. Wilkinson used laboratory experiments to show that hungry bats were not fed at random, but that individual bats were more likely to be fed by those to whom they had previously provided the same service. These results confirm that vampire bats recognize each other as individuals, remember which individuals have performed altruistic behaviors, and then return favors. Despite such encouraging results, reciprocal altruism is difficult to study in nature because it generally requires long-term observations of marked individuals and may occur simultaneously with kin selection.

Animal Communication

Only through communication can one animal influence behavior of another. Compared to the enormous communicative potential of human speech, however, nonhuman communication is severely restricted. Animals may communicate by sounds, scents, touch (including electric and thermal signals), pheromones (as sensed by antennae in insects and vomeronasal glands in mammals), and movement. Indeed any sensory channel or combination of channels may be used, giving animal communication richness and variety.

Phylogenetic studies are important for testing hypotheses of the evolution of mating-behavioral and morphological characters by sexual selection. Phylogenetic and behavioral studies of swordtail fish species by Alexandra Basolo show that female preference for the male sword evolved prior to evolution of the sword, consistent with the hypothesis that initial formation and elongation of the sword in males occurred through sexual selection by female preference. Another important phylogenetic study of mating behavior examines evolution of male displays in leks of neotropical bowerbirds. Richard Prum identified 44 behavioral characters that have been used to evolve species-specific behavioral displays in these birds, and he reconstructed the historical sequence by which these displays evolved. His results show a general evolutionary trend toward increased complexity of displays and a tendency for behavioral changes to precede changes in plumage. A new behavioral display that highlights a particular area of plumage subjects that plumage to sexual selection for morphological elaboration. These studies suggest that behavioral evolution may be a major factor determining the action of selection on morphological characters. Some evolutionists have proposed that behavioral evolution generally accelerates morphological evolution, and that a change in behavior is often a critical factor permitting evolution of new adaptive zones.

Unlike our language, which is composed of words with definite meanings that may be rearranged to generate an almost infinite array of new meanings and images, communication of other animals consists of a limited repertoire of signals. Typically, each signal conveys one and only one message. These messages are not divided or rearranged to construct *new kinds* of information. A single message from a sender may, however, contain several bits of relevant information for a receiver.

A cricket's song announces to an unfertilized female the species of the sender (males of different species have different songs), his sex (only males sing), his location (source of the song), and social status (only a male able to defend the area around his burrow sings from one location). This information is crucial to a female and accomplishes a biological function. However, there is no way for a male to alter his song to provide additional information concerning food, predators, or habitat, which might improve his mate's chances for survival and thus enhance his own fitness.

Chemical Sex Attraction in Moths

Mate attraction in silkworm moths illustrates an extreme case of stereotyped, single-message communication that has evolved to serve a single biological function: mating. Virgin female silkworm moths have special glands that produce a chemical sex attractant to which males are sensitive. Adult males use their large bushy antennae, covered with thousands of sensory hairs that function as receptors (Figure 36.22), to detect the chemical attractant **bombykol** (a complex alcohol named after the silkworm, *Bombyx mori*).

To attract males, females sit quietly and emit a minute amount of bombykol, which is carried downwind. When a few molecules reach a male's antennae, he is stimulated to fly upwind

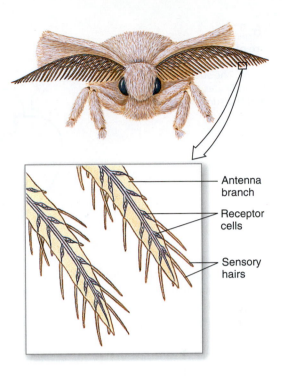

Figure 36.22
Large antennae of a male silkworm moth, *Bombyx mori;* these are especially sensitive to the sex attractant (pheromone) released by the female moth.

Antenna branch

Receptor cells

Sensory hairs

in search of the female. His search is at first random, but, when by chance he approaches within a few hundred yards of the female, he encounters a concentration gradient of the attractant. Guided by the gradient, he flies toward the female, finds her, and copulates with her.

In this example of chemical communication the attractant bombykol, a *pheromone* (p. 741), serves as a signal to bring the sexes together. Its effectiveness is ensured because natural selection favors evolution of males with antennal receptors sensitive enough to detect the attractant at great distances (several miles). Males whose genotype produces a less sensitive sensory system fail to locate a female and thus are reproductively eliminated from the population.

Language of Honey Bees

Among the most sophisticated and complex of all nonhuman communication systems is the symbolic language of bees. Honey bees communicate the location of food resources when these sources are too distant to be located easily by individual bees. They communicate by dances, which have important accompanying sounds and are mainly of two forms. The form having the most informational richness is the **waggle dance** (Figure 36.23). Bees most commonly execute these dances when a forager has returned from a rich source, carrying either nectar in her stomach or pollen grains packed in basketlike spaces formed by hairs on her legs. The waggle dance is roughly in the pattern of a figure-eight made against the vertical surface on the comb inside the hive. One cycle of the dance consists of three components: (1) a circle with a diameter

Figure 36.23

Waggle dance of honey bees used to communicate both the direction and distance of a food source. The straight run of the waggle dance indicates direction according to the position of the sun (angles X and Y).

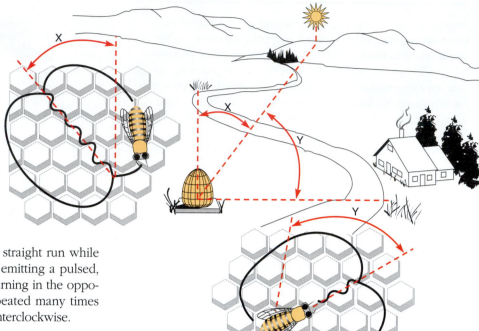

about three times the length of the bee, (2) a straight run while waggling the abdomen from side to side and emitting a pulsed, low-frequency sound, and (3) another circle, turning in the opposite direction from the first. This dance is repeated many times with the circling alternating clockwise and counterclockwise.

Studies of foraging communication in bees provide an outstanding example of how active controversy drives scientific discovery. The precise information communicated by bee dances on location of food sources was discovered in 1943 by Karl von Frisch, a recipient of the 1973 Nobel Prize in Physiology or Medicine. This discovery was strongly challenged beginning in 1967 by Adrian Wenner and Patrick Wells, whose experiments indicated that a bee's dance was not needed to communicate position of food sources, and that odors could best explain a bee's foraging communications. The resulting conflict caused the experiments of both sides to be scrutinized for sources of ambiguity and uncontrolled variables. New experiments were conducted by many researchers to resolve this conflict, ultimately revealing that different sides of the original controversy had revealed different components of a large repertoire of communication mechanisms. Among the experiments demonstrating that bees communicate precise information through dances are ones in which a robot bee with vibrating metal wings was moved through the waggle dance and its associated sounds. When operated in a hive, the computer-directed robot successfully recruited attending bees to visit preselected food dishes outside the hive that had not been visited previously. This work showed that the sounds associated with the dance were as important as its visual appearance. The dance is used, however, mainly when food near the hive is limited and distant food sources are transient; when a scented food source is readily and continuously available nearby, bees communicate primarily using the odor cues studied by Wenner and Wells. A major message of James and Carol Gould's work on honey bees is that bees use several sensory systems to navigate, and that a primary system has backups in case it fails. Discovery of this full range of sensory communication mechanisms, and the circumstances conducive to a bee's use of each one, owes a great debt to the controversy sparked by the apparently conflicting, but ultimately consolidated, findings of von Frisch, Wenner, and Wells.

The straight, waggle run is the important informational component of the dance. Waggle dances are performed almost always in clear weather, and the direction of the straight run is related to the position of the sun. If the forager has located food directly toward the sun, she will make her waggle run straight upward over the vertical surface of the comb. If food was located 60 degrees to the right of the sun, her waggle run is 60 degrees right of vertical. We see then that the waggle run points at the same angle relative to the vertical as the food is located relative to the sun.

Distance of the food source is also coded into bee dances. If food is close to the hive (less than 50 m), the forager employs a simpler dance called the **round dance.** The forager simply turns a complete clockwise circle, then turns, and completes a counterclockwise circle, a performance repeated many times. Other workers cluster around the scout and become stimulated by the dance as well as by chemicals of the nectar and pollen grains from flowers she has visited. The recruits then fly out and search all directions but do not stray far.

If the food source is farther away, round dances become waggle dances, which provide both distance and directional information. The tempo of the waggle dance is related inversely to the food's distance. If food is about 100 m away, each figure-eight cycle lasts about 1.25 seconds; if 1000 m away, it lasts about 3 seconds; and if about 8 km (5 miles) away, it lasts 8 seconds. When food is plentiful, the bees may not dance at all. But when food is scarce, the dancing becomes intense, and other workers cluster around returning scouts and follow them through the dance patterns.

Communication by Displays

A display is a kind of behavior or series of behaviors that serves communication. The release of sex attractant by a female moth and the dances of bees just described are examples of displays; so are alarm calls of herring gulls, songs of the white-crowned sparrow, courtship dances of the sage grouse, and "eyespots" on the hind wings used by certain moths to startle potential predators.

The elaborate pair-bonding displays of blue-footed boobies (Figure 36.24) are performed with maximum intensity when the birds come together after a period of separation. The male at right in the illustration is sky pointing: the head and tail are pointed skyward and the wings are swiveled forward to display their glossy upper surfaces to the female. This display is accompanied by a high-piping whistle. The female at left is parading. She steps with exaggerated slow deliberation, lifting each brilliant blue foot in turn, as though holding it aloft momentarily for the male to admire. Such highly personalized displays, performed with droll solemnity, appear comical, even inane to an observer. Indeed the boobies, whose name is derived from the Spanish word "bobo" meaning clown, presumably were so designated for their amusing antics.

The exaggerated nature of the displays ensures that the message is not missed or misunderstood. Such displays are essential to establish and to maintain a strong pair bond between male and female. This requirement also explains the repetitious nature of displays that follow one another throughout courtship and until laying of eggs. Repetitious displays maintain a state of mutual stimulation between male and female, ensuring the cooperation necessary for copulation and subsequent incubation and care of young.

Communication Between Humans and Other Animals

One uncertainty in studies of animal communication is understanding what sensory channel an animal is using. Signals may be visual displays, odors, vocalizations, tactile vibrations, or electrical currents (as, for example, among certain fishes). Even more difficult is establishing two-way communication between humans and other animals since an investigator must translate meanings into symbols that the animal can understand. Furthermore, people are poor social partners for most other animals.

Animal Cognition

One of the most fascinating subjects in animal behavior is animal intelligence and awareness. Animal cognition is a general term for mental function, including perception, thinking, and memory. Many biologists consider some mental processes of animals similar to those of humans. Recent studies on animal cognition with nonhuman primates and African grey parrots have yielded fascinating results.

In the late 1960s Beatrix and Allen Gardner of the University of Nevada in Reno began using American Sign Language (ASL) to train a chimpanzee named Washoe to communicate with her hands as deaf people do. By age five Washoe could sign 132 words, which she could put into strings forming sentences and phrases. She could answer questions, make suggestions, and convey moods. In one session, when asked what a swan was, Washoe answered "water bird." Washoe also taught signs to other chimpanzees. At first, signs were used as play but soon the chimpanzees used them to make spontaneous requests to trainers such as "drink," "tickle," and "hug." Similar work has been done with other primates including gorillas, orangutans, and bonobos.

Irene Pepperberg of the University of Arizona has worked for years with an African grey parrot named Alex. Because parrots can vocalize like humans, Pepperberg could communicate with Alex using human vocal language. Over the years Alex learned a number of attributes including colors, shapes, and materials for more than 100 objects. Alex can identify objects by colors, size, and shape. Thus, if Alex is given two objects of the same color but one larger than the other, he could state that the difference between them was "size." Alex also can count and relate to the trainer how many objects of each particular category are present.

Conscious awareness is also part of cognition. Donald Griffin wrote two books suggesting that many animals are self-aware and can think and reason. The ability of apes, parrots, and other animals to use language-related skills is significant because it shows us their cognitive abilities so that we can begin to communicate with them. The possibility that animals may think and have a conscious awareness has shed new light on animal behavior studies and added new significance to our studies of animals in general. Studies of animal cognition remain highly controversial.

The animal behaviorist Irven DeVore reported how choosing the proper channel for dialogue can have more than academic interest:[1]

> One day on the savanna I was away from my truck watching a baboon troop when a young juvenile came and picked up my binoculars. I knew if the glasses disappeared into the troop they'd be lost, so I grabbed them back. The juvenile screamed. Immediately every adult male in the troop rushed at me—I realized what a cornered leopard must feel like. The truck was 30 or 40 feet away. I had to face the males. I started smacking my lips very loudly, a gesture that says as strongly as a baboon can, "I mean you no harm." The males came charging up, growling, snarling, showing their teeth. Right in front of me they halted, cocked their heads to one side—and started lip-smacking back to me. They lip-smacked. I lip-smacked, "I mean you no harm." "I mean *you* no harm." It was, in retrospect, a marvelous conversation. But while my lips talked baboon, my feet edged toward the truck until I could leap inside and close the door.

[1]DeVore, Irven. The marvels of animal behavior. 1972. Washington, D.C., National Geographic Society.

Figure 36.24

A pair of Galápagos blue-footed boobies, *Sula nebouxii,* display to each other. A male (*right*) is sky pointing; a female (*left*) is parading. Such vivid, stereotyped, communicative displays serve to maintain reciprocal stimulation and cooperative behavior during courtship, mating, nesting, and care of the young.

The study of animal communication has made great strides, buoyed by the assimilation of a wealth of facts and information about communication in many species. The animal world is filled with communication. In recognizing that reasoning and insight are not required for effective, highly organized behavior, we should not conclude that other animals are, as Descartes proclaimed in the seventeenth century, nothing more than machines.

SUMMARY

Animal behavior has emerged as a scientific discipline from four different approaches. Comparative psychology emphasizes the identification of mechanisms controlling behavior, using relatively few species, with the intent that these mechanisms might have wide applicability among animals. Ethology is the study of behavior, both innate and learned, of animals in their natural habitats. Behavioral ecologists have shown that behavioral traits have evolutionary histories and may evolve by natural selection. Sociobiology aims to understand how and why social behavior in animals has evolved. Both ethology and sociobiology distinguish between studies that focus on the mechanisms of behavior (proximate causation) and those that focus on function or evolution of behavior (ultimate causation).

Students of animal behavior have observed and cataloged many behavioral patterns of animals that are highly predictable and almost invariable in performance. Often these patterns are triggered, or "released," by specific, and usually simple, environmental stimuli, called sign stimuli. Although such formalized behaviors may be released inappropriately at times, they are efficient and enable the animal to respond rapidly. Development of behavioral patterns depends on an interaction between an organism and its environment. For this reason, behavioral scientists prefer not to describe behaviors that are largely invariable in their performance as "instinctive" or "innate."

Behavior may be modified by learning through experience. Two simple kinds of learning behavior are habituation, which is reduction or elimination of a behavioral response in the absence of any reward or punishment; and sensitization, in which a repeated stimulus increases the strength of a behavioral response. The gill-withdrawal reflex of the marine mollusc, *Aplysia,* is described as a protective response that can be modified experimentally to show either habituation or sensitization. The modification of the alarm response of herring-gull chicks is another example of habituation. Another form of learning is imprinting, the lasting recognition bond that forms early in life between the young of many social animals and their mothers.

Social behavior is behavior arising from interactions of members of a species with one another. In social organizations, animals tend to remain together, communicate with each other, and usually resist intrusions by "outsiders." Advantages of sociality include cooperative defense from predators, cooperative searching for food, improved reproductive performance and parental care of young, and transmission of useful information through the society. Because social animals compete with one another for resources (such as food, sexual mates, and shelter), conflicts are often resolved by a form of overt hostility called aggression. Most aggressive encounters between conspecifics are stylized bouts involving more bluff than intent to injure or to kill. Dominance hierarchies, in which a priority of access to common resources is established by aggression, are common in social organizations. Territoriality is an alternative to dominance. A territory is a defended area from which intruders of the same species are excluded.

Mating systems include monogamy, the mating of an individual with only one partner of the opposite sex each breeding season, and polygamy, the mating of an individual with two or more partners in a breeding season. Two forms of polygamy are polygyny, the mating of a male with more than one female, and polyandry, the mating of a female with more than one male. Several types of polygyny are recognized.

A behavior in which one animal may reduce its own fitness to increase fitness of others is called altruistic behavior. Examples of risky behaviors include one member of a social group warning others of a predator, and cooperative behavior among social insects

in which an individual may forego reproduction to raise its younger siblings. The favored explanation of altruism is kin selection, in which the recipient of an altruistic act is sufficiently closely related to the altruist that the recipient's survival would benefit the genes shared with the altruist.

Communication, often considered the essence of social organization, is the means by which animals influence behavior of other animals, using sounds, chemical signals, visual displays, touch, or other sensory signals. Compared to the richness of human language, animals communicate with a very limited repertoire of signals. One of the most famous examples of animal communication is that of the symbolic dances of honey bees. Birds communicate by calls and songs and, especially, by visual displays. By ritualization, simple movements have evolved to become conspicuous signals having definite meanings.

REVIEW QUESTIONS

1. How do experimental approaches of comparative psychology and ethology differ? Comment on the aims and methods employed by each.
2. Egg-retrieval behavior of greylag geese is an excellent example of a highly predictable behavior. Interpret this behavior within the framework of classical ethology, using these terms: releaser, sign stimulus, and stereotyped behavior. Interpret the territorial defense behavior of male three-spined sticklebacks in the same context.
3. The idea that behavior must be *either* innate or learned has been called a "nature versus nurture" controversy. What reasons are there for believing that such a strict dichotomy does not exist?
4. Two kinds of simple learning are habituation and imprinting. Distinguish these two types of learning, and offer an example of each.
5. Some strains of bees show hygienic behavior by uncapping cells containing larvae infected with a bacterial disease called foulbrood and removing the dead larvae from the hive. What evidence shows that this behavior is transmitted by two independently segregating genes?
6. Discuss some advantages of sociality for animals. If social living has so many advantages, why do many animals successfully live alone?
7. Suggest why aggression, which might seem counterproductive, exists among social animals.
8. What is the selective advantage to winners, as well as to losers, that aggressive encounters within species for social dominance are usually ritualized displays or symbolic fights rather than unrestrained fights to death?

9. Of what use is a territory to an animal, and how is a territory established and kept? What is the difference between territory and home range?
10. Polygyny is a form of polygamy in which a male mates with more than one female. Explain how three forms of polygyny differ from each other: resource-defense polygyny, female-defense polygyny, and male-dominance polygyny.
11. Give an example of an altruistic behavior and explain how such behavior conflicts with Darwin's expectation that animals should act selfishly to produce as many offspring as possible.
12. Earlier explanations of altruistic behavior as a form of group selection have been supplanted by Hamilton's hypothesis of kin selection. What distinguishes kin selection and how does it accord with the notion of inclusive fitness, the relative number of an individual's alleles that pass to the next generation?
13. Comment on limitations of animal communication compared to those of human communication.
14. Dance language used by returning forager honey bees to specify location of food is an example of remarkably complex communication among "simple" animals. How is direction and distance information coded into the waggle dance of the bees? What is the purpose of the round dance?
15. What is meant by "ritualization" in display communication? What is the adaptive significance of ritualization?
16. Early efforts by humans to communicate vocally with chimpanzees were almost total failures; however, researchers have learned how to communicate successfully with apes. How was this task accomplished?

SELECTED REFERENCES

Alcock, J. 2005. Animal behavior: an evolutionary approach, ed. 8. Sunderland, Massachusetts, Sinauer Associates, Inc. *Clearly written and well-illustrated discussion of genetics, physiology, ecology, and history of behavior in an evolutionary perspective.*

Attenborough, D. 1990. The trials of life: a natural history of animal behavior. Boston, Little, Brown and Company. *Superb photographs and flowing text describe the life cycles of organisms, often focusing on unusual and exciting patterns of behavior.*

Basolo, A. L. 1996. The phylogenetic distribution of a female preference. Syst. Biol. **45**:290–307. *A phylogentic analysis of behavioral evolution.*

Bekoff, M., and D. Jamieson (eds.). 1996. Readings in animal cognition. Cambridge, Massachusetts, The MIT Press. *Selected readings of papers in animal cognition by authors in the field.*

Bradbury, J. W., and S. L. Vehrencamp. 1998. Principles of animal communication. Sunderland, Massachusetts, Sinauer Associates. *A comprehensive text on animal communication.*

Clutton-Brock, T. 2002. Breeding together: kin selection and mutualism in cooperative vertebrates. Science **296**:69–72. *A good update on selective explanations of cooperative behavior.*

Drickamer, L. C., S. H. Vessey, and D. Meikle. 1996. Animal behavior: mechanisms, ecology, and evolution, ed. 3. Dubuque, William C. Brown Publishers. *Comprehensive, with helpful discussions on the methods and experimental approaches used to answer behavioral questions.*

Gould, J. L., and C. G. Gould. 1994. The animal mind. New York, Scientific American Library. *Attractively illustrated, engagingly written exploration of animal behavior and efforts of researchers to determine what happens inside minds of animals.*

Greenspan, R. J. 1995. Understanding the genetic construction of behavior. Sci. Am. **272**:72–78 (Apr.). *Studies of courtship and mating in fruit flies indicate that behavior is regulated by many multipurpose genes, each of which handles diverse responsibilities in the body.*

Houck, L. D., and L. C. Drickamer (eds.). 1996. Foundations of animal behavior. Chicago, The University of Chicago Press. *Classic papers in animal behavior with commentaries.*

Kirchner, W. H., and W. F. Towne. 1994. The sensory basis of the honeybee's dance language. Sci. Am. **270**:74–80 (June). *Experiments with a robotic honey bee that can dance and emit sounds similar to living bees show conclusively that the dance language successfully recruits foragers to food outside the hive.*

Lorenz, K. Z. 1952. King Solomon's ring. New York, Thomas Y. Crowell Company, Inc. *One of the most delightful books ever written about behavior of animals.*

Manning, A., and M. S. Dawkins. 1998. An introduction to animal behaviour, ed. 5. Cambridge, England, Cambridge University Press. *Survey of animal behavior, drawing from ethology, physiology, and comparative psychology.*

Preston-Mafham, R., and K. Preston-Mafham. 1993. The encyclopedia of land invertebrate behavior. Cambridge, Massachusetts, The MIT Press. *Numerous examples of fascinating invertebrate behavior in a series of informative and beautifully illustrated essays. Highly recommended.*

Prum, R. O. 1990. Phylogenetic analysis of the evolution of display behavior in the neotropical manakins (Aves: Pipridae). *Ethology* **84**:202–231. *A phylogenetic analysis of behavioral evolution.*

Queller, D. C., and J. E. Strassmann. 1998. Kin selection and social insects. BioScience **48**(3):165–175. *How kin selection operates in social insects, and why most cases of altruism are found in social insects.*

Ridley, M. 1995. Animal behavior: an introduction to behavioral mechanisms, development, and ecology, ed. 2. Oxford, Blackwell Scientific Publications. *The principles of animal behavior presented with well-chosen examples and clear illustrations.*

Savage-Rumbaugh, S., S. G. Shanker, and T. J. Taylor. 1998. Ape language: from conditioned response to symbol. New York, Oxford University Press. *Details the general area of ape language.*

ONLINE LEARNING CENTER

Visit www.mhhe.com/hickmanipz14e for chapter quizzing, key term flash cards, web links, and more!

PART FIVE

Animals and Their Environments

5

Grizzly bear.

The Biosphere and Animal Distribution

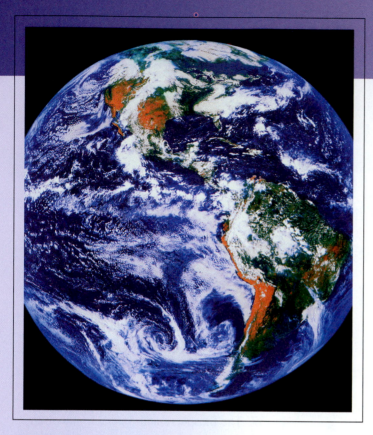

Spaceship earth.

Spaceship Earth

All life is confined to the **biosphere,** a thin veneer on the earth's surface. From the first remarkable photographs of earth taken from the *Apollo* spacecraft, revealing a beautiful blue and white globe lying against a limitless backdrop of space, viewers were struck and perhaps humbled by our isolation and insignificance in the universe. The phrase "spaceship earth" became part of our vocabulary as we realized that all the resources for sustaining life except for solar energy are restricted to a thin layer of land and sea and a narrow veil of atmosphere above it. We could better appreciate just how thin the biosphere is if we could shrink the earth and all of its dimensions to a 1 m sphere. We would no longer perceive vertical dimensions on the earth's surface. The highest mountains would fail to penetrate a thin coat of paint applied to our shrunken earth; a fingernail's scratch on the surface would exceed the depth of an ocean's deepest trenches.

Earth's biosphere and the organisms in it evolve together. A living form is a transient link built of environmental materials, which are later returned to the environment to be used again in the recreation of new life. Life, death, decay, and re-creation have been the cycle of existence since life began.

The primitive earth of approximately 5 billion years ago, barren, stormy, and volcanic with a reducing atmosphere of ammonia, methane, and water (Figure 37.1), supported the prebiotic syntheses that led to life's beginnings (p. 28), but it was totally unsuited, indeed lethal, for the living organisms that inhabit the earth today. The appearance of free oxygen in the atmosphere, produced largely if not almost entirely by life, is an example of the reciprocity between organism and environment. Although oxygen was poisonous to early forms of life, its gradual accumulation from photosynthesis led some forms to evolve oxygen metabolism, on which most organisms now depend. As living organisms adapt and evolve, they change their environments. In so doing they must themselves change.

The many properties that make the earth wonderfully suitable for life were first recognized and examined in detail by Lawrence J. Henderson (1878 to 1942) in his book *The Fitness of the Environment,* published in 1913. The profound insights of this distinguished Harvard biochemist and physiologist were remarkable, appearing as they did before ecology had become a science. His insightful understanding of reciprocity between organism and environment has become a principle that underlies all ecological science. Henderson's book deserves a broader appreciation than it has received; it is, for example, seldom mentioned in ecology textbooks.

DISTRIBUTION OF LIFE ON EARTH

Biosphere and Its Subdivisions

The biosphere as usually defined is the thin outer layer of the earth capable of supporting life. It is probably best viewed as a global system that includes all life on earth and the physical environments in which living organisms exist and interact. The physical subdivisions of the biosphere include the lithosphere, hydrosphere, and atmosphere.

The **lithosphere** is the rocky material of the earth's outer shell and is the ultimate source of all mineral elements required by living organisms. The **hydrosphere** is the water on or near the earth's surface, and it extends into the lithosphere and the atmosphere. Water is distributed over the earth by a global hydrological cycle of evaporation, precipitation, and runoff. Five-sixths of the evaporation is from oceans, and more water is evaporated from oceans than is returned to them by precipitation. Oceanic evaporation therefore provides much of the rainfall that supports life on land. The gaseous component of the biosphere, the **atmosphere,** extends to some 3500 km above the earth's surface, but all life is confined to the lowest 8 to 15 km (troposphere). The screening layer in the atmosphere of oxygen-ozone is concentrated mostly between 20 and 25 km. The main gases in the troposphere are (by volume) nitrogen, 78%; oxygen, 21%; argon, 0.93%; carbon dioxide, 0.03%; and variable amounts of water vapor.

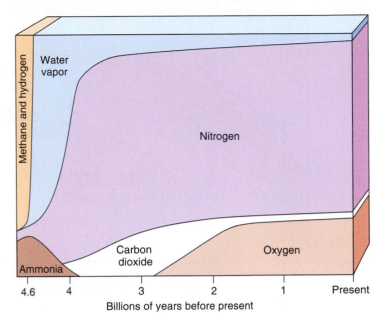

Figure 37.1

Changing composition of earth's atmosphere over time. The percentage of each gas in the atmosphere is proportional to its band width on the vertical axis. The primitive atmosphere formed as hydrogen (H_2), methane (CH_4), and ammonia (NH_3). Hydrogen gas, too light to be held by the earth's gravitational field, was lost to space. Nitrogen (N_2), carbon dioxide (CO_2), sulfur dioxide (SO_2), and water vapor (H_2O) emitted from volcanoes replaced the remaining primitive gases. The first free oxygen (O_2) was formed by solar radiation acting on water molecules (photochemical dissociation) in the atmosphere. When oxygen-producing plants appeared 3 to 3.5 billion years ago, atmospheric oxygen gradually rose to its present level approximately 400 million years ago.

Atmospheric oxygen originates almost entirely from photosynthesis. Since the mid-Paleozoic era, oxygen consumed by living organisms for respiration has approximately equaled oxygen production. Earth's surplus of free oxygen is unlikely to be depleted because oxygen reserves in the atmosphere and in the oceans are so enormous that the supply could last thousands of years even if all photosynthetic replenishment suddenly would cease.

The rapid input of carbon dioxide into the atmosphere from burning fossil fuels may significantly affect the earth's heat budget. Much of the sun's short-wave light energy absorbed by earth's surface reradiates as long-wave infrared heat energy (Figure 37.2). Materials in the atmosphere, especially carbon dioxide and water vapor, impede this heat loss and raise atmospheric temperature. This heating of the atmosphere is called the "greenhouse effect," since the atmosphere traps reradiated heat from the earth much as the glass of a greenhouse traps heat reradiated by plants and soil inside. Although the greenhouse effect provides conditions essential for all life on earth, the gradual accumulation of carbon dioxide could increase the temperature of the biosphere as a whole and raise sea level by melting polar ice (Figure 37.3).

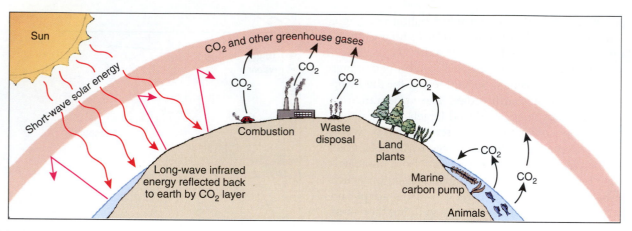

Figure 37.2

"Greenhouse effect." Carbon dioxide and water vapor in the atmosphere are transparent to sunlight but absorb heat energy reradiated from the earth, leading to warming of atmospheric air. In the marine carbon pump, carbon dioxide fixed by marine plants (especially phytoplankton) and carried in ionic form (HCO_3^-) by cold water sinks to great depths and is thereby removed from the atmosphere until deep-sea currents move it to the ocean surface in tropical regions.

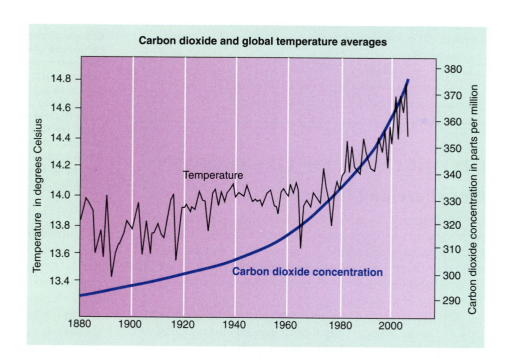

Figure 37.3

Rise in global atmospheric carbon dioxide and global temperature averages for the past 126 years. Data points before 1958 come from analysis of air trapped in bubbles in glacial ice from sites around the world. Atmospheric carbon dioxide has climbed steadily for more than a century while earth's temperature has followed a more erratic upward trend.

Atmospheric carbon dioxide increased from about 280 parts per million (ppm) before the Industrial Revolution to an average 377 ppm today. In the past century global temperature has increased 0.4° C and most experts agree that it will have increased 2° to 6° C when carbon dioxide and other heat-trapping greenhouse gases have doubled in the next century. Atmospheric carbon dioxide and temperatures have been much higher than these projected levels at various times in earth's long history, such as parts of the Paleozoic era (Devonian through Carboniferous) during which the entire world was warm and humid.

Terrestrial Environments: Biomes

A biome is a major biotic unit bearing a characteristic and easily recognized array of plant life. Botanists long ago recognized that the earth's terrestrial environment could be divided into large units having a distinctive vegetation, such as forests, prairies, and deserts. Animal distribution has always been more difficult to map, because plant and animal distributions do not exactly coincide. Zoogeographers use plant distributions as the basic biotic units and recognize biomes as distinctive combinations of plants and animals. A biome is therefore identified by its dominant plant formation (Figure 37.4), but because animals depend on plants, each biome supports a characteristic fauna.

Each biome is distinctive, but its borders are not. Anyone who has traveled across North America knows that plant communities blend into one another over broad areas. Moist deciduous

Figure 37.4

Major biomes of North America. Boundaries between biomes are not distinct as shown but grade into one another over broad areas.

forests of the Appalachians change gradually to drier oak forests of the upper Mississippi Valley, and then to oak woodlands with grassy understory, which yield to tall and mixed prairies (now corn and wheatlands), then to desert grasslands, and finally to desert shrublands. The indistinct boundaries where the dominant plants of adjacent biomes are mixed together form an almost continuous gradient called an **ecocline.** Thus biomes are abstractions, a convenient way to organize our concepts about different communities. Nevertheless, anyone can distinguish a grassland, deciduous forest, coniferous forest, or shrub desert by the dominant plants in each, and we can make reasonable assumptions about the animals that live in each biome.

A biome's distinctiveness is determined mainly by climate, characteristic rainfall and temperature of each region, and amount of solar radiation. Global variation in climate arises from uneven heating of the atmosphere by the sun. Because the sun's rays strike higher latitudes at a lower angle, atmospheric heating is less there than at the equator (Figure 37.5). Air warmed at the equator rises and moves toward the poles. It is replaced by cold air moving away from the poles at lower levels. Earth's rotation complicates this pattern, producing a Coriolis effect that deflects moving air

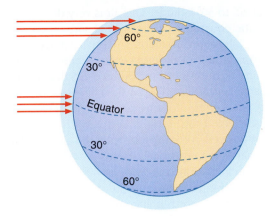

Figure 37.5

Earth's climate is determined by differential solar radiation between the higher latitudes and equator. Solar energy is spread across a much larger, slanting surface area at high latitudes than is an equivalent amount of energy at the equator.

to the right in the Northern Hemisphere and to the left in the Southern Hemisphere. Air circulation in each hemisphere forms three latitudinal zones, called cells (Figure 37.6). In the Northern Hemisphere, hot moist air at the equator cools and condenses as it rises, providing rainfall for the lush vegetation of equatorial rain forests. Warm air then flows northward at high levels, cools,

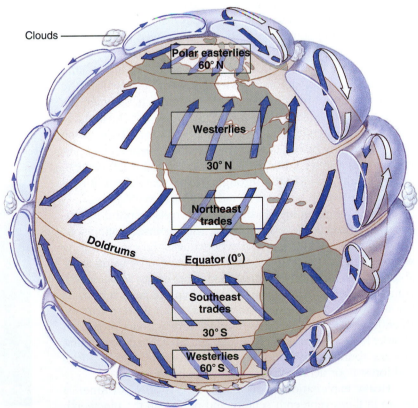

Figure 37.6

Earth as a heat engine. As the result of unequal heating across earth's surface, together with other factors such as earth's rotation, circulation of the oceans, and presence of landmasses, the earth acts like a giant heat engine that imposes a complex patchwork of climates on earth. See text for explanation.

and sinks at 20° to 30° latitude. This air is very dry, having lost its moisture at the equator. As air heats it absorbs even more moisture, causing intense evaporation at earth's surface and producing a subtropical belt of deserts centered between 15° and 30° north (deserts of the American southwest, Saharan Africa, Arabian Peninsula, and India). The air then flows southward toward the equator, gathering moisture as it moves across the ocean, and being deflected to the right as the northeast trade winds. The cycle in this cell is completed when the air, now laden with moisture, reaches the equator.

A second circulation cell between 30° and 60° north arises when cool air sinking at around 30° moves northward at the surface. At 50° to 60° north it encounters cold air moving south from the North Pole, producing an unstable stormy area with abundant precipitation. Warmer air from the south is deflected upward and turns south at high altitude to complete the second cell. A third polar cell forms when cold, southward-moving Arctic air returns to the pole at high altitude.

The principal terrestrial biomes are temperate deciduous forest, temperate coniferous forest, tropical forest, grassland, tundra, and desert. In this brief survey, we refer especially to biomes of North America and consider predominant features of each.

Temperate Deciduous Forest

Temperate deciduous forest, best developed in eastern North America, encompasses several forest types that change gradually from northeast to south. Deciduous, broad-leaved trees such as oak, maple, and beech predominate. Seasonal aspects are better defined in this biome than in any other. The deciduous habit is an adaptation of dormancy for low-energy levels from the sun in winter and freezing winter temperatures. In summer, the relatively dense forests form a closed canopy and deep shade below. Consequently, understory plants grow rapidly in spring and flower early before the canopy develops. Mean annual precipitation is relatively high (75 to 125 cm, or 30 to 50 inches), and rain falls periodically throughout the year. Mean annual temperatures range between 5° and 18° C (41° to 65° F).

Animal communities in deciduous forests respond to seasonal change in various ways. Some, such as insect-eating warblers, migrate. Others, such as woodchucks, hibernate during winter months. Others unable to escape survive by using available food (as do deer) or stored food supplies (as do squirrels). Hunting and habitat loss have largely eliminated large carnivores (mountain lions, bobcats, and wolves) from eastern forests. Deer, however, thrive in second-growth forests under protection of strict hunting management. Insect and invertebrate communities are abundant in deciduous forests because decaying logs and forest-floor debris provide excellent shelter.

Heavy exploitation of deciduous forests of North America began in the seventeenth century and peaked in the nineteenth century. Logging removed the once-magnificent stands of temperate hardwoods. With the opening of the prairie for agriculture, many eastern farms were abandoned and returned gradually to deciduous forests.

Coniferous Forest

In North America coniferous forests form a broad, continuous, continent-wide belt stretching across Canada and Alaska, and south through the Rocky Mountains into Mexico. This biome continues across northern Eurasia, making it one of the largest plant formations on earth. It is dominated by evergreens—pine, fir, spruce, and cedar—which are adapted to withstand freezing and to exploit short summer growing seasons. Conical trees with flexible branches shed snow easily. The northern area is the **boreal** (northern) **forest,** often called **taiga** (a Russian word, pronounced "tie-ga"). Taiga is dominated by white and black spruce, balsam, subalpine fir, larch, and birch. Mean annual precipitation is less than 100 cm (40 inches) and average temperatures range from −5° to +3° C (23° to 37° F).

In central North America, the taiga merges into **mixed temperate forest,** dominated by sugar maple, white pine, red pine, and eastern hemlock. Much of this forest was destroyed by logging and replaced by second growth, which has a higher proportion of deciduous trees, including hard maples, beech, and yellow birch. Oaks and hickories dominate the mixed temperate forests farther south and occur less frequently in the pine-dominated **southern evergreen forests,** which occupy much of the southeastern United States. The last old-growth coniferous forests of the Pacific northwest are rapidly falling to commercial logging.

Mammals of boreal forests include deer, moose (Figure 37.7), elk, snowshoe hares, various rodents, carnivores such as wolves, foxes, wolverines, lynxes, weasels, martins, and omnivorous bears. They are adapted physiologically or behaviorally for long, cold, snowy winters. Common birds are chickadees, nuthatches, warblers, and jays. One bird, the red crossbill, has a beak specialized for picking seeds from cones. Mosquitos and flies pester animals and people in this biome. Southern coniferous forests lack many mammals found in the north, but they have more snakes, lizards, and amphibians.

Figure 37.7

A bull moose browses on dwarf birch in the boreal coniferous forest or taiga biome. Note shedding of antler skin ("velvet"), signifying that antler growth is complete and that breeding season is approaching.

Tropical Forest

A worldwide equatorial belt of tropical forests has high rainfall (more than 200 cm [80 inches] per year), high humidity, relatively high and constant temperatures averaging over 17° C (63° F), and little seasonal variation in day length. These conditions nurture luxurious, uninterrupted growth that reaches its greatest intensity in rain forests. In sharp contrast to temperate deciduous forests, dominated by relatively few tree species, tropical forests contain thousands of species, none of which dominates. A single hectare typically contains 50 to 70 tree species versus 10 to 20 tree species in an equivalent area of hardwood forest in the eastern United States. Climbing plants and epiphytes occur among the trunks and limbs. A distinctive feature of tropical forests is stratification of life into six to eight feeding strata (Figure 37.8).

Insectivorous birds and bats occupy air above the canopy; below the canopy birds, fruit bats and other mammals feed on leaves and fruit. In the middle zones are arboreal mammals (such as monkeys and sloths), numerous birds, insectivorous bats, insects, and amphibians. Climbing animals, such as squirrels and civets, move along tree trunks, feeding from all strata. On the ground are large mammals such as the large rodents of South America (for example, capybara, paca, and agouti) and members of the pig family. Finally, small insectivorous, carnivorous, and herbivorous animals search the litter and lower tree trunks for food. No other biome matches tropical forests in incredible variety of animal species. Food webs (p. 834) are intricate and notoriously difficult to unravel.

Tropical forests, especially the enormous expanse centered in the Amazon Basin, are the most seriously threatened forest ecosystems. Large areas are cleared for agriculture by "slash-and-burn" methods, but farms are soon abandoned because soil fertility is low. It may seem paradoxical that a biome as luxuriant as a tropical forest should have poor soil; however, nutrients released by decomposition are rapidly recycled by plants, microbes, and fungi, leaving no reservoir of humus. In many areas, the soil rapidly becomes a hard, bricklike crust called **laterite** after plants are removed. Tropical plants cannot recolonize such areas. Other pressures on tropical forests include logging by multinational timber companies and clearing of land for cattle ranching.

Grassland

The North American prairie biome is one of the world's most extensive grasslands, extending from the Rocky Mountain edge on the west to eastern deciduous forest on the east, and from northern Mexico in the south to the Canadian provinces of Alberta, Saskatchewan, and Manitoba in the north. Original grassland associations of plants and animals have been widely transformed into the most productive agricultural region in the world, dominated by monocultures of cereal grains. In grazing lands, virtually all major native grasses have been replaced by alien species. Of the once dominant herbivore, bison (Figure 37.9), very few survive, but jackrabbits, prairie dogs, ground squirrels, and antelope remain. Mammalian predators include coyotes, ferrets, and badgers, although, of these, only coyotes are common. Vast tracts of open tallgrass prairie remain in the Flint Hills of Kansas and northern Oklahoma, and large areas of short-grass prairie occur in western Kansas and Nebraska. These areas maintain native vegetation and predatory animals, including raptors, cougars, and bobcats. Rainfall on the North American prairie ranges from about 80 cm (31 inches) in the east to 40 cm (16 inches) in the west. Average annual temperatures range between 10° and 20° C (50° to 68° F).

Tundra

Tundra is characteristic of severe, cold climatic regions, especially treeless Arctic regions and high mountaintops. Plant life must adapt to a short growing season of about 60 days and to

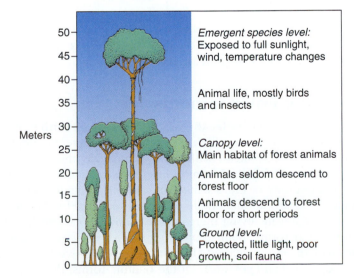

Figure 37.8

Profile of tropical forest, showing stratification of animal and plant life into six strata. The animal biomass is small compared with the biomass of plants.

Figure 37.9

Bison grazing on a short-grass prairie.

Figure 37.10

A large male caribou on the Alaskan tundra. The gregarious caribou travel in large herds, feeding in summer on grasses, dwarf willow, and birch, but in winter almost exclusively on lichen.

a soil that remains frozen for most of the year. Average annual precipitation is usually less than 25 cm (10 inches) and annual temperature averages about −10° C (14° F).

Most tundra regions are covered with bogs, marshes, ponds, and a spongy mat of decayed vegetation, although high tundras may be covered only with lichens and grasses. Despite thin soil and a short growing season, vegetation of dwarf woody plants, grasses, sedges, and lichens may be profuse. Characteristic animals of the Arctic tundra are lemmings, caribou (Figure 37.10), musk-oxen, arctic foxes, arctic hares, ptarmigans, and (during the summer) many migratory birds.

Desert

Deserts are arid regions where rainfall is low (less than 25 cm [10 inches] a year), and water evaporation is high. The North American desert has two parts, the hot deserts of the southwest (Mojave, Sonoran, and Chihuahuan) and the cool, high desert in the rain shadow of the High Sierras and Cascade Mountains. Desert plants, such as thorny shrubs and cacti, have reduced foliage, drought-resistant seeds, and other adaptations for conserving water. Many large desert animals have remarkable anatomical and physiological adaptations for keeping cool and conserving water (p. 681). Most smaller animals avoid severe conditions by living in burrows or using nocturnal habits. Mammals include mule deer, peccaries, cottontails, jackrabbits, kangaroo rats, and ground squirrels. Typical birds are roadrunners, cactus wrens, turkey vultures, and burrowing owls. Lizards, snakes, and tortoises are numerous, and a few species of toads are common. Arthropods include a great variety of insects and arachnids.

Inland Waters

Of all the world's water, 2.5% is freshwater. Most freshwater exists in polar ice caps or underground in aquifers and soil, leaving only 0.01% of the world's inland waters as habitat for aquatic life. A quarter of the world's vertebrates and nearly half of its fishes live in these fragile "islands" of water, which must also supply human needs for irrigation, drinking water, hydroelectric power, and waste disposal.

Inland waters exist as running-water, or **lotic** (L. *lotus,* action of washing) habitats, and standing-water, or **lentic** (L. *lentus,* slow) habitats. Lotic habitats follow a gradient from mountain brooks to streams and rivers. Brooks and streams with high-velocity water flow contain much dissolved oxygen because of their turbulence. Nutrients are chiefly from organic detritus washed from adjacent terrestrial areas. More slowly moving rivers have less dissolved oxygen and more floating algae and plants. Their fauna tolerates lower oxygen concentration.

Lentic habitats, such as ponds and lakes, have still lower concentrations of oxygen, particularly in the deeper areas. Animals living on underwater substrates or submerged vegetation **(benthos)** include snails and mussels, crustaceans, and a wide variety of insects. Many swimming forms, called **nekton,** occur in lakes and larger ponds. Depending on the nutrients available, a large contingent of small floating plants and animals **(plankton)** may occur. Ponds and lakes have short life spans, a few hundred to many thousands of years depending on size and rate of sedimentation, and undergo great physical change as they age. For example, the Great Lakes of North America, which occupy depressions gouged by glacial advances of the Pleistocene epoch, became ice free about 5000 years ago.

A striking exception to the short lifetimes of most lakes is Lake Baikal in southern Siberia. This enormous lake, 1741 m deep (more than 1 mile), is by far the oldest lake in the world, dating from at least the Paleocene—more than 60 million years BP. The speciation of sculpins in Lake Baikal is illustrated in Figure 6.21, page 120.

Many freshwater habitats are severely damaged by human pollution such as dumping of toxic industrial wastes and enormous quantities of sewage. Of the Great Lakes, Lake Erie is the most seriously affected by inflow of nitrates and phosphates. These nutrients fertilize the lake, creating huge blooms of algae that sink and decompose at the bottom of the lake, producing anoxic conditions that harm aquatic life.

Oceanic Environments

Oceans represent by far the largest portion of earth's biosphere, covering 71% of earth's surface to an average depth of 3.75 km (2.3 miles), with their greatest depths reaching more than 11.5 km (7.2 miles) below sea level. The evident monotony of the ocean's surface belies a variety of life below. Oceans are the cradle of life, as reflected by the variety of organisms living there—more than 200,000 species of unicellular forms, plants, and animals. About 98% of these forms live on the seabed **(benthic);** only 2% live freely in the open ocean **(pelagic).** Of the benthic forms, the largest biomass occurs in the intertidal zone or shallow depths of oceans, but species diversity increases from shallow waters to a maximum at 2000 to 3000 m depth and then declines at greater depths.

Figure 37.11

Major marine zones. The continental shelf, slope, and rise collectively form the continental margin.

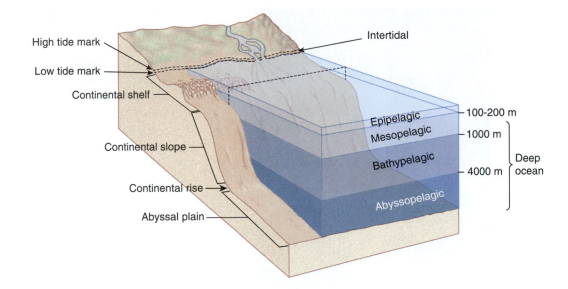

The most productive areas are concentrated along continental margins and a few areas where waters are enriched by organic nutrients and debris lifted by upwelling currents into the sunlit, or **photic,** zone, where photosynthetic activity occurs. With certain notable exceptions (see note in Chapter 38, p. 839), all life below the photic zone is supported by a light "rain" of organic particles from above.

An **estuary** is a semienclosed transition zone where freshwater enters the sea. Despite unstable salinity caused by the variable entry of freshwater, an estuary is a nutrient-rich habitat that supports a diverse fauna.

Benthic communities of the ocean floor occupy geological provinces categorized by topography, substrate, and distance from shore (Figure 37.11). Closest to shore are an ocean's **continental margins,** which contain (1) a **continental shelf** extending from shallow nearshore waters outward to a depth of 120 to 400 m, (2) a **continental slope** marking a sharp descent from the outer shelf to a depth of 3000 to 5000 m, and (3) a **continental rise,** thick sediments piled at the base of the continental slope. Beyond this continental margin is the deep-sea basin or **abyssal plain,** an offshore plain with submarine channels and hills averaging 4000 m in depth but descending to 11,000 m. The abyssal plain has little seasonal variation in temperature and illumination and therefore is relatively stable despite considerable spatial heterogeneity.

Rocky Intertidal Zone

The **intertidal zone** is the portion of the continental shelf exposed to air during low tides; animals of intertidal communities experience daily fluctuations between marine and terrestrial environments. Attached to rocky intertidal substrates are periwinkles, barnacles, mussels, and other forms whose exoskeleton protects them from desiccation and physical abrasion by waves (Figure 37.12). These attached forms are consumed by marine gastropods and seastars. Interactions between physical stresses, predation, and interspecific competition (p. 831) often produce

visibly distinct bands; periwinkles dominate on the most exposed rocks, barnacles in intermediate areas, and mussels on the most submerged surfaces. Depressions in the rocky surface often produce tidal pools isolated on an otherwise exposed shore. These tidal pools support anemones, corals, tunicates, and other forms ill-suited to the fully exposed habitat. Attached seaweeds are often interspersed among rocky intertidal faunas. Rocky intertidal faunas are abundant on the northern Atlantic and Pacific coasts of North America.

Rocky Subtidal Zone

Kelp forests (Figure 37.13) dominated by brown seaweeds occupy shallow subtidal waters worldwide, reaching even the Arctic and Antarctic Circles. Kelps attach to a firm substrate by holdfasts and grow upward, some reaching the sea surface and

Figure 37.12

A rocky intertidal community typically contains attached bivalve molluscs and barnacles, whose exoskeleton provides protection from desiccation and wave action, and predatory seastars.

A

B

Figure 37.13

Kelp forests of the rocky subtidal zone **(A)** are grazed by sea urchins, which are consumed by sea otters **(B)**. A large population of sea otters keeps a kelp forest dense by removing sea urchins.

forming a canopy analogous to the canopy of a tropical forest. Grazing by sea urchins and damage by storms greatly alter the structure of a kelp forest. Numerous species of abalones, sea urchins, and limpets graze kelp forests of the Pacific Coast of North America. These kelp forests support diverse animal life including suspension-feeding mussels and their crustacean predators. Populations of sea otters, which consume molluscs, sea urchins, and fishes of kelp forests, increase a kelp forest's density by removing sea urchins, a powerful grazer.

Coral reefs occur off the coasts of continents and volcanic islands and include atolls, a series of reefs encircling a submerged volcanic island. Reefs protect from wave damage a particularly diverse subtidal community (Figure 37.14), whose substrate is a topographically complex structure formed by mutualistic growths of corals and single-celled algae (pp. 279–282). A single reef may contain 50 or more species of coral with different species dominant at different depths. The complex topography of a reef divides its surface into numerous subcommunities associated with varying levels of illumination and physical orientation, thereby supporting hundreds of species of fishes and snails, plus other cnidarians, crustaceans, sponges, polychaetes, molluscs, echinoderms, tunicates, and other invertebrates. Because complex symbiotic relationships characterize a coral-reef community, no single species dominates its structure. Reefs are therefore less disturbed by changes in single species than are Pacific kelp forests, whose community structure changes greatly with local density of otter populations. Reef communities nonetheless are very susceptible to damage by chemical pollution and increases in water temperature.

Many species competing for limited space on a coral reef show agonistic interactions. Clonal ectoprocts compete by overgrowing others, causing some groups to evolve rapid clonal budding and erect structures resistant to overgrowth. Some slowly growing corals destroy their neighbors with stinging tentacles and digestive secretions. Mutualistic relationships also exist, including protection of clown fishes by otherwise predatory sea anemones attached to reef corals (p. 277). Thus, the apparent stability of coral reefs is achieved by dynamic interactions among many species.

Nearshore Soft Sediments

Intertidal and nearshore subtidal environments with soft sediments support a number of marine biomes including beaches, mudflats, **salt marshes, sea-grass beds,** and **mangrove communities.** An intertidal sand flat is initially colonized by grasses, followed by marsh mussels, burrowing crabs and shrimp, and deposit-feeding polychaetes, forming the characteristics of a salt marsh (Figure 37.15). Small creeks in salt marshes are particularly favorable habitats for many polychaetes, mussels, periwinkles, crustaceans, and fishes. Small fishes, such as killifishes, attract predatory terns and kingfishers. These marshes are important sources of organic matter and provide nurseries for many marine fish species.

Shallow coastal subtidal environments include beds of sea grasses, which often colonize newly deposited sediments and

Figure 37.14
A coral reef's topographic complexity permits it to support diverse subtidal communities with complex symbiotic relationships among species.

become dense along the Atlantic coasts of Europe and North America. Hydroids, sponges, and ectoprocts occur among the grasses, which also support larvae of scallops.

In calm waters of tropical and subtropical marine coasts, mangrove trees grow in submerged soft sediments, forming thick forests along the shoreline. Submerged mangrove roots support a rich marine community of detritus feeders, including oysters, crabs, and shrimp. Fishes also abound. Mangroves are perhaps unique in supporting a marine community at their roots and simultaneously a terrestrial community in their exposed branches.

Deep-Sea Sediments

The deep sea includes the continental slope, continental rise, and abyssal plain. These regions contain mostly soft sediments, with clean sands predominating where strong currents flow and fine muds where currents are weak. Suspension-feeding invertebrates (p. 709) dominate sandy substrates but are rare on muddy ones. Experiments on suspension-feeding clams show that high turbidity in muddy areas damages suspension-feeding systems. Deposit feeders abound in muddy substrates, producing a patchiness of animal communities on the abyssal plain corresponding to substrate type. Deposit-feeding sea cucumbers, polychaetes, and echiurid worms produce fecal mounds that provide local substrates for smaller suspension-feeding bivalves, polychaetes, and crustaceans in an area otherwise unsuited for suspension-feeding. Dead fish and plants fall to the seafloor, supporting bacteria and deposit feeders.

A

B

Figure 37.15
Nearshore marine environments with soft sediments include salt marshes (A) and mangroves (B).

Hydrothermal Vents

Sporadic occurrence of hydrothermal vents (p. 839) contributes further patchiness to the environments and animal communities of the deep sea. Hydrothermal vents occur on the abyssal plain in areas of submarine volcanic activity, which produces a hard substrate and hot, sulfide-rich water. Archaebacteria that derive their energy by oxidizing sulfide form mats on the rocky surfaces near the vents, where they are grazed by bivalves, limpets, and crabs. Other bivalves have sulfur-oxidizing archaebacteria as symbionts in their gills. Giant pogonophoran or siboglinid worms (p. 370) also harbor symbiotic archaebacteria to obtain nutrition. Hydrothermal vents are ephemeral; repeated colonization of newly formed vents propagates these communities.

Pelagic Realm

The vast open ocean is called the **pelagic** realm (Figure 37.16). Despite its size (90% of the total oceanic area), the pelagic realm is relatively impoverished biologically because, as organisms die, they sink from the photic zone, carrying nutrients into the bathypelagic zone.

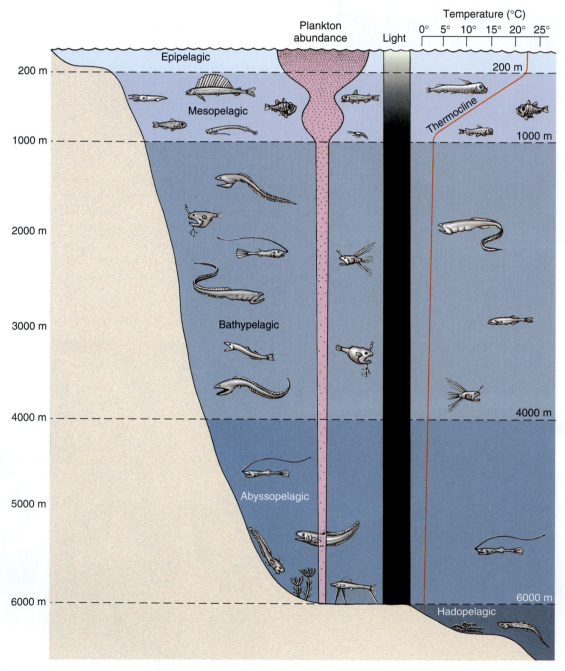

Figure 37.16

Life in pelagic zones. Each zone supports a distinct community of organisms. Animals in zones below the mesopelagic depend on the meager rain of food that sinks from the epipelagic and mesopelagic zones.

Areas of **upwelling** and convergence of ocean currents are vital sources of nutrient renewal for the surface photic zone. The enormously productive polar seas are an example. Before their populations were exploited by humans, baleen whales probably consumed around 77 million tons of krill (a shrimplike animal, Figure 20.27, p. 436) per year, exceeding the entire catch of all fishes, crustaceans, and molluscs taken by the world's fishing fleet in a year. The enormous krill population was sustained by phytoplankton, the base of the food web (p. 834), which in turn flourished because of abundant nutrients in the Antarctic sea.

The world's most productive fisheries are centered on upwelling regions. Before its collapse in 1972, the Peruvian anchovy fishery, which depended on the Peru Current, provided 22% of all fish caught in the world! Earlier, the California sardine fishery and Japanese herring fishery, both fisheries of upwelling regions, were intensively harvested to collapse and never recovered. The world's fisheries today are seriously imperiled by overfishing, degradation of fish habitats by trawling, wasteful fishing methods, and marine pollution. Some of the world's greatest fishing grounds, such as the Grand Banks and Georges Banks of eastern North America, have been destroyed.

Below the surface, or **epipelagic** zone, layers of the pelagic realm are the great ocean depths, characterized by enormous pressure, perpetual darkness, and a constant temperature near 0° C. It remained a world unknown to humans until recently, when baited cameras, bathyscaphs, and deep-water trawls have been lowered to view and to sample the ocean floor. There are several distinct habitats in the ocean depths (Figure 37.16). The **mesopelagic** zone is the "twilight zone," which receives dim light and supports a varied community of animals. Below the mesopelagic zone is a world of perpetual darkness, divided into three depth zones as shown in Figure 37.16: bathypelagic, abyssopelagic, and hadopelagic. Deep-sea forms depend on a meager rain of organic debris from above that escapes consumption by organisms in the water column.

ANIMAL DISTRIBUTION (ZOOGEOGRAPHY)

Zoogeography describes patterns of animal distribution and species diversity and seeks to explain why species and species diversity are distributed as they are. Zoogeography is a subset of biogeography, which seeks evolutionary and ecological explanation for the spatial distributions of all living forms. Most species occupy limited geographic areas. Why animals are distributed as they are is not always obvious because similar habitats on separate continents may contain quite different kinds of animals. A particular species may be absent from a region that supports similar animals because barriers to dispersal prevent it from getting there or because established populations of other animals prevent it from colonizing. Thus we would like to discover why animals occur where they are and not where one thinks they ought to be.

Explanations for geographical distributions of animals are found by studying the history of nature. The fossil record plainly shows that animals once flourished in regions that they no longer occupy. Extinction has a major role, but many groups left descendants that migrated to other regions and survived. For example, camels originated in North America, where their oldest fossils occur. Camels spread during the Pleistocene epoch by way of Alaska to Eurasia and Africa, where true camels live today, and to South America, where their living descendants include llamas, alpacas, guanacos, and vicuñas. (The Pleistocene began about 1.7 million years BP and ended about 11 thousand years BP; see the geological timetable on the back inside-cover.) Then camels went extinct in North America about 10,000 years BP at the end of the Ice Age. Thus the history of an animal species and its ancestors must be documented before one can understand why it lives where it does. Earth's surface undergoes enormous change. Many land areas were once covered by seas. Fertile plains have been claimed by advancing desert; impassable mountain barriers have formed where none existed before; and inhospitable ice fields have retreated in a warming climate to be replaced by forests. Geological change is responsible for much of the alteration in animal (and plant) distribution and is powerful in shaping organic evolution.

Phylogenetic systematics permits us to reconstruct histories of animal distributions (see Chapter 10). A cladogram presents the structure of common evolutionary descent among species. Geographical distributions of closely related species are mapped onto a cladogram to generate hypotheses of those species' geographic histories. The large aquatic hellbenders of eastern North America are unlike any other salamanders except for two species of East Asia. Molecular calibration of their phylogeny (p. 113) suggests that the hellbenders separated from their East Asian relatives about 28 million years ago, when a temporary connection joined forests and streams of eastern Asia and eastern North America across Alaska and northern Canada, areas that later became inhospitable for salamanders. Because the closest living relatives of hellbenders and Asian giant salamanders are the smaller hynobiid salamanders of Asia, the best hypothesis is that the giant salamanders originated in Asia and that a hellbender lineage dispersed to North America 28 million years ago. Many other plant and animal populations moved between eastern Asia and eastern North America at about the same time, giving a common historical explanation for multiple disjunct distributions.

Disjunct Distributions

Zoogeographers are challenged to explain numerous discontinuous or **disjunct distributions:** closely related species living in widely separated areas of a continent, or even the world (Figure 37.17). How could a group of animals become so dispersed geographically? Either a population moves from its place of origin to a new location (**dispersal**), traversing intervening territory unsuited for long-term colonization, or the environment changes, breaking a once continuously distributed species into geographically separated populations (**vicariance**). Climatic changes may contract and fragment areas of habitat favorable for a species, or physical movement of landmasses or waterways may carry different populations of a species away from each other.

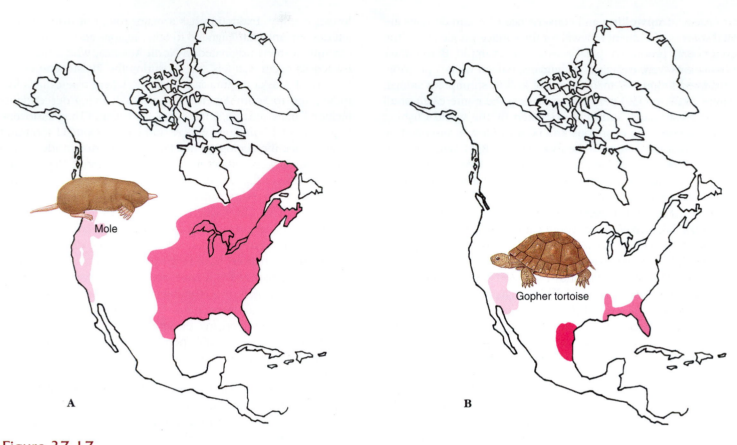

Figure 37.17

Disjunct distributions in North America. **A,** Moles of the family Talpidae probably entered North America across the Bering land bridge that once joined North America and Asia during the Tertiary period. Eastern and western populations are now separated by the Rocky Mountains. **B,** Gopher tortoises of the genus *Gopherus* occupy three fully isolated desert areas.

Distribution by Dispersal

By dispersal, animals enter new localities from their places of origin. Dispersal involves *emigration* from one region and *immigration* into another. Dispersal is a *one-way,* outward movement distinguished from *periodic* movement back and forth between two localities, such as seasonal migration of many birds (p. 601). Dispersing animals may move actively by their own power or be passively dispersed by wind, by floating or rafting on rivers, lakes, or the sea, or by hitching rides on other animals. Animal species should expand their geographic distributions in this manner across all favorable habitat accessible to them. As the last Pleistocene glaciers retreated northward, habitats favorable for many temperate species became available on formerly glaciated territory in North America, Europe, and Asia. Species that originated immediately south of the glaciated territory prior to glacial retreat then expanded northward as new habitats appeared. Because the reproductive rate of animal populations is great, a continuous pressure drives populations to expand across all favorable habitats.

Dispersal easily explains movement of animal populations into favorable habitats that are geographically adjacent to their places of origin. This movement produces an expanded but geographically continuous distribution. Can dispersal also explain the origins of geographically disjunct distributions? For example, flightless ratite birds (p. 116) inhabit disjunct landmasses primarily of the Southern Hemisphere including Africa, Australia, Madagascar, New Guinea, New Zealand, and South America. These landmasses are separated from each other by ocean, a very strong barrier to ratite dispersal. To explain this distribution by dispersal, one must postulate a **center of origin** from which the group dispersed to reach all of the widely separated landmasses on which it now occurs. Because ratites do not fly, a dispersalist hypothesis requires intermittent, passive rafting of individuals across the ocean. Is this hypothesis reasonable? We know from studies of the Galápagos Islands and Hawaii (see Chapter 6) that occasional, long-distance dispersal of terrestrial animals and plants across oceans does occur. This is the only way that terrestrial animals could colonize islands produced by oceanic volcanoes. For flightless birds and many other discontinuously distributed animals, however, an alternative explanation of disjunct distributions is the hypothesis of vicariance (L. *vicarius,* a substitute).

Distribution by Vicariance

Disjunct distributions of animals may be created by physical changes in the environment that cause formerly continuous habitats to become disjunct. Areas once joined may become separated by barriers impenetrable for many animals. The study of fragmentation of biotas in this manner is called **vicariance biogeography.** At the species level, "vicariance" is often used as a synonym of "allopatry," which is simply a distribution of populations in geographically separated areas (p. 118). Lava flows from a volcano may cause a formerly continuous forest to separate into geographically discontinuous patches, thereby breaking many species of plants and animals into geographically isolated populations.

Perhaps the most dramatic vicariant phenomenon is continental drift, through which a once continuous landmass is sequentially broken into continents and islands separated by ocean. All terrestrial and freshwater animal species that had spread across the initially continuous landmass would be sequentially fragmented into many populations on different continents and islands separated by ocean. Vicariance by continental drift gives us another hypothesis for explaining the disjunct distribution of flightless birds; they may descend from an ancestral species that was widespread in the Southern Hemisphere when Africa, Australia, Madagascar, New Guinea, New Zealand, and South America were in closer contact (Figure 37.18) than they are today. When these landmasses moved apart across the ocean, ancestral species would have fragmented into disjunct populations that evolved independently, producing the diversity of forms observed today.

Suppose that different species of flightless birds evolved allopatrically as continental drifting sequentially broke their terrestrial environment into isolated pieces. If we construct a cladogram or phylogenetic tree of these birds as shown in Figure 37.19, the earliest divergence should correspond to the first vicariant event that fragmented their common ancestral species. All subsequent branching events on the tree should correspond sequentially to subsequent vicariant events that fragmented major lineages further. Our tree hypothetically reconstructs the history of vicariant events for the group. If we erase the names of species from terminal branches of the tree and replace them with geographic areas in which each species is found, we have a hypothesis for the sequential separation of different geographic areas. We can test this vicariant hypothesis further by identifying other groups of terrestrial organisms that have different species in each of the same geographic areas as flightless birds. If our hypothesis is correct, these groups were fragmented geographically by the same vicariant events that fragmented flightless birds. We therefore predict that the cladograms or phylogenetic trees constructed for species in the other groups will show the same branching pattern as the flightless-bird tree when we replace the species names with those of the areas they inhabit. If this hypothesis is confirmed, we have a **general area cladogram** that depicts the history of fragmentation of the different geographic areas studied. This general hypothesis of vicariance can be investigated further using geological and climatic studies.

In most groups of organisms, recurring vicariance and dispersal have contributed to evolution of disjunct distributional patterns. Methods of vicariance biogeography are useful for finding such cases. Indeed, the cladogram of flightless birds is not just a simple grouping of birds that inhabit nearby areas. We can ask whether any branch on the cladogram for a particular group of species is inconsistent with the general area cladogram for geographic areas that these species inhabit. Suppose that the cladogram for a particular taxon is consistent with the area cladogram except for placement of a single branch. We explain most geographic disjunctions within the taxon by vicariance but look for dispersal to explain the single branch not compatible with the general area cladogram. In this way, we can focus our study of dispersal on specific cases in which it is most likely to have occurred.

Continental-Drift Theory

It is no accident that enthusiasm for vicariance biogeography accompanied acceptance of continental-drift theory by geologists. Continental-drift theory is not new (it was proposed in 1912 by German meteorologist Alfred Wegener), but it remained largely disfavored until the theory of **plate tectonics** provided a mechanism to explain drifting continents. According to the theory of plate tectonics (tectonics means "deforming movement"), earth's surface is composed of 6 to 10 rocky plates, about 100 km thick, which shift position on a more malleable underlying layer. Wegener proposed that earth's continents had been drifting like rafts following breakup of a single great landmass called Pangaea ("all land"). The original breakup of Pangaea occurred approximately 200 million years ago. Two great supercontinents formed: a northern Laurasia and a southern Gondwana, separated from each other by the Tethys Sea (see Figure 37.18). At the end of the Jurassic period, some 135 million years ago, the supercontinents began to fragment and to drift apart. Laurasia split into North America, most of Eurasia, and Greenland. Gondwana split into South America, Africa, Madagascar, Arabia, India, Australia, New Guinea, Antarctica, and numerous smaller fragments that now form Southeast Asia. The Arabian, Indian, and Southeast Asian fragments gradually moved across the Tethys Sea and eventually collided with Laurasia, to which they are now attached. This theory is supported by the appearance of fit between the continents, by airborne paleomagnetic surveys, by seismographic studies, by the presence of mid-ocean ridges where the tectonic plates arise, and by a wealth of biological data.

Continental drift explains several otherwise puzzling distributions of animals, such as the similarity of invertebrate fossils in Africa and South America, as well as certain similarities in present-day faunas at the same latitudes on the two continents. However, the continents have been separated for all of the Cenozoic era and probably for much of the Mesozoic era as well, much too long to explain the distributions of some modern organisms such as placental mammals. Continental-drift theory is, nevertheless, enormously useful in explaining interconnections between floras and faunas of the past.

The present distribution of marsupial mammals is an excellent example of the influence of continental breakup. Marsupials appeared in the Middle Cretaceous period, about 100 million years BP, probably in South America. Because South America was then connected to Australia through Antarctica (then much warmer

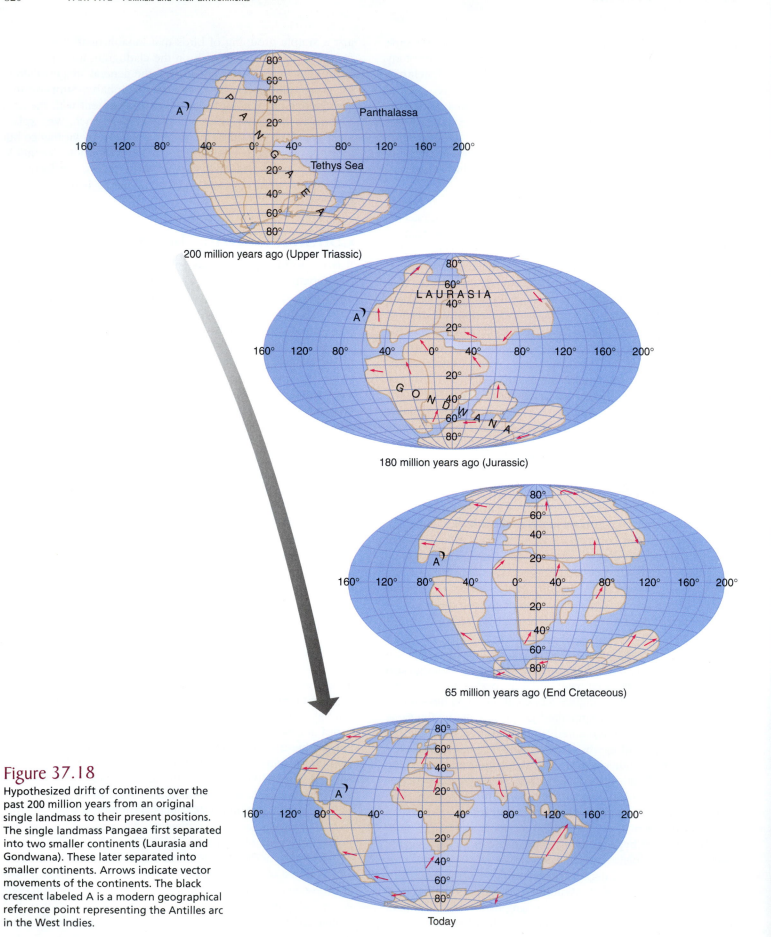

200 million years ago (Upper Triassic)

180 million years ago (Jurassic)

65 million years ago (End Cretaceous)

Today

Figure 37.18

Hypothesized drift of continents over the past 200 million years from an original single landmass to their present positions. The single landmass Pangaea first separated into two smaller continents (Laurasia and Gondwana). These later separated into smaller continents. Arrows indicate vector movements of the continents. The black crescent labeled A is a modern geographical reference point representing the Antilles arc in the West Indies.

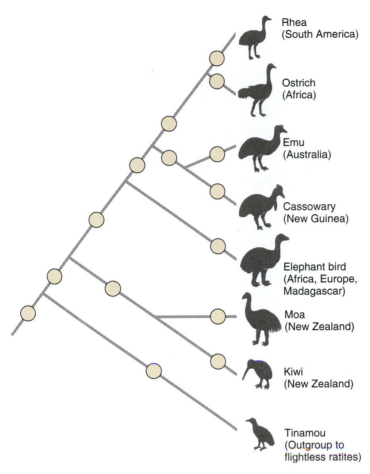

Figure 37.19

Phylogenetic relationships inferred for flightless birds (see Chapter 6, p. 116). Vicariance biogeography proposes that these flightless species descended from an ancestral species that was widespread in the Southern Hemisphere when Africa, Australia, Madagascar, New Guinea, New Zealand, and South America were connected. By moving apart, these landmasses fragmented both themselves and the flightless populations they contained. If the vicariance hypothesis is correct, the phylogenetic branching sequence inferred for allopatric flightless species reflects the sequence by which their landmasses broke apart from each other. This hypothesis is tested by looking for similar phylogenetic patterns in other groups of animals and plants whose ancestral populations would have been fragmented by the same geological events. Widespread geographic distribution of the elephant bird suggests that it has dispersed following the fragmentation of landmasses.

than it is today), marsupials spread through all three continents. They also moved into North America, but there they encountered placental mammals, which had dispersed to that continent from Asia. Marsupials evidently could not coexist with placentals, and so became extinct in North America. (North American marsupials today, opossums, are relatively recent arrivals from South America.) Placentals followed marsupials into South America, but by that time marsupials had expanded and were too firmly established to be displaced. In the meantime, about 50 million years BP, Australia drifted apart from Antarctica, barring entrance to placentals. Australia remained in isolation, allowing marsupials to diversify into the present rich and varied faunas.

Many people consider Alfred Russel Wallace (see Figure 6.1) the founder of modern historical biogeography. Wallace conducted extensive field studies in the Malay Archipelago, where he discovered an abrupt faunal change between Asian faunal elements and Australian/New Guinean ones. Pheasants, parrots, monkeys, numerous lizard groups, and even marine invertebrates are among the faunal elements whose geographic distributions show abrupt boundaries at this line. This biogeographic boundary is called "Wallace's Line," and bisects the current country of Indonesia (Figure 37.20). Wallace's Line has been a biogeographic mystery since he described it because there are no obvious environmental changes or barriers that would explain the abrupt displacement of faunas across the line. Plate tectonics offers the best explanation of Wallace's Line. Although currently close, the Southeast Asian plates separated from the Australian/New Guinean plate during the breakup of Gondwana, and these different plates spent many millions of years crossing the Tethys Sea in isolation from each other before finding their present locations. Wallace's Line marks the approximate boundary between the Southeast Asian and Australian/New Guinean plates, whose faunas diverged greatly during their long evolutionary separation. Molecular phylogenetic studies comparing lizard groups distributed on opposite sides of Wallace's Line support the interpretation that these groups were formerly isolated by a large expanse of ocean and only recently became geographic neighbors in the Malay Archipelago.

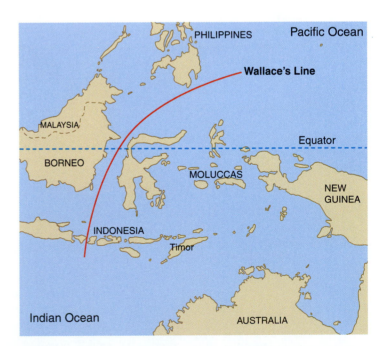

Figure 37.20

The enigmatic borderline described by Alfred Russel Wallace as marking the geographic separation of Asian (upper left) and Australian (center to lower right) faunas in the Malay Archipelago. A collision of tectonic plates brought formerly distant landmasses and their faunas in close geographic proximity, thereby forming "Wallace's Line."

Isthmus
of Panama

Figure 37.21

The Great American Interchange. The Isthmus of Panama emerged approximately 3 million years ago, permitting the extensive interchange of many families of mammals. At top are representatives of 38 South American genera that walked north across the isthmus. At bottom are representatives of 47 North American genera that migrated to South America. The North American immigrants diversified rapidly after entering South America. South American immigrants to North America diversified little and most became extinct.

Temporary Land Bridges

Temporary land bridges also have been important pathways of dispersal. An important and well-established land bridge that no longer exists connected Asia and North America across the Bering Strait. It was across this corridor that the placentals moved from Asia into North America.

Today a land bridge connects North and South America at the Isthmus of Panama, but from the mid-Eocene epoch (50 million years BP) to the end of the Pliocene epoch (3 million years BP), water completely separated the two continents. During this long period, the major groups of mammals evolved in distinctive directions on each continent. When the land bridge was reestablished at the end of the Pliocene epoch, a tide of mammals moved in both directions (Figure 37.21). This dispersal is called the "Great American Interchange," one of the most important minglings of distinct continental faunas in earth's history.

For a period both continents gained mammalian diversity, but extinction of large numbers of mammals on both continents soon followed. North American carnivores such as raccoons, weasels, foxes, dogs, cats (including sabercats), and bears began preying on South American mammals. Other North American invaders included hoofed mammals (horses, tapirs, peccaries, camelids, deer, antelopes, and mastodonts), rabbits, and several families of rodents. These mammals displaced many South American residents occupying similar habitats. Today nearly half of South America's mammals descend from recent North American invaders. Only a few South American invaders survived in North America: porcupines, armadillos, and opossums. Several other South American groups, including giant ground sloths, glyptodonts, anteaters, giant aquatic capybaras, toxodonts (rhino-sized plant eaters), and giant armadillos, entered North America but subsequently went extinct there.

SUMMARY

The biosphere is a thin life-containing blanket surrounding the earth. Life on earth is possible because of a steady supply of energy from the sun, presence of water, a suitable range of temperatures, the correct proportion of major and minor elements, and the screening of lethal ultraviolet radiation by atmospheric ozone. Earth's environment and living organisms have evolved together, each deeply marking the other.

The biosphere comprises lithosphere, the earth's rocky shell; hydrosphere, the global distribution of water; and atmosphere, the blanket of gas surrounding the earth.

Earth's terrestrial environment comprises various biomes, each one bearing a distinctive array of plant life and associated animal life. Eastern deciduous forest has distinct seasons and autumn leaf fall. North of the deciduous forest is coniferous forest, whose northern range is called taiga, an area dominated by needle-leafed trees adapted for heavy snowfall. Animals of the taiga are adapted for long, snowy winters.

The tropical forest is the richest biome, characterized in part by a great diversity of plant species and the vertical stratification of animal habitats. Most tropical forest soils rapidly deteriorate when forest is removed.

The most modified biome is grassland, or prairie, which has been converted largely to agriculture and grazing. The tundra biome of the far north and the desert biome are both severe environments for animal life, but they are populated nevertheless with organisms that have evolved appropriate adaptations.

Freshwater habitats include rivers and streams (lotic habitats) and ponds and lakes (lentic habitats). All are geologically ephemeral habitats strongly influenced by nutrient input.

Oceans occupy 71% of earth's surface. The photic, or sunlit, zone supports photosynthetic activity by phytoplankton. Oceanic animal communities are classified according to topographic features, substrate, and distance from the shore. Benthic communities with rocky substrates include intertidal regions with tide pools, subtidal coral reefs and kelp forests, and hydrothermal vent communities of the deep sea. Benthic communities on soft sediments include nearshore salt marshes, sea-grass beds, and mangrove communities. Benthic deep-sea communities form a patchwork of suspension feeders on sandy substrates and deposit feeders on muddy ones. Coral reefs are the most ecologically diverse of these benthic communities. Pelagic communities include a shallow-water zone overlying the continental shelf. This zone is the locus of the world's great fisheries, which are especially productive in areas of upwelling where nutrients are constantly renewed. The deeper waters of open ocean occupy most of the ocean's area but have low biological productivity.

Zoogeography is the study of animal distribution on earth and its history using evidence from current distributions of animal species, fossils, and phylogenetic systematic analyses. Animals have become distributed by dispersal, the spread of populations from their centers of origin, and by vicariance, the separation of populations by barriers. Continental drift, now strongly supported by plate tectonic theory, explains how some animal groups become geographically separated so that evolutionary diversification could occur. It also explains how certain groups, such as marsupial mammals, became isolated from others. Temporary land bridges have served as important pathways for animal dispersal.

REVIEW QUESTIONS

1. What special conditions on earth make this planet especially favorable for life?
2. How have the earth and life on it evolved together with each deeply influencing the other?
3. What is the biosphere? How would you distinguish between the following subdivisions of the biosphere: lithosphere, hydrosphere, atmosphere?

4. What is the origin of oxygen on earth? What would happen to earth's supply of oxygen if photosynthesis suddenly were to cease?

5. What is the evidence that increasing carbon dioxide levels in the atmosphere are responsible for an increase in the "greenhouse effect"?

6. What is a biome? Briefly describe six examples of biomes.

7. Describe three kinds of benthic marine communities that use hard substrates and three that use soft substrates. What are the main physical factors that separate community types within each substrate category?

8. What are some very productive marine environments, and why are they so productive?

9. What is the source of nutrients for animals living in the deep-sea habitat?

10. What are some reasons why a species may be absent from a habitat or region to which it should adapt well?

11. Define and distinguish the alternative explanations for disjunct distributions among animals: dispersal and vicariance.

12. Who first proposed the continental-drift theory? What three sources of evidence convinced geologists that the theory was correct?

13. How does continental-drift theory help to explain the disjunct distribution of marsupial mammals in Australia and South America?

14. What was the Great American Interchange? When did it occur, and what were the results?

SELECTED REFERENCES

Berner, E. K., and R. A. Berner. 1996. Global environment: water, air, and geochemical cycles. Upper Saddle River, New Jersey, Prentice-Hall, Inc. *A geochemistry textbook with good coverage of global water and air circulation, greenhouse effect, acid rain, and geochemistry of rivers, lakes, and oceans.*

Castro, P., and M. E. Huber. 2005. Marine biology, ed. 5. Boston, McGraw-Hill Higher Education. *A global coverage of marine biology designed for nonmajors.*

Cox, C. B., and P. D. Moore. 2005. Biogeography: an ecological and evolutionary approach, ed. 7. Boston, Blackwell Science Ltd. *Highly readable account with a strong ecological emphasis.*

Henderson, L. J. 1913. The fitness of the environment. New York, Macmillan, Inc. *This short but influential book, one of the great classics of biological literature, explains how conditions on our planet made life possible.*

Lieberman, B. S. 2000. Paleobiogeography. New York, Kluwer Academic/Plenum Publishers. *A current coverage of historical biogeography with emphasis on use of fossils to study global change, plate tectonics, and evolution.*

Levinton, J. S. 2001. Marine biology, ed. 2. Oxford, U.K., Oxford University Press. *A thorough coverage of oceanic ecosystems with photographs of diverse communities.*

Lomolino, M. V., B. R. Riddle, and J. H. Brown. 2006. Biogeography, ed. 3. Sunderland, Massachusetts, Sinauer Associates. *A current general textbook of biogeography.*

Lomolino, M. V., D. F. Sax, and J. H. Brown. 2004. Foundations of biogeography. Chicago, University of Chicago Press. *Collects classic papers of biogeography from 1700s to 1975 with new commentary.*

MacDonald, G. 2003. Biogeography: introduction to space, time and life. New York, John Wiley and Sons, Inc. *A recent introductory textbook of biogeography.*

Marshall, R. 2005. Alaska wilderness: exploring the central Brooks Range, ed. 3. Berkeley, California, University of California Press. *A chronicle of the original exploration of Alaskan tundra, with updates on its conservation status.*

Rothschild, L. J., and A. M. Lister. 2003. Evolution on planet earth: the impact of the physical environment. San Diego, Academic Press. *A geological and climatic perspective on life's evolutionary history.*

Van Oosterzee, P. 1997. Where worlds collide: the Wallace Line. Ithaca, New York, Cornell University Press. *An entertaining account of the founding of biogeography by Alfred Russel Wallace with emphasis on the paradox of an abrupt faunal change in the Malay Archipelago.*

Whitfield, P., P. D. Moore, and B. Cox. 2002. Biomes and habitats. New York, Macmillan Reference USA. *A recent coverage of the earth's biomes.*

ONLINE LEARNING CENTER

Visit www.mhhe.com/hickmanipz14e for chapter quizzing, key term flash cards, web links, and more!

Animal Ecology

The ecological niche of a praying mantis includes other insects as a food resource.

Every Species Has Its Niche

The lavish richness of earth's biomass is organized into a hierarchy of interacting units: an individual organism, a population, a species, a community, and finally an ecosystem, that most bewilderingly complex of all natural systems. Central to ecological study is the habitat, the spatial location where an animal lives. An animal's use of resources and the conditions it tolerates in its habitat constitute its niche: how it gets its food, how it achieves reproductive perpetuity—in short, how it survives and propagates. The concept of niche applies also to populations and species and is usually studied by ecologists at these levels. The niche of a species, for example, comprises the resources collectively used and conditions tolerated by its members to sustain the species in its ecosystem.

The niche of a species is a product of its evolution, and once it is established, no other species in the community can evolve to exploit exactly the same resources. The "competitive exclusion principle" states that no two species occupy the same niche in the same geographic area. Different species are therefore able to form

an ecological community in which each one has a different role in their shared environment.

In the mid-nineteenth century, the German zoologist Ernst Haeckel introduced the term **ecology,** defined as the "relation of the animal to its organic as well as inorganic environment." Environment here includes everything external to the animal but most importantly its immediate surroundings. Although we no longer restrict ecology to animals alone, Haeckel's definition is still sound. Animal ecology is now a highly synthetic science that incorporates everything we know about behavior, physiology, genetics, and evolution of animals to study interactions between populations of animals and their environments. The major goal of animal ecology is to understand how these diverse interactions determine geographical distributions and abundances of animal populations. Such knowledge is crucial for ensuring continued survival of many populations when their natural environments are altered.

THE HIERARCHY OF ECOLOGY

Ecology is studied as a hierarchy of biological systems in interaction with their environments. At the base of the ecological hierarchy is an **organism.** To understand why animals live where they do, ecologists must examine the varied physiological and behavioral mechanisms that animals use to survive, to grow, and to reproduce. A near-perfect physiological balance between production and loss of heat is required for success of certain endothermic species (such as birds and mammals) under extreme temperatures, as found in the Arctic or a desert. Other species succeed in these situations by escaping the most extreme conditions by migration, hibernation, or torpidity. Insects, fishes, and other ectotherms (animals whose body temperature depends on environmental heat) compensate for fluctuating temperatures by altering behavior and biochemical and cellular processes. Thus an animal's physiological capacities permit it to live under changing and often adverse environmental conditions. Behavioral responses are important also for obtaining food, finding shelter, escaping enemies and unfavorable environments, finding a mate, and caring for offspring. Physiological mechanisms and behaviors that improve adaptability to environments assist survival of organisms. Ecologists who focus their studies at the organismal level are called physiological ecologists or behavioral ecologists.

Animals in nature coexist with others of the same species; these groups are called **populations.** Populations have properties that cannot be discovered by studying individual animals alone. These properties include genetic variability among individuals (polymorphism), growth in numbers over time, and factors that limit the density of individuals in each area. Ecological studies at the population level help us to predict the future success of endangered species and to discover controls for pest species.

Just as individuals do not exist alone in nature, populations of different species co-occur in more complex associations called **communities.** The complexity of a community is measured as **species diversity,** the number of different species that coexist to form the community. The populations of species in a community interact with each other in many ways, the most prevalent of which are **predation, parasitism,** and **competition. Predators** obtain energy and nutrients by killing and eating other animals, called prey. **Parasites** derive similar benefits from their host organisms, but usually do not kill the hosts. A **parasitoid** is a parasite that kills its host organism. Competition occurs when food or space are limited, and members of the same or different species interfere with each other's use of shared resources. **Mutualism** occurs when both members of a pair of species benefit from their interactions, usually by avoiding negative interactions with other species. Communities are complex because all of these interactions occur simultaneously, and their individual effects on the community often cannot be isolated.

Most people know that lions, tigers, and wolves are predators, but the world of invertebrates also includes numerous predaceous animals. These predators include unicellular organisms, jellyfish and their relatives, various worms, predaceous insects, sea stars, and many others.

Ecological communities are biological components of even larger, more complex entities called **ecosystems.** An ecosystem consists of all populations in a community together with their physical environments. The study of ecosystems reveals two key processes in nature, the flow of energy and the cycling of materials through biological channels. The largest ecosystem is the **biosphere,** the thin veneer of land, water, and atmosphere that envelops the planet and supports all life on earth (see Chapter 37).

Environment and Niche

An animal's environment consists of all conditions that directly affect its survival and reproduction. These factors include space, forms of energy such as sunlight, heat, wind and water currents, and also materials such as soil, air, water, and numerous chemicals. The environment also includes other organisms, which can be an animal's food, or its predators, competitors, hosts, or parasites. The environment thus includes both abiotic (nonliving) and biotic (living) factors. Some environmental factors, such as space and food, are utilized directly by an animal and are called **resources.**

A resource may be expendable or nonexpendable, depending on how an animal uses it. Food is expendable, because once eaten it is no longer available. Food therefore must be continuously replenished in the environment. Space, whether total living area or a subset such as the number of suitable nesting sites, is not exhausted by being used, and thus is nonexpendable.

The physical space where an animal lives, which contains its environment, is its **habitat.** Size of a habitat is variable. A rotten log is a normal habitat for carpenter ants. Such logs occur in larger habitats called forests where deer also live. However, deer forage in open meadows, so their habitat is larger than the forest. On a larger scale, some migratory birds occupy forests of the north temperate region during summer and move to the tropics during winter. Thus, habitat is defined by an animal's normal activity rather than by arbitrary physical boundaries.

Animals of any species experience environmental limits of temperature, moisture, and food within which they can grow, reproduce, and survive. A suitable environment therefore must meet all requirements for life. A freshwater clam living in a tropical lake could tolerate the temperature of a tropical ocean, but would be killed by the ocean's salinity. A brittle star living in the Arctic Ocean could tolerate the salinity of the tropical ocean but not its temperature. Thus temperature and salinity are two separate dimensions of an animal's environmental limits. If we add another variable, such as pH (p. 24), we increase our description to three dimensions (Figure 38.1). If we consider all environmental conditions that permit members of a species to survive and to multiply, we define the role of that species in nature as distinct from all others. This unique, multidimensional relationship of a species with its environment is called its **niche** (see the chapter-opening essay, p. 825). Dimensions of the niche vary among members of a species, making the niche subject to evolution by natural selection. The niche of a species undergoes evolutionary changes over successive generations.

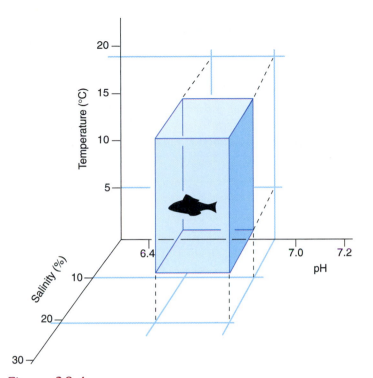

Figure 38.1

Three-dimensional niche volume of a hypothetical animal showing three tolerance ranges. This graphic representation is one way to show a portion of the multidimensional nature of environmental relations.

Animals may be generalists or specialists with respect to tolerance of environmental conditions. For example, most fish are adapted to either freshwater or seawater, but not both. However, those that occupy marshes, such as the killifish, *Fundulus heteroclitus,* easily tolerate changes in salinity during tidal cycles in these estuarine habitats as freshwater from land mixes with seawater. Similarly, although most snakes can eat a wide variety of animal prey, others have narrow dietary requirements; for example, the African snake, *Dasypeltis scaber,* is specialized to eat bird eggs (see Figure 32.3, p. 711).

However broad may be the tolerance limits of an animal, it experiences only a single set of conditions at a time. An animal probably will not experience in its lifetime all environmental conditions that it might tolerate. Thus, we must distinguish an animal's **fundamental niche,** which describes its potential role, and its **realized niche,** the subset of potentially suitable environments that an animal actually experiences. Likewise, we must distinguish fundamental niche from realized niche at the population and species levels, at which this phenomenon is mainly studied. For example, competition within a community may limit the realized niche of a species to a much smaller range of conditions than predicted by its fundamental niche.

Populations

An animal exists in nature as a member of a population, a reproductively interactive group of animals of a single species (p. 118). A species may be a single, cohesive population or may contain many geographically disjunct populations, often called **demes.**

Because members of a deme interbreed, they share a common gene pool (p. 126).

Movement of individuals among demes within a species can impart some evolutionary cohesion to the species as a whole. Local environments may change unpredictably, sometimes causing a local deme to become severely depleted or eliminated. Immigration is therefore a crucial source of replacement among demes within a region. Extinction of a species may be avoided if risk of extinction is spread among many demes, because simultaneous destruction of the environments of all demes in this manner is unlikely unless a catastrophe is widespread. Interaction among demes in this manner is called **metapopulation dynamics,** with "metapopulation" referring to a population subdivided into multiple genetically interacting demes. In some species, gene flow and recolonization among demes may be nearly symmetrical. If some demes are stable and others more susceptible to extinction, the more stable ones, termed **source demes,** differentially supply emigrants to the less stable ones, called **sink demes.**

Each population or deme has a characteristic **age structure, sex ratio,** and **growth rate.** The study of these properties and the factors that influence them is called demography. Demographic characteristics vary according to lifestyles of the species under study. For example, some animals (and most plants) are **modular.** Modular animals, such as sponges, corals, and ectoprocts, consist of colonies of genetically identical organisms. Reproduction is by asexual **cloning,** as described for hydrozoans in Chapter 13 (p. 270). Most colonies also have distinct periods of gamete formation and sexual reproduction. Colonies propagate also by fragmentation, as seen on coral reefs during severe storms. Pieces of coral may scatter by wave action on a reef, forming propagules for new reefs. For these modular animals, age structure and sex ratio are difficult to determine. Changes in colony size are used to measure growth rate, but counting individuals is more difficult and less meaningful than in **unitary** animals, which are independently living organisms.

Most animals are unitary. However, even some unitary species reproduce by **parthenogenesis** (p. 140). Parthenogenetic species occur in many animal taxa, including insects, fishes, salamanders, and lizards. Such groups contain only females, which lay unfertilized eggs that hatch into daughters whose genotype comes entirely from their mothers. The praying mantid, *Bruneria borealis,* common in the southeastern United States, is a parthenogenetic unitary animal.

Most metazoans are biparental (p. 139), and reproduction follows a period of organismal growth and maturation. Each new generation begins with a **cohort** of individuals born at the same time. Of course, individuals of any cohort do not all survive to reproduce. For a population to retain constant size from generation to generation, each adult female must replace herself on average with one daughter that survives to reproduce. If females produce on average more than one viable daughter, the population will grow; if fewer than one, the population will decline.

Animal species have different characteristic patterns of **survivorship** from birth until death of the last member of a cohort. The three principal types of survivorship are illustrated in Figure 38.2. Curve I, in which all individuals die at the same time, probably occurs only rarely in nature. Curve II, in which rate of mortality

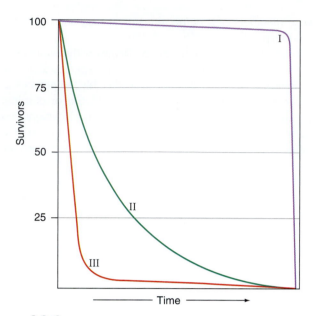

Figure 38.2

Three principal types of theoretical survivorship curves. See text for explanation.

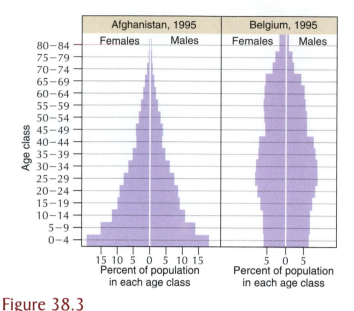

Figure 38.3

Age structure profiles of the human populations of Afghanistan and Belgium in 1995 contrast the rapidly growing, youthful population of Afghanistan with the stable population of Belgium, where the fertility rate is below replacement.

as a proportion of survivors is constant over all ages, is characteristic of some animals that care for their young, as do many birds. Human populations generally fall somewhere between curves I and II, depending on nutrition and medical care.

Survivorship of most invertebrates, and of vertebrates like fish that produce great numbers of offspring, resembles curve III. For example, a mature female marine prosobranch snail, *Ilyanassa obsoleta,* produces thousands of eggs each reproductive period. Zygotes become free-swimming planktonic veliger larvae, which are scattered far from the mother's habitat by oceanic currents. They form part of the plankton and experience high mortality from numerous animals that consume plankton. Furthermore, the larvae require a specific, sandy substrate on which to settle and to metamorphose into adult snails. The probability of a larva surviving long enough to find a suitable habitat is very low, and most of the cohort dies during the veliger stage. We therefore see a rapid drop in survivorship in the first part of the curve. The few larvae that do survive to become snails have improved odds of surviving further, as shown by the more gentle slope of the curve for older snails. Thus, high reproductive output balances high juvenile mortality.

Most animals do not survive to reach reproductive age; however, those that do survive often reproduce only once before they die, as do many insect species of the temperate zone. Here, adults reproduce before the onset of winter and die, leaving only their eggs to overwinter and to repopulate their habitat the following spring. Similarly, Pacific salmon after several years return from the ocean to freshwater to spawn only once, after which all adults of a cohort die. However, other animals survive long enough to produce multiple cohorts of offspring that may mature and reproduce while their parents are still alive and reproductively active.

Populations of animals containing multiple cohorts, such as robins, box turtles, and humans, exhibit **age structure.** Analysis of age structure reveals whether a population is actively growing, stable, or declining. Figure 38.3 shows age profiles of two populations. On a global scale, humans exhibit an age structure similar to curve I in Figure 38.2, although age structures vary among regions.

Population Growth and Intrinsic Regulation

Population growth is the difference between rates of birth and death. As Darwin recognized from an essay by Thomas Malthus (p. 108), all populations have an inherent ability to grow exponentially. This ability is called the **intrinsic rate of increase,** denoted by the symbol **r.** The steeply rising curve in Figure 38.4 shows this kind of growth. If species continually grew in this fashion, earth's resources soon would be exhausted and mass extinction would follow. A bacterium dividing three times per hour could produce a colony a foot deep over the entire earth after 36 hours, and this mass would be over our heads only one hour later. Animals have much lower potential growth rates than bacteria, but they could achieve the same kind of result over a longer period of time, given unlimited resources. Many insects lay thousands of eggs each year. A single codfish may spawn 6 million eggs in a season, and a field mouse can produce 17 litters of five to seven offspring each year. Obviously, unrestricted growth is uncommon in nature.

Even in the most benign environment, a growing population eventually exhausts food or space. Exponential increases such as locust outbreaks or planktonic blooms in lakes must end when food or space is expended. Actually, among all resources that could limit a population, the one in shortest supply relative to the needs of the population is depleted before others. This one is termed the **limiting resource.** The largest population that can be supported by the limiting resource in a habitat is called the **carrying capacity** of that environment, symbolized **K.** Ideally, a population would slow its growth rate in response to diminishing resources until it just reaches K, as represented by the sigmoid curve in Figure 38.4. The mathematical expressions of exponential and sigmoid (or logistic) growth curves are compared in the box on page 831. Sigmoid growth occurs when negative feedback exists between growth rate and population density. This phenomenon is called density dependence, and it is the mechanism for intrinsic regulation of populations. We can compare density dependence by negative feedback to the way endothermic animals regulate their body temperatures when the environmental temperature exceeds an optimum. If the resource is expendable, as is food, carrying capacity is reached when rate of resource replenishment equals rate of depletion by the population; the population is then at K for that limiting resource. According to the logistic model, when population density reaches K, rates of birth and death are equal and growth of the population ceases. If food is being replaced at a rate that supports the current population, but no more, a population of grasshoppers in a green meadow may be at carrying capacity even though we see plenty of unconsumed food.

Although experimental populations of protozoa may fit the logistic growth curve closely, most populations in nature fluctuate above and below carrying capacity. For example, after sheep were introduced to Tasmania around 1800, their numbers changed logistically with small oscillations around an average population size of about 1.7 million; we thereby infer the carrying capacity of the environment to be 1.7 million sheep (Figure 38.5A). Ring-necked pheasants introduced on an island in Ontario, Canada exhibited wider oscillations (Figure 38.5B).

Why do intrinsically regulated populations oscillate this way? First, the carrying capacity of an environment can change over time, causing a change of population density as dictated by a limiting resource. Second, animals always experience a lag between the time that a resource becomes limiting and the time that the population responds by reducing its rate of growth. Third, **extrinsic** factors occasionally may limit a population's growth below carrying capacity. We consider extrinsic factors in the next section.

On a global scale, humans have the longest record of exponential population growth (Figure 38.5C). Although famine and war have restrained growth of populations locally, the only dip in global human growth resulted from bubonic plague ("black death"), which decimated much of Europe during the fourteenth

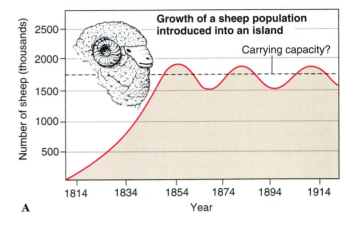

A

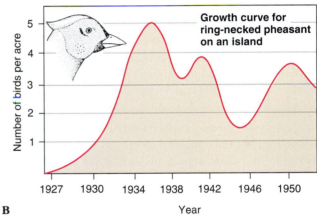

B

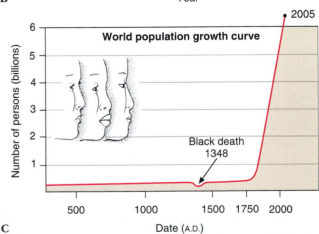

C

Figure 38.5

Growth curves for sheep (**A**), ring-necked pheasant (**B**), and world human populations (**C**) throughout history. Note that the sheep population on an island is stable because of human control of the population, but the ring-necked pheasant population oscillates greatly, probably because of large changes in carrying capacity. Where would you place the carrying capacity for the human population?

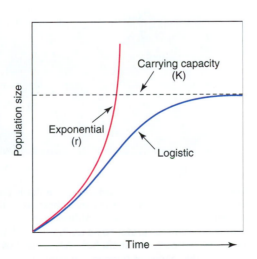

Figure 38.4

Population growth, showing exponential growth of a species in an unlimited environment, and logistic growth in a limited environment. K = population size at carrying capacity; r = intrinsic rate of population growth.

century. What then is the carrying capacity for the human population? The answer is far from simple, and several important factors must be considered when estimating the human K.

With development of agriculture, the carrying capacity of the environment increased, and the human population grew steadily from 5 million around 8000 B.C., when agriculture was introduced, to 16 million around 4000 B.C. Despite the toll of terrible famines, disease, and war, the population reached 500 million by 1650. With the Industrial Revolution in Europe and England in the eighteenth century, followed by a medical revolution, discovery of new lands for colonization, and better agricultural practices, the human carrying capacity increased dramatically. The population doubled to 1 billion around 1850. It doubled again to 2 billion by 1927, to 4 billion in 1974, passed 6 billion in October 1999, and is expected to reach 8 billion by the year 2040. Thus, the growth has been exponential and remains high (Figure 38.5C).

Between 1970 and 2004 the annual growth rate of the human population decreased from 1.9% to 1.23%. At 1.23%, it will take nearly 57 years for the world population to double rather than 36.5 years at the higher annual growth rate. The decrease is credited to better family planning. Nevertheless, half the global population is under 25 years old and most live in developing countries where reliable contraception is limited or nonexistent. Thus, despite a drop in growth rate, the greatest surge in population lies ahead, with a projected 3 billion people added within the next five decades, the most rapid increase ever in human numbers.

To estimate carrying capacity for the human species, we must consider not only quantity of resources, but quality of life. Many of the 6.4 billion people alive today are malnourished. At present 99% of our food comes from the land, and the tiny fraction that we derive from the sea is decreasing due to exploitation of fish stocks (p. 817). Although what would constitute the maximum sustainable agricultural output is uncertain, food production cannot sustain exponential population growth indefinitely.

Extrinsic Limits to Growth

We have seen that the intrinsic carrying capacity of a population for an environment prevents unlimited exponential growth of the population. Population growth also can be limited by extrinsic biotic factors, including predation, parasitism (including disease-causing pathogens), and interspecific competition, or by abiotic influences such as floods, fires, and storms. Although abiotic factors certainly reduce populations in nature, they cannot truly regulate population growth because their effect is wholly independent of population size; abiotic limiting factors are **density-independent.** A single hailstorm may kill most young of wading bird populations, and a forest fire may eliminate entire populations of many animals, regardless of how many individuals exist.

In contrast, biotic factors act in a **density-dependent** manner. Predators and parasites respond to changes in density of their prey and host populations, respectively, to maintain those populations at fairly constant sizes. These sizes are below carrying capacity, because populations regulated by predation or parasitism are

not limited by their resources. Competition between species for a common limiting resource lowers the effective carrying capacity for each species below that of either one alone.

Community Ecology
Interactions among Populations in Communities

Populations of animals that form a **community** interact in ways that can be detrimental (−), beneficial (+), or neutral (0) to each species, depending on the interaction. For instance, we consider a predator's effect on prey as (−), because survival of the prey animal is reduced. However, the same interaction benefits the predator (+) because food obtained from prey increases a predator's ability to survive and to reproduce. Thus, the predator-prey interaction is + − . Ecologists use this shorthand notation to characterize interspecific interactions because it shows the direction in which the interaction affects each species.

We see other kinds of + − interactions. One is **parasitism,** in which the parasite benefits by using the host as a home and source of nutrition, and the host is harmed. **Herbivory,** in which an animal eats a plant, is another + − relationship. **Commensalism** is an interaction that benefits one species and neither harms nor benefits the other (+0). Most bacteria that normally inhabit our intestinal tracts do not affect us (0), but the bacteria benefit (+) by having food and a place to live. Some harmless bacteria in our intestinal tracts prevent entry of more harmful bacterial species, causing this commensalism to become a mutualism. A classic example of commensalism is the association of pilot fishes and remoras with sharks (Figure 38.6). These fishes get the "crumbs" remaining when the host shark makes its kill, but we now know that some remoras also feed on ectoparasites of sharks. Therefore, this commensalism also grades into mutualism.

Organisms engaged in mutalism have a friendlier arrangement than commensalistic species, because the fitness of both is enhanced (++). Biologists are finding mutualistic relationships far more common in nature than previously believed

Figure 38.6
Four remoras, *Remora* sp., attached to a shark. Remoras feed on fragments of food left by their shark host, as well as on pelagic invertebrates and small fishes. Although they actually are good swimmers, remoras prefer to be pulled through the water by marine creatures or boats. The shark host may benefit by having embedded copepod skin parasites removed by remoras.

Exponential and Logistic Growth

We describe the sigmoid growth curve (see Figure 38.4) by a simple model called the logistic equation. The slope at any point on the growth curve is the growth rate, how rapidly population size is changing with time. If N represents the number of organisms and t the time, we can, in the language of calculus, express growth as an instantaneous rate:

$$dN/dt = \text{rate of change in number of organisms per time}$$
$$\text{at a particular instant in time.}$$

When populations are in an environment of unlimited resources (unlimited food and space, and no competition from other organisms), growth is limited only by the inherent capacity of the population to reproduce itself. Under these ideal conditions growth is expressed by the symbol r, which is defined as the intrinsic rate of population growth per capita. The index r is actually the difference between the birth rate and death rate per individual in the population per unit time. The growth rate of the population as a whole is then:

$$dN/dt = rN$$

This expression describes the rapid **exponential growth** illustrated by the early upward-curving portion of the sigmoid growth curve (see Figure 38.4).

Growth rate for populations in the real world slows as the upper limit is approached, and eventually stops. At this point N has reached its maximum density because the space being studied is "saturated" with animals. This limit is called the carrying capacity of the environment and is expressed by the symbol K. The sigmoid population growth curve now is described by the logistic equation, which is written as:

$$dN/dt = rN([K - N]/K)$$

This equation states that the rate of increase per unit of time (dN/dt) = rate of growth per capita (r) × population size (N) × unutilized freedom for growth ($[K - N]/K$). We see from the equation that when the population approaches carrying capacity, $K - N$ approaches 0, dN/dt also approaches 0, and the curve flattens.

Populations occasionally overshoot the carrying capacity of the environment so that N exceeds K. The population then exhausts a resource (usually food or shelter). The rate of growth, dN/dt, then becomes negative and the population declines.

(Figure 38.7). Some mutualistic relationships are not only beneficial but necessary for survival of one or both species. An example is the relationship between a termite and protozoans inhabiting its gut. Bacterial symbionts of the protozoans digest wood eaten by the termite because they produce an enzyme, lacking in the termite, that digests cellulose; the termite lives on waste products of protozoan-bacterial metabolism. In return, the protozoans and their bacteria gain a habitat and food supply. Such absolute interdependence among species is a liability if one participant is lost. *Calvaria* trees native to the island of Mauritius have not reproduced successfully for over 300 years, because their seeds germinate only after being eaten and passed through the gut of a dodo bird, now extinct.

Competition between species reduces fitnesses of both (−−). Many biologists, including Darwin, considered competition the most common and important interaction in nature. Ecologists have constructed most of their theories of community structure from the premise that competition is the chief organizing factor in species assemblages. Sometimes the effect on one species in a competitive relationship is negligible. This condition is called **amensalism, or asymmetric competition** (0−). For example, two species of barnacles in rocky intertidal habitats, *Chthamalus stellatus* and *Balanus balanoides,* compete for space. A famous experiment by Joseph Connell[1] demonstrated that *B. balanoides* excluded *C. stellatus* from a portion of the habitat, whereas *C. stellatus* had no effect on *B. balanoides.*

We have treated interactions as occurring between pairs of species. However, in natural communities containing populations of many species, a predator may have more than one prey and several animals may compete for the same resource (p. 826). Thus, ecological communities are quite complex and dynamic, a challenge to ecologists who study this level of natural organization.

Competition and Character Displacement

Competition occurs when two or more species share a **limiting resource.** Simply sharing food or space with another species does not produce competition unless the resource is in short supply relative to the needs of the species that share it. Thus, we cannot demonstrate competition in nature based solely on sharing of resources. However, we find evidence of competition by investigating the different ways that species exploit a resource.

Competing species may reduce conflict by reducing overlap of their niches. **Niche overlap** is the portion of resources shared by the niches of two or more species. For example, if two species of birds eat seeds of exactly the same size, competition eventually excludes the species less able to exploit the resource. This example illustrates the principle of **competitive exclusion:** strongly competing species cannot coexist indefinitely. To coexist in the same habitat, species must specialize by partitioning a shared resource and using different portions of it. Specialization of this kind is called **character displacement.**

Character displacement usually appears as differences in organismal morphology or behavior related to exploitation of a resource. For example, in his classic study of Galápagos finches (p. 122), English ornithologist David Lack noticed that bill sizes of these birds depended on whether they occurred together on the same island (Figure 38.8). On the islands Daphne and Los Hermanos, where *Geospiza fuliginosa* and *G. fortis* occur separately and

[1]Connell, J. H. 1961. The influence of interspecific competition and other factors on the distribution of the barnacle *Chthamalus stellatus.* Ecology 42:710–723.

Figure 38.7

Among the many mutualisms that abound in nature is the whistling thorn acacia of the African savanna and the ants that make their homes in the acacia's swollen galls. The acacia provides both protection for the ants' larvae *(photograph of opened gall, right)* and honeylike secretions used by the ants as food. In turn, the ants protect the tree from herbivores by swarming out as soon as the tree is touched. Giraffes, which love the tender acacia leaves, seem immune to the ants' fiery stings.

therefore do not compete with each other, bill sizes are nearly identical; on the island Santa Cruz, where both *G. fuliginosa* and *G. fortis* coexist, their bill sizes do not overlap. These results suggest resource partitioning, because bill size determines the size of seeds eaten. Work by American ornithologist Peter Grant confirms what Lack suspected: *G. fuliginosa* with its smaller bill selects smaller seeds than does *G. fortis* with its larger bill. Where these two species coexist, competition between them has led to evolutionary displacement of bill sizes to diminish competition. Absence of competition today has been called appropriately "the ghost of competition past."

Character displacement promotes coexistence by reducing niche overlap. When several species share the same general resources by such partitioning, they form a **guild.** Just as a guild in medieval times constituted a brotherhood of men sharing a common trade, species in an ecological guild share a common livelihood. The term guild was introduced to ecology by Richard Root in his 1967 paper on niche patterns of the blue-gray gnatcatcher.[2] A classic example of a bird guild is Robert MacArthur's study of a feeding guild comprising five species of warblers in spruce woods of the northeastern United States.[3] At first glance, we might ask how five birds, very similar in size and appearance, could coexist by feeding on insects in the same tree. However, on close inspection MacArthur found subtle differences among these birds in sites of foraging (Figure 38.9). One species searched only on outer branches of spruce crowns; another species used the top 60% of the tree's outer and inner branches away from the trunk; another species concentrated on inner branches near the trunk; another species used the midsection from the periphery to the trunk; and still another species foraged in the bottom 20% of the tree. These observations suggest that each warbler's niche within this guild is formed by structural differences in habitat.

Guilds are not limited to birds. For example, a study done in England on insects associated with Scotch broom plants revealed nine different guilds of insects, including three species of stem miners, two gall-forming species, two that fed on seeds and five that fed on leaves. Another insect guild consists of three species of praying mantids that avoid both competition and predation by differing in sizes of their prey, timing of hatching, and height of vegetation in which they forage.

[2]Root, R. B. 1967. The niche exploitation pattern of the blue-gray gnatcatcher. Ecological Monographs 37:317–350.
[3]MacArthur, R. H. 1958. Population ecology of some warblers of northeastern coniferous forests. Ecology 39:599–619.

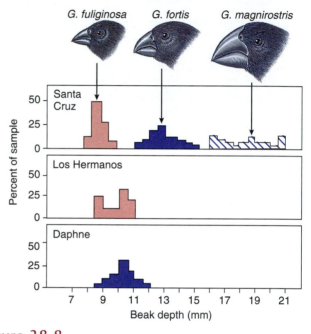

Figure 38.8

Displacement of beak sizes in Darwin's finches from the Galápagos Islands. Beak depths are given for the ground finches *Geospiza fuliginosa* (pink bars) and *G. fortis* (blue bars) where they occur together (sympatric) on Santa Cruz Island and where they occur alone on the islands Daphne and Los Hermanos. *Geospiza magnirostris* is another large ground finch that lives on Santa Cruz.

Predators and Parasites

Ecological warfare waged by predators against their prey causes coevolution: predators get better at catching prey, and prey get better at escaping predators. This is an evolutionary race that predators cannot afford to win. If a predator became so efficient that it exterminated its prey, the predator species would become extinct. Because most predators feed on more than a single species, specialization on a single prey species to the point of extermination is uncommon.

However, when a predator does rely primarily on a single prey species, both populations tend to fluctuate cyclically. First prey density increases, then that of the predator until prey become scarce. At that point, predators must adjust their population size downward by leaving the area, lowering reproduction, or dying. When density of the predator population falls enough to allow reproduction by prey to outpace mortality from predation, the cycle begins again. Thus, populations of both predators

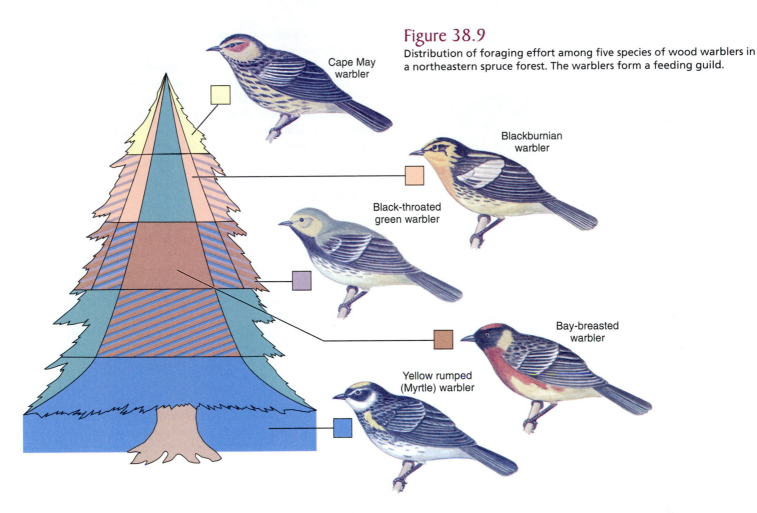

Figure 38.9

Distribution of foraging effort among five species of wood warblers in a northeastern spruce forest. The warblers form a feeding guild.

Cape May warbler

Blackburnian warbler

Black-throated green warbler

Bay-breasted warbler

Yellow rumped (Myrtle) warbler

and prey show cycles of abundance, but increases and decreases in predator abundance are slightly delayed relative to those of prey because of the time lag in a predator's response to changing prey density. We can illustrate this process in the laboratory with protozoa (Figure 38.10). Perhaps the longest documented natural example of a predator-prey cycle is between Canadian populations of snowshoe hares and lynxes (see Figure 28.28, p. 631).

The war between predators and prey reaches high art in the evolution of defenses by potential prey. Many palatable animals escape detection by matching their background, or by resembling some inedible feature of the environment (such as a stick). Such defenses are called cryptic. In contrast to cryptic defenses, toxic or distasteful animals advertise their strategy with bright colors and conspicuous behavior. Such defenses are termed aposematic. These species are protected because predators learn to recognize and to avoid them after distasteful encounters.

When distasteful prey adopt warning coloration, advantages of deceit arise for palatable prey. Palatable prey can deceive potential predators by mimicking distasteful prey, a phenomenon called **Batesian mimicry.** Coral snakes and monarch butterflies are both brightly colored, noxious prey. Coral snakes have a venomous bite, and monarch butterflies are poisonous because caterpillars store poison (cardiac glycoside) from milkweeds that they eat. Both species serve as **models** for other species, called **mimics,** that do not possess toxins of their own but look like the model species that do (Figure 38.11A and B).

In another form of mimicry termed **Müllerian mimicry,** two or more toxic species resemble each other (Figure 38.11C). What may an animal that has its own poison gain by evolving resemblance to another poisonous animal? The answer is that a predator needs only to experience the toxicity of one species to avoid all similar prey. A predator can learn one warning signal more easily than many! The benefits that two distasteful species derive from mutual mimicry are not always equal, for example, when a mildly distasteful species mimics a highly poisonous one. Such cases illustrate a continuum between Batesian and Müllerian mimicry.

Sometimes the influence of one population on others is so pervasive that its absence drastically changes an entire community. We call such a population a **keystone species.**[4] On the rocky intertidal shores of western North America, the sea star, *Pisaster ochraceous,* is a keystone species. Sea stars are a major predator of the mussel, *Mytilus californianus.* When sea stars were removed experimentally from a patch of Washington State coastline, mussels expanded in numbers, occupying all space previously used by 25 other invertebrate and algal species (Figure 38.12). Keystone predators act by reducing prey populations below the level at which resources, such as space, are limiting. The original notion that all keystone species were predators has been broadened to include any species whose removal causes extinction of others.

[4]Paine, R. T. 1969. A note on trophic complexity and community stability. American Naturalist **103:**91–93.

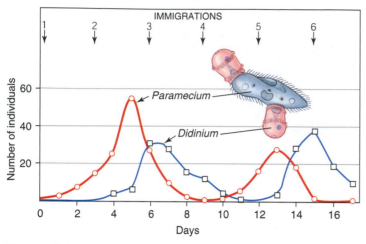

Figure 38.10

Classic predator-prey experiment by Russian biologist G. F. Gause in 1934 shows the cyclic interaction between predator *(Didinium)* and prey *(Paramecium)* in laboratory culture. When the *Didinium* find and eat all the *Paramecium,* the *Didinium* themselves starve. Gause could keep the two species coexisting only by occasionally introducing one *Didinium* and one *Paramecium* to the culture *(arrows);* these introductions simulated migration from an outside source.

By reducing competition, keystone species may allow more species to coexist on the same resource. Consequently they contribute to maintaining diversity in a community. Keystone species illustrate a more general phenomenon, disturbance. Periodic natural disturbances such as fires and hurricanes also may prevent monopolization of resources and competitive exclusion by a few broadly adapted competitors. Disturbances may permit more species to coexist in such highly diverse communities as coral reefs and rain forests.

Parasites are often considered freeloaders because they appear to benefit from their hosts at no expense. Virulence is correlated, at least in part, with availability of new hosts. Coevolution between parasite and host is expected to generate an increasingly benign, less virulent relationship if host organisms are uncommon and/or difficult for a parasite to infect. Selection favors a benign relationship, because a parasite's fitness is diminished if its host dies. When alternative hosts are common and transmission rates high, continued colonization of new hosts makes any particular host's life less valuable to the parasite, so that high virulence may have no disadvantage.

Ecosystems

Transfer of energy and materials among organisms within ecosystems is the ultimate level of organization in nature. Energy and materials are required to construct and to maintain life, and their incorporation into biological systems is called **productivity.** Productivity is divided into component **trophic levels** based on how organisms obtain energy and materials. Trophic levels are linked together into **food chains,** which denote movement of energy from plant compounds to organisms that eat plants, then to other organisms that eat the plant feeders, and possibly further through a linear series of organisms that feed and are then eaten by others. Food chains connect to form **food webs** (Figure 38.13), pathways for transfer of energy and materials among organisms within an ecosystem.

Primary producers are organisms that begin productivity by fixing and storing energy from outside the ecosystem. Primary producers usually are green plants that capture solar energy through **photosynthesis** (but see an exception in the box titled "Life without the Sun"). Powered by solar energy, plants assimilate and organize minerals, water, and carbon dioxide into living tissue. All other organisms survive by consuming this tissue, or by consuming organisms that consumed this tissue. **Consumers** include **herbivores,** which eat plants directly, and **carnivores,** which eat other animals. The most

Figure 38.11

Artful guises abound in the tropics.
A, A palatable viceroy butterfly *(top photograph)* mimics a distasteful monarch butterfly *(lower photograph).*
B, A harmless clearwing moth *(top photograph)* mimics a yellowjacket wasp, which is armed with a stinger *(lower photograph).* Both **A** and **B** illustrate Batesian mimicry.
C, Two unpalatable tropical butterflies of different taxonomic families resemble one another, an example of Müllerian mimicry.

A

B

C

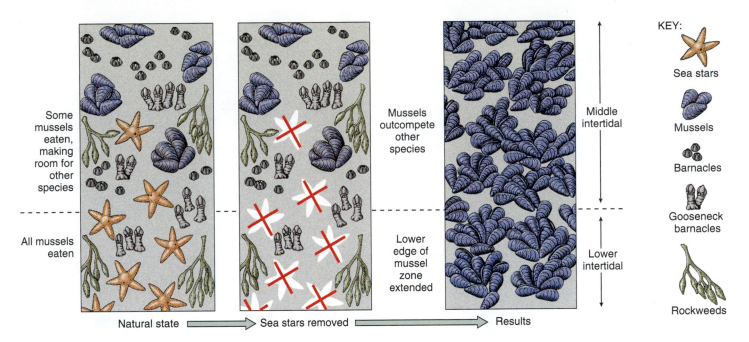

Figure 38.12

The experimental removal of a keystone species, the predatory sea star, *Pisaster ochraceus*, from an intertidal community completely changes the structure of the community. With their principal predator missing, mussels form dense beds by outcompeting and replacing other intertidal species.

important consumers are **decomposers,** mainly bacteria and fungi that break dead organic matter into its mineral components, returning it to a soluble form again available to plants at the base of the nutrient cycle (p. 838). Although important chemicals such as nitrogen and carbon are reused endlessly through biological cycling, all energy ultimately is lost from the ecosystem as heat and not recycled. Thus, no ecosystems are truly closed; all require new energy input from the sun or hydrothermal vents (p. 839).

Energy Flow and Productivity

Every organism in nature has an **energy budget.** Just as we each must partition our income for housing, food, utilities, and taxes, each organism must obtain enough energy to meet its metabolic costs, to grow, and to reproduce.

Ecologists divide the budget into three main components: **gross productivity, net productivity,** and **respiration.** Gross productivity is like gross income; it is the total energy assimilated, analogous to your pay before deductions. When an animal eats, food passes through its gut and nutrients are absorbed. Most energy assimilated from these nutrients serves the animal's metabolic demands, which include cellular metabolism and regulation of body heat in endotherms. Energy required for metabolic maintenance is respiration, which is deducted from gross productivity to yield net productivity, an animal's take-home pay. Net productivity is energy stored by an animal in its tissues as **biomass.** This energy is available for growth, and also for reproduction, which is population growth.

An animal's energy budget is expressed by a simple equation, in which gross and net productivity are represented by P_g and P_n, respectively, and respiration is R:

$$P_n = P_g - R$$

This equation states the first law of thermodynamics (p. 10) in the context of ecology. Its important messages are that the energy budget of every animal is finite and may be limiting, and that energy is available for growth of individuals and populations only after maintenance is satisfied.

The second law of thermodynamics, which states that total disorder or randomness of a system always increases, is important when we study energy transfers between trophic levels in food webs. Energy for maintenance, R, usually constitutes more than 90% of the assimilated energy (P_g) for animal consumers. More than 90% of the energy in an animal's food is lost as heat, and less than 10% is stored as biomass. Each succeeding trophic level therefore contains only 10% of the energy in the next lower trophic level. Most ecosystems are thereby limited to five or fewer trophic levels.

"R.F.D.2."

SORRY. I KNOW IT SEEMS UNFAIR.

YOU SEE, THIS EGG IS A LINK IN THE FOOD CHAIN

AND I, A HUMAN, AM AT THE END OF THE CHAIN.

I HOPE.

Steve Stinson and Roanoke Times and World-News

Figure 38.13

Midwinter food web in *Salicornia* salt marsh of San Francisco Bay area.

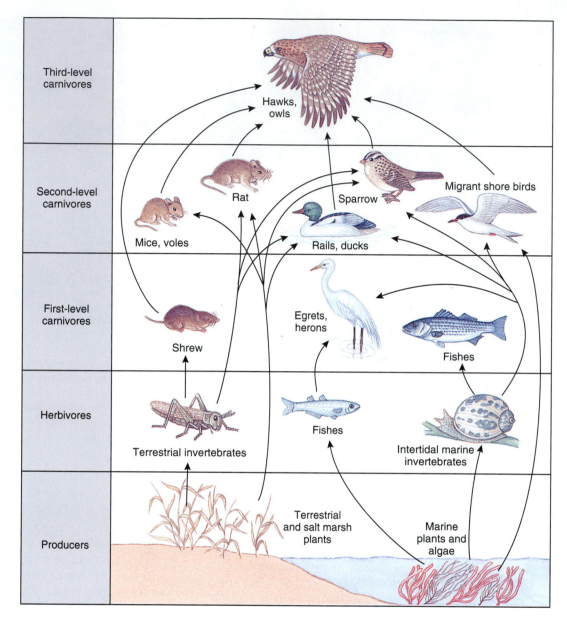

Third-level carnivores — Hawks, owls

Second-level carnivores — Rat, Sparrow, Migrant shore birds, Mice, voles, Rails, ducks

First-level carnivores — Shrew, Egrets, herons, Fishes

Herbivores — Terrestrial invertebrates, Fishes, Intertidal marine invertebrates

Producers — Terrestrial and salt marsh plants, Marine plants and algae

Our ability to feed a growing human population is influenced profoundly by the second law of thermodynamics (p. 10). People, who are at the end of the food chain, may eat the grain, fruits, and vegetables of plants that fix the sun's energy in chemical bonds; this very short chain represents an efficient use of potential energy. People also may eat beef from animals that eat grass that fixes the sun's energy; the addition of a trophic level decreases available energy by a factor of 10. Ten times as much plant biomass is needed to feed humans as meat eaters as to feed humans as grain eaters. Consider a person who eats a bass that eats a sunfish that eats zooplankton that eats phytoplankton that fixes the sun's energy. The tenfold loss of energy at each trophic level in this five-step chain requires that the pond must produce 5 tons of phytoplankton for a person to gain a pound by eating bass. If the human population depended on bass for survival, we would quickly exhaust this resource.

These figures must be considered as we look to the sea for food. Productivity of oceans is very low and limited largely to estuaries, marshes, reefs, and upwellings where nutrients are available to phytoplankton producers (p. 817). Such areas constitute a small part of the ocean. The rest is a watery void.

Marine fisheries supply about one-fifth of the world's dietary protein, but much of this protein is used to feed livestock and poultry. If we remember the rule of 10-to-1 loss in energy with each transfer of material between trophic levels, then use of fish as food for livestock rather than humans is poor use of a valuable resource in a protein-deficient world. Fishes that we prefer to eat include flounder, tuna, and halibut, which are three or four levels up the food chain. Every 125 g of tuna requires one metric ton of phytoplankton to produce. If humans are to derive greater benefit from oceans as a food source, we must eat more of the less desirable fishes from lower trophic levels.

When we examine the food chain in terms of biomass at each level, we can construct **ecological pyramids** either of numbers, energy, or biomass. A pyramid of numbers (Figure 38.14A), also called an **Eltonian pyramid,** depicts numbers of organisms transferred between each trophic level. This pyramid provides a vivid impression of the great difference in numbers of organisms

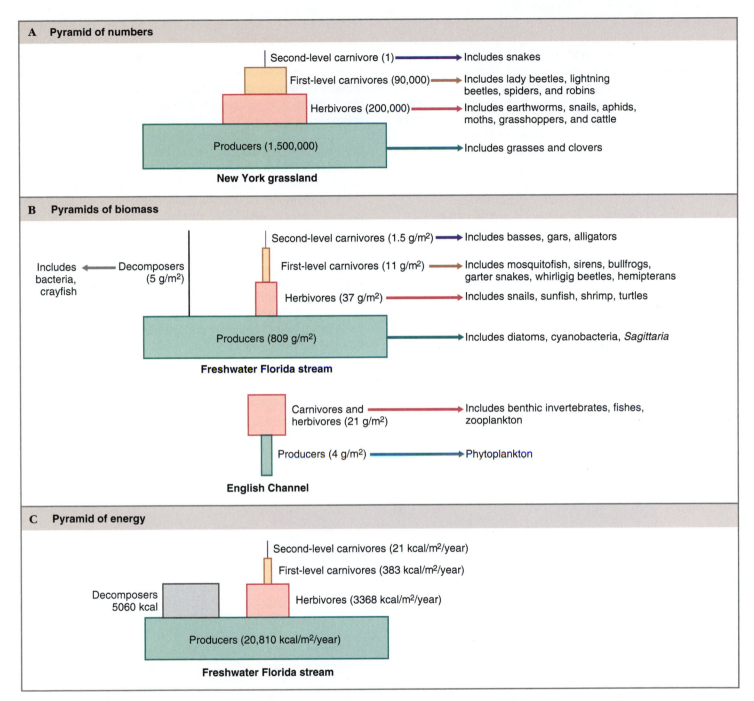

Figure 38.14

Ecological pyramids with representative organisms from each trophic level. **A,** Pyramid of numbers of organisms counted for 100 square feet of grassland in New York. **B,** Pyramids of biomass for a freshwater Florida stream (top) and plankton of the English Channel (inverted pyramid at bottom). **C,** Pyramid of energy for the same Florida stream shown in B. *(From the work of E. P. Odum and H. T. Odum.)*

involved in each step of the chain, and supports the observation that large predatory animals are rarer than the small animals on which they feed. However, a pyramid of numbers does not indicate actual mass of organisms at each level.

The concepts of food chains and ecological pyramids were invented and first explained in 1923 by Charles Elton, a young ecologist at Oxford University. Working for a summer on a treeless arctic island,

Elton watched arctic foxes as they roamed, noting what they ate and, in turn, what their prey had eaten, until he was able to trace the complex cycling of nitrogen in food throughout the animal community. Elton realized that life in a food chain comes in discrete sizes, because each form evolved to be much bigger than the thing it eats. He thus explained the common observation that large animals are rare while their smaller prey are common; the ecological pyramids illustrating this phenomenon now bear Elton's name.

More instructive are pyramids of biomass (Figure 38.14B), which depict the total bulk, or "standing crop," of organisms at each trophic level. Such pyramids usually slope upward because mass and energy are lost at each transfer. However, in some aquatic ecosystems in which the producers are algae, which have short life spans and rapid turnover, the pyramid is inverted. Algae tolerate heavy exploitation by zooplankton consumers. Therefore, the base of the pyramid (biomass of phytoplankton) is smaller than the biomass of zooplankton it supports. This inverted pyramid is analogous to a person who weighs far more than the food in a refrigerator, but who is sustained from the refrigerator because food is constantly replenished.

A third type of pyramid is a pyramid of energy, which shows rate of energy flow between levels (Figure 38.14C). An energy pyramid is never inverted because energy transferred from each level is less than what entered it. A pyramid of energy gives the best overall picture of community structure because it is based on production. In the English Channel, productivity of phytoplankton exceeds that of zooplankton, even though biomass of phytoplankton is less than biomass of zooplankton (because of heavy grazing by zooplankton consumers).

Nutrient Cycles

All elements essential for life are derived from environmental air, soil, rocks, and water. When plants and animals die and their bodies decay, or when organic substances are burned or oxidized, elements and inorganic compounds essential for life processes (nutrients) are released and returned to the environment. Decomposers fulfill an essential role in this process by feeding on remains of plants and animals and on fecal material. The result is that nutrients flow in a perpetual cycle between biotic and abiotic components of an ecosystem. Nutrient cycles are often called **biogeochemical cycles** because they involve exchanges between living organisms (bio-) and rocks, air, and water of the earth's crust (geo-). Continuous input of energy from the sun keeps nutrients flowing and the ecosystem functioning (Figure 38.15).

Our synthetic compounds challenge nature's nutrient cycling because decomposers have not evolved ways to degrade them. Probably most harmful of these materials to ecosystemic processes are pesticides. Pesticides in natural food webs can be insidious for three reasons. First, many pesticides become concentrated as they travel up succeeding trophic levels. The highest concentrations occur in the biomass of top carnivores such as hawks and owls, diminishing their ability to reproduce. Second, many species killed by pesticides are not pests but merely innocent bystanders, called nontarget species. Nontarget effects happen when pesticides move outside the agricultural field to which they were applied, through rainwater runoff, leaching through soil, or being dispersed by wind. The third problem is persistence; some chemical pesticides have great longevity in the environment, so that nontarget effects persist long after pesticide application.

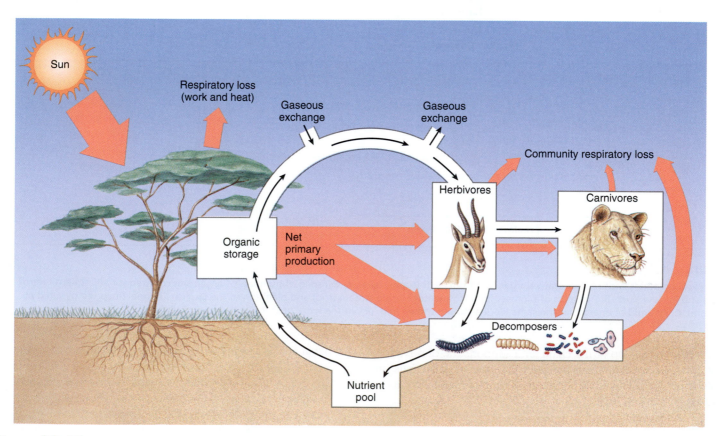

Figure 38.15

Nutrient cycles and energy flow in a terrestrial ecosystem. Note that nutrients are recycled, whereas energy flow (orange) is one way.

Life without the Sun

For many years, ecologists thought that all animals depended directly or indirectly on primary production from solar energy. However, in 1977 and 1979, dense communities of animals were discovered living on the seafloor adjacent to vents of hot water issuing from rifts (Galápagos Rift and East Pacific Rise) where tectonic plates on the seafloor slowly spread apart. These communities (see photo) include several species of molluscs, some crabs, polychaete worms, enteropneusts (acorn worms), and giant pogonophoran or siboglinid worms. The temperature of seawater above and immediately around vents is 7° to 23° C where it is heated by basaltic intrusions, whereas the surrounding normal seawater is 2° C.

   The producers in these vent communities are chemoautotrophic bacteria that derive energy from oxidation of hydrogen sulfide abundant in the vent water and fix carbon dioxide into organic carbon. Some animals in the vent communities—for example, bivalve molluscs—are filter feeders that ingest bacteria. Others, such as giant pogonophoran tubeworms (see p. 370), which lack mouths and digestive tracts, harbor colonies of symbiotic bacteria in their tissues and use the organic carbon that these bacteria synthesize.

A population of giant pogonophoran or siboglinid tubeworms grows in dense profusion near a Galápagos Rift thermal vent, photographed at 2800 m (about 9000 feet) from the deep submersible *Alvin.* Also visible in the photograph are mussels and crabs.

Genetic engineering of crop plants aims to improve their resistance toward pests to lessen the need for chemical pesticides.

EXTINCTION AND BIODIVERSITY

Biodiversity exists because rates of speciation on average slightly exceed rates of extinction in earth's evolutionary history. Approximately 99% of all species that have lived are extinct. Speciation rates represent an ongoing process of geographic expansion of populations by dispersal followed by geographic fragmentation (p. 818), producing a multiplication of species. Speciation rates vary enormously among animal taxa and geographic areas; typical rates include 0.2 to 0.4 speciation events per species per million years, as measured for Cretaceous marine gastropods of the Atlantic coast; average duration of these gastropod species was 2 to 6 million years.

   Extinction rates show episodic peaks and valleys throughout earth's evolutionary history. Paleontologist David Raup analyzed episodicity of extinction peaks by dividing the 600-million-year history of the marine fossil record into successive intervals of 1 million years duration; he then measured the percentage of existing species that suffered extinction during each interval. Rates of species extinction in the 600 intervals range from near zero to 96%, averaging about 25% (Figure 38.16A). We measure episodicity of extinction by asking the question, How long must we wait on average for an extinction peak killing at least 30% of existing species, or perhaps 65%? Answers to these questions are shown by Raup's "kill curve" (Figure 38.16B). Extinction events killing at least 5% of the existing species occur almost continuously throughout geological time. Extinction events killing at least 30% of existing species occur every 10 million years on average, whereas extinction events killing at least 65% of existing species occur on average every 100 million years; the latter episodes clearly qualify as mass extinctions

(p. 133). Figure 38.16A reveals a continuous distribution of extinction rates from very high to very low; a common dichotomy made between mass extinction and "background" extinction is therefore misleading. Most dramatic are the "big five" extinctions that each killed more than 75% of the existing species (Table 38.1); however, these events collectively represent only 4% of species extinctions in the past 600 million years.

   Fossil studies show that species whose geographic ranges are large have lower average rates of extinction than those with small geographic ranges. Species of Atlantic Cretaceous gastropods differed greatly in geographic range depending on modes of larval feeding. Some species had pelagic, plankton-feeding ("planktotrophic") larvae, which were carried long distances by oceanic currents; these species maintained large geographic ranges averaging 2000 km along the Atlantic Coast. Other species had heavy larvae that settled to the ocean floor as benthic feeders immediately upon hatching; these nonplanktotrophic species average less than one-fourth the geographic range of their plankton-feeding

TABLE 38.1

Comparison of Species Extinction Levels for the Big Five Mass Extinctions*

Extinction Episode	Age, Myr Before Present	Percent Extinction
Cretaceous	65	76
Triassic	208	76
Permian	245	96
Devonian	367	82
Ordovician	439	85

* After David Raup (1995).

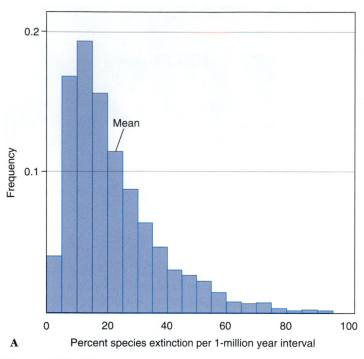

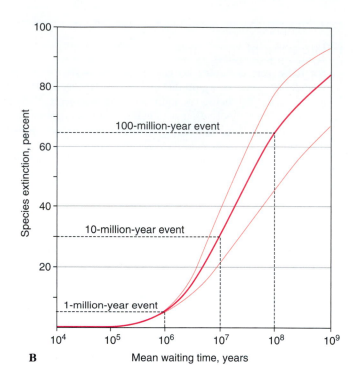

Figure 38.16

A, Variation in extinction rates for species in the marine fossil record. Data were obtained by dividing the fossil record into 600 consecutive intervals, each one having a duration of 1 million years (Myr), beginning 600 Myr ago. Percent species extinction was measured for each interval. Almost 20% of the intervals had extinction rates of 10 to 15%, the highest bar on this graph. The mean species extinction rate is 25% per million years, and the mean species duration is 4 Myr. **B,** Species "kill curve" for the data summarized in **A.** Waiting time is the average interval between events equal to or greater than a given intensity of species extinction. The kill curve demonstrates an episodic distribution of extinction peaks over the past 600 Myr. If extinction rates were constant through time, the kill curve would be indistinguishable from the x axis. The lighter curves on either side of the dark one show the statistical measurement error; the actual curve could lie anywhere between these lighter curves, but the dark line provides the best estimate.
After David Raup (1995).

counterparts. A nonplanktotrophic species is about three times more likely to suffer extinction than a plankton-feeding one, but their geographic fragmentation also makes nonplanktotrophic species twice as likely to undergo speciation.

A paradox of biodiversity is that geographic **habitat fragmentation** of a species simultaneously increases rates of both local extinction and speciation. The African antelopes shown in Figure 6.11 illustrate a similar contrast; in the past 6 million years, one group (hartebeests, blesboks, and wildebeests) experienced multiple events of speciation and extinction with seven species alive today, whereas the other lineage (impalas) persisted as a single species through the same time period. Impalas are about equal to the sum of the other seven species in number of animals alive today. This contrast shows that increased species diversity evolves at the peril of greater extinction risk for each individual species.

Higher taxa, such as orders, families, and genera (p. 200), also gain some protection from extinction by having large geographic ranges. Raup notes that higher taxa containing many species collectively distributed across a large geographic area are unlikely to go extinct. When such extinctions do happen, as seen for many dinosaur and marine ammonoid taxa at the end of the Cretaceous (Table 38.1), unusually catastrophic conditions are inferred. An asteroid bombardment at the end-Cretaceous extinction (p. 133) appears to have caused within a short time widespread fires and extreme darkness and cold followed by extreme

heat, all conditions well outside the evolutionary tolerances of many animal taxa whose members were formerly abundant. Only by chance would a taxon likely contain species able to withstand a challenge unprecedented in the group's evolutionary history.

Darwin explained extinctions of higher taxa by interspecific competition, but paleontological studies now refute this claim. Paleontologist Michael Benton estimates that less than 15% of the extinct tetrapod families could have been killed by competition with other families. Ecological and fossil studies of ectoprocts demonstrate that species of one order (Cheilostomatida) have outcompeted those of another order (Tubuliporata) ecologically by overgrowing their colonies for many millions of years without driving Tubuliporata extinct.

Decline or extinction of one taxon often frees resources to which another taxon becomes adapted at a much later time, leading to evolutionary proliferation of ecologically diverse species in the latter. Resources freed by extinction of dinosaur taxa at the end of the Cretaceous are considered important for the later proliferation of mammalian adaptive and species diversity in the Cenozoic (p. 113).

Fossil studies of extinction help us to put into evolutionary perspective the consequences of human-mediated ecological changes for biodiversity. Fragmentation of populations as seen especially on islands (p. 122) produces locally high rates of species formation and endemism (p. 202), but these young species are unusually prone to extinction because their geographic ranges are small. About half of all areas in the world that contain at least two endemic bird species are islands, even though islands form less than 10% of the earth's terrestrial habitats. Island species are often particularly prone to destruction by introduction of invasive exotic species. For example, land snails of the genus *Partula* on the Tahitian island of Moorea were a major study system for island speciation until introduction of exotic snails displaced these native species. Mainland habitats, such as forests, are fragmented into virtual islands when development clears vast areas of habitat and when introduced species invade these areas. Because tropical regions have high species endemism, human fragmentation of these environments is particularly likely to produce species extinction.

A major challenge to animal species conservation is to obtain an inventory of the earth's species diversity. Estimates of the earth's total number of species are typically as large as 10 million, and this number may be too low by an order of magnitude. Vagueness of these estimates reflects both practical and conceptual problems. A thorough survey of geographic genetic variation among natural populations requires expensive molecular-genetic analyses (see Chapter 6) and is critical for applying any of the current species concepts (see Chapter 10). This analysis is feasible only for taxa

with relatively small numbers of large individuals, such as those at the top of a food chain or ecological pyramid. Beetles and nematodes constitute two taxa whose large numbers of small organisms challenge hopes for a comprehensive taxonomic survey. Even with appropriate data in hand, conflicting perspectives on what should constitute a species (discussed in Chapter 10) would preclude an unambiguous tally of species numbers. Such conflicts would be particularly intense in groups of animals that do not show primarily a simple bisexual mode of reproduction. Clearly, conservation efforts cannot await a thorough taxonomic inventory of all animal populations. Maintenance of diverse ecosystems as identified in Chapter 37 is an initial priority for preventing widespread species extinction.

One hopes that human disturbance of natural environments will not make the present time a competitor with the big five mass extinctions of Table 38.1. Human activity clearly has induced numerous species extinctions, and we must avoid making the present time one that climbs too high on Raup's kill curve (Figure 38.16B). Evolutionary studies suggest, however, that widely distributed higher taxa are unlikely to go extinct even during episodes of high species extinction. Lacking an inventory of species diversity in animals, we must avoid creating conditions that would selectively destroy higher taxa.

SUMMARY

Ecology is the study of relationships between organisms and their environments to explain the distribution and abundance of species on earth. The physical space that contains an animal's environment is its habitat. Within the habitat are physical and biological conditions suitable to survival and reproduction, which constitute the niche of an animal, population, or species.

Animal populations consist of demes of interbreeding members sharing a common gene pool. Cohorts of animals have characteristic patterns of survivorship that represent adaptive trade-offs between parental care and numbers of offspring. Animal populations consisting of overlapping cohorts have age structure that indicates whether they are growing, declining, or at equilibrium.

Every species in nature has an intrinsic rate of increase that provides a potential for exponential growth. The human population is growing exponentially at about 1.23% each year, and is expected to increase from 6.4 billion to 8 billion by the year 2040. Population growth may be regulated intrinsically by the carrying capacity of the environment, extrinsically by competition between species for a limiting resource, or by predators or parasites. Density-independent abiotic factors can limit, but not truly regulate, population growth.

Communities consist of populations that interact with one another through competition, predation, parasitism, commensalism, and mutualism. These relationships result from coevolution among populations within communities. Guilds of species avoid competitive exclusion by character displacement, the partitioning of limited resources by morphological specialization. Keystone predators are those that control community structure and reduce competition

among prey, which increases species diversity. Parasites and their hosts evolve a benign relationship that ensures their coexistence.

Ecosystems consist of communities and their abiotic environments. Animals occupy the trophic levels of herbivorous and carnivorous consumers within ecosystems. All organisms have an energy budget consisting of gross and net productivity, and respiration. For animals, respiration usually is at least 90% of this budget. Thus, transfer of energy from one trophic level to another is limited to about 10%, which in turn limits the number of trophic levels in an ecosystem. Ecological pyramids of energy depict how productivity decreases in successively higher trophic levels of food webs.

Ecosystem productivity is a result of energy flow and material cycles within ecosystems. All energy is lost as heat, but nutrients and other materials including pesticides are recycled. No ecosystem is closed because all depend upon exchanges of energy and materials from outside sources.

Biodiversity exists because rates of speciation on average slightly exceed rates of extinction in earth's evolutionary history. Approximately 99% of all species that have lived are now extinct. Extinction rates in the geologic past are highly episodic, with species extinction ranging from near zero to 96% in different million-year intervals. Species having large geographic ranges experience lower average rates of extinction than species with smaller ranges, and the same relationship holds for higher taxa. Paleontological studies of extinction provide an important perspective for evaluating potential evolutionary consequences of human-mediated species extinction.

REVIEW QUESTIONS

1. The term ecology is derived from the Greek meaning "house" or "place to live." However, as used by scientists, the term "ecology" is not the same as "environment." How do these terms differ?

2. How would you distinguish between ecosystem, community, and population?

3. What is the distinction between habitat and environment?

4. Define the niche concept. How does the "realized niche" of a population differ from its "fundamental niche"? How does the concept of niche differ from the concept of guild?

5. Populations of independently living (unitary) animals have a characteristic age structure, sex ratio, and growth rate. However, these properties are difficult to determine for modular animals. Why?

6. Explain which of the three survivorship curves in Figure 38.2 best fits the following: (a) a population in which mortality as a proportion of survivors is constant; (b) a population in which there is little early death and most individuals live to old age; (c) a population that experiences heavy mortality of the very young but with the survivors living to old age. Offer an example from the real world of each survivorship pattern.

7. Contrast exponential and logistic growth of a population. Under what conditions might you expect a population to exhibit exponential growth? Why cannot exponential growth be perpetuated indefinitely?

8. Growth of a population may be hindered by either density-dependent or density-independent mechanisms. Define and contrast these two mechanisms. Offer examples of how growth of the human population might be curbed by either agent.

9. Herbivory is an example of an interspecific interaction that is beneficial for the animal (+) but harmful to the plant it eats (−). What are some + − interactions among animal populations? What is the difference between commensalism and mutualism?

10. Explain how character displacement can ease competition between coexisting species.

11. Define predation. How does the predator-prey relationship differ from the parasite-host relationship? Why is the evolutionary race between predator and prey one that the predator cannot afford to win?

12. Mimicry of monarch butterflies by viceroys is an example of a palatable species resembling a toxic one. What is the advantage to the viceroy of this form of mimicry? What is the advantage to a toxic species of mimicking another toxic species?

13. A keystone species is defined as one whose removal from a community causes the extinction of other species. How does this extinction happen?

14. What is a trophic level, and how does it relate to a food web?

15. Define productivity as the word is used in ecology. What is a primary producer? What is the distinction between gross productivity, net productivity, and respiration? What is the relation of net productivity to biomass (or standing crop)?

16. What is a food chain? How does a food chain differ from a food web?

17. How is it possible to have an inverted pyramid of biomass in which the consumers have a greater biomass than the producers? Can you think of an example of an inverted pyramid of *numbers* in which there are, for example, more herbivores than plants on which they feed?

18. The pyramid of energy has been offered as an example of the second law of thermodynamics (p. 10). Why?

19. Animal communities surrounding deep-sea thermal vents apparently exist in total independence of solar energy. How is this existence possible?

20. What do paleontological studies show about the relationship between the geographic range of a species and its probability of undergoing speciation or extinction? How does this present a paradox for biodiversity?

SELECTED REFERENCES

Benton, M. J. 1996. On the nonprevalence of competitive replacement in the evolution of tetrapods. In Jablonski, D. H. Erwin, and J. H. Lipps (eds.), Evolutionary paleobiology, pp. 185–210. Chicago, University of Chicago Press. *Shows that ecological competition does not explain extinction of tetrapod taxonomic families.*

Chase, J. M., and M. A. Leibold. 2003. Ecological niches: linking classical and contemporary approaches. Chicago, University of Chicago Press. *An insightful coverage of niche concepts in community ecology.*

Krebs, C. J. 2001. Ecology: the experimental analysis of distribution and abundance, ed. 5. San Francisco, Benjamin-Cummings. *Important treatment of population ecology.*

Molles, M. C. Jr. 2005. Ecology: concepts and applications, ed. 3. New York, McGraw-Hill. *A concise and well-illustrated survey of ecology.*

Pianka, E. R. 2000. Evolutionary ecology. San Francisco, Benjamin-Cummings. *An introduction to ecology written from an evolutionary perspective.*

Raup, D. M. 1995. The role of extinction in evolution. In W. M. Fitch and F. J. Ayala (eds.), Tempo and mode in evolution, pp. 109–124. Washington, D.C., National Academy Press. *A paleontologist's insights on extinction and biodiversity.*

Ricklefs, R. E., and G. L. Miller. 2000. Ecology. New York, W. H. Freeman. *Clearly written, well-illustrated general ecology text.*

Sax, D. F., J. J. Stachowicz, and S. D. Gaines, 2005. Species invasions: insights into ecology, evolution, and biogeography. Sunderland, Massachusetts, Sinauer Associates, *The establishment of alien species in an area can reveal ecosystem properties and the structuring of ecological communities.*

Sibly, R. M., J. Hone, and T. H. Clutton-Brock. 2003. Wildlife population growth rates. Cambridge, U.K., Cambridge University Press. *Contributions by many ecologists illustrate current methods for measuring growths of natural populations.*

Smil, V. 1997. Global population and the nitrogen cycle. Sci. Am. **277:**78–81 (July). *The sudden growth of the human population worldwide in the 20th century parallels global consumption of synthetically-produced nitrogen-rich fertilizers upon which humans now heavily depend for food production, but there are adverse consequences for the environment.*

Smith, R. L., and T. M. Smith. 2001. Ecology and field biology. San Francisco, Benjamin-Cummings. *Clearly written, well-illustrated general ecology text.*

ONLINE LEARNING CENTER

Visit www.mhhe.com/hickmanipz14e for chapter quizzing, key term flash cards, web links, and more!

This glossary lists definitions, pronunciations, and derivations of the most important recurrent technical terms, units, and names (excluding taxa) used in the text.

A

abiotic (ā′bī-ät′ik) (Gr. *a,* without, + *biōtos,* life, livable). Characterized by the absence of life.

abomasum (ab′ō-mā′səm) (L. *ab,* from, + *omasum,* paunch). Fourth and last chamber of the stomach of ruminant mammals.

aboral (ab-o′rəl) (L. *ab,* from, + *os,* mouth). A region of an animal opposite the mouth.

abscess (ab′ses) (L. *abscessus,* a going away). Dead cells and tissue fluid confined in a localized area, causing swelling.

abyssal plain (ə-bis′-əl plān′). An offshore seafloor with submarine channels and hills averaging 4000 m in depth but descending to 11,000 m below the sea surface.

acanthodians (a′kan-thō′dē-əns) (Gr. *akantha,* prickly, thorny). Along with placoderms, these groups were among the earliest known true jawed fishes from Lower Silurian to Lower Permian.

acanthor (ə-kan′thor) (Gr. *akantha,* spine or thorn, + *or*). First larval form of acanthocephalans in the intermediate host.

acclimatization (ə-klī′mə-tə-zā-shən) (L. *ad,* to, + Gr. *klima,* climate). Gradual physiological adjustment of an organism in response to relatively long-lasting environmental changes.

acetabulum (as′ə-tab′ū-ləm) (L. a little saucer for vinegar). True sucker, especially in flukes and leeches; the socket in the hip bone that receives the thigh bone.

acicula (ə-sik′ū-lə) (L. *acicula,* a small needle). Needlelike supporting bristle in parapodia of some polychaetes.

acid A molecule that dissociates in solution to produce a hydrogen ion (H⁺).

acinus (as′ə-nəs), pl. **acini** (as′ə-nī) (L. grape). A small lobe of a compound gland or a saclike cavity at the termination of a passage.

acoelomate (ā-sēl′ə-māt′) (Gr. *a,* not, + *koilōma,* cavity). Without a coelom, as in flatworms and proboscis worms.

acontium (ə-kän′chē-əm), pl. **acontia** (Gr. *akontion,* dart). Threadlike structure bearing nematocysts located on mesentery of sea anemone.

acquired immune deficiency syndrome (AIDS) An eventual consequence of infection with the human immunodeficiency virus in which the immune response is severely disabled. The disease is ultimately fatal, and no cure has been discovered.

acrocentric (ak′rō-sen′trək) (Gr. *akros,* tip, + *kentron,* center). Chromosome with centromere near the end.

acron (a′crän) (Gr. *akron,* mountaintop, fr. *akros,* tip). Preoral region of an insect.

actin (Gr. *aktis,* ray, beam). A protein in all cells that forms part of the cytoskeleton and is involved in the processes of endocytosis and exocytosis. In contractile tissues it forms the thin filaments of striated muscle.

actinotroch (ək-tin′ə-trōk) (Gr. *aktis,* ray, beam, + *trochos,* wheel). Larval form found in Phoronida.

active transport Mediated transport in which a transmembrane protein transports a molecule across a plasma membrane against a concentration gradient; requires expenditure of energy; contrasts with **facilitated diffusion.**

adaptation (L. *adaptatus,* fitted). An anatomical structure, physiological process, or behavioral trait that evolved by natural selection and improves an organism's ability to survive and to leave descendants; contrasts with **exaptation.**

adaptive radiation Evolutionary diversification that produces numerous ecologically disparate lineages from a single ancestral one, especially when this diversification occurs within a short interval of geological time.

adaptive value Degree to which a characteristic helps an organism to survive and to reproduce or lends greater fitness in its environment; selective advantage.

adaptive zone A characteristic reaction and mutual relationship between environment and organism ("way of life") demonstrated by a group of evolutionarily related organisms.

adductor (ə-duk′tər) (L. *ad,* to, + *ducere,* to lead). A muscle that draws a part toward a median axis, or a muscle that draws the two valves of a mollusc shell together.

adenine (ad′nēn, ad′ə-nēn) (Gr. *adēn,* gland, + *ine,* suffix). A purine base; component of nucleotides and nucleic acids.

adenosine (ə-den′ə-sēn) **(di-, tri) phosphate** (ADP and ATP). A nucleotide composed of adenine, ribose sugar, and two (ADP) or three (ATP) phosphate units; ATP is an energy-rich compound that, with ADP, serves as a phosphate bond-energy transfer system in cells.

adhesion junction Transmembrane proteins serving as an intercellular, anchoring connection.

adipokine (ad′ə-pō′-kīn). A peptide hormone secreted by white adipose tissue. Over 50 adipokines have been isolated.

adipose (ad′ə-pōs) (L. *adeps,* fat). Fatty tissue; fatty.

adrenaline (ə-dren′ə-lən) (L. *ad,* to,+ *renalis,* pertaining to kidneys). A hormone produced by the adrenal, or suprarenal, gland; epinephrine.

adsorption (ad-sorp′shən) (L. *ad,* to, + *sorbeo,* to absorb). The adhesion of molecules to solid bodies.

aerobic (a-rō′bik) (Gr. *aēr,* air, + *bios,* life). Oxygen-dependent form of respiration.

afferent (af′ə-rənt) (L. *ad,* to, + *ferre,* to bear). Adjective meaning leading or bearing toward some organ, for example, nerves conducting impulses toward the brain or blood vessels carrying blood toward an organ; contrasts with **efferent.**

age structure An accounting of the ages of individuals in a population at a particular time and place.

aggression (ə-gres′hən) (L. *aggressus,* attack). An offensive action or procedure.

agnathan (ag-nā′-thən (Gr. *a,* without, + *gnathos,* jaw). A jawless fish of the paraphyletic superclass Agnatha of the phylum Chordata.

agonistic behavior (Gr. *agōnistēs,* combatant). An offensive action or threat directed toward another organism.

AIDS See **acquired immune deficiency syndrome.**

alate (ā′lāt) (L. *alatus,* wing). Winged.

albumin (al-byū′mən) (L. *albumen,* white of egg). Any of a large class of simple proteins that are important constituents of vertebrate blood plasma and tissue fluids and also present in milk, whites of eggs, and other animal substances.

alimentary (al′ə-men′tə-rē) (L. *alimentum,* food, nourishment). Having to do with nutrition or nourishment.

allantois (ə-lan′tois) (Gr. *allas,* sausage, + *eidos,* form). One of the extraembryonic membranes of the amniotes that functions in respiration and excretion in birds and nonavian reptiles and plays an important role in the development of the placenta in most mammals.

bat / āpe / ärmadillo / herring / fēmale / finch / līce / crocodile / crōw / cóin / duck / ūnicorn / ə indicates unaccented vowel sound "uh" as in mammal, fishes, cardinal, heron, vulture / stress as in bi-ol′o-gy, bi′o-log′i-cal

allele (ə-lēl′) (Gr. *allēlōn,* of one another). Alternative forms of genes coding for the same trait; situated at the same locus in homologous chromosomes.

allelic frequency An estimation of the proportion of gametes produced in a population (gene pool) that contains a particular allelic form of a particular gene.

allograft (a′lō-graft) (Gr. *allos,* other, + graft). A piece of tissue or an organ transferred from one individual to another individual of the same species, not identical twins; contrasts with homograft.

allometry (ə-lom′ə-trē) (Gr. *allos,* other, + *metry,* measure). Relative growth of a part in relation to the whole organism.

allopatric (Gr. *allos,* other, + *patra,* native land). In separate and mutually exclusive geographical regions.

allopatric speciation The hypothesis that new species are formed by dividing an ancestral species into geographically isolated subpopulations that evolve **reproductive barriers** between them through independent evolutionary divergence from their common ancestor.

alpha-helix (Gr. *alpha,* first, + L. *helix,* spiral). Spiral arrangement of the genetic DNA molecule; regular coiled arrangement of polypeptide chain in proteins; a secondary structure of proteins.

altricial (al-tri′shəl) (L. *altrices,* nourishers). Referring to young animals (especially birds) having the young hatched or born in an immature, dependent condition.

altruistic behavior A term used initially by Darwin to denote a behavior performed by an individual to aid others while apparently increasing its own risk. Unless such behaviors have at least indirect benefits favoring the individual performing them, evolution of such behaviors could not be explained by natural selection. Neo–Darwinian explanations of such behaviors include kin selection and reciprocal altruism.

alula (al′yə-lə) (L. dim. of *ala,* wing). The first digit or thumb of a bird's wing, much reduced in size.

alveolate (al-vē′ə-lāt) (L. dim. of *alveus,* cavity, hollow). Any member of a protozoan clade having alveolar sacs; includes ciliates, apicomplexans, and dinoflagellates.

alveolus (al-vē′ə-ləs) (L. dim. of *alveus,* cavity, hollow). A small cavity or pit, such as a microscopic air sac of the lungs, terminal part of an alveolar gland, or bony socket of a tooth. Membranous sacs under the plasma membrane of certain protozoa.

ambulacra (am′byə-lak′rə) (L. *ambulare,* to walk). In echinoderms, radiating grooves where podia of the water-vascular system characteristically project outside the organism.

amebocyte (ə-mē′bə-sīt) (Gr. *amoibē,* change, + *kytos,* hollow vessel). Cell in metazoan invertebrate often functioning in defense against invading particles.

ameboid (ə-mē′boid) (Gr. *amoibē,* change, + *oid,* like). Ameba-like in putting forth pseudopodia.

amebozoa (ə-mē′bə-zō′ə) (Gr. *amoibē,* change, + *zōon,* animal). A protozoan clade containing slime molds and amebas with lobose pseudopodia.

amensalism (ā-men′səl-iz′əm). An asymmetric competitive interaction between two species in an ecological community in which only one of the species is affected.

amictic (ə-mik′tic) (Gr. *a,* without, + *miktos,* mixed or blended). Pertaining to female rotifers, which produce only diploid eggs that cannot be fertilized, or to the eggs produced by such females; contrasts with **mictic.**

amino acid (ə-mē′nō) (amine, an organic compound). An organic acid with an amino group (—NH_2). A fundamental subunit of proteins and peptides.

amitosis (ā′mī-tō′səs) (Gr. *a,* not, + *mitos,* thread). A form of cell division in which mitotic nuclear changes do not occur; cleavage without separation of daughter chromosomes.

ammocoetes (am-mō-cēt′ēz) (Gr. *ammo,* sand, + *coet,* bed). The filter-feeding, larval stage of lampreys.

amniocentesis (am′nē-ō-sin-tē′səs) (Gr. *amnion,* membrane around the fetus, + *centes,* puncture). Procedure for withdrawing a sample of fluid around the developing embryo for examination of chromosomes in the embryonic cells and other tests.

amnion (am′nē-än) (Gr. *amnion,* membrane around the fetus). The innermost of the extraembryonic membranes forming a fluid-filled sac around the embryo in amniotes.

amniote (am′nē-ōt). Having an amnion; as a noun, an animal that develops an amnion in embryonic life; refers collectively to nonavian reptiles, birds, and mammals.

amniotic egg (am′nē-ot′ək). A shelled vertebrate egg containing four membranes (yolk sac, amnion, chorion, and allantois).

amphiblastula (am′fə-blas′chə-lə) (Gr. *amphi,* on both sides, + *blastos,* germ, + L. *ula,* small). Free-swimming larval stage of certain marine sponges; blastula-like but with cells of the animal pole flagellated and those of the vegetal pole unflagellated.

amphid (am′fəd) (Gr. *amphidea,* anything that is bound around). One of a pair of anterior sense organs in certain nematodes.

amphipathic (am-fi-pa′thək) (Gr. *amphi,* on both sides, + *pathos,* suffering, passion). Adjective to describe a molecule with one part soluble in water (polar) and another part insoluble in water (nonpolar).

amplexus (am-plek′səs) (L. embrace). The copulatory embrace of frogs or toads.

ampulla (am-pūl′ə) (L. flask). Membranous vesicle; dilation at one end of each semicircular canal containing sensory epithelium; muscular vesicle above tube foot in water-vascular system of echinoderms.

amylase (am′ə-lās′) (L. *amylum,* starch, + *ase,* suffix meaning enzyme). An enzyme that breaks starch into smaller units.

anabolism (ə-na′bə-li′zəm) (Gr. *ana,* up, + *bol,* to throw, + *ism,* suffix meaning state of condition). Constructive metabolism.

anadromous (an-ad′rə-məs) (Gr. *anadromos,* running upward). Refers to fishes that migrate up streams from the sea to spawn.

anaerobic (an′ə-rō′bik) (Gr. *an,* not, + *aēr,* air, + *bios,* life). Not dependent on oxygen for respiration.

analogy (L. *analogus,* ratio). Similarity of function but not of origin.

anamniote (an′am-nē-ōt). A vertebrate that lacks an amniotic membrane around the embryo. Includes fishes and amphibians.

anaphylaxis (an′ə-fə-lax′əs) (Gr. *ana-,* up, + *phylax,* guard). A systemic (whole body) immediate hypersensitivity reaction.

anapsid (ə-nap′səd) (Gr. *an,* without, + *apsis,* arch). Amniotes in which the skull lacks temporal openings.

anastomosis (ə-nas′tə-mō′səs) (Gr. *ana,* again, + *stoma,* mouth). A union of two or more blood vessels, fibers, or other structures to form a branching network.

ancestral A character state inferred to have been present in the most recent common ancestral population of a group of organisms.

ancestrula (an-ses′trə-lə) (NL, *ancestor*). First individual (zooid) of a bryozoan colony that arises from metamorphosis of a free-swimming larva (in marine forms) or a statoblast (in freshwater forms).

androgen (an′drə-jən) (Gr. *anēr, andros,* man, + *genēs,* born). Any of a group of steroid male sex hormones.

androgenic gland (an′drō-jen′ək) (Gr. *anēr,* male, + *gennaein,* to produce). Gland in Crustacea that causes development of male characteristics.

aneuploidy (an′ū-ploid′ē) (Gr. *an,* without, not, + *eu,* good, well, + *ploid,* multiple of). Loss or gain of a chromosome, cells of the organism have one fewer than normal chromosome number, or one extra chromosome, for example, trisomy 21 (Down syndrome).

angiotensin (an′jē-o-ten′sən) (Gr. *angeion,* vessel, + L. *tensio,* to stretch). Blood protein formed from the interaction of renin and a liver protein, causing increased blood pressure and stimulating release of aldosterone and ADH.

Angstrom (after Ångström, Swedish physicist). A unit of one ten-millionth of a millimeter (one ten-thousandth of a micrometer); it is represented by the symbol Å.

anhydrase (an-hī′drās) (Gr. *an,* not, + *hydōr,* water, + *ase,* enzyme suffix). An enzyme involved in the removal of water from a compound. Carbonic anhydrase promotes the conversion of carbonic acid into water and carbon dioxide.

anisogametes (an′īs-ō-gam′ēts) (Gr. *anisos,* unequal, + *gametēs,* spouse). Gametes of a species that differ in form or size.

anlage (än′lä-gə) (Ger. laying out, foundation). Rudimentary form; primordium.

annulus (an′yəl-əs) (L. ring). Any ringlike structure, such as superficial rings on leeches.

antenna (L. sail yard). A sensory appendage on the head of arthropods, or the second of the two such pairs of structures in crustaceans.

antennal gland Excretory gland of Crustacea located in the antennal metamere.

anterior (L. comparative of *ante,* before). The head end of an organism, or (as an adjective) toward that end.

anthracosaurs (an-thrak'ə-sors) (Gr. *anthrax,* coal, carbon, + *sauros,* lizard). A group of Paleozoic labyrinthodont tetrapods.

anthropoid (an'thrə-poyd) (Gr. *anthrōpos,* man, + *eidos,* form). Resembling humans: especially the great apes.

antibodies (an'te-bä-dēs). Protein (immunoglobulins) in cell surfaces and dissolved in blood or tissue fluid, capable of combining with the antigens that stimulated their production. Formation of antibody-antigen complexes immobilize invading organisms, which can then be removed by phagocytosis of the complexes.

anticodon (an'tī-kō'don). A sequence of three nucleotides in transfer RNA that is complementary to a codon in messenger RNA.

antigen (an'ti-jən). Any substance capable of stimulating an immune response, most often a protein.

antigenic determinant See **epitope.**

antimicrobial peptides Peptides secreted during an innate immune response in animals and plants. See **defensins.**

antiparallel In the double helix of DNA, paired strands are oriented so that the 3′ end of one strand is opposite the 5′ end of the other strand.

aperture (ap'ər-chər) (L. *apertura* from *aperire,* to uncover). An opening; the opening into the first whorl of a gastropod shell.

apex (ā'peks) (L. tip). Highest or uppermost point; the lower pointed end of the heart.

apical (ā'pə-kl) (L. *apex,* tip). Pertaining to the tip or apex.

apical complex A certain combination of organelles found in the protozoan phylum Apicomplexa.

apocrine (ap'ə-krən) (Gr. *apo,* away, + *krinein,* to separate). Applies to a type of mammalian sweat gland that produces a viscous secretion by detaching part of the cytoplasm of secreting cells.

apoptosis (a'pə-tō'səs) (Gr. *apo-,* away from, + *ptōsis,* a falling). Genetically determined cell death, "programmed" cell death.

apopyle (ap'ə-pīl) (Gr. *apo,* away from, + *pylē,* gate). In sponges, opening of the radial canal into the spongocoel.

appendicular (L. *ad,* to, + *pendere,* to hang). Pertaining to appendages; pertaining to vermiform appendix.

aquaporins (aw'kwə-pōr'ins). Water channels or pores composed of transmembrane proteins that allow water to move through plasma membranes. They may be open at all times or gated to open or close by a specific signal.

arboreal (är-bōr'ē-al) (L. *arbor,* tree). Living in trees.

archaeocytes (ärk'ē-ō-sites) (Gr. *archaios,* beginning, + *kytos,* hollow vessel). Ameboid cells of varied function in sponges.

archenteron (ärk-en'tə-rän) (Gr. *archē,* beginning, + *enteron,* gut). The main cavity of an embryo in the gastrula stage; it is lined with endoderm and represents the future digestive cavity.

archinephros (ärk'ē-nəf'rōs) (Gr. *archaois,* ancient, + *nephros,* kidney). Ancestral vertebrate kidney, existing today only in the embryo of hagfishes.

archosaur (är'kə-sor) (Gr. *archōn,* ruling, + *sauros,* lizard). A clade of diapsid vertebrates that includes the living birds and crocodiles and the extinct pterosaurs and dinosaurs.

areolar (a-rē'ə-ler) (L. *areola,* small space). A small area, such as spaces between fibers of connective tissue.

arginine phosphate Phosphate storage compound (phosphagen) found in many invertebrates and used to regenerate stores of ATP.

Aristotle's lantern Masticating apparatus of some sea urchins.

arteriole (är-tir'ē-ōl) (L. *arteria,* artery). A small arterial branch that delivers blood to a capillary network.

artery (ärt'ə-rē) (L. *arteria,* artery). A blood vessel that carries blood away from the heart and toward a peripheral cavity.

artiodactyl (är'ti-o-dak'təl) (Gr. *artios,* even, + *daktylos,* toe). One of an order of ungulate mammals with two or four digits on each foot.

asconoid (Gr. *askos,* bladder). Simplest form of sponges, with canals leading directly from the outside to the interior.

asexual Without distinct sexual organs; not involving formation of gametes.

assimilation (L. *assimilatio,* bringing into conformity). Absorption of digested nutrients to form complex organic protoplasmic materials.

asymmetric competition See **amensalism.**

athecate (ā'thē-kāt') (Gr. *a-,* an absence of something, + *thēkē,* a case, box). An organism without a theca.

atherosclerosis (a'thə-rō-sklə-rō'səs) (Gr. *athērōma,* tumor full of gruel-like material, + *sklērōs,* hard). Disease characterized by fatty plaques forming in the inner lining of arteries.

atmosphere (at'-mə-sfi(ə)r). The gaseous component of the biosphere, extending from the earth's surface to an altitude of 3500 km.

atoke (ā'tōk) (Gr. *a,* without, + *tokos,* offspring). Anterior, nonreproductive part of a marine polychaete, as distinct from the posterior, reproductive part (epitoke) during the breeding season.

atoll (ə-tōl') (Maldivian, *atolu*). A coral reef or island surrounding a lagoon.

atom The smallest unit of an element, composed of a dense nucleus of protons and (usually) neutrons surrounded by a system of electrons.

ATP Adenosine triphosphate. In biochemistry, an ester of adenosine and triphosphoric acid.

atrium (ā'trē-əm) (L. *atrium,* vestibule). One of the less muscular chambers of the heart; also, the tympanic cavity of the ear; also, the large cavity containing the pharynx in tunicates and cephalochordates.

auricle (aw'ri-kl) (L. *auricula,* a small ear). One of the less muscular chambers of the heart; atrium; the external ear, or pinna; any earlike lobe or process.

auricularia (ə-rik'u-lar'e-ə) (L. *auricula,* a small ear). A type of larva found in Holothuroidea.

autogamy (aw-täg'ə-me) (Gr. *autos,* self, + *gamos,* marriage). Condition in which the gametic nuclei produced by meiosis fuse within the same organism that produced them to restore the diploid number.

autosome (aw'tə-sōm) (Gr. *autos,* self, + *sōma,* body). Any chromosome that is not a sex chromosome.

autotomy (aw-tät'ə-me) (Gr. *autos,* self, + *tomos,* a cutting). Detachment of a part of the body by an organism.

autotroph (aw'tə-trōf) (Gr. *autos,* self, + *trophos,* feeder). An organism that makes its organic nutrients from inorganic raw materials.

autotrophic nutrition (Gr. *autos,* self, + *trophia,* denoting nutrition). Nutrition characterized by the ability to use simple inorganic substances for the synthesis of more complex organic compounds, as in green plants and some bacteria.

average effect A quantitative-genetic parameter estimating the incremental contribution of each copy of a particular allele to the mean value of a particular organismal phenotype (such as height or weight) in the population being studied. Average effect is calculated from measurements of the population frequencies of all genotypes containing the allele, and the mean deviations of each genotypic class from the mean value of the phenotype in the population as a whole.

avicularium (L. *avicula,* small bird, + *aria,* like or connected with). Modified zooid that is attached to the surface of the feeding zooid in Ectoprocta and resembles a bird's beak.

axial (L. *axis,* axle). Relating to the axis, or stem; on or along the axis.

axocoel (ak'sə-cēl) (Gr. *axon,* an axle, + *koilos,* hollow). The most anterior of three coelomic spaces that appear during larval echinoderm development.

axolotl (ak'sə-lot'l) (Nahuatl, *atl,* water, + *xolotl,* doll, servant, spirit). Larval stage of any of several species of the genus *Ambystoma* (such as *A. tigrinum*) exhibiting neotenic reproduction.

axon (ak'sän) (Gr. *axōn*). Elongate extension of a neuron that conducts signals away from the cell body and toward the synaptic terminals.

---

bat / āpe / ärmadillo / herring / fēmale / finch / līce / crocodile / crōw / cóin / duck / ūnicorn / ə indicates unaccented vowel sound "uh" as in mammal, fishes, cardinal, heron, vulture / stress as in bi-ol'o-gy, bi'o-log'i-cal

axoneme (aks′ə-nēm) (L. *axis,* axle, + Gr. *nēma,* thread). The microtubules in a cilium or flagellum, usually arranged as a circlet of nine pairs enclosing one central pair; also, the microtubules of an axopodium.

axopodium (ak′sə-pō′di-um) (Gr. *axon,* an axis, + *podion,* small foot). Long, slender, more or less permanent pseudopodium found in certain amebas. (Also **axopod.**)

axostyle (aks′ō-stil). Tubelike organelle in some flagellate protozoa, extending from the area of the kinetosomes to the posterior end, where it often protrudes.

B

B cell A type of lymphocyte that is most important in the humoral immune response.

barrier reef A coral reef that runs approximately parallel to the shore and is separated from the shore by a lagoon.

basal body Also known as kinetosome and blepharoplast, a cylinder of nine triplets of microtubules found basal to a flagellum or cilium; same structure as a centriole.

basal disc Aboral attachment site on a cnidarian polyp.

base A molecule that dissociates in solution to produce a hydroxide ion.

basis, basipodite (bā′səs, bā-si′pə-dīt) (Gr. *basis,* base, + *pous, podos,* foot). The distal or second joint of the protopod of a crustacean appendage.

Batesian mimicry A condition in which a prey species evolves close resemblance to the aposematic phenotype (such as warning coloration) of an inedible species to deceive a potential predator into avoiding it.

bathypelagic (bath′ə-pe-laj′ik) (Gr. *bathys,* deep, + *pelagos,* open sea). Relating to or inhabiting the deep sea.

behavioral ecology The study of animal behaviors as used to promote survival and reproduction in a population's natural habitat.

benthos (ben′thäs) (Gr. depth of the sea). Organisms that live on the substrate underlying seas and lakes; adj., **benthic.** Also, the submerged substrate itself.

bilateral cleavage In an early embryo, the first cell division plane divides the zygote into left and right sides which are maintained throughout subsequent cleavages.

Bilateria (bī′lə-tir′-ē-ə) (L. *bi-,* two, + *latus,* side). Bilaterally symmetrical animals.

bilirubin (bil′ə-ru-bən) (L. *bilis,* bile, + *rubeo,* to be red). A breakdown product of the heme group of hemoglobin, excreted in bile.

binary fission A mode of asexual reproduction in which an animal splits into two approximately equal offspring.

biogenesis (bī′ō-jen′ə-səs) (Gr. *bios,* life, + *genesis,* birth). The doctrine that life originates only from preexisting life.

biogenetic law See **recapitulation.**

biogeochemical cycle A description of the flow of elementary matter, such as carbon or phosphorus, through the component parts of an ecosystem and its **abiotic** environment,

including the amount of an element present at the various stages of a **food web.**

binomial nomenclature The Linnaean system in which a species is given a compound name, the first word being the genus in which the species is placed (capitalized) and the second word being the species epithet (uncapitalized) used to separate a species from other members of the same genus.

biological species concept A reproductive community of populations (reproductively isolated from others) that occupies a specific niche in nature.

bioluminescence Method of light production by living organisms in which usually certain proteins (luciferins), in the presence of oxygen and an enzyme (luciferase), are converted to oxyluciferins with the liberation of light.

biomass (Gr. *bios,* life, + *maza,* lump or mass). The weight of total living organisms or of a species population per unit of area.

biome (bī-ōm) (Gr. *bios,* life, + *ōma,* abstract group suffix). Complex of plant and animal communities characterized by climatic and soil conditions; the largest ecological unit.

biosphere (Gr. *bios,* life, + *sphaira,* globe). That part of earth containing living organisms.

biotic (bī-äd′ik) (Gr. *biōtos,* life, livable). Of or relating to life.

bipinnaria (L. *bi,* double, + *pinna,* wing, + *aria,* like or connected with). Free-swimming, ciliated, bilateral larva of the asteroid echinoderms; develops into the brachiolaria larva.

biradial symmetry A type of radial symmetry in which only two planes passing through the oral-aboral axis yield mirror images because some structure is paired.

biramous (bī-rām′əs) (L. *bi,* double, + *ramus,* a branch). Adjective describing appendages with two distinct branches, contrasted with uniramous, unbranched.

bivalent (bī-vāl′ənt) (L. *bi,* double, + *valen,* strength, worth). The pairs of homologous chromosomes at synapsis in the first meiotic division, a tetrad.

blastocoel (blas′tə-sēl) (Gr. *blastos,* germ, + *koilos,* hollow). Cavity of the blastula.

blastocyst (blast′ō-sist) (Gr. *blastos,* germ, + *kystis,* bladder). Mammalian embryo in the blastula stage.

blastomere (Gr. *blastos,* germ, + *meros,* part). An early cleavage cell.

blastopore (Gr. *blastos,* germ, + *poros,* passage, pore). External opening of the archenteron in the gastrula.

blastula (Gr. *blastos,* germ, + L. *ula,* dim.). Early embryological stage of many animals; consists of a hollow mass of cells.

blending See **polygenic inheritance.**

blepharoplast (blə-fä′rə-plast) (Gr. *blepharon,* eyelid, + *plastos,* formed). See **basal body.**

blood plasma The liquid, noncellular fraction of blood, including dissolved substances.

blood type Characteristic of human blood given by the particular antigens on the membranes of the erythrocytes, genetically determined, causing agglutination when

incompatible groups are mixed; the blood types are designated A, B, O, AB, Rh negative, Rh positive, and others.

Bohr effect A characteristic of hemoglobin that causes it to dissociate from oxygen in greater degree at higher concentrations of carbon dioxide.

bombykol (bäm′-bə-käl). A volatile chemical produced by virgin female silkworm moths to attract males.

book gill Respiratory structure of aquatic chelicerates (Arthropoda) where many thin blood-filled gills are layered like the pages of a book. Gas exchange occurs as seawater passes between each pair of gills.

book lung Respiratory structure of terrestrial chelicerates (Arthropoda) where many thin-walled air pockets extend into a blood-filled chamber in the abdomen.

boreal (bōr′ē-əl) (L. *boreas,* north wind). Relating to a northern biotic area characterized by a predominance of coniferous forests and tundra.

BP Before the present.

brachial (brak′ē-əl) (L. *brachium,* forearm). Referring to the arm.

brachiation (brak′ē-ā′shən) (L. *brachium,* arm). Locomotion by swinging by the arms from one hold to another.

brachiolaria (brak′ē-ō-lār′ē-ə) (L. *brachiola,* little arm, + *aria,* pertaining to). This asteroid larva develops from the bipinnaria larva and has three preoral holdfast processes.

brackish Water having a salinity intermediate between freshwater and seawater, ranging from 0.5 to 30 parts per thousand.

bradyzoite An individual coccidian (a single-celled parasite) such as *Toxoplasma gondii* that is encased in a tissue cyst and divides slowly.

branchial (brank′ē-əl) (Gr. *branchia,* gills). Referring to gills.

bronchiole (brän′kē-ōl) (Gr. *bronchion,* dim. of *bronchos,* windpipe). Small, thin-walled branch of the bronchus.

bronchus (brän′kəs), pl. **bronchi** (Gr. *bronchos,* windpipe). Either of two primary divisions of the trachea that lead to the right and left lung.

brown bodies Remnants of the lophophore and digestive tract of a degenerating adult ectoproct left behind in the chamber as a new lophophore and digestive tract are formed.

brown fat Mitochondria-rich, heat-generating adipose tissue of endothermic vertebrates.

bryophytes (brī′ə-fīts) (Gr. *bryō,* to sprout up, + *phyta,* plants). Nonvascular plants comprising mosses, liverworts, and hornworts.

buccal (buk′əl) (L. *bucca,* cheek). Referring to the mouth cavity.

budding Reproduction in which the offspring arises as an outgrowth from the parent and is initially smaller than the parent. Failure of the offspring to separate from the parent leads to colony formation.

buffer Any substance or chemical compound that tends to keep pH levels constant when acids or bases are added.

bursa, pl. **bursae** (M.L. *bursa,* pouch, purse made of skin). A saclike cavity. In ophiuroid echinoderms, pouches opening at bases of arms and functioning in respiration and reproduction (genitorespiratory bursae).

C

caecilian (si-sil′-yən). Denotes any member of the amphibian order Gymnophiona (also called Apoda).

caecum See **cecum.**

calciferous glands (kal-si′fə-rəs). Glands in an earthworm that secrete calcium ions into the gut.

calorie (kal′ə-rē) (L. *calere,* to be warm). Unit of heat defined as the amount of heat required to heat 1 g of water from 14.5 to 15.5° C; 1 cal = 4.184 joules in the International System of Units.

calyx (kā-′liks) (L. bud cup of a flower). Any of various cup-shaped zoological structures.

cancellous (kan′səl-əs) (L. *cancelli,* latticework, + *osus,* full of). Having a spongy or porous structure.

capillary Extremely narrow (average in mammals is about 8 μm in diameter) blood vessels linking arterial and venous systems within tissues, and composed of a single layer of endothelial cells. Capillaries provide an interface between the circulatory system and cells allowing filtration of oxygen, nutrients and signaling molecules (for example, hormones) to cells and uptake of metabolic wastes from cells.

capitulum (ka-pi′tə-ləm) (L. small head). Term applied to small, headlike structures of various organisms, including the projection from the body of ticks and mites carrying mouthparts.

captacula (kap-tak′ū-lə) (L. *captare,* to lie in wait for). Tentacles extending from the head of scaphopod molluscs, used in feeding.

carapace (kar′ə-pās) (F. from Sp. *carapacho,* shell). Shieldlike plate covering the cephalothorax of certain crustaceans; dorsal part of the shell of a turtle.

carbohydrate (L. *carbo,* charcoal, + Gr. *hydōr,* water). Compounds of carbon, hydrogen, and oxygen having the generalized formula $(CH_2O)_n$; aldehyde or ketone derivatives of polyhydric alcohols, with hydrogen and oxygen atoms attached in a 2:1 ratio.

carboxyl (kär-bäk′səl) (carbon + oxygen + yl, chemical radical suffix). The acid group of organic molecules (—COOH).

cardiac (kär′dē-ak) (Gr. *kardia,* heart). Belonging or relating to the heart.

carinate (kar′ə-nāt) (L. *carina,* keel). Having a keel, in particular the flying birds with a keeled sternum for the insertion of flight muscles; contrasts with ratite.

carnivore (kar′nə-vōr′) (L. *carnivorous,* flesh eating). One of the flesh-eating mammals of the order Carnivora. Also, any organism that eats animals. Adj., **carnivorous.**

carotene (kär′ə-tēn) (L. *carota,* carrot, + *ene,* unsaturated straight-chain hydrocarbons). A red,

orange, or yellow pigment belonging to the group of carotenoids; precursor of vitamin A.

carrier An individual heterozygous for a recessive allele, such as an allele for a genetic disease, who is phenotypically normal but may transmit the recessive allele to offspring.

carrying capacity The maximum number of individuals that can persist under specified environmental conditions.

cartilage (L. *cartilago;* akin to L. *cratis,* wickerwork). A translucent, specialized connective tissue that forms most of the skeleton of embryos, very young vertebrates, and adult cartilaginous fishes, such as sharks and rays; in other adult vertebrates much of it is converted into bone.

caste (kast) (L. *castus,* pure, separated). One of the polymorphic forms within an insect society, each caste having its specific duties, such as queen, worker, and soldier.

catabolism (Gr. *kata,* downward, + *bol,* to throw, + *ism,* suffix meaning state of condition). Destructive metabolism; process in which complex molecules are reduced to simpler ones.

catadromous (kə-tad′rə-məs) (Gr. *kata,* down, + *dromos,* a running). Refers to fishes that migrate from freshwater to the ocean to spawn.

catalyst (kad′ə-ləst) (Gr. *kata,* down, + *lysis,* a loosening). A substance that accelerates a chemical reaction but does not become a part of the end product.

catastrophic species selection Differential survival among species during a time of mass extinction based on character variation that permits some species but not others to withstand severe environmental disturbances, such as those caused by an asteroid impact.

catch collagen In echinoderms a mutable collagenous tissue under neuronal control that can be changed from a "liquid" to a "solid" form very quickly

caudal (käd′l) (L. *cauda,* tail). Constituting, belonging to, or relating to a tail.

caveolae (ka-vē′ə-lē) (L. *cavea,* a cave, + dim. suffix). The invaginated vesicles and pits in pinocytosis.

cDNA See **complementary DNA.**

cecum, caecum (sē′kəm) (L. *caecus,* blind). A blind pouch at the beginning of the large intestine; any similar pouch.

cell-mediated immune response Immune response involving cell surfaces only, not antibody production, specifically the T_H1 arm of the immune response. Contrasts with **humoral immune response.**

cellulase (sel′ū-lās) (L. *cella,* small room). An enzyme that cleaves cellulose; only synthesized by bacteria and some protists.

cellulose (sel′ū-lōs) (L. *cella,* small room). Chief polysaccharide constituent of the cell wall of green plants and some fungi; an insoluble carbohydrate $(C_6H_{10}O_5)_n$ that is converted into glucose by hydrolysis.

center of origin The geographic area occupied by a species or higher taxon during

its initial evolution; contrasts to areas colonized by dispersal following a species's evolutionary origin.

centriole (sen′trē-ol) (Gr. *kentron,* center of a circle, + L. *ola,* small). A minute cytoplasmic organelle usually found in the centrosome and considered the active division center of the animal cell; organizes spindle fibers during mitosis and meiosis. Same structure as basal body or kinetosome.

centrolecithal (sen′tro-les′ə-thəl) (Gr. *kentron,* center, + *lekithos,* yolk, + Eng. *al,* adjective). Pertaining to an insect egg with the yolk concentrated in the center.

centromere (sen′trə-mir) (Gr. *kentron,* center, + *meros,* part). A localized constriction in a characteristic position on a given chromosome, bearing the kinetochore.

centrosome (sen′trə-sōm) (Gr. *kentron,* center, + *sōma,* body). Microtubule organizing center in nuclear division in most eukaryotic cells; in animals and many unicellular organisms it surrounds the centrioles.

cephalization (sef′ə-li-zā-shən) (Gr. *kephale,* head). The evolutionary process by which sensory organs and specialized appendages become localized in the head end of animals.

cephalothorax (sef′ə-lä-thō′raks) (Gr. *kephale,* head, + thorax). A body division found in many Arachnida and higher Crustacea, in which the head is fused with some or all of the thoracic segments.

cerata (sə-ra′tə) (Gr. *keras,* a horn, bow). Dorsal processes on some nudibranchs for gaseous exchange.

cercaria (ser-kar′ē-ə) (Gr. *kerkos,* tail, + L. *aria,* like or connected with). Tadpolelike larva of trematodes (flukes).

cervical (sər′və-kəl) (L. *cervix,* neck). Relating to a neck.

channels Pores created by transmembrane proteins that allow ions and water to move through plasma membranes by diffusion. They may be open at all times or gated by specific signals to open or to close them (for example, chemically-gated or voltage-gated channels).

character (kar′ik-tər). A component of phenotype (including specific molecular, morphological, behavioral, or other features) used by systematists to diagnose species or higher taxa, or to evaluate phylogenetic relationships among different species or higher taxa, or relationships among populations within a species.

character displacement Differences in morphology or behavior within a species caused by competition with another species; characteristics typical of one species differ according to whether the other species is present or absent from a local community.

charging In protein synthesis, a reaction catalyzed by tRNA synthetase, in which an amino acid is attached to its particular tRNA molecule.

chela (kēl′ə) (Gr. *chēlē,* claw). Pincerlike claw.

chelicera (kə-lis′ə-rə), pl. **chelicerae** (Gr. *chēlē,* claw, + *keras,* horn). One of a pair of the most anterior head appendages on the members of the subphylum Chelicerata.

chelifore (kə′-li-for) (Gr. *chēlē*, claw, + OE *fore*, before). First pair of appendages on a pycnogonid; sometimes absent; if present, chelate or achelate.

chelipeds (kēl′ə-peds) (Gr. *chēlē*, claw, + L. *pes*, foot). Pincerlike first pair of legs in most decapod crustaceans; specialized for seizing and crushing.

chemoautotroph (ke′mō-aw′tō-trōf) (Gr. *chemeia*, transmutation, + *autos*, self, + *trophos*, feeder). An organism utilizing inorganic compounds as a source of energy.

chemotaxis (kē′mō-tak′səs) (Gr. *chēmeia*, an infusion, + *taxō* from *tassō*, to put in order). Orientation movement of cells or organisms in response to a chemical stimulus.

chemotroph (kem′ə-trōf) (Gr. *chēmeia*, an infusion, + *tropē*, to turn). An organism that derives nourishment from inorganic substances without using chlorophyll.

chiasma (kī-az′mə), pl. **chiasmata** (Gr. cross). An intersection or crossing, as of nerves; a connection point between homologous chromatids where crossing over has occurred at synapsis.

chitin (kī′tən) (Fr. *chitine*, from Gr. *chitōn*, tunic). A hard substance that forms part of the cuticle of arthropods and is found sparingly in certain other invertebrates; a nitrogenous polysaccharide insoluble in water, alcohol, dilute acids, and digestive juices of most animals.

chlorocruorin (klō′rō-kroo′ə-rən) (Gr. *chlōros*, light green, + L. *cruor*, blood). A greenish iron-containing respiratory pigment dissolved in the blood plasma of certain marine polychaetes.

chlorogogen cells (klōr′ə-gog-ən) (Gr. *chlōros*, light green, + *agōgos*, a leading, a guide). Modified peritoneal cells, greenish or brownish, clustered around the digestive tract of certain annelids; apparently they aid in elimination of nitrogenous wastes and in food transport.

chlorophyll (klō′rə-fil) (Gr. *chlōros*, light green, + *phyllōn*, leaf). Green pigment found in plants and in some animals; necessary for photosynthesis.

chloroplast (klō′rə-plast) (Gr. *chlōros*, light green, + *plastos*, molded). A plastid containing chlorophyll and usually other pigments, found in cytoplasm of plant cells.

choanoblast (kō-an′ə-blast) (Gr. *choanē*, funnel, + *blastos*, germ). One of several cellular elements within the syncytial tissue of a hexactinellid sponge; choanoblasts bear flagellated extensions called collar bodies.

choanocyte (kō-an′ə-sīt) (Gr. *choanē*, funnel, + *kytos*, hollow vessel). One of the flagellated collar cells that line cavities and canals of sponges.

choanoflagellate Any member of a protozoan clade having a single flagellum surrounded by a column of microvilli; some form colonies and all are included within the larger clade of opisthokonts.

cholinergic (kōl′i-nər′jik) (Gr. *chole*, bile, + *ergon*, work). A nerve fiber that releases acetylcholine from its axon terminal.

chorioallantoic membrane (kō′rē-ō-al′an-tō′-ic) (Gr. *chorion*, skin, + *allas*, sausage, + *eidos*, form). A vascular envelope around some amniote embryos formed by fusion of mesoderm from the chorion and allantois.

chorioallantoic placenta (kō′rē-ō-al′an-tō′-ic) (Gr. *chorion*, skin, + *allas*, sausage). A type of placenta that occurs in placental mammals and some marsupials in which materials are exchanged between the embryo and mother through a structure modified from the embryonic chorionic and allantoic membranes.

chorion (kō′rē-on) (Gr. *chorion*, skin). The outer of the double membrane that surrounds the embryo of nonavian reptiles, birds, and mammals; in mammals it contributes to the placenta.

choroid (kōr′oid) (Gr. *chorion*, skin, + *eidos*, form). Delicate, highly vascular membrane; in vertebrate eye; the layer between the retina and sclera.

chorionic villi (kō′rē-än′ic vil′ē) (Gr. *chorion*, skin, + L. *villi*, pl. of *villus*, shaggy hair, tuft of hair). Fingerlike extensions containing blood vessels that lie on the outer surface of a vertebrate chorionic membrane.

choriovitelline placenta (kor′ē-ō-vi′təl-ən) (Gr. *chorion*, skin, + *vittel*, yolk of an egg). An often transitory placenta that forms during the early developmental stages of marsupials and placental mammals. Also called the "yolk sac placenta," it is formed from the yolk sac and chorionic membrane of the embryo.

chromatid (krō′mə-tid) (Gr. *chromato*, from *chrōma*, color, + L. *id*, feminine stem for particle of specified kind). A replicated chromosome joined to its sister chromatid by the centromere; separates and becomes daughter chromosome at anaphase of mitosis or anaphase of the second meiotic division.

chromatin (krō′mə-tin) (Gr. *chrōma*, color). The nucleoprotein material of a chromosome; the hereditary material containing DNA.

chromatophore (krō-mat′ə-fōr) (Gr. *chrōma*, color, + *pherein*, to bear). Pigment cell, usually in the dermis, in which the pigment can be dispersed or concentrated.

chromomere (krō′mō-mir) (Gr. *chrōma*, color, + *meros*, part). One of the chromatin granules of characteristic size on the chromosome; may be identical with a gene or a cluster of genes.

chromonema (krō-mə-nē′mə) (Gr. *chrōma*, color, + *nēma*, thread). A convoluted thread in prophase of mitosis or the central thread in a chromosome.

chromoplast (krō′mə-plast) (Gr. *chrōma*, color, + *plastos*, molded). A plastid that contains pigment.

chromosomal theory of inheritance The general theory synthesizing results of Mendelian genetics and cytology to propose particulate inheritance of hereditary factors carried on chromosomes in eukaryotes.

chromosome (krō′mə-sōm) (Gr. *chrōma*, color, + *sōma*, body). A complex body, spherical or rod-shaped, that arises from the nuclear network during mitosis, splits longitudinally, and carries a part of the organism's genetic information as genes composed of DNA associated with proteins.

chrysalis (kris′ə-lis) (L. from Gr. *chrysos*, gold). The pupal stage of a butterfly.

chyme (kīm) (Gr. *chymos*, juice). Semifluid mass of partly digested food in stomach and small intestine as digestion proceeds.

cilium (sil′i-əm), pl. **cilia** (L. eyelid). A hairlike, vibratile organelle process found on many animal cells. Cilia may be used in moving particles along the cell surface or, in ciliate protozoans, for locomotion.

cinclides (sing′klid-əs), sing. **cinclis** (sing′kləs) (Gr. *kinklis*, latticed gate or partition). Small pores in the external body wall of sea anemones for extrusion of acontia.

circadian (sər-kād′ē-ən) (L. *circa*, around, + *dies*, day). Occurring at a period of approximately 24 hours.

cirrus (sir′əs) (L. curl). A hairlike tuft on an insect appendage; locomotor organelle of fused cilia; male copulatory organ of some invertebrates.

cisternae (sis-ter′nē) (L. *cista*, box). Space between membranes of the endoplasmic reticulum within cells.

cistron (sis′trən) (L. *cista*, box). A series of codons in DNA that code for an entire polypeptide chain.

clade (klād) (Gr. *klados*, branch, sprout). A taxon or other group consisting of an ancestral species and all of its descendants, forming a distinct branch on a phylogenetic tree.

cladistics (klad-is′-təks) (Gr. *klados*, branch, sprout). A system of arranging taxa by analysis of evolutionarily derived characteristics so that the arrangement will reflect phylogenetic relationships.

cladogram (klād′ə-gram) (Gr. *klados*, branch, + *gramma*, letter). A branching diagram showing the pattern of sharing of evolutionarily derived characters among species or higher taxa.

clasper (klas′-pər). Digitiform projection on the medial side of pelvic fins of male chondrichthians and some placoderms; used as an intromittent organ to transfer sperm to female reproductive tract

clathrin (kla′thrən) (L. *chathri*, latticework). A protein forming a lattice structure lining the invaginated pits during receptor-mediated endocytosis.

cleavage (O.E. *cleofan*, to cut). Process of nuclear and cell division in animal zygote.

climax (klī′maks) (Gr. *klimax*, ladder). Stage of relative stability attained by a community of organisms, often the culminating development of a natural succession. Also, orgasm.

climax community (Gr. *klimax*, ladder, staircase, climax). A self-perpetuating, more-or-less stable community of organisms that continues as long as environmental conditions under which it developed prevail.

clitellum (klĭ-tel′əm) (L. *clitellae,* packsaddle). Thickened saddlelike portion of certain midbody segments of many oligochaetes and leeches. Produces the cocoon in which eggs are deposited during sexual reproduction.

cloaca (klō-ā′kə) (L. sewer). Posterior chamber of digestive tract in many vertebrates, receiving feces and urogenital products. In certain invertebrates, a terminal portion of digestive tract that serves also as respiratory, excretory, or reproductive tract.

clone (klōn) (Gr. *klōn,* twig). All descendants derived by asexual reproduction from a single individual.

cloning (klō′ning). Production of genetically identical organisms by asexual reproduction.

cnida (nī′də) (Gr. *knidē,* nettle). Stinging or adhesive organelles formed within cnidocytes in phylum Cnidaria; nematocysts are a common type.

cnidoblast (nī′də-blast) (Gr. *knidē,* nettle, + *blastos,* germ). See **cnidocyte.**

cnidocil (nī′də-sil) (Gr. *knidē,* nettle, + L. *cilium,* hair). Modified cilium on nematocyst-bearing cnidocytes in cnidarians; triggers nematocyst.

cnidocyte (nī′də-sīt) (Gr. *knidē,* nettle, + *kytos,* hollow vessel). Modified interstitial cell that holds the nematocyst; during development of the nematocyst, the cnidocyte is a cnidoblast.

coacervate (kō′ə-sər′vət) (L. *coacervatus,* to heap up). An aggregate of colloidal droplets held together by electrostatic forces.

coagulation (kō-ag′ū-lā-shən). Process in which enzymes are activated to produce clotting of blood.

coccidian (kok-sid′ē-ən) (Gr. *kokkis,* kernel, grain). Intracellular protozoan parasite belonging to a class within phylum Apicomplexa; the organism causing malaria is an example.

cochlea (kōk′lē-ə) (L. snail, from Gr. *kochlos,* a shellfish). A tubular cavity of the inner ear containing the essential organs of hearing; occurs in crocodiles, birds, and mammals; spirally coiled in mammals.

cocoon (kə-kun′) (Fr. *cocon,* shell). Protective covering of a resting or developmental stage, sometimes used to refer to both the covering and its contents; for example, the cocoon of a moth or the protective covering for the developing embryos in some annelids.

codominance A condition in which each allele maintains its distinctive homozygous expression in the heterozygous condition, not a blending of the separate homozygous phenotypes (contrasts with intermediate

inheritance). Genes for A and B blood types show codominance (p. 126).

codon (kō′dän) (L. code, + on). In messenger RNA a sequence of three adjacent nucleotides that code for one amino acid.

coelenteron (sē-len′tər-on) (Gr. *koilos,* hollow, + *enteron,* intestine). Internal cavity of a cnidarian; gastrovascular cavity; archenteron.

coelom (sē′lōm) (Gr. *koilōma,* cavity). The body cavity in triploblastic animals, lined with mesodermal peritoneum.

coelomocyte (sē′lō′mə-sīt) (Gr. *koilōma,* cavity, + *kytos,* hollow vessel). Another name for amebocyte; primitive or undifferentiated cell of the coelom and the water-vascular system of echinoderms.

coelomoduct (sē-lō′mə-dukt) (Gr. *koilos,* hollow, + L. *ductus,* a leading). A duct that carries gametes or excretory products (or both) from the coelom to the exterior.

coenecium, coenoecium (sə-nēs[h]′ē-əm) (Gr. *koinos,* common, + *oikion,* house). The common secreted investment of an ectoproct colony; may be chitinous, gelatinous, or calcareous.

coenenchyme (sēn′ən-kīm) (Gr. *koinos,* shared in common, + *enchyma,* something poured in). Extensive mesogleal tissue between the polyps of an octocorallian (phylum Cnidaria) colony.

coenocytic (sē-nə-sit′ik) (Gr. *koinos,* common, + *kytos,* hollow vessel). A tissue in which the nuclei are not separated by cell membranes; syncytial.

coenosarc (sē′nə-särk) (Gr. *koinos,* shared in common, + *sarkos,* flesh). The inner, living part of hydrocauli in hydroids.

coenzyme (kō-en′zīm) (L. prefix, *co,* with, + Gr. *enzymos,* leavened, from *en,* in, + *zymē,* leaven). A required substance in the activation of an enzyme; a prosthetic or nonprotein constituent of an enzyme.

cohort (kō′hort). All organisms of a population born within a specified interval of time.

collagen (käl′ə-jən) (Gr. *kolla,* glue, + *genos,* descent). A structural protein of connective tissue, the most abundant protein in the animal kingdom, characterized by high content of the amino acids glycine, alanine, proline, and hydroxyproline.

collar bodies Extensions of choanoblasts bearing flagellated collars in hexactinellid sponges.

collar cells Cells having a single flagellum surrounded by a ring of microvilli. Sponge choanocytes are collar cells, as are choanoflagellates, but collar cells also occur in other taxa.

collenchyme (käl′ən-kīm) (Gr. *kolla,* glue, + *enchyma,* infusion). A gelatinous mesenchyme containing undifferentiated cells; found in cnidarians and ctenophores.

collencyte (käl′lən-sīt) (Gr. *kolla,* glue, + *en,* in, + *kytos,* hollow vessel). A type of cell in

sponges that is star-shaped and apparently contractile.

colloblast (käl′ə-blast) (Gr. *kolla,* glue, + *blastos,* germ). A glue-secreting cell on the tentacles of ctenophores.

colloid (kä′loid) (Gr. *kolla,* glue, + *eidos,* form). A two-phase system in which particles of one phase are suspended in the second phase.

columella (kä′lū-mel′ə) (L. small column). Central pillar in gastropod shells.

comb plate One of the plates of fused cilia arranged in rows for ctenophore locomotion.

commensalism (kə-men′səl-iz′əm) (L. *cum,* together with, + *mensa,* table). A relationship in which one individual lives close to or on another and benefits, and the host is unaffected; often symbiotic.

common descent Darwin's theory that all forms of life are derived from a shared ancestral population through a branching of evolutionary lineages.

community (L. *communitas,* community, fellowship). An assemblage of organisms that are associated in a common environment and interact with each other in a self-sustaining and self-regulating relation.

comparative biochemistry Studies of the structures of biological macromolecules, especially proteins and nucleic acids, and their variation within and among related species to reveal homologies of macromolecular structure.

comparative cytology Studies of the structures of chromosomes within and among related species to reveal homologies of chromosomal structure.

comparative method Characteristics of related populations or species are compared systematically to test hypotheses of common evolutionary descent and to examine ultimate causes underlying origins of biological features.

comparative morphology Studies of organismal form and its variation within and among related species to reveal homologies of organismal characters.

comparative psychology A field of study used to identify general rules of behavior that apply to humans and other animals.

competition Some degree of overlap in ecological niches of two populations in the same community, such that both depend on the same food source, shelter, or other resources, and negatively affect each other's survival.

competitive exclusion An ecological principle stating that two species whose niches are very similar cannot coexist indefinitely in the same community; one species is driven to extinction by competition between them.

complement Collective name for a series of proteins in the blood. Proteins of the complement series are activated by antibody bound to invading organisms and lead to rupture of foreign cells. Some complement proteins also bind to antibody-antigen complexes enhancing phagocytosis by phagocytic cells of the immune system.

complementary DNA (cDNA) DNA prepared by transcribing the base sequence from mRNA into DNA by reverse transcriptase; also called **copy DNA.**

compound A substance whose molecules are composed of atoms of two or more elements.

condensation reaction A chemical reaction in which reactant molecules are combined by the removal of a water molecule (a hydrogen from one and a hydroxyl from the other reactant).

condyle (kän′dīl) (Gr. *kondylos,* bump). A process on a bone used for articulation.

cone (kōn′). One of a group of cells on the retina of a vertebrate eye used to perceive color and to form images in well-illuminated settings.

conjugation (kon′ju-ga′shən) (L. *conjugare,* to yoke together). Temporary union of two ciliate protozoa while they are exchanging chromatin material and undergoing nuclear phenomena resulting in binary fission. Also, formation of cytoplasmic bridges between bacteria for transfer of plasmids.

conodonts (kon′-ə-donts) (Gr. *con,* cone + *odont,* tooth). Microscopic, toothlike fossils belonging to an extinct vertebrate-like animal known from the Cambrian to Triassic periods.

conspecific (L. *com,* together, + *species*). A member of the same species.

consumer (kən-sū′-mer). An organism whose energy and matter are acquired by eating other organisms, which may be **primary producers, herbivores,** or **carnivores.**

continental margin A portion of the ocean floor located adjacent to the shore; comprises a continental shelf, continental slope, and a continental rise.

continental rise Thick sediments piled at the base of the submarine continental slope.

continental shelf Portion of the continental margin of the ocean floor extending from shallow nearshore waters outward to a depth of 120 to 400 m below the sea surface.

continental slope A sharp descent of the ocean floor from the outer continental shelf to a depth of 3000 to 4000 m below the sea surface.

contractile vacuole A clear fluid-filled cell vacuole in protozoa and a few lower metazoa; collects water and releases it to the outside in a cyclical manner, for osmoregulation and some excretion.

control That part of a scientific experiment to which the experimental variable is not applied but which is similar to the experimental group in all other respects.

conus arteriosus (kōn′-əs är-tir′-ē-ō-səs′). An extension of the ventricle through which blood exits the heart in amphibians and some fishes; in mammals, an extension of the right ventricle serving the pulmonary circuit.

cooperative behavior Participation by an organism in a group activity that enhances each participant's genetic contribution

to future generations. Includes collective foraging and breeding behaviors.

coprophagy (kə-prä′fə-jē) (Gr. *kopros,* dung, + *phagein,* to eat). Feeding on dung or excrement as a normal behavior among animals; reinjestion of feces.

copulation (Fr. from L. *copulare,* to couple). Sexual union to facilitate the reception of sperm by the female.

copy DNA See **complementary DNA.**

coralline algae Algae that precipitate calcium carbonate in their tissues; important contributors to coral reef mass.

corium (kō′re-um) (L. *corium,* leather). The deep layer of the skin; dermis.

cornea (kor′nē-ə) (L. *corneus,* horny). The outer transparent coat of the eye.

corneum (kor′nē-əm) (L. *corneus,* horny). Epithelial layer of dead, keratinized cells. Stratum corneum.

cornified (kor′nə-fīd) (L. *corneus,* horny). Adjective for conversion of epithelial cells into nonliving, keratinized cells.

corona (kə-rō′nə) (L. crown). Head or upper portion of a structure; ciliated disc on anterior end of rotifers.

corpora allata (kor′pə-rə əl-la′tə) (L. *corpus,* body, + *allatum,* aided). Endocrine glands in insects that produce juvenile hormone.

corpora cardiaca (kor′pə-rə kar-dī′ə-cə) (L. *corpus,* body, + Gr. *kardiakos,* belonging to the heart). Paired organs behind the brain of insects, serve as storage and release organs for prothoracicotropic hormone (PTTH).

cortex (kor′teks) (L. bark). The outer layer of a structure.

cosmopolitan Used to describe a species or higher taxon that has a very large geographical range, such as the worldwide distribution of humans.

covalent bond A chemical bond in which electrons are shared between atoms.

coxa, coxopodite (kox′ə, kəx-ä′pə-dīt) (L. *coxa,* hip, + Gr. *pous, podos,* foot). The proximal joint of an insect or arachnid leg; in crustaceans, the proximal joint of the protopod.

creatine phosphate High-energy phosphate compound found in the muscle of vertebrates and some invertebrates, used to regenerate stores of ATP.

Cretaceous extinction A mass extinction that occurred 65 million years ago in which 76% of existing species, including all dinosaurs, became extinct, marking the end of the Mesozoic era.

cretin (krēt′n) (Fr. *crétin,* [dialect], fr. L. *christianus,* Christian, to indicate idiots so afflicted were also human). A human with severe mental, somatic, and sexual retardation resulting from hypothyroidism during early stages of development.

crista (kris′ta), pl. **cristae** (L. *crista,* crest). A crest or ridge on a body organ or organelle; a platelike projection formed by the inner membrane of mitochondrion.

crop (kräp). A region of the esophagus specialized for storing food.

crossing over Exchange of parts of nonsister chromatids at synapsis in the first meiotic division.

cryptobiotic (Gr. *kryptos,* hidden, + *biōticus,* pertaining to life). Living in concealment; refers to insects and other animals that live in secluded situations, such as underground or in wood; also tardigrades and some nematodes, rotifers, and others that survive harsh environmental conditions by assuming for a time a state of very low metabolism.

ctenidia (te-ni′dē-ə) (Gr. *kteis,* comb). Comblike structures, especially gills of molluscs; also applied to comb plates of Ctenophora.

ctenoid scales (ten′oid) (Gr. *kteis, ktenos,* comb). Thin, overlapping dermal scales of some fishes; exposed posterior margins have fine, toothlike spines.

cupula (ku′pū-lə) (L. little tub). Small inverted cuplike structure housing another structure; gelatinous matrix covering hair cells in lateral-line and equilibrium organs.

cutaneous respiration Use of the integument by amphibians for gas exchange between the blood and air. It is the primary respiratory structure of terrestrial lungless forms, including plethodontid salamanders.

cuticle (ku′ti-kəl) (L. *cutis,* skin). A protective, noncellular, organic layer secreted by the external epithelium (hypodermis) of many invertebrates. In higher animals the term refers to the epidermis or outer skin.

cyanobacteria (sī-an-ō-bak-ter′ē-ə) (Gr. *kyanos,* a dark-blue substance, + *bakterion,* dim. of *baktron,* a staff). Photosynthetic prokaryotes, also called blue-green algae, cyanophytes.

cyanophyte (sī-an′ō-fit) (Gr. *kyanos,* a dark-blue substance, + *phyton,* plant). A cyanobacterium, blue-green alga.

cyclin A protein important in the control of the cell division cycle and mitosis.

cycloid scales (sī′-kloid) (Gr. *kyklos,* circle). Thin, overlapping dermal scales of some fishes; posterior margins are smooth.

cydippid larva (sī-dip′pid) (Gr. *kydippe,* mythological Athenian maiden). Free-swimming larva of most ctenophores; superficially similar to the adult.

cynodonts (sīn′ə-dänts) (Gr. *kynodōn,* canine tooth). A group of mammal-like carnivorous synapsids of the Upper Permian and Triassic.

cyrtocyte (ser′tō-sīt) (Gr. *kyrtē,* a fish basket, cage, + *kytos,* hollow vessel). A pro-tonephridial cell with a single flagellum enclosed in a cylinder of cytoplasmic rods.

cyst (sist) (Gr. *kystis,* a bladder, pouch). A resistant, quiescent stage of an organism, usually with a secreted wall.

cystacanth (sis′tə-kanth) (Gr. *kystis,* bladder, pouch, + *akantha,* thorn). Juvenile stage of an acanthocephalan that is infective to the definitive host.

cysticercoid (sis′tə-ser′koyd) (Gr. *kystis,* bladder, + *kerkos,* tail, + *eidos,* form). A type of juvenile tapeworm composed of a solid-bodied **cyst** containing an invaginated scolex; contrasts with **cysticercus.**

cysticercus (sis′tə-ser′kəs) (Gr. *kystis,* bladder, + *kerkos,* tail). A type of juvenile tapeworm in which an invaginated and introverted scolex is contained in a fluid-filled bladder; contrasts with **cysticercoid.**

cystid (sis′tid) (Gr. *kystis,* bladder). In an ectoproct, the dead secreted outer parts plus the adherent underlying living layers.

cytochrome (sī′tə-krōm) (Gr. *kytos,* hollow vessel, + *chrōma,* color). Several iron-containing pigments that serve as electron carriers in aerobic respiration.

cytokine (sī′tə-kīn) (Gr. *kytos,* hollow vessel, + *kinein,* to move). A polypeptide hormone that mediates communication between cells participating in an immune response and is secreted as a neurotransmitter by the brain. Many different cytokines have been isolated and these can affect cells that secrete them, nearby cells or cells at distant locations.

cytokinesis (sī′tə-kin-ē′sis) (Gr. *kytos,* hollow vessel, + *kinesis,* movement). Division of the cytoplasm of a cell.

cytopharynx (Gr. *kytos,* hollow vessel, + *pharynx,* throat). Short tubular gullet in ciliate protozoa.

cytoplasm (sī′tə-plasm) (Gr. *kytos,* hollow vessel, + *plasma,* mold). The living matter of the cell, excluding the nucleus.

cytoproct (sī′tə-prokt) (Gr. *kytos,* hollow vessel, + *prōktos,* anus). Site on a protozoan where undigestible matter is expelled.

cytopyge (sī′tə-pīj) (Gr. *kytos,* hollow vessel, + *pyge,* rump or buttocks). In some protozoa, localized site for expulsion of wastes.

cytosol (sī′tə-sol) (Gr. *kytos,* hollow vessel, + L. *sol,* from *solutus,* to loosen). Unstructured portion of the cytoplasm in which the organelles are bathed.

cytosome (sī′tə-sōm) (Gr. *kytos,* hollow vessel, + *sōma,* body). The cell body inside the plasma membrane.

cytostome (sī′tə-stōm) (Gr. *kytos,* hollow vessel, + *stoma,* mouth). The cell mouth in many protozoa.

cytotoxic T cells (Gr. *kytos,* hollow vessel, + toxin). A special T cell activated during cell-mediated immune responses that recognizes and destroys virus-infected cells.

D

dactylozooid (dak-til′ə-zō-id) (Gr. *dakos,* bite, sting, + *tylos,* knob, + *zōon,* animal). A polyp of a colonial hydroid specialized for defense or killing food.

Darwinism Theory of evolution emphasizing common descent of all living organisms, gradual change, multiplication of species, and natural selection.

data, sing. **datum** (Gr. *dateomai,* to divide, cut in pieces). The results in a scientific experiment, or descriptive observations, upon which a conclusion is based.

deciduous (də-sij′ə-wəs) (L. *decidere,* to fall off). Shed or falling off at end of a growing period.

decomposer (dē′-kəm-pō′zər). A **consumer** that breaks organic matter into soluble components available to plants at the base of the food web; most are bacteria or fungi.

deduction (L. *deductus,* led apart, split, separated). Reasoning from the general to the particular from given premises to their necessary conclusion.

defensins Antimicrobial peptides produced abundantly by cells in and around the linings of mammalian intestine, urogenital tract, and respiratory tract and by neutrophils.

definitive host The host in which sexual reproduction of a symbiont occurs; if no sexual reproduction, then the host in which the symbiont becomes mature and reproduces; contrasts with **intermediate host.**

delayed type hypersensitivity Inflammatory reaction based primarily on cell-mediated immunity.

deletion Chromosomal material is excised and lost from a chromosome.

deme (dēm) (Gr. populace). A local population of closely related animals.

demography (də-mäg′rə-fē) (Gr. *demos,* people, + *graphy*). The properties of the rate of growth and the **age structure** of populations.

dendrite (den′drīt) (Gr. *dendron,* tree). Any of nerve cell processes that conduct impulses toward the cell body.

density-dependent Refers to biotic environmental factors, such as predators and parasites, whose effects on a population vary according to the number of organisms in the population.

density-independent Refers to abiotic environmental factors, such as fires, floods, and temperature changes, whose effects on a population are unaffected by the number of organisms in the population.

deoxyribonucleic acid (DNA) The genetic material of all organisms, characteristically organized into linear sequences of genes.

deoxyribose (dē-äk-sē-rī′bōs) (L. *deoxy,* loss of oxygen, + *ribose,* a pentose sugar). A five-carbon sugar similar to ribose but lacking one oxygen atom. It forms part of the fundamental structure of DNA nucleotides.

depolarization A change in the voltage in a positive direction recorded across a plasma membrane (see **membrane potential**). This allows a signal to be transmitted in excitable cells, such as nerve, muscle, and sensory cells.

derived A character state inferred to have arisen within a taxon being studied and not present in the most recent common ancestral population of the taxon.

dermal (Gr. *derma,* skin). Pertaining to the skin; cutaneous.

dermal ostia (Gr. *derma,* skin; L. *ostium,* door). Incurrent pores in a sponge.

dermis The inner, sensitive mesodermal layer of skin; corium.

desmosome (dez′mə-sōm) (Gr. *desmos,* bond, + *sōma,* body). Buttonlike plaque serving as an intercellular connection.

determinate cleavage The type of cleavage, usually spiral, in which the fate of the blastomeres is determined very early in development; mosaic cleavage.

detritus (də-trī′tus) (L. that which is rubbed or worn away). Any fine particulate debris of organic or inorganic origin.

Deuterostomia (dū′tə-rō-stō′mē-ə) (Gr. *deuteros,* second, secondary, + *stoma,* mouth). A group of phyla in which cleavage is indeterminate (regulative) and primitively radial. The endomesoderm is enterocoelous, and the mouth is derived away from the blastopore. Includes Echinodermata, Chordata, and Hemichordata; contrasts with **Protostomia.**

dextral (dex′trəl) (L. *dexter,* right-handed). Pertaining to the right; in gastropods, shell is dextral if opening is to right of columella when held with spire up and facing observer.

diapause (dī′ə-pawz) (Gr. *diapausis,* pause). A period of arrested development in the life cycle of insects and certain other animals in which physiological activity is very low and the animal is highly resistant to unfavorable external conditions.

diapsids (dī-ap′səds) (Gr. *di,* two, + *apsis,* arch). Amniotes in which the skull bears two pairs of temporal openings; includes nonavian reptiles (possibly excluding turtles) and birds as living representatives.

diastole (dī-as′tə-lē) (Gr. *diastolē,* dilation). Relaxation and expansion of the heart during which the chambers are filled with blood.

dictyosome (dik′tē-ə-sōm) (Gr. *diction,* to throw, + *sōma,* body). A part of the secretory system of endoplasmic reticulum in protozoans, also called Golgi bodies.

diffusion (L. *diffusus,* dispersion). The movement of particles or molecules from area of high concentration of the particles or molecules to area of lower concentration.

digestion Reduction of food by mechanical and chemical means to simple, soluble molecules that can be absorbed and transported to body cells.

digitigrade (dij′ə-tə-grād) (L. *digitus,* finger, toe, + *gradus,* step, degree). Walking on the digits with the posterior part of the foot raised; contrasts with **plantigrade.**

dihybrid (dī-hī′brəd) (Gr. *dis,* twice, + L. *hibrida,* mixed offspring). A hybrid whose parents differ in two distinct characters; an offspring having two different alleles at two different loci, for example, *A/a B/b.*

dimorphism (dī-mor′fizm) (Gr. *di,* two, + *morphē,* form). Existence within a species of two distinct forms according to color, sex, size, organ structure, or behavior. Occurrence of two kinds of zooids in a colonial organism.

---

bat / āpe / ärmadillo / herring / fēmale / finch / līce / crocodile / crōw / cóin / duck / ūnicorn / ə indicates unaccented vowel sound "uh" as in mammal, fishes, cardinal, heron, vulture / stress as in bi-ol′o-gy, bi′o-log′i-cal

dinoflagellate (dī-nō-fla′jə-lāt) (Gr. *dinos,* whirling, + L. *flagellum,* a whip). Any member of a protozoan clade having two flagella, one in the equatorial region of the body and other trailing; cells naked or with a test of cellulose plates.

dioecious (dī-ē′shəs) (Gr. *di,* two, + *oikos,* house). Having male and female gonads in separate individuals.

diphycercal (dif′i-ser′kəl) (Gr. *diphyēs,* twofold, + *kerkos,* tail). A tail that tapers to a point, as in lungfishes; vertebral column extends to tip without upturning.

diphyodont (di′fi-ə-dänt) (Gr. *diphyēs,* twofold, + *odous,* tooth). Having deciduous and permanent sets of teeth successively.

diploblastic (di′plə-blas′tək) (Gr. *diploos,* double, + *blastos,* bud). Organism with two germ layers, endoderm and ectoderm.

diploid (dip′loid) (Gr. *diploos,* double, + *eidos,* form). Having the somatic (double, or 2*n*) number of chromosomes or twice the number characteristic of a gamete of a given species.

diplomonad (dip′lō-mō′nad) (Gr. *diploos,* double, + *monos,* lone, single). Any member of a protozoan clade having four kinetosomes and lacking mitochondria.

direct development A life cycle sequence from zygote to adult without intervening larval stages.

directional selection A selective process favoring an extreme value of a quantitative phenotype in a population, potentially causing a change in the mean value of the phenotype of the population.

disaccharides (dī-sak′ə-rīds) (Gr. *dis,* twice, + L. *saccharum,* sugar). A class of sugars (such as lactose, maltose, and sucrose) that yield two monosaccharides on hydrolysis.

discoidal cleavage In an early embryo, cell division occurs within a small disc of cytoplasm atop a large mass of yolk.

disjunct distribution Denotes the geographic range of a species or group of closely related species being separated into two or more geographically isolated areas.

dispersal (dis-pər′-səl). Movement of organisms from their place of birth to a new geographic area for permanent residence. Founder events are special, rare cases of dispersal in which dispersing individuals cross a geographic barrier unsuited to permanent residence and start a new population beyond the barrier.

disruptive selection A selective process in which a population's mean value of a quantitative phenotype is disfavored over extreme values, potentially causing a bimodal distribution of phenotypes to evolve.

distal (dis′təl). Farther from the center of the body than a reference point.

disulfide bond (dī-səl′-fīd bänd). A covalent bond between the sulfur atoms of two cysteine amino acids. Formation of such bonds between nonadjacent cysteines in a polypeptide stabilizes the tertiary structure of a protein; disulfide bonds between cysteines

of different polypeptides contribute to the quaternary structure of a protein.

DNA See **deoxyribonucleic acid.**

DNA barcoding A technique for identifying organisms to species using sequence information of a standard gene present in all animals. The mitochondrial gene encoding cytochrome *c* oxidase I (*COI*) is often used.

DNA ligase (lī′gāse). An enzyme that joins the ends of two separate pieces of DNA.

dominance hierarchy A social ranking, formed through agonistic behavior, in which individuals are associated with each other so that some have greater access to resources than do others.

dominant An allele that is expressed regardless of the nature of the corresponding allele on the homologous chromosome.

dorsal (dor′səl) (L. *dorsum,* back). Toward the back, or upper surface, of an animal.

double circulation A system in tetrapod vertebrates for distributing blood through two separate circuits. One circuit carries deoxygenated blood from the heart through the lungs for oxygenation and back to the heart as oxygenated blood; a separate circuit carries oxygenated blood to tissues where oxygen is released and then back to the heart in deoxygenated form. Double circulation features various mechanisms for separating the alternative circuits.

double helix The fundamental structure of a DNA molecule consisting of two paired strands held together by complementary pairing of bases and forming the three-dimensional structure of an alpha helix. The paired strands are antiparallel because the 3′ end of one strand is opposite the 5′ end of the other one.

Down syndrome A congenital syndrome including mental retardation, caused by the cells in a person's body having an extra chromosome 21; also called trisomy 21.

dual-gland adhesive organ Organs in the epidermis of most turbellarians, with three cell types; viscid and releasing gland cells and anchor cells.

duodenum (dū-ə-dēn′əm) (L. *duodeni,* twelve each, fr. its length, about 12 fingers' width). The first and shortest portion of the small intestine lying between the pyloric end of the stomach and the jejunum.

duplication An extra copy of chromosomal material is produced and inserted within a chromosome.

dyad (dī′əd) (Gr. *dyas,* two). A product of the division of a tetrad during the first meiotic division.

E

eccrine (ek′rən) (Gr. *ek,* out of, + *krinein,* to separate). Applies to a type of mammalian sweat gland that produces a watery secretion.

ecdysis (ek-dī′-sis) (Gr. *ekdysis,* to strip off, escape). Shedding of outer cuticular layer; molting, as in insects or crustaceans.

ecdysone (ek-dī′sōn) (Gr. *ekdysis,* to strip off). Molting hormone of arthropods, stimulates growth and ecdysis, produced by prothoracic glands in insects and Y organs in crustaceans.

ecdysozoan protostome (ek′də-sō-zō′ən prō′tə-stōm) (Gr. *ekdysis,* to strip off, escape, + *zōon,* animal. Gr. *protos,* first, + *stoma,* mouth). Any member of a clade within Protostomia whose members shed the cuticle as they grow; includes arthropods, nematodes, and several smaller phyla.

ecocline (ek′ō-klīn) (Gr. *oikos,* home, + *klino,* to slope, recline). The gradient between adjacent biomes; a gradient of environmental conditions.

ecological pyramid A quantitative measurement of a **food web** in terms of amount of **biomass,** numbers of **organisms,** or energy at each of the different **trophic** levels present (**producers, herbivores,** first-level **carnivores,** higher-level carnivores).

ecology (Gr. *oikos,* house, + *logos,* discourse). Part of biology that concerns the relationship between organisms and their environment.

ecosystem (ek′ō-sis-təm) (eco[logy] from Gr. *oikos,* house, + system). An ecological unit consisting of both the biotic communities and the nonliving (abiotic) environment, which interact to produce a stable system.

ecotone (ek′ō-tōn) (eco[logy] from Gr. *oikos,* home, + *tonos,* stress). The transition zone between two adjacent ecological communities.

ectoderm (ek′tō-derm) (Gr. *ektos,* outside, + *derma,* skin). Outer layer of cells of an early embryo (gastrula stage); one of the germ layers, also sometimes used to include tissues derived from ectoderm.

ectognathous (ek′tə-nā′thəs) (Gr. *ektos,* outside, without, + *gnathos,* jaw). A derived character shared by most insects: mandibles and maxillae not in pouches.

ectolecithal (ek′tō-les′ə-thəl) (Gr. *ektos,* outside, + *lekithos,* yolk). Yolk for nutrition of the embryo contributed by cells that are separate from the egg cell and are combined with the zygote by envelopment within the eggshell.

ectoneural (ek′tə-nu-rəl) (Gr. *ektos,* outside, without, + *neuron,* nerve). Oral (chief) nervous system in echinoderms.

ectoparasite (ek′tō-par′ə-sīt). A **parasite** that resides on the outside surface of its host organism; contrasts with **endoparasite.**

ectoplasm (ek′tō-plazm) (Gr. *ektos,* outside, + *plasma,* form). The cortex of a cell or that part of cytoplasm just under the cell surface; contrasts with **endoplasm.**

ectothermic (ek′tō-therm′ic) (Gr. *ektos,* outside, + *thermē,* heat). Having a body temperature derived from heat acquired from the environment; contrasts with **endothermic.**

edema (ē-dē′mə) (Gr. *oidēma,* swelling). Buildup of fluid in interstitial spaces, causing swelling.

effect macroevolution Differential rates of speciation and/or extinction among evolving lineages attributed to interactions between their differing organismal-level emergent properties and the environments shared by the lineages; contrasts with **species selection.**

effector (L. *efficere,* bring to pass). An organ, tissue, or cell that becomes active in response to stimulation.

efferent (ef′ə-rənt) (L. *ex,* out, + *ferre,* to bear). Leading or conveying away from some organ, for example, action potentials conducted away from the brain, or blood conveyed away from an organ; contrasts with **afferent.**

egestion (ē-jes′chən) (L. *egestus,* to discharge). Act of casting out indigestible or waste matter from the body by any normal route.

electron A subatomic particle with a negative charge and a mass of 9.1066×10^{-28} gram.

eleocyte (el′ē-ə-sīt) (Gr. *elaion,* oil, + *kytos,* hollow vessel). Fat-containing cells in annelids that originate from the chlorogogen tissue.

elephantiasis (el-ə-fən-tī′ə-səs). Disfiguring condition caused by chronic infection with filarial worms *Wuchereria bancrofti* and *Brugia malayi.*

Eltonian pyramid An **ecological pyramid** showing numbers of **organisms** at each of the **trophic** levels.

embryogenesis (em′brē-ō-jen′ə-səs) (Gr. *embryon,* embryo, + *genesis,* origin). The origin and development of the embryo; embryogeny.

emergence (L. *e,* out, + *mergere,* to plunge). The appearance of properties in a biological system (at the molecular, cellular, organismal, or species levels) that cannot be deduced from knowledge of the component parts taken separately or in partial combinations; such properties are termed **emergent properties.**

emigrate (L. *emigrare,* to move out). To move *from* one area to another to establish residence.

emulsion (ə-məl′shən) (L. *emulsus,* milked out). A colloidal system in which both phases are liquids.

encystment (en-sist′-mənt). Process of cyst formation.

endemic (en-dem′ik) (Gr. *en,* in, + *demos,* populace). Peculiar to a certain region or country; native to a restricted area; not introduced.

endergonic (en-dər-gän′ik) (Gr. *endon,* within, + *ergon,* work). Used in reference to a chemical reaction that requires energy; energy absorbing.

endite (en′dīt) (Gr. *endon,* within). Medial process on an arthropod limb.

endochondral (en′dō-kän′drōl) (Gr. *endon,* within, + *chondros,* cartilage). Occurring with the substance of cartilage, especially bone formation.

endocrine (en′də-krən) (Gr. *endon,* within, + *krinein,* to separate). Refers to a gland that is without a duct and that releases its product directly into the blood or lymph.

endocytosis (en′dō-sī-tō-səs) (Gr. *endon,* within, + *kytos,* hollow vessel). The engulfment of matter by phagocytosis, potocytosis, receptor-mediated endocytosis, and by bulk-phase (nonspecific) endocytosis.

endoderm (en′də-dərm) (Gr. *endon,* within, + *derma,* skin). Innermost germ layer of an embryo, forming the primitive gut; also may refer to tissues derived from endoderm.

endognathous (en′də-nā-thəs) (Gr. *endon,* within, + *gnathous,* jaw). An ancestral character in insects, found in orders Diplura, Collembola, and Protura, in which the mandibles and maxillae are located in pouches.

endolecithal (en′də-les′ə-thəl) (Gr. *endon,* within, + *lekithos,* yolk). Yolk for nutrition of the embryo incorporated into the egg cell itself.

endolymph (en′də-limf) (Gr. *endon,* within, + *lympha,* water). Fluid that fills most of the membranous labyrinth of the vertebrate ear.

endometrium (en′də-mē′trē-əm) (Gr. *endon,* within, + *mētra,* womb). The mucous membrane lining the uterus.

endoparasite (en′dō-par′ə-sīt). A pararsite that resides inside the body of its host organism; contrasts with **ectoparasite.**

endoplasm (en′də-pla-zm) (Gr. *endon,* within, + *plasma,* mold or form). The portion of cytoplasm that immediately surrounds the nucleus; contrasts with **ectoplasm.**

endoplasmic reticulum A complex of membranes within a cell; may bear ribosomes (rough) or not (smooth).

endopod, endopodite (en′də-päd, en-dop′ə-dīt) (Gr. *endon,* within, + *pous, podos,* foot). Medial branch of a biramous crustacean appendage.

endopterygote (en′dəp-ter′i-gōt) (Gr. *endon,* within, + *pteron,* feather, wing). Insect in which the wing buds develop internally; has holometabolous metamorphosis.

endorphin (en-dor′fin) (contraction of endogenous morphine). Group of opiate-like brain neuropeptides that modulate pain perception and are implicated in many other functions.

endoskeleton (Gr. *endon,* within, + *skeletos,* hard). A skeleton or supporting framework within the living tissues of an organism; contrasts with **exoskeleton.**

endostyle (en′də-stīl) (Gr. *endon,* within, + *stylos,* a pillar). Mucus-secreting, ciliated groove(s) in the floor of the pharynx of tunicates, cephalochordates, and larval jawless fishes useful for accumulating and moving food particles to the stomach.

endosymbiosis (en′dō-sim-bī-ō-səs) (Gr. *endon,* within, + *syn,* with, + *bios,* life). Association between organisms of different species whereby one lives inside another.

endothelium (en-də-thē′lē-əm) (Gr. *endon,* within, + *thēlē,* nipple). Squamous epithelium lining internal body cavities such as heart and blood vessels. Adj., **endothelial.**

endothermic (en′də-therm′ic) (Gr. *endon,* within, + *thermē,* heat). Having a body temperature determined by heat derived from an animal's own oxidative metabolism; contrasts with **ectothermic.**

energy budget An economic analysis of the energy used by an organism, partitioned into **gross productivity, net productivity,** and **respiration.**

enkephalin (en-kef′ə-lin) (Gr. *endon,* within, + *kephale,* head). Group of small brain neuropeptides with opiate-like qualities.

enterocoel (en′tər-ō-sēl′) (Gr. *enteron,* gut, + *koilos,* hollow). A type of coelom formed by the outpouching of a mesodermal sac from the endoderm of the primitive gut.

enterocoelic mesoderm formation Embryonic formation of mesoderm by a pouchlike outfolding from the archenteron, which then expands and obliterates the blastocoel, thus forming a large cavity, the coelom, lined with mesoderm.

enterocoelomate (en′ter-ō-sēl′ō-māte) (Gr. *enteron,* gut, + *koilōma,* cavity, + Eng. *ate,* state of). An animal having an enterocoel, such as an echinoderm or a vertebrate.

enteron (en′tə-rän) (Gr. intestine). The digestive cavity.

entocodon (en′tə-kō′dän) (L. *entos,* within, + L. *codex,* tablet). One of three layers, with ectoderm and endoderm, of a developing medusa bud on a hydrozoan colony; the entocodon is derived from ectoderm and produces smooth and striated muscles in the medusa.

entomology (en′tə-mol′ə-jē) (Gr. *entoma,* an insect, + *logos,* discourse). Study of insects.

entozoic (en-tə-zō′ic) (Gr. *entos,* within, + *zōon,* animal). Living within another animal; internally parasitic (chiefly parasitic worms).

entropy (en′trə-pē) (Gr. *en,* in, on, + *tropos,* turn, change in manner). A quantity that is the measure of energy in a system not available for doing work.

enzyme (en′zīm) (Gr. *enzymos,* leavened, from *en,* in, + *zyme,* leaven). A substance, produced by living cells, that is capable of accelerating specific chemical transformations, such as hydrolysis, oxidation, or reduction, but is unaltered itself in the process; a biological catalyst.

eocytes (ē′ə-sīts) (Gr. *ēōs,* the dawn, + *kytos,* hollow vessel). A group of prokaryotes currently classified among the Archaebacteria but possibly a sister group of eukaryotes.

ephyra (ef′ə-rə) (Gr. *Ephyra,* Greek city). Refers to castlelike appearance. Medusa bud from a scyphozoan polyp.

epidermis (ep′ə-dər′məs) (Gr. *epi,* on, upon, + *derma,* skin). The outer, nonvascular layer of skin of ectodermal origin; in invertebrates, a single layer of ectodermal epithelium.

epididymis (ep′ə-did′ə-məs) (Gr. *epi,* on, upon, + *didymos,* testicle). Part of the sperm duct that is coiled and lying near the testis.

---

bat / āpe / ärmadillo / herring / fēmale / finch / līce / crocodile / crōw / côin / duck / ūnicorn / ə indicates unaccented vowel sound "uh" as in mamm**a**l, fish**e**s, cardinal, her**o**n, vult**u**re / stress as in bi-ol′o-gy, bi′o-log′i-cal

epigenesis (ep′ə-jen′ə-sis) (Gr. *epi*, on, upon, + *genesis*, birth). The embryological (and generally accepted) view that an embryo is a new creation that develops and differentiates step by step from an initial stage; the progressive production of new parts that were nonexistent as such in the original zygote.

epigenetics (ep′ə-je-net′iks) (Gr. *epi*, on, upon, + *genesis*, birth). Study of the relationship between genotype and phenotype as mediated by developmental processes.

epipelagic (ep′-ə-pə-laj′-ik). Refers to the upper zone of the marine pelagic realm, which receives greater illumination than lower zones.

epipod, epipodite (ep′ē-päd, e-pip′ə-dīt) (Gr. *epi*, on, upon, + *pous, podos*, foot). A lateral process on the protopod of a crustacean appendage, often modified as a gill.

epistasis (e-pis′tā-səs) (Gr. *epi*, on, upon, + *stasis*, standing). Prevention of expression of an allele at one locus by an allele at another locus.

epistome (ep′i-stōm) (Gr. *epi*, on, upon, + *stoma*, mouth). Flap over the mouth in some lophophorates bearing the protocoel.

epithelium (ep′ə-thē′lē-əm) (Gr. *epi*, on, upon, + *thēlē*, nipple). A cellular tissue covering a free surface or lining a tube or cavity.

epitoke (ep′ə-tōk) (Gr. *epitokos*, fruitful). Posterior part of a marine polychaete when swollen with developing gonads during the breeding season; contrasts with **atoke.**

epitope That portion of an antigen to which an antibody or T-cell receptor binds. Also called **antigenic determinant.**

erythroblastosis fetalis (ə-rith′rə-blas-tō′səs fə-tal′əs) (Gr. *erythros*, red, + *blastos*, germ, + *osis*, a disease; L. *fetalis*, relating to a fetus). A disease of newborn infants caused when Rh-negative mothers develop antibodies against the Rh-positive blood of the fetus. See **blood type.**

erythrocyte (ə-rith′rə-sīt) (Gr. *erythros*, red, + *kytos*, hollow vessel). Red blood cell; has hemoglobin to carry oxygen from lungs or gills to tissues; during formation in mammals, erythrocytes lose their nuclei; those of other vertebrates retain the nuclei.

esthete (es-thēt′) (Gr. *esthēs*, a garment). Light sensory receptor on a shell of a chiton (phylum Mollusca).

estivation (es′tə-vā′shen) (L. *aestivates*, to spend the summer). A state of dormancy during the summer when temperatures are high, food is scarce and/or dehydration threatens. Metabolism and breathing rate decrease.

estrus (es′trəs) (L. *oestrus*, gadfly, frenzy). The period of heat of the female associated with ovulation of the egg. Time of maximum sexual receptivity.

estuary (es′chə-we′rē) (L. *aestuarium*, estuary). An arm of the sea where the tide meets the current of a freshwater drainage.

ethology (e-thäl′-ə-jē) (Gr. *ethos*, character, + *logos*, discourse). The study of animal behavior in natural environments.

euchromatin (ū′krō-mə-tən) (Gr. *eu*, good, well, + *chrōma*, color). Part of the chromatin that absorbs less stain than heterochromatin; contains active genes.

eukaryote, eukaryotic, eucaryotic (ū′ka-rē-ot′ik) (Gr. *eu*, good, true, + *karyon*, nut, kernel). Organisms whose cells characteristically contain a membrane-bound nucleus or nuclei; contrasts with **prokaryotic.**

eumetazoan (ū-met-ə-zō′ən) (Gr. *eu*, good, true, + *meta*, after, + *zōon*, animal). Any multicellular animal with distinct germ layers that form true tissues; animals beyond the cellular grade of organization.

euploidy (ū′ploid′ē) (Gr. *eu*, good, well, + *ploid*, multiple of). Presence of one or more complete sets of chromosomes and no partial sets in a cell nucleus; includes **haploidy, diploidy,** and **polyploidy.**

euryhaline (ū′-rə-hā′lin) (Gr. *eurys*, broad, + *hals*, salt). Able to tolerate wide ranges of saltwater concentrations.

euryphagous (yə-rif′ə-gəs) (Gr. *eurys*, broad, + *phagein*, to eat). Eating a large variety of foods.

eurytopic (ū-rə-täp′ik) (Gr. *eurys*, broad, + *topos*, place). Refers to an organism with a wide environmental range.

eusociality (ū′sō-shē-al′-ət-ē). A reproductive division of labor among members of a population or species. Generations overlap, and nonreproductive individuals help to raise younger relatives that are not the helpers' offspring. Ants, many bees, and some wasps are eusocial.

eutely (u′te-lē) (Gr. *euteia*, thrift). Condition of a body composed of a constant number of cells or nuclei in all adult members of a species, as in rotifers, acanthocephalans, and nematodes.

evagination (ē-vaj′ə-nā′shen) (L. *e*, out, + *vagina*, sheath). An outpocketing from a hollow structure.

evolution (L. *evolvere*, to unfold). Organic evolution encompasses all changes in the characteristics and diversity of life on earth throughout its history.

evolutionary duration The length of time that a species or higher taxon exists in geological time.

evolutionary sciences Empirical investigation of ultimate causes in biology using the comparative method.

evolutionary species concept A single lineage of ancestral-descendant populations that maintains its identity from other such lineages and has its own evolutionary tendencies and historical fate; differs from the biological species concept by explicitly including a time dimension and including asexual lineages.

evolutionary stable strategy (ESS) Application of the mathematical theory of games to evaluate whether a system of social behaviors is robust to the evolution of "cheating" behaviors that would threaten its stability; denotes a social system that should persist over long periods of evolutionary time because it prevails in competition with others that might arise.

evolutionary taxonomy A system of classification, formalized by George Gaylord Simpson, that groups species into Linnaean higher taxa representing a hierarchy of distinct adaptive zones; such taxa may be monophyletic or paraphyletic but not polyphyletic.

evolvability Refers to the opportunities for morphological evolution conferred on an evolving lineage that possesses a "toolkit" of semiautonomous developmental modules that can be expressed at various stages of ontogeny and at multiple physical locations in the body. For example, evolution of tetrapod limbs featured ectopic expression of developmental genetic modules normally expressed in the vertebral column.

exaptation (ek-sap′-tā′shən). Evolutionary cooption of an organismal or molecular character for a biological role unrelated to the character's evolutionary origin. Bird feathers are considered an exaptation for flight because they originated prior to avian flight but provided utility for flight following its origin; contrasts with **adaptation;** bird feathers are considered an adaptation for the biological role of thermoregulation.

excision repair Means by which cells repair certain kinds of damage (dimerized pyrimidines) in their DNA.

exergonic (ek′sər-gon′ik) (Gr. *exō*, outside of, + *ergon*, work). An energy-yielding reaction.

exite (ex′īt) (Gr. *exō*, outside). Process from lateral side of an arthropod limb.

exocrine (ek′sə-krən) (Gr. *exō*, outside, + *krinein*, to separate). A type of gland that releases its secretion through a duct; contrasts with **endocrine.**

exocytosis (eks′ə-sī-tō′səs) (Gr. *exo*, outside, + *kytos*, hollow vessel). Transport of a substance from inside a cell to the outside.

exon (ex′ən) (Gr. *exō*, outside). Part of the mRNA as transcribed from the DNA that contains a portion of the information necessary for final gene product.

exopod, exopodite (ex′ə-päd, ex-äp′ə-dīt) (Gr. *exō*, outside, + *pous, podos*, foot). Lateral branch of a biramous crustacean appendage.

exopterygote (ek′səp-ter′i-gōt) (Gr. *exō*, without, + *pteron*, feather, wing). Insect in which the wing buds develop externally during nymphal instars; has hemimetabolous metamorphosis.

exoskeleton (ek′sō-skel′ə-tən) (Gr. *exō*, outside, + *skeletos*, hard). A supporting structure secreted by ectoderm or epidermis; external, not enveloped by living tissue; contrasts with **endoskeleton.**

experiment (L. *experiri*, to try). A trial made to support or to disprove a hypothesis.

experimental method A general procedure for testing hypotheses by predicting how a biological system will respond to a disturbance, making the disturbance under controlled conditions, and then comparing the observed results with the predicted ones.

experimental sciences Empirical investigation of proximate causes in biology using the **experimental method.**

exponential growth Refers to an increase in the numbers of individuals in a population by at least a factor of 2 each generation.

exteroceptor (ek′stər-ō-sep′tər) (L. *exter*, outward, + *capere*, to take). A sense organ excited by stimuli from the external world.

extrinsic factor An environmental variable that influences the biological properties of a population, such as observed number of individuals or rate of growth.

extrusome (eks′trə-sōm) (L. *extrusus*, driven out, + *soma*, body). Any membrane-bound organelle used to extrude something from a protozoan cell.

eyelid (ī′-lid). A thin surface of skin and muscle that can be closed to protect an eye from light, abrasion, and/or desiccation. Occurs in many but not all terrestrial vertebrates.

F

facilitated diffusion Mediated transport in which a transmembrane protein makes possible diffusion of a molecule across a plasma membrane in the direction of a concentration gradient; contrasts with **active transport.**

FAD Abbreviation for flavin adenine dinucleotide, an electron acceptor in the respiratory chain.

fascicle (fas′ə-kəl) (L. *fasciculus*, small bundle). A small bundle, usually referring to a collection of muscle fibers or nerve axons.

fatty acid Any of a series of saturated organic acids having the general formula $C_nH_{2n}O_2$; occurs in natural fats of animals and plants.

female-defense polygyny A male obtains more than one mate because multiple females aggregate and can be guarded as a unit from other males.

fermentation (L. *fermentum*, ferment). Enzymatic transformation, without oxygen, or organic substrates, especially carbohydrates, yielding products such as alcohols, acids, and carbon dioxide.

fiber (L. *fibra*, thread). A thin cell or strand of protoplasmic material produced or secreted by a cell and lying outside the cell.

fibril (L. *fibra*, thread). A strand of protoplasm produced by a cell and lying within the cell.

fibrillar (fī′brə-lər) (L. *fibrilla*, small fiber). Composed of or pertaining to fibrils or fibers.

fibrin Protein that forms a meshwork, trapping erythrocytes, to become blood clot. Precursor is fibrinogen.

fibrosis (fī-brō′səs). Deposition of fibrous connective tissue in a localized site, during process of tissue repair or to isolate a source of antigen.

filipodium (fi′li-pō′de-əm) (L. *filum*, thread, + Gr. *pous, podos*, a foot). A type of pseudopodium that is very slender and may branch but does not rejoin to form a mesh.

filter-feeding Any feeding process by which particulate food is filtered from water in which it is suspended.

first law of thermodynamics Energy is neither created nor destroyed but may be converted from one form to another.

fission (L. *fissio*, a splitting). Asexual reproduction by a division of the body into two or more parts.

fitness Degree of suitability for a particular environment. Genetic fitness is relative contribution by organisms of a particular genotype to the next generation; organisms with high genetic fitness are naturally selected and their genetic characteristics become prevalent in a population.

5′ end (five-prime end) The terminus of a nucleic acid molecule that consists of phosphate attached to the 5′ carbon of the terminal sugar (the opposite terminus is the 3′ end).

flagellum (flə-jel′em), pl. **flagella** (L. a whip). Whiplike organelle of locomotion.

flame cell Specialized hollow excretory or osmoregulatory structure of one or several small cells containing a tuft of flagella (the "flame") and situated at the end of a minute tubule; connected tubules ultimately open to the outside. See **solenocyte, protonephridium.**

fluke (O.E. *flōc*, flatfish). A member of class Trematoda or class Monogenea. Also, certain of the flatfishes (order Pleuronectiformes).

FMN Abbreviation for flavin mononucleotide, the prosthetic group of a protein (flavoprotein) and a carrier in the electron transport chain in respiration.

food chain Movement of energy from plant compounds to organisms that eat plants, then to other organisms that eat the plant feeders, and possibly further through a linear series of organisms that feed and are then eaten by others. Food chains connect and branch to form food webs.

food vacuole A digestive organelle in the cell.

food web An analysis relating species in an ecological community according to how they acquire nutrition, such as fixing atmospheric carbon **(producers)**, consuming producers **(herbivores)**, consuming herbivores (first-level **carnivores),** or consuming carnivores (higher-level carnivores).

foraminiferan (for′əm-i-nif′-ər-ən) (L. *foramin*, hole, performation, + *fero*, to bear). Granuloreticulosean amebas bearing a test with many openings.

fossil (fos′əl). Any remains or impression of an organism from a past geological age that has been preserved by natural processes, usually by mineralization in the earth's crust.

fossorial (fä-sōr′ē-əl) (L. *fossor*, digger). Characterized by digging or burrowing.

fouling Contamination of feeding or respiratory areas of an organism by excrement, sediment, or other matter. Also, accumulation of sessile marine organisms on the hull of a boat or ship so as to impede its progress through the water.

founder event Establishment of a new population by a small number of individuals (sometimes a single female carrying fertile eggs) that disperse from their parental population to a new location geographically isolated from the parental population.

fovea (fō′vē-ə) (L. small pit). A small pit or depression; especially the fovea centralis, a small pit containing only cones in the retina of some vertebrates, a point of acute vision.

free energy The energy available for doing work in a chemical system.

free-living Any organism not intimately associated with a host organism.

fringing reef A type of coral reef close to a landmass having a small lagoon or no lagoon between it and the shore.

frog (fräg′). Denotes any member of the amphibian order Anura (also called Salientia).

frontal plane A plane parallel to the main axis of the body and at right angles to the sagittal plane.

fundamental niche A variety of roles potentially performed by an organism or population in an ecological community; limits on such roles are set by the intrinsic biological attributes of an organism or population. See also **niche** and **realized niche.**

fusiform (fū′zə-form) (L. *fusus*, spindle, + *forma*, shape). Spindle-shaped; tapering toward each end.

G

gamete (ga′mēt, gə-mēt′) (Gr. *gamos*, marriage). A mature haploid sex cell; usually, male and female gametes can be distinguished. An egg or a sperm.

gametic meiosis Meiosis that occurs during formation of the gametes, as in humans and other metazoa.

gametocyte (gə-mēt′ə-sīt) (Gr. *gametēs*, spouse, + *kytos*, hollow vessel). The mother cell of a gamete; an immature gamete.

ganglion (gang′lē-ən), pl. **ganglia** (Gr. little tumor). An aggregation of nerve tissue containing nerve-cell bodies.

ganoid scales (ga′noid) (Gr. *ganos*, brightness). Thick, bony, rhombic scales of some bony fishes; not overlapping.

gap genes Genes expressed in broad regions along the anterior-posterior axis of a developing embryo (e.g., produce head, thorax, and abdomen in *Drosophila*); mutations result in gaps in formation of segments.

bat / āpe / ärmadillo / herring / fēmale / finch / līce / crocodile / crōw / cöin / duck / ūnicorn / ə indicates unaccented vowel sound "uh" as in mammal, fishes, cardinal, heron, vulture / stress as in bi-ol′o-gy, bi′o-log′i-cal

gap junction Pores formed by a ring of transmembrane proteins communicating the cytoplasm between two cells.

gastrocoel (gas'tro-cēl) (Gr. *gastēr,* stomach, + *koilos,* hollow). An embryonic cavity forming in gastrulation that becomes the adult gut; also called an **archenteron.**

gastrodermis (gas'tro-dər'mis) (Gr. *gastēr,* stomach, + *derma,* skin). Lining of the digestive cavity of cnidarians.

gastrolith (gas'trə-lith) (Gr. *gastēr,* stomach, + *lithos,* stone). Calcareous body in the wall of the cardiac stomach of crayfish and other Malacostraca, preceding the molt.

gastrovascular cavity (Gr. *gastēr,* stomach, + L. *vasculum,* small vessel). Body cavity in certain lower invertebrates that functions in both digestion and circulation and has a single opening serving as both mouth and anus.

gastrozooid (gas'trə-zō-id) (Gr. *gastēr,* stomach, + *zōon,* animal). The feeding polyp of a hydroid, a hydranth.

gastrula (gas'trə-lə) (Gr. *gastēr,* stomach, + L. *ula,* dim.). Embryonic stage, usually cap- or sac-shaped, with walls of two layers of cells surrounding a cavity (archenteron) with one opening (blastopore).

gastrulation (gas'trə-lā'shən) (Gr. *gastēr,* stomach). Process by which an early metazoan embryo becomes a gastrula, acquiring first two and then three layers of cells.

gel (jel) (from gelatin, from L. *gelare,* to freeze). That state of a colloidal system in which the solid particles form the continuous phase and the fluid medium forms the discontinuous phase.

gemmule (je'mūl) (L. *gemma,* bud, + *ula,* dim.). Asexual, cystlike reproductive unit in freshwater sponges; formed in summer or autumn and capable of overwintering.

gene (Gr. *genos,* descent). A nucleic acid sequence (usually DNA) that encodes a functional polypeptide or RNA sequence.

gene pool A collection of all of the alleles of all of the genes in a population.

general area cladogram A cladogram depicting the sequence of historical fragmentation among various geographic areas of endemism; represents the shared branching patterns of the individual cladograms of many taxa that share common areas of endemism.

genetic code The correspondence between the sequence of bases in a DNA or messenger RNA molecule and the sequence of amino acids in an encoded protein.

genetic drift Random change in allelic frequencies in a population. In small populations, genetic variation at a locus may be lost by chance fixation of a single allelic variant.

genome (je'nōm) (Gr. *genos,* offspring, + *ōma,* abstract group). All the DNA in a haploid set of chromosomes (nuclear genome), organelle (mitochondrial genome, chloroplast genome) or virus (viral genome, which in some viruses consists of RNA rather than DNA).

genomics (jē-nō'miks). Mapping and sequencing of genomes (= structural genomics). Functional genomics is the development and application of experimental approaches to assess gene function. Functional genomics uses information derived from structural genomics.

genotype (jēn'ō-tīp) (Gr. *genos,* offspring, + *typos,* form). The genetic constitution, expressed and latent, of an organism; the total set of genes present in the cells of an organism; contrasts with **phenotype.**

genus (jē-nus), pl. **genera** (L. race). A group of related species with taxonomic rank between family and species.

geographic range The specific geographic area occupied by members of a population, species or higher taxon.

germinal vesicle Mature nucleus of a primary oocyte, enlarged and filled with RNA.

germinative zone The site immediately following the scolex on the body of a mature tapeworm where new proglottids are produced.

germ layer In an animal embryo, one of three basic layers (ectoderm, endoderm, mesoderm) from which the various organs and tissues arise in the multicellular animal.

germ plasm Cell lineages giving rise to the germ cells of a multicellular organism, as distinct from the somatoplasm.

germovitellarium (jer'mō-vit-ə-lar'ē-əm) (L. *germen,* a bud, offshoot, + *vitellus,* yolk). Closely associated ovary (germarium) and yolk-producing structure (vitellarium) in rotifers.

gestation (jes-tā'shən) (L. *gestare,* to bear). The period in which offspring are carried in the uterus.

globulins (glo'bū-lənz) (L. *globus,* a globe, ball, + *-ulus,* ending denoting tendency). A large group of compact proteins with high molecular weight; includes immunoglobulins (antibodies).

glochidium (glō-kid'ē-əm) (Gr. *glochis,* point, + *idion,* dim.). Bivalved larval stage of freshwater mussels.

glomerulus (glä-mer'ū-ləs) (L. *glomus,* ball). A tuft of capillaries projecting into a renal corpuscle in a kidney. Also, a small spongy mass of tissue in the proboscis of hemichordates, presumed to have an excretory function. Also, a concentration of nerve fibers situated in the olfactory bulb.

gluconeogenesis (glū-cō-nē-ō-gən'ə-səs) (Gr. *glykys,* sweet, + *neos,* new, + *genesis,* origin). Synthesis of glucose from protein or lipid precursors.

glucose A 6-carbon sugar particularly important for cellular metabolism in living organisms (= dextrose).

glycogen (glī'kə-jən) (Gr. *glykys,* sweet, + *genēs,* produced). A polysaccharide constituting the principal form in which carbohydrate is stored in animals; animal starch.

glycolysis (glī-kol'ə-səs) (Gr. *glykys,* sweet, + *lysis,* a loosening). Enzymatic breakdown of glucose (especially) or glycogen into phosphate derivatives with release of energy.

gnathobase (nath'ō-bās') (Gr. *gnathos,* jaw, + base). A median basic process on certain appendages in some arthropods, usually for biting or crushing food.

gnathostomes (nath'ō-stōms) (Gr. *gnathos,* jaw, + *stoma,* mouth). Vertebrates with jaws.

Golgi complex (gōl'jē) (after Golgi, Italian histologist). An organelle in cells that serves as a collecting and packaging center for secreted polypeptides and proteins.

gonad (gō'nad) (N.L. *gonas,* primary sex organ). An organ that produces gametes (ovary in the female and testis in the male).

gonangium (gō-nan'jē-əm) (N.L. *gonas,* primary sex organ, + *angeion,* dim. of vessel). Reproductive zooid of hydroid colony (Cnidaria).

gonoduct (Gr. *gonos,* seed, progeny, + duct). Duct leading from a gonad to the exterior.

gonophore (gän'ə-for) (Gr. *gonos,* seed, progeny, + *phoros,* bearer). Sexual reproductive structure developing from reduced medusae in some hydrozoans; it may be retained on the colony or released.

gonopore (gän'ə-por) (Gr. *gonos,* seed, progeny, + *poros,* an opening). A genital pore found in many invertebrates.

grade (L. *gradus,* step). A level of organismal complexity or adaptive zone characteristic of a group of evolutionarily related organisms.

gradualism (graj'ə-wal-iz'əm). A component of Darwin's evolutionary theory postulating that evolution occurs by the temporal accumulation of small, incremental changes in populations, usually across very long periods of geological time; it opposes claims that evolution can occur by large, discontinuous or macromutational changes.

granular glands Integumentary structures of modern amphibians associated with secretion of defensive compounds.

granulocytes (gran'ū-lə-sīts) (L. *granulus,* small grain, + Gr. *kytos,* hollow vessel). White blood cells (neutrophils, eosinophils, and basophils) bearing in their cytoplasm "granules" (vacuoles) that stain deeply.

granuloreticulosan (gran'yə-lō-ru-tik-yə-lō'səm) (L. *granulus,* small grain, + *reticulum,* a net). Any member of a protozoan clade having branched and netlike pseudopodia; includes foraminiferans.

green gland Excretory gland of certain Crustacea; the antennal gland.

gregarine (gre-ga-rin') (L. *gregarious,* belonging to a herd or flock). Protozoan parasites belonging to class Gregarinea within phylum Apicomplexa; these organisms infect guts or body cavities of invertebrates.

gregarious (L. *grex,* herd). Living in groups or flocks.

gross productivity A measurement of the total energy assimilated by an organism.

ground substance The matrix in which connective tissue fibers are embedded.

group selection A hypothesis that selection sometimes acts on an association of individuals in a population rather than directly on the individuals themselves; proposed to explain evolution of individual behaviors that do not benefit an individual directly but may benefit an association that includes the individual performing the behavior. Critical analyses have largely discredited group-selectional hypotheses in favor of alternatives, such as kin selection and reciprocal altruism.

growth rate The proportion by which a population changes in numbers of individuals at a given time by reproduction and possibly immigration.

guanine (gwä′nēn) (Sp. from Quechua, *huanu*, dung). A white crystalline purine base, $C_5H_5N_5O$, occurring in various animal tissues and in guano and other animal excrements.

guild (gild) (M.E. *gilde*, payment, tribute). Species of a local community that partition resources through character displacement to avoid niche overlap and competition, such as Galápagos finch communities whose component species differ in beak size for specializing on different-size seeds.

gynandromorph (ji-nan′drə-mawrf) (Gr. *gyn*, female, + *andr*, male, + *morphē*, form). An abnormal individual exhibiting characteristics of both sexes in different parts of the body; for example the left side of a bilateral organism may show characteristics of one sex and the right side those of the other sex.

gynocophoric canal (gī′nə-kə-fōr′ik) (Gr. *gynē*, woman, + *pherein*, to carry). Groove in male schistosomes (certain trematodes) that carries the female.

H

habitat (L. *habitare*, to dwell). The place where an organism normally lives or where individuals of a population live.

habitat fragmentation Emergence of geographic barriers that disrupt a formerly continuous geographic distribution of populations of a species. Evolutionary rates of species formation and species extinction are increased by this occurrence.

habituation A kind of learning in which continued exposure to the same stimulus produces diminishing responses; contrasts with **sensitization.**

hair cell An important sensory component of several kinds of mechano- and auditory receptors found in both invertebrate (statocyst) and vertebrate (vestibular organ, organ of Corti) organs of equilibrium and hearing. The "hairs" are cilia, or sensory endings that project from the cell surface, and when bent by mechanical stimuli they generate nerve impulses or action potentials, communicating a signal to the central nervous system.

halter (hal′tər), pl. **halteres** (hal-ti′rēz) (Gr. leap). In Diptera, small club-shaped structure on each side of the metathorax representing the hindwings; thought to be sense organs for balancing; also called balancer.

haplodiploidy (Gr. *haploos*, single, + *diploos*, double, + *eidos*, form). Reproduction in which haploid males are produced parthenogenetically, and diploid females are from fertilized eggs.

haploid (Gr. *haploos*, single). The reduced, or *n*, number of chromosomes, typical of gametes, as opposed to the diploid, or *2n*, number found in somatic cells. In certain groups, mature organisms may have a haploid number of chromosomes.

Hardy-Weinberg equilibrium Mathematical demonstration that the Mendelian hereditary process does not change the populational frequencies of alleles or genotypes across generations, and that change in allelic or genotypic frequencies requires factors such as natural selection, genetic drift in finite populations, recurring mutation, migration of individuals among populations, and nonrandom mating.

hectocotylus (hek-tə-kät′ə-ləs) (Gr. *hekaton*, hundred, + *kotylē*, cup). Specialized, and sometimes autonomous, arm that serves as a male copulatory organ in cephalopods.

heliozoan (hēl′ē-ō-zō′ən) (Gr. *hēlios*, sun, + *zōon*, animal). Descriptive term for a freshwater ameba, naked or testate.

hemal system (hē′məl) (Gr. *haima*, blood). System of small vessels in echinoderms; function is probably distribution of nutrients to specific body regions.

hemerythrin (hē′mə-rith′rin) (Gr. *haima*, blood, + *erythros*, red). A red, iron-containing respiratory pigment found in the blood of some polychaetes, sipunculids, priapulids, and brachiopods.

hemidesmosome (he-mē-dez′mə-sōm) (Gk. *hēmi*, half, + *desmos*, bond, + *sōma*, body). Buttonlike plaque composed of transmembrane proteins at the base of cells that anchor them to underlying connective tissue layers.

hemimetabolous (he′mi-mə-ta′bə-ləs) (Gr. *hēmi*, half, + *metabolē*, change). Refers to gradual metamorphosis during development of insects, without a pupal stage.

hemizygous (he′mē-zī′gəs) (Gr. *hēmi*, half, + *zygōtos*, yoked). For animals having chromosomal sex determination in which one sex (termed the heterogametic sex) has only one copy of a particular sex chromosome, genotypes of heterogametic individuals consist of a single copy of all genes located on that sex chromosome.

hemocoel (hēm′ə-sēl) (Gr. *haima*, blood, + *koiloma*, cavity). Major body space in arthropods formed by fusion of embryonic coelom with blastocoel; contains blood (hemolymph).

hemoglobin (Gr. *haima*, blood, + L. *globulus*, globule). An iron-containing respiratory pigment occurring in vertebrate red blood cells and in blood plasma of many invertebrates; a compound of an iron porphyrin heme and globin proteins.

hemolymph (hē′mə-limf) (Gr. *haima*, blood, + L. *lympha*, water). Fluid in the coelom or hemocoel of some invertebrates that represents the blood and lymph of vertebrates.

hemozoin (hē-mə-zo′ən) (Gr. *haima*, blood, + *zōon*, an animal). Insoluble digestion product of malarial parasites produced from hemoglobin.

hepatic (hə-pat′ik) (Gr. *hēpatikos*, of the liver). Pertaining to the liver.

herbivore ([h]ərb′ə-vōr′) (L. *herba*, green crop, + *vorare*, to devour). Any organism subsisting on plants. Adj., **herbivorous.**

herbivory [h]ər-biv′-ə-rē′). Denotes the condition of animals feeding on plants or the destruction of plant biomass by such feeding.

heredity (L. *heres*, heir). The faithful transmission of biological traits from parents to their offspring.

hermaphrodite (hə[r]-maf′rə-dīt) (Gr. *hermaphroditos*, containing both sexes; from Greek mythology, Hermaphroditos, son of Hermes and Aphrodite). An organism with both male and female functional reproductive organs. **Hermaphroditism** may refer to an aberration in unisexual animals; **monoecy** implies that this is the normal condition for the species.

hermatypic (hər-mə-ti′pik) (Gr. *herma*, reef, + *typos*, pattern). Relating to reef-forming corals.

heterocercal (het′ər-o-sər′kəl) (Gr. *heteros*, different, + *kerkos*, tail). In some fishes, a tail with the upper lobe larger than the lower, and the end of the vertebral column somewhat upturned in the upper lobe, as in sharks.

heterochromatin (het′ə-rō-krōm′ə-tən) (Gr. *heteros*, different, + *chrōma*, color). Chromatin that stains intensely and appears to represent genetically inactive areas.

heterochrony (het′ə-rō-krōn-y) (Gr. *heteros*, different, + *chronos*, time). Evolutionary change in the relative time of appearance or rate of development of characteristics from ancestor to descendant.

heterodont (het′ə-ro-dänt) (Gr. *heteros*, different, + *odous*, tooth). Having teeth differentiated into incisors, canines, and molars for different purposes.

heterokont (het′ə-rō-känt) (Gr. *heteros*, other, different, + *kont*, pole). Refers to flagellated cells with two different anterior flagella, one long, hairy, and forward directed, and the other short, smooth, and trailing posteriorly.

heterolobosea (het′ə-rō-lō-bō′sē-ə) (Gr. *heteros*, other, different, + *lobos*, lobe). A protozoan clade in which most members can assume both ameboid and flagellate forms.

heterostracans (Gr. *heteros*, different, + *ostrakon*, shell). A group of extinct fishes with dermal armor and no jaws or paired fins; known from the Ordovician to Devonian periods.

heterotopy (het′ə-rō-tō′-pē). An evolutionary change in the physical location of a structure or developmental process in the organismal body plan.

heterotroph (het′ə-rō-trōf) (Gr. *heteros*, different, + *trophos*, feeder). An organism that obtains both organic and inorganic raw materials from the environment in order to live; includes most animals and those plants that do not have photosynthesis.

heterozygote (het′ə-rō-zī′gōt) (Gr. *heteros*, different, + *zygotos*, yoked). An organism in which homologous chromosomes contain different allelic forms (often dominant and recessive) of a locus; derived from a zygote formed by union of gametes of dissimilar allelic constitution.

hexamerous (hek-sam′ər-əs) (Gr. *hex*, six, + *meros*, part). Six parts, specifically, symmetry based on six or multiples thereof.

hibernation (L. *hibernus*, wintry). Condition, especially of mammals, of passing the winter in a torpid state in which the body temperature drops nearly to freezing and the metabolism drops close to zero.

hierarchical system A scheme arranging organisms into a series of taxa of increasing inclusiveness, as illustrated by Linnaean classification.

high heat of vaporization Denotes the large amount of energy needed to convert liquid water at 100° C to gaseous phase (more than 500 kilocalories per gram).

high specific heat capacity Denotes the large amount of energy needed to raise the temperature of liquid water (1° C per gram).

high surface tension Denotes the large pressure needed to break the surface of liquid water relative to all other liquids except mercury.

histogenesis (his-tō-jen′ə-sis) (Gr. *histos*, tissue, + *genesis*, descent). Formation and development of tissue.

histology (hi-stäl′-ə-jē) (Gr. *histos*, web, tissue, + *logos*, discourse). The study of the microscopic anatomy of tissues.

histone (hi′stōn) (Gr. *histos*, tissue). Any of several simple proteins found in cell nuclei and complexed with DNA. Histones yield a high proportion of basic amino acids on hydrolysis; characteristic of eukaryotes.

holoblastic cleavage (Gr. *holo*, whole, + *blastos*, germ). Complete and approximately equal division of cells in early embryo; found in mammals, amphioxus, and many aquatic invertebrates that have eggs with a small amount of yolk.

holometabolous (hō′lō-mə-ta′bə-ləs) (Gr. *holo*, complete, + *metabolē*, change). Complete metamorphosis during development.

holophytic nutrition (hōl′ō-fit′ik) (Gr. *holo*, whole, + *phyt*, plant). Occurs in green plants and certain protozoa and involves synthesis of carbohydrates from carbon dioxide and water in the presence of light, chlorophyll, and certain enzymes.

holozoic nutrition (hōl′ō-zō′ik) (Gr. *holo*, whole, + *zoikos*, of animals). Type of nutrition involving ingestion of liquid or solid organic food particles.

home range The area over which an animal moves in its activities. Unlike territories, home ranges are not defended.

homeobox (hō′mē-ō-box) (Gr. *homoios*, like, resembling, + L. *buxus*, boxtree [used in the sense of enclosed, contained]). A highly conserved 180-base-pair sequence found in homeotic genes, regulatory sequences of protein-coding genes that regulate development.

homeostasis (hō′mē-ō-stā′sis) (Gr. *homeo*, alike, + *stasis*, state or standing). Maintenance of an internal steady state by means of self-regulation.

homeothermic (hō′mē-ō-thər′mik) (Gr. *homeo*, alike, + *thermē*, heat). Having a nearly uniform body temperature, regulated independent of the environmental temperature.

homeotic genes (hō-mē-ät′ik) (Gr. *homoios*, like, resembling). Genes, identified through mutations, that give developmental identity to specific body segments.

hominid (häm′ə-noid) (L. *homo, hominis,* man). A member of the family Hominidae, which includes chimpanzees, gorillas, humans, orangutans, and extinct forms descended from their most recent common ancestor.

hominoid (häm′ə-noid). Relating to the Hominoidea, a superfamily of primates to which apes and humans are assigned.

homocercal (hō′mə-ser′kəl) (Gr. *homos*, same, common, + *kerkos*, tail). A tail with the upper and lower lobes symmetrical and the vertebral column ending near the middle of the base, as in most telost fishes.

homodont (hō′mō-dänt) (Gr. *homos*, same, + *odous*, tooth). Having all teeth similar in form.

homograft See **allograft.**

homolog, homologue (hōm′ə-log) One member of a set of homologous structures or one of a pair of homologous chromosomes.

homology (hō-mäl′ə-jē) (Gr. *homologos*, agreeing). Similarity of parts or organs of different organisms caused by evolutionary derivation from a corresponding part or organ in a remote ancestor, and usually having a similar embryonic origin. May also refer to a matching pair of chromosomes. Serial homology is the correspondence in the same individual of repeated structures having the same origin and development, such as the appendages of arthropods. Adj., **homologous.**

homoplasy (hō′mō′plā′sē). Phenotypic similarity among characteristics of different species or populations (including molecular, morphological, behavioral, or other features) that does not accurately represent patterns of common evolutionary descent (= non-homologous similarity); it is produced by evolutionary parallelism, convergence and/or reversal, and is revealed by incongruence

among different characters on a cladogram or phylogenetic tree.

homozygote (hō-mō-zī′gōt) (Gr. *homos*, same, + *zygotos*, yoked). An organism having identical alleles at one or more genetic loci. Adj., **homozygous.**

humoral (hū′mər-əl) (L. *humor*, a fluid). Pertaining to an endocrine secretion.

humoral immune response Immune response involving production of antibodies, specifically the T_H2 arm of the immune response. Contrasts with **cell-mediated immune response.**

hyaline (hī′ə-lən) (Gr. *hyalos*, glass). Adj., glassy, translucent. Noun, a clear, glassy, structureless material occurring, for example, in cartilage, vitreous body, mucin, and glycogen.

hybrid Refers to offspring of crosses between genetically distinct populations, such as those recognized as different varieties or species.

hybridization Natural or artificial genetic crossing between genetically distinct populations, such as ones considered different varieties or species.

hybridoma (hī-brid-ō′mah) (contraction of hybrid + myeloma). Fused product of a normal and a myeloma (cancer) cell, which has some of the characteristics of the normal cell.

hydatid cyst (hī-da′təd) (Gr. *hydatis*, watery vesicle). A type of cyst formed by juveniles of certain tapeworms (*Echinococcus*) in their vertebrate hosts.

hydranth (hī′dranth) (Gr. *hydōr*, water, + *anthos*, flower). Nutritive zooid of hydroid colony.

hydrocaulus (hī′drə-kä′ləs) (Gr. *hydōr*, water, + *kaulos*, stem of a plant). Stalks or "stems" of a hydroid colony, the parts between the hydrorhiza and the hydranths.

hydrocoel (hī′-drə-sēl) (Gr. *hydōr*, water, + *koilos*, hollow). Second or middle coelomic compartment in echinoderms; left hydrocoel gives rise to water-vascular system.

hydrocorals Members of phylum Cnidaria, class Hydrozoa, with massive calcareous skeletons.

hydrogen bond A relatively weak chemical bond resulting from unequal charge distribution within molecules, in which a hydrogen atom covalently bonded to another atom is attracted to the electronegative portion of another molecule.

hydrogenosome (hī-drō-gen′ə-sōm) (Gr. *hydōr*, water, + *genos*, kind, + *sōma*, body). An anaerobic cellular organelle presumed to be derived from a mitochondrion.

hydroid The polyp form of a cnidarian as distinguished from the medusa form. Any cnidarian of the class Hydrozoa, order Hydroida.

hydrolysis (Gr. *hydōr*, water, + *lysis*, a loosening). The decomposition of a chemical compound by the addition of water; the splitting of a molecule into its groupings so that the split products acquire hydrogen and hydroxyl groups.

hydrorhiza (hī′drə-rī′zə) (Gr. *hydōr*, water, + *rhiza*, a root). Rootlike stolon that attaches a hydroid to its substrate.

hydrosphere (Gr. *hydōr*, water, + *sphaira*, ball, sphere). Aqueous envelope of the earth.

hydrostatic pressure The pressure exerted by a fluid (gas or liquid), defined as force per unit area. For example, the hydrostatic pressure of one atmosphere (1 atm) is 14.7 lb/in².

hydrostatic skeleton A mass of fluid or plastic parenchyma enclosed within a muscular wall to provide the support necessary for antagonistic muscle action; for example, parenchyma in acoelomates and perivisceral fluids in pseudocoelomates serve as hydrostatic skeletons.

hydrothermal vent A submarine hot spring; seawater seeping through the sea floor is heated by magma and expelled back into the sea through the hydrothermal vent.

hydroxyl (hydrogen + oxygen, + yl). Containing an OH⁻ group, a negatively charged ion formed by alkalies in water.

hyomandibular (hī-ō-mən-dib′yə-lər) (Gr. *hyoeides* [shaped like the Gr. letter upsilon Υ, + *eidos*, form], + L. *mandere*, to chew). Skeleton derived from the hyoid gill arch, forming part of articulation of the lower jaw of fishes, and forming the stapes of the ear of amniotic vertebrates.

hyperosmotic (Gr. *hyper*, over, + *ōsmos*, impulse). Refers to a solution whose osmotic pressure is greater than that of another solution to which it is compared; contains a greater concentration of dissolved particles and gains water through a selectively permeable membrane from a solution containing fewer particles; contrasts with **hypoosmotic.**

hyperparasitism (hī′pər-par′ə-sit-iz-əm) (Gr. *hyper*, over, + *para*, beside, + *sitos*, food). Parasitism of a parasite by another parasite.

hyperpolarization A change in the voltage in a negative direction recorded across a plasma membrane (see **membrane potential**). This allows a signal to be transmitted in excitable cells, such as nerve, muscle, and sensory cells.

hypertrophy (hī-pər′trə-fē) (Gr. *hyper*, over, + *trophē*, nourishment). Abnormal increase in size of a part or organ.

hypodermis (hī′pə-dər′mis) (Gr. *hypo*, under, + L. *dermis*, skin). The cellular layer lying beneath and secreting the cuticle of annelids, arthropods, and certain other invertebrates.

hypoosmotic (Gr. *hypo*, under, + *ōsmos*, impulse). Refers to a solution whose osmotic pressure is less than that of another solution with which it is compared or taken as a standard; contains a lesser concentration of dissolved particles and loses water during osmosis; contrasts with **hyperosmotic.**

hypophysis (hī-pof′ə-sis) (Gr. *hypo*, under, + *physis*, growth). Pituitary body.

hypostome (hī′pə-stōm) (Gr. *hypo*, under, + *stoma*, mouth). Name applied to a structure in various invertebrates (such as mites and ticks), located at the posterior or ventral area of the mouth.

hypothalamus (hī-pō-thal′ə-mis) (Gr. *hypo*, under, + *thalamos*, inner chamber). A ventral part of the forebrain beneath the thalamus; one of the centers of the autonomic nervous system and neuroendocrine regulation.

hypothesis (Gr. *hypothesis*, foundation, supposition). A statement or proposition that can be tested by observation or experiment.

hypothetico-deductive method (Gr. *hypotithenai*, to suppose, + L. *deducere*, to lead). Scientific process of making a conjecture and then seeking empirical tests that potentially lead to its rejection.

I

ichthyosaur (ik′thē-ō-sor) (Gr. *ichthyo*, fish, + *saur*, lizard). Aquatic, Mesozoic reptiles characterized by a porpoiselike body, but with a vertical tail and large eyes.

imago (ə-mā′gō). The adult and sexually mature insect.

immediate cause See **proximate cause.**

immediate hypersensitivity Inflammatory reaction based primarily on humoral immunity.

immunity Ability of tissues in an organism to recognize and to defend against nonself invaders. **Innate immunity** is a mechanism of defense that does not depend on prior exposure to the invader; **acquired immunity** is specific to a nonself material, requires time for development, and occurs more quickly and vigorously on secondary response.

immunoglobulin (im′yə-nə-glä′byə-lən) (L. *immunis*, free, + *globus*, globe). Any of a group of plasma proteins, produced by B cells and plasma cells, that participates in the immune response by combining with the antigen that stimulated its production. Antibody.

imprinting (im′print-ing) (L. *imprimere*, to impress, imprint). Rapid and usually stable learning pattern appearing early in the life of a member of a social species and involving recognition of its own species; may involve attraction to the first moving object seen.

inbreeding The tendency among members of a population to mate preferentially with close relatives.

inclusive fitness A modification of the genetic concept of fitness to consider not only the average numbers of offspring produced by organisms of a particular genotype but also the consequences of those organisms for the reproductive outputs of their close relatives. Because close relatives have genes that are identical by descent, an organism that foregoes reproduction to enhance the reproductive success of numerous close relatives nonetheless contributes copies of its genes to future generations; such an organism could have high inclusive fitness despite a fitness of zero.

incomplete dominance See **intermediate inheritance.**

incus (in′kəs) (L. *incus*, anvil). The middle of a chain of three bones of the mammalian middle ear; homologous with the quadrate bone of early vertebrates.

indeterminate cleavage A type of embryonic development in which the fate of the blastomeres is not determined very early as to tissues or organs, for example, in echinoderms and vertebrates; regulative cleavage.

indigenous (ən-dij′ə-nəs) (L. *indigena*, native). Pertains to organisms that are native to a particular region; not introduced.

indirect development A life cycle sequence from zygote to adult with intervening larval stages.

induction (L. *inducere, inductum*, to lead). Reasoning from the particular to the general; deriving a general statement (hypothesis) based on individual observations. In embryology, the alteration of cell fates as the result of interaction with neighboring cells.

inductor (in-duk′ter) (L. *inducere*, to introduce, lead in). In embryology, a tissue or organ that causes the differentiation of another tissue or organ.

inflammation (in′fləm-mā′shən) (L. *inflammare*, from *flamma*, flame). A part of the innate immune response at a site of antigen invasion or tissue injury. A cascade of events causes phagocytic immune cell activation and damage repair in the affected region. The area becomes swollen, red, and painful during this process.

infraciliature (in-frə-sil′e-ə-tūr) (L. *infra*, below, + *cilia*, eyelashes). The organelles just below the cilia in ciliate protozoa.

infundibulum (in′fun-dib′u-ləm) (L. funnel). Stalk of the neurohypophysis linking the pituitary to the diencephalon.

ingression (en-gres′ən) (L. *ingressus*, to go into, enter). Migration of individual cells from the embryo surface to the interior of the embryo during development.

inheritance of acquired characteristics The discredited Lamarckian notion that organisms, by striving to meet the demands of their environments, obtain new adaptations and pass them by heredity to their offspring.

innate (i-nāt′) (L. *innatus*, inborn). A characteristic based partly or wholly on genetic or epigenetic constitution.

instar (inz′tär) (L. form). Stage in the life of an insect or other arthropod between molts.

instinct (L. *instinctus*, impelled). Stereotyped, predictable, genetically programmed behavior. Learning may or may not be involved.

integument (ən-teg′ū-mənt) (L. *integumentum*, covering). An external covering or enveloping layer.

interbreeding population The most inclusive grouping of organisms for which sexual reproduction and associated genetic recombination occur freely across generations; implies an absence of biological

---

bat / āpe / ärmadillo / herring / fēmale / finch / līce / crocodile / crōw / cóin / duck / ūnicorn / ə indicates unaccented vowel sound "uh" as in mammal, fishes, cardinal, heron, vulture / stress as in bi-ol′o-gy, bi′o-log′i-cal

barriers to fertilization of gametes between females and males included in the grouping. A criterion of the biological species concept.

intercellular (in-tər-sel′yə-lər) (L. *inter*, among, + *cellula*, chamber). Occurring between body cells.

interferons Several cytokines encoded by different genes, important in mediation of innate immunity and inflammation.

interleukin-1 A cytokine produced by macrophages that stimulates T cells, B cells, and macrophages.

interleukin-2 A cytokine produced by T helper lymphocytes that leads to proliferation of B and T cells and enhanced activity of natural killer cells.

interleukins A series of cytokines produced primarily by cells of the immune system, such as macrophages, T and B cells, and also by endothelial cells and fibroblasts. Target cells are various leukocytes and other cells primarily involved with enhancing an immune response. Given the name "interleukins" when it was believed that they were produced only by leukocytes and their target cells were limited to leukocytes.

intermediary meiosis Meiosis that occurs neither during gamete formation nor immediately after zygote formation, resulting in both haploid and diploid generations, such as in foraminiferan protozoa.

intermediate host A host in which some development of a symbiont occurs, but in which maturation and sexual reproduction do not occur.

intermediate inheritance The alternative alleles of a gene do not have a simple dominance relationship, so that the heterozygote shows a condition intermediate between or different from homozygotes for each allele.

internal nostrils Palatal structures connecting the nasal cavity and throat in lungfishes and tetrapod vertebrates; used for olfaction and/ or breathing when the mouth is closed.

interstitial (in-tər-sti′shəl) (L. *inter*, among, + *sistere*, to stand). Situated in the interstices or spaces between structures such as cells, organs, or grains of sand.

intertidal zone A portion of an ocean's continental shelf that gets exposed to air during low tides and submerged at high tides.

intracellular (in-trə-sel′yə-lər) (L. *intra*, inside, + *cellula*, chamber). Occurring within a body cell or within body cells.

intrinsic growth rate Exponential growth rate of a population; the difference between the density-independent components of the birth and death rates of a natural population with stable age distribution.

intrinsic rate of increase See **intrinsic growth rate.**

intron (in′trän) (L. *intra*, within). Portion of mRNA as transcribed from DNA that will not form part of mature mRNA, and therefore does not encode an amino-acid sequence in the protein product.

introvert (L. *intro*, inward, + *vertere*, to turn). The anterior narrow portion that can be withdrawn (introverted) into the trunk of a sipunculid worm.

invagination (in-vaj′ə-nā′shən) (L. *in*, in, + *vagina*, sheath). An infolding of a layer of tissue to form a saclike structure.

inversion (L. *invertere*, to turn upside down). A turning inward or inside out, as in embryogenesis of sponges; also, reversal in order of genes or reversal of a chromosome segment.

ion An atom or group of atoms with a net positive or negative electrical charge because of the loss or gain of electrons.

ionic bond A chemical bond formed by transfer of one or more electrons from one atom to another; characteristic of salts.

iridophore (ī-rid′ə-fōr) (Gr. *iris*, rainbow, or iris of eye). Iridescent or silvery chromatophores containing crystals or plates of guanine or other purine.

irritability (L. *irritare*, to provoke). A general property of all organisms involving the ability to respond to stimuli or changes in the environment.

isogametes (īs′o-gam′ēts) (Gr. *isos*, equal, + *gametēs*, spouse). Gametes of a species in which gametes of both sexes are alike in size and appearance.

isolecithal (ī′sə-les′ə-thəl) (Gr. *isos*, equal, + *lekithos*, yolk, + *al*). Pertaining to a zygote (or ovum) with yolk evenly distributed. Homolecithal.

isosmotic A liquid having the same osmotic pressure as another, reference liquid.

isotonic (Gr. *isos*, equal, + *tonikos*, tension). Pertaining to solutions having the same or equal osmotic pressure; isosmotic.

isotope (Gr. *isos*, equal, + *topos*, place). One of several different forms (species) of a chemical element, differing from each other in atomic mass but not in atomic number.

iteroparity (i′tər-o-pa′ri-tē′). A life history in which individual organisms of a population normally reproduce more than one time before dying; contrasts with **semelparity.**

J

Jacobson's organ (*Jacobson*, nineteenth-century Danish surgeon and anatomist). Also called the vomeronasal organ. A chemosensory organ in the roof of the mouth of many terrestrial vertebrates; pheromones are transferred to this organ by the tongue.

juvenile hormone Hormone produced by the corpora allata of insects; its effects include maintenance of larval or nymphal characteristics during development.

juxtaglomerular apparatus (jək′stə-glä-mer′yə-lər) (L. *juxta*, close to, + *glomus*, ball). Complex of sensory cells located in the afferent arteriole adjacent to the glomerulus and a loop of the distal tubule, which produces the enzyme renin.

K

kentrogon (ken′trə-gən) (Gr. *kentron*, a point, spine, + *gonos*, progeny, generation). A larva of the cirripede order Rhizocephala (subphylum Crustacea) that functions to inject the parasite cells into the host hemocoel.

keratin (ker′ə-tən) (Gr. *kera*, horn, + *in*, suffix of proteins). A scleroprotein found in epidermal tissues and modified into hard structures such as horns, hair, nails, and reptile scales.

keystone species A species (typically a predator) whose removal leads to reduced species diversity within the community.

kinesis (kə-nē′səs) (Gr. *kinēsis*, movement). Movements by an organism in response to stimulus.

kinetochore (kī-nēt′ə-kōr) (Gr. *kinein*, to move, + *choris*, asunder, apart). A disc of proteins located on the centromere, specialized to interact with the spindle fibers during mitosis.

kinetodesma (kə-nē′tə-dez′mə), pl. **kinetodesmata** (Gr. *kinein*, to move, + *desma*, bond). Fibril arising from the kinetosome of a cilium in a ciliate protozoan, and passing along the kinetosomes of cilia in that same row.

kinetoplast (kə-nēt′ə-plast) (Gr. *kinatos*, moving, + *plastos*, molded, formed). Cellular organelle that functions in association with a kinetosome at the base of a flagellum; presumed to be derived from a mitochondrion.

kinetosome (kən-ēt′ə-sōm) (Gr. *kinētos*, moving, + *sōma*, body). The self-duplicating granule at the base of the flagellum or cilium; similar to centriole, also called basal body or blepharoplast.

kinety (kə-nē′tē) (Gr. *kinein*, to move). All the kinetosomes and kinetodesmata of a row of cilia.

kinin (kī′nin) (Gr. *kinein*, to move, + *in*, suffix of hormones). A type of local hormone that acts near its site of origin; also called **parahormone** or **tissue hormone.**

kin selection An extension of the genetical theory of natural selection to explain altruistic behaviors that preferentially benefit close relatives; an individual's genetic contribution to future generations is enhanced by promoting survival of relatives because their shared genes are identical by descent.

Kupffer cells Phagocytic cells in the liver, part of the reticuloendothelial system.

kwashiorkor (kwash-ē-or′kər) (from Ghana). Malnutrition caused by diet high in carbohydrate and extremely low in protein.

L

labium (lā′bē-əm) (L. a lip). The lower lip of the insect formed by fusion of the second pair of maxillae.

labrum (lā′brəm) (L. a lip). The upper lip of insects and crustaceans situated above or in front of the mandibles; also refers to the outer lip of a gastropod shell.

labyrinth (L. *labyrinthus,* labyrinth). Vertebrate internal ear, composed of a series of fluid-filled sacs and tubules (membranous labyrinth) suspended within bone cavities (osseous labyrinth).

labyrinthodont (lab′ə-rin′thə-dänt) (Gr. *labyrinthos,* labyrinth, + *odous, odontos* tooth). A group of Paleozoic tetrapods containing the temnospondyls and the anthracosaurs.

lachrymal glands (lak′rə-məl) (L. *lacrimia,* tear). Structures in terrestrial vertebrates that secrete tears to lubricate the eyes.

lacteal (lak′te-əl) (L. *lacteus,* of milk). Noun, one of the lymph vessels in the villus of the intestine. Adj., relating to milk.

lacuna (lə-kū′nə), pl. **lacunae** (L. pit, cavity). A sinus; a space between cells; a cavity in cartilage or bone.

lacunar system A netlike set of circulatory canals filled with fluid in an acanthocephalan.

lagena (lə-jē′nə) (L. large flask). Portion of the primitive ear in which sound is translated into nerve impulses; evolutionary beginning of cochlea.

Lamarckism Hypothesis, as expounded by Jean Baptiste de Lamarck, of evolution by the acquisition during an organism's lifetime of characteristics that are transmitted to offspring.

lamella (lə-mel′ə) (L. dim. of *lamina,* plate). One of the two plates forming a gill in a bivalve mollusc. One of the thin layers of bone laid concentrically around an osteon (Haversian canal). Any thin, platelike structure.

lappets Lobes around the margin of scyphozoan medusae (phylum Cnidaria).

larva (lar′və), pl. **larvae** (L. a ghost). An immature stage that is quite different from the adult.

larynx (lar′inks) (Gr. the larynx, gullet). Modified upper portion of respiratory tract of air-breathing vertebrates, bounded by the glottis above and the trachea below; voice box. Adj., **laryngeal** (lə-rin′j(ē)əl), relating to the larynx.

lateral (L. *latus,* the side, flank). Of or pertaining to the side of an animal; a *bilateral* animal has two sides.

lateral-line system Sensory organ, consisting of neuromast organs in canals and grooves on the head and sides of the body of fishes and some amphibians, that detects water vibrations.

laterite (lat′ə-rīt) (L. *later,* brick). Group of hard, red soils from tropical areas that show intense weathering and leaching of bases and silica, leaving aluminum hydroxides and iron oxides. Adj., **lateritic.**

law of independent assortment Also called **Mendel's second law.** Segregation of allelic forms of a gene into gametes occurs at random with respect to the segregation of alleles of a second gene located on a different pair of homologous chromosomes.

Genes located very far apart on a single chromosome also sometimes show independent assortment because crossovers occur between them at a high rate.

law of segregation Mendel's first law of inheritance, in which the paired particulate factors influencing variation of a trait separate from each other in gamete formation so that each gamete carries only one of the factors.

lecithotrophy (le′sə-thə-trō′fē) (Gr. *lekithos,* yolk of egg, + *trophos,* one who feeds). Nutrition of an embryo directly from the yolk of an ovum.

leishmaniasis (līsh′mə-nī′ə-səs) (Fr. Sir W. B. Leishman, 1926, British medical officer). Disease caused by infection with protozoans of the genus *Leishmania.*

lek (lek) (Sw. play, game). An area where animals assemble for communal courtship display and mating.

lemniscus (lem-nis′kəs) (L. ribbon). One of a pair of internal projections of the epidermis from the neck region of Acanthocephala, which functions in fluid control in the protrusion and invagination of the proboscis.

lentic (len′tik) (L. *lentus,* slow). Of or relating to standing water such as swamp, pond, or lake.

lepidosaurs (lep′ə-dō-sors) (L. *lepidos,* scale, + *sauros,* lizard). A lineage of diapsid reptiles that appeared in the Permian and that includes the modern snakes, lizards, amphisbaenids, and tuataras, and the extinct ichthyosaurs.

lepospondyls (lep′ə-spänd′ls) (Gr. *lepos,* scale, + *spondylos,* vertebra). A group of Paleozoic amphibians distinguished by the possession of spool-shaped vertebral centra.

leptocephalus (lep′tə-sef′ə-ləs), pl. **leptocephali** (Gr. *leptos,* thin, + *kephalē,* head). Transparent, ribbonlike migratory larva of eels and related teleosts.

leuconoid (lū′kə-noid) (Gr. *leukon,* white, + *eidos,* like). A type of canal system in sponges where choanocytes reside in chambers.

leukemism (lū′kə-mi-zəm) (Gr. *leukos,* white, + *ismos,* condition of). Presence of white pelage or plumage in animals with normally pigmented eyes and skin.

leukocyte (lū′kə-sīt) (Gr. *leukos,* white, + *kytos,* hollow vessel). Any of several kinds of white blood cells (for example, granulocytes, lymphocytes, monocytes), so called because they bear no hemoglobin, as do red blood cells.

library In molecular biology, a set of clones containing recombinant DNA. Obtained from and representing the genome of the organism.

ligament (lig′ə-mənt) (L. *ligamentum,* bandage). A tough, dense band of connective tissue connecting one bone to another.

ligand (lī′gənd) (L. *ligo,* to bind). A molecule that specifically binds to a receptor; for example, a hormone (ligand) binds specifically to its receptor on the cell surface.

ligation (lī-gā′shən) (L. *ligo,* to bind). End-to-end joining of two pieces of DNA.

limax form (lī′məx) (L. *limax,* slug). Form of pseudopodial movement in which entire organism moves without extending a discrete pseudopodium.

limiting resource A particular source of nutrition, energy, or living space whose scarcity is causally associated with a population having fewer individuals than otherwise expected in a particular environment.

linkage, linked Genes carried on the same chromosome whose allelic variants tend to sort together in gamete formation (violates Mendel's second law of independent assortment).

lipase (lī′pās) (Gr. *lipos,* fat, + *ase,* enzyme suffix). An enzyme that accelerates the hydrolysis or synthesis of fats.

lipid, lipoid (li′pid) (Gr. *lipos,* fat). Certain fatlike substances such as triglycerides composed of fatty acids and glycerol, often containing other groups such as phosphoric acid. Lipids combine with proteins and carbohydrates to form principal structural components of cells such as the plasma membrane.

Lissamphibia (lis′-am-fi′-bē-ə). A tetrapod clade comprising modern amphibians (caecilians, frogs, and salamanders) and all descendants of their most recent common ancestor.

lithosphere (lith′ə-sfir) (Gr. *lithos,* rock, + *sphaira,* ball). The rocky component of the earth's surface layers.

littoral (lit′ə-rəl) (L. *litoralis,* seashore). Adj., pertaining to the shore. Noun, that portion of the seafloor between the extent of high and low tides, intertidal; in lakes, the shallow part from the shore to the lakeward limit of aquatic plants.

lobopodium (lō′bə-pō′de-əm) (Gr. *lobos,* lobe, + *pous, podos,* foot). Blunt, lobelike pseudopodium.

lobosea (lō-bō′sē-ə) (Gr. *lobos,* lobe). A protozoan clade comprising amebas with lobopodia.

locus (lō′kəs), pl. **loci** (lō′sī) (L. place). Position of a gene in a chromosome.

logistic equation A mathematical expression describing an idealized sigmoid curve of population growth.

lophocyte (lō′fə-sīt) (Gr. *lophos,* crest, + *kytos,* hollow vessel). Type of sponge amebocyte that secretes bundles of fibrils.

lophophore (lōf′ə-fōr) (Gr. *lophos,* crest, + *phoros,* bearing). Tentacle-bearing ridge or arm within which is an extension of the coelomic cavity in lophophorate animals (ectoprocts, brachiopods, and phoronids).

lophotrochozoan protostome (lō′fə-trō′kə-zō′ən) (Gr. *lophos,* crest, + *trochos,* wheel,

+ *zōon,* animal). Any member of a clade within Protostomia whose members generally possess either a trochophore larva or a lophophore; examples are annelids, molluscs, and ectoprocts.

lorica (lo′rə-kə) (L. corselet). Protective external case found in some protozoa, rotifers, and others.

lotic (lō′tik) (L. *lotus,* action of washing or bathing). Of or pertaining to running water, such as a brook or river.

low viscosity Refers to the ability of liquid water to flow quickly and easily through containers of varying spatial dimensions, such as vessels of animal circulatory systems.

lumbar (lum′bär) (L. *lumbus,* loin). Relating to or near the loins or lower back.

lumen (lū′mən) (L. light). The cavity of a tube or organ.

lymph (limf) (L. *lympha,* water). The fluid in the lymphatic system formed from excess fluid that would otherwise accumulate in interstitial spaces between cells during capillary exchange.

lymphocyte (lim′fō-sīt) (L. *lympha,* water, goddess of water, + Gr. *kytos,* hollow vessel). Cell in blood and lymph that has a central role in immune responses. See **T cell** and **B cell.**

lysosome (lī′sə-sōm) (Gr. *lysis,* loosing, + *sōma,* body). Intracellular organelle consisting of a membrane enclosing several digestive enzymes that are released when the lysosome fuses with vescicles or endosomes produced as a result of endocytosis.

M

macroevolution (L. *makros,* long, large, + *evolvere,* to unfold). Evolutionary change on a grand scale, encompassing the origin of novel designs, evolutionary trends, adaptive radiation, and mass extinction.

macrogamete (mak′rə-gam′ēt) (Gr. *makros,* long, large, + *gamos,* marriage). The larger of the two gamete types in a heterogametic organism, considered the female gamete.

macromere (mak′rə-mer′) (Gr. *makros,* long, large, + *meros,* part). The largest size class of blastomeres in a cleaving embryo when the blastomeres differ in size from one another.

macromolecule A very large molecule, such as a protein, polysaccharide, or nucleic acid.

macronucleus (ma′krō-nū′klē-əs) (Gr. *makros,* long, large, + *nucleus,* kernel). The larger of the two kinds of nuclei in ciliate protozoa; controls all cell functions except reproduction.

macrophage (mak′rə-fāj) (Gr. *makros,* long, large, + *phagō,* to eat). A phagocytic cell type in vertebrates that performs crucial functions in the immune response and inflammation, such as presenting antigenic epitopes to T cells and producing several cytokines.

madreporite (ma′drə-pōr′īt) (Fr. *madrépore,* reef-building coral, + *ite,* suffix for some body parts). Sievelike structure that is the intake for the water-vascular system of echinoderms.

major histocompatibility complex (MHC) Complex of genes coding for proteins inserted in the cell membrane; the proteins are the basis of self-nonself recognition by the immune system.

malacostracan (mal′ə-käs′trə-kən) (Gr. *malako,* soft, + *ostracon,* shell). Any member of the crustacean subclass Malacostraca, which includes both aquatic and terrestrial forms of crabs, lobsters, shrimps, pillbugs, sand fleas, and others.

malaria (mə-lar′-ē-ə) (It. *malaria,* bad air). A disease marked by periodic chills, fever, anemia, and other symptoms, caused by *Plasmodium* spp.

male-dominance polygyny A male obtains more than one mate because females choose him over other males in an aggregation.

malleus (mal′ē-əs) (L. hammer). The ossicle attached to the tympanum in middle ears of mammals.

Malpighian tubules (mal-pig′ē-ən) (Marcello Malpighi, Italian anatomist, 1628–1694). Blind tubules opening into the hindgut of nearly all insects and some myriapods and arachnids, and functioning primarily as excretory organs.

mandible (L. *mandibula,* jaw). One of the lower jawbones in vertebrates; one of the head appendages in arthropods.

mangrove community A rich submarine community of deposit feeders (crabs, oysters, and shrimp) associated with the submerged roots of mangrove trees along some tropical coasts.

mantle Soft extension of the body wall in certain invertebrates, for example, brachiopods and molluscs, which usually secretes a shell; thin body wall of tunicates.

manubrium (man-ū′bri-əm) (L. handle). The portion projecting from the oral side of a jellyfish medusa, bearing the mouth; oral cone; presternum or anterior part of sternum; handlelike part of malleus of ear.

marasmus (mə-raz′məs) (Gr. *marasmos,* to waste away). Malnutrition, especially of infants, caused by a diet deficient in both calories and protein.

marsupial (mär-sū′pē-əl) (Gr. *marsypion,* little pouch). One of the pouched mammals of the subclass Metatheria.

mass extinction A relatively short interval of geological time in which a large portion (75% to 95%) of existing species or higher taxa are eliminated nearly simultaneously.

mast cells Inflammatory cells primarily in connective tissue. Upon activation by an antigen they release pharmacologically active compounds leading to redness and swelling.

mastax (mas′təx) (Gr. jaws). Pharyngeal mill of rotifers.

matrix (mā′triks) (L. *mater,* mother). The intercellular substance of a tissue, or that part of a tissue into which an organ or process is set.

maturation (L. *maturus,* ripe). The process of ripening; the final stages in the preparation of gametes for fertilization.

maxilla (mak-sil′ə) (L. dim. of *mala,* jaw). One of the upper jawbones in vertebrates; one of the head appendages in arthropods.

maxilliped (mak-sil′ə-ped) (L. *maxilla,* jaw, + *pes,* foot). One of the pairs of head appendages located just posterior to the maxilla in crustaceans, a thoracic appendage that has become incorporated into the feeding mouthparts.

medial (mē′dē-əl). Situated, or occurring, in the middle.

mediated transport Transport of a substance across a cell membrane mediated by a protein carrier molecule in the membrane.

medulla (mə-dul′ə) (L. marrow). The inner portion of an organ in contrast to the cortex or outer portion. Also, part of hindbrain.

medusa (mə-dū′sə) (Gr. mythology, female monster with snake-entwined hair). A jellyfish, or the free-swimming stage in the life cycle of cnidarians.

Mehlis′ gland (me′ləs). Glands of uncertain function surrounding the ootype of trematodes and cestodes.

meiofauna (mī′ō-faw-nə) (Gr. *meion,* smaller, + L. *faunus,* god of the woods). Small invertebrates found in the interstices between sand grains.

meiosis (mī-ō′səs) (Gr. from *mieoun,* to make small). The nuclear changes by means of which the chromosomes are reduced from the diploid to the haploid number; in animals, usually occurs in the last two divisions in the formation of the mature egg or sperm.

melanin (mel′ə-nin) (Gr. *melas,* black). Black or dark-brown pigment found in plant or animal structures.

melanophore (mel′ə-nə-fōr, mə-lan′ə-fōr) (Gr. *melania,* blackness, + *pherein,* to bear). Black or brown chromatophore containing melanin.

membranelle (mem-bra-nel′). A tiny membranelike structure, may be formed by fused cilia.

membrane potential The voltage recorded across a plasma membrane due to an unequal distribution of ions and charged molecules on either side of the membrane. Such an unequal distribution of charge is caused by selective permeability of plasma membranes to specific ions and charged molecules.

memory cells Population of long-lived T and B lymphocytes remaining after an initial immune response and providing for the secondary response.

meninges (mə-nin′jez), sing. **meninx** (Gr. *mēninx,* membrane). Any of three membranes (arachnoid, dura mater, pia mater) that envelop the vertebrate brain and spinal cord. Also, solid connective tissue sheath enclosing the central nervous system of some vertebrates.

menopause (men′ō-pawz) (Gr. *men,* month, + *pauein,* to cease). In the human female, that time of life when ovulation ceases; cessation of the menstrual cycle.

menstruation (men′stroo-ā′shən) (L. *menstrua,* the menses, from *mensis,* month). The discharge of blood and uterine tissue from the vagina at the beginning of a menstrual cycle. It occurs during the first few days of the ovarian cycle.

meroblastic (mer-ə-blas′tik) (Gr. *meros*, part, + *blastos*, germ). Partial cleavage occurring in zygotes having a large amount of yolk at the vegetal pole; cleavage restricted to a small area on the surface of the egg.

merozoite (me′rə-zō′īt) (Gr. *meros*, part, + *zōon*, animal). A very small trophozoite at the stage just after cytokinesis has been completed in multiple fission of a protozoan.

mesenchyme (me′zən-kīm) (Gr. *mesos*, middle, + *enchyma*, infusion). Embryonic connective tissue; irregular or amebocytic cells often embedded in gelatinous matrix.

mesentery (mes′ən-ter′ē) (L. *mesenterium*, mesentery). Peritoneal fold serving to hold the viscera in position.

mesocoel (mez′ō-sēl) (Gr. *mesos*, middle, + *koilos*, hollow). Middle body coelomic compartment in some deuterostomes, anterior in lophophorates, corresponds to hydrocoel in echinoderms.

mesoderm (me′zə-dərm) (Gr. *mesos*, middle, + *derm*, skin). The third germ layer, formed in the gastrula between the ectoderm and endoderm; gives rise to connective tissues, muscle, urogenital and vascular systems, and the peritoneum.

mesoglea (mez′ō-glē′ə) (Gr. *mesos*, middle, + *glia*, glue). The layer of jellylike or cement material between the epidermis and gastrodermis in cnidarians and ctenophores.

mesohyl (mez′ō-hīl) (Gr. *mesos*, middle, + *hylē*, a wood). Gelatinous matrix surrounding sponge cells; mesoglea, mesenchyme.

mesolecithal (mezō-ləs′ə-thəl) (Gr. *mesos*, middle, + *lekithos*, yolk). Pertaining to a zygote (or ovum) having a moderate amount of yolk concentrated in the vegetal pole.

mesonephros (me-zō-nef′rōs) (Gr. *mesos*, middle, + *nephros*, kidney). The middle of three pairs of embryonic renal organs in vertebrates. Functional kidney of embryonic amniotes; its collecting duct is a Wolffian duct. Adj., **mesonephric.**

mesopelagic (me-zō-pə-laj′-ik). Refers to the "twilight zone" of open oceanic waters marking the transition from the well-lit epipelagic zone above it to a zone of total darkness below it.

mesosome (mez′ə-sōm) (Gr. *mesos*, middle, + *sōma*, body). The portion of the body in lophophorates and some deuterostomes that contains the mesocoel.

messenger RNA (mRNA) A form of ribonucleic acid that carries genetic information from the gene to the ribosome, where it determines the order of amino acids as a polypeptide is formed.

metabolism (Gr. *metabolē*, change). A group of processes that includes digestion, recovery of energy (respiration), and synthesis of molecules and structures by organisms; the sum of the constructive (anabolic) and destructive (catabolic) processes.

metacentric (me′tə-sen′trək) (Gr. *meta*, between, among, after, + *kentron*, center). Chromosome with centromere at or near the middle.

metacercaria (me′tə-sər-ka′rē-ə) (Gr. *meta*, between, among, after, + *kerkos*, tail, + L. *aria*, connected with). Fluke juvenile (cercaria) that has lost its tail and has become encysted.

metacoel (met′ə-sēl) (Gr. *meta*, between, among, after, + *koilos*, hollow). Posterior coelomic compartment in some deuterostomes and lophophorates; corresponds to somatocoel in echinoderms.

metamere (met′ə-mēr) (Gr. *meta*, after, + *meros*, part). A repeated body unit along the longitudinal axis of an animal, a somite, or segment.

metamerism (mə-ta′-mə-ri′zəm) (Gr. *meta*, between, after, + *meros*, part). Condition of being composed of serially repeated parts (metameres); serial segmentation.

metamorphosis (Gr. *meta*, between, among, after, + *morphē*, form, + *osis*, state of). Sharp change in form during postembryonic development, for example, tadpole to frog or larval insect to adult.

metanephridium (me′tə-nə-fri′dē-əm) (Gr. *meta*, between, among, after, + *nephros*, kidney). A type of tubular nephridium with the inner open end draining the coelom and the outer open end discharging to the exterior.

metanephros (me′tə-ne′frōs) (Gr. *meta*, between, among, after, + *nephros*, kidney). Vertebrate kidney formed from most posterior of three embryonic regions capable of forming renal organs; functional kidney of adult amniotes; drained by ureter. Adj., **metanephric.**

metapopulation dynamics The structure of a large population that comprises numerous semiautonomous subpopulations, termed demes, with some limited movement of individuals among demes. Demes of a metapopulation are often geographically distinct.

metasome (met′ə-sōm) (Gr. *meta*, after, behind, + *sōma*, body). The portion of the body in lophophorates and some deuterostomes that contains the metacoel.

metazoa (met-ə-zō′ə) (Gr. *meta*, after, + *zōon*, animal). Multicellular animals.

MHC See **major histocompatibility complex.**

microevolution (mī-krō-ev-ə-lü′shən). (L. *mikros*, small, + *evolvere*, to unfold). A change in the gene pool of a population across generations.

microfilament (mī′krō-fil′ə-mənt) (Gr. *mikros*, small, + L. *filum*, a thread). A thin, linear structural protein forming a part of the cytoskeleton in all cells. Part of contractile protein machinery of muscle cells.

microfilariae (mīk′rə-fil-ar′ē-ē) (Gr. *mikros*, small, + L. *filum*, a thread). Partially developed juveniles borne alive by filarial worms (phylum Nematoda).

microgamete (mīk′rə-gam′ēt) (Gr. *mikros*, small, + *gamos*, marriage). The smaller of the two gamete types in a heterogametic organism, considered the male gamete.

microglial cells Phagocytic cells in the central nervous system, part of the reticuloendothelial system.

micromere (mīk′rə-mēr′) (Gr. *mikros*, small, + *meros*, part). The smallest size class of blastomeres in a cleaving embryo when the blastomeres differ in size from one another.

micron (μ) (mī′krän) (Gr. neuter of *mikros*, small). One one-thousandth of a millimeter; about 1/25,000 of an inch. Now largely replaced by micrometer (μm).

microneme (mī′krə-nēm) (Gr. *mikros*, small, + *nēma*, thread). One type of structure forming the apical complex in the phylum Apicomplexa, slender and elongate, leading to the anterior and thought to function in host-cell penetration.

micronucleus A small nucleus found in ciliate protozoa; controls the reproductive functions of these organisms.

micropyle (mīk′rə-pīl) (Gr. *mikros*, small, + *pileos*, a cap). The small opening through which the cells emerge from a gemmule (phylum Porifera).

microsporidian (mī′krə-spō-rid′ē-ən) (Gr. *micros*, small, + *spora*, seed, + *idion*, dim. suffix). Any member of a protozoan clade comprising intracellular parasites with a distinctive spore morphology.

microthrix See **microvillus.**

microtubule (Gr. *mikros*, small, + L. *tubule*, pipe). A long, tubular cytoskeletal element with an outside diameter of 20 to 27 nm. Microtubules influence cell shape and play important roles during cell division.

microvillus (Gr. *mikros*, small, + L. *villus*, shaggy hair). Narrow, cylindrical cytoplasmic projection from epithelial cells; microvilli form the brush border of several types of epithelial cells. Also, microvilli with unusual structure cover the surface of cestode tegument (also called **microthrix** [pl. **microtriches**]).

mictic (mik′tik) (Gr. *miktos*, mixed or blended). Pertaining to haploid egg of rotifers or the females that lay such eggs.

mimic (mim′ik) (Gr. *mimicus*, imitator). A species whose morphological or behavioral characteristics copy those of another species because those characteristics deter shared predators.

mimicry Evolution by natural selection of similar forms by different species, such as the sharing of warning signals that discourage common predators. In Batesian mimicry, a species tasteful to a predator deceptively copies warning signals of a distasteful species. In Müllerian mimicry, common warning signals are evolved by two or more distasteful species to avoid a common predator.

mineralocorticoids (min(ə)rəl-ō-kort′ə-koids) (M. E. *minerale,* ore, + L. *cortex,* bark, + *oid,* suffix denoting likeness of form). Steroid hormones of the adrenal cortex, especially aldosterone, that regulate salt balance.

miracidium (mīr′ə-sid′ē-əm) (Gr. *meirakidion,* youthful person). A minute ciliated larval stage in the life of flukes.

mitochondrion (mīt′ə-kän′drē-ən) (Gr. *mitos,* a thread, + *chondrion,* dim. of *chondros,* corn, grain). An eukaryotic cellular organelle in which aerobic metabolism occurs.

mitosis (mī-tō′səs) (Gr. *mitos,* thread, + *osis,* state of). Nuclear division in which there is an equal qualitative and quantitative division of the chromosomal material between the two resulting nuclei; ordinary cell division.

mixed temperate forest Central North American woodlands containing evergreen trees such as white pine, red pine, and hemlock mixed with various deciduous trees, including maples, oaks, and hickories.

model (mod′l) (Fr. *modèle,* pattern). A species whose morphological or behavioral characteristics are copied by another species because those characteristics deter shared predators.

modular (mäj′ə-lər). Describes the structure of a colony of genetically identical organisms that are physically associated and produced asexually by cloning.

molecule A configuration of atomic nuclei and electrons bound together by chemical bonds.

molting (mōl′-ting). Shedding of the outer cuticular layer; see **ecdysis.**

monocyte (mon′ə-sīt) (Gr. *monos,* single, + *kytos,* hollow vessel). A type of leukocyte that becomes a phagocytic cell (macrophage) after moving into tissues.

monoecious (mə-nē′shəs) (Gr. *monos,* single, + *oikos,* house). Having both male and female gonads in the same organism; hermaphroditic.

monogamy (mə-näg′ə-mē) (Gr. *monos,* single, + *gamos,* marriage). The condition of having a single mate at any one time. Adj., **monogamous**.

monohybrid cross (Gr. *monos,* single, + L. *hybrida,* mongrel). Production of hybrid offspring from parents different in one specified character.

monomer (mä′nə-mər) (Gr. *monos,* single, + *meros,* part). A molecule of simple structure, but capable of linking with others to form polymers.

monophyly (män′ə-fī-lē) (Gr. *monos,* single, + *phyle,* tribe). The condition that a taxon or other group of organisms contains the most recent common ancestor of the group and all of its descendants; contrasts with **polyphyly** and **paraphyly.**

monosaccharide (män′nə-sa′kə-rīd) (Gr. *monos,* one, + *sakcharon,* sugar, from Sanskrit *sarkarā,* gravel, sugar). A simple sugar that cannot be decomposed into smaller sugar molecules; the most common are pentoses (such as ribose) and hexoses (such as glucose).

monosomy, monosomic (mä′nə-sō′mē) (Gr. *monos,* one, + *sōmē,* body). The chromosomal constitution of an otherwise diploid organism in which a single chromosome is missing (chromosome number = 2n − 1).

monotreme (mä′nō-trēm) (Gr. *monas,* single, + *trēmatos,* hole). Any member of an order of egg-laying (oviparous) mammals; duck-billed platypus and echidnas.

monozoic (mo′nə-zō′ik) (Gr. *monos,* single, + *zōon,* animal). Tapeworms with a single proglottid, do not undergo strobilation to form chain of proglottids.

morphogen (morf′ə-gən) (Gr. *morphē,* form, + *genesis,* origin). Soluble molecule acting on target cells, or forming a gradient from producing cells to target cells, to specify cell fates; an agent of embryonic induction and epistasis.

morphogenesis (mor′fə-je′nə-səs) (Gr. *morphē,* form, + *genesis,* origin). Development of the architectural features of organisms; formation and differentiation of tissues and organs.

morphogenetic determinant Certain protein or messenger RNA in egg cytoplasm distributed among descendant cells during cleavage to direct later gene expression and to specify cell fate; the basis of mosaic development.

morphological species concept See **typological species concept.**

morphology (Gr. *morphē,* form, + L. *logia,* study, from Gr. *logos,* work). The science of structure. Includes cytology, the study of cell structure; histology, the study of tissue structure; and anatomy, the study of gross structure.

morula (mär′u-lə) (L. *morum,* mulberry, + *ula,* dim.). Solid ball of cells in early stage of embryonic development.

mosaic cleavage Embryonic development characterized by independent differentiation of each part of the embryo; determinate cleavage.

mucin (mū′sən) (L. *mucus,* nasal mucus). Any of a group of glycoproteins secreted by certain cells, especially those of salivary glands.

mucus (mū′kəs) (L. *mucus,* nasal mucus). Viscid, slippery secretion rich in mucins produced by secretory cells such as those in mucous membranes. Adj., **mucous.**

Müllerian mimicry A condition in which two inedible species evolve similar aposematic phenotypes (such as warning coloration) to discourage potential predators from attempting to eat either one.

Müller's larva Free-swimming ciliated larva that resembles a modified ctenophore, characteristic of certain marine polyclad turbellarians.

multiple fission A mode of asexual reproduction in some single-celled eukaryotes in which the nuclei divide more than once before cytokinesis occurs.

multiplication of species The Darwinian theory that the evolutionary process generates new species through a branching of evolutionary lineages derived from an ancestral species.

mushroom bodies Region of the protocerebrum of the insect brain associated with learning.

mutation (mū-tā′shən) (L. *mutare,* to change). A stable and abrupt change of a gene; the heritable modification of a characteristic.

mutualism (mū′chə-wə-li′zəm) (L. *mutuus,* lent, borrowed, reciprocal). A type of interaction in which two different species derive benefit from their association and in which the association is necessary to both; often symbiotic.

mycetozoa (mī-sē′-tə-zō′ə) (Gr. *mykētos,* a fungus, + *zōon,* animal). A eukaryotic clade containing cellular, acellular, and protostelid slime molds.

myelin (mī′ə-lən) (Gr. *myelos,* marrow). A fatty material forming the medullary sheath of nerve fibers.

myocyte (mī′ə-sīt) (Gr. *mys,* muscle, + *kytos,* hollow vessel). Contractile cell (pinacocyte) in sponges.

myofibril (Gr. *mys,* muscle, + L. dim. of *fibra,* fiber). A contractile filament within muscle or muscle fiber.

myogenic (mī′o-jen′ik) (Gr. *mys,* muscle, + N.L., *genic,* giving rise to). Originating in muscle, such as heartbeat arising in vertebrate cardiac muscle because of inherent rhythmical properties of muscle rather than because of neural stimuli.

myomere (mī′ə-mēr) (Gr. *mys,* muscle, + *meros,* part). A muscle segment of successive segmental trunk musculature.

myosin (mī′ə-sin) (Gr. *mys,* muscle, + *in,* suffix, belonging to). A large, actin-binding protein of contractile tissue that forms the thick filaments of striated muscle. During contraction it combines with actin to form actomyosin.

myotome (mī′ə-tōm) (Gr. *mys,* muscle, + *tomos,* cutting). That part of a somite destined to form muscles; the muscle group innervated by a single spinal nerve.

N

nacre (nā′kər) (F. mother-of-pearl). Innermost lustrous layer of mollusc shell, secreted by mantle epithelium. Adj., **nacreous.**

NAD Abbreviation of nicotinamide adenine dinucleotide, an electron acceptor or donor in many metabolic reactions.

nares (na′rēz), sing. **naris** (L. nostrils). Openings into the nasal cavity, both internally and externally, in the head of a vertebrate.

natural killer cells Lymphocyte-like cells that can kill virus-infected cells and tumor cells in the absence of antibody.

natural selection A nonrandom reproduction of varying organisms in a population that results in the survival of those best adapted to their environment and elimination of those less well adapted; leads to evolutionary change if the variation is heritable.

nauplius (naw′plē-əs) (L. a kind of shellfish). A free-swimming microscopic larval stage of certain crustaceans, with three pairs of appendages (antennules, antennae, and mandibles) and median eye. Characteristic of ostracods, copepods, barnacles, and some others.

nekton (nek′tən) (Gr. neuter of *nēktos,* swimming). Term for actively swimming organisms, essentially independent of wave and current action; contrasts with **plankton.**

nematocyst (ne-ma′-tə-sist′) (Gr. *nēma,* thread, + *kystis,* bladder). Stinging organelle of cnidarians.

neo-Darwinism (nē′ō′ där′wə-niz′əm). A modified version of Darwin's evolutionary theory that eliminates elements of the Lamarckian inheritance of acquired characteristics and pangenesis that were present in Darwin's formulation; this theory originated with August Weismann in the late nineteenth century and, after incorporating Mendelian genetic principles, has become the currently favored version of Darwinian evolutionary theory.

neoplastic growth (nē-ə-plas′-tik grōth′). Proliferation of cells at an uncharacteristically high rate within the body of a multicellular organism, leading to cancerous tumors and metastases.

neopterygian (nē-äp′tə-rij′ē-ən) (Gr. *neos,* new, + *pteryx,* fin). Any of a large group of bony fishes that includes most modern species.

neotenine See **juvenile hormone.**

neoteny (nē′ə-tē′nē, nē-ot′ə-nē) (Gr. *neos,* new, + *teinein,* to extend). An evolutionary process by which organismal development is retarded relative to sexual maturation; produces a descendant that reaches sexual maturity while retaining a morphology characteristic of the preadult or larval stage of an ancestor.

nephridiopore (nə-frid′ē-ə-pōr′) (Gr. *nephros,* kidneys, + *porus,* pore). An external excretory opening in invertebrates.

nephridium (nə-frid′ē-əm) (Gr. *nephridios,* of the kidney). One of the segmentally arranged, paired excretory tubules of many invertebrates, notably the annelids. In a broad sense, any tubule specialized for excretion and/or osmoregulation; with an external opening and with or without an internal opening.

nephron (ne′frän) (Gr. *nephros,* kidney). Functional unit of kidney structure of vertebrates, consisting of a Bowman's capsule, an enclosed glomerulus, and the attached uriniferous tubule.

nephrostome (nef′rə-stōm) (Gr. *nephros,* kidney, + *stoma,* mouth). Ciliated, funnel-shaped opening of a nephridium.

neritic (nə-rit′ik) (Gr. *nērites,* a mussel). Portion of the sea overlying the continental shelf, specifically from the subtidal zone to a depth of 200 m.

nested hierarchy A pattern in which species are ordered into a series of increasingly more inclusive clades according to the taxonomic distribution of synapomorphies.

net productivity The energy stored by an organism, equal to the energy assimilated (**gross productivity**) minus the energy used for metabolic maintenance (**respiration**).

neural crest Populations of ectodermally derived embryonic cells that differentiate into many skeletal, neural, and sensory structures unique to vertebrates.

neurochord (nur′ə-kord) (L. *nervus,* sinew, tendon, + Gr. *chorda,* cord). Longitudinal nerve chord of hemichordates that lies above the notochord.

neurogenic (nū-rä-jen′ik) (Gr. *neuron,* nerve, + N.L. *genic,* give rise to). Originating in nervous tissue, as does the rhythmical beat of some arthropod hearts.

neuroglia (nū-ra-glē′-ə) (Gr. *neuron,* nerve, + *glia,* glue). Tissue supporting and filling the spaces between the nerve cells of the central nervous system.

neurolemma (nū-rə-lem′ə) (Gr. *neuron,* nerve, + *lemma,* skin). Delicate nucleated outer sheath of a nerve cell; sheath of Schwann.

neuromast (Gr. *neuron,* sinew, nerve, + *mastos,* knoll). Cluster of sense cells on or near the surface of a fish or amphibian that is sensitive to vibratory stimuli and water.

neuron (Gr. nerve). A nerve cell.

neuropodium (nū′rə-pō′de-əm) (Gr. *neuron,* nerve, + *pous, podos,* foot). Lobe of parapodium nearer the ventral side in polychaete annelids.

neurosecretory cell (nu′rō-sə-krēt′o-rē). Any cell (neuron) of the nervous system that produces a hormone.

neutron A subatomic particle lacking an electrical charge and having a mass 1839 times that of an electron and found in the nucleus of atoms.

niche (nich′). The role of an organism in an ecological community; its unique way of life and its relationship to other biotic and abiotic factors.

niche overlap A comparison of two species quantifying the proportion of each species' resources that are utilized also by the other species.

nictitating membrane (nik′tə-tā-ting) (L. *nicto,* to wink). Third eyelid, a transparent membrane of birds and many nonavian reptiles and mammals, that can be pulled across the eye.

nitrogen fixation (Gr. *nitron,* soda, + *gen,* producing). Reduction of molecular nitrogen to ammonia by some bacteria and cyanobacteria, often followed by

nitrification, the oxidation of ammonia to nitrites and nitrates by other bacteria.

nitrogenous base The molecular subunit of a nucleotide attached at the 1′ carbon of deoxyribose or ribose and which participates in hydrogen bonding between nucleotide strands. Includes adenine, cytosine, guanine, thymine, and uracil.

nondisjunction Failure of a pair of homologous chromosomes to separate during meiosis, leading to one gamete with $n + 1$ chromosomes (see **trisomy**) and another gamete with $n - 1$ chromosomes.

notochord (nō′tə-kord′) (Gr. *nōtos,* back, + *chorda,* cord). An elongated cellular cord, enclosed in a sheath, which forms the axial skeleton of chordate embryos, jawless fishes, and adult cephalochordates.

notopodium (nō′tə-pō′de-əm) (Gr. *nōtos,* back, + *pous, podos,* foot). Lobe of parapodium nearer the dorsal side in polychaete annelids.

nucleic acid (nu′klē′ik) (L. *nucleus,* kernel). One of a class of molecules composed of joined nucleotides; chief types are deoxyribonucleic acid (DNA), found in cell nuclei (chromosomes) and mitochondria, and ribonucleic acid (RNA), found both in cell nuclei (chromosomes and nucleoli) and in cytoplasmic ribosomes.

nucleoid (nu′klē-oid) (L. *nucleus,* kernel, + *oid,* like). The region in a prokaryotic cell where the chromosome is found.

nucleolus (nu-klē′ə-ləs) (dim. of L. *nucleus,* kernel). A deeply staining body within the nucleus of a cell and containing RNA; nucleoli are specialized portions of certain chromosomes that carry multiple copies of the information to synthesize ribosomal RNA.

nucleoplasm (nu′klē-ō-plazm′) (L. *nucleus,* kernel, + Gr. *plasma,* mold). Protoplasm of nucleus, as distinguished from cytoplasm.

nucleoprotein A molecule composed of nucleic acid and protein; occurs in the nucleus and cytoplasm of all cells.

nucleosome (nu′klē-ə-som) (L. *nucleus,* kernel, + *sōma,* body). A repeating subunit of chromatin in which one and three-quarter turns of the double-helical DNA are wound around eight molecules of histones.

nucleotide (nu′klē-ə-tīd) A molecule consisting of phosphate, 5-carbon sugar (ribose or deoxyribose), and a purine or a pyrimidine; the purines are adenine and guanine, and the pyrimidines are cytosine, thymine, and uracil.

nucleus (nū′klē-əs) (L. *nucleus,* a little nut, the kernel). The organelle in eukaryotes that contains the chromatin and which is bounded by a double membrane (nuclear envelope).

nuptial flight (nup′shəl). The mating flight of insects, especially that of the queen with male or males.

bat / āpe / ärmadillo / herring / fēmale / finch / līce / crocodile / crōw / cȯin / duck / ūnicorn / ə indicates unaccented vowel sound "uh" as in mammal, fishes, cardinal, heron, vulture / stress as in bi-ol′o-gy, bi′o-log′i-cal

nurse cells Single cells or layers of cells surrounding or adjacent to other cells or structures for which the nurse cells provide nutrient or other molecules (for example, for insect oocytes or *Trichinella* spp. juveniles).

nymph (L. *nympha*, nymph, bride). An immature stage (following hatching) of a hemimetabolous insect that lacks a pupal stage.

O

ocellus (ō-sel′əs) (L. dim. of *oculus*, eye). A simple eye or eyespot in many types of invertebrates.

octomerous (ok-tom′ər-əs) (Gr. *oct*, eight, + *meros*, part). Eight parts, specifically, symmetry based on eight.

odontophore (ō-don′tə-fōr′) (Gr. *odous*, tooth, + *pherein*, to carry). Tooth-bearing organ in molluscs, including the radula, radular sac, muscles, and cartilages.

olfactory (äl-fakt′(ə)-rē) (L. *olor*, smell, + *factus*, to bring about). Pertaining to the sense of smell.

olfactory epithelium A specialized chemosensory surface tissue inside the nasal cavities of aquatic and terrestrial vertebrates.

omasum (ō-mā′səm) (L. paunch). The third compartment of the stomach of a ruminant mammal.

ommatidium (ä′mə-tid′ē-əm) (Gr. *omma*, eye, + *idium*, small). One of the optical units of the compound eye of arthropods.

omnivore (äm′nə-vōr) (L. *omnis*, all, + *vorare*, to devour). An animal that uses a variety of animal and plant material in its diet.

oncogene (än′kə-jēn) (Gr. *onkos*, protuberance, tumor, + *genos*, descent). Any of a number of genes associated with neoplastic growth (cancer). The gene in its benign state, either inactivated or performing its normal role, is a **proto-oncogene.**

oncomiracidium (än′kō-mīr′ə-sid′ē-əm) (Gr. *onkos*, barb, hook, + *meirakidion*, youthful person). A ciliated larva of a monogenetic trematode.

oncosphere (än′kəs-fēr) (Gr. *onkinos*, a hook, + *sphaira*, ball). Rounded larva common to all cestodes, bears hooks.

ontogeny (än-tä′jə-nē) (Gr. *ontos*, being, + *geneia*, act of being born, from *genēs*, born). The course of development of an individual from egg to senescence.

oocyst (ō′ə-sist) (Gr. *ōion*, egg, + *kystis*, bladder). Cyst formed around zygote of malaria and related organisms.

oocyte (ō′ə-sīt) (Gr. *ōion*, egg, + *kytos*, hollow). Stage in formation of ovum, just preceding first meiotic division (primary oocyte) or just following first meiotic division (secondary oocyte).

ooecium (ō-ēs′ē-əm) (Gr. *ōion*, egg, + *oikos*, house, + L. *ium*, from). Brood pouch; compartment for developing embryos in ectoprocts.

oogenesis (ō-ə-jen′ə-səs) (Gr. *ōion*, egg, + *genesis*, descent). Formation, development, and maturation of a female gamete or ovum.

oogonium (ō′ə-gōn′ē-əm) (Gr. *ōion*, egg, + *gonos*, offspring). A cell that, by continued division, gives rise to oocytes; an ovum in a primary follicle immediately before the beginning of maturation.

ookinete (ō-ə-kī′nēt) (Gr. *ōion*, egg, + *kinein*, to move). The motile zygote of malarial parasites.

ootid (ō-ə-tid′) (Gr. *ōion*, egg, + *idion*, dim.). Stage of formation of ovum after second meiotic division following expulsion of second polar body.

ootype (ō′ə-tīp) (Gr. *ōion*, egg, + *typos*, mold). Part of oviduct in flatworms that receives ducts from vitelline glands and Mehlis' gland.

operculum (ō-per′kū-ləm) (L. cover). The gill cover in bony fishes; protective plate in some snails.

operon (äp′ə-rän). A genetic unit consisting of a cluster of genes under the control of other genes, found in prokaryotes.

ophthalmic (äf-thal′mik) (Gr. *ophthalamos*, an eye). Pertaining to the eye.

opisthaptor (ä′pəs-thap′tər) (Gr. *opisthen*, behind, + *haptein*, to fasten). Posterior attachment organ of a monogenetic trematode.

opisthokont (ō-pis′thə-känt) (G. *opisthen*, behind, + *kontos*, a pole). Any member of a eukaryotic clade comprising fungi, microsporidians, choanoflagellates, and metazoans; if present, flagellated cells possess a single posterior flagellum.

opisthonephros (ō-pisth′ō-nef-rōs) (Gr. *opisth*, back + *nephros*, kidney). A kidney that develops from the middle and posterior portions of the nephrogenic region of vertebrates and is drained by the Wolffian duct or accessory ducts. Functional kidney of most adult anamniotes (fishes and amphibians). Adj., **opisthonephric.**

opisthosoma (ō-pis′thə-sō′mə) (Gr. *opisthe*, behind, + *sōma*, body). Posterior body region in arachnids and pogonophorans.

opsonization (op′sən-i-zā′shən) (Gr. *opsonein*, to buy victuals, to cater). The facilitation of phagocytosis of antigens by phagocytes in the blood or tissues. Mediated by antibody bound to the particles to form antibody-antigen complexes, or by complement proteins (vertebrates) or complement-like proteins (invertebrates) bound to antigen.

oral disc The end of a cnidarian polyp bearing the mouth.

oral lobe A flaplike extension of the mouth of a scyphozoan medusa that aids in feeding.

orbit (L. circle). The cavity of the skull housing the eyeball.

organelle (Gr. *organon*, tool, organ, + L. *ella*, dim.). Specialized part of a cell; literally, a small organ that performs functions analogous to organs of multicellular animals.

organism (or′-gə-niz′-əm). A biological individual composed of one or more cells, tissues, and/or organs whose parts are interdependent in producing a collective physiological system. Organisms of the same species may form **populations.**

organizer (or′gan-ī-zer) (Gr. *organos*, fashioning). Area of an embryo that directs subsequent development of other parts.

orthogenesis (ōr′thō-jen′ə-səs). A unidirectional trend in the evolutionary history of a lineage as revealed by the fossil record; also, a now discredited, anti-Darwinian evolutionary theory, popular around 1900, postulating that genetic momentum forced lineages to evolve in a predestined linear direction that was independent of external factors and often led to decline and extinction.

osculum (os′kū-ləm) (L. *osculum*, a little mouth). Excurrent opening in a sponge.

osmole Molecular weight of a solute, in grams, divided by the number of ions or particles into which it dissociates in solution. Adj., **osmolar.**

osmoregulation Maintenance of proper internal salt and water concentrations in a cell or in the body of a living organism; active regulation of internal osmotic pressure.

osmosis (oz-mō′sis) (Gr. *ōsmos*, act of pushing, impulse). The flow of solvent (usually water) through a semipermeable membrane.

osmotic potential Osmotic pressure.

osmotroph (oz′mə-trōf) (Gr. *ōsmos*, a thrusting, impulse, + *trophē*, to eat). A heterotrophic organism that absorbs dissolved nutrients.

osphradium (äs-frā′dē-əm) (Gr. *osphradion*, small bouquet, dim. of *osphra*, smell). A chemoreceptive sense organ in aquatic snails and bivalves that tests incoming water.

ossicles (L. *ossiculum*, small bone). Small separate pieces of echinoderm endoskeleton. Also, tiny bones of the middle ear of vertebrates.

osteoblast (os′tē-ō-blast) (Gr. *osteon*, bone, + *blastos*, bud). A bone-forming cell.

osteoclast (os′tē-ō-clast) (Gr. *osteon*, bone, + *klan*, to break). A large, multinucleate cell that functions in bone dissolution.

osteocyte (os′tē-ə-sīt) (Gr. *osteon*, bone, + *kytos*, hollow). A bone cell that is characteristic of adult bone, has developed from an osteoblast, and is located in a lacuna of the bone substance.

osteoderm (äs′tē-ə-dərm) (Gr. *osteon*, bone, + *derma*, skin). A bony, dermal plate located under and supporting an epidermal scale.

osteon (os′tē-on) (Gr. bone). Unit of bone structure; Haversian system.

osteostracans (os-tē-os′trə-kəns) (Gr. *osteon*, bone, + *ostrakon*, shell). A group of jawless, extinct fishes with dermal armor and pectoral fins known from the Silurian and Devonian periods.

ostium (L. door). Opening.

otolith (ō′-tō-lith) (Gr. *ous, otos*, ear, + *lithos*, stone). Calcareous concretions in the membranous labyrinth of the inner ear of some vertebrates, or in the auditory organ of certain invertebrates.

outgroup In phylogenetic systematic studies, a species or group of species closely related to but not included within a taxon whose phylogeny is being studied, and used to polarize variation of characters and to root the phylogenetic tree.

outgroup comparison A method for determining the polarity of a character in cladistic analysis of a taxonomic group. Character states found within the group being studied are judged ancestral if they occur also in related taxa outside the study group (= outgroups); character states that occur only within the taxon being studied but not in outgroups are judged to have been derived evolutionarily within the group being studied.

oviger (ō′vi-jər) (L. *ovum*, egg, + *gerere*, to bear). Leg that carries eggs in pycnogonids.

oviparity (ō′və-pa′rə-tē) (L. *ovum*, egg, + *parere*, to bring forth). Reproduction in which eggs are released by the female; development of offspring occurs outside the maternal body. Adj., **oviparous** (ō-vip′ə-rəs).

ovipositor (ō′və-päz′ə-tər) (L. *ovum*, egg, + *positor*, builder, placer, + *or*, suffix denoting agent or doer). In many female insects a structure at the posterior end of the abdomen for laying eggs.

ovoviviparity (ō′vo-vī-və-par′ə-tē) (L. *ovum*, egg, + *vivere*, to live, + *parere*, to bring forth). Reproduction in which eggs develop within the maternal body without additional nourishment from the parent and hatch within the parent, or immediately after laying. Adj., **ovoviviparous** (ō′vo-vī-vip′ə-rəs).

ovum (L. *ovum*, egg). Mature female germ cell (egg).

oxidation (äk′sə-dā-shən) (Fr. *oxider*, to oxidize, from Gr. *oxys*, sharp, + *ation*). The loss of an electron by an atom or molecule; sometimes addition of oxygen chemically to a substance. Opposite of reduction, in which an electron is accepted by an atom or molecule.

oxidative metabolism Cellular respiration using molecular oxygen as the final electron acceptor.

oxidative phosphorylation (äk′sə-dād′iv fäs′fər-i-lā′shən). The conversion of inorganic phosphate to energy-rich phosphate of ATP, involving electron transport through a respiratory chain to molecular oxygen.

P

p53 protein A tumor-suppressor protein with critical functions in normal cells. A mutation in the gene that encodes it, *p53*, can cause loss of control over cell division and thus may lead to cancer.

paedogenesis (pē-dō-jen′ə-sis) (Gr. *pais*, child, + *genēs*, born). Reproduction by immature or larval animals caused by acceleration of maturation. Progenesis.

paedomorphosis (pē-dō-mor′fə-səs) (Gr. *pais*, child, + *morphē*, form). Retention of ancestral juvenile features in later stages of the ontogeny of descendants.

pair bond An affiliation between an adult male and an adult female for reproduction. Characteristic of monogamous species.

pangenesis (pan-jen′ə-sis) (Gr. *pan*, all, + *genesis*, descent). Darwin's hypothesis that hereditary characteristics are carried by individual body cells that produce particles that collect in the germ cells.

papilla (pə-pil′ə) (L. nipple). A small nipplelike projection. A vascular process that nourishes the root of a hair, feather, or developing tooth.

papula (pa′pū-lə) (L. pimple). Respiratory processes on skin of sea stars; also, pustules on skin.

parabasalid (pa′rə-bā′sə-lid) (Gr. *para*, beside, + *basis*, body). Any member of a protozoan clade having a flagellum and parabasal bodies.

parabiosis (pa′rə-bī-ō′sis) (Gr. *para*, beside, + *biosis*, mode of life). The fusion of two individuals, resulting in mutual physiological intimacy.

parabronchi (par-ə-brong′kī) (Gr. *para*, beside, + *bronchos*, windpipe). Fine air-conduction pathways of the bird lung.

paradigm A powerful scientific theory that explains diverse observations and guides active research, such as Darwin's theory of common descent of life.

paramylon granules (par′ə-mī-lən) (Gr. *para*, beside, + *mylos*, mill, grinder). Organelles containing the starchlike substance paramylon; in some algae and flagellates.

parapatric speciation A branching of population lineages to form separate species in which the diverging lineages are mostly nonoverlapping in geographic distribution but make contact along a narrow borderline. This controversial mode of speciation contrasts with allopatric speciation and sympatric speciation.

paraphyly (par′ə-fī-lē) (Gr. *para*, beside, + *phyle*, tribe). The condition that a taxon or other group of organisms contains the most recent common ancestor of all members of the group but excludes some descendants of that ancestor; contrasts with **monophyly** and **polyphyly.**

parapodium (pa′rə-pō′dē-əm) (Gr. *para*, beside, + *pous, podos*, foot). One of the paired lateral processes on each side of most segments in polychaete annelids; variously modified for locomotion, respiration, or feeding.

parasite (par′ə-sīt). An organism that lives physically on or in and at the expense of another organism.

parasitism (par′ə-sīt′izəm) (Gr. *parasitos*, from *para*, beside, + *sitos*, food). The condition of an organism living in or on another organism (host) at whose expense the parasite is maintained; destructive symbiosis.

parasitoid (par′ə-si′-toid). An organism that is a typical parasite early in its development but that finally kills the host during or at the completion of development; used to reference many insect parasites of other insects.

parasympathetic (par′ə-sim-pə-thet′ik) (Gr. *para*, beside, + *sympathes*, sympathetic, from *syn*, with, + *pathos*, feeling). One of the subdivisions of the autonomic nervous system, whose neurons originate in the brain and enter the periphery through the brain stem and posterior part of the spinal cord.

parenchyma (pə-ren′kə-mə) (Gr. anything poured in beside). In lower animals, a spongy mass of vacuolated mesenchyme cells filling spaces between viscera, muscles, or epithelia; in some, cell bodies of muscle cells. Also, the specialized tissue of an organ as distinguished from the supporting connective tissue.

parenchymula (pa′rən-kī′mū-lə) (Gr. *para*, beside, + *enchyma*, infusion). Flagellated, solid-bodied larva of some sponges.

parietal (pä-rī-ə-təl) (L. *paries*, wall). Something next to, or forming part of, a wall of a structure.

parsimony (pär′sə-mō-nē) (L. *parsus*, to spare). A general methodological principle that the simplest hypothesis capable of explaining observations is the best working hypothesis and should be tested first before investigating more complex hypotheses. In phylogenetic systematics, this principle involves using the phylogenetic tree that requires the smallest amount of evolutionary change as the best working hypothesis of phylogenetic relationships.

parthenogenesis (pär′thə-nō-gen′ə-sis) (Gr. *parthenos*, virgin, + L. from Gr. *genesis*, origin). Unisexual reproduction involving the production of young by females not fertilized by males; common in rotifers, cladocerans, aphids, bees, ants, and wasps. A parthenogenetic egg may be diploid or haploid.

particulate inheritance Theories of heredity in which hereditary factors are discrete entities that do not blend when transmitted through the same organism, such as the paired factors identified in Mendel's experiments.

pathogenic (path′ə-jen′ik) (Gr. *pathos*, disease, + N.L. *genic*, giving rise to). Producing or capable of producing disease.

PCR See **polymerase chain reaction.**

peck order A hierarchy of social privilege in a flock of birds.

pecten (L. comb). Any of several types of comblike structures on various organisms, for example, a pigmented, vascular, and comblike process that projects into the vitreous humor from the retina at a point of entrance of the optic nerve in the eyes of all birds and many reptiles.

pectines (pek′tīnz) (L. comb, pl. of **pecten**). Sensory appendage on abdomens of scorpions.

pectoral (pek′tə-rəl) (L. *pectoralis*, from *pectus*, the breast). Of or pertaining to the breast or

---

bat / āpe / ärmadillo / herring / fēmale / finch / līce / crocodile / crōw / cŏin / duck / ūnicorn / ə indicates unaccented vowel sound "uh" as in mammal, fishes, cardinal, heron, vulture / stress as in bi-ol′o-gy, bi′o-log′i-cal

chest; to the pectoral girdle; or to a pair of keratinized shields of the plastron of certain turtles.

pedal laceration Asexual reproduction in sea anemones, a form of fission.

pedalium (pə-dal′ē-əm) (L. *pedalis,* of or belonging to the foot). Flattened blade at the base of the tentacles in cubozoan medusae (Cnidaria).

pedicel (ped′ə-sel) (L. *pediculus,* little foot). A small or short stalk or stem. In insects, the second segment of an antenna or the waist of an ant.

pedicellaria (ped′ə-sə-lar′ē-ə) (L. *pediculus,* little foot, + *aria,* like or connected with). One of many minute pincerlike organs on the surface of certain echinoderms.

pedipalps (ped′ə-palps′) (L. *pes, pedis,* foot, + *palpus,* stroking, caress). Second pair of appendages of arachnids.

pedogenesis See **paedogenesis.**

peduncle (pē′dun-kəl) (L. *pedunculus,* dim. of *pes,* foot). A stalk. Also, a band of white matter joining different parts of the brain.

pelage (pel′ij) (Fr. fur). Hairy covering of mammals.

pelagic (pə-laj′ik) (Gr. *pelagos,* the open sea). Occupying or moving through water rather than the underlying substrate; contrasts with **benthic** (see **benthos**).

pellicle (pel′ə-kəl) (L. *pellicula,* dim. of *pellis,* skin). Thin, translucent, secreted envelope covering many protozoa.

pelvic (pel′vik) (L. *pelvis,* a basin). Situated at or near the pelvis, as applied to girdle, cavity, fins, and limbs.

pelycosaur (pel′ə-kō-sor) (Gr. *pelyx,* basin, + *sauros,* lizard). Any of a group of Permian synapsids characterized by homodont dentition and sprawling limbs.

pen A flattened flexible internal support in a squid; a reminant of the ancestral shell.

pentadactyl (pen-tə-dak′təl) (Gr. *pente,* five, + *daktylos,* finger). With five digits, or five fingerlike parts, to the hand or foot.

pentamerous symmetry (pen-tam′ər-əs) (Gr. *pente,* five, + *meros,* part). A radial symmetry based on five or multiples thereof.

peptidase (pep′tə-dās) (Gr. *peptein,* to digest, + *ase,* enzyme suffix). An enzyme that splits peptides, releasing smaller peptides or amino acids.

peptide bond A bond that binds amino acids together into a polypeptide chain, formed by removing an OH from the carboxyl group of one amino acid and an H from the amino group of another to form an amide bond—CO—NH—.

perennibranchiate (pə-ren′ə-brank′ē-āt) (L. *perennis,* throughout the year, + Gr. *branchia,* gills). Having permanent gills, relating especially to certain paedomorphic salamanders.

pericardium (pə-ri-kär′dē-əm) (Gr. *peri,* around, + *kardia,* heart). Area around heart; membrane around heart.

periostracum (pe-rē-äs′trə-kəm) (Gr. *peri,* around, + *ostrakon,* shell). Outer layer of a mollusc shell.

peripheral (pə-ri′fər-əl) (Gr. *peripherein,* to move around). Structure or location distant from center, near outer boundaries.

periproct (per′ə-präkt) (Gr. *peri,* around, + *prōktos,* anus). Region of aboral plates around the anus of echinoids.

perisarc (per′ə-särk) (Gr. *peri,* around, + *sarx,* flesh). Sheath covering the stalk and branches of a hydroid.

perissodactyl (pə-ris′ə-dak′təl) (Gr. *perissos,* odd, + *daktylos,* finger, toe). Pertaining to an order of ungulate mammals with an odd number of digits.

peristalsis (per′ə-stal′səs) (Gr. *peristaltikos,* compressing around). The series of alternate relaxations and contractions that force food through the alimentary canal.

peristomium (per′ə-stō′mē-əm) (Gr. *peri,* around, + *stoma,* mouth). Foremost true segment of an annelid; it bears the mouth.

peritoneum (per′ə-tə-nē′əm) (Gr. *peritonaios,* stretched around). The membrane that lines the coelom and covers the coelomic viscera.

Permian extinction A mass extinction that occurred 245 million years ago in which 96% of existing species became extinct, marking the end of the Paleozoic era.

perpetual change The most basic theory of evolution, that the living world is neither constant nor cycling, but is always undergoing irreversible modification through time.

petaloids (pe′tə-loids) (Gr. *petalon,* leaf, + *eidos,* form). Describes flowerlike arrangement of respiratory podia in irregular sea urchins.

pH (*potential of* **h**ydrogen). A symbol referring to the relative concentration of hydrogen ions in a solution; pH values are from 0 to 14, and the lower the value, the more acid or hydrogen ions in the solution. Equal to the negative logarithm of the hydrogen ion concentration.

phagocyte (fag′ō-sīt) (Gr. *phagein,* to eat, + *kytos,* hollow vessel). Any cell that engulfs and devours microorganisms or other particles.

phagocytosis (fag′ō-sī-tō-səs) (Gr. *phagein,* to eat, + *kytos,* hollow vessel). The engulfment of a particle by a phagocyte or a protozoan.

phagosome (fa′gō-sōm) (Gr. *phagein,* to eat, + *sōma,* body). Membrane-bound vesicle in cytoplasm containing food material engulfed by phagocytosis.

phagotroph (fag′ō-trōf) (Gr. *phagein,* to eat, + *trophē,* food). A heterotrophic organism that ingests solid particles for food.

pharynx (far′inks), pl. **pharynges** (Gr. *pharynx,* gullet). The part of the digestive tract between the mouth cavity and the esophagus that in vertebrates, is common to both digestive and respiratory tracts. In cephalochordates the gill slits open from it.

phasmid (faz′mid) (Gr. *phasma,* apparition, phantom, + *id*). One of a pair of glands or sensory structures in the posterior end of certain nematodes.

phenetic taxonomy (fə-ne′tik) (Gr. *phaneros,* visible, evident). Refers to the use of a criterion of overall similarity to classify organisms into taxa; contrasts with classifications based explicitly on a reconstruction of phylogeny.

phenotype (fē′nō-tīp) (Gr. *phainein,* to show). The visible or expressed characteristics of an organism, influenced by the genotype, but not all genes in the genotype are expressed.

phenotypic gradualism The hypothesis that new traits, even those that are strikingly different from ancestral ones, evolve by a long series of small, incremental steps.

pheromone (fer′ō-mōn) (Gr. *pherein,* to carry, + *hormōn,* exciting, stirring up). Chemical substance released by one organism that influences the behavior or physiological processes of another organism.

phosphagen (fäs′fə-jən) (phosphate + gen). A term for creatine phosphate and arginine phosphate, which store and may be sources of high-energy phosphate bonds.

phosphatide (fäs′fə-tīd′) (phosphate + ide). A complex phosphoric ester lipid, such as lecithin, found in all cells. Phospholipid.

phosphorylation (fäs′fə-rə-lā′shən). The addition of a phosphate group, that is, —PO₃, to a compound.

photic (fō′-tik). Sunlit portions of oceanic waters inhabited by photosynthetic organisms.

photoautotroph (fōt-ō-aw′-tō-trōf) (Gr. *photōs,* light, + *autos,* self, + *trophos,* feeder). An organism requiring light as a source of energy for making organic nutrients from inorganic raw materials.

photosynthesis (fōt-ō-sin′thə-sis) (Gr. *phōs,* light, + *synthesis,* action or putting together). The synthesis of carbohydrates from carbon dioxide and water in chlorophyll-containing cells exposed to light.

phototaxis (fōt′ō-tak′sis) (Gr. *phōs,* light, + *taxis,* arranging, order). A taxis in which light is the orienting stimulus. An involuntary tendency for an organism to turn toward (positive) or away from (negative) light.

phototrophs (fōt′-ō-trōfs) (Gr. *phōs, phōtos,* light, + *trophē,* nourishment). Organisms capable of using CO₂ in the presence of light as a source of metabolic energy.

phyletic gradualism A model of evolution in which morphological evolutionary change is continuous and incremental and occurs mainly within unbranched species or lineages over long periods of geological time; contrasts with **punctuated equilibrium.**

phyllopodium (fī′lə-pō′dē-əm) (Gr. *phyllon,* leaf, + *pous, podos,* foot). Leaflike swimming appendage of branchiopod crustaceans.

phylogenetic species concept An irreducible (basal) cluster of organisms, diagnosably distinct from other such clusters, and within which there is a parental pattern of ancestry and descent.

phylogenetic systematics See **cladistics.**

phylogenetic tree A tree diagram whose branches represent current or past evolutionary lineages and which shows the hypothesized patterns of common descent among those lineages.

phylogeny (fī-loj′ə-nē) (Gr. *phylon*, tribe, race, + *geneia*, origin). The origin and diversification of any taxon, or the evolutionary history of its origin and diversification, usually presented in the form of a dendrogram.

phylum (fī′ləm), pl. **phyla** (N.L. from Gr. *phylon*, race, tribe). A chief category, between kingdom and class, of taxonomic classifications into which are grouped organisms of common descent that share a fundamental pattern of organization.

physiology (L. *physiologia*, natural science). A branch of biology covering the organic processes and phenomena of an organism or any of its parts or of a particular bodily process.

phytoflagellates (fī-tə-fla′jə-lāts). Members of the former class Phytomastigophorea, plantlike flagellates.

phytophagous (fī-täf′ə-gəs) (Gr. *phyton*, plant, + *phagein*, to eat). Organisms that feed on plants.

pinacocyte (pin′ə-kō-sīt′) (Gr. *pinax*, tablet, + *kytops*, hollow vessel). Flattened cells composing dermal epithelium in sponges.

pinacoderm (pə-nak′ə-dərm) (Gr. *pinax*, plank, tablet, + *derma*, skin). The layer of pinacocytes in sponges.

pinna (pin′ə) (L. feather, sharp point). The external ear. Also a feather, wing, or fin or similar part.

pinocytosis (pin′o-sī-tō′sis, pīn′o-sī-tō′sis) (Gr. *pinein*, to drink, + *kytos*, hollow vessel, + *osis*, condition). Fluid acquisition by a cell in which specific receptors bind ions/molecules present on plasma membranes, which are invaginated and pinch off to form small vesicles. See **caveolae.**

placenta (plə-sen′tə) (L. flat cake). The vascular structure, embryonic and maternal, through which the embryo and fetus are nourished while in the uterus.

placentotrophy (plə-sent′ō-trō′fē) (L. *placenta*, flat cake, + *trophos*, one who feeds). Nutrition of an embryo from a placenta.

placode (pla′kōd) (Gr. *plakos*, flat round plate). Localized, platelike thickening of vertebrate head ectoderm from which a specialized structure develops; such structures include eye lens, special sense organs, and certain neurons.

placoderms (plak′ə-dərm) (Gr. *plax*, plate, + *derma*, skin). A group of heavily armored jawed fishes of the Silurian and Devonian periods.

placoid scale (pla′kòid) (Gr. *plax*, *plakos*, tablet, plate). Type of scale found in cartilaginous fishes, with basal plate of dentin embedded in the skin and a backward-pointing spine tipped with enamel.

plankton (plank′tən) (Gr. neuter of *planktos*, wandering). The passively floating animal and plant life of a body of water; contrasts with **nekton.**

plantigrade (plan′tə-grād′) (L. *planta*, sole, + *gradus*, step, degree). Pertaining to animals that walk on the whole surface of the foot (for example, humans and bears); contrasts with **digitigrade.**

planula (plan′yə-lə) (N.L. dim. from L. *planus*, flat). Free-swimming, ciliated larval type of cnidarians; usually flattened and ovoid, with an outer layer of ectodermal cells and an inner mass of endodermal cells.

planuloid ancestor (plan′yə-lòid) (L. *planus*, flat, + Gr. *eidos*, form). Hypothetical form representing ancestor of Cnidaria and Platyhelminthes.

plasma cell (plaz′mə) (Gr. *plasma*, a form, mold). A descendant cell of a B cell, functions to secrete antibodies.

plasma membrane (plaz′mə) (Gr. *plasma*, a form, mold). A living, external, limiting, protoplasmic structure that functions to regulate exchange of nutrients across the cell surface.

plasmalemma (plaz′mə-lem-ə) (Gr. *plasma*, a form, mold, + *lemma*, rind, sheath). The cell membrane or plasma membrane.

plasmid (plaz′məd) (Gr. *plasma*, a form, mold). A small circle of DNA that may be carried by a bacterium in addition to its genomic DNA.

plasmodium (plaz-mō′dē-əm) (Gr. *plasma*, a form, mold, + *eidos*, form). Multinucleate ameboid mass, syncytial.

plastid (plas′təd) (Gr. *plast*, formed, molded, + L. *id*, feminine stem for particle of specified kind). A membranous organelle in plant cells functioning in photosynthesis and/or nutrient storage, for example, chloroplast.

plastron (plast′trən) (Fr. *plastron*, breast plate). Ventral body shield of turtles; structure in corresponding position in certain arthropods; thin film of gas retained by epicuticle hairs of aquatic insects.

platelet (plāt′lət) (Gr. dim. of *plattus*, flat). A tiny cell fragment in the blood that releases substances initiating blood clotting.

plate tectonics Geological shifting of position by rocky plates of the earth's crust relative to underlying layers. This theory explains the changing positions of continents through geological time, formation of mountain ranges, and patterns in the formation of volcanic-island archipelagoes.

pleiotropy, pleiotropic (plī-ə-trō′pic) (Gr. *pleiōn*, more, + *tropos*, to turn). Pertaining to a gene producing more than one effect; affecting multiple phenotypic characteristics.

pleopod (plē′ə-päd) (Gr. *plein*, to sail, + *pous*, *podos*, foot). One of the swimming appendages on the abdomen of a crustacean.

plesiomorphic (plē′sē-ə-mōr′fik). An ancestral condition of a variable character.

pleura (plu′rə) (Gr. side, rib). The membrane that lines each half of the thorax and covers the lungs.

plexus (plek′səs) (L. network, braid). A network, especially of nerves or blood vessels.

pluteus (plü′tē-əs), pl. **plutei** (L. *pluteus*, movable shed, reading desk). Echinoid or ophiuroid larva with elongated processes like the supports of a desk; originally called "painter's easel larva."

pneumatophore (nū-ma′-tə-fōr) (Gr. *pneumatos*, wind, breathing, + *phōros*, bearing). A gas-filled float of the Portuguese man-of-war and some other siphonophores, which are specialized hydrozoan colonies (phylum Cnidaria).

pneumostome (nū′mə-stōm) (Gr. *pneuma*, breathing, + *stoma*, mouth). The opening of the mantle cavity (lung) of pulmonate gastropods to the outside.

podium (pō′de-əm) (Gr. *pous*, *podos*, foot). A footlike structure, for example, the tube foot of echinoderms.

poikilothermic (pòi-ki′lə-thər′mik) (Gr. *poikilos*, variable, + thermal). Pertaining to animals whose body temperature is variable and fluctuates with that of the environment; cold blooded; contrasts with **ectothermic.**

polarity (Gr. *polos*, axis). In systematics, the ordering of alternative states of a taxonomic character from evolutionarily ancestral to derived conditions. In developmental biology, the tendency for the axis of an ovum to orient corresponding to the axis of the mother. Also, condition of having opposite poles; differential distribution of gradation along an axis.

polarization (L. *polaris*, polar, + Gr. *iz*, make). The arrangement of positive electrical charges on one side of a surface membrane and negative electrical charges on the other side (in nerves and muscles).

Polian vesicles (pō′le-ən) (from G. S. Poli, Italian naturalist). Vesicles opening into ring canal in most asteroids and holothuroids.

polyandry (pol′ē-an′drē) (Gr. *polys*, many, + *anēr*, man). Condition of having more than one male mate at one time.

polyembryony Asexual proliferation of a single fertilized egg to produce many embryos.

polygamy (pə-lig′ə-mē) (Gr. *polys*, many, + *gamos*, marriage). Condition of having more than one mate at a time.

polygenic inheritance Inheritance of traits influenced by multiple alleles; traits show continuous variation between extremes; offspring are usually intermediate between the two parents; also known as **blending** and **quantitative inheritance.**

polygyny (pə-lij′ə-nē) (Gr. *polys*, many, + *gynē*, woman). Condition of having more than one female mate at one time.

polymer (pä′lə-mər) (Gr. *polys*, many, + *meros*, part). A chemical compound composed of repeated structural units called monomers.

polymerase chain reaction (PCR) A technique for preparing large quantities of DNA from tiny samples, making it easy to clone

---

bat / āpe / ärmadillo / herring / fēmale / finch / līce / crocodile / crōw / cóin / duck / ūnicorn / ə indicates unaccented vowel sound "uh" as in mammal, fishes, cardinal, heron, vulture / stress as in bi-ol′o-gy, bi′o-log′i-cal

a specific gene as long as part of the sequence of the gene is known.

polymerization (pə-lim′ər-ə-zā′shən). The process of forming a polymer or polymeric compound.

polymorphism (pä′lē-mor′fi-zəm) (Gr. *polys*, many, + *morphē*, form). The presence in a species of more than one structural type of individual; genetic variation in a population.

polynucleotide (poly + nucleotide). A nucleic acid having many nucleotides combined in a linear chain.

polyp (pä′lip) (Gr. *polypous*, many-footed). Individual of the phylum Cnidaria, generally adapted for attachment to the substratum at the aboral end, often forms colonies.

polypeptide (pä-lē-pep′tīd) (Gr. *polys*, many, + *peptein*, to digest). A molecule consisting of a single linear chain of amino acids.

polyphyly (päl′ē-fī′lē) (Gr. *polys*, many, + *phylon*, tribe). The condition that a taxon or other group of organisms does not contain the most recent common ancestor of all members of the group, implying that it has multiple evolutionary origins; such groups are not valid as formal taxa and are recognized as such only through error. Contrasts with **monophyly** and **paraphyly.**

polyphyodont (pä′lē-fī′ə-dänt) (Gr. *polyphyes*, manifold, + *odous*, tooth). Having several sets of teeth in succession.

polypide (pä′li-pīd) (L. *polypus*, polyp). An individual or zooid in a colony, specifically in ectoprocts, which has a lophophore, digestive tract, muscles, and nerve centers.

polyploid (pä′lə-ploid′) (Gr. *polys*, many, + *ploidy*, number of chromosomes). An organism possessing more than two full homologous sets of chromosomes.

polysaccharide (pä′lē-sak′ə-rid, -rīd). (Gr. *polys*, many, + *sakcharon*, sugar, from Sanskrit *sarkarā*, gravel, sugar). A carbohydrate composed of many monosaccharide units, for example, glycogen, starch, and cellulose.

polysome (polyribosome) (Gr. *polys*, many, + *sōma*, body). Two or more ribosomes connected by a molecule of messenger RNA.

polyspermy (pä′lē-spər′mē) (Gr. *polys*, many, + *sperma*, seed). Entrance of more than one sperm during fertilization of an egg.

polytene chromosomes (pä′li-tēn) (Gr. *polys*, many, + *tainia*, band). Chromosomes in some somatic cells of some insects in which the chromatin replicates repeatedly without undergoing mitosis.

pongid (pän′jəd) (L. *Pongo*, type genus of orangutan). Of or relating to the formerly recognized but now discontinued family Pongidae, comprising the anthropoid apes (gorillas, chimpanzees, gibbons, orangutans).

population (L. *populus*, people). A group of organisms of the same species inhabiting a specific geographical locality.

populational gradualism The observation that new genetic variants become established in a population by increasing their frequencies across generations incrementally, initially from one or a few individuals and eventually characterizing a majority of the population.

porocyte (pō′rə-sīt) (Gr. *porus*, passage, pore, + *kytos*, hollow vessel). Type of cell found in asconoid sponges through which water enters the spongocoel.

portal system (L. *porta*, gate). System of large veins beginning and ending with a bed of capillaries; for example, hepatic portal and renal portal system in vertebrates.

positive assortative mating A tendency for individuals to mate with others that resemble themselves for one or more varying traits in a population.

posterior (L. latter). Situated at or toward the rear of the body; in bilateral forms, the end of the main body axis opposite the head.

preadaptation The possession of a trait that coincidentally predisposes an organism for survival in an environment different from those encountered in its evolutionary history.

prebiotic synthesis The chemical synthesis that occurred before the emergence of life.

precocial (prē-kō′shəl) (L. *praecoquere*, to ripen beforehand). An organism capable of independent living at or shortly after birth, requiring little parental care; contrasts with **altricial.**

predaceous, predacious (prē-dā′shəs) (L. *praedator*, a plunderer, *praeda*, prey). Living by killing and consuming other animals; predatory.

predation (prə-dā′shən). An interaction between species in an ecological community in which members of one species (prey) serve as food for another species **(predator).**

predator (pred′ə-tər) (L. *praedator*, a plunderer, *praeda*, prey). An organism that preys on other organisms for its food.

preformation Discredited concept that gametes contained already-formed young that unfolded or expanded during development.

prehensile (prē-hen′səl) (L. *prehendere*, to seize). Adapted for grasping.

premunition A resistance to reinfection by an animal (host) when some infective organisms remain in the host's body.

primary bilateral symmetry Usually applied to a radially symmetrical organism descended from a bilateral ancestor and developing from a bilaterally symmetrical larva.

primary heterotroph Denotes the hypothesis that the earliest microorganisms to evolve obtained nutrients from an environment that did not contain autotrophic organisms.

primary organizer Region of an embryo near the dorsal lip of the blastopore capable of inducing embryo development when transplanted to a host embryo; host forms two embryos after transplantation.

primary producer A species whose members begin **productivity** by acquiring energy and matter from **abiotic** sources, such as plants that synthesize sugars from water and carbon dioxide using solar energy (see **photosynthesis**).

primary radial symmetry Usually applied to a radially symmetrical organism that did not have a bilateral ancestor or larva, in contrast to a secondarily radial organism.

primary structure For a protein, the linear sequence of amino acids in a polypeptide chain. For a nucleic acid, the linear sequence of bases in the molecule.

primate (prī′-māt) (L. *primus*, first). Any mammal of the order Primates, which includes the tarsiers, lemurs, marmosets, monkeys, apes, and humans.

primitive (L. *primus*, first). Primordial; ancient; little evolved; characteristics closely approximating those possessed by an early ancestor.

prion (prī′-än). An infectious protein that causes proteins of the host organism to assume a nonstandard and often pathological spatial conformation, as in "mad cow" disease.

proboscis (prō-bäs′əs) (Gr. *pro*, before, + *boskein*, feed). A snout or trunk. Also, tubular sucking or feeding organ with the mouth at the end as in planarians, leeches, and insects. Also, the sensory and defensive organ at the anterior end of certain invertebrates.

producers (L. *producere*, to bring forth). Organisms, such as plants, able to produce their own food from inorganic substances.

production In ecology, the energy accumulated by an organism that becomes incorporated into new biomass.

productivity (prō′dak-tiv′-ət-ē). A property of a biological system measured by the amount of energy and/or materials that it incorporates.

product rule The probability of independent events occurring together is the product of the probabilities of the events occurring separately.

progesterone (prō-jes′tə-rōn′) (L. *pro*, before, + *gestare*, to carry). Hormone secreted by the corpus luteum and the placenta; prepares the uterus for the fertilized egg and maintains the capacity of the uterus to hold the embryo and fetus.

proglottid (prō-glät′əd) (Gr. *proglōttis*, tongue tip, from *pro*, before, + *glōtta*, tongue, + *id*, suffix). Portion of a tapeworm containing a set of reproductive organs; usually corresponds to a segment.

prohormone (prō′hor-mōn) (Gr. *pro*, before, + *hormaein*, to excite). A precursor of a hormone, especially a peptide hormone.

prokaryote, prokaryotic, procaryotic (pro-kar′ē-ät′ik) (Gr. *pro*, before, + *karyon*, kernel, nut). Not having a membrane-bound nucleus or nuclei. Prokaryotic cells characterize the bacteria and cyanobacteria; contrasts with **eukaryotic.**

promoter A region of DNA to which the RNA polymerase must have access for transcription of a structural gene to begin.

pronephros (prō-nef′rōs) (Gr. *pro*, before, + *nephros*, kidney). Most anterior of three pairs of embryonic renal organs of vertebrates, functional only in adult hagfishes and larval fishes and amphibians, and vestigial in amniote embryos. Adj., **pronephric.**

proprioceptor (prō′prē-ə-sep′tər) (L. *proprius*, own, particular, + receptor). Sensory receptor located deep within the tissues, especially muscles, tendons, and joints, that is responsive to changes in muscle stretch, body position, and movement.

prosimian (prō-sim′ē-ən) (Gr. *pro*, before, + L. *simia*, ape). Any member of a group of arboreal primates including lemurs, tarsiers, and lorises, but excluding monkeys, apes, and humans.

prosoma (prō-sōm′ə) (Gr. *pro*, before, + *sōma*, body). Anterior part of an invertebrate in which primitive segmentation is not visible; fused head and thorax of arthropod; cephalothorax.

prosopyle (prōs′ō-pīl) (Gr. *prosō*, forward, + *pyle*, gate). Connections between the incurrent and radial canals in some sponges.

prostaglandins (präs′tə-glan′dəns). A family of fatty-acid hormones, originally discovered in semen, known to have powerful effects on smooth muscle, nerves, circulation, and reproductive organs.

prostomium (prō-stōm′ē-əm) (Gr. *protos*, first, + *stoma*, mouth, + *-idion*, dim. ending). Anterior closure of a metameric animal, anterior to the mouth.

protandrous (prō-tan′drəs) (Gr. *prōtos*, first, + *anēr*, male). Condition of hermaphroditic animals and plants in which male organs and their products appear before the corresponding female organs and products, thus preventing self-fertilization.

protease (prō′tē-ās) (Gr. *protein*, + *ase*, enzyme). An enzyme that digests proteins; includes proteinases and peptidases.

protein (prō′tēn, prō′tē-ən) (Gr. *protein*, from *proteios*, primary). A macromolecule of carbon, hydrogen, oxygen, nitrogen and usually sulfur; composed of chains of amino acids joined by peptide bonds; present in all cells.

proteome (prō′tē-ōm) (Gr. *protein*, + L. suffix *-oma*, group). The set of protein molecules produced by an organism during its lifetime. Scientific study of this phenomenon is called **proteomics.**

prothoracic glands Glands that secrete the hormone ecdysone, or molting hormone, in the prothorax of insects.

prothoracicotropic hormone Hormone secreted by the brain of insects that stimulates the prothoracic gland to secrete ecdysone, or molting hormone.

prothrombin (prō-thräm′bən) (Gr. *pro*, before, + *thrombos*, clot). A constituent of blood plasma that is changed to thrombin by a catalytic sequence that includes thromboplastin, calcium, and plasma globulins; involved in blood clotting.

protist (prō′tist) (Gr. *protos*, first). A member of the paraphyletic kingdom Protista, generally considered to include the protozoan groups and eukaryotic algae.

protocoel (prō′tə-sēl) (Gr. *protos*, first, + *koilos*, hollow). The anterior coelomic compartment in some deuterostomes, corresponds to the axocoel in echinoderms.

protocooperation A mutually beneficial interaction between organisms in which the interaction is not physiologically necessary to the survival of either.

proton A subatomic particle with a positive electrical charge and having a mass of 1836 times that of an electron; found in the nucleus of atoms.

protonephridium (prō′tə-nə-frid′ē-əm) (Gr. *protos*, first, + *nephros*, kidney). Primitive osmoregulatory or excretory organ consisting of a tubule terminating internally with flame bulb or solenocyte; the unit of a flame bulb system.

proto-oncogene See **oncogene.**

protoplasm (prō′tə-plazm) (Gr. *protos*, first, + *plasma*, form). Organized living substance; cytoplasm and nucleoplasm of the cell.

protopod, protopodite (prō′tə-päd, prō-top′ə-dīt) (Gr. *protos*, first, + *pous, podos*, foot). Basal portion of crustacean appendage, containing coxa and basis.

Protostomia (prō′tə-stō′mē-ə) (Gr. *protos*, first, + *stoma*, mouth). A group of phyla in which cleavage is determinate, the coelom (in coelomate forms) is formed by proliferation of mesodermal bands (schizocoelic formation), the mesoderm is formed from a particular blastomere (called 4d), and the mouth is derived from or near the blastopore. Includes the Annelida, Arthropoda, Mollusca, and a number of minor phyla; contrasts with **Deuterostomia.**

proventriculus (prō′ven-trik′ū-ləs) (L. *pro*, before, + *ventriculum*, ventricle). In birds the glandular stomach between the crop and gizzard. In insects, a muscular dilation of the foregut armed internally with chitinous teeth.

proximal (L. *proximus*, nearest). Situated toward or near the point of attachment; opposite of distal, distant.

proximate cause (L. *proximus*, nearest, + *causa*). Explanations of the functioning of a biological system at a particular time and place, such as how an animal performs its metabolic, physiological, and behavioral activities. (=immediate cause)

pseudocoel (sū′do-sēl) (Gr. *pseudēs*, false, + *koilōma*, cavity). A body cavity not lined with peritoneum and not a part of the blood or digestive systems, embryonically derived from the blastocoel.

pseudocoelomate (sū′də-sē′lə-māt) (Gr. *pseudēs*, false, + *koilōma*, cavity, + *ate*, suffix). Having a body cavity formed from a persistent blastocoel and lined with mesoderm on only one side.

pseudopodium (sū′də-pō′dē-əm) (Gr. *pseudēs*, false, + *podion*, small foot, + *eidos*, form). A temporary cytoplasmic protrusion extended out from a protozoan or ameboid cell, and serving for locomotion or for engulfing food.

puff Strands of DNA spread apart at certain locations on giant chromosomes of some flies where that DNA is being transcribed.

pulmonary (pul′mən-ner-ē) (L. *pulmo*, lung, + *aria*, suffix denoting connected to). Relating to or associated with lungs.

punctuated equilibrium A model of evolution in which morphological evolutionary change is discontinuous, being associated primarily with discrete, geologically instantaneous events of speciation leading to phylogenetic branching; morphological evolutionary stasis characterizes species between episodes of speciation; contrasts with **phyletic gradualism.**

pupa (pū′pə) (L. girl, doll, puppet). Inactive quiescent stage of the holometabolous insects. It follows the larval stages and precedes the adult stage.

purine (pū′rēn) (L. *purus*, pure, + *urina*, urine). Organic base with carbon and nitrogen atoms in two interlocking rings. The parent substance of adenine, guanine, and other naturally occurring bases.

pygidium (pī-jid′e-əm) (Gr. *pygē*, rump, buttocks, + *-idion*, dim. ending). Posterior closure of a segmented animal, bearing the anus.

pyrimidine (pī-rim′ə-dēn) (alter. of pyridine, from Gr. *pyr*, fire, + *id*, adj. suffix, + *ine*). An organic base composed of a single ring of carbon and nitrogen atoms; parent substance of several bases found in nucleic acids.

Q

quaternary structure For a protein that contains more than one polypeptide chain, the three-dimensional configuration formed by bonds between side groups of amino acids located on different polypeptide chains.

quantitative inheritance See **polygenic inheritance.**

queen In entomology, the single fully developed female in a colony of social insects such as bees, ants, and termites, distinguished from workers, nonreproductive females, and soldiers.

R

radial canals Canals along the ambulacra radiating from the ring canal of echinoderms; also choanocyte-lined canals in syconoid sponges.

radial cleavage Embryonic development in which early cleavage planes are symmetrical to the polar axis, each blastomere of one tier lying directly above the corresponding blastomere of the next layer; indeterminate cleavage.

radial symmetry A morphological condition in which the parts of an animal are arranged concentrically around an oral-aboral axis, and more than one imaginary plane through this axis yields halves that are mirror images of each other.

Radiata (rä′dē-ä′tə) (L. *radius,* ray). Phyla showing radial symmetry, specifically Cnidaria and Ctenophora.

radiolarian (rā′dē-ə-la′rē-ən) (L. *radius,* ray, spoke of a wheel, + *Lar,* tutelary god of house and field). Amebas with actinopodia and beautiful tests.

radioles (rā′dē-ōlz) (L. *radius,* ray, spoke of a wheel). Featherlike processes from the head of many tubicolous polychaete worms (phylum Annelida), used primarily for feeding.

radula (ra′jə-lə) (L. scraper). Rasping tongue found in most molluscs.

ramicristate (rä-mē-kris′tāt) (L. *ramus,* branch, + *cristatus,* crested). Any member of a protozoan clade having branched tubular cristae in mitochondria; typically ameboid forms, naked or testate, including true slime molds.

Ras protein A protein that initiates a cascade of reactions leading to cell division when a growth factor is bound to the cell surface. The gene encoding Ras becomes an oncogene when a mutation produces a form of Ras protein that initiates the cascade even in the absence of the growth factor.

ratite (ra′tīt) (L. *ratis,* raft). Referring to birds having an unkeeled sternum; contrasts with **carinate.**

realized niche The role actually performed by an organism or population in its ecological community at a particular time and place as constrained by both its intrinsic biological attributes and particular environmental conditions. See also **niche** and **fundamental niche.**

recapitulation Summarizing or repeating; hypothesis that an individual repeats its phylogenetic history in its development.

receptor-mediated endocytosis Endocytosis of large molecules, which are bound to surface receptors in clathrin-coated pits.

recessive An allele that must be homozygous to influence a phenotype.

recombinant DNA DNA from two different species, such as a virus and a mammal, combined into a single molecule.

reciprocal altruism Evolution of a behavioral repertoire in which an organism performs behaviors benefiting other members of the population, possibly at its own increased peril, because such behaviors are returned by individuals that receive its benefits.

redia (rē′dē-ə), pl. **rediae** (rē′dē-ē) (from Redi, Italian biologist). A larval stage in the life cycle of flukes; it is produced by a sporocyst larva, and in turn gives rise to many cercariae.

reduction In chemistry, the gain of an electron by an atom or molecule of a substance; also the addition of hydrogen to, or the removal of oxygen from, a substance.

regulative cleavage See **radial cleavage.**

regulative development Embryonic development determined by interactions among neighboring cells; cell fates are not fixed early in development.

relative fitness A comparison of two or more different genotypes for the average numbers of offspring produced per individual in a population. Based on measurements of relative fitnesses of alternative diploid genotypes, relative fitnesses can be assigned analytically to individual alleles.

releaser (L. *relaxare,* to unloose). Simple stimulus that elicits an innate behavior pattern.

renin (rē′nin) (L. *ren,* kidney). An enzyme produced by the kidney juxtaglomerular apparatus that initiates changes leading to increased blood pressure and increased sodium reabsorption.

rennin (re′nən) (M.E. *renne,* to run). A milk-clotting endopeptidase secreted by the stomach of some young mammals, including bovine calves and human infants.

replication (L. *replicatio,* a folding back). In genetics, the duplication of one or more DNA molecules from the preexisting molecule.

reproductive barrier (L. *re,* + *producere,* to lead forward; M.F. *barriere,* bar). The factors that prevent one sexually propagating population from interbreeding and exchanging genes with another population.

reproductive community A general criterion for the species category shared to some degree by all formal species concepts is that species constitute a reproductively bounded population or lineage of populations that does not freely merge with others in nature.

repugnatorial glands (L. *repugnare,* to resist). Glands secreting a noxious substance for defense or offense, for example, as in the millipedes.

resource (rē′so(ə)rs). An available source of nutrition, energy, or space in which to live.

resource-defense polygyny A male gains reproductive access to multiple females indirectly by holding a critical resource.

respiration (L. *respiratio,* breathing). Gaseous interchange between an organism and its surrounding medium. In the cell, the release of energy by the oxidation of food molecules.

restriction endonuclease An enzyme that cleaves a DNA molecule at a particular base sequence.

rete mirabile (rē′tē mə-rab′ə-lē) (L. wonderful net). A network of small blood vessels so arranged that the incoming blood runs countercurrent to the outgoing blood and thus makes possible efficient exchange between the two bloodstreams. Such a mechanism serves to maintain the high concentration of gases in the fish swim bladder.

reticular (rə-tik′ū-lər) (L. *reticulum,* small net). Resembling a net in appearance or structure.

reticuloendothelial system (rə-tic′ū-lō-en-dō-thēl′ē-əl) (L. *reticulum,* dim. of net, + Gr. *endon,* within, + *thele,* nipple). The fixed phagocytic cells in the tissues, especially the liver, lymph nodes, spleen, and others; also called the mononuclear phagocyte system.

reticulopodia (rə-tik′ū-lə-pō′dē-ə) (L. *reticulum,* dim. of *rete,* net, + *podos, pous,* foot). Pseudopodia that branch and rejoin extensively.

reticulum (rə-tik′yə-ləm) (L. *rete,* dim. *reticulum,* a net). Second stomach of ruminants; a netlike structure.

retina (ret′nə, ret′ən-ə) (L. *rete,* net). The posterior sensory membrane of the eye that receives images.

retortamonad (rē-tort′ə-mō′nəd) (L. *retro,* bend backward, + *monas,* single). Any member of a protozoan clade comprising certain heterotrophic flagellates.

rhabdite (rab′dīt) (Gr. *rhabdos,* rod). Rodlike structures in the cells of the epidermis or underlying parenchyma in certain turbellarians. They are discharged in mucous secretions.

rheoreceptor (rē′ə-rē-cep′tər) (Gr. *rheos,* a flowing, + receptor). A sensory organ of aquatic animals that responds to water current.

rhinarium (rī-na′rē-əm) (Gr. *rhis,* nose). Hairless area surrounding the nose of a mammal.

rhinophore (rī′nə-fōr) (Gr. *rhis,* nose, + *pherein,* to carry). Chemoreceptive tentacles in some molluscs (opisthobranch gastropods).

rhipidistian (rip-ə-dis′tē-ən) (Gr. *rhipis,* fan, + *histion,* sail, web). Member of a group of Paleozoic lobe-finned fishes.

rhizopodia (rī′zə-pō′dē-ə) (Gr. *rhiza,* root, + *podos,* foot). Branched filamentous pseudopodia made by some amebas.

rhopalium (rō-pā′lē-əm) (N.L. from Gr. *rhopalon,* a club). One of the marginal, club-shaped sense organs of certain jellyfishes; tentaculocyst.

rhoptries (rōp′trēz) (Gr. *rhopalon,* club, + *tryō,* to rub, wear out). Club-shaped bodies in Apicomplexa composing one of the structures of the apical complex; open at anterior and apparently functioning in penetration of host cell.

rhynchocoel (ring′kō-sēl) (Gr. *rhynchos,* snout, + *koilos,* hollow). In nemerteans, the dorsal tubular cavity that contains the inverted proboscis. It has no opening to the outside.

ribonucleic acid A linear polymer of nucleotides containing the sugar ribose, often bending to form complex spatial structures stabilized by hydrogen bonds between bases of nonadjacent nucleotides. It functions in gene expression and protein synthesis in all living forms and is hypothesized to have been the basis of early, precellular life.

ribose (rī'bōs) A five-carbon sugar forming part of the fundamental structure of RNA nucleotides, including ATP molecules used to store chemical energy in cellular metabolism.

ribosomal RNA (rRNA) Ribonucleic acids that form the physical structures of ribosomes in association with ribosomal proteins.

ribosome (rī'bə-sōm). Subcellular structure composed of protein and ribonucleic acid. May be free in the cytoplasm or attached to the membranes of the endoplasmic reticulum; functions in protein synthesis.

ritualization In ethology, the evolutionary modification, usually intensification, of a behavior pattern to serve communication. Includes symbolic **ritualized threat displays** that carry mutually understood meanings to establish a dominance hierarchy within a population.

RNA Ribonucleic acid, of which there are several different kinds, such as messenger RNA, ribosomal RNA, and transfer RNA (mRNA, rRNA, tRNA), as well as many structural and regulatory RNAs.

RNA polymerase One of three kinds of enzymes that synthesize RNA using ribonucleotide triphosphates (ATP, CTP, GTP, UTP) and a DNA template. RNA polymerase I synthesizes ribosomal RNA; RNA polymerase II synthesizes messenger RNA, and RNA polymerase II synthesizes transfer RNA.

RNA world Hypothetical stage in the evolution of life on earth in which both catalysis and replication were performed by RNA, not protein enzymes and DNA.

rod One of a group of cells on the retina of a vertebrate eye serving vision in low light.

rostellum (räs'tel'ləm) (L. small beak). Projecting structure on scolex of tapeworm, often with hooks.

rostrum (räs'trəm) (L. ship's beak). A snoutlike projection on the head.

rotational cleavage Unique pattern of early embryo formation in most mammals whereby cells at the second cell division appear to have turned relative to each other.

round dance A behavior performed by a hive bee to notify other members of the hive population that a food source is nearby.

rumen (rū'mən) (L. cud). The large first compartment of the stomach of ruminant mammals. Serves as a fermentation chamber in which bacteria degrade cellulose.

ruminant (rūm'ə-nənt) (L. *ruminare,* to chew the cud). Artiodactyl mammals with a complex four-chambered stomach including a foregut that contains bacteria and serves as a fermentation chamber.

S

saccule (sa'kūl) (L. *sacculus,* small bag). Small chamber of the membranous labyrinth of the inner ear.

sacrum (sā'krəm, sā'krəl) (L. *sacer,* sacred). Bone formed by fused vertebrae to which pelvic girdle is attached; pertaining to the sacrum. Adj., **sacral**.

sagittal (saj'ə-təl) (L. *sagitta,* arrow). Pertaining to the median anteroposterior plane that divides a bilaterally symmetrical organism into right and left halves.

salamander (sal'-ə-man'-der). Denotes any member of the amphibian order Urodela (also called Caudata, sometimes with a different content of fossil forms).

salt (L. *sal,* salt). The reaction product of an acid and a base; dissociates in water solution to negative and positive ions, but not H⁺ or OH⁻.

salt marsh A marine community located on intertidal sand flats typically containing grasses, marsh mussels, burrowing crabs and shrimp, and deposit-feeding polychaetes; provides nursery areas for many marine fishes.

saprophagous (sə-präf'ə-gəs) (Gr. *sapros,* rotten, + *phagos,* from *phagein,* to eat). Feeding on decaying matter; saprobic; saprozoic.

saprophyte (sap'rə-fīt) (Gr. *sapros,* rotten, + *phyton,* plant). A plant living on dead or decaying organic matter.

saprozoic nutrition (sap-rə-zō'ik) (Gr. *sapros,* rotten, + *zōon,* animal). Animal nutrition by absorption of dissolved salts and simple organic nutrients from surrounding medium; also refers to feeding on decaying matter.

sarcolemma (sär'kə-lem'ə) (Gr. *sarx,* flesh, + *lemma,* rind). The thin, noncellular sheath that encloses a striated muscle fiber.

sarcomere (sär'kə-mir) (Gr. *sarx,* flesh, + *meros,* part). Transverse segment of striated muscle forming the fundamental contractile unit.

sarcoplasm (sär'kə-plaz'əm) (Gr. *sarx,* flesh, + *plasma,* mold). The clear, semifluid cytoplasm between the fibrils of muscle fibers.

saturated fatty acid A subunit of fats containing a carboxyl group and a carbon chain in which all carbons are joined by single bonds and each carbon in the chain is bonded to two hydrogen atoms.

sauropterygians (so-räp'tə-rij'ē-əns) (Gr. *sauros,* lizard, + *pteryginos,* winged). Marine Mesozoic reptiles often with paddlelike limbs and elongate necks; includes plesiosaurs, pliosaurs, and placodonts.

scalids (skä-lədz) (Gr. *skalis,* hoe, mattock). Recurved spines on the head of kinorhynchs.

schistosomiasis (shis'tō-sō-mī'-ə-sis) (Gr. *schistos,* divided, + *soma,* body, + *lasis,* a diseased condition). Infection with blood flukes of the genus *Schistosoma.*

schizocoel (skiz'ō-sēl) (Gr. *schizo,* from *schizein,* to split, + *koilōma,* cavity). A coelom formed by the splitting of embryonic mesoderm. Noun, **schizocoelomate,** an animal with a schizocoel, such as an arthropod or mollusc. Adj., **schizocoelous,** the process of forming a coelom by mesodermal splitting.

schizocoelous mesoderm formation (skiz'ō-sēl-ləs). Embryonic formation of the mesoderm as cords of cells between ectoderm and endoderm; splitting of these cords produces the coelomic space.

schizogony (skə-zä'gə-nē) (Gr. *schizein,* to split, + *gonos,* seed). Multiple asexual fission.

scientific revolution Philosopher Thomas Kuhn's term for a phase of scientific discovery in which research reveals flaws in an existing paradigm causing it to be discarded in favor of an alternative one.

sclerite (skler'īt) (Gr. *sklēros,* hard). A hard chitinous or calcareous plate or spicule; one of the plates forming the exoskeleton of arthropods, especially insects.

scleroblast (skler'ə-blast) (Gr. *sklēros,* hard, + *blastos,* germ). An amebocyte specialized to secrete a spicule; found in sponges.

sclerocyte (skler'ə-sīt) (Gr. *sklēros,* hard, + *kytos,* hollow vessel). An amebocyte in sponges that secretes spicules.

sclerotic (skler-äd'ik) (Gr. *sklēros,* hard). Pertaining to the tough outer coat of the eyeball.

sclerotin (skler'-ə-tən) (Gr. *sklērotēs,* hardness). Insoluble, tanned protein permeating the cuticle of arthropods.

sclerotization (sklər-ə-tə-zā'shən). Process of hardening of the cuticle of arthropods by the formation of stabilizing cross linkages between peptide chains of adjacent protein molecules.

scolex (skō'leks) (Gr. *skōlēx,* worm, grub). The holdfast, or so-called head, of a tapeworm; bears suckers and, in some, hooks, and posterior to it new proglottids are differentiated.

scrotum (skrō'təm) (L. bag). The pouch that contains the testes in most mammals.

scyphistoma (sī-fis'tə-mə) (Gr. *skyphos,* cup, + *stoma,* mouth). A stage in the development of scyphozoan jellyfish just after the larva becomes attached, the polyp form of a scyphozoan.

sea-grass bed A coastal subtidal marine community comprising a thicket of sea grasses and associated invertebrates, including hydroids, sponges, ectoprocts, and larvae of sea scallops.

sebaceous (sə-bāsh'əs) (L. *sebaceus,* made of tallow). A type of mammalian epidermal gland that produces a fatty substance.

---

bat / āpe / ärmadillo / herring / fēmale / finch / līce / crocodile / crōw / cóin / duck / ūnicorn / ə indicates unaccented vowel sound "uh" as in mammal, fishes, cardinal, heron, vulture / stress as in bi-ol'o-gy, bi'o-log'i-cal

sebum (sē′bəm) (L. grease, tallow). Oily secretion of the sebaceous glands of the skin.

secondary induction Specification of cell fates due to interactions with cells not belonging to the primary organizer region of the embryo.

secondary structure For a protein, the three-dimensional configuration formed by bond angles between adjacent amino acids in the linear polypeptide chain. A common secondary structure is an alpha helix, which makes helical turns in a clockwise direction like a screw.

second law of thermodynamics Physical systems tend to proceed to a state of increased disorder termed entropy.

sedentary (sed′ən-ter-ē). Stationary, sitting, inactive; staying in one place.

segment-polarity genes Genes active during development to specify anterior-posterior structures within a segment.

segmentation Division of the body into discrete segments or metameres; also called **metamerism.**

selectively permeable Permeable to small particles, such as water and certain inorganic ions, but not to larger molecules.

semelparity (se′məl-pā′ri-tē′). A life history in which individual organisms of a population normally reproduce only one time before dying, although numerous offspring may be produced at the time of reproduction; contrasts with **iteroparity.**

seminiferous (sem-ə-nif′rəs) (L. semen, semen, + ferre, to bear). Pertains to the tubules that produce or carry semen in the testes.

semipermeable (L. semi, half, + permeabilis, capable of being passed through). Permeable to small particles, such as water and certain inorganic ions, but not to larger molecules.

sensillum, pl. **sensilla** (sen-si′ləm) (L. sensus, sense). A small sense organ, especially in the arthropods.

sensitization (sen′-sət-ə-zā′-shən). A kind of learning in which an animal acquires a characteristic response to a particular stimulus; contrasts with **habituation.**

septal filament Unattached edge of an internal partition (septum) in a sea anemone gastrovascular cavity that extends into the cavity and is armed with nematocysts and gland cells.

septum, pl. **septa** (L. fence). A wall between two cavities.

serial homology See **homology.**

serosa (sə-rō′sə) (N.L. from L. serum, serum). The outer embryonic membrane of birds and reptiles; chorion. Also, the peritoneal lining of the body cavity.

serotonin (sir′ə-tōn′ən) (L. serum, serum). A phenolic amine, found in the serum of clotted blood and in many other tissues, that possesses several poorly understood metabolic, vascular, and neural functions; 5-hydroxytryptamine.

serous (sir′əs) (L. serum, serum). Watery, resembling serum; applied to glands, tissue, cells, fluid.

serum (sir′əm) (L. whey, serum). The liquid that separates from the blood after coagulation; blood plasma from which fibrinogen has been removed. Also, the clear portion of a biological fluid separated from its particulate elements.

sessile (ses′əl) (L. sessilis, low, dwarf). Attached at the base; fixed to one spot.

seta (sē′tə), pl. **setae** (sē′tē) (L. bristle). A needlelike chitinous structure of the integument of annelids, arthropods, and others.

sex cell A haploid cell (ovum or sperm) whose fertilization by one of the opposite kind produces a diploid zygote; also called a **gamete.**

sex chromosomes Chromosomes that determine gender of an animal. They may bear a few or many other genes.

sex ratio An accounting of the proportion of males versus females in a population at a particular time and place.

sexual selection Differential propagation among varying organisms caused by the greater success of some forms in the reproductive process (mating success and fertility). A characteristic favored by sexual selection may be detrimental to survival and disfavored by natural selection.

sibling species Reproductively isolated species that are so similar morphologically that they are difficult or impossible to distinguish using morphological characters.

sickle cell anemia A condition that causes the red blood cells to collapse (sickle) under oxygen stress. The condition becomes manifest when an individual is homozygous for the gene for hemoglobin-S (HbS).

sign stimulus An ethological term denoting an entity (such as a particular sound, shape, or color) whose perception by an animal elicits a stereotypical behavioral pattern.

siliceous (sə-li′shəs) (L. silex, flint). Containing silica.

simian (sim′ē-ən) (L. simia, ape). Pertaining to monkeys or apes.

sinistral (si′nə-strəl, sə-ni′stral) (L. sinister, left). Pertaining to the left; in gastropods, shell is sinistral if opening is to left of columella when held with spire up and facing observer.

sink deme A subpopulation (deme) whose members are drawn disproportionately from other subpopulations of the same species (see **metapopulation dynamics**); for example, a deme occupying an environmentally unstable area whose members are periodically destroyed by climatic changes and then replenished by colonists from other demes when favorable conditions are restored.

sinoatrial node (si′nō-āt′rē-əl) (L. sinus, curved, + atrium, hall). Specialized cardiac muscle cells that serve as the pacemaker in a tetrapod heart.

sinus (sī′nəs) (L. curve). A cavity or space in tissues or in bone.

siphon (sī′-fən). A tube for directing water flow.

siphonoglyph (sī′fän′ə-glif′) (Gr. siphōn, reed, tube, siphon, + glyphē, carving). Ciliated furrow in the gullet of sea anemones.

siphuncle (sī′fun-kəl) (L. siphunculus, small tube). Cord of tissue running through the shell of a nautiloid, connecting all chambers with the animal's body.

sister group The relationship between a pair of species or higher taxa that are each other's closest phylogenetic relatives.

smallest distinct groupings A general criterion for the species category shared to some degree by all formal species concepts is that species be the least inclusive population or lineage of populations having a unique history of common descent. Violation of this criterion would blur the distinction between species and higher taxa.

socially coordinated behavior Any activity in which an organism adjusts its actions to the presence of others to increase its own reproductive success. Includes cooperative behaviors as well as agonistic ones.

sociobiology Ethological study of social behavior in humans or other animals.

solenia (sō-len′ē-ə) (Gr. sōlēn, pipe). Channels through the coenenchyme connecting the polyps in an octocorallian colony (phylum Cnidaria).

solenocyte (sō-len′ə-sīt) (Gr. sōlēn, pipe, + kytos, hollow vessel). Special type of flame bulb in which the bulb bears a flagellum instead of a tuft of flagella. See **flame cell, protonephridium.**

solvent (säl-vent′). A liquid into which a substance is dissolved.

soma (sō′mə) (Gr. body). The whole of an organism except the germ cells (germ plasm).

somatic (sō-mat′ik) (Gr. sōma, body). Refers to the body, for example, somatic cells in contrast to germ cells.

somatocoel (sə-mat′ə-sēl) (Gr. sōma, the body, + koilos, hollow). Posterior coelomic compartment of echinoderms; left somatocoel gives rise to oral coelom, and right somatocoel becomes aboral coelom.

somatoplasm (sō′ma-tō-pla′zm) (Gr. sōma, body, + plasma, anything formed). The living matter that forms the mass of the body as distinguished from germ plasm, which comprises the reproductive cells. The protoplasm of body cells.

somite (sō′mīt) (Gr. soma, body). One of the blocklike masses of mesoderm arranged segmentally (metamerically) in a longitudinal series beside the neural tube of the embryo; metamere.

sorting Differential survival and reproduction among varying individuals; often confused with natural selection which is one possible cause of sorting.

southern evergreen forests Pine-dominated woodlands of the southeastern United States.

source deme A stable subpopulation (deme) that serves differentially as a source of colonists for establishing, joining, or replacing other such subpopulations of the same species (see **metapopulation dynamics**); for example, a deme inhabiting an environmentally stable area whose members routinely establish transitory populations in environmentally unstable nearby areas.

speciation (spē′sē-ā′shən) (L. *species,* kind). The evolutionary process or event by which new species arise.

species (spē′shez, spē′sēz) sing. and pl. (L. particular kind). A group of interbreeding individuals of common ancestry that are reproductively isolated from all other such groups; a taxonomic unit ranking below a genus and designated by a binomen consisting of its genus and the species name.

species diversity The number of different **species** that coexist at a given time and place to form an ecological **community.**

species epithet The second (uncapitalized) word in the Linnaean binomial nomenclature of species used to separate an individual species from other members of the same genus.

species selection Differential rates of speciation and/or extinction among varying evolutionary lineages caused by interactions among species-level emergent characteristics and the environment; contrasts with **effect macroevolution.**

Spemann organizer A region of an embryo acting as the primary organizer (see **primary organizer**).

spermatheca (spər′mə-thē′kə) (Gr. *sperma,* seed, + *thēkē,* case). A sac in the female reproductive organs for the reception and storage of sperm.

spermatid (spər′mə-təd) (Gr. *sperma,* seed, + *eidos,* form). A growth stage of a male reproductive cell arising by division of a secondary spermatocyte; gives rise to a spermatozoon.

spermatocyte (spər-ma′-tō-sīt) (Gr. *sperma,* seed, + *kytos,* hollow vessel). A growth stage of a male reproductive cell; gives rise to a spermatid.

spermatogenesis (spər-ma′-tō-jen′-ə-səs) (Gr. *sperma,* seed, + *genesis,* origin). Formation and maturation of spermatozoa.

spermatogonium (spər′ma′-tō-gō′nē-əm) (Gr. *sperma,* seed, + *gonē,* offspring). Precursor of mature male reproductive cell; gives rise directly to a spermatocyte.

spermatophore (spər-ma′-tō-fōr′) (Gr. *sperma, spermatos,* seed, + *pherein,* to bear). Capsule or packet enclosing sperm, produced by males of several invertebrate groups and a few vertebrates.

sphincter (sfingk′tər) (Gr. *sphinkter,* band, sphincter, from *sphingein,* to bind tight). A ring-shaped muscle capable of closing a tubular opening by constriction.

spicule (spi′kyul) (L. dim. *spica,* point). One of the minute calcareous or siliceous skeletal bodies found in sponges, radiolarians, soft corals, and sea cucumbers.

spiracle (spi′rə-kəl) (L. *spiraculum,* from *spirare,* to breathe). External opening of a trachea in arthropods. One of a pair of openings on the head of elasmobranchs for passage of water. Exhalent aperture of tadpole gill chamber.

spiral cleavage A type of embryonic cleavage in which cleavage planes are diagonal to the polar axis and unequal cells are produced by the alternate clockwise and counterclockwise cleavage around the axis of polarity; determinate cleavage.

spiral valve A thin surface within the conus arteriosus of an amphibian heart that directs oxygenated blood toward the lungs and deoxygenated blood to other body organs.

spongin (spun′jin) (L. *spongia,* sponge). Fibrous, collagenous material forming the skeletal network of demosponges.

spongioblast (spun′je-o-blast) (Gr. *spongos,* sponge, + *blastos,* bud). Cell in a sponge that secretes spongin, a protein.

spongocoel (spun′jō-sēl) (Gr. *spongos,* sponge, + *koilos,* hollow). Central cavity in sponges.

spongocyte (spun′jō-sīt) (Gr. *spongos,* sponge, + *kytos,* hollow vessel). A cell in sponges that secretes spongin.

sporocyst (spō′rə-sist) (Gr. *sporos,* seed, + *kystis,* pouch). A larval stage in the life cycle of flukes; it originates from a miracidium.

sporogony (spor-äg′ə-nē) (Gr. *sporos,* seed, + *gonos,* birth). Multiple fission to produce sporozoites after zygote formation.

sporozoite (spō′rə-zō′īt) (Gr. *sporos,* seed, + *zōon,* animal, + *ite,* suffix for body part). A stage in the life history of many sporozoan protozoa; released from oocysts.

squalene (skwā′lēn) (L. *squalus,* a kind of fish). A liquid acyclic triterpene hydrocarbon found especially in the liver oil of sharks.

squamous epithelium (skwā′məs) (L. *squama,* scale, + *osus,* full of). Simple epithelium of flat, nucleated cells.

stabilizing selection A selective process in which the mean value of a quantitative phenotype is favored over extreme values in a population, potentially stabilizing the mean value.

stapes (stā′pēz) (L. stirrup). Stirrup-shaped innermost bone of the middle ear.

statoblast (stad′ə-blast) (Gr. *statos,* standing, fixed, + *blastos,* germ). Biconvex capsule containing germinative cells and produced by most freshwater ectoprocts by asexual budding. Under favorable conditions it germinates to give rise to new zooid.

statocyst (Gr. *statos,* standing, + *kystis,* bladder). Sense organ of equilibrium; a fluid-filled cellular cyst containing one or more granules (statoliths) used to sense direction of gravity.

statolith (Gr. *statos,* standing, + *lithos,* stone). Small calcareous body resting on tufts of cilia in the statocyst.

stenohaline (sten-ə-hā′līn, -lən) (Gr. *stenos,* narrow, + *hals,* salt). Pertaining to aquatic organisms that have restricted tolerance to changes in environmental saltwater concentration.

stenophagous (stə-näf′ə-gəs) (Gr. *stenos,* narrow, + *phagein,* to eat). Eating few kinds of foods.

stenotopic (sten-ə-tō′pik) (Gr. *stenos,* narrow, + *topos,* place). Refers to an organism with a narrow range of adaptability to environmental change; having a restricted environmental distribution.

stereogastrula (ste′rē-ə-gas′trə-lə) (Gr. *stereos,* solid, + *gastēr,* stomach, + L. *ula,* dim.). A solid type of gastrula, such as the planula of cnidarians.

stereom (ster′ē-ōm) (Gr. *stereos,* solid, hard, firm). Meshwork structure of endoskeletal ossicles of echinoderms.

stereotyped behavior A pattern of behavior repeated with little variation in performance.

sternum (ster′nəm) (L. breastbone). Ventral plate of an arthropod body segment; breastbone of vertebrates.

sterol (ste′rōl), steroid (ste′roid) (Gr. *stereos,* solid, + L. *ol,* from *oleum,* oil). One of a class of organic compounds containing a molecular skeleton of four fused carbon rings; it includes cholesterol, sex hormones, adrenocortical hormones, and vitamin D.

stigma (Gr. *stigma,* mark, tatoo mark). Eyespot in certain protozoa. Spiracle of certain terrestrial arthropods.

stolon (stō′lən) (L. *stolō, stolonis,* a shoot, or sucker of a plant). A rootlike extension of the body wall giving rise to buds that may develop into new zooids, thus forming a compound animal in which the zooids remain united by the stolon. Found in some colonial anthozoans, hydrozoans, ectoprocts, and ascidians.

stoma (stō′mə) (Gr. mouth). A mouthlike opening.

stomochord (stō′mə-kord) (Gr. *stoma,* mouth, + *chordē,* cord). Anterior evagination of the dorsal wall of the buccal cavity into the proboscis of hemichordates; the buccal diverticulum.

stramenopile (strə-men′ə-pīl) (L. *stramen,* straw, + *pilus,* a hair). Any member of a protozoan clade with tubular mitochondrial cristae, and typically having 3-part tubular hairs on a long anterior flagellum.

strobila (strō′bə-lə) (Gr. *strobilē,* lint plug like a pine cone [*strobilos*]). A stage in the development of the scyphozoan jellyfish. Also, the chain of proglottids of a tapeworm.

strobilation (strō′bə-lā′shən) (Gr. *strobilos,* a pine cone). Repeated, linear budding of individuals, as in scyphozoans (phylum Cnidaria), or sets of reproductive organs, as in tapeworms (phylum Platyhelminthes).

stroma (strō′mə) (Gr. *strōma,* bedding). Supporting connective tissue framework of an animal organ; filmy framework of red blood corpuscles and certain cells.

structural gene A gene carrying the information to construct a protein.

subnivean (sub-ni′vē-ən) (L. *sub,* under, below, + *nivis,* snow). Applied to environments beneath snow, in which snow insulates against a colder atmospheric temperature.

substrate The substance upon which an enzyme acts; also, a base or foundation (substratum); and the substance or base on which an organism grows.

sugar The fundamental subunit of carbohydrates having carbon, hydrogen, and oxygen in a usual ratio of 1:2:1, respectively. Two sugars (deoxyribose, ribose) form part of the nucleotide structure of nucleic acids.

survivorship (sər-vī′vər-ship). The proportion of individuals of a cohort or population that persist from one point in their life history (such as birth) to another one (such as reproductive maturity or a specified age).

suspension feeder Aquatic organisms that collect suspended food particles from the surrounding water; particles may be filtered or taken by other methods.

swim bladder Gas-filled sac of many bony fishes used in buoyancy and in some cases, respiratory gas exchange.

sycon (sī′kon) (Gr. *sykon,* fig). A type of canal system in certain sponges. Sometimes called syconoid.

symbiosis (sim′bī-ōs′əs, sim′bē-ōs′əs) (Gr. *syn,* with, + *bios,* life). The living together of two different species in an intimate relationship. At least one species benefits; the other species may benefit, may be unaffected, or may be harmed (mutualism, commensalism, and parasitism).

sympatric (sim′pa′-trik) (Gr. *syn,* with, + *patra,* native land). Having the same or overlapping regions of geographical distribution. Noun, **sympatry.**

sympatric speciation A branching of population lineages to form separate species during which the diverging lineages co-occupy a geographic area. A controversial mode of speciation that contrasts with allopatric speciation and parapatric speciation.

symplesiomorphy (sim-plē′zē-ə-mōr′fē). Sharing among species of ancestral characteristics, not indicative that the species form a monophyletic group.

synapomorphy (sin-ap′o-mor′fē) (Gr. *syn,* together with, + *apo,* of, + *morphe,* form). Shared, evolutionarily derived character states that are used to recover patterns of common descent among two or more species.

synapse (si′naps, si-naps′) (Gr. *synapsis,* contact, union). The place at which an action potential passes between neuronal processes, typically from an axon of one nerve cell to a dendrite of another nerve cell.

synapsids (si-nap′sədz) (Gr. *synapsis,* contact, union). An amniote group comprising the mammals and the ancestral mammal-like reptiles, having a skull with a single pair of temporal openings.

synapsis (si-nap′səs) (Gr. *synapsis,* contact, union). The time when the pairs of homologous chromosomes lie alongside each other in the first meiotic division.

synaptonemal complex (sin-ap′tə-nē′məl) (Gr. *synapsis,* a joining together, + *nēma,* thread). The structure that holds homologous chromosomes together during synapsis in prophase of meiosis I.

syncytium (sən-sish′e-əm) (Gr. *syn,* with, + *kytos,* hollow). A multinucleated cell. Adj., **syncytial.**

syndrome (sin′drōm) (Gr. *syn,* with, + *dramein,* to run). A group of symptoms characteristic of a particular disease or abnormality.

syngamy (sin′gə-mē) (Gr. *syn,* with, + *gamos,* marriage). Fertilization of one gamete with another individual gamete to form a zygote, found in most animals with sexual reproduction.

synkaryon (sin-ker′e-on) (Gr. *syn,* with, + *karyon,* nucleus). Zygote nucleus resulting from fusion of pronuclei.

syrinx (sir′inks) (Gr. shepherd's pipe). The vocal organ of birds located at the base of the trachea.

systematics (sis-tə-mat′iks). Science of classification and reconstruction of phylogeny.

systole (sis′tə-lē) (Gr. *systolē,* drawing together). Contraction of heart.

T

tactile (tak′til) (L. *tactilis,* able to be touched, from *tangere,* to touch). Pertaining to touch.

taenidia (tə′nid′-ē-ə) (Gr. *tainia,* ribbon). Spiral thickenings of the cuticle that support tracheae (phylum Arthropoda).

tagma, pl. **tagmata** (Gr. *tagma,* arrangement, order, row). A compound body section of an arthropod resulting from embryonic fusion of two or more segments; for example, head, thorax, abdomen.

tagmatization, tagmosis Organization of the arthropod body into tagmata.

taiga (tī′gä) (Russ.). Habitat zone characterized by large tracts of coniferous forests, long, cold winters, and short summers; most typical in Canada and Siberia.

tantulus (tan′tə-ləs) (Gr. *tantulus,* so small). Larva of a tantulocaridan (subphylum Crustacea).

taxis (tak′sis), pl. **taxes** (Gr. *taxis,* arrangement). An orientation movement by a (usually) simple organism in response to an environmental stimulus.

taxon (tak′son), pl. **taxa** (Gr. *taxis,* arrangement). Any taxonomic group or entity.

taxonomic rank The Linnaean category (kingdom, phylum, class, order, family, genus, species and variations thereof) into which a recognized taxon is placed.

taxonomy (tak-sän′ə-mi) (Gr. *taxis,* arrangement, + *nomos,* law). Study of the principles of scientific classification; systematic ordering and naming of organisms.

T cell A type of lymphocyte important in cellular immune response and in regulation of most immune responses.

T-cell receptors Receptors born on surfaces of T cells. The variable region of a T-cell receptor binds with a specific antigen.

tectorial membrane (tek-to′-rē-əl mem′-brān). A structure in the anuran inner ear that detects low-frequency sounds.

tectum (tek′təm) (L. roof). A rooflike structure, for example, dorsal part of capitulum in ticks and mites.

tegmen (teg′mən) (L. *tegmen,* a cover). External epithelium of crinoids (phylum Echinodermata).

tegument (teg′ū-ment) (L. *tegumentum,* from *tegere,* to cover). An integument: specifically external covering in cestodes and trematodes, formerly considered a cuticle.

telencephalon (tel′en-sef′ə-lon) (Gr. *telos,* end, + *encephalon,* brain). The most anterior vesicle of the brain; the anterior-most subdivision of the prosencephalon that becomes the cerebrum and associated structures.

teleology (tē′lē-äl′ə-jē) (Gr. *telos,* end, + L. *logia,* study of, from Gr. *logos,* word). The philosophical view that natural events are goal directed and are preordained, as opposed to the scientific view of mechanical determinism.

teleost (tē′lē-ost). A clade of ray-finned fishes characterized by a humocercal caudal fin.

telocentric (tē′lō-sen′trək) (Gr. *telos,* end, + *kentron,* center). Chromosome with centromere at the end.

telolecithal (te-lō-les′ə-thəl) (Gr. *telos,* end, + *lekithos,* yolk, + *al*). Having the yolk concentrated at one end of an egg.

telson (tel′sən) (Gr. *telson,* extremity). Posterior projection of the last body segment in many crustaceans.

temnospondyls (tem-nō-spän′dəls) (Gr. *temnō,* to cut, + *spondylos,* vertebra). A large group of early tetrapods that lived from the Carboniferous to the Triassic.

template (tem′plāt). A pattern or mold guiding the formation of a duplicate; often used with reference to gene duplication.

tendon (ten′dən) (L. *tendo,* tendon). Fibrous band connecting muscle to bone or other movable structure.

tentaculocyst (ten-tak′u-lō-sist) (L. *tentaculum,* feeler, + *kystis,* pouch). One of the sense organs along the margin of medusae; a rhopalium.

tergum (ter′gəm) (L. back). Dorsal part of an arthropod body segment.

territory (L. *territorium,* from *terra,* earth). A restricted area preempted by an animal or pair of animals, usually for breeding purposes, and guarded from other individuals of the same species.

tertiary structure For a protein, the three-dimensional configuration formed by bonds between side groups of amino acids located in different regions of a polypeptide chain. The disulfide bond between two cysteine amino acids is a common example.

test (L. *testa,* shell). A shell or hardened outer covering.

testate The condition of having a test.

testcross A genetic cross used to determine the genotype (homozygous versus heterozygous) of an individual showing a genetically dominant phenotype. The individual tested is crossed to a homozygous recessive individual. Homozygous individuals tested will produce only offspring with the dominant phenotype, whereas heterozygous individuals will produce offspring with approximately even numbers of dominant and recessive phenotypes.

tetrad (te′trad) (Gr. *tetras,* four). Group of two pairs of chromatids at synapsis and resulting from the replication of paired homologous chromosomes; the bivalent.

tetrapods (te′trə-päds) (Gr. *tetras,* four, + *pous, podos,* foot). Four-footed vertebrates; the group includes amphibians, nonavian reptiles, birds, and mammals.

theca (thē′kə) (Gr. *thēkē,* a case for something, a box). A protective covering for an organism or an organ.

thecate (thē′kāt) (Gr. *thēkē,* a case, box). An organism bearing a theca.

theory (thē′-ə-rē). A scientific hypothesis or set of related hypotheses that offer very powerful explanations for a wide variety of related phenomena and serve to organize scientific investigation of those phenomena.

therapsids (thə-rap′sidz) (Gr. *theraps,* an attendant). Extinct amniotes, from the Permian to Triassic, from which true mammals evolved.

thermocline (thər′mō-klīn) (Gr. *thermē,* heat, + *klinein,* to swerve). Layer of water separating upper warmer and lighter water from lower colder and heavier water in a lake or sea; a stratum of abrupt change in water temperature.

thoracic (thō-ra′sək) (L. *thōrax,* chest). Pertaining to the thorax or chest.

3′ end (three-prime end) The terminus of a nucleic acid molecule that consists of a hydroxyl group attached to the 3′ carbon of the terminal sugar. Nucleic-acid synthesis consists of adding nucleotides to this end of the molecule. (The opposite terminus is the 5′ end.)

thrombin Enzyme catalyzing fibrinogen transformation into fibrin. Percursor is **prothrombin.**

Tiedemann's bodies (tēd′ə-mənz) (from F. Tiedemann, German anatomist). Four or five pairs of pouchlike bodies attached to the ring canal of sea stars, apparently functioning in production of coelomocytes.

tight junction Region of actual fusion of cell membranes between two adjacent cells.

tissue (ti′shu) (M.E. *tissu,* tissue). An aggregation of cells, usually of the same kind, organized to perform a common function.

titer (ti′tər) (Fr. *titrer,* to titrate). Concentration of a substance in a solution as determined by titration.

Toll-like receptors (TLRs) Named for the Toll protein family discovered in *Drosophila,* Toll-like receptors are located on cell membranes of vertebrates. When activated by binding a microbe, they signal the cell to synthesize an appropriate antimicrobial peptide. Recognizing patterns rather than specific molecular configurations, they are a vital part of innate immune defenses.

tornaria (tor-na′rē-ə) (L. *tornare,* to turn). A free-swimming larva of enteropneusts that rotates as it swims; resembles somewhat the bipinnaria larva of echinoderms.

torsion (L. *torquere,* to twist). A twisting phenomenon in gastropod development that alters the position of the visceral and pallial organs by 180 degrees.

toxicyst (tox′i-sist) (Gr. *toxikon,* poison, + *kystis,* bladder). Structures possessed by predatory ciliate protozoa, which on stimulation expel a poison to subdue the prey.

trabecular reticulum (trə-bek′ū-lər rə-tik′ū-ləm) (L. *trabecula,* a small beam; *reticulum,* a net). A bilayered, syncytial tissue forming the main body structure of hexactinellid sponges (phylum Porifera).

trachea (trā′kē-ə) (M.L. windpipe). The cartilage-lined tube used to conduct air between the pharynx and lungs of tetrapods. Also, any of the air tubes of insects.

tracheal system (trāk′ē-əl) (L. *trachia,* windpipe). A network of thin-walled tubes that branch throughout the entire body of terrestrial insects; used for respiration.

tracheole (trāk′ē-ōl) (L. *trachia,* windpipe). Fine branches of the tracheal system, filled with fluid, but not shed at ecdysis.

trachyline-line (trak′ā-līn-līn) (Gr. trachys, rough, + *linum,* thread). Descriptive term for an unusual hydrozoan life cycle (phylum Cnidaria) where a larva metamorphoses directly to a medusa stage without an intervening polyp stage.

transcription Formation of messenger RNA from the coded DNA.

transcription factor A steriod or protein molecule that binds to a chromosome at the locus of a gene either to activate or to repress synthesis of RNA complementary to the gene's sense strand.

transduction Condition in which bacterial DNA (and the genetic characteristics it bears) is transferred from one bacterium to another by the agent of viral infection.

transfer RNA (tRNA) A form of RNA of about 70 or 80 nucleotides, which is an adapter

molecule in the synthesis of proteins. A specific amino-acid molecule is carried by transfer RNA to a ribosome-messenger RNA complex for incorporation into a polypeptide.

transformation Condition in which DNA in the environment of bacteria somehow penetrates them and is incorporated into their genetic complement, so that their progeny inherit the genetic characters so acquired.

transformational theory of evolution Any evolutionary hypothesis in which change occurs via the restructuring of individual organisms during their ontogenies and transmission of such phenotypic alterations to offspring, as in Lamarck's theory. Contrasts with variational theories of evolution, such as Darwinian natural selection.

translation (L. a transferring). The process in which the genetic information present in messenger RNA is used to direct the order of specific amino acids during protein synthesis.

translocation Chromosomal material moved from one chromosome to a nonhomologous chromosome, often in a reciprocal exchange.

transporter A transmembrane protein transport molecule in the plasma membrane that permits otherwise impermeable ions and/ or molecules to be transported across the membrane, called carrier-mediated transport.

transverse plane (L. *transversus,* across). A plane or section that lies or passes across a body or structure.

trend A directional change in the characteristic features or patterns of diversity in a group of organisms when viewed over long periods of evolutionary time in the fossil record.

trichinosis (trik-ən-ō′səs). Disease caused by infection with the nematode *Trichinella spiralis.*

trichocyst (trik′ə-sist) (Gr. *thrix,* hair, + *kystis,* bladder). Saclike protrusible organelle in the ectoplasm of ciliates, which discharges as a threadlike weapon of defense.

triglyceride (trī-glis′ə-rīd) (Gr. *tria,* three, + *glykys,* sweet, + *ide,* suffix denoting compound). A triester of glycerol with one, two, or three kinds of fatty acids.

trimerous (trī′mə-rəs) (Gr. *treis,* three, + *meros,* a part). Body in three main divisions, as in lophophorates and some deuterostomes.

tripartite (trī-par′tīt). See **trimerous.**

triploblastic (trip′lō-blas′tik) (Gr. *triploos,* triple, + *blastos,* germ). Pertaining to metazoa in which the embryo has three primary germ layers—ectoderm, mesoderm, and endoderm.

trisomy The chromosomal constitution of an otherwise diploid organism in which a single extra chromosome is present (chromosome number = $2n + 1$).

trisomy 21 See **Down syndrome.**

trochophore (trōk′ə-fōr) (Gr. *trochos,* wheel, + *pherein,* to bear). A free-swimming ciliated marine larva characteristic of most molluscs and certain ectoprocts, brachiopods, and marine worms; an ovoid or pyriform body with preoral circlet of cilia and sometimes a secondary circlet behind the mouth.

trophallaxis (trōf′ə-lak′səs) (Gr. *trophē,* food, + *allaxis,* barter, exchange). Exchange of food between young and adults, especially certain social insects.

trophi (trō′fī) (Gr. *trophos,* one who feeds). Jaw-like structures in the mastax of rotifers.

trophic (trō′fək) (Gr. *trophē,* food). Pertaining to feeding and nutrition.

trophic level Position of a species in a **food web,** such as producer, herbivore, first-level carnivore, or higher-level carnivore.

trophoblast (trōf′ə-blast) (Gr. *trephein,* to nourish, + *blastos,* germ). Outer ectodermal nutritive layer of blastodermic vesicle; in mammals it is part of the chorion and attaches to the uterine wall.

trophosome (trōf′ə-sōm) (Gr. *trophē,* food, + *sōma,* body). Organ in poganophorans or siboglinids bearing mutualistic bacteria, derived from midgut.

trophozoite (trōf′ə-zō′īt) (Gr. *trophē,* food, + *zōon,* animal). Adult stage in the life cycle of a protozoan in which it is actively absorbing nourishment.

tropic (trä′pic) (Gr. *tropē,* to turn toward). Related to the tropics (tropical); in endocrinology, a hormone that influences the action of another hormone or endocrine gland (usually pronounced trō′pic).

tropomyosin (trōp′ə-mī′ə-sən) (Gr. *tropos,* turn, + *mys,* muscle). Low-molecular weight, actin-binding protein surrounding the actin filaments of striated muscle. It works with troponin to regulate muscle contraction.

troponin (trə-pōn′in). Complex of globular, actin-binding proteins positioned at intervals along the actin filament of skeletal muscle; serves as a calcium-dependent switch in muscle contraction.

tube feet (podia) Numerous small, muscular, fluid-filled tubes projecting from an echinoderm; part of water-vascular system; used in locomotion, clinging, food handling, and respiration.

tubercle (tū′bər-kəl) (L. *tuberculum,* small hump). Small protuberance, knob, or swelling.

tubulin (tū′bū-lən) (L. *tubulus,* small tube, + *in,* belonging to). Globular protein forming the hollow cylinder of microtubules.

tumor necrosis factor A cytokine, the most important source of which is macrophages and T helper lymphocytes, that is a major mediator of inflammation.

tumor-suppressor gene A gene whose products act as restraints on cell division by triggering apoptosis, controlling transcription of other genes, restraining progression in phases of the cell cycle, or by other means.

tundra (tun′drə) (Russ. from Lapp, *tundar,* hill). Terrestrial habitat zone, located between taiga and polar regions; characterized by absence of trees, short growing season, and mostly frozen soil during much of the year.

tunic (L. *tunica,* tunic, coat). In tunicates, a cuticular, cellulose-containing covering of the body secreted by the underlying body wall.

turbinates (tər′bin-āts) (L. *turbin,* whirling). Highly convoluted bones covered in mucous membrane in the nasal cavity of endotherms that serve to reduce heat and water lost during respiration.

tympanic membrane (tim-pan′ik) (Gr. *tympanon,* drum). The surface that separates the outer and middle ear (eardrum).

type specimen A specimen deposited in a museum that formally defines the name of the species that it represents.

typhlosole (tif′lə-sōl′) (Gr. *typhlos,* blind, + *sōlēn,* channel, pipe). A longitudinal fold projecting into the intestine in certain invertebrates such as the earthworm.

typological species concept The discredited, pre-Darwinian notion that species are classes defined by the presence of fixed, unchanging characters (=“essence”) shared by all members.

typology (tī-päl′ə-jē) (L. *typus,* image). A classification of organisms in which members of a taxon are perceived to share intrinsic, essential properties, and variation among organisms is regarded as uninteresting and unimportant.

U

ulcer (ul-sər) (L. *ulcus,* ulcer). An abscess that opens through the skin or a mucous surface.

ultimate cause (L. *ultimatus,* last, + *causa).* The evolutionary factors responsible for the origin, state of being, or role of a biological system.

umbilical (L. *umbilicus,* navel). Refers to the navel, or umbilical cord.

umbo (um′bō), pl. **umbones** (əm-bō′nēz) (L. boss of a shield). One of the prominences on either side of the hinge region in a bivalve mollusc shell. Also, the “beak” of a brachiopod shell.

undulating membrane A membranous structure on a protozoan associated with a flagellum; on other protozoa may be formed from fused cilia.

ungulate (un′gū-lət) (L. *ungula,* hoof). Hooved. Noun, any hooved mammal.

uniformitarianism (ū′nə-fōr′mə-ter′ē-ə-niz′əm). Methodological assumptions that the laws of chemistry and physics have remained constant throughout the history of the earth, and that past geological events occurred by processes that can be observed today.

unique density behavior Refers to greater density of water in liquid than in solid phase.

uniramous (ū′nə-rām′əs) (L. *unus,* one, + *ramus,* a branch). Adjective describing unbranched appendages (phylum Arthropoda).

unitary (ū′nə-ter′ē). Describes the structure of a population in which reproduction is strictly sexual and each organism is genetically distinct from others.

unsaturated fatty acid A subunit of fat molecules containing a carboxyl group and a carbon chain in which two or more carbons are joined by double bonds and therefore capable of forming additional bonds with other atoms, such as hydrogen.

ureter (ūr′ə-tər) (Gr. *ouētēr,* ureter). Duct carrying urine from metanephric kidney to bladder or cloaca.

urethra (ū-rē′thrə) (Gr. *ourethra,* urethra). The tube from the urinary bladder to the exterior in both sexes.

uropod (ū′rə-pod) (Gr. *oura,* tail, + *pous, podos,* foot). Posteriormost appendage of many crustaceans.

urostyle (ū′-rō-stīl). A rodlike structure comprising fused vertebrae at the posterior end of the anuran vertebral column; homologous to tail vertebrae of other tetrapods.

utricle (ū′trə-kəl) (L. *utriculus,* little bag). That part of the inner ear containing the receptors for dynamic body balance; the semicircular canals lead from and to the utricle.

V

vacuole (vak′yə-wōl) (L. *vacuus,* empty, + Fr. *ole,* dim.). A membrane-bound, fluid-filled space in a cell.

valence (vā′ləns) (L. *valere,* to have power). Degree of combining power of an element as expressed by the number of atoms of hydrogen (or its equivalent) that the element can hold (if negative) or displace in a reaction (if positive). The oxidation state of an element in a compound. The number of electrons gained, shared, or lost by an atom when forming a bond with one or more other atoms.

valve (L. *valva,* leaf of a double door). One of the two shells of a typical bivalve mollusc or brachiopod. In cardiovascular and lymphatic systems, valves allow one-way flow of blood or lymph.

variation (L. *varius,* various). Differences among individuals of a group or species that cannot be ascribed to age, sex, or position in the life cycle.

variational theory of evolution An evolutionary hypothesis such as Darwinian natural selection in which change occurs in the frequencies of alternative genetic characteristics across generations in a population rather than by modification of heredity by characteristics acquired by an organism during its ontogeny. Contrasts with transformational theories of evolution, such as Lamarckism.

vector (L. a bearer, carrier, from *vehere, vectum,* to carry). Any agent that carries and transmits pathogenic microorganisms from one host to another host. Also, in molecular biology, an agent such as bacteriophage or plasmid that carries recombinant DNA.

vegetal plate Region formed by flattening the embryonic vegetal pole at the beginning of gastrulation.

vegetal pole Region of an egg with a high concentration of yolk; this region is opposite the animal pole where cytoplasm is concentrated.

veins (vānz) (L. *vena,* a vein). Blood vessels that carry blood toward the heart; in insects, fine extensions of the tracheal system that support the wings.

velarium (və-laʹrē-əm) (L. *velum,* veil, covering). Shelflike extension of the subumbrellar edge in cubozoans (phylum Cnidaria).

veliger (vēlʹə-jər) (L. *velum,* veil, covering). Larval form of certain molluscs; develops from the trochophore and has the beginning of a foot, mantle, and shell.

velum (vēʹləm) (L. veil, covering). A membrane on the subumbrellar surface of jellyfish of class Hydrozoa. Also, a ciliated swimming organ of the veliger larva.

ventral (venʹtrəl) (L. *venter,* belly). Situated on the lower or abdominal surface.

ventricle (venʹ-tri-kəl). A chamber of the vertebrate heart that receives blood from an atrium (a separate heart chamber) and pumps blood from the heart.

venule (venʹūl) (L. *venula,* dim. of *vena,* vein). Small vessel conducting blood from capillaries to vein; small vein of insect wing.

vermiform (verʹmə-form) (L. *vermis,* worm, + *forma,* shape). Adjective to describe any wormlike organism; an adult (nematogen) rhombozoan (phylum Mesozoa).

vesicular (ve-sikʹyə-ler) (L. *vesicula,* small bladder, blister). Descriptive term for a granular appearance of the nucleus in many protozoans due to clumping of chromatin; also composed of vesicle-like cavities; also bladderlike.

vestige (vesʹtij) (L. *vestigium,* footprint). A rudimentary organ that may have been well developed in some ancestor or in the embryo.

vibraculum (vī-brakʹyə-ləm) (N.L.). Modified bryozoan individual (zooid) in which the operculum is highly elongated to act as a freely-moving whip.

vibrissa (vī-brisʹə), pl. **vibrissae** (L. nostril hair). Stiff hairs that grow from the nostrils or other parts of the face of many mammals and that serve as tactile organs; "whiskers."

vicariance (vī-karʹē-ənts) (L. *vicarius,* a substitute). Geographical separation of populations, especially as imposed by discontinuities in the physical environment that fragmented populations that were formerly geographically continuous.

vicariance biogeography An approach to historical biogeography that emphasizes locating physical barriers that simultaneously fragment codistributed species into local areas of geographic endemism; explains shared patterns of cladogenesis among geographically codistributed taxa.

vicariant speciation Formation of species in allopatry initiated by intrusion of a physical barrier that fragments a species into geographically isolated populations. Contrasts with speciation by a founder event, which requires establishment of a new population by a rare movement of individuals across an otherwise strong geographic barrier into territory not occupied by the parent population.

villus (vilʹəs), pl. **villi** (L. tuft of hair). A small fingerlike process on the wall of the small intestine that increases the surface area for absorption of digested nutrients. Also one of the branching, vascular processes on the embryonic portion of the placenta.

virus (vīʹrəs) (L. slimy liquid, poison). A submicroscopic noncellular particle composed of a nucleoprotein core and a protein shell; parasitic; will grow and reproduce only in a host cell.

viscera (visʹər-ə) (L. pl. of *viscus,* internal organ). Internal organs in the body cavity.

visceral (visʹər-əl). Pertaining to viscera.

vitalism (L. *vita,* life). The discredited viewpoint that natural processes are controlled by supernatural forces and cannot be explained through the laws of physics and chemistry and mechanistic processes alone.

vitamin (L. *vita,* life, + *amine,* from former supposed chemical origin). An organic substance required in small amounts for normal metabolic function; must be supplied in the diet or by intestinal flora because the organism cannot synthesize it. An exception is vitamin D_3, which is manufactured in the skin in the presence of sunlight.

vitellaria (viʹtəl-larʹē-ə) (L. *vitellus,* yolk of an egg). Structures in many flatworms that produce vitelline cells, which provide eggshell material and nutrient for the embryo.

vitelline gland See **vitellaria.**

vitelline membrane (və-telʹən, vīʹtəl-ən) (L. *vitellus,* yolk of an egg). The noncellular membrane that encloses the egg cell.

viviparity (vīʹvə-parʹə-tē) (L. *vivus,* alive, + *parere,* to bring forth). Reproduction in which eggs develop within the female body, which supplies nutritional aid as in therian mammals, many nonavian reptiles, and some fishes; offspring are born as juveniles. Adj., **viviparous** (vī-vipʹə-rəs).

vocal cords Paired muscles whose vibration in the larynx (voice box) of many terrestrial vertebrates produces sound.

W

waggle dance A complex behavior performed by a hive bee to direct other members of the hive population toward a distant food source.

water-vascular system System of fluid-filled closed tubes and ducts peculiar to echinoderms; used to move tentacles and tube feet that serve variously for clinging, food handling, locomotion, and respiration.

weir (wer) (Old English *wer,* a fence placed in a stream to catch fish). Interlocking extensions of a flame cell and a collecting tubule cell in some protonephridia.

X

xanthophore (zanʹthə-fōr) (Gr. *xanthos,* yellow, + *pherein,* to bear). A chromatophore containing yellow pigment.

xenograft (zēʹnə-graft). Graft of tissue from a species different from the recipient.

X-organ Neurosecretory organ in eyestalk of crustaceans that secretes molt-inhibiting hormone.

Y

Y-organ Gland in the antennal or maxillary segment of some crustaceans that secretes molting hormone.

Z

zoecium, zooecium (zō-ēʹshē-əm) (Gr. *zōon,* animal, + *oikos,* house). Cuticular sheath or shell of Ectoprocta.

zoochlorella (zōʹə-klōr-elʹə) (Gr. *zōon,* life, + *Chlorella*). Any of various minute green algae (usually *Chlorella*) that live symbiotically within the cytoplasm of some protozoa and other invertebrates.

zooflagellates (zōʹə-flaʹjə-lāts). Members of the former Zoomastigophorea, animal-like flagellates (former phylum Sarcomastigophora).

zooid (zō-id) (Gr. *zōon,* life). An individual member of a colony of animals, such as colonial cnidarians and ectoprocts.

zooxanthella (zōʹə-zan-thəlʹə) (Gr. *zōon,* animal, + *xanthos,* yellow). A minute dinoflagellate alga living in the tissues of many types of marine invertebrates.

zygote (Gr. *zygōtos,* yoked). A fertilized egg.

zygotic meiosis Meiosis that occurs within the first few divisions after zygote formation; thus all stages in the life cycle other than the zygote are haploid.

---

bat / āpe / ärmadillo / herring / fēmale / finch / līce / crocodile / crōw / cȯin / duck / ūnicorn / ə indicates unaccented vowel sound "uh" as in mammal, fishes, cardinal, heron, vulture / stress as in bi-olʹo-gy, biʹo-logʹi-cal

CREDITS

PHOTOS

Part Openers:

One: Cleveland P. Hickman, Jr.; Two: © Tom Tietz/Stone Images/Getty; Three: © Frank & Joyce Burek/Getty Images; Four & Five: Cleveland P. Hickman, Jr.

Chapter 1

Opener: Cleveland P. Hickman, Jr.;
1.1a: © Dave Fleetham/Visuals Unlimited;
1.1b: © Steve McCutcheon/Visuals Unlimited;
1.1c: © Peter Ziminski/Visuals Unlimited;
1.1d: © Link/Visuals Unlimited;
1.1D(inset): © T.E. Adams/Visuals Unlimited; **1.1e:** Courtesy Duke University Marine Laboratory; **1.2a:** Courtesy of IBM U.K. Scientific Centre; **1.3:** © John D. Cunningham/Visuals Unlimited; **1.4:** © David M. Phillips/Visuals Unlimited; **1.5a:** © N. P. Salzman;
1.5b: © Ed Reschke; **1.5c:** © Ken Highfill/Photo Researchers, Inc.; **1.5d(both):** © William C. Ober; **1.6a:** © A.C. Barrington Brown/Photo Researchers, Inc.; **1.7a:** © M. Abbey/Visuals Unlimited; **1.7b:** © S. Dalton/National Audubon Society/Photo Researchers, Inc.;
1.8a,b: © D. Kline/Visuals Unlimited;
1.11a,b: © Michael Tweedie/Photo Researchers, Inc.; **1.12:** Courtesy American Museum of Natural History, Neg. #326669; **1.16a,b:** Courtesy Gregor Mendel Museum, Brno. Czechoslavakia; **1.18:** Pre'vost and Dumas; **1.19:** © Carolina Biological Supply/Phototake

Chapter 2

Opener: Larry S. Roberts; **2.4:** © G.I. Bernard/Animals Animals/Earth Scenes; **2.16:** Courtesy Kevin Walsh; **2.17:** Courtesy R.M. Syren and S.W. Rox, Institute of Molecular Evolution/University of Miami, Coral Gables, Florida; **2.18:** Cleveland P. Hickman, Jr.; **2.21:** Ben Barnhart

Chapter 3

Opener: © William S. Ober; **3.1:** © Russell Illig/Getty Images; **3.5:** Courtesy A. Wayne Vogl; **3.7:** Courtesy of G.E. Palade, University of California School of Medicine; **3.8b:** Courtesy Richard Rodewald; **3.9b & 3.11b:** Courtesy of Charles Flickinger; **3.12:** Courtesy A. Wayne Vogl; **3.13a:** © K.G. Murti/Visuals Unlimited; **3.14b:** Courtesy Kent McDonald; **3.16:** Courtesy Susumu Ito; **3.24(all):** © Times Mirror Higher Education Group, Inc./Kingsley Stern, photographer

Chapter 4

Opener: © Gary W. Carter/Visuals Unlimited

Chapter 5

Opener: Larry S. Roberts; **5.1:** Courtesy Gregor Mendel Museum, Brno, Czechoslovakia; **5.8A:** © Peter J. Bryant/Biological Photo Service

Chapter 6

Opener: © Siede Preis/Getty Images; **6.1a:** Courtesy American Museum of Natural History, New York, Neg. #32662; **6.1b, 6.2, 6.3:** The Natural History Museum, London; **6.5A:** © The Bridgeman Art Library International; **6.5B:** © Stock Montage; **6.6 & 6.7:** Cleveland P. Hickman, Jr.; **6.8A:** © Ken Lucas/Biological Photo Service; **6.8B:** © A.J. Copley/Visuals Unlimited; **6.8C:** © Roberta Hess Poinar; **6.8D:** Courtesy G.O. Poinar, University of California at Berkeley; **6.9A:** Courtesy W. Boehm; **6.10:** Cleveland P. Hickman, Jr.; **Page 115:** Dr. Mariano Rocchi; **6.15:** Courtesy Library of Congress; **6.19:** Courtesy M.K. Kelley, Courtesy of Harvard University Press; **6.23B:** Cleveland P. Hickman, Jr.; **6.24:** Courtesy of Storrs Agricultural Experiment Station, University of Connecticut at Storrs; **6.27:** Fritz Goro; **6.29:** © Timothy W. Ranson/Biological Photo Service; **6.30:** © Krasemann/Photo Researchers, Inc.; **6.31B:** Courtesy Dr. Robert K. Selander; **6.34:** Courtesy of the Canada Center for Remote Sensing, Energy, Mines, and Resources, Canada

Chapter 7

Opener: © Francis Leroy, Biocosmos/SPL/Photo Researchers, Inc.; **7.3:** © Robert Humbert/Biological Photo Service; **7.7:** From R.G. Kessel and R.H. Kardon, Tissues and Organs: A Text-Atlas of Scanning Electron Microscopy, 1979, W.H. Freeman and Co.

Chapter 8

Opener: A. Vodicka & John C. Gerhart/University of California, Berkeley; **8.5:** Courtesy G. Schatten; **8.17:** © F.R. Turner/Biological Photo Service

Chapter 9

Opener: Larry S. Roberts; **9.8A-C, 9.9(both) & 9.10A:** © Ed Reschke; **9.10B:** Cleveland P. Hickman, Jr.; **9.10C,D & 9.11(all):** © Ed Reschke

Chapter 10

Opener: Cleveland P. Hickman, Jr.; **10.1:** Courtesy Library of Congress; **10.3:** © Kjell Sandved; **10.7:** Courtesy American Museum of Natural History, Neg. #334101; **10.8A:** © M. Coe/OSF/Animals Animals/Earth Scenes; **10.8B:** © D. Allen/OSF/Animals Animals/Earth Scenes; **10.10:** Courtesy of Dr. George W. Byers, University of Kansas

Chapter 11

Opener: © M. Abbey/Visuals Unlimited; **11.3:** © Dr. David M. Phillips/Visuals Unlimited; **11.4:** L. Tetley; **11.5(all):** © M. Abbey/Visuals Unlimited; **11.8:** © Manfred Kage/Peter Arnold; **11.9B:** Courtesy Dr. Ian R. Gibbons; **11.11A:** Courtesy L. Evans Roth; **11.33A:** © Manfred Kage/Peter Arnold; **11.33B:** © A.M. Siegelman/Visuals Unlimited; **11.34:** Courtesy J. and M. Cachon. From Lee, J.J. S.H. Hutner and E.C. Bovee (editors). 1985. An Illustrated Guide to the Protozoa, Society of Protozoologists, Allen Press

Chapter 12

Opener: © William C. Ober; **12.6:** Larry S. Roberts; **12.8:** © William C. Ober; **12.15A:** Larry S. Roberts; **12.15B:** © William C. Ober; **12.15C:** Larry S. Roberts

Chapter 13

Opener & 13.1A: © William C. Ober; **13.6:** Thien T. Mai; **13.8A:** © William C. Ober; **13.9:** © R. Harbo; **13.10:** © William C. Ober; **13.13:** © Cabisco/Visuals Unlimited; **13.14:** © Carolina Biological Supply/Phototake; **13.15:** © Peter Parks/OSF/Animals Animals/Earth Scenes; **13.16A:** Larry S. Roberts; **13.16B & (inset):** © William C. Ober; **13.17:** © R. Harbo; **13.18:** Larry S. Roberts; **13.20:** © R. Harbo; **13.22:** © D.W. Gotshall; **13.23A:** © Jeff Rotman Photography; **13.23B:** Larry S. Roberts; **13.25A,B:** © R. Harbo; **13.26:** Larry S. Roberts; **13.27A-C:** © William C. Ober; **13.29:** Larry S. Roberts; **13.30 & 13.31A:** © William C. Ober; **13.33 & 13.34A-C:** Larry S. Roberts; **13.35B:** Cleveland P. Hickman, Jr.; **13.36A:** © William C. Ober; **13.36B:** © Kjell Sandved; **13.38:** © Dr. Ronald L. Shimek, 2004

Chapter 14

Opener: © L. Newman and A. Flowers; **14.4A:** © John D. Cunningham/Visuals Unlimited, Inc.; **14.13:** H. Zaiman, M.D./Fropm H. Zaiman.

A Pictorial Presentation of Parasites; **14.16**: Courtesy of A.W. Cheever/From H. Zaiman. A Pictorial Presentation of Parasites; **14.17**: © Arthur M. Seigelman/Visuals Unlimited; **14.22**: © Cabisco/Visuals Unlimited; **14.23**: Ana Flisser; **14.27**: © Stan Elems/Visuals Unlimited; **14.29**: Cleveland P. Hickman, Jr.

Chapter 15

Opener: Courtesy D. Despommier/From H. Zaiman. A Pictorial Presentation of Parasites; **15.3B**: Reinhardt M. Kristensen; **15.4C**: Courtesy to Martin V. Sørensen; **15.5**: © John Walsh/Photo Researchers, Inc.; **15.8C**: © Larry Stepanowicz/Visuals Unlimited; **15.9**: Courtesy of Dr. Peter Funch, University of Aarhus; **15.10A**: © Larry Stepanowicz/Visuals Unlimited; **15.12C**: Image courtesy of Dr. Kerstin Wasson; **15.13A**: © William Ober; **15.14A,B**: Cleveland P. Hickman, Jr.; **15.15A,B**: ©Robert Brons/Biological Photo Service; **15.16**: From R.M. Sayre, Trans. Am. Micros. 88: 266-274, 1969

Chapter 16

Opener: Larry S. Roberts; **16.1A**: Kjell B. Sandved/Visuals Unlimited; **16.1B**: © R. Harbo; **16.1C**: Larry S. Roberts; **16.1D**: © R. Harbo; **16.1E & 16.4B**: Larry S. Roberts; **16.7B**: © Dr. Thurston C. Lacalli, Biology Department, University of Victoria; **16.8**: Kjell B. Sandved/Visuals Unlimited; **16.11**: © Daniel Gotshall/Visuals Unlimited; **16.16A**: © Gerald and Buff Corsi/Visuals Unlimited; **16.16B**: © David Wrobel/Visuals Unlimited; **16.17A,B**: A. Kerstitch/Visuals Unlimited; **16.20A**: © R. Harbo; **16.20B**: © Tom Phillipp; **16.21A**: © R. Harbo; **16.21B**: Larry S. Roberts; **16.22A,B**: Cleveland P. Hickman, Jr.; **16.23 & 16.24A**: Larry S. Roberts; **16.24B**: Cleveland P. Hickman, Jr.; **16.25A**: © R. Harbo; **16.25B**: © D.P. Wilson/Frank Lane Picture Agency Ltd.; **16.28A**: © R. Harbo; **16.29, 16.32A,B & 16.33**: Larry S. Roberts; **16.36B & 16.37A**: Richard J. Neves; **16.38**: Larry S. Roberts; **16.39B**: © Dave Fleetham/Tom Stack & Associates

Chapter 17

Opener: Photo gear, #CRAB 02.TIF; **17.2E**: General Biological Supply; **17.4A**: © William Ober; **17.4B**: Larry S. Roberts; **17.9**: Cleveland P. Hickman, Jr.; **17.10**: Larry S. Roberts; **17.15**: Courtesy J.F. Grassle/Woods Hole Oceanographic Institution; **17.21**: © G.L. Twiest/Visuals Unlimited; **17.23**: Photograph by T. Branning; **17.25, 17.26B & 17.29B**: Cleveland P. Hickman, Jr.

Chapter 18

Opener: Courtesy D. Despommier/From H. Zaiman. A Pictorial Presentation of Parasites; **18.5A**: M. Hickman; **18.5B**: © Dr. M.A. Ansary/Science Photo Library; **18.6A**: © Dr. Dennis Kunkel/Visuals Unlimited; **18.6B**: E. Pike/From H. Zaiman. A Pictorial Presentation of Parasites;

18.8: Larry S. Roberts; **18.9A**: © R. Calentin/Visuals Unlimited; **18.9B**: H. Zaiman, M.D./From H. Zaiman. A Pictorial Presentation of Parasites; **18.11**: Contributed by E.L. Schiller, AFIP; **18.12**: Sharon Patton; **18.13**: Dr. Andreas Schmidt-Rhaesa; **18.14B**: © David Scharf/ Photo Researchers, Inc.; **18.15B**: © Erling Svensen; **18.16B**: Reinhard Kristensen in Kopenhagen; **18.17B**: © Dr. James Castner; **18.19 & 18.21**: Courtesy D.R. Nelson; **18.22**: R.M. Sayre, Trans. Am. Micros. 88: 266-274, 1969

Chapter 19

Opener: Cleveland P. Hickman, Jr.; **19.4A,B**: © A.J. Copley/Visuals Unlimited; **19.7B**: Cleveland P. Hickman, Jr; **19.10 & 19.11**: © J.H. Gerard/Nature Press; **19.12**: © Todd Zimmerman/Natural History Museum of Los Angeles County; **19.13A,B**: © J.H. Gerard/Nature Press; **19.14A**: © Todd Zimmerman/Natural History Museum of Los Angeles County; **19.14B**: Cleveland P. Hickman, Jr.; **19.15A,B**: © J.H. Gerard/Nature Press; **19.16**: From G.W. Wharton, "Mites and commercial extracts of house dust" in Science 167: 1382-1383. Copyright © 1970 by AAAS; **19.17**: S. Roberts; **19.18**: © D.S. Snyder/Visuals Unlimited; **19.19**: © A.M. Siegelman/Visuals Unlimited; **19.20**: © John D. Cunningham/Visuals Unlimited; **19.22A & 19.23A**: © Dr. James L. Castner; **19.24B**: © Dan Kline/Visuals Unlimited

Chapter 20

Opener: Cleveland P. Hickman, Jr.; **20.20**: Courtesy J. Ubelaker; **20.21A**: © William C. Ober; **20.21B**: © R. Harbo; **20.23A**: Cleveland P. Hickman, Jr.; **20.24**: Larry S. Roberts; **20.25A**: © Kjell Sandved/Visuals Unlimited; **20.25B,C**: © Kjell Sandved; **20.26A,B**: © Larry S. Roberts; **20.28A**: © Cleveland P. Hickman, Jr.; **20.28B**: © R. Harbo; **20.28C**: © Cleveland P. Hickman, Jr.; **20.28D,E & 20.29**: Larry S. Roberts

Chapter 21

Opener: 21.2B: © Larry S. Roberts; **21.5A,B**: © Ron West/Nature Photography; **21.7A,B**: © Kjell Sandved; **21.8**: Cleveland P. Hickman, Jr.; **21.9**: © J.H. Gerard/Nature Press; **21.10**: © Dr. James Castner; **21.13**: © John D. Cunningham/Visuals Unlimited; **21.14**: Courtesy Jay Georgi; **21.15**: © Dr. James L. Castner; **21.16A**: Cleveland P. Hickman, Jr.; **21.16B**: J.H. Gerard/Nature Press; **21.21A,B**: Cleveland P. Hickman, Jr.; **21.23A,B**: © Robert Brons/Biological Photo Service; **21.24A,B**: © Dr. James L. Castner; **21.26A**: Cleveland P. Hickman, Jr.; **21.26B**: © J. H. Gerard/Nature Press; **21.26C**: © Carolina Biological Supply/Phototake; **21.27A,B**: © J. H. Gerard/Nature Press; **21.28A-C**: © Kjell Sandved; **21.29**: © J. H. Gerard/Nature Press; **21.30**: Courtesy J.E. Lloyd; **21.31**: © K. Lorenzen/Andromeda/Educational Images; **21.32A**: © J.H. Gerard/Nature Press; **21.32B & 21.33A**: © Dr. James L. Castner; **21.33B**: Larry S. Roberts;

21.34A-C: © Dr. James L. Castner; **21.35A**: © L.L. Rue, III; **21.35B & 21.36**: © Dr. James L. Castner; **21.37**: © Kjell Sandved; **21.38B, 21.39 & 21.40**: © Dr. James L. Castner; **21.41A,B**: © Kjell Sandved; **21.41C**: Cleveland P. Hickman, Jr.; **21.41D**: © Kjell Sandved

Chapter 22

Opener: © Ken Lucas/Visuals Unlimited; **22.2B**: Thuesen, E.V., and R. Bieri, 1987; **22.4A-C**: Larry S. Roberts; **22.4D**: © Godfrey Merlin; **22.7F & 22.8A**: © R. Harbo; **22.8B**: © D.W. Gotshall; **22.10**: Larry S. Roberts; **22.14A**: © R. Harbo; **22.14B**: Larry S. Roberts; **22.17A,B**: © R. Harbo; **22.18A**: © Jeffrey L. Rotman/Corbis; **22.18B,C**: © R. Harbo; **22.18D**: © William C. Ober; **22.19**: © R. Harbo; **22.20A,B**: © A. Kerstitch/Visuals Unlimited; **22.21A,B**: Larry S. Roberts; **22.24A-C & 22.27A,B**: © R. Harbo; **22.27C & 22.29**: Larry S. Roberts

Chapter 23

Opener: © Heather Angel; **23.5A**: © www.deepseaimages.com; **23.5B**: Larry S. Roberts; **23.7**: Andrew J. Martinez/Photo Researchers, Inc.; **23.9A**: Cleveland P. Hickman, Jr.

Chapter 24

Opener: © Scott Henderson; **24.4**: © Berthoule-Scott/Jacana/Photo Researchers; **24.9 & 24.12A,B**: © Jeff Rotman Photography; **24.19A,B**: Courtesy John G. Shedd Aquarium/Patrice Ceisel; **24.20A**: © James D. Watt/Animals Animals; **24.20B**: © Biological Photo Service; **24.20C**: © Jeff Rotman Photography; **24.20D**: © Fred McConnaughey/Photo Researchers, Inc.; **24.31**: © D.W. Gotshall; **24.32**: © Mary Beth Angelo/Photo Researchers, Inc.; **24.34**: © Will Troyer/Visuals Unlimited; **24.35**: © D.W. Gotshall; **24.36**: Courtesy of F. McConnaughey

Chapter 25

Opener: Cleveland P. Hickman, Jr.; **25.4B**: Smith/Photo Researchers, Inc.; **25.5A,B**: Courtesy L. Houck; **25.8**: © Michael Redmer/Visuals Unlimited; **25.10**: Allan Larson; **25.11A**: © Ken Lucas/Biological Photo Service; **25.11B & 25.12**: Cleveland P. Hickman, Jr.; **25.13**: Courtesy American Museum of Natural History, Neg. #125617; **25.14**: Cleveland P. Hickman, Jr.; **25.17**: © E.R. Degginger/Animals Animals/Earth Scenes; **25.24**: Cleveland P. Hickman, Jr.

Chapter 26

Opener: Courtesy of Ron Magill/Miami Metrozoo; **26.7 & 26.8**: Cleveland P. Hickman, Jr.; **26.10**: Digital Vision/Getty Images; **26.11**: © OSF LTD; **26.13**: © John Mitchell/Photo Researchers, Inc.; **26.14**: Cleveland P. Hickman, Jr.; **26.17**: Craig K. Lorenz/Photo Researchers, Inc.; **26.15**: © Stephen Dalton/Photo Researchers, Inc.; **26.16**: © L.L. Rue, III; **26.18**: © Paul Freed/Animals Animals/Earth Scenes;

26.19B: © Austin J. Stevens/Animals Animals/ Earth Scenes; **26.20:** Cleveland P. Hickman, Jr.; **26.21:** © Joe McDonald/Visuals Unlimited; **26.23:** Cleveland P. Hickman, Jr.; **26.24:** © Renee Lynn/Stone Images/Getty; **26.27:** © Zig Leszczynski/Animals Animals; **26.28A:** Cleveland P. Hickman, Jr.; **26.28B:** © George McCarthy/ Corbis

Chapter 27

Opener: William J. Weber/Visuals Unlimited; **27.1A:** Courtesy American Museum of Natural History, Neg. #125065; **27.4:** Cleveland P. Hickman, Jr.; **27.6:** © Corbis; **27.22:** © D. Poe/Visuals Unlimited; **27.23A,B:** © L.L. Rue, III; **27.24:** Cleveland P. Hickman, Jr.; **27.26:** © John Gerland/Visuals Unlimited; **27.27:** © Richard R. Hansen/Photo Researchers, Inc.; **27.29:** © L.L. Rue, III; **27.30:** Courtesy of Culver Pictures; **27.31-27.34:** Cleveland P. Hickman, Jr.

Chapter 28

Opener & 28.7: © L.L. Rue, III; **28.8A:** © PhotoDisc; **28.8B:** © Corbis; **28.9:** R.E. Treat; **28.13:** © L.L. Rue, III; **28.14 & 28.16:** © Gerlach/Visuals Unlimited; **28.17A:** Cleveland P. Hickman, Jr.; **28.19:** S. Malowski/Visuals Unlimited; **28.21:** © Kjell Sandved/Visuals Unlimited; **28.24:** © L.L. Rue, III; **28.26:** © M.H. Tierney, Jr./Visuals Unlimited; **28.27:** © G. Herben/Visuals Unlimited; **28.29:** Cleveland P. Hickman, Jr.; **28.30:** © L.L. Rue, III; **28.31:** Cleveland P. Hickman, Jr.; **28.32A:** Courtesy of Zoological Society of San Diego; **28.32B:** © Timothy Ransom/Biological Photo Services; **28.33:** © Milton H. Tierney, Jr./Visuals Unlimited; **28.35:** © John Reader; **28.37:** Photodisc/Getty Images; **28.38-28.40:** Cleveland P. Hickman, Jr.; **28.42:** © William C. Ober

Chapter 29

Opener: © Eric Soder/Photo Researchers, Inc.; **29.11B:** Courtesy Dr. Ian R. Gibbons; **29.13A,B:** Ed Reschke; **29.13C:** © G.W. Willis M.D./ Biological Photo Service

Chapter 30

Opener: Cleveland P. Hickman, Jr.; **30.1:** From J.F. Fulton and L.G. Wilson, Selected Reading in the History of Physiology, 1966. Courtesy of Charles C. Thomas, Publisher, Springfield, Illinois; **30.11:** From R.G. Kessel and R.H. Kardon, Tissues and Organs: A Text-Atlas of Scanning Electron Microscopy, 1979 W.H. Freeman and Co.; **30.21:** © L.L. Rue, III

Chapter 31

Opener: © Andrew Syred/Photo Researchers, Inc.; **31.2:** From J.F. Fulton and L.G. Wilson, Selected Readings in the History of Physiology, 1966. Courtesy of Charles C. Thomas, Publisher, Springfield, Illinois; **31.4A,B:** Courtesy P.P.C. Graziadei; **31.5:** © David M. Phillips/Visuals Unlimited

Chapter 32

Opener: Cleveland P. Hickman, Jr.; **32.3(both):** Courtesy Carl Gans; **32.6:** Cleveland P. Hickman, Jr.; **32.11:** Courtesy of Wyeth-Ayerst Laboratories; **32.12:** From R.G. Kessel and R.H. Kardon, Tissues and Organs: A Text-Atlas of Scanning Electron Microscopy, 1979 W.H. Freeman and Co.; **32.13D:** Courtesy of J.D. Berlin; **32.17:** Hospital Tribune 8:1, 1974

Chapter 33

Opener: © D.H. Ellis/Visuals Unlimited

Chapter 34

Opener: © Ed Reschke; **34.1A,B:** From J.F. Fulton and L.G. Wilson, Selected Readings in the History of Physiology, 1966. Courtesy of Charles C. Thomas, Publisher, Springfield, IL.; **34.10:** From J. A. Prior, et al., Physical Diagnosis, 1951 Mosby-Year Book, Inc.; **34.16:** From J.F. Fulton and L.G. Wilson, Selected Readings in the History of Physiology, 1966. Courtesy of Charles C. Thomas Publisher, Springfield, IL.

Chapter 35

Opener: Dr. Kari Lounatmaa/Photo Researchers, Inc.; **35.2:** From Van der Knapp, W.P.W., and E.S. Loker, "Immune mechanisms in trematode-snail interactions," Parasit. Today, 6:175-182, 1990; **35.8:** Courtesy of H. Zaiman, M.D.; **35.9:** © SUI/Visuals Unlimited

Chapter 36

Opener: © John Gerlach/Visuals Unlimited; **36.1A:** Thomas McAvoy/Life Magazine 1995. Time Inc./Getty Images; **36.1B,C:** Time & Life Pictures/Getty Images; **36.7:** Cleveland P. Hickman, Jr.; **36.9:** Nina Leen/Life Magazine 1995. Time Inc./Getty Images; **36.11 & 36.12:** Cleveland P. Hickman, Jr.; **36.14:** © Michele Westmoreland/Corbis; **36.15:** Corbis; **36.17:** Cleveland P. Hickman, Jr.; **36.18:** © Tom McHugh/Photo Researchers, Inc.; **36.19:** © Ray Richardson/Animals Animals; **36.21:** © Richard R. Hansen/Photo Researchers, Inc.

Chapter 37

Opener: StockTrek/Getty Images; **37.7:** Cleveland P. Hickman, Jr.; **37.9:** © F. Gohier/ Photo Researchers, Inc.; **37.10:** Cleveland P. Hickman, Jr.; **37.12:** © Doug Wechsler/Animals Animals/Earth Scenes; **37.13A:** © Gregory Ochocki/Photo Researchers, Inc.; **37.13B:** © Stephen J. Krasemann/Photo Researchers, Inc.; **37.14:** © Frank & Joyce Burek/Getty Images; **37.15A:** © Raymond Gehman/Corbis; **37.15B:** © Robert Lubeck/Animals Animals/Earth Scenes

Chapter 38

Opener: © Larry Hurd; **38.6:** © Noble Proctor/ Photo Researchers, Inc.; **38.7(both):** Cleveland P. Hickman, Jr.; **38.11A(top):** © Patti Murray/ Animals Animals/Earth Scenes; **38.11A(bottom):** © Wild & Natural/Animals Animals/Earth Scenes;

38.11B(both),C: © James L. Castner; **Page 839:** © D. Foster/WHOI/Visuals Unlimited

LINE ART AND TEXT

Figure 1.13 Source: From S. Gould, *Ontogeny and Phylogeny.* Harvard University Press, 1977.
Figure 6.16 Source: After J. Cracraft, *Ibis* 116:294-521 (1974).
Figure 6.22 Source: After P.R. Grant, Speciation and adaptive radiation of Darwin's finches, *American Scientist* 69:653-663, 1981.
Figure 8.1 Source: From N. Hartsoeker, *Essai de deoprique,* 1964.
Figure 8.18: Redrawn from *From Egg to Adult,* A report from Howard Hughes Medical Institute, © 1992.
Figure 8.19 Source: After E.M. De Robertis, O. Guillermo, and C.V.E. Wright, Homeobox genes and the vertebrate body plan, *Sci. Am.* 263: 46-52, July 1990.
Figure 10.5 Source: After E.O. Wiley, *Phylogenetics,* John Wiley & Sons, New York, 1981.
Figure 11.12 Redrawn from T.P. Stossel. The machinery of cell crawling. *Sci. Am.* 271:54-63, September, 1994.
Figure 22.7 Source: Courtesy of Tim Doyle.
Figure 22.13 Source: After A.N. Baker, F.W.E. Row, and H.E.S. Clark. A new class of Ecinodermata from New Zealand, *Nature* 321:862-864, 1986.
Figure 23.10 Source: After S. J. Gould, *Wonderful Life.* W.W. Norton, New York, 1989.
Figure 27.12b Source: After K. Schmidt-Nielsen, *Animal Physiology,* 4e. Cambridge University Press, 1990
Figure 28.20 Source: After N. Sugar, Biosonar and neural computation in bats, *Sci. Am* 262: 60-68, June 1990.
Figure 36.2 Source: After K. Lorenz and N. Tinbergen, *Zeit. Tierpsycho* 2:1-29, 1938.
Figure 36.10 Source: After J. Alcock, *Animal Behavior: An Evolutionary Approach,* 3e., Sinauer Associates, Sunderland, MA, 1984, from a photography by Masakasu Konishi.
Figure 36.16 Source: From C. Darwin, *Expression of the Emotions in Man and Animals.* Appleton and Co., New York, 1872.
Figure 37.3 With permission from *Natural History,* March 1990, Copyright the American Museum of Natural History.
Figure 37.18 Adapted from "The breakup of Pangaea" by Robert S. Dietz and John C. Holden. Copyright © October 1970 by Scientific American, Inc. All rights reserved. Adapted by permission.
Figure 38.3 Source: Data from E. Bos, et al. 1994. *World Population Projections 1994-95.* Baltimore, Johns Hopkins University Press for the World Bank.

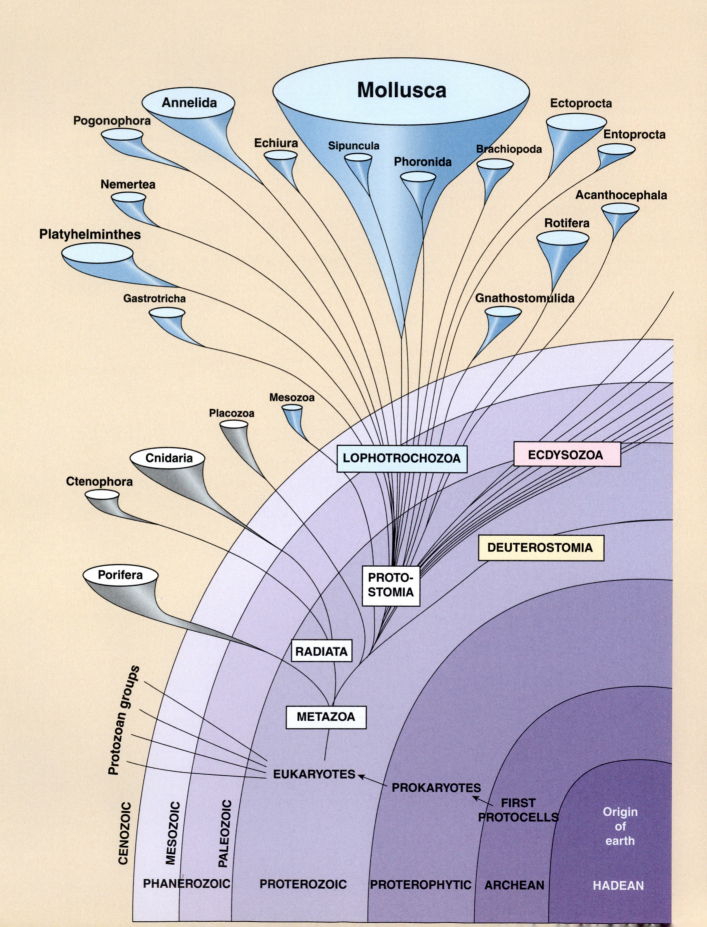

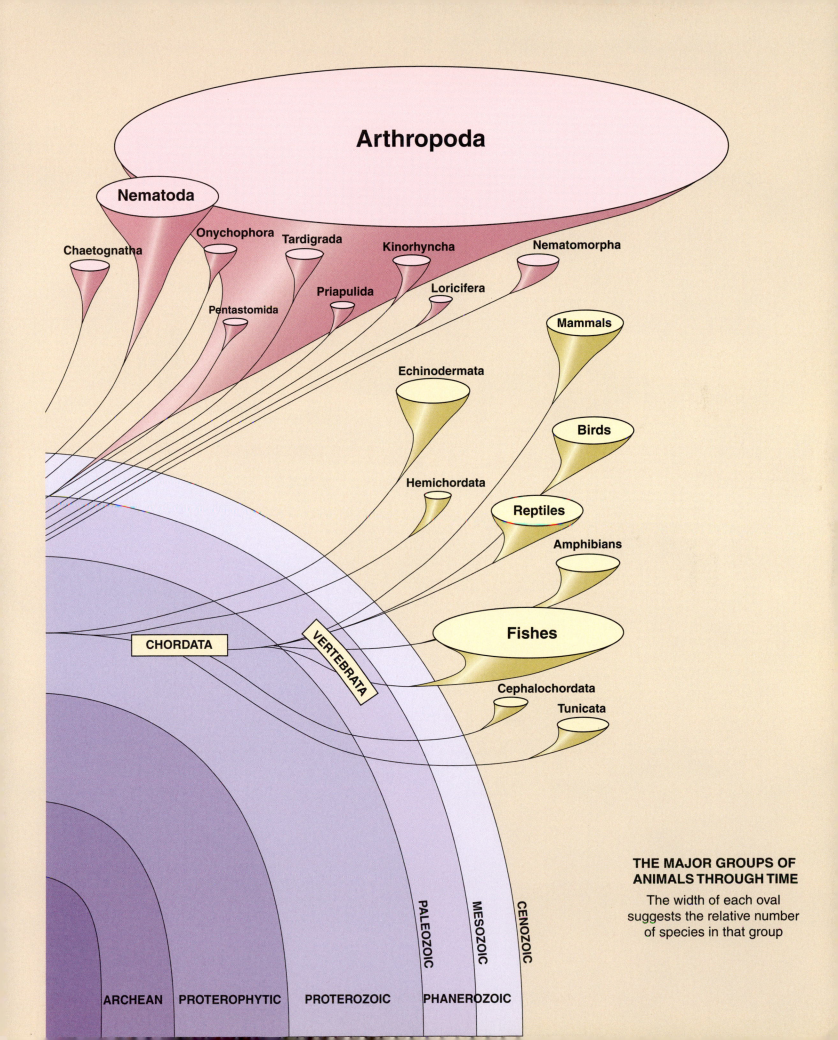

Arthropoda

Nematoda

Chaetognatha

Onychophora

Tardigrada

Kinorhyncha

Nematomorpha

Pentastomida

Priapulida

Loricifera

Mammals

Echinodermata

Birds

Hemichordata

Reptiles

Amphibians

CHORDATA

VERTEBRATA

Fishes

Cephalochordata

Tunicata

PALEOZOIC

MESOZOIC

CENOZOIC

ARCHEAN PROTEROPHYTIC PROTEROZOIC PHANEROZOIC

THE MAJOR GROUPS OF ANIMALS THROUGH TIME

The width of each oval suggests the relative number of species in that group